SELECTED UNITS AND CONVERSION FACTORS

SI UNITS

	Entity	Abbreviation	Name
Fundamental units	Length	m	meter
	Mass	kg	kilogram
	Time	s	second
	Charge	C	coulomb
Supplementary unit	Angle	rad	radian
Derived units	Force	$N = kg \cdot m/s^2$	newton
	Energy	$J = kg \cdot m^2/s^2$	joule
	Power	$W = J/s$	watt
	Pressure	$Pa = N/m^2$	pascal
	Frequency	$Hz = 1/s$	hertz
	Electric potential	$V = J/C$	volt
	Capacitance	$F = C/V$	farad
	Current	$A = C/s$	ampere
	Resistance	$\Omega = V/A$	ohm
	Magnetic field	$T = N/(A \cdot m)$	tesla
	Nuclear decay rate	$Bq = 1/s$	becquerel

SELECTED BRITISH UNITS

Length	1 inch (in.) = 2.54 cm (exactly)
	1 foot (ft) = 0.3048 m
	1 mile (mi) = 1.609 km
Force	1 pound (lb) = 4.448 N
Energy	1 British thermal unit (Btu) = 1.055×10^3 J
Power	1 horsepower (hp) = 746 W
Pressure	1 lb/in^2 = 6.895×10^3 Pa

OTHER UNITS

Length	1 light year (ly) = 9.46×10^{15} m
	1 astronomical unit (au) = 1.50×10^{11} m
	1 nautical mile = 1.852 km
	1 angstrom (Å) = 10^{-10} m
Area	1 acre (ac) = 4.05×10^3 m^2
	1 square foot (ft^2) = 9.29×10^{-2} m^2
	1 barn (b) = 10^{-28} m^2
Volume	1 liter (L) = 10^{-3} m^3
	1 U.S. gallon (gal) = 3.785×10^{-3} m^3
Mass	1 solar mass = 1.99×10^{30} kg
	1 metric ton = 10^3 kg
	1 atomic mass unit (u) = 1.6605×10^{-27} kg
Time	1 year (y) = 3.16×10^7 s
	1 day (d) = 86,400 s
Speed	1 mile per hour (mph) = 1.609 km/h
	1 nautical mile per hour (naut) = 1.852 km/h
Angle	1 degree (°) = 1.745×10^{-2} rad
	1 minute of arc (′) = 1/60 degree
	1 second of arc (″) = 1/60 minute of arc
	1 grad = 1.571×10^{-2} rad

Energy	1 kiloton T.
	1 kilowatt·h
	1 food calorie
	1 calorie (cal)
	1 electron volt ($= 1.60 \times 10^{-19}$ J
Pressure	1 atmosphere (atm) = 1.013×10^5 Pa
	1 millimeter of mercury (mm Hg) = 133.3 Pa
	1 torricelli (torr) = 1 mm Hg = 133.3 Pa
Nuclear decay rate	1 curie (Ci) = 3.70×10^{10} Bq

USEFUL FORMULAE

Circumference of a circle of radius r or diameter d: $C = 2\pi r = \pi d$

Area of a circle of radius r or diameter d: $A = \pi r^2 = \pi d^2/4$

Area of a sphere of radius r: $A = 4\pi r^2$

Volume of a sphere of radius r: $V = (4/3)(\pi r^3)$

Other formulae (such as the Pythagorean theorem) appear in the readily accessible math review in Chapter 1.

COLLEGE PHYSICS

COLLEGE PHYSICS

PAUL PETER URONE
California State University, Sacramento

Brooks/Cole Publishing Company
ITP An International Thomson Publishing Company

Pacific Grove • Albany • Belmont • Bonn • Boston • Cincinnati
Detroit • Johannesburg • London • Madrid • Melbourne • Mexico City
New York • Paris • Singapore • Tokyo • Toronto • Washington

Sponsoring Editor: *Keith Dodson*
Editorial Assistants: *Georgia Jurickovich and Leigh Hamilton*
Production Coordinator: *Marjorie Z. Sanders*
Production: *Electronic Publishing Services Inc., NYC*
Marketing Team: *Margaret Parks, Caroline Croley, and Michele Mootz*
Interior Design: *Electronic Publishing Services Inc., NYC*
Cover Design: *Cheryl Carrington*

Interior Illustration: *Electronic Publishing Services Inc., NYC*
Cover Photo: *© 1997/Thomas T. Duncan/Adventure Photo & Film*
Photo Researcher: *Electronic Publishing Services Inc., NYC*
Typesetting: *Electronic Publishing Services Inc., NYC*
Printing and Binding: *Von Hoffmann Printing Company*

COPYRIGHT © 1998 by Brooks/Cole Publishing Company
A Division of International Thomson Publishing Inc.
I(T)P The ITP logo is a registered trademark under license.

For more information, contact:

BROOKS/COLE PUBLISHING COMPANY
511 Forest Lodge Road
Pacific Grove, CA 93950
USA

International Thomson Publishing Europe
Berkshire House 168-173
High Holborn
London WC1V 7AA
England

Thomas Nelson Australia
102 Dodds Street
South Melbourne, 3205
Victoria, Australia

Nelson Canada
1120 Birchmount Road
Scarborough, Ontario
Canada M1K 5G4

International Thomson Editores
Seneca 53
Col. Polanco
11560 México, D. F., México

International Thomson Publishing GmbH
Königswinterer Strasse 418
53227 Bonn
Germany

International Thomson Publishing Asia
221 Henderson Road
#05-10 Henderson Building
Singapore 0315

International Thomson Publishing Japan
Hirakawacho Kyowa Building, 3F
2-2-1 Hirakawacho
Chiyoda-ku, Tokyo 102
Japan

Printed in the United States of America.

10 9 8 7 6 5 4 3 2 1

Library of Congress Cataloging-in-Publication Data

Urone, Paul Peter.
    College physics / Paul Peter Urone.
        p. cm.
    Includes index.
    ISBN 0-534-35190-5
    1. Physics. I. Title.
QC23.U76   1998
    530—dc21
                                                    97-46326
                                                     CIP

This book is dedicated to Patty, my friend, my lover, my wife.
Without her neither this book nor my heart would exist.

PREFACE

TO THE STUDENT

This book is written for you. It is based on a quarter-century of teaching college-level physics and influenced by a strong recollection of my own struggles as a student. I have spent thousands of hours in the classroom, in the laboratory, and in tutoring centers talking and *listening* to students. To the nearly three thousand students I have known over the years, I say this: You are a part of this book. Your ideas, your reactions to concepts, your needs as well as your wishes have influenced every aspect of this text. To the students now using this text: I hope the story of physics told in this book will make the subject understandable, useful, and enjoyable.

A good story is important, and physics is rich in stories. This book tells some of those stories and laces them with applications that are convincing evidence that physics is visible everywhere. Applications range from driving a car to launching a rocket, from a skater whirling on ice to a neutron star spinning in space, and from taking your temperature to taking a chest x ray. In spite of having more years behind me than most of you, I share many of your interests—a skier on the slopes is much more illuminating than a block on an incline.

But it isn't enough for you to read a story or listen to an entertaining lecture. You must take action—active inquiry, exploration, and extrapolation are the next steps to useful understanding. One activity you will undoubtedly have in your physics course is homework. At the end of each chapter in this text, you will find questions and problems your instructor may assign for homework. While that may seem burdensome at times, it is crucial to real understanding. This is where you really learn—in doing things yourself and questioning the world around you. Learning to solve physics problems is much like learning to ride a bicycle. *Your action* is necessary in addition to watching how someone else does it.

This book contains many features that are designed to help you learn the concepts of physics, use those concepts to solve problems, and also recognize how the solution fits nature (is correct). To learn more about the features in this text, consult the material titled *Special Features*. I hope you will find much in this book that interests you, helps you to succeed in your courses, and allows you to apply physics outside the classroom. The most powerful form of knowledge is that which enables you to go beyond what you are told.

TO THE INSTRUCTOR

This text is intended for one-year introductory courses requiring algebra and some trigonometry, but no calculus. It is written for both the students and instructors of today. After describing the general features of the text, I elaborate on the special features of the text designed to achieve its goals.

Goals

For a subject as sophisticated as physics and a process as complex as learning, our goals are as common as they are fundamental: to help transmit the concepts of physics and the analytical skills to apply them; to build an appreciation of the underlying simplicity and unity in nature; to impart a sense of the beauty and wonder found in the physical universe.

General Approach

Topics are introduced conceptually with a steady progression to precise definitions and analytical applications. The analytical aspect (problem solving) is tied back to the conceptual before moving on to another topic. Each chapter, for example, opens with an attractive photograph relevant to the subject of the chapter. The introduction to the chapter is made with interesting applications that are easy to visualize and that most students should have observed for themselves. The conceptual foundation of the chapter is then laid with increasing precision. Definitions and quantitative relationships follow. Worked examples not only illustrate the analytical aspects of a problem, but always tie the strategy and results back into the conceptual aspect of the topic (the *Discussion* section of each worked example). This forms a cycle: Starting with familiar applications, there is a progression from conceptual to analytical, which is tied back to the conceptual at the end.

Organization, Level, and Content

This book is innovative, yet practical for most potential users. Topics ordinarily covered in a one-year course at the algebra-trig level are included in a relatively traditional sequence. Certain highly sophisticated topics (such as the use of phasors in AC circuits) are omitted to allow sufficient treatment of the topics that remain. Thus, the length of the text has been controlled without eliminating those topics most often included in introductory courses. This does not mean that the text is written at a low level. It means that there is a uniformity of treatment that does not introduce a special technique for a single topic.

There is considerable latitude on the part of the instructor regarding level and content. By choosing the types of problems assigned (from three denoted levels of difficulty), the instructor can determine the level of sophistication required of the student. Likewise, by choosing the topics covered in depth, the instructor can determine both the level and the topical emphasis of the course.

Concepts and Calculations

In recent years there has been a growing awareness that conceptual skills need to be addressed as directly as analytical skills—that an ability to calculate does not guarantee conceptual understanding. It has always been my belief that the issues of conceptual and analytical skills should be addressed simultaneously. In order to unify conceptual, analytical, and calculational skills within the learning process, I have integrated them throughout the text. This can be seen in a series of features described in the section *Special Features*. It is particularly easy to see how this is built into the *worked examples* and *problem-solving strategies* in the text. But it is also fundamental to the general approach of conceptual to analytical tied back to conceptual described previously.

Modern Perspective

Professional study groups and surveys have identified a need for expanded emphasis of modern physics. This is met in this text by the early incorporation of familiar ideas from modern physics, such as the existence of atoms and the conversion of mass into energy. The chapters on modern physics are more complete than most, with an entire chapter devoted to applications of nuclear physics and another to particle physics. The final chapter of the text, *Frontiers of Physics,* is devoted to the most current and exciting endeavors in physics. It ends with a section titled *Questions We Know to Ask.*

Accuracy

Every effort has been made not only to keep the text free from error but to have it accurately portray the physical universe. My goal has always been to produce a first edition text of second edition quality. Thorough peer reviews performed at many stages of the writing and editing of the text have been carefully taken into account. Also considered were the results of selected student reviews and extensive classroom testing. During the final stages of production, every equation, calculation, and number in the text was checked yet again by the author and by another qualified physicist for accuracy and consistency in significant figures. Two physicists each solved all of the end-of-chapter problems in detail. Many of those problems have been solved by other individuals as well. This has eliminated ambiguities in the statements of problems, ensured that the situations posed were reasonable, and made the number of significant figures used consistent with the simple rules elaborated in the text. The answers to odd-numbered problems that appear at the end of the book are thus felt to be nearly error free, as are the solutions to problems found in the *Instructor's Solution Manual*. The art program was rendered with the advice of a physicist and was checked for accuracy and effectiveness by the author and other individuals during production.

SUPPLEMENTS

The text is supported by a complete package of text supplements:

Student's Solutions Manual, prepared by Steven Yount of the United States Patent Office and Gary Shoemaker of California State University, Sacramento. Provides carefully worked-out solutions to the odd-numbered problems, answers to which appear in the back of the book.

Instructor's Solutions Manual, also by Steven Yount and Gary Shoemaker. This manual includes detailed solutions for all problems in the end-of-chapter sections.

Instructor's Solutions Disk, a computer disk version of the *Instructor's Solutions Manual*. This is available to instructors in Microsoft Word for Windows and Macintosh.

Test Item File, by Frank D. Stekel of University of Wisconsin–Whitewater. The test bank contains more than 1600 questions of varying formats and level of difficulty.

Computerized Testing A computerized version of the printed test items is available to DOS, Windows, and Macintosh platforms.

Transparencies Approximately 115 four-color transparency acetates of key figures from the text are included in this package.

CNN Physics Today This 50 minute video contains numerous segments highlighting everyday physics—from spiderwebs to Cirque du Soleil. Available to adopters of the text.

SPECIAL FEATURES

The following briefly describe special features of this text.

Applications Applications *of physics relevant* and *visible* to students are an integral part of making the physics interesting. I revel in finding physics everywhere and have incorporated that passion in this text. This is present in the body of each chapter and continues in the worked examples and end-of-chapter conceptual questions and problems. There is a discernible point to most of what the students read and do. When the results of an application of physics give a new insight, the physics sings as it does to those of us who have always found it inherently interesting.

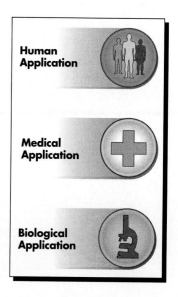

Throughout the book, you will find icons for significant instances of *human, medical,* and *biological applications*. Many students in algebra-trig based physics courses are majoring in related disciplines and find these applications stimulating. *Human, medical,* and *biological* applications are an integral part of the prose of the text and are found in worked examples and in end-of-chapter conceptual questions and problems, but those that obscure the exposition of the physics are avoided. Certain sections are devoted to human, medical, and biological topics, including: forces and torques in muscles and joints (Section 5.7), fluids and pressures in the human body (Section 10.8), nerve conduction (Section 19.7), color and color vision (Section 24.10), and biological effects of ionizing radiation (Section 30.5). The table of contents includes many others.

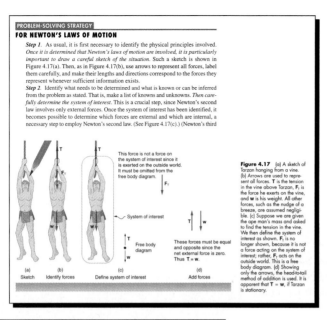

Step 1. As usual, it is first necessary to identify the physical principles involved. *Once it is determined that Newton's laws of motion are involved, it is particularly important to draw a careful sketch of the situation.* Such a sketch is shown in Figure 4.17(a). Then, as in Figure 4.17(b), use arrows to represent all forces, label them carefully, and make their lengths and directions correspond to the forces they represent whenever sufficient information exists.

Step 2. Identify what needs to be determined and what is known or can be inferred from the problem as stated. That is, make a list of knowns and unknowns. *Then carefully determine the system of interest.* This is a crucial step, since Newton's second law involves only external forces. Once the system of interest has been identified, it becomes possible to determine which forces are external and which are internal, a necessary step to employ Newton's second law. (See Figure 4.17(c).) (Newton's third

Figure 4.17 (a) A sketch of Tarzan hanging from a vine. (b) Arrows are used to represent all forces. **T** is the tension in the vine above Tarzan, **F₁** is the force he exerts on the vine, and **w** is his weight. All other forces, such as the nudge of a breeze, are assumed negligible. (c) Suppose we are given the ape man's mass and asked to find the tension in the vine. We then define the system of interest as shown. **F₁** is no longer shown, because it is not a force acting on the system of interest; rather, **F₁** acts on the outside world. This is a free body diagram. (d) Showing only the arrows, the head-to-tail method of addition is used. It is apparent that **T** = **w**, if Tarzan is stationary.

Problem-solving strategies *Problem-solving strategies* are first presented in a special section and subsequently appear at crucial points in the text where students can benefit most from them. Problem-solving strategies have a logical structure that is reinforced in the *worked examples* and supported in certain places by line drawings that illustrate various steps. There are three types of problem-solving strategies. The largest number are found in the main body of the text. The other two types are those for the *integrated concepts* problems and those for the *unreasonable results* problems. These help the student to accomplish the goals of those two different types of problems. All problem-solving strategies contain conceptual elements as well as analytical.

EXAMPLE 3.8   VELOCITY RELATIVE TO . . .?

An airline passenger drops a quarter in change while the plane is moving at 260 m/s in level flight. What is the velocity of the quarter when it strikes the floor 1.50 m below its point of release: (a) Measured relative to the plane? (b) Measured relative to the earth?

Strategy Both problems can be solved with the techniques for falling objects and projectiles. In part (a), the initial velocity of the quarter is zero relative to the plane, making the motion that of a falling object (one-dimensional). In part (b), the initial velocity is 260 m/s horizontal relative to the earth and gravity is vertical, making this a projectile motion. In both parts it is best to use a coordinate system with vertical and horizontal axes.

Solution for (a) With the given information, we note that the initial velocity and position are zero, and the final

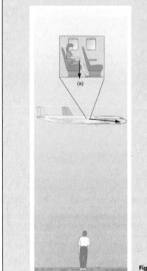

Figure 3.19 More classical relativity. The motion of a quarter dropped inside an airplane as viewed by two different observers. (a) An observer in the plane sees the quarter fall straight down. (b) An observer on the ground sees the quarter move almost horizontally.

position is −1.50 m. The final velocity can be found using Equation 3.10d:

$$v^2 = v_0^2 - 2gy$$

Substituting known values, we get

$$v^2 = 0 - 2(9.8 \text{ m/s}^2)(-1.50 \text{ m}) = 29.4 \text{ m}^2/\text{s}^2$$

yielding

$$v = -5.42 \text{ m/s}$$

The minus sign is *chosen* when the square root is taken; it means the velocity is downward. There is no initial horizontal velocity relative to the plane and no horizontal acceleration, and so the motion is straight down relative to the plane.

Solution for (b) Since the initial vertical velocity is zero relative to the ground and since vertical motion is independent of horizontal motion, the final vertical velocity for the quarter relative to the ground is $v_y = -5.42$ m/s, the same as found in part (a). There is no horizontal acceleration, and so the initial and final horizontal velocities are the same and $v_x = 260$ m/s. The x- and y-components of velocity can be combined, using Equation 3.16, to find the magnitude of the final velocity:

$$v = \sqrt{v_x^2 + v_y^2}$$

Thus,

$$v = \sqrt{(260 \text{ m/s})^2 + (-5.42 \text{ m/s})^2}$$

yielding

$$v = 260.06 \text{ m/s}$$

The direction is given by Equation 3.17:

$$\theta = \tan^{-1}(v_y/v_x) = \tan^{-1}(-5.42/260)$$

so that

$$\theta = \tan^{-1}(-0.0208) = -1.19°$$

Discussion In part (a), the final velocity relative to the plane is the same as it would be if the quarter were dropped from rest on earth and fell 1.50 m. This fits our experience; objects fall the same way when the plane is flying horizontally as when it is at rest on the ground. This is also true in moving cars. In part (b), an observer on the ground sees a much

(continued)

Worked examples *Worked examples have four distinct parts* to promote both analytical and conceptual skills. Worked examples are *introduced in words*, always using some application that should be of interest. This is followed by a clearly labeled *Strategy* section that emphasizes the concepts involved and how solving the problem relates to those concepts. Next comes the mathematical *Solution* to the problem. Finally, there is a *Discussion* section. This last step ties the analytical back to the conceptual by discussing the answer of the worked example—its meaning, magnitude, further implications, and so on are used to give the analytical process conceptual meaning and to link topics together.

Many worked examples contain *multiple-part problems* to help the students learn how to approach normal situations, in which problems tend to have multiple parts. Finally, worked examples *employ the techniques of the problem-solving strategies* so that students can see how those strategies succeed in practice as well as in theory.

Integrated concept problems Many would say that a central tenet of physics is its underlying unity and connections. Although topics are necessarily introduced separately, I have written numerous *integrated concept problems*, to encourage students to apply principles broadly. In many chapters, there are *integrated concept worked examples* that guide the student in approaching situations that have more than one physical principle involved. These are augmented by *integrated concept problem-solving strategies* that explicitly help the student draw ideas from more than one area and apply them conceptually and analytically to solving a problem. While some texts contain problems drawing from many areas, very few aid students in mastering such problems. In this text, such problem-solving support has been a central idea in the development of these problems.

These problems involve physical principles from more than one chapter. Integrated knowledge from a broad range of topics is much more powerful than a narrow application of physics. Energy, for example, is involved in kinematics, whence the relevance of Chapter 2. The following topics are involved in some or all of the problems in this section:

Topics	Location
Kinematics	Chapter 2
Two-dimensional kinematics	Chapter 3
Dynamics	Chapter 4
Statics, elasticity	Chapter 5

PROBLEM-SOLVING STRATEGY

Use the following strategy to solve integrated concept problems.

Step 1. *Identify which physical principles are involved.* This is done by following the problem-solving strategies found in this and previous chapters (see the table of contents or look in the index). Listing the givens and the quantities to be calculated will allow you to identify the principles involved.

Step 2. *Solve the problem using strategies outlined in the text.* If these are available for the specific topic, you should refer to them. You should also refer to the sections of the text that deal with a particular topic.

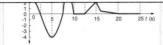

Newton's universal law of gravitation is modified by Einstein's general theory of relativity, as we shall see in Chapter 32. Newton's gravity is not seriously in error—it was and still is an extremely good approximation for most situations. Einstein's modification is most noticeable in extremely large gravitational fields, such as near black holes. However, general relativity also explains such things as small but long-known deviations of the orbit of the planet Mercury from classical predictions.

Connections boxes To illustrate the underlying unity of apparently different topics, *Connections boxes* are placed in the margin where appropriate. They contain brief discussions of conceptual connections not necessarily obvious at the introductory level. Connections boxes augment the integrated concepts problems and their supporting materials, adding emphasis to the most powerful aspects of physics—its underlying unity, breadth of applicability, and fundamental simplicity.

Unreasonable results *Unreasonable result problems* are unique to this text. They are designed to further emphasize that properly applied physics must describe nature accurately and is not simply the process of solving equations. These problems contain a premise that produces an unreasonable answer. For example, if the heat generated by metabolizing an average day's food is retained, a person's body temperature will rise to a lethal level. Thus, physics correctly applied produces in this case a result that is never observed—the student must recognize that the premise of complete heat retention is at fault. These problems are clearly labeled and are found at the very end of the end-of-chapter problems—all other problems in the text produce reasonable results and often contain discussion to emphasize that physics must fit nature. Taken with the careful accuracy of the text and the discussions at the end of *worked examples*, unreasonable results problems can help students examine the concepts of a problem as well as the mechanics of solving it.

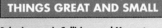

Figure 2.22 (a) Graph of displacement versus time for a jogger who starts on a run, stops, and goes back to pick up his keys at 2.5 m before continuing. (b) Graph of velocity versus time produced by finding the slope of the displacement graph. (c) Graph of acceleration versus time produced by finding the slope of the velocity graph. Problems 54 and 55.

UNREASONABLE RESULTS

Physics must describe nature accurately. The following problems have results that are unreasonable because one premise is unreasonable or because certain of its premises are inconsistent with one another. The physical principle applied correctly then produces an unreasonable result. For example, if a person starting a foot race accelerates at 0.40 m/s² for 100 s, his final speed will be 40 m/s (about 90 miles per hour)—clearly unreasonable because the time of 100 s is an unreasonable premise. The physics is correct in a sense, but there is more to describing nature than just manipulating equations correctly. Checking the result of a problem to see if it is reasonable does more than help uncover errors in problem solving—it also builds intuition in judging whether nature is being accurately described.

PROBLEM-SOLVING STRATEGIES

Use the following strategies to determine if an answer is reasonable and, if it is not, to determine what is the cause.

Step 1. Solve the problem using strategies as outlined in Section 2.5 and in the format followed in the worked examples in the text. In the

example given in the preceding paragraph, you would identify the givens as the acceleration and time and use Equation 2.9 to find the unknown final velocity. That is,

$$v = v_0 + at = 0 + (0.40 \text{ m/s}^2)(100 \text{ s}) = 40 \text{ m/s}$$

Step 2. Check to see if the answer is reasonable. Is it too large or too small, or does it have the wrong sign, improper units, . . .? In the present case, the velocity is about four times higher than a person can run—so it is too large.

Step 3. If the answer is unreasonable, look for what specifically could cause the identified difficulty. In the example of the runner, there are only two assumptions that are suspect. The acceleration could be too great or the time too long. People can easily accelerate at 0.40 m/s²; thus, the time must be too long.

2.58 Suppose you drop a rock from a cliff and you observe that it falls 39.2 m in 2.00 s. (a) What is the acceleration of the rock? (b) What is unreasonable about the result? (c) Which premise is unreasonable, or are the premises inconsistent?

2.59 An advertisement claims that a certain automobile can stop from a speed of 35.0 m/s in a distance of 20.0 m. (a) What is the average deceleration? (b) What is unreasonable about the result? (c) Which premise is unreasonable?

Things Great and Small In these special topic essays, Macroscopic Phenomena (such as air pressure) are explained with Submicroscopic Phenomena (such as atoms bouncing off walls). These essays support the modern perspective by describing aspects of modern physics before they are formally treated in later chapters. Connections are also made between apparently disparate phenomena.

Summary Chapter summaries are thorough and functional and present all important definitions and numbered equations. Students are able to find the definitions of all terms and symbols as well as their physical relationships. The structure of the summary makes plain the fundamental principles of the chapter and will serve as useful study guides.

Index The *index* of this book is unique. It is more detailed and sophisticated, yet easier to use, than any other. There is significant cross-referencing, making it easy for readers to find what they seek. Index entries often tell a small story—for example, the name of a famous scientist is listed with a sublisting of what he or she is known for.

THINGS GREAT AND SMALL

Submicroscopic Collisions and Momentum

The conservation of momentum principle not only applies to the macroscopic world, it is also essential to our explorations of the substructure of matter. Giant machines hurl submicroscopic particles at one another and evaluate the results by assuming conservation of momentum (among other things). Figure 7.A shows a detector used in such experiments.

On the small scale, we find that particles and details of their properties invisible to the naked eye can be measured with our instruments, and models of the submicroscopic world can be constructed to describe the results. Momentum is found to be a property of all submicroscopic particles *including massless particles*, such as photons composing light. This hints that momentum may have an identity beyond mass times velocity. Furthermore, we find that the conservation of momentum principle is valid. We use it to analyze the mass and other properties of previously undetected particles, such as the nucleus of the atom and the existence of quarks inside the nucleus itself. Figure 7.B illustrates how a particle scattering backward from another implies that its target is massive and dense. Experiments seeking evidence of quarks inside nuclei scattered high-energy electrons from protons (nuclei of hydrogen atoms). Occasionally, electrons scattered straight backward in a manner that implied a very small and very dense particle *inside* the proton—this is considered nearly direct evidence of quarks. The analysis, again, was based partly on the same conservation of momentum principle that works so well on the large scale.

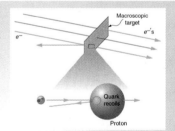

Figure 7.B A submicroscopic particle scatters straight backward from a target particle. In experiments seeking evidence for quarks, electrons were observed to occasionally scatter straight backward from a proton.

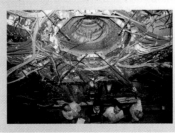

Figure 7.A A huge accelerator hurls tiny particles at one another to explore the submicroscopic world. The results of the collisions are measured with huge detectors such as this, and they are analyzed with the conservation of momentum principle, among others.

Glossary The *glossary* contains all relevant symbols, their meaning, and where they are defined in the text. The glossary thus enables students to take the first step in organizing their material by defining symbols and tracing their first use to help locate basic definitions.

ACKNOWLEDGMENTS

It is a pleasure to acknowledge the contributions of those who have helped shape this book. As mentioned earlier, the influence of my students has been profound. I have also benefited from interactions with many colleagues over many years. I thank all of you for the very important role you have played in the development of this book.

Most of the writing process is incredibly solitary, but there are junctures at which the direct input and work of others is not only enjoyed but absolutely crucial. There were many rounds of reviews during the writing of this book, producing invaluable advice ranging from general comments on organization to specific aspects of physics and modern teaching theory. A partial list of those who reviewed part or all of this text includes:

Barron Arenson
Mesa Community College

Mark John Badway

Elizabeth Behrman
Wichita State University

Treasure Brasher
West Texas A & M University

Charles Brient
Ohio University

Michael Browne
University of Idaho

Ron Canterna
University of Wyoming

Neal Cason
University of Notre Dame

Carlo Catucci

Lowell Christinen
American River College

Robert Cole
University of Southern California

Lawrence B. Coleman
University of California–Davis

Steven Davis
University of Arkansas–Little Rock

Miles J. Dresser
Washington State University

Timothy Duncan

A. Gordon Emslie
University of Alabama–Huntsville

Donald Francheschetti
Memphis State University

Simon George
California State University–Long Beach

Grant W. Hart
Brigham Young University

Tom R. Herrmann
Eastern Oregon University

David Jenkins
Virginia Polytechnic Institute

Shiva Kumar
Yale University

Clement Lam
North Harris College

Michael Lieber
University of Arkansas

Bo Lou
Ferris State University

Lloyd Makarowitz
State University of New York
   at Farmingdale

Gerardo Muñoz
California State University–Fresno

Lawrence S. Pinsky
University of Houston

Dinko Pocanic
University of Virginia

James Rittenbach
Yakima Valley Community College

Dennis Ross
Iowa State University

Roy Rubins
University of Texas at Arlington

Alvin Rusk
Southwest State University

Kandula S. R. Sastry
University of Massachusetts

Arthur G. Schmidt
Northwestern University

Wesley Shanholtzer
Marshall University

Joseph Shinar
Iowa State University

Mark Sprague
East Carolina University

Frank D. Stekel
University of Wisconsin–Whitewater

James F. Sullivan
University of Cincinnati

Paul Varlashkin
East Carolina University

Jane Webb
Christopher Newport University

Robert Wild
University of California–Riverside

Rand Worland
University of Puget Sound

To all of you I give my sincere thanks. Those errors that remain are entirely my own responsibility, and I will be happy to be informed about them.

Certain individuals merit special mention. Steve Yount solved almost every problem in the text, starting as a motivated student and finishing some years later as a physicist. Professor Gary Shoemaker of my own department also solved all of these problems giving them an added degree of accuracy. Also in my department, Professor Duane Aston classroom-tested part of the manuscript. The support and influence of my former editor, Rick Mixter, contributed to the successful completion of this text.

My deepest debt of gratitude is to my friends and family. All of them tolerated me and my preoccupation with this project over many years, gave me support in times of despair, and shared the moments of joy. In ways they may not realize, my parents, my wife's parents, and my children have carried me along. But my wife, Patty, has done more than anyone should expect or hope. I will be forever grateful.

It is my good fortune to have this book published by Brooks/Cole ITP. Marjorie Sanders as production editor coordinated the production in a cheerful, even handed, and considerate manner. Kelly Shoemaker produced the cover of this text based on a photograph I wish I could claim I had taken myself. A special thanks to Keith Dodson. Your work has not only gotten me through, it has had an impact on the entire book. You have been steadfast in your faith in this project and your commitment to its careful and imaginative execution. Keith also spent innumerable hours managing the production of the supplements for the text. I don't know that I will miss those e-mails on weekends, but I certainly did appreciate them.

Electronic Publishing Services Inc. designed the text, performed the final copyediting and proofreading, rendered the art, found the photographs, and put everything together in a highly professional manner. As production editor at EPS, Rob Anglin did a fine job of overseeing the work. Francis Hogan, an artist in his own right, did a wonderful job of photo research and encouraged me in making the final, necessary cuts. Francis also supplied several splendid photographs of his own. Andrew Schwartz performed the copyediting with a precision worthy of his moniker, "Eagle Eye," but with an intelligence beyond the need for another comma. Many individuals were involved in rendering the art—it did not escape my notice that they thought carefully about each piece, carrying them beyond the minimum that might have sufficed.

Paul Peter Urone

ABOUT THE AUTHOR

Paul Peter Urone (Peter, as he calls himself) is a professor of physics at California State University, Sacramento. Peter is a product of the Sputnik era, being in ninth grade when the space race began. Encouraged by the rush toward high technology, he received a bachelor's degree in physics from the University of Colorado in Boulder, where he later went on to earn his Ph.D. Following graduate school, he performed basic research in nuclear physics in The Netherlands, England, and at Stony Brook in New York before joining the faculty at CSU, Sacramento in 1973. After some years there he turned his attention to medical and biological applications of physics and authored the text *Physics with Health Science Applications*, now in its sixteenth printing.

A dedicated teacher, Peter receives high marks from his students and considers himself a compulsive story teller. He firmly believes that the role of an educator is to inspire, motivate, and guide as well as to explain. Peter's greatest intellectual joy is in the beauty of physics and especially in being able to communicate that appreciation to others in a manner they find useful.

When not writing or teaching, Peter pursues his interests in ancient Greek coins, folk art, and Native American art. His true passion, however, is spending as much time as possible with his wife, Patty, and their two sons, Chris and Dustin.

BRIEF CONTENTS

CONTENTS

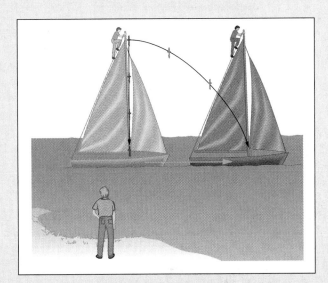

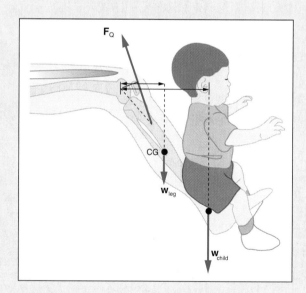

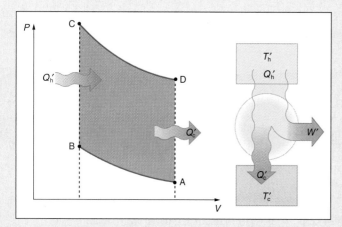

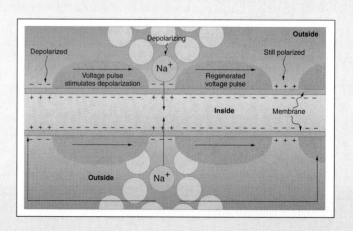

28 ATOMIC PHYSICS   741

29 RADIOACTIVITY AND NUCLEAR PHYSICS   779

30 APPLICATIONS OF NUCLEAR PHYSICS   811

31 PARTICLE PHYSICS   840

SPECIAL APPLICATIONS FEATURES

PROBLEM-SOLVING STRATEGIES

INTRODUCTION

1

Physical Quantities, Units, and Mathematical Overview

Galaxies are as immense as atoms are small. Yet the same laws of physics govern both, and all the rest of nature, an indication of the underlying unity in the universe. The laws of physics are surprisingly few in number, implying an underlying simplicity to nature's apparent complexity.

1.1 PHYSICS: AN INTRODUCTION

The physical universe is enormously complex in its detail. Every day, each of us observes a great variety of objects and phenomena. And over the centuries, the curiosity of the human race has led us collectively to explore and catalog a wealth of information vastly beyond the capacity of any single person to assimilate. From the flight of birds to the colors of flowers, from lightning to gravity, from quarks to clusters of galaxies, from the flow of time to the mystery of the creation of the universe, we have asked questions and assembled huge arrays of facts. And in the face of all these details, we have discovered that a surprisingly small and unified set of physical laws can explain what we observe. As humans, we make generalizations and seek order. We have found that nature is remarkably cooperative—it exhibits the *underlying order and simplicity* we so value.

It is the underlying order of nature that makes science in general, and physics in particular, so enjoyable to study. For example, what do a bag of French fries and a car battery have in common? Both contain energy that can be converted to other forms. The law of conservation of energy (which says that energy can change form but is never lost) ties together such topics as food calories, batteries, heat, light, and watch springs. Understanding this law makes it easier to learn about the various forms that energy takes and how they relate to one another. Apparently unrelated topics are connected through broadly applicable physical laws, permitting a comprehension superior to the memorization of lists of facts.

The unifying aspect of physical laws and the basic simplicity of nature form the underlying themes of this text. In learning to apply these laws, you will, of course, study the most important topics in physics. More importantly, you will gain analytical abilities that enable you to apply these laws far beyond the scope of what can be included in a single book. This first section discusses the realm of physics (to help define what physics is), some applications of physics (to help illustrate its relevance to other disciplines), and more precisely what constitutes a physical law (to help illuminate the importance of experimentation to theory).

Science and the Realm of Physics

Science consists of the laws that are the general truths of nature and the body of knowledge they encompass. Scientists are continually trying to expand that body of knowledge and to perfect the human understanding and expression of the laws that govern it. **Physics** is the most basic of the sciences, concerning itself with the interactions of energy, matter, space, and time, and especially with questions of what underlies every phenomenon. A physicist is not content to know that energy is conserved (a fact of major importance), but also wants to know *why* it is conserved. It is this almost childlike curiosity that has produced major advances in our understanding of the universe. A question such as, "Why does an apple fall to the ground?" may seem frivolous, but such a question reputedly led Isaac Newton to an understanding of gravity and the ability to calculate the orbits of the moon and planets. The concern for basics essentially defines the *realm of physics*.

Because of this concern for basics, physics has been the home of some of the most important and exciting discoveries in science. Questions such as, "What are the basic

Figure 1.1 To better understand nature, physicists perform experiments and develop theories. (a) The laboratory pictured here explores the substructure of matter in highly technical efforts mounted by large groups of experimentalists. (b) Albert Einstein was the consummate theoretician—he not only explained experimental results, but also developed theories that correctly predicted previously unknown characteristics of nature. He is shown here with Wolfgang Pauli, another important theoretician.

(a)

(b)

building blocks of matter?" have led to the discoveries of atoms, nuclei, and the likelihood that quarks are the underlying substructure of nuclei. The question, "Is time the same for all objects?" led to the discovery that a twin traveling to a distant star at high speed would age less than her sister who remained on earth. Some of these discoveries have drastically changed the way we view the universe and have even affected our concept of reality. For example, it appears that when an electron moves from one orbit to another in an atom it does not exist between orbits.

Applications of Physics

You need not be a scientist to use physics. On the contrary, a knowledge of physics is useful in everyday situations as well as in nonscientific professions. It can help you understand how microwave ovens work, why metals should not be put into them, and why they might affect pacemakers. (See Figure 1.2.) You can understand the hazards of radiation and rationally evaluate these hazards more easily when you know some physics. The reason a car's radiator should be black to get rid of heat while a roof should be white to reflect heat also is supplied by physics. The operation of a car's ignition system, a ground fault interrupter, and the three-wire electrical power distribution system are much easier to understand when you think about them in terms of basic physics. This list can be extended unendingly.

Physics is the foundation of many important disciplines and contributes directly to others. Chemistry, for example—since it deals with the interactions of atoms and molecules—is rooted in atomic and molecular physics. Most branches of engineering are applied physics. Others, such as chemical engineering, contain central elements of applied physics. Parts of geology rely heavily on physics, such as radioactive dating of rocks, earthquake analysis, and heat transfer in the earth. In architecture, physics is at the heart of structural stability, and is involved in the acoustics, heating, lighting, and cooling of buildings. Some disciplines, such as biophysics and geophysics, are hybrids of physics.

Physics finds many applications in the biological sciences from the microscopic level, where it helps describe membrane properties (see Figure 1.3), to the macroscopic level, where it can explain the heat, work, and power associated with the human body. Physics has numerous applications in medical diagnostics, such as x rays, nuclear magnetic resonance imaging, and ultrasonic blood flow measurements. Medical therapy sometimes directly involves physics, such as cancer radiotherapy, which depends on the physics of ionizing radiation. Applications of physics also can be found in how musical instruments make sound, how the eye detects color, and why lasers are well suited for transmitting information.

(a)

(b)

Figure 1.2 These two applications of physics have more in common than meets the eye—both utilize electromagnetic waves. (a) The microwave oven uses electromagnetic waves to heat food. (b) This laser-produced hologram is an entertaining application of electromagnetic waves.

Human & Medical Application

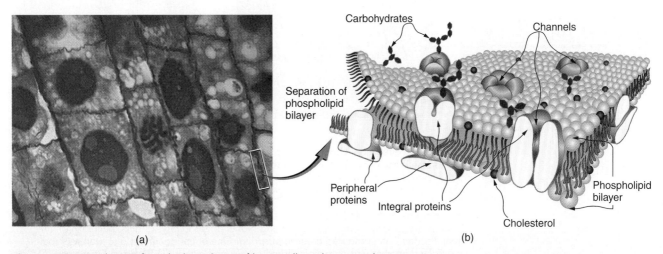

(a)   (b)

Figure 1.3 Membranes form the boundaries of living cells and are complex in structure and behavior. The disciplines of biology, chemistry, and physics are needed to understand membranes. Many of the most fundamental properties of life, such as the firing of nerve cells, are related to membranes. (a) This photograph shows cells with membranes as boundaries. (b) An artist's rendition of the membrane's structure.

It is not necessary to study all applications of physics formally. What is most useful is a knowledge of the basic laws of physics and a skill in the analytical methods for applying them. Furthermore, since physics has retained the most basic aspects of science, it is used by all of the sciences, and the study of physics makes other sciences easier to understand.

Models, Theories, and Laws; The Role of Experimentation

Just what are laws, and how are they discovered? The **laws of nature** are concise descriptions of the universe around us; they are human statements of the underlying laws or rules that all natural processes follow. Such laws are intrinsic to the universe and are not subject to human whim, only to human discovery and understanding, however imperfect. Their discovery is a very human endeavor, with all the elements of mystery, imagination, struggle, triumph, and disappointment inherent in any creative effort. (See Figure 1.4.) The cornerstone of discovering natural laws is observation; science must describe the universe as it is, not as we may imagine it to be.

We all are curious to some extent. We look about us, make generalizations, and try to understand what we see—for example, what type of clouds signal an oncoming storm. As we become serious about exploring nature, we become more organized and formal in collecting and analyzing data. We attempt greater precision, perform controlled experiments, if we can, and write down ideas about how the data may be organized and unified. Models, theories, and laws are formulated to generalize the results of these experiments.

A **model** is a mental image or analogy to objects or phenomena that we can experience directly. An example is the planetary model of the atom in which electrons are pictured as orbiting the nucleus, analogous to the way planets orbit the sun. (See Figure 1.5.) We cannot observe electron orbits directly, but the mental image helps explain the observations we can make, such as the emission of light from hot gases (atomic spectra).

A **theory** is usually a larger-scale and more broadly applicable generalization than a model and often seeks to describe nature with mathematical precision. Some theories

(a)

(b)

(c)

Figure 1.4   Newton, Curie, and Feynman are three famous physicists who illustrate how human the pursuit of physics is. (a) **Isaac Newton** (1642–1727) was very reluctant to publish his revolutionary work and had to be convinced to do so. In his later years he stepped down from his academic post and became exchequer of the Royal Mint, a post he took seriously, inventing reeding on coins to prevent trimming. (b) **Marie Curie** (1867–1934) sacrificed monetary assets to help finance her early research and damaged her physical well-being with radiation exposure. In addition to winning Nobel prizes in physics and chemistry (an unmatched feat), she produced creative offspring. One of her daughters also won a Nobel prize, and another was an accomplished pianist and author. (c) **Richard Feynman** (1918–1988) questioned authority and convention—for example, during the investigation of the Challenger disaster and in fighting war-time censorship of letters to his dying wife. His impressive contributions to theoretical physics were all the more valuable for his brilliant ability to lecture and express concepts with clarity and wit.

include models to help visualize phenomena, whereas others do not. Newton's theory of gravity, for example, does not require a model or mental image, because we can observe the objects directly with our own senses. The kinetic theory of gases, on the other hand, is a model in which a gas is viewed as being composed of atoms and molecules. Atoms and molecules are too small to be observed directly with our senses—thus, we picture them mentally to understand what our instruments tell us about the behavior of gases.

The designation **law** is reserved for a concise and very general statement, such as the law that energy is conserved in any process, or Newton's second law of motion, which relates force, mass, and acceleration via the simple equation $F = ma$. Einstein's theory of relativity, although universally applicable, is not concise enough to be called a law. Less broadly applicable statements are usually called **principles** (such as Pascal's principle, which is applicable only in fluids), but the distinction between laws and principles often is not carefully made.

The models, theories, and laws we devise sometimes *imply the existence of objects or phenomena as yet unobserved*. This is a remarkable triumph and a tribute to the power of science. The law of conservation of energy implied the existence of particles called neutrinos years before they were actually observed. Einstein's general theory of relativity predicted that light would bend noticeably when passing a massive body, years before this was observed for starlight passing the sun during a solar eclipse. In each case, the law or theory was designed to explain previously observed phenomena but had implications beyond what was known.

It is the *underlying order* in the universe that enables scientists to make such spectacular predictions. However, if *experiment* does not verify our predictions, then the theory or law is wrong, no matter how dear it is. Laws can never be known with absolute certainty, since it is impossible to perform every imaginable experiment. If a good-quality, verifiable experiment contradicts a well-established law, then the law must be modified. Since perhaps billions of observations may have been found to be consistent with a law, an experiment showing a violation of that law will be subject to greater scrutiny than normal, but observation is the absolute determinant of truth. A law may be completely overthrown, as were certain parts of classical physics; or it may be modified, as the law of conservation of energy was modified when it was discovered that energy can be converted into matter and matter into energy. Einstein's famous equation $E = mc^2$ precisely states the connection between mass and energy. Thus the law of conservation of energy was modified to be the law of conservation of mass plus energy.

The study of science in general and physics in particular is an adventure much like the exploration of uncharted ocean. Discoveries are made; models, theories, and laws formulated; and the beauty of the physical universe is made more sublime for the insights gained.

The Evolution of Natural Philosophy into Modern Physics

Physics was not always a separate and distinct discipline and is not now isolated from other sciences. The word *physics* comes from Greek, meaning nature. The study of nature came to be called "natural philosophy." From ancient times through the Renaissance, natural philosophy encompassed many fields, including astronomy, biology, chemistry, physics, mathematics, and medicine. Over the last few centuries, the growth of knowledge has resulted in ever-increasing specialization and branching of natural philosophy into separate fields, with physics retaining the most basic facets. (See Figure 1.6.) Physics as it developed from the Renaissance to the end of the 19th century is called **classical physics**. It was transformed into modern physics by the revolutionary discoveries made starting at the beginning of the 20th century.

Classical physics is not an exact description of the universe, but it is an excellent approximation under the following conditions: matter must be moving at speeds less than about 1% of the speed of light,* the objects dealt with must be large enough to be seen with a microscope, and only weak gravitational fields, such as on earth, can be involved. Since humans live under such circumstances, classical physics seems intuitively reasonable, while many aspects of modern physics seem bizarre. This is why models are so

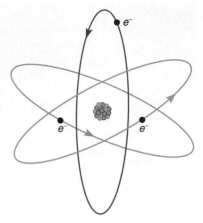

Figure 1.5 What is a model? This is the planetary model of the atom showing electrons orbiting the nucleus. It is a not a photograph, but a drawing that we use to form a mental image of the actual atom, which we cannot see directly with our eyes because it is too small. A good model is consistent with all the experimental observations we can make with instruments, such as the emission of light when an electron drops to a lower orbit. The planetary model is partially, though not completely, satisfactory in this respect—electron orbits are not as definite as pictured.

---

*The speed of light is 3.00×10^8 meters per second; thus, 1% is 3000 km/s or 1865 miles per second—still faster than most motions.

(a)

(b)

(c)

Figure 1.6 Over the centuries natural philosophy has evolved to more specialized disciplines, as illustrated by the contributions of some of the greatest minds in history. (a) The Greek philosopher **Aristotle** (384–322 B.C.) wrote on a broad range of topics including physics, animals, the soul, politics, and poetry. (b) **Galileo Galilei** (1564–1642) laid the foundation of modern experimentation and made contributions in mathematics, physics, and astronomy. (c) **Niels Bohr** (1885–1962) made fundamental contributions to the development of quantum mechanics, one part of modern physics.

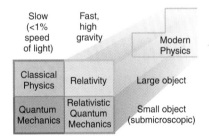

Figure 1.7 The realms of physics are the regions where the major theories apply.

useful in modern physics—it deals with phenomena we do not ordinarily experience. We can relate to models in human terms and visualize what happens when objects move at high speeds or imagine what objects too small to observe with our senses might be like. For example, although we have never seen an atom with our eyes, we can understand its properties because we can picture it in our minds.

Some of the most spectacular advances in science have been made in modern physics. Many of the laws of classical physics have been modified or overthrown, and revolutionary changes in technology, society, and our very thinking have resulted. Like science fiction, modern physics is filled with fascinating objects beyond our normal experiences, but it has the advantage over science fiction of being very real. Why, then, is a majority of this text devoted to topics of classical physics? There are two main reasons: a knowledge of classical physics is necessary to understand modern physics, and classical physics gives an extremely accurate description of the universe under a wide range of everyday circumstances.

Modern physics itself consists of the two revolutionary theories, relativity and quantum mechanics. These theories deal with the fast and the small, respectively. **Relativity** must be used whenever an object is traveling at greater than about 1% of the speed of light or experiences a strong gravitational field such as that near the sun. **Quantum mechanics** must be used for objects smaller than can be seen with a microscope. The combination of these two theories is **relativistic quantum mechanics**, and it must be used when small objects travel at high speeds or experience a strong gravitational field. Relativistic quantum mechanics is the best universally applicable theory we have. Because of its mathematical complexity, it is used only when necessary, and the other theories are used whenever they will produce sufficiently accurate results. We will find, however, that we can do a great deal of modern physics with the algebra and trigonometry used in this text. Figure 1.7 illustrates the realms of the various theories of physics.

The next chapter begins the actual study of the fascinating discoveries of physics, and of the monumental unifying principles that give them meaning. The remainder of this chapter helps lay a foundation for our trek through physics.

1.2 PHYSICAL QUANTITIES AND UNITS

The range of objects and phenomena studied in physics is immense. From the incredibly short lifetime of an exotic nucleus to the age of the earth, from the tiny size of subnuclear particles to the vast distance to the edge of the known universe, from the force exerted by a jumping flea to the force between the earth and the sun, there are enough factors of 10 to challenge the imagination of even the most experienced scientist. Giving numerical values for physical quantities and equations for physical principles allows us to understand nature much more deeply than does qualitative description alone. To comprehend these vast ranges, we must also have accepted units in which to express them. And we shall find that (even in the potentially mundane discussion of meters, kilograms, and seconds) a profound simplicity of nature appears—all physical quantities can be expressed as combinations of only four fundamental physical quantities.

The goal of physics is to describe the physical universe—both to gather knowledge about it and to discover the physical laws that rule it. The only way to accomplish these goals is through observation and measurement of physical quantities. The relationship between physical quantities and measurement is fundamental: we define a **physical quantity** either by *specifying how it is measured* or by *stating how it is calculated* from other measurements. For example, we define distance and time by specifying methods for measuring them, while we define average speed by stating that it is calculated as distance traveled divided by time of travel. When this is done explicitly, such as Einstein did when developing modern relativity or you may do in the laboratory component of your physics course, it is often referred to as an **operational definition**.

To make any sense, numerical measurements must be made relative to some *standard*, or **unit**. For example, a distance could be measured to be 80 meters, 0.080 kilometers, 260 feet, or 0.050 miles, but the numbers alone would be meaningless. (See Figure 1.8.)

Unfortunately, there are many systems of units to contend with. Just as jargon develops for every field, it seems that every field also develops its own system of units. This may be convenient for people in each field, but it can make life difficult. In particular, it hampers communication between fields because conversions must always be made from one unit system to another. In recent decades, the situation has improved with international agreements to use a standard system of units called **SI units**, named for the French *Système International*.

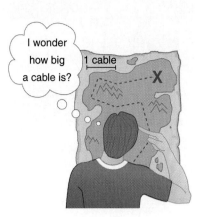

Figure 1.8 Distances given in unknown units are maddeningly useless.

SI Units: Fundamental and Derived Units

Table 1.1 gives the fundamental SI units that are used throughout this text. This text uses non-SI units in a few applications where they are in very common use, such as the measurement of blood pressure in millimeters of mercury (mm Hg). Whenever non-SI units are discussed, they will be tied to SI units through conversions.

It is an intriguing fact that some physical quantities are more fundamental than others and that the most fundamental physical quantities can be defined *only* in terms of the procedure used to measure them. The units in which they are measured are thus called **fundamental units**. In this text the fundamental physical quantities are taken to be length, mass, time, and charge. All other physical quantities, such as force and electric current, can be expressed as algebraic combinations of length, mass, time, and charge (for example, speed is length divided by time); their units are called **derived units**.*

Units of Time, Length, and Mass: The Second, Meter, and Kilogram

The SI unit for time, the **second** (abbreviated s), has a long history. For many years it was defined as 1/86,400 of a mean solar day. More recently, a new standard was adopted to gain greater accuracy and to define the second in terms of a nonvarying physical phe-

---

*If you look up fundamental units in other references, you may find that *seven* fundamental physical quantities are listed rather than four. You should not worry about this, since only four are really *basic and independent*. Three more are sometimes listed for convenience. There are also alternative groupings of fundamental physical quantities (for example, length, mass, time, and *temperature*), but each grouping contains only *four* physical quantities. Another such grouping is time, mass, the speed of light, and temperature. It is a characteristic of nature, and not an invention of humans, that only four physical quantities are needed to derive all others.

TABLE 1.1

FUNDAMENTAL SI UNITS			
Length	*Mass*	*Time*	*Charge*
meter (m)	kilogram (kg)	second (s)	coulomb (C)

Figure 1.9 An atomic clock such as this one uses the vibrations of cesium atoms to keep time to a precision of better than a microsecond per year. The fundamental unit of time, the second, is based on such clocks.

CONNECTIONS

Relativity

There is another reason, in addition to greater accuracy, for basing the meter on the speed of light. In Chapter 26 on relativity, we will see that the speed of light is a constant independent of the relative motion of the source of light and the observer. This implies that the speed of light is a very basic physical quantity. The constancy of *c* led to questions about other fundamental physical quantities. For example, it was found that different observers can measure different times for the same event.

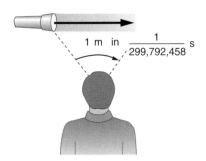

Figure 1.10 The meter is defined to be the distance light travels in 1/299,792,458 of a second in a vacuum. Distance traveled is speed multiplied by time.

nomenon (the day is getting longer since the earth's rotation is slowing). Cesium atoms can be made to vibrate in a very steady way, and these vibrations can be readily observed and counted. In 1967 the second was redefined as the time required for 9,192,631,770 of these vibrations. (See Figure 1.9.) Accuracy in the fundamental units is essential, since all measurements are ultimately expressed in terms of fundamental units and can be no more accurate than they are.*

The **meter** (abbreviated m) is the SI unit for length; it also has a long history, and the method of determining it has evolved to improve its accuracy. The meter was first defined in 1791 as 1/10,000,000 of the distance from the equator to the north pole. This was improved in 1889 by redefining the meter to be the distance between two engraved lines on a platinum-iridium bar now kept near Paris. By 1960 it had become possible to define the meter even more accurately in terms of the wavelength of light; it was again redefined, as 1,650,763.73 wavelengths of an orange light emitted by krypton atoms. In 1983, partly for greater accuracy, the meter was given its present definition as the distance light travels in a vacuum in 1/299,792,458 of a second. (See Figure 1.10.) This is a more radical change than the others, since it has the effect of defining the speed of light to be exactly 299,792,458 meters per second and forcing the meter to be consistent with this speed and the definition of the second. The value of the meter will change if the speed of light is someday measured with greater accuracy.

The meter is officially no longer a fundamental unit; but the speed of light (denoted *c*) now is. The four fundamental physical quantities in the SI system are officially time, mass, the speed of light (rather than length), and temperature. This is an example of how a different set of fundamental physical quantities can and is now being used. It does not affect the way we handle our fundamental units in this text—*we shall use length, mass, time, and charge as our fundamental units*.

The SI unit for mass is the **kilogram** (abbreviated kg); it is defined to be the mass of a platinum-iridium cylinder kept with the old meter standard at the International Bureau of Weights and Measures near Paris. Careful copies of the standard kilogram are also kept at the United States' National Institute of Standards and Technology and at other locations around the world. The determination of all other masses can be ultimately traced to a comparison with the standard mass. Present techniques for measuring masses are most accurate on the macroscopic scale (rather than on an atomic scale), where one kilogram can be measured with an accuracy of one part in 10^9. An atomic standard for mass may someday be adopted if one can be developed with a greater accuracy than the macroscopic standard.

Charge will be introduced in the second half of the text when electricity and magnetism are covered. The first half of this book is concerned with mechanics, fluids, heat, and waves. In these subjects all pertinent physical quantities can be expressed as algebraic combinations of length, mass, and time.

Metric Prefixes

SI units form one type of **metric system**, meaning that factors of 10 are used to span the ranges of values (rather than a factor of 12 for inches to feet, for example). Table 1.2 gives metric prefixes and symbols used to denote various factors of 10. Some of the units in Table 1.2 have not yet been discussed (they will be defined in later chapters), but examples you can relate to are given.

Metric systems have the advantage that conversions of units involve only powers of 10. There are 100 centimeters in a meter, 1000 meters in a kilometer, and so on. In nonmetric systems, such as the British system, the relationships are not as simple—there are 12 inches in a foot, 5280 feet in a mile, and so on. Another advantage to the metric system is that the same unit can be used over extremely large ranges of values simply by using an appropriate metric prefix. For example, distances in meters are suitable in construction, while distances in kilometers are appropriate for air travel, and nanometers are convenient in optical design. With a metric system there is no need for inventing new units for particular applications.

---

*Measurement, accuracy, and significant figures are discussed in depth in Section 1.3.

THINGS GREAT AND SMALL

This is the first of a series of vignettes called **Things Great and Small** that introduce you to aspects of modern physics throughout the text by illustrating explanations of the large based on the small. Over and over again, we have found that the characteristics of nature on the microscopic or small scale explain phenomena known on the macroscopic or GREAT scale. The existence and characteristics of tiny atoms, for example, explain large-scale properties of matter such as the transparency of diamond and the opacity of charcoal. Small-scale phenomena cannot be properly treated without quantum mechanics and other aspects of modern physics, but we can gain a greater appreciation of nature from a qualitative examination of the small scale before studying these topics. It would be terribly dull to pretend that atoms do not exist until we get to the last part of our studies, and it would be obtuse to pretend that none of modern physics is familiar to you before you study physics.

Figure 1.A An atomic clock is a large-scale object the operation of which is based on the microscopic oscillation of atoms. Other fundamental units may someday be based on small-scale phenomena.

Quest for Microscopic Standards for Basic Units

The fundamental units described in this chapter are those that produce the greatest accuracy and precision in measurement. There is a sense among physicists that, since there is an underlying microscopic substructure to matter, it would be most satisfying to base our standards of measurement on microscopic objects and phenomena. This has been accomplished for the standard of time, which is based on the oscillations of cesium atoms. (See Figure 1.A.)

The standard for length was once based on the wavelength of light (a small-scale length) emitted by a certain type of atom, but it has been supplanted by the more precise measurement of the speed of light. If it becomes possible to measure the mass of atoms to greater precision than the kilogram standard, it may become possible to base mass measurements on the small scale. There are also possibilities that electrical phenomena on the small scale may someday allow us to base the unit of charge on that of electrons and protons, but at present charge is related to large-scale currents and forces between wires.

TABLE 1.2

METRIC PREFIXES FOR POWERS OF 10 AND THEIR SYMBOLS

Prefix	Symbol	Value*	Example (some are approximate)			
exa	E	10^{18}	exameter	$= $ Em	$= 10^{18}$ m	distance light travels in a century
peta	P	10^{15}	petasecond	$= $ Ps	$= 10^{15}$ s	30 million years
tera	T	10^{12}	terawatt	$= $ TW	$= 10^{12}$ W	powerful laser output
giga	G	10^{9}	gigahertz	$= $ GHz	$= 10^{9}$ Hz	a microwave frequency
mega	M	10^{6}	megacurie	$= $ MCi	$= 10^{6}$ Ci	high radioactivity
kilo	k	10^{3}	kilometer	$= $ km	$= 10^{3}$ m	about 6/10 mile
hecto	h	10^{2}	hectoliter	$= $ hL	$= 10^{2}$ L	26 gallons
deka	da	10^{1}	dekagram	$= $ dag	$= 10^{1}$ g	teaspoon of butter
—	—	$10^{0}(= 1)$				
deci	d	10^{-1}	deciliter	$= $ dL	$= 10^{-1}$ L	less than half a soda
centi	c	10^{-2}	centimeter	$= $ cm	$= 10^{-2}$ m	fingertip thickness
milli	m	10^{-3}	millimeter	$= $ mm	$= 10^{-3}$ m	flea at its shoulders
micro	μ	10^{-6}	micrometer	$= $ μm	$= 10^{-6}$ m	detail in microscope
nano	n	10^{-9}	nanogram	$= $ ng	$= 10^{-9}$ g	small speck of dust
pico	p	10^{-12}	picofarad	$= $ pF	$= 10^{-12}$ F	small capacitor in radio
femto	f	10^{-15}	femtometer	$= $ fm	$= 10^{-15}$ m	size of a proton
atto	a	10^{-18}	attosecond	$= $ as	$= 10^{-18}$ s	time light crosses an atom

*See Section 1.5 for a discussion of powers of 10.

Figure 1.11 The tremendous range of observable phenomena in nature challenges the imagination.

Known Ranges of Length, Mass, and Time

The vastness of the universe and the breadth over which physics applies are illustrated by Table 1.3—in it interesting examples of known lengths, masses, and times are given. Examination of this table will give you some feeling for the range of possible topics and numerical values. (See Figure 1.11.)

Similar ranges exist for other physical quantities, such as force, and are similarly difficult to fully comprehend. In our normal everyday lives we seldom experience such tremendous ranges. For example, an hour is 3 factors of 10 longer than a heartbeat, a year 7 factors of 10 longer, and a lifetime 9 factors of 10 longer. Even this 9 factors of 10 in time is hard to grasp, let alone the 41 factors of 10 in the time table or the *83* factors of 10 in the mass table! Yet vast as these ranges may be, the same laws of physics govern on all scales.

Unit Conversion

We often find it useful or necessary to convert from one type of unit to another. We may wish to do something as simple as converting 80 meters to kilometers (km), which is done as follows:

$$80 \text{ m} \times \frac{1 \text{ km}}{1000 \text{ m}} = 0.080 \text{ km}$$

Note that the unwanted m unit cancels, leaving only the wanted km unit. The factor 1 km/1000 m is an example of a **conversion factor**. Some examples of conversion factors are

$$1 \text{ km} = 1000 \text{ m}$$

$$1 \text{ in.} = 2.54 \text{ cm}$$

A more complete list of conversion factors is given at the beginning of the book.

TABLE 1.3

APPROXIMATE VALUES OF LENGTH, MASS, AND TIME

Lengths in meters	Masses in kilograms (more precise values in parentheses)	Times in seconds (more precise values in parentheses)
10^{-17} Present experimental limit to smallest observable detail	10^{-30} Mass of an electron (9.11×10^{-31} kg)	10^{-23} Time for light to cross a proton
10^{-15} Diameter of proton	10^{-27} Mass of a hydrogen atom (1.67×10^{-27} kg)	10^{-22} Mean life of an extremely unstable nucleus
10^{-14} Diameter of uranium nucleus	10^{-15} Mass of a bacterium	10^{-15} Time for one oscillation of visible light
10^{-10} Diameter of hydrogen atom	10^{-5} Mass of a mosquito	10^{-13} Time for one vibration of an atom in a solid
10^{-8} Thickness of membranes in cells of living organisms	10^{-2} Mass of a hummingbird	10^{-8} Time for one oscillation of an FM radio wave
10^{-6} Wavelength of visible light	1 Mass of a liter of water (about a quart)	10^{-3} Duration of a nerve impulse
10^{-3} Size of grain of sand	10^{2} Mass of a person	1 Time for one heartbeat
1 Height of a 4-year-old child	10^{3} Mass of a car	10^{5} One day (8.64×10^4 s)
10^{2} Length of a football field	10^{8} Mass of a large ship	10^{7} One year (3.16×10^7 s)
10^{4} Greatest ocean depth	10^{12} Mass of a large iceberg	10^{9} About half the life expectancy of a human
10^{7} Diameter of the earth	10^{15} Mass of the nucleus of a comet	10^{11} Recorded history
10^{11} Distance from the earth to the sun	10^{23} Mass of the moon (7.36×10^{22} kg)	10^{17} Age of the earth
10^{16} Distance traveled by light in 1 year (a light year)	10^{25} Mass of the earth (5.98×10^{24} kg)	10^{18} Age of the universe
10^{21} Diameter of the Milky Way galaxy	10^{30} Mass of the sun (1.99×10^{30} kg)	Range of times $= 10^{18}/10^{-23} = \mathbf{10^{41}}$
10^{22} Distance from earth to nearest large galaxy (Andromeda)	10^{42} Mass of the Milky Way galaxy (current upper limit)	
10^{26} Distance from earth to edge of the known universe	10^{53} Mass of the known universe (current upper limit)	
Range of lengths $= 10^{26}/10^{-17} = \mathbf{10^{43}}$	Range of masses $= 10^{53}/10^{-30} = \mathbf{10^{83}}$	

EXAMPLE 1.1   HOW DO YOU MAKE UNIT CONVERSIONS?

Suppose you drive the 10.0 km from home to your university in 20.0 minutes. Calculate your average speed (a) in kilometers per hour (km/h) and (b) in meters per second (m/s).

Strategy First we calculate the average speed using the given units. Then we can get the average speed into the desired units by picking the correct conversion factor and multiplying by it. The correct conversion factor is the one that cancels the unwanted unit and leaves the desired unit in its place.

Solution for (a) Average speed is distance traveled divided by time of travel. (Take this as a given for now—average speed and other motion concepts will be formally defined in Chapter 2.) In equation form,

$$\text{average speed} = \frac{\text{distance}}{\text{time}}$$

Now substitute the given values for distance and time, which yields

$$\text{average speed} = \frac{10.0 \text{ km}}{20.0 \text{ min}} = 0.500 \frac{\text{km}}{\text{min}}$$

To convert km/min to km/h, we multiply by the conversion factor that will cancel minutes and leave hours. That conversion factor is 60 min/h. Thus,

$$\text{average speed} = 0.500 \frac{\text{km}}{\text{min}} \times \frac{60 \text{ min}}{1 \text{ h}} = 30.0 \frac{\text{km}}{\text{h}}$$

Discussion for (a) The answer seems reasonable—if you travel 10 km in a third of an hour (20 min), you would travel three times that far in an hour. How can you be certain that the conversion factor, which is a ratio (equivalent to 1), is not upside down? The answer is that *you must check to see that the units cancel as they should*. If you accidentally get the ratio upside down, then the units do not cancel; rather, they give you the wrong units as follows:

$$\frac{\text{km}}{\text{min}} \times \frac{1 \text{ h}}{60 \text{ min}} = \frac{1}{60} \frac{\text{km} \cdot \text{h}}{\text{min}^2}$$

which are obviously not the desired km/h.

Solution for (b) There are several ways to get the average speed into meters per second. We choose to start with the answer to the previous part and convert km/h to m/s. Two conversion factors are needed—one to convert hours to seconds, and another to convert kilometers to meters. Multiplying by these gives

$$\text{average speed} = 30.0 \frac{\text{km}}{\text{h}} \times \frac{1 \text{ h}}{3600 \text{ s}} \times \frac{1000 \text{ m}}{1 \text{ km}}$$
$$= 8.33 \frac{\text{m}}{\text{s}}$$

Discussion for (b) If we had started with 0.500 km/min, we would have needed different conversion factors, but the answer would have been the same 8.33 m/s (left as an end-of-chapter problem).

Unit conversions have real-world applications along with those in physics and other disciplines. (See Figure 1.12.)

The remaining sections of this chapter are devoted to an overview of the mathematics needed for success in using this text—both in understanding the physics presented and in working problem assignments. If you are concerned about your mathematical skills, or simply curious about what mathematics is required in this text, you will find the following sections useful and perhaps even reassuring. For example, you may have noted that the answers in the worked example just above were given to three digits. Why is this? When do you need to be concerned about the number of digits in something you calculate? Why not write down all the digits your calculator produces? Take a look at the next section to find out.

1.3   MEASUREMENT, ACCURACY, AND UNCERTAINTY; SIGNIFICANT FIGURES

Science is based on observation and experiment—that is, on *measurements*. Our lists of facts, and the physical laws that govern them, are all verified by measurement. **Accuracy** is how close a measurement is to the "true" value—and no measurement is exact. For example, you may measure the time for a sprinter in the 100 meter dash to be 9.84 s, but the measurement may only be accurate to plus or minus 0.05 s, meaning the actual time

Figure 1.12 Not everyone does unit conversions correctly. To see what is wrong, convert 30 miles per hour to kilometers per hour—the answer is *not* 40 km/h. These speed limit signs once appeared in the state of Florida.

can be anything from 9.79 to 9.89 s. The **uncertainty** in a measurement is an estimate of the amount it can be off from the "true" value—the greater the accuracy, the smaller the uncertainty. Here the uncertainty is 0.05 s. To represent these ideas in symbols, we call the time t, and its uncertainty δt. (δ is the lower case Greek letter delta. δt is pronounced "delta t.") The measurement with its uncertainty is written as

$$t \pm \delta t = (9.84 \pm 0.05) \text{ s}$$

where the symbol $\pm$ means plus or minus. More generally, if we represent a measurement by a symbol A, then its uncertainty is given the symbol δA ("delta A"), and they are written together as $A \pm \delta A$. The uncertainty in a measurement is as important as the measurement itself. For example, if a medical thermometer had an uncertainty of 3.0°C, it would be useless in determining whether a person has a fever. In scientific work models, theories, and laws need only describe nature within experimental uncertainties to be considered valid.

How do we estimate the uncertainty in a measurement? We often assume a certain accuracy for the measurements we make, such as when reading the speedometer in our car, or taking a person's temperature. To be more precise we must consider the factors that cause uncertainty in a measurement (this is done in all scientific work and is a favorite activity in the laboratory component of physics courses). Those factors contributing to uncertainty in a measurement include

1. Limitations of the measuring device
2. The skill of the person making the measurement
3. Irregularities in the object being measured
4. Any other factors that affect the outcome (highly dependent on the situation)

In our example of measuring the time of a sprinter, such factors contributing to the uncertainty could be the following: the stopwatch has 0.01 s as its smallest division; the person using the stopwatch has trouble starting and stopping it (reaction time); and it is unclear which part of the runner crossed the finish line first. Other factors might be whether temperature affects the stopwatch and whether the stopwatch is calibrated. At any rate, the uncertainty in a measurement must be based on a careful consideration of all the factors that might contribute and their possible effects. This is emphasized in the laboratory component of most physics courses, and it is one of the most important aspects of scientific research, since the impact of research depends a great deal on its accuracy.

Percent uncertainty   One of the more useful ways of expressing an uncertainty is as a percent of the measured value. Considering our runner, for example, we ask what percent of 9.84 s is the uncertainty of 0.05 s? It is

$$\frac{0.05 \text{ s}}{9.84 \text{ s}} \times 100 = 0.5\%$$

The **percent uncertainty** (*% unc*) is defined to be

$$\% \ unc = \frac{\delta A}{A} \times 100 \qquad \textbf{(1.1)}$$

We quote the measurement and its percent uncertainty as $A \pm \% \ unc$. For example, the runner's time is 9.84 s $\pm$ 0.5%. Or we may simply characterize our measurement as being good to 0.5%.

Uncertainties in calculations   There is an uncertainty in anything calculated from measured quantities. For example, the area of a floor calculated from measurements of its length and width has an uncertainty because the length and width have uncertainties. How big is the uncertainty in something you calculate by multiplication or division? If the measurements going into the calculation have small uncertainties (a few percent or less), then the **method of adding percents** can be used for multiplication or division. This method says that *the percent uncertainty in a quantity calculated by multiplication or division is the sum of the percent uncertainties in the items used to make the calcula-*

tion. For example, if a floor has a length of 4.00 m and a width of 3.00 m, with uncertainties of 2% and 1% respectively, then the area of the floor is 12.0 m² and has an uncertainty of 3%. (Expressed as an area this is 0.36 m², which we round to 0.4 m² since the percent uncertainty is only a one-digit number.)

Significant Figures

An approximate method to keep track of the accuracy of numbers is to write only those figures that are significant. The last digit in a number is considered to have some significance, but it may not be exact. Where our time measurement gave (9.84 ± 0.05) s, for example, the last digit in 9.84 was significant, but not exact. It is incorrect to write this time as 9.8 s, because it implies less accuracy than is present. It is also incorrect to write it as 9.840 s, because the 0 has no significance if the digit preceding it has an uncertainty. Using the **method of significant figures**, the rule is that *the last digit written down is the first digit with some uncertainty*. The number of digits is then the number of significant figures—our example has three significant figures.

"I asked you a question, buddy ... What's the square root of 5,248?"

Zeros Special consideration is given to zeros when counting significant figures. The zeros in 0.053 are not significant, because they are only placekeepers that locate the decimal point. There are two significant figures in 0.053. The zeros in 10.053 are significant—this number has five significant figures. The zeros in 1300 may or may not be significant depending on the style of writing numbers. They could mean the number is known to the last digit, or they could be placekeepers. So 1300 could have two, three, or four significant figures. (To avoid this ambiguity, write 1300 in scientific notation. See Section 1.5.) *Zeros are significant except when they serve only as placekeepers.*

Significant figures in calculations A calculated number has a limited number of significant figures. Even if a pocket calculator gives an answer to nine digits, not all of them are necessarily significant since the numbers that went into the calculation had limited accuracy. So the calculated number has an uncertainty. There are two approximate* rules for finding the number of significant figures in a calculated number:

1. *Multiplication and division*. Here the rule is that *the result has the same number of significant figures as the quantity having the least significant figures entering into the calculation*. For example, the area of a circle can be calculated from its radius using $A = \pi r^2$. Let us see how many significant figures the area has if the radius has only two—say, $r = 1.2$ m. Then,

$$A = \pi r^2 = (3.1415927\ldots)(1.2 \text{ m})^2 = 4.5238934 \text{ m}^2$$

is what you would get using a calculator that has an eight-digit output. But since the radius has only two significant figures, it limits the calculated quantity to two significant figures or

$$A = 4.5 \text{ m}^2$$

even though π is good to at least eight digits.

2. *Addition and subtraction*. Suppose you buy 7.56 kg of potatoes in a grocery store as measured with a scale accurate to 0.01 kg. Then you drop off 6.052 kg of potatoes at your laboratory as measured by a scale accurate to 0.001 kg. Finally, you go home and add 13.7 kg of potatoes you already have. This final mass is measured by a bathroom scale accurate to only 0.1 kg. How many kilograms of potatoes do you now have, and how many significant figures are appropriate in the answer? The mass is found by simple addition and subtraction:

$$
\begin{array}{r}
7.56 \text{ kg} \\
-6.052 \text{ kg} \\
\underline{13.7 \quad \text{ kg}} \\
15.208 \text{ kg} = 15.2 \text{ kg}
\end{array}
$$

---

*The entire approach of significant figures is an approximate technique for tracking the accuracy of calculated numbers. The method of adding percents is slightly better. More accurate error analysis, including statistical analysis, is used in research and is beyond the scope of this text.

Here we choose the least accurate number, which is 13.7, to determine the last digit we can write in the answer. The result cannot have any accuracy beyond the tenths place. Stated more generally, the *last significant figure in addition and subtraction is in the same column as the last significant figure in the least accurate number.*

Significant figures in this text   Since the rules for significant figures are approximate, most numbers in this text are assumed to be good to three significant figures. Furthermore, consistent numbers of significant figures are used in all worked examples and in answers to odd-numbered problems. You will note that an answer given to three digits is based on input good to at least three digits, for example. If the input has fewer significant figures, the answer will also have fewer significant figures. Care is also taken that the number of significant figures is reasonable for the situation posed. In some topics, particularly in optics, more accurate numbers are needed and more than three significant figures will then be used. Finally, if a number is exact, such as the 2 in the formula for the circumference of a circle, $c = 2\pi r$, it does not affect the number of significant figures in a calculation.

1.4*   EXPONENTS

When we write 2^4, we mean 2 multiplied by itself 4 times—that is,

$$2^4 = 2 \times 2 \times 2 \times 2 = 16$$

The superscript 4 is the *exponent*, 2 is the *base*, and in words 2^4 is called 2 to the 4th power, or 4 factors of 2. More generally, a^n is a times a taken n times, n is the **exponent**, a is the **base**, and in words a^n is called a to the nth power, or n factors of a. *Two special cases* are

$$a^1 = a \tag{1.2}$$

and, by definition,

$$a^0 = 1 \tag{1.3}$$

The *rule for multiplying two numbers with the same base a* is

$$(a^n)(a^m) = a^{n+m} \tag{1.4}$$

For example,

$$(10^2)(10^3) = (10 \times 10)(10 \times 10 \times 10) = 10^5$$

Similarly, the *rule for division of one number by another with the same base* is

$$\frac{a^n}{a^m} = a^{n-m} \tag{1.5}$$

For example,

$$\frac{2^6}{2^2} = \frac{2 \times 2 \times 2 \times 2 \times 2 \times 2}{2 \times 2} = 2^4$$

There is no simplifying rule for multiplication or division of numbers with different bases, such as $(3^2)(5^7)$ or $(6^3)/(7^2)$. (Or more generally, you cannot simplify numbers such as $(a^n)(b^m)$ or $(a^n)/(b^m)$.)

The division rule, Equation 1.5, implies that

$$\frac{1}{a^m} = a^{-m} \tag{1.6}$$

since $1 = a^0$. Equation 1.6 is very often useful for powers of 10. For example, $1/10^5 = 10^{-5}$. The equation is also very useful in algebraic manipulations.

If numbers having the *same exponent* are multiplied or divided, it is possible to simplify expressions. The *rule for multiplying two numbers with the same exponent* is

$$(a^n)(b^n) = (ab)^n \tag{1.7}$$

For example, $(4^3)(2^3) = (8)^3$. (Prove this for yourself by writing it out.) Another example is $50^4 = (5^4)(10^4)$. The *rule for division of one number by another with the same exponent* is

$$\frac{a^n}{b^n} = \left(\frac{a}{b}\right)^n \tag{1.8}$$

For example, $(3^4)/(4^4) = (3/4)^4 = (0.75)^4$.

A **root** is a fractional exponent. For example, $4^{1/2}$ is the square root of 4, or $4^{1/2} = \pm 2$. Note that if you multiply $4^{1/2}$ times itself you get 4. This is really what is meant by a root—if you square $4^{1/2}$, you get 4. More generally, $a^{1/n}$ is the nth root of a, and if you multiply $a^{1/n}$ by itself n times you get a. That is,

$$(a^{1/n})^n = a \qquad \qquad \textbf{(1.9)}$$

since

$$(a^{1/n})^n = a^{n/n} = a^1 = a$$

For example, $(4^{1/2})^2 = (\pm 2)^2 = 4$.

REVIEW EXERCISES

1   (a) What is $10^0 = ?$ (b) What is $10^1 = ?$ (c) What is $x^0 = ?$
2   (a) What is $3^0 = ?$ (b) What is $3^1 = ?$ (c) What is $x^1 = ?$
3   (a) What is $10^{-1} = ?$ (b) What is $(10^2)^{1/2} = ?$ (c) What is $(3^2)^0 = ?$

Simplify the following:

4	$(x^3)(x^8)$	**12**	$(x^8)/(x^{10})$	**20**	$(10^{-19})/(10^6)$
5	$(10^5)(10^{-12})$	**13**	$(7^3)/(7^5)$	**21**	$1/x^{18}$
6	$(x^{-3})(y^{-3})$	**14**	$(7^3)/(12^3)$	**22**	$1/10^{-34}$
7	$(x^{1/5})(y^{1/5})$	**15**	$(8^{1/3})/(27^{1/3})$	**23**	$(x^5)(y^5)/x^3$
8	$(5^6)(10^6)$	**16**	$(4^7)(15^7)$	**24**	$(x^6y^3)/(xy)^2$
9	$(x^{-7})/(x^{-15})$	**17**	$(3^{1/2})(5^{1/2})$	**25**	$(xy)^2/(xy)^{-9}$
10	$(x^3)/(y^3)$	**18**	$(10^4)(10^{-9})$	**26**	$x^4/(xy)^{-3}$
11	$(x^5)/(x^{-3})$	**19**	$(10^4)/(10^{-9})$	**27**	$(x^8)^{1/8}(y)/(x^2)(y^2)$

Answers to Odd-Numbered Review Exercises

1	(a) 1 (b) 10 (c) 1	**11**	x^8	**21**	x^{-18}
3	(a) 1/10 (b) 10 (c) 1	**13**	7^{-2}	**23**	x^2y^5
5	10^{-7}	**15**	$(8/27)^{1/3} = 2/3$	**25**	$(xy)^{11}$
7	$(xy)^{1/5}$	**17**	$(15)^{1/2}$	**27**	$(xy)^{-1}$
9	x^8	**19**	10^{13}		

1.5* SCIENTIFIC NOTATION AND POWERS OF 10

Scientific notation is a convenient way of writing large or small numbers and also helps with significant figures. For example, the distance between the earth and the sun averages about 150,000,000 km. As written, this number is awkward and may have anywhere between two and nine significant figures. In scientific notation we would write this as 1.50×10^8 km, if it was known to have three significant figures. The smallest detail ordinarily seen in a microscope is about 0.000001 m, and this number is more conveniently written as 1×10^{-6} m.

In standard **scientific notation** a quantity is rewritten as a number between 1 and 10 multiplied by the correct power of 10. For example, 150,000,000 is written as 1.50 (a number between 1 and 10) times 10^8, a power of 10. By **power of 10** we mean the base 10 with an exponent.[†] In words, we say that 10^8 is eight powers of 10. We also commonly say that 10^8 is 10 to the eighth.

To put a quantity into scientific notation, we count the number of places we must move its decimal to get a number between 1 and 10. For example, the decimal following 150,000,000 must be moved eight places to the left to give 1.50. That is, 1.50,000,000. Hence 150,000,000 = 1.50×10^8. Similarly, the decimal in 0.000001

[†]Most numbers we work with are in the base 10 (decimal) system. See Section 1.4 for a discussion of exponents.

must be moved six places to the right to give 1. That is, 0.000001. Hence $0.000001 = 1 \times 10^{-6}$. On a handheld calculator you enter the power-of-10 exponent with the "EXP" or "EE" key. You can also put any number into scientific notation with a "SCI" or similar key.

Since numbers in scientific notation all have the same base, the rules for adding, subtracting, multiplying, and dividing them are simple. For example, $(1.5 \times 10^8) + (2.3 \times 10^8) = 3.8 \times 10^8$. Note that for addition and subtraction the numbers must have the same power of 10. The *rule for addition and subtraction of numbers in scientific notation* is

$$(a \times 10^n) \pm (b \times 10^n) = (a \pm b) \times 10^n \tag{1.10}$$

This is two rules in one—the plus is used for addition and the minus for subtraction. If the numbers do not have the same power of 10, one must be converted to the other's power. For example, if we wish to subtract 6.8×10^7 from 1.5×10^8, we first convert $1.5 \times 10^8 = 15 \times 10^7$ (since the first factor is multiplied by 10, the second is divided by 10, making it 10^7. Now $(1.5 \times 10^8) - (6.8 \times 10^7) = (15 \times 10^7) - (6.8 \times 10^7) = (15 - 6.8) \times 10^7 = 8.2 \times 10^7$.

The rules for multiplication and division do not require the numbers to have the same power of 10. For example, $(3.0 \times 10^2) \cdot (2.5 \times 10^7) = 7.5 \times 10^9$. The *rule for multiplication of numbers in scientific notation* is

$$(a \times 10^n) \cdot (b \times 10^m) = (ab) \times 10^{n+m} \tag{1.11}$$

The *rule for division of numbers in scientific notation* is

$$\frac{a \times 10^n}{b \times 10^m} = \frac{a}{b} \times 10^{n-m} \tag{1.12}$$

So that, for example, $(7.5 \times 10^9)/(2.5 \times 10^7) = 3.0 \times 10^2$.

If we square a number, we are raising it to the second power, and the rule for this is also simple. For example, $(3.0 \times 10^8)^2 = 9.0 \times 10^{16}$. The *rule for raising a number in scientific notation to a power* is

$$(a \times 10^n)^m = a^m \times 10^{nm} \tag{1.13}$$

which means that each part of the number is raised to the mth power. Equation 1.13 is also valid for roots; in this case, m is simply a fraction. For example, $(6.25 \times 10^4)^{1/2} = (6.25)^{1/2} \times (10^4)^{1/2} = 2.50 \times 10^2$.

REVIEW EXERCISES

Express the following in standard scientific notation:

28   47,032.102
29   0.00000105

30   215.1
31   6,835,000.0

32   0.0000000000000016
33   1,470,000,000.0

Express the following in ordinary decimal notation:

34 1.6×10^{-19}
35 8.85×10^{-12}

36 3.0×10^8
37 1.523×10^1

38 0.8032×10^3
39 480.5×10^{-9}

Simplify the following to a single number in standard scientific notation:

40 $(1.5 \times 10^4) + (0.84 \times 10^4)$
41 $(1.5 \times 10^4) - (0.84 \times 10^4)$
42 $(-3 \times 10^5)(7 \times 10^3)$
43 $(1.05 \times 10^{-34})(3 \times 10^8)$
44 $(6.02 \times 10^{23})(4.7 \times 10^{-3})$
45 $(-1.01 \times 10^5)(-6.0 \times 10^3)$
46 $(-1.01 \times 10^5)/(-6.0 \times 10^3)$
47 $(6.02 \times 10^{23})/(4.7 \times 10^{-3})$

48 $(1.05 \times 10^{-34})/(3 \times 10^8)$
49 $(-3 \times 10^5)/(7 \times 10^3)$
50 $(-5 \times 10^4)^{-3}$
51 $(6.63 \times 10^{-34})^{-2}$
52 $(1.6 \times 10^{-19})^2$
53 $(3.0 \times 10^8)^2$
54 $\dfrac{(3.16 \times 10^7) \cdot (3.0 \times 10^8)}{(5.36 \times 10^4) \cdot (9.7 \times 10^{-2})}$

55 $\dfrac{(6.63 \times 10^{-34}) \cdot (1.602 \times 10^{-19})^2}{(8.85 \times 10^{-12})^3 \cdot (9.11 \times 10^{-31})}$

56 $(6.09 \times 10^8)^3/(1.01 \times 10^5)^{1/2}$
57 $(-2.1 \times 10^{-3})/(7.55 \times 10^8)^2$

Answers to Odd-Numbered Review Exercises

29	1.05×10^{-6}	**39**	0.0000004805	**49**	-4.29×10^{1}
31	6.8350000×10^{6}	**41**	6.6×10^{3}	**51**	2.28×10^{66}
33	$1.4700000000 \times 10^{9}$	**43**	3.15×10^{-26}	**53**	9.0×10^{16}
35	0.00000000000885	**45**	6.06×10^{8}	**55**	2.69×10^{-8}
37	15.23	**47**	1.28×10^{26}	**57**	-3.68×10^{-21}

1.6* LOGARITHMS

Logarithms are needed to solve certain problems that contain an exponent. For example, suppose you know that $10^{y} = 83.0$ and you want to find what y is. These days it is easy to find y using an inexpensive calculator. Simply enter 83.0 and press the log key, and the calculator will read $1.91907 \ldots$ If you then enter this number and press the key labeled 10^{y} (often labeled *inverse log* or *antilog*), the calculator will give 83.0. In equation form, this is expressed as $y = \log_{10} 83.0 = 1.91907 \ldots$, and $10^{1.91907 \cdots} = 83.0$, or antilog$(1.91907 \ldots) = 83.0$. Logarithms are useful in certain parts of physics, such as sound intensity, circuits, and radioactive decay, and so we briefly examine them here.

The general definition of a logarithm is

$$\text{if } x = a^{y}, \quad \text{then } y = \log_{a} x \qquad \text{(1.14)}$$

This defines y to be the **logarithm of x to the base a**, and x is the **antilog of y**. The base a can be any number; 10 is one of the most common, since the decimal system is base 10 and we express so many numbers in scientific notation with powers of 10. Thus a common logarithm is defined to be base 10, having $a = 10$, so that

$$\text{if } x = 10^{y}, \quad \text{then } y = \log_{10} x \qquad \text{(1.15)}$$

which defines y to be the **common logarithm of x** (to the base 10); x is the **antilog of y**. It is customary to write log for $\log_{10}$.

Another frequently used base is the **natural number** $e = 2.71828 \ldots,$[†] so that a natural logarithm is defined by

$$\text{if } x = e^{y}, \quad \text{then } y = \ln x \qquad \text{(1.16)}$$

which defines y to be the **natural logarithm of x** (to the base e), and x is the **antiln of y**. It is customary to write ln for $\log_{e}$, and most scientific calculators have both a log and a ln key. Now, if we have $e^{y} = 83$ and want to know what y is, we can find it on a calculator by entering 83 and hitting the ln key, which will produce the number $4.41884 \ldots$; that is, ln $83 = 4.41884 \ldots$ If we enter that number and press the e^{y} key (often labeled *inverse ln* or *antiln*), the calculator will give 83 back. Thus $e^{4.41884 \cdots} = 83$, or antiln$(4.41884 \ldots) = 83$.

Although calculators have made handling logarithms easier (tables must be used if no calculator is available), the following rules can still be useful in problem solving, following textbook derivations, and simplifying equations. The abbreviation log is used for both $\log_{10}$ and ln, since the rules are identical for each (and, in fact, for logarithms to any base).

The first rule is the *rule for taking the logarithm of a product*:

$$\log(AB) = \log A + \log B \qquad \text{(1.17)}$$

This rule is particularly useful for common logarithms of numbers in scientific notation, since

$$\log_{10}(A \times 10^{n}) = \log_{10} A + \log_{10} 10^{n}$$

and since $\log_{10} 10^{n} = n$,

$$\log_{10}(A \times 10^{n}) = \log_{10} A + n$$

[†]The natural number e has special properties that make it useful in certain topics, such as electrical circuits and radioactive decay, where we will explore its use further.

This is very useful if you have no calculator. You only need a log table for numbers 1 through 10, since A is between 1 and 10 in scientific notation. The *rule for taking the logarithm of a ratio* can also be useful. This is

$$\log(A/B) = \log A - \log B \tag{1.18}$$

Also useful is the *rule for taking the logarithm of a number expressed as a power*:

$$\log A^n = n \log A \tag{1.19}$$

As noted, all of the numbered rules are valid for $\log_{10}$ and ln, as well as logarithms to any other base.

REVIEW EXERCISES

Find the values of the following:

58	$\log 10^{-8}$	**66**	antilog 8.477	**74**	$\ln(5.4 \times 10^{-3})$
59	$\log 10^{22}$	**67**	antilog -33.18	**75**	$\ln(6.02 \times 10^{23})$
60	$\log(3.7 \times 10^{10})$	**68**	antilog -18.8	**76**	antiln 6
61	$\log(1.6 \times 10^{-19})$	**69**	antilog 23.78	**77**	antiln -2
62	$\log(5.4 \times 10^{-3})$	**70**	$\ln 10^{-8}$	**78**	antiln 8.477
63	$\log(6.02 \times 10^{23})$	**71**	$\ln 10^{22}$	**79**	antiln -33.18
64	antilog 6	**72**	$\ln(3.7 \times 10^{10})$	**80**	antiln -18.8
65	antilog -2	**73**	$\ln(1.6 \times 10^{-19})$	**81**	antiln 23.78

Simplify the following:

82	$\log(xy)$	**90**	$\log(x^3/y)$	**98**	$\ln(x^5 y^3)$
83	$\log(xy^2)$	**91**	$\log(1/xy)$	**99**	$\ln(x^{-2} y^7)$
84	$\log(x^3 y)$	**92**	$\log(x \times 10^4)$	**100**	$\ln(x/y)$
85	$\log(x^{-3} y^{-2})$	**93**	$\log(x \times 10^{-7})$	**101**	$\ln(y/x)$
86	$\log(x^5 y^3)$	**94**	$\ln(xy)$	**102**	$\ln(x^3/y)$
87	$\log(x^{-2} y^7)$	**95**	$\ln(xy^2)$	**103**	$\ln(1/xy)$
88	$\log(x/y)$	**96**	$\ln(x^3 y)$	**104**	$\ln 10$
89	$\log(y/x)$	**97**	$\ln(x^{-3} y^{-2})$	**105**	$\ln(x \times 10^n)$

Answers to Odd-Numbered Review Exercises

59	22	**75**	54.8	**91**	$-\log x - \log y$
61	-18.8	**77**	0.135	**93**	$\log x - 7$
63	23.8	**79**	3.89×10^{-15}	**95**	$\ln x + 2 \ln y$
65	0.010	**81**	2.13×10^{10}	**97**	$-3 \ln x - 2 \ln y$
67	6.61×10^{-34}	**83**	$\log x + 2 \log y$	**99**	$-2 \ln x + 7 \ln y$
69	6.03×10^{23}	**85**	$-3 \log x - 2 \log y$	**101**	$\ln y - \ln x$
71	50.7	**87**	$-2 \log x + 7 \log y$	**103**	$-\ln x - \ln y$
73	-43.3	**89**	$\log y - \log x$	**105**	$\ln x + 2.30n$

1.7*   THE RULES OF ALGEBRA

Algebra is a generalized form of arithmetic in which symbols are used for numbers and physical quantities. Algebra is used extensively in physics and many other disciplines. In physics, algebraic expressions are very often used to give the relationship between physical quantities, such as in $E = mc^2$, and also for solving problems in our attempts to understand and describe nature. The rules of algebra utilized in this text are straightforward and logical and will help you to follow the text as well as solve problems.

One of the fundamental rules in algebra is that whatever you do to one side of an equation you must do to the other side. For example, you can divide each side of an equation by 2 and it remains true; so if $a = b$, then $a/2 = b/2$ as well. This helps solve simple problems such as $2x = 6$, where we divide each side by 2 to get $x = 3$. More formally:

An equation remains true if the same operation is performed on both sides.   (1.20)

These operations include *addition*, *subtraction*, *multiplication*, *division*, *taking the power or root*, and *taking the logarithm or antilog*—as long as the same thing is done to both sides, the equation remains true. Let us take a look at a few of these operations.

Addition and Subtraction

Consider the equation $x - 5 = 7$, which has the obvious solution $x = 12$. To get this via algebraic manipulation, as might be necessary in a more difficult problem, just add 5 to each side of the equation, so that the equation

$$x - 5 = 7$$

becomes

$$x - 5 + 5 = 7 + 5 = 12$$

and

$$x = 12$$

Note that this is equivalent to moving -5 to the other side of the equation and changing its sign to $+5$. That is,

$$x - 5 = 7$$

so that

$$x = 7 + 5 = 12$$

The *rule for moving a quantity from one side of an equation to the other is that the quantity changes sign* in the process:

$$\text{if } a + b = c, \quad \text{then } a = c - b \tag{1.21}$$

In our example, b was -5, so that it changed to $+5$ when moved across the equal sign. This simple type of algebraic manipulation is very common in the text and is crucial to solving problems.

Multiplication and Division

Suppose we want to solve for x in the following equation:

$$\frac{x}{4} = \frac{5}{8}$$

We multiply each side by 8:

$$8 \cdot \frac{x}{4} = 8 \cdot \frac{5}{8}$$

On the left side we divide 8 by 4, and the 8s on the right side cancel. Thus,

$$2x = 5$$

Now divide both sides by 2:

$$x = \frac{5}{2} = 2.5$$

Cross-multiplication can be useful in solving algebraic problems. The *rule for cross-multiplication* is

$$\text{if } \frac{a}{b} = \frac{c}{d}, \quad \text{then } ad = bc \tag{1.22}$$

which is another very useful algebraic manipulation. Using this on the previous problem would produce $8x = 4 \cdot 5 = 20$. Then dividing both sides by 8 gives $x = 20/8 = 2.5$.

Powers and Roots

We may need to solve the following equation for x:

$$\sqrt{\frac{x}{2}} = 5$$

Simply square both sides (take each side to the power of 2), yielding

$$\frac{x}{2} = 25$$

Then multiply each side by 2, obtaining

$$x = 50$$

as the solution.

An even simpler problem is

$$x^2 = 25$$

We take the square root of each side, which gives

$$(x^2)^{1/2} = (25)^{1/2}$$

Thus,

$$x = \pm 5$$

are the solutions. Note that there are two solutions because squaring either $+5$ or -5 produces 25. In general, when the unknown (x here) is squared in an equation, there will be two solutions (they are sometimes identical).

The Quadratic Formula

There are many equations in physics in which one of the quantities is squared, such as $E = mc^2$. Many of these equations are in a form like $x^2 + 3x + 2 = 0$, which cannot be solved for x by simply taking the square root.

For a **quadratic equation**, which is an equation in the form

$$ax^2 + bx + c = 0$$

where x is the unknown and a, b, and c are numbers, there is a general solution. This solution is given by the **quadratic formula**:

$$x = \frac{-b \pm \sqrt{b^2 - 4ac}}{2a} \tag{1.23}$$

The formula gives two values for x, one for the plus sign and the other for the minus.

If, for example, we wanted to solve

$$x^2 + 3x + 2 = 0$$

we first determine the values of $a = 1$, $b = 3$, and $c = 2$. Then we enter these into the quadratic formula given by Equation 1.23, which yields

$$x = \frac{-3 \pm \sqrt{3^2 - 4 \cdot 1 \cdot 2}}{2 \cdot 1} = \frac{-3 \pm \sqrt{1}}{2}$$

Thus,

$$x = \frac{-3 \pm 1}{2} = \begin{cases} -1 \\ -2 \end{cases}$$

are the two solutions for x. Let us check that the solutions are correct. For $x = -1$, we see that $x^2 + 3x + 2 = 1 - 3 + 2 = 0$. For $x = -2$, we see that $x^2 + 3x + 2 = 4 - 6 + 2 = 0$.

Simultaneous Equations

All of our examples thus far have had a single unknown; usually we have called it x. A single **linear equation** (one that has no powers or roots other than 1) can be solved to give a unique value for a single unknown. But if there are, say, two unknowns in an equation, it cannot be solved to give unique values of each. For example, if we know that $x + 2y = 3$, there is not a unique pair of values for x and y that satisfy (solve) the equation—there are an infinite number of pairs of x and y that work. To determine unique values for x and y, we need a second independent equation (one that cannot be obtained simply by manipulating the first). The general rule is

> *n* **independent equations are needed to produce unique values for**
> *n* **unknowns.** **(1.24)**

In solving physics problems, you may need to determine more than one unknown at a time. The number of equations needed is the same as the number of unknowns, and you must then *solve simultaneous equations*. The question now is how to do this.

Let us take the equation $x + 2y = 3$, and say that we also know that $x - y = 9$. We manipulate these equations to eliminate one unknown (either x or y) and solve for the other. For easy reference we label the equations (a) and (b):

$$x + 2y = 3 \quad \text{(a)}$$
$$x - y = 9 \quad \text{(b)}$$

We can eliminate y by multiplying (b) by 2 and adding the result to (a):

$$
\begin{array}{ll}
x + 2y = 3 & \text{(a)} \\
\underline{2x - 2y = 18} & \text{(b}') \\
3x \quad\;\; = 21 &
\end{array}
$$

so that

This is a *single equation in a single unknown*. (That is the goal of the manipulations.) It can be solved in this case by dividing each side by 3, yielding

$$x = 7$$

Now any of the equations (a), (b), or (b$'$) can be used to find y. Choosing (a), and substituting 7 for x, we have another single equation in a single unknown:

$$7 + 2y = 3$$

By moving the 7 across, we get

$$2y = 3 - 7 = -4$$

Dividing by 2 gives

$$y = -2$$

The solution to our two simultaneous equations is $x = 7$ and $y = -2$. To check that these values satisfy equations (a) and (b), we substitute them for x and y, respectively. In equation (a), this yields $7 + 2(-2) = 3$. In equation (b), this yields $7 - (-2) = 9$. So both equations are satisfied, and the answers for x and y are correct.

The same technique works for solving any number of simultaneous equations. The more unknowns, the more time consuming the process may be, but it is no more difficult mathematically than manipulating two equations in two unknowns. Finally, note that we can add or subtract various equations because the sides are equal—thus, the same quantity is being added to or subtracted from each side.

REVIEW EXERCISES

Solve for the unknown in each equation. That is, perform algebraic manipulations to obtain an equation with the unknown on one side and all other terms on the other side. Identify the operation you perform in each step.

106 $v^2 = v_0^2 + 2ax$; find v_0

107 $v = v_0 + at$; find t

108 $\dfrac{1}{p} + \dfrac{1}{q} = \dfrac{1}{f}$; find q

109 $x = v_0 t + (1/2)at^2$; find t

110 $t = (2x/a)^{1/2}$; find x

111 $\dfrac{h'}{h} = -\dfrac{p}{q}$; find q

112 $\dfrac{P_1 - P_2}{\dfrac{8\eta L}{\pi r^4}} = \dfrac{V}{t}$; find r

113 $mg - F + T = ma$; find F

114 $KE = (1/2)I\omega^2$, find ω

115 $Fd \cos \theta = W$; find d

116 $c = f\lambda$; find λ

117 $F = qvB \sin \theta$; find θ

Solve the following simultaneous equations for x and y:

118 $x + y^2 = 7, x - y = 5$

119 $x + 5y = 47, 15x + y = -35$

120 $-x + y = -3, x - y^2 = 27$

Answers to Odd-Numbered Review Exercises

107 $t = (v - v_0)/a$

109 $t = \dfrac{-v_0 \pm \sqrt{v_0^2 + 2ax}}{a}$

111 $q = -p(h/h')$

113 $F = T + m(g - a)$

115 $d = W/(F \cos \theta)$

117 $\theta = \sin^{-1}(F/qvB)$

119 $x = -3, y = 10$

1.8* ANGLES, TRIANGLES, AND SIMPLE TRIGONOMETRY

In our efforts to describe nature in this text, we often employ geometry and simple trigonometry. Both come into play, for example, when we trace the path taken by light going through a camera lens. This last section in our mathematics overview presents the important fundamentals of geometry and trigonometry used in this text.

Angles

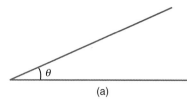

An **angle** represented by the symbol θ—such as that shown between two lines in Figure 1.13(a), can be thought of as a measure of how much the lines diverge—the greater the angle, the greater the divergence of the lines. Angles are usually measured in degrees (°) but also may be measured in radians (rad). There are 360° or 2π rad in one complete revolution. Two straight lines are defined to be **parallel** if they have the same direction. The angle between parallel lines is 0°. Two straight lines are defined to be **perpendicular** if the angle between them is 90°, as in Figure 1.13(b).

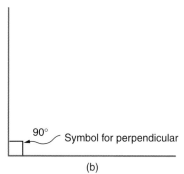

Angles are equal if:

1. They are directly across an intersection between two straight lines, called a *vertex*, as shown in Figure 1.14.
2. The sides are parallel, as shown in Figure 1.15.
3. The sides are mutually perpendicular, as shown in Figure 1.16.
4. When a straight line crosses two parallel lines, various pairs of angles are equal, as shown in Figure 1.17. (This results from statements 1 and 2, but it is useful in itself.)

Figure 1.13 (a) The angle θ between the lines is most often measured in degrees or radians. (b) Two lines are **perpendicular** if the angle between them is 90° (π radians).

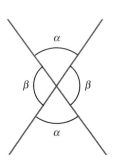

Figure 1.14 Angles across a vertex are equal. Here the angles are called α (alpha) and β (beta).

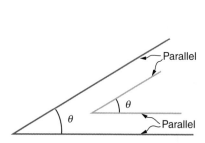

Figure 1.15 These angles are equal because the sides are parallel.

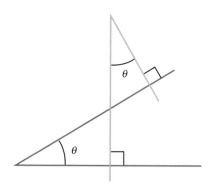

Figure 1.16 These angles are equal because the sides are mutually perpendicular.

Triangles

We often encounter triangles in physics. Here are some of their properties:

1. The sum of the angles inside a triangle is 180°, as seen in Figure 1.18.
2. An **equilateral** triangle is defined to have sides of equal length. Its angles must also be equal, so that each angle is 60°, as seen in Figure 1.19(a).
3. An **isosceles** triangle is defined to have two sides of equal length and, thus, two equal angles. (See Figure 1.19(b).)
4. If two triangles have the same three angles, they are defined to be **similar** triangles. (See Figure 1.20.) Their sides will be proportional, so that

$$\frac{a_1}{a_2} = \frac{b_1}{b_2} = \frac{c_1}{c_2} \quad \text{(similar triangles)} \quad (1.25)$$

Similar triangles that can be exactly superimposed are called **congruent** triangles.

5. A **right** triangle is one having a 90° angle. An example is shown in Figure 1.21. Right triangles are very useful in physics and are by far the most commonly encountered triangles.
6. The **Pythagorean theorem** gives the relationship among the sides of a right triangle to be

$$a^2 + b^2 = c^2 \quad \text{(right triangle)} \quad (1.26)$$

where a, b, and c are the lengths of the sides as shown in Figure 1.21. The side opposite the right angle is called the **hypotenuse**.

Simple Trigonometry

Trigonometry is very useful in representing physics. Right triangles are particularly important, and we often are concerned about the ratios of their sides and how this relates to the angles in the triangle. For example, the triangle in Figure 1.21 might represent the slope of a hill with the hypotenuse making an angle θ with the horizontal. Then the altitude gained going up the slope (b) has something to do with the angle and the distance traveled (c). The following are useful definitions and relationships:

1. The trigonometric functions of sine (**sin θ**), cosine (**cos θ**), and tangent (**tan θ**)—are defined in terms of the right triangle shown in Figure 1.22, where the sides are now labeled o, a, and h for opposite, adjacent, and hypotenuse, respectively. In equation form, sin θ, cos θ, and tan θ are

$$\left.\begin{array}{l} \sin \theta \equiv \dfrac{o}{h} \\[2mm] \cos \theta \equiv \dfrac{a}{h} \\[2mm] \tan \theta \equiv \dfrac{o}{a} \end{array}\right\} \quad (1.27)$$

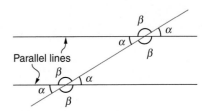

Figure 1.17 Only two different angles, here labeled α and β, are formed when a straight line crosses two parallel lines.

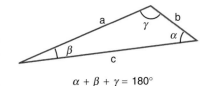

$$\alpha + \beta + \gamma = 180°$$

Figure 1.18 The angles inside a triangle always add up to 180°. Here that means that $\alpha + \beta + \gamma = 180°$.

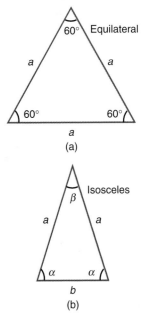

Figure 1.19 (a) An equilateral triangle has three equal-length sides and three equal angles. (b) An isosceles triangle has two equal-length sides and two equal angles.

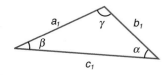

Similar triangles

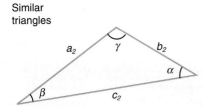

Figure 1.20 Similar triangles have the same three angles α, β, and γ. The lengths of their sides are proportional.

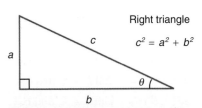

Figure 1.21 A right triangle has a 90° angle in it.

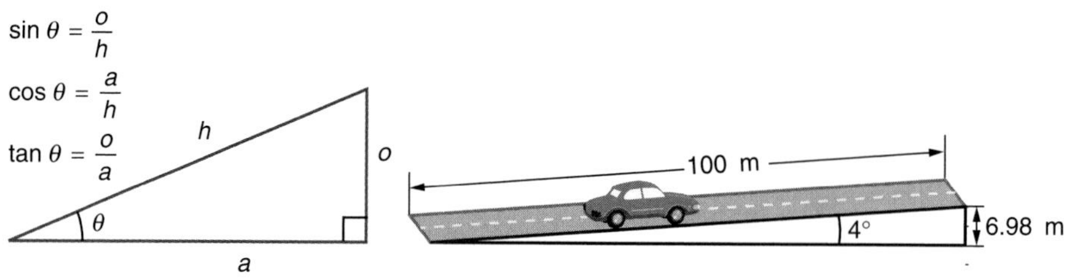

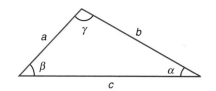

$$\sin \theta = \frac{o}{h}$$

$$\cos \theta = \frac{a}{h}$$

$$\tan \theta = \frac{o}{a}$$

Figure 1.22 A right triangle with the sides labeld o, a, and h for opposite, adjacent, and hypotenuse. The ratios of the lengths of the sides are given by various trigonometric functions.

Note that $\tan \theta = (\sin \theta)/(\cos \theta)$. Here the Pythagorean theorem is $o^2 + a^2 = h^2$.

Suppose the triangle in Figure 1.22 represents a hill with a 4.00° slope, and we would like to know how much altitude we gain for every 100 m we travel up the hill. In this case the altitude gained is represented by o, and the distance traveled by h. So we want to find o given $h = 100$ m and $\theta = 4.00°$. Since $\sin \theta = o/h$, we see that $o = h \sin \theta$. Using a pocket calculator, enter 4 and hit the sin key, getting $\sin \theta = 0.0698$. Thus $o = (100\ \text{m})(0.0698) = 6.98$ m is the gain in altitude for every 100 m traveled up the slope of a 4.00° hill.

2. Three trigonometric functions are defined to be the reciprocals of those given in Equation 1.27. They are the cosecant (**csc θ**), the secant (**sec θ**), and the cotangent (**cot θ**), defined to be $1/(\sin \theta)$, $1/(\cos \theta)$, and $1/(\tan \theta)$, respectively.

3. Given the definitions of sine and cosine, there are certain useful relationships that are true for any angles, whether in a triangle or not. Among these trigonometric identities are three that are used in this text:

$$\sin^2\theta + \cos^2\theta = 1 \tag{1.28}$$

$$\sin(\alpha \pm \beta) = \sin \alpha \cos \beta \pm \cos \alpha \sin \beta \tag{1.29}$$

$$\cos(\alpha \pm \beta) = \cos \alpha \cos \beta \mp \sin \alpha \sin \beta \tag{1.30}$$

Figure 1.23

4. For any triangle, such as the one in Figure 1.23, the following relationships are true:

$$c^2 = a^2 + b^2 - 2ab \cos \gamma \quad \text{(law of cosines)} \tag{1.31}$$

$$\frac{a}{\sin \alpha} = \frac{b}{\sin \beta} = \frac{c}{\sin \gamma} \quad \text{(law of sines)} \tag{1.32}$$

REVIEW EXERCISES

The next 4 exercises refer to similar triangles, like those in Figure 1.20.

121   If $a_1 = 3$, $b_1 = 4.5$, and $a_2 = 45$, find b_2.

122   If $b_1 = 3$ and $c_1 = 4$, what is the ratio b_2/c_2?

123   If $c_2 = 17$, $a_2 = 10$, and $a_1 = 2.5$, find c_1.

124   If $a_1 = 6$, $b_1 = 2.1$, $c_1 = 5$, and $a_2 = 0.15$, find b_2 and c_2.

The next 12 exercises refer to a right triangle, like that in Figure 1.22.

125   If $o = 3$ and $a = 4$, find h.

126   If $a = 4$ and $h = 5$, find o.

127   If $o = 3$ and $a = 4$, find $\sin \theta$.

128   If $a = 4$ and $h = 5$, find $\cos \theta$.

129   If $o = 3$ and $a = 4$, find θ.

130   If $a = 4$ and $h = 5$, find θ.

131   If $o = 3$ and $a = 4$, find $\tan \theta$.

132   If $o = 3$ and $a = 4$, find $\sin \theta$, $\cos \theta$, and $\tan \theta$. Show that $\tan \theta = (\sin \theta)/(\cos \theta)$ by direct division.

133   If $\theta = 30°$ and $h = 5$, find o.

134   If $\theta = 30°$ and $h = 5$, find a.

135   If $\theta = 30°$ and $h = 5$, show $o^2 + a^2 = h^2$ by calculating a and h.

136   If $\theta = 30°$ and $h = 5$, verify the law of cosines.

137   Show that $\sin^2\theta + \cos^2\theta = 1$, for $\theta = 60°$.

138   Show that $\sin^2\theta + \cos^2\theta = 1$, for $\theta = 5°$.

The next 4 exercises refer to a triangle like that in Figure 1.23, where $a = 2$, $b = 3$, and $\gamma = 105°$.

139 What is c?

140 What is β?

141 What is α?

142 Show that $\alpha + \beta + \gamma = 180°$ by finding α and β.

Answers to Odd-Numbered Review Exercises

121 67.5

123 4.25

125 5

127 0.6

129 36.9°

131 0.75

133 2.5

135 $2.5^2 + 4.33^2$
$= 6.25 + 18.75 = 25 = 5^2$

137 $0.866^2 + 0.5^2$
$= 0.75 + 0.25 = 1$

139 4.01

141 28.8°

SUMMARY

There is underlying order and simplicity in nature, an attribute we value highly. **Science** consists of the laws that are the general truths of nature and the body of knowledge they encompass. These **laws of nature** are rules that all natural processes follow. **Physics** is the most basic of the sciences, concerning itself with energy, matter, space, and time, and their interactions. A **model** is a mental image or analogy to objects or phenomena that we can experience directly; it can be crucial to visualization of indirectly observed phenomena. A **theory** is usually a larger-scale and more broadly applicable generalization than a model and often describes nature with mathematical precision. The word **law** is reserved for a concise and very general human statement of a rule of nature. Less broadly applicable statements are usually called **principles**. The validity of models, theories, and laws is determined by observation and experiment. Physics can be divided into four realms called **classical physics**, **relativity**, **quantum mechanics**, and **relativistic quantum mechanics**. The last three make up **modern physics**.

Physical quantities are defined by how they are measured or how they are calculated from measurements. **Units** are standards for expressing the measurement of physical quantities. All units can be expressed as combinations of four **fundamental units**. The particular group of four fundamental units used in this text are the **meter** (m) for length, the **kilogram** (kg) for mass, the **second** (s) for time, and the **coulomb** (C) for charge. These units are part of a **metric system** where powers of 10 are used to relate quantities over the vast ranges encountered in nature.

Accuracy is how close a measurement is to the "true" value. The **uncertainty** in a measurement is an estimate of the amount by which it may differ from the "true" value. Quantities calculated from measurements have uncertainties that can be estimated with significant figures or by adding percent uncertainties as discussed in the text.

CONCEPTUAL QUESTIONS

1.1 Models are particularly useful in relativity and quantum mechanics, where conditions are outside those normally encountered by humans. What is a model?

1.2 How does a model differ from a theory?

1.3 If two different theories describe experimental observations equally well, can one be said to be more valid than the other (assuming both use accepted rules of logic)?

1.4 What determines the validity of a theory?

1.5 Certain criteria must be satisfied if a measurement or observation is to be believed. What are some of these criteria? Will the criteria necessarily be as strict for an expected result as for an unexpected result?

1.6 Can the validity of a model be limited, or must it be universally valid? A theory? A law?

1.7 Classical physics is a good approximation to modern physics under certain circumstances. What are they?

1.8 When is it *necessary* to use relativistic quantum mechanics? Can it be used accurately in all circumstances?

1.9 Identify some advantages and disadvantages of metric units.

1.10 Give an example of a physical quantity the unit of which is not stated but implied by context. For example, "I turned 23 last week" implies *23 years*.

1.11 Look up a unit, such as a barleycorn, not listed in the text (the more obscure, the better). Define the unit and state its relationship to SI units. (Hint: Look in a dictionary or encyclopedia under "weights and measures").

1.12 The four fundamental units used in this text, and from which all other units can be expressed, are the length, mass, time, and charge units of m, kg, s, and C, respectively. What are two other groupings of four units that can serve as the fundamental units from which all others are derived?

1.13 What is the relationship between the accuracy and uncertainty of a measurement?

PROBLEMS

Note: See the Review Exercises at the ends of Sections 1.4 through 1.8 for problems relating to the mathematical overview.

Sections 1.1 An Introduction to Physics and 1.2 Physical Quantities and Units

1.1 Can classical physics be used to accurately describe a satellite moving at a speed of 7500 meters per second? Explain why or why not.

Problems 1.2 through 1.15 should be solved using data from Table 1.3.

1.2 How many heartbeats are there in a lifetime?

1.3 A generation is about one-third of a lifetime. Approximately how many generations have passed since the beginning of recorded history?

1.4 What fraction of the age of the earth is encompassed by recorded history?

1.5 How many times longer than the mean life of an extremely unstable nucleus is the lifetime of a human?

1.6 Calculate the approximate distance to the Andromeda galaxy in light years.

1.7 (a) What is the ratio of the distance to the nearest galaxy to the size of the Milky Way galaxy? This is typical of the distances between galaxies as compared with their sizes. (b) What is the ratio of the size of the Milky Way to the distance to the sun? (c) What is the ratio of the distance to Andromeda to the distance to the sun?

1.8 Calculate the approximate number of atoms in a bacterium, assuming the average mass of an atom is ten times the mass of a hydrogen atom.

1.9 Approximately how many atoms thick is a cell membrane, assuming all atoms average about twice the size of a hydrogen atom?

1.10 If the sun's mass is about average, how many stars are there in the Milky Way galaxy?

1.11 (a) What fraction of the earth's diameter is the greatest ocean depth? (b) The greatest mountain height?

1.12 What is the approximate number of galaxies in the known universe, assuming that the Milky Way has an average mass and that all of the mass in the known universe is in galaxies?

1.13 If the mass of the sun is about average and if half of the matter in the universe is in stars, approximately how many stars are there in the known universe?

• 1.14 (a) Calculate the number of cells in a hummingbird assuming the mass of an average cell is ten times the mass of a bacterium. (b) Making the same assumption, how many cells are there in a human?

• 1.15 Assuming one nerve impulse must end before another can begin, what is the maximum firing rate of a nerve in impulses per second?

Problems 1.16 through 1.26 are unit conversion problems that can be solved using data from tables and appendices in this text.

1.16 The speed limit on certain interstate highways is 65 miles per hour. (a) What is this in feet per second? (b) How many km/h is this?

1.17 A car is traveling at a speed of 88 feet per second. (a) What is its speed in km/h? (b) Is it exceeding the 55 mile per hour speed limit?

1.18 Show that 1.0 m/s = 3.6 km/h.

1.19 American football is played on a field 100 yards long, excluding the end zones. How long is the field in meters?

1.20 Soccer fields vary in size. A large soccer field is 115 meters long and 85 meters wide. What are its dimensions in feet and inches?

1.21 What is the height in meters of a 6 foot 1.0 inch tall person?

1.22 Mount Everest, at 29,028 feet, is the tallest mountain on earth. What is its height in kilometers?

1.23 The speed of sound is measured to be 342 meters per second on a certain day. What is this in miles per hour?

• 1.24 Tectonic plates are large segments of the earth's crust that move slowly. Suppose one such plate has an average speed of 4.0 cm per year. (a) What distance does it move in one second at this speed? (b) What is its speed in miles per million years?

• 1.25 (a) Using data from the appropriate tables for the size of the earth's orbit around the sun, calculate the average speed of the earth in its orbit in kilometers per second. (b) What is this in feet per second?

• 1.26 Solve part (b) of Example 1.1 by starting with the average speed as 0.500 km/min and convert this to m/s. Do you get the same result as in the text?

Section 1.3 Measurement, Accuracy and Uncertainty; Significant Figures

Express your answers to problems in this section to the correct number of significant figures.

1.27 Suppose your bathroom scale reads your mass as 65 kg, with a 3% uncertainty. What is the uncertainty in your mass?

1.28 A good-quality measuring tape can be off by 0.50 cm over a distance of 20 m. What is its percent uncertainty?

1.29 (a) A car speedometer has a 5% uncertainty. What is the range of possible speeds when it reads 90 km/h? (b) Convert this to miles per hour.

1.30 An infant's pulse rate is measured to be 130 ± 5 beats per minute. What is the percent uncertainty in this measurement?

1.31 (a) Suppose a person has an average heart rate of 72.0 beats per minute. How many beats does she have in 2.0 y? (b) In 2.00 y? (c) In 2.000 y?

1.32 An Australian beer can contains 708.7 mL. How much is left after 498 mL is removed?

1.33 State how many significant figures are proper in the results of the following calculations: (a) $(106.7)(98.2)/(46.210)(1.01)$ (b) $(18.7)^2$ (c) $(1.60 \times 10^{-19})(3712)$.

• 1.34 (a) How many significant figures are in the numbers 99 and 100? (b) If the uncertainty in each number is 1, what is the percent uncertainty in each? (c) Which is a more meaningful way to express the accuracy of these two numbers, significant figures or percent uncertainties?

• 1.35 (a) If your speedometer has an uncertainty of 2.0 miles per hour at a speed of 60 miles per hour, what is the percent uncertainty? (b) If it has the same percent uncertainty when it reads 45 miles per hour, what is the range of speeds you could be going?

• **1.36** (a) A blood pressure is measured to be 120 ± 2 mm Hg. What is its percent uncertainty? (b) Assuming the same percent uncertainty, what is the uncertainty in a blood pressure measurement of 80 mm Hg?

• **1.37** A person measures her heart rate by counting the number of beats in 30 seconds. If 40 ± 1 beats are counted in 30.0 ± 0.5 s, what is the heart rate and its uncertainty in beats per minute?

• **1.38** What is the area of a circle 3.102 cm in diameter?

• **1.39** If a marathon runner averages 9.5 miles per hour, how long does it take him to run a 26.22 mile marathon?

⦂ **1.40** The sides of a small rectangular box are measured to be 1.80 ± 0.01 cm, 2.05 ± 0.02 cm, and 3.1 ± 0.1 cm long. Calculate its volume and uncertainty in cubic centimeters.

2 KINEMATICS

This bird's flight is a beautiful example of one-dimensional motion.

You can find motion anywhere you look. Even when you are resting, for example, your heart moves blood through your veins. Everything from a tennis game to a space probe flyby of the planet Neptune involves motion. Questions about motion are interesting in themselves (such as, "Where will that football come down if it is thrown at a certain angle and speed?"). But there is another reason for studying motion before you study other subjects in physics. Certain concepts related to motion, such as acceleration, are basic to the study of later material, such as force. There is a logical sequence of topics in physics that is similar to, but not nearly as rigid as, that found in mathematics.

Our formal study of physics begins with **kinematics**, which is defined to be the *study of motion without regard to mass or force*. The word *kinematics* comes from a Greek term, meaning motion, and is related to other English words such as *cinema* (movies!) and *kinesiology* (the study of human motion). The study of motion goes back to ancient times and has obviously interested humans since before recorded history. The first recorded studies of motion concerned movement of the sun and planets across the sky. Interest in their motions led many centuries later to the discovery of gravity and the revolutionary idea that the earth is not the center of the universe.

In this and the next chapter we will study only the *motion* of a football, for example, without worrying about what forces cause it to have that particular path. Such considerations will come in later chapters. In this chapter, we examine the simplest type of motion—namely, motion along a straight line, or one-dimensional motion. In the next chapter, we consider motion along curved paths (two- and three-dimensional motions), such as a car rounding a curve, by applying the concepts developed here.

2.1 DISPLACEMENT

To be able to describe the motion of an object, you must first be able to say where it is. More precisely, you need to specify its position relative to a convenient reference frame. The earth is very often used as a reference frame. But you might also use another reference frame, such as an airplane when you fly. (See Figure 2.1.)

If an object moves relative to a reference frame (such as the professor moves right relative to the earth and the passenger moves toward the rear of a plane), then position changes. We define **displacement** to be the *change in position* of an object. (The word *displacement* implies that an object has moved, or been displaced.) The SI unit for displacement is the meter (m), discussed in Section 1.2, but we sometimes use km, miles, feet, and other units of length. In symbols, displacement Δx is defined to be

$$\Delta x = x_2 - x_1 \qquad\qquad (2.1)$$

where x_1 is the initial position and x_2 is the final position. In this text the upper case Greek letter Δ (delta) always means "change in" whatever quantity follows it; thus, Δx means *change in position*.

Note that displacement has a direction as well as a magnitude. The professor's displacement is 2.0 m to the right, and the passenger's displacement is 10 m to the rear. In one-dimensional motion the direction can be specified with a plus or minus sign. The professor's position is initially 1.5 m and finally 3.5 m; thus, her displacement is $\Delta x = x_2 - x_1 = 3.5\,\text{m} - 1.5\,\text{m} = +2.0\,\text{m}$. So movement to the right is positive here, whereas motion to the left is negative. Similarly, the airplane passenger's initial position is $x_1 = 12\,\text{m}$ and his final position is $x_2 = 2.0\,\text{m}$, and so his displacement is $\Delta x = x_2 - x_1 = 2.0\,\text{m} - 12\,\text{m} = -10\,\text{m}$. His motion toward the rear of the plane is signified by a negative displacement.

Distance is defined to be *the magnitude or size of displacement*. By magnitude, we mean a number with a unit. Distance has no direction and, hence, no sign. For example, the distance the professor walks is 2.0 m and the distance the airplane passenger walks is 10 m. Neither distance has a sign. It is important to note that the *total distance traveled* can be greater than distance. For example, the professor could pace back and forth many times, perhaps stepping off 150 m during a lecture (part of her path is shown as the curvy black line beneath the graph), yet still end up only 2.0 m to the right of her starting point. Then the displacement would be +2.0 m, the distance 2.0 m, but the total distance traveled 150 m. In kinematics we nearly always deal with displacement and distance between starting and ending points, and almost never with total distance traveled.

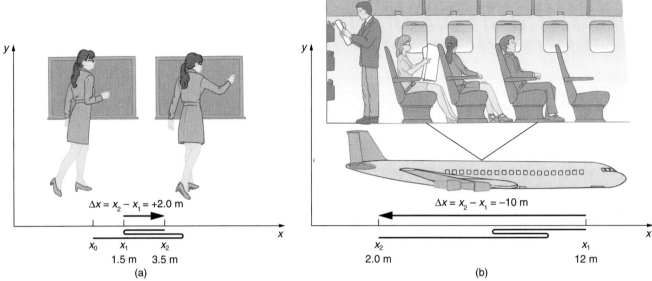

Figure 2.1 (a) A professor paces left and right while lecturing. Her position *relative to the earth* is given by x. The +2.0 m displacement of the professor relative to the earth is represented by an arrow pointing to the right. (b) A passenger moves from his seat to the magazine bin. His location *relative to the airplane* is given by x. The −10 m displacement of the passenger relative to the plane is represented by an arrow toward the rear of the plane that is five times the length of the one for the professor.

Vectors and Scalars

Any quantity with both *magnitude and direction* is defined to be a **vector**. Displacement is thus one type of vector. Other examples of vectors include a velocity 90 km/h east and a force 500 newtons (about 112 pounds) straight down. (The direction of a vector in one-dimensional motion is given by a plus or minus sign.) Figure 2.1 shows how vectors can be represented by arrows. An arrow used to represent a vector has a length proportional to the vector's magnitude and points in the same direction as the vector.

Some physical quantities either have no direction or none is specified. Any quantity with a *magnitude only*, but no direction, is defined to be a **scalar**. For example, a 20°C temperature, the 125 kcal of energy in a candy bar, a 90 kilometer per hour (km/h) speed limit, a person's 1.8 m height, and a distance of 2.0 m are all scalars with no specified direction. Note, however, that a scalar can be negative, such as a −20°C temperature. The minus indicates a point on a scale rather than a direction.

2.2 TIME, VELOCITY, AND SPEED

There is obviously more to motion than distance and displacement. Questions such as, "How long does a foot race take?" and "What was the runner's speed?" cannot be answered without other concepts. In this section we add definitions of time, velocity, and speed to expand our description of motion.

Time

There are many fascinating but unanswered questions about time. Can it be reversed? Is it continuous? Does it have an absolute beginning or end? The lack of answers to these questions does not affect our ability to work with time in a practical sense. As discussed in Section 1.2, the most fundamental physical quantities are defined by how they are measured. So it is with time. Every measurement of time involves measuring a change in some physical quantity. It may be a number on a digital clock, a heartbeat, or the position of the sun in the sky. In physics the definition of time is simple—**time** is *change*. It is impossible to know that time has passed unless something changes. The amount of time or change is calibrated by comparison with a standard. The SI unit for time is the second, abbreviated s (as defined in Section 1.2). We might, for example, observe that a certain pendulum makes one full swing each 0.75 s, thereby calibrating it. We could then use the pendulum to measure time by counting its swings or, of course, by connecting the pendulum to a clock mechanism that registers time on a dial. This allows not only the amount of time to be measured, but a sequence of events to be determined.

How does time relate to motion? We are usually interested in elapsed time for a particular motion, such as how long it takes an airplane passenger to get from his seat to the magazine bin. To find elapsed time, we note the time at the beginning and end of the motion and subtract. For example, a lecture may start at 11:00 A.M. and end at 11:50 A.M., so that the elapsed time was 50 minutes. More formally, we use the symbol t for time, and we define **elapsed time** Δt, to be the difference between the ending time and beginning time:

$$\Delta t = t_2 - t_1 \qquad \textbf{(2.2)}$$

where t_1 is the time at the beginning of the motion, t_2 is the time at the end of the motion, and Δt means change in time or elapsed time. (As usual, Δ means change in the quantity following.) Life is simpler if the beginning time t_1 is taken to be zero, as when we use a stopwatch. Then, for example, the stopwatch would simply read zero at the start of lecture and 50 min at the end. If $t_1 = 0$, then $\Delta t = t_2 = t$. In this text, for simplicity's sake, motion *starts at time equal to zero and the symbol t is used for elapsed time unless otherwise specified.*

Velocity

Your intuitive notion of velocity is basically the same as its scientific definition. You know that if you have a large displacement in a small amount of time you have a large velocity, and that velocity has units of distance divided by time, such as miles per hour. **Average velocity** is precisely defined to be *displacement (change in position) divided by the time of travel and to have the same direction as displacement.* This definition implies that *velocity is a vector.* In symbols, average velocity is

$$\left. \begin{aligned} \bar{v} &= \frac{\Delta x}{\Delta t} \\[2ex] &= \frac{x_2 - x_1}{t_2 - t_1} \end{aligned} \right\} \qquad \textbf{(2.3)}$$

where $\bar{v}$ is the *average* (indicated by the bar over the v) velocity, Δx is the change in position, and x_1 and x_2 are the beginning and ending positions at times at times t_1 and t_2, respectively. If the starting time t_1 is taken to be zero, then the average velocity is simply $\bar{v} = \Delta x/t$. Suppose, for example, the airplane passenger considered earlier took 20 seconds to get to the magazine bin; then his average velocity would be $\bar{v} = \Delta x/t = (-10 \text{ m})/(20 \text{ s}) = -0.50$ m/s. The minus sign indicates the velocity is toward the rear of the plane. The SI unit for velocity is m/s, but many other units, such as km/h, mi/h, and cm/s, are in common use.

The average velocity of some object does not tell us anything about what happens to it between the starting point and ending point, such as whether the airplane passenger stops momentarily or backs up. To get more details, we must consider smaller segments of the trip over smaller time intervals. (See Figure 2.2.)

The smaller the time intervals considered in a motion, the more detailed the information. When we carry this process to its logical conclusion, we consider an infinitesimal interval. If that is done, then the average velocity becomes the *instantaneous velocity* or *the velocity at a specific instant.* We read the magnitude of instantaneous velocity on our car's speedometer, for example. More precisely, **instantaneous velocity** v is the average velocity at a specific instant in time (or over an infinitesimally small time interval). Mathematically, finding instantaneous velocity v at a precise instant t can involve taking a limit, a calculus operation beyond the scope of this text. However, under many circumstances (elaborated in later sections), we can find precise values for instantaneous velocity without calculus.

Speed

In everyday language, speed and velocity mean the same thing; they are, however, distinct concepts in physics. One major difference is that speed has no direction. Thus *speed is a scalar.* **Instantaneous speed** is simply defined to be *the magnitude of instantaneous*

Figure 2.2 A more detailed record of the airplane passenger heading for the magazine bin showing smaller segments of his trip. Each of the smaller segments has its own average velocity.

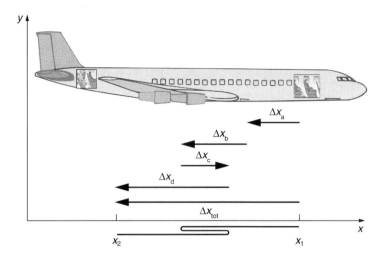

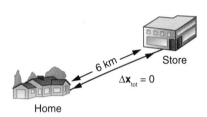

Figure 2.3 During a 30 minute trip to the store, the total distance traveled is 6.0 km. The average speed is 12 km/h. The displacement for the round trip is zero, since there was no net change in position. Thus the average velocity is zero.

velocity. So, for example, suppose the airplane passenger at one instant had an instantaneous velocity of −3.0 m/s (the minus meaning toward the rear of the plane). At that same time his instantaneous speed was 3.0 m/s. Or suppose that at one time during a shopping trip your instantaneous velocity is 40 km/h due north. Your instantaneous speed at that instant is 40 km/h—the same magnitude but without a direction.

Average speed is quite distinct from average velocity. **Average speed** is defined to be *total distance traveled divided by elapsed time.* We have noted that total distance traveled can be greater than displacement. So average speed can be greater than average velocity, which is displacement divided by time. For example, if you drive to a store and return home in half an hour, and your car's odometer shows the total distance traveled was 6.0 km, then your average speed was 12 km/h. But your average velocity was *zero* since your displacement for the round trip is zero. (Displacement is change in position and, thus, is zero for a round trip.) This shows that average speed is *not* simply the magnitude of average velocity. (See Figure 2.3.)

2.3 ACCELERATION

In everyday conversation, to accelerate means to speed up. The accelerator in a car can in fact cause it to speed up. The greater the acceleration, the greater the change in velocity in a given time. The formal definition of acceleration is consistent with these notions, but more inclusive. **Acceleration** is *the rate at which velocity changes.* In symbols, **acceleration $\bar{a}$** is defined to be

$$\left. \begin{array}{l} \bar{a} = \dfrac{\Delta v}{\Delta t} \\[2mm] = \dfrac{v_2 - v_1}{t_2 - t_1} \end{array} \right\} \tag{2.4}$$

(The bar over the *a* means *average* acceleration.) The *units for acceleration* come directly from its definition and are distance divided by time squared—m/s^2 in SI units. (See the following example for some elaboration on the meaning of these units.) *Acceleration is a vector* in the *same direction as the change in velocity* Δv. Now, since velocity is a vector, it can change either in magnitude or in direction. This means, for example, that if a car turns a corner at constant speed, there is still an acceleration because direction is changing. The quicker you turn, the greater the acceleration. So there is an acceleration when velocity changes either in magnitude (an increase or decrease in speed) or in direction, or both.

Figure 2.4 Racehorses leaving gate.

EXAMPLE 2.1 A HORSE ACCELERATES

A racehorse coming out of the gates, as in Figure 2.4, accelerates from rest to a velocity of 15.0 m/s due west, in 1.80 s. Calculate its average acceleration.

Strategy We can solve this problem by identifying Δv and Δt from the given information and then calculating the acceleration directly from its definition in Equation 2.4.

(continued)

(continued)

Solution Since the horse is going from 0 to 15.0 m/s due west, its change in velocity equals its final velocity. That is, $\Delta v = v_2 - 0 = 15.0$ m/s due west. The change in time or elapsed time Δt is given to be 1.80 s. The definition of average acceleration is

$$\bar{a} = \frac{\Delta v}{\Delta t}$$

Entering the values of Δv and Δt gives

$$\bar{a} = \frac{15.0 \text{ m/s due west}}{1.80 \text{ s}}$$

so that

$$\bar{a} = 8.33 \text{ m/s}^2 \text{ due west}$$

Discussion An acceleration of 8.33 m/s² due west means that the horse increases its velocity by 8.33 m/s due west each second. (That is, (8.33 m/s)/s, which we write as 8.33 m/s².) This is truly an average acceleration, because the ride is not smooth. We shall see later that an acceleration of this magnitude would require the rider to hang on with a force nearly equal to his weight.

Figure 2.5 Graphs of instantaneous acceleration versus time for two different one-dimensional motions. (a) Here acceleration varies only slightly and is always in the same direction since it is positive. The average over the interval is nearly the same as the acceleration at any given time. (b) Here the acceleration varies greatly, perhaps representing a package on a post office conveyor belt that is accelerated forward and backward as it bumps along. It is necessary to consider small time intervals (such as from 0 to 1.0 s) with constant or nearly constant acceleration in such a situation.

t (s)	a (m/s²)
0–1.0	3.0
1.0–3.0	–2.0
3.0–4.2	0.0
4.2–4.8	1.5
4.8–5.5	5.0
5.5–6.0	–4.0

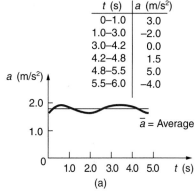

Instantaneous acceleration *a*, or the *acceleration at a specific instant in time*, is obtained by the same process as discussed for instantaneous velocity in the previous section—that is, by considering an infinitesimally small interval of time. How do we find instantaneous acceleration using only algebra? The answer is that we choose an average acceleration that is representative of the motion. Figure 2.5 shows graphs of instantaneous acceleration versus time for two very different motions. In Figure 2.5(a), the acceleration varies slightly and the average over the entire interval is nearly the same as the instantaneous acceleration at any time. It would be accurate to treat this motion as if it had a constant acceleration equal to the average. In Figure 2.5(b), the acceleration varies drastically over time. In such situations it is best to consider smaller time intervals and choose an average acceleration for each. For example, we could consider motion over the time intervals from 0 to 1 s and from 1 to 3 s as separate motions with accelerations of +3.0 m/s² and −2.0 m/s², respectively.

The next several examples consider the motion of the subway shuttle train shown in Figure 2.6. In Figure 2.6(a) the shuttle moves to the right, and in Figure 2.6(b) it moves to the left. The examples are designed to further illustrate aspects of motion and to illustrate some of the reasoning that goes into solving problems.

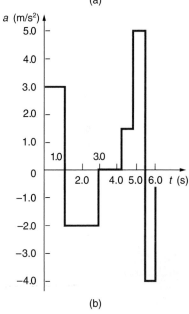

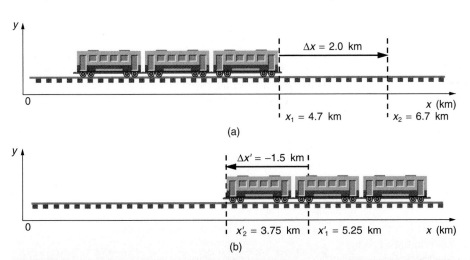

Figure 2.6 One-dimensional motion of a subway shuttle train considered in Examples 2.2 through 2.7. Here we have chosen the x-axis so that + means to the right and − means to the left for displacements, velocities, and accelerations, as discussed in the examples. (a) The subway shuttle train moves to the right from x_1 to x_2. Its displacement Δx is +2.0 km. (b) The shuttle moves to the left from x_1' to x_2'. Its displacement $\Delta x'$ is −1.5 km.

EXAMPLE 2.2   WHAT IS THE SIZE AND SIGN OF DISPLACEMENT FOR A SUBWAY SHUTTLE?

What are the displacements for the motions shown in parts (a) and (b) of Figure 2.6?

Strategy To find displacement, we use its definition from Equation 2.1. This is straightforward since the initial and final positions are given.

Solution Equation 2.1 defines displacement as the change in position, which is the difference between the final and initial positions:

$$\Delta x = x_2 - x_1$$

In Figure 2.6(a), we see that $x_2 = 6.7$ km and $x_1 = 4.7$ km;

thus,

$$\Delta x = 6.7 \text{ km} - 4.7 \text{ km} = +2.0 \text{ km}$$

Similarly, in Figure 2.6(b), we see that $x_2' = 3.75$ km and $x_1' = 5.25$ km; thus,

$$\Delta x' = 3.75 \text{ km} - 5.25 \text{ km} = -1.50 \text{ km}$$

Discussion The direction of the motion in (a) is to the right and therefore its displacement has a positive sign, whereas motion in (b) is to the left and thus has a negative sign.

EXAMPLE 2.3   DISTANCE AND DISPLACEMENT ARE NOT ALWAYS THE SAME THING

What are the distances for the motions shown in parts (a) and (b) of Figure 2.6?

Concept Distance is defined to be the magnitude of displacement, which was found in the previous example.

Solution The displacements for Figures 2.6(a) and (b) are $\Delta x = +2.0$ km and $\Delta x' = -1.50$ km. The distances are thus 2.0 and 1.50 km, respectively.

Discussion Distance is a scalar and has magnitude but no sign.

EXAMPLE 2.4   ACCELERATION TO START THE TRAIN

Suppose the train in Figure 2.6(a) accelerates from rest to 30.0 km/h in the first 20.0 s of its motion. What is its average acceleration during that time interval?

Strategy As in Example 2.1, we must identify the change in velocity and the change in time to calculate the acceleration using Equation 2.4. Since the train starts from rest, its change in velocity is $\Delta v = +30.0$ km/h, where the plus sign means velocity to the right. The acceleration takes place over the first 20.0 s, and so $\Delta t = 20.0$ s.

Solution Now using the definition of average acceleration in Equation 2.4,

$$\bar{a} = \frac{\Delta v}{\Delta t}$$

Entering the values of Δv and Δt gives

$$\bar{a} = \frac{+30.0 \text{ km/h}}{20.0 \text{ s}}$$

which has mixed units for time (hours and seconds). Let us get everything into meters and seconds by doing a unit conversion as discussed in Section 1.2:

$$\bar{a} = \frac{+30.0 \text{ km/h}}{20.0 \text{ s}} \times \frac{10^3 \text{ m}}{1 \text{ km}} \times \frac{1 \text{ h}}{3600 \text{ s}}$$

which yields

$$\bar{a} = +0.417 \text{ m/s}^2$$

Discussion The plus sign means that acceleration is to the right. This is reasonable because the train starts from rest and ends up with a velocity to the right (also positive). So acceleration is in the same direction as the *change* in velocity, as is always the case.

EXAMPLE 2.5   ACCELERATION TO SLOW THE TRAIN

Now suppose that at the end of its trip the train in Figure 2.6(a) slows to a stop from a speed of 30.0 km/h in 8.00 s. What is its average acceleration while stopping?

Strategy As in the previous example, we must find the

change in velocity and the change in time.

Solution Using the definition of change in velocity, we see that it is negative in this problem:

$$\Delta v = v_2 - v_1 = 0 - 30.0 \text{ km/h} = -30.0 \text{ km/h}$$

(continued)

(continued)

The change in time is simply

$$\Delta t = 8.00 \text{ s}$$

Now using the definition of average acceleration in Equation 2.4,

$$\bar{a} = \frac{\Delta v}{\Delta t}$$

we get

$$\bar{a} = \frac{-30.0 \text{ km/h}}{8.00 \text{ s}}$$

Doing a unit conversion as before,

$$\bar{a} = \frac{-30.0 \text{ km/h}}{8.00 \text{ s}} \times \frac{10^3 \text{ m}}{1 \text{ km}} \times \frac{1 \text{ h}}{3600 \text{ s}}$$

which yields

$$\bar{a} = -1.04 \text{ m/s}^2$$

Discussion The minus sign means that acceleration is to the left. This is reasonable; the train initially has a positive velocity in this problem, and a negative acceleration would oppose the motion. Again acceleration is in the same direction as the *change* in velocity, which is negative here.

EXAMPLE 2.6 AVERAGE VELOCITY OF THE TRAIN

What is the average velocity of the train in Figure 2.6(b) if it takes 5.0 min to make its trip?

Concept Average velocity is displacement divided by time. It will be negative here, since the train moves to the left and has a negative displacement.

Solution Displacement was found to be $\Delta x' = -1.5$ km in Example 2.2 and time is given as $\Delta t = 5.0$ min. Thus,

from the definition of average velocity in Equation 2.3, and converting units to km/h,

$$\bar{v} = \frac{\Delta x'}{\Delta t} = \frac{-1.5 \text{ km}}{5.0 \text{ min}} \times \frac{60 \text{ min}}{1 \text{ h}} = -18 \text{ km/h}$$

Discussion Negative velocity indicates motion to the left.

EXAMPLE 2.7 ACCELERATION MORE SUBTLE

Finally, suppose the train in Figure 2.6(b) slows to a stop from a velocity of −20.0 km/h in 10.0 s. What is its average acceleration?

Strategy As before, we must find the change in velocity and the change in time to calculate average acceleration.

Solution The change in velocity here is actually positive, since

$$\Delta v = v_2 - v_1 = 0 - (-20.0 \text{ km/h}) = +20.0 \text{ km/h}$$

The change in time is simply

$$\Delta t = 10.0 \text{ s}$$

The definition of average acceleration in Equation 2.4 is

$$\bar{a} = \frac{\Delta v}{\Delta t}$$

so that

$$\bar{a} = \frac{+20.0 \text{ km/h}}{10.0 \text{ s}}$$

We make a unit conversion as before:

$$\bar{a} = \frac{+20.0 \text{ km/h}}{10.0 \text{ s}} \times \frac{10^3 \text{ m}}{1 \text{ km}} \times \frac{1 \text{ h}}{3600 \text{ s}}$$

which yields

$$\bar{a} = +0.556 \text{ m/s}^2$$

Discussion The plus sign means that acceleration is to the right. This is reasonable because the train initially has a negative velocity (to the left) in this problem and a positive acceleration opposes the motion (and so is to the right). Again, acceleration is in the same direction as the *change* in velocity, which is positive here.

Perhaps the most important thing to note about these examples are the signs of the answers. Plus means the quantity is to the right and minus means to the left. This is easy to imagine for displacement and velocity. But it is a little less obvious for acceleration. Most people interpret negative acceleration as slowing of an object. This was not the case in Example 2.7, where a positive acceleration slowed a negative velocity. What was crucial was that the acceleration was in the opposite direction of the velocity. In fact, a negative acceleration will *increase* a negative velocity. For example, the train moving to the left in Figure 2.6(b) is speeded up by an acceleration to the left. In that case both v and a are negative. The plus and minus signs give the directions of the accelerations. If acceleration has the same sign as velocity, the object speeds up. If acceleration has the opposite sign of velocity, the object slows down.

2.4 MOTION EQUATIONS FOR CONSTANT ACCELERATION IN ONE DIMENSION

There are additional motion relationships we have yet to explore. For example, we know intuitively that the greater the acceleration, say of a car moving away from a stop sign, the greater displacement in a given time. But we have not developed a specific equation that relates acceleration and displacement. In this section we develop some convenient equations for this and other kinematic relationships.

We can derive some very useful equations starting from the definitions of displacement, velocity, and acceleration already covered. First, let us make some simplifications in notation. As mentioned in Section 2.2, taking the initial time to be zero, as if time is measured with a stopwatch, is a great simplification. Since elapsed time is $\Delta t = t_2 - t_1$, taking $t_1 = 0$ means that $\Delta t = t_2$ = the final time on the stopwatch. When initial time is taken to be zero, we use the subscript 0 to denote initial values of position and velocity. That is, x_0 is *the initial position* and v_0 is *the initial velocity*. We put no subscripts on the final values. That is, t *is the final time*, x *is the final position*, and v *is the final velocity*. This gives a simpler expression for elapsed time—now, $\Delta t = t$. It also simplifies the expression for displacement, which is now $\Delta x = x - x_0$. And it simplifies the expression for change in velocity, which is now $\Delta v = v - v_0$. To summarize, using the simplified notation, with the initial time taken to be zero,

$$\left. \begin{aligned} \Delta t &= t \\ \Delta x &= x - x_0 \\ \Delta v &= v - v_0 \end{aligned} \right\} \tag{2.5}$$

where *the subscript* 0 *denotes an initial value and the absence of a subscript denotes a final value* in whatever motion is under consideration.

We now make the important assumption that *acceleration is a constant*. This assumption allows us to avoid using calculus to find instantaneous acceleration. Since acceleration is constant, the average and instantaneous accelerations are equal. That is,

$$\bar{a} = a = \text{constant} \tag{2.6}$$

so we use the symbol a for acceleration at all times. Assuming acceleration to be constant does not seriously limit the situations we can study nor degrade the accuracy of our treatment. For one thing, acceleration *is* constant in a great number of situations. Furthermore, in many other situations we can accurately describe motion by assuming a constant acceleration equal to the average acceleration for that motion. Finally, in motions where acceleration changes drastically, such as a car accelerating to top speed and then braking to a stop, the motion can be considered in separate parts, each of which has its own constant acceleration.

To get our first two new equations, we start with the definition of average velocity given in Equation 2.3,

$$\bar{v} = \frac{\Delta x}{\Delta t}$$

Substituting the simplified notation for Δx and Δt as given in Equation 2.5 yields

$$\bar{v} = \frac{x - x_0}{t}$$

Solving for x yields

$$x = x_0 + \bar{v}t \qquad [\text{constant } a] \tag{2.7}$$

where the average velocity is

$$\bar{v} = \frac{v_0 + v}{2} \qquad \text{[constant } a]\qquad\qquad \textbf{(2.8)}$$

Equation 2.8 reflects the fact that when acceleration is constant, $\bar{v}$ is just the simple average of initial and final velocities. For example, if you steadily increase your velocity (that is, with constant acceleration) from 30 to 60 km/h, then your average velocity during this steady increase is 45 km/h. Using Equation 2.8 to confirm this, we see that

$$\bar{v} = \frac{30 \text{ km/h} + 60 \text{ km/h}}{2} = 45 \text{ km/h}$$

as seemed logical.

EXAMPLE 2.8 HOW FAR DOES THE JOGGER RUN?

A jogger runs down a straight stretch of road with an average velocity of 4.00 m/s for 2.00 min. What is his final position, taking his initial position to be zero?

Strategy The final position x is given by Equation 2.7:

$$x = x_0 + \bar{v}t$$

To find x, we identify the values of $x_0 = 0$, $\bar{v} = 4.00$ m/s, and $t = 120$ s from the statement of the problem and substitute them into the equation.

Solution Entering the known values into Equation 2.7 gives

$$x = x_0 + \bar{v}t = 0 + (4.00 \text{ m/s})(120 \text{ s}) = 480 \text{ m}$$

Discussion Velocity and final displacement are both positive, which means they are in the same direction.

We could have solved Example 2.8 starting with the definition of average velocity $\bar{v} = \Delta x/\Delta t$, but it seems easier to use Equation 2.7, and that is reason enough. Equation 2.7 gives additional insight into the relationship of displacement, velocity, and time. It shows, for example, that displacement is a linear function of average velocity. (By linear function, we mean that displacement depends on $\bar{v}$ rather than some other power, such as $\bar{v}^2$.) On an automobile trip, for example, we will get twice as far in a given time if we average 90 km/h than if we average 45 km/h.

We can derive another useful equation by manipulating the definition of acceleration given in Equation 2.4,

$$a = \frac{\Delta v}{\Delta t}$$

Substituting the simplified notation for Δv and Δt as given in Equation 2.5 gives

$$a = \frac{v - v_0}{t}$$

Solving for v yields

$$v = v_0 + at \qquad \text{[constant } a]\qquad\qquad \textbf{(2.9)}$$

which is another useful kinematic relationship.

EXAMPLE 2.9 SLOWING AN AIRPLANE AFTER LANDING

An airplane lands with an initial velocity of 70.0 m/s and then decelerates at 1.50 m/s^2 for 40.0 s. What is its final velocity? (See Figure 2.7.)

Strategy The final velocity v is given by Equation 2.9:

$$v = v_0 + at$$

To find v, we identify the values of $v_0 = 70.0$ m/s, $a =$ -1.50 m/s^2 (negative because it is in the opposite direction of the velocity), and $t = 40.0$ s.

Solution Entering the known values into Equation 2.9 gives

$$v = v_0 + at = 70.0 \text{ m/s} + (-1.50 \text{ m/s}^2)(40.0 \text{ s})$$

(continued)

(continued)

which yields

$$v = 10.0 \text{ m/s}$$

as desired when slowing down, but still positive. With jet engines, reverse thrust could be maintained long enough to stop the plane and start moving it backward. That would be indicated by a negative final velocity, which is not the case here.

Discussion The final velocity is much less than the initial,

$v_0 = 70$ m/s

$a = -1.5$ m/s^2

$t_0 = 0$

$v = 10$ m/s

$a = -1.5$ m/s^2

$t = 40$ s

Figure 2.7 The airplane in Example 2.9 lands with an initial velocity of 70.0 m/s and slows to a final velocity of 10.0 m/s before heading for the terminal. The acceleration is negative since its direction is opposite to velocity.

In addition to being useful in problem solving, Equation 2.9 gives us insight into the relationship among velocity, acceleration, and time. From it we can see, for instance, that final velocity clearly depends on how large acceleration is and how long it lasts. For example, an MX ballistic missile has a larger average acceleration than the space shuttle and achieves a greater velocity in the first minute or two of flight (the actual MX burn time is classified—short burn time missiles are more difficult for an enemy to destroy). But the space shuttle obtains a greater final velocity, so that it can orbit the earth rather than come directly back down as the MX does. The space shuttle does this by accelerating for a longer time. Another insight provided by Equation 2.9 is that if acceleration is zero, then final velocity equals the initial velocity ($v = v_0$), as expected. If a is negative, then the final velocity is less than the initial velocity, as in the previous example. All of these observations fit our intuition, and it is always useful to examine basic equations in light of our intuition and experiences to check that they do indeed describe nature accurately.

A third equation can be obtained by combining Equations 2.7, 2.8, and 2.9. Equation 2.7 states

$$x = x_0 + \bar{v}t$$

We use Equation 2.9 to find an expression for $\bar{v}$; the equation gives

$$v = v_0 + at$$

Adding v_0 to each side of this equation and dividing by 2 gives

$$\frac{v_0 + v}{2} = v_0 + \tfrac{1}{2}at$$

But the expression $(v_0 + v)/2$ is defined to be $\bar{v}$ in Equation 2.8. Thus,

$$\bar{v} = v_0 + \tfrac{1}{2}at$$

Now we substitute this expression for $\bar{v}$ into Equation 2.7, yielding

$$x = x_0 + v_0t + \tfrac{1}{2}at^2 \qquad \text{[constant } a\text{]} \qquad \textbf{(2.10)}$$

This is another useful kinematic relationship.

EXAMPLE 2.10 DRAGSTERS COVER A LOT OF GROUND FAST

Dragsters such as the one shown in Figure 2.8 can achieve accelerations of 15.0 m/s^2. Suppose such a dragster accelerates from rest at this rate for 7.30 s. How far does it travel in this time?

Strategy We are asked to find displacement, which is x if we take x_0 to be zero. Equation 2.10 can be used to find x once we identify v_0, a, and t from the statement of the problem. Starting from rest means that $v_0 = 0$, a is given as 15.0 m/s^2, and t is also given, as 7.30 s.

(continued)

(continued)

Solution Equation 2.10 is

$$x = x_0 + v_0 t + \tfrac{1}{2}at^2$$

Since the initial position and velocity are both zero, this simplifies to

$$x = \tfrac{1}{2}at^2$$

Substituting the identified values of a and t gives

$$x = \tfrac{1}{2}(15.0 \text{ m/s}^2)(7.30 \text{ s})^2$$

yielding

$$x = 400 \text{ m}$$

Discussion If we convert 400 m to miles, we find that the distance covered is approximately one-quarter of a mile, the standard distance for drag racing. So the answer is reasonable. This is an impressive displacement in only 7.30 s, but top notch dragsters can do a quarter mile in even less time than this.

Figure 2.8 A dragster during a run. Notice that the background is blurred because the photographer followed the motion of the dragster.

What else can we learn by examining Equation 2.10? We see, for one thing, that displacement depends on elapsed time squared when acceleration is not zero. In this example, then, the dragster only covers one-fourth of the total distance in the first half of the elapsed time. Further, if acceleration is zero, then the initial velocity equals average velocity ($v_0 = \bar{v}$) and Equation 2.10 becomes the same as Equation 2.7.

A fourth useful equation can be obtained from another algebraic manipulation of Equations 2.7–2.9. Equation 2.9 is solved for t, and the result is substituted into Equation 2.7. Then the definition of $\bar{v}$ from Equation 2.8 is substituted and terms are gathered, yielding

$$v^2 = v_0^2 + 2a(x - x_0) \qquad \text{[constant } a\text{]} \qquad \textbf{(2.11)}$$

This is yet another useful kinematic relationship.

EXAMPLE 2.11 DRAGSTERS REACH GREAT SPEEDS, TOO

Calculate the final velocity of the dragster in Example 2.10, without using information about time.

Strategy Equation 2.11 is ideally suited to this task; it relates velocities, acceleration, and displacement, and no time information is required. First, we identify that $v_0 = 0$, since the dragster starts from rest. Then we note that $x - x_0 = 400$ m was the answer to Example 2.10. Finally, the acceleration was given to be $a = 15.0$ m/s^2.

Solution Equation 2.11 is

$$v^2 = v_0^2 + 2a(x - x_0)$$

Substituting the known values gives

$$v^2 = 0 + 2(15.0 \text{ m/s}^2)(400 \text{ m})$$

Thus,

$$v^2 = 1.20 \times 10^4 \text{ m}^2/\text{s}^2$$

To get v, we take the square root:

$$v = (1.20 \times 10^4 \text{ m}^2/\text{s}^2)^{1/2}$$

Thus,

$$v = 110 \text{ m/s}$$

Discussion 110 m/s is about 245 miles per hour, but even this breakneck speed is short of the record for the quarter mile! Note that square roots have two values; we took the positive value to indicate a velocity in the same direction as the acceleration.

TABLE 2.1

SUMMARY OF KINEMATIC EQUATIONS

$x = x_0 + \bar{v}t$ [constant a]		(2.7)
$\bar{v} = \dfrac{v_0 + v}{2}$ [constant a]		(2.8)
$v = v_0 + at$ [constant a]		(2.9)
$x = x_0 + v_0 t + \frac{1}{2}at^2$ [constant a]		(2.10)
$v^2 = v_0^2 + 2a(x - x_0)$ [constant a]		(2.11)

Once again an examination of Equation 2.11 produces insight into the general relationships among physical quantities. For example, the final velocity depends on how large the acceleration is and the distance over which it acts. Table 2.1 gives a list of useful motion equations for your convenience.

In the following examples, we further explore one-dimensional motion, but in situations requiring slightly more algebraic manipulation. The examples also give insight into problem-solving techniques.

EXAMPLE 2.12 HOW FAR DOES YOUR CAR GO WHEN COMING TO A HALT?

On dry concrete a car can decelerate at a rate of 7.00 m/s², whereas on wet concrete it can decelerate at only 5.00 m/s². Find the distances necessary to stop a car moving at 30.0 m/s (about 67 miles per hour) (a) on dry concrete and (b) on wet concrete. (c) Repeat both calculations, finding the displacement from the point where the driver sees a traffic light turn red, taking into account his reaction time of 0.500 s to get his foot on the brake. (See Figure 2.9.)

Strategy and Concept In all but the simplest problems, it is useful to list known values and to identify exactly what needs to be determined. We shall do this explicitly in the next several examples, using a table to set them off.

Solution to (a) By examining the statement of the problem, we identify what is to be determined and what is given or can be surmised. Note that a deceleration is an acceleration in the opposite direction of velocity. So, here, if v is positive, then a is negative.

Known	To be determined
$v_0 = 30.0$ m/s	$x - x_0$
$v = 0$	
$a = -7.00$ m/s²	

If x_0 is assumed to be zero, then x is the quantity to be determined. Note that we have no information about time; thus, we need an equation without time in it. Examining Table 2.1, we find that only Equation 2.11 does not involve

time. Furthermore, all quantities in Equation 2.11, except $x - x_0$, are known. This equation therefore provides the easiest way to find $x - x_0$. First, it must be manipulated algebraically to get $x - x_0$ on one side and everything else on the other. We start with Equation 2.11:

$$v^2 = v_0^2 + 2a(x - x_0)$$

We know the final velocity is $v = 0$, and substituting this into the equation simplifies it to

$$0 = v_0^2 + 2a(x - x_0)$$

so that

$$x - x_0 = \frac{-v_0^2}{2a}$$

Entering known values yields

$$x - x_0 = \frac{-(30.0 \text{ m/s})^2}{2(-7.00 \text{ m/s}^2)}$$

Thus,

$$\boxed{x - x_0 = 64.3 \text{ m} \qquad \text{when dry (about 211 ft)}}$$

Solution to (b) This part can be solved in exactly the same manner as part (a). The only difference is that the deceleration is 5.00 m/s². The result, without repeating all the steps, is

$$\boxed{x' - x_0' = 90.0 \text{ m} \qquad \text{when wet (about 295 ft)}}$$

Solution to (c) It is reasonable to assume that the velocity remains constant during the driver's reaction time.

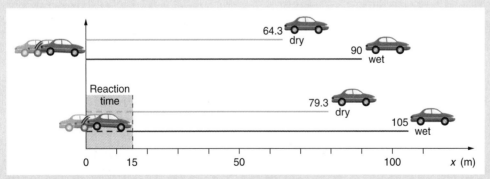

Figure 2.9 The distance necessary to stop a car varies greatly depending on road conditions and driver reaction time. Shown here are the braking distances for dry and wet pavement as calculated in Example 2.12, for a car initially traveling at 30 m/s. Also shown are the total distances traveled from the point where the driver first sees a light turn red, assuming a 0.500 s reaction time.

(continued)

(continued)

Therefore, total displacement is what was found above plus the distance traveled by the car during the reaction time.

Known	To be determined
$\bar{v} = 30.0$ m/s	$x'' - x_0''$
$t' = 0.500$ s	
$a'' = 0$	

With these knowns, Equation 2.7 is the easiest to use:

$$x'' = x_0'' + \bar{v}t$$

Solving for $x'' - x_0''$ gives

$$x'' - x_0'' = \bar{v}t''$$

and substituting the known values yields

$$x'' - x_0'' = (30.0 \text{ m/s})(0.500 \text{ s}) = 15.0 \text{ m}$$

This means the car drifts 15.0 m while the driver reacts, making the total displacements in the two cases of dry and wet concrete 15.0 m greater than if he reacted instantly.

Thus, including reaction time:

$$x - x_0 = 64.3 \text{ m} + 15.0 \text{ m}$$
$$= 79.3 \text{ m} \qquad \text{when dry (about 260 ft)}$$

and

$$x' - x_0' = 90.0 \text{ m} + 15.0 \text{ m}$$
$$= 105 \text{ m} \qquad \text{when wet (about 344 ft)}$$

Discussion The displacements found in this example seem reasonable for stopping a fast moving car. It is interesting that reaction time adds significantly to the displacements. But more important is the general approach to solving problems. We identify the knowns and the quantities to be determined and then find an appropriate equation. There is often more than one way to solve a problem. The various parts of this example can in fact be solved by other methods, but the solutions presented above are the shortest.

EXAMPLE 2.13 HOW MUCH TIME IS NEEDED FOR A CAR TO MERGE INTO TRAFFIC?

Suppose a car merges into freeway traffic on a 200 meter long ramp. If its initial velocity is 10.0 m/s and it accelerates at 2.00 m/s^2, how long does it take to travel the 200 m up the ramp? (Such information might be useful to a traffic engineer.)

Strategy and Concept We are asked to solve for the time t. As before, we identify the known quantities in order to choose a convenient physical relationship (that is, an equation with one unknown t). It is clearly stated that the initial velocity is $v_0 = 10.0$ m/s, and that the acceleration is $a = 2.00$ m/s^2. The 200 m displacement is $x - x_0$, so our table is

Known	To be determined
$v_0 = 10.0$ m/s	t
$x - x_0 = 200$ m	
$a = 2.00$ m/s^2	

Equation 2.10 is the only one that we can use to solve this problem without having to first solve for the final velocity v. Furthermore, all physical quantities in Equation 2.10, except t, are known.

Solution Equation 2.10 is

$$x = x_0 + v_0 t + \tfrac{1}{2}at^2$$

Rearranging terms gives

$$x - x_0 = v_0 t + \tfrac{1}{2}at^2$$

The only unknown in this equation is t. The quadratic formula, discussed in Section 1.7, can be used to solve for t. First, let us enter the knowns into the equation and simplify it. This gives

$$200 \text{ m} = (10.0 \text{ m/s})t + \tfrac{1}{2}(2.00 \text{ m/s}^2)t^2$$

The units of meters (m) cancel because they are in each term. We can get the units of seconds (s) to cancel by taking $t = t$ s, where t is the magnitude of time and s is the unit. Doing so leaves

$$200 = 10t + t^2$$

To use the quadratic formula, it is necessary to rearrange this to get 0 on one side of the equation. We get

$$t^2 + 10t - 200 = 0$$

This is a quadratic equation of the form

$$at^2 + bt + c = 0$$

where the constants are $a = 1$, $b = 10$, and $c = -200$. Its solutions are given by Equation 1.23, the quadratic formula:

$$t = \frac{-b \pm \sqrt{b^2 - 4ac}}{2a}$$

This yields two solutions for t, which are

$$t = 10.0 \quad \text{and} \quad -20.0$$

(It is left as an exercise for the reader to verify these solutions.) In this case, then, the time is $t = t$ s, or

$$t = 10.0 \text{ s} \quad \text{and} \quad -20.0 \text{ s}$$

A negative value for time is unreasonable, since it would mean that the event happened 20 s before the motion began. We can discard that solution. Thus,

$$t = 10.0 \text{ s}$$

Discussion Whenever an equation contains an unknown squared, there will be two solutions. In some problems both solutions are meaningful, but in others such as the above, only one solution is reasonable. The 10.0 s answer seems reasonable for a typical freeway on-ramp.

With the basics of kinematics established, we can go on to many other interesting examples and applications. In the process of developing kinematics, we have also glimpsed a general approach to problem solving that produces both correct answers and insights into physical relationships. The next section discusses problem-solving basics and outlines an approach that will help you succeed.

2.5 PROBLEM-SOLVING BASICS

CONNECTIONS

This is the first of many problem-solving strategies. Others follow in later chapters and build on these basics. The strategies are designed for clarity and ease of use. The main idea is stated in italics with elaboration following. The steps in all strategies are similar and involve the conceptualization of the problem, followed by analysis, and ending with another look at the concepts. Worked examples also follow this approach, which is designed to help you both solve and conceptualize (understand) problems. Such problem-solving techniques are useful in many fields outside physics.

Problem-solving skills are obviously essential to success in a quantitative course in physics. More importantly, the ability to apply broad physical principles, usually represented by equations, to specific situations is a very powerful form of knowledge. It is much more powerful than memorizing a list of facts. Analytical skills and problem-solving abilities can be applied to new situations, whereas a list of facts cannot be made long enough to contain every possible circumstance. Such skills are useful both for solving the end-of-chapter problems in this text and for applying physics in everyday and professional life.

PROBLEM-SOLVING STRATEGIES

While there is no simple step-by-step method that works for every problem, the following general procedures facilitate problem solving and make it more meaningful. (The methods used to solve examples in preceding sections already have utilized this general approach.) A certain amount of creativity and insight is required as well.

Step 1. Examine the situation to determine which physical principles are involved. It often helps to *draw a simple sketch* at the outset. Once you have identified the physical principles, it is much easier to find and apply the equations representing those principles. Although finding the correct equation is essential, keep in mind that equations represent physical principles, laws of nature, and relationships among physical quantities. Without a conceptual understanding of a problem, a numerical solution is meaningless.

Step 2. Identify exactly what needs to be determined in the problem (identify the unknowns). In complex problems, especially, it is not always obvious what needs to be found or in what sequence. Making a list can help.

Step 3. Make a list of what is given or can be inferred from the problem as stated (identify the knowns). Many problems are stated very succinctly and require some inspection to determine what is known. A sketch can also be very useful at this point. Formally identifying the knowns is of particular importance in applying physics in real-world situations.

Step 4. Solve the appropriate equation or equations for the quantity to be determined (the unknown). If the equation contains more than one unknown, then an additional equation is needed to solve the problem. In some problems, several unknowns must be determined to get at the one needed most. In such problems it is especially important to keep physical principles in mind to avoid going astray in a sea of equations.

Step 5. Substitute the knowns along with their units into the appropriate equation, and obtain numerical solutions complete with units. This step produces the numerical answer; it also provides a check on units that can help you find errors. If the units of the answer are incorrect, then an error has been made. However, be warned that correct units do not guarantee that the numerical part of the answer is also correct.

Step 6. Check the answer to see if it is reasonable: Does it make sense? This final step is extremely important—the goal of physics is to accurately describe nature. To see if the answer is reasonable, check both its magnitude and its sign, in addition to its units. Your intuition will improve as you solve more and more physics problems, and it will become possible for you to make finer and finer judgments regarding whether nature is adequately described by the answer to a problem. This step brings the problem back to its conceptual meaning. If you can judge whether the answer is reasonable, you have a deeper understanding of physics than just being able to mechanically solve a problem.

When solving problems, we often perform these steps in different order, and we also tend to do several steps simultaneously. There is no rigid procedure that will work every time. Creativity and insight grow with experience, and the basics of problem solving become almost automatic. One way to get practice is to work out the text's examples for yourself as you read. Another is to work as many end-of-chapter problems as possible, starting with the easiest to build confidence and progressing to the more difficult. Once you become involved in physics, you will see it all around, and you can begin to apply it to situations you encounter outside the classroom, just as is done in many of the applications in this text.

2.6 FALLING OBJECTS

Falling objects form an interesting class of motion problems. For example, we can estimate the depth of a vertical mine shaft by dropping a rock into it and listening for it to hit bottom. By applying the kinematics developed so far to falling objects, we can examine some interesting situations and learn much about gravity in the process.

The most remarkable and unexpected fact about falling objects is that, if air resistance and friction are negligible, then in a given location all objects fall toward the center of the earth with the *same constant acceleration, independent of their mass*. (See Figure 2.10.) This experimentally determined fact is unexpected, because we are so accustomed to the effects of air resistance and friction that we expect light objects to fall slower than heavy ones. Air resistance opposes the motion of an object through the air, while friction between objects—such as between clothes and a laundry chute—also opposes motion. An object *falling without air resistance or friction* is defined to be in **free-fall**.

It is now common knowledge that the force of gravity causes objects to fall. The acceleration of free-falling objects is therefore called the **acceleration due to gravity**. The acceleration due to gravity is *constant*, which means we can apply the kinematics equations developed in Section 2.4 to any falling object where air resistance and friction are negligible. This opens a broad class of interesting situations to us. The acceleration due to gravity is so important that it is given its own symbol—g; it is constant at any given location on the earth and has the following average value:

$$g = 9.80 \text{ m/s}^2 \qquad \textbf{(2.12)}$$

Although g varies from 9.78 to 9.83 m/s^2, depending on latitude, altitude, underlying geological formations, and local topography, the average value of 9.80 m/s^2 will be used unless otherwise specified. The direction of g is *downward*. In fact, its direction *defines* what we call vertical.

The first person to demonstrate convincingly that all free-falling objects fall at the same rate was Galileo Galilei (1564–1642), one of the greatest scientists of all time. (His image appears in Figure 1.6.) Legend has it that Galileo demonstrated this characteristic of gravity by dropping various masses from the Leaning Tower of Pisa, but there is no historical evidence that he actually used the famed tower. His recorded laboratory experiments on gravity settled some very old controversies, proving incorrect the then-popular "logical" notion that heavier objects fall faster. What he demonstrated was that there are causes—namely, air resistance and friction—for different fall rates, and in free-fall all objects fall with the same acceleration.

More important than Galileo's discoveries about gravity was his influence in establishing experimentation and observation as the basis for understanding nature's laws. He was instrumental in ending older methods of determining "truth" that relied primarily on argument and dogmatic adherence to certain beliefs independent of detailed observation.

In air | In a vacuum | In a vacuum (the hard way)

Figure 2.10 A hammer and a feather will fall with the same constant acceleration if air resistance is made negligible. This is a general characteristic of gravity not unique to the earth, as astronaut David R. Scott demonstrated on the moon, where the acceleration due to gravity is only 1.67 m/s^2.

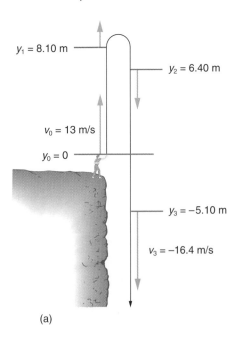

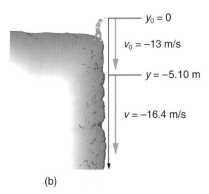

Figure 2.11 (a) A person throws a rock straight up as explored in Example 2.14. The arrows are velocity vectors at 0, 1.00, 2.00, and 3.00 s. (b) A person throws a rock straight down from a cliff with the same initial speed as before, as in Example 2.15. Note that at the same distance below the point of release, the rock has the same velocity in both cases.

Galileo is often credited as being the father of modern science, for observation and experimentation are its very core.

One-Dimensional Motion Involving Gravity

The best way to see the basic features of motion involving gravity is to start with the simplest situations and then progress toward more complex ones. So we start by considering straight up and down motion with no air resistance or friction. This means that the initial velocity is vertical (or zero if an object is dropped), and the objects are in free-fall. Under these circumstances the motion is one-dimensional and has constant acceleration of magnitude g. The kinematic equations, 2.7–2.11, listed in Table 2.1 can be used with $-g$ substituted for a. Using $-g$ is equivalent to defining up to be positive and down negative. We will also represent vertical displacement with the symbol y and use x for horizontal displacement.

EXAMPLE 2.14   A ROCK THROWN STRAIGHT UP EVENTUALLY COMES DOWN

A person standing on the edge of a high cliff throws a rock straight up with an initial velocity of 13.0 m/s. The rock misses the edge of the cliff as it falls back to earth. Calculate the position and velocity of the rock 1.00, 2.00, and 3.00 s after it is thrown, neglecting the effects of air resistance. (See Figure 2.11(a).)

Strategy and Concept We are asked to determine the position y at various times. It is reasonable to take the initial position y_0 to be zero. This is one-dimensional motion along the vertical direction. We use plus and minus signs to indicate direction, with up positive and down negative (we could just as easily do it the other way around). Since up is positive, the initial velocity, $v_0 = +13.0$ m/s, must be positive, too. The acceleration due to gravity is downward, so that $a = -g = -9.80$ m/s^2. It is crucial that the initial velocity and the acceleration due to gravity have opposite signs. This indicates that gravity opposes the velocity and will slow and eventually reverse it. Our table of values for the entire example, where the subscripts 1, 2, and 3 denote the three times, is

Known	To be determined
$y_0 = 0$ (chosen)	y_1, y_2, y_3
$v_0 = 13.0$ m/s	v_1, v_2, v_3
$a = -g = -9.80$ m/s^2	
$t = 1.00, 2.00, 3.00$ s	

Solution for y_1 First, let us find the position at $t = 1.00$ s. The table for this is

Known	To be determined
$y_0 = 0$ (chosen)	y_1
$v_0 = 13.0$ m/s	
$a = -g = -9.80$ m/s^2	
$t = 1.00$ s	

Examining the kinematic equations in Table 2.1, we see that Equation 2.10 has the final position (y_1 here) on one side (no algebraic manipulation required) and all the other quantities in the equation are known. It is thus the easiest to use. Equation 2.10 is (with y_1 for the final position)

$$y_1 = y_0 + v_0 t + \tfrac{1}{2}at^2$$

(continued)

(continued)

Substituting $-g$ for a yields
$$y_1 = y_0 + v_0 t - \tfrac{1}{2}gt^2$$
Entering the knowns yields
$$y_1 = 0 + (13.0 \ \text{m/s})(1.00 \ \text{s}) - \tfrac{1}{2}(9.80 \ \text{m/s}^2)(1.00 \ \text{s})^2$$
so that
$$y_1 = 8.10 \ \text{m}$$

Discussion The rock is 8.10 m *above its starting point* at $t = 1.00$ s, since y_1 is positive. It could be moving up or down; the only way to tell is to calculate v_1 and find out if it is positive or negative. A positive value for v_1 would mean that it was moving up. Note that according to the rules for subtraction, this answer ($13.0 - 4.90 = 8.10$) should only be given to 2 digits. It is expressed to three digits, however, since it is close to 10 and significant figures are approximate.

Solution for v_1 To find v_1, it is useful to make another table of knowns, adding the now-known value of position y_1:

Known	To be determined
$y_0 = 0$ (chosen)	v_1
$y_1 = 8.10$ m	
$v_0 = 13.0$ m/s	
$a = -g = -9.80$ m/s^2	
$t = 1.00$ s	

Again by examining Table 2.1 we see that Equation 2.9 is the most straightforward to find v_1—it is
$$v = v_0 + at$$
Substituting $-g$ for a and v_1 for v yields
$$v_1 = v_0 - gt$$

and entering the knowns gives
$$\begin{aligned} v_1 &= 13.0 \ \text{m/s} - (9.80 \ \text{m/s}^2)(1.00 \ \text{s}) \\ &= 3.20 \ \text{m/s} \end{aligned}$$

Discussion The positive value for v_1 means that the rock is still heading upward at $t = 1.00$ s. It has slowed from its original 13.0 m/s, as expected.

Solution for remaining times The calculations for the position and velocity at $t = 2.00$ and 3.00 s are identical in nature to those already done. The results are summarized in the following table and are illustrated in Figure 2.11(a).

$t = 1.00$ s	$y_1 = 8.10$ m	$v_1 = 3.20$ m/s
$t = 2.00$ s	$y_2 = 6.40$ m	$v_2 = -6.60$ m/s
$t = 3.00$ s	$y_3 = -5.10$ m	$v_3 = -16.40$ m/s

Discussion The interpretation of these results is important. At 1.00 s, the rock is above its starting point and heading upward, since y_1 and v_1 are both positive. At 2.00 s, the rock is still above its starting point, but the negative velocity means it is moving downward. At 3.00 s, both y_3 and v_3 are negative, meaning the rock is below its starting point and continuing to move downward. Note that the values for y are the *positions* (or displacements) of the rock, *not* the total distances traveled. Finally, note that free-fall applies to upward motion as well as downward. Both have the same acceleration. Astronauts training in the famous Vomit Comet, for example, experience free-fall while arcing up as well as down, as we will discuss in more detail later.

EXAMPLE 2.15 A ROCK THROWN DOWN STILL ACCELERATES

What happens if the person on the cliff throws the rock straight down, as in Figure 2.11(b), instead of straight up? To explore this question, calculate the velocity of the rock when it is 5.10 m below the starting point if it is thrown downward with an initial speed of 13.0 m/s.

Strategy and Concept Since up is positive, the final position of the rock is negative ($y = -5.10$ m) because it is below the starting point. Similarly, the initial velocity is downward and therefore negative, as is the acceleration due to gravity. We expect the final velocity to turn out negative since the rock will continue to move downward.

Solution Examining the table of known values, we see that time is not known, and so we seek an equation for v that does not involve time.

Known	To be determined
$y_0 = 0$ (chosen)	v
$y = -5.10$ m	
$v_0 = -13.0$ m/s	
$a = -g = -9.80$ m/s^2	

Of the kinematic equations in Table 2.1, Equation 2.11 seems well suited, since the only unknown in it is v:
$$v^2 = v_0^2 + 2a(y - y_0)$$
Substituting $-g$ for a,
$$v^2 = v_0^2 - 2g(y - y_0)$$
Entering known values, we obtain
$$\begin{aligned} v^2 &= (-13.0 \ \text{m/s})^2 - 2(9.80 \ \text{m/s}^2)(-5.10 \ \text{m}) \\ &= 169 \ \text{m}^2/\text{s}^2 + 99.96 \ \text{m}^2/\text{s}^2 = 268.96 \ \text{m}^2/\text{s}^2 \end{aligned}$$
Taking the square root, noting that a square root can be positive or negative, gives
$$v = \pm 16.4 \ \text{m/s}$$
The negative root is chosen to indicate that the rock is still heading down. Thus,
$$v = -16.4 \ \text{m/s}$$

Discussion Note that this is exactly the same velocity the rock had at this position when it was thrown straight upward with the same initial speed. (See Example 2.14 and Figure 2.11(a).) This is not an accidental result. When all forces but

(continued)

(continued)

gravity are negligible, the *speed* of a falling object depends only on its initial speed and its vertical position relative to the starting point. For example, if the velocity of the rock is calculated at a height of 8.10 m above the starting point (using the method of the previous example) when the initial velocity is 13.0 m/s straight up, a result of ±3.20 m/s is obtained.

Here both signs are meaningful; the positive value occurs when the rock is at 8.10 m and heading up, and the negative value occurs when the rock is at 8.10 m and heading back down. It has the same *speed* but the opposite direction. There is a symmetry here that will be mentioned again when conservation of energy is considered in Chapter 6.

EXAMPLE 2.16   FIND *g* FROM DATA ON A FALLING OBJECT

The acceleration due to gravity can be calculated from data taken in an introductory physics laboratory course. An object, usually a metal ball for which air resistance is negligible, is dropped and the time it takes to fall a known distance is measured. See, for example, Figure 2.12. Very accurate results can be produced with this method if sufficient care is taken in measuring the distance fallen and the elapsed time. Suppose the distance fallen is measured to be 1.0000 m and the time 0.45173 s, with a small uncertainty in the last digit of each number. Calculate the acceleration due to gravity.

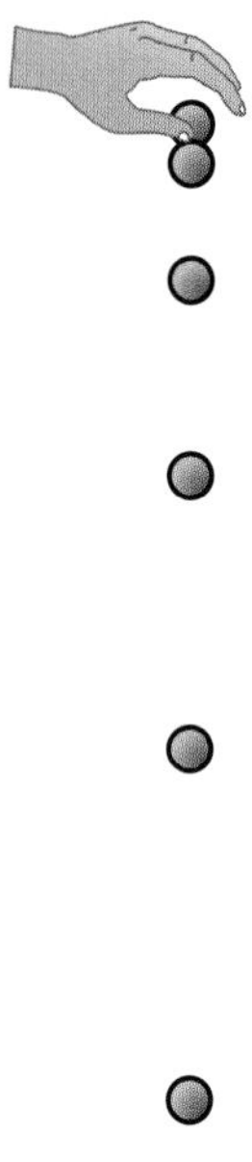

	y (m)	*v* (m/s)	*t* (s)
	0	0	0
	−0.049	−0.98	0.1
	−0.196	−1.96	0.2
	−0.441	−2.94	0.3
	−0.784	−3.92	0.4
	−1.225	−4.90	0.5

Figure 2.12   Positions and velocities of a metal ball released from rest when air resistance is negligible. Velocity is seen to increase linearly with time while displacement increases with time squared, as expected for motion with a constant acceleration.

Strategy   Up is positive, and so the displacement $(y - y_0)$ is negative. Time is given and the initial velocity is surmised to be zero since the ball is dropped and not thrown down. Since the ball is in free-fall, its acceleration is that due to gravity; thus

Known	*To be determined*
$y - y_0 = -1.0000$ m	$a \ (= -g)$
$t = 0.45173$ s	
$v_0 = 0$	

Solution   Searching for the most straightforward method for finding acceleration leads us to Equation 2.10, since we know everything in it except acceleration. It is

$$y = y_0 + v_0 t + \tfrac{1}{2}at^2$$

Substituting 0 for v_0 yields

$$y = y_0 + \tfrac{1}{2}at^2$$

Solving for *a* gives

$$a = \frac{2(y - y_0)}{t^2}$$

Substituting known values yields

$$a = \frac{2(-1.0000 \text{ m})}{(0.45173 \text{ s})^2}$$

Thus,

$$a = -9.801 \text{ m/s}^2 = -g$$

So that

$$\boxed{g = 9.801 \text{ m/s}^2}$$

Discussion   The negative value for *a* indicates it is downward, as expected. Since the data going into the calculation is relatively precise, this value for *g* is more precise than the average value of 9.80 m/s^2; it represents the local value for the acceleration due to gravity.

2.7*   GRAPHICAL ANALYSIS OF ONE-DIMENSIONAL MOTION

A graph, like a picture, is worth a thousand words. Graphs such as those in preceding sections not only contain numerical information, they also reveal the relationship of one physical quantity to another. This section uses graphs of displacement, velocity, and acceleration versus time to illustrate one-dimensional kinematics.

Slopes and General Relationships

First note that graphs in this text have perpendicular axes, one horizontal on the page and the other vertical. When two physical quantities are plotted against one another in such a graph, the horizontal axis is usually considered to be an independent variable and the vertical axis a dependent variable. If we call the horizontal axis the x-axis and the vertical axis the y-axis, as in Figure 2.13, a straight-line graph has the general form

$$y = mx + b$$

Here m is the **slope**, defined to be the rise divided by the run (as seen in the figure) of the straight line. The letter b is used for the **y-intercept**, which is the point at which the line crosses the vertical axis.

Time is usually an independent variable that other quantities, such as displacement, depend upon. A graph of displacement versus time would, thus, have x on the vertical axis and t on the horizontal axis. Figure 2.14 is just such a straight-line graph. Using the relationship between dependent and independent variables, we see that the slope in Figure 2.14 is average velocity $\bar{v}$ and the intercept is displacement at time zero—that is, x_0. Substituting these symbols into $y = mx + b$ gives

$$x = \bar{v}t + x_0$$

or

$$x = x_0 + \bar{v}t$$

which is the kinematic relation in Equation 2.7. Thus a graph of displacement versus time gives a general relationship among displacement, velocity, and time, as well as giving detailed numerical information about a specific situation.

Figure 2.14 shows a graph of displacement versus time for a jet-powered car on the Bonneville Salt Flats. From the figure you can see that the car has a displacement of 400 m at time 0, 650 m at $t = 1.0$ s, and so on. Its displacement at times other than those listed in the table can be read from the graph; furthermore, information about its velocity and acceleration can also be obtained from the graph.

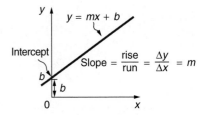

Figure 2.13 A straight-line graph. The equation for the straight line is $y = mx + b$.

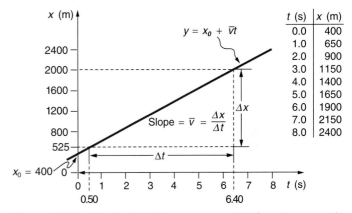

t (s)	x (m)
0.0	400
1.0	650
2.0	900
3.0	1150
4.0	1400
5.0	1650
6.0	1900
7.0	2150
8.0	2400

Figure 2.14 Graph of displacement versus time for a jet-powered car on the Bonneville Salt Flats.

EXAMPLE 2.17 AVERAGE VELOCITY FROM A GRAPH OF DISPLACEMENT VERSUS TIME

Find the average velocity of the car the position of which is graphed in Figure 2.14.

Strategy and Concept The slope of a graph of x vs. t is average velocity, since slope equals rise over run. In this case, rise = Δx, and run = Δt, so that

$$\text{slope} = \frac{\Delta x}{\Delta t} = \bar{v}$$

Since the slope is constant here, any two points on the graph can be used to find the slope. (Generally speaking, it is most accurate to use two widely separated points on the straight line. This is because any error in reading data from the graph is proportionally smaller if the interval is larger.)

Solution Using the two points shown, we get

$$\bar{v} = \frac{\Delta x}{\Delta t} = \frac{2000 \text{ m} - 525 \text{ m}}{6.4 \text{ s} - 0.50 \text{ s}}$$

yielding

$$\boxed{\bar{v} = 250 \text{ m/s}}$$

Discussion This is an impressively large land speed (900 km/h, or about 560 mi/h). But it is considerably shy of the record of 283 m/s (1020 km/h or 633 mi/h) set in 1983.

The graphs in Figure 2.15 represent the motion of the jet-powered car as it accelerates toward its top speed, but only during the time when its acceleration is constant. Time starts at zero for this motion (as if measured with a stopwatch), and the displacement and velocity are initially 200 m and 15 m/s, respectively.

The graph of displacement versus time in Figure 2.15(a) is a curve rather than a straight line. The slope of the curve becomes steeper as time progresses, implying that the velocity is increasing. The slope at any point on a displacement versus time graph is the instantaneous velocity at that point. It is found by drawing a straight line tangent to the curve at the point of interest and taking the slope of this straight line. This is shown for two points in Figure 2.15(a). If this is done at every point on the curve and the values plotted against time, then the graph of velocity versus time shown in Figure 2.15(b) is obtained.

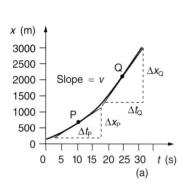

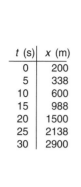

t (s)	x (m)
0	200
5	338
10	600
15	988
20	1500
25	2138
30	2900

(a)

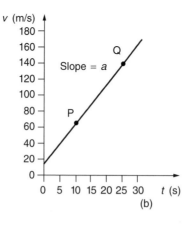

t (s)	v (m/s)
0	15
5	40
10	65
15	90
20	115
25	140
30	165

(b)

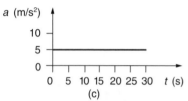

(c)

Figure 2.15 Motion of a jet-powered car during the time span when its acceleration is constant. (a) The slope of an x vs. t graph is velocity. This is shown at two points, and the instantaneous velocities obtained are plotted in the next graph. Instantaneous velocity at any point is the slope of the tangent at that point. (b) The slope of the v vs. t graph is constant for this part of the motion, indicating constant acceleration. (c) Acceleration has the constant value of $5.0 \ \text{m/s}^2$ over the time interval plotted.

EXAMPLE 2.18 INSTANTANEOUS VELOCITY FROM THE SLOPE AT A POINT

Calculate the velocity of the jet car at a time of 25 s by finding the slope of the x vs. t graph in Figure 2.15(a).

Strategy and Concept The slope of a curve at a point is equal to the slope of a straight line tangent to the curve at that point. This is illustrated in Figure 2.15(a), where Q labels the point at $t = 25$ s.

Solution Data from Figure 2.15(a) yield

$$\text{slope} = v_Q = \frac{\Delta x_Q}{\Delta t_Q} = \frac{(3120 - 1300) \ \text{m}}{(32 - 19) \ \text{s}}$$

where the endpoints of the tangent at Q correspond to a position of 1300 m at time 19 s and a position of 3120 m at time 32 s. Thus,

$$v_Q = \frac{1820 \ \text{m}}{13 \ \text{s}}$$
$$= 140 \ \text{m/s}$$

Discussion This is the value given in the table for v at $t = 25$ s. The value of 140 m/s for v_Q is plotted in Figure 2.15(b). The entire graph of v vs. t can be obtained in this fashion.

Carrying this one step further, we note that the slope of a velocity versus time graph is acceleration. Slope is rise divided by run; on a v vs. t graph, rise = Δv and run = Δt, so that

$$\text{slope} = \frac{\Delta v}{\Delta t} = a$$

Since the velocity versus time graph in Figure 2.15(b) is a straight line, its slope is the same everywhere, implying that acceleration is constant. Acceleration versus time is graphed in Figure 2.15(c).

Additional general information can be obtained from Figure 2.15(b) and the expression for a straight line,

$$y = mx + b$$

In this case the vertical axis (y) is v, the intercept b is v_0, the slope (m) is a, and the horizontal axis (x) is t. Substituting these symbols yields

$$v = v_0 + at$$

This is Equation 2.9. A general relationship for velocity, acceleration, and time has again been obtained from a graph.

It is not accidental that the same equations are obtained by graphical analysis as by algebraic techniques. In fact, an important way to *discover* physical relationships is to measure various physical quantities and then make graphs of one quantity against another to see if they are correlated in any way. Correlations imply physical relationships and might be evidenced by smooth graphs such as those above. From such graphs, mathematical

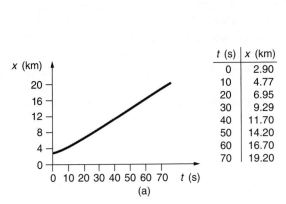

(a)

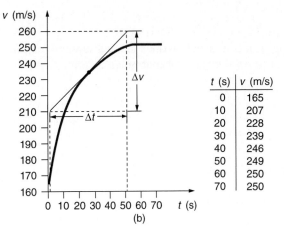

(b)

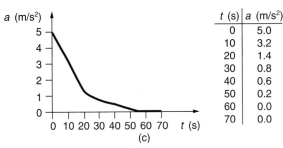

(c)

Figure 2.16   Motion of a jet-powered car as it reaches its top velocity. This motion begins where the motion in Figure 2.15 ends. (a) The slope of this graph is velocity; it is plotted in the next graph. (b) The velocity gradually approaches its top value. The slope of this graph is acceleration; it is plotted in the final graph. (c) Acceleration gradually declines to zero when velocity becomes constant.

relationships can sometimes be postulated. Further experiments are performed to determine the validity of the hypothesized relationships.

Now consider the motion of the jet car as it goes from 165 m/s to its top velocity of 250 m/s, graphed in Figure 2.16. Time again starts at zero, and the initial displacement and velocity are 2900 m and 165 m/s, respectively. (These were the final displacement and velocity of the car in the motion graphed in Figure 2.15.) Acceleration gradually decreases from 5.0 m/s^2 to zero when the car hits 250 m/s. The slope of the x vs. t graph increases until $t = 55$ s, after which time the slope is constant. Similarly, velocity increases until 55 s and then becomes constant, since acceleration decreases to zero at 55 s and remains zero afterward.

EXAMPLE 2.19   ACCELERATION FROM A GRAPH OF VELOCITY VERSUS TIME

Calculate the acceleration of the jet car at a time of 25 s by finding the slope of the v vs. t graph in Figure 2.16(b).

Strategy and Concept   The slope of the curve at $t = 25$ s is equal to the slope of the line tangent at that point, as illustrated in Figure 2.16(b).

Solution   Taking numbers from the figure gives values of v and t that we can subtract to find Δv and Δt:

$$\text{slope} = \frac{\Delta v}{\Delta t} = \frac{(260 - 210) \text{ m/s}}{(51 - 1.0) \text{ s}}$$

where the endpoints of the tangent to the curve at 25 s, as drawn in Figure 2.16(b), correspond to a velocity of 210 m/s at 1.0 s, and 260 m/s at 51 s. Thus,

$$a = \frac{50 \text{ m/s}}{50 \text{ s}} = 1.0 \text{ m/s}^2$$

Discussion   Note that this value for a is consistent with the value plotted in Figure 2.16(c) at $t = 25$ s.

A graph of displacement versus time can be used to generate a graph of velocity versus time, and a graph of velocity versus time can be used to generate a graph of acceleration versus time. We do this by finding the slope of the graphs at every point. If acceleration is constant, we can determine equations that describe the graphs. These equations were previously included in Table 2.1. Graphical analysis of motion can be used to describe both specific and general characteristics of kinematics. This is true in other topics in physics. An important aspect of exploring physical relationships is to graph them and look for underlying relationships.

SUMMARY

Kinematics is the study of motion without regard to mass or force. In this chapter, it is limited to motion along a straight line, called one-dimensional motion. **Displacement** is defined to be the change in position of an object. In symbols, displacement Δx is defined to be

$$\Delta x = x_2 - x_1 \tag{2.1}$$

where x_1 is the initial position and x_2 is the final position. In this text, the Greek letter Δ (delta) always means "change in" whatever quantity follows it. The SI unit for displacement is the meter. Displacement has a direction as well as a magnitude. **Distance** is defined to be the magnitude of displacement.

Time is measured in terms of change, and its SI unit is the second (s). Elapsed time for an event is

$$\Delta t = t_2 - t_1 \tag{2.2}$$

where t_1 is the initial time and t_2 is the final time. The initial time is often taken to be zero, as if measured with a stopwatch; the elapsed time is then just t.

Average velocity $\bar{v}$ is defined to be displacement divided by the time of travel. In symbols, average velocity is

$$\left. \begin{aligned} \bar{v} &= \frac{\Delta x}{\Delta t} \\ &= \frac{x_2 - x_1}{t_2 - t_1} \end{aligned} \right\} \tag{2.3}$$

Its SI unit is m/s. Velocity has a direction specified. **Instantaneous velocity** v is the velocity at a specific instant or the average velocity for an infinitesimal or zero time interval. **Instantaneous speed** is defined to be the magnitude of instantaneous velocity. **Average speed** is the total distance traveled divided by elapsed time. (Average speed is *not* the magnitude of the average velocity.) Speed has no direction specified.

Acceleration is the rate at which velocity changes. In symbols, **average acceleration** $\bar{a}$ is

$$\left. \begin{aligned} \bar{a} &= \frac{\Delta v}{\Delta t} \\ &= \frac{v_2 - v_1}{t_2 - t_1} \end{aligned} \right\} \tag{2.4}$$

The SI unit for acceleration is m/s². **Instantaneous acceleration** a is the acceleration at a specific instant in time. Acceleration has a direction specified.

A **vector** is any quantity that has magnitude and direction. A **scalar** is any quantity that has magnitude but no direction. Displacement, velocity, and acceleration are vectors, whereas distance and speed are scalars. In one-dimensional motion, direction is specified by a plus or minus sign to signify left, right, up, down, and the like.

To simplify calculations we take acceleration to be constant, so that $\bar{a} = a$ at all times. We also take initial time to be zero. Initial position and velocity are given a subscript 0; final values have no subscript. Thus,

$$\left. \begin{aligned} \Delta t &= t \\ \Delta x &= x - x_0 \\ \Delta v &= v - v_0 \end{aligned} \right\} \tag{2.5}$$

The following kinematic equations for motion with constant a are useful:

$$x = x_0 + \bar{v}t \quad \text{[constant } a] \tag{2.7}$$

$$\bar{v} = \frac{v_0 + v}{2} \quad \text{[constant } a] \tag{2.8}$$

$$v = v_0 + at \quad \text{[constant } a] \tag{2.9}$$

$$x = x_0 + v_0 t + \tfrac{1}{2}at^2 \quad \text{[constant } a] \tag{2.10}$$

$$v^2 = v_0^2 + 2a(x - x_0) \quad \text{[constant } a] \tag{2.11}$$

where $\bar{v}$ is the average velocity. In vertical motion, y is substituted for x.

An object in **free-fall** has only gravity acting on it (no air resistance or friction). On earth, all free-falling objects have an **acceleration due to gravity** g, which averages

$$g = 9.80 \text{ m/s}^2 \tag{2.12}$$

Since acceleration is constant, the kinematic equations above can be applied with $-g$ substituted for a. Up is positive for displacement, velocity, and acceleration.

Graphs can be used to analyze motion. The slope of an x vs. t graph is velocity, and the slope of a v vs. t graph is acceleration. Average velocity, instantaneous velocity, and acceleration can all be obtained in this manner.

CONCEPTUAL QUESTIONS

2.1 According to a highway map, the driving distance from San Francisco to Sacramento is 138 km. The distance between these cities, as measured by a reconnaissance satellite, is 123 km. Can these two numbers be consistent? If so, explain how. If not, explain why not.

2.2 Give an example in which there are clear distinctions among total distance traveled, distance, and displacement. Specifically identify each quantity in your example.

2.3 Under what circumstances does total distance traveled equal distance? What is the only case in which distance and displacement are exactly the same?

2.4 Give an example (but not one from the text) of a device used to measure time, and identify what change in that device indicates a change in time.

2.5 There is a distinction between average speed and the magnitude of average velocity. Give an example that illustrates the difference between these two quantities.

2.6 If you divide the total distance traveled on a car trip (as determined by the odometer) by the time for the trip, are you calculating the average speed or the magnitude of the average velocity? Under what circumstances are these two quantities the same?

2.7 How are instantaneous velocity and instantaneous speed related to one another? How do they differ?

2.8 What is the average acceleration of the earth relative to the sun over one year's time? What is its average velocity over the same time period?

2.9 It is possible for speed to be constant while acceleration is not zero? Give an example of such a situation.

2.10 Is it possible for velocity to be constant while acceleration is not zero? Explain.

2.11 Give an example in which velocity is zero yet acceleration is not.

2.12 The subway shuttle car shown in Figure 2.6(b) is moving to the left (a negative velocity). Suppose it comes to a stop. What then is the direction of its acceleration? Is the acceleration positive or negative?

2.13 Plus and minus signs are used in one-dimensional motion to indicate direction. What is the sign of an acceleration that reduces the magnitude of a negative velocity? Of a positive velocity?

2.14 What is the acceleration of a rock thrown straight upward on the way up? At the top of its flight? On the way down?

2.15 An object that is thrown straight up falls back to earth. This is a one-dimensional motion. When is its velocity zero? Does its velocity change direction? Does the acceleration due to gravity have the same sign on the way up as on the way down?

2.16 Suppose you throw a rock nearly straight up at a coconut in a palm tree, and the rock misses on the way up but hits the coconut on the way down. Neglecting air resistance, how does the speed of the rock when it hits the coconut on the way down compare with what it would have been if it had hit the coconut on the way up? Is it more likely to dislodge the coconut on the way up or down? Explain.

2.17 If an object is thrown straight up and air resistance is negligible, then its speed when it returns to the starting point is the same as when it was released. If air resistance were *not* negligible, how would its speed upon return compare with its initial speed? How would the maximum height to which it rises be affected?

2.18 The severity of a fall depends on your speed when you strike the ground. All factors but gravity being the same, how many times higher could a safe fall on the moon be than on the earth?

2.19 How many times higher could an astronaut jump on the moon than on earth if his takeoff speed is the same as on earth?

2.20* Sketch a graph of velocity versus time corresponding to the graph of displacement versus time given in Figure 2.17. Identify the time (t_a, t_b, t_c, t_d, or t_e) at which the instantaneous velocity is greatest, the time at which it is zero, and the time at which it is negative.

2.21* Sketch a graph of velocity versus time corresponding to the graph of displacement versus time given in Figure 2.18. Identify the time or times (t_a, t_b, t_c, etc.) at which the instantaneous velocity is greatest. At which times is it zero? At which times is it negative?

2.22* Sketch a graph of acceleration versus time corresponding to the graph of velocity versus time given in Figure 2.19.

2.23* Sketch a graph of acceleration versus time corresponding to the graph of velocity versus time given in Figure 2.20. Identify the time or times (t_a, t_b, t_c, etc.) at which the acceleration is greatest. At which times is it zero? At which times is it negative?

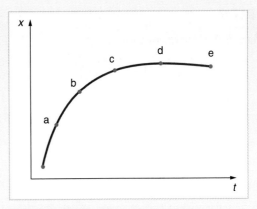

Figure 2.17 Question 20.

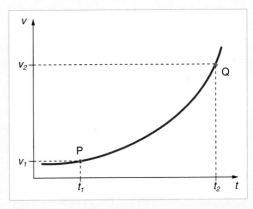

Figure 2.18 Graph of displacement versus time for the motion of an object. Question 21.

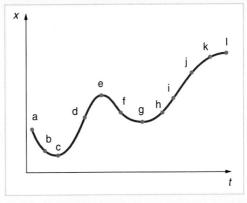

Figure 2.19 Graph of velocity versus time for a rocket launch. Question 22.

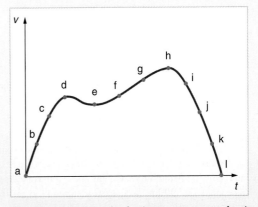

Figure 2.20 Graph of velocity versus time for the motion of an object. Question 23.

PROBLEMS

Section 2.1 Displacement

2.1 Find the following for path A in Figure 2.21: (a) The total distance traveled. (b) The distance from start to finish. (c) The displacement from start to finish.

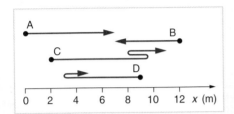

Figure 2.21 Various one-dimensional paths. Problems 1 through 4.

2.2 Find the following for path B in Figure 2.21: (a) The total distance traveled. (b) The distance from start to finish. (c) The displacement from start to finish.

2.3 Find the following for path C in Figure 2.21: (a) The total distance traveled. (b) The distance from start to finish. (c) The displacement from start to finish.

2.4 Find the following for path D in Figure 2.21: (a) The total distance traveled. (b) The distance from start to finish. (c) The displacement from start to finish.

Sections 2.2–2.4 Displacement, Velocity, Acceleration, and Time

2.5 (a) Calculate the earth's average speed relative to the sun. (b) What is its average velocity over a period of one year?

2.6 A helicopter blade is spinning at 100 revolutions per minute. Its tip is 5.00 m from the center of rotation. (a) Calculate the average speed of the blade tip in the helicopter's frame of reference. (b) What is its average velocity over one revolution?

• 2.7 The planetary model of the atom pictures electrons orbiting the atomic nucleus much as planets orbit the sun. In this model one can view hydrogen, the simplest atom, as having a single electron in a circular orbit 1.06×10^{-10} m in diameter. (a) If the average speed of the electron in this orbit is known to be 2.20×10^{6} m/s, calculate the number of revolutions per second it makes about the nucleus. (b) What is the electron's average velocity?

2.8 With the ability to measure the motion of land masses, continental drift has become an established fact. The North American and European continents are drifting apart at a rate of about 3 cm/y. At this rate how long will it take them to drift 500 km further apart than at present?

2.9 Land west of the San Andreas Fault is moving at an average velocity of about 6 cm/y northwest relative to land east of the fault. Los Angeles is west of the fault and may thus someday be at the same latitude as San Francisco, which is east of the fault. How far in the future will this occur if the displacement to be made is 590 km northwest, assuming the motion remains constant?

2.10 On May 26, 1934, a streamlined, stainless steel diesel train called the Zephyr set the world's nonstop long-distance speed record for trains. Its run from Denver to Chicago took 13 h 4 min 58 s and was witnessed by more than a million people along the route. The total distance traveled was 1633.8 km. What was its average speed in km/h and m/s?

2.11 Tidal friction is slowing the rotation of the earth. As a result, the orbit of the moon is increasing in radius (the reasons for this are discussed in Chapter 7) at a rate of approximately 4 cm/y. Assuming this to be a constant rate, how many years will pass before the radius of the moon's orbit increases by 3.84×10^{6} m (1%)?

• 2.12 A student drove to the university from her home and noted that the odometer on her car increased by 12.0 km. The trip took 18.0 min. (a) What was her average speed? (b) If the straight-line distance from her home to the university is 10.3 km in a direction 25.0° south of east, what was her average velocity? (c) If she returned home by the same path 7 h 30 min after she left, what were her average speed and velocity for the entire trip?

• 2.13 A football quarterback runs 15.0 m straight down the playing field in 2.50 s. He is then hit and pushed 3.00 m straight backward in 1.75 s. He breaks the tackle and runs straight forward another 21.0 m in 5.20 s. Calculate his average velocity for each of the three intervals and for the entire motion.

⁞ 2.14 Conversations with astronauts on the lunar surface were characterized by a kind of echo in which the earthbound person's voice was so loud in the astronaut's space helmet that it was picked up by his microphone and transmitted back to earth. It is reasonable to assume that the echo time equals the time necessary for the radio wave to travel from earth to the moon and back (that is, neglecting any time delays in the electronic equipment). Calculate the distance from the earth to the moon given that the echo time was 2.56 s and that radio waves travel at the speed of light, 3.00×10^{8} m/s.

2.15 A properly motivated cheetah can accelerate from rest to a speed of 30.0 m/s in 7.00 s. What is its acceleration?

2.16 On December 10, 1954, Dr. John Paul Stapp rode a rocket sled to a speed of 282 m/s (632 mi/h), enduring extreme accelerations. He was accelerated from rest to his top speed in 5.00 s and brought jarringly back to rest in only 1.40 s! Calculate his acceleration and deceleration. Express each in multiples of g by taking its ratio to the acceleration of gravity.

2.17 An Olympic-class sprinter starts a race with an acceleration of 4.50 m/s². What is her speed 2.40 s later?

2.18 A commuter backs her car out of her garage with an acceleration of 1.40 m/s². (a) How long does it take her to reach a speed of 2.00 m/s? (b) If she then brakes to a stop in 0.800 s, what is her deceleration?

2.19 A pitched baseball is caught in a well-padded mitt. If the deceleration of the ball is 2.10×10^{4} m/s² and 1.85 ms elapse from the time the ball first touches the mitt until it stops, what was the initial velocity of the ball?

2.20 Assume that an MX missile goes from rest to a suborbital speed of 6.50 km/s in 60.0 s (the actual speed and time are classified). What is its average acceleration in m/s² and in multiples of g?

2.21 A bullet in a gun is accelerated from the firing chamber to the end of the barrel at an average rate of 6.20×10^5 m/s^2 for 8.10×10^{-4} s. What is its muzzle velocity (that is, its final velocity)?

•**2.22** (a) A light-rail commuter train accelerates at a rate of 1.35 m/s^2. How long does it take it to reach its top speed of 80.0 km/h starting from rest? (b) The same train ordinarily decelerates at a rate of 1.65 m/s^2. How long does it take to come to a stop from its top speed? (c) In emergencies the train can decelerate more rapidly, coming to rest from 80.0 km/h in 8.30 s. What is its emergency deceleration in m/s^2?

2.23 While entering a freeway, a car accelerates from rest at a rate of 2.40 m/s^2 for 12.0 s. (a) Draw a sketch of the situation. (b) List the knowns in this problem. (c) How far does the car travel in those 12.0 s? To solve this part, first identify the unknown, and then discuss how you chose the appropriate equation to solve for it. After choosing the equation, show your steps in solving for the unknown, checking units, and discuss if the answer is reasonable. (d) What is the car's final velocity? Solve for this unknown in the same manner as in part (c), showing all steps explicitly.

2.24 At the end of a race a runner decelerates from a velocity of 9.00 m/s at a rate of 0.400 m/s^2. (a) How far does she travel in the next 15.0 s? (b) What is her final velocity?

2.25 Blood is accelerated from rest to 30.0 cm/s in a distance of 1.80 cm by the left ventricle of the heart. (a) Make a sketch of the situation. (b) List the knowns in this problem. (c) How long does the acceleration take? To solve this part, first identify the unknown, and then discuss how you chose the appropriate equation to solve for it. After choosing the equation, show your steps in solving for the unknown, checking units. (d) Is the answer reasonable when compared with the time for a heartbeat?

2.26 In a slap shot a hockey player accelerates the puck from a velocity of 8.00 m/s to 40.0 m/s in the same direction. If this takes 3.33×10^{-2} s, calculate the distance over which the acceleration acts.

2.27 A powerful motorcycle can accelerate from rest to 26.8 m/s (60 mi/h) in only 3.90 s. (a) What is its average acceleration? (b) How far does it travel in that time?

2.28 Freight trains can produce only relatively small accelerations and decelerations. (a) What is the final velocity of a freight train that accelerates at a rate of 0.0500 m/s^2 for 8.00 min, starting with an initial velocity of 4.00 m/s? (b) If the train can decelerate at a rate of 0.550 m/s^2, how long will it take to come to a stop from this velocity? (c) How far will it travel in each case?

2.29 A fireworks shell is accelerated from rest to a velocity of 65.0 m/s over a distance of 0.250 m. (a) How long did the acceleration last? (b) Calculate the acceleration.

•**2.30** A swan on a lake gets airborne by flapping its wings and running on top of the water. (a) If the swan must reach a velocity of 6.00 m/s to take off and it accelerates from rest at an average rate of 0.350 m/s^2, how far will it travel before becoming airborne? (b) How long does this take?

•**2.31** A woodpecker's brain is specially protected from large decelerations by tendonlike attachments inside the skull. While pecking on a tree the woodpecker's head comes to a stop from an initial velocity of 0.600 m/s in a distance of only 2.00 mm. (a) Find the deceleration in m/s^2 and in multiples of g. (b) Calculate the stopping time. (c) The tendons cradling the brain stretch, making its stopping distance 4.50 mm (greater than the head and, hence, less deceleration of the brain). What is the brain's deceleration, expressed in multiples of g?

• **2.32** The light created by a TV, computer, or oscilloscope screen is caused by electrons striking the phosphor-coated interior surface. These electrons are easily accelerated by voltages inside the tube. Typically, an electron might be accelerated at 8.00×10^{16} m/s^2 over a distance of 2.10 cm. (a) List the knowns in this problem. (b) What is its final velocity if the electron starts from rest? To solve this part, first identify the unknown, then discuss how you chose the appropriate equation to solve for it. After choosing the equation, show your steps in solving for the unknown, checking units, and discuss if the answer is reasonable. (c) How long does it take to reach this velocity? (d) Once accelerated, the electron travels another 18.0 cm to the front screen at constant velocity. How long does this take? (e) Do the last two answers seem reasonable? If not, say why not. If so, state why.

• **2.33** An unwary football player collides with a padded goalpost while running at a velocity of 7.50 m/s and comes to a full stop after compressing the padding and his body 0.350 m. (a) How long does the collision last? (b) What is his deceleration?

⁞**2.34** An express subway train passes through an underground station. It enters with an initial velocity of 22.0 m/s and decelerates at a rate of 0.150 m/s^2 as it goes through. The station in 210 m long. (a) How long is the nose of the train in the station? (b) How fast is it going when the nose leaves the station? (c) If the train is 130 m long, when does the end of the train leave the station? (d) What is the velocity of the end of the train as it leaves?

⁞**2.35** The Concorde is a supersonic transport (SST) that has been in regular commercial use since 1976. This problem considers a typical flight from Washington, D.C., to Paris, a total distance of 6180 km. (a) The plane accelerates from rest to 295 m/s (the speed of sound at cruising altitude) in the first 16.0 min of flight. Calculate its average acceleration. (b) In the following 38.0 min, the plane accelerates to its top speed of 596 m/s (Mach 2.02, or 2.02 times the speed of sound). What is its average acceleration for this part of the flight? (c) For the next 2.00 h the plane cruises at its top speed. How far has it traveled in the first 2 h 54 min of flight? (d) The plane now decelerates and lands after decelerating for 39 min. What is the average speed for the last part of the flight? (Note that deceleration is not constant for this last leg, since the plane lands at a speed of about 100 m/s and comes quickly to a stop.) (e) Calculate the average speed for the entire flight.

⁞**2.36** A bicycle racer sprints at the end of a race to clinch a victory. The racer has an initial velocity of 11.5 m/s and accelerates at the rate of 0.500 m/s^2 for 7.00 s. (a) What is his final velocity? (b) The racer continues at this velocity to the finish line. If he was 300 m from the finish line when he started to accelerate, how much time did he save? (c) One other racer was 5.00 m ahead when the winner started to accelerate, but he was unable to accelerate and traveled at 11.8 m/s until the finish line. How far ahead of him (in meters and in seconds) did the winner finish?

Section 2.6 Falling Objects

Assume air resistance is negligible unless otherwise stated.

2.37 Calculate the displacement and velocity at times of 0.500, 1.00, 1.50, and 2.00 s for a snowball thrown straight up with an initial velocity of 15.0 m/s. Take the point of release to be $y_0 = 0$.

2.38 Calculate the displacement and velocity at times of 0.500, 1.00, 1.50, 2.00, and 2.50 s for a rock thrown straight down with an initial velocity of 14.0 m/s from the Verrazano Narrows bridge in New York City. The underside of this bridge is 70.0 m above the water at its center.

2.39 A basketball referee tosses the ball straight up for the starting tipoff. At what velocity must a basketball player leave the ground to rise 1.25 m above the floor in an attempt to get the ball?

2.40 A rescue helicopter is hovering over a person whose boat has sunk. One of the rescuers throws a life preserver straight down to the victim with an initial velocity of 1.40 m/s and observes that it takes 1.8 s to reach the water. (a) List the knowns in this problem. (b) How high above the water was the preserver released? Note that the downdraft of the helicopter reduces the effects of air resistance on the falling preserver, so that an acceleration equal to that of gravity is reasonable.

• 2.41 A dolphin in an aquatic show jumps straight up out of the water at a velocity of 13.0 m/s. (a) List the knowns in this problem. (b) How high does his body rise above the water? To solve this part, first note that the final velocity is now a known and identify its value. Then identify the unknown, and discuss how you chose the appropriate equation to solve for it. After choosing the equation, show your steps in solving for the unknown, checking units, and discuss if the answer is reasonable. (c) How long is the dolphin in the air? Neglect any effects due to his size or orientation.

• 2.42 A swimmer bounces straight up from a diving board and falls feet first into a pool. She starts with a velocity of 4.00 m/s, and her takeoff point is 1.80 m above the pool. (a) How long are her feet in the air? (b) What is her highest point above the board? (c) What is her velocity when her feet hit the water?

• 2.43 (a) Calculate the height of a cliff if it takes 2.35 s for a rock to hit the ground when it is thrown straight up from the cliff with an initial velocity of 8.00 m/s. (b) How long would it take to reach the ground if it is thrown straight down with the same speed?

• 2.44 A very strong, but inept, shot putter puts the shot straight up with an initial velocity of 11.0 m/s. How long does he have to get out of the way if the shot was released at a height of 2.20 m, and he is 1.80 m tall (he could end up shorter)?

⁞ 2.45 There is a 250 m high cliff at Half Dome in Yosemite National Park. Suppose a boulder breaks loose from the top of this cliff. (a) How fast will it be going when it strikes the ground? (b) Assuming a reaction time of 0.300 s, how long will a tourist at the bottom have to get out of the way after hearing the sound of the rock breaking loose (neglecting the height of the tourist, which would become negligible anyway if hit)? The speed of sound is 335 m/s on this day.

⁞ 2.46 Suppose you drop a rock into a dark well and, using precision equipment, you measure the time for the sound of a splash to return. (a) Neglecting the time required for sound to travel up the well, calculate the distance to the water if the sound returns in 2.0000 s. (b) Now calculate the distance taking into account the time for sound to travel up the well. The speed of sound is 332.00 m/s in this well.

⁞ 2.47 A steel ball is dropped onto a hard floor from a height of 1.50 m and rebounds to a height of 1.45 m. (a) Calculate its velocity just before it strikes the floor. (b) Calculate its velocity just after it leaves the floor on its way back up. (c) Calculate its acceleration during contact with the floor if that contact lasts 0.0800 ms. (d) How much did the ball compress during its collision with the floor, assuming the floor is absolutely rigid?

⁞ 2.48 Solve Problem 2.47 for a soft tennis ball that is dropped from a height of 1.50 m and rebounds to a height of 1.10 m, and for which the collision with the floor lasts 3.50 ms.

Section 2.7* Graphical Analysis of One-Dimensional Motion

Note: There is always uncertainty in numbers taken from graphs. If your answers differ from expected values, examine them to see if they are within data extraction uncertainties estimated by you.

2.49* (a) By taking the slope of the curve in Figure 2.15(a), verify that the velocity of the jet car is 115 m/s at $t = 20$ s. (b) By taking the slope of the curve at any point in Figure 2.15(b), verify that the jet car's acceleration is 5.0 m/s².

2.50* Take the slope of the curve in Figure 2.16(a) to verify that the velocity at $t = 10$ s is 207 m/s.

2.51* Take the slope of the curve in Figure 2.16(a) to verify that the velocity at $t = 30$ s is 239 m/s.

2.52* (a) By taking the slope of the curve in Figure 2.16(a), verify that the velocity is 249 m/s at $t = 50$ s. (b) By taking the slope of the curve in Figure 2.16(b), verify that the acceleration is 3.2 m/s² at $t = 10$ s.

2.53* (a) By taking the slope of the curve in Figure 2.16(b), verify that the acceleration is 1.4 m/s² at $t = 20$ s. (b) Repeat to verify that the acceleration is 0.8 m/s² at $t = 30$ s.

2.54* (a) Take the slope of the curve in Figure 2.22(a) to find the velocity at $t = 2.5$ s. (b) Repeat at 7.5 s. These values must be consistent with the graph in Figure 2.22(b).

2.55* (a) Take the slope of the curve in Figure 2.22(b) to find the acceleration at $t = 2.5$ s. (b) Repeat at 5.0 and 15 s. Check for consistency with the graph in Figure 2.22(c).

⁞ 2.56* Take the slope of the displacement graph in Figure 2.23, at $t = 10$, 30, and 50 s, to find the velocities at those times. Do these agree with the values read from the velocity versus time graph?

⁞ 2.57* Take the slope of the velocity graph in Figure 2.23, at $t = 5.0$, 10, and 30 s, to find the accelerations at those times. Do these agree with the values read from the acceleration versus time graph?

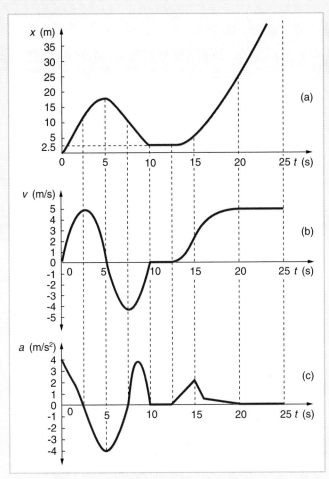

Figure 2.22 (a) Graph of displacement versus time for a jogger who starts on a run, stops, and goes back to pick up his keys at 2.5 m before continuing. (b) Graph of velocity versus time produced by finding the slope of the displacement graph. (c) Graph of acceleration versus time produced by finding the slope of the velocity graph. Problems 54 and 55.

UNREASONABLE RESULTS

Physics must describe nature accurately. The following problems have results that are unreasonable because one premise is unreasonable or because certain of its premises are inconsistent with one another. The physical principle applied correctly then produces an unreasonable result. For example, if a person starting a foot race accelerates at 0.40 m/s² for 100 s, his final speed will be 40 m/s (about 90 miles per hour)—clearly unreasonable because the time of 100 s is an unreasonable premise. The physics is correct in a sense, but there is more to describing nature than just manipulating equations correctly. Checking the result of a problem to see if it is reasonable does more than help uncover errors in problem solving—it also builds intuition in judging whether nature is being accurately described.

PROBLEM-SOLVING STRATEGIES

Use the following strategies to determine if an answer is reasonable and, if it is not, to determine what is the cause.

Step 1. *Solve the problem using strategies as outlined in Section 2.5 and in the format followed in the worked examples in the text.* In the

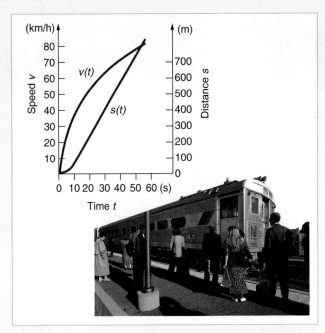

Figure 2.23 Graphs of displacement, velocity, and acceleration versus time for this light-rail vehicle, which is used in metropolitan transit systems. Problems 56 and 57.

example given in the preceding paragraph, you would identify the givens as the acceleration and time and use Equation 2.9 to find the unknown final velocity. That is,

$$v = v_0 + at = 0 + (0.40 \text{ m/s}^2)(100 \text{ s}) = 40 \text{ m/s}$$

Step 2. *Check to see if the answer is reasonable.* Is it too large or too small, or does it have the wrong sign, improper units, . . .? In the present case, the velocity is about four times higher than a person can run—so it is too large.

Step 3. *If the answer is unreasonable, look for what specifically could cause the identified difficulty.* In the example of the runner, there are only two assumptions that are suspect. The acceleration could be too great or the time too long. People can easily accelerate at 0.40 m/s²; thus, the time must be too long.

2.58 Suppose you drop a rock from a cliff and you observe that it falls 39.2 m in 2.00 s. (a) What is the acceleration of the rock? (b) What is unreasonable about the result? (c) Which premise is unreasonable, or are the premises inconsistent?

2.59 An advertisement claims that a certain automobile can stop from a speed of 35.0 m/s in a distance of 20.0 m. (a) What is the average deceleration? (b) What is unreasonable about the result? (c) Which premise is unreasonable?

TWO-DIMENSIONAL KINEMATICS

You may have experienced vigorous curved path motions like this

theme park ride. Curved-path motion is not one-dimensional, but is

described in a similar fashion.

The arc of a football, the orbit of a satellite, a bicycle rounding a curve, a swimmer diving into a pool, and a puppy chasing its tail are but a few examples of motions along curved paths. In fact, most motions in nature follow curved paths rather than straight lines. Motion along a curved path on a flat surface (a plane), such as that of a ball on a pool table or a skater on a rink, is two-dimensional and, thus, described by two-dimensional kinematics. Motion not confined to a plane, such as a car following a winding mountain road, is described by three-dimensional kinematics. Both two- and three-dimensional kinematics are simple extensions of the one-dimensional kinematics developed for straight-line motion in the previous chapter. This simple extension will allow us to apply physics to many more situations, and it will also yield unexpected insights about nature.

3.1 KINEMATICS IN TWO DIMENSIONS: AN INTRODUCTION

What are some of the important differences between one- and two-dimensional motion? To answer this question, let us examine some examples of two-dimensional motion. First, suppose you want to walk from one point to another in a city with uniform square blocks, as pictured in Figure 3.1. The straight-line path that a helicopter might fly is blocked to you as a pedestrian, and so you are forced to take a two-dimensional path, such as the one shown. You walked 14 blocks in all, 9 east followed by 5 north. What is the straight-line distance? There is an old adage that the shortest distance between two points is a straight line. The two legs of the trip and the straight-line path form a right triangle, and so the Pythagorean theorem* can be used to find the straight-line distance. The hypotenuse of the triangle is the straight-line path, and so its length in units of city blocks is $\sqrt{9^2 + 5^2} = 10.3$, considerably shorter than the 14 blocks you walked.

The fact that the straight-line distance (10.3 blocks) in Figure 3.1 is less than the total distance walked (14 blocks) is one example of a general characteristic of vectors. (**Vectors** were defined in Section 2.1 to be any quantity that has magnitude and direction.) *Vectors do not add like ordinary numbers unless they are along the same direction.*

As in Chapter 2, we use arrows to represent vectors. The length of the arrow is proportional to the vector's magnitude. This is shown by hash marks in Figure 3.1. The arrow points in the same direction as the vector. The three vectors in Figure 3.1 all have different directions. The first represents a 9 block displacement east. The second represents a 5 block displacement north. These are added to give the third vector, the 10.3 block total displacement. In this example the magnitude of the total displacement can be found using the Pythagorean theorem, since the first and second vectors are perpendicular. In following sections, we will develop techniques for adding vectors having any direction, not just those perpendicular to one another.

Now let us compare the motions of two balls. One ball is dropped from rest. At the same instant, another is thrown horizontally from the same height and follows a curved path. (See Figure 3.2.) A stroboscope has captured the positions of the balls at fixed time

---

*See Section 1.8 for a review of angles, triangles, and simple trigonometry, including the Pythagorean theorem.

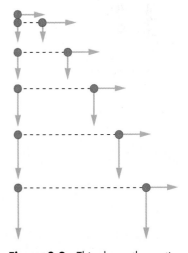

Figure 3.2 This shows the motions of two identical balls—one falls from rest, the other has an initial horizontal velocity. Each subsequent position is an equal time interval. Arrows represent horizontal and vertical velocities at each position. The vertical velocities and positions are identical for both balls and the horizontal velocity is constant, indicating that the vertical and horizontal motions are independent.

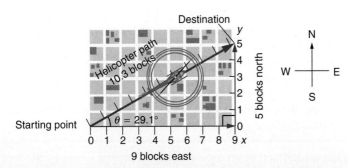

Figure 3.1 A pedestrian walks a two-dimensional path between two points in a city. The straight-line path followed by a helicopter between the two points is shorter than the 14 blocks walked by the pedestrian. All blocks are square and the same size.

intervals as they fell. It is remarkable that the vertical positions of the two balls are the same for each flash of the strobe. This implies that the vertical motion is independent of whether or not the ball is moving horizontally. Furthermore, careful examination of the ball thrown horizontally shows it travels the same horizontal distance between flashes. This means that the horizontal velocity is constant, and affected neither by vertical motion nor by gravity (which is vertical). It has been found that

Motions along perpendicular directions, such as horizontal and vertical, are independent.

The two-dimensional curved path of the horizontally thrown ball is composed of two independent one-dimensional motions. The key to analyzing such motion, called projectile motion, is to break it into motions along perpendicular directions. We shall see how to do this in the sections that follow, and in subsequent chapters we will find such techniques to be useful in many areas of physics.

There are two commonly employed techniques for dealing with the special characteristics of vectors in two and three dimensions. Section 3.2 introduces graphical techniques, and Section 3.3 covers analytical (mathematical) techniques.

3.2 VECTOR DEFINITIONS AND GRAPHICAL METHODS OF VECTOR ADDITION AND SUBTRACTION

A vector was defined in Section 2.1 to be any quantity that has magnitude and direction. Displacement and velocity are two examples of vectors; there are many others, though, such as acceleration and force. In one-dimensional, or straight-line, motion, the direction of a vector can be given simply by a plus or minus sign. In two dimensions, we specify the direction of a vector relative to some reference frame, using an arrow having length proportional to the vector's magnitude and pointing in the direction of the vector. Figure 3.3 shows such a **graphical representation of a vector**, using as an example the total displacement for the person walking in a city considered in the previous section. We shall use the notation that a boldface symbol, such as **D**, stands for a vector. Its magnitude is represented by the symbol in italics, D, and its direction by θ.

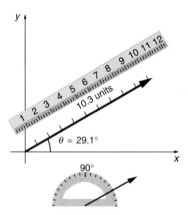

Figure 3.3 An arrow is drawn to represent the total displacement vector **D** for the person walking in a city considered in Figure 3.1. Using a protractor, a line is drawn at an angle θ relative to the east–west axis. The length of the arrow, D, is proportional to the vector's magnitude and is measured along the line with a ruler.

PROBLEM-SOLVING STRATEGY

FOR THE GRAPHICAL ADDITION OF VECTORS

The **graphical addition of vectors** uses the **head-to-tail method** illustrated in Figure 3.4. To use this method:

Step 1. *Draw an arrow to represent the first vector using a ruler and protractor*, as was done in Figure 3.3.
Step 2. Now draw an arrow to represent the second vector. *Place the tail of the second vector at the head of the first* as shown in Figure 3.4. (Movement of a vector without changing its direction or magnitude is called *translation*.)
Step 3. *Continue this process for each vector to be added.* (In Figure 3.4 only two vectors are added, whereas in Figure 3.5 three are added.)
Step 4. *Draw an arrow from the tail of the first vector to the head of the last.* This is the **resultant**, or the sum, of the other vectors.
Step 5. To get the **magnitude of the resultant**, *measure its length with a ruler.*
Step 6. To get the **direction of the resultant**, *measure the angle it makes with the reference frame using a protractor.*

The graphical addition of vectors is limited in accuracy only by the precision with which the drawings can be made. It is valid for any number of vectors. Figure 3.5 shows how the three vectors given in Example 3.1 are added graphically.

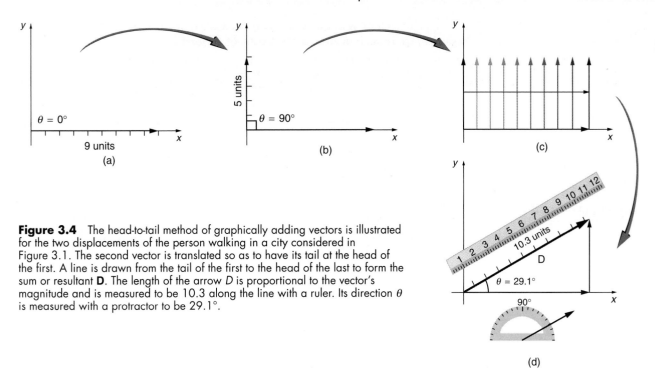

Figure 3.4 The head-to-tail method of graphically adding vectors is illustrated for the two displacements of the person walking in a city considered in Figure 3.1. The second vector is translated so as to have its tail at the head of the first. A line is drawn from the tail of the first to the head of the last to form the sum or resultant **D**. The length of the arrow D is proportional to the vector's magnitude and is measured to be 10.3 along the line with a ruler. Its direction θ is measured with a protractor to be 29.1°.

EXAMPLE 3.1   ADDING VECTORS GRAPHICALLY (HEAD-TO-TAIL)

Use the graphical technique for adding vectors to find the total displacement of a person who walks the following three paths (displacements) on a flat field. First, she walks 25.0 m in a direction 49.0° north of east. Then, she walks 23.0 m heading 15.0° north of east. Finally, she turns and walks 32.0 m in a direction 68.0° south of east.

Strategy Represent each displacement vector graphically with an arrow, labeling the first **A**, the second **B**, and the third **C**, making the lengths proportional to the distance and the directions as specified relative to an east–west line. The head-to-tail method outlined above will give a value for the magnitude and direction of the resultant displacement, denoted **R**.

Solution Figure 3.5 shows how the three displacement

vectors are drawn and placed head to tail, and how the resultant is measured. The total displacement **R** is seen to have a magnitude of 50.0 m and to lie in a direction 7.0° south of east. In symbols, this is

$$R = 50.0 \text{ m}$$

and

$$\theta = 7.0° \text{ south of east}$$

Discussion The head-to-tail graphical method of vector addition works for any number of vectors. It is illustrated for two vectors in Figure 3.4 and for three in Figure 3.5. The resultant is also independent of the order in which the vectors are added, as seen in Figure 3.5(c).

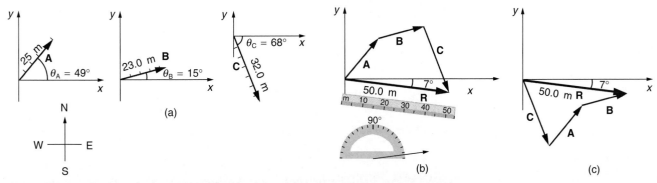

Figure 3.5 (a) The three displacements walked by the woman in Example 3.1 are represented as arrows with appropriate lengths and directions. (b) The arrows are placed head to tail. Their sum is the resultant **R** drawn from the tail of the first to the head of the last vector. Its length R is measured with a ruler, and its direction θ is measured with a protractor. (c) The same resultant is obtained when the vectors are added in a different order. This is true for all vector addition.

In Figure 3.5(c) the same vectors are added in a different order; this yields the same resultant. This is an important characteristic of vectors:

> **Vectors can be added in any order.**

Although illustrated for only one case in Figure 3.5, it is true in every case. Vector addition is thus **commutative**—that is, it can be performed in any order (just like the addition of ordinary numbers—you get the same result whether you add 2 + 3 or 3 + 2, for example). In symbols, this can be stated as **A** + **B** = **B** + **A**.

Vector subtraction is a straightforward extension of vector addition. To define subtraction (say we want to subtract **B** from **A**, written **A** − **B**), we must first define what we mean by −**B**. The **negative** of a vector **B** is defined to be −**B**; that is, graphically *the negative of any vector has the same magnitude but the opposite direction*, as shown in Figure 3.6. In other words, −**B** has the same length as **B**, but points in the reverse direction.

The **subtraction** of vector **B** from vector **A** is defined to be the addition of −**B** to **A**. That is, **A** − **B** ≡ **A** + (−**B**). Thus vector subtraction is the addition of a negative vector, analogous to the subtraction of scalars (where, for example, 5 − 2 ≡ 5 + (−2)). The result is independent of the order in which the subtraction is made. When vectors are subtracted graphically, the techniques outlined above are used, as the following example illustrates.

Figure 3.6 The negative of a vector is just another vector of the same magnitude but pointing in the opposite direction. So −**B** is the negative of **B**—it has the same length but opposite direction.

EXAMPLE 3.2 SUBTRACTING VECTORS GRAPHICALLY

Use graphical techniques to subtract the vector **B** from the vector **A** shown in Figure 3.7(a). These vectors again represent displacements made by a person walking on flat ground. The vector **A** represents the first leg of the trip, in which a person walks 27.5 m in a direction 66.0° north of east. The vector **B** represents a displacement 30.0 m in a direction 112° north of east (or 22.0° west of north).

Strategy and Concept Subtracting **B** from **A** means that the person walks the second leg in the opposite direction of **B**, which is seen in Figure 3.7(a) to be a distance of 30.0 m at an angle 68.0° south of east.

Solution Figure 3.7(b) shows how the total displacement (resultant) **R** is obtained graphically by placing the vectors **A** and −**B** head to tail and then drawing a line from the tail of **A** to the head of −**B**. The magnitude and direction of **R** are determined with ruler and protractor to be

$$R = 23.0 \text{ m}$$

and

$$\theta = 7.5° \text{ south of east}$$

Discussion Since subtraction of a vector is the same as addition of one with the opposite direction, the graphical method of subtracting vectors works the same as for addition.

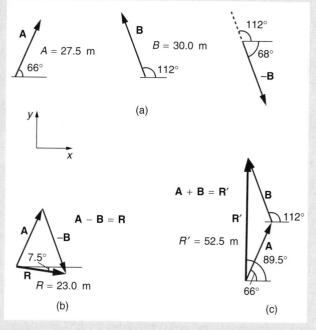

Figure 3.7 Graphical subtraction of vectors discussed in Example 3.2, where **A** − **B** is found. (a) The vectors **A**, **B**, and −**B** represent displacements for a person walking on level ground. (b) To subtract **B** from **A**, the arrows **A** and −**B** are placed head to tail. Their sum is the resultant **R** drawn from the tail of the first to the head of the last vector. Its length *R* is measured with a ruler, and its direction *θ* is measured with a protractor. (c) The addition of the vectors, **A** + **B** = **R′**, is shown for contrast. It represents a different second leg on a walk, and so the person ends up at a different location.

If we decided to walk three times as far on the first leg of the trip considered in the preceding example, then we would walk 3 times 27.5 m, or 82.5 m, in a direction 66.0° north of east. This is an example of multiplying a vector by a positive scalar—the magnitude is changed, but the direction stays the same. If the scalar is negative, then multiplying a vector by it changes the vector's magnitude and gives the new vector the opposite direction. The definition of the **multiplication of a vector A by a scalar** c is that the magnitude of the vector becomes the absolute value of cA, and its direction is not changed if c is positive and is reversed if c is negative. In our case, $c = 3$ and $A = 27.5$ m. Vectors are multiplied by scalars in many situations.

Given a vector, we often want to find what other vectors added together produce it; this is the reverse of vector addition and subtraction. For example, we may know that the total displacement of a person walking in a city is 10.3 blocks in a direction 29.1° north of east and want to find out how many blocks east and north had to be walked. This is called *finding the components (or parts)* of the displacement in the east and north directions, and it is the inverse of the process followed in Figure 3.4 to find the total displacement. It is one example of finding the components of a vector. There are many situations in physics where this is a useful thing to do. Most of these involve finding components along perpendicular axes (such as north and east), so that right triangles are involved. The analytical techniques in the next section are ideal for finding components.

3.3   ANALYTICAL METHODS OF VECTOR ADDITION AND SUBTRACTION

Analytical methods of vector addition and subtraction employ geometry and simple trigonometry rather than the ruler and protractor of graphical methods. Part of the graphical technique is retained, since vectors are still represented by arrows for easy visualization. However, analytical methods are more concise and accurate than graphical methods, which are limited by the accuracy with which a drawing can be made. Analytical methods are limited only by the accuracy with which physical quantities are known. These methods have been made very convenient with the advent of inexpensive handheld calculators.

Analytical techniques and right triangles go hand in hand in physics. This is because, among other things, motions along perpendicular directions are independent. We very often need to add and subtract vectors that are perpendicular, as we did for the person taking a walk in a city. We will also often find doing the reverse very useful—given a vector like **A** in Figure 3.8, we may wish to find what two perpendicular vectors, $\mathbf{A}_x$ and $\mathbf{A}_y$, add to produce it. $\mathbf{A}_x$ and $\mathbf{A}_y$ are defined to be the **components of A along the x- and y-axes**. The three vectors **A**, $\mathbf{A}_x$, and $\mathbf{A}_y$ form a right triangle.

If the vector **A** is known, then its magnitude A (its length) and its angle θ (its direction) are known. **To find A_x and A_y, its x- and y-components**, we use the following relationships for a right triangle*:

$$A_x = A \cos \theta \qquad (3.1)$$

and

$$A_y = A \sin \theta \qquad (3.2)$$

Suppose, for example, that **A** is the vector representing the total displacement of the person walking in a city considered in the preceding two sections. Then $A = 10.3$ blocks and $\theta = 29.1°$, so that $A_x = A \cos \theta = (10.3 \text{ blocks})(\cos 29.1°) = (10.3 \text{ blocks})(0.874) = 9.00$ blocks. This is consistent with the original situation, a 9 block eastward displacement. Similarly, $A_y = A \sin \theta = (10.3 \text{ blocks})(\sin 29.1°) = (10.3 \text{ blocks})(0.486) = 5.00$ blocks, or a 5 block northward displacement as before.

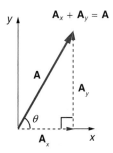

Figure 3.8   The vector **A**, with its tail at the origin of an x,y-coordinate system, is shown together with its x- and y-components, $\mathbf{A}_x$ and $\mathbf{A}_y$. These vectors form a right triangle. The analytical relationships among these vectors are summarized in Equations 3.1 through 3.4.

---

*See Section 1.8 for a review of simple trigonometry including relationships for right triangles. Note that Equations 3.1 and 3.2 are for θ measured relative to the positive x-axis.

If the perpendicular components A_x and A_y are known, then **A** can be found analytically. **To find the magnitude A and direction θ of a vector from its perpendicular components A_x and A_y,** we use the following relationships:

$$A = \sqrt{A_x^2 + A_y^2} \tag{3.3}$$

$$\theta = \tan^{-1}(A_y/A_x) \tag{3.4}$$

Equation 3.3 is just the Pythagorean theorem for the right triangle in Figure 3.8. For example, if A_x and A_y are 9 and 5 blocks, respectively, then $A = \sqrt{9^2 + 5^2} = 10.3$, again consistent with the example of the person walking in a city. Finally, the direction is $\theta = \tan^{-1}(5/9) = 29.1°$, as before.

Equations 3.1 and 3.2 are used to find the perpendicular components of a vector—that is, to go from A and θ to A_x and A_y. Equations 3.3 and 3.4 are used to find a vector from its perpendicular components—that is, to go from A_x and A_y to A and θ. Both processes are crucial to analytical methods of vector addition and subtraction.

To see how to add vectors using perpendicular components, consider Figure 3.9, in which the vectors **A** and **B** are added to produce the resultant **R**. If **A** and **B** represent two legs of a walk (two displacements), then **R** is the total displacement. The person taking the walk ends up at the tip of **R**. There are many ways to arrive at the same point. In particular, the person could have walked first in the x-direction and then in the y-direction. Those paths are the x- and y-components of the resultant, R_x and R_y. If we know R_x and R_y, we can find R and θ using Equations 3.3 and 3.4. This is what is done when the analytical method of vector addition is used.

PROBLEM-SOLVING STRATEGY

FOR THE ANALYTICAL ADDITION OF VECTORS

The **addition of vectors using perpendicular components** is illustrated in Figure 3.9. To use this analytical method:

Step 1. *Find the components of each vector to be added along the chosen perpendicular axes.* Use Equations 3.1 and 3.2 to find the components. In Figure 3.9, these components are A_x, A_y, B_x, and B_y. The angles **A** and **B** make with the x-axis are θ_A and θ_B, respectively.

Step 2. *Find the components of the resultant along each axis by adding the components of the individual vectors along that axis.* That is, as shown in Figure 3.9,

$$R_x = A_x + B_x \tag{3.5}$$

and

$$R_y = A_y + B_y \tag{3.6}$$

Components along the same axis, say the x-axis, are vectors along the same line and, thus, can be added to one another like ordinary numbers. The same is true for components along the y-axis. (For example, a 9 block eastward walk could be taken in two legs, the first 3 blocks east and the second 6 blocks east, for a total of 9, since they are along the same direction.) So breaking vectors into components along common axes makes it easier to add them. Now that the components of **R** are known, its magnitude and direction can be found.

Step 3. *To get the* **magnitude of the resultant** R, *use the Pythagorean theorem as stated in Equation 3.3.* That is,

$$R = \sqrt{R_x^2 + R_y^2} \tag{3.7}$$

Step 4. *To get the* **direction of the resultant** θ, *use Equation 3.4. That is,*

$$\theta = \tan^{-1}(R_y/R_x) \tag{3.8}$$

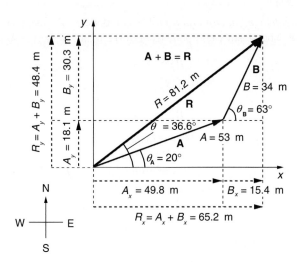

Figure 3.9 The addition of two vectors using perpendicular components. Here **A** and **B** are two legs of a walk, and **R** is the resultant or total displacement. The components of **A** and **B** are found along the perpendicular axes. The components along each axis are added to obtain the components of **R**. Then R and its direction θ are calculated.

The following example illustrates this technique for adding vectors using perpendicular components.

EXAMPLE 3.3   ADD VECTORS BY ADDING COMPONENTS

Add the vector **A** to the vector **B** shown in Figure 3.9, using perpendicular components along the x- and y-axes. The x- and y-axes are along the east–west and north–south directions, respectively. Vector **A** represents the first leg of a walk in which a person walks 53.0 m in a direction 20.0° north of east. Vector **B** represents the second leg, a displacement of 34.0 m in a direction 63.0° north of east.

Strategy and Concept The components of **A** and **B** along the x- and y-axes represent walking due east and due north to get to the same ending point. Once found, they are combined to produce the resultant.

Solution Following the method outlined above, we first find the components of **A** and **B** along the x- and y-axes. Note that $A = 53.0$ m, $\theta_A = 20.0°$, $B = 34.0$ m, and $\theta_B = 63.0°$. We find the x-components by using Equation 3.1, which gives

$$A_x = A \cos \theta_A = (53.0 \text{ m})(\cos 20.0°)$$
$$= (53.0 \text{ m})(0.940) = 49.8 \text{ m}$$

and

$$B_x = B \cos \theta_B = (34.0 \text{ m})(\cos 63.0°)$$
$$= (34.0 \text{ m})(0.454) = 15.4 \text{ m}$$

Similarly, the y-components are found using Equation 3.2:

$$A_y = A \sin \theta_A = (53.0 \text{ m})(\sin 20.0°)$$
$$= (53.0 \text{ m})(0.342) = 18.1 \text{ m}$$

and

$$B_y = B \sin \theta_B = (34.0 \text{ m})(\sin 63.0°)$$
$$= (34.0 \text{ m})(0.891) = 30.3 \text{ m}$$

The x- and y-components of the resultant are thus

$$R_x = A_x + B_x = 49.8 \text{ m} + 15.4 \text{ m} = 65.2 \text{ m}$$

and

$$R_y = A_y + B_y = 18.1 \text{ m} + 30.3 \text{ m} = 48.4 \text{ m}$$

Now we can find the magnitude of the resultant by using the Pythagorean theorem as in Equation 3.7:

$$R = \sqrt{R_x^2 + R_y^2} = \sqrt{(65.2)^2 + (48.4)^2} \text{ m}$$

so that

$$R = \sqrt{6601} \text{ m} = 81.2 \text{ m}$$

Finally, we find the direction of the resultant by using Equation 3.8:

$$\theta = \tan^{-1}(R_y/R_x) = +\tan^{-1}(48.4/65.2)$$

Thus,

$$\theta = \tan^{-1}(0.742) = 36.6°$$

Discussion This example illustrates the addition of vectors using perpendicular components as shown in Figure 3.9. Vector subtraction using perpendicular components is very similar—it is just the addition of a negative vector.

Subtraction of vectors is accomplished by the addition of a negative vector. That is, $\mathbf{A} - \mathbf{B} \equiv \mathbf{A} + (-\mathbf{B})$. Thus, *the method for the subtraction of vectors using perpendicular*

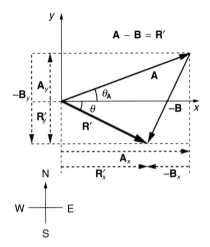

Figure 3.10 The subtraction of the two vectors shown in Figure 3.9. The components of −**B** are the negatives of the components of **B**. The method of subtraction is the same as that for addition.

components is identical to that for addition. The components of −**B** are the negatives of the components of **B**. The *x*- and *y*-components of the resultant **R**′ = **A** − **B** are thus

$$R'_x = A_x - B_x$$

and

$$R'_y = A_y - B_y$$

and the rest of the method outlined above is identical to that for addition. (See Figure 3.10.)

Analyzing vectors using perpendicular components is very useful in many areas of physics, since perpendicular quantities are often independent of one another. The next section, on projectile motion, is the first of many in which using perpendicular components helps make the picture clear and simplifies the physics.

3.4 PROJECTILE MOTION

Projectile motion is the motion of an object moving through the air, subject to the acceleration of gravity. The object is called a *projectile*, and its path is called its *trajectory*. The motion of falling objects, as covered in Section 2.6, is a simple one-dimensional type of projectile motion in which there is no horizontal movement. In this section we consider two-dimensional projectile motion, such as that of a shot put or other object for which *air resistance is negligible*.

The most important fact to remember here is that *motions along perpendicular axes are independent*. This was discussed in Section 3.1 and illustrated in Figure 3.2, where vertical and horizontal motions were seen to be independent. The key to analyzing two-dimensional projectile motion is to break it into two motions, one along the horizontal axis and the other along the vertical. (This is the most sensible choice of axes, since gravity is vertical—thus, there will be no acceleration along the horizontal axis when air resistance is negligible.) As is customary, we call the horizontal axis the *x*-axis and the vertical axis the *y*-axis. Figure 3.11 illustrates the notation for displacement, where **s** is defined to be the total displacement and **x** and **y** are its components along the horizontal and vertical axes, respectively. The magnitudes of these vectors are *s*, *x*, and *y*.

Of course, to describe motion we must deal with velocity and acceleration, as well as with displacement. We must find their components along the *x*- and *y*-axes, too. The components of acceleration are very simple. Since we assume all forces except gravity are negligible, $a_y = -g = -9.80 \text{ m/s}^2$. Since gravity is vertical, $a_x = 0$. Both accelerations are constant, and so the kinematic equations, 2.7–2.11, can be used. Given these assumptions, the following steps are then used to analyze projectile motion.

Figure 3.11 The total displacement **s** of a football at a point along its path. **s** has components **x** and **y** along the horizontal and vertical axes. Its magnitude is *s*, and it makes an angle θ with the horizontal.

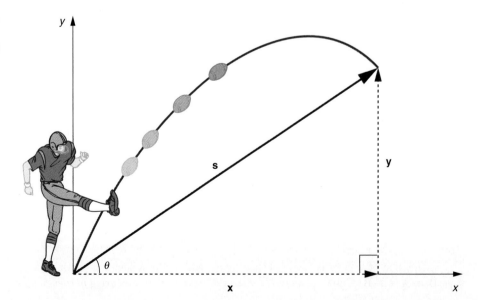

PROBLEM-SOLVING STRATEGY

FOR PROJECTILE MOTION

Step 1. *Break the motion into horizontal and vertical components along the x- and y-axes.* These are perpendicular axes, and so Equations 3.1 and 3.2 are used. The components of displacement **s** along these axes are x and y. The components of the velocity **v** are $v_x = v \cos \theta$ and $v_y = v \sin \theta$, where v is the magnitude of the velocity and θ is its direction, as shown in Figure 3.12. Initial values are denoted with a subscript 0, as usual.

Step 2. *Treat the motion as two independent one-dimensional motions, one horizontal and the other vertical.* The kinematic equations for horizontal and vertical motion take the following forms:

Horizontal motion ($a_x = 0$)

$$x = x_0 + v_x t \tag{3.9a}$$

$$v_x = v_{0x} = \bar{v}_x = \text{a constant} \tag{3.9b}$$

Vertical motion ($a_y = -g = -9.80 \ m/s^2$)

$$y = y_0 + \bar{v}_y t \tag{3.10a}$$

$$v_y = v_{0y} - gt \tag{3.10b}$$

$$y = y_0 + v_{0y} t - \tfrac{1}{2} g t^2 \tag{3.10c}$$

$$v_y^2 = v_{0y}^2 - 2g(y - y_0) \tag{3.10d}$$

Step 3. *Solve for whatever the unknowns are in the two separate motions—one horizontal and one vertical.* Note that the motions have time t in common. The problem solving procedures here are the same as for one-dimensional kinematics and are illustrated in the solved examples below.

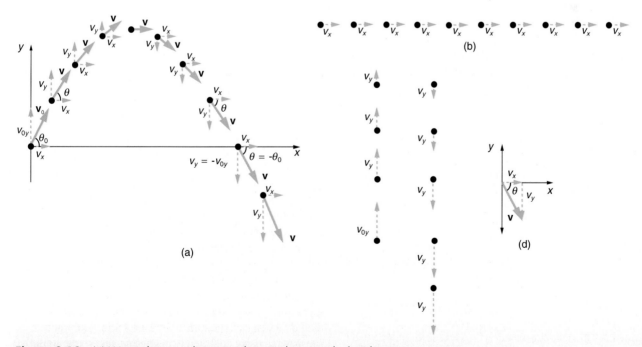

Figure 3.12 (a) We analyze two-dimensional projectile motion by breaking it into two independent one-dimensional motions along the vertical and horizontal axes. (b) The horizontal motion is simple, since $a_x = 0$ and v_x is thus constant. (c) The vertical motion is the same as for falling bodies. (d) The x- and y-motions are recombined to give the total velocity.

Step 4. *Recombine the two motions to find the total displacement* **s** *and velocity* **v**. Since the *x*- and *y*-motions are perpendicular, we do this using the techniques outlined in the previous section and employing Equations 3.3 and 3.4 in the following form, where θ is the direction of the displacement **s** and θ_v is the direction of the velocity **v**:

Total displacement and velocity

$$s = \sqrt{x^2 + y^2} \qquad \text{(3.11a)}$$

$$\theta = \tan^{-1}(y/x) \qquad \text{(3.11b)}$$

$$v = \sqrt{v_x^2 + v_y^2} \qquad \text{(3.11c)}$$

$$\theta_v = \tan^{-1}(v_y/v_x) \qquad \text{(3.11d)}$$

EXAMPLE 3.4   A FIREWORKS PROJECTILE EXPLODES HIGH AND AWAY

During a fireworks display, a shell is shot into the air with an initial speed of 70.0 m/s at an angle of 75° above the horizontal, as illustrated in Figure 3.13. The fuse is timed to ignite the shell just as it reaches its highest point above the ground. (a) Calculate the height at which the shell explodes. (b) How much time passed between the launch of the shell and the explosion? (c) How far downrange did the shell explode?

Strategy Since air resistance is negligible for the unexploded shell, the analysis method outlined above can be used. The motion can be broken into horizontal and vertical motions in which $a_x = 0$ and $a_y = -g$, and Equations 3.9 and 3.10 can be used to solve for the desired quantities. For simplicity, we take x_0 and y_0 to be zero.

Solution for (a) By "height" we mean the altitude or vertical position *y* above the starting point. The highest point in any trajectory is reached when $v_y = 0$. We use Equation 3.10d to find *y*:

$$v_y^2 = v_{0y}^2 - 2g(y - y_0)$$

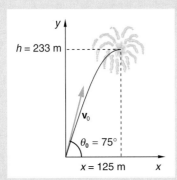

Figure 3.13 The trajectory of a fireworks shell as described in Example 3.4. The fuse is set to explode the shell at the highest point in its trajectory, which is found to be at a height of 233 m and 125 m downrange.

Because y_0 and v_y are both zero, the equation simplifies to

$$0 = v_{0y}^2 - 2gy$$

Solving for *y* gives

$$y = \frac{v_{0y}^2}{2g}$$

Now we must find v_{0y}, the component of the initial velocity in the *y*-direction. It is given by $v_{0y} = v_0 \sin \theta_0$, where v_0 is the initial velocity of 70.0 m/s, and $\theta_0 = 75°$ is the initial angle. Thus,

$$v_{0y} = v_0 \sin \theta_0 = (70.0 \text{ m/s})(\sin 75°) = 67.6 \text{ m/s}$$

And *y* is

$$y = \frac{(67.6 \text{ m/s})^2}{2(9.80 \text{ m/s}^2)}$$

so that

$$y = 233 \text{ m}$$

Discussion for (a) Note that since up is positive, the initial velocity is positive, as is the maximum height, but the acceleration of gravity is negative. Note also that the maximum height depends only on the vertical component of the initial velocity, so that any projectile with a 67.6 m/s initial vertical velocity will reach a maximum height of 233 m (neglecting air resistance). The numbers in this example are reasonable for large fireworks displays, the shells of which do reach such heights before exploding. In practice, air resistance is not completely negligible, and so the initial velocity would have to be somewhat larger than that given to reach the same height.

(continued)

(continued)

Solution for (b) As in many physics problems, there is more than one way to solve for the time to the highest point. In this case, the easiest method is to use Equation 3.10a. Since y_0 is zero, this is simply

$$y = \bar{v}_y t$$

where the average vertical velocity is $\bar{v}_y = (v_{0y} + v_y)/2 = (67.6 \text{ m/s} + 0)/2 = 33.8$ m/s, so that

$$t = \frac{y}{\bar{v}_y} = \frac{233 \text{ m}}{33.8 \text{ m/s}}$$

$$= 6.90 \text{ s}$$

Discussion for (b) This time is also reasonable for large fireworks. When you are able to see the launch of fireworks, you will notice several seconds pass before the shell explodes.

Solution for (c) The distance downrange is the horizontal displacement x. Since air resistance is negligible,

$a_x = 0$ and the horizontal velocity is constant, as discussed above. The horizontal displacement is horizontal velocity × time as given by Equation 3.9a, with x_0 equal to zero:

$$x = v_x t$$

where v_x is the x-component of the velocity, which is given by $v_x = v_0 \cos \theta_0$. Now,

$$v_x = v_0 \cos \theta_0 = (70.0 \text{ m/s})(\cos 75°) = 18.1 \text{ m/s}$$

The time t for both motions is the same, and so x is

$$x = (18.1 \text{ m/s})(6.90 \text{ s}) = 125 \text{ m}$$

Discussion for (c) The horizontal motion is a very simple constant velocity in the absence of air resistance. The horizontal displacement found here could be useful in keeping the fireworks fragments from falling on spectators. Once the shell explodes, air resistance has a major effect, and many fragments will land directly below.

In solving part (a) of the preceding example, the expression we found for y is valid for any projectile motion where air resistance is negligible. Call the maximum height h; then,

$$h = \frac{v_{0y}^2}{2g} \tag{3.12}$$

This equation defines the **maximum height of a projectile** and depends only on the vertical component of the initial velocity.

EXAMPLE 3.5 HOT ROCK PROJECTILE

Kilauea in Hawaii is the world's most continuously active volcano. Very active volcanoes characteristically eject red hot rocks and lava rather than smoke and ash. Suppose a large rock is ejected from the volcano with a speed of 25.0 m/s and at an angle 35° above the horizontal, as shown in Figure 3.14. The rock strikes the side of the volcano at an altitude 20.0 m lower than its starting point. (a) Calculate the time it takes the rock to follow this path. (b) What are the magnitude and direction of the rock's velocity at impact?

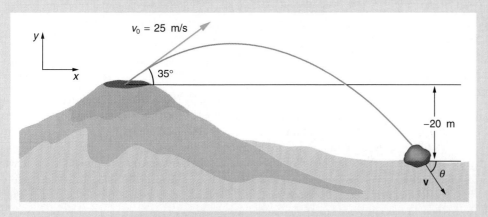

Figure 3.14 The trajectory of a rock ejected from the Kilauea volcano as described in Example 3.5.

(continued)

(continued)

Strategy Again, breaking this two-dimensional motion into two independent one-dimensional motions will allow us to solve for the desired quantities. The time a projectile is in the air is governed by its vertical motion alone. We will solve for t first. While the rock is rising and falling vertically, the horizontal motion continues at a constant velocity. This example asks for the final velocity. Thus the vertical and horizontal results will be recombined to obtain v and θ_v at the final time t determined in the first part of the example.

Solution for (a) While the rock is in the air, it rises and then falls to a final position 20.0 m lower than its starting altitude. We can find the time for this by using Equation 3.10c, which is

$$y = y_0 + v_{0y}t - \tfrac{1}{2}gt^2$$

If we take the initial position y_0 to be zero, then the final position is $y = -20.0$ m. Now the initial vertical velocity is the vertical component of the initial velocity, found from $v_{0y} = v_0 \sin \theta_0 = (25.0 \text{ m/s})(\sin 35°) = 14.3$ m/s. Substituting known values and writing the time as $t = t$ s (this makes the units cancel as previously done in Example 2.13) yields

$$-20.0 = 14.3t - 4.90t^2$$

Rearranging terms gives a quadratic equation in t:

$$4.90t^2 - 14.3t - 20.0 = 0$$

This is a quadratic equation of the form $at^2 + bt + c = 0$, where the constants are $a = 4.90, b = -14.3$, and $c = -20.0$. Its solutions are given by Equation 1.23, the quadratic formula:

$$t = \frac{-b \pm \sqrt{b^2 - 4ac}}{2a}$$

This yields two solutions: t = 3.96 and −1.03. (It is left as an exercise for the reader to verify these solutions.) The time is $t = 3.96$ s or −1.03 s. The negative value of time implies an event before the start of motion, and so we discard it. Thus,

$$t = 3.96 \text{ s}$$

Discussion for (a) The time for projectile motion is completely determined by the vertical motion. So any projectile that has an initial vertical velocity of 14.3 m/s and lands 20.0 m below its starting altitude will spend 3.96 s in the air.

Solution for (b) From the information now in hand, we can find the final horizontal and vertical velocities v_x and v_y, and combine them to find the total velocity v and the angle θ_v it makes with the horizontal. Of course, v_x is constant and is given by

$$v_x = v_0 \cos \theta_0 = (25.0 \text{ m/s})(\cos 35°) = 20.5 \text{ m/s}$$

The final vertical velocity is given by Equation 3.10b:

$$v_y = v_{0y} - gt$$

where v_{0y} was found in part (a) to be 14.3 m/s. Thus,

$$v_y = 14.3 \text{ m/s} - (9.80 \text{ m/s}^2)(3.96 \text{ s})$$

so that

$$v_y = -24.5 \text{ m/s}$$

To find the magnitude of the final velocity v, we combine its perpendicular components, using Equation 3.11c:

$$v = \sqrt{v_x^2 + v_y^2} = \sqrt{(20.5 \text{ m/s})^2 + (-24.5 \text{ m/s})^2}$$

which gives

$$v = 31.9 \text{ m/s}$$

The direction θ_v is found from Equation 3.11d:

$$\theta_v = \tan^{-1}(v_y/v_x)$$

so that

$$\theta_v = \tan^{-1}(-24.5/20.5) = \tan^{-1}(-1.19)$$

Thus,

$$\theta_v = -50.1°$$

Discussion for (b) The negative angle means that the velocity is 50.1° below the horizontal. This is consistent with the fact that the final vertical velocity is negative and hence downward—as you would expect since the final altitude is 20.0 m lower than the initial altitude. (See Figure 3.14.)

One of the most important things illustrated by projectile motion is that vertical and horizontal motions are independent and have only time in common. Galileo was the first person to fully comprehend this characteristic. He used it to predict the range of a projectile. On level ground, we define **range** to be the horizontal distance R traveled by a projectile. Galileo and many others were interested in the range of projectiles primarily for military purposes—such as aiming cannons. However, investigating the range of projectiles can shed light on other interesting phenomena, such as the orbits of earth satellites. Let us consider projectile range further.

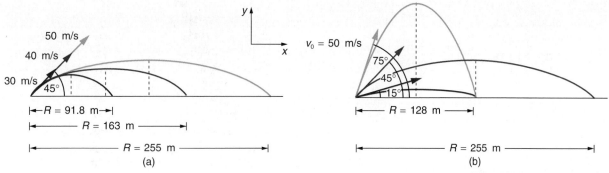

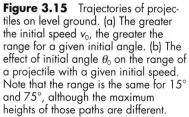

Figure 3.15 Trajectories of projectiles on level ground. (a) The greater the initial speed v_0, the greater the range for a given initial angle. (b) The effect of initial angle θ_0 on the range of a projectile with a given initial speed. Note that the range is the same for 15° and 75°, although the maximum heights of those paths are different.

How does the initial velocity of a projectile affect its range? Obviously, the greater the initial speed v_0, the greater the range, as shown in Figure 3.15(a). The initial angle θ_0 also has a dramatic effect on the range, as illustrated in Figure 3.15(b). For a fixed initial speed, such as might be produced by a cannon, the maximum range is obtained with $\theta_0 = 45°$. Interestingly, for every initial angle except 45°, there are two angles that give the same range—the sum of those angles is 90°. The range also depends on the strength of the acceleration of gravity g. The lunar astronaut Alan Shepherd was able to drive a golf ball a great distance on the moon because gravity is weaker there. The range R of a projectile on *level ground* for which air resistance is negligible is given by

$$R = \frac{v_0^2 \sin 2\theta_0}{g} \tag{3.13}$$

where v_0 is the initial speed and θ_0 is the initial angle relative to the horizontal. The proof of this equation is left as an end-of-chapter problem (hints are given), but it does fit the major features of projectile range as described.

When we speak of the range of a projectile on level ground, we assume that R is very small compared with the circumference of the earth. If, however, the range is large, the earth curves away below the projectile and gravity changes direction along the path. The range is larger than predicted by Equation 3.13 because the projectile has farther to fall than it would on level ground. (See Figure 3.16.) If the initial speed is great enough, the projectile goes into orbit. This possibility was recognized centuries before it could be accomplished. When an object is in orbit, the earth curves away from underneath it at the same rate as it falls. It thus falls continuously but never hits the surface. These and other aspects of orbital motion, such as the rotation of the earth, will be covered analytically and in greater depth in Chapter 8.

Once again we see that thinking about one topic, such as the range of a projectile, can lead us to others, such as earth orbits. In the next section, we will examine the addition of velocities, another important aspect of two-dimensional kinematics. This will also yield insights beyond the immediate topic.

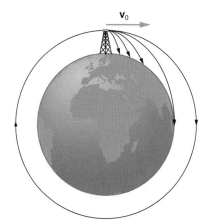

Figure 3.16 Projectile to satellite. In each case shown here, a projectile is launched from a very high tower to avoid air resistance. With increasing initial speed, the range increases and becomes longer than it would be on level ground because the earth curves away underneath its path. With a great enough initial speed, orbit is achieved.

3.5 ADDITION OF VELOCITIES

Relative Velocity

If a person rows a boat across a rapidly flowing river and tries to head directly for the other shore, the boat instead moves *diagonally* relative to the shore, as in Figure 3.17(a). The boat does not move in the direction in which it is pointed. The reason, of course, is that the river carries the boat downstream. Similarly, if a small airplane flies overhead in a strong crosswind, you can sometimes see that the plane is not moving in the direction in which it is pointed, as illustrated in Figure 3.17(b). The plane is moving straight ahead relative to the air, but it is carried sideways by the movement of the air mass relative to the ground.

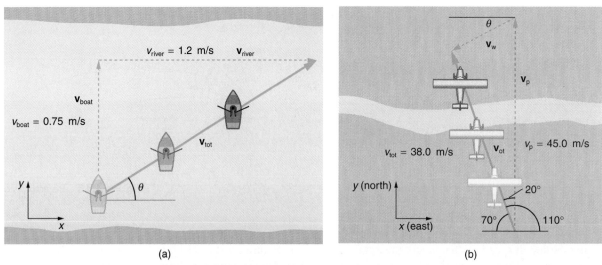

Figure 3.17 Examples of addition of velocities. (a) A boat trying to head straight across a river will actually move diagonally relative to the shore as shown. Its total velocity (solid arrow) relative to the shore is the sum of its velocity relative to the river plus the velocity of the river relative to the shore. (b) An airplane heading straight north is instead carried to the west and slowed down by wind. The plane does not move relative to the ground in the direction it points; rather, it moves in the direction of its total velocity (solid arrow). These figures are the subjects of Examples 3.6 and 3.7.

In each of these situations, an object has a velocity relative to a medium (such as a river) and that medium has a velocity relative to an observer on solid ground. The velocity of the object *relative to the observer* is the sum of these velocity vectors, as indicated in Figure 3.17. These are only two of many situations in which it is useful to add velocities. We actually did add velocities in Example 3.5 (in which horizontal and vertical velocity components were added to obtain the magnitude and direction of the total velocity). In this section, we first reexamine how to add velocities and then consider certain aspects of what relative velocity means.

How do we add velocities? Velocity is a vector (it has both magnitude and direction); the rules of vector addition discussed in Sections 3.2 and 3.3 apply to the addition of velocities, just as they do for any other vectors. In one-dimensional motion, the addition of velocities is simple—they add like ordinary numbers. For example, if a hockey player is moving at 8 m/s straight toward the goal and hits the puck in the same direction with a velocity of 34 m/s relative to his body, then the velocity of the puck is 42 m/s relative to the stationary, profusely sweating goalie standing in front of the goal.

In two-dimensional motion, either graphical or analytical techniques can be used to add velocities. We will concentrate on analytical techniques. The following equations give the relationships between the magnitude and direction of velocity (v and θ) and its components (v_x and v_y) along the x- and y-axes of an appropriately chosen coordinate system:

$$v_x = v \cos \theta \tag{3.14}$$

$$v_y = v \sin \theta \tag{3.15}$$

$$v = \sqrt{v_x^2 + v_y^2} \tag{3.16}$$

$$\theta = \tan^{-1}(v_y/v_x) \tag{3.17}$$

These are Equations 3.1–3.4, valid for any vectors, adapted specifically for velocity. The first two equations are used to find the components of a velocity when its magnitude and direction are known. The last two are used to find the magnitude and direction of velocity when its components are known.

EXAMPLE 3.6 ADDING VELOCITIES OF BOAT AND RIVER

Referring again to Figure 3.17(a), calculate the magnitude and direction of the boat's velocity relative to an observer on the shore, if the velocity of the boat is 0.750 m/s relative to the river and the speed of the river is 1.20 m/s to the right.

Strategy We start by choosing a coordinate system with its x-axis parallel to the velocity of the river, as shown in Figure 3.17(a). Since the boat is directed straight toward the other shore, its velocity relative to the water is parallel to the y-axis and perpendicular to the velocity of the river. Thus we can add the two velocities by using Equations 3.16 and 3.17 directly.

Solution From Equation 3.16, the magnitude of the total velocity v_{tot} is

$$v_{tot} = \sqrt{v_x^2 + v_y^2}$$

where

$$v_x = v_{river} = 1.20 \ \text{m/s}$$

and

$$v_y = v_{boat} = 0.750 \ \text{m/s}$$

Thus,

$$v_{tot} = \sqrt{(1.20 \ \text{m/s})^2 + (0.750 \ \text{m/s})^2}$$

yielding

$$v_{tot} = 1.42 \ \text{m/s}$$

The direction of the total velocity θ is given by Equation 3.17:

$$\theta = \tan^{-1}(v_y/v_x) = \tan^{-1}(0.750/1.20)$$

This gives

$$\theta = 32.0°$$

Discussion Both the magnitude v and the direction θ are consistent with Figure 3.17(a). Note that since the velocity of the river is large compared with the velocity of the boat, it is swept rapidly downstream. This is evidenced by the small angle (only 32.0°) the total velocity has relative to the river bank.

EXAMPLE 3.7 WHAT WIND VELOCITY CAUSES AN AIRPLANE TO DRIFT?

Calculate the wind velocity for the situation shown in Figure 3.17(b). The plane is known to be moving at 45.0 m/s due north relative to the air mass, while its velocity relative to the ground is 38.0 m/s in a direction 20.0° west of north.

Strategy In this problem we know the total velocity $\mathbf{v}_{tot}$ and that it is the sum of two other velocities, $\mathbf{v}_w$ (the wind) and $\mathbf{v}_p$ (the plane relative to the air mass). $\mathbf{v}_p$ is known, and we are asked to find $\mathbf{v}_w$. None of the velocities are perpendicular, but it is possible to find their components along perpendicular axes. If we can find the components of $\mathbf{v}_w$, then we can combine them to solve for its magnitude and direction. As shown in Figure 3.17(b), we choose a coordinate system with its x-axis due east and its y-axis due north (parallel to $\mathbf{v}_p$). (You may wish to look back at the discussion of the addition of vectors using perpendicular components in Section 3.3.)

Solution Since $\mathbf{v}_{tot}$ is the vector sum of the $\mathbf{v}_w$ and $\mathbf{v}_p$, its x- and y-components are the sums of the x- and y-components of the wind and plane velocities. That is,

$$v_{tot\,x} = v_{wx}$$

and

$$v_{tot\,y} = v_{wy} + v_p$$

We can use the first of these two equations to find v_{wx}:

$$v_{wx} = v_{tot\,x} = v_{tot} \cos 110°$$

Since $v_{tot} = 38.0$ m/s and $\cos 110° = -0.342$, we have

$$v_{wx} = (38.0 \ \text{m/s})(-0.342) = -13.0 \ \text{m/s}$$

The minus sign indicates motion west.

Now, to find v_{wy}, we note that

$$v_{wy} = v_{tot\,y} - v_p$$

Here $v_{tot\,y} = v_{tot} \sin 110°$; thus,

$$v_{wy} = (38.0 \ \text{m/s})(0.940) - 45.0 \ \text{m/s} = -9.29 \ \text{m/s}$$

This minus sign indicates motion south.

Now that the perpendicular components of the wind velocity $\mathbf{v}_{wx}$ and $\mathbf{v}_{wy}$ are known, we can find the magnitude and direction of $\mathbf{v}_w$ using Equations 3.16 and 3.17. First, the magnitude, from Equation 3.16, is

$$v_w = \sqrt{v_{wx}^2 + v_{wy}^2}$$

$$= \sqrt{(-13.0 \ \text{m/s})^2 + (-9.29 \ \text{m/s})^2}$$

(continued)

(continued)
so that

$$v_w = 16.0 \text{ m/s}$$

The direction is given by Equation 3.17:

$$\theta = \tan^{-1}(v_{wy}/v_{wx}) = \tan^{-1}(-9.29/-13.0)$$

giving

$$\theta = 35.6°$$

Discussion The wind's speed and direction are consistent with the significant effect it has on the total velocity of the plane, as seen in Figure 3.17(b). Since the plane is fighting a strong combination of crosswind and headwind, it ends up with a total velocity significantly less than its velocity relative to the air mass as well as heading in a different direction.

Note that in both of the last two examples, we were able to make the mathematics easier by choosing a coordinate system with one axis parallel to one of the velocities. We will repeatedly find that choosing an appropriate coordinate system makes problem solving easier. For example, in projectile motion we always use a coordinate system with one axis parallel to gravity.

Relative Velocities and Classical Relativity

When adding velocities, we have been careful to specify that the velocity is relative to some reference frame. These are called *relative velocities*. For example, the velocity of an airplane relative to an air mass is different from its velocity relative to the ground. Both are quite different from the velocity of an airplane relative to its passengers (which should be close to zero). Relative velocities are one aspect of **relativity**, which is defined to be the study of how different observers moving relative to each other measure the same phenomenon.

Nearly everyone has heard of relativity and immediately associates it with Albert Einstein (1879–1955), the greatest physicist of the 20th century. Einstein revolutionized our view of nature with his *modern* theory of relativity, which we shall study in later chapters. The relative velocities in this section are actually aspects of classical relativity, first discussed correctly by Galileo and Isaac Newton. **Classical relativity** is limited to situations where velocities are less than about 1% of the speed of light—that is, less than 3000 km/s. Most things we encounter move slower than this.

Let us consider an example of what two different observers see in a situation analyzed long ago by Galileo. Suppose a sailor at the top of a mast on a moving ship drops his knife. Where will it hit the deck? Will it hit at the base of the mast, or will it hit behind the mast because the ship is moving forward? The answer is that if air resistance is negligible, the knife will hit at the base of the mast at a point directly below its point of release. Now let us consider what two different observers see when the knife drops. One observer is on the ship and the other on shore. The knife has no horizontal velocity relative to the observer on the ship, and so he sees it fall straight down the mast. (See Figure 3.18.) To the observer on shore, the knife and the ship have the *same* horizontal velocity, so both move the same distance forward while the knife is falling. This observer sees the curved path shown in Figure 3.18. Although the paths look different to the different observers, the same result is seen by each—the knife hits at the base of the mast and not behind it. To get the correct description, it is crucial to correctly specify the velocities relative to the observer.

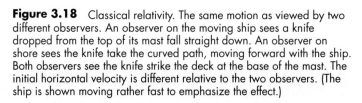

Figure 3.18 Classical relativity. The same motion as viewed by two different observers. An observer on the moving ship sees a knife dropped from the top of its mast fall straight down. An observer on shore sees the knife take the curved path, moving forward with the ship. Both observers see the knife strike the deck at the base of the mast. The initial horizontal velocity is different relative to the two observers. (The ship is shown moving rather fast to emphasize the effect.)

EXAMPLE 3.8 VELOCITY RELATIVE TO . . .?

An airline passenger drops a quarter in change while the plane is moving at 260 m/s in level flight. What is the velocity of the quarter when it strikes the floor 1.50 m below its point of release: (a) Measured relative to the plane? (b) Measured relative to the earth?

Strategy Both problems can be solved with the techniques for falling objects and projectiles. In part (a), the initial velocity of the quarter is zero relative to the plane, making the motion that of a falling object (one-dimensional). In part (b), the initial velocity is 260 m/s horizontal relative to the earth and gravity is vertical, making this a projectile motion. In both parts it is best to use a coordinate system with vertical and horizontal axes.

Solution for (a) With the given information, we note that the initial velocity and position are zero, and the final position is −1.50 m. The final velocity can be found using Equation 3.10d:

$$v^2 = v_0^2 - 2gy$$

Substituting known values, we get

$$v^2 = 0 - 2(9.8 \text{ m/s}^2)(-1.50 \text{ m}) = 29.4 \text{ m}^2/\text{s}^2$$

yielding

$$v = -5.42 \text{ m/s}$$

The minus sign is *chosen* when the square root is taken; it means the velocity is downward. There is no initial horizontal velocity relative to the plane and no horizontal acceleration, and so the motion is straight down relative to the plane.

Solution for (b) Since the initial vertical velocity is zero relative to the ground and since vertical motion is independent of horizontal motion, the final vertical velocity for the quarter relative to the ground is $v_y = -5.42$ m/s, the same as found in part (a). There is no horizontal acceleration, and so the initial and final horizontal velocities are the same and $v_x = 260$ m/s. The x- and y-components of velocity can be combined, using Equation 3.16, to find the magnitude of the final velocity:

$$v = \sqrt{v_x^2 + v_y^2}$$

Thus,

$$v = \sqrt{(260 \text{ m/s})^2 + (-5.42 \text{ m/s})^2}$$

yielding

$$v = 260.06 \text{ m/s}$$

The direction is given by Equation 3.17:

$$\theta = \tan^{-1}(v_y/v_x) = \tan^{-1}(-5.42/260)$$

so that

$$\theta = \tan^{-1}(-0.0208) = -1.19°$$

Discussion In part (a), the final velocity relative to the plane is the same as it would be if the quarter were dropped from rest on earth and fell 1.50 m. This fits our experience; objects fall the same way when the plane is flying horizontally as when it is at rest on the ground. This is also true in moving cars. In part (b), an observer on the ground sees a much

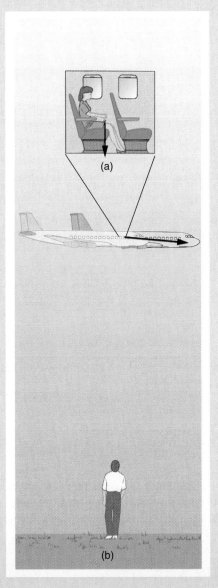

Figure 3.19 More classical relativity. The motion of a quarter dropped inside an airplane as viewed by two different observers. (a) An observer in the plane sees the quarter fall straight down. (b) An observer on the ground sees the quarter move almost horizontally.

(continued)

(continued)

different motion for the quarter. It is moving so fast horizontally to begin with that its final velocity is barely greater than the initial velocity. Once again, we see that in two dimensions vectors do not add like ordinary numbers—the final velocity v in part (b) is *not* $(260 - 5.42)$ m/s; rather, it is 260.06 m/s. The velocity's magnitude had to be calculated to five digits to see any difference from that of the airplane. The motions as seen by different observers (one in the plane and one on the ground) in this example are analogous to those discussed for the knife dropped from the mast of a moving ship, except that the velocity of the plane is much larger, so that the two observers see *very* different paths. (See Figure 3.19.) Additionally, both observers see the quarter fall 1.50 m vertically, but the one on the ground also sees it move forward 144 m (this calculation is left for the reader). Thus one observer sees a vertical path, the other a nearly horizontal path.

CONNECTIONS

Modern relativity is covered in Chapters 26 and 32. Like classical relativity, it is concerned with how observers moving relative to one another measure the same physical phenomena. But because Einstein was able to clearly define how measurements are made (some involve light) and because the speed of light is the same for all observers, the outcomes are spectacularly unexpected. Time varies with observer, energy is stored as increased mass, and more surprises await.

Classical relativity does not contain great surprises, though it clarifies the description of nature. In contrast, the modern (or Einstein's) theory of relativity contains immense surprises. Einstein predicted that time, length, and mass are not absolute but depend on the motion of the observer. He also predicted that mass and energy could be converted to one another (his famous equation $E = mc^2$) and that very large velocities do not add as simply as in classical physics, because nothing can exceed the speed of light. These and all of the other seemingly bizarre predictions made by Einstein have been verified by experiment and have altered our view of nature. Einstein's tremendous contributions came in part because he carefully took into account that all observations are made relative to a particular frame of reference.

SUMMARY

Two-dimensional kinematics differs from one-dimensional kinematics in two essential ways. First, vectors do not add as simply in two dimensions as in one; they require mathematical techniques different from adding simple scalars. Second, motions along perpendicular directions are independent, so that two-dimensional motion is composed of two independent one-dimensional motions.

A **vector** is any quantity that has magnitude and direction. If **A** is the symbol for a vector, then A is the symbol for its magnitude. The **graphical addition of vectors** uses the **head-to-tail method**. An arrow is drawn for each vector with arrow length proportional to the vector magnitude and pointing in the vector's direction. These arrows are placed head to tail. The **resultant**, or sum, is an arrow drawn from the tail of the first to the head of the last. The **magnitude** and **direction** of the resultant are determined by measurements of its length and direction with a ruler and protractor. Vector addition is **commutative**—that is, vectors can be added in any order. **Vector subtraction** is the addition of a negative vector—that is, **A** − **B** ≡ **A** + (−**B**). The **negative** of a vector **B** is written −**B** and is defined to be a vector of the same length having the opposite direction. **Multiplication of a vector A by a positive scalar** changes its magnitude, but does not change its direction.

The **analytical method** of vector addition and subtraction uses vector components. To find the **components** A_x and A_y of a vector **A** along perpendicular x- and y-axes, we use

$$A_x = A \cos \theta \qquad (3.1)$$

and

$$A_y = A \sin \theta \qquad (3.2)$$

where θ is the direction of **A** relative to the x-axis. To find the magnitude and direction of **A** from its perpendicular components, we use

$$A = \sqrt{A_x^2 + A_y^2} \qquad (3.3)$$

and

$$\theta = \tan^{-1}(A_y/A_x) \qquad (3.4)$$

If vectors **A** and **B** are added to give a resultant **R**, the components of **R** are given by

$$R_x = A_x + B_x \qquad (3.5)$$

and

$$R_y = A_y + B_y \qquad (3.6)$$

and the magnitude and direction of **R** are found by combining its components using equations such as 3.3 and 3.4. If **B** is subtracted from **A**, then Equations 3.5 and 3.6 become

$$R_x = A_x - B_x$$

and

$$R_y = A_y - B_y$$

Projectile motion is the motion of an object projected into the air and subjected to the acceleration of gravity. We handle projectile motion by **breaking it into components** along vertical and horizontal axes. Those motions are independent but share a common time. The solutions of this pair of one-dimensional motions are combined to give two-dimensional quantities, such as velocity and displacement. The **maximum height h reached by a projectile**, neglecting air resistance, is

$$h = \frac{v_{0y}^2}{2g} \tag{3.12}$$

where v_{0y} is the vertical component of its initial velocity. The **range R** of a projectile is the horizontal distance it travels. On *level ground*, the range is given by

$$R = \frac{v_0^2 \sin 2\theta_0}{g} \tag{3.13}$$

where v_0 is the initial speed and θ_0 is the initial angle relative to the horizontal.

Velocities in two dimensions are added using the same analytical vector techniques, which are rewritten as Equations 3.14–3.17. Relative velocity varies dramatically with reference frame. **Relativity** is the study of how different observers measure the same phenomenon, particularly when the observers move relative to one another. **Classical relativity** is limited to situations where speed is less than about 1% of the speed of light.

CONCEPTUAL QUESTIONS

3.1 Which of the following is a vector: a person's height, the altitude in Death Valley, the age of the earth, a nonzero force, the boiling point of water, the cost of this book, the earth's population?

3.2 Give a specific example of a vector, stating its magnitude, units, and direction.

3.3 What do vectors and scalars have in common? How do they differ?

3.4 What frame or frames of reference do you instinctively use when driving a car? When flying in a commercial jet airplane?

3.5 There is a passage in the Beatles' song *The Fool on the Hill* that says, "The fool on the hill sees the sun going down and the eyes in his head see the world spinning round." Write down how you would explain to the fool that both points of view are valid (earth stationary, earth spinning round) depending on your choice of frame of reference. Define both frames for him.

3.6 Two campers in a national forest hike from their cabin to a lake, each taking a different path, as illustrated in Figure 3.20. The total distance traveled along path 1 is 7.5 km, and that along path 2 is 8.2 km. What is the final displacement of each camper?

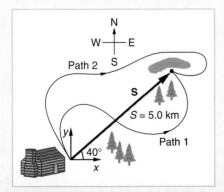

Figure 3.20 Different paths are taken by two campers from the same starting point to the same destination. What is their final displacement? Question 6 and Problem 3.

3.7 If an airplane pilot is told to fly 123 km in a straight line to get from San Francisco to Sacramento, explain why she could end up anywhere on the circle shown in Figure 3.21. What other information would she need to get to Sacramento?

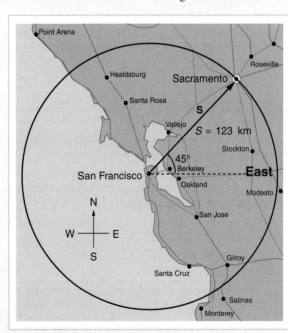

Figure 3.21 A map of the region surrounding San Francisco and a displacement vector to Sacramento. Question 7 and Problem 13.

3.8 Suppose you make two perpendicular steps **A** and **B** (that is, two nonzero displacements), say one south and one east. Is it possible for these two steps to return to your original position? More generally, can two nonzero perpendicular vectors ever add to zero?

3.9 Suppose you take two steps **A** and **B** (that is, two nonzero displacements). Under what circumstances can you end up at your starting point? More generally, under what circumstances can two nonzero vectors add to give zero? Is the maximum distance you can end up from the starting point $A + B$, the sum of the lengths of the two steps?

3.10 Suppose you add two vectors **A** and **B**. What relative direction between them produces the resultant with the greatest magnitude? What is the maximum magnitude? What relative direction between them produces the resultant with the smallest magnitude? What is the minimum magnitude?

3.11 Give an example of a nonzero vector that has a component of zero.

3.12 Explain why a vector cannot have a component greater than its own magnitude.

3.13 If the vectors **A** and **B** are perpendicular, what is the component of **A** along the direction of **B**? What is the component of **B** along the direction of **A**?

3.14 Explain why it is not possible to add a scalar to a vector.

3.15 If you take two steps of different sizes, can you end up at your starting point? More generally, can two vectors with different magnitudes ever add to zero? Can three or more?

3.16 How does air resistance affect the range and maximum height of projectile motion?

3.17 Answer the following questions for projectile motion on level ground assuming negligible air resistance (the initial angle being neither 0° nor 90°): (a) Is the velocity ever zero? (b) When is the velocity a minimum? A maximum? (c) Can the velocity ever be the same as the initial velocity at a time other than at $t = 0$? (d) Can the speed ever be the same as the initial speed at a time other than at $t = 0$?

3.18 For a fixed initial speed, the range of a projectile is determined by the angle at which it is fired. For all but the maximum, there are two angles that give the same range. Considering factors that might affect the ability of an archer to hit a target, such as wind, explain why the smaller angle (closer to the horizontal) is preferable. When would it be necessary for the archer to use the larger angle? Why does the punter in a football game use the higher trajectory?

3.19 During a lecture demonstration, a professor places two quarters on the edge of a table. She then flicks one of the quarters horizontally off the table, simultaneously nudging the other over the edge. Describe the subsequent motion of the two quarters, in particular noting whether they hit the floor at the same time.

3.20 If someone is riding in the back of a pickup truck and throws a baseball straight backward, is it possible for the ball to fall straight down as viewed by a person standing at the side of the road? Under what condition would this occur? How would the motion of the ball appear to the person who threw it?

3.21 The hat of a jogger running at constant velocity drops off the back of his head. Draw a sketch showing the path of the hat in the jogger's frame of reference. Draw its path as viewed by a stationary observer.

3.22 A clod of dirt falls from the bed of a moving truck. It strikes the ground directly below the end of the truck. What is the direction of its velocity relative to the truck just before it hits? Is this the same as the direction of its velocity relative to ground just before it hits? Explain.

PROBLEMS

Section 3.2 Graphical Methods of Vector Addition and Subtraction

Use graphical methods to solve these problems. You may assume data taken from graphs is accurate to three digits.

3.1 Find the following for path A in Figure 3.22: (a) The total distance traveled. (b) The magnitude and direction of the displacement from start to finish.

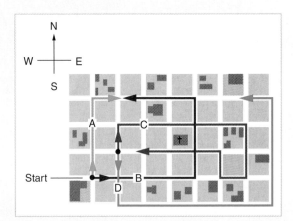

Figure 3.22 The various lines represent paths taken by different people walking in a city. All blocks are 120 m on a side. Problems 1, 2, 13, and 14.

3.2 Find the following for path B in Figure 3.22: (a) The total distance traveled. (b) The magnitude and direction of the displacement from start to finish.

3.3 Find the north and east components of the displacement for the hikers shown in Figure 3.20.

3.4 Suppose you walk 18.0 m straight west and then 25.0 m straight north. How far are you from your starting point, and what is the compass direction of a line connecting your starting point to your final position? (If you represent the two legs of the walk as vector displacements **A** and **B**, as in Figure 3.23, then this problem asks you to find their sum **R** = **A** + **B**.)

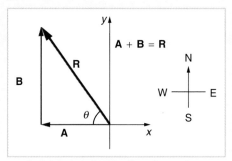

Figure 3.23 The two displacements **A** and **B** add to give a total displacement **R** having magnitude R and direction θ. Problems 4, 16, and 17.

3.5 Suppose you first walk 12.0 m in a direction 20° west of north and then 20.0 m in a direction 40° south of west. How far are you from your starting point, and what is the compass direction of a line connecting your starting point to your final position? (If you represent the two legs of the walk as vector displacements **A** and **B**, as in Figure 3.24, then this problem finds their sum **R** = **A** + **B**.)

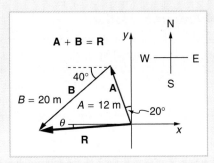

Figure 3.24 The two displacements **A** and **B** add to give a total displacement **R** with magnitude *R* and direction *θ*. Problems 5, 6, and 7.

• **3.6** Repeat Problem 3.5, but reverse the order of the two legs of the walk; show that you get the same final result. That is, you first walk leg **B**, which is 20.0 m in a direction 40° south of west, and then leg **A**, which is 12.0 m in a direction 20° west of north. (This problem shows that **B** + **A** = **A** + **B**.)

• **3.7** (a) Repeat Problem 3.5, but for the second leg you walk 20.0 m in a direction 40° north of east. (This is equivalent to subtracting **B** from **A**—that is, to finding **R′** = **A** − **B**.) (b) Repeat Problem 3.5, but now you first walk 20.0 m in a direction 40° south of west and then 12.0 m in a direction 20° east of south. (This is equivalent to subtracting **A** from **B**—that is, to finding **R″** = **B** − **A** = −**R′**. Is that consistent with your result?)

• **3.8** Show that the order of addition of three vectors does not affect their sum. Do this by choosing any three vectors **A**, **B**, and **C**, all having different lengths and directions. Find the sum **A** + **B** + **C**; then find their sum when added in a different order (there are five other orders in which **A**, **B**, and **C** can be added; choose only one), and show the result is the same.

• **3.9** Show that the sum of the vectors discussed in Example 3.2 gives the result shown in Figure 3.7(c).

: **3.10** Find the magnitudes of velocities v_A and v_B in Figure 3.25.

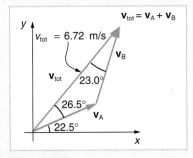

Figure 3.25 The two velocities v_A and v_B add to give a total v_{tot}. Problems 10, 11, 12, 25, 26, and 27.

: **3.11** Find the components of v_{tot} along the *x*- and *y*-axes in Figure 3.25.

: **3.12** Find the components of v_{tot} along a set of perpendicular axes rotated 30° counterclockwise relative to those in Figure 3.25.

Section 3.3 Analytical Methods of Vector Addition and Subtraction
Use analytical techniques to solve these problems.

3.13 Find the following for path C in Figure 3.22: (a) The total distance traveled. (b) The magnitude and direction of the displacement from start to finish.

3.14 Find the following for path D in Figure 3.22: (a) The total distance traveled. (b) The magnitude and direction of the displacement from start to finish.

3.15 Find the north and east components of the displacement from San Francisco to Sacramento shown in Figure 3.21.

3.16 Repeat Problem 3.4 using analytical techniques. (See Figure 3.23.)

3.17 Repeat Problem 3.4 using analytical techniques, but reverse the order of the two legs of the walk and show that you get the same final result. (This problem shows that adding them in reverse order gives the same result—that is, **B** + **A** = **A** + **B**.)

3.18 You drive 7.50 km in a straight line in a direction 15° east of north. (a) Find the distances you would have to drive straight east and then straight north to arrive at the same point. (This is equivalent to finding the components of the displacement along the east and north directions.) (b) Show that you still arrive at the same point if the east and north legs are reversed in order.

• **3.19** Do Problem 3.4 using analytical techniques and change the second leg of the walk to 25.0 m straight south. (This is equivalent to subtracting **B** from **A**—that is, to finding **R′** = **A** − **B**.) (b) Repeat again, but now you first walk 25.0 m north and then 18.0 m east. (This is equivalent to subtracting **A** from **B**—that is, to finding **R″** = **B** − **A** = −**R′**. Is that consistent with your result?)

• **3.20** Show that the order of addition of three vectors does not affect their sum. Do this by choosing any three vectors **A**, **B**, and **C**, all having different lengths and directions. Find the sum **A** + **B** + **C**; then find their sum when added in a different order (there are five other orders in which **A**, **B**, and **C** can be added; choose only one), and show the result is the same.

• **3.21** A new landowner has a triangular piece of flat land she wishes to fence. Starting at the west corner, she measures the first side to be 80.0 m long and the next to be 105 m—these sides are represented as displacement vectors **A** and **B** in Figure 3.26. She then correctly calculates the length and orientation of the third side **C**. What is her result?

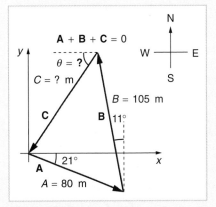

Figure 3.26 Problem 21.

: **3.22** You fly 32.0 km in a straight line in still air a direction 35° south of west. (a) Find the distances you would have to fly straight south and then straight west to arrive at the same point. (This is equivalent to finding the components of the displacement along the south and west directions.) (b) Find the distances you would have to fly first in a direction 45° south of west and then in a direction 45° west of north.

These are the components of the displacement along a different set of axes—one rotated 45°.

⋮ 3.23 A Texas rancher wants to fence off his four-sided plot of flat land. He measures the first three sides, shown as **A**, **B**, and **C** in Figure 3.27, and then correctly calculates the length and orientation of the fourth side **D**. What is his result?

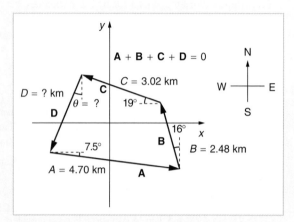

Figure 3.27 Problem 23.

⋮ 3.24 In an attempt to escape his island, Gilligan builds a raft and sets to sea. The wind shifts a great deal during the day, and he is blown along the following straight lines: 2.50 km 45° north of west; then 4.70 km 60° south of east; then 1.30 km 25° south of west; then 5.10 km straight east; then 1.70 km 5° east of north; then 7.20 km 55° south of west; and finally 2.80 km 10° north of east. What is his final position relative to the island?

⋮ 3.25 Find the magnitudes of velocities v_A and v_B in Figure 3.25.

⋮ 3.26 Find the components of v_{tot} along the *x*- and *y*-axes in Figure 3.25.

⋮ 3.27 Find the components of v_{tot} along a set of perpendicular axes rotated 30° counterclockwise relative to those in Figure 3.25.

⋮ 3.28 Suppose a pilot flies 40.0 km in a direction 60° north of east and then flies 30.0 km in a direction 15° north of east as shown in Figure 3.28. Find her total distance *R* from the starting point and the direction *θ* of the straight-line path to the final position.

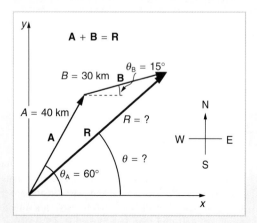

Figure 3.28 The paths flown by the pilot in Problem 28.

Section 3.4 Projectile Motion

3.29 A daredevil is attempting to jump her motorcycle over a line of buses parked end to end by driving up a 32° ramp at a speed of 40.0 m/s (90 mi/h). How many buses can she clear if the top of the takeoff ramp is at the same height as the bus tops and the buses are 20.0 m long? (Not a good thing to try, since the answer is based on the assumption of negligible air resistance and she does not clear the last bus by much.)

3.30 An archer shoots an arrow at a 75.0 m distant target, the bull's-eye of which is at same height as the release height of the arrow. (a) At what angle must the arrow be released to hit the bull's-eye if its initial speed is 35.0 m/s? (Although neglected here, the atmosphere provides significant lift to real arrows.) (b) There is a large tree halfway between the archer and the target with an overhanging branch 3.50 m above the release height of the arrow. Will the arrow go over or under the branch?

3.31 A football quarterback throws a pass 25.0 m straight downfield, where it is caught at the same height as it left the quarterback's hand. (a) At what angle was the ball thrown if its initial speed was 20.0 m/s, assuming that the smaller of the two possible angles was used? (b) What other angle gives the same range, and why would it not be used?

3.32 Verify the ranges for the projectiles in Figure 3.15(a) for $\theta = 45.0°$ and the given initial velocities.

3.33 Verify the ranges shown for the projectiles in Figure 3.15(b) for an initial velocity of 50.0 m/s at the given initial angles.

• 3.34 The cannon on a battleship can fire a shell a maximum distance of 32.0 km. (a) Calculate the initial velocity of the shell. (b) What maximum height does it reach? (At its highest, the shell is above 60% of the atmosphere—but air resistance is not really negligible as assumed to make this problem easier.) (c) The ocean is not flat, since the earth is curved. How many meters lower will its surface be 32.0 km from the ship along a horizontal line parallel to the surface at the ship? (This gives some idea of the error introduced by the assumption of a flat earth in projectile motion.)

• 3.35 A football quarterback is moving straight backward at a speed of 2.00 m/s when he throws a pass to a player 18.0 m straight downfield. (a) If the ball is thrown at an angle of 25.0° relative to the ground and is caught at the same height as it is released, what is its initial speed relative to the ground? (b) How long does it take to get to the receiver? (c) What is its maximum height above its point of release?

• 3.36 Gun sights are adjusted to aim high to compensate for the effect of gravity, effectively making the gun accurate only for a specific range. If a gun is sighted to hit targets that are at the same height as the gun and 100 m away, how low will the bullet hit if aimed directly at a target 150 m away? The muzzle velocity of the bullet is 275 m/s.

• 3.37 An eagle is flying horizontally at a speed of 3.00 m/s when the fish in her talons wiggles loose and falls into the lake 5.00 m below. Calculate the velocity of the fish relative to the water when it hits the water.

• 3.38 An owl is carrying a mouse to the chicks in its nest. It is 4.00 m west and 12.0 m above the center of the 30 cm diameter nest and is flying east at 3.50 m/s at an angle 30° below the horizontal when it accidentally drops the mouse. Is the owl lucky enough to have the mouse hit the nest? To answer this question, calculate the horizontal position of the mouse when it has fallen 12.0 m.

: 3.39 Suppose a football player kicks a successful field goal from a distance of 36.6 m (40 yd) from the goalpost. Find the initial speed of the football if it just clears the crossbar 3.05 m (10 ft) above the ground, given the initial direction to be 40° above the horizontal.

: 3.40 By what height would the ball in Problem 3.39 clear the crossbar if it is kicked with the same initial speed (20.1 m/s) but at an angle of 45° above the horizontal?

: 3.41 The free throw line in basketball is 4.57 m (15 ft) from the basket, which is 3.05 m (10 ft) above the floor. A player standing on the free throw line throws the ball with an initial speed of 7.15 m/s, releasing it at a height of 2.44 m (8 ft) above the floor. At what angle above the horizontal must the ball be thrown to exactly hit the basket? Note that most players will use a large initial angle rather than a flat shot because it allows for a larger margin of error.

: 3.42 On August 12, 1987, Alessandro Andrei set a world record in the shot put with a throw of 22.9 m (75 ft 2 in.). What was the initial speed of the shot if he released it at a height of 2.10 m and threw it at an angle of 38° above the horizontal? (Although the maximum distance for a projectile on level ground is achieved at 45°, the ideal angle is smaller when the projectile lands at an altitude below its point of release; thus, 38° will give a longer range than 45° in the shot put.)

: 3.43 Prove that the trajectory of a projectile is parabolic, having the form $y = ax + bx^2$. To do this, solve the expression $x = v_{0x}t$ for t and substitute it into the expression for $y = v_{0y}t - (1/2)gt^2$. (These are the equations for the x and y positions of a projectile that starts at the origin.) You should obtain an equation of the form $y = ax + bx^2$ where a and b are constants.

: 3.44 Derive Equation 3.13 for the range of a projectile on level ground by finding the time t at which y becomes zero and substituting this value of t into the expression for $x - x_0$, noting that $R = x - x_0$.

Section 3.5 Addition of Velocities

3.45 Bryan Allen pedaled a human-powered aircraft across the English Channel from the cliffs of Dover to Cap Gris-Nez on June 12, 1979. (a) He flew for 169 min at an average velocity of 3.53 m/s in a direction 45° south of east. What was his total displacement? (b) Allen encountered a headwind averaging 2.00 m/s almost precisely in the opposite direction of his motion relative to the earth. What was his average velocity relative to the air? (c) What was his total displacement relative to the air mass?

3.46 A seagull flies at a velocity of 9.00 m/s straight into the wind. (a) If it takes the bird 20.0 min to travel 6.00 km relative to the earth, what is the velocity of the wind? (b) If the bird turns around and flies with the wind, how long will he take to return 6.00 km?

3.47 Near the end of a marathon race, the first two runners are separated by a distance of 45.0 m. The front runner has a velocity of 3.50 m/s, and the second a velocity of 4.20 m/s. (a) What is the velocity of the second runner relative to the first? (b) If the front runner is 250 m from the finish line, who will win the race, assuming they run at constant velocity? (c) What distance ahead will the winner be when she crosses the finish line?

3.48 Verify that the quarter dropped by the airline passenger in Example 3.8 travels 144 m horizontally while falling 1.50 m in the frame of reference of the earth.

• 3.49 Calculate the initial velocity of the ball *relative to the quarterback* in Problem 3.35. Use the answer to Problem 3.35(a) as the starting point of this problem.

• 3.50 A ship sets sail from Rotterdam, The Netherlands, heading due north at 7.00 m/s relative to the water. The local ocean current is 1.50 m/s in a direction 40° north of east. What is the velocity of the ship relative to the earth?

• 3.51 A jet airplane flying over Cheyenne, Wyoming, has an air speed of 260 m/s in a direction 5.0° south of west. It is in the jet stream, which is blowing at 45.0 m/s in a direction 20° south of east. What is the velocity of the airplane relative to the earth?

• 3.52 A knife is dropped from the top of a 15.0 m high mast on a ship moving at 1.75 m/s due south. Calculate the velocity of the knife when it hits the deck of the ship: (a) relative to the ship and (b) relative to a stationary observer on shore.

• 3.53 The velocity of the wind relative to the water is crucial to sailboats. Suppose a sailboat is in an ocean current that has a velocity of 2.20 m/s in a direction 30° east of north relative to the earth. It encounters a wind that has a velocity of 4.50 m/s in a direction of 50° south of west relative to the earth. What is the velocity of the wind relative to the water?

• 3.54 The great astronomer Edwin Hubble discovered that all distant galaxies are receding from our Milky Way galaxy with velocities proportional to their distances. It appears to an observer on the earth that we are at the center of an expanding universe. Figure 3.29 illustrates this for five galaxies lying along a straight line, with the Milky Way galaxy at the center. Using the data from the figure, calculate the velocities: (a) relative to galaxy 2 and (b) relative to galaxy 5. The results mean that observers on all galaxies will see themselves at the center of the expanding universe, and they would likely be aware of relative velocities, concluding that it is not possible to locate the center of expansion with the given information.

: 3.55 (a) In what direction would the ship in Problem 3.50 have to travel in order to have a velocity straight north relative to the earth, assuming its speed relative to the water remains 7.00 m/s? (b) What would its speed be relative to the earth?

: 3.56 (a) Another airplane is flying in a jet stream that is blowing at 45.0 m/s in a direction 20° south of east (as in

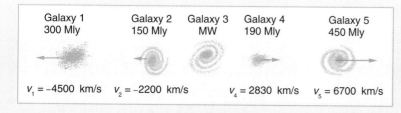

Galaxy 1	Galaxy 2	Galaxy 3	Galaxy 4	Galaxy 5
300 Mly	150 Mly	MW	190 Mly	450 Mly

$v_1 = -4500$ km/s $v_2 = -2200$ km/s $v_4 = 2830$ km/s $v_5 = 6700$ km/s

Figure 3.29 Five galaxies on a straight line, showing their distances and velocities relative to the Milky Way (MW) galaxy. The distances are in millions of light years (Mly), where a light year is the distance light travels in 1 y. The velocities are nearly proportional to the distances. The sizes of the galaxies are greatly exaggerated; an average galaxy is about 0.1 Mly across. Problem 54.

Problem 3.51). Its direction of motion relative to the earth is 45° south of west, while its direction of travel relative to the air is 5° south of west. What is the airplane's speed relative to the air mass? (b) What is the airplane's speed relative to the earth?

: 3.57 A ship sailing in the Gulf Stream is heading 25° west of north at a speed of 4.00 m/s relative to the water. Its velocity relative to the earth is 4.80 m/s 5° west of north. What is the velocity of the Gulf Stream? (The velocity obtained is typical for the Gulf Stream a few hundred kilometers off the east coast of the United States.)

: 3.58 A hockey player is moving at 8.00 m/s when he hits the puck toward the goal. The speed of the puck relative to the player is 29.0 m/s. The line between the center of the goal and the player makes a 90° angle relative to his path as shown in Figure 3.30. What angle must the puck's velocity make relative to the player (in his frame of reference) to hit the center of the goal?

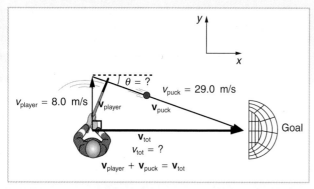

Figure 3.30 A hockey player moving across the rink must shoot backward to give the puck a velocity toward the goal. Problem 58.

UNREASONABLE RESULTS

Physics must describe nature accurately. The following problems have results that are unreasonable because one premise is unreasonable or because certain of its premises are inconsistent with one another. The physical principle applied correctly then produces an unreasonable result. The physics is correct in a sense, but there is more to describing nature than just manipulating equations correctly. Checking the result of a problem to see if it is reasonable not only helps uncover errors in problem solving—it also builds intuition in judging whether nature is being accurately described.

Note: Strategies for determining if an answer is reasonable appear in Chapter 2. They can be found with the section of problems labeled *Unreasonable Results* at the end of that chapter.

3.59 (a) Find the maximum range of a super cannon that has a muzzle velocity of 4.0 km/s. (b) What is unreasonable about the range you found? (c) Is the premise unreasonable, or is the available equation inapplicable? Explain.

3.60 Suppose you wish to shoot supplies straight up to astronauts in an orbit 36,000 km above the surface of the earth. (a) At what velocity must the supplies be launched? (b) What is unreasonable about this velocity? (c) Is there a problem with the relative velocity between the supplies and the astronauts when the supplies reach their maximum height? (d) Is the premise unreasonable, or is the available equation inapplicable? Explain.

• 3.61 A commercial airplane has an air speed of 280 m/s due east and flies with a strong tailwind. It travels 3000 km in a direction 5° south of east in 1.50 h. (a) What was the velocity of the plane relative to the ground? (b) Calculate the magnitude and direction of the tailwind's velocity. (c) What is unreasonable about both of these velocities? (d) Which premise is unreasonable?

DYNAMICS

Newton's Laws of Motion

This athlete is exerting large forces with many of his muscles at this point in his vault. The effect of all forces on his path is described by Newton's laws of motion.

Motions attract our attention. We appreciate the motion itself, and marvel at the forces that must be necessary to achieve certain motions. Motion is the subject of kinematics, but kinematics only describes the *way* objects move. **Dynamics** extends the study of motion to consider mass and the forces that affect motion. Newton's laws of motion are the foundation upon which dynamics is built. These laws provide a first example of the breadth and simplicity of nature's laws.

Newton's laws were developed by the legendary Isaac Newton. (His image appears in Figure 1.6.) The development of Newton's laws marks the transition from the Renaissance into the modern era. This transition was characterized by a revolutionary change in the way of thinking about the physical universe. Among the many great people who contributed to this change were Newton and Galileo. For many centuries natural philosophers had debated the nature of the universe based on certain rules of logic and gave great weight to the thoughts of earlier classical philosophers such as Aristotle. Galileo was instrumental in establishing *observation* as the absolute determinant of truth, rather than "logical" argument. Most notable among Galileo's achievements in demonstrating the importance of observation was his use of the telescope. He discovered moons around Jupiter and made other observations that were inconsistent with certain ancient ideas and religious dogma. For this, and because of the manner in which he dealt with those in authority, Galileo was tried by the Inquisition and punished. He spent the final years of his life under a form of house arrest. Because others before Galileo had also made progress in *observing* the nature of the universe, and because repeated observations verified those of Galileo, his work could not be suppressed or denied.

Galileo also contributed to the formation of what is now called *Newton's* first law of motion. Galileo, along with the French philosopher Rene Descartes (1596–1650) among many others, developed older ideas and made contributions upon which Newton built. Newton himself acknowledged the work of his predecessors, which enabled him to develop laws of motion, discover the law of gravity, invent calculus, and make great contributions to the theories of light and color.

It was not until the advent of modern physics early in the 20th century that Newton's laws had to be modified. At that time, Albert Einstein (1879–1955) developed the theory of relativity and, along with many other individuals, contributed greatly to quantum theory. (The successes of those theories are based on observation.) Newton's laws are an extremely good approximation to the more accurate relativistic and quantum mechanical theories *under the following conditions*: Objects dealt with must be moving at much less than the speed of light (less than about 10^6 m/s), and they must be larger than molecules (larger than about 10^{-9} m). These constraints define the realm of classical mechanics, as was discussed in Chapter 1. All situations we consider in this chapter, and all those preceding the introduction of relativity in Chapter 26, are in the realm of classical physics.

4.1 FORCE: THE CONCEPT

To understand Newton's laws, we need a working definition of force. Our intuitive definition of **force**—that it is a push or a pull—is a good place to start. We know that a push

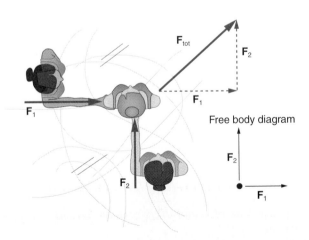

Figure 4.1 Forces add like other vectors, as illustrated in this overhead view of two ice skaters pushing on a third. As expected, the total force is in the direction shown. Forces (vectors) represented by arrows can be added using the familiar head-to-tail method or by trigonometric methods, as before. (See Chapter 3.) The total force on any body is the vector sum of all the individual forces acting on it.

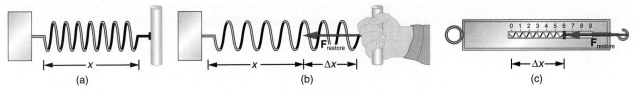

Figure 4.2 A standard unit of force can be based on some reproducible force, such as that exerted by a stretched spring. (a) This spring has a length x when undistorted. (b) When stretched a distance Δx, the spring exerts a restoring force, **F**$_{restore}$, that is reproducible. (c) One device that uses a spring to measure force is the spring scale. The force **F**$_{restore}$ is exerted on whatever is attached to the hook. **F**$_{restore}$ has a magnitude of 6 units of the force standard being used.

or pull has magnitude and direction and can vary considerably in both regards. For example, rocket engines can exert a tremendous upward push, while gravity may exert a tiny downward pull on a flea. Our everyday experiences also give us a good idea of how multiple forces add. If two people push in different directions on a third, as illustrated in Figure 4.1, we intuitively expect the total force to be in the direction shown. Force is, in fact, another type of vector. It adds like other vectors, as illustrated in Figure 4.1.

A more quantitative definition of force can be based on some standard force, just as distance is measured in units relative to a standard distance. One possibility is to stretch a spring a fixed distance, as illustrated in Figure 4.2, and use the force it exerts to pull itself back to its relaxed shape—called a *restoring force*—as a standard. The magnitude of all other forces can be stated as multiples of this standard unit of force. Many other possibilities exist for standard forces. (One that we will encounter in Chapter 21 is the magnetic force between two wires carrying electric current.) An alternative definition of force will be given in Section 4.2, and there will be detailed discussion of the types of forces in Sections 4.5 and 4.6.

4.2 NEWTON'S FIRST LAW OF MOTION: MASS

Experience suggests that an object at rest will remain at rest if left alone, and that an object in motion tends to slow down and stop unless some effort is made to keep it moving. What **Newton's first law of motion** states, however, is the following:

> **A body at rest remains at rest, or if in motion remains in motion at constant velocity, unless acted on by a net external force.**

This does not really contradict our experience. Rather it states that there is a *cause* (net external force) *for any change in velocity*, such as the slowing of an object due to friction.*

The idea of cause and effect is crucial in describing accurately what happens in various situations. For example, consider what happens if you have an object sliding along a rough horizontal surface. The object quickly grinds to a halt. On a smoother surface, it slides farther. On a smooth, lubricated surface, it slides even farther. Extrapolating to a frictionless surface, we can imagine the object sliding in a straight line indefinitely. Friction is thus viewed (consistent with Newton's first law) as the *cause* of slowing, and the object would not slow down at all if friction were eliminated. If we know enough about friction, we can accurately predict how fast the object will slow down.

Newton's first law is completely general and can be applied to anything from an object sliding on a table to a satellite in orbit to blood pumped from the heart. Experiments have thoroughly verified that any change in velocity (speed or direction) is caused by an external force. The idea of generally applicable laws is important not only here—it is a basic feature of all laws of physics. The genius of Galileo, who first developed the idea for the first law, and Newton, who clarified it, was to ask the very fundamental question,

---

*Friction is a commonly encountered force that opposes motion between systems in contact with one another. It is discussed in some detail in Section 4.5.

"What is the cause?" Thinking in terms of cause and effect is a worldview fundamentally different from the typical earlier attitude. Earlier a question such as, "Why does a tiger have stripes?" would have been answered, "That is the nature of the beast." True perhaps, but not a useful insight.

Mass

The property by which a body remains at rest or remains in motion with constant velocity is called **inertia**, and Newton's first law is often called the **law of inertia**. Experience tells us that some objects have more inertia than others. It is obviously more difficult to change the motion of a boulder than that of a basketball, for example. The inertia of an object is measured by its **mass**—the quantity of matter it contains. The quantity of matter in an object is determined by the numbers of atoms and molecules of various types in it. Thus, unlike weight, mass does not vary with location. The mass of an object is the same on the earth, in orbit, or on the surface of the moon. In practice, it is very difficult to count and identify all of the atoms and molecules in an object, and so operationally masses are determined by comparison with the standard kilogram, as first discussed in Section 1.2.

4.3 NEWTON'S SECOND LAW OF MOTION: CONCEPT OF A SYSTEM

Newton's first law of motion states the cause and effect relationship between force and changes in motion. *Newton's second law of motion* is more quantitative and is used extensively to calculate what happens in situations involving force and mass. Before we can write down Newton's second law as a simple equation giving the exact relationship of force, mass, and acceleration, we need to sharpen some ideas that have already been mentioned.

First, what do we mean by a change in motion? The answer is that a change in motion is equivalent to a change in velocity. A change in velocity means, by definition, there is an acceleration. Newton's first law says that net external force causes a change in motion; thus, we see that *net external force causes acceleration.*

Another question immediately arises. What do we mean by external force? Your intuitive notion of external is correct—an **external force** acts from outside the system of interest. For example, in Figure 4.3(a) the system of interest is the wagon plus the child in it. The two forces exerted by the other children are external forces. An internal force acts between elements of the system. Again looking at Figure 4.3(a), the force the child in the wagon exerts to hang onto it is an internal force between elements of the system of interest. Only external forces affect the motion of a system, according to Newton's first law. (The internal forces actually cancel, as we shall see in the next section.) You must choose the system before you can determine which forces are external. Sometimes the system is obvious, whereas other times choosing a system is more subtle. The **concept of a system** is crucial to many areas of physics as well as to the correct application of Newton's laws, and we shall discuss it specifically many times on our journey through physics.

Now, it seems reasonable that acceleration should be directly proportional to and in the same direction as the net or total external force acting on a system. This has been verified experimentally and is illustrated in Figure 4.3. In part (a), a smaller force causes a smaller acceleration than the larger force illustrated in part (c). For completeness, the vertical forces are also shown; they are assumed to cancel. (The vertical forces are the weight **w** and support of the ground **N**, and they will be discussed in more detail in later sections.) Part (b) shows how vectors representing the external forces add.

To get an equation for Newton's second law, we first write the relationship of acceleration and external force as the proportionality

$$\mathbf{a} \propto \text{net } \mathbf{F}$$

where the symbol $\propto$ means "proportional to," and net **F** is the **net external force**. (The net external force is the vector sum of all external forces and can be determined graphically, using the head-to-tail method, or analytically, using components, and so on. The

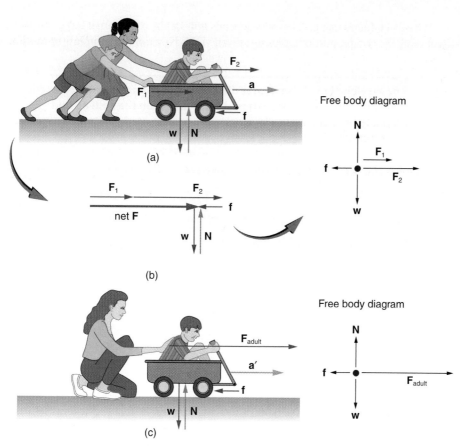

Figure 4.3 Different forces exerted on the same mass produce different accelerations. (a) Two children push a wagon with a child in it. Arrows representing all external forces are shown. The system of interest is the wagon and its rider. The weight **w** of the system and the support of the ground **N** are also shown for completeness and are assumed to cancel. (b) Vector addition of the forces using the head-to-tail method shows that the net external force is horizontal, as is the acceleration of the system. (c) A larger net external force produces a larger acceleration ($a' > a$) when an adult pushes the child.

techniques are the same as for the addition of other vectors, covered in Chapter 3.) This proportionality states what we have said in words—*acceleration is directly proportional to the net external force*. It is a tremendous simplification not to have to consider the numerous internal forces acting between objects within the system, such as muscular forces within the child's body, let alone the myriad of forces between atoms in the objects. Once the system of interest is chosen, it is straightforward to identify the external forces and ignore the internal ones.

Now, it also seems reasonable that acceleration should be inversely proportional to the mass of the system. In other words, the larger mass, the smaller the acceleration produced by a given force. And indeed, as illustrated in Figure 4.4, the same net external force applied to a car produces a much smaller acceleration than when applied to a basketball. The proportionality is written as

$$a \propto \frac{1}{m}$$

where m is the mass of the system. Experiments have shown that acceleration is exactly inversely proportional to mass, just as it is exactly linearly proportional to the net external force.

Figure 4.4 The same force exerted on systems of different masses produces different accelerations. (a) A basketball player pushes on a basketball to make a pass. (The effect of gravity on the ball is ignored.) (b) The same player exerts an identical force on a stalled car and produces a far smaller acceleration (even if friction is negligible).

It has been found that acceleration depends *only* on the net external force and mass. Combining the two proportionalities just given yields **Newton's second law of motion**:

The acceleration of a system is directly proportional to and in the same direction as the net external force acting on the system, and inversely proportional to its mass.

In equation form, Newton's second law of motion is

$$\mathbf{a} = \frac{\text{net } \mathbf{F}}{m} \tag{4.1}$$

This is often written in a more familiar form:

$$\text{net } \mathbf{F} = m\mathbf{a} \tag{4.2}$$

Although these last two equations are really the same, the first gives more insight into what Newton's second law means. It is a cause and effect relationship among three quantities that is not simply based on their definitions. Its validity is completely based on experimental verification.

Units of Force

Equation 4.2 is used to define the units of force in terms of the three basic units for mass, length, and time. The SI unit of force is called the **newton** (abbreviated N) and is the force needed to accelerate a 1 kilogram system at the rate of 1 meter per second squared. That is, since $\mathbf{F} = m\mathbf{a}$,

$$1 \text{ N} = 1 \text{ kg} \cdot \frac{\text{m}}{\text{s}^2}$$

In the United States, the most familiar unit of force is the British pound (lb). But in this text the newton, like SI units in general, will be used almost exclusively. When a comparison with more familiar units of force is useful, it will be made through conversion coefficients. For example, 1 N = 0.225 lb.

EXAMPLE 4.1   WHAT ACCELERATION CAN A PERSON PUSHING A CAR PRODUCE?

Suppose that the net external force exerted on the car (push minus friction) in Figure 4.4 is 350 N (about 79 lb) parallel to the ground, and the mass of the car is 1800 kg. What is its acceleration?

Strategy Since net F and m are given, the acceleration can be calculated directly from Newton's second law as stated in Equation 4.1, $a = (\text{net } F)/m$.

Solution The magnitude of the acceleration $\mathbf{a}$ is $a = (\text{net } F)/m$. Entering known values gives

$$a = \frac{350 \text{ N}}{1800 \text{ kg}}$$

Substituting the units kg · m/s² for N yields

$$a = \frac{350 \text{ kg} \cdot \text{m/s}^2}{1800 \text{ kg}} = 0.194 \text{ m/s}^2$$

Discussion The direction of the acceleration is the same as that of the net (total) force (given to be parallel to the ground). There is no information given in this example about the individual external forces acting on the system, but we can say something about their relative magnitudes. For example, the force exerted by the person pushing the car must be greater than the friction opposing the motion, and the vertical forces must cancel if there is to be no acceleration in the vertical direction.

EXAMPLE 4.2   WHAT ROCKET THRUST ACCELERATES THIS SLED?

Calculate the force exerted by each rocket, called its thrust T, for the situation shown in Figure 4.5. The sled's initial acceleration is 49.0 m/s², the mass of the system is 2100 kg, and the force of friction opposing the motion is known to be 600 N.

Strategy Although there are forces acting vertically and horizontally, we assume the vertical forces cancel since there is no vertical acceleration. This leaves us with only horizontal forces and a simpler one-dimensional problem. Directions are indicated with plus or minus signs, with right taken as the positive direction.

Solution Since acceleration, mass, and force of friction are given, we start with Newton's second law and look for ways to find the thrust of the engines. Hence we begin with

$$a = \frac{\text{net } F}{m}$$

where net F is the net force along the horizontal direction. We can see from Figure 4.5 that the engine thrusts add, while friction opposes the thrust. In equation form, the net external force is

$$\text{net } F = 4T - f$$

Substituting this into Newton's second law gives

$$a = \frac{4T - f}{m}$$

Using a little algebra, we solve for the total thrust $4T$:

$$4T = ma + f$$

Substituting known values yields

$$4T = (2100 \text{ kg})(49.0 \text{ m/s}^2) + 600 \text{ N}$$

So the total thrust is

$$4T = 1.04 \times 10^5 \text{ N}$$

and the individual thrusts are

$$T = 2.59 \times 10^4 \text{ N}$$

Discussion This is about 5800 pounds of thrust each, realistic for rocket sled experiments performed in the 1950s. Note that in this example, as in the preceding one, the system of interest is obvious. We will see in later examples that choosing the system of interest is crucial—and the choice is not always obvious.

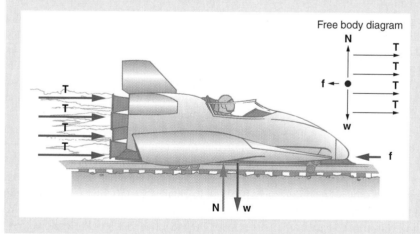

Free body diagram

Figure 4.5 The rocket sled considered in Example 4.2. Each rocket creates an identical thrust **T**. As in other examples where there is no vertical motion, the ground exerts an upward force **N** on the system that is equal in magnitude and opposite in direction to its weight **w**. The system here is the sled, its rockets, and rider, and so none of the forces between these objects are considered. The arrow representing friction is drawn larger than scale.

Newton's laws of motion are among the greatest insights into motion in history, relating force, mass, and acceleration. Those physical quantities can be defined independently from Newton's laws, and so the laws tell us something basic and universal about nature. The next section introduces the third and final law of motion.

4.4   NEWTON'S THIRD LAW OF MOTION: SYMMETRY

There is a passage in the musical *Man of la Mancha* that relates to Newton's third law of motion. Sancho, in describing a fight with his wife to Don Quixote, says, "Of course I hit her back, Your Grace, but she's a lot harder than me and you know what they say,

'Whether the stone hits the pitcher or the pitcher hits the stone, it's going to be bad for the pitcher.'" This is exactly what happens whenever one body exerts a force on another—the first also experiences a force. Numerous common experiences, such as stubbing a toe, confirm this. It is precisely stated in **Newton's third law of motion**:

> **Whenever one body exerts a force on a second body, the first body experiences a force that is equal in magnitude and opposite in direction to the one it exerts.**

This law represents a certain *symmetry in nature*: forces always occur in pairs, and one body cannot exert a force on another without experiencing a force itself. The law has practical uses in analyzing the origin of forces and understanding which forces are external to a system. We sometimes refer to this law loosely as "action-reaction," where the force exerted is the action and the force experienced as a consequence is the reaction.

We can readily see Newton's third law at work if we consider how people move about. Consider a swimmer pushing off from the side of a pool, as illustrated in Figure 4.6. She pushes against the pool wall with her feet and accelerates in the direction *opposite* to that of her push. The wall has exerted an equal and opposite force back on the swimmer. You might think that two equal and opposite forces would cancel, but they do not *because they act on different systems*. If we select the swimmer to be the system of interest, as in the figure, then $\mathbf{F}_{wall}$ is an external force on this system and affects its motion. The swimmer moves in the direction of $\mathbf{F}_{wall}$. In contrast, the force $\mathbf{F}_{feet}$ acts on the wall and not on the system of interest. Thus $\mathbf{F}_{feet}$ does not directly affect the motion of the system and does not cancel $\mathbf{F}_{wall}$. Note that the swimmer pushes in the direction opposite to that in which she wishes to move. The reaction to her push is thus in the desired direction.

Other examples of Newton's third law are easy to find. As a professor paces in front of a chalkboard, she exerts a force backward on the floor. The floor exerts a reaction force forward on the professor that actually causes her to move. Similarly, a car accelerates due to the ground pushing forward in reaction to the drive wheels pushing backward on the ground. You can see evidence of the wheels pushing backward when tires spin on a gravel road and throw rocks backward. Rockets move forward by expelling gas backward at high velocity. This means the rocket exerts a large backward force on the gas in its combustion chamber, and the gas therefore exerts a large reaction force (called thrust) forward on the rocket. It is a common misconception that rockets propel themselves by pushing on the ground or on the air behind them. They actually work better in a vacuum, where they can more readily expel the exhaust gases. Helicopters push down on the air, thereby experiencing an upward reaction force. Birds and airplanes also fly by exerting force on air in a direction opposite to that of the needed force. For example, the wings of a bird force air downward and backward in order to get lift and move forward. Professional boxers sometimes break a hand by hitting an opponent's jaw.

Figure 4.6 When the swimmer exerts a force $\mathbf{F}_{feet}$ on the wall, she accelerates in the direction opposite to that of her push. This means the net external force on her is in the direction opposite to $\mathbf{F}_{feet}$. This is because, in accordance with Newton's third law of motion, the wall exerts a force $\mathbf{F}_{wall}$ on her in the direction opposite to the one she exerts on it. The line around the swimmer indicates the system of interest. Note that $\mathbf{F}_{feet}$ does not act on this system and, thus, does *not* cancel $\mathbf{F}_{wall}$. Not shown are the force of gravity and the buoyant force of the water, which do cancel.

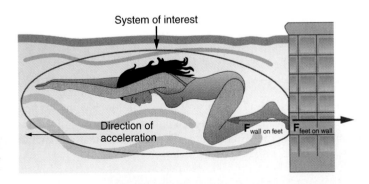

EXAMPLE 4.3 CHOOSE THE CORRECT SYSTEM TO SOLVE THIS PROBLEM

A physics professor pushes a cart of demonstration equipment to a lecture hall, as seen in Figure 4.7. The professor's mass is 85.0 kg, the cart's is 20.0 kg, and the equipment's is 15.0 kg. Calculate the acceleration produced when the professor exerts a backward force of 150 N on the floor. All forces opposing the motion, such as friction and air resistance, total 4.00 N.

Strategy Define the system to be the professor, cart, and equipment, since they accelerate as a unit. This is System 1 in the figure. The professor pushes backward with a force F_{foot} of 150 N. According to Newton's third law, the floor exerts a forward reaction force F_{floor} of 150 N on System 1. We can assume there is no net force in the vertical direction since there is no vertical motion, and so the problem is one-dimensional along the horizontal direction. As noted, **f** opposes the motion and is thus in the opposite direction of $\mathbf{F}_{floor}$. There are no other significant forces acting on System 1. If the net external force can be found from all this information, we can use Newton's second law to find the acceleration as requested.

Solution Newton's second law is given by Equation 4.2 as

$$a = \frac{net\ F}{m}$$

The net external force on System 1 is deduced from the figure and the discussion above to be

$$net\ F = F_{floor} - f = 150\ N - 4.00\ N = 146\ N$$

The mass of System 1 is

$$m = (85.0 + 20.0 + 15.0)\ kg = 120\ kg$$

These values of net F and m give an acceleration of

$$a = \frac{146\ N}{120\ kg} = 1.22\ m/s^2$$

Discussion None of the forces between components of System 1, such as between the professor's hands and the cart, contribute to the net external force because they are internal to System 1. Another way to look at this is to note that forces between components of a system cancel because they are equal in magnitude and opposite in direction. For example, the force exerted by the professor on the cart results in an equal and opposite force back on her. In this case both forces act on the same system and, therefore, cancel. Thus internal forces (between components of a system) cancel. Choosing System 1 was crucial to solving this problem.

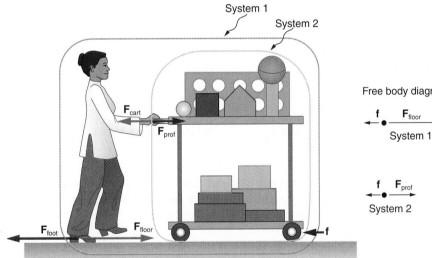

System 1
System 2

Free body diagrams

f $\mathbf{F}_{floor}$
System 1

f $\mathbf{F}_{prof}$
System 2

$\mathbf{F}_{cart}$
$\mathbf{F}_{prof}$
$\mathbf{F}_{foot}$ $\mathbf{F}_{floor}$ **f**

Figure 4.7 Examples 4.3 and 4.4 consider this professor as she pushes a cart of demonstration equipment. The lengths of the arrows are proportional to the magnitudes of the forces (except for **f**, since it is too small to draw to scale). Different questions are asked in each example; thus, the system of interest must be defined differently for each. System 1 is appropriate for Example 4.3, since it asks for the acceleration of the entire group of objects. Only $\mathbf{F}_{floor}$ and **f** are external forces acting on System 1. All other forces either cancel or act on the outside world. System 2 is chosen for Example 4.4 so that $\mathbf{F}_{prof}$ will be an external force and enter into Newton's second law. The weight of the systems and the supporting forces of the floor cancel and, for simplicity, are not shown.

EXAMPLE 4.4 A DIFFERENT SYSTEM MUST BE CHOSEN TO SOLVE THIS PROBLEM

Calculate the force the professor exerts on the cart in Figure 4.7 using data from Example 4.3 if needed.

Strategy If we now define the system of interest to be the cart plus equipment (System 2 in the figure), then the net ex-

ternal force on System 2 is the force the professor exerts on the cart minus friction. The force the professor exerts on the cart, $\mathbf{F}_{prof}$, is an external force acting on System 2. $\mathbf{F}_{prof}$ was internal to System 1, but it is external to System 2 and will enter Newton's second law for this system.

(continued)

(continued)

Solution Newton's second law can be used to find F_{prof}. Starting with

$$a = \frac{net\ F}{m}$$

and noting that the net external force on System 2 is

$$net\ F = F_{prof} - f$$

we solve for F_{prof}, the desired quantity:

$$F_{prof} = net\ F + f$$

The value of f is given, and so we must calculate net F. That can be done since both the acceleration and mass of System 2 are known. Using Newton's second law in the form of Equation 4.2, we see that

$$net\ F = ma$$

where the mass of System 2 is 35.0 kg ($m = 20.0$ kg + 15.0 kg) and its acceleration was found to be $a = 1.22$ m/s² in the previous example. Thus,

$$net\ F = (35.0\ kg)(1.22\ m/s^2) = 42.6\ N$$

Now we can find the desired force:

$$F_{prof} = 42.6\ N + 4.0\ N = 46.6\ N$$

Discussion It is interesting that this force is significantly less than the 150 N force the professor exerted backward on the floor. Not all of that 150 N force is transmitted to the cart; some of it is used to accelerate the professor. The choice of System 2 as illustrated in the figure is crucial to solving this problem.

The choice of a system is an important analytical step both in solving problems and in thoroughly understanding the physics of the situation (not necessarily the same thing).

4.5 WEIGHT, FRICTION, TENSION, AND OTHER CLASSES OF FORCES

Forces are given many names, such as push, pull, thrust, lift, weight, friction, and tension. Traditionally, forces have been grouped into several categories and given names relating to their source, how they are transmitted, or their effects. The most important of these categories are discussed in this section, together with some interesting applications.

Weight and the Force of Gravity

When an object is dropped, it accelerates toward the center of the earth. Newton's second law states that the net force on the object is responsible for its acceleration. If air resistance is negligible, the net force on a falling object is the **force of gravity**, commonly called its **weight** w. Galileo was instrumental in showing that, in the absence of air resistance, all objects fall with the same acceleration g. We can use Newton's second law as stated in Equation 4.2 to derive an equation for weight. The magnitude of the net external force is

$$net\ F = ma$$

For an object in free-fall, the net external force is its weight—or net $F = w$—and its acceleration is that of gravity—or $a = g$. Substituting these into Newton's second law gives

$$w = mg \qquad \textbf{(4.4)}$$

This is the equation for *weight*—the force of gravity on a mass m. Since $g = 9.8$ m/s², the weight of a 1.0 kg object on earth is 9.8 N, as we see using Equation 4.4 to calculate w:

$$w = mg = (1.0\ kg)(9.8\ m/s^2) = 9.8\ N$$

The acceleration of gravity g varies slightly over the surface of the earth, so that the weight of an object depends on location and is not an intrinsic property of the object. Weight varies dramatically if one leaves the surface of the earth. On the moon, for example, the acceleration of gravity is only 1.67 m/s². Using the conversion factor that 1.0 N = 0.225 lb, a 1.0 kg mass is seen to have a weight of 2.2 lb on earth and only about 0.4 lb on the moon.

It is important to be aware that weight and mass are very different physical quantities, although they are intimately related. Mass is the quantity of matter and does not vary in classical physics, whereas weight is the force of gravity and does vary. It is tempting to equate the two, since most of our examples take place on earth, where the weight of

CONNECTIONS

The acceleration due to gravity g is the same for all objects in free-fall in a given location. This was first discussed for the kinematics of falling objects in Section 2.6. We will find this fact useful in problems involving weight or the force of gravity. When we study gravitational force as described by Newton's universal law of gravity, we will finally see *why* the acceleration due to gravity is the same for all objects.

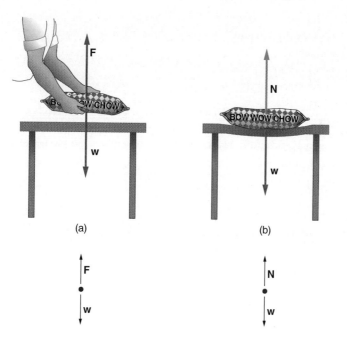

Figure 4.8 (a) The person holding the bag of dog food must supply an upward force **F** equal in magnitude and opposite in direction to the weight of the food **w**. (b) The table sags when the dog food is placed on it until it supplies a force **N** equal in magnitude and opposite in direction to the weight of the load.

an object varies only slightly with location. Furthermore, the terms *mass* and *weight* are used interchangeably in everyday language; for example, our medical records often show our "weight" in kilograms* but never in the correct units of newtons.

Weight is a pervasive force that acts at all times and so must be counteracted to keep an object from falling. You definitely notice that you must support the weight of a heavy object by pushing up on it when you hold it as illustrated in Figure 4.8(a). But how do inanimate objects like a table support the weight of a mass placed on them, such as shown in Figure 4.8(b)? When the bag of dog food is placed on the table, the table actually sags slightly under the load. This would be noticeable if the load were placed on a card table, but even rigid objects deform when a force is applied to them. Unless the object is deformed beyond its limit, it will exert a restoring force much like a deformed spring. The greater the deformation, the greater the restoring force. So when the load is placed on the table, the table sags until the restoring force becomes as large as the weight of the load. At this point the net external force on the load is zero. That is the situation when the load is stationary on the table.

The conclusion is that whatever supports a load must supply an upward force equal to the weight of the load, as we assumed in a few of the previous examples. If the force supporting a load is perpendicular to the surface of contact between the load and its support, this force is defined to be a **normal force** and given the symbol **N**. The word *normal* means perpendicular to a surface.

EXAMPLE 4.5 WEIGHT ON AN INCLINE, A TWO-DIMENSIONAL PROBLEM

Consider the skier on a slope shown in Figure 4.9. Her mass including equipment is 60.0 kg. (a) What is her acceleration if friction is negligible? (b) What is her acceleration if friction is known to be 45.0 N?

Strategy and Concept This is a two-dimensional problem, since the forces on the skier (the system of interest) are not parallel. The approach we have used in two-dimensional kinematics also works very well here. Choose a convenient coordinate system and project the vectors onto its axes, giving *two* connected *one*-dimensional problems to solve. The most convenient coordinate system for motion on an incline is one that has one coordinate parallel to the slope and one perpendicular to the slope. (Remember that motions along

(continued)

---

*Since the variation in gravity is slight, the conversion of 1.0 kg to 2.2 lb is often employed, taking g as 9.8 m/s^2. Conversion of kg to lb is technically incorrect, since units of mass cannot be converted to units of force. The correct conversion is 9.8 N = 2.2 lb.

(continued)

mutually perpendicular axes are independent.) We use the symbols ⊥ and ∥ to represent perpendicular and parallel, respectively. This choice of axes simplifies this type of problem, because there is no motion perpendicular to the slope and because friction is always parallel to the surface between two objects. The only external forces acting on the system are the skier's weight, friction, and the support of the slope, respectively labeled **w**, **f**, and **N** in Figure 4.9. **N** is perpendicular to the slope and **f** is parallel to it. But **w** is not in the direction of either axis, and so the first step we take is to project it into components along the chosen axes, defining $w_∥$ to be the component of weight parallel to the slope and $w_⊥$ the component of weight perpendicular to the slope. Once this is done, we can consider the two separate problems of forces parallel to the slope and forces perpendicular to the slope.

Solution The magnitude of the component of the weight parallel to the slope is

$$w_∥ = w \sin 25° = mg \sin 25°$$

and the magnitude of the component of the weight perpendicular to the slope is

$$w_⊥ = w \cos 25° = mg \cos 25°$$

(a) Since the acceleration is parallel to the slope, we need only consider forces parallel to the slope. (Forces perpendicular to the slope add to zero, since there is no acceleration in that direction.) The forces parallel to the slope are $w_∥$ and friction f. Using Newton's second law, with subscripts to denote quantities parallel to the slope,

$$a_∥ = \frac{\text{net } F_∥}{m}$$

where net $F_∥ = w_∥ = mg \sin 25°$, assuming no friction, so that

$$a_∥ = \frac{mg \sin 25°}{m} = g \sin 25°$$

Thus,

$$a_∥ = (9.80 \text{ m/s}^2)(0.423) = 4.14 \text{ m/s}^2$$

is the acceleration.

(b) Here we have a given value for friction, and we know its direction is parallel to the slope and it opposes motion between surfaces in contact. So the net external force is now

$$\text{net } F_∥ = w_∥ - f$$

and substituting this into Newton's second law, $a_∥ = \text{net } F_∥/m$ gives

$$a_∥ = \frac{w_∥ - f}{m} = \frac{mg \sin 25° - f}{m}$$

We substitute known values to obtain

$$a_∥ = \frac{(60.0 \text{ kg})(9.80 \text{ m/s}^2)(0.423) - 45.0 \text{ N}}{60.0 \text{ kg}}$$

which yields

$$a_∥ = 3.39 \text{ m/s}^2$$

as the acceleration when there is opposing friction.

Discussion Since friction opposes motion between surfaces, the acceleration is smaller when there is friction than when there is none. In fact, it is a general result that if friction on an incline is negligible, then the acceleration down the incline is $a = g \sin θ$, *regardless of mass*. This is related to the previously discussed fact that all objects fall with the same acceleration in the absence of air resistance. Similarly, all objects slide down a frictionless incline with the same acceleration.

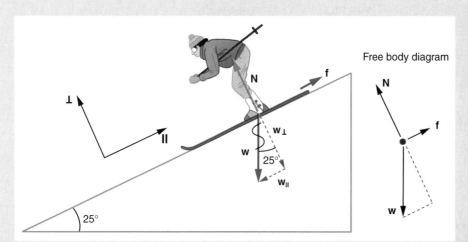

Figure 4.9 The skier in Example 4.5. Since motion and friction are parallel to the slope, it is most convenient to project all forces onto a coordinate system where one axis is parallel to the slope and the other is perpendicular (axes shown to left of skier). **N** is perpendicular to the slope and **f** is parallel to the slope, but **w** has components along both axes, namely $w_⊥$ and $w_∥$. **N** is equal in magnitude to $w_⊥$, so that there is no motion perpendicular to the slope, but f is less than $w_∥$, so that there is a downslope acceleration (along the ∥-axis).

Friction: A Brief Treatment

Although **friction** is a common force *opposing motion between systems in contact*, it is actually very complicated and is not completely understood even today. However, we can deal with its more elementary general characteristics and understand the circumstances in which it behaves simply.

One of the simpler characteristics of friction is that it is parallel to the contact surface between systems. It also varies depending on whether there is motion between the systems or not. If two systems are in contact and moving relative to one another, friction between them is called **kinetic friction**. For example, there is friction slowing a hockey puck sliding on ice. But when objects are stationary, **static friction** can act between them; static friction is potentially greater than kinetic friction between the objects.

Imagine, for example, trying to slide a heavy crate across a concrete floor—you can push hard on the crate and not move it at all. This means that static friction responds to what you do—it increases to be equal to and in the opposite direction of your push. But if you push hard enough, the crate seems to slip suddenly and to be easier to keep in motion than it was to get it started. If you add mass to the crate, say by having another box on top of it, you need to push even harder to get it started and to keep it moving. Furthermore, if you oiled the concrete you would find it to be easier to get the crate started and keep it going, as you might expect.

Figure 4.10 is a crude pictorial representation of how friction occurs at the interface between two objects. Close-up inspection of the surfaces shows them to be rough. So when you push to get the object moving, you must raise the object until it can skip along with just the tips hitting, break off the points, or do both. A considerable force can be resisted with no apparent motion. The harder the surfaces are pushed together (such as when another box is placed on the crate), the more force is needed to move them. Part of the friction is due to adhesive forces between the surface molecules of the two objects, which explains the dependence of friction on substances. Adhesion varies with substances in contact and is a complicated aspect of surface physics. Once the object is moving, there are fewer points of contact (fewer molecules adhering); hence, less force is required to keep it moving. At small but nonzero speeds, friction is nearly independent of speed.

The magnitude of the frictional force has two forms; one for static situations, the other for when there is motion. When there is no motion between the objects, the **magnitude of static friction** f_s is

$$f_s \leq \mu_s N \tag{4.5}$$

where μ_s is the **coefficient of static friction**, and N is the **magnitude of the normal force** (the force perpendicular to the surface, as defined above). The symbol $\leq$ means *less than or equal to*, implying that static friction can have a maximum value of $\mu_s N$. Static friction is a responsive force that increases to be equal and opposite to whatever force is exerted to start motion, up to its maximum limit.

The **magnitude of kinetic friction** f_k is given by

$$f_k = \mu_k N \tag{4.6}$$

where μ_k is the **coefficient of kinetic friction**. As seen in Table 4.1, the coefficients of kinetic friction are less than their static counterparts. Equations 4.5 and 4.6 include the

Figure 4.10 Frictional forces, such as **f**, always oppose motion or attempted motion between objects in contact. Friction arises in part because of the roughness of the surfaces in contact, as seen in the expanded view. In order for the object to move, it must rise to where the peaks can skip along. Thus a force is needed just to get started. Some of the peaks will be broken off, also requiring a force to maintain motion. Much of the friction is actually due to attractive forces between molecules of the two objects, so that even perfectly smooth surfaces are not friction-free. Such adhesive forces also depend on substances, explaining, for example, why rubber-soled shoes slip less than those with leather soles.

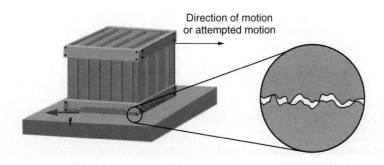

Direction of motion
or attempted motion

TABLE 4.1

COEFFICIENTS OF STATIC AND KINETIC FRICTION*

System	Static friction (μ_s)	Kinetic friction (μ_k)
Rubber on dry concrete	1.0	0.7
Rubber on wet concrete	0.7	0.5
Wood on wood	0.5	0.3
Waxed wood on wet snow	0.14	0.1
Metal on wood	0.5	0.3
Steel on steel (dry)	0.6	0.3
Steel on steel (oiled)	0.05	0.03
Teflon on steel	0.04	0.04
Bone lubricated by synovial fluid	0.016	0.015
Shoes on wood	0.9	0.7
Shoes on ice	0.1	0.05
Ice on ice	0.1	0.03
Steel on ice	0.4	0.02

*Values vary with circumstances.

dependence of friction on materials and the normal force. The direction of friction is parallel to the surface between objects and perpendicular to the normal force. For example, if the crate you are trying to push has a mass of 100 kg, then the normal force would be equal to its weight of $w = mg = (100 \text{ kg})(9.80 \text{ m/s}^2) = 980 \text{ N}$, perpendicular to the floor. If the coefficient of static friction is 0.5, you would have to exert a force parallel to the floor greater than $f_s = \mu_s N = 0.5 \times 980 \text{ N} = 490 \text{ N}$ to move the crate. Once there is motion, friction is less and the coefficient of kinetic friction might be 0.3, so that a force of only 294 N ($f_k = \mu_k N = 0.3 \times 980 \text{ N}$) would keep it moving at a constant speed. If the floor is lubricated, both coefficients are considerably less.

EXAMPLE 4.6 FRICTION ON A SLOPE

Find the coefficient of kinetic friction for the skier described in Example 4.5(b).

Strategy The magnitude of kinetic friction was given in Example 4.5 to be 45.0 N. Kinetic friction is related to the normal force by Equation 4.6, $f_k = \mu_k N$; thus, the coefficient of kinetic friction can be found if we can find the normal force of the skier on a slope. The normal force is perpendicular to the slope, and since there is no motion parallel to the slope, the normal force should equal the component of the skier's weight perpendicular to the slope. (See the free body diagram* in Figure 4.9.) That is,

$$N = w_\perp = w \cos 25° = mg \cos 25°$$

Combining these two equations, we get

$$f_k = \mu_k mg \cos 25°$$

which can be solved for the coefficient of kinetic friction μ_k.

Solution Solving the last equation for μ_k gives

$$\mu_k = \frac{f_k}{N} = \frac{f_k}{w \cos 25°} = \frac{f_k}{mg \cos 25°}$$

Substituting known values on the right-hand side of the equation,

$$\mu_k = \frac{45.0 \text{ N}}{(60 \text{ kg})(9.8 \text{ m/s}^2)(0.906)} = 0.084$$

Discussion This is a little smaller than the coefficient listed in Table 4.1 for waxed wood on snow, but it is still reasonable since values of the coefficients of friction can vary greatly. (This is also why the answer is given to only two significant figures.) In situations like this, where an object of mass m slides down a slope that makes an angle θ with the horizontal, friction is given by $f_k = \mu_k mg \cos \theta$. All objects will slide down a slope with constant acceleration under these circumstances. The proof of this is left as an end-of-chapter exercise.

---

*A free body diagram reduces the object to a point and shows all external forces acting on the system of interest. Free body diagrams are discussed in Section 4.7 on problem-solving strategies for Newton's laws.

THINGS GREAT AND SMALL

Submicroscopic Explanations of Friction

The simpler aspects of friction dealt with in this section are its macroscopic (large-scale) characteristics. Great strides have been made in the atomic-scale explanation of friction during the past two decades. It is being found that friction is indeed very complex, but that fundamentally correct ideas about its atomic nature seem to exist. These ideas not only explain some of the simpler aspects of friction—they also hold hope for developing nearly friction-free environments that could save hundreds of billions of dollars in energy converted unnecessarily to heat.

Figure 4.A illustrates one macroscopic characteristic of friction that is explained by microscopic research. We have noted that friction is proportional to the normal force, but not to the area in contact, a somewhat counterintuitive notion. When two rough surfaces are in contact, the *actual* contact area is a tiny fraction of the total area since only high spots touch. When a greater normal force is exerted,

the *actual* contact area increases, and it is found that the friction is proportional to this area.

But the atomic-scale view promises to explain far more than the simpler features of friction. The mechanism for how heat is generated is now being determined. Surface atoms adhere and cause atomic lattices to vibrate—essentially creating sound waves that are eventually damped into heat. Chemical reactions that are related to frictional wear can also occur between atoms and molecules on the surfaces. Figure 4.B shows how the tip of a probe drawn across another material is deformed by atomic-scale friction. The force needed to drag the tip can be measured and is found to be related to shear stress, which will be discussed in the next chapter. The variation in shear stress is remarkable (more than a factor of 10^{12}) and difficult to predict theoretically, but shear stress is yielding a fundamental understanding of an ancient large-scale phenomenon—friction.

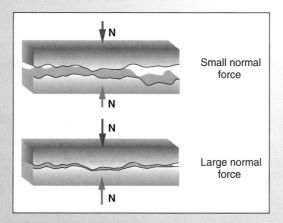

Figure 4.A Two rough surfaces in contact have a much smaller area of actual contact that their total area. When a greater normal force is exerted the area of actual contact increases as does friction.

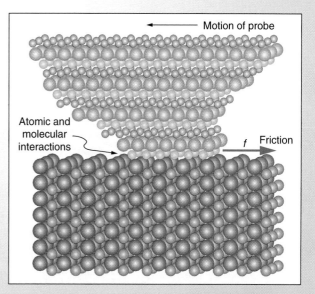

Figure 4.B The tip of a probe is deformed sideways by frictional force as the probe is dragged across a surface. Measurements of how the force varies for different materials are yielding fundamental insights into the atomic nature of friction.

Tension

A **tension** is a force along the length of a medium, especially a force carried by a flexible medium, such as a rope or cable. The word *tension* comes from a Latin word meaning "to stretch." Not coincidentally, the flexible cords that carry muscle forces to other parts of the body are called *tendons*. Any flexible connector, such as a string, rope, chain, wire, or cable, can only exert pulls parallel to its length; thus, a force carried by a flexible connector is a tension with direction parallel to the connector. Consider a person holding a mass on a rope as shown in Figure 4.11. Tension in the rope must equal the weight of the supported mass,

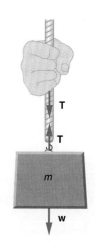

Figure 4.11 When a perfectly flexible connector (one requiring no force to bend it) such as this rope transmits a force **T**, that force must be parallel to the length of the rope, as shown. The pull such a flexible connector exerts is a *tension*. Note that the rope pulls with equal force but in *opposite directions* on the hand and the supported mass (neglecting the weight of the rope). This is an example of Newton's third law. The rope is the medium that carries the equal and opposite forces between the two objects.

as we can prove using Newton's second law. If the 5.00 kg mass in the figure is stationary, then its acceleration is zero, and thus net **F** = 0. The only external forces acting on the mass are its weight **w** and the tension **T** supplied by the rope. Thus,

$$\text{net } F = T - w = 0$$

where T and w are the magnitudes of the tension and force and their signs indicate direction, with up being positive here. Thus, just as you would expect, the tension equals the weight of the supported mass:

$$T = w = mg$$

For a 5.00 kg mass, then (neglecting the mass of the rope), we see T = (5.00 kg)(9.80 m/s^2) = 49.0 N.

Flexible connectors are often used to transmit forces around corners, such as in a hospital traction system, a finger joint, and a bicycle brake cable. If there is no friction, the tension is transmitted undiminished. Only its direction changes, and it is always parallel to the flexible connector, as illustrated in Figure 4.12.

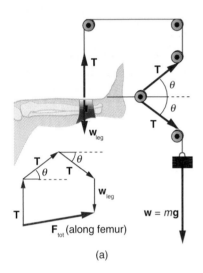

(a)

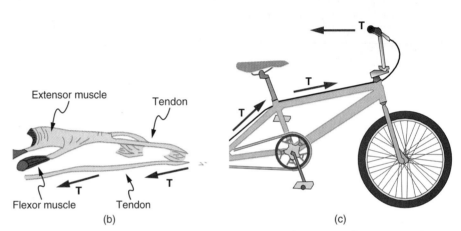

(b)

(c)

Figure 4.12 (a) A traction system in which wires are used to transmit forces. Frictionless pulleys change the direction of the force T without changing its magnitude. The system may look complicated, but the direction of T is always parallel to the wire. (b) Tendons in the finger carry force T from the muscles to other parts of the finger, usually changing the force's direction, but not its magnitude (the tendons are relatively friction-free). (c) The brake cable on a bicycle carries the tension T from the handlebars to the brake mechanism. Again, the direction but not the magnitude of T is changed.

EXAMPLE 4.7 WHAT IS THE TENSION IN A TIGHTROPE?

Calculate the tension in the wire supporting the 70.0 kg tightrope walker shown in Figure 4.13.

Strategy As shown in the figure, the wire is not perfectly horizontal, but is bent under the person's weight. Tension along its length, thus, has an upward component that can support the weight. As usual, forces are vectors represented pictorially by arrows having the same directions as the forces and lengths proportional to their magnitudes. The system is the tightrope walker, and the only external forces acting on him are his weight **w**, and the two tensions **T**$_L$ and **T**$_R$ (L and R for left and right), as illustrated. It is reasonable to neglect the weight of the wire itself. The net external force is zero since the system is stationary. When the

head-to-tail method of vector addition is used in Figure 4.13(b), the arrows must form a closed triangle to add to zero. A little trigonometry can now be used to find the tensions. One conclusion is possible at the start—we can see from part (b) of the figure that the magnitudes of the tensions T_L and T_R must be equal. This is because **w** is vertical and the angles are the same on either side, forming an isosceles triangle.

As in previous two-dimensional problems where no two vectors are parallel, the easiest method of solution is to pick a convenient coordinate system and project the vectors onto its axes. In this case the best coordinate system has one axis horizontal and the other vertical. We call those the x- and y-axes, respectively.

(continued)

(continued)

Solution First consider the horizontal components of the forces (denoted with a subscript x):

$$\text{net } F_x = T_{Lx} - T_{Rx}$$

The net external horizontal force net $F_x = 0$, since the person is stationary. Thus,

$$T_{Lx} = T_{Rx}$$

Since $T_{Lx} = T_L \cos 5.0°$ and $T_{Rx} = T_R \cos 5.0°$, we see that

$$T_L \cos 5.0° = T_R \cos 5.0°$$

Thus,

$$T_L = T_R = T$$

as predicted. Now, considering the vertical components (denoted by a subscript y), we can solve for T. Again, Newton's second law implies that net $F_y = 0$, since the person is stationary. Thus, as illustrated in part (c) of the figure,

$$\text{net } F_y = T_y + T_y - w = 0$$

or

$$2T_y = w$$

where $T_y = T \sin 5.0°$. Hence,

$$2T \sin 5.0° = w$$

and

$$T = \frac{w}{2 \sin 5.0°} = \frac{mg}{2 \sin 5.0°}$$

so that

$$T = \frac{(70.0 \text{ kg})(9.80 \text{ m/s}^2)}{2(0.0872)} = \frac{686 \text{ N}}{0.174}$$

and the tension is

$$\boxed{T = 3.94 \times 10^3 \text{ N}}$$

Discussion The tension is almost six times the 686 N weight of the tightrope walker. Since the wire is nearly horizontal, the vertical component of its tension is only a small fraction of the tension in the wire. The large horizontal components are in opposite directions and cancel, and so most of the tension in the wire is not used to support the weight of the tightrope walker.

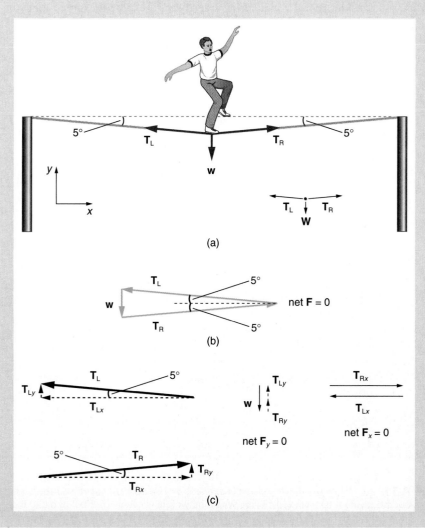

Figure 4.13 (a) The tightrope walker is the system of interest in Example 4.7. The tensions acting on the person are in the direction of the wire carrying them, as shown. (b) The head-to-tail method of vector addition. $\mathbf{T}_L$ and $\mathbf{T}_R$ must be equal in magnitude, since an isosceles triangle is formed. (c) When the vectors are projected onto vertical and horizontal axes, their components along those axes must add to zero, since the tightrope walker is stationary. The small angle results in T being much greater than w.

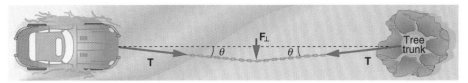

Figure 4.14 A very large tension is created in the chain by pushing on it perpendicular to its length, used in this illustration to pull a car out of the mud when no tow truck is available. Each time the car moves forward, the chain is tightened to keep it as nearly straight as possible. The tension in the chain is given by $T = F_\perp/(2 \sin \theta)$; since θ is small, T is very large. This situation is analogous to the tightrope walker shown in Figure 4.13, except that the tensions shown here are those transmitted to the car and the tree rather than those acting at the point where $\mathbf{F}_\perp$ is applied.

In fact, if we wish to *create* a very large tension, all we have to do is exert a force perpendicular to a flexible connector, as illustrated in Figure 4.14. The general expression for the tension created when a force ($\mathbf{F}_\perp$) is exerted perpendicular to and at the middle of a flexible connector is $T = F_\perp/(2 \sin \theta)$, so that T becomes very large as θ approaches zero. Even the relatively small weight of any flexible connector will cause it to sag, since an infinite tension would result if it were horizontal (i.e., θ and $\sin \theta = 0$). (See Figure 4.15.)

Real Forces and Inertial Frames*

There is another distinction among forces in addition to the types already mentioned. Some forces are real, whereas others are not. **Real forces** are those that have some physical origin, such as the pull of gravity. **Fictitious forces** are those that arise simply because an observer is in a frame of reference that rotates (like a merry-go-round) or is accelerated (like a car slowing down). For example, if a satellite is heading due north above the earth's northern hemisphere, then to an observer on earth it will appear to experience a force to the west that has no physical origin.[†] Of course, what is happening here is that the earth is rotating toward the east and moves east under the satellite. In the earth's frame this looks like a westward force on the satellite, or it can be interpreted as a violation of Newton's first law (the law of inertia) in the earth's frame. An **inertial frame of reference** is one in which all forces are real and, equivalently, one in which Newton's laws have the simple forms given in this chapter.

The rotation of the earth is slow enough that the earth is nearly an inertial frame. You ordinarily must perform precise experiments to observe fictitious forces and the slight

[†]This force is called the Coriolis force. The interested reader can explore this and other types of fictitious forces in an optional section of Chapter 8.

(a)

(b)

Figure 4.15 (a) Unless an infinite tension is exerted, any flexible connector will sag under its own weight, giving a characteristic curve when the weight is evenly distributed along the length. Shown is a chain suspended between posts. (b) Suspension bridges are essentially very heavy flexible connectors. The weight of the bridge is evenly distributed along the length of flexible connectors, usually cables, which take on the characteristic shape.

departures from Newton's laws, such as the effect just described. On the large scale, such as for the rotation of weather systems and ocean currents, the effects can be observed. The crucial factor in determining whether a frame of reference is inertial or not is whether it accelerates or rotates relative to a known inertial frame. Unless stated otherwise, all phenomena discussed in this text are considered in inertial frames.

All the forces discussed above are real forces, but there are a number of other real forces, such as drag, lift, and thrust, that are not discussed in this section. They are more specialized, and it is not necessary to discuss every type of force. It is natural, however, to ask where the basic simplicity is in the long list of forces. Are some more basic than others? Are some different manifestations of the same underlying force? The answer to both questions is yes, as will be seen in the next (optional) section and in the treatment of modern physics later in the text.

4.6* THE FOUR BASIC FORCES: AN INTRODUCTION

One of the most remarkable simplifications in physics is that only four distinct forces account for all known phenomena. In fact, nearly all of the forces we experience directly are due to only one basic force, called the electromagnetic force. (Gravity is the only force we experience directly that is not electromagnetic.) This is a tremendous simplification of the myriad *apparently* different forces we can list, only a few of which were discussed in the previous section.

The four basic forces are gravity, the electromagnetic force, the weak nuclear force, and the strong nuclear force. Their properties are summarized in Table 4.2. Since the weak and strong nuclear forces act over an extremely short range, we do not experience them directly, although they are crucial to the very structure of matter. These forces determine which nuclei are stable and which decay, and they are the basis of the release of energy in certain nuclear reactions. Nuclear forces determine not only the stability of nuclei, but also the relative abundance of elements in nature. The nucleus of an atom determines the number of electrons it has and, hence, indirectly determines the chemistry of the atom. More will be said of all of these topics in later chapters.

Gravity is surprisingly weak—it is only because gravity is always attractive that we notice it at all. Our weight is the force of gravity due to the *entire* earth acting on us. On the very large scale, as in astronomical systems, gravity is the dominant force determining the motions of moons, planets, stars, and galaxies. Gravity also affects the nature of space and time. As we shall see later in the study of general relativity, space is curved in the vicinity of very massive bodies, such as the sun, and time actually slows down near massive bodies.

Electromagnetic forces can be either attractive or repulsive, and they nearly cancel for macroscopic objects. (Remember that it is the *net* external force that is important.) If they did not cancel, electromagnetic forces would completely overwhelm gravity. The electromagnetic force is a combination of electrical forces (such as evidenced by static electricity) and magnetic forces (such as affect a compass needle). These two forces were thought to be quite distinct until early in the 19th century, when it began to be discovered that they are different manifestations of the same force. This is a classical case of the *unification of forces*. Similarly, friction, tension, and all of the other classes of forces we experience directly (except gravity, of course) are due to electromagnetic interactions

CONNECTIONS

The four basic forces will be encountered in more detail as you progress through the text. Gravity is defined in Chapter 8, electric force in Chapter 17, magnetic force in Chapter 21, and nuclear forces in Chapter 29. On a macroscopic scale, electromagnetism and gravity are the basis for all forces. The nuclear forces are vital to the substructure of matter, but they are not directly experienced on the macroscopic scale.

CONNECTIONS

Attempts to unify the four basic forces are discussed in relation to elementary particles in Section 31.6. By "unify," we mean finding connections between the forces that make them different manifestations of a single force. Even if such unification is achieved, the forces will retain their separate characteristics on the macroscopic scale and may be identical only under extreme conditions such as those existing in the early universe.

TABLE 4.2

PROPERTIES OF THE FOUR BASIC FORCES				
Force	*Approximate relative strengths*	*Range*	*+/−* [†]	*Carrier particle*
Gravity	10^{-38}	∞	+ only	Graviton (conjectured)
Electromagnetic	10^{-2}	∞	+/−	Photon (observed)
Weak nuclear	10^{-13}	$<10^{-18}$ m	+/−	W^+, W^-, Z^0 (observed)[††]
Strong nuclear	1	$<10^{-15}$ m	+/−	Gluons (conjectured)[§]

[†] + attractive; − repulsive; +/− both.
[††] Called vector bosons—predicted by theory and first observed in 1983.
[§] Eight proposed—evidenced by meson exchange in nuclei.

of atoms and molecules. It is still convenient to consider them separately, however, in specific applications because of the ways they manifest themselves.

Physicists are now exploring whether the four basic forces are in some way related. Attempts to unify all forces into one come under the rubric of Grand Unified Theories (GUTs), with which there has been some success in recent years. It is now known that under conditions of extremely high density and temperature, such as existed in the early universe, the electromagnetic and weak nuclear forces are indistinguishable. They can now be considered to be different manifestations of one force, called the *electroweak* force. So the list of four has been reduced in a sense to only three. Further progress in unifying all forces is proving difficult—especially the inclusion of gravity, which has the special characteristics of affecting the space and time in which the other forces exist. While the unification of forces will not affect how we discuss forces in this text, it is fascinating that such underlying simplicity exists in the face of the overt complexity of the universe. There is no reason that nature must be simple—it simply is.

Action at a Distance: Concept of a Field

All forces act at a distance. This is obvious for gravity. The earth and moon, for example, interact without coming into contact. It is also true for all other forces. Friction, for example, is electromagnetic force between atoms that may not actually touch. What is it that carries forces between objects? One way to answer this question is to imagine that a **force field** surrounds whatever object creates the force. A second object (often called a test object) placed in this field will experience a force that is a function of location and other variables. The field itself is the "thing" that carries the force from one object to another. The field is defined so as to be a characteristic of the object creating it; the field does not depend on the test object placed in it. The gravitational field of the earth, for example, is a function of the mass of the earth and distance from its center, independent of the presence of other masses. The concept of a field is useful in that equations can be written for force fields surrounding objects (for gravity, this yields $w = mg$ at the earth's surface), and motions can be calculated from these equations.

The field concept has been applied very successfully; we can calculate motions and describe nature to high precision using field equations. As useful as the field concept is, however, it leaves unanswered the question of what carries the force. It has been proposed in recent decades, starting in 1935 with Hideki Yukawa's (1907–1981) work on the strong nuclear force, that all forces are transmitted by the exchange of elementary particles. We can visualize particle exchange as analogous to macroscopic phenomena such as two people passing a basketball back and forth, thereby exerting a repulsive force without touching one another. (See Figure 4.16.) This idea deepens rather than contradicts field concepts. It is more satisfying philosophically to think of something physical actually moving between objects acting at a distance. Table 4.2 lists the exchange or carrier

CONNECTIONS

The concept of a force field is also used in connection with electric charge and is presented in Section 17.4. It is also a useful idea for all the basic forces, as will be seen in Chapter 31. Fields help us to visualize forces and how they are transmitted, as well as to describe them with precision *and* to link forces with subatomic carrier particles.

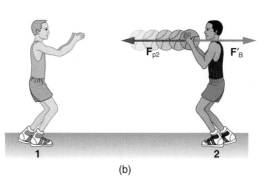

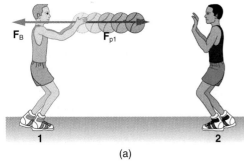

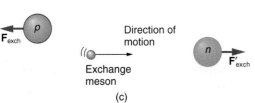

Figure 4.16 The exchange of masses resulting in repulsive forces. (a) The person throwing the basketball exerts a force F_{p1} on it toward the other person and feels a reaction force F_B away from the second person. (b) The person catching the basketball exerts a force F_{p2} on it to stop it and feels a reaction force F_B' away from the first person. (c) The analogous exchange of a meson between a proton and a neutron carries the strong nuclear forces F_{exch} and F_{exch}' between them. An attractive force can also be exerted by the exchange of a mass—if person 2 pulled the basketball away from the first person as he tried to retain it, then the force between them would be attractive.

particles, both observed and proposed, that carry the four forces. But the real fruit of the particle exchange proposal is that searches for Yukawa's particle found it *and* a number of others that were completely unexpected, stimulating yet more research. All of this eventually led to the proposal of quarks as the underlying substructure of matter, which is a basic tenet of GUTs. If successful, these theories will explain not only forces, but also the structure of matter itself.

The ideas presented in this section are but a glimpse into topics of modern physics that will be covered in much greater depth in later chapters.

4.7 FURTHER APPLICATIONS OF NEWTON'S LAWS OF MOTION: PROBLEM-SOLVING STRATEGIES

Success in problem solving is obviously necessary in understanding and applying physical principles, not to mention the more immediate need of passing exams. The basics of problem solving, presented in Section 2.5, are followed here, but specific strategies useful in applying Newton's laws of motion are emphasized. These techniques also reinforce concepts that are useful in many other areas of physics. Many problem solving strategies are stated outright in worked examples, and so the following techniques should reinforce skills you have already begun to develop.

PROBLEM-SOLVING STRATEGY

FOR NEWTON'S LAWS OF MOTION

Step 1. As usual, it is first necessary to identify the physical principles involved. *Once it is determined that Newton's laws of motion are involved, it is particularly important to draw a careful sketch of the situation.* Such a sketch is shown in Figure 4.17(a). Then, as in Figure 4.17(b), use arrows to represent all forces, label them carefully, and make their lengths and directions correspond to the forces they represent whenever sufficient information exists.

Step 2. Identify what needs to be determined and what is known or can be inferred from the problem as stated. That is, make a list of knowns and unknowns. *Then carefully determine the system of interest.* This is a crucial step, since Newton's second law involves only external forces. Once the system of interest has been identified, it becomes possible to determine which forces are external and which are internal, a necessary step to employ Newton's second law. (See Figure 4.17(c).) (Newton's third

CONNECTIONS

This problem-solving strategy builds on those first discussed in Chapter 2, but this one is more specific to Newton's laws of motion. There are also problem-solving strategies at the ends of chapters for the Integrated Concepts and Unreasonable Results problems.

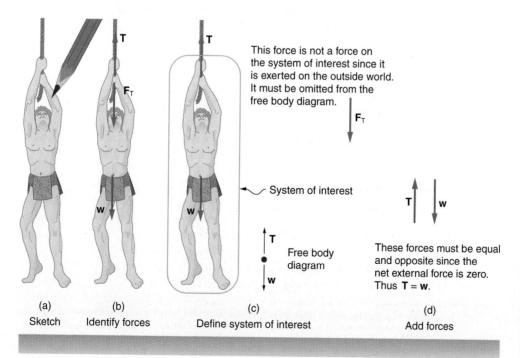

This force is not a force on the system of interest since it is exerted on the outside world. It must be omitted from the free body diagram.

System of interest

Free body diagram

These forces must be equal and opposite since the net external force is zero. Thus **T** = **w**.

(a) Sketch   (b) Identify forces   (c) Define system of interest   (d) Add forces

Figure 4.17 (a) A sketch of Tarzan hanging from a vine. (b) Arrows are used to represent all forces. **T** is the tension in the vine above Tarzan, $\mathbf{F_T}$ is the force he exerts on the vine, and **w** is his weight. All other forces, such as the nudge of a breeze, are assumed negligible. (c) Suppose we are given the ape man's mass and asked to find the tension in the vine. We then define the system of interest as shown. $\mathbf{F_T}$ is no longer shown, because it is not a force acting on the system of interest; rather, $\mathbf{F_T}$ acts on the outside world. This is a free body diagram. (d) Showing only the arrows, the head-to-tail method of addition is used. It is apparent that **T** = **w**, if Tarzan is stationary.

law may be used to identify whether forces are exerted between components of a system (internal) or between the system and something outside (external).) As illustrated in Examples 4.3 and 4.4, the system of interest depends on what question we need to answer, and trial and error may be necessary to choose the correct system for the problem at hand. This becomes easier with practice, eventually developing into an almost unconscious process. Skill in clearly defining systems will be beneficial in later chapters as well.

A diagram showing the system of interest and all of the external forces is called a **free body diagram**. We have drawn several of these in worked examples. Figure 4.17(c) shows a free body diagram for the system of interest. Note that no internal forces are shown.

Step 3. Once a free body diagram is drawn, *Newton's second law can be applied to solve the problem.* This is done in Figure 4.17(d) for a particular situation. In general, once external forces are clearly identified in free body diagrams, it should be a straightforward task to put them into equation form and solve for the unknown, as done in all previous examples. If the problem is one-dimensional—that is, if all forces are parallel—then they add like scalars. If the problem is two-dimensional, then it must be broken down into a pair of one-dimensional problems. This is done by projecting the force vectors onto a set of axes chosen for their convenience. As seen in previous examples, the choice of axes can simplify the problem. For example, when an incline is involved, a set of axes with one axis parallel to the incline and one perpendicular is most convenient. It is almost always convenient to make one axis parallel to the direction of motion, if this is known.*

Step 4. As always, *check the solution to see if it is reasonable.* In some cases, this is obvious. For example, it is reasonable to find that friction causes an object to slide down an incline more slowly than when no friction exists. In practice, intuition develops gradually through problem solving, and with experience it becomes progressively easier to judge whether an answer is reasonable or not.

Further Applications of Newton's Laws of Motion

There are many interesting applications of Newton's laws of motion, a few more of which are presented in this section. These serve also to illustrate some further subtleties of physics and to help build problem solving skills using the techniques explicitly outlined in Sections 2.5.

We have discussed that when an object rests on a horizontal surface, there is a normal force supporting it equal in magnitude to its weight. Furthermore, simple friction is always proportional to the normal force. What happens if we pull up on an object as in Figure 4.18, or push down on an object as we would push on a grocery cart? Consider the following example.

EXAMPLE 4.8 FRICTION WHEN YOU PULL UP

Suppose Mother Goose pulls with a force of 80.0 N on the leash of her dog Grimm. The dog is seen at rest in Figure 4.18. Calculate Grimm's *initial* acceleration, given that his mass is 12.0 kg and that the coefficient of static friction between his back and the ground is 0.700.

Strategy The dog needs to be the system of interest, since we want to find his acceleration. Grimm's free body diagram is shown in Figure 4.18(b). Notice that the normal force N is less than the weight w. Mother Goose supports part of the dog's weight, since she is pulling at an angle above the horizontal. This has the effect of reducing friction compared with what it would be if she pulled horizontally on his collar. (The opposite would be true if she pushed down on him.) We have a two-dimensional problem, since not all of the forces are parallel to one another. Figure 4.18(c)

(continued)

---

*The technique of projecting vectors onto carefully chosen axes is useful not only in problem solving, but also in describing nature. This will be noted during the development of certain new principles in later chapters.

(continued)

shows the external forces projected onto a set of axes with the x-axis horizontal and the y-axis vertical. The vertical forces add to zero (net $F_y = 0$). Since the vertical component of Mother Goose's pull ($F_y = 18.0$ N) is much less than Grimm's weight, the ground supports the rest of his weight with **N**. There is no motion in the vertical direction. We need to find friction before we can calculate the acceleration in the horizontal direction.

Solution Friction is proportional to the normal force. We find the normal force by considering the net external force in the vertical direction, which is

$$\text{net } F_y = N + F_y - w = 0$$

Solving this equation for N gives

$$N = w - F_y = mg - F_y$$

The force of friction (when he first begins to slide) has the maximum static value of $f = \mu_s N$. Substituting the expression just obtained for N yields

$$f = \mu_s(mg - F_y)$$

Noting that $F_y = F \sin 13° = 18.0$ N, we find

$$f = 0.700[(12.0 \text{ kg})(9.80 \text{ m/s}^2) - 18.0 \text{ N}]$$

which yields

$$f = 69.7 \text{ N}$$

Now consider the net external force in the horizontal direction:

$$\text{net } F_x = F_x - f$$

The horizontal acceleration is found using Newton's second law:

$$a_x = \frac{\text{net } F_x}{m} = \frac{F_x - f}{m}$$

where $F_x = F \cos 13° = 77.9$ N. Thus,

$$a_x = \frac{(77.9 - 69.7) \text{ N}}{12.0 \text{ kg}} = 0.686 \text{ m/s}^2$$

Discussion The answer is calculated using unrounded values of F_x and f. The answer is itself correct to three significant figures. If you enter the numbers in the last equation into your calculator, you will get an answer of 0.683 N. This is an example of how rounding off before the end of a problem can introduce errors. Note that static friction is considered here, since the problem is to find Grimm's *initial* acceleration. Once moving, he will accelerate more rapidly (much like the crate discussed earlier) since kinetic friction is less than static.

(a)

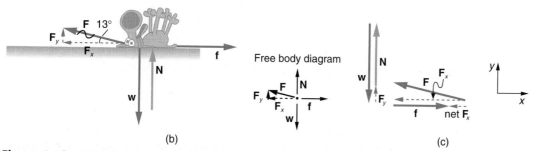

(b)

(c)

Figure 4.18 (a) Mother Goose accelerates her dog Grimm by pulling on his leash as examined in Example 4.8. (b) The free body diagram for the system of interest. (c) The horizontal and vertical components of the forces. The vertical forces add to zero, while the horizontal forces nearly cancel, resulting in a small net external force. Note that the normal force N is less than Grimm's weight, supporting only that part of his weight not supported by the leash. That is, $N = w - F_y$.

EXAMPLE 4.9 FRICTION OPPOSES MOTION OR ATTEMPTED MOTION

Suppose two tugboats push on a barge, as shown in Figure 4.19. If the mass of the barge is 5.00×10^6 kg and its acceleration is observed to be 7.50×10^{-2} m/s^2 in the direction shown, what is the drag of the water on the barge resisting the motion?

Strategy The directions and magnitudes of acceleration and the applied forces are given in Figure 4.19(a). Since the barge is flat bottomed, the drag of the water, $\mathbf{F}_D$, will be in the direction opposite to $\mathbf{F}_x + \mathbf{F}_y$, as shown in the free body diagram in Figure 4.19(b). The system of interest here is the barge, since the forces on *it* are given as well as its acceleration. In this problem the best approach is to calculate the direction and magnitude of $\mathbf{F}_x + \mathbf{F}_y$ (called $\mathbf{F}_{app}$) and then apply Newton's second law to find $\mathbf{F}_D$.

Solution Since $\mathbf{F}_x$ and $\mathbf{F}_y$ are perpendicular, the magnitude and direction of $\mathbf{F}_{app}$ are easily found. First, the magnitude is given by the Pythagorean theorem:

$$F_{app} = (F_x^2 + F_y^2)^{1/2} = 4.50 \times 10^5 \text{ N}$$

The angle is given by

$$\theta = \tan^{-1} \frac{F_y}{F_x} = 53.1°$$

which is the same direction as that of the acceleration. This means that the net external force is in the same direction as $\mathbf{F}_{app}$, and $\mathbf{F}_D$ must be in the opposite direction of $\mathbf{F}_{app}$ as claimed. The problem is now one-dimensional. From

Figure 4.19(b), we can see that

$$\text{net } F = F_{app} - F_D$$

But Newton's second law states that

$$\text{net } F = ma$$

Thus,

$$ma = F_{app} - F_D$$

This can be solved for the drag of the water F_D in terms of known quantities:

$$F_D = F_{app} - ma$$

Substituting known values gives

$$F_D = (4.50 \times 10^5 \text{ N})$$
$$- (5.00 \times 10^6 \text{ kg})(7.50 \times 10^{-2} \text{ m/s}^2)$$

Thus the magnitude of the drag of the water is

$$\boxed{F_D = 7.50 \times 10^4 \text{ N}}$$

The direction of $\mathbf{F}_D$ has already been determined to be in the direction opposite to $\mathbf{F}_{app}$, or at an angle of 53.1°.

Discussion Numbers used in this example are reasonable for a moderately large barge. It is certainly difficult to obtain larger accelerations with tugboats, and small speeds are desirable to avoid running the barge into the docks. Drag is relatively small for a well-designed hull at low speeds, consistent with the answer to this example, where F_D is less than 1/600th of the weight of the ship.

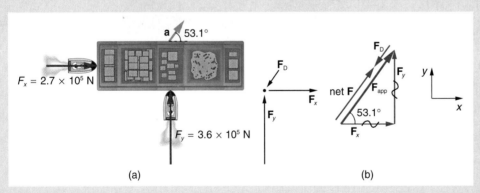

Figure 4.19 (a) A view from above of two tugboats pushing on a barge as discussed in Example 4.9. (b) The free body diagram* for the ship with the two vertical forces omitted—the weight of the barge and the buoyant force of the water supporting it cancel and are not shown. Since the applied forces are perpendicular, the x- and y-axes are in the same direction as $\mathbf{F}_x$ and $\mathbf{F}_y$. The problem quickly becomes a one-dimensional problem along the direction of $\mathbf{F}_{app}$, since friction is in the opposite direction of $\mathbf{F}_{app}$.

In the earlier example of a tightrope walker (Example 4.7), we noted that the tensions in wires supporting a mass were equal only because the angles on either side were equal. Consider the following example, where the angles are not equal—slightly more trigonometry is involved.

---

*A free body diagram reduces the object to a point and shows all external forces acting on the system of interest. Free body diagrams are discussed in Section 4.7 on problem-solving strategies for Newton's laws.

EXAMPLE 4.10   DIFFERENT TENSIONS AT DIFFERENT ANGLES

Consider the traffic light (of mass 15.0 kg) suspended from two wires as shown in Figure 4.20. Find the tension in each wire, neglecting the masses of the wires.

Strategy   The system of interest is the traffic light, and its free body diagram is shown in Figure 4.20(c). The three forces involved are not parallel, and so they must be projected onto a coordinate system. The most convenient one has one axis vertical and one horizontal, and the vector projections on it are shown in part (d) of the figure. There are two unknowns in this problem (T_1 and T_2), and so two equations are needed to find them. These two equations come from applying Newton's second law along the vertical and horizontal axes, noting that the net external force is zero along each axis because acceleration is zero.

Solution   First consider the horizontal or x-axis:

$$\text{net } F_x = T_{2x} - T_{1x} = 0$$

Thus, as you might expect,

$$T_{1x} = T_{2x}$$

This gives us the following relationship between T_1 and T_2:

$$T_1 \cos 30° = T_2 \cos 45°$$

Thus,

$$T_2 = 1.225T_1$$

Note that T_1 and T_2 are not equal in this case, because the angles on either side are not equal. It is reasonable that T_2 ends up being greater than T_1, because it is exerted more vertically than T_1.

Now consider the force components along the vertical or y-axis:

$$\text{net } F_y = T_{1y} + T_{2y} - w = 0$$

This implies

$$T_{1y} + T_{2y} = w$$

Substituting the expressions for the vertical components gives

$$T_1 \sin 30° + T_2 \sin 45° = w$$

There are two unknowns in this equation, but substituting the expression for T_2 in terms of T_1 reduces this to one equation in one unknown:

$$T_1(0.500) + (1.225T_1)(0.707) = w = mg$$

which yields

$$1.366T_1 = (15.0 \text{ kg})(9.80 \text{ m/s}^2)$$

Solving this last equation gives the magnitude of T_1 to be

$$\boxed{T_1 = 108 \text{ N}}$$

Finally, the magnitude of T_2 is determined using the relationship between them, $T_2 = 1.225T_1$, found above. Thus we obtain

$$\boxed{T_2 = 132 \text{ N}}$$

Discussion   Both tensions would be larger if both wires were more horizontal, and they will be equal if and only if the angles on either side are the same (as they were in the earlier example of a tightrope walker).

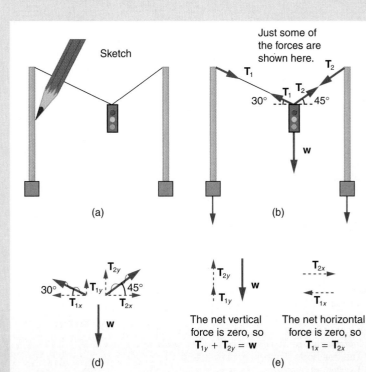

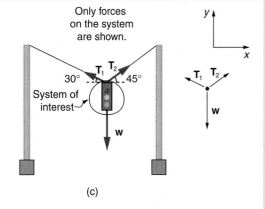

Figure 4.20   (a) A traffic light suspended from two wires. See Example 4.10. (b) Some of the forces involved. (c) Only forces acting on the system are shown here. The free body diagram for the traffic light is also shown. (d) The forces projected onto vertical (y) and horizontal (x) axes. The horizontal components of the tensions must cancel, and the sum of the vertical components of the tensions must equal the weight of the traffic light. (e) Head-to-tail addition of the vertical and horizontal forces.

Riding elevators is an everyday experience. But can you predict what you would see on the dial of a bathroom scale if you stood on it during an elevator ride? Will you see a value greater than your weight when the elevator starts up? What about when the elevator moves upward at a constant speed: will the scale still read more than your weight at rest? Consider the following example.

EXAMPLE 4.11   WHAT DOES THE BATHROOM SCALE READ IN AN ELEVATOR?

Figure 4.21 shows a 75.0 kg man (weight about 165 lb) standing on a bathroom scale in an elevator. Calculate the scale reading: (a) if the elevator accelerates upward at a rate of 1.20 m/s², and (b) if the elevator moves upward at a constant speed.

Strategy If the scale is accurate, its reading will equal F_p, the force the person exerts downward on it. Figure 4.21(a) shows the numerous forces acting on the elevator, scale, and person. It makes this one-dimensional problem look much more formidable than if the person is chosen to be the system of interest and a free body diagram is drawn as in Figure 4.21(b). Analysis of the free body diagram using Newton's laws can produce answers to both parts (a) and (b) of this example, as well as some other questions that might arise. The only forces acting on the person are his weight **w** and the upward force of the scale **F**$_s$. According to Newton's third law **F**$_p$ and **F**$_s$ are equal in magnitude and opposite in

direction, so that we need to find F_s in order to find what the scale reads. We can do this, as usual, by applying Newton's second law,

$$\text{net } F = ma$$

From the free body diagram we see that net $F = F_s - w$, so that

$$F_s - w = ma$$

Solving for F_s gives an equation with only one unknown:

$$F_s = ma + w$$

or simply

$$F_s = ma + mg$$

No assumptions were made about the acceleration, and so this solution should be valid for a variety of accelerations in addition to the ones in this exercise. However, it cannot be used if the elevator accelerates downward faster than g, because the person would then be in contact with the ceiling rather than the scale.

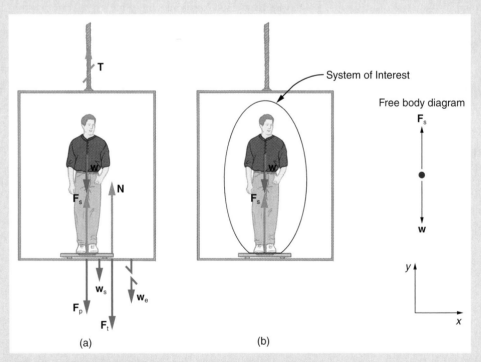

Figure 4.21 (a) The various forces acting when a person stands on a bathroom scale in an elevator. The arrows are approximately correct for when the elevator is accelerating upward—broken arrows represent forces that are too large to draw to scale. **T** is the tension in the supporting cable, **w** is the weight of the person, **w**$_s$ the weight of the scale, **w**$_e$ the weight of the elevator, **F**$_s$ the force of the scale on the person, **F**$_p$ the force of the person on the scale, **F**$_t$ the force of the scale on the floor of the elevator, and **N** the force of the floor upward on the scale. (b) The free body diagram for Example 4.11, showing only the external forces acting on the designated system of interest.

(continued)

(continued)

Solution for (a) In this part of the problem $a = 1.20$ m/s^2, so that

$$F_s = (75.0 \text{ kg})(1.20 \text{ m/s}^2) + (75.0 \text{ kg})(9.80 \text{ m/s}^2)$$

yielding

$$F_s = 825 \text{ N}$$

Discussion for (a) This is about 185 lb—the scale reading is greater than his 735 N (165 lb) weight. This means that the scale is pushing up on the person with a force greater than his weight, as it must in order to accelerate him upward. Clearly, the greater the acceleration of the elevator, the greater the scale reading, consistent with what you feel in rapidly accelerating versus slowly accelerating elevators.

Solution for (b) Now, what happens when the elevator reaches a constant upward velocity? Will the scale still read more than his weight? For any constant velocity—up, down, or stationary—acceleration is zero because $a = \Delta v / \Delta t$ and $\Delta v = 0$ for any constant velocity. Thus,

$$F_s = ma + mg = 0 + mg$$

Now

$$F_s = (75.0 \text{ kg})(9.80 \text{ m/s}^2)$$

which gives

$$F_s = 735 \text{ N}$$

Discussion for (b) This force in newtons is equivalent to 165 lb. The scale reading equals the person's weight whenever the elevator is moving at a constant velocity—up, down, or stationary.

The solution to the previous example also applies to an elevator accelerating downward, as mentioned. When an elevator accelerates downward, a is negative, and the scale reading is *less* than the weight of the person, until a constant downward velocity is reached, at which time the scale reading again becomes equal to the person's weight.

The applications in this section are but a further sampling of the power of Newton's laws of motion. We will find reasons to apply Newton's laws throughout the text and particularly in the next several chapters on mechanics.

SUMMARY

Dynamics is the study of motion including mass and the forces that affect motion. **Mass** is the quantity of matter. (Units for mass and methods of measuring it were discussed in Chapter 1.) **Force** is a push or pull that can be defined in terms of various standards, and it is a vector having both magnitude and direction.

Newton's three laws of motion are the foundation of dynamics. **Newton's first law of motion** states that

> **A body at rest remains at rest, or if in motion remains in motion at constant velocity, unless acted on by a net external force.**

A change in velocity means a change in its magnitude or direction, or both. An **external force** is one acting on a system from outside the system, as opposed to internal forces that act between components of the system. **Newton's second law of motion** states that:

> **The acceleration of a system is directly proportional to and in the same direction as the net external force acting on the system, and inversely proportional to its mass.**

Newton's second law is written in equation form as

$$\mathbf{a} = \frac{\text{net } \mathbf{F}}{m} \qquad (4.1)$$

or

$$\text{net } \mathbf{F} = m\mathbf{a} \qquad (4.2)$$

where net $\mathbf{F}$ is the **net external force**, or the vector sum of all external forces acting on the system. The SI unit of force is the **newton** (abbreviated N), given by

$$1 \text{ N} = 1 \text{ kg} \cdot \frac{\text{m}}{\text{s}^2}$$

equivalent to about 0.225 pounds of force.

Newton's third law of motion represents a basic symmetry in nature. It states:

> **Whenever one body exerts a force on a second body, the first body experiences a force that is equal in magnitude and opposite in direction to the one it exerts.**

Several types of forces are important. **Weight** w is the force of gravity on a mass m, and it is given by

$$w = mg \qquad (4.4)$$

where g is the acceleration of gravity. Weight depends on location, whereas mass does not. **Friction** is a contact force between systems opposing motion or attempted motion between them. Simple friction is proportional to the **normal force** N pushing the systems together (normal force is perpendicular to the contact surface between systems), and it depends on the materials involved. **Static friction** f_s between systems stationary relative to one another is given by

$$f_s \leq \mu_s N \qquad (4.5)$$

where μ_s is the **coefficient of static friction**, which depends on the materials in contact. **Kinetic friction** f_k between systems moving relative to one another is given by

$$f_k = \mu_k N \qquad (4.6)$$

where μ_k is the **coefficient of kinetic friction**, which also depends on materials. Representative values for μ_s and μ_k are given in Table 4.1.

Other types of force include **tension**, especially lengthwise in a flexible medium like a wire, and the **thrust** or push of a rocket. In any **inertial frame of reference** (one that is not accelerated or rotated), Newton's laws have the simple forms given in this chapter and all forces are **real forces** having a physical origin.

The various types of forces that are placed into categories useful in many applications are all manifestations of the **four basic forces** in nature. The properties of these forces are summarized in Table 4.2. Everything experienced directly is due to either electromagnetic forces or gravity. The nuclear forces are responsible for the submicroscopic structure of matter, but they are not directly sensed because of their short ranges. Attempts are being made to show all four forces are different manifestations of a single unified force. A **force field** surrounds an object creating a force and is the carrier of that force.

Problem-solving strategies are developed and discussed in the chapter's worked examples and are summarized in Section 4.7. These are aimed at situations involving Newton's laws of motion and build on general techniques presented in Chapter 2.

CONCEPTUAL QUESTIONS

4.1 Which statement is correct? (a) Net force causes motion. (b) Net force causes change in motion. Explain your answer.

4.2 How are inertia and mass related?

4.3 When you take off in a jet aircraft, there is a sensation of being pushed back into the seat. Explain why you move backward in the seat—is there really a force backward on you? (The same reasoning explains whiplash injuries, in which the head is apparently thrown backward.)

4.4 Propose a force standard different from the example of a stretched spring discussed in the text. Your standard must be capable of producing the same force repeatedly.

4.5 What properties do forces have that allow us to classify them as vectors?

4.6 What is the relationship between weight and mass? Which is an intrinsic unchanging property of a body?

4.7 Why can we neglect forces such as those holding a body together when we apply Newton's second law of motion?

4.8 Explain how the choice of the "system of interest" affects which forces must be considered when applying Newton's second law of motion.

4.9 Describe a situation in which the net external force on a system is not zero yet its speed remains constant.

4.10 A system can have a nonzero velocity while the net external force on it *is* zero. Describe such a situation.

4.11 What is the net external force acting on a rock thrown straight up when it is at the top of its trajectory?

4.12 (a) Give an example of different net external forces acting on the same system to produce different accelerations. (See Figure 4.3, for example.) (b) Give an example of the same net external force acting on systems of different masses, producing different accelerations. (See Figure 4.4, for example.) (c) What law accurately describes both effects? State it in words and as an equation.

4.13 If the acceleration of a system is zero, are no external forces acting on it? What about internal forces? Explain your answers.

4.14 Describe a situation in which one system exerts a force on another and, as a consequence, experiences a force that is equal in magnitude and opposite in direction. (This will be an example of Newton's third law of motion.)

4.15 Why does an ordinary rifle recoil (kick backward) when fired? The barrel of a recoilless rifle is open at both ends—describe how Newton's third law applies when one is fired. Can you safely stand close behind one when it is fired?

4.16 A football lineman reasons that it is senseless to try to outpush the opposing player, since no matter how hard he pushes he will experience an equal and opposite force from the other player. Use Newton's laws and draw a free body diagram of an appropriate system to explain how he can still outpush the opposition if he is strong enough.

4.17 Newton's third law of motion tells us that forces always occur in pairs of equal and opposite magnitude. Explain how the choice of the "system of interest" affects whether or not one such pair of forces cancels.

4.18 The effect of gravity on the basketball in Figure 4.4(a) is ignored. When gravity *is* taken into account, what is the direction of the net external force on the basketball—above horizontal, below horizontal, or still horizontal?

4.19 Define normal force. What is its relationship to friction when friction behaves simply?

4.20 The glue on tape can exert forces. Can these forces be a type of simple friction? Explain, considering especially that tape can stick to vertical walls and even to ceilings.

4.21 When you push a piece of chalk across a chalkboard, it sometimes screeches because it rapidly alternates between slipping and sticking to the board. Describe this process in more detail, in particular explaining how it is related to the fact that kinetic friction is less than static friction. (The same slip-grab process occurs when tires screech on pavement.)

4.22 Explain, in terms of the properties of the four basic forces, why people notice gravity acting on their bodies if it is such a comparatively weak force.

4.23 What is the dominant force between astronomical objects? Why are the other three basic forces less significant over these very large distances?

4.24 Give a detailed example of how the exchange of a particle can result in an *attractive* force. (For example, consider one child pulling a toy out of the hands of another.)

4.25 To simulate the apparent weightlessness of space orbit, astronauts are trained in the hold of a cargo aircraft that is accelerating downward at g. Why will they appear to be weightless, as measured by standing on a bathroom scale, in this accelerated frame of reference? Is there any difference between their apparent weightlessness in orbit and in the aircraft?

4.26 A cartoon shows the toupee coming off the head of an elevator passenger when the elevator rapidly stops an upward ride. Can this really happen without the person being tied to the floor of the elevator? Explain your answer.

PROBLEMS

Sections 4.3 and 4.4 Newton's Laws of Motion

You may assume data taken from illustrations is accurate to three digits.

4.1 A 63.0 kg sprinter starts a race with an acceleration of 4.20 m/s^2. What is the net external force on him?

4.2 A grocery bagger pushes a 4.50 kg bag of potatoes in such a way that the net external force on it is 60.0 N. Calculate its acceleration.

4.3 Since astronauts in orbit are apparently weightless, a clever method of measuring their masses is needed to monitor their mass gains or losses to adjust diets. One way to do this is to exert a known force on an astronaut and measure the acceleration produced. Suppose a net external force of 50.0 N is exerted and the astronaut's acceleration is measured to be 0.893 m/s^2. Calculate her mass.

4.4 What net external force is exerted on a 1100 kg artillery shell fired from a battleship if the shell is accelerated at 2.40×10^4 m/s^2? What force is exerted on the ship by the artillery shell?

4.5 In Example 4.1 the net external force on the car is stated to be 350 N. If the force of friction opposing the motion is 240 N, what force F (in newtons) is the person exerting on the car?

4.6 The same rocket sled drawn in Figure 4.5 is decelerated at a rate of 196 m/s^2. What force is necessary to produce this deceleration? Assume that the rockets are off. The mass of the system is 2100 kg.

4.7 If the rocket sled discussed in Example 4.2 starts with only one rocket burning, what is its acceleration? Assume all other factors remain as before. Why is the acceleration not one-fourth of what it is with all rockets burning?

• **4.8** Suppose two children push horizontally, but in exactly opposite directions, on a third child in a wagon. The first child exerts a force of 75.0 N, the second a force of 90.0 N, friction is 12.0 N, and the mass of the third child plus wagon is 23.0 kg. (a) What is the system of interest if the acceleration of the child in the wagon is to be calculated? (b) Draw a free body diagram, including the weight and all other forces acting on the system. (c) Calculate the acceleration. (d) What would the acceleration be if friction is 15.0 N?

• **4.9** A powerful motorcycle can produce an acceleration of 3.50 m/s^2 while traveling at 90.0 km/h. At that speed the forces resisting motion, including friction and air resistance, total 400 N. What force does the motorcycle exert backward on the ground to produce its acceleration if its mass with rider is 245 kg?

• **4.10** A brave but inadequate football player is being pushed backward by an opposing player who is exerting a force of 800 N on him. The mass of the losing player plus equipment and ball is 90.0 kg, and he is accelerating at 1.20 m/s^2 backward. (a) What is the force of friction between the losing player's feet and the Astroturf? (b) What force does the winning player exert on the ground to move forward if his mass plus equipment is 110 kg? (c) Draw a sketch of the situation showing the system of interest used to solve each part.

• **4.11** A 5.00×10^5 kg rocket is accelerating straight up. Its engines produce 1.50×10^7 N of thrust, and air resistance is 4.50×10^6 N. What is the rocket's acceleration?

• **4.12** The wheels of a midsize car exert a force of 2100 N backward on the road to accelerate the car in the forward direction. If the force of friction including air resistance is 250 N and the acceleration of the car is 1.80 m/s^2, what is the mass of the car plus its occupants?

⁝ **4.13** A freight train consists of two 80,000 kg engines and 45 cars of average mass 55,000 kg. (a) What force must each engine exert backward on the track to accelerate the train at a rate of 5.00×10^{-2} m/s^2 if the force of friction is 7.50×10^5 N, assuming the engines exert identical forces? This is not a large frictional force for such a massive system. Rolling friction for trains is small, and consequently trains are very energy efficient transportation systems. (b) What is the force in the coupling between the 37th and 38th cars (this is the force each exerts on the other), assuming all cars have the same mass and that friction is evenly distributed among all of the cars and engines?

⁝ **4.14** Commercial airplanes are sometimes pushed out of the passenger loading area by a tractor. (a) The 1800 kg tractor exerts a force of 1.75×10^4 N backward on the pavement, and the system experiences forces resisting motion that total 2400 N. If the acceleration is 0.150 m/s^2, what is the mass of the airplane? (b) Calculate the force exerted by the tractor on the airplane, assuming 2200 N of the friction is experienced by the airplane. (c) Draw two sketches showing the systems of interest used to solve each part, including the free body diagrams for each.

⁝ **4.15** A 1100 kg car pulls a boat on a trailer. (a) What total force resists the motion of the car, boat, and trailer, if the car exerts a 1900 N force on the road and produces an acceleration of 0.550 m/s^2? The mass of the boat plus trailer is 700 kg. (b) What is the force in the hitch between the car and the trailer if 80% of the resisting forces are experienced by the boat and trailer?

‡4.16 Two teams of nine members each engage in a tug of war. The first team's members have average masses of 68 kg and exert average forces of 1350 N horizontally. The second team's members have average masses of 73 kg and exert average forces of 1365 N horizontally. (a) What is the acceleration of the two teams? (b) What is the tension in the section of rope between the teams?

Sections 4.5–4.7 Types of Forces and More Applications of Newton's Laws

Note: In problems utilizing coefficients of friction, you may assume three-digit accuracy for these coefficients. Rounding to one or two digits may mask the effect being explored in the problem.

• 4.17 (a) Find the magnitudes of the forces F_1 and F_2 that add to give the total force F shown in Figure 4.22. This may be done either graphically or using trigonometry. (b) Show graphically that the same total force is obtained independent of the order of addition of F_1 and F_2. (c) Find the direction and magnitude of some other pair of vectors that add to give F. Draw these to scale on the same drawing used in part (b) or a similar one.

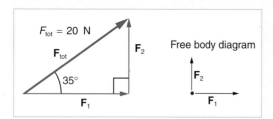

Figure 4.22   Problem 17.

• 4.18 Two children pull a snow saucer exerting forces as shown from above in Figure 4.23. Find the acceleration of the 9.00 kg sled. Note that the direction of the frictional force is unspecified; it will be in the opposite direction of the sum of F_1 and F_2.

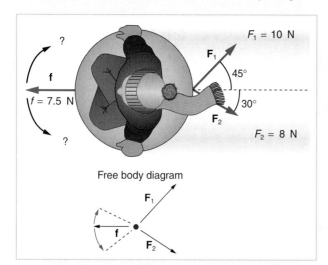

Figure 4.23   An overhead view of the horizontal forces acting on a child's snow saucer. Problem 18.

• 4.19 A flea jumps by exerting a force of 1.20×10^{-5} N straight down on the ground. A breeze blowing on the flea parallel to the ground exerts a force of 0.500×10^{-6} N on the flea.

Find the direction and magnitude of the acceleration of the flea if its mass is 6.00×10^{-7} kg. Do not neglect the force of gravity.

‡4.20 The rocket sled shown in Figure 4.5 accelerates at a rate of 49.0 m/s^2. Its passenger has a mass of 75.0 kg. (a) Calculate the horizontal component of the force the seat exerts against his body. Compare this with his weight by taking a ratio. (b) Calculate the direction and magnitude of the total force the seat exerts against his body.

‡4.21 Repeat Problem 4.20 for the situation in which the rocket sled decelerates at a rate of 201 m/s^2. In this problem the forces are exerted by the seat and restraining belts.

4.22 Verify the statement in the text that a 1.00 kg mass weighs approximately 2.2 lb on the earth and approximately 0.40 lb on the moon. Use the most accurate information in the text to get your answer to the greatest possible number of significant figures.

4.23 The weight of an astronaut plus his space suit on the moon was only 60.0 lb. How much do they weigh (in pounds) on the earth?

4.24 Suppose the mass of a fully loaded module in which astronauts take off from the moon is 10,000 kg. The thrust of its engines is 30,000 N. (a) Calculate its acceleration in a vertical takeoff from the moon. (b) Could it lift off from the earth? If not, why not? If it could, calculate its acceleration.

• 4.25 What force does a trampoline have to apply to a 45.0 kg gymnast to accelerate her straight up at 7.50 m/s^2? Note that the answer is independent of the velocity of the gymnast— she can be moving either up or down or be stationary.

4.26 A physics major is cooking breakfast when he notices that the frictional force between the steel spatula and the Teflon frying pan is only 0.200 N. Knowing the coefficient of kinetic friction between the two materials, he quickly calculates the normal force. What is it?

4.27 (a) When rebuilding her car's engine, a physics major must exert 300 N of force to insert a dry steel piston into a steel cylinder. What is the normal force between the piston and cylinder? (b) What force would she have to exert if the steel parts are oiled?

4.28 (a) What is the maximum frictional force in the knee joint of a person who supports 66.0 kg of her mass on that knee? (b) During strenuous exercise it is possible to exert forces to the joints easily ten times greater than the weight being supported. What is the maximum force of friction under such conditions? The frictional forces in joints are relatively small in all circumstances except when the joints deteriorate, such as from injury or arthritis. Increased frictional forces can cause further damage and pain.

4.29 If you consider the drag of water opposing the motion of a ship to be similar to friction, what is the effective coefficient of friction between the barge and water in Example 4.9?

• 4.30 (a) Repeat Example 4.8 for the situation in which the dog Grimm is moving, assuming a kinetic coefficient of friction of 0.500. (b) What is the ratio of the acceleration found here to the initial acceleration found in the example? Why is this ratio not the same as the ratio of the coefficients of friction, 0.7/0.5?

• 4.31 Suppose you have a 120 kg wooden crate resting on a wood floor. (a) What maximum force can you exert horizontally on the crate without moving it? (b) If you continue to exert this force once the crate starts to slip, what will its acceleration then be?

• **4.32** (a) If half of the weight of a flatbed truck is supported by its two drive wheels, what is the maximum acceleration it can achieve on dry concrete? (b) Will a metal cabinet lying on the wooden bed of the truck slip if it accelerates at this rate? (c) Solve both problems assuming the truck has four-wheel drive.

: **4.33** A team of eight dogs pulls a sled with waxed wood runners on wet snow (mush). The dogs have average masses of 19.0 kg, and the loaded sled with its rider has a mass of 210 kg. (a) Calculate the acceleration starting from rest if each dog exerts an average force of 185 N backward on the snow. (b) What is the acceleration once the sled starts to move? (c) For both situations calculate the force in the coupling between the dogs and the sled.

: **4.34** Consider the 65.0 kg ice skater being pushed by two others shown in Figure 4.1. (a) Find the direction and magnitude of $\mathbf{F}_{tot}$, the total force exerted on her by the others, given that the magnitudes F_1 and F_2 are 26.4 and 18.6 N, respectively. (b) What is her initial acceleration if she is initially stationary and wearing steel-bladed skates that point in the direction of $\mathbf{F}_{tot}$? (c) What is her acceleration assuming she is already moving in the direction of $\mathbf{F}_{tot}$? Remember that friction is always in the opposite direction of motion or attempted motion between surfaces in contact.

• **4.35** Show that the acceleration of any object down a frictionless incline that makes an angle θ with the horizontal is $a = g \sin \theta$. (Note that this acceleration is independent of mass.)

: **4.36** Show that the acceleration of any object down an incline where friction behaves simply (that is, where $f_k = \mu_k N$) is $a = g(\sin \theta - \mu_k \cos \theta)$. Note that the acceleration is independent of mass and reduces to the expression found in Problem 4.35 when friction becomes negligibly small ($\mu_k = 0$).

: **4.37** If an object is to rest on an incline without slipping, then friction must equal the component of the weight of the object parallel to the incline. This requires greater and greater friction for steeper slopes. Show that the maximum angle of an incline above the horizontal for which an object will not slide down is $\theta = \tan^{-1} \mu_s$. You may use the result of the previous problem. Assume that $a = 0$ and that static friction has reached its maximum value.

: **4.38** Calculate the maximum deceleration of a car that is heading down a 6° slope (one that makes an angle of 6° with the horizontal) under the following road conditions. You may assume that the weight of the car is evenly distributed on all four tires and that the static coefficient of friction is involved—that is, the tires are not allowed to slip during the deceleration. (a) On dry concrete. (b) On wet concrete. (c) On ice, assuming that $\mu_s = 0.100$, the same as for shoes on ice.

: **4.39** Calculate the maximum acceleration of a car that is heading up a 4° slope (one that makes an angle of 4° with the horizontal) under the following road conditions. Assume that only half the weight of the car is supported by the two drive wheels and that the static coefficient of friction is involved—that is, the tires are not allowed to slip during the acceleration. (a) On dry concrete. (b) On wet concrete.

(c) On ice, assuming that $\mu_s = 0.100$, the same as for shoes on ice.

: **4.40** Repeat Problem 4.39 for a car with four wheel drive.

: **4.41** Repeat Problem 4.13 for the same train going up a 2.00° slope (one that makes an angle of 2.00° with the horizontal).

• **4.42** (a) Calculate the tension in a vertical strand of spiderweb if a spider of mass 8.00×10^{-5} kg hangs motionless on it. (b) Calculate the tension in a horizontal strand of spiderweb if the same spider sits motionless in the middle of it much like the tightrope walker in Figure 4.13. The strand sags at an angle of 12.0° below the horizontal. Compare this with the tension in the vertical strand (find their ratio).

• **4.43** Suppose a 60.0 kg gymnast climbs a rope. (a) What is the tension in the rope if he climbs at a constant speed? (b) What is the tension in the rope if he accelerates upward at a rate of 1.50 m/s²?

• **4.44** Suppose your car was mired deeply in the mud and you wanted to use the method illustrated in Figure 4.14 to pull it out. (a) What force would you have to exert perpendicular to the center of a rope to produce a force of 12,000 N on the car if the angle is 2.00°? (b) Real ropes stretch under such forces. What force would be exerted on the car if the angle increases to 7.00° and you still apply the force found in part (a) to its center?

• **4.45** What force is exerted on the tooth in Figure 4.24 if the tension in the wire is 25.0 N? Note that the force applied to the tooth is smaller than the tension in the wire, but this is necessitated by practical considerations of how force can be applied in the mouth.

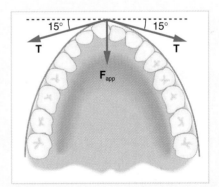

Figure 4.24 Braces are used to apply forces to teeth to realign them. Shown in this figure are the tensions applied by the wire to the protruding tooth. The total force applied to the tooth by the wire, $\mathbf{F}_{app}$, points straight toward the back of the mouth. Problem 45.

: **4.46** Show that, as stated in the text, a force $\mathbf{F}_\perp$ exerted on a flexible medium at its center and perpendicular to its length (such as on the tightrope wire in Figure 4.13) gives rise to a tension of magnitude $T = F_\perp/(2 \sin \theta)$.

4.47 (a) What is the strength of the weak nuclear force relative to the strong nuclear force? (b) What is the strength of the weak nuclear force relative to the electromagnetic force? Since the weak nuclear force only acts at very short distances, such as inside nuclei, where the strong and electromagnetic forces also act, it is surprising that we have any knowledge of it at all. We have such knowledge because the weak nuclear force is responsible for β decay, a type of nuclear decay not explained by other forces.

4.48 (a) What is the ratio of the strength of gravity to that of the strong nuclear force? (b) What is the ratio of the strength of gravity to that of the weak nuclear force? (c) What is the ratio of the strength of gravity to that of the electromagnetic force? The correct implication of the small answers is that gravity is negligible in describing the nucleus, where all three of the other forces exist.

4.49 What is the ratio of the strength of the strong nuclear to that of the electromagnetic force? Based on this, one expects that the strong force dominates the nucleus, which is true for small nuclei. Large nuclei, however, have sizes greater than the range of the strong nuclear force, and the electromagnetic force begins to affect nuclear stability. These facts will be used to explain nuclear fusion and fission in a later chapter.

• **4.50** Consider the baby being weighed in Figure 4.25. (a) What is the mass of the child and basket if a scale reading of 102 N is observed? (b) What is the tension T in the cord attaching the child to the scale? (c) What is the tension T' in the cord attaching the scale to the ceiling, if the scale has a mass of 0.500 kg? (d) Draw a sketch of the situation indicating the system of interest used to solve each part. The masses of the cords are negligible.

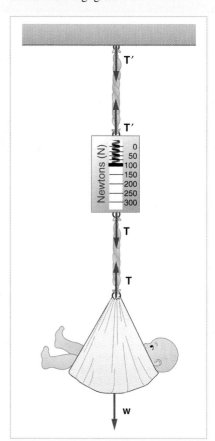

Figure 4.25 Problem 50.

• **4.51** Figure 4.26 shows Batman and Robin hanging motionless from a bat rope. Batman's mass is 90.0 kg, while Robin's is 55.0 kg, and the mass of the rope is negligible. (a) Draw your own sketch of the situation showing all forces acting on Batman, Robin, and the rope. (b) Find the tension in the rope above Batman. (c) Find the newton in the rope between Batman and Robin. Indicate on your sketch the system of interest used to solve each part.

Figure 4.26 Batman and Robin hang motionless from a bat rope as they try to figure out what to do next. Will the tension be the same everywhere in the rope? Problem 51.

• **4.52** A 76.0 kg person is being pulled away from a burning building as shown in Figure 4.27. Calculate the tension in the two ropes if the person is momentarily motionless. Include a free body diagram in your solution.

• **4.53** A shopper pushes a grocery cart by exerting a force on the handle at a downward angle 35.0° below the horizontal. The loaded cart has a mass of 28.0 kg, and the force of friction is 60.0 N. (a) Draw a free body diagram for the system of interest. (b) What force must the shopper exert to move at a constant velocity? Note that this situation is very similar to that shown in Figure 4.29(a).

⁞ **4.54** Consider the 52.0 kg mountain climber in Figure 4.28. (a) Find the tension in the rope and the force that the mountain climber must exert with her feet on the vertical rock face

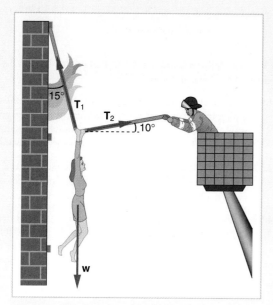

Figure 4.27 The force T_2 needed to hold steady the person being rescued from the fire is less than her weight and less than the force T_1 in the other rope, since the more vertical rope supports a greater part of her weight (a vertical force). Problem 52.

Figure 4.28 Part of the climber's weight is supported by her rope, and part by friction between her feet and the rock face. Problem 54.

to remain stationary. Assume that the force is exerted parallel to her legs. Also, assume negligible force exerted by her arms. (b) What is the minimum coefficient of friction between her shoes and the cliff?

4.55 A contestant in a winter games event pushes a 45.0 kg block of ice across a frozen lake as shown in Figure 4.29(a). (a) Calculate the minimum force F he must exert to get the block moving. (b) What is its acceleration once it starts to move, if that force is maintained?

4.56 Repeat Problem 4.55 with the contestant pulling the block of ice with a rope over his shoulder at the same angle *above* the horizontal as shown in Figure 4.29(b).

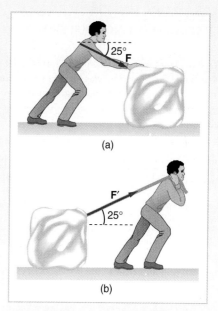

Figure 4.29 Which method of sliding a block of ice requires less force—(a) pushing or (b) pulling at the same angle above the horizontal? Problems 55 and 56.

4.57 Calculate the force a mother must exert to hold her 12.0 kg child in an elevator under the following conditions. (a) The elevator accelerates upward at 0.850 m/s^2. (b) The elevator moves upward at a constant speed. (c) The upward bound elevator decelerates at 2.30 m/s^2. (d) Calculate the ratio of the force to the weight of the child in each case and show the free body diagram used (same for all parts).

INTEGRATED CONCEPTS

The following problems involve concepts from this chapter and several others. Physics is at its most interesting and most powerful when applied to general situations that involve more than a narrow set of physical principles. For example, forces produce accelerations, a topic of kinematics, and hence the relevance of Chapters 2 and 3. The following topics are involved in the problems in this section:

Topics	*Location*
Kinematics	Chapter 2
Two-dimensional kinematics	Chapter 3
Dynamics	Chapter 4

PROBLEM-SOLVING STRATEGY

Step 1. *Identify which physical principles are involved*. This is done by following the problem-solving strategies as given in Section 2.5 and other places. Listing the givens and the quantities to be calculated will allow you to identify the principles involved.

Step 2. *Solve the problem using strategies outlined in the text*. If these are available for the specific topic, you should refer to them. You should also refer to the sections of the text that deal with a particular topic. The following worked example illustrates how these strategies are applied to an integrated concept problem.

EXAMPLE WHAT FORCE MUST A SOCCER PLAYER
EXERT TO REACH TOP SPEED?

A soccer player starts from rest and accelerates forward, reaching a velocity of 8.00 m/s in 2.50 s. (a) What was his average acceleration? (b) What average force did he exert backward on the ground to achieve this acceleration? The player's mass is 70.0 kg, and air resistance is negligible.

Strategy Step 1 To solve an *integrated concept problem*, such as those following this example, we must first identify the physical principles involved and identify the chapters in which they are found. Part (a) of this example considers *acceleration* along a straight line. This is a topic of *kinematics* covered in Chapter 2. Part (b) deals with *force*, a topic of *dynamics* found in this chapter.

Strategy Step 2 The following solutions to each part of the example illustrate how the specific problem solving strategies are applied. These involve identifying knowns and unknowns, checking to see if the answer is reasonable, and so forth.

Solution for (a): Kinematics (Chapter 2) We are given the initial and final velocities (zero and 8.00 m/s forward); thus, the change in velocity is $\Delta v = 8.00$ m/s. We are given the elapsed time, and so $\Delta t = 2.50$ s. The unknown is acceleration, which can be found from its definition in Equation 2.4:

$$a = \frac{\Delta v}{\Delta t}$$

Substituting the known values yields

$$a = \frac{8.00 \text{ m/s}}{2.50 \text{ s}}$$

$$= 3.20 \text{ m/s}^2$$

Discussion for (a) This is an attainable acceleration for an athlete in good condition.

Solution for (b): Force (Chapter 4) Here we are asked to find the average force the player exerts backward to achieve this forward acceleration. Neglecting air resistance, this would be equal in magnitude to the net external force on the player, since this force causes his acceleration. Since we now know the player's acceleration and are given his mass, we can use Newton's second law to find the force exerted. That is, from Equation 4.2,

$$\text{net } F = ma$$

Substituting the known values of m and a gives

$$\text{net } F = (3.20 \text{ m/s}^2)(70.0 \text{ kg})$$

$$= 224 \text{ N}$$

Discussion for (b): This is about 50 pounds, a reasonable average force.

This worked example illustrates how to apply problem-solving strategies to situations that include topics from different chapters. The first step is to identify the physical principles involved in the problem. The second step is to solve for the unknown using familiar problem-solving strategies. These strategies are found throughout the text, and many worked examples show how to use them for single topics. You will find these techniques for integrated concept problems useful in applications of physics outside of a physics course, such as in your profession, in other science disciplines, and in every-

day life. The following problems will build your skills in the broad application of physical principles.

4.58 A 35.0 kg dolphin decelerates from 12.0 to 7.50 m/s in 2.30 s to join another dolphin in play. What average force was exerted to slow him if he was moving horizontally? (Gravity is balanced by the buoyant force of the water.)

4.59 When starting a foot race, a 70.0 kg sprinter exerts an average force of 650 N backward on the ground for 0.800 s. (a) What is his final speed? (b) How far does he travel?

• **4.60** A large rocket has a mass of 2.00×10^6 kg at takeoff, and its engines produce a thrust of 3.50×10^7 N. (a) Find its initial acceleration if it takes off vertically. (b) How long does it take to reach a velocity of 120 km/h straight up, assuming constant mass and thrust? (In reality, the mass decreases as fuel is consumed; thus, the acceleration increases.)

• **4.61** A basketball player jumps straight up for a ball. To do this, he lowers his body 0.300 m and then accelerates through this distance by forcefully straightening his legs. This player leaves the floor with a vertical velocity sufficient to carry him 0.900 m above the floor. (a) Calculate his velocity when he leaves the floor. (b) Calculate his acceleration while he is straightening his legs. He goes from zero to the velocity found in part (a) in a distance of 0.300 m. (c) Calculate the force he exerts on the floor to do this, given his mass is 110 kg.

• **4.62** A 2.50 kg fireworks shell is fired straight up from a mortar and reaches a height of 110 m. (a) Neglecting air resistance (a bad approximation, but make it anyway), calculate the shell's velocity when it leaves the mortar. (b) The mortar itself is a tube 0.450 m long. Calculate the average acceleration of the shell in the tube to go from zero to the velocity found in (a). (c) What is the average force on the shell in the mortar? Express in newtons and as a ratio to the weight of the shell.

• **4.63** Some cars are equipped with an antilocking device on their brakes to prevent the tires from slipping in an emergency stop, thus taking advantage of the larger static coefficient of friction. (Static friction also allows you to steer.) (a) Calculate the maximum stopping force (friction) on dry concrete for a 800 kg car if the tires do not slip. (b) Calculate the maximum deceleration in this case. (c) How far will the car travel while stopping from an initial speed of 90.0 km/h? (d) Repeat each part for tires that slip. (Although the mass of the car is given in this problem, the stopping distance is actually independent of mass.)

• **4.64** (a) Calculate the maximum acceleration of a car on dry concrete that supports half of its weight on its drive wheels. (b) How much time will it take such a car to accelerate from rest to 90.0 km/h at maximum acceleration? (c) What is the maximum deceleration of the car, noting that all four tires are used for braking? (d) What minimum distance will the car travel while stopping from a speed of 90.0 km/h, measuring from the point where the brakes are first applied? Note that dragsters are able to accelerate at a much greater rate than indicated here by creating a coefficient of friction greater that 1.

: 4.65 Repeat Problem 4.64 for a car going up a slope that makes a 2.50° angle with the horizontal.

: 4.66 Repeat Problem 4.62 for a shell fired at an angle 10.0° from the vertical.

: 4.67 An elevator filled with passengers has a mass of 1700 kg. (a) The elevator accelerates upward from rest at a rate of 1.20 m/s^2 for 1.50 s. Calculate the tension in the cable supporting the elevator. (b) The elevator continues upward at constant velocity for 8.50 s. What is the tension in the cable during this time? (c) The elevator decelerates at a rate of 0.600 m/s^2 for 3.00 s. What is the tension in the cable during deceleration? (d) How high has the elevator moved above its original starting point, and what is its final velocity?

UNREASONABLE RESULTS

Physics must describe nature accurately. The following problems have results that are unreasonable because one premise is unreasonable or because certain premises are inconsistent with one another. The physical principle applied correctly then produces an unreasonable result. For example, if a person starting a foot race accelerates at 0.400 m/s^2 for 100 s, her final speed will be 40.0 m/s (about 90 miles per hour)—clearly unreasonable because the time of 100 s is an unreasonable premise. The physics is correct in a sense, but there is more to describing nature than just manipulating equations correctly. Checking the result of a problem to see if it is reasonable not only helps uncover errors in problem solving—it also builds intuition in judging whether nature is being accurately described.

Note: Strategies for determining if an answer is reasonable appear in Chapter 2. They can be found with the section of problems labeled *Unreasonable Results* at the end of that chapter.

- **4.68** (a) Repeat Problem 4.15(a) but assuming an acceleration of 1.20 m/s^2 is produced. (b) What is unreasonable about the result? (c) Which premise is unreasonable (pretty obvious since only one has been changed), and why is it unreasonable?

- **4.69** (a) What is the final velocity of a car originally traveling at 50.0 km/h that decelerates at a rate of 0.400 m/s^2 for 50.0 s? (b) What is unreasonable about the result? (c) Which premise is unreasonable, or which premises are inconsistent?

- **4.70** (a) What is the initial acceleration of a rocket that has a mass of 1.50×10^6 kg at takeoff, the engines of which produce a thrust of 2.00×10^6 N? Do not neglect gravity. (b) What is unreasonable about the result? (This result has been unintentionally achieved by real rockets at times.) (c) Which premise is unreasonable, or which premises are inconsistent?

- **4.71** A 75.0 kg man stands on a bathroom scale in an elevator that accelerates from rest to 30.0 m/s in 2.00 s. (a) Calculate the scale reading in newtons and compare it with his weight. (The scale exerts a upward force on him equal to its reading.) (b) What is unreasonable about the result? (c) Which premise is unreasonable, or which premises are inconsistent?

A boulder in such a precarious position is usually called a balancing rock.

What might desks, bridges, buildings, trees, and mountains have in common—at least in the eyes of a physicist? The answer is that they are ordinarily motionless relative to the earth. Furthermore, their acceleration is zero, since they remain motionless. But that means they also have something in common with a car moving at a constant velocity. Anything with a constant velocity has an acceleration of zero, too. Now, the important part—Newton's second law states that net $\mathbf{F} = m\mathbf{a}$, and so the net external force is zero for all stationary objects *and* for all objects moving at constant velocity. There are forces acting, but they are balanced. That is, they are in *equilibrium*. **Statics** is the study of forces in equilibrium, a large group of situations that makes up a special case of Newton's second law. We have already considered a few such situations; in this chapter, we cover the topic more thoroughly, including consideration of such possible effects as rotation and deformation of an object by the forces acting on it. This leads to two important new concepts, torque and elasticity, which are also introduced in this chapter.

How can we guarantee a body is in equilibrium, and what can we learn from systems that are in equilibrium? There are actually two conditions that must be satisfied to achieve equilibrium. They are the topics of the first two sections of this chapter.

5.1 THE FIRST CONDITION FOR EQUILIBRIUM

The first condition necessary to achieve equilibrium is the one already mentioned: *the net external force on the system must be zero*. Expressed as an equation, this is simply

$$\text{net } \mathbf{F} = 0 \qquad\qquad (5.1)$$

Note that if net $\mathbf{F}$ is zero, then the net external force in *any* direction is zero. For example, the net external forces along the typical x- and y-axes are zero. This is written as

$$\text{net } F_x = \text{net } F_y = 0$$

Figure 5.1 illustrates situations where net $\mathbf{F} = 0$ for both **static equilibrium** (motionless), and **dynamic equilibrium** (constant velocity).

But it is not enough just to have the net external force be zero for a system to be in equilibrium. Consider the two situations illustrated in Figure 5.2. The net external force is zero in both situations shown in the figure; but in one case, equilibrium is achieved, whereas in the other, it is not. In part (a), the stick remains motionless. But in part (b), with the same forces applied in different places, the stick rotates. Apparently, the point at which a force is applied is another factor in determining whether equilibrium is achieved. This will be explored further in the next section.

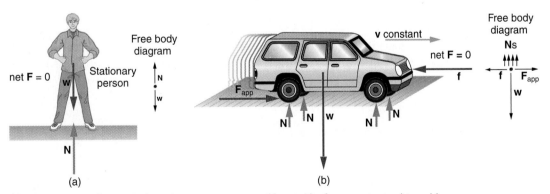

Figure 5.1 (a) This motionless person is in *static equilibrium*. The forces acting on him add up to zero. Both forces are vertical in this case. (b) This car is in *dynamic equilibrium* because it is moving at constant velocity. There are horizontal and vertical forces, but the net external force in any direction is zero. The applied force $\mathbf{F}_{app}$ is balanced by friction, and the weight of the car is supported by the normal forces, here shown to be equal for all four tires.

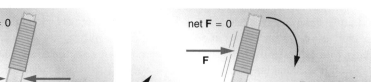

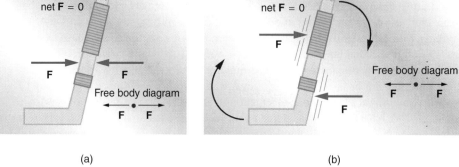

(a)   (b)

Figure 5.2 A hockey stick on ice with two equal and opposite horizontal forces applied to it. Friction is negligible, and the force of gravity is balanced by the support of the ice (a normal force). Thus net **F** = 0 in both cases. (a) Equilibrium is achieved, in this case static equilibrium. (b) The same forces are applied at other points and the stick rotates—in fact, it experiences an accelerated rotation. Thus net **F** = 0 is a necessary—but not sufficient—condition for achieving equilibrium.

5.2 THE SECOND CONDITION FOR EQUILIBRIUM

Torque

The second condition necessary to achieve equilibrium has to do with avoiding accelerated rotation. A body or system in equilibrium can rotate provided that its rate of rotation is constant and not changed by the forces acting on it. To see what factors affect rotation, let us think about what happens when you open an ordinary door by rotating it on its hinges.

Several familiar factors determine how effective you are in opening the door. (See Figure 5.3.) First of all, the larger force, the more effective it is in opening the door—obviously, the harder you push, the more rapidly the door opens. Second, the point at which you push is crucial. If you apply your force too close to the hinges, the door will open slowly if at all. Most people have been embarrassed by making this mistake and bumping up against a door when it did not open as quickly as expected. Finally, the direction you push is also important. The most effective direction is perpendicular to the door—we push in this direction almost instinctively.

These three factors—the magnitude, direction, and point of application of the force—are incorporated into the definition of the physical quantity called torque. **Torque** is the effectiveness of a force in changing or accelerating a rotation.* In equation form, the magnitude of torque is defined to be

$$\tau = rF \sin \theta \tag{5.2}$$

where τ (the Greek letter tau) is the symbol for torque, r is the distance from the pivot point to the point where the force is applied, F is the magnitude of the force, and θ is the angle between r and F, as seen in Figures 5.3(e) and 5.4. An alternative expression for

---

*Analogous to force changing a velocity.

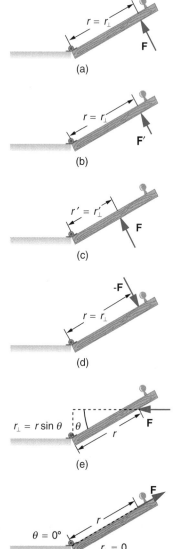

Figure 5.3 Torque is the effectiveness of a force in changing a rotation, illustrated here for rotating a door on its hinges (as viewed from overhead). Torque has both magnitude and direction. (a) Counterclockwise torque is produced by this force, meaning that the door will rotate in a counterclockwise direction if **F** is the only force acting. (b) A smaller counterclockwise torque is produced by a smaller force **F′** acting at the same distance from the hinges (the pivot point). (c) The same force as in (a) produces a smaller counterclockwise torque when applied at a smaller distance from the hinges. (d) A clockwise torque is produced by a force in the opposite direction. Its magnitude is the same as in (a). (e) A smaller counterclockwise torque is produced by the same magnitude force acting in a different direction. Here θ is less than 90°. (f) Torque is zero here since the force just pulls on the hinges, producing no rotation. $\theta = 0°$ in this case.

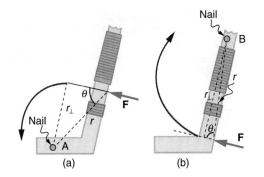

Figure 5.4 A force applied to an object creates a torque, which depends on the location of the pivot point. (a) Shown here are the three factors r, F, and θ for pivot point A on a body—r is the distance from the chosen pivot point to the point where the force F is applied, and θ is the angle between **r** and **F** as shown. If the object can rotate about point A, it will rotate counterclockwise. This means torque is counterclockwise relative to pivot A. (b) Relative to point B, torque is greater since the perpendicular lever arm $r'_\perp$ is larger than $r_\perp$. The torque from the applied force will cause a clockwise rotation around point B, and so it is a clockwise torque relative to B.

torque is given in terms of the **perpendicular lever arm** $r_\perp$ shown in Figures 5.3 and 5.4, which is defined as

$$r_\perp = r \sin \theta \qquad (5.3)$$

so that

$$\tau = r_\perp F \qquad (5.4)$$

The perpendicular lever arm $r_\perp$ is the shortest distance from the pivot point to the line along which **F** lies; it is shown as a dashed line in the figures. Note that $r_\perp$ is perpendicular to **F**, as its name implies. It is sometimes easier to find or visualize $r_\perp$ than to find both r and θ. In such cases, it may be more convenient to use Equation 5.4 rather than 5.2 for torque, but both are equally valid.*

The **SI unit of torque** is newtons times meters, usually written as N·m. For example, if you push perpendicular to the door with a force of 40 N at a distance of 0.800 m from the hinges, you exert a torque of 32 N·m (0.800 m × 40 N × sin 90°) relative to the hinges. If you reduce the force to 20 N, the torque is reduced to 16 N·m, and so on.

The direction and magnitude of a torque depend on where the pivot is, such as the hinges on a door. Torque is generally not the same for different points, since both r and θ depend on the location of the pivot. Note that for rotation in a plane, torque has two possible directions. Torque is either clockwise or counterclockwise relative to the chosen pivot point, as illustrated for points A and B in Figure 5.4. If the object can rotate about point A, it will rotate counterclockwise. This means the torque for the force shown is counterclockwise relative to A. But if the object can rotate about point B, it will rotate clockwise, which means the torque for the force shown is clockwise relative to B. The magnitude of the torque is also different for the two points. It is greater for point B, because the lever arm relative to B is longer.

Now, **the second condition necessary to achieve equilibrium** is that *the net external torque on a system must be zero.* (An external torque is one created by an external force.) You can choose the point about which the torques are calculated. It can be the physical pivot point of a system or any other point in space—but it must be the same point for all torques. If the second condition (net external torque is zero) is satisfied for one point in space, it will also hold for *any* point in space, in or out of the system of interest.[†] The second condition necessary to achieve equilibrium is stated in equation form as

$$\text{net } \tau = 0 \qquad (5.5)$$

where net means total. Torques in opposite directions are assigned opposite signs. A common convention is to call counterclockwise (ccw) torques positive and clockwise (cw) torques negative. Both r and F are always positive. Equation 5.5 can then be written as

$$\text{net } \tau = \text{net } \tau_{cw} + \text{net } \tau_{ccw} = 0$$

---

*Another alternative is to use the perpendicular component of the force, which is given by $F_\perp = F \sin \theta$. If that is done, then $\tau = rF_\perp$, and the effectiveness of a torque is determined by the magnitude of its component perpendicular to r.

[†]This very useful fact is only true in inertial frames of reference.

Figure 5.5 Two children balancing a seesaw satisfy both conditions for equilibrium. The lighter child sits farther from the pivot to create a torque equal in magnitude to that of the heavier child. See Example 5.1.

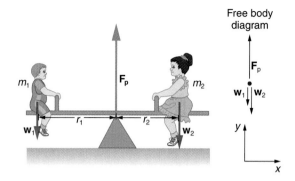

In this text we consider the magnitudes (no plus or minus signs) of the sums and state the second condition for equilibrium as

$$\text{net } \tau_{cw} = \text{net } \tau_{ccw} \tag{5.6}$$

When using this expression, we think of the sums of the clockwise and counterclockwise torques as being equal to achieve equilibrium. This can be useful if you think of the torques balancing one another in equilibrium.

When two children balance a seesaw as shown in Figure 5.5, they satisfy the two conditions for equilibrium. Most people have perfect intuition about seesaws, knowing that the lighter child must sit farther from the pivot and that a heavy child can keep a lighter one off the ground indefinitely.

EXAMPLE 5.1   SHE SAW TORQUES ON A SEESAW

The two children shown in Figure 5.5 are balanced on a seesaw of negligible mass. (This assumption is made to keep the example simple—more involved examples will follow.) The first child has a mass of 26.0 kg and sits 1.60 m from the pivot. (a) If the second child has a mass of 32.0 kg, how far is she from the pivot? (b) What is F_p, the supporting force exerted by the pivot?

Strategy Both conditions for equilibrium must be satisfied. In part (a), we are asked for a distance; thus, the second condition (regarding torques) must be used, since the first (regarding only forces) has no distances in it. To apply the second condition for equilibrium, we first identify the system of interest to be the seesaw plus the two children. We take the supporting pivot to be the point about which the torques are calculated. We then identify all external forces acting on the system.

Solution for (a) The three external forces acting on the system are the weights of the two children and the supporting force of the pivot. Let us examine the torque produced by each. Torque is defined to be

$$\tau = rF \sin \theta$$

Here $\theta = 90°$, so that $\sin \theta = 1$ for all three forces. That means $r_\perp = r$ for all three. The torques exerted by the three forces are first,

$$\tau_1 = r_1 w_1$$

second,

$$\tau_2 = r_2 w_2$$

and third,

$$\tau_p = r_p F_p = 0 \cdot F_p = 0$$

Since F_p acts directly on the pivot point, the distance r_p is zero. A force acting on the pivot cannot cause a rotation, just as pushing directly on the hinges of a door will not cause it to rotate. Now, τ_1 is a counterclockwise torque and τ_2 is a clockwise torque, so that the second condition for equilibrium

$$\text{net } \tau_{cw} = \text{net } \tau_{ccw}$$

is simply

$$\tau_2 = \tau_1$$

Thus,

$$r_2 w_2 = r_1 w_1$$

Weight is mass times the acceleration of gravity. Entering mg for w, we get

$$r_2 m_2 g = r_1 m_1 g$$

Solve this for the unknown r_2:

$$r_2 = r_1 \frac{m_1}{m_2}$$

The quantities on the right side of the equation are known; thus, r_2 is

$$r_2 = (1.60 \text{ m}) \frac{26.0 \text{ kg}}{32.0 \text{ kg}} = 1.30 \text{ m}$$

(continued)

(continued)

As expected, the heavier child must sit closer to the pivot (1.30 vs. 1.60 m) to balance the seesaw.

Solution for (b) This part asks for a force F_p. The easiest way to find it is to use the first condition for equilibrium:

$$\text{net } \mathbf{F} = 0$$

The forces are all vertical, so that we are dealing with a one-dimensional problem along the vertical axis; hence, the condition can be written as

$$\text{net } F_y = 0$$

where we again call the vertical axis the y-axis. Using plus and minus signs to indicate the directions of the forces, we see

$$F_p - w_1 - w_2 = 0$$

This yields what might have been guessed at the start:

$$F_p = w_1 + w_2$$

So the pivot supplies a supporting force equal to the total weight of the system:

$$F_p = m_1 g + m_2 g$$

Entering known values gives

$$F_p = (26.0 \text{ kg})(9.80 \text{ m/s}^2) + (32.0 \text{ kg})(9.80 \text{ m/s}^2)$$
$$= 568 \text{ N}$$

Discussion The two results make intuitive sense. The heavier child sits closer to the pivot. The pivot supports the weight of the two children. Part (b) can also be solved using the second condition for equilibrium, since both distances are known. This is left as an end-of-chapter problem.

Several aspects of the preceding example have broad implications. First, the choice of the pivot as the point about which torques are calculated simplified the problem. Since F_p is exerted on the pivot point, its lever arm is zero. Thus the torque exerted by the supporting force F_p is zero relative to that pivot point. The second condition for equilibrium holds for any point, and so we choose the pivot to simplify the solution of the problem.

Second, the acceleration of gravity canceled in this problem, and we were left with a ratio of masses. *This will not always be the case.* Always enter the correct forces—do not jump ahead to some ratio of masses.

Third, the weight of each child is distributed over an area of the seesaw, yet we treated it as if the force were exerted at a single point. This is not an approximation—the distances r_1 and r_2 are the distances to points directly below the center of gravity of each child. As we shall see in the next section, the mass and weight of a system can act as if they are located at a single point.

Finally, note that the concept of torque has an importance beyond static equilibrium. Torque plays the same role in rotational motion that force plays in linear motion.

5.3 CENTER OF MASS; CENTER OF GRAVITY

Most real objects and all systems are spread out in space rather than being point particles. Yet here again nature exhibits simplicity. Certain characteristics of a system act as if its mass is concentrated at a single point, called the **center of mass** (CM). The center of mass lies at a position that is a weighted average of the mass distribution in a body. It will be closer to the more massive part of a system—for example, nearer the thicker end of a baseball bat—and in the exact geometric center of a symmetrical system, such as a ball. (See Figure 5.6.) The position of the center of mass in a three-dimensional system is defined to be

$$\mathbf{r}_{CM} = \frac{m_1 \mathbf{r}_1 + m_2 \mathbf{r}_2 + m_3 \mathbf{r}_3 + \cdots}{m_1 + m_2 + m_3 + \cdots} = \frac{m_1 \mathbf{r}_1 + m_2 \mathbf{r}_2 + m_3 \mathbf{r}_3 + \cdots}{M} \qquad (5.7)$$

where $\mathbf{r}_{CM}$ is the position of the CM; $\mathbf{r}_1$ is the position of mass m_1, $\mathbf{r}_2$ is the position of mass m_2, and so on; and M is the total mass of the system. It is easiest to visualize the CM of a simple system, and so we consider a system of just two masses located on an x-axis. Then the vector $\mathbf{r}$ becomes x and the CM lies at $x_{CM} = (m_1 x_1 + m_2 x_2)/(m_1 + m_2)$. If one mass is larger than the other, the CM will lie closer to it. If the two masses are equal, then the CM will lie halfway between them at $x_{CM} = (x_1 + x_2)/2$. (See Figure 5.7.)

Figure 5.6 The center of mass (CM) of an object or system is a weighted average of its mass distribution. (a) The CM of a baseball bat lies closer to the thick end. (b) The CM of any symmetrical object, such as this baseball, lies at its geometrical center.

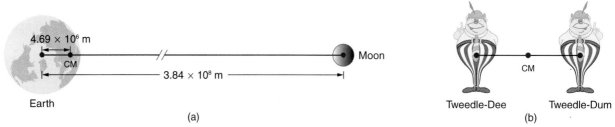

Figure 5.7 (a) The CM of the earth-moon system lies inside the earth, since the earth is 81 times more massive than the moon. The top drawing shows the system to scale. Note how large the separation of the earth and moon is compared with their diameters (about 30 times the diameter of the earth). (b) Tweedle-Dee and Tweedle-Dum have identical masses. When they are considered as a simple system, their CM lies halfway between them. The dots in them are their individual CMs.

EXAMPLE 5.2 WHERE IS THE CM ON A BALANCED SEESAW?

Calculate the position of the CM of the two children playing on the seesaw in Figure 5.5.

Strategy To solve this problem, we must locate the origin of our coordinate system. To make the calculation easier, we take the origin to be at the CM of the 26.0 kg child (the one on the left in Figure 5.5). Thus $x_1 = 0$ and $x_2 = 1.60$ m + 1.30 m = 2.90 m. The location of the CM is then found using Equation 5.7 for the two masses in our system.

Solution For two masses on the x-axis, Equation 5.7 is

$$x_{CM} = \frac{m_1 x_1 + m_2 x_2}{m_1 + m_2}$$

Entering the known values,

$$x_{CM} = \frac{(26.0 \text{ kg})(0) + (32.0 \text{ kg})(2.90 \text{ m})}{58.0 \text{ kg}}$$

$$= 1.60 \text{ m}$$

Discussion As expected, the CM is closer to the larger child: it is 1.30 m from him versus 1.60 m from the smaller child. Note that the CM of the two children is positioned exactly over the pivot. This is anything but accidental. The system is balanced (in equilibrium) *because* its CM is over the point of support.

The weight of a system also acts as if it is concentrated at a single point, called the **center of gravity** (CG). The center of gravity is at the same place as the CM provided gravity is uniform (a very good approximation in most circumstances). For instance, in the preceding example the center of gravity of the two children is at the same location as their center of mass. Thus the combined weight of the two children was exerted straight down on the pivot, causing no torque. The position of the center of gravity in a three-dimensional system is defined to be

$$\mathbf{r}_{CG} = \frac{m_1 g \mathbf{r}_1 + m_2 g \mathbf{r}_2 + m_3 g \mathbf{r}_3 + \cdots}{m_1 g + m_2 g + m_3 g + \cdots} = \frac{w_1 \mathbf{r}_1 + w_2 \mathbf{r}_2 + w_3 \mathbf{r}_3 + \cdots}{w_{tot}}$$

In most cases g is the same for each mass and therefore cancels, since it appears in the numerator and denominator. This leaves us with the same equation as for the center of mass. Thus,

$$\mathbf{r}_{CG} = \mathbf{r}_{CM} \qquad \text{(in a uniform gravitational field)} \qquad (5.8)$$

Ordinarily, there is little difference between the locations of CG and CM. One implication of this is that the torque exerted by the weight of a system is the same as if its total weight were located at the CG. This is another tremendous simplification, for even the smallest macroscopic object has huge numbers of atoms, which we need not consider individually. Consider Figure 5.8, for example.

We will find the concepts of CM and CG very useful in this and other chapters.

5.4 STABILITY

It is one thing to have a system in equilibrium; it is quite another for it to be stable. The child perched precariously on her father's finger in Figure 5.8, for example, is not in stable equilibrium. There are *three types of equilibrium—stable, unstable,* and *neutral.* Figures 5.9, 5.10, and 5.11 illustrate various examples.

A system is said to be in **stable equilibrium** if, when displaced from equilibrium, it experiences a net force or torque in a direction opposite to the direction of the displacement. For example, a marble in the bottom of a bowl will experience a *restoring* force when displaced from its equilibrium position. This force moves it back toward the equilibrium position. Most systems are in stable equilibrium, especially for small displacements. For another example of stable equilibrium, see Figure 5.9.

A system is in **unstable equilibrium** if, when displaced, it experiences a net force or torque in the *same* direction as the displacement from equilibrium. A system in unstable equilibrium accelerates away from its equilibrium position if displaced even slightly. An obvious example is a ball resting on top of a hill. Once displaced, it accelerates away from the crest. See Figure 5.10 for another example of unstable equilibrium.

A system is in **neutral equilibrium** if its equilibrium is independent of displacements from its original position. A marble on a flat horizontal surface is an example. Combinations of these situations are possible. For example, a marble on a saddle is stable for displacements toward the front or back of the saddle and unstable for displacements to the side. Figure 5.11 shows another example of neutral equilibrium.

When we consider how far a system in stable equilibrium can be displaced before it becomes unstable, we find that some systems in stable equilibrium are more stable than others. Both the pencil in Figure 5.9 and the person in Figure 5.12 are in stable equilibrium, but become unstable for relatively small displacements to the side. The critical point is reached when the CG is no longer above the base of support. Additionally, since the CG of a person's body is above the pivots in the hips, displacements must be quickly controlled. This is a central nervous system function that is developed when we learn to hold our bodies erect as infants. An animal like a chicken has an easier system to control.

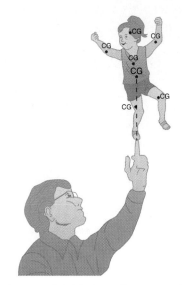

Figure 5.8 A balanced system, such as the child on her father's finger, has its CG directly over the pivot, so that the torque of the total weight is zero. This is equivalent to having the torques of the individual parts balanced about the pivot point, in this case the fingertip. The CGs of the arms, legs, head, and torso are labeled with smaller type.

Human & Biological Application

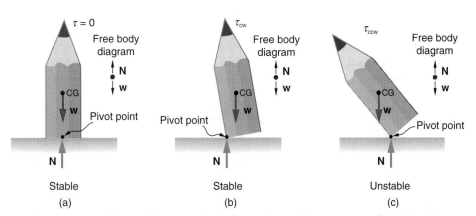

Figure 5.9 (a) Stable equilibrium. Both conditions for equilibrium are satisfied. The total torque about any pivot, including one just beneath the CG as shown, is zero. (b) If the pencil is displaced slightly to the side (counterclockwise), it is no longer in equilibrium. Its weight creates a *clockwise* torque that returns the pencil to its equilibrium position. (c) If the pencil is displaced too far, the torque caused by its weight changes direction to counterclockwise and causes the displacement to increase.

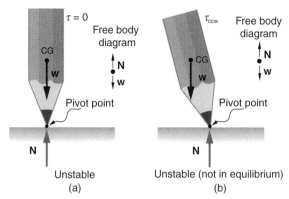

Figure 5.10 (a) Unstable equilibrium, although both conditions for equilibrium are satisfied. (b) If the pencil is displaced even slightly, a torque is created by its weight that is in the same direction as the displacement, causing it to increase.

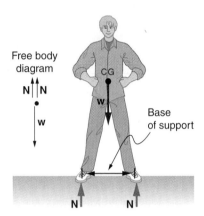

Figure 5.12 The CG of a person is above the hip joints (one of the main pivots in the body) and lies between two narrowly separated feet. Like a pencil standing on its eraser, the person is in stable equilibrium in relation to sideways displacements, but relatively small displacements take his CG outside the base of support and make him unstable. Humans are less stable relative to forward and backward displacements because the feet are not very long. Muscles are used extensively to balance the body in the front-to-back direction.

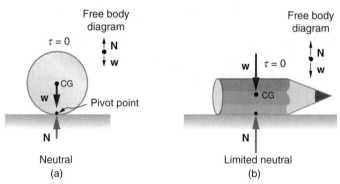

Figure 5.11 (a) Neutral equilibrium. The CG of a sphere on a flat surface lies directly above the point of support independent of position on the surface. It is therefore in equilibrium in any location, and if displaced it will *not* tend to move. (b) Because it has a circular cross section, the pencil is in neutral equilibrium for displacements perpendicular to its length.

Figure 5.13 shows that the CG of a chicken lies below its hip joints and between its widely separated and broad feet. Even relatively large displacements of the chicken's CG are stable and result in restoring forces and torques that return it to its equilibrium position with little effort on the chicken's part. Not all birds are like chickens, of course. Some, such as the flamingo, have balance systems at least as sophisticated as humans.

Engineers and architects strive to achieve very stable equilibrium for buildings and other systems that must withstand not only gravity but wind, earthquakes, and other forces that displace them from equilibrium. Although the examples in this section emphasize gravitational forces, the basic conditions for equilibrium are the same for all types of forces. The net external force must be zero, and the net torque must also be zero.

5.5 APPLICATIONS OF STATICS, INCLUDING PROBLEM-SOLVING STRATEGIES

Statics can be applied to a variety of situations, ranging from raising a drawbridge to bad posture and back strain. We begin with a discussion of problem solving strategies specifically useful for statics. Since statics is a special case of Newton's laws, both the general problem-solving strategies and the special strategies for Newton's laws, discussed in Sections 2.5 and 4.7, still apply.

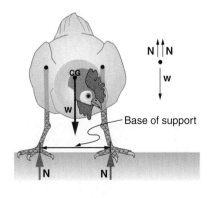

Figure 5.13 The CG of a chicken is below the hip joints and lies above a broad base of support formed by widely separated and large feet. The chicken is thus in very stable equilibrium, since a relatively large displacement is needed to render it unstable. The body of the chicken is supported from above by the hips and acts like a pendulum between them. The chicken is thus stable for front-to-back displacements as well as for side-to-side displacements.

PROBLEM-SOLVING STRATEGY

FOR STATIC EQUILIBRIUM SITUATIONS

Step 1. The first step is to determine that the system is in static equilibrium. This is always the case when the *acceleration of the system is zero and accelerated rotation does not occur*.

Step 2. Identify the knowns and unknowns and then *determine the system of interest*, just as for any problem involving Newton's laws of motion. Try to choose a system that eliminates unknown forces and torques by making them internal to the system, and involves the known forces and torques by making them external to the system.

Step 3. It is particularly important to *draw a free body diagram for the system of interest*. Carefully label all forces, and note their relative magnitudes, directions, and points of application whenever these are known.

Step 4. Solve the problem by applying either or both of the conditions for equilibrium (represented by Equations 5.1 and 5.6), depending on the list of knowns and unknowns. If the second condition is involved, *choose the pivot point to simplify the solution*. Any pivot point can be chosen, but the most useful ones cause torques by unknown forces to be zero. (Torque is zero if the force is applied at the pivot (then $r = 0$), or along a line through the pivot point (then $\theta = 0$)). Always choose a convenient coordinate system for projecting forces.

Step 5. *Check the solution to see if it is reasonable*, by examining the magnitude, direction, and units of the answer. The importance of this last step never diminishes, although in unfamiliar applications it is usually more difficult to judge reasonableness. This becomes progressively easier with experience.

CONNECTIONS

Basic problem-solving strategies were presented in Section 2.5, and other techniques have been presented for specific topics. Consult the index to locate particular problem-solving strategies.

Now let us apply this problem-solving strategy for the pole vaulter shown in Figure 5.14. The pole is uniform and has a mass of 5.00 kg. In Figure 5.14(a), the pole's CG lies halfway between the vaulter's hands. It seems reasonable that the force exerted by each hand is equal to half the weight of the pole, or 24.5 N. This obviously satisfies the first

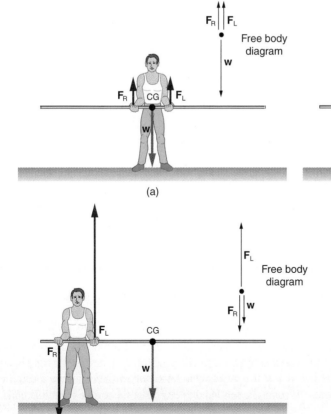

(a)

(b)

(c)

Figure 5.14   (a) A pole vaulter holding a pole with its CG halfway between his hands. Each hand exerts a force equal to half the weight of the pole, $F_R = F_L = w/2$. (b) The pole vaulter moves the pole to the right, and the forces the hands exert are no longer equal. See Example 5.3. (c) If the pole is held with its CG to the left of the person, then he must push down with his right hand and up with his left. The forces he exerts are larger here, since they are in opposite directions and since the CG is a long distance from either hand. See Example 5.4.

condition for equilibrium (net $F = 0$). The second condition (net $\tau = 0$) is also satisfied, as we can see by choosing the CG to be the pivot point. The weight exerts no torque about a pivot point located at the CG, since it is applied at that point and its lever arm is zero. The equal forces exerted by the hands are equidistant from the chosen pivot, and so they exert equal and opposite torques. Similar arguments hold for other systems where supporting forces are exerted symmetrically about the CG. For example, the four legs of a uniform table each support one-fourth of its weight.

If the pole vaulter holds the pole as shown in Figure 5.14(b), things are not as simple. The total force he exerts is still equal to the weight of the pole, but it is not evenly divided between the hands. (If $F_L = F_R$, then the torques about the CG would not be equal since the lever arms are different.) Logically, the right hand should support more of the weight, since it is closer to the CG. In fact, if the right hand is moved directly under the CG it will support all of the weight. This situation is exactly analogous to two people carrying a load; the one closer to the CG carries more of its weight. Finding the forces F_L and F_R is straightforward, as the next example shows.

EXAMPLE 5.3   WHAT FORCE IS NEEDED TO SUPPORT A WEIGHT HELD NEAR ITS CG?

For the situation shown in Figure 5.14(b), calculate: (a) F_R, the force exerted by the right hand, and (b) F_L, the force exerted by the left hand. The hands are 0.900 m apart, and the CG of the pole is 0.600 m from the left hand.

Strategy Figure 5.14(b) includes a free body diagram for the pole, the system of interest. There is not enough information to use the first condition for equilibrium (net $F = 0$), since two of the three forces are unknown and the hand forces cannot be assumed to be equal in this case. There is enough information to use the second condition for equilibrium (net $\tau = 0$) if the pivot point is chosen to be at either hand, thereby making the torque from that hand zero. We choose to locate the pivot at the left hand in this part of the problem, to eliminate the torque from the left hand.

Solution for (a) There are now only two nonzero torques, those from gravity (τ_w) and from the right hand (τ_R). Stating the second condition in terms of clockwise and counterclockwise torques,

$$\text{net } \tau_{cw} = \text{net } \tau_{ccw}$$

Here this is

$$\tau_R = \tau_w$$

since the weight creates a counterclockwise torque and the right hand a clockwise toque. Using the definition of torque, $\tau = rF \sin \theta$, noting that $\theta = 90°$, and substituting known values, we obtain

$$(0.900 \text{ m})(F_R) = (0.600 \text{ m})(mg)$$

Thus,

$$F_R = (0.667)(5.00 \text{ kg})(9.80 \text{ m/s}^2) = 32.7 \text{ N}$$

Solution for (b) The first condition for equilibrium is based on the free body diagram in the figure. This implies that

$$F_L + F_R = w = mg$$

Solving for F_L, we obtain

$$F_L = mg - F_R = 0.333mg$$

$$= (0.333)(5.00 \text{ kg})(9.80 \text{ m/s}^2) = 16.3 \text{ N}$$

Discussion F_L is seen to be exactly half of F_R, as might be guessed, since F_L is applied twice as far from the CG as F_R.

Now the pole vaulter holds the pole as he might at the start of a run, shown in Figure 5.14(c), and the forces change again. Both are considerably greater, and one force reverses direction.

EXAMPLE 5.4   WHAT FORCES MUST BE EXERTED IF THEY ARE TO ONE SIDE OF THE CG?

In Figure 5.14(c) the CG of the pole is 2.00 m from the left hand, and the hands are 0.700 m apart. Calculate the force exerted by (a) the right hand and (b) the left hand of the pole vaulter.

Strategy Figure 5.14(c) includes a free body diagram for the pole, and we can use methods very similar to those in the

previous example. First, note the directions of the hand forces. The directions shown in the figure are based on intuition. If either direction is incorrect, then its solution will be negative, indicating the true direction to be opposite of the one assumed. To first find the force exerted by the right hand, we use the second condition for equilibrium; we locate the pivot at the left hand to eliminate the torque from this hand.

(continued)

(continued)

Solution for (a) The two nonzero torques are those from gravity (τ_w) and from the right hand (τ_R), but now the weight creates a clockwise torque and the right hand a counterclockwise torque. So the second condition for equilibrium

$$\text{net } \tau_{cw} = \text{net } \tau_{ccw}$$

becomes

$$\tau_w = \tau_R$$

This becomes

$$(2.00 \text{ m})(mg) = (0.700 \text{ m})(F_R)$$

Solving for the unknown and substituting known values yields

$$F_R = \frac{2.00 \text{ m}}{0.700 \text{ m}} (5.00 \text{ kg})(9.80 \text{ m/s}^2)$$

$$= 140 \text{ N}$$

Discussion for (a) This result for the force exerted by the right hand is considerably greater than the weight of the pole (49.0 N). The positive value for F_R confirms its direction is downward as assumed.

Solution for (b) Given the value of F_R found above, there are several ways to solve for F_L. Perhaps the easiest is to use the first condition for equilibrium,

$$\text{net } F = 0$$

In the present case, this gives

$$F_L - F_R - mg = 0$$

Solving for the unknown force yields

$$F_L = F_R + mg$$

In other words, the left hand supports the weight of the pole and counteracts the downward force of the right hand. Entering known values gives

$$F_L = 140 \text{ N} + 49.0 \text{ N} = 189 \text{ N}$$

Discussion for (b) Again, F_L is much larger than the weight of the pole, because the two hands must exert forces in opposite directions to hold the pole as shown. In fact, it becomes very difficult to hold the pole if the hands are placed closer together, as you can easily imagine.

The following two sections explore other applications of statics, introducing additional aspects and complexities, and employing similar approaches to those used above.

5.6* SIMPLE MACHINES

Simple machines are devices—including levers, gears, pulleys, wedges, and screws— used to apply forces and torques, often increasing them dramatically. We can analyze and understand simple machines with the principles of statics, since in many uses these devices are stationary or move at constant velocity.

Figure 5.15 shows one type of lever used as a nail puller. Crowbars, seesaws, and other such levers are all analogous to this one. What interests us most is that the force exerted by the nail puller, F_o, is much greater than the force applied to it at the other end, F_i. The ratio of output (F_o) to input (F_i) for any simple machine is called its **mechanical advantage (MA)**[†]:

$$\text{MA} = \frac{F_o}{F_i} \tag{5.9}$$

Torques are involved in levers, since there is rotation about a pivot point. Distances from the physical pivot of the lever are crucial, and we can obtain a useful expression for the MA in terms of them. There are three vertical forces acting on the nail puller

[†]This is actually the *ideal* mechanical advantage. In practice, friction may reduce the output force, and the reduced ratio is called the *actual* mechanical advantage.

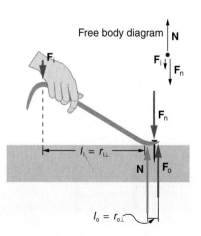

Figure 5.15 The nail puller is a lever with a large mechanical advantage. The external forces on the nail puller are represented by solid arrows. The force the nail puller applies to the nail (F_o) is shaded to show it is not a force on the nail puller. The reaction force the nail exerts back on the puller (F_n) is an external force and is equal and opposite to F_o. The perpendicular lever arms of the input and output forces are l_i and l_o.

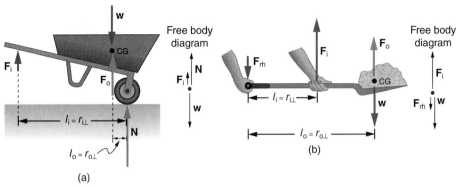

Figure 5.16 Both of these lever systems are simple machines, with their pivots in different places relative to the forces applied to and by them. They differ from each other, and they differ from the nail puller in that both the input and output forces are on the same side of the pivot. Solid arrows represent external forces on the system of interest, while shaded arrows represent output forces. (a) A wheelbarrow enables you to lift much heavier loads that you could with your body alone. The pivot is the wheel's axle. (b) The force output of this lever (supporting the shovel's load) is less than the force input (from the hand nearest the load), because the input is exerted closer to the pivot than the output is. The pivot here is at the handle held by the right hand.

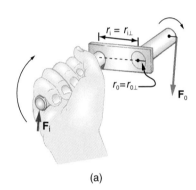

(a)

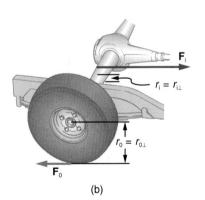

(b)

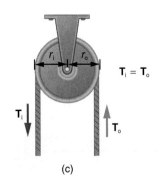

(c)

(the system of interest)—these are F_i, F_n, and N. F_n is the reaction force back on the system, equal and opposite to F_o. (Note that F_o is not a force on the system.) N is the normal force up on the lever, and its torque is zero since it is exerted at the pivot. The torques due to F_i and F_n must thus be equal to satisfy the second condition for equilibrium (net $\tau = 0$). Hence,

$$l_i F_i = l_o F_o$$

where l_i and l_o are the perpendicular lever arms of the input and output forces. Rearranging the last equation gives $F_o/F_i = l_i/l_o$. Thus,

$$\text{MA} = \frac{l_i}{l_o} \qquad (5.10)$$

This equation is true in general for levers, and it matches our experience. For example, the longer the handle on a nail puller, the greater the force you can exert with it—since the greater the distance from the pivot, the greater torque a given force creates. You might exert a force on the handle of a typical nail puller 50.0 cm from the pivot and have a nail 1.50 cm on the other side, creating an MA = 50.0 cm/1.50 cm = 33.3. So, from Equation 5.9, you create a force $F_o = \text{MA} \cdot F_i = 33.3 F_i$—that is, 33.3 times greater than the input.

Two other types of levers that differ slightly from the nail puller are shown in Figure 5.16. All these lever types are similar in that only three forces are involved—the input, the output, and the force on the pivot—and their MAs are given by Equation 5.10 with distances measured relative to the physical pivot. (More complicated types of levers, of course, can also be analyzed using the principles of statics.)

A crank is a lever that can be rotated 360° about its pivot, as shown in Figure 5.17. It may not look like a lever, but its physics is the same. The MA for a crank is simply r_i/r_o. Wheels and gears have this simple expression for their MAs, too. The MA can be greater than 1, as it is for the crank, or less than 1, as it is for the simplified car axle driving the wheels. If the axle's radius is 2.00 cm and the wheel's radius is 24.0 cm, then MA = 1/12 = 0.0833, and the axle would have to exert a force of 12,000 N on the wheel

Figure 5.17 Solid arrows represent input forces, while shaded arrows represent output forces. (a) A crank acts just like a lever that can be rotated continuously about its pivot. Cranks are usually designed to have a large MA. (b) A very simplified automobile axle drives wheels, which have a larger diameter, yielding an MA less than 1. (c) This pulley changes the direction of the force T exerted by the cord without changing its magnitude. It thus has an MA of 1.

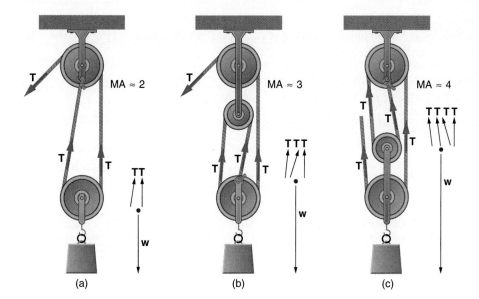

Figure 5.18 The only forces shown in this figure are those each pulley system exerts on the load (the system of interest) and the input force T. (a) This pulley system has two cables attached to its load, thus applying a force of approximately $2T$. It has an MA ≈ 2. (b) Similarly, this pulley system has three cables attached to its load and has an MA ≈ 3. (c) This pulley system applies a force of $4T$, so that it has MA ≈ 4. Effectively it has four cables pulling on the system of interest. The MAs will be integers if the cables all pull exactly vertically.

for it to exert a force of 1000 N on the ground. (Real cars have differentials that are much more complex and have a different MA.) An ordinary pulley has an MA of 1; it only changes the direction of the force and not its magnitude. Once again we see underlying connections in physics. Levers seem to be very specialized, but they are exactly analogous to cranks, wheels, gears, and pulleys.

Combinations of pulleys, such as illustrated in Figure 5.18, are used to multiply force. If the pulleys are friction-free, the force output is approximately an integral multiple of the tension in the cable. The number of cables pulling directly upward on the system of interest, as illustrated in Figure 5.18, is approximately the MA of the pulley. Since each attachment applies an external force in approximately the same direction as the others, they add, producing a total force that is nearly an integral multiple of T.

5.7* FORCES AND TORQUES IN MUSCLES AND JOINTS

Muscles, bones, and joints are some of the most interesting applications of statics. There are some surprises. Muscles, for example, exert far greater forces than we might think. Figure 5.19 shows a forearm holding a book and a schematic diagram of

Human & Biological Application

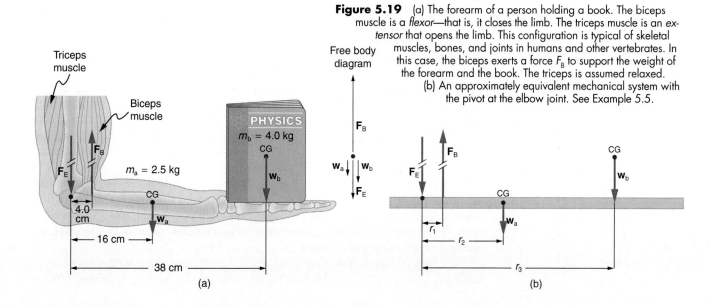

Figure 5.19 (a) The forearm of a person holding a book. The biceps muscle is a *flexor*—that is, it closes the limb. The triceps muscle is an *extensor* that opens the limb. This configuration is typical of skeletal muscles, bones, and joints in humans and other vertebrates. In this case, the biceps exerts a force F_B to support the weight of the forearm and the book. The triceps is assumed relaxed.
(b) An approximately equivalent mechanical system with the pivot at the elbow joint. See Example 5.5.

an analogous lever system. The schematic is a good approximation to the forearm, which looks more complicated than it is, and we can get some insight into the way typical muscle systems function by analyzing it.

EXAMPLE 5.5   MUSCLES EXERT BIGGER FORCES THAN YOU MIGHT THINK

Calculate the force the biceps muscle must exert to hold the forearm and its load as shown in Figure 5.19, and compare this force with the weight of the forearm plus its load. You may take the data in the figure to be accurate to three digits.

Strategy There are four forces acting on the forearm and its load (the system of interest). They are the force of the biceps F_B, the force at the elbow joint F_E, and the weights of the forearm w_a and its load w_b. Two of these are unknown (F_B and F_E), so that the first condition for equilibrium cannot by itself yield F_B. But if we use the second condition and choose the pivot to be at the elbow, then the torque due to F_E is zero, and the only unknown becomes F_B.

Solution The torques created by the weights are clockwise relative to the pivot, while that created by the biceps is counterclockwise; thus, the second condition for equilibrium (net τ_{cw} = net τ_{ccw}) becomes

$$r_2 w_a + r_3 w_b = r_1 F_B$$

Note that $\sin \theta = 1$ for all forces, since $\theta = 90°$ for all forces. This equation can easily be solved for F_B in terms of known quantities, yielding

$$F_B = \frac{r_2 w_a + r_3 w_b}{r_1}$$

Entering knowns,

$$F_B = \frac{(16.0 \text{ cm})(2.50 \text{ kg})(9.80 \text{ m/s}^2)}{4.00 \text{ cm}}$$
$$+ \frac{(38.0 \text{ cm})(4.00 \text{ kg})(9.80 \text{ m/s}^2)}{4.00 \text{ cm}}$$

yielding

$$F_B = 470 \text{ N}$$

Now the combined weight of the arm and its load is (6.50 kg)(9.80 m/s²) = 63.7 N, so that the ratio of the force exerted by the biceps to the total weight is

$$\frac{F_B}{w_a + w_b} = \frac{470}{63.7} = 7.38$$

Discussion In more familiar units, the biceps is exerting a force of 106 lb to support a total weight of 14.3 lb, that is, a force 7.38 times the weight supported.

Like the biceps, most skeletal muscles exert much larger forces within the body than limbs apply to the outside world. The reason is clear once we realize that most muscles are attached close to joints, causing these systems to have mechanical advantages much less than 1. This is true both for muscles that close limbs (such as the biceps), called *flexors*, and those that open limbs (such as the triceps), called *extensors*. Viewing them as simple machines, the input force is much greater than the output force.

Very large forces are also created in the joints. In the previous example, the downward force F_E exerted by the humerus at the joint equals 407 N, or 6.38 times the total weight supported. (The calculation of F_E is straightforward and is left as an end-of-chapter problem.) Because of the way things are put together, joints and muscles often exert forces in opposite directions and thus subtract (here the upward force of the muscle minus the downward force of the joint equals the weight supported—i.e., 470 N − 407 N ≈ 63.7 N = weight supported). Forces in muscles and joints are largest when their load is a long distance from the joint, as the book is in the previous example. Racket sports, such as tennis, exaggerate this effect by extending the arm. It is no wonder that joint deterioration, such as tennis elbow, results from extreme effort in such sports.

The leg works much as the arm does, except that the knee is more complicated than the elbow. Connecting tissues in the limbs, such as tendons and cartilage, as well as the joints are sometimes damaged by the large forces they carry. Often this is due to accidents, but very heavily muscled athletes, such as weightlifters, can tear muscles and connecting tissue through effort alone.

The back is considerably more complicated than the arm or leg, with many muscles and many joints between vertebrae, all having mechanical advantages less than 1. Back

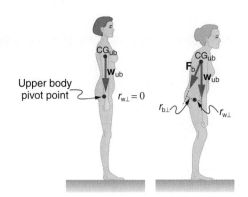

Figure 5.20 (a) Good posture places the upper body's CG over the pivots in the hips, eliminating the need for muscle action to balance the body. (b) Poor posture requires exertion by the back muscles to counteract the clockwise torque produced about the pivot by the upper body's weight. The back muscles have a small effective perpendicular lever arm, $r_{b\perp}$, and must therefore exert a large force $\mathbf{F}_b$. Note that the legs lean backward to keep the CG of the entire body above the base of support in the feet.

muscles must therefore exert very large forces, which are borne by the spinal column. Disks crushed by mere exertion are very common. The jaw is somewhat exceptional—the masseter muscles that close the jaw have a mechanical advantage greater than 1 for the back teeth, allowing us to exert very large forces with them.

Figure 5.20 shows how bad posture causes back strain. In part (a), we see a person with good posture. Note that her upper body's CG is directly above the pivot point in the hips, which in turn is directly above the base of support at her feet. Because of this, her upper body's weight exerts no torque about the hips. The only force needed is a vertical force at the hips equal to the weight supported. No muscle action is required, since the bones are rigid and transmit this force from the floor. This is a position of unstable equilibrium, but only small forces are needed to bring the upper body back to vertical if it is slightly displaced. Bad posture is shown in part (b); we see that the upper body's CG is in front of the pivot in the hips. This creates a clockwise torque about the hips that is counteracted by muscles in the lower back. These muscles must exert large forces, since they have typically small mechanical advantages. (Alternatively, the perpendicular lever arm for the muscles is much smaller than for the CG.) Prolonged muscle action produces muscle strain. Note that the CG of the entire body is still directly above the base of support in part (b) of Figure 5.20—it must be or the person would not be in equilibrium. We lean forward for the same reason when carrying a load on our backs, to the side when carrying a load in one arm, and backward when carrying a load in front of us, as seen in Figure 5.21.

There is a familiar admonition against lifting with our backs. This, even more than bad posture, can cause muscle strain and damage disks and vertebrae, since abnormally large forces are created in the back muscles and spine.

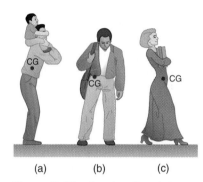

(a)   (b)   (c)

Figure 5.21 People adjust their stance to maintain balance. (a) A father carrying his son piggyback leans forward to position their overall CG above the base of support at his feet. (b) A student carrying a shoulder bag leans to the side to keep the overall CG over his feet. (c) Another student carrying a load of books in her arms leans backward for the same reason.

EXAMPLE 5.6   DO NOT LIFT WITH YOUR BACK

Consider the person lifting a heavy box with his back, shown in Figure 5.22. (a) Calculate the force in the back muscles F_B needed to support the upper body plus the box and compare this with their weight. The mass of the upper body is 55.0 kg and that of the box is 30.0 kg. (b) Calculate the magnitude and direction of the force $\mathbf{F}_V$ exerted by the vertebrae on the spine at the indicated pivot point. Again, data in the figure is taken to be accurate to three digits.

Strategy By now, we sense that the second condition for equilibrium is a good place to start, and inspection of the knowns confirms that it can be used to solve for F_B if the pivot is chosen to be at the hips. The torques created by w_{ub} and w_{box} are clockwise, while that created by F_B is counterclockwise.

Solution for (a) Using the perpendicular lever arms

given in the figure, the second condition (net τ_{cw} = net τ_{ccw}) becomes

$$(35.0 \text{ cm})(55.0 \text{ kg})(9.80 \text{ m/s}^2)$$

$$+ (50.0 \text{ cm})(30.0 \text{ kg})(9.80 \text{ m/s}^2) = (8.00 \text{ cm}) \cdot F_B$$

Solving for F_B yields

$$F_B = 4200 \text{ N}$$

The ratio of the force the back muscles exert to the weight of the upper body plus its load is

$$\frac{F_B}{w_{ub} + w_{box}} = \frac{4200 \text{ N}}{833 \text{ N}} = 5.04$$

(continued)

(continued)

This force is considerably larger than it would be if the load were not present.

Solution for (b) More important in terms of its damage potential is the force on the vertebrae $\mathbf{F}_V$. The first condition for equilibrium (net $\mathbf{F} = 0$) can be used to find its magnitude and direction. Using y for vertical and x for horizontal, the condition for the net external forces along those axes to be zero is written

$$\text{net } F_y = \text{net } F_x = 0$$

Starting with the vertical (y-) components, this yields

$$F_{Vy} - w_{ub} - w_{box} - F_B \sin 29.0° = 0$$

Thus,

$$F_{Vy} = w_{ub} + w_{box} + F_B \sin 29.0°$$

$$= 833 \text{ N} + (4200 \text{ N}) \sin 29.0°$$

yielding

$$F_{Vy} = 2870 \text{ N}$$

Similarly, for the horizontal (x-) components,

$$F_{Vx} - F_B \cos 29.0° = 0$$

yielding

$$F_{Vx} = 3670 \text{ N}$$

The magnitude of F_V is given by the Pythagorean theorem:

$$F_V = (F_{Vy}^2 + F_{Vx}^2)^{1/2} = 4660 \text{ N}$$

The direction of F_V is

$$\theta = \tan^{-1}(F_{Vy}/F_{Vx}) = 38.0°$$

Note that the ratio of F_V to the weight supported is

$$\frac{F_V}{w_{ub} + w_{box}} = \frac{4660 \text{ N}}{833 \text{ N}} = 5.60$$

Discussion This force is 5.60 times greater than it would be if the person were standing erect. The trouble with the back is not so much that the forces are large—because similar forces are created in our hips, knees, and ankles—but that our spines are relatively weak. Proper lifting, performed with the back erect and using the legs to raise the body and load, creates much smaller forces in the back—in this case, 5.60 times smaller.

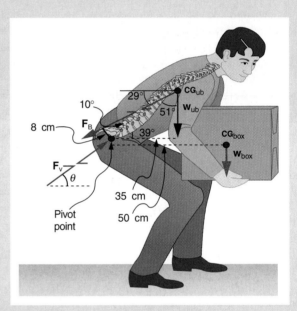

Figure 5.22 Large forces are exerted by the back muscles and experienced in the vertebrae when a person lifts with his back, since these muscles have a small effective perpendicular lever arm. The data shown here are analyzed in Example 5.6.

What are the benefits of having most skeletal muscles attached so close to joints? One is speed. A small muscle contraction produces a large movement of the limb. Others are flexibility and agility, made possible by the large numbers of joints and the ranges over which they function. It is difficult to imagine a system with biceps muscles attached at the wrist, for example, that would be capable of the broad range of movement we vertebrates possess.

There are some interesting complexities in real systems of muscles, bones, and joints. One is that the pivot point in many joints changes location as the joint is flexed, so that the perpendicular lever arms and the mechanical advantage of the system change, too. Thus, for example, the force the biceps muscle must exert to hold up a book varies as the forearm is flexed. Similar things happen in the legs, explaining, for example, why there is less leg strain when a bicycle seat is set at the proper height. The methods employed in this section give a reasonable description of real systems provided enough is known about the dimensions of the system. There are many other interesting examples of force and torque in the body—a few of these are the subject of end-of-chapter problems.

5.8 ELASTICITY: STRESS AND STRAIN

Is there any difference between a situation in which very small forces are in equilibrium and one where very large forces are in equilibrium? Intuitively, it seems obvious that large forces must have some effect on an object, even in equilibrium. For example, if two people push in opposite directions on a car with equal force, the car will not move. If two large bulldozers do the same thing, the car will not move, but it will noticeably change shape. In this section we consider some of the effects of forces on the shapes of objects.

Even very small forces are known to cause some deformation. Moreover, for small deformations, two important characteristics are observed. First, the object returns to its original shape when the force is removed—that is, the deformation is **elastic** for small deformations. Second, the size of the deformation is proportional to the force—that is, for small deformations, **Hooke's law*** is obeyed. In equation form, Hooke's law is

$$F = k \, \Delta L \qquad \qquad \textbf{(5.11)}$$

where ΔL is the amount of deformation (the change in length, for example) produced by the force F, and k is a proportionality constant that depends on the shape and composition of the object and the direction of the force. Rearranging this to

$$\Delta L = \frac{F}{k}$$

makes it clear that the deformation is proportional to the applied force. For example, a guitar string made of nylon stretches when it is tightened, and the elongation ΔL is proportional to the force applied (at least for small deformations). Thicker nylon strings and ones made of steel stretch less for the same applied force, implying they have a larger k. (See Figure 5.23.) Finally, all three strings return to their normal lengths when the force is removed, provided the deformation was small. (See Figure 5.24.) Most materials will behave in this manner if the deformation is less than about 0.1%, or about 1 part in 10^3.

We now consider three specific types of deformations—changes in length, shear (to the side), and changes in volume. All deformations are assumed to be small unless otherwise stated.

Tension and Compression; Elastic Modulus

A change in length ΔL is produced when a force is applied to a wire or rod parallel to its length L_0, either stretching it (a tension) or compressing it. (See Figure 5.25.)

---

*Traditionally called a law, although doing so is a misuse of the term *law* since the relationship is approximate and not universally applicable. It was first carefully studied by and is named for the English scientist Robert Hooke (1635–1703).

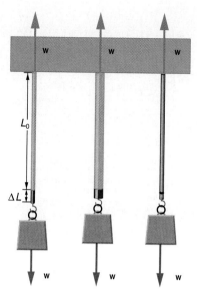

Figure 5.23 The same force, in this case a weight, applied to three different guitar strings produces the three different deformations shown as shaded segments. The string on the left is thin nylon, the one in the middle is thicker nylon, and the one on the right is steel. Also shown are the supporting forces, which are equal to the supported weight since the net external force on each string is zero.

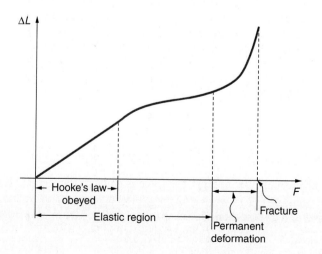

Figure 5.24 A graph of deformation ΔL versus applied force F. The straight segment is the linear region where Hooke's law is obeyed. The slope of the straight region is $1/k$. For larger forces, the graph is curved but the deformation is still elastic— ΔL will return to zero if the force is removed. Still greater forces permanently deform the object until it finally fractures. The shape of the curve near fracture depends on several factors, including how the force F is applied. Note that in this graph the slope increases just before fracture, indicating that a small increase in F is producing a large increase in L near fracture.

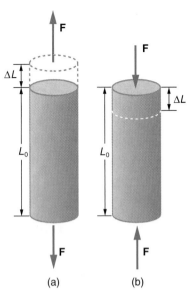

Figure 5.25 (a) Tension. The rod is stretched a length ΔL when a force is applied parallel to its length. (b) Compression. The same rod is compressed by forces in the opposite direction. For very small deformations and uniform materials, ΔL is approximately the same for tension and compression. For larger deformations, the cross-sectional area changes as the rod is compressed or stretched.

TABLE 5.1

ELASTIC MODULI*

Material	Young's modulus (tension–compression) γ (10^9 N/m^2)	Shear modulus S (10^9 N/m^2)	Bulk modulus B (10^9 N/m^2)
Aluminum	70	25	75
Bone	16	80	8
Brass	90	35	75
Brick	15		
Concrete	20		
Glass	70	20	30
Granite	45	20	45
Hair (human)	10		
Hardwood	15	10	
Iron, cast	100	40	90
Lead	16	5	50
Marble	60	20	70
Nylon	5		
Polystyrene	3		
Silk	6		
Spider thread	3		
Steel	210	80	130
Tendon	1		
Acetone			0.7
Ethanol			0.9
Glycerin			4.5
Mercury			25
Water			2.2

*Approximate and average values. Young's moduli (γ) for tension and compression sometimes differ but are averaged here.

Experiments have shown that the change in length (ΔL) depends on only a few variables. As already noted, ΔL is proportional to the force F and depends on the substance from which the object is made. Additionally, the change in length is proportional to the original length L_0, and inversely proportional to the cross-sectional area of the wire or rod. For example, a long guitar string will stretch more than a short one, and a thick string will stretch less than a thin one. We can combine all these factors into one equation for ΔL:

$$\Delta L = \frac{1}{\gamma} \frac{F}{A} L_0 \qquad (5.12)$$

Human Application

where A is the cross-sectional area and γ is a factor, called the **elastic modulus** or **Young's modulus**, that depends on the substance. Table 5.1 lists values of γ for several materials—those with a large γ are said to have a large *tensile strength* because they deform less for a given tension or compression. Young's moduli are not listed for liquids and gases because they cannot be stretched or compressed in only one direction. Note that there is an assumption that the object does not accelerate, so that there are actually *two* applied forces of magnitude F acting in opposite directions. For example, the strings in Figure 5.23 are being pulled down by a force of magnitude w and held up by the ceiling, which also exerts a force of magnitude w.

EXAMPLE 5.7 HOW MUCH DOES YOUR LEG SHORTEN WHEN YOU STAND ON IT?

Calculate the change in length of the upper leg bone (the femur) when a 70.0 kg man supports 62.0 kg of his mass on it, assuming it to be equivalent to a uniform rod that is 40.0 cm long and 2.00 cm in radius.

Strategy The force is equal to the weight supported, or $F = mg = (62.0 \text{ kg})(9.80 \text{ m/s}^2) = 607.6$ N, and the cross-sectional area is $\pi r^2 = 1.257 \times 10^{-3} \text{ m}^2$. Equation 5.12 can be used to find the change in length.

(continued)

(continued)

Solution All quantities except ΔL in Equation 5.12 are known. Entering them gives

$$\Delta L = \frac{1}{1.6 \times 10^{10} \ \text{N/m}^2} \cdot \frac{607.6 \ \text{N}}{1.257 \times 10^{-3} \ \text{m}^2} \cdot 0.400 \ \text{m}$$
$$= 1.2 \times 10^{-5} \ \text{m}$$

Discussion This small change in length seems reasonable, consistent with our experience that bones are rigid. In fact, even the rather large forces encountered during strenuous physical activity do not compress or bend bones by large amounts. Although bone is rigid compared with fat or muscle, note that several of the substances listed in Table 5.1 have larger γ. In other words, they are more rigid and have greater tensile strength.

Equation 5.12 is traditionally rearranged and written in the following form:

$$\frac{F}{A} = \gamma \frac{\Delta L}{L_0}$$

The ratio of force to area, F/A, is defined to be **stress**, and the ratio of the change in length to length, $\Delta L/L_0$, is defined to be **strain**. In other words,

$$\text{stress} = \gamma \times \text{strain} \qquad (5.13)$$

In this form, the equation is analogous to Hooke's law, with stress analogous to force and strain analogous to deformation. This general idea—that force and the deformation it causes are proportional for small deformations—applies to changes in length, sideways bending, and changes in volume.

Sideways Stress; Shear Modulus

Figure 5.26 illustrates what is meant by a sideways stress or a *shearing force*. Here the deformation is called Δx and it is *perpendicular* to L_0, rather than parallel as with tension and compression. Shear deformation behaves very similarly to tension and compression and can be described with very similar equations. The expression for shear deformation is

$$\Delta x = \frac{1}{S} \frac{F}{A} L_0 \qquad (5.14)$$

where S is the **shear modulus** (see Table 5.1) and F is the force applied perpendicular to L_0 and parallel to the cross-sectional area A. Again, to keep the object from accelerating, there are actually two equal and opposite forces F applied across opposite faces as illustrated in Figure 5.26. The equation is logical—for example, it is easier to bend a long thin pencil (small A) than a short thick one, and both are more easily bent than similar steel rods (large S).

Examination of the shear moduli in Table 5.1 reveals some telling patterns. For example, shear moduli are less than Young's moduli for most materials. Bone is a remarkable exception. Its shear modulus is not only greater than its Young's modulus, it is as large as steel's. This is one reason that bones can be long and relatively thin. The shear moduli for concrete and brick are very small; they are too highly variable to be listed. Concrete used in buildings can withstand compression, as in pillars and arches, but is very poor against shear, as might be encountered in heavily loaded floors or during earthquakes. Modern structures were made possible by the advent of steel and steel-reinforced concrete. Almost by definition, liquids and gases have shear moduli near zero, because they flow in response to shearing forces.

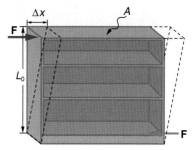

Figure 5.26 Shearing forces are applied perpendicular to the length L_0 and parallel to the area A, producing a deformation Δx. Vertical forces are not shown, but it should be kept in mind that in addition to the two shearing forces, **F**, there must be supporting forces to keep the object from rotating. The distorting effects of these supporting forces are ignored in this treatment. The weight of the object also is not shown, since it is usually negligible compared with forces large enough to cause significant deformations.

EXAMPLE 5.8 THAT NAIL DOES NOT BEND MUCH UNDER A LOAD

Find the mass of the picture hanging from a steel nail as shown in Figure 5.27, given that the nail bends only 1.80 μm.

Strategy The force F on the nail (neglecting the nail's own weight) is the weight of the picture w. If we can find w, then the mass of the picture is just w/g. Equation 5.14 can be solved for F.

(continued)

(continued)

Solution Solving Equation 5.14 for F, we see that all other quantities can be found:

$$F = \frac{SA}{L_0} \Delta x$$

S is found in Table 5.1; r is 0.750 mm, and so the cross-sectional area is $A = \pi r^2 = 1.77 \times 10^{-6} \text{ m}^2$; and the value for L_0 is shown in the figure. Thus,

$$F = \frac{(8.0 \times 10^{10} \text{ N/m}^2)(1.77 \times 10^{-6} \text{ m}^2)}{(5.00 \times 10^{-3} \text{ m})}$$
$$\times (1.80 \times 10^{-6} \text{ m}) = 51 \text{ N}$$

This 51 N force is the weight w of the picture, and so the picture's mass is

$$m = \frac{w}{g} = \frac{F}{g} = 5.2 \text{ kg}$$

Discussion This is a fairly massive picture, and it is impressive that the nail flexes only 1.80 µm—an amount undetectable to the unaided eye.

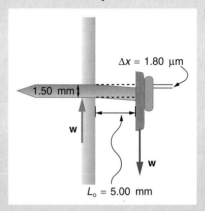

Figure 5.27 Side view of a nail with a picture hung from it. The nail flexes very slightly (shown much larger than actual) because of the shearing effect of the supported weight. Also shown is the upward force of the wall on the nail, illustrating that there are equal and opposite forces applied across opposite cross sections of the nail. See Example 5.8 for a calculation of the mass of the picture.

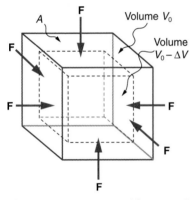

Figure 5.28 An inward force on all surfaces compresses this cube. Its change in volume is proportional to the force per unit area and its original volume, and is related to the compressibility of the substance.

Volume Deformations; Bulk Modulus

An object will be compressed in all directions if inward forces are applied evenly on all its surfaces as in Figure 5.28. It is relatively easy to compress gases and extremely difficult to compress liquids and solids. For example, air in a wine bottle is compressed when it is corked. But if you try corking a brim-full bottle, you cannot compress the wine—some must be removed if the cork is to be inserted. The reason for these different compressibilities is that atoms and molecules are separated by large empty spaces in gases, but packed close together in liquids and solids. To compress a gas, you must force its atoms and molecules closer together. To compress liquids and solids, you must actually compress their atoms and molecules, and very strong electromagnetic forces in them oppose this.

We can describe the compression or volume deformation of an object with an equation. First, we note that a force "applied evenly" is defined to have the same ratio of force to area (F/A) on all surfaces. The deformation produced is a change in volume ΔV, which is found to behave very similarly to shear, tension, and compression previously discussed. (This is not surprising, since a compression of the entire object is equivalent to compressing each of its three dimensions.) The relationship of the change in volume to other physical quantities is given by

$$\Delta V = \frac{1}{B} \frac{F}{A} V_0 \tag{5.15}$$

where B is the **bulk modulus** (see Table 5.1), V_0 is the original volume, and F/A is the force per unit area applied uniformly inward on all surfaces. Note that no bulk moduli are given for gases—they are treated as a separate subject in later chapters.

What are some examples of bulk compression of solids and liquids? One practical example is the manufacture of industrial-grade diamonds by compressing carbon with an extremely large force per unit area. The carbon atoms rearrange their crystalline structure into the more tightly packed pattern of diamond. In nature, a similar process occurs deep underground, where extremely large forces result from the weight of overlying material. Another natural source of large compressive forces is the pressure created by the weight of water, especially in deep parts of the oceans. Water exerts an inward force on all surfaces of a submerged object, and even on the water itself. At great depths, water is measurably compressed, as the following example illustrates.

EXAMPLE 5.9 HOW MUCH IS WATER COMPRESSED AT GREAT OCEAN DEPTHS?

Calculate the fractional decrease in volume ($\Delta V/V_0$) for sea water at about 5 km depth, where the force per unit area is 5.00×10^7 N/m^2.

Strategy Equation 5.15 is the correct physical relationship. All quantities in the equation except $\Delta V/V_0$ are known.

Solution Solving Equation 5.15 for the unknown $\Delta V/V_0$ gives

$$\frac{\Delta V}{V_0} = \frac{1}{B}\frac{F}{A}$$

Substituting known values,

$$\frac{\Delta V}{V_0} = (2.2 \times 10^9 \text{ N/m}^2)(5.00 \times 10^7 \text{ N/m}^2)$$
$$= 0.023 = 2.3\%$$

Discussion Although measurable, this is not a very great decrease in volume considering that the force per unit area is about 1 million pounds per square foot! Liquids and solids are extraordinarily difficult to compress.

Conversely, very large forces are created by liquids and solids when they try to expand but are constrained from doing so—which is equivalent to compressing them to less than their normal volume. This often occurs when a contained material warms up, since most materials expand when their temperature increases. If the materials are tightly constrained, they deform or break their container. Another very common example occurs when water freezes. Water, unlike most materials, expands when it freezes, and it can easily fracture a boulder, rupture a biological cell, or crack an engine block that gets in its way.

Other types of deformations, such as torsion or twisting, behave analogously to the tension, shear, and bulk deformations considered here.

SUMMARY

Statics is the study of forces in equilibrium. Two conditions must be met to achieve **equilibrium**, which is defined to be motion without linear or rotational acceleration. **The first condition necessary to achieve equilibrium** is that *the net external force on the system must be zero*, so that

$$\text{net } \mathbf{F} = 0 \tag{5.1}$$

The second condition assures torques are also balanced. **Torque** is *the effectiveness of a force in producing a rotation* and is defined to be

$$\tau = rF \sin \theta \tag{5.2}$$

where τ is torque, r is the distance from the pivot point to the point where the force is applied, F is the magnitude of the force, and θ is the angle between r and F. The **perpendicular lever arm** $r_\perp$ is defined to be

$$r_\perp = r \sin \theta \tag{5.3}$$

so that

$$\tau = r_\perp F \tag{5.4}$$

The perpendicular lever arm $r_\perp$ is the shortest distance from the pivot point to the line along which $\mathbf{F}$ lies. The **SI unit for torque** is the newton-meter (N·m). **The second condition necessary to achieve equilibrium** is that *the net external torque on a system must be zero*:

$$\text{net } \tau = 0 \tag{5.5}$$

also written as

$$\text{net } \tau_{cw} = \text{net } \tau_{ccw} \tag{5.6}$$

where the directions are relative to any point you choose.

A system acts as if its mass is concentrated at a single point, called the **center of mass** (CM). The position of the center of mass is

$$\mathbf{r}_{CM} = \frac{m_1\mathbf{r}_1 + m_2\mathbf{r}_2 + m_3\mathbf{r}_3 + \cdots}{m_1 + m_2 + m_3 + \cdots}$$
$$= \frac{m_1\mathbf{r}_1 + m_2\mathbf{r}_2 + m_3\mathbf{r}_3 + \cdots}{M} \tag{5.7}$$

where $\mathbf{r}_{CM}$ is the position of the CM; $\mathbf{r}_1$ is the position of mass m_1, $\mathbf{r}_2$ is the position of mass m_2, and so on; and M is the total mass of the system. The weight of a system acts as if it is concentrated at a single point, called the **center of gravity** (CG), where the location $\mathbf{r}_{CG}$ is

$$\mathbf{r}_{CG} = \mathbf{r}_{CM} \quad \text{(in a uniform gravitational field)} \tag{5.8}$$

Objects can be treated like points with their masses at CM and weight acting at CG.

A system is said to be in **stable equilibrium** if, when displaced from equilibrium, it experiences a net force or torque in a direction *opposite* the direction of the displacement. It is in **unstable equilibrium** if, when displaced from equilibrium, it experiences a net force or torque in the *same* direction as the displacement from equilibrium. It is in **neutral equilibrium** if its equilibrium is independent of displacements from its original position.

A **simple machine** is a device used to apply forces and torques. Examples include levers, gears, and pulleys. **Mechanical advantage (MA)** is defined to be the ratio of output force (F_o) to input force (F_i):

$$MA = \frac{F_o}{F_i} \qquad (5.9)$$

In terms of perpendicular lever arms, this is

$$MA = \frac{l_i}{l_o} \qquad (5.10)$$

where l_i is the lever arm of the input force and l_o is the lever arm of the output force.

A body is **elastic** if it can be deformed and return to its original shape when force is removed. Most objects obey Hooke's law,

$$F = k \, \Delta L \qquad (5.11)$$

for small deformations. ΔL is the amount of deformation produced by the force F, and k is a proportionality constant that depends on the shape and composition of the object and the direction of the force. A force exerted along the length of an ob-

ject is called a **tension** and will stretch or compress the object an amount ΔL given by

$$\Delta L = \frac{1}{\gamma} \frac{F}{A} L_0 \qquad (5.12)$$

where A is the cross-sectional area and γ is a factor, called the **elastic modulus** or **Young's modulus**, that depends on the substance. This can also be expressed in terms of **stress**, defined to be F/A, and **strain**, defined to be $\Delta L/L_0$. Then,

$$\text{stress} = \gamma \times \text{strain} \qquad (5.13)$$

A **shear** force is perpendicular to the length L_0 and produces a shear deformation Δx given by

$$\Delta x = \frac{1}{S} \frac{F}{A} L_0 \qquad (5.14)$$

where S is the **shear modulus** (see Table 5.1) and F is the force applied perpendicular to L_0 and parallel to the cross-sectional area A. Force applied to all sides of an object changes its volume an amount ΔV given by

$$\Delta V = \frac{1}{B} \frac{F}{A} V_0 \qquad (5.15)$$

where B is the **bulk modulus**, V_0 is the original volume, and F/A is the force per unit area applied uniformly inward on all surfaces. Table 5.1 gives representative values for γ, S, and B—the tensile, shear, and bulk moduli of various substances.

CONCEPTUAL QUESTIONS

5.1 What can you say about the velocity of a moving body that is in dynamic equilibrium? Draw a sketch of such a body using clearly labeled arrows to represent all external forces on it.

5.2 Under what conditions can a rotating body be in equilibrium? Give an example.

5.3 What three factors affect the torque created by a force relative to a specific pivot point?

5.4 A wrecking ball is being used to knock down a building. One tall unsupported concrete wall remains standing. If the wrecking ball hits the wall near the top, is the wall more likely to fall over by rotating at its base or by falling straight down? Explain your answer. How is it most likely to fall if it is struck with the same force at its base? Note that this depends on how tightly the wall is attached at its base.

5.5 Weekend mechanics sometimes put a length of pipe over the handle of a wrench when trying to remove a very tight bolt. How does this help? (It is also hazardous since it can break the bolt.)

5.6 An aspiring scientist holds a meter stick on his forefingers as shown in Figure 5.29(a). He then slides his hands together and finds that they meet at the center of the stick, as in Figure 5.29(c). Explain why this happens independent of the starting positions of the fingers. Also explain why the process is not reversible.

5.7 Can the center of mass of a body be located outside of it? If so, give an example. If not, explain why not.

5.8 A round pencil lying on its side as in Figure 5.11(b) is in neutral equilibrium relative to displacements perpendicular to its length. What is its stability relative to displacements parallel to its length?

5.9 A tightrope walker is in unstable equilibrium. Explain why by discussing what happens if there is a displacement to the side.

5.10 Explain why a tightrope walker can survive better on a slightly loose wire than a very tight one by examining the stability of her equilibrium on the wire and the possibility of shifting her base of support.

5.11 Figure 5.30 shows a motorcycle high wire circus act. The system is actually in stable equilibrium relative to the wire. Explain how this is achieved.

5.12 Explain the need for tall towers on a suspension bridge to ensure stable equilibrium.

5.13 Why does the person in Figure 5.31 carry the load with its center of mass directly above her neck vertebrae?

5.14 Certain types of dinosaurs were bipedal, although they had tiny brains. What is a good reason that these creatures invariably had long tails if they had long necks?

5.15 The first condition for equilibrium is used to solve part (b) of Example 5.4. Where would you locate the pivot point if you wished to solve that problem using the *second* condition for equilibrium (assuming you had already solved part (a))? Can you solve the problem with the pivot located anywhere else?

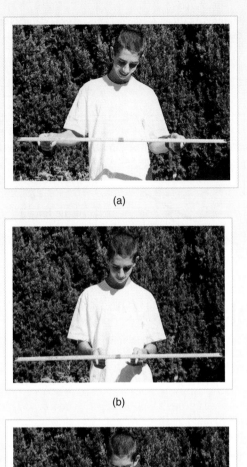

(a)

(b)

(c)

Figure 5.29 It is possible to go from (a) to (b) to (c) by just sliding the fingers. The reverse is not possible. Question 6.

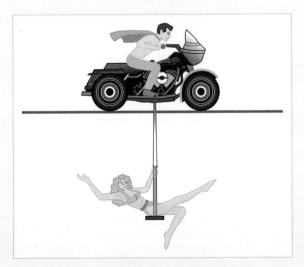

Figure 5.30 How precarious is this system's balance? Question 11 and Problem 12.

Figure 5.31 Question 13.

5.16 A common type of nutcracker is shown in Figure 5.32. Which of the simple machines in Figures 5.15 and 5.16 is most similar to it?

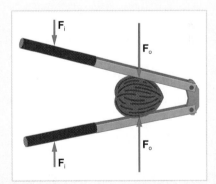

Figure 5.32 This common form of nutcracker is a double lever system. Question 16.

5.17 Scissors are like a double lever system. Which of the simple machines in Figures 5.15 and 5.16 is most analogous to scissors?

5.18 Suppose you pull a nail at a constant rate using a nail puller. Is the nail puller in equilibrium? What if you pull the nail with some acceleration (an increasing rate)—is the nail puller in equilibrium then? In which case is the force applied to the nail puller larger and why?

5.19 The MAs of the pulley systems in Figure 5.18 are slightly less than integers because not all the forces are parallel. What else would reduce their MAs in practical applications?

5.20 What are the MAs of the two gear systems shown in Figure 5.33 assuming no friction? Identify which system in Figure 5.17 is most analogous to each gear system.

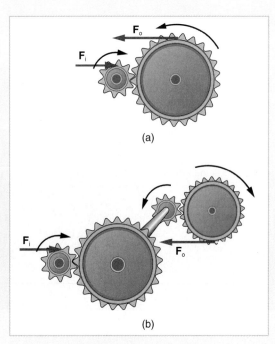

(a)

(b)

Figure 5.33 What are the MAs of these gear systems? The small gears each have 10 teeth, while the large ones have 25. The effective radius of each large gear is 2.1 cm and that of each shaft is 0.20 cm. Question 20.

5.21 Why are the forces exerted on the outside world by the limbs of our bodies usually much smaller than the forces exerted by muscles inside the body?

5.22 Explain why pregnant women often suffer from back strain late in their pregnancy.

5.23 Young's modulus is the same for tension and compression if the strain is so small that the cross-sectional area of the object does not change appreciably. Will Young's modulus be greater for tension or compression if the strain is large? Explain your answer. (Ignore effects due to composite materials or odd shapes.)

5.24 Figure 5.34 shows a graph of stress versus strain. Demonstrate that the slope in the linear region (where Equations 5.12 and 5.13 are valid) equals Young's modulus.

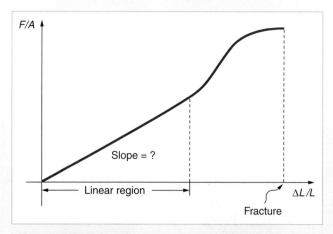

Figure 5.34 A graph of stress versus strain for a rod under tension. Question 24.

5.25 Are the two forces shown in Figure 5.27 the only external forces acting on the nail (ignoring the nail's weight)? Explain.

5.26 An old carpenter's trick to keep nails from bending when they are pounded into hard materials is to grip the center of the nail firmly with pliers. Why does this help?

5.27 When a glass bottle full of vinegar warms up, both the vinegar and the glass expand, but vinegar expands significantly more with temperature than glass. The bottle will break if it was filled to its tightly capped lid. Explain why, and also explain how a pocket of air above the vinegar would prevent the break. (This is the function of the air above liquids in glass containers.)

5.28 A coil spring is a torsion device. Explain how compressing or stretching a spring twists the wire or rod making up its coils. How does this design allow us to compress or stretch the spring by a large fraction of its length without producing a torsional strain large enough to fracture it?

PROBLEMS

Sections 5.1–5.5 Statics in Various Situations

5.1 When opening a door, you push on it perpendicularly with a force of 55.0 N at a distance of 0.850 m from the hinges. What torque are you exerting relative to the hinges? (Does it matter if you push at the same height as the hinges?)

5.2 When tightening a bolt, you push perpendicularly on a wrench with a force of 165 N at a distance of 0.140 m from the center of the bolt. (a) How much torque are you exerting in N·m (relative to the center of the bolt)? (b) Convert this torque to ft·lb.

5.3 Two children push on opposite sides of a door during play. Both push horizontally and perpendicular to the door. One child pushes with a force of 175 N at a distance of 0.600 m from the hinges, and the second pushes at a distance of

0.450 m. What force must the second exert to keep the door from moving? Assume friction is negligible.

5.4 Use the second condition for equilibrium (net $\tau = 0$) to calculate F_p in Example 5.1, employing any data given or solved for in part (a) of the example.

•5.5 Repeat the seesaw problem in Example 5.1 with the center of mass of the seesaw 0.160 m to the left of the pivot (on the side of the lighter child) and assuming a mass of 12.0 kg for the seesaw. The other data given in the example remain unchanged.

•5.6 Engines are often rated by their torque output. A certain engine produces a maximum torque of 150 ft·lb on its 3.00 cm diameter crank shaft. (a) What is this torque in N·m? (b) Calculate the maximum force it can produce at the edge of the shaft in pounds and newtons.

•**5.7** Suppose a horse leans against a wall as in Figure 5.35. Calculate the force exerted on the wall assuming that force is horizontal and using the data in the schematic representation of the situation. Note that the force exerted on the wall is equal and opposite to the force exerted on the horse, keeping it in equilibrium. The total mass of the horse and rider is 500 kg. Take the data to be accurate to three digits.

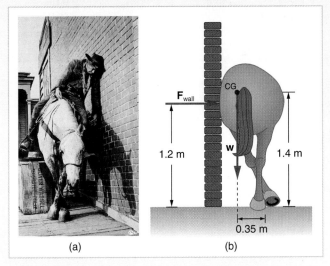

(a)   (b)

Figure 5.35 (a) Lee Marvin and horse leaning against a wall in the movie *Cat Ballou*. (b) Schematic representation showing forces, dimensions, mass, and location of the center of mass. Problems 7 and 8.

•**5.8** (a) Calculate the magnitude and direction of the force on each foot of the horse in Figure 5.35 (two are on the ground), assuming the center of mass of the horse is midway between the feet. The total mass of the horse and rider is 500 kg. (b) What is the minimum coefficient of friction between the hooves and ground? Note that the force exerted by the wall is horizontal.

‡**5.9** A 17.0 m high and 11.0 m long wall under construction and its bracing are shown in Figure 5.36. The wall is in stable equilibrium without the bracing but can pivot at its base. Calculate the force exerted by each of the 10 braces if a strong wind exerts a horizontal force of 650 N on each square meter of the wall. Assume all braces exert equal forces parallel to their lengths and neglect the thickness of the wall.

5.10 Verify the position of the center of mass of the earth-moon system shown in Figure 5.7(a).

5.11 Find the center of mass of the earth-sun system.

5.12 Find the center of mass of the motorcycle plus its riders shown in Figure 5.30, given the following information: the rider has a mass of 65.0 kg and his center of mass is 1.10 m above the wire; the motorcycle's mass is 125 kg and its center of mass is 0.300 m above the wire; and the passenger and swing have a mass of 70.0 kg and their center of mass is 1.75 m below the wire. The CMs are all on the same vertical line.

•**5.13** A 12.5 cm tall soft drink can has a mass of 17.2 g and contains 362 g of soda when full. Assuming the can is symmetric, its center of mass is obviously 6.25 cm above its base when full and when empty. (a) Where is its center of mass when half full? (b) When one-tenth full?

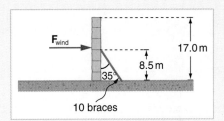

Figure 5.36 Side view of a uniform concrete wall 11.0 m wide by 17.0 m tall. There are ten braces. Problem 9.

•**5.14** (a) What force must be exerted by the wind to support a 2.50 kg chicken in the position shown in Figure 5.37? (b) What is the ratio of this force to the chicken's weight? Does this support the contention that the chicken has a relatively stable construction?

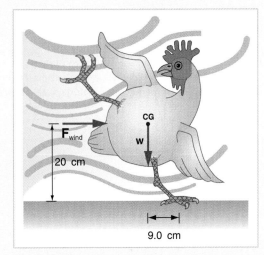

Figure 5.37 Wind does not easily topple the stable chicken. Problem 14.

•**5.15** Solve Example 5.3(b) using the second condition for equilibrium (net $\tau = 0$), and employing any data given or solved for in part (a) of the example.

•**5.16** Given that each hand supports half the weight of the pole in Figure 5.14(a), show that the second condition for equilibrium (net $\tau = 0$) is satisfied for a pivot other than the one located at the center of gravity of the pole.

•**5.17** Suppose the weight of the drawbridge in Figure 5.38 is supported entirely by its hinges and the opposite shore, so that its cables are slack. (a) What fraction of the weight is supported by the opposite shore if the point of support is directly beneath the cable attachments? (b) What is the direction and magnitude of the force the hinges exert on the bridge under these circumstances? The mass of the bridge is 2500 kg.

‡**5.18** Suppose a 900 kg car is on the bridge in Figure 5.38 with its center of mass halfway between the hinges and the cable attachments. (The bridge is supported by the cables and hinges only.) (a) Find the force in the cables. (b) Find the direction and magnitude of the force exerted by the hinges on the bridge.

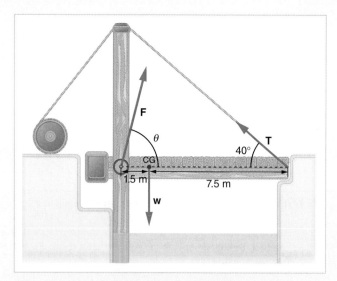

Figure 5.38   A small drawbridge, showing the forces on the hinges (**F**), its weight (**w**), and the tension in its wires (**T**). Problems 17 and 18.

⁞ **5.19**   A sandwich board advertising sign is constructed as shown in Figure 5.39. Its mass is 8.00 kg. (a) Calculate the tension in the chain assuming no friction between the legs and the sidewalk. (b) What force is exerted by each side on the hinge?

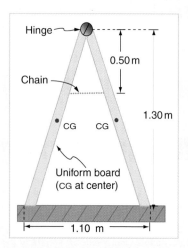

Figure 5.39   An advertising sign. Problems 19 and 20.

⁞ **5.20**   (a) What minimum coefficient of friction is needed between the legs and the ground to keep the sign in Figure 5.39 in the position shown if the chain breaks? (b) What force is exerted by each side on the hinge?

⁞ **5.21**   A 40.0 kg cattle gate is hung on a fence post as shown in Figure 5.40. Calculate the magnitude and direction of the force exerted on the gate by each hinge, assuming these forces to be equal in magnitude. The gate's center of mass is at its geometric center.

Section 5.6* Simple Machines

5.22*   (a) What is the mechanical advantage of a nail puller—similar to the one shown in Figure 5.15—where you exert a force 45.0 cm from the pivot and the nail is 1.80 cm on the other side? (b) What minimum force must you exert to apply a force of 1250 N to the nail?

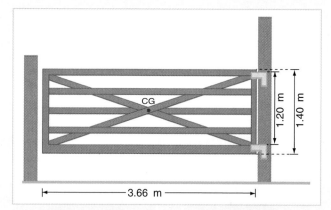

Figure 5.40   A long gate supported by hinges. Problem 21.

5.23*   (a) What is the mechanical advantage of a wheelbarrow, such as the one in Figure 5.16(a), if the center of gravity of the wheelbarrow and its load has a perpendicular lever arm of 5.50 cm, while the hands have a perpendicular lever arm of 1.02 m? (b) What upward force must you exert to support the wheelbarrow and its load if their combined mass is 55.0 kg? (c) What force does the wheel exert on the ground?

5.24*   Consider the shovel shown in Figure 5.16(b). (a) What is its mechanical advantage if the left hand is 65.0 cm from the handle while the center of mass of the shovel and its load is 95.0 cm from the handle? (b) What force must you exert to support the shovel and its load if their combined mass is 11.0 kg?

5.25*   An antique wind-up record player has a crank in which the force is exerted 7.50 cm from the axis and the radius of the internal shaft is 0.480 cm. What is its mechanical advantage?

5.26*   A typical automobile has an axle with a 1.10 cm radius driving a tire with a radius of 27.5 cm. What is its mechanical advantage assuming the very simplified model in Figure 5.17(b)?

• **5.27***   What force does the nail puller in Problem 5.22 exert on the supporting surface? The nail puller has a mass of 2.10 kg.

• **5.28***   If you used an ideal pulley of the type shown in Figure 5.18(a) to support a car engine of mass 115 kg: (a) What would the tension in the rope be? (b) What force must the ceiling supply, assuming you pull straight down on the rope? Neglect the pulley system's mass.

• **5.29***   Repeat Problem 5.28 for the pulley shown in Figure 5.18(c), assuming you pull straight up on the rope. The pulley system's mass is 7.00 kg.

• **5.30***   Suppose you are using an ideal pulley system of the type shown in Figure 5.18(b) and find it necessary to exert a force of 250 N to support a load. (a) What is the load's mass? (b) What force is exerted on the ceiling? Neglect the pulley system's mass.

• **5.31***   (a) What force does each side of the nutcracker shown in Figure 5.32 exert on a nut 3.20 cm from the pivot? The hand exerts a perpendicular force of 50.0 N on each side at a distance of 11.0 cm from the pivot. (b) What is the total force on the nut, and what keeps it from slipping out? (No calculations are necessary for this part.)

• **5.32*** A supertanker uses a windlass (a type of winch) like the one shown in Figure 5.41 to hoist its 19,000 kg anchor. What force must be exerted on the outside wheel to lift the anchor at constant speed, neglecting friction and assuming the anchor is out of the water?

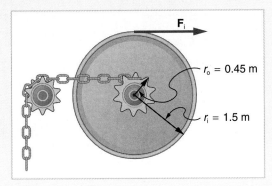

Figure 5.41 Schematic of a ship's windlass showing radii. Problem 32.

Section 5.7* Forces and Torques in Muscles and Joints

5.33* Two muscles in the back of the leg pull on the Achilles tendon as shown in Figure 5.42. What total force do they exert?

5.34* The upper leg muscle (quadriceps) exerts a force of 1250 N, which is carried by a tendon over the kneecap (the patella) at the angles shown in Figure 5.43. Find the direction and magnitude of the force exerted by the kneecap on the upper leg bone (the femur).

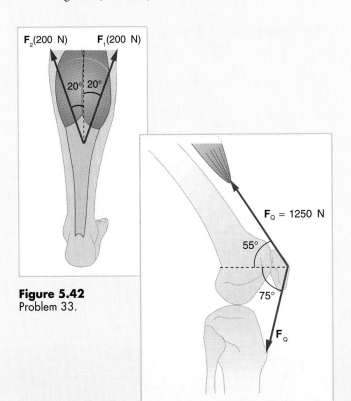

Figure 5.42
Problem 33.

Figure 5.43 Problem 34.

5.35* A device for exercising the upper leg muscle is shown in Figure 5.44, together with a schematic representation of an equivalent lever system. Calculate the force exerted by the upper leg muscle to lift the mass at a constant speed.

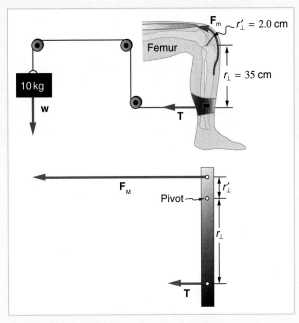

Figure 5.44 A mass is connected by pulleys and wires to the ankle in this exercise device. Problem 35.

• **5.36*** A person working at a drafting board may hold her head as shown in Figure 5.45, requiring muscle action to support the head. The three major forces acting are shown. Calculate the direction and magnitude of the force supplied by the upper vertebrae F_V to hold the head stationary, assuming that it acts along a line through the center of mass as do the weight and muscle force. Is the direction of F_V likely to produce minimal strain in the upper vertebrae?

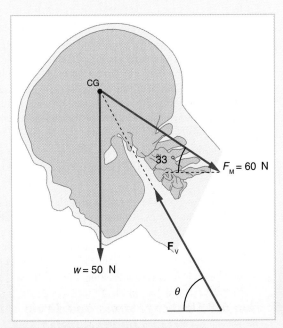

Figure 5.45 Problem 36.

• **5.37*** Even when the head is held erect, as in Figure 5.46, its center of mass is not directly over the principal point of support (the atlanto-occipital joint). The muscles in the back of the neck must therefore exert a force to keep it erect. That is why your head falls forward when you fall asleep in class. (a) Calculate the force exerted by these muscles using the information in the figure. (b) What is the force on the pivot?

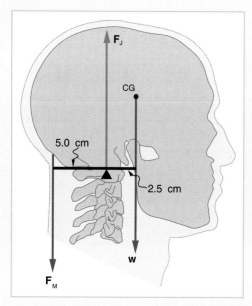

Figure 5.46 The center of mass of the head lies in front of its major point of support, requiring muscle action to hold the head erect. A simplified lever system is shown. Problem 37.

• **5.38*** Verify that the force in the elbow joint in Example 5.5 is 407 N, as stated in the text.

• **5.39*** A 75.0 kg man stands on his toes by exerting an upward force through the Achilles tendon, as in Figure 5.47. (a) What is the force in the Achilles tendon if he stands on one foot? (b) Calculate the force at the pivot of the simplified lever system shown—that force is representative of forces in the ankle joint.

‡ **5.40*** (a) What force must the woman in Figure 5.48 exert on the floor with each hand to do a push-up? Assume that she moves up at a constant speed. (b) The triceps muscle in the back of her upper arm has an effective lever arm of 1.75 cm, and she exerts force on the floor a horizontal distance of 20.0 cm from the elbow joint. Calculate the force in the triceps muscle, and take its ratio to her weight.

‡ **5.41*** A father lifts his child as shown in Figure 5.49. What force must the upper leg muscle exert to lift the child at a constant speed?

‡ **5.42*** Unlike most of the other muscles in our bodies, the masseter muscle in the jaw, as illustrated in Figure 5.50, is attached relatively far from the joint, enabling very large forces to be exerted by the back teeth. (a) Using the information in the figure, calculate the force exerted by the teeth on the bullet. (b) Calculate the force on the joint.

Section 5.8 Elasticity: Stress and Strain

5.43 The "lead" in pencils is a graphite composition with a Young's modulus of about 1×10^9 N/m^2. Calculate the

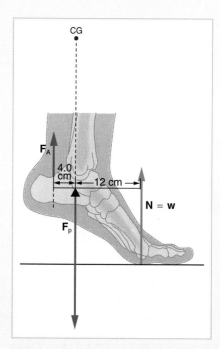

Figure 5.47 The muscles in the back of the leg pull on the Achilles tendon when a person stands on his toes. A simplified lever system is shown. Problem 39.

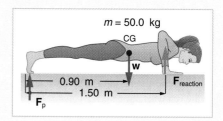

Figure 5.48 Problem 40.

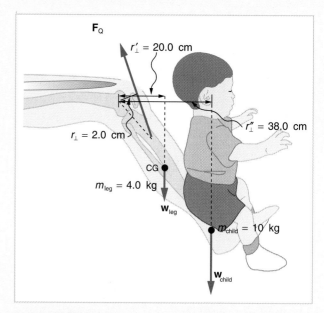

Figure 5.49 Problem 41.

change in length of the lead in an automatic pencil if you tap it straight into the pencil with a force of 4.0 N. The lead is 0.50 mm in diameter and 60 mm long.

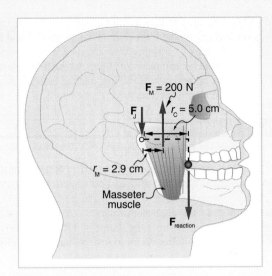

Figure 5.50 Problem 42.

5.44 TV broadcast antennas are the tallest artificial structures on earth. In 1987 a 72.0 kg physicist placed himself and 400 kg of equipment at the top of one 610 m high antenna to perform gravity experiments. How much was the antenna compressed, if we consider it to be equivalent to a steel cylinder 0.150 m in radius?

5.45 How much does a 65.0 kg mountain climber stretch her 0.800 cm diameter nylon rope when she hangs 35.0 m below a rock outcropping?

5.46 A 20.0 m tall hollow aluminum flagpole is equivalent in strength to a solid cylinder 4.00 cm in diameter. A strong wind bends the pole much as a horizontal force of 900 N exerted at the top would. How far to the side does the top of the pole flex?

5.47 As an oil well is drilled, each new section of drill pipe supports its own weight and that of the pipe and drill bit beneath it. Calculate the stretch in a new 6.00 m length of steel pipe that supports 3.00 km of pipe having a mass of 20.0 kg/m and a 100 kg drill bit. The pipe is equivalent in strength to a solid cylinder 5.00 cm in diameter.

5.48 Calculate the force a piano tuner applies to stretch a steel piano wire 8.00 mm, if the wire is originally 0.850 mm in diameter and 1.35 m long.

• 5.49 When using a pencil eraser, you exert a vertical force of 6.00 N at a distance of 2.00 cm from the hardwood-eraser joint. The pencil is 6.00 mm in diameter and is held at an angle of 20.0° to the horizontal. (a) How much does the wood flex perpendicular to its length? (b) How much is it compressed lengthwise?

• 5.50 To consider the effect of wires hung on poles we take data from Example 4.10, in which tensions in wires supporting a traffic light were calculated. The left wire made an angle 30.0° below the horizontal with the top of its pole and carried a tension of 108 N. The 12.0 m tall hollow aluminum pole is equivalent in strength to a 4.50 cm diameter solid cylinder. (a) How far is it bent to the side? (b) By how much is it compressed?

• 5.51 A moonshiner makes the error of filling a glass jar to the brim and capping it tightly. The moonshine expands more than the glass when it warms up, in such a way that the vol-

ume increases by 0.2% (that is, $\Delta V/V_0 = 2 \times 10^{-3}$) relative to the space available. Calculate the force exerted by the moonshine per square centimeter if its bulk modulus is 1.8×10^9 N/m², assuming the jar does not break. In view of your answer, do you think the jar survives?

• 5.52 When water freezes, its volume increases by 9.05% (that is, $\Delta V/V_0 = 9.05 \times 10^{-2}$). What force per unit area is water capable of exerting on a container when it freezes? (It is acceptable to use the bulk modulus of water in this problem.) Is it surprising that such forces can fracture engine blocks, boulders, and the like?

• 5.53 This problem returns to the tightrope walker studied in Example 4.7, who created a tension of 3.94×10^3 N in a wire making an angle 5° below the horizontal with each supporting pole. Calculate how much this tension stretches the steel wire if it was originally 15 m long and 0.50 cm in diameter.

⁞ 5.54 The pole in Figure 5.51 is at a 90° bend in a power line and is therefore subjected to more shear force than ones in straight parts of the line. The tension in each line is 4.00×10^4 N, at the angles shown. The pole is 15.0 m tall, has an 18.0 cm diameter, and can be considered to have half the strength of hardwood. (a) Calculate the compression of the pole. (b) Find how much it bends and in what direction. (c) Find the tension in a guy wire used to keep the pole straight if it is attached to the top of the pole at an angle of 30.0° with the vertical. (Clearly, the guy wire must be in the opposite direction of the bend.)

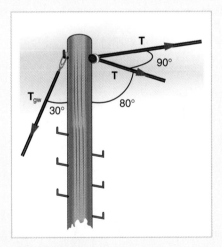

Figure 5.51 Problem 54.

INTEGRATED CONCEPTS

The following problems involve material from more than one chapter. Integrated knowledge from a broad range of topics is much more powerful than a narrow application of physics. Here you will need to refer to material in Chapters 2, 3, and 4 as well as in Chapter 5.

PROBLEM-SOLVING STRATEGY

Use the following strategy to solve integrated concept problems.

Step 1. *Identify which physical principles are involved.* This is done by following the problem-solving strategies as given in Section 2.5

and in other places. Listing the givens and the quantities to be calculated will allow you to identify the principles involved.

Step 2. *Solve the problem using strategies outlined in the text.* If these are available for the specific topic, you should refer to them. You should also refer to the sections of the text that deal with a particular topic.

Note: A worked example that can help you solve Integrated Concept problems appears in Chapter 4. Consult the section of problems labeled *Integrated Concepts* at the end of that chapter.

: 5.55 A fully loaded elevator has a mass of 2500 kg and is accelerated upward at 0.750 m/s^2 by a 2.20 cm diameter steel cable. Calculate the stretch in the 35.0 m length of cable between the elevator and the reel pulling it up.

: 5.56 The force applied to a trailer hitch is exerted on a metal sphere atop a vertical cylinder that consequently experiences shear stresses. Calculate the shear deformation Δx in the 2.00 cm long, 0.950 cm diameter steel cylinder, if the 850 kg trailer it pulls is accelerated up a 2.00° incline at 1.10 m/s^2.

: 5.57 A baseball with a mass of 0.140 kg is hit by a hardwood bat that can be reasonably approximated by a 5.00 cm diameter cylinder. Calculate the shear deformation Δx of the bat given the ball is hit 40.0 cm from the hands, has an initial velocity of 60.0 km/h, a final velocity of 80.0 km/h in the opposite direction, and the collision lasts 0.500 ms.

: 5.58 A 2.00 cm diameter, 5.00 cm long brass cylinder is dropped on its end from a height of 1.50 m above a hard surface. Assuming that the cylinder rebounds with the same speed as it hit the floor and given that the collision lasts 0.200 ms, calculate the average deformation of the 133 g cylinder.

: 5.59 Calculate the speed of a bullet leaving a rifle (its muzzle velocity) given the following information. The lead bullet has a length of 1.40 cm, a diameter of 0.500 cm, a mass of 3.11 g, and is compressed 38.0 μm during its acceleration. The rifle barrel is 0.650 m long, and the bullet is assumed to have the same acceleration for the length of the barrel.

UNREASONABLE RESULTS

The following problems have results that are unreasonable because some premise is unreasonable or because certain of the premises are inconsistent with one another. Physical principles applied correctly then produce unreasonable results. The purpose of these problems is to give practice in assessing whether nature is being accurately described, and if it is not, to trace the source of difficulty. This is very much like the process followed in original research when physical principles as well as faulty premises are tested.

Note: Strategies for determining if an answer is reasonable appear in Chapter 2. They can be found with the section of problems labeled *Unreasonable Results* at the end of that chapter.

• 5.60 (a) Calculate the force needed to stretch a 1.50 mm diameter nylon guitar string 3.00 cm when tuning if the string is originally 0.800 m long. (b) What is unreasonable about the result? (c) Which premise is unreasonable, or which premises are inconsistent?

• 5.61 Suppose two children are using a uniform seesaw that is 3.00 m long and has its center of mass over the pivot. The first child has a mass of 30.0 kg and sits 1.40 m from the pivot. (a) Calculate where the second 18.0 kg child must sit to balance the seesaw. (b) What is unreasonable about the result? (c) Which premise is unreasonable, or which premises are inconsistent?

WORK, ENERGY, AND POWER

How many forms of energy can you identify in this photograph?

Energy is one of the great catchwords of modern times. You can no doubt name many forms of energy, from that we consume in foods, to the energy we use to run our cars, to the sunlight that warms us on the beach. You can also cite examples of what people call energy that may not be scientific, such as someone having an energetic personality. But in fact, many of the common uses of the word energy *are* scientifically meaningful. Energy is one of the most important concepts in physics. Not only does energy have many interesting forms, it is involved in almost all phenomena. What makes it even more important is that the total amount of energy is constant. It may move about and change forms, but it cannot appear from nothing or disappear without a trace. Energy is one of a handful of physical quantities that is *conserved*. True without exception, conservation of energy (as physicists like to call this fact) is based solely on experiment. Even when we learn about new forms of energy, conservation of energy is always found to be true. Perhaps the most dramatic example of this was supplied by Einstein when he suggested that mass is interchangeable with energy (his famous equation $E = mc^2$) and, in this sense, is a form of energy.

There is no simple, yet accurate, definition for energy. Energy is characterized by its many forms and the fact that it is conserved. We can loosely define **energy** as the ability to do work, admitting that in some circumstances not all energy is available to do work. Because of the association of energy with work, we begin the chapter with a discussion of work. Work is intimately related to energy and how energy may move from one system to another or change form.

6.1 WORK: THE SCIENTIFIC DEFINITION

The scientific definition of work differs in some ways from its everyday meaning. Certain things we think of as hard work, such as writing a term paper or carrying a heavy load on level ground, are not work as defined by the scientist. The scientific definition of work reveals its relationship to energy—whenever work is done, energy is transferred or changes form.

For work, in the scientific sense, to be done, a force must be exerted and there must be motion or displacement in the direction of the force. Formally, the **work** done on a system by a constant force is defined to be *the product of the component of the force in the direction of motion times the displacement*. In equation form, this is

$$W = Fd \cos \theta \tag{6.1}$$

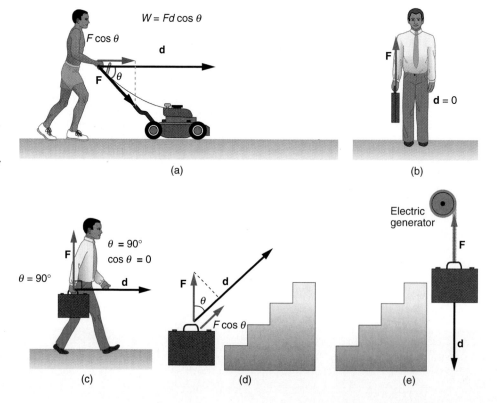

Figure 6.1 (a) The work done by the force **F** on this lawn mower is $Fd \cos \theta$. Note that $F \cos \theta$ is the component of the force in the direction of motion. (b) A person holding a briefcase does no work on it, since there is no motion. No energy is transferred to or from the briefcase. (c) No work is done on the briefcase, and no energy is transferred to it, when it is carried horizontally at a constant speed. (d) Work is done when a person carries the briefcase upstairs, since there is motion in the direction of the force. Energy is transferred to the briefcase and could in turn be used to do work. (e) Energy is transferred out of the briefcase and into an electric generator. Here the work done on the briefcase by the generator is negative, removing energy from the briefcase, since **F** and **d** are in opposite directions.

where W is work, d is the displacement or distance moved, and θ is the angle between the force vector **F** and the displacement vector **d**, such as in Figure 6.1(a).

To examine what the definition of work means, let us consider the other situations shown in Figure 6.1. The person holding the briefcase in Figure 6.1(b) does no work, for example. Here $d = 0$, and so $W = 0$. Why is it you get tired just holding a load? The answer is that your muscles are doing work against one another, *but they are doing no work on the system of interest* (the briefcase). There must be motion for work to be done, and there must be a component of the force in the direction of the motion. For example, the person carrying the briefcase on level ground in Figure 6.1(c) does no work on it, since the force is perpendicular to the motion. That is, cos 90° = 0, and so $W = 0$. In contrast, when a force exerted on the system has a component in the direction of motion, such as in Figure 6.1(d), work *is* done—energy is transferred to the briefcase. Finally, in Figure 6.1(e), energy is taken from the briefcase to run a generator. There are two good ways to interpret this energy transfer. One interpretation is that the briefcase does work on the generator, giving it energy. The other interpretation is that the generator does negative work on the briefcase, thus removing energy from it. The drawing shows the latter, with the force from the generator upward on the briefcase, and the displacement downward. This makes $\theta = 180°$, and cos 180° = −1; therefore, W is negative.

Work and energy have the same units. From the definition of work, we see that those units are force times distance. Thus, in SI units, work and energy are measured in **newton-meters**. A newton-meter is given the special name **joule** (J), and $1 \text{ J} = 1 \text{ N·m} = 1 \text{ kg·m}^2/\text{s}^2$.* One joule is not a large amount of energy. One *calorie* of heat (the amount required to warm 1 g of water by 1°C) is equivalent to 4.186 J, while one *food calorie* (1 kcal) is equivalent to 4186 J.

EXAMPLE 6.1   HOW MUCH WORK DO YOU DO TO PUSH A LAWN MOWER ACROSS A LARGE LAWN?

How much work is done on the lawn mower by the person in Figure 6.1(a) if he exerts a force of 75.0 N at an angle 35.0° below the horizontal and pushes the mower 25.0 m on level ground? Convert the amount of work from joules to kcal and compare it with this person's average daily intake of 2500 kcal of food energy.

Strategy We can solve this problem by substituting the given values into the definition of work done on a system, stated in Equation 6.1. The force, angle, and displacement are given, so that only the work W is unknown.

Solution Equation 6.1 is

$$W = Fd \cos \theta$$

Substituting the known values gives

$$W = (75.0 \text{ N})(25.0 \text{ m}) \cos 35.0°$$
$$= 1536 \text{ J} = 1.54 \times 10^3 \text{ J}$$

Converting the work in joules to kcal yields $W = 1536$ J × (1 kcal/4186 J) = 0.367 kcal. The ratio of the work done to the daily consumption is thus

$$\frac{W}{2500 \text{ kcal}} = 1.47 \times 10^{-4}$$

Discussion This is a tiny fraction of what the person consumes, but it is typical. Very little of the food energy we consume is used to do work. Even when we work all day long, less than 10% of our food energy intake is used to do work and more than 90% is converted to heat or stored in the form of fat.

6.2   KINETIC ENERGY AND THE WORK-ENERGY THEOREM

What happens to the work done on a system? Energy is transferred into the system, but in what form? Does it remain in the system or move on? The answers depend on the situation. For example, if the lawn mower in Figure 6.1(a) is pushed just hard enough

---

*Note that torque, defined in Section 5.2, also has units of N·m, but torque is definitely neither energy nor work. The reason is that for all torques, we deal with the distance of the force from a pivot, rather than the distance the force moves.

Figure 6.2 (a) A graph of $F \cos \theta$ vs. d. The area under the curve equals the work done by the force. (b) A graph of $F \cos \theta$ vs. d in which the force varies. The work done for each interval is the area of each strip; thus, the total area under the curve equals the total work done.

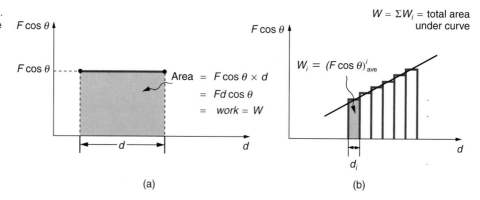

(a)

(b)

to keep it going at a constant speed, then energy put into the mower by the person is removed continuously by friction and eventually leaves the system in the form of heat. In contrast, work done on the briefcase by the person carrying it up stairs in Figure 6.1(d) is stored in the briefcase and can be recovered at any time, such as shown in Figure 6.1(e). In fact, the energy imparted in lifting stone blocks during construction of the pyramids remains in them and has the potential to do work.

In this section we begin the study of various types of work and forms of energy. We will find that some types of work leave the energy of a system constant, for example, whereas others change the system in some way, such as making it move. We will also develop definitions of important forms of energy, such as the energy of motion.

Let us start by considering the total, or net, work done on a system. Net work is defined to be the sum of work done by all external forces—that is, **net work** is the work done by the net external force, net F.* In equation form, this is net $W = (\text{net } F)d \cos \theta$. Figure 6.2(a) shows a graph of force versus displacement for the component of the force in the direction of the displacement—that is, an $F \cos \theta$ vs. d graph. In this case, $F \cos \theta$ is a constant. You can see that the area under the graph is $Fd \cos \theta$, or the work done. Figure 6.2(b) shows a more general process where the force varies. The area under the curve is divided into strips, each having an average force $(F \cos \theta)_{i(\text{ave})}$. The work done is $W_i = (F \cos \theta)_{i(\text{ave})}d_i$ for each strip, and the total work done is the sum of the W_i. Thus the total work done is the total area under the curve, a useful property to which we shall later refer.

This will be simpler to examine if we consider a one-dimensional situation where a force is used to accelerate an object in a direction parallel to its initial velocity. In such a situation, such as the package on the roller-type conveyor system shown in Figure 6.3, $\theta = 0$ and $\cos \theta = 1$. The net work (net W) is thus

$$\text{net } W = (\text{net } F)d$$

By using Newton's second law, and doing some algebra, we can take this to an interesting conclusion. Substituting net $F = ma$ from Newton's second law gives

$$\text{net } W = mad$$

To get some relationship between net work and the velocity given to a system by the net force acting on it, we take $d = x - x_0$ and use Equation 2.11, $v^2 = v_0^2 + 2ad$ (note that a appears

---

*We know from the study of Newton's laws in Chapter 4 that net force causes an acceleration. The work done by the net force will be seen to give a system energy of motion and in the process we will also find an expression for the energy of motion.

Figure 6.3 A package on a roller belt is pushed horizontally through a distance d. The package is accelerated from v_0 to v by the net force, net F. The kinetic energy of the package is increased, indicating the net work done on the system is positive. See Example 6.2.

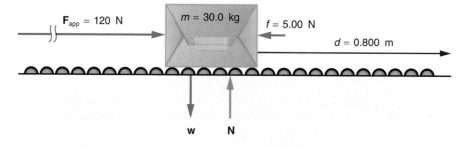

in the expression for the net work). Solving for acceleration gives $a = (v^2 - v_0^2)/2d$. When a is substituted into the preceding expression for net W, we obtain

$$\text{net } W = m \cdot \frac{v^2 - v_0^2}{2d} \cdot d$$

The d cancels, and we rearrange this to

$$\text{net } W = \tfrac{1}{2}mv^2 - \tfrac{1}{2}mv_0^2 \qquad \textbf{(6.2)}$$

This expression is called the **work-energy theorem**, and it is actually *true in general* (even for forces that vary in direction and magnitude), although we have derived it for the special case of a constant force parallel to the displacement. The theorem implies that the net work on a system equals the change in the quantity $(1/2)mv^2$. Just what is this quantity? It is our first example of a form of energy.

The quantity $(1/2)mv^2$ in the work-energy theorem is defined to be the **transla-tional* kinetic energy** (KE) of a mass m moving at a velocity v. That is,

$$\text{KE} = \tfrac{1}{2}mv^2 \qquad \textbf{(6.3)}$$

is the energy associated with motion. Kinetic energy is a highly visible form of energy associated with the motion of a particle, single body, or system of objects moving together. (The kinetic energy of atoms and molecules inside objects is hidden and is called thermal energy—it will be discussed later on.) We are intuitively aware that it takes energy to get an object, like a car or the package in Figure 6.3, up to speed; but it may be a bit surprising that kinetic energy is proportional to speed squared. This means, for example, that a car traveling at 100 km/h has four times the kinetic energy it has at 50.0 km/h, helping explain why high-speed collisions are so devastating. We will now consider a series of examples to illustrate various aspects of work and energy.

EXAMPLE 6.2 WHAT IS THE KINETIC ENERGY OF A PACKAGE?

Suppose a 30.0 kg package on the roller-type conveyor system in Figure 6.3 is moving at 0.500 m/s. What is its kinetic energy?

Strategy Since the mass m and velocity v are given, the kinetic energy can be calculated from its definition as given in Equation 6.3.

Solution Equation 6.3 states that kinetic energy is

$$\text{KE} = \tfrac{1}{2}mv^2$$

Entering known values gives

$$\text{KE} = 0.5(30.0 \text{ kg})(0.500 \text{ m/s})^2$$

which yields

$$\text{KE} = 3.75 \text{ kg} \cdot \text{m}^2/\text{s}^2 = 3.75 \text{ J}$$

Discussion Note that the units of kinetic energy are joules, the same as the units of work, as mentioned when work was first defined. It is also interesting that although this is a fairly massive package, its kinetic energy is not large at this relatively low velocity. This is consistent with the fact that people can move packages like this without exhausting themselves.

EXAMPLE 6.3 HOW MUCH WORK IS DONE TO ACCELERATE A PACKAGE?

Suppose that you push on the 30.0 kg package in Figure 6.3 with a force of 120 N through a distance of 0.800 m, and that opposing friction averages 5.00 N. Calculate the net work done on the package.

Strategy and Concept This is a one-dimensional problem, since the applied force, friction, and the displacement are all horizontal. (See Figure 6.3.) The net work is

thus the net force times distance.

Solution The net force is the push minus friction, or net $F = 120 \text{ N} - 5.00 \text{ N} = 115 \text{ N}$. Thus the net work is

$$\begin{aligned}\text{net } W = \text{net } Fd &= (115 \text{ N})(0.800 \text{ m}) \\ &= 92.0 \text{ N} \cdot \text{m} = 92.0 \text{ J}\end{aligned}$$

(continued)

---

*Translational kinetic energy is distinct from *rotational* kinetic energy, which is considered separately in Chapter 9.

(continued)

Discussion This is the net work done on the package. The person actually does more work than this, because friction opposes the motion. Friction does negative work and removes some of the energy the person expends and converts it to heat. The net work equals the sum of the work done by each individual force.

EXAMPLE 6.4   WORK AND ENERGY CAN REVEAL SPEED

Find the velocity of the package in Figure 6.3 just after the push, using work and energy concepts.

Strategy Here the work-energy theorem can be used, because we have just calculated the net work, net W, and the initial kinetic energy, $(1/2)mv_0^2$. This allows us to find the final kinetic energy $(1/2)mv^2$ and, thus, the final velocity.

Solution The work-energy theorem is Equation 6.2:

$$\text{net } W = \tfrac{1}{2}mv^2 - \tfrac{1}{2}mv_0^2$$

Solving for $(1/2)mv^2$ gives

$$\tfrac{1}{2}mv^2 = \text{net } W + \tfrac{1}{2}mv_0^2$$

Thus,

$$\tfrac{1}{2}mv^2 = 92.0 \text{ J} + 3.75 \text{ J} = 95.75 \text{ J}$$

Solving for the final velocity as requested and entering known values gives

$$v = \left[\frac{2(95.75 \text{ J})}{m}\right]^{1/2} = \left[\frac{191.5 \text{ kg}\cdot\text{m}^2/\text{s}^2}{30.0 \text{ kg}}\right]^{1/2}$$
$$= 2.53 \text{ m/s}$$

Discussion Using work and energy, we not only arrive at an answer, we see that the final kinetic energy is the sum of its initial kinetic energy and the net work done on the package. This means that the work indeed adds to the energy of the package.

EXAMPLE 6.5   WORK AND ENERGY CAN REVEAL DISTANCE, TOO

How far does the package in Figure 6.3 coast after the push, assuming friction remains constant? Use work and energy considerations.

Strategy We know that once the person stops pushing, friction brings the package to rest. In terms of energy, what happens is that friction does negative work until it has removed all of the package's kinetic energy. The work done by friction is the force of friction times the distance traveled; hence, this gives us a way of finding the distance.

Solution The amount of work done by friction, or W_f, must equal the kinetic energy just after the push. Thus $W_f = 95.75$ J. Furthermore, $W_f = fd'$, where d' is the

distance it takes to stop. Thus,

$$d' = \frac{W_f}{f} = \frac{95.75 \text{ J}}{5.00 \text{ N}}$$

and so

$$d' = 19.2 \text{ m}$$

Discussion This is a reasonable distance for a package to coast on this relatively friction-free conveyor system. Note that the work done by friction is negative (the force is in the opposite direction of motion) and, hence, removes the kinetic energy.

All of the examples in this section can be solved without considering energy—but also without gaining insight about what work and energy are doing in this situation. Moreover, solutions involving energy are generally shorter and easier than those using kinematics and dynamics alone.

6.3   GRAVITATIONAL POTENTIAL ENERGY

Climbing stairs and lifting objects is work in both the scientific and everyday sense—it is work done against gravity. When work is performed, there is a transformation of energy. The work done against gravity goes into an important form of stored energy that we will explore in this section.

Let us calculate the work done in lifting a mass m a height h, such as in Figure 6.4(a). If the mass is lifted straight up at constant speed, then the force needed to lift it is equal to its weight mg. The work done on the mass is then $W = Fd = mgh$. We define this to be the

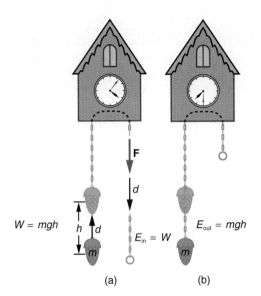

Figure 6.4 (a) The work done to lift the mass is stored in it as gravitational potential energy. (b) As the mass moves downward, gravitational potential energy is removed from it to run the cuckoo clock.

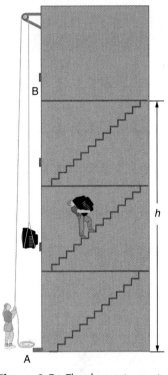

Figure 6.5 The change in gravitational potential energy (ΔPE_g) between points A and B is independent of path. $\Delta PE_g = mgh$ for any path between the two points. Gravity is one of a small class of forces where the work done by or against the force depends only on the starting and ending points, not on the path between them.

gravitational potential energy (PE_g) put into the mass. This form of energy has the potential to be converted to other forms of energy, such as kinetic energy. If we release the mass, gravity will do an amount of work mgh on it, thereby increasing its kinetic energy by that same amount (by the work-energy theorem). We will find it more useful to consider just the conversion of PE to KE without explicitly considering the intermediate step of work.

More precisely, we define the *change* in gravitational potential energy ΔPE_g to be

$$\Delta PE_g = mgh \qquad (6.4)$$

where h is the change in vertical position.* For example, if a 0.500 kg mass hung from a cuckoo clock is raised 1.00 m, then its increase in gravitational potential energy is $mgh = (0.500 \text{ kg})(9.80 \text{ m/s}^2)(1.00 \text{ m}) = 4.90 \text{ kg} \cdot \text{m}^2/\text{s}^2 = 4.90 \text{ J}$. (Note that the units of gravitational potential energy turn out to be joules, the same as for work and other forms of energy.) As the clock runs, the mass is lowered and we think of it as gradually giving up its 4.90 J of gravitational potential energy *without directly considering the force of gravity doing work.*

Equation 6.4 is true for any path that has a change in height of h, not just one where the mass is lifted straight up. (See Figure 6.5.) It is much easier to calculate mgh (a simple multiplication) than it is to calculate the work done along a complicated path. The idea of gravitational potential energy has the double advantage that it is very broadly applicable and it makes calculations easier. From now on, we will consider that any change in vertical position h of a mass m is accompanied by a change in gravitational potential energy mgh, and we will avoid the equivalent but more difficult task of calculating work done by or against gravity.

EXAMPLE 6.6   ENERGY CONSIDERATIONS MAKE A COMPLEX PATH EASY TO CALCULATE

(a) What is the final speed of the roller coaster in Figure 6.6 if it starts from rest and friction is negligible? (b) What is its final speed (again assuming negligible friction) if its initial speed is 5.00 m/s?

Strategy The roller coaster loses potential energy as it goes downhill, all of which is converted to kinetic energy, assuming friction is negligible. That is, the loss of gravitational potential energy equals the gain in kinetic energy.

Writing this in equation form, $\Delta PE_g = \Delta KE$. Using the equations for PE_g and KE, we can solve for the final velocity v, which is the desired quantity.

Solution for (a) Here the initial kinetic energy is zero, so that $\Delta KE = (1/2)mv_f^2$. Equation 6.4 states that $\Delta PE_g = mgh$. Thus,

$$\Delta PE_g = \Delta KE$$

(continued)

---

*Ordinarily, we would call the change in vertical position Δh; but for simplicity in notation we take our initial position to be zero, so that $\Delta h = h$.

(continued)

becomes

$$mgh = \tfrac{1}{2}mv_f^2$$

Solving for v_f, we find that mass cancels and that

$$v_f = (2gh)^{1/2}$$

Substituting known values,

$$v_f = [2(9.80 \text{ m/s}^2)(20.0 \text{ m})]^{1/2}$$
$$= 19.8 \text{ m/s}$$

Solution for (b) Again $\Delta PE_g = \Delta KE$, but in this case there is initial kinetic energy; thus,

$$mgh = \tfrac{1}{2}mv_f^2 - \tfrac{1}{2}mv_0^2$$

Rearranging gives

$$\tfrac{1}{2}mv_f^2 = mgh + \tfrac{1}{2}mv_0^2$$

This means that the final kinetic energy is the sum of the initial kinetic energy and the gravitational potential energy. Mass again cancels, and

$$v_f = (2gh + v_0^2)^{1/2}$$

This equation is very similar to the kinematics equation

$v = (v_0^2 + 2ad)^{1/2}$, but it is more general—the kinematics equation is only valid for constant acceleration, and our equation above is valid for any path and need not have a constant acceleration. Now, substituting known values,

$$v_f = [2(9.80 \text{ m/s}^2)(20.0 \text{ m}) + 25.0 \text{ m}^2/\text{s}^2]^{1/2}$$
$$= 20.4 \text{ m/s}$$

Discussion and Implications First, note that mass cancels when friction is negligible. This is quite consistent with observations made in earlier chapters that all objects fall at the same rate if friction is negligible. Second, only the speed of the roller coaster is considered; there is no information about its direction at any point. This reveals another general truth. When friction is negligible, the speed of a falling body depends only on its initial speed and altitude, and not on its mass or the path taken. For example, the roller coaster will have the same final speed whether it falls 20.0 m straight down or takes a complicated path like the one in the figure. Third, and perhaps unexpectedly, the final speed in part (b) is greater than in part (a), but by far less than 5.00 m/s. Finally, note that speed can be found at *any* altitude along the way by simply using the appropriate value of h at the point of interest.

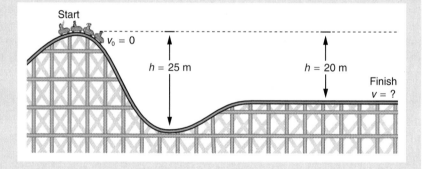

Figure 6.6 The speed of a roller coaster increases as gravity pulls it downhill and is greatest at its lowest point. Viewed in terms of energy, the roller coaster's gravitational potential energy is converted to kinetic energy. If friction is negligible, all ΔPE_g is converted to KE.

We have seen that work done by or against gravity depends only on the starting and ending points, and not on the path between, allowing us to define the simplifying concept of gravitational potential energy. We can do the same thing for a few other forces, and we will see that this leads to a formal definition of the conservation of energy law.

6.4 CONSERVATIVE FORCES AND POTENTIAL ENERGY

Work is done by a force; and some forces, such as gravity, have special characteristics. A **conservative force** is one, like gravity, for which work done by or against it depends only on the starting and ending points of a motion and not on the path taken. We can define a potential energy (PE) for any conservative force, just as we did for gravity. For example, when you wind a watch, you do work against its spring and store energy in it. This energy is fully recoverable as work, and it is useful to think of it as potential energy contained in the spring. Indeed, the reason that the spring has this characteristic is that its force is *conservative*. That is, a conservative force results in stored or potential energy. Gravitational potential energy is an example, as is the energy stored in a spring. We will also see how conservative forces are related to the conservation of energy.

First, we will obtain an expression for the potential energy stored in a spring (PE_s). We calculate the work done to compress or stretch a spring that obeys Hooke's law (Equation 5.11). (See Figure 6.7.) Here a spring is stretched a distance x along its length. The force needed to stretch the spring is $F = kx$, where k is the spring's force constant. The force increases linearly from 0 at the start to kx in the fully stretched position. The average force is $kx/2$. Thus the work done against the spring's force is $W_s = Fd = (kx/2) \cdot x = (1/2)kx^2$. Alternatively, we noted in Section 6.2 that the area under a graph of F vs. x is the work done by the force. In Figure 6.7(c) we see this area is also $(1/2)kx^2$. We therefore define the **potential energy PE_s of a spring** to be

$$PE_s = \tfrac{1}{2}kx^2 \tag{6.5}$$

where k is the spring's force constant and x is the displacement from its undeformed position. The potential energy represents the work done *on* the spring and the energy stored in it as a result of stretching it a distance x. PE_s does not depend on the path taken; it depends only on the stretch x in the final configuration.

Equation 6.5 has general validity beyond the special case for which it was derived. Potential energy can be stored in any elastic medium by deforming it. Indeed, the general definition of **potential energy** is energy due to position, shape, or configuration. For shape or position deformations, stored energy is $PE_s = (1/2)kx^2$, where k is the force constant of the particular system and x is its deformation. Another example is seen in Figure 6.8.

Let us now consider what form the work-energy theorem takes when only conservative forces are involved. This will lead us to the conservation of energy principle. The work-energy theorem states that the net work done by all forces acting on a system equals its change in kinetic energy. In equation form, this is

$$\text{net } W = \tfrac{1}{2}mv^2 - \tfrac{1}{2}mv_0^2 = \Delta KE$$

If only conservative forces act, then

$$\text{net } W = W_c$$

where W_c is the total work done by all conservative forces. Thus,

$$W_c = \Delta KE$$

Now, if a conservative force, such as gravity or a spring force, does work, the system loses potential energy. That is, $W_c = -\Delta PE$. Therefore,

$$-\Delta PE = \Delta KE$$

or

$$\Delta KE + \Delta PE = 0$$

This means that the total kinetic and potential energy is constant for any process involving only conservative forces. That is,

$$\left.\begin{array}{r} KE + PE = \text{constant} \\[1em] KE_i + PE_i = KE_f + PE_f \end{array}\right\} \quad \text{(conservative forces only)} \tag{6.6}$$

where i and f denote initial and final values. This is a form of the work-energy theorem for conservative forces; it is known as the **conservation of mechanical energy** principle. The total kinetic plus potential energy of a system is defined to be its **mechanical energy** (KE + PE). A system that experiences only conservative forces is called a **closed system** because energy never enters or leaves it, but only changes form among KE and the various types of PE, with the total energy constant.

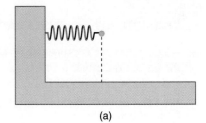

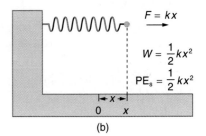

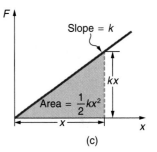

(c)

Figure 6.7 (a) An undeformed spring has no PE_s stored in it. (b) The force needed to stretch the spring a distance x is $F = kx$, and the work done to stretch it is $(1/2)kx^2$. Since the force is conservative, this work is stored as potential energy (PE_s) in the spring. It can be fully recovered. (c) A graph of F vs. x has a slope of k, and the area under the graph is $(1/2)kx^2$. Thus the work done or potential energy stored is $(1/2)kx^2$.

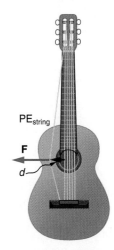

Figure 6.8 Work is done to deform the guitar string, giving it potential energy. When released, the potential energy is converted to kinetic energy and back to potential as the string oscillates back and forth. A very small fraction is dissipated as sound energy, slowly removing energy in a nonconservative process.

EXAMPLE 6.7   CONSERVATION OF MECHANICAL ENERGY USED TO CALCULATE SPEED

A 100 g toy car is propelled by a compressed spring, as shown in Figure 6.9. It follows a track that rises 0.180 m above the starting point. The spring is compressed 4.00 cm and has a force constant of 250 N/m. Assuming negligible friction, find (a) how fast the car is going before it starts up the slope and (b) how fast it is going at the top of the slope.

Strategy Spring force and gravity are conservative forces, and so conservation of mechanical energy can be used. Thus,

$$KE_i + PE_i = KE_f + PE_f$$

or

$$\tfrac{1}{2}mv_i^2 + mgh_i + \tfrac{1}{2}kx_i^2 = \tfrac{1}{2}mv_f^2 + mgh_f + \tfrac{1}{2}kx_f^2$$

where h is the height (vertical position) and x is the compression of the spring. This general statement looks complex but becomes much simpler when we start considering specific situations. First, we must identify the initial and final conditions in a problem; then, we enter them into the last equation to solve for an unknown.

Solution for (a) This part of the problem is limited to conditions just before the car is released and just after it leaves the spring. Take the initial height to be zero, so that both h_i and h_f are zero. Furthermore, the initial speed v_i is zero and the final compression of the spring x_f is zero, and so several terms in the conservation of mechanical energy are zero and it simplifies to

$$\tfrac{1}{2}kx_i^2 = \tfrac{1}{2}mv_f^2$$

In other words, the initial potential energy in the spring is converted completely to kinetic energy in the absence of friction. Solving for the final velocity and entering known values yields

$$v_f = \left(\frac{k}{m}\right)^{1/2} x_i = \left(\frac{250 \text{ N/m}}{0.100 \text{ kg}}\right)^{1/2} (0.0400 \text{ m})$$

$$= 2.00 \text{ m/s}$$

Solution for (b) One method of finding the speed at the top of the slope is to consider conditions just before the car is released and just after it reaches the top of the slope, completely ignoring everything in between. Doing the same type of analysis to find which terms are zero, the conservation of mechanical energy becomes

$$\tfrac{1}{2}kx_i^2 = \tfrac{1}{2}mv_f^2 + mgh_f$$

This means that the spring's initial potential energy is converted partly to gravitational potential energy and partly to kinetic energy. The final speed at the top of the slope will be less than at the bottom. Solving for v_f and substituting known values gives

$$v_f = \left(\frac{k}{m} x_i^2 - 2gh_f\right)^{1/2}$$

$$= \left[\frac{250 \text{ N/m}}{0.100 \text{ kg}} \cdot (0.0400 \text{ m})^2\right.$$

$$\left. -2(9.80 \text{ m/s}^2)(0.180 \text{ m})\right]^{1/2}$$

$$= 0.690 \text{ m/s}$$

Discussion Another way to solve this problem is to realize that the car's kinetic energy before it goes up the slope is converted partly to potential energy—that is, to take the final conditions in part (a) to be the initial conditions in part (b). This is left as an end-of-chapter problem.

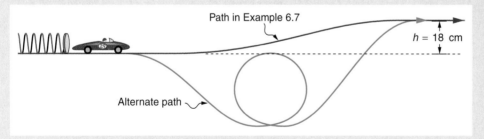

Path in Example 6.7

$h = 18$ cm

Alternate path

Figure 6.9 A toy car is pushed by a compressed spring and coasts up a slope. Assuming negligible friction, the potential energy in the spring is first completely converted to kinetic energy, and then to a combination of kinetic and gravitational potential energy as the car rises. The details of the path are unimportant since all forces are conservative—the car would have the same final speed if it took the alternate path shown.

Note that for conservative forces, we do not directly calculate the work they do; rather, we consider their effects through their corresponding potential energies, just as we did in the preceding example. Note also that we do not consider details of the path taken—only the starting and ending points are important (as long as the path is not impossible). This is usually a tremendous simplification, since the path may be complicated and forces may vary along the way.

6.5 NONCONSERVATIVE FORCES: OPEN SYSTEMS

Forces are either conservative or nonconservative. A **nonconservative force** is one for which work depends on the path taken. Friction is a good example of a nonconservative force. As illustrated in Figure 6.10, work done against friction depends on the length of the path between the starting and ending points. Because of this dependence on path, there is no potential energy associated with nonconservative forces. More important is that the work done by a nonconservative force *adds or removes mechanical energy from a system*. Friction, for example, creates thermal energy that dissipates, removing energy from the system. Furthermore, even if the thermal energy is retained or captured, it cannot be fully converted back to work, so that it is lost in that sense, too.

A system subjected to nonconservative forces is an **open system**, because energy may enter or leave it as well as change forms within it. We often say that energy is not conserved for such systems, but this should not be taken to imply that the energy disappears or appears from nowhere. It simply enters or leaves the system being considered, and its source or destination can be identified. An exact statement is that *mechanical energy may* not be conserved when nonconservative forces act. For example, when a car is brought to a stop by friction on level ground, it loses kinetic energy, which is dissipated as heat, reducing its mechanical energy. Figure 6.11 compares the effects of conservative and nonconservative forces.

Now let us consider what form the work-energy theorem takes when both conservative and nonconservative forces act. This will lead us to see that the work done by nonconservative forces equals the change in the mechanical energy of a system. As noted in Section 6.2, the work-energy theorem states that the net work on a system equals the change in its kinetic energy, or net $W = \Delta KE$. The net work is the sum of the work by nonconservative forces plus the work by conservative forces. That is,

$$\text{net } W = W_{nc} + W_c$$

so that

$$W_{nc} + W_c = \Delta KE$$

where W_{nc} is the *total work done by all nonconservative forces* and W_c is the *total work done by all conservative forces*.

Consider Figure 6.12, where a person pushes a crate up a ramp and is opposed by friction. As in the previous section, we note that work done by a conservative force comes from a loss of potential energy, so that $W_c = -\Delta PE$. Substituting this into the previous equation and solving for W_{nc} gives

$$W_{nc} = \Delta KE + \Delta PE \qquad \textbf{(6.7a)}$$

This means that the total mechanical energy (KE + PE) changes by exactly the amount of work done by nonconservative forces. In Figure 6.12, this is the work done by the person minus the work done by friction. So even if energy is not conserved for the system of interest (like the crate), we know that an equal amount of work was done to cause the change.

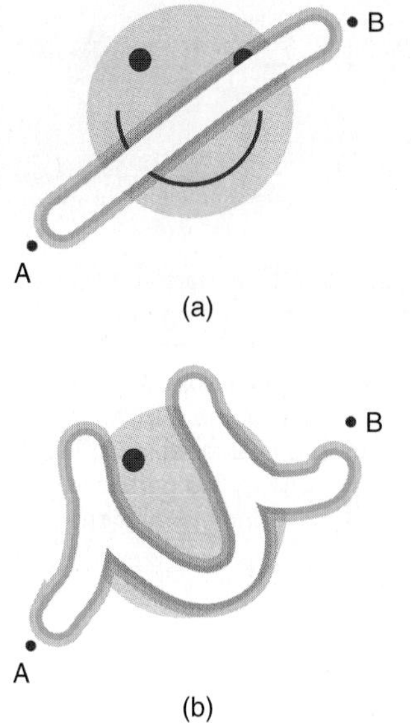

Figure 6.10 The amount of the irritating happy face erased depends on the path taken by the eraser between points A and B, as does the work done against friction. Less work is done and less of the face is erased for the path in (a) than for the path in (b). The force here is friction, and most of the work goes into thermal energy that subsequently leaves the system. The energy expended cannot be fully recovered, but it's worth it.

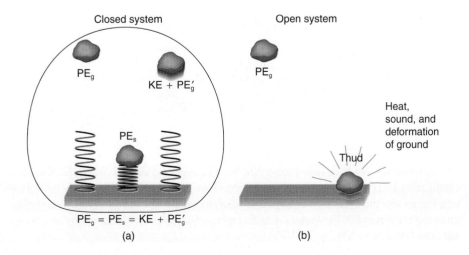

Closed system | Open system

PE_g

$KE + PE_g'$

PE_g

Heat, sound, and deformation of ground

PE_s

Thud

$PE_g = PE_s = KE + PE_g'$

(a) | (b)

Figure 6.11 Comparison of the effects of conservative and nonconservative forces on the mechanical energy of a system. (a) Closed system. When a rock is dropped onto a spring, its mechanical energy remains constant (neglecting air resistance) because the force in the spring is conservative. The spring can propel the rock back to its original height, where it once again has only potential energy due to gravity. (b) Open system. When the same rock is dropped onto the ground, it is stopped by nonconservative forces that dissipate its mechanical energy as heat, sound, and surface distortion. The rock has lost mechanical energy.

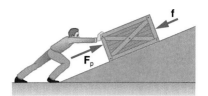

Figure 6.12 A person pushes a crate up a ramp, doing work on the crate. Friction and gravity also do work on the crate, both opposing the person. As the crate is pushed up the ramp, it gains mechanical energy, implying that the work done by the person is greater than the work done by friction.

We rearrange Equation 6.7a to obtain

$$KE_i + PE_i + W_{nc} = KE_f + PE_f \qquad \textbf{(6.7b)}$$

This means that the amount of work done by nonconservative forces adds to the mechanical energy of a system. If W_{nc} is positive, then mechanical energy is increased, such as when the man pushes the crate up the ramp in Figure 6.12. If W_{nc} is negative, then mechanical energy is decreased, such as when the rock hits the ground in Figure 6.11(b). If W_{nc} is zero, then we have conservation of mechanical energy, such as when nonconservative forces balance. For example, when you push a lawn mower at constant speed on level ground, your work is removed by friction and the mower has a constant energy.

EXAMPLE 6.8   HOW FAR MUST A BASEBALL PLAYER SLIDE TO COME TO REST?

Consider the situation shown in Figure 6.13, where a baseball player slides to a stop on level ground. Using energy considerations, calculate the distance the 85.0 kg baseball player slides, given his initial speed is 7.00 m/s and the force of friction against him is a constant 450 N.

Strategy Friction stops the player by converting his kinetic energy into other forms, including heat. Friction must do an amount of work equal to the initial kinetic energy of the player. This means that friction times distance slid equals the initial kinetic energy, or $f \cdot d = (1/2)mv_i^2$. This can be proven more rigorously by applying the relationship stated in Equation 6.7b:

$$KE_i + PE_i + W_{nc} = KE_f + PE_f$$

The form of potential energy here is gravitational (mgh). Since h is constant, PE is the same before and after; thus, PE subtracts out. The final kinetic energy is zero. The work done by friction is negative, since f is in the opposite direc-

tion of motion (that is, $\theta = 180°$, and so $\cos \theta = -1$). Thus $W_{nc} = -f \cdot d$. The equation simplifies to

$$\tfrac{1}{2}mv_i^2 - f \cdot d = 0$$

or

$$f \cdot d = \tfrac{1}{2}mv_i^2$$

This can now be solved for the distance d.

Solution Solving the previous equation for d and substituting known values yields

$$d = \frac{\tfrac{1}{2}mv_i^2}{f} = \frac{0.5(85.0 \text{ kg})(7.00 \text{ m/s})^2}{450 \text{ N}}$$

$$= 4.63 \text{ m}$$

Discussion This is a distance of about 15.0 ft, reasonable for a slide in baseball. The most important point of the example is that the amount of nonconservative work equals the change in mechanical energy. You must work harder to stop a truck, with its large mechanical energy, than to stop a mosquito, for example.

$$KE = \tfrac{1}{2}mv^2 = f \cdot d$$

d

$f = 450 \text{ N}$

Figure 6.13 The baseball player slides to a stop in a distance d. In the process, friction removes the player's kinetic energy by doing an amount of work $f \cdot d$ equal to the initial kinetic energy.

In summary, we have now progressed from considering only conservative forces to considering all types of forces, conservative and nonconservative. When only conservative forces act, the system is closed, and mechanical energy is conserved. When nonconservative forces act, the system is open, and mechanical energy changes by an amount equal to W_{nc}.

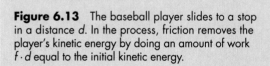

6.6 CONSERVATION OF ENERGY

Energy, as we have noted, is conserved, making it one of the most important physical quantities in nature. The **conservation of energy law** can be stated as follows:

Total energy is constant in any process. It may change in form or be transferred from one system to another, but the total remains the same.

We have explored some forms of energy and some ways it can be transferred from one system to another. This led to the definition of two major types of energy—mechanical energy (KE + PE), and energy transferred via work done by nonconservative forces (W_{nc}). But energy takes *many* other forms, manifesting itself in *many* different ways, and we need to be able to deal with all these before we can write an equation for the above general statement of the conservation of energy.

At this point, we deal with all other forms of energy by lumping them into a single group called **other energy** (OE). Then we can state the conservation of energy in equation form as

$$KE_i + PE_i + W_{nc} + OE_i = KE_f + PE_f + OE_f \qquad (6.8)$$

All types of energy and work can be included in this very general statement of conservation of energy. Kinetic energy is KE, work done by a conservative force is represented by PE, work done by nonconservative forces is W_{nc}, and all other energies are included as OE. Equation 6.8 applies to all previous examples; in those situations OE was constant, and so it subtracted out and was not directly considered.

When does OE play a role? One example occurs when fuel is loaded into an airplane. This fuel is converted to kinetic energy when the plane moves, to potential energy when it gains altitude, to heat (another form of OE) in the engines, and work against friction as it flies (W_{nc}).

What are some other forms of energy? You can probably name a number of forms of energy not yet discussed. Many of these will be covered in later chapters, but it is interesting to name a few. Electrical energy is a common form that is converted to many other forms and does work in a wide range of practical situations. Chemical fuels, such as gasoline and food, carry energy that can be transferred to a system without doing work on it. Chemical fuel can also produce electrical energy, such as in batteries. Batteries can in turn produce light, which is a very pure form of energy. Most energy sources on earth are in fact stored energy from sunlight. This includes food energy and hydroelectric energy. Nuclear energy comes from processes that convert measurable amounts of mass into energy. Nuclear energy creates sunlight, electricity in power plants, and heat and blast in weapons. Atoms and molecules inside all objects are in random motion. This internal kinetic energy is called thermal energy, since it is related to the temperature of the object. These and all other forms of energy can be converted into one another and can do work, and the total is always conserved.

Table 6.1 gives the amount of energy in various situations. The range of energies and the variety of types and situations is impressive.

CONNECTIONS

Energy
The fact that energy is conserved and has many forms makes it very important. You will find that energy is discussed in nearly every remaining chapter of the text, since it is involved in all processes. It will also become apparent that many situations are best understood in terms of energy and that problems are often most easily conceptualized and solved by considering energy.

PROBLEM-SOLVING STRATEGIES

FOR ENERGY

You will find the following problem-solving strategies useful whenever you deal with energy. The strategies help in organizing and reinforcing energy concepts. In fact, they are used in the examples presented in this chapter. The familiar general problem-solving strategies presented in Section 2.5—involving identifying physical principles, knowns, and unknowns, checking units, and so on—continue to be relevant here.

Step 1. Determine the system of interest. A sketch may help.
Step 2. Analyze the situation to *determine the types of work and energy involved in both the initial and final states*. Make a list of the various types. This will allow you

to *judge whether the system of interest is open or closed*. It is open if energy enters or leaves it. If not, it is closed. A system is open if nonconservative forces, such as friction, act, or if there are any changes in OE, such as the conversion of fuel to another form of energy. Once it is determined if the system is closed or open, use step 3 or step 4, respectively.

Step 3. For *closed systems (conservative forces only), mechanical energy is conserved.* That is,

$$KE_i + PE_i = KE_f + PE_f$$

This is the conservation of energy law for closed systems—in this case, W_{nc} is zero and all forms of OE are zero or constant. Closed systems are precisely described by considering only kinetic and potential energies. Nothing else need be considered.

Step 4. For *open systems*, mechanical energy may not be conserved. Work may be done by nonconservative forces, and other forms of energy (OE) may enter or leave the system. *The conservation of energy law in its most general form must be used.* This is

$$KE_i + PE_i + W_{nc} + OE_i = KE_f + PE_f + OE_f$$

In most problems, one or more of the terms is zero, simplifying its solution. Do not calculate W_c, the work done by conservative forces. It is incorporated in the PE terms.

Step 5. You have already identified the types of work and energy involved (in step 2). Before solving for the unknown, *eliminate terms wherever possible* to simplify the algebra. For example, choose $h = 0$ at either the initial or final point, so that PE_g is zero there. Then solve for the unknown in the customary manner.

Step 6. *Check the answer to see if it is reasonable.* Once you have solved a problem, reexamine the forms of work and energy to see if you have set up the conservation of energy equation correctly. For example, work done against friction should be negative, potential energy at the bottom of a hill should be less than at the top, and so on. Also check to see that the numerical value obtained is reasonable. For example, the final speed of a skateboarder who coasts down a 3 m high ramp could reasonably be 20 km/h, but *not* 80 km/h.

TABLE 6.1

VARIOUS ENERGIES	
Object/phenomenon	*Energy in joules*
Big bang	10^{68}
Energy released in a supernova	10^{44}
Hydrogen fusion energy in the oceans	10^{34}
Annual U.S. energy use	8×10^{19}
Large fusion bomb (9 megaton)	3.8×10^{16}
1 kg hydrogen (fusion to helium)	6.4×10^{14}
1 kg uranium (nuclear fission)	8.0×10^{13}
Hiroshima-size fission bomb (10 kiloton)	4.2×10^{13}
90,000 ton aircraft carrier at 30 knots	1.1×10^{10}
1 barrel crude oil	5.9×10^9
1 ton TNT	4.2×10^9
1 gallon gasoline	1.3×10^8
Daily adult food intake	1.2×10^7
1 ton car at 90 km/h	3.1×10^5
1 g fat (9.3 kcal)	3.9×10^4
1 g carbohydrate (4.1 kcal)	1.7×10^4
1 g protein (4.1 kcal)	1.7×10^4
Baseball at 100 mph	1.5×10^2
Mosquito (10^{-2} g) at 0.5 m/s	1.3×10^{-6}
Single electron in a TV tube	4.0×10^{-15}
Single electron in 120 V line	1.9×10^{-17}
Energy to break one DNA strand	10^{-19}

6.7 POWER

Power—the word conjures up many images. A professional football lineman muscling aside his opponent. A dragster roaring away from the starting line. Mount St. Helens blowing its flank into the atmosphere. A rocket blasting off, as in Figure 6.14.

These images of power have in common the rapid performance of work, consistent with the scientific definition of **power** (*P*) as the rate at which work is done. In equation form, this is

$$P = \frac{W}{t} \tag{6.9}$$

The SI unit for power is the familiar **watt** (W), where 1 watt = 1 joule/second (1 W = 1 J/s). Since work represents energy transfer, power is also the rate at which energy is expended. A 60 watt light bulb, for example, expends 60 joules of energy per second. Great power means a large amount of work or energy in a short time. For example, when a powerful car accelerates rapidly, it does a large amount of work and expends a large amount of fuel in a short time.

Figure 6.14 This powerful rocket does work and consumes energy at a very high rate.

EXAMPLE 6.9 POWER WHEN RUNNING UP STAIRS

What is the power output of mechanical energy for a 60.0 kg woman who runs up a 3.00 m high flight of stairs in 2.50 s, starting from rest but having a final speed of 2.00 m/s? (See Figure 6.15.)

Strategy and Concept The work going into mechanical energy is $W = \Delta KE + \Delta PE$. If we take $h = 0$ at the bottom of the stairs, then both KE and PE_g are initially zero; thus, $W = KE + PE_g = (1/2)mv^2 + mgh$. Since all terms are given, we can calculate W and then divide it by time to get power.

Solution Substituting the expression for W into the definition of power given in Equation 6.9 yields

$$P = \frac{W}{t} = \frac{\frac{1}{2}mv^2 + mgh}{t}$$

Entering known values yields

$$P = \frac{0.5(60.0 \text{ kg})(2.00 \text{ m/s})^2 + (60.0 \text{ kg})(9.80 \text{ m/s}^2)(3.00 \text{ m})}{2.50 \text{ s}}$$

$$= \frac{120 \text{ J} + 1764 \text{ J}}{2.50 \text{ s}}$$

$$= 754 \text{ W}$$

Discussion Getting the person up the stairs requires 1764 J compared with only 120 J needed to increase kinetic energy; thus, most of her power output is required for climbing rather than accelerating.

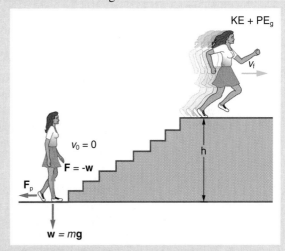

Figure 6.15 When this woman runs upstairs starting from rest, she converts food energy into kinetic energy, gravitational potential energy, and waste heat. Her power output depends on how fast she does this.

It is impressive that this woman's useful power output is slightly greater than 1 horsepower (1 hp = 746 W)! People can generate more than a horsepower with their leg muscles for short periods of time by rapidly converting available blood sugar and oxygen into work output. (A horse can put out 1 hp for hours on end.) Once oxygen is depleted, power output decreases and the person begins to breathe rapidly to obtain oxygen to metabolize more food—this is known as the *aerobic* stage of exercise. If the woman climbed the stairs slowly, then her power output would be much less although the amount of work would be the same.

Examples of power are limited only by the imagination, since there are as many types as there are forms of work and energy. (See Table 6.2 for some examples.) Sunlight reaching the earth's surface carries a maximum power of about 1.3 kilowatts per square

TABLE 6.2

VARIOUS POWERS	
Object/phenomenon	*Power in watts*
Supernova (at peak)	5×10^{37}
Milky Way galaxy	10^{37}
Crab nebula pulsar	10^{28}
Sun	4×10^{26}
Mount St. Helens eruption (maximum)	4×10^{15}
Lightning bolt	2×10^{12}
Nuclear power plant (total electric and heat)	3×10^{9}
Aircraft carrier (total useful and heat)	10^{8}
Dragster (total useful and heat)	2×10^{6}
Automobile (total useful and heat)	8×10^{4}
Football lineman (total useful and heat)	5×10^{3}
Clothes dryer	4×10^{3}
Person at rest (all heat)	100
Typical incandescent light (total useful and heat)	60
Heart, person at rest (total useful and heat)	8
Electric clock	3
Pocket calculator	10^{-3}

meter (kW/m²). A tiny fraction of this is retained by the earth long term. Our consumption rate of fossil fuels is far greater than the rate at which they are stored, and so it is inevitable that they will be depleted. Power implies that energy is transferred, perhaps changing form. It is rarely possible to change one form completely into another. For example, a 60 W incandescent bulb converts only 5 W of electrical power to light, with 55 W going into heat. Furthermore, the typical electric power plant converts only 35 to 40% of its fuel consumption into electric power. The remainder becomes a huge amount of waste heat that must be dispersed as rapidly as created. A nuclear power plant may produce 1000 MW of electric power, but it consumes nuclear energy at a rate of about 2800 MW, creating waste heat at a rate of 1800 MW. (See Figure 6.16.)

We usually have to pay for the energy we use. It is interesting and easy to estimate the cost of energy for an electrical appliance if its power consumption rate and time used are known. The higher the power consumption rate and the longer it is used, the greater the cost. The power consumption rate is $P = W/t = E/t$, where E is the energy supplied by the electric company, and so the energy consumed over a time t is

$$E = Pt$$

Electric bills state the energy used in units of kilowatt-hours (kW·h) or power times time. This is a convenient unit, since electrical power consumption at the kilowatt level for hours of time is typical.

Figure 6.16 Tremendous amounts of electric power are generated by nuclear power plants such as this one, but an even larger amount of power goes into waste heat. The large cooling towers are needed to disperse waste heat as rapidly as it is produced. The reactors are in the two domed buildings to the right of the cooling towers. This is not unique to nuclear plants but is an unavoidable consequence of generating electric power from heat produced by any fuel—nuclear, coal, oil, natural gas, or the like.

EXAMPLE 6.10 ENERGY COST FROM POWER AND TIME

What is the cost of running a 200 W stereo 6.00 h per day for 30.0 days if the cost of electricity is 9.00¢ per kW·h?

Strategy Cost is based on energy consumed; thus, we must find E from $E = Pt$ and then calculate the cost.

Solution The energy consumed in kW·h is

$$E = Pt = (0.200 \ \text{kW})(6.00 \ \text{h/day})(30.0 \ \text{days})$$
$$= 36.0 \ \text{kW·h}$$

and the cost is simply given by

$$\text{cost} = (36.0 \ \text{kW·h})(9.00 \ \text{¢/kW·h}) = \$3.24$$

Discussion The cost of using the stereo in this example is neither exorbitant nor negligible. It is clear that the cost is a combination of power and time. When both are high, such as for an air conditioner in the summer, the cost is high.

The motivation to save energy costs has become more compelling with its ever-increasing price. Armed with the knowledge that energy consumed is power times time, you can estimate costs for yourself and make the necessary value judgments about where to save energy. Either power or time must be limited. It is most cost-effective to limit the use of high-power devices that normally operate for long periods of time, such as water heaters and air conditioners. This would not include relatively high power devices like toasters, since they are on only a few minutes per day. It would also not include electric clocks, in spite of their 24-hour-per-day usage, since they are very low power devices. It is sometimes possible to use devices that have greater efficiencies—that is, devices that consume less power to accomplish the same task. One example is the fluorescent light bulb, which produces over six times as much light per watt of power consumed than its incandescent cousin.

6.8* WORK, ENERGY, AND POWER IN HUMANS; INTRODUCTION TO EFFICIENCY

Our own bodies, like all living organisms, are energy conversion machines. We convert food energy into three major forms—work, thermal energy, and chemical energy stored as fat, with less than 5% excreted unused. (See Figure 6.17.) The fraction going into each form depends both on how much we eat and on our level of physical activity. If we eat more than is needed to do work and stay warm, the remainder goes into body fat.

Work done by a person is sometimes called *useful work*, which is *work done on the outside world*, such as lifting weights. Useful work requires a force exerted through a distance on the outside world, and so it excludes internal work, such as that done by the heart when pumping blood. But useful work does include that done in climbing stairs or accelerating to a full run, since these are accomplished by exerting forces on the outside world. Forces exerted by the body are nonconservative, so that they can change the mechanical energy (KE + PE) of the system worked upon, and this is often the goal. A softball player throwing a ball into the air, for example, increases both its kinetic and potential energy.

The body is incapable of converting all food energy to work. Most food energy is actually converted to thermal energy. When you do useful work, you also produce thermal energy in proportion to the work done. This can be viewed as an *inefficient* conversion of energy to work. We define **efficiency** *Eff* to be the useful work output divided by the energy input. In equation form, this is

$$Eff = \frac{W_{out}}{E_{in}} \qquad \textbf{(6.10a)}$$

where W_{out} is useful work output and E_{in} is the energy consumed (food digested, in humans). Since W_{out} can end up as any form of energy, efficiency can also be written as

$$Eff = \frac{E_{out}}{E_{in}} \qquad \textbf{(6.10b)}$$

where E_{out} is some desired energy output. This is a slightly different definition of efficiency, since the work output does not necessarily go entirely into the desired energy

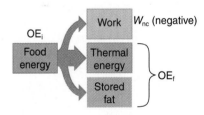

Figure 6.17 Energy consumed by humans is converted to work, thermal energy, and stored fat. By far the largest fraction goes to thermal energy, although the fraction varies depending on the type of physical activity.

TABLE 6.3

EFFICIENCY OF THE HUMAN BODY AND MECHANICAL DEVICES	
Activity/device	Efficiency* (%)
Body	
Cycling and climbing	20
Swimming, surface	2
Swimming, submerged	4
Shoveling	3
Weightlifting	9
Steam engine	17
Gasoline engine	30
Diesel engine	35
Nuclear power plant	35
Coal power plant	42

*Representative values

form. Furthermore, since power is energy divided by time, if we divide the numerator and denominator by time, we get an expression for efficiency in terms of power:

$$Eff = \frac{P_{out}}{P_{in}} \qquad (6.10c)$$

where P_{in} is the power consumption or input, and P_{out} is the useful power output going into useful work or a desired form of energy. For our purposes, it is acceptable to choose whichever definition of efficiency is most convenient.*

Table 6.3 lists efficiencies of certain human activities and those of various other systems. Because energy is conserved, efficiency cannot be greater than 1, but it is remarkable that all efficiencies in the table are *significantly* less than 1 (that is, much less than 100%). The reason is that all these systems convert thermal energy or stored chemical energy into useful work.

Thermal energy involves random motions of atoms and molecules, whereas work involves a highly coordinated motion in which a force is exerted through a distance. It is not only difficult to coordinate a system's random thermal motion into an organized motion in a single direction, it is not even possible—as will be discussed in more detail in the chapter on thermodynamics.

The body's efficiency is limited by the efficiency of muscles in converting food calories[†] into useful work. Single muscles have maximum efficiencies of about 25 or 30%. But overall efficiency includes energy to operate the entire body. Thus the body's efficiency is always less than that of an individual muscle.

EXAMPLE 6.11 WHAT IS THE EFFICIENCY OF THE BODY IN LIFTING A MASS?

Find the efficiency of the person in Figure 6.18 who metabolizes 8.00 kcal of food energy while lifting 120 kg to a height of 2.30 m above its starting point.

Strategy Here the work output goes into increasing the gravitational potential energy of the weights; thus, $W_{out} = mgh$. Since work output and energy input are given, Equation 6.10a can be used to calculate the efficiency.

Solution In this situation Equation 6.10a becomes

$$Eff = \frac{W_{out}}{E_{in}} = \frac{mgh}{E_{in}}$$

Entering known values gives

$$Eff = \frac{(120 \text{ kg})(9.80 \text{ m/s}^2)(2.30 \text{ m})}{(8.00 \text{ kcal})(4186 \text{ J/kcal})} = \frac{2705 \text{ J}}{33,488 \text{ J}}$$

$$= 0.0833 = 8.33\%$$

Discussion This efficiency is consistent with the 9% value given in Table 6.3. Note that the 120 kg mass is that of the object lifted and does not include the mass of the arms, because we are interested only in work on the outside world.

Figure 6.18 This weightlifter does work on the barbell when he lifts it. The efficiency of the arms is small, so that he consumes about 12 times as much food energy as he does work. The excess food energy is converted to thermal energy.

Food energy is taken in as a form of OE. When digested or metabolized, it is converted into other forms. Since food is metabolized in a process that amounts to oxidation, we can measure the energy people use during various activities by measuring their oxygen use. (See Figure 6.19.) Since metabolization of food is similar to other oxidation processes, the energy content of food is determined by burning it and measuring the kilocalories of heat produced.

---

*The first expression for efficiency is the most scientifically rigorous, whereas the next two can contain elements of subjectivity regarding what the desired energy or power output may be. We shall not be overly concerned with the distinction until we discuss heat engines in some detail in Chapter 14.
[†]Recall that a food calorie is equal to one kilocalorie (kcal).

Speed is important in many physical activities, such as foot races. The faster the task is performed, the greater the power output. For example, if you run up stairs rather than walk, your power output ($P = mgh/t$) will be greater because t is smaller. You do the same amount of work (mgh) in both cases, but the effects on your body differ if you run rather than walk. Vigorous work or exercise causes immediate increases in heart and respiration rates and can have long-term effects on your metabolic rate. Table 6.4 gives the energy and oxygen consumption rates for a number of activities.

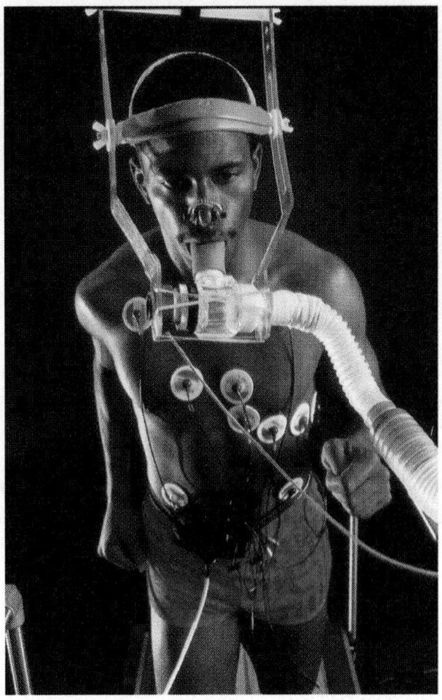

TABLE 6.4

ENERGY AND OXYGEN CONSUMPTION RATES* (POWER)

| | Energy consumption | | Oxygen consumption |
Activity	kcal/min	watts	in liters O_2/min
Sleeping	1.2	83	0.24
Sitting at rest	1.7	120	0.34
Standing relaxed	1.8	125	0.36
Sitting in class	3.0	210	0.60
Walking slowly (4.8 km/h)	3.8	265	0.76
Cycling (13–18 km/h)	5.7	400	1.14
Shivering	6.1	425	1.21
Playing tennis	6.3	440	1.26
Swimming breaststroke	6.8	475	1.36
Ice skating (14.5 km/h)	7.8	545	1.56
Climbing stairs (116/min)	9.8	685	1.96
Cycling (21 km/h)	10.0	700	2.00
Running cross-country	10.6	740	2.12
Playing basketball	11.4	800	2.28
Cycling, professional racer	26.5	1855	5.30
Sprinting	34.5	2415	6.90

*Normal 76 kg male.

Figure 6.19 The person on this treadmill is breathing through an apparatus that measures the amount of oxygen used. This provides a direct measure of the person's metabolic rate, which is the rate at which food energy is converted to another form. Such measurements can indicate level of athletic conditioning as well as certain medical problems.

EXAMPLE 6.12 RECREATIONAL ENERGY USED BY THE BODY

Suppose that during one day of a camping trip, you walk slowly for 7.00 h, spend 1.00 h climbing vigorously to the top of a mesa, make camp for 1.50 h (this requires the same amount of power as slow cycling), stand relaxed for 2.00 h around camp, sit and talk while restful for 4.50 h, and sleep for 8.00 h. You eat 3000 kcal (an average amount) of food on this day. (a) How many kcal of energy do you metabolize? (b) How many grams of fat do you gain or lose, given that fat has an energy content of 9.3 kcal/g? (c) How many kcal of body heat did you generate, assuming that the climbing was done with an efficiency of 20%, and all other activities produced no useful work?

Strategy and Solution for (a) Here we are asked for how much energy your body consumed. Energy is power times time, and so we find it for each activity using powers from Table 6.4 and the given times. Thus food energy metabolized is

$$E = P_1 t_1 + P_2 t_2 + \cdots$$

$$= (3.8 \text{ kcal/min})(420 \text{ min}) + (9.8 \text{ kcal/min})(60 \text{ min})$$
$$+ (5.7 \text{ kcal/min})(90 \text{ min}) + (1.8 \text{ kcal/min})(120 \text{ min})$$
$$+ (1.7 \text{ kcal/min})(270 \text{ min}) + (1.2 \text{ kcal/min})(480 \text{ min})$$

$$= (1596 \text{ kcal}) + (588 \text{ kcal}) + (513 \text{ kcal}) + (216 \text{ kcal})$$
$$+ (459 \text{ kcal}) + (576 \text{ kcal})$$

Thus,

$$E = 3948 \text{ kcal} = 3.9 \times 10^3 \text{ kcal}$$

Over half of this was used for walking and climbing.

Strategy and Solution for (b) If you use more energy than you consume, your body will obtain the needed energy by metabolizing body fat. You metabolized 3948 kcal but only consumed 3000 kcal, so that your body will get 948 kcal from fat. The amount lost will be

$$\text{fat lost} = (948 \text{ kcal}) \frac{1.0 \text{ g fat}}{9.3 \text{ kcal}} = 102 \text{ g}$$

This is less than 4 ounces. Most people have plenty of body fat to spare, but if none is readily available the body will break down muscle tissue. This is a hazard during extreme exertion, such as running marathons, and for anorexics or persons on unsupervised starvation diets.

Strategy and Solution for (c) The body produces additional heat when useful work is done. In the first part of this example, we found that climbing required 588 kcal of food energy. The efficiency for climbing is 20%,

(continued)

(continued)

meaning that 20% of this goes to useful work and 80% to heat—that is,

$$W_{out} = Eff \cdot E_{in} = (0.20)(588 \text{ kcal}) = 118 \text{ kcal}$$

Climbing is the *only* useful work for the day; everything else goes to heat, and so the amount of heat is

$$\text{heat} = (3948 - 118) \text{ kcal} = 3.8 \times 10^3 \text{ kcal}$$

Discussion This is a rather large amount of heat when you consider that 1 kcal of heat will increase the temperature of 1 kg of water 1°C—enough heat was generated to raise the temperature of a person about 60°C to 97°C, or nearly to the boiling point! Clearly, little or none of the heat is retained.

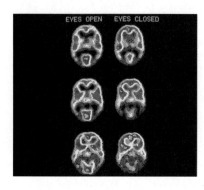

Figure 6.20 These brain scans use nuclear tracing techniques to show the level of energy consumption in the brain. The vision center in the back of the head shows the relative lack of energy consumption in a person whose eyes are closed. Increased energy utilization is apparent in the vision centers when the same person's eyes are open.

All bodily functions, from thinking to lifting weights, require energy. (See Figure 6.20.) The many small muscle actions accompanying all quiet activity, from sleeping to head scratching, ultimately become thermal energy, as do less visible muscle actions by the heart, lungs, and digestive tract. Shivering, in fact, is an involuntary response to low body temperature that pits muscles against one another to produce body heat and no work. The kidneys and liver consume a surprising amount of energy, but the biggest surprise of all it that a full 25% of all energy consumed by the body is used to maintain electrical potentials in all living cells. (Nerve cells use this electrical potential in nerve impulses.) This bioelectrical energy ultimately becomes mostly thermal energy, but some is utilized to power chemical processes such as in the kidneys and liver, and in fat production.

SUMMARY

Work, energy, and power are among the most fundamental and commonly encountered concepts in physics. **Energy** is loosely defined to be the ability to do work, although not all energy is available to do work. **Work** is the product of the component of the force in the direction of motion with the displacement. In equation form,

$$W = Fd \cos \theta \qquad (6.1)$$

where W is work and θ is the angle between the force **F** and the displacement **d**. The SI unit for work and energy is the **joule** (J), where $1 \text{ J} = 1 \text{ N} \cdot \text{m} = 1 \text{ kg} \cdot \text{m}^2/\text{s}^2$.

The **work-energy theorem** states that the net work, net W, on a system changes its kinetic energy. This is written

$$\text{net } W = \tfrac{1}{2}mv^2 - \tfrac{1}{2}mv_0^2 \qquad (6.2)$$

where **kinetic energy** is defined to be

$$\text{KE} = \tfrac{1}{2}mv^2 \qquad (6.3)$$

for a mass m having a velocity v. Work and energy related to gravity are treated by considering the **gravitational potential energy** (PE_g), in which the change in gravitational potential energy, ΔPE_g, is

$$\Delta\text{PE}_g = mgh \qquad (6.4)$$

with h being the change in vertical position and g the acceleration of gravity.

A **conservative force** is one for which work depends only on the starting and ending points of a motion, not on the path taken. We can define a potential energy (PE) for any conservative force, just as we defined PE_g for gravity. The **potential energy PE_s of a spring** is

$$\text{PE}_s = \tfrac{1}{2}kx^2 \qquad (6.5)$$

where k is the spring's force constant and x is the displacement from its undeformed position. **Mechanical energy** is defined to be KE + PE and, for a conservative force,

$$\left.\begin{array}{l} \text{KE} + \text{PE} = \text{constant} \\[4pt] \text{or} \\[4pt] \text{KE}_i + \text{PE}_i = \text{KE}_f + \text{PE}_f \end{array}\right\} \begin{array}{l}\text{(conservative} \\ \text{forces only)} \end{array} \qquad (6.6)$$

where i and f denote initial and final values. This is known as the **conservation of mechanical energy**. A system that experiences only conservative forces is called a **closed system**.

A **nonconservative force** is one for which work depends on the path. Work done by a nonconservative force W_{nc} changes the mechanical energy of a system. A system subjected to nonconservative forces is an **open system**. It can be shown that

$$W_{nc} = \Delta\text{KE} + \Delta\text{PE} \qquad (6.7a)$$

or

$$\text{KE}_i + \text{PE}_i + W_{nc} = \text{KE}_f + \text{PE}_f \qquad (6.7b)$$

When all forms of energy are considered, conservation of energy is written

$$KE_i + PE_i + W_{nc} + OE_i = KE_f + PE_f + OE_f \quad \textbf{(6.8)}$$

where **OE** is all **other forms of energy** besides KE and PE. Table 6.1 lists various energies.

Power (P) is the rate at which work is done, or

$$P = \frac{W}{t} \quad \textbf{(6.9)}$$

The SI unit for power is the familiar **watt** (W), where 1 W = 1 J/s. Table 6.2 lists various powers.

Energy is often utilized to do work, but seldom can all energy be converted completely to work. The efficiency *Eff* of a machine or human is defined to be

$$Eff = \frac{W_{out}}{E_{in}} \quad \textbf{(6.10a)}$$

where W_{out} is work output and E_{in} is the energy consumed. Another way to express this is

$$Eff = \frac{E_{out}}{E_{in}} \quad \textbf{(6.10b)}$$

where E_{out} is some desired *energy* output. In terms of power output and input, efficiency is

$$Eff = \frac{P_{out}}{P_{in}} \quad \textbf{(6.10c)}$$

where P_{in} is the power consumption or input, and P_{out} is the power output going into work or a desired form of energy. Table 6.3 lists representative efficiencies, and Table 6.4 gives power consumption rates in humans.

CONCEPTUAL QUESTIONS

6.1 Give an example of something we think of as work in everyday circumstances that is *not* work in the scientific sense. Is energy transferred or changed in form in your example? If so, explain how this is accomplished without doing work.

6.2 Give an example of a situation in which there is a force and a displacement, but the force does no work. Explain why it does no work.

6.3 Describe a situation in which a force is exerted for a long time but does no work. Explain why.

6.4 The person in Figure 6.1(a) does work on the lawn mower. Under what conditions would the mower gain energy? Under what conditions would it lose energy?

6.5 Work done on a system puts energy into it. Work done by a system removes energy from it. Give an example of each.

6.6 When solving for velocity in Example 6.4, we kept only the positive root. Why?

6.7 In Example 6.6(b), we calculated the final speed of a roller coaster that descended 20 m in altitude and had an initial speed of 5 m/s downhill. If the roller coaster had had an initial speed of 5 m/s *uphill* instead, and it coasted uphill, stopped, and then rolled back down to a final point 20 m below the start, we would find it had the same final speed. Explain in terms of conservation of energy.

6.8 What is a conservative force?

6.9 The force exerted by a diving board is conservative providing that internal friction is negligible. Assuming this is so, describe changes in the potential energy of a diving board as a swimmer dives from it, starting just before the swimmer steps on the board until just after his feet leave it.

6.10 What is a closed system? Can energy change form in it? Give an example.

6.11 Define mechanical energy. What is the relationship of mechanical energy to nonconservative forces? What happens to mechanical energy if only conservative forces act?

6.12 What is the relationship of potential energy to conservative force?

6.13 What is an open system? Does the mechanical energy of an open system change in every process? Explain.

6.14 Consider the following scenario. A car for which friction is *not* negligible accelerates from rest down a hill, running out of gasoline after a short distance. The driver lets the car coast farther down the hill, then up and over a small crest. He then coasts down that hill into a gas station, where he brakes to a stop and fills the tank with gasoline. Identify the forms of energy the car has, and how they are changed and transferred in this series of events.

6.15 Describe the energy transfers and transformations for a javelin, starting from the point at which an athlete picks up the javelin and ending when the javelin is stuck into the ground after being thrown.

6.16 Do devices with efficiencies less than 1 violate the conservation of energy principle? Explain.

6.17 Most electrical appliances are rated in watts. Does this rating depend on how long the appliance is on? (When off it is a zero watt device.) Explain in terms of the definition of power.

6.18 Explain, in terms of the definition of power, why energy consumption is sometimes listed in kilowatt-hours rather than joules. What is the relationship between these two energy units?

6.19 A spark of static electricity, such as that you might receive from a doorknob on a cold dry day, may carry a few hundred watts of power. Explain why you are not injured by such a spark.

6.20 Does the work you do on a book when you lift it onto a shelf depend on the path taken? On the time taken? On the height of the shelf? On the mass of the book?

6.21[*] Explain why it is easier to climb a mountain on a zigzag path rather than one straight up. Is your increase in gravitational potential energy the same in both cases? Is your energy consumption the same in both?

6.22 Do you do work on the outside world when you rub your hands together to warm them? What is the efficiency of this activity?

6.23 Shivering is an involuntary response to lowered body temperature. What is the efficiency of the body when shivering, and is this a desirable value?

6.24 Discuss the relative effectiveness of dieting and exercise in losing weight, noting that most athletic activities consume food energy at a rate of 300 to 400 kcal/h, while a single cup of yogurt can contain 325 kcal. Specifically, is it likely that exercise alone will be sufficient to lose weight? You may wish to consider that regular exercise may increase the metabolic rate, whereas protracted dieting may reduce it.

PROBLEMS

Section 6.1 Work

6.1 How much work does a grocery clerk do on a can of soup he pushes 0.600 m horizontally with a force of 5.00 N? Express your answer in joules and kilocalories.

6.2 Find the work done by a 75.0 kg person in climbing a 2.50 m flight of stairs at a constant speed.

• 6.3 (a) Calculate the work done on a 1500 kg elevator by its cable to lift it 40.0 m at constant speed, assuming friction averages 100 N. (b) What is the work done on the elevator by gravity in this process? (c) What is the total work done on the elevator?

• 6.4 Suppose a car travels 108 km at a speed of 30.0 m/s, and uses 2.00 gallons of gasoline. Only 30% of the gasoline goes into work done against friction. (See Table 6.1 for the energy content of gasoline.) (a) What is the force of friction? (b) If friction is directly proportional to speed, how many gallons will be used to drive 108 km at a speed of 28.0 m/s?

• 6.5 Calculate the work done by an 85.0 kg trucker who pushes a crate 4.00 m up along a ramp that makes an angle of 20.0° with the horizontal. He exerts a force of 500 N on the crate parallel to the ramp and moves at a constant speed. Be certain to include the work he does on the crate *and* on his body to get up the ramp.

• 6.6 How much work is done by the boy pulling his sister 30.0 m in a wagon as shown in Figure 6.21?

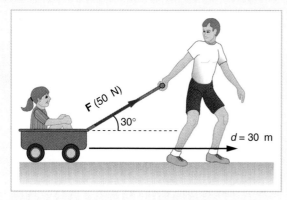

Figure 6.21 The boy does work on the wagon plus child when he pulls it as shown. Problem 6.

‡ 6.7 A shopper pushes a grocery cart 20.0 m at constant speed on level ground, against a 35.0 N frictional force. He pushes in a direction 25.0° below the horizontal. (a) What is the work done on the cart by friction? (b) What is the work done on the cart by gravity? (c) What is the work done on the cart by the shopper? (d) Find the force the shopper exerts, using energy considerations. (e) What is the total work done on the cart?

‡ 6.8 Suppose the ski patrol lowers a rescue sled and victim, having a total mass of 90.0 kg, down a 60.0° slope at constant speed, as shown in Figure 6.22. The coefficient of friction between the sled and the snow is 0.100. (a) How much work is done by friction as the sled moves 30.0 m along the hill? (b) How much work is done by the rope on the sled in this distance? (c) What is the work done by gravity on the sled? (d) What is the total work done?

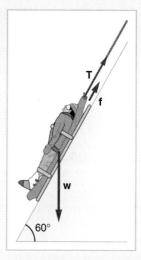

Figure 6.22 A rescue sled and victim are lowered down a steep slope. Problem 8.

Section 6.2 Kinetic Energy

6.9 Compare the kinetic energy of a 20,000 kg truck moving at 110 km/h with that of an 80.0 kg astronaut in orbit moving at 27,500 km/h.

6.10 How fast must a 3000 kg elephant move to have the same kinetic energy as a 65.0 kg sprinter running at 10.0 m/s?

• 6.11 Confirm the value given for the kinetic energy of an aircraft carrier in Table 6.1. You will need to look up the definition of a nautical mile (1 knot = 1 nautical mile/h).

• 6.12 (a) Calculate the force needed to bring a 950 kg car to rest from a speed of 90.0 km/h in a distance of 120 m (a fairly typical distance for a nonpanic stop). (b) Suppose instead the car hits a concrete abutment at full speed and is brought to a stop in 2.00 m. Calculate the force exerted on the car and compare it with the force found in part (a).

• 6.13 Federal statute requires a car's bumper be designed to withstand a 2.50 mph (1.12 m/s) collision with an immovable object without damage to the body of the car. The bumper cushions the shock by absorbing the force over a distance. Calculate the average force on a bumper that collapses 0.200 m while bringing a 900 kg car to rest from an initial speed of 1.12 m/s.

• **6.14** Boxing gloves are padded to lessen the force of a blow. (a) Calculate the force exerted by a boxing glove on an opponent's face, if the glove and face compress 7.50 cm during a blow in which the 7.00 kg arm and glove are brought to rest from an initial speed of 10.0 m/s. (b) Calculate the force exerted by an identical blow in the gory old days when no gloves were used and the knuckles and face would compress only 2.00 cm.

⁞ **6.15** Using energy considerations, calculate the average force a 60.0 kg sprinter exerts backward on the track to accelerate from 2.00 to 8.00 m/s in a distance of 25.0 m, if he encounters a headwind that exerts an average force of 30.0 N against him.

Sections 6.3–6.6 Conservation of Energy

6.16 Using values from Table 6.1, how many DNA molecules could be broken in half by the energy carried by a single electron in a TV tube? (These electrons are not dangerous in themselves, but they do create dangerous x rays. All modern TVs have shielding that absorbs x rays before they can escape and expose viewers.)

6.17 If the energy in fusion bombs were used to supply the energy needs of the United States, how many of the 9 megaton variety would be needed for a year's supply (using data from Table 6.1)? This is not as far-fetched as it may sound—there are thousands of nuclear bombs, and their energy can be trapped in underground explosions and converted to electricity, as natural geothermal energy is.

6.18 Use of hydrogen fusion to supply energy is a dream that may be realized in the next century. It would be a relatively clean and almost limitless supply of energy, as can be seen from Table 6.1. To illustrate this, calculate how many years the present energy needs of the United States could be supplied by one millionth of the oceans' hydrogen fusion energy.

6.19 A hydroelectric power facility converts gravitational potential energy of water behind a dam to electric energy. (a) What is the gravitational potential energy relative to the generators of a lake of volume 50.0 km^3 (mass = 5.00×10^{13} kg), given the lake has an average height of 40.0 m above the generators? (b) Compare this with the energy stored in a 9 megaton fusion bomb.

6.20 How much gravitational potential energy (relative to the ground on which it is built) is stored in the Great Pyramid of Cheops, given its mass is about 7×10^9 kg and its center of mass is 36.5 m above the surrounding ground? How does this compare with the daily food intake of a person?

6.21 Suppose a 50.0 kg teenager picks up an 11.0 kg stereo and raises it 0.750 m from the floor to a tabletop. (a) How much work did he do on the stereo? (b) How much work did he do to raise his own center of mass 0.400 m in the process?

• **6.22** A pogo stick has a spring with a force constant of 2.50×10^4 N/m, which can be compressed 12.0 cm. To what maximum height can a child jump on the stick using only the energy in the spring, if the child and stick have a total mass of 40.0 kg?

• **6.23** In Example 6.6 we found the speed of a roller coaster that had descended 20.0 m was only slightly greater when it had an initial speed of 5.00 m/s than when it started from rest. This implies that $\Delta PE \gg KE_i$. Confirm this statement by taking the ratio of ΔPE to KE_i. (Note that mass cancels.)

• **6.24** A 5.00×10^5 kg subway train is brought to a stop from a speed of 0.500 m/s in 0.400 m by a large spring bumper at the end of its track. What is the force constant k of the spring?

• **6.25** (a) Find the speed of the roller coaster considered in Example 6.6(b) at its lowest point, 25.0 m below the starting point. Assume the initial speed is 5.00 m/s and friction is negligible. (b) Show that the final speed will be 20.4 m/s at 20.0 m below the starting point, taking the initial conditions to be those in the previous part. (c) At what height below the starting point will the roller coaster have a speed of 10.0 m/s?

• **6.26** Solve Example 6.7(b) by taking the initial conditions to be those just before the toy starts up the slope (as suggested in the last paragraph there). That is, show the final speed of the toy car is 0.690 m/s if its initial speed is 2.00 m/s and it coasts up a frictionless slope gaining 0.180 m in altitude.

⁞ **6.27** In a downhill ski race surprisingly little advantage is gained by getting a running start. (This is because the initial kinetic energy is small compared with the gain in gravitational potential energy on even small hills.) To demonstrate this, find the final speed and the time taken for a skier who skies 70.0 m along a 30° slope neglecting friction: (a) Starting from rest. (b) Starting with an initial speed of 2.50 m/s.

⁞ **6.28** A 60.0 kg skier with an initial speed of 12.0 m/s coasts up a 2.50 m high rise as shown in Figure 6.23. Find his final speed at the top, given that the coefficient of friction between her skis and the snow is 0.0800. (Hint: Find the distance traveled up the incline assuming a straight-line path as shown in the figure.)

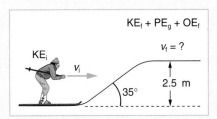

Figure 6.23 The skier's initial kinetic energy is partially used in coasting to the top of a rise. Problem 28.

⁞ **6.29** Using energy considerations and assuming negligible air resistance, show that a rock thrown from a bridge 20.0 m above water with an initial speed of 15.0 m/s strikes the water with a speed of 24.8 m/s independent of the direction thrown.

⁞ **6.30** (a) How high a hill can a car coast up (engine disengaged) if friction is negligible and its initial speed is 110 km/h? (b) If, in actuality, a 750 kg car with an initial speed of 110 km/h is observed to coast up a hill to a height 22.0 m above its starting point, how much thermal energy was generated by friction? (c) What is the average force of friction if the hill has a slope 2.5° above the horizontal?

Section 6.7 Power

6.31 The Crab nebula pulsar is the remnant of a supernova that occurred in A.D. 1054. Using data from Table 6.2, calculate the approximate factor by which the power output of this astronomical object has declined since its explosion.

6.32 Suppose a star 1000 times brighter than our sun (that is, emitting 1000 times the power) suddenly supernovas. Using data from Table 6.2: (a) By what factor does its power output increase? (b) How many times brighter than our entire Milky Way galaxy is it?

6.33 A person in good physical condition can put out 100 W of useful power for several hours at a stretch, perhaps by pedaling a mechanism that drives an electric generator. Neglecting any problems of generator efficiency and practical considerations such as resting time: (a) How many people would it take to run a 4.00 kW electric clothes dryer? (b) How many people would it take to replace a large electric power plant that generates 800 MW?

6.34 What is the cost of operating a 3.00 W electric clock for a year if the cost of electricity is 9.00¢ per kW·h?

6.35 A large household air conditioner may consume 15.0 kW of power. What is the cost of operating this air conditioner 3.00 hours per day for 30.0 days if the cost of electricity is 11.0¢ per kW·h?

6.36 (a) What is the average power consumption in watts of an appliance that uses 5.00 kW·h of energy per day? (b) How many joules of energy does it consume in a year?

• 6.37 (a) What is the average useful power output of a person who does 6.00×10^6 J of useful work in 8.00 h? (b) Working at this rate, how long will it take this person to lift 2000 kg of bricks 1.50 m to a platform? (Work done to lift his body can be omitted since it is not considered useful output here.)

• 6.38 A 500 kg dragster accelerates from rest to a final speed of 110 m/s in 400 m (about a quarter of a mile) and encounters an average frictional force of 1200 N. What is its average power output in watts and horsepower if this takes 7.30 s?

• 6.39 (a) How long will it take an 850 kg car with a useful power output of 40.0 hp (1 hp = 746 W) to reach a speed of 15.0 m/s, neglecting friction? (b) How long will this take if the car also climbs a 3.00 m high hill in the process?

• 6.40 (a) Find the useful power output of an elevator motor that lifts a 2500 kg load a height of 35.0 m in 12.0 s, if it also increases the speed from rest to 4.00 m/s. Note that the total mass of the counterbalanced system is 10,000 kg—so that only 2500 kg is raised in height, but the full 10,000 kg is accelerated. (b) What does it cost, if electricity is 9.00¢ per kW·h?

• 6.41 (a) What is the available energy content in joules of a battery that operates a 2.00 W electric clock for 18 months? (b) How long can a battery that can supply 8.00×10^4 J run a pocket calculator that consumes energy at the rate of 1.00×10^{-3} W?

‡ 6.42 Calculate the minimum safe distance you can be from a speaker that puts out 2.00 W of sound, using the following assumptions: The sound power spreads uniformly over a sphere of radius r and area $4\pi r^2$, your eardrum's area is 0.800 cm², and it cannot tolerate more than 1.00×10^{-6} W. Damage is done by this small amount of power after prolonged exposure and can be repaired by the body, but damage done by 10^{-4} W (found at many rock concerts) occurs almost instantaneously and prolonged exposure at that level produces permanent damage.

‡ 6.43 Calculate the power output needed for a 950 kg car to climb a 2.00° slope at a constant 30.0 m/s while encountering wind resistance and friction totaling 600 N.

‡ 6.44 (a) How long would it take a 1.50×10^5 kg airplane with engines that produce 100 MW of power to reach a speed of 250 m/s and an altitude of 12.0 km if air resistance were negligible? (b) If it actually takes 900 s, what is the power? (c) Given this power, what is the average force of air resistance if the airplane takes 1200 s? (Hint: You must find the distance the plane travels in 1200 s assuming constant acceleration.)

‡ 6.45 (a) Calculate the power per square meter reaching the upper atmosphere from the sun. (Take the power output of the sun to be 4.00×10^{26} W.) (b) Part of this is absorbed and reflected by the atmosphere, so that a maximum of 1.30 kW/m² reaches the earth's surface. Calculate the area in km² of solar energy collectors needed to replace an electric power plant that generates 750 MW if the collectors convert an average of 2.00% of the maximum power into electricity. (This small conversion efficiency is due to the devices themselves, and the fact that the sun is directly overhead only briefly.)

Section 6.8* Work, Energy and Power in Humans

6.46 How long can you play tennis on the 200 kcal of energy in a candy bar?

6.47 How long can you rapidly climb stairs (116/min) on the 93.0 kcal of energy in a 10.0 g pat of butter? How many flights is this if each flight has 16 stairs?

6.48 What is the power output in watts and horsepower of a 70.0 kg sprinter who accelerates from rest to 10.0 m/s in 3.00 s?

6.49 Calculate the power output in watts and horsepower of a shot-putter who takes 1.20 s to accelerate the 7.27 kg shot from rest to 14.0 m/s, while raising it 0.800 m. (Do not include the power produced to accelerate his body.)

6.50 (a) What is the efficiency of an out-of-condition professor who does 2.10×10^5 J of useful work while metabolizing 500 kcal of food energy? (b) How many food calories would a well-conditioned athlete metabolize in doing the same work with an efficiency of 20%?

6.51 Energy that is not utilized for work or heat is converted to body fat containing about 9.30 kcal/g. How many grams of fat will you gain if you eat 2500 kcal one day and do nothing but sit relaxed for 16.0 h and sleep for the other 8.00 h? Use data from Table 6.4 for the energy consumption rates of these activities.

6.52 Using data from Table 6.4, calculate the daily caloric needs of a person who sleeps for 7.00 h, walks for 2.00 h, attends classes for 4.00 h, cycles for 2.00 h, sits relaxed for 3.00 h, and studies for 6.00 h. (Studying consumes energy at the same rate as sitting in class.)

• 6.53 What is the efficiency of a subject on a treadmill who puts out work at the rate of 100 W while consuming oxygen at the rate of 2.00 L/min? (Hint: See Table 6.4.)

• 6.54 Shoveling snow can be extremely taxing since the arms have such a low efficiency in this activity. Suppose a person shoveling a sidewalk metabolizes food at the rate of 800 W. (a) What is her useful power output? (b) How long will it take her to lift 3000 kg of snow 1.20 m? (This could be the amount of heavy snow on 60.0 ft of sidewalk.) (c) How much waste heat in kcal will she generate in the process?

• 6.55 Very large forces are produced in joints when a person jumps from some height to the ground. (a) Calculate the force produced if an 80.0 kg person jumps from a 0.600 m high ledge and lands stiffly, compressing joint material 1.50 cm as a result. (Be certain to include the weight of the person.) (b) In practice the knees bend almost involuntarily to help extend the distance over which you stop. Calculate the force produced if the stopping distance is 0.300 m. (c) Compare both forces with the weight of the person.

· 6.56* Jogging on hard surfaces with insufficiently padded shoes produces large forces in the feet and legs. (a) Calculate the force needed to stop the downward motion of a jogger's leg, if his leg has a mass of 13.0 kg, a speed of 6.00 m/s, and stops in a distance of 1.50 cm. (Be certain to include the weight of the 75.0 kg jogger's body.) (b) Compare this force with the weight of the jogger.

· 6.57* (a) Calculate the energy in kcal used by a 55.0 kg woman who does 50 deep knee bends in which her center of mass is lowered and raised 0.400 m. (She does work in both directions.) You may assume her efficiency is 20%. (b) What is the average power consumption rate in watts if she does this in 3.00 min?

· 6.58* Kanellos Kanellopoulos flew 119 km (74 miles) from Crete to Santorini on April 23, 1988, in the Daedalus 88, an aircraft powered by a bicycle-type drive mechanism. His useful power output for the 234 min trip was about 350 W. Using the efficiency for the legs from Table 6.3, calculate the food energy in kcal he metabolized during the flight.

· 6.59* The swimmer shown in Figure 6.24 exerts an average horizontal backward force of 80.0 N with his arm during each 1.80 m long stroke. (a) What is his work output in each stroke? (b) Calculate the power output of his arms if he does 120 strokes per minute.

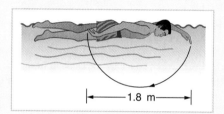

Figure 6.24 The work done by this swimmer's arms is considered in Problem 59.

⁝ 6.60* Mountain climbers carry bottled oxygen when at very high altitudes. (a) Assuming that a mountain climber uses oxygen at twice the rate for climbing 116 stairs per minute (because of low air temperature and winds), calculate how many liters of oxygen a climber would need for 10.0 h of climbing. (These are liters at sea level). Note that only 40% of the inhaled oxygen is utilized; the rest is exhaled. (b) How much useful work does the climber do if he and his equipment have a mass of 90.0 kg and he gains 1000 m of altitude? (c) What is his efficiency for the 10.0 h climb?

⁝ 6.61* The awe-inspiring Great Pyramid of Cheops was built more than 4500 years ago. Its square base, originally 230 m on a side, covered 13.1 acres, and it was 146 m high, with a mass of about 7×10^9 kg. (Its dimensions are slightly different today due to quarrying and some sagging.) Estimates are that 20,000 workers spent 20 years to construct it, working 12-hour days, 330 days per year. (a) Calculate the gravitational potential energy stored in the pyramid, given its center of mass is at one-fourth its height. (b) Only a fraction of the workers lifted blocks; most were involved in support services such as building ramps, bringing food and water, and hauling blocks to the site. Calculate the efficiency of the workers who did the lifting, assuming there were 1000 of them and they consumed food energy at the rate of 300 kcal/h. (The small answer means most of their work went into friction and lifting and lowering their own bodies.) (c) Calculate the mass of food that had to be supplied each day, assuming that the average worker required 3600 kcal per day and that their diet was 5% protein, 60% carbohydrate, and 35% fat. (This neglects the mass of bulk and nondigestible materials consumed.)

INTEGRATED CONCEPTS

These problems involve physical principles from more than one chapter. Integrated knowledge from a broad range of topics is much more powerful than a narrow application of physics. Energy, for example, is involved in kinematics, whence the relevance of Chapter 2. The following topics are involved in some or all of the problems in this section:

Topics	Location
Kinematics	Chapter 2
Two-dimensional kinematics	Chapter 3
Dynamics	Chapter 4
Statics, elasticity	Chapter 5

PROBLEM-SOLVING STRATEGY

Use the following strategy to solve integrated concept problems.

Step 1. *Identify which physical principles are involved.* This is done by following the problem-solving strategies found in this and previous chapters (see the table of contents or look in the index). Listing the givens and the quantities to be calculated will allow you to identify the principles involved.

Step 2. *Solve the problem using strategies outlined in the text.* If these are available for the specific topic, you should refer to them. You should also refer to the sections of the text that deal with a particular topic.

The following worked example illustrates how this strategy is applied to an integrated concept problem.

EXAMPLE **WHAT FORCE IS NEEDED TO TOSS A BALL 1 m INTO THE AIR?**

Topics	Location
Work, energy, and power	Chapter 6
Dynamics	Chapter 4

Suppose you toss a ball straight up with your hand. You start from rest and move it upward through 0.100 m before it leaves your hand at a speed of 3.00 m/s. (a) Using energy considerations, find the maximum height the ball reaches, assuming negligible air resistance. (b) What force did you exert on the ball if it had a mass of 0.400 kg?

Strategy Step 1 To solve an *integrated concept problem*, such as this example and the problems following it, we must first identify the physical principles involved and identify the chapters in which they are found. Part (a) of this example specifically asks you to use *energy*, which is covered in this chapter. Part (b) deals with *force*, a topic of Chapter 4.

Strategy Step 2 The following solutions to each part of the example illustrate how the specific problem-solving strategies are applied. These involve identifying knowns and unknowns, checking to see if the answer is reasonable, and so on.

Solution for (a): Energy (Chapter 6) Just as the ball leaves your hand it has kinetic energy, which is converted completely to gravitational potential energy when it reaches its maximum height. Neglecting air resistance, we can equate these two forms of energy and try to find the maximum height. That is,

$$KE_i = PE_f$$

or

$$\tfrac{1}{2}mv_i^2 = mgh$$

Noting that mass m cancels, we solve for the unknown, which is h. This yields

$$h = \frac{v_i^2}{2g}$$

Substituting knowns,

$$h = \frac{(3.00 \text{ m/s})^2}{2(9.80 \text{ m/s}^2)}$$

$$= 0.459 \text{ m}$$

Discussion for (a) This is a reasonable height to be casually tossing a ball.

Solution for (b): Dynamics (Chapter 4) We can find the net force by equating the work done with the kinetic energy. That is,

$$\text{net } Fd = \tfrac{1}{2}mv_i^2$$

so that net F is

$$\text{net } F = \frac{mv_i^2}{2d}$$

Substituting knowns yields

$$\text{net } F = \frac{(0.400 \text{ kg})(3.00 \text{ m/s})^2}{2(0.100 \text{ m})}$$

$$= 18.0 \text{ N}$$

But the net force equals the force you exerted (F) minus the weight of the ball. That is,

$$\text{net } F = F - w$$

so that

$$F = \text{net } F + w = \text{net } F + mg$$

$$= 18.0 \text{ N} + (0.400 \text{ kg})(9.80 \text{ m/s}^2)$$

$$= 21.9 \text{ N}$$

Discussion for (b) This is force you must exert on the ball as you lift it the 0.100 m before it leaves your hand. This force is greater than the weight of the ball (which is 3.92 N), as it must be in order to accelerate it upward.

⋮6.62 A 105 kg basketball player crouches down 0.400 m while waiting to jump for a ball. After exerting a force on the floor through this 0.400 m, his feet leave the floor and his center of gravity rises 0.950 m above its normal standing erect position. (a) Using energy considerations, calculate his velocity when he leaves the floor. (b) What average force did he exert on the floor? (Do not neglect the force to support his weight as well as that to accelerate him.) (c) What was his power output during the acceleration phase?

⋮6.63 (a) Calculate the force the woman in Figure 6.25 exerts to do a push-up at constant speed, taking all data to be known to three digits. (b) How much work does she do if her center of mass rises 0.240 m? (c) What is her useful power output if she does 25 push-ups in 1 min? (Should work done lowering her body be included? See the discussion of useful work in Section 6.8.)

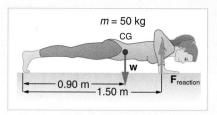

Figure 6.25 Push-ups done by this woman are the subject of Problem 63.

⋮6.64 Solve Examples 6.4 and 6.5 using only kinematics and dynamics. Are the solutions lengthier than those using energy considerations?

⋮6.65 Repeat Problem 6.28 using only kinematics and dynamics. Is the solution lengthier than that using energy considerations?

⋮6.66 A toy gun uses a spring with a force constant of 300 N/m to propel a 10.0 g steel ball. If the spring is compressed 7.00 cm and friction is negligible: (a) How much force is needed to compress the spring? (b) To what maximum height can the ball be shot? (c) At what angles above the horizontal may a child aim to hit a target 3.00 m away at the same height as the gun? (d) What is the gun's maximum range on level ground?

⋮6.67 (a) What force must be supplied by an elevator cable to produce an acceleration of 0.800 m/s² against a 200 N frictional force, if the mass of the loaded elevator is 1500 kg? (b) How much work is done by the cable in lifting the elevator 20.0 m? (c) What is the final speed of the elevator if it starts from rest? (d) How much work went into thermal energy?

UNREASONABLE RESULTS

As in previous chapters, problems with unreasonable results are included to give practice in assessing whether nature is being accurately described, and to trace the source of difficulty if it is not. This is very much like the process followed in original research when physical principles as well as faulty premises are tested. The following problems give results that are unreasonable even when the physics is correctly applied, because some premise is unreasonable or because certain premises are inconsistent.

Note: Strategies for determining if an answer is reasonable appear in Chapter 2. They can be found with the section of problems labeled *Unreasonable Results* at the end of that chapter.

⋮6.68 A car advertisement claims that its 900 kg car accelerated from rest to 30.0 m/s and drove 100 km, gaining 3.00 km in altitude, on 1.00 gallons of gasoline. The average force of friction including air resistance was 700 N. (a) Calculate the car's efficiency. (b) What is unreasonable about the result? (c) Which premise is unreasonable, or which premises are inconsistent?

⋮6.69 Body fat is metabolized, supplying 9.30 kcal/g, when dietary intake is less than need. The manufacturers of an exercise bicycle claim that you can lose 1.00 lb of fat per day by vigorously exercising for 2.00 h per day on their machine. (a) How many kcal are supplied by the metabolization of 1.00 lb of fat? (b) Calculate the kcal/min that you would have to utilize to metabolize fat at the rate of 1.00 lb in 2.00 h. (c) What is unreasonable about the results? (d) Which premise is unreasonable, or which premises are inconsistent?

LINEAR MOMENTUM

The airborne football player has great momentum,

which will soon affect his target.

We use the term *momentum* with various shades of meaning in everyday language, and most of them are consistent with its precise scientific definition. We speak of sports teams or politicians gaining and maintaining the momentum to win. We also recognize that momentum has something to do with collisions. For example, looking at the massive football player in the photograph hurtling through midair toward an opponent, we expect his momentum will have a great effect. Generally, momentum implies a tendency to continue on course—to move in the same direction—and is associated with great mass and speed.

Momentum—like energy—is important because it is conserved. Only a few physical quantities are conserved in nature, and their study yields fundamental insight into the workings of nature, as we shall see in our study of momentum.

7.1 LINEAR MOMENTUM AND FORCE

The scientific definition of linear momentum is consistent with our intuitive feeling that a large, fast-moving object has great momentum. **Linear momentum*** is the product of a system's mass times its velocity. In symbols, it is expressed as

$$\mathbf{p} = m\mathbf{v} \tag{7.1}$$

Momentum **p** is a vector having the same direction as the velocity **v**. The SI unit for momentum is kg·m/s.

EXAMPLE 7.1 HOW GREAT IS THE MOMENTUM OF A FOOTBALL PLAYER?

(a) Calculate the momentum of a 110 kg football player running at 8.00 m/s. (b) Compare the player's momentum with that of a hard-thrown 0.410 kg football that has a speed of 25.0 m/s.

Strategy No information is given regarding direction, and so we can calculate only the magnitude of the player's momentum p. (As usual a symbol in italics is a magnitude, whereas one in boldface is a vector.) In both parts of this example, the magnitude of momentum can be calculated directly from the definition of momentum given in Equation 7.1, which becomes

$$p = mv$$

when only magnitudes are involved.

Solution for (a) Substituting known values for mass

and speed gives

$$p = (110 \text{ kg})(8.00 \text{ m/s}) = 880 \text{ kg·m/s}$$

Solution for (b) Similarly, for the football,

$$p_{\text{fb}} = (0.410 \text{ kg})(25.0 \text{ m/s}) = 10.3 \text{ kg·m/s}$$

where p_{fb} is the momentum of the football. The ratio of the player's momentum to that of the football is

$$\frac{p}{p_{\text{fb}}} = \frac{880}{10.3} = 85.9$$

Discussion The momentum of the player is much greater than that of the football, as you might guess. This is why the player's motion is only slightly affected if he catches the ball. We shall quantify what happens in such collisions in terms of momentum a little later on in this chapter.

Momentum and Newton's Second Law

The importance of momentum, unlike that of energy, was recognized early in the development of classical physics. It was deemed so important that it was called the "quantity of motion." Newton actually stated his **second law of motion in terms of momentum**: the net external force equals the change in momentum of a system divided by the time over which it changes. In symbols, this is

$$\text{net } \mathbf{F} = \frac{\Delta \mathbf{p}}{\Delta t} \tag{7.2}$$

where net **F** is the net external force, $\Delta \mathbf{p}$ is the change in momentum, and Δt is the change in time.

---

*Simply referred to as *momentum* from now on. In Chapter 9 we will consider the separate topic of angular momentum associated with rotation.

This statement of Newton's second law of motion includes the more familiar net $\mathbf{F} = m\mathbf{a}$ as a special case. We can derive this form as follows. First, note that the change in momentum $\Delta\mathbf{p}$ is

$$\Delta\mathbf{p} = \Delta(m\mathbf{v})$$

If mass is constant, then

$$\Delta(m\mathbf{v}) = m\,\Delta\mathbf{v}$$

so that for constant mass, Newton's second law of motion becomes

$$\text{net } \mathbf{F} = \frac{\Delta\mathbf{p}}{\Delta t} = \frac{m\,\Delta\mathbf{v}}{\Delta t}$$

Since $\Delta\mathbf{v}/\Delta t = \mathbf{a}$, we get the familiar

$$\text{net } \mathbf{F} = m\mathbf{a}$$

when the mass of the system is constant.

Newton's second law of motion stated in terms of momentum, as in Equation 7.2, is more generally applicable because it can be applied to systems the mass of which is changing, such as a rocket, as well as to systems of constant mass. We will consider systems with varying mass in some detail in Section 7.7, but we will still find the relationship of momentum to force to be useful when mass is constant, such as in the following example.

EXAMPLE 7.2   WHAT AVERAGE FORCE DOES A BAT EXERT TO CHANGE THE MOMENTUM OF A BALL?

What is the average force exerted on a 0.140 kg baseball by a bat, given that the ball's initial velocity is 45.0 m/s and that its final velocity, after a 1.30 ms impact, is 65.0 m/s in exactly the opposite direction?

Strategy This is a one-dimensional problem (with directions indicated by plus and minus signs), since the ball exactly reverses its direction. Newton's second law stated in terms of momentum is then written

$$\text{net } F = \frac{\Delta p}{\Delta t}$$

As noted above, when mass is constant the change in momentum is

$$\Delta p = m\,\Delta v = m(v_{\text{f}} - v_{\text{i}})$$

In this example, the initial and final velocities are given, as is the change in time; thus, once Δp is calculated, Equation 7.2 can be used to find the force.

Solution Substituting the given values for the initial and final velocities,

$$\Delta p = m(v_{\text{f}} - v_{\text{i}})$$
$$= (0.140 \text{ kg})[65.0 \text{ m/s} - (-45.0 \text{ m/s})]$$
$$= 15.4 \text{ kg}\cdot\text{m/s}$$

Now the net external force is given by Equation 7.2:

$$\text{net } F = \frac{\Delta p}{\Delta t} = \frac{15.4 \text{ kg}\cdot\text{m/s}}{1.30 \times 10^{-3} \text{ s}}$$
$$= 1.18 \times 10^{4} \text{ N}$$

Discussion This force is more than 2600 lb! This is the average force the bat exerts on the ball during its brief impact (neglecting the 1.37 N force of gravity). Figure 7.1 shows the effect of a golf club on a golf ball, similar to what happens when a baseball bat hits a baseball. This problem can also be solved by finding the acceleration and using net $F = ma$. One additional step is required.

Figure 7.1   The tremendous force exerted on a golf ball by a golf club is evident in this photograph.

7.2 IMPULSE

The effect of a force on an object depends on how long it acts, as well as how great the force is. In Example 7.2, a very large force acting for a short time had a great effect on the momentum of a baseball. A small force could cause the same change in momentum, but it would have to act for a much longer time. For example, if the ball were thrown upward, the force of gravity (which is much smaller than the bat's force) would eventually reverse the momentum of the ball.

Quantitatively, the effect we are talking about is the *change in momentum* $\Delta \mathbf{p}$. By rearranging Equation 7.2 to be

$$\Delta \mathbf{p} = \text{net } \mathbf{F} \cdot \Delta t$$

we can see how the change in momentum equals the average net external force multiplied by the time this force acts. The quantity net $\mathbf{F} \cdot \Delta t$ is given the name **impulse**. Impulse is the same as the change in momentum.

EXAMPLE 7.3   IMPULSE OF A GOALPOST ON A FOOTBALL PLAYER

Calculate the final speed of a 110 kg football player initially running at 8.00 m/s who collides head on with a padded goalpost and experiences a backward force of 1.76×10^4 N for 5.50×10^{-2} s.

Strategy We can calculate the impulse $\Delta p = \text{net } F \cdot \Delta t$ on the player, since the force and time are given. Once the change in momentum Δp is known, we can then get the final velocity, since the player's mass and initial velocity are also given.

Solution Entering the given information, the impulse given to the player is

$$\Delta p = \text{net } F \cdot \Delta t = (-1.76 \times 10^4 \text{ N})(5.50 \times 10^{-2} \text{ s})$$
$$= -968 \text{ kg} \cdot \text{m/s}$$

The minus sign indicates that the force is in the direction opposite to the initial momentum. (So is Δp.) We can find the final velocity by noting that the change in momentum is

$$\Delta p = mv_{\mathrm{f}} - mv_{\mathrm{i}}$$

We solve this for v_{f}:

$$v_{\mathrm{f}} = \frac{\Delta p}{m} + v_{\mathrm{i}}$$

Then we enter known values:

$$v_{\mathrm{f}} = \frac{-968 \text{ kg} \cdot \text{m/s}}{110 \text{ kg}} + 8.00 \text{ m/s}$$
$$= -0.800 \text{ m/s}$$

Discussion The minus sign indicates the player bounces backward (and probably slumps to the ground).

CONNECTIONS

The assumption of a constant force in the definition of impulse is analogous to the assumption of a constant acceleration in kinematics. In both cases nature is adequately described without the use of calculus.

Our definition of impulse includes an assumption that the force is constant over the time interval Δt. Usually, this is not the case. Forces vary considerably even during the brief time intervals considered. It is, however, possible to find an average effective force that produces the same result as the time-varying one. Figure 7.2 shows a graph of what an actual force looks like as a function of time. The area under the curve has units of momentum and is equal to the impulse or change in momentum between times t_1 and t_2. That area is equal to the area inside the rectangle bounded by F_{eff}, t_1, and t_2. Thus the impulses and their effects are the same for both the actual and effective forces. Note that in Section 6.2 we found that the area under the graph of force versus *distance* is the *work* done by the force.

Figure 7.2   (a) A graph of force versus time for an actual force and an equivalent effective force. The areas under the two curves are equal. (b) A piano hammer striking a string would generate a force similar to F_{actual}, but its impulse might be the same as that of F_{eff}. Experiments would determine whether the two different forces produce different overtones.

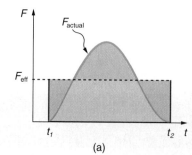

(a)

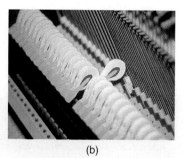

(b)

7.3 CONSERVATION OF MOMENTUM

We have said that momentum is important because it is conserved. Yet it was not conserved in preceding examples, where large changes in momentum were produced by forces acting on the system of interest. Under what circumstances is momentum conserved?

The answer to this question entails considering a large enough system. It is always possible to find a larger system in which total momentum is constant, even if momentum changes for components of the system. In the case of the football player hitting the goalpost, for example, the earth actually recoils—conserving momentum—because of the force applied to it through the goalpost. Since the earth is many orders of magnitude more massive that the player, its recoil is unmeasurably small and can be neglected in any practical sense, but it is real nevertheless.

Consider what happens if the masses of two colliding objects are more similar than those of the football player and the earth—for example, one car bumping into another, as shown in Figure 7.3. Both cars are coasting in the same direction when the lead car (labeled 2) is bumped by the following car (labeled 1). The only unbalanced force on each car is the force of the collision. Car 1 slows down as a result of the collision, losing some momentum, while car 2 is sped up and gains some momentum. We shall now show that the total momentum of the two-car system remains constant.

Using the definition of impulse, the change in momentum of car 1 is

$$\Delta p_1 = F_1 \Delta t$$

where F_1 is the force on car 1 due to car 2, and Δt is the time the force acts (the duration of the collision). Similarly, the change in momentum of car 2 is

$$\Delta p_2 = F_2 \Delta t$$

where F_2 is the force on car 2 due to car 1, and we assume the duration of the collision Δt is the same for both cars.* We know from Newton's third law that $F_2 = -F_1$, and so

$$\Delta p_2 = -F_1 \Delta t = -\Delta p_1$$

Thus the changes in momentum are equal and opposite, and

$$\Delta p_1 + \Delta p_2 = 0$$

---

*Intuitively, it seems obvious that the duration of the collision is the same for both cars, but it is only true for macroscopic objects traveling at ordinary speeds. This assumption is modified in modern physics, without affecting the result that momentum is conserved.

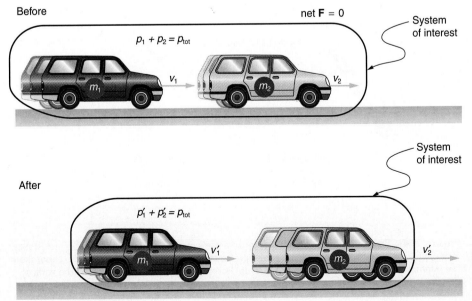

Figure 7.3 A car bumps into the one it is following. The first car slows down and the second speeds up. The momentum of each car is changed, but, assuming negligible friction, the total momentum p_{tot} of the two cars is the same before and after the collision.

Since the changes in momentum add to zero, the total momentum of the two-car system is constant. That is,

$$p_1 + p_2 = \text{constant}$$

or

$$p_1 + p_2 = p_1' + p_2'$$

where p_1' and p_2' are the momenta of cars 1 and 2 after the collision. (We often use primes to denote the final state.)

This result—that momentum is conserved—has a validity far beyond the preceding one-dimensional case. It can be similarly shown that total momentum is conserved for any isolated system, with any number of bodies in it. In equation form, the **conservation of momentum principle** is written

$$\left. \begin{array}{c} \mathbf{p}_{\text{tot}} = \text{constant} \\[1em] \mathbf{p}_{\text{tot}} = \mathbf{p}_{\text{tot}}' \end{array} \right\} \quad \text{(isolated system)} \qquad \textbf{(7.3)}$$

or

where $\mathbf{p}_{\text{tot}}$ is the total momentum (the sum of the momenta of the individual bodies in the system) and $\mathbf{p}_{\text{tot}}'$ is the total momentum some time later. (The total momentum can be shown to be the momentum of the center of mass of the system.) An **isolated system** is defined to be one for which the net external force is zero (net $\mathbf{F} = 0$).

Perhaps an easier way to see that momentum is conserved for an isolated system is to consider Newton's second law in terms of momentum, net $\mathbf{F} = \Delta \mathbf{p}_{\text{tot}}/\Delta t$. For an isolated system, net $\mathbf{F} = 0$; thus, $\Delta \mathbf{p}_{\text{tot}} = 0$, and $\mathbf{p}_{\text{tot}}$ is constant.

We have noted that the three dimensions in nature are independent, and it is interesting to note that momentum can be conserved along one direction and not another. For example, in projectile motion where air resistance is negligible, momentum is conserved in the horizontal direction but not along the vertical, since horizontal forces are zero while the net vertical force is not. (See Figure 7.4.)

The conservation of momentum principle can be applied to systems as different as a comet striking the earth and a gas containing huge numbers of atoms and molecules. Conservation of momentum is violated only when the net external force is not zero. But another larger system can always be considered in which momentum is conserved by simply including the source of the external force. For example, in the collision of two cars considered above, the two-car system conserves momentum while each one-car system does not.

CONNECTIONS

Conservation of momentum is quite useful in describing collisions. We shall see in later chapters that momentum is crucial to our understanding of the submicroscopic world because much of what we know about it comes from collision experiments. For an example, see the box on the next page.

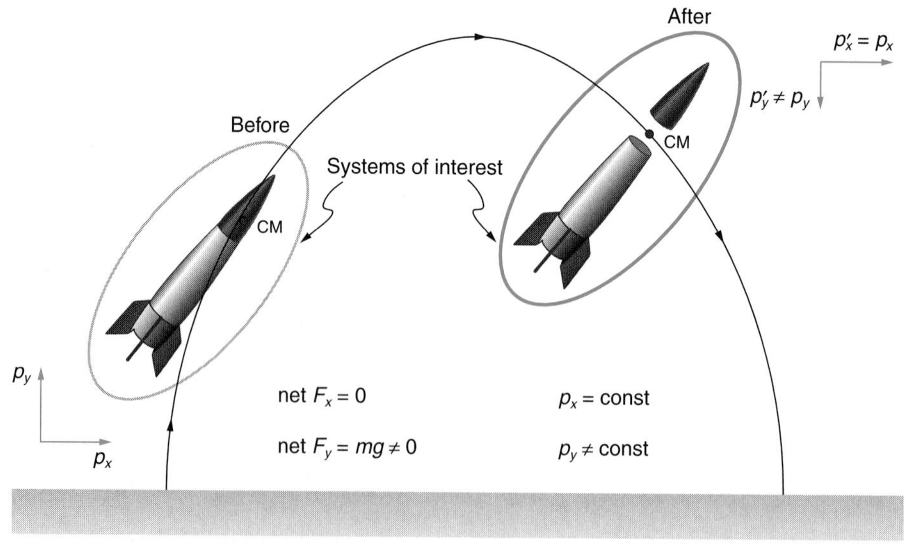

Figure 7.4 The horizontal component of a projectile's momentum is conserved if air resistance is negligible, even in this case where a space probe separates. The forces causing the separation are internal to the system, so that the net external horizontal force net F_x is still zero. The vertical component of the momentum is not conserved, since the net vertical force net F_y is not zero. The center of mass of the space probe takes the same path it would if the separation did not occur.

THINGS GREAT AND SMALL

Submicroscopic Collisions and Momentum

The conservation of momentum principle not only applies to the macroscopic world, it is also essential to our explorations of the substructure of matter. Giant machines hurl submicroscopic particles at one another and evaluate the results by assuming conservation of momentum (among other things). Figure 7.A shows a detector used in such experiments.

On the small scale, we find that particles and details of their properties invisible to the naked eye can be measured with our instruments, and models of the submicroscopic world can be constructed to describe the results. Momentum is found to be a property of all submicroscopic particles *including massless particles*, such as photons composing light. This hints that momentum may have an identity beyond mass times velocity. Furthermore, we find that the conservation of momentum principle is valid. We use it to analyze the mass and other properties of previously unde-

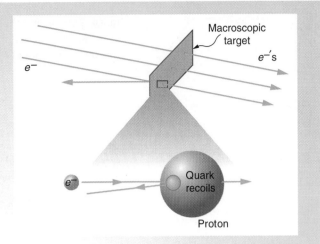

Figure 7.B A submicroscopic particle scatters straight backward from a target particle. In experiments seeking evidence for quarks, electrons were observed to occasionally scatter straight backward from a proton.

tected particles, such as the nucleus of the atom and theexistence of quarks inside the nucleus itself. Figure 7.B illustrates how a particle scattering backward from another implies that its target is massive and dense. Experiments seeking evidence of quarks inside nuclei scattered high-energy electrons from protons (nuclei of hydrogen atoms). Occasionally, electrons scattered straight backward in a manner that implied a very small and very dense particle *inside* the proton—this is considered nearly direct evidence of quarks. The analysis, again, was based partly on the same conservation of momentum principle that works so well on the large scale.

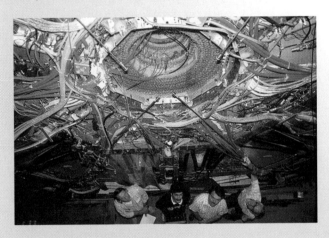

Figure 7.A A huge accelerator hurls tiny particles at one another to explore the submicroscopic world. The results of the collisions are measured with huge detectors such as this, and they are analyzed with the conservation of momentum principle, among others.

7.4 ELASTIC COLLISIONS IN ONE DIMENSION

In this and the next two sections, we consider various types of two-body collisions. These are the easiest to analyze, and they illustrate many of the physical principles involved in collisions. The conservation of momentum principle is very useful here, and it can be used whenever the net external force on a system is zero.

We start with the elastic collision of two bodies moving along the same line—a one-dimensional problem. An **elastic collision** is one that conserves internal kinetic energy. Internal kinetic energy is the sum of the kinetic energies of the bodies in the system. Figure 7.5 illustrates an elastic collision in which internal kinetic energy and momentum are conserved.

Figure 7.5 An elastic one-dimensional two-body collision. Momentum *and* internal kinetic energy are conserved. Truly elastic collisions can only be achieved with submicroscopic particles, such as electrons striking nuclei. Macroscopic collisions can be very nearly, but not quite, elastic—some kinetic energy is always converted into other forms such as heat and sound. One macroscopic collision that is nearly elastic is that of two steel blocks on ice. Another nearly elastic collision is that between two carts with spring bumpers on an air track.

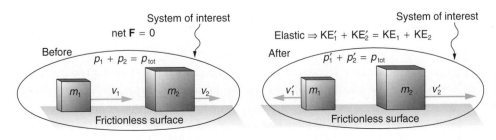

Now, to solve problems involving one-dimensional elastic collisions between two masses, we can use the equations for conservation of momentum and conservation of internal kinetic energy. First, the equation for conservation of momentum for two bodies in a one-dimensional collision is

$$p_1 + p_2 = p_1' + p_2'$$

or

$$m_1 v_1 + m_2 v_2 = m_1 v_1' + m_2 v_2' \qquad \text{(net } F = 0\text{)} \qquad \textbf{(7.4)}$$

where prime ($'$) indicates values after the collision. By definition, an elastic collision conserves internal kinetic energy, and so the sum of kinetic energies before the collision equals the sum after the collision. Thus,

$$\tfrac{1}{2}m_1 v_1^2 + \tfrac{1}{2}m_2 v_2^2 = \tfrac{1}{2}m_1 v_1'^2 + \tfrac{1}{2}m_2 v_2'^2 \qquad \text{(two-body elastic collision)} \qquad \textbf{(7.5)}$$

is the equation for conservation of internal kinetic energy in a one-dimensional collision.

EXAMPLE 7.4 AN ELASTIC COLLISION

Calculate the velocities of two masses following an elastic collision, given that $m_1 = 0.500$ kg, $m_2 = 3.50$ kg, $v_1 = 4.00$ m/s, and $v_2 = 0$.

Strategy and Concept First visualize what the initial conditions mean—a small mass strikes a larger mass that is initially at rest. This is slightly simpler than the situation shown in Figure 7.5, where both masses are initially moving. We are asked to find two unknowns (the final velocities v_1' and v_2'). To find two unknowns, we must use two independent equations. Since this is an elastic collision, we can use Equations 7.4 and 7.5. Both are simplified by the fact that mass 2 is initially at rest, and thus $v_2 = 0$. Once we simplify these equations, we combine them algebraically to solve for the unknowns.

Solution Noting that $v_2 = 0$, conservation of momentum stated in Equation 7.4 simplifies to

$$p_1 = p_1' + p_2'$$

or

$$m_1 v_1 = m_1 v_1' + m_2 v_2'$$

Conservation of internal kinetic energy, stated in Equation 7.5, is

$$\tfrac{1}{2}m_1 v_1^2 = \tfrac{1}{2}m_1 v_1'^2 + \tfrac{1}{2}m_2 v_2'^2$$

Solving the first (conservation of momentum) for v_2', we obtain

$$v_2' = \frac{m_1}{m_2}(v_1 - v_1')$$

Substituting this into the equation for conservation of internal kinetic energy eliminates the variable v_2', leaving only

v_1' as an unknown (the algebra is left as an exercise for the reader). There are two solutions to any quadratic equation; in this example, they are

$$v_1' = 4.00 \text{ m/s}$$

and

$$v_1' = -3.00 \text{ m/s}$$

As noted when quadratic equations were encountered in earlier chapters, both solutions may or may not be meaningful. In this case, the first solution is the same as the initial condition. The first solution thus represents the situation before the collision and is discarded. The second solution ($v_1' = -3.00$ m/s) is negative, meaning that the first mass bounces backward. When this negative value of v_1' is used to find the velocity of the second mass after the collision, we get

$$v_2' = \frac{m_1}{m_2}(v_1 - v_1') = \frac{0.500 \text{ kg}}{3.50 \text{ kg}}[4.00 - (-3.00)] \text{ m/s}$$

or

$$v_2' = 1.00 \text{ m/s}$$

Discussion The result of this example is intuitively reasonable. A small mass strikes a larger one at rest and bounces backward. The larger one is knocked forward, but with a low speed. (This is like a Honda bouncing backward off a Cadillac that is initially at rest.) As a check, try calculating the internal kinetic energy before and after the collision. You will see that it is unchanged at 4.00 J. Also check the total momentum before and after the collision; you will find it, too, is unchanged.

The equations for conservation of momentum and internal kinetic energy as written above can be used to describe any one-dimensional elastic collision of two bodies. These equations can be extended to more bodies if needed.

7.5 INELASTIC COLLISIONS IN ONE DIMENSION

We have seen that in an elastic collision internal kinetic energy is conserved. An **inelastic collision** is one in which internal kinetic energy changes (it is *not* conserved). This means the forces between colliding objects may remove or add internal kinetic energy. Work done by internal forces may change the forms of energy within a system. For inelastic collisions, this internal work may transform some internal kinetic energy into heat, such as when colliding objects stick together. Or it may convert stored energy into internal kinetic energy, such as when exploding bolts separate a satellite from its launch vehicle.

Figure 7.6 shows an example of an inelastic collision. Two equal masses head toward one another at equal speeds and then stick together. Their total internal kinetic energy is initially mv^2 (twice $(1/2)mv^2$). The two masses come to rest after sticking together, conserving momentum. But the internal kinetic energy is zero after the collision. A collision in which the objects stick together is sometimes called perfectly inelastic, since it reduces internal kinetic energy more than does any other type of inelastic collision. In fact, it reduces internal kinetic energy to the minimum it can while still conserving momentum.

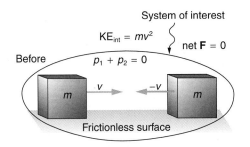
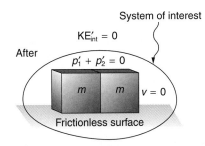

Figure 7.6 An inelastic one-dimensional two-body collision. Momentum is conserved, but internal kinetic energy is *not* conserved. (a) Two equal masses initially head directly toward one another at the same speed. (b) The masses stick together (a perfectly inelastic collision), and so their final velocity is zero. The internal kinetic energy of the system changes in any inelastic collision and is reduced to *zero* in this example.

EXAMPLE 7.5 INELASTIC COLLISION OF PUCK AND GOALIE

(a) Find the recoil velocity of a 70.0 kg hockey goalie, originally at rest, who catches a 0.150 kg hockey puck slapped at him at a velocity of 35.0 m/s. (b) How much kinetic energy is lost in the collision? Assume friction between the ice and the puck-goalie system is negligible. (See Figure 7.7.)

Strategy Momentum is conserved because the net external force on the puck-goalie system is zero. We can thus use conservation of momentum to find the final velocity. Note that the initial velocity of the goalie is zero and that the final velocity of the puck and goalie are the same. Once the final velocity is found, the kinetic energies can be calculated before and after the collision and compared as requested.

Solution for (a) Momentum is conserved because the net external force on the puck-goalie system is zero.

Conservation of momentum, given by Equation 7.4, is

$$p_1 + p_2 = p_1' + p_2'$$

or

$$m_1 v_1 + m_2 v_2 = m_1 v_1' + m_2 v_2'$$

Since the goalie is initially at rest, we know $v_2 = 0$; and since the goalie catches the puck, the final velocities are equal, or $v_1' = v_2' = v'$. Thus the conservation of momentum equation simplifies to

$$m_1 v_1 = (m_1 + m_2)v'$$

Solving for v',

$$v' = \frac{m_1}{m_1 + m_2} v_1$$

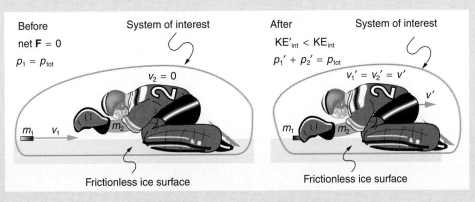

Figure 7.7 A goalie catches a hockey puck and recoils backward. The initial kinetic energy of the puck is almost entirely converted to heat and sound in this inelastic collision.

(continued)

(continued)

Entering known values,

$$v' = \frac{0.150 \text{ kg}}{70.15 \text{ kg}} \, 35.0 \text{ m/s} = 7.48 \times 10^{-2} \text{ m/s}$$

Discussion for (a) This is a small recoil velocity (about 3 inches per second) in the same direction as the puck's original velocity, as we might expect.

Solution for (b) Before the collision, the internal kinetic energy KE_{int} of the system is that of the hockey puck, since the goalie is initially at rest. KE_{int} is, thus, initially

$$\text{KE}_{\text{int}} = \tfrac{1}{2}mv^2 = \tfrac{1}{2}(0.150 \text{ kg})(35.0 \text{ m/s})^2$$
$$= 91.9 \text{ J}$$

After the collision, the internal kinetic energy is

$$\text{KE}'_{\text{int}} = \tfrac{1}{2}(m + M)v^2 = \tfrac{1}{2}(70.15 \text{ kg})(7.48 \times 10^{-2} \text{ m/s})^2$$
$$= 0.196 \text{ J}$$

The loss of internal kinetic energy is thus

$$\text{KE}'_{\text{int}} - \text{KE}_{\text{int}} = 0.196 \text{ J} - 91.9 \text{ J}$$
$$= -91.7 \text{ J}$$

Discussion for (b) Nearly all of the initial internal kinetic energy is lost in this perfectly inelastic collision. It is mostly converted to heat and sound.

In some collisions, the objects do not stick together and less of the internal kinetic energy is removed—such as happens in most automobile accidents. Alternatively, stored energy may be converted into internal kinetic energy in a collision. Figure 7.8 shows a one-dimensional example in which two carts on an air track collide, releasing potential energy from a compressed spring. The following example deals with data from such a collision.

EXAMPLE 7.6 A COLLISION THAT RELEASES STORED ENERGY IS INELASTIC

In the collision pictured in Figure 7.8, two carts collide inelastically, and a spring releases its potential energy and converts it to internal kinetic energy. The mass of the cart on the left is 0.350 kg and its initial velocity is 2.00 m/s. The cart on the right has a mass of 0.500 kg and an initial velocity of −0.500 m/s. After the collision, the first cart is observed to recoil with a velocity of −4.00 m/s. (a) What is the final velocity of the 0.500 kg cart? (b) How much energy was released by the spring (assuming all of it was converted into internal kinetic energy)?

Strategy We can use conservation of momentum to find the final velocity of cart 2, since net $F = 0$ (the track is frictionless). Once this is done, we can compare internal kinetic energy before and after the collision to see how much energy was released by the spring.

Solution for (a) As before, the equation for conservation of momentum in a two-body system is

$$m_1v_1 + m_2v_2 = m_1v'_1 + m_2v'_2$$

The only unknown in this equation is v'_2. Solving for v'_2 and substituting known values yields

$$v'_2 = \frac{m_1v_1 + m_2v_2 - m_1v'_1}{m_2}$$

$$= \frac{(0.350 \text{ kg})(2.00 \text{ m/s}) + (0.500 \text{ kg})(-0.500 \text{ m/s})}{0.500 \text{ kg}}$$

$$- \frac{(0.350 \text{ kg})(-4.00 \text{ m/s})}{0.500 \text{ kg}}$$

$$= 3.70 \text{ m/s}$$

Solution for (b) The internal kinetic energy before the collision is

$$\text{KE}_{\text{int}} = \tfrac{1}{2}m_1v_1^2 + \tfrac{1}{2}m_2v_2^2$$
$$= \tfrac{1}{2}(0.350 \text{ kg})(2.00 \text{ m/s})^2$$
$$+ \tfrac{1}{2}(0.500 \text{ kg})(-0.500 \text{ m/s})^2$$
$$= 0.763 \text{ J}$$

After the collision, the internal kinetic energy is

$$\text{KE}'_{\text{int}} = \tfrac{1}{2}m_1v_1'^2 + \tfrac{1}{2}m_2v_2'^2$$
$$= \tfrac{1}{2}(0.350 \text{ kg})(-4.00 \text{ m/s})^2$$
$$+ \tfrac{1}{2}(0.500 \text{ kg})(3.70 \text{ m/s})^2$$
$$= 6.22 \text{ J}$$

The change in internal kinetic energy is thus

$$\text{KE}'_{\text{int}} - \text{KE}_{\text{int}} = 6.22 \text{ J} - 0.763 \text{ J}$$
$$= 5.46 \text{ J}$$

Discussion The final velocity of cart 2 is large and positive, meaning that it is moving to the right after the collision. The internal kinetic energy increases by 5.46 J. That energy was released by the spring.

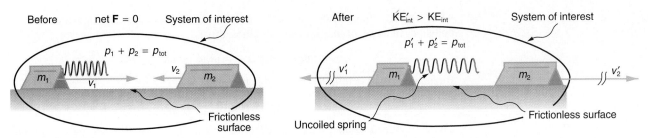

Figure 7.8 An air track is nearly frictionless, so that momentum is conserved. Motion is one-dimensional. In this collision, examined in Example 7.6, potential energy of a compressed spring is released during the collision and is converted to internal kinetic energy.

7.6* COLLISIONS OF POINT MASSES IN TWO DIMENSIONS

In the previous two sections, we considered only one-dimensional collisions; in such collisions, the incoming and outgoing velocities are all along the same line. But what about collisions, such as those between billiard balls, in which objects scatter to the side? (See Figure 7.9.) These are two-dimensional collisions, and we shall see that their study is an extension of the one-dimensional analysis already presented. The approach taken, just as was done in two-dimensional kinematics and dynamics, is to choose a convenient coordinate system and break the motion into components along perpendicular axes. This yields a pair of one-dimensional problems to be solved simultaneously.

One complication arising in two-dimensional collisions is that the objects might rotate before or after their collision. For example, if two skaters hook arms as they pass by one another, they will spin in circles. We will not consider rotation until the next chapter, and so for now, we arrange things so that no rotation is possible. (You may recall we avoided rotation in the material on statics, too.) To avoid rotation, we consider only the scattering of *point masses*—that is, structureless particles that cannot rotate or spin.

We start by assuming that net **F** = 0, so that momentum **p** is conserved. The simplest collision is one in which one of the masses is initially at rest. (See Figure 7.10.) The best choice for a coordinate system is one with an axis parallel to the velocity of the incoming mass, as shown in Figure 7.10. Since momentum is conserved, the components of momentum along the x- and y-axes (p_x and p_y) will also be conserved. But with the chosen coordinate system, p_y is initially zero and p_x is the momentum of the incoming mass. Both facts simplify the analysis.[†]

Along the x-axis, the equation for conservation of momentum is

$$p_{1x} + p_{2x} = p'_{1x} + p'_{2x}$$

Figure 7.9 The collision of these billiard balls is two-dimensional, confined to the horizontal plane

---

[†] Even with the simplifying assumptions of point masses, one mass initially at rest, and a convenient coordinate system, we still gain new insights into nature from the analysis of two-dimensional collisions.

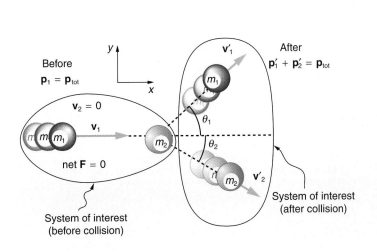

Figure 7.10 A two-dimensional collision with the coordinate system chosen so that m_2 is initially at rest and v_1 is parallel to the x-axis. This is sometimes called the laboratory coordinate system, since many scattering experiments have a target mass that is stationary in the laboratory, while particles are scattered from it to determine its character. The masses may not be observed directly, but their initial and final velocities are. This type of experiment has revealed much of what we know about the submicroscopic world.

where the subscripts denote the masses and axes and prime denotes the situation after the collision. In terms of masses and velocities, this is

$$m_1 v_{1x} + m_2 v_{2x} = m_1 v'_{1x} + m_2 v'_{2x}$$

But since mass 2 is initially at rest, this becomes

$$m_1 v_{1x} = m_1 v'_{1x} + m_2 v'_{2x}$$

The components of the velocities along the x-axis are $v \cos \theta$. Since mass 1 initially moves along the x-axis, we see $v_{1x} = v_1$. Thus conservation of momentum along the x-axis gives the following equation:

$$m_1 v_1 = m_1 v'_1 \cos \theta_1 + m_2 v'_2 \cos \theta_2 \qquad \textbf{(7.6a)}$$

where θ_1 and θ_2 are shown in Figure 7.10.

Along the y-axis, the equation for conservation of momentum is

$$p_{1y} + p_{2y} = p'_{1y} + p'_{2y}$$

or

$$m_1 v_{1y} + m_2 v_{2y} = m_1 v'_{1y} + m_2 v'_{2y}$$

But v_{1y} is zero, since mass 1 initially moves along the x-axis. And since mass 2 is initially at rest, v_{2y} is also zero. The equation for conservation of momentum along the y-axis becomes

$$0 = m_1 v'_{1y} + m_2 v'_{2y}$$

The components of the velocities along the y-axis are $v \sin \theta$. Thus conservation of momentum along the y-axis gives the following equation:

$$0 = m_1 v'_1 \sin \theta_1 + m_2 v'_2 \sin \theta_2 \qquad \textbf{(7.6b)}$$

Equations 7.6a and b are very useful in analyzing two-dimensional collisions of point masses, where one is originally stationary (a common laboratory situation). But two equations can only be used to find two unknowns, and so other data may be necessary when collision experiments are used to explore nature at the submicroscopic level.

EXAMPLE 7.7   DETERMINE THE FINAL VELOCITY OF AN UNSEEN MASS FROM THE SCATTERING OF ANOTHER MASS

Suppose the following experiment is performed. A 0.250 kg mass is slid on a frictionless surface into a dark room, where it strikes an initially stationary object with mass 0.400 kg. The 0.250 kg mass emerges from the room at an angle of 45.0° with its incoming direction. (See Figure 7.11.) The speed of the 0.250 kg mass was originally 2.00 m/s, and it is 1.50 m/s after the collision. Calculate the magnitude and direction of the velocity (v'_2 and θ_2) of the 0.400 kg mass after the collision.

Strategy Momentum is conserved since the surface is frictionless. The coordinate system shown in the figure is one in which m_2 is originally at rest and the initial velocity is parallel to the x-axis, so that conservation of momentum as stated in Equations 7.6a and b is applicable. Everything is known in these equations except v'_2 and θ_2, which are precisely the quantities we wish to find. We can find two unknowns because we have two independent equations: conservation of momentum in the x- and y-directions.

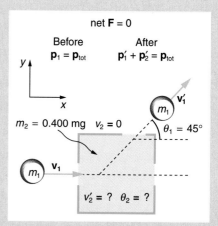

Figure 7.11 A collision taking place in a dark room is explored in Example 7.7. The incoming mass m_1 is scattered by an initially stationary object. Only the stationary object's mass m_2 is known. By measuring the angle and speed at which m_1 emerges from the room, it is possible to calculate the magnitude and direction of m_2's velocity after the collision.

(continued)

(continued)

Solution Solving Equation 7.6b for $v_2' \sin \theta_2$ and 7.6a for $v_2' \cos \theta_2$, and taking the ratio yields an equation in which all but one quantity is known:

$$\tan \theta_2 = \frac{v_1' \sin \theta_1}{v_1' \cos \theta_1 - v_1}$$

Entering known values gives

$$\tan \theta_2 = \frac{(1.50 \text{ m/s})(0.707)}{(1.50 \text{ m/s})(0.707) - 2.00 \text{ m/s}} = -1.129$$

Thus,

$$\theta_2 = \tan^{-1}(-1.129) = -48.5°$$

The minus sign indicates that m_2 is scattered to the right, as expected (since m_1 scatters to the left). Either Equation 7.6a or b can now be used to solve for v_2', but the latter is easiest

since it has fewer terms. It yields

$$v_2' = -\frac{m_1}{m_2} \cdot v_1' \cdot \frac{\sin \theta_1}{\sin \theta_2}$$

Entering known values gives

$$v_2' = -\frac{0.250 \text{ kg}}{0.400 \text{ kg}} \cdot 1.50 \text{ m/s} \cdot \frac{0.707}{-0.749}$$

Thus,

$$v_2' = 0.886 \text{ m/s}$$

Discussion It is instructive to calculate the internal kinetic energy of this two-mass system before and after the collision. (This is left as an end-of-chapter problem.) If you do this, you will find that the internal kinetic energy is less after the collision, and so it is an inelastic collision. This type of result makes a physicist want to explore the system further.

Elastic Collisions of Two Equal Masses

Some interesting situations arise when the two colliding masses are equal and the collision is elastic. This is nearly the case with billiard balls, and precisely the case with some subatomic collisions. We can thus get a mental image of a subatomic collision by thinking about billiards or pool. (Refer to Figure 7.10 for masses and angles.) First, an elastic collision conserves internal kinetic energy. Again, let us assume mass 2 is initially at rest. Then the internal kinetic energy before and after the collision of two equal masses is

$$\tfrac{1}{2}mv_1^2 = \tfrac{1}{2}mv_1'^2 + \tfrac{1}{2}mv_2'^2$$

Because the masses are equal, $m_1 = m_2 = m$. Algebraic manipulation (left to the reader) of Equations 7.6a and b for conservation of momentum in the *x*- and *y*-directions can show that

$$\tfrac{1}{2}mv_1^2 = \tfrac{1}{2}mv_1'^2 + \tfrac{1}{2}mv_2'^2 + mv_1'v_2' \cos(\theta_1 + \theta_2)$$

The two preceding equations can both be true only if

$$mv_1'v_2' \cos(\theta_1 + \theta_2) = 0$$

There are three ways that this term can be zero. They are

1. $v_1' = 0$	Head-on collision; incoming ball stops.
2. $v_2' = 0$	No collision; incoming ball continues unaffected.
3. $\cos(\theta_1 + \theta_2) = 0$	Angle of separation $(\theta_1 + \theta_2)$ is 90° after the collision.

All three of these are familiar occurrences in billiards and pool, although most of us try to avoid the second. If you play enough pool, you will notice that the angle between the balls is very close to 90° after the collision, although it will vary from this value if a great deal of spin is placed on the ball. (Large spin carries in extra energy and a quantity called angular momentum, which must also be conserved.) (See Figures 7.9 and 7.12.) The assumption that the scattering of billiard balls is elastic is reasonable based on the correctness of the three results it produces. This also implies that, to a good approximation, momentum is conserved in billiards and pool. The end-of-chapter problems explore these and other characteristics of two-dimensional collisions.

CONNECTIONS

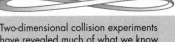

Two-dimensional collision experiments have revealed much of what we know about the subatomic world, as we shall see in the chapters on nuclear and particle physics. Rutherford, for example, discovered the nature of the atomic nucleus from such experiments.

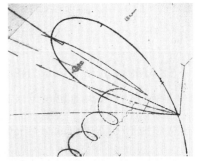

Figure 7.12 A collision taking place between two protons is recorded by the lines of bubbles in this bubble chamber photograph. Since the protons have equal masses and scatter elastically, the angle between their outgoing paths is 90°. The curvature in the paths is due to a magnetic field. The angle of separation will differ from 90° at very high velocities because of relativistic effects.

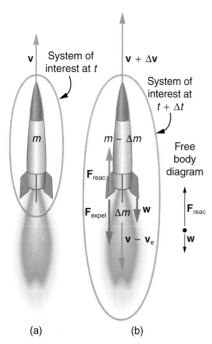

Figure 7.13 (a) This rocket has a mass m and an upward velocity v. The net external force on the system is $-mg$ if air resistance is neglected. (b) A time Δt later the system has two main parts, the ejected gas and the remainder of the rocket. The reaction force on the rocket is what overcomes gravity and accelerates it upward.

7.7* INTRODUCTION TO ROCKET PROPULSION

Rockets range in size from fireworks so small that ordinary people hazard their use to immense Saturn Vs that once propelled massive payloads toward the moon. The propulsion of all rockets, jet engines, deflating balloons, and even squid and octopuses is explained by the same physical principle—Newton's third law of motion. Matter is forcefully ejected from a system, producing an equal and opposite reaction on what remains. Another common example is the recoil of a gun. The gun exerts a force on a bullet to accelerate it and consequently experiences an equal and opposite force, causing the gun's recoil or kick.

Figure 7.13 shows a rocket accelerating straight up. In part (a), the rocket has a mass m and a velocity v relative to the earth, and hence a momentum mv. In part (b), a time Δt has elapsed in which the rocket has ejected a mass Δm of hot gas at a velocity v_e relative to the rocket. The remainder of the mass $(m - \Delta m)$ now has a greater velocity $(v + \Delta v)$. The momentum of the entire system (rocket plus expelled gas) has actually decreased, because the force of gravity has acted for a time Δt, producing a negative impulse $\Delta p = -mg \, \Delta t$. (Remember that impulse is the net external force on a system multiplied by the time it acts, and it equals the change in momentum of the system.) So the center of mass of the system is in free fall but, by rapidly expelling mass, part of the system can accelerate upward. It is a commonly held misconception that the rocket exhaust pushes on the ground. In fact, gases are easier to expel into a vacuum, and a rocket's thrust is greater there than in the atmosphere or on the launch pad.

By calculating the change in momentum for the entire system over Δt, and equating it to the impulse, the following expression can be shown to be a good approximation for the acceleration of the rocket:

$$a = \frac{v_e}{m} \frac{\Delta m}{\Delta t} - g \qquad (7.7)$$

"The rocket" is that part of the system remaining after the gas is ejected, and g is the acceleration of gravity.

A rocket's acceleration depends on three major factors, consistent with Equation 7.7. First, the greater the exhaust velocity of the gases relative to the rocket, v_e, the greater the acceleration. The practical limit for v_e is about 2.5×10^3 m/s for conventional (nonnuclear) hot gas propulsion systems. The second factor is the rate at which mass is ejected from the rocket. This is the term $\Delta m / \Delta t$ in the equation. The faster the rocket burns its fuel, the greater its acceleration. The third factor is the mass m of the rocket. The smaller the mass, all other factors being the same, the greater the acceleration. Rocket mass m decreases dramatically during flight since most of the rocket is fuel to begin with, so that acceleration grows continuously, reaching a maximum just before the fuel is exhausted.

EXAMPLE 7.8 FIND THE INITIAL ACCELERATION OF A MOON LAUNCH

A Saturn V's mass at liftoff was 2.80×10^6 kg, its fuel burn rate was 1.40×10^4 kg/s, and the exhaust velocity was 2.40×10^3 m/s. Calculate its initial acceleration.

Strategy This is a straightforward application of the expression for acceleration given in Equation 7.7, since a is the unknown and all of the terms on the right side of the equation are given.

Solution Substituting the given values into Equation 7.7 yields

$$a = \frac{v_e}{m} \frac{\Delta m}{\Delta t} - g$$

$$= \frac{2.40 \times 10^3 \text{ m/s}}{2.80 \times 10^6 \text{ kg}} \cdot (1.40 \times 10^4 \text{ kg/s}) - 9.80 \text{ m/s}^2$$

$$= 2.20 \text{ m/s}^2$$

(continued)

(continued)

Discussion This is fairly small even for an initial acceleration. The acceleration does increase steadily as the rocket burns fuel, since m decreases while v_e and $\Delta m/\Delta t$ remain constant. Knowing this acceleration and the mass of the rocket, you can show that the thrust of engines was 3.36×10^7 N, or about 7.5 million pounds.

To achieve the high speeds needed to hop continents, obtain orbit, or escape the earth's gravity altogether, the mass of the rocket other than fuel must be as small as possible. It can be shown that, in the absence of air resistance and neglecting gravity, the final velocity of a one-stage rocket initially at rest is

$$v = v_e \ln \frac{m_0}{m_r} \tag{7.8}$$

where $\ln(m_0/m_r)$ is the natural logarithm of the ratio of the initial mass of the rocket (m_0) to what is left (m_r) after all of the fuel is exhausted.* For example, let us calculate the mass ratio needed to escape the earth's gravity starting from rest, given the escape velocity from the earth is about 11.2×10^3 m/s, and assuming an exhaust velocity $v_e = 2.5 \times 10^3$ m/s. Then,

$$\ln \frac{m_0}{m_r} = \frac{v}{v_e} = \frac{11.2 \times 10^3 \text{ m/s}}{2.5 \times 10^3 \text{ m/s}} = 4.48$$

Solving for m_0/m_r gives

$$\frac{m_0}{m_r} = \text{antiln } 4.48 = 88$$

Thus the mass of the rocket is

$$m_r = \frac{m_0}{88}$$

This result means that only 1/88 of the mass is left, and 87/88 was fuel. Expressed as percentages, 98.9% of the rocket is fuel, while payload, engines, fuel tanks, and other components make up only 1.10%. Taking air resistance and gravity into account, the mass remaining m_r can only be about $m_0/180$. It is difficult to build a rocket in which the fuel has a mass 180 times *everything* else. The solution is multistage rockets. Each stage only needs to achieve part of the final velocity and is discarded after it burns its fuel. The result is that each successive stage can have smaller engines and more payload relative to its fuel. Once out of the atmosphere, the ratio of payload to fuel becomes more favorable, too.

The space shuttle is an attempt at an economical vehicle with some reusable parts, such as the solid fuel boosters and the craft itself. (See Figure 7.14.) Its manned nature, however, makes it at *least* as costly for launching satellites as unmanned expendable rockets. Ideally, the shuttle would only be used when human activities are required for the success of a mission, such as the repair of the Hubble Space Telescope. Satellites can also be launched from airplanes. This has the double advantage that the initial velocity is significantly above zero and the rocket can avoid most of the atmosphere's resistance.

Figure 7.14 The space shuttle has a number of reusable parts. Solid fuel boosters on either side are recovered and refueled after each flight, and the entire orbiter also returns to earth for use in subsequent flights. The large liquid fuel tank is expended. The space shuttle is a complex assemblage of technologies, employing both solid and liquid fuel and pioneering ceramic tiles as reentry heat shields. As a result, it permits multiple launches as opposed to single-use rockets.

---

*Note that v in Equation 7.8 is actually the *change* in velocity, and so the equation can be used for any segment of the flight. If we start from rest, the change in velocity equals the final velocity.

SUMMARY

Linear momentum (*momentum* for brevity) is the product of a system's mass times its velocity, expressed as

$$\mathbf{p} = m\mathbf{v} \qquad (7.1)$$

and having units of kg·m/s. **Newton's second law of motion in terms of momentum** is

$$\text{net } \mathbf{F} = \frac{\Delta \mathbf{p}}{\Delta t} \qquad (7.2)$$

where net $\mathbf{F}$ is the net external force, $\Delta \mathbf{p}$ is the change in momentum, and Δt is the change in time. This becomes net $\mathbf{F} = m\mathbf{a}$ when mass is constant.

Impulse is defined to be the change of momentum $\Delta \mathbf{p}$ of a system and is given by

$$\Delta \mathbf{p} = \text{net } \mathbf{F} \cdot \Delta t$$

where net $\mathbf{F}$ acts on the system for a time Δt. If net $\mathbf{F} = 0$, then $\Delta \mathbf{p} = 0$, and momentum is conserved. The **conservation of momentum principle** is written

$$\left.\begin{array}{l} \mathbf{p}_{\text{tot}} = \text{constant} \\[1ex] \text{or} \\[1ex] \mathbf{p}_{\text{tot}} = \mathbf{p}'_{\text{tot}} \end{array}\right\} \quad \text{(isolated system)} \qquad (7.3)$$

where $\mathbf{p}_{\text{tot}}$ is the initial total momentum and $\mathbf{p}'_{\text{tot}}$ is the total momentum some time later. An **isolated system** is defined to be one for which the net external force is zero (net $\mathbf{F} = 0$).

An **elastic collision** is one that conserves internal kinetic energy. For two bodies in a one-dimensional collision, conservation of momentum is

$$\left.\begin{array}{l} p_1 + p_2 = p'_1 + p'_2 \\[1ex] \text{or} \\[1ex] m_1 v_1 + m_2 v_2 = m_1 v'_1 + m_2 v'_2 \end{array}\right\} \quad (\text{net } F = 0) \quad (7.4)$$

and conservation of internal kinetic energy in a one-dimensional collision is

$$\tfrac{1}{2} m_1 v_1^2 + \tfrac{1}{2} m_2 v_2^2 = \tfrac{1}{2} m_1 v_1'^2 + \tfrac{1}{2} m_2 v_2'^2 \qquad (7.5)$$
$$\text{(two-body elastic collision)}$$

An **inelastic collision** does not conserve internal kinetic energy, although it will conserve momentum if net $\mathbf{F} = 0$.

Two-dimensional collisions of point masses where mass 2 is initially at rest conserve momentum along the initial direction of mass 1 (the x-axis), stated by

$$m_1 v_1 = m_1 v'_1 \cos \theta_1 + m_2 v'_2 \cos \theta_2 \qquad (7.6a)$$

and along the direction perpendicular to the initial velocity (the y-axis), stated by

$$0 = m_1 v'_1 \sin \theta_1 + m_2 v'_2 \sin \theta_2 \qquad (7.6b)$$

See Figure 7.10 for explication of velocities, angles, and the coordinate system. Elastic collisions of two equal masses with one originally at rest produce a final state where the angle between the final velocities is 90°.

The vertical acceleration of a rocket is

$$a = \frac{v_e}{m} \frac{\Delta m}{\Delta t} - g \qquad (7.7)$$

where v_e is the exhaust velocity, m is the mass of the rocket, $\Delta m/\Delta t$ is the rate at which fuel is burnt, and g is the acceleration of gravity. The final velocity of a one-stage rocket initially at rest is

$$v = v_e \ln \frac{m_0}{m_r} \qquad (7.8)$$

where m_0 is the rocket's initial mass and m_r its remaining mass after the fuel is burnt.

CONCEPTUAL QUESTIONS

7.1 A small mass and a large mass each have the same momentum. Which mass has the largest kinetic energy?

7.2 A small mass and a large mass each have the same kinetic energy. Which mass has the largest momentum?

7.3 Football coaches advise players to block, hit, and tackle with their feet on the ground rather than by leaping through the air. Using the concepts of momentum, work, and energy, explain how a football player can be more effective with his feet on the ground.

7.4 How can a small force impart the same momentum to an object as a large force?

7.5 Explain in terms of impulse how padding reduces forces in a collision. State in terms of a real example, such as a car hitting a padded barricade.

7.6 Under what circumstances is momentum conserved?

7.7 Can momentum be conserved if there are external forces acting? If so, under what conditions? If not, why not?

7.8 Momentum can be conserved in one direction while not being conserved in another. What is the angle between the directions? Give an example.

7.9 Explain in terms of momentum and Newton's laws how a car's air resistance is due in part to the fact that it pushes air in its direction of motion.

7.10 Can objects in a system have momentum while the momentum of the system is zero? Explain.

7.11 Must the total energy of a system be conserved whenever its momentum is conserved? Explain why or why not.

7.12 What is an elastic collision?

7.13 What is an inelastic collision? What is a perfectly inelastic collision?

7.14 A mixed pair of ice skaters in a show are standing motionless at arms length just before starting a routine. They reach out, clasp hands, and pull themselves together with arm action only. Assuming no friction with the ice, what is their velocity after their bodies meet?

7.15 A small pickup with a camper shell slowly coasts with negligible friction toward a red light. Two children are playing in the back making various inelastic collisions with each other and the walls. What is the effect of the children on the motion of the center of mass of the system (truck plus entire load)? What is their effect on the motion of the truck?

7.16 Suppose a fireworks shell explodes, breaking into three large pieces for which air resistance is negligible. How is the motion of the center of mass affected by the explosion? How would it be affected if the pieces experienced significantly more air resistance than the intact shell?

7.17 During one Skylab flight, an astronaut was positioned motionless in the center of the orbiter, out of reach of any solid object on which he could exert a force. Suggest a method by which he could move himself away from this position, and explain the physics involved.

7.18 It is possible for the velocity of a rocket to be greater than the exhaust velocity of the gases it ejects. When that is the case, the gas velocity and momentum are in the same direction as the rocket's. How does the rocket still obtain thrust by ejecting the gases?

7.19 Figure 7.15 shows a cube at rest with a small mass heading toward it. (a) Describe the directions (angle θ_1) at which the small mass can emerge after colliding elastically with

the cube. How does θ_1 depend on b, the so-called impact parameter? Ignore any effects that might be due to rotation after the collision, and assume that the cube is much more massive than the small mass. (b) Answer the same questions if the small mass instead collides with a massive sphere.

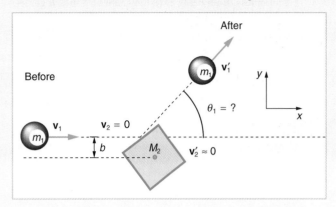

Figure 7.15 A small mass approaches a collision with a much more massive cube, after which its velocity has the direction θ_1. The angles to which the small mass can be scattered are determined by the shape of the object it strikes and the impact parameter b. Question 19.

PROBLEMS

Sections 7.1 and 7.2 Momentum, Force, and Impulse

7.1 (a) Calculate the momentum of a 2000 kg elephant charging a hunter at a speed of 7.50 m/s. (b) Compare the elephant's momentum with that of a 0.0400 kg bullet fired at a speed of 600 m/s. (c) What is the momentum of the 90.0 kg hunter running at 7.40 m/s after missing the elephant?

7.2 (a) What is the mass of a large ship that has a momentum of 1.25×10^6 kg·m/s, when moving at a speed of 48.0 km/h? (b) Compare the ship's momentum with that of a 1100 kg artillery shell fired at a speed of 630 m/s.

7.3 At what speed would a 5.00×10^5 kg airplane have to fly to have a momentum of 1.25×10^6 kg·m/s (the same as the ship in Problem 7.2)?

7.4 (a) What is the momentum of a 1.20×10^4 kg garbage truck moving at 30.0 m/s? (b) At what speed would an 8.00 kg trash can have the same momentum?

7.5 A bullet is accelerated down the barrel of a gun by hot gases produced in the combustion of gun powder. What is the average force exerted on a 0.0300 kg bullet to accelerate it to a speed of 600 m/s in a time of 2.00 ms?

•7.6 Using data in the text on the mass of the earth and the size of its orbit, calculate its linear momentum.

•7.7 A person slaps his leg with his hand, bringing it to rest in 2.50 ms from an initial speed of 4.00 m/s. (a) What is the average force exerted on the leg, taking the effective mass of the hand and forearm to be 1.50 kg? (b) Would the force be any different if the man clapped his hands together at the same speed and brought them to rest in the same time? Explain why or why not.

•7.8 A professional boxer hits his opponent with a 1000 N (225 lb) horizontal blow that lasts for 0.150 s. (a) Calculate the impulse imparted by this blow. (b) What is the opponent's final velocity, if his mass is 105 kg and he is motionless in midair when struck near his center of mass?

•7.9 Suppose a child drives a bumper car into the side rail, which exerts a force of 4000 N on the car for 0.200 s. (a) What impulse is imparted by this force? (b) Find the final velocity of the bumper car if its initial velocity was 2.80 m/s and the car plus driver have a mass of 200 kg. You may neglect friction between the car and floor.

•7.10 One hazard of space travel is debris left by previous missions. There are several thousand masses large enough to detect by radar orbiting the earth, but there are far greater numbers of very small masses, such as flakes of paint. Calculate the force exerted by a 0.100 mg chip of paint that strikes a space shuttle window at a relative speed of 4.00×10^3 m/s, given the collision lasts 6.00×10^{-8} s. Such a collision chipped the window of the ill-fated Challenger in June 1983, causing $50,000 of damage.

⁑7.11 A cruise ship with a mass of 1.00×10^7 kg strikes a pier at a speed of 0.750 m/s, and comes to rest 6.00 m later, damaging the ship, the pier, and the tugboat captain's finances. Calculate the average force exerted on the pier using the concept of impulse. (Hint: First calculate the time it took to bring the ship to rest.)

⁑7.12 A 0.450 kg hammer is moving horizontally at 7.00 m/s when it strikes a nail and comes to rest after driving it 1.00 cm into a board. (a) Calculate the duration of the impact. (b) What was the average force exerted on the nail?

⁑7.13 Water from a fire hose is directed horizontally against a wall at a rate of 50.0 kg/s and a speed of 42.0 m/s. Calculate the force exerted on the wall, assuming the water's horizontal momentum is reduced to zero.

⁑7.14 When serving a tennis ball, the player hits it when its velocity is zero (at the highest point of a vertical toss). The racket exerts a force of 540 N on the ball for 5.00 ms, giving it a final velocity of 45.0 m/s. Using these data, find the mass of the ball.

⁑7.15 Starting with the definitions of momentum and kinetic energy, derive an equation for the kinetic energy of a particle expressed as a function of its momentum.

Sections 7.3–7.5 Conservation of Momentum; One-Dimensional Collisions

7.16 Suppose a 0.200 kg Gumby slides on ice at a speed of 0.750 m/s, running into 0.350 kg Pokey, who was initially motionless. Both being clay, they naturally stick together. What is their final velocity?

7.17 Train cars are coupled together by being bumped into one another. Suppose two loaded train cars are moving toward one another, the first having a mass of 150,000 kg and a velocity of 0.300 m/s, and the second having a mass of 110,000 kg and a velocity of −0.120 m/s. (The minus indicates direction of motion.) What is their final velocity?

7.18 What is the velocity of a 900 kg car initially moving at 30.0 m/s, just after it hits a 150 kg deer initially running at 12.0 m/s in the same direction? Assume the deer remains on the car.

7.19 A 1.80 kg falcon catches a 0.650 kg dove from behind in midair. What is their velocity after impact if the falcon's velocity is initially 28.0 m/s and the dove's 7.00 m/s in the same direction?

7.20 Two football players collide head-on in midair while chasing a pass. The first player has a 95.0 kg mass and an initial velocity of 6.00 m/s, while the second player has a 115 kg mass and initial velocity of −3.50 m/s. What is their velocity just after impact if they cling together?

7.21 What is the speed of a 1.20×10^4 kg trash truck initially moving at 25.0 m/s just after it hits and adheres to an 80.0 kg trash can initially at rest?

• **7.22** Two identical masses (such as billiard balls) have a one-dimensional collision in which one is initially motionless. After the collision, the moving mass is stationary and the other moves with the same speed as the other originally had. Show that both momentum and kinetic energy are conserved.

• **7.23** The moon's craters are remnants of meteor collisions occurring over the last 3 or 4 billion years and continuing today. Suppose a fairly large asteroid having a mass of 5.00×10^{12} kg (about a kilometer across) strikes the moon at a speed of 15.0 km/s. (a) At what speed does the 7.36×10^{22} kg moon recoil after the perfectly inelastic collision? (b) How much kinetic energy is lost in the collision? Such an event may have been observed by medieval English monks who reported observing a red glow and subsequent haze about the moon.

• **7.24** Suppose the goalie in Example 7.5 had elastically reflected the puck back in the direction from which it had originated. What would the final velocities be in that case?

• **7.25** A 0.240 kg billiard ball moving at 3.00 m/s strikes the bumper and bounces straight back at 2.40 m/s (80% of its original speed). The collision lasts 0.0150 s. (a) Calculate the average force exerted on the ball by the bumper. (b) How much kinetic energy in joules is lost during the collision? (c) What percent of the original energy is left?

• **7.26** In an ice show, a 60.0 kg skater leaps into the air and is caught by an initially stationary 75.0 kg skater. (a) What is their final velocity assuming negligible friction and that the leaper's original horizontal velocity was 4.00 m/s? (b) How much kinetic energy is lost?

• **7.27** Using mass and speed data from Example 7.1, and assuming that the player catches the ball with his feet off the ground with both of them moving horizontally, calculate: (a) The final velocity if the ball and player are going in the same direction. (b) The loss of kinetic energy in this case. (c) Repeat both parts for the ball and the player going in opposite directions. Note that the loss of kinetic energy is related to how much it stings to catch the pass.

• **7.28** A 6.00×10^7 kg battleship originally at rest fires a 1100 kg artillery shell horizontally with a velocity of 575 m/s. (a) If the shell if fired straight aft, there will be negligible friction opposing the ship's recoil. Calculate its recoil velocity. (b) Calculate the increase in internal kinetic energy (that is, for the ship and the shell). This is less than the energy released by the gun powder—significant heat is generated.

• **7.29** Two manned satellites approaching one another at a relative speed of 0.250 m/s intend to dock. The first has a mass of 4.00×10^3 kg, and the second a mass of 7.50×10^3 kg. (a) Calculate the final velocity after docking in the frame of reference in which the first satellite was originally at rest. (b) What is the loss of kinetic energy in this inelastic collision? (c) Repeat both parts in the frame of reference in which the second satellite was originally at rest. Explain why the change in velocity is different in the two frames, whereas the change in kinetic energy is the same in both.

‡ **7.30** If the two satellites in Problem 7.29 collide elastically rather than dock, what is their final relative velocity?

‡ **7.31** Space probes may be separated from their launchers by exploding bolts. (They bolt away from one another.) Suppose a 4800 kg satellite uses this method to separate from the 1500 kg remains of its launcher, and that 5000 J of kinetic energy is supplied to the two parts. What are their subsequent velocities in the frame of reference in which they were at rest before separation?

• **7.32** A 30,000 kg freight car is coasting at 0.850 m/s with negligible friction under a hopper that dumps 110,000 kg of scrap metal into it. (a) What is the final velocity of the loaded freight car? (b) How much kinetic energy is lost?

‡ **7.33** A 0.0250 kg bullet is accelerated from rest to a speed of 550 m/s in a 3.00 kg rifle. The pain of the rifle's kick is much worse if you hold the gun loosely a few centimeters from your shoulder rather than holding it tightly against your shoulder. (a) Calculate the recoil velocity of the rifle if it is held loosely away from the shoulder. (b) How much kinetic energy does the rifle gain? (c) What is the recoil velocity if the rifle is held tightly against the shoulder, making the effective mass 28.0 kg? (d) How much kinetic energy is transferred to the rifle-shoulder combination? The pain is related to the amount of kinetic energy, which is significantly less in this latter situation.

‡ **7.34** One of the most worrisome waste products of a nuclear reactor is plutonium 239 (^{239}Pu). This nucleus is radioactive and decays by splitting into a helium-4 nucleus and a uranium-235 nucleus (^{4}He + ^{235}U), the latter of which is also radioactive and will itself decay some time later. The energy emitted in the plutonium decay is 8.40×10^{-13} J and is entirely converted to kinetic energy of the helium and uranium nuclei. The mass of the helium nucleus is 6.68×10^{-27} kg, while that of the uranium is 3.92×10^{-25} kg (note that the

ratio of the masses is 4 to 235). (a) Calculate the velocities of the two nuclei, assuming the plutonium nucleus is originally at rest. (b) How much kinetic energy does each nucleus carry away? Note that the data given here are accurate to three digits only.

Section 7.6* Two-Dimensional Collisions

• **7.35*** Two identical pucks collide on an air hockey table. One was originally at rest. (a) If the incoming puck has a speed of 6.00 m/s and scatters to an angle of 30.0°, what is the velocity (magnitude and direction) of the second puck? (You may use the result that $\theta_1 + \theta_2 = 90°$ for elastic collisions of identical masses.) (b) Confirm that the collision is elastic.

• **7.36*** Confirm that the results of Example 7.7 do conserve momentum in both the x- and y-directions.

• **7.37*** A 3000 kg cannon is mounted so that it can recoil only in the horizontal direction. (a) Calculate its recoil velocity when it fires a 15.0 kg shell at 480 m/s at an angle of 20.0° above the horizontal. (b) What is the kinetic energy of the cannon? This energy is dissipated as heat in the shock absorbers that stop its recoil.

• **7.38*** A 5.50 kg bowling ball moving at 9.00 m/s collides with a 0.850 kg bowling pin, which is scattered at an angle of 85.0° to the initial direction of the bowling ball, with a speed of 15.0 m/s. (a) Calculate the final velocity (magnitude and direction) of the bowling ball. (b) Is the collision elastic?

⁞**7.39*** Rutherford demonstrated that nuclei were very small and dense by scattering helium-4 nuclei (^{4}He) from gold-197 nuclei (^{197}Au). The energy of the incoming helium nucleus was 8.00×10^{-13} J, and the masses of the helium and gold nuclei were 6.68×10^{-27} and 3.29×10^{-25} kg, respectively (note that their mass ratio is 4 to 197). (a) If the helium scatters to an angle of 120° in an elastic collision, calculate its final speed, and the final velocity (magnitude and direction) of the gold nucleus. (b) What is the final kinetic energy of the helium nucleus?

⁞**7.40*** Two cars collide at an icy intersection and stick together afterward. The first car has a mass of 1200 kg and was approaching at 8.00 m/s due south. The second car has a mass of 850 kg and was approaching at 17.0 m/s due west. (a) Calculate the final velocity (magnitude and direction) of the cars. (b) How much kinetic energy is lost in the collision? (This energy goes into deformation of the cars.) Note that since both cars have an initial velocity, you cannot use Equations 7.6a and b; rather, you must look for other simplifying aspects.

⁞**7.41*** Starting with Equations 7.6a and b for conservation of momentum in the x- and y-directions assuming that one mass is originally stationary, prove that for an elastic collision of equal masses,

$$\tfrac{1}{2}mv_1^2 = \tfrac{1}{2}mv_1'^2 + \tfrac{1}{2}mv_2'^2 + mv_1'v_2' \cos(\theta_1 + \theta_2)$$

as discussed in the text.

⁞**7.42*** Show that the collision in Example 7.7 is not elastic by calculating the internal kinetic energy before and after the collision.

Section 7.7* Rocket Propulsion

7.43* Antiballistic missiles (ABMs) are designed to have very large accelerations so that they may intercept fast-moving incoming missiles in the short time available. What is the take-off acceleration of a 10,000 kg ABM that expels 196 kg of gas per second at an exhaust velocity of 2.50×10^3 m/s?

7.44* What is the acceleration of a 5000 kg rocket taking off from the moon, where the acceleration of gravity is only 1.6 m/s², if the rocket expels 8.00 kg of gas per second at an exhaust velocity of 2.20×10^3 m/s?

7.45* Calculate the increase in velocity of a 4000 kg space probe that expels 3500 kg of its mass at an exhaust velocity of 2.00×10^3 m/s. You may assume gravity is negligible at the probe's location.

7.46* Ion-propulsion rockets have been proposed for use in space. They employ atomic ionization techniques and nuclear energy sources to produce extremely high exhaust velocities, perhaps as great as 8.00×10^6 m/s. This allows a much more favorable payload to fuel ratio. To illustrate this: (a) Calculate the increase in velocity of a 20,000 kg space probe that expels only 40.0 kg of its mass at the given exhaust velocity. (b) These engines are usually designed to produce a very small thrust for a very long time—the type of engine that might be useful on a trip to the outer planets, for example. Calculate the acceleration of such an engine if it expels 4.50×10^{-6} kg/s at the given velocity, assuming the acceleration of gravity is negligible.

• **7.47*** (a) Calculate the thrust of the rocket in Problem 7.43. (b) What is the rocket's acceleration 46.0 s after liftoff, neglecting any change in the acceleration of gravity with altitude?

• **7.48*** Calculate the maximum rate at which a rocket can expel gases if its acceleration cannot exceed seven times that of gravity (this is a consideration on all manned flights and for some equipment). The mass of the rocket just as it runs out of fuel is 75,000 kg, and its exhaust velocity is 2.40×10^3 m/s. Assume that the acceleration of gravity is the same as on the earth's surface (9.80 m/s²).

• **7.49*** How much of a 100,000 kg single-stage rocket can be anything but fuel if the rocket is to have a final speed of 8.00 km/s, given that it expels gases at an exhaust velocity of 2.20×10^3 m/s?

• **7.50*** Given the following data for a fire extinguisher–toy wagon rocket experiment (see Figure 7.16), calculate the average exhaust velocity of the gases expelled from the extinguisher. Starting from rest, the final velocity is 10.0 m/s. The total mass is initially 75.0 kg and is 70.0 kg after the extinguisher is fired.

INTEGRATED CONCEPTS

These problems involve physical principles from more than one chapter. Integrated knowledge from a broad range of topics is much more powerful than a narrow application of physics. Here you will need to refer to material in Chapters 3, 4, 5, and 6 as well as in Chapter 7.

Note: Strategies for dealing with problems having concepts from more than one chapter appear in Chapter 6, among

Figure 7.16 Problem 50.

others. They can be found with the section of problems labeled *Integrated Concepts* **at the end of that chapter.**

‡ 7.51 Suppose a 12.0 kg fireworks shell is shot into the air with an initial velocity of 65.0 m/s at an angle of 80.0° above the horizontal. At the highest point of its trajectory, a small explosive charge separates it into two pieces, neither of which ignite (two duds). One 9.00 kg piece falls straight down, having zero velocity just after the explosion. Neglect air resistance (a poor approximation, but do it anyway). (a) At what horizontal distance from the starting point does the 9.00 kg piece hit the ground? (b) Calculate the velocity of the 3.00 kg piece just after the separation. (c) At what horizontal distance from the starting point does the 3.00 kg piece hit the ground?

‡ 7.52 A highly motivated Indiana Jones swings on a vine to rescue a heroine from an unspeakable death, scooping her into his arm at the bottom of his swing and taking her up to temporary safety on a ledge. Indiana, being a college professor with some knowledge of physics, calculated whether the maneuver was possible before he attempted it—a calculation we shall repeat using the following data. Indiana's mass, including his glued-on hat, is 70.0 kg, the heroine's is 48.0 kg, and the ledge from which he starts is 8.00 m above the low point of his swing. (a) What is his velocity just before he scoops her up, assuming he starts from rest? (b) What is their velocity just after scooping her up? (c) How much kinetic energy is lost in their collision? (d) What is the highest ledge they can swing to?

‡ 7.53 A 0.0250 kg bullet moving horizontally at 450 m/s strikes a stationary 0.500 kg block. (a) What is their velocity just after the collision? (b) The bullet-embedded block slides 8.0 m on a horizontal surface with a 0.30 kinetic coefficient of friction. Now what is its velocity? (c) The bullet-embedded block now strikes and sticks to a stationary 2.00 kg block. How far does this combination travel before stopping?

‡ 7.54 When a hammer strikes a nail, it is compressed along its length. Suppose that a steel nail 8.00 cm long and 3.00 mm in diameter is compressed an average of 65.0 μm by a

0.450 kg hammer during a 1.20 ms collision. (a) What average force was exerted on the nail by the hammer? (b) What is the impulse imparted to the hammer? (c) If the hammer is moving at 7.00 m/s just before the collision, what is its final velocity?

‡ 7.55 Professional golfers can drive the ball over 200 m. While air resistance is significant, its effects are nearly canceled because lift is gained from backspin on the ball, and so a reasonable description of the ball's trajectory is obtained if air resistance is neglected. Suppose a golfer hits a ball 200 m on level ground, giving it an initial velocity at an angle of 45.0°. The duration of the impact of the club and ball has been measured to be 1.00 ms. (a) Calculate the velocity of the ball just after it is struck. (b) What is the 0.0450 kg ball's change in momentum? (c) Find the average force exerted by the club on the ball. (d) Find the average shear deformation of the club's 0.800 m long and 1.00 cm diameter steel shaft.

‡ 7.56 A 4000 kg lunar probe is moving at 1.50 km/s in a region where the earth's and moon's gravitational attractions cancel. It is necessary to change its direction of motion by 1.00° without changing its speed. (a) Calculate how long it must fire its engines, given that they expel 6.00 kg of gas per second at an exhaust velocity of 2.00×10^3 m/s. (b) How long would it fire its engines to change direction by 1.00° if the thrust was directed perpendicular to its original motion? (c) What would its final speed be?

‡ 7.57 Using data from Example 7.8, calculate the acceleration of a Saturn V rocket 20.0 s after launch, assuming that it is heading straight up and given that an air resistance of 20,000 N opposes the motion. You may neglect any change in the acceleration of gravity with altitude, but you must take into account that the rocket has consumed part of its fuel.

‡ 7.58 Calculate the upward thrust generated by a Saturn V rocket using data from Example 7.8, and verify that it is 3.36×10^7 N as stated.

UNREASONABLE RESULTS

As in previous chapters, problems with unreasonable results are included to give practice in assessing whether nature is being accurately described, and to trace the source of difficulty if it is not. This is very much like the process followed in original research when physical principles as well as faulty premises are tested. The following problems give results that are unreasonable even when the physics is correctly applied, because some premise is unreasonable or because certain premises are inconsistent.

PROBLEM-SOLVING STRATEGY

Use the following strategy to determine if an answer is reasonable and, if it is not, to determine what is the cause.

Step 1. *Solve the problem using strategies as outlined in Section 2.5 and in other places in the text.* See the table of contents to locate problem-solving strategies. Use the format followed in the worked examples in the text to solve the problem as usual.

Step 2. *Check to see if the answer is reasonable.* Is it too large or too small, or does it have the wrong sign, improper units, . . .?

Step 3. *If the answer is unreasonable, look for what specifically could cause the identified difficulty.*

•**7.59** A 1000 kg car moving at 30.0 m/s hits a padded barricade at a freeway off-ramp. The barricade is designed to bring the car to a stop more gently than a concrete wall and exerts a 200 N force on the car. (a) Calculate the duration Δt of the collision. (b) What is unreasonable about this result? (c) Which premise is unreasonable?

•**7.60** Consider a one-dimensional elastic collision between two masses on a frictionless surface. Mass 1 is 5.00 g and has an initial velocity of 2.00 m/s. After the collision, its velocity is −10.0 m/s. (a) What is the recoil velocity of mass 2 if it is initially at rest and has a mass of 10.0 kg? (b) What is unreasonable about this result? (c) Which premise is unreasonable, or which premises are inconsistent?

•**7.61** Consider a one-dimensional perfectly inelastic collision between two masses on a frictionless surface. Mass 1 is 4.00 kg and has an initial velocity of 3.00 m/s, while mass 2 is initially at rest. After the collision, the velocity of the two masses stuck together is 3.50 m/s. (a) What is the mass of mass 2? (b) What is unreasonable about this result? (c) Which premise is unreasonable?

⁝**7.62*** Squid have been reported to jump from the ocean and travel 30.0 m (measured horizontally) before reentering the water. (a) Calculate the initial speed of the squid if it leaves the water at an angle of 20.0°, assuming negligible lift from the air and negligible air resistance. (b) The squid propels itself by squirting water. What fraction of its mass would it have to eject in order to achieve the speed found in the previous part? The water is ejected at 12.0 m/s; gravity and friction are neglected. (c) What is unreasonable about the results? (d) Which premise is unreasonable, or which premises are inconsistent?

8 UNIFORM CIRCULAR MOTION AND GRAVITATION

The Indianapolis race car moves in a circular path as it makes the turn. Its wheels also spin rapidly—the latter completing many revolutions, the former only part of one (a circular arc). The same physical principles are involved for each.

Many motions, such as the arc of a bird's flight or the earth's path around the sun, are curved. Recall that Newton's first law tells us that motion is along a straight line at constant speed unless there is a net external force. We will therefore study not only motion along curves, but also the forces that cause it, including gravity.

This chapter deals with the simplest form of curved motion, **uniform circular motion**—motion in a circular path at constant speed. Its study illustrates most concepts associated with rotational motion and leads to the study of many new topics we group under the name *rotation*. Pure *rotational motion* occurs when points in an object move in circular paths centered on one point. *Translational motion* is that with no rotation.

8.1 ROTATION ANGLE AND ANGULAR VELOCITY

We begin the study of uniform circular motion by defining two angular quantities needed to describe rotational motion.

Rotation Angle

When objects rotate about some axis—for example, when the CD (compact disc) in Figure 8.1 rotates about its center—each point in the object follows a circular arc. Consider a line from the center of the CD to its edge. Each pit (used to record sound) along this line moves through the same angle in the same time. The rotation angle is the amount of rotation and is analogous to linear distance. We define the **rotation angle** $\Delta\theta$ to be the ratio of the arc length to the radius of curvature:

$$\Delta\theta = \frac{\Delta s}{r} \tag{8.1}$$

The **arc length** Δs is the distance traveled along a circular path. (See Figure 8.2.) Note that r is the **radius of curvature** of the circular path.[*]

We know that for one complete revolution the arc length is the circumference of a circle of radius r. The circumference of a circle is $2\pi r$. Thus for one complete revolution the rotation angle is

$$\Delta\theta = \frac{2\pi r}{r} = 2\pi$$

This is the basis for defining the units used to measure rotation angles $\Delta\theta$ to be **radians** (rad), defined so that

$$2\pi \text{ rad} = 1 \text{ revolution} \tag{8.2}$$

If $\Delta\theta = 2\pi$ rad, then the CD has made one complete revolution, and every point on the CD is back at its original position. You are no doubt familiar with degrees, another unit for rotation angle, and know that there are 360° in a circle or one revolution. The relationship between radians and degrees is thus

$$2\pi \text{ rad} = 360°$$

so that

$$1 \text{ rad} = \frac{360°}{2\pi} = 57.3°$$

Angular Velocity

How fast is an object rotating? We define **angular velocity** ω as the rate of change of an angle. In symbols, this is

$$\omega = \frac{\Delta\theta}{\Delta t} \tag{8.3}$$

Figure 8.1 All points on a CD travel in circular arcs. The pits along a line from the center to the edge all move through the same angle $\Delta\theta$ in a time Δt.

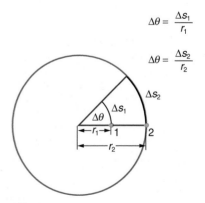

$$\Delta\theta = \frac{\Delta s_1}{r_1}$$

$$\Delta\theta = \frac{\Delta s_2}{r_2}$$

Figure 8.2 Points 1 and 2 rotate through the same angle, but point 2 moves through a greater arc length because it is a greater distance from the center of rotation.

[*]For a path that is not a complete circle, the radius of curvature is the radius of the arc or of the circle you obtain by completing the arc.

where a rotation $\Delta\theta$ takes place in a time Δt. The greater the rotation angle in a given amount of time, the greater the angular velocity. Its units are radians per second (rad/s).

Angular velocity ω is analogous to linear velocity v. To get the precise relationship between angular and linear velocity, we again consider a pit on the rotating CD. It moves an arc length Δs in a time Δt, and so it has a linear velocity

$$v = \frac{\Delta s}{\Delta t}$$

From Equation 8.1 we see that $\Delta s = r\,\Delta\theta$. Substituting this into the expression for v gives

$$v = \frac{r\,\Delta\theta}{\Delta t} = r\omega$$

We write this relationship two different ways and gain two different insights:

$$\left.\begin{array}{c} v = r\omega \\[2mm] \omega = \dfrac{v}{r} \end{array}\right\} \qquad (8.4)$$

or

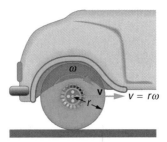

Figure 8.3 A car moving at a velocity v to the right has a tire rotating with an angular velocity ω. The speed of the tread of the tire relative to the axle is v, the same as if the car were jacked up. Thus the car moves forward at $v = r\omega$, where r is the tire radius. A larger angular velocity for the tire means a greater velocity for the car.

The first relationship in Equation 8.4 states that the linear velocity v is proportional to the distance from the center of rotation; thus, it is largest for a point on the rim (largest r), as you might expect. The second relationship in Equation 8.4 can be illustrated by considering the tire of a moving car. The speed of a point on the rim of the tire is the same as the speed v of the car. (See Figure 8.3.) So the faster the car moves, the faster the tire spins—large v, large ω, since $v = r\omega$. Similarly, a larger-radius tire rotating at the same angular velocity would produce a greater speed for the car.

EXAMPLE 8.1 HOW FAST DOES A CAR TIRE SPIN?

Calculate the angular velocity of a 0.300 m radius car tire when the car travels at 15.0 m/s (about 33 mph). (See Figure 8.3.)

Strategy Since the speed of the tire rim is the same as the speed of the car, we have $v = 15.0$ m/s. The radius of the tire is given to be $r = 0.300$ m. Knowing v and r, we can use the second relationship in Equation 8.4 to calculate the angular velocity.

Solution Equation 8.4 states

$$\omega = \frac{v}{r}$$

Substituting the knowns,

$$\omega = \frac{15.0 \text{ m/s}}{0.300 \text{ m}} = 50.0 \text{ rad/s}$$

Discussion When we cancel units in the above calculation, we get 50.0/s. But the angular velocity must have units of rad/s. Since radians are actually unitless (radians are defined as a ratio of distance to distance), we can simply insert them into the answer for the angular velocity. Also note that if an earth mover with much larger tires, say 1.20 m in radius, were moving at the same speed of 15.0 m/s, its tires would rotate more slowly. They would have an angular velocity $\omega = (15.0 \text{ m/s})/(1.20 \text{ m}) = 12.5$ rad/s.

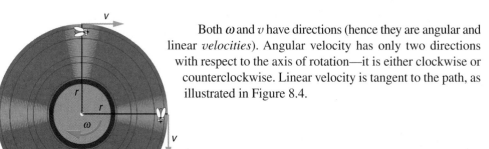

Both ω and v have directions (hence they are angular and linear *velocities*). Angular velocity has only two directions with respect to the axis of rotation—it is either clockwise or counterclockwise. Linear velocity is tangent to the path, as illustrated in Figure 8.4.

Figure 8.4 As an object moves in a circle, here a fly on the edge of an old-fashioned vinyl record, its instantaneous velocity is always tangent to the circle. The direction of the angular velocity is clockwise in this case.

8.2 CENTRIPETAL ACCELERATION

We know from kinematics that acceleration changes velocity, either in magnitude or direction, or both. In uniform circular motion, velocity changes direction, and so there is an associated acceleration. You experience this acceleration yourself when you turn a corner in your car. (If you hold the wheel steady and move at constant speed, you are in uniform circular motion.) What you notice is a sideways acceleration because you and the car are changing direction. The sharper the curve and the greater your speed, the more noticeable this acceleration is. In this section we examine the direction and magnitude of that acceleration.

Figure 8.5 shows an object moving in a circular path at constant speed. The direction of the velocity is shown at two points along the path. Acceleration is in the direction of the change in velocity, which points directly toward the center of rotation (the center of the circular path). This is shown with the vector diagram in the figure. We call the acceleration needed to make an object move in uniform circular motion the **centripetal acceleration*** (a_c); *centripetal* means "toward the center."

The direction of centripetal acceleration is toward the center of curvature, but what is its magnitude? Note that the triangle formed by the velocity vectors and the one formed by the radii and Δr are similar (see Section 1.8 for a mathematical review of triangles). Both triangles are isosceles triangles (two equal sides) and both are labeled ABC. The two equal sides of the velocity vector triangle are the speeds $v_1 = v_2 = v$. Taking the ratio of BC to BA in each triangle, we obtain

$$\frac{\Delta v}{v} = \frac{\Delta r}{r}$$

Acceleration is $\Delta v/\Delta t$, and so we solve this expression for Δv:

$$\Delta v = \frac{v}{r}\,\Delta r$$

Then we divide this by Δt, yielding

$$\frac{\Delta v}{\Delta t} = \frac{v}{r}\frac{\Delta r}{\Delta t}$$

Finally, noting that $\Delta v/\Delta t = a_c$ and that $\Delta r/\Delta t = v$, we see that the magnitude of centripetal acceleration is

$$a_c = \frac{v^2}{r}$$

which is the acceleration needed to make you go in a circle of radius r at a speed v. So centripetal acceleration is greater at high speed, as you have noticed in driving a car. But it is a bit surprising that a_c is proportional to speed squared, implying, for example, that it is four times as hard to take a curve at 60 mph than at 30 mph. A sharp corner has a small radius, so that a_c is greater, as you have noticed.

It is also useful to express a_c in terms of angular velocity. Substituting $v = r\omega$ into the above expression, we find $a_c = (r\omega)^2/r = r\omega^2$. In equation form, the magnitude of centripetal acceleration is

or
$$\left.\begin{array}{l} a_c = \dfrac{v^2}{r} \\[2mm] a_c = r\omega^2 \end{array}\right\} \tag{8.5}$$

You may use whichever expression is more convenient, as the next two examples illustrate.

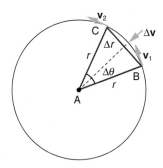

Figure 8.5 The directions of the velocity at two different points are shown, and the change in velocity $\Delta\mathbf{v}$ is seen to point directly toward the center of curvature. Since $\mathbf{a}_c = \Delta\mathbf{v}/\Delta t$, it is also toward the center; $\mathbf{a}_c$ is called centripetal acceleration.

---

*Centripetal acceleration is sometimes called radial acceleration.

EXAMPLE 8.2 HOW DOES THE CENTRIPETAL ACCELERATION OF A CAR AROUND A CURVE COMPARE WITH GRAVITY?

What is the magnitude of the centripetal acceleration needed to make a car follow a curve of radius 500 m at a speed of 25.0 m/s (about 56 mph)? Compare the acceleration with that of gravity for this fairly gentle curve taken at highway speed. (See Figure 8.6(a).)

Strategy Since v and r are given, the first expression in Equation 8.5 is the most convenient to use.

Solution Entering the given values of $v = 25.0$ m/s and

$r = 500$ m into the first expression for a_c gives

$$a_c = \frac{v^2}{r} = \frac{(25.0 \text{ m/s})^2}{500 \text{ m}} = 1.25 \text{ m/s}^2$$

Discussion To compare this with the acceleration of gravity, we take the ratio of $a_c/g = (1.25 \text{ m/s}^2)/(9.80 \text{ m/s}^2) = 0.128$. Thus $a_c = 0.128g$ and is quite noticeable.

EXAMPLE 8.3 HOW BIG IS THE CENTRIPETAL ACCELERATION IN AN ULTRACENTRIFUGE?

Calculate the centripetal acceleration of a point 7.50 cm from the axis of an ultracentrifuge spinning at 75,000 rpm. Determine the ratio of this acceleration to that of gravity for comparison's sake. (See Figure 8.6(b).)

Strategy The term rpm stands for revolutions per minute. By converting this to radians per second, we obtain the angular velocity ω. Since r is given, we can then use the second expression in Equation 8.5 to calculate the centripetal acceleration.

Solution To convert 75,000 rpm to rad/s, we use the facts that one revolution is 2π radians and a minute is 60.0 s. Thus,

$$\omega = 75,000 \frac{\text{rev}}{\text{min}} \cdot \frac{2\pi \text{ rad}}{1 \text{ rev}} \cdot \frac{1 \text{ min}}{60.0 \text{ s}} = 7854 \text{ rad/s}$$

Now the centripetal acceleration is given by the second expression in Equation 8.5 as

$$a_c = r\omega^2$$

Substituting known values gives

$$a_c = (0.0750 \text{ m})(7854 \text{ rad/s})^2 = 4.63 \times 10^6 \text{ m/s}^2$$

Note that the unitless radians are discarded in order to get the correct units for centripetal acceleration. Taking the ratio of a_c to g yields

$$\frac{a_c}{g} = 4.72 \times 10^5$$

Discussion This last result means that centripetal acceleration is 472,000 gs! It is no wonder that such high-ω centrifuges are called ultracentrifuges. The extremely large accelerations needed to move in circles at this rotation rate greatly decrease the time needed to cause sedimentation of blood cells or other materials.

Of course, a net external force is needed to cause any acceleration, just as Newton proposed in his second law of motion. So a net external force is needed to cause a centripetal acceleration. The next section considers the forces involved in circular motion.

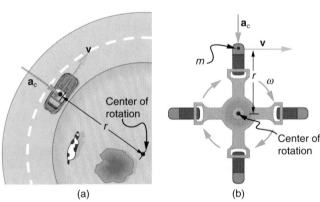

Figure 8.6 (a) The car following a circular path at constant speed is accelerated perpendicular to its velocity, as shown. The magnitude of this centripetal acceleration is found in Example 8.2. (b) A particle of mass m in a centrifuge is rotating at constant angular velocity ω. It must be accelerated perpendicular to its velocity or it would continue in a straight line. The magnitude of the necessary acceleration is found in Example 8.3.

(a)   (b)

8.3 CENTRIPETAL FORCE

Any force or combination of forces can cause a centripetal or radial acceleration. Just a few examples are the tension in the rope on a tether ball, the force of earth's gravity on the moon, friction between roller skates and the rink floor, a banked roadway's force on a car, and elastic forces in the wall of a spinning washing machine.

Any net force causing uniform circular motion is called a **centripetal force**. The direction of a centripetal force is toward the center of curvature, the same as the direction of centripetal acceleration. According to Newton's second law of motion, net force is mass times acceleration: net $F = ma$. For uniform circular motion, the acceleration is the centripetal acceleration—$a = a_c$. Thus the magnitude of centripetal force F_c is

$$F_c = ma_c \qquad (8.6)$$

By using the expressions for centripetal acceleration a_c from Equation 8.5, we get two expressions for the centripetal force F_c in terms of mass, velocity, angular velocity, and radius of curvature:

or
$$\left.\begin{array}{l} F_c = m \dfrac{v^2}{r} \\[2mm] F_c = mr\omega^2 \end{array}\right\} \qquad (8.7)$$

You may use whichever expression for centripetal force is more convenient. Centripetal force F_c is always perpendicular to the path, since a_c is perpendicular to the velocity.

Note that if you solve the first expression for r, you get

$$r = \frac{mv^2}{F_c}$$

This implies that for a given mass and velocity, a large centripetal force causes a small radius of curvature—that is, a tight curve. (See Figure 8.7.)

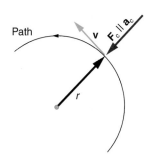

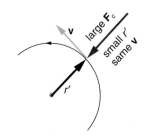

Figure 8.7 Centripetal force is perpendicular to velocity and causes uniform circular motion. The larger the F_c, the smaller the radius of curvature r and the sharper the curve. The second curve has the same v, but a larger F_c produces a smaller r'.

EXAMPLE 8.4 WHAT COEFFICIENT OF FRICTION DO CAR TIRES NEED ON A FLAT CURVE?

(a) Calculate the centripetal force exerted on a 900 kg car that negotiates a 500 m radius curve at 25.0 m/s. (Just as the car in Example 8.2 did.) (b) Assuming an unbanked curve, find the minimum static coefficient of friction between the tires and the road. (See Figure 8.8.)

Strategy and Solution for (a) We know that $F_c = ma_c$, and we have a value for a_c from Example 8.2. Thus,

$$F_c = ma_c$$
$$= (900 \text{ kg})(1.25 \text{ m/s}^2) = 1125 \text{ N}$$

Strategy for (b) Figure 8.8 shows the forces acting on the car on an unbanked (level ground) curve. Friction is to the left, causing the car to turn; thus, friction is the centripetal force in this case. We know from Section 4.5 that the maximum static friction (the tires roll but do not slip) is $\mu_s N$, where μ_s is the static coefficient of friction and N is the normal force. Thus the centripetal force in this situation is

$$F_c = f = \mu_s N = \mu_s mg$$

The normal force equals the car's weight on level ground, so that $N = mg$. Now we have a relationship between centripetal force and the coefficient of friction. Using the first expression for F_c from Equation 8.7,

$$m \frac{v^2}{r} = \mu_s mg$$

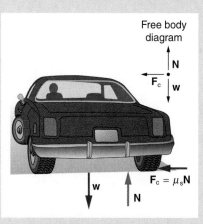

Figure 8.8 This car on level ground is moving away and turning to the left. The centripetal force causing the car to turn in a circular path is due to friction between the tires and the road. A minimum coefficient of friction is needed, or the car will move in a larger-radius curve and leave the roadway.

(continued)

(continued)

We solve this for μ_s, noting that mass cancels, and obtain

$$\mu_s = \frac{v^2}{rg}$$

Solution for (b) Substituting the knowns,

$$\mu_s = \frac{(25.0 \text{ m/s})^2}{(500 \text{ m})(9.80 \text{ m/s}^2)} = 0.13$$

Discussion We could also solve part (a) using the first expression in Equation 8.7 since m, v, and r are given.

The coefficient of friction found in part (b) is much smaller than typical between tires and roads. (Since coefficients of friction are approximate, the answer is given to only two digits.) The car will still negotiate the curve if the coefficient is greater than 0.13, since static friction is a responsive force that will grow only to the needed amount ($f = \mu_s N$). It is notable that mass cancels, implying, for example, that it does not matter how heavily loaded the car is. Mass cancels because friction is assumed proportional to the normal force, which in turn is proportional to mass.

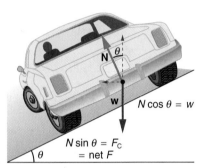

Figure 8.9 The car on this banked curve is moving away and turning to the left. For ideal banking, the net external force equals a horizontal centripetal force in the absence of friction. The components of the normal force N in the vertical and horizontal directions must equal the centripetal force and the weight of the car, respectively. In cases in which forces are not parallel, it is most convenient to consider components along perpendicular axes—in this case, the vertical and horizontal directions.

Let us now consider banked curves, where the slope of the road helps you negotiate the curve. (See Figure 8.9.) The greater the angle θ, the faster you can take the curve. Race tracks, for example, often have steeply banked curves. In an ideally banked curve, the angle θ is such that you can negotiate the curve at a certain speed without the aid of friction between the tires and the road. We will derive an expression for θ for an ideally banked curve and consider an example related to it.

Figure 8.9 shows a free body diagram for a car on a frictionless banked curve. If the angle θ is ideal for the speed and radius, then the net external force will equal the necessary centripetal force. The only two external forces acting on the car are its weight **w** and the normal force of the road **N**. (A frictionless surface can only exert force perpendicular to the surface—that is, a normal force.) These two forces must add to give a net external force that is horizontal toward the center of curvature and has magnitude mv^2/r. Since this is the crucial force and it is horizontal, we use a coordinate system with vertical and horizontal axes. Only the normal force has a horizontal component, and so this must equal the centripetal force—that is,

$$N \sin \theta = \frac{mv^2}{r}$$

The net vertical force must be zero, meaning that the vertical components of the two external forces must be equal in magnitude and opposite in direction. From the figure, we see that the vertical component of the normal force is $N \cos \theta$, and the only other vertical force is the car's weight. These must be equal in magnitude; thus,

$$N \cos \theta = mg$$

Now we can combine the last two equations to eliminate N and get an expression for θ, as desired. Solving the second equation for $N = mg/(\cos \theta)$, and substituting this into the first yields

$$mg \frac{\sin \theta}{\cos \theta} = \frac{mv^2}{r}$$

Since $(\sin \theta)/(\cos \theta) = \tan \theta$, we obtain

$$\theta = \tan^{-1} \frac{v^2}{rg} \qquad \text{(ideally banked curve)}$$

This expression can be understood by considering how θ depends on v and r. Large θ will be obtained for large v and small r. That is, roads must be steeply banked for high speeds and sharp curves. Friction helps, since it allows you to take the curve at greater or lower speed than if the curve is frictionless. Note that θ does not depend on the mass of the vehicle.

EXAMPLE 8.5   WHAT IS THE IDEAL SPEED TO TAKE A STEEPLY BANKED TIGHT CURVE?

Curves on some test tracks and race courses, such as Daytona, are very steeply banked. This, with the aid of tire friction and very stable car configurations, allows the curves to be taken at very high speed. To illustrate, calculate the speed at which a 100 m radius curve banked at 65.0° should be driven if the road is frictionless.

Strategy We first note that all terms in the expression for the ideal angle of a banked curve except for speed are known; thus, we need only rearrange it so that speed appears on the left-hand side and then substitute known quantities.

Solution Starting with

$$\tan \theta = \frac{v^2}{rg}$$

we get

$$v = (rg \tan \theta)^{1/2}$$

Noting that $\tan 65.0° = 2.14$, we obtain

$$v = [(100 \text{ m})(9.80 \text{ m/s}^2)(2.14)]^{1/2}$$

$$= 45.8 \text{ m/s}$$

Discussion This is just over 100 mph, consistent with a very steeply banked and rather sharp curve. Tire friction enables a vehicle to take the curve at significantly higher speeds.

Calculations similar to those in the preceding examples can be performed for a host of interesting situations in which centripetal force is involved—a number of these are posed in the end-of-chapter problems.

8.4*   FICTITIOUS FORCES AND NONINERTIAL FRAMES: THE CORIOLIS FORCE

What do taking off in a jet airplane, turning a corner in a car, riding a merry-go-round, and the counterclockwise motion of a hurricane have in common? Each exhibits fictitious forces—unreal forces that arise, and may *seem* real, because the observer's frame of reference is accelerating or rotating.

When taking off in a jet, most people would agree it feels as if you are being pushed back into the seat as the airplane accelerates down the runway. Yet a physicist would say that *you* tend to remain stationary while the *seat* pushes forward on you, and there is no real force backward on you. An even more common experience occurs when you make a tight curve in your car—say, to the right. You feel as if you are thrown (i.e., *forced*) toward the left relative to the car. Again, a physicist would say that *you* are going in a straight line but the *car* moves to the right, and there is no real force on you to the left. (See Figure 8.10.)

We can reconcile these points of view by examining the frames of reference used. Let us concentrate on people in a car. Passengers instinctively use the car as a frame of reference, whereas a physicist uses the earth. The physicist chooses the earth because it is very nearly an inertial frame of reference[†]—one in which all forces are real (that is, in which all forces have an identifiable physical origin). In such a frame of reference, Newton's laws of motion take the form given in Chapter 4. The car is a **noninertial frame of reference** because it is accelerated to the side. The force to the left sensed by car passengers is a **fictitious force** having no physical origin. There is nothing real pushing them left—the car is actually accelerating to the right.

Let us now take a mental ride on a merry-go-round—specifically, a rapidly rotating playground merry-go-round. You take the merry-go-round to be your frame of reference because you rotate together. In that noninertial frame, you feel a fictitious force, named **centrifugal force**, trying to throw you off. You must hang on tightly to counteract the centrifugal force. In the earth's frame of reference, there is no force trying to throw you off.

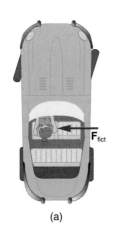

(a)

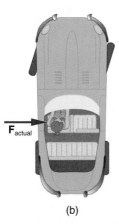

(b)

Figure 8.10 (a) The car driver feels herself forced to the left relative to the car when she makes a right turn. This is a fictitious force arising from the use of the car as a frame of reference. (b) In the earth's frame of reference, the driver moves in a straight line, obeying Newton's first law, and the car moves to the right. There is no real force to the left on the driver relative to the earth. There *is* a real force to the right on the car to make it turn.

[†]Inertial frames of reference and real forces were discussed briefly in Section 4.5.

Figure 8.11 (a) A rider on a merry-go-round feels as if he is being thrown off. This fictitious force is called centrifugal force—it explains the rider's motion in the rotating frame of reference. (b) In an inertial frame of reference and according to Newton's laws, it is his inertia that carries him off and *not* a real force (shaded rider has net **F** = 0 and heads straight). A real force, **F**$_{centripetal}$, is needed to cause a circular path.

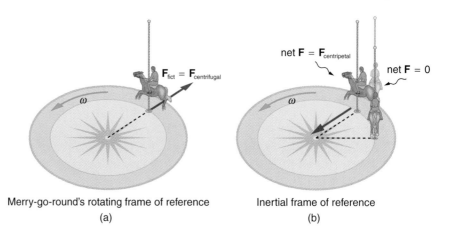

Merry-go-round's rotating frame of reference
(a)

Inertial frame of reference
(b)

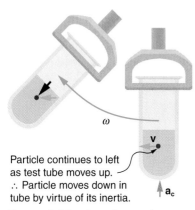

Particle continues to left as test tube moves up. ∴ Particle moves down in tube by virtue of its inertia.

Figure 8.12 Centrifuges utilize inertia to perform their task. Particles in the fluid sediment out because their inertia carries them away from the center of rotation. The large angular velocity of the centrifuge quickens the sedimentation. Ultimately, the particles will come into contact with the test tube walls, which will then supply the centripetal force needed to make them go in a circle of constant radius.

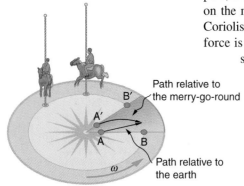

Path relative to the merry-go-round

Path relative to the earth

Figure 8.13 Looking down on the counterclockwise rotation of a merry-go-round, we see that a ball slid straight toward the edge follows a path curved to the right. The person slides the ball toward point B, starting at point A. Both points rotate to the shaded positions shown in the time that the ball follows the curved path in the rotating frame and a straight path in the earth's frame.

Rather you must hang on to force yourself go in a circle because otherwise you would go in a straight line, right off the merry-go-round. (See Figure 8.11.)

This effect (that inertia carries you away from the center of rotation if there is no centripetal force to cause circular motion) is put to good use in centrifuges. (See Figure 8.12.) A centrifuge spins a sample very rapidly. Viewed from the rotating frame of reference, the fictitious centrifugal force throws particles outward, hastening their sedimentation. The greater the angular velocity, the greater the centrifugal force. But what really happens is that the inertia of the particles carries them along a line tangent to the circle while the test tube is forced in a circular path by a centripetal force.

Let us now consider what happens if something moves in a frame of reference that rotates. For example, what if you slide a ball directly away from the center of the merry-go-round, as shown in Figure 8.13? The ball follows a straight path relative to the earth (assuming negligible friction) and a path curved to the right on the merry-go-round's surface. A person standing next to the merry-go-round sees the ball moving straight and the merry-go-round rotating underneath it. In the merry-go-round's frame of reference, we explain the apparent curve to the right by using a fictitious force, called the **Coriolis force**, that causes the ball to curve to the right. The fictitious Coriolis force can be used by anyone in that frame of reference to explain why objects follow curved paths.

We have up to now considered the earth to be an inertial frame of reference with little or no worry about effects due to its rotation. Yet such effects *do* exist—in the rotation of weather systems, for example. Most consequences of the earth's rotation can be qualitatively understood by analogy with the merry-go-round. Viewed from above the north pole, the earth rotates counterclockwise as does the merry-go-round in Figure 8.13. As on the merry-go-round, any motion in the earth's northern hemisphere experiences a Coriolis force to the right. Since the angular velocity of the earth is small, the Coriolis force is usually negligible, but for large-scale motions, such as wind patterns, it has substantial effects.

The Coriolis force causes hurricanes in the northern hemisphere to rotate in the counterclockwise direction, while those in the southern hemisphere rotate in the clockwise direction. (The earth's rotation is clockwise when viewed from above the south pole, and the Coriolis force is consequently to the left in the southern hemisphere.) Figure 8.14 helps show how these rotations take place. Air flows toward any region of low pressure, and hurricanes contain particularly low pressures. Thus winds flow toward the center of a hurricane or a low-pressure weather system. In the northern hemisphere, these inward winds are deflected to the right, as shown in the figure, producing a counterclockwise circulation for low-pressure zones of any type. Low pressure is caused by rising air, which also produces cooling and cloud formation, making the low-pressure patterns quite visible from space. Conversely, wind circulation around high-pressure zones is clockwise in the northern hemisphere but is less visible because high pressure produces fewer clouds.

The rotation of hurricanes and the path of a ball on a merry-go-round can just as well be explained by inertia and the rotation of the system underneath. When noninertial

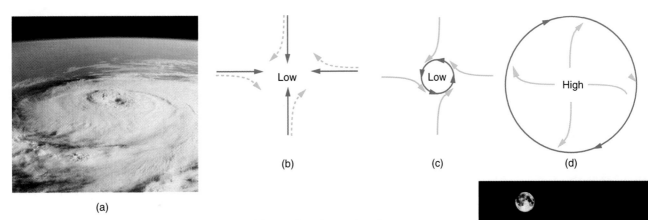

(a)

(b)　　　　(c)　　　　(d)

Figure 8.14 (a) The counterclockwise rotation of this northern hemisphere hurricane is a major consequence of the Coriolis force. (b) Without the Coriolis force, air would flow straight into a low-pressure zone, such as that found in hurricanes. (c) The Coriolis force deflects the winds to the right, producing a counterclockwise rotation. (d) Wind flowing away from a high-pressure zone is also deflected to the right, producing a clockwise rotation. (e) The opposite direction of rotation is produced by the Coriolis force in the southern hemisphere.

frames are used, fictitious forces, such as the Coriolis force, must be invented to explain the curved path. There is no identifiable physical source for these fictitious forces. In an inertial frame, inertia explains the path, and no force is found to be without an identifiable source. Either view allows the description of nature, but that in an inertial frame is simplest in the sense that all forces have real origins and explanations.

8.5 NEWTON'S UNIVERSAL LAW OF GRAVITATION

What do aching feet, a falling apple, and the orbit of the moon have in common? Each is caused by the force of gravity. Our feet are strained by supporting our weight—the force of earth's gravity on us. An apple falls from a tree because of the same force acting a few meters above the earth's surface. And the moon orbits the earth because gravity is able to supply the necessary centripetal force at a distance of hundreds of millions of meters. In fact, the same force causes planets to orbit the sun, stars to orbit the galaxy, and galaxies to cluster together. Gravity is another example of underlying simplicity in nature. This universal force works the same way on all distance scales and describes a vast number of phenomena.

Newton first precisely defined the force of gravity and showed that it could explain both falling bodies and astronomical motions. (See Figure 8.15.) But Newton was not the first to suspect that the same force caused both our weight and the motion of planets. His predecessor Galileo had contended that falling bodies and planetary motions had the same cause. Some of Newton's contemporaries, such as Robert Hooke, Christopher Wren, and Edmund Halley, had also made some progress toward understanding gravitation. But Newton was the first to propose gravity's exact mathematical form and to use that form to show that the motion of heavenly bodies should be conic sections—circles, ellipses, parabolas, and hyperbolas. This theoretical prediction was a major triumph—it had been known for some time that moons, planets, and comets follow such paths, but no one had been able to propose a mechanism that caused them to follow these paths and not others.

The force of gravity is relatively simple. It is always attractive, and it depends only on the masses involved and the distance between them. (See Figure 8.16.) Stated in modern language, **Newton's universal law of gravitation** is

> **Every particle in the universe attracts every other particle with a force along a line joining them. The force is directly proportional to the product of their masses and inversely proportional to the square of the distance between them.**

(e)

Figure 8.15 According to early accounts, Newton was inspired to make the connection between falling bodies and astronomical motions when he saw an apple fall from a tree and realized that if gravity could extend above the ground to a tree, it might also reach the moon. The inspiration of Newton's apple is a part of worldwide folklore and may even be based in fact. Great importance is attached to it because Newton's universal law of gravitation and his laws of motion answered very old questions about nature and gave tremendous support to the notion of underlying simplicity and unity in nature. Scientists still expect underlying simplicity to emerge from their ongoing inquiries into nature.

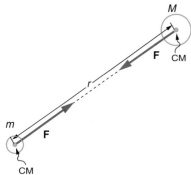

Figure 8.16 Gravitational attraction is along a line joining the centers of mass of these two bodies. The magnitude of the force is the same on each, consistent with Newton's third law.

For two bodies having masses m and M with a distance r between their centers of mass, the equation for Newton's universal law of gravitation is

$$F = G \frac{mM}{r^2} \tag{8.8}$$

where F is the magnitude of the gravitational force and G is a proportionality factor called the **gravitational constant**. G is a universal constant—that is, it is thought to be the same everywhere in the universe. It has been measured experimentally to be

$$G = 6.673 \times 10^{-11} \frac{\text{N} \cdot \text{m}^2}{\text{kg}^2}$$

in SI units. Note that the units of G are such that a force in newtons is obtained from Equation 8.8 when considering masses in kilograms and distance in meters. For example, two 1.000 kg masses separated by 1.000 m will experience a gravitational attraction of 6.673×10^{-11} N. This is an extraordinarily small force. The small magnitude of the gravitational force is consistent with everyday experience. We are unaware that even large objects like mountains exert gravitational forces on us. In fact, our body weight is the force of attraction of the *entire earth* on us.

Recall that the acceleration of gravity g is about 9.8 m/s² on earth. We can now determine why this is so. The weight of an object mg is the gravitational force between it and the earth. Substituting mg for F in Newton's universal law of gravitation

$$mg = G \frac{mM}{r^2}$$

gives where m is the mass of the object, M is the mass of the earth, and r is the distance to the center of the earth (the distance between the centers of mass of the object and earth). (See Figure 8.17.) The mass m of the object cancels, leaving an equation for g:

$$g = G \frac{M}{r^2}$$

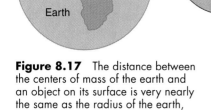

Figure 8.17 The distance between the centers of mass of the earth and an object on its surface is very nearly the same as the radius of the earth, since the earth is so much larger than the object.

Substituting known values for the mass and radius of the earth,

$$g = \left(6.67 \times 10^{-11} \frac{\text{N} \cdot \text{m}^2}{\text{kg}^2}\right) \cdot \frac{5.98 \times 10^{24} \text{ kg}}{(6.38 \times 10^6 \text{ m})^2}$$

and we obtain a value for the acceleration of a falling body:

$$g = 9.80 \text{ m/s}^2$$

This is the expected value *and is independent of the body's mass.** Newton's law of gravitation takes Galileo's observation that all masses fall with the same acceleration a step further, explaining the observation in terms of a force that causes objects to fall—in fact, in terms of a universally existing force of attraction between masses.

In the following example, we make a comparison similar to one made by Newton himself. He noted that if gravity caused the moon to orbit the earth, then the acceleration of gravity should equal the centripetal acceleration of the moon in its orbit. Newton found the two accelerations agree "pretty nearly."

CONNECTIONS

Attempts are still being made to understand the force of gravity. As we shall see in Chapters 31 and 32, modern physics is exploring the connections of gravity to other forces, space, and time. General relativity alters our view of gravity, inducing us to think of gravity as bending space and time.

---

*This calculation of g neglects air resistance (necessary for objects to fall at g) and the small variations in the earth's radius. There are effects due to the earth's rotation that cause g, as measured relative to the earth, to be smaller at the equator than at the poles. Nevertheless, the value calculated *is* the correct value of g in an inertial frame of reference.

EXAMPLE 8.6 EARTH'S GRAVITY IS THE CENTRIPETAL FORCE MAKING THE MOON MOVE IN A CURVED PATH

(a) Find the acceleration of the earth's gravity at the distance of the moon. (b) Calculate the centripetal acceleration needed to keep the moon in its orbit (assuming a circular orbit about a fixed earth), and compare it with the acceleration of gravity just found.

(continued)

(continued)

Strategy for (a) This calculation is the same as the one finding the acceleration of gravity at the earth's surface, except that r is the distance to the moon. The radius of the moon's nearly circular orbit is 3.84×10^8 m.

Solution for (a) Substituting known values into the expression for g found just above yields

$$g = G\frac{M}{r^2} = \left(6.67 \times 10^{-11}\ \frac{\text{N·m}^2}{\text{kg}^2}\right) \cdot \frac{5.98 \times 10^{24}\ \text{kg}}{(3.84 \times 10^8\ \text{m})^2}$$

$$= 2.70 \times 10^{-3}\ \text{m/s}^2$$

Strategy for (b) Centripetal acceleration can be calculated using either form of Equation 8.5. We choose to use the second form:

$$a_c = r\omega^2$$

where ω is the angular velocity of the moon about the earth.

Solution for (b) Given that the period of the moon's orbit is 27.3 days, we see ω is

$$\omega = \frac{\Delta\theta}{\Delta t} = \frac{2\pi\ \text{rad}}{(27.3\ \text{d})(86,400\ \text{s/d})} = 2.66 \times 10^{-6}\ \frac{\text{rad}}{\text{s}}$$

The centripetal acceleration is

$$a_c = r\omega^2 = (3.84 \times 10^8\ \text{m})(2.66 \times 10^{-6}\ \text{rad/s})^2$$

$$= 2.72 \times 10^{-3}\ \text{m/s}^2$$

Discussion The centripetal acceleration of the moon found in (b) differs by less than 1% from the acceleration of earth's gravity found in (a). The agreement is approximate since the moon's orbit is slightly elliptical, and the earth is not stationary (rather the earth-moon system rotates about its center of mass, which is located some 1700 km below the earth's surface). The clear implication is that the earth's gravitational force causes the moon to orbit the earth.

Why does the earth not remain stationary as the moon orbits it? Because, as expected from Newton's third law, if the earth exerts a force on the moon, then the moon should exert an equal and opposite force on the earth. (See Figure 8.18.) We do not sense the moon's effect on the earth's motion, because the moon's gravity moves our bodies right along with the earth.

Tides

Ocean tides are one very observable result of the moon's gravity acting on the earth. Figure 8.19 is a simplified drawing of the moon's position relative to the tides. Since water can flow, a high tide is created on the side of the earth nearest the moon, where the moon's gravitational pull is strongest. Why is there also a high tide on the opposite side of the earth? The answer is that the earth is pulled toward the moon more than the water on the far side, because the earth is closer to the moon. So water on the moon's side is pulled away from the earth, and the earth is pulled away from water on the far side. As the earth rotates, the tidal bulge keeps its orientation with the moon. Thus there are two tides per day (actually two per 25 h since the moon moves in its orbit each day, too).

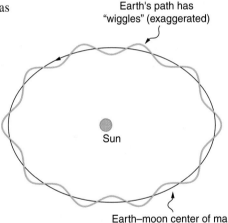

(a)   (not to scale)

(b)

Earth's path has "wiggles" (exaggerated)

Sun

Earth–moon center of mass follows elliptical path

Figure 8.18 (a) The earth and moon rotate approximately once a month around their common center of mass. (b) Their center of mass orbits the sun in an elliptical orbit, but the earth's path around the sun has wiggles in it. Similar wiggles in the paths of stars have been observed and are considered direct evidence of planets orbiting the stars. The planets' reflected light is too dim to be observed.

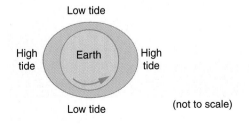

Low tide

High tide   Earth   High tide

Low tide   (not to scale)

Moon

Figure 8.19 The moon causes ocean tides by attracting the water on the near side more than the earth, and by attracting the earth more than the water on the far side. The distances and sizes are not to scale. There are two high and two low tides per day at any location, since the earth rotates under the tidal bulge.

Figure 8.20 (a, b) The highest tides occur when the earth, moon, and sun are aligned. (c) The lowest tides occur when the sun lies at 90° to the earth-moon alignment. Note that figure is not drawn to scale.

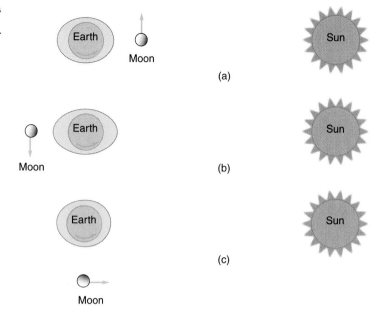

The sun also affects tides, though it has about half the effect of the moon. However, the largest tides, called spring tides, occur when the earth, moon, and sun are aligned. The smallest tides, called neap tides, occur when the sun is at a 90° angle to the earth-moon alignment. (See Figure 8.20.)

Tides are not unique to the earth but occur in many astronomical systems. The most extreme occur where gravity is strongest and varies most rapidly, such as near black holes. (See Figure 8.21.) A few likely candidates for black holes have been observed in our galaxy. These have masses greater than the sun's but have diameters only a few kilometers across. The tidal forces near them are so great that they can actually tear matter from a companion star.

The Cavendish Experiment Then and Now

As previously noted, the gravitational constant G is determined experimentally. This was first accurately done by Henry Cavendish (1731–1810), an English scientist, in 1798, more than 100 years after Newton published his universal law of gravitation. The

Figure 8.21 A black hole is an object with such strong gravity that not even light can escape it. This black hole was created by the supernova of one star in a two-star system. The tidal forces created by the black hole are so great that they tear matter from the companion star. This matter is compressed and heated as it is sucked into the black hole, creating light and x rays observable from earth.

measurement of G is very basic and important because it determines the strength of one of the four forces in nature. Cavendish's experiment was very difficult because he measured the tiny gravitational attraction between two ordinary-sized masses (tens of kilograms at most), employing apparatus like that in Figure 8.22. Remarkably, his value for G differs by less than 1% from the best modern value.

One important consequence of knowing G was that an accurate value for the mass of the earth could finally be obtained. This was done by measuring the acceleration of gravity as accurately as possible and then calculating the mass of the earth M from the relationship

$$g = G\frac{M}{r^2}$$

Rearranging to solve for M yields

$$M = \frac{gr^2}{G}$$

So M can be calculated because all quantities on the right, including the radius of the earth r, are known from direct measurements. We shall see in the next section that knowing G also allows the determination of astronomical masses.

The Cavendish experiment is also used to explore other aspects of gravity. One of the most interesting questions is whether the force of gravity depends on substance as well as mass—for example, whether one kilogram of lead exerts the same gravitational pull as one kilogram of water. A Hungarian scientist named Roland von Eötvös pioneered this inquiry early in this century. He found, with an accuracy of 5 parts per billion, that the force of gravity does not depend on substance. Such experiments continue today, and their interpretations are at times the source of considerable controversy. Only with the advent of general relativity has it become necessary to alter Newton's law of gravity proposed so long ago. But that alteration does not add a new force and is only noticeable for very high gravity or extremely precise measurements.

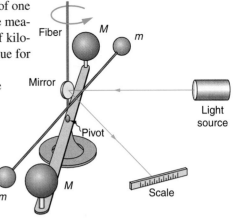

Figure 8.22 Cavendish used apparatus like this to measure the gravitational attraction between the two suspended spheres (*m*) and the two on the stand (*M*) by observing the amount of torsion (twisting) created in the fiber. Distance between the masses can be varied to check the dependence of the force on distance. Modern experiments of this type continue to explore gravity.

8.6 SATELLITES AND KEPLER'S LAWS: AN ARGUMENT FOR SIMPLICITY

Examples of gravitational orbits abound. Hundreds of artificial satellites orbit the earth together with thousands of pieces of debris. The moon's orbit about the earth has intrigued humans from time immemorial. The orbits of planets, asteroids, meteors, and comets about the sun are no less interesting. If we look further, we see almost unimaginable numbers of stars, galaxies, and other celestial objects orbiting one another and interacting through gravity.

All these motions are governed by gravitational force, and it is possible to describe them to various degrees of precision. Precise descriptions of complex systems must be made with large computers. However, we can describe an important class of orbits without the use of computers, and we shall find it instructive to study them. These orbits have the following characteristics:

1. *A small mass m orbits a much larger mass M.* This allows us to view the motion as if M were stationary—in fact, as if from an inertial frame of reference placed on M—without significant error. Mass m is the satellite of M, if the orbit is closed.
2. *The system is isolated from other masses.* This allows us to neglect the small effects due to outside masses.

The conditions are satisfied, to good approximation, by earth satellites (including the moon), by objects orbiting the sun, and by the satellites of other planets. Historically,

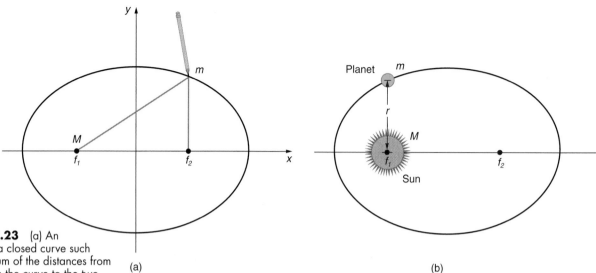

Figure 8.23 (a) An ellipse is a closed curve such that the sum of the distances from a point on the curve to the two foci (f_1 and f_2) is a constant. One can be drawn as shown by putting a pin at each focus, and then placing a string around a pencil and the pins and tracing a line on paper. A circle is a special case of an ellipse in which the two foci coincide (thus any point on the circle is the same distance from the center). (b) For any closed gravitational orbit, m follows an elliptical path with M at one focus. Kepler's first law states this fact for planets orbiting the sun.

planets were studied first, and there is a classical set of three laws, called Kepler's laws of planetary motion, that describe the orbits of all bodies satisfying the two previous conditions (not just planets in our solar system). These descriptive laws are named for the German astronomer Johannes Kepler (1571–1630), who devised them after careful study (over some 20 years) of a large amount of preexisting data. **Kepler's laws of planetary motion** are as follows:

Kepler's first law: **The orbit of each planet about the sun is an ellipse with the sun at one focus.* (See Figure 8.23.)**

Kepler's second law: **Each planet moves so that an imaginary line drawn from the sun to the planet sweeps out equal areas in equal times. (See Figure 8.24.)**

Kepler's third law: **The ratio of the squares of the periods of any two planets about the sun is equal to the ratio of the cubes of their average distances from the sun. In equation form, this is**

$$\frac{T_1^2}{T_2^2} = \frac{r_1^3}{r_2^3} \qquad (8.9)$$

where T is the period (time for one orbit) and r is the average radius. This equation is valid only for comparing two small masses orbiting the same large one. Most importantly, this is a descriptive equation, giving no information as to the cause of the equality.

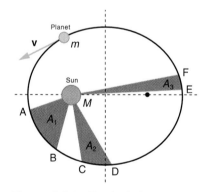

Figure 8.24 The shaded regions have equal areas. It takes equal times for m to go from A to B, from C to D, and from E to F. The mass m moves fastest when it is closest to M, since it gains kinetic energy as it loses gravitational potential energy. Kepler's second law was originally devised for planets orbiting the sun, but it has broader validity.

Note again that while, for historical reasons, Kepler's laws are stated for planets orbiting the sun, they are actually valid for all bodies satisfying the two previously stated conditions.

---

*The path can be any conic section—that is, a circle, ellipse, parabola, or hyperbola—with the largest mass at one focus. Conic sections have foci, with a circle's focus at its center and an ellipse having two foci. Most objects in the solar system are in bound orbits (where m cannot escape the gravity of M)—these are always elliptical.

EXAMPLE 8.7 FIND THE TIME FOR ONE ORBIT OF AN EARTH SATELLITE

Given that the moon orbits the earth each 27.3 days and that it is an average distance of 3.84×10^8 m from the center of the earth, calculate the period of an artificial satellite orbiting at an average altitude of 1500 km above the surface of the earth.

Strategy The period, or time for one orbit, is related to the radius of the orbit by Kepler's third law, given in mathematical form in Equation 8.9. Let us use the subscript 1 for the moon and the subscript 2 for the satellite. We are asked to find T_2. The given information tells us that the orbital radius of the moon is $r_1 = 3.84 \times 10^8$ m, and the period of the moon is $T_1 = 27.3$ days. The height of the artificial satellite above the earth's surface is given, and so we must add the radius of the earth to get $r_2 = (1500 + 6380)$ km = 7880 km, where the radius of the earth is 6380 km. Now all quantities are known, and so T_2 can be found.

Solution Kepler's third law is

$$\frac{T_1^2}{T_2^2} = \frac{r_1^3}{r_2^3}$$

To solve for T_2, we cross-multiply and take the square root, yielding

$$T_2 = \left(\frac{r_2}{r_1}\right)^{3/2} T_1$$

Substituting known values yields

$$T_2 = \left(\frac{7880 \text{ km}}{3.84 \times 10^5 \text{ km}}\right)^{3/2} 27.3 \text{ d} \cdot \frac{24.0 \text{ h}}{\text{d}}$$
$$= 1.93 \text{ h}$$

Discussion This is a reasonable period for a satellite in a fairly low orbit. It is interesting that any satellite at this altitude will orbit in the same amount of time. This fact is related to the condition that the satellite's mass is small compared with that of the earth. It is also connected to the fact that all masses fall at the same rate in the absence of air resistance.

People immediately search for deeper meaning when broadly applicable laws, like Kepler's, are discovered. It was Newton who took the next giant step when he proposed the law of gravity. While Kepler was able to discover *what* was happening, Newton discovered the force of gravity to be the *cause*.

Derivation of Kepler's Third Law for Circular Orbits

We shall derive Kepler's third law, starting with Newton's laws of motion and his universal law of gravity. The point is to demonstrate that the force of gravity is the cause for Kepler's laws (although we will only derive the third one).

Let us consider a circular orbit of a small mass m around a large mass M, satisfying the two conditions stated at the beginning of this section. Gravity supplies the centripetal force to m. Starting with Newton's second law applied to circular motion (see Equation 8.7),

$$\text{net } F = ma_c = m\frac{v^2}{r}$$

The net external force on m is gravity, and so we substitute the force of gravity for net F:

$$G\frac{mM}{r^2} = m\frac{v^2}{r}$$

The mass m cancels, yielding

$$G\frac{M}{r} = v^2$$

The fact that m cancels is another aspect of the oft-noted fact that at a given location all masses fall with the same acceleration. Here we see that at a given orbital radius r, all masses orbit at the same speed. (This was implied by the result of the preceding worked example, too.) Now, to get at Kepler's third law, we must get the period T into the equation. By definition, period T is the time for one complete orbit. Now the average speed v is the circumference divided by the period—that is,

$$v = \frac{2\pi r}{T}$$

Substituting this into the previous equation gives

$$G\,\frac{M}{r} = \frac{4\pi^2 r^2}{T^2}$$

Solving for T^2 yields

$$T^2 = \frac{4\pi^2}{GM}\,r^3 \tag{8.10}$$

Using subscripts 1 and 2 to denote two different satellites, and taking the ratio of the last equation for satellite 1 to satellite 2 yields

$$\frac{T_1^2}{T_2^2} = \frac{r_1^3}{r_2^3}$$

This is Kepler's third law. Note that Kepler's third law is valid only for comparing satellites of the same parent body, since only then does the mass of parent body M cancel.

Now consider what we get if we solve Equation 8.10 for the ratio r^3/T^2. We obtain a relationship that can be used to determine the mass M of a parent body from the orbits of its satellites:

$$\frac{r^3}{T^2} = \frac{G}{4\pi^2}\,M \tag{8.11}$$

If r and T are known for a satellite, then the mass M of the parent can be calculated. This principle has been used extensively to find the masses of heavenly bodies that have satellites. Furthermore, the ratio r^3/T^2 should be a constant for all satellites of the same parent body (since $r^3/T^2 = GM/4\pi^2$). (See Table 8.1.)

It is clear from Table 8.1 that the ratio of r^3/T^2 is constant, at least to the third digit, for all listed satellites of the sun and Jupiter. Small variations in that ratio have two causes—uncertainties in the r and T data, and perturbations of the orbits due to other bodies. Interestingly, those perturbations can be—and have been—used to predict the location of new planets and moons. This is another verification of Newton's universal law of gravitation.

The Case for Simplicity

The development of the universal law of gravity by Newton played a pivotal role in the history of ideas. While it is beyond the scope of this text to cover that history in any detail, we note some important points.

CONNECTIONS

Newton's universal law of gravitation is modified by Einstein's general theory of relativity, as we shall see in Chapter 32. Newton's gravity is not seriously in error—it was and still is an extremely good approximation for most situations. Einstein's modification is most noticeable in extremely large gravitational fields, such as near black holes. However, general relativity also explains such things as small but long-known deviations of the orbit of the planet Mercury from classical predictions.

TABLE 8.1

ORBITAL DATA AND KEPLER'S THIRD LAW

Parent	Satellite	Average orbital radius r (km)	Period T (y)	r^3/T^2 (km³/y²)
Earth	Moon	3.84×10^5	0.07481	1.01×10^{19}
Sun	Mercury	5.79×10^7	0.2409	3.34×10^{24}
	Venus	1.082×10^8	0.6150	3.35×10^{24}
	Earth	1.496×10^8	1.000	3.35×10^{24}
	Mars	2.279×10^8	1.881	3.35×10^{24}
	Jupiter	7.783×10^8	11.86	3.35×10^{24}
	Saturn	1.427×10^9	29.46	3.35×10^{24}
	Uranus	2.870×10^9	84.00	3.35×10^{24}
	Neptune	4.497×10^9	164.8	3.35×10^{24}
	Pluto	5.90×10^9	248.3	3.33×10^{24}
Jupiter	Io	4.22×10^5	0.00485 (1.77 d)	3.19×10^{21}
	Europa	6.71×10^5	0.00972 (3.55 d)	3.20×10^{21}
	Ganymede	1.07×10^5	0.0196 (7.16 d)	3.19×10^{21}
	Callisto	1.88×10^5	0.0457 (16.19 d)	3.20×10^{21}

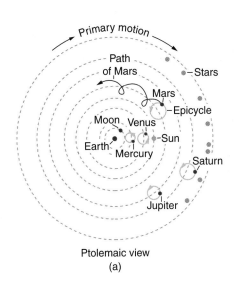

Primary motion

Path of Mars

Stars

Mars

Epicycle

Moon Venus

Sun

Earth

Mercury

Saturn

Jupiter

Ptolemaic view
(a)

Copernican view
(b)

Figure 8.25 (a) The Ptolemaic model of the universe has the earth at the center with the moon, planets, sun, and stars revolving about it in complex superpositions of circular paths. This geocentric model, which can be made progressively more accurate by adding more circles, is purely descriptive, containing no hints as to what causes motions might have. (b) The Copernican model has the sun at the center of the solar system. It is fully explained by a small number of laws of physics, including Newton's universal law of gravitation.

The universal law of gravity is a good example of a physical principle that is very broadly applicable. That single equation for the force of gravity describes all situations in which gravity acts. It gives a cause for a vast number of effects, such as the orbits of the planets and moons in the solar system. It epitomizes the underlying unity and simplicity of physics.

Before the discoveries of Kepler, Copernicus, Galileo, Newton, and others, the solar system was thought to revolve around the earth as shown in Figure 8.25(a). This is called the Ptolemaic view, for the Greek philosopher who lived in the second century A.D. It is characterized by a list of facts for the motions of planets with no cause and effect explanation. There tended to be a different rule for each heavenly body and a general lack of simplicity.

Figure 8.25(b) represents the modern or Copernican model. There is a small set of rules and a single underlying force that explain not only all motions in the solar system, but all other situations involving gravity. The breadth and simplicity of the laws of physics are compelling. As our knowledge of nature has grown, the basic simplicity of its laws has become ever more evident.

SUMMARY

Uniform circular motion is motion in a circle at constant speed. The **rotation angle** $\Delta\theta$ is defined to be the ratio of the arc length to the radius of curvature:

$$\Delta\theta = \frac{\Delta s}{r} \qquad (8.1)$$

where **arc length** Δs is distance traveled along a circular path and r is the **radius of curvature** of the circular path. $\Delta\theta$ is measured in units of **radians** (rad), for which

$$2\pi \text{ rad} = 1 \text{ revolution} \qquad (8.2)$$

The conversion between radians and degrees is 1 rad = 57.3°.

Angular velocity ω is the rate of change of an angle,

$$\omega = \frac{\Delta\theta}{\Delta t} \qquad (8.3)$$

where a rotation $\Delta\theta$ takes place in a time Δt. The units of an-gular velocity are radians per second (rad/s). Linear velocity v and angular velocity ω are related by

$$\left. \begin{array}{c} v = r\omega \\[6pt] \omega = \dfrac{v}{r} \end{array} \right\} \qquad (8.4)$$

Centripetal acceleration a_c is the acceleration needed for uniform circular motion. It is perpendicular to the velocity and has the magnitude

$$\left. \begin{array}{c} a_c = \dfrac{v^2}{r} \\[6pt] a_c = r\omega^2 \end{array} \right\} \qquad (8.5)$$

The units of centripetal acceleration are m/s^2.

Centripetal force F_c is any force causing uniform circular motion. It is perpendicular to velocity and has magnitude

$$F_c = ma_c \qquad (8.6)$$

which can also be expressed as

or

$$\left.\begin{array}{l} F_c = m \dfrac{v^2}{r} \\[2mm] F_c = mr\omega^2 \end{array}\right\} \qquad (8.7)$$

Rotating and accelerated frames of reference are **noninertial**. **Fictitious forces**, such as the Coriolis force, are needed to explain motion in such frames.

Newton's universal law of gravity is stated in words as follows:

> Every particle in the universe attracts every other particle with a force along a line joining them. The force is directly proportional to the product of their masses and inversely proportional to the square of the distance between them.

In equation form, this is

$$F = G \dfrac{mM}{r^2} \qquad (8.8)$$

where F is the magnitude of the gravitational force. G is the **gravitational constant**, given by $G = 6.673 \times 10^{-11}\,\text{N}\cdot\text{m}^2/\text{kg}^2$. Newton's law of gravity applies universally.

Kepler's laws are stated for a small mass m orbiting a larger mass M in near isolation. Kepler's laws of planetary motion are then as follows:

> *Kepler's first law:* **The orbit of each planet about the sun is an ellipse with the sun at one focus.**
>
> *Kepler's second law:* **Each planet moves so that an imaginary line drawn from the sun to the planet sweeps out equal areas in equal times.**
>
> *Kepler's third law:* **The ratio of the squares of the periods of any two planets about the sun is equal to the ratio of the cubes of their average distances from the sun:**
>
> $$\dfrac{T_1^2}{T_2^2} = \dfrac{r_1^3}{r_2^3} \qquad (8.9)$$
>
> **where T is the period (time for one orbit) and r is the average radius of the orbit.**

The period and radius of a satellite's orbit about a larger body M are related by

$$T^2 = \dfrac{4\pi^2}{GM} r^3 \qquad (8.10)$$

or

$$\dfrac{r^3}{T^2} = \dfrac{G}{4\pi^2} M \qquad (8.11)$$

Table 8.1 explores these relationships for satellites of the sun and the planet Jupiter.

CONCEPTUAL QUESTIONS

8.1 There is an analogy between rotational and linear physical quantities. What rotational quantities are analogous to distance and velocity?

8.2 Can centripetal acceleration change the speed of circular motion? Explain.

8.3 If you wish to reduce the stress (which is related to centripetal force) on high-speed tires, would you use large- or small-diameter tires? Explain.

8.4 Define centripetal force. Can any type of force (for example, tension, gravity, friction, and so on) be a centripetal force? Can any combination of forces be a centripetal force?

8.5 Can a centripetal force do work? Explain.

8.6 Identify the force or combination of forces that causes centripetal acceleration for the following: a tether ball, one of Saturn's moons, a car on an unbanked curve, a car on a banked curve (not necessarily ideally banked), a test tube in a centrifuge, an airplane making a turn.

8.7 Race car drivers routinely cut corners as shown in Figure 8.26. Explain how this allows the curve to be taken at the greatest speed.

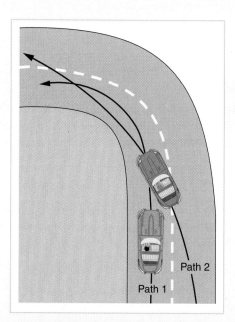

Figure 8.26 Two paths around a race track curve are shown. Race car drivers will take the inside path (called cutting the corner) whenever possible because it allows them to take the curve at the highest speed. Question 7.

8.8 A number of amusement parks have rides that make vertical loops like the one shown in Figure 8.27. For safety, the cars are attached to the rails in such a way that they cannot fall off. If the car goes over the top at just the right speed, gravity alone will supply the centripetal force. What other force acts and what is its direction if: (a) The car goes over the top at faster than this speed? (b) Slower?

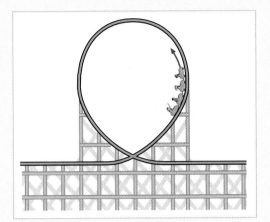

Figure 8.27 Amusement rides with a vertical loop are an example of a form of curved motion. Questions 8 and 9.

8.9 What is the direction of the force exerted by the car on the passenger as the car goes over the top of the amusement ride pictured in Figure 8.27 under the following circumstances: (a) The car goes over the top at such a speed that gravity is the only force acting? (b) The car goes over the top faster than this speed? (c) The car goes over the top slower than this speed?

8.10 Suppose a child is riding on a merry-go-round at a distance about halfway between its center and edge. She has a lunch box resting on wax paper, so that there is very little friction between it and the merry-go-round. Which path shown in Figure 8.28 will the lunch box take when she lets go? The lunch box leaves a trail in the dust on the merry-go-round. Is that trail straight, curved to the left, or curved to the right?

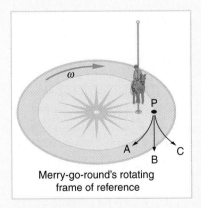

Merry-go-round's rotating frame of reference

Figure 8.28 A child riding on a merry-go-round releases her lunch box at point P. This is a view from above the clockwise rotation. Assuming it slides with negligible friction, will it follow path A, B, or C, as viewed from the earth's frame of reference? What will be the shape of the path it leaves in the dust on the merry-go-round? Question 10.

8.11* Do you feel yourself thrown to either side when you negotiate a curve that is ideally banked for your car's speed? What is the direction of the force exerted on you by the car seat?

8.12* Carefully observe a wet dog shake off water. Describe the physics the dog instinctively uses and, in particular, discuss if rotational motion is crucial to the process.

8.13* Suppose a mass is moving in a circular path on a frictionless table as shown in Figure 8.29. In the earth's frame of reference, there is no centrifugal force pulling the mass away from the center of rotation, yet there is a very real force stretching the string attaching the mass to the nail. Using concepts related to centripetal force and Newton's third law, explain what force stretches the string, identifying its physical origin.

Figure 8.29 A mass attached to a nail on a frictionless table moves in a circular path. The force stretching the string is real and not fictional. What is the physical origin of the force on the string? Question 13.

8.14* When a toilet is flushed or a sink is drained, the water (and other stuff) begins to rotate about the drain on the way down. Assuming no initial rotation and a flow initially directly straight toward the drain, explain what causes the rotation and which direction it has in the northern hemisphere. (Note that this is a small effect and in most toilets the rotation is caused by directional jets of water from the tank.) Would the direction of rotation reverse if water were forced up the drain?

8.15* Is there a real force that throws water from clothes during the spin cycle of a washing machine? Explain how the water is removed.

8.16* There is an amusement park ride in which riders enter a large vertical barrel and stand against the wall on its horizontal floor. The barrel is spun up and the floor drops away. Riders feel as if they are pinned to the wall by a force something like gravity. This is a fictitious force sensed and used by the riders to explain events in the rotating frame of reference of the barrel. Explain in an inertial frame of reference (the earth is nearly so) what pins the riders to the wall, and identify all of the real forces acting on them.

8.17 Action at a distance, such as is the case for gravity, was once thought to be illogical and therefore untrue. What is the ultimate determinant of the truth in physics?

8.18 A nonrotating frame of reference placed at the center of the sun is very nearly an inertial one. Why is it not exactly an inertial frame?

8.19 Draw a free body diagram for a satellite in an elliptical orbit showing why its speed increases as it approaches the parent body and decreases as it moves away.

8.20 Newton's laws of motion and gravity were the first to convincingly demonstrate the underlying simplicity and unity in nature. Many other examples have since been discovered, and we now expect to find such underlying order in complex situations. Is there proof that such order will always be found in new explorations?

8.21 In what frame(s) of reference are Kepler's laws valid? Are Kepler's laws purely descriptive, or do they contain causal information?

PROBLEMS

Sections 8.1–8.3 Uniform Circular Motion, Centripetal Acceleration, and Centripetal Force

8.1 Semitractor trucks have an odometer on one hub of a trailer wheel. The hub is weighted so that it does not rotate, but it contains gears to count the number of wheel revolutions— it then calculates the distance traveled. If the wheel has a 1.15 m diameter and goes through 200,000 rotations, how many kilometers should the odometer read?

8.2 An automobile with 0.260 m radius tires travels 80,000 km before wearing them out. How many revolutions do the tires make, neglecting any backing up and any change in radius due to wear?

8.3 Vinyl record albums revolve $33\frac{1}{3}$ times per minute. What is this in revolutions per second? In radians per second?

8.4 A truck with 0.420 m radius tires travels at 32.0 m/s. At how many radians per second are the tires rotating? What is this in rpm?

8.5 Taking the age of the earth to be about 4×10^9 y and assuming the radius of its orbit has not changed and is circular, calculate the approximate total distance the earth has traveled since its birth (in a frame of reference stationary with respect to the sun).

8.6 The propeller of a World War II fighter plane is 2.30 m in diameter. (a) What is its angular velocity in rad/s if it spins at 1200 rpm? (b) What is the linear speed of its tip at this angular velocity if the plane is stationary on the tarmac? (c) What is the centripetal acceleration of the propeller tip under these conditions? Calculate it in m/s^2 and convert to multiples of g.

8.7 Helicopter blades withstand tremendous stresses. In addition to supporting the weight of the helicopter, they are spun at rapid rates and experience large centripetal accelerations, especially at the tip. (a) Calculate the centripetal acceleration at the tip of a 4.00 m long helicopter blade that rotates at 300 rpm. (b) Compare the linear speed of the tip with the speed of sound (taken to be 340 m/s on this day).

8.8 An ordinary workshop grindstone has a radius of 7.50 cm and rotates at 6500 rpm. (a) Calculate the centripetal acceleration at its edge in m/s^2 and convert it to multiples of g. (b) What is the linear speed of a point on its edge?

8.9 Olympic class ice skaters are able to spin at about 5 revolutions per second. (a) What is a 5.00 rev/s angular velocity in rad/s? (b) What is the centripetal acceleration of the skater's nose if it is 0.120 m from the axis of rotation? (c) An exceptional skater named Dick Button was able to spin much faster in the 1950s than anyone since—at about 9 revolutions per second! What was the centripetal acceleration of the tip of his nose, assuming it is at 0.120 m radius?

8.10 Verify the linear speed of an ultracentrifuge is about 0.50 km/s, and the earth in its orbit is about 30 km/s by calculating: (a) The linear speed of a point on an ultracentrifuge

0.100 m from its center, rotating at 50,000 rpm. (b) The linear speed of the earth in its orbit about the sun (use data from the text on the radius of the earth's orbit and approximate it as being circular).

8.11 A rotating space station is said to create "artificial gravity"—a loose term used for an acceleration that would be crudely similar to gravity. The outer wall of the rotating space station would become a floor for the astronauts, and centripetal acceleration supplied by the floor would allow astronauts to exercise and maintain muscle and bone strength more naturally than in nonrotating space environments. If the space station is 200 m in diameter, what angular velocity would produce an "artificial gravity" of 9.80 m/s^2 at the rim?

8.12 At takeoff a commercial jet has a 60.0 m/s speed. Its tires have a diameter of 0.850 m. (a) At how many rpm are the tires rotating? (b) What is the centripetal acceleration at the edge of the tire? (c) With what force must a determined 10^{-15} kg bacterium cling to the rim? (d) Take the ratio of this force to the bacterium's weight.

8.13 (a) A 22.0 kg child is riding a playground merry-go-round that is rotating at 40.0 rpm. What centripetal force must she exert to stay on if she is 1.25 m from its center? (b) What centripetal force does she need to stay on an amusement park merry-go-round that rotates at 3.00 rpm if she is 8.00 m from its center? (c) Compare each force with her weight.

8.14 An Indianapolis 500 race car takes a 350 m radius curve at 300 km/h. (a) What centripetal acceleration is needed? (b) How many gs is this?

8.15 What is the ideal banking angle for a gentle turn of 1.20 km radius on a highway with a 105 km/h speed limit (about 65 mph), assuming everyone travels at the limit?

8.16 What is the ideal speed to take a 100 m radius curve banked at a 20.0° angle?

8.17 What is the radius of a bobsled turn banked at 85.0° and taken at 30.0 m/s, assuming it is ideally banked?

8.18 (a) A jet fighter flying at 300 m/s (just below the speed of sound) makes a turn of radius 1.85 km. What is its centripetal acceleration in gs? (b) Suppose the pilot makes an emergency turn to avoid an approaching missile, subjecting himself to a centripetal acceleration of 10 gs, while flying at 450 m/s (supersonic). What is the radius of his turn? (This must be short-lived because fighter planes can only briefly endure such large accelerations without serious damage, and the pilot will soon black out at 10 gs.)

8.19 At equinox (around March 21 and September 23) day and night have equal lengths everywhere on earth. Along the earth's equator, the sun passes directly overhead in its semicircular arc across the sky on those days. Neglecting any effects due to the movement of the earth in its orbit about the sun: (a) Calculate the angular velocity of the sun across the equatorial sky in degrees per minute. (b) Express the time it takes the sun to move across the sky in units of minutes per solar diameter. You can use this to estimate the approximate time until sunset—a good way to win bets near dusk. (Note that you must find data on the size of the earth's orbit and the diameter of the sun to solve this problem.)

⁚ 8.20 A large centrifuge, like the one shown in Figure 8.30, is used to expose aspiring astronauts to accelerations similar to those experienced in rocket launches and atmospheric reentries. (a) At what angular velocity is the centripetal acceleration 10 *g*s if the rider is 15.0 m from the center of rotation? (b) The rider's cage hangs on a pivot at the end of the arm, allowing it to swing outward during rotation as shown in Figure 8.30(b). At what angle θ below the horizontal will the cage hang when the centripetal acceleration is 10 *g*s? (Hint: The arm supplies centripetal force and supports the weight of the cage. Draw a free body diagram of the forces to see what the angle θ should be.)

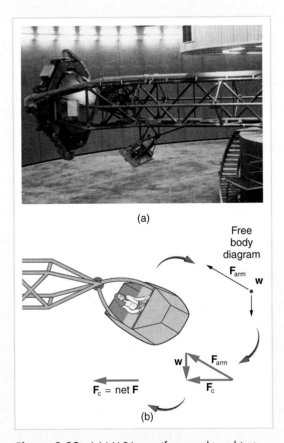

(a)

(b)

Figure 8.30 (a) NASA centrifuge used to subject trainees to accelerations similar to those experienced in rocket launches and reentries. (b) Rider in cage showing how the cage pivots outward during rotation. This allows the total force exerted on the rider by the cage to be along its axis at all times. Problem 20.

⁚ 8.21 Part of riding a bicycle involves leaning at the correct angle when making a turn, as seen in Figure 8.31. To be stable, the force exerted by the ground must be on a line going through the center of gravity. The force on the bicycle wheel can be resolved into two perpendicular components—friction parallel to the road (this must supply the centripetal force), and the vertical normal force (which must equal the system's weight). (a) Show that θ (as defined in the figure) is related to the speed v and radius of curvature r of the turn in the

same way as for an ideally banked roadway—that is,

$$\theta = \tan^{-1} \frac{v^2}{rg}$$

(b) Calculate θ for a 12.0 m/s turn of radius 30.0 m (as in a race).

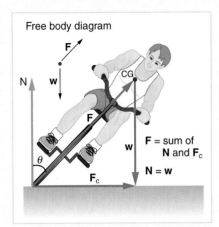

Figure 8.31 A bicyclist negotiating a turn on level ground must lean at the correct angle—the ability to do this becomes instinctive. It is necessary for the force of the ground on the wheel to be on a line through the center of gravity. The net external force on the system is the centripetal force. The vertical component of the force on the wheel cancels the weight of the system while its horizontal component must supply the centripetal force. This produces a relationship among the angle θ, the speed v, and the radius of curvature r of the turn similar to that for the ideal banking of roadways. Problem 21.

Section 8.5 Law of Gravity
Data on the sun and planets is found in Table 8.1 and the appendices.

8.22 (a) Calculate the mass of the earth given the acceleration of gravity at the north pole is 9.830 m/s² and the radius of the earth is 6371 km from pole to pole. (b) Compare this with the accepted value of 5.979×10^{24} kg.

8.23 (a) Calculate the acceleration of gravity at the earth due to the moon. (b) Calculate the acceleration of gravity at the earth due to the sun. (c) Take the ratio of the moon's acceleration to the sun's and comment on why the tides are predominantly due to the moon in spite of this number.

8.24 (a) What is the acceleration of gravity on the surface of the moon? (b) On the surface of Mars? The mass of Mars is 6.418×10^{33} kg and its radius is 3.38×10^6 m.

8.25 (a) Calculate the acceleration of gravity on the surface of the sun. (b) By what factor would your weight increase if you could stand on the sun? (Never mind that you can't.)

• 8.26 The moon and earth rotate about their common center of mass, which is located about 4700 km from the center of the earth. (This is 1690 km below the surface. See Figure 5.7(a).) (a) Calculate the acceleration of the moon's gravity at that point. (b) Calculate the centripetal acceleration of the center of the earth as it rotates about that point once each lunar month (about 27.3 days) and compare it with the acceleration found in part (a). Comment on whether or not they are equal and why they should or should not be.

• **8.27** Solve part (b) of Example 8.6 using $a_c = v^2/r$.

• **8.28** Astrology, that unlikely and vague pseudoscience, makes much of the position of the planets at the moment of birth. The only known force a planet exerts on earth is gravitational. (a) Calculate the gravitational force exerted on a 4.20 kg baby by a 100 kg father 0.200 m away at birth (assisting so he is close). (b) Calculate the force on the baby due to Jupiter if it is at its closest to the earth, some 6.29×10^{11} m away, showing it to be comparable to that of the father. Other objects in the room and the hospital building also exert similar gravitational forces. (Of course, there could be an unknown force acting, but scientists first need to be convinced that there is even an effect, much less that an unknown force causes it.)

• **8.29** The existence of the planet Pluto was proposed based on irregularities in Neptune's orbit. Pluto was subsequently discovered near its predicted position. But it now appears that the discovery was fortuitous, since Pluto is small and the irregularities in Neptune's orbit were not well known. To illustrate that Pluto has a minor effect on the orbit of Neptune compared with other planets: (a) Calculate the acceleration of gravity at Neptune due to Pluto when they are 4.50×10^{12} m apart, as they are at present. The mass of Pluto is 1.4×10^{22} kg. (b) Calculate the acceleration of gravity at Neptune due to Uranus, presently about 2.50×10^{12} m apart, and compare it with that due to Pluto. The mass of Uranus is 8.62×10^{25} kg.

⁑ **8.30** (a) The sun orbits the Milky Way galaxy once each 2.60×10^8 y, with a roughly circular orbit averaging 3.00×10^4 light years in radius. (A light year is the distance traveled by light in 1 y.) Calculate the centripetal acceleration of the sun in its galactic orbit. Does your result support the contention that a nearly inertial frame of reference can be located at the sun? (b) Calculate the average speed of the sun in its galactic orbit. Does the answer surprise you?

Section 8.6 Satellites and Kepler's Laws

8.31 (a) Calculate the ratio r^3/T^2 for the moon's orbit about the earth, where r is the average radius and T the period. (b) Calculate the same ratio for any moon of Jupiter. (c) Why are the answers not equal?

8.32 An isochronous earth satellite is one that has an orbital period of precisely 1 day. Such orbits are useful for communication and weather observation because the satellite remains above the same point on the earth (provided they orbit in the equatorial plane in the same direction as the earth's rotation). Calculate the radius of such an orbit based on the data for the moon in Table 8.1.

• **8.33** Calculate the mass of the sun based on data for the earth's orbit and compare the value obtained with the sun's actual mass.

• **8.34** Find the mass of Jupiter based on data for the orbit of one of its moons, and compare your result with its actual mass.

• **8.35** Find the ratio of the mass of Jupiter to that of the earth based on data in Table 8.1.

⁑ **8.36** Astronomical observations of our Milky Way galaxy indicate that it has a mass of about 8×10^{11} solar masses. A star orbiting on the galaxy's periphery is about 6×10^4 light years from its center. (a) What should the orbital period of that star

be? (b) If its period is 6.0×10^7 y instead, what is the mass of the galaxy? Such calculations are used to imply the existence of "dark matter" in the universe and have indicated, for example, the existence of very massive black holes at the centers of some galaxies.

INTEGRATED CONCEPTS

These problems involve physical principles from more than one chapter. Integrated knowledge from a broad range of topics is much more powerful than a narrow application of physics. Here you will need to refer to material in Chapters 2, 3, 4, and 6 as well as in Chapter 7.

Note: Strategies for dealing with problems having concepts from more than one chapter appear in Chapter 6, among others. Those strategies are accompanied by a worked example that can help you see how to apply them. Consult the section of problems labeled *Integrated Concepts* **at the end of Chapter 6. The table of contents and the index list the locations and topics of all problem-solving strategies.**

• **8.37** If a car takes a banked curve at less than the ideal speed, friction is needed to keep it from sliding toward the inside of the curve (a real problem on icy mountain roads). (a) Calculate the ideal speed to take a 100 m radius curve banked at 15.0°. (b) What is the minimum coefficient of friction needed for a frightened driver to take the same curve at 20.0 km/h?

⁑ **8.38** Space debris left from old satellites and their launchers is becoming a hazard to other satellites. (a) Calculate the speed of a satellite in an orbit 900 km above the earth's surface. (b) Suppose a loose rivet is in an orbit of the same radius that intersects the satellite's orbit at an angle of 90° relative to the earth. What is the velocity of the rivet relative to the satellite just before striking it? (c) Given the rivet is 3.00 mm in size, how long will its collision with the satellite last? (d) If its mass is 0.500 g, what is the average force it exerts on the satellite? (e) How much energy in joules is generated by the collision? (The satellite's velocity does not change appreciably, because its mass is much greater than the rivet's.)

⁑ **8.39** Modern roller coasters have vertical loops like the one shown in Figure 8.32. The radius of curvature is smaller at the top than on the sides so that the downward centripetal acceleration at the top will be greater than the acceleration due to gravity, keeping the passengers pressed firmly into their seats. (a) What is the speed of the roller coaster at the top of the loop if the radius of curvature there is 15.0 m and the downward acceleration of the car is $1.50g$? (b) How high above the top of the loop must the roller coaster start from rest, assuming negligible friction? (c) If it actually starts 5.00 m higher than your answer to the previous part, how much energy did it lose to friction? Its mass is 1500 kg.

⁑ **8.40** A children's ride at a fair is shown in Figure 8.33. The seats pivot outward on metal rods and hang vertically at rest. (a) Calculate the angle made with the vertical if the rider's center of mass is initially 3.00 m from the center of rotation and 3.20 m below the pivot point. The angular velocity is 10.0 rpm. (b) How much does the steel rod stretch (loaded at full speed as compared with loaded at rest) if it is 1.00 cm in diameter and 3.20 m in length, given that the combined mass of the chair and child is 25.0 kg?

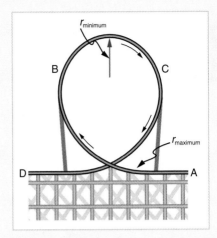

Figure 8.32 Teardrop-shaped loops are used in the latest roller coasters so that the radius of curvature gradually decreases to a minimum at the top. This means that the centripetal acceleration builds from zero to a maximum at the top and gradually decreases again. A circular loop would cause a jolting change in acceleration at entry, a disadvantage discovered long ago in railroad curve design. With a small radius of curvature at the top, the centripetal acceleration can more easily be kept greater than g so that the passengers do not lose contact with their seats nor do they need seat belts to keep them in place. Problem 39.

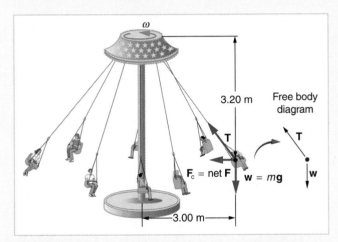

Figure 8.33 Riders on this children's merry-go-round swing outward to ever larger angles as their angular velocity increases. The net force on the rider is the needed centripetal force, supplied by the suspension rod, which also supports their weight. Some riders enjoy the accelerations experienced. Problem 40.

UNREASONABLE RESULTS

As in previous chapters, problems with unreasonable results are included to give practice in assessing whether nature is being accurately described, and to trace the source of difficulty if it is not. This is very much like the process followed in original research when physical principles as well as faulty premises are tested. The following problems give results that are unreasonable even when the physics is correctly applied, because some premise is unreasonable or because certain premises are inconsistent.

PROBLEM-SOLVING STRATEGY

To determine if an answer is reasonable, and to determine the cause if it is not, do the following.

Step 1. *Solve the problem using strategy as outlined in Section 2.5 and in other places in the text.* See the table of contents to locate problem-solving strategies. Use the format followed in the worked examples in the text to solve the problem as usual.

Step 2. *Check to see if the answer is reasonable.* Is it too large or too small, or does it have the wrong sign, improper units, . . .?

Step 3. *If the answer is unreasonable, look for what specifically could cause the identified difficulty.*

8.41 (a) Calculate the minimum coefficient of friction needed for a car to negotiate an unbanked 50.0 m radius curve at 30.0 m/s. (b) What is unreasonable about the result? (c) Which premises are unreasonable or inconsistent?

• **8.42** A father pushes his child on a swing so that his speed is 9.00 m/s at the lowest point of his path. The swing is suspended 2.00 m above the child's center of mass. (a) What is the centripetal acceleration of the child at the low point? (b) What force does the child exert on the seat if his mass is 18.0 kg? (c) What is unreasonable about these results? (d) Which premises are unreasonable or inconsistent?

• **8.43** A mountain 10.0 km from a person exerts a gravitational force on him equal to 2.00% of his weight. (a) Calculate the mass of the mountain. (b) Compare the mountain's mass with that of the entire earth. (c) What is unreasonable about these results? (d) Which premises are unreasonable or inconsistent? (Note that accurate gravitational measurements *can* easily detect the effect of nearby mountains and variations in local geology.)

• **8.44** (a) Based on Kepler's laws and information on the orbital characteristics of the moon, calculate the orbital radius for an earth satellite having a period of 1.00 h. (b) What is unreasonable about this result? (c) What is unreasonable or inconsistent about the premise of a 1.00 h orbit?

9 ROTATIONAL MOTION AND ANGULAR MOMENTUM

The mention of a tornado conjures up images of raw destructive power.
Tornadoes blow houses away as if they were made of paper and have been
known to pierce tree trunks with pieces of straw. They descend from clouds in
funnel-like shapes that spin violently, particularly at the bottom, producing
winds as high as 500 km/h.

Why do tornadoes spin at all? And why so rapidly? The answer is that air masses that produce tornadoes are themselves rotating; and when the radius of rotation narrows, the rate of rotation increases. An ice skater increases her spin in an exactly analogous manner. (See Figure 9.1.) The skater starts her rotation with outstretched limbs and increases her spin by pulling them in toward her body. The same physics describes the exhilarating spin of a skater and the wrenching force of a tornado.

Clearly, force, energy, and power are associated with rotational motion. These and other aspects of rotational motion are covered in this chapter. We shall see that all important aspects of rotational motion either have already been defined for linear motion or have exact analogues in linear motion. We look first at angular acceleration, the rotational analogue of linear acceleration.

Figure 9.1 This skater increases her rate of spin by pulling in her arms and her free leg.

9.1 ANGULAR ACCELERATION

Chapter 8 discussed only uniform circular motion, which is motion in a circle at constant speed and, hence, constant angular velocity. Recall that angular velocity ω was defined as the time rate of change of angle θ:

$$\omega = \frac{\Delta\theta}{\Delta t} \tag{9.1}$$

where θ is the angle of rotation. (See Figure 9.2.) The relationship between angular velocity ω and linear velocity v was also defined in Chapter 8 as

$$\left. \begin{aligned} v &= r\omega \\ \text{or} \\ \omega &= \frac{v}{r} \end{aligned} \right\} \tag{9.2}$$

where r is the radius of curvature, also as seen in Figure 9.2.

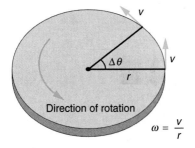

Direction of rotation

$$\omega = \frac{v}{r}$$

Figure 9.2 This shows uniform circular motion and some of the quantities defined for it.

Angular velocity is not constant when a skater pulls in her arms, when a child starts up a merry-go-round from rest, or when a computer's hard disk slows to a halt when switched off. In each of these cases there is an *angular acceleration*, in which ω changes. The faster the change occurs, the greater the angular acceleration. **Angular acceleration** α is defined as the rate of change of angular velocity. In equation form, angular acceleration is expressed as follows:

$$\alpha = \frac{\Delta\omega}{\Delta t} \tag{9.3}$$

where $\Delta\omega$ is the **change in angular velocity** and Δt is the change in time. The units of angular acceleration are (rad/s)/s, or rad/s^2. If ω increases, then α is positive. If ω decreases, then α is negative.*

EXAMPLE 9.1 ANGULAR ACCELERATION AND DECELERATION OF A BIKE WHEEL

Suppose a youngster puts her bicycle on its back and starts the rear wheel spinning from rest to a final angular velocity of 250 rpm in 5.00 s. (a) Calculate the angular acceleration in rad/s^2. (b) If she now slams on the brakes, causing an angular acceleration of −87.3 rad/s^2, how long does it take the wheel to stop?

Strategy for (a) The angular acceleration can be found directly from its definition in Equation 9.3, since the final angular velocity and time are given. We see that $\Delta\omega$ is 250 rpm and Δt is 5.00 s.

Solution for (a) Entering known information into the definition of angular acceleration gives

$$\alpha = \frac{\Delta\omega}{\Delta t} = \frac{250 \text{ rpm}}{5.00 \text{ s}}$$

Since $\Delta\omega$ is in revolutions per minute (rpm) and we want the standard units of rad/s^2 for angular acceleration, we must convert $\Delta\omega$ from rpm to rad/s:

$$\Delta\omega = 250 \, \frac{\text{rev}}{\text{min}} \cdot \frac{2\pi \text{ rad}}{1 \text{ rev}} \cdot \frac{1.00 \text{ min}}{60.0 \text{ s}} = 26.2 \, \frac{\text{rad}}{\text{s}}$$

(continued)

From here to Section 9.7, we will consider only motions in which the axis of rotation (the axle of a wheel, for example) does not change in orientation. This makes the direction of α simple—it is either positive (in the same direction as positive ω) or negative (in the direction opposite to positive ω).

(continued)

Entering this into the expression for α gives

$$\alpha = \frac{\Delta\omega}{\Delta t} = \frac{26.2 \text{ rad/s}}{5.00 \text{ s}} = 5.24 \text{ rad/s}^2$$

Strategy for (b) In this part we know the angular acceleration and the initial angular velocity. We can find the time to stop by using the definition of angular acceleration and solving for Δt, yielding

$$\Delta t = \frac{\Delta\omega}{\alpha}$$

Solution for (b) Here the angular velocity decreases from 26.2 rad/s (250 rpm) to zero, so that $\Delta\omega$ is -26.2 rad/s,

and α is given to be -87.3 rad/s^2. Thus,

$$\Delta t = \frac{-26.2 \text{ rad/s}}{-87.3 \text{ rad/s}^2} = 0.300 \text{ s}$$

Discussion Note that the angular acceleration when the girl starts the wheel spinning is small and positive, taking a few seconds to produce a positive angular velocity. When she hits the brake, the angular acceleration is large and negative. This quickly reduces the angular velocity to zero. In both cases the relationships are analogous to what happens with linear motion. For example, a large acceleration only takes a small time to change velocity, such as when a jet is launched from an aircraft carrier.

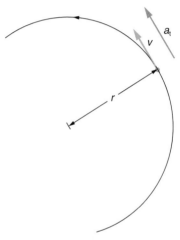

Figure 9.3 Linear acceleration causes the magnitude of velocity to change and is tangent to the motion. Linear acceleration is also called tangential acceleration a_t.

If the bicycle in the preceding example had been on its wheels instead of upside-down, it would first have accelerated along the ground and then come to a stop. This connection between circular motion and linear motion needs to be explored. We would like to know, for example, how linear and angular acceleration are related. In circular motion, linear acceleration is *tangent* to the circle at the point of interest. (See Figure 9.3.) Linear acceleration is thus called **tangential acceleration** a_t.

Linear or tangential acceleration changes the magnitude of velocity but not its direction. We know from Chapter 8 that centripetal acceleration a_c changes the direction of the velocity but not its magnitude. Centripetal acceleration causes circular motion. (See Figure 9.4.) Thus a_t and a_c are perpendicular and independent of one another.* Tangential acceleration a_t is directly related to the angular acceleration α and can increase or decrease the velocity, but does not change its direction.

Now we can find the exact relationship between linear acceleration a_t and angular acceleration α. Since linear acceleration changes the magnitude of velocity, it is defined (as it was in Chapter 2) to be

$$a_t = \frac{\Delta v}{\Delta t}$$

For circular motion, note that $v = r\omega$, so that

$$a_t = \frac{\Delta(r\omega)}{\Delta t}$$

The radius r is constant for circular motion, and so $\Delta(r\omega) = r\,\Delta\omega$. Thus,

$$a_t = r\frac{\Delta\omega}{\Delta t}$$

By definition (see Equation 9.3), $\Delta\omega/\Delta t = \alpha$. Thus,

$$a_t = r\alpha$$

or

$$\alpha = \frac{a_t}{r}$$

(9.4)

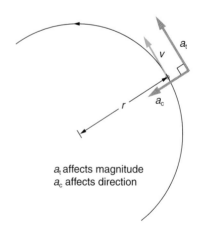

a_t affects magnitude
a_c affects direction

Figure 9.4 Centripetal acceleration causes the direction of velocity to change and is perpendicular to the motion. Centripetal and tangential acceleration are thus perpendicular.

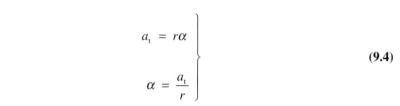

*Since a_t and a_c are mutually independent, they can be added to give any acceleration for a particle on a general curved path. We are limiting ourselves to circular motion, so that a_c must always equal v^2/r—otherwise, the radius of curvature will change. If a_c is larger than v^2/r, the radius of curvature will decrease; if smaller, then r will increase.

This means that linear and angular acceleration are directly proportional. The greater the angular acceleration, the larger the linear (tangential) acceleration, and vice versa. For example, the greater the angular acceleration of a car's drive wheels, the greater the acceleration of the car. The radius also matters. For example, the smaller a wheel, the smaller its linear acceleration for a given angular acceleration α.

EXAMPLE 9.2 WHAT IS THE ANGULAR ACCELERATION OF A MOTORCYCLE WHEEL?

A powerful motorcycle can accelerate from 0 to 30.0 m/s (about 67 mph) in 4.20 s. What is the angular acceleration of its 0.320 m radius wheels? (See Figure 9.5.)

Strategy We are given information about the linear velocities of the motorcycle. Thus we can find its linear acceleration

Figure 9.5 The linear acceleration of a motorcycle is accompanied by an angular acceleration of its wheels.

a_t. Then the second expression in Equation 9.4 can be used to find the angular acceleration.

Solution Linear acceleration is

$$a_t = \frac{\Delta v}{\Delta t} = \frac{30.0 \text{ m/s}}{4.20 \text{ s}} = 7.14 \text{ m/s}^2$$

We also know the radius of the wheels. Entering the values for a_t and r into Equation 9.4,

$$\alpha = \frac{a_t}{r} = \frac{7.14 \text{ m/s}^2}{0.320 \text{ m}}$$

$$= 22.3 \text{ rad/s}^2$$

Discussion Units of radians are automatically involved in any relationship between angular and linear quantities.

So far, we have defined three rotational quantities—θ, ω, and α. These quantities are analogous to the translational quantities x, v, and a. Table 9.1 displays the rotational quantities, the analogous translational quantities, and the relationships between them.

9.2 KINEMATICS OF ROTATIONAL MOTION

Just by using our intuition, we can begin to see how rotational quantities like θ, ω, and α are related to one another. If a motorcycle wheel, for example, has a large angular acceleration for a fairly long time, it ends up spinning rapidly and rotates through many revolutions. In more technical terms, this means that if α is large for a long t, then the final ω and θ are large. This is exactly analogous to the fact that the motorcycle's large translational acceleration produces a large final velocity, and the distance traveled will be large, too.

Kinematics is the description of motion. The **kinematics of rotational motion** describes the relationships among rotation angle, angular velocity, angular acceleration, and time. Let us start by finding an equation relating ω, α, and t. To do this, we recall a familiar kinematic equation for translational, or straight-line, motion:

$$v = v_0 + at \qquad [\text{constant } a]$$

Note that in rotational motion $a = a_t$, and we shall use the symbol a for tangential or linear acceleration from now on. As in linear kinematics, we assume a is constant. This means that angular acceleration α is also a constant, since $\alpha = a/r$. Now let us substitute $v = r\omega$ and $a = r\alpha$ into the linear equation above:

$$r\omega = r\omega_0 + r\alpha t$$

The radius cancels, yielding

$$\omega = \omega_0 + \alpha t \qquad [\text{constant } \alpha]$$

TABLE 9.1

ROTATIONAL AND TRANSLATIONAL QUANTITIES

Rotational	Translational	Relationship
θ	x	$\theta = \frac{x}{r}$
ω	v	$\omega = \frac{v}{r}$
α	a	$\alpha = \frac{a_t}{r}$

CONNECTIONS

Kinematics for rotational motion is completely analogous to translational kinematics, first presented in Chapter 2. Kinematics is concerned with the description of motion without regard to force or mass. We will find that translational kinematic quantities such as displacement, velocity, and acceleration have direct analogues in rotational motion.

where ω_0 is the initial angular velocity. This last equation is a *kinematical relationship* among ω, α, and t—that is, it describes their relationship without reference to forces or masses that may affect rotation. It is also precisely analogous in form to its translational counterpart.

Starting with the four kinematical equations we developed in Chapter 2, we can derive the following four rotational kinematical equations (presented together with their translational counterparts):

Rotational	Translational		
$\theta = \bar{\omega}t$	$x = \bar{v}t$	(constant α, a)	(9.5a)
$\omega = \omega_0 + \alpha t$	$v = v_0 + at$	(constant α, a)	(9.5b)
$\theta = \omega_0 t + \frac{1}{2}\alpha t^2$	$x = v_0 t + \frac{1}{2}at^2$	(constant α, a)	(9.5c)
$\omega^2 = \omega_0^2 + 2\alpha\theta$	$v^2 = v_0^2 + 2ax$	(constant α, a)	(9.5d)

In these equations, the subscript 0 denotes initial values (θ_0, x_0, and t_0 are taken to be zero), and the average angular velocity $\bar{\omega}$ and average velocity $\bar{v}$ are as follows:

$$\bar{\omega} = \frac{\omega_0 + \omega}{2} \quad \text{and} \quad \bar{v} = \frac{v_0 + v}{2} \quad \text{(constant } \alpha, \ a) \tag{9.5e}$$

Equations 9.5a–e can be used to solve any rotational or translational kinematics problem in which α and a are constant.

PROBLEM-SOLVING STRATEGY

FOR ROTATIONAL KINEMATICS*

Step 1. *Examine the situation to determine that rotational kinematics is involved.* Rotation must be involved, but without the need to consider forces or masses that affect the motion.

Step 2. *Identify exactly what needs to be determined in the problem (identify the unknowns).* A sketch of the situation is useful.

Step 3. *Make a list of what is given or can be inferred from the problem as stated (identify the knowns).*

Step 4. *Solve the appropriate equation or equations for the quantity to be determined (the unknown).* It can be useful to think in terms of a translational analogue since you are familiar by now with such motion.

Step 5. *Substitute the knowns along with their units into the appropriate equation, and obtain numerical solutions complete with units.* Be certain to use units of radians for angles.

Step 6. *Check the answer to see if it is reasonable: Does it make sense?*

EXAMPLE 9.3   AN ACCELERATING FISHING REEL

A deep-sea fisherman hooks a big one that pulls line from his fishing reel, which is initially at rest. The reel is given an angular acceleration of 110 rad/s² for 2.00 s, and line unwinds from the reel at a radius of 4.50 cm from its axis of rotation. (See Figure 9.6.) (a) What is the final angular velocity of the reel? (b) At what speed is line leaving the reel after 2.00 s elapses? (c) How many revolutions does the reel make? (d) How many meters of line come off the reel in this time?

Strategy In each part of this example, the strategy is the same as it was for solving problems in linear kinematics. In particular, knowns are identified and a relationship is then sought that can be used to solve for the unknown.

(continued)

---

*You will note a similarity in approach for the problem-solving strategies in the various sections of this text. This is because the same logic applies to most topics in physics and to other disciplines as well. It is likely that the general problem-solving skills you develop in using this text will be useful to you in many other areas. There is evidence, for example, that the best medical diagnosticians did well in their physics courses!

(continued)

Solution for (a) Here α and t are given and ω is wanted. The most straightforward equation to use is 9.5b, since the unknown is already on one side and all other terms are known. That equation states that

$$\omega = \omega_0 + \alpha t$$

We are also given that $\omega_0 = 0$ (it starts from rest), so that

$$\omega = 0 + (110 \text{ rad/s}^2)(2.00 \text{ s}) = 220 \text{ rad/s}$$

Solution for (b) Now that ω is known, the speed v can most easily be found using the relationship

$$v = r\omega$$

The radius r of the reel is given to be 4.50 cm; thus,

$$v = (0.0450 \text{ m})(220 \text{ rad/s}) = 9.90 \text{ m/s}$$

Note again that radians must always be used in any calculation relating linear and angular quantities.

Solution for (c) Here we are asked to find the number of revolutions. Since 1 rev = 2π rad, we can find the number of revolutions by finding θ in radians. We are given α and t, and we know ω_0 is zero, so that θ can be obtained using Equation 9.5c:

$$\theta = \omega_0 t + \tfrac{1}{2}\alpha t^2$$

$$= 0 + (0.500)(110 \text{ rad/s}^2)(2.00 \text{ s})^2 = 220 \text{ rad}$$

Converting radians to revolutions gives

$$\theta = (220 \text{ rad})\frac{1 \text{ rev}}{2\pi \text{ rad}} = 35.0 \text{ revolutions}$$

Solution for (d) The number of meters of line is x, which can be obtained through its relationship with θ:

$$x = r\theta = (0.0450 \text{ m})(220 \text{ rad}) = 9.90 \text{ m}$$

Discussion This example illustrates that relationships among rotational quantities are highly analogous to those among linear quantities. We also see in this example how linear and rotational quantities are connected. The answers to the questions are realistic. The reel is found to spin at 220 rad/s, which is 2100 rpm! (No wonder reels sometimes make high-pitched sounds.) The amount of line played out is 9.90 m, about right for when the big one bites.

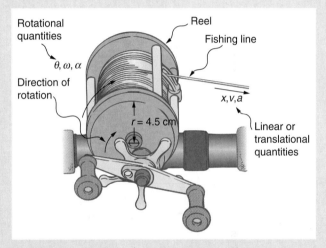

Figure 9.6 Fishing line coming off a rotating reel moves linearly. Examples 9.3 and 9.4 consider relationships between rotational and linear quantities associated with a fishing reel.

EXAMPLE 9.4 THE REEL SLOWS DOWN

Now let us consider what happens if the fisherman applies a brake to the spinning reel, achieving an angular acceleration of −300 rad/s². How long does it take the reel to come to a stop?

Strategy We are asked to find the time t for the reel to come to a stop. The initial and final conditions are different from those in the previous problem, which involved the same fishing reel. Now we see the initial angular velocity is $\omega_0 = 220$ rad/s and the final angular velocity ω is zero. The angular acceleration is given to be $\alpha = -300$ rad/s². Examining the available equations, we see all quantities but t are known in Equation 9.5b, making it easiest to use.

Solution Equation 9.5b states

$$\omega = \omega_0 + \alpha t$$

We solve the equation algebraically for t, and then substitute the knowns as usual, yielding

$$t = \frac{\omega - \omega_0}{\alpha} = \frac{0 - 220 \text{ rad/s}}{-300 \text{ rad/s}^2} = 0.733 \text{ s}$$

Discussion Note that care must be taken with the signs that indicate the directions of various quantities. Note also that the time to stop the reel is fairly small because the acceleration is rather large.

EXAMPLE 9.5 TRAINS ACCELERATE SLOWLY AND SO DO THEIR WHEELS

Large freight trains accelerate very slowly. Suppose one such train accelerates from rest, giving its 0.350 m radius wheels an angular acceleration of 0.250 rad/s². After the wheels have revolved 200 times, assuming no slippage: (a) How far has the train moved down the track? (b) What are the final angular velocity of the wheels and the linear velocity of the train?

Strategy In part (a) we are asked to find x, and in (b) we are asked to find ω and v. We are given the number of revolutions θ, the radius of the wheels r, and the angular acceleration α.

Solution for (a) The distance x is very easily found from the relationship between distance and rotation angle:

$$\theta = \frac{x}{r}$$

Solving this for x yields

$$x = r\theta$$

Before using this equation, we must convert the number of revolutions into radians, since we are dealing with a relationship between linear and rotational quantities:

$$\theta = (200 \text{ rev})\frac{2\pi \text{ rad}}{1 \text{ rev}} = 1257 \text{ rad}$$

Now we can substitute the knowns, yielding

$$x = r\theta = (0.350 \text{ m})(1257 \text{ rad}) = 440 \text{ m}$$

Solution for (b) We cannot use any equation with t in it to find ω, since the equation would have at least two unknowns. Equation 9.5d will work, since we know everything in it but ω:

$$\omega^2 = \omega_0^2 + 2\alpha\theta$$

Taking the square root and entering known values gives

$$\omega = [0 + 2(0.250 \text{ rad/s}^2)(1257 \text{ rad})]^{1/2}$$
$$= 25.1 \text{ rad/s}$$

We can find v through its relationship to ω:

$$v = r\omega = (0.350 \text{ m})(25.1 \text{ rad/s}) = 8.77 \text{ m/s}$$

Discussion The distance traveled is fairly large and the final velocity is fairly slow (just under 20 mph).

There is translational motion even for something spinning in place, as the following example illustrates. Figure 9.7 shows a fly on the edge of a record merrily spinning along to a 1960s rock album. The following example calculates the total distance he travels.

EXAMPLE 9.6 HOW FAR DOES THE FLY ON THE EDGE OF A RECORD TRAVEL?

Calculate the total distance traveled by the fly during the 20.0 min the 0.150 m radius record in Figure 9.7 spins at 33.3 rpm. (Do not worry about the start-up and slow-down times and motions.)

Strategy First find the total number of revolutions θ, and then the linear distance x. Equation 9.5a can be used to find θ since $\bar{\omega}$ is given to be 33.3 rpm.

Solution Entering known values into Equation 9.5a gives

$$\theta = \bar{\omega}t = (33.3 \text{ rpm})(20.0 \text{ min}) = 666 \text{ rev}$$

As always, it is necessary to convert revolutions to radians before calculating a linear quantity like x from an angular

one like θ:

$$\theta = 666 \text{ rev} \cdot \frac{2\pi \text{ rad}}{1 \text{ rev}} = 4185 \text{ rad}$$

Now, using the relationship between x and θ,

$$x = r\theta = (0.150 \text{ m})(4185 \text{ rad}) = 628 \text{ m}$$

Discussion Quite a trip! Note that this is the total distance traveled. Displacement is actually zero for complete revolutions since they bring the fly back to its original position. The distinction between total distance traveled and displacement was first noted in Chapter 2.

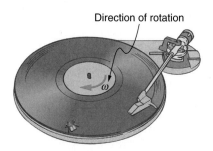

Figure 9.7 The fly makes many revolutions while the album plays. Example 9.6 finds his total distance traveled.

9.3 DYNAMICS OF ROTATIONAL MOTION: ROTATIONAL INERTIA

If you have ever spun a bike wheel or pushed a merry-go-round, you know that force is needed to change angular velocity. (See Figure 9.8.) In fact, your intuition is reliable in predicting many of the factors that are involved. For example, we know that a door opens slowly if we push too close to its hinges. Furthermore, we know that the more massive the door, the more slowly it opens. This implies that the farther the force is applied from the pivot, the greater the angular acceleration; another implication is that angular acceleration is inversely proportional to mass. This all should seem very similar to the familiar relationship among force, mass, and acceleration embodied in Newton's second law of motion. There are, in fact, precise rotational analogues to both force and mass.

To develop the precise relationship among force, mass, radius, and angular acceleration, consider what happens if we exert a force F on a point mass m that is at a distance r from a pivot point, as shown in Figure 9.9. Since the force is perpendicular to r, an acceleration $a = F/m$ is obtained in the direction of F. We can rearrange this to

$$F = ma$$

and then look for ways to relate this to rotational quantities. We note that $a = r\alpha$, and we substitute this into $F = ma$, yielding

$$F = mr\alpha$$

Recall from Chapter 5 that **torque** is the effectiveness of a force in producing a rotation. In this case, since F is perpendicular to r, torque is simply $\tau = rF$. So if we multiply both sides of the equation by r, we get torque on the left-hand side. That is,

$$rF = mr^2\alpha$$

or

$$\tau = mr^2\alpha$$

This last equation is the rotational analogue of Newton's second law ($F = ma$), where torque is analogous to force, angular acceleration is analogous to translational acceleration, and mr^2 is analogous to mass (or inertia). The quantity mr^2 is called the **rotational inertia** or **moment of inertia** of a point mass m a distance r from the center of rotation.

Rotational Inertia and Moment of Inertia

Before we can consider the rotation of anything other than a point mass like the one in Figure 9.9, we must extend the idea of rotational inertia to all types of bodies. To do this, we define the **moment of inertia** I of a body to be the sum of mr^2 for all the point masses* of which it is composed. That is,

$$I = \sum mr^2 \tag{9.6}$$

Here I is analogous to m in translational motion. Actually, calculating I is beyond the scope of this text except for one simple case—that of a hoop, which has all its mass at the same distance from its axis. A hoop's moment of inertia about its axis is therefore MR^2, where M is its total mass and R its radius. (We use M and R for an entire body to distinguish them from m and r for point masses.) In all other cases, we must consult Table 9.2 for formulas for I. Note that I has units of mass times distance squared (kg·m²), as we might expect from its definition.

The general relationship among torque, moment of inertia, and angular acceleration is

$$\left. \begin{array}{c} \text{net } \tau = I\alpha \\[1em] \alpha = \dfrac{\text{net } \tau}{I} \end{array} \right\} \tag{9.7}$$

or

CONNECTIONS

Dynamics for rotational motion is completely analogous to linear or translational dynamics, first presented in Chapter 4. Dynamics is concerned with force and mass and their effect on motion. For rotational motion we will find direct analogues to force and mass that behave just as we would expect from our earlier experiences.

Figure 9.8 Force is required to spin the bike wheel. The greater the force, the greater the angular acceleration produced. The more massive the wheel, the smaller the angular acceleration. If you push on a spoke closer to the axle, the angular acceleration will be smaller.

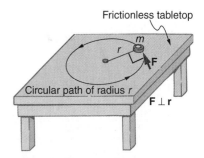

Figure 9.9 A mass is supported by a horizontal frictionless table and is attached to a pivot point by a cord that supplies centripetal force. A force F is applied to the mass perpendicular to the radius r, causing it to accelerate about the pivot point. The force is kept perpendicular to r.

CONNECTIONS

Torque is defined and its directions are discussed in Chapter 5. In statics, the net torque is zero, and there is no angular acceleration. In rotational motion, net torque is the cause of angular acceleration, exactly as in Newton's second law of motion for rotation.

TABLE 9.2

SOME ROTATIONAL INERTIAS

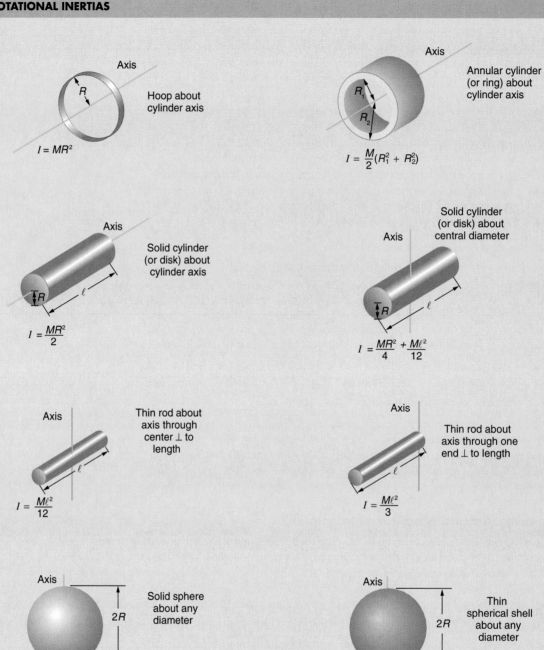

Hoop about cylinder axis

$$I = MR^2$$

Annular cylinder (or ring) about cylinder axis

$$I = \frac{M}{2}(R_1^2 + R_2^2)$$

Solid cylinder (or disk) about cylinder axis

$$I = \frac{MR^2}{2}$$

Solid cylinder (or disk) about central diameter

$$I = \frac{MR^2}{4} + \frac{M\ell^2}{12}$$

Thin rod about axis through center $\perp$ to length

$$I = \frac{M\ell^2}{12}$$

Thin rod about axis through one end $\perp$ to length

$$I = \frac{M\ell^2}{3}$$

Solid sphere about any diameter

$$I = \frac{2MR^2}{5}$$

Thin spherical shell about any diameter

$$I = \frac{2MR^2}{3}$$

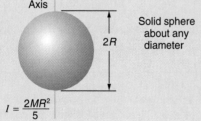

Hoop about any diameter

$$I = \frac{MR^2}{2}$$

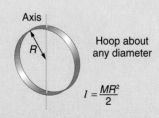

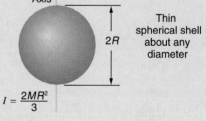

Slab about $\perp$ axis through center

$$I = \frac{M(a^2 + b^2)}{12}$$

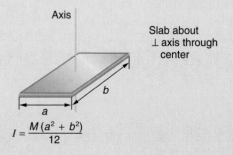

where net τ is the total torque from all forces relative to a chosen axis. For simplicity, we will only consider torques exerted by forces in the plane of the rotation. Such torques are either positive or negative and add like ordinary numbers. The relationship in Equation 9.7 is the rotational analogue to Newton's second law and is very generally applicable. It is actually valid for *any* torque, applied to *any* object, relative to *any* axis.

As we would intuitively expect, the larger the torque, the larger the angular acceleration. The harder a child pushes on a merry-go-round, for example, the faster it accelerates. Furthermore, the more massive a merry-go-round is, the slower it accelerates. This is the effect of the moment of inertia—the larger the moment of inertia, the smaller the angular acceleration. But there is an additional twist. The moment of inertia depends not only on the mass of an object, but also on its *distribution* of mass relative to the line about which it rotates. For example, it will be much easier to accelerate a merry-go-round full of children if they stand close to its axis than if they all stand at the outer edge. The mass is the same in both cases; but the moment of inertia is much larger when the children are at the edge.

PROBLEM-SOLVING STRATEGY

FOR ROTATIONAL DYNAMICS

Step 1. *Examine the situation to determine that torque and mass are involved in the rotation. Draw a careful sketch of the situation.*

Step 2. *Determine the system of interest.*

Step 3. *Draw a free body diagram.* That is, draw and label all external forces acting on the system of interest. We worked with free body diagrams in Chapters 4 and 5.

Step 4. *Apply Equation 9.7, the rotational equivalent of Newton's second law, to solve the problem.* Care must be taken to use the correct moment of inertia and to consider the torque about the point of rotation.

Step 5. As always, *check the solution to see if it is reasonable.*

EXAMPLE 9.7   THE EFFECT OF MASS DISTRIBUTION ON A MERRY-GO-ROUND

Consider the father pushing a playground merry-go-round in Figure 9.10. He exerts a force of 250 N at the edge of the 50.0 kg merry-go-round, which has a 1.50 m radius. Calculate the angular acceleration produced: (a) when no one is on the merry-go-round; (b) when the 18.0 kg child sits 1.25 m from the center. Consider the merry-go-round itself to be a uniform disk with negligible retarding friction.

Strategy Angular acceleration is given directly by the second expression of Equation 9.7:

$$\alpha = \frac{\tau}{I}$$

To solve for α, we must first calculate the torque τ (which is the same in both cases) and moment of inertia I (which is greater in the second case). To find the torque, we note that the applied force is perpendicular to the radius and friction is negligible, so that

$$\tau = rF \sin \theta = rF = (1.50 \text{ m})(250 \text{ N}) = 375 \text{ N} \cdot \text{m}$$

Solution for (a) The moment of inertia of a disk is given in Table 9.2 to be

$$\tfrac{1}{2}MR^2$$

Here $M = 50.0$ kg and $R = 1.50$ m, so that

$$I = 0.5(50.0 \text{ kg})(1.50 \text{ m})^2 = 56.25 \text{ kg} \cdot \text{m}^2$$

Now we can substitute known values; we find the angular acceleration to be

$$\alpha = \frac{\tau}{I} = \frac{375 \text{ N} \cdot \text{m}}{56.25 \text{ kg} \cdot \text{m}^2} = 6.67 \frac{\text{rad}}{\text{s}^2}$$

Solution for (b) We expect the angular acceleration to be less in this part, because the moment of inertia is greater with the child on the merry-go-round. To find the new I, we first find the child's moment of inertia I_c by considering the

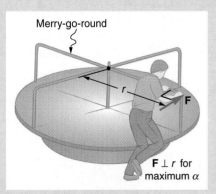

Merry-go-round

r

F

F ⊥ r for maximum α

Figure 9.10   A father pushes a playground merry-go-round at its edge and perpendicular to its radius to achieve maximum torque. Example 9.7 calculates the angular accelerations produced when the merry-go-round is empty and when a child is on it.

(continued)

(continued)

child to be equivalent to a point mass at a distance of 1.25 m from the axis. Then,

$$I_c = MR^2 = (18.0 \text{ kg})(1.25 \text{ m})^2 = 28.13 \text{ kg} \cdot \text{m}^2$$

The total moment of inertia is the sum of those of the merry-go-round and the child (to justify this to yourself, examine the definition of *I*):

$$I = 28.13 \text{ kg} \cdot \text{m}^2 + 56.25 \text{ kg} \cdot \text{m}^2 = 84.38 \text{ kg} \cdot \text{m}^2$$

Substituting known values into the equation for α gives

$$\alpha = \frac{\tau}{I} = \frac{375 \text{ N} \cdot \text{m}}{84.38 \text{ kg} \cdot \text{m}^2} = 4.44 \frac{\text{rad}}{\text{s}^2}$$

Discussion The angular acceleration is less with a child on the merry-go-round than when it is empty, as expected. The angular accelerations found are reasonable. If, for example, the father kept pushing perpendicularly for 2.00 s, he would give the merry-go-round an angular velocity of 13.3 rad/s when empty and 8.88 rad/s with the child on it. These correspond to 2.12 and 1.41 rev/s, respectively. Confirmation of these numbers is left as an exercise for the reader.

9.4 ROTATIONAL KINETIC ENERGY: WORK-ENERGY REVISITED

We turn next to a consideration of work and energy associated with rotational motion. Figure 9.11 shows a worker using an electric grindstone propelled by a motor. Sparks are flying, and noise and vibration are created as layers of steel are pared from the pole. The stone continues to turn even after the motor is turned off, but it is eventually brought to a stop by friction. Clearly, the motor had to do work to get the stone spinning. This work went into heat, light, sound, vibration, and considerable *rotational kinetic energy*.

Work must be done to rotate objects like grindstones or merry-go-rounds. Work was defined in Chapter 6 for translational motion, and we can build on that knowledge when considering work done in rotational motion. The simplest rotational situation is one in which the net force is exerted perpendicular to the radius of a disk (as shown in Figure 9.12) and remains perpendicular as the disk starts to rotate. The force is parallel to the displacement, and so the net work done is the product of the force times the arc length traveled:

$$\text{net } W = (\text{net } F) \cdot \Delta s$$

To get torque and other rotational quantities into the equation, we multiply and divide the right-hand side of the equation by *r*, and gather terms:

$$\text{net } W = (r \text{ net } F) \frac{\Delta s}{r}$$

We recognize that $r \text{ net } F = \text{net } \tau$ and $\Delta s / r = \theta$, so that

$$\text{net } W = (\text{net } \tau)\theta \tag{9.8}$$

This is the expression for rotational work. It is very similar to the familiar definition of translational work as force times distance. Here torque is analogous to force, and angle is analogous to distance. Equation 9.8 is valid in general, even though it was derived for a special case.

To get an expression for rotational kinetic energy, we must again perform some algebraic manipulations. The first step is to note that net $\tau = I\alpha$, so that

$$\text{net } W = I\alpha\theta$$

Now we solve one of the kinematics equations for $\alpha\theta$. We start with Equation 9.5d:

$$\omega^2 = \omega_0^2 + 2\alpha\theta$$

Next we solve for $\alpha\theta$:

$$\alpha\theta = \frac{\omega^2 - \omega_0^2}{2}$$

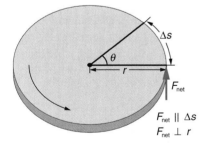

Figure 9.11 The motor does work in spinning the grindstone, giving it rotational kinetic energy. That energy is then converted to heat, light, sound, and vibration.

Figure 9.12 The net force on this disk is kept perpendicular to its radius as the force causes the disk to rotate. The net work done is thus (net *F*) · Δ*s*. The net work goes into rotational kinetic energy.

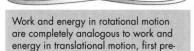

CONNECTIONS

Work and energy in rotational motion are completely analogous to work and energy in translational motion, first presented in Chapter 6.

Substituting this into the equation for net W and gathering terms yields

$$\text{net } W = \tfrac{1}{2}I\omega^2 - \tfrac{1}{2}I\omega_0^2 \qquad (9.9)$$

This is the **work-energy theorem** for rotational motion only.* As you may recall, net work changes the kinetic energy of a system. By analogy with translational motion, we define the term $(1/2)I\omega^2$ to be **rotational kinetic energy** KE_{rot} for a body with a moment of inertia I and an angular velocity ω:

$$KE_{rot} = \tfrac{1}{2}I\omega^2 \qquad (9.10)$$

The expression for rotational kinetic energy is exactly analogous to translational kinetic energy, with I being analogous to m and ω to v. Rotational kinetic energy has important effects. Flywheels, for example, can be used to store large amounts of rotational kinetic energy. (See Figure 9.13.) The rotational kinetic energy of steam turbines can be used as an emergency source of energy in the event of a loss of steam pressure.

Figure 9.13 Experimental vehicles like this bus have been constructed in which rotational kinetic energy is stored in a large flywheel. When the bus goes down a hill, its transmission converts its gravitational potential energy into KE_{rot}. It can also convert translational kinetic energy, when the bus must stop, into KE_{rot}. The flywheel's energy can then be used to accelerate, to go up another hill, or to keep the bus going against friction.

EXAMPLE 9.8   WORK AND ENERGY TO SPIN A GRINDSTONE

Consider a person who spins a large grindstone by placing her hand on its edge and exerting a force through part of a revolution as shown in Figure 9.14. In this example we verify that the work done by the torque she exerts equals the change in rotational energy. (a) How much work is done if she exerts a force of 200 N through a rotation of 1.00 rad (57.3°)? The force is kept perpendicular to the grindstone's 0.320 m radius at the point of application, and friction is negligible. (b) What is the final angular velocity if the grindstone has a mass of 85.0 kg? (c) What is the final rotational kinetic energy? (It should equal the work done.)

Strategy To find the work, we can use Equation 9.8, net $W = (\text{net } \tau)\theta$. We have enough information to calculate the torque and are given the rotation angle. In the second part, we can find the final angular velocity using one of the kinematical relationships. In the last part, we can calculate the rotational kinetic energy from its expression in Equation 9.10.

Solution for (a) The net work is expressed in Equation 9.8 as

$$\text{net } W = (\text{net } \tau)\theta$$

where net τ is the applied force times the radius (rF) since there is no retarding friction, and the force is perpendicular to r. The angle θ is given. Substituting values yields

$$\text{net } W = rF\theta = (0.320 \text{ m})(200 \text{ N})(1.00 \text{ rad})$$
$$= 64.0 \text{ N} \cdot \text{m}$$

Noting that $1 \text{ N} \cdot \text{m} = 1 \text{ J}$,

$$\boxed{\text{net } W = 64.0 \text{ J}}$$

Solution for (b) To find ω from the given information requires more than one step. We start with the kinematic

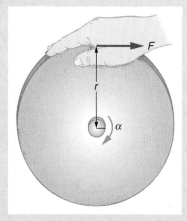

Figure 9.14 A large grindstone is given a spin by a person grasping its outer edge.

relationship in Equation 9.5d,

$$\omega^2 = \omega_0^2 + 2\alpha\theta$$

Note that $\omega_0 = 0$ since we start from rest. Taking the square root gives

$$\omega = (2\alpha\theta)^{1/2}$$

Now we need to find α. One possibility is

$$\alpha = \frac{\text{net } \tau}{I}$$

Here the torque is

$$\text{net } \tau = rF = (0.320 \text{ m})(200 \text{ N}) = 64.0 \text{ N} \cdot \text{m}$$

The formula for the moment of inertia for a disk is found in Table 9.2:

$$I = \tfrac{1}{2}MR^2 = 0.5(85.0 \text{ kg})(0.320 \text{ m})^2 = 4.352 \text{ kg} \cdot \text{m}^2$$

(continued)

---

*If there is both rotation and translation, then net $W = \Delta KE_{rot} + \Delta KE_{trans}$, where $KE_{trans} = (1/2)mv^2$.

(continued)

Substituting into the expression for α, we obtain

$$\alpha = \frac{64.0 \ \text{N} \cdot \text{m}}{4.352 \ \text{kg} \cdot \text{m}^2} = 14.7 \ \frac{\text{rad}}{\text{s}^2}$$

Now substitute this and the given value for θ into the expression for ω above:

$$\omega = (2\alpha\theta)^{1/2} = \left[2\left(14.7 \ \frac{\text{rad}}{\text{s}^2} \right)(1.00 \ \text{rad}) \right]^{1/2} = 5.42 \ \frac{\text{rad}}{\text{s}}$$

Solution for (c) The final rotational kinetic energy is

$$KE_{rot} = \tfrac{1}{2}I\omega^2$$

Both I and ω were found above. Thus,

$$KE_{rot} = 0.5(4.352 \ \text{kg} \cdot \text{m}^2)(5.42 \ \text{rad/s})^2 = 64.0 \ \text{J}$$

Discussion The final rotational kinetic energy does equal the work done by the torque. This confirms that the work done went into rotational kinetic energy. We could, in fact, have used energy instead of a kinematical relation to solve part (b). We will do this in later examples.

Helicopter pilots are quite familiar with rotational kinetic energy. They know, for example, that a point of no return will be reached if they allow their blades to slow below a critical angular velocity during flight. They lose lift, and it is impossible to immediately get the blades spinning fast enough to regain it. Rotational kinetic energy must be supplied to the blades to get them to rotate faster, and this cannot be supplied in time to avoid a crash. Because of weight limitations, helicopter engines are too small to supply both the energy needed for lift and to replenish the rotational kinetic energy of the blades once they have slowed down. The rotational kinetic energy is put into them before takeoff and must not be allowed to drop below the crucial level.

PROBLEM-SOLVING STRATEGY

FOR ROTATIONAL ENERGY

> **Step 1.** *Determine that energy or work is involved in the rotation.*
> **Step 2.** *Determine the system of interest.* A sketch usually helps.
> **Step 3.** Analyze the situation to *determine the types of work and energy involved.*
> **Step 4.** For *closed systems, mechanical energy is conserved.* That is,
>
> $$KE_i + PE_i = KE_f + PE_f$$
>
> Note that KE_i and KE_f may each include translational and rotational contributions.
> **Step 5.** For *open systems* mechanical energy may not be conserved, and other forms of energy (OE such as heat) may enter or leave the system. Determine what they are, and calculate them as necessary.
> **Step 6.** *Eliminate terms wherever possible* to simplify the algebra.
> **Step 7.** *Check the answer to see if it is reasonable.*

EXAMPLE 9.9 HELICOPTER ENERGIES

A typical small helicopter, like the one in Figure 9.15, has four blades, each 4.00 m long with a mass of 50.0 kg. The blades can be approximated as thin rods that rotate about one end of an axis perpendicular to their length. The helicopter has a total loaded mass of 1000 kg. (a) Calculate the rotational kinetic energy in the blades when they rotate at 300 rpm. (b) Calculate the translational kinetic energy of the helicopter when it flies at 20.0 m/s, and compare it with the rotational energy in the blades. (c) To what height could the helicopter be raised if all of the rotational kinetic energy could be used to lift it?

Strategy and Concept Rotational and translational kinetic energies can be calculated from their definitions.

The last part of the problem relates to the idea that energy can change form, in this case from rotational kinetic energy to gravitational potential energy.

Solution for (a) The rotational kinetic energy has been defined in Equation 9.10 as

$$KE_{rot} = \tfrac{1}{2}I\omega^2$$

We must convert the angular velocity to radians per second and calculate the moment of inertia before we can find KE_{rot}. The angular velocity ω is

$$\omega = \frac{300 \ \text{rev}}{1.00 \ \text{min}} \cdot \frac{2\pi \ \text{rad}}{1 \ \text{rev}} \cdot \frac{1.00 \ \text{min}}{60.0 \ \text{s}} = 31.4 \ \frac{\text{rad}}{\text{s}}$$

(continued)

(continued)

The moment of inertia of one blade will be that of a thin rod rotated about its end, found in Table 9.2. The total I is four times this, since there are four blades. Thus,

$$I = 4\frac{M\ell^2}{3} = 4\frac{(50.0 \text{ kg})(4.00 \text{ m})^2}{3} = 1067 \text{ kg}\cdot\text{m}^2$$

Entering ω and I into the expression for kinetic energy gives

$$KE_{rot} = 0.5(1067 \text{ kg}\cdot\text{m}^2)(31.4 \text{ rad/s})^2$$
$$= 5.26 \times 10^5 \text{ J}$$

Solution for (b) Translational kinetic energy was defined in Chapter 6. Entering the given values of mass and velocity, we obtain

$$KE_{trans} = \tfrac{1}{2}mv^2 = 0.5(1000 \text{ kg})(20.0 \text{ m/s})^2 = 2.00 \times 10^5 \text{ J}$$

To compare kinetic energies, we take the ratio of translational kinetic energy to rotational kinetic energy. This ratio is

$$\frac{2.00 \times 10^4 \text{ J}}{5.26 \times 10^5 \text{ J}} = 0.380$$

Solution for (c) At the maximum height, all rotational kinetic energy will have been converted to gravitational energy. To find this height, we equate those two energies:

$$KE_{rot} = PE_{grav}$$

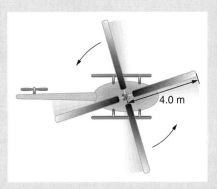

Figure 9.15 Helicopters store large amounts of rotational kinetic energy in their blades. This energy must be put into the blades before takeoff and maintained until the end of the flight. The engines do not have enough power to simultaneously provide lift and put significant rotational energy into the blades.

or

$$\tfrac{1}{2}I\omega^2 = mgh$$

We now solve for h and substitute known values

$$h = \frac{\tfrac{1}{2}I\omega^2}{mg} = \frac{5.26 \times 10^5 \text{ J}}{(1000 \text{ kg})(9.80 \text{ m/s}^2)} = 53.7 \text{ m}$$

Discussion The ratio of translational to rotational kinetic energy is only 0.380. This tells us that most of the kinetic energy of the helicopter is in its spinning blades—something you probably would not suspect. The 53.7 m height to which the helicopter could be raised with the rotational kinetic energy is also impressive, again emphasizing the amount of rotational kinetic energy involved.

How Thick Is the Soup? Or Why Don't All Objects Roll Downhill at the Same Rate?

One of the quality controls in a tomato soup factory consists of rolling filled cans down a ramp. If they roll too fast, the soup is too thin. Why should cans of identical size and mass roll down an incline at different rates? And why should the thickest soup roll the slowest?

The easiest way to answer these questions is to consider energy. Suppose each can starts down the ramp from rest. This means each starts with the same gravitational potential energy PE_{grav}, which is converted entirely to KE, provided each rolls without slipping. KE, however, can take the form of KE_{trans} or KE_{rot}, and total KE is the sum of the two. If a can *rolls* down a ramp, it puts part of its energy into rotation, leaving less for translation. Thus it goes slower than it would if it slid down. Furthermore, the thin soup does not rotate, whereas the thick soup does, because it sticks to the can. The thick soup thus puts more energy into rotation than the thin soup, and it rolls more slowly. (See Figure 9.16.)

The general principle here is, of course, conservation of energy. Assuming no losses to friction, conservation of energy is expressed in equation form as

$$KE_i + PE_i = KE_f + PE_f$$

In the soup can situation we are considering, the initial kinetic energy and the final potential energy are zero, and so

$$PE_i = KE_f$$

More specifically,

$$PE_{grav} = KE_{trans} + KE_{rot}$$

CONNECTIONS

Conservation of energy includes rotational motion, since rotational kinetic energy is another form of KE. Chapter 6 has a detailed treatment of conservation of energy.

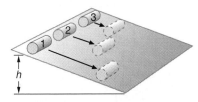

Figure 9.16 Three cans of soup with identical masses race down an incline. The first does not roll but slides frictionlessly. It wins because it converts all its PE into translational KE. The second contains thin soup and comes in second because part of its initial PE goes to rotating the can (but not the thin soup). The third contains thick soup. It comes in third because the soup rotates along with the can, taking even more of the initial PE for rotational KE and leaving less for translational KE.

or

$$mgh = \tfrac{1}{2}mv^2 + \tfrac{1}{2}I\omega^2$$

So the initial mgh is divided between translational kinetic energy and rotational kinetic energy; and the greater I is, the less energy goes into translation. If the object slides down frictionlessly, then $\omega = 0$, and all the energy goes into translation; thus, the can goes faster.

EXAMPLE 9.10   HOW FAST DOES A CYLINDER ROLL DOWN AN INCLINE?

Calculate the final speed of a solid cylinder that rolls down a 2.00 m high incline starting from rest, given the cylinder has a mass of 0.750 kg and a radius of 4.00 cm.

Strategy We can solve for the final velocity using conservation of energy, but we must first express rotational quantities in terms of translational quantities to end up with v as the only unknown.

Solution Conservation of energy for this situation is written as described above:

$$mgh = \tfrac{1}{2}mv^2 + \tfrac{1}{2}I\omega^2$$

Before we can solve for v, we must get an expression for I from Table 9.2. Since v and ω are related, we must also substitute the relationship $\omega = v/R$ into the expression. This yields

$$mgh = \tfrac{1}{2}mv^2 + \tfrac{1}{2}\left(\tfrac{1}{2}mR^2\right)\left(\frac{v^2}{R^2}\right)$$

Interestingly, the radius of the cylinder R and its mass m cancel, yielding

$$gh = \tfrac{1}{2}v^2 + \tfrac{1}{4}v^2 = \tfrac{3}{4}v^2$$

Solving algebraically for v gives

$$v = \left(\frac{4gh}{3}\right)^{1/2}$$

Substituting known values,

$$v = \left[\frac{4(9.80 \text{ m/s}^2)(2.00 \text{ m})}{3}\right]^{1/2} = 5.11 \text{ m/s}$$

Discussion Since m and R cancel, the result $v = (4gh/3)^{1/2}$ is valid for any solid cylinder, implying that all solid cylinders will roll down an incline at the same rate independent of their mass and size. (Rolling cylinders down inclines is what Galileo actually did to show that objects fall at the same rate independent of mass.) Note that if the cylinder slid frictionlessly down the incline without rolling, then all of the gravitational potential energy would go into translational kinetic energy. Thus $(1/2)mv^2 = mgh$ and $v = (2gh)^{1/2}$, which is greater than $(4gh/3)^{1/2}$. That is, the cylinder would go faster at the bottom.

CONNECTIONS

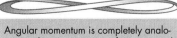

Angular momentum is completely analogous to linear momentum, first presented in Chapter 7. It has the same implications in terms of carrying rotation forward, and it is conserved when the net external torque is zero. Angular momentum, like linear momentum, is also a property of the submicroscopic world and its particles.

9.5   ANGULAR MOMENTUM AND ITS CONSERVATION

Why does the earth keep on spinning? What started it spinning to begin with? And how does an ice skater manage to spin faster and faster simply by pulling her arms in? Why does she not have to exert a torque to spin faster? Questions like these have answers based in angular momentum, the rotational analogue to linear momentum.

By now the pattern is clear—every rotational phenomenon has a direct translational analogue. It seems quite reasonable, then, to define **angular momentum** L as

$$L = I\omega \qquad\qquad (9.11)$$

This is an exact analogue to the definition of linear momentum as $p = mv$. As we would expect, a body with a large moment of inertia I, like the earth, has a very large angular momentum. One with a large angular velocity ω, like a centrifuge, also has a rather large angular momentum.

EXAMPLE 9.11   ANGULAR MOMENTUM OF THE EARTH

Calculate the angular momentum of the earth.

Strategy No information is given in the statement of the problem; so we must look up pertinent data before we can calculate $L = I\omega$. First, according to Table 9.2, the formula for the moment of inertia of a sphere is

$$I = \frac{2MR^2}{5}$$

so that

$$L = I\omega = \frac{2MR^2\omega}{5}$$

On the inside front cover, the mass of the earth M is found to be 5.979×10^{24} kg and its radius R is given as 6.376×10^6 m. The earth's angular velocity ω is, of course, exactly one revolution per day, but we must convert ω to radians per second to do the calculation in SI units.

(continued)

(continued)

Solution Substituting known information into the expression for L and converting ω to radians per second gives

$$L = 0.4(5.979 \times 10^{24} \text{ kg})(6.376 \times 10^6 \text{ m})^2 \left(\frac{1 \text{ rev}}{\text{d}} \right)$$

$$= 9.72 \times 10^{37} \text{ kg} \cdot \text{m}^2/\text{d}$$

Substituting 2π rad for 1 rev, and 8.64×10^4 s for 1 day gives

$$L = (9.72 \times 10^{37} \text{ kg} \cdot \text{m}^2) \left(\frac{2\pi \text{ rad}}{8.64 \times 10^4 \text{ s}} \right)$$

$$= 7.07 \times 10^{33} \text{ kg} \cdot \text{m}^2/\text{s}$$

Discussion This is a large number, demonstrating that the earth, as expected, has a tremendous angular momentum. The answer is approximate, since we have assumed a constant density for the earth in order to estimate its moment of inertia.

When you push a merry-go-round, spin a bike wheel, or open a door, you exert a torque. If the torque you exert is greater than opposing torques, then the rotation accelerates, and angular momentum increases. The greater the net torque, the more rapid the increase in L. The relationship between torque and angular momentum is

$$\text{net } \tau = \frac{\Delta L}{\Delta t} \qquad \textbf{(9.12)}$$

This is exactly analogous to the relationship between force and linear momentum, net $F = \Delta p/\Delta t$. Equation 9.12 is very fundamental and broadly applicable.* It is, in fact, the rotational form of Newton's second law.

EXAMPLE 9.12 TORQUE PUTS ANGULAR MOMENTUM INTO A LAZY SUSAN

Figure 9.17 shows a lazy Susan being rotated by a person in quest of sustenance. Suppose the person exerts a 2.50 N force perpendicular to the lazy Susan's 0.260 m radius for 0.150 s. (a) What is the final angular momentum of the lazy Susan if it starts from rest, assuming friction is negligible? (b) What is the final angular velocity of the lazy Susan, given that its mass is 4.00 kg and assuming its moment of inertia is that of a disk?

Strategy We can find the angular momentum by solving Equation 9.12 for ΔL, and using the given information to calculate the torque. The final angular momentum equals the change in angular momentum, since the lazy Susan starts from rest. That is, $\Delta L = L$. To find the final velocity, we must calculate ω from the definition of L in Equation 9.11.

Figure 9.17 A party-goer exerts a torque on a lazy Susan to make it rotate. Example 9.12 examines the relationship between torque and the angular momentum produced.

Solution for (a) Solving Equation 9.12 for ΔL gives

$$\Delta L = (\text{net } \tau) \cdot \Delta t$$

Since the force is perpendicular to r, we see that net $\tau = rF$, so that

$$L = rF \cdot \Delta t = (0.260 \text{ m})(2.50 \text{ N})(0.150 \text{ s})$$

$$= 9.75 \times 10^{-2} \text{ kg} \cdot \text{m}^2/\text{s}$$

Solution for (b) The final angular velocity can be calculated from the definition of angular momentum,

$$L = I\omega$$

Solving for ω and substituting the formula for the moment of inertia of a disk gives

$$\omega = \frac{L}{I} = \frac{L}{\frac{1}{2} M R^2}$$

And substituting values yields

$$\omega = \frac{9.75 \times 10^{-2} \text{ kg} \cdot \text{m}^2/\text{s}}{(0.500)(4.00 \text{ kg})(0.260 \text{ m})^2} = 0.721 \text{ rad/s}$$

Discussion Note that the imparted angular momentum does not depend on any property of the object but only on torque and time. The final angular velocity is equivalent to one revolution in 8.71 s (it is left as an exercise for the reader to show this), which is about right for a lazy Susan.

---

*For example, if I is constant, then net $\tau = \Delta(I\omega)/\Delta t = I \Delta(\omega)/\Delta t = I\alpha$. So Equation 9.12 contains the earlier relationship net $\tau = I\alpha$ as a special case.

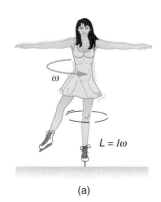

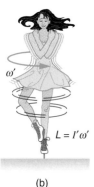

(a)

(b)

Figure 9.18 (a) An ice skater spinning on the tip of her skate with her arms extended. Her angular momentum is conserved because the net torque on her is negligibly small. (b) Her rate of spin increases greatly when she pulls in her arms, decreasing her moment of inertia. The work she does to pull in her arms goes into an increase in rotational kinetic energy.

Conservation of Angular Momentum

We can now understand why the earth keeps on spinning. As we saw in the previous example, $\Delta L = (\text{net } \tau) \cdot \Delta t$. This means that to change angular momentum, a torque must act over some period of time. Since the earth has a large angular momentum, a large torque acting over a long time is needed to change its rate of spin. Tidal friction exerts torque that is slowing the earth's rotation, but tens of millions of years must pass before the change is very significant. Recent research indicates the length of the day was 18 hours some 900 million years ago. Only the tides exert significant retarding torques on the earth, and so it will continue to spin, although ever more slowly, for many billions of years.

What we have here is, in fact, another conservation law. If the net torque is *zero*, then angular momentum is constant or *conserved*. We can see this rigorously by considering Equation 9.12 for the situation in which the net torque is zero. In that case,

$$\text{net } \tau = 0$$

implying that

$$\frac{\Delta L}{\Delta t} = 0$$

If the change in angular momentum ΔL is zero, then the angular momentum is constant; thus,

$$\left. \begin{aligned} L &= \text{constant} \\ \text{or} \\ L &= L' \end{aligned} \right\} \quad (\text{net } \tau = 0) \qquad \textbf{(9.13)}$$

This is the **law of conservation of angular momentum**. Conservation laws are as scarce as they are important.

An example of conservation of angular momentum is seen in Figure 9.18, in which a world-class ice skater is executing a spin. The net torque on her is very close to zero, since there is relatively little friction between her skates and the ice and because the friction is exerted very close to the pivot point. (Both F and r are small, and so τ is negligibly small.) Consequently, she can spin for quite some time. She can do something else, too. She can increase her rate of spin by pulling her arms and legs in. Why does this increase her rate of spin? The answer is that her angular momentum is constant, so that

$$L = L'$$

Expressing this in terms of the moment of inertia,

$$I\omega = I'\omega'$$

where the primed quantities refer to conditions after she has pulled in her arms and reduced her moment of inertia. Since I' is smaller, the angular velocity ω' must increase to keep the angular momentum constant. The change can be dramatic, as the following example shows.

EXAMPLE 9.13 ANGULAR MOMENTUM AND SPINNING SKATER

Suppose an ice skater like the one in Figure 9.18 is spinning at 0.800 revolutions per second with her arms extended. She has a moment of inertia of 2.34 kg·m² with her arms extended and of 0.363 kg·m² with her arms close to her body. (These moments of inertia are based on reasonable assumptions about a 60.0 kg skater. Their calculation is left as an end-of-chapter problem.) (a) What is her angular velocity in revolutions per second after she pulls in her arms? (b) What is her rotational kinetic energy before and after she does this?

Strategy In the first part of the problem, we are looking for the skater's angular velocity ω' after she has pulled in her arms. To find this, we use the conservation of angular momentum and note that the moments of inertia and initial angular velocity are given. To find the initial and final kinetic energies, we use the definition of rotational kinetic energy given by Equation 9.10.

Solution for (a) Since torque is negligible (as discussed above), the conservation of angular momentum given in

(continued)

(continued)

Equation 9.13 is applicable. Thus,

$$L = L'$$

or

$$I\omega = I'\omega'$$

Solving for ω' and substituting known values gives

$$\omega' = \frac{I}{I'}\omega = \frac{2.34 \text{ kg}\cdot\text{m}^2}{0.363 \text{ kg}\cdot\text{m}^2}\cdot 0.800 \text{ rev/s}$$

$$= 5.16 \text{ rev/s}$$

Solution for (b) Equation 9.10 gives rotational kinetic energy to be

$$\text{KE}_{\text{rot}} = \tfrac{1}{2}I\omega^2$$

The initial value is found by substituting known values and converting the angular velocity to rad/s:

$$\text{KE}_{\text{rot}} = 0.5(2.34 \text{ kg}\cdot\text{m}^2)[(0.800 \text{ rev/s})(2\pi \text{ rad/rev})]^2$$

$$= 29.6 \text{ J}$$

The final rotational kinetic energy is

$$\text{KE}'_{\text{rot}} = \tfrac{1}{2}I'\omega'^2$$

Substituting known values gives

$$\text{KE}'_{\text{rot}} = 0.5(0.363 \text{ kg}\cdot\text{m}^2)[(5.16 \text{ rev/s})(2\pi \text{ rad/rev})]^2$$

$$= 191 \text{ J}$$

Discussion In both parts there is an impressive increase. First, the final angular velocity is large, although most world-class skaters can achieve spin rates about this great. Second, the final kinetic energy is much greater than the initial. The increase in rotational kinetic energy comes from work done by the skater in pulling in her arms. This is internal work that depletes some of the skater's food energy.

There are many other examples of objects that increase their rate of spin because something reduced their moment of inertia. Tornadoes are one example. Storm systems that create tornadoes are slowly rotating. When the radius of rotation narrows, even in a local region, angular velocity increases, sometimes to the furious level of a tornado. The earth is another example. Our planet was born from a huge cloud of gas and dust, the rotation of which came from turbulence in an even larger cloud. Gravity caused the cloud to contract, and the rotation rate increased as a result. (See Figure 9.19.)

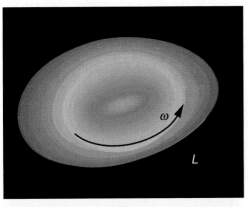

(a)

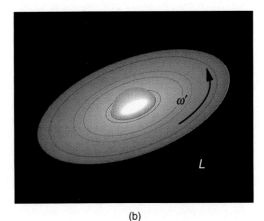

(b)

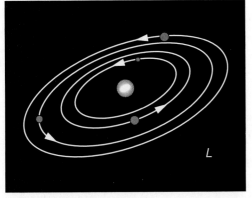

(c)

Figure 9.19 The solar system coalesced from a cloud of gas and dust that was originally rotating. The orbital motions and spins of the planets are in the same direction as the original spin and conserve the angular momentum of the parent cloud.

Figure 9.20 The bowling ball causes the pins to fly, some of them spinning violently.

9.6* COLLISIONS OF EXTENDED BODIES IN TWO DIMENSIONS: A BRIEF TREATMENT OF THINGS THAT GO BUMP AND SPIN

Bowling pins are sent flying and spinning when hit by a bowling ball—angular momentum as well as linear momentum and energy have been imparted to the pins. (See Figure 9.20.) Many collisions involve angular momentum. Cars, for example, may spin and collide on ice. An adept pool player may put "English" on the ball—that is, make the cue ball spin in order to make certain shots. We now take a brief look at what happens when bodies that can rotate collide.

Consider the relatively simple collision shown in Figure 9.21, in which a disk strikes and adheres to an initially motionless stick nailed at one end to a frictionless surface. After the collision, the two rotate about the nail. There is one unbalanced external force on the system at the nail. This force exerts no torque, since its lever arm r is zero. Angular momentum is therefore conserved in the collision. Kinetic energy is not conserved, since the collision is inelastic. It is possible that momentum is not conserved either, since the force at the nail may have a component in the direction of the disk's initial velocity.

Figure 9.21 (a) A disk slides toward a motionless stick on a frictionless surface. (b) The disk hits the stick at one end and adheres to it, and they rotate together, pivoting about the nail. Angular momentum is conserved for this inelastic collision, because the surface is frictionless and the unbalanced external force at the nail exerts no torque.

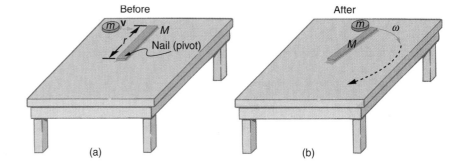

(a)   (b)

EXAMPLE 9.14 ROTATION IN A COLLISION

Suppose the disk in Figure 9.21 has a mass of 50.0 g and an initial velocity of 30.0 m/s when it strikes the 1.20-m-long 2.00 kg stick. (a) What is the angular velocity of the two after the collision? (b) What is the kinetic energy before and after the collision? (c) What is the total linear momentum before and after the collision?

Strategy for (a) We can answer the first question using conservation of angular momentum as noted. Since angular momentum is $I\omega$, we can solve for angular velocity.

Solution for (a) Letting primed quantities stand for conditions after the collision, conservation of angular momentum states

$$L = L'$$

where both are calculated relative to the pivot point. The initial angular momentum is that of the disk just before it strikes the stick. That is,

$$L = I\omega$$

where I is the moment of inertia of the disk and ω is its angular velocity about the pivot point. Now $I = mr^2$ (taking the disk to be approximately a point mass) and $\omega = v/r$, so that

$$L = mr^2 \frac{v}{r} = mvr$$

After the collision,

$$L' = I'\omega'$$

It is ω' that we wish to find. Conservation of angular momentum gives

$$I'\omega' = mvr$$

Rearranging,

$$\omega'_z = \frac{mvr}{I'}$$

where I' is the moment of inertia of the stick and disk stuck together, which is the sum of their individual moments of inertia about the nail. Table 9.2 gives the formula for a rod rotating about one end to be $I = Mr^2/3$. Thus,

$$I' = mr^2 + \frac{Mr^2}{3} = \left(m + \frac{M}{3}\right)r^2$$

Entering known values yields

$$I' = (0.0500 \text{ kg} + 0.667 \text{ kg})(1.20 \text{ m})^2 = 1.032 \text{ kg} \cdot \text{m}^2$$

The value of I' is now entered into the expression for ω':

$$\omega' = \frac{mvr}{I'} = \frac{(0.0500 \text{ kg})(30.0 \text{ m/s})(1.20 \text{ m})}{1.032 \text{ kg} \cdot \text{m}^2}$$

$$= 1.744 \frac{\text{rad}}{\text{s}} = 1.74 \frac{\text{rad}}{\text{s}}$$

(continued)

(continued)

Strategy for (b) The kinetic energy before the collision is the incoming disk's translational kinetic energy, and after the collision it is the rotational kinetic energy of the two stuck together.

Solution for (b) First, we calculate the translational kinetic energy by entering given values for the mass and speed of the incoming disk. This gives

$$KE = \tfrac{1}{2}mv^2 = 0.500(0.0500 \text{ kg})(30.0 \text{ m/s})^2 = 22.5 \text{ J}$$

After the collision, the rotational kinetic energy can be found because we now know the final angular velocity and the final moment of inertia. Thus, entering the values, we have

$$KE' = \tfrac{1}{2}I'\omega'^2 = 0.5(1.032 \text{ kg·m}^2)\left(1.744 \ \frac{\text{rad}}{\text{s}}\right)^2$$
$$= 1.57 \text{ J}$$

Strategy for (c) The linear momentum before the collision is that of the disk. After the collision, it is the sum of the disk's momentum and that of the center of mass of the stick.

Solution for (c) Before the collision, then, linear momentum is

$$p = mv = (0.0500 \text{ kg})(30.0 \text{ m/s}) = 1.50 \text{ kg·m/s}$$

After the collision, the disk and the stick's center of mass move in the same direction. The total linear momentum is

that of the disk moving at a new velocity $v' = r\omega'$ plus that of the stick's center of mass, which moves at half this speed since $v_{CM} = (r/2)\omega' = v'/2$. Thus,

$$p' = mv' + Mv_{CM} = mv' + \frac{Mv'}{2}$$

Gathering terms,

$$p' = \left(m + \frac{M}{2}\right)v'$$

so that

$$p' = \left(m + \frac{M}{2}\right)r\omega'$$

Substituting known values,

$$p' = (1.050 \text{ kg})(1.20 \text{ m})(1.744 \text{ rad/s}) = 2.20 \text{ kg·m/s}$$

Discussion First note that the kinetic energy is less after the collision, as predicted, since the collision is inelastic. More surprising is that the momentum after the collision is actually greater than before the collision! This can be understood if you consider how the nail affects the stick and vice versa. Apparently, the stick pushes backward on the nail when first struck by the disk. The nail's reaction (consistent with Newton's third law) is to push forward on the stick, imparting momentum to it in the same direction in which the disk was initially moving, thereby increasing the momentum of the system.

Example 9.14 has other implications. For example, what would happen if the disk hit very close to the nail? Obviously, a force would be exerted on the nail in the forward direction. So when the stick is struck at the end farthest from the nail, a backward force is exerted on the nail, and when it is hit at the end nearest the nail, a forward force is exerted on the nail. Thus striking it at a certain point in between produces *no* force on the nail. This intermediate point is known as the *percussion point*.

An analogous situation occurs in tennis. (See Figure 9.22.) If you hit a ball with the end of your racket, the handle is pulled away from your hand. If you hit a ball much farther down, say on the shaft of the racket, the handle is pushed into your palm. And if you

Figure 9.22 A disk hitting a stick is compared to a tennis ball being hit by a racket. (a) When the ball strikes the racket near the end, a backward force is exerted on the hand. (b) When the racket is struck much farther down, a forward force is exerted on the hand. (c) When the racket is struck at the percussion point, no force is delivered to the hand.

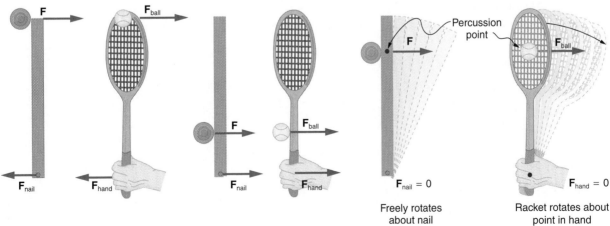

(a) (b) (c)

hit the ball at the racket's percussion point (what some people call the "sweet spot"), then *no* force is exerted on your hand. The same effect occurs for a baseball bat.

9.7* GYROSCOPIC EFFECTS: VECTOR ASPECTS OF ANGULAR MOMENTUM

Gyroscopes do strange things. The one in Figure 9.23, for example, slowly turns in a circle (called *precessing*) instead of falling over as it would if it were not spinning. What holds it up? And why does it precess? The answers to these questions are based on the fact that angular momentum *has direction as well as magnitude*—it is a vector—and that torque affects both the direction and the magnitude of angular momentum.

What is the direction of the angular momentum of a rotating object like the disk in Figure 9.24? The figure shows the right-hand rule used to find the direction of both angular momentum and angular velocity. Both **L** and **ω** are vectors—each has direction and magnitude. Both can be represented by arrows. The right-hand rule defines both to be perpendicular to the plane of rotation in the direction shown. Since angular momentum is related to angular velocity by $L = I\omega$, the direction of **L** is the same as the direction of **ω**. Notice in the figure that both point along the axis of rotation.[†]

Now recall that torque changes angular momentum as expressed by

$$\text{net } \tau = \frac{\Delta \mathbf{L}}{\Delta t}$$

This means that the direction of $\Delta \mathbf{L}$ is the same as the direction of the torque τ that creates it. This is illustrated in Figure 9.25, which shows the direction of torque and the angular momentum it creates.

Let us now consider a bicycle wheel with a couple of handles attached to it, as shown in Figure 9.26. (This is a popular demonstration device among physicists, because it does unexpected things.) The wheel rotates as shown, giving it an angular momentum to the right. Suppose the person holding the wheel tries to rotate it as shown. Her natural expectation is that the wheel will rotate in the direction she pushes it—but

---

[†]Prior to this section, we considered only cases where this axis was fixed. In such cases, the directions of θ, ω, α, τ, and L were denoted by $+$ and $-$ signs corresponding to clockwise and counterclockwise rotation.

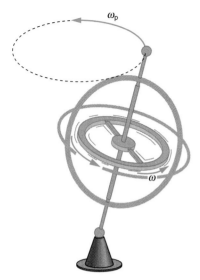

Figure 9.23 Torque can change the direction of angular momentum, the magnitude, or both. This gyroscope slowly precesses in a circle instead of falling on its side. Gravity exerts a torque on the gyroscope that causes the direction of its angular momentum to change.

Figure 9.24 (a) This disk is rotating counterclockwise when viewed from above. (b) Right-hand rule. The direction of angular velocity **ω** and angular momentum **L** are defined to be the direction in which the thumb of your right hand points when you curl your fingers in the direction of the disk's rotation as shown.

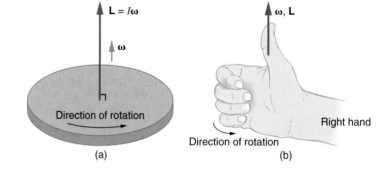

Figure 9.25 (a) Torque is perpendicular to the plane formed by *r* and *F* and is in the direction your right thumb would point if you curled your fingers in the direction of *F*. (b) The direction of the torque is thus the same as that of the angular momentum it creates.

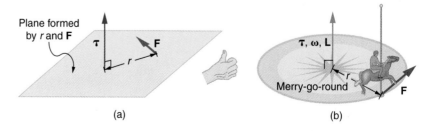

what happens is quite different. The forces exerted create a torque that is horizontal toward the person, as shown in Figure 9.26(a). This torque creates a change in angular momentum $\Delta \mathbf{L}$ in the same direction, perpendicular to the original angular momentum $\mathbf{L}$ and, thus, changing its direction. Figure 9.26(b) shows how $\Delta \mathbf{L}$ and $\mathbf{L}$ add, giving a new angular momentum with direction more toward the person than before. The axis of the wheel has thus moved *perpendicular to the forces exerted on it*, instead of in the expected direction.

This same logic explains the precession of a gyroscope. Figure 9.27 shows the forces acting on a spinning gyroscope. The torque produced is perpendicular to the angular momentum and, thus, changes its direction but not its magnitude. The gyroscope precesses about a vertical axis, since the torque is always horizontal and perpendicular to $\mathbf{L}$. If the gyroscope is *not* spinning, it acquires angular momentum in the direction of the torque ($\mathbf{L} = \Delta \mathbf{L}$), and it rotates about a horizontal axis, falling over just as we would expect.

The earth itself acts like a gigantic gyroscope. Its angular momentum is along its axis and points at Polaris, the North Star. But the earth is slowly precessing (once in about 26,000 years) due to the torque of the sun and moon on its nonspherical shape. The details are beyond the scope of this text, but the interested reader can pursue the topic with the knowledge that it is an aspect of the vector nature of angular momentum.

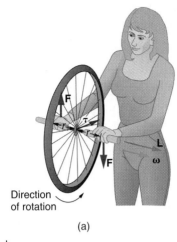

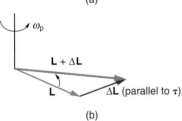

Figure 9.26 (a) A person holding the spinning bike wheel lifts with her right hand and pushes down with her left hand in an attempt to rotate the wheel. This creates a torque directly toward her. This torque causes a change in angular momentum $\Delta \mathbf{L}$ in exactly the same direction. (b) This vector diagram shows how $\Delta \mathbf{L}$ and $\mathbf{L}$ add, producing a new angular momentum pointing more toward the person. The wheel moves toward the person, perpendicular to the forces she exerts on it.

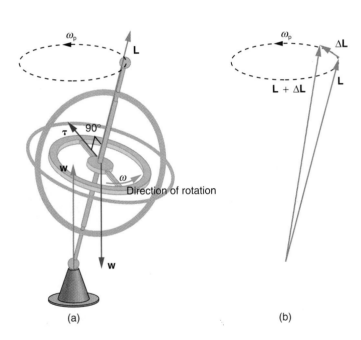

Figure 9.27 (a) The forces on a spinning gyroscope are its weight and the supporting force from the stand. These create a horizontal torque on the gyroscope, which creates a change in angular momentum $\Delta \mathbf{L}$ that is also horizontal. (b) $\Delta \mathbf{L}$ and $\mathbf{L}$ add to produce a new angular momentum with the same magnitude, but different direction, so that the gyroscope precesses in the direction shown instead of falling over.

SUMMARY

Although defined in Chapter 8, it is useful to recall the definition of angular velocity,

$$\omega = \frac{\Delta \theta}{\Delta t} \tag{9.1}$$

and the relationship between angular velocity and linear velocity in rotational motion:

$$v = r\omega$$

or

$$\omega = \frac{v}{r} \tag{9.2}$$

Angular acceleration α is the rate of change of angular velocity:

$$\alpha = \frac{\Delta \omega}{\Delta t} \tag{9.3}$$

where $\Delta \omega$ is the **change in angular velocity** and Δt is the change in time. The units of angular acceleration are rad/s^2. Linear acceleration a is also called **tangential acceleration** a_t, and it is related to angular acceleration by

$$a_t = r\alpha$$

or

$$\alpha = \frac{a_t}{r} \tag{9.4}$$

Rotational and translational kinematics are analogous, as described by the following equations:

Rotational and Translational Quantities

Rotational	Translational	
$\theta = \bar{\omega}t$	$x = \bar{v}t$	(constant α, a) **(9.5a)**
$\omega = \omega_0 + \alpha t$	$v = v_0 + at$	(constant α, a) **(9.5b)**
$\theta = \omega_0 t + \frac{1}{2}\alpha t^2$	$x = v_0 t + \frac{1}{2}at^2$	(constant α, a) **(9.5c)**
$\omega^2 = \omega_0^2 + 2\alpha\theta$	$v^2 = v_0^2 + 2ax$	(constant α, a) **(9.5d)**

The **moment of inertia** I for a body is the rotational analogue to mass and is defined by

$$I = \sum mr^2 \qquad (9.6)$$

Formulas for I for common shapes are found in Table 9.2. The relationship between torque (defined in Chapter 5) and angular acceleration is analogous to Newton's second law:

$$\left.\begin{array}{l} \text{net } \tau = I\alpha \\[12pt] \alpha = \dfrac{\text{net } \tau}{I} \end{array}\right\} \qquad (9.7)$$

or

Work done by a torque is given by

$$\text{net } W = (\text{net } \tau)\theta \qquad (9.8)$$

and the **work-energy theorem** in rotational motion is

$$\text{net } W = \tfrac{1}{2}I\omega^2 - \tfrac{1}{2}I\omega_0^2 \qquad (9.9)$$

The **rotation kinetic energy** of an object having an angular velocity ω is

$$\text{KE}_{\text{rot}} = \tfrac{1}{2}I\omega^2 \qquad (9.10)$$

Angular momentum L is analogous to linear momentum and is given by

$$L = I\omega \qquad (9.11)$$

Angular momentum is changed by torque, following the relationship

$$\text{net } \tau = \frac{\Delta L}{\Delta t} \qquad (9.12)$$

Angular momentum is conserved if the net torque is zero:

$$\left.\begin{array}{l} L = \text{constant} \\[12pt] L = L' \end{array}\right\} \quad (\text{net } \tau = 0) \qquad (9.13)$$

or

This is known as the **law of conservation of angular momentum**. Angular momentum may be conserved in collisions.

Rotational and translational dynamics are analogous, as summarized in Table 9.3.

TABLE 9.3

ROTATIONAL AND TRANSLATIONAL QUANTITIES AND THEIR RELATIONSHIPS		
Rotational	*Translational*	*Relationship*
θ	x	$\theta = \dfrac{x}{r}$
ω	v	$\omega = \dfrac{v}{r}$
α	a	$\alpha = \dfrac{a_t}{r}$
I	m	$I = \sum mr^2$
τ	F	$\tau = rF \sin \theta$
L	p	$L = rp$

CONCEPTUAL QUESTIONS

9.1 There is an analogy between rotational and translational physical quantities. Identify the rotational term analogous to each of the following: acceleration, force, mass, work, translational kinetic energy, linear momentum, impulse.

9.2 Explain why centripetal acceleration changes the direction of velocity in circular motion but not its magnitude.

9.3 In circular motion, a tangential acceleration can change the magnitude of the velocity but not its direction. Explain.

9.4 Suppose a bug is on the edge of a record. Does it experience nonzero tangential acceleration, centripetal acceleration, or both when: (a) The record starts to spin? (b) The record rotates at constant angular velocity? (c) The record slows to a halt?

9.5 The moment of inertia of a long rod spun about an axis through one end perpendicular to its length is $ML^2/3$. Why is this greater than it would be if you spun a point mass M at the location of the center of mass of the rod (at $L/2$)? (That would be $ML^2/4$.)

9.6 Why is the moment of inertia of a hoop of mass M and radius R greater than that of a disk having the same mass and radius?

Why is the moment of inertia of a spherical shell of mass M and radius R greater than that of a solid sphere having the same mass and radius?

9.7 Give an example in which a small force exerts a large torque. Give another example in which a large force exerts a small torque.

9.8 When reducing the mass of a racing bike, the greatest benefit is realized from reducing the mass of the tires and wheel rims. Why does this allow a racer to achieve greater accelerations than would an identical reduction in the mass of the bicycle's frame?

9.9 A ball slides up a ramp without friction. It is then rolled without slipping and with the same initial velocity up another ramp. In which case does it reach a greater height, and why?

9.10 Describe the energy transformations involved when a yo-yo is thrown downward and then climbs back up its string to be caught in the user's hand.

9.11 What energy transformations are involved when a dragster engine is revved, its clutch let out rapidly, its tires spun, and it starts to accelerate forward? Describe the source and transformation of energy at each step.

9.12 The earth has more rotational kinetic energy now than did the cloud of gas and dust from which it formed. Where did this energy come from?

9.13 When you rev the engine of your car with the transmission in neutral, you notice that the car rocks in the opposite sense of the engine's rotation. Explain in terms of conservation of angular momentum. Is the angular momentum of the car conserved for long (for more than a few seconds)?

9.14 During a daredevil show, a car is driven up a ramp and jumped across a river. Although the ramp is straight, the car rotates as it flies through the air, with its front end moving down and its rear moving up. What causes this rotation, assuming no acceleration of the engine or wheels in flight?

9.15 Suppose a child gets off a rotating merry-go-round. Does the angular velocity of the merry-go-round increase, decrease, or remain the same if: (a) She jumps off radially? (b) She jumps backward to land motionless? (c) She jumps straight up and hangs onto an overhead tree branch? (d) She jumps off forward, tangential to the edge? Explain each response.

9.16 Helicopters have a small propeller on their tail to keep them from rotating in the opposite direction of their main lifting blades. Explain in terms of Newton's third law why the helicopter body rotates in the opposite direction of the blades.

9.17 Whenever a helicopter has two sets of lifting blades, they rotate in opposite directions (and there will be no tail propeller). Explain why it is best to have the blades rotate in opposite directions.

9.18 Describe how work is done by a skater pulling in her arms during a spin. In particular, identify the force exerted and the distance moved, noting that a component of the force is in the direction moved. Why is angular momentum not increased by this action?

9.19 When there is a global heating trend, the atmosphere expands and the length of the day increases. Explain why it increases.

9.20 Nearly all conventional piston engines have flywheels on them to smooth out engine vibrations caused by the thrust of individual piston firings. Why does the flywheel have this effect?

9.21 Jet turbines spin rapidly. They are designed to fly apart if something makes them seize suddenly, rather than transfer angular momentum to the plane's wing, possibly tearing it off. Explain how flying apart conserves angular momentum without transferring it to the wing.

9.22 An astronaut tightens a bolt on a satellite in orbit. He rotates in the opposite direction that the bolt rotates, and the satellite

rotates in the same direction as the bolt. Explain why. If a handhold is available on the satellite, can this counterrotation be prevented? Explain.

9.23 Competitive divers pull their limbs in and curl up their bodies when they do flips. Just before entering the water, they fully extend their limbs to enter straight down. Explain the effect of both actions on their angular velocity. Also explain the effect on their angular momentum. (See Figure 9.28.)

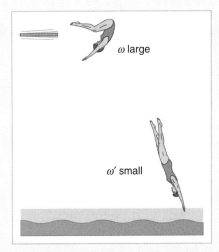

Figure 9.28 The diver spins rapidly when curled up and slows when she extends her limbs before entering the water. Question 23.

9.24 Draw a free body diagram to show how a diver gains angular momentum when leaving the diving board.

9.25* Describe two different collisions—one in which angular momentum is conserved, and the other in which it is not. What condition determines if angular momentum is conserved in a collision?

9.26* Suppose a hockey puck strikes a hockey stick that lies flat on the ice and is free to move in any direction. Which quantities are likely to be conserved: Angular momentum? Linear momentum? Kinetic energy (assuming the puck and stick are very resilient)?

9.27* While driving his motorcycle at highway speed, a physics student notices that pulling back lightly on the right handlebar tips the cycle to the left and produces a left turn. Explain why this happens.

9.28* Gyroscopes used in guidance systems to indicate directions in space must have an angular momentum that does not change in direction. Yet they are often subjected to large forces and accelerations. How can the direction of their angular momentum be constant when they are accelerated?

PROBLEMS

Sections 9.1 and 9.2 Rotational Kinematics

9.1 The terminus of a tornado is 60.0 m in diameter and carries 500 km/h winds. What is its angular velocity in revolutions per second?

9.2 (a) Calculate the linear acceleration of a car, the 0.300 m radius tires of which have an angular acceleration of 10.0 rad/s^2. Assume no slippage. (b) How many revolutions do the tires make in 2.50 s if they start from rest? (c) What is

their final angular velocity? (d) What is the final velocity of the car?

9.3 A phonograph record accelerates from rest to 33.3 rpm in 1.50 s. (a) What is its angular acceleration in rad/s^2? (b) How many revolutions does it go through in the process?

9.4 A phonograph slows from an initial 45.0 rpm at a rate of 0.0500 rad/s^2. (a) How long does it take to come to rest? (b) How many revolutions does it make before stopping?

•**9.5** During a panic stop, a car decelerates at 7.00 m/s^2. (a) What is the angular acceleration of its 0.280 m radius tires, assuming they do not slip on the pavement? (b) How many revolutions do the tires make before coming to rest, given their initial angular velocity is 95.0 rad/s? (c) How long does it take? (d) What distance does the car travel in this time? (e) What was its initial velocity?

‡**9.6** Suppose a yo-yo has a center shaft 0.250 cm in radius and that string is being pulled from it. (a) If the string is stationary and the yo-yo accelerates away from it at a rate of 1.50 m/s^2, what is its angular acceleration? (b) What is the angular velocity after 0.750 s if it starts from rest? (c) The outside radius of the yo-yo is 3.50 cm. What is the tangential acceleration of a point on its edge?

Section 9.3 Rotational Dynamics

9.7 This problem considers additional aspects of Example 9.7. (a) How long does it take the father to get the merry-go-round and child up to an angular velocity of 1.50 rad/s? (b) How many revolutions must he go through to do this? (c) If he exerts a slowing force of 300 N at a radius of 1.35 m, how long would it take him to stop them?

•**9.8** Calculate the moment of inertia of a skater—similar to the one considered in Example 9.13—given the following information. (a) The 60.0 kg skater is approximated as a cylinder 0.110 m in radius. (b) The skater with arms extended is approximately a 52.5 kg cylinder, 0.110 m in radius, having two 0.900 m long arms of 3.75 kg each, extending straight out from her body like rods rotated about their ends.

•**9.9** Suppose you exert a force of 180 N tangential to a 0.280-m-radius 75.0 kg grindstone (a solid disk). (a) What torque is exerted? (b) What is the angular acceleration assuming negligible opposing friction? (c) What is the angular acceleration if there is an opposing frictional force of 20.0 N exerted 1.50 cm from the axis?

•**9.10** Consider the 12.0 kg motorcycle wheel shown in Figure 9.29. Assume it to be approximately an annular ring with an inner radius of 0.280 m and an outer radius of 0.330 m. The motorcycle is on its center stand, so that the wheel can spin freely. (a) If the drive chain exerts a force of 2200 N at a radius of 5.00 cm, what is the angular acceleration of the wheel? (b) What is the tangential acceleration of a point on the outer edge of the tire? (c) How long, starting from rest, does it take to reach an angular velocity of 80.0 rad/s?

•**9.11** Zorch, an archenemy of Superman, decides to slow the earth's rotation to once per 28.0 h by exerting an opposing force at the equator and parallel to it. Superman is not immediately concerned, because he knows Zorch can only exert a force of 4.00×10^7 N (a little greater than a Saturn V rocket's thrust). How long must Zorch push with this force to accomplish his goal? (This gives Superman time to devote to other villains.)

‡**9.12** An automobile engine can produce 200 N·m of torque. (a) What is this torque in ft·lb? (This is consistent with an engine torque quoted in Problem 5.6.) (b) Calculate the angular acceleration produced if 95.0% of this torque is applied to the drive shaft, axle, and rear wheels of a car, given the following information. The car is suspended so that the wheels can turn freely. Each wheel acts like an 0.180-m-radius 15.0 kg disk. The walls of each tire act like a 2.00 kg annular ring of inside radius 0.180 m and outside radius 0.320 m. The tread of each tire acts like a 10.0 kg hoop of radius 0.330 m. The 14.0 kg axle acts like a 2.00 cm radius rod. The 30.0 kg drive shaft acts like a 3.20 cm radius rod.

Section 9.4 Rotational Energy and Work

9.13 This problem considers energy and work aspects of Example 9.7—use data from that example as needed. (a) Calculate the rotational kinetic energy in the merry-go-round plus child when they have an angular velocity of 20.0 rpm. (b) Using energy considerations, find the number of revolutions the father will have to push to achieve this angular velocity starting from rest. (c) Again using energy considerations, calculate the force the father must exert to stop the merry-go-round in two revolutions.

9.14 What is the final velocity of a hoop that rolls without slipping down a 5.00 m high hill, starting from rest?

9.15 (a) Calculate the rotational kinetic energy of the earth on its axis. (b) What is the rotational kinetic energy of the earth in its orbit around the sun?

9.16 Calculate the rotational kinetic energy in the motorcycle wheel described in Problem 9.10 (see also Figure 9.29) if its angular velocity is 120 rad/s.

•**9.17** A bus contains a 1500 kg, 0.600 m radius flywheel (a disk) and has a total mass of 10,000 kg. (a) Calculate the angular velocity the flywheel must have to contain enough energy to take the bus from rest to a speed of 20.0 m/s, assuming 90.0% of the rotational kinetic energy can be transformed into translational energy. (b) How high a hill can the bus climb with this stored energy and still have a speed of 3.00 m/s at the top of the hill?

•**9.18** (a) A ball with an initial velocity of 8.00 m/s rolls up a hill without slipping. Treating the ball as a spherical shell, calculate the vertical height it reaches. (b) Repeat the calculation for the same ball if it slides up the hill without rolling.

‡**9.19** Consider two cylinders that start down an incline from rest. One rolls without slipping, while the other slides frictionlessly without rolling. They both travel a short distance at the bottom and then start up another incline. (a) Show that they

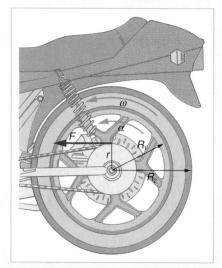

Figure 9.29 A motorcycle wheel has a moment of inertia approximately that of an annular ring. Problems 10, 16, 22, and 35.

both reach the same height on the other incline, and that this height is equal to their original height. (b) Find the ratio of the times taken to do this.

⁝ 9.20 What is the moment of inertia of an object that rolls without slipping down a 2.00 m high incline starting from rest, and has a final velocity of 6.00 m/s? Express the moment of inertia as a multiple of MR^2, where M is the mass of the object and R is its radius.

⁝ 9.21 Repeat Problem 9.11 using energy considerations.

⁝ 9.22 Suppose a 200 kg motorcycle has two wheels like the one described in Problem 9.10 (see also Figure 9.29) and is heading toward a hill at a speed of 30.0 m/s. (a) How high can it coast up the hill, neglecting friction? (b) How much energy is lost to friction if it only gains an altitude of 35.0 m before coming to rest?

Section 9.5 Angular Momentum and Its Conservation

9.23 (a) Calculate the angular momentum of the earth in its orbit around the sun. (b) Compare this with the angular momentum of the earth on its axis.

9.24 (a) What is the angular momentum of the moon in its orbit around the earth? (b) How does this compare with the angular momentum of the moon on its axis? Remember that the moon keeps one side toward the earth at all times.

• 9.25 Suppose you start an antique car by exerting a force of 300 N on its crank for 0.250 s. What angular momentum is given to the engine if the handle of the crank is 0.300 m from the pivot and the force is exerted to create maximum torque the entire time?

• 9.26 A 1.80 m radius playground merry-go-round has a mass of 120 kg and is rotating with an angular velocity of 0.500 rev/s. What is its angular velocity after a 22.0 kg child gets onto it by grabbing its outer edge? The child is initially at rest.

• 9.27 Three children are riding on the edge of a 1.60-m-radius 100 kg merry-go-round, spinning at 20.0 rpm. The children have masses of 22.0, 28.0, and 33.0 kg. If the child with a mass of 28.0 kg moves to the center of the merry-go-round, what is the new angular velocity in rpm?

Section 9.6* Collisions Involving Rotation

• 9.28* Repeat Example 9.14 with the disk striking and adhering to the stick 0.100 m from the nail.

• 9.29* Repeat Example 9.14 with the disk originally spinning clockwise at 1000 rpm and having a radius of 1.50 cm.

• 9.30* Twin skaters approach one another as shown in Figure 9.30 and lock hands. (a) Calculate their final angular velocity, given each had an initial speed of 2.50 m/s relative to the ice. Each has a mass of 70.0 kg, and their centers of mass are 0.800 m from their locked hands. You may approximate their moments of inertia to be that of point masses at this radius. (b) Compare the initial and final kinetic energy.

⁝ 9.31* Suppose a 0.250 kg ball is thrown at 15.0 m/s to a motionless person standing on ice who catches it with an outstretched arm as shown in Figure 9.31. (a) Calculate the final linear velocity of the person, given his mass is 70.0 kg. (b) What is his angular velocity if each arm has a 5.00 kg mass? You may treat his arms as uniform rods of length

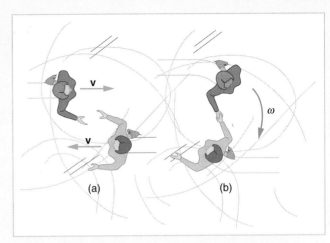

Figure 9.30 (a) Twin skaters approach each other with identical speeds. (b) The skaters lock hands and spin. Problem 30.

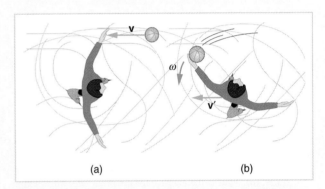

Figure 9.31 (a) Overhead view of a person standing motionless on ice about to catch a ball. Both arms are outstretched. (b) After catching the ball, the skater recoils and rotates. Problem 31.

0.900 m and the rest of his body as a uniform cylinder of radius 0.180 m. Neglect the effect of the ball on his center of mass, so that it remains in his geometrical center. (c) Compare the initial and final total kinetic energy.

⁝ 9.32* Repeat Example 9.14 with the stick free to move translationally as well as rotationally.

INTEGRATED CONCEPTS

These problems involve physical principles from more than one chapter. Integrated knowledge from a broad range of topics is much more powerful than a narrow application of physics. Here you will need to refer to material in Chapters 2, 3, 4, 6, 7, and 8, as well as in Chapter 9.

PROBLEM-SOLVING STRATEGY

Use the following strategy to solve integrated concept problems.

Step 1. *Identify which physical principles are involved.* Chapter 9 deals with aspects of kinematics, dynamics, energy, and momentum in situations involving rotation.

Step 2. *Solve the problem using strategies outlined in the text.* These can be found both in Chapter 9 as well as in earlier chapters on kinematics, dynamics, and energy.

Note: Worked examples that can help you see how to apply these strategies to integrated concept problems appear in a number of chapters, the most recent being Chapter 6. (See the section of problems labeled *Integrated Concepts* at the end of that chapter and consult the index for others.)

•**9.33** An ultracentrifuge accelerates from rest to 100,000 rpm in 2.00 min. (a) What is its angular acceleration in rad/s^2? (b) What is the tangential acceleration of a point 9.50 cm from the axis of rotation? (c) What is the radial acceleration in m/s^2 and *g*s of this point at full rpm?

•**9.34** You have a 90.0 kg, 0.340 m radius grindstone (a disk) turning at 90.0 rpm, and you press a steel ax against it with a radial force of 20.0 N. (a) Assuming the kinetic coefficient of friction between steel and stone is 0.20, calculate the angular acceleration of the grindstone. (b) How many turns will the stone make before coming to rest?

‡**9.35** If you get a tire spinning and then drop it onto the ground, it will accelerate in the forward direction until the tangential velocity of the tire's edge equals the translational velocity of the entire wheel. Suppose you do this with a motorcycle wheel like the one in Problem 9.10 and Figure 9.29. (a) What will its final translational velocity be if its initial angular velocity is 90.0 rad/s? (b) What is the angular acceleration of the tire, assuming the coefficient of friction between it and the road is 0.70? (c) How many revolutions does it make before it stops slipping?

‡**9.36** The axis of the earth makes a 23.5° angle with a perpendicular to the plane of the earth's orbit. As shown in Figure 9.32 this axis precesses, making one complete rotation in 25,780 y. (a) Calculate the change in angular momentum in half this time. (b) What is the average torque producing this change in angular momentum? (c) If this torque were created by a single force (it is not) acting at the most effective point on the equator, what would its magnitude be?

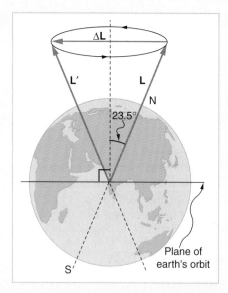

Figure 9.32 The earth's axis slowly precesses, always making an angle of 23.5° with a perpendicular to the plane of the earth's orbit. The change in angular momentum for the two positions shown is quite large, although the magnitude *L* is unchanged. Problem 36.

UNREASONABLE RESULTS

As in previous chapters, problems with unreasonable results are included to give practice in assessing whether nature is being accurately described, and to trace the source of difficulty if it is not. This is very much like the process followed in original research when physical principles as well as faulty premises are tested. The following problems give results that are unreasonable even when the physics is correctly applied, because some premise is unreasonable or because certain premises are inconsistent.

PROBLEM-SOLVING STRATEGY

To determine if an answer is reasonable, and to determine the cause if it is not, do the following.

Step 1. Solve the problem using strategies as outlined in several places in this chapter. See the table of contents to locate all problem-solving strategies. Use the format followed in the text's worked examples to solve the problem as usual.

Step 2. Check to see if the answer is reasonable. Is it too large or too small, or does it have the wrong sign, improper units, . . .?

Step 3. If the answer is unreasonable, look for what specifically could cause the identified difficulty.

9.37 You are told that a professional basketball player spins the ball with an angular acceleration of 100 rad/s^2. (a) What is the ball's final angular velocity, if the ball starts from rest and the acceleration lasts 2.00 s? (b) What is unreasonable about the result? (c) Which premises are unreasonable or inconsistent?

•**9.38** A gymnast doing a forward flip lands on the mat and exerts a 500 N · m torque to slow and then reverse her angular velocity. Her initial angular velocity is 10.0 rad/s, and her moment of inertia is 0.050 kg · m^2. (a) What time is required for her to exactly reverse her spin? (b) What is unreasonable about the result? (c) Which premises are unreasonable or inconsistent?

•**9.39** An advertisement claims that a 800 kg car is aided by its 20.0 kg flywheel, which can accelerate the car from rest to a speed of 30.0 m/s. The flywheel is a disk with a 0.150 m radius. (a) Calculate the angular velocity the flywheel must have if 95.0% of its rotational energy is used to get the car up to speed. (b) What is unreasonable about the result? (c) Which premise is unreasonable, or which premises are inconsistent?

FLUID STATICS

The fluid essential to all life has a beauty of its own.

It also helps support the weight of this swimmer.

Much of what we value in life is fluid: that breath of fresh winter air; the hot blue flame in our furnace; the water we drink, swim in, and bathe in; the blood in our veins. What exactly is a fluid? Can we understand fluids with the laws already presented, or will new laws emerge from their study? The physical characteristics of static or stationary fluids and some of the laws that govern their behavior are the topics of this chapter. The chapter following this one explores aspects of fluid flow.

10.1 WHAT IS A FLUID?

We define a fluid by how it deforms when it is stressed or subjected to force. As discussed in Section 5.8, there are three types of stress. These are tension (force along the length), shear (sideways force), and bulk (compression from all sides). Most of what we have studied so far has involved solid objects, which deform very little when stressed. Most fluids, however, deform easily and do not spring back to their original shape—that is, *fluids flow*. More precisely, a **fluid** is a state of matter that yields to sideways or shearing forces.*

Matter most commonly exists in three major states that we call **phases of matter**. These are solids, liquids, and gases. We can understand the phases of matter and their properties by considering the forces between atoms contained in them. (See Figure 10.1.) Atoms in solids are in close contact, with forces between them that allow the atoms to vibrate but not to change neighbors. (These forces can be thought of as springs that can be stretched or compressed, but are not easily broken.) Thus a solid resists all types of stress. A solid cannot be greatly deformed, because the forces do not allow its atoms to move about freely. Solids also resist compression, because the atoms are essentially in contact and would have to be forced into one another. Simply put, a **solid** retains its shape without the aid of a container.

Atoms in liquids, like those in solids, are also essentially in contact, but they can slide about and change neighbors freely. Thus a liquid resists compression just like a solid, but a liquid can easily flow and be deformed. The forces between atoms in a liquid are attractive and do not allow the atoms to easily break away. A **liquid** remains in a

*Because fluids flow, their shear modulus is zero.

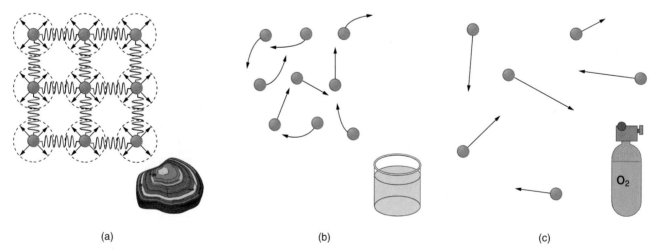

(a)   (b)   (c)

Figure 10.1 (a) Atoms in a solid always have the same neighbors, held near home by forces represented here by springs. The atoms are essentially in contact with one another. A rock is an example of a solid. It can stand alone because of the forces holding its atoms together. (b) Atoms in a liquid are also in close contact but can slide over one another. Forces between them strongly resist attempts to push them closer together and also hold them in close contact. Water is an example of a liquid. It can flow, but it also remains in an open container because of the forces between its atoms. (c) Atoms in a gas are separated by distances that are considerably larger than their diameters. They freely move about. Gas must be held in a closed container to prevent it from moving out freely.

container with an open top. Atoms in gases are separated by distances that are large compared with the atom. The forces between gas atoms are very weak except when they collide. Gases thus not only flow but are relatively easy to compress and expand because there is much space and little force between gas atoms. A **gas** will escape an open container. Since gases and liquids both flow, they are both fluids. The major distinction is that gases are easily compressed and expanded, whereas liquids are not. We shall generally refer to both gases and liquids simply as fluids, and make a distinction between them only when they behave differently.

10.2 DENSITY

Which weighs more, a ton of feathers or a ton of bricks? This old riddle plays with the distinction between mass and density. A ton is a ton, of course; but bricks have much greater density than feathers, and so we are tempted to think of them as heavier. (See Figure 10.2.)

Figure 10.2 A ton of feathers and a ton of bricks have the same mass, but the feathers make a much bigger pile because they have a lower density.

Density, as you will see, is an important characteristic of substances. It is crucial, for example, in determining whether an object sinks or floats in a fluid. **Density** is the mass per unit volume of a substance or object. In equation form, density is

$$\rho = \frac{m}{V} \tag{10.1}$$

where the Greek letter ρ (rho) is the symbol for density, m is the mass, and V is the volume occupied. In the riddle regarding the feathers and bricks, the masses are the same, but the volume occupied by the feathers is much greater, since their density is much lower. The SI unit of density is kg/m^3; representative values are given in Table 10.1. The metric system was originally devised so that water would have a density of 1 g/cm^3, which is equivalent to $10^3 \, kg/m^3$. Thus the basic mass unit, the kilogram, was first devised to be the mass of $10^3 \, cm^3$ of water.

As you can see by examining Table 10.1, the density of an object may help identify what it is. The density of gold, for example, is much greater than that of brass. Density also reveals something about the phase of matter and its substructure. Notice that the densities of liquids and solids are roughly comparable, consistent with the fact that their atoms are in close contact. The densities of gases are much smaller than those of liquids and solids, because the atoms in gases are separated by large amounts of empty space.

EXAMPLE 10.1   FIND THE MASS OF A RESERVOIR FROM ITS VOLUME

What mass of water is held behind a dam, the reservoir of which has a surface area of 50.0 km^2 and an average depth of 40.0 m?

Strategy We can calculate the volume V of the reservoir from its dimensions, and we can find the density of water ρ in Table 10.1. Then the mass m can be found from the definition of density ($\rho = m/V$).

Solution Solving Equation 10.1 for m gives

$$m = \rho V$$

The volume V of the reservoir is its surface area A times its average depth h:

$$V = Ah = (50.0 \text{ km}^2)(40.0 \text{ m})$$

$$= \left[(50.0 \text{ km}^2)\left(\frac{10^3 \text{ m}}{1 \text{ km}}\right)^2 \right](40.0 \text{ m}) = 2.00 \times 10^9 \text{ m}^3$$

The density of water ρ is determined from Table 10.1 to be $1.000 \times 10^3 \, kg/m^3$. Substituting V and r into the expression for mass gives

$$m = (1.00 \times 10^3 \text{ kg/m}^3)(2.00 \times 10^9 \text{ m}^3)$$
$$= 2.00 \times 10^{12} \text{ kg}$$

Discussion A large reservoir contains a very large mass of water. Its weight is $mg = 1.96 \times 10^{13}$ N. It is reasonable to ask whether the dam must supply a force equal to this tremendous weight. The answer is no. As we shall see in the following sections, the force the dam must supply can be much smaller than the weight of the water it holds back.

TABLE 10.1

DENSITIES OF VARIOUS SUBSTANCES*

Substance	ρ $(10^3$ kg/m³ or g/cm³)	Substance	ρ $(10^3$ kg/m³ or g/cm³)	Substance	ρ $(10^3$ kg/m³ or g/cm³)
Solids		Granite	2.7	**Gases**	
Aluminum	2.7	Earth's crust	3.3	Air	1.29×10^{-3}
Brass	8.44	Wood	0.3–0.9	Carbon dioxide	1.98×10^{-3}
Copper (average)	8.8	Ice (0°C)	0.917	Carbon monoxide	1.25×10^{-3}
Gold	19.32	Bone	1.7–2.0	Hydrogen	0.090×10^{-3}
Iron or steel	7.8			Helium	0.18×10^{-3}
Lead	11.3	**Liquids**		Methane	0.72×10^{-3}
Silver (average)	10.1	Water (4°C)	1.000	Nitrogen	1.25×10^{-3}
Tungsten	19.30	Blood	1.05	Nitrous oxide	1.98×10^{-3}
Uranium	18.70	Sea water	1.025	Oxygen	1.43×10^{-3}
Concrete	2.30–3.0	Mercury	13.6	Steam (100°C)	0.60×10^{-3}
Cork	0.24	Ethyl alcohol	0.79		
Glass, common		Gasoline	0.68		
(average)	2.6	Glycerin	1.26		
		Olive oil	0.92		

*At 0°C and standard atmospheric pressure, unless otherwise specified.

10.3 PRESSURE

You have no doubt heard of blood pressure. Atmospheric pressure is discussed regularly on weather broadcasts. These are only two of many examples of pressures in fluids. Pressure is related to, but not the same as, force. Sometimes the same force has very different effects, as in Figure 10.3. The area to which a force is applied can make a great deal of difference. The relevant physical quantity is **pressure** P, defined to be

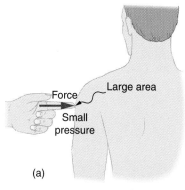

$$P = \frac{F}{A} \tag{10.2}$$

where F is a force applied perpendicularly to area A. The SI unit for pressure is 1 N/m², called a **pascal** (1 Pa $= 1$ N/m²). You may not have noticed it, but car tires have recommended pressures in kilopascals (kPa) printed on their sidewalls. There are a host of other pressure units in common use. Among them are pounds per square inch (lb/in² or psi), and millimeters of mercury (mm Hg). Pressure is defined for all states of matter, but we shall find it extremely important in fluids.

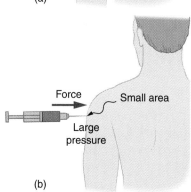

(a)

(b)

Figure 10.3 (a) While the person being poked might be irritated, the force has little lasting effect. (b) The same force applied to a small area has a decidedly different effect. The pressure exerted by the needle is great enough to break the skin.

EXAMPLE 10.2   WHAT FORCE DOES A PRESSURE EXERT?

A space shuttle astronaut is working outside the shuttle. The pressure gauge on her air tank reads 6.90×10^6 N/m² (about 1000 lb/in²). What force does the air exert on the flat end of the cylindrical tank, which is a disk 0.150 m in diameter?

Strategy We can find the force exerted from the definition of pressure given in Equation 10.2, $P = F/A$, provided we can find the area A acted upon.

Solution Rearranging the definition of pressure to solve for force, we see that

$$F = PA$$

Here pressure P is given, and A is the area of the end of the cylinder, which is given by $A = \pi r^2$. Thus,

$$F = (6.90 \times 10^6 \text{ N/m}^2)(3.14)(0.0750 \text{ m})^2$$
$$= 1.22 \times 10^5 \text{ N}$$

Discussion This is more than 27,000 lb! No wonder the tank must be strong. Since we found $F = PA$, we see that the force exerted by a pressure is directly proportional to the area acted upon as well as the pressure itself.

The force exerted on the end of the tank is perpendicular to its inside surface. This is because the force is exerted by a static or stationary fluid. We have already seen that fluids cannot *withstand* shearing (sideways) forces; they cannot *exert* shearing forces, either. Fluid pressure and the forces it creates thus have well-defined directions: they are always perpendicular to any surface. (See Figure 10.4, for example.) Finally, note that pressure is exerted on all surfaces. Swimmers, as well as the tire in Figure 10.4, feel pressure on all sides. (See Figure 10.5.)

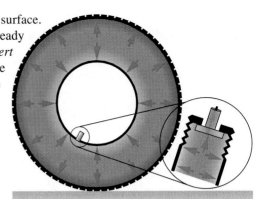

10.4 VARIATION OF PRESSURE WITH DEPTH IN A FLUID

If your ears have ever popped on a plane flight or ached during a deep dive in a swimming pool, you have experienced the effect of depth on pressure in a fluid. The higher the altitude, the less atmospheric pressure; and the greater the depth in a pool, the greater the water pressure. In both cases, pressure is greater at greater depth; but the effect is much more pronounced in water than in air. You may notice an air pressure change on an elevator ride that takes you many stories, but you need only dive a meter or so below the surface of a pool to feel the pressure increase. The difference is that water is far denser than air.

Figure 10.4 Pressure inside this tire exerts force perpendicular to all surfaces it contacts. Arrows give representative directions and magnitudes of the forces exerted at various points. Static fluids cannot exert sideways (shearing) forces.

One very important cause of pressure in a fluid is its own weight. The more dense a fluid, the more weight, all other things being the same. Consider the container in Figure 10.6. Its bottom supports the weight of the fluid in it. Let us calculate the pressure exerted on the bottom by the weight of the fluid. That pressure is the weight of the fluid mg divided by the area A supporting it:

$$P = \frac{mg}{A}$$

We can find the mass of the fluid from its volume and density:

$$m = \rho V$$

The volume of the fluid V is related to the dimensions of the container. It is

$$V = Ah$$

where A is the cross-sectional area and h the depth. Combining the last two equations gives

$$m = \rho Ah$$

If we enter this into the expression for pressure, we obtain

$$P = \frac{(\rho Ah)g}{A}$$

The area cancels, yielding

$$P = h\rho g \qquad (10.3)$$

This is the **pressure due to the weight of a fluid**. The equation has general validity beyond the special conditions under which it is derived here. If the container were not there, the surrounding fluid would exert this pressure, keeping the fluid static. Thus Equation 10.3 represents the pressure due to the weight of any fluid of average density ρ at any depth h below its surface.

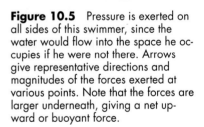

Figure 10.5 Pressure is exerted on all sides of this swimmer, since the water would flow into the space he occupies if he were not there. Arrows give representative directions and magnitudes of the forces exerted at various points. Note that the forces are larger underneath, giving a net upward or buoyant force.

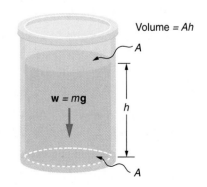

Figure 10.6 The bottom of this container supports the entire weight of the fluid in it. The vertical sides can exert no upward force on the fluid (since it can withstand no shearing force), and so the bottom must support it all.

EXAMPLE 10.3 WHAT FORCE MUST A DAM WITHSTAND?

In Example 10.1, we calculated the mass of water in a large reservoir. We will now consider the pressure and force acting on the dam retaining the water. (See Figure 10.7.) The dam is 500 m wide, and the water is 80.0 m deep at the dam. (a) What is the average pressure on the dam due to the water? (b) Calculate the force exerted against the dam, and compare it with the weight of the water the dam retains (previously found to be 1.96×10^{13} N).

Strategy for (a) The average pressure created by the water is that at the average depth of 40.0 m, since pressure increases linearly with depth.

Solution for (a) Equation 10.3 gives the pressure due to the weight of a fluid as

$$P = h\rho g$$

Entering the density of water from Table 10.1 and taking h to be the average depth of 40.0 m, we obtain

$$\overline{P} = (40.0 \text{ m})\left(10^3 \frac{\text{kg}}{\text{m}^3}\right)\left(9.80 \frac{\text{m}}{\text{s}^2}\right)$$

$$= 3.92 \times 10^5 \frac{\text{N}}{\text{m}^2} = 392 \text{ kPa}$$

Strategy for (b) The force exerted on the dam by the water is the average pressure times the area of contact:

$$F = \overline{P}A$$

Solution for (b) We have already found the value for $\overline{P}$. The area of the dam is $A = 80.0 \text{ m} \times 500 \text{ m} = 4.00 \times 10^4 \text{ m}^2$,

so that

$$F = (3.92 \times 10^5 \text{ N/m}^2)(4.00 \times 10^4 \text{ m}^2)$$
$$= 1.57 \times 10^{10} \text{ N}$$

Discussion Although this is a large force, it is small compared with the 1.96×10^{13} N weight of the water in the reservoir—in fact, it is only $8.00 \times 10^{-2}\%$ of the weight. Note that the pressure found in part (a) is completely independent of the size of the lake—it depends only on its depth. Thus the force depends only on the water's depth and the dimensions of the dam, *not* on the dimensions of the reservoir.

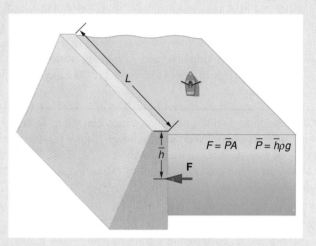

Figure 10.7 The dam must withstand the force exerted against it by the water it retains. This force is small compared with the weight of the water behind the dam.

Atmospheric pressure is a prime example of pressure due to the weight of a fluid. It may vary due to large-scale flow, but we shall consider only its static pressure, which is due to its weight. **Standard atmospheric pressure** P_{atm} is the average value at sea level, measured to be

$$1 \text{ atmosphere (atm)} = P_{atm} = 1.01 \times 10^5 \text{ N/m}^2 = 101 \text{ kPa} \qquad \textbf{(10.4)}$$

This means that on a windless average day at sea level, the column of air above 1.00 m² of the earth's surface has a weight of 1.01×10^5 N. (See Figure 10.8.)

EXAMPLE 10.4 HOW DENSE IS THE AIR?

Calculate the average density of the atmosphere, given that it extends to an altitude of 120 km. Compare this density with that listed in Table 10.1.

Strategy If we solve Equation 10.3 for density, we see that

$$\overline{\rho} = \frac{P}{hg}$$

We then take P to be atmospheric pressure, h is given, and g is known, and so we can use this to calculate $\overline{\rho}$.

Solution Entering known values into the expression for $\overline{\rho}$ yields

$$\overline{\rho} = \frac{1.01 \times 10^5 \text{ N/m}^2}{(120 \times 10^3 \text{ m})(9.80 \text{ m/s}^2)} = 8.59 \times 10^{-2} \text{ kg/m}^3$$

Discussion The result is the average density of air between the earth's surface and the top of the earth's atmosphere, which essentially ends at 120 km. The density of air at sea level is given in Table 10.1 as 1.29 kg/m³—about 15 times its average value. Because air is so compressible, it is this dense only near the surface of the earth. Air density declines rapidly with altitude.

EXAMPLE 10.5 WHAT DEPTH OF WATER CREATES THE SAME PRESSURE AS THE ENTIRE ATMOSPHERE?

Calculate the depth in water at which the pressure due to the weight of the water equals 1.00 atm.

Strategy We begin by solving Equation 10.3 for depth h:

$$h = \frac{P}{\rho g}$$

Then we take P to be 1.00 atm and ρ to be the density of the water that creates the pressure.

Solution Entering the known values into the expression for h gives

$$h = \frac{1.01 \times 10^5 \text{ N/m}^2}{(1.00 \times 10^3 \text{ kg/m}^3)(9.80 \text{ m/s}^2)} = 10.3 \text{ m}$$

Discussion Just 10.3 m of water creates the same pressure as 120 km of air. Since water is nearly incompressible, we can neglect any change in its density over this depth.

What do you suppose the *total* pressure is at a depth of 10.3 m in a swimming pool? Does the atmospheric pressure on the water's surface affect the pressure below? The answer is yes. This seems only logical, since both the water's weight and the atmosphere's weight must be supported. So the *total* pressure at 10.3 m is 2 atm, half from the water and half from the air above. We shall see in the next section that fluid pressures always add in this way.

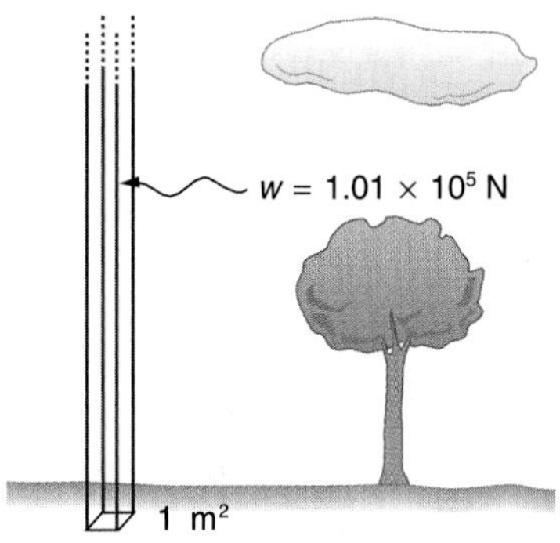

10.5 PASCAL'S PRINCIPLE

Pressure is force per unit area. Can pressure be created in a fluid by pushing directly on it? The answer to this question is yes, but it is much easier if the fluid is enclosed. The heart, for example, creates blood pressure by pushing directly on the blood in an enclosed system. If you try to push on a fluid in an open system, such as a river, the fluid flows away. An enclosed fluid cannot flow away, and so pressure is more easily created by an applied force.

What happens to a pressure exerted on an enclosed fluid? Since atoms in a fluid are free to move about, they transmit the pressure to all parts of the fluid and to the walls of the container. Remarkably, the pressure is transmitted *undiminished*. This is called **Pascal's principle**, because it was first clearly stated by the French philosopher and scientist Blaise Pascal (1623–1662):

> **A change in pressure applied to an enclosed fluid is transmitted undiminished to all portions of the fluid and to the walls of its container.**

Pascal's principle, an experimentally verified fact, is what makes pressure so important in fluids. Since a change in pressure is transmitted undiminished in an enclosed fluid, we often know more about pressure than other physical quantities in fluids. Moreover, Pascal's principle implies that *the total pressure in a fluid is the sum of the pressures from different sources.* We shall find this fact—that pressures add—to be very useful.

One of the most important applications of Pascal's principle is found in a **hydraulic system**, which is an enclosed fluid system used to exert forces. The most common hydraulic systems are those that operate car brakes. Let us first consider the simple hydraulic system shown in Figure 10.9.

We can derive a relationship between the forces in the simple hydraulic system shown in Figure 10.9 by applying Pascal's principle. Note first that the two pistons in the system are at the same height, and so there will be no difference in pressure due to a difference in depth. Now the pressure created by F_1 acting on area A_1 is simply $P_1 = F_1/A_1$, as defined by Equation 10.2. According to Pascal's principle, this pressure is transmitted undiminished throughout the fluid and to all walls of the container. Thus a pressure P_2 is created at the other piston that is equal to P_1. That is,

$$P_1 = P_2$$

Figure 10.8 Atmospheric pressure at sea level averages 1.01×10^5 N/m^2, since the column of air over this 1 m^2, extending to the top of the atmosphere, has a weight of 1.01×10^5 N.

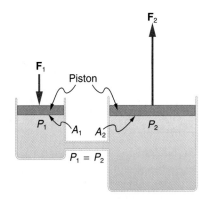

Figure 10.9 A typical hydraulic system with two fluid-filled cylinders, capped with pistons and connected by a tube called a hydraulic line. A downward force F_1 on the left piston creates a pressure that is transmitted undiminished to all parts of the enclosed fluid. This results in an upward force F_2 on the right piston that is larger than F_1 because the right piston has a larger area.

But since $P_2 = F_2/A_2$, we see that

$$\frac{F_1}{A_1} = \frac{F_2}{A_2}$$ (10.5)

This equation relates the ratios of force to area in any hydraulic system, provided that the pistons are at the same vertical height and that friction in the system is negligible. Hydraulic systems can increase or decrease the force applied to them. To make the force larger, the pressure is applied to a larger area. For example, if a 100 N force is applied to the left cylinder in Figure 10.9 and the right one has an area 5 times greater, then the force out is 500 N. Hydraulic systems are analogous to simple levers, but they have the advantage that pressure can be sent through tortuously curved lines to several places at once.

EXAMPLE 10.6   PASCAL PUTS ON THE BRAKES

Consider the automobile hydraulic system shown in Figure 10.10. A force of 100 N is applied to the brake pedal, which acts on the cylinder—called the master—through a lever. A force of 500 N is exerted on the master cylinder. (The reader can verify the force is 500 N using techniques of statics from Chapter 5.) Pressure created in the master cylinder is transmitted to four so-called slave cylinders. The master cylinder has a diameter of 0.500 cm, while each slave cylinder has a diameter of 2.50 cm. Calculate the force F_2 created at each of the slave cylinders.

Strategy We are given the force F_1 that is applied to the master cylinder. The cross-sectional areas A_1 and A_2 can be calculated from their given diameters. Then Equation 10.5 can be used to find the force F_2.

Solution Pascal's principle applied to hydraulic systems is given by Equation 10.5:

$$\frac{F_1}{A_1} = \frac{F_2}{A_2}$$

Manipulate this algebraically to get F_2 on one side and substitute known values:

$$F_2 = \frac{A_2}{A_1} F_1 = \frac{\pi r_2^2}{\pi r_1^2} F_1 = \frac{(1.25 \text{ cm})^2}{(0.250 \text{ cm})^2} \cdot 500 \text{ N}$$

$$= 1.25 \times 10^3 \text{ N}$$

Discussion This is the force exerted by each of the four slave cylinders. Note that we can add as many slave cylinders as we wish. If each has a 2.50 cm diameter, each will exert a 1.25×10^3 N force.

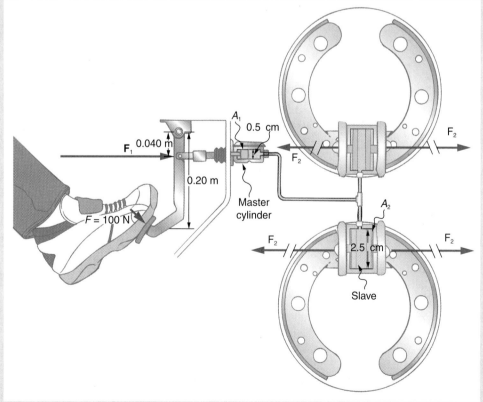
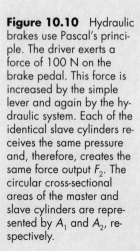

Figure 10.10 Hydraulic brakes use Pascal's principle. The driver exerts a force of 100 N on the brake pedal. This force is increased by the simple lever and again by the hydraulic system. Each of the identical slave cylinders receives the same pressure and, therefore, creates the same force output F_2. The circular cross-sectional areas of the master and slave cylinders are represented by A_1 and A_2, respectively.

A simple hydraulic system, like a simple machine, can increase force but cannot do more work than done on it. Work is force times distance moved, and the slave cylinder moves through a smaller distance than the master cylinder. Furthermore, the more slaves added, the smaller the distance each moves. Many hydraulic systems—such as power brakes and those in bull-dozers and so on—have a motorized pump that actually does most of the work in the system.

10.6   GAUGE PRESSURE, ABSOLUTE PRESSURE, AND PRESSURE MEASUREMENT

If you limp into a gasoline station with a nearly flat tire, you will notice the tire gauge on the air line reads nearly zero when you begin to fill it. In fact, if there were a gaping hole in your tire, the gauge would read zero, even though atmospheric pressure exists in the tire. Why does the gauge read zero? There is no mystery here. To be most easily interpreted, tire gauges are simply designed to read zero at atmospheric pressure and positive when pressure is greater than atmospheric.

Similarly, atmospheric pressure adds to blood pressure in every part of the circulatory system. (As noted in the previous section, the total pressure in a fluid is the sum of the pressures from different sources—here, the heart and the atmosphere.) But atmospheric pressure has no effect on blood flow since it adds to the pressure coming out of the heart and going back into it, too. What is important is how much *greater* blood pressure is than atmospheric pressure. Blood pressure measurements, like tire pressures, are thus made relative to atmospheric pressure.

In brief, it is very common for pressure gauges to ignore atmospheric pressure—that is, to read zero at atmospheric pressure. We therefore define **gauge pressure** to be the amount of pressure above or below atmospheric pressure. Gauge pressure is positive for pressures above atmospheric pressure, and negative for pressures below it.

In fact, atmospheric pressure does add to the pressure in any fluid not enclosed in a rigid container. This is true because of Pascal's principle. The total pressure, or **absolute pressure**, is thus the sum of gauge pressure and atmospheric pressure:

$$P_{abs} = P_g + P_{atm} \qquad (10.6)$$

where P_{abs} is absolute pressure, P_g is gauge pressure, and P_{atm} is atmospheric pressure. For example, if your tire gauge reads 34 psi (pounds per square inch), then the absolute pressure is 34 psi plus 14.7 psi (P_{atm} in psi), or 48.7 psi.

For reasons we will explore later, in most cases absolute pressure in fluids cannot be negative. Fluids push rather than pull. This means that the smallest absolute pressure is zero. (A negative absolute pressure is a pull.) Thus the smallest possible gauge pressure is $P_g = -P_{atm}$ (this makes P_{abs} zero). There is no theoretical limit to how large a gauge pressure can be.

There are a host of devices for measuring pressure, ranging from tire gauges to blood pressure cuffs. Pascal's principle is of major importance in these devices. The undiminished transmission of pressure through a fluid allows precise remote sensing of pressures. Remote sensing is often more convenient than putting a measuring device into a system, such as a person's artery.

Figure 10.11 shows one of the many types of mechanical pressure gauges in use today. In all mechanical pressure gauges, pressure creates a force that is converted (or transduced) into some type of readout.

An entire class of gauges uses the property that pressure due to the weight of a fluid is $P = h\rho g$. Consider the U-shaped tube shown in Figure 10.12, for example. This simple tube is called a *manometer*. In Figure 10.12(a), both sides of the tube are open to the atmosphere. Atmospheric pressure therefore pushes down on each side and is transmitted undiminished to the opposite side, and its effect cancels. If the fluid is deeper on one side, a greater pressure is created on the deeper side, and the fluid flows away from that side until the depths are equal.

Let us examine how a manometer is used to measure pressure. Suppose one side of the U-tube is connected to some source of pressure P_{abs}, such as the toy balloon in Figure 10.12(b) or the vacuum-packed peanut jar shown in Figure 10.12(c). Pressure is transmitted undiminished to the manometer, and the fluid levels are no longer equal. In Figure 10.12(b), P_{abs} is greater than atmospheric pressure, whereas in Figure 10.12(c) P_{abs} is less

CONNECTIONS

Conservation of energy applied to a hydraulic system tells us that the system cannot do more work than is done on it. Work transfers energy, and so the work output cannot exceed the work input. Power brakes and other similar hydraulic systems use pumps to supply extra energy when needed.

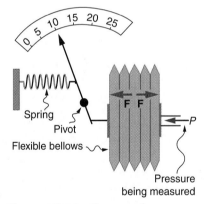

Figure 10.11 This aneroid gauge utilizes flexible bellows connected to a mechanical indicator to measure pressure.

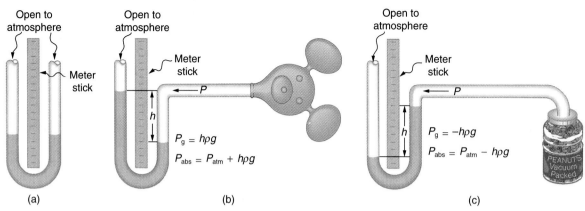

Figure 10.12 An open-tube manometer has one side open to the atmosphere. (a) Fluid depth must be the same on both sides, or the pressure each side exerts at the bottom will be unequal, and there will be flow from the deeper side. (b) A positive gauge pressure $P_g = h\rho g$ transmitted to one side of the manometer can support a column of fluid of height h. (c) Similarly, atmospheric pressure is greater than a negative gauge pressure P_g by an amount $h\rho g$. The jar's rigidity prevents atmospheric pressure from being transmitted to the peanuts.

than atmospheric pressure. In both cases, P_{abs} differs from atmospheric pressure by an amount $h\rho g$, where ρ is the density of the fluid in the manometer. In Figure 10.12(b), P_{abs} can support a column of fluid of height h, and so it must exert a pressure $h\rho g$ greater than atmospheric pressure (the gauge pressure P_g is positive). In Figure 10.12(c), atmospheric pressure can support a column of fluid of height h, and so P_{abs} is less than atmospheric pressure by an amount $h\rho g$ (the gauge pressure P_g is negative). A manometer with one side open to the atmosphere is an ideal device for measuring gauge pressures. The gauge pressure is $P_g = h\rho g$, and it is found by measuring h.

EXAMPLE 10.7 NATURAL GAS PRESSURE MEASURED BY A MANOMETER

A water-filled manometer is connected to a residential natural gas line to test the gas pressure. The water rises on the side that is open to the atmosphere, as shown in Figure 10.12(b). What is the natural gas gauge pressure if the high side is 18.0 cm above the low side? (That is, h is 18.0 cm.)

Strategy and Concept As discussed above, gauge pressure measured by a manometer is

$$P_g = h\rho g$$

where ρ is the density of the fluid in the manometer (water in this case, so that $\rho = 1.000 \times 10^3$ kg/m^3). We are given that h is 18.0 cm or 0.180 m, and g is the acceleration of gravity.

Solution Entering values into the expression for P_g, we obtain

$$P_g = (0.180 \text{ m})(1.000 \times 10^3 \text{ kg/m}^3)(9.80 \text{ m/s}^2)$$
$$= 1.76 \times 10^3 \text{ N/m}^2$$

Discussion The total pressure is greater than atmospheric pressure, and so natural gas will flow from the line. The gauge pressure is not very large, but it is sufficient to supply the fuel. Mercury is sometimes used in manometers. If it were used in this example, then the ρ of the fluid in the manometer would be 13.6 times greater and the height h would be 1/13.6 of that found for a water manometer. This would be small enough to make the measurement difficult. Mercury manometers are used to measure larger pressures, such as blood pressure. Finally, note that the density of water is known to four digits (because of its role in the early metric system), but the final answer is given to three since the other data is only specified to three digits.

Medical Application

Mercury manometers are often used to measure arterial blood pressure. An inflatable cuff is placed on the upper arm as shown in Figure 10.13. By squeezing the bulb, the person making the measurement creates pressure, which is transmitted undiminished to both the main artery in the arm and the manometer. When this applied pressure exceeds blood pressure, blood flow below the cuff is cut off. The person making the measurement then slowly lowers the applied pressure and listens for blood flow to resume. Blood pressure pulsates because of the pumping action of the heart, reaching a maximum, called systolic pressure, and a minimum, called diastolic pressure, with each heartbeat. Systolic pressure is measured by noting h when blood flow first begins as cuff pressure is lowered. Diastolic pressure is measured by noting h when blood flows without interruption. The typical blood pressure of a young adult raises the mercury to a height of 120 mm at systolic and 80 mm at diastolic. This is commonly quoted as 120 over 80, or 120/80. The first pressure is representative of the maximum output of the heart; the second is due to the elasticity of the arteries in maintaining the pressure between beats.

A **barometer** is a device that measures atmospheric pressure. A mercury barometer is shown in Figure 10.14. This device measures atmospheric pressure, rather than gauge pressure, because there is a nearly pure vacuum above the mercury in the tube. The height of the mercury is such that $h\rho g = P_{atm}$. When atmospheric pressure varies, the mercury rises or falls, giving important clues to weather forecasters. The barometer can also be used as an altimeter, since average atmospheric pressure varies with altitude. Mercury barometers and manometers are so common that units of mm Hg are often quoted for atmospheric pressure and blood pressures. Table 10.2 gives conversion factors for some of the more commonly used units of pressure.

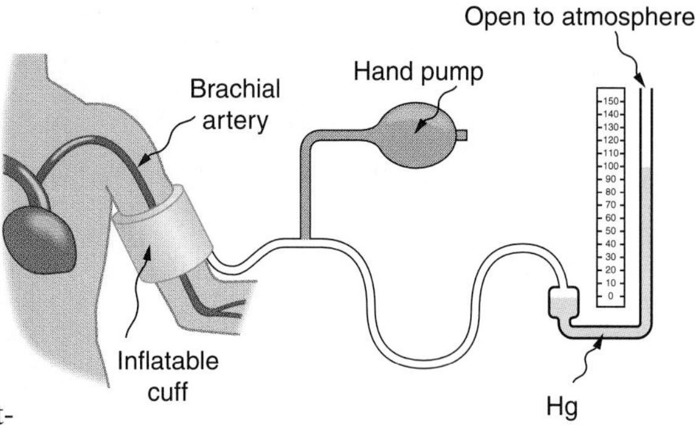

Figure 10.13 In routine blood pressure measurements, an inflatable cuff is placed on the upper arm at the same level as the heart. Blood flow is detected just below the cuff, and corresponding pressures are transmitted to a mercury-filled manometer.

10.7 ARCHIMEDES' PRINCIPLE: BUOYANT FORCE, DENSITY MEASUREMENT, AND WHY SOME THINGS FLOAT

When you rise from lounging in a warm bath, your arms may feel strangely heavy. That is because you no longer have the buoyant support of the water. Where does this buoyant force come from? Why is it that some things float and others do not? Do objects that sink get any support at all from the fluid? Is your body buoyed by the atmosphere, or are only helium balloons affected? (See Figure 10.15.)

Answers to all these questions, and many others, are based on the fact that pressure increases with depth in a fluid. This means that the upward force on the bottom of an object in a fluid is greater than the downward force on the top of the object. There is a net upward, or *buoyant*, force on any object in any fluid. (See Figure 10.16.) If the buoyant force is greater than the object's weight, the object will rise to the surface and float. If the buoyant force is less than the object's weight, the object will sink. But the buoyant force exists whether the object floats or sinks.

Just how great is this buoyant force? To answer this question, think about what happens when a submerged object is removed from a fluid, as in Figure 10.17. The space it occupied is filled by fluid having a weight w_{fl}. This weight is supported by the surrounding fluid, and so the buoyant force must equal w_{fl}, the weight of the fluid displaced by the object. It is a tribute to the genius of the Greek mathematician and inventor Archimedes (ca. 287–212 B.C.) that he stated this principle long before concepts of force were well established. Stated in words, **Archimedes' principle** is as follows:

The buoyant force on an object equals the weight of the fluid it displaces.

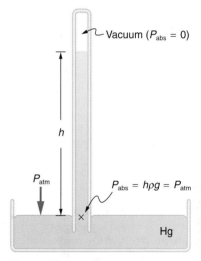

Figure 10.14 A mercury barometer measures atmospheric pressure. The pressure created by the mercury's weight, $h\rho g$, equals atmospheric pressure. The atmosphere is able to force mercury in the tube to a height h because the pressure above the mercury is zero.

TABLE 10.2

CONVERSION FACTORS FOR VARIOUS PRESSURE UNITS*	
Conversion to N/m² (Pa)	*Conversion from atm*
1.0 atm = 1.013×10^5 N/m²	1.0 atm = 1.013×10^5 N/m²
1.0 dyn/cm² = 0.10 N/m²	1.0 atm = 1.013×10^6 dyn/cm²
1.0 kg/cm² = 9.8×10^4 N/m²	1.0 atm = 1.03 kg/cm²
1.0 lb/in² = 6.90×10^3 N/m²	1.0 atm = 14.7 lb/in²
1.0 mm Hg = 133 N/m²	1.0 atm = 760 mm Hg
1.0 cm Hg = 1.33×10^3 N/m²	1.0 atm = 76.0 cm Hg
1.0 cm water = 98.1 N/m²	1.0 atm = 1.03×10^3 cm water
1.0 bar = 1.000×10^5 N/m²	1.0 atm = 1.013 bar

*Note that 1 Pa = 1 N/m²; 1 Torr = 1 mm Hg.

Figure 10.15 (a) Even objects that sink, like this anchor, are partly supported by water when submerged. (b) Submarines have adjustable density so that they may float or sink as desired. (c) Helium-filled balloons tug upward on their strings, demonstrating air's buoyant effect.

(a)

(b)

(c)

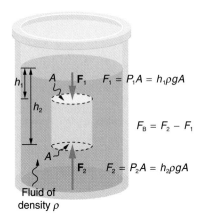

Figure 10.16 Pressure due to the weight of a fluid increases with depth since $P = h\rho g$. This pressure thus creates an upward force on the bottom of the cylinder that is greater than the downward force on its top. Their difference is the buoyant force F_B. (Horizontal forces cancel.)

In the figure: A, $\mathbf{F}_1$, $F_1 = P_1A = h_1\rho gA$; $F_B = F_2 - F_1$; A, $\mathbf{F}_2$, $F_2 = P_2A = h_2\rho gA$; Fluid of density ρ.

In equation form, Archimedes' principle is

$$F_B = w_{fl} \tag{10.7}$$

where F_B is the buoyant force and w_{fl} is the weight of the fluid displaced by the object. Archimedes' principle is valid in general, for any object in any fluid, whether partially or totally submerged.

Floating and Sinking

Drop a lump of clay in water. It will sink. Then mold the lump of clay into the shape of a boat, and it will float. Because of its shape, the boat displaces more water and experiences a greater buoyant force. The same is true of steel ships.

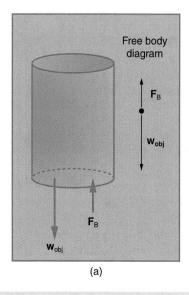

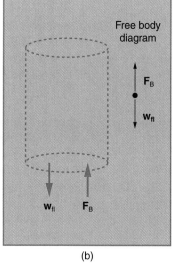

Figure 10.17 (a) An object submerged in a fluid experiences a buoyant force F_B. If F_B is greater than the weight of the object, the object will rise. If F_B is less than the weight of the object, the object will sink. (b) If the object is removed, it is replaced by fluid having weight w_{fl}. Since this weight is supported by surrounding fluid, the buoyant force must equal the weight of the fluid displaced. That is, $F_B = w_{fl}$, a statement of Archimedes' principle.

Free body diagram — $\mathbf{F}_B$, $\mathbf{w}_{obj}$, $\mathbf{F}_B$, $\mathbf{w}_{obj}$

(a)

Free body diagram — $\mathbf{F}_B$, $\mathbf{w}_{fl}$, $\mathbf{w}_{fl}$, $\mathbf{F}_B$

(b)

EXAMPLE 10.8 BUOYANT FORCE MAY DEPEND ON SHAPE

(a) Calculate the buoyant force on 10,000 tons (1.00×10^7 kg) of steel completely submerged in water, and compare this with the steel's weight. (b) What is the maximum buoyant force that water could exert on this same steel if it were shaped into a boat that could displace 1.00×10^5 m^3 of water?

Strategy for (a) To find the buoyant force, we must find the weight of water displaced. We can do this by using the densities of water and steel given in Table 10.1. We note that since the steel is completely submerged, its volume and the water's volume are the same. Once we know the volume of water, we can find its mass and weight.

Solution for (a) First, we solve the definition of density $\rho = m/V$ for the steel's volume, and then we substitute values for mass and density. This gives

$$V_{st} = \frac{m_{st}}{\rho_{st}} = \frac{1.00 \times 10^7 \text{ kg}}{7.8 \times 10^3 \text{ kg/m}^3} = 1.28 \times 10^3 \text{ m}^3$$

As noted, this is also the volume of water displaced, V_w. We can now find the mass of water displaced from the relationship between its volume and density. We solve the definition of density for the mass of water, and substitute

(continued)

(continued)

known values. This gives

$$m_w = \rho_w V_w = (1.000 \times 10^3 \text{ kg/m}^3)(1.28 \times 10^3 \text{ m}^3)$$
$$= 1.28 \times 10^6 \text{ kg}$$

The weight of water displaced is $m_w g$, and so the buoyant force is

$$F_B = w_w = m_w g = (1.28 \times 10^6 \text{ kg})(9.80 \text{ m/s}^2)$$
$$= 1.3 \times 10^7 \text{ N}$$

The steel's weight is $m_w g = 9.80 \times 10^7$ N, much greater than the buoyant force, and so the steel will remain submerged. Note that the buoyant force is rounded to two digits because the density of steel is given to only two digits.

Strategy for (b) Here we are given the maximum volume of water the steel boat can displace. The buoyant force

is the weight of this volume of water.

Solution for (b) The mass of water displaced is found from its relationship to density and volume, both of which are known. That is,

$$m_w = \rho_w V_w = (1.000 \times 10^3 \text{ kg/m}^3)(1.00 \times 10^5 \text{ m}^3)$$
$$= 1.00 \times 10^8 \text{ kg}$$

The maximum buoyant force is the weight of this much water, or

$$F_B = w_w = m_w g = (1.00 \times 10^8 \text{ kg})(9.80 \text{ m/s}^2)$$
$$= 9.80 \times 10^8 \text{ N}$$

Discussion The maximum buoyant force is ten times the weight of the steel, meaning the ship can carry a load nine times its own weight.

Density and Archimedes' Principle

Density has a great deal to do with Archimedes' principle. The average density of an object is what ultimately determines whether it floats. If its average density is less than that of the surrounding fluid, it will float. This is because the fluid, having a higher density, contains more mass and hence more weight in the same volume. The buoyant force, which equals the weight displaced, is thus greater than the weight of the object.

How much of a floating object is submerged depends on how the object's density is related to that of the fluid. In Figure 10.18, for example, the unloaded ship has a lower density and less of it is submerged compared with the same ship loaded. We can derive a quantitative expression for the fraction submerged by considering density. The fraction submerged is the ratio of the volume submerged to the volume of the object, or

$$\text{fraction submerged} = \frac{V_{sub}}{V_{obj}} = \frac{V_{fl}}{V_{obj}}$$

since the volume submerged equals the volume of fluid displaced. Now we can get a relationship between the densities by substituting $V = m/\rho$ into the expression. This gives

$$\frac{V_{fl}}{V_{obj}} = \frac{m_{fl}/\rho_{fl}}{m_{obj}/\overline{\rho}_{obj}}$$

where $\overline{\rho}_{obj}$ is the average density of the object and ρ_{fl} is the density of the fluid. Since the

(a)

(b)

Figure 10.18   An unloaded ship (a) floats higher in the water than it does when loaded (b).

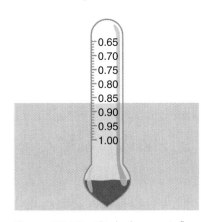

Figure 10.19 This hydrometer is floating in a fluid of specific gravity 0.87. The glass hydrometer is filled with air and weighted with lead at the bottom. It floats highest in the densest fluids and has been calibrated and labeled so that specific gravity can be read from it directly.

object floats, its mass and that of the displaced fluid are equal, and so they cancel from the equation, leaving

$$\text{fraction submerged} = \frac{\overline{\rho}_{obj}}{\rho_{fl}} \tag{10.8}$$

We use this last relationship to measure densities. This is done by measuring the fraction of a floating object that is submerged—for example, with a hydrometer. It is useful to define the ratio of the density of an object to a fluid (usually water) as **specific gravity**:

$$\text{specific gravity} = \frac{\rho}{\rho_w} \tag{10.9}$$

where ρ is the average density of the object or substance and ρ_w is the density of water at 4.00°C. Specific gravity is unitless, independent of whatever units are used for ρ. If an object floats, its specific gravity is less than 1. If it sinks, its specific gravity is greater than 1. Moreover, the fraction of a floating object that is submerged equals its specific gravity. We measure the specific gravity of fluids, such as battery acid, radiator fluid, and urine, as an indicator of their condition. One device for measuring specific gravity is shown in Figure 10.19.

EXAMPLE 10.9   WHAT IS THE AVERAGE DENSITY OF THE FLOATING WOMAN?

Suppose a 60.0 kg woman floats in fresh water with 97.0% of her volume submerged when her lungs are full of air. What is her average density?

Strategy We can find the woman's density by solving Equation 10.8 for the density of the object. This yields

$$\overline{\rho}_{obj} = \overline{\rho}_{person} = (\text{fraction submerged}) \cdot \rho_{fl}$$

We know both the fraction submerged and the density of water, and so we can calculate the woman's density.

Solution Entering the known values into the expression for her density, we obtain

$$\overline{\rho}_{person} = 0.970 \cdot \left(10^3 \, \frac{\text{kg}}{\text{m}^3}\right) = 970 \, \frac{\text{kg}}{\text{m}^3}$$

Discussion Her density is less than the fluid density. We expect this since she floats. Body density is one indicator of a person's percent body fat, of interest in medical diagnostics and athletic training. (See Figure 10.20.)

Human Application

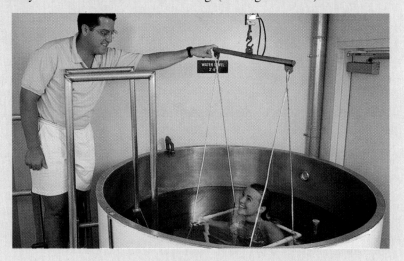

Figure 10.20 Subject in "fat tank," where she is weighed while completely submerged as part of a body density determination. The subject must completely empty her lungs and hold a metal weight in order to sink. Corrections are made for the residual air in her lungs (measured separately) and the metal weight. Her corrected submerged weight, her weight in air, and pinch tests of strategic fatty areas are used to calculate her percent body fat.

There are many obvious examples of lower-density objects or substances floating in higher-density fluids—oil on water, a hot air balloon, a bit of cork in wine, an iceberg, and hot wax in a "lava lamp," to name a few. Less obvious examples include real lava rising in a volcano and mountain ranges floating on the higher-density crust and mantle beneath them. Even the seemingly solid earth has fluid characteristics.

More Density Measurements

One of the most common techniques for determining density is shown in Figure 10.21. An object, here a coin, is weighed in air and then weighed again while submerged

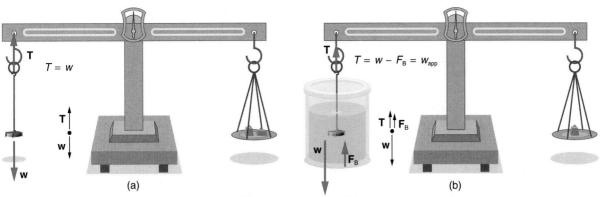

Figure 10.21 (a) A coin is weighed in air. (b) The apparent weight of the coin is determined while it is completely submerged in a fluid of known density. These two measurements are used to calculate the density of the coin.

in a liquid. The density of the coin, an indication of its authenticity, can be calculated if the fluid density is known. This same technique can also be used to determine the density of the fluid if the density of the coin is known. All this is based on Archimedes' principle.

Archimedes' principle states that the buoyant force on the object equals the weight of the fluid displaced. This, in turn, means that the object *appears* to weigh less when submerged; we call this measurement the object's *apparent weight*. The object suffers an *apparent weight loss* equal to the weight of the fluid displaced. Alternatively, on balances that measure mass, the object suffers an *apparent mass loss* equal to the mass of fluid displaced. That is,

> **apparent weight loss = weight of fluid displaced**

or

> **apparent mass loss = mass of fluid displaced**

The next example illustrates the use of this technique.

EXAMPLE 10.10 IS THE COIN AUTHENTIC?

An ancient Greek coin's mass is determined in air to be 8.630 g. When the coin is submerged in water as shown in Figure 10.21, its apparent mass is 7.800 g. Calculate its density, given that water has a density of 1.000 g/cm^3 and that effects caused by the wire suspending the coin are negligible.

Strategy To calculate the coin's density, we need its mass (which is given) and its volume. The volume of the coin equals the volume of water displaced. The volume of water displaced V_w can be found by solving the equation for density, $\rho = m/V$, for V.

Solution The volume of water is

$$V_w = \frac{m_w}{\rho_w}$$

where m_w is the mass of water displaced. As noted, the mass of the water displaced equals the apparent mass loss, which is

$$m_w = (8.630 - 7.800)\ g = 0.830\ g$$

Thus the volume of water is

$$V_w = \frac{0.830\ g}{1.000\ g/cm^3} = 0.830\ cm^3$$

This is also the volume of the coin, since it is completely submerged. We can now find the density of the coin using the definition of density:

$$\rho_c = \frac{m_c}{V_c} = \frac{8.630\ g}{0.830\ cm^3} = 10.4\ g/cm^3$$

Discussion You can see from Table 10.1 that this density is very close to that of pure silver, appropriate for this type of ancient coin. Most modern counterfeits are not pure silver.

This brings us back to Archimedes' principle and how it came into being. As the story goes, the king of Syracuse gave Archimedes the task of determining whether the royal crown maker was supplying a crown of pure gold. The purity of gold is difficult to determine by color (it can be diluted with other metals and still look as yellow as pure gold), and other analytical techniques had not yet been conceived. Even in ancient times, however, it had been realized that the density of gold was greater than that of any other then-known substance. Archimedes purportedly agonized over his task and had his inspiration one day while at the public baths, pondering the support the water gave his body. He came up with his now-famous principle, saw how to apply it to determine density, and ran naked down the streets of Syracuse crying "Eureka!" (Greek for "I have found it"). Similar behavior can be observed in contemporary physicists from time to time.

10.8* COHESION AND ADHESION IN LIQUIDS: SURFACE TENSION AND CAPILLARY ACTION

Children blow soap bubbles and play in the spray of a sprinkler on a hot summer day. (See Figure 10.22.) An underwater spider keeps his air supply in a shiny bubble he carries wrapped around him. A technician draws blood into a small-diameter tube just by touching it to a drop on a pricked finger. A premature infant struggles to inflate her lungs. What is the common thread? All these activities are dominated by the attractive forces between atoms and molecules in liquids—both within a liquid and between the liquid and its surroundings.

Attractive forces between molecules of the same type are called **cohesive forces**. Liquids can, for example, be held in open containers because cohesive forces hold the molecules together. Attractive forces between molecules of different types are called **adhesive forces**. Such forces cause liquid drops to cling to window panes, for example. In this section we examine effects directly attributable to cohesive and adhesive forces in liquids.

Surface Tension

Cohesive forces between molecules cause the surface of a liquid to act like a stretched rubber sheet. It contracts to the smallest possible surface area—perhaps forming a droplet. This general effect is called *surface tension*. Figure 10.23 shows how cohesive forces cause liquids to reduce surface area whenever possible. Molecules on the surface are pulled inward by cohesive forces. This is seen to reduce the surface area. Molecules inside the liquid experience zero net force, since they have neighbors on all sides.

The model of a liquid surface acting like a stretched elastic sheet effectively explains surface tension effects. For example, some insects can walk on water (as opposed to floating in it) as we would walk on a trampoline—they dent the surface as shown in Figure 10.24(a). Figure 10.24(b) shows another example, where a needle rests on a

Figure 10.22 The soap bubbles in this photograph are caused by cohesive forces among molecules in liquids.

CONNECTIONS

Forces between atoms and molecules underlie the macroscopic effect called surface tension. These attractive forces pull the molecules closer together and tend to minimize the surface area. This is another example of a submicroscopic explanation for a macroscopic phenomenon.

Figure 10.23 Surface tension results from cohesive forces and acts to make surface area as small as possible. The forces on a surface molecule are shown—note the direction of the net force in each case. (a) A nonspherical drop becomes spherical, reducing its surface area. (b) A bump on the surface of a liquid is flattened, also reducing surface area.

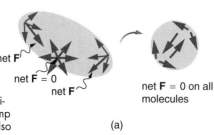

(a)

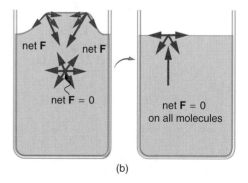

(b)

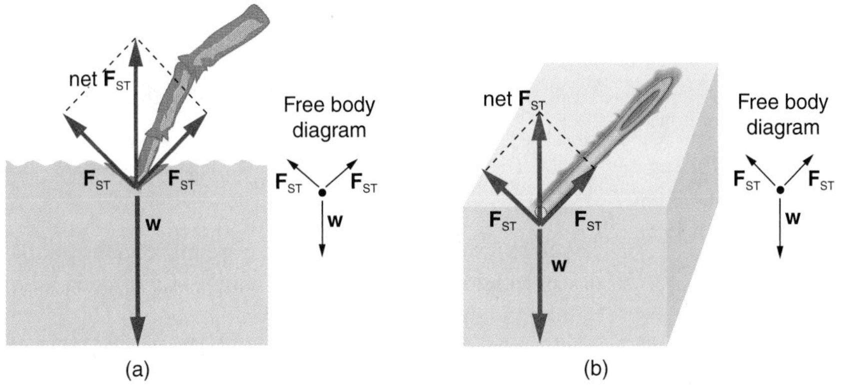

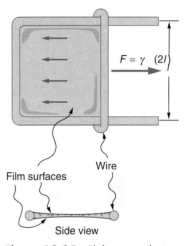

Figure 10.24 Surface tension supporting the weight of an insect and an iron needle, both of which rest on the surface without penetrating it. They are not floating; rather, they are supported by the surface of the liquid. (a) An insect leg dents the water surface. F_{ST} is a restoring force (surface tension) parallel to the surface. (b) An iron needle similarly dents a water surface until the restoring force (surface tension) grows to equal its weight.

water surface. It cannot, and does not, float, because its density is greater than that of water. Rather, its weight is supported by forces in the stretched surface that try to make the surface smaller or flatter. If the needle were placed point down on the surface, its weight acting on a smaller area would break the surface, and it would sink.

Surface tension is proportional to the strength of the cohesive force, which varies with the type of liquid. Figure 10.25 shows one way to measure surface tension. The liquid film exerts a force on the movable wire, trying to reduce its surface area. This force depends on surface tension and can be accurately measured. **Surface tension** γ is defined to be the force F per unit length L exerted by a stretched liquid membrane:

$$\gamma = \frac{F}{L} \tag{10.10}$$

Table 10.3 lists values of γ for some liquids. Suppose, for example, that the device in Figure 10.25 has a film of soapy water in it that is 2.00 cm across. Then the force would be $F = \gamma L = (0.037 \text{ N/m})(0.0400 \text{ m}) = 1.48 \times 10^{-3} \text{ N}$. It is possible to measure such small forces with very sensitive and friction-free devices.

Surface tension creates pressure inside bubbles. The contraction of the film places pressure on gas trapped inside. Think about what happens when air is released from a balloon. The rubber contracts, forcing air out, propelling the balloon like a rocket and causing the greatest acceleration just before the balloon runs out of air. This behavior implies that the pressure inside a bubble increases as the bubble gets smaller. In fact, the gauge pressure P inside a spherical bubble is

$$P = \frac{4\gamma}{r} \tag{10.11}$$

where r is the radius of the bubble. Thus the pressure inside a bubble is greatest when the bubble is the smallest. Another bit of evidence for this is illustrated in Figure 10.26, where two bubbles collide. The smaller bubble forces air into the larger, and they combine to form a single bubble.

Figure 10.25 Sliding wire device used for measuring surface tension; the device exerts a force to reduce the film's surface area. The force needed to hold the wire in place is $F = \gamma L = \gamma(2l)$, since there are *two* liquid surfaces attached to the wire. This force remains nearly constant as the film is stretched, until the film approaches its breaking point.

TABLE 10.3

SURFACE TENSION OF SOME LIQUIDS*

Liquid	Surface tension γ (N/m)
Water	
at 0°C	0.0756
at 20°C	0.0728
at 100°C	0.0589
Soapy water (typical)	0.0370
Ethyl alcohol	0.0223
Glycerin	0.0631
Mercury	0.465
Olive oil	0.032
Tissue fluids (typical)	0.050
Blood, whole at 37°C	0.058
Blood plasma at 37°C	0.073
Gold at 1070°C	1.000
Oxygen at −193°C	0.0157
Helium at −269°C	0.00012

*At 20°C unless otherwise stated.

Figure 10.26 When two bubbles combine, the smaller always forces air into the larger. Pressure in a bubble is inversely proportional to its radius, so that the smaller bubble has a greater pressure than the larger.

EXAMPLE 10.11 PRESSURE INSIDE A BUBBLE

Calculate the gauge pressure inside a soap bubble 2.00×10^{-4} m in radius using the surface tension for soapy water in Table 10.3. Convert this pressure to mm Hg.

Strategy The radius is given and surface tension can be found in Table 10.3, and so P can be found directly from Equation 10.11.

Solution Substituting r and γ into Equation 10.11, we obtain

$$P = \frac{4\gamma}{r} = \frac{4(0.037 \text{ N/m})}{2.00 \times 10^{-4} \text{ m}} = 740 \text{ N/m}^2 = 740 \text{ Pa}$$

We use a conversion factor to get this into units of mm Hg:

$$P = (740 \text{ N/m}^2)\frac{1.00 \text{ mm Hg}}{133 \text{ N/m}^2} = 5.56 \text{ mm Hg}$$

Discussion Note that if a hole were made in the bubble, the air would be forced out, the bubble would decrease in radius, and the pressure inside would *increase*. This is why a balloon deflates most vigorously just before running out of air.

Human & Medical Application

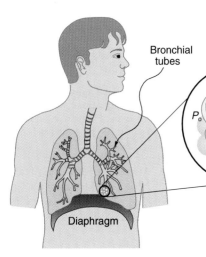

Figure 10.27 Bronchial tubes in the lungs branch into ever-smaller structures, finally ending in alveoli. The alveoli act like tiny bubbles. The surface tension of their mucous lining aids in exhalation and can prevent inhalation if too great.

Our lungs contain hundreds of millions of mucus-lined sacs called *alveoli*, which are very similar in size and surface tension to the bubble in the example. (See Figure 10.27.) You can exhale without any muscle action by allowing surface tension to contract these sacs. Surface tension in the alveoli needs to remain within a certain range. If it is too great, as it is if water gets into the lungs, then you cannot inhale. This is a severe problem in resuscitating drowning victims. Another problem occurs in newborn infants born without a surface-tension-reducing substance called a *surfactant*. Without this surfactant, the lungs are very difficult to inflate. This condition is known as hyaline membrane disease and is a leading cause of death, particularly in premature infants. Emphysema produces yet another problem with the alveoli. Alveolar walls deteriorate, and the sacs combine to form larger sacs that produce smaller pressure, reducing the ability of emphysema victims to exhale.

Adhesion and Capillary Action

Why is it that water beads up on a waxed car but does not on bare paint? The answer is that the adhesive forces between water and wax are much smaller than those between water and paint. (See Figure 10.28.) The **contact angle** θ is directly related to the relative strength of the cohesive and adhesive forces. *The larger the strength of the cohesive force relative to the adhesive force, the larger θ is*, and the more the liquid tends to form a droplet. The smaller θ is, the smaller the relative strength, so

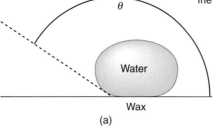

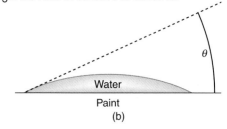

Figure 10.28 In the photograph, water beads on the waxed car paint and flattens on the unwaxed paint. (a) Water forms beads on the waxed surface because the cohesive forces responsible for surface tension are larger than the adhesive forces, which tend to flatten the drop. (b) Water beads on bare paint are flattened considerably because the adhesive forces between water and paint are strong, overcoming surface tension. The contact angle θ is directly related to the relative strengths of the cohesive and adhesive forces. The larger θ is, the larger the ratio of cohesive to adhesive.

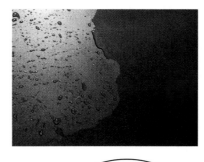

that the adhesive force is able to flatten the drop. Table 10.4 lists contact angles for several combinations of liquids and solids.

One important phenomenon related to the relative strength of cohesive and adhesive forces is **capillary action**—the tendency of a fluid to be raised or suppressed in a narrow tube, or *capillary tube*. This is what causes blood to be drawn into a small-diameter tube when the tube touches a drop.

If a capillary tube is placed vertically into a liquid, as shown in Figure 10.29, capillary action will raise or suppress the liquid inside the tube depending on the combination of substances. The actual effect depends on the relative strength of the cohesive and adhesive forces and, thus, the contact angle θ given in the table. If θ is less than 90°, then the fluid will be raised; if θ is greater than 90°, it will be suppressed. Mercury, for example, has a very large surface tension and a large contact angle with glass. Its surface curves downward in the tube something like a drop.* The tendency of surface tension is always to reduce the surface area. Surface tension thus flattens the curved liquid surface in a capillary tube. This results in a downward force in mercury and an upward force in water, as seen in Figure 10.29.

Capillary action can move liquids horizontally over very large distances, but the height to which it can raise or suppress a liquid in a tube is limited by gravity. This height, h, is

$$h = \frac{2\gamma \cos \theta}{\rho g r} \qquad (10.12)$$

The expression makes good sense intuitively. The height is directly proportional to surface tension γ, which is its direct cause. Furthermore, the height is inversely proportional to tube radius—the smaller the radius, the higher the fluid can be raised, since a smaller tube holds less mass. The height is also inversely proportional to fluid density, since a larger density means a greater mass in the same volume. (See Figure 10.30.)

---

*The curved surface of a fluid in a tube is called a **meniscus**.

Figure 10.29 (a) Mercury is suppressed in a glass tube because its contact angle is greater than 90.0°. Surface tension exerts a downward force as it flattens the mercury, suppressing it in the tube. The dashed line shows the shape the mercury surface would have without the flattening effect of surface tension. (b) Water is raised in a glass tube because its contact angle is nearly 0°. Surface tension therefore exerts an upward force when it flattens the surface to reduce its area.

TABLE 10.4

CONTACT ANGLES OF SOME SUBSTANCES*

Interface	Contact angle θ
Mercury–glass	140°
Water–glass	0°
Water–paraffin	107°
Water–silver	90°
Organic liquids (most)–glass	0°
Ethyl alcohol–glass	0°
Kerosene–glass	26°

*At 20°C.

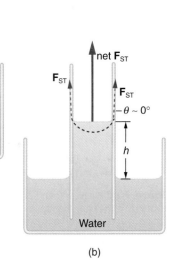

(a)

(b)

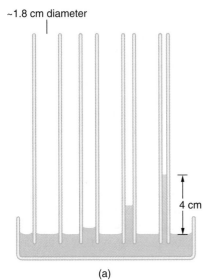

(a)

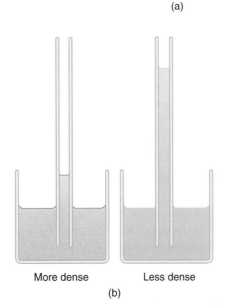

More dense   Less dense

(b)

Figure 10.30 (a) Capillary action depends on the radius of a tube. The smaller the tube, the greater the height reached. The height is negligible for large-radius tubes. (b) A denser fluid in the same tube rises to a smaller height, all other factors being the same.

EXAMPLE 10.12 CAPILLARY ACTION AND SAP

Can capillary action be solely responsible for sap rising in trees? To answer this question, calculate the radius of a capillary tube that would raise sap 100 m to the top of a giant redwood, assuming that sap's density is 1050 kg/m³, its contact angle is zero, and its surface tension is the same as water's at 20.0°C.

Strategy The height of capillary action is given in Equation 10.12, and every quantity in that equation is known except for r.

Solution Solving for r and substituting known values produces

$$r = \frac{2\gamma \cos\theta}{\rho g h} = \frac{2(0.0728 \text{ N/m})(1)}{(1050 \text{ kg/m}^3)(9.80 \text{ m/s}^2)(100 \text{ m})}$$

$$= 1.42 \times 10^{-7} \text{ m}$$

Discussion This result is unreasonable. Sap in trees moves through the *xylem*, which forms tubes with radii as small as 2.5×10^{-5} m. This is about 180 times as large as the radius found necessary here to raise sap 100 m. This means that capillary action alone cannot be solely responsible for sap getting to the tops of trees.

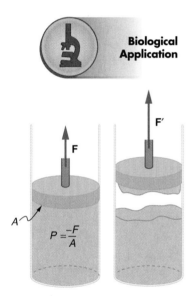

Biological Application

$$P = \frac{-F}{A}$$

Figure 10.31 (a) When the piston is raised, it stretches the liquid slightly, putting it under tension and creating a negative absolute pressure $P = -F/A$. (b) The liquid eventually separates, giving an experimental limit to negative pressure in this liquid.

Human & Medical Applications

How *does* sap get to the tops of trees? The question has not been completely resolved, but it appears that it is pulled up like a chain held together by cohesive forces. As each molecule of sap enters a leaf or evaporates, the entire chain is pulled up a notch. A pull in a fluid implies negative absolute pressure, something not ordinarily encountered. In most situations, *fluids can push but can exert only negligible pull*, because the cohesive forces seem to be too small to hold the molecules tightly together. It has proved difficult to test this in laboratory studies, apparently because of great sensitivity to any impurities. Figure 10.31 shows one device for studying negative pressure. Some experiments have demonstrated that negative pressures sufficient to pull sap to the tops of the tallest trees *can* be achieved.

10.9* PRESSURES IN THE BODY AND THEIR MEASUREMENT

Next to taking a person's temperature, measuring blood pressure is the most common of all medical examinations. Control of high blood pressure is largely responsible for significant decreases in heart attack and stroke fatalities achieved in the last two decades. Other pressures in the body can be measured and may provide valuable medical indicators. In this section, we consider a few examples together with some of the physics that accompanies them. Table 10.5 lists some of the measured pressures in mm Hg, the units most commonly quoted.

Blood Pressures

Common arterial blood pressure measurements, discussed in Section 10.6, produce values for systolic and diastolic pressures, both of which have health implications. When systolic pressure is chronically high, the risk of stroke and heart attack is increased. If, however, it is too low, fainting is a problem. Systolic pressure increases dramatically during exercise and returns to normal afterward. This produces no ill effects and may, in fact, be beneficial to the tone of the circulatory system. Diastolic pressure can be an indicator of fluid balance. When low, it may indicate that a person is hemorrhaging internally and needs a transfusion. Conversely, high diastolic pressure indicates a ballooning of the blood vessels, which may be due to the transfusion of too much fluid into the circulatory system. This can seriously strain the heart in its attempt to pump blood.

Blood pressure is also measured in the major veins, the heart chambers, arteries to the brain, and the lungs. But these pressures are usually only monitored during surgery or for patients in intensive care. To obtain the measurements, qualified health care workers thread thin tubes called catheters into appropriate locations to transmit pressures to external measuring devices. The effects of some of these pressures on flow in the circulatory system will be discussed in more detail in the next chapter.

TABLE 10.5

TYPICAL PRESSURES IN HUMANS*	
Body system	*Gauge pressure in mm Hg*
Blood pressures in large arteries (resting)	
Maximum (systolic)	100–140
Minimum (diastolic)	60–90
Blood pressure in large veins	4–15
Eye	12–24
Brain and spinal fluid (lying down)	5–12
Bladder	
While filling	0–25
When full	100–150
Chest cavity between lungs and ribs	−8 to −4
Inside lungs	−2 to +3
Digestive tract	
Esophagus	−2
Stomach	0–20
Intestines	10–20
Middle ear	<1

*See also Chapter 11 for other important pressures in humans.

Pressure in the Eye

The shape of the eye is maintained by fluid pressure, called *intraocular pressure*, which is normally in the range of 12.0 to 24.0 mm Hg. When circulation of fluid in the eye is blocked, it can lead to a buildup in pressure, a condition called *glaucoma*. Pressure can become as great as 85.0 mm Hg. The force exerted by this abnormally large pressure can permanently damage the optic nerve. To get an idea of the force involved, suppose the back of the eye has an area of 6 cm^2, and the pressure is 85.0 mm Hg. The force is $F = PA$. To get F in newtons, we convert the area to m^2 and the pressure to N/m^2. Thus $F = h\rho gA = (85 \times 10^{-3} \text{ m})(13.6 \times 10^3 \text{ kg/m}^3)(9.80 \text{ m/s}^2)(6 \times 10^{-4} \text{ m}^2) = 6.8 \text{ N}$, or about 1.5 lb. If you imagine 1.5 lb resting on your eye, you realize this is enough to cause damage.

Human & Medical Applications

People over 40 are at greatest risk of having glaucoma and should routinely have their intraocular pressure tested. Most measurements involve exerting a force on the eye over some area (a pressure) and observing the eye's response. If intraocular pressure is high, the eye will deform less and rebound more vigorously than normal. Excessive intraocular pressure can be reliably detected and sometimes controlled.

Pressures Associated with the Lungs

Pressure inside the lungs increases and decreases with each breath. It drops below atmospheric pressure (negative gauge pressure) when you inhale, and the atmosphere forces its way in. It increases above atmospheric pressure (positive gauge pressure) when you exhale, forcing air out.

Lung pressure has several causes. Muscle action in the diaphragm and rib cage is necessary for inhalation; this muscle action pulls the lungs outward and reduces the pressure within them. (See Figure 10.32.) Surface tension in the alveoli creates a positive pressure opposing inhalation. You can exhale without muscle action by letting surface tension in the alveoli create its positive pressure. People on respirators exhale in this manner. Muscle action can add to this positive pressure to aid exhalation, such as when you blow up a balloon or when you cough.

Figure 10.32 (a) During inhalation, muscles expand the chest, and the diaphragm moves downward, reducing pressure inside the lungs to less than atmospheric (negative gauge pressure). Pressure between the lungs and chest wall is even lower to overcome the positive pressure created by surface tension in the lungs. (b) During gentle exhalation, the muscles simply relax and surface tension in the alveoli creates a positive pressure inside the lungs, forcing air out. Pressure between the chest wall and lungs remains negative to keep them attached to the chest wall, but it is less negative than during inhalation.

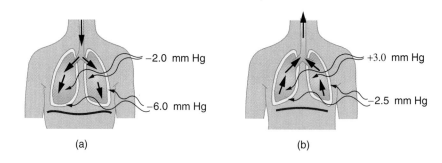

(a)   (b)

The lungs, in fact, would collapse due to the surface tension in the alveoli, if they were not attached to the inside of the chest wall by liquid adhesion. Pressure in the liquid attaching the lungs to the inside of the chest wall is thus negative, ranging from −4 to −8 mm Hg during exhalation and inhalation, respectively. If air is allowed to enter the chest cavity, it breaks the attachment, and one or both lungs may collapse. Suction is applied to the chest cavity of surgery patients and trauma victims to reestablish negative pressure and inflate the lungs.

Other Pressures in the Body

Human & Medical Applications

Spinal column and skull   Normally, there is a 5 to 12 mm Hg pressure in the fluid surrounding the brain and filling the spinal column. This cerebrospinal fluid serves many purposes, one of which is to supply flotation to the brain. The buoyant force supplied by the fluid nearly equals the weight of the brain, since their densities are nearly equal. If there is a loss of fluid, the brain rests on the inside of the skull, causing severe headaches. Spinal fluid pressure is measured by means of a needle inserted between vertebrae that transmits the pressure to a suitable measuring device.

Bladder pressure   This bodily pressure is one of which we are often aware. In fact, there is a relationship between our awareness of this pressure and a subsequent increase in it. Bladder pressure climbs steadily from zero to about 25 mm Hg as the bladder fills to its normal capacity of 500 cm^3. This pressure triggers the *micturition reflex*, which stimulates the feeling of needing to urinate. What is more, it also causes muscles around the bladder to contract, raising the pressure to over 100 mm Hg, accentuating the sensation. Coughing, straining, tensing in cold weather, wearing tight clothes, and experiencing simple nervous tension all can increase bladder pressure and trigger this reflex. So can the weight of a pregnant woman's fetus. Bladder pressure can be measured by a catheter or by inserting a needle through the bladder wall and transmitting the pressure to an appropriate measuring device. One hazard of high bladder pressure, perhaps created by an obstruction, is that it can force urine back into the kidneys, damaging them severely.

Pressures in the skeletal system   These are the largest pressures in the body, because forces in the skeletal system can be large, while the areas to which they are applied, such as in the joints, are small. For example, when a person lifts improperly, a force of 5000 N may be created between vertebrae in the spine, and this may be applied to an area as small as 10 cm^2. The pressure created is $P = F/A = (5000 \text{ N})/(10^{-3} \text{ m}^2) = 5.0 \times 10^6$ N/m^2 or about 50 atm! This pressure can damage both disk and bone. Even under normal circumstances, forces between vertebrae in the spine are large enough to create pressures of several atmospheres. Most causes of excessive pressure in the skeletal system can be avoided by lifting properly and avoiding extreme physical activity.

There are many other interesting and medically significant pressures in the body. For example, pressure caused by various muscle actions drives food and waste through the digestive system. Stomach pressure behaves much like bladder pressure and is tied to the sensation of hunger. Pressure in the relaxed esophagus is normally negative because pressure in the chest cavity is normally negative. Positive pressure in the stomach may thus force acid into the esophagus, causing "heartburn." Pressure in the middle ear can result in significant force on the eardrum if it differs greatly from atmospheric pressure. This is most noticeable in plane flights and is very important in scuba diving, in which external pressure can be quite large. The eustachian tubes connect the middle ear to the throat and allow us to equalize pressure in the middle ear to avoid an imbalance of force on the eardrum.

Many bodily pressures are associated with the flow of fluids. Fluid flow will be discussed in the next chapter.

SUMMARY

A **fluid** is a state of matter that yields to sideways or shearing forces. Liquids and gases are both fluids. **Fluid statics** is the physics of stationary fluids.

Density is the mass per unit volume of a substance or object,

$$\rho = \frac{m}{V} \tag{10.1}$$

where ρ is the density, m is the mass, and V is the volume occupied. Table 10.1 lists the densities, in the SI unit of kg/m^3, of a number of substances.

Pressure is the force applied per unit area, given by

$$P = \frac{F}{A} \tag{10.2}$$

where F is a force applied perpendicularly to area A. The SI unit for pressure is $1\ N/m^2$, called a **pascal** (Pa). Table 10.2 lists conversion factors for various pressure units. The **pressure due to the weight of a fluid** is given by

$$P = h\rho g \tag{10.3}$$

where h is the depth below the surface of the fluid, ρ is its average density, and g is the acceleration of gravity. Standard atmospheric pressure is due to the weight of the atmosphere and has the value $1.01 \times 10^5\ N/m^2$.

Pascal's principle is the basis for many important effects in fluids. It states:

> **A change in pressure applied to an enclosed fluid is transmitted undiminished to all portions of the fluid and to the walls of its container.**

Pascal's principle implies that *the total pressure in a fluid is the sum of the pressures from different sources.* Pascal's principle applied to hydraulic systems results in the following:

$$\frac{F_1}{A_1} = \frac{F_2}{A_2} \tag{10.5}$$

where the forces F_1 and F_2 are applied to pistons of areas A_1 and A_2, respectively.

Gauge pressure P_g is the amount of pressure above or below atmospheric pressure P_{atm}. Absolute pressure P_{abs} is the total pressure, given by

$$P_{abs} = P_g + P_{atm} \tag{10.6}$$

Most gauges register gauge pressure P_g.

Archimedes' principle states that for any object in any fluid:

> **The buoyant force on an object equals the weight of the fluid it displaces.**

In equation form, Archimedes' principle is

$$F_B = w_{fl} \tag{10.7}$$

where F_B is the buoyant force and w_{fl} is the weight of the fluid displaced by the object. The fraction of a floating object that is submerged is given by

$$\text{fraction submerged} = \frac{\bar{\rho}_{obj}}{\rho_{fl}} \tag{10.8}$$

The **specific gravity** of a substance is defined to be

$$\text{specific gravity} = \frac{\rho}{\rho_w} \tag{10.9}$$

where ρ_w is the density of water. When an object's weight or mass is determined while submerged, it appears to be less, or

apparent weight loss	=	weight of fluid displaced

or

apparent mass loss	=	mass of fluid displaced

There are forces between molecules. Attractive forces between molecules of the same type are **cohesive forces**, whereas those between different types are **adhesive forces**. **Surface tension** γ is a force per unit length in a liquid that causes its surface to contract to the smallest possible area; it is given by

$$\gamma = \frac{F}{L} \tag{10.10}$$

Table 10.3 gives values of γ for some liquids. Surface tension causes the gauge pressure inside a bubble with radius r to be

$$P = \frac{4\gamma}{r} \tag{10.11}$$

The combination of surface tension and adhesion affects the contact angle θ of a liquid on a surface, representative values of which are given in Table 10.4. The combination of surface tension and adhesion also causes **capillary action**, which raises or suppresses fluids in a tube to a height

$$h = \frac{2\gamma \cos \theta}{\rho g r} \tag{10.12}$$

where r is the tube radius, ρ the fluid density, and θ the contact angle.

Pressures are important in the body; values for pressures at various sites in the body are given in Table 10.5.

CONCEPTUAL QUESTIONS

10.1 What physical characteristic distinguishes a fluid from a solid?

10.2 Why are gases easier to compress than liquids and solids?

10.3 How do gases differ from liquids?

10.4 Approximately how does the density of air vary with altitude?

10.5 Give an example in which density is used to identify the substance composing an object. Would information in addition to average density be needed to identify the substances in an object composed of more than one material?

10.6 How is pressure related to the sharpness of a knife and its ability to cut?

10.7 Why does a dull hypodermic needle hurt more than a sharp one?

10.8 The outward force on one end of an air tank was calculated in Example 10.2. How is this force balanced? (The tank does not accelerate, and so the force must be balanced.)

10.9 Why is force exerted by static fluids always perpendicular to a surface?

10.10 Toe dancing (as in ballet) is much harder on toes than normal dancing or walking. Explain in terms of pressure.

10.11 How do jogging on soft ground and wearing padded shoes reduce the pressures to which the feet and legs are subjected?

10.12 Why does atmospheric pressure decrease more rapidly than linearly with altitude?

10.13 Atmospheric pressure exerts a large force (equal to the weight of the atmosphere above your body—about 10 tons) on the top of your body when you are sunbathing. Why are you able to get up?

10.14 What are two reasons that mercury rather than water is used in barometers?

10.15 You can break a strong bottle by pounding a cork into it with your fist, but the cork must press directly against liquid filling the bottle—there can be no air between the cork and liquid. Explain why the bottle breaks, and why it will not if there is air between the cork and liquid.

10.16 Suppose the master cylinder in a hydraulic system is at a greater height than the slave. Explain how this will affect the force produced at the slave.

10.17 Figure 10.33 shows how sandbags placed around a leak outside a river levee can effectively stop the flow of water under the levee. Explain how the small amount of water inside the column formed by the sandbags is able to balance the much larger body of water behind the levee.

10.18 Is there a net force on a dam due to atmospheric pressure? Explain.

10.19 Does atmospheric pressure add to the gas pressure in a rigid tank? In a toy balloon? When, in general, does atmospheric pressure *not* affect the total pressure in a fluid?

10.20 Explain why the fluid reaches equal levels on either side of a manometer if both sides are open to the atmosphere, even if the tubes are of different diameters.

10.21 Figure 10.13 shows how a common measurement of arterial blood pressure is made. Is there any effect on the measured pressure if the manometer is lowered? What is the effect of raising the arm above the shoulder? What is the effect of placing the cuff on the upper leg with the person standing? Explain your answers in terms of pressure created by the weight of a fluid.

10.22 Considering the magnitude of typical arterial blood pressures, why are mercury rather than water manometers used for these measurements?

10.23 How do you convert pressure units like millimeters of mercury, centimeters of water, and inches of mercury into units like newtons per meter squared without resorting to a table of pressure conversion factors?

10.24 Why is it difficult to swim under water in the Great Salt Lake?

10.25 More force is required to pull the plug in a full bathtub than when it is empty. Does this contradict Archimedes' principle? Explain.

10.26 Do fluids exert buoyant forces in a "weightless" environment, such as in the space shuttle? Explain.

10.27 Will the same ship float higher in salt water than in fresh water? Explain.

10.28 Figure 10.34 shows a glass of ice water filled to the brim. Will the water overflow when the ice melts? Explain.

Figure 10.34 What happens to the water level when the ice in this glass melts? Question 28.

10.29 Marbles dropped into a partially filled bathtub sink to the bottom. Part of their weight is supported by buoyant force, yet the downward force on the bottom of the tub increases by exactly the weight of the marbles. Explain why.

10.30 The density of oil is less than that of water, yet a loaded oil tanker sits lower in the water than an empty one. Why?

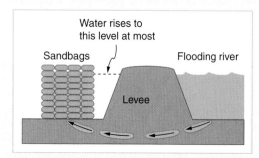

Figure 10.33 Because the river level is very high, it has started to leak under the levee. Sandbags are placed around the leak, and the water held by them rises until it is the same level as the river, at which point the water there stops rising. Question 17.

10.31*Is surface tension due to cohesive or adhesive forces, or both?

10.32*Is capillary action due to cohesive or adhesive forces, or both?

10.33*Birds such as ducks, geese, and swans have greater densities than water, yet they are able to sit on its surface. Explain this ability, noting that water does not wet their feathers and that they cannot sit on soapy water.

10.34*Water beads up on an oily sunbather, but not on her neighbor, whose skin is not oiled. Explain in terms of cohesive and adhesive forces.

10.35*Could capillary action be used to move fluids in a "weightless" environment, such as in a space probe?

10.36*What effect does capillary action have on the reading of a manometer with uniform diameter? Explain.

10.37*Pressure between the inside chest wall and the outside of the lungs normally remains negative. Explain how pressure inside the lungs can become positive (to cause exhalation) without muscle action.

10.38*Why are teeth with points better for cutting and flat teeth better for grinding food?

PROBLEMS

Note: When considering significant figures in these problems, you may assume all densities to have three significant figures so that certain effects are not rounded off or obscured.

Section 10.2 Density

10.1 Gold is sold by the troy ounce (31.103 g). What is the volume of 1 troy ounce of pure gold?

10.2 Mercury is commonly supplied in flasks containing 34.5 kg (about 76 lb). What is the volume in liters of this much mercury?

10.3 What is the mass of a deep breath of air having a volume of 2.00 L?

10.4 A straightforward method of finding the density of an object is to measure its mass and then measure its volume by submerging it in a graduated cylinder. What is the density of a 240 g rock that displaces 89.0 cm^3 of water? (Note that the accuracy and practical applications of this technique are more limited than a variety of others that are based on Archimedes' principle.)

10.5 Suppose you have a coffee mug with a circular cross section and vertical sides (uniform radius). What is its inside radius if it holds 375 g of coffee when filled to a depth of 7.50 cm? Assume coffee has the same density as water.

10.6 Phone booth cramming periodically gains popularity among certain groups. (a) What is the volume of a 65.0 kg person whose density is 980 kg/m^3? (b) What is the absolute maximum number of such people that could be crammed into a phone booth with inside measurements of 0.800 m depth, 0.800 m width, and 2.15 m height? Neglect space in the booth occupied by the telephone, clothing, and so on. Note that it would be in poor taste not to round your answer to the next lowest integer. The taste of the crammers can be neglected.

10.7 A rectangular gasoline tank can hold 50.0 kg of gasoline when full. What is the depth of the tank if it is 0.500 m wide by 0.900 m long?

10.8 A trash compactor can reduce the volume of its contents to 0.350 their original value. Neglecting the mass of air expelled, by what factor is the density of the trash increased?

10.9 A 2.50 kg steel gas can holds 20.0 L of gasoline when full. What is the average density of the full gas can, taking into account the volume occupied by steel as well as by gasoline?

10.10 What is the density of 18.0 karat gold that is a mixture of 18 parts gold, 5 parts silver, and 1 part copper? (These are parts by mass, not volume.) Assume that this is a simple mixture having an average density equal to the weighted densities of its constituents.

10.11 There is relatively little empty space between atoms in solids and liquids, so that the *average* density of an atom is about the same as matter on a macroscopic scale—approximately 10^3 kg/m^3. The nucleus of an atom has a radius about 10^{-5} that of the atom and contains nearly all the mass of the entire atom. (a) What is the approximate density of a nucleus? (b) One remnant of a supernova, called a neutron star, can have the density of a nucleus. What would be the radius of a neutron star with a mass 10 times that of our sun?

Sections 10.3 and 10.4 Pressure and Pressure Due to the Weight of a Fluid

10.12 As a woman walks, her entire weight is momentarily placed on one heel of her high-heeled shoes. Calculate the pressure exerted on the floor by the heel if it has an area of 1.50 cm^2 and the woman's mass is 55.0 kg. Express the force in N/m^2 and lb/in^2. (In the early days of commercial flight, women were not allowed to wear high-heeled shoes because aircraft floors were too thin to resist such large pressures.)

10.13 The pressure exerted by a phonograph needle on a record is surprisingly large. If the equivalent of 1.00 g of mass is supported by a needle the tip of which is a circle 0.200 mm in radius, what pressure is exerted on the record in N/m^2?

10.14 Nail tips exert tremendous pressures because they exert a large force on a small area. What force must be exerted on a nail with a 1.00 mm diameter circular tip to create a pressure of 3.00 × 10^9 N/m^2? (This is possible because the hammer striking the nail is brought to rest is such a short distance.)

10.15 What depth of mercury creates a pressure of 1.00 atm?

10.16 The greatest ocean depths on earth are found in the Marianas Trench near the Philippines. Calculate the pressure due to the ocean at the bottom of this trench, given its depth to be 11.0 km and assuming sea water's density is constant all the way down. (The validity of the assumption of constant density is examined in one of the integrated concept problems.)

10.17 Water towers store water above the level of consumers for times of heavy use, eliminating the need for high-speed pumps. How high above the user must the water level be to create a gauge pressure of 3.00 × 10^5 N/m^2?

10.18 Verify that the SI unit of $h\rho g$ is N/m^2.

• **10.19** How much force is exerted on one side of an 8.50 by 11.0 in. sheet of paper by the atmosphere? How can the paper withstand such a force?

• **10.20** What pressure is exerted on the bottom of the gasoline tank in Problem 10.7 by the weight of the gasoline in it when full?

• **10.21** Calculate the average pressure exerted on the palm of a shot-putter's hand by the shot, if the area of contact is 50.0 cm^2 and he exerts a force of 800 N on it. Express the pressure in N/m^2 and compare it with the 1.00×10^6 N/m^2 pressures sometimes encountered in the skeletal system.

• **10.22** The left side of the heart creates a pressure of 120 mm Hg by exerting a force directly on the blood over an effective area of 15.0 cm^2. What force does it exert to accomplish this?

⁑ **10.23** Show that the total force on a rectangular dam due to the water behind it increases with the *square* of the water depth. In particular, show that this force is $F = \rho g h^2 L/2$, where ρ is the density of water, h is its depth at the dam, and L is the length of the dam. You may assume the face of the dam is vertical. (Hint: Calculate the average pressure exerted and multiply this by the area in contact with the water. See Figure 10.7.)

Sections 10.5 and 10.6 Pascal's Principle

10.24 How much pressure is transmitted in the hydraulic system considered in Example 10.6? Express in N/m^2 and atmospheres.

10.25 What force must be exerted on the master cylinder of a hydraulic lift to support the weight of a 2000 kg car (a big car) resting on the slave cylinder? The master cylinder has a 2.00 cm diameter, while the slave's is 24.0 cm.

10.26 A philistine pours the remnants of several bottles of wine into a jug after a party. He then inserts the 2.00 cm diameter cork into the bottle, placing it in direct contact with the wine. He is amazed when he pounds the cork into place and the jug's 14.0 cm diameter bottom breaks away. Calculate the extra force exerted against the bottom if he pounded the cork with a 120 N force.

10.27 Find the gauge and absolute pressures in the balloon and peanut jar shown in Figure 10.12, assuming the manometer connected to the balloon utilizes water while that connected to the jar contains mercury. Express in units of cm water for the balloon and mm Hg for the jar, taking $h = 0.0500$ m for each.

10.28 (a) Convert normal blood pressure readings of 120 over 80 mm Hg to N/m^2 using the relationship for pressure due to the weight of a fluid ($P = h\rho g$) rather than a conversion factor. (b) Convert these pressures to lb/in^2.

• **10.29** A certain hydraulic system is designed to exert a force 100 times as large as the one put into it. (a) What must the ratio of the area of the slave cylinder to the area of the master cylinder be? (b) What must the ratio of their diameters be? (c) By what factor is the distance through which the output force moves reduced relative to the distance through which the input force moves? Assume no losses to friction.

• **10.30** How tall must a water-filled manometer be to measure blood pressures as high as 300 mm Hg?

• **10.31** Pressure cookers have been around for more than 300 years, although the early ones had a nasty habit of exploding. How much force must the latches holding the lid onto a pressure cooker be able to withstand if the circular lid is 25.0 cm in diameter and the gauge pressure inside is 3.00 atm? Neglect the weight of the lid.

• **10.32** Suppose you measure a standing person's blood pressure by placing the cuff on his leg 0.500 m below the heart. Calculate the pressures you would observe (in units of mm Hg) if the pressures at the heart are 120 over 80 mm Hg. Assume there is no loss of pressure due to resistance in the circulatory system (a reasonable assumption, since major arteries are large).

⁑ **10.33** (a) Show that the maximum height to which water can be raised in a syringe or similar device (like a barometer) is 10.3 m, assuming normal atmospheric pressure. (b) To what maximum height can gasoline be raised in such devices?

⁑ **10.34** A submarine is stranded on the bottom of the ocean with its hatch 25.0 m below the surface. Calculate the force needed to open the hatch from the inside, given it is circular and 0.450 m in diameter. Air pressure inside the submarine is 1.00 atm.

⁑ **10.35** Verify that work input equals work output for a hydraulic system assuming no losses to friction. Do this by showing that the distance the output force moves is reduced by the same factor that the output force is increased. Assume the volume of the fluid is constant.

⁑ **10.36** Assuming bicycle tires are perfectly flexible and support the weight of bicycle and rider by pressure alone, calculate the total area of the tires in contact with the ground. The bicycle plus rider has a mass of 80.0 kg, and the gauge pressure in the tires is 3.50×10^5 N/m^2.

Section 10.7 Archimedes' Principle

10.37 What fraction of ice is submerged when it floats in fresh water, given the density of water at 0°C is very close to 1000 kg/m^3?

10.38 Logs sometimes float vertically in a lake because one end has become water-logged and denser than the other. What is the average density of a uniform-diameter log that floats with 20.0% of its length above water?

10.39 Find the density of a fluid in which a hydrometer having a density of 0.750 g/cm^3 floats with 92.0% of its volume submerged.

10.40 If your body has a density of 995 kg/m^3, what fraction of you will be submerged when floating quietly in: (a) Fresh water? (b) The Great Salt Lake, which has a density of 1030 kg/m^3?

• **10.41** Bird bones have air pockets in them to reduce their weight—this also gives them an average density significantly less than that of the bones of other animals. Suppose an ornithologist weighs a bird bone in air and in water and finds its mass is 45.0 g and its apparent mass when submerged is 3.60 g (it is watertight). (a) What mass of water is displaced? (b) What is the volume of the bone? (c) What is its average density?

• **10.42** A rock with a mass of 540 g in air is found to have an apparent mass of 342 g when submerged in water. (a) What mass of water is displaced? (b) What is the volume of the rock? (c) What is its average density? Is this consistent with the value for granite?

• **10.43** Archimedes' principle can be used to calculate the density of a fluid as well as that of a solid. Suppose a chunk of iron with a mass of 390.0 g in air is found to have an apparent mass of 350.5 g when completely submerged in an unknown liquid. (a) What mass of fluid does the iron displace? (b) What is the volume of iron, using its density as given in Table 10.1? (c) Calculate the fluid's density and identify it.

• **10.44** In a fat tank measurement of a woman's density, she is found to have a mass of 62.0 kg in air and an apparent mass of 0.850 kg when completely submerged with lungs empty. (a) What mass of water does she displace? (b) What is her volume? (c) Calculate her density. (d) If her lung capacity is 1.75 L, is she able to float without treading water with her lungs filled with air?

• **10.45** Some fish have a density slightly less than that of water and must exert a force (swim) to stay submerged. What force must a 85.0 kg grouper exert to stay submerged in salt water if its body density is 1015 kg/m³?

• **10.46** (a) Calculate the buoyant force on a 2.00 L helium balloon. (b) Given the mass of the rubber in the balloon is 1.50 g, what is the net vertical force on the balloon if it is let go? You can neglect the volume of the rubber.

• **10.47** (a) What is the density of a person who floats in fresh water with 4.00% of her volume above the surface? This could be measured by placing her in a tank with marks on the side to measure how much water she displaces when floating and when held under water (briefly). (b) What percent of her volume is above the surface when she floats in sea water?

• **10.48** A certain man has a mass of 80.0 kg and a density of 995 kg/m³ (exclusive of the air in his lungs). (a) Calculate his volume. (b) Find the buoyant force air exerts on him. (c) What is the ratio of the buoyant force to his weight?

⁞ **10.49** A simple compass can be made by placing a small bar magnet on a cork floating in water. (a) What fraction of a plain cork will be submerged when floating in water? (b) If the cork has a mass of 10.0 g and a 20.0 g bar magnet is placed on it, what fraction of the cork will be submerged? (c) Will the bar magnet and cork float in ethyl alcohol?

⁞ **10.50** What fraction of an iron anchor's weight will be supported by buoyant force when submerged in salt water?

⁞ **10.51** Referring to Figure 10.16, prove that the buoyant force on the cylinder is equal to the weight of the fluid displaced (Archimedes' principle). You may assume that the buoyant force is $F_2 - F_1$ and that the ends of the cylinder have equal areas A. Note that the volume of the cylinder (and that of the fluid it displaces) equals $(h_2 - h_1)A$.

⁞ **10.52** A 75.0 kg man floats in fresh water with 3.00% of his volume above water when his lungs are empty, and 5.00% above water when his lungs are full. Calculate the volume of air he inhales—called his lung capacity—in liters.

⁞ **10.53** Scurrilous con artists have been known to represent gold-plated tungsten ingots as pure gold and sell them to the greedy at prices much below gold value but deservedly far above the cost of tungsten. With what accuracy must you be able to measure the mass of such an ingot in and out of water to tell that it is almost pure tungsten rather than pure gold?

Section 10.8* Surface Tension and Capillary Action

10.54* What is the gauge pressure in mm Hg inside a soap bubble 0.100 m in diameter?

10.55* Calculate the force on the slide wire in Figure 10.25 if it is 3.50 cm long and the fluid is ethyl alcohol.

10.56* Figure 10.30(a) shows the effect of tube radius on the height to which capillary action can raise a fluid. (a) Calculate h for water in a 0.900 cm radius glass tube—a rather large tube like the one on the left. (b) What is the radius of the glass tube on the right if it raises water to 4.00 cm?

10.57* We stated in Example 10.12 that a xylem tube is of radius 2.50×10^{-5} m. Verify that such a tube raises sap less than a meter by finding h for it, making the same assumptions as in the example.

• **10.58*** What fluid is in the device shown in Figure 10.25 if the force is 3.16×10^{-3} N and the length of the wire is 2.50 cm? Calculate γ and find a likely match from Table 10.3.

• **10.59*** (a) If the gauge pressure inside a 10.0 cm radius rubber balloon is 15.0 cm of water, what is the effective surface tension of the balloon? (b) What will the gauge pressure inside the balloon be when its radius is 5.00 cm, assuming the rubber acts like a liquid surface?

• **10.60*** Calculate the gauge pressures inside 2.00 cm radius bubbles of water, alcohol, and soapy water. Which liquid forms the most stable bubbles, neglecting any effects of evaporation?

• **10.61*** Suppose water is raised by capillary action to a height of 5.00 cm in a glass tube. (a) To what height will it be raised in a paraffin tube of the same radius? (b) In a silver tube of the same radius?

• **10.62*** Calculate the contact angle θ for olive oil if capillary action raises it to a height of 7.07 cm in a 0.100 mm radius glass tube. Is this value consistent with that for most organic liquids?

⁞ **10.63*** When two soap bubbles touch, the larger is inflated by the smaller until they form a single bubble. (a) What is the gauge pressure inside a 1.50 cm radius soap bubble? (b) Inside a 4.00 cm radius soap bubble? (c) Inside the single bubble they form if no air is lost when they touch?

⁞ **10.64*** Calculate the ratio of the heights to which water and mercury are raised by capillary action in the same glass tube.

⁞ **10.65*** What is the ratio of heights to which ethanol and water are raised by capillary action in the same glass tube?

Section 10.9* Pressures in the Body

Note that although this section is optional, many of its associated problems are straightforward applications of fluid pressure and require no previous knowledge of the body.

10.66* One way to force air into an unconscious person's lungs is to squeeze on a balloon appropriately connected to the subject. What force must you exert on the balloon with your hands to create a gauge pressure of 4.00 cm water, assuming you squeeze on an effective area of 50.0 cm²?

10.67* Heroes in movies hide beneath water and breathe through a hollow reed (villains never catch on to this trick). In practice, you cannot inhale in this manner if your lungs are more than 60.0 cm below the surface. What maximum negative gauge pressure can you create in your lungs on dry land, assuming you can achieve −3.00 cm water pressure with your lungs 60.0 cm below the surface?

• **10.68** Gauge pressure in the fluid surrounding an infant's brain may rise as high as 85.0 mm Hg (5 to 12 mm Hg is normal), creating an outward force large enough to make the skull grow abnormally large. (a) Calculate this outward force in newtons on each side of an infant's skull if the effective area of each side is 70.0 cm². (b) What is the net force on the skull?

• **10.69** A full-term fetus typically has a mass of 3.50 kg. (a) What pressure does the weight of such a fetus create if it rests on the mother's bladder, supported on an area of 90.0 cm²? (b) Convert this pressure to mm Hg and determine if it alone is great enough to trigger the micturition reflex (it will add to any pressure already existing in the bladder).

• **10.70** Calculate the maximum force in newtons exerted by the blood on an aneurysm, or ballooning, in a major artery, given the maximum blood pressure for this person is 150 mm Hg and the effective area of the aneurysm is 20.0 cm². Note that this force is great enough to cause further enlargement and subsequently greater force on the ever-thinner vessel wall.

• **10.71** If the pressure in the esophagus is −2.00 mm Hg while that in the stomach is +20.0 mm Hg, to what height could stomach fluid rise in the esophagus, assuming a density of 1.10 g/cm³? (This will not occur if the muscle closing the lower end of the esophagus is working properly.)

• **10.72** Suppose a 3.00 N force can rupture an eardrum having an area of 1.00 cm². (a) Calculate the maximum tolerable gauge pressure inside the eardrum (in the middle ear) in N/m² and convert it to mm Hg. (Pressures in the middle ear may rise when an infection causes a fluid buildup.) (b) At what depth in fresh water would this person's eardrum rupture, assuming the gauge pressure in the middle ear is zero?

‡ **10.73** Pressure in the spinal fluid is measured as shown in Figure 10.35. If the pressure in the spinal fluid is 10.0 mm Hg: (a) What is the reading of the water manometer in cm water? (b) What is the reading if the person sits up, placing the top of the fluid 60.0 cm above the tap? The fluid density is 1.05 g/cm³.

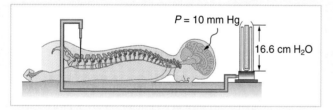

P = 10 mm Hg

16.6 cm H₂O

Figure 10.35 A water manometer used to measure pressure in the spinal fluid. The height of the fluid in the manometer is measured relative to the spinal column, and the manometer is open to the atmosphere. The measured pressure will be considerably greater if the person sits up. Problem 73.

INTEGRATED CONCEPTS

These problems involve physical principles from more than one chapter. Integrated knowledge from a broad range of topics is much more powerful than a narrow application of physics. Here you will need to refer to material in Chapters 4, 5, and 6, as well as in Chapter 10.

PROBLEM-SOLVING STRATEGY

Use the following strategy to solve integrated concept problems.

Step 1. *Identify which physical principles are involved.*

Step 2. *Solve the problem using strategies outlined in the text.*

Note: Worked examples that can help you see how to apply this strategy to integrated concept problems appear in a number of chapters, the most recent being Chapter 6. (See the section of problems labeled *Integrated Concepts* at the end of that chapter and consult the index for others.)

• **10.74** (a) How high will water rise in a 0.500 mm radius glass capillary tube? (b) How much gravitational potential energy does the water gain? (c) Discuss possible sources of this energy.

• **10.75** A negative pressure of 25.0 atm can sometimes be achieved with the device in Figure 10.31 before the water separates. (a) To what height could such a negative gauge pressure raise water? (b) How much would a steel wire of the same diameter and length as this capillary stretch if suspended from above?

‡ **10.76** Suppose you hit a steel nail with a 0.500 kg hammer, initially moving at 15.0 m/s and brought to rest in 2.80 mm. (a) What average force is exerted on the nail? (b) How much is the nail compressed if it is 2.50 mm in diameter and 6.00 cm long? (c) What pressure is created on the 1.00 mm diameter tip of the nail? (It will be similar to the pressure given in Problem 10.14.)

‡ **10.77** (a) As done in Problem 10.16, calculate the pressure due to the ocean at the bottom of the Marianas Trench near the Philippines, given its depth to be 11.0 km and assuming the density of sea water is constant all the way down. (b) Calculate the percent decrease in volume of sea water due to such a pressure, assuming its bulk modulus is the same as water and is constant. (c) What would be the percent increase in its density? Is the assumption of constant density valid? Will the actual pressure be greater or smaller than that calculated under this assumption?

‡ **10.78** The hydraulic system of a back hoe is used to lift a load as shown in Figure 10.36. (a) Calculate the force F the slave cylinder must exert to support the 400 kg load and the 150 kg brace and shovel. (b) What is the pressure in the hydraulic fluid if the slave cylinder is 2.50 cm in diameter? (c) What force would you have to exert on a lever with a mechanical advantage of 5.00 acting on a master cylinder 0.800 cm in diameter to create this pressure?

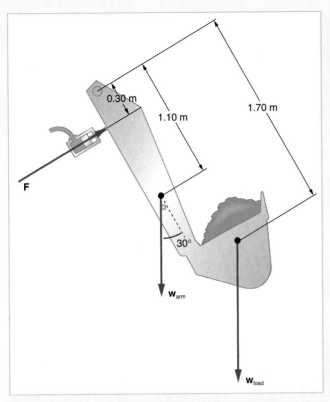

0.30 m

1.10 m

1.70 m

F

30°

W_{arm}

W_{load}

Figure 10.36 Hydraulic and mechanical lever systems are used in heavy machinery such as this back hoe. Problem 78.

As in previous chapters, problems with unreasonable results are included to give practice in assessing whether nature is being accurately described, and to trace the source of difficulty if it is not. This is very much like the process followed in original research when physical principles as well as faulty premises are tested. The following problems give results that are unreasonable even when the physics is correctly applied, because some premise is unreasonable or because certain premises are inconsistent.

PROBLEM-SOLVING STRATEGY

To determine if an answer is reasonable, and, to determine the cause if it is not, do the following.

Step 1. *Solve the problem using strategies as outlined in several places in this chapter.* See the table of contents to locate all problem-solving strategies. Use the format followed in the worked examples in the text to solve the problem as usual.

Step 2. *Check to see if the answer is reasonable.* Is it too large or too small, or does it have the wrong sign, improper units, . . .?

Step 3. *If the answer is unreasonable, look for what specifically could cause the identified difficulty.*

• **10.79** Some miners wish to remove water from a mine shaft. A pipe is lowered to the water 90.0 m below, and a negative pressure is applied to raise the water. (a) Calculate the pressure needed to raise the water. (b) What is unreasonable about this pressure? (c) What is unreasonable about the premise?

• **10.80** You are pumping up a bicycle tire with a hand pump, the piston of which has a 2.00 cm radius. (a) What force in newtons must you exert to create a pressure of 6.90×10^5 N/m^2 (100 lb/in^2)? (b) What is unreasonable about this result? (c) Which premises are unreasonable or inconsistent?

11 FLUID DYNAMICS

Many fluids are flowing in this scene. Water from the hoses and smoke from the fire are visible flows. Less visible is the flow of air and the flow of fluids on the ground and within the people fighting the fire.

We have dealt with many situations in which fluids are static. But by their very definition, fluids flow. Examples come easily—a column of smoke rises from a camp fire, water streams from a fire hose, blood courses through your veins. Why does rising smoke curl and twist? How does a nozzle increase the speed of water emerging from a hose? How does the body regulate blood flow? The physics of fluids in motion—**fluid dynamics**—allows us to answer these and many other questions.

11.1 FLOW RATE AND ITS RELATION TO VELOCITY

Flow rate $\mathcal{F}$ is defined to be the volume per unit time flowing past a point (really through an area A, as seen in Figure 11.1). In symbols, this is

$$\mathcal{F} = \frac{V}{t} \qquad (11.1)$$

where V is volume and t time. The SI unit for flow rate is m³/s, but a number of other units for $\mathcal{F}$ are in common use. A swimming pool pump, for example, may be rated at 100 gallons per minute (gal/min), and the heart of a resting adult pumps blood at a rate of 5.00 liters per minute (L/min). Note that a **liter** (L) is 1/1000 of a cubic meter or 1000 cubic centimeters (10^{-3} m³ or 10^3 cm³) and just over a quart in size. In this text we shall use whatever metric units are most convenient for a given situation.

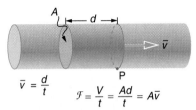

Figure 11.1 Flow rate is the volume of fluid per unit time flowing past a point through the area A. Here the shaded cylinder of fluid flows past point P in a uniform pipe in time t. The volume of the cylinder is Ad and the average velocity is $\bar{v} = d/t$, so that the flow rate is $\mathcal{F} = Ad/t = A\bar{v}$.

In the figure:
$$\bar{v} = \frac{d}{t} \qquad \mathcal{F} = \frac{V}{t} = \frac{Ad}{t} = A\bar{v}$$

EXAMPLE 11.1 THE HEART PUMPS A LOT OF BLOOD IN A LIFETIME

How many cubic meters of blood does the heart pump in a 75.0 year lifetime, assuming the average flow rate is 5.00 L/min?

Strategy Time and flow rate $\mathcal{F}$ are given, and so the volume can be calculated from the definition of flow rate.

Solution Solving Equation 11.1 for volume gives

$$V = \mathcal{F}t$$

Substituting known values yields

$$V = \left(\frac{5.00 \text{ L}}{1 \text{ min}}\right)(75.0 \text{ y})\left(\frac{1 \text{ m}^3}{10^3 \text{ L}}\right)\left(5.26 \times 10^5 \frac{\text{min}}{\text{y}}\right)$$
$$= 1.97 \times 10^5 \text{ m}^3$$

Discussion This is about 200,000 tons of blood, an impressive figure.

Flow rate and velocity are related, but quite different, physical quantities. To make the distinction clear, think about the flow rate of a river. The greater the velocity of the water, the greater the flow rate of the river. But flow rate also depends on the size of the river. A rapid mountain stream carries far less water than the lazy Mississippi. The precise relationship between flow rate $\mathcal{F}$ and velocity $\bar{v}$ is

$$\mathcal{F} = A\bar{v} \qquad (11.2)$$

where A is the cross-sectional area and $\bar{v}$ is the average velocity. This seems logical enough. It tells us that flow rate is directly proportional to both the velocity and the size of a river, pipe, or other conduit. The larger the conduit, the greater its cross-sectional area. Figure 11.1 illustrates how this relationship is obtained. The shaded cylinder has a volume

$$V = Ad$$

which flows past the point P in a time t. Dividing both sides of this relationship by t gives

$$\frac{V}{t} = \frac{Ad}{t}$$

We note that $\mathcal{F} = V/t$ and the average velocity is $\bar{v} = d/t$. Thus the equation becomes $\mathcal{F} = A\bar{v}$, as stated in Equation 11.2.

What happens when a tube with a fluid moving through it changes in size? When water flows from a hose into a narrow spray nozzle, it emerges with a large velocity—

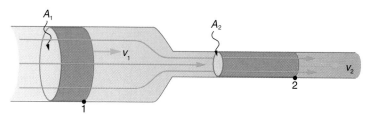

Figure 11.2 When a tube narrows, the same volume occupies a greater length. For the same volume to pass points 1 and 2 in a given time, the velocity must be greater at point 2. The process is exactly reversible. If the fluid flows in the opposite direction, its velocity will decrease when the tube widens.

that is the purpose of the nozzle. Conversely, when a river empties into one end of a reservoir the water slows considerably, perhaps picking up speed again when it leaves the other end of the reservoir. In other words, velocity increases when cross-sectional area decreases, and velocity decreases when cross-sectional area increases. Figure 11.2 shows this relationship for a tube with a varying radius.

If the fluid in Figure 11.2 is incompressible, then the same amount of fluid flows past any point in the tube in a given time. That is, the flow rate is the same at all points. In particular, for points 1 and 2,

or
$$\left.\begin{array}{c} \mathcal{F}_1 = \mathcal{F}_2 \\ A_1\bar{v}_1 = A_2\bar{v}_2 \end{array}\right\} \tag{11.3}$$

Equation 11.3 is called the **equation of continuity**, and it is valid for any incompressible fluid. Since liquids are incompressible, the equation can be used for them. However, gases are compressible, and so the equation must be applied with caution to gases if they are subjected to compression or expansion. The equation states that cross-sectional area times average velocity is a constant. Thus, when the cross-sectional area decreases, the velocity must increase to keep the flow rate constant. Conversely, the velocity must decrease when the area increases.

EXAMPLE 11.2   VELOCITY INCREASES WHEN A TUBE NARROWS

A nozzle with a radius of 0.250 cm is attached to a garden hose with a radius of 0.900 cm. The flow rate through hose and nozzle is 0.500 liters per second. Calculate the velocity of the water (a) in the hose and (b) in the nozzle.

Strategy We can use the relationship between flow rate and velocity to find both velocities. We will use the subscript 1 for the hose and 2 for the nozzle.

Solution for (a) First, we solve Equation 11.2 for $\bar{v}_1$ and note that the cross-sectional area is $A = \pi r^2$, yielding

$$\bar{v}_1 = \frac{\mathcal{F}}{A_1} = \frac{\mathcal{F}}{\pi r_1^2}$$

Substituting known values and making appropriate unit conversions yields

$$\bar{v}_1 = \frac{(0.500 \text{ L/s})(10^{-3} \text{ m}^3/\text{L})}{\pi(9.00 \times 10^{-3} \text{ m}^3)^2} = 1.96 \text{ m/s}$$

Solution for (b) We could repeat this calculation to find the velocity in the nozzle $\bar{v}_2$, but we will use the equation of continuity to give a somewhat different insight. Equation 11.3 states

$$A_1\bar{v}_1 = A_2\bar{v}_2$$

Solving for $\bar{v}_2$ and substituting πr^2 for the cross-sectional area yields

$$\bar{v}_2 = \frac{A_1}{A_2}\bar{v}_1 = \frac{\pi r_1^2}{\pi r_2^2}\bar{v}_1 = \frac{r_1^2}{r_2^2}\bar{v}_1$$

Substituting known values,

$$\bar{v}_2 = \frac{(0.900 \text{ cm})^2}{(0.250 \text{ cm})^2} \cdot 1.96 \text{ m/s} = 25.5 \text{ m/s}$$

Discussion A velocity of 1.96 m/s is about right for water emerging from a nozzleless hose. The nozzle produces a considerably faster stream merely by constricting the flow to a narrower tube.

The solution to the last part of the example shows that velocity is inversely proportional to the *square* of the radius of the tube, making for large effects when radius varies. We can blow out a candle at quite a distance, for example, by pursing our lips, whereas blowing on a candle with our mouth wide open is quite ineffective. (See Figure 11.3.)

Figure 11.3 Children quickly learn that blowing air through a narrow pucker produces a high-velocity stream.

11.2 BERNOULLI'S EQUATION

When a fluid flows into a narrower channel, its velocity increases. That means its kinetic energy also increases. Where does that energy come from? In fact, the fluid's pressure does work to increase its kinetic energy: pressure times area equals force, and so pressure can do work and increase kinetic energy. As a result, *pressure drops in a rapidly moving fluid*, whether or not the fluid is confined to a tube.

There are a number of common examples of pressure dropping in rapidly moving fluids. Shower curtains have a disagreeable habit of bulging into the shower stall when the shower is on. The high-velocity stream of water and air creates a negative gauge pressure inside the shower, and standard atmospheric pressure on the other side pushes the curtain in. You may also have noticed that when passing a truck on the highway, your car tends to veer toward it. The reason is the same—the high velocity of the air between the car and the truck creates a negative pressure, and the vehicles are pushed together by greater pressure on the outside. (See Figure 11.4.) This effect was observed as far back as the mid-1800s, when it was found that trains passing in opposite directions tipped precariously toward one another.

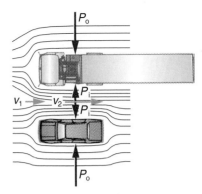

Figure 11.4 An overhead view of a car passing a truck on a highway. Air passing between the vehicles flows in a narrower channel and must increase its velocity (v_2 is greater than v_1), causing the pressure between them to drop (P_i is less than P_o). Greater pressure on the outside pushes the car and truck together.

Bernoulli's Equation

The relationship between pressure and velocity in fluids is described quantitatively by **Bernoulli's equation**, named after its discoverer, the Swiss scientist Daniel Bernoulli (1700–1782). Bernoulli's equation states that for an incompressible, frictionless fluid, the following sum is constant:

$$P + \tfrac{1}{2}\rho v^2 + \rho g h = \text{constant}$$

where P is the absolute pressure, ρ is the fluid density, h is the height above some reference point, and g is the acceleration of gravity. If we follow a small volume of fluid along its path, various quantities in the sum may change, but the total is constant. Let the subscripts 1 and 2 refer to any two points along the path that the bit of fluid follows; Bernoulli's equation becomes

$$P_1 + \tfrac{1}{2}\rho v_1^2 + \rho g h_1 = P_2 + \tfrac{1}{2}\rho v_2^2 + \rho g h_2 \qquad \textbf{(11.4)}$$

Bernoulli's equation is directly related to the conservation of energy principle. Note that the second and third terms look a lot like kinetic and potential energy, with ρ instead of m. In fact, each term in the equation has units of energy per unit volume. We can prove this for the second term by substituting $\rho = m/V$ into it and gathering terms:

$$\tfrac{1}{2}\rho v^2 = \frac{\tfrac{1}{2}mv^2}{V} = \frac{\text{KE}}{V}$$

So $(1/2)\rho v^2$ is the kinetic energy per unit volume. Making the same substitution into the third term in the equation, we find

$$\rho g h = \frac{mgh}{V} = \frac{\text{PE}_g}{V}$$

So $\rho g h$ is the gravitational potential energy per unit volume. Note that pressure P has units of energy per unit volume, too. Since $P = F/A$, its units are N/m². If we multiply these by m/m, we obtain $\text{N} \cdot \text{m/m}^3 = \text{J/m}^3$, or energy per unit volume. Bernoulli's equation is, in fact, just a convenient statement of conservation of energy for an incompressible fluid in the absence of friction.

Bernoulli's equation has many terms in it, and it is broadly applicable. To understand it better, we will look at a number of specific situations that simplify and illustrate its use and meaning.

CONNECTIONS

Conservation of energy applied to fluid flow produces Bernoulli's equation. The terms in Bernoulli's equation resemble KE and PE$_g$ in form, but they are actually energy per unit volume. If other forms of energy are involved in fluid flow, Bernoulli's equation can be modified (just as conservation of energy is modified) to take these forms into account. Such forms of energy include thermal energy dissipated because of fluid viscosity.

Bernoulli's Equation for Static Fluids

Let us first consider the very simple situation where the fluid is static—that is, $v_1 = v_2 = 0$. Bernoulli's equation in that case is

$$P_1 + \rho g h_1 = P_2 + \rho g h_2$$

We can further simplify the equation by taking $h_2 = 0$ (we can always choose some height to be zero, just as we often have done for other situations involving gravity). In that case, we get

$$P_2 = P_1 + \rho g h_1$$

This tells us that, in static fluids, pressure increases with depth. As we go from point 1 to point 2 in the fluid, the depth increases by h_1, and consequently, P_2 is greater than P_1 by an amount $\rho g h_1$. In the very simplest case, P_1 is zero at the top of the fluid, and we get the familiar relationship $P = \rho g h$. The fact that the pressure due to the weight of a fluid is $\rho g h$ is included in Bernoulli's equation! Although we introduce Bernoulli's equation for fluid flow, it includes much of what we studied for static fluids in the preceding chapter.

Bernoulli's Principle—Bernoulli's Equation at Constant Depth

Another important situation is one in which the fluid moves but its depth is constant—that is, $h_1 = h_2$. Under that condition, Bernoulli's equation becomes

$$P_1 + \tfrac{1}{2}\rho v_1^2 = P_2 + \tfrac{1}{2}\rho v_2^2 \tag{11.5}$$

Situations in which fluid flows at a constant depth are so important that this equation is often called **Bernoulli's principle**. It is Bernoulli's equation for fluids at constant depth. (Note again that this applies to a small volume of fluid as we follow it along its path.) As we have just discussed, pressure drops in a rapidly moving fluid. We can see this from Bernoulli's principle. For example, if v_2 is greater than v_1 in the equation, then P_2 must be less than P_1 for the equality to hold.

EXAMPLE 11.3 PRESSURE DROPS AS A FLUID SPEEDS UP

In a previous example, we found that water's velocity increased from 1.96 m/s in a certain hose to 25.5 m/s in its nozzle. Calculate the pressure in the nozzle, given the absolute pressure in the hose is 4.00×10^5 N/m² (nearly 60 lb/in²) and assuming level, frictionless flow.

Strategy Level flow means constant depth, and so Bernoulli's principle applies. We use the subscript 1 for values in the hose and 2 for those in the nozzle. We are thus asked to find P_2.

Solution Solving Bernoulli's principle for P_2 yields

$$P_2 = P_1 + \tfrac{1}{2}\rho v_1^2 - \tfrac{1}{2}\rho v_2^2 = P_1 + \tfrac{1}{2}\rho(v_1^2 - v_2^2)$$

Substituting known values,

$$
\begin{aligned}
P_2 &= 4.00 \times 10^5 \text{ N/m}^2 \\
&\quad + \tfrac{1}{2}(10^3 \text{ kg/m}^3)[(1.96 \text{ m/s})^2 - (25.5 \text{ m/s})^2] \\
&= 0.777 \times 10^5 \text{ N/m}^2
\end{aligned}
$$

Discussion This absolute pressure is less than atmospheric pressure! If we were to drill a hole in the side of the nozzle, air would be forced into the side of the nozzle by the atmosphere instead of water squirting out as it does from the end of the nozzle. This can have many uses, as we shall see.

Applications of Bernoulli's Principle

There are a number of devices and situations in which fluid flows at a constant height and, thus, can be analyzed with Bernoulli's principle.

Entrainment People have long put the Bernoulli principle to work by using reduced pressure in high-velocity fluids to move things about. Higher pressure outside the high-velocity fluid forces other fluids into the stream as shown in Figure 11.5. This process is called *entrainment*. Entrainment devices have been in use since ancient times, particularly as pumps to raise water small heights, as in draining swamps, fields, or other low-lying areas.

Wings and sails The wing is a beautiful example of Bernoulli's principle in action. Figure 11.6(a) shows the characteristic shape of a wing. The upper surface is longer, thereby causing air to flow faster over it. The pressure on top of the wing is reduced, creating a net upward force or lift. (Wings can also gain lift by pushing air downward, utilizing the conservation of momentum principle.) Sails also have the characteristic shape of a wing. (See Figure 11.6(b).) Pressure on the front side of the sail, P_{front}, is lower

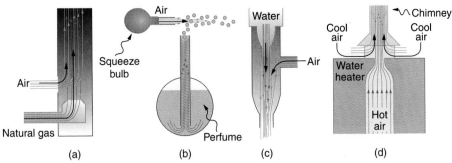

(a)   (b)   (c)   (d)

Figure 11.5 Examples of entrainment devices that use increased fluid velocity to create low pressures, which then *entrain* one fluid into another. (a) A Bunsen burner uses an adjustable gas nozzle, entraining air for proper combustion. (b) An atomizer uses a squeeze bulb to create a jet of air that entrains drops of perfume. Paint sprayers and carburetors use very similar techniques to move their respective liquids. (c) A common aspirator uses a high-velocity stream of water to create a negative pressure. Aspirators may be used as suction pumps in dental and surgical situations or for draining a flooded basement. (d) The chimney of a water heater is designed to entrain cool air into the pipe leading through the ceiling.

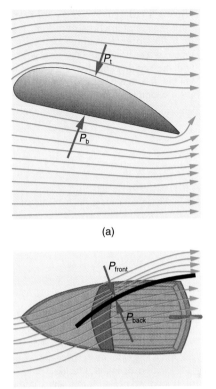

(a)

(b)

Figure 11.6 (a) The Bernoulli principle helps explain lift generated by a wing. (b) Sails use the same technique to generate part of their thrust.

than P_{back} on the back of the sail. This results in a forward force and even allows you to sail into the wind.

Velocity measurement Figure 11.7 shows two devices that measure fluid velocity based on Bernoulli's principle. The manometer in Figure 11.7(a) is connected to two tubes that are small enough not to appreciably disturb the flow. The tube facing the oncoming fluid creates a dead spot having zero velocity ($v_1 = 0$) in front of it, while fluid passing the other tube has velocity v_2. This means that Bernoulli's principle as stated in Equation 11.5 becomes

$$P_1 = P_2 + \tfrac{1}{2}\rho v_2^2$$

Thus pressure P_2 over the second opening is reduced by $(1/2)\rho v_2^2$, and so the fluid in the manometer rises by h on the side connected to the second opening, where

$$h \propto \tfrac{1}{2}\rho v_2^2$$

(Recall from Chapter 4 that the symbol $\propto$ means "proportional to.") Solving for v_2, we see that

$$v_2 \propto \sqrt{h}$$

Figure 11.7(b) shows a version of this device that is in common use for measuring various fluid velocities; such devices are frequently used as air speed indicators in aircraft.

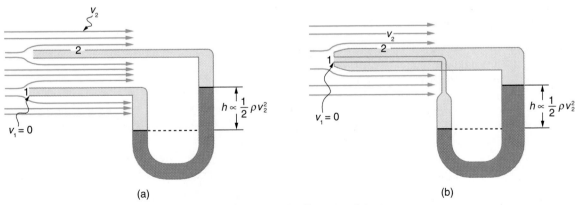

(a)   (b)

Figure 11.7 Measurement of fluid velocity based on Bernoulli's principle. (a) A manometer is connected to two tubes that are close together and small enough not to disturb the flow. Tube 1 is open at the end facing the flow. A dead spot having zero velocity is created there. Tube 2 has an opening on the side, and so the fluid has a velocity v across the opening; thus, pressure there drops. The difference in pressure at the manometer is $(1/2)\rho v_2^2$, and so h is proportional to $(1/2)\rho v_2^2$. (b) This variety of velocity measuring device is a *Prandtl tube*, but it is usually referred to as a Pitot tube.

Figure 11.8 (a) Water gushes from a large tube at the base of a dam. (b) In the absence of significant resistance, water flows from the reservoir with the same speed it would have if it fell the distance h without friction. This is an example of Torricelli's theorem.

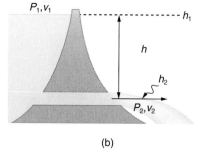

(a)   (b)

11.3 THE MOST GENERAL APPLICATIONS OF BERNOULLI'S EQUATION

Torricelli's Theorem

Figure 11.8 shows water gushing from a large tube through a dam. What is its velocity as it emerges? Interestingly, if resistance is negligible, the speed is just what it would be if the water fell a distance h from the surface of the reservoir; the water's speed is independent of the size of the opening. Bernoulli's equation must be used since the depth is not constant. We consider water flowing from the surface (point 1) to the tube's outlet (point 2). Bernoulli's equation as stated in Equation 11.4 is

$$P_1 + \tfrac{1}{2}\rho v_1^2 + \rho g h_1 = P_2 + \tfrac{1}{2}\rho v_2^2 + \rho g h_2$$

Both P_1 and P_2 equal atmospheric pressure* and subtract out of the equation, leaving

$$\tfrac{1}{2}\rho v_1^2 + \rho g h_1 = \tfrac{1}{2}\rho v_2^2 + \rho g h_2$$

Solving this equation for v_2^2, noting that the density ρ cancels, yields

$$v_2^2 = v_1^2 + 2g(h_1 - h_2)$$

We let $h = h_1 - h_2$; the equation then becomes

$$v_2^2 = v_1^2 + 2gh$$

where h is the height dropped by the water. This is an equation we had in kinematics for any object falling a distance h with negligible resistance. In fluids, the last equation is called *Torricelli's theorem*. Note that the result is independent of the velocity's direction, just as we found when applying conservation of energy to falling objects.

All preceding applications of Bernoulli's equation involved simplifying conditions, such as constant height or constant pressure. The next example is a more general application of Bernoulli's equation in which pressure, velocity, and height all change. (See Figure 11.9.)

Figure 11.9 Pressure in the nozzle of this fire hose is less than at ground level for two reasons: the water has to go uphill to get to the nozzle, and velocity increases in the nozzle. In spite of its lowered pressure, the water can exert a large force on anything it strikes. It does this by virtue of its kinetic energy. Pressure in the water stream becomes equal to atmospheric pressure once it emerges into the air.

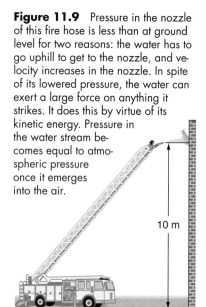

10 m

EXAMPLE 11.4 PRESSURE IN ANOTHER NOZZLE

Fire hoses used in major structure fires have inside diameters of 6.40 cm (about 2.5 in.). Suppose such a hose carries a flow of 40.0 liters per second (over 10 gallons per second) starting at a gauge pressure of 1.75×10^6 N/m² (about 250 psi). The hose goes 10.0 m up a ladder to a nozzle having an inside diameter of 3.00 cm. Assuming negligible resistance, what is the pressure in the nozzle?

Strategy Here we must use Bernoulli's equation to solve for the pressure, since depth is not constant. Otherwise, this example is solved in a manner similar to Example 11.3.

Solution Bernoulli's equation states

$$P_1 + \tfrac{1}{2}\rho v_1^2 + \rho g h_1 = P_2 + \tfrac{1}{2}\rho v_2^2 + \rho g h_2$$

where the subscripts 1 and 2 refer to the initial conditions at ground level and final conditions inside the nozzle, respectively. We must first find the velocities v_1 and v_2. Since $\mathcal{F} = A_1 v_1$, we get

$$v_1 = \frac{\mathcal{F}}{A_1} = \frac{40.0 \times 10^{-3}\ \text{m}^3/\text{s}}{\pi (3.20 \times 10^{-2}\ \text{m})^2} = 12.4\ \text{m/s}$$

Similarly, we find

$$v_2 = 56.6\ \text{m/s}$$

(continued)

---

*P_1 is atmospheric pressure because it is the pressure at the top of the reservoir. P_2 must be atmospheric pressure, since the emerging water is surrounded by the atmosphere and cannot have a pressure different from atmospheric pressure.

(continued)

(This rather large velocity is helpful in reaching the fire.) Now, taking h_1 to be zero, we solve Bernoulli's equation for P_2:

$$P_2 = P_1 + \tfrac{1}{2}\rho(v_1^2 - v_2^2) - \rho g h_2$$

Substituting known values yields

$$
\begin{aligned}
P_2 &= 1.75 \times 10^6 \ \text{N/m}^2 \\
&\quad + (500 \ \text{kg/m}^3)[(12.4 \ \text{m/s})^2 - (56.6 \ \text{m/s})^2] \\
&\quad - (10^3 \ \text{kg/m}^3)(9.80 \ \text{m/s}^2)(10.0 \ \text{m}) \\
&= 1.28 \times 10^5 \ \text{N/m}^2
\end{aligned}
$$

Discussion This is a gauge pressure, since the initial pressure was given as a gauge pressure. Note that because of this pressure, there is a large force backward on the hose. Usually, the hose is braced against the ladder or more than one firefighter holds the hose.

Power in Fluid Flow

Power (as defined in Chapter 6) is the *rate* at which work is done or energy in any form is used or supplied. To see the relationship of power to fluid flow, consider Bernoulli's equation:

$$P + \tfrac{1}{2}\rho v^2 + \rho g h = \text{constant}$$

All three terms have units of energy per unit volume, as discussed in the previous section. Now, considering units, if we multiply energy per unit volume by flow rate (volume per unit time), we get units of power. That is, $(E/V) \cdot (V/t) = E/t$. This means that if we multiply Bernoulli's equation by flow rate $\mathcal{F}$, we get power. In equation form, this is

$$(P + \tfrac{1}{2}\rho v^2 + \rho g h) \cdot \mathcal{F} = \text{power} \qquad \textbf{(11.6)}$$

Each term has a clear physical meaning. For example, $P \cdot \mathcal{F}$ is the power supplied to a fluid, by a pump perhaps, to give it its pressure P. Similarly, $(1/2)\rho v^2 \cdot \mathcal{F}$ is the power supplied to a fluid to give it its kinetic energy. And $\rho g h \cdot \mathcal{F}$ is the power going to gravitational potential energy.

CONNECTIONS

Power was defined in Chapter 6 as the rate of energy transferred per unit time, or E/t. Fluid flow involves several types of power. Each type of power is identified with a specific type of energy being expended or changed in form.

EXAMPLE 11.5 POWER IN A MOVING FLUID

Suppose the fire hose in the previous example is fed by a pump that receives water through a 6.40 cm diameter hose coming from a hydrant with a pressure of $0.700 \times 10^6 \ \text{N/m}^2$ (about 100 psi). What power does the pump supply to the water?

Strategy and Concept Here we must consider energy forms as well as how they relate to fluid flow. Since the input and output hoses have the same diameters and are at the same height, the pump does not change the velocity of the water nor its height, and so the water's kinetic energy and gravitational potential energy are unchanged. That means the pump only supplies power to increase water pressure by $1.05 \times 10^6 \ \text{N/m}^2$ (from 0.700 to $1.75 \times 10^6 \ \text{N/m}^2$).

Solution As discussed above, the power associated with pressure is

$$
\begin{aligned}
\text{power} &= P \cdot \mathcal{F} \\
&= (1.05 \times 10^6 \ \text{N/m}^2)(40.0 \times 10^{-3} \ \text{m}^3/\text{s}) \\
&= 4.20 \times 10^4 \ \text{W} = 42.0 \ \text{kW}
\end{aligned}
$$

Discussion This is about 56 horsepower. Such a substantial amount of power requires a large pump, such as is found on some fire trucks. The pump in this example increases only the water's pressure. If a pump—such as the heart—directly increases velocity and height as well as pressure, we would have to calculate all three terms to find the power it supplies.

11.4 VISCOSITY AND LAMINAR FLOW: POISEUILLE'S LAW

When you pour yourself a glass of juice, it flows freely and quickly. But when you pour syrup on your pancakes, it flows slowly and sticks to the pitcher. The difference is fluid friction, both within the fluid itself and between the fluid and its surroundings. We call this property of fluids *viscosity*. Juice has low viscosity, whereas syrup has high viscosity. In the previous sections we have considered ideal fluids with little or no viscosity. In this section we will investigate what factors, including viscosity, affect the rate of fluid flow.

The precise definition of viscosity is based on laminar, or nonturbulent, flow. Before we can define viscosity, then, we need to define laminar flow and turbulent flow. Figure 11.10 shows both types of flow. **Laminar** flow is characterized by smooth flow of the fluid in layers that do not mix. Turbulent flow, or **turbulence**, is characterized by eddies and swirls that mix layers of fluid together.

Figure 11.10 Smoke rises smoothly for a while and then begins to form swirls and eddies. The smooth flow is called laminar, whereas the swirls and eddies typify turbulent flow. If you watch the smoke (being careful not to breathe), you will notice that it rises more rapidly when flowing smoothly than after it becomes turbulent, implying that turbulence poses more resistance to flow.

Figure 11.11 shows schematically how laminar and turbulent flow differ. Layers flow without mixing when flow is laminar. When there is turbulence, the layers mix, and there are significant velocities in directions other than the overall direction of flow. The lines that are shown in many illustrations are the paths followed by small volumes of fluids. These are called *streamlines*. Streamlines are smooth and continuous when flow is laminar, but break up and mix when flow is turbulent. Turbulence has two important causes. First, any obstruction or sharp corner, such as in a faucet, creates turbulence by imparting velocities perpendicular to the flow. Second, high velocity causes turbulence. Drag between adjacent layers of fluid and with the surroundings forms swirls and eddies if the velocity is great enough. We shall concentrate on laminar flow for the remainder of this section, leaving certain aspects of turbulence for later sections.

Figure 11.12 shows how viscosity is measured for a fluid. Two parallel plates have the specific fluid between them. The bottom plate is held fixed, while the top plate is moved to the right, dragging fluid with it. The layer (or lamina) of fluid in contact with either plate does not move relative to the plate, and so the top layer moves at v while the bottom layer remains at rest. Each successive layer from the top down exerts a force on the one below it, trying to drag it along, producing a continuous variation in velocity from v to 0 as shown. Care is taken that the flow is laminar; that is, the layers do not mix.

A force F is required to keep the top plate in Figure 11.12 moving at a constant velocity v, and experiments have shown that this force depends on four factors. First, F is directly proportional to v (until you go so fast that turbulence occurs—then a much larger force is needed, and it has a more complicated dependence on v). Second, F is proportional to the area of the plate A. This seems reasonable, since A is directly proportional to the amount of fluid being moved. Third, F is inversely proportional to the distance between the plates L. This is also reasonable; L is like a lever arm, and the greater the lever arm, the less force that is needed.* Fourth, F is directly proportional to *the coefficient of viscosity*, η (the Greek letter eta). The greater the viscosity, the greater the force required. These dependencies are combined into the equation

$$F = \eta \frac{vA}{L} \tag{11.7}$$

which gives us a working definition of fluid **viscosity** η. Solving for η gives

$$\eta = \frac{FL}{vA}$$

which defines viscosity in terms of how it is measured. The SI unit of viscosity is $N \cdot m/[(m/s)m^2] = (N/m^2) \cdot s$. Table 11.1 lists coefficients of viscosity for various fluids.

Viscosity varies from one fluid to another by several orders of magnitude. As you might expect, the viscosities of gases are much less than those of liquids.

Laminar Flow Confined to Tubes—Poiseuille's Law

What causes flow? The answer, not surprisingly, is pressure. In fact, there is a very simple relationship between horizontal flow and pressure. Flow rate $\mathcal{F}$ is in the direction from high to low pressure. The greater the pressure difference between two points, the greater the flow rate. This is precisely stated as

$$\mathcal{F} = \frac{P_2 - P_1}{R} \tag{11.8}$$

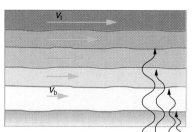

Friction between layers

(a)

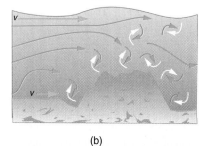

(b)

(c)

Figure 11.11 (a) Laminar flow occurs in layers without mixing. Notice that viscosity causes drag between layers as well as with the fixed surface. (b) Turbulent flow mixes the fluid. There is more interaction, greater heating, and more resistance than in laminar flow. (c) Here laminar flow becomes turbulent.

---

*The motion in Figure 11.12 is like a continuous shearing motion. As noted previously, fluids have zero shear strength, but the *rate* at which they are sheared is related to the same geometrical factors A and L as is shear deformation for solids. See Section 5.8.

where P_1 and P_2 are the pressures at two points, such as at either end of a tube, and R is the resistance to flow. Resistance R includes everything, except pressure, that affects flow rate. For example, R is greater for a long tube than a short one. The greater the viscosity of a fluid, the greater R is. Turbulence greatly increases R, whereas increasing the diameter of a tube decreases R.

If viscosity is zero, the fluid is frictionless and resistance to flow is zero, too. Comparing frictionless flow in a tube to viscous flow, as in Figure 11.13, we see that for a viscous fluid velocity is greatest at midstream because of drag at the boundaries. We can see the effect of viscosity in a Bunsen burner flame, even though the viscosity of natural gas is small.

The resistance R to laminar flow of an incompressible fluid having viscosity η through a horizontal tube of uniform radius r and length l, such as the one in Figure 11.14, is given by

$$R = \frac{8\eta l}{\pi r^4} \tag{11.9}$$

This equation is called **Poiseuille's law for resistance** after the French scientist J. L. Poiseuille (1799–1869), who derived it in an attempt to understand the flow of blood, an often turbulent fluid.

Let us examine Poiseuille's expression for R to see if it makes good intuitive sense. We see that resistance is directly proportional to both fluid viscosity η and the length l of a tube. After all, both of these directly affect the amount of friction encountered—the greater either is, the greater the resistance and the smaller the flow. The radius r of a tube affects the resistance, which again makes sense, since the greater the radius, the greater the flow, all other factors being the same. But it is surprising that r is raised to the *fourth* power in Poiseuille's law. This means that any change in the radius of a tube has a very large effect on resistance. For example, doubling the radius of a tube decreases resistance by a factor of $2^4 = 16$!

Taken together, Equations 11.8 and 11.9 give the following expression for flow rate:

$$\mathcal{F} = \frac{(P_2 - P_1)\pi r^4}{8\eta l} \tag{11.10}$$

This equation describes laminar flow through a tube. It is sometimes called Poiseuille's law for laminar flow, or simply **Poiseuille's law**.

There are a host of situations in which Poiseuille's law either exactly describes flow or gives the correct sense of its dependence on the various factors involved. For example, if your car has an oil pressure gauge, you may notice

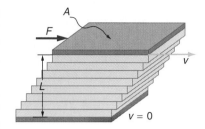

Figure 11.12 Laminar flow of fluid between two plates of area A. The bottom plate is fixed. When the top plate is pushed to the right, it drags the fluid along with it.

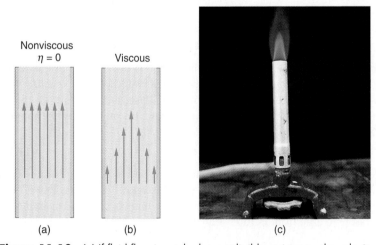

Figure 11.13 (a) If fluid flow in a tube has negligible resistance, the velocity is the same all across the tube. (b) When a viscous fluid flows through a tube, its velocity at the walls is zero, increasing steadily to its maximum at the center of the tube. (c) The shape of the Bunsen burner flame is due to the velocity profile across the tube.

TABLE 11.1

COEFFICIENTS OF VISCOSITY OF VARIOUS FLUIDS

Fluid	Temperature (°C)	Viscosity η 10^{-3} $(N/m^2) \cdot s$
Gases		
Air	0	0.0171
	20	0.0181
	40	0.0190
	100	0.0218
Ammonia	20	0.00974
Carbon dioxide	20	0.0147
Helium	20	0.0196
Hydrogen	0	0.0090
Mercury	20	0.0450
Oxygen	20	0.0203
Steam	100	0.0130
Liquids		
Water	0	1.792
	20	1.005
	37	0.6947
	40	0.656
	100	0.284
Whole blood*	20	3.015
	37	2.084
Blood plasma*	20	1.810
	37	1.257
Ethyl alcohol	20	1.20
Methanol	20	0.584
Oil (heavy machine)	20	661
Oil (SAE 10)	30	200
Oil (olive)	20	138
Glycerin	20	1500

*The ratios of the viscosities of blood to water are nearly constant between 0°C and 37°C.

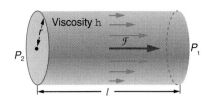

Figure 11.14 Poiseuille's law applies to laminar flow of an incompressible fluid of viscosity η through a tube of length l and radius r. The direction of flow is from greater to lower pressure. Flow rate $\mathcal{F}$ is directly proportional to the pressure difference $P_2 - P_1$, and inversely proportional to the length l of the tube and viscosity η of the fluid. Flow rate increases with r^4, the *fourth* power of the radius.

Human & Biological Applications

that oil pressure is high when the engine is cold. This is because motor oil has greater viscosity when cold than when warm, and so pressure must be greater to pump the same amount of cold oil. The circulatory system provides many examples—blood flow is regulated by changes in vessel size and blood pressure. During vigorous exercise, blood vessels are selectively dilated to important muscles and organs and blood pressure increased. This creates both greater overall blood flow and increased flow to specific areas. Conversely, decreases in vessel radii, perhaps from plaques in the arteries, can greatly reduce blood flow. If a vessel's radius is reduced by only 5% to 0.95 of its original value, the flow rate is reduced to about $(0.95)^4 = 0.81$ of its original value. A 19% decrease in flow is caused by a 5% decrease in radius. The body may compensate by increasing blood pressure by 19%, but this presents hazards to the heart and any vessel that has weakened walls.

EXAMPLE 11.6 WHAT PRESSURE PRODUCES THIS FLOW RATE?

An intravenous (IV) system is supplying saline solution to a patient at the rate of 0.120 cm³/s through a needle of radius 0.150 mm and length 2.50 cm. What pressure is needed at the entrance of the needle to cause this flow, assuming the viscosity of the saline solution to be the same as that of water? Blood pressure in the patient's vein is 8.00 mm Hg.

Strategy Assuming laminar flow, Poiseuille's law as expressed in Equation 11.10 applies. This is

$$\mathcal{F} = \frac{(P_2 - P_1)\pi r^4}{8\eta l}$$

where P_2 is the pressure at the entrance of the needle and P_1 is the pressure in the vein. The only unknown is P_2.

Solution Solving for P_2 yields

$$P_2 = \frac{8\eta l}{\pi r^4}\mathcal{F} + P_1$$

P_1 is given as 8.00 mm Hg, which converts to 1.066×10^3 N/m². Substituting this and the other known values yields

$$P_2 = \left[\frac{8(1.00 \times 10^{-3}\ \text{N·s/m}^2)(2.50 \times 10^{-2}\ \text{m})}{\pi(0.150 \times 10^{-3}\ \text{m})^4}\right]$$
$$\times (1.20 \times 10^{-7}\ \text{m}^3/\text{s}) + 1.066 \times 10^3\ \text{N/m}^2$$
$$= 1.62 \times 10^4\ \text{N/m}^2$$

Discussion This pressure could be supplied by an IV bottle with the surface of the saline solution 1.61 m above the entrance to the needle (left as an end-of-chapter problem), assuming that there is negligible pressure drop in the tubing leading to the needle.

Flow and Resistance as Causes of Pressure Drops

You may have noticed that water pressure in your home is lower than normal on hot summer days. This pressure drop occurs in the water main before it reaches your home. Let us consider flow through the water main as illustrated in Figure 11.15. We can understand why the pressure P_1 drops during times of heavy use by rearranging Equation 11.8 to

$$P_2 - P_1 = R \cdot \mathcal{F} \tag{11.11}$$

where, in this case, P_2 is the pressure at the water works and R is the resistance of the water main. During times of heavy use, the flow rate $\mathcal{F}$ is large. This means that $P_2 - P_1$ must also be large. Thus P_1 must decrease. It is correct to think of flow and resistance as causing the pressure to drop from P_2 to P_1. Equation 11.11 is valid for both laminar and turbulent flows.

We can use Equation 11.11 to analyze pressure drops occurring in more complex systems in which the tube radius is not the same everywhere. Resistance will be much greater in narrow places, such as an obstructed coronary artery. For a given flow rate $\mathcal{F}$, the pressure drop will be greatest where the tube is most narrow. This is how faucets con-

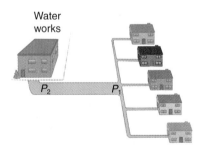

Figure 11.15 During times of heavy use, there is a significant pressure drop in a water main, and P_1 supplied to users is significantly less than P_2 created at the water works. If the flow is very small, then the pressure drop is negligible, and $P_1 \approx P_2$.

trol flow. Additionally, R is greatly increased by turbulence, and so a constriction that creates turbulence greatly reduces the pressure downstream. Plaque in an artery reduces pressure and hence flow, both by its resistance and by the turbulence it creates.

11.5*   THE ONSET OF TURBULENCE

Can we predict if flow will be laminar or turbulent? Sometimes. We know that flow in a very smooth tube or around a smooth, streamlined object will be laminar at low velocity. We also know that at a high velocity even flow in a smooth tube or around a smooth object will experience turbulence. In between, it is more difficult to predict. In fact, at intermediate velocities, flow may oscillate back and forth indefinitely between laminar and turbulent. (See Figure 11.16.)

An indicator called the **Reynolds number** N_R can reveal whether flow is laminar or turbulent. For flow in a tube of uniform diameter, the Reynolds number is defined as

$$N_R = \frac{2\rho vr}{\eta} \qquad \text{(flow in tube)} \qquad \textbf{(11.12)}$$

where ρ is the fluid density, v its velocity, η its viscosity, and r the tube radius. The Reynolds number is a unitless quantity. Experiments have revealed N_R is related to the onset of turbulence. For N_R below about 2000, flow is laminar. For N_R above about 3000, flow is turbulent. For values of N_R between 2000 and 3000, flow is unstable—that is, it can be laminar, but small obstructions and surface roughness can make it turbulent, and it may oscillate randomly between being laminar and turbulent.

EXAMPLE 11.7   IS THIS FLOW LAMINAR OR TURBULENT?

Calculate the Reynolds number for flow in the needle considered in Example 11.6 to verify the assumption that the flow is laminar. Assume the density of the saline solution is 1025 kg/m³.

Strategy We have all of the information needed except the fluid velocity v, which can be calculated from $v = \mathcal{F}/A = 1.70$ m/s (verification left as an end-of-chapter problem).

Solution Entering the known values into Equation 11.12

for N_R gives

$$\begin{aligned} N_R &= \frac{2\rho vr}{\eta} \\ &= \frac{2\,(1025\ \text{kg/m}^3)(1.70\ \text{m/s})(0.150 \times 10^{-3}\ \text{m})}{(1.00 \times 10^{-3}\ \text{N}\cdot\text{s/m}^2)} \\ &= 523 \end{aligned}$$

Discussion Since N_R is well below 2000, the flow should indeed be laminar.

The topic of chaos has become quite popular in the last few years. A system is defined to be **chaotic** when its behavior is so sensitive to some factor that it is extremely difficult to predict. The field of **chaos** is the study of chaotic behavior. A good example of chaotic behavior is the flow of a fluid with a Reynolds number between 2000 and 3000. Whether or not the flow is turbulent is difficult, but not impossible, to predict—the difficulty lies in the extremely sensitive dependence on factors like roughness and obstructions. A tiny variation in one factor has an exaggerated (or nonlinear) effect on the

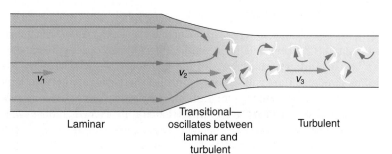

Laminar

Transitional—
oscillates between
laminar and
turbulent

Turbulent

Figure 11.16 Flow is laminar in the largest part of this tube and turbulent in the narrowest part, where velocity is high. In the transition region, the flow can oscillate chaotically between laminar and turbulent flow.

flow. Things as disparate as turbulence, the orbit of Pluto, and the onset of irregular heartbeat are chaotic and can be analyzed with similar techniques.

11.6* MOTION OF AN OBJECT IN A VISCOUS FLUID

A moving object in a viscous fluid is equivalent to a stationary object in a fluid stream. (For example, when you ride a bicycle at 10 m/s in still air, you feel the air in your face exactly as if you were stationary in a 10 m/s wind.) Flow of the stationary fluid around a moving object may be laminar, turbulent, or a combination of the two. Just as with flow in tubes, it is possible to predict when a moving object creates turbulence. We use another form of the Reynolds number N_R', defined for an object moving in a fluid to be

$$N_R' = \frac{\rho v L}{\eta} \quad \text{(object in fluid)} \quad \textbf{(11.13)}$$

where L is a characteristic length of the object (a sphere's diameter, for example), ρ the fluid density, η its viscosity, and v the object's velocity in the fluid. If N_R' is less than about 1, flow around the object can be laminar, particularly if the object has a smooth shape. The transition to turbulent flow occurs for N_R' between 1 and about 10, depending on surface roughness and so on. Depending on the surface, there can be a turbulent wake behind the object with some laminar flow over its surface. For an N_R' between 10 and 10^6, the flow may be either laminar or turbulent and may oscillate between the two. For N_R' greater than about 10^6, the flow is entirely turbulent, even at the surface of the object. (See Figure 11.17.) Laminar flow occurs mostly when the objects in the fluid are small, such as raindrops, pollen, and blood cells in a plasma.

Biological Application

EXAMPLE 11.8 DOES A BASEBALL HAVE A TURBULENT WAKE?

Calculate the Reynolds number N_R' for a baseball pitched at 40.0 m/s (about 90 mph—a fastball). The ball's diameter is 7.40 cm.

Strategy We can use Equation 11.13 to calculate N_R', since all values in it are either given or can be found in tables of density and viscosity.

Solution Substituting values into the equation for N_R' yields

$$N_R' = \frac{\rho v L}{\eta} = \frac{(1.29 \text{ kg/m}^3)(40.0 \text{ m/s})(0.0740 \text{ m})}{1.81 \times 10^{-5} \text{ N} \cdot \text{s/m}^2}$$

$$= 2.11 \times 10^5$$

Discussion This value is high enough to imply a turbulent wake. Most large objects, such as airplanes and sailboats, create significant turbulence as they move. As noted before, the Bernoulli principle gives only qualitatively correct results in such situations.

One of the consequences of viscosity is a resistance force called **viscous drag** F_v that is exerted on a moving object. It typically depends strongly on the object's velocity (in contrast with simple friction). Experiments have shown that for laminar flow (N_R' less than about 1) viscous drag is proportional to velocity, whereas for N_R' between about 1 and 10^6 viscous drag is proportional to velocity squared. (This is a strong dependence and is pertinent to bicycle racing, where even a small headwind causes significantly increased drag on the racer.) For N_R' greater than 10^6, drag increases dramatically and behaves with greater complexity. For laminar flow around a sphere, F_v is proportional to fluid viscosity η, the object's characteristic size L, and its velocity v. All of which makes sense—the more viscous the fluid and the larger the object, the more drag we expect.

An interesting consequence of the increase in F_v with velocity is that an object falling through a fluid will not continue to accelerate indefinitely (as it would if we neglect air resistance, for example). Instead, viscous drag grows, slowing acceleration, until a critical velocity, called the **terminal velocity**, is reached and the acceleration of the object becomes zero. Once this happens, the object continues to fall at constant velocity

Biological Application

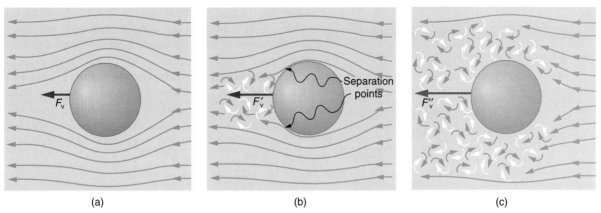

Figure 11.17 (a) Motion of this sphere to the right is equivalent to fluid flow to the left. Here the flow is laminar with N'_R less than 1. There is a force, called *viscous drag F_v*, to the left on the ball due to the fluid's viscosity. (b) At a higher velocity the flow becomes partially turbulent, creating a wake starting where the flow lines separate from the surface. Pressure in the wake is less than in front of the sphere, because fluid velocity is less, creating a net force to the left F'_v that is significantly greater than for laminar flow. Here N'_R is greater than 10. (c) At much higher velocities, where N'_R is greater than 10^6, flow becomes turbulent everywhere on the surface and behind the sphere. Drag increases dramatically.

(the terminal velocity). This is true for particles of sand falling in the ocean, cells falling in a centrifuge, and sky divers falling through the air. Figure 11.18 shows some of the factors that affect terminal velocity. There is a viscous drag on the object that depends on the viscosity of the fluid and the size of the object. But there is also a buoyant force that depends on the density of the object relative to the fluid. Terminal velocity will be greatest for low-viscosity fluids and objects with high densities and small sizes. Thus a sky diver falls slower with outspread limbs than when he is in a cannonball shape.

Knowledge of terminal velocity is useful for estimating sedimentation rates of small particles. We know from watching mud settle out of dirty water that these are usually slow processes. The centrifuge is used to speed sedimentation by creating an accelerated frame in which gravity is replaced by centripetal acceleration, which can be much greater, increasing the terminal velocity.

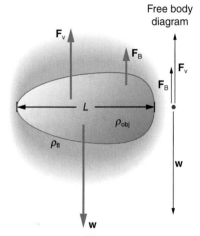

Figure 11.18 There are three forces acting on an object falling through a viscous fluid: its weight w, the viscous drag F_v, and the buoyant force F_B.

11.7* MOLECULAR TRANSPORT PHENOMENA: DIFFUSION, OSMOSIS, AND RELATED PROCESSES

There is something fishy about the ice cube from your freezer—how did it pick up those food odors? How does soaking a sprained ankle in Epsom salt reduce swelling? The answer to these questions are related to atomic and molecular transport phenomena—another mode of fluid motion.

Atoms and molecules are in constant motion at any temperature above absolute zero. In fluids they move about randomly even in the absence of macroscopic flow. This motion is called a random walk and is illustrated in Figure 11.19. **Diffusion** is the movement of substances due to random thermal molecular motion. Fluids, like fish fumes, can even diffuse through solids.

Diffusion is a slow process over macroscopic distances. Common densities are great enough that molecules cannot travel very far before having a collision that can scatter them in any direction, including straight backward. It can be shown that the average distance x_{rms} a molecule travels is proportional to the square root of time:

$$x_{rms} = \sqrt{2Dt} \tag{11.14}$$

where x_{rms} stands for the root mean square distance and is the statistical average for the process. D is the diffusion constant for the particular molecule in a specific medium. Table 11.2 lists representative values of D.

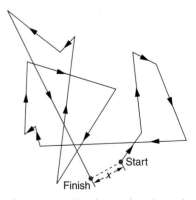

Figure 11.19 The random thermal motion of a molecule in a fluid in time t. This type of motion is called a *random walk*.

TABLE 11.2

DIFFUSION CONSTANTS FOR VARIOUS MOLECULES*		
Diffusing molecule	*Medium*	*D (m²/s)*
Hydrogen (H_2)	Air	6.4×10^{-5}
Oxygen (O_2)	Air	1.8×10^{-5}
Oxygen (O_2)	Water	1.0×10^{-9}
Glucose ($C_6H_{12}O_6$)	Water	6.7×10^{-10}
Hemoglobin	Water	6.9×10^{-11}
DNA	Water	1.3×10^{-12}

*At 20°C and 1 atm.

Note that D gets progressively smaller for more massive molecules. This is because the average molecular speed at a given temperature is inversely proportional to molecular mass. Thus the more massive molecules diffuse more slowly. Another interesting point is that D for oxygen in air is much greater than the D for oxygen in water. In water, the oxygen makes many more collisions in its random walk and is slowed considerably. Finally, note that diffusion constants increase with temperature, because average molecular speed increases with temperature.*

EXAMPLE 11.9 HOW LONG DOES DIFFUSION TAKE?

Calculate the average time it takes a glucose molecule to move 1.0 cm in water.

Strategy We can use Equation 11.14, the expression for the average distance moved in time t, and solve it for t. All other quantities are known.

Solution Solving for t and substituting known values yields

$$t = \frac{x_{rms}^2}{2D} = \frac{(0.010\ \text{m})^2}{2(6.7 \times 10^{-10}\ \text{m}^2/\text{s})}$$
$$= 7.5 \times 10^4\ \text{s} = 21\ \text{h}$$

Discussion This is a remarkably long time for glucose to move a mere centimeter! This is why we stir sugar into water rather than waiting for it to diffuse.

Biological Application

Because diffusion is typically very slow, its most important effects occur over small distances. For example, the cornea of the eye gets most of its oxygen by diffusion through the thin tear layer covering it.

The Rate and Direction of Diffusion

If you very carefully place a drop of food coloring in a still glass of water, it will slowly diffuse into the colorless surroundings until its concentration is the same everywhere. This type of diffusion is called free diffusion, because there are no barriers inhibiting it. Let us examine its direction and rate. Diffusion always moves toward regions of lower concentration. Molecular motion is random in direction, and so simple chance dictates that more molecules will move out of a region of high concentration than into it. The rate of diffusion is higher initially than after the process is partially completed. (See Figure 11.20.)

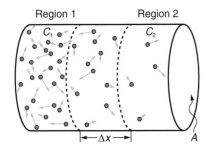

Figure 11.20 Diffusion proceeds from a region of higher concentration to a lower one. The rate of movement is proportional to the difference in concentration.

The rate of diffusion is proportional to the concentration difference. Many more molecules will leave a region of high concentration than enter it from a region of low concentration. In fact, if the concentrations were the same, there would be no *net* movement. The rate of diffusion is also proportional to the diffusion constant D, which is determined experimentally. The farther a molecule can diffuse in a given time, the more likely it is to leave the region of high concentration. Many of the factors that affect the rate are hidden in the diffusion constant D. For example, temperature and cohesive and adhesive forces all affect values of D.

Osmosis and Dialysis—Diffusion across Membranes

Some of the most interesting examples of diffusion occur through barriers that affect the rates of diffusion. For example, when you soak a swollen ankle in Epsom salt, water diffuses through your skin. Many substances regularly move through cell membranes; oxygen moves in, carbon dioxide out, nutrients in, and wastes out, for example. Because membranes are thin structures (typically 65×10^{-10} to 100×10^{-10} m across) diffusion rates through them can be high. Diffusion through membranes is an important method of transport.

Biological Application

Membranes are generally selectively permeable, or **semipermeable**. (See Figure 11.21.) One type of semipermeable membrane has small pores that allow only small molecules to pass

---

*The average kinetic energy of molecules, $(1/2)mv^2$, is proportional to absolute temperature. The precise relationship is explored in the Chapter 12 on Temperature and Kinetic Theory.

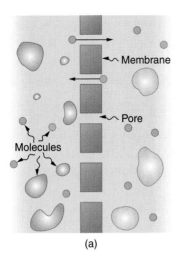
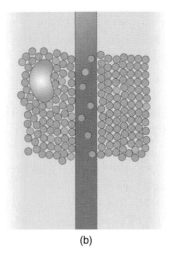

(a)   (b)

Figure 11.21   (a) A semipermeable membrane with small pores that allow only small molecules to pass through. (b) Certain molecules dissolve in this membrane and diffuse across it.

through. In other types of membranes, the molecules may actually dissolve in the membrane or react with molecules in it while moving across. Membrane function, in fact, is the subject of much current research, involving not only physiology but also chemistry and physics.

Osmosis is the transport of water through a semipermeable membrane from a region of high concentration to a region of low concentration. Osmosis is driven by the imbalance in water concentration. For example, water is more concentrated in your body than in Epsom salt. When you soak a swollen ankle in Epsom salt, the water moves out of your body into the lower-concentration region in the salt. Similarly, **dialysis** is the transport of any other molecule through a semipermeable membrane due to its concentration difference. Both osmosis and dialysis are used by the kidneys to cleanse the blood.

Osmosis can create a substantial pressure. Consider what happens if osmosis continues for some time, as illustrated in Figure 11.22. Water passes by osmosis from the left into the region on the right, where it is less concentrated, causing the solution on the right to rise. This movement will continue until the pressure $h\rho g$ created by the extra height of fluid on the right is large enough to stop further osmosis. This pressure is called a *back pressure*. The back pressure $h\rho g$ that stops osmosis is called the **relative osmotic pressure** if neither solution is pure water, and it is called the **osmotic pressure** if one solution is pure water. Osmotic pressure can be large, depending on how great the concentration difference is. For example, if pure water and sea water are separated by a semipermeable membrane that passes no salt, osmotic pressure will be 25.9 atm. This means that water will diffuse through the membrane until the salt water surface rises 261 m above the pure water surface! One example of pressure created by osmosis is turgor in plants (many wilt when too dry). Dialysis can similarly cause substantial pressures.

Reverse osmosis and **reverse dialysis** (also called filtration) are processes that occur when back pressure is sufficient to reverse their normal direction. Back pressure can be created naturally as on the right side of Figure 11.22. (This can also be done with a piston.) Reverse osmosis can be used to desalinate water by simply forcing it through a membrane that will not pass salt. Similarly, reverse dialysis can be used to filter out any substance that a given membrane will not pass.

One further example of the movement of substances through membranes deserves mention. We sometimes find that substances pass in the direction opposite to what we expect. Cypress tree roots, for example, extract pure water from salt water, though osmosis would move it in the opposite direction. This is not reverse osmosis, because there is no back pressure to cause it. What is happening is called **active transport**, a process in which a living membrane expends energy to move substances across it. Many living membranes move water and other substances by active transport. The kidneys, for example, not only use osmosis and dialysis—they also employ significant active transport to move substances into and out of blood. In fact, it is estimated that at least 25% of the body's energy is expended on active transport of substances at the cellular level. The study of active transport carries us into the realms of microbiology, biophysics, and biochemistry and is a fascinating application of the laws of nature to living structures.

Biological Application

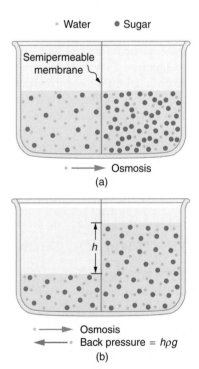

• Water   • Sugar

Semipermeable membrane

→ Osmosis
(a)

h

→ Osmosis
← Back pressure = $h\rho g$
(b)

Figure 11.22   (a) Two sugar-water solutions of different concentrations, separated by a semipermeable membrane that passes water but not sugar. Osmosis will be to the right, since water is less concentrated there. (b) The fluid level rises until the back pressure $h\rho g$ equals the relative osmotic pressure; then, the net transfer of water is zero.

SUMMARY

Fluid dynamics is the physics of moving fluids. **Flow rate** $\mathcal{F}$ is defined to be the volume V flowing past a point in time t, or

$$\mathcal{F} = \frac{V}{t} \qquad (11.1)$$

where V is volume and t time. The SI unit of volume is m³. Another common unit is the **liter** (L), which is 10^{-3} m³. Flow rate and velocity are related by

$$\mathcal{F} = A\bar{v} \qquad (11.2)$$

where A is the cross-sectional area of the flow and $\bar{v}$ is its average velocity. For incompressible fluids, flow rate at various points is constant. That is,

$$\left.\begin{array}{c} \mathcal{F}_1 = \mathcal{F}_2 \\ A_1\bar{v}_1 = A_2\bar{v}_2 \end{array}\right\} \qquad (11.3)$$

or

Bernoulli's equation states that the sum on each side of the following equation is constant, or the same at any two points in an incompressible, frictionless fluid:

$$P_1 + \tfrac{1}{2}\rho v_1^2 + \rho g h_1 = P_2 + \tfrac{1}{2}\rho v_2^2 + \rho g h_2 \qquad (11.4)$$

Bernoulli's principle is Bernoulli's equation applied to situations in which depth is constant. The terms involving height subtract out, yielding

$$P_1 + \tfrac{1}{2}\rho v_1^2 = P_2 + \tfrac{1}{2}\rho v_2^2 \qquad (11.5)$$

Power in fluid flow is given by

$$(P + \tfrac{1}{2}\rho v^2 + \rho g h)\cdot \mathcal{F} = \text{power} \qquad (11.6)$$

where the first term is power associated with pressure, the second is power associated with velocity, and the third is power associated with height.

Laminar flow is characterized by smooth flow of the fluid in layers that do not mix. **Turbulence** is characterized by eddies and swirls that mix layers of fluid together. Fluid **viscosity** η is due to friction within a fluid. Representative values are given in Table 11.1.

Flow is proportional to pressure difference, and inversely proportional to resistance:

$$\mathcal{F} = \frac{P_2 - P_1}{R} \qquad (11.8)$$

For laminar flow in a tube, **Poiseuille's law for resistance** states that

$$R = \frac{8\eta l}{\pi r^4} \qquad (11.9)$$

Poiseuille's law for flow in a tube is

$$\mathcal{F} = \frac{(P_2 - P_1)\pi r^4}{8\eta l} \qquad (11.10)$$

Pressure drop caused by flow and resistance is given by

$$P_2 - P_1 = R\cdot \mathcal{F} \qquad (11.11)$$

The **Reynolds number** N_R can reveal whether flow is laminar or turbulent. It is

$$N_R = \frac{2\rho vr}{\eta} \quad \text{(flow in tube)} \qquad (11.12)$$

For N_R below about 2000, flow is laminar. For N_R above about 3000, flow is turbulent. For values of N_R between 2000 and 3000, it may be either or both. When an object moves in a fluid, a different form of the Reynolds number

$$N_R' = \frac{\rho vL}{\eta} \quad \text{(object in fluid)} \qquad (11.13)$$

indicates whether flow is laminar or turbulent. For N_R' less than about 1, flow is laminar. For N_R' greater than 10^6, it is entirely turbulent.

Diffusion is the movement of substances due to random thermal molecular motion. The average distance x_{rms} a molecule travels by diffusion is given by

$$x_{rms} = \sqrt{2Dt} \qquad (11.14)$$

where D is the diffusion constant, representative values of which are found in Table 11.2. **Osmosis** is the transport of water through a semipermeable membrane from a region of high concentration to a region of low concentration. **Dialysis** is the transport of any other molecule through a semipermeable membrane due to its concentration difference. Both processes can be reversed by back pressure. **Active transport** is a process in which a living membrane expends energy to move substances across it.

CONCEPTUAL QUESTIONS

11.1 What is the difference between flow rate and fluid velocity? How are they related?

11.2 Many figures in the text show streamlines. Explain why fluid velocity is greatest where streamlines are closest together. (Hint: Consider the relationship between fluid velocity and the cross-sectional area through which it flows.)

11.3 You can squirt water a considerably greater distance by placing your thumb over the end of a garden hose than by leaving it completely uncovered. Explain how this works.

11.4 Water is shot nearly vertically upward in a decorative fountain and the stream is observed to broaden as it rises. Conversely, a stream of water falling straight down from a faucet narrows. Explain why, and discuss whether surface tension enhances or reduces the effect in each case.

11.5 Look back to Figure 11.4. Explain two things: Why is P_o less than atmospheric? Why is P_o greater than P_i?

11.6 Based on Bernoulli's equation, what are three forms of energy in a fluid? (Note that these forms are conservative, unlike heat and other dissipative forms not included in Bernoulli's equation.)

11.7 Give an example of entrainment not mentioned in the text.

11.8 Many entrainment devices have a constriction, called a Venturi, such as shown in Figure 11.23. How does this bolster entrainment?

11.9 Some chimney pipes have a T-shape, with a crosspiece on top that helps draw up gases whenever there is even a slight breeze. Explain how this works in terms of Bernoulli's principle.

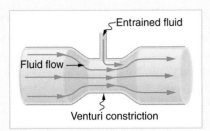

Figure 11.23 A tube with a narrow segment designed to enhance entrainment is called a Venturi. These are very commonly used in carburetors and aspirators. Question 8.

11.10 Is there a limit to the height to which an entrainment device can raise a fluid? Explain.

11.11 Why is it preferable for airplanes to take off into the wind rather than with the wind?

11.12 Roofs are sometimes pushed off vertically during a hurricane, and buildings sometimes explode outward when hit by a tornado. Use Bernoulli's principle to explain.

11.13 Why does a sailboat need a keel?

11.14 It is dangerous to stand close to railroad tracks when a rapidly moving commuter train passes. Explain why atmospheric pressure would push you toward the moving train.

11.15 Water that has emerged from a hose into the atmosphere has a gauge pressure of zero. Why? When you put your hand in front of the emerging stream you feel a force, yet the water's gauge pressure is zero. Explain where the force comes from in terms of energy.

11.16 The old rubber boot shown in Figure 11.24 has two leaks. To what maximum height can the water squirt from leak 1? How does the velocity of water emerging from leak 2 differ from that of leak 1? Explain your responses in terms of energy.

Figure 11.24 Water emerges from two leaks in an old boot. Question 16.

11.17 Explain why the viscosity of a liquid decreases with temperature—that is, how might increased temperature reduce the effects of cohesive forces in a liquid? Also explain why the viscosity of a gas increases with temperature—that is, how does increased gas temperature create more collisions between atoms and molecules?

11.18 Water pressure inside a hose nozzle can be less than atmospheric due to the Bernoulli effect. Explain in terms of energy how the water can emerge from the nozzle against the opposing atmospheric pressure.

11.19 When paddling a canoe upstream, it is wisest to travel as near shore as possible. When canoeing downstream, it may be best to stay near the middle. Explain why.

11.20 Why does flow decrease in your shower when someone flushes the toilet?

11.21 Plumbing usually includes air-filled tubes near faucets, as shown in Figure 11.25. Explain why they are needed and how they work.

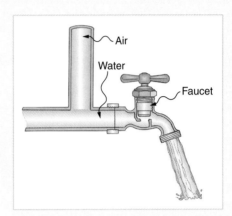

Figure 11.25 The vertical tube near the faucet remains full of air and serves a useful purpose. Question 21.

11.22 Ultrasound can be used to measure blood velocity in the body. If there is a partial constriction of an artery, where would you expect blood velocity to be greatest, at or nearby the constriction? What are the two distinct causes of higher resistance in the constriction?

11.23 Sink drains often have a device such as that shown in Figure 11.26 to help speed the flow of water. How does this work?

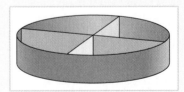

Figure 11.26 You will find devices such as this in many drains. They significantly increase flow rate. Question 23.

11.24 Some ceiling fans have decorative wicker reeds on their blades. Discuss whether these fans are as quiet and efficient as those with smooth blades.

11.25* What direction will a helium balloon move inside a car that is slowing down—toward the front or back? Explain.

11.26* Will identical raindrops fall more rapidly in 5°C air or 25°C air, neglecting any differences in air density? Explain your answer.

11.27* Why would you expect the rate of diffusion to increase with temperature? Can you give an example, such as the fact that you can dissolve sugar more rapidly in hot water?

11.28* How are osmosis and dialysis similar? How do they differ?

PROBLEMS

Section 11.1 Flow Rate and Velocity

11.1 What is the average flow rate in cm^3/s of gasoline to the engine of a car traveling at 60.0 miles per hour if it averages 30.0 miles per gallon?

11.2 The heart of a resting adult pumps blood at a rate of 5.00 L/min. (a) Convert this to cm^3/s. (b) What is this in m^3/s?

11.3 (a) During drought conditions, the American River is 200 ft across and averages 10.0 ft in depth, carrying a flow of 500 ft^3/s. What is the average velocity of the river? (b) At flood stage, the river carries 100,000 ft^3/s and is 350 ft across at one point. What is its average depth if it flows at an average velocity of 15.0 ft/s?

• 11.4 A supertanker can hold 3.00×10^5 m^3 of liquid (nearly 300,000 tons of crude oil). (a) How long would it take to fill the tanker if you could divert a small river flowing at 2500 ft^3/s into it? (b) How long for the same river at a flood stage flow of 100,000 ft^3/s?

• 11.5 A new residential swimming pool with a capacity of 20,000 gallons is being filled with garden hoses delivering a total flow of 15.0 gallons per minute. (a) How long does it take to fill? (b) How long would it take to fill if you could divert a moderate size river, flowing at 50,000 ft^3/s, into it?

• 11.6 The flow rate of blood through a 2.00×10^{-6} m radius capillary is 3.80×10^{-9} cm^3/s. (a) What is the blood velocity? (This small velocity allows time for diffusion of materials to and from the blood.) (b) Assuming all the blood in the body passes through capillaries, how many of them must there be to carry a total flow of 90.0 cm^3/s? (The large number obtained is an overestimate, but it is still reasonable.)

• 11.7 (a) What is the fluid velocity in a 9.00 cm diameter fire hose carrying 80.0 liters of water per second? (b) What is the flow rate in gallons per minute?

• 11.8 The main uptake air duct of a forced air furnace is 0.300 m in diameter. What is the average speed of air in the duct if it carries a volume equal to that of the house's interior every 15 min? The inside volume of the house is equivalent to a rectangular solid 13.0 m wide by 20.0 m long by 2.75 m high.

• 11.9 Verify that the fluid velocity in Example 11.7 actually is 1.70 m/s as stated.

• 11.10 Water is moving at a velocity of 2.00 m/s through a hose with an internal diameter of 1.60 cm. (a) What is the flow rate in liters per second? (b) Fluid velocity in this hose's nozzle is 15.0 m/s. What is the nozzle's inside diameter?

‡ 11.11 Prove that the speed of an incompressible fluid through a constriction, such as in a Venturi tube, increases by a factor equal to the square of the factor by which the diameter decreases. (The converse applies for flow out of a constriction into a larger-diameter region.)

‡ 11.12 Water emerges straight down from a 1.80 cm diameter faucet at a speed of 0.500 m/s. (Because of the construction of the faucet, there is no variation in speed across the stream.) (a) What is the flow rate in cm^3/s? (b) What is the diameter of the stream 0.200 m below the faucet? Neglect any effects due to surface tension.

Sections 11.2 and 11.3 Bernoulli's Equation

11.13 Verify that pressure has units of energy per unit volume.

11.14 Suppose you have a wind speed gauge like the Pitot tube shown in Figure 11.7(b). By what factor must wind speed increase to double h on the manometer? Is this independent of the moving fluid and the fluid in the manometer?

11.15 If the pressure reading of your Pitot tube is 15.0 mm Hg at a speed of 200 km/h, what will it be at 700 km/h at the same altitude?

11.16 Calculate the maximum height to which water could be squirted with the hose in Example 11.2 if it: (a) Emerges from the nozzle. (b) Emerges with the nozzle removed, assuming the same flow rate.

• 11.17 Every few years, winds in Boulder, Colorado, attain sustained speeds of 45.0 m/s (about 100 miles per hour) when the jet stream descends during early spring. Approximately what is the force due to the Bernoulli effect on a roof having an area of 220 m^2? Typical air density in Boulder is 1.14 kg/m^3, and the corresponding atmospheric pressure is 8.89×10^4 N/m^2. (Bernoulli's principle as stated in the text assumes laminar flow. Using the principle here produces only an approximate result, because there is significant turbulence.)

• 11.18 Calculate the approximate force on a square meter of sail, given the horizontal velocity of the wind is 6.00 m/s parallel to its front surface and 3.50 m/s along its back surface. Take the density of air to be 1.29 kg/m^3. (The calculation, based on Bernoulli's principle, is approximate due to the effects of turbulence.)

• 11.19 (a) What is the pressure drop due to the Bernoulli effect as water goes into a 3.00 cm diameter nozzle from a 9.00 cm diameter fire hose while carrying a flow of 40.0 L/s? (b) To what maximum height above the nozzle can this water rise? (The actual height will be significantly smaller due to air resistance.)

• 11.20 Hoover Dam generates electricity with water taken from a depth of 150 m and with an average flow rate of 650 m^3/s. (a) Calculate the power in this flow. Note that the velocity and height of the water are not changed appreciably by the generators, but its pressure is reduced to atmospheric. (b) What is the ratio of this power to the facility's average of 450 MW?

‡ 11.21 A frequently quoted rule of thumb in aircraft design is that wings should produce about 1000 N of lift per square meter of wing. (The fact that a wing has a top and bottom surface does not double its area.) (a) At takeoff the aircraft travels at 60.0 m/s, so that the air velocity relative to the bottom of the wing is 60.0 m/s. Given the sea level density of air to be 1.29 kg/m^3, how fast must it move over the upper surface to create the ideal lift? (b) How fast must air move over the upper surface at a cruising speed of 245 m/s and at an altitude where air density is one-fourth that at sea level? Note that this is not all of the aircraft's lift—some comes from the body of the plane, some from engine thrust, and so on. Furthermore, Bernoulli's principle gives an approximate answer because flow over the wing creates turbulence.

‡ 11.22 (a) Using Bernoulli's equation, show that the measured fluid velocity v for a Pitot tube, like the one in Figure 11.7(b), is given by

$$v = \left(\frac{2\rho' g h}{\rho} \right)^{1/2}$$

where h is the height of the manometer fluid, ρ' the density of the manometer fluid, ρ the density of the moving fluid, and g the acceleration of gravity. (Note that v is indeed proportional to the square root of h, as stated in the text.) (b) Calculate v for moving air if a mercury manometer's h is 0.200 m.

11.23 The left ventricle of a resting adult's heart pumps blood at a flow rate of 83.0 cm³/s, increasing its pressure by 110 mm Hg, its velocity from zero to 30.0 cm/s, and its height by 5.00 cm. (All numbers are averaged over the entire heartbeat.) Calculate the total power output of the left ventricle. Note that most of the power is used to increase blood pressure.

11.24 A sump pump is draining a flooded basement at the rate of 0.750 L/s, with an output pressure of 3.00×10^5 N/m². (a) The water enters a hose with a 3.00 cm inside diameter and rises 2.50 m above the pump. What is its pressure at this point? (b) The hose goes over the foundation wall, losing 0.500 m in height, and widens to 4.00 cm diameter. What is the pressure now? You may neglect frictional losses in both parts of the problem.

Section 11.4 Viscosity and Poiseuille's Law

11.25 (a) Calculate the retarding force due to the viscosity of the air layer between a cart and a level air track given the following information—air temperature is 20°C, the cart is moving at 0.400 m/s, its surface area is 2.50×10^{-2} m², and the thickness of the air layer is 6.00×10^{-5} m. (b) What is the ratio of this force to the weight of the 0.300 kg cart?

11.26 What force is needed to pull one microscope slide over another at a speed of 1.00 cm/s, if there is a 0.500 mm thick layer of 20°C water between them and the contact area is 8.00 cm²?

11.27 A glucose solution being administered with an IV has a flow rate of 4.00 cm³/min. What will the new flow rate be if the glucose is replaced by whole blood having the same density but a viscosity 2.50 times that of the glucose? All other factors remain constant.

11.28 Fluid originally flows through a tube at a rate of 100 cm³/s. To illustrate the sensitivity of flow rate to various factors, calculate the new flow rate for the following changes with all other factors remaining the same as in the original conditions. (a) Pressure difference increases by a factor of 1.50. (b) A new fluid with 3.00 times greater viscosity is substituted. (c) The tube is replaced by one having 4.00 times the length. (d) Another tube is used with a radius 0.100 times the original. (e) Yet another tube is substituted with a radius 0.100 times the original and half the length, *and* the pressure difference is increased by a factor of 1.50.

11.29 Suppose a blood vessel's radius is decreased to 90.0% of its original value by plaque deposits and the body compensates by increasing the pressure difference along the vessel to keep the flow rate constant. By what factor must the pressure difference increase?

11.30 The flow rate of blood in a coronary artery has been reduced to half its normal value by plaque deposits. By what factor has the radius of the artery been reduced?

11.31 A 1.50 mm thick layer of oil is placed between two microscope slides. It is found that a force of 5.50×10^{-4} N is required to glide one over the other at a speed of 1.00 cm/s when their contact area is 6.00 cm². What is the oil's viscosity? What type of oil might it be?

11.32 (a) Verify that a 19.0% decrease in laminar flow through a tube is caused by a 5.00% decrease in radius, assuming that all other factors remain constant, as stated in the text. (b) What increase in flow is obtained from a 5.00% increase in radius, again assuming all other factors remain constant?

11.33 Example 11.6 dealt with the flow of saline solution in an IV system. (a) Verify that a pressure of 1.62×10^4 N/m² is created at a depth of 1.61 m in a saline solution, assuming its density to be that of sea water. (b) Calculate the new flow rate if the height of the saline solution is decreased to 1.50 m. (c) At what height would the direction of flow be reversed? (This can be a problem when patients stand up.)

11.34 Water supplied to a house by a water main has a pressure of 3.00×10^5 N/m² early on a summer day when neighborhood use is low. This pressure produces a flow of 20.0 L/min through a garden hose. Later in the day, pressure at the exit of the water main and entrance to the house drops, and a flow of only 8.00 L/min is obtained through the same hose. (a) What pressure is now being supplied to the house, assuming resistance is constant? (b) By what factor did the flow rate in the water main increase in order to cause this decrease in delivered pressure? The pressure at the entrance of the water main is 5.00×10^5 N/m², and the original flow rate was 200 L/min. (c) How many more users are there, assuming each would consume 20.0 L/min in the morning?

11.35 An oil gusher shoots crude oil 25.0 m into the air through a 0.100 m diameter pipe. Neglecting air resistance but not the resistance of the pipe, and assuming laminar flow, calculate the pressure at the entrance of the 50.0 m long vertical pipe. Take the density of the oil to be 900 kg/m³ and its viscosity to be 1.00 (N/m²)·s. Note that you must take into account the pressure due to the 50.0 m column of oil in the pipe.

11.36 Concrete is pumped from a cement mixer to the place it is being laid, instead of being carried in wheelbarrows. The flow rate is 200 L/min through a 50.0 m long, 8.00 cm diameter hose, and the pressure at the pump is 8.00×10^6 N/m² (about 1200 psi). (a) Calculate the resistance of the hose. (b) What is the viscosity of the concrete, assuming the flow is laminar? (c) How much power is being supplied, assuming the point of use is at the same level as the pump? You may neglect the power supplied to increase the concrete's velocity.

Sections 11.5* and 11.6* Turbulence

11.37 Verify that the flow of oil is laminar (barely) for the oil gusher in Problem 11.35.

11.38 Show that the Reynolds number N_R is unitless by plugging in units for all the quantities in its definition and canceling.

11.39 Calculate the Reynolds numbers for flow in the hose and nozzle considered in Example 11.2 to determine if the flow in either can possibly be laminar.

11.40 Calculate the Reynolds numbers for flow in the fire hose and nozzle considered in Example 11.4 to show that the flow in each must be turbulent.

11.41 Verify that the flow of concrete is laminar for the situation considered in Problem 11.36, taking concrete's viscosity to be 48.0 (N/m²)·s, and given its density is 2300 kg/m³.

11.42 At what flow rate might turbulence begin to develop in a 0.200 m diameter water main? Assume a 20°C temperature.

: 11.43* Gasoline is piped underground from refineries to major users. The flow rate is 3.00×10^{-2} m³/s (about 500 gal/min), the viscosity of gasoline is 1.00×10^{-3} (N/m²)·s, and its density is 680 kg/m³. (a) What minimum diameter must the pipe have if the Reynolds number is to be less than 1000? (b) What pressure difference must be maintained along each kilometer of the pipe to maintain this flow rate?

: 11.44* Assuming that blood is an ideal fluid, calculate the average flow rate at which turbulence is a certainty in the aorta. Take the diameter of the aorta to be 2.50 cm. (Turbulence will actually occur at lower average flow rates, because blood is not an ideal fluid. Furthermore, since blood flow is pulsatile, turbulence may occur during only the high-velocity part of each heartbeat.)

Section 11.7* Diffusion

11.45* You can smell perfume very shortly after opening the bottle. To show that it is not reaching your nose by diffusion, calculate the average distance a perfume molecule moves in one second in air, given its diffusion constant D to be 1.00×10^{-6} m²/s.

11.46* What is the ratio of the average distances that oxygen will diffuse in a given time in air and water? Why is the distance less in water (equivalently, why is D less in water)?

• 11.47* Oxygen reaches the veinless cornea of the eye by diffusing through its 0.500 mm thick tear layer. How long does it take the average oxygen molecule to do this?

UNREASONABLE RESULTS

The following problems have results that are unreasonable because some assumption or premise is unreasonable, or because certain of the premises are inconsistent with one another. Physical principles applied correctly then produce unreasonable results. The purpose of these problems is to give practice in assessing whether nature is being accurately described, and to trace the source of difficulty if it is not.

Note: Strategies for dealing with problems having unreasonable results appear in Chapter 10, among others. They can be found with the section of problems labeled *Unreasonable Results* at the end of that chapter.

11.48 A mountain stream is 10.0 m wide and averages 2.00 m in depth. During the spring runoff, the flow in the stream reaches 100,000 m³/s. (a) What is the average velocity of the stream under these conditions? (b) What is unreasonable about this velocity? (c) What is unreasonable or inconsistent about the premises?

: 11.49 A fairly large garden hose has an internal radius of 0.600 cm (about 5/8 in. outside diameter) and a length of 23.0 m (about 75 ft). The nozzleless horizontal hose is attached to a faucet, and it delivers 5.00 L/s. (a) What water pressure is supplied by the faucet? Convert the pressure to lb/in². (b) What is unreasonable about this pressure? (c) What is unreasonable about the premise? (d) What is the Reynolds number for the given flow?

TEMPERATURE, KINETIC THEORY, AND THE GAS LAWS

12

The welder's electric arc creates enough heat to melt the rod, spray sparks, and burn the retina of an unprotected eye. Its heat can be felt on exposed skin a few meters away, and its light can be seen for kilometers.

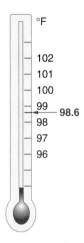

Figure 12.1 A typical mercury thermometer. The mercury expands and contracts more rapidly than the glass containing it. When the thermometer's temperature increases, mercury from the bulb is forced into the narrow tube, producing a large change in the length of the column for a small change in temperature.

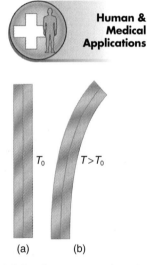

Figure 12.2 The curvature of a bimetallic strip depends on temperature. (a) The strip is straight at the starting temperature, where its two components have the same length. (b) At a higher temperature, this strip bends to the right because the metal on the left has expanded more than its companion.

Figure 12.3 (a) The thermistor (shown exposed in this enlargement) has replaced the mercury thermometer at many medical facilities. Its electrical properties depend sensitively and reproducibly on temperature. (b) Liquid crystals in the numbers on this plastic strip change color at certain temperatures, in this case indicating a normal internal temperature.

Heat is a form of energy transfer very familiar to us. We feel the warmth of the summer sun and the chill of a clear summer night; we heat coffee and drink it to warm up after a winter stroll; we sweat to keep cool. Manifestations of heat are apparent throughout the universe. The earth's slowly cooling interior sometimes creates lava flows, suggesting it is still hot within. The sun warms the earth's surface and is the source of much of the energy we find on it. Increasing atmospheric carbon dioxide threatens to trap more of the sun's heat, perhaps fundamentally altering the ecosphere. In space, supernovas explode, briefly radiating more heat than an entire galaxy.

Just what is heat? How is it related to temperature? How is it transferred, and what are its effects? Can it be converted freely and completely to other forms of energy and to work? Three chapters are devoted to these and many other questions. Again, we shall find in answering these questions that in spite of the many and varied phenomena involved, there is a small set of underlying physical principles that unite the subjects and tie them to other fields.

12.1 TEMPERATURE

The concept of temperature has evolved from the common concepts of hot and cold. Our human perception of what feels hot or cold is a relative one. For example, suppose you place one hand in hot water and the other in cold, and after a while place both in tepid water. The tepid water will feel cool to the hand that was in hot water and warm to the hand that was in cold water. The scientific definition of temperature is less ambiguous than your senses of hot and cold. **Temperature** is operationally defined to be what we measure with a thermometer.* Two thermometers placed in the hot and cold water would show the hot to have a higher temperature. If both were then placed in the tepid water, both would give identical readings (within measurement uncertainties). In this section, we discuss temperature, its measurement by thermometers, and its relationship to thermal equilibrium.

Any physical property that depends on temperature, and whose response to temperature is reproducible, can be used as the basis of a thermometer. Since many physical properties depend on temperature, the variety of thermometers is truly remarkable. For example, volume increases with temperature for most substances. The common mercury thermometer is based on this property. (See Figure 12.1.) So is the bimetallic strip. (See Figure 12.2.) Other properties used to measure temperature include electrical resistance and color, as shown in Figure 12.3, and the emission of infrared radiation, as shown in Figure 12.4.

---

*This sounds like a joke, but it is a serious statement. Many physical quantities are defined solely in terms of how they are measured. We shall see later how temperature is related to the kinetic energies of atoms and molecules, a more physical explanation.

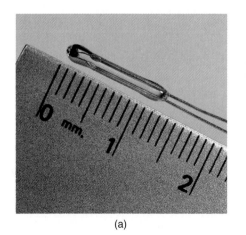

(a)

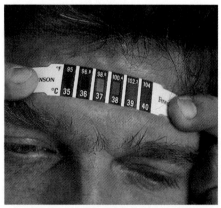

(b)

Human & Medical Applications

Temperature Scales

The idea of a thermometer raises the issue of temperature units and scales. A temperature scale can be created by identifying two easily reproducible temperatures, commonly the freezing and boiling temperatures of water at standard atmospheric pressure, and assigning values to those two points. For example, the **Fahrenheit** scale (still the most frequently used in the United States) has the freezing and boiling points of water at 32°F and 212°F, respectively. The unit of temperature on this scale is the *Fahrenheit degree* (°F). The rest of the world uses the **Celsius** scale (which replaced the slightly different *centigrade* scale), which has the freezing and boiling points of water at 0°C and 100°C. Its unit is the *Celsius degree* (°C). Note that the Celsius degree is larger than the Fahrenheit degree by a factor of 9/5, since only 100 Celsius degrees span the same range as 180 Fahrenheit degrees (180/100 = 9/5).

A temperature scale in common scientific use is the **Kelvin** scale. The Kelvin scale is an *absolute temperature* scale defined to have 0 K at the lowest possible temperature, called **absolute zero**. The official temperature unit on this scale is the *kelvin*, abbreviated K, with no degree sign. The kelvin is exactly the same size as the Celsius degree. Note that the freezing and boiling points of water are 273.15 K and 373.15 K, respectively. Those points are 100 kelvin apart and 100 Celsius degrees apart.

The concept of absolute zero arose from the behavior of gases. Figure 12.5 shows how the volume of gases decreases as temperature is lowered. When this behavior was first accurately measured, it was not yet possible to produce extremely low temperatures. It was noted that the volumes of all gases extrapolate to zero at the same temperature, −273.15°C. This implied that there is a temperature below which it is impossible to go. This temperature was called absolute zero. Today we know that gases first liquefy and then freeze before reaching absolute zero, so that their volume does not vanish. But we have found that there is an absolute lowest temperature. At that temperature, all the energy it is possible to remove from the atoms and molecules in a system has been extracted. It is the same temperature for all systems. Absolute zero is −273.15°C or 0 K.

The relationships between the three common temperature scales are shown in Figure 12.6, and a few interesting temperatures are shown in Figure 12.7. Temperatures on these three scales can be converted to one another using the following equations:

$$T \,(°F) = \tfrac{9}{5}T \,(°C) + 32 \tag{12.1a}$$

$$T \,(°C) = \tfrac{5}{9}[T \,(°F) - 32] \tag{12.1b}$$

$$T \,(K) = T \,(°C) + 273.15 \tag{12.1c}$$

where *T* is the symbol for temperature. The Kelvin scale, the only absolute scale of the

Figure 12.4 This patient is having her temperature taken by a device inserted into the ear. Infrared radiation (the emission of which varies with temperature) from the ear canal is measured and a temperature readout is quickly produced.

CONNECTIONS

Atomic and molecular characteristics explain and underlie the macroscopic characteristics of materials, as noted in other chapters. We shall explore these submicroscopic explanations in more detail later. See, for example, the discussion of kinetic theory in Section 12.4.

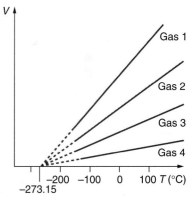

Figure 12.5 Graph of volume versus temperature for various gases kept at a constant pressure. Note that all of the graphs extrapolate to zero at the same temperature.

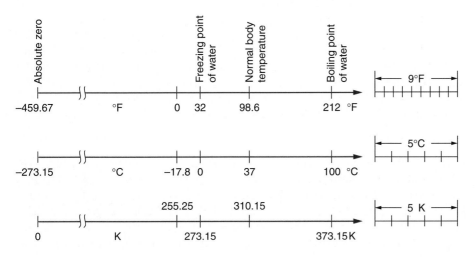

Figure 12.6 Relations among the Fahrenheit, Celsius, and Kelvin temperature scales, rounded to the nearest degree. The relative sizes of the degrees are also shown.

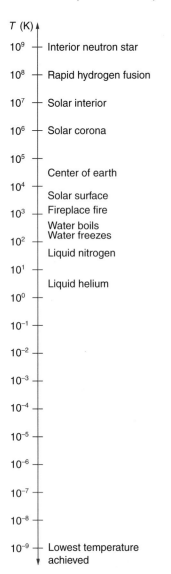

T (K)

10^9	Interior neutron star
10^8	Rapid hydrogen fusion
10^7	Solar interior
10^6	Solar corona
10^5	
	Center of earth
10^4	
	Solar surface
10^3	Fireplace fire
	Water boils
10^2	Water freezes
	Liquid nitrogen
10^1	
	Liquid helium
10^0	
10^{-1}	
10^{-2}	
10^{-3}	
10^{-4}	
10^{-5}	
10^{-6}	
10^{-7}	
10^{-8}	
10^{-9}	Lowest temperature achieved

Figure 12.7 This logarithmic scale illustrates the tremendous range of temperatures in nature. Note that zero on a logarithmic scale would occur off the bottom of the page at infinity.

three, is used extensively in scientific work since a number of physical quantities, such as the volume of an ideal gas, are directly related to absolute temperature.

Thermal Equilibrium and the Zeroth Law of Thermodynamics

Can we take it for granted that a thermometer measures the temperature of any object it contacts? Thermometers really take their own temperature—whatever property we measure to determine temperature is a property of the thermometer itself. But we can be confident that the thermometer measures the temperature of, say, the inside of your mouth, because of a basic law verified by experiment. Any two systems placed in thermal contact (meaning heat can be transferred between them) will reach the same temperature. That is, if one is hotter, it will cool, and the other will warm until they have exactly the same temperature. The objects are then in **thermal equilibrium**, and no further changes will take place. Something profound is happening here: the systems interact and change because their temperatures differ, and the changes stop once their temperatures are the same. Thus the temperature a thermometer registers *does* represent the system with which it is in thermal equilibrium.

Furthermore, experiment has shown that if two systems, say A and B, are in equilibrium with one another, and B is in equilibrium with a third system, C, then A is also in equilibrium with C. This may seem obvious, since all three have the same temperature, but it is also an experimental fact basic to all of thermodynamics. It is called the **zeroth law of thermodynamics**. The odd name for this law came about because it was accepted as a law after the first and second laws of thermodynamics had been developed and given their numbers. Yet this law on equilibrium logically comes before the first and second laws, and it is thus called the zeroth law.

12.2 THERMAL EXPANSION OF SOLIDS AND LIQUIDS

The expansion of mercury in a thermometer is but one of many commonly encountered examples of **thermal expansion**, the change in size with temperature. Hot air rises because its volume has increased, lowering its density compared with that of surrounding air. The same happens in all liquids and gases, driving natural convection in homes, oceans, and weather systems. Solids also undergo thermal expansion. Railroad tracks and bridges, for example, have expansion joints to allow them to freely expand and contract with temperature changes. (See Figure 12.8.)

What are the basic properties of thermal expansion? First, it is clearly related to temperature change. The greater the temperature change, the more a bimetallic strip will bend, for example. Second, it depends on the material. Mercury, for example, expands more as a function of temperature than does glass.

Why does thermal expansion occur at all? The explanation is found on the submicroscopic scale. As noted in Section 10.1, atoms and molecules vibrate and move within a substance. They thus have kinetic energy. With increasing temperature, the atoms and molecules gain kinetic energy, colliding with their neighbors more forcibly, and stretch the average distance between them. (Something like stretching the springs in Figure 10.1.) For a solid, this implies that size will increase by a certain amount per atom along all dimensions. Thus a large object will have a greater change in length than a small one, because it has more atoms. That is, the change in length ΔL is proportional to length L. The dependence of thermal expansion on temperature, substance, and length is summarized in the equation for the amount of **linear thermal expansion**:

$$\Delta L = \alpha L \, \Delta T \qquad (12.2)$$

Here ΔL is the change in length L, ΔT is the change in temperature, and α is the **coefficient of linear expansion**, which varies slightly with temperature. Table 12.1 lists representative values of α, which may have units of 1/°C or 1/K. (Since the size of the kelvin and the degree Celsius are the same, both α and ΔT can be expressed in terms of either kelvins or degrees Celsius.) Equation 12.2 is quite accurate for small changes in temperature and can be used for large changes in temperature if an average value of α is used.

Figure 12.8 Thermal expansion joints on bridges allow them to change length without buckling.

TABLE 12.1

THERMAL EXPANSION COEFFICIENTS AT 20°C

Material	Coefficient of linear expansion α (1/°C)	Coefficient of volume expansion β (1/°C)
Solids		
Aluminum	25×10^{-6}	75×10^{-6}
Brass	19×10^{-6}	56×10^{-6}
Copper	17×10^{-6}	51×10^{-6}
Gold	14×10^{-6}	42×10^{-6}
Iron or steel	12×10^{-6}	35×10^{-6}
Invar	0.9×10^{-6}	2.7×10^{-6}
Lead	29×10^{-6}	87×10^{-6}
Silver	18×10^{-6}	54×10^{-6}
Glass (ordinary)	9×10^{-6}	27×10^{-6}
Glass (Pyrex)	3×10^{-6}	9×10^{-6}
Quartz	0.4×10^{-6}	1×10^{-6}
Concrete, brick	$\sim 12 \times 10^{-6}$	$\sim 36 \times 10^{-6}$
Marble (average)	2.5×10^{-6}	7.5×10^{-6}
Liquids*		
Ether		1650×10^{-6}
Ethyl alcohol		1100×10^{-6}
Gasoline		950×10^{-6}
Glycerin		500×10^{-6}
Mercury		180×10^{-6}
Water		210×10^{-6}
Gases*		
Air and most other gases at atmospheric pressure		3400×10^{-6}

*Values for liquids and gases are approximate.

EXAMPLE 12.1   A VISIBLE THERMAL EXPANSION

The main span of the Golden Gate bridge is 1275 m long when at its coldest. The bridge is subjected to temperature extremes ranging from −15°C to 40°C. What is its change in length between these two temperatures?

Strategy The change in length is given by Equation 12.2. We use the coefficient of linear expansion for steel from Table 12.1, and we note that the change in temperature is 55°C.

Solution Substituting α for steel and the given informa- tion for length and temperature into Equation 12.2 yields

$$\Delta L = \alpha L \, \Delta T$$
$$= (12 \times 10^{-6}/°C)(1275 \text{ m})(55°C)$$
$$= 0.84 \text{ m}$$

Discussion While not large compared with the length of the bridge, this change in length is observable. Generally it is spread over many expansion joints, so that the movement at each is small.

Objects expand along all dimensions, as illustrated in Figure 12.9. That is, their areas and volumes increase with temperature as well as their lengths. Interestingly, this means that *holes* also get larger with temperature! If you cut a hole in a metal plate, the re- maining material will expand exactly as it would if the plug were still in place. The plug

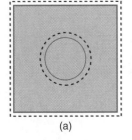
(a)

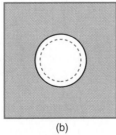
(b)

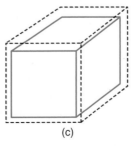
(c)

Figure 12.9 Objects expand in all directions as temperature increases. (a) Area increases because both length and width increase. The area of a cir- cular plug also increases. (b) If the plug is removed, the hole it leaves gets larger with increasing temperature, just as if the expanding plug were still in place. (c) Volume also increases, since all three dimensions increase.

would get bigger, and so the hole must get bigger, too. For small temperature changes, the change in area ΔA is given by

$$\Delta A = 2\alpha A \, \Delta T \tag{12.3}$$

The change in volume ΔV is very nearly $\Delta V = 3\alpha V \, \Delta T$. It is traditional to write this as

$$\Delta V = \beta V \, \Delta T \tag{12.4}$$

where β is the **volume coefficient of expansion** and $\beta \approx 3\alpha$. (You will note that the values of β in Table 12.1 are very nearly, but not always exactly, equal to 3α.)

There are interesting effects that occur because different materials expand differently with temperature. If you have ever noticed gasoline dripping from a freshly filled gas tank on a hot day, you have observed the effects of differential expansion. Gasoline starts out at ground temperature and cools the steel tank when it is filled. Both expand as they warm to air temperature, but gasoline expands much more than steel and so it may overflow. Just how much do you lose? The next example gives some idea.

EXAMPLE 12.2 HOW MUCH MORE DOES THE GASOLINE EXPAND THAN THE TANK?

Suppose your 60.0 L steel gas tank is full of gasoline, both having a temperature of 15.0°C. How much gasoline has spilled by the time they warm to 35.0°C?

Strategy and Concept Both the tank and the gasoline increase in volume, but the gasoline increases more. So the amount spilled is the difference in their changes in volume. We treat the gas tank as if it were solid steel, because it expands as if it were.

Solution The increase in volume of the steel tank is

$$\Delta V_{\text{S}} = \beta_{\text{S}} V_{\text{S}} \, \Delta T$$

The increase in volume of the gasoline is

$$\Delta V_{\text{gas}} = \beta_{\text{gas}} V_{\text{gas}} \, \Delta T$$

The amount spilled is the difference in volumes:

$$V_{\text{spill}} = \Delta V_{\text{gas}} - \Delta V_{\text{S}}$$

Each of these changes is given by Equation 12.4. Noting that the original volumes are equal and gathering terms gives

$$
\begin{aligned}
V_{\text{spill}} &= (\beta_{\text{gas}} - \beta_{\text{S}}) V \, \Delta T \\
&= [(950 - 35) \times 10^{-6}/\text{°C}](60.0 \text{ L})(20.0\text{°C}) \\
&= 1.10 \text{ L}
\end{aligned}
$$

Discussion This is a significant amount, particularly for a 60.0 L tank.

If you try to just cap the tank tightly so it cannot overflow, you will find that it leaks anyway, either around the gas cap or by bursting the tank! Tightly constricting the expanding gasoline is equivalent to compressing it, and both liquids and solids resist being compressed with extremely large forces. That is why an air gap is left over liquids in rigid containers—it allows them to expand and contract without stressing the container.

Thermal Stress

Thermal stress is stress created by thermal expansion or contraction. (See Section 5.8 for a discussion of stress and strain.) Thermal stress can be destructive, as when expanding gasoline ruptures a gas tank. It can instead be useful, such as in manufacturing when two parts are joined together by heating one, slipping it over the other, and allowing it to cool. Thermal stress can in fact explain many phenomena, such as the weathering of rocks and pavement by the expansion of ice when it freezes.

EXAMPLE 12.3 HOW LARGE IS THIS THERMAL STRESS?

What pressure would be created in the gas tank considered in Example 12.2 if the gasoline increases in temperature from 15.0°C to 35.0°C without being allowed to expand?

Strategy To solve this problem we must refer back to Equation 5.15, which relates a change in volume ΔV to pressure:

$$\Delta V = \frac{1}{B}\frac{F}{A} V_0$$

where F/A is pressure, V_0 is the original volume, and B is the bulk modulus of the material involved.

Solution We solve the previous equation for pressure:

$$P = \frac{F}{A} = \frac{\Delta V}{V_0} B$$

Then we substitute known values to answer the question

(continued)

(continued)

posed. The bulk modulus for gasoline is found from Table 5.1 to be 1.00×10^9 N/m². The change in volume is the amount that would spill, namely $\Delta V = 1.10$ L, as found in the previous example. The original volume V_0 is what the gasoline would become if allowed to expand freely—this turns out to be $V_0 = 61.14$ L (verification left as an end-of-chapter problem). Substituting these

values, we obtain

$$P = \frac{1.10 \text{ L}}{61.14 \text{ L}}(1.00 \times 10^9 \text{ N/m}^2) = 1.8 \times 10^7 \text{ N/m}^2$$

Discussion This pressure is about 2500 lb/in², *much* more than a gas tank can handle.

Forces and pressures created by thermal stress are typically as great as the one in the example. Railroad tracks and roadways can buckle in the heat if they lack sufficient expansion joints. (See Figure 12.10.) Power lines sag more in the summer than in the winter and will snap in the winter if there is insufficient slack. Cracks open and close in plaster walls as a house warms and cools. Glass cooking pans will crack if cooled rapidly and unevenly because of differential contraction and the stresses it creates. (Pyrex is less susceptible because of its small coefficient of thermal expansion.) More ominously, nuclear reactor pressure vessels are threatened by overly rapid cooling. None has failed, but several have been cooled faster than is desirable. Cells are ruptured when foods are frozen, detracting from their taste. Repeated thawing and freezing accentuate the damage. Even the oceans can be affected. A portion of the rise in sea level during global warming is due to the thermal expansion of sea water.

Biological Application

Figure 12.10 This crack in a concrete sidewalk was created by thermal stress, an indication of how great such stress can be.

12.3 THE IDEAL GAS LAW

In this section we continue to explore the thermal behavior of gases. In particular, we will examine the submicroscopic* behavior of the atoms and molecules that compose gases.

We noted in Section 10.1 that gases are easily compressed. We can see evidence of this in Table 12.1, where you will note that gases have the largest βs in the table. This means that they expand and contract very rapidly with temperature changes. Additionally, you will note that most gases expand at the *same* rate, having the same β. Why should gases all behave in nearly the same way, when liquids and solids have widely varying expansion rates?

The best answers are found on the submicroscopic scale. Atoms and molecules in gases are typically separated by distances that are large compared with their sizes. At typical separation distances in gases, as illustrated in Figure 12.11, the forces between atoms and molecules are weak, and they essentially do not interact except during collisions. Thus gases expand rapidly because the atomic attraction is weak. Furthermore, their behavior is relatively independent of the details of the forces between atoms (since the forces are weak) and, therefore, independent of the type of gas. In short, the thermal expansion of gases is simple; it depends mostly on temperature and the number of atoms and molecules per unit volume. In contrast, in liquids and solids, atoms and molecules are closer together and are quite sensitive to the details of the forces between them.

To get some idea of how the pressure, temperature, and volume of a gas are related to one another, let us consider what happens when you put air into an initially deflated tire. The tire's volume first increases in direct proportion to the amount of air injected without increasing tire pressure. Once the tire has expanded to nearly full size, the walls of the tire limit its volume from increasing much more. If we continue to put air in, the volume does not increase much, but the pressure does. The more air, the greater the pressure. Once inflated, the pressure also increases with temperature—most manufacturers specify optimal tire pressure when tires are cold. (See Figure 12.12.)

The following expression describes the behavior of the tire—and much more. The **ideal gas law** states that

$$PV = NkT \tag{12.5}$$

Figure 12.11 Atoms and molecules in a gas are typically widely separated as shown. Since the forces between them are quite weak at these distances, the properties of a gas depend more on the number of atoms per unit volume and on temperature than they do on the type of atom.

---

*Since atoms and molecules are too small to be observed with a microscope, we refer to objects on this scale as submicroscopic.

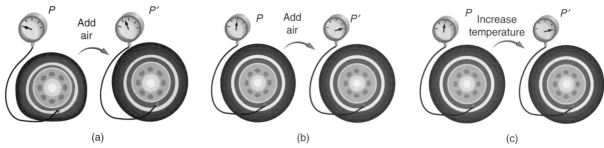

Figure 12.12 (a) When air is pumped into a deflated tire, its volume first increases without much increase in pressure. (b) When the tire is filled to a certain point, the tire walls resist further expansion and the pressure increases with more air. (c) Once the tire is inflated, its pressure increases with temperature.

where P is the absolute pressure* in a gas, V the volume it occupies, N the number of atoms and molecules in the gas, and T its absolute temperature. The constant k is called the **Boltzmann constant** and has the value

$$k = 1.38 \times 10^{-23} \text{ J/K} \tag{12.6}$$

The ideal gas law is so named because it describes the behavior of an **ideal gas**—defined to be one for which the volume occupied by its atoms is a negligible fraction of V, and for which the interactions of its atoms are negligible except for the exchange of energy and momentum during collisions. The remarkable thing about the ideal gas law is that it works very well for most real gases under most conditions,[†] because their atoms and molecules are separated by distances large compared with their sizes, as discussed.

Let us see how the ideal gas law is consistent with filling the tire just described. At first, tire pressure P is constant and equal to atmospheric, and k and T are also constant, so that from Equation 12.5 we expect V to increase in direct proportion to the number of atoms and molecules N put into the tire. (It is a reasonable assumption that the temperature is constant if the tire is pumped up slowly.) Once V is constrained to be constant, the equation predicts that P should rise in proportion to N. Later on when the temperature T does increase, while V and N stay the same, the equation implies P should increase.

EXAMPLE 12.4 TIRE PRESSURE INCREASES WITH TEMPERATURE

Suppose your bicycle tire is fully inflated with an absolute pressure of 7.00×10^5 N/m² (a gauge pressure just under 90.0 lb/in²) at a temperature of 18.0°C. What will its pressure be after its temperature has risen to 35.0°C? There are no appreciable leaks nor changes in volume.

Strategy We are asked to find the pressure P, which can be calculated from Equation 12.5 given enough information. At first, it may seem that not enough information is given since the volume V and number of atoms N are not specified. But these remain constant. If we take a ratio of the ideal gas law before and after the temperature increase, we get them to cancel:

$$\frac{P'\cancel{V'}}{P\cancel{V}} = \frac{\cancel{N'}k T'}{\cancel{N}kT}$$

Primed quantities represent conditions after the temperature

has risen. This leaves us with

$$P' = P\frac{T'}{T}$$

where the temperatures must be in kelvins, since T and T' are absolute temperatures.

Solution Substituting $T = (18.0 + 273)$ K = 291 K, and $T' = (35.0 + 273)$ K = 308 K into the last equation yields

$$P' = (7.00 \times 10^5 \text{ N/m}^2)\frac{308 \text{ K}}{291 \text{ K}} = 7.41 \times 10^5 \text{ N/m}^2$$

Discussion The final temperature is about 6% greater than the original temperature, and so the final pressure is about 6% greater. Note that *absolute* pressure and *absolute* temperature *must* be used in the ideal gas law.

Just how many atoms and molecules are there in a typically encountered body, such as the gas in a tire or the water in a drink? We can use the ideal gas law to give us an idea of just how large N typically is.

---

*See Section 10.6 for definitions of absolute and gauge pressures.
[†]Note, for example, that N is the total number of atoms and molecules, completely independent of the type of gas.

EXAMPLE 12.5 HOW MANY ATOMS AND MOLECULES ARE THERE IN A VOLUME OF GAS?

Calculate the number of atoms and molecules in a cubic meter of gas at standard temperature and pressure (STP), defined to be 0°C and atmospheric pressure.

Strategy Since pressure, volume, and temperature are all specified, we can use the ideal gas law to find N.

Solution Solving Equation 12.5 for N and substituting known values gives

$$N = \frac{PV}{kT} = \frac{(1.01 \times 10^5 \text{ N/m}^2)(1.00 \text{ m}^3)}{(1.38 \times 10^{-23} \text{ J/K})(273 \text{ K})}$$

$$= 2.68 \times 10^{25}$$

Discussion This is an undeniably large number, especially considering that a gas is mostly empty space. N is huge even in small volumes. For example, 1 cm³ has 2.68×10^{19} atoms and molecules in it. Once again, note that N is the same for all types or mixtures of gases.

Moles and Avogadro's Number

It is sometimes convenient to work with another unit for counting the number of atoms and molecules, the mole. A **mole** (abbreviated mol) is defined to be the amount of a substance that contains as many elementary particles (atoms or molecules) as there are atoms in exactly 0.012 kg of carbon-12. The actual number of atoms or molecules in one mole is called **Avogadro's number** (N_A) in recognition of Amedeo Avogadro (1776–1856), an Italian scientist. Avogadro developed the concept of the mole based on his hypothesis that equal volumes of gas at the same pressure and temperature contain equal numbers of molecules. Today, the value of Avogadro's number is known to be

$$N_A = 6.02 \times 10^{23}/\text{mol} \tag{12.7}$$

One mole always contains 6.02×10^{23} particles (atoms or molecules), whatever the element or substance. Figure 12.13 illustrates how big a mole is. A mole of any substance has a mass in grams equal to its molecular mass, which can be determined from the periodic table of elements.

CONNECTIONS

A precise value for Avogadro's number was not determined until early this century, when a theory of Einstein's was used by others to determine the size and masses of atoms. This will be discussed in Chapter 28.

EXAMPLE 12.6 MOLES PER CUBIC METER AND LITERS PER MOLE

Calculate: (a) the number of moles in 1.00 m³ of gas at STP; (b) the number of liters of gas per mole.

Strategy and Solution for (a) We are asked to find the number of moles per cubic meter, and we know from the previous example that the number of atoms or molecules per cubic meter is 2.68×10^{25}. So the number of moles can be found by dividing this number by Avogadro's number. We let n stand for the number of moles; thus,

$$n/\text{m}^3 = (2.68 \times 10^{25}/\text{m}^3)\left(\frac{1.00 \text{ mol}}{6.02 \times 10^{23}}\right)$$

$$= 44.5 \text{ mol/m}^3$$

Strategy and Solution for (b) Using the value obtained for the number of moles in a cubic meter, inverting it, and converting cubic meters to liters, we obtain

$$\frac{1.00 \text{ m}^3}{44.5 \text{ mol}} (10^3 \text{ L/m}^3) = 22.5 \text{ L/mol}$$

Discussion This value is very close to the accepted value of 22.4 L/mol. The slight difference is due to rounding errors caused by using three-digit input.

The Ideal Gas Law Restated Using Moles

A very common expression of the ideal gas law uses the number of moles, n, rather than the number of atoms and molecules, N. It is easily obtained starting with the ideal gas law,

$$PV = NkT$$

and converting N into the number of moles. To do this, multiply and divide the equation

Figure 12.13 How big is a mole? (a) One mole of water is only 18 mL, less than an ounce, but contains 6.02×10^{23} water molecules. (b) Even a tiny bacterium contains 10^{11} to 10^{12} molecules, but this is only a fraction of a mole. (c) One mole of anything macroscopic, like this mole of table tennis balls covering the earth to a depth of about 35 km, is a huge amount.

(a)

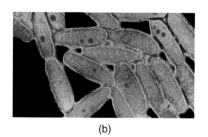

(b)

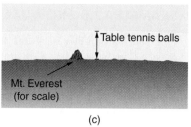

(c)

by Avogadro's number N_A, yielding

$$PV = \frac{N}{N_A} N_A kT$$

Note that $N/N_A = n$, the number of moles, since N is the number of atoms and N_A is the number of atoms per mole. We define a new quantity, $R = N_A k$, and we obtain a common expression of the **ideal gas law in terms of moles**:

$$PV = nRT \qquad \qquad \textbf{(12.8)}$$

where n is the number of moles and R is the **universal gas constant**—so called because its value is the same for all gases. To obtain a numerical value for R, we multiply N_A by k:

$$R = N_A k = (6.02 \times 10^{23}/\text{mol})(1.38 \times 10^{-23} \text{ J/K})$$

This gives R in SI units as

$$R = 8.31 \text{ J/mol} \cdot \text{K}$$

In other units, R is

$$R = 1.99 \text{ cal/mol} \cdot \text{K}$$

and

$$R = 0.0821 \text{ L} \cdot \text{atm/mol} \cdot \text{K}$$

You can use whichever value of R is most convenient. We shall find many uses for the ideal gas law in both its forms.

EXAMPLE 12.7   MOLES IN A BIKE TIRE

How many moles of gas are in the bike tire considered in Example 12.4, given that its volume is $2.00 \times 10^{-3} \text{ m}^3$ (2.00 L)?

Strategy We can solve the ideal gas law, as stated in Equation 12.8, for the number of moles.

Solution Solving for n and substituting known values yields

$$n = \frac{PV}{RT} = \frac{(7.00 \times 10^5 \text{ N/m}^2)(2.00 \times 10^{-3} \text{ m}^3)}{(8.31 \text{ J/mol} \cdot \text{K})(291 \text{ K})}$$

$$= 0.579 \text{ mol}$$

Discussion The most convenient value for R in this case is 8.31 J/mol · K because our knowns are in SI units. The pressure and temperature were obtained from the initial conditions in Example 12.4, but we would get the same answer if we used the final values.

CONNECTIONS

The ideal gas law can be considered to be another manifestation of the law of conservation of energy. Work done on a gas results in an increase in its energy, increasing pressure, volume, or temperature (internal kinetic energy).

The Ideal Gas Law and Energy

Let us now examine the role of energy in the behavior of gases. When you pump up a bike tire by hand, you do noticeable work, exerting a force through a distance repeatedly. Where does that energy go? Part of it can go to heating the pump and air, and part to increasing pressure.

The ideal gas law is closely related to energy. Note that the units on both sides are joules. The right-hand side of the ideal gas law in Equation 12.5 is NkT. This term is roughly the amount of kinetic energy of N atoms or molecules at an absolute temperature T. The left-hand side of the ideal gas law is PV, which also has units of joules. (We know from our study of fluids that pressure is one type of potential energy per unit volume, and so pressure times volume is energy.) The meaning is that there is energy in a gas related to both its pressure and volume. That energy can be changed in various ways including removing it to do work—something we will explore in detail in Chapter 14—such as in gasoline or steam engines, turbines, and so on.

FOR THE IDEAL GAS LAW

Step 1. *Examine the situation to determine that an ideal gas is involved.* Most gases are nearly ideal in behavior. If the situation involves a phase change, the gas is usually *not* ideal.

Step 2. *Identify exactly what needs to be determined in the problem (identify the unknowns).* A written list is useful.

Step 3. *Determine whether the number of molecules is known or the number of moles, in order to decide which form of the ideal gas law to use.* The first form is Equation 12.5 and involves *N*, the number of atoms or molecules. The second is Equation 12.8 and involves *n*, the number of moles.

Step 4. *Make a list of what is given or can be inferred from the problem as stated (identify the knowns).*

Step 5. *Solve the ideal gas law for the quantity to be determined (the unknown).* You may need to take a ratio of final to initial states to get rid of unknown but unchanging variables.

Step 6. *Substitute the knowns along with their units into the appropriate equation, and obtain numerical solutions complete with units.* Be certain to use absolute temperature and absolute pressure.

Step 7. *Check the answer to see if it is reasonable: Does it make sense?*

12.4 KINETIC THEORY: MOLECULAR EXPLANATION OF PRESSURE AND TEMPERATURE

We have developed macroscopic definitions of pressure and temperature. Pressure is simply force divided by the area on which the force is exerted. Temperature is what we measure with a thermometer. It is interesting and useful to investigate both pressure and temperature on a submicroscopic scale as well, by considering the action of atoms or molecules in a gas. We will, for example, find that pressure and temperature are in fact intimately related. The model we use in investigating the submicroscopic view is **kinetic theory**, which takes matter to be made of atoms and molecules in continuous random motion.

Figure 12.14 shows an elastic collision of a gas molecule with the wall of a container, creating a force on the wall. Huge numbers of molecules will collide with the wall in a short time, creating an average force per unit area. This is the source of pressure in a gas. The greater the number of molecules in the container, the higher the pressure. Similarly, the greater the average velocity of the molecules, the greater the pressure. The actual relationships are derived in the accompanying Things Great and Small essay by looking at the small scale in some detail. The following relationship results:

$$PV = \tfrac{1}{3}Nm\overline{v^2} \tag{12.9}$$

where *P* is the pressure (average force per unit area), *V* is the volume of gas in the container, *N* is the number of molecules in the container, *m* is the mass of the molecule, and $\overline{v^2}$ is the average of the molecular velocity squared. Equation 12.9 is another expression for the ideal gas law, but now in terms of submicroscopic properties.

What can we learn from this submicroscopic version of the ideal gas law? One thing we can now do is to derive a precise relationship between temperature and the average kinetic energy of molecules in a gas. Recall the previous expression of the ideal gas law in Equation 12.5:

$$PV = NkT$$

Equating the right-hand side of this equation with the right-hand side of Equation 12.9 gives

$$\tfrac{1}{3}Nm\overline{v^2} = NkT$$

CONNECTIONS

As noted in previous chapters, atomic and molecular characteristics explain and underlie the macroscopic world. This entire section is devoted to such explanations.

Figure 12.14 When a molecule collides with a rigid wall, the component of its momentum perpendicular to the wall is reversed. A force is thus exerted on the wall, creating pressure.

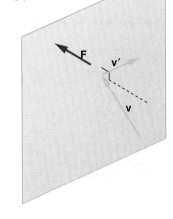

THINGS GREAT AND SMALL

Submicroscopic Origin of Pressure in a Gas

Figure 12.A shows a box with a gas exerting a pressure on the walls. We know from our previous discussions that putting more gas in the box means greater pressure, and that increasing the temperature also increases pressure. But why should this be so? A look on the submicroscopic scale gives us some answers and produces an alternative expression for the ideal gas law.

Also shown in Figure 12.A is an expanded view of an elastic collision of a gas molecule with the wall of a container. Calculating the average force exerted by such molecules will lead us to the ideal gas law and to the precise connection between temperature and molecular kinetic energy. The assumption is made (consistent with all previous discussions and supported by experimental evidence) that the molecule is small compared with the average separation of molecules in the gas, and that its interaction with other molecules and the walls is negligible except during direct collisions. We also assume the wall is rigid and that the molecule's direction changes but its speed remains constant (and hence its kinetic energy and the magnitude of its momentum remain constant). This is not always a good assumption, but a more detailed consideration of the molecule's average exchange of energy and momentum with wall molecules gives the identical result as this greatly simplifying assumption.

If the molecule's velocity changes in the x-direction, so does its momentum in that direction—it goes from $-mv_x$ to $+mv_x$. Thus its change in momentum is $\Delta(mv) = +mv_x - (-mv_x) = 2mv_x$. The force exerted on it is given by Newton's second law stated in terms of momentum:

$$F = \frac{\Delta p}{\Delta t} = \frac{2mv_x}{\Delta t}$$

This force will be very large for a very short time, and then zero until the next collision of this molecule as it bounces around in the box. We are looking for an average force; to get it, we take Δt to be the average time between collisions of this molecule with this wall. This average time is just the time it would take the molecule to go across and back (a distance $2l$) at a speed of v_x. Thus $\Delta t = 2l/v_x$, and the expression for the force becomes

$$F = \frac{2mv_x}{2l/v_x} = \frac{mv_x^2}{l}$$

This is the force due to one molecule. To get the total force, we multiply by the number of molecules N and use their average squared velocity, yielding

$$F = N\frac{m\overline{v_x^2}}{l}$$

where the bar over a quantity means its average value. We would like to have the force in terms of the total velocity v,

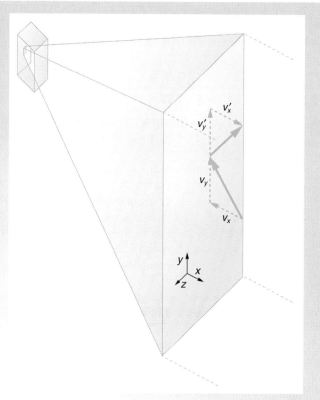

Figure 12.A Gas in a box exerts an outward pressure on its walls. A molecule colliding with a rigid wall has the direction of its velocity and momentum in the x-direction reversed. This direction is perpendicular to the wall. The components of its velocity momentum in the y- and z-directions are not changed. This means there is no force parallel to the wall.

rather than just its x-component. This is easily accomplished by noting that the total velocity squared is the sum of the squares of its components, so that

$$\overline{v^2} = \overline{v_x^2} + \overline{v_y^2} + \overline{v_z^2}$$

Since the velocities are random, their average components in all directions are the same:

$$\overline{v_x^2} = \overline{v_y^2} = \overline{v_z^2}$$

Thus,

$$\overline{v^2} = 3\overline{v_x^2}$$

or

$$\overline{v_x^2} = \tfrac{1}{3}\overline{v^2}$$

Substituting this into the expression for F gives

$$F = N\frac{m\overline{v^2}}{3l}$$

Pressure is F/A, so that we will divide this by A, and noting that $Al = V$, the volume of the box, we obtain

$$P = \frac{F}{A} = N\frac{m\overline{v^2}}{3Al} = \frac{1}{3}\frac{Nm\overline{v^2}}{V}$$

giving the important result

$$PV = \tfrac{1}{3}Nm\overline{v^2} \qquad (12.9)$$

This is another expression of the **ideal gas law**.

We can get the average kinetic energy of a molecule, $(1/2)m\overline{v^2}$, from the left-hand side of the equation by canceling N and multiplying by $3/2$. This produces the result that the average kinetic energy of a molecule is directly related to absolute temperature. The resulting equation is

$$\overline{\text{KE}} = \tfrac{1}{2}m\overline{v^2} = \tfrac{3}{2}kT \tag{12.10}$$

The average kinetic energy of a molecule, $\overline{\text{KE}}$, is called **thermal energy**. Equation 12.10 is a submicroscopic interpretation of temperature, and it has been found to be valid for gases and reasonably accurate in liquids and solids. It is another definition of temperature related to submicroscopic energy.

It is sometimes useful to rearrange Equation 12.10, solving for the average speed of molecules in a gas in terms their temperature. This yields

$$v_{\text{rms}} = \sqrt{\frac{3kT}{m}} \tag{12.11}$$

where v_{rms} stands for root mean square (rms) speed. This is a particular type of average for a statistical process.

EXAMPLE 12.8 ENERGY AND SPEED OF A GAS MOLECULE

(a) What is the average kinetic energy of a gas molecule at 20.0°C (room temperature)? (b) Find the rms speed of a nitrogen molecule (N_2) at this temperature.

Strategy for (a) The average kinetic energy is given in Equation 12.10. Before substituting values into this equation, we must convert the given temperature to kelvins. This gives $T = (20.0 + 273)$ K $= 293$ K.

Solution for (a) The temperature alone is sufficient to find the average kinetic energy. Substituting the temperature gives

$$\overline{\text{KE}} = \tfrac{3}{2}kT = \tfrac{3}{2}(1.38 \times 10^{-23} \text{ J/K})(293 \text{ K})$$
$$= 6.07 \times 10^{-21} \text{ J}$$

Strategy for (b) Finding the rms speed of a nitrogen molecule is a straightforward calculation using Equation 12.11, but we must first find the mass of a nitrogen molecule. Using the molecular mass of nitrogen from the periodic table,

$$m = \frac{2(14.0067) \times 10^{-3} \text{ kg/mol}}{6.02 \times 10^{23}/\text{mol}} = 4.65 \times 10^{-26} \text{ kg}$$

Solution for (b) Substituting this mass and the value for k into Equation 12.11 yields

$$v_{\text{rms}} = \sqrt{\frac{3kT}{m}} = \sqrt{\frac{3(1.38 \times 10^{-23} \text{ J/K})(293 \text{ K})}{4.65 \times 10^{-26} \text{ kg}}}$$
$$= 511 \text{ m/s}$$

Discussion Note that the average kinetic energy of the molecule is completely independent of the type of mole-

cule. Average kinetic energy depends only on the absolute temperature. The kinetic energy in joules is very small, not enough to trigger any of our senses. The rms velocity of the nitrogen molecule is surprisingly large. It is difficult to imagine that air molecules are moving nearly this fast when we watch still air or perhaps wait for the smell of perfume to waft our way. The mean free path (the distance a molecule can move on average between collisions) of molecules in air is very small, and so the molecules bounce madly about, not getting very far per second. (See the discussion of diffusion in Section 11.7, for example.) The high value for rms speed is reflected in the speed of sound, however, which is about 340 m/s at room temperature. The faster the rms speed of air molecules, the faster sound vibrations can be transferred through the air. Consistently, the speed of sound increases with temperature and is greater in gases with small molecular masses, like helium. (See Figure 12.15.)

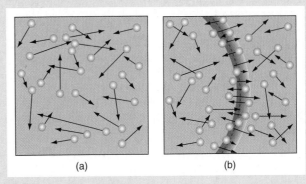

(a)   (b)

Figure 12.15 (a) There are so many molecules moving so fast in an ordinary gas that they bounce furiously about, engaging in billions of collisions per second. (b) Individual molecules do not move very far in a small time, but disturbances such as sound waves are transmitted at speeds related to the molecular speeds.

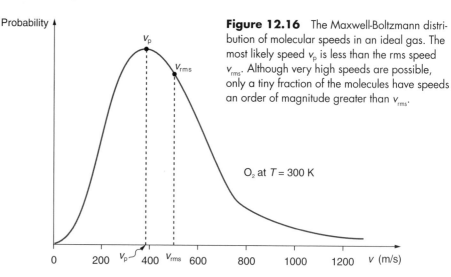

Figure 12.16 The Maxwell-Boltzmann distribution of molecular speeds in an ideal gas. The most likely speed v_p is less than the rms speed v_{rms}. Although very high speeds are possible, only a tiny fraction of the molecules have speeds an order of magnitude greater than v_{rms}.

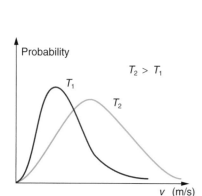

Figure 12.17 The Maxwell-Boltzmann distribution is shifted to higher speeds and is broadened at higher temperatures.

Distribution of Molecular Speeds

The motion of molecules in a gas is random in magnitude and direction for individual molecules, but a gas of many molecules has a predictable orderly distribution of molecular speeds. This is called the Maxwell-Boltzmann distribution after its originators, who calculated it based on kinetic theory. (It has since been confirmed experimentally.) (See Figure 12.16.) The distribution has a long tail on the high-speed side, since a few molecules may go several times the rms speed. The most probable speed v_p is less than the rms speed v_{rms}. Figure 12.17 shows the effect of temperature on the distribution. The curve is shifted to higher speeds at higher temperatures, with a broader range of speeds.

EXAMPLE 12.9   FAST ENOUGH TO ESCAPE?

The escape velocity from the earth's gravity is 11.1 km/s at an altitude of 100 km, near the top of the atmosphere. Anything having this speed in an upward direction will escape the earth's gravity entirely. At what temperature would helium atoms have an rms speed equal to the escape velocity?

Strategy We can solve either Equation 12.10 or 12.11 for T, yielding

$$T = \frac{m\overline{v^2}}{3k}$$

where v is the escape velocity, 11.1 km/s, k is the Boltzmann constant, and m is the mass of a helium atom, which we must calculate. As in the previous example, we find m by using the molecular mass of helium from the periodic table. This gives

$$m = \frac{4.0026 \times 10^{-3} \text{ kg/mol}}{6.02 \times 10^{23}/\text{mol}} = 6.65 \times 10^{-27} \text{ kg}$$

Solution Entering these values into the expression for T yields

$$T = \frac{(6.65 \times 10^{-27} \text{ kg})(11.1 \times 10^{3} \text{ m/s})^2}{3(1.38 \times 10^{-23} \text{ J/K})}$$

$$= 1.98 \times 10^{4} \text{ K}$$

Discussion This temperature is much higher than atmospheric temperature, which is on the order of 250 K at high altitude. There are very few helium atoms left in the atmosphere, but it originally had many. The reason is that even at normal temperatures there are a small number of helium atoms with speeds higher than the earth's escape velocity. These escape and others that gain such speeds also escape, eventually removing nearly all light atoms and molecules (such as hydrogen) from the atmosphere. Heavier molecules, like oxygen, nitrogen, and water (very little of which reaches very high altitude), have smaller rms speeds, and so it is less likely that any of them will escape. In fact, so few have speeds above the escape velocity that billions of years are required to lose significant amounts of the atmosphere. Figure 12.18 shows the lack of an atmosphere on the moon. Since gravity is weaker on the moon, the escape velocity is less. This is the topic of an end-of-chapter problem.

Figure 12.18 Unlike the earth, there is no atmosphere on the moon. This is because gravity is less on the moon and escape velocity is consequently lower there. The black sky on the moon shows there is no gas to scatter sunlight.

12.5 PHASE CHANGES

To this point we have considered the behavior of ideal gases. Real gases behave like ideal gases provided they are not close to the temperature where they change phase,* perhaps condensing to a liquid or freezing to a solid. Figure 12.19 shows how a real gas behaves like an ideal gas, until its temperature is lowered to the region where it begins to condense to a liquid. When a gas liquefies, its molecules become very close together in a liquid state. There is thus a dramatic decrease in volume, as seen in the figure. Cooled further, it goes from liquid to solid, and its volume continues to decrease with temperature, but it never reaches zero because of the finite volume of its molecules.

High pressure may also cause a gas to change phase to a liquid. Carbon dioxide, for example, is a gas at room temperature and atmospheric pressure, but it becomes a liquid under sufficiently high pressure. We can examine aspects of the behavior of a substance by plotting a graph of pressure versus volume, called a **PV diagram**. When the substance behaves like an ideal gas, the ideal gas law describes the relationship between its pressure and volume. That is,

$$PV = NkT \qquad \text{(ideal gas)}$$

Now, assuming the number of molecules is fixed, at a given temperature,

$$PV = \text{constant} \qquad \text{(ideal gas, constant temperature)}$$

So, for example, the volume of the gas will decrease as the pressure increases. If you plot the relationship $PV = \text{constant}$ on a PV diagram, you get a hyperbolic curve. Figure 12.20 shows a graph of pressure versus volume. The hyperbolic curves represent ideal gas behavior at various fixed temperatures and are thus called *isotherms*. At lower temperatures, the curves begin to look less like hyperbolas—the gas is not behaving ideally and may even contain liquid. There is a **critical point**—that is, a **critical temperature**—above which liquid cannot exist. At sufficiently high pressure above the critical point, the gas will have the density of a liquid but will not exhibit surface tension and will not remain in an open container. Carbon dioxide, for example, cannot be liquefied at a temperature above 31.0°C. **Critical pressure** is the minimum pressure needed for liquid to exist at the critical temperature. Table 12.2 lists representative critical temperatures and pressures.

*The three major phases of matter and their submicroscopic structure were first discussed in Section 10.1.

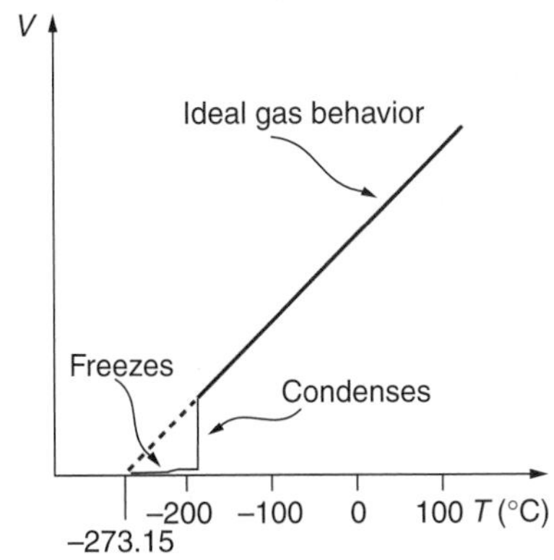

Figure 12.19 A graph of volume versus temperature for a real gas at constant pressure. The straight part of the graph represents ideal gas behavior—it extrapolates to zero volume at −273.15°C, or absolute zero. The volume actually drops precipitously at the liquefaction point and decreases slightly after that, but it never becomes zero.

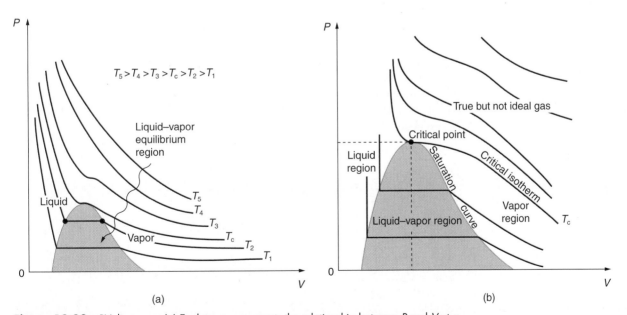

Figure 12.20 PV diagrams. (a) Each curve represents the relationship between P and V at a fixed temperature. The upper curves are at higher temperatures. The lower curves are not hyperbolic, because the gas is no longer an ideal gas. (b) An expanded portion of the PV diagram for low temperatures where there is the possibility of a partial or total phase change to liquid.

TABLE 12.2

CRITICAL TEMPERATURES AND PRESSURES				
	Critical temperature		*Critical pressure*	
Substance	*K*	*°C*	*N/m²*	*atm*
Water	647.4	374.3	22.12×10^6	219.0
Sulfur dioxide	430.7	157.6	7.88×10^6	78.0
Ammonia	405.5	132.4	11.28×10^6	111.7
Carbon dioxide	304.2	31.1	7.39×10^6	73.2
Oxygen	154.8	−118.4	5.08×10^6	50.3
Nitrogen	126.2	−146.9	3.39×10^6	33.6
Hydrogen	33.3	−239.9	1.30×10^6	12.9
Helium	5.3	−267.9	0.229×10^6	2.27

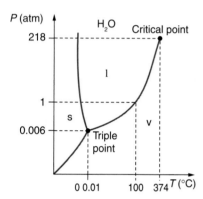

Figure 12.21 The phase diagram (*PT* graph) for water. Note that the axes are nonlinear and the graph is not to scale. This graph is simplified— there are several other exotic phases of ice at higher pressures.

Phase Diagrams

Another type of graph that provides considerable insight is a plot of pressure versus temperature. There are well-defined regions on such a graph that correspond to various phases of matter, and so *PT* graphs are called **phase diagrams**. Figure 12.21 shows the phase diagram for water, with various shaded areas indicating different phases. If you know the pressure and temperature, you can determine from the graph what the phase of water will be. The solid lines between phases are points where the phases are in equilibrium. For example, at 1.00 atm the boiling point of water is 100°C. The boiling temperature rises steadily to 374°C at a pressure of 218 atm. The curve ends at that point, the *critical point*, because at higher temperatures the liquid phase does not exist at any pressure. (The critical point occurs at the critical temperature, as you can see for water from Table 12.2.) The critical temperature for oxygen is −118°C, and so any attempt to liquefy oxygen above this temperature is futile.

Similarly, the curve between the solid and liquid regions in Figure 12.21 gives the melting temperature at various pressures. For example, the melting point is 0°C at 1.00 atm, as expected. Note that at a fixed temperature you can change the phase from ice to water by increasing pressure. Water melts from pressure in the hands of a snowball maker, for example. It is interesting that at low pressure there is no liquid phase, but water can exist as either gas or solid. (For water this is true at any pressure below 0.00600 atm.) The phase change from solid directly to gas is called **sublimation**. Sublimation accounts for large losses of snow pack that never make it into a river, for the routine automatic defrosting of a freezer, and the freeze-drying process applied to so many foods.

All three curves on the phase diagram meet at a single point called the **triple point**, where all three phases exist in equilibrium. This is called a point because the three phases are in equilibrium only in a very narrow range about it. For water, the triple point occurs at 273.16 K and is a more accurate calibration temperature than the melting point of water at 1.00 atm, or 273.15 K. See Table 12.3 for triple point values of other substances.

Equilibrium

Liquid and gas phases are in equilibrium at the boiling temperature. (See Figure 12.22.) If a single substance is in a closed container at the boiling point, then the liquid is boiling and the gas is condensing at the same rate, so that there is no net change in their relative amounts. Molecules in the liquid escape to become a gas at the same rate that molecules in the gas strike the liquid and stick, or form drops and stay liquid. The combination of temperature and pressure has to be just right; they must be at a point on the curve between liquid and gas in a phase diagram. For example, if the temperature and pressure are both increased by just the right amount, then equilibrium is maintained because both boiling and condensation rates increase by the same amount.

TABLE 12.3

TRIPLE POINT TEMPERATURES AND PRESSURES				
	Temperature		*Pressure*	
Substance	*K*	*°C*	*N/m²*	*atm*
Water	273.16	0.01	6.10×10^2	0.00600
Carbon dioxide	216.55	−56.60	5.16×10^5	5.11
Sulfur dioxide	197.68	−75.47	1.67×10^3	0.0167
Ammonia	195.40	−77.75	6.06×10^3	0.0600
Nitrogen	63.18	−210.0	1.25×10^4	0.124
Oxygen	54.36	−218.8	1.52×10^2	0.00151
Hydrogen	13.84	−259.3	7.04×10^3	0.0697

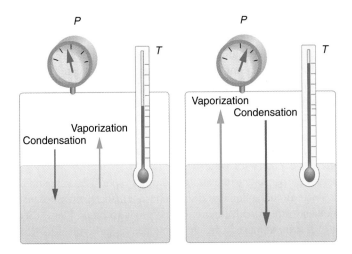

Figure 12.22 Equilibrium between liquid and gas at two different boiling points. (a) The rates of boiling and condensation are equal at this combination of temperature and pressure, and so the liquid and gas phases are in equilibrium. (b) At a higher temperature, the boiling rate increases, but increased pressure can compensate by increasing the condensation rate a like amount. Equilibrium still exists at this higher boiling point.

One example of equilibrium between liquid and gas is that of water and steam at 100°C and 1.00 atm. This is the boiling point, so that they should exist in equilibrium. Why is it that an open pot of water at 100°C boils completely away? One condition mentioned above is not met. The gas surrounding an open pot is not pure water: It is mixed with air. If pure water and steam were in a closed container at 100°C and 1.00 atm, they would coexist. But with air over the pot, there are fewer molecules of water to condense, and so boiling dominates. What about water at 20.0°C and 1.00 atm? This is well into the liquid region, yet an open glass of water at this temperature will completely evaporate. Again, the reason is that the gas around it is air and not pure water vapor, so that even the reduced evaporation rate is greater than the condensation rate of water from dry air. If the glass is sealed, then the liquid phase remains. We call the gas phase a **vapor**, when it exists, as it does for water at 20.0°C, at a temperature below the boiling temperature. (See Figure 12.23.)

Vapor Pressure, Partial Pressure, and Dalton's Law

We define **vapor pressure** to be the gas pressure created by the liquid and solid phases of a substance. Vapor pressure is created by the faster molecules breaking away from the liquid or solid. Vapor pressure depends on both the substance and its temperature— the greater the temperature, the greater the vapor pressure.

Liquid and solid phases are not necessarily stable. If the surrounding gas is not the same substance as the liquid or solid, then molecules that escape the liquid or solid are unlikely to be replaced.

Partial pressure is defined to be the pressure a gas would create if it alone occupied the total volume available. *The total pressure is the sum of partial pressures of the component gases*, assuming ideal gas behavior and no chemical reactions. This is known as **Dalton's law of partial pressures** after the English scientist John Dalton (1766–1844), who first proposed it. Dalton's law is based on kinetic theory—each gas creates its pressure by atomic collisions, independent of the other gases present. (It is also consistent with the fact that pressures add, as implied by Pascal's principle.) Now we can see that water evaporates and ice sublimates when their vapor pressures exceed the partial pressure of water vapor in the surrounding mixture of gases. If their vapor pressures are less than the partial pressure of water vapor in the surrounding gas, then condensation or frost forms.

Figure 12.23 These formations are the result of water evaporating from this lake. Great amounts of water change from liquid to vapor even though the lake's temperature is well below the boiling point. The reason is that the vapor pressure of water is greater than the partial pressure of water in the air over the lake.

12.6 HUMIDITY, EVAPORATION, AND BOILING

"It's not the heat, it's the humidity," is a hackneyed expression, but it makes a valid point. We keep cool in hot weather by evaporating sweat from our skin and water from our breathing passages. Since evaporation is inhibited by high humidity, we feel hotter at a given temperature when humidity is high. Low humidity can also cause discomfort from excessive drying of mucous membranes.

TABLE 12.4

SATURATION VAPOR DENSITY OF WATER

Temperature (°C)	Vapor pressure (N/m^2)	Saturation vapor density (g/m^3)
−50	4.0	0.039
−20	1.04×10^2	0.89
−10	2.60×10^2	2.36
0	6.10×10^2	4.84
5	8.68×10^2	6.80
10	1.19×10^3	9.40
15	1.69×10^3	12.8
20	2.33×10^3	17.2
25	3.17×10^3	23.0
30	4.24×10^3	30.4
37	6.31×10^3	44.0
40	7.34×10^3	51.1
50	1.23×10^4	82.4
60	1.99×10^4	130
70	3.12×10^4	197
80	4.73×10^4	294
90	7.01×10^4	418
95	8.59×10^4	505
100	$\mathbf{1.01 \times 10^5}$	**598**
120	1.99×10^5	1095
150	4.76×10^5	2430
200	1.55×10^6	7090
220	2.32×10^6	10,200

When we say humidity, we really mean *relative humidity*. Relative humidity tells us how much water vapor is in the air compared with the maximum possible. At its maximum, the relative humidity is 100%, and evaporation is inhibited. The amount of water vapor the air can hold depends on the temperature. For example, relative humidity rises in the evening as air temperature declines, sometimes reaching the *dew point*. At the dew point, relative humidity is 100%, and fog may result from the condensation of water droplets. Conversely, if you wish to dry something (perhaps your hair), it is more effective to blow hot air on it than cold. The hot air has a greater capacity for water vapor.

What underlies the capacity of the air to hold water vapor is the vapor pressure of water and ice. Both the liquid and solid phases emanate vapor because there are molecules in them with high enough speeds to break loose. (See Figure 12.24(a).) If a lid is placed over the container, as in Figure 12.24(b), evaporation continues, building up pressure, until there is enough vapor for the reverse process of condensation to balance the evaporation. Then equilibrium would be achieved, and the vapor pressure would equal the partial pressure in the container. Vapor pressure increases with temperature because molecular speeds shift higher with increasing temperature. Table 12.4 gives representative values of water vapor pressure over a range of temperatures.

Relative humidity is related to the *partial pressure** of water vapor in the air. At 100% humidity, the partial pressure is equal to the vapor pressure and there can be no more water in the vapor phase. If the partial pressure is less than the vapor pressure, then evaporation will take place—humidity is less than 100%. If the partial pressure is greater than the vapor pressure, then condensation takes place. The capacity of air to "hold" water vapor is determined by the vapor pressure of water and ice and has nothing to do with the properties of air itself.

EXAMPLE 12.10 DENSITY AND VAPOR PRESSURE

Table 12.4 gives the vapor pressure of water at 20.0°C to be 2.33×10^3 N/m². Using the ideal gas law, calculate the density of water vapor in g/m³ that would create a partial pressure equal to this vapor pressure. Compare the result with the saturation vapor density given in the table.

Strategy and Concept The ideal gas law stated in terms of moles is

$$PV = nRT$$

If we can solve this for n/V, we can calculate the number of moles per cubic meter and convert this to grams per cubic meter as requested.

Solution Solving the ideal gas law for n/V and substituting known values gives

$$\frac{n}{V} = \frac{P}{RT} = \frac{2.33 \times 10^3 \text{ N/m}^2}{(8.31 \text{ J/mol·K})(293 \text{ K})} = 0.957 \frac{\text{mol}}{\text{m}^3}$$

The molecular mass of water is 18.0, and so there are 18.0 g/mol. Thus the density ρ in g/m³ is

$$\rho = \left(0.957 \frac{\text{mol}}{\text{m}^3}\right)\left(\frac{18.0 \text{ g}}{\text{mol}}\right) = 17.2 \text{ g/m}^3$$

Discussion The density found here is obtained by assuming a pressure equal to the vapor pressure of water at 20.0°C. The density found is identical to the value in Table 12.4. That means that a vapor density of 17.2 g/m³ at 20.0°C creates a partial pressure of 2.33×10^3 N/m², equal to the vapor pressure of water at that temperature. If those two pressures are equal, then the liquid and vapor phases are in equilibrium, and the humidity is 100%. Thus at 20.0°C there can be no more than 17.2 g of water vapor per m³, and so this is the saturation vapor density at that temperature. This example also confirms that water vapor behaves like an ideal gas, since the pressure and density are consistent with the ideal gas law (assuming the density in the table is correct). The saturation vapor densities listed in Table 12.4 are the maximum amounts of water vapor that air can hold at various temperatures. Again, these have nothing to do with the properties of air but are directly related to the vapor pressures of water given in the table and can be calculated from them using the ideal gas law.

---

*Defined in Section 12.5 as the pressure a gas would create if it alone occupied the volume.

Figure 12.24 (a) Because of the distribution of speeds and kinetic energies, some water molecules can break away to the vapor phase even at temperatures below the ordinary boiling point. (b) If the container is sealed, evaporation will continue until there is enough vapor density for the condensation rate to equal the evaporation rate. This vapor density and the partial pressure it creates are the saturation values. They increase with temperature and are independent of the presence of other gases, such as air. They depend only on the vapor pressure of water.

(a)   (b)

We define **percent relative humidity** to be the ratio of vapor density to saturation vapor density, or

$$\text{percent relative humidity} = \frac{\text{vapor density}}{\text{saturation vapor density}} \times 100 \quad \textbf{(12.12)}$$

We can use this and the data in Table 12.4 to do a variety of interesting calculations, keeping in mind that relative humidity is based on the comparison of the partial pressure of water vapor in the air and the vapor pressure of water and ice.

EXAMPLE 12.11 HUMIDITY AND DEW POINT

(a) Calculate the percent relative humidity on a day when the temperature is 25.0°C and the air contains 9.40 g of water vapor per m³. (b) At what temperature will this air reach 100% relative humidity? This is the dew point. (c) What is the humidity when the air temperature is 25.0°C and the dew point is −10.0°C?

Strategy and Solution for (a) Percent relative humidity is defined by Equation 12.12 to be the ratio of vapor density to saturation vapor density. The first is given to be 9.40 g/m³, and the second is found in Table 12.4 to be 23.0 g/m³. Thus,

$$\text{percent relative humidity} = \frac{9.40 \text{ g/m}^3}{23.0 \text{ g/m}^3} \times 100 = 40.9\%$$

Strategy and Solution for (b) The air contains 9.40 g/m³ of water vapor. The relative humidity will be 100% at a temperature where 9.40 g/m³ is the saturation density. Inspection of Table 12.4 reveals this to be the case at 10.0°C. So the relative humidity will be 100% at 10.0°C,

and that temperature is called the dew point for this air.

Strategy and Solution for (c) Here the dew point temperature is given to be −10.0°C. From Table 12.4 we see that the vapor density is 2.36 g/m³, because this is the saturation vapor density at −10.0°C. The saturation vapor density at 25.0°C is seen to be 23.0 g/m³. Thus the relative humidity at 25.0°C is

$$\text{percent relative humidity} = \frac{2.36 \text{ g/m}^3}{23.0 \text{ g/m}^3} \times 100 = 10.3\%$$

Discussion The importance of the dew point is that air temperature cannot drop below 10.0°C in part (b), or −10.0°C in part (c), without water vapor condensing out of the air. Considerable heat is released, and this can prevent the temperature from dropping. When dew points are below 0°C, freezing temperatures are a greater possibility, and so farmers and others often keep track of the dew point. This is also the reason that deserts have much greater day to night temperature swings than humid regions.

Why does water boil at 100°C? You will note from Table 12.4 that the vapor pressure of water at 100°C is 1.01×10^5 N/m², or 1.00 atm. Thus it can evaporate without limit at this temperature and pressure. This is why the boiling point is 100°C at atmospheric pressure. But why does it form bubbles when it boils? Interestingly, this is because water ordinarily contains significant amounts of dissolved air and other impurities—attested to by the small bubbles of air you can usually observe in a glass of water. If one of these bubbles starts out attached to the bottom of the container at 20.0°C, then it has some water vapor in it (about 2.30%). The pressure in the bubble is 1.00 atm and must

Biological Application

Figure 12.25 (a) An air bubble in water starts out saturated with water vapor at 20.0°C. (b) As the temperature rises, water vapor enters the bubble because its vapor pressure increases. The bubble expands to keep its pressure at 1.00 atm. (c) At 100°C, water vapor enters the bubble continuously because water's vapor pressure exceeds its partial pressure in the bubble, which must be less than 1.00 atm. The bubble thus grows and rises to the surface.

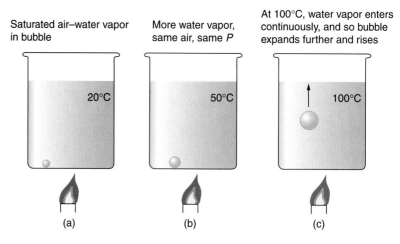

Saturated air–water vapor in bubble

More water vapor, same air, same P

At 100°C, water vapor enters continuously, and so bubble expands further and rises

(a) (b) (c)

remain so (ignoring the slight pressure due to the depth of water). As the bubble's temperature rises, the amount of air in it stays the same but the water vapor increases, and so the bubble must expand to keep its pressure at 1.00 atm. At 100°C, water vapor will enter the bubble continuously, trying to make its partial pressure equal to 1.00 atm. It cannot reach this pressure, since the bubble also contains air and has a total pressure of 1.00 atm. The bubble therefore grows in size, the buoyant force on it increases until it breaks away, and it rises rapidly to the surface—we call this boiling! (See Figure 12.25.)

SUMMARY

Temperature is measured by a thermometer and is related to the average kinetic energy of atoms and molecules in a system (see Equation 12.10 below). **Fahrenheit**, **Celsius**, and **Kelvin** temperature scales are related by

$$T\ (°F) = \tfrac{9}{5}T\ (°C) + 32 \qquad \textbf{(12.1a)}$$

$$T\ (°C) = \tfrac{5}{9}[T\ (°F) - 32] \qquad \textbf{(12.1b)}$$

$$T\ (K) = T\ (°C) + 273.15 \qquad \textbf{(12.1c)}$$

Systems are in **thermal equilibrium** when they have the same temperature. The **zeroth law of thermodynamics** states that if systems A and B are in thermal equilibrium with each other, and a third system C is in equilibrium with A, then C is also in equilibrium with B.

Thermal expansion is the change in the size of an object with temperature. The length of an object changes by

$$\Delta L = \alpha L\ \Delta T \qquad \textbf{(12.2)}$$

where ΔL is the change in length L, ΔT is the change in temperature, and α is the **coefficient of linear expansion**. Table 12.1 lists representative values of α. The area of an object changes by

$$\Delta A = 2\alpha A\ \Delta T \qquad \textbf{(12.3)}$$

where ΔA is the change in area A. The volume changes by

$$\Delta V = \beta V\ \Delta T \qquad \textbf{(12.4)}$$

where ΔV is the change in volume V, and β is the **volume coefficient of expansion**, which can also be found in Table 12.1. **Thermal stress** is created when thermal expansion is constrained.

The **ideal gas law** states that

$$PV = NkT \qquad \textbf{(12.5)}$$

where P is the absolute pressure in a gas, V the volume it occupies, N the number of atoms and molecules in the gas, T its absolute temperature, and k is the **Boltzmann constant**,

$$k = 1.38 \times 10^{-23}\ \text{J/K} \qquad \textbf{(12.6)}$$

An **ideal gas** is described by the ideal gas law. Real gases often behave like ideal gases.

A **mole** of a substance contains **Avogadro's number** (N_A) of atoms or molecules, where

$$N_A = 6.02 \times 10^{23}/\text{mol} \qquad \textbf{(12.7)}$$

A mole of any substance has a mass in grams equal to its molecular weight, which can be determined from the periodic table of elements. The ideal gas law stated in terms of moles is

$$PV = nRT \qquad \textbf{(12.8)}$$

where n is the number of moles and R is the **universal gas constant**. Three values for R are 8.31 J/mol·K, 1.99 cal/mol·K, and 0.0821 L·atm/mol·K.

Kinetic theory takes matter to be made of atoms and molecules in continuous random motion. The ideal gas law stated in terms of molecular mass and velocity is

$$PV = \tfrac{1}{3}Nm\overline{v^2} \qquad \textbf{(12.9)}$$

where N is the number of molecules in the container, m is the mass of the molecule, and $\overline{v^2}$ is the average of the molecular velocity squared. **Thermal energy** is defined to be the average kinetic energy $\overline{KE}$ of a molecule:

$$\overline{KE} = \tfrac{1}{2}m\overline{v^2} = \tfrac{3}{2}kT \qquad \textbf{(12.10)}$$

This equation gives a precise definition of absolute temperature. The root mean square speed of atoms or molecules at temperature T is

$$v_{\text{rms}} = \sqrt{\frac{3kT}{m}} \qquad \textbf{(12.11)}$$

Phase changes among the various phases of matter are affected by temperature and pressure. There is a **critical temperature** above which a liquid cannot exist. **Critical pressure** is the minimum pressure needed for a liquid to exist. Representative values of both are found in Table 12.2. **Phase diagrams** are graphs of pressure versus temperature that show conditions under which various phases exist. There is a **triple point** at which all three phases exist in equilibrium. See Table 12.3 for values. A gas at a temperature below its boiling point is called a **vapor**. **Vapor pressure** is the gas pressure created by a substance. **Partial pressure** is the pressure a gas would create if it existed alone. **Dalton's law** states that total pressure is the sum of partial pressures of all gases present.

Percent relative humidity is defined to be

percent relative humidity

$$= \frac{\text{vapor density}}{\text{saturation vapor density}} \times 100 \qquad \textbf{(12.12)}$$

Saturation vapor density and vapor pressure for water are found in Table 12.4.

CONCEPTUAL QUESTIONS

12.1 What does it mean to say that two bodies are in thermal equilibrium?

12.2 Give an example of a physical property that varies with temperature and describe how it is used to measure temperature.

12.3 What will the bimetallic strip in Figure 12.2(a) do if you lower its temperature to less than its original value?

12.4 A constant-volume gas thermometer contains a fixed amount of gas. What property of the gas is measured to indicate its temperature?

12.5 When a cold mercury thermometer is placed in a hot liquid, the column of mercury goes *down* slightly before going up. Explain why.

12.6 Thermal stresses caused by uneven cooling can easily break glass cookware. Explain why Pyrex, a glass with a small coefficient of linear expansion, is less susceptible.

12.7 One method of getting a tight fit, say of a metal peg in a hole in a metal block, is to manufacture the peg slightly larger than the hole. It is then inserted when at a different temperature than the block. Should the block be hotter or colder than the peg at insertion? Explain.

12.8 Does it really help to run hot water over a tight metal lid on a glass jar before trying to open it? Explain.

12.9 What does the ideal gas law predict about the volume of a gas at 0 K? Explain why this will not actually happen.

12.10 Why doesn't the ideal gas law depend on the relative sizes of atoms and molecules?

12.11 Under what circumstances would you expect a gas to behave significantly different than predicted by the ideal gas law?

12.12 How is momentum related to the pressure exerted by a gas? Explain on the submicroscopic level, considering the behavior of atoms and molecules.

12.13 A pressure cooker contains water and steam in equilibrium at a pressure greater than atmospheric. How does this increase cooking speed?

12.14 Why does condensation form most rapidly on the coldest object in a room—for example, on a glass of ice water?

12.15 Is the vapor pressure of solid carbon dioxide (dry ice) equal to 1.00 atm at −78.5°C? (See Figure 12.26.)

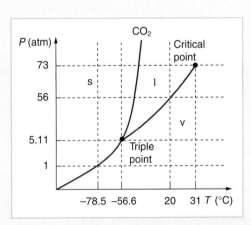

Figure 12.26 The phase diagram for carbon dioxide. The axes are nonlinear, and the graph is not to scale. Dry ice is solid carbon dioxide and has a sublimation temperature of −78.5°C. Questions 15 and 16.

12.16 Can carbon dioxide be liquefied at room temperature? If so, how? If not, why not? (See Figure 12.26.)

12.17 Why can oxygen not be liquefied at room temperature by placing it under a large enough pressure to force its molecules together?

12.18 What is the distinction between gas and vapor?

12.19 Since humidity depends only on water's vapor pressure and temperature, are the saturation vapor densities listed in Table 12.4 valid in an atmosphere of helium at a pressure of $1.01 \times 10^5 \text{ N/m}^2$, rather than air? Are those values affected by altitude on earth?

12.20 Why does a beaker of 40.0°C water placed in a vacuum chamber start to boil as the chamber is evacuated? At what pressure does the boiling begin?

12.21 Why do liquids like alcohol and ether evaporate much more rapidly than water at STP?

PROBLEMS

Section 12.1 Temperature Scales

12.1 What is the Fahrenheit temperature of a person with a 39.0°C fever?

12.2 Frost damage to most plants occurs at temperatures of 28.0°F or lower. What is 28.0°F on the Celsius scale?

12.3 To conserve energy, room temperatures are kept at 68.0°F in the winter and 78.0°F in the summer. What are these temperatures on the Celsius scale?

12.4 A tungsten light bulb filament may operate at 2900 K. What is its Fahrenheit temperature?

12.5 The surface temperature of the sun is about 5750 K. What is this on the Fahrenheit scale?

12.6 One of the hottest temperatures ever recorded on the surface of the earth was 134°F in Death Valley. What is this on the Celsius scale?

• 12.7 (a) Suppose a cold front blows into your locale and drops the temperature by 40.0 Fahrenheit degrees. How much of a drop is this in Celsius degrees? (b) Show that any change in temperature in Fahrenheit degrees is nine-fifths the change in Celsius degrees.

• 12.8 (a) At what temperature do the Fahrenheit and Celsius scales have the same numerical value? (b) At what temperature do the Fahrenheit and Kelvin scales have the same numerical value?

Section 12.2 Thermal Expansion

12.9 The height of the Washington Monument is measured to be 170 m on a day when its temperature is 35.0°C. What will its height be on a day when its temperature falls to −10.0°C? Although the monument is made of limestone, take the thermal coefficient of expansion to be that of marble.

12.10 How large an expansion gap should be left between steel railroad rails if they may reach a maximum temperature 35.0°C greater than when they were laid? Their original length is 10.0 m.

12.11 What is the change in length of a 3.00 cm long column of mercury if its temperature changes from 37.0°C to 40.0°C, assuming it is unconstrained?

12.12 Global warming will produce rising sea levels partly due to melting ice caps but also due to the expansion of water as average ocean temperatures rise. To get some idea of the size of this effect, calculate the change in length of a column of water 1.00 km high for a temperature increase of 1.00°C. Note that this is only an approximate calculation since ocean warming is not uniform with depth.

12.13 Show that 60.00 L of gasoline originally at 15.0°C will expand to 61.14 L when it warms to 35.0°C, as claimed in Example 12.3.

• 12.14 (a) Suppose a meter stick made of steel and one made of invar are the same length at 0°C. What is their difference in length at 22.0°C? (b) Repeat the calculation for two 30.0 m long surveyor's tapes.

• 12.15 (a) If a 500 mL glass beaker is filled to the brim with ethyl alcohol at a temperature of 5.00°C, how much will overflow when their temperature reaches 22.0°C? (b) How much *less* water would overflow under the same conditions?

• 12.16 Most automobiles have a coolant reservoir to catch radiator fluid that may overflow when the engine is hot. The radiator is made of copper and is filled to its 16.0 L capacity when at 10.0°C. What volume of radiator fluid will overflow when the radiator and fluid reach their 95.0°C operating temperature, given the fluid's volume coefficient of expansion is $\beta = 400 \times 10^{-6}$/°C? Note that this is approximate, since most car radiators have operating temperatures greater than 95.0°C.

‡ 12.17 A physicist makes a cup of instant coffee and notices that as the coffee cools its level drops 3.00 mm in the glass cup. Show that this cannot be due to thermal contraction by calculating the decrease in level if the 350 cm³ of coffee is in a 7.00 cm diameter cup and decreases in temperature from 95.0°C to 45.0°C. (Most of the drop in level is actually due to escaping bubbles of air.)

‡ 12.18 The density of 0°C water is very nearly 1000 kg/m³ (it is actually 999.84 kg/m³), whereas that of 0°C ice is 917 kg/m³. Calculate the pressure necessary to keep ice from expanding when it freezes, neglecting the effect such a large pressure would have on the freezing temperature. (This problem gives you only an indication of how large the forces associated with freezing water might be.)

‡ 12.19 Show that $\beta \approx 3\alpha$, by calculating the change in volume ΔV of a cube with sides of length L.

Section 12.3 The Ideal Gas Law

12.20 The gauge pressure in your car tires is 2.50×10^5 N/m² at a temperature of 35.0°C when you drive it onto a ferry boat to Alaska. What is their gauge pressure later, when their temperature has dropped to −40.0°C?

12.21 Convert an absolute pressure of 7.00×10^5 N/m² to gauge pressure in lb/in². (This was stated to be just less than 90.0 lb/in² in Example 12.4. Is it?)

12.22 Suppose a gas-filled incandescent light bulb is manufactured to have atmospheric pressure in it at 20.0°C. Find the gauge pressure inside such a bulb when it is hot, assuming its average temperature is 60.0°C (an approximation) and neglecting any change in volume due to thermal expansion or gas leaks.

12.23 Large helium-filled balloons are used to lift scientific experiments to high altitudes. (a) What is the pressure inside such a balloon if it starts out at sea level with a temperature of 10.0°C and rises to an altitude where its volume is eight times the original volume and its temperature is −50.0°C? (b) What is the gauge pressure?

12.24 Confirm that the units of nRT are those of energy for each value of R—8.31 J/mol·K, 1.99 cal/mol·K, and 0.0821 L·atm/mol·K.

12.25 In the text it was shown that $N/V = 2.68 \times 10^{25}$/m³ for gas at STP. (a) Show that this is equivalent to $N/V = 2.68 \times 10^{19}$/cm³, as stated. (b) How many atoms are there in one μm³ (a cubic micrometer) at STP?

• 12.26 (a) What is the volume in km³ of one mole of sand grains if each grain is a cube 1.00 mm on a side? (b) How many kilometers of beaches would this cover if the beach averages 100 m in width and 10.0 m in depth? Neglect air spaces between grains.

• **12.27** An expensive vacuum system can achieve a pressure as low as 1.00×10^{-7} N/m² at 20°C. How many atoms are there in a cubic centimeter at this pressure and temperature?

• **12.28** The density of gas atoms at a certain location in the space above our planet is about 1.00×10^{11}/m³, and the pressure is 2.75×10^{-10} N/m². What is the temperature there?

• **12.29** The bicycle tire considered in Examples 12.4 and 12.7 has a pressure of 7.00×10^5 N/m² at a temperature of 18.0°C, and contains 2.00 L of gas. What will its pressure be if you let out an amount of air that has a volume of 100 cm³ at atmospheric pressure? Assume tire temperature and volume remain constant.

• **12.30** A high-pressure gas cylinder contains 50.0 L of toxic gas at a pressure of 1.40×10^7 N/m² and a temperature of 25.0°C. Its valve leaks after the cylinder is dropped. The cylinder is cooled to dry ice temperatures (−78.5°C), to reduce the leak rate and pressure so that it can be safely repaired. (a) What is the final pressure in the tank, assuming a negligible amount of gas leaks while being cooled and that there is no phase change? (b) What is the final pressure if one-tenth of the gas escapes? (c) To what temperature must the tank be cooled to reduce the pressure to 1.00 atm (assuming the gas does not change phase and that there is no leakage during cooling)? (d) Does cooling the tank appear to be a practical solution?

• **12.31** Find the number of moles in 2.00 L of gas at 35.0°C and under 7.41×10^7 N/m² of pressure. (Should the answer be the same as obtained in Example 12.7?)

⁑**12.32** Calculate the depth to which one mole of 3.75 cm diameter table tennis balls would cover the earth if the space between balls adds an extra 25.0% to their volume and they are not crushed by their own weight.

⁑**12.33** (a) What is the gauge pressure in a 25.0°C car tire containing 3.60 mol of gas in a 30.0 L volume? (b) What will its gauge pressure be if you add 1.00 L of gas originally at atmospheric pressure and 25.0°C? Assume the temperature returns to 25.0°C and the volume remains constant.

⁑**12.34** The actual final pressure for the light bulb in Problem 12.22 will be less than calculated because the glass bulb will expand. What will the actual final pressure be, taking this into account? Is this a negligible difference?

⁑**12.35** (a) In the deep space between galaxies the density of atoms is as low as 10^6/m³, and the temperature is a frigid 2.7 K. What is the pressure? (b) What volume in m³ is occupied by 1 mol of gas? (c) If this volume is a cube, what is the length of its sides in kilometers?

Section 12.4 Kinetic Theory

12.36 Some incandescent light bulbs are filled with argon gas. What is v_{rms} for argon atoms near the filament, assuming their temperature is 2500 K?

12.37 Average atomic and molecular speeds (v_{rms}) are large even at low temperatures. What is v_{rms} for helium atoms at 5.00 K, just one degree above helium's liquefaction temperature?

12.38 (a) What is the average kinetic energy in joules of hydrogen atoms on the 5500°C surface of the sun? (b) What is the average kinetic energy of helium atoms in a region of the solar corona where the temperature is 6.00×10^5 K?

12.39 At what temperature would oxygen molecules (molecular mass 32.0) have an average velocity v_{rms} equal to the earth's escape velocity of 11.1 km/s?

12.40 The escape velocity from the moon is much smaller than from the earth, only 2.38 km/s. At what temperature would hydrogen molecules (molecular mass 2.016) have an average velocity v_{rms} equal to the moon's escape velocity?

• **12.41** Nuclear fusion, the energy source of the sun and hydrogen bombs, occurs much more readily when the average kinetic energy of the atoms is high—that is, at high temperatures. Suppose you want the atoms in your fusion experiment to have average kinetic energies of 6.40×10^{-14} J. What temperature is needed?

• **12.42** Suppose that the average velocity (v_{rms}) of carbon dioxide molecules (molecular mass 44.0) in a flame is found to be 1.05×10^5 m/s. What temperature does this represent?

• **12.43** Hydrogen molecules (molecular mass 2.016) have an average velocity v_{rms} equal to 193 m/s. What is the temperature?

• **12.44** Much of the gas near the sun is atomic hydrogen. Its temperature would have to be 1.50×10^7 K for the average velocity v_{rms} to equal the escape velocity from the sun. What is that velocity?

⁑**12.45** There are two important isotopes of uranium—^{235}U and ^{238}U; these are nearly identical chemically but have different atomic masses. Only ^{235}U is very useful in nuclear reactors. One of the techniques for separating them is based on the different average velocities v_{rms} of uranium hexafluoride gas, UF_6. (a) The molecular masses for $^{235}UF_6$ and $^{238}UF_6$ are 349.0 and 352.0, respectively. What is the ratio of their average velocities? (b) At what temperature would their average velocities differ by 1.00 m/s?

Sections 12.5 and 12.6 Phase Changes and Humidity

12.46 Dry air is 78.1% nitrogen. What is the partial pressure of nitrogen when atmospheric pressure is 1.01×10^5 N/m²?

12.47 (a) What is the vapor pressure of water at 20.0°C? (b) What percentage of atmospheric pressure is this? (c) What percent of 20.0°C air is water vapor if it has 100% relative humidity? (The density of dry air at 20.0°C is 1.20 kg/m³.)

12.48 Pressure cookers increase cooking speed by raising the boiling temperature of water above its value at atmospheric pressure. (a) What pressure is necessary to raise the boiling point to 120°C? (b) What gauge pressure is this?

12.49 (a) At what temperature does water boil at an altitude of 1500 m (about 5000 ft) on a day when atmospheric pressure is 8.59×10^4 N/m²? (b) What about at an altitude of 3000 m (about 10,000 ft) when atmospheric pressure is 7.00×10^4 N/m²?

12.50 What is atmospheric pressure on top of Mt. Everest on a day when water boils there at a temperature of 70.0°C?

12.51 At a spot in the high Andes, water boils at 80.0°C, greatly reducing the cooking speed of potatoes, for example. What is atmospheric pressure at this location?

12.52 What is the relative humidity on a 25.0°C day when the air contains 18.0 g/m³ of water vapor?

12.53 What is the density of water vapor in g/m³ on a hot dry day in the desert when the temperature is 40.0°C and the relative humidity is 6.00%?

• **12.54** A deep sea diver should breath a gas mixture that has the same oxygen partial pressure as at sea level, where dry air contains 20.9% oxygen and has a total pressure of $1.01 \times 10^5 \text{ N/m}^2$. (a) What is the partial pressure of oxygen at sea level? (b) If the diver breathes a gas mixture at a pressure of $2.00 \times 10^6 \text{ N/m}^2$, what percent oxygen should it be to have the same oxygen partial pressure as at sea level?

• **12.55** Table 12.4 gives the vapor pressure of water at 40.0°C to be $7.34 \times 10^3 \text{ N/m}^2$. Using the ideal gas law, calculate the density of water vapor in g/m^3 that creates a partial pressure equal to this vapor pressure. The result should be the same as the saturation vapor density at that temperature (51.1 g/m^3).

• **12.56** Air in human lungs has a temperature of 37.0°C and a saturation vapor density of 44.0 g/m^3 (see Table 12.4). (a) If 2.00 L of air is exhaled and very dry air inhaled, what is the maximum loss of water vapor by the person? (b) Calculate the partial pressure of water vapor having this density, and compare it with the vapor pressure in Table 12.4 ($6.31 \times 10^3 \text{ N/m}^2$).

• **12.57** If the relative humidity is 90.0% on a muggy summer morning when the temperature is 20.0°C, what will it be later in the day when the temperature is 30.0°C, assuming the water vapor density remains constant?

• **12.58** Late on an autumn day, the relative humidity is 45.0% and the temperature is 20.0°C. What will the relative humidity be that evening when the temperature has dropped to 10.0°C, assuming constant water vapor density?

‡ **12.59** Atmospheric pressure atop Mt. Everest is $3.30 \times 10^4 \text{ N/m}^2$. (a) What is the partial pressure of oxygen there if it is 20.9% of the air? (b) What percent oxygen should a mountain climber breathe so that its partial pressure is the same as at sea level, where atmospheric pressure is $1.01 \times 10^5 \text{ N/m}^2$?

‡ **12.60** What is the dew point (the temperature at which 100% relative humidity would occur) on a day when relative humidity is 39.0% at a temperature of 20.0°C?

‡ **12.61** On a certain day, the temperature is 25.0°C and the relative humidity is 90.0%. How many grams of water must condense out of each cubic meter of air if the temperature falls to 15.0°C? Such a drop in temperature can, thus, produce heavy dew or fog.

INTEGRATED CONCEPTS

These problems involve physical principles from more than one chapter. Integrated knowledge from a broad range of topics is much more powerful than a narrow application of physics. Here you will need to refer to material in Chapters 4, 6, and perhaps others, as well as in Chapter 12.

PROBLEM-SOLVING STRATEGY

Use the following strategy to solve integrated concept problems.

Step 1. *Identify which physical principles are involved.*

Step 2. *Solve the problem using strategies outlined in the text.*

Note: Worked examples that can help you see how to apply this strategy to integrated concept problems appear in a number of chapters, the most recent being Chapter 6. (See the section of problems labeled *Integrated Concepts* at the end of that chapter and consult the index for others.)

• **12.62** The boiling point of water increases with depth because pressure increases with depth. At what depth will fresh water

have a boiling point of 150°C, if the surface of the water is at sea level?

• **12.63** (a) At what depth in fresh water is the critical pressure of water reached, given that the surface is at sea level? (b) At what temperature will this water boil? (c) Is a significantly higher temperature needed to boil water at a greater depth?

‡ **12.64** To get an idea of the small effect that temperature has on Archimedes' principle, calculate the fraction of a copper block's weight that is supported by buoyant force in 0°C water and compare this with the fraction supported in 95.0°C water.

‡ **12.65** If you want to cook in water at 150°C, you need a pressure cooker that can withstand the necessary pressure. (a) What pressure is required for the boiling point of water to be this high? (b) If the lid of the pressure cooker is a disk 25.0 cm in diameter, what force must it be able to withstand at this pressure?

UNREASONABLE RESULTS

As in previous chapters, problems with unreasonable results are included to give practice in assessing whether nature is being accurately described, and to trace the source of difficulty if it is not. This is very much like the process followed in original research when physical principles as well as faulty premises are tested. The following problems give results that are unreasonable even when the physics is correctly applied, because some premise is unreasonable or because certain premises are inconsistent.

PROBLEM-SOLVING STRATEGY

To determine if an answer is reasonable, and to determine the cause if it is not, do the following.

Step 1. *Solve the problem using strategies as outlined in this chapter.* See the table of contents to locate problem-solving strategies in other chapters. Use the format followed in the worked examples in the text to solve the problem as usual.

Step 2. *Check to see if the answer is reasonable.* Is it too large or too small, or does it have the wrong sign, improper units, . . .?

Step 3. *If the answer is unreasonable, look for what specifically could cause the identified difficulty.*

12.66 (a) How many moles per cubic meter of an ideal gas are there at a pressure of $1.00 \times 10^{14} \text{ N/m}^2$, and at 0°C? (b) What is unreasonable about this result? (c) Which premise or assumption is responsible?

• **12.67** (a) An automobile mechanic claims that an aluminum rod fits loosely into its hole on an aluminum engine block because the engine is hot and the rod is cold. If the hole is 10.0% bigger in diameter than the 22.0°C rod, at what temperature will the rod be the same size as the hole? (b) What is unreasonable about this temperature? (c) Which premise is responsible?

• **12.68** The temperature inside a supernova explosion is said to be 2.00×10^{13} K. (a) What would the average velocity v_{rms} of hydrogen atoms be? (b) What is unreasonable about this velocity? (c) Which premise or assumption is responsible?

• **12.69** Suppose the relative humidity is 80% on a day when the temperature is 30.0°C. (a) What will the relative humidity be if the air cools to 25.0°C and the vapor density remains constant? (b) What is unreasonable about this result? (c) Which premise is responsible?

HEAT AND
HEAT TRANSFER METHODS

Great heat in motion is evident in this lava flow.

(a)

(b)

Figure 13.1 (a) The chilling effect of a clear breezy night is produced by the wind and by the less obvious radiation of body heat to cold outer space. (b) Great controversy once raged about the age of the earth, now generally accepted to be about 4.5 billion years. Much of the debate centered on the earth's molten interior. According to our understanding of heat transfer, the earth's center should have long since cooled off if the earth is really that old. The discovery of radioactivity in rocks revealed the source of energy that keeps the interior of the earth molten despite heat lost to the surface and from there to cold, dark outer space.

CONNECTIONS

The now-familiar law of **conservation of energy** did not arise from some theory. On the contrary, it slowly dawned upon scientists that not only are energy's many forms interconvertible, their total is conserved, too. All this was based on *observations and experiment*. Heat is yet another form of energy—specifically energy in transit due to a temperature difference.

It is fascinating that energy takes so many different forms. Heat is one of its most intriguing. Heat is often hidden, it only exists when in transit, and it is transferred by a number of distinctly different methods. Heat touches every aspect of our lives and enters into our understanding of the functioning of the universe. From comprehending the chill we feel on a clear breezy night to explaining why the earth's center has yet to cool, we find heat intimately involved. (See Figure 13.1.) This chapter defines and explores heat, its effects, and the methods by which it is transferred.

13.1 HEAT

The word *heat* has numerous meanings in everyday language, many of which are different from its precise scientific definition. Scientifically, heat is a particular form of energy transfer. You may recall from Chapter 6 that work also transfers energy. In fact, heat and work are the only two distinct methods of energy transfer.

Work is an organized transfer of energy in which a force acts through a distance. Heat involves the random thermal motions of atoms and molecules and, thus, is less organized in a sense. We know from experience that if we place two objects having different temperatures in contact, perhaps as shown in Figure 13.2, they will interact until their temperatures are the same. Temperature is directly related to the average kinetic energy of atoms and molecules in the objects. It is apparent that some form of energy is transferred. No work is done on or by either object, because no force acts through a distance. The transfer of energy is caused by the temperature difference and ceases once the temperatures are equal. This leads to the following definition of **heat**:

Heat is energy transferred solely due to a temperature difference.

As noted in Chapter 12, heat is easily confused with temperature. For example, we may say the heat was unbearable, when we mean the temperature was high. Heat is a form of energy, whereas temperature is not. There *is* a cause and effect relationship between temperature and heat. Temperature differences *cause* heat to flow.

Naturally, since heat is energy, it has units of energy (the *joule* in SI units). The *calorie* is another commonly employed unit of heat. The calorie is defined by one effect heat can have. A **calorie** (cal) is the amount of energy needed to change the temperature of 1.00 g of water by 1.00°C—specifically, between 14.5°C and 15.5°C, since there is a slight temperature dependence. Perhaps the most common unit of heat is the **kilocalorie** (kcal), the amount of energy needed to change the temperature of 1.00 kg of water by 1.00°C, sometimes called a big calorie (Cal). Since mass is most often specified in kilograms, kilocalories are very commonly used. Food calories are actually kilocalories, a fact not easily determined from package labeling.

It is also possible to change the temperature of water, or any other substance, by doing work, since work can transfer energy into or out of a system. This realization was crucial to establishing that heat is a form of energy. One of those most responsible was the English physicist James Prescott Joule (1818–1889). Joule performed myriad ex-

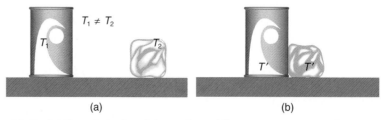

(a)   (b)

Figure 13.2 (a) The soft drink and the ice have different temperatures and are not in equilibrium. (b) When allowed to interact, energy is transferred until the two achieve equilibrium when they reach the same temperature. Heat is the form of energy transferred. No work is done by one object on the other.

periments to establish the **mechanical equivalent of heat**—*the amount of work needed to produce the same effects as a given amount of heat*. The best modern value for this equivalence is

$$1.000 \text{ kcal} = 4186 \text{ J} \qquad (13.1)$$

where the kilocalorie is the amount of heat defined above. Figure 13.3 shows one of Joule's most famous experimental setups for demonstrating the mechanical equivalent of heat. Joule's experiments not only demonstrated that work and heat can produce the same effects, they also helped establish the conservation of energy principle. So significant were his contributions that the SI unit of energy was named after him.

When heat enters a system, it is transformed into some other form of energy. It may become increased kinetic energy of atoms and molecules. That is, heat entering a system can increase the system's temperature, such as when a stove heats food. Heat can also change the phase of matter, such as when heat melts ice. Both effects involve the *internal energy* of the system. The same is true of work—once it enters a system, it becomes some form of energy. And work can produce the same two effects (temperature change and phase change) as heat. Since these two final forms of energy give no clue as to whether heat or work created them, it is incorrect to refer to such energy as "heat content" or "work content."

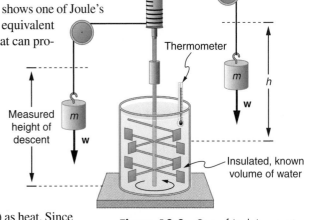

Figure 13.3 One of Joule's experiments for measuring the mechanical equivalent of heat. Gravitational PE (work done by gravity) is converted into KE and then randomized by viscosity and turbulence into increased average kinetic energy of atoms and molecules in the system, producing a temperature increase.

13.2 TEMPERATURE CHANGE AND SPECIFIC HEAT

One of the major effects of heat is temperature change. Clearly, heat input increases temperature, whereas heat output decreases it. But how much heat is required to change temperature by a degree, and what other factors are involved? To explore these questions without complications, we first assume there is no phase change. (Phase changes will be explored separately in the next section.) Second, we assume there is no work done on or by the system, so that the only energy entering or leaving it is heat.

Experiments have shown that the amount of heat required to produce a given temperature change depends primarily on three factors—the size of the temperature change, the mass of the system, and the substance and phase involved. (See Figure 13.4.)

The first two factors, temperature change and mass, are the most easily understood. Intuitively, it seems reasonable that the amount of heat is directly proportional to the temperature change. More concretely, on the submicroscopic scale, we know that the average kinetic energy of atoms and molecules is directly proportional to absolute temperature, and so we would expect twice the energy to be needed for twice the temperature change. Similarly, it seems reasonable that the amount of heat required to create a given temperature change is directly proportional to the system's mass. The greater the mass, the more atoms and molecules, and the more heat required to change their average kinetic energy.

The third important factor is the substance involved, including the phase it is in. For example, very different amounts of heat are required to change the temperature of 1.00 kg of copper and the same mass of water by 1.00°C (copper takes less, by a factor of 10.8).

CONNECTIONS

As noted many times and in many places, explanations on the submicroscopic scale provide profound insight. At this stage, we are considering systems with large numbers of atoms and molecules. At a later stage—in the chapters on modern physics—we consider *individual* atoms and molecules, introducing concepts of quantum mechanics, and so gain further insight.

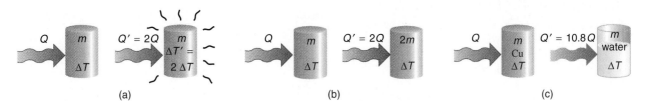

Figure 13.4 The amount of heat Q required to cause a temperature change depends on the size of the temperature change, the mass of the system, and the substance and phase involved. (a) The amount of heat is directly proportional to temperature change. (b) The amount of heat is also directly proportional to mass. (c) The amount of heat depends on the substance and its phase. For example, it takes 10.8 times as much heat to produce the same temperature increase in water than in an equal mass of copper.

Furthermore, different amounts of heat are required to change the temperature of 1.00 kg of *ice* and 1.00 kg of water by 1.00°C (ice takes less, by about a factor of 2).

The quantitative relationship between heat and temperature change contains all three factors just mentioned. In equation form, it is

$$Q = mc \, \Delta T \tag{13.2}$$

where Q is the symbol for heat, m is the mass of the system, and ΔT is the change in temperature. The symbol c stands for **specific heat**, which is the amount of heat required to change the temperature of 1.00 kg of mass by 1.00°C. Specific heat c depends on substance; its SI unit is J/kg·K or J/kg·°C (recall that kelvins and degrees Celsius are the same size). Thus Q is in joules when mass is in kg and temperature change is in °C. If heat in kcal is desired, then c may be expressed in kcal/kg·°C.

Biological Application

Values of specific heat c must generally be looked up in tables, because there is no simple way to calculate them. Specific heat depends on the substance (reflecting its substructure), its phase, and its temperature. Table 13.1 lists representative values of specific heat c for temperature changes made with volume held constant.*

EXAMPLE 13.1   HOW MUCH HEAT IS THERE IN FOOD?

The body uses food energy in three ways. It can store the energy as fat or use it to do work, but most of it goes to producing body heat—that is, to increasing the temperature of the entire body. We must get rid of this body heat to maintain a normal temperature. How much will the temperature of a 75.0 kg man increase if he eats a cup of yogurt containing 350 kcal, given he retains all the energy but does no work?

Strategy We are asked to find the temperature increase ΔT, which depends on the amount of heat, the mass of the system, and the specific heat. All the needed information is given except the specific heat, which is listed in Table 13.1.

Solution Solving the relationship between heat and temperature increase, $Q = mc \, \Delta T$, for ΔT and substituting

known values yields

$$\Delta T = \frac{Q}{mc} = \frac{350 \text{ kcal}}{(75.0 \text{ kg})(0.83 \text{ kcal/kg·°C})}$$
$$= 5.60°C$$

Discussion and Implication This is an unreasonable temperature increase, since it would raise normal body temperature to 42.6°C—over 108°F! The implication is that the body does not retain all of this heat. We will explore the ways the body gains or loses heat to the environment in later sections.

EXAMPLE 13.2   TRUCK BRAKES OVERHEAT ON DOWNHILL RUNS

Truck brakes used to control speed on a downhill run do work, converting gravitational potential energy into increased internal energy (higher temperature) in the brake material. This prevents the gravitational potential energy from going into kinetic energy of the truck. The problem is that the mass of the truck is large compared with the mass of the brake material absorbing the energy, and the temperature increase may occur too fast for heat to be carried away from the brakes. Calculate the temperature increase of 100 kg of brake material having an average specific heat of 800 J/kg·°C, if it retains all the energy from a 10,000 kg truck descending 75.0 m at constant speed.

Strategy and Concept Here there is a conversion of gravitational potential energy into internal kinetic energy. We first calculate the gravitational potential energy (Mgh) that the entire truck loses in its descent, and then we find the temperature increase produced in the brake material alone.

Solution The gravitational potential energy going into increasing the temperature of the brakes is

$$Mgh = (10,000 \text{ kg})(9.80 \text{ m/s}^2)(75.0 \text{ m})$$
$$= 7.35 \times 10^6 \text{ J}$$

(continued)

---

*This is sometimes written as c_v. Specific heat measured at constant pressure, c_p, differs from c_v, especially for gases. It is beyond the scope of this text to deal with this distinction in detail. We shall use values of c_v, unless otherwise stated.

(continued)
Solving Equation 13.2 for the change in temperature yields

$$\Delta T = \frac{Q}{mc}$$

where Q is 7.35×10^6 J, and m is the mass of the brake material. Substituting these values and the given specific heat

of the brake material gives

$$\Delta T = \frac{7.35 \times 10^6 \text{ J}}{(100 \text{ kg})(800 \text{ J/kg}\cdot{}^\circ\text{C})} = 92.0{}^\circ\text{C}$$

Discussion This temperature increase would likely bring the brake material temperature above the boiling point of water.

TABLE 13.1

SPECIFIC HEATS* OF VARIOUS SUBSTANCES

Substance	Specific heat c	
	J/kg·°C	kcal/kg·°C†
Solids		
Aluminum	900	0.215
Asbestos	800	0.19
Concrete, granite (average)	840	0.20
Copper	387	0.0924
Glass	840	0.20
Gold	129	0.0308
Human body (average at 37°C)	3500	0.83
Ice (average, −50°C to 0°C)	2090	0.50
Iron, steel	452	0.108
Lead	128	0.0305
Silver	235	0.0562
Wood	1700	0.4
Liquids		
Benzene	1740	0.415
Ethanol	2450	0.586
Glycerin	2410	0.576
Mercury	139	0.0333
Water (15.0°C)	4186	1.000
Gases††		
Air (dry)	721 (1015)	0.172 (0.242)
Ammonia	1670 (2190)	0.399 (0.523)
Carbon dioxide	638 (833)	0.152 (0.199)
Nitrogen	739 (1040)	0.177 (0.248)
Oxygen	651 (913)	0.156 (0.218)
Steam (100°C)	1520 (2020)	0.363 (0.482)

*Values for solids and liquids at constant volume and at 25.0°C, except as noted.
†These values are identical in units of cal/g·°C.
††c_v at constant volume and at 15.0°C, except as noted, and at 1.00 atm average pressure. Values in parentheses are c_p at a constant pressure of 1.00 atm.

Note that in Example 13.2, it is *not* correct to say that heat was put into the truck's brakes. Rather, work was done to increase their temperature. Once their temperature is up, heat will flow from them into the cooler surroundings, but no heat is put into them in this situation. The example is another illustration of the mechanical equivalent of heat. (See Figure 13.5.) The same temperature increase *could* be produced by putting 7.35×10^6 J of heat into the brakes—say, with a blow torch.

Figure 13.5 The smoking brakes on this truck are visible evidence of the mechanical equivalent of heat. See Example 13.2.

EXAMPLE 13.3 HEAT EXCHANGED BETWEEN WATER AND PAN

Suppose you pour 0.250 kg of 20.0°C water (about a cup) into a 0.500 kg aluminum pan fresh off the stove with a temperature of 150°C. Assuming that the pan is on an insulated pad and that a negligible amount of water boils off, what temperature do the water and pan reach a short time later?

(continued)

(continued)

Strategy The exchange of heat between the pan and water will take place rapidly. The pan is on an insulated pad, and so very little heat will be exchanged with the surroundings (such as its plastic handle, which is not included in the given 0.500 kg mass). Thus the heat lost by the pan equals the heat gained by the water. Heat transfer stops once both reach the same temperature. In equation form, this is

$$Q_{hot} = Q_{cold}$$

More specifically, in this case

$$Q_{Al} = Q_{w}$$

where Q_{Al} and Q_{w} are the heat lost by the aluminum and gained by the water, respectively. Using Equation 13.2 to express the heat needed for a temperature change, we see that

$$m_{Al}c_{Al}\,\Delta T_{Al} = m_{w}c_{w}\,\Delta T_{w}$$

Remember that the pan and water reach the same final temperature, which we call T_f. Thus the change in temperature of the pan and water are $\Delta T_{Al} = 150°C - T_f$ and $\Delta T_{w} = T_f - 20°C$. Entering these expressions into the previous equation gives

$$m_{Al}c_{Al}(150°C - T_f) = m_{w}c_{w}(T_f - 20.0°C)$$

which is an equation in just one unknown (T_f), since the specific heats are listed in Table 13.1—and T_f is exactly what we want to find.

Solution Solving the last equation for T_f yields

$$T_f = \frac{m_{Al}c_{Al}(150°C) + m_{w}c_{w}(20.0°C)}{m_{Al}c_{Al} + m_{w}c_{w}}$$

Next we substitute known values:

$$
\begin{aligned}
T_f &= \big[(0.500 \text{ kg})(0.215 \text{ kcal/kg}\cdot°\text{C})(150°C) \\
&\quad + (0.250 \text{ kg})(1.00 \text{ kcal/kg}\cdot°\text{C})(20.0°C)\big] \\
&\quad \div \big[(0.500 \text{ kg})(0.215 \text{ kcal/kg}\cdot°\text{C}) \\
&\quad + (0.250 \text{ kg})(1.00 \text{ kcal/kg}\cdot°\text{C})\big] \\
&= \frac{21.125 \text{ kcal}}{0.3575 \text{ kcal/°C}} \\
&= 59.1°C
\end{aligned}
$$

Discussion Why is the final temperature so much closer to 20.0°C than 150°C? The reason is that water has a greater specific heat than most common substances and, thus, changes temperature more slowly for a given heat input. A large body of water, such as a lake, requires so much heat input to increase its temperature, and can transfer so much heat into a colder environment, that it reduces temperature extremes in its vicinity. This example (and several end-of-chapter problems) are typical of what is called a *calorimetry* problem—that is, one in which heat is exchanged between two systems until they reach a common temperature.

Figure 13.6 Heat from the higher-temperature roof is causing a phase change.

13.3 PHASE CHANGE AND LATENT HEAT

We have discussed one major effect of heat—temperature change. Now let us look at the other—phase change. In Figure 13.6, for example, we see water dripping from icicles melting on a roof warmed by the sun. Conversely, water freezes in an ice tray cooled by lower-temperature surroundings.

Energy is required to melt a solid, because the cohesive forces between molecules must be partially overcome to allow the molecules to move about. Similarly, energy is needed to vaporize a liquid, because in so doing the molecules are separated and molecular attractive forces are overcome. Note that there is no temperature change until a phase change is complete. The energy for a phase change can come from heat transferred into the system from one at higher temperature, like the sun. It can also come from work done on the system, as when sliding friction melts tire rubber in a panic stop.

Conversely, energy is released in the reverse processes of freezing and condensation, usually in the form of thermal energy. The cohesive forces do work on molecules in bringing them together, and the energy produced must be dissipated to allow them to stay together. (See Figure 13.7.)

The amount of energy involved in a phase change depends on two major factors. Both are quite logical in view of the submicroscopic picture of phase changes just described. First, the energy is proportional to the mass that changes phase, because a certain amount of energy is required for each molecule and hence the total energy needed is directly proportional to the number of molecules and, thus, to mass. Second, the energy required depends on the substance, because the strength of molecular forces is

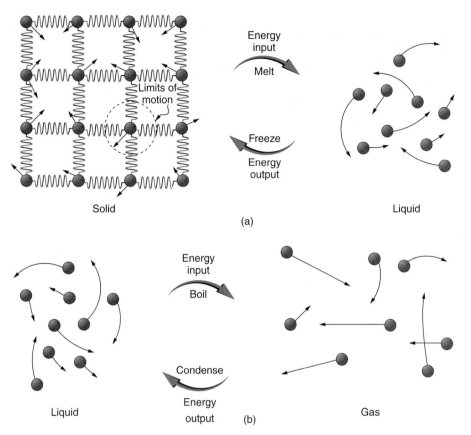

Figure 13.7 (a) Energy is required to partially overcome the attractive forces between molecules in a solid to form a liquid. That same energy must be removed for freezing to take place. (b) Molecules are separated by large distances when going from liquid to vapor, requiring significant energy to overcome molecular attraction. The same energy must be removed for condensation to take place. There is no temperature change until a phase change is complete.

substance related. These two factors are represented in equation form by

and

$$Q = mL_f \qquad \text{(melting–freezing)}$$

$$Q = mL_v \qquad \text{(vaporization–condensation)}$$

(13.3)

In these equations, Q is the amount of energy required to change the phase of a mass m. Q is referred to as **latent heat**. L_f, in the first equation, is the **heat of fusion**—the energy required to change 1.00 kg of a substance from the solid to liquid phase. L_v, in the second equation, is the **heat of vaporization**—the energy required to change 1.00 kg of a substance from the liquid to vapor phase. Both L_f and L_v are measured in units of J/kg. Both depend on the substance, being representative of the strength of its molecular forces, as noted earlier. L_f and L_v are collectively called **latent heat coefficients**. They are *latent*, or hidden, because in phase changes energy enters or leaves a system without changing temperature; so, in effect, the energy is hidden. Table 13.2 lists representative values of L_f and L_v together with melting and boiling points.

The table shows that significant amounts of energy are involved with phase changes. Let us look, for example, at how much energy is needed to melt a kilogram of ice at 0°C to produce a kilogram of water at 0°C. Using Equation 13.3 and the value for water from Table 13.2, we find $Q = mL_f = (1.00 \text{ kg})(79.8 \text{ kcal/kg}) = 79.8 \text{ kcal}$ is the amount required just to melt a kilogram of ice. This 79.8 kcal of energy would raise the temperature of the water from 0°C to 79.8°C if it were already melted. Even more energy is required to vaporize water; some 539 kcal/kg at the normal boiling point, 100°C, produces 100°C steam.

Why is there no temperature change during a phase change? To answer this, let us consider ice melting. Ice cannot exist at temperatures greater than the melting point, and so if any of the surrounding water is heated to a higher temperature, it transfers its heat to the remaining lower-temperature ice. This will continue until there is no ice left. So all of the energy input goes to melting, and none to temperature increase, until the melting is complete. The same is true for all substances when they melt. It is also true at the

CONNECTIONS

Latent heat is another form of potential energy. In Chapter 6, potential energy was defined for gravity, springs, and conservative forces in general.

TABLE 13.2

HEATS OF FUSION AND VAPORIZATION*

Substance	Melting point (°C)	L_f kJ/kg	kcal/kg	Boiling point (°C)	L_v kJ/kg	kcal/kg
Helium	−269.7	5.23	1.25	−268.9	20.9	4.99
Hydrogen	−259.3	58.6	14.0	−252.9	452	108
Nitrogen	−210.0	25.5	6.09	−195.8	201	48.0
Oxygen	−218.8	13.8	3.30	−183.0	213	50.9
Ethanol	−114	104	24.9	78.3	854	204
Ammonia	−75	453	108	−33.4	1370	327
Mercury	−38.9	11.8	2.82	357	272	65.0
Water	0.00	334	79.8	100.0	2256†	539†
Sulfur	119	38.1	9.10	444.6	326	77.9
Lead	327	24.5	5.85	1750	871	208
Antimony	631	165	39.4	1440	561	134
Aluminum	660	380	90	2450	11,400	2720
Silver	961	88.3	21.1	2193	2336	558
Gold	1063	64.5	15.4	2660	1578	377
Copper	1083	134	32.0	2595	5069	1211
Uranium	1133	84	20	3900	1900	454
Tungsten	3410	184	44	5900	4810	1150

*Values quoted at the normal freezing and boiling points at 1.00 atm.
†At 37.0°C (body temperature), the heat of vaporization L_v for water is 2430 kJ/kg or 580 kcal/kg.

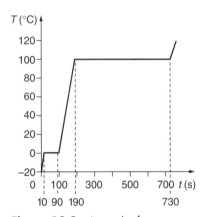

Figure 13.8 A graph of temperature versus time for 1.00 kg of ice as 1.00 kcal/s goes into it. The energy is spread very uniformly throughout the sample, so that there are no local hot spots. The system is constructed so there is no loss of water vapor as the ice warms, melts to water, and then boils to steam. The long periods of constant temperature during phase changes occur when large amounts of energy input are going into latent heat rather than causing a temperature increase.

Human & Medical Applications

boiling point. If there are any local temperature increases above the boiling point, heat is transferred to the lower-temperature liquid until all of it is vaporized. Phase changes thus have a stabilizing effect on temperature, tending to prevent it from changing until the phase change is complete.

Phase changes can have a tremendous stabilizing effect even on temperatures that are not near the melting and boiling points. This is because vaporization and liquefaction can occur at temperatures below the boiling point. For example, air temperatures in humid climates rarely go above 35.0°C, because so much energy goes into evaporating water into the air. Similarly, temperatures in humid weather rarely fall below the dew point, because so much heat is released by condensation as the air goes below that temperature. But in some circumstances, phase change can have a *de*stabilizing effect. For example, it supplies the energy to drive thunderstorms, as will be discussed in Section 13.6 on convective heat transfer.

Let us examine the effect of phase change more precisely by considering the steady input of heat into some water that starts out as −20.0°C ice. (See Figure 13.8.) The ice's temperature rises steadily to 0°C and then remains constant until all the ice is melted. Once this has occurred, the water temperature rises steadily, but it rises at a slower rate than the ice did, because it has a greater specific heat than ice. At 100°C, boiling commences, and the temperature does not change again until all the water is vaporized. Note also that the time to vaporize the water is significantly longer than the melting time, since the latent heat of vaporization is greater than the latent heat of fusion. The resulting steam then increases in temperature steadily and at a faster rate than water, because steam has a lower specific heat.

Of course, water evaporates at temperatures less than the boiling point. As you might expect, more energy is required than at the boiling point, because less energy is available from random thermal motions. For example, at body temperature, 580 kcal/kg is needed to evaporate water, such as perspiration from the skin. This energy loss reduces the temperature of the remaining water and skin, an effective cooling mechanism in hot weather. To get rid of the heat from 350 kcal of yogurt, you would have to evaporate $m = Q/L_v = (350 \text{ kcal})/(580 \text{ kcal/kg}) = 0.603$ kg of sweat. High humidity will inhibit evaporation, and so body temperature might rise, leaving unevaporated sweat on your brow.

EXAMPLE 13.4 PHASE CHANGE AND TEMPERATURE CHANGE

An alcohol rub can quickly reduce a fever victim's temperature by taking away body heat, which is used to evaporate the alcohol from the skin. The evaporation is rapid, because air is normally devoid of alcohol vapor, and so the vapor pressure of alcohol is greater than its near zero partial pressure in air. (Relative humidity has no effect, because it describes the partial pressure of *water* vapor.) How many degrees is the temperature of a 55.0 kg person reduced by the evaporation of 500 cm³ (half a liter) of ethyl alcohol from her skin?

Strategy and Concept If the evaporation takes place very rapidly, as it should, then all of the heat for the phase change comes from the person's body, and negligible body heat is produced during the process. We will assume that the decreased temperature is spread evenly through her body. Thus the heat going into the alcohol, Q_{alc}, equals the heat coming out of the person, Q_p. That is,

$$Q_{alc} = Q_p$$

The heat that goes into the alcohol causes it to change phase, and so we express it with Equation 13.3 as $Q_{alc} = m_{alc}L_v$. The heat leaving the person reduces temperature, and so we express it with Equation 13.2 as $Q_p = m_p c_p \Delta T$. Thus,

$$m_{alc}L_v = m_p c_p \Delta T$$

We can solve this for the unknown ΔT, giving

$$\Delta T = \frac{m_{alc}L_v}{m_p c_p}$$

Solution We know or can look up all quantities on the right, except for the mass of alcohol. Its density is found in Table 10.1 to be 790 kg/m³, or 0.790 kg/L; thus, half a liter has a mass of 0.395 kg. Now, substituting numerical values into the preceding equation, we get

$$\Delta T = \frac{(0.395 \text{ kg})(204 \text{ kcal/kg})}{(55.0 \text{ kg})(0.83 \text{ kcal/kg} \cdot °C)} = 1.8°C$$

Discussion The result of 1.8°C is roughly 3°F, a significant decrease. In practice, the actual temperature change will differ somewhat from this value because of the assumptions made. For example, the heat of vaporization of alcohol is greater than 204 kcal/kg, because the temperature is below alcohol's boiling point. Additionally, the cooling effect must be transferred throughout the body, giving it time to produce more heat. Furthermore, some heat for evaporation may come from the environment. Nevertheless, the effect is very real, since phase changes in general involve large amounts of energy compared with that in temperature changes.

We have seen that vaporization uses heat. Condensation is the reverse process, and it releases heat. This may seem surprising, since we associate condensation with cold objects—the glass in Figure 13.9, for example. But heat must be removed to make a vapor condense. The amount of heat is exactly the same as that required to make the phase change in the other direction, from liquid to vapor, and so it can be calculated from Equation 13.3.

Heat is also released when a liquid freezes. (A freezer must remove this heat to make ice cubes.) The heat released in this phase change can actually be used to prevent frost damage in orchards. This process, surprisingly, involves spraying the trees with water. (See Figure 13.10.) Most plants can tolerate temperatures a few degrees below 0°C. If significant amounts of water are sprayed, then a large amount of heat is emitted at 0°C as the water freezes, and spraying can prevent temperatures from dropping to lower, more damaging levels.

Figure 13.9 Condensation forms on a glass of ice water because the temperature of nearby air is reduced below the dew point. The air cannot hold as much water as it did at room temperature, and so water condenses. Heat of vaporization is released when the water condenses, speeding the melting of the ice in the glass.

Figure 13.10 The ice on these trees released large amounts of heat when it froze, helping to prevent the temperature of the trees from dropping below 0°C. Water is intentionally sprayed on orchards to help prevent hard frosts.

Human Application

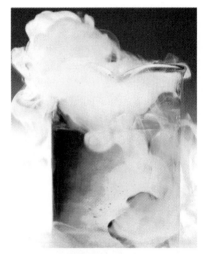

Figure 13.11 Direct transitions between solid and vapor are common, sometimes useful and even beautiful. (a) Dry ice sublimates directly to carbon dioxide gas. The visible vapor is made of water droplets. (b) Frost forms patterns on a very cold window, an example of a solid formed directly from a vapor.

Have you ever noticed that snow can disappear into thin air without a trace of liquid water? Or seen ice cubes slowly wasting away in a freezer? These are examples of sublimation, the direct transition from solid to vapor phase, first defined in Section 12.5. Water is not the only substance that can sublimate. As shown in Figure 13.11, frozen carbon dioxide, known as dry ice, does the same thing. (In fact, carbon dioxide cannot exist as a liquid at atmospheric pressure.) Certain air fresheners use the sublimation of a solid to inject a perfume into the room. Moth balls are a slightly toxic example of a phenol that sublimates, while some solids, such as osmium tetroxide, are so toxic they must be kept in sealed containers to prevent human exposure to their sublimation-produced vapors. As noted in the discussion of phase changes in Section 12.5, direct transitions from solid to vapor occur because the vapor pressure of solids is not zero. The reverse process, solids forming from vapors, also takes place. Frost can form on very cold windows without first going through the liquid stage, for example.

All phase transitions involve energy. In the case of direct solid–vapor transitions, the energy required is given by the equation $Q = mL_s$, where L_s is the **heat of sublimation**, the energy required to change 1.00 kg of a substance from the solid to vapor phase. L_s is analogous to L_f and L_v, and it depends on substance. Sublimation requires energy input, so that dry ice is an effective coolant, whereas frost releases energy. The amount of energy is on the same order of magnitude as that for other phase transitions.

The material presented in this section and the preceding one allows us to calculate any number of effects related to temperature and phase change. In each case, it is necessary to identify which temperature and phase changes are taking place and then to apply the appropriate equation. It should always be kept in mind that either heat or work can cause both temperature and phase changes.

PROBLEM-SOLVING STRATEGIES

FOR THE EFFECTS OF HEAT

Step 1. Examine the situation to determine that there is a change in the temperature or phase caused by heat entering or leaving a system. When the presence or absence of a phase change is not obvious, you may wish to first solve the problem as if there is no phase change and examine the temperature change obtained. If it is great enough to take you past a boiling or melting point, you would then go back and do the problem in steps—temperature change, phase change, subsequent temperature change, and so on.

Step 2. Identify and list all objects that change temperature and phase.

Step 3. Identify exactly what needs to be determined in the problem (identify the unknowns). A written list is useful.

Step 4. Make a list of what is given or can be inferred from the problem as stated (identify the knowns).

Step 5. Solve the appropriate equation for the quantity to be determined (the unknown). For a temperature change, Equation 13.2 is appropriate. Table 13.1 lists information relevant to temperature change. For a phase change, Equation 13.3 is appropriate. Table 13.2 lists information relevant to phase change.

Step 6. Substitute the knowns along with their units into the appropriate equation, and obtain numerical solutions complete with units. You will need to do this in steps if there is more than one stage to the process (such as a temperature change followed by a phase change).

Step 7. Check the answer to see if it is reasonable: Does it make sense? For example, be certain that the temperature change does not also cause a phase change that you have not taken into account.

13.4 INTRODUCTION TO HEAT TRANSFER METHODS

Equally interesting as the effects of heat on a system are the methods by which heat flows or is transferred. Whenever there is a temperature difference, heat will flow. It may flow rapidly, such as through a cooking pan, or slowly, such as through the walls of a picnic ice chest. We can manipulate heat flow rates by choosing materials (such as thick wool clothing for the winter), controlling air movement (such as by using weather stripping around doors), or by choice of color (such as a white roof to reflect the summer sun). So many processes involve heat flow that it is hard to imagine a situation where no heat at all flows. Yet every process involving heat flow proceeds by only three methods:

1. **Conduction** is the transfer of heat through stationary matter by physical contact. (The matter is stationary on a macroscopic scale—we know there is thermal motion of the atoms and molecules at any temperature above absolute zero.) Heat transferred between the electric burner on a stove and the bottom of a pan is transferred by conduction.
2. **Convection** is the transfer of heat by the macroscopic movement of mass. This is what takes place in a home with a forced-air furnace, for example.
3. Heat transfer by **radiation** occurs when microwaves, infrared radiation, visible light, or another form of electromagnetic radiation is emitted or absorbed. An obvious example is the warming of the earth by the sun. Less obvious is the radiation of infrared from an automobile radiator.

We examine these methods in some detail in the three following sections. Each method has unique and interesting characteristics, but all three do have this in common: they transfer energy solely because of temperature difference. By definition, such energy is heat. (See Figure 13.12.)

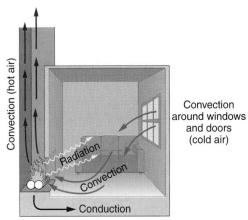

Figure 13.12 A fireplace transfers heat by all three methods: conduction, convection, and radiation. Radiation is responsible for most of the heat entering the room. Conduction also moves heat into the room, but at a much slower rate. Convection removes heat from the room through the cold air entering around windows and the hot air rising up the chimney.

13.5 CONDUCTION

As you walk barefoot across the living room carpet in your cold house and step onto the tile floor in the kitchen, you expect the cold sensation that follows. What is intriguing is that the carpet and tile are at the same temperature. Why should one feel colder than the other? The answer is that tile conducts heat away from your feet much more rapidly than carpet. In fact, the *rate at which heat is transferred* is at least as important as the amount transferred.

Obviously, some materials conduct heat faster than others. What else besides the choice of material affects the rate of heat conduction? Taking a look at what happens at the submicroscopic level, as in Figure 13.13, helps us to see the major factors involved. In addition to the choice of material, its thickness affects the rate of heat transfer, as do the temperature difference and the contact area.

Imagine what happens when molecules collide at the surface between two bodies at different temperatures, as in Figure 13.13. Molecules in the higher-temperature object have greater average kinetic energy. The most probable result of a collision between a low-energy and a high-energy molecule is the transfer of energy from high to low. This is the mechanism via which the net transfer of heat from hot to cold takes place by conduction. We can tell this energy transfer is heat because it is random in character.

The mechanism by which conduction proceeds has clear implications. First, the greater the temperature difference, the greater the difference in average molecular kinetic

CONNECTIONS

Heat transfer by conduction is due to atomic and molecular collisions at the contact surface between the two bodies. This means that the laws of conservation of momentum and energy are involved, even though the two macroscopic bodies are at rest. Heat conduction through a body takes place in part by the same collisions, but it can also be carried by free electrons. As we shall see in later chapters, the same materials that conduct heat rapidly also tend to conduct electricity well. Much of both types of conduction is carried by free electrons.

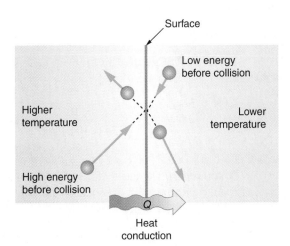

Figure 13.13 The molecules in two bodies at different temperatures have different average kinetic energies. Collisions occurring at the contact surface tend to transfer energy from high- to low-temperature regions.

Figure 13.14 Heat is conducted through any material, here represented by a rectangular bar, be it window glass or walrus blubber. The rate of heat transfer by conduction is directly proportional to the surface area A, the temperature difference $T_2 - T_1$, and the substance's conductivity k. The rate of heat transfer is inversely proportional to the thickness d.

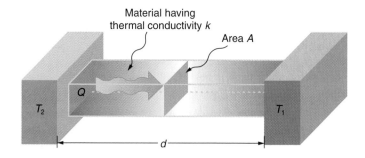

energies, and so the faster the rate of heat transfer. Thus you will get a more severe burn from boiling water than from hot tap water. Conversely, if the temperatures are the same, the net heat transfer rate falls to zero, and equilibrium is achieved.

Second, the greater the contact area, the greater the number of collisions per unit time and the greater the rate of heat transfer. For example, if you touch a cold wall with your palm, your hand cools faster than if you just touch it with a fingertip.

A third factor is the thickness of the material through which the heat moves. Figure 13.14 shows a slab of material with different temperatures on either side. Suppose that T_2 is greater than T_1, so that heat moves from left to right. Heat entering the left side is transferred to the right by submicroscopic collisions. The thicker the material, the more time it takes for this process to carry the heat to the cold side. This confirms the common knowledge that thick clothing is warmer in winter than thin clothing and explains the thick blubber on Arctic animals.

Fourth is the more complex dependence on the substructure of matter. Some substances transfer heat more effectively than others. This is the characteristic we refer to as good or poor *conductivity*—some substances, like most metals (which have many free electrons), transfer heat very rapidly, whereas others, like rubber, do not.

All four factors are included in a simple equation that was deduced from and is confirmed by experiments. The **rate of conductive heat transfer** through a slab of material such as the one in Figure 13.14 is given by

$$\frac{Q}{t} = \frac{kA(T_2 - T_1)}{d} \tag{13.4}$$

where Q/t is the rate of heat transfer in watts or kcal/s, k is the **thermal conductivity** of the material, A and d are its surface area and thickness as shown in Figure 13.14, and $T_2 - T_1$ is the temperature difference across the slab. Table 13.3 gives representative values of thermal conductivity.

EXAMPLE 13.5 CONDUCTION RATE THROUGH AN ICE CHEST

A certain Styrofoam ice chest has a total area of 0.950 m^2 and walls that average 2.50 cm thick. The contents—ice, water, and canned beverages at 0°C—are kept cold by melting ice. How much ice melts in one day if the ice chest is kept in the 35.0°C trunk of a car?

Strategy and Concept This question involves more than one concept. To find the amount of ice melted, we must find the total heat transferred. This can be obtained by calculating the rate of heat transfer by conduction and multiplying by time.

Solution The rate of heat transfer by conduction is given by Equation 13.4:

$$\frac{Q}{t} = \frac{kA(T_2 - T_1)}{d}$$

Substituting known values yields

$$\frac{Q}{t} = \frac{(0.010 \text{ J/s} \cdot \text{m} \cdot °\text{C})(0.950 \text{ m}^2)(35.0°\text{C})}{0.0250 \text{ m}} = 13.3 \text{ J/s}$$

This is a small rate of heat flow, only 13.3 W. The *total* amount of heat transferred in 1 day (86,400 s) is the rate of heat transfer multiplied by the number of seconds in one day. That is,

$$Q = (Q/t) \cdot t = (13.3 \text{ J/s})(86,400 \text{ s}) = 1.15 \times 10^6 \text{ J}$$

The amount of ice that this heat will melt is found by solving $Q = mL_f$ for m:

$$m = \frac{Q}{L_f} = \frac{1.15 \times 10^6 \text{ J}}{334 \times 10^3 \text{ J/kg}} = 3.44 \text{ kg}$$

Discussion The result of 3.44 kg, or about 7.6 lb, seems about right. Based on experience, you might expect to use a 7 to 10 lb bag of ice a day, with a little extra required if you add any warm food or beverages.

Inspecting the conductivities in Table 13.3 reveals that Styrofoam is a very poor conductor. A poor conductor is the same as a good insulator. Other good insulators include glass wool, rock wool, wool itself, and goose down. Like Styrofoam, these all incorporate many pockets of air, taking advantage of air's poor conductivity. Nature was first here, with eons of feathers and fur.

A combination of material and thickness is often manipulated to get good insulators—the smaller the conductivity k and the larger the thickness d, the better. The ratio of $R = d/k$ will thus be large for a good insulator. The ratio d/k is called the **R factor**. The rate of conductive heat transfer is inversely proportional to R. The larger R is, the better the insulation. R factors are most commonly quoted for household insulation, refrigerators, and the like—unfortunately, still in British units of $ft^2 \cdot {}°F \cdot h/Btu$, although the units usually go unstated. A couple of representative values are an R factor of 11 for 3.5 in. batts of insulation and an R factor of 19 for 6.5 in. batts. Walls are usually insulated with 3.5 in. batts, while ceilings are usually insulated with 6.5 in. batts. In cold climates, thicker batts may be used in ceilings and walls.

Note that in Table 13.3 the best thermal conductors—silver, copper, gold, and aluminum—are also the best electrical conductors, again related to the density of free electrons in them. A cooking utensil typically is made from a good conductor.

TABLE 13.3

THERMAL CONDUCTIVITIES OF COMMON SUBSTANCES*

Substance	Thermal conductivity k $(J/s \cdot m \cdot °C)$
Silver	420
Copper	390
Gold	318
Aluminum	220
Steel, iron	80
Steel (stainless)	14
Ice	2.2
Glass (average)	0.84
Concrete, brick	0.84
Water	0.6
Fatty tissue, without blood	0.2
Asbestos	0.16
Plasterboard	0.16
Wood	0.08–0.16
Snow (dry)	0.10
Cork	0.042
Glass wool, rock wool	0.042
Wool	0.04
Down	0.025
Air	0.023
Styrofoam	0.010

*At temperatures near 0°C

EXAMPLE 13.6 CONDUCTION THROUGH AN ALUMINUM PAN

A cast aluminum sauce pan with a 0.800 cm thick bottom that is 14.0 cm in diameter is sitting on an electrical stove and is boiling water at the rate of 1.00 g/s. What is the temperature difference across (through) the bottom of the pan?

Strategy Conduction through the aluminum is the primary method of heat transfer here, and so we use Equation 13.4 and solve for the temperature difference $T_2 - T_1$:

$$\frac{Q}{t} = \frac{kA(T_2 - T_1)}{d}$$

Thus,

$$T_2 - T_1 = \frac{Q}{t}\left(\frac{d}{kA}\right)$$

Values for d, k, and A can be found from the given information and from Table 13.3. But we must calculate Q/t from the information that 1.00 g of water is boiling away each second. The heat goes to phase change, so that $Q = mL_v$; thus,

$$\frac{Q}{t} = \frac{mL_v}{t}$$

$$= \frac{(1.00 \times 10^{-3}\ kg)(2256 \times 10^3\ J/kg)}{1.00\ s} = 2256\ J/s$$

This is about 2.26 kW and is typical for an electric stove.

Solution Now d, k, and $A = \pi r^2 = 1.54 \times 10^{-2}\ m^2$ can all be entered into the equation for the temperature difference, yielding

$$T_2 - T_1 = (2256\ J/s)\ \frac{8.00 \times 10^{-3}\ m}{(220\ J/s \cdot m \cdot °C)(1.54 \times 10^{-2}\ m^2)}$$

$$= 5.33°C$$

Discussion This is a remarkably small temperature difference. Consider that the stove burner is red hot while the inside of the pan is near 100°C because of its contact with boiling water. This contact effectively cools the bottom of the pan in spite of its proximity to the very hot stove burner. Aluminum is such a good conductor that it only takes this small temperature difference to produce a heat flow of 2.26 kW into the pan.

Of course, conduction is not always the only method of heat transfer, or even the dominant method as it was in the previous examples. Indeed, it may even be absent, as it is in a vacuum. But whenever conduction takes place, it has the characteristics described in this section.

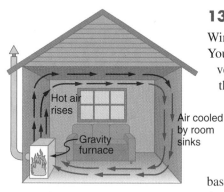

Figure 13.15 Air heated by the so-called gravity furnace expands and rises, forming a convective loop that transfers heat to other parts of the room. As the air is cooled on the ceiling and outside wall, it contracts, eventually becoming more dense than room air and sinking to the floor. A properly designed heating system using natural convection, like this one, can be quite efficient in uniformly heating a home.

Human & Biological Application

13.6 CONVECTION

Winds, ocean currents, and the plume of a fire all transfer heat by natural convection. Your circulatory system and an automobile air conditioner transfer heat by forced convection. Since convection transfers heat by the macroscopic movement of mass through macroscopic distances, it is generally a much faster method of heat transfer than conduction. Convection is also much more complicated than conduction, especially when the flow is turbulent, as it usually is. Nevertheless, we can describe convection rather well and do some straightforward and realistic calculations of its effects.

Natural convection is driven by buoyant force. The familiar adage that heat rises is based on the fact that most substances expand when they warm up, so that their density decreases, and they rise. The house in Figure 13.15 is quietly heated in this manner, as is the pot of water on the stove in Figure 13.16. Ocean currents and large-scale atmospheric circulation carry huge quantities of heat from one part of the globe to another. Both are examples of natural convection.

Forced convection is driven by mechanical forces, such as exerted by fans or the lungs. The circulatory system is another example of heat transfer by forced convection. When overheated, the body dilates blood vessels to the skin, sending more blood there to be cooled. Of course, if the skin is hotter than internal temperature, as when you are in a hot tub, the effect may be to put more heat into the body rather than remove it. This can dangerously increase core temperature, causing vessels to further dilate and the heart to pump more blood and more heat at an even faster rate. Each breath of air we take can also transfer heat into or, more commonly, out of the body.

There are many other examples of heat transfer by forced convection. Forced-air heating and cooling systems are mainstays in our homes, workplaces, and schools. Automobile water pumps and hair dryers also use forced convection. Whether forced or natural, convection often transfers heat very rapidly.

EXAMPLE 13.7 CONVECTION OF AIR THROUGH THE WALLS OF A HOUSE

Most houses are anything but airtight. Air goes in and out around doors and windows, through cracks and crevices, following wiring to switches and outlets, and so on. The air in a typical house is completely replaced in less than an hour. Suppose that a moderate-sized house has inside dimensions equivalent to a rectangle 12.0 m wide by 18.0 m long and 3.00 m high, and that all its air is replaced in 30.0 min. Calculate the amount of heat per unit time in watts that is needed to warm the air by 10.0°C, thus replacing the heat lost by convection alone.

Strategy We are asked to find the amount of heat per unit time, and the time of complete turnover is given to be 30.0 min. First, then, we must find the amount of heat needed to warm the air, which is

$$Q = mc\,\Delta T$$

We are given that ΔT is 10.0°C, but we must still find values for the mass of air and its specific heat before we can calculate Q. The specific heat of air is a weighted average of the specific heats of nitrogen and oxygen, which gives $c = c_p \approx 1000\ \text{J/kg}\cdot\text{°C}$ from Table 13.1 (note that the specific heat at constant pressure must be used for this process). The mass of air can be determined from its density (found in Table 10.1) and the given volume of the house. That is,

$$m = \rho V = (1.29\ \text{kg/m}^3)[(12.0 \times 18.0 \times 3.00)\ \text{m}^3] = 836\ \text{kg}.$$

Solution We substitute the values just found into the equation giving the heat needed for a temperature change:

$$Q = mc\,\Delta T$$
$$= (836\ \text{kg})(1000\ \text{J/kg}\cdot\text{°C})(10.0\text{°C}) = 8.36 \times 10^6\ \text{J}$$

The heat needed per unit time is thus

$$\frac{Q}{t} = \frac{8.36 \times 10^6\ \text{J}}{1800\ \text{s}} = 4.6\ \text{kW}$$

Discussion This rate of heat production in watts (expressed to two digits, since c_p is approximate) is seen to be equivalent to the power consumed by about forty-six 100 W light bulbs. Newly constructed homes are designed for a turnover time of 2 hours, rather than the half-hour for the house of this example. Weather stripping, caulking, and improved window seals are commonly employed. More extreme measures are sometimes taken in very cold (or hot) climates to achieve a tight standard of more than 6 hours for one air turnover. Still longer turnover times are unhealthy, because a minimum amount of fresh air is necessary to supply oxygen for breathing and to dilute household pollutants.

TABLE 13.4

WIND-CHILL FACTORS*

Moving air temperature (°C)	Wind Speed (m/s)				
	2	5	10	15	20
5	3	−1	−8	−10	−12
2	0	−7	−12	−16	−18
0	−2	−9	−15	−18	−20
−5	−7	−15	−22	−26	−29
−10	−12	−21	−29	−34	−36
−20	−23	−34	−44	−50	−52
−40	−44	−59	−73	−82	−84

*Approximate temperature in °C of still air that produces the same rate of cooling as winds at the speeds and temperatures listed.

Figure 13.16 Convection plays an important role in transferring heat inside this pot of water. Once conducted to the inside, heat is transferred to other parts of the pot mostly by convection. The hotter water expands, decreases in density, and rises to transfer heat to other parts of the pot, while colder water sinks to be heated at the bottom.

A cold wind is much more chilling than still air because convection aids conduction in removing heat. Table 13.4 gives approximate wind-chill factors, which are the temperatures of still air that produce the same rate of cooling as air of a given temperature and speed. Wind-chill factors are a dramatic reminder of convection's ability to move heat faster than conduction. For example, a 15.0 m/s wind at 0°C has the chilling equivalent of still air at about −18°C.

Although air can carry heat rapidly by convection, we have seen in the previous section that air is actually a poor conductor. Of course, a poor conductor is a good insulator. The key to whether air will act as an insulator or transfer heat by convection is the amount of space it has to move around in. The space between the inside and outside walls of a house, for example, is about 3.5 in. (or about 9 cm)—large enough for convection to work very effectively. This is why we need to fill that air gap with insulation. In contrast, the gap between the two panes of a double-paned window is about 1 cm. This space is about ideal for preventing heat transfer. It is small enough to prevent convection and large enough to take advantage of air's low conductivity and prevent conduction, too. Fur also takes advantage of the low conductivity of air by trapping it in spaces too small to support convection, as shown in Figure 13.17. Fur enjoys the added advantage, for an insulator, of having low mass. Feathers and glass wool insulation also use air as an effective low-mass insulator.

Some very interesting things happen *when convection is accompanied by a phase change*. The energy involved in the phase change can be taken from one system to another. In some cases, that energy can even be *transferred from a cool system to a warmer one*. For example, energy is required to evaporate sweat from our skin. Some of that heat is removed from the skin, thus reducing its temperature. If there is no convection in the surrounding air, the evaporation will quickly cease as the air around the body becomes saturated. But if there is convection, the saturated air is carried away, taking with it the energy used to evaporate the sweat and supplying less humid air—thus allowing more evaporation to occur. This takes heat out of the body even when surrounding temperatures are greater than body temperature. At the same time, all other methods of heat transfer are putting heat into the body; thus, convection accompanied by this phase change is our only way of cooling off.

Human Application

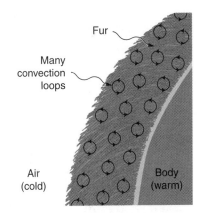

Figure 13.17 Fur is filled with air, breaking it up into many small pockets. Convection is very slow here because the loops are so small. The low conductivity of air thus makes fur a very good lightweight insulator.

Human & Biological Applications

EXAMPLE 13.8 SWEAT RIDS US OF BODY HEAT

The average person produces body heat at the rate of 120 W when sitting at rest. At what rate must water evaporate from the body to get rid of all this body heat? (This might occur when a person is sitting in the shade and surrounding temperatures are the same as skin temperature, eliminating heat transfer by other methods.)

Strategy All of the body heat goes into the energy needed for a phase change ($Q = mL_v$). Thus the amount of

heat per unit time is

$$\frac{Q}{t} = \frac{mL_v}{t} = 120 \text{ W} = 120 \frac{\text{J}}{\text{s}}$$

We divide both sides of the equation by L_v to find that the mass evaporated per unit time is

$$\frac{m}{t} = \frac{120 \text{ J/s}}{L_v}$$

(continued)

(continued)

Solution The problem can be solved by entering the value of L_v, which is seen in Table 13.2 to be 2430 kJ/kg, or 2430 J/g, at body temperature. This yields

$$\frac{m}{t} = \frac{120 \text{ J/s}}{2430 \text{ J/g}} = 0.0494 \text{ g/s} = 2.96 \text{ g/min}$$

Discussion Evaporating about 3 g/min seems reasonable. This would be about 180 g (about 7 ounces) per hour. If the air is very dry, the sweat may evaporate without even being noticed. A significant amount of the evaporation also takes place in the lungs and breathing passages.

Figure 13.18 Here convection accompanied by a phase change releases the heat needed to drive this thunderhead into the stratosphere.

Figure 13.19 The phase change that occurs when this iceberg melts involves a tremendous amount of heat.

Another important example of the combination of phase change and convection occurs when water evaporates from the oceans. The evaporation removes heat from the oceans, releasing that heat when condensation takes place. This is the driving power behind thunderheads, those great cumulus clouds that rise dramatically as much as 20.0 km into the stratosphere. Water vapor carried in by convection condenses, releasing tremendous amounts of heat. This heat causes the air to expand and rise to greater heights, where it is colder. More condensation occurs in these colder regions, which in turn drives the cloud even higher. This is called positive feedback, since the process reinforces and accelerates itself. (See Figure 13.18.) Such systems sometimes produce violent storms, replete with lightning and hail, and constitute the mechanism driving hurricanes. Heat can be released from condensation in regions warmer or colder than where the evaporation took place, and so it is possible again to transfer heat from a cool environment to a warmer one. In the case of hurricanes, the heat can be converted into work in a giant heat engine not unlike the smaller ones we will discuss in the next chapter.

The movement of icebergs is another example of convection accompanied by a phase change. (See Figure 13.19.) Suppose an iceberg is carried south from Greenland into warmer Atlantic waters. The ice absorbs a great amount of heat when it melts, effectively taking heat out of the warmer water. When the ice formed, it released the same amount of heat in Greenland. This particular convection, which takes ice southward, transfers heat from south to north (or cold from north to south, loosely speaking).

13.7 RADIATION

You can feel the heat from a fireplace without being close enough for either convection or conduction to bring it to you. (See Figure 13.20.) The same is true of an oven. You can

Figure 13.20 Most of the heat from this fire reaches the observers by infrared radiation. The visible light, although dramatic, carries relatively little energy. Convection takes heat away from the observer as hot air rises, while conduction is negligibly slow here. Skin is very sensitive to infrared radiation, so that you can sense the presence of a fire without looking at it directly.

sometimes tell that an oven is hot without touching its door or looking inside—it may just warm you as you walk by. These are examples of heat transfer by radiation, a process distinctly different from both conduction and convection. Radiation is the transfer of energy by electromagnetic waves, such as radio waves, microwaves, infrared radiation, visible light, and ultraviolet radiation. While conduction and convection obviously must travel through matter, radiation can go through either vacuum or matter, and a vacuum transmits radiation without absorption.

Electromagnetic radiation takes place for many reasons. In this section, we concentrate on radiation caused by a temperature difference—that is, heat transferred by radiation. We will refer to this simply as radiation. (Other types of radiation will be covered in other chapters.) In particular, we will explore how radiation is created and how the rate of heat transfer by radiation depends on temperature, color, and surface area.

The thermal motion of atoms and molecules in any object at a temperature above absolute zero causes them to radiate electromagnetic waves.* Thermal motion increases with temperature, and so, as you would expect, the amount of radiation increases with temperature. An object radiates more when it is hot than when it is cool. Temperature change also shifts the type of radiation emitted. For example, the radiated heat you feel from an oven is mostly in the infrared, while the hotter electrical element inside glows red to orange. At higher temperatures, such as that of steel in a blast furnace, the glow becomes yellow-white. Bluish light with some ultraviolet is emitted at even higher temperatures, such as found on the surfaces of some stars. Figure 13.21 shows how the spectra of electromagnetic waves vary with temperature.

All objects absorb radiation as well as radiate it. Of course, some things absorb radiation faster than others. It is the *color* of the object that affects its rate of heat transfer by radiation. Black is most effective. Most of us avoid wearing black clothing in the hot summer sun, for instance. Similarly, black asphalt in a parking lot will be hotter than the adjacent gray sidewalk on a sunny summer day, because black absorbs better than gray. The reverse is also true—that is, black radiates better than gray. Thus, on a clear summer night, the asphalt will be colder than the gray sidewalk, because black radiates the heat away more rapidly. An *ideal radiator* is the same color as an *ideal absorber*—jet black. More precisely, an **ideal absorber** captures all the radiation that falls on it. In contrast, white is a poor absorber and it is also a poor radiator. A black object absorbs all radiation falling on it, whereas a white object reflects all radiation, like a perfect mirror. (A perfect, polished white surface is mirrorlike in appearance, and a crushed mirror looks white.) (See Figures 13.22 and 13.23.)

---

*The reasons for this involve the acceleration of charges, which we will discuss in later chapters.

CONNECTIONS

When electromagnetic radiation acts to transfer heat, it is a perfect illustration of how *heat can exist only in transit.* Visible light, for example, can carry heat from the sun to the earth, but it is converted to some other form of energy when absorbed by the earth and is not stored as light. This is true of all heat transfer, whether it is by conduction, convection, or radiation.

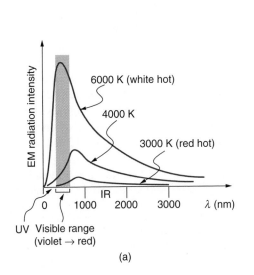

(a)

(b)

Figure 13.21 (a) A graph of the spectra of electromagnetic waves emitted from an ideal radiator at three different temperatures. The intensity or rate of radiation emission increases dramatically with temperature, and the spectrum shifts toward the visible and ultraviolet parts of the spectrum. The shaded portions denote the visible parts of the spectra. From these, it is apparent that the shift toward the ultraviolet with temperature makes the visible part go from red to white to blue in appearance as temperature increases. (b) Note the variations in color corresponding to variations in flame temperature.

Figure 13.22 This photograph clearly shows that the darker pavement is hotter than the lighter pavement (much more of the ice on the right has melted), although both have been in the sun for the same time.

Absorb **Radiate**

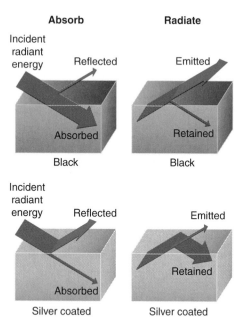

Figure 13.23 A black object both absorbs and radiates well. A white object is a very poor absorber and a very poor radiator. It is as if radiation from the inside is reflected back into the white object, whereas radiation from the inside of the black object is "absorbed" when it hits the surface and finds itself on the outside.

Human Applications

Shadings of white through gray to black make up the simplest colors, having a uniform ability to absorb all parts of the spectrum. Colored objects behave in similar but more complex ways, preferentially absorbing and radiating various types of electromagnetic radiation. This gives them color in the visible range, and it may make them special in other ranges. For example, the strong absorption of infrared by the skin allows us to be very sensitive to it.

What affects the rate of heat transfer by radiation emitted from some object? There is a relatively simple equation for this. Before writing it down, let us re-examine the factors mentioned above. First, we know the rate of radiation depends on temperature. It is an experimental fact that it is directly proportional to the *fourth power* of the absolute temperature—a remarkably strong temperature dependence. Second, the rate depends on color. It ranges from rapid for black to slow for white. Third, the greater the surface area of the object, the more rapidly it radiates. If you knock the coals of a fire apart, there can be a noticeable increase in radiation due to an increase in radiating surface area. All of this is expressed in the **Stefan-Boltzmann law of radiation** for the **rate of radiation** from an object:

$$\frac{Q}{t} = \sigma e A T^4 \tag{13.5}$$

where $\sigma = 5.67 \times 10^{-8}$ J/s·m²·K⁴ is the **Stefan-Boltzmann constant** and is determined from experiment, A is the surface area of the object, and T is its absolute temperature in kelvins. The symbol e stands for the **emissivity** of the object, a measure of how well it radiates. An ideal jet-black radiator has $e = 1$, whereas a perfect reflector has $e = 0$. Real objects fall between these two values. For example, tungsten light bulb filaments have an e of about 0.5, and lampblack has the greatest emissivity known, about 0.99.

Skin is a remarkably good absorber and emitter of infrared radiation, having an emissivity of 0.97 in the infrared. This means that we are all nearly jet black in the infrared, in spite of the obvious variations in skin color in the visible part of the spectrum. This is why we can so easily feel heat falling on our skin. Our ability to emit infrared is the basis of night scopes used militarily to detect body heat. We absorb *and* radiate very well in the infrared. Even small temperature variations can be detected because of the T^4 dependence of radiation on absolute temperature. Images, caled *thermographs,* can be used medically to detect fevered regions of the body perhaps indicative of disease. Similar techniques can be used to detect heat leaks in homes (see Figure 13.24) or even the warm exhaust of an enemy submarine.

All objects both emit and absorb radiation. The *net* rate of heat transfer by radiation (absorption minus emission) is related to both the temperature of the object and that of

Figure 13.24 Thermograph of a home shows temperature variations indicative of where heat loss is most severe. The windows are seen to be a major source of heat loss.

its surroundings. Assuming that an object with a temperature T_1 is surrounded by an environment with uniform temperature T_2, the **net rate of heat transfer by radiation** is

$$\frac{Q}{t} = \sigma e A (T_2^4 - T_1^4) \qquad (13.6)$$

where e is the emissivity of the object alone. In other words, there is no dependency on whether the surroundings are white, gray, or black; the balance of radiation into and out of the object depends on how well *it* emits and absorbs radiation. When Q/t is positive, there is a net influx of radiation; as in all simple processes, the net flow of heat is from hot to cold.

EXAMPLE 13.9 HOW MUCH DOES A PERSON RADIATE?

What is rate of heat transfer by radiation from an unclothed person standing in a dark room with a temperature of 22.0°C, if the person has a normal 33.0°C skin temperature and a surface area of 1.50 m²? The emissivity of skin is 0.97 in the infrared, where the radiation takes place.

Strategy We can solve this by using Equation 13.6, since all factors are given and only heat transfer by radiation is asked for.

Solution Recall that absolute temperature in kelvins must be used; thus, $T_2 = 295$ K and $T_1 = 306$ K, so that

$$\frac{Q}{t} = \sigma e A (T_2^4 - T_1^4)$$
$$= (5.67 \times 10^{-8} \text{ J/s} \cdot \text{m}^2 \cdot \text{K}^4)(0.97)(1.50 \text{ m}^2)$$
$$\times [(295 \text{ K})^4 - (306 \text{ K})^4]$$
$$= -99 \text{ J/s} = -99 \text{ W}$$

Discussion This is a significant rate of heat loss (note the minus sign), considering that a person at rest may produce only 125 W of body heat and that conduction and convection will also be removing heat. And, indeed, we would probably expect to get cold if we did this. Clothing would significantly reduce heat loss by all methods, because it slows both conduction and convection, and because it has a lower emissivity (especially if it is white) than skin.

Radiation is almost exclusively the method by which the earth receives heat from the sun and gives it up to outer space. Of course, the sun is much hotter than the earth, and so the net transfer is to the earth. The rate of transfer is less than Equation 13.6 would predict though, because the sun does not fill the sky. This effectively gives the daytime sky a lower average temperature than the sun's. The average emissivity of the earth is about 0.65, but this is complicated by the fact that cloud cover, which is highly reflective, varies. There is a negative feedback (one in which a change produces an effect that opposes that change) between clouds and heat; greater temperatures evaporate more water to form more clouds, which reflect heat back into space, reducing temperature. The oft-mentioned **greenhouse effect** is directly related to the variation of the earth's emissivity with radiation type. (See Figure 13.25.) Certain atmospheric gases, carbon dioxide

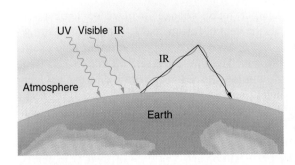

Figure 13.25 The greenhouse effect is a name given to the trapping of heat in the earth's atmosphere by a process very similar to that used in greenhouses. The atmosphere, like window glass, is transparent to incoming visible radiation and some of the sun's infrared. These are absorbed by the earth, and reemitted as infrared. The atmosphere, like glass, traps these rays, keeping the earth warmer than it would otherwise be. The amount of trapping depends on trace gases like carbon dioxide; changes in the quantities of these gases could affect the entire earth's temperatures.

Figure 13.26 This simple but effective solar cooker uses the greenhouse effect and insulation to trap solar energy. Made of inexpensive, durable materials, it is a great money and labor saver and is of particular economic value in energy-poor Third World countries.

among them, absorb visible radiation from the sun much better than they emit infrared from the earth, effectively increasing the earth's temperature, in the same way glass increases temperatures in a greenhouse. (See Figure 13.26.) There has been a tremendous increase in carbon dioxide, and other so-called greenhouse gases, due to fossil fuel burning in recent centuries. This will affect the earth's climate in ways we do not yet know.

Conversely, dark space is very cold, about 3 K, and so the earth radiates heat into the dark sky. Clouds reflect the radiated heat back to the surface, greatly reducing heat loss, just as they greatly reduce heat gain during the day. In other words, clouds have a much lower emissivity than the oceans and land masses. The rate of heat loss from soil can be so rapid that radiative frost may occur on clear summer evenings, even in warm latitudes. A tiny amount of the heat absorbed on the day side of the earth is stored for long periods of time and is found in many of our sources of energy, such as fossil fuels, as well as wood, hydroelectric power, and wind energy. Since carbon dioxide is not the only greenhouse gas, nor even the primary one (water vapor is), its effect is not as dramatic as its increase in the atmosphere.

FOR THE METHODS OF HEAT TRANSFER

Step 1. *Examine the situation to determine that the rate of heat transfer is involved.*

Step 2. *Identify the type of heat transfer—conduction, convection, or radiation.*

Step 3. *Identify exactly what needs to be determined in the problem (identify the unknowns). A written list is useful.*

Step 4. *Make a list of what is given or can be inferred from the problem as stated (identify the knowns).*

Step 5. *Solve the appropriate equation for the quantity to be determined (the unknown). For conduction, Equation 13.4 is appropriate. Table 13.3 lists thermal conductivities. For convection, determine the amount of matter moved and use Equation 13.2 to calculate the heat involved in the temperature change. If a phase change accompanies convection, Equation 13.3 is appropriate to find the heat involved in the phase change. Table 13.2 lists information relevant to phase change. For radiation, Equation 13.6 gives the net heat transfer rate.*

Step 6. *Substitute the knowns along with their units into the appropriate equation, and obtain numerical solutions complete with units.*

Step 7. *Check the answer to see if it is reasonable: Does it make sense?*

SUMMARY

Heat and work are the two distinct methods of energy transfer. **Heat** is energy transferred solely because of a temperature difference. Any energy unit can be used for heat, but the most common is the **kilocalorie** (kcal), defined to be the amount of energy needed to change the temperature of 1.00 kg of water between 14.5°C and 15.5°C. The **mechanical equivalent** of this amount of heat is

$$1.000 \text{ kcal} = 4.186 \text{ J} \qquad (13.1)$$

The two distinct effects of heat on a system are temperature change and phase change. (Work can produce identical effects.)

The amount of heat involved in a temperature change is

$$Q = mc \, \Delta T \qquad (13.2)$$

where Q is heat, m is the mass of a system, and ΔT is its change in temperature. **Specific heat** c is the amount of energy needed per unit mass per degree of temperature change. Specific heats depend primarily on substance; representative values are given in Table 13.1.

The amount of heat involved in phase changes is

$$
\left.
\begin{aligned}
Q &= mL_f \qquad \text{(melting–freezing)} \\
\text{and} & \\
Q &= mL_v \qquad \text{(vaporization–condensation)}
\end{aligned}
\right\} \quad (13.3)
$$

where Q is the amount of energy required to change the phase of a mass m. Energy involved in phase changes is called **latent heat**, since phase change involves no temperature change. L_f is the **heat of fusion**, and L_v is the **heat of vaporization**. L_f and L_v are collectively called **latent heat coefficients**; representative values are given in Table 13.2. L_s is the **heat of sublimation**, the heat required for direct solid–vapor phase changes.

The three methods of heat transfer are conduction, convection, and radiation.

Conduction is the transfer of heat through stationary matter by physical contact. The **rate of heat transfer by**

conduction through a slab of material is

$$\frac{Q}{t} = \frac{kA(T_2 - T_1)}{d} \qquad \text{(13.4)}$$

where Q/t is the rate of heat transfer in watts or kcal/s, k the **thermal conductivity** of the material, A the surface area, d the thickness of the slab of material, and $T_2 - T_1$ the temperature difference across the slab. Table 13.3 gives representative values of thermal conductivity k.

Convection is the transfer of heat by the macroscopic movement of mass. Convection can be natural or forced and generally transfers heat faster than conduction. Table 13.4 gives wind-chill factors, indicating that moving air has the same chilling effect of much colder stationary air. *Convection that occurs along with a phase change* can carry heat from cold regions to warm ones.

Heat transfer by **radiation** is energy transferred by electromagnetic waves, directly as a result of a temperature dif-

ference. All bodies above absolute zero radiate at a rate given by the **Stefan-Boltzmann law of radiation**:

$$\frac{Q}{t} = \sigma e A T^4 \qquad \text{(13.5)}$$

where $\sigma = 5.67 \times 10^{-8} \text{ J/s} \cdot \text{m}^2 \cdot \text{K}^4$ is the **Stefan-Boltzmann constant**, A the surface area of the object, and T its absolute temperature. The symbol e stands for the **emissivity** of the object; a blackbody or ideal radiator has $e = 1$, whereas a shiny white or perfect reflector has $e = 0$, with real objects having es between 1 and 0. The **net rate of heat transfer by radiation** is

$$\frac{Q}{t} = \sigma e A (T_2^4 - T_1^4) \qquad \text{(13.6)}$$

where T_1 is the temperature of an object surrounded by an environment with uniform temperature T_2, and e is the emissivity of the *object*.

CONCEPTUAL QUESTIONS

13.1 How is heat related to temperature?

13.2 Describe a situation in which heat is converted to another form of energy. Give another example in which some form of energy is converted into heat.

13.3 When heat enters a system, is the energy stored as heat? Explain briefly.

13.4 What three factors affect the amount of heat needed to change an object's temperature?

13.5 Heat can cause temperature and phase changes. What else can?

13.6 The brakes in a car increase in temperature by ΔT when bringing the car to rest from a speed v. How much greater would ΔT be if the car initially had twice the speed? You may assume the car stops so fast that no heat is transferred out of the brakes.

13.7 How does the latent heat of fusion of water help slow the decrease of air temperatures, perhaps preventing them from falling significantly below 0°C, in the vicinity of large bodies of water?

13.8 What is the temperature of ice right after it is formed by freezing water?

13.9 If you place 0°C ice into 0°C water in an insulated container, what will happen? Will some ice melt, will more water freeze, or will neither take place?

13.10 What effect does condensation on a glass of ice water have on the rate at which the ice melts? Will the condensation speed the melting process or slow it?

13.11 In very humid climates where there are numerous bodies of water, such as in Florida, it is unusual for temperatures to rise above about 35°C (95°F). In deserts, however, temperatures can rise far above this. Explain how the evaporation of water helps limit high temperatures in humid climates.

13.12 In the winter, it is often warmer in San Francisco than in nearby Sacramento, 150 km inland. In the summer, it is nearly always hotter in Sacramento. Explain how the bodies of water surrounding San Francisco moderate its temperature extremes.

13.13 Why does an ice bag filled with 0°C ice stay cold for a much longer time than one filled with the same amount of water at 0°C?

13.14 Putting a lid on a boiling pot greatly reduces the amount of heat needed to keep it boiling. Explain why.

13.15 Freeze-dried foods have been dehydrated in a vacuum. During the process, the food freezes and must be heated to facilitate dehydration. Explain both how the vacuum speeds dehydration and why the food freezes as a result.

13.16 When still air cools by radiating at night, it is unusual for temperatures to fall below the dew point. Explain why.

13.17 In a physics classroom demonstration, the instructor inflates a balloon by mouth and then cools it in liquid nitrogen. When cold, the shrunken balloon has a small amount of light blue liquid in it and some snowlike crystals, too. As it warms up, the liquid boils, and then part of the crystals sublimate, with some crystals lingering for awhile and then producing a liquid. Identify the blue liquid and the two solids in the cold balloon. Justify your identifications using data from Table 13.2.

13.18 Some electric stoves have a flat ceramic surface with heating elements hidden beneath. A pot placed over a heating element will be heated, while it is safe to touch the surface only a few centimeters away. Why is a ceramic, with a conductivity less than that of a metal but greater than that of a good insulator, an ideal choice for the stove top?

13.19 What are the major methods of heat transfer from the hot core of the earth to its surface? From its surface into outer space?

13.20 Loose-fitting white clothing covering most of the body is ideal for desert dwellers, both in the hot sun and during cold evenings. Explain how such clothing is advantageous both day and night.

13.21 What methods of heat transfer do we slow by closing curtains on a cold night?

13.22 One way to make a fireplace more energy efficient is to have an external air supply for the combustion of its fuel. Another is to have room air circulate around the outside of the fire box and back into the room. Detail the methods of heat transfer involved in each.

13.23 When our bodies get too warm, they respond by sweating and increasing blood circulation to the surface to carry away heat from the core. What effect will this have on you if you are in a 40.0°C hot tub?

13.24 Figure 13.27 is a cut-away drawing of a thermos bottle, a device designed specifically to slow all forms of heat transfer. Explain the functions of the various parts, such as the vacuum, the silvering, the thin-walled long glass neck, the rubber support, the air layer, and the stopper.

13.25 When watching a daytime circus in a large, dark-colored tent, you sense significant heat coming from the tent. Explain.

13.26 Satellites designed to observe the radiation from cold (3 K) dark space have sensors that are shaded from the sun, earth, and moon and are cooled to very low temperatures. Why must the sensors be at low temperature?

13.27 Why are cloudy nights generally warmer than clear ones?

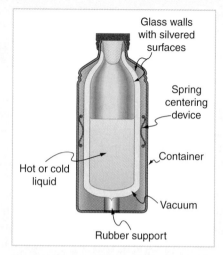

Figure 13.27 Thermos bottle construction is designed to inhibit all methods of heat transfer. Question 24.

PROBLEMS

Sections 13.2 and 13.3 Temperature Change and Phase Change

13.1 On a hot day, an 80,000 L home swimming pool's temperature increases by 1.50°C. What was the net heat input? Ignore any complications, such as loss of water by evaporation.

13.2 Show that 1 cal/g·°C = 1 kcal/kg·°C.

13.3 To sterilize a 50.0 g glass baby bottle, we must raise its temperature from 22.0°C to 95.0°C. How much heat is required?

13.4 The same amount of heat entering identical masses of different substances produces different temperature changes. Calculate the final temperature when 1.00 kcal of heat enters 1.00 kg of the following, originally at 20.0°C: (a) water; (b) concrete; (c) steel; (d) mercury.

13.5 How many kilocalories of heat are required to thaw a 0.450 kg package of frozen vegetables originally at 0°C if their heat of fusion is the same as that of water?

13.6 Some gun fanciers make their own bullets, which involves melting and casting the lead slugs. How much heat is needed to raise the temperature and melt 0.500 kg of lead, starting from 25.0°C?

13.7 An ice bag containing 0°C ice is much more effective in absorbing heat than one containing the same amount of 0°C water. (a) How much heat in kcal is required to raise the temperature of 0.800 kg of water from 0°C to 30.0°C? (b) How much heat is required to first melt 0.800 kg of 0°C ice and then raise its temperature?

•13.8 A 0.250 kg block of a pure material is heated from 20.0°C to 65.0°C by the addition of 1.04 kcal of heat. Calculate its specific heat, and identify the substance of which it is most likely composed.

•13.9 (a) How much heat is required to raise the temperature of a 0.750 kg aluminum pot containing 2.50 kg of water from 30.0°C to the boiling point and then boil away 0.750 kg of water? (b) How long does this take if the rate of heat input is 500 W?

•13.10 A 3.00 g lead bullet traveling at 500 m/s strikes a target, converting its kinetic energy into thermal energy. Show that if half this energy goes into the bullet, it will melt, assuming its initial temperature is 40.0°C.

•13.11 Suppose identical amounts of heat entering different masses of copper and water cause identical changes in temperature. What is the ratio of the mass of copper to water?

•13.12 The formation of condensation on a glass of ice water causes the ice to melt faster than it would otherwise. If 8.00 g of condensation forms on a glass containing water and 200 g of ice, how many grams will melt as a result? Assume no other heat transfer occurs.

•13.13 A large power plant heats 1000 kg of water per second to high-temperature steam to run its electrical generators. (a) How much heat is needed each second to raise the water temperature from 30.0°C to 100°C, boil it, and raise the resulting steam from 100°C to 450°C? (b) How much power is needed in megawatts? (Note: In real power plants, this process occurs under high pressure, altering the boiling point and specific heats. The results of this problem are only approximate.)

•13.14 Burns produced by steam at 100°C are much more severe than those produced by the same mass of 100°C water. To verify this: (a) Calculate the heat that must be removed from 5.00 g of 100°C water to lower its temperature to 50.0°C. (b) Calculate the heat that must be removed from 5.00 g of 100°C steam to condense it and lower its temperature to 50.0°C. (c) Calculate the mass of human flesh that the heat produced in each case can raise from the normal 37.0°C to 50.0°C. (Flesh is badly damaged at 50.0°C.)

•13.15 On a trip, you notice that a 3.50 kg bag of ice lasts an average of one day in your cold chest. What is the average power in watts entering the ice if it starts at 0°C and completely melts to 0°C water in exactly one day?

• **13.16** On a certain dry sunny day, a swimming pool's temperature would rise by 1.50°C if not for evaporation. What fraction of the water must evaporate to carry away precisely enough heat to keep the temperature constant?

• **13.17** (a) How much heat is required to raise the temperature of a 0.200 kg piece of ice from −20.0°C to 130°C, including the heat needed for phase changes? (b) How much time is required for each stage, assuming a constant 5.00 kcal/s rate of heat input? (c) Make a graph of temperature versus time for this process.

• **13.18** Verify the accuracy of the graph in Figure 13.8 by calculating the temperature at times of 10, 50, 90, 150, 500, and 750 s.

• **13.19** How many grams of coffee must evaporate from 350 g of coffee in a 100 g glass cup to cool them from 95.0°C to 45.0°C? You may assume the coffee has the same thermal properties as water and that the average heat of vaporization is 560 cal/g. (You may neglect the change in mass of the coffee as it cools. This gives an answer that is slightly larger than correct.)

• **13.20** It is difficult to extinguish a fire on a crude oil tanker, because each liter of crude oil releases 2.80×10^7 J when burned. To illustrate the difficulty, calculate the number of liters of water that must be expended to absorb the energy released by burning 1.00 L of crude oil, if the water has its temperature raised from 20.0°C to 100°C, it boils, and the resulting steam is raised to 300°C.

• **13.21** The number of kilocalories in food is determined by calorimetry techniques in which the food is burned and the amount of heat released is measured. How many kilocalories per gram are there in a 5.00 g peanut, if the energy from burning it is trapped in 0.500 kg of water held in a 0.100 kg aluminum cup, causing a 54.9°C temperature increase?

• **13.22** To help prevent frost damage, 4.00 kg of 0°C water is sprayed onto a fruit tree. (a) How much heat will be released by the water as it freezes? (b) How much would the temperature of the 200 kg tree decrease if it lost this amount of heat, given its specific heat is 0.800 kcal/kg · °C, and no phase change occurs?

• **13.23** A 0.250 kg aluminum bowl holding 0.800 kg of soup at 25.0°C is placed in a freezer. What is the final temperature if 90.0 kcal of heat is removed from the bowl and soup, assuming the soup's thermal properties as the same as water's?

‡ **13.24** A 0.0500 kg ice cube at −30.0°C is placed in 0.400 kg of 35.0°C water in a very well insulated container. What is the final temperature?

‡ **13.25** If you pour 0.0100 kg of 20.0°C water onto a 1.20 kg block of ice initially at −15.0°C, what is the final temperature? You may assume the water cools so rapidly that effects of the surroundings are negligible.

‡ **13.26** Indigenous peoples sometimes cooked in watertight baskets by placing enough hot rocks into the water to bring it to a boil. What mass of 500°C rock must be placed in 4.00 kg of 15.0°C water to bring its temperature to 100°C, if 0.0250 kg of water escapes as vapor from the initial sizzle? You may neglect the effects of the surroundings and take the average specific heat of the rocks to be that of granite.

‡ **13.27** What would be the final temperature of the pan and water in Example 13.3 if 0.260 kg of water were placed in the pan and

0.0100 kg of the water evaporated immediately, leaving the remainder to come to a common temperature with the pan?

‡ **13.28** Even when shut down after a period of normal use, a large commercial nuclear reactor produces heat at the rate of 150 MW by the radioactive decay of fission products. This causes a rapid increase in temperature if the cooling system fails. (a) Calculate the rate of temperature increase in degrees Celsius per second (°C/s), if the mass of the reactor core is 1.60×10^5 kg and has an average specific heat of 0.0800 kcal/kg · °C. (b) How long would it take to obtain a temperature increase of 2000°C? (The initial rate of temperature increase would be greater than calculated here, because the heat is concentrated in a smaller mass, but later, the temperature increase would slow because the 5×10^5 kg steel containment vessel would begin to be heated, too.)

‡ **13.29** In some countries, liquid nitrogen is used on dairy trucks instead of mechanical refrigerators. A 3.00 h delivery trip requires 200 L of liquid nitrogen, which has a density of 808 kg/m³. (a) Calculate the heat needed to evaporate this amount of liquid nitrogen and raise its temperature to 3.00°C. (You must use c_p, and you can assume it is constant over the temperature range.) This is the amount of cooling the liquid nitrogen supplies. (b) What is this in kilowatt-hours? (c) Compare the amount of cooling obtained from melting an identical mass of 0°C ice with that from the liquid nitrogen.

‡ **13.30** In 1986 a gargantuan iceberg broke away from the Ross Ice Shelf in Antarctica. It was approximately a rectangle 160 km long, 40.0 km wide, and 250 m thick. (a) What is the mass of this iceberg, given that the density of ice is 917 kg/m³? (b) How much heat in joules is required to melt it? (c) Compare this amount of energy with the annual U.S. energy use given in Table 6.1—that is, what is the ratio of these two amounts? (d) How many years would it take sunlight alone to melt ice this thick, if the ice absorbs an average of 100 W/m², 12.00 h per day?

‡ **13.31** Following vigorous exercise, an 80.0 kg person's body temperature is 40.0°C. At what rate in watts must he dissipate heat to reduce his temperature to 37.0°C in 30.0 min, assuming his body continues to produce heat at the rate of 150 W?

‡ **13.32** Rubbing your hands together warms them by converting work into thermal energy. If a woman rubs her hands back and forth for a total of 20 rubs a distance of 7.50 cm each and with a frictional force averaging 40.0 N, what is the temperature increase? The mass of tissue warmed is only 0.100 kg, mostly in the palms and fingers.

Sections 13.5–13.7 Heat Transfer Methods

13.33 Calculate the rate of heat conduction through 13.0 cm thick house walls that have an average thermal conductivity twice that of glass wool. The surface area of the walls is 120 m² and their inside surface is at 18.0°C, while the outside is at 5.00°C.

13.34 The rate of heat conduction out of a window on a winter day is rapid enough to chill the air next to it. To see just how rapidly windows conduct heat, calculate the rate of conduction in watts through a 3.00 m² window that is 0.635 cm thick (1/4 in.) if the temperatures of the inner and outer surfaces are 5.00°C and

−10.0°C, respectively. This rapid rate will not be maintained—the inner surface will cool, and frost may even form.

13.35 Calculate the rate of heat conduction out of the human body, assuming that the core internal temperature is 37.0°C, the skin temperature is 34.0°C, the thickness of the tissues between averages 1.00 cm, and the surface area is 1.40 m².

13.36 Suppose you stand with one foot on ceramic flooring and one on a wool carpet, making contact over an 80.0 cm² area with each foot. Both the ceramic and the carpet are 2.00 cm thick and are 10.0°C on their bottoms. At what rate must each foot supply heat to keep the top of the ceramic and carpet at 33.0°C?

13.37 A man consumes 3000 kcal of food in one day, converting most of it to body heat. If he loses half this energy by evaporating water (through breathing and sweating), how many kilograms of water evaporate?

13.38 At what temperature does still air cause the same chill factor as −5°C air moving at 15 m/s?

13.39 At what wind speed does −10°C air cause the same chill factor as still air at −29°C?

13.40 At what rate does heat radiate from a 275 m² black roof on a night when the roof's temperature is 30.0°C and the surrounding temperature is 15.0°C? The emissivity of the roof is 0.900.

13.41 Cherry-red embers in a fireplace are at 850°C and have an exposed area of 0.200 m² and an emissivity of 0.980. The surrounding room has a temperature of 18.0°C. If 50.0% of the radiant energy enters the room, what is the net rate of radiant heat transfer in kilowatts?

13.42 Radiant heat makes it impossible to stand close to a hot lava flow. Calculate the rate of heat loss by radiation from 1.00 m² of 1200°C fresh lava into 30.0°C surroundings, assuming lava's emissivity is 1.

13.43 Calculate the rate of heat transfer by radiation from a car radiator at 110°C into a 50.0°C environment, if the radiator has an emissivity of 0.750 and a 1.20 m² surface area.

13.44 Find the net rate of heat loss by radiation from a skier standing in the shade, given the following. She is completely clothed in white (head to foot, including a ski mask), the clothes have an emissivity of 0.200 and a surface temperature of 10.0°C, the surroundings are at −15.0°C, and her surface area is 1.60 m².

• **13.45** (a) What is the rate of heat conduction through the 3.00 cm thick fur of a large animal having a 1.40 m² surface area? Assume that the animal's skin temperature is 32.0°C, that the air temperature is −5.00°C, and that fur has the same thermal conductivity as air. (b) What food intake in kcal will the animal need in one day to replace this heat?

• **13.46** A walrus loses heat by conduction through its blubber at the rate of 150 W when immersed in −1.00°C water. Its internal core temperature is 37.0°C, and it has a surface area of 2.00 m². What is the average thickness of its blubber, which has the conductivity of fatty tissue without blood?

• **13.47** Compare the rate of heat conduction through a 13.0 cm thick wall that has an area of 10.0 m² and a thermal conductivity twice that of glass wool with the rate of heat conduction through a 0.750 cm thick window that has an area of 2.00 m², assuming the same temperature difference across each.

• **13.48** Suppose a person is covered head to foot by wool clothing with average thickness 2.00 cm and is losing heat by conduction through the clothing at the rate of 50.0 W. What is the temperature difference across the clothing, given the surface area is 1.40 m²?

• **13.49** Some stove tops are smooth ceramic for easy cleaning. If the ceramic is 0.600 cm thick and conducts heat through the same area and at the same rate as computed in Example 13.6, what is the temperature difference across it? Ceramic has the same thermal conductivity as glass and brick.

• **13.50** The "steam" above a freshly made cup of instant coffee is really water vapor droplets condensing after evaporating from the hot coffee. What is the final temperature of 250 g of hot coffee initially at 90.0°C if 2.00 g evaporates from it? The coffee is in a Styrofoam cup, and so other methods of heat transfer can be neglected.

• **13.51** How many kilograms of water must evaporate from a 60.0 kg woman to lower her body temperature by 0.750°C?

• **13.52** On a hot dry day, evaporation from a lake takes away just enough heat to balance the 1.00 kW/m² of incoming sunlight. What mass of water evaporates in 1.00 h from each square meter?

• **13.53** One winter day, the climate control system of a large university classroom building malfunctions. As a result, 500 m³ of excess air is brought in each minute. At what rate in kilowatts must heat be supplied to warm this air by 10.0°C to normal inside temperature?

• **13.54** The Kilauea volcano in Hawaii is the world's most active, disgorging about 5×10^5 m³ of 1200°C lava per day. What is the rate of heat transfer out of the earth by convection if this lava has a density of 2700 kg/m³ and eventually cools to 30°C? Assume that the specific heat of lava is the same as that of granite.

• **13.55** During heavy exercise, the body pumps 2.00 L of blood per minute to the surface, where it is cooled by 2.00°C. What is the rate of heat loss from this forced convection alone, assuming blood has the same specific heat as water and its density is 1050 kg/m³?

• **13.56** A person inhales and exhales 2.00 L of 37.0°C air, evaporating 4.00×10^{-2} g of water from the lungs and breathing passages with each breath. (a) How much heat is lost in each breath from this evaporation? (b) What is the rate of heat loss in watts if the person is breathing at a moderate rate of 18.0 breaths per minute?

• **13.57** Thermography is a technique for measuring radiant heat and detecting variations in surface temperatures that may be medically or militarily meaningful. (a) What is the percent increase in the rate of heat transfer by radiation from a given area at a temperature of 34.0°C compared with that at 33.0°C, such as on a person's skin? (b) What is the percent increase in the rate of heat transfer by radiation from a given area at a temperature of 34.0°C compared with that at 20.0°C, such as for warm and cool jeep hoods?

‡ **13.58** A glass coffee pot has a 9.00 cm diameter circular bottom in contact with a heating element that keeps the coffee warm with a continuous 50.0 W heat input. (a) What is the temperature of the bottom of the pot, if it is 3.00 mm thick and the inside temperature is 60.0°C? (b) If the temperature of

the coffee remains constant and all of the heat input is removed by evaporation, how many grams per minute evaporate? Take the heat of vaporization to be 560 cal/g.

⁝13.59 Lava beds and slag heaps cool very slowly, sometimes remaining hot inside for years. To illustrate: (a) Calculate the rate of heat conduction per square meter through a 30.0 m thickness of granite if the surface is 20.0°C and the interior is 50.0°C, assuming the granite has the same thermal conductivity as brick. (b) At this rate, how long does it take for enough heat to flow to cool 1.00 m³ (2700 kg) of granite by 1.00°C?

⁝13.60 (a) Calculate the rate of heat conduction through a double-paned window that has a 1.50 m² area and is made of two panes of 0.800 cm thick glass separated by a 1.00 cm air gap. The inside surface temperature is 15.0°C, while that on the outside is −10.0°C. (Hint: There are identical temperature drops across the two glass panes. First find these and then the temperature drop across the air gap.) (b) Calculate the rate of heat conduction through a 1.60 cm thick window of the same dimensions and with the same temperatures. Compare your answer with that for part (a).

⁝13.61 What is the temperature increase of air that is blown through a hair dryer at the rate of 2.00 L per second (cold volume) and carries 1200 W from the heating element?

⁝13.62 Forced convection of air through a car radiator at the rate of 1.00 m³/s carries away heat at the rate of 75.0 kW. What is the temperature increase of the air?

⁝13.63 A large body of lava from a volcano has stopped flowing and is slowly cooling. The interior of the lava is at 1200°C, its surface is at 450°C, and the surroundings are at 27.0°C. (a) Calculate the rate at which heat is transferred by radiation from 1.00 m² of surface lava into the surroundings, assuming the emissivity is 1.00. (b) Suppose heat is conducted to the surface at the same rate. What is the thickness of the lava between the 450°C surface and the 1200°C interior, assuming the conductivity is the same as that of brick?

⁝13.64 The sun radiates like a perfect blackbody with an emissivity of exactly 1. (a) Calculate the surface temperature of the sun, given it is a sphere with a 7.00×10^8 m radius that radiates 3.80×10^{26} W into 3 K space. (b) How much power does the sun radiate per square meter of its surface? (c) How much power in watts per square meter is this at the distance of the earth, 1.50×10^{11} m away? (This number is called the solar constant.)

⁝13.65 A bareback rider under a circus tent feels the heat radiating from the sunlit portion of the tent. Calculate the temperature of the tent canvas based on the following information. The bareback rider's skin temperature is 34.0°C and has an emissivity of 0.970 and an exposed area of 0.400 m². He receives radiant heat at the rate of 20.0 W—half what you would calculate if the entire region behind him was hot. The rest of the surroundings are at 34.0°C.

⁝13.66 Calculate the temperature the entire sky would have to be in order to transfer heat by radiation at 1000 W/m²—about the rate at which the sun transfers heat when it is directly overhead on a clear day. This is the effective temperature of the sky, a kind of average that takes account of the fact that the sun occupies only a small part of the sky but is much hotter than the rest. Assume that the body receiving the heat has a temperature of 27.0°C.

INTEGRATED CONCEPTS

Physics is most interesting when applied to general situations involving more than a narrow set of physical principles. The following problems involve physical principles from more than one chapter. Material from Chapters 6, 10, 11, and 12 is directly involved. Heat is a form of energy, hence the relevance of Chapter 6, Work, Energy, and Power. Convection involves Archimedes' principle, which is introduced in Chapter 10, Fluid Statics. Convection also involves fluid flow—hence the connection with Chapter 11, Fluid Dynamics. Finally, heat is intimately related to the material in Chapter 12, Temperature, Kinetic Theory, and the Gas Laws. You will need to refer to these and perhaps other chapters to work the following problems.

Note: Strategies for dealing with problems having concepts from more than one chapter appear in Chapter 12, among others. There are also worked examples, which can help you see how to apply these strategies to integrated concept problems, in a number of chapters, the most recent being Chapter 6. (See the section of problems labeled *Integrated Concepts* at the ends of those chapters and consult the index for others.)

13.67 One 30.0°C day the relative humidity is 75.0%, and that evening the temperature drops to 20.0°C, well below the dew point. (a) How many grams of water condense from each cubic meter of air? (b) How much heat is released by this condensation? (c) What temperature increase could this cause in dry air?

•13.68 Large meteors sometimes strike the earth, converting most of their kinetic energy into heat. (a) What is the kinetic energy of a 10^9 kg meteor moving at 25.0 km/s? (b) If this meteor lands in a deep ocean and 80% of its kinetic energy goes into heating water, how many kilograms of water could it raise by 5.0°C?

•13.69 Frozen waste from airplane toilets is routinely ejected at high altitude. Ordinarily it breaks up and disperses over a large area, but sometimes it holds together and strikes the ground. Calculate the mass of 0°C ice that can be melted by the conversion of kinetic and gravitational potential energy to heat when a 20.0 kg piece of frozen waste is released at 12.0 km altitude while moving at 250 m/s and strikes the ground at 100 m/s. (Since less than 20.0 kg melts, a significant mess results.)

•13.70 A large electrical power facility produces 1600 MW of waste heat, which is dissipated in cooling towers by warming air flowing through them by 5.00°C. What is the necessary flow rate of air in m³/s?

•13.71 In certain large geographic regions, the underlying rock is hot. Wells can be drilled and water circulated through the rock to extract heat for the generation of electricity. (a) Calculate the amount of heat that can be extracted by cooling 1.00 km³ of granite by 100°C. (b) How long will it take to remove this amount of heat at the rate of 300 MW, assuming no heat is put back into the 1.00 km³ of rock by its surroundings?

•13.72 You lose heat by evaporating water from your lungs and breathing passages. (a) Calculate the maximum number of grams of water that can be evaporated when you inhale 1.50 L of 37.0°C air with an original relative humidity of 40.0%. (Body temperature is also 37.0°C.) (b) How many joules of energy are required to evaporate this amount? (c) What is the rate of heat loss in watts from this method, if you breathe at a normal resting rate of 10.0 breaths per minute?

13.73 (a) What is the temperature increase of water falling 50.0 m over Niagara Falls? (b) What fraction must evaporate to keep the temperature constant?

13.74 Hot air rises because it has expanded. It then displaces a greater volume of cold air, which increases the buoyant force on it. (a) Calculate the ratio of the buoyant force to the weight of 50.0°C air surrounded by 20.0°C air. (b) What energy is needed to heat 1.00 m^3 from 20.0°C to 50.0°C? (c) What gravitational potential energy is gained by this volume of air if it rises 1.00 m? Will this cause a significant cooling of the air?

UNREASONABLE RESULTS

The following problems have results that are unreasonable because some premise is unreasonable or because certain of the premises are inconsistent with one another. Physical principles applied correctly then produce unreasonable results. The purpose of these problems is to give practice in assessing whether nature is being accurately described, and if it is not to trace the source of difficulty.

Note: Strategies for dealing with problems having unreasonable results appear in Chapter 12, among others. They can be found with the section of problems labeled *Unreasonable Results* at the end of that chapter.

13.75 (a) What is the temperature increase of an 80.0 kg person who consumes 2500 kcal of food in one day and retains 95.0% of the energy as body heat? (b) What is unreasonable about this result? (c) Which premise or assumption is responsible?

13.76 A slightly deranged Arctic inventor surrounded by ice thinks it would be much less mechanically complex to cool a car engine by melting ice on it than by having a water-cooled system with a radiator, water pump, antifreeze, and so on. (a) If 80.0% of the energy in 1.00 gal of gasoline goes into waste heat in a car engine, how many kilograms of 0°C ice could it melt? (b) Is this a reasonable amount of ice to carry around to cool the engine for 1.00 gal of gasoline consumption? (c) What premises or assumptions are unreasonable?

13.77 (a) Calculate the rate of heat transfer by conduction through a 1.00 m^2 window that is 0.750 cm thick, if its inner surface is at 22.0°C and its outer surface is at 35.0°C. (b) What is unreasonable about this result? (c) Which premise or assumption is responsible?

13.78 A meteorite 1.20 cm in diameter is so hot immediately after penetrating the atmosphere that it radiates 20.0 kW of power. (a) What is its temperature, if the surroundings are at 20.0°C and it has an emissivity of 0.800? (b) What is unreasonable about this result? (c) Which premise or assumption is responsible?

THERMODYNAMICS

A heat engine converts some heat energy into work. It is apparent that this locomotive not only produces work, but also disperses large amounts of wasted heat energy.

Since heat is a form of energy, it can be converted to work or to any other form of energy. An automobile engine, for example, produces heat by burning gasoline. It can convert the heat into work by exerting a force through a distance. That work can in turn be converted into a variety of other forms: into the car's kinetic or gravitational potential energy; into electrical energy to run the spark plugs, radio, and lights; and back into stored energy in the car's battery. But most of the heat produced from burning gasoline in the engine is not converted to work. Rather, it is dumped into the environment, implying that the engine is quite inefficient.

Why is it often said that modern gasoline engines cannot be made to be significantly more efficient? We hear the same about the conversion of heat to electrical energy in large power stations, be they coal, oil, natural gas, or nuclear fired. Why is that the case? Is the inefficiency caused by design problems that could be solved with better engineering and superior materials? Is it part of some money-making conspiracy by those who sell energy? The truth is actually more interesting and reveals much about the nature of heat.

Basic physical laws govern the conversion of heat into work and place insurmountable limits onto its efficiency. This chapter will explore these laws as well as many applications and concepts associated with them. These topics are part of **thermodynamics**—the study of heat, its movement, and its relationship to work. (See Figure 14.1.)

14.1 THE FIRST LAW OF THERMODYNAMICS

If we are interested in the conversion of heat into work, then the conservation of energy principle is bound to be important. The first law of thermodynamics applies the conservation of energy principle to systems where heat and work are the methods of transferring energy into and out of the system. The **first law of thermodynamics** states that the change in internal energy of a system equals the net heat input minus the net work output. In equation form, the first law of thermodynamics is

$$\Delta U = Q - W \qquad\qquad (14.1)$$

Here ΔU is the **change in internal energy** U of the system. Q is the **net heat transferred into the system**—that is, Q is the sum of all heat transferred into and out of the system. W is the **net work done by the system**—that is, W is the sum of all work done on or by the system. We use the following sign conventions: if Q is positive, then there is a net transfer of heat into the system; if W is positive, then there is net work done by the system. So positive Q adds energy to the system and positive W takes energy from the system. Thus $\Delta U = Q - W$. Note also that if more heat is added than work done, the difference is stored as internal energy. Heat engines are a good example of this—heat is added to them so that they can do work. (See Figure 14.2.) We will now examine Q, W, and ΔU further.

Heat Q and Work W

Heat and work (Q and W) are two of the most important ways of bringing energy into or taking it out of a system. The processes are quite different. Heat transfer, a less organized

Figure 14.1 Like the steam engine at the opening of this chapter, this is a device for converting heat into work.

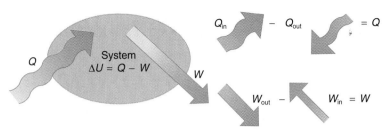

Figure 14.2 The first law of thermodynamics is the conservation of energy principle stated for a system where heat and work are the methods of transferring energy. Q represents the net heat transfer—it is the sum of all heat transferred into and out of the system. Q is positive for net heat transfer into the system. W is the total work done on and by the system. W is positive when more work is done by the system than on it. The change in the internal energy of the system, ΔU, is related to heat and work by the first law of thermodynamics, $\Delta U = Q - W$.

process, is driven by temperature differences. Work, a quite organized process, involves a macroscopic force exerted through a distance. Nevertheless, heat and work can produce identical results.* For example, both can cause temperature increase. Heat transferred into a system, as when the sun warms the air in a bicycle tire, can increase its temperature, and so can work done on the system, as when work is done on the air by pumping it into the tire. Once the temperature increase has occurred, it is impossible to tell whether it was caused by heat or work. This is an important point. Heat and work are energy in transit—neither is stored as such in a system. Both can change the internal energy U of a system. Internal energy is a form of energy completely different from either heat or work.

Internal Energy U

We can think about the internal energy of a system in two different but consistent ways. The first is the submicroscopic view of the atoms and molecules in a system. The **internal energy** U of a system is the sum of the kinetic and potential energies of its atoms and molecules. Recall that kinetic plus potential energy is called mechanical energy. Thus internal energy is the sum of submicroscopic mechanical energy—a quite logical explanation of how energy is stored in a system, since all matter is composed of atoms and molecules. Since it is impossible to keep track of all individual atoms and molecules, we must deal with averages and distributions. These are more like overall or macroscopic characteristics. In fact, a second way to view the internal energy of a system is in terms of its macroscopic characteristics.

Macroscopically, we define the change in internal energy ΔU to be that given by Equation 14.1, the first law of thermodynamics. It has been verified by many detailed experiments that $\Delta U = Q - W$, where ΔU is the change in total kinetic and potential energy of all atoms and molecules in a system. It has also been experimentally determined that the internal energy U of a system depends only on the state of the system and not how it reached that state. More specifically, U is found to be a function of a few macroscopic quantities (pressure, volume, and temperature, for example), independent of past history such as whether there has been heat transfer or work done. This means that if we know the state of a system, we can calculate its internal energy U from a few macroscopic quantities.

To get a better idea of how to think about the internal energy of a system, let us examine a system going from state 1 to state 2. It has internal energy U_1 in state 1, and it has internal energy U_2 in state 2, no matter how it got to either state. This means that the change in internal energy $\Delta U = U_2 - U_1$ is independent of what caused the change. In other words, ΔU is independent of path. By path, we mean the method of getting from the starting point to the ending point. Why is this of any interest? Note that $\Delta U = Q - W$. Both Q and W *depend on path*, but ΔU does not. That means, for one thing, that internal energy is easier to consider than either heat or work.

CONNECTIONS

In thermodynamics, we often use the macroscopic picture when making calculations, while the submicroscopic picture gives underlying explanations in terms of averages and distributions. We shall again see this in later sections of this chapter. In some thermodynamic topics, such as entropy, calculations are also made using the submicroscopic view.

---

*This was first noted when the mechanical equivalent of heat was discussed in Section 13.1. Furthermore, in Sections 13.2 and 13.3, when temperature and phase changes were investigated, it was noted that the necessary energy could come from heat or work.

EXAMPLE 14.1 THE SAME CHANGE IN *U* IS PRODUCED BY TWO DIFFERENT PROCESSES

(a) Suppose 40.00 J of heat flows into a system while it does 10.00 J of work. Later, 25.00 J of heat flows out of the system while 4.00 J of work is done on the system. What is the change in internal energy? (b) What is the change in internal energy of a system that loses a total of 150.00 J of heat and has 159.00 J of work done on it? (See Figure 14.3.)

Strategy In part (a), we must first find the net heat and net work from the given information. Then the first law of thermodynamics (Equation 14.1) can be used to find the change in internal energy. In part (b), the net heat and work are given, so the equation can be used directly.

Solution for (a) The net heat transfer is the heat input minus the heat output, or

$$Q = 40.00 \text{ J} - 25.00 \text{ J} = 15.00 \text{ J}$$

Similarly, the total work is the work done by the system minus the work done on the system, or

$$W = 10.00 \text{ J} - 4.00 \text{ J} = 6.00 \text{ J}$$

Thus the change in internal energy is given by Equation 14.1:

$$\Delta U = Q - W = 15.00 \text{ J} - 6.00 \text{ J} = 9.00 \text{ J}$$

We can also find the change in internal energy for each of the two steps. First, consider 40.00 J of heat in and 10.00 J of work out, or

$$\Delta U_1 = Q_1 - W_1 = 40.00 \text{ J} - 10.00 \text{ J} = 30.00 \text{ J}$$

Now consider 25.00 J of heat out and 4.00 J of work in, or

$$\Delta U_2 = Q_2 - W_2 = -25.00 \text{ J} - (-4.00 \text{ J}) = -21.00 \text{ J}$$

The total change is the sum of these two steps, or

$$\Delta U = \Delta U_1 + \Delta U_2 = 30.00 \text{ J} - 21.00 \text{ J} = 9.00 \text{ J}$$

Discussion for (a) No matter if you look at the overall process or take it in steps, the change in internal energy is the same.

Solution for (b) Here the net heat transfer and total work are given directly to be $Q = -150.00$ J and $W = -159.00$ J, so that

$$\Delta U = Q - W = -150.00 \text{ J} - (-159.00 \text{ J}) = 9.00 \text{ J}$$

Discussion A very different process in part (b) produces the same 9.00 J change in internal energy as in part (a). Note that the change in the system in both parts is related to ΔU and not to the individual Qs or Ws involved. The system ends up in the same state in both (a) and (b). Parts (a) and (b) present two different paths for the system to follow between the same starting and ending points, and the change in internal energy for each is the same—it is independent of path.

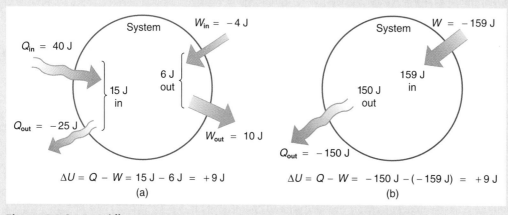

Figure 14.3 Two different processes produce the same change in a system. (a) Heat transfers a total of 15.00 J into the system, while work takes out a total of 6.00 J. The change in internal energy is $\Delta U = Q - W = 9.00$ J. (b) Heat removes 150.00 J from the system while work puts 159.00 J into it, producing an increase of 9.00 J in internal energy. If the system starts out in the same state in (a) and (b), it will end up in the same final state in either case—its final state is related to internal energy, not how it got it.

Human Metabolism and the First Law of Thermodynamics

Human & Biological Applications

Human metabolism (the conversion of food into heat, work, and fat) is an interesting example of the first law of thermodynamics in action. Section 6.8 treated work, energy, and power in human activity in some detail. We now take another look at these topics—this time via the first law of thermodynamics. Considering the body as the system of interest, we can use the first law to examine heat, work, and internal energy in activities ranging from sleep to heavy work.*

What are some of the major characteristics of heat, work, and energy in the body? For one, body temperature is normally kept constant by transfers of heat to surroundings. This means Q is negative. Another fact is that the body usually does work on the outside world. This means W is positive. In such situations, then, the body loses internal energy since $\Delta U = Q - W$ is negative.

Now consider the effects of eating. Eating increases the internal energy of the body by adding chemical potential energy (an unromantic view of a good steak). The body *metabolizes* all the food we consume. Basically, metabolism is an oxidation process in which the chemical potential energy of food is released. There are three places this internal energy can go—to heat, to work, and to stored fat (a tiny fraction may go to cell repair and growth). Heat and work take internal energy out of the body, and food puts it back. If you eat just the right amount of food, then your average internal energy remains constant. Whatever you lose to heat and work is replaced by food, so that, in the long run, $\Delta U = 0$. If you overeat repeatedly, then ΔU is always positive, and your body stores this extra internal energy as fat. The reverse is true if you eat too little. If ΔU is negative for a few days, then the body metabolizes its own fat to produce the heat and work that take energy from the body. This is how dieting produces weight loss.

Life isn't always this simple, as any dieter knows. The body stores fat or metabolizes it only if caloric intake changes for a period of several days. Once you have been on a major diet, the next one is less successful because your body alters the way it responds to low caloric intake. Your basal metabolic rate is your average rate of food conversion into heat and work. The body adjusts its basal metabolic rate to partially compensate for overeating or undereating. It decreases the metabolic rate rather than eliminate its own fat to replace lost food intake. You will chill more easily and feel less energetic as a result of the lower metabolic rate, and you will not lose weight as fast as before. Exercise helps to lose weight, because it forces you to produce both heat and work and raises your metabolic rate even when you are at rest. Weight loss is also aided by the fact that the efficiency of the body in converting internal energy to work is quite low, so that the loss of internal energy resulting from doing work is much greater than the work done.†

The body provides us with an excellent indication that many thermodynamic processes are *irreversible*. An irreversible process can go in one direction but not the reverse. For example, although body fat can be converted to work and heat, work done on the body and heat put into it cannot be converted to body fat. Otherwise, we could skip lunch by sunning ourselves or by walking down stairs. Another example of an irreversible thermodynamic process is photosynthesis. This is the intake of one form of heat—light—by plants and its conversion to chemical potential energy. Both applications of the first law of thermodynamics are illustrated in Figure 14.4. One great advantage of conservation laws such as the first law of thermodynamics is that they accurately describe the beginning and ending points of complex processes, such as metabolism and photosynthesis, without regard to the complications in between. Table 14.1 presents a summary of terms relevant to the first law of thermodynamics.

14.2 THE FIRST LAW AND SOME SIMPLE PROCESSES

One of the most important things we can do with heat is to use it to do work for us. Such a device is called a **heat engine**. Car engines and steam turbines that generate electricity are examples of heat engines. Figure 14.5 shows schematically how the first law of thermodynamics applies to the typical heat engine.

One of the ways in which heat is converted to work is illustrated in Figure 14.6. Heat is added to a gas in a cylinder, increasing its pressure and thereby the force it exerts on

---

*The conversion of heat to work by machines is treated in detail in following sections.
†See Table 6.3 for the efficiencies of the body in various activities. See also Table 6.4 for the rates in kcal/min at which various physical activities consume the body's internal energy.

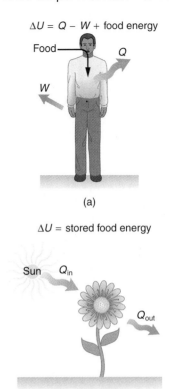

$\Delta U = Q - W$ + food energy

(a)

ΔU = stored food energy

Q_{in}

Sun

Q_{out}

(b)

Figure 14.4 (a) The first law of thermodynamics applied to metabolism. Heat lost and work done remove internal energy, while food intake replaces it. (b) Plants convert part of the radiant heat transfer in sunlight to stored chemical energy, a process called photosynthesis.

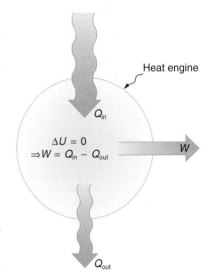

Heat engine

Q_{in}

$\Delta U = 0$
$\Rightarrow W = Q_{in} - Q_{out}$

W

Q_{out}

Figure 14.5 Schematic representation of a heat engine, governed, of course, by the first law of thermodynamics. It is impossible to devise a system where $Q_{out} = 0$, so that some heat is always wasted.

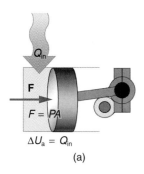

F = PA

$\Delta U_a = Q_{in}$

(a)

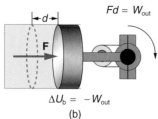

$Fd = W_{out}$

$\Delta U_b = -W_{out}$

(b)

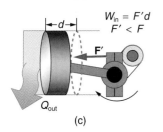

$W_{in} = F'd$
$F' < F$

Q_{out}

(c)

Figure 14.6 (a) Heat is added to the gas in a cylinder. The heat increases the internal energy of the gas, creating higher pressure and temperature. (b) The force exerted on the movable cylinder does work as the gas expands. Gas pressure and temperature decrease when it expands, indicating that its internal energy has been decreased by doing work. (c) Heat is removed to further reduce pressure in the gas so that the piston can be more easily returned to its starting position.

TABLE 14.1

SUMMARY OF TERMS FOR THE FIRST LAW OF THERMODYNAMICS
U Internal energy—the sum of the kinetic and potential energies of a system's atoms and molecules. Can be divided into many subcategories, such as thermal and chemical energy. Depends only on the state of a system (such as its P, V, and T) not on how the energy entered the system. Path independent.
Q Heat—energy transferred because of a temperature difference. Characterized by random molecular motion. Highly dependent on path. Q entering a system is positive.
W Work—energy transferred by a force moving through a distance. An organized, orderly process. Path dependent. W done by a system is positive.

a movable piston. The gas does work on the outside world, as this force moves the piston through some distance. Heat input has resulted in work output. To repeat this process, the piston is returned to its starting point. Heat is removed from the gas so that its pressure decreases, lessening the amount of work needed to return the piston. Variations of this process are employed daily in hundreds of millions of heat engines. We will examine heat engines in detail in the next section. In this section, we consider some of the simpler underlying processes on which heat engines are based.

PV Diagrams and Their Relationship to Work Done on or by a Gas

Figure 14.7 shows an **isobaric process**, in which a gas does work on a piston at constant pressure. Since the pressure is constant, the force exerted is constant and the work done is*

$$W = Fd$$

Now $F = PA$, and so

$$W = PAd$$

Since the volume of a cylinder is its cross-sectional area A times its length d, we see that $Ad = \Delta V$, the change in volume; thus,

$$W = P\,\Delta V \qquad \text{(isobaric process)} \qquad \textbf{(14.2)}$$

Note that if ΔV is positive, then W is positive, meaning that work is done *by* the gas on the outside world.

It is not surprising that $W = P\,\Delta V$, since we have already noted in our treatment of fluids that pressure is a type of potential energy per unit volume and that pressure in fact has units of energy divided by volume. We also noted in our discussion of the ideal gas law that PV has units of energy. In this case, some of the energy associated with pressure becomes work.

Figure 14.8 shows a graph of pressure versus volume (that is, a PV diagram as introduced in Section 12.5) for an isobaric process. You can see in the figure that the work

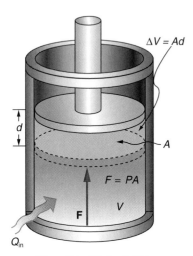

$\Delta V = Ad$

A

$F = PA$

V

F

Q_{in}

$W_{out} = Fd = PAd = P\,\Delta V$

Figure 14.7 An isobaric expansion of a gas requires heat input to keep the pressure constant. Since pressure is constant, the amount of work done is $P\,\Delta V$.

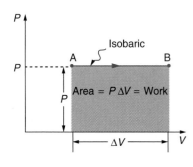

Isobaric

Area = $P\,\Delta V$ = Work

---

*We know from Chapter 6 that work is $W = Fd\cos\theta$. Here $\cos\theta = 1$, because the force and displacement are in the same direction.

Figure 14.8 A graph of pressure versus volume for a constant-pressure, or isobaric, process, such as the one shown in Figure 14.7. The area under the curve equals the work done by the gas, since $W = P\,\Delta V$.

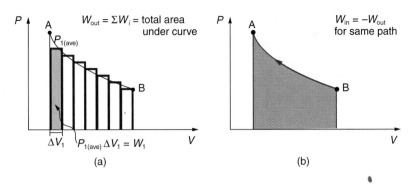

(a)

(b)

Figure 14.9 (a) A *PV* diagram in which pressure varies as well as volume. The work done for each interval is its average pressure times the change in volume, or the area under the curve over that interval. Thus the total area under the curve equals the total work done. (b) Work must be done *on* the system to follow the reverse path. This is interpreted as a negative area under the curve.

done is the area under the graph. This is a very useful and broadly applicable property of *PV* diagrams: *the work done on or by a system in going from one state to another equals the area under the curve on a PV diagram.*

We can see where this leads by considering Figure 14.9(a), which shows a more general process in which both pressure and volume change. The area under the curve is closely approximated by dividing it into strips, each having an average constant pressure $P_{i(\text{ave})}$. The work done is $W_i = P_{i(\text{ave})} \Delta V_i$ for each strip, and the total work done is the sum of the W_i. Thus the total work done is the total area under the curve. If the path is reversed, as in Figure 14.9(b), then work is done on the system. The area under the curve in that case is negative, because ΔV is negative.

PV diagrams clearly illustrate that *the amount of work depends on the path taken and not just the endpoints.* This is seen in Figure 14.10(a), where more work is done in going from A to C by the upper path than by the lower one. The vertical paths, where volume is constant, are called **isochoric** processes. Since volume is constant, $\Delta V = 0$, and no work is done in an isochoric process. Now, if the system follows the cyclical path ABCDA, as in Figure 14.10(b), then the total work done is the area inside the loop. The negative area below path CD subtracts, leaving only the area inside the rectangle. In fact, the work done in any cyclical process (one that returns to its starting point) is the area inside the loop it forms on a *PV* diagram, as Figure 14.10(c) illustrates for a general cyclical process. Note that the loop must be traversed in the clockwise direction for work to be positive—that is, for there to be a net work output.

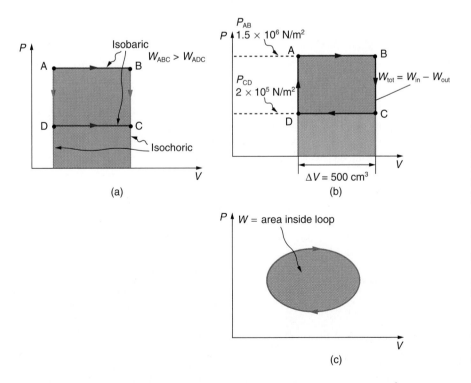

(a)

(b)

(c)

Figure 14.10 (a) The work done in going from A to C depends on path. It is greater for the path ABC than for the path ADC, because the former is at higher pressure. In both cases, the work done is the area under the path. This area is greater for path ABC. (b) The total work done in the cyclical process ABCDA is the area inside the loop, since the negative area below CD subtracts, leaving just the area inside the rectangle. (The values given for the pressures and the change in volume are intended for use in Example 14.2.) (c) The area inside any closed loop is the work done in the cyclical process. If the loop is traversed in the counterclockwise direction, net work is done on the system rather than by it.

EXAMPLE 14.2 TOTAL WORK DONE IN A CYCLICAL PROCESS EQUALS THE AREA INSIDE THE CLOSED LOOP ON A *PV* DIAGRAM

Calculate the total work done in the process ABCDA shown in Figure 14.10(b) by the following two methods to verify that work equals the area inside the closed loop on the *PV* diagram. (a) Calculate the work done along each segment of the path and add these values to get the total work. (b) Calculate the area inside the rectangle ABCDA.

Strategy To find the work along any path on a *PV* diagram, you use the fact that work is pressure times change in volume, or $W = P \Delta V$. So in part (a) this is calculated for each leg of the path around the closed loop.

Solution for (a) The work along path AB is

$$W_{AB} = P_{AB} \Delta V_{AB}$$
$$= (1.50 \times 10^6 \text{ N/m}^2)(5.00 \times 10^{-4} \text{ m}^3) = 750 \text{ J}$$

Since the path BC is isochoric, $\Delta V_{BC} = 0$, and so $W_{BC} = 0$. The work along path CD is negative, since ΔV_{CD} is negative (the volume decreases). The work is

$$W_{CD} = P_{CD} \Delta V_{CD}$$
$$= (2.00 \times 10^5 \text{ N/m}^2)(-5.00 \times 10^{-4} \text{ m}^3) = -100 \text{ J}$$

Again, since the path DA is isochoric, $\Delta V_{DA} = 0$, and so $W_{DA} = 0$. Now the total work is

$$W = W_{AB} + W_{BC} + W_{CD} + W_{DA}$$
$$= 750 \text{ J} + 0 + (-100 \text{ J}) + 0 = 650 \text{ J}$$

Solution for (b) The area inside the rectangle is its height times its width, or

$$\text{area} = (P_{AB} - P_{CD}) \cdot \Delta V$$
$$= [(1.50 \times 10^6 \text{ N/m}^2) - (2.00 \times 10^5 \text{ N/m}^2)]$$
$$\times (5.00 \times 10^{-4} \text{ m}^3)$$
$$= 650 \text{ J}$$

Thus,

$$\text{area} = 650 \text{ J} = W$$

Discussion The result, as anticipated, is that the area equals the work done. The area is often easier to calculate than is the work done along each path. It is also convenient to visualize the area inside different curves on *PV* diagrams in order to see which processes might produce the most work.

Figure 14.11(a) shows two other important processes on a *PV* diagram. For comparison, both are shown starting from the same point A. The upper curve ending at point B is an **isothermal** process—that is, one in which temperature is kept constant. If the gas behaves like an ideal gas, as is often the case, and if no phase change occurs, then $PV = nRT$. Since T is constant, PV is a constant for an isothermal process. We ordinarily expect the temperature of a gas to decrease as it expands, and so we correctly suspect that heat must be added to keep the temperature constant during an isothermal expansion. To show this more rigorously for the special case of a monatomic ideal gas, we note that the average kinetic energy of an atom in such a gas is given by Equation 12.10 to be

$$\tfrac{1}{2}m\overline{v^2} = \tfrac{3}{2}kT$$

The kinetic energy of the atoms in a monatomic ideal gas is its only form of internal energy, and so its total internal energy U is

$$U = N \cdot \tfrac{1}{2}m\overline{v^2} = \tfrac{3}{2}NkT \qquad \text{(monatomic ideal gas)} \qquad \textbf{(14.3)}$$

where N is the number of atoms in the gas. This means that the internal energy of an ideal monatomic gas is constant during an isothermal process—that is, $\Delta U = 0$. If the internal energy does not change, then the net heat input must equal the net work output. That

Figure 14.11 (a) The upper curve is an isothermal process ($\Delta T = 0$), whereas the lower curve is an adiabatic process ($Q = 0$). Both start from the same point A, but the isothermal process does more work than the adiabatic because heat must be added to keep its temperature constant. This keeps the pressure higher all along the isothermal path than along the adiabatic path, producing more work. The adiabatic path thus ends up with a lower pressure and temperature at point C, even though the final volume is the same as for the isothermal process. (b) The cycle ABCA produces a net work output.

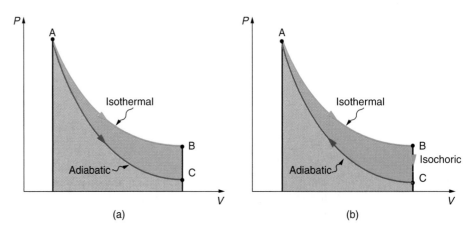
(a) (b)

is, since $\Delta U = Q - W = 0$ here, $Q = W$. We must add just enough heat to replace the work done. An isothermal process is inherently slow, because heat must be added continuously to keep the gas temperature constant at all times and must be allowed to spread through the gas so that there are no hot or cold regions.

Also shown in Figure 14.11(a) is a curve AC for an **adiabatic** process, defined to be one in which there is no heat transfer—that is, $Q = 0$. Processes that are nearly adiabatic can be achieved either by using very effective insulation or by performing the process so fast that there is little time for heat to flow. Temperature must decrease during an adiabatic process, since work is done at the expense of internal energy and $U = (3/2)NkT$. In fact, since $Q = 0$, we see $\Delta U = -W$ for an adiabatic process. Lower temperature results in lower pressure along the way, so that curve AC is lower than curve AB, and less work is done. If the path ABCA could be followed by cooling the gas from B to C isochorically, as in Figure 14.11(b), there would be a net work output.

Reversible Processes

Both isothermal and adiabatic processes such as shown in Figure 14.11 are reversible in principle. A **reversible process** is one in which both the system and its environment can return to exactly the states they were in by following the reverse path. The reverse isothermal and adiabatic paths are BA and CA, respectively. Real macroscopic processes are never exactly reversible. Our system is a gas (like that in Figure 14.7), and its environment is the piston, cylinder, and the rest of the universe. If there are any energy dissipating mechanisms, such as friction or turbulence, then heat is transferred to the environment for either direction of the piston. So, for example, if the path BA is followed and there is friction, then the gas will be returned to its original state but the environment will not—it will have been heated in both directions. Reversibility requires the direction of heat transfer to reverse for the reverse path. Since dissipative mechanisms cannot be completely eliminated, real processes cannot be reversible.

There must be reasons that real macroscopic processes cannot be reversible. We can imagine them going in reverse. For example, heat flows spontaneously from hot to cold and never spontaneously the reverse. Yet it would not violate the first law of thermodynamics for this to happen. In fact, all spontaneous processes, such as a bubble bursting, never go in reverse. There is a second thermodynamic law that forbids them from going in reverse. When we study this law, we will learn something about nature and also find that the law limits the efficiency of heat engines. We will find that heat engines with the greatest possible theoretical efficiency would have to use reversible processes, and even they cannot convert all heat into work. Table 14.2 summarizes the simpler thermodynamic processes and their definitions.

TABLE 14.2

SUMMARY OF SIMPLE THERMODYNAMIC PROCESSES	
Isobaric	*Constant pressure* $W = P\,\Delta V$
Isochoric	*Constant volume* $W = 0$
Isothermal	*Constant temperature* $Q = W$
Adiabatic	*No heat transfer* $Q = 0$

14.3 INTRODUCTION TO THE SECOND LAW OF THERMODYNAMICS: HEAT ENGINES AND THEIR EFFICIENCY

The second law of thermodynamics has to do with the direction taken by spontaneous processes. Many processes occur spontaneously in one direction only—that is, they are irreversible. More precisely, an **irreversible process** is one that depends on path. If the process can only go in one direction, then the reverse path differs fundamentally and the process cannot be reversible. For example, as just noted in the previous section, heat always flows from higher to lower temperature. A cold object in contact with a hot one never gets colder by transferring heat to the hot object and making it hotter. Another example is that mechanical energy, such as kinetic energy, can be completely converted to heat by friction, but the reverse is impossible. A hot stationary object never spontaneously cools off and starts moving. Yet another example is the expansion of a puff of gas introduced into one corner of a vacuum chamber. The gas expands to fill the chamber, but it never regroups in the corner. The random motion of the gas molecules could take them all back to the corner, but this is never observed to happen. (See Figure 14.12.)

The fact that certain processes never occur suggests that there is a law forbidding them to occur. The first law of thermodynamics would allow them to occur—none violates conservation of energy. The law that forbids these processes is called the second law of thermodynamics. We shall see that the second law can be stated in many ways that may seem different, but which in fact are equivalent. Like all natural laws, the second law of

Figure 14.12 Examples of one-way processes in nature. (a) Heat always flows spontaneously from hot to cold and not from cold to hot. (b) The brakes of this car convert its kinetic energy to heat transferred to the environment. The reverse process is impossible. (c) The burst of gas let into this vacuum chamber quickly expands to uniformly fill every part of the chamber. The random motions of the gas molecules will never return them to the corner.

thermodynamics supplies insights into nature, and its several statements imply that it is broadly applicable, fundamentally affecting many apparently disparate processes.

The already familiar direction of heat flow from hot to cold is the basis of our first version of the **second law of thermodynamics**:

> **Heat flows spontaneously from higher- to lower-temperature bodies and never spontaneously in the reverse direction.**

Another way of stating this: it is impossible for any process to have as its sole result the transfer of heat from a cooler to a hotter object.

Heat Engines

Now let us consider a device that can divert some of the natural flow of heat into work. As noted in the previous section, such a device is called a heat engine, and one is shown schematically in Figure 14.13(b). Gasoline and diesel engines, jet engines, and steam turbines are all heat engines that produce work by utilizing part of the heat from some source. Heat flowing from the hot object (or hot reservoir) is Q_h, while waste heat flowing into the cold object (or cold reservoir) is Q_c, and the work output of the engine is W. The temperatures of the hot and cold reservoirs are T_h and T_c, respectively.

Heat sources are usually expensive. Gasoline, for example, is increasingly expensive. Thus we would like to convert as much of the heat to work as possible. In fact, we would like W to equal Q_h, and for there to be no heat lost to the environment ($Q_c = 0$). Unfortunately, this is impossible. The **second law of thermodynamics in terms of the conversion of heat to work** (the second form of the second law) is as follows:

> **It is impossible for any system to absorb heat from a reservoir and convert it completely to work in a cyclical process in which the system returns to its initial state.**

A **cyclical process** brings a system, such as the gas in a cylinder, back to its original condition at the end of every cycle. Most heat engines, such as reciprocating piston engines and rotating turbines, utilize cyclical processes. The second law, just stated in its second form, clearly states that such engines cannot be perfect converters of heat into work. But even one-shot

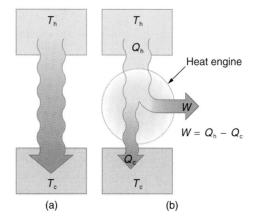

Figure 14.13 (a) Heat flows spontaneously from a hot object to a cold one, consistent with the second law of thermodynamics. (b) A heat engine, represented here by a circle, converts part of the heat flow to work. The hot and cold objects are called the hot and cold reservoirs. Q_h is the heat flowing out of the hot reservoir, W is the work output, and Q_c is the waste heat flowing into the cold reservoir.

(noncyclical) devices, like rocket engines, do not convert all heat into work. Before going into the underlying reasons for the limits on the conversion of heat into work, we need to explore the relationships among W, Q_h, and Q_c and to define the efficiency of a cyclical heat engine.

As noted, a cyclical process brings the system back to its original condition at the end of every cycle. This means that the system's internal energy U is the same at the beginning and end of every cycle—that is, $\Delta U = 0$. The first law of thermodynamics states that

$$\Delta U = Q - W$$

where Q is the *net* heat transferred during the cycle ($Q = Q_h - Q_c$)* and W is the net work done by the system. Since $\Delta U = 0$, we have

$$0 = Q - W$$

so that

$$W = Q$$

This means the net work output equals the net heat input, or

$$W = Q_h - Q_c \quad \text{(cyclical process)} \tag{14.4}$$

just as shown schematically in Figure 14.13(b). The problem is that in all processes, there is some waste heat Q_c—and usually a very significant amount at that.

In the conversion of energy to work, we are always faced with getting less out than we put in. We usually define *efficiency* to be the ratio of what we get to what we spend, or $Eff = W/E_{in}$. In that spirit, we define the efficiency of a heat engine to be its net work output W divided by its heat input Q_h; that is,

$$Eff = \frac{W}{Q_h} \tag{14.5a}$$

Since $W = Q_h - Q_c$ in a cyclical process, we can also express this as

$$Eff = \frac{Q_h - Q_c}{Q_h} = 1 - \frac{Q_c}{Q_h} \quad \text{(cyclical process)} \tag{14.5b}$$

making it clear that an efficiency of 1, or 100%, is possible only if there is no waste heat ($Q_c = 0$).

EXAMPLE 14.3   DAILY WORK DONE BY A NUCLEAR POWER PLANT AND ITS EFFICIENCY

In a single day, a large nuclear power plant—a huge heat engine that converts heat from nuclear processes into work used to generate electricity—produces 2.50×10^{14} J of heat in its reactor and throws away 1.65×10^{14} J of waste heat. (a) How many joules of work are put out? (b) What is the efficiency of the plant?

Strategy for (a) We can use Equation 14.4 to find the work output W, since a cyclical process is used in nuclear power plants. In this process, water is boiled under pressure to form high-temperature steam, which is used to run electric turbines, and then condensed back to water to start the cycle again.

Solution for (a) Equation 14.4 gives the work output to be

$$W = Q_h - Q_c$$

Substituting the given values,

$$W = 2.50 \times 10^{14} \text{ J} - 1.65 \times 10^{14} \text{ J}$$
$$= 8.50 \times 10^{13} \text{ J}$$

Strategy for (b) The efficiency can be calculated with Equation 14.5a or b. Since Q_h is given and work W was found in the first part of this example, it is easiest to use Equation 14.5a.

Solution for (b) Using Equation 14.5a to find efficiency,

$$Eff = \frac{W}{Q_h}$$

W was just found to be 8.50×10^{13} J, and Q_h is given; thus,

$$Eff = \frac{8.50 \times 10^{13} \text{ J}}{2.50 \times 10^{14} \text{ J}}$$
$$= 0.340, \text{ or } 34.0\%$$

Discussion If all the work is converted to electricity in a period of one day, the average power output is 984 MW (left as an exercise for the reader to verify), typical for a large nuclear power plant. The efficiency found is acceptably close to the value of 35.0% given for nuclear power plants in

(continued)

---

*Note that all Qs are positive. The direction of heat flow is indicated by a plus or minus sign. For example, Q_c flows out of the system and so is preceded by a minus sign.

(continued)

Table 6.3. It is distressing that fully 66.0% of the energy produced goes to waste heat, which is simply dumped into the environment. While the laws of thermodynamics limit the efficiency of such plants—including plants fired by coal, oil, and natural gas—the waste heat could be, and sometimes is, used for heating homes or for industrial processes. The generally low cost of energy has not made it economical to use the waste heat from most heat engines, however.

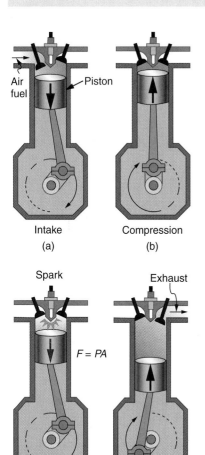

Figure 14.14 The four-stroke internal combustion gasoline engine converts heat into work in the cyclical process shown here. The piston is connected to a rotating crankshaft, which both takes work out of and does work on the gas in the cylinder. (a) Air is mixed with fuel during the intake stroke. (b) During the compression stroke, the air-fuel mixture is rapidly compressed in a nearly adiabatic process, as the piston rises with the valves closed. Work is done on the gas. (c) The power stroke has two distinct parts. First, the air-fuel mixture is ignited, converting chemical potential energy into thermal energy almost instantaneously, causing a great increase in pressure. Then the piston descends, and the gas does work by exerting a force through a distance in a nearly adiabatic process. (d) The exhaust stroke expels the hot gas to prepare the engine for another cycle, starting again with the intake stroke.

With the information given in the preceding example, we can find such things as the efficiency of a heat engine without any knowledge of how it operates. But looking further produces greater insight. Figure 14.14 illustrates the operation of the common four-stroke gasoline engine. The four steps shown complete this heat engine's cycle, bringing the gasoline-air mixture back to its original condition. The *PV* diagram in Figure 14.15 shows an analogous process, less complex but similar enough to the internal combustion engine to shed light on its operation.

The Otto cycle shown in Figure 14.15 is analogous to the cycle used in the four-stroke internal combustion engine, illustrating its most important aspects. The adiabatic process AB corresponds to the nearly adiabatic compression stroke of the gasoline engine. In both cases, work is done on the system (the gas mixture in the cylinder), increasing its temperature and pressure. Along path BC of the Otto cycle, heat Q_h is added at constant volume, causing a further increase in pressure and temperature. This corresponds to the effect of thermal energy created by burning fuel in an internal combustion engine; it takes place so rapidly that the volume is nearly constant there, too. Path CD in the Otto cycle is an adiabatic expansion that does work on the outside world, just as the power stroke of an internal combustion engine does in its nearly adiabatic expansion. The work done by the system along path CD is greater that the work done on the system along path AB, because the pressure is greater, and so there is a net work output. Along path DA in the Otto cycle, heat Q_c is removed from the system at constant volume to reduce its temperature and pressure, returning it to its original state. In an internal combustion engine, this corresponds to the exhaust of hot gases and the intake of an air-gasoline mixture at a considerably lower temperature. In both cases, heat is dissipated into the environment along this final path.

The net work done by a cyclical process is the area inside the closed path on a *PV* diagram, such as that inside path ABCDA in Figure 14.15. Note that in every imaginable cyclical process, it is absolutely necessary to remove heat from the system in order to get a net work output. In the Otto cycle, heat is removed along path DA. If no heat is removed, then the return path is the same, and the net work output is zero. The lower the temperature on the path AB, the less work has to be done to compress the gas. The area

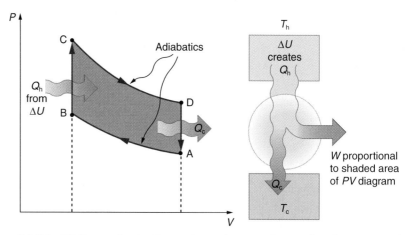

Figure 14.15 *PV* diagram for the Otto cycle, analogous to that employed in an internal combustion engine. Point A corresponds to the start of the compression stroke of an internal combustion engine. Paths AB and CD are adiabatic and correspond closely to the compression and power strokes of an internal combustion engine, respectively. Paths BC and DA are isochoric and accomplish similar results to the ignition and exhaust-intake portions, respectively, of the internal combustion engine's cycle. Work is done on the gas along path AB, but more work is done by the gas along path CD, so that there is a net work output.

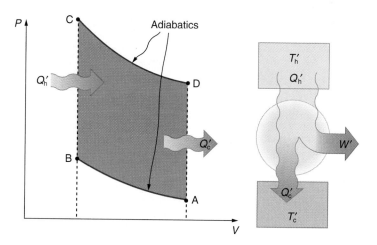

Figure 14.16 This Otto cycle produces a greater work output than the one in the Figure 14.15, because the starting temperature of path CD is higher and the starting temperature of path AB is lower. The area inside the loop is greater, corresponding to greater net work output.

inside the closed path is greater, and so the engine puts out more work and is thus more efficient. Similarly, the higher the temperature along path CD, the more work output. (See Figure 14.16.) So efficiency is related to the temperatures of the hot and cold reservoirs. In the next section, we shall see what the absolute limit to the efficiency of a heat engine is, and how it is related to temperature.

14.4   CARNOT'S PERFECT HEAT ENGINE: THE SECOND LAW OF THERMODYNAMICS RESTATED

We know from the second law of thermodynamics that a heat engine cannot be 100% efficient, since there must always be some waste heat Q_c. How efficient, then, can a heat engine be? This question was answered at a theoretical level in 1824 by a young French engineer, Sadi Carnot (1796–1832), in his study of the then-emerging heat engine technology crucial to the Industrial Revolution. He devised a theoretical cycle, now called the **Carnot cycle**, that is the most efficient cyclical process possible. The second law of thermodynamics can be restated in terms of the Carnot cycle, and so what Carnot actually discovered was this fundamental law. Any heat engine employing the Carnot cycle is called a **Carnot engine**.

What is crucial to the Carnot cycle—and, in fact, defines it—is that only reversible processes are used. Irreversible processes involve dissipative factors, such as friction and turbulence, which convert energy and work to heat. This increases waste heat Q_c and reduces the efficiency of the engine. Obviously, then, reversible processes are superior. Stated in terms of reversible processes, the **second law of thermodynamics** has a third form:

> **A Carnot engine operating between two given temperatures has the greatest possible efficiency. Furthermore, all engines employing only reversible processes have this same maximum efficiency when operating between the same given temperatures.**

Figure 14.17 shows the *PV* diagram for a Carnot cycle. The cycle comprises two isothermal and two adiabatic processes. Recall that both isothermal and adiabatic processes are, in principle, reversible.

Carnot also determined just what the efficiency of a perfect heat engine—that is, a Carnot engine—would be. It is always true that the efficiency of a cyclical heat engine is given by Equation 14.5b as

$$Eff = \frac{Q_h - Q_c}{Q_h} = 1 - \frac{Q_c}{Q_h}$$

What Carnot found was that for a perfect heat engine the ratio of the heats, Q_c/Q_h, equals the ratio of the absolute temperatures of the heat reservoirs. That is, $Q_c/Q_h = T_c/T_h$ for a

Figure 14.17 The *PV* diagram for a Carnot cycle, employing only reversible isothermal and adiabatic processes. Heat Q_h flows in during the isothermal path AB, which takes place at constant temperature T_h. Heat Q_c flows out during the isothermal path CD, which takes place at constant temperature T_c. The net work output *W* equals the area inside the path ABCDA. Also shown is a schematic of a Carnot engine operating between hot and cold reservoirs at temperatures T_h and T_c. Any heat engine using reversible processes and operating between these two temperatures will have the same maximum efficiency as the Carnot engine.

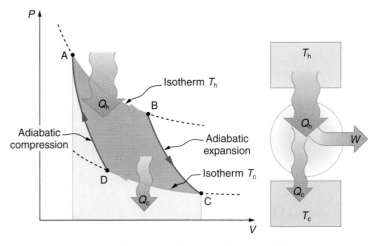

Carnot engine, so that the maximum or **Carnot efficiency** Eff_C is given by

$$Eff_C = 1 - \frac{T_c}{T_h} \tag{14.6}$$

where T_h and T_c are in kelvins (or any other absolute temperature scale). No real heat engine can do as well as the Carnot efficiency—an actual efficiency about 0.7 of this maximum is usually the best that can be accomplished.

Carnot's interesting result implies that 100% efficiency would only be possible if $T_c = 0$ K—that is, only if the cold reservoir were at absolute zero, a practical and theoretical impossibility. But the physical implication is this—the only way to convert all heat to work is to remove *all* thermal energy, and this requires a cold reservoir at absolute zero.

It is also apparent that the greatest efficiencies are obtained when T_c/T_h is as small as possible. Just as discussed for the Otto cycle in the previous section, this means that efficiency is best for the greatest hot temperature and smallest low temperature possible for the hot and cold reservoirs. (This increases the area inside the closed loop on the *PV* diagram; also, it seems reasonable that the greater the temperature difference, the easier it is to divert the heat flow to work.) The actual reservoir temperatures of a heat engine are usually related to the type of heat source and the temperature of the environment into which waste heat is dissipated. Consider the following example.

EXAMPLE 14.4 MAXIMUM THEORETICAL EFFICIENCY FOR A NUCLEAR REACTOR

A nuclear power reactor heats pressurized water to 300°C, being limited by materials from using higher temperatures. Heat from this water is transferred in a complex process (see Figure 14.18); eventually, water is cooled to 27.0°C before

Figure 14.18 Schematic diagram of a pressurized water reactor and the steam turbines that convert work into electrical energy. Heat exchange is used to generate steam, in part to avoid contamination of the generators with radioactivity. Two turbines are used, because this is less expensive than operating a single generator that produces the same amount of electrical energy. The steam is condensed to liquid before being returned to the heat exchange, to keep exit steam pressure low and aid the flow of steam through the turbines (equivalent to utilizing a lower-temperature cold reservoir). The considerable condensation energy must be dissipated into the local environment.

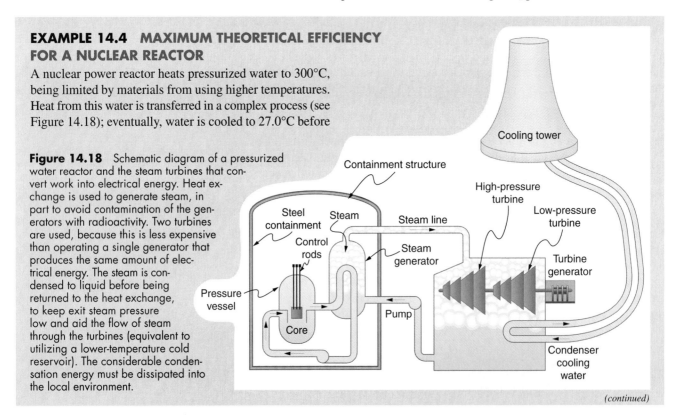

(continued)

(continued)

being heated again to start the cycle over. Calculate the maximum theoretical efficiency for a heat engine operating between these two temperatures.

Strategy Since temperatures are given for the hot and cold reservoirs of this heat engine, Equation 14.6 can be used to calculate the desired Carnot (maximum theoretical) efficiency. Those temperatures must first be converted to kelvins.

Solution The hot and cold reservoir temperatures are given as 300°C and 27.0°C, respectively. In kelvins, then, $T_h = 573$ K and $T_c = 300$ K, so that the maximum efficiency is

$$Eff_C = 1 - \frac{T_c}{T_h}$$

Thus,

$$Eff_C = 1 - \frac{300 \text{ K}}{573 \text{ K}}$$
$$= 0.476, \text{ or } 47.6\%$$

Discussion A nuclear power plant's actual efficiency is about 35%, a little better than 0.7 times the maximum possible, a tribute to superior engineering. Electrical power plants fired by coal, oil, and natural gas have greater actual efficiencies (about 42%), because their simpler boilers can go to higher temperatures and pressures. The cold reservoir temperature in any of these plants is limited by the local environment, but those dumping waste heat into the air can produce temperatures lower than surrounding air if cooling towers using water evaporation are employed.

Since all real processes are irreversible, the actual efficiency of a heat engine can never be as great as that of a Carnot engine, as illustrated in Figure 14.19(a). Even with the best heat engine possible, there are always dissipative processes in peripheral equipment, such as electrical transformers or automobile transmissions. These further reduce the overall efficiency by converting some of the engine's work output back into heat, as shown in Figure 14.19(b).

14.5* HEAT PUMPS AND REFRIGERATORS

Heat pumps, air conditioners, and refrigerators move heat from cold to hot. They are heat engines run backward.[†] (See Figure 14.20.) Heat Q_c is removed from a cold reservoir and moved into a hot one. This requires work input W, which is also converted to heat. Thus

---

[†]We say *backward* rather than *in reverse*, because real heat engines use irreversible processes; they can be run backward but not truly reversed. Only Carnot engines are precisely reversible.

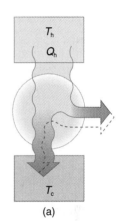

(a)

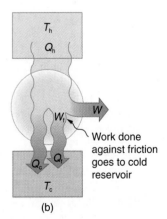

(b)

Figure 14.19 Real heat engines are less efficient than Carnot engines. (a) Real engines use irreversible processes, reducing the amount of heat converted to work. Solid lines represent the actual process; the dashed lines are what a Carnot engine would do between the same two reservoirs. (b) Friction and other dissipative processes in the output mechanisms of a heat engine convert some of its work output into waste heat.

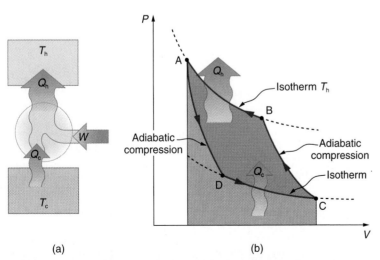

(a)   (b)

Figure 14.20 Heat pumps, air conditioners, and refrigerators are heat engines operated backward. The one shown here is based on a Carnot (reversible) engine. (a) Schematic diagram showing a heat pump moving heat from a cold reservoir to a hot one. The directions of W, Q_h, and Q_c are opposite what they would be in a heat engine. (b) PV diagram for a Carnot cycle similar to that in Figure 14.17 but reversed, following path ADCBA. The area inside the loop is negative, meaning there is a net work input. Heat Q_c flows *into* the system from a cold reservoir along path DC, and heat Q_h flows *out* of the system into a hot reservoir along path BA.

the heat transferred to the hot reservoir is $Q_h = Q_c + W$. (Note that Q_h, Q_c, and W are positive, with their directions indicated on schematics rather than by sign.) A heat pump's mission is to add heat to a warm environment, such as a home in the winter, putting heat Q_h into it. The mission of air conditioners and refrigerators is to chill a cold environment by removing heat Q_c from it.

Heat Pumps

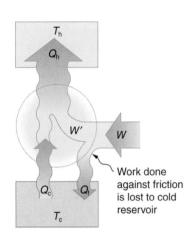

Figure 14.21 When a real heat engine is run backward, some of the intended work input is converted into heat before it gets into the heat engine, thereby reducing its coefficient of performance COP_{hp}. If all of W had gone into the heat pump, then Q_h would have been greater. The best heat pump is one based on a Carnot engine, since there are no dissipative processes to reduce the heat delivered to the hot reservoir.

The great advantage of using a heat pump to move heat into your home, rather than just burning fuel to produce heat, is that a heat pump supplies $Q_h = Q_c + W$. You only pay for W, and you get an additional amount of heat Q_c from the outside free. When you burn fuel to produce heat, you pay for all of it. The disadvantage is that the work input (required by the second law of thermodynamics) is sometimes more expensive than simply burning fuel. The quality of a heat pump is therefore judged by how much heat Q_h is put into the hot reservoir compared with how much work input W is required. In the spirit of taking the ratio of what you get to what you spend, we define a **heat pump's coefficient of performance** (COP_{hp}) to be

$$COP_{hp} = \frac{Q_h}{W} \qquad (14.7)$$

Note that in Equation 14.5a we had $Eff = W/Q_h$; thus, we see that $COP_{hp} = 1/Eff$, an important and interesting fact. First, since the efficiency of any heat engine is less than 1, it means that COP_{hp} is always greater than 1—that is, a heat pump always puts out more heat Q_h than work put into it. Second, it means that heat pumps work best when temperature differences are small. The efficiency of a perfect, or Carnot, engine is $Eff_C = 1 - (T_c/T_h)$; thus, the smaller the temperature difference, the smaller the efficiency and the greater $COP_{hp} = 1/Eff$ is.

Friction and other irreversible processes reduce heat engine efficiency, but they do *not* benefit the operation of a heat pump—instead, they reduce the amount of work input by converting part of it to heat before it gets into the heat pump and dispersing heat back into the cold reservoir. (See Figure 14.21.)

EXAMPLE 14.5 THE BEST COP_{hp} OF A HEAT PUMP FOR HOME USE

A heat pump used to heat a home must employ a cycle that produces a hot temperature greater than typical indoor temperature so that it can transfer heat to the inside. Similarly, it must produce a cold temperature that is colder than outdoor temperature so that it can absorb heat from outside. Its hot and cold reservoir temperatures therefore cannot be too close, placing a limit on its COP_{hp}. (See Figure 14.22.) What is the best coefficient of performance possible for such a heat pump, if it has a hot reservoir temperature of 45.0°C and a cold reservoir temperature of −15.0°C?

Strategy A Carnot engine reversed will give the best possible performance as a heat pump. As noted above, $COP_{hp} = 1/Eff$, so that we need to first calculate the Carnot efficiency to solve this problem.

Solution Carnot efficiency in terms of absolute temperature is given by Equation 14.6 to be

$$Eff_C = 1 - \frac{T_c}{T_h}$$

The temperatures in kelvins are $T_h = 318$ K and $T_c = 258$ K, so that

$$Eff_C = 1 - \frac{258}{318} = 0.1887$$

Thus, from the discussion above,

$$COP_{hp} = \frac{1}{Eff} = \frac{1}{0.1887} = 5.30$$

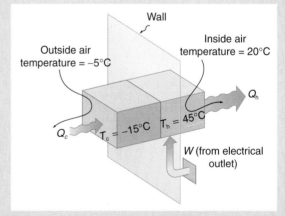

Figure 14.22 The heat pump of Example 14.5 absorbs heat from the outside and transfers it to the inside along with the work done to run the pump. Note that the cold temperature produced by the heat pump is lower than the outside temperature, so that heat is transferred into the pump. The pump produces a hot temperature greater than the indoor temperature in order to transfer heat into the house.

(continued)

(continued)

or

$$COP_{hp} = \frac{Q_h}{W} = 5.30$$

so that

$$Q_h = 5.30W$$

Discussion This result means that the heat pump delivers 5.30 times as much heat as the work put into it. It would cost 5.30 times as much for the same heat produced by an electric room heater as it does for that produced by a heat pump.

Real heat pumps do not perform quite as well as the ideal one in the previous example; their COP_{hp}s range from about 2 to 4. This means that the heat Q_h real heat pumps deliver is 2 to 4 times as great as the work W put into them. Their economical feasibility is still limited, however, since W is usually supplied by electrical energy that costs more per joule than heat produced by burning fuels like natural gas. Furthermore, the initial cost of a heat pump is greater than that of many furnaces, so that a heat pump must last longer for its cost to be recovered. Heat pumps are most likely to be economically superior where winter temperatures are mild, electricity is relatively cheap, and other fuels are relatively expensive.

Air Conditioners and Refrigerators

Air conditioners and refrigerators have the mission of cooling something down in a warm environment. As with heat pumps, work input is required to move heat from cold to hot, and this is expensive. The quality of air conditioners and refrigerators is judged by how much heat Q_c is removed from a cold environment compared with how much work input W is required. We thus define the **coefficient of performance** (COP_{ref}) of an air conditioner or refrigerator to be

$$COP_{ref} = \frac{Q_c}{W} \tag{14.8}$$

Noting again that $Q_h = Q_c + W$, we can see that an air conditioner will have a lower coefficient of performance than a heat pump, because $COP_{hp} = Q_h/W$ and Q_h is greater than Q_c. It is left as an end-of-chapter problem to show that

$$COP_{ref} = COP_{hp} - 1$$

for a heat engine used as either an air conditioner or a heat pump operating between the same two temperatures. Real air conditioners and refrigerators typically do remarkably well, having COP_{ref}s ranging from 2 to 6. These are better than the COP_{hp} values for the heat pumps mentioned above, because the temperature differences are smaller, but they are less than those for Carnot engines operating between the same two temperatures.

A type of COP rating system called the *energy efficiency rating* (*EER*) has been developed. *EER*s are expressed in mixed units of British thermal units (Btu) per hour of heating or cooling divided by the power input in watts. Room air conditioners are readily available with *EER*s ranging from 6 to 12. Although not the same as the *COP*s just described, these *EER*s are good for comparison purposes—the greater the *EER*, the cheaper an air conditioner is to operate (but the higher its purchase price is likely to be).

PROBLEM-SOLVING STRATEGIES

FOR THERMODYNAMICS

Step 1. Examine the situation to determine if heat, work, and internal energy are involved. In particular, this may concern any system where the primary methods of transferring energy are heat and work. Heat engines, heat pumps, refrigerators, and air conditioners are examples of such systems.

Step 2. Identify the system of interest.

Step 3. Identify exactly what needs to be determined in the problem (identify the unknowns). A written list is useful. Maximum efficiency means a Carnot engine is involved. Efficiency is not the same as the coefficient of performance.

Step 4. Make a list of what is given or can be inferred from the problem as stated (identify the knowns). It is important to distinguish heat input from heat output, as well as work input from work output. In many situations, it is useful to determine the type of process, such as isothermal or adiabatic.

Step 5. Solve the appropriate equation for the quantity to be determined (the unknown).

Step 6. Substitute the knowns along with their units into the appropriate equation, and obtain numerical solutions complete with units.

Step 7. Check the answer to see if it is reasonable: Does it make sense? For example, efficiency is always less than 1, whereas coefficients of performance can be greater than 1.

14.6 ENTROPY AND THE SECOND LAW OF THERMODYNAMICS: DISORDER AND THE UNAVAILABILITY OF ENERGY

There is yet another way of expressing the second law of thermodynamics. This version relates to a concept called entropy. By examining it, we shall see that the directions associated with the second law—heat flowing from hot to cold, for example—are related to the tendency in nature for systems to become disordered and for less energy to be available for use as work. The entropy of a system can in fact be shown to be a measure of its disorder and of the unavailability of energy to do work.

We can see how entropy is defined by recalling our discussion of the Carnot engine. We noted that for a Carnot cycle, and hence for any reversible processes, $Q_c/Q_h = T_c/T_h$. Rearranging terms yields

$$\frac{Q_c}{T_c} = \frac{Q_h}{T_h}$$

for any reversible process. Q_c and Q_h are absolute values of the heat transferred at temperatures T_c and T_h, respectively. This ratio of Q/T is defined to be the **change in entropy** ΔS for a reversible process:

$$\Delta S = \left(\frac{Q}{T}\right)_{rev} \tag{14.9}$$

where Q is the heat flow, which is positive for heat input and negative for heat output, and T is the absolute temperature at which the reversible process takes place. (The SI unit for entropy is J/K.) If temperature changes during the process, then it is usually a good approximation to take T to be the average temperature, avoiding the need to use integral calculus to find ΔS.

The definition of ΔS is strictly valid only for reversible processes, such as used in a Carnot engine. However, we can find ΔS precisely even for real, irreversible processes. The reason is that the entropy S of a system, like internal energy U, depends only on the condition of the system and not how it reached that condition. Thus the change in entropy ΔS of a system between state 1 and state 2 is the same no matter how the change occurs. We just need to find or imagine a reversible process that takes us from state 1 to state 2 and calculate ΔS for that process. That will be the change in entropy for any process going from state 1 to state 2. (See Figure 14.23.)

Now let us take a look at the change in entropy of a Carnot engine and its heat reservoirs for one full cycle. The hot reservoir has a loss of entropy $\Delta S_h = -Q_h/T_h$, because heat flows out of it. The cold reservoir has a gain of entropy $\Delta S_c = Q_c/T_c$, because heat flows into it. (We assume the reservoirs are large enough that their temperatures are constant.) So the total change in entropy is

$$\Delta S_{tot} = \Delta S_h + \Delta S_c$$

Thus, since we know that $Q_h/T_h = Q_c/T_c$ for a Carnot engine,

$$\Delta S_{tot} = -\frac{Q_h}{T_h} + \frac{Q_c}{T_c} = 0$$

This result, which has general validity, means that

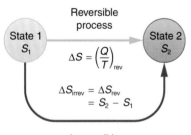

Figure 14.23 When a system goes from state 1 to state 2, its entropy changes by the same amount ΔS, whether a hypothetical reversible path is followed or a real irreversible path is taken. We can therefore calculate ΔS for the reversible path using its definition in Equation 14.9.

The total change in entropy for a system in any reversible process is zero.

The entropy of various parts of the system may change, but the total change is zero. Furthermore, the system does not affect the entropy of its surroundings, since it does not exchange heat with them. Thus the reversible process changes neither the total entropy of the system nor the entropy of its surroundings. Sometimes this is stated more grandiosely as follows: *reversible processes do not affect the total entropy of the universe*. Real processes are not reversible, though, and they do change total entropy, as the following example illustrates.

EXAMPLE 14.6 ENTROPY INCREASES IN AN IRREVERSIBLE (REAL) PROCESS

Spontaneous heat transfer from hot to cold is an irreversible process. Calculate the total change in entropy if 4000 J of heat is transferred from a hot reservoir at $T_h = 600$ K (327°C) to a cold reservoir at $T_c = 250$ K (−23.0°C), assuming there is no temperature change in either reservoir. (See Figure 14.24.)

Strategy How can we calculate the change in entropy for an irreversible process when Equation 14.9 is valid only for reversible processes? Remember that the total change in entropy of the hot and cold reservoirs will be the same whether a reversible or irreversible process moves the heat from hot to cold. So we can calculate the change in entropy of the hot reservoir for a hypothetical reversible process that removes 4000 J of heat from it; then we do the same for a hypothetical reversible process that adds 4000 J of heat to the cold reservoir. This produces the same changes in the hot and cold reservoirs that letting the heat flow irreversibly between them would, and so it also produces the same changes in entropy.

Solution We now calculate the two changes in entropy using Equation 14.9. First, for the hot reservoir losing heat,

$$\Delta S_h = \frac{-Q_h}{T_h} = \frac{-4000 \text{ J}}{600 \text{ K}} = -6.67 \text{ J/K}$$

And for the cold reservoir,

$$\Delta S_c = \frac{Q_c}{T_c} = \frac{4000 \text{ J}}{250 \text{ K}} = 16.0 \text{ J/K}$$

Thus the total is

$$\begin{aligned}\Delta S_{tot} &= \Delta S_h + \Delta S_c \\ &= (-6.67 + 16.0) \text{ J/K} \\ &= 9.33 \text{ J/K}\end{aligned}$$

Discussion There is an increase in entropy for the system of two heat reservoirs undergoing this irreversible transfer of heat. We will see that this means there is a loss of ability to do work with this transferred energy. Entropy has increased, and energy has become unavailable to do work.

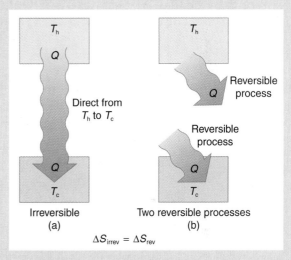

Figure 14.24 (a) Heat transfer from a hot object to a cold one is an irreversible process that produces an overall increase in entropy. (b) The same final state and, thus, the same change in entropy is achieved for the objects if reversible processes remove and add the same amount of heat to each.

It is reasonable that entropy increases for heat transfer from hot to cold. Since the change in entropy is Q/T, there is a larger change at lower temperature. The decrease in entropy of the hot object is therefore less than the increase in entropy of the cold object, producing an overall increase, just as in the previous example. This result is very general:

> **There is an increase in entropy for any system undergoing an irreversible process.**

With respect to entropy, there are only two possibilities: entropy is constant for a reversible process, and it increases for an irreversible process. There is a fourth version of **the second law of thermodynamics stated in terms of entropy**:

> **Total entropy either increases or remains constant in any process; it never decreases.**

So, for example, heat cannot flow spontaneously from cold to hot, because entropy would decrease.

Entropy is very different from energy. Entropy is *not* conserved but increases in all real processes. Noting that reversible processes (such as in Carnot engines) are the ones that convert the most heat to work, and are also the ones that keep entropy constant, leads us to make a connection between entropy and the availability of energy to do work.

Entropy and the Unavailability of Energy to Do Work

What does a change in entropy mean, and why should we be interested in it? One reason is that entropy is directly related to the fact that not all heat can be converted into work. The next example gives some indication of how an increase in entropy results in less heat being converted to work.

EXAMPLE 14.7 LESS WORK IS PRODUCED BY A GIVEN AMOUNT OF HEAT WHEN ENTROPY IS GREATER

(a) Calculate the work output of a Carnot engine operating between temperatures of 600 K and 100 K for 4000 J of heat input. (b) Now suppose that the 4000 J of heat first flows from the 600 K reservoir to a 250 K reservoir (this produces the increase in entropy calculated in the previous example) before flowing into a Carnot engine operating between 250 K and 100 K. What work output is produced? (See Figure 14.25.)

Strategy In both parts, we must first calculate the Carnot efficiency and then the work output.

Solution for (a) The Carnot efficiency is given by Equation 14.6:

$$Eff_C = 1 - \frac{T_c}{T_h}$$

Substituting the given temperatures yields

$$Eff_C = 1 - \frac{100 \text{ K}}{600 \text{ K}} = 0.833$$

Now the work output can be calculated using the definition of efficiency for any heat engine as given by Equation 14.5a:

$$Eff = \frac{W}{Q_h}$$

Solving for W and substituting known terms gives

$$W = Eff_c \cdot Q_h$$
$$= (0.833)(4000 \text{ J}) = 3333 \text{ J}$$

Solution for (b) Similarly,

$$Eff'_c = 1 - \frac{T_c}{T'_h} = 1 - \frac{100 \text{ K}}{250 \text{ K}} = 0.600$$

so that

$$W = Eff'_c \cdot Q_h$$
$$= (0.600)(4000 \text{ J}) = 2400 \text{ J}$$

Discussion There is 933 J less work from the same amount of heat in the second process. This is an important result. The same amount of heat put into two perfect engines produces different amounts of work, because entropy differs in the two cases. In the second case, entropy is greater and less work is produced. Entropy is associated with the *unavailability* of energy to do work.

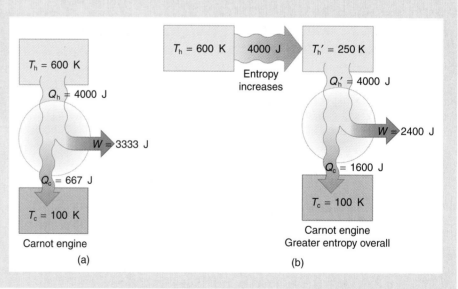

Figure 14.25 (a) A Carnot engine working between 600 K and 100 K has 4000 J of heat input and performs 3333 J of work. (b) The 4000 J of heat is first transferred irreversibly to a 250 K reservoir and then goes into a Carnot engine. The increase in entropy caused by the transfer of the heat to a colder reservoir results in a smaller work output of 2400 J. There is a permanent loss of 933 J of energy for the purpose of doing work.

When entropy increases, a certain amount of energy becomes *permanently* unavailable to do work. The energy is not lost, but its character is changed, so that some of it can never be converted to work—that is, to an organized force acting through a distance.

For instance, in the previous example, 933 J less work was done after an increase in entropy of 9.33 J/K. It can be shown that the amount of energy that becomes unavailable for work is

$$W_{unavail} = \Delta S \cdot T_0 \qquad \textbf{(14.10)}$$

where T_0 is the lowest temperature utilized. In the previous example, $W_{unavail} = (9.33$ J/K)(100 K) = 933 J, as found.

Heat Death of the Universe: An Overdose of Entropy

This leads us to a disconcerting conclusion about the future. As entropy increases, less and less energy in the universe is available to do work. On earth, we have great stores of energy such as fossil and nuclear fuels, and large-scale temperature differences. But as these are used, a certain fraction of the energy they contain can never be converted into work. Eventually, all fuels will be exhausted, all temperatures will be the same, and it will be impossible for heat engines to function, and for work to be done.

Will energy coming in from outer space, such as solar energy, replenish our stores? In human terms, yes—the sun will continue to supply us with energy for several billion years. But in terms of the universe, and the very long term, very large scale picture, the entropy of the universe is increasing, and so the availability of energy to do work is decreasing. Eventually, when all stars have died, all forms of potential energy have been utilized, and all temperatures have become the same, there will be no possibility of doing work.

This sorry fate is known as the *heat death of the universe*, the end of all activity. It lies very far in the future, many factors of 10 more remote than the beginning of the universe some 15 or 20 billion years ago.

Order to Disorder

Entropy is related not only to the unavailability of energy to do work—it is also a measure of disorder. For example, melting a block of ice takes a highly structured and orderly system of water molecules and converts it into a disorderly liquid in which molecules have no fixed positions. (See Figure 14.26.) There is a large increase in entropy in the process, as seen in the following example.

EXAMPLE 14.8 ENTROPY ASSOCIATED WITH DISORDER

Find the increase in entropy of 1.00 kg of ice originally at 0°C that is melted to form 0°C water.

Strategy As before, the change in entropy can be calculated from the definition of ΔS once we find the amount of heat Q needed to melt the ice.

Solution The change in entropy is defined in Equation 14.9 to be

$$\Delta S = \frac{Q}{T}$$

Here Q is the amount of heat needed to melt 1.00 kg of ice and is given by Equation 13.3:

$$Q = mL_f$$

where m is the mass and L_f is the latent heat of fusion. From Table 13.2, we see that $L_f = 334$ kJ/kg for water, so that

$$Q = (1.00 \text{ kg})(334 \text{ kJ/kg}) = 3.34 \times 10^5 \text{ J}$$

Now the change in entropy is positive, since heat is added to cause the phase change; thus,

$$\Delta S = \frac{Q}{T} = \frac{3.34 \times 10^5 \text{ J}}{T}$$

T is the melting temperature of ice. That is, $T = 0°C = 273$ K. So the change in entropy is

$$\Delta S = \frac{3.34 \times 10^5 \text{ J}}{273 \text{ K}}$$
$$= 1.22 \times 10^3 \text{ J/K}$$

Discussion This is a significant increase in entropy accompanying an increase in disorder.

In another easily imagined example, suppose we mix equal masses of water originally at two different temperatures, say 20.0°C and 40.0°C. We get water at an intermediate temperature of 30.0°C. Three things have happened: Entropy has increased, some energy has become unavailable to do work, and the system has become less orderly. Let us think about each of them.

Order

Disorder

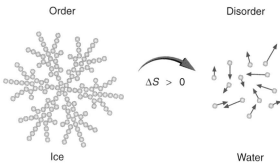

$\Delta S > 0$

Ice

Water

Figure 14.26 When ice melts, it becomes more disordered and less structured. The systematic arrangement of molecules in a crystal structure is replaced by a more random and less orderly movement of molecules without fixed locations or orientations. Its entropy increases since heat enters it. Entropy is a measure of disorder.

First, entropy has increased for the same reason that it did in Example 14.6. Mixing the two bodies of water has the same effect as removing heat from the hot one and adding the same amount of heat to the cold one. It decreases the entropy of the hot water but increases the entropy of the cold water by a greater amount, producing an overall increase in entropy.

Second, once the two masses of water are mixed, there is only one temperature—you cannot run a heat engine with them. The energy that could have been used to run a heat engine is now unavailable to do work.

Third, the mixture is less orderly, or to use another word, less structured. Rather than having two masses at different temperatures and with different distributions of molecular speeds, and so on, we now have a single mass with a uniform temperature.

These three things—entropy, unavailability of energy, and disorder—are not only related but are in fact essentially equivalent.

Life, Evolution, and the Second Law of Thermodynamics

Some people misuse the second law of thermodynamics, stated in terms of entropy, to say that the existence and evolution of life violate the law and thus require divine intervention. It is a fact that living organisms are highly structured, and much lower in entropy than the substances from which they grow. It is also true that the evolution of life from inert matter to its present forms represents a large decrease in entropy for living systems. But it is *always* possible for the entropy of one part of the universe to decrease, provided the total change in entropy of the universe increases. In equation form, we can write this as follows:

$$\Delta S_{tot} = \Delta S_{syst} + \Delta S_{envir} > 0$$

Thus ΔS_{syst} can be negative as long as ΔS_{envir} is positive and greater in magnitude.

How is it possible for a system to decrease its entropy? Energy transfer is necessary. Energy coming from the sun can decrease the entropy of local systems on earth—that is, ΔS_{syst} is negative. But the overall entropy of the rest of the universe increases by a greater amount—that is, ΔS_{envir} is positive and greater in magnitude. Thus $\Delta S_{tot} = \Delta S_{syst} + \Delta S_{envir} > 0$, and the second law of thermodynamics is *not* violated.

Every time a plant stores some solar energy in the form of chemical potential energy, or an updraft of warm air lifts a soaring bird, the earth can be viewed as a heat engine operating between a hot reservoir supplied by the sun and a cold reservoir supplied by dark outer space—a heat engine of high complexity, causing local decreases in entropy as it utilizes part of the heat flowing from the sun into deep space. There is a large total increase in entropy resulting from this massive heat transfer. A small part of this heat is stored in structured systems on earth, producing much smaller local decreases in entropy on earth. (See Figure 14.27.)

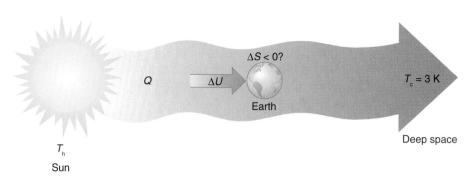

$\Delta S < 0$?

Q

ΔU

Earth

$T_c = 3$ K

Deep space

T_h

Sun

Figure 14.27 The entropy of the earth may decrease in the process of intercepting a small part of the heat flowing from the sun into deep space. Entropy for the entire process increases greatly while the earth becomes more structured with living systems and stored energy in various forms.

14.7 STATISTICAL INTERPRETATION OF ENTROPY AND THE SECOND LAW: THE UNDERLYING EXPLANATION

The various statements of the second law of thermodynamics state what happens rather than why it happens. Why should heat flow only from hot to cold? Why should energy become ever less available to do work? Why should the universe become increasingly disorderly? The answer is that it is a matter of overwhelming probability. Disorder is simply vastly more likely than order.

When you watch an emerging rain storm begin to wet the ground, you will notice that the drops fall in a disorganized manner both in time and in space. Some fall close together, some far apart, but they never fall in straight, orderly rows. It is not impossible for rain to fall in an orderly pattern, just highly unlikely, because there are many more disorderly ways than orderly ones. To illustrate this fact, we will examine some random processes, starting with coin tosses.

Coin Tosses

What are the possible outcomes of tossing 5 coins? Each coin can land either heads or tails.* On the large scale, we are concerned only with the total heads and tails and not with the order in which heads and tails appear. There are the following possibilities:

> 5 heads, 0 tails
> 4 heads, 1 tail
> 3 heads, 2 tails
> 2 heads, 3 tails
> 1 head,   4 tails
> 0 heads, 5 tails

These are what we call macrostates. A **macrostate** is an overall property of a system. It does not specify the details of the system, such as the order in which heads and tails occur or which coins are heads or tails.

Using this nomenclature, a system of 5 coins has the 6 possible macrostates just listed. Some macrostates are more likely to occur than others. For instance, there is only one way to get 5 heads, but there are several ways to get 3 heads and 2 tails, making that macrostate more probable. Table 14.3 lists of all the ways in which 5 coins can be tossed, taking into account the order in which heads and tails occur. Each

---

*There are actually four possibilities when a coin is tossed—heads, tails, lands on edge, does not come down; the last two are so exceedingly unlikely that they are omitted from our consideration.

TABLE 14.3

5-COIN TOSS		
Macrostate	*Individual microstates*	*Number of microstates*
5 heads, 0 tails	HHHHH	1
4 heads, 1 tail	HHHHT, HHHTH, HHTHH, HTHHH, THHHH	5
3 heads, 2 tails	HHHTT, HHTTH, HTTHH, TTHHH, HHTHT, HTHTH, THTHH, HTHHT, THHTH, THHHT	10
2 heads, 3 tails	TTTHH, TTHHT, THHTT, HHTTT, TTHTH, THTHT, HTHTT, THTTH, HTTHT, HTTTH	10
1 head, 4 tails	TTTTH, TTTHT, TTHTT, THTTT, HTTTT	5
0 heads, 5 tails	TTTTT	1
		Total: 32

TABLE 14.4

100-COIN TOSS		
Macrostate		Number of microstates
Heads	Tails	W
100	0	1
99	1	1.0×10^2
95	5	7.5×10^7
90	10	1.7×10^{13}
75	25	2.4×10^{23}
60	40	1.4×10^{28}
55	45	6.1×10^{28}
51	49	9.9×10^{28}
50	50	1.0×10^{29}
49	51	9.9×10^{28}
45	55	6.1×10^{28}
40	60	1.4×10^{28}
25	75	2.4×10^{23}
10	90	1.7×10^{13}
5	95	7.5×10^7
1	99	1.0×10^2
0	100	1

Total: 1.27×10^{30}

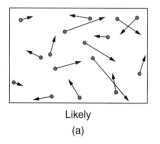

Likely

(a)

Highly unlikely

(b)

Figure 14.28 (a) The ordinary state of gas in a container is a disorderly, random distribution of atoms with a Maxwell-Boltzmann distribution of speeds. It is so unlikely that they would ever end up in one corner of the container that it might as well be impossible. (b) With energy transfer, the gas can be forced into one corner and its entropy greatly reduced. But left alone, it will spontaneously increase its entropy and return to the normal conditions, because they are immensely more likely.

individual possibility is called a **microstate**—a detailed description of every element of a system.

The macrostate of 3 heads and 2 tails can be achieved in 10 ways and is thus 10 times more probable than the one having 5 heads. Not surprisingly, it is equally probable to have the reverse, 2 heads and 3 tails. Similarly, it is equally probable to get 5 tails as it is to get 5 heads. Note that all of these conclusions are based on the crucial assumption that each microstate is equally probable. With coin tosses, this requires that the coins not be asymmetric in a way that favors one side over the other, as with loaded dice. With any system, the assumption that all microstates are equally probable must be valid, or the analysis will be erroneous.

The two most orderly possibilities are 5 heads or 5 tails. (They are more structured than the others.) They are also the least likely, only 2 out of 32 possibilities. The most disorderly possibilities are 3 heads and 2 tails and its reverse. (They are the least structured.) The most disorderly possibilities are also the most likely, with 20 out of 32 possibilities for the 3 heads and 2 tails and its reverse. If we start with an orderly array like 5 heads and toss the coins, it is very likely that we will get a less orderly array as a result, since 30 out of the 32 possibilities are less orderly. So even if you start with an orderly state, there is a strong tendency to go from order to disorder, from low entropy to high entropy. The reverse can happen, but it is unlikely.

This interesting result becomes dramatic and compelling for larger systems. Consider what happens if you have 100 coins instead of just 5. The most orderly arrangements (most structured) are 100 heads or 100 tails. The least orderly (least structured) is that of 50 heads and 50 tails. There is only 1 way (1 microstate) to get the most orderly arrangement of 100 heads. There are 100 ways (100 microstates) to get the next most orderly arrangement of 99 heads and 1 tail (also 100 to get its reverse). And there are 1.0×10^{29} ways to get 50 heads and 50 tails, the least orderly arrangement. Table 14.4 is an abbreviated list of the various macrostates and the number of microstates in each. The total number of microstates—the total number of different ways 100 coins can be tossed—is an impressively large 1.27×10^{30}. Now, if we start with an orderly macrostate like 100 heads and toss the coins, there is a virtual certainty that we will get a less orderly macrostate. If we keep tossing the coins, it is possible, but exceedingly unlikely, that we will ever get back to the most orderly macrostate. If you tossed the coins once each second, you could expect to get either 100 heads or 100 tails once in 2×10^{22} years! This is 1 trillion (10^{12}) times longer than the age of the universe, and so the chances are essentially zero. In contrast, there is an 8% chance of getting 50 heads, a 73% chance of getting from 45 to 55 heads, and a 96% chance of getting from 40 to 60 heads. Disorder is highly likely.

Disorder in a Gas

The fantastic growth in the odds favoring disorder that we see in going from 5 to 100 coins continues as the number of entities in the system increases. Let us now imagine applying this approach to perhaps a small sample of gas. Because counting microstates and macrostates involves statistics, this is called **statistical analysis**. The macrostates of a gas correspond to its macroscopic properties, such as volume, temperature, and pressure; and its microstates correspond to the detailed description of the positions and velocities of its atoms. Even a small amount of gas has a huge number of atoms: $1.0 \, \text{cm}^3$ of an ideal gas at 1.0 atm and 0°C has 2.7×10^{19} atoms. So each macrostate has an immense number of microstates. In plain language, this means that there are an immense number of ways in which the atoms in a gas can be arranged and still give it the same pressure, temperature, and so on.

The most likely conditions (or macrostates) for a gas are those we see all the time—a random distribution of atoms in space with a Maxwell-Boltzmann distribution of speeds in random directions. (See Figures 12.16 and 12.17 in Section 12.4, on kinetic theory.) This is the most disorderly and least structured condition we can imagine. In contrast, one type of very orderly and structured macrostate has all of the atoms in one corner of a container with identical velocities. There are very few ways to accomplish this (very few microstates corresponding to it), and so it is exceedingly unlikely ever to occur. (See Figure 14.28(b).) Indeed, it is so unlikely that we have a law saying that it is impossible, which has never been observed to be violated—namely, the second law of thermodynamics.

The disordered condition is one of high entropy, and the ordered one has low entropy. With a transfer of energy from another system, we could force all of the atoms into one corner and have a local decrease in entropy, but at the cost of an overall increase in entropy of the universe. If the atoms start out in one corner, they will quickly disperse and become uniformly distributed and will never return to the orderly original state. (See Figure 14.28(a).) Entropy will increase. With such a large sample of atoms, it is possible—but unimaginably unlikely—for entropy to decrease. Disorder is simply vastly more likely than order.

The arguments that disorder and high entropy are the most probable states are quite convincing. The great Austrian physicist Ludwig Boltzmann (1844–1906)—who, along with Maxwell, made so many contributions to kinetic theory—proved that the entropy of a system in a given state (a macrostate) can be written

$$S = k \ln W \qquad (14.11)$$

where $k = 1.38 \times 10^{-23}$ J/K is Boltzmann's constant, and $\ln W$ is the natural logarithm of the number of microstates W corresponding to the given macrostate. W is proportional to the probability that the macrostate will occur. Thus entropy is directly related to the probability of a state—the more likely the state, the greater its entropy. Boltzmann proved that this expression for S is equivalent to the definition $\Delta S = Q/T$, which we have used so extensively.

Thus the second law of thermodynamics is explained on a very basic level. Entropy either remains the same or increases in every process because of the extraordinarily small probability of a decrease, based on the extraordinarily larger number of microstates in systems with greater entropy. There is a slight modification: entropy *can* decrease, but for any macroscopic system, it is so unlikely that it will never be observed.

EXAMPLE 14.9 ENTROPY INCREASES IN A COIN TOSS

Suppose you toss 100 coins starting with 60 heads and 40 tails, and you get the most likely result, 50 heads and 50 tails. What is the change in entropy?

Strategy Noting that the number of microstates is labeled W in the Table 14.4 for the 100-coin toss, we can use Equation 14.11 to calculate the change in entropy.

Solution The change in entropy is

$$\Delta S = S_f - S_i = k \ln W_f - k \ln W_i$$

where i stands for the initial 60 heads and 40 tails state, and f for the final 50 heads and 50 tails state. Substituting the Ws from Table 14.4 gives

$$\Delta S = (1.38 \times 10^{-23} \text{ J/K})[\ln(1.0 \times 10^{29}) \\ - \ln(1.4 \times 10^{28})]$$
$$= 2.7 \times 10^{-23} \text{ J/K}$$

Discussion This increase in entropy means we have gone to a less orderly situation. It is not impossible for further tosses to take you back to 60 heads and 40 tails, but it is less likely. There is about a 1 in 90 chance for that decrease in entropy (-2.7×10^{-23} J/K) to occur. If we calculate the decrease in entropy to go the most orderly state, we get $\Delta S = -92 \times 10^{-23}$ J/K. There is about a 1 in 10^{30} chance of this. So very small decreases in entropy are unlikely, then, but slightly greater decreases are impossibly unlikely. This implies, again, that for a macroscopic system, a decrease in entropy is impossible. For example, for 1.00 kg of 0°C ice to spontaneously give up heat to its 0°C environment, reversing the process in Example 14.8, there would be a decrease in entropy of 1.22×10^3 J/K. Given that a ΔS of 10^{-21} J/K corresponds to a 1 in 10^{30} chance, a decrease of this size (10^3 J/K) is an *utter* impossibility. Even for a milligram of ice to spontaneously refreeze is impossible.

PROBLEM-SOLVING STRATEGY

FOR ENTROPY

Step 1. Examine the situation to determine if entropy is involved.

Step 2. Identify the system of interest.

Step 3. Identify exactly what needs to be determined in the problem (identify the unknowns). A written list is useful.

Step 4. Make a list of what is given or can be inferred from the problem as stated (identify the knowns). You must carefully identify the heat transferred, if any, and the temperature at which the process takes place. It is also important to identify the initial and final states.

Step 5. Solve the appropriate equation for the quantity to be determined (the unknown). Note that the change in entropy can be determined between any states by calculating it for a reversible process.

Step 6. Substitute the knowns along with their units into the appropriate equation, and obtain numerical solutions complete with units.

Step 7. Check to see if it is reasonable: Does it make sense? For example, total entropy should increase for any real process or be constant for a reversible process. Disordered states should be more probable and have greater entropy than ordered states.

SUMMARY

Thermodynamics is the study of heat, its movement, and its relationship to work. The **first law of thermodynamics** states that the change in internal energy of a system equals the net heat input minus the net work output, or

$$\Delta U = Q - W \qquad (14.1)$$

where ΔU is the **change in internal energy** U, Q is the **net heat transferred**, and W is the **net work done**. The first law of thermodynamics is based on the law of conservation of energy. Table 14.1 presents a more detailed summary of these terms.

A **heat engine** is a device that converts heat to work. Several simple processes involving gases may be used by heat engines. These are summarized in Table 14.2 and are **isobaric** (constant-pressure) processes, **isochoric** (constant-volume) processes, **isothermal** (constant-temperature) processes, and **adiabatic** (no-heat-transfer) processes. All these processes are shown as paths on various PV diagrams. The work done in any process shown on a PV diagram equals the area under the curve, and the work done in a **cyclical process** (one that returns the system to its original conditions) equals the area inside the loop on a PV diagram. A **reversible** process is one that returns to its initial state by following a reverse path. An **irreversible** process is one that depends on path, including the direction along a path. Real processes are irreversible and less efficient in heat engines.

The **second law of thermodynamics** stated in its first form is

---

Heat flows spontaneously from higher- to lower-temperature bodies and never spontaneously in the reverse direction.

The second form of the **second law of thermodynamics** is

It is impossible for any system to absorb heat from a reservoir and convert it completely to work in a cyclical process in which the system returns to its initial state.

The work output of a heat engine is

$$W = Q_h - Q_c \qquad \text{(cyclical process)} \qquad (14.4)$$

where Q_h is the heat input from a hot reservoir and Q_c is the

waste heat that flows into a cold reservoir. The *efficiency* of a heat engine is

$$Eff = \frac{W}{Q_h} \qquad (14.5a)$$

or

$$Eff = \frac{Q_h - Q_c}{Q_h} = 1 - \frac{Q_c}{Q_h} \qquad \text{(cyclical process)} \qquad (14.5b)$$

A **Carnot engine** is a heat engine that utilizes the most efficient (reversible) process, called the **Carnot cycle**. The third form of the **second law of thermodynamics** is

---

A Carnot engine operating between two given temperatures has the greatest possible efficiency. Furthermore, all engines employing only reversible processes have this same maximum efficiency when operating between the same given temperatures.

The maximum or **Carnot efficiency** Eff_C is given by

$$Eff_C = 1 - \frac{T_c}{T_h} \qquad (14.6)$$

where T_h and T_c are the temperatures in kelvins of the hot and cold reservoirs.

Heat engines can be run backward to move heat from cold to hot, as in heat pumps, refrigerators, and air conditioners. A **heat pump's coefficient of performance** (COP_{hp}) is defined to be

$$COP_{hp} = \frac{Q_h}{W} \qquad (14.7)$$

where Q_h is the heat moved to a hot reservoir and W is the work input. Since $COP_{hp} = 1/Eff$, it is greatest for small temperature differences. The **coefficient of performance** (COP_{ref}) of an air conditioner or refrigerator is

$$COP_{ref} = \frac{Q_c}{W} \qquad (14.8)$$

where Q_c is the heat removed from a cold reservoir and W is the work input.

The **change in entropy** ΔS for a reversible process is

$$\Delta S = \left(\frac{Q}{T}\right)_{rev} \qquad (14.9)$$

where Q is the heat flow (positive for heat input, negative for heat output) and T is the absolute temperature. The entropy of individual systems may increase or decrease, but the total change in entropy is stated in the fourth version of the **second law of thermodynamics:**

> **Total entropy either increases or remains constant in any process; it never decreases.**

Total entropy remains constant in reversible processes and increases in irreversible (real) processes. An increase in entropy ΔS results in energy becoming unavailable to do work in the amount

$$W_{unavail} = \Delta S \cdot T_0 \qquad \textbf{(14.10)}$$

where T_0 is the lowest temperature utilized.

Entropy is associated with disorder and increases because disorder is vastly more probable than order. The **macrostate** of a system is its overall or large-scale condition. Each individual possibility for a given macrostate is called a **microstate**, a detailed description of every element of the system. The probability of a macrostate is determined by how many different microstates W it has. More microstates are possible for disorderly macrostates, and the entropy of a macrostate is

$$S = k \ln W \qquad \textbf{(14.11)}$$

where k is Boltzmann's constant, and $\ln W$ is the natural logarithm of the number of microstates W corresponding to the macrostate. Entropy increases in real processes, because systems move toward more probable conditions.

CONCEPTUAL QUESTIONS

14.1 The first law of thermodynamics and the earlier statements of conservation of energy in Chapter 6 are clearly related. How do they differ in the categories of energy considered?

14.2 Heat Q and work W are always energy in transit, whereas internal energy U is energy stored in a system. Give an example of each type of energy, and state specifically how it is either in transit or resides in a system.

14.3 How do heat and thermal energy differ? In particular, which can be stored as such in a system and which cannot?

14.4 If you run down some stairs and stop, what happens to your kinetic energy and your initial gravitational potential energy?

14.5 Give an explanation of how food energy can be viewed as molecular potential energy (consistent with the microscopic definition of internal energy).

14.6 Identify the type of energy transferred to your body in each of the following as either internal energy, heat, or work: (a) basking in sunlight; (b) eating food; (c) riding an elevator to a higher floor.

14.7 Identify the type of energy transferred out of your body in each of the following as either internal energy, heat, or work: (a) floating quietly in 25.0°C water; (b) lifting a barbell; (c) liposuction.

14.8 Explain in terms of the first law of thermodynamics how a decrease in your body's metabolic rate could cause you to chill more easily and feel less energetic than normal.

14.9 One method of converting heat to work is to transfer heat into a gas, which expands, doing work on a piston (see Figure 14.6, for example). (a) Is the heat converted directly to work in an isobaric process, or does it go through another form first? Explain. (b) What about an isothermal process? (c) An adiabatic process (where heat was transferred in prior to the adiabatic process)?

14.10 Would Question 14.9 make any sense for an isochoric process? Explain.

14.11 We ordinarily say that $\Delta U = 0$ for an isothermal process. Does this assume no phase change takes place? Explain.

14.12 The temperature of a rapidly expanding gas decreases. Explain why in terms of the first law of thermodynamics. (Hint: Consider whether the gas does work and whether heat can be rapidly conducted into the gas.)

14.13 Which cyclical process represented by the two closed loops, ABCFA and ABDEA, on the PV diagram in Figure 14.29 produces the greatest *net* work? Is it also the one with the smallest work input required to return it to point A? Explain your responses.

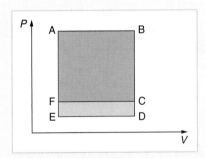

Figure 14.29 The two cyclical processes shown on this PV diagram start with and return the system to the conditions at point A, but they follow different paths and produce different amounts of work. Question 13.

14.14 A real process may be nearly adiabatic if it occurs over a very short time. How does the short time span help the process to be adiabatic?

14.15 It is unlikely that a process can be isothermal unless it is a very slow process. Explain why. Is the same true for isobaric and isochoric processes? Explain.

14.16 Is a temperature difference necessary to operate a heat engine? State why or why not.

14.17 Definitions of efficiency vary depending on how energy is being converted. Compare the definitions of efficiency for the human body and heat engines. How does the definition of efficiency in each relate to the type of energy being converted into work?

14.18 Why—other than the fact that the second law of thermodynamics says reversible engines are the most efficient—should heat engines employing reversible processes be more efficient than those employing irreversible processes? Consider that dissipative mechanisms are one cause of irreversibility.

14.19 Can improved engineering and materials be employed in heat engines to reduce waste heat? Can they eliminate waste heat entirely?

14.20 Does the second law of thermodynamics alter the conservation of energy principle?

14.21 Consider a system with a certain energy content, from which we wish to extract as much work as possible. Should the system's entropy be high or low? Is this orderly or disorderly? Structured or uniform? Explain briefly.

14.22 Does a gas become more orderly when it liquefies? Does its entropy change? If so, does it increase or decrease? Explain.

14.23 Explain how water's entropy can decrease when it freezes without violating the second law of thermodynamics. Specifically, explain what happens to the entropy of its surroundings.

14.24 Is a uniform-temperature gas more or less orderly than one with several different temperatures? Which is more structured? Which can convert heat to work without heat transfer from another system?

14.25 Give an example of a spontaneous process in which a system becomes less ordered and energy becomes less available to do work. What happens to the system's entropy in this process?

14.26 What is the change in entropy in an adiabatic process? Does this imply that adiabatic processes are reversible? Can a process be precisely adiabatic for a macroscopic system?

14.27 Does the entropy of a star increase or decrease as it radiates? Does the entropy of the space into which it radiates (which has a temperature of about 3 K) increase or decrease? What does this do to the entropy of the universe?

14.28 Explain why a building made of bricks has smaller entropy than the same bricks in a disorganized pile. Do this by considering the number of ways that each could be formed (number of microstates in each macrostate).

14.29 Use the statistical interpretation of entropy to explain how you know that a videotape is running backward if it shows a cigarette getting longer while smoke sinks to and enters its tip and ashes occasionally jump up to it from below. What law does this process violate? Does it necessarily violate any others?

14.30 If you mix equal amounts of 20.0°C and 40.0°C water, as discussed in the text, you will obtain 30.0°C water, provided there is no interaction with surroundings. Has energy been lost to this system? Has its entropy increased, decreased, or remained constant? Random molecular motion can separate the water into two equal masses of 20.0°C and 40.0°C again—how is the likelihood of this related to the mass of water involved?

14.31[*] How are the statistical interpretation of entropy and the second law of thermodynamics related to the direction taken by diffusion and osmosis? (See Section 11.7 for discussion of these and related processes.)

14.32[*] What is the advantage of a heat pump over a direct source of heat, such as burning fuels (converting internal energy from stored chemical into thermal)? Why, then, is it still sometimes cheaper to produce heat by burning fuel?

14.33[*] Why do refrigerators, air conditioners, and heat pumps operate most cost-effectively for cycles with a small difference between T_h and T_c? (Note that the temperatures of the cycle employed are crucial to its *COP*.)

14.34[*] Grocery store managers contend that there is *less* total energy consumption in the summer if the store is kept at a *low* temperature. Make arguments to support or refute this claim, taking into account that there are numerous refrigerators and freezers in the store.

PROBLEMS

Section 14.1 First Law of Thermodynamics

14.1 What is the change in internal energy of a car if you put 12.0 gallons of gasoline into its tank? All other factors, such as the car's temperature, are constant. (See Table 6.1 for useful information.)

14.2 How much heat was lost by a system, if its internal energy decreased by 150 J while it was doing 30.0 J of work?

14.3 A system does 1.80×10^8 J of work while dissipating 7.50×10^8 J of heat. What is the change in internal energy of the system assuming no other changes (such as in temperature or by the addition of fuel)?

14.4 What is the change in internal energy of a system that takes in 3.00×10^6 J of heat, and does 4.50×10^5 J of work while dissipating 8.00×10^6 J of heat?

• 14.5 Suppose a woman does 500 J of work and dissipates 9500 J of heat in the process. (a) What is the decrease in her internal energy, assuming no change in temperature or consumption of food? (That is, there is no other energy transfer.) Express your answer in kcal. (b) What is her efficiency?

• 14.6 (a) How much food energy in kcal will a man metabolize in the process of doing 35.0 kJ of work with an efficiency of 5.00%? (b) How much heat must be dissipated to keep his temperature constant?

• 14.7 (a) What is the average metabolic rate in watts of the typical man who metabolizes 2500 kcal of food energy in one day? (b) What is the maximum amount of work in joules he can do without breaking down fat, assuming a maximum efficiency of 20.0%? (c) Compare his work output with the daily output of a 0.250 horsepower output motor.

• 14.8 How long will the energy in a 350 kcal cup of yogurt last in a woman doing work at the rate of 150 W with an efficiency of 20.0% (such as in leisurely climbing stairs)?

: 14.9 A woman climbing the Washington Monument metabolizes 140 kcal of food energy. If the her efficiency is 18.0%, how much waste heat had to be dissipated to keep her temperature constant?

Section 14.2 The First Law and Simple Processes

14.10 Show that the units of pressure times volume (PV) are energy.

14.11 (a) What would be the net work done by the system whose PV diagram is shown in Figure 14.10(b), if it followed the path ABCA with a straight line from C to A? (b) What is the net heat transfer into the system?

14.12 An automobile tire contains 0.0380 m³ of air at a pressure of 2.20×10^5 N/m² (about 32 psi). How much more internal energy does this gas have than the same volume has at zero gauge pressure?

• **14.13** A helium-filled toy balloon has a gauge pressure of 0.200 atm and a volume of 10.0 liters. How much greater is the internal energy of the helium in the balloon than it would be at zero gauge pressure?

• **14.14** Steam to drive a locomotive is supplied at a constant gauge pressure of 1.75×10^6 N/m² (about 250 psi) to a 0.200 m radius piston. (a) By calculating $P \Delta V$, find the amount of work done by the steam when the piston moves 0.800 m. Note that this is the net work output, since gauge pressure is used. (b) Now find the amount of work by calculating the force exerted times the distance traveled. Is the answer the same as in part (a)?

• **14.15** A hand-driven tire pump has a 2.50 cm diameter piston and a maximum stroke of 30.0 cm. (a) How much work do you do in one stroke if the average gauge pressure is 2.40×10^5 N/m² (about 35 psi)? (b) What average force do you exert on the piston, neglecting friction and gravity?

• **14.16** Calculate the net work output of a heat engine following path ABCDA in Figure 14.30.

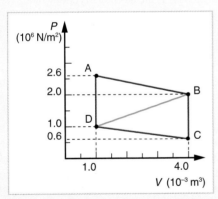

Figure 14.30 The PV diagram considered for heat engines in Problems 16 and 17.

• **14.17** What is the net work output of a heat engine that follows path ABDA in Figure 14.30, with a straight line from B to D? Why is the work output less than for path ABCDA?

Section 14.3 The Second Law of Thermodynamics and Heat Engines

14.18 A certain heat engine does 10.0 kJ of work and dissipates 8.50 kJ of waste heat in a cyclical process. (a) What was the heat input to this engine? (b) What was its efficiency?

14.19 With 2.50×10^6 J of heat input, a given cyclical heat engine can do only 1.50×10^5 J of work. (a) What is its efficiency? (b) How much waste heat is produced?

14.20 (a) What is the work output of a cyclical heat engine having a 22.0% efficiency and 6.00×10^9 J of heat input? (b) How much waste heat is produced?

14.21 (a) What is the efficiency of a cyclical heat engine that wastes 75.0 kJ of heat for every 95.0 kJ of heat input? (b) How much work does it produce for 100 kJ of heat input?

• **14.22** The engine of a large ship does 2.00×10^8 J of work with an efficiency of 5.00%. (a) How much waste heat is produced? (b) How many barrels of fuel are consumed, if each barrel produces 6.00×10^9 J of heat when burned?

• **14.23** (a) How much waste heat is produced by an electrical power station that uses 1.25×10^{14} J of heat input with an efficiency of 42.0%? (b) What is the ratio of waste heat to work output? (c) How much work is done?

⁞ **14.24** This problem compares the energy output and waste heat produced by two different types of nuclear power plants—one with the normal efficiency of 34.0% considered in Example 14.3, and another with an improved efficiency of 40.0%. Suppose both produce the same amount of heat in one day, 2.50×10^{14} J. (a) How much more electrical energy is produced by the more efficient plant? (b) How much less waste heat is generated by the more efficient plant? (One type of more efficient nuclear power plant, the gas-cooled reactor, has been too undependable to be economically feasible in spite of its greater efficiency.)

Section 14.4 Carnot Engines

14.25 There have been proposals to extract energy from the oceans by operating a heat engine between the warm upper layers, which may be at 20.0°C, and the cold deep water that is at 5.00°C. What is the maximum efficiency of a heat engine operating between these two temperatures? (In practice, you cannot even do this well. The best you can expect to get is an efficiency 0.700 times the maximum.)

14.26 A gas-cooled nuclear reactor operates between hot and cold reservoir temperatures of 700°C and 27.0°C (compared with a standard power reactor that has a hot temperature of only 300°C). (a) What is the maximum efficiency of a heat engine operating between 700°C and 27.0°C? (b) Find the ratio of this efficiency to the Carnot efficiency of a standard nuclear reactor (found in Example 14.4).

• **14.27** A certain gasoline engine has an efficiency of 30.0%. What would the hot reservoir temperature be for a Carnot engine having that efficiency, if it operates with a cold reservoir temperature of 200°C?

• **14.28** Steam locomotives have an efficiency of 17.0% and operate with a hot steam temperature of 425°C. (a) What would the cold reservoir temperature be if this were a Carnot engine? (b) What would the maximum efficiency of this steam engine be if its cold reservoir temperature were 150°C?

• **14.29** (a) What is the hot reservoir temperature of a Carnot engine that has an efficiency of 42.0% and a cold reservoir temperature of 27.0°C? (b) What must the hot reservoir temperature be for a real heat engine that achieves 0.700 of the maximum efficiency, but still has an efficiency of 42.0% (and a cold reservoir at 27.0°C)?

⁞ **14.30** Practical steam engines utilize 450°C steam, which is later exhausted at 270°C. (a) What is the maximum efficiency that such a heat engine can have? (b) Since 270°C steam is still quite hot, a second steam engine is sometimes operated using the exhaust of the first. What is the maximum efficiency of the second engine if its exhaust has a temperature of 150°C? (c) What is the overall efficiency of the two engines? (d) Show that this is the same efficiency as a single Carnot engine operating between 450°C and 150°C.

14.31 It is possible to operate a second heat engine using the exhaust of a first heat engine—the two engines are said to be operated in series. For example, the first engine could operate between temperatures T_1 and T_2, while the second would operate between T_2 and a lower temperature T_3. Prove that the overall efficiency of two Carnot engines run in series is the same as a single Carnot engine operating between the highest and lowest temperatures, T_1 and T_3.

Section 14.5* Heat Pumps, Air Conditioners, and Refrigerators

14.32 What is the coefficient of performance of an ideal heat pump that pumps heat from a cold temperature of −25.0°C to a hot temperature of 40.0°C?

14.33 Suppose you have an ideal refrigerator that cools an environment at −20.0°C and exhausts its heat at 50.0°C. What is its coefficient of performance?

14.34 What is the best coefficient of performance possible for a hypothetical refrigerator that could make liquid nitrogen at −200°C and exhaust its heat at 35.0°C?

14.35 In a very mild winter climate, a heat pump could extract heat from an environment at −5.00°C and exhaust it at 35.0°C. What is the best possible coefficient of performance for these temperatures?

• 14.36 (a) What is the best coefficient of performance for a heat pump that has a hot reservoir temperature of 50.0°C and a cold reservoir temperature of −20.0°C? (b) How much heat in kilocalories would it pump into the warm environment if 3.60×10^7 J of work (10.0 kW · h) is put into it? (c) If the cost of this work input is 10.0¢/kW · h, how does its cost compare with the direct production of heat by burning natural gas at a cost of 85.0¢ per therm (a common unit of energy for natural gas), where a therm equals 1.055×10^8 J?

• 14.37 (a) What is the best coefficient of performance for a refrigerator that cools an environment at −30.0°C and exhausts its heat at 45.0°C? (b) How much work in joules must be done in order to remove 1000 kcal of heat from the cold environment? (c) What is the cost of doing this if the work costs 10.0¢ per 3.60×10^6 J (a kilowatt-hour)? (c) How many kcal of heat are dumped into the warm environment?

• 14.38 Suppose you want to operate an ideal refrigerator with a cold temperature of −10.0°C, and you would like it to have a coefficient of performance of 7.00. What is the hot reservoir temperature for such a refrigerator?

• 14.39 An ideal heat pump is being considered for use in heating an environment with a temperature of 22.0°C. What is the cold reservoir temperature if the pump is to have a coefficient of performance of 12.0?

• 14.40 A 4 ton air conditioner removes 5.06×10^7 J (48,000 British thermal units) from a cold environment in 1.00 h. What energy input in joules is necessary to do this if the air conditioner has an *EER* of 12.0?

14.41 Show that the coefficients of performance of refrigerators and heat pumps are related by

$$COP_{ref} = COP_{hp} - 1$$

Start with the definitions of the *COP*s and the conservation of energy relationship between Q_h, Q_c, and W.

14.42 The energy efficiency rating (*EER*) of an air conditioner or refrigerator is defined to be the number of British thermal units of heat removed from a cold environment per hour divided by the watts of power input—that is,

$$EER = \frac{Q_c/t_1}{W/t_2} \propto \frac{Q_c}{W} = COP_{ref}$$

but this is expressed in different mixed units (t_1 is in hours, t_2 is in seconds, Q_c is in British thermal units, and W is in joules). (a) What is the relationship between *EER* and COP_{ref}, given that a British thermal unit is 1054 J, and a watt-hour is 3.60×10^3 J? That is, what is the constant of proportionality? (b) What COP_{ref}s then correspond to commercially available air conditioners, the *EER*s of which range from 6.00 to 12.0?

Section 14.6 Entropy

Note that several of the problems in this section require information from Chapter 13 regarding the effects of heat on temperature and phase.

14.43 On a winter day, a certain house loses 5.00×10^8 J heat to the outside (about 500,000 Btu). What is the total change in entropy due to this heat transfer alone, assuming an average indoor temperature of 21.0°C and an average outdoor temperature of 5.00°C?

14.44 A car parked in the sun on a hot summer day absorbs 4.00×10^6 J of heat, increasing its temperature from 35.0°C to 45.0°C. What is the increase in entropy of the car due to this heat transfer alone?

14.45 A hot rock ejected from a lava fountain cools from 1100°C to 40.0°C, and its entropy decreases by 950 J/K. How much heat in joules left the rock?

• 14.46 When 1.60×10^5 J of heat enters a cherry pie initially at 20.0°C, its entropy increases by 480 J/K. What is its final temperature?

• 14.47 The sun radiates heat at the rate of 3.80×10^{26} W from its 5500°C surface into dark empty space (a negligible fraction falls on the earth and other planets). The effective temperature of deep space is −270°C. (a) What is the increase in entropy in one day due to this heat transfer? (b) How much work is made unavailable?

• 14.48 (a) How much heat flows from 1.00 kg of water at 40.0°C when it is placed in contact with 1.00 kg of 20°C water in reaching equilibrium? (b) What is the change in entropy due to this heat transfer? (c) How much work in made unavailable, taking the lowest temperature to be 20.0°C?

• 14.49 What is the decrease in entropy of 25.0 g of water that condenses on a bathroom mirror at a temperature of 35.0°C, assuming no change in temperature and given the latent heat of vaporization to be 2450 kJ/kg?

• 14.50 Find the increase in entropy of 1.00 kg of liquid nitrogen that starts at its boiling temperature, boils, and warms to 20.0°C at constant pressure. See Tables 13.1 and 13.2 for useful information.

14.51 A large electrical power plant generates 1000 MW of electricity with an efficiency of 35.0%. (a) Calculate the heat input to the plant, Q_h, in one day. (b) How much heat Q_c is exhausted in one day? (c) If the cooling towers exhaust the heat in the form of 35.0°C air into the local air mass, which increases in temperature from 18.0°C to 20.0°C, what is the total increase in entropy due to this heat transfer? (d) How much energy becomes unavailable to do work because of this increase in entropy, assuming an 18.0°C lowest temperature? (Part of Q_c could, of course, be utilized to operate heat engines or for simple space heating, but it rarely is.)

14.52 (a) How much heat flows from 20.0 kg of 90.0°C water placed in contact with 20.0 kg of 10.0°C water, producing a final temperature of 50.0°C? (b) How much work could a Carnot engine do with this amount of heat input, assuming it operates between two reservoirs at constant temperatures of 90.0°C and 10.0°C? (c) What increase in entropy is produced by mixing 20.0 kg of 90.0°C water with 20.0°C of 10.0°C water? (d) Calculate the amount of work made unavailable by this mixing using a low temperature of 10.0°C, and compare it with the work done by the Carnot engine.

Section 14.7 Statistical Interpretation of Entropy and the Second Law of Thermodynamics

14.53 Using Table 14.4, verify the contention that if you toss 100 coins each second, you can expect to get 100 heads or 100 tails once in 2×10^{22} years, but calculate the time to two-digit accuracy.

14.54 What percent of the time will you get something in the range from 60 heads and 40 tails through 40 heads and 60 tails when tossing 100 coins? The total number of microstates in that range is 1.22×10^{30}.

14.55 (a) If tossing 100 coins, how many ways (microstates) are there to get the three most likely macrostates of 49 heads and 51 tails, 50 heads and 50 tails, and 51 heads and 49 tails? (b) What percent of the total possibilities is this?

• **14.56** (a) What is the change in entropy if you start with 100 coins in the 45 heads and 55 tails macrostate, toss them, and get 51 heads and 49 tails? (b) What if you get 75 heads and 25 tails? (c) How much more likely is 51 heads and 49 tails than 75 heads and 25 tails? (d) Does either outcome violate the second law of thermodynamics?

• **14.57** (a) What is the change in entropy if you start with 10 coins in the 5 heads and 5 tails macrostate, toss them, and get 2 heads and 8 tails? (b) How much more likely is 5 heads and 5 tails than 2 head and 8 tails? (Take the ratio of the number of microstates to find out.)

10-COIN TOSS

Macrostate		Number of
Heads	Tails	Microstates W
10	0	1
9	1	10
8	2	45
7	3	120
6	4	210
5	5	252
4	6	210
3	7	120
2	8	45
1	9	10
0	10	1
		Total: 1024

Problems 57 and 58 refer to this table.

• **14.58** (a) If you toss 10 coins, what percent of the time will you get the three most likely macrostates (6 heads and 4 tails, 5 heads and 5 tails, 4 heads and 6 tails)? (b) You can realistically toss 10 coins and count the number of heads and tails about twice a minute. At that rate, how long will it take on average to get either 10 heads and 0 tails or 0 heads and 10 tails?

⁞ **14.59** (a) Construct a table showing the macrostates and all of the individual microstates for tossing 6 coins. (Use Table 14.3 as a guide.) (b) How many macrostates are there? (c) What is the total number of microstates? (d) What percent chance is there of tossing 5 heads and 1 tail? (e) How much more likely are you to toss 3 heads and 3 tails than 5 heads and 1 tail? (Take the ratio of the number of microstates to find out.)

INTEGRATED CONCEPTS

The following problems involve concepts from this chapter and several others. Physics is most interesting when applied to general situations involving more than a narrow set of physical principles. For example, heat is a form of energy, and hence the relevance of Chapter 6, Work, Energy, and Power. The following topics are involved in some or all of the problems in this section:

Topics	Location
Dynamics	Chapter 4
Stress and strain	Chapter 5
Work, energy, and power	Chapter 6
Momentum	Chapter 7
Flow rate	Chapter 11
Thermal expansion	Chapter 12
Temperature change and phase change	Chapter 13
Thermodynamics	Chapter 14

PROBLEM-SOLVING STRATEGY

Step 1. Identify which physical principles are involved.

Step 2. Solve the problem using strategies outlined in the text.

The following worked example illustrates how this strategy is applied to an integrated concept problem.

EXAMPLE **WASTE HEAT PRODUCED BY A NUCLEAR POWER PLANT AND THE AMOUNT OF WATER NEEDED TO CARRY IT AWAY**

Topics	Location
Thermodynamics	Chapter 14
Temperature change	Chapter 13
Flow rate	Chapter 11
Density (Fluid statics)	Chapter 10

In one day the reactors of a large nuclear power plant generate 5.00×10^{14} J of heat, which it converts to electrical energy with an efficiency of 35.0%. (a) How many joules of waste heat does the plant create in one day? (b) Ocean water with a specific heat of 4186 J/kg·°C is used to carry away the heat. How many kilograms of water are required if the water temperature is increased from 5.00°C to 25.0°C? (c) What is the flow rate of the water in m³/s, given its density is 1050 kg/m³?

Strategy Step 1 To solve an *integrated concept problem*, such as those following this example, we must first identify the physical principles involved and identify the chapters in which they are found.

Part (a) of this example considers efficiency and waste heat. These are topics of *thermodynamics* and are covered in this chapter. Part (b) deals with *temperature change*, a topic of Chapter 13. Part (c) asks for *flow*, a topic of Chapter 11, and requires us to use *density*, a topic of Chapter 10, to get volume from mass.

Strategy Step 2 The following solutions to each part of the example illustrate how the specific problem-solving strategies are applied. These involve identifying knowns and unknowns, checking to see if the answer is reasonable, and so on.

Solution for (a): Thermodynamics (Chapter 14) A nuclear power plant is a heat engine. We are asked to find its waste heat Q_c. Efficiency is given by Equation 14.5a to be

$$Eff = \frac{W}{Q_h}$$

In this example, we are given $Eff = 35.0\%$ and $Q_h = 5.00 \times 10^{14}$ J, and so we can find the work done:

$$W = Eff \cdot Q_h = (0.350)(5.00 \times 10^{14} \text{ J}) = 1.75 \times 10^{14} \text{ J}$$

From Equation 14.4 we know that $W = Q_h - Q_c$, and so the waste heat is

$$Q_c = Q_h - W = (5.00 \times 10^{14} - 1.75 \times 10^{14}) \text{ J} = 3.25 \times 10^{14} \text{ J}$$

Discussion for (a) This is a very large amount of heat, but that is typical for large electrical generating stations.

Solution for (b): Temperature change (Chapter 13) Here we are asked to find the mass of water that this amount of heat can raise by a given temperature. The relationship between change in temperature and heat is given by Equation 13.2:

$$Q_c = mc \, \Delta T$$

Solving for m as requested, and substituting known values, we obtain

$$m = \frac{Q_c}{c \, \Delta T}$$

$$= \frac{3.25 \times 10^{14} \text{ J}}{(4186 \text{ J/kg} \cdot {}^\circ\text{C})(25.0{}^\circ\text{C} - 5.00{}^\circ\text{C})}$$

$$= 3.88 \times 10^9 \text{ kg}$$

Discussion for (b) This is a very large amount of water, consistent with the fact that the amount of waste heat is large, and is also typical of all electrical generating facilities. Their environmental impacts can be great.

Solution for (c): Flow rate and volume (Chapters 10 and 11) The volume of sea water can be found from density and mass using the definition of density found in Equation 10.1:

$$\rho = \frac{m}{V}$$

Solving for V and substituting known values gives

$$V = \frac{m}{\rho} = \frac{3.88 \times 10^9 \text{ kg}}{1050 \text{ kg/m}^3}$$

$$= 3.70 \times 10^6 \text{ m}^3$$

Flow rate is defined by Equation 11.1 to be V/t, or volume per unit time. Thus,

$$\mathcal{F} = \frac{3.70 \times 10^6 \text{ m}^3}{8.64 \times 10^4 \text{ s}}$$

$$= 42.8 \text{ m}^3/\text{s}$$

Discussion for (c) This is a flow rate of a small river. It is great enough to fill a backyard swimming pool in 1 or 2 seconds.

This worked example illustrates how to apply problem-solving strategies to situations that include topics in different chapters. The first step is to identify the physical principles involved in the problem. The second step is to solve for the unknowns using familiar problem-solving strategies. These are found throughout the text, and many worked examples show how to use them for single topics. In this integrated concepts example, you can see how to apply them across many topics. You will find these techniques useful in applications of physics outside a physics course, such as in your profession, in other science disciplines, and in everyday life. The following problems will build your skills in the broad application of physical principles.

14.60 The cooling capacities of air conditioners are often rated in tons, where a 1.00 ton air conditioner removes 12.65 MJ of heat from a cold environment in 1.00 h (this odd amount of heat is 12,000 Btu). (a) What mass of ice melting would remove the same amount of heat? (b) How many hours of operation would be equivalent to melting 2000 lb of ice (a short ton)?

14.61 The waste heat produced by the ship in Problem 14.22 is dissipated into the water by heating it 5.00°C (this warmed water is later mixed with ocean water at the ambient temperature). (a) How many kilograms of water does this take? (b) What is the flow of cooling water in m³/s if the ship does its work in 2.00 min?

•**14.62** What is the mass of the person in Problem 14.9 if the height climbed is 169 m?

•**14.63** In a drive to a nearby ski resort, a 1000 kg car goes from rest to a speed of 90.0 km/h and gains 1.00 km in altitude, using 1.00 gal of gasoline in the process. (Burning 1.00 gal of gasoline produces 1.30×10^8 J of heat.) (a) What was the overall efficiency for the trip, and why is it so much lower than that given for gasoline engines in Table 6.3? (b) How much waste heat was produced?

•**14.64** A 500 kg dragster reaches a speed of 107 m/s on level ground, starting from rest. Its efficiency is only 8.00% in producing kinetic energy, because of friction and its design for high power without regard to efficiency. (a) How many gallons of fuel does it use during this acceleration, given there are 1.30×10^8 J of energy per gallon? (b) How much waste heat is produced? (c) Once it is brought back to rest (assuming the engine shuts off at the end of the race), how much waste heat has been produced?

•**14.65** While taking a plane flight to a meeting, a physicist decides to do a mental calculation of how much fuel the plane would consume in the absence of friction, assuming its engines have an efficiency of 20.0% and the mass of the plane is 200,000 kg. Each gallon of jet fuel produces 1.30×10^8 J of heat when burned. (a) How many gallons of fuel are used to get the plane from rest to the takeoff speed of 60.0 m/s? (b) How many gallons to get to a cruising speed of 250 m/s and an altitude of 10.0 km? (c) How much waste heat is produced in each case? (Both the actual fuel consumption and heat produced will be greater because of air resistance.)

:14.66 A granite column in a building is 8.00 m high and has a mass of 2.16×10^4 kg. (a) How much heat is needed to increase its temperature by 10.0°C, given its specific heat is 0.200 kcal/kg·°C? (b) What is its change in length if its coefficient of linear expansion is 2.50×10^{-6}/°C? (c) Calculate the force it would exert if not allowed to expand, given its cross-sectional area to be 1.00 m² and the Young's modulus for granite to be 45.0×10^9 N/m². Note that this force is much larger than the weight it supports, so that it will expand almost freely. (d) How much work does it perform when it expands, if it supports a mass of 1.00×10^4 kg? (e) What is its change in internal energy?

:14.67 An electrical power plant is constructed to extract energy from the ocean; it utilizes warm water at 20.0°C and cold water at 5.00°C. (a) Calculate its efficiency assuming it is 0.700 times the Carnot efficiency. (b) What is the rate of heat input needed to produce 500 MW of electrical power? (c) What flow of water in m³/s must it take in, given the ocean water's density is 1050 kg/m³ and its specific heat is 1.00 kcal/kg·°C?

:14.68 Using the information from Problem 14.14: (a) Calculate the power output of the steam engine if there are four cylinders, each making 2.00 strokes (of 0.800 m) per second. (b) What is the rate of heat input if the efficiency is 17.0%? (c) Assuming constant power output, how long would it take to increase the 3.00×10^6 kg train's speed from 10.0 to 11.0 m/s on level ground? (d) Is this possible if the maximum tractive force is 0.350 times the weight of the 1.50×10^5 kg engine?

UNREASONABLE RESULTS

The following problems have results that are unreasonable because some premise is unreasonable or because certain of the premises are inconsistent with one another. Physical principles applied correctly then produce unreasonable results. The purpose of these problems is to give practice in assessing whether nature is being accurately described, and if it is not to trace the source of difficulty.

Note: Strategies for dealing with problems having unreasonable results appear in Chapter 12, among others. They can be found with the section of problems labeled *Unreasonable Results* at the end of that chapter.

14.69 What is wrong with the claim that a cyclical heat engine does 4.00 kJ of work on an input of 24.0 kJ of heat while dissipating 16.0 kJ of waste heat?

•14.70 (a) Calculate the cold reservoir temperature of a steam engine that utilizes hot steam at 450°C and has a Carnot efficiency of 0.700. (b) What is unreasonable about the temperature? (c) Which premise is unreasonable?

•14.71 (a) Suppose you want to design a steam engine that exhausts at 270°C and has a Carnot efficiency of 0.800. What temperature of hot steam must you utilize? (b) What is unreasonable about the temperature? (c) Which premise is unreasonable?

•14.72 (a) A cyclical heat engine, operating between temperatures of 450°C and 150°C, produces 4.00 MJ of work on a heat input of 5.00 MJ. How much waste heat does it produce? (b) What is unreasonable about the engine? (c) Which premise is unreasonable?

15 OSCILLATORY MOTION AND WAVES

There are at least four types of waves in this picture—only the water waves

are evident. There are also sound waves, light waves,

and waves on the guitar strings.

What do an ocean buoy, a child in a swing, a guitar string, and atoms in a crystal have in common? They all **oscillate**—that is, they move back and forth between two points. Many systems oscillate, and they have certain characteristics in common. All oscillations involve force and energy, for example. You push a child in a swing to get the motion started. The energy of atoms vibrating in a crystal can be increased with heat. You put energy into a guitar string when you pluck it. In this chapter, we will study oscillatory motion and the forces and energy involved with it.

Some oscillations create waves. A vibrating guitar string creates sound waves, for example. You can make water waves in a swimming pool by slapping the water repeatedly with your hand. You can no doubt think of other types of waves. Some, like water waves, are visible and some, like sound waves, are not. But every wave is a disturbance that moves from its source and carries energy. Other examples of waves include earthquakes and visible light. Even subatomic particles, like electrons, behave like waves.

By studying oscillatory motion and waves, we shall find once again that a small number of underlying principles describe all of them and that wave phenomena are more common than you have ever imagined. We begin by studying the type of force that underlies the simplest oscillations and waves.

15.1 HOOKE'S LAW: STRESS AND STRAIN RECALLED

Newton's first law implies that an object that oscillates back and forth is experiencing forces. Without force, the object would move in a straight line at constant speed rather than oscillate. Consider, for example, plucking a plastic ruler to the left as shown in Figure 15.1. This deformation is opposed by a force in the opposite direction—a **restoring force**. Once released, the restoring force causes the ruler to move back toward its stable equilibrium position, where the net force on it is zero. But by the time the ruler gets there, it has momentum and continues to move to the right, producing the opposite deformation. It is then forced to the left, back through equilibrium, and the process is repeated until dissipative forces dampen the motion. (These forces remove mechanical energy from the system, gradually reducing the motion until the ruler comes to rest.)

The simplest oscillations occur when the restoring force is directly proportional to the displacement. Interestingly, this is the rule rather than the exception. In fact, for small deformations, the restoring force is almost always directly proportional to the deformation. When stress and strain were covered in Section 5.8, a name was given to this relationship between force and displacement—it is called Hooke's law*:

$$F = -kx \qquad (15.1)$$

Here F is the restoring force, x is the displacement from equilibrium, or **deformation**, and k is a constant related to how hard it is to deform the system. The minus sign indicates the restoring force is in the direction opposite to the displacement. (See Figure 15.2.)

The **force constant** k is related to the rigidity or stiffness of a system—the larger the force constant, the greater the restoring force, and the stiffer the system. (The units of k are newtons per meter (N/m).) For example, k is directly related to Young's modulus when we stretch a string. Figure 15.3 shows a graph of the absolute value of the restoring force versus the displacement for a system that obeys Hooke's law—a simple spring in this case. The slope of the graph equals the force constant k in newtons per meter. A favorite physics laboratory exercise is to measure restoring forces created by springs, determine if they follow Hooke's law, and calculate their force constants if they do.

---

*Although Hooke's law holds for small displacements and many other cases, it is not universally valid and is thus not a "law" in the strictest sense. This is of no great concern, since other laws, such as Ampere's law, also suffer from being limited or being only classically relevant.

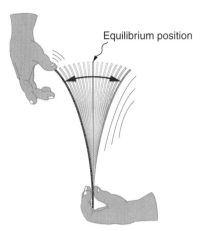

Figure 15.1 When displaced from its vertical equilibrium position, this plastic ruler oscillates back and forth because of the restoring force opposing displacement. When the ruler is on the left, there is a force to the right, and vice versa.

CONNECTIONS

Hooke's law was first introduced when elastic deformations were considered in Chapter 5. It is obeyed by most systems for small deformations or displacements from equilibrium. We will see that this means almost all systems in equilibrium will oscillate in a simple way if slightly disturbed.

Figure 15.2 (a) The plastic ruler has been released, and the restoring force is returning it to its equilibrium position. (b) The net force is zero at the equilibrium position, but the ruler has momentum and continues to move to the right. (c) The restoring force is in the opposite direction. It stops the ruler and moves it back toward equilibrium again. (d) Now the ruler has momentum to the left. (e) In the absence of damping, the ruler reaches its original position. From there, the motion will repeat itself.

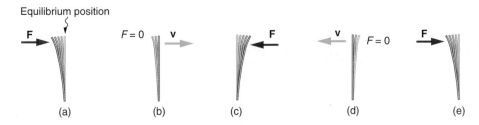

Figure 15.3 (a) A graph of absolute value of the restoring force versus displacement. The fact that the graph is a straight line means that the system obeys Hooke's law. The slope of the graph is the force constant k. (b) The data in the graph were generated by measuring the displacement of a spring from equilibrium when supporting various weights. The restoring force equals the weight supported, if the mass is stationary.

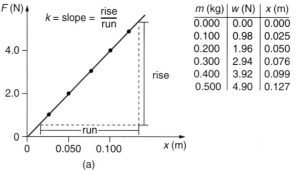

m (kg)	w (N)	x (m)
0.000	0.00	0.000
0.100	0.98	0.025
0.200	1.96	0.050
0.300	2.94	0.076
0.400	3.92	0.099
0.500	4.90	0.127

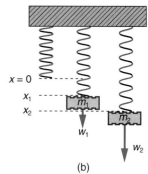

(a)   (b)

EXAMPLE 15.1   HOW STIFF ARE CAR SPRINGS?

What is the force constant for the suspension system of a car that settles 1.20 cm when an 80.0 kg person gets in?

Strategy and Concept Consider the car to be in its equilibrium position $x = 0$ before the person gets in. It then settles down 1.20 cm, which means it is displaced to a position $x = -1.20 \times 10^{-2}$ m. At that point, the springs supply a restoring force F equal to the person's weight $w = mg = (80.0 \text{ kg})(9.80 \text{ m/s}^2) = 784$ N. We take this force to be F in Hooke's law. Knowing F and x, we can then solve for the force constant k.

Solution Solving Hooke's law, $F = -kx$, for k and sub-

stituting known values, we obtain

$$k = -\frac{F}{x} = -\frac{784 \text{ N}}{-1.20 \times 10^{-2} \text{ m}} = 6.53 \times 10^4 \text{ N/m}$$

Discussion Note that F and x have opposite signs because they are in opposite directions—the restoring force is up, and the displacement down. Note also that the car would oscillate up and down when the person got in if it were not for damping provided by shock absorbers. Bouncing cars are a sure sign of bad shocks.

Energy in Hooke's Law Deformations

Work must be done to produce a deformation. That is, a force must be exerted through a distance, whether you pluck a guitar string or compress a car spring. If the only result is deformation and no work goes into frictional heating or kinetic energy, then all the work is initially stored in the deformed object as potential energy. As noted in Section 6.4, the potential energy stored in a spring is $\text{PE}_s = (1/2)kx^2$. Here we generalize the idea to elastic potential energy for a deformation of any system obeying Hooke's law. Thus,

$$\text{PE}_{el} = \tfrac{1}{2}kx^2 \tag{15.2}$$

where PE_{el} is the **elastic potential energy** stored in any deformed system that has a displacement x from equilibrium and a force constant k, and that obeys Hooke's law.

We would like to find the work done in deforming a system, in order to find the energy stored. This work is done by an applied force F_{app}. The applied force is exactly opposite of the restoring force (action-reaction), and so $F_{app} = kx$. Figure 15.4 shows a graph of the applied force versus deformation x for a system that obeys Hooke's law. Work done on the system is force times distance, which equals the area under the curve, or $(1/2)kx^2$.

Figure 15.4 A graph of applied force versus distance for the deformation of a system obeying Hooke's law. The work done on the system equals the area under the graph. This is the area of the triangle, which is one-half its base times its height, or $(1/2)kx^2$.

Method A
$$W = \frac{1}{2}bh = \frac{1}{2}kxx$$
$$W = \frac{1}{2}kx^2$$

Method B
$$W = f \cdot x = \left(\frac{1}{2}kx\right)(x)$$
$$W = \frac{1}{2}kx^2$$

(This is Method A in the figure.) Another way to look at this is to note that the force increases linearly from 0 to kx, so that the average force is $(1/2)kx$, the distance moved is x, and thus $W = F_{app} \cdot d = [(1/2)kx](x) = (1/2)kx^2$. (This is Method B in the figure.)

EXAMPLE 15.2 ENERGY IN A TOY GUN'S SPRING

(a) How much energy is stored in the spring of a toy gun that has a force constant of 50.0 N/m and is compressed 0.150 m? (b) In the absence of friction and neglecting the mass of the spring, at what speed will a 2.00 g plastic bullet be ejected from the gun? (See Figure 15.5.)

Strategy for (a) The energy stored in the spring can be found directly from Equation 15.2, since k and x are given.

Solution for (a) Entering the given values for k and x yields

$$PE_{el} = \tfrac{1}{2}kx^2 = \tfrac{1}{2}(50.0 \ N/m)(0.150 \ m)^2 = 0.563 \ N \cdot m$$
$$= 0.563 \ J$$

Strategy for (b) Since there is no friction, the potential energy is converted entirely into kinetic energy. The expression for kinetic energy can be solved for the plastic bullet's speed.

Solution for (b) Without friction, conservation of energy implies

$$KE_f = PE_{el}$$

or

$$\tfrac{1}{2}mv^2 = \tfrac{1}{2}kx^2 = PE_{el} = 0.563 \ J$$

Solving for v yields

$$v = \left[\frac{2\,PE_{el}}{m}\right]^{1/2} = \left[\frac{2(0.563 \ J)}{0.002 \ kg}\right]^{1/2} = 23.7 \ m/s$$

Discussion This is an impressive projectile speed for a toy gun (more than 50 miles per hour). The numbers in this problem seem reasonable. The force needed to compress the spring is small enough for a child to manage, and the energy imparted to the plastic bullet is small enough to limit the damage it might do. Yet the speed of the bullet is great enough for it to travel an acceptable (to a child) distance.

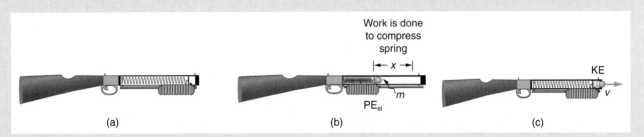

Work is done to compress spring

$\leftarrow x \rightarrow$

KE

PE_{el}

(a)   (b)   (c)

Figure 15.5 The toy gun in Example 15.2. (a) Before being cocked, the spring is uncompressed. (b) The spring has been compressed a distance x, and the projectile is in place. (c) When released, the spring converts elastic potential energy PE_{el} into kinetic energy.

15.2 PERIOD AND FREQUENCY IN OSCILLATIONS

When you pluck a guitar string, the resulting sound has a steady tone and lasts a long time. Each successive vibration of the string takes the same time as the previous one. We define **periodic motion** to be a repetitious oscillation, such as exhibited by the guitar string. The time to complete one oscillation remains constant and is called the **period** T; its units are usually seconds, but may be any convenient period of time. The word *period* refers to the time for some event whether repetitious or not; but we shall be primarily interested in periodic motion, which is by definition repetitive.

A concept closely related to period is the frequency of an event. For example, if you get a paycheck twice a month, the frequency of payment is two per month and the period between checks is half a month. **Frequency** f is defined to be the number of events per unit time. For periodic motion, frequency is the number of oscillations per unit time. The relationship between frequency and period is

$$f = \frac{1}{T} \tag{15.3}$$

The SI unit for frequency is the cycle per second, which is defined to be a **hertz** (Hz):

$$1 \ Hz = 1 \frac{cycle}{s}$$

A cycle is one complete oscillation. Note that a vibration can be a single or multiple event, whereas oscillations are usually repetitive for a significant number of cycles.

EXAMPLE 15.3 FREQUENCY OF MEDICAL ULTRASOUND AND PERIOD OF MIDDLE C

(a) A medical imaging device creates ultrasound by oscillating with a period of 0.400 μs. What is the frequency of this oscillation? (b) The frequency of middle C on a typical musical instrument is 264 Hz. What is the time for one complete oscillation?

Strategy Both questions can be answered using the relationship between period and frequency. In part (a), the period T is given and we are asked to find frequency f. In part (b), the frequency f is given and we are asked to find the period T.

Solution for (a) The frequency f can be found using Equation 15.3, since T is given. Substituting 0.400 μs for T, the frequency is

$$f = \frac{1}{T} = \frac{1}{0.400 \times 10^{-6} \text{ s}} = 2.50 \times 10^6 \text{/s} = 2.50 \text{ MHz}$$

Oscillations at this frequency create ultrasound—that is, sound with a frequency far above the highest frequency humans can hear.

Solution for (b) The time for one complete oscillation is the period T. Since we know $f = 1/T$, we can rearrange this to find T:

$$T = \frac{1}{f}$$

Now we substitute the given value for the frequency and find that

$$T = \frac{1}{f} = \frac{1}{264 \text{ Hz}} = \frac{1}{264 \text{ cycles/s}}$$
$$= 3.79 \times 10^{-3} \text{ s} = 3.79 \text{ ms}$$

Discussion The frequency of sound found in (a) is higher than humans can hear and, therefore, is called ultrasound. The period found in (b) is the time per cycle, but this is often quoted as simply the time in convenient units (ms or milliseconds in this case).

15.3 SIMPLE HARMONIC MOTION: A SPECIAL PERIODIC MOTION

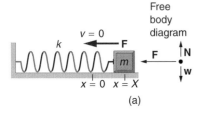

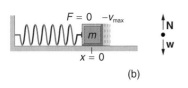

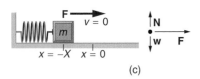

The oscillations of a system in which the net force obeys Hooke's law are of special importance, because they are very common. They are also the simplest oscillatory motions. **Simple harmonic motion** is the name given to oscillatory motion for a system where the net force obeys Hooke's law, and such a system is called a **simple harmonic oscillator**. If the net force obeys Hooke's law, there is no damping (by friction or other nonconservative forces). Thus a simple harmonic oscillator oscillates with equal displacement on either side of the equilibrium position, as shown for a mass on a spring in Figure 15.6. The maximum displacement from equilibrium is called the **amplitude** X. The units for amplitude and displacement are the same, but depend on the type of oscillation. For the mass on the spring the units of amplitude and displacement are meters, whereas for sound oscillations they have units of pressure, and yet other units for other types of oscillations. Since amplitude is the maximum displacement, it is related to the amount of energy in the oscillation.

What is so significant about simple harmonic motion? One special thing is that the period T and frequency f of a simple harmonic oscillator are independent of amplitude. The string of a guitar, for example, will oscillate with the same frequency whether plucked gently or hard. Since the period is constant, a simple harmonic oscillator can be used as a clock.

Two important factors do affect the period of a simple harmonic oscillator. The period is related to how stiff the system is. A very stiff object has a large force constant k, which causes the system to have a smaller period. For example, you can adjust a diving board's stiffness—the stiffer it is, the faster it vibrates, and the shorter its period. Period

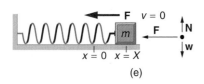

Figure 15.6 A mass attached to a spring sliding on a frictionless surface is an uncomplicated simple harmonic oscillator. When displaced from equilibrium, it performs simple harmonic motion with amplitude X and period T. Its maximum speed occurs as it passes through equilibrium. The stiffer the spring, the smaller the period T. The greater the mass, the greater the period T.

also depends on the mass of the oscillating system. The more massive the system, the longer the period. For example, a heavy person on a diving board bounces up and down more slowly than a light one.

In fact, the mass m and the force constant k are the *only* factors that affect the period and frequency of simple harmonic motion. In a later optional section, it will be shown that the **period of a simple harmonic oscillator** is

$$T = 2\pi \sqrt{\frac{m}{k}} \qquad (15.4a)$$

and, since $f = 1/T$, the **frequency of a simple harmonic oscillator** is

$$f = \frac{1}{2\pi} \sqrt{\frac{k}{m}} \qquad (15.4b)$$

Note that neither T nor f has any dependence on amplitude.

EXAMPLE 15.4 A CAR BOUNCES WITH A CHARACTERISTIC FREQUENCY

If the shock absorbers in a car like the one in Example 15.1 go bad, then it will oscillate at the least provocation, such as when going over bumps in the road and after stopping. (See Figure 15.7.) Calculate the frequency and period of these oscillations for the car considered in Example 15.1, if its mass including its load is 900 kg.

Strategy The frequency of the car's oscillations will be that of a simple harmonic oscillator as given in Equation 15.4b. The mass is given, and the force constant k of the suspension system was found in Example 15.1 to be 6.53×10^4 N/m.

Solution Entering the known values of k and m into Equation 15.4b yields

$$f = \frac{1}{2\pi} \sqrt{\frac{k}{m}} = \frac{1}{2\pi} \sqrt{\frac{6.53 \times 10^4 \text{ N/m}}{900 \text{ kg}}} = \frac{1}{2\pi} \sqrt{72.6/\text{s}^2}$$
$$= 1.36/\text{s} = 1.36 \text{ Hz}$$

The period can be calculated using Equation 15.4a, but it is simpler to use the relationship $T = 1/f$ and substitute the value just found for f:

$$T = \frac{1}{f} = \frac{1}{1.36 \text{ Hz}} = 0.737 \text{ s}$$

Discussion The values of T and f both seem about right for a bouncing car.

The Link between Simple Harmonic Motion and Waves

If a time-exposure photograph of the bouncing car in Example 15.4 were taken as it drives by, the headlight would make a wavelike streak, as shown in Figure 15.7. Similarly, Figure 15.8 shows a mass bouncing on a spring as it leaves a wavelike trace of its position on a moving strip of paper. Both waves are sine functions. All simple harmonic motion is intimately related to sine and cosine waves.

The displacement in any simple harmonic motion—that is, one in which the net restoring force follows Hooke's law—is given by

$$x = X \cos \frac{2\pi t}{T} \qquad (15.5)$$

where t is time and X is amplitude.* At $t = 0$ the initial position is $x_0 = X$, and the displacement oscillates back and forth with a period T (when $t = T$, we get $x = X$ again). Furthermore, the velocity v is

$$v = -v_{max} \sin \frac{2\pi t}{T}$$

where

$$v_{max} = \sqrt{\frac{k}{m}} X$$

$\qquad (15.6)$

The mass has zero velocity at maximum displacement—for example, $v = 0$ when $t = 0$,

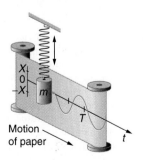

Figure 15.7 The bouncing car makes a wavelike motion. If the restoring force in the suspension system follows Hooke's law, then the wave is a sine function. (The wave is a trace produced by the headlight as the car moves to the right.)

Figure 15.8 The vertical position of a mass bouncing on a spring is recorded on a strip of moving paper, leaving a sine wave.

---

*The proof of this requires differential calculus.

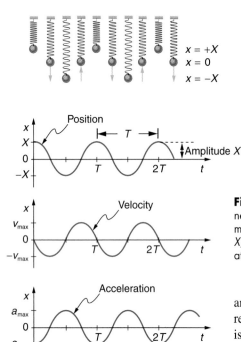

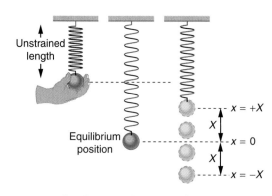

Figure 15.9 Graphs of x, v, and a versus t for the motion of a mass on a spring. The net force on the mass follows Hooke's law, and so the mass undergoes simple harmonic motion. Note that the initial position has the vertical displacement at its maximum value X; v is initially zero and then negative as the mass moves down; and the initial acceleration is negative, back toward the equilibrium position, becoming zero at that point.

and at that time $x = X$. The minus sign in the first equation gives the correct direction for the velocity. Just after the start of the motion, for instance, the velocity is negative since the system is moving back toward the equilibrium point. Finally, we can get an expression for a using Newton's second law. (Then we have x, v, t, and a, the quantities needed for kinematics and a description of simple harmonic motion.) According to Newton's second law, the acceleration is $a = F/m = -kx/m$, so that a is also a cosine function:

$$a = -\frac{kX}{m} \cos \frac{2\pi t}{T} \tag{15.7}$$

Hence a is directly proportional to, and in the opposite direction of, x. Figure 15.9 shows the simple harmonic motion of a mass on a spring and presents graphs of x, v, and a versus time.

The most important point here is that these equations are mathematically straightforward and are valid for all simple harmonic motion. They are very useful in visualizing waves associated with simple harmonic motion, including visualizing how waves add to one another.

15.4 THE SIMPLE PENDULUM

There are many pendulums around. Some have uses, such as in clocks, some are for fun, such as a child's swing, and some are just there, such as the sinker on a fish line. For small displacements, a pendulum is a simple harmonic oscillator. A **simple pendulum** is defined to have a small mass, also known as the pendulum bob, which is suspended from a light wire or string, such as shown in Figure 15.10. Exploring the simple pendulum a bit further, we can discover the conditions under which it performs simple harmonic motion, and we can derive an interesting expression for its period.

We begin by defining the displacement to be the arc length s. We see from Figure 15.10 that the net force on the bob is tangent to the arc and equals $-mg \sin \theta$. The weight mg is broken into components. Tension in the string exactly cancels the component parallel to the string. This leaves a *net* restoring force back toward the equilibrium position at $\theta = 0$.

Now, if we can show that the restoring force is directly proportional to the displacement, then we have a simple harmonic oscillator. With that in mind, note that for small angles (less than about 15°), $\sin \theta \approx \theta$, since they differ by about 1% or less at smaller angles.* Thus, for angles less than 15°, the restoring force F is

$$F = -mg \sin \theta \approx -mg\theta$$

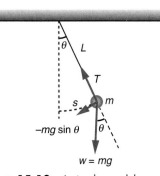

Figure 15.10 A simple pendulum has a small-diameter bob and a string of very small mass but strong enough not to stretch appreciably. The linear displacement from equilibrium is s, the length of the arc. Also shown are the forces on the bob, which result in a net force of −mg sin θ toward the equilibrium position—that is, a restoring force.

---

*To verify this, you must express θ in radians. Then, using a calculator, you will find that $\sin \theta \approx \theta$ for small angles.

The displacement s is also directly related to θ. When θ is expressed in radians, the arc length in a circle is related to its radius (L in this instance) by[†]

$$s = L\theta$$

so that

$$\theta = \frac{s}{L}$$

For small angles, then, the expression for the restoring force is

$$F \approx -\frac{mg}{L}\, s$$

This is of the form

$$F = -kx$$

where the force constant $k = mg/L$ and the displacement $x = s$. Thus, for angles less than about 15°, the restoring force is directly proportional to the displacement, and the simple pendulum is a simple harmonic oscillator.

Using Equation 15.4a, we can find the period of a pendulum for amplitudes less than about 15°. For the simple pendulum,

$$T = 2\pi\sqrt{\frac{m}{k}} = 2\pi\sqrt{\frac{m}{mg/L}}$$

Thus,

$$T = 2\pi\sqrt{\frac{L}{g}} \tag{15.8}$$

for the period of a simple pendulum. This result is interesting because of its simplicity. The only things that affect the period of a simple pendulum are its length and the acceleration of gravity. The period is completely independent of other factors, such as mass. As with simple harmonic oscillators, T for a pendulum is nearly independent of amplitude, especially if θ less than 15°. Even simple pendulum clocks can be finely adjusted and accurate.

Note the dependence on T on g. If the length of a pendulum is precisely known, it can actually be used to measure the acceleration of gravity. Consider the following example.

EXAMPLE 15.5 THE PERIOD OF A PENDULUM IS USED TO MEASURE THE ACCELERATION OF GRAVITY

What is the acceleration of gravity in a region where a simple pendulum having a length 75.000 cm has a period of 1.7357 s?

Strategy We are asked to find g given the period T and the length L of a pendulum. We can solve Equation 15.8 for g, assuming only that the angle of deflection is less than 15°.

Solution Squaring Equation 15.8, solving for g, and substituting known values gives

$$g = 4\pi^2\frac{L}{T^2} = 4\pi^2\frac{0.75000 \text{ m}}{(1.7357 \text{ s})^2}$$

yielding

$$g = 9.828 \text{ m/s}^2$$

Discussion This method for determining g can be very accurate. (This is why length and period are given to five digits in this example.) Knowing g can be important in geological research, for example. A map of g over large geographical regions aids the study of plate tectonics and helps in the search for oil fields.

15.5* ENERGY AND THE SIMPLE HARMONIC OSCILLATOR

To study the energy of a simple harmonic oscillator, we first consider all the forms its energy can have. We know from Section 15.1 that the energy stored in the deformation of a simple harmonic oscillator is a form of potential energy given by

$$PE_{el} = \tfrac{1}{2}kx^2$$

Since a simple harmonic oscillator has no dissipative forces, the other important form of

---

[†]This was first defined and discussed in Section 8.1.

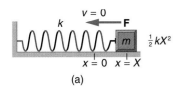

(a)

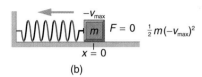

(b)

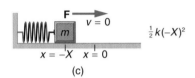

(c)

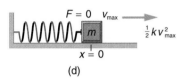

(d)

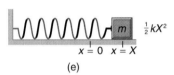

(e)

Figure 15.11 The transformation of energy in simple harmonic motion is illustrated for a mass attached to a spring on a frictionless surface.

energy is kinetic energy KE. Conservation of energy for these two forms is

$$KE + PE_{el} = \text{constant}$$

or

$$\tfrac{1}{2}mv^2 + \tfrac{1}{2}kx^2 = \text{constant} \tag{15.9}$$

This statement of conservation of energy is valid for *all* simple harmonic oscillators, including ones where gravity plays a role. The effect of gravity for, say, a mass suspended on a vertical spring is to shift the equilibrium position downward. But it does not alter the fact that the *net* force is $-kx$ relative to the equilibrium position, so that the total potential energy relative to that position is $\tfrac{1}{2}kx^2$. (See Figure 15.9, for example.) (This means that the effect of gravity is included in PE_{el}, and the term mgh does not enter the conservation of energy expression explicitly.)

In the case of undamped simple harmonic motion, the energy oscillates back and forth between kinetic and potential, going completely from one to the other as the system oscillates. So for the simple example of a mass on a frictionless surface attached to a spring, as shown again in Figure 15.11, the motion starts with all of the energy stored in the spring. As the mass starts to move, the elastic potential energy is converted to kinetic energy, becoming entirely kinetic at the equilibrium position. It then is converted back into elastic potential energy by the spring, the velocity becomes zero when the kinetic energy is completely converted, and so on. This concept provides extra insight here and in later applications of simple harmonic motion, such as alternating current circuits.

The conservation of energy principle can be used to derive an expression for velocity v. If we start our simple harmonic motion with zero velocity and maximum displacement ($x = X$), then the total energy is $(1/2)kX^2$. This total is constant and is shifted back and forth between kinetic and potential energy, at most times being part each. The conservation of energy for this system in equation form is thus

$$\tfrac{1}{2}mv^2 + \tfrac{1}{2}kx^2 = \tfrac{1}{2}kX^2$$

Solving this for v yields

$$v = \pm\sqrt{\frac{k}{m}(X^2 - x^2)}$$

Manipulating this expression algebraically gives

$$v = \pm\sqrt{\frac{k}{m}}X\sqrt{1 - \frac{x^2}{X^2}}$$

From this expression, we see that at $x = 0$ the velocity is a maximum v_{max}, and so

$$v = \pm v_{max}\sqrt{1 - \frac{x^2}{X^2}} \tag{15.10}$$

where

$$v_{max} = \sqrt{\frac{k}{m}}X$$

as stated earlier in Equation 15.6. Notice that the maximum velocity depends on three factors. Maximum velocity is directly proportional to amplitude. As you might guess, the greater the maximum displacement, the greater the maximum velocity. Maximum velocity is also greater for stiffer systems, because they exert greater force for the same displacement. This is seen in the expression for v_{max}—it is proportional to the square root of the force constant k. Finally, the maximum velocity is smaller for larger masses, being inversely proportional to the square root of m. Large masses accelerate more slowly for a given force.

EXAMPLE 15.6 MAXIMUM SPEED OF AN OSCILLATING SYSTEM

Suppose that the car considered in previous examples—it has a 900 kg mass and a suspension system with a force constant $k = 6.53 \times 10^4$ N/m—hits a bump and bounces with an amplitude of 0.100 m. What is its maximum vertical velocity assuming no damping?

Strategy Given m and k, and identifying X as 0.100 m, we can use the expression for v_{max} given in Equation 15.10.

(continued)

(continued)

Solution Substituting known values into Equation 15.10 gives

$$v_{max} = \sqrt{\frac{k}{m}}X = \sqrt{\frac{6.53 \times 10^4 \text{ N/m}}{900 \text{ kg}}} \cdot 0.100 \text{ m}$$

$$= 0.852 \text{ m/s}$$

Discussion This seems reasonable for a bouncing car. There are other ways to use conservation of energy to find v_{max}. We could use it directly, as was done in Example 15.2, for example.

15.6* UNIFORM CIRCULAR MOTION AND SIMPLE HARMONIC MOTION

There is an easy way to produce simple harmonic motion by using uniform circular motion. Figure 15.12 shows one way of doing this. A ball is attached to a uniformly rotating turntable, and its shadow is projected as shown. The shadow undergoes simple harmonic motion. It is usually easier to get uniform circular motion (ω constant) than it is to get a large system to follow Hooke's law for large visible displacements. So observing the projection of uniform circular motion, as in Figure 15.12, is often easier than observing a precise large-scale simple harmonic oscillator. If studied in sufficient depth, simple harmonic motion produced in this manner can give considerable insight into many aspects of oscillations and waves and is very useful mathematically. In our brief treatment, we shall indicate some of the major features of this relationship and how they might be useful.

Figure 15.13 shows the basic relationship between uniform circular motion and simple harmonic motion. The point P travels around the circle at constant angular velocity ω. The point P is analogous to the ball on the turntable. The projection of the position of P onto a fixed axis undergoes simple harmonic motion and is analogous to the shadow of the ball. At the time shown in the figure, the projection has position x and moves to the left with velocity v. The velocity of the point P around the circle equals $\mathbf{v}_{max}$. The projection of $\mathbf{v}_{max}$ on the x-axis is the velocity v of the simple harmonic motion along the x-axis.

To see that the projection undergoes simple harmonic motion, note that its position x is given by

$$x = X \cos \theta$$

where $\theta = \omega t$, with ω being the constant angular velocity and X the radius of the circular path. Thus,

$$x = X \cos \omega t$$

The angular velocity ω is in radians per unit time, in this case 2π radians in the time for one revolution T. That is, $\omega = 2\pi/T$. Substituting this expression for ω, we see that the position x is given by

$$x = X \cos \frac{2\pi t}{T}$$

This is the same expression we had for the position of a simple harmonic oscillator in Section 15.3. If we make a graph of position versus time as in Figure 15.14, we see again the wavelike character (typical of simple harmonic motion) of the projection of uniform circular motion onto the x-axis.

Now let us use Figure 15.13 to do some further analysis of uniform circular motion as it relates to simple harmonic motion. The triangle formed by the velocities in the figure and that formed by the displacements (X, x, and $\sqrt{X^2 - x^2}$) are similar right triangles. Taking ratios of similar sides, we see that

$$\frac{v}{v_{max}} = \frac{\sqrt{X^2 - x^2}}{X} = \sqrt{1 - \frac{x^2}{X^2}}$$

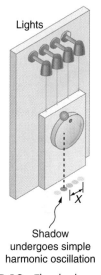

Lights

Shadow undergoes simple harmonic oscillation

Figure 15.12 The shadow of a ball rotating at constant angular velocity ω on a turntable goes back and forth in precise simple harmonic motion.

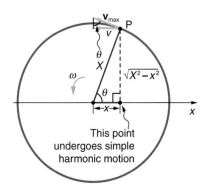

This point undergoes simple harmonic motion

Figure 15.13 A point P moving on a circular path with a constant angular velocity ω is undergoing uniform circular motion. Its projection on the x-axis undergoes simple harmonic motion. Also shown is the velocity of this point around the circle, $\mathbf{v}_{max}$, and its projection, which is v. Note that these velocities form a similar triangle to the displacement triangle.

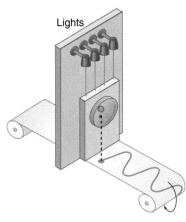

Lights

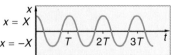

Figure 15.14 The position of the projection of uniform circular motion performs simple harmonic motion, as this wavelike graph of x vs. t indicates.

We can solve this for the speed v, yielding

$$v = v_{max}\sqrt{1 - \frac{x^2}{X^2}}$$

This expression for the speed of a simple harmonic oscillator is exactly the same as was obtained from conservation of energy considerations in the previous section. You can begin to see that it is possible to get all of the characteristics of simple harmonic motion from the analysis of the projection of uniform circular motion.

Finally, let us consider the period T of the motion of the projection. This is the time it takes the point P to complete one revolution. That time is the circumference of the circle $2\pi X$ divided by the velocity around the circle, v_{max}. Thus the period T is

$$T = \frac{2\pi X}{v_{max}}$$

We know from conservation of energy considerations in the previous section that

$$v_{max} = \sqrt{\frac{k}{m}}X$$

Solving this for X/v_{max} gives

$$\frac{X}{v_{max}} = \sqrt{\frac{m}{k}}$$

Substituting this into the expression for T yields

$$T = 2\pi\sqrt{\frac{m}{k}}$$

Thus the period of the motion is the same as we had in Equation 15.4a for a simple harmonic oscillator. We have proved this is the period using the relationship between uniform circular motion and simple harmonic motion.

Future sections occasionally refer to the connection between uniform circular and simple harmonic motion. Moreover, if you carry your study of physics and its applications to greater depths, you will find this relationship useful. It can, for example, help to analyze how waves add when they are superimposed.

15.7 DAMPED HARMONIC MOTION

A guitar string stops oscillating a few seconds after being plucked. To keep a child happy on a swing, you must keep pushing. Although we can often make friction and other nonconservative forces negligibly small, completely undamped motion is rare. In fact, we may even *want* to damp oscillations, such as with automobile shock absorbers.

For a system with a small amount of damping, the period and frequency are nearly the same as for simple harmonic motion, but the amplitude gradually decreases as shown in Figure 15.15. What happens is that the nonconservative damping force removes energy from the system, usually in the form of thermal energy. Equation 6.7a states this in general for nonconservative forces as $W_{nc} = \Delta KE + \Delta PE$, or

$$W_{nc} = \Delta(KE + PE)$$

Figure 15.15 A graph of displacement versus time for a harmonic oscillator with a small amount of damping. The amplitude slowly decreases, but the period and frequency are nearly the same as if the system were completely undamped.

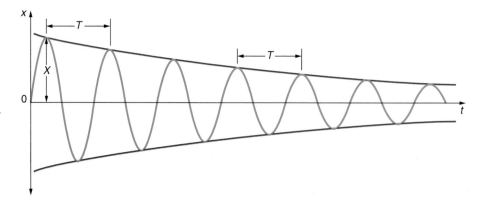

where W_{nc} is work done by a nonconservative force (here the damping force). For a damped harmonic oscillator, W_{nc} is negative, removing mechanical energy (KE + PE) from the system.

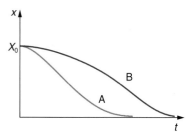

Figure 15.16 Displacement versus time for a critically damped harmonic oscillator (A) and an overdamped harmonic oscillator (B). The critically damped oscillator returns to equilibrium at $x = 0$ in the smallest time possible without overshooting.

If you gradually increase the amount of damping on a system, the period and frequency begin to be affected, because damping opposes and hence slows the back and forth motion. (The net force is smaller in both directions.) For very large damping, the system does not even oscillate—it slowly moves toward equilibrium. Figure 15.16 shows the displacement of a harmonic oscillator for different amounts of damping. When we want to damp out oscillations, such as in the needle in a gas gauge or the suspension of a car, we may want the system to return to equilibrium as quickly as possible without overshooting through the equilibrium position. **Critical damping** is defined to be the condition in which the damping of an oscillator results in it returning as quickly as possible to its equilibrium position without oscillating back and forth through it. Critical damping is represented by curve A in Figure 15.16. With less than critical damping, the system will return to equilibrium faster but will overshoot and cross over a few times. Such a system is **underdamped**; its displacement is represented by the curve in Figure 15.15. Curve B in Figure 15.16 represents an **overdamped** system. It moves more slowly toward equilibrium than the critically damped system.

Critical damping is often desired, because such a system returns to and remains at equilibrium rapidly. Additionally, a constant force applied to a critically damped system, such as the force on the needle of a gauge, moves the system to a new equilibrium position in the shortest time possible without overshooting or oscillating about the new position. This is very handy for gauges, bathroom scales, and many other systems that we wish to move about without having them oscillate when they get to their new position. Damping forces can vary greatly in character. Friction, for example, is sometimes independent of velocity (as assumed in most places in this text). But many damping forces depend on velocity—sometimes in complex ways, sometimes being simply proportional to v.

EXAMPLE 15.7 SPRING ENERGY DISSIPATED BY FRICTION

Suppose a 0.200 kg mass is connected to a spring as in Figure 15.11, but with simple friction between it and the surface having a coefficient of friction $\mu_k = 0.0800$. What total distance does the mass travel if it is released 0.100 m from equilibrium starting at $v = 0$? The force constant of the spring is $k = 50.0$ N/m.

Strategy Since the motion starts from rest, the energy in the system is initially $PE_{el} = (1/2)kX^2$. This energy is removed by work done against friction $W_{nc} = fd$, where d is the total distance traveled and $f = \mu_k mg$ is the force of friction. By equating the work done to the energy removed, we can solve for the distance d.

Solution The work done equals the initial stored elastic potential energy. That is,

$$W_{nc} = \Delta(KE + PE) = \tfrac{1}{2}kX^2$$

But we know $W_{nc} = fd$, and we enter the friction as $f = \mu_k mg$; thus,

$$W_{nc} = \mu_k mgd$$

Combining these two equations, we see that

$$\tfrac{1}{2}kX^2 = \mu_k mgd$$

Solving for d and entering known values yields

$$d = \frac{\tfrac{1}{2}kX^2}{\mu_k mg} = \frac{\tfrac{1}{2}(50.0 \text{ N/m})(0.100 \text{ m})^2}{(0.0800)(0.200 \text{ kg})(9.80 \text{ m/s}^2)}$$

$$= 1.59 \text{ m}$$

Discussion This is the total distance traveled back and forth across $x = 0$, the undamped equilibrium position. This system will not return to $x = 0$ for this type of damping force, since static friction will exceed the restoring force at the end of the motion. This system is underdamped. In contrast, an overdamped system with a simple constant damping force never gets back to $x = 0$. For example, if this system had a damping force 20 times greater, it would only move 0.0795 m back from its original 0.100 m position.

15.8 FORCED OSCILLATIONS AND RESONANCE

Sit in front of a piano sometime and sing a loud brief note at it with the dampers off its strings. It will sing the same note back at you—the strings having the same frequencies as your voice are resonating in response to the forces in the sound waves you sent to them.

Figure 15.17 The paddle ball on its rubber band moves in response to the finger supporting it. If the finger moves with the natural frequency f_0 of the ball on the rubber band, then a resonance is achieved, and the amplitude of the ball's oscillations increases dramatically. At higher and lower driving frequencies, energy is transferred to the ball less efficiently, and it responds with lower-amplitude oscillations.

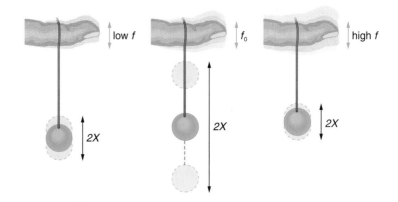

This is a good example of the fact that objects—in this case, piano strings—can be forced to oscillate but oscillate best at their natural frequency. In this section, we shall briefly explore applying a *periodic driving force* to a simple harmonic oscillator. The driving force puts energy into the system at a certain frequency, not necessarily the same as the natural frequency of the system. The **natural frequency** is the frequency at which a system would oscillate if there were no driving and no damping force.

Most of us have played with toys involving a mass supported on an elastic band, something like the paddle ball suspended from a finger in Figure 15.17. Imagine this is your finger. At first you hold it steady, and the ball bounces up and down with a small amount of damping. If you move your finger up and down slowly, the ball will follow along without bouncing much on its own. As you increase the frequency at which you move your finger up and down, the ball will respond by oscillating with an increasing amplitude. When you drive the ball at its natural frequency, the ball's oscillations increase in amplitude with each oscillation for as long as you drive it.* The phenomenon of driving a system with a frequency equal to its natural frequency is called **resonance**. A system being driven at its natural frequency is said to **resonate**. As the driving frequency gets progressively higher than the resonant or natural frequency, the amplitude of the oscillations becomes smaller, until the oscillations nearly disappear and your finger simply moves up and down with little effect on the ball.

Figure 15.18 shows a graph of the amplitude of a damped harmonic oscillator as a function of the frequency of the periodic force driving it. There are three curves on the graph, each representing a different amount of damping. All three curves peak at the point where the frequency of the driving force equals the natural frequency of the harmonic oscillator. The highest peak, or greatest response, is for the least amount of damping, because less energy is removed by the damping force.

---

*When the amplitude becomes very large, the restoring force departs from Hooke's law, affecting the resonant frequency. This is called the nonlinear region, or a distorted response, such as sometimes is intentionally produced in rock music.

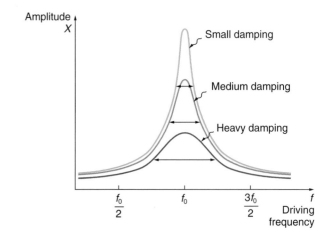

Figure 15.18 Amplitude of a harmonic oscillator as a function of the frequency of the driving force. The curves represent the same oscillator with the same natural frequency but with different amounts of damping. Resonance occurs when the driving frequency equals the natural frequency, and the greatest response is for the least amount of damping. The narrowest response is also for the least damping.

It is interesting that the widths of the resonance curves shown in Figure 15.18 depend on damping. The less damping, the narrower the resonance. The message is that if you want a driven oscillator to resonate at a very specific frequency, you need as little damping as possible. This is the case for piano strings and many other musical instruments. Conversely, if you want small-amplitude oscillations, like in a car's suspension system, then you want heavy damping. This reduces the amplitude but also makes the system respond at more frequencies, a trade-off.

These features of driven harmonic oscillators apply to a huge variety of systems. When you tune your radio, for example, you are adjusting its resonant frequency so that it only oscillates to the desired station's broadcast (driving) frequency. The more selective the radio is in discriminating between stations, the smaller its damping. Magnetic resonance imaging is a widely used medical diagnostic tool in which atomic nuclei, mostly of hydrogen, are made to resonate by incoming microwaves. A child on a swing is driven at the pendulum's natural frequency to achieve maximum amplitude. In all of these cases, the efficiency of energy transfer from the driving force into the oscillator is best at resonance. Speed bumps prove that even a car's suspension system is not immune to resonance. In spite of finely engineered shock absorbers, which ordinarily convert driving energy to heat almost as fast as it comes in, speed bumps still cause a large-amplitude oscillation. Figure 15.19 shows a photograph of a famous example of the destructive effects of a driven harmonic oscillation.

15.9 WAVES

What do we mean when we say something is a wave? The most intuitive and easiest wave to imagine is the familiar water wave. More precisely, a **wave** is a disturbance that propagates or moves from the place it was created. For water waves, the disturbance is in the surface of the water, perhaps created by a rock thrown into a pond or by a swimmer splashing the surface repeatedly. For sound waves, the disturbance is a change in air pressure, perhaps created by a vibrating string. For earthquakes, there are several types of disturbances, including disturbance of the earth's surface and pressure disturbances under the surface. Even radio waves are most easily understood using an analogy with water waves. Visualizing water waves is useful because there is more to it than just a mental image. Water waves exhibit characteristics common to all waves, such as amplitude, period, and frequency. All wave characteristics can be described by a small set of underlying principles.

The simplest waves repeat for several cycles and are associated with simple harmonic motion. Let us start by considering the simplified water wave in Figure 15.20. The wave is an up and down disturbance of the water surface. It causes a seagull to move up and down in simple harmonic motion as the wave crests and troughs (peaks and valleys) pass under the bird. The time for one complete up and down motion is the wave's period T. The wave's frequency is $f = 1/T$, as usual. The wave itself moves to the right in the figure. This movement of the wave is actually the disturbance moving to the right, not the water itself (or the bird would move to the right). We define **wave velocity** v_w to be the speed at which the disturbance moves. Wave velocity is sometimes also called the

Figure 15.19 In 1940, the Tacoma Narrows bridge collapsed. Heavy cross winds drove the bridge into oscillations at its resonant frequency. Damping decreased when support cables broke loose and started to slip over the towers, allowing increasingly greater amplitude until the structure failed.

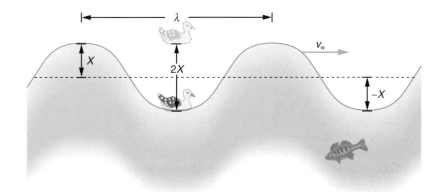

Figure 15.20 An idealized ocean wave passes under a seagull that bobs up and down in simple harmonic motion. The wave has a wavelength λ, which is the distance between adjacent identical parts of the wave. The up and down disturbance of the surface propagates parallel to the surface at a speed v_w.

propagation velocity or propagation speed, because the disturbance propagates from one location to another.

The water wave in the figure also has a length associated with it, called its **wavelength** λ, the distance between adjacent identical parts of a wave. (λ is the distance parallel to the direction of propagation.) The speed of propagation v_w is the distance the wave travels in a given time, which is one wavelength in the time of one period. In equation form, that is

or

$$v_w = \frac{\lambda}{T}$$

$$v_w = f\lambda$$

(15.11)

This fundamental relationship holds for all types of waves. For water waves, v_w is the speed of a surface wave; for sound, v_w is the speed of sound; and for visible light, v_w is the speed of light, for example. All waves have period, frequency, and wavelength as well as wave velocity.

EXAMPLE 15.8   HOW FAST DOES A WAVE PROPAGATE?

Calculate the wave velocity of the ocean wave in Figure 15.20, if the distance between wave crests is 10.0 m and the time for the gull to bob up and down is 5.00 s.

Strategy We are asked to find v_w. The given information tells us that $\lambda = 10.0$ m and $T = 5.00$ s, and so we can use the first expression in Equation 15.11 to find the wave velocity.

Solution Entering the known values into Equation 15.11,

we obtain

$$v_w = \frac{\lambda}{T} = \frac{10.0 \text{ m}}{5.00 \text{ s}} = 2.00 \text{ m/s}$$

Discussion This slow speed seems reasonable for an ocean wave. Note that this is the speed at which the wave moves to the right in the figure, not the varying speed at which the seagull moves up and down.

Transverse and Longitudinal Waves

A simple wave consists of a periodic disturbance that propagates from one place to another. The water wave in Figure 15.20 propagates in the horizontal direction while the surface is disturbed in the vertical direction. Such a wave is called a **transverse wave** or shear wave; in such a wave, the disturbance is perpendicular to the direction of propagation. In contrast, in a **longitudinal wave** or compressional wave, the disturbance is parallel to the direction of propagation. Figure 15.21 shows an example of each for disturbances on a Slinky. The size of the disturbance is its amplitude X and is completely independent of the speed of propagation v_w.

Waves may be transverse, longitudinal, or a combination of the two. (Water waves are actually a combination of transverse and longitudinal. The simplified water wave illustrated in Figure 15.20 shows no longitudinal motion of the bird.) The waves on the strings of musical instruments are transverse. So are electromagnetic waves, such as visible light.

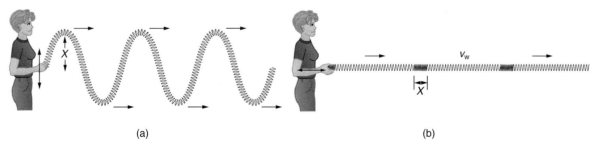

(a)   (b)

Figure 15.21 Waves on a Slinky propagate parallel to its length. (a) A transverse wave is created when the Slinky is shaken perpendicular to its length. (b) A longitudinal wave is created when the Slinky is shaken parallel to its length.

Sound waves in air and water are longitudinal. Their disturbances are periodic variations in pressure that are transmitted in fluids. Fluids do not have appreciable shear strength, and thus sound waves in them must be longitudinal or compressional. Sound in solids can be both longitudinal and transverse. (See Figure 15.22.)

Earthquake waves under the earth's surface also have both longitudinal and transverse components (called compressional or P-waves, and shear or S-waves, respectively) with important individual characteristics—they propagate at different speeds, for example. Earthquakes also have surface waves that are similar to surface waves on water.

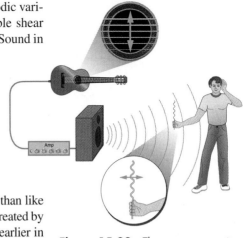

Figure 15.22 The wave on a guitar string is transverse. The sound wave rattles a sheet of paper in a direction that shows the sound wave is longitudinal.

15.10 SUPERPOSITION AND INTERFERENCE

Most waves do not look very simple. They look more like those in Figure 15.23 than like the simple water wave considered in the previous section. (Simple waves may be created by a simple harmonic oscillation and, thus, have a sinusoidal shape like that shown earlier in Figure 15.14.) Certainly, complex waves are more interesting, even beautiful, but they look formidable. Most waves appear complex because they result from several simple waves adding together. Luckily, perhaps surprisingly, the rules for adding waves are quite simple.

When two or more waves arrive at the same point, they superimpose themselves on one another. More specifically, the disturbances of waves are superimposed when they come together—a phenomenon called **superposition**. Each disturbance corresponds to a force, and forces add. If the disturbances are along the same line, then the resulting wave is a simple addition of the disturbances of the individual waves—that is, their amplitudes add. Figure 15.24 illustrates this in two special cases, both of which produce simple results. Part (a) shows two identical waves that arrive at the same point exactly in phase—that is, precisely aligned crest to crest and trough to trough—producing pure **constructive interference**. Since the disturbances add, pure constructive interference produces a wave with twice the amplitude of the individual waves, but with the same wavelength. Part (b) shows two identical waves that arrive exactly out of phase—that is, precisely aligned crest to trough—producing pure **destructive interference**. Since the disturbances are in the opposite direction for this superposition, the resulting amplitude is zero for pure destructive interference—the waves completely cancel.

While pure constructive and pure destructive interference do occur, they require precisely aligned identical waves. The superposition of most waves produces a combination of constructive and destructive interference and can vary from place to place and time to time. Sound from a stereo, for example, can be loud in one spot and quiet in another. This means the sound waves add partially constructively and partially destructively at different locations. A stereo has at least two speakers creating sound waves, and waves can reflect from walls. All these superimpose. An example of sounds that vary over time from constructive to destructive is found in the combined whine of airplane jets heard by a

Figure 15.23 These waves result from the superposition of several waves from different sources, producing a complex pattern.

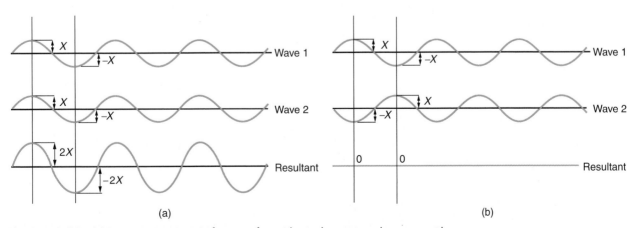

(a) (b)

Figure 15.24 (a) Pure constructive interference of two identical waves produces one with twice the amplitude, but the same wavelength. (b) Pure destructive interference of two identical waves produces zero amplitude, or complete cancellation.

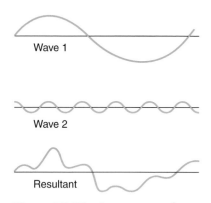

Figure 15.25 Superposition of non-identical waves exhibits both constructive and destructive interference. The amplitudes simply add.

stationary passenger. The combined sound can fluctuate up and down in volume as the sound from the two engines varies in time from constructive to destructive. These examples are waves that are similar.

An example of the superposition of two dissimilar waves is shown in Figure 15.25. Here again, the disturbances add and subtract, producing a more complicated looking wave. If more waves are added, the resultant wave becomes even more complex in appearance, as seen in Figure 15.26.

Standing Waves

Sometimes waves do not seem to move; rather, they just vibrate in place. This can be seen on the surface of a glass of milk in a refrigerator, for example. Vibrations from the refrigerator motor create waves on the milk that oscillate up and down but do not seem to move across the surface. This is one example of a standing wave. There are other standing waves, such as on guitar strings and in organ pipes. Standing waves are formed by the superposition of two or more moving waves, such as illustrated in Figure 15.27 for two identical waves moving in opposite directions. The waves move through each other with their disturbances adding as they go by. If the two waves have the same amplitude and wavelength, then they alternate between constructive and destructive interference. The resultant looks like a wave standing in place and, thus, is called a **standing wave**. With the glass of milk, the two waves that produce standing waves may come from reflections from the side of the glass.

Standing waves are also found on the strings of musical instruments and are due to reflections of waves from the ends of the string. Figure 15.28 shows three standing waves that can be created on a string that is fixed at both ends. **Nodes** are the points where the string does not move; more generally, nodes are where the wave disturbance is zero in a standing wave. The fixed ends of strings must be nodes, too, since the string cannot move there. **Loops** are the regions of greatest amplitude in a standing wave. Standing waves on strings have a frequency that is related to the propagation speed v_w of the disturbance on the string. The wavelength λ is determined by the distance between the points where the string is fixed in place. The lowest frequency, called the **fundamental frequency**, is thus for the longest wavelength, which is seen to be $\lambda_1 = 2L$ in the figure. The fundamental frequency is $f_1 = v_w/\lambda_1 = v_w/2L$. As seen in the figure, the first overtone has $f_2 = 2f_1$, and the second overtone has $f_3 = 3f_1$. All of these frequencies can be changed by adjusting the tension in the string. The greater the tension, the greater v_w is and the higher the frequencies. This is familiar to anyone who has ever observed a string instrument being tuned. We will see in later chapters that standing waves are crucial to many resonance phenomena, such as in sounding boxes on string instruments.

Figure 15.26 Sounds from various musical instruments are superpositions of unique sets of waves. The resulting total waves may be complex in appearance, but they result from the simple addition of the disturbances (or amplitudes) of all the individual waves.

Beats

Striking two adjacent keys on a piano produces a warbling combination usually considered to be unpleasant. The superposition of two waves of similar but not identical frequencies is the culprit. A version of this is often noticeable in jet aircraft, particularly

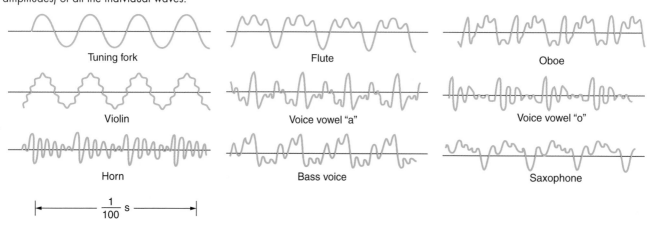

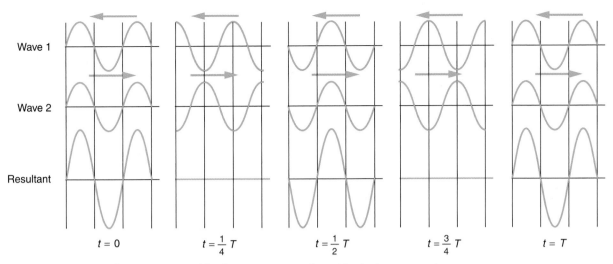

Figure 15.27 Standing wave created by the superposition of two identical waves moving in opposite directions. The oscillations are at fixed locations in space and result from alternately constructive and destructive interference.

the two-engine variety, while taxiing. The combined sound of the engines goes up and down in loudness. This happens because the sound waves have similar but not identical frequencies. The discordant warbling of the piano and the fluctuating loudness of the jet engine noise are both due to alternately constructive and destructive interference as the two waves get in and out of phase. Figure 15.29 illustrates this graphically.

The wave resulting from the superposition of two similar frequency waves has a frequency that is the average of the two. This wave fluctuates in amplitude, or beats, with a frequency called the **beat frequency**. We can see how this works by adding two waves together mathematically. Note that a wave can be represented at one point in space as

$$x = X \cos \frac{2\pi t}{T} = X \cos 2\pi f t$$

where $f = 1/T$ is the frequency of the wave. Adding two waves that have different frequencies but identical amplitudes produces a resultant

$$x = x_1 + x_2$$

More specifically,

$$x = X \cos 2\pi f_1 t + X \cos 2\pi f_2 t$$

Using a trigonometric identity, it can be shown that

$$x = 2X \cos \pi f_B t \cos 2\pi f_{ave} t$$

where

$$f_B = |f_1 - f_2| \qquad \textbf{(15.12)}$$

is the beat frequency, and f_{ave} is the average of f_1 and f_2. This means that the resultant wave has twice the amplitude and the average frequency of the two superimposed waves, but it also fluctuates in overall amplitude at the beat frequency f_B. The first cosine term in the

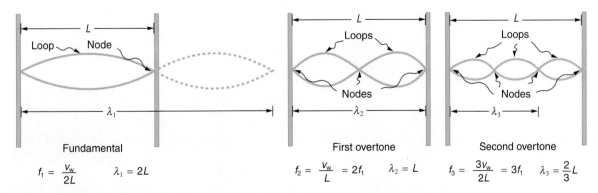

Figure 15.28 Standing waves on a string.

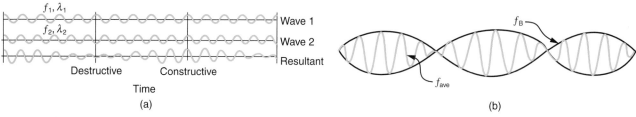

Figure 15.29 (a) Beats are produced by the superposition of two waves of slightly different frequencies but identical amplitudes. The waves alternate in time between constructive and destructive interference, giving the resulting wave a time-varying amplitude. (b) A longer-term view shows the amplitude envelope with a frequency f_B and the wave itself with a frequency f_{ave}.

expression effectively causes the amplitude to go up and down as seen in Figure 15.29(b). The second cosine term is the wave with frequency f_{ave}, also seen in the figure. This result is valid for all types of waves, but if it is a sound wave then what we hear (providing the two frequencies are similar) is an average frequency that gets louder and softer (warbles) at the beat frequency.

Piano tuners use beats routinely in their work. When comparing a note with a tuning fork, they listen for beats and adjust the string until the beats go away (to zero frequency). For example, if the tuning fork has a 256 Hz frequency and 2 beats per second are heard, then the other frequency is either 254 or 258 Hz. Most keys hit multiple strings, and these are actually adjusted until they have nearly the same frequency and give a slow beat for richness. Twelve-string guitars and mandolins are also tuned using beats.

While they may sometimes be annoying in audible sounds, we will find that beats have many applications. Observing beats is a very useful way to compare similar frequencies. There are applications of beats as apparently disparate as in ultrasonic imaging and radar speed traps, as we will see in later chapters.

15.11 ENERGY IN WAVES: INTENSITY

All waves carry energy. The energy of some waves can be directly observed. Earthquakes shake whole cities to the ground, performing the work of thousands of wrecking balls. (See Figure 15.30.) Loud sounds pulverize nerve cells in the inner ear, causing permanent hearing loss. Ultrasound is used for deep-heat treatment of muscle strains. A laser beam can burn away a malignancy. Water waves chew up beaches and levees and are experimentally utilized to generate electricity in specially designed devices.

The amount of energy in a wave is clearly related to its amplitude. Large-amplitude earthquakes produce large ground displacements. Loud sounds have higher pressures and come from larger-amplitude source vibrations than soft ones. Large ocean breakers churn up the shore more than small ones.

More quantitatively, a wave is a displacement that is resisted by a restoring force. The larger the displacement x, the larger the force $F = kx$ needed to create it. Since work W

Figure 15.30 The destructive effect of an earthquake is palpable evidence of the energy carried in these waves. The Richter scale (and its more modern versions) rating of earthquakes is related to both their amplitude and the energy they carry.

is related to force times distance, Fx, and energy is put into the wave by the work done to create it, the energy in a wave is related to amplitude. In fact, a wave's energy is directly proportional to its amplitude squared, since $W \propto Fx = kx^2$.*

The energy effects of a wave depend on time as well as amplitude. For example, the longer deep-heat ultrasound is applied, the more energy it transfers. Waves can also be concentrated or spread out. Sunlight, for example, can be focused to burn wood. Earthquakes spread out, doing less damage farther from the source. In both cases, changing the area the waves cover has important effects. All the pertinent factors are included in the definition of **intensity** I as power per unit area:

$$I = \frac{P}{A} \qquad (15.13)$$

where P is the power carried by the wave through area A. The definition of intensity is valid for any energy in transit, including that carried by waves. The SI unit for intensity is watts per square meter (W/m^2). For example, infrared and visible energy from the sun impinge on the earth at an intensity of 1300 W/m^2 just above the atmosphere. There are other intensity-related units in use, too. The most common are decibels—for example, a 90 decibel sound level corresponds to an intensity of 10^{-3} W/m^2. (This is not much power per unit area considering that 90 decibels is such a high sound level. Decibels will be discussed in some detail in the next chapter.)

EXAMPLE 15.9 ENERGY AND POWER IN SUNLIGHT

The average intensity of sunlight on the earth's surface is about 700 W/m^2. (a) Calculate the amount of energy that falls on a solar collector having an area of 0.500 m^2 in 4.00 h. (b) What intensity would such sunlight have if concentrated by a magnifying glass onto an area 200 times smaller than its own?

Strategy for (a) Since power is energy per unit time, or $P = E/t$, the definition of intensity can be written as $I = P/A = (E/t)/A$, and this can be solved for E with the given information.

Solution for (a) Solving the definition of intensity given in Equation 15.13 for E gives

$$I = \frac{P}{A} = \frac{E/t}{A}$$

Thus,

$$E = IAt$$

Substituting known values,

$$E = (700 \text{ W/m}^2)(0.500 \text{ m}^2)[(4.00)(3600 \text{ s})]$$
$$= 5.04 \times 10^6 \text{ J}$$

or about 1200 kcal.

Strategy for (b) Taking a ratio of new intensity to old, using primes for new quantities, we will find that it depends on the ratio of the areas. All other quantities will cancel.

Solution for (b) Taking the ratio of intensities gives

$$\frac{I'}{I} = \frac{P'/A'}{P/A} = \frac{A}{A'}$$

The powers cancel, since $P' = P$. Noting that $A = 200A'$, we see

$$\frac{I'}{I} = 200$$

Thus,

$$I' = 200I = 200(700 \text{ W/m}^2) = 1.40 \times 10^5 \text{ W/m}^2$$

Discussion The energy falling on the solar collector in 4 hours in part (a) is enough to be useful—for example, to heat a significant amount of water. The intensity of the concentrated sunlight found in (b) could even start fires.

EXAMPLE 15.10 CONSTRUCTIVE INTERFERENCE INCREASES INTENSITY

If two identical waves, each having an intensity of 1.00 W/m^2, interfere perfectly constructively, what is the intensity of the resulting wave?

Strategy We know from the previous section that two identical waves, with equal amplitudes X, add with amplitude $2X$ in the resulting wave when they interfere perfectly

(continued)

---

*Recall that the potential energy in an elastic deformation is $PE_{el} = (1/2)kx^2$. A wave carries this energy and can transfer it to other systems and into other forms.

(continued)

constructively. Since a wave's intensity is proportional to amplitude squared, this means the intensity is four times as great as in the individual waves.

Solution Intensity is proportional to amplitude squared. For the new amplitude, this yields

$$I' \propto (X')^2 = (2X)^2 = 4X^2$$

For the old amplitude, the intensity was

$$I \propto X^2$$

Taking the ratio of new to old gives

$$\frac{I}{I'} = 4$$

Thus,

$$I' = 4I = 4.00 \ \text{W/m}^2$$

Discussion The intensity goes up by a factor of 4 when the amplitude doubles. This is a little disquieting. The two individual waves each have intensities of 1.00 W/m², yet their sum has an intensity of 4.00 W/m², which may appear to violate conservation of energy. This, of course, cannot happen. What does happen is intriguing. The area over which the intensity is 4.00 W/m² is much less than the area covered by the two waves before they interfered. There are other areas where the intensity is zero. The addition of waves is not as simple as our first look in Section 15.10 suggested. We ac-

tually will get a pattern of both constructive and destructive interference whenever two waves are added. For example, if we have two stereo speakers putting out 1.00 W/m² each, there will be places in the room where the intensity is 4.00 W/m², other places where it is zero, and others in between. Figure 15.31 shows what this might look like. We will pursue interference patterns further in later chapters, finding them to be a general feature of wave superposition.

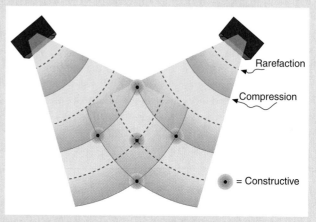

Figure 15.31 These stereo speakers produce both constructive and destructive interference in the room, a property common to the superposition of all types of waves. The shading is proportional to intensity.

SUMMARY

An **oscillation** is a back and forth motion between two points. It may create a **wave**, which is a disturbance that propagates from where it was created. The simplest type of oscillations and waves are related to systems that obey **Hooke's law**:

$$F = -kx \tag{15.1}$$

where F is the **restoring force**, x is the displacement from equilibrium or **deformation**, and k is the **force constant** of the system. **Elastic potential energy** PE_{el} stored in the deformation of a system that obeys Hooke's law is given by

$$PE_{el} = \tfrac{1}{2}kx^2 \tag{15.2}$$

This energy is converted to kinetic energy when the system is allowed to oscillate. **Periodic motion** is a repetitious oscillation. The time for one oscillation is the **period** T, and the number of oscillations per unit time is the **frequency** f. These are related by

$$f = \frac{1}{T} \tag{15.3}$$

The units of T are seconds, and the units of f are hertz (Hz) or cycles/second.

 Simple harmonic motion is oscillatory motion for a system that obeys Hooke's law. Such a system is also called a **simple harmonic oscillator**. Maximum displacement is

the **amplitude** X. The period T and frequency f of a simple harmonic oscillator are

$$T = 2\pi\sqrt{\frac{m}{k}} \tag{15.4a}$$

and

$$f = \frac{1}{2\pi}\sqrt{\frac{k}{m}} \tag{15.4b}$$

where m is the mass of the system. Displacement in simple harmonic motion as a function of time is

$$x = X\cos\frac{2\pi t}{T} \tag{15.5}$$

The velocity is

$$v = -v_{max}\sin\frac{2\pi t}{T}$$

where

$$v_{max} = \sqrt{\frac{k}{m}}X \tag{15.6}$$

The acceleration is

$$a = -\frac{kX}{m}\cos\frac{2\pi t}{T} \tag{15.7}$$

A mass m suspended by a wire of length L is a **simple pendulum** and undergoes simple harmonic motion for amplitudes less than about 15°. Its period is

$$T = 2\pi\sqrt{\frac{L}{g}} \qquad (15.8)$$

where g is the acceleration of gravity.

Energy in the simple harmonic oscillator is shared between elastic potential energy and kinetic energy, with the total being constant:

$$\tfrac{1}{2}mv^2 + \tfrac{1}{2}kx^2 = \text{constant} \qquad (15.9)$$

Uniform circular motion produces simple harmonic motion if the position of a point so moving is projected along the axis of the circle.

Damped harmonic oscillators have nonconservative forces that dissipate their energy. **Critical damping** returns the system to equilibrium as fast as possible without overshooting. An **underdamped** system will oscillate through equilibrium, whereas an **overdamped** system moves more slowly toward equilibrium than one that is critically damped.

A periodic driving force applied to a simple harmonic oscillator produces a **resonance** if it drives the system at its **natural frequency**. The system is said to **resonate**. Large amplitude and energy infusion occur. The more damping a system has, the broader response it has to driving frequencies.

A **wave** is a disturbance that moves from the point of creation with a **wave velocity** v_w. It has a **wavelength** λ, which is the distance between adjacent identical parts of the wave. These are related to the wave's frequency and period by

$$\left.\begin{aligned} v_w &= \frac{\lambda}{T} \\[2mm] v_w &= f\lambda \end{aligned}\right\} \qquad (15.11)$$

A **transverse** wave has a disturbance perpendicular to its direction of propagation, whereas a **longitudinal** wave has a disturbance parallel to its direction of propagation.

Superposition is the combination of two waves at the same location. **Constructive interference** occurs when two identical waves are superimposed in phase. **Destructive interference** occurs when two identical waves are superimposed exactly out of phase. A **standing wave** is one in which two waves superimpose to produce a wave that varies in amplitude but does not propagate. **Nodes** are points of no motion in standing waves, whereas **loops** are regions of motion. Waves on a string are resonant standing waves with a **fundamental frequency** and can occur at higher multiples of the fundamental.

Beats occur when waves of similar frequencies f_1 and f_2 are superimposed. The resulting amplitude oscillates with a **beat frequency** given by

$$f_B = |f_1 - f_2| \qquad (15.12)$$

Intensity is defined to be power per unit area:

$$I = \frac{P}{A} \qquad (15.13)$$

and has units of W/m^2.

CONCEPTUAL QUESTIONS

15.1 How is the force constant k in Hooke's law related to the elastic moduli γ, S, and B discussed in Section 5.8? Specifically explain whether k should be directly or inversely proportional to them.

15.2 Describe a system in which elastic potential energy is stored.

15.3 Describe a situation in which stored elastic potential energy is converted to another form of energy. Identify the other form of energy.

15.4 What conditions must be met to produce simple harmonic motion?

15.5 If frequency is not constant for some oscillation, can the oscillation be simple harmonic motion?

15.6 Give an example of a simple harmonic oscillator, specifically noting how its frequency is independent of amplitude.

15.7 Explain why you expect an object made of a stiff material to vibrate at a higher frequency than a similar object made of a spongy material.

15.8 As you pass a semitractor-trailer on an interstate, you notice that its trailer is bouncing up and down slowly. Is it more likely that the trailer is heavily loaded or nearly empty? Explain.

15.9 Low riders modify cars to be much closer to the ground than when manufactured. Should they install stiffer springs? Explain.

15.10 At what point in its motion is the acceleration of a simple harmonic oscillator zero? Is the velocity zero at the same point?

15.11 A person standing on a diving board undergoes simple harmonic motion. At what point in the motion is speed the greatest? Is the board curved or straight at this point? How is this point related to where the board would be if the person stood motionless on it? How does the maximum speed depend on the mass of the person, stiffness of the board, and amplitude of the bounce?

15.12 Pendulum clocks are made to run at the correct rate by adjusting the pendulum's length. Suppose you move from one city to another where the acceleration of gravity is slightly greater, taking your pendulum clock with you. Will you have to lengthen or shorten the pendulum to keep the correct time, other factors remaining constant? Explain.

15.13 How will the rate at which a pendulum clock runs depend on temperature? Should this be a large effect? Explain.

15.14 Explain in terms of energy how dissipative forces like friction reduce the amplitude of a harmonic oscillator. Also explain how a driving mechanism can compensate. (A pendulum clock is such a system.)

15.15 Give an example of a damped harmonic oscillator. (They are more common than undamped or simple harmonic oscillators.)

15.16 What are critical damping, overdamping, and underdamping? How would a car bounce after a bump under each of these conditions?

15.17 Most harmonic oscillators are damped and, if undriven, eventually come to a stop. How is this related to the second law of thermodynamics?

15.18 Give an example of resonance in which you make it clear how the driving frequency is related to the system's natural frequency.

15.19 Draw a graph of displacement versus time, comparing underdamped motion to critically damped motion. Illustrate that underdamped motion returns to equilibrium sooner than critically damped motion, but continues to oscillate.

15.20 Does the equilibrium position change for damped harmonic motion like that in Example 15.7? Is there a range of equilibrium positions? Explain.

15.21 Give one example of a transverse wave and another of a longitudinal wave, being careful to note the relative directions of the disturbance and wave propagation in each.

15.22 What is the difference between propagation speed and the frequency of a wave? Does one or the other (or both) affect wavelength? If so, how?

15.23 Two identical waves undergo pure constructive interference. Is the resultant intensity twice that of the individual waves? Explain.

15.24 Speakers in stereo systems have two color-coded terminals in back to indicate how to hook up the wires. If the wires are reversed, the speaker moves in a direction opposite that of a properly connected speaker. Explain why it is important to have both speakers connected the same way.

15.25 Would the amplitudes of two superimposed waves necessarily add if they were in a medium that did not follow Hooke's law?

15.26 Circular water waves decrease in amplitude as they move away from where a rock was dropped. Explain why.

15.27 We have noted that the energy and intensity of a wave is proportional to its amplitude squared. How is this related to elastic potential energy? Would intensity necessarily be proportional to amplitude squared if the medium did not follow Hooke's law?

15.28* The piston in an engine is attached by a rod to a rotating crankshaft as shown in Figure 14.6. The piston does not undergo simple harmonic motion when the engine runs at a steady speed. Explain why not.

PROBLEMS

Section 15.1 Hooke's Law Deformations

15.1 A bathroom spring scale obeys Hooke's law and is depressed 0.750 cm by its maximum load of 120 kg. What is the spring's effective force constant?

15.2 Fish are hung on a spring scale to determine their mass (most fishermen feel no obligation to report the mass truthfully). (a) What is the force constant of the spring in such a scale if it stretches 8.00 cm for a 10.0 kg load? (b) What is the mass of a fish that stretches the spring 5.50 cm? (c) How far apart are the half-kilogram marks on the scale?

15.3 The springs of a pickup truck act like a single spring with a force constant of 1.30×10^5 N/m. How much will the truck be depressed by its maximum load of 1000 kg?

15.4 What energy is put into the car springs in Example 15.1 when the 80.0 kg person gets into the car?

•15.5 One type of BB gun uses a spring-driven plunger to blow the BB from its barrel. (a) Calculate the force constant of its spring if you must compress it 0.150 m to drive the 0.0500 kg plunger to a top speed of 20.0 m/s. (b) What force must be exerted to compress the spring?

•15.6 The barricade at the end of a subway line has a large spring designed to compress 2.00 m when stopping a 1.00×10^5 kg train moving at 0.500 m/s. (a) What is the force constant of the spring? (b) What speed would the train be going if it only compressed the spring 0.500 m? (c) What force does the spring exert when compressed 0.500 m?

•15.7 When an 80.0 kg man stands on a pogo stick, the spring is compressed 0.120 m. (a) What is the force constant of the spring? (b) Will the spring be compressed more when he hops down the road?

•15.8 A 0.500 kg mass hung from a spring stretches it 0.180 m. (a) How much does it stretch two such identical springs if hung as in Figure 15.32(a)? (b) How much does it stretch the two springs if hung as in Figure 15.32(b)? (c) Calculate the effective force constant of a single replacement spring in each case.

•15.9 If the pickup truck in Problem 15.3 has four identical springs, what is the force constant of each?

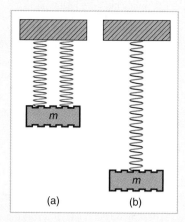

Figure 15.32 (a) A mass hung from two springs in parallel. (b) A mass hung from two springs in series. Problem 8.

⦂15.10 A spring has a length of 0.200 m when a 0.300 kg mass hangs from it, and a length of 0.750 m when a 1.95 kg mass hangs from it. (a) What is the force constant of the spring? (b) What is the unloaded length of the spring?

:15.11 Suppose you have three springs with force constants of 2.00×10^3, 5.00×10^2, and 6.00×10^5 N/m. What is their effective force constant if one is hung from another in series, similar to the two springs in Figure 15.32(b)? You may assume the springs have negligible mass. (Hint: Consider what happens if a mass is hung from them—each spring is stretched as if the mass were hung on it alone.)

:15.12 Show that when springs having force constants $k_1, k_2, k_3, \ldots$ are hung in series, as are the two springs in Figure 15.32(b), their effective force constant k_{eff} is given by the following equation:

$$\frac{1}{k_{eff}} = \frac{1}{k_1} + \frac{1}{k_2} + \frac{1}{k_3} + \cdots$$

(Hint: Consider that the total deformation of the system from equilibrium when a force is applied equals the sum of the deformations of the individual springs, each of which experiences the full force. Enter this total into Hooke's law to obtain an effective force constant. Note that k_{eff} is less than that of the weakest of the individual springs.)

Sections 15.2 and 15.3 Period and Frequency; Simple Harmonic Motion

15.13 What is the period of 60.0 Hz electrical power?

15.14 If your heart rate is 150 beats per minute during strenuous exercise, what is the time per beat in units of seconds?

15.15 Find the frequency of a tuning fork that takes 2.50×10^{-3} s to complete one oscillation.

15.16 A stroboscope is set to flash every 8.00×10^{-5} s. What is the frequency of the flashes?

15.17 A tire has a tread pattern with a crevice every 2.00 cm. Each crevice makes a single vibration as the tire moves. What is the frequency of these vibrations if the car moves at 30.0 m/s?

•15.18 Each piston of an engine makes a sharp sound every other revolution of the engine. (a) How fast is a race car going if its eight-cylinder engine emits a sound of frequency 750 Hz, given that the engine makes 2000 revolutions per kilometer? (b) At how many revolutions per minute is the engine rotating?

•15.19 Near the top of the Citicorp Bank building in New York City, there is a 4.00×10^5 kg mass on springs having adjustable force constants. Its function is to dampen wind-driven oscillations of the building by oscillating at the same frequency as the building is being driven—the driving force is transferred to the mass, which oscillates instead of the entire building. (a) What effective force constant should the springs have to make the mass oscillate with a period of 2.00 s? (b) What energy is stored in the springs for a 2.00 m displacement from equilibrium?

•15.20 The device pictured in Figure 15.33 entertains infants while keeping them from wandering. The child bounces in a harness suspended from a door frame by elastic bands. (a) If the elastic stretches 0.250 m while supporting a 12.0 kg child, what is its force constant? (b) What is the time for one complete bounce of this child? (c) What is the child's maximum velocity if the amplitude of her bounce is 0.200 m?

Figure 15.33 A child suspended in a harness in a doorway. Problems 20 and 54.

•15.21 A 90.0 kg skydiver hanging from a parachute bounces up and down with a period of 1.50 s. What is the new period of oscillation when a second skydiver, whose mass is 60.0 kg, hangs from the legs of the first?

•15.22 One type of cuckoo clock keeps time by having a mass bouncing on a spring, usually something cute like a cherub in a chair. What force constant is needed to produce a period of 0.500 s for a 0.0150 kg mass?

•15.23 At what frequency would a 60.0 kg person bounce on the bathroom scale in Problem 15.1? You may neglect the mass of the scale's moving parts.

:15.24 The length of nylon rope from which a mountain climber is suspended has a force constant of 1.40×10^4 N/m. (a) What is the frequency at which he bounces, given his mass plus equipment to be 90.0 kg? (b) How much would this rope stretch to break the climber's fall, if he free-falls 2.00 m before the rope runs out of slack? (c) Repeat both parts of this problem in the situation where twice this length of nylon rope is used.

:15.25 A 0.500 kg mass suspended from a spring oscillates with a period of 1.50 s. How much mass must be added to change the period to 2.00 s?

:15.26 Suppose you attach a mass to a vertical spring originally at rest, and let it bounce up and down. You release the mass from rest at the spring's original rest length. (a) Show that the spring exerts an upward force of $2.00mg$ on the mass at its lowest point. (b) If the spring has a force constant of 10.0 N/m and a 0.250 kg mass is set in motion as described, find the amplitude of the oscillations. (c) Find the maximum velocity.

Section 15.4 The Simple Pendulum

As usual, the acceleration of gravity in these problems is taken to be $g = 9.80$ m/s², unless otherwise specified.

15.27 What is the length of a pendulum that has a period of 0.500 s?

15.28 Some people think a pendulum with a period of 1.00 s can be driven with "mental energy" or psycho kinetically, because its period is the same as an average heartbeat. True or not, what is the length of such a pendulum?

15.29 What is the period of a 1.00 m long pendulum?

15.30 How long does it take a child on a swing to complete one swing if her center of gravity is 4.00 m below the pivot?

15.31 The pendulum on a cuckoo clock is 5.00 cm long. What is its frequency?

15.32 Two parakeets sit on a swing with their combined center of mass 10.0 cm below the pivot. At what frequency do they swing?

• **15.33** A pendulum with a period of 3.00000 s, in a location where the acceleration of gravity is 9.79 m/s^2, is moved to a location where it is 9.82 m/s^2. What is its new period?

• **15.34** When a pendulum with a period of 2.00000 s is moved to a new location from one where the acceleration of gravity was 9.80 m/s^2, its new period becomes 1.99796 s. What is the acceleration of gravity at its new location?

• **15.35** (a) What is the effect on the period of a pendulum of doubling its length? (b) Of decreasing its length by 5.00%?

• **15.36** Find the ratio of the new period of a pendulum to its old period, if transported from the earth to the moon, where the acceleration of gravity is 1.63 m/s^2.

• **15.37** At what rate will a pendulum clock run on the moon, where the acceleration of gravity is 1.63 m/s^2, if it keeps time accurately on earth? That is, find the time in hours it takes the clock's hour hand to make one revolution on the moon.

⁝ **15.38** The length of a clock's pendulum can be adjusted so that it keeps time accurately. With what precision must the length be known for such a clock to have an accuracy of 1.00 second in a year, all other variables being neglected?

⁝ **15.39** Suppose the length of a clock's pendulum is changed by 1.000%, exactly at noon one day. What time will it read 24.00 hours later, assuming it kept perfect time before the change? Note that there are two answers, and perform the calculation to four-digit precision.

⁝ **15.40** If a pendulum-driven clock gains 5.00 s/day, what fractional change in pendulum length must be made for it to keep perfect time?

Sections 15.5* and 15.6* The Simple Harmonic Oscillator's Energy and Connection to Uniform Circular Motion

15.41* What is the maximum velocity of a 55.0 kg person bouncing on a bathroom scale having a force constant of 1.50×10^5 N/m, if the amplitude of the bounce is 0.200 cm?

15.42* A novelty clock has a 0.0100 kg mass bouncing on a spring that has a force constant of 1.25 N/m. What is the maximum velocity of the mass if it bounces 3.00 cm above and below its equilibrium position?

• **15.43*** What is the maximum energy stored in the spring in Problem 15.41?

• **15.44*** How many joules of kinetic energy does the mass in Problem 15.42 have at its maximum velocity?

• **15.45*** What is the velocity of the car in Example 15.6 when it is 0.0500 m above its equilibrium position?

⁝ **15.46*** At what positions is the speed of a simple harmonic oscillator half its maximum? That is, what values of x/X give $v = \pm v_{max}/2$, where X is the amplitude of the motion?

⁝ **15.47*** A steam locomotive has a 1.20 m diameter drive wheel connected to a piston by a rod attached 0.450 m from the wheel's center. What is the maximum velocity of the piston relative to the locomotive when it is traveling at 40.0 m/s? Assume simple harmonic motion for the piston (an approximation).

⁝ **15.48*** A ladybug sits 12.0 cm from the center of a Beatles album spinning at 33.33 rpm. What is the maximum velocity of its shadow on the wall behind the turntable, if illuminated parallel to the record by the parallel rays of the setting sun?

Sections 15.7 and 15.8 Damped and Driven Oscillations

15.49 How much energy must the shock absorbers of a 1200 kg car dissipate in order to damp a bounce that initially has a velocity of 0.800 m/s at the equilibrium position? Assume the car returns to its original vertical position.

15.50 What distance would the mass in Example 15.7 travel if it were released 0.200 m from equilibrium with zero initial velocity?

• **15.51** If a car has a suspension system with a force constant of 5.00×10^4 N/m, how much energy must its shocks remove to dampen an oscillation starting with a maximum displacement of 0.0750 m?

• **15.52** (a) How much will a spring that has a force constant of 40.0 N/m be stretched by a 0.500 kg mass hung motionless from it? (b) Calculate the decrease in gravitational potential energy of the 0.500 kg mass when it descends this distance. (c) Part of this gravitational energy goes into the spring. Calculate the energy stored in the spring by this stretch, and compare it with the gravitational potential energy. Explain where the rest of the energy might go.

⁝ **15.53** Suppose you have a 0.750 kg mass on a horizontal surface connected to a spring that has a force constant of 150 N/m. There is simple friction between the mass and surface with a static coefficient of friction $\mu_s = 0.100$. (a) How far can the spring be stretched without moving the mass? (b) If the mass is set into oscillation with an amplitude twice the distance found in part (a), and the kinetic coefficient of friction is $\mu_k = 0.0850$, what total distance does it travel before stopping? Assume it starts at the maximum amplitude.

⁝ **15.54** A 10.0 kg child bounces on an elastic band, as in Figure 15.33, with an amplitude of 0.200 m. The band has a force constant of 500 N/m. Each time the child's feet touch the ground, she pushes and increases her mechanical energy by 0.100 J. (a) How much does her amplitude increase after 20.0 bounces? (b) If she then stops driving the oscillation and damping is equivalent to simple friction with $\mu_k = 0.0500$, what total distance does she travel before stopping?

⁝ **15.55** A suspension bridge oscillates with an effective force constant of 1.00×10^8 N/m. (a) How much energy is needed to make it oscillate with an amplitude of 0.100 m? (b) If soldiers march across the bridge with a cadence equal to the bridge's natural frequency and impart 1.00×10^4 J of energy each second, how long does it take for the bridge's oscillations to go from 0.100 to 0.500 m amplitude?

Section 15.9 Waves

15.56 Storms in the South Pacific can create waves that travel all the way to the California coast, 12,000 km away. How long does it take them if they travel at 15.0 m/s?

15.57 Waves on a swimming pool propagate at 0.750 m/s. You splash the water at one end of the pool and observe the wave go to the opposite end, reflect, and return in 30.0 s. How far away is the other end?

15.58 Wind gusts create ripples on the ocean that have a wavelength of 5.00 cm and propagate at 2.00 m/s. What is their frequency?

15.59 How many times a minute does a boat bob up and down on ocean waves that have a wavelength of 40.0 m and a propagation speed of 5.00 m/s?

15.60 Suppose you have a Slinky stretched out on a nearly frictionless floor and you shake it back and forth 4.00 times per second. You note the crests are 0.600 m apart. What is the speed of propagation?

15.61 Scouts at summer camp shake the rope bridge they have just crossed and observe the wave crests to be 8.00 m apart. If they shake it twice per second, what is the propagation speed of the waves?

15.62 What is the wavelength of the waves you create in a swimming pool if you splash your hand at a rate of 2.00 Hz and the waves propagate at 0.800 m/s?

15.63 Radio waves transmitted through space at 3.00×10^8 m/s by the Voyager spacecraft have a wavelength of 0.120 m. What is their frequency?

• **15.64** What is the wavelength of an earthquake that shakes you with a frequency of 10.0 Hz and gets to another city 84.0 km away in 12.0 s?

• **15.65** Your ear is capable of noticing that sounds arrive as little as 1.00 ms apart. What is the minimum distance between objects that produce echoes that will arrive at noticeably different times on a day where the speed of sound is 340 m/s?

⁝ **15.66** Seismographs measure the arrival times of earthquakes with a precision of 0.100 s. To get the distance to the epicenter of the quake, they compare the arrival times of S- and P-waves, which travel at different speeds. If S- and P-waves travel at 4.00 and 7.20 km/s, respectively, in the region considered, how precisely can the distance to the source of the earthquake be determined?

Section 15.10 Superposition

15.67 A car has two horns, one emitting a frequency of 199 Hz and the other 203 Hz. What beat frequency do they produce?

15.68 The middle-C hammer of a piano hits two strings, producing beats of 1.50 Hz. One of the strings is tuned to 260.00 Hz. What frequencies could the other string have?

15.69 Two tuning forks having frequencies of 460 and 464 Hz are struck simultaneously. What average frequency will you hear, and what will the beat frequency be?

• **15.70** Twin jet engines on an airplane are producing an average sound frequency of 4100 Hz with a beat frequency of 0.500 Hz. What are their individual frequencies?

• **15.71** Three adjacent keys on a piano (F, F-sharp, and G) are struck simultaneously, producing frequencies of 349, 370, and 392 Hz. What beat frequencies are produced by this discordant combination?

Section 15.11 Wave Intensity

15.72 A device used to measure the intensity of sunlight has an area of 100 cm² and registers 6.50 W. What is the intensity in W/m²?

15.73 What is the power delivered by a typical helium-neon laser to an area of 1.00 mm² if the intensity is 250 W/m²?

15.74 Ultrasound of intensity 1.50×10^{-2} W/m² is produced by the rectangular head of a medical imaging device measuring 3.00 by 5.00 cm. What is its power output?

15.75 What increase in amplitude of a wave is needed to increase intensity by a factor of 50?

15.76 The amplitude of echoes in an anechoic chamber is decreased by a factor of 30.0. By what factor does the intensity decrease?

• **15.77** Ultrasonic deep heating is accomplished by applying ultrasound of intensity 5.00×10^3 W/m² with a circular transducer having a radius of 3.00 cm. How long must this be applied for 1.00 kcal to be emitted?

• **15.78** If the ultrasound in Problems 15.74 and 15.77 is identical except for amplitude, calculate the ratio of amplitudes of 5.00×10^3 W/m² and 1.50×10^{-2} W/m² ultrasound.

• **15.79** Energy from the sun arrives at the top of the earth's atmosphere with an intensity of 1.30 kW/m². How long does it take for 1.80×10^9 J to arrive on an area of 1.00 m²? This is the average monthly electrical energy consumption of a family of four in the United States.

• **15.80** Suppose you have a device that extracts energy from ocean breakers in direct proportion to their intensity. If the device produces 10.0 kW of power on a day when the breakers are 1.20 m high, how much will it produce when they are 0.600 m high?

• **15.81** A microphone receiving a pure sound tone feeds an oscilloscope, producing a wave on its screen. If the sound intensity is originally 2.00×10^{-5} W/m², but is turned up until the amplitude increases by 30.0%, what is the new intensity?

• **15.82** A solar collector is 10.0% efficient in gathering solar energy and converting it to heat. If the average intensity of sunlight on one day is 700 W/m², what area should your collector have to gather energy at the rate of 100 W?

• **15.83** What is the intensity in W/m² of a laser beam used to burn away cancerous tissue that, when 90.0% absorbed, puts 500 J of energy into a circular spot 2.00 mm in diameter in 4.00 s?

INTEGRATED CONCEPTS

Physics is most interesting when applied to general situations involving more than a narrow set of physical principles. The following problems involve physical principles from more than one chapter. Material from Chapters 3, 5, 6, 12, and 13 is directly involved. For example, waves carry energy, and hence the relevance of Chapter 6, Work, Energy, and Power. You will need to refer to these and perhaps other chapters to work the following problems.

Note: Strategies for dealing with problems having concepts from more than one chapter appear in Chapter 14, among others. Worked examples that can help you see how to apply these strate-

gies to integrated concept problems appear in a number of chapters, the most recent being Chapter 14. (See the section of problems labeled *Integrated Concepts* at the end of that chapter, and consult the index for others.)

• **15.84** A 0.800 m long brass pendulum experiences a 20.0°C temperature increase. (a) Calculate the original period. (b) What is the new period? (c) Find the ratio of the change in period to the original period, and calculate how long a clock run by this pendulum would take to fall 1.00 s behind the correct time.

• **15.85** The period of a pendulum is independent of mass if its length remains constant. But if a larger mass stretches the support, the period will increase. Suppose a 1.20 m long steel wire has a 1.00 mm diameter and originally has a 0.500 kg mass, which is replaced by a 5.00 kg mass. (a) Calculate the new length of the wire. (b) Show that this makes a very small difference in the period of the pendulum by comparing the periods of a 1.20 m long pendulum and one of the new length. (Assume each mass is suspended at its center of mass at the end of the wire.)

⁚15.86 A mass will bounce up and down on the end of a wire by stretching it along its length. Suppose in a physics demonstration a 10.0 kg mass is hung from the end of a 30.0 m long steel wire that is 0.750 mm in diameter. (a) Calculate the change in length of the wire. (b) What is the effective force constant of the wire? (c) At what frequency will the mass bounce?

⁚15.87 A physics major who runs away to join the circus is asked to design a spring-loaded cannon to shoot a performer safely across center ring. (a) Calculate the speed at which the performer must leave the cannon to land 20.0 m away (horizontally) and 10.0 m below, if the cannon is inclined at an angle of 42.0° above the horizontal. (b) What force constant must the spring have, if the barrel is 15.0 m long and the mass to be accelerated is 75.0 kg? Note that both kinetic and gravitational potential energy must be supplied. (c) Calculate the maximum acceleration in *g*s, and assess whether a person can withstand it.

⁚15.88 The spring over one wheel of a car has a force constant of 1.50×10^4 N/m, and there is a shock absorber that damps the wheel's motion. The wheel hits a bump, rises 0.120 m, and returns to its original height. (a) How much heat is generated by the shock absorber? You may assume the same amount of heat is generated on the way up as on the way down. (b) What is the temperature increase of the 1.50 kg shock absorber if its average specific heat is 1100 J/kg·°C?

UNREASONABLE RESULTS

The following problems have results that are unreasonable because some premise is unreasonable or because certain of the premises are inconsistent with one another. Physical principles applied correctly then produce unreasonable results. The purpose of these problems is to give practice in assessing whether nature is being accurately described, and if it is not to trace the source of difficulty.

15.89 (a) Calculate the propagation speed of waves on the surface of the ocean that have crests 2.00 m apart and arrive at the rate of 30.0 per second. (b) What is unreasonable about this result? (c) Which premise or assumption is responsible?

15.90 (a) What force constant must a spring have for a 20.0 kg mass to oscillate on it with a 1.00 ms period? (b) What is unreasonable about this result? (c) Which premise or assumption is responsible?

15.91 (a) If the speed of sound in air is 340 m/s, what frequency would your voice be emitting to produce a sound of wavelength 1.70 mm? (b) What is unreasonable about this result? (c) Which premise or assumption is responsible?

• **15.92** (a) What decrease in temperature would cause the period of a brass pendulum to decrease by 5.00%? (b) What is unreasonable about this result? (c) Which premise or assumption is responsible?

SOUND AND HEARING

16

*This glass has been shattered by a high-intensity sound wave having its
resonant frequency. While the sound is not visible, and the event took place
where no one could hear it, the effects of the sound prove its existence.*

Figure 16.1

If a tree falls in the forest and no one is there to hear it, does it make a sound? (See Figure 16.1.) The answer to this old philosophical question depends on how you define sound. If it is defined to be the human perception, then there was no sound. If sound is defined to be a disturbance of the atoms in matter, transmitted from its origin outward—that is, a wave—then there was a sound. This wave is the physical phenomenon called sound. Its perception is hearing. Both the physical phenomenon and its perception are interesting and will be considered in this chapter. We shall make a clear distinction between sound and hearing; they are related, but not identical.

16.1 SOUND

Sound was discussed a number of times in the previous chapter as an illustration of waves. Since hearing is one of our most important senses, it is interesting to see how the physical properties of sound correspond to our perceptions of it. But sound has important applications beyond hearing. Ultrasound, for example, is not heard but can be employed to form medical images.

The physical phenomenon of **sound** is defined to be a disturbance of matter that is transmitted from its source outward. Sound is a wave. On the atomic scale, it is a disturbance of atoms that is far more ordered than their thermal motions. In many instances, sound is a periodic wave, and the atoms undergo simple harmonic motion. We shall begin with such periodic sound waves and eventually will consider more complex sounds, such as sonic booms. All of the wave concepts developed in the previous chapter apply to sound. We will study them in some depth for sound and will also include the perception of sound, or hearing.

A vibrating string creates a sound wave as illustrated in Figure 16.2. As the string oscillates back and forth, it transfers energy to the air, mostly as thermal energy created by turbulence. But a small part of the string's energy goes into compressing and expanding surrounding air, creating slightly higher and lower local pressures. These compressions and rarefactions move out as longitudinal pressure waves having the same frequency as the string—they are the disturbance that is a sound wave. (Sound waves in air and most fluids are longitudinal, since fluids have almost no shear strength. In solids, sound waves can be both transverse and longitudinal.) Figure 16.2 shows a graph of pressure versus distance from the source.

The amplitude of a sound wave decreases with distance from its source, because the energy of the wave is spread over a larger and larger area. But it is also absorbed by ob-

Figure 16.2 (a) A vibrating string moving to the right compresses the air in front of it and expands the air behind it. (b) As it moves to the left, it creates another compression and rarefaction as the ones on the right move away from the string. (c) After many vibrations, there are a series of compressions and rarefactions moving out from the string as a sound wave. The graph shows gauge pressure versus distance from the source. Pressures vary only slightly from atmospheric for ordinary sounds.

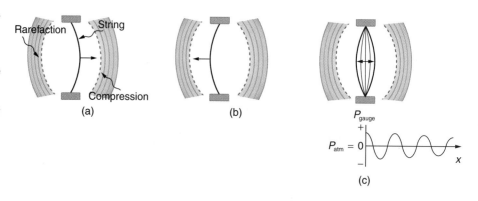

Figure 16.3 Sound wave compressions and rarefactions travel up the ear canal and force the eardrum to vibrate. There is a net force on the eardrum, since the sound wave pressures differ from the atmospheric pressure found behind the eardrum. A complicated mechanism converts the vibrations to nerve impulses, which are perceived by the person.

jects, such as the eardrum in Figure 16.3, and converted to thermal energy by viscosity in the air. Additionally, each compression heats the air slightly and each rarefaction cools it, so that there is a flow of heat that reduces the organized disturbance into random thermal motions. (These processes can be viewed as a manifestation of the second law of thermodynamics.) Whether the heat flow from compression to rarefaction is significant depends on how far apart they are—that is, it depends on wavelength. Wavelength, frequency, amplitude, and speed of propagation are important for sound, as they are for all waves. We begin the study of the wave properties of sound in the next section.

16.2 SPEED OF SOUND, FREQUENCY, AND WAVELENGTH

Sound, like all waves, travels at a certain speed and has the properties of frequency and wavelength, defined in the previous chapter. You can observe direct evidence of the speed of sound while watching a fireworks display. The flash of an explosion is seen well before its sound is heard, implying both that sound travels at a finite speed and that it is slower than light. You can also directly sense the frequency of a sound. Perception of frequency is called **pitch**. The wavelength of sound is not directly sensed, but indirect evidence is found in the correlation of the size of musical instruments with their pitch. Small instruments, like a piccolo, typically make high-pitch sounds, while large instruments, like a tuba, typically make low-pitch sounds. High pitch means small wavelength, and the size of a musical instrument is directly related to the wavelengths of sound it produces. So a small instrument creates a short-wavelength sound. Similar arguments hold that a large instrument creates long-wavelength sounds.

The relationship of the speed of sound, its frequency, and wavelength is the same as given for all waves in Equation 15.11, repeated here with a new equation number:

$$v_w = f\lambda \qquad (16.1)$$

where v_w is the speed of sound, f is its frequency, and λ its wavelength. The wavelength of a sound is the distance between adjacent identical parts of a wave—for example, between adjacent compressions as illustrated in Figure 16.4. The frequency is the same as that of the source and is the number of waves that pass a point per unit time.

Table 16.1 makes it apparent that the speed of sound varies greatly in different media. The speed of sound in a medium is determined by a combination of its rigidity (or compressibility in gases) and its density. The more rigid (or less compressible) the medium, the faster the speed of sound. This is analogous to the fact that the frequency of simple harmonic motion is directly proportional to the stiffness of the oscillating

Human Application

TABLE 16.1

SPEED OF SOUND IN VARIOUS MEDIA	
Medium	v_w *(m/s)*
Gases at 0°C	
Air	331
Carbon dioxide	259
Oxygen	316
Helium	965
Hydrogen	1290
Liquids at 20°C	
Ethanol	1160
Mercury	1450
Water, fresh	1480
Sea water	1540
Human tissue	1540
Solids (longitudinal or bulk)	
Vulcanized rubber	54
Polyethylene	920
Marble	3810
Glass, Pyrex	5640
Lead	1960
Aluminum	5120
Steel	5960

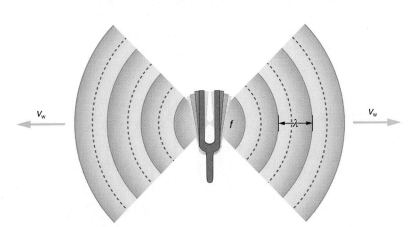

Figure 16.4 A sound wave emanates from a source vibrating at a frequency f, propagates at v_w, and has a wavelength λ.

object. The greater the density of a medium, the slower the speed of sound, analogous to the fact that the frequency of simple harmonic motion is inversely proportional to mass. The speed of sound in air is low, because air is compressible. Because liquids and solids are relatively rigid and very difficult to compress, the speed of sound in them is generally greater than in gases.

Earthquakes, essentially sound waves in the earth, are an interesting example of how the speed of sound depends on the rigidity of the medium. Earthquakes have both longitudinal and transverse components, and these travel at different speeds. Table 5.1 in Section 5.8 shows that the bulk modulus of granite is greater than its shear modulus. For that reason, the speed of longitudinal or pressure waves (P-waves) in earthquakes in granite is significantly higher than the speed of transverse or shear waves (S-waves). Both components of earthquakes travel slower in less rigid material, such as sediments. P-waves have speeds of 4.00 to 7.00 km/s, and S-waves correspondingly range in speed from 2.00 to 5.00 km/s, both being faster in more rigid material. The P-wave gets progressively farther ahead of the S-wave as they travel through the earth. The time between the P- and S-waves is routinely used to determine the distance to their source, the epicenter of the earthquake.

The speed of sound is affected by temperature in a given medium. For air at sea level, the speed of sound is given by

$$v_{\text{w}} = (331 \text{ m/s})\sqrt{\frac{T \text{ (K)}}{273 \text{ K}}} \qquad \textbf{(16.2)}$$

In Section 12.4, it was noted that the speed of sound in gases is related to the average speed of molecules in the gas, v_{rms}, and that

$$v_{\text{rms}} = \sqrt{\frac{3kT}{m}}$$

so that it is reasonable that the speed of sound in air and other gases should depend on the square root of temperature.* While not negligible, this is not a strong dependence. At 0°C, the speed of sound is 331 m/s, whereas at 20.0°C it is 343 m/s, less than a 4% increase. Figure 16.5 shows a use of the speed of sound to sense distances. Echoes are also used in medical imaging, as we shall see.

One of the more important properties of sound is that its speed is usually nearly independent of frequency. This is certainly true in open air for sounds in the audible range of 20 to 20,000 Hz. If this were not true, you would certainly notice it for music played by a marching band in a football stadium, for example. Suppose that high-frequency sounds traveled faster—then the farther you were from the band, the more the low-pitch instruments would lag the high-pitch ones. But the music from all instruments arrives in cadence independent of distance, and so all frequencies must travel at nearly the same speed. According to Equation 16.1,

$$v_{\text{w}} = f\lambda$$

---

*The exact dependence of the speed of sound on physical properties such as temperature, density, and bulk modulus can be derived considering stress and strain, and the thermodynamics of ideal gases, but this is beyond the scope of this text.

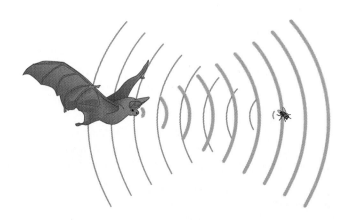

Figure 16.5 A bat uses sound echoes to find its way about and to catch prey. The time for the echo to return is directly proportional to the distance.

In a given medium under fixed conditions, v_w is constant, so that there is a relationship between f and λ; the higher the frequency, the smaller the wavelength. See Figure 16.6 and consider the following example.

EXAMPLE 16.1 WHAT ARE THE WAVELENGTHS OF AUDIBLE SOUNDS?

Calculate the wavelengths of sounds at the extremes of the audible range, 20 and 20,000 Hz, in 30.0°C air.

Strategy To find wavelength from frequency, we can use $v_w = f\lambda$, but we first need a value for v_w. That is given by Equation 16.2.

Solution Equation 16.2 is

$$v_w = (331 \text{ m/s})\sqrt{\frac{T \text{ (K)}}{273 \text{ K}}}$$

Entering the temperature in K,

$$v_w = (331 \text{ m/s})\sqrt{\frac{303 \text{ K}}{273 \text{ K}}} = 348.7 \text{ m/s}$$

Now, solving the relationship between speed and wavelength for λ,

$$\lambda = \frac{v_w}{f}$$

Entering the minimum frequency gives the maximum wavelength:

$$\lambda_{max} = \frac{348.7 \text{ m/s}}{20 \text{ Hz}} = 17 \text{ m}$$

Similarly, the maximum frequency gives the minimum wavelength:

$$\lambda_{min} = \frac{348.7 \text{ m/s}}{20,000 \text{ Hz}} = 0.0174 \text{ m} = 1.74 \text{ cm}$$

Discussion Since the product of f times λ equals a constant, the smaller f is the larger λ must be, and vice versa.

The speed of sound can change when sound travels from one medium to another. The frequency usually remains the same, since it is like a driven oscillation and has the frequency of the original source. If v_w changes and f remains the same, then the wavelength λ must change. That is, since $v_w = f\lambda$, the higher the speed the greater the wavelength for a given frequency.

16.3 SOUND INTENSITY AND SOUND LEVEL

In a quiet forest, you can sometimes hear a single leaf fall to the ground. After settling into bed, you may hear your blood pulsing through your ears. But when a passing motorist has his boom box turned up, you cannot even hear what the person next to you in your car is saying. We are all very familiar with the loudness of sounds and aware that they are related to how energetically the source is vibrating. We often hear about the hazards of high decibel levels and the fact that some famous rock musicians can no longer hear well enough to perform. The relevant physical quantity is sound intensity, a concept that is valid for all sounds whether or not in the audible range.

Intensity was defined in Section 15.11 to be the power per unit area carried by a wave. In equation form, **intensity** I is

$$I = \frac{P}{A} \qquad (16.3)$$

where P is the power crossing area A. The SI unit for I is W/m². The intensity of any wave is related to its amplitude squared. For sound waves, the amplitude is related to the maximum pressure in the sound wave, so that high-intensity sounds have relatively large gauge pressures (above and below atmospheric pressure). This is consistent with the fact that the sound wave is created by some vibration; the greater its amplitude, the more the air is compressed in the sound it creates. (See Figure 16.7.)

Sound levels are quoted in decibels (dB) much more often than sound intensities in watts per meter squared. Decibels are the unit of choice in the scientific literature as well as the popular media. The reasons for this choice of units are related to how we perceive sounds. Our ear seems to respond proportionally to the logarithm* of the intensity rather

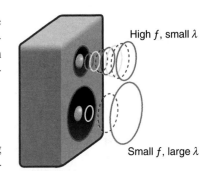

High f, small λ

Small f, large λ

Figure 16.6 Low-frequency sounds have a greater wavelength than the high-frequency sounds, since they travel at the same speed. Here the lower-frequency sounds are emitted by the large speaker, called a woofer, while the higher-frequency sounds are emitted by the small speaker, called a tweeter.

Human Application

---

*Section 1.6 gives a review of logarithms.

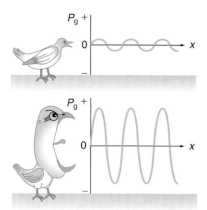

Figure 16.7 Graphs of the pressures in two sound waves of different intensities. The more intense sound is created by a source that has larger-amplitude oscillations and has greater pressure maxima and minima. Since pressures are higher in the greater-intensity sound, it can exert larger forces on the objects it encounters.

TABLE 16.2

SOUND LEVELS AND INTENSITIES

Sound level β (dB)	Intensity I (W/m^2)	Example/effect
0	1×10^{-12}	Threshold of hearing at 1000 Hz
10	1×10^{-11}	Rustle of leaves
20	1×10^{-10}	Whisper at 1 m distance
30	1×10^{-9}	Quiet home
40	1×10^{-8}	Average home
50	1×10^{-7}	Average office, soft music
60	1×10^{-6}	Normal conversation
70	1×10^{-5}	Noisy office, busy traffic
80	1×10^{-4}	Loud radio, classroom lecture
90	1×10^{-3}	Inside subway train
		Damage from prolonged exposure*
100	1×10^{-2}	Noisy factory, siren at 30 m
		Damage from 8 h per day exposure
110	1×10^{-1}	Damage from 30 min per day exposure
120	1	Loud rock concert, pneumatic chipper at 2 m, threshold of pain
		Damage in seconds
140	1×10^{2}	Jet airplane at 30 m, severe pain
160	1×10^{4}	Bursting of eardrums

*Several government agencies and health-related professional associations recommend 85 dB not be exceeded for 8 h daily exposures in the absence of hearing protection.

than directly to the intensity. The **sound level** β in decibels of a sound having an intensity I in W/m^2 is defined to be

$$\beta \text{ (dB)} = 10 \log_{10}\left(\frac{I}{I_0}\right) \qquad \textbf{(16.4)}$$

where $I_0 = 10^{-12}\ W/m^2$ is a reference intensity. In particular, I_0 is the lowest or threshold intensity of sound a person with normal hearing can perceive at a frequency of 1000 Hz. Sound level is not really the same as an intensity. Since β is defined in terms of a ratio, it is really a unitless quantity telling you the *level* of the sound relative to a fixed standard ($10^{-12}\ W/m^2$ in this case). The units of **decibels** (dB) are used to indicate this ratio is multiplied by 10 in its definition. The bel, upon which the decibel is based, is named for Alexander Graham Bell, the inventor of the telephone.

The decibel level of a sound having the threshold intensity of $10^{-12}\ W/m^2$ is $\beta = 0$ dB, because $\log_{10} 1 = 0$. That is, the threshold of hearing is 0 decibels. Table 16.2 gives levels in decibels and intensities in W/m^2 for some familiar sounds.

One of the more striking things about the intensities in Table 16.2 is that the intensity in W/m^2 is quite small for most sounds. The ear is sensitive to as little as a trillionth of a watt per meter squared—even more impressive when you realize that the area of the eardrum is only about 1 cm^2, so that only $10^{-16}\ W$ falls on it at the threshold of hearing! Air molecules in a sound wave of this intensity vibrate over a distance of less than one molecular diameter, and the gauge pressures involved are less than 10^{-9} atm.

Another impressive feature of the sounds in the table is their range. Sound intensity varies by a factor of 10^{12} from threshold to a sound that causes damage in seconds. You are unaware of this tremendous range in sound intensity, because your ear responds approximately as the logarithm of intensity. Thus sound levels in decibels fit your experience better than intensities in W/m^2. The decibel scale is also easier to relate to because we are more accustomed to dealing with numbers like 0, 53, or 120 than numbers like 1.00×10^{-11}.

Human Application

EXAMPLE 16.2 HOW TO CALCULATE A SOUND LEVEL IN DECIBELS

Calculate the sound level in decibels for a sound having an intensity of $5.00 \times 10^{-4}\ W/m^2$ (five times as intense as an 80.0 dB sound).

Strategy We are given I, and so we can calculate β

(continued)

(continued)
straight from its definition in Equation 16.4.

Solution Entering known values of I and I_0 into Equation 16.4 gives

$$\beta \ (\text{dB}) = 10 \ \log_{10}\left(\frac{I}{I_0}\right) = 10 \ \log_{10}\left(\frac{5.00 \times 10^{-4} \ \text{W/m}^2}{1.00 \times 10^{-12} \ \text{W/m}^2}\right)$$

$$= 10 \ \log_{10}(5.00 \times 10^8) = 10 \cdot (8.70) \ \text{dB}$$

$$= 87.0 \ \text{dB}$$

Discussion This 87.0 dB sound has an intensity five times as great as an 80.0 dB sound. So a factor of five in intensity corresponds to a difference of 7.00 dB in sound level. This is true for any intensities differing by a factor of five.

EXAMPLE 16.3 HOW DOES DOUBLING INTENSITY AFFECT THE DECIBEL LEVEL?

Show that if one sound is twice as intense as another, it has a sound level about 3 dB higher.

Strategy We are given the ratio of two intensities is 2 to 1, and we are then asked to find the difference in their sound levels in decibels. This can be done using some of the properties of logarithms as reviewed in Section 1.6.

Solution We are given the ratio of two intensities is 2 to 1, or

$$\frac{I_2}{I_1} = 2.00$$

and we wish to show that the difference in sound level is about 3 dB. That is, we want to show

$$\beta_2 - \beta_1 = 3 \ \text{dB}$$

Now, noting from Equation 1.18 in Section 1.6 that

$$\log_{10}b - \log_{10}a = \log_{10}\left(\frac{b}{a}\right)$$

and using the definition of β, we get

$$\beta_2 - \beta_1 = 10 \ \log_{10}\left(\frac{I_2}{I_1}\right) = 10 \ \log_{10} 2.00$$

$$= 10 \cdot (0.301) \ \text{dB}$$

Thus,

$$\beta_2 - \beta_1 = 3.01 \ \text{dB}$$

Discussion This means that the two sound levels differ by 3.01 dB, or about 3 dB, as advertised. Note that since only the ratio I_2/I_1 is given, and not the actual intensities, this result is true for any intensities that differ by a factor of two. For example, a 56.0 dB sound is twice as intense as a 53.0 dB sound, a 97.0 dB sound is half as intense as a 100 dB sound, and so on.

One more observation readily verified by examining Table 16.2 or using Equation 16.4 is that each factor of 10 in intensity corresponds to 10 dB. For example, a 90 dB sound compared with a 60 dB sound is 30 dB greater, or three factors of 10 (that is, 10^3 times) as intense. Another example is that if one sound is 10^7 as intense as another, it is 70 dB higher. Equation 16.4 can be used to convert from I to β and vice versa for any numerical values, but there are some handy numbers in 3.0, 7.0, and 10.0 dB *differences*—they correspond to factors of 2.0, 5.0, and 10.0 in intensity, respectively. This is summarized in Table 16.3.

TABLE 16.3

RATIOS OF INTENSITIES AND CORRESPONDING DIFFERENCES IN SOUND LEVELS

I_2/I_1	$\beta_2 - \beta_1$
2.0	3.0 dB
5.0	7.0 dB
10.0	10.0 dB

16.4 DOPPLER EFFECT AND SONIC BOOMS

The characteristic sound of a motorcycle buzzing by is an example of the Doppler effect. The high-pitch scream shifts dramatically to a lower-pitch roar as the motorcycle passes by a stationary observer. The closer the motorcycle brushes by, the more abrupt the shift. The faster the motorcycle moves, the greater the shift. We also hear this characteristic sliding frequency for passing race cars, airplanes, and trains. It is so familiar that it is used to imply motion, such as in Road Runner cartoons, and children often mimic it in play.

The **Doppler effect** is an alteration in the observed frequency of a sound due to motion of either the source or the observer. Although less familiar, this effect is easily noticed for a stationary source and moving observer. For example, if you ride a train past a stationary warning bell, you will hear the bell's frequency shift from high to low as you pass by. The actual change in frequency or shift due to relative motion of source and observer is called a **Doppler shift**. The Doppler effect and Doppler shift are named for the Austrian physicist and mathematician Christian Johann Doppler (1803–1853), who did experiments with both moving sources and moving observers. Doppler, for example, had musicians play on a moving open train car and also play standing next to the train tracks as a train passed by. Their music was observed both on and off the train, and changes in frequency were measured.

What causes the Doppler shift? Figure 16.8 compares sound waves emitted by stationary and moving sources in a stationary air mass. Each disturbance spreads out spherically from the point where the sound was emitted. If the source is stationary, then all of the spheres are centered on the same point, and the stationary observers on either side see the same wavelength and frequency as emitted by the source, as in Figure 16.8(a). If the source is moving, as in Figure 16.8(b), then the situation is different. Each disturbance moves out in a sphere from the point where it was emitted, but the point of emission moves. This causes the waves to be closer together on one side and farther apart on the other. Thus the wavelength is shorter in the direction the source is moving (on the right in Figure 16.8(b)), and longer in the opposite direction (on the left). Finally, if the observers move, as in Figure 16.8(c), the frequency at which they receive the disturbances changes. The observer moving toward the source receives them at a higher frequency, and the person moving away from the source receives them at a lower frequency.

We know that wavelength and frequency are related by $v_w = f\lambda$, where v_w is the fixed speed of sound. The sound moves in a medium and has the same speed v_w in that medium whether the source is moving or not. Thus f times λ is a constant. Since the observer on the right in Figure 16.8(b) receives a shorter wavelength, the frequency she receives must be higher. Similarly, the observer on the left receives a longer wavelength, and hence he observes a lower frequency. The same thing happens in Figure 16.8(c). A higher frequency is received by the observer moving toward the source, and a lower frequency is received by an observer moving away from the source. In general, then, relative motion of source and observer toward one another increases the received frequency. Relative motion apart decreases frequency. The greater the relative speed, the greater the effect.

For a stationary observer and a moving source, the frequency received by the observer f_{obs} can be shown to be

$$f_{obs} = f_s\left(\frac{v_w}{v_w \mp v_s}\right) \qquad \textbf{(16.5)}$$

where f_s is the frequency of the source, v_s is the speed of the source along a line joining the source and observer, and v_w is the speed of sound. The minus sign is used for motion toward the observer and the plus for motion away from the observer, producing the ap-

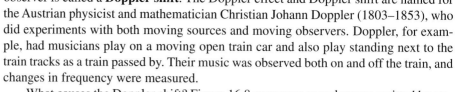
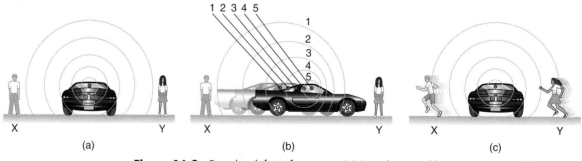

Figure 16.8 Doppler shifts in frequency. (a) Sounds emitted by a source spread out in spherical waves. Since the source, observers, and air are stationary, the wavelength and frequency are the same in all directions and to all observers. (b) Sounds emitted by a source moving to the right spread out from the points at which they were emitted. The wavelength is reduced and, consequently, the frequency is increased in the direction of motion, so that observer hears a higher-pitch sound. The opposite is true on the left, where the wavelength is increased and the frequency is reduced. (c) The same effect is produced when the observers move relative to the source. Motion toward increases frequency. Motion apart decreases frequency.

propriate shifts up and down in frequency. Note that the greater the speed of the source, the greater the effect. Similarly, for a stationary source and moving observer,

$$f_{obs} = f_s\left(\frac{v_w \pm v_{obs}}{v_w}\right) \tag{16.6}$$

where v_{obs} is the speed of the observer along a line joining the source and observer.* Here the plus sign is for motion toward the source, and the minus is for motion away from the source.

EXAMPLE 16.4   DOPPLER SHIFT OF A TRAIN HORN

Suppose a train with a 150 Hz horn is moving at 35.0 m/s in still air on a day when the speed of sound is 340 m/s. (a) What frequencies are observed by a stationary person at the side of the tracks as the train approaches and after it passes? (b) What frequency is observed by the train's engineer?

Strategy To find the observed frequency in part (a), Equation 16.5 must be used because the source is moving. The minus sign is used for the approaching train, and the plus sign for the receding train. In part (b), there are two Doppler shifts—one for a moving source and the other for a moving observer.

Solution for (a) As the train approaches, we use the minus sign in Equation 16.5. Entering known values then gives

$$f_{obs} = f_s\left(\frac{v_w}{v_w - v_s}\right) = (150\text{ Hz})\left(\frac{340\text{ m/s}}{340\text{ m/s} - 35.0\text{ m/s}}\right)$$
$$= (150\text{ Hz})(1.11) = 167\text{ Hz}$$

Similarly, as the train recedes, we use the plus sign and find that

$$f'_{obs} = f_s\left(\frac{v_w}{v_w + v_s}\right) = (150\text{ Hz})\left(\frac{340\text{ m/s}}{340\text{ m/s} + 35\text{ m/s}}\right)$$
$$= (150\text{ Hz})(0.907) = 136\text{ Hz}$$

Discussion for (a) The numbers calculated are valid when the train is far enough away that the motion is nearly along the line joining train and observer. In both cases, the shift is significant and easily noticed. Note that the shift is 17.0 Hz for motion toward and 14.0 Hz for motion away. It is not symmetric.

Solution for (b) It seems reasonable that the engineer would receive the same frequency as emitted by the horn, since the relative velocity between them is zero. Relative to the medium (air), the speeds are $v_s = v_{obs} = 35.0$ m/s. The first Doppler shift is for the moving observer; the second is for the moving source. Thus,

$$f_{obs} = \left[f_s\left(\frac{v_w + v_{obs}}{v_w}\right)\right]\left(\frac{v_w}{v_w + v_s}\right)$$

The quantity in brackets is the Doppler-shifted frequency due to a moving observer. The factor on the right is the effect of the moving source. Since $v_s = v_{obs}$ in this case, everything but f_s cancels, yielding

$$f_{obs} = f_s$$

Discussion for (b) We may expect that there is no change in frequency when source and observer move together because it fits your experience. For example, there is no Doppler shift in the frequency of conversations between driver and passenger on a motorcycle. People talking when a wind moves the air between them also observe no Doppler shift in their conversation. The crucial point is that source and observer are not moving relative to one another.

Sonic Booms to Bow Wakes

What happens to the sound produced by a moving source, such as a jet airplane, that approaches or even exceeds the speed of sound? The answer to this question applies not only to sound but to all other waves as well.

Suppose a jet airplane is coming nearly straight at you, emitting a sound of frequency f_s. The greater the plane's speed v_s, the greater the Doppler shift and the greater f_{obs} is. Now, as v_s approaches the speed of sound, f_{obs} approaches infinity, since the denominator in Equation 16.5 approaches zero. Physically (at the speed of sound), this means that in front of the source each successive wave is superimposed on the previous one, because the source moves forward at the speed of sound. The observer gets them all at the same

---

*Equations 16.5 and 16.6 assume relative motion along a straight line directly toward or directly away from each other. For motion along any other line, take v_s and v_w to be the components of the speeds along a line joining source and observer.

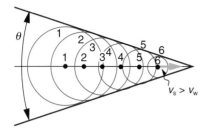

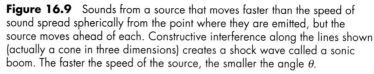

Figure 16.9 Sounds from a source that moves faster than the speed of sound spread spherically from the point where they are emitted, but the source moves ahead of each. Constructive interference along the lines shown (actually a cone in three dimensions) creates a shock wave called a sonic boom. The faster the speed of the source, the smaller the angle θ.

Human Application

instant, and so the frequency is infinite. (Before airplanes exceeded the speed of sound, some people argued it would be impossible because such constructive superposition would produce pressures great enough to destroy the airplane.) If the source exceeds the speed of sound, no sound is received by the observer until the source has passed, so that the sounds from the approaching source are mixed with those from it when receding. This appears messy, but something interesting happens as part of this—a sonic boom is created. (See Figure 16.9.)

There is constructive interference along the lines shown (a cone in three dimensions) from similar sound waves arriving there simultaneously. This forms a disturbance called a **sonic boom**, a constructive interference of sound created by an object moving faster than sound. Inside the cone, the interference is mostly destructive, and so sound intensity there is much less than on the shock wave. An aircraft creates two sonic booms, one from its nose and one from its tail. (See Figure 16.10.) During television coverage of space shuttle landings, two distinct booms can often be heard. These are separated by exactly the time it would take the shuttle to pass by a point. Observers on the ground often do not see the aircraft creating the sonic boom, because it has passed by before the shock wave reaches them, as seen in Figure 16.10. If the aircraft flies close by at low altitude, pressures in the sonic boom can be destructive and break windows as well as rattle nerves. Because of this, supersonic flights are banned over populated areas of the United States.

Sonic booms are one example of a broader phenomenon called bow wakes. A **bow wake**, such as the one in Figure 16.11, is created when the wave source moves faster than the wave propagation speed. Water waves spread out in circles from the point where created, and the bow wake is the familiar V-shaped wake trailing the source. A more exotic bow wake is created when a subatomic particle travels through a substance faster than the speed of light in that medium. If the particle creates light in its passage, that light spreads on a cone with an angle indicative of the speed of the particle, as illustrated in Figure 16.12. Such a bow wake is called Cerenkov radiation and is commonly observed in particle physics.

Doppler shifts and sonic booms are interesting sound phenomena that occur in all types of waves. They can be of considerable use. For example, the Doppler shift in ultrasound can be used to measure blood velocity, while police use the Doppler shift in radar (a microwave) to measure car velocities. In most of this chapter, the general properties of waves previously developed are applied to sound. In this section, however, a property of sound was found to have general applicability to waves.

Figure 16.10 Two sonic booms, created by the nose and tail of an aircraft, are observed on the ground after the plane has passed by.

Figure 16.11 Bow wake created by a duck. Constructive interference produces the rather structured wake, while there is relatively little wave action inside the wake, where interference is mostly destructive.

Figure 16.12 (a) A subatomic particle creating light as it passes through a substance faster than the speed of light in that medium ($v_s > v_l$). The light spreads on a cone trailing the particle exactly analogous to a sonic boom. This characteristic light is called Cerenkov radiation. (b) The glow in this reactor pool is Cerenkov radiation caused by subatomic particles traveling faster than the speed of light in water.

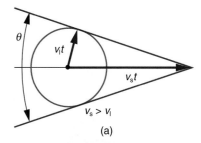

(a)

(b)

16.5   SOUND INTERFERENCE AND RESONANCE: STANDING WAVES IN AIR COLUMNS

Interference is the hallmark of waves, all of which exhibit constructive and destructive interference exactly analogous to that seen for water waves. In fact, one way to prove something "is a wave" is to observe interference effects. So, sound being a wave, we expect it to exhibit interference; we have already mentioned a few such effects, like the beats from two similar notes played simultaneously.

Figure 16.13 shows a clever use of sound interference to cancel noise. Larger-scale applications of active noise reduction by destructive interference are contemplated for entire passenger compartments in commercial aircraft. To obtain destructive interference, a fast electronic analysis is performed, and a second sound is introduced with its maxima and minima exactly reversed from the incoming noise. Sound waves in fluids are pressure waves, and consistent with Pascal's principle, pressures from two different sources add and subtract like simple numbers—that is, positive and negative gauge pressures add to a much smaller pressure, producing a lower-intensity sound. While completely destructive interference is possible only under the simplest conditions, it is possible to reduce noise levels by 30 dB or more using this technique.

Where else can we observe sound interference? All sound resonances, such as in musical instruments, are due to constructive and destructive interference. Only the resonant frequencies interfere constructively to form standing waves, while others interfere destructively and are absent. From the toot made by blowing over a bottle, to the characteristic flavor of a violin's sounding box, to the recognizability of a great singer's voice, resonance and standing waves play a vital role.

Suppose we hold a tuning fork near the end of a tube that is closed at the other end, as shown in Figure 16.14. If the tuning fork has just the right frequency, the air column in the tube resonates loudly, but at most frequencies it vibrates very little. This just means that the air column has only certain natural frequencies. The figure shows how a resonance at the lowest of these natural frequencies is formed. A disturbance travels down the tube at the speed of sound and bounces off the closed end. If the tube is just the right length, the reflected sound arrives back at the tuning fork exactly half a cycle later, and it interferes constructively with the continuing sound produced by the tuning fork. The incoming and reflected sounds form a standing wave in the tube as shown.

The standing wave formed in the tube has its maximum air displacements (a loop) at the open end, where motion is unconstrained, and no displacement (a node) at the

CONNECTIONS

Interference is such a fundamental aspect of waves that observing interference is proof that something is a wave. The wave nature of light was established by experiments showing interference. Similarly, when electrons scattered from crystals exhibited interference, their wave nature was confirmed to be exactly as predicted by symmetry with certain wave characteristics of light. Interference was defined in Section 15.10.

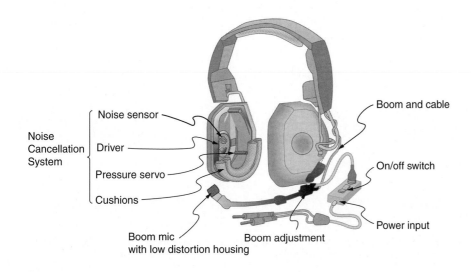

Noise sensor
Boom and cable
Noise Cancellation System
Driver
On/off switch
Pressure servo
Cushions
Power input
Boom mic with low distortion housing
Boom adjustment

Figure 16.13 Headphones designed to cancel noise with destructive interference create a sound wave exactly opposite to incoming sound. These can be more effective than the simple passive attenuation used in most ear protection. Such headphones were used on the record-setting, around the world nonstop flight of the Voyager aircraft to protect the pilots' hearing from engine noise.

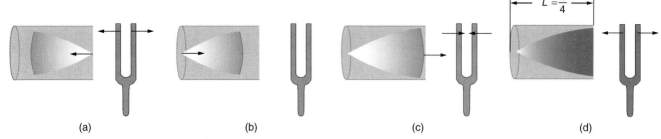

(a)   (b)   (c)   (d)

Figure 16.14 Resonance of air in a tube closed at one end, caused by a tuning fork. (a) A disturbance moves down the tube. (b) The disturbance reflects from the closed end of the tube. (c) If the length of the tube L is just right, the disturbance gets back to the tuning fork half a cycle later and interferes constructively with the continuing sound from the tuning fork. This forms a standing wave, and the air column resonates. (d) A graph of air displacement along the length of the tube shows none at the closed end, where the motion is constrained, and a maximum at the open end. This standing wave has one-fourth of its wavelength in the tube, so that $\lambda = 4L$.

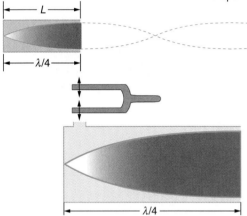

Figure 16.15 The same standing wave is created in the tube by a vibration introduced near its closed end.

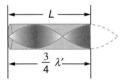

Figure 16.16 Another resonance for a tube closed at one end. This has maximum air displacements at the open end, and none at the closed end. The wavelength is shorter, with three-fourths λ' equaling the length of the tube, so that $\lambda' = 4L/3$. This higher-frequency vibration is the first overtone.

closed end, where air movement is halted. The distance from a node to a loop is one-fourth of a wavelength, and this equals the length of the tube; thus, $\lambda = 4L$. This same resonance can be produced by a vibration introduced at or near the closed end of the tube, as shown in Figure 16.15. It is best to consider this a natural vibration of the air column independent of how it is induced.

Given that maximum air displacements are possible at the open end and none at the closed end, there are other, shorter wavelengths that can resonate in the tube, such as the one shown in Figure 16.16. Here the standing wave has three-fourths of its wavelength in the tube, or $L = (3/4)\lambda'$, so that $\lambda' = 4L/3$.

Continuing this process reveals a whole series of shorter-wavelength and higher-frequency sounds that resonate in the tube. We use specific terms for the resonances in any system. The lowest resonant frequency is called the **fundamental**, while all higher resonant frequencies are called **overtones**. All resonant frequencies are integral multiples of the fundamental, and they are collectively called **harmonics**. The fundamental is the first harmonic, the first overtone is the second harmonic, and so on. Figure 16.17 shows the fundamental and the first three overtones (the first four harmonics) in a tube closed at one end.

The fundamental and overtones can be present simultaneously in a variety of combinations. For example, middle C on a trumpet has a sound distinctively different from middle C on a clarinet, both instruments being modified versions of a tube closed at one end. The fundamental frequency is the same (and usually the most intense), but the overtones and their mix of intensities are different and subject to shading by the musician. This mix is what gives various musical instruments (and human voices) their distinctive characteristics, whether they have air columns, strings, sounding boxes, or

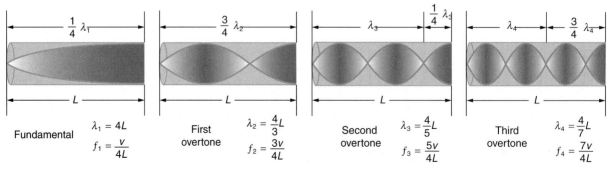

Fundamental	First overtone	Second overtone	Third overtone
$\lambda_1 = 4L$	$\lambda_2 = \frac{4}{3}L$	$\lambda_3 = \frac{4}{5}L$	$\lambda_4 = \frac{4}{7}L$
$f_1 = \frac{v}{4L}$	$f_2 = \frac{3v}{4L}$	$f_3 = \frac{5v}{4L}$	$f_4 = \frac{7v}{4L}$

Figure 16.17 The fundamental and three lowest overtones for a tube closed at one end. All have maximum air displacements at the open end and none at the closed end.

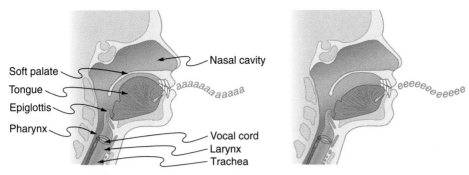

Figure 16.18   The throat and mouth form an air column closed at one end that resonates in response to vibrations in the voice box. The spectrum of overtones and their intensities vary with mouth shaping and tongue position to form different sounds. The voice box can be replaced with a mechanical vibrator, and understandable speech is still possible. Variations in basic shapes make different voices recognizable.

drumheads. In fact, much of what we vocalize is determined by shaping the throat and mouth and positioning the tongue to adjust the fundamental and combination of overtones. Simple resonant cavities can be made to resonate with the sound of the vowels, for example. (See Figure 16.18.)

Human Application

Now let us look for a pattern in the resonant frequencies for a simple tube that is closed at one end. The fundamental has $\lambda = 4L$, and frequency is related to wavelength and the speed of sound as given by

$$v_w = f\lambda$$

Solving for f gives

$$f = \frac{v_w}{\lambda} = \frac{v_w}{4L}$$

where v_w is the speed of sound in air. Similarly, the first overtone has $\lambda' = 4L/3$ (see Figure 16.17), so that

$$f' = 3\frac{v_w}{4L} = 3f$$

Since $f' = 3f$, we call the first overtone the third harmonic. Continuing this process, we see a pattern that can be generalized in a single expression. The **resonant frequencies of a tube closed at one end** are

$$f_n = n\frac{v_w}{4L}, \qquad n = 1, 3, 5, \ldots \qquad \textbf{(16.7)}$$

where f_1 is the fundamental, f_3 is the first overtone, and so on. It is interesting that the resonant frequencies depend on the speed of sound and, hence, on temperature. This poses a noticeable problem for organs in old unheated cathedrals, and it is also the reason that musicians commonly bring their wind instruments to room temperature before playing them.

EXAMPLE 16.5   HOW LONG IS A TUBE WITH A 128 HZ FUNDAMENTAL?

(a) What length should a tube closed at one end have on a day when air temperature is 22.0°C, if its fundamental frequency is to be 128 Hz (C below middle C)? (b) What is the frequency of its fourth overtone?

Strategy   Given the frequency, the length L can be found from the relationship in Equation 16.7, but we will first need to find the speed of sound v_w.

Solution for (a)   From Equation 16.7, the fundamental

frequency is $f_1 = v_w/4L$. Solving this for the length gives

$$L = \frac{v_w}{4f_1}$$

To find the speed of sound, we use Equation 16.2:

$$v_w = (331 \text{ m/s})\sqrt{\frac{T \text{ (K)}}{273 \text{ K}}} = (331 \text{ m/s})\sqrt{\frac{295 \text{ K}}{273 \text{ K}}}$$

$$= 344 \text{ m/s}$$

(continued)

(continued)

Entering the value of into the expression for L gives

$$L = \frac{v_w}{4f_1} = \frac{344 \text{ m/s}}{4(128 \text{ Hz})} = 0.672 \text{ m}$$

Discussion for (a) Many wind instruments are modified tubes that have finger holes, valves, and other devices for changing the length of the resonating air column and, hence, the frequency of the note played. Horns producing very low frequencies, like tubas, require tubes so long that they are coiled into loops.

Solution for (b) Since the first overtone has $n = 3$, and

the second has $n = 5$, we see that the fourth overtone has $n = 9$. Entering this into Equation 16.7 gives

$$f_9 = 9\frac{v_w}{4L} = 9f_1 = 1152 \text{ Hz}$$

Discussion for (b) Whether this overtone occurs in a simple tube or a musical instrument depends on how it is stimulated to vibrate and the details of its shape. The trombone, for example, does not produce its fundamental frequency and only makes overtones.

Another type of tube is one that is open at both ends. Its resonances can be analyzed in a very similar fashion to those for the tube closed at one end. The air columns in tubes open at both ends have maximum air displacements at both ends, as illustrated in Figure 16.19. Standing waves form as shown.

Based on the fact that a tube open at both ends has maximum air displacements at both ends, and using Figure 16.19 as a guide, we can see that the **resonant frequencies of a tube open at both ends** are

$$f_n = n\frac{v_w}{2L}, \qquad n = 1, 2, 3, \dots \tag{16.8}$$

CONNECTIONS

Resonance occurs in many different systems, including strings, air columns, and atoms. Resonance is the driven or forced oscillation of a system at its natural frequency. At resonance, energy is transferred rapidly to the oscillating system, and the amplitude of its oscillations grow until the system no longer obeys Hooke's law. An example of this is the distorted sound intentionally produced in certain types of rock music.

where f_1 is the fundamental, f_2 is the first overtone, f_3 is the second overtone, and so on. Note a tube open at both ends has a fundamental frequency twice what it would have if closed at one end. It also has a different spectrum of overtones than a tube closed at one end. So if you had two tubes with the same fundamental frequency but one was open at both ends and the other was closed at one end, they would sound different when played because they have different overtones. Middle C, for example, would sound richer played on an open tube, because it has even multiples of the fundamental as well as odd. A closed tube has only odd multiples.

Wind instruments use resonance in air columns to amplify tones made by lips or vibrating reeds. Other instruments also use air resonance in clever ways to amplify sound. Figure 16.20 shows a violin and a guitar, both of which have sounding boxes but with different shapes, resulting in different overtone structures. The vibrating string creates a sound that resonates in the sounding box, greatly amplifying the sound and creating

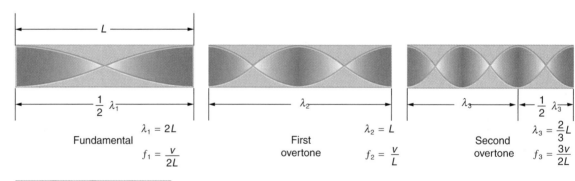

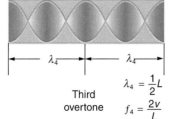

Figure 16.19 The resonant frequencies of a tube open at both ends are shown, including the fundamental and the first three overtones. In all cases the maximum air displacements occur at both ends of the tube, giving it different natural frequencies than a tube closed at one end.

(a)   (b)

Figure 16.20 These string instruments use resonance in their sounding boxes to amplify and enrich the sound created by their vibrating strings. The bridge and supports couple the string vibrations to the sounding boxes and air within.

overtones that give the instrument its characteristic flavor. The more complex the shape of the sounding box, the greater its ability to resonate over a wide range of frequencies. The primitive xylophone shown in Figure 16.21 uses pots below the wooden slats to amplify their tones. The resonance of the pot can be adjusted by adding water.

We have emphasized sound applications in our discussions of resonance and standing waves, but these ideas apply to any system that has wave characteristics. Vibrating strings, for example, are actually resonating and have fundamentals and overtones similar to those for air columns. More subtle are the resonances in atoms due to the wave character of their electrons. Their orbits can be viewed as standing waves, which have a fundamental (ground state) and overtones (excited states). We will delve into this in later chapters, but it is fascinating that wave characteristics apply to such a tremendous range of physical systems.

Figure 16.21 Resonance has been used in musical instruments since prehistoric times. This primitive xylophone uses pots as resonance chambers to amplify its sound.

16.6 HEARING

The human ear has a tremendous range and sensitivity. It can give us a wealth of simple information—such as pitch, loudness, and direction. And from its input we can detect musical quality and nuances of voiced emotion. How is our hearing related to the physical qualities of sound, and how does the hearing mechanism work? This section will explore these topics briefly.

Human & Biological Application

Hearing is the perception of sound. (Perception is commonly defined to be awareness through the senses, a typically circular definition of higher-level processes in living organisms.) Normal human hearing encompasses frequencies from 20 to 20,000 Hz, an impressive range. Sounds below 20 Hz are called **infrasound**, whereas those above 20,000 Hz are **ultrasound**. Neither is perceived. When we do hear low-frequency vibrations, like those of a diving board, we hear the individual vibrations only because there are higher-frequency sounds in each. Other animals have hearing ranges different from that of humans. Dogs can hear sounds as high as 30,000 Hz, whereas bats and dolphins can hear up to 100,000 Hz sounds. Elephants are known to respond to frequencies below 20 Hz.

The perception of frequency is called pitch, as defined in Section 16.2. Most of us have excellent relative pitch, which means that we can tell whether one sound has a different frequency from another. Typically, we can discriminate between two sounds if their frequencies differ by 0.3% or more. For example, 500.0 and 501.5 Hz are noticeably different. Pitch perception is directly related to frequency and is not greatly affected by other physical quantities such as intensity. Pitch perception is the most directly related to a physical quantity of all our hearing senses. Some people can identify musical notes, such as A-sharp, C, or E-flat, just by listening to them. This uncommon ability is called perfect pitch.

Human Applications

TABLE 16.4

SOUND PERCEPTIONS	
Perception	*Physical quantity*
Pitch	Frequency
Loudness	Intensity and frequency
Timbre	Number and relative intensity of multiple frequencies

The ear is remarkably sensitive to low-intensity sounds. The lowest audible intensity or threshold is about 10^{-12} W/m^2 or 0 dB, as discussed in Section 16.3. Sounds as much as 10^{12} more intense can be briefly tolerated without damage. Very few measuring devices are capable of observations over a range of a trillion. The perception of intensity is called **loudness**. At a given frequency, it is possible to discern differences of about 1 dB, and a change of 3 dB is easily noticed. But loudness is not related to intensity alone. Frequency has a major effect on how loud a sound seems. The ear has its maximum sensitivity to frequencies in the 2000 to 5000 Hz range, so that sounds in this range are perceived as being louder than, say, those at 500 or 10,000 Hz, even when they all have the same intensity. Sounds near the high- and low-frequency extremes of the hearing range seem even less loud, because the ear is even less sensitive at those frequencies. Table 16.4 gives the dependence of certain human hearing perceptions on physical quantities.

When a violin plays middle C, there is no mistaking it for a piano playing the same note. The reason is that each instrument produces a distinctive set of frequencies and intensities. We call our perception of these combinations of frequencies and intensities tone quality, or more commonly the **timbre** of the sound. It is more difficult to correlate timbre perception to physical quantities than it is for loudness or pitch perception. Timbre is more subjective. Terms such as dull, brilliant, warm, cold, pure, and rich are employed to describe the timbre of a sound. So the consideration of timbre takes us into the realm of perceptual psychology, where higher-level processes in the brain are dominant. This is true for other perceptions of sound, such as music and noise. We shall not delve further into them; rather, we concentrate on the question of loudness perception.

A unit called a **phon** is used to express loudness numerically. Phons differ from decibels because the phon is a unit of loudness perception, whereas the decibel is a unit of physical intensity. Figure 16.22 shows the relationship of loudness to intensity and frequency in some detail. The curved lines are equal-loudness curves. Each curve is labeled with its loudness in phons. Any sound along a given curve will be perceived as equally loud by the average person. The curves were determined by having large numbers of people compare the loudness of sounds at different frequencies and intensities. At a frequency of 1000 Hz, phons are taken to be numerically equal to decibels. The following example helps illustrate how to use the graph.

EXAMPLE 16.6 LOUDNESS VERSUS INTENSITY AND FREQUENCY

(a) What is the loudness in phons of a 100 Hz sound that has an intensity of 80 dB? (b) What is the intensity in decibels of a 4000 Hz sound having a loudness of 70 phons? (c) At what intensity will a 8000 Hz sound have the same loudness as a 200 Hz sound at 60 dB?

Strategy and Solution for (a) The square grid of the graph relating phons and decibels in Figure 16.22 is a plot of intensity versus frequency, both physical quantities. To find the loudness of a given sound, you must know its frequency and intensity and locate that point on the square grid. Then interpolate between loudness curves to get the loudness in phons. In this case, 100 Hz at 80 dB lies halfway between the curves marked 70 and 80 phons, and so its loudness is *75 phons.*

Strategy and Solution for (b) To find the intensity

of a sound, you must have its frequency and loudness. Once that point is located, the intensity can be determined from the vertical axis. In this case, these are given to be 4000 Hz at 70 phons. Follow the 70 phon curve until it reaches 4000 Hz. At that point, it is below the 70 dB line at about *67 dB*, so that is its intensity.

Strategy and Solution for (c) First, locate the point for a 200 Hz and 60 dB sound. It lies just slightly above the 50 phon curve, and so its loudness is 51 phons. Then see where the 51 phon level is at 8000 Hz. It lies at *63 dB*.

Discussion These answers, like all information extracted from Figure 16.22, have uncertainties of several phons or several decibels, partly due to difficulties in interpolation, but mostly related to uncertainties in the equal-loudness curves.

Further examination of the graph in Figure 16.22 reveals some interesting facts about human hearing. First, sounds below the 0 phon curve are not perceived by most people. So, for example, a 60 Hz sound at 40 dB is inaudible. The 0 phon curve represents the threshold of normal hearing.* We can hear some sounds at intensities below 0

---

*It is actually the threshold for the one percentile of the population with the best hearing. But most young people can hear sounds at or very near 0 phons, while 50% of all people can hear sounds above 15 phons.

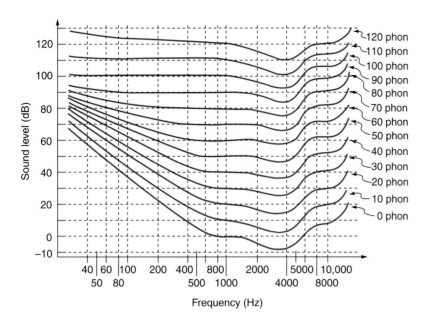

Figure 16.22 The relationship of loudness in phons to intensity in decibels for persons with normal hearing. The curved lines are equal-loudness curves—all sounds on a given curve are perceived as equally loud. Phons and decibels are defined to be the same at 1000 Hz.

dB. For example, a −3 dB, 5000 Hz sound is audible, since it lies above the 0 phon curve. The loudness curves all have dips in them between about 2000 and 5000 Hz. This means the ear is most sensitive to frequencies in that range. For example, a 15 dB sound at 4000 Hz has a loudness of 20 phons, the same as a 20 dB sound at 1000 Hz. The curves rise at both extremes of the frequency range, indicating that a greater-intensity sound is needed at those frequencies to be perceived to be as loud as at middle frequencies. For example, a sound at 10,000 Hz must have an intensity of 30 dB to seem as loud as a 20 dB sound at 1000 Hz. Sounds above 120 phons are painful as well as damaging.

We do not often utilize our full range of hearing. This is particularly true for frequencies above 8000 Hz, which are rare in the environment and are unnecessary for understanding conversation or appreciating music. In fact, people who have lost the ability to hear such high frequencies are usually unaware of their loss until tested. The shaded region in Figure 16.23 is the frequency and intensity region where most conversational sounds fall. The curved lines indicate what effect hearing losses of 40 and 60 phons will have. A 40 phon hearing loss at all frequencies still allows a person to understand conversation, although it will seem very quiet. A person with a 60 phon loss at all frequencies will hear only the lowest frequencies and will not be able to understand speech unless it is much louder than normal. Even so, speech may seem indistinct, because higher frequencies are not as well perceived.

Human Application

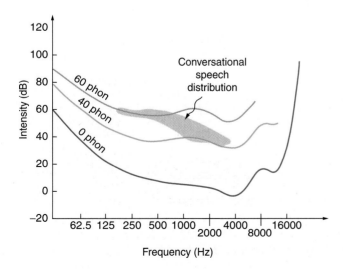

Figure 16.23 The shaded region represents frequencies and intensities found in normal conversational speech. The 0 phon line represents the normal hearing threshold, while those at 40 and 60 represent thresholds for people with 40 and 60 phon hearing losses.

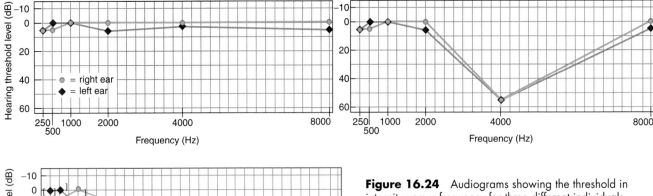

Figure 16.24 Audiograms showing the threshold in intensity versus frequency for three different individuals. Intensity is measured relative to the normal threshold. The top left graph is that of a person with normal hearing. The graph to its right has a dip at 4000 Hz and is that of a child who suffered hearing loss due to a cap gun. The third graph is typical of presbycusis, the progressive loss of higher frequency hearing with age. Tests performed by bone conduction (brackets) can distinguish nerve damage from middle ear damage.

Hearing tests are performed over a range of frequencies, usually from 250 to 8000 Hz, and can be displayed graphically in an audiogram like that in Figure 16.24. The hearing threshold is measured in dB *relative to the normal threshold*, so that normal hearing registers as 0 dB at all frequencies. Hearing loss caused by noise typically shows a dip near the frequency that caused the loss. The most common form of hearing loss comes with age and is called *presbycusis*—literally elder ear. Such loss is increasingly severe at higher frequencies, and interferes with music appreciation and speech recognition.

The Hearing Mechanism

Human & Biological Application

The hearing mechanism involves some interesting physics. The ear is a transducer that converts sound waves into nerve impulses in a manner much more sophisticated than, but analogous to, a microphone. Figure 16.25 shows the gross anatomy of the ear with its division into three parts: the outer ear or ear canal; the middle ear, which runs from the eardrum to the cochlea; and the inner ear, which is the cochlea itself. The thing you normally call an ear is technically called the pinna.

The outer ear, or ear canal, carries sound to the recessed protected eardrum. The air column in the ear canal resonates and is partially responsible for the sensitivity of the ear

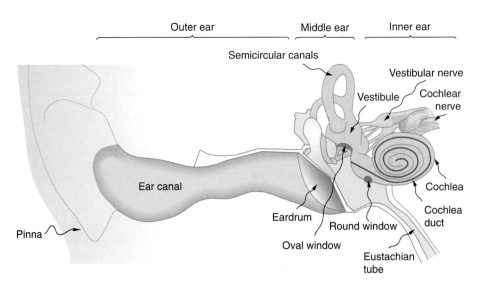

Figure 16.25 Gross structure of the human ear.

Figure 16.26 Schematic of the middle ear's system for converting sound pressure into force, increasing that force through a lever system, and applying the increased force to a small area of the cochlea, thereby creating a pressure about 40 times that in the original sound wave. A protective muscle reaction to intense sounds greatly reduces the mechanical advantage of the lever system.

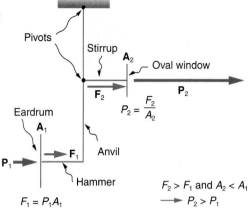

to sounds in the 2000 to 5000 Hz range. The middle ear converts sound into mechanical vibrations and applies these to the cochlea. There is a lever system that takes the force exerted on the eardrum by sound pressure variations and increases it. This increased force is then applied to an area of the cochlea smaller than the eardrum, creating pressure waves in the cochlea some 40 times greater than those impinging on the eardrum. (See Figure 16.26.) Two muscles in the middle ear (not shown) protect the inner ear from very intense sounds. They react to intense sound in a few milliseconds and reduce the force transmitted to the cochlea. This protective reaction can also be triggered by your own voice, so that humming while shooting a gun, for example, can reduce noise damage.

Figure 16.27 shows the middle and inner ear in greater detail. Pressure waves moving through the cochlea cause the tectorial membrane to vibrate, rubbing cilia (sometimes called hairs), which stimulate nerves that send signals to the brain. The membrane resonates at different positions for different frequencies, with high frequencies stimulating nerves at the near end and low frequencies at the far end. The complete operation of the cochlea is still not understood, but several mechanisms for sending information to the brain are known to be involved. For sounds below about 1000 Hz, the nerves send signals at the same frequency as the sound. For frequencies greater than about 1000 Hz, the nerves signal frequency by position. There is a structure to the cilia, and there are connections between nerve cells that perform signal processing before information is sent to the brain. Intensity information is partly indicated by the number of nerve signals and by volleys of signals. The brain processes the cochlear nerve signals to provide additional information such as source direction (based on time and intensity comparisons of sounds from both ears). Higher-level processing produces many nuances, such as music appreciation.

Hearing losses can occur because of problems in the middle or inner ear. Conductive losses in the middle ear can be partially overcome by sending sound vibrations to the cochlea through the skull. Hearing aids for this purpose usually press against the bone behind the ear, rather than simply amplifying the sound sent into the ear canal as many hearing aids do. Damage to the nerves in the cochlea is not repairable, but amplification can partially compensate. There is a risk that amplification will produce further damage. Another common failure in the cochlea is damage or loss of the cilia but with nerves remaining functional. Cochlear implants that stimulate the nerves directly are being developed and tested with some success.

Biological Application

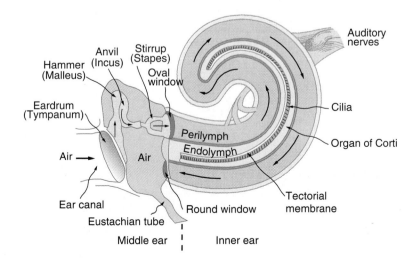

Figure 16.27 Schematic of the middle and inner ears. The inner ear, or cochlea, is a coiled tube about 3 mm in diameter and 3 cm in length if uncoiled. When the oval window is forced inward, as shown, the pressure wave travels through the perilymph in the direction of the arrows, stimulating nerves at the base of cilia in the organ of Corti.

Medical Applications

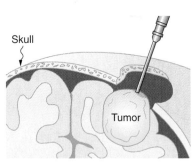

Skull

Tumor

Figure 16.28 The tip of this small probe oscillates at 23 kHz with such a large amplitude that it pulverizes tissue on contact. The debris is then aspirated. The speed of the tip may exceed the speed of sound in tissue, thus creating shock waves and cavitation, rather than a smooth simple harmonic oscillator–type wave.

16.7 ULTRASOUND

Any sound with a frequency above 20,000 Hz (or 20 kHz)—that is, above the highest audible frequency—is defined to be ultrasound. In practice, it is possible to create ultrasound frequencies up to more than a gigahertz. (Higher frequencies are difficult to create; furthermore, they propagate poorly because they are very strongly absorbed.) Ultrasound has a tremendous number of applications, which range from burglar alarms to the guidance systems of bats. We begin our discussion of ultrasound with some of its applications in medicine.

Ultrasound in Medical Therapy

Ultrasound, like any wave, carries energy that can be absorbed by the medium carrying it, producing effects that vary with intensity. Focused to intensities of 10^3 to 10^5 W/m^2, ultrasound can shatter gallstones or pulverize cancerous tissue in surgical procedures. (See Figure 16.28.) Intensities this great can damage individual cells, variously causing their protoplasm to stream inside them, altering their permeability, or rupturing their walls through cavitation. Cavitation is the creation of vapor cavities in a fluid—the longitudinal vibrations in ultrasound alternatively compress and expand the medium, and at sufficient amplitudes the expansion separates molecules. Most cavitation damage is done when the cavities collapse, producing even greater shock pressures.

Most of the energy carried by high-intensity ultrasound in tissue is converted to thermal energy. In fact, intensities of 10^3 to 10^4 W/m^2 are commonly used for deep-heat treatments called **ultrasound diathermy**. Frequencies of 0.8 to 1 MHz are typical. In both athletics and physical therapy, ultrasound diathermy is most often applied to injured or overworked muscles to relieve pain and improve flexibility. Skill is needed by the therapist to avoid "bone burns" and other tissue damage caused by overheating and cavitation, sometimes made worse by reflection and focusing of the ultrasound by joint and bone tissue.

Ultrasound in Medical Diagnostics

The applications of ultrasound in medical diagnostics have produced untold benefits with no known risks. Diagnostic intensities are too low (about 10^{-2} W/m^2) to cause thermal damage. More significantly, ultrasound has been in use for several decades and detailed follow-up studies do not show evidence of ill effects, quite unlike the case for x rays.

Even simple ultrasonic echoes provide useful information. Echoes are produced when ultrasound is partially reflected by surfaces between different tissues, especially between those having different densities. (See Figure 16.29.) The speed of sound in different tissues is nearly constant. Thus the time for each echo to return to the source is directly proportional to the distance of the surface from the source. Such simple distance information can reveal the presence of a foreign object in the eye, or the distance to and thickness of various tissue layers and organs.

The most common ultrasound applications produce an image like that in Figure 16.30(b). The speaker-microphone broadcasts a directional beam, sweeping the beam across the area of interest. This is accomplished by having multiple ultrasound sources in the probe's head, which are phased to interfere constructively in a given, adjustable direction.

Figure 16.29 (a) An ultrasound speaker doubles as a microphone. Brief bleeps are broadcast, and echoes are recorded from various depths. (b) Graph of echo intensity versus time. The time for echoes to return is directly proportional to the distance of the reflector, yielding this information noninvasively.

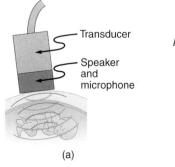

Transducer

Speaker and microphone

(a)

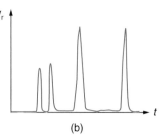

I_r

t

(b)

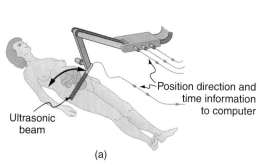

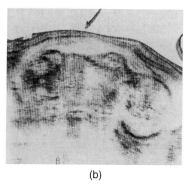

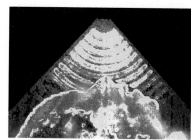

(a)

(b)

Figure 16.31 A state of the art ultrasonic image shows the eyelashes of a fetus. Cleft palates, spinal defects, and other information can be reliably obtained from such images.

Figure 16.30 (a) An ultrasonic image is produced by sweeping the ultrasonic beam across the area of interest, in this case the woman's abdomen. Data is recorded and analyzed in a computer, providing a two-dimensional image. (b) Ultrasonic scan of Christopher M. Urone, age –3 weeks.

Echoes are measured as a function of position as well as depth. A computer constructs an image that reveals the shape and density of internal structures.

How much detail can ultrasound reveal? The image in Figure 16.30(b) is typical of low-cost systems, but that in Figure 16.31 shows the remarkable detail possible with more advanced systems. Whenever a wave is used as a probe, it is very difficult to detect details smaller than its wavelength λ. Indeed, current technology cannot do quite this well. Abdominal scans may use a 7 MHz frequency, and the speed of sound in tissue is about 1540 m/s—so the wavelength limit to detail would be $\lambda = v_w/f = (1540 \text{ m/s})/(7 \times 10^6$ Hz$) = 0.22$ mm. In practice, 1 mm detail is attainable, which is sufficient for many purposes. Higher-frequency ultrasound would allow greater detail, but it does not penetrate as well as lower frequencies do. The accepted rule of thumb is that you can effectively scan to a depth of about 500λ into tissue. For 7 MHz, this penetration limit is 500×0.22 mm, which is 0.11 m. Higher frequencies may be employed in smaller organs, such as the eye, but are not practical for looking deep into the body.

In addition to shape information, ultrasonic scans can produce density information superior to that found in x rays. This is because the intensity of a reflected sound is related to changes in density. Sound is most strongly reflected at places where density changes are greatest.

Another major use of ultrasound in medical diagnostics is to detect motion and determine velocity through the Doppler shift of an echo, known as **Doppler-shifted ultrasound**. It is used to monitor fetal heartbeat, measure blood velocity, and detect occlusions in blood vessels, for example. (See Figure 16.32.) The magnitude of the Doppler shift in an echo is directly proportional to the velocity of whatever reflects the sound. Since an echo is involved, there is actually a double shift. The first occurs because the reflector (say a fetal heart) is a moving observer and receives a Doppler-shifted frequency. The reflector then acts as a moving source, producing a second Doppler shift.

A clever technique is used to measure the Doppler shift in an echo. The frequency of the echoed sound is superimposed on the broadcast frequency, producing beats. Beat frequency is $f_B = |f_1 - f_2|$, and so it is directly proportional to the Doppler shift $(f_1 - f_2)$ and, hence, the reflector's velocity. The advantage in this technique is that the Doppler shift is small, since the reflector's velocity is small, so that great accuracy would be needed to measure the shift directly. But measuring the beat frequency is easy, and it is not affected if the broadcast frequency varies somewhat. Furthermore, the beat frequency is in the audible range and can be amplified for audio feedback to the medical observer.

Medical Applications

Figure 16.32 This Doppler-shifted ultrasonic image of a partially occluded artery uses color to indicate velocity. The highest velocities are in red, while the lowest are blue. The blood must move faster through the constriction to carry the same flow.

CONNECTIONS

Doppler-shifted radar echoes are used to measure wind velocities in storms as well as aircraft and automobile speeds. The principle is the same as for Doppler-shifted ultrasound. There is evidence that bats and dolphins may also sense the velocity of the object (such as prey) reflecting their ultrasound signals by observing its Doppler shift.

EXAMPLE 16.7 DOPPLER-SHIFTED ULTRASOUND REVEALS BLOOD VELOCITY

Ultrasound with frequency 2.50 MHz is sent toward blood in an artery that is moving toward the source at 20.0 cm/s, as illustrated in Figure 16.33. (a) What frequency does the blood receive? (b) What frequency returns to the source?

(c) What beat frequency is produced if the source and returning frequencies are mixed?

Strategy The first two questions can be answered using

(continued)

(continued)

Equations 16.5 and 16.6 for the Doppler shift. The last question asks for beat frequency, which is the difference of the original and returning frequencies.

Solution for (a) The blood is a moving observer, and so the frequency it receives is given by Equation 16.6:

$$f_{obs} = f_s\left(\frac{v_w + v_b}{v_w}\right)$$

where v_b is the blood velocity (v_{obs}, here), and the plus sign is chosen because the motion is toward the source. Entering given values,

$$f_{obs} = (2,500,000 \text{ Hz})\left(\frac{1540 \text{ m/s} + 0.2 \text{ m/s}}{1540 \text{ m/s}}\right)$$

$$= 2,500,325 \text{ Hz}$$

Solution for (b) Now the blood acts as a moving source and the microphone as a stationary observer. The frequency leaving the blood is 2,500,325 Hz, but it is shifted upward as given by Equation 16.5:

$$f'_{obs} = f_{obs}\left(\frac{v_w}{v_w - v_b}\right)$$

where f'_{obs} is the frequency received by the speaker-microphone, the source velocity is v_b, and the minus sign is used because the motion is toward the observer. Now,

$$f'_{obs} = (2,500,325 \text{ Hz})\left(\frac{1540 \text{ m/s}}{1540 \text{ m/s} - 0.200 \text{ m/s}}\right)$$

$$= 2,500,649 \text{ Hz}$$

Solution for (c) The beat frequency is simply the absolute value of the difference between f_s and f'_{obs}, as stated in Equation 15.12. Thus,

$$f_B = \left|f'_{obs} - f_s\right| = \left|2,500,649 \text{ Hz} - 2,000,000 \text{ Hz}\right|$$

$$= 649 \text{ Hz}$$

Discussion The Doppler shifts are quite small compared with the original frequency of 2.50 MHz. It is far easier to measure the beat frequency than it is to measure the echo frequency with an accuracy great enough to see shifts of a few hundred hertz out of a couple of megahertz. Furthermore, variations in the source frequency do not greatly affect the beat frequency, since both f_s and f'_{obs} would increase or decrease. Those changes subtract out in $f_B = \left|f'_{obs} - f_s\right|$.

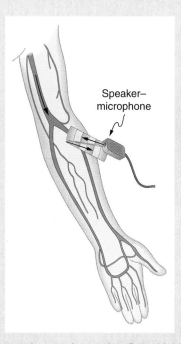

Figure 16.33 Ultrasound is partly reflected by blood cells and plasma back toward the speaker-microphone. Since the cells are moving, two Doppler shifts are produced—one for blood as a moving observer, and the other for the reflected sound coming from a moving source. The magnitude of the shift is directly proportional to blood velocity.

Industrial and Other Applications of Ultrasound

Industrial, retail, and research applications of ultrasound are common. A few are discussed here. Ultrasonic cleaners have many uses. Jewelry, machined parts, and other objects with odd shapes and crevices are immersed in a cleaning fluid that is agitated with ultrasound typically about 40 kHz in frequency. The intensity is great enough to cause cavitation, which is responsible for most of the cleansing action. Since cavitation-produced shock pressures are large and well transmitted in a fluid, they reach into small crevices where even a low surface tension cleaning fluid might not penetrate.

Sonar is a familiar application of ultrasound. Sonar typically employs ultrasonic frequencies in the range from 30.0 to 100 kHz. Bats, dolphins, submarines, and even some birds use ultrasonic sonar. Echoes are analyzed to give distance and size information both for guidance and finding prey. In most sonar applications, the sound reflects quite well because the objects of interest have significantly different density than the medium traveled in. When the Doppler shift is observed, velocity information can also be obtained. Submarines sonar does this, and there is evidence that some bats also sense velocity from their echoes. (See Figure 16.34.)

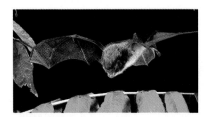

Figure 16.34 This bat has used ultrasound to locate the insect it is about to catch and eat.

Similarly, there are a range of relatively inexpensive devices that measure distance by timing ultrasonic echoes. Many cameras, for example, use such information to focus automatically. Some doors open when their ultrasonic ranging devices detect a close enough object, and certain home security lights turn on when their ultrasonic rangers observe motion. Ultrasonic "measuring tapes" also exist to measure such things as room dimensions. Sinks in public restrooms are sometimes automated with ultrasound devices to turn on and off when people wash their hands. This avoids the spread of germs and can conserve water.

Ultrasound is used for nondestructive testing in industry and by the military. Since ultrasound reflects well from any large change in density, it can reveal cracks and voids in solids, such as aircraft wings, that are too small to be seen with x rays. For similar reasons, ultrasound is also good for measuring the thickness of coatings, particularly where there are several layers involved.

Basic research in solid state physics employs ultrasound. Its attenuation is related to a number of physical characteristics, making it a useful probe. Among these are structural changes like those found in liquid crystals, the transition of a material to a superconducting phase, as well as density and other properties.

These examples of the uses of ultrasound are meant to whet the appetites of the curious, as well as to illustrate the underlying physics of ultrasound. There are many more applications, as you can easily discover for yourself.

SUMMARY

Sound is a disturbance of matter that is transmitted from its source outward and is one type of wave. **Hearing** is the perception of sound.

The relationship of the speed of sound v_w, its frequency f, and its wavelength λ is

$$v_w = f\lambda \tag{16.1}$$

which is the same relationship given for all waves in Equation 15.11. Table 16.1 lists the speed of sound in various media. In air, the speed of sound is related to air temperature T by

$$v_w = (331 \text{ m/s})\sqrt{\frac{T \text{ (K)}}{273 \text{ K}}} \tag{16.2}$$

v_w is the same for all frequencies and wavelengths.

Intensity is the same for a sound wave as was defined in Equation 15.13 for all waves; it is

$$I = \frac{P}{A} \tag{16.3}$$

where P is the power crossing area A. The SI unit for I is W/m^2. **Sound level** in units of **decibels** is

$$\beta \text{ (dB)} = 10 \log_{10}\left(\frac{I}{I_0}\right) \tag{16.4}$$

where $I_0 = 10^{-12} \text{ W/m}^2$ is the threshold intensity of hearing. Table 16.2 relates sound levels and intensities for familiar sounds. Table 16.3 gives relationships between ratios of intensities and differences in sound levels.

The **Doppler effect** is an alteration in the observed frequency of a sound due to motion of either the source or the observer. The actual change in frequency is called the **Doppler shift**. For a stationary observer and a moving source, the observed frequency f_{obs} is

$$f_{obs} = f_s\left(\frac{v_w}{v_w \mp v_s}\right) \tag{16.5}$$

where f_s is the frequency of the source, v_s is the speed of the source, and v_w is the speed of sound. The minus sign is used for motion toward the observer and the plus for motion away. Similarly, for a stationary source and moving observer,

$$f_{obs} = f_s\left(\frac{v_w \pm v_{obs}}{v_w}\right) \tag{16.6}$$

where v_{obs} is the speed of the observer.

A **sonic boom** is a constructive interference of sound created by an object moving faster than sound. It is a type of **bow wake** created when any wave source moves faster than the wave propagation speed.

Sound interference and resonance have the same properties as defined for all waves in Chapter 15. In air columns, the lowest-frequency resonance is called the **fundamental**, whereas all higher resonant frequencies are called **overtones**. Collectively, they are called **harmonics**. The **resonant frequencies of a tube closed at one end** are

$$f_n = n\frac{v_w}{4L}, \qquad n = 1, 3, 5, \ldots \tag{16.7}$$

where f_1 is the fundamental and L is the length of the tube. The **resonant frequencies of a tube open at both ends** are

$$f_n = n\frac{v_w}{2L}, \qquad n = 1, 2, 3, \ldots \tag{16.8}$$

The range of audible frequencies is 20 to 20,000 Hz.

Those above 20,000 Hz are **ultrasound**, whereas those below 20 Hz are **infrasound**. The perception of frequency is **pitch**, and the perception of intensity is **loudness**. Loudness has units of **phons**. Table 16.4 gives relationships between sound perceptions and physical characteristics. The relationship between loudness, intensity, and frequency is given in graphical form in Figure 16.22.

Applications of ultrasound are explored in Section 16.7.

CONCEPTUAL QUESTIONS

16.1 How do sound vibrations of atoms differ from thermal motion?

16.2 When sound passes from one medium to another where its speed is different, does its frequency or wavelength change? Explain briefly.

16.3 The speed of sound in human tissue is the same as in sea water. Why would you expect this? Would you expect the speed of sound to be the same in bone?

16.4 If a sound source broadcasts uniformly in all directions and there is no absorption or reflection of the sound, its intensity will decrease with distance from the source squared. Explain why.

16.5 Is the Doppler shift real, or just a sensory illusion?

16.6 Doppler shifts vary slightly with temperature. Explain why.

16.7 Due to efficiency considerations related to its bow wake, the supersonic transport aircraft must maintain a cruising speed that is a constant ratio to the speed of sound (a constant Mach number). If it flies from warm air into colder air, should it increase or decrease its speed? Explain.

16.8 When you hear a sonic boom, you often cannot see the plane that made it. Why is that?

16.9 While driving a car with a broken muffler to the repair shop, you notice that the noise inside the car is particularly loud at certain engine speeds. Explain what is happening to cause this effect.

16.10 While consuming a soft drink from a bottle, you pause to blow over the top to make it toot. Explain why the frequency of the toot changes as the bottle is emptied. How is blowing over the top able to cause the bottle to resonate at so many different frequencies? (Hint: Turbulence is noisy.)

16.11 How does an unamplified guitar produce sounds so much more intense than those of a plucked string held taut by a simple stick?

16.12 Sound system speakers are designed *not* to resonate at any audible frequency. Explain why.

16.13 Operatic sopranos are reputed to be able to break a glass by singing at it. What is the mechanism, and what sound characteristics—such as frequency, sound velocity, and intensity—are crucial to this task?

16.14 What is the difference between an overtone and a harmonic? Are all harmonics overtones? Are all overtones harmonics?

16.15 You are given two wind instruments of identical length. One is open at both ends, whereas the other is closed at one end. Which is able to produce the lowest frequency?

16.16 Suppose music is recorded at a certain intensity and the recording is later played at lower intensity. Why is it desirable to amplify the low-frequency sounds more than others? Note that most CD and tape players have a "loudness" adjustment that does this for you.

16.17 Why can a hearing test show that your threshold of hearing is 0 dB at 250 Hz, when Figure 16.22 implies that no one can hear such a frequency at less than 20 dB?

16.18 Give an example of a sound that is intense but not loud.

16.19 How are pitch and frequency related? How do they differ?

16.20 What physical characteristics of sound affect loudness? Is loudness a physical characteristic of sound?

16.21 The ear is not very sensitive to low-frequency sounds, such as those made by the heart. How does a stethoscope, like the one shown in Figure 16.35, both increase the amount of sound coming out of the chest and increase its intensity as it passes from the bell to the ear?

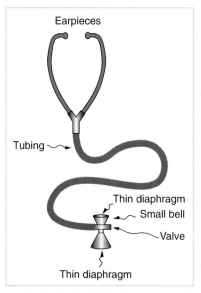

Figure 16.35 A stethoscope is used to improve transmission of sound out of the body and to amplify it. Question 21.

16.22 If audible sound follows a rule of thumb similar to that for ultrasound, in penetrating a certain number of wavelengths before being absorbed, would you expect the high or low frequencies from your neighbor's stereo to penetrate into your house? How does this compare with your experience?

16.23 When using Doppler-shifted ultrasound to locate partial occlusions in blood vessels, you observe the highest shift at the occlusion. Explain in terms of the relationship between flow rate and fluid velocity.

16.24 Submarine ultrasound sonar produces strong echoes from rocks, ships, and other submarines, but weak echoes from thermal currents. Explain why.

16.25 Ultrasonic cleaners have their greatest effect on the outside of the object, with relatively little ultrasound getting into the object to strain it internally. Why is this?

16.26 The frequencies used in ultrasound diathermy are low enough that the sound penetrates to the other side of the body. Explain why most of the ultrasound is reflected back into the body and does not escape.

16.27 Why is it that sound waves in a loudspeaker are so poorly transmitted to the air? (This explains why loudspeakers have such a low efficiency in converting their vibrational energy into sound in air.)

PROBLEMS

Section 16.2 Speed of Sound, Frequency, and Wavelength

16.1 When poked by a spear, an operatic soprano lets out a 1200 Hz shriek. What is its wavelength if the speed of sound is 345 m/s?

16.2 What frequency sound has a 0.100 m wavelength when the speed of sound is 340 m/s?

16.3 Calculate the speed of sound on a day when a 1500 Hz frequency has a wavelength of 0.221 m.

16.4 (a) What is the speed of sound in a medium where a 100 kHz frequency produces a 5.96 cm wavelength? (b) Which substance in Table 16.1 is this likely to be?

16.5 Show that the speed of sound in 20.0°C air is 343 m/s, as claimed in the text.

16.6 Air temperature in the Sahara desert can reach 56.0°C (about 134°F). What is the speed of sound in air at that temperature?

• **16.7** Dolphins make sounds in air and water. What is the ratio of the wavelength of a sound in air to its wavelength in water? Assume air temperature is 20.0°C.

• **16.8** A sonar echo returns to a submarine 1.20 s after being emitted. What is the distance to the object creating the echo?

• **16.9** If a submarine's sonar can measure echo times with a precision of 0.0100 s, what is the smallest difference in distances it can detect?

⁞ **16.10** A physicist at a fireworks display times the lag between seeing an explosion and hearing its sound, and finds it to be 0.400 s. (a) How far away is the explosion if air temperature is 24.0°C, neglecting the time taken for light to reach the physicist? (b) Calculate the distance to the explosion taking the speed of light into account. Note that this is negligibly greater.

Section 16.3 Sound Intensity and Sound Level

16.11 The warning tag on a lawn mower states that it produces noise at a level of 91.0 dB. What is this in W/m^2?

16.12 What sound level in dB is produced by earphones that create an intensity of 4.00×10^{-2} W/m^2?

16.13 Show that an intensity of 10^{-12} W/m^2 is the same as 10^{-16} W/cm^2, as stated following Table 16.2.

16.14 (a) What is the decibel level of a sound that is twice as intense as a 90.0 dB sound? (b) One that is one-fifth as intense?

16.15 (a) What is the intensity of a sound that has a level 7.00 dB lower than a 4.00×10^{-9} W/m^2 sound? (b) One that is 3.00 dB higher?

• **16.16** (a) How much more intense is a sound that has a level 17.0 dB higher than another? (b) If one sound has a level 23.0 dB less than another, what is the ratio of their intensities?

• **16.17** People with good hearing can perceive sounds as low in level as −8.00 dB at a frequency of 3000 Hz. What is the intensity of this sound in W/m^2?

• **16.18** If a large housefly 3.0 m away from you makes a noise of 40.0 dB, what is the noise level of 1000 flies at that distance, assuming interference has a negligible effect?

• **16.19** Ten cars in a circle at a boom box competition produce a 120 dB sound level at the center of the circle. What is the average sound level produced there by each, assuming interference effects can be neglected?

• **16.20** The amplitude of a sound wave is measured in terms of its maximum gauge pressure. By what factor does the amplitude of a sound wave increase if the sound level goes up by 40.0 dB?

• **16.21** If a sound level of 0 dB at 1000 Hz corresponds to a maximum gauge pressure (sound amplitude) of 10^{-9} atm, what is the maximum gauge pressure in a 60 dB sound? A 120 dB sound?

⁞ **16.22** An 8 hour exposure to a sound level of 90.0 dB may cause hearing damage. What energy in joules falls on a 0.800 cm diameter eardrum so exposed?

⁞ **16.23** Ear trumpets were never very common, but they did aid people with hearing losses by gathering sound over a large area and concentrating it on the smaller area of the eardrum. What decibel increase does an ear trumpet produce if its sound gathering area is 900 cm^2, the area of the eardrum is 0.500 cm^2, but it only has an efficiency of 5.00% in transmitting the sound to the eardrum?

⁞ **16.24** Sound is more effectively transmitted into a stethoscope by direct contact than through the air, and it is further intensified by being concentrated on the smaller area of the eardrum. It is reasonable to assume that sound is transmitted into a stethoscope 100 times as effectively compared with transmission though the air. What, then, is the gain in decibels produced by a stethoscope that has a sound gathering area of 15.0 cm^2, and concentrates the sound onto two eardrums with a total area of 0.900 cm^2 with an efficiency of 40.0%?

⁞ **16.25** Loudspeakers can produce intense sounds with surprisingly small energy input in spite of their low efficiencies. Calculate the power input needed to produce a 90.0 dB sound level for a 12.0 cm diameter speaker that has an efficiency of 1.00%. (This is the sound level right at the speaker.)

Section 16.4 Doppler Effect

16.26 (a) What frequency is received by a person watching an oncoming ambulance moving at 110 km/h and emitting a steady 800 Hz sound from its siren? The speed of sound on this day is 345 m/s. (b) What frequency does she receive after the ambulance has passed?

16.27 (a) At an air show a jet flies directly toward the stands at a speed of 1200 km/h, emitting a frequency of 3500 Hz, on a day when the speed of sound is 342 m/s. What frequency is received by the observers? (b) What frequency do they receive as the plane flies directly away from them?

16.28 What frequency is received by a mouse just before being dispatched by a hawk flying at it at 25.0 m/s and emitting a screech of frequency 3500 Hz? Take the speed of sound to be 331 m/s.

• **16.29** A spectator at a parade receives an 888 Hz tone from an oncoming trumpeter who is playing an 880 Hz note. At what speed is the musician approaching if the speed of sound is 338 m/s?

• **16.30** A light-rail commuter train blows its 200 Hz horn as it approaches a crossing. The speed of sound is 335 m/s. (a) An observer waiting at the crossing receives a frequency of 208 Hz. What is the speed of the train? (b) What frequency does the observer receive as the train moves away?

• **16.31** Can you perceive the shift in frequency produced when you pull a tuning fork toward you at 10.0 m/s on a day when the speed of sound is 344 m/s? To answer this, calculate the factor by which the frequency shifts and see if it is greater than 0.300%.

⁞ **16.32** Two eagles fly directly toward one another, the first at 15.0 m/s and the second at 20.0 m/s. Both screech, one emitting a frequency of 3200 Hz and the other one of 3800 Hz. What frequencies do they receive if the speed of sound is 330 m/s?

⁞ **16.33** What is the minimum speed at which a source must travel toward you for you to be able to hear that its frequency is Doppler shifted? That is, what speed produces a shift of 0.300% on a day when the speed of sound is 331 m/s?

Section 16.5 Resonances and Beats

16.34 A custom car built for pretentious people has two brass horns that are supposed to produce the same frequency but actually emit 263.8 and 264.5 Hz. What beat frequency is produced?

16.35 What beat frequencies will be present: (a) If the musical notes A and C are played together (frequencies of 220 and 264 Hz)? (b) If D and F are played together (frequencies of 297 and 352 Hz)? (c) If all four are played together?

16.36 What beat frequencies result if a piano hammer hits three strings that emit frequencies of 127.8, 128.1, and 128.3 Hz?

16.37 A piano tuner hears a beat every 2.00 s when listening to a 264.0 Hz tuning fork and a single piano string. What are the two possible frequencies of the string?

16.38 (a) What is the fundamental frequency of a 0.672 m long tube, open at both ends, on a day when the speed of sound is 344 m/s? (b) What is the frequency of its second harmonic?

16.39 If a wind instrument, like a tuba, has a fundamental frequency of 32.0 Hz, what are its first three overtones? It is closed at one end.

16.40 What are the first three overtones of a bassoon that has a fundamental frequency of 90.0 Hz? It is open at both ends.

16.41 How long must a flute be in order to have a fundamental frequency of 262 Hz (this is middle C on the evenly tempered chromatic scale) on a day when air temperature is 20.0°C? It is open at both ends.

16.42 What length should an oboe have in order to produce a fundamental frequency of 110 Hz on a day when the speed of sound is 343 m/s? It is open at both ends.

• **16.43** What is the length of a tube that has a fundamental frequency of 176 Hz and a first overtone of 352 Hz if the speed of sound is 343 m/s?

• **16.44** (a) Find the length of an organ pipe closed at one end that produces a fundamental frequency of 256 Hz when air temperature is 18.0°C. (b) What is its fundamental frequency at 25.0°C?

• **16.45** By what fraction will the frequencies produced by a wind instrument change when air temperature goes from 10.0°C to 30.0°C? That is, find the ratio of the frequencies at those temperatures.

• **16.46** The ear canal resonates like a tube closed at one end. If ear canals range in length from 1.80 to 2.60 cm in an average population, what is the range of resonant frequencies? Take air temperature to be 37.0°C, the same as body temperature.

• **16.47** Calculate the first overtone in an ear canal, which resonates like a 2.40 cm long tube closed at one end, taking air temperature to be 37.0°C. Is the ear particularly sensitive to such a frequency?

⁞ **16.48** A crude approximation of voice production is to consider the breathing passages and mouth to be a resonating tube closed at one end. (a) What is the fundamental frequency if the tube is 0.240 m long, taking air temperature to be 37.0°C? (b) What would this frequency become if the person replaced the air with helium? Assume the same temperature dependence for helium as air.

⁞ **16.49** Students in a physics lab are asked to find the length of a tube that has a fundamental frequency of 256 Hz. They ring a 256 Hz tuning fork and hold it over a vertical tube partially filled with water. The level of the water is varied until resonance occurs. What is the air temperature if the resonance occurs for a length of 0.336 m?

⁞ **16.50** What frequencies will a 1.80 m long tube produce in the audible range at 20.0°C if: (a) The tube is closed at one end? (b) It is open at both ends?

Section 16.6 Hearing

16.51 The factor of 10^{12} in the range of intensities to which the ear can respond, from threshold to that causing damage after brief exposure, is truly remarkable. If you could measure distances over the same range with a single instrument and the smallest distance you could measure was 1 mm, what would the largest be?

16.52 The frequencies to which the ear responds vary by a factor of 10^3. Suppose the speedometer on your car measured speeds differing by the same factor of 10^3, and the greatest speed it reads is 90.0 mi/h. What would be the slowest nonzero speed it could read?

16.53 What are the closest frequencies to 500 Hz that an average person can clearly distinguish as being different in frequency from 500 Hz? The sounds are not present simultaneously.

16.54 Can the average person tell that a 2002 Hz sound has a different frequency than a 1999 Hz sound without playing them simultaneously?

16.55 If your radio is producing an average sound level of 85 dB, what is the next lowest sound level that is clearly less intense?

16.56 Can you tell that your roommate turned up the TV if its average sound level goes from 70 to 73 dB?

16.57 Based on the graph in Figure 16.22, what is the threshold of hearing in decibels for frequencies of 60, 400, 1000, 4000, and 15,000 Hz? Note that many AC electrical appliances produce 60 Hz, 400 Hz is common in music, 1000 Hz is a reference frequency, your maximum sensitivity is near 4000 Hz, and many older TVs produce a 15,750 Hz whine.

16.58 What sound levels must sounds of frequencies 60, 3000, and 8000 Hz have in order to have the same loudness as a 40 dB sound of frequency 1000 Hz (that is, to have a loudness of 40 phons)?

16.59 What is the sound level in decibels of a 600 Hz tone if it has a loudness of 20 phons? If it has a loudness of 70 phons?

16.60 (a) What are the loudnesses in phons of sounds having frequencies of 200, 1000, 5000, and 10,000 Hz, if they are all at the same 60.0 dB sound level? (b) If they are all at 110 dB? (c) At 20.0 dB?

16.61 Suppose a person has a 50 dB hearing loss at all frequencies. By how many factors of 10 will low-intensity sounds need to be amplified to seem normal to this person? Note that smaller amplification is appropriate for more intense sounds to avoid further hearing damage.

16.62 If a woman needs an amplification of 5.0×10^5 times the threshold intensity to enable her to hear at all frequencies, what is her overall hearing loss in dB? Note that smaller amplification is appropriate for more intense sounds to avoid further damage to her hearing from levels above 90 dB.

• **16.63** (a) What is the intensity in W/m^2 of a just barely audible 200 Hz sound? (b) A barely audible 4000 Hz sound?

• **16.64** (a) Find the intensity in W/m^2 of a 60.0 Hz sound having a loudness of 60 phons. (b) Do the same for a 10,000 Hz sound.

⁝ **16.65** A person has a hearing threshold 10 dB above normal at 100 Hz and 50 dB above normal at 4000 Hz. How much more intense must a 4000 Hz tone be than a 100 Hz tone if they are both barely audible to this person?

⁝ **16.66** A child has a hearing loss of 60 dB near 5000 Hz, due to noise exposure, and normal hearing elsewhere. How much more intense is a 5000 Hz tone than a 400 Hz tone if they are both barely audible to the child?

⁝ **16.67** What is the ratio of intensities of two sounds if the first is just barely discernible as louder to a person than the second?

Section 16.7 Ultrasound

16.68 What is the sound level in dB of ultrasound of intensity 10^5 W/m^2, used to pulverize tissue during surgery?

16.69 Is 155 dB ultrasound in the range of intensities utilized for deep heating? Calculate the intensity, and compare this with values quoted in the text.

16.70 Find the sound level in dB of 2.00×10^{-2} W/m^2 ultrasound used in medical diagnostics.

16.71 (a) Calculate the minimum frequency of ultrasound that will allow you to see details as small as 0.250 mm in human tissue. (b) What is the effective depth to which this sound is effective as a diagnostic probe?

• **16.72** (a) Find the size of the smallest detail observable in human tissue with 20.0 MHz ultrasound. (b) Is its effective penetration depth great enough to examine the entire eye (about 3.00 cm is needed)? (c) What is the wavelength of such ultrasound in 0°C air?

• **16.73** Echo times are measured by diagnostic ultrasound scanners to determine distances to reflecting surfaces in a patient. What is the difference in echo times for tissues that are 3.50 and 3.60 cm beneath the surface? (This is the minimum resolving time for the scanner to see details as small as 0.100 cm, or 1.00 mm. Discrimination of smaller time differences is needed to see smaller details.)

• **16.74** (a) How far apart are two layers of tissue that produce echoes having round-trip times (used to measure distances) that differ by 0.750 μs? (b) What minimum frequency must the ultrasound have to see detail this small?

⁝ **16.75** A diagnostic ultrasound echo is reflected from moving blood and returns with a frequency 500 Hz higher than its original 2.00 MHz. What is the velocity of the blood?

⁝ **16.76** Ultrasound reflected from an oncoming bloodstream that is moving at 30.0 cm/s is mixed with the original frequency of 2.50 MHz to produce beats. What is the beat frequency?

⁝ **16.77** A beat frequency of 150 Hz is produced by an ultrasonic echo mixed with its original frequency of 2.50 MHz after being reflected from the heart wall of a patient. What is the speed of the wall?

INTEGRATED CONCEPTS

Physics is most interesting when applied to general situations involving more than a narrow set of physical principles. The following problems involve physical principles from more than one chapter. Material from Chapters 5, 10, and 13 is directly involved. You will need to refer to these and perhaps other chapters to work the following problems.

Note: Problem-solving strategies and a worked example that can help you solve integrated concept problems appear in Chapter 14. They can be found with the section of problems labeled *Integrated Concepts* **at the end of that chapter. Consult the index for the locations of other problem-solving strategies and worked examples for integrated concept problems.**

16.78 The lever system in the middle ear, shown in Figure 16.26, transmits sound waves to the inner ear. If the lever system has a mechanical advantage of 2, and the area the increased force is exerted upon is 0.050 that of the eardrum, show that the pressure created in the cochlear fluid is 40 times that of the sound wave on the eardrum.

⁝ **16.79** A deep-heat treatment using an ultrasound diathermy unit consists of sending ultrasound with intensity 1.00×10^4 W/m^2 into a patient for 5.00 min. (a) Calculate the energy transmitted if the ultrasound is applied with a circular head of diameter 3.50 cm. (b) What temperature increase is produced if the energy is totally absorbed in 0.300 kg of tissue, assuming no other heat transfer during the treatment?

UNREASONABLE RESULTS

The following problems have results that are unreasonable because some premise is unreasonable or because certain of the premises are inconsistent with one another. Physical principles applied correctly then produce unreasonable results. The purpose of these problems is to give practice in assessing whether nature is being accurately described, and if it is not to trace the source of difficulty.

16.80 (a) What is the speed of sound in air on a day when a 264 Hz sound has a wavelength of 2.75 m? (b) What is unreasonable about this result? (c) Which premise or assumption is responsible?

16.81 (a) What is the fundamental frequency of a child's whistle that has a length of 3.40 mm and is closed at one end? Take the speed of sound to be 343 m/s. (b) What is unreasonable about this result? (c) Which premise or assumption is responsible?

16.82 Fire alarms in public buildings are designed to be so loud that they force people to evacuate—for example, making ordinary conversation impossible. (a) A safety engineer wants an alarm that creates a 10^2 W/m^2 sound intensity. What sound level is this in dB? (b) What is unreasonable about this result? (c) Which premise or assumption is responsible?

• 16.83 While crossing the finish line of a 100 m dash, the winner lets out a 1000 Hz whoop that is received as 1001 Hz by spectators directly in front of her. (a) Calculate her speed if the speed of sound is 345 m/s that day. (b) What is unreasonable about this result? (c) Which premise or assumption is responsible?

ELECTRIC CHARGE AND ELECTRIC FIELD

Static electricity, perhaps created while combing dry hair,

attracts these pieces of paper to the comb.

Figure 17.1 When Benjamin Franklin demonstrated that lightning was related to static electricity, he made a connection that is now part of the evidence that all directly experienced forces are manifestations of the electromagnetic force.

The image of Benjamin Franklin (1706–1790) flying a kite in a thunderstorm is familiar to every schoolchild. (See Figure 17.1.) In this experiment, Franklin demonstrated a connection between lightning and static electricity. Sparks were drawn from a key hung on a kite string during an electrical storm. These sparks were like those produced by static electricity, such as the spark that jumps from your finger to a metal doorknob after you walk across a wool carpet. What Franklin demonstrated in his dangerous experiment was a connection between phenomena on two different scales: one the grand power of an electrical storm, the other an effect of more human proportions. Connections like this one reveal the underlying unity of the laws of nature, an aspect we humans find particularly appealing.

Much has been written about Franklin. His experiments were only part of the life of a man who was a scientist, inventor, revolutionary, statesman, and writer. Franklin's experiments were not performed in isolation, nor were they the only ones to reveal connections. For example, the Italian scientist Luigi Galvani (1737–1798) performed a series of experiments in which static electricity was used to stimulate contractions of frog leg muscles, an effect already known in humans subjected to static discharges. But Galvani also found that certain combinations of metals produced the same effect in frogs as static discharge. Allesandro Volta (1745–1827), partly inspired by Galvani's work, experimented with various combinations of metals and developed the battery.

During the same era, other scientists were making progress in discovering fundamental connections. The periodic table was being developed as the systematic properties of the elements were being discovered. This influenced the development and refinement of the concept of atoms as the basis of matter. Such submicroscopic descriptions of matter also help explain a great deal more. Atomic and molecular interactions, such as the forces of friction, cohesion, and adhesion, are now known to be electromagnetic. Furthermore, all the macroscopic forces that we experience directly, such as the support of the ground beneath our feet and the tension in a rope, are different manifestations of electromagnetic force. Even the force of gravity is sensed through the electromagnetic interaction of molecules, such as between those in our feet and those on the top of a bathroom scale.

This chapter begins the study of electromagnetic phenomena at a fundamental level. The next several chapters will cover static electricity, moving electricity, and magnetism—collectively known as electromagnetism. In this chapter, we begin with the study of electric phenomena due to charges that are at least temporarily stationary, called **electrostatics**, or static electricity.

17.1 STATIC ELECTRICITY AND CHARGE: CONSERVATION OF CHARGE

What makes plastic wrap cling? Static electricity. (See Figure 17.2(a).) Not only are applications of static electricity common these days, its existence has been known since ancient times. The first record of its effects dates to ancient Greeks who noted more than

Figure 17.2 (a) The force holding plastic wrap onto a bowl is the static electrical force. In fact, the quality of plastic wrap is judged by how large this force is and how long it lasts—that is, how static it is. (b) These clothes cling together because of static electricity generated by rubbing during the drying process.

(a)

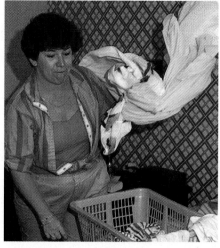

(b)

500 years B.C. that polishing amber temporarily enabled it to attract bits of straw. The very word *electric* derives from the Greek word for amber (*elektron*).

Many of the characteristics of static electricity can be explored by rubbing things together. Rubbing creates the spark you get from walking across a wool carpet, for example. Static cling generated in a clothes dryer and the attraction of straw to recently polished amber also result from rubbing. (See Figure 17.2(b).) Similarly, lightning results from air movements under certain weather conditions. You can also rub a balloon on your hair, and the static electricity created can then make the balloon cling to a wall. Static electricity can be explored with simple equipment, as we shall see. First, we list some of the most basic characteristics of static electricity:

1. The effects of static electricity are explained by a physical quantity not previously introduced, called **electric charge**.
2. There are only two types of charge, one called **positive** and the other called **negative**.
3. Like charges repel, whereas unlike charges attract.
4. The force between charges decreases with distance.

How do we know there are two types of charge? When various materials are rubbed together in controlled ways, certain combinations of materials always produce positive charge on one material and negative on the other. For example, when glass is rubbed with silk, the glass becomes positively charged and the silk negatively charged. Since the glass and silk have opposite charges, they attract one another like clothes that have rubbed together in a dryer. Two glass rods rubbed in this manner will repel one another, since each rod has positive charge on it. Similarly, two silk cloths so rubbed will repel, since both cloths have negative charge. Figure 17.3 shows how these simple materials can be used to explore the nature of the force between charges.

More sophisticated questions arise. Where do these charges come from? Can you create or destroy charge? Is there a smallest unit of charge? Exactly how does the force depend on the amount of charge and the distance between charges? Such questions obviously occurred to Franklin and the other early researchers, and they interest us even today.

Charge Carried by Electrons and Protons

Today we have the advantage of knowing that normal matter is made of atoms, and that atoms contain positive and negative charge, usually in equal amounts. Figure 17.4 shows a simple model of an atom with negative electrons orbiting its positive nucleus. The nucleus is positive due to the presence of positive protons. Nearly all charge in nature is due to electrons and protons, which are two of the three building blocks of most matter. (The third is the neutron, the characteristics of which, along with those of electrons, protons, atoms, and molecules, will be discussed in some detail in later chapters.) Other charge carrying particles are observed in cosmic rays and nuclear decay, and are created in particle accelerators. All but the electron and proton survive only a short time and are quite rare by comparison.

The charges of electrons and protons are identical in magnitude but opposite in sign. Furthermore, all charges in nature are integral multiples of this basic quantity of charge, meaning that all charges are made of combinations of a basic unit of charge. Usually, charges are formed by combinations of electrons and protons. The magnitude of this basic charge is

$$|q_e| = 1.60 \times 10^{-19} \text{ C} \qquad (17.1)$$

The symbol q is commonly used for charge. The SI unit of charge is the **coulomb** (C), a quantity of charge that is much greater than that on the elementary building blocks of matter. For example, the number of protons needed to make a charge of 1.00 C is 1.00 C × (1 proton)/(1.60 × 10^{-19} C) = 6.25 × 10^{18} protons. Similarly, 6.25 × 10^{18} electrons have a combined charge of −1.00 C. Just as there is a smallest bit of an element (an atom), there is a smallest bit of charge. There is no directly observed charge smaller than $|q_e|$,* and all observed charges are integral multiples of $|q_e|$.

---

*See the Things Great and Small essay. Quarks with charges of one-third and two-thirds $|q_e|$ are thought to be the substructure of a broad class of particles that includes protons.

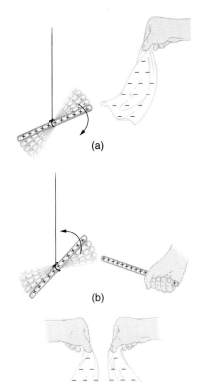

(a)

(b)

(c)

Figure 17.3 A glass rod becomes positively charged when rubbed with silk, while the silk becomes negatively charged. (a) The glass rod is attracted to the silk. (b) Two similarly charged glass rods repel. (c) Two similarly charged silk cloths repel.

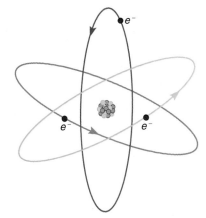

Figure 17.4 This simplified view of an atom is called the planetary model of the atom. Negative electrons orbit a much heavier positive nucleus, as the planets orbit the much heavier sun. The forces in the atom are electromagnetic, whereas those in the planetary system are gravitational. Normal macroscopic amounts of matter contain immense numbers of atoms and molecules and, hence, even greater numbers of individual negative and positive charges.

THINGS GREAT AND SMALL

The Submicroscopic Origin of Charge

With the exception of exotic, short-lived particles, all charge in nature is carried by electrons and protons. Electrons carry the charge we have named negative. Protons carry an equal-magnitude charge that we call positive. (See Figure 17.A.) Electron and proton charge are considered fundamental building blocks, since all other charges are integral multiples of those carried by electrons and protons. Electrons and protons are also two of the three fundamental building blocks of ordinary matter. The neutron is the third and has zero total charge.

Figure 17.A shows a person touching a Van de Graaff generator and receiving excess positive charge. The repul-sion of like charges causes the person's hair to stand up; the expanded view of a hair shows the existence of both types of charges, but an excess of positive. The further blowup shows an artist's conception of an electron and a proton perhaps found in an atom in a strand of hair.

The electron seems to have no substructure; in con-trast, there is strong evidence that quarks are a substructure of protons and certain other particles. (This will be ex-plained in some detail in later chapters.) Although quarks have never been directly observed, they are believed to carry fractional charges as seen in Figure 17.B. Charges on electrons and protons and all other directly observable particles are unitary, but these quark substructures carry fractional charges. There are continuing attempts to ob-serve fractional charge directly and to learn of the proper-ties of quarks, which are perhaps the ultimate substructure of matter.

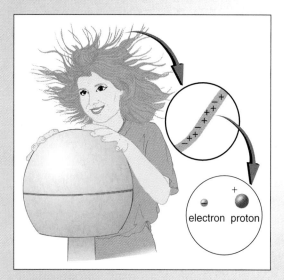

Figure 17.A When this person touches a Van de Graaff gen-erator, she receives an excess of positive charge, causing her hair to stand on end. The charges in one hair are shown. An artist's conception of an electron and a proton illustrate the par-ticles carrying the negative and positive charges. We cannot really see these particles with visible light because they are so small (the electron seems to be an infinitesimal point), but we know a great deal about their measurable properties, such as the charges they carry.

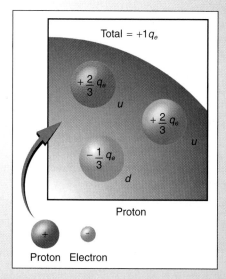

Figure 17.B Artist's conception of fractional quark charges in-side a proton. Those quark charges add up to the single posi-tive charge on the proton. When the substructure of protons is explored by scattering extremely energetic electrons from them, it appears that there are pointlike particles inside the proton. Attempts to isolate quarks have always failed for reasons that are not completely understood.

Separation of Charge in Atoms

Charges in atoms and molecules can be separated—for example, by rubbing materials to-gether. Some atoms and molecules have a greater affinity for electrons than others and will become negatively charged by close contact in rubbing, leaving the other material positively charged. (See Figure 17.5.) (Positive charge can similarly be transferred by rubbing.) Methods other than rubbing can also separate charges. Batteries, for example, use combinations of substances that interact in such a way as to separate charges. Chemical interactions may transfer negative charge from one substance to the other, mak-ing one battery terminal negative and leaving the other positive.

No charge is actually created or destroyed when charges are separated as we have been discussing. Rather, existing charges are moved about. In fact, in all situations the

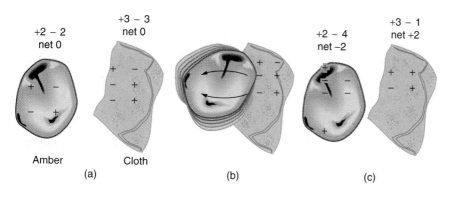

Amber   Cloth
(a)   (b)   (c)

+2 − 2
net 0

+3 − 3
net 0

+2 − 4
net −2

+3 − 1
net +2

Figure 17.5 When materials are rubbed together, charges can be separated, particularly if one material has a greater affinity for electrons than another (this affinity is complex; it is related to chemical properties that are also explained by the greater affinity for electrons of some atoms than others). This separation of charge leaves one substance with a net positive charge and the other with a net negative, causing them to be attracted to one another. (a) Both the amber and cloth are originally neutral. Only a tiny fraction of the charges are involved, and only a few of them are shown here. (b) When rubbed together, some negative charge is transferred to the amber, leaving the cloth with a net positive charge. (c) When separated, the amber and cloth now have net charges.

total amount of charge is always constant. This universally obeyed law of nature is called the **law of conservation of charge** and is simply stated:

Total charge is constant in any process.

In more exotic situations, such as in particle accelerators, mass can be created from energy in the amount $\Delta m = E/c^2$. Sometimes, the created mass is charged, such as when an electron is created. Whenever a charged particle is created, another having an opposite charge is always created along with it, so that the total charge created is zero. Usually, the two particles are matter-antimatter counterparts. For example, an antielectron would usually be created at the same time as an electron. The antielectron has a positive charge (it is called a positron), and so the total charge created is zero. (See Figure 17.6.) All particles have antimatter counterparts with opposite signs. When matter and antimatter counterparts are brought together, they completely annihilate one another. By annihilate, we mean that the mass of the two particles is converted to energy, again obeying the relationship $\Delta m = E/c^2$. Since the two particles have equal and opposite charges, the total charge is zero before and after the annihilation; thus, total charge is conserved.

The law of conservation of charge is absolute—it has never been observed to be violated. Charge, then, is a special physical quantity, joining a very short list of other quantities that are always conserved. Those conserved quantities discussed to this point in the text are energy, momentum, angular momentum, and charge. There are a few others that we will encounter later on, but the number of conserved quantities in nature is small.

CONNECTIONS

Conservation Laws

Only a limited number of physical quantities are universally conserved. Charge is one—energy, momentum, and angular momentum are the only others covered so far. Because they are conserved, these physical quantities are used to explain more phenomena and form more connections than other, less basic quantities. We will find that conserved quantities give us great insight into the rules followed by nature and hints to the organization of nature. Discoveries of conservation laws have led to further discoveries, such as the weak nuclear force and the quark substructure of protons and other particles.

17.2 CONDUCTORS AND INSULATORS: CHARGING BY CONTACT AND BY INDUCTION

Some substances, such as metals and salty water, allow charges to move through them with relative ease. Some of the electrons in metals and similar conductors are not bound to individual atoms or sites in the material. These are called free electrons. They can move through the material much as air moves through loose sand. Any substance that has free electrons and allows charge to move relatively freely through it is called a **conductor**. The electrons may collide with fixed atoms and molecules, losing some energy, but they can move in a conductor. **Superconductors** allow the movement of charge without any loss of energy at all. Salty water and other similar conducting materials contain free ions that can move through them. An **ion** is an atom or molecule having a positive or negative (nonzero) total charge.

Other substances, such as glass, do not allow charges to move through them. These are called insulators. Electrons and ions in **insulators** are bound in place and cannot move easily.* Pure water and dry table salt are insulators, for example, whereas molten salt and salty water are conductors.

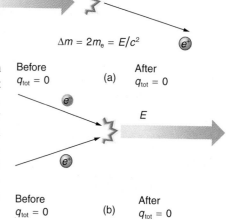

$\Delta m = 2m_e = E/c^2$

Before   After
$q_{tot} = 0$ (a) $q_{tot} = 0$

E

Before   After
$q_{tot} = 0$ (b) $q_{tot} = 0$

Figure 17.6 (a) When enough energy is present, it can be converted into matter. Here the matter created is an electron–antielectron pair. The total charge before and after this event is zero. (b) When matter and antimatter annihilate, the total charge is conserved at zero before and after the annihilation.

*The charges may move, but very slowly compared with conductors—perhaps slowed by a factor of as much as 10^{23}.

Figure 17.7 An electroscope is a favorite instrument in physics demonstrations and student laboratories. It is typically made with gold foil leaves hung from a metal stem and is insulated from the air in a glass-walled container. (a) A positively charged glass rod is brought near the electroscope, attracting electrons to the top and leaving a net positive charge on the leaves. Like charges in the light flexible gold leaves repel, separating them. (b) When the rod is touched against the ball, electrons are attracted and transferred, reducing the net charge on the glass rod but leaving the electroscope positively charged. (c) The excess charges are evenly distributed in the stem and leaves of the electroscope once the glass rod is removed.

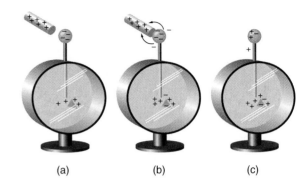

(a)   (b)   (c)

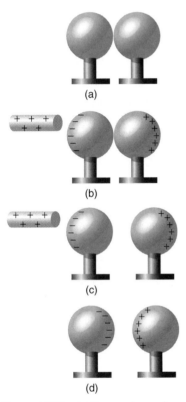

(a)

(b)

(c)

(d)

Figure 17.8 (a) Two uncharged or neutral metal spheres are in contact with each other but insulated from the rest of the world. (b) A positively charged glass rod is brought near the sphere on the left, attracting negative charge and leaving the other sphere positively charged. (c) The spheres are separated before the rod is removed, thus separating negative and positive charge. (d) The spheres retain net charges after the inducing rod is removed—without ever having been touched by a charged object.

Charging by Contact

Figure 17.7 shows an electroscope being charged by touching a positively charged glass rod. The glass rod is an insulator, and so it must actually touch the electroscope to transfer charge to or from it. (Note that the extra positive charges reside on the surface of the glass rod as a result of rubbing it with silk.) Since only electrons move in metals, we see that they are attracted to the top of the electroscope and some are transferred to the positive rod by touch, leaving the electroscope with a net positive charge. Electrostatic repulsion in the leaves of the charged electroscope separates them against the force of gravity. Similarly, the electroscope can be negatively charged by contact with a negatively charged object.

Charging by Induction

It is not necessary to transfer excess charge directly to an object in order to charge it. Figure 17.8 shows a method of inducing a charge, as opposed to charging by direct contact. Here we see two neutral metal spheres in contact with one another but insulated from the rest of the world. A positively charged rod is brought near one of them, attracting negative charge to that side, leaving the other sphere positively charged. This is an example of induced polarization of neutral objects. **Polarization** is the separation of charges in an object that remains neutral. Note that the object closest to the charged rod receives an opposite charge when charged by induction. Note also that no charge is removed from the charged rod, so that this process can be repeated without depleting the supply of excess charge.

Another method of charging by induction is shown in Figure 17.9. The neutral metal sphere is polarized when a charged rod is brought near it. The sphere is then **grounded**, meaning that a conducting wire is run from the sphere to the ground. Since the earth is large and most ground is a good conductor, it can supply or accept excess charge easily. In this case, electrons are attracted to the sphere through a wire called the

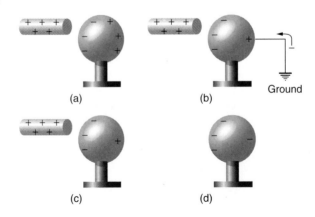

(a)   (b)   Ground

(c)   (d)

Figure 17.9 Charging by induction, using a ground connection. (a) A positively charged rod is brought near a neutral metal sphere, polarizing it. (b) The sphere is grounded, allowing electrons to be attracted from the earth's ample supply. (c) The ground connection is broken. (d) The positive rod is removed, leaving the sphere with an induced negative charge.

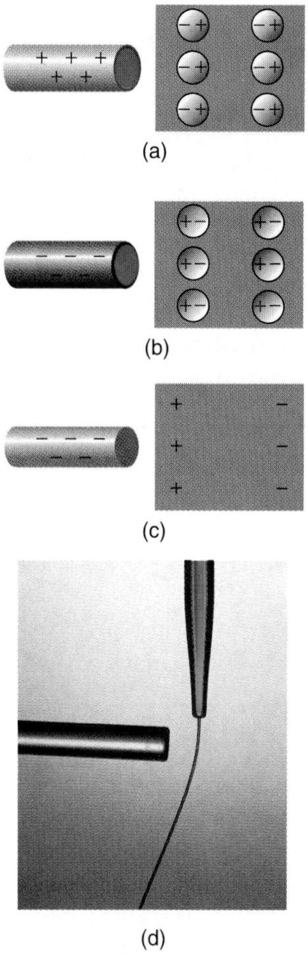

Figure 17.10 Both positive and negative objects attract a neutral object by polarizing its molecules. (a) A positive object brought near a neutral insulator polarizes its molecules. There is a slight shift in the distribution of the electrons orbiting the molecule, with unlike charges being brought nearer and like charges moved away. Since the electrostatic force decreases with distance, there is a net attraction. (b) A negative object produces the opposite polarization, but again attracts the neutral object. (c) The same effect occurs for a conductor; since the unlike charges are closer, there is a net attraction. (d) Can you explain the attraction of water to the charged rod?

ground wire, because it supplies a conducting path to the ground. The ground connection is broken before the charged rod is removed, leaving the sphere with an excess charge opposite to that of the rod. Again, an opposite charge is achieved when charging by induction and the charged rod loses none of its excess charge.

Neutral objects can be attracted to any charged object. The pieces of straw attracted to polished amber are neutral, for example. If you run a plastic comb through your hair, the charged comb can pick up neutral pieces of paper. Figure 17.10 shows how the polarization of atoms and molecules in neutral objects results in their attraction to a charged object. When a charged rod is brought near a neutral substance, an insulator in this case, the distribution of charge in atoms and molecules is shifted slightly. Opposite charge is attracted nearer the external charged rod, while like charge is repelled. Since the electrostatic force decreases with distance, the repulsion of like charges is weaker than the attraction of unlike charges, and so there is a net attraction. Thus a positively charged glass rod attracts neutral pieces of paper, as will a negatively charged rubber rod. Some molecules, like water, are **polar molecules**. Polar molecules have a natural or inherent separation of charge, although they are neutral overall. Polar molecules are particularly affected by other charged objects and show greater polarization effects than molecules with naturally uniform charge distributions.

17.3 COULOMB'S LAW

The main features of the electrostatic force—the existence of two types of charge, likes repel, unlikes attract, and the decrease of force with distance—were eventually refined, and expressed as a mathematical formula. The mathematical formula for the electrostatic force is called **Coulomb's law** after the French physicist Charles Coulomb (1736–1806), who performed experiments and first proposed the formula

$$F = k \frac{q_1 q_2}{r^2} \qquad (17.2)$$

for the force between two point charges, q_1 and q_2, separated by a distance r. In SI units, the constant $k = 8.988 \times 10^9$ N·m^2/C$^2 \approx 9.00 \times 10^9$ N·m^2/C^2.* The force is understood to be along the line joining the two charges. (See Figure 17.11.) Although the formula is simple, it was no mean task to prove it. The experiments Coulomb did, with the primitive equipment then available, were difficult. Modern experiments have verified Coulomb's law to great precision. For example, it has been shown that the force is inversely proportional to distance squared ($F \propto 1/r^2$) to an accuracy of 1 part in 10^{16}.

---

*We will find in later chapters that k is related to the speed of light, since light is an electromagnetic wave and involves the Coulomb force.

CONNECTIONS

The Four Basic Forces
It is remarkable that the Coulomb force between two charges, $F = kq_1q_2/r^2$, is so mathematically similar to the gravitational force between two masses, $F = GmM/r^2$. See Section 4.6 for an introductory discussion of other ties among the four basic forces. These will also be explored in more detail in later chapters. We will find, for example, that the $1/r^2$ dependence of the force is related to the particles that carry it.

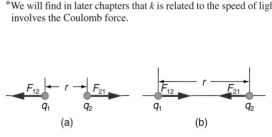

(a)

(b)

Figure 17.11 The magnitude of the electrostatic force F between point charges q_1 and q_2 separated by a distance r is given by Coulomb's law. Note that Newton's third law applies as usual—the force on q_1 is equal in magnitude and opposite in direction to the force it exerts on q_2. (a) Like charges. (b) Unlike charges.

EXAMPLE 17.1 HOW STRONG IS THE COULOMB FORCE RELATIVE TO GRAVITY?

Compare the electrostatic force between an electron and proton separated by 0.530×10^{-10} m with the gravitational force between them. This distance is their average separation in a hydrogen atom.

Strategy To compare the two forces, we first compute the electrostatic force using Coulomb's law, stated in Equation 17.2. We then calculate the gravitational force using Newton's universal law of gravitation, stated in Equation 8.8. Finally, we take a ratio to see how the forces compare in magnitude.

Solution Entering the given and known information about the charges and separation of the electron and proton into the expression of Coulomb's law yields

$$F = k \frac{q_1 q_2}{r^2}$$

$$= (9.00 \times 10^9 \ \text{N} \cdot \text{m}^2/\text{C}^2)$$

$$\times \frac{(-1.60 \times 10^{-19} \ \text{C})(1.60 \times 10^{-19} \ \text{C})}{(0.530 \times 10^{-10} \ \text{m})^2}$$

Thus the Coulomb force is

$$F = -8.20 \times 10^{-8} \ \text{N}$$

The minus sign indicates an attractive force. This is a very large force for an electron—it would cause an acceleration of 9.00×10^{22} m/s^2 (verification is left as an end-of-chapter problem).

The gravitational force is given by Equation 8.8:

$$F_G = G \frac{mM}{r^2}$$

where $G = 6.67 \times 10^{-11}$ N·m^2/kg^2. Here m and M represent the electron and proton masses, which can be found in the appendices. Entering values for the knowns yields

$$F_G = (6.67 \times 10^{-11} \ \text{N} \cdot \text{m}^2/\text{kg}^2)$$

$$\times \frac{(9.11 \times 10^{-31} \ \text{kg})(1.67 \times 10^{-27} \ \text{kg})}{(0.530 \times 10^{-10} \ \text{m})^2}$$

$$= 3.61 \times 10^{-47} \ \text{N}$$

This is also an attractive force, although it is traditionally shown as positive since gravity is always attractive. The ratio of the magnitude of the electrostatic force to gravitational force in this case is, thus,

$$\boxed{\frac{F}{F_G} = 2.27 \times 10^{39}}$$

Discussion This is a remarkably large ratio! Note that this will be the ratio of electrostatic force to gravitational force for an electron and a proton at any distance (taking the ratio before entering numerical values shows that the distance cancels). This ratio gives some indication of just how much larger the Coulomb force is than the gravitational force between two of the most common particles in nature.

As the previous example implies, gravitational force is completely negligible on a small scale, where the interactions of individual charged particles are important. On a large scale, such as between the earth and a person, the reverse is true. Most objects are nearly electrically neutral, and so attractive and repulsive Coulomb forces nearly cancel. Gravity on a large scale dominates interactions between large objects because gravitational force is always attractive, while Coulomb forces tend to cancel.

EXAMPLE 17.2 WHAT FORCE IS NEEDED TO SEPARATE ALL THE POSITIVE AND NEGATIVE CHARGE IN A GRAM OF HYDROGEN?

Suppose you could separate all of the protons and electrons in a mole of hydrogen gas (about a gram) to a distance of 1.00 m. What would be the force of attraction?

Strategy The separated charges will strongly attract because of the Coulomb force. To answer the question, we must therefore calculate that force. First, we must find the amount of charge involved. One mole contains Avogadro's number, or 6.02×10^{23} atoms. Each hydrogen atom has a single proton and a single electron. Thus the total charge of the separated protons is $(6.02 \times 10^{23}) \cdot (1.60 \times 10^{-19} \ \text{C}) = 9.63 \times 10^4$ C, and the total charge of the electrons is equal and opposite, or -9.63×10^4 C. The magnitude of the attractive Coulomb force can now be found from Equation 17.2.

Solution Entering the values determined for charge into

the equation for the Coulomb force yields

$$F = k \frac{q_1 q_2}{r^2}$$

$$= (9.00 \times 10^9 \ \text{N} \cdot \text{m}^2/\text{C}^2)$$

$$\times \frac{(9.63 \times 10^4 \ \text{C})^2}{(1.00 \ \text{m})^2}$$

$$= 8.35 \times 10^{19} \ \text{N}$$

Discussion This force is what would be needed to separate the electrons and protons in 1 g of hydrogen to a meter apart, but it is unreasonably large. The implication is that it would be impossible to separate these charges in the manner suggested. And that is the point—the electrostatic force is large by any measure.

17.4 ELECTRIC FIELD: CONCEPT OF A FIELD REVISITED

Contact forces, such as between a baseball and a bat, are explained on the small scale by the interaction of the charges in atoms and molecules in close proximity. They interact through forces that include the Coulomb force. Action at a distance is a force between objects that are not close enough for their atoms to "touch." That is, they are separated by more than a few atomic diameters. Action at a distance can also be through the Coulomb force. For example, a charged rubber comb attracts neutral bits of paper from a distance via the Coulomb force. It is very useful to think of an object being surrounded in space by a **force field**. The force field carries the force to another object (called a test object) some distance away.

The Coulomb force field surrounding any charge extends throughout space. Its magnitude is given by Equation 17.2, $F = kqQ/r^2$, for a point particle having a charge Q acting on a test charge q at a distance r. (See Figure 17.12.) We see from this equation that both the magnitude and direction of the Coulomb force field depend on Q and the test charge q.

To simplify things, we would prefer to have a field that depends only on Q and not on the test charge q. The electric field is defined in such a manner that it represents only the charge creating it and is unique at every point in space. Specifically, the **electric field E** is defined to be the ratio of the Coulomb force to the test charge:

$$\mathbf{E} = \frac{\mathbf{F}}{q} \qquad (17.3)$$

where $\mathbf{F}$ is the electrostatic force (or Coulomb force) exerted on a positive test charge q. It is understood that $\mathbf{E}$ is in the same direction as $\mathbf{F}$. It is also assumed that q is so small that it does not alter the charge distribution creating the electric field. The units of electric field are newtons per coulomb (N/C). If the electric field is known, then the electrostatic force on any charge q is simply obtained by multiplying charge times electric field, or

$$\mathbf{F} = q\mathbf{E}$$

which is another way of writing Equation 17.3.

Consider the electric field created by a point charge Q. The force it exerts on a test charge q is $F = kqQ/r^2$. Thus the magnitude of the electric field, E, for a point charge is

$$E = \frac{F}{q} = k\frac{qQ}{qr^2} = k\frac{Q}{r^2}$$

Since the test charge cancels, we see that

$$E = k\frac{Q}{r^2} \qquad \text{(point charge)} \qquad (17.4)$$

The electric field is thus seen to depend only on the charge Q creating the field and the distance r; it is completely independent of the test charge q.

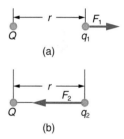

Figure 17.12 The Coulomb force field created by a positive charge Q is shown acting on two different charges. Both charges are the same distance from Q. (a) Since q_1 is positive, the force F_1 acting on it is repulsive. (b) The charge q_2 is negative and greater in magnitude than q_1, and so the force F_2 acting on it is attractive and stronger than F_1. The Coulomb force field is thus not unique at any point in space, because it depends on the test charges q_1 and q_2 as well as the charge Q.

EXAMPLE 17.3   THE ELECTRIC FIELD OF A POINT CHARGE

(a) Calculate the electric field E due to a point charge of 2.00 nC at a distance of 5.00 mm from the charge.
(b) What force does this electric field exert on a charge of $-0.250\ \mu C$?

Strategy In part (a), we find the electric field for a point charge by using Equation 17.4. Once the electric field is found, the force on a charge can be calculated for part (b) by using $F = qE$.

Solution for (a) The electric field for a point charge is given by Equation 17.4. Here $Q = 2.00 \times 10^{-9}$ C and

$r = 5.00 \times 10^{-3}$ m. Entering those values gives

$$E = k\frac{Q}{r^2}$$

$$= (9.00 \times 10^9\ \text{N·m}^2/\text{C}^2)$$

$$\times \frac{(2.00 \times 10^{-9}\ \text{C})}{(5.00 \times 10^{-3}\ \text{m})^2}$$

$$= 7.20 \times 10^5\ \text{N/C}$$

Discussion for (a) This electric field strength is the same at any point 5.00 mm away from the charge Q that

(continued)

(continued)

creates the field. It is positive, meaning that it has a direction pointing away from the charge Q.

Solution for (b) The force on a charge $q = -0.250 \, \mu C$ is then obtained by multiplying charge times electric field strength:

$$F = qE = (-0.250 \times 10^{-6} \text{ C})(7.20 \times 10^5 \text{ N/C})$$

$$= -0.180 \text{ N}$$

Discussion for (b) The minus sign means the force is attractive, as expected for unlike charges. The charges in this example are typical of common static electricity, and the modest attractive force obtained is similar to forces experienced in static cling and similar situations.

17.5 ELECTRIC FIELD LINES: MULTIPLE CHARGES

Drawings using lines to represent electric fields around charged objects are very useful in visualizing field strength and direction. Since the electric field has both magnitude and direction, it is a vector. Like all vectors, the electric field can be represented by an arrow that has length proportional to its magnitude and that points in the correct direction. (We have used arrows extensively to represent force vectors, for example.) Figure 17.13 shows two pictorial representations of the same electric field created by a positive point charge Q. The standard representation uses continuous lines as in Figure 17.13(b), rather than numerous individual arrows as in Figure 17.13(a).

Note that the electric field is defined for a *positive* test charge q, so that the field lines point away from a positive charge and toward a negative charge. (See Figure 17.14.) The electric field strength is exactly proportional to the number of field lines per unit area, since the electric field for a point charge is $E = kQ/r^2$ and area is proportional to r^2. This pictorial representation, in which field lines represent the direction and their closeness

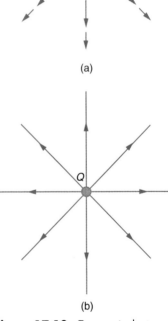

(a)

(b)

Figure 17.13 Two equivalent representations of the electric field due to a positive charge Q. (a) Arrows representing the electric field's magnitude and direction. (b) In the *standard representation*, the arrows are replaced by continuous field lines having the same direction at any point as the electric field. The closeness of the lines is directly related to the strength of the electric field. A test charge placed anywhere will feel a force in the direction of the field line; this force will have a strength proportional to the density of the lines (being greater near the charge, for example).

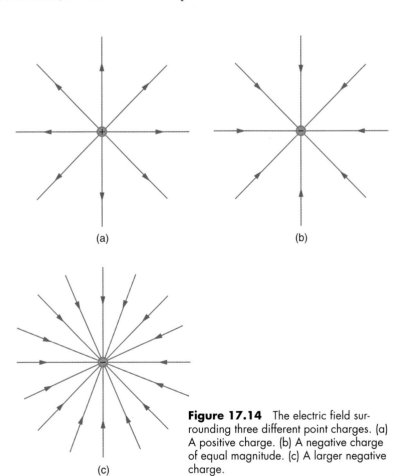

(a)

(b)

(c)

Figure 17.14 The electric field surrounding three different point charges. (a) A positive charge. (b) A negative charge of equal magnitude. (c) A larger negative charge.

(that is, their areal density or the number of lines crossing a unit area) represents strength, is used for all fields: electrostatic, gravitational, magnetic, and others.

In many situations, there are multiple charges. The total electric field created by multiple charges is the vector sum of the individual fields created by each charge.* Figure 17.15 shows how electric field vectors add, and it shows how the field for two positive charges results from this vector addition.

While the electric fields from multiple charges are more complex than those of single charges, some simple features are easily noticed. For example, the field is weaker between like charges, as shown by the lines being farther apart in that region. (This is because the fields from each charge exert opposing forces on any charge placed between them.) (See Figures 17.15 and 17.16(a).) Furthermore, at a great distance from two like charges, the field becomes identical to the field from a single, larger charge. Figure 17.16(b) shows the electric field of two unlike charges. The field is stronger between the charges. In that region, the fields from each charge are in the same direction, and so their strengths add. The field of two unlike charges is weak at large distances, because the fields of the individual charges are in opposite directions and so their strengths subtract. At very large distances, the field of two unlike charges looks like that of a smaller single charge.

We use electric field lines to visualize and analyze electric fields (the lines are a pictorial tool, not a physical entity in themselves). The **properties of electric field lines** for any charge distribution can be summarized as follows:

1. Field lines must begin on positive charges *and* terminate on negative charges, or at infinity in the hypothetical case of isolated charges.
2. The number of field lines leaving a positive charge or entering a negative charge is proportional to the magnitude of the charge.
3. The strength of the field is proportional to the closeness of the field lines—more precisely, it is proportional to the number of lines per unit area perpendicular to the lines.
4. The direction of the electric field is tangent to the field line at any point in space.
5. Field lines can never cross.

The last property means that the field is unique at any point. The field line represents the direction of the field; so if they crossed, the field would have two directions at that location (an impossibility if the field is unique).

---

*This statement is more profound than it may appear. It means that the fields are independent of one another and add. This is modified by modern quantum field theory in extreme cases.

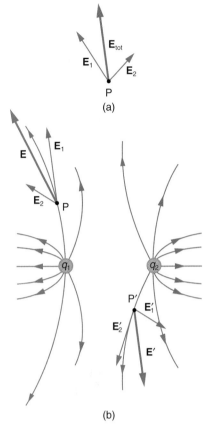

Figure 17.15 (a) The electric fields E_1 and E_2 at point P add to E_{tot}. (b) Two positive point charges q_1 and q_2 produce the resultant electric field shown. The field is tangent to the lines at any given point, as illustrated at points P and P′.

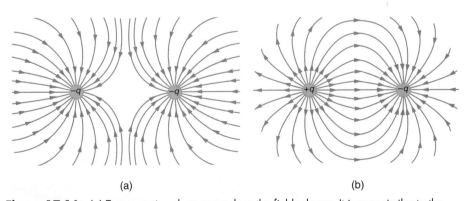

Figure 17.16 (a) Two negative charges produce the fields shown. It is very similar to the field produced by two positive charges, except that the directions are reversed. The field is clearly weaker between the charges. The individual forces on a test charge in that region are in opposite directions. (b) Two opposite charges produce the field shown, which is stronger in the region between the charges.

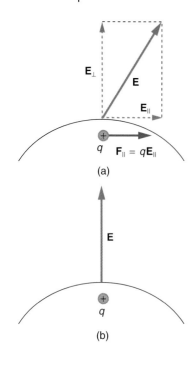

(a)

(b)

Figure 17.17 When an electric field is applied to a conductor, free charges move until the field is perpendicular to the surface. (a) The component of the electric field parallel to the surface exerts a force on the charge, which moves until that component is zero. (b) The field is perpendicular to the surface in electrostatic equilibrium. The free charge is brought to the surface, where electrostatic forces are in equilibrium.

17.6 CONDUCTORS AND ELECTRIC FIELDS IN STATIC EQUILIBRIUM

Conductors contain charges that move easily. When excess charge is placed on a conductor or the conductor is put into a static electric field, charges in the conductor quickly respond to reach a steady state called **electrostatic equilibrium**. Figure 17.17 shows the effect of an electric field on free charges in a conductor. The free charges move until the field is perpendicular to the conductor's surface. There can be no component of the field parallel to the surface in electrostatic equilibrium, since, if there were, it would produce further movement of charge. A positive free charge is shown, but free charges can be either positive or negative and are, in fact, negative in metals. The motion of a positive charge is equivalent to the motion of a negative charge in the opposite direction.

A conductor placed in an electric field will be polarized. (See also Section 17.2, where metal spheres were polarized by a charged rod and the polarization of atoms and molecules was discussed.) Figure 17.18 shows the result of placing a neutral conductor in an originally uniform electric field. The field becomes stronger near the conductor but entirely disappears inside it.

Excess charges placed on a spherical conductor repel and move until they are evenly distributed, as shown in Figure 17.19. Excess charge is forced to the surface until the field inside the conductor is zero. Outside the conductor, the field is exactly the same as if the conductor were replaced by a point charge at its center equal to the excess charge.

A **conductor in electrostatic equilibrium** has the following properties:

1. The electric field is zero inside a conductor.
2. Just outside a conductor, the electric field lines are perpendicular to its surface, ending or beginning on charges on the surface.
3. Any excess charge resides entirely on the surface or surfaces of a conductor.

These properties are consistent with the situations already discussed and can be used to analyze any conductor in electrostatic equilibrium. This can lead to some interesting new insights, such as the following.

How can a very uniform electric field be created? Consider a system of two metal plates with opposite charges on them, as shown in Figure 17.20. The properties of conductors in electrostatic equilibrium indicate that the electric field between the plates will

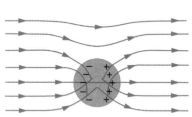

Figure 17.18 A spherical conductor in static equilibrium with an originally uniform electric field. Free charges move, polarizing the conductor, until the electric field lines are perpendicular to the surface. The field lines end on excess negative charge on one surface and begin again on excess positive charge on the other surface. There can be no electric field inside the conductor, since free charges move in response and neutralize it.

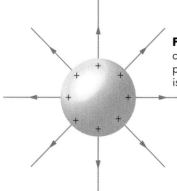

Figure 17.19 The mutual repulsion of excess positive charges on a spherical conductor distributes them uniformly on its surface. The resulting electric field is perpendicular to the surface and zero inside. Outside the conductor, the field is identical to that of a point charge at the center equal to the excess charge.

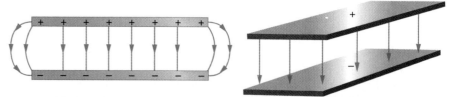

Figure 17.20 Two views of two metal plates with equal but opposite excess charges. The field between them is uniform in strength and direction except near the edges. One use of such a field is to produce uniform acceleration of charges between the plates, such as in the electron gun of a TV tube.

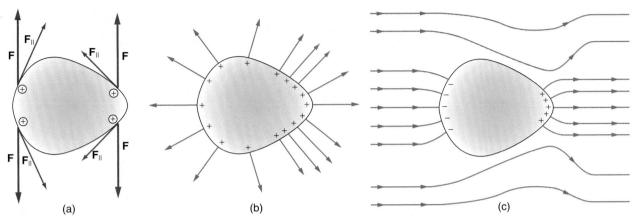

Figure 17.21 Excess charge on a nonuniform conductor becomes most concentrated at the location of greatest curvature. (a) The forces between identical pairs of charges at either end of the conductor are identical, but the components of the forces parallel to the surface are different. It is $F_\parallel$ that moves the charges apart once they have reached the surface. (b) Since $F_\parallel$ is smallest at the more pointed end, the charges are left closer together, producing the electric field shown. (c) An uncharged conductor in an originally uniform electric field is polarized, with the most concentrated charge at its most pointed end.

be uniform in strength and direction as shown. Except near the edges, the excess charges distribute themselves uniformly, producing field lines that are uniformly spaced (hence uniform in strength) and perpendicular to the surfaces (hence uniform in direction, since the plates are flat). The edge effects are less important when the plates are close together.

What happens if a conductor has sharp corners or is pointed? Excess charges on a nonuniform conductor become concentrated at the sharpest points. Additionally, excess charge may move on or off the conductor at the sharpest points. To see how and why this happens, consider the charged conductor in Figure 17.21. The electrostatic repulsion of like charges is most effective in moving them apart on the flattest surface, and so they become least concentrated there. This is because the forces between identical pairs of charges at either end of the conductor are identical, but the components of the forces parallel to the surfaces are different. The component parallel to the surface is greatest on the flattest surface and, hence, more effective in moving the charge. The same effect is produced on a conductor by an externally applied electric field, as seen in Figure 17.21(c). Since the field lines must be perpendicular to the surface, more of them are concentrated on the most curved parts.

On a very sharply curved surface, such as shown in Figure 17.22, the charges are so concentrated at the point that the resulting electric field can be great enough to remove them from the surface. This can be useful. Lightning rods work best when they are most pointed. The large charges created in storm clouds induce an opposite charge on a building that can result in a lightning bolt. The induced charge is bled away continually by a lightning rod, preventing the more dramatic lightning strike. Of course, we sometimes wish to prevent the transfer of charge rather than to facilitate it. In that case, the conductor should be very smooth and have as large a radius of curvature as possible. (See Figure 17.23.) Smooth surfaces are used on high-voltage transmission lines, for example, to avoid leakage of charge into the air.

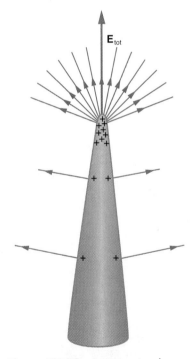

Figure 17.22 A very pointed conductor has a large charge concentration at the point. The electric field is very strong at the point and can exert a force large enough to transfer charge on or off the conductor. Lightning rods are used to prevent the buildup of large excess charges on structures and, thus, are pointed.

PROBLEM-SOLVING STRATEGIES
FOR ELECTROSTATICS

Step 1. Examine the situation to determine if static electricity is involved. This may concern separated stationary charges, the forces among them, and the electric fields they create.

Step 2. Identify the system of interest. This includes noting the number, locations, and types of charges involved.

Step 3. Identify exactly what needs to be determined in the problem (identify the unknowns). A written list is useful. Determine whether the Coulomb force is to be

Figure 17.23 (a) A lightning rod is pointed to facilitate the transfer of charge. (b) This Van de Graaff generator has a smooth surface with a large radius of curvature to prevent the transfer of charge and allow a large voltage to be generated. The mutual repulsion of like charges is evident in the woman's hair.

(a)   (b)

considered directly—if so, it may be useful to *draw a free body diagram*. If electric field lines are to be drawn, *follow the numbered rules given in Sections 17.5 and 17.6.*

Step 4. *Make a list of what is given or can be inferred from the problem as stated (identify the knowns).* It is important to distinguish the Coulomb force F from the electric field E, for example.

Step 5. *Solve the appropriate equation for the quantity to be determined (the unknown) or draw the field lines as requested.*

Step 6. *Examine the answer to see if it is reasonable: Does it make sense?* Are units correct and the numbers involved reasonable?

17.7 APPLICATIONS OF ELECTROSTATICS

The Van de Graaff Generator

Van de Graaff generators (or Van de Graaffs) are not only spectacular devices used to demonstrate high voltage due to static electricity—they are also used for serious research. The first was built by Robert Van de Graaff in 1931 (based on original suggestions by Lord Kelvin) for use in nuclear physics research. Figure 17.24 shows a schematic along with a photograph of a large research version. Van de Graaffs utilize both smooth and pointed surfaces, and conductors and insulators to generate large static charges and, hence, large voltages. (The connection between voltage and charge will be made in the next chapter.) A very large excess charge can be deposited on the sphere, because it moves quickly to the outer surface. Practical limits arise because the large electric fields polarize and eventually ionize surrounding materials, creating free charges that neutralize excess charge or allow it to escape. Nevertheless, voltages of 50 million volts are well within practical limits.

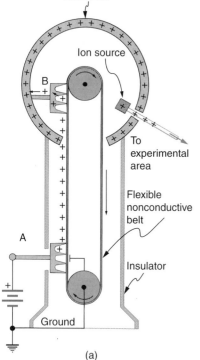

Conductor

Ion source

B

To experimental area

Flexible nonconductive belt

A

Insulator

Ground

(a)

(b)

Figure 17.24 (a) A battery supplies excess positive charge to a pointed conductor, the points of which spray the charge onto a moving insulating belt near the bottom. The pointed conductor in the large sphere picks up the charge (the induced electric field at the points is so large it removes the charge from the belt). This can be done because the charge does not remain inside the conducting sphere but moves to its outside surface. An ion source inside the sphere produces positive ions, which are accelerated away to high velocities. (b) A large research Van de Graaff lies within tanks containing a pressurized insulating gas. Evacuated beam pipes carry accelerated charged ions to experimental areas.

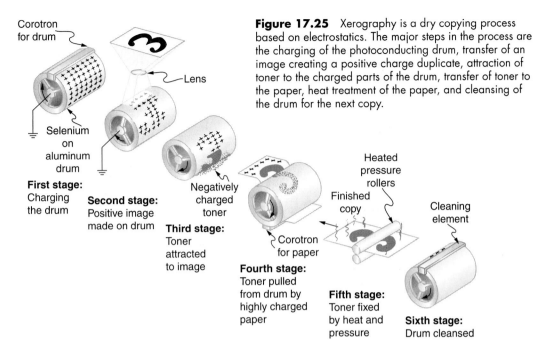

Figure 17.25 Xerography is a dry copying process based on electrostatics. The major steps in the process are the charging of the photoconducting drum, transfer of an image creating a positive charge duplicate, attraction of toner to the charged parts of the drum, transfer of toner to the paper, heat treatment of the paper, and cleansing of the drum for the next copy.

Xerography

Most copy machines use an electrostatic process called *xerography*—a word coined from the Greek words *xeros* for dry and *graphos* for writing. The heart of the process is shown in simplified form in Figure 17.25. A selenium-coated aluminum drum is sprayed with positive charge from points on a device called a corotron. Selenium is a substance with an interesting property—it is a *photoconductor*. That is, selenium is an insulator when in the dark and a conductor when exposed to light. In the first stage of the process, the conducting aluminum drum is grounded, so that a negative charge is induced under the thin layer of uniformly positively charged selenium. In the second stage, the surface of the drum is exposed to the image of whatever is to be copied. Where the image is light, the selenium becomes conducting, and the positive charge is neutralized. In dark areas, the positive charge remains, and so the image has been transferred to the drum. The third stage takes a dry black powder, called the *toner*, and sprays it with a negative charge so that it will be attracted to the positive regions of the drum. Next, a blank piece of paper is given a greater positive charge than on the drum so that it will pull the toner from the drum. Finally, the paper and electrostatically held toner are passed through heated pressure rollers, which melt and permanently adhere the toner within the fibers of the paper.

Laser Printers

Laser printers use the xerographic process to make high-quality images on paper, employing a laser to produce an image on the photoconducting drum as shown in Figure 17.26.

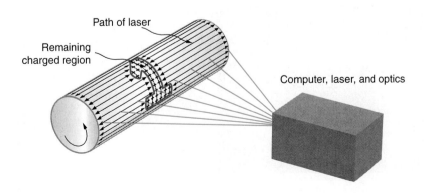

Figure 17.26 In a laser printer, a laser beam is scanned across a photoconducting drum, leaving a positive charge image. The other steps for charging the drum and transferring the image to paper are the same as in xerography. Laser light can be very precisely controlled, enabling laser printers to produce high-quality images.

Figure 17.27 The nozzle of an ink jet printer produces small ink droplets, which are sprayed with electrostatic charge. Various computer-driven devices are then used to direct the droplets to the correct positions on a page.

In its most common application, the laser printer receives output from a computer, and it can achieve high-quality output because of the precision with which laser light can be controlled. Many laser printers do significant information processing, such as making sophisticated letters or fonts, and may contain a computer more powerful than the one giving them the raw data to be printed.

Ink Jet Printers and Electrostatic Painting

The ink jet printer, most commonly used to print computer-generated text and graphics, also employs electrostatics. A nozzle makes a fine spray of tiny ink droplets, which are then given an electrostatic charge. (See Figure 17.27.) Once charged, the droplets can be directed, using charged plates, with great precision to form letters and images on paper. Ink jet printers can produce color images by using a black jet and three other jets with primary colors, usually cyan, magenta, and yellow, much as a color television produces color. (This is more difficult with xerography, requiring multiple drums and toners.)

Electrostatic painting employs electrostatic charge to spray paint onto odd-shaped surfaces. Mutual repulsion of like charges causes the paint to fly away from its source. Surface tension forms drops, which are then attracted by unlike charges to the surface to be painted. Electrostatic painting can reach those hard to get at places, applying an even coat in a controlled manner. (See Figure 17.28.) If the object is a conductor, the electric field is perpendicular to the surface, tending to bring the drops in perpendicularly. Corners and points on conductors will receive extra paint. Felt can similarly be applied.

Figure 17.28 (a) Electrostatic painting in progress. (b) Paint given a large excess charge flies outward, forming droplets. The object to be painted is given the charge opposite to the paint and attracts it perpendicularly to its surfaces.

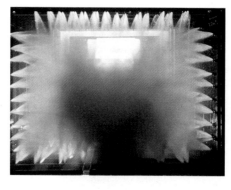

(a)

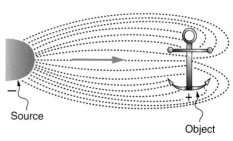

(b)

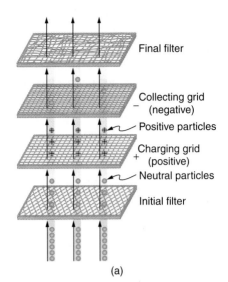

Final filter

Collecting grid
(negative)

Positive particles

Charging grid
(positive)

Neutral particles

Initial filter

(a)

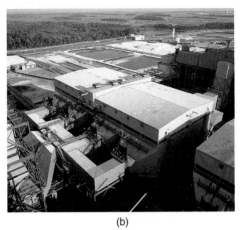

(b)

Figure 17.29 (a) Schematic of an electrostatic precipitator. Air is passed through grids of opposite charge. The first grid charges airborne particles, while the second attracts and collects them. (b) The dramatic effect of electrostatic precipitators is seen by the *absence* of smoke from this power plant.

Smoke Precipitators and Electrostatic Air Cleaning

Another important application of electrostatics is found in air cleaners, both large and small. The electrostatic part of the process places excess (usually positive) charge on smoke, dust, pollen, and other particles in the air and then passes the air through an oppositely charged grid that attracts and retains the charged particles. (See Figure 17.29.) Large electrostatic precipitators are used industrially to remove all manner of particles from air discharged into the environment. Home precipitators are very effective in removing polluting particles, irritants, and allergenics, often in conjunction with the home heating and air conditioning system.

SUMMARY

Electrostatics or static electricity, common and easily observed, results from stationary **electric charge**. There are only two types of charge, **positive** and **negative**. Like charges repel, unlike charges attract, and the force between charges decreases with the square of the distance. The vast majority of positive charge in nature is carried by protons, while the vast majority of negative charge is carried by electrons. An **ion** is an atom or molecule that has nonzero total charge due to having unequal numbers of electrons and protons. The SI unit for charge is the **coulomb** (C), with protons and electrons having charges of opposite sign but equal magnitude; that basic magnitude $|q_e|$ is

$$|q_e| = 1.60 \times 10^{-19} \text{ C} \qquad (17.1)$$

Total charge is constant in any process—that is, **total charge is conserved**. Whenever charge is created or destroyed, equal amounts of positive and negative are involved. Most often, existing charges are separated from neutral objects to obtain some net charge.

A **conductor** is a substance that allows charge to move through it with relative ease, whereas an **insulator** does not. A conducting object is said to be **grounded** if it is connected to the earth through a conductor. Grounding allows transfer of charge to and from the earth's large reservoir. Objects can be **charged by contact** with another, charged object and obtain the same sign charge. If an object is temporarily grounded, it

can be charged by **induction**, and obtains the opposite sign charge. **Polarization** is the separation of charges in a neutral object. **Polar molecules** have an inherent separation of charge.

Coulomb's law gives the force between point charges. It is

$$F = k \frac{q_1 q_2}{r^2} \qquad (17.2)$$

where q_1 and q_2 are two point charges separated by a distance r, and $k \approx 9.00 \times 10^9$ N·m²/C². This *Coulomb force* is extremely basic, since most charges are due to pointlike particles. It is responsible for all electrostatic effects and underlies most macroscopic forces. The Coulomb force is extraordinarily strong compared with gravity, another basic force—but unlike gravity it can cancel, since it can be either attractive or repulsive.

The **electric field E** is defined to be

$$\mathbf{E} = \frac{\mathbf{F}}{q} \qquad (17.3)$$

where **F** is the Coulomb force exerted on a small positive test charge q. **E** is in the same direction as **F**, and has units of N/C, but unlike the Coulomb force, the electric field **E** is unique at every point and represents only the charge creating it. The electric field E created by a point charge Q is

$$E = k \frac{Q}{r^2} \qquad \text{(point charge)} \qquad (17.4)$$

where r is the distance from Q.

Drawings of electric field lines are useful visual tools. The **properties of electric field lines** for any charge distribution are:

1. Field lines must begin on positive charges *and* terminate on negative charges, or at infinity in the hypothetical case of isolated charges.
2. The number of field lines leaving a positive charge or entering a negative charge is proportional to the magnitude of the charge.
3. The strength of the field is proportional to the closeness of the field lines—more precisely, it is proportional to the number of lines per unit area perpendicular to the lines.

4. The direction of the electric field is tangent to the field line at any point in space.
5. Field lines can never cross.

A **conductor in electrostatic equilibrium** has the following properties:

1. The electric field is zero inside a conductor.
2. Just outside a conductor, the electric field lines are perpendicular to its surface, ending or beginning on charges on the surface.
3. Any excess charge resides entirely on the surface or surfaces of a conductor.

CONCEPTUAL QUESTIONS

17.1 There are very large numbers of charged particles in most objects. Why, then, don't most objects exhibit static electricity?

17.2 Why do most objects tend to contain nearly equal numbers of positive and negative charges?

17.3 An eccentric inventor attempts to levitate by first placing a large negative charge on himself and then putting a large positive charge on the ceiling of his workshop. Instead, while attempting to place a large negative charge on himself, his clothes fly off. Explain.

17.4 When you bring your finger very near the knob of a charged electroscope, the leaves drop; they then rise again when you take your finger away. Explain.

17.5 If you have charged an electroscope by contact with a positively charged object, describe how you could use it to determine the charge of other objects. Specifically, what would the leaves of the electroscope do if other charged objects were brought near its knob?

17.6 When a glass rod is rubbed with silk, it becomes positive and the silk becomes negative—yet both attract dust. Does the dust have a third type of charge that is attracted to both positive and negative? Explain.

17.7 Why does a car always attract dust right after it is polished? (Note that car wax and car tires are insulators.)

17.8 Describe how a positively charged object can be used to give another object a negative charge. What is the name of this process?

17.9 What is grounding? What effect does it have on a charged conductor? On a charged insulator?

17.10 Look back at Figure 17.8. Explain, in terms of Coulomb's law, why the separated charges are closer together in part (d) than in part (c).

17.11 Figure 17.30 shows the charge distribution in a water molecule, which is called a polar molecule because it has an inherent separation of charge. Given water's polar character, explain what effect humidity has on removing excess charge from objects.

17.12 Explain, in terms of Coulomb's law, why a polar molecule (such as in Figure 17.30) is attracted by both positive and negative charges.

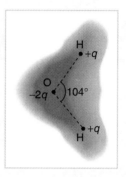

Figure 17.30 Schematic representation of the outer electron cloud of a neutral water molecule. The electrons spend more time near the oxygen than the hydrogens, giving a permanent charge separation as shown. Water is thus a *polar molecule*. It is more easily affected by electrostatic forces than molecules with uniform charge distributions. Questions 11 and 12.

17.13 Given the polar character of water molecules, explain how ions in the air form nucleation centers for rain droplets.

17.14 Why must the test charge q in the definition of the electric field be vanishingly small?

17.15 Are the direction and magnitude of the Coulomb force unique at a given point in space? What about the electric field?

17.16 The purpose of this question is to compare and contrast the Coulomb force field and the electric field. To do this, make a list of five properties for the Coulomb force field analogous to the five properties listed for electric field lines. Compare each item in your list of Coulomb force field properties with those of the electric field—are they the same or different? (For example, electric field lines cannot cross. Is the same true for Coulomb field lines?)

17.17 Figure 17.31 shows an electric field extending over three regions, labeled I, II, and III. Answer the following questions. (a) Are there any isolated charges? If so, in what region and what are their signs? (b) Where is the field strongest? (c) Where is it weakest? (d) Where is the field the most uniform?

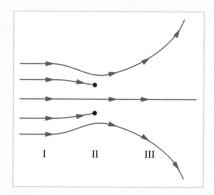

Figure 17.31 Question 17.

17.18 Is the object in Figure 17.32 a conductor or an insulator? Justify your answer.

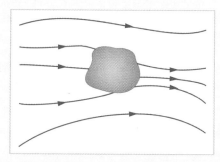

Figure 17.32 The electric field in the vicinity of an object. Questions 18 and 19.

17.19 If the electric field lines in Figure 17.32 were perpendicular to the object, would it necessarily be a conductor? Explain.

17.20 The discussion of the electric field between two parallel conducting plates, shown in Figure 17.20, states that edge effects are less important if the plates are close together. What does close mean? That is, is the actual plate separation crucial, or is the ratio of plate separation to plate area crucial?

17.21 Would the self-created electric field in Figure 17.22 remove positive or negative charge from the pointed conductor? Would the same sign charge be removed from a neutral pointed conductor by the application of a similar externally created electric field? (The answers to both questions have implications for charge transfer utilizing points.)

17.22 Why is a golfer with a metal club over her shoulder vulnerable to lightning in an open fairway? Would she be any safer under a tree?

17.23 Can the belt of a Van de Graaff be a conductor? Explain.

17.24 Are you relatively safe from lightning inside an automobile? Give two reasons.

17.25 Discuss pros and cons of a lightning rod being grounded versus simply being attached to a building.

PROBLEMS

Sections 17.1 and 17.2 Static Electricity and Charge

17.1 Common static electricity involves charges ranging from nanocoulombs to microcoulombs. (a) How many electrons are needed to form a charge of -2.00 nC? (b) How many electrons must be removed from a neutral object to leave a net charge of 0.500 μC?

17.2 If 1.80×10^{20} electrons move through a pocket calculator during a full day's operation, how many coulombs of charge moved through it?

17.3 To start a car engine, the car battery moves 3.75×10^{21} electrons through the starter motor. How many coulombs of charge was moved?

17.4 A certain lightning bolt moves 40.0 C of charge. How many fundamental units of charge $|q_e|$ is this?

• 17.5 Suppose a speck of dust in an electrostatic precipitator has 1.0000×10^{12} protons in it and has a net charge of -5.00 nC (a very large charge for a small speck). How many electrons does it have?

• 17.6 An amoeba has 1.00×10^{16} protons and a net charge of 0.300 pC. (a) How many fewer electrons are there than protons? (b) If you paired them up, what fraction of the protons would have no electrons?

: 17.7 A 50.0 g ball of copper has a net charge of 2.00 μC. What fraction of the copper's electrons have been removed? (Each copper atom has 29 protons, and copper has an atomic mass of 63.5.)

: 17.8 What net charge would you place on a 100 g piece of sulfur if you put an extra electron on 1 in 10^{12} of its atoms? (Sulfur has an atomic mass of 32.1.)

: 17.9 How many coulombs of positive charge are there in 4.00 kg of plutonium, given its atomic mass is 244 and that each plutonium atom has 94 protons?

Section 17.3 Coulomb's Law

17.10 What is the repulsive force between two pith balls that are 8.00 cm apart and have equal charges of -30.0 nC?

17.11 How strong is the attractive force between a glass rod with a 0.700 μC charge and a silk cloth with a -0.600 μC charge, which are 12.0 cm apart, assuming they act like point charges?

17.12 Two point charges exert a 5.00 N force on each other. What will the force become if the distance between them is increased by a factor of three?

17.13 Two point charges are brought closer together, increasing the force between them by a factor of 25. By what factor was their separation decreased?

17.14 How far apart must two point charges of 75.0 nC (typical of static electricity) be to have a force of 1.00 N between them?

• **17.15** Bare free charges do not remain stationary when close together. To illustrate this, calculate the acceleration of two isolated protons separated by 2.00 nm (a typical distance between gas atoms).

• **17.16** (a) Verify the acceleration for the electron in Example 17.1 is 9.00×10^{22} m/=s², as stated. (b) What is the proton's acceleration?

• **17.17** By what factor must you change the distance between two point charges to change the force between them by a factor of 10?

• **17.18** Suppose you have a total charge q_{tot} that you can split in any manner. Once split, the separation distance is fixed. How do you split the charge to achieve the greatest force?

• **17.19** Common transparent tape becomes charged when pulled from a dispenser. If one piece is placed above another, the repulsive force can be great enough to support the top piece's weight. Assuming equal point charges (only an approximation), calculate the magnitude of the charge if electrostatic force is great enough to support the weight of a 10.0 mg piece of tape held 1.00 cm above another.

• **17.20** Show that the ratio of the electrostatic to gravitational force between an electron and proton is independent of distance, as stated in Example 17.1.

• **17.21** (a) Find the ratio of the electrostatic to gravitational force between two electrons. (b) What is this ratio for two protons? (c) Are these ratios consistent with the approximate relative strengths given in Table 4.2?

‡ **17.22** At what distance is the electrostatic force between two protons equal to the weight of one?

‡ **17.23** A certain five cent coin contains 5.00 g of nickel. What fraction of the nickel atoms' electrons, removed and placed 1.00 m above it, would support the weight of this coin? The atomic mass of nickel is 58.7, and each nickel atom contains 28 electrons and 28 protons.

‡ **17.24** (a) Two point charges totaling 8.00 µC exert a repulsive force of 0.150 N on one another when separated by 0.500 m. What is the charge on each? (b) What is the charge on each if the force is attractive?

‡ **17.25** Point charges of 5.00 and −3.00 µC are placed 0.250 m apart. (a) Where can a third charge be placed so that the net force on it is zero? (b) What if both charges are positive?

Sections 17.4 and 17.7 Electric Fields and Their Applications

17.26 What is the magnitude and direction of an electric field that exerts a 2.00×10^{-5} N upward force on a −1.75 µC charge?

17.27 What is the magnitude and direction of the force exerted on a 3.50 µC charge by a 250 N/C electric field that points due east?

17.28 Calculate the magnitude of the electric field 2.00 m from a point charge of 5.00 mC (such as found on the terminal of a Van de Graaff).

17.29 (a) What magnitude point charge creates a 10,000 N/C electric field at a distance of 0.250 m? (b) How large is the field at 10.0 m?

• **17.30** Calculate the initial (from rest) acceleration of a proton in a 5.00×10^6 N/C electric field (such as created by a research Van de Graaff).

• **17.31** (a) Find the direction and magnitude of an electric field that exerts a 4.80×10^{-17} N westward force on an electron. (b) What magnitude and direction force does this field exert on a proton?

• **17.32** (a) What is the electric field 5.00 m from the center of the terminal of a Van de Graaff with a 3.00 mC charge, noting that the field is equivalent to that of a point charge at the center of the terminal? (b) Find the force this field exerts on 2.00 µC of charge on the belt at that distance.

• **17.33** What is the direction and magnitude of an electric field that supports the weight of a free electron near the surface of the earth?

• **17.34** A simple and common technique for accelerating electrons is shown in Figure 17.33, where a uniform electric field is created between two plates. Electrons are released, usually from a hot filament, near the negative plate, and there is a small hole in the positive plate that allows the electrons to continue moving. (a) Calculate the acceleration of the electron if the field strength is 2.50×10^4 N/C. (b) Explain why the electron will not be pulled back to the positive plate once it moves through the hole.

‡ **17.35** The earth has a net charge that produces an electric field of approximately 150 N/C downward at its surface. (a) What is

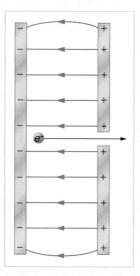

Figure 17.33 Parallel conducting plates with opposite charges on them create a relatively uniform electric field used to accelerate electrons to the right. Those that go through the hole can be used to make a TV or computer screen glow, or to produce x rays. Problems 34 and 49.

the magnitude and sign of the excess charge, noting the electric field of a conducting sphere is equivalent to a point charge at its center? (b) What acceleration will the field produce on a free electron near the earth's surface? (c) What mass object with a single extra electron will have its weight supported by this field?

‡ **17.36** Point charges of 25.0 and 45.0 µC are placed 0.500 m apart. (a) At what point along the line between them is the electric field zero? (b) What is the electric field halfway between them?

‡ **17.37** What can you say about two charges q_1 and q_2, if the electric field one-fourth of the way from q_1 to q_2 is zero?

Sections 17.5 and 17.6 Electric Field Lines

17.38 (a) Sketch the electric field lines near a point charge $+q$.
(b) Do the same for a point charge $-3.00q$.

17.39 Sketch the electric field lines a long distance from the charge distributions shown in Figure 17.16(a) and (b).

17.40 (a) Figure 17.34 shows the electric field lines near two charges q_1 and q_2. What is the ratio of their magnitudes?
(b) Sketch the electric field lines a long distance from the charges shown in Figure 17.34.

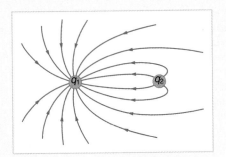

Figure 17.34 The electric field near two charges. Problem 40.

•**17.41** Sketch the electric field lines in the vicinity of two opposite charges, where the negative charge is three times greater in magnitude than the positive. See Figure 17.34 for a similar situation.

•**17.42** Sketch the electric field lines in the vicinity of the conductor in Figure 17.35, given the field was originally uniform and parallel to the object's long axis. Is the resulting field small near the long side of the object?

Figure 17.35 Problem 42.

•**17.43** Sketch the electric field lines in the vicinity of the conductor in Figure 17.36, given the field was originally uniform and parallel to the object's long axis. Is the resulting field small near the long side of the object?

Figure 17.36 Problem 43.

•**17.44** Sketch the electric field between the two conducting plates shown in Figure 17.37, given the top plate is positive and an equal amount of negative charge is on the bottom plate. Be certain to indicate the distribution of charge on the plates.

Figure 17.37 Problem 44.

⁝**17.45** Sketch the electric field lines in the vicinity of the charged insulator in Figure 17.38 noting its nonuniform charge distribution.

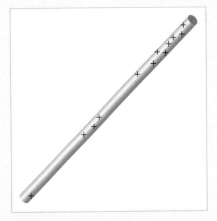

Figure 17.38 A charged insulating rod such as might be used in a classroom demonstration. Problem 45.

INTEGRATED CONCEPTS

The following problems involve concepts from this chapter and several others. Physics is most interesting when applied to general situations involving more than a narrow set of physical principles. The electric field exerts force on charges, for example, and hence the relevance of Chapter 4, Dynamics. The following topics are involved in some or all of the problems in this section:

Topics	*Location*
Kinematics	Chapter 2
Two-dimensional kinematics	Chapter 3
Dynamics	Chapter 4
Statics	Chapter 5
Uniform circular motion	Chapter 8
Fluid statics	Chapter 10

PROBLEM-SOLVING STRATEGY

Step 1. *Identify which physical principles are involved.*

Step 2. *Solve the problem using strategies outlined in the text.*

The following worked example illustrates how this strategy is applied to an integrated concept problem.

EXAMPLE ACCELERATION OF A CHARGED DROP OF GASOLINE

The following topics are involved in this integrated concepts worked example:

Topics	Location
Electric charge and field	Chapter 17
Dynamics	Chapter 4

If steps are not taken to ground a gas pump, static electricity can be placed on gasoline when filling your car's tank. Suppose a tiny drop of gasoline has a mass of 4.00×10^{-15} kg and is given a positive charge of 3.20×10^{-19} C. (a) Find the weight of the drop. (b) Calculate the electric force on the drop if there is an upward electric field of strength 3.00×10^5 N/C due to other static electricity in the vicinity. (c) Calculate the drop's acceleration.

Strategy Step 1 To solve an *integrated concept problem,* such as those following this example, we must first identify the physical principles involved and identify the chapters in which they are found. Part (a) of this example asks for weight. This is a topic of *dynamics* and defined in Chapter 4. Part (b) deals with electric force on a charge, a topic of the current Chapter 17 on *electric charge and field.* Part (c) asks for acceleration, knowing forces and mass. These are part of Newton's laws, also found in Chapter 4.

Strategy Step 2 The following solutions to each part of the example illustrate how the specific problem-solving strategies are applied. These involve identifying knowns and unknowns, checking to see if the answer is reasonable, and so on.

Solution to (a): Dynamics (Chapter 4) Weight is mass times the acceleration of gravity, as first expressed in Equation 4.4:

$$w = mg$$

Entering the given mass and the average acceleration of gravity yields

$$w = (4.00 \times 10^{-15} \text{ kg})(9.80 \text{ m/s}^2) = 3.92 \times 10^{-14} \text{ N}$$

Discussion for (a) This is a small weight, consistent with the small mass of the drop.

Solution for (b): Electric force (Chapter 17) The force an electric field exerts on a charge is given by rearranging Equation 17.3:

$$F = qE$$

Here we are given the charge (3.20×10^{-19} C is twice the fundamental unit of charge) and the electric field strength, and so the electric force is found to be

$$F = (3.20 \times 10^{-19} \text{ C})(3.00 \times 10^5 \text{ N/C}) = 9.60 \times 10^{-14} \text{ N}$$

Discussion for (b) While this is a small force, it is greater than the weight of the drop.

Solution for (c): Acceleration (Chapter 4) The acceleration can be found using Newton's second law, provided we can identify all of the external forces acting on the drop. We assume only the drop's weight and the electric force are significant. Since the drop has a positive charge and the electric field is given to be upward, the electric force is upward. We thus have a one-dimensional (vertical direction) problem, and we can state Newton's second law as

$$a = \frac{\text{net } F}{m}$$

where net $F = F - w$. Entering this and the known values into the

expression for Newton's second law yields

$$a = \frac{F - w}{m}$$
$$= \frac{9.60 \times 10^{-14} \text{ N} - 3.92 \times 10^{-14} \text{ N}}{4.00 \times 10^{-15} \text{ kg}}$$
$$= 14.2 \text{ m/s}^2$$

Discussion for (c) This is an upward acceleration great enough to carry the drop to places where you might not wish to have gasoline.

This worked example illustrates how to apply problem-solving strategies to situations that include topics in different chapters. The first step is to identify the physical principles involved in the problem. The second step is to solve for the unknown using familiar problem-solving strategies. These are found throughout the text, and many worked examples show how to use them for single topics. In this integrated concepts example, you can see how to apply them across several topics. You will find these techniques useful in applications of physics outside a physics course, such as in your profession, in other science disciplines, and in everyday life. The following problems will build your skills in the broad application of physical principles.

• **17.46** Calculate the angular velocity ω of an electron orbiting a proton in the hydrogen atom, given the radius of the orbit is 0.530×10^{-10} m. You may assume that the proton is stationary and the centripetal force is supplied by Coulomb attraction.

• **17.47** An electron has an initial velocity of 5.00×10^6 m/s in a uniform 2.00×10^5 N/C strength electric field. The field accelerates the electron in the direction opposite to its initial velocity. (a) What is the direction of the electric field? (b) How far does the electron travel before coming to rest? (c) How long does it take the electron to come to rest? (d) What is the electron's velocity when it returns to its starting point?

• **17.48** The practical limit to an electric field in air is about 3.00×10^6 N/C. Above this strength, sparking takes place because air begins to ionize and charges flow, reducing the field. (a) Calculate the distance a free proton must travel in this field to reach 3.00% of the speed of light, starting from rest. (b) Is this practical in air, or must it occur in a vacuum?

• **17.49** (a) Repeat Problem 17.34, but, in addition, find the speed of the electron after it has been accelerated through a distance of 2.00 cm. (b) How long does this take?

• **17.50** A 5.00 g charged insulating ball hangs on a 30.0 cm long string in a uniform horizontal electric field as shown in Figure 17.39. Given the charge on the ball is 1.00 μC, find the strength of the field.

‡ **17.51** Two identical 0.500 g insulating balls with equal charges hang on 20.0 cm long strings as shown in Figure 17.40. Find their charge.

‡ **17.52** Figure 17.41 shows an electron passing between two charged metal plates that create an 100 N/C vertical electric field perpendicular to the electron's original horizontal velocity. (These can be used to change the electron's direction, such as in an oscilloscope.) The initial speed of the electron is 3.00×10^6 m/s, and the horizontal distance it travels in the uniform field is 4.00 cm. (a) What is its vertical deflection?

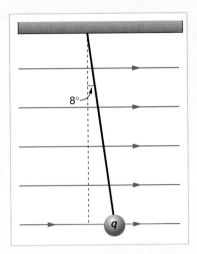

Figure 17.39 A horizontal electric field causes the charged ball to hang at an angle of 8.00°. Problem 50.

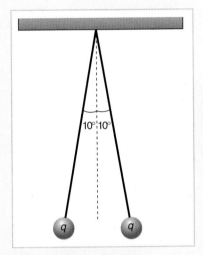

Figure 17.40 Two charged balls repel to an angle of 10.0°. Problem 51.

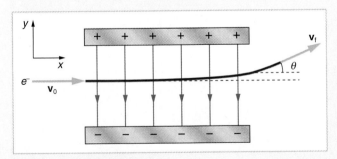

Figure 17.41 Problem 52.

(b) What is the vertical component of its final velocity? (c) At what angle does it exit? Neglect any edge effects.

‡ **17.53** The classic Millikan oil drop experiment was the first to obtain an accurate measurement of the charge on an electron. In it, oil drops were suspended against gravity by a vertical electric field. (See Figure 17.42.) Given the oil drop to be 1.00 μm in radius and have a density of 920 kg/m³:

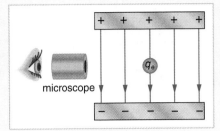

Figure 17.42 In the Millikan oil drop experiment, small drops can be suspended in an electric field by the force exerted on a single excess electron. Classically, this experiment was used to determine the electron charge q_e by measuring the electric field and mass of the drop. Problem 53.

(a) Find the weight of the drop. (b) If the drop has a single excess electron, find the electric field strength needed to balance its weight.

UNREASONABLE RESULTS

The following problems have results that are unreasonable because some premise is unreasonable or because certain of the premises are inconsistent with one another. Physical principles applied correctly then produce unreasonable results. The purpose of these problems is to give practice in assessing whether nature is being accurately described, and if it is not to trace the source of difficulty.

PROBLEM-SOLVING STRATEGY

To determine if an answer is reasonable, and to determine the cause if it is not, do the following.

Step 1. *Solve the problem using strategies as outlined in Section 17.6 and in several other places in the text.* See the table of contents to locate problem-solving strategies. Use the format followed in the worked examples in the text to solve the problem as usual.

Step 2. *Check to see if the answer is reasonable.* Is it too large or too small, or does it have the wrong sign, improper units, . . .?

Step 3. *If the answer is unreasonable, look for what specifically could cause the identified difficulty.* Usually, the manner in which the answer is unreasonable is an indication of the difficulty. For example, an extremely large Coulomb force could be due to the assumption of an excessively large separated charge.

17.54 (a) Calculate the electric field strength near a 10.0 cm diameter conducting sphere that has 1.00 C of excess charge on it. (b) What is unreasonable about this result? (c) Which assumptions are responsible?

• **17.55** (a) Two 0.500 g raindrops in a thunderhead are 1.00 cm apart when they each acquire 1.00 mC charges. Find their acceleration. (b) What is unreasonable about this result? (c) Which premise or assumption is responsible?

• **17.56** A wrecking yard inventor wants to pick up cars by charging a 0.400 m diameter ball and inducing an equal and opposite charge on the car. If a car has a 1000 kg mass and the ball is to be able to lift it from a distance of 1.00 m: (a) What charge must be used? (b) What is the electric field near the surface of the ball? (c) Why are these results unreasonable? (d) Which premise or assumption is responsible?

18 ELECTRIC POTENTIAL AND ELECTRIC ENERGY

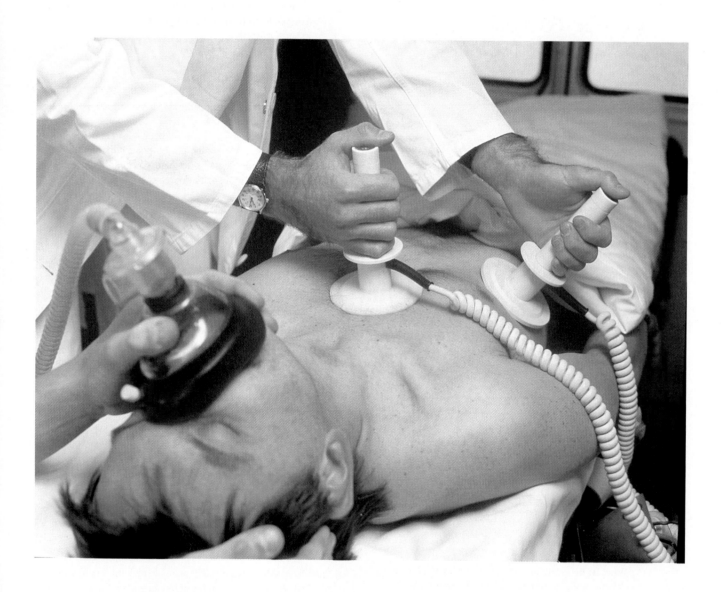

Electric potential energy.

The previous chapter just scratched the surface (or at least rubbed it) of electrical phenomena. Two of the most familiar aspects of electricity are its energy and voltage. We know, for example, that great amounts of electrical energy can be stored in batteries, are transmitted cross-country through power lines, and may jump from clouds to explode the sap of trees. We also know about voltages associated with electricity. Batteries are typically a few volts, the outlet in our home produces 120 volts, and power lines can be as high as hundreds of thousands of volts. But energy and voltage are not the same thing. A motorcycle battery, for example, is small and would not be very successful in replacing the much larger car battery, yet each has the same voltage. In this chapter, we shall examine the relationship between voltage and electrical energy and begin to explore some of the many applications of electricity.

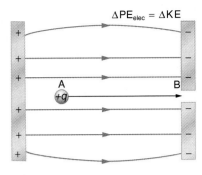

Figure 18.1 A charge accelerated by an electric field is analogous to a mass going down a hill. In both cases potential energy is converted to another form. Work is done by a force, but since this force is conservative we treat the work as potential energy.

18.1 ELECTRIC POTENTIAL ENERGY: POTENTIAL DIFFERENCE

When a free positive charge q is accelerated by an electric field, such as shown in Figure 18.1, it is given kinetic energy. The process is analogous to gravity. It is as if the charge is going down an electrical hill where its electric potential energy is converted to kinetic energy. Let us explore the work done on a charge q by the electric field in this process, so that we may develop a definition of electric potential energy.

The electrostatic or Coulomb force is conservative,* which means that the work done on q while moving from point A to point B depends only on the initial and final positions, not on the path taken. This is exactly analogous to gravity in the absence of dissipative forces such as friction. When a force is conservative, it is possible to define a potential energy associated with the force, and it is usually easier to deal with the potential energy (because it depends only on position) than to calculate the work directly.

We use the letters PE to denote **electric potential energy**, which has units of joules. Change in potential energy, ΔPE, is crucial, since the work done *by* a conservative force is the negative of the change in potential energy; that is, $W = -\Delta$PE. For example, work W done to accelerate a positive charge from rest is positive and results from a *loss* in PE, or a negative ΔPE. There must be a minus sign in front of ΔPE to make W positive. PE can be found at any point by taking one point as a reference and calculating the work needed to move a charge to the other point.

Calculating the work directly is generally difficult, since $W = Fd \cos \theta$, and the direction and magnitude of F can be complex for multiple charges, for odd-shaped objects, and along arbitrary paths. But we do know that, since $F = qE$, the work and, hence, ΔPE is proportional to the test charge q. To have a physical quantity that is independent of test charge, we define **electric potential** V (or simply **potential**, since electric is understood) to be the potential energy per unit charge:

$$V = \frac{\text{PE}}{q}$$

Since PE is proportional to q, this cancels the dependence on q. Thus V does not depend on q. The change in potential energy ΔPE is crucial, and so we are concerned with the **difference in potential** or **potential difference** ΔV between two points, where

$$\Delta V = V_{\text{B}} - V_{\text{A}} = \frac{\Delta \text{PE}}{q} \qquad \textbf{(18.1)}$$

The potential difference between points A and B, $V_{\text{B}} - V_{\text{A}}$, is thus defined to be the change in potential energy of a charge q moved from A to B, divided by the charge. Units of potential difference are joules per coulomb, given the name **volt** (V) after Allesandro Volta:

$$1 \text{ V} = 1 \text{ J/C} \qquad \textbf{(18.2)}$$

The familiar term **voltage** is the common name for potential difference. Keep in mind that whenever a voltage is quoted, it is understood to be the potential difference between two

CONNECTIONS

Potential Energy

For a detailed treatment of work, energy, and potential energy, see Chapter 6. In particular, Section 6.4—Conservative Forces and Potential Energy—is directly relevant. Gravitational potential energy and electric potential energy are quite analogous. Potential energy accounts for work done by a conservative force and gives added insight regarding energy and energy transformation without the necessity of dealing with the force directly. It is much more common, for example, to use the concept of voltage (related to electric potential energy) than to deal with the Coulomb force directly.

---

*Processes in which an accelerated charge radiates, or where electrical energy is converted to thermal energy, must be considered separately. These are dissipative processes analogous to friction.

points. For example, every battery has two terminals, and its voltage is the potential difference between them. More fundamentally, the point you choose to be zero volts is arbitrary. This is analogous to the fact that gravitational potential energy has an arbitrary zero, such as sea level or perhaps a lecture hall floor.

In summary, the important relationship between potential difference (or voltage) and electrical potential energy is given by

and
$$\left.\begin{aligned} \Delta V &= \frac{\Delta PE}{q} \\ \Delta PE &= q\,\Delta V \end{aligned}\right\} \qquad (18.3)$$

The second expression is equivalent to the first.

Voltage V is *not* the same as energy. Voltage is the energy per unit charge. Thus a motorcycle battery and a car battery can both have the same voltage (more precisely, the same potential difference between battery terminals), yet one holds much more energy than the other since $\Delta PE = q\,\Delta V$. The car battery can separate and move more charge than the motorcycle battery, although both are 12 V batteries.

EXAMPLE 18.1 TWO DIFFERENT 12.0 V BATTERIES STORE DIFFERENT AMOUNTS OF ENERGY

Suppose you have a 12.0 V motorcycle battery that can move 5000 C of charge, and a 12.0 V car battery that can move 60,000 C of charge. How much energy does each deliver?

Strategy To say we have a 12.0 V battery means that its terminals have a 12.0 V potential difference. When such a battery moves charge, it puts the charge through a potential difference of 12.0 V, and the charge is given a change in potential energy equal to

$$\Delta PE = q\,\Delta V$$

So to find the energy output, we multiply the charge moved by the potential difference.

Solution For the motorcycle battery, $q = 5000$ C and $\Delta V = 12.0$ V. The total energy delivered by the motorcycle battery is

$$\begin{aligned} \Delta PE_{cycle} &= (5000 \text{ C})(12.0 \text{ V}) \\ &= (5000 \text{ C})(12.0 \text{ J/C}) \\ &= 6.00 \times 10^4 \text{ J} \end{aligned}$$

Similarly, for the car battery, $q = 60,000$ C and

$$\begin{aligned} \Delta PE_{car} &= (60,000 \text{ C})(12.0 \text{ V}) \\ &= 7.20 \times 10^5 \text{ J} \end{aligned}$$

Discussion While voltage and energy are related, they are not the same thing. The voltages of the batteries are identical, but the energy supplied by each is quite different. Note also that as a battery is discharged, some of its energy is used internally and its terminal voltage drops, such as when headlights dim for a low car battery. The energy supplied by the battery is still calculated as in this example, but not all of the energy is available for external use.

Note that the energies calculated in the previous example are absolute values. The change in potential energy for the battery is negative, since it loses energy. These batteries, like many electrical systems, actually move negative charge—electrons in particular. The batteries repel electrons from their negative terminals through whatever circuitry is involved and attract them to their positive terminals as shown in Figure 18.2. The change in potential is $\Delta V = V_B - V_A = +12$ V and the charge q is negative, so that $\Delta PE = q\,\Delta V$ is negative, meaning the potential energy of the battery has decreased when q has moved from A to B.

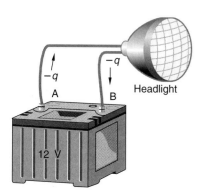

Figure 18.2 A battery moves negative charge from its negative terminal through a headlight to its positive terminal. Appropriate combinations of chemicals in the battery separate charges so that the negative terminal has an excess of negative charge, which is repelled by it and attracted to the excess positive charge on the other terminal. In terms of potential, the positive terminal is at a higher voltage than the negative. Inside the battery, both positive and negative charges move.

EXAMPLE 18.2 HOW MANY ELECTRONS MOVE THROUGH A HEADLIGHT EACH SECOND?

When a 12.0 V car battery runs a single 30.0 W headlight, how many electrons pass through it each second?

Strategy To find the number of electrons, we must first find the charge that moved in 1.00 s. The charge moved is related to voltage and energy through Equation 18.3, $\Delta PE = q \, \Delta V$. A 30.0 W lamp uses 30.0 joules per second, so that $\Delta PE = -30.0$ J here, and $\Delta V = +12.0$ V.

Solution To find the charge q moved, we solve Equation 18.3 for q. Starting with

$$\Delta PE = q \, \Delta V$$

q is then

$$q = \frac{\Delta PE}{\Delta V}$$

Now, since the electrons are going from the negative terminal to the positive, we see that $\Delta V = +12.0$ V. Since the battery loses energy, we have $\Delta PE = -30.0$ J. Entering these

values into the expression for q yields

$$q = \frac{-30.0 \text{ J}}{+12.0 \text{ V}} = \frac{-30.0 \text{ J}}{+12.0 \text{ J/C}} = -2.50 \text{ C}$$

The number of electrons, n_e, is the total charge divided by the charge per electron. That is,

$$n_e = \frac{2.50 \text{ C}}{-1.60 \times 10^{-19} \text{ C}/e^-} = 1.56 \times 10^{19} \text{ electrons}$$

Discussion This is a very large number. It is no wonder that we do not ordinarily observe individual electrons with so many being present in ordinary systems. In fact, electricity had been in use for many decades before it was determined that the moving charges in many circumstances were negative. Positive charge moving in the opposite direction of negative charge often produces identical effects; this makes it difficult to determine which is moving or whether both are moving.

The Electron Volt

The energy per electron is very small in macroscopic situations like that in the previous example—a tiny fraction of a joule. But on a submicroscopic scale, such energy per particle (electron, proton, or ion) can be of great importance. For example, even a tiny fraction of a joule can be great enough for these particles to destroy organic molecules and harm living tissue. The particle may do its damage by direct collision, or it may create harmful x rays, which can also inflict damage. It is useful to have an energy unit related to submicroscopic effects. Figure 18.3 shows a situation related to the definition of such an energy unit. An electron is accelerated between two charged metal plates as it might be in a TV tube, computer terminal, or x-ray tube. The electron is given kinetic energy that is later converted to another form—light in a TV, for example. (Note that downhill for the electron is uphill for a positive charge.) Since energy is related to voltage by $\Delta PE = q \, \Delta V$, we can think of the joule as a coulomb-volt. On the submicroscopic scale, it is more convenient to define an energy unit called the **electron volt** (eV) to be *the energy given a fundamental charge accelerated through a potential difference of 1 V.* In equation form,

$$1 \text{ eV} = (1.60 \times 10^{-19} \text{ C})(1 \text{ V}) = (1.60 \times 10^{-19} \text{ C})(1.00 \text{ J/C})$$

or

$$1 \text{ eV} = 1.60 \times 10^{-19} \text{ J} \tag{18.4}$$

An electron accelerated through a potential difference of 1 V is given an energy of 1 eV. It follows that an electron accelerated through 50 V is given 50 eV. A potential difference of 100,000 V (100 kV) will give an electron an energy of 100,000 eV (100 keV), and so on. Similarly, an ion with a double positive charge accelerated through 100 V will be given 200 eV of energy. These simple relationships between accelerating voltage and particle charges make the electron volt a simple and convenient energy unit in such circumstances.

The electron volt is commonly employed in submicroscopic processes—chemical valence energies and molecular and nuclear binding energies are among the quantities often expressed in electron volts. For example, about 5 eV of energy is required to break up certain organic molecules. If a proton is accelerated from rest through a potential difference of 30 kV, it is given an energy of 30 keV (30,000 eV) and it can break up as many as 6000 of these molecules (30,000 eV/5 eV per molecule = 6000 molecules). Nuclear decay energies are on the order of 1 MeV (1,000,000 eV) per event and can, thus, produce significant biological damage.

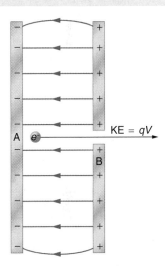

Figure 18.3 A typical electron gun accelerates electrons using a potential difference between two metal plates. The energy of the electron in electron volts is numerically the same as the voltage between the plates. For example, a 5000 V potential difference produces 5000 eV electrons.

CONNECTIONS

Energy Units

The electron volt (eV) is the most common energy unit for submicroscopic processes. This will be particularly noticeable in the chapters on modern physics. Energy is so important to so many subjects that there is a tendency to define a special energy unit for each major topic. There are, for example, calories for food energy, kilowatt-hours for electrical energy, and therms for natural gas energy.

Conservation of Energy

For conservative forces, such as the electrostatic force, conservation of energy states that mechanical energy is a constant. Mechanical energy is the sum of kinetic and potential energy; that is, KE + PE. Here PE is the electric potential energy. Conservation of energy is stated in equation form as

$$KE + PE = \text{constant}$$

or

$$KE_i + PE_i = KE_f + PE_f$$

where i and f stand for initial and final conditions. As we have found many times before, considering energy can give us insights and facilitate problem solving.

EXAMPLE 18.3 ELECTRICAL POTENTIAL ENERGY CONVERTED TO KINETIC ENERGY

Calculate the final speed of a free electron accelerated from rest through a potential difference of 100 V, perhaps as in Figure 18.3.

Strategy Assuming the electron is accelerated in a vacuum and neglecting gravity (we will check on this assumption later), all of the electrical potential energy is converted into kinetic energy. We can identify the initial and final forms of energy to be $KE_i = 0$, $KE_f = (1/2)mv^2$, $PE_i = qV$, and $PE_f = 0$.

Solution Conservation of energy states that

$$KE_i + PE_i = KE_f + PE_f$$

Entering the forms identified above, we obtain

$$qV = \tfrac{1}{2}mv^2$$

We solve this for v:

$$v = \left(\frac{2qV}{m}\right)^{1/2}$$

Entering values for q, V, and m gives

$$v = \left[\frac{2(-1.60 \times 10^{-19}\ \text{C})(-100\ \text{J/C})}{9.11 \times 10^{-31}\ \text{kg}}\right]^{1/2}$$
$$= 5.93 \times 10^6\ \text{m/s}$$

Discussion Note that both the charge and the initial voltage are negative, as in Figure 18.3. From the discussions in Chapter 17, we know that electrostatic forces on small particles are generally very large compared with gravity. The large final speed confirms that gravity is indeed negligible here. The large speed also indicates how easy it is to accelerate electrons with small voltages, because of their very small mass. Voltages higher than the 100 V in this problem are typically used in electron guns. Those higher voltages produce electron speeds so great that relativistic effects must be taken into account. That is why a low voltage is considered (accurately) in this example.

18.2 ELECTRIC POTENTIAL IN A UNIFORM ELECTRIC FIELD

In the previous section, we explored the relationship between voltage and energy. In this section, we will explore the relationship between voltage and electric field (defined in Section 17.4). For example, a uniform electric field E can be created by placing a potential difference (or voltage) V across two parallel metal plates. (See Figure 18.4.) Examining this will tell us what voltage is needed to create a certain electric field strength; it will also reveal a more fundamental relationship between electric potential and electric field. From a physicist's point of view, either V or E can be used to describe any charge distribution. V is most closely tied to energy, whereas E is most closely related to force. The relationship between V and E is revealed by calculating the work done by the force in moving a charge from point A to point B. But, as noted in the previous section, this is complex for arbitrary charge distributions, requiring calculus. We therefore look at a uniform field as an interesting special case.

The work done by the electric field in Figure 18.4 to move a positive charge q from A to B is

$$W = -\Delta PE = -q\,\Delta V$$

The potential difference between points A and B is

$$-\Delta V = V_A - V_B = V_{AB}$$

Entering this into the expression for work yields

$$W = qV_{AB}$$

Work is $W = Fd \cos \theta$; here $\cos \theta = 1$, since the path is parallel to the field, and so

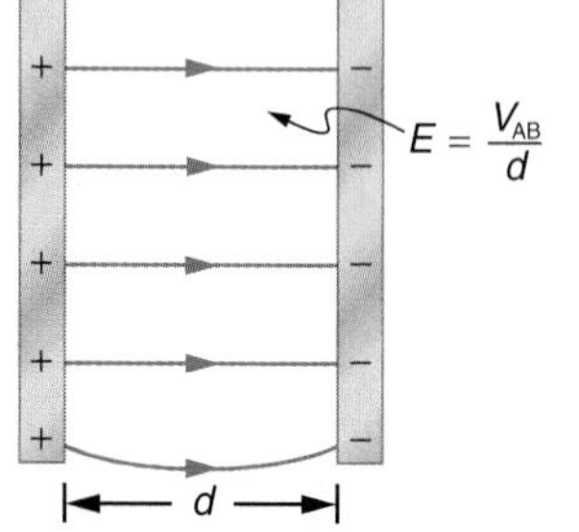

Figure 18.4 The relationship of V and E between parallel conducting plates is $E = V/d$.

$W = Fd$. Since $F = qE$, we see that $W = qEd$. Substituting this expression for work into the previous equation gives

$$qEd = qV_{AB}$$

The charge cancels, and so the voltage between points A and B is seen to be

and
$$\left.\begin{array}{l} V_{AB} = Ed \\ E = \dfrac{V_{AB}}{d} \end{array}\right\} \quad \text{(uniform } E\text{-field only)} \qquad \textbf{(18.5)}$$

d is the distance from A to B, or the distance between the plates in Figure 18.4. Note that Equation 18.5 implies the units for electric field are volts per meter. We already know the units for electric field are newtons per coulomb; thus the following relation among units is valid:

$$1 \ \text{N/C} = 1 \ \text{V/m} \qquad \textbf{(18.6)}$$

EXAMPLE 18.4 WHAT IS THE HIGHEST VOLTAGE POSSIBLE BETWEEN TWO PLATES?

Dry air will support a maximum electric field strength of about 3×10^6 V/m. Above that value, the field creates enough ionization in the air to make the air a conductor. This allows a discharge or spark that reduces the field. What, then, is the maximum voltage between two parallel conducting plates separated by 2.5 cm of dry air?

Strategy We are given the maximum electric field E between the plates and the distance d between them. Equation 18.5 can thus be used to calculate the maximum voltage.

Solution The potential difference or voltage between the plates is given by Equation 18.5:

$$V_{AB} = Ed$$

Entering the given values for E and d gives

$$V_{AB} = (3 \times 10^6 \ \text{V/m})(0.025 \ \text{m}) = 7.5 \times 10^4 \ \text{V}$$

or

$$\boxed{V_{AB} = 75 \ \text{kV}}$$

Discussion The answer is quoted to only two digits, since the maximum field strength is approximate. One of the implications of this is that it takes about 75 kV to make a spark jump a 2.5 cm (1 in.) gap, or 150 kV for a 5 cm spark. This limits the voltages that can exist between conductors, perhaps on a power transmission line. Smaller voltage will cause a spark if there are points on the surface, since points create greater fields than smooth surfaces. Humid air breaks down at a lower field strength, meaning that a smaller voltage will make a spark jump through

humid air. The largest voltages can be built up, say with static electricity, on dry days.

Figure 18.5 A spark chamber is used to trace the paths of high-energy particles. Ionization created by the particles as they pass through the gas between the plates allows a spark to jump. The sparks are perpendicular to the plates, following electric field lines between them. The potential difference between adjacent plates is not high enough to cause sparks without the ionization produced by cosmic rays or particles from accelerator experiments.

EXAMPLE 18.5 FIELD AND FORCE INSIDE AN ELECTRON GUN

(a) An electron gun (like the one in Figure 18.3) has parallel plates separated by 4.00 cm and gives electrons an energy of 25.0 keV. What is the electric field strength between the plates? (b) What force would this field exert on a piece of plastic with a 0.500 μC charge that gets between the plates?

Strategy Since voltage and plate separation are given, the electric field strength can be calculated directly from the second expression in Equation 18.5. Once the electric field strength is known, the force on a charge is found using $F = qE$.

(continued)

(continued)

Solution for (a) The expression for the electric field between two uniform metal plates is given by Equation 18.5:

$$E = \frac{V_{AB}}{d}$$

Since the electron is a single charge and is given 25.0 keV, the potential difference must be 25.0 kV. Entering this value for V_{AB} and the plate separation of 0.0400 m, we obtain

$$E = \frac{25.0 \text{ kV}}{0.0400 \text{ m}} = 6.25 \times 10^5 \text{ V/m}$$

Solution for (b) The force on a charge in an electric field is obtained from Equation 17.3:

$$F = qE$$

Substituting known values gives

$$F = (0.500 \times 10^{-6} \text{ C})(6.25 \times 10^5 \text{ V/m}) = 0.310 \text{ N}$$

Discussion Note that the units are newtons, since 1 V/m = 1 N/C. The force on the charge is the same no matter where the charge is located between the plates. This is because the E-field is uniform between the plates.

In more general situations, whether or not the E-field is uniform, the electric field E points in the direction of decreasing potential. That is because the force on a positive charge is in the direction of E and also in the direction of lower V. Furthermore, the magnitude of E equals the rate of decrease of V with distance. The faster V decreases over distance, the greater the E-field. In equation form, the general relationship between voltage and electric field is

$$E = -\frac{\Delta V}{\Delta s} \tag{18.7}$$

where Δs is the distance over which the change in potential, ΔV, takes place. (This is consistent with Equation 18.5, but with $\Delta s = d$ and $-\Delta V = V_{AB}$.) The electric field is said to be the *gradient* (as in grade or slope) of the electric potential. For continually changing potentials, ΔV and Δs become infinitesimals—differential calculus must be employed.

18.3 ELECTRIC POTENTIAL DUE TO A POINT CHARGE

Point charges, such as electrons, are among the fundamental building blocks of matter. Furthermore, spherical charge distributions (like on a metal sphere) create external electric fields exactly like a point charge. The electric potential due to a point charge is, thus, a case we need to consider. Using calculus to find the work needed to move a test charge q from a large distance away to a distance of r from a point charge Q, and noting the connection between work and potential ($W = -q\,\Delta V$), it can be shown that the **electric potential V of a point charge** is

$$V = k\frac{Q}{r} \qquad \text{(point charge)} \tag{18.8}$$

The potential at infinity is chosen to be zero.* Thus V for a point charge decreases with distance, whereas E for a point charge decreases with distance squared ($E = F/q = kQ/r^2$).

There is a major difference between V and E. The electric potential V is a scalar and has no direction, whereas the electric field E is a vector. To find the voltage due to a combination of point charges, you add the individual voltages as numbers. To find the total electric field, you must add the individual fields as vectors, taking magnitude and direction into account. This is consistent with the fact that V is closely associated with energy, a scalar, whereas E is closely associated with force, a vector.

EXAMPLE 18.6 WHAT VOLTAGE IS PRODUCED BY A SMALL CHARGE ON A METAL SPHERE?

Charges in static electricity are typically in the nanocoulomb (nC) to microcoulomb (μC) range. What is the voltage 5.00 cm away from the center of a metal sphere that has a -3.00 nC static charge?

Strategy As we have discussed in the previous chapter, charge on a metal sphere spreads out uniformly and produces a field like that of a point charge located at its center. Thus Equation 18.8 can be used to find the voltage V.

(continued)

---

*V can be thought of as the absolute potential at a distance r from Q. Strictly speaking, V is the *potential difference* between r and infinity with $V = 0$ at infinity as a reference voltage.

(continued)

Solution Entering known values into the expression for the potential of a point charge, we obtain

$$V = k\frac{Q}{r} = (9.00 \times 10^9 \text{ N} \cdot \text{m}^2/\text{C}^2)\frac{-3.00 \times 10^{-9} \text{ C}}{5.00 \times 10^{-2} \text{ m}}$$

$$= -540 \text{ V}$$

Discussion The negative value for voltage means a positive charge would be attracted from a larger distance, since the potential is lower (more negative) than at larger distances. Conversely, a negative charge would be repelled, as expected.

EXAMPLE 18.7   WHAT IS THE EXCESS CHARGE ON THE VAN DE GRAAFF?

A demonstration Van de Graaff generator has a 25.0 cm diameter metal sphere that produces a voltage of 100 kV near its surface. (See Figure 18.6.) What excess charge resides on the sphere?

Strategy The potential will be the same as that of a point charge 12.5 cm away. We can thus solve Equation 18.8 to find the excess charge.

Solution Solving Equation 18.8 for Q and entering known values gives

$$Q = \frac{rV}{k} = \frac{(0.125 \text{ m})(100 \text{ kV})}{9.00 \times 10^9 \text{ N} \cdot \text{m}^2/\text{C}^2}$$

$$= 1.39 \text{ }\mu\text{C}$$

Discussion This is a relatively small charge, but it produces a rather large voltage. We have another indication here that it is difficult to store isolated charges.

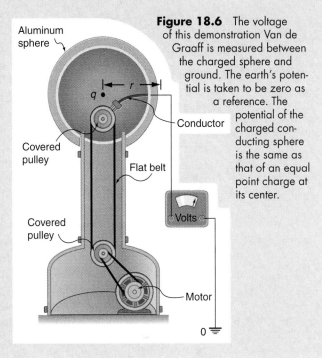

Figure 18.6 The voltage of this demonstration Van de Graaff is measured between the charged sphere and ground. The earth's potential is taken to be zero as a reference. The potential of the charged conducting sphere is the same as that of an equal point charge at its center.

The voltages in both of these examples could be measured with a meter that compares the measured potential with ground potential. Ground potential is often taken to be zero (instead of taking the potential at infinity to be zero). It is the potential difference between two points that is of importance, and very often there is a tacit assumption that some reference point, such as the earth or a very distant point, is at zero potential. As noted in Section 18.1, this is analogous to taking sea level as $h = 0$ when considering gravitational potential energy $\text{PE}_g = mgh$.

18.4   EQUIPOTENTIAL LINES

We can represent electric potentials (voltages) pictorially just as we drew pictures to illustrate electric fields. Of course, the two are related. Consider Figure 18.7, which shows an isolated positive point charge and its electric field lines. While we use blue arrows to represent the magnitude and direction of the electric field, we use green lines to represent places where the electric potential is constant. These are called **equipotential lines** in two dimensions, or *equipotential surfaces* in three dimensions. The potential for a point charge is the same anywhere on an imaginary sphere of radius r surrounding the charge. This is true since the potential for a point charge is given by $V = kQ/r$ and, thus, has the same value at any point that is a given distance r from the charge. An equipotential sphere is a circle in the two-dimensional view of Figure 18.7. Since the electric field lines point radially away from the charge, they are perpendicular to the equipotential lines.

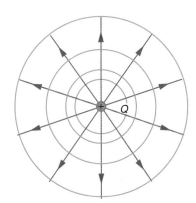

Figure 18.7 An isolated point charge Q, with its electric field lines in blue and equipotential lines in green. The potential is the same along each equipotential line, meaning that no work is required to move a charge anywhere along one of those lines. Work is needed to move a charge from one equipotential to another. Equipotential lines are perpendicular to electric field lines in every case.

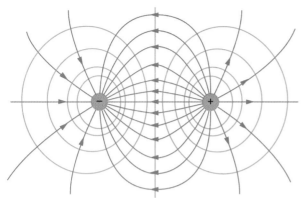

Figure 18.8 The electric field lines and equipotential lines for two equal but opposite charges. The equipotential lines can be drawn by making them perpendicular to the electric field lines, if those are known. Note that the potential is greatest (most positive) near the positive charge and least (most negative) near the negative charge.

It is important to note that *equipotential lines are always perpendicular to electric field lines.* No work is required to move a charge along an equipotential, since $\Delta V = 0$. Thus the work is

$$W = -\Delta PE = -q \, \Delta V = 0$$

Work is zero if force is perpendicular to motion. Force is in the same direction as E, so that motion along an equipotential must be perpendicular to E. More precisely, work is related to the electric field by

$$W = Fd \cos \theta = qEd \cos \theta = 0$$

Neither q nor E nor d is zero, and so $\cos \theta$ must be 0, meaning θ must be 90°. In other words, motion along an equipotential is perpendicular to E.

One of the rules for static electric fields and conductors is that the electric field must be perpendicular to the surface of any conductor. This implies that a *conductor is an equipotential surface in static situations.* There can be no voltage across a conductor, or charges will flow. One of the uses of this fact is that a conductor can be fixed at zero volts by connecting it to the earth with a good conductor—a process called **grounding**. Grounding can be a useful safety tool. For example, grounding the metal case of an electrical appliance ensures that it is at zero volts relative to the earth.

Because a conductor is an equipotential, it can replace any equipotential surface. For example, in Figure 18.7 a charged spherical conductor can replace the point charge, and the electric field and potential surfaces outside of it will be unchanged, confirming the contention that a spherical charge distribution is equivalent to a point charge at its center.

Figure 18.8 shows the electric field and equipotential lines for two equal and opposite charges. Given the electric field lines, the equipotential lines can be drawn simply by making them perpendicular to the electric field lines. Conversely, given the equipotential lines, as in Figure 18.9(a), the electric field lines can be drawn by making them perpendicular to the equipotentials, as in Figure 18.9(b).

One of the more important cases is that of the by now familiar parallel conducting plates shown in Figure 18.10. Between the plates, the equipotentials are evenly spaced and parallel. The same field could be maintained by placing conducting plates at the equipotential lines and having them at the potentials listed.

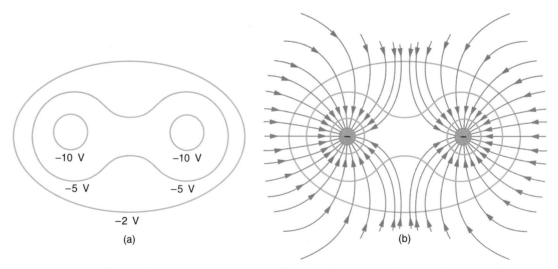

Figure 18.9 (a) These equipotential lines might be measured with a voltmeter in a laboratory experiment. (b) The corresponding electric field lines are found by drawing them perpendicular to the equipotentials. Note that these fields are consistent with two equal negative charges.

18.5 CAPACITORS AND DIELECTRICS

A **capacitor** is a device used to store charge. Capacitors have applications ranging from filtering static out of radio reception to energy storage in heart defibrillators. Typically, commercial capacitors have two conducting parts close to one another, but not touching, such as those in Figure 18.11. When battery terminals are connected to an initially uncharged capacitor, equal amounts of positive and negative charge, $+Q$ and $-Q$, are separated into its two parts. The capacitor remains neutral overall, but we refer to it as storing a charge Q in this circumstance.

The amount of charge Q a capacitor can store depends on two major factors—the voltage applied and the capacitor's physical characteristics, such as its size.

It is easy to see the relationship of voltage to stored charge for a parallel plate capacitor, such as shown in Figure 18.12. Each electric field line starts on an individual positive charge and ends on a negative one, so that there will be more field lines if there is more charge. (Drawing a single field line per charge is a convenience, only. There can be many field lines for each charge, but the total number is proportional to the number of charges.) The electric field strength is, thus, directly proportional to Q. That is,

$$E \propto Q$$

where the symbol $\propto$ means "proportional to." From the discussion in Section 18.2, we know that the voltage across parallel plates is $V = Ed$. Thus,

$$V \propto E$$

It follows, then, that $V \propto Q$, and conversely,

$$Q \propto V$$

This is true in general: The greater the voltage applied to any capacitor, the greater the charge stored in it.

Different capacitors will store different amounts of charge for the same applied voltage, depending on their physical characteristics. We define their **capacitance** C to be such that the charge Q stored in a capacitor is proportional to C. The charge stored in a capacitor is given by

$$Q = CV \qquad \textbf{(18.9)}$$

This equation expresses the two major factors affecting the amount of charge stored. Those factors are the physical characteristics of the capacitor, C, and the voltage, V. Rearranging the equation, we see that *capacitance C is the amount of charge stored per volt*, or

$$C = \frac{Q}{V}$$

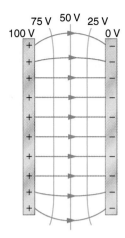

Figure 18.10 The electric field and equipotential lines of two metal plates.

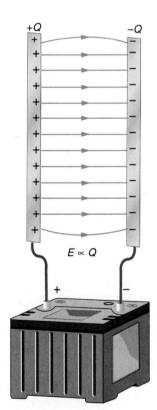

Figure 18.12 Electric field lines in this parallel plate capacitor, as always, start on positive charges and end on negative charges. Since the electric field strength is proportional to the density of field lines, it is also proportional to the amount of charge on the capacitor.

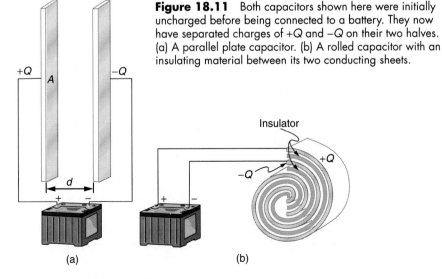

Figure 18.11 Both capacitors shown here were initially uncharged before being connected to a battery. They now have separated charges of $+Q$ and $-Q$ on their two halves. (a) A parallel plate capacitor. (b) A rolled capacitor with an insulating material between its two conducting sheets.

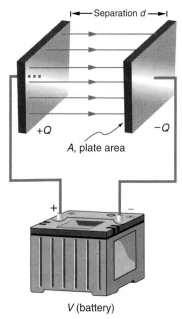

|←— Separation d —→|

+Q   −Q

A, plate area

+   −

V (battery)

Figure 18.13 Parallel plate capacitor with plates separated by a distance d; each plate has area A.

The unit of capacitance is the **farad** (F), named for Michael Faraday, the English scientist whose influence will be discussed in later sections. Since capacitance is charge per unit voltage, we see that a farad is a coulomb per volt, or

$$1 \text{ F} = \frac{1 \text{ C}}{1 \text{ V}} \tag{18.10}$$

A 1 farad capacitor would be able to store 1 coulomb (a very large amount of charge) with the application of only 1 volt. One farad is, thus, a very large capacitance. Typical capacitors range from fractions of a picofarad (pF) to millifarads (mF).

Parallel Plate Capacitors

The parallel plate capacitor shown in Figure 18.13 has two identical conducting plates, each having a surface area A, separated by a distance d. When a voltage V is applied to the capacitor, it stores a charge Q as shown. We can see how its capacitance depends on A and d by considering the characteristics of the Coulomb force. We know that like charges repel, unlike charges attract, and the force between charges decreases with distance. So it seems quite reasonable that the bigger the plates are, the more charge they can store—because the charges can spread out more. Thus C should be greater for larger A. Similarly, the closer the plates are together, the greater the attraction of the opposite charges on them. So C should be greater for smaller d.

It can be shown that for a parallel plate capacitor there are only two factors (A and d) that affect its capacitance C. The **capacitance of a parallel plate capacitor** is given by

$$C = \varepsilon_0 \frac{A}{d} \qquad \text{(parallel plate capacitor)} \tag{18.11}$$

A is the area of one plate in square meters, and d is the distance between plates in meters. The constant $\varepsilon_0 = 8.85 \times 10^{-12}$ F/m is the **permittivity of free space**; this is its numerical value in SI units.* (The units of F/m are equivalent to $C^2/N \cdot m^2$.) The small numerical value of ε_0 is related to the large size of the farad. A capacitor must have a large area to have a capacitance approaching a farad.

EXAMPLE 18.8   CAPACITANCE AND CHARGE STORED IN A PARALLEL PLATE CAPACITOR

(a) What is the capacitance of a parallel plate capacitor with metal plates, each of area 1.00 m^2, separated by 1.00 mm? (b) What charge is stored in this capacitor if a voltage of 3000 V is applied to it?

Strategy Finding the capacitance C is a straightforward application of Equation 18.11. Once C is found, the charge stored can be found using Equation 18.9.

Solution for (a) Entering the given values into Equation 18.11 for the capacitance of a parallel plate capacitor yields

$$C = \varepsilon_0 \frac{A}{d} = \left(8.85 \times 10^{-12} \frac{\text{F}}{\text{m}}\right) \frac{1.00 \text{ m}^2}{1.00 \times 10^{-3} \text{ m}}$$

$$= 8.85 \times 10^{-9} \text{ F} = 8.85 \text{ nF}$$

Discussion for (a) This small value for the capacitance indicates how difficult it is to make a device with a large capacitance. Special techniques help, such as using very large area thin foils placed close together.

Solution for (b) The charge stored in any capacitor is given by Equation 18.9. Entering the known values into that equation gives

$$Q = CV = (8.85 \times 10^{-9} \text{ F})(3.00 \times 10^3 \text{ V})$$
$$= 26.6 \text{ μC}$$

Discussion for (b) This charge is only slightly greater than those found in typical static electricity. Since air breaks down at about 3.00×10^3 V/mm, more charge cannot be stored on this capacitor by increasing the voltage.

Dielectrics

The previous example highlights the difficulty of storing large charges in capacitors. If d is made smaller to produce a larger capacitance, then the maximum voltage must be reduced

---

*The permittivity of free space is related to the constant k in the Coulomb force by $\varepsilon_0 = 1/(4\pi k)$. This indicates a direct connection between the behavior of capacitors and the strength of the Coulomb force.

proportionally to avoid breakdown (which depends on $E = V/d$). An important solution to this difficulty is to put an insulating material, called a **dielectric**, between the plates of a capacitor and make d as small as possible. Not only does the smaller d make the capacitance greater, but many insulators can withstand greater electric fields than air before breaking down.

There is another benefit to using a dielectric in a capacitor. Depending on the material used, the capacitance is *greater* than given by Equation 18.11 by a factor κ, called the **dielectric constant**. A parallel plate capacitor with a dielectric between its plates has a capacitance given by

$$C = \kappa \varepsilon_0 \frac{A}{d} \quad \text{(parallel plate capacitor—with dielectric)} \quad \textbf{(18.12)}$$

Values of the dielectric constant κ for various materials are given in Table 18.1.* Note that κ for vacuum is exactly 1, and so Equation 18.12 is valid in that case, too. If a dielectric is used, perhaps by placing Teflon between the plates of the capacitor in Example 18.8, then the capacitance is greater by the factor κ, which for Teflon is 2.1.

Note also that the dielectric constant for air is very close to 1, so that air-filled capacitors act much like those with vacuum between their plates *except* that the air can become conducting if the electric field strength (recall that $E = V/d$ for a parallel plate capacitor) becomes too great. Also shown in Table 18.1 are maximum electric field strengths in V/m, called **dielectric strengths**, for several materials. These are the fields above which the material begins to break down and conduct. The dielectric strength imposes a limit on the voltage that can be applied for a given plate separation. For instance, in Example 18.8, the separation is 1.00 mm, and so the voltage limit for air is 3000 V (3×10^6 V/m $\cdot 10^{-3}$ m). However, the limit for a 1.00 mm separation filled with Teflon is 60,000 V, since the dielectric strength of Teflon is 60×10^6 V/m. So the same capacitor filled with Teflon has a greater capacitance and can be subjected to a much greater voltage, allowing it to store a maximum charge of $Q = CV = (2.10)(8.85 \text{ nF})(60{,}000 \text{ V}) = 1.12 \text{ mC}$. This is 42 times the charge of the same air-filled capacitor.

How does a dielectric increase capacitance? Polarization of the insulator is responsible. The more easily it is polarized, the greater its dielectric constant κ. Water, for example, is a polar molecule, and water has a relatively large dielectric constant of 80. The effect of polarization can be best explained in terms of the characteristics of the Coulomb force. Figure 18.14 shows the separation of charge schematically in the molecules of a dielectric material placed between the charged plates of a capacitor. The Coulomb force between the closest ends of the molecules and the charge on the plates is attractive and very strong, since they are very close together. This attracts more charge onto the plates than if the space were empty and the opposite charges were a distance d away.

*Sometimes $\kappa \varepsilon_0$ is written as ε, and ε is called the *permittivity of the material*.

TABLE 18.1

DIELECTRIC CONSTANTS AND DIELECTRIC STRENGTHS FOR VARIOUS MATERIALS AT 20°C

Material	Dielectric constant κ	Dielectric strength (V/m)
Vacuum	1.00000	—
Air	1.00059	3×10^6
Bakelite	4.9	24×10^6
Fused quartz	3.78	8×10^6
Neoprene rubber	6.7	12×10^6
Nylon	3.4	14×10^6
Paper	3.7	16×10^6
Polystyrene	2.56	24×10^6
Pyrex glass	5.6	14×10^6
Silicon oil	2.5	15×10^6
Strontium titanate	233	8×10^6
Teflon	2.1	60×10^6
Water	80	—

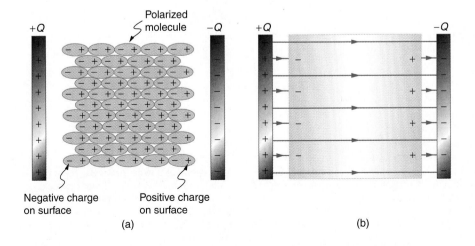

Negative charge on surface Positive charge on surface

(a) (b)

Figure 18.14 (a) The molecules in the insulating material between the plates of a capacitor are polarized by the charged plates. This produces a layer of opposite charge on the surface of the dielectric that attracts more charge onto the plate, increasing its capacitance. (b) The dielectric reduces the electric field strength inside the capacitor, resulting in a smaller voltage between the plates for the same charge. The capacitor stores the same charge for a smaller voltage, implying that it has a larger capacitance because of the dielectric.

Another way to understand how a dielectric increases capacitance is to consider its effect on the electric field inside the capacitor. Figure 18.14(b) shows the electric field lines with a dielectric in place. Since the field lines end on charges in the dielectric, there are fewer of them going from one side of the capacitor to the other. So the electric field strength is less than if there was a vacuum between the plates, even though the same charge is on the plates. Now, the voltage between the plates is $V = Ed$, and so it, too, is reduced by the dielectric. What this all means is that there is a smaller voltage V for the same charge Q; since $C = Q/V$, the capacitance C is greater.

Those who pursue the topic beyond the scope of this text will find that the dielectric constant is generally *defined* to be $\kappa = E_0/E$, or the ratio of the electric field in a vacuum to that in the dielectric material, and is intimately related to the polarizability of the material.

THINGS GREAT AND SMALL

The Submicroscopic Origin of Polarization

Polarization is a separation of charge within an atom or molecule. As has been noted, the planetary model of the atom pictures it as having a positive nucleus orbited by negative electrons, analogous to the planets orbiting the sun. This model is very helpful in explaining a vast range of phenomena and will be refined in the chapters on modern physics. The submicroscopic origin of polarization can be modeled as shown in Figure 18.A.

We will find in later chapters that the orbits of electrons are more properly viewed as electron clouds with the density of the cloud related to the probability of finding an electron in that location (as opposed to the definite locations and paths of planets in their orbits around the sun). This cloud is shifted by the Coulomb force so that the atom on average has a separation of charge. Although the atom remains neutral, it can now be the source of a Coulomb force, since a charge brought near the atom will be closer to one type of charge than the other.

Some molecules, such as those of water, have an inherent separation of charge and are thus called polar molecules. Figure 18.B illustrates the separation of charge in a water molecule, which has two hydrogen atoms and one oxygen atom (H_2O). The water molecule is not symmetric—it has the hydrogens on one side in a boomerang shape. The electrons in a water molecule are more concentrated around the more highly charged oxygen nucleus than around the hydrogen nuclei. This makes the oxygen end of the molecule slightly negative and leaves the hydrogens slightly positive. The inherent separation of charge in polar molecules makes it easier to align them with external fields and charges. Polar molecules therefore exhibit greater polarization effects and have greater dielectric constants. Those who study chemistry will find that the polar nature of water has many effects. For example, water molecules gather ions much more effectively because they have an electric field and a separation of charge to attract charges of both signs.

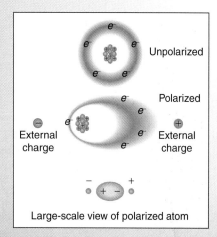

Figure 18.A Artist's conception of a polarized atom. The orbits of electrons around the nucleus are shifted slightly by the external charges (shown exaggerated). The resulting separation of charge within the atom means that it is polarized. Note that unlike charge is now closer to the external charges causing the polarization.

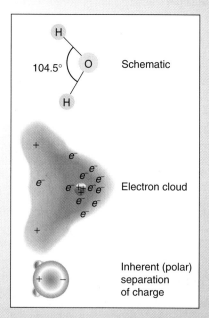

Figure 18.B Artist's conception of a water molecule. There is an inherent separation of charge, and so water is a polar molecule. Electrons in the molecule are attracted to the oxygen nucleus and leave an excess of positive charge near the two hydrogen nuclei

18.6 CAPACITORS IN SERIES AND PARALLEL

Several capacitors may be connected together in a variety of applications. Multiple connections of capacitors act like a single equivalent capacitor. The total capacitance of this equivalent single capacitor depends both on the individual capacitors and how they are connected. There are two simple and common types of connections, called *series* and *parallel*, for which we can easily calculate the total capacitance. Certain more complicated connections can also be related to combinations of series and parallel.

Capacitors in Series

Figure 18.15(a) shows a series connection of three capacitors with a voltage applied. As for any capacitor, the capacitance of the combination is related to charge and voltage by

$$C = \frac{Q}{V}$$

Note in the figure that opposite charges of magnitude Q flow to either side of the originally uncharged combination of capacitors when the voltage V is applied. Conservation of charge requires that equal-magnitude charges be created on the plates of the individual capacitors, since charge is only being separated in these originally neutral devices. The end result is that the combination resembles a single capacitor with an effective plate separation greater than that of the individual capacitors alone. (See Figure 18.15(b).) Larger plate separation means smaller capacitance. It is a general feature of series connections of capacitors that the total capacitance is less than any of the individual capacitances.

We can find an expression for the total capacitance by considering the voltage across the individual capacitors shown in Figure 18.15(a). Solving $C = Q/V$ for V gives

$$V = \frac{Q}{C}$$

The voltages across the individual capacitors are thus $V_1 = Q/C_1$, $V_2 = Q/C_2$, and $V_3 = Q/C_3$. The total voltage is the sum of the individual voltages:

$$V = V_1 + V_2 + V_3$$

Now, calling the total capacitance C_s for series capacitance, consider that

$$V = \frac{Q}{C_s} = V_1 + V_2 + V_3$$

Entering the expressions for V_1, V_2, and V_3, we get

$$\frac{Q}{C_s} = \frac{Q}{C_1} + \frac{Q}{C_2} + \frac{Q}{C_3}$$

Canceling the Qs, we obtain the equation for the **total capacitance in series** C_s to be

$$\frac{1}{C_s} = \frac{1}{C_1} + \frac{1}{C_2} + \frac{1}{C_3} + \cdots \qquad \textbf{(18.13)}$$

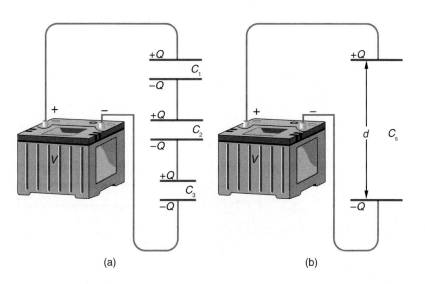

(a)   (b)

Figure 18.15 (a) Capacitors connected in series. The magnitude of the charge on all plates is Q. (b) An equivalent capacitor has a large plate separation d. Series connections produce a total capacitance that is less than that of any of the individual capacitors.

where $\cdots$ indicates that the expression is valid for any number of capacitors connected in series. An expression of this form always results in a total capacitance C_s that is less than any of the individual capacitances C_1, C_2, . . ., as the next example illustrates.

EXAMPLE 18.9 WHAT IS THE SERIES CAPACITANCE?

Find the total capacitance for three capacitors connected in series, given their individual capacitances are 1.00, 5.00, and 8.00 μF.

Strategy With the given information, the total or series capacitance can be found using Equation 18.13.

Solution Entering the given capacitances into the expression for $1/C_s$ found in Equation 18.13 gives

$$\frac{1}{C_s} = \frac{1}{C_1} + \frac{1}{C_2} + \frac{1}{C_3}$$

$$= \frac{1}{1.00\ \mu F} + \frac{1}{5.00\ \mu F} + \frac{1}{8.00\ \mu F} = \frac{1.325}{\mu F}$$

Inverting to find C_s yields

$$C_s = \frac{\mu F}{1.325} = 0.755\ \mu F$$

Discussion The total series capacitance C_s is less than the smallest individual capacitance, as promised. In series connections of capacitors, the sum is less than the parts. In fact, it is less than any individual. Note that it is sometimes possible, and more convenient, to solve an equation like the above by finding the least common denominator, which in this case is 40. Thus,

$$\frac{1}{C_s} = \frac{40}{40\ \mu F} + \frac{8}{40\ \mu F} + \frac{5}{40\ \mu F} = \frac{53}{40\ \mu F}$$

so that $C_s = 40\ \mu F/53 = 0.755\ \mu F$.

Capacitors in Parallel

Figure 18.16(a) shows a parallel connection of three capacitors with a voltage applied. Here the total capacitance is easier to find than in the series case. To find the equivalent total capacitance C_p, we first note that the voltage across each capacitor is V, the same as that of the source, since they are connected directly to it through a conductor. (Conductors are equipotentials, and so the voltage across the capacitors is the same as that across the voltage source.) Thus the capacitors have the same charges on them as they would have if connected individually to the voltage source. The total charge Q is the sum of the individual charges:

$$Q = Q_1 + Q_2 + Q_3$$

Using the relationship $Q = CV$, we see that the total charge is $Q = C_pV$, and the individual charges are $Q_1 = C_1V$, $Q_2 = C_2V$, and $Q_3 = C_3V$. Entering these into the previous equation gives

$$C_pV = C_1V + C_2V + C_3V$$

Canceling V from the equation, we obtain the equation for the **total capacitance in parallel** C_p:

$$C_p = C_1 + C_2 + C_3 + \cdots \qquad (18.14)$$

Total capacitance in parallel is the simple sum of the individual capacitances. (Again the $\cdots$ indicates the expression is valid for any number of capacitors connected in parallel.) So, for example, if the capacitors in Example 18.9 were connected in parallel,

Figure 18.16 (a) Capacitors in parallel. Each is connected directly to the voltage source just as if it were all alone, and so the total capacitance in parallel is just the sum of the individual capacitances. (b) The equivalent capacitor has a larger plate area and can therefore hold more charge than the individual capacitors.

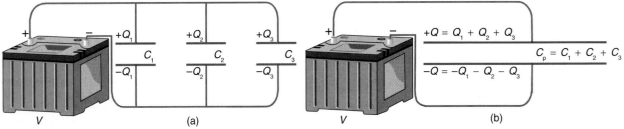

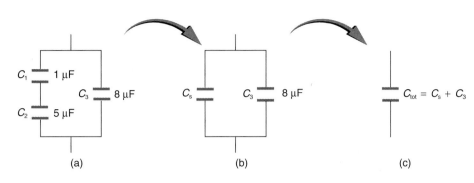

Figure 18.17 (a) There are both series and parallel connections of capacitors here. See Example 18.10. (b) C_1 and C_2 are in series; their equivalent capacitance C_s is less than either of them. (c) Note that C_s is in parallel with C_3. The total capacitance is, thus, the sum of C_s and C_3.

their capacitance would simply be $C_p = 1.00 \, \mu\text{F} + 5.00 \, \mu\text{F} + 8.00 \, \mu\text{F} = 14.0 \, \mu\text{F}$. The equivalent capacitor for a parallel connection has an effectively larger plate area and, thus, a larger capacitance, as illustrated in Figure 18.16(b).

More complicated connections of capacitors can sometimes be combinations of series and parallel. (See Figure 18.17.) To find the total capacitance of such combinations, we identify series and parallel parts, compute their capacitances, and then find the total.

EXAMPLE 18.10 A MIXTURE OF SERIES AND PARALLEL CAPACITANCE

Find the total capacitance of the combination of capacitors shown in Figure 18.17, taking their values to three digits.

Strategy To find the total capacitance, we first identify which capacitors are in series and which are in parallel. Capacitors C_1 and C_2 are in series. Their combination, labeled C_s in the figure, is in parallel with C_3.

Solution Since C_1 and C_2 are in series, their total capacitance is given by Equation 18.13. Entering their values into the equation gives

$$\frac{1}{C_s} = \frac{1}{C_1} + \frac{1}{C_2} = \frac{1}{1.00 \, \mu\text{F}} + \frac{1}{5.00 \, \mu\text{F}} = \frac{1.20}{\mu\text{F}}$$

Inverting gives

$$C_s = 0.833 \, \mu\text{F}$$

This equivalent series capacitance is in parallel with the third capacitor; thus, the total is the sum

$$C_{tot} = C_s + C_3 = 8.83 \, \mu\text{F}$$

Discussion This technique of analyzing the combinations of capacitors piece by piece until a total is obtained can be applied to larger combinations of capacitors.

18.7 ENERGY STORED IN CAPACITORS

Most of us have seen dramatizations of heart defibrillation in which medical personnel pass an electric current through a patient's heart to get it to beat normally. Often realistic in detail, the person applying the shock directs another person to "make it 400 joules this time." The energy delivered by the defibrillator is stored in a capacitor and can be adjusted to fit the situation. SI units of joules are often employed. Less dramatic is the use of capacitors in microelectronics, such as certain handheld calculators, to supply energy when batteries are changed. (See Figure 18.18.) Capacitors are also used to supply energy for flash lamps on cameras.

Energy stored in a capacitor is electrical potential energy, and it is thus related to the charge Q and voltage V on the capacitor. We must be careful when applying the equation for electrical potential energy ($\Delta\text{PE} = q \, \Delta V$) to a capacitor. Remember that ΔPE is the potential energy of a charge q going through a voltage ΔV. But the capacitor starts with zero voltage and gradually comes up to its full voltage as it is charged. The first charge placed on a capacitor sees $\Delta V = 0$, since the capacitor has zero voltage when uncharged. The final charge placed on a capacitor sees $\Delta V = V$, since the capacitor now has its full voltage V on it. The average voltage on the capacitor during the charging process is $V/2$, and so the average voltage experienced by the full charge Q is $V/2$. Thus the energy stored in a capacitor, E_{cap}, is

$$E_{cap} = Q \frac{V}{2}$$

Figure 18.18 Energy stored in the large capacitor is used to preserve the memory of an electronic calculator when its batteries are changed.

where Q is the charge on a capacitor with a voltage V applied. (Note that the energy is *not* QV, but $QV/2$.) Charge and voltage on a capacitor are related by $Q = CV$, and so the expression for E_{cap}, **the energy stored in a capacitor**, can be algebraically manipulated into three equivalent expressions:

Medical Application

$$E_{cap} = \frac{QV}{2} = \frac{CV^2}{2} = \frac{Q^2}{2C} \qquad \textbf{(18.15)}$$

where Q is the charge and V the voltage on a capacitor C. The energy is in joules for a charge in coulombs, voltage in volts, and capacitance in farads.

EXAMPLE 18.11 CAPACITANCE IN A HEART DEFIBRILLATOR

A heart defibrillator delivers 400 J of energy by discharging a capacitor initially at 10,000 V. What is its capacitance?

Strategy We are given E_{cap} and V, and we are asked to find the capacitance C. Of the three expressions in Equation 18.15, the most convenient relationship is

$$E_{cap} = \frac{CV^2}{2}$$

Solution Solving this expression for C and entering the

given values yields

$$C = \frac{2E_{cap}}{V^2} = \frac{800 \text{ J}}{(1.00 \times 10^4 \text{ V})^2} = 8.00 \times 10^{-6} \text{ F}$$

or

$$\boxed{C = 8.00 \text{ } \mu\text{F}}$$

Discussion This is a fairly large, but manageable, capacitance at 10,000 V.

SUMMARY

The electrostatic or Coulomb force is conservative; thus, it is most convenient to define an **electric potential energy** PE and to consider ΔPE when moving a test charge q between two points. Since ΔPE is proportional to the test charge q, we define the **potential difference** ΔV between two points A and B to be

$$\Delta V = V_B - V_A = \frac{\Delta PE}{q} \qquad \textbf{(18.1)}$$

Potential difference is commonly called **voltage** and has units of **volts** (V), where

$$1 \text{ V} = 1 \text{ J/C} \qquad \textbf{(18.2)}$$

The relationship between energy and voltage is

and $$\left. \begin{array}{l} \Delta V = \dfrac{\Delta PE}{q} \\[2mm] \Delta PE = q \, \Delta V \end{array} \right\} \qquad \textbf{(18.3)}$$

More precisely, Equation 18.3 is the relationship between potential difference and electric potential energy—the first has units of volts and the latter has units of joules.

The **electron volt** (eV) is defined to be

$$1 \text{ eV} = 1.60 \times 10^{-19} \text{ J} \qquad \textbf{(18.4)}$$

An electron volt is the energy given a fundamental charge when it is accelerated through a potential difference of 1 V.

Since the electrostatic force is conservative, the familiar conservation of mechanical energy,

$$KE_i + PE_i = KE_f + PE_f$$

can be freely employed. PE is the sum of all potential energies involved (PE can be electric potential energy alone).

The relationship between electric potential (voltage) and

the electric field is simplest *for a uniform electric field*:

or $$\left. \begin{array}{l} V_{AB} = Ed \\[2mm] E = \dfrac{V_{AB}}{d} \end{array} \right\} \qquad \text{(uniform E-field only)} \qquad \textbf{(18.5)}$$

V_{AB} is the voltage between points A and B, and d is the component of the distance between points A and B parallel to E. The following relationship among units is useful:

$$1 \text{ N/C} = 1 \text{ V/m} \qquad \textbf{(18.6)}$$

In all situations, the electric field E points in the direction of decreasing potential. The magnitude E is related to how fast the potential changes over distance:

$$E = -\frac{\Delta V}{\Delta s} \qquad \textbf{(18.7)}$$

where Δs is the distance. (This is consistent with Equation 18.5 with $\Delta s = d$ and $-\Delta V = V_{AB}$.)

Uniform distributions of charge create a potential outside that is the same as that of a point charge at their center. The **electric potential V of a point charge Q** at a distance r is

$$V = k\frac{Q}{r} \qquad \text{(point charge)} \qquad \textbf{(18.8)}$$

The potential at infinity is zero.

Locations where the electric potential (voltage) is constant are called **equipotential lines** in two dimensions, or equipotential *surfaces* in three dimensions. Equipotential lines are *always* perpendicular to electric field lines. A conductor in static equilibrium is an equipotential surface. Since conductors are equipotentials, an object can be forced to have the same potential as the earth if it is connected to it by a conductor—a process called **grounding**.

A **capacitor** is a device used to store charge. The amount of charge Q it can store is directly proportional to the voltage V applied and to its capacitance C:

$$Q = CV \qquad \textbf{(18.9)}$$

The **capacitance** C is determined by the physical characteristics of the capacitor and has units of **farads** (F); a farad is one coulomb of stored charge per volt applied:

$$1\ F = \frac{1\ C}{1\ V} \qquad \textbf{(18.10)}$$

Parallel plate capacitors with plates of area A separated by a distance d have capacitance

$$C = \varepsilon_0\,\frac{A}{d} \qquad \text{(parallel plate capacitor)} \qquad \textbf{(18.11)}$$

The constant $\varepsilon_0 = 8.85 \times 10^{-12}$ F/m is the **permittivity of free space**. A **dielectric** is an insulating material placed between the plates of a capacitor, which increases its capacitance by a factor of κ, called the **dielectric constant**. Thus,

$$C = \kappa\varepsilon_0\,\frac{A}{d} \qquad \text{(parallel plate capacitor—with dielectric)}$$

$$\textbf{(18.12)}$$

The **dielectric strength** of a material (in V/m) determines how large a voltage can be applied to a capacitor before it breaks down, a larger dielectric strength generally enabling a greater charge to be stored. Table 18.1 gives values of the dielectric constants and dielectric strengths of selected materials.

Two types of simple connections of multiple capacitors are called *series* and *parallel*. The **total capacitance in series** C_s is

$$\frac{1}{C_s} = \frac{1}{C_1} + \frac{1}{C_2} + \frac{1}{C_3} + \cdots \qquad \textbf{(18.13)}$$

for the individual capacitances C_1, C_2, C_3, ... Series connections have less total capacitance than any of the individual capacitances. The **total capacitance in parallel** C_p is

$$C_p = C_1 + C_2 + C_3 + \cdots \qquad \textbf{(18.14)}$$

Energy stored in capacitors is related to the capacitance C, the applied voltage V, and the charge Q stored. Three expressions for E_{cap}, **the energy stored in a capacitor**, are

$$E_{cap} = \frac{QV}{2} = \frac{CV^2}{2} = \frac{Q^2}{2C} \qquad \textbf{(18.15)}$$

CONCEPTUAL QUESTIONS

18.1 Voltage is the common word for potential difference. Which term is more descriptive, voltage or potential difference?

18.2 If the voltage between two points is zero, can a test charge be moved between them with zero net work being done? Can this necessarily be done without exerting a force? Explain.

18.3 What is the relationship between voltage and energy? More precisely, what is the relationship between potential difference and electric potential energy?

18.4 Voltages are always measured between two points. Why?

18.5 How are units of volts and electron volts related? How do they differ?

18.6 Discuss how potential difference and electric field strength are related. Give an example.

18.7 What is the strength of the electric field in a region where the electric potential is constant?

18.8 Will a negative charge, initially at rest, move toward higher or lower potential? Explain why.

18.9 Is potential difference a scalar or a vector?

18.10 In what region of space is the potential due to a uniformly charged sphere the same as that of a point charge? In what region does it differ from that of a point charge?

18.11 Can the potential of a nonuniformly charged sphere be the same as that of a point charge? Explain.

18.12 What is an equipotential line or surface?

18.13 Explain in your own words why equipotential lines and surfaces must be perpendicular to electric field lines.

18.14 Can different equipotential lines cross? Explain.

18.15 Does the capacitance of a device depend on the applied voltage? Does the charge stored in it?

18.16 Use the characteristics of the Coulomb force to explain why capacitance should be proportional to the plate area of a capacitor. Similarly, explain why capacitance should be inversely proportional to the separation between plates.

18.17 Give the reason that a dielectric material increases capacitance compared with what it would be with air between the plates of a capacitor. What is the independent reason that a dielectric material also allows a greater voltage to be applied to a capacitor? (The dielectric thus increases C and permits a greater V.)

18.18 How does the polar character of water molecules help to explain water's relatively large dielectric constant?

18.19 Sparks will occur between the plates of an air-filled capacitor at lower voltage when the air is humid than when dry. Explain why, considering the polar character of water molecules.

18.20 Water has a large dielectric constant, but it is rarely used in capacitors. Explain why.

18.21 If you wish to store a large amount of energy in a capacitor bank, would you connect capacitors in series or parallel? Explain.

18.22 How does the energy contained in a charged capacitor change when a dielectric is inserted, assuming the capacitor is isolated and its charge is constant? Does this imply that work was done?

18.23 What happens to the energy stored in a capacitor connected to a battery when a dielectric is inserted? Was work done in the process?

PROBLEMS

Section 18.1 Electric Potential Energy and Voltage

18.1 What energy in joules is involved in a spark from a doorknob to your finger, if a charge of 125 nC moved through a potential difference of 5000 V?

18.2 The spark from a demonstration Van de Graaff generator moves 0.500 μC of charge through a potential difference of 100,000 V. How much energy is dissipated?

18.3 What is the energy content of a 3.00 V calculator battery that has the ability to move 200 C of charge over its lifetime?

18.4 When removing a wool sweater, a static discharge of 1.00 μC dissipates 5.00×10^{-3} J of energy. What voltage was involved?

18.5 A large lightning bolt dissipates 3.00×10^9 J of energy and moves 30.0 C of charge. What was its voltage?

18.6 A spark plug uses a voltage of 25.0 kV to turn 7.50 J of electrical energy into thermal energy to ignite fuel. How much charge moves in the process?

• 18.7 Find the power output in watts of a 9.00 V battery that moves 2.00 C of charge in 1.00 h.

• 18.8 A 1.50 V battery supplies 0.300 W of power to a small flashlight for 10.0 min. (a) How much charge does it move? (b) How many electrons must move to carry this charge?

• 18.9 What is the voltage of the batteries supplying a flashlight with 0.900 W of power if they move 270 C in 15.0 min in the process?

• 18.10 (a) Calculate the power output in watts of a 12.0 V truck battery that moves 800 C of charge in 4.00 s. (b) How many electrons move to carry this energy? (c) Find the energy per electron in joules and convert it to electron volts.

• 18.11 If an electron travels 0.200 m from an electron gun to a TV screen in 25.0 ns, what voltage was used to accelerate it? (Note that the voltage you obtain here is lower than actually used in TVs to avoid the necessity of relativistic corrections.)

⁝18.12 Find the ratio of speeds of an electron and a negative hydrogen ion (one having an extra electron) accelerated through the same voltage, assuming nonrelativistic final speeds. Take the mass of the hydrogen ion to be 1.67×10^{-27} kg.

⁝18.13 What voltage is needed to accelerate a negative hydrogen ion (one having an extra electron) from rest to 1.00% of the speed of light?

⁝18.14 A bare helium nucleus has two positive charges and a mass of 6.64×10^{-27} kg. (a) Calculate its kinetic energy in joules at 2.00% of the speed of light. (b) What is this in electron volts? (c) What voltage would be needed to obtain this energy?

Section 18.2 Uniform Electric Field and Voltage

18.15 Show that units of V/m and N/C for electric field strength are indeed equivalent.

18.16 What is the strength of the electric field between two parallel conducting plates separated by 1.00 cm and having a potential difference (voltage) between them of 1.50×10^4 V?

18.17 The electric field strength between two parallel conducting plates separated by 4.00 cm is 7.50×10^4 V/m. (a) What is the potential difference between the plates? (b) The plate with the lowest potential is taken to be at zero volts. What is the potential 1.00 cm from that plate (and 3.00 cm from the other)?

18.18 How far apart are two conducting plates that have an electric field strength of 4.50×10^3 V/m between them, if their potential difference is 15.0 kV?

• 18.19 (a) Will the electric field strength between two parallel conducting plates exceed the breakdown strength for air (3×10^6 V/m) if the plates are separated by 2.00 mm and a potential difference of 5000 V is applied? (b) How close together can the plates be with this applied voltage?

• 18.20 Two parallel conducting plates are separated by 10.0 cm, and one of them is taken to be at zero volts. (a) What is the electric field strength between them, if the potential 8.00 cm from the zero volt plate (and 2.00 cm from the other) is 450 V? (b) What is the voltage between the plates?

• 18.21 Find the maximum potential difference between two parallel conducting plates separated by 0.500 cm of air, given the maximum sustainable electric field strength in air to be 3×10^6 V/m.

⁝18.22 A doubly charged ion is accelerated to an energy of 32.0 keV by the electric field between two parallel conducting plates separated by 2.00 cm. What is the electric field strength between the plates?

⁝18.23 An electron is to be accelerated in a uniform electric field having a strength of 2.00×10^6 V/m. (a) What energy in keV is given to the electron if it is accelerated through 0.400 m? (b) Over what distance would it have to be accelerated to increase its energy by 50.0 GeV (as in the Stanford Linear Accelerator, which is actually smaller than this)?

Section 18.3 Electric Potential of a Point Charge

18.24 A 0.500 cm diameter plastic sphere, used in a static electricity demonstration, has a uniformly distributed 40.0 pC charge on its surface. What is the potential near its surface?

18.25 What is the potential 0.530×10^{-10} m from a proton (the average distance between the proton and electron in a hydrogen atom)?

18.26 A sphere has a surface uniformly charged with 1.00 C. At what distance from its center is the potential 5.00 MV?

18.27 How far from a 1.00 μC point charge will the potential be 100 V?

18.28 What is the sign and magnitude of a point charge that produces a potential of −2.00 V at a distance of 1.00 mm?

18.29 If the potential due to a point charge is 500 V at a distance of 15.0 m, what is the sign and magnitude of the charge?

• 18.30 In nuclear fission, a nucleus splits roughly in half. (a) What is the potential 2.00×10^{-14} m from a fragment that has 46 protons in it? (b) What is the potential energy in MeV of a similarly charged fragment at this distance?

• 18.31 A research Van de Graaff generator has a 2.00 m diameter metal sphere with a charge of 5.00 mC on it. (a) What is the potential near its surface? (b) At what distance from its center is the potential 1.00 MV? (c) An oxygen atom with three missing electrons is released near the Van de Graaff. What is its energy in MeV at this distance?

: 18.32 An electrostatic paint sprayer has a 0.200 m diameter metal sphere at a potential of 25.0 kV that repels paint droplets onto a grounded object. (a) What charge is on the sphere? (b) What charge must a 0.100 mg drop of paint have to arrive at the object with a speed of 10.0 m/s?

Section 18.4 Equipotential Lines

18.33 (a) Sketch the equipotential lines near a point charge $+q$. Indicate the direction of increasing potential. (b) Do the same for a point charge $-3.00q$.

18.34 Sketch the equipotential lines for the two equal positive charges shown in Figure 18.19. Indicate the direction of increasing potential.

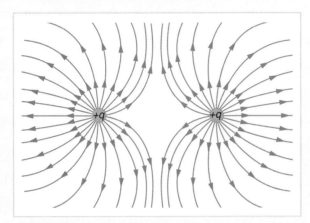

Figure 18.19 The electric field near two equal positive charges. Problem 34.

18.35 Figure 18.20 shows the electric field lines near two charges q_1 and q_2, the first having a magnitude four times that of the second. Sketch the equipotential lines for these two charges, and indicate the direction of increasing potential.

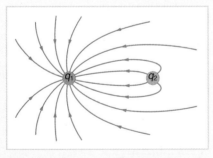

Figure 18.20 The electric field near two charges. Problems 35 and 36.

18.36 Sketch the equipotential lines a long distance from the charges shown in Figure 18.20. Indicate the direction of increasing potential.

• 18.37 Sketch the equipotential lines in the vicinity of two opposite charges, where the negative charge is three times as great in magnitude as the positive. See Figure 18.20 for a similar situation. Indicate the direction of increasing potential.

• 18.38 Sketch the equipotential lines in the vicinity of the negatively charged conductor in Figure 18.21. How will these equipotentials look a long distance from the object?

Figure 18.21 Problem 38.

: 18.39 Sketch the equipotential lines surrounding the two conducting plates shown in Figure 18.22, given the top plate is positive and the bottom plate has an equal amount of negative charge. Be certain to indicate the distribution of charge on the plates.

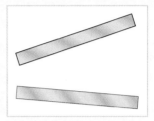

Figure 18.22 Problem 39.

: 18.40 (a) Sketch the electric field lines in the vicinity of the charged insulator in Figure 18.23. Note its nonuniform charge distribution. (b) Sketch equipotential lines surrounding the insulator. Indicate the direction of increasing potential.

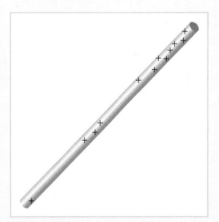

Figure 18.23 A charged insulating rod such as might be used in a classroom demonstration. Problem 40.

Sections 18.5–18.7 Capacitors

Take the data in figures to be good to three significant digits.

18.41 What charge is stored in a 180 μF capacitor when 120 V is applied to it?

18.42 Find the charge stored when 5.50 V is applied to an 8.00 pF capacitor.

18.43 What charge is stored in the capacitor in Example 18.11?

18.44 Calculate the voltage applied to a 2.00 μF capacitor when it holds 3.10 μC of charge.

18.45 What voltage must be applied to an 8.00 nF capacitor to store 0.160 mC of charge?

18.46 What capacitance is needed to store 3.00 μC of charge at a voltage of 120 V?

18.47 What is the capacitance of a large Van de Graaff generator's terminal, given it stores 8.00 mC of charge at a voltage of 12.0 MV?

18.48 Find the capacitance of a parallel plate capacitor having plates of area 5.00 m² that are separated by 0.100 mm of Teflon.

18.49 (a) What is the capacitance of a parallel plate capacitor having plates of area 1.50 m² that are separated by 0.0200 mm of neoprene rubber? (b) What charge does it hold when 9.00 V is applied to it?

18.50 Suppose you want a capacitor bank with a total capacitance of 0.750 F and you possess numerous 1.50 mF capacitors. What is the smallest number you could hook together to achieve your goal, and how would you connect them?

18.51 What total capacitances can you make by connecting a 5.00 and 8.00 μF capacitor together?

18.52 A 165 μF capacitor is used in conjunction with a motor. How much energy is stored in it when 118 V is applied?

• **18.53** A nervous physicist worries that the two metal shelves of his wood frame bookcase might obtain a high voltage if charged by static electricity, perhaps produced by friction. (a) What is the capacitance of the empty shelves if they have area 0.100 m² and are 0.200 m apart? (b) What is the voltage between them if opposite charges of magnitude 2.00 nC are placed on them? (c) To show that this voltage poses a small hazard, calculate the energy stored.

• **18.54** If the energy stored in the capacitor in Example 18.11 is only 300 J (rather than 400 J) but the charge the same, what is the applied voltage?

• **18.55** Suppose you have a 9.00 V battery, a 2.00 μF capacitor, and a 7.40 μF capacitor. (a) Find the charge and energy stored if the capacitors are connected to the battery in series. (b) Do the same for a parallel connection.

• **18.56** Find the total capacitance of the combination of capacitors shown in Figure 18.24.

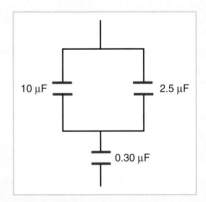

Figure 18.24 A combination of series and parallel connections of capacitors. Problems 56 and 60.

• **18.57** Find the total capacitance of the combination of capacitors shown in Figure 18.25.

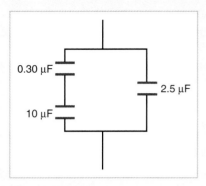

Figure 18.25 A combination of series and parallel connections of capacitors. Problems 57 and 61.

• **18.58** A large capacitance of 1.00 mF is needed for a certain application. (a) Calculate the area the parallel plates of such a capacitor must have if they are separated by 3.00 μm of Teflon. (b) What is the maximum voltage that can be applied? (c) Find the maximum charge that can be stored? (d) Calculate the volume of Teflon alone in the capacitor.

• **18.59** A high technology 0.100 F capacitor is used to store energy to operate a microelectronics device during a power failure. (a) What energy is stored in the capacitor when 3.00 V is applied to it? (b) How long can this capacitor supply energy at the rate of 2.00 mW?

⁞ **18.60** (a) Find the charge stored on each capacitor in Figure 18.24 when a 1.55 V battery is connected to the combination. (b) What energy is stored in each?

⁞ **18.61** What total energy is stored in the capacitors in Figure 18.25 if 1.80×10^{-4} J is stored in the 2.50 μF capacitor?

⁞ **18.62** Using $Q = CV$ and the expression for the capacitance of a parallel plate capacitor with a dielectric, show that the maximum charge that can be stored is $\kappa \varepsilon_0 A$ times the dielectric strength of the dielectric material used. Thus, for a given dielectric, the maximum charge depends on plate area but not plate separation.

⁞ **18.63** Show that for a given dielectric material, the maximum energy a parallel plate capacitor can store is directly proportional to the volume of dielectric (Volume = $A \cdot d$). Note that the applied voltage is limited by the dielectric strength.

⁞ **18.64** Find the total capacitance of the combination of capacitors shown in Figure 18.26.

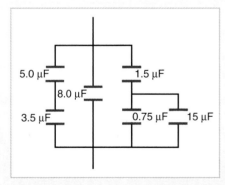

Figure 18.26 A combination of series and parallel connections of capacitors. Problem 64.

INTEGRATED CONCEPTS

Physics is most interesting when applied to general situations involving more than a narrow set of physical principles. These general problems involve physical principles from more than one chapter—especially those involving energy related to static electricity. Material from the chapters on work, energy and power, and thermodynamics is directly involved. You may wish to refer to Chapters 6, 12, 13, 17, and perhaps others to solve the following problems.

Note: Problem-solving strategies and a worked example that can help you solve integrated concept problems appear in Chapter 17. They can be found with the section of problems labeled *Integrated Concepts* at the end of that chapter. Consult the index for the locations of other problem-solving strategies and worked examples for integrated concept problems.

18.65 Singly charged gas ions are accelerated from rest through a voltage of 13.0 V. At what temperature will the average kinetic energy of gas molecules be the same as that given these ions?

18.66 The temperature near the center of the sun is thought to be 30 million degrees Celsius ($3.0 \times 10^7 °C$). Through what voltage must a singly charged ion be accelerated to have the same energy as the average kinetic energy of ions at this temperature?

• 18.67 A lightning bolt strikes a tree, moving 20.0 C of charge through a potential difference of 100 MV. (a) What energy was dissipated? (b) What mass of water could be raised from 15.0°C to the boiling point and then boiled by this energy? Note that the expansion of the steam upon boiling can literally blow the tree apart.

• 18.68 A 12.0 V battery-operated bottle warmer heats 50.0 g of glass, 250 g of baby formula, and 200 g of aluminum from 20.0°C to 90.0°C. (a) How much charge is moved by the battery? (b) How many electrons per second flow if it takes 5.00 min?

• 18.69 A prankster applies 450 V to an 80.0 µF capacitor and then tosses it to an unsuspecting victim. The victim's finger is burned by the discharge of the capacitor through 0.200 g of flesh. What is the temperature increase of the flesh? Is it reasonable to assume no phase change?

• 18.70 A battery-operated car utilizes a 12.0 V system. Find the charge the batteries must be able to move in order to accelerate the 750 kg car from rest to 25.0 m/s, make it climb a 200 m high hill, and then cause it to travel at a constant 25.0 m/s by exerting a 500 N force for an hour.

⁑ 18.71 Fusion probability is greatly enhanced when appropriate nuclei are brought close together, but mutual Coulomb repulsion must be overcome. This can be done using the kinetic energy of high-temperature gas ions or by accelerating the nuclei toward one another. (a) Calculate the potential energy of two singly charged nuclei separated by 1.00×10^{-12} m by finding the voltage of one at that distance and multiplying by the charge of the other. (b) At what temperature will atoms of a gas have an average kinetic energy equal to this needed electrical potential energy?

⁑ 18.72 Consider Problem 17.53 on the Millikan oil drop experiment. Find the voltage between two metal plates separated by 2.50 cm that would produce the necessary electric field.

UNREASONABLE RESULTS

The following problems have results that are unreasonable because some premise is unreasonable or because certain of the premises are inconsistent with one another. Physical principles applied correctly then produce unreasonable results. The purpose of these problems is to give practice in assessing whether nature is being accurately described, and to trace the source of difficulty if it is not.

PROBLEM-SOLVING STRATEGY

To determine if an answer is reasonable, and to determine the cause if it is not, do the following.

Step 1. *Solve the problem using strategies as outlined in several places in the text.* See the table of contents to locate problem-solving strategies. Use the format followed in the worked examples in this chapter to solve the problem as usual.

Step 2. *Check to see if the answer is reasonable.* Is it too large or too small, or does it have the wrong sign, improper units, . . .?

Step 3. *If the answer is unreasonable, look for what specifically could cause the identified difficulty.* Usually, the manner in which the answer is unreasonable is an indication of the difficulty. For example, an extremely large voltage could be due to the assumption of an excessively large separated charge.

18.73 (a) An 8.00 µF capacitor is connected in parallel to another capacitor, producing a total capacitance of 5.00 µF. What is the capacitance of the second capacitor? (b) What is unreasonable about this result? (c) Which assumptions are unreasonable or inconsistent?

18.74 (a) On a particular day, it takes 9600 J of electric energy to start a truck's engine. Calculate the capacitance of a capacitor that could store that amount of energy at 12.0 V. (b) What is unreasonable about this result? (c) Which assumptions are responsible?

18.75 (a) Find the voltage near a 10.0 cm diameter metal sphere that has 8.00 C of excess positive charge on it. (b) What is unreasonable about this result? (c) Which assumptions are responsible?

• 18.76 (a) What is the final speed of an electron accelerated from rest through a voltage of 25.0 MV by a negatively charged Van de Graaff terminal? (b) What is unreasonable about this result? (c) Which assumptions are responsible?

⁑ 18.77 (a) A certain parallel plate capacitor has plates of area 4.00 m^2, separated by 0.0100 mm of nylon, and stores 0.170 C of charge. What is the applied voltage? (b) What is unreasonable about this result? (c) Which assumptions are responsible or inconsistent?

19 ELECTRIC CURRENT, RESISTANCE, AND OHM'S LAW

Electric energy in massive quantities is transmitted from this hydroelectric facility by the movement of charge—that is, by electric current. The rectangular structure at the base of the dam houses the generators.

The flicker of numbers on a handheld calculator, nerve impulses carrying signals of vision to the brain, an electric train pulling its load over a mountain pass, a hydroelectric plant sending energy to metropolitan and rural users—these and many other examples of electricity involve *electric current*, the *movement of charge*. Whereas the previous two chapters concentrated on static electricity and the fundamental force underlying its behavior, the next few chapters will be devoted to electric and magnetic phenomena involving current. In addition to exploring applications of electricity, we shall gain new insights into nature—in particular, the fact that all magnetism results from electric current.

19.1 CURRENT

Electric current is defined to be the rate at which charge flows. A large current, such as that used to start a truck engine, moves a large amount of charge in a small time, whereas a small current, such as that used to operate a handheld calculator, moves a small amount of charge over a long period of time. In equation form, **electric current *I*** is defined to be

$$I = \frac{\Delta Q}{\Delta t} \qquad (19.1)$$

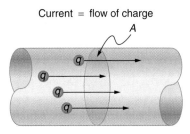

Current = flow of charge

Figure 19.1 The rate of flow of charge is current. An ampere is the flow of one coulomb through an area in one second.

where ΔQ is the amount of charge passing through a given area in time Δt. (As in previous chapters, initial time is often taken to be zero, in which case $\Delta t = t$.) (See Figure 19.1.) The SI unit for current is the familiar **ampere** (A), named for the French physicist Andre Ampere (1775–1836). Since $I = \Delta Q/\Delta t$, we see that an ampere is one coulomb per second:

$$1 \text{ A} = 1 \text{ C/s} \qquad (19.2)$$

Not only are fuses and circuit breakers rated in amperes (or amps), so are many electrical appliances.

EXAMPLE 19.1 CURRENTS LARGE AND SMALL

(a) How much current is put out by a truck battery that moves 720 C of charge in 4.00 s while starting an engine? (b) How long does it take 1.00 C of charge to flow through a handheld calculator if a 0.300 mA current is flowing?

Strategy We can use the definition of current in Equation 19.1 to find the current in part (a), since charge and time are given. In part (b), we rearrange the definition of current and use the given values of charge and current to find the time required.

Solution for (a) Entering the given values for charge and time into the definition of current gives

$$I = \frac{\Delta Q}{\Delta t} = \frac{720 \text{ C}}{4.00 \text{ s}} = 180 \text{ C/s}$$
$$= 180 \text{ A}$$

Discussion for (a) This is a large current, consistent with the large power needs of the truck starter, and the fact that a large charge is moved in a small amount of time.

Solution for (b) Solving the relationship $I = \Delta Q/\Delta t$ for time Δt, and entering the known values for charge and current gives

$$\Delta t = \frac{\Delta Q}{I} = \frac{1.00 \text{ C}}{0.300 \times 10^{-3} \text{ C/s}}$$
$$= 3.33 \times 10^3 \text{ s}$$

Discussion for (b) This time is just shy of an hour. The small current used by the handheld calculator takes a much longer time to move a smaller charge than the large current of the truck starter. Such small current and energy demands allow handheld calculators to operate from solar cells or to get many hours of use out of small batteries.

Figure 19.2 shows a simple circuit and the standard schematic representation of a battery, conducting path, and load (a resistor). Schematics are very useful in visualizing the main features of a circuit. A single schematic can represent a wide variety of situations. The schematic in Figure 19.2(b), for example, can represent anything from a truck battery connected to a headlight to a AAA cell connected to a penlight.

Note that the direction of current flow in Figure 19.2 is from positive to negative. *The direction of conventional current is the direction that positive charge would flow.* Depending on the situation, positive charges, negative charges, or both may move. In metal wires, for example, current is carried by electrons—that is, negative charges move.

Figure 19.2 (a) A simple electric circuit. A closed path for current to flow through is supplied by conducting wires connecting a load to the terminals of a battery. (b) In this schematic, the battery is represented by the symbol ⊣⊢, conducting wires are shown as straight lines, and the symbol -\/\/\/- represents the load. The schematic represents a wide variety of similar circuits.

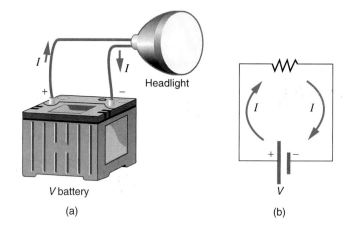

(a)

(b)

In ionic solutions, such as salt water, both positive and negative charges move. This is also true in nerve cells. A Van de Graaff generator used for nuclear research can produce a current of pure positive charges, such as protons. Figure 19.3 illustrates the movement of charged particles that compose a current. The fact that conventional current is taken to be in the direction that positive charge would flow can be traced back to Benjamin Franklin. He named the type of charge associated with electrons negative, long before they were known to carry current in so many situations. Franklin, in fact, was totally unaware of the small-scale structure of electricity.

It is important to realize that there is an electric field in conductors carrying current, as illustrated in Figure 19.3. Unlike static electricity, where a conductor in equilibrium cannot have an electric field in it, conductors carrying a current have an electric field and are not in static equilibrium. An electric field is needed to supply energy to move the charges.

EXAMPLE 19.2 ELECTRONS THROUGH A CALCULATOR—HOW MANY?

If the 0.300 mA current through the handheld calculator mentioned in Example 19.1 is carried by electrons, how many electrons per second pass through it?

Strategy Since each electron has a charge of 1.60×10^{-19} C, we can convert the current in coulombs per second to electrons per second.

Solution Starting with the definition of current, we have

$$I = \frac{\Delta Q}{\Delta t} = \frac{0.300 \times 10^{-3} \text{ C}}{\text{s}}$$

We divide this by the charge per electron, so that

$$\frac{e}{s} = \frac{0.300 \times 10^{-3} \text{ C}}{\text{s}} \times \frac{1 e}{1.60 \times 10^{-19} \text{ C}}$$

$$= 1.88 \times 10^{15} \frac{e}{\text{s}}$$

Discussion There are so many charged particles moving, even in small currents, that individual charges are not noticed, just as individual water molecules are not noticed in water flow.

Drift Velocity

Electrical signals are known to move very rapidly. Telephone conversations carried by currents in wires cover large distances without noticeable delays. Lights come on as soon as a switch is thrown. Most electrical signals carried by currents travel at speeds on the

Figure 19.3 Current I is the rate at which charge moves through an area A, such as the cross section of a wire. Conventional current is defined to move in the direction of the electric field. (a) Positive charges move in the direction of the electric field and the same direction as conventional current. (b) Negative charges move in the opposite direction of the electric field. Conventional current is in the direction opposite to the movement of negative charge.

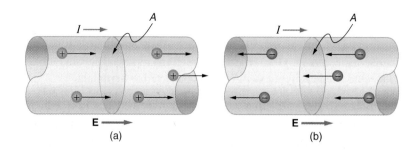

(a)

(b)

order of 10^8 m/s, a significant fraction of the speed of light. Interestingly, the individual charges that make up the current move *much* more slowly on average, typically drifting at speeds on the order of 10^{-4} m/s. How do we reconcile these two speeds, and what does it tell us about standard conductors?

The high speed of electrical signals results from the fact that the force between charges acts rapidly at a distance. Thus, when a free charge is forced into a wire, as in Figure 19.4, the incoming charge pushes other charges ahead of it, which in turn push on charges farther down the line. The density of charge in a system cannot easily be increased, and so the signal is passed on rapidly. The resulting electrical shock wave moves through the system at nearly the speed of light. To be precise, this rapidly moving signal or shock wave is a rapidly propagating change in electric field.

Good conductors have large numbers of free charges in them. In metals, the free charges are free electrons. Figure 19.5 shows how free electrons move through an ordinary conductor. The distance that an individual electron can move between collisions with atoms or other electrons is quite small. The electron paths, thus, appear nearly random, like the motion of atoms in a gas. But there is an electric field in the conductor that causes the electrons to drift in the direction shown (opposite to the field, since they are negative). The **drift velocity** v_d is the average velocity of the free charges. Drift velocity is quite small, since there are so many free charges. If we have an estimate of the density of free electrons in a conductor, we can calculate the drift velocity for a given current. The larger the density, the lower the velocity required for a given current.

The free electron collisions transfer energy to the atoms of the conductor. The electric field does work in moving the electrons through a distance, but that work does not increase the kinetic energy (nor speed, therefore) of the electrons. The work is transferred to the conductor's atoms, possibly increasing temperature. Thus a continuous power input is required to keep a current flowing. An exception, of course, is found in superconductors; for reasons we shall explore in a later chapter, superconductors do not absorb energy from the electron collisions. Superconductors can have a steady current without a continual supply of energy, a great energy savings. In contrast, the supply of energy can be useful, such as in a light bulb filament.

We can obtain an expression for the relationship between current and drift velocity by considering the number of free charges in a segment of wire, as illustrated in Figure 19.6. The *number of free charges per unit volume* is given the symbol n and depends on the material. The shaded segment has a volume Ax, so that the number of free charges in it is nAx. The charge ΔQ in this segment is thus $qnAx$, where q is the amount of charge on each carrier (for electrons, q is -1.60×10^{-19} C). Current is charge moved per unit time; thus, if all the original charges move out of this segment in time Δt, the current is

$$I = \frac{\Delta Q}{\Delta t} = \frac{qnAx}{\Delta t}$$

Note that $x/\Delta t$ is the drift velocity, since the charges move an average distance x in a time Δt. Rearranging terms gives

$$I = nqAv_d \qquad (19.3)$$

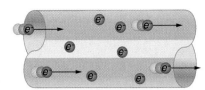

Figure 19.4 When charged particles are forced into this volume of a conductor, an equal number are quickly forced to leave. The repulsion between like charges makes it difficult to increase the number of charges in a volume. Thus, as one charge enters, another leaves almost immediately, carrying the signal rapidly forward.

CONNECTIONS

Good electrical conductors are often good heat conductors, too. This is because large numbers of free electrons can carry electrical current and can transport thermal energy. Compare data from Tables 19.1 and 13.3 and you can verify for yourself that materials that are good electrical conductors are generally good thermal conductors.

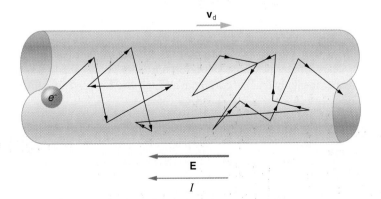

Figure 19.5 Free electrons moving in a conductor make many collisions with other electrons and atoms. The path of one electron is shown. The average velocity of the free charges is called the drift velocity, v_d, and it is in the direction opposite to the electric field for electrons. The collisions normally transfer energy to the conductor, requiring a constant supply of energy to maintain a steady current.

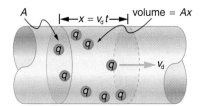

Figure 19.6 All the charges in the shaded volume of this wire move out in a time t, having a drift velocity $v_d = x/t$. See text for further discussion.

where I is the current through a wire of cross-sectional area A made of a material with a free charge density n. The carriers of the current each have charge q and move with a drift velocity v_d.

EXAMPLE 19.3   DRIFT VELOCITY IN A COMMON WIRE

Calculate the drift velocity of electrons in a 12 gauge copper wire (which has a diameter of 2.053 mm) carrying a 20.0 A current, given there is one free electron per copper atom. Household wiring often contains 12 gauge copper wire, and the maximum current allowed in such wire is usually 20 A.

Strategy We can find the drift velocity using Equation 19.3 if we can identify all of the other terms in the equation.

Solution Rearranging Equation 19.3 to isolate drift velocity gives

$$v_d = \frac{I}{nqA}$$

The current I is given to be 20.0 A, $q = -1.60 \times 10^{-19}$ C is the charge of an electron, and $A = \pi r^2 = 3.31 \times 10^{-6}$ m^2. The density of free charges per unit volume, n, is not given directly, but it can be determined from the knowledge that there is one free electron per copper atom. n is thus the same as the number of copper atoms per m^3. The periodic table gives the atomic mass of copper to be 63.5, and so there are 6.02×10^{26} atoms (Avogadro's number for 1 kg mol) per 63.5 kg of copper. Table 10.1 gives the density of

copper to be 8.80×10^3 kg/m^3, and so the number of copper atoms per cubic meter is

$$n = \frac{6.02 \times 10^{26}}{63.5 \text{ kg}} \cdot 8.80 \times 10^3 \text{ kg/m}^3$$
$$= 8.34 \times 10^{28}/\text{m}^3$$

Substituting these values into the expression for drift velocity gives

$$v_d = \frac{I}{nqA}$$
$$= \frac{20.0 \text{ A}}{(8.34 \times 10^{28}/\text{m}^3)(-1.60 \times 10^{-19} \text{ C})(3.31 \times 10^{-6} \text{ m}^2)}$$
$$= -4.53 \times 10^{-4} \text{ m/s}$$

Discussion The minus sign indicates that the negative charges are moving in the direction opposite to conventional current. The small value for drift velocity (on the order of 10^{-4} m/s) confirms that the signal moves on the order of 10^{12} times faster (about 10^8 m/s) than the charges that carry it, archetypal for a shock wave.

Note that simple drift velocity is not the entire story. In some metals, for example, not all free electrons move, and those that do thus move somewhat faster than the drift velocity.

19.2   OHM'S LAW: RESISTANCE AND SIMPLE CIRCUITS

What drives current? We can think of various devices—such as batteries, generators, wall outlets, and so on—which are necessary to maintain a current. All such devices create a potential difference and are loosely referred to as voltage sources.* When a voltage source is connected to a conductor, it applies a potential difference V that creates an electric field. The electric field in turn exerts force on charges, causing current.

Ohm's Law

The current that flows through most substances is directly proportional to the voltage V applied to it. The German physicist Georg Simon Ohm (1787–1854) was the first to demonstrate experimentally that the *current* in a metal wire *is directly proportional to the voltage applied*:

$$I \propto V$$

---

*Voltage output varies with the device and its load, being lower when the device is supplying current. This is discussed in detail in Section 20.2.

This important relationship is known as **Ohm's law**. It can be viewed as a cause and effect relationship, with voltage the cause and current the effect.

Resistance and Simple Circuits

If voltage drives current, what impedes it? The electric property that impedes current (crudely similar to friction and air resistance) is called **resistance** R. Collisions of moving charges with atoms and molecules in a substance transfer energy to the substance and limit current. Resistance is defined to have current inversely proportional to it, or

$$I \propto \frac{1}{R}$$

Thus, for example, current is cut in half if resistance doubles. Combining the relationships of current to voltage and current to resistance gives

$$I = \frac{V}{R} \qquad (19.4)$$

This relationship is also called **Ohm's law**. Ohm's law in this form really defines resistance for certain materials. Ohm's law (like Hooke's law) is not universally valid. The many substances for which Ohm's law holds are called *ohmic*. These include good conductors like copper and aluminum, and some poor conductors under certain circumstances.* An object that has simple resistance is called a *resistor*, even if its resistance is small. The unit for resistance is an **ohm** and given the symbol Ω (upper case Greek omega). Rearranging Equation 19.4 gives $R = V/I$, and so the units of resistance are 1 ohm = 1 volt per ampere:

$$1 \ \Omega = 1 \ V/A \qquad (19.5)$$

Figure 19.7 shows the schematic for a simple circuit. A **simple circuit** has a single voltage source and a single resistor. The wires connecting the voltage source to the resistor can be assumed to have negligible resistance, or their resistance can be included in R.

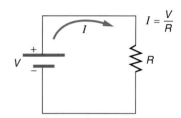

Figure 19.7 A simple electric circuit in which a closed path for current to flow is supplied by conductors (usually metal wires) connecting a load to the terminals of a battery, represented by the symbol ⊣⊢. The symbol ⌁ represents the single resistor and includes any resistance in the connections to the voltage source.

EXAMPLE 19.4   HEADLIGHT RESISTANCE

What is the resistance of an automobile headlight through which 2.50 A flows when 12.0 V is applied to it?

Strategy We can rearrange Ohm's law as stated in Equation 19.4 and use it to find the resistance.

Solution Rearranging Equation 19.4 and substituting known values gives

$$R = \frac{V}{I} = \frac{12.0 \ V}{2.50 \ A} = 4.80 \ \Omega$$

Discussion This is a relatively small resistance, but it is larger than the cold resistance of the headlight. As we shall see in the next section, resistance usually increases with temperature, and so the bulb has a lower resistance when it is first switched on and draws considerably more current during its brief warm-up period.

Resistances range over many orders of magnitude. Some ceramic insulators, such as those used to support power lines, have resistances of $10^{12} \ \Omega$ or more. A dry person may have a hand to foot resistance of $10^5 \ \Omega$, whereas the resistance of the human heart is about $10^3 \ \Omega$. A meter long piece of large-diameter copper wire may have a resistance of $10^{-5} \ \Omega$, and superconductors have no resistance at all (they are nonohmic). Resistance is related to the shape of an object and the material of which it is composed, as will be seen in the next section.

---

*Ohmic materials have a resistance R that is independent of voltage V and current I.

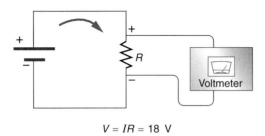

$V = IR = 18$ V

Figure 19.8 The voltage drop across a resistor in a simple circuit equals the voltage output of the battery.

CONNECTIONS

Conservation of Energy

In a simple electrical circuit, the sole resistor converts energy supplied by the source into another form. Conservation of energy is evidenced here by the fact that all of the energy supplied by the source is converted to another form by the resistor alone. We will find that conservation of energy has other important applications in circuits and is a powerful tool in circuit analysis.

TABLE 19.1

RESISTIVITIES ρ OF VARIOUS MATERIALS AT 20°C

Material	Resistivity ρ ($\Omega \cdot m$)
Conductors	
Silver	1.59×10^{-8}
Copper	1.72×10^{-8}
Gold	2.44×10^{-8}
Aluminum	2.65×10^{-8}
Tungsten	5.6×10^{-8}
Iron	9.71×10^{-8}
Platinum	10.6×10^{-8}
Steel	20×10^{-8}
Lead	22×10^{-8}
Manganin	
(Cu, Mn, Ni alloy)	44×10^{-8}
Constantin	
(Cu, Ni alloy)	49×10^{-8}
Mercury	96×10^{-8}
Nichrome	
(Ni, Fe, Cr alloy)	100×10^{-8}
Semiconductors*	
Carbon (pure)	3.5×10^{-5}
Carbon	$(3.5–60) \times 10^{-5}$
Germanium (pure)	600×10^{-3}
Germanium	$(1–600) \times 10^{-3}$
Silicon (pure)	2300
Silicon	0.1–2300
Insulators	
Amber	5×10^{14}
Glass	$10^9–10^{14}$
Lucite	$>10^{13}$
Mica	$10^{11}–10^{15}$
Quartz (fused)	75×10^{16}
Rubber (hard)	$10^{13}–10^{16}$
Sulfur	10^{15}
Teflon	$>10^{13}$
Wood	$10^8–10^{11}$

*Values depend strongly on amounts and types of impurities.

Additional insight is gained by solving Equation 19.4 for V, yielding

$$V = IR$$

This expression for V can be interpreted as the *voltage drop across a resistor produced by the flow of current I*. The phrase *IR drop* is often used for this voltage. For instance, the headlight in Example 19.4 has an *IR* drop of 12.0 V. If voltage is measured at various points in a circuit, it will be seen to increase at the voltage source and decrease at the resistor. Voltage is similar to fluid pressure. The voltage source is like a pump, creating pressure, causing current—the flow of charge. The resistor is like a pipe that reduces pressure and limits flow because of its resistance. Conservation of energy has important consequences here. The voltage source supplies energy, and the resistor converts it to another form (such as thermal energy). In a simple circuit (one with a single simple resistor), the voltage supplied by the source equals the voltage drop at the resistor, since PE $= q\,\Delta V$ and the same q flows through each. Thus the energy supplied by the voltage source and the energy converted by the resistor are equal. (See Figure 19.8.)

19.3 RESISTANCE AND RESISTIVITY

The resistance of an object depends on its shape and the material of which it is composed. The cylindrical resistor in Figure 19.9 is easy to analyze, and, by so doing, we can gain insight into the resistance of more complicated shapes. As you might expect, the cylinder's electric resistance R is directly proportional to its length L, similar to the resistance of a pipe to fluid flow. The longer the cylinder is, the more collisions charges will make with its atoms. The greater the diameter of the cylinder, the more current it can carry (again similar to the flow of fluid through a pipe). In fact, R is inversely proportional to the cylinder's cross-sectional area A.

For a given shape, the resistance depends on the material of which the object is composed. Different materials offer different resistance to the flow of charge. We define the **resistivity** ρ of a substance so that the resistance R of an object is directly proportional to ρ. Resistivity ρ is an *intrinsic* property of a material, independent of its shape or size. The resistance R of a uniform cylinder of length L, of cross-sectional area A, and made of a material with resistivity ρ is

$$R = \frac{\rho L}{A} \qquad (19.6)$$

Table 19.1 gives representative values of ρ. In Table 19.1, materials are separated into categories of conductors, semiconductors, and insulators, based on broad groupings of resistivities. Conductors have the smallest resistivities, insulators the largest, and semiconductors are intermediate. Conductors have varying but large free charge densities, whereas most charges in insulators are bound to atoms and are not free to move.

Figure 19.9 A uniform cylinder of length L and cross-sectional area A. Its resistance to the flow of current is similar to the resistance posed by a pipe to fluid flow. The longer the cylinder, the greater its resistance. The larger its cross-sectional area A, the smaller its resistance.

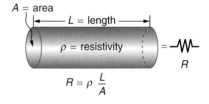

A = area
L = length
ρ = resistivity
$R = \rho \dfrac{L}{A}$

Semiconductors are intermediate, having far fewer free charges than conductors, and having properties that make the number of free charges depend strongly on the type and amount of impurities in the semiconductor. These unique properties of semiconductors are put to use in modern electronics, as will be explored in later chapters.

EXAMPLE 19.5   DIAMETER OF A HEADLIGHT FILAMENT

A car headlight filament is made of tungsten and has a cold resistance of 0.350 Ω. If the filament is a cylinder 4.00 cm long (it may be coiled to save space), what is its diameter?

Strategy We can rearrange Equation 19.6 to find the cross-sectional area A of the filament from the given information. Then its diameter can be found by assuming it has a circular cross section.

Solution The cross-sectional area, found by rearranging the expression for the resistance of a cylinder given in Equation 19.6, is

$$A = \frac{\rho L}{R}$$

Substituting the given values, and taking ρ from Table 19.1, yields

$$A = \frac{(5.6 \times 10^{-8} \ \Omega \cdot m)(4.00 \times 10^{-2} \ m)}{0.350 \ \Omega}$$

$$= 6.40 \times 10^{-9} \ m^2$$

The area of a circle is related to its diameter D by

$$A = \frac{\pi D^2}{4}$$

Solving for the diameter D, and substituting the value found for A, gives

$$D = 2\left(\frac{A}{\pi}\right)^{1/2} = 2\left(\frac{6.40 \times 10^{-9} \ m^2}{3.14}\right)^{1/2}$$

$$= 9.0 \times 10^{-5} \ m$$

Discussion The diameter is just under a tenth of a millimeter. It is quoted to only two digits, because ρ is known to only two digits.

Temperature Variation of Resistance

The resistivity of all materials depends on temperature. Some even become superconductors (zero resistivity) at low temperatures. (See Figure 19.10.) Conversely, the resistivity of conductors increases with increasing temperature. Since the atoms vibrate more rapidly and over larger distances at higher temperatures, the electrons moving through a metal make more collisions, effectively making the resistivity higher. Over relatively small temperature changes (about 100°C or less), resistivity ρ varies with temperature change ΔT as expressed in the following equation:

$$\rho = \rho_0(1 + \alpha \Delta T) \tag{19.7}$$

where ρ_0 is the original resistivity, and α is the **temperature coefficient of resistivity**. (See Table 19.2.) For larger temperature changes, α may vary or a nonlinear equation may

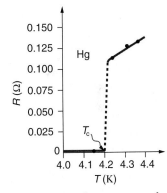

Figure 19.10 The resistance of a sample of mercury is zero at very low temperatures—it is a superconductor up to about 4.2 K. Above that critical temperature, its resistance makes a sudden jump and then increases nearly linearly with temperature.

TABLE 19.2

TEMPERATURE COEFFICIENTS OF RESISTIVITY α			
Material	*Coefficient* α *(1/°C)**	*Material*	*Coefficient* α *(1/°C)**
Conductors		*Conductors*	
Silver	3.8×10^{-3}	Constantin	
Copper	3.9×10^{-3}	(Cu, Ni alloy)	0.002×10^{-3}
Gold	3.4×10^{-3}	Mercury	0.89×10^{-3}
Aluminum	3.9×10^{-3}	Nichrome	
Tungsten	4.5×10^{-3}	(Ni, Fe, Cr alloy)	0.4×10^{-3}
Iron	5.0×10^{-3}		
Platinum	3.93×10^{-3}	*Semiconductors*	
Lead	3.9×10^{-3}	Carbon (pure)	-0.5×10^{-3}
Manganin		Germanium (pure)	-50×10^{-3}
(Cu, Mn, Ni alloy)	0.000×10^{-3}	Silicon (pure)	-70×10^{-3}

*Values at 20°C.

be needed to find ρ. Note that α is positive for metals, meaning their resistivity increases with temperature. Some alloys have been developed specifically to have a small temperature dependence. Manganin, for example, has α close to zero (to three digits on the scale in Table 19.2), and so its resistivity varies only slightly with temperature. This is useful for making a temperature-independent resistance standard, for example.

Note also that α is negative for the semiconductors listed in Table 19.2, meaning that their resistivity decreases with increasing temperature. They become better conductors at higher temperature, because increased thermal agitation increases the number of free charges available to carry current. This property of decreasing ρ with temperature is also related to the type and amount of impurities present in the semiconductors.

The resistance of an object also depends on temperature since R is directly proportional to ρ. For a cylinder we know $R = \rho L/A$, and so, if L and A do not change greatly with temperature, R will have the same temperature dependence as ρ. (Examination of the coefficients of linear expansion in Table 12.1 shows them to be about two orders of magnitude less than typical temperature coefficients of resistivity, and so the effect of temperature on L and A is about two orders of magnitude less than on ρ.) Thus,

$$R = R_0(1 + \alpha\,\Delta T) \tag{19.8}$$

is the temperature dependence of the resistance of an object, where R_0 is the original resistance and R is the resistance after a temperature change ΔT. Numerous thermometers are based on the effect of temperature on resistance. (See Figure 19.11.) One of the most common is the thermistor, a semiconductor crystal with a strong temperature dependence, the resistance of which is measured to obtain its temperature. The device is small, so that it quickly comes into thermal equilibrium with the part of a person it touches.

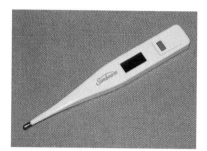

Figure 19.11 These familiar thermometers are based on the automated measurement of a thermistor's temperature-dependent resistance.

EXAMPLE 19.6   HOT FILAMENT RESISTANCE

Although caution must be used in applying Equations 19.7 and 19.8 for temperature changes greater than 100°C, for tungsten the equations work reasonably well for very large temperature changes. What, then, is the resistance of the tungsten filament in the previous example if its temperature is increased to a typical operating temperature of 2850°C?

Strategy This is a straightforward application of Equation 19.8, since the original resistance of the filament was given to be $R_0 = 0.350\ \Omega$, and the temperature change is $\Delta T = 2830$°C.

Solution Thus the hot resistance R is obtained by entering known values into Equation 19.8:

$$
\begin{aligned}
R &= R_0(1 + \alpha\,\Delta T) \\
&= (0.350\ \Omega)[1 + (4.5 \times 10^{-3}/°C)(2830°C)] \\
&= 4.8\ \Omega
\end{aligned}
$$

Discussion This value is consistent with the headlight resistance in Example 19.4.

19.4   ELECTRIC POWER AND ENERGY

Power is commonly associated with electricity. Power transmission lines immediately come to mind. We also think of light bulbs in terms of their power ratings in watts. Let us compare a 25 W bulb with a 60 W bulb. (See Figure 19.12.) Since both operate on the same voltage, the 60 W bulb must draw more current to have a greater power rating. Thus the 60 W bulb's resistance must be lower than that of a 25 W bulb. If we increase voltage, we also increase power. For example, when a 25 W bulb that is designed to operate on 120 V is connected to 240 V, it briefly glows very brightly and then burns out. Precisely how are voltage, current, and resistance related to electric power?

We already know from the discussions in Chapter 18 that electric energy depends on both the voltage involved and the charge moved. This is expressed most simply as $PE = qV$, where q is the charge moved and V is the voltage (or more precisely, the potential difference the charge moves through). Power is the rate at which energy is moved, and so electric power is

$$P = \frac{PE}{t} = \frac{qV}{t}$$

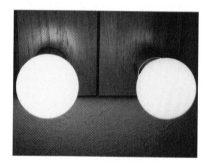

Figure 19.12 Which of these light bulbs, the 25 or the 60 W, has the higher resistance? Which draws more current? Which uses the most energy? Can you tell from the color that the 25 W filament is cooler?

Recognizing that current is $I = q/t$ (note that $\Delta t = t$ here), the expression for power becomes

$$P = IV \tag{19.9}$$

Electric power (P) is simply the product of current times voltage. Power has familiar units of watts. Since the SI unit for potential energy (PE) is the joule, power has units of joules per second, or watts. Thus $1 \text{ A} \cdot \text{V} = 1 \text{ W}$. For example, the fuse for a car's cigarette lighter may be rated at 20 A, so that the circuit can deliver a maximum power $P = IV = (20 \text{ A})(12 \text{ V}) = 240 \text{ W}$. In some applications, electric power may be expressed as volt-amperes or even kilovolt-amperes ($1 \text{ kV} \cdot \text{A} = 1 \text{ kW}$).

To see the relationship of power to resistance, we combine Ohm's law with Equation 19.9. Substituting $I = V/R$ gives $P = (V/R)V = V^2/R$. Similarly, substituting $V = IR$ gives $P = I(IR) = I^2R$. Three expressions for electric power are listed together here for convenience:

$$P = IV \tag{19.9a}$$

$$P = \frac{V^2}{R} \tag{19.9b}$$

$$P = I^2R \tag{19.9c}$$

Note that the first equation is always valid, whereas the other two can only be used for resistors. In a simple circuit, with one voltage source and a single resistor, the power supplied by the voltage source and that dissipated by the resistor are identical. (In more complicated circuits, considered in the next few chapters, P can be the power dissipated by a single device and not the total power in the circuit.)

Different insights can be gained from the three different expressions for electric power. For example, Equation 19.9b implies that the lower the resistance connected to a given voltage source, the greater the power delivered. Furthermore, since voltage is squared in Equation 19.9b, the effect of applying a higher voltage is perhaps greater than expected. Thus, when the voltage is doubled to a 25 W bulb, its power nearly quadruples to about 100 W, burning it out. If the bulb's resistance remained constant, its power would be exactly 100 W, but at the higher temperature its resistance is higher, too.

EXAMPLE 19.7 HOT AND COLD POWER

(a) Find the power dissipated by the car headlight considered in Examples 19.4–19.6, both when it is hot and when it is cold. (b) What current does it draw when cold?

Strategy and Solution for (a) For the hot headlight, we know voltage and current, so that we can use $P = IV$ to find the power. For the cold headlight, we know the voltage and resistance, so that we can use $P = V^2/R$ to find the power.

Entering the known values of current and voltage for the hot headlight, we obtain

$$P = IV = (2.50 \text{ A})(12.0 \text{ V}) = 30.0 \text{ W}$$

The cold resistance was $0.350 \ \Omega$, and so the power it uses when first switched on is

$$P = \frac{V^2}{R} = \frac{(12.0 \text{ V})^2}{0.350 \ \Omega}$$
$$= 411 \text{ W}$$

Discussion for (a) The 30 W dissipated by the hot headlight is typical. But the 411 W when cold is surprisingly higher. The initial power quickly decreases as the bulb heats up and its resistance increases.

Strategy and Solution for (b) The current when the bulb is cold can be found several different ways. We rearrange one of the power equations, $P = I^2R$, and enter known values, obtaining

$$I = \sqrt{\frac{P}{R}} = \sqrt{\frac{411 \text{ W}}{0.350 \ \Omega}} = 34.3 \text{ A}$$

Discussion for (b) The cold current is remarkably higher than the steady state value of 2.50 A, but the current will quickly decline to that value as the bulb heats up. Most fuses and circuit breakers (used to limit the current in a circuit) are designed to tolerate very high currents briefly as a device comes on. In some cases, such as with electric motors, the current remains high for several seconds, necessitating special "slow blow" fuses.

Energy, Power, and Time

The relationship $E = Pt$ is one that you will find useful in many different contexts. The energy your body uses in exercise is related to the power level and duration of your activity, for example. The amount of heating by a power source is related to the power level and time it is applied. Even the radiation dose of an x-ray image is related to the power and time of exposure.

The Cost of Electricity

The more electric appliances you use and the longer they are left on, the higher your electric bill. This familiar fact is based on the relationship between energy and power. You pay for the energy used. Since $P = E/t$, we see that

$$E = Pt$$

is the energy used by a device using power P for a time interval t. For example, the more light bulbs burning, the greater P used; the longer they are on, the greater t is. The energy unit on electric bills is the kilowatt-hour (kW·h), consistent with the relationship $E = Pt$. It is easy to estimate the cost of operating electric appliances if you have some idea of their power consumption rate in watts or kilowatts, the time they are on in hours, and the cost per kilowatt-hour for your electric utility.

EXAMPLE 19.8 ANOTHER REASON TO TURN OFF THE TV

If the cost of electricity in your district is 9.50¢ per kW·h, what is the cost of operating a 200 W television 4.00 h per day for 30.0 days?

Strategy To find the cost, we first find the energy used in kilowatt-hours and then multiply by the cost per kilowatt-hour.

Solution The energy used in kilowatt-hours is found by entering the power and time into the expression for energy:

$$E = Pt = (200 \text{ W})(120 \text{ h}) = 24,000 \text{ W·h}$$

In kilowatt-hours, this is

$$E = 24.0 \text{ kW·h}$$

Now the cost is

$$\text{cost} = (24.0 \text{ kW·h})(9.50 \text{ ¢/kW·h}) = \$2.28$$

Discussion While cheap, the cost is not just a few pennies. When applying this idea to other electrical devices, it is apparent that low-power appliances, like electric clocks, cost almost nothing to operate. Conversely, high-power appliances, such as electric water heaters, can be expensive, particularly if they are on for many hours. Trimming power consumption by choosing smaller or more efficient appliances and limiting the time they are on saves money, not to mention energy.

Kilowatt-hours, like all other specialized energy units such as food calories, can be converted to joules. You can prove to yourself that $1 \text{ kW·h} = 3.6 \times 10^6 \text{ J}$.

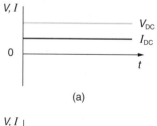

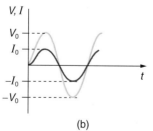

Figure 19.13 (a) DC voltage and current are constant in time, once the current is established. (b) A graph of voltage and current versus time for 60 Hz AC power such as is commonly used in the United States. The voltage and current are sinusoidal and are in phase for a simple resistance circuit. The frequencies and peak voltages of AC sources differ greatly.

19.5 ALTERNATING CURRENT VERSUS DIRECT CURRENT

Most of the examples dealt with so far, and particularly those utilizing batteries, have constant voltage sources. Once the current is established, it is thus also a constant. **Direct current** (DC) is defined to be the steady state of a constant-voltage circuit. But there are also many well-known applications that use a time-varying voltage source. If the source varies periodically, particularly sinusoidally, the circuit is known as an **alternating current** (AC) circuit. Examples include the commercial and residential power that serves so many of our needs. Figure 19.13 shows graphs of voltage and current versus time for typical DC and AC power.

Figure 19.14 shows a schematic of a simple circuit with an AC voltage source represented by the symbol ⊘. The voltage between the terminals fluctuates as shown, with the **AC voltage** given by

$$V = V_0 \sin 2\pi ft \qquad (19.10)$$

where V is the voltage at time t, V_0 is the peak voltage, and f is the frequency in hertz. For this simple resistance circuit, $I = V/R$, and so the **AC current** is

$$I = I_0 \sin 2\pi ft \qquad (19.11)$$

where I is the current at time t, and $I_0 = V_0/R$ is the peak current. The voltage and current are said to be in phase, as seen in Figure 19.13.

Current in the resistor alternates back and forth just like the driving voltage, since $I = V/R$. If the resistor is a fluorescent light bulb, for example, it brightens and dims 120 times per second as the current repeatedly goes through zero. A 120 Hz flicker is too rapid for your eyes to detect, but if you wave your hand back and forth between your face and a fluorescent light, you will see a stroboscopic effect evidencing AC. The fact that the light output fluctuates means that the power is fluctuating. The power supplied is $P = IV$. Using the expressions for I and V above, we see that the time dependence of power is $P = I_0 V_0 \sin^2 2\pi ft$, as shown in Figure 19.15.

We are most often concerned with average power rather than its fluctuations—that 60 W light bulb in your desk lamp has an *average* power consumption of 60 W, for example. As illustrated in Figure 19.15, the average power P_{ave} is*

$$P_{ave} = \tfrac{1}{2}I_0 V_0 \qquad (19.12)$$

Similarly, we define an average or **rms current** I_{rms} and average or **rms voltage** V_{rms} to be, respectively,

$$I_{rms} = \frac{I_0}{\sqrt{2}} \qquad (19.13a)$$

and

$$V_{rms} = \frac{V_0}{\sqrt{2}} \qquad (19.13b)$$

where **rms** stands for **root mean square**, a particular kind of average.[†] Now,

$$P_{ave} = I_{rms} V_{rms} \qquad (19.14)$$

which gives

$$P_{ave} = \frac{I_0}{\sqrt{2}} \cdot \frac{V_0}{\sqrt{2}} = \tfrac{1}{2}I_0 V_0$$

consistent with Equation 19.12. It is standard practice to quote I_{rms}, V_{rms}, and P_{ave} rather than the peak values. For example, most household electricity is 120 V AC, which means that V_{rms} is 120 V. The common 20 A circuit breaker will interrupt a sustained I_{rms} greater than 20 A. Your 1.0 kW microwave oven consumes $P_{ave} = 1.0$ kW, and so on. You can think of these rms and average values as the equivalent DC values for a simple resistive circuit.

To summarize, when dealing with AC, Ohm's law and the equations for power are completely analogous to those for DC, but rms and average values are used for AC. Thus, for AC, Ohm's law is written

$$I_{rms} = \frac{V_{rms}}{R} \qquad (19.15)$$

The various expressions for AC power P_{ave} are

$$P_{ave} = I_{rms} V_{rms} \qquad (19.16a)$$

$$P_{ave} = \frac{V_{rms}^2}{R} \qquad (19.16b)$$

and

$$P_{ave} = I_{rms}^2 R \qquad (19.16c)$$

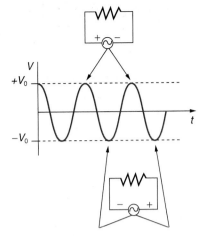

Figure 19.14 The potential difference V between the terminals of an AC voltage source ⊘ fluctuates as shown. The mathematical expression for V is given in Equation 19.10.

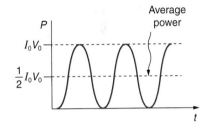

Figure 19.15 AC power as a function of time. Since the voltage and current are in phase here, their product is nonnegative and fluctuates between zero and $I_0 V_0$. Average power is $(1/2)I_0 V_0$.

EXAMPLE 19.9   PEAK VOLTAGE AND POWER FOR AC

(a) What is the value of the peak voltage for 120 V AC power? (b) What is the peak power consumption rate of a 60.0 W AC light bulb?

We can use Equation 19.13b to find the peak voltage, and we can manipulate the definition of power to find the peak power from the given average power.

Strategy We are given V_{rms} is 120 V and P_{ave} is 60.0 W.

Solution for (a) Solving Equation 19.13b for the peak
(continued)

---

*This is evident from the graph, since the areas above and below the $(1/2)I_0 V_0$ line are equal, but it can also be proven using trigonometric identities.

[†]To obtain a root mean square, the particular quantity is squared, its mean (or average) is found, and the square root taken. This is useful for AC, since the average value is zero.

(continued)

voltage V_0 and substituting the known value for V_{rms} gives

$$V_0 = \sqrt{2}\,V_{rms} = 1.414(120\ \text{V}) = 170\ \text{V}$$

Discussion for (a) This means that the AC voltage swings from +170 to –170 V and back 60 times every second. An equivalent DC voltage is a constant 120 V.

Solution for (b) Peak power is peak current times peak

voltage. Thus,

$$P_0 = I_0 V_0 = 2(\tfrac{1}{2}I_0 V_0) = 2P_{ave}$$

We know the average power is 60.0 W, and so

$$P_0 = 2(60.0\ \text{W}) = 120\ \text{W}$$

Discussion for (b) So the power swings from zero to 120 W one hundred and twenty times per second (twice each cycle), and the power averages 60 W.

Why Use AC for Power Distribution?

Most large power distribution systems are AC. Moreover, the power is transmitted at much higher voltages than the 120 V or 220 V AC we use in homes and on the job. Economies of scale make it cheaper to build a few very large electric power generation plants than to build numerous small ones. This necessitates sending power long distances, and it is obviously important that energy losses en route be minimized. High voltages can be transmitted with much smaller power losses than low voltages, as we shall see. (See Figure 19.16.) For safety reasons, the voltage at the user is reduced to familiar values. The crucial factor is that it is much easier to increase and decrease AC voltages than DC, so that AC is used in most large power distribution systems.

Figure 19.16 Power is distributed over large distances at high voltage to reduce power loss in the transmission lines. These large transformers reduce the voltage sent from the power plant for local distribution. Smaller transformers again reduce the voltage for safe residential and commercial use.

EXAMPLE 19.10 POWER LOSSES ARE LESS FOR HIGH TRANSMISSION VOLTAGES

(a) What current is needed to transmit 100 MW of power at 200 kV? (b) What is the power dissipated by the transmission lines if they have a resistance of 1.00 Ω? (c) What percentage of the power is lost in the transmission lines?

Strategy We are given $P_{ave} = 100$ MW, $V_{rms} = 200$ kV, and the resistance of the lines is $R = 1.00$ Ω. Using these givens, we can find the current flowing (from $P = IV$) and then the power dissipated in the lines (I^2R), and we take the ratio to the total power transmitted.

Solution for (a) To find the current, we rearrange the relationship $P_{ave} = I_{rms}V_{rms}$ and substitute known values. This gives

$$I_{rms} = \frac{P_{ave}}{V_{rms}} = \frac{100 \times 10^6\ \text{W}}{200 \times 10^3\ \text{V}} = 500\ \text{A}$$

Solution for (b) Knowing the current and given the resistance of the lines, the power dissipated in them is found from Equation 19.16c. Substituting the known values gives

$$P_{ave} = I_{rms}^2 R = (500\ \text{A})^2(1.00\ \Omega) = 250\ \text{kW}$$

Solution for (c) The percent loss is the ratio of this lost power to the total or input power, multiplied by 100:

$$\%\ \text{loss} = \frac{250\ \text{kW}}{100\ \text{MW}} \times 100 = 0.250\%$$

Discussion One-fourth of a percent is an acceptable loss. Note that if 100 MW of power had been transmitted at 25 kV, then a current of 4000 A would have been needed. This would result in a power loss in the lines of 16.0 MW, or 16.0% rather than 0.250%. The lower the voltage, the more current is needed, and the greater the power loss in the fixed-resistance transmission lines. Of course, lower-resistance lines can be built, but this requires larger and more expensive wires. If superconducting lines can be economically produced, there would be no loss in the transmission lines at all. But, as we shall see in a later chapter, there is a limit to current in superconductors, too. In short, high voltages are more economical for transmitting power, and AC voltage is much easier to raise and lower, so that AC is used in most large-scale power distribution systems.

It is widely recognized that high voltages pose greater hazards than low voltages. But, in fact, some high voltages, such as those associated with common static electricity, can be harmless. So it is not voltage alone that determines a hazard. It is not so widely recognized that AC shocks are often more harmful than similar DC shocks. Thomas Edison thought this was the case and engaged in bitter fights, in particular with George Westinghouse and Nikola Tesla, over the use of AC in early power distribution systems.

19.6* ELECTRIC HAZARDS, THERMAL AND SHOCK

There are two known hazards of electricity—thermal and shock. A **thermal hazard** is one where excessive electric power causes undesired thermal effects, such as starting a fire in the wall of a house. A **shock hazard** occurs when electric current passes through a person. Shocks range in severity from painful, but otherwise harmless, to heart-stopping lethality. This section considers these hazards and the various factors affecting them in a quantitative manner. Section 22.8 will consider systems and devices for preventing electrical hazards.

Human & Biological Application

Thermal Hazards

Electric power causes undesired thermal effects whenever electric energy is converted to heat at a rate faster than it can be safely dissipated. A classic example of this is the **short circuit**, a low-resistance path between terminals of a voltage source. An example of a short circuit is shown in Figure 19.17. Insulation on wires leading to an appliance has worn through, allowing the wires to come into contact. Such an undesired contact with a high voltage is called a **short**. Since the resistance of the short, r, is very small, the power dissipated in the short, $P = V^2/r$, is very large. For example, if V is 120 V and r is 0.100 Ω, then the power is 144 kW, much greater than that used by a typical household appliance. Thermal energy delivered at this rate will very quickly raise the temperature of surrounding materials, melting or perhaps igniting them.

One particularly insidious aspect of a short circuit is that its resistance may actually be decreased by the heat generated. This can happen if the short creates ionization. These charged atoms and molecules are free to move and, thus, lower the resistance r. Since $P = V^2/r$, the power dissipated in the short rises, possibly causing more ionization, more power, and so on. High voltages, such as the 480 V AC used in some industrial applications, lend themselves to this hazard, because higher voltages create higher initial power production in a short.

Another serious, but less dramatic, thermal hazard occurs when wires supplying power to a user are overloaded with too great a current. As discussed in the previous section, the power dissipated in the supply wires is $P = I^2R_w$, where R_w is the resistance of the wires and I the current flowing through them. If either I or R_w is too large, the wires overheat. For example, a worn appliance cord (with some of its braided wires broken) may have $R_w = 2.00\ \Omega$ rather than the 0.100 Ω it should be. If 10.0 A of current passes through the cord, then $P = I^2R_w = 200$ W is dissipated in the cord—much more than is safe. Similarly, if a wire with a 0.100 Ω resistance is meant to carry a few amps, but is instead carrying 100 A, it will severely overheat. The power dissipated in the wire will in that case be $P = 1000$ W. Fuses and circuit breakers are used to limit

Figure 19.17 A short circuit is an undesired low-resistance path across a voltage source. (a) Worn insulation on the wires of a toaster allow them to come into contact with a low resistance r. Since $P = V^2/r$, thermal power is created so rapidly that the cord melts or burns. (b) A schematic of the short circuit.

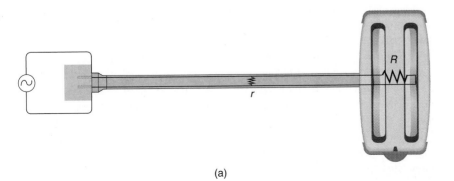

(a)

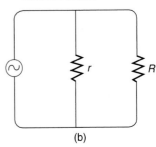

(b)

Figure 19.18 (a) A fuse has a metal strip with a low melting point that, when overheated by an excessive current, permanently breaks the connection of a circuit to a voltage source. (b) A circuit breaker is an automatic but restorable electric switch. The one shown here has a bimetallic strip that bends to the right and into the notch if overheated. The spring then forces the metal strip downward, breaking the electrical connection at the points.

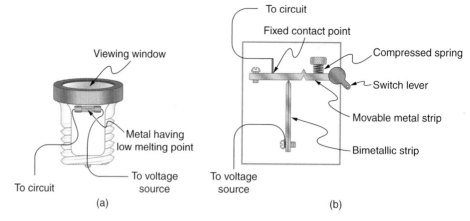

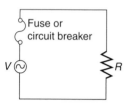

Figure 19.19 Schematic of a circuit with a fuse or circuit breaker in it. Fuses and circuit breakers act like automatic switches that open when sustained current exceeds desired limits.

excessive currents. (See Figures 19.18 and 19.19.) Each device opens the circuit automatically when a sustained current exceeds safe limits.

Fuses and circuit breakers for typical household voltages and currents are relatively simple to produce, but those for large voltages and currents experience special problems. For example, when a circuit breaker tries to interrupt the flow of high-voltage electricity, a spark can jump across its points that ionizes the air in the gap and allows the current to continue flowing. Large circuit breakers found in power distribution systems employ insulating gas and even use jets of gas to blow out such sparks. Here AC is safer than DC, since AC current goes through zero 120 times per second, giving a quick opportunity to extinguish these arcs.

Shock Hazards

Electrical currents through people produce tremendously varied effects. An electrical current can be used to block back pain. The possibility of using electrical current to stimulate muscle action in paralyzed limbs, perhaps allowing paraplegics to walk, is under study. TV dramatizations in which electrical shocks are used to bring a heart attack victim out of ventricular fibrillation (a massively irregular, often fatal, beating of the heart) are more than common. Yet most electrical shock fatalities occur because a current put the heart *into* fibrillation. A pacemaker uses electrical shocks to stimulate the heart to beat properly. Some fatal shocks do not produce burns, but warts can be safely burned off with electric current. Of course, there are consistent explanations for these disparate effects. The major factors upon which the effects of electrical shock depend are

1. The amount of current I
2. The path taken by the current
3. The duration of the shock
4. The frequency f of the current ($f = 0$ for DC)

Table 19.3 gives the effects of electrical shocks as a function of current for a typical accidental shock. The effects are for a shock that passes through the trunk of the body, has a duration of 1 s, and is caused by 60 Hz power.

Very small currents pass harmlessly and unfelt through the body. This happens to you regularly without your knowledge. The threshold of sensation is only 1 mA and, although unpleasant, shocks are apparently harmless for currents less than 5 mA. A great number of safety rules take the 5 mA value for the maximum allowed shock. At 10 to 20 mA and above, the current can stimulate sustained muscular contractions much as regular nerve impulses do. People sometimes say they were knocked across the room by a shock, but what really happened was that certain muscles contacted, propelling them in a manner not of their own choosing. (See Figure 19.20(a).) More frightening, and potentially more dangerous, is the *"can't let go" effect* illustrated in Figure 19.20(b). The muscles that close the fingers are stronger than those that open them, so that the hand closes involuntarily on the wire shocking it. This can prolong the shock indefinitely. It can also be a danger to a person trying to rescue the victim, because the rescuer's hand may close about the victim's wrist.

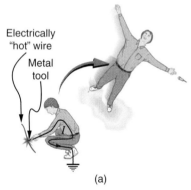

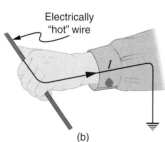

Figure 19.20 An electric current can cause muscular contractions with varying effects. (a) The victim is "thrown" backward by involuntary muscle contractions that extend the legs and torso. (b) The victim can't let go of the wire that is stimulating all the muscles in the hand. Those that close the fingers are stronger than those that open them.

TABLE 19.3

EFFECTS OF ELECTRICAL SHOCK AS A FUNCTION OF CURRENT*

Current (mA)	Effect
1	Threshold of sensation
5	Maximum harmless current
10–20	Onset of sustained muscular contraction; cannot let go for duration of shock; contraction of chest muscles may stop breathing during shock
50	Onset of pain
100–300+	Ventricular fibrillation possible; often fatal
300	Onset of burns depending on concentration of current
6000 (6 A)	Onset of sustained ventricular contraction and respiratory paralysis; both cease when shock ends; heartbeat may return to normal; used to defibrillate the heart

*For an average male shocked through trunk of body for 1 s by 60 Hz AC. Values for females are 60–80% of those listed.

Modern electric fences, used in animal enclosures, are now pulsed on and off to allow people who touch them to get free, rendering them less lethal than in the past.

Human & Biological Application

Greater currents may affect the heart. Its electrical patterns can be disrupted, so that it beats irregularly and ineffectively in a condition called *ventricular fibrillation*. This condition often lingers after the shock and is fatal due to a lack of blood circulation. The threshold for ventricular fibrillation is between 100 and 300 mA. At about 300 mA and above, the shock can cause burns, depending on the concentration of current—the more concentrated, the greater the likelihood of burns.

Very large currents cause the heart and diaphragm to contract for the duration of the shock. Both the heart and breathing stop. Interestingly, both often return to normal following the shock. The electrical patterns on the heart are completely erased in a manner that the heart can start afresh with normal beating, as opposed to the permanent disruption caused by smaller currents than can put the heart into ventricular fibrillation. The latter is something like scribbling on a blackboard, whereas the former completely erases it. TV dramatizations of electric shock used to bring a heart attack victim out of ventricular fibrillation also show large paddles. These are used to spread out current passed through the victim to reduce the likelihood of burns.

Current is the major factor determining shock severity (given other conditions such as path, duration, and frequency are fixed, such as in the table and preceding discussion). A larger voltage is more hazardous, but since $I = V/R$, the severity of the shock depends on the *combination* of voltage and resistance. For example, a person with dry skin has a resistance of about 100 kΩ. If he comes into contact with 120 V AC, a current $I = (120 \text{ V})/(100 \text{ k}\Omega) = 1.2$ mA passes harmlessly through him. The same person soaking wet may have a resistance of 5000 Ω and the same 120 V will produce a current of 24 mA—above the "can't let go" threshold and potentially dangerous.

Most of the body's resistance is in its dry skin. When wet, salts go into ion form, lowering the resistance significantly. The interior of the body has a much lower resistance than dry skin because of all the ionic solutions and fluids it contains. If skin resistance is bypassed, such as by an intravenous infusion, a catheter, or exposed pacemaker leads, a person is rendered **microshock sensitive**. In this condition, currents about 1/1000 those listed in Table 19.3 produce similar effects. During open heart surgery, currents as small as 20 μA can be used to still the heart. Stringent electrical safety requirements in hospitals, particularly in surgery and intensive care, are related to the doubly disadvantaged microshock sensitive patient. The break in the skin has reduced his resistance, and so the same voltage causes a greater current, and a much smaller current has a greater effect.

Factors other than current that affect the severity of a shock are its path, duration, and AC frequency. Path has obvious consequences. For example, the heart is unaffected by an electric shock through the brain, such as may be used to treat manic depression. And it is a general truth that the longer the duration of a shock, the greater its effects. Figure 19.21 presents a graph that illustrates the effects of frequency on a shock. The curves show the minimum current for two different effects, as a function of frequency. The lower the

Figure 19.21 Graph of average values for the threshold of sensation and the "can't let go" current as a function of frequency. The lower the value, the more sensitive the body is at that frequency.

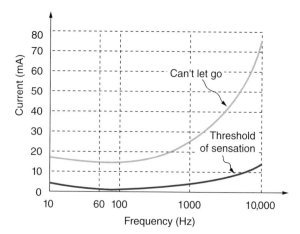

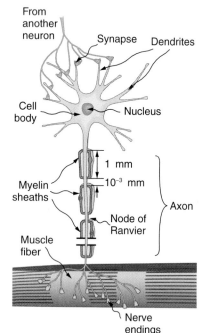

Figure 19.22 Are these dangerous? The answer depends on AC frequency and the power involved.

current needed, the more sensitive the body is at that frequency. Ironically, the body is most sensitive to frequencies near the 60 Hz frequency in common use. The body is slightly less sensitive for DC ($f = 0$), mildly confirming Edison's claims that AC presents a greater hazard. At higher and higher frequencies, the body becomes progressively less sensitive to any effects that involve nerves. This is related to the maximum rates at which nerves can fire or be stimulated. At very high frequencies, electrical current travels only on the surface of a person. Thus a wart can be burned off with very high frequency current without causing the heart to stop. (Do not try this at home with 60 Hz AC!) Some of the spectacular demonstrations of electricity, in which high voltage arcs are passed through the air and over people's bodies, employ high frequencies and low currents. (See Figure 19.22.)

19.7* NERVE CONDUCTION

Electric currents in the vastly complex system of billions of nerves in our body allow us to sense the world, control parts of our body, and think. These are representative of the three major functions of nerves. First, nerves carry messages from our sensory organs and others to the central nervous system, consisting of the brain and spinal cord. Second, nerves carry messages from the central nervous system to muscles and other organs. Third, nerves transmit and process signals within the central nervous system. The sheer number of nerve cells and the incredibly greater number of connections between them makes this system the subtle wonder that it is. **Nerve conduction** is a general term for electrical signals carried by nerve cells. It is one aspect of **bioelectricity**, or electrical effects in and created by biological systems.

Nerve cells, properly called **neurons**, look different from other cells—they have tendrils, some of them many centimeters long, connecting them with other cells. (See Figure 19.23.) Signals arrive at the cell body across *synapses* or through *dendrites*, stimulating the neuron to generate its own signal, sent along its long *axon* to other nerve or muscle cells. Signals may arrive from many other locations and be transmitted to yet others, conditioning the synapses by use, giving the system its complexity and its ability to learn.

The method by which these electric currents are generated and transmitted is more complex than the simple movement of free charges in a conductor, but it can be understood with principles already discussed in this text. The most important of these are the Coulomb force and diffusion.

Figure 19.24 illustrates how a voltage (potential difference) is created across the cell wall of a neuron in its resting state. This thin membrane separates electrically neutral fluids having differing concentrations of ions, the most important varieties being Na^+, K^+, and Cl^- (these are sodium, potassium, and chlorine ions with single plus or minus charges as indicated). As discussed in Section 11.7, free ions will diffuse from a region of high concentration to one of low concentration. But the cell wall is **semipermeable**, meaning that some ions may cross it while others cannot. In its resting state, the wall is permeable to K^+ and Cl^-, and impermeable to Na^+. Diffusion of K^+ and Cl^- thus creates the layers of positive and negative charge on the outside and inside of the wall. The Coulomb force prevents the ions from diffusing across in their entirety. Once the charge layer has built

Figure 19.23 A neuron with its dendrites and long axon. Signals in the form of electric currents reach the cell body through dendrites and across synapses, stimulating the neuron to generate its own signal sent down the axon. The number of interconnections can be far greater than shown here.

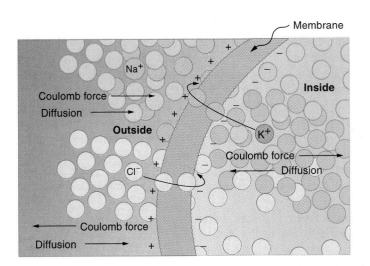

Figure 19.24 The semipermeable membrane of a cell wall has different concentrations of ions inside and out. Diffusion moves the K⁺ and Cl⁻ ions in the direction shown, until the Coulomb force halts further transfer. This results in a layer of positive charge on the outside, a layer of negative charge on the inside, and thus a voltage across the cell wall. The membrane is normally impermeable to Na⁺.

up, the repulsion of like charges prevents more from moving across, and the attraction of unlike charges prevents more from leaving either side. The result is two layers of charge right on the wall, with diffusion being balanced by the Coulomb force. A tiny fraction of the charges move across and the fluids remain neutral (other ions are present), while a separation of charge and a voltage have been created across the membrane.

The separation of charge creates a potential difference of 70 to 90 mV across the cell wall. While this is a small voltage, the resulting electric field ($E = V/d$) across the only 8 nm thick wall is immense (on the order of 11 MV/m!) and has fundamental effects on its structure and permeability. Now, if the exterior of a neuron is taken to be at 0 V, then the interior has a *resting potential* of about −90 mV. Such voltages are created across the walls of almost all types of animal cells but are largest in nerve and muscle cells. In fact, fully 25% of the energy used by cells goes toward creating and maintaining these potentials.

Electric currents along the cell wall are created by any stimulus that changes the wall's permeability. The wall temporarily becomes permeable to Na⁺, which then rushes in, driven both by diffusion and the Coulomb force. This inrush of Na⁺ first neutralizes the inside wall, or *depolarizes* it, and then makes it slightly positive. The depolarization causes the membrane to again become impermeable to Na⁺, and the movement of K⁺ quickly returns the cell to its resting potential, or *repolarizes* it. The sequence of events results in a voltage pulse, called the **action potential**. (See Figure 19.25.) Only small

Biological Application

Figure 19.25 An action potential is the pulse of voltage inside a nerve cell graphed here. It is caused by movements of ions across the cell wall as shown. Depolarization occurs when a stimulus makes the wall permeable to Na⁺ ions. Repolarization follows as the wall again becomes impermeable to Na⁺, and K⁺ moves from high to low concentration. In the long term, active transport slowly maintains the concentration differences, but the cell may fire hundreds of times in rapid succession without seriously depleting them.

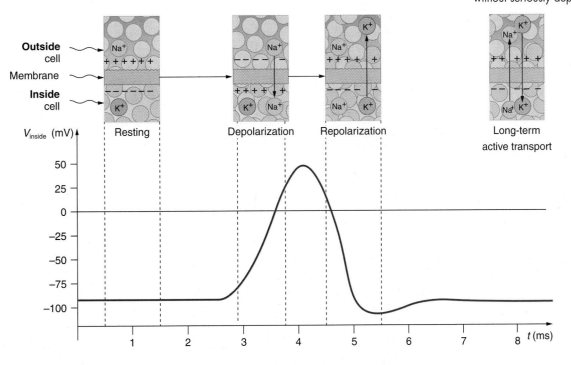

Biological Application

fractions of the ions move, so that the cell can fire many hundreds of times without depleting the excess concentrations of Na$^+$ and K$^+$. Eventually, the cell must replenish these ions to maintain the concentration differences that create bioelectricity. This sodium-potassium pump is an example of *active transport*, wherein cell energy is used to move ions across membranes against diffusion gradients and the Coulomb force.

The action potential is a voltage pulse at one location on a cell wall. How does it get transmitted along the cell wall, and in particular down an axon, as a nerve impulse? The answer is that the changing voltage and electric fields affect the permeability of the adjacent cell wall, so that the same process takes place there. The adjacent wall depolarizes, affecting the wall further down, and so on, as illustrated in Figure 19.26. Thus the action potential stimulated at one location triggers a *nerve impulse* that moves slowly (about 1 m/s) along the cell wall.

Some axons, like that in Figure 19.23, are sheathed with *myelin*. Figure 19.27 shows an enlarged view of an axon having myelin sheaths characteristically separated by unmyelinated gaps. This arrangement gives the axon a number of interesting properties. Since myelin is an insulator, it prevents signals from jumping between adjacent nerves (cross talk). Additionally, the myelinated regions transmit electric signals at a very high speed, as an ordinary conductor or resistor would. There is no action potential in the myelinated regions, so that no cell energy is used in them. There is an *IR* signal loss in

Figure 19.26 A nerve impulse is the propagation of an action potential along a cell wall. A stimulus causes an action potential at one location, which changes the permeability of the adjacent membrane, causing an action potential there. This in turn affects the wall further down, so that the action potential moves slowly (in electrical terms) along the cell wall. Although the impulse is due to Na$^+$ and K$^+$ going *across* the wall, it is equivalent to a wave of charge moving *along* the outside and inside of the wall.

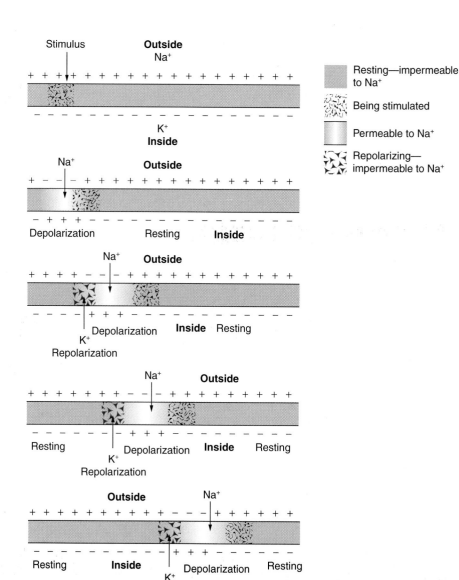

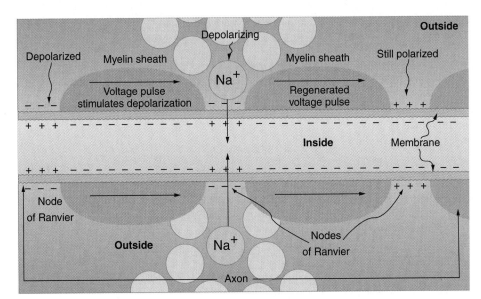

Figure 19.27 Propagation of a nerve impulse down a myelinated axon. The signal travels very fast and without energy input in the myelinated regions, but it loses voltage. It is regenerated in the gaps. The signal moves faster than in unmyelinated axons and is insulated from signals in other nerves, limiting cross talk.

the myelin, but the signal is regenerated in the gaps, where the voltage pulse triggers the action potential at full voltage. So a myelinated axon transmits a nerve impulse faster, with less energy consumption, and is better protected from cross talk than an unmyelinated one. Not all axons are myelinated, so that cross talk and slow signal transmission are a characteristic of the normal operation of these axons, another variable in the nervous system.

Most animal cells can fire or create their own action potential. Muscle cells contract when they fire and are often induced to do so by a nerve impulse. In fact, nerve and muscle cells are physiologically similar, and there are even hybrid cells, such as in the heart, that have characteristics of both nerves and muscles. Some animals, like the infamous electric eel (see Figure 19.28), use muscles ganged so that their voltages add in order to create a shock great enough to stun prey.

Figure 19.28 An electric eel flexes its muscles to create a voltage that stuns prey.

SUMMARY

Electric current I is the rate at which charge flows, given by

$$I = \frac{\Delta Q}{\Delta t} \qquad (19.1)$$

where ΔQ is the amount of charge passing through an area in time Δt. The direction of conventional current is taken as the direction that positive charge moves. The SI unit for current is the **ampere** (A), where

$$1 \text{ A} = 1 \text{ C/s} \qquad (19.2)$$

Current is the flow of free charges, such as electrons and ions. **Drift velocity** v_d is the average speed at which these charges move. Current I is proportional to drift velocity v_d, as expressed in the relationship

$$I = nqAv_d \qquad (19.3)$$

Here I is the current through a wire of cross-sectional area A.

The wire's material has a free charge density n, and each carrier has charge q and a drift velocity v_d. Electrical signals travel at speeds about 10^{12} greater than drift velocity.

A *simple circuit* is one in which there is a single voltage source and a single resistance. One statement of **Ohm's law** gives the relationship between current, voltage, and resistance in a simple circuit to be

$$I = \frac{V}{R} \qquad (19.4)$$

R is the **resistance**, with units of **ohms** (Ω), related to volts and amperes by

$$1 \text{ } \Omega = 1 \text{ V/A} \qquad (19.5)$$

There is a voltage or *IR drop* across a resistor, caused by the current flowing through it, given by $V = IR$.

The resistance R of a cylinder of length L and cross-sectional area A is

$$R = \frac{\rho L}{A} \quad (19.6)$$

where ρ is the **resistivity** of the material. Values of ρ in Table 19.1 show that materials fall into three groups—*conductors*, *semiconductors*, and *insulators*. Temperature affects resistivity; for relatively small temperature changes ΔT, resistivity is

$$\rho = \rho_0(1 + \alpha\,\Delta T) \quad (19.7)$$

where ρ_0 is the original resistivity. Table 19.2 gives values for α, the **temperature coefficient of resistivity**. The resistance R, too, of an object varies with temperature:

$$R = R_0(1 + \alpha\,\Delta T) \quad (19.8)$$

where R_0 is the original resistance.

Electric power P is the rate (in watts) that energy is supplied by a source or dissipated by a device. Three expressions for electrical power are

$$P = IV \quad (19.9a)$$

$$P = \frac{V^2}{R} \quad (19.9b)$$

and

$$P = I^2R \quad (19.9c)$$

The energy used by a device with a power P over a time t is $E = Pt$.

Direct current (DC) refers to systems where the source voltage is constant. The voltage source of an **alternating current** (AC) system puts out

$$V = V_0 \sin 2\pi ft \quad (19.10)$$

where V is the voltage at time t, V_0 is the peak voltage, and f is the frequency in hertz. In a simple circuit, $I = V/R$ and **AC current** is

$$I = I_0 \sin 2\pi ft \quad (19.11)$$

where I is the current at time t, and $I_0 = V_0/R$ is the peak current. The average AC power P_{ave} is

$$P_{ave} = \tfrac{1}{2}I_0V_0 \quad (19.12)$$

Average or **rms current** I_{rms} and average or **rms voltage** V_{rms} are

$$I_{rms} = \frac{I_0}{\sqrt{2}} \quad (19.13a)$$

and

$$V_{rms} = \frac{V_0}{\sqrt{2}} \quad (19.13b)$$

where *rms* stands for **root mean square**. Thus,

$$P_{ave} = I_{rms}V_{rms} \quad (19.14)$$

Ohm's law for AC is

$$I_{rms} = \frac{V_{rms}}{R} \quad (19.15)$$

Expressions for P_{ave} are

$$P_{ave} = I_{rms}V_{rms} \quad (19.16a)$$

$$P_{ave} = \frac{V_{rms}^2}{R} \quad (19.16b)$$

and

$$P_{ave} = I_{rms}^2 R \quad (19.16c)$$

analogous to DC.

The two types of electric hazards are **thermal** (excessive power) and **shock** (current through a person). Shock severity is determined by current, path, duration, and AC frequency. Table 19.3 lists shock hazards as a function of current, and Figure 19.21 graphs the threshold current for two hazards as a function of frequency.

Electric potentials in **neurons** and other cells are created by ionic concentration differences across **semipermeable** membranes. Stimuli change the permeability and create **action potentials** that propagate along neurons. Myelin sheaths speed this process and reduce the needed energy input.

CONCEPTUAL QUESTIONS

19.1 Can a wire carry a current and still be neutral—that is, have a total charge of zero? Explain.

19.2 Car batteries are rated in ampere-hours (A · h). To what physical quantity do ampere-hours correspond (voltage, charge, . . .), and what relationship do ampere-hours have to energy content?

19.3 If two different wires having identical cross-sectional areas carry the same current, will the drift velocity be higher or lower in the better conductor? Explain in terms of Equation 19.3, by considering how the density of charge carriers n relates to whether a material is a good conductor or not.

19.4 Why are *two* conducting paths from a voltage source to an electrical device needed to operate the device?

19.5 In cars, one battery terminal is connected to the metal body. How does this allow a single wire to supply current to electrical devices rather than two?

19.6 Why isn't a bird sitting on a high-voltage power line electrocuted? Contrast this with the situation in which a large bird hits *two* wires simultaneously with its wings.

19.7 The *IR* drop across a resistor means that there is a change in potential or voltage across the resistor. Is there any change in current as it passes through a resistor? Explain.

19.8 How is the *IR* drop in a resistor similar to the pressure drop in a fluid flowing through a pipe?

19.9 In which of the three semiconducting materials listed in Table 19.1 do impurities supply free charges? (Hint: Examine the range of resistivity for each and determine whether the pure semiconductor has the higher or lower conductivity.)

19.10 Does the resistance of an object depend on the path current takes through it? Consider, for example, a rectangular bar—is its resistance the same along its length as across its width? (See Figure 19.29.)

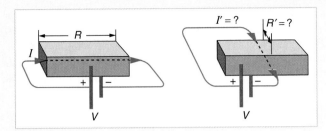

Figure 19.29 Does current taking two different paths through the same object encounter different resistance? Question 10.

19.11 If aluminum and copper wires of the same length have the same resistance, which has the largest diameter? Why?

19.12 Explain why Equation 19.8 for the temperature variation of the resistance R of an object is not as accurate as Equation 19.7, which gives the temperature variation of resistivity ρ.

19.13 Why do incandescent light bulbs grow dim late in their lives, particularly just before their filaments break?

19.14 The power dissipated in a resistor is given by $P = V^2/R$, which means power decreases if resistance increases. Yet this power is also given by $P = I^2R$, which means power increases if resistance increases. Explain why there is no contradiction here.

19.15 Give an example of a use of AC power other than in the household. Similarly, give an example of a use of DC power other than that supplied by batteries.

19.16 Why do voltage, current, and power go through zero 120 times per second for 60 Hz AC electricity?

19.17 You are riding in a train, gazing into the distance through its window. As close objects streak by, you notice that the nearby fluorescent lights make *dashed* streaks. Explain.

19.18* What are the two major hazards of electricity?

19.19* Why isn't a short circuit a shock hazard?

19.20* What determines the severity of a shock? Can you say that a certain voltage is hazardous without further information?

19.21* An electrified needle is used to burn off warts, with the circuit being completed by having the patient sit on a large butt plate. Why is this plate large?

19.22* Some surgery is performed with high-voltage electricity passing from a metal scalpel through the tissue being cut. Considering the nature of electric fields at the surface of conductors, why would you expect most of the current to flow from the sharp edge of the scalpel? Do you think high- or low-frequency AC is used?

19.23* Before working on a power transmission line, linemen will touch the line with the back of the hand as a final check that the voltage is zero. Why the back of the hand?

19.24* Why is the resistance of wet skin so much smaller than dry, and why do blood and other bodily fluids have low resistances?

19.25* Could a person on intravenous infusion (an IV) be microshock sensitive?

19.26* In view of the small currents that cause shock hazards and the larger currents that circuit breakers and fuses interrupt, how do they play a role in preventing shock hazards?

19.27* Note that in Figure 19.24, both the concentration gradient and the Coulomb force tend to move Na^+ ions into the cell. What prevents this?

19.28* Define depolarization, repolarization, and the action potential.

19.29* Explain the properties of myelinated nerves in terms of the insulating properties of myelin.

PROBLEMS

Section 19.1 Current and Drift Velocity

19.1 What is the current in milliamperes put out by the solar cells of a pocket calculator through which 4.00 C of charge passes in 4.00 h?

19.2 600 C passes through a flashlight in 0.500 h. What is the average current?

19.3 What is the current when a typical static charge of 0.250 μC moves from your finger to a metal doorknob in 1.00 μs?

19.4 Find the current when 2.00 nC jumps between your comb and hair over a 0.500 μs time interval.

19.5 A large lightning bolt had a 20,000 A current and moved 30.0 C of charge. What was its duration?

19.6 The 200 A current through a spark plug moves 0.300 mC of charge. How long does the spark last?

• **19.7** (a) A defibrillator passes 12.0 A of current through the torso of a person for 0.0100 s. How much charge moves? (b) How many electrons pass through the wires connected to the patient? (See Figure 19.30.)

• **19.8** A clock battery wears out after moving 10,000 C of charge through the clock at a rate of 0.500 mA. (a) How long did the clock run? (b) How many electrons per second flowed?

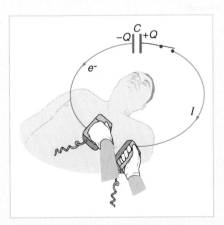

Figure 19.30 The capacitor in a defibrillation unit drives a current through the heart of a patient. Problem 7.

• **19.9** The batteries of a submerged nonnuclear submarine supply 1000 A at full speed ahead. How long does it take to move Avogadro's number (6.02×10^{23}) of electrons at this rate?

• **19.10** Electron guns are used in x-ray tubes. The electrons are accelerated through a relatively large voltage and directed onto a metal target, producing x rays. (a) How many electrons per second strike the target if the current is 0.500 mA? (b) What charge strikes the target in 0.750 s?

• **19.11** A large cyclotron directs a beam of He^{++} nuclei onto a target with a beam current of 0.250 mA. (a) How many He^{++} nuclei per second is this? (b) How long does it take for 1.00 C to strike the target? (c) How long before 1.00 mol of He^{++} nuclei strike the target?

• **19.12** Repeat Example 19.3, but for a wire made of silver and given there is one free electron per silver atom.

• **19.13** Using the results of Example 19.3, find the drift velocity in a copper wire of twice the diameter and carrying 20.0 A.

• **19.14** A 14 gauge copper wire has a diameter of 1.628 mm. What magnitude current flows when the drift velocity is 1.00 mm/s? (See Example 19.3 for useful information.)

⁞ **19.15** SPEAR, a storage ring about 72.0 m in diameter at the Stanford Linear Accelerator, has a 20.0 A circulating beam of electrons that are moving at nearly the speed of light. (See Figure 19.31.) How many electrons are in the beam?

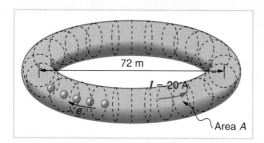

Figure 19.31 Electrons circulating in the storage ring called SPEAR constitute a 20.0 A current. Because they travel close to the speed of light, each electron completes many orbits in each second. Problem 15.

Sections 19.2 and 19.3 Ohm's Law and Resistance

19.16 What current flows through the bulb of a 3.00 V flashlight when its hot resistance is 3.60 Ω?

19.17 Calculate the effective resistance of a pocket calculator that has a 1.35 V battery and through which 0.200 mA flows.

19.18 What is the effective resistance of a car's starter motor when 150 A flows through it as the car battery applies 11.0 V to the motor?

19.19 How many volts are supplied to operate an indicator light on a VCR that has a resistance of 140 Ω, given 25.0 mA passes through it?

19.20 (a) Find the voltage drop in an extension cord having a 0.0600 Ω resistance and through which 5.00 A is flowing. (b) A cheaper cord utilizes thinner wire and has a resistance of 0.300 Ω. What is the voltage drop in it when 5.00 A flows? (The voltage to whatever appliance is supplied by the cords is reduced by these amounts.)

19.21 A power transmission line is hung from metal towers with glass insulators having a resistance of 1.00×10^9 Ω. What current flows through the insulator if the voltage is 200 kV? (Some high-voltage lines are DC.)

19.22 What is the resistance of a 20.0 m long piece of 12 gauge copper wire having a 2.053 mm diameter?

19.23 The diameter of 0 gauge copper wire is 8.252 mm. Find the resistance of a 1.00 km length of such wire used for power transmission.

19.24 If the 0.100 mm diameter tungsten filament in a light bulb is to have a resistance of 0.200 Ω at 20.0°C, how long should it be?

• **19.25** Find the ratio of the diameter of aluminum to copper wire, if they have the same resistance per unit length (as they might in household wiring).

• **19.26** What current flows through a 2.54 cm diameter rod of pure silicon that is 20.0 cm long, when 1000 V is applied to it? (Such a rod may be used to make nuclear particle detectors, for example.)

• **19.27** To what temperature must you raise a copper wire, originally at 20.0°C, to double its resistance, neglecting any changes in dimensions?

• **19.28** A resistor made of Nichrome wire is used in an application where its resistance cannot change more than 1.00% from its value at 20.0°C. Over what temperature range can it be used?

• **19.29** Of what material is a resistor made if its resistance is 40.0% greater at 100°C than at 20.0°C?

• **19.30** An electronic device designed to operate at any temperature in the range from −10.0°C to 55.0°C contains pure carbon resistors. By what factor does their resistance increase over this range?

• **19.31** (a) Of what material is a wire made, if it is 25.0 m long with a 0.100 mm diameter and has a resistance of 77.7 Ω at 20.0°C? (b) What is its resistance at 150°C?

⁞ **19.32** Assuming a constant temperature coefficient of resistivity, what is the maximum percent decrease in the resistance of a Constantin wire starting at 20.0°C?

⁞ **19.33** A wire is drawn through a die, stretching it to four times its original length. By what factor does its resistance increase?

⁞ **19.34** A copper wire has a resistance of 0.500 Ω at 20.0°C, and an iron wire has a resistance of 0.525 Ω at the same temperature. At what temperature are their resistances equal?

⁞ **19.35** Digital medical thermometers determine temperature by measuring the resistance of a semiconductor device called a thermistor (which has $\alpha = -0.0600/°C$) when it is at the same temperature as the patient. What is a patient's temperature if the thermistor's resistance at that temperature is 82.0% of its value at 37.0°C (normal body temperature)?

Sections 19.4 and 19.5 Electrical Power and Energy; Alternating Current

19.36 What is the power of a 100 MV lightning bolt having a current of 20,000 A?

19.37 What power is supplied to the starter motor of a large truck that draws 250 A of current from a 24.0 V battery hookup?

19.38 What is the power output of the pocket calculator's solar cells in Problem 19.1, given their voltage output is 3.00 V? (See Figure 19.32.)

19.39 How many watts does the flashlight in Problem 19.2 use if its voltage is 3.00 V?

Figure 19.32 The strip of solar cells just above the keys of this calculator convert light to electricity to supply its energy needs. Problem 38.

19.40 Find the power dissipated in each of the extension cords in Problem 19.20.

19.41 Verify that the units of a volt-ampere are watts, as implied by the equation $P = IV$.

19.42 Show that the units $1 \text{ V}^2/\Omega = 1 \text{ W}$, as implied by the equation $P = V^2/R$.

19.43 Show that the units $1 \text{ A}^2 \cdot \Omega = 1 \text{ W}$, as implied by the equation $P = I^2R$.

19.44 Verify the energy unit equivalence that $1 \text{ kW} \cdot \text{h} = 3.60 \times 10^6 \text{ J}$.

19.45 Electrons in an x-ray tube are accelerated through 100 kV and directed toward a target to produce x rays. Calculate the power of the electron beam in this tube if it has a current of 15.0 mA.

19.46 An electric water heater consumes 5.00 kW for 2.00 h per day. What is the cost of running it for one year if electricity costs 9.00 ¢/kW·h?

19.47 Certain heavy industrial equipment uses AC power that has a peak voltage of 679 V. What is the rms voltage?

19.48 A certain circuit breaker trips when the rms current is 15.0 A. What is the corresponding peak current?

19.49 Military aircraft use 400 Hz AC power, because it is possible to design lighter weight equipment at this higher frequency. What is the time for one complete cycle of this power?

• **19.50** Certain older cars have 6.00 V electrical systems. (a) What is the hot resistance of a 30.0 W headlight in such a car? (b) What current flows through it?

• **19.51** Alkaline batteries have the advantage of putting out constant voltage until very nearly the end of their life. How long will an alkaline battery rated at 1.00 A·h and 1.58 V keep a 1.00 W flashlight bulb burning?

• **19.52** A tourist takes his 25.0 W, 120 V AC razor to Europe, finds a special adapter, and plugs it into 220 V AC. Assuming

constant resistance, what power does the razor consume as it is ruined?

• **19.53** A cauterizer, used to stop bleeding in surgery, puts out 2.00 mA at 15.0 kV. (a) What is its power output? (b) What is the resistance of the path?

• **19.54** In this problem, you will verify statements made at the end of Example 19.10. (a) What current is needed to transmit 100 MW of power at a voltage of 25.0 kV? (b) Find the power loss in a 1.00 Ω transmission line. (c) What percent loss does this represent?

• **19.55** Power companies decrease voltage output to reduce power output in times of extremely high usage. What percent power reduction is achieved by a voltage decrease from 120 to 110 V?

• **19.56** (a) What is the hot resistance of a 25 W light bulb that runs on 120 V AC? (b) If the bulb's operating temperature is 2700°C, what is its resistance at 2600°C?

• **19.57** A small office building air conditioner operates on 408 V AC and consumes 50.0 kW. (a) What is its effective resistance? (b) What is the cost of running the air conditioner during a hot summer month when it is on 8.00 h per day for 30 days and electricity costs 9.00 ¢/kW·h?

• **19.58** The average TV in the United States is said to be on 6 hours per day. Estimate the yearly cost of electricity to operate the approximately 100 million TVs, assuming their power consumption is 150 W and the cost of electricity averages 9.00 ¢/kW·h.

• **19.59** What is the peak power consumption of a 120 V AC microwave oven that draws 10.0 A?

• **19.60** What is the peak current through a 500 W room heater that operates on 120 V AC power?

⁝ **19.61** An old light bulb draws only 50.0 W, rather than its original 60.0 W, due to evaporative thinning of its filament. By what factor is its diameter reduced, assuming uniform thinning along its length? Neglect any effects caused by temperature differences.

⁝ **19.62** Two different electrical devices have the same power consumption, but one is meant to be operated on 120 V AC and the other on 220 V AC. (a) What is the ratio of their resistances? (b) What is the ratio of their currents? (c) Assuming its resistance is unaffected, by what factor will the power increase if a 120 V AC device is connected to 220 V AC?

⁝ **19.63** 00 gauge copper wire has a diameter of 9.266 mm. Calculate the power loss in a kilometer of such wire when it carries 100 A.

⁝ **19.64** Nichrome wire is used in some radiative heaters. (a) Find the resistance needed if the average power output is to be 1.00 kW utilizing 120 V AC. (b) What length of Nichrome wire, having a cross-sectional area of 5.00 mm², is needed if the operating temperature is 500°C? (c) What power will it draw when first switched on?

⁝ **19.65** Find the time after $t = 0$ when the instantaneous voltage of 60 Hz AC first reaches the following values: (a) $V_0/2$ (b) V_0 (c) 0.

⁝ **19.66** (a) At what two times in the first period following $t = 0$ does the instantaneous voltage in 60 Hz AC equal V_{rms}? (b) $-V_{rms}$?

Section 19.6* Electrical Hazards

19.67* (a) How much power is dissipated in a short circuit of 220 V AC through a resistance of 0.250 Ω? (b) What current flows?

19.68* What voltage is involved in a 1.44 kW short circuit through a 0.100 Ω resistance?

19.69* Find the current through a person and identify the likely effect on her if she touches a 120 V AC source: (a) if she is standing on a rubber mat and offers a total resistance of 300 kΩ; (b) if she is standing barefoot on wet grass and has a resistance of only 4500 Ω.

19.70* While taking a bath, a person touches the metal case of a radio. The path through the person to the drainpipe and ground has a resistance of 4000 Ω. What is the smallest voltage on the case of the radio that could cause ventricular fibrillation?

19.71* Foolishly trying to fish a burning piece of bread from a toaster with a metal butter knife, a man comes into contact with 120 V AC. He does not even feel it since, luckily, he is wearing rubber-soled shoes. What is the minimum resistance of the path the current follows through the person?

19.72* During surgery, a current as small as 20.0 μA applied directly to the heart may cause ventricular fibrillation. If the resistance of the exposed heart is 300 Ω, what is the smallest voltage that poses this danger?

• **19.73*** (a) What is the resistance of a 220 V AC short circuit that generates a peak power of 96.8 kW? (b) What would the average power be if the voltage was 120 V AC?

INTEGRATED CONCEPTS

The following problems involve concepts from this chapter and several others. Physics is most interesting when applied to general situations involving more than a narrow set of physical principles. For example, electric current carries energy, hence the relevance of Chapter 6, Work, Energy, and Power. The following topics are involved in some or all of the problems in this section:

Topics	Location
Kinematics	Chapter 2
Work, energy, and power	Chapter 6
Fluid statics	Chapter 10
Temperature, gases	Chapter 12
Heat and heat transfer	Chapter 13
Electric potential and energy	Chapter 18
Current, resistance, Ohm's law	Chapter 19

PROBLEM-SOLVING STRATEGY

Step 1. Identify which physical principles are involved.

Step 2. Solve the problem using strategies outlined in the text.

The following worked example illustrates how these strategies are applied to an integrated concept problem.

EXAMPLE   YOUR STARTER MOTOR CAN GET HOT

The following topics are involved in this integrated concepts worked example:

Topics	Location
Current, resistance, Ohm's law	Chapter 19
Heat and heat transfer	Chapter 13
Work, energy, and power	Chapter 6

Part of the electrical energy you use to start your car goes into heating the starter motor. Suppose your 12.0 V system supplies a 200 A current to the starter motor for 4.00 s. (a) How much energy is supplied? (b) Suppose 20.0% of this energy goes into heating 1.00 kg of copper wire in the motor. Calculate the temperature increase of the copper.

Strategy Step 1 To solve an **integrated concept problem**, such as those following this example, we must first identify the physical principles involved and identify the chapters in which they are found. Part (a) of this example asks for *energy*, which can be found from *power* and *time*. These are topics of Chapter 6 and the present Chapter 19. Part (b) considers a *temperature change*, a topic of Chapter 13.

Strategy Step 2 The following solutions to each part of the example illustrate how specific problem-solving strategies are applied. These involve identifying knowns and unknowns, checking to see if the answer is reasonable, and so on.

Solution for (a): Energy (Chapter 6), Electric power (Chapter 19) Energy is power times time, or

$$E = Pt$$

which is a rearrangement of the definition of power from Chapter 6. Electrical power can be expressed as current times voltage, or

$$P = IV$$

which is Equation 19.9a. Combining these two expressions and entering known values yields

$$E = IVt$$
$$= (200 \text{ A})(12.0 \text{ V})(4.00 \text{ s}) = 9600 \text{ J}$$

Discussion for (a) This is a moderate amount of energy to be delivered in a short time, consistent with the large current supplied to start the car.

Solution for (b): Temperature change (Chapter 13) We are asked to find the temperature increase of a given mass of copper. The energy that goes into increasing the temperature is 20.0% of that found in the previous part, or $Q = (0.200)(9600 \text{ J}) = 1920 \text{ J}$. The relationship of energy and temperature change is given by Equation 13.2, $Q = mc \, \Delta T$, where c is the specific heat of copper. Rearranging to solve for temperature change gives

$$\Delta T = \frac{Q}{mc}$$

The specific heat of copper is found in Table 13.1 and is 387 J/kg·°C. Substituting known values yields

$$\Delta T = \frac{1920 \text{ J}}{(1.00 \text{ kg})(387 \text{ J/kg·°C})} = 4.96°C$$

Discussion for (b) This is a tolerable temperature change (about 9°F), but it occurs in only 4 s. If your car does not start easily, you may have to crank for many more seconds and the starter motor can become quite hot.

This worked example illustrates how to apply problem-solving strategies to situations that include topics in different chapters. The first step is to identify the physical principles involved in the problem. The second step is to solve for the unknown using familiar problem-solving strategies. These are found throughout the text. Moreover, many worked examples show how to use problem-solving strategies for single topics. In this integrated concepts example, you can see how to apply them across several topics. You will find these techniques useful in applications of physics outside a physics course, such as in your profession, in other science disciplines, and in every-day life. The following problems will build your skills in the broad application of physical principles.

- **19.74** A heart defibrillator passes 10.0 A through a patient's torso for 5.00 ms in an attempt to restore normal beating. (a) How much charge passed? (b) What voltage was applied if 500 J of energy was dissipated? (c) What was the path's resistance? (d) Find the temperature increase caused in the 8.00 kg of affected tissue.

- **19.75** Cold vaporizers pass a current through water, evaporating it with only a small increase in temperature. One such home device is rated at 3.50 A and utilizes 120 V AC with 95.0% efficiency. (a) What is the vaporization rate in grams per minute? (b) How much water must you put into the vaporizer for 8.00 h of overnight operation? (See Figure 19.33.)

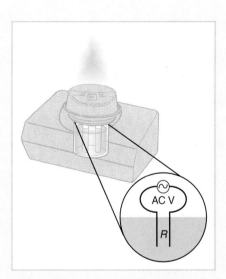

Figure 19.33 This cold vaporizer passes current directly through water, vaporizing it directly with relatively little temperature increase. Problem 75

- **19.76** (a) What energy is dissipated by a lightning bolt having a 20,000 A current, a voltage of 100 MV, and a length of 1.00 ms? (b) What mass of tree sap could be raised from 18.0°C to its boiling point and then evaporated by this energy, assuming sap has the same thermal characteristics as water?

- **19.77** What current must be produced by a 12.0 V battery-operated bottle warmer in order to heat 75.0 g of glass, 250 g of baby formula, and 300 g of aluminum from 20.0°C to 90.0°C in 5.00 min?

- **19.78** How much time is needed for a surgical cauterizer to raise the temperature of 1.00 g of tissue from 37.0°C to 100°C and then boil away 0.500 g of water, if it puts out 2.00 mA at 15.0 kV? Ignore heat losses to the surroundings.

- **19.79** Hydroelectric generators at Hoover Dam produce a maximum current of 8000 A at 250 kV. (a) What is the power output? (b) The water that powers the generators enters and leaves the system at low speed (thus its kinetic energy does not change) but loses 160 m in altitude. How many cubic meters per second are needed, assuming 85.0% efficiency?

- **19.80** Assuming 95.0% efficiency for the conversion of electrical power by the motor, what current must the 12.0 V batteries of a 750 kg electric car be able to supply: (a) To accelerate from rest to 25.0 m/s in 1.00 min? (b) To climb a 200 m high hill in 2.00 min at a constant 25.0 m/s speed while exerting 500 N of force to overcome air resistance and friction? (c) To travel at a constant 25.0 m/s speed, exerting a 500 N force to overcome air resistance and friction?

- **19.81** A light-rail commuter train draws 630 A of 650 V DC electricity when accelerating. (a) What is its power consumption rate in kilowatts? (b) How long does it take to reach 20.0 m/s starting from rest if its loaded mass is 5.30×10^4 kg, assuming 95.0% efficiency and constant power? (c) Find its average acceleration.

- **19.82** (a) An aluminum power transmission line has a resistance of 0.0580 Ω/km. What is its mass per kilometer? (b) What is the mass per kilometer of a copper line having the same resistance?

- **19.83** (a) Redo Example 19.6 taking into account the thermal expansion of the tungsten filament. You may assume a thermal expansion coefficient of 12×10^{-6}/°C. (b) By what fraction does your answer differ from that in the example?

- **19.84** An immersion heater utilizing 120 V can raise the temperature of a 100 g aluminum cup containing 350 g of water from 20.0°C to 95.0°C in 2.00 min. Find its resistance, assuming it is constant during the process.

- **19.85** A short circuit in a 120 V appliance cord has a 0.500 Ω resistance. Calculate the temperature rise of the 2.00 g of surrounding materials, assuming their specific heat capacity is 0.200 cal/g·°C and that it takes 0.0500 s for a circuit breaker to interrupt the current. Is this likely to be damaging?

- **19.86** (a) What is the cost of heating a hot tub containing 1500 kg of water from 10.0°C to 40.0°C, assuming 75.0% efficiency to take heat loss to surroundings into account? The cost of electricity is 9.00 ¢/kW·h. (b) What current was used by the 220 V AC electric heater, if this took 4.00 h?

UNREASONABLE RESULTS

The following problems have results that are unreasonable because some premise is unreasonable or because certain of the premises are inconsistent with one another. Physical principles applied correctly then produce unreasonable results. The purpose of these problems is to give practice in assessing whether nature is being accurately described, and if it is not to trace the source of difficulty.

Note: Strategies for determining if an answer is reasonable appear in Chapters 17 and 18, among others. Look at the section of problems labeled *Unreasonable Results* at the ends of those chapters.

• **19.87** (a) To what temperature must you raise a resistor made of Constantin to double its resistance, assuming a constant temperature coefficient of resistivity? (b) To cut it in half? (c) What is unreasonable about these results? (d) Which assumptions are unreasonable, or which premises are inconsistent?

• **19.88** (a) What current is needed to transmit 100 MW of power at 480 V? (b) What power is dissipated by the transmission lines if they have a 1.00 Ω resistance? (c) What is unreasonable about this result? (d) Which assumptions are unreasonable, or which premises are inconsistent?

⫶ **19.89** (a) What current is needed to transmit 100 MW of power at 10.0 kV? (b) Find the resistance of 1.00 km of wire that would cause a 0.0100% power loss. (c) What is the diameter of a 1.00 km long copper wire having this resistance? (d) What is unreasonable about these results? (e) Which assumptions are unreasonable, or which premises are inconsistent?

*The complexity of the electric circuits in the computer
is surpassed by those in the human brain.*

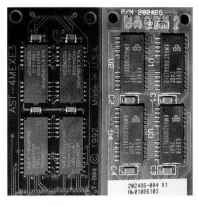

Figure 20.1 This circuit is even more complex than it appears. The principles developed in this text are the basis of the most complex circuits. We shall apply them to simpler devices.

Electric circuits are commonplace. Some are simple, such as those in flashlights. Others, such as the innards of supercomputers, are extremely complex. (See Figure 20.1.) This chapter takes the topic of electric circuits a step beyond the simple circuits covered in the previous chapter. When the circuit is purely resistive, everything in this chapter applies to both DC and AC. Matters become more complex when capacitance is involved—this chapter does consider what happens when capacitors are connected to DC voltage sources. But the interaction of capacitors and other nonresistive devices with AC is left for a later chapter. Finally, a number of important DC instruments, such as meters that measure voltage and current, are covered in this chapter.

20.1 RESISTORS IN SERIES AND PARALLEL

Most circuits have more than one resistor. The simplest combinations of resistors are the series and parallel connections illustrated in Figure 20.2. Note that these look just like the series and parallel connections of capacitors presented in Section 18.6. Like capacitors, the total resistance of a combination of resistors depends on both their values and how they are connected.

Resistors in Series

When are resistors in series? The answer is, **resistors are in series** whenever current must flow through devices sequentially. For example, if current flows through a person holding a screwdriver and into the earth, then R_1 could be the resistance of the screwdriver's shaft, R_2 the resistance of its handle, R_3 the person's resistance, and R_4 the resistance of her shoes. Figure 20.3 shows resistors in series connected to a voltage source. It seems reasonable that the total resistance is the sum of the individual resistances, considering that the current has to pass through each resistor in sequence. (This would be an advantage to a person avoiding shock, who could reduce the current by wearing high-resistance rubber-soled shoes. It could be a disadvantage if one of the resistances is a faulty high-resistance cord to an appliance that reduces operating current.) To verify that resistances in series add, let us consider the voltage drop in each resistor.

We know from the discussion of Ohm's law in Section 19.2 that there is a voltage drop ($V = IR$) across a resistor when a current flows through it.* So the voltage drop across R_1 is $V_1 = IR_1$, that across R_2 is $V_2 = IR_2$, and that across R_3 is $V_3 = IR_3$. The sum of these voltages equals the voltage output of the source; that is,

$$V = V_1 + V_2 + V_3$$

This is based on conservation of energy and conservation of charge. Since electrical potential energy is PE $= qV$, the energy supplied by the source is qV, while that dissipated by the resistors is $qV_1 + qV_2 + qV_3$. These energies must be equal, because there is no other source and no other destination for energy in the circuit. Thus $qV = qV_1 + qV_2 + qV_3$. The charge q cancels, yielding $V = V_1 + V_2 + V_3$, as stated. (The same charge q passes through the battery and each resistor in a given amount of time, since there is no capacitance to store charge, there is no place for charge to leak,

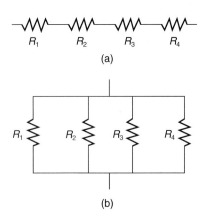

Figure 20.2 (a) A series connection of resistors. (b) A parallel connection of resistors.

CONNECTIONS

Conservation Laws

The derivations of the expressions for series and parallel resistance are based on **conservation of energy** and **conservation of charge**, introduced in Chapters 6 and 17, respectively. These two conservation laws are directly involved in all electrical phenomena and will be invoked repeatedly to explain both specific effects and general behavior of electricity.

*You may also think of this as the voltage necessary to make a current I flow through the resistor.

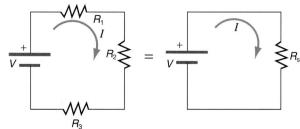

Figure 20.3 Three resistors connected in series to a battery and the equivalent single or series resistance.

and charge is conserved.) Now substituting the values for the individual voltages gives

$$V = IR_1 + IR_2 + IR_3 = I(R_1 + R_2 + R_3)$$

Note that for the equivalent single resistance R_s, we have

$$V = IR_s$$

This implies that the total or equivalent series resistance R_s of three resistors is $R_s = R_1 + R_2 + R_3$. This logic is valid in general for any number of resistors in series; thus, the **total resistance R_s of a series connection** is

$$R_s = R_1 + R_2 + R_3 + \cdots \qquad (20.1)$$

as proposed. Since all of the current must pass through each resistor, it experiences the resistance of each, and resistances in series simply add up.

EXAMPLE 20.1 ANALYSIS OF A SERIES CIRCUIT

Given the voltage output of the battery in Figure 20.3 is 12.0 V, and that the resistances are $R_1 = 1.00\ \Omega$, $R_2 = 6.00\ \Omega$, and $R_3 = 13.0\ \Omega$: (a) What is the total resistance? (b) Find the current. (c) Calculate the voltage drop in each resistor, and show these add to equal the voltage output of the source. (d) Calculate the power dissipated by each resistor. (e) Find the power output of the source, and show it equals the total power dissipated by the resistors.

Strategy and Solution for (a) The total resistance is simply the sum of the individual resistances, as given by Equation 20.1. Thus,

$$R_s = R_1 + R_2 + R_3 = 20.0\ \Omega$$

Strategy and Solution for (b) The current is found using Ohm's law. Entering the value of the applied voltage and the total resistance yields the current for the circuit to be

$$I = \frac{V}{R_s} = \frac{12.0\ \text{V}}{20.0\ \Omega} = 0.600\ \text{A}$$

Strategy and Solution for (c) The voltage or IR drop in a resistor is given by $V = IR$. Entering the current and the value of the first resistance yields

$$V_1 = IR_1 = (0.600\ \text{A})(1.0\ \Omega) = 0.600\ \text{V}$$

Similarly,

$$V_2 = IR_2 = (0.600\ \text{A})(6.0\ \Omega) = 3.60\ \text{V}$$

and

$$V_3 = IR_3 = (0.600\ \text{A})(13.0\ \Omega) = 7.80\ \text{V}$$

Discussion for (c) The three IR drops add to 12.0 V, as predicted:

$$V_1 + V_2 + V_3 = (0.600 + 3.60 + 7.80)\ \text{V} = 12.0\ \text{V}$$

Strategy and Solution for (d) The easiest way to calculate power is to use $P = I^2R$, since each resistor has the same full current flowing through it. Thus the power dissipated by the first resistor is

$$P_1 = I^2R_1 = (0.600\ \text{A})^2(1.00\ \Omega) = 0.360\ \text{W}$$

Similarly,

$$P_2 = I^2R_2 = (0.600\ \text{A})^2(6.00\ \Omega) = 2.16\ \text{W}$$

and

$$P_3 = I^2R_3 = (0.600\ \text{A})^2(13.0\ \Omega) = 4.68\ \text{W}$$

Discussion for (d) Power can also be calculated using $P = IV$ or $P = V^2/R$, where V is the voltage drop across the resistor (not the full voltage of the source). The same values will be obtained.

Strategy and Solution for (e) The easiest way to calculate power output of the source is to use $P = IV$, where V is the source voltage. This gives

$$P = (0.600\ \text{A})(12.0\ \text{V}) = 7.20\ \text{W}$$

Discussion for (e) Not coincidentally, total power dissipated by the resistors is also 7.20 W, the same as the power put out by the source. That is,

$$P_1 + P_2 + P_3 = (0.360 + 2.16 + 4.68)\ \text{W} = 7.20\ \text{W}$$

Power is energy per unit time, and so conservation of energy requires the power output of the source be equal to the total power dissipated by the resistors.

Briefly, the **major features of resistors in series** are

1. Series resistances add: $R_s = R_1 + R_2 + R_3 + \cdots$.
2. The same current flows through each resistor in series.
3. Series resistors do not get the total source voltage, but divide it.

Figure 20.4 Three resistors connected in parallel to a battery and the equivalent single or parallel resistance.

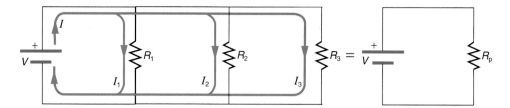

Resistors in Parallel

Figure 20.4 shows resistors in parallel, wired to a voltage source. **Resistors are in parallel** when each resistor is connected directly to the voltage source by connecting wires having negligible resistance. Each resistor thus has the full voltage of the source applied to it. Each resistor draws the same current it would if it alone were connected to the voltage source (provided the voltage source is not overloaded—a topic we will address in the next section). For example, an automobile's headlights, radio, and so on, are wired in parallel, so that they utilize the full voltage of the source and can operate completely independently.

To find an expression for the equivalent parallel resistance R_p, let us consider the currents that flow and how they are related to resistance. Since each resistor has the full voltage, the currents flowing through the individual resistors are $I_1 = V/R_1$, $I_2 = V/R_2$, and $I_3 = V/R_3$. Conservation of charge implies that the total current I produced by the source is the sum of these currents:

$$I = I_1 + I_2 + I_3$$

Substituting the expressions for the individual currents gives

$$I = \frac{V}{R_1} + \frac{V}{R_2} + \frac{V}{R_3} = V\left(\frac{1}{R_1} + \frac{1}{R_2} + \frac{1}{R_3}\right)$$

Note that Ohm's law for the equivalent single resistance gives

$$I = \frac{V}{R_p} = V\left(\frac{1}{R_p}\right)$$

The terms inside the parentheses in the last two equations must be equal. Generalizing to any number of resistors, the **total resistance R_p of a parallel connection** is related to the individual resistances by

$$\frac{1}{R_p} = \frac{1}{R_1} + \frac{1}{R_2} + \frac{1}{R_3} + \cdots \qquad \textbf{(20.2)}$$

This relationship results in a total resistance R_p that is less than the smallest of the individual resistances. When resistors are connected in parallel, more current flows from the source than would for any of them individually, and so the total resistance is lower.

EXAMPLE 20.2   ANALYSIS OF A PARALLEL CIRCUIT

Given the voltage output of the battery and resistances in the parallel connection in Figure 20.4 are the same as the previously considered series connection ($V = 12.0$ V, $R_1 = 1.00\ \Omega$, $R_2 = 6.00\ \Omega$, and $R_3 = 13.0\ \Omega$): (a) What is the total resistance? (b) Find the total current. (c) Calculate the currents in each resistor, and show these add to equal the total current output of the source. (d) Calculate the power dissipated by each resistor. (e) Find the power output of the source, and show it equals the total power dissipated by the resistors.

Strategy and Solution for (a) The total resistance for a parallel combination of resistors is found using

Equation 20.2. Entering known values gives

$$\frac{1}{R_p} = \frac{1}{R_1} + \frac{1}{R_2} + \frac{1}{R_3} = \frac{1}{1.00\ \Omega} + \frac{1}{6.00\ \Omega} + \frac{1}{13.0\ \Omega}$$

Thus,

$$\frac{1}{R_p} = \frac{1.00}{\Omega} + \frac{0.167}{\Omega} + \frac{0.0769}{\Omega} = \frac{1.244}{\Omega}$$

We must invert this to find the total resistance R_p. This yields

$$R_p = \frac{1}{1.244}\ \Omega = 0.804\ \Omega$$

(continued)

(continued)

Discussion for (a) R_p is, as predicted, less than the smallest individual resistance.

Strategy and Solution for (b) The total current can be found from Ohm's law, substituting R_p for the total resistance. This gives

$$I = \frac{V}{R_p} = \frac{12.0 \text{ V}}{0.804 \text{ }\Omega} = 14.92 \text{ A}$$

Discussion for (b) I is much larger than for the same devices connected in series (see the previous example), since a parallel connection reduces the total resistance.

Strategy and Solution for (c) The individual currents are easily calculated from Ohm's law, since each resistor gets the full voltage. Thus,

$$I_1 = \frac{V}{R_1} = \frac{12.0 \text{ V}}{1.00 \text{ }\Omega} = 12.0 \text{ A}$$

Similarly,

$$I_2 = \frac{V}{R_2} = \frac{12.0 \text{ V}}{6.00 \text{ }\Omega} = 2.00 \text{ A}$$

and

$$I_3 = \frac{V}{R_3} = \frac{12.0 \text{ V}}{13.0 \text{ }\Omega} = 0.92 \text{ A}$$

Discussion for (c) The total current is the sum of the individual currents:

$$I_1 + I_2 + I_3 = 14.92 \text{ A}$$

This is consistent with conservation of charge.

Strategy and Solution for (d) The power dissipated by each resistor can be found using any of the equations relating power to current, voltage, and resistance, since all

three are known. Let us use $P = V^2/R$, since each resistor gets full voltage. Thus,

$$P_1 = \frac{V^2}{R_1} = \frac{(12.0 \text{ V})^2}{1.00 \text{ }\Omega} = 144 \text{ W}$$

Similarly,

$$P_2 = \frac{V^2}{R_2} = \frac{(12.0 \text{ V})^2}{6.00 \text{ }\Omega} = 24.0 \text{ W}$$

and

$$P_3 = \frac{V^2}{R_3} = \frac{(12.0 \text{ V})^2}{13.0 \text{ }\Omega} = 11.1 \text{ W}$$

Discussion for (d) Each power is considerably higher in parallel than when connected in series to the same voltage source.

Strategy and Solution for (e) The total power can also be calculated in several ways. Choosing $P = IV$, and entering the total current, yields

$$P = IV = (14.92 \text{ A})(12.0 \text{ V}) = 179.1 \text{ W}$$

Discussion for (e) Total power dissipated by the resistors is also 179.1 W:

$$P_1 + P_2 + P_3 = (144 + 24.0 + 11.1) \text{ W} = 179.1 \text{ W}$$

This is consistent with the law of conservation of energy.

Overall Discussion Some answers are given to four digits in this example (although only three-digit input is used) in order to show the answers are consistent. Finally, note that both the currents and powers in parallel connections are greater than for the same devices in series.

Briefly, the **major features of resistors in parallel** are

1. Parallel resistance is found from $1/R_p = 1/R_1 + 1/R_2 + 1/R_3 + \cdots$, and it is smaller than any individual resistance in the combination.
2. Each resistor in parallel has the same full voltage of the source applied to it.*
3. Parallel resistors do not get the total current, but divide it.

Combinations of Series and Parallel

More complex connections of resistors are sometimes just combinations of series and parallel. These are commonly encountered, especially when wire resistance is considered. Combinations of series and parallel can be reduced to a single equivalent resistance using the technique illustrated in Figure 20.5. Various parts are identified as either series or parallel, reduced to their equivalents, and further reduced until a single resistance is left. The process is more time consuming than difficult.

---

*Power distribution systems most often use parallel connections to supply the myriad devices served with the same voltage and to allow them to operate independently.

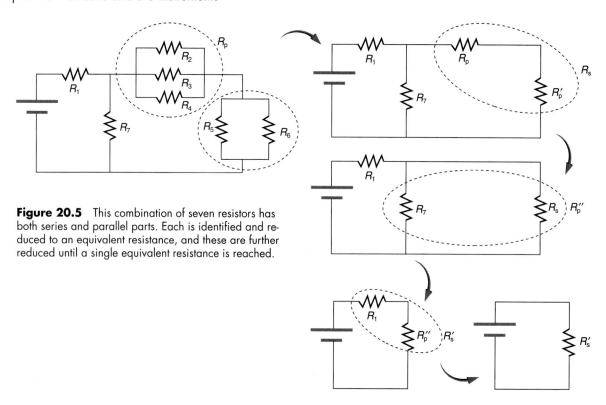

Figure 20.5 This combination of seven resistors has both series and parallel parts. Each is identified and reduced to an equivalent resistance, and these are further reduced until a single equivalent resistance is reached.

The simplest combination of series and parallel resistance, shown in Figure 20.6, is also the most instructive, since it is found in many applications. For example, R_1 could be the resistance of wires from a car battery to its electrical devices, which are in parallel. R_2 and R_3 could be the starter motor and a passenger compartment light. We have previously assumed that wire resistance is negligible, but, when it is not, it has important effects, as the next example indicates.

EXAMPLE 20.3 A COMBINATION OF SERIES AND PARALLEL

Figure 20.6 shows the resistors from the previous two examples wired in a different way—a combination of series and parallel. We can consider R_1 to be the resistance of wires leading to R_2 and R_3. With that in mind: (a) Find the total resistance. (b) What is the IR drop in R_1? (c) Find the current I_2 through R_2. (d) What power is dissipated by R_2?

Strategy and Solution for (a) To find the total resistance, we note that R_2 and R_3 are in parallel and their combination R_p is in series with R_1. Thus the total or equivalent resistance of this combination is

$$R_{tot} = R_1 + R_p$$

First, we find R_p using Equation 20.2 and entering known values:

$$\frac{1}{R_p} = \frac{1}{R_2} + \frac{1}{R_3} = \frac{1}{6.00 \ \Omega} + \frac{1}{13.0 \ \Omega} = \frac{0.2436}{\Omega}$$

Inverting gives

$$R_p = \frac{1}{0.2436} \ \Omega = 4.11 \ \Omega$$

So the total resistance is

$$R_{tot} = R_1 + R_p = 1.00 \ \Omega + 4.11 \ \Omega = 5.11 \ \Omega$$

Discussion for (a) The total resistance of this combination is intermediate between the pure series and pure parallel values (20.0 and 0.804 Ω, respectively) found for the same resistors in the two previous examples.

Strategy and Solution for (b) To find the IR drop in R_1, we note that the full current I flows through R_1. Thus its IR drop is

$$V_1 = IR_1$$

We must find I before we can calculate V_1. The total current I is found using Ohm's law for the circuit. That is,

$$I = \frac{V}{R_{tot}} = \frac{12.0 \ V}{5.11 \ \Omega} = 2.35 \ A$$

Entering this into the expression above, we get

$$V_1 = IR_1 = (2.35 \ A)(1.00 \ \Omega) = 2.35 \ V$$

Discussion for (b) The voltage applied to R_2 and R_3 is less that the total voltage by an amount V_1. When wire resistance is large, it can significantly affect the operation of the devices represented by R_2 and R_3.

(continued)

(continued)

Strategy and Solution for (c) To find the current through R_2, we must first find the voltage applied to it. We call this voltage V_p, because it is applied to a parallel combination of resistors. The voltage applied to both R_2 and R_3 is reduced by the amount V_1, and so it is

$$V_p = V - V_1 = 12.0 \text{ V} - 2.35 \text{ V} = 9.65 \text{ V}$$

Now the current I_2 through R_2 is found using Ohm's law:

$$I_2 = \frac{V_p}{R_2} = \frac{9.65 \text{ V}}{6.00 \text{ }\Omega} = 1.61 \text{ A}$$

Discussion for (c) The current is less than the 2.00 A that flowed through R_2 when it was connected in parallel to the battery in Example 20.2.

Strategy and Solution for (d) The power dissipated by R_2 is given by

$$P_2 = (I_2)^2 R_2 = (1.61 \text{ A})^2(6.00 \text{ }\Omega) = 15.5 \text{ W}$$

Discussion for (d) The power is less than the 24.0 W this resistor dissipated when connected in parallel to the 12.0 V source.

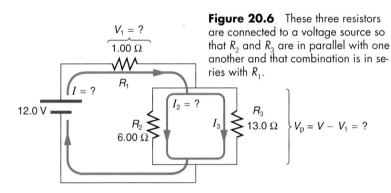

Figure 20.6 These three resistors are connected to a voltage source so that R_2 and R_3 are in parallel with one another and that combination is in series with R_1.

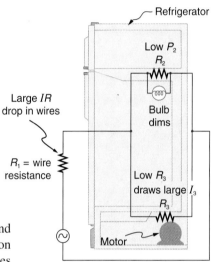

Figure 20.7 Why do lights dim when a large appliance is switched on? The answer is that the large current the appliance draws causes a significant *IR* drop in the wires and reduces the voltage received by the light.

One implication of this last example is that resistance in wires reduces the current and power delivered to a resistor. If wire resistance is relatively large, as in a worn extension cord, then this loss can be significant. If a large current is drawn, the *IR* drop in the wires can also be significant. For example, when you are rummaging in the refrigerator and the motor comes on, the refrigerator light dims momentarily. Similarly, you can see the passenger compartment light dim when you start the engine of your car (although this may be due to resistance inside the battery itself, as we shall see in the next section). What is happening in these high-current situations is illustrated in Figure 20.7. The device represented by R_3 has a very low resistance, and so when it is switched on, a large current flows. This increased current causes a larger *IR* drop in the wires represented by R_1, reducing the voltage received by the light bulb (which is R_2), which then dims noticeably.

PROBLEM-SOLVING STRATEGIES

FOR SERIES AND PARALLEL RESISTORS

Step 1. *Draw a clear circuit diagram, labeling all resistors and voltage sources.* This step includes a list of the knowns for the problem, since they are labeled in your circuit diagram.

Step 2. *Identify exactly what needs to be determined in the problem (identify the unknowns).* A written list is useful.

Step 3. *Determine whether resistors are in series, parallel, or a combination of both series and parallel.* Examine the circuit diagram to make this assessment. Resistors are in series if the same current must pass sequentially through them.

Step 4. *Use the appropriate list of major features for series or parallel to solve for the unknowns.* There is one list for series and another for parallel. If your problem has a combination of series and parallel, reduce it in steps by considering individual groups of series or parallel connections as done in this section and the previous worked example. *Special note: When finding R_p, the reciprocal must be taken with care.*

Step 5. Check to see if the answers are reasonable and consistent. Units and numerical results must be reasonable. Total series resistance should be greater, whereas total parallel resistance should be smaller, for example. Power should be greater for the same devices in parallel compared with series, and so on.

20.2 ELECTROMOTIVE FORCE: TERMINAL VOLTAGE

When you forget to turn off your car lights, they slowly dim as the battery runs down. Why don't they simply blink off when the battery's energy is gone? Their gradual dimming implies that battery output voltage decreases as the battery is depleted. Furthermore, if you connect an excessive number of 12 V lights in parallel to a car battery, they will be dim even when the battery is fresh and even if the wires to the lights have very low resistance. This implies that the battery's output voltage is reduced by the overload. The reason for the decrease in output voltage for depleted or overloaded batteries is that *all voltage sources have two fundamental parts*—a source of electrical energy and an internal resistance. Let us examine both.

Electromotive Force

You can think of many different types of voltage sources. Batteries themselves come in many varieties. There are many types of mechanical generators, driven by many different energy sources, ranging from nuclear to wind. Solar cells create voltages directly from light, while thermoelectric devices create voltage from temperature differences. A few voltage sources are shown in Figure 20.8. All such devices create a potential difference and can supply current if connected to a resistance. On the small scale, the potential difference creates an electric field that exerts force on charges, causing current. We thus use the name *electromotive force*, abbreviated *emf* and given the symbol $\mathcal{E}$. $\mathcal{E}$ is not a force at all; it is a special type of potential difference. To be precise, the **electromotive force** (emf or $\mathcal{E}$) *is the potential difference of a source when no current is flowing.* Units of $\mathcal{E}$ are volts.

(a)

(b)

(c)

(d)

Figure 20.8 A variety of voltage sources. The voltage output of each depends on its construction and load, and equals $\mathcal{E}$ only if there is no load.

$\mathcal{E}$ is directly related to the source of potential difference, such as the particular combination of chemicals in a battery. $\mathcal{E}$ differs from the voltage output of the device when current flows. The voltage across the terminals of a battery, for example, is less than $\mathcal{E}$ when the battery supplies current, and it declines further as the battery is depleted or loaded down. However, if the device's output voltage can be measured without drawing current, then output voltage will equal $\mathcal{E}$ (even for a very depleted battery).

Internal Resistance

As noted before, a 12 V truck battery is physically larger, contains more charge and energy, and can deliver a larger current than a 12 V motorcycle battery. Both are lead-acid batteries with identical emfs, but, because of its size, the truck battery has a smaller **internal resistance** r. Internal resistance is the inherent resistance to the flow of current within the source itself. Figure 20.9 is a schematic representation of the two fundamental parts of any voltage source. The emf $\mathcal{E}$ and internal resistance r are in series. The smaller the internal resistance r, for a given $\mathcal{E}$, the more current and the more power the source can supply.

Note that internal resistance r can behave in complex ways. As noted, internal resistance increases as a battery is depleted. But internal resistance may also depend on the magnitude and direction of the current through a voltage source, its temperature, and even its history. The internal resistance of rechargeable nickel-cadmium cells, for example, depends on how many times and how deeply they have been depleted.

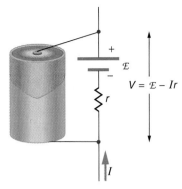

Figure 20.9 Any voltage source (in this case, a carbon-zinc dry cell) has an emf $\mathcal{E}$ related to its source of potential difference, and an internal resistance r related to its construction. Also shown are the output terminals across which the terminal voltage V is measured. Since $V = \mathcal{E} - Ir$, terminal voltage equals $\mathcal{E}$ only if there is no current flowing.

THINGS GREAT AND SMALL

The Submicroscopic Origin of Battery Potential

Various types of batteries are available, with emfs ($\mathcal{E}$s) determined by the combination of chemicals involved. We can view this as a molecular reaction (what much of chemistry is about) that separates charge. The lead-acid battery used in cars and other vehicles is one of the most common types. A single cell of this battery is seen in Figure 20.A. The positive or cathode terminal of the cell is connected to a lead oxide plate, while the negative or anode terminal is connected to a lead plate. Both plates are immersed in sulfuric acid, the electrolyte for the system.

The details of the chemical reaction are left to the reader to pursue in a chemistry text, but their results at the molecular level help explain the potential created by the battery. Figure 20.B shows the result of a single chemical reaction. Two electrons are placed on the anode, making it negative, provided that the cathode supplied two electrons. This leaves the cathode positively charged, because it has lost two electrons. In short, a separation of charge has been driven by a chemical reaction. Note that the reaction will not take place unless there is a complete circuit to allow two electrons to be supplied to the cathode. Under many circumstances, these electrons come from the anode, flow through a resistance, and return to the cathode. Note also that since the chemical reactions involve substances with resistance, it is not possible to create the emf without an internal resistance.

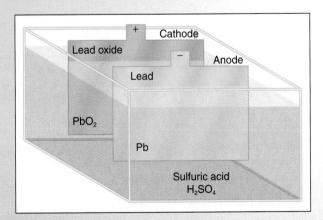

Figure 20.A Artist's conception of a lead-acid cell. Chemical reactions in a lead-acid cell separate charge, sending negative charge to the anode, which is connected to the lead plates. The lead oxide plates are connected to the positive or cathode terminal of the cell. Sulfuric acid conducts the charge as well as participating in the chemical reaction.

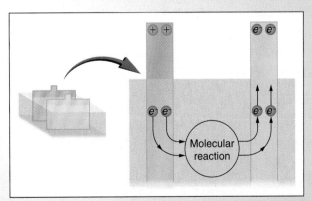

Figure 20.B Artist's conception of two electrons being forced onto the anode of a cell and two electrons being removed from the cathode of the cell. The chemical reaction in a lead-acid battery places two electrons on the anode and removes two from the cathode. It requires a closed circuit to proceed, since the two electrons must be supplied to the cathode.

What is responsible for the unique potential of the particular combination of chemicals? Quantum mechanical descriptions of molecules, which take into account the types of atoms and numbers of electrons in them, are able to predict the energy states they can have and the energies of reactions between them. In the lead-acid battery, an energy of 2 eV is given to each electron sent to the anode. As discussed in Chapter 18 (and elsewhere), voltage is electrical potential energy divided by charge: $V = PE/q$. An electron volt is the energy given to a single electron by a voltage of 1 V. So the voltage here is 2 V, since 2 eV is given to each electron. It is the energy produced in each molecular reaction that produces the voltage. A different reaction produces a different energy and, hence, a different voltage.

Terminal Voltage

The voltage output of a device is measured across its terminals and, thus, is called its **terminal voltage** V. Terminal voltage is given by

$$V = \mathcal{E} - Ir \qquad (20.3)$$

where $\mathcal{E}$ is the emf, r is the internal resistance, and I is the current flowing at the time of the measurement. I is positive if current flows away from the positive terminal, as shown in Figure 20.9. You can see that the larger the current, the smaller the terminal voltage. Similarly, the larger the internal resistance, the smaller the terminal voltage.

Suppose a load resistance R_{load} is connected to a voltage source, as in Figure 20.10. Since the resistances are in series, the total resistance in the circuit is $R_{load} + r$. Thus the current is given by Ohm's law to be

$$I = \frac{\mathcal{E}}{R_{load} + r}$$

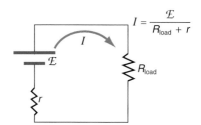

Figure 20.10 Schematic of a voltage source and its load R_{load}. Since the internal resistance r is in series with the load, it can significantly affect the terminal voltage and current delivered to the load.

We see from this expression that the smaller the internal resistance r, the greater the current the voltage source supplies to its load R_{load}. As batteries are depleted, r increases. If r becomes a significant fraction of the load resistance, then the current is significantly reduced, as the following example illustrates.

EXAMPLE 20.4 TERMINAL VOLTAGE AND LOAD

A certain battery has a 12.0 V emf and an internal resistance of 0.100 Ω. (a) Calculate its terminal voltage when connected to a 10.0 Ω load. (b) What is the terminal voltage when connected to a 0.500 Ω load? (c) What power does the 0.500 Ω load dissipate? (d) If the internal resistance grows to 0.500 Ω, find the current, terminal voltage, and power dissipated by a 0.500 Ω load.

Strategy The analysis above gave an expression for current when internal resistance is taken into account. Once the current is found, the terminal voltage can be calculated using Equation 20.3. Once current is found, the power dissipated by a resistor can also be found.

Solution for (a) Entering the given values for the emf, load resistance, and internal resistance into the expression above yields

$$I = \frac{\mathcal{E}}{R_{load} + r} = \frac{12.0 \text{ V}}{10.1 \text{ Ω}} = 1.188 \text{ A}$$

The terminal voltage is found by entering the known values into Equation 20.3:

$$V = \mathcal{E} - Ir = 12.0 \text{ V} - (1.188 \text{ A})(0.100 \text{ Ω})$$
$$= 11.9 \text{ V}$$

Discussion for (a) The terminal voltage here is only slightly lower than the emf, implying that 10.0 Ω is a light load for this particular battery.

Solution for (b) Similarly, with R_{load} = 0.500 Ω, the current is

$$I = \frac{\mathcal{E}}{R_{load} + r} = \frac{12.0 \text{ V}}{0.600 \text{ Ω}} = 20.0 \text{ A}$$

The terminal voltage is now

$$V = \mathcal{E} - Ir = 12.0 \text{ V} - (20.0 \text{ A})(0.100 \text{ Ω})$$
$$= 10.0 \text{ V}$$

Discussion for (b) This terminal voltage exhibits a more significant reduction compared with $\mathcal{E}$, implying 0.500 Ω is a heavy load for this battery.

Solution for (c) The power dissipated by the 0.500 Ω load can be found using I^2R. Entering the known values gives

$$P_{load} = I^2 R_{load} = (400 \text{ A}^2)(0.500 \text{ Ω}) = 200 \text{ W}$$

Discussion for (c) Note that this power can also be obtained using the expressions V^2/R or IV, where V is the

(continued)

(continued)
terminal voltage (10.0 V in this case).

Solution for (d) Here the internal resistance has increased, perhaps due to the depletion of the battery, to the point where it is as great as the load resistance. As before, we first find the current by entering the known values into the expression, yielding

$$I = \frac{\mathcal{E}}{R_{load} + r} = \frac{12.0 \text{ V}}{1.00 \text{ }\Omega} = 12.0 \text{ A}$$

Now the terminal voltage is

$$V = \mathcal{E} - Ir = 12.0 \text{ V} - (12.0 \text{ A})(0.500 \text{ }\Omega)$$
$$= 6.00 \text{ V}$$

And the power dissipated by the load is

$$P_{load} = I^2 R_{load} = (144 \text{ A}^2)(0.500 \text{ }\Omega) = 72.0 \text{ W}$$

Discussion for (d) We see that the increased internal resistance has significantly decreased terminal voltage, current, and power delivered to a load.

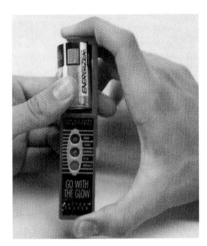

Figure 20.11 These two battery testers measure terminal voltage under a load to determine the condition of a battery. The large device is used to test car batteries and has a small resistance that can dissipate large amounts of power. The small device is used on small batteries and has colored lights to indicate the acceptability of their terminal voltage.

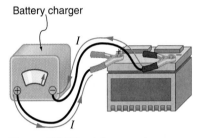

Figure 20.12 A battery charger reverses the normal direction of current through a battery, reversing its chemical reaction and replenishing its chemical potential.

Battery testers, such as those in Figure 20.11, intentionally draw current to determine whether the terminal voltage drops below an acceptable level. They really test the internal resistance of the battery. If internal resistance is high, the battery is weak, as evidenced by its low terminal voltage.

Some batteries can be recharged by passing a current through them in the direction opposite to the current they supply to a resistance. This is done routinely in cars, and it is represented pictorially in Figure 20.12 for a battery charger. The voltage output of the battery charger must be greater than the emf of the battery to reverse current through it. This will cause the terminal voltage of the battery to be greater than $\mathcal{E}$, since $V = \mathcal{E} - Ir$, and I is now negative.

Multiple Voltage Sources

There are two voltage sources when a battery charger is used. Voltage sources connected in series are relatively simple. *When voltage sources are in series, their internal resistances add and their emfs add algebraically.* (See Figure 20.13.) Series connections of voltage sources are common, for example, in flashlights, toys, and other appliances. Usually, the cells are in the same sense in order to produce a larger total emf. But if the cells oppose one another, such as when one is put into an appliance backward, the total emf is less, since it is the *algebraic* sum of the individual emfs. A battery is a multiple connection of voltaic cells, as shown in Figure 20.14. The

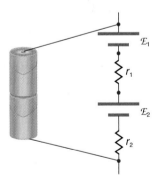

Figure 20.13 A series connection of two voltage sources. The emfs and internal resistances add, giving a total emf of $\mathcal{E}_1 + \mathcal{E}_2$ and a total internal resistance of $r_1 + r_2$.

Figure 20.14 Batteries are multiple connections of individual cells, as shown in this modern rendition of an old print. Single cells, such as AA or C cells, are commonly called batteries, although this is technically incorrect.

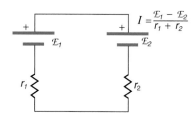

Figure 20.15 These two voltage sources are connected in series with their emfs in opposition. Current flows in the direction of the greater emf and is limited to $I = (\mathcal{E}_1 - \mathcal{E}_2)/(r_1 + r_2)$ by the sum of the internal resistances. A battery charger connected to a battery is an example of such a connection. The charger must have a larger emf than the battery to reverse current through it.

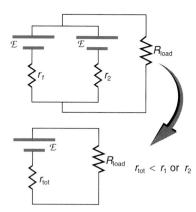

Figure 20.17 Two voltage sources with identical emfs connected in parallel produce the same emf but have a smaller total internal resistance than the individual sources. Parallel combinations are often used to deliver more current. Here $I = \mathcal{E}/(r_{tot} + R_{load})$ flows through the load.

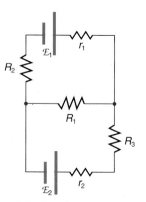

Figure 20.18 This circuit cannot be reduced to a combination of series and parallel connections. Kirchhoff's rules, special applications of the laws of conservation of charge and energy, can be used to analyze it.

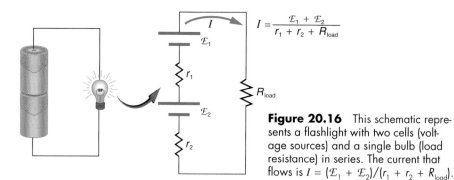

Figure 20.16 This schematic represents a flashlight with two cells (voltage sources) and a single bulb (load resistance) in series. The current that flows is $I = (\mathcal{E}_1 + \mathcal{E}_2)/(r_1 + r_2 + R_{load})$.

disadvantage of series connections of cells is that their internal resistances add. The author once owned a 1957 MGA that had two 6 V batteries in series, rather than a single 12 V battery. This arrangement produced a large internal resistance that caused him many problems in starting the engine.

If the series connection of two voltage sources is made into a complete circuit with the emfs in opposition, then a current of magnitude $I = (\mathcal{E}_1 - \mathcal{E}_2)/(r_1 + r_2)$ flows. See Figure 20.15, for example, which shows a circuit exactly analogous to the battery charger discussed above. If two voltage sources in series with emfs in the same sense are connected to a load R_{load}, as in Figure 20.16, then $I = (\mathcal{E}_1 + \mathcal{E}_2)/(r_1 + r_2 + R_{load})$ flows.

Figure 20.17 shows two voltage sources with identical emfs in parallel and connected to a load resistance. In this simple case, the total emf is the same as the individual emfs. But the total internal resistance is reduced, since the internal resistances are in parallel. The parallel connection thus can produce a larger current. Here $I = \mathcal{E}/(r_{tot} + R_{load})$ flows through the load, and r_{tot} is less than those of the individual batteries. For example, some diesel-powered cars use two 12 V batteries in parallel; they produce a total emf of 12 V but can deliver the larger current needed to start a diesel engine.

20.3 KIRCHHOFF'S RULES

Many complex circuits, such as the one in Figure 20.18, cannot be analyzed with the series-parallel techniques developed above. There are, however, two circuit analysis rules that can be used to analyze any circuit, simple or complex. These rules are special cases of the laws of conservation of charge and conservation of energy. The rules are known as **Kirchhoff's rules**, after their inventor Gustav Kirchhoff (1824–1887):

> **Kirchhoff's first rule—the junction rule. The sum of all currents entering a junction must equal the sum of all currents leaving the junction.**
>
> **Kirchhoff's second rule—the loop rule. The algebraic sum of changes in potential around any closed circuit path (loop) must be zero.**

Explanations of the two rules will now be given, followed by problem-solving hints for applying Kirchhoff's rules, and a worked example that uses them.

Kirchhoff's first rule (the junction rule) is an application of the conservation of charge to a junction; it is illustrated in Figure 20.19. Current is the flow of charge, and charge is conserved; thus, whatever charge flows into the junction must flow out.* (We, in fact, used this argument in Section 20.1 when deriving the equation for resistances in parallel, and we saw one verification of it in part (c) of Example 20.2.) In Figure 20.19,

---

*The junction rule needs to be modified where current leaks from the wires, but this does not occur in the applications we consider.

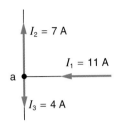

Figure 20.19 The junction rule. An example of Kirchhoff's first rule where the sum of the currents into a junction equals the sum of the currents out of a junction. In this case, the current going into the junction splits and comes out as two currents, so that $I_1 = I_2 + I_3$. Here I_1 must be 11 A, since I_2 is 7 A and I_3 is 4 A.

CONNECTIONS

Conservation Laws

Kirchhoff's rules for circuit analysis are applications of conservation laws to circuits. The first rule is the application of conservation of charge, while the second rule is the application of conservation of energy. Conservation laws, even used in a specific application, such as circuit analysis, are so basic as to form the foundation of that application.

Kirchhoff's first rule requires that $I_1 = I_2 + I_3$. Equations like this can and will be used to analyze circuits and to solve circuit problems.

Kirchhoff's second rule (the loop rule) is an application of conservation of energy. The loop rule is stated in terms of potential V rather than energy, but the two are related since $PE_{elec} = qV$. In a closed loop, whatever energy is supplied by emfs must be used by devices in the loop, since there are no outside sources of energy and no other destinations for energy. (We also used this reasoning in Section 20.1 when deriving the equation for series resistances, and we saw one verification of it in part (c) of Example 20.1.) Figure 20.20 illustrates the changes in potential in a simple series circuit loop. Kirchhoff's second rule requires $\mathcal{E} - Ir - IR_1 - IR_2 = 0$. Rearranged, this is $\mathcal{E} = Ir + IR_1 + IR_2$, which means the emf equals the sum of the IR (voltage) drops in the loop.

By applying Kirchhoff's rules, we generate equations that allow us to find the unknowns in circuits. The unknowns may be currents, emfs, or resistances. Each time a rule is applied, an equation is produced. If there are as many independent equations as unknowns, then the problem can be solved. *There are two decisions you must make when applying Kirchhoff's rules.* These decisions determine the signs of various quantities in the equations you obtain from applying the rules.

1. *When applying Kirchhoff's first or junction rule, you must label the current in each branch and decide in what direction it is going.* For example, in Figures 20.18, 20.19, and 20.20, currents are labeled I_1, I_2, I_3, and I, and arrows indicate their directions. There is no risk here, for if you choose the wrong direction, the current will be of the correct magnitude but negative.

2. *When applying Kirchhoff's second or loop rule, you must identify a closed loop and decide in which direction to go around it, clockwise or counterclockwise.* For example, in Figure 20.20 the loop was traversed in the same direction as the current (clockwise). Again, there is no risk; going around the circuit in the opposite direction reverses the sign of every term in the equation, which is like multiplying both sides of the equation by −1. Figure 20.21 and the following points will help you get the plus or minus signs right when applying the loop rule. Note that the resistors and emfs are traversed by going from a to b.

(a) When a resistor is traversed in the same direction as the current, the change in potential is −IR. (See Figure 20.21(a).)

(b) When a resistor is traversed in the direction opposite to the current, the change in potential is +IR. (See Figure 20.21(b).)

(c) When an emf is traversed from − to + (the same direction it moves positive charge), the change in potential is +$\mathcal{E}$. (See Figure 20.21(c).)

(d) When an emf is traversed from + to − (opposite to the direction it moves positive charge), the change in potential is −$\mathcal{E}$. (See Figure 20.21(d).)

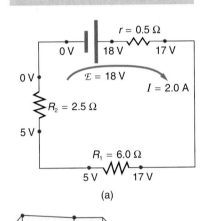

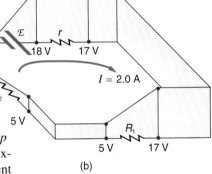

(a)

(b)

Figure 20.20 The loop rule. An example of Kirchhoff's second rule where the sum of the changes in potential *around a closed loop* must be zero. (a) In this standard schematic of a simple series circuit, the emf supplies 18 V, which is reduced to zero by the resistances, with 1 V across the internal resistance, and 12 and 5 V across the two load resistances, for a total of 18 V. (b) This perspective view represents the potential as something like a roller coaster, where charge is raised in potential by the emf and lowered by the resistances.

Direction of traverse a ⟶ b

$\Delta V = V_b - V_a = -IR$
(a)

$\Delta V = V_b - V_a = +IR$
(b)

$\Delta V = V_b - V_a = +\mathcal{E}$
(c)

$\Delta V = V_b - V_a = -\mathcal{E}$
(d)

Figure 20.21 Each of these resistors and emfs is traversed from a to b. The potential changes are shown beneath each element and are explained in the text.

EXAMPLE 20.5   USING KIRCHHOFF'S RULES TO FIND CURRENTS

Find the currents flowing in the circuit in Figure 20.22.

Strategy This circuit is complex enough that the currents cannot be found using Ohm's law and the series-parallel techniques—it is necessary to use Kirchhoff's rules. Currents have been labeled I_1, I_2, and I_3 in the figure and assumptions made on their directions. Locations have been labeled with letters a through h. In the solution we will apply the junction and loop rules, seeking three independent equations to allow us to solve for the three unknown currents.

Solution We begin by applying Kirchhoff's first or junction rule at point a. This gives

$$I_1 = I_2 + I_3 \tag{i}$$

since I_1 flows into the junction, while I_2 and I_3 flow out. Applying the junction rule at e produces exactly the same equation, so that no new information is obtained. Equation i is a single equation with three unknowns—three independent equations are needed, and so the loop rule must be applied.

Now we consider the loop abcdea. Going from a to b, we traverse R_2 in the same (assumed) direction of the current I_2, and so the change in potential is $-I_2 R_2$. Then going from b to c, we go from − to +, so that the change in potential is $+\mathcal{E}_1$. Traversing the internal resistance r_1 from c to d gives $-I_2 r_1$. Completing the loop by going from d to a again traverses a resistor in the same direction as its current, giving a change in potential of $-I_1 R_1$. The loop rule states that the changes in potential sum to zero. Thus,

$$-I_2 R_2 + \mathcal{E}_1 - I_2 r_1 - I_1 R_1 = -I_2(R_2 + r_1) + \mathcal{E}_1 - I_1 R_1 = 0$$

Substituting values for the resistances and emf, and canceling the ampere unit gives

$$-3I_2 + 18 - 6I_1 = 0 \tag{ii}$$

Now applying the loop rule to aefgha (we could have chosen abcdefgha as well) similarly gives

$$+I_1 R_1 + I_3 R_3 + I_3 r_2 - \mathcal{E}_2 = +I_1 R_1 + I_3(R_3 + r_2) - \mathcal{E}_2 = 0$$

Note that the signs are reversed compared with the other loop, because elements are traversed in the opposite direction. With values entered, this becomes

$$+6I_1 + 2I_3 - 45 = 0 \tag{iii}$$

Equations i, ii, and iii are sufficient to solve for the three unknown currents. First, solve Equation ii for I_2:

$$I_2 = 6 - 2I_1 \tag{iv}$$

Now solve Equation iii for I_3:

$$I_3 = 22.5 - 3I_1 \tag{v}$$

Substituting Equations iv and v into Equation i allows us to find a value for I_1:

$$I_1 = I_2 + I_3 = (6 - 2I_1) + (22.5 - 3I_1) = 28.5 - 5I_1$$

So $6I_1 = 28.5$, and

$$I_1 = 4.75 \text{ A}$$

Substituting this value for I_1 into Equation iv gives $I_2 = 6 - 2I_1 = 6 - 9.50$, so that

$$I_2 = -3.50 \text{ A}$$

The minus sign means I_2 flows in the direction opposite to that assumed in Figure 20.22. Finally, substituting the value for I_1 into Equation v gives $I_3 = 22.5 - 3I_1 = 22.5 - 14.25$, so that

$$I_3 = 8.25 \text{ A}$$

Discussion Just as a check, we note that indeed $I_1 = I_2 + I_3$.

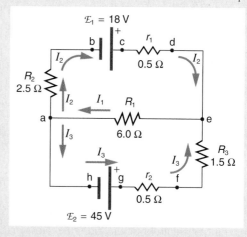

Figure 20.22   This circuit is similar to that in Figure 20.18, but the resistances and emfs are specified. The currents in each branch are labeled and assumed to move in the directions shown. This example uses Kirchhoff's rules to find these currents.

PROBLEM-SOLVING STRATEGIES

FOR KIRCHHOFF'S RULES

Step 1. Make certain there is a clear circuit diagram on which you can *label all known and unknown* resistances, emfs, and currents. If a current is unknown, you must *assign it a direction*. This is necessary for determining the signs of potential changes. If you assign the direction incorrectly, the current will be found to have a negative value—no harm done.

Step 2. *Apply the junction rule* to any junction in the circuit. Each time the junction rule is applied, you should get an equation with a current that does not appear in a previous application—if not, then the equation is redundant.

Step 3. *Apply the loop rule* to as many loops as needed to solve for the unknowns in the problem. (There must be as many independent equations as unknowns.) To apply the loop rule, you must choose a direction to go around the loop. Then carefully and consistently determine the signs of the potential changes for each element using points (a) through (d) discussed above in conjunction with Figure 20.21.

Step 4.   *Solve the simultaneous equations for the unknowns.* This may involve many algebraic steps, requiring careful checking and rechecking.

Step 5.   *Check to see if the answers are reasonable and consistent.* The numbers should be of the correct order of magnitude, neither exceeding large nor vanishingly small. The signs should be reasonable—for example, no resistance should be negative. Check to see that the values obtained satisfy the various equations obtained from applying the rules. The currents should satisfy the junction rule, for example.

The material in this section is correct in theory. We should be able to verify it by making measurements of current and voltage. In fact, some of the devices used to make such measurements are straightforward applications of the principles covered so far and are explored in the next sections. As we shall see, a very basic, even profound, fact results—making a measurement alters the quantity being measured.

Figure 20.23   These gauges are voltmeters that measure the voltage output of "sender" units, which are hopefully proportional to the amount of gasoline in the tank, engine temperature, and speed of the car.

20.4   DC VOLTMETERS AND AMMETERS

Voltmeters measure voltage, whereas **ammeters** measure current. Some of the meters you see on automobile dashboards, tape recorders, and tuner-amplifiers are voltmeters or ammeters. (See Figure 20.23.) The internal construction of the simplest of these meters and how they are connected to the system they monitor give further insight into applications of series and parallel connections.

Voltmeters are connected in parallel with whatever device's voltage is to be measured. A parallel connection is used, because objects in parallel experience the same potential difference. (See Figure 20.24, where the voltmeter is represented by the symbol —Ⓥ—.) Ammeters are connected in series with whatever device's current is to be measured. A series connection is used because objects in series have the same current passing through them. (See Figure 20.25, where the ammeter is represented by the symbol —Ⓐ—.)

Analog meters have a needle that swivels to point at numbers on a scale, as opposed to digital meters, which have direct numerical readouts. The heart of most analog meters is a device called a **galvanometer**, denoted by —Ⓖ—. Current flow through a galvanometer produces a proportional needle deflection. The two crucial characteristics of a given galvanometer are its resistance and current sensitivity. *Current sensitivity* is defined to be the current that gives a full-scale deflection of the galvanometer's needle. For example, a galvanometer with a current sensitivity of 50 μA has a maximum deflection of its needle when 50 μA flows through it, reads half-scale when 25 μA flows through it, and so on. If such a galvanometer has a 25 Ω resistance, then a voltage of only $V = IR = (50 \text{ μA})(25 \text{ Ω}) = 1.25$ mV produces a full-scale reading. By connecting resistors to this galvanometer, you can use it as either a voltmeter or ammeter that can measure a broad range of voltages or currents.

Figure 20.26 shows how a galvanometer can be used as a voltmeter by connecting it in series with a large resistance. The value of the resistance R is determined by the maximum voltage to be measured. Suppose you want 10 V to produce a full-scale deflection

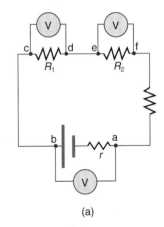

(a)

(b)

Figure 20.24   (a) To measure potential differences in this series circuit, the voltmeter —Ⓥ— is placed in parallel with the voltage source or either of the resistors. Note that *terminal voltage* is measured between points a and b. It is not possible to connect the voltmeter directly across the emf. (b) An analog voltmeter in use.

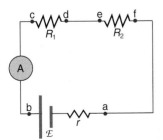

Figure 20.25   An ammeter is placed in series to measure current. All of the current in this circuit flows through the meter. The ammeter would have the same reading if located between points d and e or between points f and a as it does in the position shown.

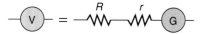

Figure 20.26 A large resistance placed in series with a galvanometer produces a voltmeter, the full-scale deflection of which depends on the choice of R. The larger the voltage to be measured, the larger R must be.

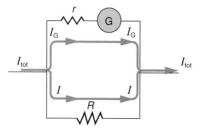

Figure 20.27 A small shunt resistance R placed in parallel with a galvanometer produces an ammeter, the full-scale deflection of which depends on the choice of R. The larger the current to be measured, the smaller R must be. Most of the current flowing through the meter is shunted through R to protect the galvanometer.

of a voltmeter containing a 25 Ω galvanometer with a 50 μA sensitivity. Then 10 V applied to the meter must produce a current of 50 μA. The total resistance must be $R_{tot} = R + r = V/I = (10 \text{ V})/(50 \text{ }\mu\text{A}) = 200$ kΩ, or $R = R_{tot} - r = 200$ k$\Omega - 25 \text{ }\Omega \approx 200$ kΩ. (R is so large that the galvanometer resistance is nearly negligible.) Note that 5 V applied to this voltmeter produces a half-scale deflection by producing a 25 μA current through the meter, and so the voltmeter's reading is proportional to voltage as desired. This voltmeter would not be useful for voltages less than about half a volt, because the meter deflection would be small and difficult to read accurately. For other voltage ranges, other resistances are placed in series with the galvanometer. Many meters have a choice of scales. That choice involves switching an appropriate resistance into series with the galvanometer.

The same galvanometer can also be made into an ammeter by placing it in parallel with a small resistance R, often called the *shunt resistance*, as shown in Figure 20.27. Since the shunt resistance is small, most of the current passes through it, allowing an ammeter to measure currents much greater than those producing a full-scale deflection of the galvanometer. Suppose, for example, an ammeter is needed that gives a full-scale deflection for 1.0 A, and contains the same 25 Ω galvanometer with its 50 μA sensitivity. Since R and r are in parallel, the voltage across them is the same. These IR drops are $IR = I_G r$, so that $I_G/I = R/r$. Solving for R, and noting that I_G is 50 μA and I is 0.999950 A,

$$R = r\frac{I_G}{I} = (25 \text{ }\Omega)\frac{50 \text{ }\mu\text{A}}{0.999950 \text{ A}} = 1.25 \times 10^{-3} \text{ }\Omega$$

Ammeters may also have multiple scales for greater flexibility in application. The various scales are achieved by switching various shunt resistances in parallel with the galvanometer—the greater the maximum current to be measured, the smaller the shunt resistance must be.

When you use a voltmeter or ammeter, you are connecting another resistor to an existing circuit and, thus, altering the circuit. Ideally, voltmeters and ammeters do not appreciably affect the circuit, but it is instructive to examine the circumstances under which they do or do not interfere. First, consider the voltmeter, which is always placed in parallel with the device being measured. Very little current flows through the voltmeter if its resistance is a few orders of magnitude greater than the device, and so the circuit is not appreciably affected. (See Figure 20.28(a).) (A large resistance in parallel with a small one has a combined resistance essentially equal to the small one.) If, however, the voltmeter's resistance is comparable to that of the device being measured, then the two in parallel have a smaller resistance, appreciably affecting the circuit. (See Figure 20.28(b).) The voltage across the device is not the same as when the voltmeter is out of the circuit.

An ammeter is placed in series in the branch of the circuit being measured, so that its resistance adds to that branch. Normally, the ammeter's resistance is very small compared with devices in the circuit, and so the extra resistance is negligible. (See Figure 20.29(a).) However, if very small resistances are involved, or if the ammeter is not as low in resistance as it should be, then the total series resistance is significantly greater, and the current in the branch being measured is reduced. (See Figure 20.29(b).)

Figure 20.28 (a) A voltmeter having a resistance much larger than the device with which it is in parallel produces a parallel resistance essentially the same as the device and does not appreciably affect the circuit being measured. (b) Here the voltmeter has the same resistance as the device, so that the parallel resistance is half of what it is when the voltmeter is not connected. This is an example of a significant alteration of the circuit and is to be avoided.

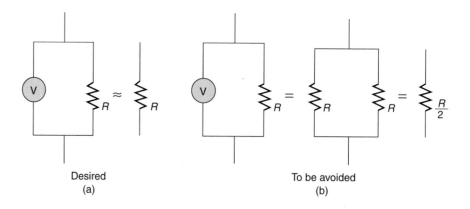

Desired
(a)

To be avoided
(b)

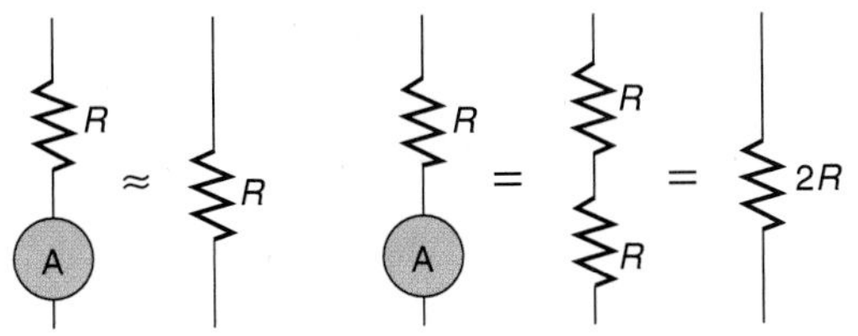

Figure 20.29 (a) An ammeter normally has such a small resistance that the total series resistance in the branch being measured is not appreciably increased. The circuit is essentially unaltered compared with when the ammeter is absent. (b) Here the ammeter's resistance is the same as that of the branch, so that the total resistance is doubled and the current is half what it is without the ammeter. This significant alteration of the circuit is to be avoided.

One solution to the problem of voltmeters and ammeters interfering with the circuits being measured is to use galvanometers with greater sensitivity. This allows construction of voltmeters with greater resistance and ammeters with smaller resistance than when less sensitive galvanometers are used. There are practical limits to galvanometer sensitivity, but it is possible to get analog meters that make measurements accurate to a few percent. Note that the inaccuracy comes from altering the circuit, not from a fault in the meter.

There is another measurement technique based on drawing no current at all and, hence, not altering the circuit at all. These are called null measurements and are the topic of the next section. Digital meters that employ solid state electronics and null measurements can attain accuracies of one part in 10^6.

20.5 NULL MEASUREMENTS

As noted in the preceding section, standard measurements of voltage and current alter the circuit being measured, introducing uncertainties in the measurements. Voltmeters draw some extra current, whereas ammeters reduce current flow. **Null measurements** balance voltages so that there is *no current flowing through the measuring device* and, therefore, no alteration of the circuit being measured. Null measurements are generally more accurate but are also more complex than the use of standard voltmeters and ammeters, and they still have limits to their precision. We shall consider a few specific types of null measurements, because they are common, interesting, and they further illuminate principles of electric circuits.

The Potentiometer

Suppose you wish to measure the emf $\mathcal{E}$ of a battery. Consider what happens if you connect the battery directly to a standard voltmeter as shown in Figure 20.30. (Once we note the problems with this measurement, we will examine a null measurement that improves accuracy.) As discussed before, the actual quantity measured is the terminal voltage V, which is related to the emf $\mathcal{E}$ of the battery by

$$V = \mathcal{E} - Ir$$

where I is the current that flows and r is the internal resistance of the battery. The emf could be accurately calculated if r were very accurately known, but it usually is not. If the current I could be made zero, then $V = \mathcal{E}$, and so emf could be directly measured. Standard voltmeters need a current to operate; thus, another technique is needed.

A **potentiometer** is a null measurement device for measuring potentials (voltages). (See Figure 20.31.) A voltage source is connected to a resistor R, say a long wire, and passes a constant current through it. There is a steady drop in potential (an IR drop) along

CONNECTIONS

Limits to Knowledge

Making a measurement alters the system being measured in a manner that produces uncertainty in the measurement. For macroscopic systems, such as the circuits discussed in this section, the alteration can usually be made negligibly small, but it cannot be eliminated entirely. For submicroscopic systems, such as atoms, nuclei, and smaller particles, measurement alters the system in a manner that cannot be made arbitrarily small. This actually limits knowledge of the system—even limiting what nature can know about itself. We shall see profound implications of this when the Heisenberg uncertainty principle is discussed in the chapters on quantum mechanics.

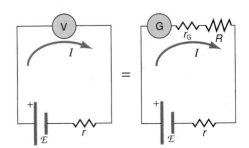

Figure 20.30 An analog voltmeter attached to a battery draws a small but nonzero current and measures a terminal voltage that differs from the emf of the battery. Since the internal resistance of the battery is not precisely known, it is not possible to calculate the emf precisely.

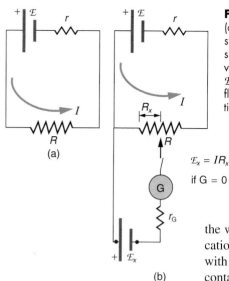

Figure 20.31 The potentiometer, a null measurement device. (a) A voltage source connected to a long wire resistor passes a constant current I through it. (b) An unknown emf is connected as shown, and the point of contact along R is adjusted until the galvanometer reads zero. The segment of wire has a resistance R_x and $\mathcal{E}_x = IR_x$, where I is unaffected by the connection since no current flows through the galvanometer. The unknown emf is thus proportional to the resistance of the wire segment.

$$\mathcal{E}_x = IR_x$$

if G = 0

the wire, so that a variable potential can be obtained by making contact at varying locations along the wire. Figure 20.31(b) shows an unknown emf $\mathcal{E}_x$ connected in series with a galvanometer. Note that $\mathcal{E}_x$ opposes the other voltage source. The location of the contact point (see arrow on drawing) is adjusted until the galvanometer reads zero. When the galvanometer reads zero, $\mathcal{E}_x = IR_x$, where R_x is the resistance of the section of wire up to the contact point. Since no current flows through the galvanometer, none flows through the unknown emf, and so $\mathcal{E}_x$ is directly sensed. Now, a very precisely known standard emf $\mathcal{E}_s$ is substituted for $\mathcal{E}_x$, and the contact point is adjusted until the galvanometer again reads zero, so that $\mathcal{E}_s = IR_s$. In both cases, no current passes through the galvanometer, and so the current I through the long wire is the same. Upon taking the ratio $\mathcal{E}_x/\mathcal{E}_s$, I cancels, giving

$$\frac{\mathcal{E}_x}{\mathcal{E}_s} = \frac{IR_x}{IR_s} = \frac{R_x}{R_s}$$

Solving for $\mathcal{E}_x$ gives

$$\mathcal{E}_x = \mathcal{E}_s \frac{R_x}{R_s}$$

Since a long uniform wire is used for R, the ratio of resistances R_x/R_s is the same as the ratio of the lengths of wire that zero the galvanometer for each emf. The three quantities on the right-hand side of the equation are now known or measured, and $\mathcal{E}_x$ can be calculated. The uncertainty in this calculation can be considerably smaller than when using a voltmeter directly, but it is not zero. There is always some uncertainty in the ratio of resistances R_x/R_s and in the standard emf $\mathcal{E}_s$. Furthermore, it is not possible to tell when the galvanometer reads exactly zero, which introduces error into both R_x and R_s, and may also affect the current I.

Resistance Measurements and the Wheatstone Bridge

There are a whole variety of so-called ohm meters that purport to measure resistance. What the most common ohm meters actually do is to apply a voltage to a resistance, measure the current, and calculate the resistance using Ohm's law. Their readout is this calculated resistance. Two configurations for ohm meters using standard voltmeters and ammeters are shown in Figure 20.32. Such configurations are limited in accuracy, because the meters alter both the voltage applied to the resistor and the current that flows through it.

The **Wheatstone bridge** is a null measurement device for calculating resistance by balancing potential drops in a circuit. (See Figure 20.33.) (The device is called a *bridge* because the galvanometer forms a bridge between two branches. A variety of bridge devices are used to make null measurements in circuits.) Resistors R_1 and R_2 are precisely known, while the arrow through R_3 indicates that it is a variable resistance. The value of R_3 can be precisely read. With the unknown resistance R_x in the circuit, R_3 is adjusted until the galvanometer reads zero. The potential difference between points b and d is then zero, meaning that b and d are at the *same* potential. With no current running through the galvanometer, it has no effect on the rest of the circuit. So the branches abc and adc are in parallel, and each branch has the full voltage of the source. That is, the *IR* drops along

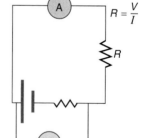

$$R = \frac{V}{I}$$

V assumed constant

(a)

$$R = \frac{V}{I}$$

V measured

(b)

Figure 20.32 Two methods for measuring resistance with standard meters. (a) Assuming a known voltage for the source, an ammeter measures current, and resistance is calculated as $R = V/I$. (b) Since the terminal voltage V varies with current, it is better to measure it. V is most accurately known when I is small, but I itself is most accurately known when it is large.

abc and adc are the same. Since b and d are at the same potential, the *IR* drop along ad must equal the *IR* drop along ab. Thus,

$$I_1 R_1 = I_2 R_3$$

Again since b and d are at the same potential, the *IR* drop along dc must equal the *IR* drop along bc. Thus,

$$I_1 R_2 = I_2 R_x$$

Taking the ratio of these last two expressions gives

$$\frac{I_1 R_1}{I_1 R_2} = \frac{I_2 R_3}{I_2 R_x}$$

Canceling the currents and solving for R_x yields

$$R_x = R_3 \frac{R_2}{R_1}$$

This equation is used to calculate the unknown resistance when current through the galvanometer is zero. This method can be very accurate (often to four significant digits), but it is limited by two factors. First, it is not possible to get the current through the galvanometer to be exactly zero. Second, there are always uncertainties in R_1, R_2, and R_3, which contribute to the uncertainty in R_x.

Figure 20.33 The Wheatstone bridge is used to calculate unknown resistances. The variable resistance R_3 is adjusted until the galvanometer reads zero with the switch closed. This simplifies the circuit, allowing R_x to be calculated based on the *IR* drops as discussed in the text.

20.6   DC CIRCUITS CONTAINING RESISTORS AND CAPACITORS

When you use a flash camera, it takes a few seconds to charge the capacitor that powers the flash. The light flash discharges the capacitor in a tiny fraction of a second. Why does one take longer than the other? This question and a number of other phenomena that involve charging and discharging capacitors are discussed in this section.[*]

An **RC circuit** is one containing a resistor *R* and a capacitor *C*. Figure 20.34 shows a simple *RC* circuit that employs a DC voltage source. The capacitor is initially uncharged. As soon as the switch is closed, current flows to and from the initially uncharged capacitor. As charge increases on the capacitor plates, there is increasing opposition to the flow of charge by the repulsion of like charges on each plate. In terms of voltage, this is because voltage across the capacitor is given by $V = Q/C$, and it opposes the battery, growing from zero to $\mathcal{E}$ when fully charged. The current thus decreases from its initial value of $I_0 = \mathcal{E}/R$ to zero as the voltage on the capacitor reaches $\mathcal{E}$.[†] The initial current is $I_0 = \mathcal{E}/R$, because all of the *IR* drop is in the resistance. So the smaller the resistance, the faster a given capacitor will be charged. Note that the internal resistance of the voltage source is included in *R*. When the batteries powering a flash camera begin to wear out, their internal resistance rises, reducing the current and lengthening the time it takes to get ready for the next flash.

---

[*]When capacitors were first discussed in Chapter 18, we ignored the time it took to charge and discharge them, concentrating on their functions as charge and energy storage devices.
[†] This can also be explained with Kirchhoff's second (loop) rule. When there is no current, there is no *IR* drop, and so the voltage on the capacitor must then equal the emf of the voltage source, $\mathcal{E}$.

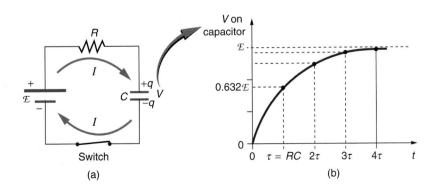

(a)   (b)

Figure 20.34 (a) An *RC* circuit with an initially uncharged capacitor. Current flows in the direction shown (opposite of electron flow) as soon as the switch is closed. Mutual repulsion of like charges in the capacitor progressively slows the flow as the capacitor is charged, stopping the current when the capacitor is fully charged and $Q = C\mathcal{E}$. (b) A graph of voltage across the capacitor versus time, with the switch closing at $t = 0$.

Voltage on the capacitor is initially zero and rises rapidly at first, since the initial current is a maximum. Figure 20.34(b) shows a graph of capacitor voltage versus time starting when the switch is closed at $t = 0$. The voltage approaches $\mathcal{E}$ asymptotically, since the closer it gets to $\mathcal{E}$ the less current flows. The equation for voltage versus time when charging a capacitor C through a resistor R, derived using calculus, is

$$V = \mathcal{E}(1 - e^{-t/RC}) \qquad \text{(charging)} \qquad \textbf{(20.4)}$$

where V is the voltage across the capacitor, $\mathcal{E}$ is the emf of the DC voltage source, and the exponential $e = 2.718...$ is the base of the natural logarithm. Note that the units of RC are seconds. We define

$$\tau = RC \qquad \textbf{(20.5)}$$

where τ is a **characteristic time for an RC circuit**. As noted before, a small resistance R allows the capacitor to charge faster. This is reasonable, since a larger current flows through a smaller resistance. It is also reasonable that the smaller the capacitor C, the less time needed to charge it. Both factors are contained in $\tau = RC$.

More quantitatively, consider what happens when $t = \tau = RC$. Then the voltage on the capacitor is $V = \mathcal{E}(1 - e^{-1}) = \mathcal{E}(1 - 0.368) = 0.632\mathcal{E}$.* This means that in the characteristic time $\tau = RC$, the voltage rises to 0.632 of its final value. The voltage will rise 0.632 of the remainder in the next time τ. It is a characteristic of the exponential that the final value is never reached, but 0.632 of the remainder to that value is achieved in every characteristic time τ. In just a few multiples of the time τ, then, the final value is very nearly achieved, as the graph in Figure 20.34(b) illustrates.

Discharging a capacitor through a resistor proceeds in a similar fashion, as Figure 20.35 illustrates. Initially, the current is $I_0 = V_0/R$, driven by the initial voltage V_0 on the capacitor. As the voltage decreases, the current and hence the rate of discharge decreases, implying another exponential formula for V. Using calculus, the voltage V on a capacitor C being discharged through a resistor R is found to be

$$V = V_0 e^{-t/RC} \qquad \text{(discharging)} \qquad \textbf{(20.6)}$$

The graph in Figure 20.35(b) is an example of this exponential decay. Again the characteristic time is $\tau = RC$. A small resistance R allows the capacitor to discharge in a small time, since the current is larger. Similarly, a small capacitance requires less time to discharge, since less charge is stored. In the first time interval $\tau = RC$ after the switch is closed, the voltage falls to 0.368 of its initial value, since $V = V_0 e^{-1} = 0.368V_0$. In each successive time τ, the voltage falls to 0.368 of its preceding value. In a few multiples of τ, the voltage becomes very close to zero, as indicated by the graph in Figure 20.35(b).

Now we can explain why the flash camera takes so much longer to charge than discharge; the resistance while charging is significantly greater than while discharging. The internal resistance of the battery accounts for most of the resistance while charging. As the battery ages, the increasing internal resistance makes the charging process even slower (you may have noticed this). The flash discharge is through a low-resistance ionized gas

---

*Using a calculator with an e^x key, you can verify that $e^{-1} = 0.368$ and $1 - e^{-1} = 0.632$, to three-digit accuracy.

Figure 20.35 (a) Closing the switch discharges the capacitor C through the resistor R. Mutual repulsion of like charges on each plate drives the current. (b) A graph of voltage across the capacitor versus time, with $V = V_0$ at $t = 0$. The voltage decreases exponentially, falling a fixed fraction of the way to zero in each subsequent characteristic time τ.

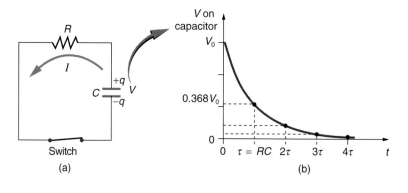

in the flash tube and proceeds very rapidly. Flash photographs, such as in Figure 20.36, can capture a brief instant of a rapid motion because the flash can be less than a microsecond in duration. Such flashes can be made extremely intense. During World War II, nighttime reconnaissance photographs were made from the air with a single flash illuminating more than a square kilometer of enemy territory. The brevity of the flash eliminated blurring due to the surveillance aircraft's motion. A modern use of intense flash lamps is to pump energy into a laser in a short, high-intensity flash.

RC circuits are commonly used for timing purposes. A mundane example of this is found in the ubiquitous intermittent wiper systems of modern cars. The time between wipes is varied by adjusting the resistance in an *RC* circuit. Another example of an *RC* circuit is found in novelty jewelry, Halloween costumes, and various toys that have battery-powered flashing lights. (See Figure 20.37.)

Figure 20.36 This photograph was obtained with an extremely brief and intense flash of light powered by the discharge of a capacitor through a gas.

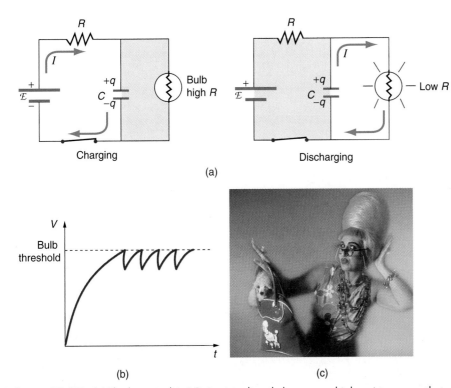

Figure 20.37 (a) The lamp in this *RC* circuit ordinarily has a very high resistance, so that the battery charges the capacitor as if the lamp were not there. When the voltage reaches a threshold value, a current flows through the lamp that dramatically reduces its resistance, and the capacitor discharges through the lamp as if the battery and charging resistor were not there. Once discharged, the process starts again, with the flash period determined by the *RC* constant τ. (b) A graph of voltage versus time for this circuit. (c) This woman's flashing light jewelry is part of an *RC* circuit.

EXAMPLE 20.6 *RC* CIRCUIT IN A HEART DEFIBRILLATOR

A heart defibrillator is used to resuscitate an accident victim by discharging a capacitor through the trunk of her body. A simple version of the circuit is seen in Figure 20.35. (a) What is the characteristic time if an 8.00 μF capacitor is used and the path resistance through her body is 1000 Ω? (b) If the initial voltage is 10.0 kV, how long does it take to decline to 500 V?

Strategy Since the resistance and capacitance are given, it is straightforward to multiply them to give the time con-

stant asked for in part (a). To find the time for the voltage to decline to 500 V, we repeatedly multiply the initial voltage by 0.368 until a voltage less than or equal to 500 V is obtained. Each multiplication corresponds to a time τ.

Solution for (a) The characteristic time is given by Equation 20.5. Entering the given values for resistance and capacitance gives

$$\tau = RC = (1000 \ \Omega)(8.00 \ \mu F) = 8.00 \ ms$$

(continued)

(continued)

Solution for (b) In the first 8.00 ms, the voltage declines to 0.368 of its initial value. This is

$$V = 0.368 V_0 = 3680 \text{ V} \quad \text{at } t = 8.00 \text{ ms}$$

After another 8.00 ms, we multiply by 0.368 again, and the voltage is

$$V' = 0.368 V = (0.368)(3680 \text{ V})$$
$$= 1350 \text{ V} \quad \text{at } t = 16.0 \text{ ms}$$

Similarly, after another 8.00 ms, the voltage is

$$V'' = 0.368 V' = (0.368)(1350 \text{ V})$$
$$= 498 \text{ V} \quad \text{at } t = 24.0 \text{ ms}$$

Discussion So after only 24.0 ms the voltage is down to 498 V, or 4.98% of its original value.* Such brief times are useful in heart defibrillation, because the brief but intense current causes a brief but effective contraction of the heart. The actual circuit in a heart defibrillator is slightly more complex than the one in Figure 20.35, to compensate for magnetic and AC effects that will be covered in the next chapter.

---

*The exact solution is obtained using the full exponential treatment and, in this case, does not differ greatly from the approximate treatment in the example. Solving $V = V_0 e^{-t/RC}$ for t gives $t = -RC \ln(V/V_0) = -RC \ln(500/10,000) = -(8.0 \text{ ms})(-2.996) = 23.97$ ms.

SUMMARY

The two simplest ways of connecting resistors are in *series* and *parallel*. Resistors are in **series** whenever current must flow through them sequentially. Resistors are in **parallel** when each resistor is connected directly to the voltage source by connecting wires having negligible resistance.

The **major features of resistors in series** are:

1. Series resistances add:
$$R_s = R_1 + R_2 + R_3 + \cdots \quad \textbf{(20.1)}$$
2. The same current flows through each resistor in series.
3. Series resistors do not get the total source voltage, but divide it.

The **major features of resistors in parallel** are:

1. Parallel resistance is found from
$$\frac{1}{R_p} = \frac{1}{R_1} + \frac{1}{R_2} + \frac{1}{R_3} + \cdots \quad \textbf{(20.2)}$$
R_p is smaller than any individual resistance in the combination.
2. Each resistor in parallel has the same full voltage of the source.
3. Parallel resistances do not get the total current, but divide it.

If a more cFomplex connection of resistors is a *combination of series and parallel*, it can be reduced to a single equivalent resistance by identifying its various parts as series or parallel, reducing each to its equivalent, and continuing until a single resistance eventually is reached.

All voltage sources have two fundamental parts—a source of electrical energy that has a characteristic emf $\mathcal{E}$, and an internal resistance r. The **electromotive force** (emf or $\mathcal{E}$) *is the potential difference of a source when no current is flowing*. The numerical value of $\mathcal{E}$ depends on the source of potential dif-

ference. The **internal resistance** r of a voltage source affects the output voltage when a current flows. The voltage output of a device is called its **terminal voltage** V and is given by

$$V = \mathcal{E} - Ir \quad \textbf{(20.3)}$$

where I is current and is positive when flowing away from the positive terminal of the voltage source. When multiple voltage sources are in series, their internal resistances add and their emfs add algebraically.

Kirchhoff's rules can be used to analyze any circuit, simple or complex:

Kirchhoff's first rule—the junction rule. The sum of all currents entering a junction must equal the sum of all currents leaving the junction.

Kirchhoff's second rule—the loop rule. The algebraic sum of changes in potential around any closed circuit path (loop) must be zero.

The two rules are based, respectively, on the laws of conservation of charge and energy. Conventions for determining the correct signs of various terms, and detailed problem-solving hints for the application of Kirchhoff's rules, can be found in Section 20.3. The simpler series and parallel rules are special cases of Kirchhoff's rules.

Voltmeters measure voltage, and **ammeters** measure current. A voltmeter is placed in parallel with the voltage source to receive full voltage and must have a large resistance to limit its effect on the circuit. An ammeter is placed in series to get the full current flowing through a branch and must have a small resistance to limit its effect on the circuit. Both can be based on combination of a resistor and a **galvanometer**, a device that gives an analog reading of current. Standard voltmeters and ammeters alter the circuit being measured and are thus limited in accuracy.

Null measurement techniques achieve greater accuracy by balancing a circuit so that no current flows through the measuring device. One such device, for determining voltage, is a **potentiometer**. Another null measurement device, for determining resistance, is the **Wheatstone bridge**. Other physical quantities can also be measured with null measurement techniques.

An *RC* circuit is one that has both resistance and capacitance. The **characteristic time constant** τ **for an** *RC* **circuit** is

$$\tau = RC \qquad \text{(20.5)}$$

When an initially uncharged ($V_0 = 0$ at $t = 0$) capacitor in series with a resistor is charged by a DC voltage source, the voltage rises, asymptotically approaching the emf $\mathcal{E}$ of the voltage source; as a function of time,

$$V = \mathcal{E}(1 - e^{-t/RC}) \qquad \text{(charging)} \qquad \text{(20.4)}$$

In each characteristic time τ, the voltage rises 0.632 of the remaining distance to the final voltage. If a capacitor with an initial voltage V_0 is discharged through a resistor starting at $t = 0$, then its voltage decreases exponentially as given by

$$V = V_0 e^{-t/RC} \qquad \text{(discharging)} \qquad \text{(20.6)}$$

In each characteristic time τ, the voltage falls to 0.368 of its initial value, approaching zero asymptotically.

CONCEPTUAL QUESTIONS

20.1 A switch has a variable resistance that is nearly zero when closed and extremely large when open, and it is placed in series with the device it controls. Explain the effect the switch in Figure 20.38 has on current when open and when closed.

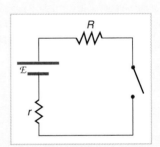

Figure 20.38 A switch is ordinarily in series with a resistance and voltage source. Ideally, the switch has nearly zero resistance when closed but has an extremely large resistance when open. Questions 1, 2, 3, and 4.

20.2 What is the voltage across the open switch in Figure 20.38?

20.3 There is a voltage across an open switch, such as in Figure 20.38. Why, then, is the power dissipated by the open switch small?

20.4 Why is the power dissipated by a closed switch, such as in Figure 20.38, small?

20.5 A student in a physics lab mistakenly wired a light bulb, battery, and switch as shown in Figure 20.39. Explain why the bulb is on when the switch is open, and off when the switch is closed. (Do not try this—it is hard on the battery!)

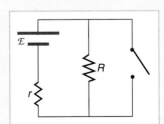

Figure 20.39 A wiring mistake put this switch in parallel with the device represented by *R*. Question 5.

20.6 Knowing that the severity of a shock depends on the magnitude of the current through your body, would you prefer to be in series or parallel with a resistance, such as the heating element of a toaster, if shocked by it? Explain.

20.7 Would your headlights dim when you start your car's engine if the wires in your automobile were superconductors? (Do not neglect the battery's internal resistance.) Explain.

20.8 Some strings of Christmas tree lights are wired in series to save wiring costs. An old version utilized bulbs that break the electrical connection, like an open switch, when they burn out. If one such bulb burns out, what happens to the others? If such a string operates on 120 V and has 20 identical bulbs, what is the normal operating voltage of each? Newer versions use bulbs that short circuit, like a closed switch, when they burn out. If one such bulb burns out, what happens to the others? If such a string operates on 120 V and has 19 remaining identical bulbs, what is then the operating voltage of each?

20.9 If two household light bulbs rated 60 and 100 W are connected in series to household power, which will be brighter? Explain.

20.10 Suppose you are doing a physics lab that asks you to put a resistor into a circuit, but all the resistors supplied have a larger resistance than the requested value. How would you connect the available resistances to attempt to get the smaller value asked for?

20.11 Before World War II, some radios got power through a "resistance cord" that had a significant resistance. A resistance cord reduces the voltage to a desired level for the radio's tubes and the like, and it saves the expense of a transformer. Explain why such cords become warm and waste energy when the radio is on.

20.12 Some light bulbs have three power settings (not including zero), obtained from multiple filaments that are individually switched and wired in parallel. What is the minimum number of filaments needed for three power settings?

20.13 Is every emf a potential difference? Is every potential difference an emf? Explain.

20.14 Explain which battery is doing the charging and which is being charged in Figure 20.40.

20.15 Given a battery, an assortment of resistors, and a variety of voltage and current measuring devices, describe how you would determine the internal resistance of the battery.

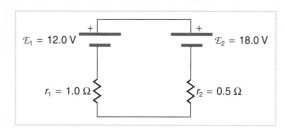

Figure 20.40 Question 14 and Problem 18.

20.16 Two different 12 V automobile batteries on a store shelf are rated at 600 and 850 "cold cranking amps." Which has the smallest internal resistance?

20.17 What are the advantages and disadvantages of connecting batteries in series? In parallel?

20.18 Semitractor trucks use four large 12 V batteries. The starter system requires 24 V, while normal operation of the truck's other electrical components utilizes 12 V. How could the four batteries be connected to produce 24 V? To produce 12 V? Why is 24 V better than 12 V for starting the truck's engine (a very heavy load)?

20.19 Can all of the currents going into the junction in Figure 20.41 be positive? Explain.

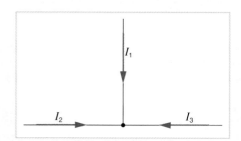

Figure 20.41 Question 19.

20.20 Apply the junction rule to junction b in Figure 20.42. Is any new information gained by applying the junction rule at e?

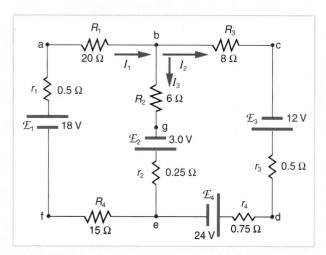

Figure 20.42 Questions 20, 21, 22, and 23 and Problem 34.

20.21 (a) What is the potential difference going from point a to point b in Figure 20.42? (b) What is the potential difference going from c to b? (c) From e to g? (d) From e to d?

20.22 Apply the loop rule to loop afedcba in Figure 20.42.

20.23 Apply the loop rule to loops abgefa and cbgedc in Figure 20.42.

20.24 Apply the loop rule to loop abcdefgha in Figure 20.22.

20.25 Why should you not connect an ammeter directly across a voltage source as shown in Figure 20.43?

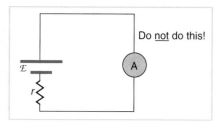

Figure 20.43 Question 25.

20.26 Suppose you are using a multimeter (one designed to measure a range of voltages, currents, and resistances) to measure current in a circuit and you inadvertently leave it in a voltmeter mode. What effect will the meter have on the circuit? What would happen if you were measuring voltage but accidentally put the meter in the ammeter mode?

20.27 Specify the points to which you could connect a voltmeter to measure the following potential differences in Figure 20.44: (a) the potential difference of the voltage source; (b) the potential difference across R_1; (c) across R_2; (d) across R_3; (e) across R_2 and R_3. Note that there may be more than one answer to each part.

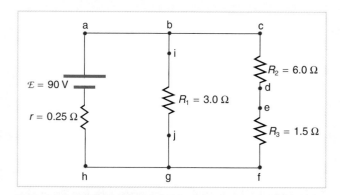

Figure 20.44 Questions 27 and 28.

20.28 To measure currents in Figure 20.44, you would replace a wire between two points with an ammeter. Specify the points between which you would place an ammeter to measure the following: (a) the total current; (b) the current flowing through R_1; (c) through R_2; (c) through R_3. Note that there may be more than one answer to each part.

20.29 Why can a null measurement be more accurate than one using standard voltmeters and ammeters? What factors limit the accuracy of null measurements?

20.30 If a potentiometer is used to measure cell emfs on the order of a few volts, why is it most accurate for the standard emf $\mathcal{E}_s$ to be the same order of magnitude and the resistances to be in the range of a few ohms?

20.31 Regarding the units involved in the relationship $\tau = RC$, verify that the units of resistance times capacitance are time—that is, $\Omega \cdot F = s$.

20.32 Draw two graphs of charge versus time on a capacitor. Draw one for charging an initially uncharged capacitor in series with a resistor, as in the circuit in Figure 20.34, starting from

$t = 0$. Draw the other for discharging a capacitor through a resistor, as in the circuit in Figure 20.35, starting at $t = 0$, with an initial charge Q_0. Show at least two intervals of τ.

20.33 When charging a capacitor, as discussed in conjunction with Figure 20.34, how long does it take for the voltage on the capacitor to reach $\mathcal{E}$? Is this a problem?

20.34 When discharging a capacitor, as discussed in conjunction with Figure 20.35, how long does it take for the voltage on the capacitor to reach zero? Is this a problem?

20.35 Referring to Figure 20.34, draw a graph of potential difference across the resistor versus time, showing at least two intervals of τ. Also draw a graph of current versus time for this situation.

20.36 Referring to Figure 20.35, draw a graph of potential difference across the resistor versus time, showing at least two intervals of τ. Also draw a graph of current versus time for this situation. (Be careful of the signs.)

20.37 In Figure 20.37, does the graph indicate the time constant is shorter for discharging than for charging? Would you expect ionized gas to have low resistance? How would you adjust R to get a longer time between flashes? Would adjusting R affect the discharge time?

20.38 An electronic apparatus may have large capacitors at high voltage in the power supply section, presenting a shock hazard even when the apparatus is switched off. A "bleeder resistor" is therefore placed across such a capacitor, as shown schematically in Figure 20.45, to bleed the charge from it after the apparatus is off. Why must the bleeder resistance be much greater than the effective resistance of the rest of the circuit? How does this affect the time constant for discharging the capacitor?

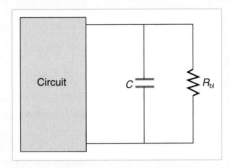

Figure 20.45 A bleeder resistor R_{bl} discharges the capacitor in this electronic device once it is switched off. Question 38.

PROBLEMS

Section 20.1 Resistors in Series and Parallel

Data taken from figures can be assumed to be good to three significant digits.

20.1 (a) What is the resistance of ten 275 Ω resistors connected in series? (b) In parallel?

20.2 (a) What is the resistance of a 100 Ω, a 2.50 kΩ, and a 4.00 kΩ resistor connected in series? (b) In parallel?

20.3 What are the largest and smallest resistances you can obtain by connecting a 36.0, a 50.0, and a 700 Ω resistor together?

• **20.4** What resistance do you need to connect to a 137 Ω resistor to get a total of 53.0 Ω?

• **20.5** Your car's 30.0 W headlight and 2.40 kW starter are ordinarily connected in parallel in a 12.0 V system. What power would one headlight and the starter consume if connected in series to a 12.0 V battery? (Neglect any other resistance in the circuit and any change in resistance in the two devices.)

• **20.6** (a) Given a 48.0 V battery, and 24.0 and 96.0 Ω resistors, find the current and power for each when connected in series. (b) Repeat when the resistances are in parallel.

• **20.7** Referring to Example 20.3 and Figure 20.6, calculate I_3 in the following two different ways: (a) from the known values of I and I_2; (b) using Ohm's law for R_3.

• **20.8** Referring to Example 20.3 and Figure 20.6: (a) Calculate P_3 and note how it compares with P_3 found in Examples 20.1 and 20.2. (b) Find the total power supplied by the source and compare it with the sum of the powers dissipated by the resistors.

⁞ **20.9** Refer to Figure 20.7 and the discussion of lights dimming when a heavy appliance comes on. (a) Given the voltage source is 120 V, the wire resistance is 0.400 Ω, and the bulb is nominally 75.0 W, what power will the bulb dissipate if a total of 30.0 A passes through the wires when the motor comes on? Assume negligible change in bulb resistance. (b) What power is consumed by the motor?

⁞ **20.10** A 240 kV power transmission line carrying 500 A is hung from grounded metal towers by ceramic insulators, each having a 1.00×10^9 Ω resistance. (a) What is the resistance to ground of 100 of these insulators? (b) Calculate the power dissipated by 100 of them. (c) What fraction of the power carried by the line is this?

⁞ **20.11** Show that if two resistors R_1 and R_2 are combined and one is much greater than the other ($R_1 \gg R_2$): (a) Their series resistance is very nearly equal to the greater resistance R_1. (b) Their parallel resistance is very nearly equal to smaller resistance R_2. (These results are mentioned in conjunction with the discussion of voltmeters and ammeters in Section 20.4).

Section 20.2 Emf and Terminal Voltage

20.12 Standard automobile batteries have six lead-acid cells in series, creating a total emf of 12.0 V. What is the emf of an individual lead-acid cell?

20.13 Carbon-zinc dry cells (sometimes referred to as nonalkaline cells) have an emf of 1.54 V, and they are produced as single cells or in various combinations to form other voltages. (a) How many 1.54 V cells are needed to make the common 9 V battery used in many small electronic devices? (b) What is the actual emf of the approximately 9 V battery?

20.14 What is the output voltage of a 3.0000 V lithium cell in a digital wristwatch that draws 0.300 mA, if the cell's internal resistance is 2.00 Ω?

20.15 (a) What is the terminal voltage of a large 1.54 V carbon-zinc dry cell used in a physics lab to supply 2.00 A to a circuit, if the cell's internal resistance is 0.100 Ω? (b) How much electrical power does the cell produce? (c) What power goes to its load?

20.16 What is the internal resistance of an automobile battery that has an emf of 12.0 V and a terminal voltage of 15.0 V while a current of 8.00 A is charging it?

20.17 (a) Find the terminal voltage of a 12.0 V motorcycle battery having a 0.600 Ω internal resistance, if it is being charged by a current of 10.0 A. (b) What is the output voltage of the battery charger?

20.18 (a) Calculate the magnitude and indicate the direction of flow of current in Figure 20.40. (b) Find the terminal voltage of each battery. (Are their values consistent with the direction of current flow?)

20.19 Referring to Example 20.4, part (c), verify that the power dissipated by the 0.500 Ω load is 200 W when calculated using V^2/R or IV, where V is the terminal voltage.

20.20 The hot resistance of a flashlight bulb is 2.30 Ω, and it is run by a 1.58 V alkaline cell having a 0.100 Ω internal resistance. (a) What current flows? (b) Calculate the power supplied to the bulb using I^2R_{bulb}. (c) Is this power the same as calculated using V^2/R_{bulb}?

20.21 The label on a portable radio recommends the use of rechargeable nickel-cadmium cells (nicads), although they have a 1.25 V emf while alkaline cells have a 1.58 V emf. Given that the radio has a 3.20 Ω effective resistance and uses four cells in series, calculate the power delivered to the radio: (a) When using nicad cells each having an internal resistance of 0.0400 Ω. (b) When using alkaline cells each having an internal resistance of 0.200 Ω.

20.22 An automobile starter motor has an equivalent resistance of 0.0500 Ω and is supplied by a 12.0 V battery with a 0.0100 Ω internal resistance. (a) What is the current to the motor? (b) What voltage is applied to it? (c) What power is supplied to the motor? (d) Repeat these calculations for when the battery connections are corroded and add 0.0900 Ω to the circuit. (Significant problems are caused by even small amounts of unwanted resistance in low-voltage, high-current applications.)

20.23 A child's electronic toy is supplied by three 1.58 V alkaline cells having internal resistances of 0.0200 Ω in series with a 1.53 V carbon-zinc dry cell having a 0.100 Ω internal resistance. The load resistance is 10.0 Ω. (a) What current flows? (b) How much power is supplied to the load? (c) What is the internal resistance of the dry cell if it goes bad, resulting in only 0.500 W being supplied to the load?

20.24 (a) What is the internal resistance of a voltage source if its terminal voltage drops by 2.00 V when the current supplied increases by 5.00 A? (b) Can the emf of the voltage source be found with the information supplied?

Section 20.3 Kirchhoff's Rules

20.25 Apply the loop rule to loop abcdefgha in Figure 20.22.

20.26 Apply the loop rule to loop aedcba in Figure 20.22.

20.27 Verify Equation ii in Example 20.5 by substituting the values found for the currents I_1 and I_2.

20.28 Verify Equation iii in Example 20.5 by substituting the values found for the currents I_1 and I_3.

20.29 Apply the junction rule at point a in Figure 20.46.

20.30 Apply the loop rule to loop abcdefghija in Figure 20.46.

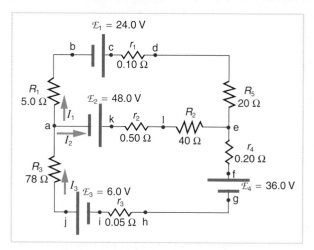

Figure 20.46 Problems 29, 30, 31, and 32.

20.31 Apply the loop rule to loop akledcba in Figure 20.46.

20.32 Find the currents flowing in the circuit in Figure 20.46.

20.33 Solve Example 20.5, but use loop abcdefgha instead of loop abcdea.

20.34 Find the currents flowing in the circuit in Figure 20.42.

Section 20.4 Standard Voltmeters and Ammeters

20.35 What is the sensitivity of the galvanometer (that is, what current gives a full-scale deflection) inside a voltmeter that has a 1.00 MΩ resistance on its 30.0 V scale?

20.36 What is the sensitivity of the galvanometer (that is, what current gives a full-scale deflection) inside a voltmeter that has a 25.0 kΩ resistance on its 100 V scale?

20.37 Find the resistance that must be placed in series with a 25.0 Ω galvanometer having a 50.0 μA sensitivity (the same as the one discussed in the text) to allow it to be used as a voltmeter with a 0.100 V full-scale reading.

20.38 Find the resistance that must be placed in series with a 25.0 Ω galvanometer having a 50.0 μA sensitivity (the same as the one discussed in the text) to allow it to be used as a voltmeter with a 3000 V full-scale reading.

20.39 Find the resistance that must be placed in parallel with a 25.0 Ω galvanometer having a 50.0 μA sensitivity (the same as the one discussed in the text) to allow it to be used as an ammeter with a 10.0 A full-scale reading.

20.40 Find the resistance that must be placed in parallel with a 25.0 Ω galvanometer having a 50.0 μA sensitivity (the same as the one discussed in the text) to allow it to be used as an ammeter with a 300 mA full-scale reading.

20.41 Find the resistance that must be placed in series with a 10.0 Ω galvanometer having a 100 μA sensitivity to allow it to be used as a voltmeter with: (a) A 300 V full-scale reading. (b) A 0.300 V full-scale reading.

20.42 Find the resistance that must be placed in parallel with a 10.0 Ω galvanometer having a 100 μA sensitivity to allow it to be used as an ammeter with: (a) A 20.0 A full-scale reading. (b) A 100 mA full-scale reading.

20.43 Suppose you measure the terminal voltage of a 1.585 V alkaline cell having an internal resistance of 0.100 Ω by placing a

1.00 kΩ voltmeter across its terminals. (a) What current flows? (b) Find the terminal voltage. (c) To see how close the measured terminal voltage is to the emf, calculate their ratio.

• **20.44** Suppose you measure the terminal voltage of a 3.200 V lithium cell having an internal resistance of 5.00 Ω by placing a 1.00 kΩ voltmeter across its terminals. (a) What current flows? (b) Find the terminal voltage. (c) To see how close the measured terminal voltage is to the emf, calculate their ratio.

• **20.45** A certain ammeter has a resistance of 5.00×10^{-5} Ω on its 3.00 A scale and contains a 10.0 Ω galvanometer. What is the sensitivity of the galvanometer?

⁞ **20.46** A 1.00 MΩ voltmeter is placed in parallel with a 75.0 kΩ resistor in a circuit. (a) What is the resistance of the combination? (b) If the voltage across the combination is kept the same as it was across the 75.0 kΩ resistor alone, what is the percent increase in current? (c) If the current through the combination is kept the same as it was through the 75.0 kΩ resistor alone, what is the percentage decrease in voltage?

⁞ **20.47** A 0.0200 Ω ammeter is placed in series with a 10.00 Ω resistor in a circuit. (a) Calculate the resistance of the combination. (b) If the voltage is kept the same across the combination as it was through the 10.00 Ω resistor alone, what is the percent decrease in current? (b) If the current is kept the same through the combination as it was through the 10.00 Ω resistor alone, what is the percent increase in voltage?

Section 20.5 Null Measurements

20.48 What is the emf $\mathcal{E}_x$ of a cell being measured in a potentiometer, if the standard cell's emf is 12.0 V and the potentiometer balances for $R_x = 5.000$ Ω and $R_s = 2.500$ Ω?

20.49 Calculate the emf $\mathcal{E}_x$ of a dry cell for which a potentiometer is balanced when $R_x = 1.200$ Ω, while an alkaline standard cell with an emf of 1.600 V requires $R_s = 1.247$ Ω to balance the potentiometer.

20.50 When an unknown resistance R_x is placed in a Wheatstone bridge, it is possible to balance the bridge by adjusting R_3 to be 2500 Ω. What is R_x if $R_2/R_1 = 0.625$?

20.51 To what value must you adjust R_3 to balance a Wheatstone bridge, if the unknown resistance R_x is 100 Ω, R_1 is 50.0 Ω, and R_2 is 175 Ω?

• **20.52** (a) What is the unknown emf $\mathcal{E}_x$ in a potentiometer that balances when R_x is 10.0 Ω, and balances when R_s is 15.0 Ω for a standard 3.000 V emf? (b) The same emf $\mathcal{E}_x$ is placed in the same potentiometer, which now balances when R_s is 15.0 Ω for a standard emf of 3.100 V. At what resistance R_x will the potentiometer balance?

• **20.53** A potentiometer is set up to use a standard 2.000 V cell and balance when $R_s = 1.000$ Ω. What range of resistances will balance the potentiometer, if you use it to measure the emfs of lithium cells that range from 3.000 to 3.200 V?

• **20.54** Suppose you want to measure resistances in the range from 10.0 Ω to 10.0 kΩ using a Wheatstone bridge that has $R_2/R_1 = 2.000$. Over what range should R_3 be adjustable?

Section 20.6 RC Circuits

20.55 The timing device in an automobile's intermittent wiper system is based on an *RC* time constant and utilizes a 0.500 μF capacitor and a variable resistor. Over what range must *R* be made to vary to achieve time constants from 2.00 to 15.0 s?

20.56 A heart pacemaker fires 72 times a minute, each time a 25.0 nF capacitor is charged (by a battery in series with a resistor) to 0.632 of its full voltage. What is the value of the resistance?

20.57 The duration of a photographic flash is related to an *RC* time constant, which is 0.100 μs for a certain camera. (a) If the resistance of the flash lamp is 0.0400 Ω during discharge, what is the size of the capacitor supplying its energy? (b) What is the time constant for charging the capacitor, if the charging resistance is 800 kΩ?

• **20.58** A 2.00 and a 7.50 μF capacitor can be connected in series or parallel, as can a 25.0 and a 100 kΩ resistor. Calculate the four *RC* time constants possible from connecting the resulting capacitance and resistance in series.

• **20.59** After two time constants, what percentage of the final voltage $\mathcal{E}$ is on an initially uncharged capacitor C, charged through a resistance R?

• **20.60** A 500 Ω resistor, an uncharged 1.50 μF capacitor, and a 6.16 V emf are connected in series. (a) What is the initial current? (b) What is the *RC* time constant? (c) What is the current after one time constant? (d) What is the voltage on the capacitor after one time constant?

⁞ **20.61** Figure 20.47 shows how a bleeder resistor is used to discharge a capacitor after an electronic device is shut off, allowing a person to work on the electronics with less risk of shock. (a) What is the time constant? (b) How long will it take to reduce the voltage on the capacitor to 0.250% (5% of 5%) of its full value once discharge begins? (c) If the capacitor is charged to a voltage V_0 through a 100 Ω resistance, calculate the time it takes to rise to $0.865V_0$ (this is about two time constants).

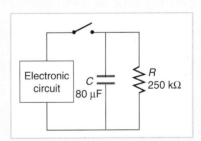

Figure 20.47 Problem 61.

⁞ **20.62** Using the exact exponential treatment, find how much time is required to discharge a 250 μF capacitor through a 500 Ω resistor down to 1.00% of its original voltage.

⁞ **20.63** Using the exact exponential treatment, find how much time is required to charge an initially uncharged 100 pF capacitor through a 75 MΩ resistor to 90% of its final voltage.

INTEGRATED CONCEPTS

Physics is most interesting when applied to general situations involving more than a narrow set of physical principles. These general problems involve physical principles from more than one chapter. Since this chapter is highly applied, however, there are fewer general problems presented than in Chapter 19 and others.

Note: Problem-solving strategies and a worked example that can help you solve integrated concept problems appear in Chapter 19. They can be found with the section of problems labeled

Integrated Concepts at the end of that chapter. Consult the index for the locations of other problem-solving strategies and worked examples for integrated concept problems.

• **20.64** A 12.0 V emf automobile battery has a terminal voltage of 16.0 V when being charged by a current of 10.0 A. (a) What is the battery's internal resistance? (b) What power is dissipated inside the battery? (c) At what rate (in °C/min) will its temperature increase if its mass is 20.0 kg and it has a specific heat of 0.300 kcal/kg · °C, assuming no heat escapes?

• **20.65** If you wish to take a picture of a bullet traveling at 500 m/s, then a very brief flash of light produced by an *RC* discharge through a flash tube can limit blurring. Assuming 1.00 mm of motion during one *RC* constant is acceptable, and given that the flash is driven by a 600 μF capacitor, what is the resistance in the flash tube?

• **20.66** A lamp in a child's Halloween costume flashes based on an *RC* discharge of a capacitor through its resistance. The effective duration of the flash is 0.250 s, during which it produces an average 0.500 W from an average 3.00 V. (a) What energy does it dissipate? (b) How much charge moves through the lamp? (c) Find the capacitance. (d) What is the resistance of the lamp?

• **20.67** A 160 μF capacitor charged to 450 V is discharged through a 31.2 kΩ resistor. (a) Find the time constant. (b) Calculate the temperature increase of the resistor, given its mass is 2.50 g and its specific heat is 0.400 cal/g · °C, noting that most of the heat is retained in the short time of the discharge.

UNREASONABLE RESULTS

The following problems have results that are unreasonable because some premise is unreasonable or because certain of the premises are inconsistent with one another. Physical principles applied correctly then produce unreasonable results. The purpose of these problems is to give practice in assessing whether nature is being accurately described, and if it is not to trace the source of difficulty.

PROBLEM-SOLVING STRATEGY

To determine if an answer is reasonable, and to determine the cause if it is not, do the following.

Step 1. Solve the problem using strategies as outlined in Sections 20.1 and 20.3 and in several other places in the text. See the table of contents to locate problem-solving strategies. Use the format followed in the worked examples in this chapter to solve the problem as usual.

Step 2. Check to see if the answer is reasonable. Is it too large or too small, or does it have the wrong sign, improper units, . . .?

Step 3. If the answer is unreasonable, look for what specifically could cause the identified difficulty. Usually, the manner in which the answer is unreasonable is an indication of the difficulty. For example, a negative terminal voltage may imply an assumption of too great a current or too great an internal resistance.

20.68 Two resistors, one having a resistance of 145 Ω, are connected in parallel to produce a total resistance of 150 Ω. (a) What is the value of the second resistance? (b) What is unreasonable about this result? (c) Which assumptions are unreasonable or inconsistent?

20.69 Two resistors, one having a resistance of 900 kΩ, are connected in series to produce a total resistance of 0.500 MΩ. (a) What is the value of the second resistance? (b) What is unreasonable about this result? (c) Which assumptions are unreasonable or inconsistent?

• **20.70** A 1.58 V alkaline cell with a 0.200 Ω internal resistance is supplying 8.50 A to a load. (a) What is its terminal voltage? (b) What is the value of the load resistance? (c) What is unreasonable about these results? (d) Which assumptions are unreasonable or inconsistent?

• **20.71** (a) What is the internal resistance of a 1.54 V dry cell that supplies 1.00 W of power to a 15.0 Ω bulb? (b) What is unreasonable about this result? (c) Which assumptions are unreasonable or inconsistent?

• **20.72** Consider the circuit in Figure 20.22, and suppose that the emfs are unknown and the currents are given to be $I_1 = 5.00$ A, $I_2 = 3.00$ A, and $I_3 = -2.00$ A. (a) Could you find the emfs? (b) What is wrong with the assumptions?

• **20.73** Suppose you have a 40.0 Ω galvanometer with a 25.0 μA sensitivity. (a) What resistance would you put in series with it to allow it to be used as a voltmeter that has a full-scale deflection for 0.500 mV? (b) What is unreasonable about this result? (c) Which assumptions are responsible?

• **20.74** (a) What resistance would you put in parallel with a 40.0 Ω galvanometer having a 25.0 μA sensitivity to allow it to be used as an ammeter that has a full-scale deflection for 10.0 μA? (b) What is unreasonable about this result? (c) Which assumptions are responsible?

• **20.75** (a) Calculate the capacitance needed to get an *RC* time constant of 1000 s with a 0.100 Ω resistor. (b) What is unreasonable about this result? (c) Which assumptions are responsible?

MAGNETISM

The magnificent spectacle of the Aurora Borealis glows in the northern sky.
Shaped by the earth's magnetic field, this light is produced
by radiation spewed from solar storms.

Figure 21.1 Magnets come in various shapes, sizes, and strengths. All have both a north and a south pole.

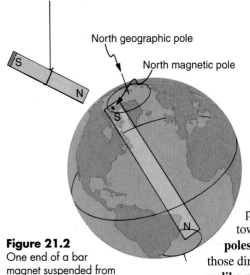

North geographic pole

North magnetic pole

Figure 21.2
One end of a bar magnet suspended from a thread points north. The magnet's two poles are labeled N and S for north seeking and south seeking.

One evening, an Alaskan sticks a note to his refrigerator with a small magnet. Through the kitchen window, the Aurora Borealis glows in the night sky. This grand spectacle is shaped by the same force that holds the note to the refrigerator.

People have been aware of magnets and magnetism for thousands of years. The earliest records date to well before the time of Christ, particularly in a region of Asia Minor called Magnesia (the name of this region is the source of words like *magnetic*). Magnetic rocks found in Magnesia, which is now part of western Turkey, stimulated interest during ancient times. A practical application for magnets was found much later, when they were employed as navigational compasses. The use of magnets in compasses resulted not only in improved long-distance sailing, but also in the names of *north* and *south* being given to the two types of magnetic poles.

Today magnetism plays many important roles in our lives. All electric motors, with uses as diverse as powering refrigerators, starting cars, and moving elevators, contain magnets. Generators, whether producing hydroelectric power or running bicycle lights, use magnetic fields. Recycling facilities employ magnets to separate iron from other refuse. Hundreds of millions of dollars are spent annually on magnetic containment of fusion as a future energy source. Medically, magnetic resonance imaging (MRI) has become an important diagnostic tool, and the use of magnetism to explore brain activity is a subject of contemporary research and development. The list of applications also includes tape recording, detection of inhaled asbestos, and levitation of high-speed trains. Magnetism is used to explain atomic energy levels, cosmic rays, and charged particles trapped in the van Allen belts. Once again, we will find all these disparate phenomena are linked by a small number of underlying physical principles.

21.1 MAGNETS

All magnets attract iron, such as that in a refrigerator door. But magnets may attract or repel other magnets. Experimentation shows that all magnets have two poles. (See Figure 21.1.) If freely suspended, as in Figure 21.2, one pole will point toward the north. The two poles are thus named the **north** and **south magnetic poles** (or more properly, north seeking and south seeking poles, for the attractions in those directions). It is a universal characteristic of all magnets that **like poles repel and unlike poles attract**. (See Figure 21.3.)

Further experimentation shows that it is **impossible to separate north and south poles** in the manner that + and − charges can be separated. (See Figure 21.4.) The fact that

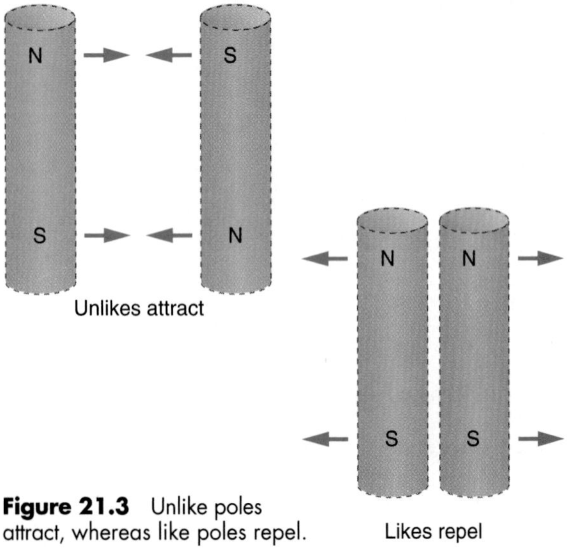

Unlikes attract

Likes repel

Figure 21.3 Unlike poles attract, whereas like poles repel.

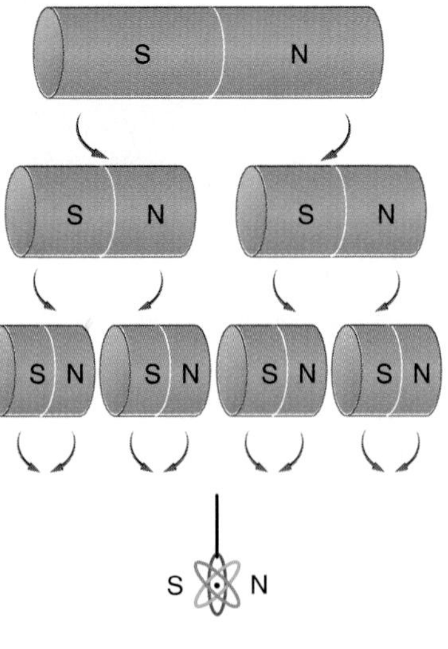

Figure 21.4 North and south poles always occur in pairs. Attempts to separate them result in more pairs of poles. If we continue to split the magnet, we will eventually get down to an iron atom with a north and south pole—these, too, cannot be separated.

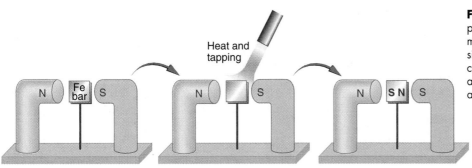

Figure 21.5 An unmagnetized piece of iron is placed between two magnets, heated, and then cooled, or simply tapped when cold. The iron becomes a permanent magnet with poles aligned as shown. Note that there are attractive forces between the magnets.

magnetic poles always occur in pairs of north and south is true from the very large scale—for example, sunspots always occur in pairs that are north and south magnetic poles—all the way down to the very small scale. Magnetic atoms have both a north and a south pole, as do many types of subatomic particles, such as electrons, protons and neutrons.

21.2 FERROMAGNETS AND ELECTROMAGNETS: UNDERLYING CURRENTS

Ferromagnets

Only certain materials, such as iron, cobalt, nickel, and gadolinium, exhibit strong magnetic effects. Such materials are called **ferromagnetic**, after the Latin word for iron, *ferrum*. Other materials exhibit weak magnetic effects, detectable only with sensitive instruments. Not only do ferromagnetic materials respond strongly to magnets (the way iron is attracted to magnets), they can be **magnetized** themselves—that is, be induced to be magnetic, or made into permanent magnets. (See Figure 21.5.)

When a magnet is brought near a previously unmagnetized ferromagnetic material, it causes local magnetization of the material with unlike poles closest, as in Figure 21.5. (This results in the attraction of the previously unmagnetized material to the magnet.) What happens on a microscopic scale is illustrated in Figure 21.6. Regions called **domains** act like small bar magnets. Within domains, the poles of individual atoms are aligned. Each atom acts like a tiny bar magnet. Domains are small and randomly oriented in an unmagnetized ferromagnetic object. In response to an external magnetic field, the domains may grow to millimeter size, aligning themselves as shown in Figure 21.6(b). This induced magnetization can be made permanent if the material is heated and then cooled, or simply tapped in the presence of other magnets.

Conversely, a permanent magnet can be demagnetized by hard blows or by heating it in the absence of another magnet. Increased thermal motion at higher temperature can disrupt and randomize the orientation and size of the domains. There is a well-defined temperature for ferromagnetic materials, called the **Curie temperature**, above which they cannot be magnetized. The Curie temperature for iron is 1043 K, which is well above room temperature. There are several elements and alloys that have Curie temperatures much lower than room temperature and are ferromagnetic only below those temperatures.

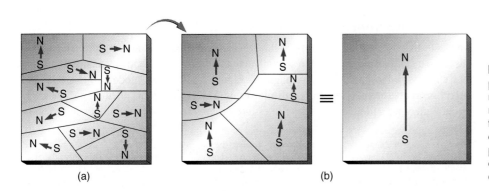

(a)   (b)

Figure 21.6 (a) An unmagnetized piece of iron (or other ferromagnetic material) has randomly oriented domains. (b) When magnetized by an external field, the domains show greater alignment, and some grow at the expense of others. Individual atoms are aligned within domains; each atom acts like a tiny bar magnet.

Figure 21.7 Iron filings near (a) a current-carrying coil and (b) a magnet act like tiny compass needles, showing the shape of their fields. Their response to a current-carrying coil and a permanent magnet are seen to be very similar, especially near the ends of the coil and magnet.

 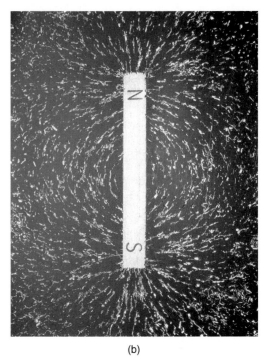

(a)   (b)

Figure 21.8 An electromagnet with a ferromagnetic core can produce very strong magnetic effects. Alignment of domains in the core produces a magnet, the poles of which are aligned with the electromagnet.

Electromagnets

Early in the 19th century, it was discovered that electrical currents cause magnetic effects. The first significant observation was by the Danish scientist Hans Christian Oersted (1777–1851), who found that a compass needle was deflected by a current-carrying wire. This was the first solid evidence that the movement of charges had any connection with magnets. **Electromagnetism** is the use of electric current to make magnets, called **electromagnets**. Electromagnets are employed for everything from a wrecking yard crane that lifts scrapped cars to controlling the beam of a 90 km circumference particle accelerator.

Figure 21.7 shows that the response of iron filings to a current-carrying coil and to a permanent bar magnet. The patterns are similar. In fact, electromagnets and ferromagnets have the same basic characteristics—for example, they have north and south poles that cannot be separated and for which like poles repel and unlike poles attract.

Combining a ferromagnet with an electromagnet can produce particularly strong magnetic effects. (See Figure 21.8.) Whenever strong magnetic effects are needed, such as lifting scrap metal, or in particle accelerators, electromagnets are enhanced by ferromagnetic materials. Limits to how strong the magnets can be made are imposed by coil resistance (it will overheat and melt at sufficiently high current), and so superconducting magnets may be employed. These are still limited, because superconducting properties are destroyed by too great a magnetic field.

Figure 21.9 shows a few uses of combinations of electromagnets and ferromagnets. Ferromagnetic materials can act as memory devices, because the direction of polarization can be reversed or erased. Magnetic information storage on audiotapes, videotapes, and computer disks are among the most common applications.

Current: The Source of All Magnetism

An electromagnet creates magnetism with an electric current. In later sections we explore this more quantitatively, finding the strength and direction of magnetic fields created by various currents. But what about ferromagnets? Figure 21.10 shows models of how **electric currents create magnetism at the submicroscopic level.**[*] Currents, including those associated with other submicroscopic particles like protons and neutrons, allow us to explain

---

[*] As we shall see in Chapter 27, we cannot directly observe the paths of individual electrons about atoms, and so a model or visual image, consistent with all direct observations, is made. We *can* directly observe the electron's orbital angular momentum, its spin momentum, and subsequent magnetic moments, all of which are explained with electric current creating subatomic magnetism.

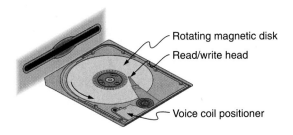

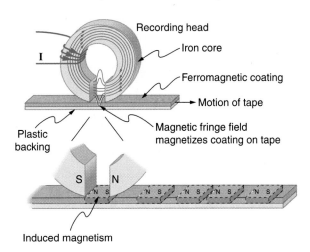

Figure 21.9 An electromagnet induces regions of permanent magnetism on a floppy disk coated with a ferromagnetic material. The information stored here is digital (a region is either magnetic or not); in other applications, it can be analog (with a varying strength), such as on audiotapes.

ferromagnetism and all other magnetic effects. Ferromagnetism, for example, results from an internal cooperative alignment of electron spins, possible in some materials but not in others.

Crucial to the statement that **electric current is the source of all magnetism** is the fact that it is impossible to separate north and south magnetic poles. (This is far different from the case of positive and negative charges, which are easily separated.) A current loop always produces a magnetic dipole—that is, a magnetic field that acts like a north and south pole pair. Since isolated north and south magnetic poles, called **magnetic monopoles**, are not observed, currents are used to explain all magnetic effects. If magnetic monopoles did exist, then we would have to modify this underlying connection that all magnetism is due to electrical current. There is no known reason that magnetic monopoles should not exist—they are simply never observed—and so searches at the subnuclear level continue. If they do *not* exist, we would like to find out why not. If they *do* exist, we would like to see evidence of them.

21.3 MAGNETIC FIELDS AND MAGNETIC FIELD LINES

Einstein is said to have been fascinated by a compass as a child, perhaps musing on how the needle felt a force without direct physical contact. His ability to think deeply and clearly about action at a distance, particularly for gravitational, electric, and magnetic forces, later enabled him to create his revolutionary theory of relativity. Since magnetic forces act at a distance, we define a **magnetic field** to represent magnetic forces. The pictorial representation of **magnetic field lines** is very useful in visualizing the strength and direction of the magnetic field. As shown in Figure 21.11, the **direction of magnetic field**

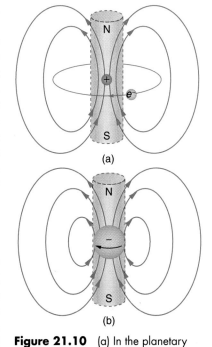

Figure 21.10 (a) In the planetary model of the atom, an electron orbits a nucleus, forming a closed current loop and creating a magnetic field with a north and south pole. (b) Electrons have spin and can be crudely pictured as rotating charge, forming a current that produces a magnetic field with a north and south pole. Neither the planetary model nor the image of a spinning electron is completely consistent with modern physics, as we shall explore in later chapters.

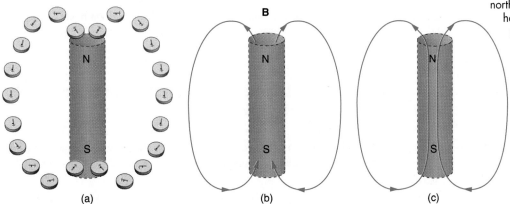

Figure 21.11 Magnetic field lines are defined to have the direction that a small compass points when placed at a location. (a) If small compasses are used to map the magnetic field around a bar magnet, they will point in the directions shown. (b) Connecting the arrows gives continuous magnetic field lines. The strength of the field is proportional to the closeness (or density) of the lines. (c) If the interior of the magnet could be probed, the field lines would be found to form continuous closed loops.

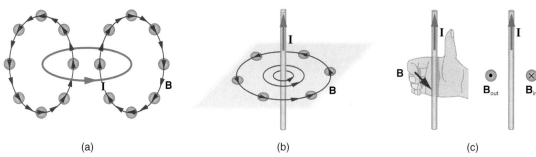

(a)   (b)   (c)

Figure 21.12 Small compasses could be used to map the fields shown here. (a) The magnetic field of a circular current loop is similar to that of a bar magnet. (b) A long straight wire creates a field with circular loops. (c) When the wire is in the plane of the paper, the field is perpendicular to the paper. Note the symbols used for field in (like the tail of an arrow) and field out (like the tip of an arrow).

CONNECTIONS

Concept of a Field

A field surrounds any object that can act on another at a distance without apparent physical connection. The field represents the object creating it, and it is discussed here for the magnetic field. See also Sections 17.4 and 17.5 for a discussion of the electric field. Section 4.6 has a general discussion of how fields relate to forces.

lines is defined to be the direction that the north end of a compass needle points. The magnetic field is traditionally called the ***B*-field**.

Small compasses used to test the magnetic field will not disturb it. (This is analogous to the way we tested electric fields with a small test charge. In both cases, the fields represent only the object creating them and not the probe testing them.) Figure 21.12 shows how the magnetic field appears for a current loop and a long straight wire, as could be explored with small compasses. A small compass placed in these fields will align itself parallel to the field line at its location, with its north pole pointing in the direction of **B**. Note the symbols used for field into and out of the paper.

Extensive exploration of magnetic fields has revealed a number of hard and fast rules. We use magnetic field lines to represent the field (the lines are a pictorial tool, not a physical entity in themselves). The **properties of magnetic field lines** can be summarized by these rules:

1. The direction of the magnetic field is tangent to the field line at any point in space. A small compass will point in the direction of the field line.
2. The strength of the field is proportional to the closeness of the lines. It is exactly proportional to the number of lines per unit area perpendicular to the lines (called the areal density).
3. Magnetic field lines can never cross, meaning the field is unique at any point in space.
4. Magnetic field lines are continuous, forming closed loops without beginning or end.

The last property is related to the fact that north and south poles cannot be separated. It is a distinct difference from electric field lines, which begin and end on positive and negative charges. If magnetic monopoles existed, then magnetic field lines would begin and end on them.

CONNECTIONS

Charges and Magnets

There is no magnetic force on static charges. But there is a magnetic force on moving charge. When charges are stationary, their electric fields do not affect magnets. But, when charges move, they create magnetic fields that exert forces on other magnets. When there is relative motion, a connection between electric and magnetic fields emerges—each affects the other. We shall explore this connection further in later chapters.

21.4 MAGNETIC FIELD STRENGTH *B*: FORCE ON A MOVING CHARGE IN A MAGNETIC FIELD

What is the mechanism by which one magnet exerts a force on another? The answer is related to the fact that all magnetism is caused by current, the flow of charge. *Magnetic fields exert forces on moving charges*, and so they exert forces on other magnets, all of which have moving charges.

Right Hand Rule-1

The magnetic force on a moving charge is one of the most fundamental known. Magnetic force is as fundamental as the electrostatic or Coulomb force. Magnetic force is more complex, in both the number of factors that affect it and in its direction, than the relatively simple Coulomb force. The *direction* of the magnetic force **F** is perpendicular to the plane formed by **v** and **B**, as determined by **right hand rule–1** (or RHR-1), which is illustrated in Figure 21.13. RHR-1 states that to determine the direction of the magnetic force on a positive moving charge, you point the thumb of the right hand in the direction of **v**, the fingers in the direction of **B**, and a perpendicular to the palm points in the direction of **F**. One way to remember this is that there is one velocity, and so the thumb represents it. There are many field lines, and so the fingers represent them. The force is in the direction you would push with your palm. The force on a negative charge is in exactly the opposite direction to that on a positive charge. The **magnitude of the magnetic force** *F* on a charge *q* moving at a speed *v* in a magnetic field of strength *B* is given by

$$F = qvB \sin \theta \qquad (21.1)$$

where θ is the angle between the directions of **v** and **B**. This force is often called the **Lorentz force**. The SI unit for magnetic field strength *B* is called the **tesla** (T) after the eccentric but brilliant inventor Nikola Tesla (1856–1943). To determine how the tesla relates to other SI units, we solve Equation 21.1 for *B*:

$$B = \frac{F}{qv \sin \theta}$$

Since $\sin \theta$ is unitless, the tesla is

$$1 \text{ T} = \frac{1 \text{ N}}{\text{C} \cdot \text{m/s}} = \frac{1 \text{ N}}{\text{A} \cdot \text{m}} \qquad (21.2)$$

noting that C/s = A. Another smaller unit, called the **gauss** (G), where $1 \text{ G} = 10^{-4} \text{ T}$, is sometimes used. The strongest permanent magnets have fields near 2 T; superconducting electromagnets may attain 10 T or more. The earth's magnetic field on its surface is only about 5×10^{-5} T, or 0.5 G.

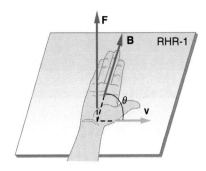

$F = qvB \sin \theta$

F ⊥ plane of **v** and **B**

Figure 21.13 Magnetic fields exert forces on moving charges. This force is one of the most basic known. The direction of the magnetic force on a moving charge is perpendicular to the plane formed by **v** and **B** and follows right hand rule–1 (RHR-1) as shown. The magnitude of the force is proportional to *q*, *v*, *B*, and the sine of the angle between **v** and **B**.

EXAMPLE 21.1 MAGNETIC FORCE DUE TO THE EARTH'S FIELD IS USUALLY SMALL

With the exception of compasses, you seldom see or personally experience forces due to the earth's small magnetic field. To illustrate this, suppose that in a physics lab you rub a glass rod with silk, placing a 20 nC positive charge on it. Calculate the force on the rod due to the earth's magnetic field, if you throw it with a horizontal velocity of 10 m/s due west in a place where the earth's field is due north parallel to the ground. (The direction of the force is determined with right hand rule–1 as shown in Figure 21.14.)

Strategy We are given the charge, its velocity, and the magnetic field strength and direction. We can thus use Equation 21.1 to find the force.

Solution Equation 21.1 states that magnetic force is

$$F = qvB \sin \theta$$

We see that $\sin \theta = 1$, since the angle between the velocity and the direction of the field is 90°. Entering the other given quantities yields

$$F = (20 \times 10^{-9} \text{ C})(10 \text{ m/s})(5 \times 10^{-5} \text{ T})$$
$$= 1 \times 10^{-11} (\text{C} \cdot \text{m/s}) \frac{\text{N}}{\text{C} \cdot \text{m/s}} = 1 \times 10^{-11} \text{ N}$$

Discussion This force is completely negligible on any macroscopic object, consistent with experience. (It is calculated to only one digit, since the earth's field varies with location and is given to only one digit.) The earth's magnetic field, however, does produce very important effects, particularly on submicroscopic particles. Some of these are explored in the next section.

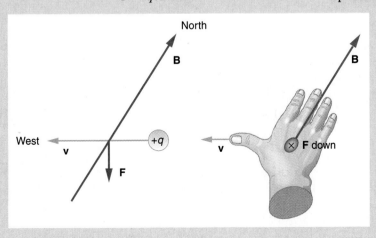

Figure 21.14 A positively charged object moving due west in a region where the earth's field is due north experiences a force that is straight down as shown. A negative charge moving in the same direction would feel a straight upward force.

Figure 21.15 Trails of bubbles are created by high-energy charged particles moving through the super-heated liquid hydrogen in this bubble chamber. There is a strong magnetic field perpendicular to the page that causes the curved paths of the particles. The radius of the path can be used to find the mass, charge, and energy of the particle.

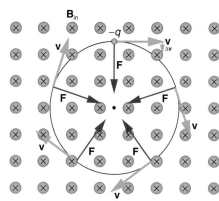

Figure 21.16 A negatively charged particle moves in the plane of the page in a region where the magnetic field is perpendicular into the page (represented by ⊗s—like the tails of arrows). The magnetic force is perpendicular to the velocity, and so velocity changes in direction but not magnitude. Uniform circular motion results.

21.5 FORCE ON A MOVING CHARGE IN A MAGNETIC FIELD: EXAMPLES AND APPLICATIONS

Magnetic force can cause a charged particle to move in a circular or spiral path. Cosmic rays are energetic charged particles in outer space, some of which approach the earth. They can be forced into spiral paths by the earth's magnetic field. Protons in giant accelerators are kept in a circular path by magnetic force. The bubble chamber photograph in Figure 21.15 shows charged particles moving in such curved paths. The curved paths of charged particles in magnetic fields are the basis of a number of phenomena and can even be used analytically, such as in a mass spectrometer.

How does the magnetic force cause circular motion? Magnetic force is always perpendicular to velocity, so that it does no work on the charged particle. The particle's kinetic energy and speed thus remain constant. The direction of motion is affected, but not the speed. This is typical of uniform circular motion (first introduced in Chapter 8). The simplest case occurs when a charged particle moves perpendicular to a uniform B-field, such as shown in Figure 21.16.* Here the magnetic force supplies the centripetal force $F_c = mv^2/r$. Noting that $\sin \theta = 1$, we see that $F = qvB$. Since the magnetic force F equals the centripetal force F_c, we have

$$qvB = \frac{mv^2}{r}$$

Solving for r yields

$$r = \frac{mv}{qB} \qquad \qquad \textbf{(21.3)}$$

Here r is the radius of curvature of the path of a charged particle with mass m and charge q, moving at a speed v perpendicular to a magnetic field of strength B. (If the velocity is not perpendicular to the magnetic field, then v is the component of the velocity perpendicular to the field. The component of the velocity parallel to the field is unaffected, since the magnetic force is zero for motion parallel to the field. This produces a spiral motion rather than a circular one.)

EXAMPLE 21.2 DON'T TRY THIS AT HOME

A magnet brought near a TV screen (or computer monitor), as in Figure 21.17, severely distorts its picture by altering the path of the electrons that make its phosphors glow. (*This permanently magnetizes and ruins the TV.*) To illustrate this, calculate the radius of curvature of the path of an electron having a velocity of 6.00×10^7 m/s (corresponding to the accelerating voltage of about 10.0 kV used in some TVs) perpendicular to a magnetic field of strength $B = 0.500$ T (obtainable with permanent magnets).

Strategy We can find the radius of curvature r directly from Equation 21.3, since all other quantities in it are given or known.

Solution Using known values for the mass and charge of
(continued)

---

*If this takes place in a vacuum, the magnetic field dominates the motion. There are many such instances.

(continued)

an electron, and entering them and the given values of v and   or
B into Equation 21.3 gives

$$r = \frac{mv}{qB} = \frac{(9.11 \times 10^{-31}\text{ kg})(6.00 \times 10^{7}\text{ m/s})}{(1.60 \times 10^{-19}\text{ C})(0.500\text{ T})}$$

$$= 6.83 \times 10^{-4}\text{ m}$$

$$r = 0.683 \text{ mm}$$

Discussion The small radius indicates a large effect. The electrons in the TV tube are made to move in very tight circles, greatly altering their paths and distorting the image.

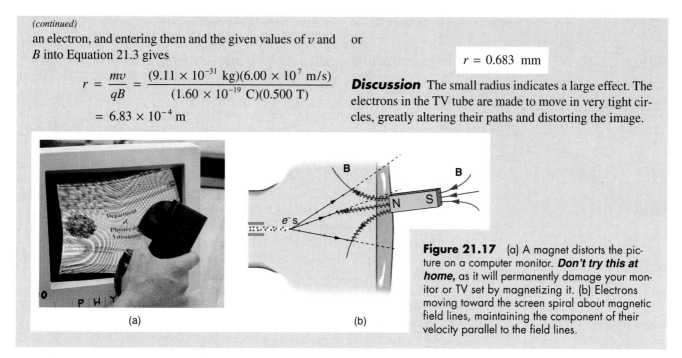

(a)   (b)

Figure 21.17 (a) A magnet distorts the picture on a computer monitor. **Don't try this at home,** as it will permanently damage your monitor or TV set by magnetizing it. (b) Electrons moving toward the screen spiral about magnetic field lines, maintaining the component of their velocity parallel to the field lines.

Figure 21.17(b) shows how electrons not moving perpendicular to magnetic field lines follow the field lines. The component of velocity parallel to the lines is unaffected, and so the charges spiral along the field lines. If field strength increases in the direction of motion, the field will exert a force to slow the charges, forming a kind of magnetic mirror. (See Figure 21.18.)

The properties of charged particles in magnetic fields are related to such things as the Aurora Borealis and particle accelerators. *Charged particles approaching magnetic field lines may get trapped in spiral orbits about the lines rather than crossing them,* as seen above. Some cosmic rays, for example, follow the earth's magnetic field lines, entering the atmosphere near the magnetic poles and causing the Aurora Borealis in the northern hemisphere. This glow of energized atoms and molecules is seen in Figure 21.19. Those particles that approach middle latitudes must cross magnetic field lines, and many are prevented from penetrating the atmosphere. Cosmic rays are a component of background radiation, which is consequently higher at the poles than at the equator.

Some incoming charged particles are trapped in the earth's magnetic field, forming two belts above the atmosphere known as the van Allen belts after their discoverer, an American physicist. (See Figure 21.20.) Particles trapped in these belts form radiation (similar to nuclear radiation) so intense that manned space flights avoid them, and satellites with sensitive electronics are kept out of them. In the few minutes it took lunar missions to cross the van Allen belts, astronauts received radiation doses more than twice the

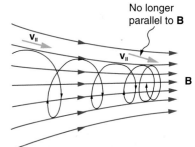

Figure 21.18 When a charged particle moves along a magnetic field line into a region where the field becomes stronger, the particle experiences a force that reduces the component of velocity parallel to the field. This force slows the motion along the field line and here reverses it, forming a magnetic mirror.

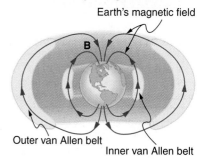

Figure 21.20 The van Allen belts are two regions in which energetic charged particles are trapped in the earth's magnetic field. One belt lies about 300 km above the surface, the other about 16,000 km. Charged particles in these belts migrate along magnetic field lines and are partially reflected away from the poles by the stronger fields there. The charged particles that enter the atmosphere are replenished by the sun and sources in deep outer space.

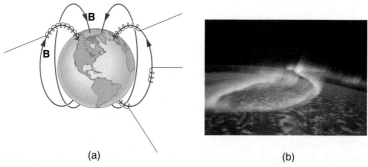

(a)   (b)

Figure 21.19 (a) Energetic electrons and protons, components of cosmic rays from the sun and deep outer space, often follow the earth's magnetic field lines rather than cross them. (b) The Aurora Borealis is one result. There is also greater background radiation near the poles than at the equator, where the cosmic rays must cross field lines to enter the atmosphere.

Figure 21.21   The Fermilab facility has a large particle accelerator that employs magnetic fields (magnets seen here in blue) to contain and direct its beam. This and other accelerators have been in use for several decades and have allowed us to discover some of the laws underlying all matter.

allowed annual exposure for radiation workers. Other planets have similar belts, especially those having strong magnetic fields like Jupiter.

Back on earth, we have devices that employ magnetic fields to contain charged particles. Among them are the giant particle accelerators that have been used to explore the substructure of matter. (See Figure 21.21.) Magnetic fields not only control the direction of the charged particles, they also are used to focus particles into beams and overcome the repulsion of like charges in these beams.

Controlled thermonuclear fusion (like that contained by gravity in the sun) is a hope for a future clean energy source. One of the most promising devices is the *tokamak*, which uses magnetic fields to contain and direct the reactive charged particles. (See Figure 21.22.) Less exotic, but more immediately practical, amplifiers in microwave ovens use a magnetic field to contain oscillating electrons. These oscillating electrons generate the microwaves sent into the oven.

Mass spectrometers have a variety of designs, and many use magnetic fields to measure mass. The curvature of a charged particle's path in the field is related to mass, as seen above, and is measured to obtain mass information. Historically, such techniques were employed in the first direct observations of electron charge and mass.

21.6   THE HALL EFFECT

We have seen effects of a magnetic field on free charges. The magnetic field also affects charges moving in a conductor. One result is the Hall effect, which has important implications and applications.

Figure 21.23 shows what happens to charges moving through a conductor in a magnetic field. The field is perpendicular to the electron drift velocity and to the width of the conductor. Note that conventional current is to the right in both parts of the figure. In part (a), electrons carry the current and move to the left. In part (b), positive charges carry the current and move to the right. Moving electrons feel a magnetic force toward one side of the conductor, leaving a net positive charge on the other side. This separation of charge *creates a voltage* $\mathcal{E}$, known as the **Hall emf**, *across* the conductor. The creation

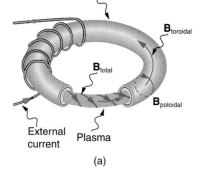

Toroidal vacuum chamber

$\mathbf{B}_{toroidal}$

$\mathbf{B}_{total}$

$\mathbf{B}_{poloidal}$

External current   Plasma

(a)

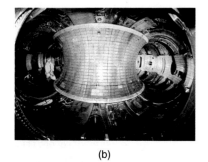

(b)

Figure 21.22   (a) Tokamaks such as this one are being improved with the goal of economical production of energy by nuclear fusion. Magnetic fields in the doughnut-shaped device contain and direct the reactive charged particles. (b) Interior view.

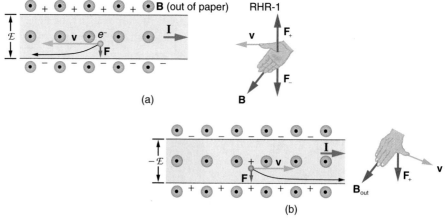

(a)

(b)

Figure 21.23   The Hall effect. (a) Electrons move to the left in this flat conductor (conventional current to the right). The magnetic field is directly out of the page, represented by circled dots; it exerts a force on the moving charges, causing a voltage $\mathcal{E}$, the Hall emf, across the conductor. (b) Positive charges moving to the right (conventional current also to the right) are moved to the side, creating a Hall emf of the opposite sign, $-\mathcal{E}$. Thus, if the direction of the field and current are known, the sign of the charge carriers can be determined from the Hall effect.

of a voltage *across* a current-carrying conductor by a magnetic field is known as the **Hall effect** after the American physicist who discovered it in 1879.

One very important use of the Hall effect is to determine whether positive or negative charges carry a current. Note that in Figure 21.23(b), where positive charges carry the current, the Hall emf has the sign opposite to when negative charges carry the current. Historically, the Hall effect was used to show that electrons carry current in metals, and it also shows that positive charges carry current in some semiconductors. The Hall effect is used today as a research tool to probe the movement of charges, their drift velocities and densities, and so on, in materials. In 1980 it was discovered that the Hall effect is quantized, an example of quantum behavior in a macroscopic object.

The Hall effect has other uses that range from determination of blood flow rate to precision measurement of magnetic field strength. To examine these quantitatively, we need an expression for the Hall emf $\mathcal{E}$ across a conductor. Consider the balance of forces on a moving charge in a situation where B, v, and ℓ are mutually perpendicular, such as shown in Figure 21.24. Although the magnetic force moves negative charges to one side, they cannot build up without limit. The electric field caused by their separation opposes the magnetic force, $F = qvB$, and the electric force, $F_e = qE$, eventually grows to equal it. That is,

$$qE = qvB$$

or

$$E = vB$$

Note that the electric field E is uniform across the conductor because the magnetic field B is uniform, as is the conductor. For a uniform electric field, Equation 18.5 gives the relationship between electric field and voltage. Here it is $E = \mathcal{E}/\ell$, where ℓ is the width of the conductor and $\mathcal{E}$ is the Hall emf. Entering this into the last expression gives

$$\frac{\mathcal{E}}{\ell} = vB$$

Solving this for the Hall emf yields

$$\mathcal{E} = B\ell v \qquad (B, v, \text{ and } \ell \text{ perpendicular}) \qquad \textbf{(21.4)}$$

where $\mathcal{E}$ is the Hall effect voltage across a conductor of width ℓ through which charges move at a speed v.

One of the most common uses of the Hall effect is in the measurement of magnetic field strength B. Such devices, called *Hall probes*, can be made very small, allowing fine position mapping. Hall probes can also be made very accurate, usually accomplished by careful calibration. Another application of the Hall effect is to measure fluid flow in any fluid that has free charges (most do). (See Figure 21.25.) A magnetic field applied perpendicular to the flow direction produces a Hall emf $\mathcal{E}$ as shown. Note that the sign of $\mathcal{E}$ depends not on the sign of the charges, but only on the directions of B and v. The magnitude of the Hall emf is $\mathcal{E} = B\ell v$, where ℓ is the fluid diameter, so that the average velocity v can be determined from $\mathcal{E}$ providing the other factors are known.

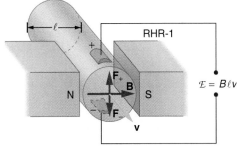

Figure 21.24 The Hall emf $\mathcal{E}$ creates an electric force that balances the magnetic force on the moving charges. The magnetic force creates the separation of charge, which builds up until it is balanced by the electric force, an equilibrium that is quickly reached.

Figure 21.25 The Hall effect can be used to measure fluid flow in any fluid having free charges, such as blood. The Hall emf $\mathcal{E}$ is measured across the tube perpendicular to the applied magnetic field and is proportional to the average velocity v. See Example 21.3.

Medical Application

EXAMPLE 21.3 HALL EFFECT FOR BLOOD FLOW

A Hall effect flow probe is placed on an artery, applying a 0.100 T magnetic field across it, in a setup similar to that in Figure 21.25. What is the Hall emf, given the vessel's inside diameter is 4.00 mm and the average blood velocity is 20.0 cm/s?

Strategy Since B, v, and ℓ are mutually perpendicular, Equation 21.4 can be used to find $\mathcal{E}$.

Solution Entering the given values for B, ℓ, and v into Equation 21.4 gives

$$\mathcal{E} = B\ell v = (0.100 \text{ T})(4.00 \times 10^{-3} \text{ m})(0.200 \text{ m/s})$$
$$= 80.0 \text{ μV}$$

Discussion This is the average voltage output. Instantaneous voltage varies with pulsating blood flow. The voltage is small in this type of measurement. $\mathcal{E}$ is particularly difficult to measure, because there are voltages associated with heart action (ECG voltages) that are on the order of millivolts. In practice, this difficulty is overcome by applying an AC magnetic field so that the Hall emf is AC with the same frequency. An amplifier can be very selective in picking out only the appropriate frequency, eliminating signals and noise at other frequencies.

Figure 21.26 The magnetic field exerts a force on a current-carrying wire in a direction given by right hand rule–1 (the same direction as that on the individual moving charges). This force can easily be large enough to move the wire, since typical currents consist of very large numbers of moving charges.

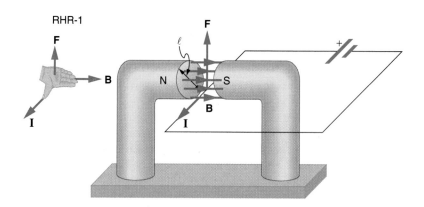

21.7 MAGNETIC FORCE ON A CURRENT-CARRYING CONDUCTOR

Since charges ordinarily cannot escape a conductor, the magnetic force on charges moving in a conductor is transmitted to the conductor itself. (See Figure 21.26.)

We can derive an expression for the magnetic force on a current by taking a sum of the magnetic forces on individual charges.* The force on an individual charge moving at the drift velocity v_d is given by Equation 21.1 to be $F = qv_dB \sin \theta$. Taking B to be uniform over a length of wire ℓ and zero elsewhere, the total magnetic force on the wire is then $F = (qv_dB \sin \theta)(N)$, where N is the number of charge carriers in the section of wire of length ℓ. Now $N = nV$, where n is the number of charge carriers per unit volume and V is the volume of wire in the field. Noting that $V = A\ell$, where A is the cross-sectional area of the wire, the force on the wire is $F = (qv_dB \sin \theta)(nA\ell)$. Gathering terms,

$$F = (nqAv_d)\ell B \sin \theta$$

In Section 19.1, where drift velocity was discussed, the quantity in parentheses was found to be the current I (see Equation 19.3). Thus,

$$F = I\ell B \sin \theta \qquad (21.5)$$

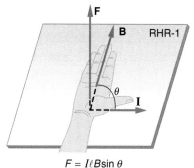

$F = I\ell B\sin\theta$

$F \perp$ plane of I and B

Figure 21.27 The force on a current-carrying wire in a magnetic field is $F = I\ell B \sin \theta$. Its direction is given by RHR-1.

is the **magnetic force on a length ℓ of wire carrying a current I in a uniform magnetic field B** as shown in Figure 21.27. If we divide both sides of this expression by ℓ, we find that the magnetic force per unit length of wire in a uniform field is $F/\ell = IB \sin \theta$. The direction of this force is given by RHR-1, with the thumb in the direction of the current I. Then with the fingers in the direction of B, a perpendicular to the palm points in the direction of F, as in Figure 21.27.

EXAMPLE 21.4 FORCE ON A WIRE IN A STRONG MAGNETIC FIELD

Calculate the force on the wire shown in Figure 21.26, given $B = 1.50$ T, $\ell = 5.00$ cm, and $I = 20.0$ A.

Strategy The force can be found with the given information by using Equation 21.5 and noting that the angle θ between I and B is 90°, so that $\sin \theta = 1$.

Solution Entering given values into Equation 21.5 yields

$$F = I\ell B \sin \theta = (20.0 \text{ A})(0.0500 \text{ m})(1.50 \text{ T})(1)$$

The units for tesla are 1 T = N/(A·m); thus,

$$\boxed{F = 1.50 \text{ N}}$$

Discussion This large magnetic field creates a significant force on a small length of wire.

Magnetic force on current-carrying conductors is used to convert electric energy to work. (Motors are a prime example—they employ loops of wire and are considered in the next section.) Magneto-hydrodynamics (MHD) is the technical name given to a clever application where magnetic force pumps fluids without moving mechanical

---

*The forces add because they are in the same direction.

parts. (See Figure 21.28.) A strong magnetic field is applied across a tube and a current is passed through the fluid at right angles to the field, resulting in a force on the fluid parallel to the tube axis as shown. The absence of moving parts makes this attractive for moving a hot, chemically active substance, such as the liquid sodium employed in some nuclear reactors. Experimental artificial hearts are testing this technique for pumping blood, perhaps circumventing the adverse effects of mechanical pumps. (Cell membranes, though, are affected by the large fields needed in MHD, delaying its practical application in humans.) MHD propulsion for nuclear submarines has been proposed, because it could be considerably quieter than conventional propeller drives. The deterrent value of nuclear submarines is based on their ability to hide and survive a first or second nuclear strike. As we slowly disassemble our nuclear weapons arsenals, the submarine branch will be the last to be decommissioned because of this ability. (See Figure 21.29.) Existing MHD drives are heavy and inefficient—much development work is needed.

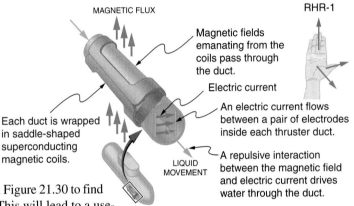

Figure 21.28 Magneto-hydrodynamics. The magnetic force on the current passed through this fluid can be used as a nonmechanical pump.

21.8 TORQUE ON A CURRENT LOOP: MOTORS AND METERS

Motors are the most common application of magnetic force on current-carrying wires. Motors have loops of wire in a magnetic field. When current is passed through the loops, the magnetic field exerts torque on the loops, which rotates a shaft. Electrical energy is converted to mechanical work in the process. (See Figure 21.30.)

Let us examine the force on each segment of the loop in Figure 21.30 to find the torques produced about the axis of the vertical shaft. (This will lead to a useful equation for the torque on the loop.) We take the field to be uniform over the rectangular loop, which has width w and height ℓ. First, we note that the forces on the top and bottom segments are vertical and, therefore, parallel to the shaft, producing no torque. Those vertical forces are equal in magnitude and opposite in direction, so that they also produce no net force on the loop. Figure 21.31 shows views of the loop from above. Torque was defined in Section 5.2 to be $\tau = rF \sin \phi$, where F is the force, r is the distance from the pivot that the force is applied, and ϕ is the angle between r and F. As seen in Figure 21.31(a), right hand rule–1 gives the forces on the sides to be equal in magnitude and opposite in direction, so that the net force is again zero. But each force produces a clockwise torque. Since $r = w/2$, the torque on each vertical segment is $(w/2)F \sin \phi$, and the two add to give a total torque

$$\tau = \frac{w}{2}F \sin \phi + \frac{w}{2}F \sin \phi = wF \sin \phi$$

Now, each vertical segment has a length ℓ that is perpendicular to B, so that the force on each is $F = I\ell B$. Entering F into the expression for torque yields

$$\tau = wI\ell B \sin \phi$$

If we have a multiple loop of N turns, we get N times the torque of one loop. Finally, note that the area of the loop is $A = w\ell$; the expression for the torque becomes

$$\tau = NIAB \sin \phi \qquad \qquad \textbf{(21.6)}$$

This is the torque on a current-carrying loop in a uniform magnetic field. Equation 21.6 can be shown to valid for a loop of any shape. The loop carries a current I, has N turns, each of area A, and the perpendicular to the loop makes an angle ϕ with the field B. The net force on the loop is zero.

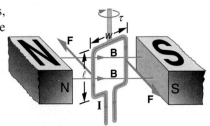

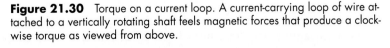

Figure 21.30 Torque on a current loop. A current-carrying loop of wire attached to a vertically rotating shaft feels magnetic forces that produce a clockwise torque as viewed from above.

MAGNETIC FLUX

RHR-1

Magnetic fields emanating from the coils pass through the duct.

Electric current

An electric current flows between a pair of electrodes inside each thruster duct.

Each duct is wrapped in saddle-shaped superconducting magnetic coils.

LIQUID MOVEMENT

A repulsive interaction between the magnetic field and electric current drives water through the duct.

Figure 21.29 An MHD propulsion system in a nuclear submarine could produce significantly less turbulence than propellers and allow it to run more silently. The development of a silent drive submarine was dramatized in the book and the film *The Hunt for Red October*.

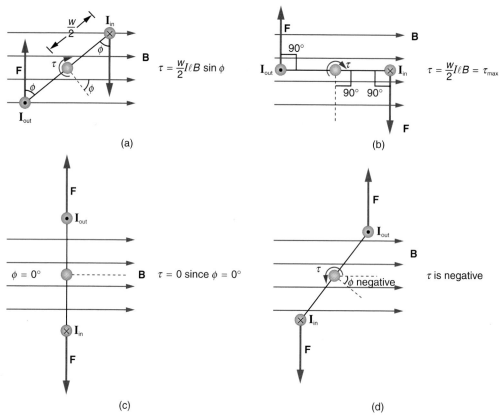

Figure 21.31 Top views of a current-carrying loop in a magnetic field. (a) The equation for torque is derived using this view. Note that the perpendicular to the loop makes an angle ϕ with the field that is the same as the angle between $w/2$ and F. (b) The maximum torque occurs when ϕ is a right angle and sin $\phi = 1$. (c) Zero (minimum) torque occurs when ϕ is zero and sin $\phi = 0$. (d) The torque reverses once the loop rotates past $\phi = 0$.

EXAMPLE 21.5 TORQUE ON A CURRENT-CARRYING LOOP IN A STRONG MAGNETIC FIELD

Find the maximum torque on a 100-turn square loop of wire 10.0 cm on a side that carries 15.0 A of current in a 2.00 T field.

Strategy Torque on the loop can be found using Equation 21.6. Maximum torque occurs when $\phi = 90°$ and sin $\phi = 1$.

Solution For sin $\phi = 1$, the maximum torque is

$$\tau_{max} = NIAB$$

Entering known values yields

$$\tau_{max} = (100)(15.0 \text{ A})(0.0100 \text{ m}^2)(2.00 \text{ T})$$
$$= 30.0 \text{ N} \cdot \text{m}$$

Discussion This torque is large enough to be useful in a motor.

The torque found in the preceding example is the maximum. As the coil rotates, torque decreases to zero at $\phi = 0$. Torque then *reverses* its direction once the coil rotates past $\phi = 0$. (See Figure 21.31(d).) This means that, unless we do something, the coil will oscillate back and forth about equilibrium at $\phi = 0$. To get the coil to continue rotating in the same direction, we can reverse the current as it passes through $\phi = 0$ with automatic switches called *brushes*. (See Figure 21.32.) Other types of motors will be seen in Chapter 22.

Meters, such as those in analog gas gauges, are another common application of magnetic torque on a current-carrying loop. Figure 21.33 shows that a meter is very similar in construction to a motor. The meter in the figure has its magnets shaped to limit the effect of ϕ by making B perpendicular to the loop over a large angular range. Thus the torque is proportional to I and not ϕ. A linear spring exerts a counter-torque that balances the current-produced torque. This makes the needle deflection proportional to I. If an exact proportionality cannot be achieved, the gauge reading can be calibrated. (Gas

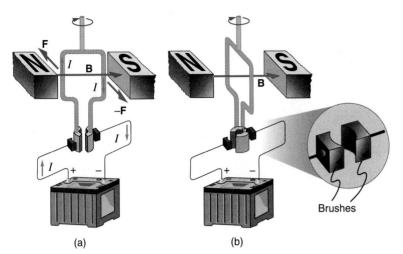

Brushes

gauges are often intentionally inaccurate to motivate a pit stop, but this is not common in other gauges.) To produce a galvanometer (see Section 20.4) that has a low resistance and responds to small currents, we use large loop area A, high magnetic field B, and low-resistance coils.

21.9 MAGNETIC FIELDS PRODUCED BY CURRENTS: AMPERE'S LAW

How much current is needed to create a significant magnetic field, perhaps as strong as the earth's field? Surveyors will tell you that electric power lines create magnetic fields that interfere with their compass readings. Indeed, when Oersted discovered in 1820 that a current in a wire affected a compass needle, he was not dealing with extremely large currents. How does the shape of wires carrying current affect the shape of the magnetic field created? We noted earlier that a current loop created a field similar to that of a bar magnet. But what about a straight wire or a toroid (doughnut)? How is the direction of a current-created field related to the direction of the current? Answers to these questions are explored in this section, together with a brief discussion of the law governing the fields created by currents.

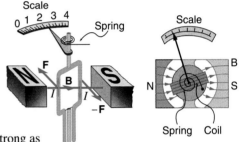

Figure 21.33 Meters are very similar to motors but only rotate through a part of a revolution. The magnetic poles of this meter are shaped to keep the component of B perpendicular to the loop constant, so that the torque does not depend on ϕ and the deflection against the return spring is proportional only to the current I.

Magnetic Field Created by a Long Straight Current-Carrying Wire: Right Hand Rule–2

Magnetic fields have both direction and magnitude. As noted before, one way to explore the direction of a magnetic field is with compasses, as shown for a long straight current-carrying wire in Figure 21.34. Hall probes can determine the magnitude of the field. The field around a long straight wire is found to be in circular loops. **Right hand rule–2** (RHR-2) emerges from this exploration and is valid for any current segment—*point the thumb in the direction of current, and the fingers curl in the direction of the magnetic field loops* created by it.

The **magnetic field strength** (magnitude) **created by a long straight current-carrying wire** is found by experiment to be

$$B = \frac{\mu_0 I}{2\pi r} \quad \text{(long straight wire)} \quad (21.7)$$

where I is the current, r is the shortest distance to the wire, and the constant

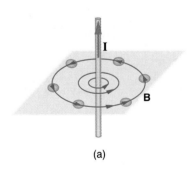

(a)

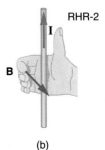

RHR-2

(b)

Figure 21.34 (a) Compasses placed near a long straight current-carrying wire indicate the field lines form circular loops centered on the wire. (b) Right hand rule–2 states that if the thumb points in the direction of the current, the fingers curl in the direction of the field. This rule is consistent with the field mapped for the long straight wire and is valid for any current segment.

$\mu_0 = 4\pi \times 10^{-7}$ T·m/A is the **permeability of free space**. (μ_0 is one of the basic constants in nature. We will see later that μ_0 is related to the speed of light.) Since the wire is very long, the magnitude of the field depends only on distance from the wire r, not on position along the wire.

EXAMPLE 21.6   WHAT CURRENT CREATES THIS FIELD STRENGTH?

Find the current in a long straight wire that would create a magnetic field twice the strength of the earth's at a distance of 5.0 cm from the wire.

Strategy The earth's field is about 5.0×10^{-5} T, and so here B due to the wire is taken to be 1.0×10^{-4} T. Equation 21.7 can be used to find I, since all other quantities are known.

Solution Solving Equation 21.7 for I, and entering known values, gives

$$I = \frac{2\pi r B}{\mu_0} = \frac{2\pi (5.0 \times 10^{-2} \text{ m})(1.0 \times 10^{-4} \text{ T})}{4\pi \times 10^{-7} \text{ T·m/A}}$$
$$= 25 \text{ A}$$

Discussion So a moderate current creates a significant magnetic field at a distance of 5.0 cm from a long straight wire. Note that the answer is stated to only two digits, since the earth's field is specified to only two digits in this example.

Ampere's Law and Others

The magnetic field of a long straight wire has more implications than you might at first suspect. *Each segment of current creates a magnetic field like that of a long straight wire, and the total field of any shape current is the vector sum of the fields due to each segment.* The formal statement of the direction and magnitude of the field due to each segment is called the **Biot-Savart law**. Integral calculus is needed to sum the field for an arbitrary shape current. This results in a more complete law, called **Ampere's law**, which relates magnetic field and current in a general way. Ampere's law in turn is a part of **Maxwell's equations**, which give a complete theory of all electromagnetic phenomena. Considerations of how Maxwell's equations appear to different observers led to the **modern theory of relativity**, and the realization that electric and magnetic fields are different manifestations of the same thing. Most of this is beyond the scope of this text in both mathematical level, requiring calculus, and in the amount of space that can be devoted to it. But for the interested student, and particularly for those who continue in physics, engineering, or similar pursuits, delving into these matters further will reveal descriptions of nature that are elegant as well as profound. In this text, we shall keep the general features in mind, such as RHR-2 and the rules for magnetic field lines listed in Section 21.3, while concentrating on the fields created in certain important situations.

CONNECTIONS

Relativity

Hearing all we do about Einstein, we sometimes get the impression he invented relativity out of nothing. On the contrary, one of Einstein's motivations was to solve difficulties in knowing how different observers see magnetic and electric fields.

Magnetic Field Created by a Current-Carrying Circular Loop

The magnetic field near a current-carrying loop of wire is shown in Figure 21.35. Both the direction and magnitude of the magnetic field created by a current-carrying loop are complex. RHR-2 can be used to give the direction of the field near the loop, but mapping with compasses and the rules about field lines given in Section 21.3 are needed for more detail. There is a simple formula for the **magnetic field strength at the center of a circular loop**. It is

$$B = \frac{\mu_0 I}{2R} \qquad \text{(at center of loop)} \tag{21.8}$$

where R is the radius of the loop. This equation is very similar to that for a straight wire, but it is valid *only* at the center of a loop of wire. The similarity of the equations does indicate that similar field strength can be obtained at the center of a loop. One way to get a larger field is to have N loops; then, the field is $B = N\mu_0 I/2R$. Note that the larger the loop, the smaller the field at its center, because the current is farther away.

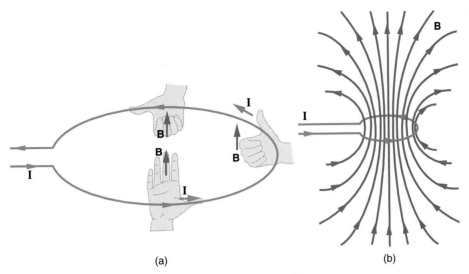

Figure 21.35 (a) RHR-2 gives the direction of the magnetic field inside and outside a current-carrying loop. (b) More detailed mapping with compasses or with a Hall probe completes the picture. The field is similar to that of a bar magnet.

Magnetic Field Created by a current-carrying Solenoid

A solenoid is a long coil of wire (as opposed to a flat loop). Because of its shape, the field inside a solenoid can be very uniform, and also very strong. The field just outside the coils is nearly zero. Figure 21.36 shows how the field looks and how its direction is given by RHR-2.

The magnetic field inside of a current-carrying solenoid is very uniform in direction and magnitude. Only near the ends does it begin to weaken and change direction. The field outside has similar complexities to flat loops and bar magnets, but the **magnetic field strength inside a solenoid** is simply

$$B = \mu_0 nI \qquad \text{(inside a solenoid)} \qquad \text{(21.9)}$$

where n is the number of loops per unit length of the solenoid ($n = N/\ell$, with N being the number of loops and ℓ the length). Note that B is the field strength anywhere in the uniform region of the interior and not just at the center. Large uniform fields spread over a large volume are possible with solenoids, as the following example implies.

EXAMPLE 21.7 FIELD STRENGTH INSIDE A SOLENOID

What is the field inside a 2.00 m long solenoid that has 2000 loops and carries a 1600 A current?

Strategy To find the field strength inside a solenoid, we use Equation 21.9. First, we note the number of loops per unit length is $n = N/\ell = 2000/(2.00 \text{ m}) = 1000/\text{m}$.

Solution Substituting known values into Equation 21.9 gives

$$B = \mu_0 nI = (4\pi \times 10^{-7} \text{ T·m/A})(1000/\text{m})(1600 \text{ A})$$
$$= 2.01 \text{ T}$$

Discussion This is a large field strength that could be established over a large-diameter solenoid, such as in medical uses of magnetic resonance imaging (MRI). The very large current is an indication that fields of this strength are not easily achieved, however. Such a large current through 1000 loops squeezed into a meter's length would produce significant heating. Current limits imposed by overheating of normal wires can be exceeded by superconducting wires, although this is expensive. There is still an upper limit to the current, since the superconducting state is disrupted by very large magnetic fields.

There are interesting variations of the flat coil and solenoid. For example, the toroidal coil used to confine the reactive particles in tokamaks is much like a solenoid bent into a circle (see Figure 21.22 in Section 21.5). The field inside a toroid is very strong but circular. Charged particles travel in circles, following the field lines, and collide with one another, perhaps inducing fusion. But the charged particles do not cross field lines and escape the toroid. A whole range of coil shapes are used to produce all sorts of magnetic field shapes. Adding ferromagnetic materials produces greater field strengths and can have a significant effect on the shape of the field. Ferromagnetic materials tend to trap magnetic fields (the field lines bend into the ferromagnetic material,

Figure 21.36 (a) Because of its shape, the field inside a solenoid is remarkably uniform in magnitude and direction, as indicated by the straight, uniformly spaced field lines. The field outside the coils is nearly zero. (b) This cutaway shows the solenoid's field.

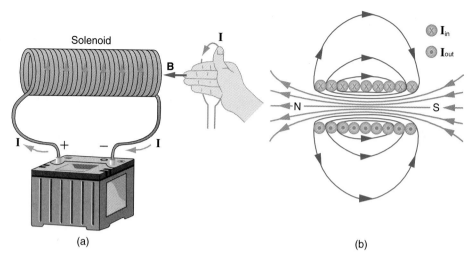

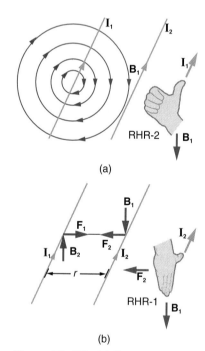

Figure 21.37 (a) The magnetic field created by a long straight conductor is perpendicular to a parallel conductor as indicated by RHR-2. (b) RHR-1 shows that the force between the parallel conductors is attractive when the currents are in the same direction. A similar analysis shows the force is repulsive between currents in opposite directions.

leaving weaker fields outside it) and are used as shields for devices that are adversely affected by magnetic fields.

21.10 MAGNETIC FORCE BETWEEN TWO PARALLEL CONDUCTORS

You might expect that there are significant forces between current-carrying wires, since ordinary currents create significant magnetic fields and these fields exert significant forces on ordinary currents. But you might not expect that the force between wires is used to define the ampere. It might also surprise you to learn that this force has something to do with why large circuit breakers burn up when they attempt to interrupt large currents.

The force between two long straight and parallel conductors can be found by applying what we have developed in preceding sections. Figure 21.37 shows the wires, their currents, the fields they create, and the subsequent forces they exert on one another. Let us consider the field created by wire 1 and the force it exerts on wire 2 (call the force F_2). The field due to I_1 at a distance r is given by Equation 21.7 to be

$$B_1 = \frac{\mu_0 I_1}{2\pi r}$$

This field is uniform along wire 2 and perpendicular to it, and so the force F_2 it exerts on wire 2 is given by Equation 21.5 with $\sin \theta = 1$:

$$F_2 = I_2 \ell B_1$$

By Newton's third law, the forces on the wires are equal in magnitude, and so we just write F for the magnitude of F_2. (Note that $F_1 = -F_2$.) Since the wires are very long, it is convenient to think in terms of F/ℓ, the force per unit length. Substituting the expression for B_1 into the last equation and rearranging terms gives

$$\frac{F}{\ell} = \frac{\mu_0 I_1 I_2}{2\pi r} \qquad \textbf{(21.10)}$$

F/ℓ is the force per unit length between two parallel currents I_1 and I_2, separated by a distance r. The force is attractive if the currents are in the same direction, repulsive if they are in opposite directions.

This force is responsible for the *pinch effect* in electric arcs and plasmas. (See Figure 21.38.) The force exists whether the currents are in wires or not. In an electric arc, where currents are moving parallel to one another, there is an attraction that squeezes currents into a smaller tube. In large circuit breakers, like those used in neighborhood power distribution systems, the pinch effect can concentrate an arc between plates of a switch trying to break a large current, burn holes, and even ignite the equipment. Another example of the pinch effect is found in the solar plasma, where jets of ionized material, such as solar flares, are shaped by magnetic forces.

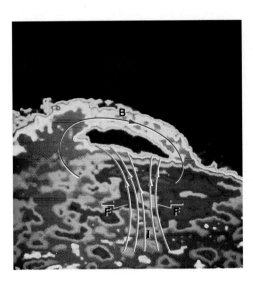

Figure 21.38 The pinch effect is due to the attraction between currents in the same direction. On the sun, surface currents of ionized gas may be pinched by this effect, forming a very large current with magnetic fields that trap glowing ions in spectacular arcs.

The *operational definition of the ampere* is based on the force between current-carrying wires. Note that for parallel wires separated by 1 meter with each carrying 1 ampere, the force per meter is

$$\frac{F}{\ell} = \frac{(4\pi \times 10^{-7} \text{ T}\cdot\text{m/A})(1 \text{ A})^2}{(2\pi)(1 \text{ m})} = 2 \times 10^{-7} \text{ N/m}$$

Since μ_0 is exactly $4\pi \times 10^{-7}$ T·m/A by definition, the force per meter is exactly 2×10^{-7} N/m. This is the basis of the operational definition of the ampere. The official SI definition of the ampere is

> **One ampere of current through each of two parallel conductors of infinite length, separated by one meter in empty space free of other magnetic fields, causes a force of exactly 2×10^{-7} N/m on each conductor.**

Infinite-length straight wires are impractical, and so, in practice, a current balance is constructed with coils of wire separated by a few centimeters. Force is measured to determine current. This also provides us with a method for measuring the coulomb. We measure the charge that flows for a current of one ampere in one second. That is, 1 C = 1 A·s. For both the ampere and the coulomb, the method of measuring force between conductors is the most accurate in practice.

21.11 EVEN MORE APPLICATIONS OF MAGNETISM

Mass Spectrometry

The curved paths followed by charged particles in magnetic fields can be put to use. As shown in Section 21.5, a charged particle moving perpendicular to a magnetic field travels in a circular path having a radius r given by Equation 21.3 to be

$$r = \frac{mv}{qB}$$

It was noted that this relationship could be used to measure the mass of charged particles such as ions. A mass spectrometer is a device that measures such masses. Most mass spectrometers use magnetic fields for this purpose, although some of them have extremely sophisticated designs. Since there are five variables in the relationship, there are many possibilities. But if v, q, and B can be fixed, then the radius of the path r is simply proportional to the mass m of the charged particle. Let us examine one such mass spectrometer that has a relatively simple design. (See Figure 21.39.) The process begins with an ion source, a device like an electron gun. The ion source gives ions their charge, accelerates them to some velocity v, and directs a beam of them into the next stage of the

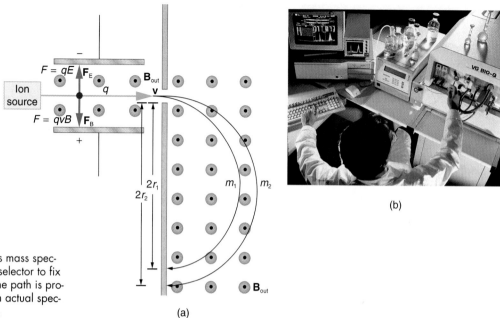

Figure 21.39 (a) This mass spectrometer uses a velocity selector to fix v, so that the radius of the path is proportional to mass. (b) An actual spectrometer and its controls.

spectrometer. This next region is a *velocity selector* that only allows particles with a particular value of v to get through.

The velocity selector has both an electric and a magnetic field, perpendicular to one another, producing forces in opposite directions on the ions. Only those ions for which the forces balance travel in a straight line into the next region. If the forces balance, then the electric force $F = qE$ equals the magnetic force $F = qvB$, so that $qE = qvB$. Noting that q cancels, we see that

$$v = \frac{E}{B} \tag{21.11}$$

is the velocity particles must have to make it through the velocity selector, and further that v can be selected by varying E and B. In the final region, there is only a uniform magnetic field, and so the charged particles move in circular arcs with radii proportional to particle mass. The paths also depend on charge q, but since q is in multiples of electron charges, it is easy to determine and to discriminate between ions in different charge states.

Cathode Ray Tubes—CRTs—and the Like

What do TVs, computer monitors, x-ray machines, and the 2 mile long Stanford Linear Accelerator have in common? All of them accelerate electrons, making them different versions of the electron gun. Many of these devices use magnetic fields to steer the accelerated electrons. Figure 21.40 shows the construction of the type of cathode ray tube (CRT) found in TVs, oscilloscopes, and computer monitors. Two pairs of coils are used to steer the electrons, one vertically and one horizontally, to their desired destination.

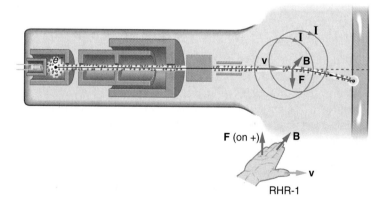

Figure 21.40 The cathode ray tube (CRT) is so named because rays of electrons originate at the cathode in the electron gun. Magnetic coils are used to steer the beam in many CRTs. In this case, the beam is moved down. Another pair of horizontal coils would steer the beam horizontally.

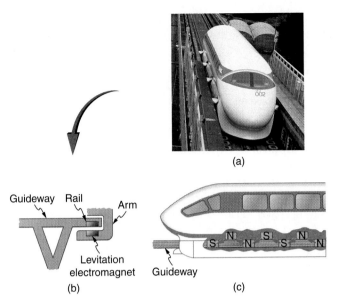

Figure 21.41 (a) A maglev train. (b) The support for the maglev is obtained from electromagnets. (c) A separate system of magnets is arranged to drive and stop the train. The magnets are timed as shown to drive the train with attraction between unlike poles and repulsion between like poles.

Maglev Trains

Some of the fastest trains in the world, such as the one shown in Figure 21.41, are driven and supported by magnetic forces. Their great speed capabilities are made possible by low friction compared with that in conventional wheeled trains. Magnetic levitation or maglev (so called because the train appears to float on air) eliminates mechanical friction in wheels and bearings and also eliminates the need to supply and remove associated rotational kinetic energy as the train accelerates and decelerates. In addition to levitating the train, magnetic forces drive it, reducing local pollution such as that emitted by diesel-electric trains.

There are many other applications of magnetism. They range from magnets used to hold an artificial jaw together after reconstructive surgery to the possibility that birds use the earth's field as a navigational guide. One use of magnetism that has become familiar in the last two decades is magnetic resonance imaging (MRI). MRI is now a highly developed diagnostic tool that is particularly well suited for images of the central nervous system and is free of the adverse effects of x rays. MRI utilizes the effect of a magnetic field in aligning magnetic atoms. There are interesting quantum effects here; thus, MRI will be discussed in some detail in a later chapter.

SUMMARY

Magnetism is a subject that includes the properties of magnets, the effect of the magnetic force on moving charges and currents, and the creation of magnetic fields by currents. There are two types of *magnetic poles*, called **north** and **south magnetic poles**. North magnetic poles are those attracted toward the earth's north pole. **Like poles repel and unlike poles attract**. Magnetic poles always occur in pairs of north and south—**it is not possible to isolate north and south poles**. This is directly attributable to the fact that **all magnetism is created by electric current**.

Ferromagnetic materials, such as iron, are those that exhibit strong magnetic effects. The atoms in ferromagnetic materials act like small magnets (due to currents within the atoms) and can be aligned, usually in millimeter-sized regions called **domains**. These domains can grow and align on a larger scale, producing permanent magnets. Such a material is

magnetized, or induced to be magnetic. Above a material's **Curie temperature**, thermal agitation destroys the alignment of atoms, and ferromagnetism disappears. **Electromagnets** employ electric currents to make magnetic fields, often aided by induced fields in ferromagnetic materials.

Magnetic fields can be pictorially represented by **magnetic field lines**, the properties of which are:

1. The field is tangent to the magnetic field line.
2. Field strength is proportional to line density.
3. Field lines cannot cross.
4. Field lines are continuous loops.

Magnetic fields exert a **force on a moving charge** q, the *magnitude* of which is

$$F = qvB \sin \theta \qquad \textbf{(21.1)}$$

where θ is the angle between the directions of v and B. The SI unit for magnetic field strength B is the **tesla** (T), which is related to other units by

$$1 \text{ T} = \frac{1 \text{ N}}{\text{C} \cdot \text{m/s}} = \frac{1 \text{ N}}{\text{A} \cdot \text{m}} \quad \textbf{(21.2)}$$

The *direction* of the force on a moving charge is given by **right hand rule–1** (RHR-1): *Point the thumb of the right hand in the direction of v, the fingers in the direction of B, and a perpendicular to the palm points in the direction of F.* The force is perpendicular to the plane formed by v and B. Since the force is zero if v is parallel to B, charged particles often follow magnetic field lines rather than cross them. Magnetic force can supply centripetal force and cause a charged particle to move in a circular path of radius

$$r = \frac{mv}{qB} \quad \textbf{(21.3)}$$

where v is component of the velocity perpendicular to B for a charged particle with mass m and charge q.

The **Hall effect** is the creation of voltage $\mathcal{E}$, known as the **Hall emf**, *across* a current-carrying conductor by a magnetic field. The Hall emf is given by

$$\mathcal{E} = Bℓv \quad (B, v, \text{ and } ℓ \text{ perpendicular}) \quad \textbf{(21.4)}$$

for a conductor of width $ℓ$ through which charges move at a speed v.

The **magnetic force on current-carrying conductors** is given by

$$F = IℓB \sin \theta \quad \textbf{(21.5)}$$

where I is the current, $ℓ$ is the length of a straight conductor in a uniform magnetic field B, and θ is the angle between I and B. The force follows RHR-1 with the thumb in the direction of I. The **magnetic torque τ on a current loop** is given by

$$\tau = NIAB \sin \phi \quad \textbf{(21.6)}$$

for a loop that carries a current I, that has N turns of area A, and such that a perpendicular to the loop makes an angle ϕ with the field B. The net force on the loop is zero if the field is uniform.

The **strength of the magnetic field created by current in a long straight wire** is given by

$$B = \frac{\mu_0 I}{2\pi r} \quad \text{(long straight wire)} \quad \textbf{(21.7)}$$

where I is the current, r is the shortest distance to the wire, and the constant $\mu_0 = 4\pi \times 10^{-7}$ T·m/A is the **permeability of free space**. The direction of the magnetic field created by a long straight wire is given by **right hand rule–2** (RHR-2): *Point the thumb of the right hand in the direction of current, and the fingers curl in the direction of the magnetic field loops* created by it. The magnetic field created by current following any path is the sum (or integral) of the fields due to segments along the path (magnitude and direction as for a straight wire), resulting in a general relationship between current and field known as **Ampere's law**.

The **magnetic field strength at the center of a circular loop** is given by

$$B = \frac{\mu_0 I}{2R} \quad \text{(at center of loop)} \quad \textbf{(21.8)}$$

where R is the radius of the loop. This equation becomes $B = N\mu_0 I/2R$ for a flat coil of N loops. RHR-2 gives the direction of the field about the loop. A long coil is called a *solenoid*. The **magnetic field strength inside a solenoid** is

$$B = \mu_0 nI \quad \text{(inside a solenoid)} \quad \textbf{(21.9)}$$

where n is the number of loops per unit length of the solenoid. The field inside is very uniform in magnitude and direction.

The **force between two parallel currents** I_1 and I_2, separated by a distance r, has a magnitude per unit length given by

$$\frac{F}{ℓ} = \frac{\mu_0 I_1 I_2}{2\pi r} \quad \textbf{(21.10)}$$

The force is attractive if the currents are in the same direction, repulsive if they are in opposite directions.

Crossed (perpendicular) electric and magnetic fields act as a velocity filter, giving equal and opposite forces on any charge with velocity perpendicular to the fields and of magnitude

$$v = \frac{E}{B} \quad \textbf{(21.11)}$$

CONCEPTUAL QUESTIONS

21.1 Volcanic and other such activity at the mid-Atlantic ridge extrudes material to fill the gap between separating tectonic plates associated with continental drift. The magnetization of rocks is found to reverse in a coordinated manner with distance from the ridge. What does this imply about the earth's magnetic field, and how could knowledge of the spreading rate be used to give its historical record?

21.2 Explain why the magnetic field would not be unique (that is, not have a single value) at a point in space where magnetic field lines cross. (Consider the direction of the field at such a point.)

21.3 List the ways in which magnetic field lines and electric field lines are similar. For example, the field direction is tangent to the line at any point in space. Also list the ways in which they differ. For example, electric force is parallel to electric field lines, whereas magnetic force on moving charges is perpendicular to magnetic field lines.

21.4 Noting that the magnetic field lines of a bar magnet resemble the electric field lines of a pair of equal and opposite charges, do you expect the magnetic field to rapidly decrease in strength with distance from the magnet? Is this consistent with your experience with magnets?

21.5 Is the earth's magnetic field parallel to the ground at all locations? If not, where is it parallel to the surface? Is its strength the same at all locations? If not, where is it greatest?

21.6 High-velocity charged particles can damage biological cells and are a component of radiation exposure in a variety of locations ranging from research facilities to natural background. Describe how you could use a magnetic field to shield yourself.

21.7 If a cosmic ray proton approaches the earth from outer space along a line toward the center of the earth that lies in the plane of the equator, in what direction will it be deflected by the earth's magnetic field? What about an electron? A neutron?

21.8 What is the direction of the magnetic field in Figure 21.15?

21.9 Discuss how the Hall effect could be used to obtain information on free charge density in a conductor. (Hint: Consider how drift velocity and current are related.)

21.10 What are the signs of the charges on the particles in Figure 21.42?

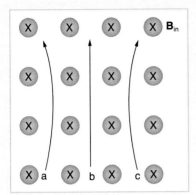

Figure 21.42 Charged particles moving in the plane of the page under the influence of a magnetic field only. Questions 10 and 11.

21.11 What happens to the paths of the particles in Figure 21.42 if an electric field is established with its direction across the page to the left?

21.12 Which of the particles in Figure 21.43 has the greatest velocity, assuming they have identical charges and masses?

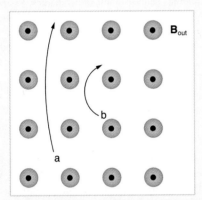

Figure 21.43 Charged particles moving in the plane of the page under the influence of a magnetic field only. Questions 12 and 13.

21.13 Which of the particles in Figure 21.43 has the greatest mass, assuming all have identical charges and velocities?

21.14 Draw a sketch of the situation in Figure 21.26 showing the direction of electrons carrying the current, and use RHR-1 to verify the direction of the force on the wire.

21.15 Verify that the direction of the force in an MHD drive, such as that in Figure 21.28, does not depend on the sign of the charges carrying the current across the fluid.

21.16 Why would a magneto-hydrodynamic drive work better in ocean water than in fresh water? Also, why would superconducting magnets be desirable?

21.17 Make a drawing and use RHR-2 to find the direction of the magnetic field of a current loop in a motor (such as in Figure 21.30). Then show that the direction of the torque on the loop is the same as produced by like poles repelling and unlike poles attracting.

21.18 Draw a diagram and use RHR-1 to show that the forces on the top and bottom segments of the motor's current loop in Figure 21.30 are vertical and produce no torque about the axis of rotation.

21.19 Which is more likely to interfere with compass readings, AC current in your refrigerator or DC current when you start your car? Explain.

21.20 Measurements of the weak and fluctuating magnetic fields associated with brain activity are called magneto-encephalograms (MEGs). Do the brain's magnetic fields imply coordinated or uncoordinated nerve impulses? Explain.

21.21 Is the force attractive or repulsive between the hot and neutral lines hung from power poles? Why?

21.22 If you have three parallel wires in the same plane, as in Figure 21.44, with currents in the outer two running in opposite directions, is it possible for the middle wire to be repelled by both? Attracted by both? Explain.

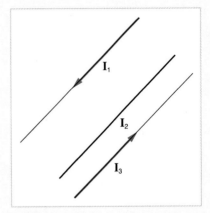

Figure 21.44 Three parallel coplanar wires with currents in the outer two in opposite directions. Question 22.

21.23 Suppose two long straight wires run perpendicular to one another without touching. Does one exert a net force on the other? If so, what is its direction? Does one exert a net torque on the other? If so, what is its direction? Justify your responses by using the right hand rules.

21.24 Use the right hand rules to show that the force between the two loops in Figure 21.45 is attractive if the currents are in the same direction, repulsive if they are in opposite directions. Is this consistent with like poles of the loops repelling and unlike poles of the loops attracting? Draw sketches to justify your answers.

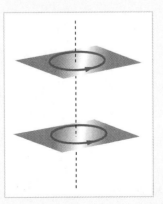

Figure 21.45 Two loops of wire carrying currents can exert forces and torques on one another. Questions 24 and 25.

21.25 If one of the loops in Figure 21.45 is tilted slightly relative to the other and their currents are in the same direction, what are the directions of the torques they exert on each other?

Does this imply that the poles of the bar magnet–like fields they create will line up with each other if the loops are allowed to rotate?

PROBLEMS

Section 21.4 Magnetic Force on a Moving Charge and RHR-1

21.1 What is the direction of the magnetic force on a positive charge that moves as shown in each of the six cases shown in Figure 21.46?

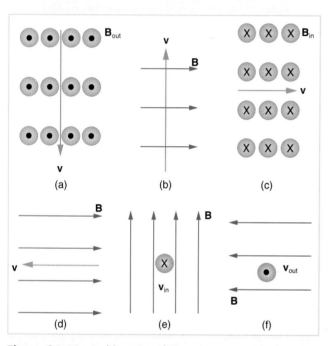

Figure 21.46 Problems 1 and 2.

21.2 Repeat Problem 21.1 for a negative charge.

21.3 What is the direction of the velocity of a negative charge that experiences the magnetic force shown in each of the three cases in Figure 21.47, assuming it moves perpendicular to **B**?

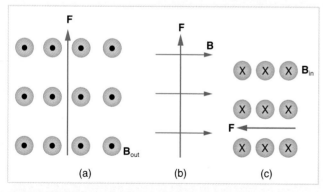

Figure 21.47 Problems 3 and 4.

21.4 Repeat Problem 21.3 for a positive charge.

21.5 What is the direction of the magnetic field that produces the magnetic force on a positive charge as shown in each of the three cases in Figure 21.48, assuming **B** is perpendicular to **v**?

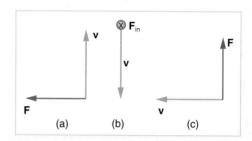

Figure 21.48 Problems 5 and 6.

21.6 Repeat Problem 21.5 for a negative charge.

21.7 What is the maximum force on a rod with a 0.100 μC charge that you pass between the poles of a 1.50 T strength permanent magnet at a speed of 5.00 m/s?

21.8 Aircraft sometimes acquire small static charges. Suppose a supersonic jet has a 0.500 μC charge and flies due west at a speed of 660 m/s over the earth's north magnetic pole, where the 8.00×10^{-5} T magnetic field points straight down. What are the direction and magnitude of the magnetic force on the plane?

•**21.9** A cosmic ray proton moving toward the earth at 5.00×10^{7} m/s experiences a magnetic force of 1.70×10^{-16} N. What is the strength of the magnetic field if there is a 45° angle between it and the proton's velocity?

•**21.10** An electron moving at 4.00×10^{3} m/s in a 1.25 T magnetic field experiences a magnetic force of 1.40×10^{-16} N. What angle does the velocity of the electron make with the magnetic field? There are two answers.

•**21.11** A physicist performing a sensitive measurement wants to limit the magnetic force on a moving charge in her equipment to less than 1.00×10^{-12} N. What is the greatest the charge can be if it moves at a maximum speed of 30.0 m/s in the earth's field?

Section 21.5 Motion of a Charge Particle in a Magnetic Field and Section 21.11 Applications

21.12 A cosmic ray electron moves at 7.50×10^{6} m/s perpendicular to the earth's magnetic field at an altitude where field strength is 1.00×10^{-5} T. What is the radius of the circular path the electron follows?

21.13 A proton moves at 7.50×10^{7} m/s, perpendicular to a magnetic field. The field causes the proton to travel in a circular path of radius 0.800 m. What is the field strength?

21.14 Viewers of *Star Trek* hear of an antimatter drive on the Starship Enterprise. One possibility for such a futuristic energy source is to store antimatter charged particles in a vacuum chamber, circulating in a magnetic field, and then extract them as needed. Antimatter annihilates with normal matter, producing pure energy. What strength magnetic field is needed to hold antiprotons, moving at 5.00×10^7 m/s, in a circular path 2.00 m in radius? Antiprotons have the same mass as protons but the opposite (negative) charge.

21.15 An oxygen-16 ion with a mass of 2.66×10^{-26} kg travels at 5.00×10^6 m/s perpendicular to a 1.20 T magnetic field, which makes it move in a circular arc with a 0.231 m radius. What positive charge is on the ion?

• **21.16** What radius circular path does an electron travel, if it moves at the same speed and in the same magnetic field as the proton in Problem 21.13?

• **21.17** A velocity selector in a mass spectrometer, like that shown in Figure 21.39, uses a 0.100 T magnetic field. (a) What electric field strength is needed to select a speed of 4.00×10^6 m/s? (b) What is the voltage between the plates if they are separated by 1.00 cm?

• **21.18** An electron in a television tube moves with a speed of 6.00×10^7 m/s, in a direction perpendicular to the earth's field, which has a strength of 5.00×10^{-5} T. (a) What strength electric field must be applied perpendicular to the earth's field to make the electron move in a straight line? (b) If this is done between plates separated by 1.00 cm, what is the voltage applied? (Note that TVs are usually surrounded by a ferromagnetic material to shield against external magnetic fields and avoid the need for such a correction.)

⁞ **21.19** (a) At what speed will a proton move in a circular path of the same radius as the electron in Problem 21.12? (b) What would the radius of the path be if the proton had the same speed as the electron? (c) What would the radius be if the proton had the same kinetic energy as the electron? (d) The same momentum?

⁞ **21.20** A mass spectrometer, like that shown in Figure 21.40, is being used to separate common oxygen-16 from the much rarer oxygen-18, taken from a sample of old glacial ice. (The relative abundance of these oxygen isotopes is related to climatic temperature at the time the ice was deposited.) The ratio of the masses of these two ions is 16 to 18, the mass of oxygen-16 is 2.66×10^{-26} kg, and they are singly charged and travel at 5.00×10^6 m/s in a 1.20 T magnetic field. What is the separation between their paths when they hit a target after traversing a semicircle?

⁞ **21.21** Triply charged uranium-235 and uranium-238 ions are being separated in a mass spectrometer, like that shown in Figure 21.39. (The much rarer uranium-235 is used for reactor fuel.) The masses of the ions are 3.90×10^{-25} kg and 3.95×10^{-25} kg, respectively, and they travel at 3.00×10^5 m/s in a 0.250 T field. What is the separation between their paths when they hit a target after traversing a semicircle?

Section 21.6 The Hall Effect

21.22 Use the information in Example 19.3 to find the Hall voltage produced by a 2.00 T field applied across a 12 gauge copper wire carrying a 20.0 A current.

21.23 What Hall voltage is produced by a 0.200 T field applied across a 2.60 cm diameter aorta when blood velocity is 60.0 cm/s?

21.24 A large water main is 2.50 m in diameter, and the average water velocity is 6.00 m/s. Find the Hall voltage produced if the pipe runs perpendicular to the earth's 5.00×10^{-5} T field.

• **21.25** A nonmechanical water meter could utilize the Hall effect by applying a magnetic field across a metal pipe and measuring the Hall voltage produced. What is the average fluid velocity in a 3.00 cm diameter pipe, if a 0.500 T field across it creates a 60.0 mV Hall voltage?

• **21.26** What is the speed of a supersonic aircraft with a 17.0 m wingspan, if it experiences a 1.60 V Hall voltage between its wing tips when in level flight over the north magnetic pole, where the earth's field strength is 8.00×10^{-5} T?

• **21.27** A Hall probe calibrated to read 1.00 µV when placed in a 2.00 T field is placed in a 0.150 T field. What is its output voltage?

⁞ **21.28** Using information in Example 19.3, what would the Hall voltage be if a 2.00 T field is applied across a 10 gauge copper wire (2.588 mm in diameter) carrying a 20.0 A current?

⁞ **21.29** Show that the Hall voltage across wires made of the same material, carrying identical currents, and subjected to the same magnetic field is inversely proportional to their diameters. (Hint: Consider how drift velocity depends on wire diameter.)

Section 21.7 Magnetic Force on a Current-Carrying Conductor

21.30 What is the direction of the magnetic force on the current in each of the six cases shown in Figure 21.49?

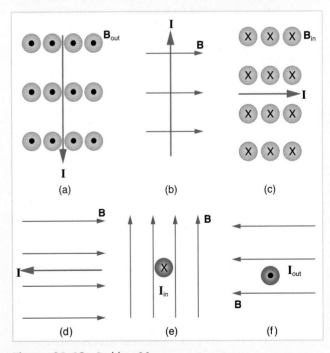

Figure 21.49 Problem 30.

21.31 What is the direction of a current that experiences the magnetic force shown in each of the three cases in Figure 21.50, assuming the current runs perpendicular to **B**?

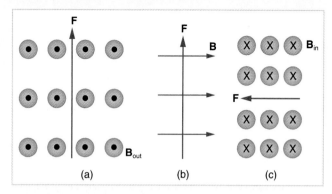

Figure 21.50 Problem 31.

21.32 What is the direction of the magnetic field that produces the magnetic force shown on the currents in each of the three cases in Figure 21.51, assuming **B** is perpendicular to **I**?

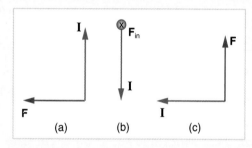

Figure 21.51 Problem 32.

21.33 (a) What is the force per meter on a lightning bolt at the equator that carries 20,000 A perpendicular to the earth's 3.00×10^{-5} T field? (b) What is the direction of the force if the current is straight up and the earth's field direction is due north, parallel to the ground?

21.34 A DC power line for a light-rail system carries 1000 A at a $30.0°$ angle to the earth's 5.00×10^{-5} T field. What is the force on a 100 m section of this line?

21.35 What force is exerted on the water in an MHD drive utilizing a 25.0 cm diameter tube, if 100 A current is passed across the tube perpendicular to a 2.00 T magnetic field? (The relatively small size of this force indicates the need for very large currents and magnetic fields to make practical MHD drives.)

21.36 A wire carrying a 30.0 A current passes between the poles of a strong magnet perpendicular to its field and experiences a 2.16 N force on the 4.00 cm of wire in the field. What is the average field strength?

• **21.37** (a) A 0.750 m long section of cable carrying current to a car starter motor makes an angle of $60.0°$ with the earth's 5.50×10^{-5} T field. What is the current when the wire experiences a force of 7.00×10^{-3} N? (b) If you run the wire between the poles of a strong horseshoe magnet, subjecting 5.00 cm of it to a 1.75 T field, what force is exerted on this segment of wire?

• **21.38** (a) What is the angle between a wire carrying 8.00 A and the 1.20 T field it is in, if 50.0 cm of the wire experiences a magnetic force of 2.40 N? (b) What is the force on the wire if it is rotated to make a 90° angle with the field?

: 21.39 The force on the rectangular loop of wire in the magnetic field shown in Figure 21.52 can be used to measure field strength. The field is uniform, and the loop is perpendicular to the field. (a) What is the direction of the magnetic force on the loop? Justify the claim that the forces on the sides of the loop are equal and opposite, independent of how much of the loop is in the field, and do not affect the net force on the loop. (b) If a current of 5.00 A is used, what is the force per tesla on the 20.0 cm wide loop?

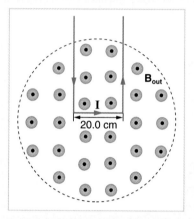

Figure 21.52 A rectangular loop of wire carrying a current is perpendicular to a magnetic field. The field is uniform in the region shown and is zero outside that region. Problems 39 and 60.

Section 21.8 Torque on a Current Loop

21.40 (a) By how many percent is the torque of a motor decreased if its permanent magnets lose 5.00% of their strength? (b) How many percent would the current need to be increased to return the torque to original values?

21.41 (a) What is the maximum torque on a 150-turn square loop of wire 18.0 cm on a side that carries a 50.0 A current in a 1.60 T field? (b) What is the torque when ϕ is $20.0°$?

21.42 Find the current through a loop needed to create a maximum torque of 9.00 N·m. The loop has 50 square turns that are 15.0 cm on a side and is in a uniform 0.800 T magnetic field.

21.43 Calculate the magnetic field strength needed on a 200-turn square loop 20.0 cm on a side to create a maximum torque of 300 N·m, if the loop is carrying 25.0 A.

21.44 Since the equation for torque on a current-carrying loop is $\tau = NIAB \sin \phi$, the units of N·m must equal units of A·m²T. Verify this.

• **21.45** (a) At what angle ϕ is the torque on a current loop 90.0% of maximum? (b) 50.0% of maximum? (c) 10.0% of maximum?

• **21.46** A proton has a magnetic field due to its spin on its axis. The field is similar to that created by a circular current loop 0.650×10^{-15} in radius with a current of 1.05×10^{4} A (no kidding). Find the maximum torque on a proton in a 2.50 T field. (This is a significant torque on a small particle.)

• **21.47** A 200-turn circular loop of radius 50.0 cm is vertical, with its axis on an east–west line. A current of 100 A circulates clockwise in the loop when viewed from the east. The earth's field here is due north, parallel to the ground, with a strength of 3.00×10^{-5} T. What are the direction and magnitude of the torque on the loop?

21.48 Repeat Problem 21.47, but with the loop lying flat on the ground with its current circulating counterclockwise in a location where the earth's field is north, but at an angle 45.0° below the horizontal and with a strength of 6.00×10^{-5} T.

Section 21.9 Magnetic Fields Produced by Currents and RHR-2

21.49 Indicate whether the magnetic field created in each of the three situations shown in Figure 21.53 is into or out of the page on the left and right of the current.

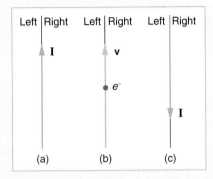

Figure 21.53 Problem 49.

21.50 What are the directions of the fields in the center of the loop and coils shown in Figure 21.54?

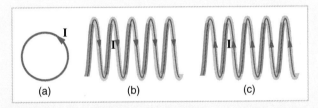

Figure 21.54 Problem 50.

21.51 What are the directions of the currents in the loop and coils shown in Figure 21.55?

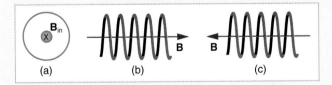

Figure 21.55 Problem 51.

21.52 To see why a cyclotron (a particle accelerator) utilizes iron to increase the magnetic field created by a coil, calculate the current needed in a 400-loop circular coil 0.660 m in radius to create a 1.20 T field at its center with no iron present.

21.53 The magnetic field of a proton is approximately like that of a circular current loop 0.650×10^{-15} m in radius carrying 1.05×10^4 A. What is the field at the center of such a loop?

21.54 Inside a motor, 30.0 A passes through a 250-turn circular loop that is 10.0 cm in radius. What is the magnetic field strength created at its center?

21.55 Nonnuclear submarines use batteries for power when submerged. (a) Find the magnetic field 50.0 cm from a straight wire carrying 1200 A from the batteries to the drive mechanism of a submarine. (b) What is the field if the wires to and from the drive mechanism are side by side?

21.56 How strong is the magnetic field inside a solenoid with 10,000 turns per meter that carries 20.0 A?

21.57 What current is needed in the solenoid described in Problem 21.56 to produce a magnetic field 10^4 times the earth's (0.500 T)?

21.58 How far from the starter cable of a car, carrying 150 A, must you be to experience a field less than the earth's (5.00×10^{-5} T)? Assume a long straight wire carries the current. (In practice, the body of your car shields the dashboard compass.)

21.59 What is the maximum current in a 1.00 m radius loop, if the field it creates at the center must be less than 5.00×10^{-7} T (1% of the earth's field)?

21.60 Measurements affect the system being measured, such as the current loop in Figure 21.52 (Problem 21.39). (a) Estimate the field the loop creates by calculating the field at the center of a circular loop 20.0 cm in diameter carrying 5.00 A. (b) What is the smallest field strength this loop can be used to measure, if its field must alter the measured field by less than 0.0100%?

21.61 Figure 21.56 shows a long straight wire just touching a loop carrying a current I_1. Both lie in the same plane. (a) What direction must the current I_2 in the straight wire have to create a field at the center of the loop in the direction opposite to that created by the loop? (b) What is the ratio of I_1/I_2 that gives zero field strength at the center of the loop? (c) What is the direction of the field directly above the loop under this circumstance?

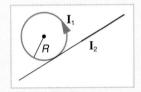

Figure 21.56 Problem 61.

21.62 Find the magnitude and direction of the magnetic field at the point equidistant from the wires in Figure 21.58(a), using the rules of vector addition to sum the contributions from each wire.

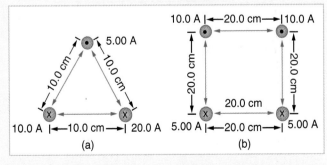

Figure 21.57 Problems 62, 63, 64, 71, and 72.

21.63 Find the magnitude and direction of the magnetic field at the point equidistant from the wires in Figure 21.57(b), using the rules of vector addition to sum the contributions from each wire.

21.64 What current is needed in the top wire in Figure 21.57(a) to produce a field of zero at the point equidistant from the wires, if the currents in the bottom two wires are both 10.0 A into the page?

Section 21.10 Magnetic Force between Two Parallel Conductors

21.65 The hot and neutral wires supplying DC power to a light-rail commuter train carry 800 A and are separated by 75.0 cm. What is the magnitude and direction of the force between 50.0 m of these wires?

21.66 The force per meter between the two wires of a jumper cable being used to start a stalled car is 0.225 N/m. (a) What is the current in the wires, given they are separated by 2.00 cm? (b) Is the force attractive or repulsive?

21.67 A 2.50 m segment of wire supplying current to the motor of a submerged submarine carries 1000 A and feels a 4.00 N repulsive force from a parallel wire 5.00 cm away. What is the direction and magnitude of the current in the other wire?

21.68 The wire carrying 400 A to the motor of a commuter train feels an attractive force of 4.00×10^{-3} N/m due to a parallel wire carrying 5.00 A to a headlight. (a) How far apart are the wires? (b) Are the currents in the same direction?

21.69 An AC appliance cord has its hot and neutral wires separated by 3.00 mm and carries a 5.00 A current. (a) What is the average force per meter between the wires in the cord? (b) What is the maximum force per meter between the wires? (c) Are the forces attractive or repulsive?

21.70 Figure 21.58 shows a long straight wire near a rectangular current loop. (a) Find the force on each segment of the loop. (b) What is the direction and magnitude of the total force on the loop?

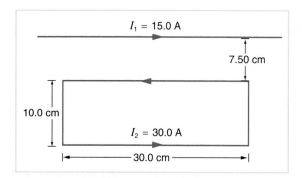

Figure 21.58 Problem 70.

21.71 Find the direction and magnitude of the force that each wire experiences in Figure 21.57(a), using vector addition.

21.72 Find the direction and magnitude of the force that each wire experiences in Figure 21.57(b), using vector addition.

INTEGRATED CONCEPTS

The following problems involve concepts from this chapter and several others. Physics is most interesting when applied to general situations involving more than a narrow set of physical principles. Magnetic fields, for example, exert forces on moving charges, hence the relevance of Chapter 4, Dynamics. The following topics are involved in some or all of the problems in this section:

Topics	Location
Two-dimensional kinematics	Chapter 3
Dynamics	Chapter 4
Statics	Chapter 5
Work, energy, and power	Chapter 6
Uniform circular motion	Chapter 8
Fluid statics	Chapter 10
Fluid dynamics	Chapter 11
Temperature, gases	Chapter 12
Oscillatory motion and waves	Chapter 15
Electric potential and energy	Chapter 18
Magnetism	Chapter 21

PROBLEM-SOLVING STRATEGY

Step 1. *Identify which physical principles are involved.*

Step 2. *Solve the problem using strategies outlined in the text.*

The following worked example illustrates how these strategies are applied to an integrated concept problem.

EXAMPLE   VOLTAGE TO ACCELERATE AND MAGNETIC FIELD TO CONFINE A PROTON

The following topics are involved in this integrated concepts worked example:

Topics	Location
Magnetism	Chapter 21
Electric potential and energy	Chapter 18
Work, energy, and power	Chapter 6

Suppose a nuclear physicist wants to accelerate protons to 10.0% of the speed of light and then confine them by having a magnetic field make them move in a circular path. (a) What voltage will accelerate protons to 3.00×10^7 m/s (10.0% of *c*)? (b) What magnetic field strength will cause the protons to move in a 0.500 m radius circular path? Assume the field is perpendicular to the proton velocity.

Strategy Step 1 To solve an **integrated concept problem,** such as those following this example, we must first identify the physical principles involved and identify the chapters in which they are found. Part (a) of this example asks for *voltage* to accelerate a charged particle. This can be found from electrical energy in Chapter 18 and using principles of *conservation of energy* from Chapter 6. Part (b) asks for a *magnetic field*, a topic of the present Chapter 21.

Strategy Step 2 The following solutions to each part of the example illustrate how specific problem-solving strategies are applied. These involve identifying knowns and unknowns, checking to see if the answer is reasonable, and so on.

Solution for (a): Energy (Chapter 6), Electric potential (Chapter 18) Electrical potential energy is charge times voltage, as given in Equation 18.3, $\Delta PE = qV$. This form of energy is converted to kinetic energy, and so conservation of energy says

$$qV = \tfrac{1}{2}mv^2$$

Rearranging this to find V, and entering the value given for the velocity and the known mass and charge of a proton, yields

$$V = \frac{mv^2}{2q}$$

$$= \frac{(1.67 \times 10^{-27} \text{ kg})(3.00 \times 10^7 \text{ m/s})^2}{2(1.60 \times 10^{-19} \text{ C})}$$

$$= 4.70 \times 10^6 \text{ V}$$

Discussion for (a) This large voltage can be achieved in the laboratory with relative ease.

Solution for (b): Magnetic field strength (Chapter 21) The relationship between magnetic field strength and the radius of the orbit of a charged particle is given by Equation 21.3. This equation is valid for situations like the present one, in which the velocity of the particle is perpendicular to the field. Rearranging Equation 21.3 to solve for B yields

$$B = \frac{mv}{rq}$$

where r is the radius of the orbit, given to be 0.500 m. Entering the known values yields

$$B = \frac{(1.67 \times 10^{-27} \text{ kg})(3.00 \times 10^7 \text{ m/s})}{(0.500 \text{ m})(1.60 \times 10^{-19} \text{ C})}$$

$$= 0.626 \text{ T}$$

Discussion for (b) A field of this magnitude can be achieved with electromagnets having iron cores. Similar and greater field strengths have been used for many decades in particle accelerators to confine the charged particles they produce.

This worked example illustrates how to apply problem-solving strategies to situations that include topics in different chapters. The first step is to identify the physical principles involved in the problem. The second step is to solve for the unknown using familiar problem-solving strategies. These are found throughout the text. Moreover, many worked examples show how to use problem-solving strategies for single topics. In this integrated concepts example, you can see how to apply them across several topics. You will find these techniques useful in applications of physics outside a physics course, such as in your profession, in other science disciplines, and in everyday life. The following problems will build your skills in the broad application of physical principles.

21.73 A pendulum is set up so that its bob swings between the poles of a permanent magnet as shown in Figure 21.59. What is the magnitude and direction of the magnetic force on the bob, if it has a positive 0.250 μC charge and is released from a height of 30.0 cm above its lowest point? The magnetic field strength is 1.50 T.

• 21.74 (a) What voltage will accelerate electrons to a speed of 6.00×10^7 m/s? (Example 21.2 states that approximately 10 kV will.) (b) Find the radius of curvature of the path of a *proton* accelerated through this potential in a 0.500 T field, and compare this with the radius of curvature of an electron accelerated through the same potential (as found in Example 21.2).

• 21.75 Find the radius of curvature of the path of a 25.0 MeV proton moving perpendicularly to the 1.20 T field of a cyclotron.

• 21.76 To construct a nonmechanical water meter, a 0.500 T magnetic field is placed across the supply water pipe to a home and the Hall voltage is recorded. (a) Find the flow rate in liters per second through a 3.00 cm diameter pipe, if the Hall voltage is 60.0 mV.

(b) What would the Hall voltage be for the same flow rate through a 10.0 cm diameter pipe with the same field applied?

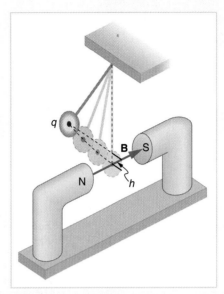

Figure 21.59 Problem 73.

• 21.77 Using the values given for an MHD drive in Problem 21.35, and assuming the force is uniformly applied to the fluid, calculate the pressure in N/m^2 created. Is this a significant fraction of an atmosphere?

• 21.78 (a) Calculate the maximum torque on a 50-turn, 1.50 cm radius circular current loop carrying 50.0 μA in a 0.500 T field. (b) If this coil is to be used in a galvanometer that reads 50.0 μA full scale, what force constant spring must be used, if it is attached 1.00 cm from the axis of rotation and is stretched by the arc moved?

• 21.79 A current balance used to define the ampere is designed so that the current through it is constant, as is the distance between wires. Even so, if the wires change length with temperature, the force between them will change. What percent change in force per degree will occur if the wires are copper?

‡ 21.80 (a) Show the period of the circular orbit of a charged particle moving perpendicularly to a uniform magnetic field is $T = 2\pi m/qB$. (b) What is the frequency f? (c) What is the angular velocity ω? Note that these results are independent of the velocity and radius of the orbit and, hence, of the energy of the particle.

‡ 21.81 A cyclotron accelerates charged particles as shown in Figure 21.60. Using the results of the previous problem, calculate the frequency of the accelerating voltage needed for a proton in a 1.20 T field.

‡ 21.82 A 0.140 kg baseball, pitched at 40.0 m/s horizontally and perpendicular to the earth's horizontal 5.00×10^{-5} T field, has a 100 nC charge on it. What distance is it deflected from its path by the magnetic force, after traveling 30.0 m horizontally?

‡ 21.83 (a) What is the direction of the force on a wire carrying a current due east in a location where the earth's field is due north? Both are parallel to the ground. (b) Calculate the force per meter if the wire carries 20.0 A and the field strength is 3.00×10^{-5} T. (c) What diameter copper wire would have its weight supported by this force? (d) Calculate the resistance per meter and the voltage per meter needed.

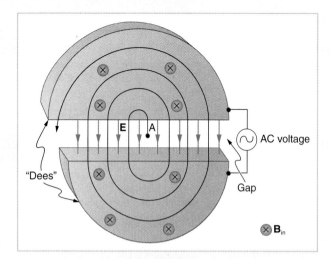

Figure 21.60 Cyclotrons accelerate charged particles orbiting in a magnetic field by placing an AC voltage on the metal Dees, between which the particles move, so that energy is added twice each orbit. The frequency is constant, since it is independent of the particle energy—the radius of the orbit simply increases with energy until the particles approach the edge and are extracted for various experiments and applications. Problems 80 and 81.

⁚21.84 One long straight wire is to be held directly above another by repulsion between their currents. The lower wire carries 100 A, and the wire 7.50 cm above it is 10 gauge (2.588 mm diameter) copper wire. (a) What current must flow in the upper wire, neglecting the earth's field? (b) What is the smallest current if the earth's 3.00×10^{-5} T field is parallel to the ground and is not neglected? (c) Is the supported wire in a stable or unstable equilibrium if displaced vertically? If displaced horizontally?

UNREASONABLE RESULTS

The following problems have results that are unreasonable because some premise is unreasonable or because certain of the premises are inconsistent with one another. Physical principles applied correctly then produce unreasonable results. The purpose of these problems is to give practice in assessing whether nature is being accurately described, and if it is not to trace the source of difficulty.

PROBLEM-SOLVING STRATEGY

To determine if an answer is reasonable and, if it is not, to determine what is the cause, do the following:

Step 1. *Solve the problem using strategies as outlined in several places in the text.* See the table of contents to locate problem-solving strategies. Use the format followed in the worked examples in this chapter to solve the problem as usual.

Step 2. *Check to see if the answer is reasonable.* Is it too large or too small, or does it have the wrong sign, improper units, . . .?

Step 3. *If the answer is unreasonable, look for what specifically could cause the identified difficulty.* Usually, the manner in which the answer is unreasonable is an indication of the difficulty.

21.85 (a) Find the charge on a baseball, thrown at 35.0 m/s perpendicular to the earth's 5.00×10^{-5} T field, that experiences a 1.00 N magnetic force. (b) What is unreasonable about this result? (c) Which assumption or premise is responsible?

21.86 A charged particle having mass 6.64×10^{-27} kg (that of a helium atom) moving at 8.70×10^{5} m/s perpendicular to a 1.50 T magnetic field travels in a circular path of radius 16.0 mm. (a) What is the charge of the particle? (b) What is unreasonable about this result? (c) Which assumptions are responsible?

21.87 An inventor wants to generate 120 V power by moving a 1.00 m long wire perpendicular to the earth's 5.00×10^{-5} T field. (a) Find the speed with which the wire must move. (b) What is unreasonable about this result? (c) Which assumption is responsible?

21.88 Frustrated by the small Hall voltage obtained in blood flow measurements, a medical physicist decides to increase the applied magnetic field strength to get a 0.500 V output for blood moving at 30.0 cm/s in a 1.50 cm diameter vessel. (a) What magnetic field strength is needed? (b) What is unreasonable about this result? (c) Which premise is responsible?

•21.89 A surveyor 100 m from a long straight 200 kV DC power line suspects that its magnetic field may equal that of the earth and affect compass readings. (a) Calculate the current in the wire needed to create a 5.00×10^{-5} T field at this distance. (b) What is unreasonable about this result? (c) Which assumption or premise is responsible?

INDUCTION

AC Circuits and Instruments

These junkyard magnets are not turned on easily or turned off easily.

The energy stored in their magnetic fields can neither be supplied instantly

nor dispersed instantly.

Figure 22.1 Physics, like this butterfly, has inherent symmetries.

Nature's displays of symmetry are beautiful and alluring. A butterfly's wings exhibit an appealing symmetry in a complex system. (See Figure 22.1.) The laws of physics display symmetries at the most basic level—these symmetries are a source of wonder and imply deeper meaning. Since we place a high value on symmetry, we look for it when we explore nature. The remarkable thing is that we find it.

The hint of symmetry between electricity and magnetism found in the preceding chapter will be elaborated upon in this chapter. Specifically, we know that a current creates a magnetic field. If nature is symmetric here, then perhaps a magnetic field can create a current. The Hall effect, discussed in the preceding chapter, is a voltage caused by the action of a magnetic field. That voltage could drive a current. Historically, it was very shortly after Oersted discovered currents cause magnetic fields that other scientists asked the following question: Can magnetic fields cause currents? The answer was soon found by experiment to be yes. In 1831, some twelve years after Oersted's discovery, the English scientist Michael Faraday (1791–1862) and the American scientist Joseph Henry (1797–1878) independently demonstrated that magnetic fields can create currents. The basic process of creating emfs and, hence, currents with magnetic fields is known as **induction**.*

Today currents induced by magnetic fields are essential to our technological society. The ubiquitous generator—found in automobiles, on bicycles, in nuclear power plants, and so on—uses magnetism to generate current. Other devices that use magnetism to induce currents include pickup coils in electric guitars, transformers of every size, certain microphones, airport security gates, and damping mechanisms on sensitive chemical balances. Not so familiar perhaps, but important nevertheless, is that the behavior of AC circuits depends strongly on the effect of magnetic fields on currents.

22.1 INDUCED EMF AND MAGNETIC FLUX

The apparatus used by Faraday[†] to demonstrate that magnetic fields can create currents is illustrated in Figure 22.2. When the switch is closed, a magnetic field is created in the coil on the top and transmitted to the one on the bottom. The galvanometer is used to detect any current induced in the coil on the bottom. It was found that each time the switch is closed, the galvanometer detects a current in one direction in the coil on the bottom. (You can also observe this in a physics lab.) Each time the switch is opened, the galvanometer detects a current in the opposite direction. Interestingly, if the switch remains closed or open for any length of time, there is no current through the galvanometer. *Closing and opening the switch* induces the current. It is the *change* in magnetic field that creates the current. More basic than the current that flows is the emf $\mathcal{E}$ that causes it. The current is a result of an *emf induced by a changing magnetic field*, whether or not there is a path for current to flow.

---

*This process is also called magnetic induction to distinguish it from charging by induction, which utilizes the Coulomb force and was discussed in Chapter 17.

[†] Although Joseph Henry was first to make the discovery, Faraday published his results first and performed more detailed experiments, and the law of induction is given Faraday's name.

Figure 22.2 Faraday's apparatus for demonstrating that a magnetic field can create a current. It is found that a *change* in the field created by the top coil induces an emf and, hence, a current in the bottom coil. When the switch is opened and closed, the galvanometer registers currents in opposite directions. No current flows through the galvanometer when the switch remains closed or open.

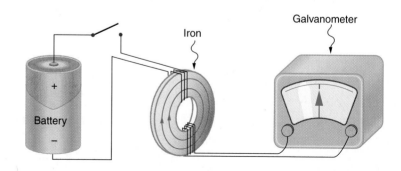

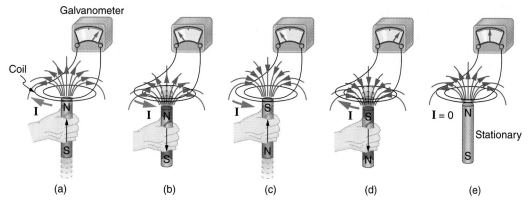

Galvanometer

Coil

(a)   (b)   (c)   (d)   (e)

Figure 22.3   Movement of a magnet relative to a coil produces emfs as shown. The same emfs are produced if the coil is moved relative to the magnet. The greater the speed, the greater the magnitude of the emf, and the emf is zero when there is no motion.

An experiment easily performed and often done in physics labs is illustrated in Figure 22.3. An emf is induced in the coil when a bar magnet is pushed in and out of it. Emfs of opposite signs are created by motion in opposite directions, and the emfs are also reversed by reversing poles. The same results are produced if the coil is moved rather than the magnet—it is the relative motion that is important. The faster the motion, the greater the emf, and there is no emf when the magnet is stationary relative to the coil.

The method of inducing an emf used in most generators is shown in Figure 22.4. A coil is rotated in a magnetic field, producing an AC emf, which depends on rotation rate and other factors that will be explored in later sections. Note that the generator is remarkably similar in construction to a motor (another symmetry).

By now it is clear that changing the magnitude or direction of a magnetic field produces an emf. Experiments revealed that there is a crucial quantity called the **magnetic flux**, Φ, given by

$$\Phi = BA \cos \theta \qquad (22.1)$$

where B is the magnetic field strength over an area A, at an angle θ with the perpendicular to the area as shown in Figure 22.5. **Any change in magnetic flux Φ induces an emf $\mathcal{E}$.** This process is defined to be **electromagnetic induction**. Units of magnetic flux Φ are $T \cdot m^2$. As seen in Figure 22.5, $B \cos \theta = B_\perp$, which is the component of B perpendicular to the area A. Thus magnetic flux is $\Phi = B_\perp A$, the product of the area and the component of the magnetic field perpendicular to it.

All induction, including the examples given so far, arises from some change in magnetic flux Φ. For example, Faraday changed B and hence Φ when opening and closing the switch in his apparatus (shown in Figure 22.2). This is also true for the bar magnet and coil shown in Figure 22.3. When rotating the coil of a generator, the angle θ and, hence, Φ is changed. Just how great an emf and what direction it takes depend on the change in Φ and how rapidly the change is made, as examined in the next section.

22.2   FARADAY'S LAW OF INDUCTION: LENZ'S LAW

Faraday's experiments showed that the emf $\mathcal{E}$ induced by a change in magnetic flux depends on only a few factors. First, $\mathcal{E}$ is directly proportional to the change in flux $\Delta\Phi$. Second, $\mathcal{E}$ is greatest when the change in time Δt is smallest—that is, $\mathcal{E}$ is inversely proportional to Δt. Finally, if a coil has N turns, an emf will be produced that is N times greater than for a single coil, so that $\mathcal{E}$ is directly proportional to N. The equation for the emf induced by a change in magnetic flux is

$$\mathcal{E} = -N \frac{\Delta\Phi}{\Delta t} \qquad (22.2)$$

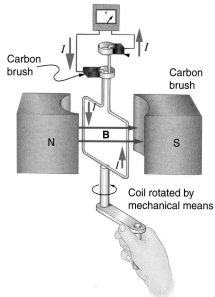

Carbon brush

Carbon brush

N   B   S

Coil rotated by mechanical means

Figure 22.4   Rotation of a coil in a magnetic field produces an emf. This is the basic construction of a generator, where work done to turn the coil is converted to electric energy. Note the generator is very similar in construction to a motor.

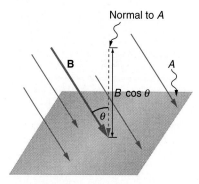

Normal to A

B

A

$B \cos \theta$

θ

$\Phi = BA \cos \theta = B_\perp A$

Figure 22.5   Magnetic flux Φ is related to the magnetic field and the area over which it exists. The flux $\Phi = BA \cos \theta$ is related to induction; any change in Φ induces an emf $\mathcal{E}$.

Figure 22.6 (a) When this bar magnet is thrust into the coil, the strength of the magnetic field increases in the coil. The current induced in the coil creates another field, in the opposite direction of the bar magnet's to oppose the increase. This is one aspect of *Lenz's law—induction opposes any change in flux.* (b) and (c) are two other situations. Verify for yourself that the direction of the induced B_{coil} shown indeed opposes the change in flux and that the current direction shown is consistent with RHR-2.

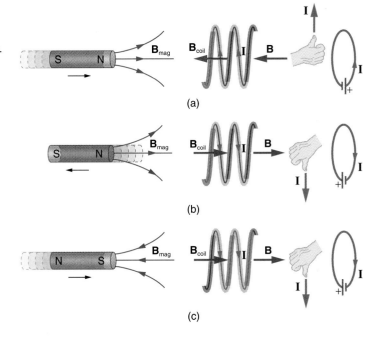

(a)

(b)

(c)

This relationship is known as **Faraday's law of induction**. The units of $\mathcal{E}$ are volts, as usual for an emf.

The minus sign in Faraday's law of induction is very important. The minus means that *the emf $\mathcal{E}$ creates a current I and magnetic field B that oppose the change in flux $\Delta\Phi$—this is known as Lenz's law.* The direction (given by the minus sign) of the emf $\mathcal{E}$ is so important that it is called **Lenz's law** after the Russian Heinrich Lenz, who, like Faraday and Henry, independently investigated aspects of induction. Faraday was aware of the direction, but Lenz stated it so clearly that he is credited for its discovery. (See Figure 22.6.)

PROBLEM-SOLVING STRATEGY

FOR LENZ'S LAW

To use Lenz's law to determine the directions of the induced magnetic fields, currents, and emfs:

> **Step 1.** *Make a sketch of the situation for use in visualizing and recording directions.*
> **Step 2.** *Determine the direction of the magnetic field B.*
> **Step 3.** *Determine whether the flux is increasing or decreasing.*
> **Step 4.** *Now determine the direction of the induced magnetic field B. It opposes the change in flux by adding or subtracting from the original field.*
> **Step 5.** *Use RHR-2 to determine the direction of the induced current I that is responsible for the induced magnetic field B.*
> **Step 6.** *The direction (or polarity) of the induced emf $\mathcal{E}$ will now drive a current in this direction and can be represented as current emerging from the positive terminal of the emf and returning to its negative terminal.*

CONNECTIONS

Conservation of Energy

Lenz's law is a manifestation of the conservation of energy. The induced emf creates a current that opposes the change in flux, because a change in flux means a change in energy. Energy can enter or leave, but not instantaneously. Lenz's law is a consequence. As the change begins, the law says induction opposes and, thus, slows the change. In fact, if the induced $\mathcal{E}$ were in the same direction as the change in flux, there would be a positive feedback that would give us free energy from no apparent source—conservation of energy would be violated.

For practice, apply these steps to the situations shown in Figure 22.6 and to others that are part of the following text material.

EXAMPLE 22.1 HOW GREAT IS THE INDUCED EMF?

Calculate the magnitude of the induced emf $\mathcal{E}$ when the magnet in Figure 22.6(a) is thrust into the coil, given the following information: The single loop coil has a radius of 6.00 cm and the average value of $B \cos \theta$ (this is given, since the bar magnet's field is complex) increases from 0.0500 to 0.250 T in 0.100 s.

(continued)

(continued)

Strategy To find the *magnitude* of $\mathcal{E}$, we use Faraday's law of induction as stated in Equation 22.2, but without the minus sign that indicates direction:

$$\mathcal{E} = N \frac{\Delta \Phi}{\Delta t}$$

Solution We are given that $N = 1$ and $\Delta t = 0.100$ s, but we must determine the change in flux $\Delta \Phi$ before we can find $\mathcal{E}$. Since the area of the loop is fixed, we see that

$$\Delta \Phi = \Delta(BA \cos \theta) = A \, \Delta(B \cos \theta)$$

Now $\Delta(B \cos \theta) = 0.200$ T, since it was given that $B \cos \theta$ changes from 0.0500 to 0.250 T. The area of the loop is

$A = \pi r^2 = (3.14)(0.060 \text{ m})^2 = 1.13 \times 10^{-2} \text{ m}^2$. Thus,

$$\Delta \Phi = (1.13 \times 10^{-2} \text{ m}^2)(0.200 \text{ T})$$

Entering the determined values into the expression for $\mathcal{E}$ gives

$$\mathcal{E} = N \frac{\Delta \Phi}{\Delta t} = \frac{(1.13 \times 10^{-2} \text{ m}^2)(0.200 \text{ T})}{0.100 \text{ s}} = 22.6 \text{ mV}$$

Discussion While this is an easily measured voltage, it is certainly not large enough for most practical applications. More loops in the coil, a stronger magnet, and faster movement make induction the practical source of voltages that it is.

22.3 MOTIONAL EMF

As we have seen, any change in magnetic flux induces an emf opposing that change—a process known as induction. Motion is one of the major causes of induction. For example, a magnet moved toward a coil induces an emf, and a coil moved toward a magnet produces a similar emf. In this section, we concentrate on motion in a magnetic field that is stationary relative to the earth, producing what is loosely called *motional emf*.

One situation where motional emf occurs is known as the Hall effect and was examined in Section 21.6. Charges moving in a magnetic field experience the magnetic force $F = qvB \sin \theta$, which moves opposite charges in opposite directions and creates an emf $\mathcal{E} = B\ell v$. We saw that the Hall effect has applications, including measurements of B and v. We will now see that the Hall effect is one aspect of the broader phenomenon of induction, and we will find that motional emf can be used as a power source.

Consider the situation shown in Figure 22.7. A rod is moved at a speed v along a pair of conducting rails separated by a distance ℓ, in a uniform magnetic field B. The rails are stationary relative to B and are connected to a stationary resistor R. The resistor could be anything from a light bulb to a voltmeter. Consider the area enclosed by the moving rod, rails, and resistor. B is perpendicular to this area, and the area is increasing as the rod moves. Thus the magnetic flux enclosed by the rails, rod, and resistor is increasing. When flux changes, an emf $\mathcal{E}$ is induced according to Faraday's law of induction.

To find the magnitude of $\mathcal{E}$ induced along the moving rod, we use Faraday's law of induction without the sign:

$$\mathcal{E} = N \frac{\Delta \Phi}{\Delta t}$$

Figure 22.7 (a) A motional emf $\mathcal{E} = B\ell v$ is induced between the rails when this rod moves to the right in the uniform magnetic field. The magnetic field **B** is into the page, perpendicular to the moving rod and rails and, hence, to the area enclosed by them. (b) Lenz's law gives the directions of the induced field and current, and the polarity of the induced emf. Since the flux is increasing, the induced field is in the opposite direction, or out of the page. RHR-2 gives the current direction shown, and the polarity of the rod will drive such a current. RHR-1 also indicates the same polarity for the rod.

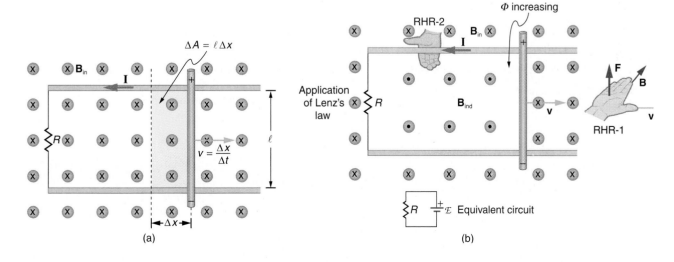

(a)   (b)

Unification of Forces

There are many connections between the electric force and the magnetic force. The fact that a moving electric field creates a magnetic field and, conversely, a moving magnetic field creates an electric field is part of why electric and magnetic forces are now considered to be different manifestations of the same force. This classic unification of electric and magnetic forces into what is called the electromagnetic force is the inspiration for contemporary efforts to unify other basic forces. See Sections 4.6 and 31.6.

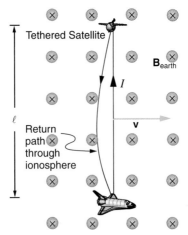

Figure 22.8 Motional emf as electrical power conversion for the space shuttle is the motivation for the Tethered Satellite experiment. A 5 kV emf is induced in the 20 km long tether while moving at orbital speed in the earth's magnetic field. The circuit is completed by a return path through the stationary ionosphere.

Here $N = 1$ and the flux $\Phi = BA \cos \theta$. We have $\theta = 0°$ and $\cos \theta = 1$, since B is perpendicular to A. Now $\Delta \Phi = \Delta(BA) = B \Delta A$, since B is uniform. Note that the area swept out by the rod is $\Delta A = \ell \Delta x$. Entering these quantities into the expression for $\mathcal{E}$ yields

$$\mathcal{E} = \frac{B \Delta A}{\Delta t} = B \frac{\ell \Delta x}{\Delta t}$$

Finally, note that $\Delta x/\Delta t = v$, the velocity of the rod. Entering this into the last expression shows that

$$\mathcal{E} = B\ell v \qquad (B, \ell, \text{ and } v \text{ perpendicular}) \qquad \text{(22.3)}$$

is the motional emf. This is the same expression given for the Hall effect in Equation 21.4.

To find the direction of the induced field, the direction of the current, and the polarity of the induced emf $\mathcal{E}$, we apply Lenz's law as explained in the preceding section. (See Figure 22.7(b).) Flux is increasing, since the area enclosed is increasing. Thus the induced field must oppose the existing one and be out of the page. RHR-2 requires that I be counterclockwise, which in turn means the top of the rod is positive as shown.

Motional emf also occurs if the magnetic field moves and the rod (or other object) is stationary relative to the earth (or some observer). We have seen an example of this in the situation where a moving magnet induces an emf in a stationary coil. It is the relative motion that is important. What is emerging in these observations is a connection between magnetic and electric fields. A moving magnetic field creates an electric field through its induced emf. We already have seen that a moving electric field creates a magnetic field—moving charge implies moving electric field and moving charge creates a magnetic field.

Motional emfs in the earth's weak field are not ordinarily very large, or we would notice voltage along metal rods, such as a screwdriver, during ordinary motions. For example, a simple calculation of the motional emf of a 1 m rod at 3.0 m/s perpendicular to the earth's field gives $\mathcal{E} = B\ell v = (5.0 \times 10^{-5}\text{ T})(1.0\text{ m})(3.0\text{ m/s}) = 150\ \mu\text{V}$. This small value is consistent with experience. There is a spectacular exception, however. In 1992 and 1996, attempts were made with the space shuttle to create large motional emfs. The Tethered Satellite was to be let out on a 20 km length of wire as shown in Figure 22.8, to create a 5 kV emf by moving at orbital speed through the earth's field. This emf could be used to convert some of the shuttle's kinetic and potential energy into electrical energy if a complete circuit could be made. To complete the circuit, the stationary ionosphere was to supply a return path for the current to flow. (The ionosphere is the rarefied and partially ionized atmosphere at orbital altitudes. It conducts because of the ionization. The ionosphere serves the same function as the stationary rails and connecting resistor in Figure 22.7, without which there would not be a complete circuit.) Drag on the current in the cable due to the magnetic force $F = I\ell B \sin \theta$ does the work that reduces the shuttle's kinetic and potential energy and allows it to be converted to electrical energy. The tests were both unsuccessful. In the first, the cable hung up and could only be extended a couple of hundred meters; in the second, the cable broke when almost fully extended. The following example indicates feasibility in principle.

EXAMPLE 22.2 LARGE MOTIONAL EMF

Calculate the motional emf induced along a 20.0 km long conductor moving at an orbital speed of 7.80 km/s perpendicular to the earth's 5.00×10^{-5} T magnetic field.

Strategy This is a straightforward application of the expression for motional emf—Equation 22.3 above.

Solution Entering the given values into Equation 22.3 gives

$$\mathcal{E} = B\ell v$$
$$= (5.00 \times 10^{-5}\text{ T})(2.0 \times 10^4\text{ m})(7.80 \times 10^3\text{ m/s})$$
$$= 7.80\text{ kV}$$

Discussion The value obtained is greater than the 5 kV design voltage for the shuttle experiment, since the actual orbital motion of the tether is not perpendicular to the earth's field. The 7.80 kV value is the maximum emf obtained when $\theta = 90°$ and $\sin \theta = 1$.

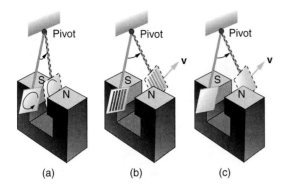

Figure 22.9 A common physics demonstration device for exploring eddy currents and magnetic damping. (a) The motion of a metal pendulum bob swinging between the poles of a magnet is quickly damped by the action of eddy currents. (b) There is little effect on the motion of a slotted metal bob, implying that eddy currents are made less effective. (c) There is also no magnetic damping on a nonconducting bob, since the eddy currents are extremely small.

22.4 EDDY CURRENTS AND MAGNETIC DAMPING

As discussed in the previous section, motional emf is induced when a conductor moves in a magnetic field or when a magnetic field moves relative to a conductor. If motional emf can cause a current loop in the conductor, we refer to that current as an **eddy current**. Eddy currents can create significant drag, called **magnetic damping**, on the motion involved. Consider the apparatus shown in Figure 22.9, which swings a pendulum bob between the poles of a strong magnet. (This is another favorite physics lab activity.) If the bob is metal, there is significant drag on the bob as it enters and leaves the field, quickly damping the motion. If, however, the bob is a slotted metal plate, as shown in Figure 22.9(b), there is a much smaller effect due to the magnet. There is no discernible effect on a bob made of an insulator. Why is there drag in both directions, and are there any uses for magnetic drag?

Figure 22.10 shows what happens to the metal plate as it enters and leaves the magnetic field. In both cases, it experiences a force opposing its motion. As it enters from the left, flux increases, and so an eddy current is set up (Faraday's law) in the counterclockwise direction (Lenz's law) as shown. Only the right-hand side of the current loop is in the field, so that there is an unopposed force on it to the left (RHR-1). When the metal plate is completely inside the field, there is no eddy current if the field is uniform, since the flux remains constant in this region. But when the plate leaves the field on the right, flux decreases, causing an eddy current in the clockwise direction that, again, experiences a force to the left, further slowing the motion. A similar analysis of what happens when the plate swings from the right toward the left shows that its motion is also damped when entering and leaving the field.

When a slotted metal plate enters the field, as shown in Figure 22.11, an emf is induced by the change in flux, but it is less effective because the slots limit the size of the current loops. Moreover, adjacent loops have currents in opposite directions, and their effects cancel. When an insulating material is used, eddy current is extremely small, and so magnetic damping on insulators is negligible. If eddy currents are to be avoided in conductors, then they can be slotted or constructed of thin layers of conducting material separated by insulating sheets.

One use of magnetic damping is found in sensitive laboratory balances. To have maximum sensitivity and accuracy, the balance must be as friction-free as possible. But if it is friction-free, then it will oscillate for a very long time. Magnetic damping is a simple and ideal solution. With magnetic damping, drag is proportional to speed and becomes zero at zero velocity. Thus the oscillations are quickly damped, after which the damping force disappears, allowing the balance to be very sensitive. (See Figure 22.12.) In most balances, magnetic damping is accomplished with a conducting disc that rotates in a fixed field.

Since eddy currents and magnetic damping occur only in conductors, recycling centers can use magnets to separate metals from other materials. Trash is dumped in

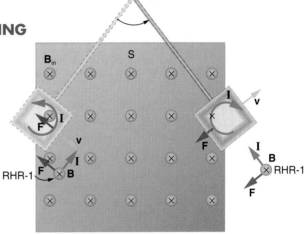

Figure 22.10 A more detailed look at the conducting plate passing between the poles of a magnet. As it enters and leaves the field, the change in flux creates an eddy current. Magnetic force on the current loop opposes the motion. There is no current and no magnetic drag when the plate is completely inside the uniform field.

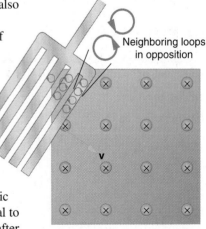

Figure 22.11 Eddy currents induced in a slotted metal plate entering a magnetic field form small loops, and the forces on them tend to cancel, thereby making magnetic drag almost zero.

Figure 22.12 Magnetic damping of this sensitive balance slows its oscillations. Since Faraday's law of induction gives the greatest effect for the most rapid change, damping is greatest for large oscillations and goes to zero as the motion stops.

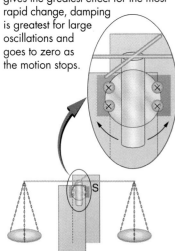

Figure 22.13 Metals can be separated from other trash by magnetic drag. Eddy currents and magnetic drag are created in the metals sent down this ramp by the powerful magnet beneath it. Nonmetals move on.

batches down a ramp, beneath which lies a powerful magnet. Conductors in the trash are slowed by magnetic damping while nonmetals in the trash move on, separating from the metals. (See Figure 22.13.) This works for all metals, not just ferromagnetic ones. A magnet can separate out the ferromagnetic materials alone by acting on stationary trash.

22.5 ELECTRIC GENERATORS

Electric generators induce an emf by rotating a coil in a magnetic field, as briefly discussed in Section 22.1. We will now explore generators in more detail. Consider the following example.

EXAMPLE 22.3 EMF INDUCED IN A GENERATOR COIL

The generator coil shown in Figure 22.14 is rotated through one-fourth of a revolution (from $\theta = 0°$ to $\theta = 90°$) in 15.0 ms. The 200-turn circular coil has a 5.00 cm radius and is in a uniform 1.25 T magnetic field. What is the average emf induced?

Strategy We use Faraday's law of induction to find the average emf induced over a time Δt:

$$\mathcal{E} = -N \frac{\Delta \Phi}{\Delta t}$$

We know that $N = 200$ and $\Delta t = 15.0$ ms, and so we must determine the change in flux $\Delta \Phi$ to find $\mathcal{E}$

Solution Since the area of the loop and the magnetic field strength are constant, we see that

$$\Delta \Phi = \Delta(BA \cos \theta) = AB \, \Delta(\cos \theta)$$

Now, $\Delta(\cos \theta) = -1.0$, since it was given that θ goes from $0°$ to $90°$. Thus $\Delta \Phi = -AB$, and

$$\mathcal{E} = N \frac{AB}{\Delta t}$$

The area of the loop is $A = \pi r^2 = (3.14)(0.0500 \text{ m})^2 = 7.85 \times 10^{-3}$ m^2. Entering this value gives

$$\mathcal{E} = 200 \frac{(7.85 \times 10^{-3} \text{ m}^2)(1.25 \text{ T})}{15.0 \times 10^{-3} \text{ s}} = 131 \text{ V}$$

Discussion This is a practical average value, similar to the 120 V used in household power.

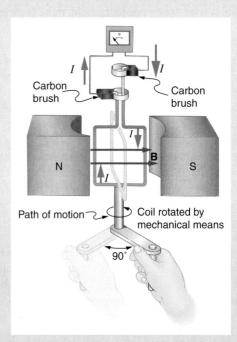

Figure 22.14 When this generator coil is rotated through one-fourth of a revolution, the magnetic flux Φ changes from its maximum to zero, inducing an emf.

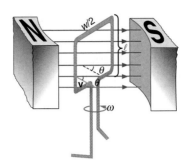

Figure 22.15 A generator with a single rectangular coil rotated at constant angular velocity in a uniform magnetic field produces an emf that varies sinusoidally in time. Note the generator is similar to a motor, except the shaft is rotated to produce a current rather than the other way around. (See Figure 21.30 in Section 21.8 on motors.)

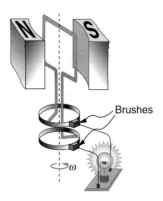

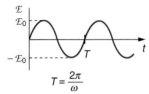

Figure 22.16 The emf of a generator is sent to a light bulb with the system of rings and brushes shown. The graph gives the emf $\mathcal{E}$ of the generator as a function of time. The period is $T = 1/f = 2\pi/\omega$, where f is the frequency.

The emf calculated in Example 22.3 is the average over one-fourth of a revolution. What is the emf at any given instant? It varies with the angle between the magnetic field and a perpendicular to the coil. We can get an expression for $\mathcal{E}$ as a function of time by considering the motional emf on a rotating rectangular coil of width w and height ℓ in a uniform magnetic field, as illustrated in Figure 22.15.

Charges in the wires of the loop experience the magnetic force, because they are moving in a magnetic field. Charges in the vertical wires experience forces parallel to the wire, causing currents. But those in the top and bottom segments feel a force perpendicular to the wire, which does not cause a current. We can thus find the induced emf by considering only the side wires. Motional emf is given by Equation 22.3 to be $\mathcal{E} = B\ell v$, where the velocity $\mathbf{v}$ is perpendicular to the magnetic field $\mathbf{B}$. Here the velocity is at an angle θ with $\mathbf{B}$, so that its component perpendicular to $\mathbf{B}$ is $v \sin \theta$ (see Figure 22.15). Thus in this case the emf induced on each side is $\mathcal{E} = B\ell v \sin \theta$, and they are in the same direction. The total emf around the loop is then

$$\mathcal{E} = 2B\ell v \sin \theta$$

This expression is valid, but it does not give $\mathcal{E}$ as a function of time. To find the time dependence of $\mathcal{E}$, we assume the coil rotates at a constant angular velocity ω (see Chapter 8, particularly Section 8.1, for definitions of angular quantities). The angle θ is related to angular velocity by $\theta = \omega t$, so that

$$\mathcal{E} = 2B\ell v \sin \omega t$$

Now, linear velocity v is related to angular velocity ω by $v = r\omega$. Here $r = w/2$, so that $v = (w/2)\omega$, and

$$\mathcal{E} = 2B\ell \frac{w}{2} \omega \sin \omega t = (\ell w) B\omega \sin \omega t$$

Noting that the area of the loop is $A = \ell w$, and allowing for N loops, we find that

$$\mathcal{E} = NAB\omega \sin \omega t \qquad \textbf{(22.4)}$$

is the **emf induced in a generator coil** of N turns and area A rotating at a constant angular velocity ω in a uniform magnetic field B. This can also be expressed as

$$\mathcal{E} = \mathcal{E}_0 \sin \omega t$$

where

$$\mathcal{E}_0 = NAB\omega \qquad \textbf{(22.5)}$$

$\mathcal{E}_0$ is the maximum (peak) emf. Note that the frequency of the oscillation is $f = \omega/2\pi$, and the period is $T = 1/f = 2\pi/\omega$. Figure 22.16 shows a graph of $\mathcal{E}$ as a function of time, and it now seems reasonable that AC voltage is sinusoidal.

The fact that the peak emf $\mathcal{E}_0 = NAB\omega$ makes good sense. The greater the number of coils, the larger their area, and the stronger the field, the greater the output voltage. It is interesting that the faster the generator is spun (greater ω), the greater the emf. This is noticeable on bicycle generators—at least the cheaper varieties. The author as a juvenile found it amusing to ride his bicycle fast enough to burn out his lights, until he had to ride home lightless one dark night.

Figure 22.17 shows a scheme by which a generator can be made to produce pulsed DC. More elaborate arrangements of multiple coils and split rings can produce smoother DC, although electronic rather than mechanical means are usually used to make ripple-free DC.

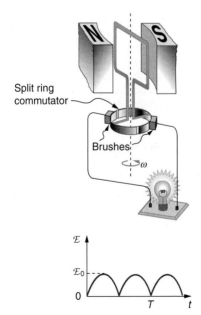

Figure 22.17 Split rings, called commutators, produce a pulsed DC emf in this configuration.

EXAMPLE 22.4 MAXIMUM EMF OF A GENERATOR

Calculate the maximum emf $\mathcal{E}_0$ of the generator that was the subject of Example 22.3.

Strategy Once ω, the angular velocity, is determined, Equation 22.5 can be used to find $\mathcal{E}_0$. All other quantities are known.

Solution Angular velocity is defined to be the change in angle per unit time:

$$\omega = \frac{\Delta\theta}{\Delta t}$$

One-fourth of a revolution is $\pi/2$ radians, and the time is 0.0150 s; thus,

$$\omega = \frac{\pi/2 \text{ rad}}{0.0150 \text{ s}}$$

$$= 104.7 \text{ rad/s} \qquad (\omega \text{ is exactly 1000 rpm})$$

We substitute this value for ω and the information from Example 22.3 into Equation 22.5, yielding

$$\mathcal{E}_0 = NAB\omega$$
$$= 200(7.85 \times 10^{-3} \text{ m}^2)(1.25 \text{ T})(104.7 \text{ rad/s})$$
$$= 206 \text{ V}$$

Discussion The maximum emf is greater than the average emf of 131 V found in Example 22.3, as it should be.

Generators illustrated in this section look very much like motors illustrated in Section 21.8. This is not coincidental. In fact, a motor becomes a generator when its shaft rotates. Certain early automobiles used their starter motor as a generator. In the next section, we shall further explore the action of a motor as a generator.

0 — 120 V
Could be greater if overspun

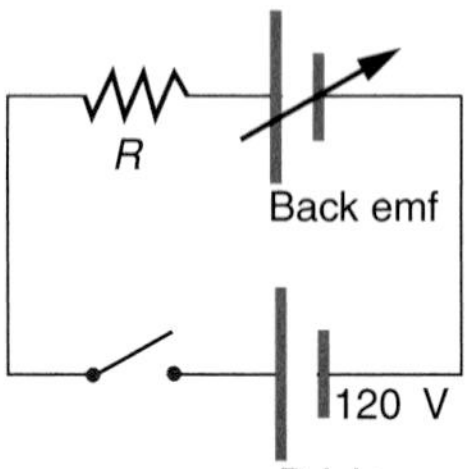

Figure 22.18 The coil of a DC motor is represented as a resistor in this schematic. The back emf is represented as a variable emf that opposes the one driving the motor. Back emf is zero when the motor is not turning, and it increases proportionally to the motor's angular velocity.

22.6 BACK EMF

It has been noted that motors and generators are very similar. Generators convert mechanical energy into electrical, whereas motors convert electrical energy into mechanical energy. Furthermore, motors and generators have the same construction. (Compare the figures of generators in the preceding section and those of motors in Section 21.8.) When the coil of a motor is turned, magnetic flux changes, and an emf (consistent with Faraday's law of induction) is induced. The motor thus acts as a generator whenever its coil rotates. This will happen whether the shaft is turned by an external input, like a belt drive, or by the action of the motor itself. That is, when a motor is doing work and its shaft is turning, an emf is generated. Lenz's law tells us the emf opposes any change, so that the input emf that powers the motor will be opposed by the motor's self-generated emf, called the **back emf** of the motor. (See Figure 22.18.)

Back emf is the generator output of a motor, and so it is proportional to the motor's angular velocity ω. It is zero when the motor is first turned on, meaning that the coil receives the full driving voltage and the motor draws maximum current when it is on but not turning. As the motor turns faster and faster, the back emf grows, always opposing the driving emf, and reduces the voltage across the coil and the amount of current it draws. This effect is noticeable in a number of situations. When a vacuum cleaner, refrigerator, or washing machine is first turned on, lights in the same circuit dim briefly due to the IR drop produced in feeder lines by the large current drawn by the motor. When a motor first comes on, it draws more current than when it runs at its normal operating speed. When a mechanical load is placed on the motor, like an electric wheelchair going up a hill, the motor slows, the back emf drops, more current flows, and more work can be done. If the motor runs at too low a speed, the larger current can overheat it (via resistive power in the coil, $P = I^2R$), perhaps even burning it out. On the other hand, if there is no mechanical load on the motor, it will increase its angular velocity ω until the back emf is nearly equal to the driving emf. Then the motor uses only enough energy to overcome friction.

Consider, for example, the motor coils represented in Figure 22.18. The coils have a 4.0 Ω equivalent resistance and are driven by a 120 V emf. Shortly after being turned on, they draw a current $I = V/R = (120 \text{ V})/(4.0 \text{ }\Omega) = 30$ A and, thus, dissipate $P = I^2R = 3.6 \text{ kW}$ of heat. Under normal operating conditions for this motor, suppose the back emf is 80 V. Then at operating speed, the total voltage across the coils is 40 V (120 V minus the 80 V back emf), and the current drawn is $I = V/R = (40 \text{ V})/(4.0 \text{ }\Omega) = 10 \text{ A}$. Under normal load, then, the power dissipated is 400 W. The latter will not cause a problem for this motor, whereas the former would burn out the coils if sustained.

22.7 TRANSFORMERS

Transformers do what their name implies—they transform voltages from one value to another.* For example, many telephones, video games, and hand vacuums have a transformer built into their plug-in unit (like that in Figure 22.19) that changes 120 V AC into whatever voltage the device uses. Transformers are also used at several points in the power distribution systems, such as illustrated in Figure 22.20. Power is sent long distances at high voltages, because less current is required for a given amount of power, and this means less line loss, as was discussed in Section 19.5. But high voltages pose greater hazards, so that transformers are employed to produce lower voltage at the user's location.

The type of transformer considered in this text—see Figure 22.21—is based on Faraday's law of induction and is very similar in construction to the apparatus Faraday used to demonstrate magnetic fields could cause currents (shown in Figure 22.2). The two coils are called the *primary* and *secondary coils*. In normal use, the input voltage is placed on the primary, and the secondary produces the transformed output voltage. Not only does the iron core trap the magnetic field created by the primary coil, its magnetization increases the field strength. Since the primary's input voltage is AC, a time-varying magnetic flux is sent to the secondary, inducing its AC output voltage.

For the simple transformer shown in Figure 22.21, the output voltage V_s depends almost entirely on the input voltage V_p and the ratio of the number of loops in the primary and secondary coils. Faraday's law of induction for the secondary coil gives its induced output voltage V_s to be

$$V_s = -N_s \frac{\Delta\Phi}{\Delta t}$$

where N_s is the number of loops in the secondary coil and $\Delta\Phi/\Delta t$ is the rate of change of magnetic flux. Note that the output voltage equals the induced emf ($V_s = \mathcal{E}_s$), provided coil resistance is small (a reasonable assumption for transformers). The cross-sectional area of the coils is the same on either side, as is the magnetic field strength, and

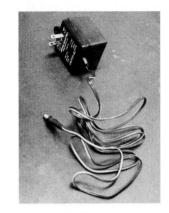

Figure 22.19 The plug-in transformer has become increasingly familiar with the proliferation of electronic devices that operate on voltages other than common 120 V AC. Most are in the 3 to 12 V range.

---

*The term voltage is used rather than emf, because transformers have internal resistance.

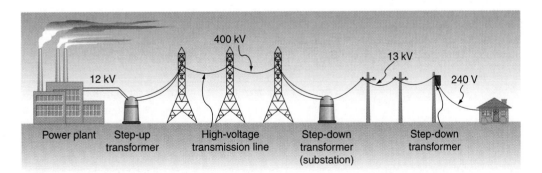

Figure 22.20 Transformers change voltages at several points in a power distribution system. Electric power is usually generated at greater than 10 kV, and transmitted long distances at voltages over 200 kV—sometimes as great as 700 kV—to limit energy losses. Local power distribution to neighborhoods or industries goes through a substation and is sent short distances at voltages ranging from 5 to 13 kV. This is reduced to 120, 240, or 480 V for safety at the individual user site.

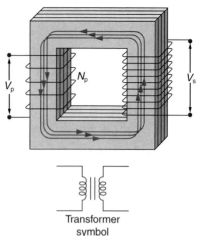

Figure 22.21 A typical construction of a simple transformer has two coils wound on a ferromagnetic core that is laminated to prevent eddy currents. The magnetic field created by the primary is mostly confined to and increased by the core, which transmits it to the secondary coil. Any change in current in the primary induces a current in the secondary.

so $\Delta\Phi/\Delta t$ is the same on either side. The input primary voltage V_p is also related to changing flux by

$$V_p = -N_p \frac{\Delta\Phi}{\Delta t}$$

The reason for this is a little more subtle. Lenz's law tells us that the primary coil opposes the change in flux caused by the input voltage V_p, hence the minus sign.* Assuming negligible coil resistance, Kirchhoff's loop rule tells us that the induced emf exactly equals the input voltage. Taking the ratio of these last two equations yields a useful relationship:

$$\frac{V_s}{V_p} = \frac{N_s}{N_p} \qquad (22.6)$$

This is known as the **transformer equation**, and it simply states that the ratio of the secondary to primary voltages in a transformer equals the ratio of the number of loops in their coils.

The output voltage of a transformer can be less than, greater than, or equal to the input voltage, depending on the number of loops in their coils. Some transformers even provide a variable output by allowing connection to be made at different points on the secondary coil. A **step-up transformer** is one that increases voltage, whereas a **step-down transformer** decreases voltage. Assuming, as we have, that resistance is negligible, the electrical power output of a transformer equals its input. This is nearly true in practice—transformer efficiency often exceeds 99%. Equating the power input and output,

$$P_p = I_p V_p = I_s V_s = P_s$$

Rearranging terms gives

$$\frac{V_s}{V_p} = \frac{I_p}{I_s}$$

Combining this with Equation 22.6, we find that

$$\frac{I_s}{I_p} = \frac{N_p}{N_s} \qquad (22.7)$$

is the relationship between the output and input currents of a transformer. So if voltage increases, current decreases. Conversely, if voltage decreases, current increases.

EXAMPLE 22.5 A STEP-UP TRANSFORMER

A portable x-ray unit has a step-up transformer, the 120 V input of which is transformed to the 100 kV output needed by the x-ray tube. The primary has 50 loops and draws a current of 10.0 A when in use. (a) What is the number of loops in the secondary? (b) Find the current output of the secondary.

Strategy and Solution for (a) We solve Equation 22.6 for N_s, the number of loops in the secondary, and enter the known values. This gives

$$N_s = N_p \frac{V_s}{V_p}$$

$$= (50)\frac{100,000 \text{ V}}{120 \text{ V}} = 4.17 \times 10^4$$

Discussion for (a) A large number of loops in the secondary (compared with the primary) is required to produce such a large voltage. This would be true for neon sign transformers and those supplying high voltage inside TVs.

Strategy and Solution for (b) We can similarly find the output current of the secondary by solving Equation 22.7 for I_s and entering known values. This gives

$$I_s = I_p \frac{N_p}{N_s}$$

$$= (10.0 \text{ A})\frac{50}{4.17 \times 10^4} = 12.0 \text{ mA}$$

Discussion for (b) As expected, the current output is significantly less than the input. In certain spectacular demonstrations, very large voltages are used to produce long arcs, but they are relatively safe because the transformer output does not supply a large current. Note that the power input here is $P_p = I_p V_p = (10.0 \text{ A})(120 \text{ V}) = 1.20 \text{ kW}$. This equals the power output $P_s = I_s V_s = (12.0 \text{ mA})(100 \text{ kV}) = 1.20 \text{ kW}$, as we assumed in the derivation of the equations used.

---

*This is an example of *self-inductance*, a topic to be explored in some detail in later sections.

The fact that transformers are based on Faraday's law of induction makes it clear why we cannot use transformers to change DC voltages. If there is no change in primary voltage, there is no voltage induced in the secondary. One possibility is to connect DC to the primary through a switch. As the switch is opened and closed, the secondary produces a voltage like that in Figure 22.22. This is not really a practical alternative, and AC is in common use wherever it is necessary to increase or decrease voltages.

EXAMPLE 22.6 A STEP-DOWN TRANSFORMER

A battery charger meant for a series connection of ten nickel-cadmium batteries (total emf of 12.5 V) needs to have a 15.0 V output to charge the batteries. It uses a step-down transformer with a 200-loop primary and a 120 V input. (a) How many loops should there be in the secondary coil? (b) If the charging current is 16.0 A, what is the input current?

Strategy and Solution for (a) You would expect the secondary to have a small number of loops. Solving Equation 22.6 for N_s and entering known values gives

$$N_s = N_p \frac{V_s}{V_p}$$

$$= (200) \frac{15.0 \text{ V}}{120 \text{ V}} = 25$$

Strategy and Solution for (b) The current input can be obtained by solving Equation 22.7 for I_p and entering

known values. This gives

$$I_p = I_s \frac{N_s}{N_p}$$

$$= (16.0 \text{ A}) \frac{25}{200} = 2.0 \text{ A}$$

Discussion The number of loops in the secondary is small, as expected for a step-down transformer. We also see that a small input current produces a larger output current in a step-down transformer. When transformers are used to operate large magnets, they sometimes have a small number of very heavy loops in the secondary. This allows the secondary to have low internal resistance and produce large currents. Note again that this solution is based on the assumption of 100% efficiency—or power out equals power in ($P_s = P_p$)—reasonable for good transformers.

Transformers have many applications in electrical safety systems, which are discussed in the next section.

22.8* ELECTRICAL SAFETY: SYSTEMS AND DEVICES

Electricity has two hazards. A *thermal hazard* occurs when there is electrical overheating. There is a *shock hazard* when electric current passes through a person. Both hazards were discussed in Section 19.6, and that material forms a useful background for this section. Here we will concentrate on systems and devices that prevent electrical hazards.

Figure 22.23 shows the schematic for a simple AC circuit with no safety features. This is not how power is distributed in practice. Modern household and industrial wiring requires the **three-wire system**, shown schematically in Figure 22.24, which has several safety features (although it is named for the three-wire cords and plugs employed). First is the familiar *circuit breaker* (or *fuse*) to prevent thermal overload. Second, there is a *case* around the appliance, such as a toaster or refrigerator. The case's safety feature is that it prevents a person from touching exposed wires and coming into electrical contact with the circuit, helping prevent shocks.

There are also *three connections to ground* shown in Figure 22.24. Recall that a ground connection is a low-resistance path directly to the earth, defined in previous chapters. The two ground connections on the *neutral wire* force it to be at zero volts relative to the earth, giving the wire its name. This wire is therefore safe to touch even if its insulation, usually white, is missing. The neutral wire is the return path for the current to follow to complete the circuit. Furthermore, the two ground connections supply an alternative path through the earth, a good conductor, to complete the circuit. The ground connection closest to the power source could be at the generating plant, while the other is at the user's location. The third ground is to the case of the appliance, through the green *ground wire*, forcing the case, too, to be at zero volts. The *hot wire* supplies voltage and current to operate the appliance and is usually insulated with black plastic. Figure 22.25 shows a more pictorial version of how the three-wire system is connected through a three-prong plug to an appliance.

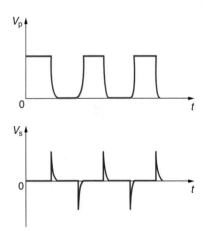

Figure 22.22 Transformers do not work for pure DC voltage input, but if it is switched on and off as on the top graph, the output will look something like that on the bottom graph. This is not the sinusoidal AC most AC appliances need.

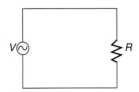

Figure 22.23 Schematic of a simple AC circuit with a voltage source and a single appliance represented by the resistance R. There are no safety features in this circuit.

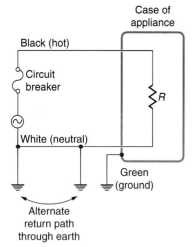

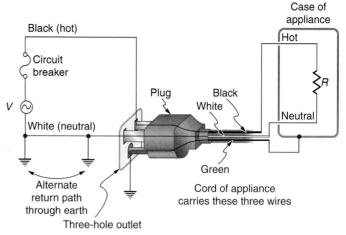

Figure 22.24 The three-wire system grounds the neutral wire at the voltage source and user location, forcing it to be at zero volts and supplying an alternative return path for the current through the earth. Also grounded to zero volts is the case of the appliance. A circuit breaker or fuse protects against thermal overload and is in series on the hot wire. The colors are those of the wire insulation.

Figure 22.25 The standard three-prong plug can only be inserted in one way, to assure proper function of the three-wire system.

Human Application

The three-wire system replaced the older two-wire system, which lacks a ground wire. Under ordinary circumstances, insulation on the hot and neutral wires prevents the case from being directly in the circuit, so that the ground wire may seem like double protection. Grounding the case solves more than one problem, however. The simplest problem is worn insulation on the hot wire that allows it to contact the case, as shown in Figure 22.26. Lacking a ground connection (some people cut the third prong off the plug because they only have two hole receptacles), a severe shock is possible. This is particularly dangerous in the kitchen, where a good connection to ground is available through a water pipe. With the ground connection intact, the circuit breaker will trip, forcing repair of the appliance. Why are some appliances still sold with two-prong plugs? These have nonconducting cases, such as power tools with impact resistant plastic cases, and are called *doubly insulated*. Modern two-prong plugs can be inserted into the asymmetric standard outlet in only one way, to ensure proper connection of hot and neutral wires.

Electromagnetic induction causes a more subtle problem that is solved by grounding the case. The AC current in appliances can induce an emf on the case. If grounded, the case voltage is kept near zero, but if the case is not grounded a shock

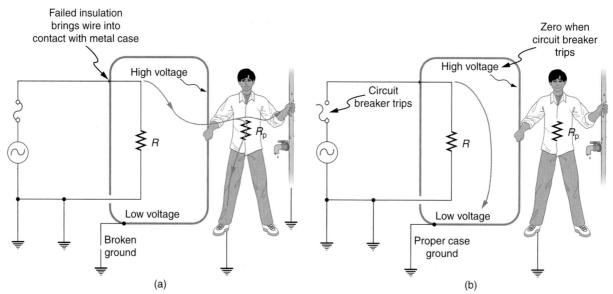

Figure 22.26 Worn insulation allows the hot wire to come into direct contact with the metal case of this appliance. (a) The ground connection being broken, the person is severely shocked. The appliance may operate normally in this situation. (b) With a proper ground, the circuit breaker trips, forcing repair of the appliance.

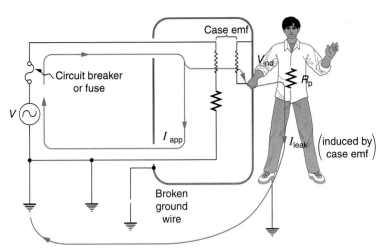

Figure 22.27 AC currents can induce an emf on the case of an appliance. The voltage can be large enough to cause a shock. If the case is grounded, the induced emf is kept near zero.

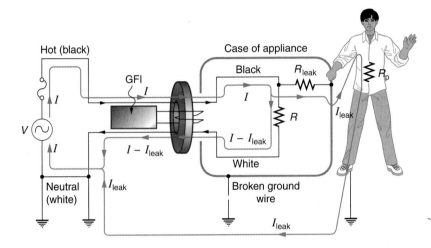

Figure 22.28 A ground fault interrupter (GFI) compares the currents in the hot and neutral wires and will trip if their difference exceeds a safe value. The leakage current here follows a hazardous path that could have been prevented by an intact ground wire.

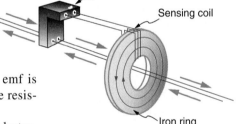

Figure 22.29 A GFI compares currents by using both to induce an emf in the same coil. If the currents are equal, they will induce equal but opposite emfs.

can occur as pictured in Figure 22.27. Current driven by the induced case emf is called a *leakage current*, although current does not necessarily pass from the resistor to the case.

A *ground fault interrupter* (GFI) is a safety device that works based on electromagnetic induction. GFIs compare the currents in the hot and neutral wires. When hot and neutral currents are not equal, it is almost always because current in the neutral is less than in the hot wire. Then some of the current, again called a leakage current, is returning to the voltage source by a path other than through the neutral wire. It is assumed that this path presents a hazard, such as shown in Figure 22.28. GFIs are usually set to interrupt the circuit if the leakage current is greater than 5 mA, the accepted maximum harmless shock. Even if the leakage current goes safely to ground through an intact ground wire, the GFI will trip, forcing repair of the leakage.

Figure 22.29 shows how a GFI works. If the currents in the hot and neutral wires are equal, then they induce equal and opposite emfs in the coil. If not, then the circuit breaker will trip. One problem with GFIs is that the currents are AC and can get out of phase. This will cause the GFI to trip even when there is not a leakage current.

Another induction-based safety device is the *isolation transformer*, shown in Figure 22.30. Most isolation transformers have equal input and output voltages. Their function is to put a large resistance between the original voltage source and the device being operated. This prevents a complete circuit between them even in the circumstance shown. There is a complete circuit through the appliance. But there is not a complete circuit for current to flow through the person in the figure, who is touching only one of the transformer's output wires, and neither output wire is grounded. The appliance is isolated from the original voltage source by the high resistance of the material between the transformer coils, hence the name isolation transformer. For current to flow through the person, it must pass through the high-resistance material between the coils, through the

Human Application

Figure 22.30 An isolation transformer puts a large resistance between the original voltage source and the device, preventing a complete circuit between them.

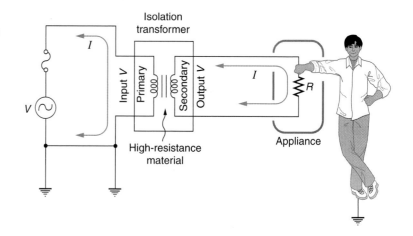

wire, the person, and back through the ground—a path with such a large resistance that the current is negligible.

The basics of electrical safety presented here help prevent many electrical hazards. Electrical safety can be pursued to greater depths. There are, for example, problems related to different ground connections for appliances in close proximity. Many other examples are found in hospitals. Microshock sensitive patients, for instance, require special protection. The interested reader can use this section as a basis for further study.

22.9 INDUCTANCE

Induction is the process in which an emf is induced by changing magnetic flux. Many examples have been discussed so far, some more effective than others. Transformers, for example, are designed to be particularly effective at inducing a desired voltage and current with very little loss of energy to other forms. Is there a useful physical quantity related to how "effective" a given device is? The answer is yes, and that physical quantity is called **inductance**.

Mutual inductance is the effect of Faraday's law of induction for one device upon another, such as the primary coil in transmitting energy to the secondary in a transformer. See Figure 22.31, where simple coils induce emfs in one another. In the many cases where the geometry of the devices is fixed, flux is changed by varying current. We therefore concentrate on the rate of change of current, $\Delta I/\Delta t$, as the cause of induction. A change in the current I_1 in one device, coil 1 in the figure, induces an emf $\mathcal{E}_2$ in the other. We express this in equation form as

$$\mathcal{E}_2 = -M \frac{\Delta I_1}{\Delta t} \qquad \textbf{(22.8a)}$$

where M is defined to be the **mutual inductance** between the two devices. The minus sign is an expression of Lenz's law. The larger the mutual inductance M, the more effective the coupling. For example, the coils in Figure 22.31 have a small M compared with the transformer coils in Figure 22.21. Units for M are $(V \cdot s)/A = \Omega \cdot s$, which is named a **henry** (H), after Joseph Henry. That is, 1 H = 1 $\Omega \cdot s$.

Nature is symmetric here. If we change the current I_2 in coil 2, we induce an emf $\mathcal{E}_1$ in coil 1, which is given by

$$\mathcal{E}_1 = -M \frac{\Delta I_2}{\Delta t} \qquad \textbf{(22.8b)}$$

where M is the same as for the reverse process. Transformers run backward with the same effectiveness, or mutual inductance M.

Mutual inductance may or may not be desirable. We want a transformer to have a large mutual inductance. But an appliance, such as an electric clothes dryer, can induce

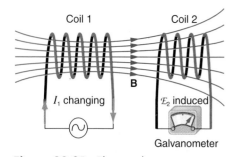

Figure 22.31 These coils can induce emfs in one another like an inefficient transformer. Their mutual inductance M indicates the effectiveness of the coupling between them. Here a change in current in coil 1 is seen to induce an emf in coil 2.

a dangerous emf on its case if the mutual inductance between its coils and the case is large. One way to reduce mutual inductance M is to counter-wind coils to cancel the magnetic field produced. (See Figure 22.32.)

Self-inductance, the effect of Faraday's law of induction of a device on itself, also exists. When, for example, current through a coil is increased, the magnetic field and flux also increase, inducing a counter emf as required by Lenz's law. Conversely, if the current is decreased, an emf is induced that opposes the decrease. Most devices have a fixed geometry, and so the change in flux is due entirely to the change in current ΔI through the device. The induced emf is related to the physical geometry of the device and the rate of change of current. It is given by

$$\mathcal{E} = -L \frac{\Delta I}{\Delta t} \qquad \textbf{(22.9)}$$

where L is the **self-inductance** of the device. A device that exhibits significant self-inductance is called an **inductor**, and given the symbol $\sim\!\!\sim\!\!\sim$. The minus sign is an expression of Lenz's law, indicating that $\mathcal{E}$ opposes the change in current. Units of self-inductance are henries (H) just as for mutual inductance. The larger the self-inductance L of a device, the greater its opposition to any change in current through it. For example, a large coil with many turns and an iron core has a large L and will not allow current to change quickly. To avoid this effect, a small L must be achieved, such as by counter-winding coils as in Figure 22.32.

A 1 H inductor is a large inductor. To illustrate this, consider a device with $L = 1.0$ H that has a 10 A current flowing through it. What happens if we try to shut off the current rapidly, perhaps in only 1.0 ms? An emf, given by Equation 22.9, will oppose the change. Thus an emf will be induced given by $\mathcal{E} = -L\,\Delta I/\Delta t = -(1.0\ \text{H})(-10\ \text{A})/(1.0\ \text{ms}) = 10{,}000$ V. The positive sign means this large voltage is in the same direction as the current, opposing its decrease. Such large emfs can cause arcs, damaging switching equipment, and so it may be necessary to change current more slowly. There are uses for such a large induced voltage. Camera flash attachments use a battery, an inductor, and a switching system to induce large voltages when the inductor is switched off and on. A capacitor is used to store the high voltage for later use in powering the flash. (See Figure 22.33.)

It is possible to calculate L for an inductor given its geometry (size and shape) and knowing the magnetic field that it creates. This is difficult in most cases, because of the complexity of the field created. So in this text the inductance L is usually a given quantity. One exception is the solenoid, because it has a very uniform field inside, a nearly zero field outside, and a simple shape. It is instructive to derive an equation for its inductance. We start by noting that the induced emf is given by Faraday's law of induction as $\mathcal{E} = -N\,\Delta\Phi/\Delta t$ and, by the definition of self-inductance, as $\mathcal{E} = -L\,\Delta I/\Delta t$. Equating these yields

$$\mathcal{E} = -N \frac{\Delta\Phi}{\Delta t} = -L \frac{\Delta I}{\Delta t}$$

Solving for L gives

$$L = N \frac{\Delta\Phi}{\Delta I} \qquad \textbf{(22.10)}$$

This equation for the self-inductance L of a device is always valid. It means that self-inductance L depends on how effective the current is in creating flux; the more effective, the greater $\Delta\Phi/\Delta I$ is.

Let us use this last equation to find an expression for the inductance of a solenoid. Since the area A of a solenoid is fixed, the change in flux is $\Delta\Phi = \Delta(BA) = A\,\Delta B$. To find ΔB, we note that the magnetic field of a solenoid is given by Equation 21.9 to be $B = \mu_0 nI = \mu_0 NI/\ell$. (Here $n = N/\ell$, where N is the number of coils and ℓ is the solenoid's length.) Only the current changes, so that $\Delta\Phi = A\,\Delta B = \mu_0 NA\,\Delta I/\ell$. Substituting $\Delta\Phi$ into Equation 22.10 gives

$$L = N \frac{\Delta\Phi}{\Delta I} = N \frac{\mu_0 NA\,\Delta I/\ell}{\Delta I}$$

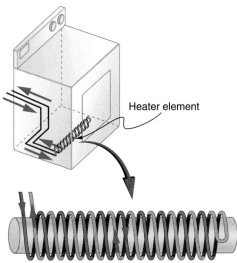

Figure 22.32 The heating coils of an electric clothes dryer can be counter-wound so that their magnetic fields cancel one another, greatly reducing the mutual inductance with the case of the dryer.

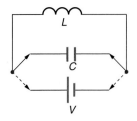

Figure 22.33 Through rapid switching of an inductor, 1.5 V batteries can be used to induce emfs of several thousand volts. This voltage can be used to store charge in a capacitor for later use, such as in a camera flash attachment.

This simplifies to

$$L = \frac{\mu_0 N^2 A}{\ell} \qquad \text{(solenoid)} \qquad \textbf{(22.11)}$$

This is the self-inductance of a solenoid of cross-sectional area A and length ℓ. Note that the inductance depends only on the physical characteristics of the solenoid, consistent with its definition.

EXAMPLE 22.7 SELF-INDUCTANCE OF A MODERATE SIZE SOLENOID

Calculate the self-inductance of a 10.0 cm long, 4.00 cm diameter solenoid that has 200 coils.

Strategy This is a straightforward application of Equation 22.11, since all quantities in the equation except L are known.

Solution Equation 22.11 gives the following expression for the self-inductance of a solenoid:

$$L = \frac{\mu_0 N^2 A}{\ell}$$

The cross-sectional area in this example is $A = \pi r^2 = (3.14)(0.0200 \text{ m})^2 = 1.26 \times 10^{-3} \text{ m}^2$, N is given to be 200,

and the length ℓ is 0.100 m. We know the permeability of free space is $m_0 = 4\pi \times 10^{-7}$ T·m/A. Substituting these into the expression for L gives

$$L = \frac{(4\pi \times 10^{-7} \text{ T·m/A})(200)^2(1.26 \times 10^{-3} \text{ m}^2)}{0.100 \text{ m}}$$

$$= 0.632 \text{ mH}$$

Discussion This solenoid is moderate in size. Its inductance of nearly a millihenry is also considered moderate.

Energy Stored in an Inductor

We know from Lenz's law that inductances oppose changes in current. There is an alternative way to look at this opposition that is based on energy. Energy is stored in a magnetic field. It takes time to build up energy, and it also takes time to deplete energy; hence, there is an opposition to rapid change. In an inductor, the magnetic field is directly proportional to current and to the inductance of the device. It can be shown that the **energy stored in an inductor** E_{ind} is given by

$$E_{ind} = \tfrac{1}{2}LI^2 \qquad \textbf{(22.12)}$$

This expression is similar to that for the energy stored in a capacitor.

EXAMPLE 22.8 ENERGY STORED IN THE FIELD OF A SOLENOID

How much energy is stored in the 0.632 mH inductor of the preceding example when a 30.0 A current flows through it?

Strategy The energy is given by Equation 22.12, and all quantities except E_{ind} are known.

Solution Substituting the value for L found in the previ-

ous example and the given current into Equation 22.12 gives

$$E_{ind} = \tfrac{1}{2}LI^2$$

$$= 0.5(0.632 \times 10^{-3} \text{ H})(30.0 \text{ A})^2 = 0.284 \text{ J}$$

Discussion This amount of energy is certainly enough to cause a spark if the current is suddenly switched off. It cannot be built up instantaneously unless the power input is infinite.

22.10 RL CIRCUITS

We know that the current through an inductor L cannot be turned on or off instantaneously. The change in current changes flux, inducing an emf opposing the change (Lenz's law). How long does the opposition last? Current *will* flow and *can* be turned off, but how long does it take? Figure 22.34 shows a switching circuit that can be used to examine current through an inductor as a function of time.

When the switch is first moved to position 1 (at $t = 0$), the current is zero and it eventually rises to $I_0 = V/R$, where R is the total resistance of the circuit. The opposition of the inductor L is greatest at the beginning, because the amount of change is greatest. The opposition it poses is in the form of an induced emf, which decreases to zero as the current approaches its final value. The opposing emf is proportional to the amount of change left. This is the hallmark of an exponential behavior, and it can be shown with calculus that

$$I = I_0(1 - e^{-t/\tau}) \qquad \text{(turning on)} \qquad \textbf{(22.13)}$$

is the current in an *RL* circuit when switched on.* The initial current is zero and approaches $I_0 = V/R$ with **a characteristic time constant τ for an *RL* circuit**, given by

$$\tau = \frac{L}{R} \qquad \textbf{(22.14)}$$

where τ has units of seconds, since 1 H = 1 $\Omega\cdot$s. In the first period of time τ, the current rises from zero to $0.632 I_0$, since $I = I_0(1 - e^{-1}) = I_0(1 - 0.368) = 0.632 I_0$. The current will go 0.632 of the remainder in the next time τ. It is a characteristic of the exponential that the final value is never reached, but 0.632 of the remainder to that value is achieved in every characteristic time τ. In just a few multiples of the time τ, the final value is very nearly achieved, as the graph in Figure 22.34(b) illustrates.

The characteristic time τ only depends on two factors, the inductance L and the resistance R. The greater the inductance L, the greater τ is, which makes sense since a large inductance is very effective in opposing change. The smaller the resistance R, the greater τ is. Again this makes sense, since a small resistance means a large final current and a greater change to get there. In both cases—large L and small R—more energy is stored in the inductor and more time is required to get it in and out.

When the switch in Figure 22.34(a) is moved to position 2 and cuts the battery out of the circuit, the current drops because of energy dissipation by the resistor. But this is also not instantaneous, since the inductor opposes the decrease in current by inducing an emf in the same direction as the battery that drove the current. Furthermore, there is a certain amount of energy, $(1/2)LI_0^2$, stored in the inductor, and it is dissipated at a finite rate. As the current approaches zero, the rate of decrease slows, since the energy dissipation rate is I^2R. Once again the behavior is exponential, and I is found to be

$$I = I_0 e^{-t/\tau} \qquad \text{(turning off)} \qquad \textbf{(22.15)}$$

(See Figure 22.34(c).) In the first period of time $\tau = L/R$ after the switch is closed, the current falls to 0.368 of its initial value, since $I = I_0 e^{-1} = 0.368 I_0$. In each successive time τ, the current falls to 0.368 of the preceding value, and in a few multiples of τ, the current becomes very close to zero as seen in the graph in Figure 22.34(c).

*Note the similarity to the exponential behavior of the voltage on a charging capacitor, given by Equation 20.4 in Section 20.6.

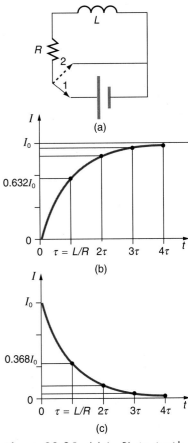

Figure 22.34 (a) An *RL* circuit with a switch to turn current on and off. When in position 1, the battery, resistor, and inductor are in series and a current is established. In position 2, the battery is removed and the current stops because of energy loss in the resistor. (b) A graph of current growth versus time when the switch is moved to position 1. (c) A graph of current decay when the switch is moved to position 2.

EXAMPLE 22.9 AN *RL* CIRCUIT

(a) What is the characteristic time constant for a 7.50 mH inductor in series with a 3.00 Ω resistor such as shown in Figure 22.34(a)? (b) Find the current 5.00 ms after the switch is moved to position 2 to disconnect the battery, if it is initially 10.0 A.

Strategy for (a) The time constant for an *RL* circuit is defined by Equation 22.14.

Solution for (a) Entering known values into the expression for τ given in Equation 22.14 yields

$$\tau = \frac{L}{R} = \frac{7.50 \text{ mH}}{3.00 \ \Omega} = 2.50 \text{ ms}$$

Discussion for (a) This is a small but definitely finite time. The coil will be very close to its full current in about ten time constants, or about 25 ms.

Strategy for (b) We can find the current by using Equation 22.15, or by considering the decline in steps. Since the time is twice the characteristic time, we consider the process in steps.

Solution for (b) In the first 2.50 ms, the current declines to 0.368 of its initial value, which is

$$I = 0.368 I_0 = (0.368)(10.0 \text{ A})$$
$$= 3.68 \text{ A} \qquad \text{at } t = 2.50 \text{ ms}$$

After another 2.50 ms, or a total of 5.00 ms, the current declines to 0.368 of the value just found. That is,

$$I' = 0.368I = (0.368)(3.68 \text{ A})$$
$$= 1.35 \text{ A} \qquad \text{at } t = 5.00 \text{ ms}$$

Discussion for (b) After another 5.00 ms has passed, the current will be 0.183 A (left as an end-of-chapter problem); so, although it does die out, the current certainly does not go to zero instantaneously.

In summary, when the voltage applied to an inductor is changed, the current also changes, *but the change in current lags the change in voltage in an RL circuit*. In this section, we looked at what happens when the voltage is simply switched on and off. In the next section, we explore how an *RL* circuit behaves when a sinusoidal AC voltage is applied.

22.11 REACTANCE, INDUCTIVE AND CAPACITIVE

Simple AC circuits containing only resistors and an AC voltage source were covered in Section 19.5. Many circuits also contain capacitors and inductors. We have seen how capacitors and inductors respond to DC voltage when it is switched on and off (Sections 20.6 and 22.10, respectively). We will now explore how inductors and capacitors react to sinusoidal AC voltage.

Inductors and Inductive Reactance

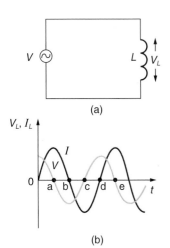

Suppose an inductor is connected directly to an AC voltage source as shown in Figure 22.35. It is reasonable to assume negligible resistance, since in practice we can make the resistance of an inductor so small that it has a negligible effect on the circuit. Also shown is a graph of voltage and current as functions of time.

The graph in Figure 22.35(b) starts with voltage at a maximum. Note that the current starts at zero and rises to its peak *after* the voltage that drives it, just as was the case when DC voltage was switched on in the preceding section. When the voltage becomes negative at point a, the current begins to decrease; it becomes zero at point b, where voltage is its most negative. The current then becomes negative, again following the voltage. The voltage becomes positive at point c and begins to make the current less negative. At point d, the current goes through zero just as the voltage reaches its positive peak to start another cycle. This behavior is summarized as follows:

> **When a sinusoidal voltage is applied to an inductor, the voltage leads the current by one-fourth of a cycle, or by a 90° phase angle.**

Figure 22.35 (a) An AC voltage source in series with an inductor *L* having negligible resistance. (b) Graph of current and voltage across the inductor as functions of time.

Current lags behind voltage, since inductors oppose change in current. Changing current induces a back emf $V = -L\,\Delta I/\Delta t$. This is considered to be an effective resistance of the inductor to AC. The rms current I through an inductor L is given by a version of Ohm's law:

$$I = \frac{V}{X_L} \tag{22.16}$$

where V is the rms voltage across the inductor and X_L is defined to be

$$X_L = 2\pi f L \tag{22.17}$$

with f the frequency of the AC voltage source in hertz.* X_L is called the **inductive reactance**, because the inductor reacts to impede the current. X_L has units of ohms (1 H = 1 Ω·s, so that frequency times inductance has units of (cycles/s)(Ω·s) = Ω), consistent with its role as an effective resistance. It makes sense that X_L is proportional to L, since the greater the induction the greater its resistance to change. It is also reasonable that X_L is proportional to frequency f, since greater frequency means greater change in current. That is, $\Delta I/\Delta t$ is large for large frequencies (large f, small Δt). The greater the change, the greater the opposition of an inductor.

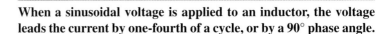

EXAMPLE 22.10 INDUCTIVE REACTANCE AND CURRENT

(a) Calculate the inductive reactance of a 3.00 mH inductor when 60.0 Hz and 10.0 kHz AC voltages are applied. (b) What is the rms current at each frequency if the applied rms voltage is 120 V?

Strategy The inductive reactance is found directly from the expression in Equation 22.17. Once X_L has been found at each frequency, Ohm's law as stated in Equation 22.16 can be used to find the current at each frequency.

Solution for (a) Entering the frequency and inductance

(continued)

*An analysis of the circuit using Kirchhoff's loop rule and calculus actually produces this expression for X_L.

(continued)

into Equation 22.17 gives

$$X_L = 2\pi fL = 6.28(60.0/s)(3.00 \text{ mH}) = 1.13 \ \Omega$$
$$\text{at } 60 \text{ Hz}$$

Similarly, at 10 kHz,

$$X_L = 2\pi fL = 6.28(1.00 \times 10^4/s)(3.00 \text{ mH}) = 188 \ \Omega$$
$$\text{at } 10 \text{ kHz}$$

Solution for (b) The rms current is now found using the version of Ohm's law in Equation 22.16, given the applied rms voltage is 120 V. For the first frequency, this yields

$$I = \frac{V}{X_L} = \frac{120 \text{ V}}{1.13 \ \Omega} = 106 \text{ A} \qquad \text{at } 60 \text{ Hz}$$

Similarly, at 10 kHz,

$$I = \frac{V}{X_L} = \frac{120 \text{ V}}{188 \ \Omega} = 0.637 \text{ A} \qquad \text{at } 10 \text{ kHz}$$

Discussion The inductor reacts very differently at the two different frequencies. At the higher frequency, its reactance is large and the current is small, consistent with how an inductor impedes rapid change. Thus high frequencies are impeded the most. Inductors can be used to filter out high frequencies; for example, a large inductor can be put in series with a sound reproduction system or in series with your home computer to reduce high-frequency sound output from your speakers or high-frequency power spikes into your computer.

Note that although the resistance in the circuit considered is negligible, the AC current is not extremely large because inductive reactance impedes its flow. With AC, there is no time for the current to become extremely large.

Capacitors and Capacitive Reactance

Consider the capacitor connected directly to an AC voltage source as shown in Figure 22.36. The resistance of a circuit like this can be made so small that it has a negligible effect compared with the capacitor, and so we can assume negligible resistance. Voltage across the capacitor and current are graphed as functions of time in the figure.

The graph in Figure 22.36 starts with voltage across the capacitor at a maximum. The current is zero at this point, because the capacitor is fully charged and halts the flow. Then voltage drops and the current becomes negative as the capacitor discharges. At point a, the capacitor has fully discharged ($Q = 0$ on it) and the voltage across it is zero. The current remains negative between points a and b, causing the voltage on the capacitor to reverse. This is complete at point b, where the current is zero and the voltage has its most negative value. The current becomes positive after point b, neutralizing the charge on the capacitor and bringing the voltage to zero at point c, which allows the current to reach its maximum. Between points c and d, the current drops to zero as the voltage rises to its peak, and the process starts to repeat. Throughout the cycle, the voltage follows what the current is doing by one-fourth of a cycle:

When a sinusoidal voltage is applied to a capacitor, the voltage follows the current by one-fourth of a cycle, or by a 90° phase angle.

The capacitor is affecting the current, having the ability to stop it altogether when fully charged. Since an AC voltage is applied, there is an rms current, but it is limited by the capacitor. This is considered to be an effective resistance of the capacitor to AC, and so the rms current I in the circuit containing only a capacitor C is given by another version of Ohm's law to be

$$I = \frac{V}{X_C} \qquad \textbf{(22.18)}$$

where V is the rms voltage and X_C is defined* to be

$$X_C = \frac{1}{2\pi fC} \qquad \textbf{(22.19)}$$

where X_C is called the **capacitive reactance**, because the capacitor reacts to impede the current. X_C has units of ohms (verification left as an exercise for the reader). X_C is inversely

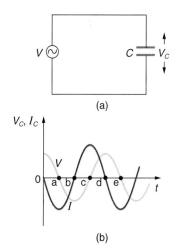

Figure 22.36 (a) An AC voltage source in series with a capacitor C having negligible resistance. (b) Graph of current and voltage across the capacitor as functions of time.

---

*As with X_L, this expression for X_C results from an analysis of the circuit using Kirchhoff's rules and calculus.

proportional to the capacitance C; the larger the capacitor, the greater the charge it can store and the greater the current that can flow. It is also inversely proportional to the frequency f; the greater the frequency, the less time there is to fully charge the capacitor, and so it impedes current less.

EXAMPLE 22.11 CAPACITIVE REACTANCE AND CURRENT

(a) Calculate the capacitive reactance of a 5.00 μF capacitor when 60.0 Hz and 10.0 kHz AC voltages are applied. (b) What is the rms current if the applied rms voltage is 120 V?

Strategy The capacitive reactance is found directly from the expression in Equation 22.19. Once X_C has been found at each frequency, Ohm's law as stated in Equation 22.18 can be used to find the current at each frequency.

Solution for (a) Entering the frequency and capacitance into Equation 22.19 gives

$$X_C = \frac{1}{2\pi f C}$$

$$= \frac{1}{6.28(60.0/\text{s})(5.00 \ \mu\text{F})} = 531 \ \Omega \quad \text{at 60 Hz}$$

Similarly, at 10 kHz,

$$X_C = \frac{1}{2\pi f C} = \frac{1}{6.28(1.00 \times 10^4/\text{s})(5.00 \ \mu\text{F})}$$

$$= 3.18 \ \Omega \quad \text{at 10 kHz}$$

Solution for (b) The rms current is now found using the version of Ohm's law in Equation 22.18, given the applied rms voltage is 120 V. For the first frequency, this yields

$$I = \frac{V}{X_C} = \frac{120 \ \text{V}}{531 \ \Omega} = 0.226 \ \text{A} \quad \text{at 60 Hz}$$

Similarly, at 10 kHz,

$$I = \frac{V}{X_C} = \frac{120 \ \text{V}}{3.18 \ \Omega} = 37.7 \ \text{A} \quad \text{at 10 kHz}$$

Discussion The capacitor reacts very differently at the two different frequencies, and in exactly the opposite way an inductor reacts. At the higher frequency, its reactance is small and the current is large. Capacitors favor change, whereas inductors oppose change. Capacitors impede low frequencies the most, since low frequency allows them time to become charged and stop the current. Capacitors can be used to filter out low frequencies. For example, a capacitor in series with a sound reproduction system rids it of 60 Hz hum.

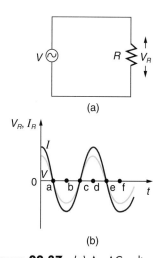

Figure 22.37 (a) An AC voltage source in series with a resistor R. (b) Graph of current and voltage across the resistor as functions of time, showing them to be exactly in phase.

Although a capacitor is basically an open circuit, there is an rms current in a circuit with an AC voltage applied to a capacitor. This is because the voltage is continually reversing, charging and discharging the capacitor. If the frequency goes to zero (DC), X_C tends to infinity, and the current is zero once the capacitor is charged. At very high frequencies, the capacitor's reactance tends to zero—it has a negligible reactance and does not impede the current (it acts like a simple wire). *Capacitors have the opposite effect on AC circuits that inductors have.*

Resistors in an AC Circuit

Just as a reminder, consider Figure 22.37, which shows an AC voltage applied to a resistor and a graph of voltage and current versus time. The voltage and current are exactly in phase in a resistor. There is no frequency dependence to the behavior of plain resistance in a circuit:

> **When a sinusoidal voltage is applied to a resistor, the voltage is exactly in phase with the current—they have a 0° phase angle.**

22.12 *RLC* SERIES AC CIRCUITS

When alone in an AC circuit, inductors, capacitors, and resistors all impede current. How do they behave when all three occur together? Interestingly, their individual resistances in ohms do not simply add. Because inductors and capacitors behave in opposite ways, they partially to totally cancel each other's effect. Figure 22.38 shows an *RLC* series circuit with an AC voltage source, the behavior of which is the subject of this section. The crux of the analysis of an *RLC* circuit is the frequency dependence of X_L and X_C, and the effect they have on the phase of voltage versus current (established in the preceding section). These give rise to the frequency dependence of the circuit, with important resonance features that are the basis of many applications, such as radio tuners.

The combined effect of resistance R, inductive reactance X_L, and capacitive reactance X_C is defined to be *impedance*, an AC analogy to resistance in a DC circuit. Current, voltage, and impedance in an *RLC* circuit are related by an AC version of Ohm's law:

$$I_0 = \frac{V_0}{Z} \quad \text{or} \quad I_{rms} = \frac{V_{rms}}{Z} \qquad \textbf{(22.20)}$$

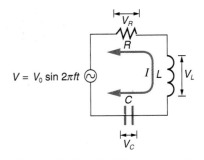

Figure 22.38 An *RLC* series circuit with an AC voltage source.

Here I_0 is the peak current, V_0 the peak source voltage, and Z is the **impedance** of the circuit. The units of impedance are ohms, and its effect on the circuit is as you might expect: the greater the impedance, the smaller the current. To get an expression for Z in terms of R, X_L, and X_C, we will now examine how the voltages across the various components are related to the source voltage. Those voltages are labeled V_R, V_L, and V_C in Figure 22.38.

Conservation of charge requires current to be the same in each part of the circuit at all times, so that we can say the currents in R, L, and C are equal and in phase. But we know from the preceding section that the voltage across the inductor V_L leads the current by one-fourth of a cycle, the voltage across the capacitor V_C follows the current by one-fourth of a cycle, and the voltage across the resistor V_R is exactly in phase with the current. Figure 22.39 shows these relationships in one graph, as well as showing the total voltage around the circuit $V = V_R + V_L + V_C$, where all four voltages are the instantaneous values. According to Kirchhoff's loop rule, the total voltage around the circuit V is also the voltage of the source.

You can see from Figure 22.39 that while V_R is in phase with the current, V_L leads by 90°, and V_C follows by 90°. Thus V_L and V_C are 180° out of phase (crest to trough) and tend to cancel, although not completely unless the have the same magnitude. Since the peak voltages are not aligned (not in phase), the peak voltage V_0 of the source does *not* equal the sum of the peak voltages across R, L, and C. The actual relationship is

$$V_0 = \sqrt{V_{0R}^2 + (V_{0L} - V_{0C})^2}$$

where V_{0R}, V_{0L}, and V_{0C} are the peak voltages across R, L, and C, respectively.* Now, using Ohm's law and definitions from the preceding section, we substitute $V_0 = I_0 Z$ into the above, as well as $V_{0R} = I_0 R$, $V_{0L} = I_0 X_L$, and $V_{0C} = I_0 X_C$, yielding

$$I_0 Z = \sqrt{I_0^2 R^2 + \left(I_0 X_L - I_0 X_C\right)^2} = I_0 \sqrt{R^2 + \left(X_L - X_C\right)^2}$$

I_0 cancels to yield an expression for Z:

$$Z = \sqrt{R^2 + \left(X_L - X_C\right)^2} \qquad \textbf{(22.21)}$$

which is the impedance of an *RLC* series AC circuit. For circuits without a resistor, take $R = 0$; for those without an inductor, take $X_L = 0$; and for those without a capacitor, take $X_C = 0$.

---

*This result, along with others, comes from an analysis made using trigonometric techniques involving rotating vectors, the projections of which are the various circuit elements' sinusoidal voltages with appropriate phase relationships. That analysis is beyond the scope of this text.

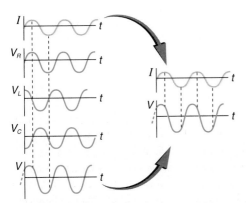

Figure 22.39 This graph shows the relationships of the voltages in an *RLC* circuit to the current. The voltages across the circuit elements add to equal the voltage of the source, which is seen to be out of phase with the current.

EXAMPLE 22.12 IMPEDANCE AND CURRENT

An *RLC* series circuit has a 40.0 Ω resistor, a 3.00 mH inductor, and a 5.00 μF capacitor. (a) Find the circuit's impedance at 60.0 Hz and 10.0 kHz, noting that these frequencies and the values for *L* and *C* are the same as in Examples 22.10 and 22.11. (b) If the voltage source has $V_{rms} = 120$ V, what is I_{rms} at each frequency?

Strategy For each frequency, we use Equation 22.21 to find the impedance and then Ohm's law to find current. We can take advantage of the results of Examples 22.10 and 22.11 rather than calculate the reactances again.

Solution for (a) At 60.0 Hz, the values of the reactances were found in Examples 22.10 and 22.11 to be $X_L = 1.13$ Ω and $X_C = 531$ Ω. Entering these and the given 40.0 Ω for resistance into Equation 22.21 yields

$$Z = \sqrt{R^2 + (X_L - X_C)^2}$$
$$= \sqrt{(40.0\ \Omega)^2 + (1.13\ \Omega - 531\ \Omega)^2}$$
$$= 531\ \Omega \quad \text{at 60.0 Hz}$$

Similarly, at 10.0 kHz, $X_L = 188$ Ω and $X_C = 3.18$ Ω, so that

$$Z = \sqrt{(40.0\ \Omega)^2 + (188\ \Omega - 3.18\ \Omega)^2}$$
$$= 190\ \Omega \quad \text{at 10.0 kHz}$$

Discussion for (a) In both cases, the result is nearly the same as the largest value, and the impedance is definitely not the sum of the individual values. It is clear that X_L dominates at high frequency and X_C dominates at low frequency.

Solution for (b) The current I_{rms} can be found using the AC version of Ohm's law in Equation 22.20:

$$I_{rms} = \frac{V_{rms}}{Z} = \frac{120\ \text{V}}{531\ \Omega} = 0.226\ \text{A} \quad \text{at 60.0 Hz}$$

Finally, at 10.0 kHz, we find

$$I_{rms} = \frac{V_{rms}}{Z} = \frac{120\ \text{V}}{190\ \Omega} = 0.633\ \text{A} \quad \text{at 10.0 kHz}$$

Discussion for (b) The current at 60.0 Hz is the same (to three digits) as found for the capacitor alone in Example 22.11. The capacitor dominates at low frequency. The current at 10.0 kHz is only slightly different from that found for the inductor alone in Example 22.10. The inductor dominates at high frequency.

Resonance in *RLC* Series AC Circuits

How does an *RLC* circuit behave as a function of the frequency of the driving voltage source? Combining Ohm's law, $I_{rms} = V_{rms}/Z$, and the expression for impedance Z from Equation 22.21 gives

$$I_{rms} = \frac{V_{rms}}{\sqrt{R^2 + (X_L - X_C)^2}}$$

The reactances vary with frequency, with X_L large at high frequencies and X_C large at low frequencies, as we have seen in three previous examples. At some intermediate frequency f_0, the reactances will be equal and cancel, giving $Z = R$—this is a minimum value for impedance, and a maximum value for I_{rms} results. We can get an expression for f_0 by taking

$$X_L = X_C$$

Substituting the definitions of X_L and X_C,

$$2\pi f_0 L = \frac{1}{2\pi f_0 C}$$

Solving this expression for f_0 yields

$$f_0 = \frac{1}{2\pi\sqrt{LC}} \tag{22.22}$$

where f_0 is the **resonant frequency** of an *RLC* series circuit. This is also the *natural frequency* at which the circuit would oscillate if not driven by the voltage source. At f_0, the effects of the inductor and capacitor cancel, so that $Z = R$, and I_{rms} is a maximum.

Resonance in AC circuits is analogous to mechanical resonance, where resonance is defined to be a forced oscillation—in this case, forced by the voltage source—at the natural frequency of the system. The receiver in a radio is an *RLC* circuit that oscillates best

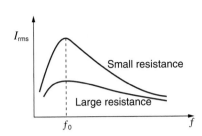

Figure 22.40 A graph of current versus frequency for two *RLC* series circuits differing only in the amount of resistance. Both have a resonance at f_0, but that for the higher resistance is lower and broader. The driving AC voltage source has a fixed amplitude V_0.

at its f_0. A variable capacitor is often used to adjust f_0 to receive a desired frequency and to reject others. Figure 22.40 is a graph of current as a function of frequency, illustrating a resonant peak in I_{rms} at f_0. The two curves are for two different circuits, which differ only in the amount of resistance in them. The peak is lower and broader for the higher-resistance circuit. Thus the higher-resistance circuit resonates less well and would not be as selective in a radio receiver, for example.

EXAMPLE 22.13 RESONANT FREQUENCY AND CURRENT

For the same *RLC* series circuit having a 40.0 Ω resistor, a 3.00 mH inductor, and a 5.00 µF capacitor: (a) Find the resonant frequency. (b) Calculate I_{rms} at resonance if V_{rms} is 120 V.

Strategy The resonant frequency is found by using the expression in Equation 22.22. The current at that frequency is the same as if the resistor alone were in the circuit.

Solution for (a) Entering the given values for L and C into the expression given for f_0 in Equation 22.22 yields

$$
\begin{aligned}
f_0 &= \frac{1}{2\pi\sqrt{LC}} \\
&= \frac{1}{2\pi\sqrt{(3.00 \times 10^{-3}\ \text{H})(5.00 \times 10^{-6}\ \text{F})}} = 1.30\ \text{kHz}
\end{aligned}
$$

Discussion for (a) We see that the resonant frequency is between 60.0 Hz and 10.0 kHz, the two frequencies chosen in earlier examples. This was to be expected, since the capacitor dominated at the low frequency and the inductor dominated at the high frequency. Their effects are the same at this intermediate frequency.

Solution for (b) The current is given by Ohm's law. At resonance, the two reactances are equal and cancel, so that the impedance equals the resistance alone. Thus,

$$
I_{rms} = \frac{V_{rms}}{Z} = \frac{120\ \text{V}}{40.0\ \Omega} = 3.00\ \text{A}
$$

Discussion for (b) At resonance, the current is greater than at the higher and lower frequencies considered for the same circuit in the preceding example.

Power in *RLC* Series AC Circuits

If current varies with frequency in an *RLC* circuit, then the power delivered to it also varies with frequency. But the average power is not simply current times voltage, as it is in purely resistive circuits. As was seen in Figure 22.39, voltage and current are out of phase in an *RLC* circuit. There is a **phase angle** ϕ between the source voltage V and the current I, which can be found from

$$
\cos \phi = \frac{R}{Z} \tag{22.23}
$$

For example, at the resonant frequency or in a purely resistive circuit $Z = R$, so that $\cos \phi = 1$. This implies that $\phi = 0°$ and that voltage and current are in phase, as expected for resistors. At other frequencies, average power is less than at resonance. This is both because voltage and current are out of phase and because I_{rms} is lower. The fact that source voltage and current are out of phase affects the power delivered to the circuit. It can be shown that the *average power* is

$$
P_{ave} = I_{rms}V_{rms} \cos \phi \tag{22.24}
$$

Thus $\cos \phi$ is called the **power factor**, which can range from 0 to 1. Power factors near 1 are desirable when designing an efficient motor, for example. At the resonant frequency, $\cos \phi = 1$.

EXAMPLE 22.14 POWER FACTOR AND POWER

For the same *RLC* series circuit having a 40.0 Ω resistor, a 3.00 mH inductor, a 5.00 μF capacitor, and a voltage source with a V_{rms} of 120 V: (a) Calculate the power factor and phase angle for f = 60.0 Hz. (b) What is the average power at 60.0 Hz? (c) Find the average power at the circuit's resonant frequency.

Strategy and Solution for (a) The power factor at 60.0 Hz is found from Equation 22.23,

$$\cos \phi = \frac{R}{Z}$$

We know Z = 531 Ω from Example 22.12, so that

$$\cos \phi = \frac{40.0 \ \Omega}{531 \ \Omega} = 0.0753 \qquad \text{at 60.0 Hz}$$

This small value indicates the voltage and current are significantly out of phase. In fact, the phase angle is

$$\phi = \cos^{-1} 0.0753 = 85.7° \qquad \text{at 60.0 Hz}$$

Discussion for (a) The phase angle is close to 90°, consistent with the fact that the capacitor dominates the

circuit at this low frequency (a pure *RC* circuit has its voltage and current 90° out of phase).

Strategy and Solution for (b) The average power at 60.0 Hz is

$$P_{ave} = I_{rms} V_{rms} \cos \phi$$

I_{rms} was found to be 0.226 A in Example 22.12. Entering the known values gives

$$P_{ave} = (0.226 \ \text{A})(120 \ \text{V})(0.0753) = 2.04 \ \text{W}$$
$$\text{at 60.0 Hz}$$

Strategy and Solution for (c) At the resonant frequency, we know $\cos \phi$ = 1, and I_{rms} was found to be 3.00 A in Example 22.13. Thus,

$$P_{ave} = (3.00 \ \text{A})(120 \ \text{V})(1) = 360 \ \text{W}$$
$$\text{at resonance (1.30 kHz)}$$

Discussion Both the current and the power factor are greater at resonance, producing significantly greater power than at higher and lower frequencies.

Power delivered to an *RLC* series AC circuit is dissipated by the resistance alone.* The inductor and capacitor have energy input and output but do not dissipate it out of the circuit. Rather they transfer energy back and forth to one another, with the resistor dissipating exactly what the voltage source puts into the circuit. The circuit is analogous to the wheel of a car driven over a washboard road as shown in Figure 22.41. The regularly spaced bumps in the road are analogous to the voltage source, driving the wheel up and down. The shock absorber is analogous to the resistance damping and limiting the amplitude of the oscillation. Energy within the system goes back and forth between kinetic (analogous to maximum current, and energy stored in an inductor) and potential energy stored in the car spring (analogous to no current, and energy stored in the electric field of a capacitor). The amplitude of the wheels' motion is a maximum if the bumps in the road are hit at the resonant frequency.

---

*This assumes no significant electromagnetic radiation from the inductor and capacitor, such as radio waves. Such radiation can happen and may even be desired, as we will see in the next chapter on electromagnetic radiation, but it can also be suppressed as is the case in this chapter.

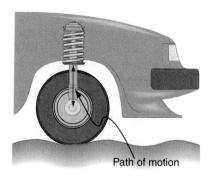

Figure 22.41 The forced but damped motion of the wheel on the car spring is analogous to an *RLC* series AC circuit. The shock absorber damps the motion and dissipates energy, analogous to the resistance in an *RLC* circuit. The mass and spring determine the resonant frequency.

Path of motion

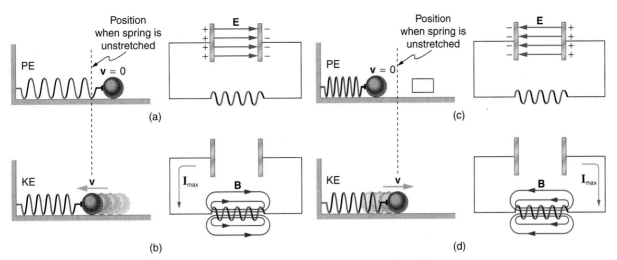

Figure 22.42 An *LC* circuit is analogous to a mass oscillating on a spring with no friction and no driving force. Energy moves back and forth between the inductor and capacitor, just as it moves from kinetic to potential in the mass-spring system.

A pure *LC* circuit with negligible resistance oscillates at f_0, the same resonant frequency as an *RLC* circuit. It can serve as a frequency standard or clock circuit—for example, in a digital wristwatch. With a very small resistance, only a very small energy input is necessary to maintain the oscillations. The circuit is analogous to a car with no shock absorbers. Once it starts oscillating, it continues at its natural frequency for some time. Figure 22.42 shows the analogy between an *LC* circuit and a mass on a spring.

SUMMARY

Induction is the creation of emfs and, hence, currents by magnetic fields. Induction is described by Faraday's and Lenz's laws.

The crucial quantity in induction is **magnetic flux Φ**, defined to be

$$\Phi = BA \cos \theta \qquad (22.1)$$

where B is the magnetic field strength over an area A at an angle θ with the perpendicular to the area. Units of magnetic flux Φ are T·m². **Any change in magnetic flux Φ induces an emf $\mathcal{E}$**—the process is defined to be **electromagnetic induction**.

Faraday's law of induction states the emf $\mathcal{E}$ induced by a change in magnetic flux is

$$\mathcal{E} = -N \frac{\Delta\Phi}{\Delta t} \qquad (22.2)$$

when flux changes by $\Delta\Phi$ in a time Δt. If $\mathcal{E}$ is induced in a coil, N is its number of turns. The minus sign means that **the emf $\mathcal{E}$ creates a current I and magnetic field B that oppose the change in flux $\Delta\Phi$**—this opposition is known as **Lenz's law**. Strategies for applying Lenz's law are given in Section 22.2.

An emf $\mathcal{E}$ induced by motion relative to a magnetic field B is called a *motional emf* and is given by

$$\mathcal{E} = B\ell v \qquad (B, \ell, \text{ and } v \text{ perpendicular}) \qquad (22.3)$$

where ℓ is the length of the object moving at v relative to the field. Currents loops induced in moving conductors are called **eddy currents**, and they can create significant drag, called **magnetic damping**.

An *electric generator* rotates a coil in a magnetic field, inducing an emf $\mathcal{E}$ given as a function of time by

$$\mathcal{E} = NAB\omega \sin \omega t \qquad (22.4)$$

where A is the area of an N-turn coil rotated at a constant angular velocity ω in a uniform magnetic field B. The **peak emf** $\mathcal{E}_0$ of a generator is

$$\mathcal{E}_0 = NAB\omega \qquad (22.5)$$

Any rotating coil will have such an induced emf—in motors, this is called **back emf**, since it opposes the emf input to the motor.

Transformers use induction to transform voltages from one value to another. For a transformer, the voltages across the primary and secondary coils are related by

$$\frac{V_s}{V_p} = \frac{N_s}{N_p} \qquad (22.6)$$

where V_p and V_s are the voltages across primary and secondary coils having N_p and N_s turns. The currents I_p and I_s in the primary and secondary coils are related by

$$\frac{I_s}{I_p} = \frac{N_p}{N_s} \qquad (22.7)$$

A *step-up transformer* increases voltage and decreases current, whereas a *step-down transformer* decreases voltage and increases current.

Electrical safety systems and devices are employed to prevent thermal and shock hazards. Circuit breakers and fuses interrupt excessive currents to prevent thermal hazards. The *three-wire system* guards against thermal and shock hazards, utilizing hot, neutral, and ground wires, and grounding the neutral wire and case of the appliance. A ground fault interrupter prevents shock by detecting the loss of current to unintentional paths. An isolation transformer insulates the device being powered from the original source, also to prevent shock. Many of these devices use induction to perform their basic function.

Inductance is the property of a device that tells how effectively it induces an emf. *Mutual inductance* is the effect of two devices on one another. A change in current $\Delta I_1/\Delta t$ in one induces an emf $\mathcal{E}_2$ in the second:

$$\mathcal{E}_2 = -M \frac{\Delta I_1}{\Delta t} \qquad \textbf{(22.8a)}$$

where M is defined to be the **mutual inductance** between the two devices, and the minus sign is due to Lenz's law. Symmetrically, a change in current $\Delta I_2/\Delta t$ through the second device induces an emf $\mathcal{E}_1$ in the first:

$$\mathcal{E}_1 = -M \frac{\Delta I_2}{\Delta t} \qquad \textbf{(22.8b)}$$

where M is the same mutual inductance as in the reverse process.

Current changes in a device induce an emf in the device itself. *Self-inductance* is the effect of the device on itself. The device is called an **inductor**, and the emf $\mathcal{E}$ induced in it by a change in current through it is

$$\mathcal{E} = -L \frac{\Delta I}{\Delta t} \qquad \textbf{(22.9)}$$

where L is the **self-inductance** of the inductor, and $\Delta I/\Delta t$ is the rate of change of current through it. The minus sign indicates that $\mathcal{E}$ opposes the change in current, as required by Lenz's law. The unit of self- and mutual inductance is the **henry** (H), where $1 \text{ H} = 1 \ \Omega \cdot \text{s}$. The self-inductance L of an inductor is proportional to how much flux changes with current. For an N-turn inductor,

$$L = N \frac{\Delta \Phi}{\Delta I} \qquad \textbf{(22.10)}$$

The self-inductance of a solenoid is

$$L = \frac{\mu_0 N^2 A}{\ell} \qquad \text{(solenoid)} \qquad \textbf{(22.11)}$$

where N is its number of turns in the solenoid, A is its cross-sectional area, ℓ is its length, and μ_0 is the permeability of free space. The **energy stored in an inductor** E_{ind} is

$$E_{\text{ind}} = \tfrac{1}{2} L I^2 \qquad \textbf{(22.12)}$$

When a series connection of a resistor and an inductor—an *RL* circuit—is connected to a voltage source, the time vari-

ation of the current is

$$I = I_0 (1 - e^{-t/\tau}) \qquad \text{(turning on)} \qquad \textbf{(22.13)}$$

where $I_0 = V/R$ is the final current. The **characteristic time constant** τ is

$$\tau = \frac{L}{R} \qquad \textbf{(22.14)}$$

In the first time constant τ, the current rises from zero to $0.632 I_0$, and 0.632 of the remainder in every subsequent time interval τ. When the inductor is shorted through a resistor, current decreases as

$$I = I_0 e^{-t/\tau} \qquad \text{(turning off)} \qquad \textbf{(22.15)}$$

Here I_0 is the initial current. Current falls to $0.368 I_0$ in the first time interval τ, and 0.368 of the remainder toward zero in each subsequent time τ.

For inductors in AC circuits, we find that **when a sinusoidal voltage is applied to an inductor, the voltage leads the current by one-fourth of a cycle, or by a 90° phase angle**. The opposition of an inductor to a change in current is expressed as a type of AC resistance. Ohm's law for an inductor is

$$I = \frac{V}{X_L} \qquad \textbf{(22.16)}$$

where V is the rms voltage across the inductor and X_L is defined to be the **inductive reactance**, given by

$$X_L = 2\pi f L \qquad \textbf{(22.17)}$$

with f the frequency of the AC voltage source in hertz. Inductive reactance X_L has units of ohms and is greatest at high frequencies.

For capacitors, we find that **when a sinusoidal voltage is applied to a capacitor, the voltage follows the current by one-fourth of a cycle, or by a 90° phase angle**. Since a capacitor can stop current when fully charged, it limits current and offers another form of AC resistance; Ohm's law for a capacitor is

$$I = \frac{V}{X_C} \qquad \textbf{(22.18)}$$

where V is the rms voltage across the capacitor and X_C is defined to be the **capacitive reactance**, given by

$$X_C = \frac{1}{2\pi f C} \qquad \textbf{(22.19)}$$

X_C has units of ohms and is greatest at low frequencies.

The AC analogy to resistance is **impedance** Z, the combined effect of resistors, inductors, and capacitors, defined by the AC version of Ohm's law:

$$I_0 = \frac{V_0}{Z} \quad \text{or} \quad I_{\text{rms}} = \frac{V_{\text{rms}}}{Z} \qquad \textbf{(22.20)}$$

where I_0 is the peak current and V_0 is the peak source voltage. Impedance has units of ohms and is given by

$$Z = \sqrt{R^2 + \left(X_L - X_C\right)^2} \qquad \textbf{(22.21)}$$

The **resonant frequency** f_0, at which $X_L = X_C$, is

$$f_0 = \frac{1}{2\pi\sqrt{LC}} \qquad \textbf{(22.22)}$$

In an AC circuit, there is a **phase angle** ϕ between source voltage V and the current I, which can be found from

$$\cos\phi = \frac{R}{Z} \qquad \textbf{(22.23)}$$

$\phi = 0°$ for a purely resistive circuit or an *RLC* circuit at resonance. The **average power delivered to an *RLC* circuit** is affected by the phase angle and is given by

$$P_{ave} = I_{rms}V_{rms}\cos\phi \qquad \textbf{(22.24)}$$

where $\cos\phi$ is called the **power factor**, which ranges from 0 to 1.

CONCEPTUAL QUESTIONS

22.1 How do the multiple-loop coils and iron ring in the version of Faraday's apparatus shown in Figure 22.2 enhance the observation of induced emf?

22.2 When a magnet is thrust into a coil as in Figure 22.3(a), what is the direction of the force exerted by the coil on the magnet? Draw a diagram showing the direction of the current induced in the coil and the magnetic field it produces, to justify your response. How does the magnitude of the force depend on the resistance of the galvanometer?

22.3 Explain how magnetic flux can be zero when the magnetic field is not zero.

22.4 Is an emf induced in the coil in Figure 22.43 when it is stretched? If so, state why and give the direction of the induced current.

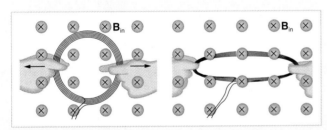

Figure 22.43 A circular coil of wire is stretched in a magnetic field. Question 4 and Problem 8.

22.5 A person who works with large magnets sometimes places her head inside a strong field. She reports feeling dizzy as she quickly turns her head. How might this be associated with induction?

22.6 Why must part of the circuit be moving relative to other parts, to have usable motional emf? Consider, for example, that the rails in Figure 22.7 are stationary relative to the magnetic field, while the rod moves.

22.7 A powerful induction cannon can be made by placing a metal cylinder inside a solenoid coil. The cylinder is forcefully expelled when solenoid current is turned on rapidly. Use Faraday's and Lenz's laws to explain how this works. Why might the cylinder get hot when the cannon is fired?

22.8 An induction stove heats a pot with an AC coil located beneath the pot (and without a hot surface). Can the stove surface be a conductor? Why won't a DC coil work?

22.9 Explain how you could thaw out a frozen water pipe by wrapping an AC coil around it. Does it matter whether or not the pipe is a conductor? Explain.

22.10 A particle accelerator sends high-velocity charged particles down an evacuated pipe. Explain how a coil of wire wrapped around the pipe could detect the passage of individual particles. Sketch a graph of the voltage output of the coil as a single particle passes through it.

22.11 Explain why magnetic damping might not be effective on an object made of several thin conducting layers separated by insulation.

22.12 Using RHR-1, show that the emfs in the sides of the generator loop in Figure 22.15 are in the same sense and thus add.

22.13 The source of a generator's electrical energy output is the work done to turn its coils. How is the work needed to turn the generator related to Lenz's law?

22.14 Suppose you find that the belt drive connecting a powerful motor to an air conditioning unit is broken and the motor is running freely. Should you be worried that the motor is consuming a great deal of energy for no useful purpose? Explain why or why not.

22.15 The primary coil of a transformer designed to use 240 V AC will burn out if 240 V DC is applied. Explain why.

22.16 Explain what causes physical vibrations in transformers at twice the frequency of the AC power involved.

22.17* Does the black insulation on hot wires prevent shock hazards, thermal hazards, or both?

22.18* Why are ordinary circuit breakers and fuses ineffective in preventing shocks?

22.19* A GFI may trip just because the hot and neutral wires connected to it are significantly different in length. Explain why.

22.20 How would you place two identical flat coils in contact so that they had the greatest mutual inductance? The least?

22.21 How would you shape a given length of wire to give it the greatest self-inductance? The least?

22.22 Verify, as was concluded without proof in Example 22.7, that units of $T\cdot m^2/A = \Omega\cdot s = H$.

22.23 Would you use a large inductance or a large capacitance in series with a system to filter out low frequencies, such as the 60 Hz hum in a sound system? Explain.

22.24 High-frequency noise in AC power can damage computers. Does the plug-in unit designed to prevent this damage use a large inductance or a large capacitance (in series with the computer) to filter out such high frequencies? Explain.

22.25 Does inductance depend on current, frequency, or both? What about inductive reactance?

22.26 Does the resonant frequency of an AC circuit depend on the peak voltage of the AC source? Explain why or why not.

22.27 Suppose you have a motor with a power factor significantly less than 1. Explain why it would be better to improve the power factor as a method of improving the motor's output, rather than to increase the voltage input.

22.28 Explain why the capacitor in Figure 22.44(a) acts as a low-frequency filter between the two circuits, whereas that in Figure 22.44(b) acts as a high-frequency filter.

22.29 If the capacitors in Figure 22.44 are replaced by inductors, which acts as a low-frequency filter and which as a high-frequency filter?

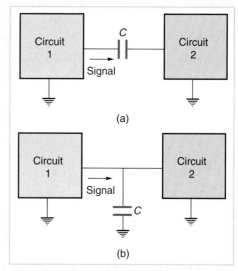

(a)

(b)

Figure 22.44 Capacitors and inductors. Questions 28 and 29. Capacitor with high frequency and low frequency. Problems 74 and 75.

PROBLEMS

Sections 22.1 and 22.2 Faraday's Law of Induction and Lenz's Law

22.1 What is the value of the magnetic flux at coil 2 in Figure 22.45(a) due to coil 1?

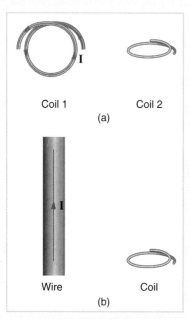

Figure 22.45 (a) The planes of the two coils are perpendicular. (b) The wire is perpendicular to the plane of the coil. Problems 1 and 2.

22.2 What is the value of the magnetic flux at the coil in Figure 22.45(b) due to the wire?

22.3 Referring to Figure 22.46(a), what is the direction of the current induced in coil 2: (a) If the current in coil 1 increases? (b) If the current in coil 1 decreases? (c) If the current in coil 1 is constant?

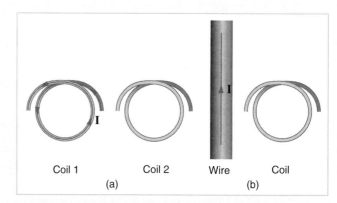

Figure 22.46 (a) The coils lie in the same plane. (b) The wire is in the plane of the coil. Problems 3, 4, 92, and 97.

22.4 Referring to Figure 22.46(b), what is the direction of the current induced in the coil: (a) If the current in the wire increases? (b) If the current in the wire decreases? (c) If the current in the wire suddenly changes direction?

22.5 Referring to Figure 22.47, what are the directions of the currents in coils 1, 2, and 3: (a) When the switch is first closed? (b) When the switch has been closed for a long time? (c) Just after the switch is opened?

22.6 Repeat Problem 22.5 with the battery reversed.

22.7 Verify that the units of $\Delta\Phi/\Delta t$ are volts. That is, show that $1 \text{ T} \cdot \text{m}^2/\text{s} = 1 \text{ V}$.

22.8 Suppose the 50-turn coil in Figure 22.43 lies in the plane of the page and originally has an area of 0.250 m². It is stretched to have no area in 0.100 s. What is the direction and magnitude of the induced emf if the uniform magnetic field is perpendicular to the page and has a strength of 1.50 T?

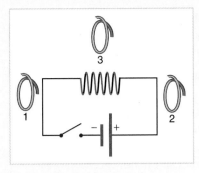

Figure 22.47 Problems 5 and 6.

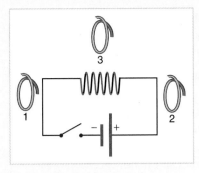

Figure 22.48 A coil is moved into and out of a region of uniform magnetic field. Problem 19.

22.9 An MRI technician moves his hand from a region of very low magnetic field strength into an MRI scanner's 2.00 T field with his fingers pointing in the direction of the field. Find the average emf induced in his wedding ring, given its diameter is 2.20 cm and assuming it takes 0.250 s to move it into the field.

• 22.10 An emf is induced by rotating a 1000-turn, 20.0 cm diameter coil in the earth's 5.00×10^{-5} T magnetic field. What average emf is induced, given the plane of the coil is originally perpendicular to the earth's field and is rotated to be parallel to the field in 10.0 ms?

• 22.11 A 0.250 m radius, 500-turn coil is rotated one-fourth of a revolution in 4.17 ms, originally having its plane perpendicular to a uniform magnetic field. (This is 60 rev/s.) Find the magnetic field strength needed to induce an average emf of 10,000 V.

Sections 22.3 and 22.4 Motional emf and Eddy Currents

22.12 Use Faraday's law, Lenz's law, and RHR-1 to show that the magnetic force on the current in the moving rod in Figure 22.7 is in the opposite direction of its velocity.

22.13 If a current flows in the Satellite Tether shown in Figure 22.8, use Faraday's law, Lenz's law, and RHR-1 to show that there is a magnetic force on the tether in the direction opposite to its velocity.

22.14 Make a drawing similar to Figure 22.10, but with the pendulum moving in the opposite direction. Then use Faraday's law, Lenz's law, and RHR-1 to show that magnetic force opposes motion.

22.15 A jet airplane with a 75.0 m wingspan is flying at 280 m/s. What emf is induced between wing tips if the vertical component of the earth's field is 3.00×10^{-5} T?

22.16 A nonferrous screwdriver is being used in a 2.00 T magnetic field. What maximum emf can be induced along its 12.0 cm length when it moves at 6.00 m/s?

• 22.17 At what speed must the sliding rod in Figure 22.7 move to produce an emf of 1.00 V in a 1.50 T field, given the rod's length is 30.0 cm?

• 22.18 The 12.0 cm long rod in Figure 22.7 moves at 4.00 m/s. What is the strength of the magnetic field if a 95.0 V emf is induced?

• 22.19 A coil is moved through a magnetic field as shown in Figure 22.48. The field is uniform inside the rectangle and zero outside. What is the direction of the induced current and what is the direction of the magnetic force on the coil at each position shown?

⦂ 22.20 Prove that when B, ℓ, and v are not mutually perpendicular, motional emf is given by $\mathcal{E} = B\ell v \sin \theta$. If v is perpendicular to B, then θ is the angle between ℓ and B. If ℓ is perpendicular to B, then θ is the angle between v and B.

⦂ 22.21 In the August 1992 space shuttle flight, only 250 m of the conducting tether considered in Example 22.2 could be let out. A 40.0 V motional emf was generated in the earth's 5.00×10^{-5} T field, while moving at 7.80×10^{3} m/s. What was the angle between the shuttle's velocity and the earth's field, assuming the conductor was perpendicular to the field?

Section 22.5 Electric Generators

22.22 Calculate the peak voltage of a generator that rotates its 200-turn, 0.100 m diameter coil at 3600 rpm in a 0.800 T field.

22.23 At what angular velocity in rpm will the peak voltage of a generator be 480 V, if its 500-turn, 8.00 cm diameter coil rotates in a 0.250 T field?

• 22.24 What is the peak emf generated by the coil considered in Problem 22.10?

• 22.25 What is the peak emf generated by the coil considered in Problem 22.11?

• 22.26 A bicycle generator rotates at 1875 rad/s, producing an 18.0 V peak emf. It has a 1.00 by 3.00 cm rectangular coil in a 0.640 T field. How many turns are in the coil?

• 22.27 An automobile generator turns at 400 rpm when the engine is idling. Its 300-turn, 5.00 by 8.00 cm rectangular coil rotates in an adjustable magnetic field so that it can produce sufficient voltage even at low rpms. What is the field strength needed to produce a 24.0 V peak emf?

⦂ 22.28 Show that if a coil rotates at an angular velocity ω, the period of its AC output is $2\pi/\omega$.

⦂ 22.29 A 75-turn, 10.0 cm diameter coil rotates at an angular velocity of 8.00 rad/s in a 1.25 T field, starting with the plane of the coil parallel to the field. (a) What is the peak emf? (b) At what time is the peak emf first reached? (c) At what time is the emf first at its most negative? (d) What is the period of the AC voltage output?

⦂ 22.30 (a) If the emf of a coil rotating in a magnetic field is zero at $t = 0$, and increases to its first peak at $t = 0.100$ ms, what is the angular velocity of the coil? (b) At what time will its next maximum occur? (c) What is the period of the output? (d) When is the output first one-fourth of its maximum? (e) When is it next one-fourth of its maximum?

Section 22.6 Back emf

22.31 Suppose a motor connected to a 120 V source draws 20.0 A when it first starts. (a) What is its resistance? (b) What current does it draw at its normal operating speed when it develops a 100 V back emf?

22.32 A motor operating on 240 V electricity has a 180 V back emf at operating speed and draws a 12.0 A current. (a) What is its resistance? (b) What current does it draw when it is first started?

• **22.33** What is the back emf of a 120 V motor that draws 12.0 A at its normal speed and 40.0 A when first starting?

• **22.34** The motor in a toy car operates on 6.00 V, developing a 4.50 V back emf at normal speed. If it draws 3.00 A at normal speed, what current does it draw when starting?

Section 22.7 Transformers

22.35 A plug-in transformer, like that in Figure 22.19, supplies 9.00 V to a video game system. (a) How many turns are in its secondary coil, if its input voltage is 120 V and the primary coil has 400 turns? (b) What is its input current when its output is 1.30 A?

22.36 An American traveler in Europe carries a transformer to convert European standard 240 V to 120 V so that she can use some small appliances on her trip. (a) What is the ratio of turns in the primary and secondary coils of her transformer? (b) What is the ratio of input to output current? (c) How could a European traveling in the United States use this same transformer to power her 240 V appliances from 120 V?

22.37 A videotape rewinder uses a plug-in transformer to convert 120 V to 12.0 V, with a maximum current output of 200 mA. (a) What is the current input? (b) What is the power input?

22.38 (a) What is the voltage output of a transformer used for rechargeable flashlight batteries, if its primary has 500 turns, its secondary 8 turns, and the input voltage is 118 V? (b) What input current is required to produce a 4.00 A output? (c) What is the power input?

• **22.39** The plug-in transformer for a laptop computer puts out 7.50 V and can supply a maximum current of 2.00 A. What is the maximum input current if the input voltage is 120 V? Assume 100% efficiency.

• **22.40** A multipurpose transformer has a secondary coil with several points at which a voltage can be extracted, giving outputs of 5.60, 12.0, and 480 V. (a) The input voltage is 120 V to a primary coil of 280 turns. What are the numbers of turns in the parts of the secondary used to produce the output voltages? (b) If the maximum input current is 10.0 A, what are the maximum output currents (each used alone)?

⁞ **22.41** A large power plant generates electricity at 12.0 kV. Its old transformer once converted the voltage to 400 kV. The secondary of this transformer is being replaced so that its output can be 500 kV for more efficient cross-country transmission on upgraded transmission lines. (a) What is the ratio of turns in the new secondary compared with the old secondary? (b) What is the ratio of new current output to old output (at 400 kV) for the same power? (c) If the upgraded transmission lines have the same resistance, what is the ratio of new line power loss to old?

⁞ **22.42** If the power output in the previous problem is 1000 MW and line resistance is 1.00 Ω, what were the old and new line losses?

Section 22.9 Induction

22.43 Two coils are placed close together in a physics lab to demonstrate Faraday's law of induction. A current of 5.00 A in one is switched off in 1.00 ms, inducing a 9.00 V emf in the other. What is their mutual inductance?

22.44 If two coils placed next to one another have a mutual inductance of 5.00 mH, what voltage is induced in one when the 2.00 A current in the other is switched off in 30.0 ms?

22.45 The 4.00 A current through a 7.50 mH inductor is switched off in 8.33 ms. What is the emf induced opposing this?

22.46 A device is turned on and 3.00 A flows through it 0.100 ms later. What is the self-inductance of the device if an induced 150 V emf opposes this?

22.47 Starting with Equation 22.8, show that the units of inductance are $(V \cdot s)/A = \Omega \cdot s$.

• **22.48** Camera flash attachments charge a capacitor to high voltage by switching the current through an inductor on and off rapidly. In what time must the 0.100 A current through a 2.00 μH inductor be switched on or off to induce a 500 V emf?

• **22.49** A large research solenoid has a self-inductance of 25.0 H. (a) What induced emf opposes shutting it off when 100 A of current through it is switched off in 80.0 ms? (b) How much energy is stored in the inductor at full current? (c) At what rate in watts must energy be dissipated to switch the current off in 80.0 ms? (d) In view of the answer to the last part, is it surprising that shutting it down this quickly is difficult?

• **22.50** (a) Calculate the self-inductance of a 50.0 cm long, 10.0 cm diameter solenoid having 1000 loops. (b) How much energy is stored in this inductor when 20.0 A of current flows through it? (c) How fast can it be turned off if the induced emf cannot exceed 3.00 V?

• **22.51** A precision laboratory resistor is made of a coil of wire 1.50 cm in diameter and 4.00 cm long, and it has 500 turns. (a) What is its self-inductance? (b) What average emf is induced if the 12.0 A current through it is turned on in 4.17 ms (one-fourth of a cycle for 60 Hz AC)? (c) What is its inductance if it is shortened to half its length and counterwound (two layers of 250 turns in opposite directions)?

• **22.52** The heating coils in a hair dryer are 0.800 cm in diameter, have a combined length of 1.00 m, and a total of 400 turns. (a) What is their total self-inductance assuming they act like a single solenoid? (b) How much energy is stored in them when 10.0 A flows? (c) What average emf opposes shutting them off if this is done in 4.17 ms (one-fourth of a cycle for 60 Hz AC)?

• **22.53** When the 20.0 A current through an inductor is turned off in 1.50 ms, an 800 V emf is induced, opposing the change. What is the value of the self-inductance?

• **22.54** How fast can the 150 A current through a 0.250 H inductor be shut off if the induced emf cannot exceed 75.0 V?

Section 22.10 RL Circuits

22.55 If you want a characteristic RL time constant of 1.00 s, and you have a 500 Ω resistor, what value of self-inductance is needed?

22.56 Your RL circuit has a characteristic time constant of 20.0 ns, and a resistance of 5.00 MΩ. (a) What is the inductance of the circuit? (b) What resistance would give you a 1.00 ns time constant, perhaps needed for quick response in an oscilloscope?

22.57 A large superconducting magnet, used for magnetic resonance imaging, has a 50.0 H inductance. If you want current through it to be adjustable with a 1.00 s characteristic time constant, what is the minimum resistance of system?

22.58 Verify that after a time of 10.0 ms, the current for the situation considered in Example 22.9 will be 0.183 A as stated.

• **22.59** Suppose you have a supply of inductors ranging from 1.00 nH to 10.0 H, and resistors ranging from 0.100 Ω to 1.00 MΩ. What is the range of characteristic RL time constants you can produce by connecting a single resistor to a single inductor?

• **22.60** (a) What is the characteristic time constant of a 25.0 mH inductor that has a resistance of 4.00 Ω? (b) If it is connected to a 12.0 V battery, what is the current after 12.5 ms?

• **22.61** What percentage of the final current I_0 flows through an inductor L in series with a resistor R, three time constants after the circuit is completed?

• **22.62** The 5.00 A current through a 1.50 H inductor is dissipated by a 2.00 Ω resistor in a circuit like that in Figure 22.34 with the switch in position 2. (a) What is the initial energy in the inductor? (b) How long will it take the current to decline to 5.00% of its initial value? (c) Calculate the average power dissipated, and compare it with the initial power dissipated by the resistor.

‡ **22.63** Use the exact exponential treatment to find how much time is required to bring the current through an 80.0 mH inductor in series with a 15.0 Ω resistor to 99.0% of its final value, starting from zero.

‡ **22.64** Using the exact exponential treatment, find the time required for the current through a 2.00 H inductor in series with a 0.500 Ω resistor to be reduced to 0.100% of its original value.

Section 22.11 Reactance

22.65 At what frequency will a 30.0 mH inductor have a reactance of 100 Ω?

22.66 What value of inductance should be used if a 20.0 kΩ reactance is needed at a frequency of 500 Hz?

22.67 What capacitance should be used to produce a 2.00 MΩ reactance at 50.0 Hz?

22.68 At what frequency will an 80.0 µF capacitor have a reactance of 0.250 Ω?

• **22.69** (a) Find the current through a 0.500 H inductor connected to a 60.0 Hz, 480 V AC source. (b) What would the current be at 100 kHz?

• **22.70** (a) What current flows when a 60.0 Hz, 480 V AC source is connected to a 0.250 µF capacitor? (b) What would the current be at 25.0 kHz?

• **22.71** A 20.0 kHz, 16.0 V source connected to an inductor produces a 2.00 A current. What is the inductance?

• **22.72** A 20.0 Hz, 16.0 V source produces a 2.00 mA current when connected to a capacitor. What is the capacitance?

• **22.73** (a) An inductor designed to filter high-frequency noise from power supplied to a personal computer is placed in series with the computer. What minimum inductance should it have to produce a 2.00 kΩ reactance for 15.0 kHz noise? (b) What is its reactance at 60.0 Hz?

• **22.74** The capacitor in Figure 22.44(a) is designed to filter low-frequency signals, impeding their transmission between circuits. (a) What capacitance is needed to produce a 100 kΩ reactance at a frequency of 120 Hz? (b) What would its reactance be at 1.00 MHz?

• **22.75** The capacitor in Figure 22.44(b) will filter high-frequency signals by shorting them to ground. (a) What capacitance is needed to produce a reactance of 10.0 mΩ for a 5.00 kHz signal? (b) What would its reactance be at 3.00 Hz?

Section 22.12 RLC Series AC Circuits

22.76 An RL circuit consists of a 40.0 Ω resistor and a 3.00 mH inductor. (a) Find its impedance at 60.0 Hz and 10.0 kHz. (b) Compare these values of Z with those found in Example 22.12, in which there was also a capacitor.

22.77 An RC circuit consists of a 40.0 Ω resistor and a 5.00 µF capacitor. (a) Find its impedance at 60.0 Hz and 10.0 kHz. (b) Compare these values of Z with those found in Example 22.12, in which there was also an inductor.

22.78 An LC circuit consists of a 3.00 mH inductor and a 5.00 µF capacitor. (a) Find its impedance at 60.0 Hz and 10.0 kHz. (b) Compare these values of Z with those found in Example 22.12, in which there was also a resistor.

22.79 What is the resonant frequency of a 0.500 mH inductor connected to a 40.0 µF capacitor?

• **22.80** To receive AM radio, you want an RLC circuit that can be made to resonate at any frequency between 500 and 1650 kHz. This is accomplished with a fixed 1.00 µH inductor connected to a variable capacitor. What range of capacitance is needed?

• **22.81** Suppose you have a supply of inductors ranging from 1.00 nH to 10.0 H, and capacitors ranging from 1.00 pF to 0.100 F. What is the range of resonant frequencies that can be achieved from combinations of a single inductor and a single capacitor?

• **22.82** What capacitance do you need to produce a resonant frequency of 1.00 GHz, when using an 8.00 nH inductor?

• **22.83** What inductance do you need to produce a resonant frequency of 60.0 Hz, when using a 2.00 µF capacitor?

• **22.84** The lowest frequency in the FM radio band is 88.0 MHz. (a) What inductance is needed to produce this resonant frequency if it is connected to a 2.50 pF capacitor? (b) The capacitor is variable, to allow the resonant frequency to be adjusted to as high as 108 MHz. What must the capacitance be at this frequency?

• **22.85** An RLC series circuit has a 2.50 Ω resistor, a 100 µH inductor, and an 80.0 µF capacitor. (a) Find the circuit's impedance at 120 Hz. (b) Find the circuit's impedance at 5.00 kHz. (c) If the voltage source has V_{rms} = 5.60 V, what is I_{rms} at each frequency? (d) What is the resonant frequency of the circuit? (e) What is I_{rms} at resonance?

• **22.86** An RLC series circuit has a 1.00 kΩ resistor, a 150 mH inductor, and a 25.0 nF capacitor. (a) Find the circuit's impedance at 500 Hz. (b) Find the circuit's impedance at 7.50 kHz. (c) If the voltage source has V_{rms} = 408 V, what is I_{rms} at each frequency? (d) What is the resonant frequency of the circuit? (e) What is I_{rms} at resonance?

22.87 For the same *RLC* series circuit as in Problem 22.85: (a) Find the power factor at $f = 120$ Hz. (b) What is the phase angle at 120 Hz? (c) What is the average power at 120 Hz? (d) Find the average power at the circuit's resonant frequency.

22.88 For the same *RLC* series circuit as in Problem 22.86: (a) Find the power factor at $f = 7.50$ kHz. (b) What is the phase angle at this frequency? (c) What is the average power at this frequency? (d) Find the average power at the circuit's resonant frequency.

22.89 An *RLC* series circuit has a 200 Ω resistor and a 25.0 mH inductor. At 8000 Hz, the phase angle is 45.0°. (a) What is the impedance? (b) Find the circuit's capacitance. (c) If $V_{rms} = 408$ V is applied, what is the average power supplied?

22.90 Referring to Example 22.14, find the average power at 10.0 kHz.

INTEGRATED CONCEPTS

Physics is most interesting and most powerful when applied to general situations involving more than a narrow set of physical principles. The integration of concepts necessary to solve problems involving several physical principles also gives greater insight into the unity of physics. You may wish to refer to Chapters 2, 4, 5, 6, 8, 13, 19, 20, 21, and perhaps others to solve the following problems.

Note: Problem-solving strategies and a worked example that can help you solve integrated concept problems appear in Chapter 21. They can be found with the section of problems labeled *Integrated Concepts* at the end of that chapter. Consult the index for the locations of other problem-solving strategies and worked examples for integrated concept problems.

22.91 The motor in a toy car is powered by four batteries in series, which produce a total emf of 6.00 V. The motor draws 3.00 A and develops a 4.50 V back emf at normal speed. Each battery has a 0.100 Ω internal resistance. What is the resistance of the motor?

22.92 Approximately how does the emf induced in the loop in Figure 22.46(b) depend on the distance of the center of the loop from the wire?

22.93 Referring to the situation in Problem 22.9: (a) What current is induced in the ring if its resistance is 0.0100 Ω? (b) What average power is dissipated? (c) What magnetic field is induced at the center of the ring?

22.94 This problem refers to the bicycle generator considered in Problem 22.26. It is driven by a 1.60 cm diameter wheel that rolls on the outside rim of the bicycle tire. (a) What is the velocity of the bicycle if the generator's angular velocity is 1875 rad/s? (b) What is the maximum emf of the generator when the bicycle moves at 10.0 m/s, noting that it was 18.0 V under the original conditions? (c) If the sophisticated generator can vary its own magnetic field, what field strength will it need at 5.00 m/s to produce a 9.00 V maximum emf?

22.95 A very large, superconducting solenoid stores 1.00 MJ of energy in its magnetic field when 100 A flows. (a) Find its self-inductance. (b) If the coils "go normal," they gain resistance and start to dissipate thermal energy. What temperature increase is produced if all the stored energy goes into heating the 1000 kg magnet, given its average specific heat is 200 J/kg·°C?

22.96 A short circuit to the grounded metal case of an appliance occurs as shown in Figure 22.49. The person touching the case is wet and only has a 3.00 kΩ resistance to ground. (a) What is the voltage on the case if 5.00 mA flows through the person? (b) What is the current in the short circuit if the resistance of the ground wire is 0.200 Ω? (c) Will this trigger the 20.0 A circuit breaker supplying the appliance?

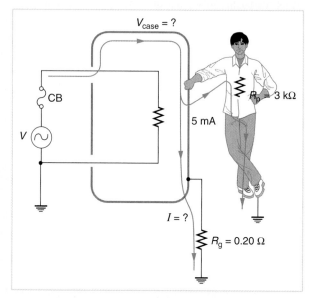

Figure 22.49 A person can be shocked even when the case of an appliance is grounded. The large short circuit current produces a voltage on the case of the appliance, since the resistance of the ground wire is not zero. Problem 96.

22.97 A lightning bolt produces a rapidly varying magnetic field. If the bolt strikes the earth vertically and acts like a current in a long straight wire, it will induce a voltage in a loop aligned like that in Figure 22.46(b). What voltage is induced in a 1.00 m diameter loop 50.0 m from a 2.00×10^6 A lightning strike, if the current falls to zero in 25.0 μs?

22.98 Derive an expression for the current in a system like that in Figure 22.7, under the following conditions. The resistance between the rails is *R*, the rails and the moving rod are identical in cross section *A* and have the same resistivity ρ. The distance between the rails is ℓ, and the rod moves at constant speed v perpendicular to the uniform field *B*. At time zero, the moving rod is next to the resistance *R*.

22.99 The Tethered Satellite in Figure 22.8 has a mass of 525 kg and is at the end of a 20.0 km long, 2.50 mm diameter cable with the tensile strength of steel. (a) How much does the cable stretch if a 100 N force is exerted to pull the satellite in? (Assume the satellite and shuttle are at the same altitude above the earth.) (b) What is the effective force constant of the cable? (c) How much energy is stored in it when stretched by the 100 N force?

22.100 The Tethered Satellite discussed in Section 22.3 is producing 5.00 kV, and a current of 10.0 A flows. (a) What magnetic drag force does this produce if the system is moving at 7.80 km/s? (b) How much kinetic energy is removed from the system in 1.00 h, neglecting any change in altitude? (c) What is the change in velocity if the mass of the system is 100,000 kg?

The following problems have results that are unreasonable because some premise is unreasonable or because certain of the premises are inconsistent with one another. Physical principles applied correctly then produce unreasonable results. The purpose of these problems is to give practice in assessing whether nature is being accurately described, and if it is not to trace the source of difficulty.

PROBLEM-SOLVING STRATEGY

To determine if an answer is reasonable, and to determine the cause if it is not, do the following.

Step 1. Solve the problem using strategies as outlined in several places in the text (such as Section 22.2). See the table of contents to locate problem-solving strategies. Use the format followed in the worked examples in this chapter to solve the problem as usual.

Step 2. Check to see if the answer is reasonable. Is it too large or too small, or does it have the wrong sign, improper units, . . .?

Step 3. If the answer is unreasonable, look for what specifically could cause the identified difficulty. Usually, the manner in which the answer is unreasonable is an indication of the difficulty.

22.101 A 500-turn coil with a 0.250 m^2 area is spun in the earth's 5.00×10^{-5} T field, producing a 12.0 kV maximum emf. (a) At what angular velocity must the coil be spun? (b) What is unreasonable about this result? (c) Which assumption or premise is responsible?

22.102 The 400 kV AC electricity from a power transmission line is fed into the primary coil of a transformer. The ratio of the number of turns in the secondary to the number in the primary is $N_s/N_p = 1000$. (a) What voltage is induced in the secondary? (b) What is unreasonable about this result? (c) Which assumption or premise is responsible?

• **22.103** A 25.0 H inductor has 100 A of current turned off in 1.00 μs. (a) What voltage is induced to oppose this? (b) What is unreasonable about this result? (c) Which assumption or premise is responsible?

• **22.104** In a recording of voltages due to brain activity (an EEG), a 10.0 mV signal with a 0.500 Hz frequency is applied to a capacitor, producing a current of 100 mA. Resistance is negligible. (a) What is the capacitance? (b) What is unreasonable about this result? (c) Which assumption or premise is responsible?

23 ELECTROMAGNETIC WAVES

Beauty brought to you courtesy of electromagnetic waves.

The beauty of a mountain vista, the warm radiance of a campfire, the sting of sunburn, the x ray revealing a broken bone, even microwave popcorn—all are brought to us by electromagnetic waves. The list of the various types of electromagnetic waves, ranging from radio transmission waves to nuclear γ-ray emissions, is interesting in itself. Even more intriguing is that all of these widely varied phenomena are different manifestations of the same thing—electromagnetic waves. (See Figure 23.1.) What are electromagnetic waves? How are they created, and how do they travel? How can we understand and organize their widely varying properties? What is their relationship to previously discussed electric and magnetic effects? These and other questions are explored in this chapter.

It is worth noting at the outset that the general phenomenon of electromagnetic waves was predicted by theory before wave aspects of all types were verified experimentally. The prediction was made by Maxwell in the mid-19th century when he formulated a single theory for all then-known electric and magnetic effects. Such a prediction of previously unobserved phenomena is an indication of the power of science in general, and physics in particular. The underlying connections and unity of physics allow certain great minds to solve puzzles without having all the pieces. The prediction of electromagnetic waves is one of most spectacular examples of this power. Certain others, such as the prediction of antimatter, will be discussed in later chapters.

Figure 23.1 The electromagnetic waves gathered by this antenna are not visible.

23.1 MAXWELL'S EQUATIONS: ELECTROMAGNETIC WAVES PREDICTED AND OBSERVED

The Scotsman James Clerk Maxwell (1831–1879) is regarded as the greatest theoretical physicist of the 19th century. (See Figure 23.2.) Although he died young, Maxwell not only formulated a complete electromagnetic theory, known as Maxwell's equations, he also developed the kinetic theory of gases and made significant contributions to the understanding of color vision and the nature of Saturn's rings.

Maxwell's theory of electromagnetism brought together all the work that had been done by Oersted, Faraday, and others, and added his own. **Maxwell's equations** are paraphrased here in words because their mathematical statement is beyond the level of this text[*]:

1. Electric field lines originate on positive charges and terminate on negative charges. The electric field is defined as the force per unit charge on a test charge, and the strength of the force is related to ε_0, the permittivity of free space. (See Chapters 17 and 18.) This first of Maxwell's equations is a generalized form of Coulomb's law known as **Gauss's law for electricity**.
2. Magnetic field lines are continuous, having no beginning or end. No magnetic monopoles are known to exist. The strength of the magnetic force is related to the permeability of free space, μ_0. (See Chapter 21.) This second of Maxwell's equations is known as **Gauss's law for magnetism**.
3. A changing magnetic field induces an emf and, hence, an electric field. The direction of the emf opposes the change. (See Chapter 22.) This third of Maxwell's equations is **Faraday's law of induction**, including **Lenz's law**.
4. Magnetic fields are generated by moving charges or by changing electric fields. This fourth of Maxwell's equations encompasses **Ampere's law** (see Chapter 21) and *adds another source of magnetism—changing electric fields*.

It is apparent that Maxwell's equations encompass the major laws of electricity and magnetism we have already covered. What is not so apparent is the symmetry that Maxwell introduced in his mathematical framework. Especially important is his addition of the hypothesis that changing electric fields create magnetic fields. This is exactly analogous (and symmetric) to Faraday's law of induction and had been suspected for some time, but it fits beautifully into Maxwell's equations.

Figure 23.2 James Clerk Maxwell.

CONNECTIONS

Unification of Forces

Maxwell's complete and symmetric theory showed that electric and magnetic forces are not separate, but different manifestations of the same thing—the electromagnetic force. This classical unification of forces is one motivation for current attempts to unify the four basic forces in nature. See the discussion in Section 4.6.

---

[*]Should you continue in physics and see the mathematical statement of his equations, you will appreciate their simplicity, unity, and elegance.

Figure 23.3 The apparatus used by Hertz to generate and detect electromagnetic waves. An *RLC* circuit connected to the first loop caused sparks across the gap and generated electromagnetic waves. Sparks across the gap in the second loop gave evidence that the waves had been received.

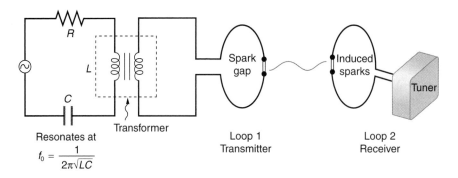

Since changing electric fields create relatively weak magnetic fields, they could not be easily detected at the time of Maxwell's hypothesis. Maxwell realized, however, that oscillating charges, like those in AC circuits, create changing fields. He predicted that these changing fields would move out from the source like waves generated on a lake by a jumping fish. The waves predicted by Maxwell would consist of oscillating electric and magnetic fields—defined to be an **electromagnetic wave** (EM wave). Electromagnetic waves would be capable of exerting forces on charges great distances from their source, and they might thus be detectable. Maxwell calculated that electromagnetic waves would propagate at a speed given by

$$c = \frac{1}{\sqrt{\mu_0 \varepsilon_0}} \tag{23.1}$$

When the values for μ_0 and ε_0 are entered into the equation for c, we find that $c = 3 \times 10^8$ m/s, *the speed of light*. In fact, Maxwell concluded that light *is* an electromagnetic wave having such wavelengths that it can be detected by the eye. Other wavelengths should exist—it remained to be seen if they did. If so, Maxwell's theory and remarkable predictions would be verified, the greatest triumph of physics since Newton. Experimental verification came within a few years, but not before Maxwell's death.

Hertz's Observations

The German physicist Heinrich Hertz (1857–1894) was the first to generate and detect certain types of electromagnetic waves in the laboratory. Starting in 1887, he performed a series of experiments that not only confirmed the existence of electromagnetic waves, but also verified that they travel at the speed of light. Hertz used an AC *RLC* circuit, like those discussed in Chapter 22, that resonates at a known frequency $f_0 = 1/(2\pi\sqrt{LC})$, and connected it to a loop of wire as shown in Figure 23.3. High voltages induced across the gap in the loop produced sparks that were visible evidence of the current in the circuit and that helped generate electromagnetic waves. Across the laboratory, Hertz had another loop attached to another *RLC* circuit, which could be tuned to the same resonant frequency as the first and could, thus, be made to receive electromagnetic waves. This loop also had a gap across which sparks were generated, giving solid evidence that electromagnetic waves had been received.

Hertz also studied the reflection, refraction, and interference patterns of the electromagnetic waves he generated, verifying their wave character. He was able to determine wavelength from the interference patterns, and knowing their frequency, he calculated propagation speed using $v = f\lambda$ (introduced as Equation 15.11 in the first discussions of waves). Hertz was thus able to prove that electromagnetic waves travel at the speed of light. The SI unit for frequency, the hertz (1 Hz = 1 cycle/s), is named in his honor.

23.2 PRODUCTION OF ELECTROMAGNETIC WAVES

We can get a good mental image of electromagnetic waves by considering how they are produced. Whenever a current varies, associated electric and magnetic fields vary, moving out like waves. Perhaps the easiest situation to visualize is a varying current in a long straight wire, produced by an AC generator at its center, as illustrated in Figure 23.4. The electric field shown next to the wire is produced by the charge distribution on the wire and varies

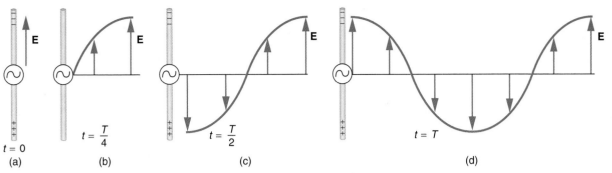

Figure 23.4 This long straight wire with an AC generator at its center becomes a broadcast antenna for electromagnetic waves. Shown here are the charge distributions at four different times. The electric field moves away from the antenna at the speed of light, forming part of an electromagnetic wave.

with it. The changing field moves out at the speed of light, becoming part of an electromagnetic wave, and making the wire into a broadcast antenna.

Closer examination of the one complete cycle shown in Figure 23.4 reveals the periodic nature of the generator-driven charges oscillating up and down in the antenna and the electric field produced. At $t = 0$, there is the maximum separation of charge with negative charges at the top and positive charges at the bottom, producing the maximum upward E-field. One-fourth of a cycle later, there is no charge separation and the field next to the antenna is zero, while the maximum in the E-field has moved away at speed c. As the process continues, the charge separation reverses and the field reaches its maximum downward value, returns to zero, and rises to its maximum upward strength at the end of one complete cycle. The outgoing wave has an amplitude proportional to the maximum separation of charge. Its wavelength is proportional to the period of the oscillation and, hence, is smaller for short periods or high frequencies. (As usual, wavelength and frequency are inversely proportional.)

Current in the antenna produces a magnetic field, as shown in Figure 23.5. The relationship between **E** and **B** is shown at one instant in Figure 23.5(a). As the current varies, the magnetic field varies in magnitude and direction. The magnetic field lines also move away from the antenna at the speed of light, forming the other part of the electromagnetic wave, as seen in Figure 23.5(b). The magnetic part of the wave has the same period and wavelength as the electric part, since they are both produced by the same movement and separation of charges in the antenna.

The electric and magnetic waves are shown together at one instant in time in Figure 23.6. The electric and magnetic fields produced by a long straight wire antenna are exactly in phase. Note that they are perpendicular to one another and to the direction of

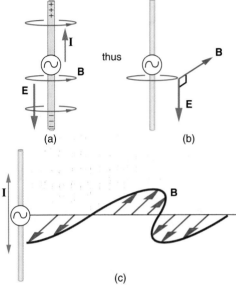

Figure 23.5 (a) The current in the antenna produces the circular magnetic field lines. The current produces the separation of charge, which in turn creates the electric field as shown. (b) The electric and magnetic fields near the wire are perpendicular; they are shown here for one point in space. (c) The magnetic field varies with current and moves away from the antenna at the speed of light.

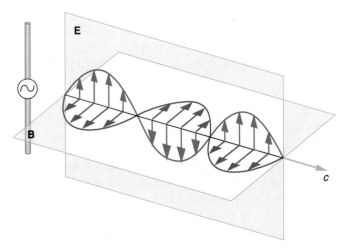

Figure 23.6 A part of the electromagnetic wave sent out from the antenna at one instant in time. The electric and magnetic fields are in phase, and they are perpendicular to one another and the direction of propagation. For clarity, the waves are shown only along one direction, but they move out in other directions, too.

propagation, making this a transverse wave. *Electromagnetic waves are transverse waves* in most cases, not just this special situation. Electromagnetic waves generally move out from a source in all directions, sometimes forming a complex radiation pattern. A linear antenna like this one will not radiate parallel to its length, for example. The wave is shown in one direction from the antenna in Figure 23.6 to illustrate its basic characteristics.

The antenna can also be driven by an AC circuit like those covered in the preceding chapter. In fact, *charges radiate whenever they are accelerated.* The electromagnetic waves produced carry energy away from the source. In the preceding chapter, it was assumed that energy did not quickly escape AC circuits, and mostly this is true. An antenna enhances the rate of electromagnetic radiation, and shielding is necessary to keep the radiation close to zero. Some familiar phenomena are based on the production of electromagnetic waves by varying currents. Your microwave oven, for example, sends electromagnetic waves, called microwaves, from a concealed antenna that has an oscillating current imposed on it.

There is a relationship between the *E*- and *B*-field strengths in an electromagnetic wave. This can be understood by again considering the antenna just described. The stronger the *E*-field created by a separation of charge, the greater the current and, hence, the greater the *B*-field created. Since current is directly proportional to voltage (Ohm's law) and voltage is directly proportional to *E*-field strength, the two should be directly proportional. It can be shown that the magnitudes of the fields do have a constant ratio, equal to the speed of light. That is,

$$\frac{E}{B} = c \qquad \qquad (23.2)$$

is the ratio of *E*-field strength to *B*-field strength in any electromagnetic wave. This is true at all times and at all locations in space.

EXAMPLE 23.1 *B*-FIELD STRENGTH IN AN ELECTROMAGNETIC WAVE

What is the maximum strength of the *B*-field in an electromagnetic wave that has a maximum *E*-field strength of 1000 V/m?

Strategy To find the *B*-field strength, we solve Equation 23.2 for *B*, yielding

$$B = \frac{E}{c}$$

Solution We are given *E*, and *c* is the speed of light.

Entering these into the expression for *B* yields

$$B = \frac{1000 \text{ V/m}}{3.00 \times 10^8 \text{ m/s}} = 3.33 \times 10^{-6} \text{ T}$$

Discussion The *B*-field strength is less than a tenth of the earth's relatively weak magnetic field. A relatively strong electric field of 1000 V/m is accompanied by a relatively weak magnetic field. Note that as this wave spreads out, say with distance from an antenna, its field strengths become progressively weaker.

The result of this example is consistent with the statement made in Section 23.1 that changing electric fields create relatively weak magnetic fields. They can be detected in electromagnetic waves, however, by taking advantage of the phenomenon of resonance, as Hertz did. A system with the same natural frequency as the electromagnetic wave can be made to oscillate. All radio and TV receivers use this principle to pick up and then amplify weak electromagnetic waves, while rejecting all others not at their resonant frequency.

23.3 THE ELECTROMAGNETIC SPECTRUM

Let us now examine how electromagnetic waves are classified into categories such as radio, infrared, ultraviolet, and so on, so that we can understand some of their similarities as well as some of their differences. We will also find that there are many connections with previously discussed topics, such as wavelength and resonance.

As noted before, an electromagnetic wave has frequency and wavelength and travels at the speed of light. The relationship among these wave characteristics was developed in

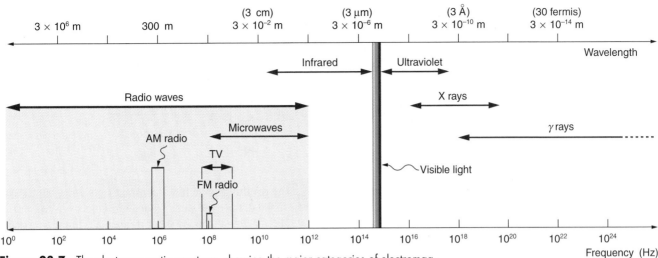

Figure 23.7 The electromagnetic spectrum, showing the major categories of electromagnetic waves. The range of frequencies and wavelengths is remarkable. The dividing line between some categories is distinct, whereas other categories overlap.

Chapter 15 and is given by Equation 15.11 as $v_w = f\lambda$, where v_w is the propagation speed of the wave. Here $v_w = c$, so that *for all electromagnetic waves,*

$$c = f\lambda \tag{23.3}$$

where f is the frequency, λ is the wavelength, and c is the speed of light. Thus, as for all waves, the greater the frequency, the smaller the wavelength. Figure 23.7 shows how the various types of electromagnetic waves are categorized according to their wavelengths and frequencies—that is, it shows the **electromagnetic spectrum**. Many of the characteristics of the various types of electromagnetic waves are related to their frequencies and wavelengths, as we shall see.

Radio Waves

The broad category of radio waves is defined to contain any electromagnetic wave produced by currents in wires and circuits. Its name derives from their most common use as a carrier of audio information (i.e., radio). The name is applied to electromagnetic waves of similar frequencies regardless of source. Radio waves from outer space, for example, do not come from alien radio stations. They are created by many astronomical phenomena, and their study has revealed much about nature on the largest scales. There are many uses for radio waves, and so the category is divided into many subcategories, including microwaves and those electromagnetic waves used for AM and FM radio, cellular telephones, and TV.

The *lowest commonly encountered radio frequencies* are produced by AC power transmission lines at frequencies of 60 Hz. (See Figure 23.8.) These extremely long wavelength electromagnetic waves are one means of energy loss in long-distance power transmission. There is an ongoing controversy regarding health hazards of these waves. Some people suspect that living near high-tension lines may cause a variety of illnesses, including cancer. But demographic data is either inconclusive or simply does not support the hazard theory.

Extremely low frequency (ELF) radio waves of about 1 kHz are used to communicate with submerged submarines. The ability of radio waves to penetrate salt water is related to their wavelength (much like ultrasound penetrating tissue)—the longer the wavelength, the farther they penetrate. Since salt water is a good conductor, radio waves are strongly absorbed by it, and very long wavelengths are needed to reach a submarine under the surface. (See Figure 23.9.)

AM radio waves are used to carry commercial radio signals in the frequency range from 540 to 1600 kHz. The abbreviation AM stands for *amplitude modulation*, which is the method for placing information on these waves. (See Figure 23.10.) A *carrier wave* having the basic frequency of the radio station, say 1530 kHz, is varied or *modulated* in

CONNECTIONS

Waves

There are many types of waves. Among the many shared attributes of waves are propagation speed, frequency, and wavelength. These are always related by the expression $v_w = f\lambda$. This chapter concentrates on EM waves, but we have seen examples of all of these characteristics for sound waves, and they also exist for water waves, earthquakes, and even for submicroscopic particles, as we shall see in later chapters.

Figure 23.8 High-tension lines radiate 60 Hz electromagnetic waves.

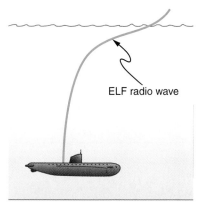

Figure 23.9 Very long wavelength radio waves are needed to reach this submarine, requiring extremely low frequency signals (ELF). Shorter wavelengths do not penetrate to any significant depth.

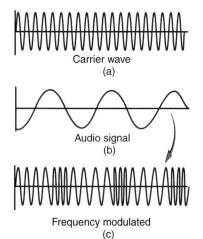

Carrier wave
(a)

Audio signal
(b)

Frequency modulated
(c)

Figure 23.11 Frequency modulation for FM radio. (a) A carrier wave at the station's basic frequency. (b) An audio signal at much lower audible frequencies. (c) The frequency of the carrier is modulated by the audio signal without changing its amplitude.

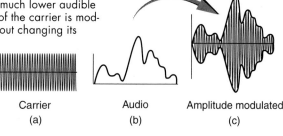

Figure 23.10 Amplitude modulation for AM radio. (a) A carrier wave at the station's basic frequency. (b) An audio signal at much lower audible frequencies. (c) The amplitude of the carrier is modulated by the audio signal without changing its basic frequency.

Carrier
(a)

Audio
(b)

Amplitude modulated
(c)

amplitude by an audio signal. The resulting wave has a constant frequency, but a varying amplitude. A radio receiver tuned to have the same resonant frequency as the carrier wave can pick up the signal, while rejecting the many other frequencies impinging on its antenna. The receiver's circuitry is designed to respond to variations in amplitude of the carrier wave to replicate the original audio signal. That audio signal is amplified to drive a speaker or perhaps to be recorded on tape.

FM radio waves are also used for commercial radio transmission, but in the frequency range of 88 to 108 MHz. FM stands for *frequency modulation*, another method of carrying information. (See Figure 23.11.) Here a carrier wave having the basic frequency of the radio station, perhaps 105.1 MHz, is modulated in frequency by the audio signal, producing a wave of constant amplitude but varying frequency. Since audible frequencies range up to 20 kHz (or 0.020 MHz) at most, the frequency of the FM radio wave can vary from the carrier by as much as 0.020 MHz. Thus the carrier frequencies of two different radio stations cannot be closer than 0.020 MHz. An FM receiver is tuned to resonate at the carrier frequency and has circuitry that responds to variations in frequency, reproducing the audio information.

FM radio is inherently less subject to noise from stray radio sources than AM radio. The reason is that amplitudes of waves add. So an AM receiver would interpret noise added onto the amplitude of its carrier wave as part of the information. An FM receiver can be made to reject amplitudes other than that of the basic carrier wave and only look for variations in frequency. It is thus easier to reject noise from FM, since noise produces a variation in amplitude.

Television is also broadcast on electromagnetic waves. Since the waves must carry a great deal of visual as well as audio information, each channel requires a larger range of frequencies than simple radio transmission. TV channels 2 through 6 utilize frequencies in the range of 54 to 88 MHz, while channels 7 through 13 lie between 174 and 216 MHz. (The entire FM radio band lies between channels 6 and 7.) TV channels 2 through 13 are called VHF (for very high frequency), while channels 14 through 84 are called UHF (for ultra high frequency) because they utilize an even higher frequency range of 470 to 890 MHz. The TV video signal is AM, while the TV audio is FM.

EXAMPLE 23.2 RADIO AND TV WAVELENGTHS

Calculate the wavelengths of a 1530 kHz AM radio signal, a 105.1 MHz FM radio signal, and a 890 MHz TV channel 84 signal.

Strategy The relationship between wavelength and frequency is given by Equation 23.3 to be $c = f\lambda$, where $c = 3.00 \times 10^8$ m/s is the speed of light (it is nearly the same in air as in a vacuum). We can rearrange this to find the wavelength for all three frequencies.

Solution Rearranging Equation 23.3 gives

$$\lambda = \frac{c}{f}$$

(a) For the $f = 1530$ kHz AM radio signal, then,

$$\lambda = \frac{3.00 \times 10^8 \text{ m/s}}{1530 \times 10^3 \text{ cycles/s}}$$
$$= 196 \text{ m}$$

(b) For the $f = 105.1$ MHz FM radio signal,

$$\lambda = \frac{3.00 \times 10^8 \text{ m/s}}{105.1 \times 10^6 \text{ cycles/s}}$$
$$= 2.85 \text{ m}$$

(continued)

(continued)

(c) And for the $f = 890$ MHz TV channel 84 signal,

$$\lambda = \frac{3.00 \times 10^8 \text{ m/s}}{890 \times 10^6 \text{ cycles/s}}$$

$$= 0.337 \text{ m}$$

Discussion These wavelengths are consistent with the spectrum in Figure 23.7. The wavelengths are also related to other properties of these electromagnetic waves, as we shall see.

The wavelengths found in the preceding example are representative of AM, FM, and TV and account for some of the differences in how they are broadcast and how well they travel. The most efficient length for a linear antenna, such as discussed in the previous section, is $\lambda/2$, half the wavelength of the electromagnetic wave. Thus a very large antenna is needed to efficiently broadcast typical AM radio with its carrier wavelengths on the order of hundreds of meters. One benefit to these long AM wavelengths is that they can go over and around rather large obstacles (like buildings and hills), just as ocean waves can go around large rocks. FM and TV are best received when there is a line of sight between the broadcast antenna and receiver, and they are often sent from very tall structures. FM and TV antennas themselves are much smaller than those used for AM, but they are elevated to achieve an unobstructed line of sight. (See Figure 23.12.)

Microwaves

Microwaves are the highest-frequency electromagnetic waves that can be produced by currents in macroscopic circuits and devices. Microwave frequencies range from about 10^9 Hz to the highest practical *LC* resonance at nearly 10^{12} Hz. Since they have high frequencies, their wavelengths are short compared with those of other radio waves—hence the name microwave. Microwaves can also be created by atoms and molecules. They are, for example, a component of electromagnetic radiation generated by thermal agitation. Since it is possible to carry more information per unit time on high frequencies, microwaves are quite suitable for communications. Most satellite-transmitted information is carried on microwaves, as are land-based long-distance transmissions. A clear line of sight between transmitter and receiver is needed because of the short wavelengths involved.

Radar is a common application of microwaves that was first developed in World War II. By detecting and timing microwave echoes, radar systems can determine the distance to objects as diverse as clouds and aircraft. A Doppler shift in the radar echo can be used to determine the speed of a car or the intensity of a rainstorm. Sophisticated radar systems are used to map the earth and other planets, with a resolution limited by wavelength. (See Figure 23.13.) The shorter the wavelength of any probe, the smaller the detail it is possible to observe.

How does the ubiquitous microwave oven create microwaves electronically, and why does food absorb them preferentially? Since microwaves can be created by thermal motions, they are both emitted and absorbed by matter. Certain molecules, such as those of water and proteins, rotate and resonate at microwave frequencies. These molecules can

Figure 23.12 A large tower is used to broadcast TV signals. The actual antennas are small structures at top of the tower—they are placed at great heights to have a clear line of sight over a large broadcast area.

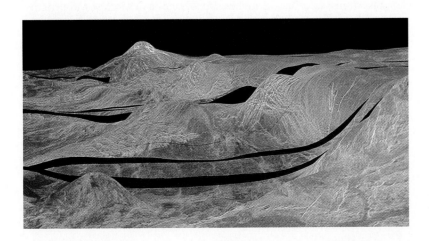

Figure 23.13 A microwave map of Venus. The Venusian atmosphere is opaque to visible light, but not to the microwaves that were used by an orbiter to create this image.

be made to rotate in resonance with incoming microwaves having their natural frequency. The energy thereby absorbed results in thermal agitation that heats food preferentially. Only later is a non-microwave-absorbing plate warmed by conduction. Hot spots in the food are related to constructive and destructive interference patterns. Rotating antennas and food turntables help spread out the hot spots.

Microwaves generated by atoms and molecules can be received and detected by electronic circuits. Deep space acts like a blackbody with a 2.7 K temperature, radiating most of its energy in the microwave frequency range. In 1964, Penzias and Wilson detected this radiation and eventually recognized that it was the radiation of the Big Bang's cooled remnants. We will discuss this in more detail in relation to cosmology in a later chapter.

Infrared Radiation

Infrared radiation is generally produced by thermal motion and the vibration and rotation of atoms and molecules. The range of infrared frequencies extends up to the lower limit of visible light, just below red. In fact, *infrared* means "below red." Frequencies at its upper limit are too high to be created in circuits, but small systems, such as atoms and molecules, can vibrate fast enough to create these waves. The lower end of the infrared frequency range overlaps the high end of the microwave range, where microwaves can also be created by atomic and molecular motions.

Human Application

Water molecules rotate and vibrate particularly well at infrared frequencies, emitting and absorbing them so efficiently that the emissivity for skin is $e = 0.97$ in the infrared.* Night vision scopes detect the infrared emitted by various warm objects, including humans, and convert it to visible light. Reconnaissance satellites can detect buildings, vehicles, and even individual humans by their infrared emissions, proportional to the fourth power of temperature. (See Figure 23.14.) More mundanely, we use infrared lamps, some of which are called quartz heaters, to preferentially warm us because we absorb infrared better than our surroundings.

The sun radiates like a nearly perfect blackbody (that is, it has $e = 1$), with a 6000 K surface temperature. About half of the solar energy arriving at the earth is in the infrared, with most of the rest in the visible part of the spectrum, and a relatively small amount in the ultraviolet. On average, 65% of the incident solar energy is absorbed by the earth, warming it during the day. At night, the earth radiates energy into cold dark space, almost entirely in the form of infrared. Over the long term, the earth reaches an average temperature such that the energy absorbed in visible and infrared in the day is radiated away at night. There are local and seasonal variations as well as changes in the amount absorbed during ice ages. But in all cases infrared plays a major role, particularly at night.

Figure 23.14 This image was made with infrared radiation captured by a satellite.

Visible Light

Visible light is the narrow segment of the electromagnetic spectrum to which the normal eye responds. Figure 23.15 shows this part of the spectrum, together with the colors associated with particular pure wavelengths. (The retina of the eye actually responds to the lowest ultraviolet frequencies, but these do not normally reach the retina because they are absorbed by the cornea and lens of the eye.) Red light has the lowest frequencies and longest wavelengths, while violet has the highest frequencies and shortest wavelengths. Blackbody radiation from the sun peaks in the visible part of the spectrum but is more intense in the red than in the violet, making the sun yellowish in appearance.

Human Application

---

*See the discussion of heat transfer by radiation in Section 13.7.

Figure 23.15 A small part of the electromagnetic spectrum that includes its visible components. The divisions between infrared, visible, and ultraviolet are not perfectly distinct, nor are those between the six rainbow colors.

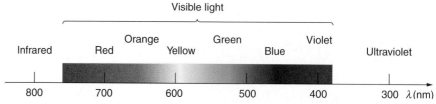

EXAMPLE 23.3 FREQUENCY OF VISIBLE LIGHT

Sunlight reaching the earth's surface peaks in intensity at a visible wavelength of about 570 nm. What is the frequency of this yellow-green light?

Strategy Equation 23.3 gives the relationship between frequency and wavelength. This can be rearranged so that the frequency can be calculated from the given information.

Solution Equation 23.3 states that $c = f\lambda$. Rearranging this to find frequency yields

$$f = \frac{c}{\lambda}$$

The speed of light is nearly the same in air as in vacuum, so that $c = 3.00 \times 10^8$ m/s. Substituting 570 nm for the wavelength of visible light yields

$$f = \frac{3.00 \times 10^8 \text{ m/s}}{570 \times 10^{-9} \text{ m}}$$
$$= 5.26 \times 10^{14} \text{ Hz}$$

Discussion As expected, this frequency is well above that of microwaves. While microwaves can be created in circuits, only small systems, such as atoms and molecules, can vibrate at the high frequencies associated with visible light. This is also true of higher-frequency electromagnetic waves, such as ultraviolet radiation, x rays, and γ rays.

Optics is the study of the behavior of visible light and other forms of electromagnetic waves. Two full chapters following this one are devoted to optics, but it is worth noting here that optics falls into two distinct categories. When electromagnetic radiation, such as visible light, interacts with objects that are large compared with its wavelength, its motion can be represented by straight lines like rays. Ray optics is the study of such situations and includes lenses and mirrors. When electromagnetic radiation interacts with objects about the same size as the wavelength or smaller, its wave nature becomes apparent. For example, observable detail is limited by the wavelength, and so visible light can never detect individual atoms, because they are so much smaller than its wavelength. Physical or wave optics is the study of such situations and includes all wave characteristics.

Ultraviolet Radiation

Ultraviolet means "above violet." The frequencies of ultraviolet electromagnetic radiation extend upward from violet, the highest-frequency visible light. Ultraviolet is also caused by atomic and molecular motions and excitations. The wavelengths of ultraviolet extend from 380 nm down to about 10 nm at its highest frequencies, which overlap with the lowest x-ray frequencies. It was recognized as early as 1801 that the solar spectrum had an invisible component beyond the violet. This part of the solar spectrum is partially absorbed by the atmosphere, particularly by ozone at high altitudes, but some ultraviolet radiation with wavelengths greater than 300 nm does reach the earth's surface.

Ultraviolet radiation has adverse effects on living cells, causing sunburn in large exposures, and skin cancer from repeated exposure. The tanning response is a defense mechanism in which the body produces pigments to absorb future exposures in inert skin layers above living cells. Even low-intensity ultraviolet can be used to sterilize hair cutting implements, implying that the energy associated with ultraviolet is deposited in a manner different from lower-frequency electromagnetic waves. (We shall see in later chapters that this is true for all electromagnetic waves with frequencies greater than visible light.) If all the sun's ultraviolet radiation reached the earth's surface, there would be extremely grave effects on the entire ecosystem from the severe cell damage it produces. Depletion of the protective ozone layer by escaped refrigerant gases called chlorofluorocarbons (CFCs) is a matter of great concern, and international attempts are being made to control these and other damaging emissions. (See Figure 23.16.)

Beneficial effects of ultraviolet radiation include sterilization, vitamin D production in the skin, and treatment of infantile jaundice. It is also used as an analytical tool to identify substances. When exposed to ultraviolet, some substances, such as minerals, glow in characteristic visible wavelengths, a process called fluorescence. So-called black lights emit ultraviolet to cause posters and clothing to fluoresce in the visible. Ultraviolet is also used in special microscopes to detect details smaller than those observable with longer-wavelength visible light microscopes.

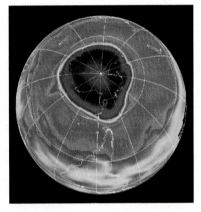

Figure 23.16 This map of ozone concentration over Antarctica shows severe depletion suspected to be caused by CFCs. Less dramatic but more general depletion has been observed over northern latitudes, suggesting the effect is global. With less ozone, more ultraviolet radiation from the sun reaches the surface, causing more damage.

Human & Medical Applications

THINGS GREAT AND SMALL

A Submicroscopic View of X-ray Production

X rays can be created in a high-voltage discharge. They are emitted in the material struck by electrons in the discharge current. There are two mechanisms by which the electrons create x rays. The first method is illustrated in Figure 23.A. Since this is a high-voltage discharge, the electron gains sufficient energy to ionize the atom. In the case shown, an inner-shell electron (one in an orbit relatively close to and tightly bound to the nucleus) is ejected. A short time later, another electron is captured and falls into the orbit in a single great plunge. The energy released by this fall is given to an EM wave known as an x ray. Since the orbits of the atom are unique to the type of atom, the energy of the x ray is characteristic of the atom hence the name *characteristic x ray*.

The second method by which an energetic electron creates an x ray when it strikes a material is illustrated in Figure 23.B. The electron interacts with charges in the material as it penetrates. These collisions transfer kinetic energy from the electron to the electrons and atoms in the material. A loss of kinetic energy implies an acceleration, in this case decreasing the electron's velocity. Whenever a charge is accelerated, it radiates EM waves. Given the high energy of the electron, these EM waves can have high energy. We call them x rays. Since the process is random, a broad spectrum of x-ray energy is emitted that is more characteristic of the electron energy than the type of material the electron encounters. Such EM radiation is called bremsstrahlung (German for "braking radiation").

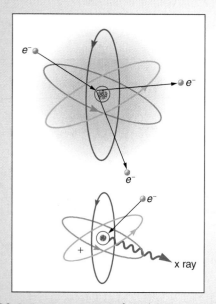

Figure 23.A Artist's conception of an electron ionizing an atom followed by the recapture of an electron and emission of an x ray. An energetic electron strikes an atom and knocks an electron out of one of the orbits closest to the nucleus. Later, the atom captures another electron, and the energy released by its fall into a low orbit generates a high-energy EM wave called an x ray.

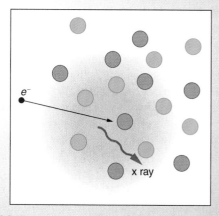

Figure 23.B Artist's conception of an electron being slowed by collisions in a material and emitting x-ray radiation. This energetic electron makes numerous collisions with electrons and atoms in a material it penetrates. An accelerated charge radiates EM waves, a second method by which x rays are created.

X Rays

Human & Medical Application

In the 1850s scientists (such as Faraday) began experimenting with high-voltage electrical discharges in tubes filled with rarefied gases. It was later found that these discharges created an invisible, penetrating form of very high frequency electromagnetic radiation. This radiation was called an *x ray*, because its identity and nature were unknown. As you can see in the Things Great and Small essay, there are two methods by which x rays are created—both are submicroscopic processes and can be caused by high-voltage discharges. While the low-frequency end of the x-ray range overlaps with the ultraviolet, x rays extend to much higher frequencies. X rays have adverse effects on living cells similar to those of ultraviolet radiation, and they have the additional liability of being more penetrating, affecting more than the surface layers of cells. Cancer and genetic defects can be induced by exposure to x rays. Because of their effect on rapidly dividing cells, x rays can also be used to treat and even cure cancer.

The widest use of x rays is for imaging objects that are opaque to visible light, such as the human body or aircraft parts. In humans, the risk of cell damage is weighed carefully against the benefit of the diagnostic information obtained. The ability of x rays to penetrate matter depends on density, and so an x-ray image can reveal very detailed density information. Figure 23.17 shows an example of the simplest type of x-ray image, an

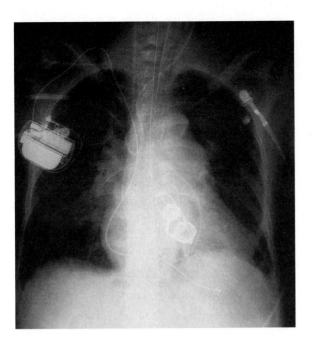

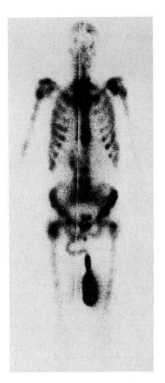

Figure 23.18 This is an image of the γ rays emitted by nuclei in a compound that is concentrated in the bones and eliminated through the kidneys. Bone cancer is evidenced by nonuniform concentration in similar structures. For example, some ribs are darker than others.

Figure 23.17 This shadow x-ray image shows many interesting features, such as artificial heart valves, a pacemaker, and the wires used to close the sternum.

x-ray shadow on film. The amount of information in a simple x-ray image is impressive, but we shall see in a later chapter that more sophisticated techniques, such as CT scans, can reveal three-dimensional information with details smaller than a millimeter.

Because they can have wavelengths less than 0.01 nm, x rays can be scattered (a process called x-ray diffraction) to detect the shape of molecules and the structure of crystals. X rays are also used as a precise analytical tool in x-ray induced fluorescence, in which the emissions are related to the types and amounts of materials present.

Gamma Rays

Soon after nuclear radioactivity was first detected in 1896, it was found that at least three distinct types of radiation were being emitted. The most penetrating nuclear radiation was called a γ *ray* (again because its identity and character were unknown), and it was later found to be an extremely high frequency electromagnetic wave. In fact, γ rays are *any electromagnetic radiation emitted by a nucleus*. This can be from natural nuclear decay or induced nuclear processes in nuclear reactors and weapons. The lower end of the γ-ray frequency range overlaps the upper end of the x-ray range, but γ rays can have the highest frequency of any electromagnetic radiation.

Gamma rays have characteristics identical to x rays of the same frequency—they differ only in source. At higher frequencies, γ rays are more penetrating and more damaging to living tissue. They have many of the same uses as x rays, including cancer therapy. Food spoilage can be greatly inhibited by exposing it to large doses of γ radiation, thereby obliterating responsible microorganisms. Damage to food cells occurs as well, and the long-term hazards of consuming radiation-preserved food are unknown and controversial. Figure 23.18 shows a medical image based on γ rays.

Human & Medical Application

23.4 ENERGY IN ELECTROMAGNETIC WAVES

Anyone who has used a microwave oven knows there is energy in electromagnetic waves. Sometimes this energy is obvious, such as in the warmth of the summer sun. Other times it is subtle, such as the unfelt energy of γ rays destroying living cells.

Electromagnetic waves can bring energy into a system by virtue of their electric and magnetic fields. These fields can exert forces and move charges in the system and, thus, do work on them. If the frequency of the electromagnetic wave is the same as the

Figure 23.19 Energy carried by a wave is proportional to its amplitude squared. With electromagnetic waves, larger E- and B-fields exert larger forces and can do more work.

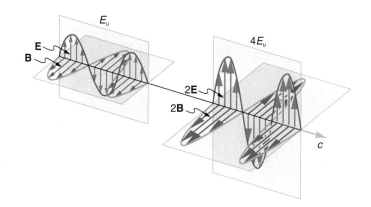

natural frequencies of the system (such as microwaves at the resonant frequency of water molecules), the transfer of energy is more efficient.

But there is energy in an electromagnetic wave whether it is absorbed or not. Once created, the fields carry energy away from a source. If absorbed, the field strengths are diminished and anything left travels on. Clearly, the larger the strength of the electric and magnetic fields, the more work they can do and the greater the energy the electromagnetic wave carries. When energy in waves was first covered in Section 15.11, it was noted that energy is proportional to the wave's amplitude squared. This is true for waves on guitar strings, for water waves, and for sound waves, where amplitude is proportional to pressure. In electromagnetic waves, the amplitude is the maximum field strength of the electric and magnetic fields. (See Figure 23.19.) Thus the energy carried and the intensity of an electromagnetic wave is proportional to E^2 and B^2. In fact, for a continuous sinusoidal electromagnetic wave, the **average intensity** I_{ave} is given by

$$I_{ave} = \frac{c\varepsilon_0 E_0^2}{2} \quad \text{(23.4a)}$$

where c is the speed of light, ε_0 the permittivity of free space, E_0 the maximum electric field strength, and intensity, as always, is power per unit area (here in W/m²).

The average intensity of an electromagnetic wave I_{ave} can also be expressed in terms of the magnetic field strength by using the relationship $B = E/c$ from Equation 23.2, and the fact that $\varepsilon_0 = 1/\mu_0 c^2$ from Equation 23.1. Algebraic manipulation produces the relationship

$$I_{ave} = \frac{cB_0^2}{2\mu_0} \quad \text{(23.4b)}$$

where μ_0 is the permeability of free space and B_0 is the maximum magnetic field strength. One more expression for I_{ave} in terms of both electric and magnetic field strengths is useful. Substituting the fact that $cB_0 = E_0$, the previous expression becomes

$$I_{ave} = \frac{E_0 B_0}{2\mu_0} \quad \text{(23.4c)}$$

Whichever of the three preceding equations is most convenient can be used, since they are really just different versions of the same principle: Energy in a wave is related to amplitude squared. Furthermore, since these equations are based on the assumption that the electromagnetic waves are sinusoidal, peak intensity is twice the average; that is, $I_0 = 2I_{ave}$.

CONNECTIONS

Waves and Particles

The behavior of electromagnetic radiation clearly exhibits wave characteristics. But we shall find in later chapters that at high frequencies, electromagnetic radiation also exhibits particle characteristics. These particle characteristics will be used to explain more of the properties of the electromagnetic spectrum and to introduce the formal study of modern physics. Another startling discovery of modern physics is that particles, such as electrons and protons, exhibit wave characteristics. This simultaneous sharing of wave and particle properties for all submicroscopic entities is one of the great symmetries in nature.

EXAMPLE 23.4 MICROWAVE INTENSITIES AND FIELDS

On its highest power setting, a certain microwave oven projects 1.00 kW of microwaves onto a 30.0 by 40.0 cm area. (a) What is the intensity in W/m²? (b) Calculate the peak electric field strength E_0 in these waves. (c) What is the peak magnetic field strength B_0?

Strategy In part (a), we can find intensity from its definition as power per unit area. Once the intensity is known, Equation 23.4 can be used to find the field strengths asked for in parts (b) and (c).

Solution for (a) Entering the given power into the def-

(continued)

(continued)

inition of intensity, and noting the area is 0.300 by 0.400 m, yields

$$I = \frac{P}{A} = \frac{1.00 \text{ kW}}{0.300 \text{ m} \times 0.400 \text{ m}}$$

Here $I = I_{ave}$, so that

$$I_{ave} = \frac{1000 \text{ W}}{0.120 \text{ m}^2} = 8.33 \times 10^3 \text{ W/m}^2$$

Note the peak intensity is twice the average: $I_0 = 2I_{ave} = 1.67 \times 10^4 \text{ W/m}^2$.

Solution for (b) To find E_0, we can rearrange Equation 23.4a to give

$$E_0 = \left(\frac{2I_{ave}}{c\varepsilon_0} \right)^{1/2}$$

Entering known values gives

$$E_0 = \sqrt{\frac{2(8.33 \times 10^3 \text{ W/m}^2)}{(3.00 \times 10^8 \text{ m/s})(8.85 \times 10^{-12} \text{ C}^2/\text{N} \cdot \text{m}^2)}}$$

$$= 2.51 \times 10^3 \text{ V/m}$$

Solution for (c) Perhaps the easiest way to find magnetic field strength, now that the electric field strength is known, is to use the relationship given by Equation 23.2:

$$B_0 = \frac{E_0}{c}$$

Entering known values gives

$$B_0 = \frac{2.51 \times 10^3 \text{ V/m}}{3.0 \times 10^8 \text{ m/s}}$$

$$= 8.35 \times 10^{-6} \text{ T}$$

Discussion As before, a relatively strong electric field is accompanied by a relatively weak magnetic field in an electromagnetic wave, since $B = E/c$, and c is a large number.

SUMMARY

Electromagnetic waves consist of oscillating electric and magnetic fields and propagate at the speed of light c. They were predicted by Maxwell, who also showed that

$$c = \frac{1}{\sqrt{\mu_0 \varepsilon_0}} \quad \textbf{(23.1)}$$

where μ_0 is the permeability of free space and ε_0 is the permittivity of free space.

Maxwell's prediction of electromagnetic waves resulted from his formulation of a complete and symmetric theory of electricity and magnetism, known as **Maxwell's equations**. These four equations are paraphrased in this text, rather than presented numerically, and encompass the major laws of electricity and magnetism. First is **Gauss's law for electricity**, second is **Gauss's law for magnetism**, third is **Faraday's law of induction**, including **Lenz's law**, and fourth is **Ampere's law** in a symmetric formulation that adds *another source of magnetism—changing electric fields*.

Electromagnetic waves are created by oscillating charges (which radiate whenever accelerated) and have the same frequency as the oscillation. Since the electric and magnetic fields in most electromagnetic wave are perpendicular to the direction the wave moves, it is ordinarily a *transverse wave*. The strengths of the electric and magnetic parts of the wave are related by

$$\frac{E}{B} = c \quad \textbf{(23.2)}$$

which implies that the magnetic field B is weak relative to the electric field E.

The relationship among the speed of propagation, wavelength, and frequency for any wave is given by $v_w = f\lambda$, so that for electromagnetic waves,

$$c = f\lambda \quad \textbf{(23.3)}$$

where f is the frequency, λ is the wavelength, and c is the speed of light.

The **electromagnetic spectrum** is separated into many categories and subcategories, based on the frequency and wavelength, source, and uses of the electromagnetic waves. Any electromagnetic wave produced by currents in wires is a *radio wave*. Radio waves are divided into many types, depending on their applications, ranging to *microwaves* at their highest frequencies. *Infrared radiation* lies below visible light in frequency and is produced by thermal motion and the vibration and rotation of atoms and molecules. Infrared's lower frequencies overlap with the highest-frequency microwaves. *Visible light* is also created by atomic and molecular motions and oscillations, and is defined as being detectable by the human eye. Its colors vary with frequency from red at the lowest to violet at the highest. *Ultraviolet* radiation starts with frequencies just above violet in the visible and is created primarily by atomic and molecular oscillations. *X rays* are created in high-voltage discharges and by other methods. Their lowest frequencies overlap the ultraviolet but extend to much higher values, overlapping at the high end with γ rays. *Gamma rays* are nuclear in origin and are defined to include the highest-frequency electromagnetic radiation of any type. Many examples and characteristics of the various components of the electromagnetic spectrum are discussed in the chapter.

The energy carried by any wave is proportional to its amplitude squared. For electromagnetic waves, this means intensity can be expressed as

$$I_{ave} = \frac{c\varepsilon_0 E_0^2}{2} \quad \textbf{(23.4a)}$$

where I_{ave} is the **average intensity** in W/m², and E_0 is the maximum electric field strength of a continuous sinusoidal wave. This can also be expressed in terms of the maximum magnetic field strength B_0 as

$$I_{ave} = \frac{cB_0^2}{2\mu_0} \quad \textbf{(23.4b)}$$

and in terms of both electric and magnetic fields as

$$I_{ave} = \frac{E_0 B_0}{2\mu_0} \quad \textbf{(23.4c)}$$

The three expressions for I_{ave} are all equivalent.

CONCEPTUAL QUESTIONS

23.1 The direction of the electric field shown in each part of Figure 23.4 is that produced by the charge distribution in the wire. Justify the direction shown in each part, using the Coulomb force law and the definition of $\mathbf{E} = \mathbf{F}/q$, where q is a *positive* test charge.

23.2 Is the direction of the magnetic field shown in Figure 23.5(a) consistent with the right hand rule for current (RHR-2) in the direction shown in the figure?

23.3 Why is the direction of the current shown in each part of Figure 23.5 opposite to the electric field produced by the wire's charge separation?

23.4 In which situation shown in Figure 23.20 will the electromagnetic wave be more successful in inducing a current in the wire? Explain.

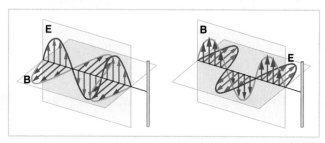

Figure 23.20 Electromagnetic waves approaching long straight wires. Question 4.

23.5 In which situation shown in Figure 23.21 will the electromagnetic wave be more successful in inducing a current in the loop? Explain.

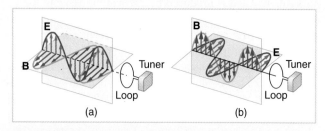

Figure 23.21 Electromagnetic waves approaching a wire loop. Question 5.

23.6 Should the straight wire antenna of a radio be vertical or horizontal to best receive radio waves broadcast by a vertical transmitter antenna? How should a loop antenna be aligned to best receive the signals? (Note that the direction of the loop that produces the best reception can be used to determine the

location of the source. It is used for that purpose in tracking tagged animals in nature studies, for example.)

23.7 Under what conditions might wires in a DC circuit emit electromagnetic waves?

23.8 Give an example of interference of electromagnetic waves.

23.9 Figure 23.22 shows the interference pattern of two radio antennas broadcasting the same signal. Explain how this is analogous to the interference pattern for sound produced by two speakers as shown in Figure 15.31. Could this be used to make a directional antenna system that broadcasts preferentially in certain directions? Explain.

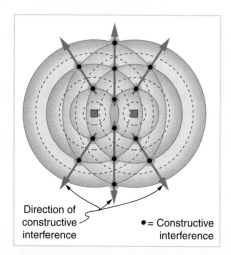

Direction of constructive interference

● = Constructive interference

Figure 23.22 An overhead view of two radio broadcast antennas sending the same signal, and the interference pattern they produce. Question 9.

23.10 If you live in an region that has a TV channel 6, you can sometimes pick up its audio portion on your FM radio receiver at the low end of the FM frequency range. Explain how this is possible. Does it imply that TV audio is broadcast as FM?

23.11 Explain why people who have the lens of their eye removed because of cataracts are able to see low-frequency ultraviolet.

23.12 How do fluorescent soap residues make clothing look "brighter and whiter" in outdoor light? Would this be effective in candlelight? (See Figure 23.23.)

23.13 Give an example of resonance in the reception of electromagnetic waves.

23.14 Illustrate that the size of details of an object that can be detected with electromagnetic waves is related to their

THE FAR SIDE By GARY LARSON

Now at your local feed store

Figure 23.23 Question 12.

wavelength, by comparing details observable with two different types (for example, radar and visible light or infrared and x rays).

23.15 Why don't buildings block radio waves as completely as they do visible light?

23.16 The rate at which information can be transmitted on an electromagnetic wave is proportional to the frequency of the wave. Is this consistent with the fact that laser telephone transmission at visible frequencies carries far more conversations per optical fiber than conventional electronic transmission in a wire? What is the implication for ELF radio communication with submarines?

23.17 Give an example of energy carried by an electromagnetic wave.

PROBLEMS

Sections 23.1–23.3 Characteristics of Electromagnetic Waves (other than intensity)

23.1 Verify that the correct value for the speed of light c is obtained when numerical values for the permeability and permittivity of free space (μ_0 and ε_0) are entered into Equation 23.1.

23.2 Show that, when SI units for μ_0 and ε_0 are entered, the units given by the right-hand side of Equation 23.1 are m/s.

23.3 What is the maximum electric field strength in an electromagnetic wave that has a maximum magnetic field strength of 5.00×10^{-4} T (about ten times the earth's)?

23.4 Verify the units obtained for magnetic field strength B in Example 23.1 (using Equation 23.2) are in fact teslas (T).

23.5 Two microwave frequencies are authorized for use in microwave ovens: 900 and 2560 MHz. Calculate the wavelength of each.

23.6 (a) Calculate the range of wavelengths for AM radio given its frequency range is 540 to 1600 kHz. (b) Do the same for the FM frequency range of 88.0 to 108 MHz.

23.7 Citizen's band (CB) radio utilizes frequencies between commercial AM and FM. What is the frequency of a 11.12 m wavelength CB channel?

23.8 Find the frequency range of visible light, given it encompasses wavelengths from 380 to 760 nm.

23.9 Electromagnetic radiation having a 15.0 μm wavelength is classified as infrared radiation. What is its frequency?

23.10 Approximately what is the smallest detail observable with a microscope that uses ultraviolet light of frequency 1.20×10^{15} Hz?

23.11 Some radar systems detect the size and shape of objects such as aircraft and geological terrain. Approximately what is the smallest observable detail utilizing 500 MHz radar?

23.12 If you wish to detect details the size of atoms (about 1×10^{-10} m) with electromagnetic radiation, it must have

a wavelength of about this size. (a) What is its frequency? (b) What type of electromagnetic radiation might this be?

23.13 If the sun suddenly turned off, we would not know it until its light stopped coming. How long would that be, given the sun is 1.50×10^{11} m away?

23.14 Distances in space are often quoted in units of light years, the distance light travels in one year. (a) How many meters is a light year? (b) How many meters is it to Andromeda, the nearest large galaxy, given it is 2.00×10^6 light years away? (c) The most distant galaxy yet discovered is 12.0×10^9 light years away. How far is this in meters?

• 23.15 A certain 60.0 Hz AC power line radiates an electromagnetic wave having a maximum electric field strength of 13.0 kV/m. (a) What is the wavelength of this very low frequency electromagnetic wave? (b) What is its maximum magnetic field strength?

• 23.16 During normal beating, the heart creates a maximum 4.00 mV potential across 0.300 m of a person's chest, creating a 1.00 Hz electromagnetic wave. (a) What is the maximum electric field strength created? (b) What is the corresponding maximum magnetic field strength in the electromagnetic wave? (c) What is the wavelength of the electromagnetic wave?

• 23.17 The ideal size (most efficient) for a broadcast antenna with one end on the ground is one-fourth the wavelength ($\lambda/4$) of the electromagnetic radiation being sent out. If a new radio station has such an antenna that is 50.0 m high, what frequency does it broadcast most efficiently? Is this in the AM or FM band?

• 23.18 TV-reception antennas for VHF (channels 2 through 13) are constructed with cross wires supported at their centers, as shown in Figure 23.24. The ideal length for the cross wires is one-half the wavelength to be received, with the more expensive antennas having one for each channel. Suppose you measure the lengths of the wires for channels 5 and 11 and find them to be 1.94 and 0.753 m long, respectively. What are the frequencies for channels 5 and 11?

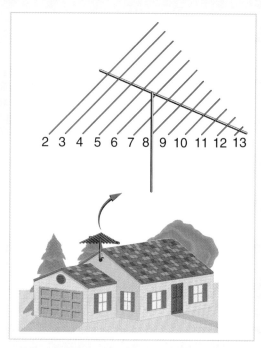

Figure 23.24 A television reception antenna has cross wires of various lengths to most efficiently receive different wavelengths. Problem 18.

• **23.19** Conversations with astronauts on lunar walks had an echo that can be used to estimate the distance to the moon. The sound spoken by the person on earth was transformed into a radio signal sent to the moon, and transformed back into sound on a speaker inside the astronaut's space suit. This sound was picked up by the microphone in the space suit (intended for the astronaut's voice) and sent back to earth as a radio echo of sorts. If the round-trip time was 2.60 s, what was the approximate distance to the moon, neglecting any delays in the electronics?

• **23.20** Lunar astronauts placed a reflector on the moon's surface, off which a laser beam is periodically reflected. The distance to the moon is calculated from the round-trip time. (a) To what accuracy in meters can the distance to the moon be determined, if this time can be measured to 0.100 ns? (b) What percent accuracy is this, given the average distance to the moon is 3.84×10^8 m?

⁝ **23.21** Radar is used to determine distances to various objects by measuring the round-trip time for an echo from the object. (a) How far away is the planet Venus if the echo time is 1000 s? (b) What is the echo time for a car 75.0 m from a Highway Patrol radar unit? (c) How accurately (in nanoseconds) must you be able to measure the echo time to an airplane 12.0 km away to determine its distance within 10.0 m?

Section 23.4 Energy in Electromagnetic Waves

23.22 What is the intensity of an electromagnetic wave with a peak electric field strength of 125 V/m?

23.23 Find the intensity of an electromagnetic wave having a peak magnetic field strength of 4.00×10^{-9} T.

• **23.24** The helium-neon lasers most commonly used in student physics laboratories have power outputs of 0.250 mW. (a) If such a laser beam is projected onto a circular spot 1.00 mm

in diameter, what is its intensity? (b) Find the peak magnetic field strength. (c) Find the peak electric field strength.

• **23.25** Find the intensity of the electromagnetic wave created by the heart, using the assumptions in Problem 23.16.

• **23.26** An AM radio transmitter broadcasts 50.0 kW of power uniformly in all directions. (a) Assuming all of the radio waves that strike the ground are completely absorbed, and that there is no absorption by the atmosphere or other objects, what is the intensity 30.0 km away? (Hint: Half the power will be spread over the area of a hemisphere.) (b) What is the maximum electric field strength at this distance?

• **23.27** Suppose the maximum safe intensity of microwaves for human exposure is taken to be 1.00 W/m². (a) If a radar unit leaks 10.0 W of microwaves (other than those sent by its antenna) uniformly in all directions, how far away must you be to be exposed to an intensity considered to be safe? Assume that the power spreads uniformly over the area of a sphere with no complications from absorption or reflection. (b) What is the maximum electric field strength at the safe intensity? (Note that early radar units leaked more than modern ones do. This caused identifiable health problems, such as cataracts, for people who worked near them.)

• **23.28** A 2.50 m diameter residential satellite dish receives TV signals that have a maximum electric field strength (for one channel) of 7.50 µV/m. (See Figure 23.25.) (a) What is the intensity of this wave? (b) What is the power received by the antenna? (c) If the orbiting satellite broadcasts uniformly over an area of 1.50×10^{13} m² (a large fraction of North America), how much power does it radiate?

Figure 23.25 Satellite dishes receive TV signals sent from orbit. Although the signals are quite weak, the receiver can detect them by being tuned to resonate at their frequency. Problem 28.

⁝ **23.29** Lasers can be constructed that produce an extremely high intensity electromagnetic wave for a brief time. They are used to ignite nuclear fusion, for example. Such a laser may produce an electromagnetic wave with a maximum electric field strength of 1.00×10^{11} V/m for a time of 1.00 ns. (a) What is the maximum magnetic field strength in the wave? (b) What is the intensity of the beam? (c) What energy does it deliver on a 1.00 mm^2 area?

23.30 Show that for a continuous sinusoidal electromagnetic wave, the peak intensity is twice the average intensity ($I_0 = 2I_{ave}$), using either the fact that $E_0 = \sqrt{2}E_{rms}$, or $B_0 = \sqrt{2}B_{rms}$, where rms means average (actually root mean square, a type of average).

23.31 Suppose a source of electromagnetic waves radiates uniformly in all directions in empty space, where there are no absorption or interference effects. (a) Show that the intensity is inversely proportional to r^2, the distance from the source squared. (b) Show that the magnitudes of the electric and magnetic fields are inversely proportional to r.

INTEGRATED CONCEPTS

The following problems involve concepts from this chapter and several others. Physics is most interesting when applied to general situations involving more than a narrow set of physical principles. For example, oscillating circuits can create electromagnetic waves, hence the relevance of Chapter 22, Induction and AC Circuits. The following topics are involved in some or all of the problems in this section:

Topics	Location
Work, energy, and power	Chapter 6
Heat and heat transfer	Chapter 13
Oscillatory motion and waves	Chapter 15
Sound	Chapter 16
Electric charge and field	Chapter 17
Magnetism	Chapter 21
Induction and AC circuits	Chapter 22

PROBLEM-SOLVING STRATEGY

Step 1. *Identify which physical principles are involved.*

Step 2. *Solve the problem using strategies outlined in the text.*

The following worked example illustrates how this strategy is applied to an integrated concept problem.

EXAMPLE WAVELENGTH OF FM AND CAPACITANCE NEEDED FOR OSCILLATOR

The following topics are involved in this integrated concepts worked example:

Topics	Location
Electromagnetic waves	Chapter 23
Induction and AC circuits	Chapter 22

The highest frequency in the FM broadcast band is 108 MHz. (a) Find the wavelength of such electromagnetic radiation. (b) If an *LC* circuit is to resonate at this frequency and has a 0.400 μH inductor, what capacitance is needed?

Strategy Step 1 To solve an **integrated concept problem,** such as those following this example, we must first identify the physical principles involved and identify the chapters in which they are found. Part (a) of this example asks for the *wavelength* of an electromagnetic wave, a topic of the present Chapter 23. Part (b) considers *resonance in an AC circuit*, a topic of Chapter 22.

Strategy Step 2 The following solutions to each part of the example illustrate how specific problem-solving strategies are applied. These involve identifying knowns and unknowns, checking to see if the answer is reasonable, and so on.

Solution for (a): EM wavelength (Chapter 23) The wavelength of an electromagnetic wave is related to its frequency by Equation 23.3:

$$c = f\lambda$$

Rearranging this to find the wavelength,

$$\lambda = \frac{c}{f}$$

Entering known values yields

$$\lambda = \frac{3.00 \times 10^8 \text{ m/s}}{108 \times 10^6 \text{ Hz}}$$
$$= 2.78 \text{ m}$$

Discussion for (a) This wavelength is consistent with the discussion of the wavelength attributes of various parts of the electromagnetic spectrum.

Solution for (b): Resonant LC circuit frequency (Chapter 22) The resonant frequency in an *LC* circuit is given by Equation 22.22:

$$f_0 = \frac{1}{2\pi\sqrt{LC}}$$

Solving this algebraically for *C* gives

$$C = \frac{1}{4\pi^2 f_0^2 L}$$

Substituting known values yields

$$C = \frac{1}{4\pi^2 (108 \times 10^6 \text{ Hz})^2 (0.400 \times 10^{-6} \text{ H})}$$
$$= 5.43 \times 10^{-12} \text{ F} = 5.43 \text{ pF}$$

Discussion for (b) /Capacitances similar to this are used in *LC* circuits, such as in your tuner-amplifier, that resonate at FM frequencies.

This worked example illustrates how to apply problem-solving strategies to situations that include topics from different chapters. The first step is to identify the physical principles involved in the problem. The second step is to solve for the unknown using familiar problem-solving strategies. These are found throughout the text. Moreover, many worked examples show how to use problem-solving strategies for single topics. In this integrated concepts example, you can see how to apply them across several topics. You will find these techniques useful in applications of physics outside a physics course, such as in your profession, in other science disciplines, and in everyday life. The following problems will build your skills in the broad application of physical principles.

23.32 (a) Calculate the ratio of the highest to lowest frequencies of electromagnetic waves the eye can see, given the wavelength range of visible light is from 380 to 760 nm. (b) Compare this with the ratio of highest to lowest frequencies the ear can hear.

23.33 (a) Calculate the rate in watts at which heat radiates (almost entirely in the infrared) from 1.0 m^2 of the earth's surface at night. Assume the emissivity is 0.90, the temperature of the earth is 15°C, and that of outer space is 2.7 K. (b) Compare the intensity of this radiation with that coming to the earth from the sun during the day, which averages about 800 W/m^2, only half of which is absorbed. (c) What is the maximum magnetic field strength in the outgoing radiation, assuming it is a continuous wave?

• **23.34** An *LC* circuit with a 5.00 pF capacitor oscillates in such a manner as to radiate at a wavelength of 3.30 m. (a) What is the resonant frequency? (b) What inductance is in series with the capacitor?

• **23.35** What capacitance in needed in series with an 800 μH inductor to form a circuit that radiates a wavelength of 196 m?

‡ 23.36 Police radar determines the speed of motor vehicles using the same Doppler shift technique employed for ultrasound in medical diagnostics. Beats are produced by mixing the double Doppler-shifted echo with the original frequency. If 1.50×10^9 Hz microwaves are used and a beat frequency of 150 Hz is produced, what is the speed of the vehicle? (Assume the same Doppler shift formulas are valid with the speed of sound replaced by the speed of light.)

‡ 23.37 Assume the mostly infrared radiation from a heat lamp acts like a continuous wave with wavelength 1.50 μm. (a) If the lamp's 200 W output is focused on a person's shoulder, over a circular area 25.0 cm in diameter, what is the intensity in W/m^2? (b) What is the peak electric field strength? (c) Find the peak magnetic field strength. (d) How long will it take to increase the temperature of the 4.00 kg shoulder by 2.00°C, assuming no other heat transfer and given its specific heat is 0.830 kcal/kg·°C?

‡ 23.38 On its highest power setting, a microwave oven increases the temperature of 0.400 kg of spaghetti by 45.0°C in 120 s. (a) What was the rate of power absorption by the spaghetti, given its specific heat is 0.900 kcal/kg·°C? (b) Find the average intensity of the microwaves, given they are absorbed over a circular area 20.0 cm in diameter. (c) What is the peak electric field strength of the microwave? (d) What is its peak magnetic field strength?

‡ 23.39 High-powered lasers can be used to vaporize neat holes through materials because they can be precisely focused on a very small area. (a) If a laser has an intensity of 2.00×10^9 W/m^2, how long would it take to raise the temperature from 20.0°C to the boiling point and then vaporize a hole through a 1.00 cm thick piece of aluminum, assuming half the energy is absorbed? (b) What is the maximum electric field strength in the laser beam?

‡ 23.40 Electromagnetic radiation from a 5.00 mW laser is concentrated on a 1.00 mm² area. (a) What is the intensity in W/m^2? (b) Suppose a 2.00 nC static charge is in the beam. What is the maximum electric force it experiences? (c) If the static charge moves at 400 m/s, what maximum magnetic force can it feel?

‡ 23.41 A 200-turn flat coil of wire 30.0 cm in diameter acts as an antenna for FM radio at a frequency of 100 MHz. The magnetic field of the incoming electromagnetic wave is perpendicular to the coil and has a maximum strength of 1.00×10^{-12} T. (a) What power is incident on the coil? (b) What average emf is induced in the coil over one-fourth of a cycle? (c) If the radio receiver has an inductance of 2.50 μH, what capacitance must it have to resonate at 100 MHz?

‡ 23.42 If electric and magnetic field strengths vary sinusoidally in time, being zero at $t = 0$, then $E = E_0 \sin 2\pi ft$ and $B = B_0 \sin 2\pi ft$. Let $f = 1.00$ GHz here. (a) When are the field strengths first zero? (b) When do they reach their most negative value? (c) How much time is needed for them to complete one cycle?

UNREASONABLE RESULTS

The following problems have results that are unreasonable because some premise is unreasonable or because certain of the premises are inconsistent with one another. Physical principles applied correctly then produce unreasonable results. The purpose of these problems is to give practice in assessing whether nature is being accurately described, and if it is not to trace the source of difficulty.

Note: Strategies for determining if an answer is reasonable appear in Chapter 22. They can be found with the section of problems labeled *Unreasonable Results* at the end of that chapter.

23.43 A researcher measures the wavelength of a 1.20 GHz electromagnetic wave to be 0.500 m. (a) Calculate the speed at which this wave propagates. (b) What is unreasonable about this result? (c) Which assumptions are unreasonable or inconsistent?

23.44 The peak magnetic field strength in a residential microwave oven is 9.20×10^{-5} T. (a) What is the intensity of the microwave? (b) What is unreasonable about this result? (c) What is wrong about the premise?

• **23.45** An *LC* circuit containing a 2.00 H inductor oscillates at such a frequency that it radiates at a 1.00 m wavelength. (a) What is the capacitance of the circuit? (b) What is unreasonable about this result? (c) Which assumptions are unreasonable or inconsistent?

• **23.46** An *LC* circuit containing a 1.00 pF capacitor oscillates at such a frequency that it radiates at a 300 nm wavelength. (a) What is the inductance of the circuit? (b) What is unreasonable about this result? (c) Which assumptions are unreasonable or inconsistent?

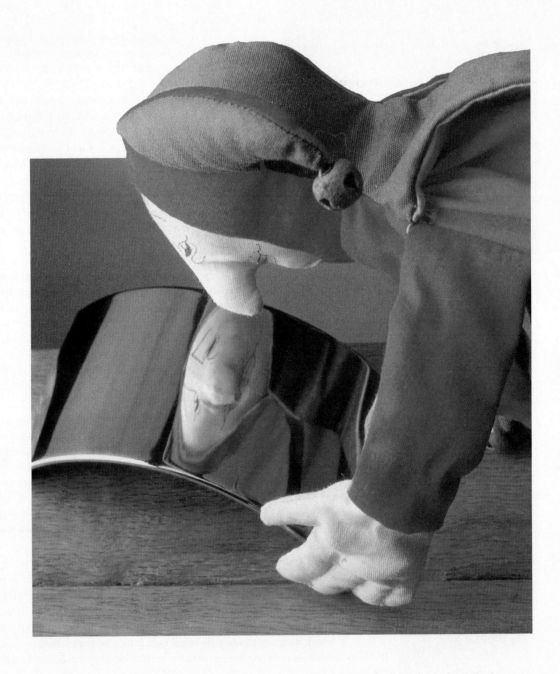

Light from this page is formed into an image by the lens of your eye, much as the lens of the camera that made this photograph. Mirrors, like lenses, can also form images that in turn are captured by your eye.

Figure 24.1 One of the joys of light.

Our lives are filled with light. Through vision, the most valued of our senses, light can evoke spiritual emotions, such as when we view a magnificent sunset. (See Figure 24.1.) Light can also simply amuse us in a theater, or warn us to stop at an intersection. It has innumerable uses beyond vision. Light can carry telephone signals through glass fibers or cook a meal in a solar oven. Life itself could not exist without light's energy. From photosynthesis in plants to the sun warming a cold-blooded animal, its supply of energy is vital.

We already know that visible light is the type of electromagnetic waves to which our eyes respond. That knowledge still leaves many questions regarding the nature of light and vision. What is color, and how do our eyes detect it? Why do diamonds sparkle? How does light travel? How do lenses and mirrors form images? These are but a few of the questions that are answered by the study of optics. **Optics** is the branch of physics that deals with the behavior of visible light and other electromagnetic waves. In particular, optics is concerned with the generation and propagation of light and its interaction with matter. What we have already learned about the generation of light in our study of heat transfer by radiation will be expanded upon in later chapters, especially those on atomic physics. In this and the next chapter, we will concentrate on the propagation of light and its interaction with matter.

It is convenient to divide optics into two major parts based on the size of objects that light encounters. When light interacts with an object that is several times as large as the light's wavelength, its observable behavior is like that of a ray; it does not prominently display its wave characteristics. This chapter will concentrate on such situations. When light interacts with smaller objects, it has very prominent wave characteristics, such as constructive and destructive interference. The next chapter, on wave optics, will concentrate on such situations.

24.1 THE RAY ASPECT OF LIGHT

There are three ways in which light can travel from a source to another location. (See Figure 24.2.) It can come directly from the source through empty space, such as from the sun to the earth. Or light can travel through various media, such as air and glass, to the person. Light can also arrive after being reflected, such as by a mirror. In all of these cases, light moves in straight lines called rays. Light may change direction when it encounters objects (such as a mirror) or in passing from one material to another (such as in passing from air to glass), but it then continues in a straight line or as a ray. The word **ray** comes from mathematics and here means a straight line that originates at some point. It is acceptable to visualize light rays as laser rays or even science fiction depictions of ray guns.

Experiments, as well as our own experiences, show that *when light interacts with objects several times as large as its wavelength, it travels in straight lines and acts like a ray*. Its wave characteristics are not pronounced in such situations. Since the wavelength of light is less than a micron (a thousandth of a millimeter), it acts like a ray in the many common situations in which it encounters objects larger than a micron. For example, when light encounters anything we can observe with unaided eyes, such as a mirror, it acts

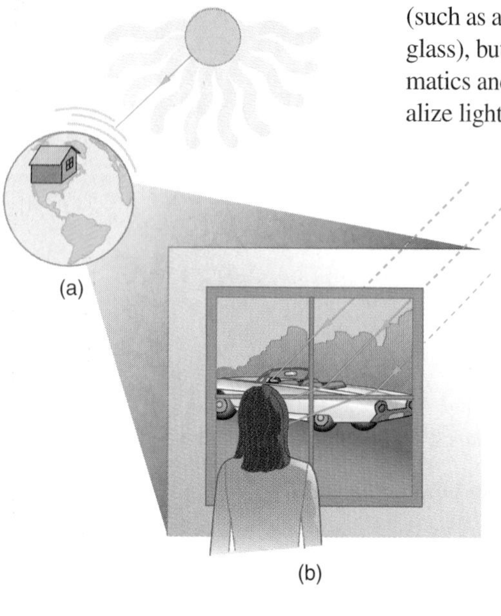

Figure 24.2 Three methods for light to travel from a source to another location. (a) Light reaches the upper atmosphere of the earth traveling through empty space directly from the source. (b) Light can reach a person in one of two ways. It can travel through media like air and glass. It can also reflect from an object like a mirror. In the situations shown here, light interacts with objects large enough that it travels in straight lines, like a ray.

like a ray with only subtle wave characteristics. We will concentrate on the ray characteristics in this chapter.

Since light moves in straight lines, changing directions when it interacts with materials, it is described by geometry and simple trigonometry. This part of optics, where the ray aspect of light dominates, is therefore called **geometric optics**. There are two laws that govern how light changes direction when it interacts with matter. These are the law of reflection, for situations in which light bounces off matter, and the law of refraction, for situations in which light passes through matter.

24.2 THE LAW OF REFLECTION

Whenever we look into a mirror, or squint at sunlight glinting from a lake, we are seeing a reflection. When you look at this page, too, you are seeing light reflected from it. Large telescopes use reflection to form an image of stars and other astronomical objects. The **law of reflection** is very simple:

> **The angle of reflection equals the angle of incidence.**

The law of reflection is illustrated in Figure 24.3, which also shows how the *angles are measured relative to the perpendicular* to the surface at the point where the light ray strikes*. We expect to see reflections from smooth surfaces, but Figure 24.4 illustrates how a rough surface reflects light. Since the light strikes different parts of the surface at different angles, it is reflected in many different directions, or **diffused**. Diffusion is what allows us to see a sheet of paper from any angle, as illustrated in Figure 24.5. Many objects, such as people, clothing, leaves, and walls, have rough surfaces and can be seen from all sides. A mirror, on the other hand, has a smooth surface (compared with the wavelength of light) and reflects light at specific angles, as illustrated in Figure 24.6. When the moon reflects from a lake, as shown in Figure 24.7, a combination of these effects takes place.

When we see ourselves in a mirror, it appears that our image is actually behind the mirror. This is illustrated in Figure 24.8. We see the light coming from a direction determined by the law of reflection. The angles are such that our image is exactly the same distance behind the mirror as we stand away from the mirror. If the mirror is on the wall of a room, the images in it are all behind the mirror, which can make the room seem bigger. Although these mirror images make objects appear to be where they cannot be (like behind a solid wall), the images are not figments of our imagination. Mirror images can be photographed and videotaped by instruments and look just as they do with our eyes (optical instruments themselves). The precise manner in which images are formed by mirrors and lenses will be treated in later sections of this chapter.

---

*Actually, a normal to the surface. Since the incident ray, reflected ray, and normal to the surface lie in a plane, the term perpendicular is used for the two-dimensional view.

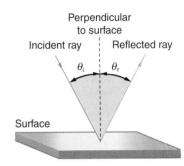

Figure 24.3 The *law of reflection* states that the angle of reflection equals the angle of incidence—$\theta_r = \theta_i$. The angles are measured relative to the perpendicular to the surface at the point where the ray strikes the surface.

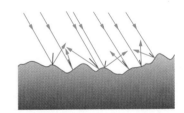

Figure 24.4 Light is diffused when it reflects from a rough surface. Here many parallel rays are incident, but they are reflected at many different angles since the surface is rough.

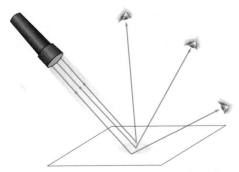

Figure 24.5 When a sheet of paper is illuminated with many parallel incident rays, it can be seen at many different angles, because its surface is rough and diffuses the light.

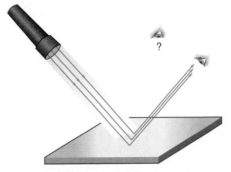

Figure 24.6 A mirror illuminated by many parallel rays reflects them in only one direction, since its surface is very smooth. Only the observer at the correct angle will see the reflected light.

Figure 24.7 Moonlight is spread out when it is reflected by the lake, since the surface is shiny but uneven.

Figure 24.8 Our image in a mirror is behind the mirror. The two rays shown are those that strike the mirror at just the correct angles to be reflected into the eyes of the person. The image appears to be in the direction the rays are coming from when they enter the eyes.

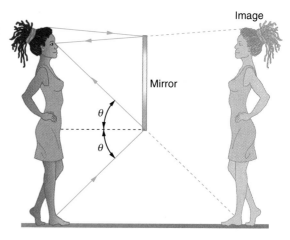

24.3   THE LAW OF REFRACTION

It is easy to notice some odd things when looking into a fish tank. For example, you may see the same fish appearing to be in two different places. (See Figure 24.9.) This is because light coming from the fish to us changes direction when it leaves the tank, and in this case, it can travel two different paths to get to our eyes. The changing of a light ray's direction (loosely called bending) when it passes through changes in matter is called **refraction**. Refraction is responsible for a tremendous range of optical phenomena, from the action of lenses to voice transmission through optical fibers.

Why does light change direction when passing from one material (medium) to another? It is because light changes speed when going from one material to another. So before we study the law of refraction, it is useful to discuss the speed of light and how it varies in different media.

The Speed of Light

Early attempts to measure the speed of light, such as those made by Galileo, determined that it moved extremely fast, perhaps instantaneously. The first real evidence that

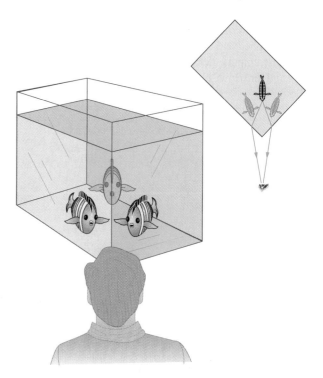

Figure 24.9 We can see the same fish in two different locations, because light changes directions when it passes from water to air. In this case, the light can reach the observer by two different paths, and so the fish seems to be in two different places. This bending of light is called refraction and is responsible for many optical phenomena.

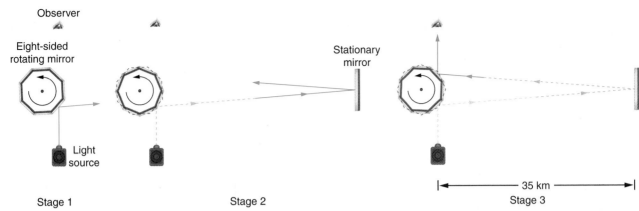

Stage 1

Stage 2

35 km

Stage 3

Figure 24.10 A schematic of early apparatus used by Michelson and others to determine the speed of light. As the mirrors rotate, the reflected ray is only briefly directed at the stationary mirror. The returning ray will be reflected into the observer's eye only if the next mirror has rotated into the correct position just as the ray returns. By measuring the correct rotation rate, the time for the round trip can be measured and the speed of light calculated.

light traveled at a finite speed came from the Danish astronomer Ole Roemer in the late 17th century. Roemer had noted that the average orbital period of one of Jupiter's moons, as measured from earth, varied depending on whether the earth was moving toward or away from Jupiter. He correctly concluded that the apparent change in period was due to the change in distance between the earth and Jupiter and the time it took light to travel this distance. In more recent times, physicists have measured the speed of light in numerous ways and with increasing accuracy. One particularly direct method, used by the American physicist Albert Michelson (1852–1931), is illustrated in Figure 24.10.* Light reflected from a rotating set of mirrors was reflected from a stationary mirror 35 km away and returned to the rotating mirrors. The time for the light to travel can be determined by how fast the mirrors must rotate for the light to be returned to the observer's eye.

The speed of light is now known to great precision. In fact, the speed of light in a vacuum c is so important that it is accepted as one of the basic physical quantities and has the fixed value[†]

$$c = 2.9972458 \times 10^8 \text{ m/s} \approx 3.00 \times 10^8 \text{ m/s} \qquad \textbf{(24.1)}$$

where the approximate value of 3.00×10^8 m/s is used whenever three-digit accuracy is sufficient. The speed of light through matter is less than it is in a vacuum, because light interacts with atoms in a material. The speed of light depends strongly on the type of material, since its interaction with different atoms, crystal lattices, and other substructures varies. We define the **index of refraction** n of a material to be

$$n = \frac{c}{v} \qquad \textbf{(24.2)}$$

where v is the speed of light in the material. Since the speed of light is always less than c in matter and equals c only in a vacuum, the index of refraction is always greater than or equal to one—that is, $n \geq 1$. Table 24.1 gives the indices of refraction for some representative substances. The values are listed for a particular wavelength of light, because they vary slightly with wavelength. (This can have important effects, such as colors produced by a prism.) Note that for gases n is close to 1.0. This seems reasonable, since atoms in gases are widely separated and light travels at c in the vacuum between atoms. It is common to take $n = 1$ for gases unless great precision is needed. Although the speed of light v in a medium varies considerably from its value c in a vacuum, it is still a large speed.

TABLE 24.1

INDEX OF REFRACTION IN VARIOUS MEDIA*	
Medium	**n**
Gases at 0°C, 1 atm	
Air	1.000293
Carbon dioxide	1.00045
Hydrogen	1.000139
Oxygen	1.000271
Liquids at 20°C	
Benzene	1.501
Carbon disulfide	1.628
Carbon tetrachloride	1.461
Ethanol	1.361
Glycerin	1.473
Water, fresh	1.333
Solids at 20°C	
Diamond	2.419
Fluorite	1.434
Glass, crown	1.52
Glass, flint	1.66
Ice at 0°C	1.309
Polystyrene	1.49
Plexiglas	1.51
Quartz, crystalline	1.544
Quartz, fused	1.458
Sodium chloride	1.544
Zircon	1.923

*For light with a wavelength of 589 nm in vacuum.

---

*Michelson later developed a more accurate, interference-based technique that produced historically significant results. Interference and other wave characteristic of optics are topics of the next chapter.

[†] As discussed in Section 1.2, the speed of light is such a basic quantity that it is taken to be *exactly* the value given in Equation 24.1, and other quantities are defined in terms of c. For example, the meter is defined to be the distance light travels in a fixed amount of time.

EXAMPLE 24.1 SPEED OF LIGHT IN MATTER

Calculate the speed of light in zircon, a material used in jewelry to imitate diamond.

Strategy The speed of light in a material, v, can be calculated from the index of refraction n of the material using Equation 24.2.

Solution Equation 24.2 states that $n = c/v$. Rearranging this to determine v gives

$$v = \frac{c}{n}$$

The index of refraction for zircon is given to be 1.923 in Table 24.1, and c is given in Equation 24.1. Entering these values in the last expression gives

$$v = \frac{3.00 \times 10^8 \text{ m/s}}{1.923}$$
$$= 1.56 \times 10^8 \text{ m/s}$$

Discussion This speed is slightly larger than half the speed of light in a vacuum and is still high compared with speeds we normally experience. The only substance listed in Table 24.1 that has a greater index of refraction than zircon is diamond. We shall see later that the large index of refraction for zircon makes it sparkle more than glass, but less than diamond.

Law of Refraction

Figure 24.11 shows how a ray of light changes directions when it passes from one medium to another. As before, the angles are measured relative to a perpendicular to the surface at the point where the light ray crosses it. (Some of the incident light will be reflected from the surface, but for now we will concentrate on the light that crosses it.) The change in direction of the light ray depends on how the speed of light changes. The change in the speed of light is related to the indices of refraction of the media involved. In the situations shown in Figure 24.11, medium 2 has a greater index of refraction than medium 1. This means that the speed of light is less in medium 2 than in medium 1. Note that as shown in Figure 24.11(a), *the direction of the ray moves closer to the perpendicular when it slows down.* Conversely, as shown in Figure 24.11(b), *the direction of the ray moves away from the perpendicular when it speeds up.* The path is exactly reversible. In both cases, you can imagine what happens by thinking about pushing a lawn mower from a sidewalk onto grass, and vice versa. Going from the sidewalk to grass, the front wheels are slowed and pulled to the side as shown. This is the same change in direction as for light when it goes from a fast medium to a slow one. When going from the grass to sidewalk, the front wheels can move faster and the mower changes direction as shown. This, too, is the same change in direction as for light going from slow to fast.

The amount that a light ray changes its direction depends both on the incident angle and the amount that the speed changes. For a ray at a given incident angle, a large change in speed causes a large change in direction. The exact mathematical relationship is the **law of refraction**, which is stated in equation form as

$$n_1 \sin \theta_1 = n_2 \sin \theta_2 \tag{24.3}$$

Here n_1 and n_2 are the indices of refraction for medium 1 and 2, and θ_1 and θ_2 are the angles between the rays and the perpendicular in medium 1 and 2 as shown in Figure 24.11. The law of refraction is also called **Snell's law** after the Dutch mathematician Willebrord

Figure 24.11 The change in direction of a light ray depends on how the speed of light changes when it crosses from one medium to another. The speed of light is greater in medium 1 than in medium 2 in the situations shown here. (a) A ray of light moves closer to the perpendicular when it slows down. This is analogous to what happens when a lawn mower goes from a sidewalk to grass. (b) A ray of light moves away from the perpendicular when it speeds up. This is analogous to what happens when a lawn mower goes from grass to sidewalk. The paths are exactly reversible.

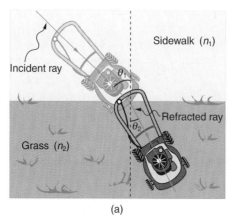
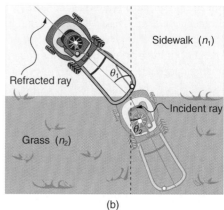

(a)   (b)

Snell (1591–1626), who discovered it in 1621. Snell's experiments showed that the law of refraction was obeyed and that a characteristic index of refraction n could be assigned to a given medium. Snell was not aware that the speed of light varied in different media, but he was able to determine indices of refraction from the way light rays changed direction.

EXAMPLE 24.2   DETERMINE THE INDEX OF REFRACTION FROM REFRACTION ITSELF

Find the index of refraction for medium 2 in Figure 24.11(a), assuming medium 1 is air and given the incident angle is 30.0° and the angle of refraction is 22.0°.

Strategy   The index of refraction for air is taken to be 1 in most cases (and to four significant figures, it is 1). Thus $n_1 = 1.00$ here. From the given information, $\theta_1 = 30.0°$ and $\theta_2 = 22.0°$. With this information, the only unknown in Snell's law is n_2, so that it can be used to find this unknown.

Solution   Snell's law is given by Equation 24.3 to be

$$n_1 \sin \theta_1 = n_2 \sin \theta_2$$

Rearranging to isolate n_2 gives

$$n_2 = n_1 \frac{\sin \theta_1}{\sin \theta_2}$$

Entering known values,

$$n_2 = 1.00 \cdot \frac{\sin 30.0°}{\sin 22.0°} = \frac{0.500}{0.375}$$
$$= 1.33$$

Discussion   This is the index of refraction for water, and Snell could have determined it by measuring the angles and performing this calculation. He would then have found 1.33 to be the appropriate index of refraction for water in all other situations, such as when a ray passes from water to glass. Today we can verify that the index of refraction is related to the speed of light in a medium by measuring that speed directly.

EXAMPLE 24.3   A LARGER CHANGE IN DIRECTION

Suppose that in a situation like that in Figure 24.11(a), light goes from air to diamond (rather than from air to water as in the preceding example) and that the incident angle is 30.0°. Calculate the angle of refraction θ_2 in the diamond.

Strategy   Again the index of refraction for air is taken to be $n_1 = 1.00$, and we are given $\theta_1 = 30.0°$. We can look up the index of refraction for diamond in Table 24.1, finding $n_2 = 2.419$. The only unknown in Snell's law is θ_2, which we wish to determine.

Solution   Solving Snell's law (Equation 24.3) for $\sin \theta_2$ yields

$$\sin \theta_2 = \frac{n_1}{n_2} \sin \theta_1$$

Entering known values,

$$\sin \theta_2 = \frac{1.00}{2.419} \cdot \sin 30.0° = 0.413 \cdot 0.500 = 0.207$$

The angle is thus

$$\theta_2 = \sin^{-1} 0.207 = 11.9°$$

Discussion   For the same 30.0° angle of incidence, the angle of refraction in diamond is significantly smaller than in water (11.9° rather than 22.0°—see the preceding example). This means there is a larger change in direction in diamond. The cause of a large change in direction is a large change in the index of refraction (or speed). In general, the larger the change in speed, the greater the effect on the direction of the ray.

24.4   TOTAL INTERNAL REFLECTION

A good-quality mirror may reflect more than 90% of the light that falls on it, absorbing the rest. But it would be useful to have a mirror that reflects all of the light that falls on it. Interestingly, we can produce *total reflection* using an aspect of *refraction*.

Consider what happens when a ray of light strikes the surface between two materials, such as is shown in Figure 24.12(a). Part of the light crosses the boundary and is refracted; the rest is reflected. If, as shown in the figure, the index of refraction for the second medium is less than for the first, the ray bends away from the perpendicular. (Since $n_1 > n_2$, the angle of refraction is greater than the angle of incidence—that is, $\theta_2 > \theta_1$.) Now imagine what happens as the incident angle is increased. This causes θ_2 to increase also. The largest the angle of refraction θ_2 can be is 90°, as shown in Figure 24.12(b). The **critical angle** θ_c for a combination of materials is defined to be the incident angle θ_1 that produces an angle of refraction of 90°. That is, θ_c is the incident angle for which $\theta_2 = 90°$. If the incident angle θ_1 is greater than the critical angle, as shown in Figure 24.12(c), then *all* of the light is reflected

Figure 24.12 (a) A ray of light crosses a boundary where the speed of light increases and the index of refraction decreases. That is, $n_2 < n_1$. The ray bends away from the perpendicular. (b) The critical angle θ_c is the one for which the angle of refraction is 90°. (c) Total internal reflection occurs when the incident angle is greater than the critical angle.

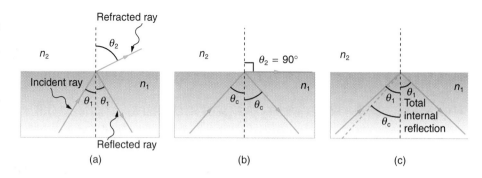

back into medium 1, a condition called **total internal reflection**. Since the angle of refraction cannot be greater than 90°, there is no refraction—there is only reflection.

Snell's law states the relationship between angles and indices of refraction. It is given by Equation 24.3:

$$n_1 \sin \theta_1 = n_2 \sin \theta_2$$

When the incident angle equals the critical angle ($\theta_1 = \theta_c$), the angle of refraction is 90° ($\theta_2 = 90°$). Noting that $\sin 90° = 1$, Snell's law in this case states

$$n_1 \sin \theta_c = n_2$$

The *critical angle* θ_c for a given combination of materials is, thus,

$$\theta_c = \sin^{-1}(n_2/n_1) \qquad \text{for } n_1 > n_2 \qquad \textbf{(24.4)}$$

Total internal reflection occurs for any incident angle greater than the critical angle θ_c, and it can only occur when the second medium has an index of refraction less than the first. Note Equation 24.4 is written for a light ray that travels in medium 1 and reflects from medium 2, as shown in the figure.

EXAMPLE 24.4   HOW BIG IS THE CRITICAL ANGLE HERE?

What is the critical angle for light leaving polystyrene (a type of plastic) and entering air?

Strategy The index of refraction for polystyrene is found to be 1.49 in Table 24.1, and the index of refraction of air can be taken to be 1.00, as before. Thus the condition that the second medium (air) has an index of refraction less than the first (plastic) is satisfied, and Equation 24.4 can be used to find the critical angle θ_c. Here, then, $n_2 = 1.00$ and $n_1 = 1.49$.

Solution The critical angle is given by Equation 24.4 to be

$$\theta_c = \sin^{-1}(n_2/n_1)$$

Substituting the identified values gives

$$\theta_c = \sin^{-1}(1.00/1.49) = \sin^{-1}0.671$$

$$= 42.2°$$

Discussion This means that any ray of light inside the plastic that strikes the surface at an angle greater than 42.2° will be totally reflected. This will make the inside surface of the clear plastic a perfect mirror for such rays without any need for the silvering used on common mirrors. Different combinations of materials have different critical angles, but any combination with $n_1 > n_2$ can produce total internal reflection. The same calculation as made here shows that the critical angle for a ray going from water to air is 48.6°, while that from diamond to air is 24.4°, and that from flint glass to crown glass is 66.3°. There is no total reflection for rays going in the other direction—for example, from air to water—since the condition that the second medium must have a smaller index of refraction is not satisfied. A number of interesting applications of total internal reflection follow.

Fiber Optics: Endoscopes to Telephones

Fiber optics is one application of total internal reflection that has grown immensely in recent years. **Fiber optics** employs the transmission of light down fibers of plastic or glass. Because the fibers are thin, light entering one is likely to strike the inside surface at an angle greater than the critical angle and, thus, be totally reflected. (See Figure 24.13.) The index of refraction outside the fiber must be smaller than inside, a condition that is easily satisfied if the fiber is surrounded by air. Rays are reflected around corners as shown, making the fibers into tiny light pipes.

Medical Application

Bundles of fibers can be used to transmit an image without a lens, as illustrated in Figure 24.14. The output of a device called an endoscope is shown in Figure 24.14(b).

Endoscopes are used to explore the body through various orifices or minor incisions. Light is transmitted down one fiber bundle to illuminate internal parts, and the reflected light is transmitted back out through another to be observed. Surgery can be performed, such as arthroscopic surgery on the knee joint, employing cutting tools attached to and observed with the endoscope. Samples can also be obtained, such as by lassoing an intestinal polyp for external examination. There are a host of other medical diagnostic and therapeutic uses for fiber optics, including transmission of an intense laser beam to burn away obstructing plaques in major arteries. With the improvement of fibers, the number and types of applications continue to increase.

Fibers in bundles are surrounded by a cladding material that has a lower index of refraction than the core. (See Figure 24.15.) The cladding prevents light from being transmitted between fibers in a bundle. Without cladding, light could pass between fibers in contact, since their indices of refraction are identical. Since no light gets into the cladding (there is total internal reflection back into the core), none can be transmitted between clad fibers that are in contact with one another.

Many telephone conversations are now carried by laser signals along optical fibers. The fibers can be made so transparent that light can travel many kilometers before it becomes dim enough to require amplification.* Lasers emit light with characteristics that allow far more conversations in one fiber than are possible with electric signals on a single conductor. We shall explore the unique characteristics of laser radiation in a later chapter.

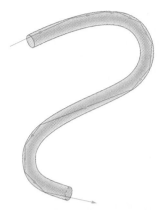

Figure 24.13 Light entering a thin fiber may strike the inside surface at large or grazing angles and is completely reflected if these angles exceed the critical angle. Such rays continue down the fiber, even following it around corners, since the angles of reflection and incidence remain large.

Corner Reflectors and the Like

A light ray that enters an object like the one in Figure 24.16(a) is reflected back exactly parallel to the direction from which it came. This is true whenever the reflecting surfaces are perpendicular, and it is independent of the angle of incidence (left as an end-of-chapter problem to prove). Such an object is called a **corner reflector**, since the light bounces from its inside corner. Many inexpensive reflector buttons on bicycles, cars, and warning signs have corner reflectors designed to return light in the direction from which it originated. It was more expensive for astronauts to place one on the moon. Laser signals are bounced from that corner reflector to measure the gradually increasing distance to the moon with great precision.

Corner reflectors are perfectly efficient when the conditions for total internal reflection are satisfied. It is not difficult to obtain total internal reflection for the situation shown in Figure 24.16(a), where the critical angle must be less than 45°. With common materials, it is easy to obtain a critical angle that is less than 45° (left as an end-of-chapter problem to verify). One use of these perfect mirrors is in binoculars, as shown in Figure 24.17.

*This remarkable feat was not easily achieved. Light can only travel a few meters through ordinary window glass before being nearly completely absorbed, for example.

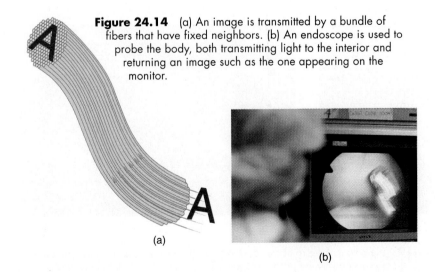

Figure 24.14 (a) An image is transmitted by a bundle of fibers that have fixed neighbors. (b) An endoscope is used to probe the body, both transmitting light to the interior and returning an image such as the one appearing on the monitor.

(a)

(b)

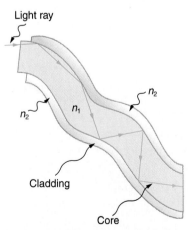

Figure 24.15 Fibers in bundles are clad by a material that has a lower index of refraction than the core to ensure total internal reflection even when fibers are in contact with one another. This shows a single fiber with its cladding.

Figure 24.16 (a) Light reflected from this corner reflector returns in the direction from which it came. (b) Astronauts placed a corner reflector on the moon to measure its gradually increasing orbital distance. (c) The light from this taillight is a reflection of the flash of the camera that took this picture on a dark night. The taillight is not on.

(a)

(b)

(c)

Figure 24.17 These binoculars employ corner reflectors with total internal reflection to get light to the observer's eyes.

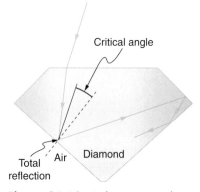

Figure 24.18 Light cannot easily escape a diamond, because its critical angle with air is so small. Most reflections are total, and the facets are placed so that light can only exit in particular ways—thus concentrating the light and making the diamond sparkle.

The Sparkle of Diamonds

Total internal reflection, coupled with a large index of refraction, explains why diamonds sparkle more than other materials. The critical angle for a diamond-to-air surface is only 24.4°, and so when light enters a diamond, it has trouble getting back out. (See Figure 24.18.) Although light freely enters the diamond, it can only exit if it makes an angle less than 24.4°. Facets on diamonds are specifically intended to make this unlikely, so that the light can only exit in certain places. Good diamonds are very clear, so that the light makes many internal reflections and is concentrated at the few places it can exit. Hence the sparkle. (Zircon is a natural gemstone that has an exceptionally large index of refraction, but not as large as diamond, so it is not as highly prized. Cubic zirconia is manufactured and has an even higher index of refraction (≈ 2.17), but still less than that of diamond.) The colors you see emerging from a sparkling diamond are not due to the diamond's color, which is usually nearly colorless. Those colors result from dispersion, the topic of the next section.

24.5 DISPERSION: THE RAINBOW AND PRISMS

Everyone enjoys the spectacle of a rainbow glimmering against a dark stormy sky. How does sunlight falling on clear drops of rain get broken into the rainbow of colors we see? The same process causes white light to be broken into colors by a clear glass prism or a diamond. (See Figure 24.19.)

We see about six colors in a rainbow. They are called red, orange, yellow, green, blue, and violet—sometimes indigo is listed, too. Those colors are associated with different wavelengths of light,* as shown in Figure 24.20. When our eye receives pure-wavelength light, we tend to see only one of the six colors, depending on wavelength. The thousands of other hues we can sense in other situations are our eye's response to various mixtures of wavelengths. White light, in particular, is a fairly uniform mixture of all visible wavelengths.

*See the discussion of visible light as a part of the electromagnetic spectrum in Section 23.3.

(a)

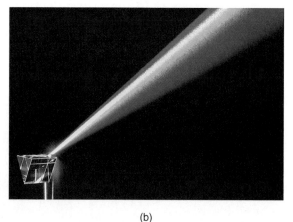

(b)

Figure 24.19 The colors of the rainbow (a) and those produced by a prism (b) are identical, although those produced by the prism are more intense and on a dark background.

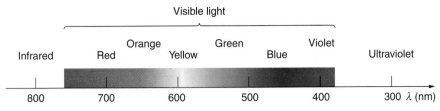

Figure 24.20 The six rainbow colors are associated with pure-wavelength light. All other hues are mixtures of wavelengths.

Sunlight, considered to be white, actually appears to be a bit yellow because of its mixture of wavelengths, but it does contain all visible wavelengths. The sequence of colors in rainbows is the same sequence as the colors plotted versus wavelength in Figure 24.20. What this implies is that white light is spread out according to wavelength in a rainbow. **Dispersion** is defined to be the spreading of white light into its full spectrum of wavelengths. More technically, dispersion occurs whenever there is a process that changes the direction of light in a manner that depends on wavelength. Dispersion, as a general phenomenon, can occur for any type of wave and always involves wavelength-dependent processes.

Refraction is responsible for dispersion in rainbows and many other situations. The angle of refraction depends on the index of refraction, as we saw in Section 24.3. We know that the index of refraction n depends on the medium. But for a given medium, n also depends on wavelength. (See Table 24.2.) Note that, for a given medium, n increases as wavelength decreases and is greatest for violet light. Thus violet light is bent more than red light, as shown for a prism in Figure 24.21, and the light is dispersed into the same sequence of wavelengths as seen in Figures 24.19 and 24.20.

Rainbows are produced by a combination of refraction and reflection. You may have noticed that you only see a rainbow when you look away from the sun. Light enters a drop of water and is reflected from the back of the drop as shown in Figure 24.22. The light is

CONNECTIONS

Dispersion

Any type of wave can exhibit dispersion. Sound waves, all types of electromagnetic waves, and water waves can be dispersed according to wavelength. Dispersion occurs whenever the speed of propagation depends on wavelength, thus separating and spreading out various wavelengths. Dispersion may require special circumstances such as in the production of a rainbow. This is also true for sound, since all frequencies ordinarily travel at the same speed. If you listen to sound through a long tube, such as a vacuum cleaner hose, you can easily hear it is dispersed by interaction with the tube. Dispersion, in fact, can reveal a great deal about what the wave has encountered that disperses its wavelengths. The dispersion of EM radiation from outer space, for example, has revealed much about the stuff between the stars.

TABLE 24.2

	Red	Orange	Yellow	Green	Blue	Violet
INDEX OF REFRACTION *n* IN SELECTED MEDIA AT VARIOUS WAVELENGTHS						
Medium	*(660 nm)*	*(610 nm)*	*(580 nm)*	*(550 nm)*	*(470 nm)*	*(410 nm)*
Water	1.331	1.332	1.333	1.335	1.338	1.342
Diamond	2.410	2.415	2.417	2.426	2.444	2.458
Glass, crown	1.512	1.514	1.518	1.519	1.524	1.530
Glass, flint	1.662	1.665	1.667	1.674	1.684	1.698
Polystyrene	1.488	1.490	1.492	1.493	1.499	1.506
Quartz, fused	1.455	1.456	1.458	1.459	1.462	1.468

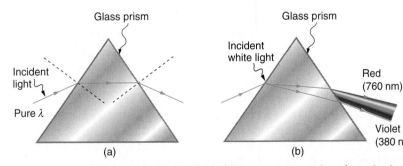

Figure 24.21 (a) A pure wavelength of light falls onto a prism and is refracted at both surfaces. (b) White light is dispersed by the prism (shown exaggerated). Since the index of refraction varies with wavelength, the angles of refraction vary with wavelength. A sequence of red to violet is produced, because the index of refraction increases steadily with decreasing wavelength.

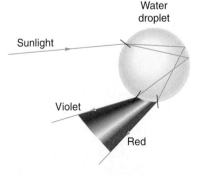

Figure 24.22 Part of the light falling on this water drop enters and is reflected from the back of the drop. This light is refracted and dispersed both as it enters and as it leaves the drop.

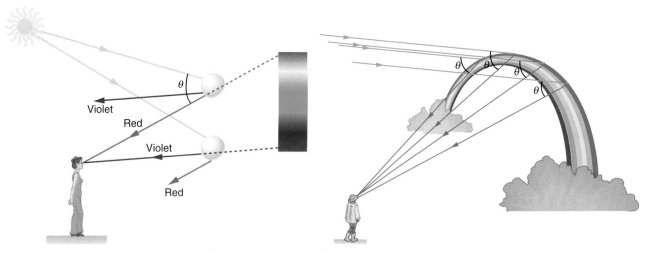

Figure 24.23 (a) Different colors emerge in different directions, and so you must look at different locations to see the various colors of a rainbow. (b) The arc of a rainbow results from the fact that a line between the observer and any point on the arc must make the correct angle with the parallel rays of sunlight to receive the refracted rays.

refracted both as it enters and as it leaves the drop. Since the index of refraction of water varies with wavelength, the light is dispersed, and a rainbow is observed as shown in Figure 24.23(a). (There is no dispersion caused by reflection at the back surface, since the law of reflection does not depend on wavelength.) The actual rainbow of colors seen by an observer depends on myriad rays being refracted and reflected toward the observer's eyes from numerous drops of water. The effect is most spectacular when the background is dark, as in stormy weather, but can be also be observed in waterfalls and lawn sprinklers. The arc of a rainbow comes from the need to look at a specific angle relative to the direction of the sun, as illustrated in Figure 24.23(b).

Dispersion may produce beautiful rainbows, but it can cause problems in optical systems. White light used to transmit messages in a fiber is dispersed, spreading out in time and eventually overlapping with other messages. Since a laser produces a nearly pure wavelength, its light experiences little dispersion, an advantage over white light for transmission of information. In contrast, dispersion of electromagnetic waves coming to us from outer space can be used to determine the amount of matter they pass through. As with many phenomena, dispersion can be useful or a nuisance, depending on the situation and our human goals.

24.6 IMAGE FORMATION BY LENSES

Lenses are found in a huge array of optical instruments, ranging from a simple magnifying glass to the eye to a camera's zoom lens. In this section, we will use the law of refraction to explore the properties of lenses and how they form images.

The word *lens* derives from the Latin word for a lentil bean, the shape of which is similar to the convex lens in Figure 24.24. The convex lens shown has been shaped so that all light rays that enter it parallel to its axis cross one another at a single point on the opposite side of the lens. (The axis is defined to be a line normal to the lens at its center, as shown in Figure 24.24.) Such a lens is called a **converging lens** for the converging effect it has on light rays. An expanded view of the path of one ray through the lens is shown, to illustrate how the ray changes direction both as it enters and as it leaves the lens. Since the index of refraction of the lens is greater than that of air, the ray moves toward the perpendicular as it enters and away from the perpendicular as it leaves. (This is in accordance with the law of refraction.) Due to the lens's shape, light is thus bent toward the axis at both surfaces. The point at which the rays cross is defined to be the **focal point** F of the lens. The distance from the center of the lens to its focal point is defined to be the **focal length** f of the lens. Figure 24.25 shows how a convex lens, such as that in a magnifying glass, can converge the nearly parallel light rays from the sun to a small spot.

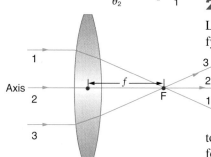

Figure 24.24 A converging lens. Rays of light entering a converging lens parallel to its axis are converged at its focal point F. (Ray 2 lies on the axis of the lens.) The distance from the center of the lens to the focal point is the lens's focal length f. An expanded view of the path taken by ray 1 shows the perpendiculars and the angles of incidence and refraction at both surfaces.

Figure 24.25 Sunlight converged by a convex magnifying glass can burn paper. Light rays from the sun are nearly parallel and cross at the focal point of the lens. The more powerful the lens, the closer to the lens the rays will cross.

The greater effect a lens has on light rays, the more powerful it is said to be. For example, a powerful converging lens will focus parallel light rays more quickly and will have a smaller focal length than a weak lens. The **power** P of a lens is defined to be the inverse of its focal length. In equation form, this is

$$P = \frac{1}{f} \qquad (24.5)$$

where f is the focal length of the lens in meters. The power of a lens P has the unit diopter (D), provided that the focal length is given in meters. That is, $1\ D = 1/m$, or $1\ m^{-1}$. (Note that this power (optical power, actually) is not the same as power in watts defined in Chapter 6. It is a concept related to the effect of optical devices on light.)

EXAMPLE 24.5 WHAT IS THE POWER OF A COMMON MAGNIFYING GLASS?

Suppose you take a magnifying glass out on a sunny day and you find that it concentrates sunlight to a small spot 8.00 cm away from the lens. What are the focal length and power of the lens?

Strategy The situation here is the same as those shown in Figures 24.24 and 24.25, since a magnifying glass is a convex or converging lens, and rays of sunlight are nearly parallel. Thus the focal length of the lens is the distance from the lens to the spot, and its power is the inverse of this distance as defined in Equation 24.5.

Solution The focal length of the lens is the distance from the center of the lens to the spot, given to be 8.00 cm. Thus,

$$f = 8.00\ \text{cm}$$

To find the power of the lens, we must first convert the focal length to meters; then, we substitute this value into Equation 24.5. This gives

$$P = \frac{1}{f} = \frac{1}{0.0800\ \text{m}} = 12.5\ \text{D}$$

Discussion This is a relatively powerful lens. The power of a lens in diopters should not be confused with the familiar concept of power in watts. It is an unfortunate fact that the word *power* is used for two completely different concepts. If you examine a prescription for eyeglasses, you will note lens powers given in diopters. If you examine the label on a motor, you will note energy consumption rate given as a power in watts.

Figure 24.26 shows a concave lens and the effect it has on rays of light that enter it parallel to its axis (the path taken by ray 2 in the figure is the axis of the lens). The concave lens is a **diverging lens,** because it causes the light rays to bend away (diverge) from its axis. In this case, the lens has been shaped so that all light rays entering it parallel to its axis appear to originate from the same point F, defined to be the focal point of a diverging lens. The distance from the center of the lens to the focal point is again called the focal length f of the lens. Note that the focal length and power of a diverging lens are *defined to be negative.* For example, if the distance to F in Figure 24.26 is 5.00 cm, then the focal length is $f = -5.00$ cm and the power of the lens is $P = -20.0$ D. An expanded view of the path of one ray through the lens is shown in the figure to illustrate how the shape of the lens, together with the law of refraction, causes the ray to follow its particular path and be diverged.

As noted in the initial discussion of the law of refraction in Section 24.3, the paths of light rays are exactly reversible. This means that the direction of the arrows could be reversed for all of the rays in Figures 24.24 and 24.26. For example, if a point light source is placed at the focal point of a convex lens, as shown in Figure 24.27, parallel light rays emerge from the other side.

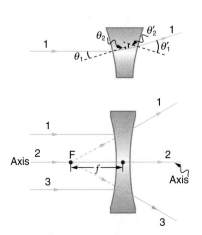

Figure 24.26 A diverging lens. Rays of light entering a diverging lens parallel to its axis are diverged, and all appear to originate at its focal point F. The dashed lines are not rays—they indicate the directions from which the rays *appear* to come. The focal length f of a diverging lens is negative. An expanded view of the path taken by ray 1 shows the perpendiculars and the angles of incidence and refraction at both surfaces.

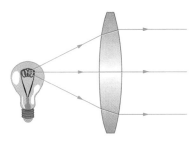

Figure 24.27 A small light source, like a light bulb filament, placed at the focal point of a convex lens results in parallel rays of light emerging from the other side. The paths are exactly the reverse of those shown in Figure 24.24. This technique is used in traffic lights and lighthouses to produce a directional beam of light from a source that emits light in all directions.

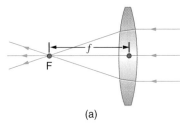

(a)

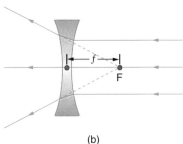

(b)

Figure 24.28 Thin lenses have the same focal length on either side. (a) Parallel light rays entering a converging lens from the right cross at its focal point on the left. (b) Parallel light rays entering a diverging lens from the right seem to come from the focal point on the right.

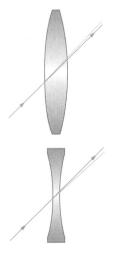

Ray Tracing and Thin Lenses

Ray tracing is the technique of determining or following (tracing) the paths that light rays take. For rays passing through matter, the law of refraction is used to trace the paths. Here we use ray tracing to help us understand the action of lenses in situations ranging from forming images on film to magnifying small print to correcting nearsightedness. While ray tracing for complicated lenses, such as in sophisticated cameras, may require computer techniques, there is a set of simple rules for tracing rays through thin lenses. A **thin lens** is defined to be one whose thickness has a negligible effect on rays passing through it. A thin symmetrical lens has two focal points, one on either side and both at the same distance from the lens. (See Figure 24.28.) Another important characteristic of a thin lens is that light rays through its center are deflected by a negligible amount, as seen in Figure 24.29.

Using paper, pencil, and a straight edge, ray tracing can accurately describe the operation of a lens. The rules for **ray tracing for thin lenses** are based on the illustrations already discussed:

1. A ray entering a converging lens parallel to its axis passes through the focal point F of the lens on the other side. (See rays 1 and 3 in Figure 24.24.)
2. A ray entering a diverging lens parallel to its axis seems to come from the focal point F. (See rays 1 and 3 in Figure 24.26.)
3. A ray passing through the center of a lens does not change direction. (See Figure 24.29, and see ray 2 in Figures 24.24 and 24.26.)
4. A ray entering a converging lens through its focal point exits parallel to its axis. (The reverse of rays 1 and 3 in Figure 24.24.)
5. A ray that enters a diverging lens by heading toward the focal point on the opposite side exits parallel to the axis. (The reverse of rays 1 and 3 in Figure 24.26.)

Image Formation by Thin Lenses

In some circumstances, a lens forms an obvious image, such as when a movie projector casts an image onto a screen. In other cases, the image is less obvious. Where, for example, is the image formed by eyeglasses? We will use ray tracing for thin lenses to illustrate how they form images, and we will develop equations to describe the image formation quantitatively.

Consider an object some distance away from a converging lens, as shown in Figure 24.30. To find the location and size of the image formed, we trace the paths of light rays originating from one point on the object, in this case the top of the person's head. The figure shows three rays from the top of the object that can be traced using the ray tracing rules given above. (Rays leave this point going in many directions, but we concentrate on only a few with paths that are easy to trace.) The first ray is one that enters the lens parallel to its axis and passes through the focal point on the other side (rule 1). The second ray passes through the center of the lens without changing direction (rule 3). The third ray passes through the nearer focal point on its way into the lens and leaves the lens parallel to its axis (rule 4). The three rays cross at the same point on the other side of the lens. The image is located at this point. All rays that come from the same point on the top of the person's head are refracted in such a way as to cross at the point shown. Rays from another point on the object, such as her belt buckle, will also cross at another common point, forming a complete image as shown. Although three rays are traced in Figure 24.30, only two are necessary to locate the image. It is best to trace rays for which there are simple ray tracing rules. Before applying ray tracing to other situations, let us consider the example shown in Figure 24.30 in more detail.

The image formed in Figure 24.30 is a **real image**, meaning that it can be projected. That is, light rays from one point on the object actually cross at the location of the image and can be projected onto a screen, a piece of film, or the retina of an eye, for example. Figure 24.31 shows how such an image would be projected onto film by a camera

Figure 24.29 The light ray through the center of a thin lens is deflected by a negligible amount and emerges parallel to its original path (shown as a shaded line).

Figure 24.30 Ray tracing is used to locate the image formed by a lens. Rays originating from the same point on the object are traced—the three chosen rays each follow one of the rules for ray tracing, so that their paths are easy to determine. The image is located at the point where the rays cross. In this case, a real image—one that can be projected on a screen—is formed. See text for further discussion.

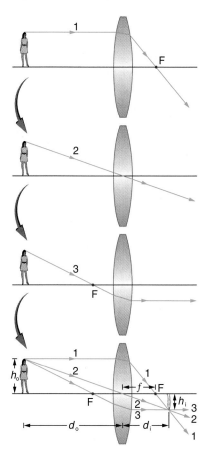

lens. Figure 24.31 also shows how a real image is projected onto the retina by the lens of an eye. Note that the image is there whether it is projected onto a screen or not.

Several important distances appear in Figure 24.30. We define d_o to be the **object distance**, the distance of an object from the center of a lens. **Image distance** d_i is defined to be the distance of the image from the center of a lens. The **height of the object** and **height of the image** are given the symbols h_o and h_i, respectively. Using the rules of ray tracing and making a scale drawing with paper and pencil, like that in Figure 24.30, we can accurately describe the location and size of an image. But the real benefit of ray tracing is in visualizing how images are formed in a variety of situations. To obtain numerical information, we use a pair of equations that can be derived from a geometric analysis of ray tracing for thin lenses. The **thin lens equations** are

$$\frac{1}{d_o} + \frac{1}{d_i} = \frac{1}{f} \tag{24.6a}$$

and

$$\frac{h_i}{h_o} = -\frac{d_i}{d_o} = m \tag{24.6b}$$

We define the ratio of image height to object height (h_i/h_o) to be the **magnification**, m. (The minus sign in Equation 24.6b will be discussed shortly.) The thin lens equations are broadly applicable to all situations involving thin lenses (and "thin" mirrors, as we will see later). We will explore many features of image formation in the following worked examples.

EXAMPLE 24.6 IMAGE OF A LIGHT BULB FILAMENT BY RAY TRACING AND BY THE THIN LENS EQUATIONS

A clear glass light bulb is placed 0.750 m from a convex lens having a 0.500 m focal length, as shown in Figure 24.32. Use ray tracing to get an approximate location for the image. Then use the thin lens equations to calculate (a) the location of the image and (b) its magnification. Verify that ray tracing and the thin lens equations produce consistent results.

Strategy and Concept Since the object is placed farther away from a converging lens than the focal length of the lens, this situation is analogous to those illustrated in Figures 24.30 and 24.31. Ray tracing to scale should produce similar results for d_i. Numerical solutions for d_i and m can be obtained using Equations 24.6a and b, noting that $d_o = 0.750$ m and $f = 0.500$ m.

Solutions *(Ray tracing)* The ray tracing to scale in Figure 24.32 shows two rays from a point on the bulb's filament crossing about 1.5 m on the far side of the lens.

(continued)

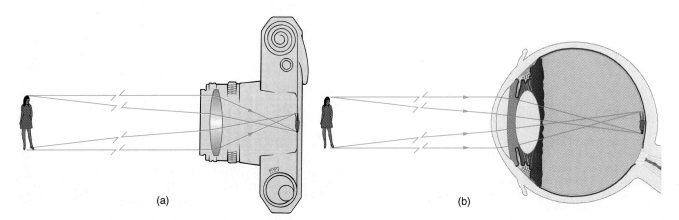

(a)

(b)

Figure 24.31 Real images can be projected. (a) A real image of the person is projected onto film. (b) The lens of an eye projects a real image on the retina.

(continued)

Thus the image distance d_i is about 1.5 m. Similarly, the image height based on ray tracing is greater than the object height by about a factor of 2, and the image is inverted. Thus m is about −2. The minus sign indicates that the image is inverted.

(*Thin lens equations*) (a) Equation 24.6a can be used to find d_i from the given information:

$$\frac{1}{d_o} + \frac{1}{d_i} = \frac{1}{f}$$

Rearranging to isolate d_i gives

$$\frac{1}{d_i} = \frac{1}{f} - \frac{1}{d_o}$$

Entering known quantities gives a value for $1/d_i$:

$$\frac{1}{d_i} = \frac{1}{0.500 \text{ m}} - \frac{1}{0.750 \text{ m}} = \frac{0.667}{\text{m}}$$

This must be inverted to find d_i:

$$d_i = \frac{\text{m}}{0.667} = 1.50 \text{ m}$$

(b) Now Equation 24.6b can be used to find the magnification m, since both d_i and d_o are known. Entering their values gives

$$m = -\frac{d_i}{d_o} = -\frac{1.50 \text{ m}}{0.750 \text{ m}} = -2.00$$

Discussion Note that the minus sign in Equation 24.6b causes the magnification to be negative when the image is inverted. Ray tracing and the use of the thin lens equations produce consistent results. The thin lens equations give the most precise results, being limited only by the accuracy of the given information. Ray tracing is limited by the accuracy with which you can draw, but it is highly useful conceptually and visually.

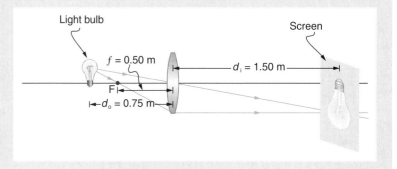

Figure 24.32 A light bulb placed 0.750 m from a lens having a 0.500 m focal length produces a real image on a poster board as discussed in Example 24.6. Ray tracing predicts the image location and size. This is a case 1 image.

Real images, such as the one considered in the previous example, are formed by converging lenses whenever an object is farther from the lens than its focal length. This is true for movie projectors, cameras, and the eye. We shall refer to these as **case 1** images. A case 1 image is formed when $d_o > f$ and f is positive. (A summary of the three cases or types of image formation appears at the end of this section.)

A different type of image is formed when an object, such as a book, is held close to a convex lens. The image is upright and larger than the object, as seen in Figure 24.33(a),

Figure 24.33 (a) A magnified image of a book page is produced by placing it closer to the converging lens than its focal length. This is a case 2 image. (b) When the same lens is held farther away from the page than the lens's focal length, an inverted image is formed. This is a case 1 image. Note that the image is in focus but the page is not, because the image is much closer to the camera taking this photograph than the page.

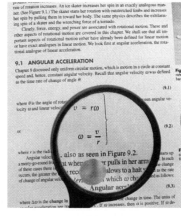

(a)   (b)

Figure 24.34 Ray tracing predicts the image location and size for an object held closer to a converging lens than its focal length. Ray 1 enters parallel to the axis and exits through the focal point on the opposite side, while ray 2 passes through the center of the lens without changing path. The two rays continue to diverge on the other side of the lens, but both appear to come from a common point, locating the upright, magnified, and virtual image. This is a case 2 image. See Example 24.7.

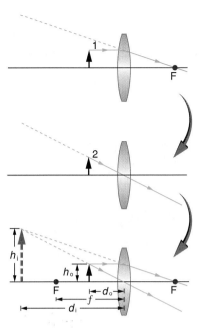

and so the lens is called a magnifier. If you slowly pull the magnifier away from the page, you will see that the magnification steadily increases until the image begins to blur. Pulling the magnifier even farther away produces an inverted image as seen in Figure 24.33(b). The distance at which the images blurs, and beyond which it inverts, is the focal length of the lens. To use a convex lens as a magnifier, the object must be closer to the converging lens than its focal length. This is called a **case 2** image. A case 2 image is formed when $d_o < f$ and f is positive.

Figure 24.34 uses ray tracing to show how an image is formed when an object is held closer to a converging lens than its focal length. Rays coming from a common point on the object continue to diverge after passing through the lens, but all appear to originate from a point at the location of the image. The image is on the same side of the lens as the object and is farther away from the lens than the object. This image, like all case 2 images, cannot be projected and, hence, is called a **virtual image**. Light rays only appear to originate at a virtual image; they do not actually pass through that location in space. A screen placed at the location of a virtual image will only receive diffuse light from the object, not focused rays from the lens. Additionally, a screen placed on the opposite side of the lens will receive rays that are still diverging, and so no image will be projected on it. We can see the magnified image with our eyes, because the lens of the eye converges the rays into a real image projected on our retina. Finally, we note that a virtual image is upright and larger than the object, meaning that the magnification is positive and greater than 1.

EXAMPLE 24.7 IMAGE PRODUCED BY A MAGNIFYING GLASS

Suppose the book page in Figure 24.33(a) is held 7.50 cm from a convex lens of focal length 10.0 cm, such as a typical magnifying glass might have. What magnification is produced?

Strategy and Concept We are given that $d_o = 7.50$ cm and $f = 10.0$ cm, so that we have a situation where the object is placed closer to the lens than its focal length. We therefore expect to get a case 2 image with a positive magnification that is greater than 1. Ray tracing produces an image like that shown in Figure 24.34, but we will use the thin lens equations to get numerical solutions in this example.

Solution To find the magnification m, we try to use Equation 24.6b, $m = -d_i/d_o$. We do not have a value for d_i, so that we must first find the location of the image using Equation 24.6a. (The procedure is the same as followed in the preceding example, where d_o and f were known.) Rearranging Equation 24.6a to isolate d_i gives

$$\frac{1}{d_i} = \frac{1}{f} - \frac{1}{d_o}$$

Entering known values, we obtain a value for $1/d_i$:

$$\frac{1}{d_i} = \frac{1}{10.0 \text{ cm}} - \frac{1}{7.50 \text{ cm}} = \frac{-0.0333}{\text{cm}}$$

This must be inverted to find d_i:

$$d_i = -\frac{\text{cm}}{0.0333} = -30.0 \text{ cm}$$

Now Equation 24.6b can be used to find the magnification m, since both d_i and d_o are known. Entering their values gives

$$m = -\frac{d_i}{d_o} = -\frac{-30.0 \text{ cm}}{10.0 \text{ cm}} = 3.00$$

Discussion A number of results in this example are true of all case 2 images, as well as being consistent with Figure 24.34. Magnification is indeed positive (as predicted), meaning the image is upright. The magnification is also greater than 1, meaning that the image is larger than the object—in this case, by a factor of 3. (It is left as an end-of-chapter problem to show the magnification increases as the

(continued)

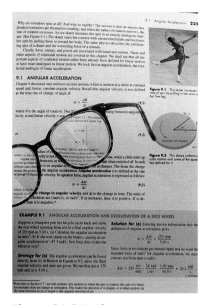

Figure 24.35 The print on a page viewed through a concave or diverging lens appears upright but smaller. This is a case 3 image.

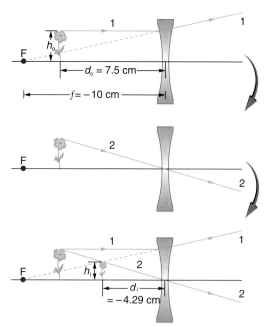

Figure 24.36 Ray tracing predicts the image location and size for a concave or diverging lens. Ray 1 enters parallel to the axis and is bent so that it appears to originate from the focal point. Ray 2 passes through the center of the lens without changing path. The two rays appear to come from a common point, locating the upright image. This is a case 3 image, which is closer to the lens than the object and smaller in height. See Example 24.8.

(continued)

object is moved away from the magnifier.) Note that the image distance is negative. This means the image is on the same side of the lens as the object. Thus the image cannot be projected and is virtual. (Negative values of d_i occur for virtual images.) The image is farther from the lens than the object, since the image distance is greater in magnitude than the object distance. The location of the image is not obvious when you look through a magnifier. In fact, since the image is bigger than the object, you may think the image is closer than the object. But the image is farther away, a fact that is useful in correcting farsightedness, as we shall see in a later section.

A third type of image is formed by a diverging or concave lens. Try looking through eyeglasses meant to correct nearsightedness. (See Figure 24.35.) You will see an image that is upright but smaller than the object. This means that the magnification is positive but less than 1. The ray diagram in Figure 24.36 shows that the image is on the same side of the lens as the object and, hence, cannot be projected—it is a virtual image. Note that the image is closer to the lens than the object. This is a **case 3** image, formed for any object by a negative focal length or diverging lens.

EXAMPLE 24.8   IMAGE PRODUCED BY A CONCAVE LENS

Suppose the book page in Figure 24.35 is held 7.50 cm from a concave lens of focal length −10.0 cm. Such a lens could be used in eyeglasses to correct pronounced nearsightedness. What magnification is produced?

Strategy and Concept This example is identical to the preceding one, except that the focal length is negative for a concave or diverging lens. The method of solution is thus the same, but the results are different in important ways.

Solution To find the magnification m, we must first find the image distance d_i using Equation 24.6a:

$$\frac{1}{d_i} = \frac{1}{f} - \frac{1}{d_o}$$

We are given that $f = -10.0$ cm and $d_o = 7.50$ cm. Entering these yields a value for $1/d_i$:

$$\frac{1}{d_i} = \frac{1}{-10.0 \text{ cm}} - \frac{1}{7.50 \text{ cm}} = \frac{-0.2333}{\text{cm}}$$

(continued)

(continued)

This must be inverted to find d_i:

$$d_i = -\frac{\text{cm}}{0.2333} = -4.29 \text{ cm}$$

Now Equation 24.6b can be used to find the magnification m, since both d_i and d_o are known. Entering their values gives

$$m = -\frac{d_i}{d_o} = -\frac{-4.29 \text{ cm}}{7.50 \text{ cm}} = 0.571$$

Discussion   A number of results in this example are true of all case 3 images, as well as being consistent with Figure 24.36. Magnification is positive (as predicted), meaning the image is upright. The magnification is also less than 1, meaning the image is smaller than the object—in this case, a little over half its size. The image distance is negative, meaning the image is on the same side of the lens as the object. (The image is virtual.) The image is closer to the lens than the object, since the image distance is smaller in magnitude than the object distance. The location of the image is not obvious when you look through a concave lens. In fact, since the image is smaller than the object, you may think it is farther away. But the image is closer than the object, a fact that is useful in correcting nearsightedness, as we shall see in a later section.

Table 24.3 summarizes the three types of images formed by single thin lenses. These are referred to as case 1, 2, and 3 images. Convex (converging) lenses can form either real or virtual images (cases 1 and 2, respectively), whereas concave (diverging) lenses can only form virtual images (always case 3). Real images are always inverted, but they can be either larger or smaller than the object. For example, a slide projector forms an image larger than the slide, whereas a camera makes an image smaller than the object being photographed. Virtual images are always upright and cannot be projected. Virtual images are larger than the object only in case 2, where a convex lens is used. The virtual image produced by a concave lens is always smaller than the object—a case 3 image. We can see and photograph virtual images only by using an additional lens to form a real image.

In the next section, we shall see that mirrors can form exactly the same types of images as lenses.

TABLE 24.3

THREE TYPES OF IMAGES FORMED BY THIN LENSES				
Type	*Formed when*	*Image type*	d_i	*m*
Case 1	f positive, $d_o > f$	Real	positive	negative
Case 2	f positive, $d_o < f$	Virtual	negative	positive $m > 1$
Case 3	f negative	Virtual	negative	positive $m < 1$

PROBLEM-SOLVING STRATEGIES

FOR LENSES

Step 1. Examine the situation to determine that image formation by a lens is involved.

Step 2. Determine whether ray tracing, the thin lens equations, or both are to be employed. A sketch is useful even if ray tracing is not specifically required by the problem.

Step 3. Identify exactly what needs to be determined in the problem (identify the unknowns).

Step 4. Make a list of what is given or can be inferred from the problem as stated (identify the knowns). It is helpful to determine whether the situation involves a case 1, 2, or 3 image. While these are just names for types of images, they have certain characteristics (given in Table 24.3) that can be of great use in solving problems.

Step 5. If ray tracing is required, use the ray tracing rules listed near the beginning of this section.

Step 6. Most quantitative problems require the use of the thin lens equations. These are solved in the usual manner by substituting knowns and solving for unknowns. Several worked examples serve as guides.

Step 7. Check to see if the answer is reasonable: Does it make sense? If you have identified the type of image (case 1, 2, or 3), you should assess whether your answer is consistent with the type of image, magnification, and so on.

24.7 IMAGE FORMATION BY MIRRORS

We only have to look as far as the nearest bathroom to find an example of an image formed by a mirror. Images in flat mirrors are the same size as the object and lie behind the mirror. Like lenses, mirrors can form a variety of images. For example, dental mirrors may produce a magnified image, just as makeup mirrors do. Security mirrors in stores, on the other hand, form images that are smaller than the object. We will use the law of reflection to understand how mirrors form images, and we will find that mirror images are analogous to those formed by lenses.

Figure 24.37 helps illustrate how a flat mirror forms an image. Two rays are shown emerging from the same point, striking the mirror, and being reflected into the observer's eye. The rays can diverge slightly, and both still get into the eye. If the rays are extrapolated backward, they seem to originate from a common point behind the mirror, locating the image. (The paths of the reflected rays into the eye are the same as if they had come directly from that point behind the mirror.) Using the law of reflection—the angle of reflection equals the angle of incidence—we can see that the image and object are the same distance from the mirror. This is a virtual image, since it cannot be projected— the rays only appear to originate from a common point behind the mirror. Obviously, if you walk behind the mirror, you cannot see the image, since the rays do not go there. But in front of the mirror, the rays behave exactly as if they had come from behind the mirror, so that is where the image lies.

Now let us consider the focal length of a mirror—for example, the concave spherical mirrors in Figure 24.38. Rays of light that strike the surface follow the law of reflection. For a mirror that is large compared with its radius of curvature, as in Figure 24.38(a), we see that the reflected rays do not cross at the same point, and the mirror does not have a well-defined focal point. If the mirror had the shape of a parabola, the rays would all cross at a single point, and the mirror would have a well-defined focal point. But parabolic mirrors are much more expensive to make than spherical mirrors. The solution is to use a mirror that is small compared with its radius of curvature, as shown in Figure 24.38(b). (This is the mirror equivalent of the thin lens approximation.) To a very good approximation, this mirror has a well-defined focal point at F that is the focal distance f from the center of the mirror. The focal length f of a concave mirror is positive, since it is a converging mirror.

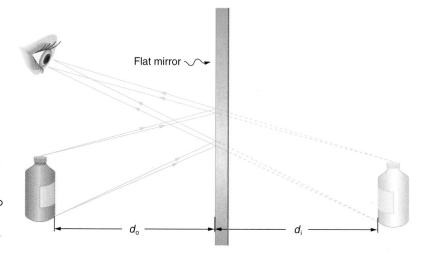

Figure 24.37 Two sets of rays from common points on a object are reflected by a flat mirror into the eye of an observer. The reflected rays seem to originate from behind the mirror, locating the virtual image.

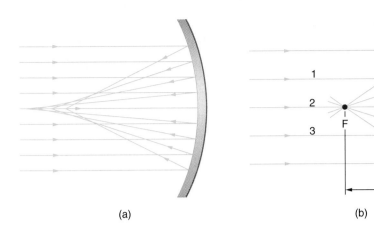

(a)   (b)

Figure 24.38 (a) Parallel rays reflected from a large spherical mirror do not all cross at a common point. (b) If a spherical mirror is small compared with its radius of curvature, parallel rays are focused to a common point. The distance of the focal point from the center of the mirror is its focal length f. Since this mirror is converging, it has a positive focal length.

Figure 24.39 *(right)* Parallel rays of light reflected from a convex spherical mirror (small in size compared with its radius of curvature) seem to originate from a well-defined focal point at the focal distance f behind the mirror. Convex mirrors diverge light rays and, thus, have a negative focal length.

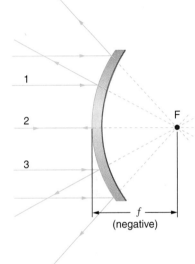

Just as for lenses, the shorter the focal length, the more powerful the mirror; thus, $P = 1/f$ for a mirror, too, as stated in Equation 24.5. A more strongly curved mirror has a shorter focal length and a greater power. Using the law of reflection and some simple trigonometry, it can be shown that the focal length is half the radius of curvature, or

$$f = \frac{R}{2} \tag{24.7}$$

where R is the radius of curvature of a spherical mirror. The smaller the radius of curvature, the smaller the focal length and, thus, the more powerful the mirror.

The convex mirror shown in Figure 24.39 also has a focal point. Parallel rays of light reflected from the mirror seem to originate from the point F at the focal distance f behind the mirror. The focal length and power of a convex mirror are negative, since it is a diverging mirror.

Ray tracing is as useful for mirrors as for lenses. The **rules for ray tracing for mirrors** are based on the illustrations just discussed:

1. A ray approaching a concave converging mirror parallel to its axis is reflected through the focal point F of the mirror on the same side. (See rays 1 and 3 in Figure 24.38(b).)
2. A ray approaching a convex diverging mirror parallel to its axis is reflected so that it seems to come from the focal point F behind the mirror. (See rays 1 and 3 in Figure 24.39.)
3. Any ray striking the center of a mirror is followed by applying the law of reflection; it makes the same angle with the axis when leaving as when approaching. (See ray 2 in Figure 24.40.)
4. A ray approaching a concave converging mirror through its focal point is reflected parallel to its axis. (The reverse of rays 1 and 3 in Figure 24.38(b).)
5. A ray approaching a convex diverging mirror by heading toward its focal point on the opposite side is reflected parallel to the axis. (The reverse of rays 1 and 3 in Figure 24.39.)

We will use ray tracing to illustrate how images are formed by mirrors, and we can use ray tracing quantitatively to obtain numerical information. But since we assume each mirror is small compared with its radius of curvature, we can use the thin lens Equations 24.6a and b for mirrors just as for we did for lenses.

Consider the situation shown in Figure 24.40, in which an object is placed farther from a concave (converging) mirror than its focal length. That is, f is positive and $d_o > f$, so that we may expect an image similar to the case 1 real image formed by a converging lens. Ray tracing in Figure 24.40 shows that the rays from a common point on the object

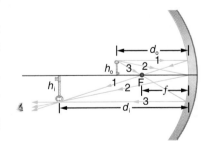

Figure 24.40 A case 1 image for a mirror. An object is farther from the converging mirror than its focal length. Rays from a common point on the object are traced using the rules in the text. Ray 1 approaches parallel to the axis, ray 2 strikes the center of the mirror, and ray 3 goes through the focal point on the way toward the mirror. All three rays cross at the same point after being reflected, locating the inverted real image. Although three rays are shown, only two of the three are needed to locate the image and determine its height.

all cross at a point on the same side of the mirror as the object. Thus a real image can be projected onto a screen placed at this location. The image distance is positive, and the image is inverted, so its magnification is negative. This is a **case 1** image for mirrors. It differs from the case 1 image for lenses only in that the image is on the same side of the mirror as the object. It is otherwise identical.

EXAMPLE 24.9   A CONCAVE REFLECTOR

Electric room heaters use a concave mirror to reflect infrared (IR) radiation from hot coils. Note that IR follows the same law of reflection as visible light. Given the mirror has a radius of curvature of 50.0 cm and produces an image of the coils 3.00 m away from the mirror, where are the coils?

Strategy and Concept We are given that the concave mirror projects a real image of the coils at an image distance $d_i = 3.00$ m. The coils are the object, and we are asked to find their location—that is, to find the object distance d_o. We are also given the radius of curvature of the mirror, so that its focal length is $f = R/2 = 25.0$ cm (positive since the mirror is concave or converging). Assuming the mirror is small compared with its radius of curvature, we can use the thin lens equations, 24.6a and b, to solve this problem.

Solution Since d_i and f are known, Equation 24.6a can be used to find d_o:

$$\frac{1}{d_o} + \frac{1}{d_i} = \frac{1}{f}$$

Rearranging to isolate d_o gives

$$\frac{1}{d_o} = \frac{1}{f} - \frac{1}{d_i}$$

Entering known quantities gives a value for $1/d_o$:

$$\frac{1}{d_o} = \frac{1}{0.250 \text{ m}} - \frac{1}{3.00 \text{ m}} = \frac{3.667}{m}$$

This must be inverted to find d_o:

$$d_o = \frac{m}{3.667} = 0.273 \text{ m} = 27.3 \text{ cm}$$

Discussion Note that the object (the filament) is farther from the mirror than the mirror's focal length. This is a case 1 image ($d_o > f$ and f positive), consistent with the fact that a real image is formed. You will get the most concentrated heat directly in front of the mirror and 3.00 m away from it. Generally, this is not desirable, since it could cause burns. Usually, you want the rays to emerge parallel, and this is accomplished by having the filament at the focal point of the mirror.

Note that the filament here is not much farther from the mirror than its focal length and that the image produced is considerably farther away. This is exactly analogous to a slide projector. Placing a slide only slightly farther away from the projector lens than its focal length produces an image significantly farther away. As the object gets closer to the focal distance, the image gets farther away. In fact, as the object distance approaches the focal length, the image distance approaches infinity and the rays are sent out parallel to one another.

What happens if an object is closer to a concave mirror than its focal length? This is analogous to a case 2 image for lenses ($d_o < f$ and f positive), which is a magnifier. In fact, this is how makeup mirrors act as magnifiers. Figure 24.41 uses ray tracing to locate the image of an object placed close to a concave mirror. Rays from a common point on the object are reflected in such a manner that they appear to be coming from

Figure 24.41 (a) A magnifying mirror. Case 2 images for mirrors are formed when a converging mirror has an object closer to it than its focal length. Ray 1 approaches parallel to the axis, ray 2 strikes the center of the mirror, and ray 3 approaches the mirror as if it came from the focal point. All three rays appear to originate from the same point after being reflected, locating the upright virtual image behind the mirror and showing it to be larger than the object. (b) Makeup mirrors are perhaps the most common use of a concave mirror to produce a larger, upright image.

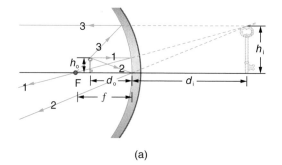

(a)

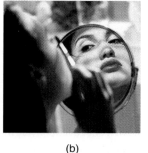

(b)

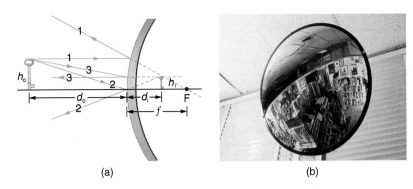

(a)

(b)

Figure 24.42 (a) Case 3 images for mirrors are formed by any convex mirror. Ray 1 approaches parallel to the axis, ray 2 strikes the center of the mirror, and ray 3 approaches toward the focal point. All three rays appear to originate from the same point after being reflected, locating the upright virtual image behind the mirror and showing it to be smaller than the object. (b) Security mirrors are convex, producing a smaller, upright image.

behind the mirror, meaning that the image is virtual and cannot be projected. As with a magnifying glass, the image is upright and larger than the object. This is a **case 2** image for mirrors and is exactly analogous to that for lenses.

A convex mirror is a diverging mirror (f is negative) and forms only one type of image. It is a **case 3** image: one that is upright and smaller than the object, just as for diverging lenses. Figure 24.42 uses ray tracing to illustrate the location and size of the case 3 image for mirrors. Since the image is behind the mirror, it cannot be projected and is thus a virtual image. It is also seen to be smaller than the object.

EXAMPLE 24.10 IMAGE IN A CONVEX MIRROR

A keratometer is a device used to measure the curvature of the cornea, particularly for fitting contact lenses. Light is reflected from the cornea, which acts like a convex mirror, and the keratometer measures the magnification of the image. The smaller the magnification, the smaller the radius of curvature of the cornea. If the light source is 12.0 cm from the cornea and the image's magnification is 0.0320, what is the cornea's radius of curvature?

Strategy If we can find the focal length of the convex mirror formed by the cornea, we can find its radius of curvature (the radius of curvature is twice the focal length of a spherical mirror). We are given that the object distance is $d_o = 12.0$ cm and that $m = 0.0320$. We first solve for the image distance d_i using Equation 24.6b, and then for f using Equation 24.6a.

Solution Equation 24.6b is $m = -d_i/d_o$. Solving this expression for d_i gives

$$d_i = -md_o$$

Entering known values yields

$$d_i = -(0.0320)(12.0 \text{ cm}) = -0.384 \text{ cm}$$

Equation 24.6a is

$$\frac{1}{f} = \frac{1}{d_o} + \frac{1}{d_i}$$

Substituting known values,

$$\frac{1}{f} = \frac{1}{12.0 \text{ cm}} + \frac{1}{-0.384 \text{ cm}} = \frac{-2.52}{\text{cm}}$$

This must be inverted to find f:

$$f = \frac{\text{cm}}{-2.52} = -0.400 \text{ cm}$$

The radius of curvature is twice the focal length, so that

$$R = 2|f| = 0.800 \text{ cm}$$

Discussion Although the focal length f of a convex mirror is defined to be negative, we take the absolute value to give us a positive value for R. The radius of curvature found here is reasonable for a cornea. In practice many corneas are not spherical, complicating the job of fitting contact lenses. Note that the image distance here is negative, consistent with the fact that the image is behind the mirror, where it cannot be projected. It is left as an end-of-chapter problem to show that for a fixed object distance, the smaller the radius of curvature, the smaller the magnification.

The three types of images formed by mirrors (cases 1, 2, and 3) are exactly analogous to those formed by lenses and are summarized in Table 24.3 at the end of the preceding section. It is easiest to concentrate on only three types of images—then remember that concave mirrors act like convex lenses, whereas convex mirrors act like concave lenses.

FOR MIRRORS

Step 1. Examine the situation to determine that image formation by a mirror is involved.

Step 2. Refer to the Problem-Solving Strategies for Lenses at the end of the previous section. The same strategies are valid for mirrors as for lenses with one qualification—*use the ray tracing rules for mirrors listed earlier in this section.*

24.8 MULTIPLE LENSES AND MIRRORS

Many optical devices are multiple-element systems having more than a single lens or mirror. (See Figure 24.43.) A microscope can be made from two convex lenses. Telescopes can also be made from two convex lenses. Many other telescopes use a mirror and a lens. High-quality camera lenses are actually composed of several individual lenses. Multiple-element systems can form useful images different from the three cases described in the preceding sections. But we shall find that it is straightforward to understand how multiple-element systems work. The image formed by the first element becomes the object for the second element. The second element forms its own image, which is the object for the third element, and so on. Ray tracing helps to visualize the image formed. If the device is composed of thin lenses and mirrors that obey the thin lens equations, then it is not difficult to describe their behavior numerically. In this section, we will examine a few of the more interesting multiple-element optical instruments.

The simplest **compound microscope** is constructed from two convex lenses and is shown schematically in Figure 24.44. The first lens is called the **objective lens**, and the second, the **eyepiece**. The purpose of a microscope is to magnify small objects, and both lenses contribute to the final magnification. Additionally, the final enlarged image is produced in a location far enough from the observer to be easily viewed, since the eye cannot focus on objects or images that are too close.

To see how the microscope in Figure 24.44 forms an image, we consider its two lenses in succession. The object is slightly farther away from the objective lens than its focal length f_o, producing a case 1 image that is larger than the object. This first image is the object for the second lens, or eyepiece. The eyepiece is intentionally located so it can further magnify the image. The eyepiece is placed so that the first image is closer to it than its focal length f_e. Thus the eyepiece acts as a magnifying glass, and the final image is made even larger. The final image remains inverted, but it is farther from the observer, making it easy to view (the eye is most relaxed when viewing distant objects and normally cannot focus closer than 25 cm). Since each lens produces a magnifica-

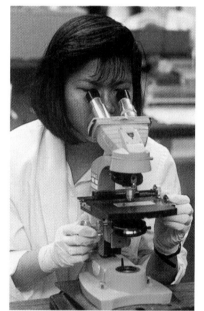

Figure 24.43 Multiple lenses and mirrors are used in this microscope.

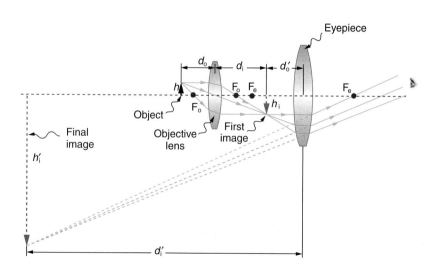

Figure 24.44 A compound microscope composed of two lenses, an objective and an eyepiece. The objective forms a case 1 image that is larger than the object. This first image is the object for the eyepiece. The eyepiece forms a case 2 final image that is further magnified.

tion that multiplies the height of the image, it is apparent that the overall magnification m is the product of the individual magnifications:

$$m = m_o m_e \qquad (24.8)$$

where m_o is the magnification of the objective and m_e is the magnification of the eyepiece. This equation can be generalized for any combination of thin lenses and mirrors that obey the thin lens equation:

> **The overall magnification of a multiple-element system is the product of the individual magnifications of its elements.**

EXAMPLE 24.11 MICROSCOPE MAGNIFICATION

Calculate the magnification of an object placed 6.20 mm from a compound microscope that has a 6.00 mm focal length objective and a 50.0 mm focal length eyepiece. The objective and eyepiece are separated by 23.0 cm.

Strategy and Concept This situation is similar to that shown in Figure 24.44. To find the overall magnification, we must find the magnification of the objective, then the magnification of the eyepiece. This involves using the thin lens equations.

Solution The magnification of the objective lens is given by Equation 24.6b as

$$m_o = -\frac{d_i}{d_o}$$

where d_o and d_i are the object and image distances, respectively, for the objective lens as labeled in Figure 24.44. The object distance is given to be $d_o = 6.20$ mm, but the image distance d_i is not known, so that it must be found using Equation 24.6a. Isolating d_i, we have

$$\frac{1}{d_i} = \frac{1}{f_o} - \frac{1}{d_o}$$

where f_o is the focal length of the objective lens. Substituting known values gives

$$\frac{1}{d_i} = \frac{1}{6.00 \text{ mm}} - \frac{1}{6.20 \text{ mm}} = \frac{0.00538}{\text{mm}}$$

We invert this to find d_i:

$$d_i = 186 \text{ mm}$$

Substituting this into the expression for m_o gives

$$m_o = -\frac{d_i}{d_o} = -\frac{186 \text{ mm}}{6.20 \text{ mm}} = -30.0$$

Now we must find the magnification of the eyepiece, which is given by

$$m_e = -\frac{d_i'}{d_o'}$$

where d_i' and d_o' are the image and object distances for the eyepiece (see Figure 24.44). The object distance is the distance

of the first image from the eyepiece. Since the first image is 186 mm to the right of the objective and the eyepiece is 230 mm to the right of the objective, the object distance is $d_o' = 230$ mm $- 186$ mm $= 44.0$ mm. This places the first image closer to the eyepiece than its focal length, so that the eyepiece will form a case 2 image as shown in the figure. We still need to find the location of the final image d_i' in order to find the magnification. This is done as before to obtain a value for $1/d_i'$:

$$\frac{1}{d_i'} = \frac{1}{f_e} - \frac{1}{d_o'} = \frac{1}{50.0 \text{ mm}} - \frac{1}{44.0 \text{ mm}} = -\frac{0.00273}{\text{mm}}$$

Inverting gives

$$d_i' = -\frac{\text{mm}}{0.00273} = -367 \text{ mm}$$

The eyepiece's magnification is thus

$$m_e = -\frac{d_i'}{d_o'} = -\frac{-367 \text{ mm}}{44.0 \text{ mm}} = 8.33$$

So the overall magnification is

$$m = m_o m_e = (-30.0)(8.33) = -250$$

Discussion Both the objective and the eyepiece contribute to the overall magnification, which is large and negative, consistent with Figure 24.44, where the image is seen to be large and inverted. In this case, the image is virtual and inverted, which cannot happen for a single element (case 2 and case 3 images for single elements are virtual and upright). The final image is 367 mm (0.367 m) to the left of the eyepiece. Had the eyepiece been placed farther from the objective, it could have formed a case 1 image to the right. Such an image can be projected on a screen, but it would be behind the head of the person in the figure and not appropriate for direct viewing. The procedure used to solve this example is applicable in any multiple-element system. Each element is treated in turn, with each forming an image that becomes the object for the next element. The process is not more difficult than for single lenses or mirrors, only lengthier.

Telescopes are meant for viewing distant objects, producing an image that is larger than would be seen with the naked eye. Telescopes gather more light than the eye, allowing dim objects to be observed. Although Galileo is often credited with inventing the telescope, he did not. What he did was more important. He constructed several early telescopes, was the first to study the heavens with them, and made monumental discoveries. Among these are the moons of Jupiter, craters and mountains on the moon, details of sunspots, and the fact that the Milky Way is composed of vast numbers of individual stars. The historic conflict between Galileo and elements of the Roman Catholic Church on some of these discoveries was a landmark event in the transition into the modern era, in which observation reigns supreme.

Figure 24.45(a) shows a telescope made of two lenses, the objective convex and the eyepiece concave, the same construction used by Galileo. Such an arrangement produces an upright image and is used in spyglasses and opera glasses, for example.

The most common two-lens telescope, like the simple microscope, uses two convex lenses and is shown in Figure 24.45(b). The object is so far away from the telescope that it is essentially at infinity compared with the focal lengths of the lenses ($d_o \approx \infty$). The first image is thus produced at $d_i = f_o$, as shown in the figure. To prove this, note that

$$\frac{1}{d_i} = \frac{1}{f_o} - \frac{1}{d_o} = \frac{1}{f_o} - \frac{1}{\infty}$$

Since $1/\infty = 0$, this simplifies to

$$\frac{1}{d_i} = \frac{1}{f_o}$$

This implies that $d_i = f_o$, as claimed. It is true that for any distant object and any lens or mirror, the image is at the focal length.

The first image formed by a telescope objective (see Figure 24.45(b)) will not be large compared with what you might see by looking at the object directly. For example, the spot formed by sunlight focused on a piece of paper by a magnifying glass is the image of the sun, and it is small. The telescope eyepiece (like the microscope eyepiece)

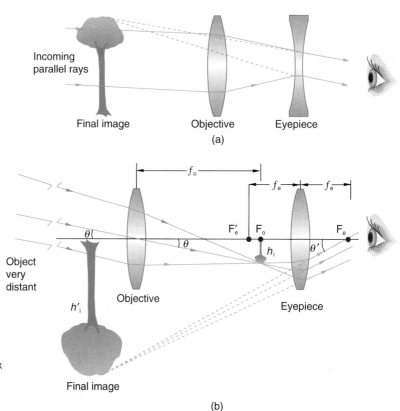

Figure 24.45 (a) Galileo made telescopes with a convex objective and a concave eyepiece. These produce an upright image and are used in spyglasses. (b) Most simple telescopes have two convex lenses. The objective forms a case 1 image that is the object for the eyepiece. The eyepiece forms a case 2 final image that is magnified.

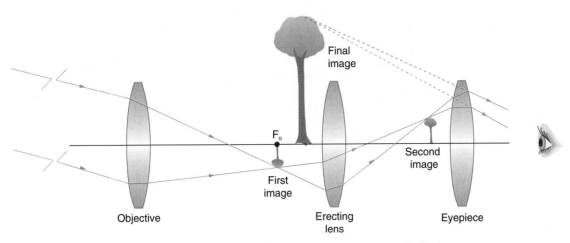

Figure 24.46 This arrangement of three lenses in a telescope produces an upright final image. The first two lenses are far enough apart that the second lens reinverts the image of the first. The third lens acts as a magnifier and keeps the image upright and in a location that is easy to view.

magnifies this first image. The distance between the eyepiece and the objective lens is made slightly less than the sum of their focal lengths so that the first image is closer to the eyepiece than its focal length. That is, d_o' is less than f_e, and so the eyepiece forms a case 2 image that is large and to the left for easy viewing. If the angle subtended by an object as viewed by the unaided eye is θ and the angle subtended by the telescope image is θ', then the **angular magnification** M is defined to be their ratio. That is, $M = \theta'/\theta$. It can be shown that the angular magnification of a telescope is related to the focal lengths of the objective and eyepiece; it is given by

$$M = \frac{\theta}{\theta'} = -\frac{f_o}{f_e} \qquad (24.9)$$

The minus sign indicates the image is inverted. To obtain the greatest angular magnification, it is best to have a long focal length objective and a short focal length eyepiece. The greater the angular magnification M, the larger an object will appear when viewed through a telescope, making more details visible. Limits to observable details are imposed by many factors, including lens quality and atmospheric disturbance.

The image in most telescopes is inverted, which is unimportant for observing the stars but a real problem for other applications, such as telescopes on ships or telescopic gun sights. If an upright image is needed, Galileo's arrangement in Figure 24.45(a) can be used. But a more common arrangement is to use a third convex lens as an eyepiece and increase the distance between the first two to reinvert the image. (See Figure 24.46.)

A telescope can also be made with a concave mirror as its first element or objective, since a concave mirror acts like a convex lens. (See Figure 24.47.) Flat mirrors are often employed in optical instruments to make them more compact or to send light to cameras and other sensing devices. There are many advantages to using mirrors rather than lenses for telescope objectives. Mirrors can be constructed much larger than lenses and can, thus, gather large amounts of light, as needed to view distant galaxies, for example. Large and relatively flat mirrors have very long focal lengths, so that great angular magnification is possible.

Additionally, lenses produce **chromatic aberrations**. An aberration is a distortion in an image, and chromatic aberrations are related to color. Since the index of refraction of lenses depends on color or wavelength, images are produced at different places and with different magnifications for different colors. (The law of reflection is independent of wavelength, and so mirrors do not have this problem, another advantage for mirrors in optical systems such as telescopes.) Figure 24.48 shows chromatic aberration for a single convex lens and its partial correction with a two-lens system. Violet rays are bent

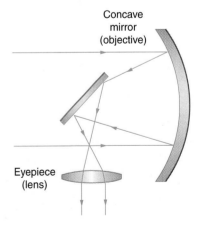

Figure 24.47 A two-element telescope composed of a mirror as the objective and a lens for the eyepiece. This telescope forms an image in the same manner as the two-convex-lens telescope already discussed, but it does not suffer from chromatic aberrations. Such telescopes can gather more light, since larger mirrors than lenses can be constructed.

Figure 24.48 (a) Chromatic aberration is caused by the dependence of a lens's index of refraction on color (wavelength). The lens is more powerful for violet (V) than for red (R), producing images with different locations and magnifications. (b) Multiple-lens systems can partially correct chromatic aberrations, but they may require lenses of different materials and add to the expense of optical systems such as cameras.

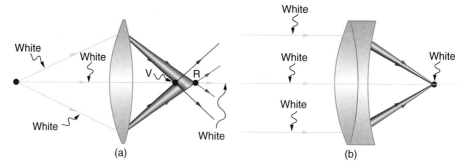

more than red, since they have a higher index of refraction and are thus focused closer to the lens. The diverging lens partially corrects this, although it is usually not possible to do so completely. Lenses of different materials and having different dispersions may be used. Lenses suffer from other aberrations due to their thickness. Expensive camera lenses are large in diameter, so that they can gather more light, and need several elements to correct for various aberrations, including chromatic aberrations. The lenses may also have specially shaped surfaces, as opposed to the simple spherical shape that is relatively easy to produce.

24.9 THE EYE AND VISION CORRECTION

The eye is perhaps the most interesting of all optical instruments. The eye is remarkable in how it forms images and in the richness of detail and color it can detect. The eye is also commonly in need of some correction, in that "normal" vision is ideal rather than normal. Image formation by the eye and common vision correction are easy to analyze with the optics already developed in this chapter.

Figure 24.49 shows the basic anatomy of the eye. The cornea and lens form a system that, to a good approximation, acts as a single thin lens. For clear vision, a real image must be projected onto the light-sensitive retina, which lies at a fixed distance from the lens. The lens of the eye adjusts its power to produce an image on the retina for objects at different distances. The center of the image falls on the fovea, which has the greatest density of light receptors and the greatest acuity in the visual field. The variable opening (or pupil) of the eye plus chemical adaptation allow the eye to detect light intensities from the lowest observable to 10^{10} times greater (without damage). The eyes sense direction, movement, sophisticated colors, and distance.

Human & Biological Application

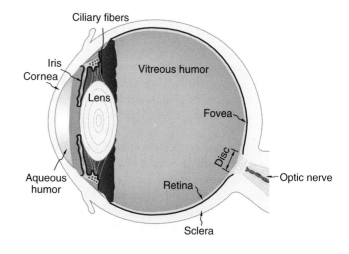

Figure 24.49 Cross-sectional anatomy of the eye. The cornea and lens of the eye act together to form a real image on the light-sensing retina, which has its densest concentration of receptors in the fovea and a blind spot over the optic nerve. The lens of the eye is adjustable in power to provide an image on the retina for varying object distances. Shown here only are the layers of tissue of varying indices of refraction in the lens.

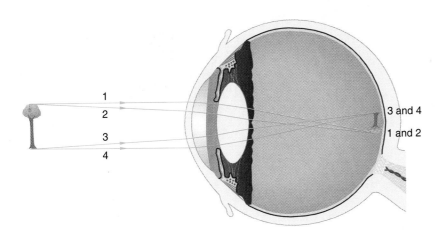

Figure 24.50 An image is formed on the retina with the light rays being converged most at the cornea and upon entering and exiting the lens. Rays from the top and bottom of the object are traced and are seen to produce an inverted real image on the retina. The distance to the object is drawn smaller than scale.

Processing of visual nerve impulses begins with interconnections in the retina and continues in the brain.

The ray diagram in Figure 24.50 shows image formation by the cornea and lens of the eye. The cornea provides about two-thirds of the power of the eye, since the speed of light changes greatly in going from air into the cornea. The lens provides the remaining power needed to produce an image on the retina. The cornea and lens can be treated as a single thin lens, even though light rays pass through many layers of material (cornea, aqueous humor, several layers in the lens, and vitreous humor), changing direction at each interface. A case 1 image is formed much like that produced by a single convex lens.

As noted, the image must fall precisely on the retina to produce clear vision—that is, the image distance d_i must equal the lens-to-retina distance. Because the lens-to-retina distance does not change, the image distance d_i must be the same for objects at all distances. The eye manages this by varying the power (and focal length) of the lens of the eye to accommodate for objects at various distances. The process of adjusting the eye's focal length is called **accommodation**. A person with normal (ideal) vision can see objects clearly at distances ranging from 25 cm to essentially infinity.

Figure 24.51 shows the accommodation of the eye for distant and near vision. Since light rays from a near object can diverge and still enter the eye, the lens must be more converging (more powerful) for close vision than for distant vision. To be more converging, the lens is made thicker by the action of the ciliary muscle surrounding it. The eye is most relaxed when viewing distant objects, one reason that microscopes and telescopes are designed to produce distant images (see Section 24.8). Vision of very distant objects is called *totally relaxed*, while close vision is termed *accommodated*, with the closest vision being *fully accommodated*.

We will use the thin lens equations, 24.6a and b, to examine image production by the eye quantitatively. First, note the power of a lens is given by Equation 24.5 as $P = 1/f$, and so we write the thin lens equations as

$$P = \frac{1}{d_o} + \frac{1}{d_i} \qquad \textbf{(24.6a)}$$

and

$$\frac{h_i}{h_o} = -\frac{d_i}{d_o} = m \qquad \textbf{(24.6b)}$$

We understand that d_i must equal the lens-to-retina distance to obtain clear vision, and that normal vision is possible for objects at distances $d_o = 25$ cm to infinity.

The eye can detect an impressive amount of detail, considering how small the image is on the retina. To get some idea of how small, consider the following example.

Human Application

Figure 24.51 Relaxed and accommodated vision for distant and close objects. (a) Light rays from the same point on a distant object must be nearly parallel to enter the eye and are more easily converged to produce an image on the retina. (b) Light rays from a close object can diverge more and still enter the eye. A more powerful lens is needed to converge them on the retina than if they were parallel.

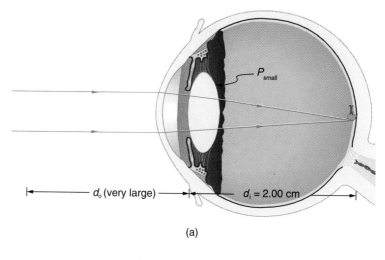

d_o (very large) $d_i = 2.00$ cm

(a)

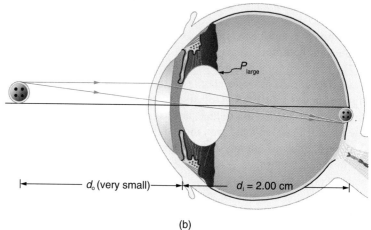

d_o (very small) $d_i = 2.00$ cm

(b)

EXAMPLE 24.12   SIZE OF IMAGE ON RETINA

What is the size of the image on the retina of a 1.20×10^{-2} cm diameter human hair, held at arm's length (60.0 cm) away? Take the lens-to-retina distance to be 2.00 cm.

Strategy We want to find the height of the image h_i, given the height of the object is $h_o = 1.20 \times 10^{-2}$ cm. We also know that the object is 60.0 cm away, so that $d_o = 60.0$ cm. For clear vision, the image distance must equal the lens-to-retina distance, and so $d_i = 2.00$ cm. Equation 24.6b can be used to find h_i with the known information.

Solution The only unknown in Equation 24.6b is h_i:

$$\frac{h_i}{h_o} = -\frac{d_i}{d_o}$$

Rearranging to isolate h_i yields

$$h_i = -h_o \cdot \frac{d_i}{d_o}$$

Substituting known values gives

$$h_i = -(1.20 \times 10^{-2} \text{ cm}) \frac{2.00 \text{ cm}}{60.0 \text{ cm}}$$

$$= -4.00 \ \mu\text{m}$$

Discussion This truly small image is not the smallest discernible—that is, the limit to visual acuity is even smaller than this. Limitations on visual acuity have to do with the wave properties of light as well as the anatomy of the eye and will be discussed in the next chapter.

EXAMPLE 24.13   POWER RANGE OF THE EYE

Calculate the power of the eye when viewing objects at the greatest and smallest distances possible with normal vision, assuming a lens-to-retina distance of 2.00 cm (a typical value).

Strategy For clear vision, the image must be on the retina, and so $d_i = 2.00$ cm here. For distant vision, $d_o \approx \infty$, and for close vision, $d_o = 25.0$ cm, as discussed above. Equation 24.6a, as written just above, can be used directly

(continued)

(continued)

to solve for P in both cases, since we know d_i and d_o. Power has units of diopters, where $1 \text{ D} = 1/\text{m}$, and so we should express all distances in meters.

Solution For distant vision,

$$P = \frac{1}{d_o} + \frac{1}{d_i} = \frac{1}{\infty} + \frac{1}{0.0200 \text{ m}}$$

Since $1/\infty = 0$, this gives

$$P = 0 + 50.0/\text{m} = 50.0 \text{ D} \qquad \text{(distant vision)}$$

Now, for close vision,

$$P = \frac{1}{d_o} + \frac{1}{d_i} = \frac{1}{0.250 \text{ m}} + \frac{1}{0.0200 \text{ m}}$$
$$= \frac{4.00}{\text{m}} + \frac{50.0}{\text{m}} = 4.00 \text{ D} + 50.0 \text{ D}$$
$$= 54.0 \text{ D} \qquad \text{(close vision)}$$

Discussion For an eye with this typical 2.00 cm lens-to-retina distance, the power ranges from 50.0 D for distant totally relaxed vision to 54.0 D for close fully accommodated vision, an 8% increase. The increase in power for close vision is consistent with the preceding discussion and the ray tracing in Figure 24.51. An 8% ability to accommodate is considered normal but is typical for people who are about 40 years old. Younger people have a greater accommodation ability, whereas older people gradually lose the ability to accommodate. The lens of the eye changes with age in ways that tend to preserve the ability to see distant objects clearly but do not allow the eye to accommodate for close vision, a condition called **presbyopia** (literally, elder eye). To correct this vision defect, we place a converging, positive power lens in front of the eye, such as found in reading glasses. Commonly available reading glasses are rated by their power in diopters, typically ranging from 1.0 to 3.5 D.

Correction of Common Vision Defects

Most people need some type of vision correction. Common vision defects are easy to understand, and some are simple to correct. Figure 24.52 illustrates two common vision defects. **Nearsightedness**, or **myopia**, is the inability to see distant objects while close objects are clear. The eye overconverges the nearly parallel rays from a distant object, and the rays cross in front of the retina. More divergent rays from a close object are converged on the retina for a clear image. The distance to the farthest object that can be seen clearly is called the **far point** of the eye (normally infinity). **Farsightedness**, or **hyperopia**, is the inability to see close objects while distant objects may be clear. A farsighted eye does not converge rays from a close object sufficiently to make the rays meet on the retina. Less diverging rays from a distant object can be converged for a clear image. The distance to the closest object that can be seen clearly is called the **near point** of the eye (normally 25 cm).

Since the nearsighted eye overconverges light rays, the correction for nearsightedness is to place a diverging spectacle lens in front of the eye. This reduces the power of an eye that is too powerful. Another way of thinking about this is that a diverging spectacle lens produces a case 3 image, which is closer to the eye than the object. (See Figure 24.53.) To determine the spectacle power needed for correction, you must know the person's far point—that is, you must know the greatest distance at which the person can see clearly. Then the image produced by a spectacle lens must be at this distance or closer for the nearsighted person to be able to see it clearly.

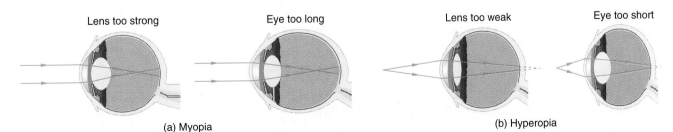

Figure 24.52 (a) The *nearsighted* (*myopic*) eye converges rays from a distant object in front of the retina; thus, they are diverging when they strike the retina, producing a blurry image. This can be caused by the lens of the eye being too powerful or the length of the eye being too great. (b) The *farsighted* (*hyperopic*) eye is unable to converge the rays from a close object by the time they strike the retina, producing blurry close vision. This can be caused by insufficient power in the lens or by the eye being too short.

Figure 24.53 Correction of nearsightedness uses a diverging lens that compensates for the overconvergence by the eye. The diverging lens produces an image closer to the eye than the object, so that the nearsighted person can see it clearly.

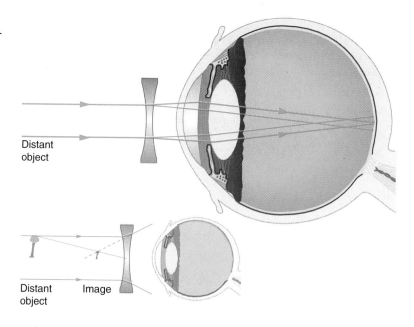

EXAMPLE 24.14 CORRECTING NEARSIGHTEDNESS

What power spectacle lens is needed to correct the vision of a nearsighted person whose far point is 30.0 cm? Assume the spectacle (corrective) lens is held 1.50 cm from the eye by eyeglass frames.

Strategy You want this nearsighted person to be able to see very distant objects clearly. That means the spectacle lens must produce an image 30.0 cm from the eye for an object very far away. An image 30.0 cm from the eye will be 28.5 cm to the left of the spectacle lens (see Figure 24.53). Thus we must get $d_i = -28.5$ cm when $d_o \approx \infty$. The image distance is negative, because it is on the same side of the spectacle as the object.

Solution Since d_i and d_o are known, the power of the spectacle lens can be found using Equation 24.6a as written above:

$$P = \frac{1}{d_o} + \frac{1}{d_i} = \frac{1}{\infty} + \frac{1}{-0.285\text{ m}}$$

Since $1/\infty = 0$, we obtain

$$P = 0 - 3.51/\text{m} = -3.51 \text{ D}$$

Discussion The negative power indicates a diverging or concave lens as expected. The spectacle produces a case 3 image closer to the eye, where the person can see it. If you examine eyeglasses for nearsighted people, you will find the lenses are thinnest in the center. Additionally, if you examine a prescription for eyeglasses for nearsighted people, you will find that the prescribed power is negative and given in units of diopters.

Since the farsighted eye underconverges light rays, the correction for farsightedness is to place a converging spectacle lens in front of the eye. This increases the power of an eye that is too weak. Another way of thinking about this is that a converging spectacle lens produces a case 2 image, which is farther from the eye than the object. (See Figure 24.54.) To determine the spectacle power needed for correction, you must know the person's near point—that is, you must know the smallest distance at which the person can see clearly. Then the image produced by a spectacle lens must be at this distance or farther for the farsighted person to be able to see it clearly.

EXAMPLE 24.15 CORRECTING FARSIGHTEDNESS

What power spectacle lens is needed to allow a farsighted person, whose near point is 1.00 m, to see an object clearly that is 25.0 cm away? Assume the spectacle (corrective) lens is held 1.50 cm from the eye by eyeglass frames.

Strategy When an object is held 25.0 cm from the person's eyes, the spectacle lens must produce an image 1.00 m away (the near point). An image 1.00 m from the eye will be 98.5 cm to the left of the spectacle lens (see Figure 24.54). Thus

(continued)

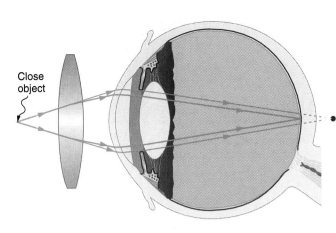

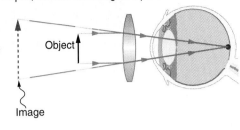

Figure 24.54 Correction of farsightedness uses a converging lens that compensates for the underconvergence by the eye. The converging lens produces an image farther from the eye than the object, so that the farsighted person can see it clearly.

(continued)
$d_i = -98.5$ cm. The image distance is negative, because it is on the same side of the spectacle as the object. The object is 23.5 cm to the left of the spectacle, so that $d_o = 23.5$ cm.

Solution Since d_i and d_o are known, the power of the spectacle lens can again be found using Equation 24.6a:

$$P = \frac{1}{d_o} + \frac{1}{d_i} = \frac{1}{0.235 \text{ m}} + \frac{1}{-0.985 \text{ m}}$$
$$= 4.26 \text{ D} - 1.02 \text{ D} = 3.24 \text{ D}$$

Discussion The positive power indicates a converging (convex) lens, as expected. The convex spectacle produces a case 2 image farther from the eye, where the person can see it. If you examine eyeglasses for farsighted people, you will find the lenses are thickest in the center. Additionally, a prescription for eyeglasses for farsighted people will have a prescribed power that is positive.

Another very common vision defect is **astigmatism**, an unevenness in the focus of the eye. This is due mostly to irregularities in the shape of the cornea but can also be due to lens irregularities or unevenness in the retina. Because of these irregularities, different parts of the lens system produce images at different locations. The eye-brain system can compensate for some of these irregularities, but they generally manifest themselves as less distinct vision or sharper images along certain axes. Figure 24.55 shows a chart used to detect astigmatism. Astigmatism can be at least partially corrected with a spectacle having the opposite irregularity of the eye. If an eyeglass prescription has a cylindrical correction, it is there to correct astigmatism. The normal corrections for near- or farsightedness are spherical corrections, uniform along all axes.

Contact lenses have advantages over glasses beyond their cosmetic aspects. One problem with glasses is that as the eye moves, it is not at a fixed distance from the spectacle lens. Contacts rest on and move with the eye, eliminating this problem. Because contacts cover a significant portion of the cornea, they provide superior peripheral vision compared with eyeglasses. Contacts also correct some corneal astigmatism caused by surface irregularities. The tear layer between the smooth contact and the cornea fills in the irregularities. Since the index of refraction of the tear layer and the cornea are very similar, you now have a regular optical surface in place of an irregular one. Contact wearers nearly always report their vision with contacts to be significantly better than with eyeglasses.

Human & Biological Application

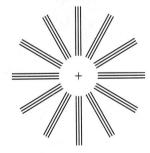

Figure 24.55 This chart can detect astigmatism, an unevenness in the focus of the eye. Check each of your eyes separately by looking at the center cross (without spectacles if you wear them). If lines along some axes appear darker or clearer than others, you have an astigmatism.

24.10* COLOR AND COLOR VISION

The gift of vision is made richer by the existence of color. Objects and lights abound with thousands of hues that stimulate our eyes, brains, and emotions. (See Figure 24.56.) Two basic questions are addressed in this brief treatment—what does color mean in scientific terms, and how do we as humans perceive it?

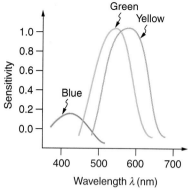

Figure 24.57 The relative sensitivity of the three types of cones, which are named for their wavelength of greatest sensitivity. Rods are about 1000 times more sensitive, and their curve peaks at about 500 nm. Evidence for the three types of cones comes from direct measurements in animal and human eyes and testing of color-blind people.

Figure 24.56 The computer-enhanced color in this image of a nebula and stars in space not only reveals more information, it also enhances the sense of beauty in the scene.

Simple Theory of Color Vision

We have already noted that color is associated with the wavelength of visible electromagnetic radiation.* When our eye receives pure-wavelength light, we tend to see only a few colors. Six of these (most often listed) are red, orange, yellow, green, blue, and violet. (Figure 24.20 shows the approximate ranges of λ that correspond to each of these colors.) These are the rainbow of colors produced when white light is dispersed according to wavelength. There are thousands of other **hues** that we can perceive but do not appear in the rainbow. These include brown, teal, gold, pink, and white. One simple theory of color vision holds that all of these hues are our eye's response to different combinations of wavelengths. This is true to an extent, but we find that color perception is even subtler than this.

There are two major types of light-sensing cells in the retina—**rods** and **cones**. Rods are more sensitive than cones by a factor of about 1000 and are solely responsible for vision in very dark environments and for peripheral vision. Rods do not yield color information. You may notice that you lose color vision when it is very dark.

Cones are concentrated in the fovea. (See Figure 24.49 for the anatomy of the eye.) There are three types of cones, and each type is sensitive to different ranges of wavelengths, as illustrated in Figure 24.57. A **simplified theory of color vision** is that there are three *primary colors* corresponding to the three types of cones. The thousands of other hues that we can distinguish are created by various combinations of stimulations of the three types of cones. Color television uses a three-color system in which the screen is covered with equal numbers of red, green, and blue phosphor dots. The broad range of hues a viewer sees is produced by various combinations of these three colors. You will perceive yellow, for example, when red and green are illuminated with the correct ratio of intensities. White may be sensed when all three are illuminated. It would seem, then, that all hues can be produced by adding three primary colors in various proportions. But there is an indication that color vision is more sophisticated. There is no unique set of three primary colors. Another set that works is yellow, green, and blue. A further indication of the need for a more complex theory of color vision is that many different combinations can produce the same hue. Yellow can be sensed with yellow light, or with a combination of red and green, and also with white light having violet removed. The three-primary-colors aspect of color vision is well established; more sophisticated theories expand on it rather than deny it.

Consider why various objects display color—that is, why is a blue jay blue and a cardinal red? The **true color of an object** is defined by its absorptive or reflective characteristics. Figure 24.58 shows white light falling on three different objects, one pure blue, one pure red, and one black, and pure red light falling on a white object. Other hues are created by more complex absorption characteristics. Pink, for example, can be due to weak absorption of all colors except red. An object can appear a different color under nonwhite illumination. For example, a pure blue object illuminated with pure red light will *appear* black, because it absorbs all the red light falling on it. But the true color of the object is blue, independent of illumination.

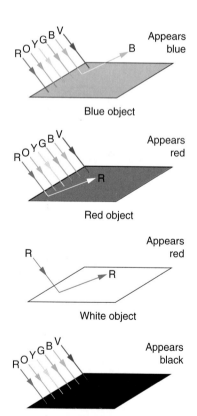

Figure 24.58 Absorption characteristics determine the true color of an object. Here three objects are illuminated by white light, and one by pure red light. White is the equal mixture of all visible wavelengths; black is the absence of light.

---

*Discussed in Section 24.5 regarding dispersion and in Section 23.3 on the visible part of the electromagnetic spectrum.

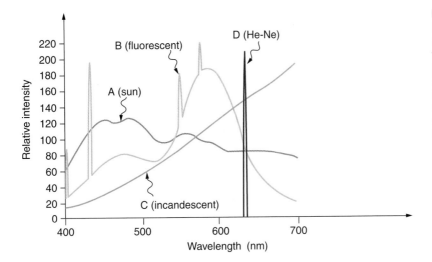

Figure 24.59 Emission spectra for various light sources. The spike for a helium-neon laser (curve D) is due to its very pure wavelength emission. Curve A is average sunlight at the earth's surface, curve B is light from a fluorescent lamp, and curve C is the output of an incandescent light. The spikes in the fluorescent output are due to atomic spectra—a topic to be explored in Chapters 27 and 28.

Similarly, **light sources have color** defined by the wavelengths they produce. A helium-neon laser emits very pure red light, the sun produces a broad yellowish spectrum, fluorescent lights emit bluish-white light, and incandescent lights emit reddish-white hues. (See Figure 24.59.) As you would expect, you sense these colors when viewing the light source directly or when illuminating a white object with them. All of this fits neatly into the simplified theory that a combination of wavelengths produces various hues.

Human Application

Color Constancy and a Modified Theory of Color Vision

The eye-brain color-sensing system can detect more information than the wavelengths of light coming from an object. It can, by comparing various objects in its view, perceive the true color of an object under varying lighting conditions—an ability that is called **color constancy**. We can sense that a white tablecloth, for example, is white whether it is illuminated by sunlight, fluorescent light, or candlelight. The wavelengths entering the eye are quite different in each case, as the graphs in Figure 24.59 imply, but the tablecloth still appears white. Our sense of color vision can detect the true color by comparing the tablecloth with its surroundings.

A modified theory of color vision must take color constancy into account. Such theories are based on a large body of anatomical evidence as well as perceptual studies. There are nerve connections among the light receptors on the retina, and there are far fewer nerve connections to the brain than there are rods and cones. This means that there is signal processing in the eye before information is sent to the brain. One result of these interconnections is that the eye makes comparisons between adjacent light receptors and is very sensitive to edges. (See Figure 24.60.) Rather than responding simply to the light entering the eye, which is uniform in the various rectangles, the eye responds to the edges and senses false darkness variations.

One theory of color vision that takes many factors into account was greatly advanced by Edwin Land, the creative founder of the Polaroid Corporation. Land proposed, based partly on his many elegant experiments, that the three types of cones are organized into systems called **retinexes**. Each retinex forms an image that is compared with the others to achieve color constancy. Thus the eye-brain system can compare a candle-illuminated white tablecloth with its generally reddish surroundings and determine that it is actually white. This **retinex theory of color vision** is an example of modified theories of color vision that attempt to account for all its subtleties. One striking experiment performed by Land demonstrates that some type of image comparison may produce color vision. Two pictures are taken of a scene on black-and-white film, one using a red filter, the other a blue filter. Resulting black-and-white slides are then projected and superimposed on a screen, producing a black-and-white image as expected. Then a red filter is placed in front of the slide taken with a red filter, and the images are again superimposed on a screen.

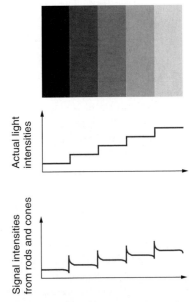

Figure 24.60 The importance of edges. Although the gray strips are uniformly shaded, as indicated by the graph immediately below them, they do not appear uniform at all. Instead, they are perceived darker on the dark side and lighter on the light side of the edge, as shown in the bottom graph. This is due to nerve impulse processing in the eye.

You would expect an image in various shades of pink, but instead the image appears to humans in full color with all the hues of the original scene. This implies that color vision can be induced by comparison of the black-and-white and red images.

Color vision is not completely understood or explained, and the retinex theory is not totally accepted. Yet it is apparent that color vision is much subtler than a first look might imply.

SUMMARY

Optics is the branch of physics that deals with the behavior of visible light and other electromagnetic waves. **Geometric optics** is the part of optics that deals with light when it moves in straight lines called **rays**, as it does when it interacts with objects that are large compared with its wavelength.

The laws of reflection and refraction govern all of geometric optics. The **law of reflection** states that *the angle of reflection equals the angle of incidence*. Angles are measured relative to a perpendicular to the surface at the point where the light strikes it.

Refraction is the change in the direction of light when it passes through matter and is related to changes in the speed of light. The speed of light in a vacuum c is a fundamental physical quantity with the fixed value

$$c = 2.9972458 \times 10^8 \text{ m/s} \approx 3.00 \times 10^8 \text{ m/s} \quad (24.1)$$

We defined the **index of refraction** n of a material to be

$$n = \frac{c}{v} \quad (24.2)$$

where v is the speed of light in the material. Values of n are given in Table 24.1. The **law of refraction**, which is also known as **Snell's law**, is given by

$$n_1 \sin \theta_1 = n_2 \sin \theta_2 \quad (24.3)$$

where n_1 and n_2 are the indices of refraction for the two media and θ_1 and θ_2 are the angles between the rays and perpendiculars to the surface at the point the ray crosses. **Total internal reflection** can occur when light goes from one medium to another with a lower index of refraction. The **critical angle** θ_c is the smallest incident angle for which there is total internal reflection and occurs when $\theta_2 = 90°$. The critical angle θ_c for a given combination of materials is

$$\theta_c = \sin^{-1}(n_2/n_1) \quad \text{for } n_1 > n_2 \quad (24.4)$$

Total internal reflection occurs for any incident angle greater than the critical angle θ_c.

Dispersion is the spreading of light, or other waves, according to wavelength. For light, it can be caused by an index of refraction that depends on λ. Dispersion is responsible for phenomena ranging from rainbows to chromatic aberrations.

A **converging lens or mirror** changes the direction of rays so that they converge more (or diverge less) than before encountering it. It has a **focal point** F located at the place where rays initially parallel to its axis are converged. Converging lenses are convex, and converging mirrors are concave. A **diverging lens or mirror** causes light rays to diverge more (or converge less) than before encountering it. The focal point of a diverging lens or mirror is the point from which rays originally parallel to its axis appear to originate after being diverged. Diverging lenses are concave, and diverging mirrors are convex. The **focal length**

f of a lens or mirror is the distance from its center to its focal point. The shorter the focal length, the more powerful its effect. The **power** P of a lens or mirror is defined to be

$$P = \frac{1}{f} \quad (24.5)$$

The unit of power is the **diopter** (D), for a focal length f in meters—1 D = 1/m, or 1 m^{-1}. Focal length and power are positive for converging lenses and mirrors, and negative for diverging lenses and mirrors.

Ray tracing is the technique of following rays that undergo reflection and refraction. There are five **rules for ray tracing for thin lenses** that are given in Section 24.6 and can be used to locate and describe image formation by a thin lens. Similarly, there are five **rules for ray tracing for thin mirrors** that are given in Section 24.7 and can be used to locate and describe image formation by a mirror having a diameter that is small compared with its radius of curvature. The important factors in image formation are the **object distance** d_o, which is the distance of an object from the center of a lens or mirror; the **image distance** d_i, which is the distance of the image from the center of the lens or mirror; the **height of the object** h_o; and the **height of the image** h_i. The **thin lens equations** relate these as follows:

$$\frac{1}{d_o} + \frac{1}{d_i} = \frac{1}{f} \quad (24.6a)$$

and

$$\frac{h_i}{h_o} = -\frac{d_i}{d_o} = m \quad (24.6b)$$

m is defined to be the **magnification**. The focal length of a spherical mirror is

$$f = \frac{R}{2} \quad (24.7)$$

where R is its radius of curvature.

A **real image** is one that can be projected, and a **virtual image** is one that cannot be projected. There are three cases for image formation by single thin lenses or mirrors that obey the thin lens equations. Refer to the summary of these three cases in Table 24.3, noting that converging lenses and mirrors (convex and concave, respectively) form the same types of images, while diverging lenses and mirrors (concave and convex, respectively) form the same types of images.

Many optical devices contain more than a single lens or mirror. These are analyzed by considering each element sequentially. The image formed by the first is the object for the second, and so on. The same ray tracing and thin lens techniques are applicable under the same provisions. The **overall magnification** of a multiple-element system is the product of the magnifications of its individual elements. For a two-element system with an objective and an eyepiece, this is

$$m = m_o m_e \qquad (24.8)$$

where m_o is the magnification of the objective and m_e is the magnification of the eyepiece, such as for a microscope. The **angular magnification** M for, say, a telescope is given by

$$M = \frac{\theta}{\theta'} = -\frac{f_o}{f_e} \qquad (24.9)$$

where θ is the angle subtended by an object viewed by the unaided eye, θ' is the angle subtended by a magnified image, and f_o and f_e are the focal lengths of the objective and eyepiece.

Image formation by the eye is adequately described by the thin lens equations. The eye produces a real image on the retina by adjusting its focal length and power in a process called **accommodation**. For close vision, the eye is fully accommodated and has its greatest power, whereas for distant vision, it is totally relaxed and has its smallest power. The loss of the ability to accommodate with age is called **presbyopia**, which is corrected by the use of a converging lens to add power for close vision. **Nearsightedness**, or **myopia**, is the in-

ability to see distant objects and is corrected with a diverging lens to reduce power. **Farsightedness**, or **hyperopia**, is the inability to see close objects and is corrected with a converging lens to increase power. In myopia and hyperopia, the corrective lenses produce images at a distance that the person can see clearly—the **far point** and **near point**, respectively.

The eye has four types of light receptors—**rods** and three types of color-sensitive **cones**. We perceive many **hues**, from light having mixtures of wavelengths. A **simplified theory of color vision** states that there are *three primary colors*, which correspond to the three types of cones, and that various combinations of the primary colors produce all the hues. The **true color of an object** is related to its relative absorption of various wavelengths of light. The **color of a light source** is related to the wavelengths it produces. **Color constancy** is the ability of the eye-brain system to discern the true color of an object illuminated by various light sources. The **retinex theory of color vision** explains color constancy by the existence of three retinexes or image systems, associated with the three types of cones, that are compared to obtain sophisticated information.

CONCEPTUAL QUESTIONS

24.1 Using the law of reflection, explain how powder takes the shine off of a person's nose. (See Figure 24.61.) What is the name of the optical effect?

(a)

(b)

Figure 24.61 A nose without (a) and with (b) powder. Question 1.

24.2 Diffusion by reflection from a rough surface is described in this chapter. Light can also be diffused by refraction. Describe how this occurs in a specific situation, such as light interacting with crushed ice.

24.3 Why is the index of refraction always greater than or equal to 1?

24.4 Does the fact that the light flash from lightning reaches you before its sound prove that the speed of light is extremely large or simply that it is greater than the speed of sound? Discuss how you could use this effect to get an estimate of the speed of light.

24.5 A ring with a colorless gemstone is dropped into water. The gemstone becomes invisible when submerged. Can it be a diamond? Explain.

24.6 Will light change direction toward or away from the perpendicular when it goes from air to water? Water to glass? Glass to air?

24.7 Explain why a person's legs appear very short when wading in a pool. Justify your explanation with a ray diagram showing the path of rays from the feet to the eye of an observer who is out of the water.

24.8 A high-quality diamond may be quite clear and colorless, transmitting all visible wavelengths with little absorption. Explain how it can sparkle with flashes of brilliant color when illuminated by white light.

24.9 Is it possible that total internal reflection plays a role in rainbows? Explain in terms of indices of refraction and angles, perhaps referring to Figure 24.22.

24.10 It can be argued that a flat piece of glass, such as in a window, is like a lens with an infinite focal length. If so, where does it form an image? That is, how are d_i and d_o related?

24.11 What are the differences between real and virtual images? How can you tell (by looking) if an image formed by a single lens or mirror is real or virtual?

24.12 Can you see a virtual image? Can you photograph one? Can one be projected onto a screen with additional lenses or mirrors? Explain your responses.

24.13 Is it necessary to project a real image onto a screen for it to exist?

24.14 At what distance is an image *always* located—at d_o, d_i, or f?

24.15 Under what circumstances will an image be located at the focal point of a lens or mirror?

24.16 What is meant by a negative magnification? What is meant by a magnification that is less than 1 in magnitude?

24.17 Can a case 1 image be larger than the object even though its magnification is always negative? Explain.

24.18 Why is the front surface of a thermometer curved as shown in Figure 24.62?

Figure 24.62 The curved surface of the thermometer serves a purpose. Question 18.

24.19 Figure 24.63 shows a light bulb between two mirrors. One mirror creates a beam of light with parallel rays; the other keeps light from escaping without being put into the beam. Where is the filament of the light in relation to the focal point or radius of curvature of each mirror?

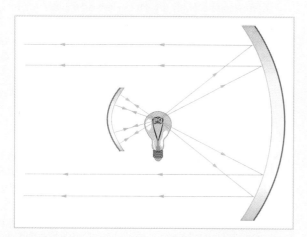

Figure 24.63 The two mirrors trap most of the bulb's light and form a directional beam as in a headlight. Question 19.

24.20 When you focus a camera, you adjust the distance of the lens from the film. If the camera lens acts like a thin lens, why can it not be a fixed distance from the film for both near and distant objects?

24.21 A thin lens has two focal points, one on either side and at equal distances from its center, and should behave the same for light entering from either side. Look through your eyeglasses (or those of a friend) backward and forward and comment on whether they are thin lenses.

24.22 It can be argued that a flat mirror has an infinite focal length. If so, where does it form an image? That is, how are d_i and d_o related?

24.23 Why does your nose look much bigger than the rest of your face when you look at your reflection in a shiny spherical Christmas tree ornament?

24.24 Geometric optics describes the interaction of light with macroscopic objects. Why, then, is it correct to use geometric optics to analyze a microscope's image?

24.25 The image produced by the microscope in Figure 24.44 cannot be projected. Could extra lenses or mirrors project it? Explain.

24.26 Why not have the objective of a microscope form a case 2 image with a large magnification? (Hint: Consider the location of that image and the difficulty that would pose for using the eyepiece as a magnifier.)

24.27 If you want your microscope or telescope to project a real image onto a screen, how would you change the placement of the eyepiece relative to the objective?

24.28 A cataract is a cloudiness in the lens of the eye. Is light dispersed or diffused by it?

24.29 If the lens of a person's eye is removed because of cataracts (as has been done since ancient times), why would you expect a spectacle lens of about 16 D to be prescribed?

24.30 It has become common to replace the cataract-clouded lens of the eye with an internal lens. This intraocular lens can be chosen so that the person has perfect distant vision. Will the person be able to read without glasses? If the person was nearsighted, is the power of the intraocular lens greater or less than the removed lens?

24.31 How does the power of a dry contact lens compare with its power when resting on the tear layer of the eye? Explain.

24.32 Why is your vision so blurry when you open your eyes while swimming under water? How does a face mask enable clear vision?

24.33 If the cornea is to be reshaped (this can be done surgically or with contact lenses) to correct myopia, should its curvature be made greater or smaller? Explain. What about for hyperopia?

24.34 When laser light is shone into a relaxed eye to spot-weld the retina to the back of the eye, the rays entering the eye must be parallel. Why?

24.35* A pure red object on a black background seems to disappear when illuminated with pure green light. Explain why.

24.36* What is color constancy, and what are its limitations?

24.37* There are different types of color blindness related to the malfunction of different types of cones. Why would it be particularly useful to study those rare individuals who are color blind only in one eye or who have a different type of color blindness in each eye?

24.38* Propose a way to study the function of the rods alone, given they can sense light about 1000 times dimmer than the cones.

PROBLEMS

Section 24.2 Law of Reflection

24.1 Suppose a man stands in front of a mirror as shown in Figure 24.64. His eyes are 1.65 m above the floor, and the top of his head is 0.13 m higher. Find the height above the floor of the top and bottom of the smallest mirror in which he can see both the top of his head and his feet. How is this distance related to the man's height?

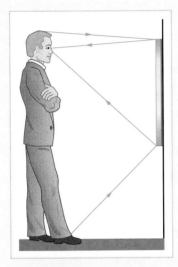

Figure 24.64 A full-length mirror is one in which you can see all of yourself. It need not be as big as you, and its size is independent of your distance from it. Problem 1.

• **24.2** Show that when light reflects from two mirrors that meet each other at a right angle, the outcoming ray is parallel to the incoming ray, as illustrated in Figure 24.65. (See also the corner reflectors in Figure 24.16.)

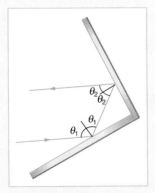

Figure 24.65 A corner reflector sends the reflected ray back in a direction parallel to the incident ray, independent of incoming direction. Problems 2 and 21.

: **24.3** Light shows staged with lasers use moving mirrors to swing beams and create colorful effects. Show that a light ray reflected from a mirror changes direction by 2ϕ when the mirror is rotated by an angle ϕ.

: **24.4** A flat mirror is neither converging nor diverging. To prove this, consider two rays originating from the same point and diverging at an angle θ. Show that after striking a plane mirror, the angle between their directions remains θ. (See Figure 24.66.)

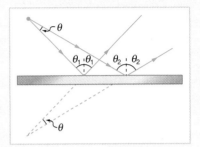

Figure 24.66 A flat mirror neither converges nor diverges light rays. Two rays continue to diverge at the same angle after reflection. Problem 4.

Section 24.3 Law of Refraction (Snell's Law)

24.5 What is the speed of light in water? In glycerin?

24.6 What is the speed of light in air? In crown glass?

24.7 Calculate the index of refraction for a medium in which the speed of light is 2.012×10^8 m/s, and identify the most likely substance based on Table 24.1.

24.8 In what substance in Table 24.1 is the speed of light 2.290×10^8 m/s?

24.9 There was a major collision of an asteroid with the moon in medieval times. It was described by monks at Canterbury Cathedral in England as a red glow on and around the moon. How long after the asteroid hit the moon, which is 3.84×10^5 km away, would the light first arrive on earth?

24.10 A scuba diver training in a pool looks at his instructor as shown in Figure 24.67. What angle does the ray from the instructor's face make with the perpendicular to the water at the point it enters? The angle in the water is 25.0°.

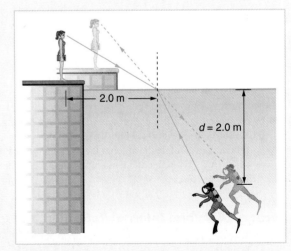

Figure 24.67 A scuba diver in a pool and his trainer look at each other. Problems 10 and 12.

- **24.11** Components of some computers communicate with each other through optical fibers having an index of refraction $n = 1.55$. What time in nanoseconds is required for a signal to travel 0.200 m through such a fiber?

- **24.12** (a) Using information in Figure 24.67, find the height of the instructor's head above the water, noting that you will first have to calculate the angle of incidence. (b) Find the apparent depth of the diver's head below water as seen by the instructor.

- **24.13** Suppose you have an unknown clear substance immersed in water, and you wish to identify it by finding its index of refraction. You arrange to have a beam of light enter it at an angle of 45.0°, and you observe the angle of refraction to be 40.3°. What is the index of refraction of the substance and its likely identification?

- **24.14** On the moon's surface, lunar astronauts placed a corner reflector, off which a laser beam is periodically reflected. The distance to the moon is calculated from the round-trip time. What percent correction is needed to account for the delay in time due to the slowing of light in the earth's atmosphere? Assume the distance to the moon is precisely 3.84×10^8 m, and the earth's atmosphere (which varies in density with altitude) is equivalent to a layer 30.0 km thick with a constant index of refraction $n = 1.000293$. (See also Problem 23.20.)

- **24.15** Figure 24.68 shows a ray of light passing from one medium into a second and then a third. Show that θ_3 is the same as it would be if the second medium were not present (provided total internal reflection does not occur).

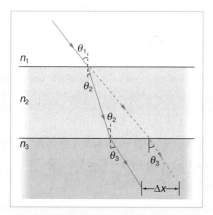

Figure 24.68 A ray of light passes from one medium to a third by traveling through a second. The final direction is the same as if the second medium were not present, but the ray is displaced by Δx (shown exaggerated). Problems 15 and 16.

- **24.16** Suppose Figure 24.68 represents a ray of light going from air through crown glass into water, such as going into a fish tank. Calculate the amount the ray is displaced by the glass (Δx), given the incident angle is 40.0° and the glass is 1.00 cm thick.

Section 24.4 Total Internal Reflection and θ_c

- **24.17** Verify that the critical angle for light going from water to air is 48.6°, as discussed at the end of Example 24.4.

- **24.18** (a) At the end of Example 24.4, it was stated that the critical angle for light going from diamond to air is 24.4°. Verify this. (b) What is the critical angle for light going from zircon to air?

- **24.19** An optical fiber uses flint glass clad with crown glass. What is the critical angle?

- **24.20** At what minimum angle will you get total internal reflection of light traveling in water and reflected from ice?

- **24.21** Suppose you are using total internal reflection to make an efficient corner reflector, such as the one in Figure 24.65. If there is air outside and the incident angle is 45.0°, what must be the minimum index of refraction of the material from which the reflector is made?

- **24.22** You can determine the index of refraction of a substance by determining its critical angle. (a) What is the index of refraction of a substance that has a critical angle of 68.4° when submerged in water? What is the substance, based on Table 24.1? (b) What would the critical angle be for this substance in air?

- **24.23** A ray of light, emitted beneath the surface of an unknown liquid with air above it, undergoes total internal reflection as shown in Figure 24.69. What is the index of refraction for the liquid and its likely identification?

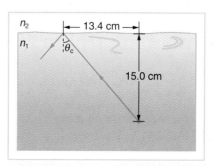

Figure 24.69 This ray strikes the surface at the critical angle and undergoes total internal reflection. Problem 23.

- **24.24** A light ray entering an optical fiber surrounded by air is first refracted and then reflected as shown in Figure 24.70. Show that if the fiber is made from crown glass, any incident ray will be totally internally reflected.

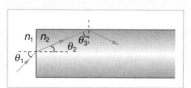

Figure 24.70 A light ray enters the end of a fiber, the surface of which is perpendicular to its sides. Problems 24 and 25 examine the conditions under which it may be totally internally reflected.

- **24.25** What is the maximum angle at which a light ray can enter a flint glass fiber clad with crown glass and undergo total internal reflection? See Figure 24.70 and note that the ray is first refracted as it enters the crown glass from air, and then reflected.

Section 24.5 Dispersion

24.26 (a) What is the ratio of the speed of red light to violet light in diamond, based on Table 24.2? (b) What is this ratio in polystyrene? (c) Which is more dispersive?

24.27 A beam of white light goes from air into water at an incident angle of 75.0°. At what angles are the red (660 nm) and violet (410 nm) parts of the light refracted?

•24.28 By how much do the critical angles for red (660 nm) and violet (410 nm) light differ in a diamond surrounded by air?

•24.29 (a) A narrow beam of light containing yellow (580 nm) and green (550 nm) wavelengths goes from polystyrene to air, striking the surface at a 30.0° incident angle. What is the angle between the colors when they emerge? (b) How far would they have to travel to be separated by 1.00 mm?

•24.30 A parallel beam of light containing orange (610 nm) and violet (410 nm) wavelengths goes from fused quartz to water, striking the surface between them at a 60.0° incident angle. What is the angle between the two colors in water?

•24.31 A ray of 610 nm light goes from air into fused quartz at an incident angle of 55.0°. At what incident angle must 470 nm light enter flint glass to have the same angle of refraction?

⁝24.32 A narrow beam of light containing red (660 nm) and blue (470 nm) wavelengths travels from air through a 1.00 cm thick flat piece of crown glass and back to air again. The beam strikes at a 30.0° incident angle. (a) At what angles do the two colors emerge? (b) By what distance are the red and blue separated when they emerge?

⁝24.33 A narrow beam of white light enters a prism made of crown glass at a 45.0° incident angle, as shown in Figure 24.71. At what angles, θ_R and θ_V, do the red (660 nm) and violet (410 nm) components of the light emerge from the prism?

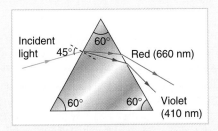

Figure 24.71 This prism will disperse the white light into a rainbow of colors. The incident angle is 45.0°, and the angles at which the red and violet light emerge are θ_R and θ_V. Problem 33.

Section 24.6 Lenses and Images

24.34 What is the power in diopters of a camera lens that has a 50.0 mm focal length?

24.35 Your camera's zoom lens has an adjustable focal length ranging from 80.0 to 200 mm. What is its range of powers?

24.36 What is the focal length of 1.75 D reading glasses found on the rack in a drugstore?

24.37 You note that your prescription for new eyeglasses is −4.50 D. What will their focal length be?

24.38 How far from the lens must the film in a camera be, if the lens has a 35.0 mm focal length and is being used to photograph a flower 75.0 cm away?

24.39 A certain slide projector has a 100 mm focal length lens. (a) How far away is the screen, if a slide is placed 103 mm from the lens and produces a sharp image? (b) If the slide is 24.0 by 36.0 mm, what are the dimensions of the image?

24.40 Repeat Problem 24.39 using ray tracing. Draw carefully to your chosen scale, and identify the rule used for each ray.

24.41 A doctor examines a mole with a 15.0 cm focal length magnifying glass held 13.5 cm from the mole. (a) Where is the image? (b) What is its magnification? (c) How big is the image of a 5.00 mm diameter mole?

24.42 Repeat Problem 24.41 using ray tracing. Draw carefully to your chosen scale, and identify the rule used for each ray.

•24.43 How far from a piece of paper must you hold your father's 2.25 D reading glasses to try to burn a hole in the paper with sunlight?

•24.44 A camera with a 50.0 mm focal length lens is being used to photograph a person standing 3.00 m away. (a) How far from the lens must the film be? (b) If the film is 36.0 mm high, what fraction of a 1.75 m tall person will fit on it?

•24.45 A camera lens used for taking close-up photographs has a focal length of 22.0 mm. The farthest it can be placed from the film is 33.0 mm. (a) What is the closest object that can be photographed? (b) What is the magnification of this closest object?

•24.46 Suppose your 50.0 mm focal length camera lens is 51.0 mm away from the film in the camera. (a) How far away is an object that is in focus? (b) What is the height of the object if its image is 2.00 cm high?

•24.47 What is the focal length of a magnifying glass that produces a magnification of 3.00 when held 5.00 cm from an object, such as a rare coin?

•24.48 What magnification will be produced by a lens of power −4.00 D (such as might be used to correct myopia) if an object is held 25.0 cm away?

•24.49 In Example 24.7, the magnification of a book held 7.50 cm from a 10.0 cm focal length lens was found to be 3.00. (a) Find the magnification for the book when it is held 8.50 cm from the magnifier. (b) Do the same for when it is held 9.50 cm from the magnifier. (c) Comment on the trend in m as the object distance increases as in these two calculations.

⁝24.50 Suppose a 200 mm focal length telephoto lens is being used to photograph mountains 10.0 km away. (a) Where is the image? (b) What is the height of the image of a 1000 m high cliff on one of the mountains?

⁝24.51 A camera with a 100 mm focal length lens is used to photograph the sun and moon. What is the height of the image of the sun on the film, given the sun is 1.40×10^6 km in diameter and is 1.50×10^8 km away?

24.52 Combine Equations 24.6a and b to show that the magnification for a thin lens is determined by its focal length and the object distance and is given by $m = f/(f - d_o)$.

24.53 (a) Using the result of Problem 24.52, show that for a case 2 image the magnification is positive and approaches infinity as the object is moved closer and closer to the focal distance. (b) What happens to the image distance in this limit?

24.54 Using the result of Problem 24.52, show that for a case 1 image the magnification is negative and approaches minus infinity as the object is moved closer and closer to the focal distance. (b) What happens to the image distance in this limit?

Section 24.7 Mirrors and Images

24.55 What is the power in diopters of a convex security mirror that has a −3.00 m focal length?

24.56 What is the power in diopters of a concave telescope mirror that has a 10.0 m radius of curvature?

24.57 What is the focal length of a makeup mirror that has a power of 1.50 D?

24.58 (a) Calculate the focal length of the mirror formed by the shiny back of a spoon that has a 3.00 cm radius of curvature. (b) What is its power in diopters?

24.59 Some telephoto cameras use a mirror rather than a lens. What radius of curvature mirror is needed to replace a 800 mm focal length telephoto lens?

24.60 Find the magnification of the heater element in Example 24.9. Note that its large magnitude helps spread out the reflected heat.

• **24.61** What is the focal length of a makeup mirror that produces a magnification of 1.50 when a person's face is 12.0 cm away?

• **24.62** A shopper standing 3.00 m from a convex security mirror sees his image with a magnification of 0.250. (a) Where is his image? (b) What is the focal length of the mirror? (c) What is its radius of curvature?

• **24.63** An object 1.50 cm high is held 3.00 cm from a person's cornea, and its reflected image is measured to be 0.167 cm high. (a) What is the magnification? (b) Where is the image? (c) Find the radius of curvature of the convex mirror formed by the cornea.

• **24.64** Do Problem 24.63 and then make a ray tracing of the object, cornea, and image. Draw carefully to your chosen scale, and identify the rule used for each ray.

24.65 Ray tracing for a flat mirror shows that the image is located a distance behind the mirror equal to the distance of the object from the mirror. This is stated $d_i = -d_o$, since this is a negative image distance (it is a virtual image). (a) What is the focal length of a flat mirror? (b) What is its power?

24.66 Show that for a flat mirror $h_i = h_o$, knowing that the image is a distance behind the mirror equal in magnitude to the distance of the object from the mirror.

24.67 Using the result of Problem 24.52, verify the statement in the discussion of Example 24.10 that the smaller the radius of curvature of a convex mirror, the smaller its magnification.

24.68 Use the law of reflection and a diagram like that in Figure 24.72 to prove that the focal length of a mirror is half its radius of curvature. That is, prove Equation 24.7, $f = R/2$.

Note this is only true for a spherical mirror if its diameter is small compared with its radius of curvature.

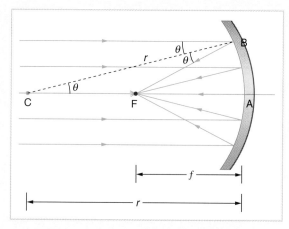

Figure 24.72 The focal length of a mirror and its radius of curvature are directly related. Problem 68.

Section 24.8 Multiple-Element Systems

24.69 A microscope with an overall magnification of 800 has an objective that magnifies by 200. (a) What is the magnification of the eyepiece? (b) If there are two other objectives that can be used, having magnifications of 100 and 400, what other total magnifications are possible?

24.70 What is the angular magnification of a telescope that has a 100 cm focal length objective and a 2.50 cm focal length eyepiece?

• **24.71** Find the distance between the objective and eyepiece lenses in the telescope in Problem 24.70 needed to produce a final image very far from the observer, where vision is most relaxed. Note that a telescope is normally used to view very distant objects.

• **24.72** (a) What magnification is produced by a 0.150 cm focal length microscope objective that is 0.155 cm from the object being viewed? (b) What is the overall magnification if an 8× eyepiece (one that produces a magnification of 8.00) is used?

• **24.73** 7.5× binoculars produce an angular magnification of −7.50, acting like a telescope. (Mirrors are used to make the image upright.) If the binoculars have objective lenses with a 75.0 cm focal length, what is the focal length of the eyepiece lenses?

• **24.74** A large reflecting telescope has an objective mirror with a 10.0 m radius of curvature. What angular magnification does it produce when a 3.00 m focal length eyepiece is used?

24.75 An amoeba is 0.305 cm away from the 0.300 cm focal length objective lens of a microscope. (a) Where is the image formed by the objective lens? (b) What is this image's magnification? (c) An eyepiece with a 2.00 cm focal length is placed 20.0 cm from the objective. Where is the final image? (d) What magnification is produced by the eyepiece? (e) What is the overall magnification?

24.76 (a) Where does an object need to be placed relative to a microscope for its 0.500 cm focal length objective to produce a magnification of −400? (b) Where should the 5.00 cm focal length eyepiece be placed to produce a further fourfold (4.00) magnification?

24.77 A small telescope has a concave mirror with a 2.00 m radius of curvature for its objective. Its eyepiece is a 4.00 cm focal length lens. (a) What is the telescope's angular magnification? (b) What angle is subtended by a 25,000 km diameter sunspot? (c) What is the angle of its telescopic image?

Section 24.9 The Eye and Vision Correction

Unless otherwise stated, the lens-to-retina distance is 2.00 cm.

24.78 What is the power of the eye when viewing an object 50.0 cm away?

24.79 Calculate the power of the eye when viewing an object 3.00 m away.

24.80 The print in many books averages 3.50 mm in height. How high is the image of the print on the retina when the book is held 30.0 cm from the eye?

24.81 What is the far point of a person whose eyes have a relaxed power of 50.5 D?

24.82 What is the near point of a person whose eyes have an accommodated power of 53.5 D?

24.83 People who do very detailed work close up, such as jewelers, often can see objects clearly at much closer than the normal 25 cm. (a) What is the power of the eyes of a woman who can see an object clearly at a distance of only 8.00 cm? (b) What is the size of an image of a 1.00 mm object, such as lettering inside a ring, held at this distance? (c) What would the size of the image be if the object were held at the normal 25.0 cm distance?

24.84 The power of a physician's eyes is 53.0 D while examining a patient. How far from her eyes is the feature being examined?

24.85 A student's eyes, while reading the blackboard, have a power of 51.0 D. How far is the board from his eyes?

24.86 Suppose a certain person's visual acuity is such that he can see objects clearly that form an image 4.00 μm high on his retina. What is the maximum distance at which he can read the 75.0 cm high letters on the side of an airplane?

24.87 A young woman with normal distant vision has a 10.0% ability to accommodate (that is, increase) the power of her eyes. What is the closest object she can see clearly?

24.88 The far point of a myopic administrator is 50.0 cm. (a) What is the relaxed power of his eyes? (b) If he has the normal 8.00% ability to accommodate, what is the closest object he can see clearly?

24.89 A very myopic man has a far point of 20.0 cm. What power contact lens (when on the eye) will correct his distant vision?

24.90 A farsighted woman has a near point of 75.0 cm. What power contact lens (when on the eye) will allow her to see objects 25.0 cm away clearly?

24.91 A myopic person sees in her medical file that her contact lens prescription is −4.00 D. What is her far point?

24.92 A mother sees that her child's contact lens prescription is +0.750 D. What is the child's near point?

24.93 Repeat Problem 24.89 for eyeglasses held 1.50 cm from the eyes.

24.94 Repeat Problem 24.90 for eyeglasses held 1.90 cm from the eyes.

24.95 Repeat Problem 24.91 for glasses that are 1.75 cm from the eyes.

24.96 Repeat Problem 24.92 for glasses that are 2.20 cm from the eyes.

24.97 (a) What power contact lens is needed to correct the vision of a person who has a far point of 1.00 m? (b) If this person has a normal 8.00% ability to accommodate (increase) the power of her eyes, what is the closest object she will be able to see clearly with contacts on?

24.98 (a) What power contact lens is needed to correct the vision of a person who has a near point of 80.0 cm and allow him to see objects 25.0 cm away? (b) If this person has a normal 8.00% ability to accommodate (increase) the power of his eyes, will they be totally relaxed when viewing very distant objects?

24.99 The lens-to-retina distance of a man is 1.98 cm, and the fully accommodated power of his eye is 54.0 D. (a) What is his near point? (b) What eyeglass power will allow him to read a book 25.0 cm from his eyes, if his glasses are 1.40 cm from his eyes?

24.100 The lens-to-retina distance of a woman is 2.03 cm, and the relaxed power of her eye is 50.0 D. (a) What is her far point? (b) What eyeglass power will allow her to see distant objects clearly, if her glasses are 1.80 cm from her eyes?

INTEGRATED CONCEPTS

Physics is most interesting and most powerful when applied to general situations involving more than a narrow set of physical principles. The integration of concepts necessary to solve problems involving several physical principles also gives greater insight into the unity of physics. You may wish to refer to Chapter 13, and perhaps others, to solve the following problems.

Note: Problem-solving strategies and a worked example that can help you solve integrated concept problems appear in Chapter 23. They can be found with the section of problems labeled *Integrated Concepts* at the end of that chapter. Consult the index for the locations of other problem-solving strategies and worked examples for integrated concept problems.

24.101 Referring to the electric room heater considered in Example 24.9, calculate the intensity of IR radiation in W/m² projected by the concave mirror on a person 3.00 m away. Assume that the heating element radiates 1500 W and has an area of 100 cm², and that half of the radiated power is reflected and focused by the mirror.

24.102 (a) Suppose you use a 15.0 cm diameter magnifying glass having a 20.0 cm focal length to focus sunlight onto a piece of coal. If the intensity of sunlight on a clear day is 1000 W/m² and it strikes the magnifying glass perpendicularly, what is the intensity of the light projected onto the coal? (b) How hot must the spot on the coal become to radiate heat away at this rate, assuming the surroundings are at 23.0°C? Neglect conduction and convection, and assume the emissivity is 1.0.

UNREASONABLE RESULTS

The following problems have results that are unreasonable because some premise is unreasonable or because certain of the premises are inconsistent with one another. Physical principles applied correctly then produce unreasonable results. The purpose of these problems is to give practice in assessing whether nature is being accurately described, and if it is not to trace the source of difficulty.

Note: Strategies for determining if an answer is reasonable appear in Chapter 22. They can be found with the section of problems labeled *Unreasonable Results* at the end of that chapter.

24.103 Suppose light travels from water to another substance, with an angle of incidence of 10.0° and an angle of refraction of 14.9°. (a) What is the index of refraction of the other substance? (b) What is unreasonable about this result? (c) Which assumptions are unreasonable or inconsistent?

24.104 Light traveling from water to a gemstone strikes the surface at an angle of 80.0° and has an angle of refraction of 15.2°. (a) What is the speed of light in the gemstone? (b) What is unreasonable about this result? (c) Which assumptions are unreasonable or inconsistent?

WAVE OPTICS

*The colors reflected by this compact disc vary with angle
and are not caused by pigments. Colors such as these
are direct evidence of the wave character of light.*

Figure 25.1 The pale yellow soap forming this bubble exhibits these brilliant colors when it is stretched thin in sunlight. How are the colors created if they are not pigments in the soap?

CONNECTIONS

Waves

The most certain indication of a wave is interference. This wave characteristic is most prominent when the wave interacts with an object that is not large compared with its the wavelength. Interference is observed for water waves, sound waves, light waves, and (as we will see in Chapter 27) for matter waves, such as electrons scattered from a crystal.

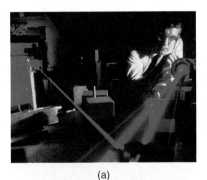

(a)

If you have ever looked at the reds, blues, and greens in a sunlit soap bubble and wondered how straw-colored soapy water could produce them, you have hit upon one of many phenomena that can only be explained by the wave character of light. (See Figure 25.1.) The same is true for the colors seen in an oil slick or in the light reflected from a compact disc. These and other interesting phenomena, such as the dispersion of white light into a rainbow of colors when passed through a narrow slit, cannot be explained fully by geometric optics. In such cases, light interacts with small objects and exhibits its wave characteristics. The branch of optics that considers the behavior of light when it exhibits wave characteristics (particularly when it interacts with small objects) is called **wave optics** (sometimes called physical optics). It is the topic of this chapter.

25.1 THE WAVE ASPECT OF LIGHT: INTERFERENCE

We know that visible light is the type of electromagnetic wave to which our eyes respond (first discussed in Section 23.3). Like all other electromagnetic waves, it obeys the equation

$$c = f\lambda \tag{25.1}$$

where $c = 3.00 \times 10^8$ m/s is the speed of light, f is the frequency of the electromagnetic wave, and λ is its wavelength. The range of visible wavelengths is approximately 380 to 760 nm. As is true of all waves, light travels in straight lines and acts like a ray when it interacts with objects several times as large as its wavelength. But when it interacts with smaller objects, it displays its wave characteristics prominently. Interference is the hallmark of a wave, and in Figure 25.2 both ray and wave characteristics can be seen. The laser beam epitomizes a ray on its way to the horizontal slits in the slide, moving in a straight line. Note that light scattered from the beam on its way to the slits is grainy, as is the light reflected from the screen. The grainy appearance is constructive and destructive interference of the laser light output. After passing through the horizontal slits, the beam spreads out vertically into a pattern of bright and dark regions caused by systematic constructive and destructive interference. A ray would move straight ahead after passing through the slits rather than spreading out.

Light has wave characteristics in various media as well as in a vacuum. When light goes from vacuum to some medium, like water, its speed and wavelength change, but its frequency f remains the same. (We can think of light as a forced oscillation that must have the frequency of the original source.) The speed of light in a medium is $v = c/n$, where n is its index of refraction. If we divide both sides of Equation 25.1 by n, we get $c/n = v = f\lambda/n$. This implies that $v = f\lambda_n$, where λ_n is the **wavelength in a medium**, and that

$$\lambda_n = \frac{\lambda}{n} \tag{25.2}$$

where λ is the wavelength in vacuum and n is the medium's index of refraction. So the wavelength of light is smaller in any medium than it is in a vacuum. For example, the range of visible wavelengths in water, which has $n = 1.333$, is (380 nm)/1.333 to (760 nm)/1.333, or $\lambda_n = 285$ to 570 nm. Although wavelengths change in going from one medium to another, colors do not, since color is associated with frequency.

25.2 HUYGENS'S PRINCIPLE: DIFFRACTION

Figure 25.3 shows how a transverse wave looks as viewed from above and from the side. A light wave can be imagined to propagate like this, although we do not actually see it wiggling through space. From above, we view the wave fronts (or wave crests) as we would by looking down on ocean waves. The side view would be a graph of the electric or magnetic field, as was shown in Figure 23.6, for example. (See also Figure 16.2 for a

(b)

Figure 25.2 A laser beam passing through two horizontal slits in the slide (a) spreads out vertically (b). The beam acts like a ray on its way to the slits, moving in a straight line. But the grainy appearance of the beam and the way it spreads out after passing through the slits are both due to interference effects—these are characteristic of waves.

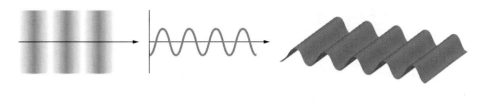

View from above   View from side   Overall view

Figure 25.3 A transverse wave, such as an electromagnetic wave like light, as viewed from above and from the side. The direction of propagation is perpendicular to the wave fronts (or wave crests) and is represented by an arrow like a ray.

similar view of how sound waves move.) The view from above is perhaps the most useful in developing concepts about wave optics.

The Dutch scientist Christian Huygens (1629–1695) developed a useful technique for determining in detail how and where waves propagate. Starting from some known position, **Huygens's principle** states that:

> **Every point on a wave front is a source of wavelets that spread out in the forward direction at the same speed as the wave itself. The new wave front is a line tangent to all of the wavelets.**

Figure 25.4 shows how Huygens's principle is applied. Each point on the wave front emits a semicircular wave that moves at the propagation speed v. These are drawn at a time t later, so that they have moved a distance $s = vt$. The new wave front is a line tangent to the wavelets and is where we would expect the wave to be a time t later. Huygens's principle works for all types of waves, including water waves, sound waves, and light waves. We will find it useful not only in describing how light waves move, but also in explaining the laws of reflection and refraction. Additionally, we will see that Huygens's principle tells us how and where light rays interfere.

Figure 25.5 shows how a mirror reflects an incoming wave at an angle equal to the incident angle, verifying the law of reflection. As the wave front strikes the mirror, wavelets are first emitted from the left part of the mirror and then the right. The wavelets closer to the left have had time to travel farther, producing a wave front moving in the direction shown.

The law of refraction can be explained by applying Huygens's principle to a wave front passing from one medium to another. (See Figure 25.6.) Each wavelet in the figure was emitted when the wave front crossed the interface between the media. Since the speed of light is smaller in the second medium, the waves do not travel as far in a given time, and the new wave front changes direction as shown. This is why a ray changes direction toward the perpendicular when light slows down. Snell's law can be derived from the geometry in Figure 25.6, but this is left as an exercise for the ambitious reader.

What happens when a wave passes through an opening, such a light shining through an open door into a dark room? For light, we expect to see a sharp shadow of the doorway on the floor of the room, and we expect no light to bend around corners into other parts of the room. When sound passes through a door, we expect to hear it everywhere in the room and, thus, expect that sound spreads out when passing through such an opening. (See Figure 25.7.) What is the difference between the behavior of sound waves and light waves in this case? The answer is that light has very short wavelengths and acts like a ray. Sound has wavelengths on the order of the size of the door and bends around corners.

If we pass light through smaller openings, often called slits, we can use Huygens's principle to see that light bends as sound does. (See Figure 25.8.) The bending of a wave

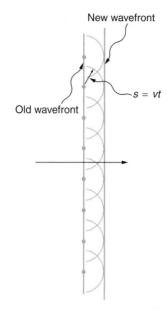

Figure 25.4 Huygens's principle applied to a straight wave front. Each point on the wave front emits a semicircular wavelet that moves a distance $s = vt$. The new wave front is a line tangent to the wavelets.

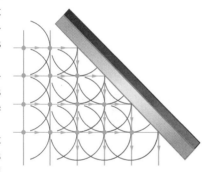

Figure 25.5 Huygens's principle applied to a straight wave front striking a mirror. The wavelets shown were emitted as each point on the wave front struck the mirror. The tangent to them shows that the new wave front has been reflected at an angle equal to the incident angle. The direction of propagation is perpendicular to the wave front, as shown.

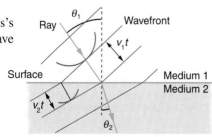

Figure 25.6 Huygens's principle applied to a straight wave front traveling from one medium to another where its speed is less. The ray bends toward the perpendicular, since the wavelets have a smaller speed in the second medium.

Figure 25.7 (a) Light passing through a doorway makes a sharp outline on the floor. Since light's wavelength is so small compared with the size of the door, it acts like a ray. (b) Sound waves bend into all parts of the room, a wave effect, because their wavelength is similar to the size of the door.

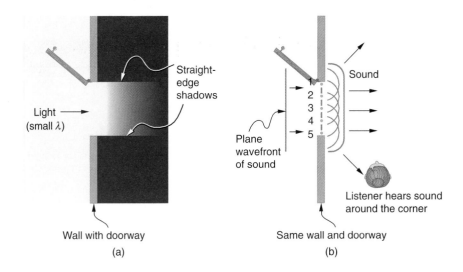

(a)

(b)

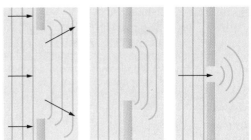

Figure 25.8 Huygens's principle applied to a straight wave front striking an opening. The edges of the wave front bend after passing through the opening, a process called *diffraction*. The amount of bending is more extreme for a small opening, consistent with the fact that wave characteristics are most noticeable for interactions with objects about the same size as the wavelength.

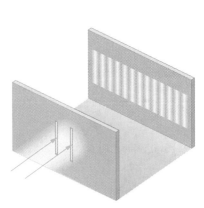

Figure 25.9 Young's double slit experiment. Here pure-wavelength light sent through a pair of vertical slits is diffracted into a pattern of numerous vertical lines spread out horizontally. Without diffraction and interference, the light would simply make two lines on the screen.

around the edges of an opening or an obstacle is called **diffraction**. Diffraction is a wave characteristic and occurs for all types of waves. If diffraction is observed for some phenomenon, it is proof that the phenomenon is a wave. Thus the horizontal diffraction of the laser beam after it passes through a slit in Figure 25.2 is evidence that light is a wave.

25.3 YOUNG'S DOUBLE SLIT EXPERIMENT

Although Christian Huygens thought that light is a wave, Isaac Newton did not. Newton felt that there were other explanations for color, and for the interference and diffraction effects that were observable at the time. Because of Newton's tremendous stature, his view generally prevailed. The fact that Huygens's principle worked was not considered direct enough evidence to prove that light is a wave. The acceptance of the wave character of light came many years later when, in 1801, the English physicist and physician Thomas Young (1773–1829) did his now classic double slit experiment. (See Figure 25.9.)

Why do we not ordinarily observe wave behavior for light, such as observed in Young's double slit experiment? First, light must interact with something small, such as the closely spaced slits used by Young, to show pronounced wave effects. Furthermore, Young used a single light source (the sun) first passed through a single slit to make the light somewhat coherent. By **coherent**, we mean waves are in phase or have a definite phase relationship. **Incoherent** means waves have random phase relationships. Why did Young then pass the light through a double slit? The answer to this question is that two slits provide two coherent light sources that then interfere constructively or destructively. Young used sunlight, where each wavelength forms its own pattern, making the effect more difficult to see. We illustrate the double slit experiment with monochromatic (pure λ) light to clarify the effect. Figure 25.10 shows the pure constructive and destructive interference of two waves having the same wavelength and amplitude. (This was first discussed and illustrated in several figures in Section 15.10.)

When light passes through narrow slits, it is diffracted into semicircular waves as shown in Figure 25.11(a) (and previously shown in Figure 25.8). Pure constructive interference occurs where the waves are crest to crest or trough to trough. Pure destructive interference occurs where they are crest to trough. The light must fall on a screen and be scattered into our eyes for us to see the pattern. An analogous pattern for water waves is shown in Figure 25.11(b). Note that regions of constructive and destructive interference move out from the slits at well-defined angles to the original beam. These angles depend on wavelength and the distance between the slits, as we shall see below.

To understand the double slit interference pattern, we consider how two waves travel from the slits to the screen as illustrated in Figure 25.12. Each slit is a different distance from a given point on the screen. Thus different numbers of wavelengths fit into each path. Waves start out from the slits in phase (crest to crest), but they may end up out of phase (crest to trough) at the screen if the paths differ in length by half a wavelength,

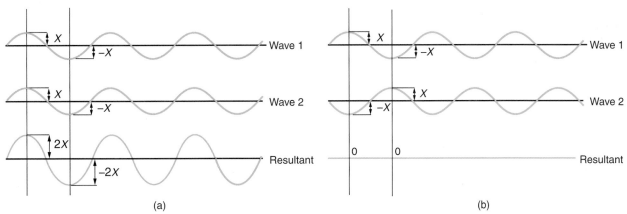

(a)

(b)

Figure 25.10 The amplitudes of waves add. Pure constructive interference is obtained when identical waves are in phase. Pure destructive interference occurs when identical waves are exactly out of phase, or shifted by half a wavelength.

interfering destructively as shown in Figure 25.12(a). If the paths differ by a whole wavelength, then the waves arrive in phase (crest to crest) at the screen, interfering constructively, as shown in Figure 25.12(b). More generally, if the paths taken by the two waves differ by any half-integral number of wavelengths $((1/2)\lambda, (3/2)\lambda, (5/2)\lambda,$ etc.), then destructive interference occurs. Similarly, if the paths taken by the two waves differ by any integral number of wavelengths ($\lambda, 2\lambda, 3\lambda$, etc.), then constructive interference occurs.

Figure 25.13 shows how to determine the path length difference for waves traveling from two slits to a common point on a screen. If the screen is a large distance away compared with the distance between the slits, then the angle θ between the path and a line from the slits to the screen (see the figure) is nearly the same for each path. The difference between the paths is shown in the figure; simple trigonometry shows it to be $d \sin \theta$, where d is the distance between the slits. To obtain **constructive interference for a double slit**, the path length difference must be an integral multiple of the wavelength, or

$$d \sin \theta = m\lambda, \qquad m = 0, 1, 2, \ldots \quad \text{(constructive)} \qquad \textbf{(25.3)}$$

Similarly, to obtain **destructive interference for a double slit**, the path length difference must be a half-integral multiple of the wavelength, or

$$d \sin \theta = (m + \tfrac{1}{2})\lambda, \qquad m = 0, 1, 2, \ldots \quad \text{(destructive)} \qquad \textbf{(25.4)}$$

where λ is the wavelength of the light, d is the distance between slits, and θ is the angle from the original direction of the beam as discussed above. We call m the **order** of the interference. For example, $m = 4$ is fourth-order interference.

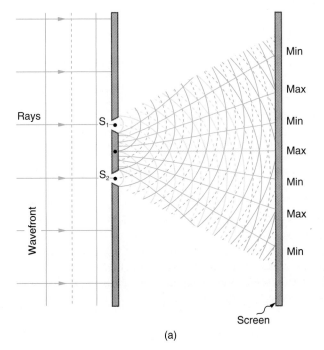

Rays

Wavefront

Screen

(a)

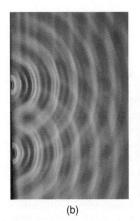

(b)

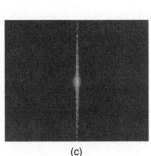

(c)

Figure 25.11 Double slits produce two coherent sources of waves that interfere. (a) Light spreads out (diffracts) from each slit, because the slits are narrow. These waves overlap and interfere constructively (bright lines) and destructively (dark regions). We can only see this if the light falls onto a screen and is scattered into our eyes. (b) Double slit interference pattern for water waves nearly identical to that for light. Wave action is greatest in regions of constructive interference and least in regions of destructive interference. (c) When light that has passed through double slits falls on a screen, we see a pattern such as this.

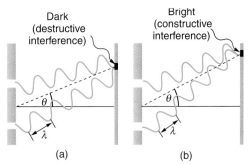

(a)   (b)

Figure 25.12 Waves follow different paths from the slits to a common point on a screen. (a) Destructive interference occurs here, because one path is a half wavelength longer than the other. The waves start in phase but arrive out of phase. (b) Constructive interference occurs here, because one path is a whole wavelength longer than the other. The waves start out and arrive in phase.

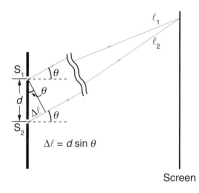

Figure 25.13 The paths from each slit to a common point on the screen differ by an amount $d \sin \theta$, assuming the distance to the screen is much greater than the distance between slits (not to scale here).

$\Delta \ell = d \sin \theta$

Screen

The equations for double slit interference imply that a series of bright and dark lines are formed. For vertical slits, the light spreads out horizontally on either side of the incident beam into a pattern called interference fringes, illustrated in Figure 25.14. The intensity of the bright fringes falls off on either side, being brightest at the center. The closer together the slits are, the more spreading there is. We can see this by examining Equation 25.3. For λ and m fixed, the smaller d is, the larger θ must be, since $\sin \theta = m\lambda/d$. This is consistent with our contention that wave effects are most noticeable when the object the wave encounters (here slits a distance d apart) is small. Small d gives large θ, hence a large effect.

EXAMPLE 25.1   FINDING WAVELENGTH FROM A DIFFRACTION PATTERN

Suppose you pass the light from a He-Ne laser through two slits separated by 0.0100 mm and find that the third bright line is formed at an angle of 10.95° relative to the incident beam. What is the wavelength of the light?

Strategy The third bright line is due to third-order constructive interference, which means that $m = 3$. We are given $d = 0.0100$ mm and $\theta = 10.95°$. The wavelength can thus be found using Equation 25.3 for constructive interference.

Solution Equation 25.3 is $d \sin \theta = m\lambda$. Solving for the wavelength λ gives

$$\lambda = \frac{d \sin \theta}{m}$$

Substituting known values yields

$$\lambda = \frac{(0.0100 \text{ mm})(\sin 10.95°)}{3}$$
$$= 6.33 \times 10^{-4} \text{ mm} = 633 \text{ nm}$$

Discussion To three digits, this is the wavelength of light emitted by the common He-Ne laser. Not by coincidence, this red color is similar to that emitted by neon lights. More important, however, is the fact that interference patterns can be used to measure wavelength. Young did this for visible wavelengths. This analytical technique is still widely used to measure electromagnetic spectra. For a given order, the angle for constructive interference increases with λ, so that spectra (measurements of intensity versus wavelength) can be obtained.

EXAMPLE 25.2   HIGHEST ORDER POSSIBLE

Interference patterns do not have an infinite number of lines, since there is a limit to how big m can be. What is the highest-order constructive interference possible with the system described in the preceding example?

Strategy and Concept Equation 25.3 describes constructive interference. For fixed values of d and λ, the larger m is, the larger $\sin \theta$ is. But the maximum value that $\sin \theta$ can have is 1, for an angle of 90° (larger angles imply light goes backward and does not reach the screen at all). Let us find which m corresponds to this maximum diffraction angle.

Solution Solving Equation 25.3 for m gives

$$m = \frac{d \sin \theta}{\lambda}$$

Taking $\sin \theta = 1$ and substituting the values of d and λ from the preceding example gives

$$m = \frac{(0.0100 \text{ mm})(1)}{633 \text{ nm}} = 15.8$$

So the largest integer m can be is 15, or

$$\boxed{m = 15}$$

Discussion The number of fringes depends on the wavelength and slit separation. The number of fringes will be very large for large slit separations. But if the slit separation becomes much greater than the wavelength, the intensity of the interference pattern changes so that the screen has two bright lines cast by the slits, as expected when light behaves like a ray.

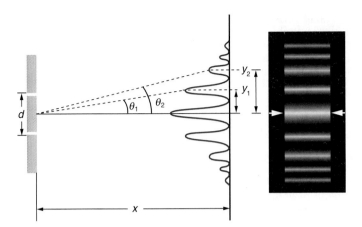

Figure 25.14 The interference pattern for a double slit has an intensity that falls off with angle. The photograph shows multiple bright and dark lines, or fringes, formed by light passing through a double slit.

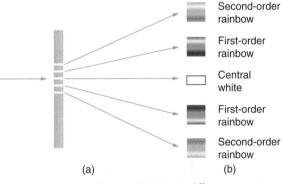

(a)   (b)

Figure 25.15 A diffraction grating is a large number of evenly spaced parallel slits. (a) Light passing through is diffracted in a pattern similar to a double slit, with bright regions at various angles. (b) The pattern obtained for white light incident on a grating. The central maximum is white, and the higher-order maxima disperse white light into their rainbow of colors.

25.4 MULTIPLE SLIT DIFFRACTION

An interesting thing happens if you pass light through a large number of evenly spaced parallel slits, called a **diffraction grating**. An interference pattern is created that is very similar to the one formed by a double slit. (See Figure 25.15.) A diffraction grating can be manufactured by scratching glass with a sharp tool in a number of precisely positioned parallel lines, with the untouched regions acting like slits. These can be photographically mass produced rather cheaply. Diffraction gratings work both for transmission of light, as in Figure 25.15, and for reflection of light, as on the bumper sticker in Figure 25.16. In addition to their use as novelty items, diffraction gratings are commonly used for spectroscopic dispersion and analysis of light. What makes them particularly useful is the fact that they form a sharper pattern than double slits do. That is, their bright regions are narrower and brighter, while their dark regions are darker. (See Figure 25.17.)

The analysis of a diffraction grating is very similar to that for a double slit. (See Figure 25.18.) As we know from our discussion of double slits in the preceding section, light is diffracted by each slit and spreads out after passing through. Rays traveling in the same direction (at an angle θ relative to the incident direction) are shown in the figure. Each of these rays travels a different distance to a common point on a screen far away. The rays start in phase, and they can be in or out of phase when they reach a screen, depending on the difference in the path lengths traveled. As seen in the figure, each ray travels a distance $d \sin \theta$ different from that of its neighbor, where d is the distance between slits. If this distance equals an integral number of wavelengths, the rays all arrive in phase, and constructive interference (a maximum) is obtained. Thus, the condition necessary to obtain **constructive interference for a diffraction grating** is

$$d \sin \theta = m\lambda, \qquad m = 0, 1, 2, \ldots \quad \text{(constructive)} \qquad \textbf{(25.5)}$$

Figure 25.16 This bumper sticker reflects different colors at different angles, because it has a diffraction grating covering it.

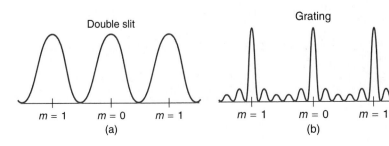

(a)   (b)

Figure 25.17 Graphs of the intensity of light passing through a double slit (a) and a diffraction grating (b) for monochromatic light. Maxima can be created at the same angles, but those for the diffraction grating are narrower and hence sharper. The maxima become narrower and the regions between darker as the number of slits is increased.

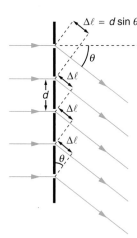

where d is the distance between slits in the grating, λ is the wavelength of light, and m is the order of the maximum. Note that this is exactly the same equation as for double slits separated by d. However, the slits are usually closer in diffraction gratings than in double slits, producing fewer maxima at larger angles.

EXAMPLE 25.3   TYPICAL DIFFRACTION GRATING EFFECTS

Diffraction gratings with 10,000 lines per centimeter are readily available. Suppose you have one, and you send a beam of white light through it to a screen 2.00 m away. (a) Find the angles for the first-order diffraction of the shortest and longest wavelengths of visible light (380 and 760 nm). (b) What is the distance between the ends of the rainbow of visible light produced on the screen? (See Figure 25.19.)

Strategy The angles can be found using Equation 25.5, once a value for the slit spacing d has been determined. Since there are 10,000 lines per centimeter, each line is separated by 1/10,000 of a centimeter. Once the angles are found, the distances along the screen can be found using simple trigonometry.

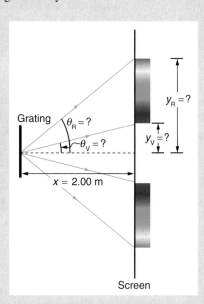

Figure 25.19 The diffraction grating considered in Example 25.3 produces a rainbow of colors on a screen a distance $x = 2.00$ m from the grating. The distances along the screen are measured perpendicular to the x-direction. Note that if the screen extends below the x-axis, another rainbow will be projected on the other side.

Solution for (a) The distance between slits is $d = (1 \text{ cm})/10,000 = 1.00 \times 10^{-4}$ cm or 1.00×10^{-6} m. Let us call the two angles θ_V for violet (380 nm) and θ_R for red (760 nm). Solving Equation 25.5 for $\sin \theta$,

$$\sin \theta_V = \frac{m\lambda_V}{d}$$

where $m = 1$ for first order and $\lambda_V = 380$ nm $= 3.80 \times 10^{-7}$ m. Substituting these values gives

$$\sin \theta_V = \frac{3.80 \times 10^{-7} \text{ m}}{1.0 \times 10^{-6} \text{ m}} = 0.380$$

Thus the angle θ_V is

$$\theta_V = \sin^{-1} 0.380 = 22.3°$$

Similarly,

$$\sin \theta_R = \frac{7.60 \times 10^{-7} \text{ m}}{1.0 \times 10^{-6} \text{ m}} = 0.760$$

Thus the angle θ_R is

$$\theta_R = \sin^{-1} 0.760 = 49.5°$$

Solution for (b) The distances on the screen are labeled y_V and y_R in Figure 25.19. Noting that $\tan \theta = y/x$, we can solve for y_V and y_R. That is,

$$y_V = x \tan \theta_V = (2.00 \text{ m}) \cdot \tan 22.3° = 0.822 \text{ m}$$

and

$$y_R = x \tan \theta_R = (2.00 \text{ m}) \cdot \tan 49.5° = 2.339 \text{ m}$$

The distance between them is therefore

$$y_R - y_V = 1.52 \text{ m}$$

(continued)

(continued)

Discussion The large distance between the red and violet ends of the rainbow produced from the white light indicates the potential this diffraction grating has as a spectroscopic tool. The more it can spread out the wavelengths (greater dispersion), the more detail can be seen in a spectrum. This depends on the quality of the diffraction grating—it must be very precisely made in addition to having closely spaced lines. Figure 25.20 shows a few spectra produced with a diffraction grating.

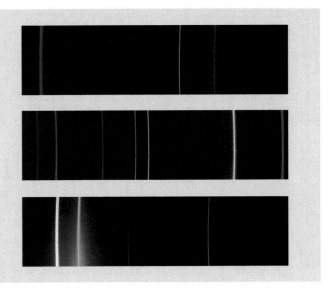

Figure 25.20 These spectra (H, He, and Hg) were produced with high-quality diffraction gratings. The amount of observable detail is related to the quality of the grating. The greater the dispersion, the more the spectrum is spread out.

25.5 SINGLE SLIT DIFFRACTION

Light passing through a single slit forms a diffraction pattern somewhat different from those formed by double slits or diffraction gratings. Figure 25.21 shows a single slit diffraction pattern. Note that the central maximum is larger than those on either side, and that the intensity decreases rapidly on either side. In contrast, a diffraction grating produces evenly spaced lines that dim slowly on either side of center.

The analysis of single slit diffraction is illustrated in Figure 25.22. Here we consider light coming from different parts of the *same* slit. According to Huygens's principle, every part of the wave front in the slit emits wavelets. These are like rays that start out in phase and head in all directions. (Each ray is perpendicular to the wave front of a wavelet.) Assuming the screen is very far away compared with the size of the slit, rays heading toward a common destination are nearly parallel. When they travel straight ahead, as in Figure 25.22(a), they remain in phase, and a central maximum is obtained. But when rays travel at an angle θ relative to the original direction of the beam, each travels a different distance to a common location, and they can arrive in or out of phase. In Figure 25.22(b), the ray from the bottom travels a distance one wavelength λ farther than the one on the top. Thus a ray from the center travels a distance $\lambda/2$ farther than the one on the left, arrives out of phase, and interferes destructively. A ray from slightly above the center and one from slightly above the bottom will also cancel one another. In fact, each ray from the slit will have another to interfere with destructively, and a minimum

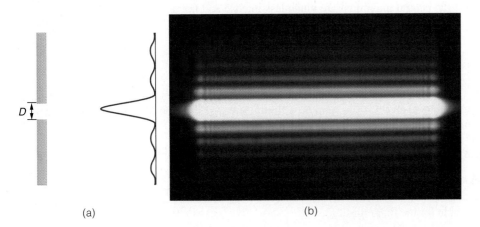

(a)

(b)

Figure 25.21 (a) Single slit diffraction pattern. Monochromatic light passing through a single slit has a central maximum and many smaller and dimmer maxima on either side. (b) The photograph shows the bright central maximum and dimmer and thinner maxima on either side.

Figure 25.22 Light passing through a single slit is diffracted in all directions and may interfere constructively or destructively, depending on the angle. The difference in path length for rays from either side of the slit is seen to be $D \sin \theta$.

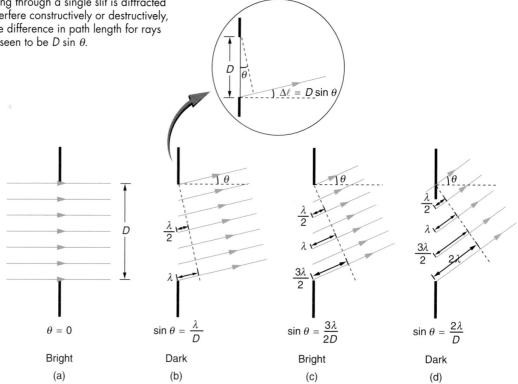

$\theta = 0$ $\sin \theta = \dfrac{\lambda}{D}$ $\sin \theta = \dfrac{3\lambda}{2D}$ $\sin \theta = \dfrac{2\lambda}{D}$

Bright   Dark   Bright   Dark

(a)   (b)   (c)   (d)

in intensity will occur at this angle. There will be another minimum at the same angle to the right of the incident direction of the light.

At the larger angle shown in Figure 25.22(c), the path lengths differ by $3\lambda/2$ for rays from the top and bottom of the slit. One ray travels a distance λ different from the one from the bottom and arrives in phase, interfering constructively. Two rays, each from slightly above those two, will also add constructively. Most rays from the slit will have another to interfere with constructively, and a maximum in intensity will occur at this angle. However, all rays do not interfere constructively for this situation, and so the maximum is not as intense as the central maximum. Finally, in Figure 25.22(d), the angle shown is large enough to create a second minimum. As seen in the figure, the difference in path length for rays from either side of the slit is $D \sin \theta$, and we see that a destructive minimum is obtained when this distance is an integral multiple of the wavelength. Thus, to obtain **destructive interference for a single slit**,

$$D \sin \theta = m\lambda, \qquad m = 1, 2, 3, \ldots \quad \text{(destructive)} \qquad \textbf{(25.6)}$$

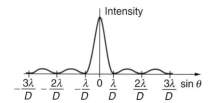

Figure 25.23 A graph of single slit diffraction intensity showing the central maximum to be wider and much more intense than those to the sides.

where D is the slit width, λ is the light's wavelength, θ is the angle relative to the original direction of the light, and m is the order of the minimum. Figure 25.23 shows a graph of intensity for single slit interference, and it is apparent that the maxima on either side of the central maximum are much less intense and not as wide. This is consistent with the photograph in Figure 25.21(b).

EXAMPLE 25.4 SINGLE SLIT DIFFRACTION

Visible light of wavelength of 550 nm falls on a single slit and produces its second diffraction minimum at an angle of 45.0° relative to the incident direction of the light. (a) What is the width of the slit? (b) At what angle is the first minimum produced? (See Figure 25.24.)

Strategy From the given information, and assuming the screen is far away from the slit, we can use Equation 25.6 first to find D, and again to find the angle for the first minimum θ_1.

Solution for (a) We are given that $\lambda = 550$ nm, $m = 2$, and $\theta_2 = 45.0°$. Solving Equation 25.6 for D and substituting known values gives

$$D = \frac{m\lambda}{\sin \theta_2} = \frac{2(550 \text{ nm})}{\sin 45.0°}$$

$$= \frac{1100 \times 10^{-9} \text{ m}}{0.707}$$

$$= 1.56 \times 10^{-6} \text{ m}$$

(continued)

(continued)

Solution for (b) Solving Equation 25.6 for $\sin\theta_1$ and substituting known values gives

$$\sin\theta_1 = \frac{m\lambda}{D} = \frac{1(550 \times 10^{-9}\ \text{m})}{1.56 \times 10^{-6}\ \text{m}} = 0.354$$

Thus the angle θ_1 is

$$\theta_1 = \sin^{-1} 0.354 = 20.7°$$

Discussion We see that the slit is narrow (it is only a few times greater than the wavelength of light). This is consistent with the fact that light must interact with an object comparable in size to its wavelength to exhibit significant wave effects such as this single slit diffraction pattern. We also see that the central maximum extends 20.7° on either side of the original beam, for a width of about 41°. The angle between the first and second minima is only about 24° (45.0° − 20.7°). Thus the second maximum is only about half as wide as the central maximum.

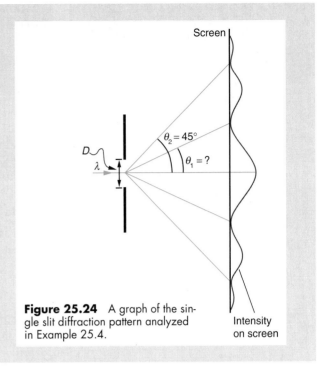

Figure 25.24 A graph of the single slit diffraction pattern analyzed in Example 25.4.

Single slit diffraction is but one example of light diffracting through openings and around edges. Figure 25.25 shows diffraction around a circular disk produced by a point-like monochromatic light source. More care is needed to observe these diffraction patterns, since the objects involved are much larger than the wavelength of light. Such diffraction patterns for large objects were observed within a few years after Young's double slit experiment, adding to the evidence that light is a wave.

25.6 LIMITS OF RESOLUTION: THE RAYLEIGH CRITERION

Light diffracts as it moves through space, bending around obstacles, interfering constructively and destructively. While this can be used as a spectroscopic tool—a diffraction grating disperses light according to wavelength, for example, and is used to produce spectra—diffraction also limits the detail we can obtain in images. Figure 25.26(a) shows the effect of passing light through a small circular aperture. Instead of a bright spot with sharp edges, a spot with a fuzzy edge surrounded by circles of light is obtained. This pattern is caused by diffraction similar to that produced by a single slit. Light from different parts of the circular aperture interferes constructively and destructively. The effect is most noticeable when the aperture is small, but the effect is there for large apertures, too.

How does diffraction affect the detail that can be observed when light passes through an aperture? Figure 25.26(b) shows the diffraction pattern produced by two point light sources that are close to one another, viewed through a circular aperture. The pattern is similar to that for a single point source, and it is just barely possible to tell that there are two light sources rather than one. If they were closer together, as in Figure 25.26(c), we could not distinguish them, thus limiting the detail or resolution we can obtain. This limit is an inescapable consequence of the wave nature of light.

There are many situations in which diffraction limits resolution. The acuity of our vision is limited because light passes through the pupil, the circular aperture of our eye.

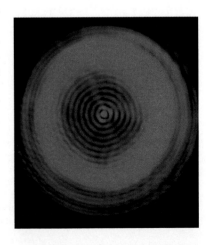

Figure 25.25 The diffraction pattern of monochromatic light produced by a disk differs from the plain shadow we might expect. The fringes and the bright dot at the center are due to interference of light diffracting around the disk.

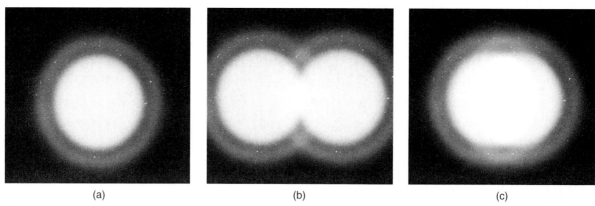

(a)   (b)   (c)

Figure 25.26 (a) Monochromatic light passed through a small circular aperture produces this diffraction pattern. (b) Two point light sources that are close to one another produce overlapping images because of diffraction. (c) If they are closer together, they cannot be resolved or distinguished.

Be aware that diffraction spreading of light is due to the limited diameter of a light beam, not the interaction with an aperture. Thus light passing through a lens with a diameter D shows this effect and spreads, blurring the image, just as light passing through an aperture of diameter D does. So diffraction limits the resolution of any system having a lens or mirror. Telescopes are also limited by diffraction, because of the finite diameter D of their primary mirror.

Just what is the limit? To answer that question, consider the diffraction pattern for a circular aperture, which has a central maximum that is wider and brighter than the maxima surrounding it (similar to a slit). (See Figure 25.27(a).) It can be shown that for a circular aperture of diameter D, the first minimum in the diffraction pattern occurs at $\theta = 1.22\lambda/D$ (providing the aperture is large compared with the wavelength of light, which is the case for most optical instruments). The accepted criterion for determining the diffraction limit to resolution based on this angle was developed by Lord Rayleigh in the 19th century. The **Rayleigh criterion** for the diffraction limit to resolution states that *two images are just resolvable when the center of the diffraction pattern of one is directly over the first minimum of the diffraction pattern of the other.* (See Figure 25.27(b).) The first minimum is at an angle of $1.22\lambda/D$, so that two point objects are just resolvable if they are separated by the angle

$$\theta = 1.22\,\frac{\lambda}{D} \qquad (25.7)$$

where λ is the wavelength of light (or other electromagnetic radiation) and D is the diameter of the aperture, lens, mirror, or the like, with which the two objects are observed. In this expression, θ has units of radians.

CONNECTIONS

Limits to Knowledge

All attempts to observe the size and shape of objects are limited by the wavelength of the probe. Even the small wavelength of light prohibits exact precision. When extremely small wavelength probes are used, the system is disturbed, still limiting our knowledge, much as making an electrical measurement alters a circuit. Heisenberg's uncertainty principle asserts that this limit is fundamental and inescapable, as we shall see in quantum mechanics.

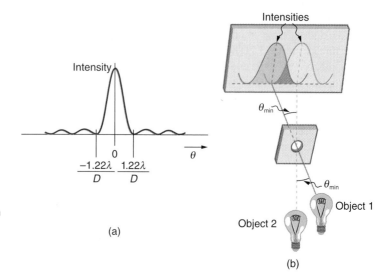

Figure 25.27 (a) Graph of intensity of the diffraction pattern for a circular aperture. Note that, similar to a single slit, the central maximum is wider and brighter than those to the sides. (b) Two point objects produce overlapping diffraction patterns. Shown here is the Rayleigh criterion for being just resolvable. The central maximum of one pattern lies on the first minimum of the other.

EXAMPLE 25.5 DIFFRACTION LIMITS OF THE HUBBLE SPACE TELESCOPE

The primary mirror of the orbiting Hubble Space Telescope has a diameter of 2.40 m. Being in orbit, this telescope avoids the degrading effects of atmospheric distortion on its resolution. (a) What is the angle between two just resolvable point light sources (perhaps two stars)? Assume an average light wavelength of 550 nm. (b) If these two stars are at the 2 million light year* distance of the Andromeda galaxy, how close together can they be and still be resolved?

Strategy The Rayleigh criterion stated in Equation 25.7 gives the smallest possible angle θ between point sources, or the best obtainable resolution. Once this angle is found, the distance between stars can be calculated, since we are given how far away they are.

Solution for (a) The Rayleigh criterion for the minimum resolvable angle is

$$\theta = 1.22 \frac{\lambda}{D}$$

Entering knowns gives

$$\theta = 1.22 \cdot \frac{550 \times 10^{-9} \text{ m}}{2.40 \text{ m}}$$
$$= 2.80 \times 10^{-7} \text{ rad}$$

Solution for (b) The distance s between two objects, a distance r away, separated by an angle θ is

$$s = r\theta$$

Substituting knowns gives

$$s = (2.0 \times 10^6 \text{ ly})(2.80 \times 10^{-7} \text{ rad})$$
$$= 0.56 \text{ ly}$$

Discussion The angle found in part (a) is extraordinarily small (less than 1/50,000 of a degree), because the primary mirror is so large compared with the wavelength of light. As noted, diffraction effects are most noticeable when light interacts with objects having sizes on the order of the wavelength of light. But the effect is still there, and there is a diffraction limit to what is observable. The actual resolution of the Hubble Telescope is not quite as good as that found here. As with all instruments, there are other effects, such as nonuniformities in mirrors or chromatic aberrations in lenses, that further limit resolution. But Figure 25.28 gives an indication of the detail observable with the Hubble because of its size and quality and especially because it is above the earth's atmosphere.

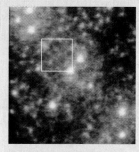

Figure 25.28 These two photographs of the same galaxy give an idea of the observable detail using the Hubble Space Telescope compared with that using a ground-based telescope. On the left is the best ground-based image. The expanded region was captured by Hubble.

The answer in part (b) indicates that two stars separated by about half a light year can be resolved. The average distance between stars in a galaxy is on the order of 5 light years in the outer parts and about 1 light year near the galactic center. So the Hubble can resolve most of the individual stars in Andromeda galaxy, even though it lies at such a huge distance that its light takes 2 million years to reach us. Figure 25.29 shows another mirror used to observe radio waves from outer space.

Figure 25.29 A 300 m diameter natural bowl at Arecibo in Puerto Rico is lined with reflective material, making it into a radio telescope. Although D for Arecibo is much larger than for the Hubble Telescope, it detects much longer wavelength radiation and its diffraction limit is significantly poorer than Hubble's. Arecibo is still very useful, because important information is carried by radio waves that is not carried by visible light.

Any beam of light having a finite diameter D and a wavelength λ exhibits diffraction spreading. The beam spreads out with an angle θ given by Equation 25.7. For example, a laser beam made of rays as parallel as possible (angles between rays as close to $\theta = 0°$ as possible) instead spreads out at an angle $\theta = 1.22\lambda/D$, where D is

*A light year is the distance light travels in 1 year.

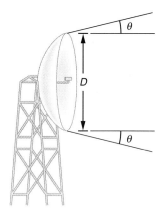

Figure 25.30 The beam produced by this microwave transmission antenna will spread out at a minimum angle $\theta = 1.22\lambda/D$ due to diffraction. It is impossible to produce a parallel beam, because the beam has a limited diameter.

the diameter of the beam and λ is its wavelength. This spreading is impossible to observe for a flashlight, because its beam is not very parallel to start with. But for long-distance transmission of laser beams or microwave signals, diffraction spreading can be significant. (See Figure 25.30.) To avoid this, we can increase D. This is done for laser light sent to the moon to measure its distance. The laser beam is expanded through a telescope to make D much larger and θ smaller.

25.7 THIN FILM INTERFERENCE

The bright colors seen in an oil slick floating on water or in a sunlit soap bubble are caused by interference. The brightest colors are those that interfere constructively. This interference is between light reflected from different surfaces of a thin film; thus, the effect is known as **thin film interference**. As noted before, interference effects are most prominent when light interacts with something having a size similar to its wavelength. A thin film is one having a thickness t smaller than a few times the wavelength of light, λ. Since color is associated with λ and since all interference depends in some way on the ratio of λ to the size of the object involved, we should expect to see different colors for different thicknesses of a film, as in Figure 25.1.

What causes thin film interference? Figure 25.31 shows how light reflected from the top and bottom surfaces of a film can interfere. Incident light is only partially reflected from the top surface of the film (ray 1). The remainder enters the film and is itself partially reflected from the bottom surface. Part of the light reflected from the bottom surface can emerge from the top of the film (ray 2) and interfere with light reflected from the top (ray 1). Since the ray that enters the film travels a greater distance, it may be in or out of phase with the ray reflected from the top. But consider for a moment, again, the bubbles in Figure 25.1. The bubbles are darkest where they are thinnest. Furthermore, if you observe a soap bubble carefully, you will note it gets dark at the point where it breaks. For very thin films, the difference in path lengths of rays 1 and 2 in Figure 25.31 is negligible; so why should they interfere destructively? The answer is that a phase change can occur upon reflection. The rule is as follows:

> **When light reflects from a medium having an index of refraction greater than that of the medium in which it is traveling, a 180° phase change (or a $\lambda/2$ shift) occurs.**

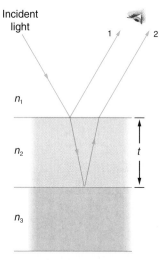

Figure 25.31 Light striking a thin film is partially reflected (ray 1) and partially refracted at the top surface. The refracted ray is partially reflected at the bottom surface and emerges as ray 2. These rays will interfere in a way that depends on the thickness of the film and the indices of refraction of the various media.

If the film in Figure 25.31 is a soap bubble (essentially water with air on both sides), then there is a $\lambda/2$ shift for ray 1 and none for ray 2. Thus when the film is very thin, the path length difference between the two rays is negligible, they are exactly out of phase, and destructive interference will occur at all wavelengths.

The thickness of the film relative to the wavelength of light is the other crucial factor in thin film interference. Ray 2 in Figure 25.31 travels a greater distance than ray 1. For light incident perpendicular to the surface, ray 2 travels a distance $2t$ farther than ray 1. When this distance is an integral or half-integral multiple of the wavelength in the medium,* constructive or destructive interference occurs, depending also on whether there is a phase change in either ray.

EXAMPLE 25.6 NONREFLECTIVE COATING USING THIN FILM INTERFERENCE

Sophisticated cameras use a series of several lenses. Light can reflect from the surfaces of these various lenses and create a haze on film that degrades image clarity. To limit these reflections, lenses are coated with a thin layer of magnesium fluoride that causes destructive thin film interference. What is the thinnest this film can be, if its index of refraction is

(continued)

---

*According to Equation 25.2, wavelength in a medium is $\lambda_n = \lambda/n$, where λ is the wavelength in vacuum and n is the index of refraction.

(continued)

1.38 and it is designed to limit the reflection of 550 nm light, normally the most intense visible wavelength? The index of refraction of glass is 1.52.

Strategy Referring to Figure 25.31, we see that $n_1 = 1.00$ for air, $n_2 = 1.38$, and $n_3 = 1.52$. Both rays 1 and 2 will have a $\lambda/2$ shift upon reflection. Thus, to obtain destructive interference, ray 2 will need to travel a half wavelength farther than ray 1. For rays incident perpendicularly, the path length difference is $2t$.

Solution To obtain destructive interference here,

$$2t = \frac{\lambda_n}{2}$$

where, according to Equation 25.2,

$$\lambda_n = \frac{\lambda}{n}$$

Thus,

$$2t = \frac{\lambda/n}{2}$$

Solving for t, and entering known values yields

$$t = \frac{\lambda/n}{4} = \frac{(550 \text{ nm})/1.38}{4}$$
$$= 99.6 \text{ nm}$$

Discussion Films such as the one in this example are most effective in producing destructive interference when the thinnest layer is used, since light over a broader range of incident angles will be reduced in intensity. These films are called nonreflective coatings; this is only an approximately correct description, though, since other wavelengths will only be partially canceled.

Thin film interference is its most constructive or most destructive when the path length difference for the two rays is an integral or half-integral wavelength, respectively. For rays incident perpendicularly, that is $2t = \lambda_n, 2\lambda_n, 3\lambda_n, \ldots$ or $2t = \lambda_n/2, 3\lambda_n/2, 5\lambda_n/2, \ldots$. To know whether interference is constructive or destructive, you must also determine if there is a phase change upon reflection. Thin film interference thus depends on film thickness, the wavelength of light, and the refractive indices. For white light incident on a film that varies in thickness, you will observe rainbows of constructive interference for various wavelengths as the thickness varies.

EXAMPLE 25.7 MORE THAN ONE THICKNESS CAN BE CONSTRUCTIVE

(a) What are the three smallest thicknesses of a soap bubble that produce constructive interference for red light having a wavelength of 650 nm? The index of refraction of soap is taken to be the same as that of water. (b) What three smallest thicknesses will give destructive interference?

Strategy and Concept Use Figure 25.31 to visualize the bubble. Note that $n_1 = n_3 = 1.00$ for air, and $n_3 = 1.333$ for soap (equivalent to water). There is thus a $\lambda/2$ shift for ray 1 reflected from the top surface of the bubble, and no shift for ray 2 reflected from the bottom surface. To get constructive interference, then, the path length difference ($2t$) must be a half-integral multiple of the wavelength—the first three being $\lambda_n/2, 3\lambda_n/2$, and $5\lambda_n/2$. To get destructive interference, the path length difference must be an integral multiple of the wavelength—the first three being 0, λ_n, and $2\lambda_n$.

Solution for (a) *Constructive interference* occurs here when

$$2t_c = \frac{\lambda_n}{2}, \frac{3\lambda_n}{2}, \frac{5\lambda_n}{2}, \ldots$$

The smallest constructive thickness t_c thus is

$$t_c = \frac{\lambda_n}{4} = \frac{\lambda/n}{4} = \frac{(650 \text{ nm})/1.333}{4}$$
$$= 122 \text{ nm}$$

The next thickness that gives constructive interference is $t'_c = 3\lambda_n/4$, so that

$$t'_c = 366 \text{ nm}$$

Finally, the third thickness producing constructive interference is $t''_c = 5\lambda_n/4$, so that

$$t''_c = 610 \text{ nm}$$

Solution for (b) For *destructive interference*, the path length difference here is an integral multiple of the wavelength. The first occurs for zero thickness, since there is a phase change at the top surface. That is,

$$t_d = 0$$

The first nonzero thickness producing destructive interference is

$$2t'_d = \lambda_n$$

Substituting known values gives

$$t'_d = \frac{\lambda_n}{2} = \frac{\lambda/n}{2} = \frac{(650 \text{ nm})/1.333}{2}$$
$$= 244 \text{ nm}$$

(continued)

(continued)

Finally, the third destructive thickness is $2t_d'' = 2\lambda_n$, so that

$$t_d'' = \lambda_n = \frac{\lambda}{n} = \frac{650 \text{ nm}}{1.333}$$

$$= 488 \text{ nm}$$

Discussion If the bubble were illuminated with pure

red light, we would see bright and dark bands at very uniform increases in thickness. First would be a dark band at 0 thickness, then bright at 122 nm thickness, then dark at 244 nm, bright at 366 nm, dark at 488 nm, and bright at 610 nm. If the bubble varied smoothly in thickness, like a smooth wedge, then the bands would be evenly spaced.

Another example of thin film interference can be seen when microscope slides are separated. (See Figure 25.32.) The slides are very flat, so that the wedge of air between them increases in thickness very uniformly. A phase change occurs at the second surface but not the first, and so there is a dark band where the slides touch. The rainbows of constructive interference repeat, going from violet to red again and again as the distance between the slides increases. As the layer of air increases, the bands become more difficult to see, because slight changes in incident angle have greater effects on path length differences. If pure-wavelength light instead of white light is used, then bright and dark bands are obtained rather than repeating rainbows.

An important application of thin film interference is found in the manufacture of optical instruments. As seen in Figure 25.33, a lens or mirror can be compared with a master as it is being ground, allowing it to be shaped to an accuracy of less than a wavelength over its entire surface. The circular bands in Figure 25.33 are called Newton's rings, because Isaac Newton described them and their use in detail. Newton did not discover them (Robert Hooke did), and Newton did not believe they were due to the wave character of light.

(a)

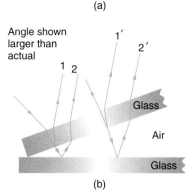

(b)

Figure 25.32 (a) The rainbow bands are produced by thin film interference in the air between the glass slides. (b) Schematic of the paths taken by rays in the wedge of air between the slides.

PROBLEM-SOLVING STRATEGIES

FOR WAVE OPTICS

Step 1. Examine the situation to determine that interference is involved. Identify whether slits or thin film interference are considered in the problem.

Step 2. If slits are involved, note that diffraction gratings and double slits produce very similar interference patterns, but that gratings have narrower (sharper) maxima. Single slit patterns are characterized by a large central maximum and smaller maxima to the sides.

Step 3. If thin film interference is involved, take note of the path length difference between the two rays that interfere. Be certain to use the wavelength in the medium involved, since it differs from the wavelength in vacuum. *Note also that there is an*

Figure 25.33 Newton's rings are observed when a lens being ground is placed over a master blank. If the lens were perfect, there would be no rings. Each successive ring of a given color indicates an increase of only one wavelength in the distance between the lens and the blank, so that great precision can be obtained.

(a)   (b)

Figure 25.34 These two photographs of a river show the effect of a polarizing filter in reducing glare in light reflected from the surface. There is less glare in (a) than in (b), since (a) was taken with a polarizing filter and (b) was not.

additional λ/2 phase shift when light reflects from a medium with a greater index of refraction.

Step 4. *Identify exactly what needs to be determined in the problem (identify the unknowns). A written list is useful.*

Step 5. *Make a list of what is given or can be inferred from the problem as stated (identify the knowns).*

Step 6. *Solve the appropriate equation for the quantity to be determined (the unknown), and enter the knowns. Slits, gratings, and the Rayleigh limit involve equations.*

Step 7. *For thin film interference, you will have constructive interference for a total shift that is an integral number of wavelengths, destructive interference for a total shift of a half-integral number of wavelengths.* Always keep in mind that crest to crest is constructive whereas crest to trough is destructive.

Step 8. *Check to see if the answer is reasonable: Does it make sense?* Angles in interference patterns cannot be greater than 90°, for example.

25.8 POLARIZATION

Polaroid sunglasses are familiar to most of us. They have a special ability to cut the glare of light reflected from water or glass. (See Figure 25.34.) Polaroids have this ability because of a wave characteristic of light called polarization. What is polarization? How is it produced? What are some of its uses? The answers to these questions are related to the wave character of light.

Light is one type of electromagnetic (EM) wave. As noted in Chapter 23, most EM waves are *transverse waves* consisting of varying electric and magnetic fields that oscillate perpendicular to the direction of propagation. (See Figure 25.35.) There are specific directions for the oscillations of the electric and magnetic fields. **Polarization** is the attribute that a wave's oscillations have a definite direction relative to the direction of propagation of the wave.* Waves having such a direction are said to be **polarized**. For an EM wave, we define the **direction of polarization** to be the direction parallel to the electric field. Thus we can think of the electric field arrows as showing the direction of polarization, as in Figure 25.35.

To examine this further, consider the transverse waves in the ropes shown in Figure 25.36. The oscillations in one rope are in a vertical plane and are said to be **vertically polarized**. Those in the other rope are in a horizontal plane and are **horizontally polarized**. If a vertical slit is placed on the first rope, the waves pass through. But a vertical slit blocks the horizontally polarized waves. For EM waves, the direction of the electric field is analogous to the disturbances on the ropes.

The sun, and many other light sources, produce waves that are randomly polarized. (See Figure 25.37.) Such light is said to be **unpolarized**, because it is composed of many

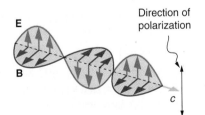

Figure 25.35 An EM wave, such as light, is a transverse wave. The electric and magnetic fields are perpendicular to the direction of propagation.

*This is not the same type of polarization as that discussed for the separation of charges in Chapter 17.

Figure 25.36 The transverse oscillations in one rope are in a vertical plane, and those in the other rope are in a horizontal plane. The first is said to be vertically polarized, and the other to be horizontally polarized. Vertical slits pass vertically polarized waves and block horizontally polarized waves.

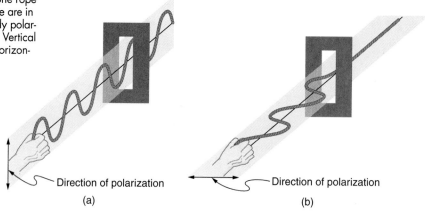

Direction of polarization

(a)

Direction of polarization

(b)

Random polarization

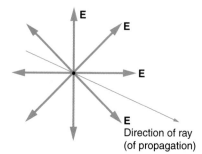

Direction of ray (of propagation)

Figure 25.37 This represents a ray of unpolarized light. The arrows represent the directions of the polarizations of the individual waves composing the ray. The light being unpolarized, the arrows point in all directions.

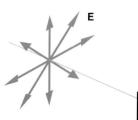

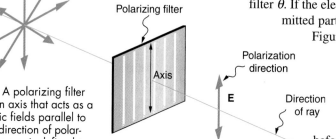

Polarizing filter

Axis

Polarization direction

E

Direction of ray

Figure 25.38 A polarizing filter has a polarization axis that acts as a slit passing electric fields parallel to its direction. The direction of polarization of an EM wave is defined to be the direction of its electric field.

waves with all possible directions of polarization. Polaroid materials, invented by the founder of Polaroid Corporation, Edwin Land, act as a *polarizing* slit for light, allowing only polarization in one direction to pass through. Polarizing filters are composed of long molecules aligned in one direction. Thinking of the molecules as many slits, analogous to those for the oscillating ropes, we can understand why only light with a specific polarization can get through. The **axis of a polarizing filter** is the direction along which the filter passes the electric field of an EM wave. (See Figure 25.38.)

Figure 25.39 shows the effect of two polarizing filters on originally unpolarized light. The first polarizes the light along its axis. When the axes of the first and second filters are aligned (parallel), then all of the polarized light passed by the first is passed by the second. If the second polarizing filter is rotated, only the component of the light parallel to the second filter's axis is passed. When the axes are perpendicular, no light is passed by the second.

Only the component of the EM wave parallel to the axis of a filter is passed. Let us call the angle between the direction of polarization and the axis of a filter θ. If the electric field has an amplitude E, then the transmitted part of the wave has an amplitude $E \cos \theta$. (See Figure 25.40.) Since the intensity of a wave is proportional to its amplitude squared, the intensity I of the transmitted wave is related to the incident wave by

$$I = I_0 \cos^2 \theta \qquad (25.8)$$

where I_0 is the intensity of the polarized wave before passing through the filter. (Equation 25.8 is known as Malus's law.)

EXAMPLE 25.8 INTENSITY REDUCTION BY A POLARIZING FILTER

What angle is needed between the direction of polarized light and the axis of a polarizing filter to reduce its intensity by 90.0%?

Strategy When the intensity is reduced by 90.0%, it is 10.0% or 0.100 times its original value. That is, $I = 0.100 I_0$. Using this information, Equation 25.8 can be used to solve for the needed angle.

Solution Equation 25.8 states that $I = I_0 \cos^2 \theta$. Solving for $\cos \theta$ and substituting the relation between I and I_0 gives

$$\cos \theta = \sqrt{\frac{I}{I_0}} = \sqrt{\frac{0.100 I_0}{I_0}} = 0.316$$

Solving for θ yields

$$\theta = \cos^{-1} 0.316 = 71.6°$$

Discussion A fairly large angle between the direction of polarization and the filter axis is needed to reduce the intensity to 10.0% of its original value. This seems reasonable based on experimenting with polarizing films. It is interesting that at an angle of 45°, the intensity is reduced to 50% of its original value (it is left as an end-of-chapter problem to show this). Note that 71.6° is 18.4° from reducing the intensity to zero, and that at an angle of 18.4° the intensity is reduced to 90.0% of its original value (also left as an end-of-chapter problem), giving evidence of symmetry.

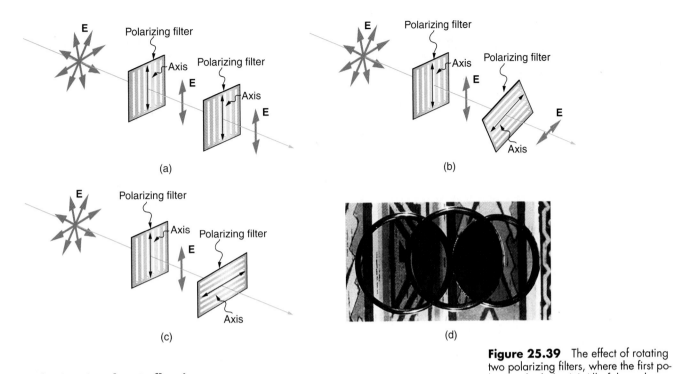

(a)

(b)

(c)

(d)

Figure 25.39 The effect of rotating two polarizing filters, where the first polarizes the light. (a) All of the polarized light is passed by the second polarizing filter, because its axis is parallel to the first. (b) As the second is rotated, only part of the light is passed. (c) When the second is perpendicular to the first, no light is passed. (d) In this photograph a polarizing filter is placed above two others. Its axis is perpendicular to the filter on the right (dark) and parallel to the filter on the left (lighter).

Polarization by Reflection

By now you can probably guess that Polaroid sunglasses cut the glare in reflected light because that light is polarized. You can check this for yourself by holding Polaroid sunglasses in front of you and rotating them while looking at light reflected from water or glass. As you rotate the sunglasses, you will notice the light gets bright and dim, but not completely black. This implies the reflected light is partially polarized and cannot be completely blocked by a polarizing filter. Figure 25.41 illustrates what happens when unpolarized light is reflected from a surface. For simplicity of drawing, only vertical and horizontal polarizations are shown (vector combinations of which can form polarizations in any direction). Vertically polarized light is preferentially refracted into the surface, so that the reflected light is left more horizontally polarized. The reasons for this phenomenon are beyond the scope of this text, but a convenient mnemonic for remembering this is to imagine the polarization direction is like an arrow. Vertical polarization would be like an arrow perpendicular to the surface and would be more likely to stick and not be reflected. Horizontal polarization is like an arrow bouncing on its side and would be more likely to be reflected. Sunglasses with vertical axes would then block more reflected light than unpolarized light from other sources.

Since the part of the light that is not reflected is refracted, the amount of polarization depends on the indices of refraction of the media involved. It can be shown that

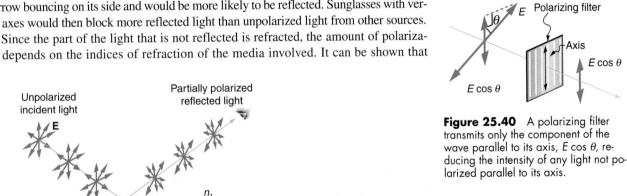

Figure 25.40 A polarizing filter transmits only the component of the wave parallel to its axis, $E \cos \theta$, reducing the intensity of any light not polarized parallel to its axis.

Unpolarized incident light

E

Partially polarized reflected light

n_1

n_2

Partially polarized refracted light

Figure 25.41 Polarization by reflection. Unpolarized light reflecting from a surface has equal amounts of vertical and horizontal polarization. The vertical components are preferentially absorbed or refracted, leaving the reflected light more horizontally polarized. This is akin to arrows striking on their sides bouncing off while arrows striking on their tips go into the surface.

THINGS GREAT AND SMALL

Atomic Explanation of Polarizing Filters

Polarizing filters have a polarization axis that acts as a slit. This slit passes electromagnetic waves (often visible light) that have an electric field parallel to the axis. This is accomplished with long molecules aligned perpendicular to the axis as shown in Figure 25.A.

Figure 25.B illustrates how the component of the electric field parallel to the long molecules is absorbed. An electromagnetic wave is composed of oscillating electric and magnetic fields. The electric field is strong compared with the magnetic (as discussed in Chapter 23) and is more effective

in exerting forces on charges in the molecules. The charged particles most affected are the electrons in the molecules, since electron masses are small. If the electron is forced to oscillate, it can absorb energy from the EM wave. This reduces the fields in the wave and, hence, reduces its intensity. In long molecules, electrons can more easily oscillate parallel to the molecule than perpendicular. The electrons are bound to the molecule and are more restricted in their movement perpendicular to the molecule. The electrons can, thus, absorb EM waves that have a component of their electric field parallel to the molecule. The electrons are much less responsive to electric fields perpendicular to the molecule and will allow those fields to pass. Thus the axis of the polarizing filter is perpendicular to the length of the molecule.

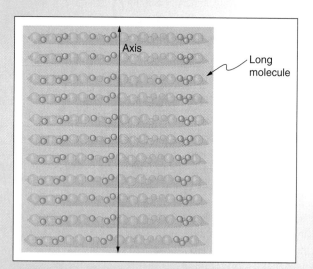

Figure 25.A Long molecules are aligned perpendicular to the axis of a polarizing filter. The component of the electric field in an EM wave perpendicular to these molecules passes through the filter, while the component parallel to the molecules is absorbed.

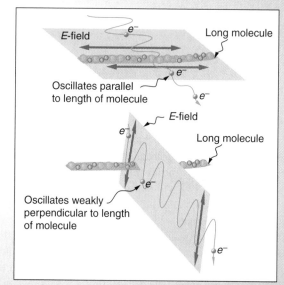

Figure 25.B Artist's conception of an electron in a long molecule being oscillated parallel to the molecule. Oscillation of the electron absorbs energy and reduces the intensity of the component of the EM wave that is parallel to the molecule.

reflected light is completely polarized at a angle of reflection θ_b, given by

$$\tan \theta_b = \frac{n_2}{n_1} \tag{25.9}$$

where n_2 is the index of refraction of the medium from which the light is reflected and n_1 is the index of refraction of the medium in which the reflected light travels. This equation is known as **Brewster's law**, and θ_b is known as **Brewster's angle** for the 19th-century Scottish physicist who discovered them.

EXAMPLE 25.9   POLARIZATION BY REFLECTION

(a) At what angle will light traveling in air be completely horizontally polarized when reflected from water? (b) From glass?

Strategy All we need to solve these problems are the indices of refraction. Air has $n_1 = 1.00$, water has $n_2 = 1.333$, and crown glass has $n_2' = 1.52$. Equation 25.9 can be directly applied to find θ_b in each case.

Solution for (a) Putting the known quantities into

Equation 25.9 gives

$$\tan \theta_b = \frac{n_2}{n_1} = 1.333$$

Solving for the angle θ_b yields

$$\theta_b = \tan^{-1} 1.333 = 53.1°$$

(continued)

(continued)

Solution to (b) Similarly,

$$\tan \theta_b' = \frac{n_2'}{n_1} = 1.52$$

Thus,

$$\theta_b' = \tan^{-1}1.52 = 56.7°$$

Discussion Light reflected at these angles could be completely blocked by a good polarizing filter held with its axis vertical. Brewster's angle for water and air are similar to those for glass and air, so that sunglasses are equally effective for light reflected from either water or glass under similar circumstances. Light not reflected is refracted into these media. So at an incident angle equal to Brewster's angle, the refracted light will be slightly vertically polarized. It will not be completely vertically polarized, because only a small fraction of the incident light is reflected, and so a significant amount of horizontally polarized light is refracted.

Polarization by Scattering

If you hold your Polaroid sunglasses in front of you and rotate them while looking at blue sky, you will see the sky get bright and dim. This is a clear indication that light scattered by air is partially polarized. Figure 25.42 helps illustrate how this happens. Since light is a transverse EM wave, it vibrates the electrons of air molecules perpendicular to the direction it is traveling. The electrons then radiate like small antennae. Since they are oscillating perpendicular to the direction of the light ray, they create EM radiation that is polarized perpendicular to the direction of the ray. When viewing the light along a line perpendicular to the original ray, as in Figure 25.42, there can be no polarization in the scattered light parallel to the original ray, because that would require the original ray to be a longitudinal wave. Along other directions, a component of the other polarization can be projected along the line of sight, and the scattered light will only be partially polarized. Furthermore, multiple scattering can bring light to your eyes from other directions and can contain different polarizations.

Photographs of the sky can be darkened by polarizing filters, a trick used by many photographers to make clouds brighter by contrast. Scattering from other particles, such as smoke or dust, can also polarize light. Detecting polarization in scattered EM waves can be a useful analytical tool in determining the scattering source.

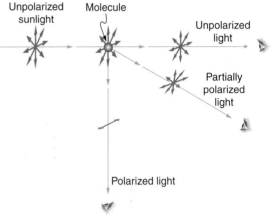

Figure 25.42 Polarization by scattering. Unpolarized light scattering from air molecules shakes their electrons perpendicular to the direction of the original ray. The scattered light therefore has a polarization perpendicular to the original direction and none parallel to the original direction.

Liquid Crystals and Other Polarization Effects in Materials

While you are undoubtedly aware of liquid crystal displays (LCDs) found in watches, calculators, laptop computer screens, and myriad other places, you are probably not aware that they are based on polarization. Liquid crystals are so named because their molecules can be aligned even though they are in a liquid. Liquid crystals have the property that they can rotate the polarization of light passing through them by 90°. Furthermore, this property can be turned off by the application of a voltage, as illustrated in Figure 25.43. It is possible to manipulate this characteristic quickly and in small well-defined regions to

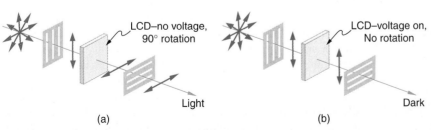 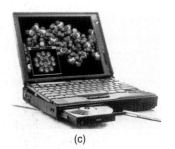

Figure 25.43 (a) Polarized light is rotated 90° by a liquid crystal and then passed by a polarizing filter that has its axis perpendicular to the original polarization direction. (b) When a voltage is applied to the liquid crystal, the polarized light is not rotated and is blocked by the filter, making the region dark in comparison with its surroundings. (c) LCDs can be made color specific, small, and fast enough to use in laptop computers.

Figure 25.44 Optical activity is the ability of some substances to rotate the plane of polarization of light passing through them. The rotation is detected with a polarizing filter or analyzer.

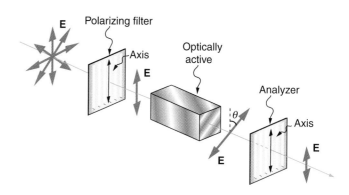

create the contrast patterns we see in so many LCD devices. There are now LCD televisions that have the advantage of a thin, low-voltage display system.

Many crystals and solutions rotate the plane of polarization of light passing through them. Such substances are said to be **optically active**. Examples include sugar water, insulin, and collagen. (See Figure 25.44.) In addition to depending on the type of substance, the amount and direction of rotation depends on a number of factors. Among these are the concentration of the substance, the distance the light travels through it, and the wavelength of light. Optical activity is due to the asymmetric shape of molecules in the substance, such as being helical. Measurements of the rotation of polarized light passing through substances can thus be used to measure concentrations, a standard technique for sugars. It can also give information on the shapes of molecules, such as proteins, and factors that affect their shapes, such as temperature and pH.

Glass and plastic become optically active when stressed; the greater the stress, the greater the effect. Optical stress analysis on complicated shapes can be performed by making plastic models of them and observing them through crossed filters, as seen in Figure 25.45. It is apparent that the effect depends on wavelength as well as stress. The wavelength dependence is sometimes also used for artistic purposes.

Another interesting phenomenon associated with polarized light is the ability of some crystals to split an unpolarized beam of light in two. Such crystals are said to be **birefringent**. (See Figure 25.46.) Each of the separated rays has a specific polarization. One behaves normally and is called the ordinary ray, whereas the other does not obey Snell's law and is called the extraordinary ray. Birefringent crystals can be used to produce polarized beams from unpolarized light. Some birefringent materials preferentially absorb one of the polarizations. These materials are called dichroic and can produce polarization by this preferential absorption. This is fundamentally how polarizing filters and other polarizers work. The interested reader is invited to further pursue the numerous properties of materials related to polarization.

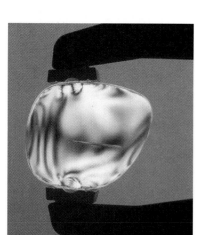

Figure 25.45 Optical stress analysis of a plastic lens placed between crossed polarizers. The polarization of light passed through the stressed lens is rotated in proportion to local stress. It is greatest where lines are closest.

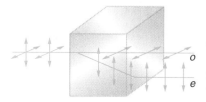

Figure 25.46 Birefringent materials, such as the common mineral calcite, split unpolarized beams of light into two. The ordinary ray behaves as expected, but the extraordinary ray does not obey Snell's law.

25.9* MICROSCOPY ENHANCED BY THE WAVE CHARACTERISTICS OF LIGHT

The use of microscopes (microscopy) to observe small details is limited by the wave nature of light. Because light diffracts significantly around small objects, it becomes impossible to observe details significantly smaller than the wavelength of light. One rule of thumb has it that details smaller than about λ are difficult to observe. Radar, for example, can detect the size of an aircraft, but not its individual rivets, since the wavelength of most radar is several centimeters or greater. Similarly, visible light cannot detect individual atoms, since they are about 0.1 nm in size and visible wavelengths range from 380 to 760 nm. Ironically, special techniques used to obtain the best possible resolution with microscopes take advantage of the same wave characteristics of light that ultimately limit the detail.

The most obvious method of obtaining better detail is to utilize shorter wavelengths. **Ultraviolet (UV) microscopes** have been constructed with special lenses that pass UV rays and utilize photographic or electronic techniques to record images. The shorter UV wavelengths allow somewhat greater detail to be observed, but drawbacks, such as the hazard of UV to living tissue and the need for special detection devices and lenses (which tend to be dispersive in the UV), severely limit the use of UV microscopes. In later chapters, we will explore practical uses of very short wavelength EM waves, such as x rays, and

CONNECTIONS

Waves

All attempts to observe the size and shape of objects are limited by the wavelength of the probe. Sonar and medical ultrasound are limited by the wavelength of sound they employ. We shall see that this is also true in electron microscopy, since electrons have a wavelength. Heisenberg's uncertainty principle asserts that this limit is fundamental and inescapable, as we shall see in quantum mechanics.

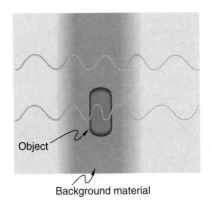

Figure 25.47 Light rays passing through a sample under a microscope will emerge with different phases depending on their paths. The object shown has a greater index of refraction than the background, and so there are more wavelengths in the ray passing through it. Superimposing these rays produces interference that varies with path, enhancing contrast between the object and background.

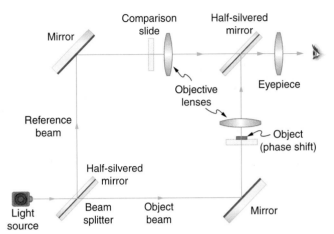

Figure 25.48 An interference microscope utilizes interference between the reference and object beam to enhance contrast. The two beams are split by a half-silvered mirror; the object beam is sent through the object, and the reference beam is sent through otherwise identical optical elements. The beams are recombined by another half-silvered mirror, and the interference depends on the various phases emerging from different parts of the object, enhancing contrast.

other short-wavelength probes, such as electrons in electron microscopes, to detect small details.

Another difficulty in microscopy is the fact that many microscopic objects do not absorb much of the light passing through them. The lack of contrast makes image interpretation very difficult. **Contrast** is the difference in intensity between objects and the background on which they are observed. Stains are commonly employed to enhance contrast, but these tend to be application specific. More general wave interference techniques can be used to produce contrast. Figure 25.47 shows the passage of light through a sample. Since the indices of refraction differ, the number of wavelengths in the paths differ. Light emerging from the object is thus out of phase with light from the background and will interfere differently, producing enhanced contrast, especially if the light is coherent and monochromatic.

Interference microscopes produce contrast between objects and background by superimposing a reference beam of light upon the light emerging from the sample. Since light from the background and objects differ in phase, there will be different amounts of constructive and destructive interference, producing the desired contrast in final intensity. Figure 25.48 shows schematically how this is done. Parallel rays of light from a source are split into two beams by a half-silvered mirror. These beams are called the object and reference beams. Each beam passes through identical optical elements, except that the object beam passes through the object we wish to observe microscopically. The light beams are recombined by another half-silvered mirror and interfere. Since the light rays passing through different parts of the object have different phases, there will be significantly different interference and, hence, greater contrast between them.

Another type of microscope utilizing wave interference and differences in phases to enhance contrast is called the **phase-contrast microscope**. While its principle is the same as the interference microscope, the phase-contrast microscope is simpler to use and construct. Its impact (and the principle upon which it is based) was so important that its developer, the Dutch physicist Frits Zernike (1888–1966), was awarded the Nobel prize in 1953. Figure 25.49 shows the enhancement in contrast obtained with a phase-contrast microscope

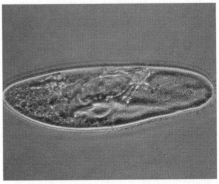

(a)

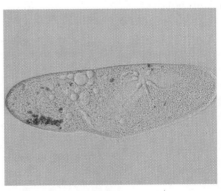

(b)

Figure 25.49 (a) An unstained image of a paramecium as seen with a conventional microscope. (b) The same object viewed with a phase-contrast microscope shows much greater contrast.

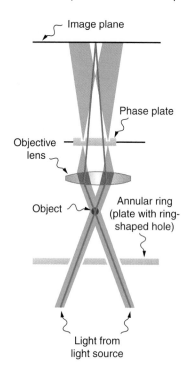

Image plane

Phase plate

Objective lens

Object

Annular ring (plate with ring-shaped hole)

Light from light source

Figure 25.50 Simplified construction of a phase-contrast microscope. Phase differences between light passing through the object and background are produced by passing the rays through different parts of a phase plate. The light rays are superimposed in the image plane, producing contrast due to their interference.

Figure 25.51 Polarizing microscopes produce brilliant color and high contrast for objects that are optically active; shown here is a photo of a superconducting ceramic.

compared with a standard microscope. Figure 25.50 shows the basic construction of a phase-contrast microscope. Phase differences between light passing through the object and background are produced by passing the rays through different parts of a phase plate (so called because it shifts the phase of the light passing through it). These two light rays are superimposed in the image plane, producing contrast due to their interference.

A **polarization microscope** also enhances contrast by utilizing a wave characteristic of light. Polarization microscopes are useful for objects that are optically active or birefringent, particularly if those characteristics vary from place to place in the object. Polarized light is sent through the object and then observed through a polarizing filter that is perpendicular to the original polarization direction. Nearly transparent objects can then appear with strong color and in high contrast. (See Figure 25.51.) Many polarization effects are wavelength dependent, producing color in the processed image. Contrast results from the action of the polarizing filter in passing only components parallel to its axis.

SUMMARY

Wave optics is the branch of optics that must be used when light interacts with small objects or whenever the wave characteristics of light are considered. Wave characteristics are those associated with interference and diffraction.

Visible light is the type of electromagnetic wave to which our eyes respond, having wavelengths in the range from 380 to 760 nm. Like all EM waves, the following relationship is valid in a vacuum:

$$c = f\lambda \qquad (25.1)$$

where $c = 3.00 \times 10^8$ m/s is the speed of light, f is the frequency of the electromagnetic wave, and λ is its wavelength in a vacuum. The **wavelength λ_n of light in a medium** having an index of refraction n is

$$\lambda_n = \frac{\lambda}{n} \qquad (25.2)$$

Its frequency is the same as in a vacuum.

An accurate technique for determining how and where waves propagate is given by **Huygens's principle**:

> **Every point on a wave front is a source of wavelets that spread out in the forward direction at the same speed as the wave itself. The new wave front is a line tangent to all of the wavelets.**

Diffraction is the bending of a wave around the edges of an opening or other obstacle.

Young's double slit experiment gave definitive proof of the wave character of light. An interference pattern is obtained by the superposition of light from two slits. There is **constructive interference** when

$$d \sin \theta = m\lambda, \qquad m = 0, 1, 2, \dots \quad \text{(constructive)} \quad (25.3)$$

where d is the distance between the slits, θ is the angle relative to the incident direction, and m is the **order**. There is **destructive interference** when

$$d \sin \theta = (m + \tfrac{1}{2})\lambda, \quad m = 0, 1, 2, \ldots \quad \text{(destructive)} \quad \textbf{(25.4)}$$

A **diffraction grating** is a large collection of evenly spaced parallel slits that produces an interference pattern similar to but sharper than that of a double slit. There is **constructive interference for a diffraction grating** when

$$d \sin \theta = m\lambda, \quad m = 0, 1, 2, \ldots \quad \text{(constructive)} \quad \textbf{(25.5)}$$

where d is the distance between slits in the grating.

A single slit produces an interference pattern characterized by a broad central maximum with narrower and dimmer maxima to the sides. There is **destructive interference for a single slit** when

$$D \sin \theta = m\lambda, \quad m = 1, 2, 3, \ldots \quad \text{(destructive)} \quad \textbf{(25.6)}$$

where D is the slit width. Note that there is no $m = 0$ minimum.

Diffraction limits resolution. For a circular aperture, lens, or mirror, the **Rayleigh criterion** states that *two images are just resolvable when the center of the diffraction pattern of one is directly over the first minimum of the diffraction pattern of the other.* This occurs for two point objects separated by the angle

$$\theta = 1.22 \frac{\lambda}{D} \quad \textbf{(25.7)}$$

where λ is the wavelength of light (or other electromagnetic radiation) and D is the diameter of the aperture, lens, mirror, or the like. This equation also gives the angular spreading of a source of light having a diameter D.

Thin film interference occurs between the light reflected from the top and bottom surfaces of a film. In addition to the path length difference, there can be a phase change:

When light reflects from a medium having an index of refraction greater than that of the medium in which it is traveling, a 180° phase change (or a $\lambda/2$ shift) occurs.

Polarization is the attribute that a wave's oscillations have a definite direction relative to the direction of propagation of the wave. EM waves are *transverse waves* that may be polarized. The **direction of polarization** is defined to be the direction parallel to the electric field of the EM wave. **Unpolarized** light is composed of many rays having random polarization directions. Light can be **polarized** by passing it through a polarizing filter or other polarizing material. The intensity I of polarized light after passing through a polarizing filter is

$$I = I_0 \cos^2\theta \quad \textbf{(25.8)}$$

where I_0 is the original intensity and θ is the angle between the direction of polarization and the axis of the filter. Polarization is also produced by reflection. Brewster's law states that **reflected light will be completely polarized** at the angle of reflection θ_b, known as **Brewster's angle**, given by a statement known as **Brewster's law**:

$$\tan \theta_b = \frac{n_2}{n_1} \quad \textbf{(25.9)}$$

where n_2 is the index of refraction of the medium from which the light is reflected and n_1 is the index of refraction of the medium in which the reflected light travels. Polarization can also be created by scattering. There are a number of types of **optically active** substances that rotate the polarization direction of light.

To improve microscope images, various techniques utilizing the wave characteristics of light have been developed. Many of these enhance contrast with interference effects.

CONCEPTUAL QUESTIONS

25.1 What type of experimental evidence indicates that light is a wave?

25.2 Give an example of a wave characteristic of light that is easily observed outside the laboratory.

25.3 How do wave effects depend on the size of the object with which the wave interacts? For example, why does sound bend around the corner of a building while light does not?

25.4 Under what conditions does light behave like a ray? Like a wave?

25.5 Go outside in the sunlight and observe your shadow. It has fuzzy edges even if you do not. Is this a diffraction effect? Explain.

25.6 Why does the wavelength of light decrease when it passes from vacuum into a medium? State which attributes change and which stay the same and, thus, require wavelength to decrease.

25.7 Does Huygens's principle apply to all types of waves?

25.8 Young's double slit experiment breaks a single light beam into two sources. Would the same pattern be obtained for two independent sources of light, such as the headlights of a distant car? Explain.

25.9 Suppose you use the same double slit to perform Young's double slit experiment in air and then repeat the experiment in water. Do the angles to the same parts of the interference pattern get larger or smaller? Does the color of the light change? Explain.

25.10 Is it possible to create a situation in which there is only destructive interference? Explain.

25.11 What is the advantage of a diffraction grating over a double slit in dispersing light into a spectrum?

25.12 What are the advantages of a diffraction grating over a prism in dispersing light for spectral analysis?

25.13 Can the lines in a diffraction grating be too close together to be useful as a spectroscopic tool for visible light? If so, what type of EM radiation would the grating be suitable for? Explain.

25.14 If a beam of white light passes through a diffraction grating with vertical lines, the light is dispersed into rainbows on the right and left. If a glass prism disperses white light to the right into a rainbow, how does the sequence of colors compare with that produced on the right by a diffraction grating?

25.15 Suppose pure-wavelength light falls on a diffraction grating. What happens to the interference pattern if the same light falls on a grating that has more lines per centimeter? What happens to the interference pattern if the same grating has a longer-wavelength light fall on it? Explain how these two effects are consistent in terms of the relationship of wavelength to the distance between slits.

25.16 It is possible for there to be no minimum in the interference pattern of a single slit. Explain why. Is the same true of double slits and diffraction gratings?

25.17 Look at a light, such as a street lamp, through the narrow gap between two fingers held close together. What type of pattern is this? Is it more distinct for a monochromatic source, such as the yellow light from a sodium vapor lamp, than for an incandescent lamp?

25.18 Figure 25.52 shows the central part of the interference pattern for a pure wavelength of red light projected onto a double slit. The pattern is actually a combination of single slit and double slit interference. Note that the bright spots are evenly spaced. Is this a double slit or single slit characteristic? Note that some of the bright spots are dim on either side of the center. Is this a single slit or double slit characteristic? Which is smaller, the slit width or the separation between slits? Explain your responses.

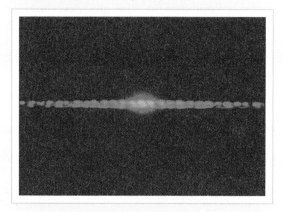

Figure 25.52 This double slit interference pattern also shows signs of single slit interference. Question 18.

25.19 Why do you expect a bright dot in the center of a symmetric shadow like that in Figure 25.25(a)?

25.20 A beam of light always spreads out. Why can't a beam be created with parallel rays to prevent spreading? Why can't lenses, mirrors, or apertures be used to correct the spreading?

25.21 How is the difference in paths taken by two originally in-phase light waves related to whether they interfere constructively or destructively? How can this be affected by reflection? By refraction?

25.22 Under what circumstances is the phase of light changed by reflection? Is the phase related to polarization?

25.23 Is there a phase change in the light reflected from either surface of a contact lens floating on a person's tear layer? The index of refraction of the lens is about 1.5, and its top surface is dry.

25.24 In placing a sample on a microscope slide, a glass cover is placed over a water drop on the glass slide. Light incident from above can reflect from the top and bottom of the glass cover and the glass slide below the water drop. At which surfaces will there be a phase change in the reflected light?

25.25 Answer Question 25.24 if the fluid between the two pieces of crown glass is carbon disulfide.

25.26 While contemplating the food value of a slice of ham, you notice a rainbow of color reflected from its moist surface. Explain its origin.

25.27 An inventor notices that a soap bubble is dark at its thinnest and realizes that destructive interference is taking place for all wavelengths. How could she use this knowledge to make a nonreflective coating for lenses that is effective at all wavelengths? That is, what limits would there be on the index of refraction and thickness of the coating? How might this be impractical?

25.28 A nonreflective coating like the one described in Example 25.6 works ideally for a single wavelength and for perpendicular incidence. What happens for other wavelengths and other incident directions? Be specific.

25.29 Why is it much more difficult to see interference fringes for light reflected from a thick piece of glass than from a thin film? Would it be easier if monochromatic light were used?

25.30 Can a sound wave in air be polarized? Explain.

25.31 No light passes through two perfect polarizing filters with perpendicular axes. However, if a third polarizing filter is placed between the original two, some light can pass. Why is this? Under what circumstance is the most light passed?

25.32 When particles scattering light are much smaller than its wavelength, the amount of scattering is proportional to $1/\lambda^4$. Does this mean there is more scattering for small λ than large λ? How does this relate to the fact that the sky is blue?

25.33 Using the information given in the preceding question, explain why sunsets are red.

PROBLEMS

Section 25.1 The Wave Aspect of Light

25.1 Show that when light passes from air to water, its wavelength decreases to 0.750 times its original value.

25.2 Find the range of visible wavelengths of light in crown glass.

25.3 What is the index of refraction of a material for which the wavelength of light is 0.671 times its value in a vacuum? Identify the likely substance.

25.4 Analysis of an interference effect in a clear solid shows that the wavelength of light in it is 329 nm. Knowing this light comes from a He-Ne laser and has a wavelength of 633 nm in air, is the substance zircon or diamond?

• **25.5** What is the ratio of thicknesses of crown glass and water that would contain the same number of wavelengths of light?

Section 25.3 The Double Slit

25.6 At what angle is the first-order maximum for 450 nm wavelength blue light falling on double slits separated by 0.0500 mm?

25.7 Calculate the angle for the third-order maximum of 580 nm wavelength yellow light falling on double slits separated by 0.100 mm.

25.8 What is the separation between two slits for which 610 nm orange light has its first maximum at an angle of 30.0°?

25.9 Find the distance between two slits that produces the first minimum for 410 nm violet light at an angle of 45.0°.

25.10 Calculate the wavelength of light that has its third minimum at an angle of 30.0° when falling on double slits separated by 3.00 μm.

25.11 What is the wavelength of light falling on double slits separated by 2.00 μm, if the third-order maximum is at an angle of 60.0°?

• **25.12** At what angle is the fourth-order maximum for the situation in Problem 25.6?

• **25.13** What is the highest-order maximum for 400 nm light falling on double slits separated by 25.0 μm?

• **25.14** Find the largest wavelength of light falling on double slits separated by 1.20 μm for which there is a first-order maximum. Is this in the visible part of the spectrum?

• **25.15** What is the smallest separation between two slits that will produce a second-order maximum for 720 nm red light?

‡ **25.16** (a) What is the smallest separation between two slits that will produce a second-order maximum for any visible light? (b) For all visible light?

‡ **25.17** (a) If the first-order maximum for pure-wavelength light falling on a double slit is at an angle of 10.0°, at what angle is the second-order maximum? (b) What is the angle of the first minimum? (c) What is the highest-order maximum possible here?

‡ **25.18** Figure 25.53 shows a double slit located a distance *x* from a screen, with the distance from the center of the screen given by *y*. When the distance *d* between the slits is relatively large, there will be numerous bright spots, called fringes. Show that, for small angles (where $\sin \theta \approx \theta$, with θ in radians), the distance between fringes is given by $\Delta y = x\lambda/d$.

‡ **25.19** Using the result of Problem 25.18, calculate the distance between fringes for 633 nm light falling on double slits separated by 0.0800 mm, located 3.00 m from a screen as in Figure 25.53.

‡ **25.20** Using the result of Problem 25.18, find the wavelength of light that produces fringes 7.50 mm apart on a screen 2.00 m from double slits separated by 0.120 mm. (See Figure 25.53.)

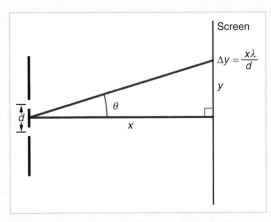

Figure 25.53 The distance between adjacent fringes is $\Delta y = x\lambda/d$, assuming the slit separation *d* is large compared with λ. Problems 18, 19, and 20.

Section 25.4 Multiple Slits (Diffraction Gratings)

25.21 A diffraction grating has 2000 lines per centimeter. At what angle will the first-order maximum be for 520 nm wavelength green light?

25.22 Find the angle for the third-order maximum for 580 nm wavelength yellow light falling on a diffraction grating having 1500 lines per centimeter.

25.23 How many lines per centimeter are there on a diffraction grating that gives a first-order maximum for 470 nm blue light at an angle of 25.0°?

25.24 What is the distance between lines on a diffraction grating that produces a second-order maximum for 760 nm red light at an angle of 60.0°?

25.25 Calculate the wavelength of light that has its second-order maximum at 45.0° when falling on a diffraction grating that has 5000 lines per centimeter.

25.26 An electric current through hydrogen gas produces several distinct wavelengths of visible light. What are the wavelengths of the hydrogen spectrum, if they form first-order maxima at angles of 24.2°, 25.7°, 29.1°, and 41.0° when projected on a diffraction grating having 10,000 lines per centimeter?

25.27 The yellow light from a sodium vapor lamp *seems* to be of pure wavelength, but it produces two first-order maxima at 36.093° and 36.129° when projected on a 10,000 line per centimeter diffraction grating. What are the two wavelengths to an accuracy of 0.1 nm?

• **25.28** At what angle does a diffraction grating produce a second-order maximum for light having a first-order maximum at 20.0°?

• **25.29** Show that a diffraction grating cannot produce a second-order maximum for a given wavelength of light unless the first-order maximum is at an angle less than 30.0°.

• **25.30** If a diffraction grating produces a first-order maximum for the shortest wavelength of visible light at 30.0°, at what angle will the first-order maximum be for the longest wavelength of visible light?

• **25.31** What do the four angles in Problem 25.26 become if a 5000 line per centimeter diffraction grating is used?

⁞ **25.32** What is the maximum number of lines per centimeter a diffraction grating can have and produce a complete first-order spectrum for visible light?

⁞ **25.33** Find the maximum number of lines per centimeter a diffraction grating can have and produce a maximum for the smallest wavelength of visible light. Would such a grating be useful for ultraviolet spectra? For infrared spectra?

⁞ **25.34** The analysis of Problem 25.18 as illustrated in Figure 25.53 also applies to diffraction gratings with lines separated by a distance d. What is the distance between fringes produced by a diffraction grating having 125 lines per centimeter for 600 nm light, if the screen is 1.50 m away?

⁞ **25.35** (a) Show that a 30,000 line per centimeter grating will not produce a maximum for visible light. (b) What is the longest wavelength for which it does produce a first-order maximum? (c) What is the greatest number of lines per centimeter a diffraction grating can have and produce a complete second-order spectrum for visible light?

Section 25.5 Single Slit Diffraction

25.36 (a) At what angle is the first minimum for 550 nm light falling on a single slit of width 1.00 μm? (b) Will there be a second minimum?

25.37 (a) Calculate the angle at which a 2.00 μm wide slit produces its first minimum for 410 nm violet light. (b) Where is the first minimum for 700 nm red light?

25.38 (a) How wide is a single slit that produces its first minimum for 633 nm light at an angle of 28.0°? (b) At what angle will the second minimum be?

25.39 (a) What is the width of a single slit that produces its first minimum at 60.0° for 600 nm light? (b) Find the wavelength of light that has its first minimum at 62.0°.

25.40 Find the wavelength of light that has its third minimum at an angle of 48.6° when it falls on a single slit of width 3.00 μm.

25.41 Calculate the wavelength of light that produces its first minimum at an angle of 36.9° when falling on a single slit of width 1.00 μm.

• **25.42** (a) Sodium vapor light averaging 589 nm in wavelength falls on a single slit of width 7.50 μm. At what angle does it produce its second minimum? (b) What is the highest-order minimum produced?

• **25.43** (a) Find the angle of the third diffraction minimum for 633 nm light falling on a slit of width 20.0 μm. (b) What slit width would place this minimum at 85.0°?

⁞ **25.44** (a) Find the angle between the first minima for the two sodium vapor lines, which have wavelengths of 589.1 and 589.6 nm, when they fall upon a single slit of width 2.00 μm. (b) What is the distance between these minima if the diffraction pattern falls on a screen 1.00 m from the slit?

⁞ **25.45** (a) What is the minimum width of a single slit (in multiples of λ) that will produce a first minimum for a wavelength λ? (b) What is its minimum width if it produces 50 minima? (c) 1000 minima?

⁞ **25.46** (a) If a single slit produces a first minimum at 14.5°, at what angle is the second-order minimum? (b) What is the angle of

the third-order minimum? (c) Is there a fourth-order minimum? (d) Use your answers to illustrate how the angular width of the central maximum is about twice the angular width of the next maximum (which is the angle between the first and second minima).

⁞ **25.47** A double slit produces a diffraction pattern that is a combination of single and double slit interference. Find the ratio of the width of the slits to the separation between them, if the first minimum of the single slit pattern falls on the fifth maximum of the double slit pattern. (This will greatly reduce the intensity of the fifth maximum.)

Section 25.6 The Rayleigh Criterion

25.48 The 300 m diameter Arecibo radio telescope pictured in Figure 25.29 detects radio waves with a 4.00 cm average wavelength. (a) What is the angle between two just resolvable point sources for this telescope? (b) How close together could these point sources be at the 2 million light year distance of the Andromeda galaxy?

25.49 Assuming the angular resolution found for the Hubble Telescope in Example 25.5, what is the smallest detail that could be observed on the moon?

25.50 Diffraction spreading for a flashlight is insignificant compared with other limitations in its optics, such as spherical aberrations in its mirror. To show this, calculate the minimum angular spreading of a flashlight beam that is originally 5.00 cm in diameter with an average wavelength of 600 nm.

• **25.51** (a) What is the minimum angular spread of a 633 nm wavelength He-Ne laser beam that is originally 1.00 mm in diameter? (b) If this laser is aimed at a mountain cliff 15.0 km away, how big will the illuminated spot be? (c) How big a spot would be illuminated on the moon, neglecting atmospheric effects? (This might be done to hit a corner reflector to measure the round-trip time and, hence, distance.)

• **25.52** A telescope can be used to enlarge the diameter of a laser beam and limit diffraction spreading. The laser beam is sent through the telescope in opposite the normal direction and can then be projected onto a satellite or the moon. (a) If this is done with the Mount Wilson telescope, producing a 2.54 m diameter beam of 633 nm light, what is the minimum angular spread of the beam? (b) Neglecting atmospheric effects, what is the size of the spot this beam would make on the moon?

• **25.53** The limit to the eye's acuity is actually related to diffraction by the pupil. (a) What is the angle between two just resolvable points of light for a 3.00 mm diameter pupil, assuming an average wavelength of 550 nm? (b) Take your result to be the practical limit for the eye. What is the greatest possible distance a car can be from you if you can resolve its two headlights, given they are 1.30 m apart? (c) What is the distance between two just resolvable points held at arm's length (0.800 m) from your eye?

• **25.54** The planet Pluto and its moon Charon are separated by 19,600 km. Neglecting atmospheric effects, should the 5.08 m diameter Mount Palomar telescope be able to resolve these bodies when they are 4.50×10^9 km from earth? Assume an average wavelength of 550 nm. (The fact that it is just barely possible to discern that these are separate bodies indicates the severity of atmospheric aberrations.)

⁞ 25.55 What is the minimum diameter mirror on a telescope that would allow you to see details as small as 5.00 km on the moon some 384,000 km away? Assume an average wavelength of 550 nm for the light received.

Section 25.7 Thin Film Interference

25.56 A soap bubble is 100 nm thick and illuminated by white light incident perpendicular to its surface. What wavelength and color of light is most constructively reflected, assuming the same index of refraction as water?

25.57 An oil slick on water is 120 nm thick and illuminated by white light incident perpendicular to its surface. What color does the oil appear (what is the most constructively reflected wavelength), given its index of refraction is 1.40?

25.58 Calculate the minimum thickness of an oil slick on water that appears blue when illuminated by white light perpendicular to its surface. Take the blue wavelength to be 470 nm and the index of refraction of oil to be 1.40.

25.59 Find the minimum thickness of a soap bubble that appears red when illuminated by white light perpendicular to its surface. Take the wavelength to be 680 nm, and assume the same index of refraction as water.

25.60 A film of soapy water ($n = 1.33$) on top of a Plexiglas cutting board has a thickness of 233 nm. What color is most strongly reflected if it is illuminated perpendicular to its surface?

25.61 What are the three smallest nonzero thicknesses of soapy water ($n = 1.33$) on Plexiglas if it appears green (constructively reflecting 520 nm light) when illuminated perpendicularly by white light?

25.62 Suppose you have a lens system that is to be used primarily for 700 nm red light. What is the second thinnest coating of fluorite that would be nonreflective for this wavelength?

• 25.63 As a soap bubble thins it becomes dark, because the path length difference becomes small compared with the wavelength of light and there is a phase shift at the top surface. If it becomes dark when the path length difference is less than one-fourth the wavelength, what is the thickest the bubble can be and appear dark at all visible wavelengths? Assume the same index of refraction as water.

• 25.64 A film of oil on water will appear dark when it is very thin, because the path length difference becomes small compared with the wavelength of light and there is a phase shift at the top surface. If it becomes dark when the path length difference is less than one-fourth the wavelength, what is the thickest the oil can be and appear dark at all visible wavelengths? Oil has an index of refraction of 1.40.

⁞ 25.65 Figure 25.33 shows two glass slides illuminated by pure-wavelength light incident perpendicularly. The top slide touches the bottom slide at one end and rests on a 0.100 mm diameter hair at the other end, forming a wedge of air. (a) How far apart are the dark bands, if the slides are 7.50 cm long and 589 nm light is used? (b) Is there any difference if the slides are made from crown or flint glass? Explain.

⁞ 25.66 Figure 25.33 shows two 7.50 cm long glass slides illuminated by pure 589 nm wavelength light incident perpendicularly. The top slide touches the bottom slide at one end and rests on some debris at the other end, forming a wedge of air. How thick is the debris, if the dark bands are 1.00 mm apart?

Section 25.8 Polarization

25.67 What angle is needed between the direction of polarized light and the axis of a polarizing filter to cut its intensity in half?

25.68 The angle between the axes of two polarizing filters is 45.0°. By how much does the second filter reduce the intensity of the light coming through the first?

25.69 If you have completely polarized light of intensity 150 W/m², what will its intensity be after passing through a polarizing filter with its axis at an 89.0° angle to the light's polarization direction?

25.70 What angle would the axis of a polarizing filter need to make with the direction of polarized light of intensity 1.00 kW/m² to reduce the intensity to 10.0 W/m²?

25.71 At the end of Example 25.8, it was stated that the intensity of polarized light is reduced to 90.0% of its original value by passing through a polarizing filter with its axis at an angle of 18.4° to the direction of polarization. Verify this statement.

• 25.72 Show that if you have three polarizing filters, with the second at an angle of 45.0° to the first and the third at an angle of 90.0° to the first, the intensity of light passed by the first will be reduced to 25.0% of its value. (This is in contrast to having only the first and third, which reduces the intensity to zero, so that placing the second between them increases the intensity of transmitted light.)

⁞ 25.73 Prove that if I is the intensity of light transmitted by two polarizing filters with axes at an angle θ, and I' is the intensity when the axes are at an angle $90.0° - \theta$, then $I + I' = I_0$, the original intensity. (Hint: Use the trigonometric identities $\cos(90.0° - \theta) = \sin\theta$ and $\cos^2\theta + \sin^2\theta = 1$.)

25.74 At what angle will light reflected from diamond be completely polarized?

25.75 What is Brewster's angle for light traveling in water that is reflected from crown glass?

25.76 A scuba diver sees light reflected from the water's surface. At what angle will this light be completely polarized?

25.77 At what angle is light inside crown glass completely polarized when reflected from water, as in a fish tank?

25.78 Light reflected at 55.6° from a window is completely polarized. What is the window's index of refraction and the likely substance of which it is made?

• 25.79 (a) Light reflected at 62.5° from a gemstone in a ring is completely polarized. Can the gem be a diamond? (b) At what angle would the light be completely polarized if the gem were in water?

⁞ 25.80 If θ_b is Brewster's angle for light reflected from the top of an interface between two substances, and θ_b' is Brewster's angle for light reflected from below, prove that $\theta_b + \theta_b' = 90.0°$.

INTEGRATED CONCEPTS

Physics is most interesting and most powerful when applied to general situations involving more than a narrow set of physical principles. The integration of concepts necessary to solve problems involving several physical principles also gives greater insight into the unity of physics. You may wish to refer to Chapters 15, 23, and perhaps others to solve the following problems.

Note: Problem-solving strategies and a worked example that can help you solve integrated concept problems appear in Chapter 23. They can be found with the section of problems labeled *Integrated Concepts* at the end of that chapter. Consult the index for the locations of other problem-solving strategies and worked examples for integrated concept problems.

25.81 A water break at the entrance to a harbor consists of a rock barrier with a 50.0 m wide opening. Ocean waves of 20.0 m wavelength approach the opening straight on. At what angle to the incident direction are the boats inside the harbor most protected against wave action?

• **25.82** If a polarizing filter reduces the intensity of polarized light to 50.0% of its original value, by how much are the electric and magnetic fields reduced?

• **25.83** Suppose you put on two pairs of Polaroid sunglasses with their axes at an angle of 15.0°. How much longer will it take the light to deposit a given amount of energy in your eye compared with a single pair of sunglasses? Assume the lenses are clear except for their polarizing characteristics.

UNREASONABLE RESULTS

The following problems have results that are unreasonable because some premise is unreasonable or because certain of the premises are inconsistent with one another. Physical principles applied correctly then produce unreasonable results. The purpose of these problems is to give practice in assessing whether nature is being accurately described, and if it is not to trace the source of difficulty.

PROBLEM-SOLVING STRATEGY

To determine if an answer is reasonable, and to determine the cause it it is not, do the following.

Step 1. Solve the problem using strategies as outlined in Section 25.7 and in several other places in the text. See the table of contents to locate problem-solving strategies. Use the format followed in the worked examples in this chapter to solve the problem as usual.

Step 2. Check to see if the answer is reasonable. Is it too large or too small, or does it have the wrong sign, improper units, . . .?

Step 3. If the answer is unreasonable, look for what specifically could cause the identified difficulty. Usually, the manner in which the answer is unreasonable is an indication of the difficulty.

25.84 Red light of wavelength of 700 nm falls on a double slit separated by 400 nm. (a) At what angle is the first-order maximum in the diffraction pattern? (b) What is unreasonable about this result? (c) Which assumptions are unreasonable or inconsistent?

25.85 (a) What visible wavelength has its fourth-order maximum at an angle of 25.0° when projected on a 25,000 line per centimeter diffraction grating? (b) What is unreasonable about this result? (c) Which assumptions are unreasonable or inconsistent?

• **25.86** An amateur astronomer wants to build a telescope with a diffraction limit that will allow him to see if there are people on the moons of Jupiter. (a) What diameter mirror is needed to be able to see 1.00 m detail on a Jovian moon at a distance of 7.50×10^8 km from earth? The wavelength of light averages 600 nm. (b) What is unreasonable about this result? (c) Which assumptions are unreasonable or inconsistent?

• **25.87** To save money on making military aircraft invisible to radar, an inventor decides to coat them with a nonreflective material having an index of refraction of 1.20, which is between that of air and the surface of the plane. This, he reasons, should be much cheaper than designing Stealth bombers. (a) What thickness should the coating be to inhibit the reflection of 4.00 cm wavelength radar? (b) What is unreasonable about this result? (c) Which assumptions are unreasonable or inconsistent?

SPECIAL RELATIVITY

Several phenomena in this photograph can be explained with the theory of relativity, but not with classical physics. The blue glow surrounding this water-immersed radioactive source is created by particles of radiation traveling faster than the speed of light in water. Nothing can travel faster than the speed of light in a vacuum.

Figure 26.1 Albert Einstein (1879–1955) was the greatest physicist of the 20th century. He not only developed modern relativity, thus revolutionizing our concept of the universe, he also made fundamental contributions to quantum mechanics.

CONNECTIONS

Realms of Physics

Classical physics is a good approximation to modern physics under conditions first discussed in Chapter 1. Modern relativity is valid in general—and it *must* be used instead of classical physics for high speeds and large gravitational fields.

Relativity. The very word instantly conjures an image of Einstein. (See Figure 26.1.) But, as we know from Chapter 3, relativity did not begin with Einstein. People have been exploring it for many centuries, with Galileo and Newton developing the first correct version of classical relativity. **Relativity** is the study of how different observers measure the same event. What Einstein developed, nearly on his own, was the modern theory of relativity. Modern relativity is divided into two parts. **Special relativity** deals with observers who are in uniform or unaccelerated motion, while **general relativity** includes accelerated relative motion and gravity. Einstein is famous because his theories of relativity made revolutionary predictions. Most importantly, his theories have been verified to great precision in a vast range of experiments, altering forever our concept of nature. Competing theories have always failed to make experimentally verifiable predictions superior to Einstein's.

It is important to note that although classical physics, in general, and classical relativity, in particular, are incorrect, they are extremely good approximations for large, slow moving objects. Otherwise, we could not use classical physics to launch satellites or build bridges. In the classical limit (objects larger than submicroscopic and moving slower than about 1% of the speed of light), relativity becomes the same as classical physics. This fact will be noted at appropriate places in our treatment.

26.1 EINSTEIN'S POSTULATES

One of the beautiful aspects of mathematics is that *all* aspects of a mathematical system can be derived from a set of axioms or postulates by using an accepted system of logic. Some systems of mathematics correspond to nature, whereas others do not, but they are all internally self-consistent and meaningful. We may attempt something like this in physics with the constraint that physics must describe nature. If we are smart enough to choose the correct postulates, then our theory will follow and will be verified by experiment. Einstein essentially did the theoretical aspect of this. With two deceptively simple postulates and a careful consideration of how measurements are actually made, he created the theory of special relativity. Of course, Einstein had reasons for his postulates that were based on past theories and experiments as well as his brilliant insight. We can understand the postulates of special relativity better if we consider what was known at the time Einstein made them.

Late in the 19th century, the major tenets of classical physics were well established. Two of the most important were Newton's laws and Maxwell's equations. As discussed in Section 23.1, Maxwell's equations describe all electromagnetic phenomena, including light. In particular, Maxwell's equations predict that light travels at $c = 3.00 \times 10^8$ m/s in a vacuum, but they do not specify the frame of reference in which light has this speed. Einstein realized there was a contradiction between this prediction and Newton's laws, in which velocities add like simple vectors. If the latter were true, then two observers moving at different speeds would see light traveling at different speeds. In fact, Einstein thought about what a light wave would look like to a person traveling along with it at a speed c. If such a motion were possible, he reasoned, then the wave would be stationary relative to the observer and have electric and magnetic fields that varied in strength at various distances from the observer but were constant in time. Einstein knew this is not allowed by Maxwell's equations. So either Maxwell's equations are wrong, or you cannot travel at c. In fact, he realized, you cannot travel at c *and* the further implication is that light *must* always travel at c relative to any observer. Maxwell's equations are correct, and Newton's addition of velocities is not correct for light.

Light had also been convincingly demonstrated to be a wave, first by Young's double slit experiment early in the 19th century, and then by many subsequent elaborations and verifications. Many types of waves were known, and all traveled in some medium, so that it was assumed that light, even in a vacuum, had some medium to carry it, and traveled at c relative to that medium. Starting in the mid-1880s, the American physicist A. A. Michelson, later aided by E. W. Morley, made a series of direct measurements of the speed of light. The startling result of their later measurements*—now known as

---

*To obtain greater precision than the experiment described in Section 24.3, Michelson and Morley split a beam of light, sent it in different directions, and observed interference effects. Their approach is now called Michelson-Morley interferometry.

the **Michelson-Morley experiment**—was that the speed of light in a vacuum is independent of the motion of the earth about the sun. The eventual conclusion derived from this result is that light, unlike mechanical waves such as sound, does not need a medium to carry it. Furthermore, the Michelson-Morley results implied that the speed of light is c independent of the motion of the source relative to the observer. (That is, everyone observes light to move at c regardless of how they move relative to the source or one another.) For a number of years, many scientists tried unsuccessfully to explain these results and still retain the general applicability of Newton's laws.

It was not until 1905, when Einstein published his first paper on special relativity, that the correct conclusion was reached. Based mostly on his analysis that Maxwell's equations would not allow another speed for light, and only slightly aware of the Michelson-Morley experiment, Einstein accepted the constancy of the speed of light and made it one of his postulates of special relativity. It is remarkable that Einstein started thinking seriously about relativity at about age 16 and later created special relativity while employed doing other work. At the time he published his results, he was a patent examiner in Bern, Switzerland. Even more incredible, in the same year he published two other important papers. One had to do with Brownian motion and helped experimentalists determine the size of atoms. The other explained the photoelectric effect and fundamentally influenced the beginnings of quantum mechanics. It was for the last of these that Einstein received the Nobel prize in 1921.

Part of Newton's laws and classical relativity is still correct. It has long been apparent that there is no such thing as absolute motion. All velocities are measured relative to some frame of reference. The simplest frames of reference are those that are not accelerated and are not rotating. Newton's first law, the law of inertia, holds exactly in such a frame, which is thus defined to be an **inertial frame of reference**. In inertial frames, a body at rest remains at rest and a body in motion moves at a constant speed in a straight line unless acted on by an outside force.

The laws of physics seem to be simplest in inertial frames. For example, the earth is only approximately an inertial frame. We observe Coriolis forces arising from our frame's rotation, complicating the description of the motion of objects relative to the earth. (Coriolis forces "cause" the rotation of hurricanes, for instance. See Section 8.4.) On earth, we sometimes notice that we cannot neglect the earth's rotation. Then net **F** is not equal to $m\mathbf{a}$, but to $m\mathbf{a}$ plus a fictitious force; this is not as simple as in an inertial frame. Not only are laws of physics simplest in inertial frames, but they should be the same in all inertial frames, since there is no preferred frame and no absolute motion.

Einstein incorporated all these ideas into two simple postulates. The **first postulate of special relativity** is:

> **The laws of physics are the same and can be stated in their simplest form in all inertial frames of reference.**

As with many fundamental statements, there is more to this postulate than meets the eye. The laws of physics include only those that satisfy this postulate. We shall find that the definitions of relativistic momentum and energy must be altered to fit. One result of these alterations is the famous equation $E = mc^2$. None of this would matter, except that experiment verifies the results of the theory.

Just as important as the first, Einstein's **second postulate of special relativity** is:

> **The speed of light c is a constant, independent of the relative motion of the source and observer.**

Deceptively simple and counterintuitive, this and the first postulate leave all else open for change. Some fundamental concepts do change. Among the changes are the loss of agreement on the elapsed time for an event, the variation of distance with speed, and the realization that matter and energy can be converted into one another.

26.2 TIME DILATION AND SIMULTANEITY

Do time intervals depend on who observes them? Intuitively, we expect the time for a process, such as the elapsed time for a foot race, to be the same for all observers. Our experience has been that disagreements over elapsed time have to do with the accuracy of measuring time. But when we carefully consider just how time is measured,* we will find that elapsed time depends on the relative motion of an observer with respect to the process being measured.

Simultaneity

We must start by considering just how we measure elapsed time. If we use a stopwatch, for example, how do we know when to start and stop the watch? (This is one of those seemingly obvious questions that bears fruit just by asking it.) One method is to use the arrival of light from the event, such as observing a light turn green to start a drag race. The timing will be more accurate if some sort of electronic detection is used, avoiding human reaction times and other complications.

Now suppose we use this method to measure the time interval between two flashes of light produced by flashlamps as shown in Figure 26.2. Two flashlamps with observer A midway between them are on a rail car that moves to the right relative to observer B. The light flashes are emitted just as A passes B, so that both A and B are equidistant from the lamps when the light is emitted. Observer B measures the time interval between the arrival of the light flashes. Light travels equal distances to him at equal speeds (according to postulate 2, the speed of light is not affected by the motion of the lamps relative to B). Thus observer B measures the flashes to be simultaneous.

Now consider what observer B sees happen to observer A. She receives the light from the right first, because she has moved toward that flash lamp, lessening the distance the light must travel and reducing the time it takes to get to her. Light travels at c relative to both observers, but observer B remains equidistant between the points where the flashes were emitted, while A gets closer to the emission point on the right. From the point of view of B, then, there is a time interval between the arrival of the flashes to

---

*The most fundamental definition of a physical quantity is made by specifying how it is measured, as discussed in Section 1.2. Measurement and uncertainty are discussed in Section 1.3. Time was first defined in Section 2.2.

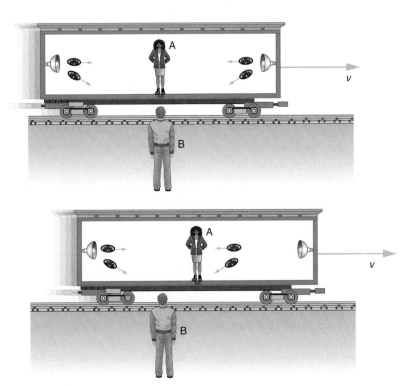

Figure 26.2 Observer B measures the elapsed time between the arrival of light flashes as described in the text. Observer A moves with the lamps on a rail car. Observer B receives the light flashes simultaneously, but he notes that observer A receives the flash from the right first. B observes the flashes to be simultaneous to him but not to A. Simultaneity is not absolute.

observer A. Observer B measures the flashes to be simultaneous relative to him but not relative to A. Here a relative velocity between observers affects whether two events are observed to be simultaneous. *Simultaneity is not absolute*.

This illustrates the power of clear thinking. We might have guessed incorrectly that if light is emitted simultaneously, then two observers halfway between the sources would see the flashes simultaneously. But careful analysis shows this not to be the case. Einstein was brilliant at this type of *thought experiment*. He very carefully considered how an observation is made and disregarded what might seem obvious. The validity of thought experiments, of course, is determined by actual observation. The genius of Einstein is evidenced by the fact that experiment always confirms his theory of relativity.

In summary: Two events are defined to be **simultaneous** if an observer measures them as occurring at the same time (such as by receiving light from the events). Two events are not necessarily simultaneous to all observers.

Time Dilation

The consideration of the measurement of elapsed time and simultaneity just above leads to an important relativistic effect called time dilation. This has to do with measurements of elapsed time by observers moving relative to one another. Suppose, for example, an astronaut measures the time it takes for light to cross her ship, bounce off a mirror, and return, as shown in Figure 26.3. How does the elapsed time the astronaut measures compare with the elapsed time measured for the same event by a person on earth, also shown in the figure? Asking this question (another thought experiment) produces a profound result. We find that the elapsed time for a process depends on who is measuring it. In this case, the time measured by the astronaut is smaller than the time measured by the earth-bound observer. This is so because the distance the light travels in the astronaut's frame is smaller than in the earth-bound frame. Light travels at the same speed in each frame, and so it will take longer to travel the greater distance in the earth-bound frame.

To quantitatively verify this contention about time depending on observer, consider the paths followed by light as seen by each observer, illustrated in Figure 26.3(c). The astronaut sees the light travel straight across and back for a total distance of $2D$, twice the width of her ship. The earth-bound observer sees the light travel a total distance $2s$. Since the ship is moving at v to the right relative to earth, light moving to the right hits the mirror in this frame. Now, time is distance divided by speed. Light travels at c in both frames, and so the time measured by the astronaut is

$$\Delta t_0 = \frac{2D}{c}$$

We define **proper time** to be the time Δt_0 measured by an observer at rest relative to the process. In this situation, the astronaut measures proper time. The time measured by

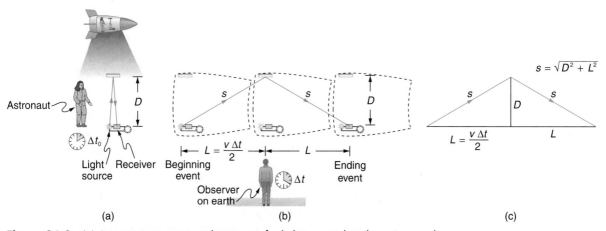

Figure 26.3 (a) An astronaut measures the time Δt_0 for light to cross her ship using an electronic timer. Light travels a distance $2D$ in the astronaut's frame. (b) A person on earth sees the light follow the longer path $2s$ and take a longer time Δt. (c) These triangles are used to find the relationship between the two distances $2D$ and $2s$.

the earth-bound observer is

$$\Delta t = \frac{2s}{c}$$

To find the relationship between Δt_0 and Δt, consider the triangles formed by D and s in Figure 26.3(c). The third side of these similar triangles is L, the distance the astronaut moves as the light goes across her ship. In the frame of the earth-bound observer,

$$L = \frac{v\,\Delta t}{2}$$

Using the Pythagorean theorem, the distance s is found to be

$$s = \sqrt{D^2 + (v\,\Delta t/2)^2}$$

Substituting s into the expression for the time interval Δt gives

$$\Delta t = \frac{2s}{c} = \frac{2\sqrt{D^2 + (v\,\Delta t/2)^2}}{c}$$

We square this equation, which yields

$$(\Delta t)^2 = \frac{4[D^2 + v^2(\Delta t)^2/4]}{c^2} = \frac{4D^2}{c^2} + \frac{v^2}{c^2}(\Delta t)^2$$

Note that if we square the first expression we had for Δt_0, we get $4D^2/c^2 = (\Delta t_0)^2$. This term appears in the preceding equation, giving us a means to relate the two time intervals. Thus,

$$(\Delta t)^2 = (\Delta t_0)^2 + \frac{v^2}{c^2}(\Delta t)^2$$

Gathering terms, we solve for Δt:

$$(\Delta t)^2 \left(1 - \frac{v^2}{c^2}\right) = (\Delta t_0)^2$$

Thus,

$$(\Delta t)^2 = \frac{(\Delta t_0)^2}{1 - \dfrac{v^2}{c^2}}$$

Taking the square root yields an important relationship between elapsed times:

$$\Delta t = \frac{\Delta t_0}{\sqrt{1 - \dfrac{v^2}{c^2}}} = \gamma\,\Delta t_0 \qquad\qquad \textbf{(26.1)}$$

where $\gamma = 1/\sqrt{1 - v^2/c^2}$.

Equation 26.1 is truly remarkable. First, as contended, elapsed time is not the same for different observers moving relative to one another, even though both are in inertial frames. Proper time Δt_0 measured by an observer, like the astronaut moving with the apparatus, is smaller than time measured by other observers. Since those other observers measure a longer time Δt, the effect is called **time dilation**. The earth-bound observer sees time dilate (get longer) for a system moving relative to the earth. Alternatively, according to the earth-bound observer, time slows in the moving frame, since less time passes there. All clocks moving relative to an observer, including biological clocks such as aging, are observed to run slow compared with a clock stationary relative to the observer.

Note that if the relative velocity is much less than the speed of light ($v \ll c$), then v^2/c^2 is extremely small, and the elapsed times Δt and Δt_0 are virtually equal. At low velocities, modern relativity approaches classical physics—our everyday experiences have very small relativistic effects. Equation 26.1 also implies that relative velocity cannot exceed the speed of light. As v approaches c, Δt approaches infinity. This would imply that time in the astronaut's frame stops at the speed of light. If v exceeded c, then we would be taking the square root of a negative number, producing an imaginary value for Δt. It is also possible that the equation is simply not valid for relative velocities greater than c.

There is considerable experimental evidence that Equation 26.1 is exactly correct. One example is found in cosmic ray particles that continuously rain down on the earth from the sun and from deep outer space. Some collisions of these particles with nuclei in the upper atmosphere result in short-lived particles called muons. The half-life (to be precisely defined in a later chapter) of a muon is 2.2 μs when it is at rest relative to the observer who measures the half-life. This would be the proper time Δt_0. Muons produced by cosmic ray particles have a range of velocities, with some moving near the speed of light. It has been found that the muon's half-life as measured by an earth-bound observer (this would be Δt) varies with velocity exactly as predicted by Equation 26.1. The faster the muon moves, the longer it lives. We on earth see the muon's half-life time dilated—as viewed from our frame, the muon's internal clock is running slower than it does when the muon is at rest relative to us.

EXAMPLE 26.1 HOW LONG DOES A SPEEDY MUON LIVE?

Suppose a muon produced as a result of a cosmic ray colliding with a nucleus in the upper atmosphere has a velocity $v = 0.950c$. Suppose it travels at constant velocity and lives 2.20 μs as measured by an observer who moves with it (this is the time on the muon's internal clock). How long does it live as measured by an earth-bound observer? (See Figure 26.4.)

Strategy Any observer moving with the system being measured observes the proper time, so that we are given $\Delta t_0 = 2.20$ μs. The earth-bound observer measures Δt as given by Equation 26.1. Since we know the velocity, the calculation is straightforward.

Solution The simplest form of Equation 26.1 is

$$\Delta t = \gamma \Delta t_0$$

where

$$\gamma = \frac{1}{\sqrt{1 - \dfrac{v^2}{c^2}}}$$

Let us first find γ and then Δt. Entering $v = 0.950c$ into the expression for γ gives

$$\gamma = \frac{1}{\sqrt{1 - \dfrac{(0.950c)^2}{c^2}}} = \frac{1}{\sqrt{1 - (0.950)^2}}$$

Note that c^2 cancels, since we express the velocity v in terms of c. Thus,

$$\gamma = \frac{1}{\sqrt{1 - 0.9025}} = \frac{1}{\sqrt{0.0975}} = \frac{1}{0.312}$$
$$= 3.20$$

Now, Δt is given by

$$\Delta t = \gamma \Delta t_0$$

Substituting known values, we get

$$\Delta t = 3.20(2.20 \text{ μs}) = 7.05 \text{ μs}$$

Discussion and Implications One implication of this example is that since $\gamma = 3.20$ at 95.0% of the speed of light ($v = 0.950c$), the relativistic effects are significant. The two time intervals differ by this factor of 3.20, where classically they would be the same. Something moving at 0.950c is said to be highly relativistic.

Figure 26.4 The muon in Example 26.1 lives longer as measured by an earth-bound observer (bottom) than as seen by its internal clock or by an observer moving with it (top).

Another implication of the preceding example is that everything an astronaut does when moving at 95.0% of the speed of light relative to the earth takes 3.20 times longer when observed from earth. Does the astronaut sense this? Not if she does not look outside her vehicle. All methods of measuring time in her frame will be affected by the same factor of 3.20. This includes her wristwatch, heart rate, cell metabolism rate, nerve impulse rate, and so on. She will have no way of telling, since all of her clocks will agree with one another because their relative velocities are zero. Motion is relative, not absolute. But what if she does look out the window?

The Twin Paradox

An intriguing consequence of time dilation is that a space traveler moving at high velocity relative to the earth would age less than her earth-bound twin. Suppose the astronaut moved at such a velocity that $\gamma = 30.0$, as in Figure 26.5. Then a trip that takes 2.00 years in her frame would take 60.0 years in her earth-bound twin's frame. Suppose her trip was 1.00 year out, a brief exploration of another star system, and then 1.00 year back. If the astronaut was 40 years old when she left, she would be 42 upon her return. But everything on earth would have aged 60.0 years, including her twin, who would be 100 if still alive. But motion is relative. To the astronaut, the spaceship is stationary and the earth moves. (This is the sensation you have when flying in a jet aircraft.) If the astronaut looks out the window of her vehicle, she will see time slow down on earth by a factor of $\gamma = 30.0$. So her earth-bound sister will have aged only 2/30 (1/15) of a year, while she aged 2.00 years. They cannot both be correct. (This contradiction might worry Doc in the "Back to the Future" movies, who warned that such inconsistencies could destroy the universe in a worst-case scenario.)

As with all paradoxes, the premise is faulty and leads to contradictory conclusions. In fact, the astronaut's motion is significantly different from that of the earth-bound twin. The astronaut is accelerated to high velocity and then decelerated to view the star system. To return to earth, she is again accelerated and decelerated. The earth-bound twin does not experience these accelerations. So the situation is not symmetric, and it is not correct to claim that the astronaut will observe the same effects as her earth-bound twin. While the twin paradox can be resolved with special relativity, that theory is expressly based on inertial frames, which by definition are not accelerated or rotating. Einstein developed general relativity to deal with accelerated frames and with gravity, a prime source of acceleration. General relativity can also be used to address the twin paradox and, according to general relativity, the astronaut will age less. We will cover some important conceptual aspects of general relativity in the final chapter of the text.

In 1971 American physicists Joseph Hafele and Richard Keating verified time dilation at low relative velocities by flying extremely accurate atomic clocks around the earth on commercial aircraft. Elapsed time was measured to an accuracy of a few nanoseconds and compared with that measured by clocks left behind. Hafele and Keating's results were within experimental uncertainties of the predictions of relativity. Both special and general relativity had to be taken into account, since gravity and accelerations were involved as well as relative motion.

Figure 26.5 The twin paradox asks why the traveling twin ages less than the earth-bound twin. That's the prediction we obtain if we consider the earth-bound twin's frame. In the astronaut's frame, however, the earth is moving and time runs slower there. Who is correct?

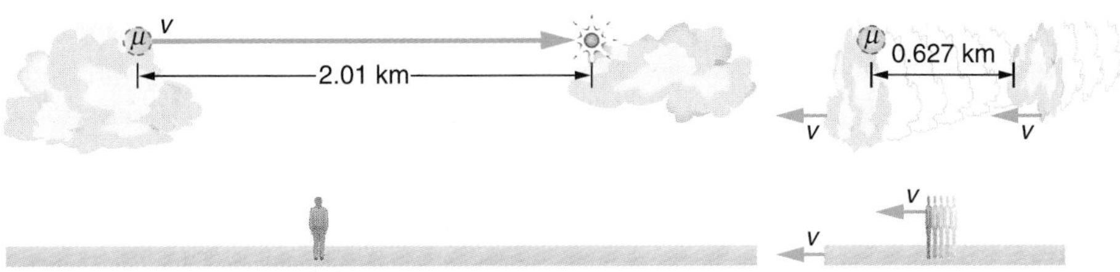

Figure 26.6 (a) The earth-bound observer sees the muon travel 2.01 km between clouds. (b) The muon sees itself travel the same path, but only a distance of 0.627 km. The earth, air, and clouds are moving relative to the muon in its frame, and all appear to have smaller lengths along the direction of travel.

26.3 LENGTH CONTRACTION

One thing all observers agree upon is relative velocity. Even though clocks measure different elapsed times for the same process, they still agree that relative velocity, which is distance divided by elapsed time, is the same. This implies that distance, too, depends on the observer's relative motion. If two observers see different times, then they must also see different distances for relative velocity to be the same to each of them.

The muon discussed in Example 26.1 illustrates this rather nicely. (See Figure 26.6.) To an observer who is on earth, the muon travels at $0.950c$ for 7.05 μs from the time it is produced until it decays. Thus it travels a distance $L_0 = v \Delta t = (0.950)(3.00 \times 10^8 \text{ m/s})(7.05 \times 10^{-6} \text{ s}) = 2.01$ km relative to the earth. To an observer moving with the muon, it lives only 2.20 μs, and so it travels $L = v \Delta t_0 = (0.950)(3.00 \times 10^8 \text{ m/s})(2.20 \times 10^{-6} \text{ s}) = 0.627$ km. So the distance between the same two events (birth and death of a muon) depends on who measures it and how they are moving relative to it. Note that the earth-bound observer measures the **proper length** L_0, because all objects except the muon are stationary relative to the earth. To the muon, the earth, air, and clouds are moving, and so the distance L it sees is not the proper length.

To get an equation relating distances measured by different observers, we note that the velocity relative to the earth-bound observer in our muon example is given by

$$v = \frac{L_0}{\Delta t}$$

The time relative to the earth-bound observer is Δt, since the object being timed is moving relative to this observer. The velocity relative to the moving observer is given by

$$v = \frac{L}{\Delta t_0}$$

The moving observer travels with the muon and, thus, observes the proper time Δt_0. The two velocities are identical; thus,

$$\frac{L_0}{\Delta t} = \frac{L}{\Delta t_0}$$

We know from Equation 26.1 that $\Delta t = \gamma \Delta t_0$. Substituting this into the above gives

$$L = \frac{L_0}{\gamma} = L_0 \sqrt{1 - \frac{v^2}{c^2}} \qquad \textbf{(26.2)}$$

This is known as **length contraction**. If we measure the length of anything moving relative to our frame, we find its length L to be smaller than the proper length L_0 that would be measured if it were stationary. For example, the muon sees a shorter path between the points where it was produced and where it decayed. Those points are fixed relative to the earth but moving relative to the muon. The muon also sees clouds and other objects as being contracted along the direction of motion.

EXAMPLE 26.2 THE DISTANCE BETWEEN STARS CONTRACTS WHEN YOU TRAVEL AT HIGH VELOCITY

Suppose an astronaut, such as the twin discussed in the preceding section, travels so fast that $\gamma = 30.0$. (a) She travels from the earth to the nearest star system, Alpha Centauri, 4.30 light years (ly) away as measured by an earth-bound observer. How far apart are the earth and Alpha Centauri as measured by the astronaut? (b) What is her velocity relative

(continued)

(continued)

to the earth? You may neglect the motion of the earth relative to the sun. (See Figure 26.7.)

Strategy and Concept   First note that a light year (ly) is a convenient unit of distance on an astronomical scale—it is the distance light travels in a year. For part (a), note that the 4.30 ly distance between Alpha Centauri and the earth is the proper distance L_0, because it is measured by an earth-bound observer to whom both stars are (approximately) stationary. To the astronaut, the earth and Alpha Centauri are moving by at the same velocity, and so the distance between them is L, contracted as expressed by Equation 26.2. In part (b), we are given γ, and so we can find v by rearranging the definition of γ to express v in terms of c.

Solution for (a)   Equation 26.2 states

$$L = \frac{L_0}{\gamma}$$

We are given $\gamma = 30.0$, and we have concluded that $L_0 = 4.30$ ly. Entering these values yields

$$l = \frac{4.30 \text{ ly}}{30.0} = 0.143 \text{ ly}$$

Solution for (b)   Substituting the value of γ into its definition gives

$$30.0 = \frac{1}{\sqrt{1 - \dfrac{v^2}{c^2}}}$$

Squaring both sides of the equation and rearranging terms gives

$$900 = \frac{1}{1 - \dfrac{v^2}{c^2}}$$

so that

$$1 - \frac{v^2}{c^2} = \frac{1}{900}$$

and

$$\frac{v^2}{c^2} = 1 - \frac{1}{900} = 0.9988 \ldots$$

Taking the square root, we find

$$\frac{v}{c} = 0.9994$$

which is rearranged to produce a value for the velocity:

$$v = 0.9994c$$

Discussion   First note that you cannot round off most calculations in special relativity until the final result is obtained, or erroneous results will be produced. The relativistic effect is large here ($\gamma = 30.0$), and we see that v is approaching (not equaling!) the speed of light. Since the distance as measured by the astronaut is so much smaller, the astronaut can travel it in much less time in her frame. People could be sent very large distances (thousands or even millions of light years) and age only a few years en route if they traveled at extremely high velocities. But, like emigrants of centuries past, they would leave the earth they know forever. Even if they returned, thousands to millions of years would have passed on earth, obliterating most of what now exists. There is a more serious practical obstacle to traveling at such velocities; immensely greater energies than classical physics predicts would be needed to achieve such high velocities. This will be discussed in Section 26.6.

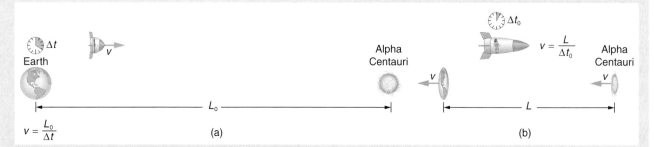

Figure 26.7   (a) The earth-bound observer measures the proper distance between the sun and Alpha Centauri. (b) The astronaut observes a length contraction, since the sun and Alpha Centauri move relative to her ship. She can travel this shorter distance in a smaller time (her proper time) without exceeding the speed of light.

Why don't we notice length contraction in everyday life? The distance to the grocery store does not seem to depend on whether we are moving or not. Examining Equation 26.2, we see that at low velocities ($v \ll c$) the lengths are nearly equal, the classical expectation. But length contraction is real, if not commonly experienced. For example, a charged particle, like an electron, traveling at relativistic velocity has its electric field lines compressed along the direction of motion as seen by a stationary observer. (See Figure 26.8.) As the electron passes a detector, such as a coil of wire, its field interacts much more briefly, an effect observed at particle accelerators such as the 3 km long Stanford Linear Accelerator

(SLAC). In fact, to an electron traveling down the beam pipe at SLAC, the accelerator and earth are all moving by and are length contracted. The relativistic effect is so great than the accelerator is only 0.5 m long to the electron. It is actually easier to get the electron beam down the pipe, since the beam does not have to be as precisely aimed to get down a short pipe as it would down one 3 km long. This, again, is an experimentally verified fact.

26.4 RELATIVISTIC ADDITION OF VELOCITIES

The classical addition of velocities was covered in depth in Section 3.5, where it was noted that velocities add as vectors. For example, the velocity of an airplane relative to the earth when there is a wind is the vector sum of its velocity relative to the air plus the air's velocity relative to the earth.

For simplicity, we restrict our consideration of velocity addition to one-dimensional motion. Classically, velocities add like regular numbers in one-dimensional motion. (See Figure 26.9.) A person riding in the back of a pickup truck throws a rock first forward, then backward at a speed of 25.0 m/s relative to the truck. The truck is moving at 20.0 m/s relative to an observer at the side of the road. We denote direction with plus and minus signs in one dimension; here forward is positive. Let v be the velocity of the truck relative to the earth, u the velocity of the rock relative to the earth-bound observer, and u' the velocity of the rock relative to the truck. Then **classical velocity addition** states that

$$u = v + u' \qquad \text{(classical velocity addition)}$$

Thus, when the rock is thrown forward, $u = 20.0 \text{ m/s} + 25.0 \text{ m/s} = 45.0 \text{ m/s}$. It makes good intuitive sense that the rock will head toward the earth-bound observer faster, because it is thrown forward from a moving vehicle. When the rock is thrown backward, $u = 20.0 \text{ m/s} + -25.0 \text{ m/s} = -5.00 \text{ m/s}$. The minus sign means the rock moves away from the earth-bound observer.

The second postulate of relativity (verified by extensive experimental observation) says that classical velocity addition does not apply to light. If it did, then the light from the truck's headlights, for example, would approach the earth-bound observer at $u = v + c$. But we know that light will move away from the truck at c relative to an observer in the truck, and it will move toward the earth-bound observer at c, too. (See Figure 26.10.)

Either light is an exception, or the classical velocity addition formula only works at low velocities. The latter is the case. The correct formula for **one-dimensional relativistic velocity addition** is

$$u = \frac{v + u'}{1 + \dfrac{vu'}{c^2}} \qquad \text{(relativistic velocity addition)} \qquad \textbf{(26.3)}$$

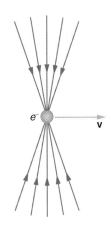

Figure 26.8 The electric field lines of a high-velocity charged particle are compressed along the direction of motion by length contraction. This produces a different signal when the particle goes through a coil, an experimentally verified effect of length contraction.

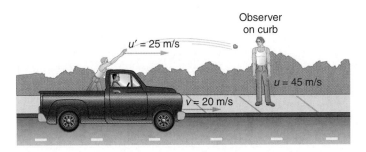

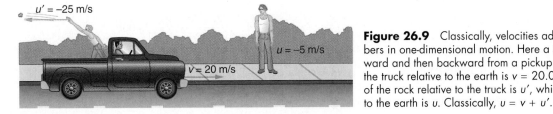

Figure 26.9 Classically, velocities add like ordinary numbers in one-dimensional motion. Here a rock is thrown forward and then backward from a pickup truck. The velocity of the truck relative to the earth is $v = 20.0$ m/s. The velocity of the rock relative to the truck is u', while its velocity relative to the earth is u. Classically, $u = v + u'$.

Figure 26.10 According to experiment and the second postulate of relativity, light from the truck's headlights moves away from the truck at c and toward the earth-bound observer at c. Classical velocity addition is not valid.

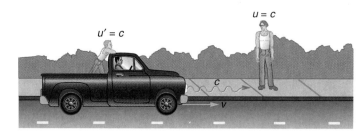

where v is the relative velocity between two observers, u is the velocity of an object relative to one observer, and u' is the velocity relative to the other observer. (For ease of visualization, we often take ourselves to measure u, while someone moving at v relative to us measures u'.) Note that the term vu'/c^2 becomes very small at low velocities, and Equation 26.3 gives a result very close to classical. As before, we see that classical velocity addition is an excellent approximation to the correct relativistic formula for small velocities. No wonder it seems correct in our experience.

EXAMPLE 26.3 THE SPEED OF LIGHT IS THE SPEED OF LIGHT

Suppose a spaceship heading directly toward the earth at half the speed of light sends a signal to us on a laser-produced beam of light. Given the light leaves the ship at c as observed from the ship, calculate the speed at which it approaches the earth.

Strategy The relative velocity is $v = 0.500c$, while the speed relative to the spaceship is $u' = c$. We are asked to find the speed relative to earth, which is u. Equation 26.3 can be used to find u with the given information.

Solution Substituting the identified values of v and u' into the relativistic velocity addition formula in Equation 26.3 gives

$$u = \frac{v + u'}{1 + \dfrac{vu'}{c^2}} = \frac{0.500c + c}{1 + \dfrac{(0.500c)(c)}{c^2}}$$

This yields

$$u = \frac{1.50c}{1 + 0.500} = c$$

Discussion Relativistic velocity addition gives the correct result. Light leaves the ship at c and approaches the earth at c. The speed of light is independent of the relative motion of source and observer, whether the observer is on the ship or earth-bound.

Velocities cannot add to greater than the speed of light, provided that v is less than c, and u' does not exceed c. The following example illustrates that relativistic velocity addition is not as symmetric as classical velocity addition.

EXAMPLE 26.4 RELATIVISTIC PACKAGE DELIVERY

Suppose the spaceship approaching the earth at half the speed of light can shoot a canister at a speed of $0.750c$. (a) At what velocity will an earth-bound observer see the canister if it is shot directly at the earth? (b) If it is shot directly away from the earth? (See Figure 26.11.)

Strategy As in the preceding example, the relative velocity of observers is $v = 0.500c$, and we are asked to find u. In part (a), we are given that $u' = +0.750c$; in (b), $u' = -0.750c$. The plus sign indicates the direction of u' is the same as v, whereas the minus indicates the direction of u' is opposite to v. Equation 26.3 can be used to calculate u in both cases.

Solution for (a) Entering the known values into Equation 26.3 gives

$$u = \frac{v + u'}{1 + \dfrac{vu'}{c^2}} = \frac{0.500c + 0.750c}{1 + \dfrac{(0.500c)(0.750c)}{c^2}}$$

This yields

$$u = \frac{1.25c}{1 + 0.375} = 0.909c$$

Solution for (b) Similarly, here

$$u = \frac{v + u'}{1 + \dfrac{vu'}{c^2}} = \frac{0.500c - 0.750c}{1 - \dfrac{(0.500c)(0.750c)}{c^2}}$$

Thus,

$$u = \frac{-0.250c}{0.625} = -0.400c$$

Discussion The minus sign indicates velocity away from the earth (in the opposite direction from v), which means the

(continued)

(continued)

canister is heading toward the earth in part (a) and away in part (b), as expected. But relativistic velocities do not add as simply as they do classically. In part (a), the canister does approach the earth faster, but not at the simple sum of 1.250c. The total velocity is less than you would get classically. And in part (b), the canister moves away from

the earth at a velocity of −0.400c, which is *faster* than the −0.250c you would expect classically. The velocities are not even symmetric. In part (a) the canister moves 0.409c faster than the ship relative to the earth, whereas in part (b) it moves −0.900c slower than the ship.

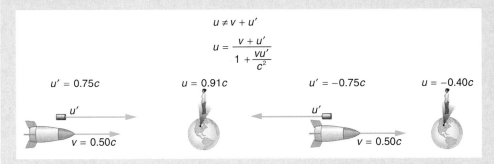

Figure 26.11 A spaceship approaches the earth at 0.500c and shoots one canister forward and another backward, both at 0.750c relative to the ship. The velocities of the canisters relative to the earth are 0.909c and −0.400c, respectively. (Classical velocity addition would give 1.250c and −0.250c, respectively.)

26.5 RELATIVISTIC MOMENTUM

Momentum is one of the most important concepts in physics. The broadest form of Newton's second law is stated in terms of momentum. And momentum is conserved whenever the net external force on a system is zero. This makes momentum conservation a fundamental tool for analyzing collisions. All of Chapter 7 was devoted to momentum, and it has been important at many subsequent points in the text, particularly where collisions were involved. We will see that momentum has the same importance in modern physics. Relativistic momentum is conserved, and much of what we know about subatomic structure comes from the analysis of collisions of accelerator-produced relativistic particles.

The first postulate of relativity states that the laws of physics are the same in all inertial frames. Does the law of conservation of momentum survive this requirement at high velocities? The answer is yes, provided that **relativistic momentum** p is defined to be classical momentum multiplied by the relativistic factor γ. In equation form, relativistic momentum p is

$$p = \gamma m u \qquad (26.4)$$

where m is the mass of the object, u is its velocity relative to an observer, and the relativistic factor $\gamma = 1/\sqrt{1 - u^2/c^2}$ here. (Note that we use u for velocity here to distinguish it from relative velocity v between observers. Only one observer is being considered here.) With p defined in this way, p_{tot} is conserved whenever the net external force is zero, just as in classical physics. Again we see that the relativistic quantity becomes virtually the same as the classical at low velocities. That is, relativistic momentum $\gamma m u$ becomes the classical $m u$ at low velocities, because γ becomes virtually 1 at low velocities.

Relativistic momentum has the same intuitive feel as classical momentum. It is greatest for large masses moving at high velocities, but, because of the factor γ, relativistic momentum approaches infinity as u approaches c. (See Figure 26.12.) This is another indication that an object with mass cannot reach the speed of light. If it did, its momentum would become infinite, an unreasonable value.

The relativistically correct definition of momentum in Equation 26.4 is sometimes taken to imply that mass varies with velocity: $m_{var} = \gamma m$. Note that m is the mass of the object as measured by a person at rest relative to the object. Thus m is defined to be the **rest mass**, which could be measured at rest, perhaps using gravity. When a mass is

CONNECTIONS

Conservation Laws

Relativistic momentum is defined in such a way that the conservation of momentum will hold in all inertial frames. Whenever the net external force on a system is zero, relativistic momentum is conserved, just as is the case for classical momentum. This has been verified in numerous experiments.

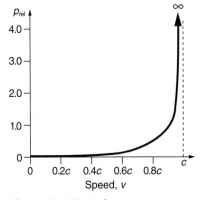

Figure 26.12 Relativistic momentum approaches infinity as the velocity of an object approaches the speed of light.

moving relative to an observer, the only way its mass can be determined is through collisions or other means in which momentum is involved. Since the mass of a moving object cannot be determined independently of momentum, the only meaningful mass is rest mass. While the idea of a mass that increases with velocity is inviting (and still appears in some textbooks), it cannot be directly confirmed; thus, we shall concentrate exclusively on rest mass. We will see that rest mass itself can vary depending on energy stored in a system ($\Delta m = \Delta E/c^2$).

In the next section, the relationship of relativistic momentum to energy is explored. That subject will produce our first inkling that objects without mass may also have momentum.

26.6 RELATIVISTIC ENERGY

Conservation of energy is one of the most important laws in physics. Not only does energy have many important forms, but each form can be converted to any other, and the total amount of energy remains constant. Since its introduction in Chapter 6, energy has played a central role in numerous topics and their applications. Relativistically, energy is still conserved, provided its definition is altered.* The altered definition of energy contains some of the most fundamental and spectacular new insights into nature found in recent history.

Total Energy and Rest Energy

The first postulate of relativity states that the laws of physics are the same in all inertial frames. Einstein showed that the law of conservation of energy is valid relativistically, if **total energy** E is defined to be

$$E = \gamma mc^2 \tag{26.5}$$

where m is rest mass, c is the speed of light, $\gamma = 1/\sqrt{1 - v^2/c^2}$, and v is the velocity of the mass relative to an observer. There are many aspects of the total energy E that we will discuss—among them are how kinetic and potential energy are included in E, and how E is related to relativistic momentum. But first, note that at rest total energy is not zero. Rather, when $v = 0$, we have $\gamma = 1$, and the **rest energy** is

$$E_0 = mc^2 \tag{26.6}$$

This is the correct form of Einstein's most famous equation, which for the first time showed that energy is related to the mass of an object at rest. For example, if energy is stored in the object, its rest mass increases. This also implies that mass can be destroyed to release energy. The implications of these first two equations regarding relativistic energy are so broad that they were not completely recognized for some years after Einstein published them in 1907, nor was the experimental proof that they are correct widely recognized at first. Einstein, it should be noted, did understand and enunciate the meanings and implications of his theory.

CONNECTIONS

Conservation Laws

Energy conservation in relativity includes the conversion between energy and mass. Relativistic energy is intentionally defined so that it will be conserved in all inertial frames, just as is the case for relativistic momentum. As a consequence, we learn that several fundamental quantities are related in ways not known in classical physics. All of these relationships are verified by experiment and have fundamental consequences.

EXAMPLE 26.5 REST ENERGY IS VERY LARGE

Calculate the rest energy of a 1.00 g mass.

Strategy and Concept According to the definition just given in Equation 26.6, the rest energy is $E_0 = mc^2$. We simply multiply mass by the speed of light squared to find rest energy.

Solution In SI units, $m = 1.00 \times 10^{-3}$ kg and $c = 3.00 \times 10^8$ m/s. Substituting these into the definition of rest energy gives

$$E_0 = mc^2 = (1.00 \times 10^{-3} \text{ kg})(3.00 \times 10^8 \text{ m/s})^2$$
$$= 9.00 \times 10^{13} \text{ kg·m}^2/\text{s}^2$$

Noting that $1 \text{ kg·m}^2/\text{s}^2 = 1$ J, we see the rest mass energy is

$$E_0 = 9.00 \times 10^{13} \text{ J}$$

Discussion This is an enormous amount of energy for a 1.00 g mass. We do not notice this energy, because it is generally not available. Rest energy is large because c is a large number and c^2 is a very large number, so that mc^2 is huge for any macroscopic mass. If you look back at some of the energies in Table 6.1, you see that the 9.00×10^{13} J rest mass energy for 1.00 g is about twice the energy released by

(continued)

---

*See the preceding section for the analogous situation with momentum.

(continued)

the Hiroshima atomic bomb and about 10,000 times the kinetic energy of a large aircraft carrier. If a way can be found to convert rest mass energy into some other form (and all forms of energy can be converted into one another), then huge amounts of energy can be obtained from the destruction of mass.

Today, the practical applications of the *conversion of mass into another form of energy*, such as in nuclear weapons and nuclear power plants, are well known. But examples also existed when Einstein first proposed the correct form of relativistic energy, and he did enunciate some of them. Nuclear radiation had been discovered in the previous decade, and it had been a mystery as to where its energy originated. The explanation was that, in certain nuclear processes, a small amount of mass is destroyed and energy is released and carried by nuclear radiation. But the amount of mass destroyed is so small that it is difficult to detect that any is missing. Although Einstein proposed this as the source of energy in the radioactive salts then being studied, it was many years before there was broad recognition that mass could be and, in fact, commonly is converted to energy. (See Figure 26.13.)

(a)

(b)

Figure 26.13 The sun (a) and the earth-bound reactors (b) convert mass into energy—the sun via fusion, the reactors via fission.

Because of the relationship of rest energy to mass, we now *consider mass to be a form of energy* rather than something separate. There had not even been a hint of this prior to Einstein's work. Such conversion is now known to be the source of the sun's energy, the energy of nuclear decay, and even the source of heat keeping the earth's interior hot.

Stored Energy and Potential Energy

What happens to energy stored in an object at rest, such as the energy put into a battery by charging it, or the energy stored in a toy gun's compressed spring? The energy input becomes part of the total energy of the object and, thus, increases its rest mass. All stored and potential energy becomes mass in a system. Why is it we don't ordinarily notice this? In fact, conservation of mass (meaning total mass is constant) was one of the great laws verified by 19th-century science. Why wasn't it noticed to be incorrect? The following example helps answer these questions.

EXAMPLE 26.6 SMALL MASS INCREASE DUE TO ENERGY INPUT

An automobile battery is rated to be able to move 600 ampere-hours (A·h) of charge at 12.0 V. (a) Calculate the increase in rest mass of such a battery when it is taken from being fully depleted to being fully charged. (b) What percent increase is this, given the battery's mass is 20.0 kg?

Strategy and Concept In part (a), we first must find the energy stored in the battery, which equals what the battery can supply in the form of electrical potential energy. Since $PE_{elec} = qV$, we have to calculate the amount of charge in 600 A·h and then multiply that by 12.0 V. We can then calculate the battery's increase in mass using $\Delta E = PE_{elec} = (\Delta m)c^2$. Part (b) is a simple ratio converted to a percent.

Solution for (a) The charge q moved by the battery can be calculated directly from the specified number of ampere-

hours, since charge is current multiplied by time. In equation form, this is

$$q = It = 600 \text{ A·h}$$

Here I is the current in amperes or coulombs per second, so that t needs to be in seconds. Thus,

$$q = (600 \text{ C/s})(3600 \text{ s}) = 2.16 \times 10^6 \text{ C}$$

Now the electric potential energy (first defined in Section 18.1) is

$$PE_{elec} = qV = (2.16 \times 10^6 \text{ C})(12.0 \text{ V})$$

Noting 1 V = 1 J/C, we find

$$PE_{elec} = 2.60 \times 10^7 \text{ J}$$

The increase in mass can now be found from $PE_{elec} = (\Delta m)c^2$. Rearranging this to find Δm and then entering knowns

(continued)

(continued)

yields

$$\Delta m = \frac{PE_{elec}}{c^2} = \frac{2.60 \times 10^7 \text{ J}}{9.00 \times 10^{16} \text{ m}^2/\text{s}^2}$$

$$= 2.88 \times 10^{-10} \text{ kg}$$

Solution for (b) The percent increase in mass is

$$\frac{\Delta m}{m} \times 100 = \left[\frac{2.88 \times 10^{-10} \text{ kg}}{20.0 \text{ kg}} \right] \times 100$$

$$= 1.44 \times 10^{-9}\%$$

Discussion Both the actual increase in mass and the percent increase are very small, since energy is divided by c^2,

a very large number. We would have to be able to measure the mass of the battery to a precision of a billionth of a percent, or 1 part in 10^{11}, to notice this increase. It is no wonder the mass variation is not readily observed. In fact, this change in mass is so small that we may question how you could verify it is real. The answer is found in nuclear processes in which the percentage of mass destroyed is large enough to be measured. The mass of the fuel of a nuclear reactor, for example, is measurably smaller when its energy has been used. In that case, stored energy has been released (converted mostly to heat and electricity) and the rest mass has decreased. This is also the case when you use the energy stored in a battery, except that the stored energy is much greater in nuclear processes, making the change in mass measurable in practice as well as in theory.

Kinetic Energy and the Ultimate Speed Limit

Kinetic energy is energy of motion. Classically, kinetic energy has the familiar expression $(1/2)mv^2$. The relativistic expression for kinetic energy is obtained from the work-energy theorem (first developed in Section 6.2), which states that the net work on a system goes into kinetic energy. If our system starts from rest, then the work-energy theorem is

$$\text{net } W = \text{KE}$$

Relativistically, at rest we have rest energy $E_0 = mc^2$. The work increases this to the total energy $E = \gamma mc^2$. Thus,

$$\text{net } W = E - E_0 = \gamma mc^2 - mc^2 = (\gamma - 1)mc^2$$

Relativistically, we have net $W = \text{KE}_{rel}$; thus,

$$\text{KE}_{rel} = (\gamma - 1)mc^2 \tag{26.7}$$

is the expression for **relativistic kinetic energy**. When motionless, we have $v = 0$ and $\gamma = 1/\sqrt{1 - v^2/c^2} = 1$, so that $\text{KE}_{rel} = 0$ at rest, as expected. But the expression for relativistic kinetic energy (like total energy and rest energy) does not look much like the classical $(1/2)mv^2$. To show that the classical expression for kinetic energy is obtained at low velocities, we note that the binomial expansion* for γ at low velocities gives

$$\gamma = 1 + \frac{1}{2}\frac{v^2}{c^2}$$

Thus, at low velocities,

$$\gamma - 1 = \frac{1}{2}\frac{v^2}{c^2}$$

Entering this into the expression for relativistic kinetic energy gives

$$\text{KE}_{rel} = \left[\frac{1}{2}\frac{v^2}{c^2}\right]mc^2 = \frac{1}{2}mv^2 = \text{KE}_{class}$$

So, in fact, relativistic kinetic energy does become the same as classical kinetic energy when $v \ll c$.

It is even more interesting to see what happens to kinetic energy when the velocity of an object approaches the speed of light. We know that γ becomes infinite as v approaches c, so that KE_{rel} also becomes infinite as the velocity approaches the speed of light. (See Figure 26.14.) An infinite amount of work (and, hence, an infinite amount of

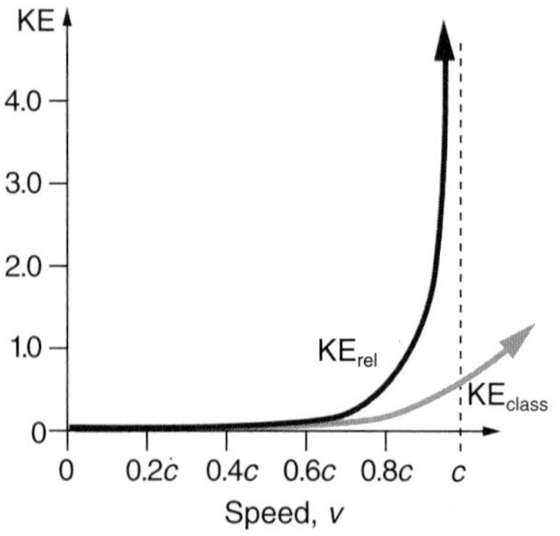

Figure 26.14 This graph of KE_{rel} versus velocity shows how kinetic energy approaches infinity as velocity approaches the speed of light. It is thus not possible for an object having mass to reach the speed of light. Also shown is KE_{class}, the classical kinetic energy, which is similar to relativistic kinetic energy at low velocities. Note that much more energy is required to reach high velocities than predicted classically.

---

*The binomial expansion is a way of expressing an algebraic quantity as a sum of an infinite series of terms. In some cases, as in the limit of small velocity here, most terms are very small. Thus the expressions derived for γ here are not exact, but they are very accurate approximations.

energy input) is required to accelerate a mass to the speed of light; thus, **no object with mass can attain the speed of light**. So the speed of light is the ultimate speed limit for any particle having mass. All of this is consistent with the fact that velocities less than c always add to less than c. Both the relativistic form for kinetic energy and the ultimate speed limit being c have been confirmed in detail in numerous experiments. No matter how much energy is put into accelerating a mass, its velocity can only approach—not reach—the speed of light.

EXAMPLE 26.7 RELATIVISTIC VERSUS CLASSICAL KINETIC ENERGY

Calculate the kinetic energy in MeV of an electron having a velocity $v = 0.990c$, and compare this with the classical value for kinetic energy at this velocity.

Strategy The expression for relativistic kinetic energy is always correct, but here it *must* be used since the velocity is highly relativistic (close to c). We will calculate the classical kinetic energy (which would be close to the relativistic value if v were less than a few percent of c) and see that it is not the same.

Solution Relativistic kinetic energy is given by Equation 26.7 to be

$$KE_{rel} = (\gamma - 1)mc^2$$

Let us first calculate γ, by substituting v into the expression for γ:

$$\gamma = \frac{1}{\sqrt{1 - \dfrac{v^2}{c^2}}} = \frac{1}{\sqrt{1 - \dfrac{(0.990)^2 c^2}{c^2}}}$$

Thus,

$$\gamma = \frac{1}{\sqrt{0.0199}} = \frac{1}{0.141} = 7.09$$

The kinetic energy is calculated by entering this value of γ and the mass of an electron into Equation 26.7:

$$KE_{rel} = (7.09 - 1)(9.11 \times 10^{-31} \text{ kg})(9.00 \times 10^{16} \text{ m}^2/\text{s}^2)$$

This gives an answer in joules, which we convert to MeV:

$$KE_{rel} = 4.99 \times 10^{-13} \text{ J} \cdot \frac{1 \text{ MeV}}{1.60 \times 10^{-13} \text{ J}}$$

so that

$$\boxed{KE_{rel} = 3.12 \text{ MeV}}$$

The classical expression for kinetic energy is

$$
\begin{aligned}
KE_{class} &= \tfrac{1}{2}mv^2 \\
&= \tfrac{1}{2}(9.11 \times 10^{-31} \text{ kg})(0.990)^2(9.00 \times 10^{16} \text{ m}^2/\text{s}^2) \\
&= 4.02 \times 10^{-14} \text{ J} \cdot \frac{1 \text{ MeV}}{1.60 \times 10^{-13} \text{ J}}
\end{aligned}
$$

Thus,

$$\boxed{KE_{class} = 0.251 \text{ MeV}}$$

Discussion As might be expected, since the velocity is 99.0% of the speed of light, the classical kinetic energy is significantly off from the correct relativistic value. Note also that the classical value is much smaller than the relativistic value. In fact, $KE_{rel}/KE_{class} = 12.4$ here. This is some indication of how difficult it is to get a mass moving close to the speed of light. Much more energy is required than predicted classically. Some people interpret this extra energy as going into increasing the mass of the system, but, as discussed in Section 26.5, this cannot be verified unambiguously. What is certain is that ever-increasing amounts of energy are needed to get the velocity of a mass a little closer to that of light.

Is there any point in getting v a little closer to c than 99.0% or 99.9%? The answer is yes. We learn a great deal by doing this. The energy that goes into a high-velocity mass can be converted to any other form, including into entirely new masses. (See Figure 26.15.) Most of what we know about the substructure of matter and the menagerie of exotic short-lived particles in nature has been learned this way. Particles are accelerated to extremely relativistic energies and made to collide with other particles, producing totally new species of particles. Patterns in the characteristics of these previously unknown particles hint at a basic substructure for all matter. These particles and some of their characteristics will be covered in Chapter 31.

Relativistic Energy and Momentum

We know classically that kinetic energy and momentum are related to each other, since $KE_{class} = p^2/2m = (mv)^2/2m = (1/2)mv^2$. Relativistically, we can obtain a relationship between energy and momentum by algebraically manipulating their definitions. This produces

$$E^2 = (pc)^2 + (mc^2)^2 \qquad \textbf{(26.8)}$$

Figure 26.15 Particle accelerators like the one at Fermilab in Batavia, Illinois, accelerate particles to extremely relativistic energies. When these collide with other particles, like nuclei, they sometimes convert kinetic energy into previously unknown particles, yielding information about the substructure of matter.

where E is the relativistic total energy and p is the relativistic momentum. This relationship between relativistic energy and relativistic momentum is more complicated than the classical, but we can gain some interesting new insights by examining it. First, total energy is related to momentum and rest mass. At rest, momentum is zero, and the equation gives the total energy to be the rest energy mc^2 (so this equation is consistent with the discussion of rest energy above). As the mass is accelerated, its rest mass m stays the same, so that the last term in the equation is constant. But as the mass is accelerated, its momentum p increases, thus increasing the total energy. At sufficiently high velocities, the rest energy term $(mc^2)^2$ becomes negligible compared with the momentum term $(pc)^2$; thus, $E \approx pc$ at extremely relativistic velocities.

If we consider momentum p to be distinct from mass, we can determine the implications of Equation 26.8 for a particle that has no mass. If we take m to be zero in Equation 26.8, then $E = pc$, or $p = E/c$. As we shall see in the next chapter, massless particles have this momentum. There are several massless particles found in nature, including photons (these are quanta of electromagnetic radiation). Another implication is that a massless particle *must* travel at c and only at c. While it is beyond the scope of this text to examine the relationship in Equation 26.8 in detail, we can see that the relationship has important implications in relativistic quantum mechanics as well as in special relativity.

PROBLEM-SOLVING STRATEGIES

FOR RELATIVITY

Step 1. Examine the situation to determine that it is necessary to use relativity. Relativistic effects are related to $\gamma = 1/\sqrt{1 - v^2/c^2}$, the quantitative relativistic factor. If γ is very close to 1, then relativistic effects are small and differ very little from the usually easier classical calculations.

Step 2. Identify exactly what needs to be determined in the problem (identify the unknowns).

Step 3. Make a list of what is given or can be inferred from the problem as stated (identify the knowns). Look in particular for information on relative velocity v.

Step 4. Make certain you understand the conceptual aspects of the problem before making any calculations. Decide, for example, which observer sees time dilated or length contracted before plugging into equations. If you have thought about who sees what, who is moving with the event being observed, who sees proper time, and so on, you will find it much easier to determine if your calculation is reasonable.

Step 5. Determine the primary type of calculation to be done to find the unknowns identified above. You will find the chapter summary helpful in determining whether a length contraction, relativistic kinetic energy, or some other concept is involved.

Step 6. Do not round off during the calculation. As noted in the text, you must often perform your calculations to many digits to see the desired effect. You may round off at the very end of the problem, but do not use a rounded number in a subsequent calculation.

Step 7. Check the answer to see if it is reasonable: Does it make sense? This may be more difficult for relativity, since we do not encounter it directly. But you can look for velocities greater than *c* or relativistic effects that are in the wrong direction (such as a time contraction where a dilation was expected).

SUMMARY

Relativity is the study of how different observers measure the same event. Modern or Einsteinian relativity is divided into two parts. **Special relativity** (the topic of this chapter) deals with observers who are in uniform or unaccelerated motion, whereas **general relativity** (covered in a later chapter) includes accelerated relative motion and gravity. Modern relativity is correct in all circumstances and, in the limit of low velocity and weak gravitation, gives the same predictions as classical relativity.

Modern relativity is based on Einstein's two postulates. The **first postulate of special relativity** is:

> **The laws of physics are the same and can be stated in their simplest form in all inertial frames of reference.**

An **inertial frame of reference** is one in which Newton's first law (the law of inertia) holds. Such frames neither accelerate nor rotate. The **second postulate of special relativity** is:

> **The speed of light *c* is a constant, independent of the relative motion of the source and observer.**

The validity of the postulates is determined by experimental verification of their predictions.

Two events are defined to be **simultaneous** if an observer measures them as occurring at the same time. They are not necessarily simultaneous to all observers—simultaneity is not absolute. Observers moving at a relative velocity *v* do not measure the same elapsed time for an event. The **proper time** for an event is the elapsed time measured by an observer at rest relative to the process; it is given the symbol Δt_0. An observer moving at *v* relative to a process sees a **time dilation**— the observer measures a longer elapsed time Δt, given by

$$\Delta t = \frac{\Delta t_0}{\sqrt{1 - \frac{v^2}{c^2}}} = \gamma \, \Delta t_0 \qquad (26.1)$$

where $\gamma = 1/\sqrt{1 - v^2/c^2}$. Distance also depends on the observer's motion. The **proper length L_0** is that measured by an observer stationary relative to the objects being measured. An observer moving at *v* relative to the system sees a **length contraction**—the observer measures a shorter length *L*, given by

$$L = \frac{L_0}{\gamma} = L_0 \sqrt{1 - \frac{v^2}{c^2}} \qquad (26.2)$$

Both observers agree that the relative velocity is *v*.

Velocities cannot add to be greater than the speed of light. The formula for **one-dimensional relativistic velocity addition** is

$$u = \frac{v + u'}{1 + \frac{vu'}{c^2}} \qquad \text{(relativistic velocity addition)} \qquad (26.3)$$

where *v* is the relative velocity between two observers, *u* is the velocity of an object relative to one observer, and *u'* is the velocity relative to the other observer.

The law of *conservation of momentum* is valid whenever the net external force is zero and for **relativistic momentum**, defined to be

$$p = \gamma m u \qquad (26.4)$$

where *m* is the object's **rest mass**, or proper mass, *u* is its velocity relative to an observer, and $\gamma = 1/\sqrt{1 - u^2/c^2}$. Except at rest, mass cannot be determined separately from momentum, and so the question of whether mass increases with velocity is left unanswered—only rest mass *m* is unambiguous.

The law of conservation of energy is also valid relativistically. **Total energy *E*** is defined to be

$$E = \gamma m c^2 \qquad (26.5)$$

and the **rest energy E_0** is

$$E_0 = m c^2 \qquad (26.6)$$

meaning that mass is a form of energy. There are processes that create and destroy mass in amounts given by this equation. In particular, stored or potential energy increases the rest mass by $\Delta m = E_{\text{stored}}/c^2$. **Relativistic kinetic energy** is the result of the net work done on a system and is given by

$$\text{KE}_{\text{rel}} = (\gamma - 1) m c^2 \qquad (26.7)$$

No object with mass can attain the speed of light, since an infinite amount of energy (or work) would be needed. The relationship between energy and momentum is given by

$$E^2 = (pc)^2 + (mc^2)^2 \qquad (26.8)$$

All of the relativistic expressions become the same as their classical counterparts in the limit that *v* is much less than *c*.

CONCEPTUAL QUESTIONS

26.1 Is classical physics exactly correct at low velocities, or just a good approximation to modern relativity?

26.2 Relativistic effects such as time dilation and length contraction are present for automobiles and airplanes. Why do these effects seem strange to us?

26.3 How has modern relativity been verified?

26.4 When you are flying in a commercial jet, it may appear to you that the airplane is stationary and the earth is moving beneath you. Is this point of view valid? Discuss briefly.

26.5 Is the earth an inertial frame of reference? Is the sun? Justify your response.

26.6 Does motion affect the rate of a clock as measured by an observer moving with it? Does motion affect how an observer moving relative to a clock measures its rate?

26.7 To whom does the elapsed time for a process seem to be longer, an observer moving relative to the process or an observer moving with the process? Which observer measures proper time?

26.8 How could you travel far into the future without aging significantly? Could this method also allow you to travel into the past?

26.9 To whom does an object seem greater in length, an observer moving with the object or an observer moving relative to the object? Which observer measures the object's proper length?

26.10 Suppose an astronaut is moving relative to the earth at a significant fraction of the speed of light. Does he observe the rate of his clocks to have slowed? What change in the rate of earth-bound clocks does he see? Does his ship seem to him to shorten? What about the distance between stars that lie on lines parallel to his motion? Do he and an earth-bound observer agree on his velocity relative to the earth?

26.11 How does modern relativity modify the law of conservation of momentum? Newton's first law? Newton's second law?

26.12 How are the classical laws of conservation of energy and conservation of mass modified by modern relativity?

26.13 What happens to the mass of water in a pot when it cools, assuming no molecules escape or are added? Is this observable in practice? Explain.

26.14 The mass of the fuel in a nuclear reactor decreases by an observable amount as it puts out energy. Is the same true for the coal and oxygen combined in a conventional power plant? If so, is this observable in practice for the coal and oxygen? Explain.

26.15 We know that the velocity of an object with mass has an upper limit of c. Is there an upper limit on its momentum? Its energy? Explain.

26.16 Given the fact that light travels at c, can it have mass?

26.17 If you use an earth-based telescope to project a laser beam onto the moon, you can move the spot across the moon's surface at a velocity greater than the speed of light. Does this violate modern relativity? (Note that light is being sent from the earth to the moon, not across the surface of the moon.)

PROBLEMS

Sections 26.2 and 26.3 Time Dilation and Length Contraction

26.1 (a) What is γ if $v = 0.250c$? (b) If $v = 0.500c$?

26.2 (a) What is γ if $v = 0.100c$? (b) If $v = 0.900c$?

26.3 Particles called π-mesons are created by accelerator beams. If these particles travel at 2.70×10^8 m/s and live 2.60×10^{-8} s when at rest relative to an observer, how long do they live as viewed in the laboratory?

26.4 A spaceship, 200 m long as seen on board, moves by the earth at $0.970c$. What is its length as measured by an earth-bound observer?

26.5 Suppose a particle called a kaon is created by cosmic radiation striking the atmosphere. It moves by you at $0.980c$, and it lives 1.24×10^{-8} s when at rest relative to an observer. How long does it live as you observe it?

• 26.6 A neutral π-meson is a particle that can be created by accelerator beams. If one such particle lives 1.40×10^{-16} s as measured in the laboratory, and 0.840×10^{-16} s when at rest relative to an observer, what is its velocity relative to the laboratory?

• 26.7 A neutron lives 900 s when at rest relative to an observer. How fast is the neutron moving relative to an observer who measures its life span to be 2065 s?

• 26.8 How fast would a 20.0 ft long sports car have to be going past you in order for it to appear only 18.0 ft long?

• 26.9 If relativistic effects are to be less than 1%, then γ must be less than 1.01. At what relative velocity is $\gamma = 1.01$?

• 26.10 If relativistic effects are to be less than 3%, then γ must be less than 1.03. At what relative velocity is $\gamma = 1.03$?

• 26.11 (a) At what relative velocity is $\gamma = 1.50$? (b) At what relative velocity is $\gamma = 100$?

• 26.12 (a) At what relative velocity is $\gamma = 2.00$? (b) At what relative velocity is $\gamma = 10.0$?

• 26.13 (a) How far does the muon in Example 26.1 travel according to the earth-bound observer? (b) How far does it travel as viewed by an observer moving with it? Base your calculation on its velocity relative to the earth and the time it lives (proper time). (c) Verify that these two distances are related through length contraction $\gamma = 3.20$.

• 26.14 (a) How long would the muon in Example 26.1 have lived as observed on earth if its velocity was $0.0500c$? (b) How far would it have traveled as observed on earth? (c) What distance is this in the muon's frame?

• 26.15 (a) How long does it take the astronaut in Example 26.2 to travel 4.30 ly at $0.9994c$ (as measured by the earth-bound observer)? (b) How long does it take according to the astronaut? (c) Verify that these two times are related through time dilation with $\gamma = 30.0$ as given.

: 26.16 How fast would a foot racer need to be running for a 100 m race to look 100 yd long?

Section 26.4 Relativistic Addition of Velocities

26.17 Suppose a spaceship heading straight toward the earth at $0.750c$ can shoot a canister at $0.500c$ relative to the ship. (a) What is the velocity of the canister relative to the earth, if it is shot directly at the earth? (b) If it is shot directly away from the earth?

26.18 Repeat Problem 26.17 with the ship heading directly away from the earth.

26.19 If a spaceship is approaching the earth at $0.100c$ and a message capsule is sent toward it at $0.100c$ relative to the earth, what is the speed of the capsule relative to the ship?

26.20 Suppose the speed of light were only 3000 m/s. A jet fighter moving toward a target on the ground at 800 m/s shoots bullets having a muzzle velocity of 1000 m/s. What is the bullets' velocity relative to the target? (If the speed of light were this small, we would experience more relativistic effects in ordinary circumstances.)

• 26.21 If two spaceships are heading directly toward each other at $0.800c$, at what speed must a canister be shot from the first ship to approach the other at $0.999c$ as seen by the second ship?

• 26.22 Two planets are on a collision course, heading directly toward each other at $0.250c$. A spaceship sent from one planet approaches the second at $0.750c$ as seen by the second planet. What is the velocity of the ship relative to the first planet?

• 26.23 When a missile is shot from one spaceship toward another, it leaves the first at $0.950c$ and approaches the other at $0.750c$. What is the relative velocity of the two ships?

• 26.24 What is the relative velocity of two spaceships, if one fires a missile at the other at $0.750c$ and the other observes it to approach at $0.950c$?

‡ 26.25 Prove that for any relative velocity v between two observers, a beam of light sent from one to the other will approach at c. (Provided v is less than c, of course.)

‡ 26.26 Show that for any relative velocity v between two observers, a beam of light projected by one directly away from the other will move away at the speed of light. (Provided v is less than c, of course.)

‡ 26.27 (a) All but the closest galaxies are receding from our own Milky Way galaxy. If a galaxy 12.0×10^9 ly away is receding from us at $0.900c$, at what velocity relative to us must we send an exploratory probe to approach the other galaxy at $0.990c$, as measured from that galaxy? (b) How long will it take the probe to reach the other galaxy as measured from earth? You may assume that the velocity of the other galaxy remains constant. (c) How long will it then take for a radio signal to be beamed back? (All of this is possible in principle, but not practical.)

Section 26.5 Relativistic Momentum

26.28 What is the momentum of an electron traveling at $0.980c$?

26.29 Find the momentum of a helium nucleus having a mass of 6.68×10^{-27} kg that is moving at $0.200c$.

• 26.30 (a) What is the momentum of a 2000 kg satellite orbiting at 4.00 km/s? (b) Find the ratio of this momentum to the classical momentum. (Hint: Use the approximation that $\gamma = 1 + (1/2)v^2/c^2$ at low velocities.)

• 26.31 (a) Find the momentum of a 1.00×10^9 kg asteroid heading toward the earth at 30.0 km/s. (b) Find the ratio of this momentum to the classical momentum. (Hint: Use the approximation that $\gamma = 1 + (1/2)v^2/c^2$ at low velocities.)

‡ 26.32 Find the velocity of a proton that has a momentum of 4.48×10^{-19} kg·m/s.

‡ 26.33 What is the velocity of an electron that has a momentum of 3.34×10^{-21} kg·m/s? Note that you must calculate the velocity to at least four digits to see the difference from c.

Section 26.6 Relativistic Energy

26.34 What is the rest energy of an electron, given its mass is 9.11×10^{-31} kg? Give your answer in joules and MeV.

26.35 Find the rest mass energy in joules and MeV of a proton, given its mass is 1.67×10^{-27} kg.

26.36 If the rest mass energies of a proton and a neutron (the two constituents of nuclei) are 938.3 and 939.6 MeV respectively, what is the difference in their masses in kilograms?

26.37 (a) Using data from Table 6.1, calculate the mass converted to energy by the fission of 1.00 kg of uranium. (b) What is the ratio of mass destroyed to the original mass, $\Delta m/m$?

26.38 A supernova explosion of a 2.00×10^{31} kg star produces 1.00×10^{44} J of energy. (a) How many kilograms of mass are converted to energy in the explosion? (b) What is the ratio $\Delta m/m$ of mass destroyed to the original mass of the star?

26.39 The Big Bang that began the universe is estimated to have released 10^{68} J of energy. How many stars could half this energy create, assuming the average star's mass is 4.00×10^{30} kg?

• 26.40 (a) Using data from Table 6.1, calculate the amount of mass converted to energy by the fusion of 1.00 kg of hydrogen. (b) What is the ratio of mass destroyed to the original mass, $\Delta m/m$? (c) How does this compare with $\Delta m/m$ for the fission of 1.00 kg of uranium?

• 26.41 There is approximately 10^{34} J of energy available from fusion of hydrogen in the world's oceans. (a) If 10^{33} J of this energy were utilized, what would be the decrease in mass of the oceans? (b) How great a volume of water does this correspond to?

• 26.42 A muon has a rest mass energy of 105.7 MeV, and it decays into an electron and a massless particle. (a) If all the lost mass is converted into the electron's kinetic energy, find γ for the electron. (b) What is the electron's velocity?

• 26.43 A π-meson is a particle that decays into a muon and a massless particle. The π-meson has a rest mass energy of 139.6 MeV, and the muon has a rest mass energy of 105.7 MeV. Suppose the π-meson is at rest and all of the missing mass goes into the muon's kinetic energy. How fast will the muon move?

• **26.44** (a) Calculate the relativistic kinetic energy of a 1000 kg car moving at 30.0 m/s if the speed of light were only 45.0 m/s. (b) Find the ratio of the relativistic kinetic energy to classical.

• **26.45** Alpha decay is nuclear decay in which a helium nucleus is emitted. If the helium nucleus has a mass of 6.80×10^{-27} kg and is given 5.00 MeV of kinetic energy, what is its velocity?

• **26.46** Beta decay is nuclear decay in which an electron is emitted. If the electron is given 0.750 MeV of kinetic energy, what is its velocity?

⁞ **26.47** A positron is an antimatter version of the electron, having exactly the same mass. When a positron and an electron meet, they annihilate, converting all of their mass into energy. (a) Find the energy released, assuming negligible kinetic energy before the annihilation. (b) If this energy is given to a proton in the form of kinetic energy, what is its velocity? (c) If this energy is given to another electron in the form of kinetic energy, what is its velocity?

⁞ **26.48** What is the kinetic energy in MeV of the π-meson in Problem 26.6, given its rest mass energy is 135 MeV?

⁞ **26.49** Find the kinetic energy in MeV of the neutron in Problem 26.7, given its rest mass energy is 939.6 MeV.

⁞ **26.50** (a) Show that $(pc)^2/(mc^2) = \gamma^2 - 1$. This means that at large velocities $pc \gg mc^2$. (b) Is $E \approx pc$ when $\gamma = 30.0$, as for the astronaut discussed in the twin paradox?

⁞ **26.51** One cosmic ray neutron has a velocity of $0.250c$ relative to the earth. (a) What is the neutron's total energy in MeV? (b) Find its momentum. (c) Is $E \approx pc$ in this situation? Discuss in terms of the equation given in part (a) of the previous problem.

INTEGRATED CONCEPTS

Physics is most interesting and most powerful when applied to general situations involving more than a narrow set of physical principles. The integration of concepts necessary to solve problems involving several physical principles also gives greater insight into the unity of physics. You may wish to refer to Chapters 6, 10, 18, and perhaps others, to solve the following problems.

Note: Problem-solving strategies and worked examples that can help you solve integrated concept problems appear in several places, the most recent being Chapter 23. Consult the section of problems labeled *Integrated Concepts* at the end of that chapter or others.

26.52 What is γ for a proton having a rest mass energy of 938.3 MeV accelerated through an effective potential of 1.0 TV (teravolt) at Fermilab outside Chicago?

26.53 (a) What is the effective accelerating potential for electrons at the Stanford Linear Accelerator, if $\gamma = 1.00 \times 10^5$ for them? (b) What is their total energy (nearly the same as kinetic in this case) in GeV?

• **26.54** (a) Using data from Table 6.1, find the mass destroyed when the energy in a barrel of crude oil is released. (b) Given these barrels contain 55.0 gal and assuming the density of crude oil is 750 kg/m³, what is the ratio of mass destroyed to original mass, $\Delta m/m$?

• **26.55** (a) Calculate the energy released by the destruction of 1.00 kg of mass. (b) How many kilograms could be lifted to a 10.0 km height by this amount of energy?

• **26.56** A Van de Graaff accelerator utilizes a 50.0 MV potential difference to accelerate charged particles such as protons. (a) What is the velocity of a proton accelerated by such a potential? (b) An electron?

• **26.57** Suppose you use an average of 500 kW·h of electric energy per month in your home. (a) How long would 1.00 g of mass converted to electric energy with an efficiency of 38.0% last you? (b) How many homes could be supplied at the 500 kW·h per month rate for one year by the energy from the described mass conversion?

⁞ **26.58** A nuclear power plant converts energy from nuclear fission into electricity with an efficiency of 35.0%. How much mass is destroyed in one year to produce a continuous 1000 MW of electric power?

⁞ **26.59** Nuclear-powered rockets were researched for some years before safety concerns became paramount. (a) What fraction of a rocket's mass would have to be destroyed to get it into a low earth orbit, neglecting the decrease in gravity? (Assume an orbital altitude of 250 km, and calculate both the kinetic (classical) and gravitational potential energy needed.) (b) If the ship has a mass of 1.00×10^5 kg (100 tons), what total yield nuclear explosion in tons of TNT is needed? (See Table 6.1.)

⁞ **26.60** The sun produces energy at a rate of 4.00×10^{26} W by the fusion of hydrogen. (a) How many kilograms of hydrogen undergo fusion each second? (b) If the sun is 90.0% hydrogen and half of this can undergo fusion before the sun changes character, how long could it produce energy at its current rate? (c) How many kilograms of mass is the sun losing per second? (d) What fraction of its mass will it have lost in the time found in part (b)?

UNREASONABLE RESULTS

The following problems have results that are unreasonable because some premise is unreasonable or because certain of the premises are inconsistent with one another. Physical principles applied correctly then produce unreasonable results. The purpose of these problems is to give practice in assessing whether nature is being accurately described, and if it is not to trace the source of difficulty.

PROBLEM-SOLVING STRATEGY

To determine if an answer is reasonable, and to determine the cause if it is not, do the following.

Step 1. *Solve the problem using strategies as outlined in Section 26.6 and in several other places in the text.* See the table of contents to locate problem-solving strategies. Use the format followed in the worked examples in this chapter to solve the problem as usual.

Step 2. *Check to see if the answer is reasonable.* Is it too large or too small, or does it have the wrong sign, improper units, . . .?

Step 3. *If the answer is unreasonable, look for what specifically could cause the identified difficulty.* Usually, the manner in which the answer is unreasonable is an indication of the difficulty.

26.61 (a) Find the value of γ for the following situation. An earth-bound observer measures 23.9 h to have passed while signals from a high-velocity space probe indicate that 24.0 h have passed on board. (b) What is unreasonable about this result? (c) Which assumptions are unreasonable or inconsistent?

26.62 (a) Find the value of γ for the following situation. An astronaut measures the length of her spaceship to be 25.0 m, while an earth-bound observer measures it to be 100 m.

(b) What is unreasonable about this result? (c) Which assumptions are unreasonable or inconsistent?

• **26.63** A spaceship is heading directly toward the earth at a velocity of $0.800c$. The astronaut on board claims that he can send a canister toward the earth at $1.20c$ relative to the earth. (a) Calculate the velocity the canister must have relative to the spaceship. (b) What is unreasonable about this result? (c) Which assumptions are unreasonable or inconsistent?

• **26.64** A proton has a mass of 1.67×10^{-27} kg. A physicist measures the proton's total energy to be 50.0 MeV. (a) What is the proton's kinetic energy? (b) What is unreasonable about this result? (c) Which assumptions are unreasonable or inconsistent?

27 INTRODUCTION TO QUANTUM MECHANICS

The Particle-Wave Duality

A deer tick imaged by an electron microscope is as monstrous

as any science fiction creature.

Quantum mechanics is the branch of physics needed to deal with submicroscopic objects. Because these objects are smaller than we can observe directly with our senses and generally must be observed with the aid of instruments, parts of quantum mechanics seem as foreign and bizarre as parts of relativity. But, like relativity, quantum mechanics is valid—truth is often stranger than fiction.

Certain aspects of quantum mechanics are familiar to us. We accept as fact that matter is composed of atoms, the smallest unit of an element, and that these atoms combine to form molecules, the smallest unit of a compound. (See Figure 27.1.) While we cannot see the individual water molecules in a stream, for example, we are aware that this is because molecules are so small and so numerous in that stream. It is commonly accepted that electrons orbit atoms in discrete shells around a tiny nucleus, itself composed of smaller particles called protons and neutrons. We are also aware that electric charge comes in tiny units carried almost entirely by electrons and protons. As with water molecules in a stream, we do not notice individual charges in the current through a light bulb, because the charges are so small and so numerous in the macroscopic situations we sense directly.

Atoms, molecules, and fundamental electron and proton charges are all examples of physical entities that are **quantized**—that is, they only appear in certain discrete values and do not have every conceivable value. *Quantized* is the opposite of *continuous*. We cannot have a fraction of an atom, or part of an electron's charge, for example. Rather, everything is built of integral multiples of these substructures. **Quantum mechanics** is the branch of physics that deals with small objects and the quantization of various entities, such as mass, charge, and energy. The **correspondence principle** states that in the classical limit (large, slow moving objects), quantum mechanics becomes the same as classical physics. In this chapter, we begin the development of quantum mechanics and its description of the strange submicroscopic world. In later chapters, we will examine many areas, such as atomic and nuclear physics, in which quantum mechanics is crucial.

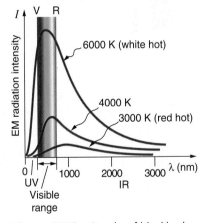

Figure 27.1 Atoms and their substructure are familiar examples of objects that require quantum mechanics to be fully explained. Certain of their characteristics, such as the discrete electron shells, cannot be explained by classical physics.

CONNECTIONS

Realms of Physics

Classical physics is a good approximation to modern physics under conditions first discussed in Chapter 1. Quantum mechanics is valid in general, and it *must* be used rather than classical physics to describe small objects, such as atoms.

27.1 QUANTIZATION OF ENERGY

Energy is quantized in some systems, meaning that the system can have only certain energies and not a continuum of energies.* We also find that some forms of energy transfer take place with discrete lumps of energy. While most of us are familiar with the quantization of matter into lumps called atoms, molecules, and the like, we are less aware that energy, too, can be quantized. Yet historically, some of the earliest clues about the necessity of quantum mechanics over classical physics came from the quantization of energy.

Where is the quantization of energy observed? Let us begin by considering the emission and absorption of electromagnetic (EM) radiation—first by solids and later by gases. Heat transfer by EM radiation was first discussed in Section 13.7, where it was noted that the EM spectrum radiated by a hot solid is linked directly to the solid's temperature. (See Figure 27.2.) An ideal radiator is one that has an emissivity of 1 at all wavelengths and, thus, is jet black. Ideal radiators are therefore called **blackbodies**, and their EM radiation is called **blackbody radiation**. It was discussed that the total intensity of the radiation varies as T^4, the fourth power of the absolute temperature of the body, and that the peak of the spectrum shifts to shorter wavelengths at higher temperatures. All of this seems quite continuous, but it was the shape of the spectrum that gave a clue that the energies of the atoms in the solid are quantized.

Using the idea that blackbody radiation emission occurs when the atoms and molecules in a body act like EM oscillators, the great German physicist Max Planck (1858–1947) struggled to describe the shape of the blackbody spectrum around 1900. (See Figure 27.3.) Planck was a consummate classical physicist, but he came to the conclusion that the energies of the EM oscillators (atoms and molecules) had to be quantized

Figure 27.2 Graphs of blackbody radiation (from an ideal radiator) at three different temperatures. The intensity or rate of radiation emission increases dramatically with temperature, and the peak of the spectrum shifts toward the visible and ultraviolet parts of the spectrum. The shape of the spectrum cannot be described with classical physics.

*Classically, this is not the case. It would be like having only certain speeds a car can travel because its kinetic energy can only have certain values.

Figure 27.3 The German physicist Max Planck had a major influence on the early development of quantum mechanics, being the first to recognize that energy is sometimes quantized. Planck also made important contributions to special relativity and classical physics.

to correctly describe the shape of the blackbody spectrum. Planck deduced that the energy of an oscillator having a frequency f is given by

$$E = nhf \qquad (27.1)$$

Here n is any nonnegative integer (0, 1, 2, 3, . . .). The symbol h stands for **Planck's constant**, one of the most important constants in modern physics, given by

$$h = 6.63 \times 10^{-34} \, \text{J·s} \qquad (27.2)$$

Equation 27.1 means that an oscillator having a frequency f (emitting and absorbing EM radiation of frequency f) can have its energy increase or decrease only in *discrete* steps of size

$$\Delta E = hf \qquad (27.3)$$

This is like a pendulum that has a characteristic oscillation frequency but which can only swing with certain amplitudes. It is also like going up and down a hill using discrete stair steps rather than being able to move up and down a continuous slope. Using the quantization of oscillators, Planck was able to correctly describe the experimentally known shape of the blackbody spectrum. This was the first indication that energy is sometimes quantized on a small scale and earned the Nobel prize for Planck in 1918. Although Planck's theory comes from observations of a macroscopic object, its analysis is based on atoms and molecules. It was such a revolutionary departure from classical physics that Planck himself was reluctant to accept his own idea that energy states are not continuous. The general acceptance of Planck's energy quantization was greatly enhanced by Einstein's explanation of the photoelectric effect (discussed in the next section), which took energy quantization a step further. Planck was fully involved in the development of both early quantum mechanics and relativity. He quickly embraced Einstein's special relativity published in 1905, and in 1906 Planck was the first to suggest the correct formula for relativistic momentum given in Equation 26.4.

Note that Planck's constant h is a very small number. So for an infrared frequency of 10^{14} Hz being emitted by a blackbody, for example, the difference between energy levels is only $\Delta E = hf = (6.63 \times 10^{-34} \, \text{J·s})(10^{14} \, \text{cycles/s}) = 6.63 \times 10^{-20}$ J, or about 0.4 eV. This 0.4 eV of energy is significant compared with typical atomic energies, which are on the order of an electron volt, or thermal energies that are typically fractions of an electron volt. But on a macroscopic or classical scale, energies are typically on the order of joules. Even if macroscopic energies are quantized, the 6.63×10^{-20} J steps are too small to be noticed. This is an example of the correspondence principle. For a large object, quantum mechanics produces results indistinguishable from those of classical physics.

Now let us turn our attention to the *emission and absorption of EM radiation by gases*. Sunlight is the most common example of gases emitting an EM spectrum that includes visible light. We also see examples in neon signs and candle flames. Studies of emissions of hot gases began more than two centuries ago, and it was soon recognized that these emission spectra contained huge amounts of information. The type of gas and its temperature, for example, could be determined. We now know that these EM emissions come from individual atoms and molecules; thus, they are called **atomic spectra**. Atomic spectra remain an important analytical tool today. Figure 27.4 shows examples of emission spectra obtained by passing an electric discharge through various materials. One of the most important characteristics of these spectra is that they are discrete. Only certain

Figure 27.4 Emission spectra of various materials. When an electrical discharge is passed through a substance, its atoms and molecules absorb energy, which is reemitted as EM radiation. The discrete nature of these emissions implies that the energy states of the atoms and molecules are quantized. Such atomic spectra were used as analytical tools for many decades before it was understood why they are quantized.

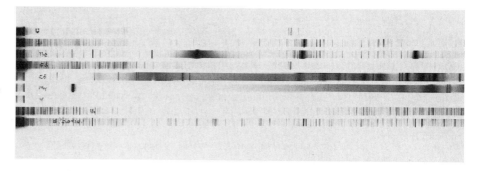

wavelengths, and hence frequencies, are emitted. If frequency and energy are associated as in Equation 27.3, the energies of the emitting atoms and molecules are quantized.

It was a major puzzle that atomic spectra are quantized. Some of the best minds of 19th-century science failed to explain why this might be. Not until the second decade of the 20th century did an answer based on quantum mechanics begin to emerge. Again a macroscopic or classical body of gas was involved, but the effect, as we shall see, is due to individual atoms and molecules.

27.2 THE PHOTOELECTRIC EFFECT

When light strikes materials, it can eject electrons from them. This is called the **photoelectric effect**, meaning that light (*photo*) creates electricity. One common use of the photoelectric effect is in light meters, such as those that adjust the automatic iris on various types of cameras. This effect has been known for more than a century and can be studied using a device such as that shown in Figure 27.5.

Figure 27.5 shows a vacuum tube with a curved metal plate and a central wire that are connected by a variable voltage source, with the wire more negative than the plate. It is a predecessor to the solid state devices used in modern applications, such as cameras. When light (or other EM radiation) strikes the curved plate in the vacuum tube, it may eject electrons. If the electrons have an energy in electron volts (eV) greater than the potential difference between the plate and the wire in volts, some electrons will be collected on the wire. Since the electron energy in eV is qV, where q is the electron charge and V the potential difference, electron energy can be measured by adjusting the voltage between the wire and the plate. The voltage that stops the electrons from reaching the wire equals the energy in eV. For example, if -3.00 V barely stops the electrons, their energy is 3.00 eV. The number of electrons ejected can also be determined by measuring the current between the wire and plate. The more light, the more electrons; a little circuitry allows this device to be used as a light meter.

What is really important about the photoelectric effect is what Einstein deduced from it. Einstein realized that there were several characteristics of the photoelectric effect that could only be explained if *EM radiation is itself quantized*. The apparently continuous stream of energy in an EM wave is actually composed of quanta of energy called photons. This is analogous to the fact that the apparently continuous stream of water from your kitchen faucet is actually composed of quanta of matter called molecules. In his explanation of the photoelectric effect, Einstein defined a quantized unit or quantum of EM energy, which we now call a **photon**, with an energy proportional to the frequency of EM radiation. In equation form, **photon energy** is

$$E = hf \tag{27.4}$$

where E is the energy of a photon of frequency f and h is Planck's constant. This revolutionary idea looks similar to Planck's quantization of energy states in blackbody oscillators, but it is quite different. It is the quantization of EM radiation itself. EM waves are composed of photons and are not continuous smooth waves. Their energy is absorbed in lumps, not continuously.* We do not observe this with our eyes, because there are so many photons in common light sources that individual photons go unnoticed. (See Figure 27.6.) The next section of the text (Section 27.3) is devoted to a discussion of photons and some of their characteristics and implications. For now, we will use the photon concept to explain the photoelectric effect, much as Einstein did.

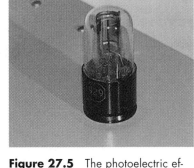

Figure 27.5 The photoelectric effect can be observed by allowing light to fall on the curved metal plate in this vacuum tube. Electrons ejected by the light are collected on the central wire. Voltage between the central wire and plate can then be adjusted to determine the energy of the ejected electrons. For example, if it is sufficiently negative, no electrons will reach the wire.

---

*This is exactly consistent with Planck's quantization of energy levels in blackbody oscillators, since these oscillators increase and decrease their energy in steps of hf by absorbing and emitting photons having $E = hf$.

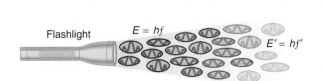

Flashlight $E = hf$ $E' = hf'$

Figure 27.6 An EM wave of frequency f is composed of photons, or individual quanta of EM radiation. The energy of each photon is $E = hf$, where h is Planck's constant and f is the frequency of the EM radiation. Higher intensity means more photons per unit area. The flashlight emits large numbers of photons of many different frequencies, hence others have energy $E' = hf'$, and so on.

The photoelectric effect has the properties discussed below. All these properties are consistent with the idea that individual photons of EM radiation are absorbed by individual electrons in a material, with the electron absorbing the photon's energy. Some of these properties are inconsistent with the idea that EM radiation is a simple wave. For simplicity, we consider what happens with monochromatic EM radiation; thus, all photons here have the same energy hf.

1. *Once EM radiation falls on a material, electrons are ejected without delay.* As soon as an individual photon is absorbed by an individual electron, the electron is ejected. If the EM radiation were a simple wave, some time would be required for sufficient energy to be deposited to eject an electron.

2. *The number of electrons ejected per unit time is proportional to the intensity of the EM radiation and to no other characteristic.* High-intensity EM radiation consists of large numbers of photons per unit area, with all photons having the same characteristic energy hf.

3. If we vary the frequency of the EM radiation falling on a material, we find the following: *For a given material, there is a threshold frequency f_0 for the EM radiation below which no electrons are ejected, regardless of intensity.* Individual photons interact with individual electrons. Thus if photon energy is too low to break an electron away, no electrons will be ejected. If EM radiation were a simple wave, sufficient energy could be obtained by increasing the intensity.

4. If we vary the intensity of the EM radiation and measure the energy of ejected electrons, we find the following: *The maximum kinetic energy of ejected electrons is independent of the intensity of the EM radiation.* Since there are so many electrons in a material, it is extremely unlikely that two photons will interact with the same electron at the same time, thereby increasing the energy given it. Instead (as noted in 2 above), increased intensity results in more electrons of the same energy being ejected. If EM radiation were a simple wave, a higher intensity could give more energy, and higher-energy electrons would be ejected.

5. *The kinetic energy of an ejected electron equals the photon energy minus the binding energy of the electron in the specific material.* An individual photon can give all of its energy to an electron. That energy is partly used to break the electron away from a material. The remainder goes into the ejected electron's kinetic energy. In equation form, this is given by

$$\text{KE}_e = hf - \text{BE} \tag{27.5}$$

where KE_e is the maximum kinetic energy of the ejected electron, hf is the photon's energy, and BE is the **binding energy** of the electron to the particular material. (BE is sometimes called the *work function* of the material.) This equation, due to Einstein in 1905, explains the properties of the photoelectric effect quantitatively. An individual photon of EM radiation (it does not come any other way) interacts with an individual electron, supplying enough energy, BE, to break it away, with the remainder going to kinetic energy. The binding energy is $\text{BE} = hf_0$, where f_0 is the threshold frequency for the particular material. Figure 27.7 shows a graph of maximum KE_e versus the frequency of incident EM radiation falling on a particular material.

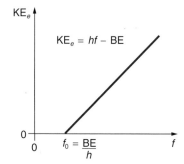

Figure 27.7 Photoelectric effect. A graph of KE_e versus the frequency of EM radiation impinging on a certain material. There is a threshold frequency below which no electrons are ejected, because the individual photon interacting with an individual electron has insufficient energy to break it away. Above the threshold energy, KE_e increases linearly with f, consistent with Equation 27.5. Note that the slope is h—the data can be used to determine Planck's constant experimentally. Einstein gave the first successful explanation of such data by proposing the idea of photons—quanta of EM radiation.

Einstein's idea that EM radiation is quantized was crucial to the beginnings of quantum mechanics. It is a far more general concept than its explanation of the photoelectric effect might imply. All EM radiation comes in the form of photons, and the characteristics of EM radiation are entirely consistent with this fact. (As we will see in the next section, many aspects of EM radiation, such as the hazards of ultraviolet (UV) radiation, can *only* be explained by photon properties.) More famous for modern relativity, Einstein planted an important seed for quantum mechanics in 1905, the same year he published his first paper on special relativity. His explanation of the photoelectric effect was the basis for the Nobel prize awarded to him in 1921. Although his other contributions to theoretical physics were also noted in that award, special and general relativity were not fully recognized in spite of having been partially verified by experiment by 1921. Einstein was world famous by the time of his Nobel prize, and he had an element of the controversial (he remained in Germany during World War I) character that remained with him all his life. Although hero-worshipped and lionized, this great man never received Nobel recognition for his most famous and important work—relativity.

EXAMPLE 27.1 PHOTON ENERGY AND THE PHOTOELECTRIC EFFECT

(a) What is the energy in joules and electron volts of a photon of 420 nm violet light? (b) What is the maximum kinetic energy of electrons ejected from calcium by 420 nm violet light, given the binding energy (or work function) of electrons for calcium metal is 2.71 eV?

Strategy and Concept To solve part (a), note that the energy of a photon is given by Equation 27.4. For part (b), once the energy of the photon is calculated, it is a straightforward application of Equation 27.5 to find the ejected electron's maximum kinetic energy, since BE is given.

Solution for (a) Photon energy is given by Equation 27.4 to be

$$E = hf$$

Since we are given the wavelength rather than the frequency, we solve the familiar relationship $c = f\lambda$ for the frequency, yielding

$$f = \frac{c}{\lambda}$$

Combining these two equations gives the useful relationship

$$E = \frac{hc}{\lambda}$$

Now substituting known values yields

$$E = \frac{(6.63 \times 10^{-34} \text{ J} \cdot \text{s})(3.00 \times 10^{8} \text{ m/s})}{420 \times 10^{-9} \text{ m}}$$
$$= 4.74 \times 10^{-19} \text{ J}$$

Converting to eV, the energy of the photon is

$$E = 4.74 \times 10^{-19} \text{ J} \cdot \frac{1 \text{ eV}}{1.6 \times 10^{-19} \text{ J}} = 2.96 \text{ eV}$$

Solution for (b) Finding the kinetic energy of the ejected electron is now a simple application of Equation 27.5. Substituting the photon energy and binding energy yields

$$\text{KE}_e = hf - \text{BE} = 2.96 \text{ eV} - 2.71 \text{ eV} = 0.25 \text{ eV}$$

Discussion The energy of this 420 nm photon of violet light is a tiny fraction of a joule, and so it is no wonder that a single photon would be difficult for us to sense directly—humans are more attuned to energies on the order of joules. But looking at the energy in electron volts, we can see that this photon has enough energy to affect atoms and molecules. A DNA molecule can be broken with about 1 eV of energy, for example, and typical atomic and molecular energies are on the order of eV, so that the UV photon in this example could have biological effects. The ejected electron (called a *photoelectron*) has a rather low energy, and it would not travel far, except in a vacuum. In fact, if the photon wavelength were longer and its energy less than 2.71 eV, then the formula would give a negative kinetic energy, an impossibility. This simply means that the 420 nm photons with their 2.96 eV energy are not much above the frequency threshold. You can show for yourself that the threshold wavelength is 459 nm (blue light). This means that if calcium metal is used in a light meter, the meter will be insensitive to wavelengths longer than those of blue light. Such a light meter would be completely insensitive to red light, for example.

27.3 PHOTONS

A photon is a quantum of EM radiation. Its energy is given by Equation 27.4 and is related to the frequency f and wavelength λ of the radiation by

$$E = hf = \frac{hc}{\lambda} \qquad \text{(energy of a photon)}$$

where E is the energy of a single photon and c is the speed of light. When working with small systems, energy in eV is often useful. Note that Planck's constant in these units is

$$h = 4.14 \times 10^{-15} \text{ eV} \cdot \text{s}$$

Since many wavelengths are stated in nanometers (nm), it is also useful to know that

$$hc = 1240 \text{ eV} \cdot \text{nm}$$

These will make many calculations a little easier.

All EM radiation is composed of photons, and some of its characteristics can be specifically attributed to photons. Figure 27.8 shows various divisions of the EM spectrum

CONNECTIONS

EM Spectrum

Characteristics of various types of EM radiation were explained in part by their wave characteristics in Chapter 23. The complete description of EM radiation is now possible with the knowledge of photon energy. You may find it useful to refer to Section 23.3 to refresh your memory about some of the characteristics of the EM spectrum.

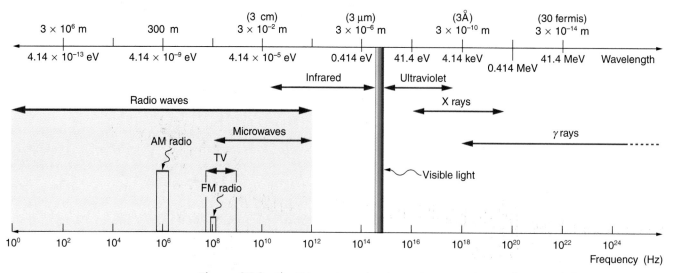

Figure 27.8 The EM spectrum, showing major categories as a function of photon energy in eV, as well as wavelength and frequency. Certain characteristics of EM radiation are directly attributable to photon energy alone.

TABLE 27.1

REPRESENTATIVE ENERGIES FOR SUBMICROSCOPIC EFFECTS (ORDER OF MAGNITUDE ONLY)	
Rotational energies of molecules	10^{-5} eV
Vibrational energies of molecules	0.1 eV
Energy between outer electron shells in atoms	1 eV
Binding energy of a weakly bound molecule	1 eV
Binding energy of a tightly bound molecule	10 eV
Energy to ionize atom or molecule	10 to 1000 eV

plotted against wavelength, frequency, and photon energy. In Chapter 23, where the EM spectrum was first discussed, photon characteristics were alluded to in the discussion of some of the characteristics of UV, x rays, and γ rays. It was noted that these types of EM radiation have properties related to their high frequencies. We can now see that such properties arise because photon energy is greatest at high frequency.

Photons act as individual quanta and interact with individual electrons, atoms, molecules, and so on. The energy a photon carries is, thus, crucial to the effects it has. Table 27.1 lists representative submicroscopic energies in eV. When we compare photon energies from the EM spectrum in Figure 27.8 with energies in the table, we can see how effects vary with the type of EM radiation.

Gamma rays, a form of nuclear and cosmic EM radiation, can have the highest frequencies and, hence, the highest photon energies in the EM spectrum. For example, a γ-ray photon with $f = 10^{21}$ Hz has an energy $E = hf = 6.63 \times 10^{-13}$ J = 4.14 MeV. This is sufficient energy to ionize thousands of atoms and molecules, since only 10 to 1000 eV are needed per ionization. In fact, γ rays are one type of *ionizing radiation*, as are x rays and UV, because they produce ionization in materials that absorb them. Because so much ionization can be produced, a single γ-ray photon can cause significant damage to biological tissue, killing cells or damaging their ability to properly reproduce. When cell reproduction is disrupted, the result can be cancer, one of the known effects of exposure to ionizing radiation. Since cancer cells are rapidly reproducing, they are exceptionally sensitive to the disruption produced by ionizing radiation. This means that ionizing radiation has uses in cancer treatment as well as risks in producing cancer.

High photon energy also enables γ rays to penetrate materials, since a collision with a single atom or molecule is unlikely to absorb all the γ-ray's energy. This can make γ rays useful as a probe, and they are sometimes used in medical imaging. One such image was shown in Figure 23.18.

X rays, as you can see in Figure 27.8, overlap with the low-frequency end of the γ-ray range. Since x rays have energies of keV and up, individual x-ray photons also can produce large amounts of ionization. At lower photon energies, x rays are not as penetrating as γ rays and are slightly less hazardous. X rays are ideal for medical imaging, their most common use, a fact that was recognized immediately upon their discovery in 1895 by the German physicist W. C. Roentgen (1845–1923). (See Figure 27.9.) Within one year of their discovery, x rays (for a time called Roentgen rays) were used for medical diagnostics. Roentgen received the 1901 Nobel prize for the discovery of x rays.

Medical Application

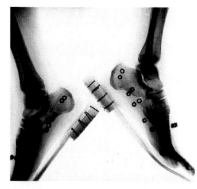

Figure 27.9 An early x-ray image showing booted feet replete with eyelets and cobbler's nails.

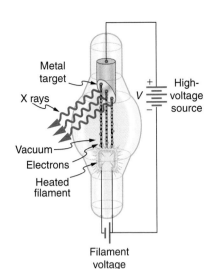

Figure 27.10 X rays are created when energetic electrons strike the anode of this cathode ray tube (CRT). Electrons interact individually with the material they strike, sometimes creating photons of EM radiation.

While γ rays originate in nuclear decay, x rays are created by the process shown in Figure 27.10. Electrons boiled from a hot filament in a vacuum tube are accelerated through a high voltage, gaining kinetic energy from the electrical potential energy. When they strike the anode, the electrons convert their kinetic energy to a variety of forms, including thermal energy. But since an accelerated charge radiates EM waves, and since the electrons act individually, photons are also created. Some of these x-ray photons obtain the full kinetic energy of the electron. The accelerated electrons originate at the cathode, so such a tube is called a cathode ray tube (CRT), and various versions of them are found in TV and computer screens as well as in x-ray machines.

CONNECTIONS

Conservation of Energy

Once again, we find that conservation of energy allows us to consider the initial and final forms that energy takes, without having to make detailed calculations of the intermediate steps. Example 27.2 is solved by considering only the initial and final forms of energy.

EXAMPLE 27.2 X-RAY PHOTON ENERGY AND X-RAY TUBE VOLTAGE

Find the maximum energy in eV of an x-ray photon created by electrons accelerated through a potential difference of 50.0 kV in a CRT like the one in Figure 27.10.

Strategy and Concept Electrons are quantized masses that can give all of their kinetic energy to a single photon when they strike the anode of a CRT. (This is something like the photoelectric effect in reverse.) The kinetic energy of the electron comes from electrical potential energy. Thus we can simply equate the photon energy to electrical potential energy—that is, $hf = qV$. (We do not have to calculate each step from beginning to end if we know that all of the starting energy qV goes to the final form hf.)

Solution The maximum photon energy is $hf = qV$, where q is the charge of the electron and V is the accelerating voltage. Thus,

$$hf = (1.60 \times 10^{-19} \text{ C})(50.0 \times 10^{3} \text{ V})$$

From the definition of the electron volt, we know 1 eV = .60 × 10⁻¹⁹ J, where 1 J = 1 C·V. Gathering terms and converting energy to eV yields

$$hf = (50.0 \times 10^{3})(1.60 \times 10^{-19} \text{ C·V})$$
$$\times \frac{1 \text{ eV}}{1.60 \times 10^{-19} \text{ C·V}}$$
$$= (50.0 \times 10^{3})(1 \text{ eV}) = 50.0 \text{ keV}$$

Discussion This example produces a result that can be applied to many similar situations. If you accelerate a single elementary charge, like that of an electron, through a potential given in volts, then its energy in eV has the same numerical value. Thus a 50.0 kV potential generates 50.0 keV electrons, which in turn can create 50.0 keV photons. Similarly, a 100 kV potential in an x-ray tube can generate 100 keV x-ray photons. Many x-ray tubes have adjustable voltages so that various energy x rays with differing abilities to penetrate can be generated.

Figure 27.11 shows the spectrum of x rays obtained from an x-ray tube. There are two distinct aspects to the spectrum. First, the smooth distribution results from electrons being decelerated in the anode material. A curve like this is obtained by detecting many photons, and it is apparent that the maximum energy is unlikely. This decelerating process produces radiation that is called **bremsstrahlung** (German for *braking radiation*). The second aspect is the existence of sharp peaks in the spectrum; these are called **characteristic x rays**, since they are characteristic of the anode material. Characteristic x rays come from atomic excitations unique to a given type of anode material. They are

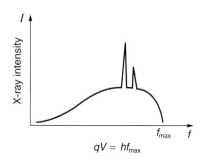

$qV = hf_{max}$

Figure 27.11 X-ray spectrum obtained when energetic electrons strike a material. The smooth part of the spectrum is bremsstrahlung, while the peaks are characteristic of the anode material. Both are atomic processes that produce energetic photons known as x-ray photons.

Human & Biological Application

akin to lines in atomic spectra, implying the energy levels of atoms are quantized. Phenomena such as discrete atomic spectra and characteristic x rays are explored further in Chapter 28, on atomic physics.

Ultraviolet radiation overlaps with the low end of the energy range of x rays, but UV is typically lower in energy. UV comes from deexcitation of atoms that may be part of a hot solid or gas. The atoms can be given energy that they later release as UV by numerous processes, including electric discharge, nuclear explosion, thermal agitation, and exposure to x rays. A UV photon has sufficient energy to ionize atoms and molecules, which makes its effects different from those of visible light. UV thus has some of the same biological effects as γ rays and x rays. For example, it can cause skin cancer and it is used as a sterilizer. The major difference is that several UV photons are required to disrupt cell reproduction or kill a bacterium, whereas single γ-ray and x-ray photons can do the same damage. But since UV does have the energy to alter molecules, it can do what visible light cannot. One of the beneficial aspects of UV is that it triggers the production of vitamin D in the skin, whereas visible light has insufficient energy per photon to alter the molecules that trigger this production. Infantile jaundice is treated by exposing the baby to UV (with eye protection), the beneficial effects of which are thought to be related to its ability to alter molecules.

EXAMPLE 27.3 PHOTON ENERGY AND EFFECTS FOR UV

Short-wavelength UV is sometimes called vacuum UV, because it is strongly absorbed by air and must be studied in a vacuum. Calculate the photon energy in eV for 100 nm vacuum UV, and estimate the number of molecules it could ionize or break apart.

Strategy Using Equation 27.4 and appropriate constants, we can find the photon energy and compare it with energy information in Table 27.1.

Solution The energy of a photon is given by

$$E = hf = \frac{hc}{\lambda}$$

Using $hc = 1240$ eV·nm, we find that

$$E = \frac{hc}{\lambda} = \frac{1240 \text{ eV·nm}}{100 \text{ nm}} = 12.4 \text{ eV}$$

Discussion and Implications According to Table 27.1, this photon energy might be able to ionize an atom or molecule, and it is about what is needed to break up a tightly bound molecule, since they are bound by approximately 10 eV. This photon energy could destroy about a dozen weakly bound molecules. Because of its high photon energy, UV disrupts atoms and molecules it interacts with. One good consequence is that all but the longest-wavelength UV is strongly absorbed and is easily blocked by sunglasses. In fact, most of the sun's UV is absorbed by a thin layer of ozone in the upper atmosphere, protecting sensitive organisms on earth.

The range of photon energies for visible light from red to violet is 1.63 to 3.26 eV, respectively (left as an end-of-chapter problem to verify). These energies are on the order of those between outer electron shells in atoms and molecules. This means that these photons can be absorbed by atoms and molecules. A *single* photon can actually stimulate the retina, for example, by altering a receptor molecule that then triggers a nerve impulse. Photons can only be absorbed or emitted by atoms and molecules that have precisely the correct energy step between levels. For example, if a red photon of frequency f encounters a molecule that has an energy between shells, ΔE, equal to hf, then the photon can be absorbed. Violet flowers absorb red and reflect violet; this implies there is no energy step between levels in the receptor molecule equal to the violet photon's energy, but there is an energy step for the red.

There are some noticeable differences in the characteristics of light between the two ends of the visible spectrum that are due to photon energies. Red light has insufficient photon energy to expose most black and white film, and it is thus used to illuminate "dark rooms" where such film is developed. Since violet light has a higher photon energy, dyes that absorb violet tend to fade more quickly than those which do not. (See Figure 27.12.) Take a look at some faded color posters in a storefront sometime, and you will notice that the blues and violets are the last to fade. This is because other dyes, such as red and green dyes, absorb blue and violet photons, the higher energies of which break up their weakly bound molecules. (Complex molecules such as those in dyes and DNA tend to be weakly bound.) Blue and violet dyes reflect those colors and, therefore, do not absorb the more energetic photons, thus suffering less molecular damage.

Transparent materials, such as some glasses, do not absorb any visible light, because there is no energy level to which the light could raise the atoms or molecules. Since individual photons interact with individual atoms, it is nearly impossible to have two photons absorbed simultaneously to reach a high level. Because of its lower photon energy, visible light can sometimes pass through many kilometers of a substance, while higher frequencies like UV, x rays, and γ rays are absorbed, because they have sufficient photon energy to ionize the material.

Figure 27.12 Why do the reds, yellows, and greens fade before the blues and violets? The answer is related to photon energy.

EXAMPLE 27.4 HOW MANY PHOTONS PER SECOND DOES A TYPICAL LIGHT BULB PRODUCE?

Assuming that 10.0% of a 100 W light bulb's energy output is in the visible range (typical for incandescent bulbs) with an average wavelength of 580 nm, calculate the number of visible photons emitted per second.

Strategy and Concept Power is energy per unit time, and so if we can find the energy per photon, we can determine the number of photons per second. This will best be done in joules, since power is given in watts, which are joules per second.

Solution The power in visible light production is 10.0% of 100 W, or 10.0 J/s. The energy of the average visible photon is found by substituting the given wavelength into the formula

$$E = \frac{hc}{\lambda}$$

This produces

$$E = \frac{(6.63 \times 10^{-34}\ \text{J·s})(3.00 \times 10^{8}\ \text{m/s})}{580 \times 10^{-9}\ \text{m}}$$

$$= 3.43 \times 10^{-19}\ \text{J}$$

The number of visible photons per second is thus

$$\text{photons/s} = \frac{10.0\ \text{J/s}}{3.43 \times 10^{-19}\ \text{J/photon}}$$

$$= 2.92 \times 10^{19}\ \text{photons/s}$$

Discussion This incredible number of photons per second is verification that individual photons are insignificant in ordinary human experience. It is also a verification of the correspondence principle—on the macroscopic scale, quantization becomes nearly continuous or classical. Finally, there are so many photons emitted by a 100 W light bulb that it can be seen by the unaided eye many miles away.

Infrared (IR) radiation has even lower photon energies than visible light and cannot significantly alter atoms and molecules. IR can be absorbed and emitted by atoms and molecules, particularly between closely spaced states. IR is extremely strongly absorbed by water, for example, because water molecules have many states separated by energies on the order of 10^{-5} to 10^{-2} eV, well within the IR and microwave energy ranges. This is why in the IR range skin is almost jet black, with an emissivity near 1—there are many states in water molecules in the skin that can absorb a large range of IR photon energies. Not all molecules have this property. Air, for example, is nearly transparent to many IR frequencies.

Microwaves are the highest frequencies that can be produced by electronic circuits, although they are also produced naturally. Thus microwaves are similar to IR but do not extend to as high frequencies. There are states in water and other molecules that have the same frequency and energy as microwaves, typically about 10^{-5} eV. This is why food absorbs microwaves more strongly than many other materials, making microwave ovens an efficient way of putting energy directly into food. Although this can be viewed as heat transfer by radiation, it is more of a resonance phenomenon.

Photon energies for both IR and microwaves are so low that huge numbers of photons are involved in any significant energy transfer by IR or microwaves (such as warming yourself with a heat lamp or cooking pizza in the microwave). Visible light, IR, microwaves, and all lower frequencies cannot produce ionization with single photons and do not ordinarily have the hazards of higher frequencies. When visible, IR, or microwave radiation *is* hazardous, such as the inducement of cataracts by microwaves, the hazard is due to huge numbers of photons acting together (not to an accumulation of photons, such as sterilization by weak UV). The negative effects of visible, IR, or microwave radiation can be thermal effects, which could be produced by any heat source. But one difference is that at very high intensity, strong electric and magnetic fields can be produced by photons acting together. Such fields can actually ionize materials. Although some people suspect that living near high-voltage power lines is hazardous to health, ongoing studies of the transient field effects produced by these lines show their strengths to be insufficient to cause damage. Demographic studies also fail to show significant correlation of ill effects with high-voltage power lines.

It is virtually impossible to detect individual photons having frequencies below microwave frequencies, because of their low photon energy. But the photons are there. A continuous EM wave, like those discussed in Chapter 23, can be constructed from the superposition of photons. At low frequencies, it is an excellent approximation to treat EM waves classically as time- and position-varying electric and magnetic fields with no discernible quantization. This is another example of the correspondence principle in situations involving huge numbers of photons.

27.4 PHOTON MOMENTUM

The quantum of EM radiation we call a photon has properties analogous to particles we can see, such as grains of sand. A photon interacts as a unit in collisions or when absorbed, rather than as an extensive wave. Quanta of mass, like electrons and protons, also act like macroscopic particles—something we expect, because they are the smallest units of matter. Particles carry momentum as well as energy. It would be intriguing if a massless particle like a photon also had momentum, since we are so accustomed to mass being part of momentum. Indeed, there has long been evidence that EM radiation carries momentum. (Maxwell and others who studied EM waves predicted that they would carry momentum.) It is now a well-established fact that photons do have momentum. In fact, photon momentum is suggested by the photoelectric effect, where photons knock electrons out of a substance. Figure 27.13 shows macroscopic evidence of photon momentum.

Figure 27.13 shows a comet with two prominent tails. What most people do not know about the tails is that they always point *away* from the sun rather than trailing behind the comet (like the tail of Bo Peep's sheep). Comet tails are composed of gases and dust evaporated from the body of the comet and ionized gas. The dust particles recoil away from the sun when photons scatter from them. Evidently, photons carry momentum in the direction of their motion (away from the sun), and some of this momentum is transferred to dust particles in collisions. Gas atoms and molecules in the blue tail are

Figure 27.13 The tails of the Hale-Bopp comet point away from the sun, evidence that light has momentum. Dust emanating from the body of the comet forms this tail. Particles of dust are pushed away from the sun by light reflecting from them. The blue ionized gas tail is also created by photons interacting with atoms in the comet material.

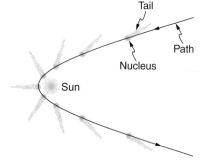

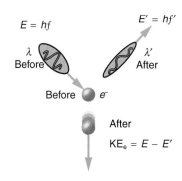

Figure 27.14 The Compton effect is the name given to the scattering of a photon by an electron. Energy and momentum are conserved, resulting in a reduction of both for the scattered photon. Studying this effect, Compton verified that photons have momentum in spite of being massless.

most affected by other particles of radiation such as protons and electrons emanating from the sun, rather than by the momentum of photons.

Momentum is conserved in quantum mechanics just as it is in relativity and classical physics. Some of the earliest direct experimental evidence of this came from scattering of x-ray photons by electrons in substances, named Compton scattering after the American physicist Arthur H. Compton (1892–1962). Around 1923, Compton observed that x rays scattered from materials had a decreased energy and correctly analyzed this as being due to scattering of photons from electrons. (See Figure 27.14.) He won a Nobel prize in 1927 for the discovery of this scattering, now called the **Compton effect**, because it helped prove that **photon momentum** is given by

$$p = \frac{h}{\lambda} \tag{27.6}$$

where h is Planck's constant and λ is the photon wavelength. Photon momentum provides another connection between mass and pure energy. First, Einstein's relativity connected them through $E = mc^2$, implying pure energy, such as massless photons, could be created from the destruction of mass. Now, we see that massless photons have momentum—another characteristic of mass. (Note that relativistic momentum given by Equation 26.4 as $p = \gamma mu$ is valid only for particles having mass.)

We can see that photon momentum is small, since $p = h/\lambda$ and h is very small. It is for this reason that we do not ordinarily observe photon momentum. Our mirrors do not tend to recoil when light reflects from them (except perhaps in cartoons). Compton saw the effects of photon momentum because he was observing x rays that have a small wavelength and a relatively large momentum, interacting with the lightest of particles, the electron.

CONNECTIONS

Conservation of Momentum

Not only is momentum conserved in all realms of physics, but all types of particles are found to have momentum. We expect particles with mass to have momentum, but now we see that massless particles including photons also carry momentum.

EXAMPLE 27.5   ELECTRON AND PHOTON MOMENTUM COMPARED

(a) Calculate the momentum of a visible photon that has a wavelength of 500 nm. (b) Find the velocity of an electron having the same momentum. (c) What is the energy of the electron, and how does it compare with the energy of the photon?

Strategy Finding the photon momentum is a straightforward application of its definition in Equation 27.6. If we find the photon momentum is small, then we can assume that an electron with the same momentum will be nonrelativistic, making it easy to find its velocity and kinetic energy from the classical formulas.

Solution for (a) Photon momentum is given by Equation 27.6:

$$p = \frac{h}{\lambda}$$

Entering the given photon wavelength yields

$$p = \frac{6.63 \times 10^{-34} \text{ J·s}}{500 \times 10^{-9} \text{ m}} = 1.33 \times 10^{-27} \text{ kg·m/s}$$

Solution for (b) Since this momentum is indeed small, we will use the classical expression $p = mv$ to

find the velocity of an electron with this momentum. Solving for v and using the known value for the mass of an electron gives

$$v = \frac{p}{m}$$
$$= \frac{1.33 \times 10^{-27} \text{ kg·m/s}}{9.11 \times 10^{-31} \text{ kg}}$$
$$= 1460 \text{ m/s}$$

Solution for (c) The electron has kinetic energy, which is classically given by

$$KE_e = \tfrac{1}{2}mv^2$$

Thus,

$$KE_e = \tfrac{1}{2}(9.11 \times 10^{-31} \text{ kg})(1460 \text{ m/s})^2 = 9.65 \times 10^{-25} \text{ J}$$

Converting this to eV, by multiplying by (1 eV)/(1.60 × 10^{-19} J), yields

$$KE_e = 6.06 \times 10^{-6} \text{ eV}$$

(continued)

(continued)

The photon energy E is

$$E = \frac{hc}{\lambda} = \frac{1240 \text{ eV} \cdot \text{nm}}{500 \text{ nm}} = 2.48 \text{ eV}$$

which is about five orders of magnitude greater.

Discussion Photon momentum is indeed small. Even if we have huge numbers of them, such as the approximately 10^{19} per second in Example 27.4, the total momentum they carry is small. An electron with the same momentum has a 1460 m/s velocity, which is clearly nonrelativistic. A more massive particle with the same momentum would have an even smaller velocity. This is borne out by the fact that it takes far less energy to give an electron the same momentum as a photon. But on a quantum mechanical scale, especially for high-energy photons interacting with small masses, photon momentum is significant. Even on a large scale, photon momentum can have an effect if there are enough of them and if there is nothing to prevent the slow recoil of mat-

ter. Comet tails are one example, but there are also proposals to build space sails that use huge low-mass mirrors to reflect sunlight. In the vacuum of space, the mirrors would gradually recoil and could actually take spacecraft from place to place in the solar system. (See Figure 27.15).

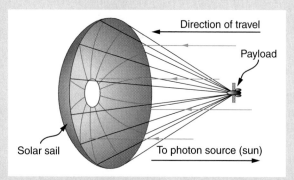

Figure 27.15 Space sails have been proposed that use the momentum of sunlight reflecting from gigantic low-mass sails to propel spacecraft about the solar system.

There is a relationship between photon momentum p and photon energy E that is consistent with the relation given in Equation 26.8 as $E^2 = (pc)^2 + (mc^2)^2$. (This was discussed briefly in Section 26.6.) We know m is zero for a photon, but p is not, so that Equation 26.8 becomes

$$E = pc$$

or

$$p = \frac{E}{c} \quad \text{(photons)} \quad \textbf{(27.7)}$$

To check the validity of Equation 27.7, note that $E = hc/\lambda$ for a photon. Substituting this into Equation 27.7 yields

$$p = \frac{hc/\lambda}{c} = \frac{h}{\lambda}$$

as determined experimentally and discussed above. Thus Equation 27.7 is consistent with Compton's result $p = h/\lambda$. For a further verification of the relationship between photon energy and momentum, see the following example.

EXAMPLE 27.6 PHOTON ENERGY AND MOMENTUM

Show that $p = E/c$ for the photon considered in the preceding example.

Strategy We will take the energy E found in the preceding example, divide it by the speed of light, and see if the same momentum is obtained as before.

Solution Given the energy of the photon is 2.48 eV and converting this to joules, we get

$$p = \frac{E}{c} = \frac{(2.48 \text{ eV})(1.60 \times 10^{-19} \text{ J/eV})}{3.00 \times 10^8 \text{ m/s}}$$

$$= 1.33 \times 10^{-27} \text{ kg} \cdot \text{m/s}$$

Discussion This value for momentum is the same as found before (note that unrounded values are used in all calculations to avoid even small rounding errors), an expected verification of the relationship $p = E/c$. This also means the relationship between energy, momentum, and mass given by $E^2 = (pc)^2 + (mc^2)^2$ applies to both matter and photons. Once again, note that p is not zero even when m is.

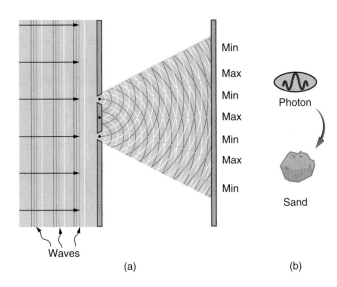

Figure 27.16 (a) The interference pattern for light through a double slit is a wave property understood by analogy to water waves. (b) The properties of photons having quantized energy and momentum and acting as a concentrated unit are understood by analogy to macroscopic particles.

27.5 THE PARTICLE-WAVE DUALITY

We have long known that EM radiation is a wave. Now we see that it is composed of photons, which are massless particles. This may seem contradictory, since we ordinarily deal with large objects that never act like both wave and particle. An ocean wave, for example, looks nothing like a rock. To understand small-scale phenomena, we make analogies with the large scale we observe directly. When we say something is a wave, we mean it shows interference effects analogous to those seen in overlapping water waves. (See Figure 27.16.) Two examples of waves are sound and EM radiation. When we say something is a particle, we mean it is quantized, has momentum, and interacts as a unit. Examples of particles include electrons, atoms, and photons of EM radiation.

There is no doubt that EM radiation interferes and has the properties of wavelength and frequency. It is a wave. There is also no doubt that it is composed of particles—photons with energy and momentum. We call this twofold nature the **particle-wave duality**, meaning that EM radiation has both particle and wave properties. This so-called duality is just a name for properties of the photon analogous to things and phenomena we can observe directly. If it seems strange, it is only because we do not ordinarily observe details on the quantum level directly.

Since we have a particle-wave duality for pure massless energy in the form of photons, and since we have seen connections between energy and mass in that both have momentum, it is reasonable to ask if there is a particle-wave duality for matter. If the EM radiation we once thought to be a pure wave has particle properties, is it possible that matter has wave properties? The answer is yes. The consequences are tremendous, as we will begin to see in the next section.

27.6 THE WAVE NATURE OF MATTER

In 1923 a French physics graduate student named Prince Louis-Victor de Broglie (1892–1987) made a radical proposal based on the hope that nature is symmetric. If EM radiation has both particle and wave properties, then nature would be symmetric if matter also had both particle and wave properties. If what we once thought of as an unequivocal wave (EM radiation) is also a particle, then what we think of as an unequivocal particle (matter) may also be a wave. De Broglie's suggestion, made as part of his doctoral thesis, was so radical that it was greeted with some skepticism. A copy of his thesis was sent to Einstein, who said it was not only probably correct, but that it might be of fundamental importance. With the support of Einstein and a few other prominent physicists, de Broglie was awarded his doctorate.

Of course, de Broglie did more than make a vague suggestion. He took both relativity and quantum mechanics into account to develop the proposal that *all particles have a wavelength* given by

$$\lambda = \frac{h}{p} \qquad \text{(matter and photons)} \qquad (27.8)$$

where h is Planck's constant and p is momentum. This is defined to be the **de Broglie wavelength**. (Note that we already have this for photons, since Equation 27.6 is $p = h/\lambda$.) The hallmark of a wave is interference. If matter is a wave, then it must exhibit constructive and destructive interference. Why isn't this ordinarily observed? The answer is that in order to see significant interference effects, a wave must interact with an object about the same size as its wavelength (as discussed extensively in Chapter 25). Since h is very small, λ is also small, especially for macroscopic objects. A 3 kg bowling ball moving at 10 m/s, for example, has $\lambda = h/p = (6.63 \times 10^{-34} \text{ J·s})/[(3 \text{ kg})(10 \text{ m/s})] = 2 \times 10^{-35}$ m. This means the bowling ball would have to interact with something about 10^{-35} m in size—far smaller than anything known. When waves interact with objects much larger than their wavelength, they move in straight lines and show negligible interference effects (such as light rays in geometric optics). To get easily observed interference effects, a larger wavelength than that of a bowling ball is needed. It would be wise to examine the smallest mass possible, the electron, since its tiny mass and momentum would produce a larger wavelength. This is exactly what was done.

American physicists Clinton J. Davisson and Lester H. Germer in 1925 and, independently, British physicist G. P. Thomson (son of J. J. Thomson, discoverer of the electron) in 1926 scattered electrons from crystals and found diffraction patterns. These patterns are exactly consistent with interference of electrons having the de Broglie wavelength and are somewhat analogous to light interacting with a diffraction grating. (See Figure 27.17.)

De Broglie's proposal of a wave nature for all particles initiated a remarkably productive era in which the foundations for quantum mechanics were laid. In 1926, the Austrian physicist Erwin Schrödinger (1887–1961) published four papers in which the wave nature of particles was treated explicitly with wave equations. At the same time, many others began important work. Among them was German physicist Werner Heisenberg (1901–1976), who, among many other contributions to quantum mechanics, formulated a mathematical treatment of the wave nature of matter that used matrices rather than wave equations. We will deal with some specifics in later sections, but it is worth noting that de Broglie's work was a watershed for the development of quantum mechanics. De Broglie was awarded the Nobel prize in 1929 for his vision, as were Davisson and G. P. Thomson in 1937 for their experimental verification of de Broglie's hypothesis.

CONNECTIONS

Waves

All microscopic particles, whether massless, like photons, or having mass, like electrons, have wave properties. The connections between mass and energy are fundamental. Not only are mass and energy interconvertible, they both have wave and particle properties.

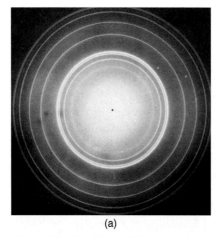
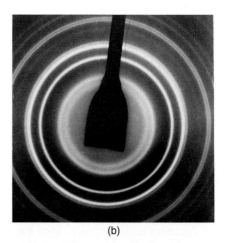

(a) (b)

Figure 27.17 These diffraction patterns were obtained for identical wavelength neutrons (a) and electrons (b) diffracted by crystalline aluminum. Bright regions are those of constructive interference, while dark regions are those of destructive interference.

EXAMPLE 27.7 ELECTRON WAVELENGTH VERSUS VELOCITY AND ENERGY

For an electron having a de Broglie wavelength of 0.167 nm (appropriate for interacting with crystal lattice structures that are about this size): (a) Calculate the electron's velocity, assuming it is nonrelativistic. (b) Calculate the electron's kinetic energy in eV.

Strategy For part (a), since the de Broglie wavelength is given, the electron's velocity can be obtained from $\lambda = h/p$ by using the nonrelativistic formula for momentum, $p = mv$. For part (b), once v is obtained (and it has been verified that v is nonrelativistic), the classical kinetic energy is simply $(1/2)mv^2$.

Solution for (a) Substituting the nonrelativistic formula for momentum ($p = mv$) into the de Broglie wavelength gives

$$\lambda = \frac{h}{p} = \frac{h}{mv}$$

Solving for v gives

$$v = \frac{h}{m\lambda}$$

Substituting known values yields

$$v = \frac{6.63 \times 10^{-34} \text{ J} \cdot \text{s}}{(9.11 \times 10^{-31} \text{ kg})(0.167 \times 10^{-9} \text{ m})}$$

$$= 4.36 \times 10^{6} \text{ m/s}$$

Solution for (b) While fast compared with an automobile, this electron's speed is not highly relativistic, and so we can comfortably use the classical formula to find the electron's kinetic energy and convert it to eV as requested:

$$\begin{aligned} \text{KE} &= \tfrac{1}{2}mv^2 \\ &= 0.5(9.11 \times 10^{-31} \text{ kg})(4.36 \times 10^{6} \text{ m/s})^2 \\ &= 8.64 \times 10^{-18} \text{ J} \cdot \frac{1 \text{ eV}}{1.60 \times 10^{-19} \text{ J}} \\ &= 54.0 \text{ eV} \end{aligned}$$

Discussion This low energy means that these 0.167 nm electrons could be obtained by accelerating them through a 54.0 V potential, an easy task. The results also confirm the assumption that the electrons are nonrelativistic, since their velocity is just over 1% of the speed of light and the kinetic energy is about 0.01% of the rest mass of an electron (0.511 MeV). If the electrons had turned out to be relativistic, we would have had to use more involved calculations employing relativistic momentum (γmv) and relativistic kinetic energy (($\gamma - 1)mc^2$).

One use of the wave nature of matter is found in the electron microscope. As we have discussed, there is a limit to the detail observed with any probe having a wavelength. Resolution, or observable detail, is limited to about one wavelength. Since a potential of only 54 V can produce electrons with subnanometer wavelengths, it is easy to get electrons with much smaller wavelengths than those of visible light (hundreds of nanometers). Electron microscopes can, thus, be constructed to detect much smaller details than optical microscopes. (See Figure 27.18.) Moreover, particle accelerators have been used and improved since the 1930s to produce even smaller wavelength probes. These particle accelerators have detected details of the nucleus and of its substructures as well.

Electrons were the first particles with mass to be directly confirmed to have the wavelength proposed by de Broglie. Subsequently, protons, helium nuclei, neutrons, and many others have been observed to exhibit

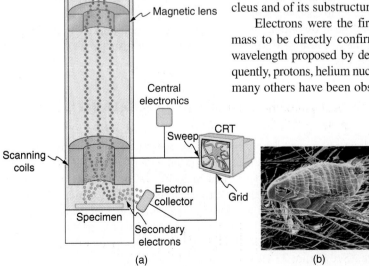

Figure 27.18 Schematic of an electron microscope (a) used to observe details as small as those in the photograph (b).

THINGS GREAT AND SMALL

A Submicroscopic Diffraction Grating

The wave nature of matter allows it to exhibit all the characteristics of other, more familiar, waves. Diffraction gratings, for example, produce diffraction patterns for light that depend on grating spacing and the wavelength of the light. This effect, as with most wave phenomena, is most pronounced when the wave interacts with objects having a size similar to its wavelength. For gratings, this is the spacing between multiple slits. (See Section 25.4 for a detailed analysis of diffraction gratings.) When electrons interact with a system having a spacing similar to the electron wavelength, they show the same types of interference patterns as light does for diffraction gratings, as shown at left in Figure 27.A.

Atoms are spaced at regular intervals in a crystal, as shown in the bottom part of Figure 27.A, and the gaps between them act like the openings in a diffraction grating. At certain incident angles, the paths of electrons differ by one wavelength and, thus, interfere constructively. At other angles, the path length differences are not an integral wavelength, and there is partial to total destructive interference. This type of scattering from a large crystal with well-defined lattice planes can produce dramatic interference patterns. It is called *Bragg reflection* for the father and son team who first explored and analyzed it in some detail. The expanded view also shows the path length differences and indicates how these depend on angle in a manner similar to the diffraction patterns for light reflecting from a diffraction grating.

The wavelength of matter is a submicroscopic characteristic that explains a macroscopic phenomenon such as

Bragg reflection. Similarly, the wavelength of light is a submicroscopic characteristic that explains the macroscopic phenomenon of diffraction patterns.

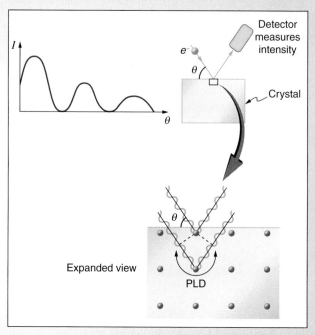

Figure 27.A This diffraction pattern at top left is produced by scattering electrons from a crystal and is graphed as a function of incident angle relative to the crystal lattice Electrons scatter from a regular array of atoms in a crystal, as shown at bottom. Electrons scattering from the second layer of atoms travel farther than those scattered from the top layer. If the path length difference (PLD) is an integral wavelength, there is constructive interference.

interference when they interact with objects having sizes similar to their de Broglie wavelength. The de Broglie wavelength for massless particles was well established in the 1920s for photons, and it has since been observed that all massless particles—and there are others such as neutrinos—have a de Broglie wavelength $\lambda = h/p$. The wave nature of all particles, having mass or not, is a universal characteristic of nature. We shall see in following sections that implications of the de Broglie wavelength include the quantization of energy in atoms and molecules, and an alteration of our basic view of nature on the small scale. The next section, for example, shows that nothing can be exactly predetermined, regardless of how hard we try. We cannot even say exactly where something is or what its energy is.

27.7 PROBABILITY: THE HEISENBERG UNCERTAINTY PRINCIPLE

Matter and photons are waves, implying they are spread out over some distance. What is the position of a particle, such as an electron? Is it at the center of the wave? The answer lies in what you observe if you measure the position of an electron. Experiments show that you will find the electron at some definite location. But if you set up exactly the same situation and measure it again, you will find the electron in a different location, often far outside any experimental uncertainty in your measurement. Repeated measurements will display a statistical distribution of locations that appears wavelike. (See Figure 27.19.)

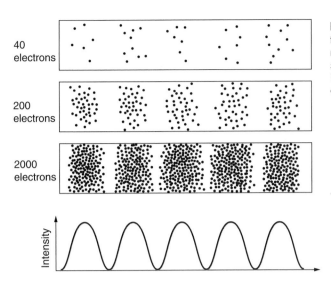

Figure 27.19 The building up of the diffraction pattern of electrons scattered from a crystal surface. Each electron arrives at a definite location, which cannot be predicted. The overall distribution shown at the bottom can be predicted as the diffraction of waves having the de Broglie wavelength of the electrons. This is the probability distribution for the electrons.

After de Broglie proposed the wave nature of matter, many physicists, including Schrödinger and Heisenberg, explored the consequences. The idea quickly emerged that, *because of its wave character, a particle's trajectory and destination cannot be predicted for each particle individually*. Each particle goes to a definite place (as illustrated in Figure 27.19). After compiling enough data, you get a distribution related to the particle's wavelength and diffraction pattern. There is a certain *probability* of finding the particle at a given location, and the overall pattern is called a **probability distribution**. Those who developed quantum mechanics devised wave equations that predicted the probability distribution in various circumstances.

It is somewhat disquieting to think that you cannot predict exactly where an individual particle will go, or even follow it to its destination. Let us explore what happens if we try to follow a particle. Consider the double slit patterns obtained for electrons and photons in Figure 27.20. First, we note that these patterns are identical, following $d \sin \theta = m\lambda$, the equation for double slit constructive interference developed in Section 25.3, where d is the slit separation and λ is the electron or photon wavelength.

Both patterns build up statistically as individual particles fall on the detector. This can be observed for photons or electrons—for now, let us concentrate on electrons. You might imagine that the electrons are interfering with one another as any waves do. To test this, you can lower the intensity until there is never more than one electron between the slits and the screen. The same interference pattern builds up. This implies that particles actually interfere with themselves. Does this also mean that the electron goes through both slits? An electron is a basic unit of matter that does not seem to be divisible. But it is a fair question, and so we should look to see if the electron traverses one slit or the other, or both. One possibility is to have coils around the slits that detect charges moving through them. What is observed is that an electron always goes through one slit or the other; it does not split to go through both. But there is a catch. If you determine that the electron went through one of the slits, you no longer get a double slit pattern—instead, you get single slit interference. There is no escape by using another method of determining which slit the electron went through. Knowing the particle went through one slit forces a single slit pattern. If you do not observe which slit the electron goes through, you obtain a double slit pattern. (See Figure 27.21.)

How does knowing which slit the electron passed through change the pattern? The answer is fundamentally important—*measurement disturbs the system being observed*. Information can be lost, and in some cases it is impossible to measure two physical quantities simultaneously to exact precision. For example, you can measure the position of a moving electron by scattering light or other electrons from it. Those probes have momentum themselves, and by scattering from the electron, they change its momentum *in a manner that loses information*. There is a limit to absolute knowledge, even in principle.

(a) Electrons

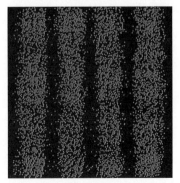

(b) Photons

Figure 27.20 Double slit interference for electrons (a) and photons (b) is identical for equal wavelengths and equal slit separations. Both patterns are probability distributions in the sense that they are built up by individual particles traversing the apparatus, the paths of which are not individually predictable.

Figure 27.21 A double slit interference pattern is obtained for particles, confirming their wave character. If the particle is observed to pass through one of the slits, a single slit pattern is obtained—here actually two superimposed single slit patterns. This is an example of observation fundamentally disturbing a system.

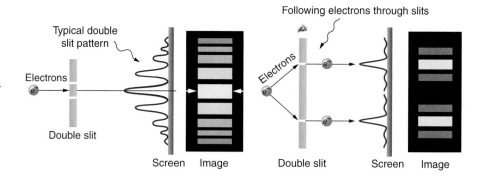

Figure 27.22 Werner Heisenberg was one of the best of those physicists who developed early quantum mechanics. Not only did his work enable a description of nature on the very small scale, it also changed our view of the availability of knowledge. Although he is universally recognized for his brilliance and the importance of his work (he received the Nobel prize in 1932, for example), Heisenberg remained in Germany during World War II and headed the German effort to build a nuclear bomb, permanently alienating himself from most of the scientific community.

It was Werner Heisenberg who first enunciated this limit to knowledge in 1927 as a result of his work on quantum mechanics and the wave characteristics of all particles. (See Figure 27.22.) Specifically, consider simultaneously measuring the position and momentum of an electron (it could be any particle). There is an **uncertainty in position** Δx that is approximately equal to the wavelength of the particle. That is,

$$\Delta x \approx \lambda$$

As discussed above, a wave is not located at one point in space. If the electron's position is measured repeatedly, a spread in locations will be observed, implying an uncertainty in position Δx. To detect the position of the particle, we must interact with it, such as having it collide with a detector. In the collision, the particle will lose momentum. This change in momentum could be anywhere from close to zero to the total momentum of the particle, $p = h/\lambda$. It is not possible to tell how much momentum will be transferred to a detector, and so there is an **uncertainty** Δp **in the momentum**, too. In fact, the uncertainty in momentum may be as large as the momentum itself, which in equation form means that

$$\Delta p \approx \frac{h}{\lambda}$$

The uncertainty in position can be reduced by using a shorter-wavelength electron, since $\Delta x \approx \lambda$. But shortening the wavelength increases the uncertainty in momentum, since $\Delta p \approx h/\lambda$. Conversely, the uncertainty in momentum can be reduced by using a longer-wavelength electron, but this increases the uncertainty in position. Mathematically, you can express this trade-off by multiplying the uncertainties. The wavelength cancels, leaving

$$\Delta x \, \Delta p \approx h$$

So if one uncertainty is reduced, the other must increase so that their product is $\approx h$.

With the use of advanced mathematics, Heisenberg showed that the best that can be done in a *simultaneous measurement of position and momentum* is

$$\Delta x \, \Delta p \geq \frac{h}{4\pi} \qquad (27.9)$$

This is known as the **Heisenberg uncertainty principle**. It is impossible to measure position x and momentum p simultaneously with uncertainties Δx and Δp that multiply to be less than $h/4\pi$. Neither uncertainty can be zero. Neither uncertainty can become small without the other becoming large. A small wavelength allows accurate position measurement, but it increases the momentum of the probe to the point that it further disturbs the momentum of a system being measured. For example, if an electron is scattered from an atom and has a wavelength small enough to detect the position of electrons in the atom, its momentum can knock the electrons from their orbits in a manner that loses information about their original motion. It is therefore impossible to follow an electron in its orbit around an atom. If you measure the electron's position, you will find it in a definite location, but the atom will be disrupted. Repeated measurements on identical atoms will produce interesting probability distributions for electrons around the atom, but they will not produce motion information.

EXAMPLE 27.8 HEISENBERG UNCERTAINTY PRINCIPLE IN POSITION AND MOMENTUM FOR AN ATOM

(a) If the position of an electron in an atom is measured to an accuracy of 0.0100 nm, what is the electron's uncertainty in velocity? (b) If the electron has this velocity, what is its kinetic energy in eV?

Strategy The uncertainty in position is the accuracy of the measurement, or $\Delta x = 0.0100$ nm. Thus the smallest uncertainty in momentum Δp can be calculated using Equation 27.9. Once the uncertainty in momentum Δp is found, the uncertainty in velocity can be found from $\Delta p = m \Delta v$.

Solution for (a) Using the equals sign in the uncertainty principle, we have

$$\Delta x \, \Delta p = \frac{h}{4\pi}$$

Solving for Δp, and substituting known values, gives

$$\Delta p = \frac{h}{4\pi \, \Delta x} = \frac{6.63 \times 10^{-34} \text{ J·s}}{4\pi(1.00 \times 10^{-11} \text{ m})}$$

Thus,

$$\Delta p = 5.28 \times 10^{-24} \text{ kg·m/s} = m \, \Delta v$$

Solving for Δv and substituting the mass of an electron gives

$$\Delta v = \frac{\Delta p}{m} = \frac{5.28 \times 10^{-24} \text{ kg·m/s}}{9.11 \times 10^{-31} \text{ kg}}$$

$$= 5.79 \times 10^6 \text{ m/s}$$

Solution for (b) Although large, this velocity is not highly relativistic, and so the electron's kinetic energy is

$$\text{KE}_e = \tfrac{1}{2}mv^2$$

$$= 0.5(9.11 \times 10^{-31} \text{ kg})(5.79 \times 10^6 \text{ m/s})^2$$

$$= 1.53 \times 10^{-17} \text{ J} \cdot \frac{1 \text{ eV}}{1.60 \times 10^{-19} \text{ J}} = 95.5 \text{ eV}$$

Discussion Since atoms are roughly 0.1 nm in size, knowing the position of an electron to 0.0100 nm localizes it reasonably well inside the atom. This would be like being able to see details one-tenth the size of the atom. But the consequent uncertainty in velocity is large. You certainly could not follow it very well if its velocity is so uncertain. To get a further idea of how large the uncertainty in velocity is, we assumed the velocity of the electron was equal to its uncertainty and found this gave a kinetic energy of 95.5 eV. This is significantly greater than the typical energy difference between levels in atoms (see Table 27.1), so that it is impossible to get a meaningful energy for the electron if we know its position even moderately well.

Why don't we notice Heisenberg's uncertainty principle in everyday life? The answer is that Planck's constant is very small. Thus the lower limit in the uncertainty of measuring the position and momentum of large objects is negligible. We can detect sunlight reflected from Jupiter, and follow the planet in its orbit around the sun. The reflected sunlight alters the momentum of Jupiter and creates an uncertainty in its momentum, but this is totally negligible compared with Jupiter's huge momentum. The correspondence principle tells us that the predictions of quantum mechanics become indistinguishable from classical physics for large objects, which is the case here.

There is another form of **Heisenberg's uncertainty principle** *for simultaneous measurements of energy and time*. In equation form,

$$\Delta E \, \Delta t \geq \frac{h}{4\pi} \qquad \textbf{(27.10)}$$

where ΔE is the uncertainty in energy and Δt is the uncertainty in time. This means that within a time interval Δt, it is not possible to measure energy precisely—there will be an uncertainty ΔE in the measurement. In order to measure energy more precisely (to make ΔE smaller), we must increase Δt. This time interval may be the amount of time we take to make the measurement, or it could be the amount of time a particular state exists, as in the next example.

EXAMPLE 27.9 HEISENBERG UNCERTAINTY PRINCIPLE IN ENERGY AND TIME FOR AN ATOM

An atom in an excited state temporarily stores energy. If the lifetime of this excited state is measured to be 10^{-10} s, what is the minimum uncertainty in the energy of the state in eV?

Strategy and Concept The minimum uncertainty in energy ΔE is found using the equals sign in Equation 27.10 and corresponds to a reasonable choice for the uncertainty in time. The largest the uncertainty in time can be is the full lifetime of the excited state, or $\Delta t = 10^{-10}$ s.

Solution Solving the uncertainty principle for ΔE and substituting known values gives

$$\Delta E = \frac{h}{4\pi \, \Delta t} = \frac{6.63 \times 10^{-34} \text{ J·s}}{4\pi(10^{-10} \text{ s})}$$
$$= 5.3 \times 10^{-25} \text{ J}$$

Now converting to eV yields

$$\Delta E = 5.3 \times 10^{-25} \text{ J} \cdot \frac{1 \text{ eV}}{1.6 \times 10^{-19} \text{ J}}$$
$$= 3 \times 10^{-6} \text{ eV}$$

Discussion The lifetime of 10^{-10} s is typical of excited states in atoms—on human time scales, they quickly emit their stored energy. An uncertainty in energy of only a few millionths of an eV results. This uncertainty is small compared with typical excitation energies in atoms, which are on the order of 1 eV. So here the uncertainty principle limits the accuracy with which we can measure the lifetime and energy of such states, but not very significantly.

The uncertainty principle for energy and time can be of great significance if the lifetime of a system is very short. Then Δt is very small, and ΔE is consequently very large. Some nuclei and exotic particles have extremely short lifetimes (as small as 10^{-25} s), causing uncertainties in energy as great as many GeV. Stored energy appears as increased rest mass, and so this means that there is significant uncertainty in the rest mass of short-lived particles. When measured repeatedly, a spread of masses or decay energies are obtained. The spread is ΔE. You might ask whether this uncertainty in energy could be avoided by not measuring the lifetime. The answer is no. Nature knows the lifetime, and so its brevity affects the energy of the particle. This is so well established experimentally that the uncertainty in decay energy is used to calculate the lifetime of short-lived states. Some nuclei and particles are so short-lived that it is difficult to measure their lifetime. But if their decay energy can be measured, its spread is ΔE, and this is used in the uncertainty principle (Equation 27.10) to calculate the lifetime Δt.

There is another consequence of the uncertainty principle for energy and time. If energy is uncertain by ΔE, then conservation of energy can be violated by ΔE for a time Δt. Neither the physicist nor nature can tell that conservation of energy has been violated, if the violation is temporary and smaller than the uncertainty in energy. While this sounds innocuous enough, we shall see in later chapters that it allows the temporary creation of matter from nothing and has implications for how nature transmits forces over very small distances.

Finally, note that in the discussion of particles and waves, we have stated that individual measurements produce precise or particle-like results. A definite position is determined each time we observe an electron, for example. But repeated measurements produce a spread in values consistent with wave characteristics. The great theoretical physicist Richard Feynman commented, "What there are, are particles." When you observe enough of them, they distribute themselves as you would expect for a wave phenomenon.

27.8 THE PARTICLE-WAVE DUALITY REVIEWED

The **particle-wave duality**, the fact that all particles have wave properties, is one of the cornerstones of quantum mechanics. We first came across it in the treatment of photons, those particles of EM radiation that also have wave properties. Later it was noted that particles of matter have wave properties as well. The dual properties of particles and waves are found for all particles, whether massless like photons, or having a rest mass like electrons. (See Figure 27.23.)

There are many submicroscopic particles in nature. Most have mass and are expected to act as particles, or the smallest units of matter. All these masses have wave properties with wavelengths given by the de Broglie relationship $\lambda = h/p$. So, too, do combinations of these particles, such as nuclei, atoms, and molecules. As a combination of masses becomes large, particularly if it is large enough to be called macroscopic, its

Figure 27.23 On a quantum mechanical scale (very small), particles with and without mass have wave properties. For example, both electrons and photons have wavelengths but come in lumps or particles.

Rock

Massive particle

Massless wave

Figure 27.24 On a classical scale (macroscopic), particles with mass behave as particles and not as waves. Particles without mass act as waves and not as particles.

wave nature becomes difficult to observe. This is consistent with our common experience with matter.

Some particles in nature are massless. We have only treated the photon so far, but all massless entities travel at the speed of light, have a wavelength, and come in lumps called particles. They have momentum given by a rearrangement of the de Broglie relationship, $p = h/\lambda$. In large combinations of these massless particles (such large combinations are common only for photons or EM waves), there is mostly wave behavior and the particle nature becomes difficult to observe. This is also consistent with experience—the wave nature of EM radiation was studied on a macroscopic scale for decades before it was discovered to be composed of photons. (See Figure 27.24.)

The particle-wave duality is a universal attribute. It is another connection between matter and energy. Not only has modern physics been able to describe nature for high speeds and small sizes, it has also discovered new connections and symmetries. There is greater unity and symmetry in nature than was known in the classical era—but they were dreamt of. A beautiful poem written by the English poet William Blake some two centuries ago contains the following four lines:

> To see the World in a Grain of Sand
> And a Heaven in a Wild Flower
> Hold Infinity in the palm of your hand
> And Eternity in an hour

SUMMARY

Quantum mechanics is the branch of physics that deals with small objects and with the quantization of various entities, such as mass, charge, and energy. The term *quantization* refers to the fact that certain physical entities are **quantized**, existing only with particular discrete values and not allowed to have every conceivable value. Quantization takes place for very small values, making macroscopic objects appear to be continuous. The **correspondence principle** states that in the classical limit (large, slow moving objects), quantum mechanics become the same as classical physics.

The first indication that energy is sometimes quantized came from **blackbody radiation**, the emission of EM radiation by an object with an emissivity of 1. Planck recognized that the energy levels of the emitting atoms and molecules were quantized, with only the allowed values of

$$E = nhf \qquad (27.1)$$

where n is any nonnegative integer (0, 1, 2, 3, . . .) and h is **Planck's constant**, given by

$$h = 6.63 \times 10^{-34} \text{ J·s} \qquad (27.2)$$

Thus the oscillator energy of atoms and molecules in a blackbody could increase or decrease only in steps of size

$$\Delta E = hf \qquad (27.3)$$

where f is the frequency of the oscillator and of the EM radiation absorbed and emitted. Another indication of energy levels being quantized in atoms and molecules comes from the lines in **atomic spectra**, which are the EM emissions of individual atoms and molecules.

The **photoelectric effect** is the name given to the process in which light or other EM radiation ejects electrons from a material. Einstein proposed **photons** to be quanta of EM radiation having energy

$$E = hf \qquad (27.4)$$

where f is the frequency of the radiation. All EM radiation is composed of photons. Einstein explained all characteristics of the photoelectric effect (elaborated on in Section 27.2) by the interaction of individual photons with individual electrons. The maximum kinetic energy of ejected electrons (photoelectrons) is given by

$$\text{KE}_e = hf - \text{BE} \qquad \text{(photoelectric effect)} \qquad (27.5)$$

where KE_e is the kinetic energy of the ejected electron, hf is the photon energy, and BE is the **binding energy** of the electron to the particular material, sometimes called the *work function* of the material.

Photon energy is responsible for many of the characteristics of various types of EM radiation, being particularly noticeable at high frequencies. At lower frequencies, multiple photons can act like the classical EM waves discussed in Chapter 23. Photons have both wave and particle characteristics. Quantized in energy, they also have **photon momentum** given by

$$p = \frac{h}{\lambda} \qquad (27.6)$$

where λ is the photon wavelength. Photon energy and momentum are related by

$$p = \frac{E}{c} \qquad \text{(photons)} \qquad (27.7)$$

where $E = hf = hc/\lambda$ for a photon.

Particles of matter also have a wavelength, called the **de Broglie wavelength**, given by

$$\lambda = \frac{h}{p} \qquad \text{(matter and photons)} \qquad (27.8)$$

where p is momentum. Our view of matter is dramatically changed by its wave character. Matter is found to have the same *interference characteristics* as any other wave and there is now a **probability distribution** for the location of a particle rather than a definite position. Another consequence of the wave character of all particles is the **Heisenberg uncertainty principle**, which limits the precision with which certain physical quantities can be known simultaneously. For position and momentum, the uncertainty principle is

$$\Delta x\, \Delta p \geq \frac{h}{4\pi} \qquad (27.9)$$

where Δx is the **uncertainty in position** and Δp is the **uncertainty in momentum**. For energy and time, the uncertainty principle is

$$\Delta E\, \Delta t \geq \frac{h}{4\pi} \qquad (27.10)$$

where ΔE is the uncertainty in energy and Δt is the uncertainty in time. While these limits are small, they are fundamentally important on the quantum mechanical scale.

The **particle-wave duality** refers to the fact that all particles—both those with and those without mass—have wave characteristics as well. This is a further connection between mass and energy.

CONCEPTUAL QUESTIONS

27.1 Give an example of a physical entity that is quantized. State specifically what the entity is and what the limits are on its values.

27.2 Give an example of a physical entity that is not quantized, in that it is continuous and may have an unbroken continuum of values.

27.3 What aspect of the blackbody spectrum forced Planck to propose quantization of energy levels in its atoms and molecules?

27.4 If Planck's constant were large, say 10^{34} times greater than it is, we would observe macroscopic entities to be quantized. Describe the motions of a child's swing under such circumstances.

27.5 Why don't we notice quantization in everyday events?

27.6 Explain how we might observe quantization by looking at a fluorescent light with a diffraction grating. Why does this macroscopic object exhibit quantization?

27.7 Is visible light the only type of EM radiation that can cause the photoelectric effect?

27.8 Which aspects of the photoelectric effect cannot be explained without photons? Which can be explained without photons? Are the latter inconsistent with the existence of photons?

27.9 Is the photoelectric effect a direct consequence of the wave character of EM radiation or of the particle character of EM radiation? Is the photoelectric effect a direct consequence of the wave character of matter or of the particle character of matter? Explain briefly.

27.10 Insulators (nonmetals) have a higher BE than metals, and it is more difficult for photons to eject electrons from insulators. Discuss how this relates to the free charges in metals that make them good conductors.

27.11 If you pick up and shake a piece of metal that has electrons in it free to move as a current, no electrons fall out. Yet if you heat the metal, electrons can be boiled off. Explain both of these facts as they relate to the amount and distribution of energy involved with shaking the object as compared with heating it.

27.12 How can a massless particle have momentum if $p = \gamma mu$ or, nonrelativistically, $p = mv$?

27.13 Why are UV, x rays, and γ rays called ionizing radiation?

27.14 How can treating food with ionizing radiation help keep it from spoiling? Would it be best to use UV or more penetrating radiation?

27.15 Television tubes are CRTs. They use an approximately 30 kV accelerating potential to send electrons to the screen, where the electrons stimulate phosphors to emit the light that forms the pictures we watch. Would you expect x rays to be created?

27.16 Tanning salons use "safe" UV with a longer wavelength than some of the UV in sunlight. This "safe" UV has enough photon energy to trigger the tanning mechanism. Is it likely to be able to cause cell damage and induce cancer with prolonged exposure?

27.17 Your pupils dilate when visible light intensity is reduced. Does wearing sunglasses that lack UV blockers increase or decrease the UV hazard to your eyes? Explain.

27.18 You can feel heat in the form of infrared radiation from a large nuclear bomb detonated in the atmosphere 75 km from you. However, none of the profusely emitted x rays or γ rays reach you. Explain.

27.19 Can a single microwave photon cause cell damage? Explain.

27.20 In an x-ray tube, the maximum photon energy is given by $hf = qV$. Would it be technically more correct to say $hf = qV + \text{BE}$, where BE is the binding energy of electrons in the target anode? Why isn't the energy stated the latter way?

27.21 What is the correct formula for the momentum of all particles, with or without mass?

27.22 Does the momentum of a photon differ in any way from the momentum of matter?

27.23 Why don't we feel the momentum of sunlight when we are on the beach?

27.24 How does the interference of water waves differ from the interference of electrons? How are they analogous?

27.25 Describe one type of evidence for the wave nature of matter.

27.26 Describe one type of evidence for the particle nature of EM radiation.

27.27 What is the Heisenberg uncertainty principle? Does it place limits on what can be known?

27.28 In what ways are matter and energy related that were not known before the development of relativity and quantum mechanics?

PROBLEMS

Section 27.1 Quantization of Energy Levels

27.1 A LiBr molecule oscillates with a frequency of 1.7×10^{13} Hz. (a) What is the difference in energy in eV between allowed oscillator states? (b) What is the approximate value of n for a state having an energy of 1.0 eV?

27.2 The difference in energy between allowed oscillator states in HBr molecules is 0.330 eV. What is the oscillation frequency of this molecule?

• 27.3 A physicist is watching an orangutan at a zoo swing lazily in a tire at the end of a rope. He (the physicist) notices that each oscillation takes 3.00 s and realizes the energy is quantized. (a) What is the difference in energy in joules between allowed oscillator states? (b) What is the value of n for a state where the energy is 5.00 J? (c) Can the quantization be observed?

Section 27.2 The Photoelectric Effect

27.4 What is the longest-wavelength EM radiation that can eject a photoelectron from silver, given the binding energy is 4.73 eV? Is this in the visible range?

27.5 Find the longest-wavelength photon that can eject an electron from potassium, given the binding energy is 2.24 eV. Is this visible EM radiation?

27.6 What is the binding energy in eV of electrons in magnesium, if the longest-wavelength photon that can eject electrons is 337 nm?

27.7 Calculate the binding energy in eV of electrons in aluminum, if the longest-wavelength photon that can eject them is 304 nm.

27.8 What is the maximum kinetic energy in eV of electrons ejected from sodium metal by 450 nm EM radiation, given the binding energy is 2.28 eV?

27.9 UV radiation having a wavelength of 120 nm falls on gold metal, to which electrons are bound by 4.82 eV. What is the maximum kinetic energy of the ejected photoelectrons?

27.10 Violet light of wavelength 400 nm ejects electrons with a maximum kinetic energy of 0.860 eV from sodium metal. What is the binding energy of electrons to sodium metal?

27.11 UV radiation having a 300 nm wavelength falls on uranium metal, ejecting 0.500 eV electrons. What is the binding energy of electrons to uranium metal?

27.12 What is the wavelength of EM radiation that ejects 2.00 eV electrons from calcium metal, given the binding energy is 2.71 eV? What type of EM radiation is this?

27.13 Find the wavelength of photons that eject 0.100 eV electrons from potassium, given the binding energy is 2.24 eV. Are these photons visible?

• 27.14 What is the maximum velocity of electrons ejected from a material by 80 nm photons, if they are bound to the material by 4.73 eV?

• 27.15 Photoelectrons from a material with a binding energy of 2.71 eV are ejected by 420 nm photons. Once ejected, how long does it take these electrons to travel 2.50 cm to a detection device?

⁞ 27.16 (a) Calculate the number of photoelectrons per second ejected from a 1.00 mm^2 area of sodium metal by 500 nm EM radiation having an intensity of 1.30 kW/m^2 (the intensity of sunlight above the earth's atmosphere). (b) Given the binding energy is 2.28 eV, what power is carried away by the electrons?

⁞ 27.17 A laser with a power output of 2.00 mW at a wavelength of 400 nm is projected onto calcium metal. (a) How many electrons per second are ejected? (b) What power is carried away by the electrons, given the binding energy is 2.28 eV?

Section 27.3 Photon Energy

27.18 What is the energy in joules and eV of a photon in a radio wave from an AM station that has a 1530 kHz broadcast frequency?

27.19 Find the energy in joules and eV of photons in radio waves from an FM station that has a 90.0 MHz broadcast frequency.

27.20 Calculate the frequency in hertz of a 1.00 MeV γ-ray photon.

27.21 (a) What is the wavelength of a 1.00 eV photon? (b) Find its frequency in hertz.

27.22 Confirm the statement in the text that the range of photon energies for visible light is 1.63 to 3.26 eV, given the range of visible wavelengths is 380 to 760 nm.

27.23 (a) Calculate the energy in eV of an IR photon of frequency 2.00×10^{13} Hz. (b) How many of these photons would need to be absorbed simultaneously by a tightly bound molecule to break it apart? (c) What is the energy in eV of a γ ray of frequency 3.00×10^{20} Hz? (d) How many tightly bound molecules could a single such γ ray break apart?

27.24 Prove that, to three-digit accuracy, $h = 4.14 \times 10^{-15}$ eV·s, as stated in the text.

27.25 Do the unit conversions necessary to show that $hc = 1240$ eV·nm, as stated in the text.

27.26 (a) What is the maximum energy in eV of photons produced in a CRT using a 25.0 kV accelerating potential, such as a color TV? (b) What is their frequency?

• 27.27 What is the accelerating voltage of an x-ray tube that produces x rays with a shortest wavelength of 0.0103 nm?

• 27.28 (a) What is the ratio of power outputs by two microwave ovens having frequencies of 950 and 2560 MHz, if they emit the same number of photons per second? (b) What is the ratio of photons per second if they have the same power output?

• 27.29 How many photons per second are emitted by the antenna of a microwave oven, if its power output is 1.00 kW at a frequency of 2560 MHz?

• **27.30** Some satellites use nuclear power. (a) If such a satellite emits a 1.00 W flux of γ rays having an average energy of 0.500 MeV, how many are emitted per second? (b) These γ rays affect other satellites. How far away must another satellite be to only receive one γ ray per second per square meter?

• **27.31** (a) If the power output of a 650 kHz radio station is 50.0 kW, how many photons per second are produced? (b) If the radio waves are broadcast uniformly in all directions, find the number of photons per second per square meter at a distance of 100 km. Assume no reflection from the ground or absorption by the air.

• **27.32** How many x-ray photons per second are created by an x-ray tube that produces a flux of x rays having a power of 1.00 W? Assume the average energy per photon in 75.0 keV.

‡ **27.33** How far away must you be from the radio station in Problem 27.31 for there to be only one photon per second per square meter? Assume no reflections or absorption, as if you were in deep outer space.

‡ **27.34** Assuming the photons produced by the light bulb in Example 27.4 spread out uniformly and are not absorbed by the atmosphere, how far away would you be if 500 photons per second enter the 3.00 mm diameter pupil of your eye? (This number easily stimulates the retina.)

Section 27.4 Photon Momentum

Note that the forms of the constants $h = 4.14 \times 10^{-15}$ eV·s and $hc = 1240$ eV·nm may be particularly useful for these problems.

27.35 Find the momentum of a 4.00 cm wavelength microwave photon.

27.36 (a) What is the momentum of a 0.0100 nm wavelength photon that could detect details of an atom? (b) What is its energy in MeV?

27.37 (a) What is the wavelength of a photon that has a momentum of 5.00×10^{-27} kg·m/s? (b) Find its energy in eV.

27.38 (a) A γ-ray photon has a momentum of 8.00×10^{-21} kg·m/s. What is its wavelength? (b) Calculate its energy in MeV.

• **27.39** (a) Calculate the momentum of a photon having a wavelength of 2.50 μm. (b) Find the velocity of an electron having the same momentum. (c) What is the kinetic energy of the electron, and how does it compare with that of the photon?

• **27.40** Repeat Problem 27.39 for a 10.0 nm wavelength photon.

• **27.41** (a) Calculate the wavelength of a photon that has the same momentum as a proton moving at 1.00% of the speed of light. (b) What is the energy of the photon in MeV? (c) What is the kinetic energy of the proton in MeV?

‡ **27.42** (a) Find the momentum of a 100 keV x-ray photon. (b) Find the equivalent velocity of a neutron with the same momentum. (c) What is the neutron's kinetic energy in keV?

‡ **27.43** Take the ratio of relativistic rest energy, $E = \gamma mc^2$, to relativistic momentum, $p = \gamma mu$, and show that in the limit that mass approaches zero, you find $E/p = c$ (consistent with Equation 27.7 for photons).

Section 27.6 The Wave Nature of Matter

27.44 At what velocity will an electron have a wavelength of 1.00 m?

27.45 What is the wavelength of an electron moving at 3.00% of the speed of light?

27.46 At what velocity does a proton have a 6.00 fm wavelength (about the size of a nucleus)? Assume the proton is nonrelativistic.

27.47 What is the velocity of a 0.400 kg billiard ball if its wavelength is 7.50 cm (large enough for it to interfere with other billiard balls)?

27.48 Find the wavelength of a proton moving at 1.00% of the speed of light.

27.49 Experiments are performed with ultracold neutrons having velocities as small as 1.00 m/s. (a) What is the wavelength of such a neutron? (b) What is its kinetic energy in eV?

• **27.50** (a) Find the velocity of a neutron that has a 6.00 fm wavelength (about the size of a nucleus). Assume the neutron is nonrelativistic. (b) What is the neutron's kinetic energy in MeV?

• **27.51** What is the wavelength of an electron accelerated through a 30.0 kV potential, as in a TV tube?

• **27.52** (a) Calculate the velocity of an electron that has a wavelength of 1.00 μm. (b) Through what voltage must the electron be accelerated to have this velocity?

• **27.53** The velocity of a proton emerging from a Van de Graaff accelerator is 25.0% of the speed of light. (a) What is the proton's wavelength? (b) What is its kinetic energy, assuming it is nonrelativistic? (c) What was the equivalent voltage through which it was accelerated?

• **27.54** The kinetic energy of an electron accelerated in an x-ray tube is 100 keV. Assuming it is nonrelativistic, what is its wavelength?

‡ **27.55** Repeat Problem 27.54 assuming the electron is relativistic.

‡ **27.56** Repeat Problem 27.53 assuming the proton is relativistic.

Section 27.7 The Heisenberg Uncertainty Principle

27.57 (a) If the position of an electron in a membrane is measured to an accuracy of 1.00 μm, what is the electron's minimum uncertainty in velocity? (b) If the electron has this velocity, what is its kinetic energy in eV?

27.58 (a) If the position of a chlorine ion in a membrane is measured to an accuracy of 1.00 μm, what is its minimum uncertainty in velocity, given its mass is 5.86×10^{-26} kg? (b) If the ion has this velocity, what is its kinetic energy in eV, and how does this compare with typical molecular binding energies?

27.59 Suppose the velocity of an electron in an atom is known to an accuracy of 2.0×10^3 m/s (reasonably accurate compared with orbital velocities). What is the electron's minimum uncertainty in position, and how does this compare with the approximate 0.1 nm size of the atom?

27.60 The velocity of a proton in an accelerator is known to an accuracy of 0.250% of the speed of light. (This could be small compared with its velocity.) What is the smallest possible uncertainty in its position?

27.61 A relatively long lived excited state of an atom has a lifetime of 3.00 ms. What is the minimum uncertainty in its energy?

27.62 (a) The lifetime of a highly unstable nucleus is 10^{-20} s. What is the smallest uncertainty in its decay energy? (b) Compare this with the rest energy of an electron.

27.63 The decay energy of a short-lived particle has an uncertainty of 1.0 MeV due to its short lifetime. What is the smallest lifetime it can have?

27.64 The decay energy of a short-lived nuclear excited state has an uncertainty of 2.0 eV due to its short lifetime. What is the smallest lifetime it can have?

• **27.65** What is the approximate uncertainty in the mass of a muon, as determined from its decay lifetime?

‡ **27.66** Derive the approximate form of Heisenberg's uncertainty principle for energy and time, $\Delta E\, \Delta t \approx h$, using the following arguments: Since the position of a particle is uncertain by $\Delta x \approx \lambda$, where λ is the wavelength of the photon used to examine it, there is an uncertainty in the time the photon takes to traverse Δx. Furthermore, the photon has an energy related to its wavelength, and it can transfer some or all of this energy to the object being examined. Thus the uncertainty in the energy of the object is also related to λ. Find Δt and ΔE; then multiply them to give the approximate uncertainty principle.

INTEGRATED CONCEPTS

The following problems involve concepts from this chapter and several others. Physics is most interesting when applied to general situations involving more than a narrow set of physical principles. For example, photons have momentum, hence the relevance of Chapter 7, Linear Momentum. The following topics are involved in some or all of the problems in this section:

Topics	Location
Dynamics	Chapter 4
Work, energy and power	Chapter 6
Linear momentum	Chapter 7
Heat and heat transfer	Chapter 13
Electric potential and energy	Chapter 18
Current and Ohm's law	Chapter 19
Wave optics	Chapter 25
Special relativity	Chapter 26

PROBLEM-SOLVING STRATEGY

Step 1. *Identify which physical principles are involved.*

Step 2. *Solve the problem using strategies outlined in the text.*

The following worked example illustrates how these strategies are applied to an integrated concept problem.

EXAMPLE   RECOIL OF A DUST PARTICLE AFTER ABSORBING A PHOTON

The following topics are involved in this integrated concepts worked example:

Topics	Location
Photons (quantum mechanics)	Chapter 27
Linear momentum	Chapter 7

A 550 nm photon (visible light) is absorbed by a 1.00 μg particle of dust in outer space. (a) Find the momentum of such a photon. (b) What is the recoil velocity of the particle of dust, assuming it is initially at rest?

Strategy Step 1 To solve an *integrated concept problem*, such as those following this example, we must first identify the physical principles involved and identify the chapters in which they are found. Part (a) of this example asks for the *momentum of a photon*, a topic of the present Chapter 27. Part (b) considers *recoil following a collision*, a topic of Chapter 7.

Strategy Step 2 The following solutions to each part of the example illustrate how specific problem-solving strategies are applied. These involve identifying knowns and unknowns, checking to see if the answer is reasonable, and so on.

Solution for (a): Photon momentum (Chapter 27) The momentum of a photon is related to its wavelength by Equation 27.6:

$$p = \frac{h}{\lambda}$$

Entering the known value for Planck's constant h and given the wavelength λ, we obtain

$$p = \frac{6.63 \times 10^{-34}\ \text{J}\cdot\text{s}}{550 \times 10^{-9}\ \text{m}}$$
$$= 1.21 \times 10^{-27}\ \text{kg}\cdot\text{m/s}$$

Discussion for (a) This momentum is small, as expected from discussions in the text and the fact that photons of visible light carry small amounts of energy and momentum compared with those carried by macroscopic objects.

Solution for (b): Recoil velocity after collision (Chapter 7) Conservation of momentum in the absorption of this photon by a grain of dust can be analyzed using Equation 7.4:

$$p_1 + p_2 = p_1' + p_2' \qquad (\text{net } F = 0)$$

The net external force is zero, since the dust is in outer space. Let 1 represent the photon and 2 the dust particle. Before the collision, the dust is at rest (relative to some observer); after the collision, there is no photon (it is absorbed). So conservation of momentum can be written

$$p_1 = p_2' = mv$$

where p_1 is the photon momentum before the collision and p_2' is the dust momentum after the collision. The mass and recoil velocity of the dust are m and v, respectively. Solving this for v, the requested quantity, yields

$$v = \frac{p}{m}$$

where p is the photon momentum found in part (a). Entering known values (noting that a microgram is 10^{-9} kg) gives

$$v = \frac{1.21 \times 10^{-27}\ \text{kg}\cdot\text{m/s}}{1.00 \times 10^{-9}\ \text{kg}}$$
$$= 1.21 \times 10^{-18}\ \text{m/s}$$

Discussion for (b) The recoil velocity of the particle of dust is extremely small. As we have noted, however, there are immense numbers of photons in sunlight and other macroscopic sources. In time, collisions and absorption of many photons could cause a significant recoil of the dust, as observed in comet tails.

This worked example illustrates how to apply problem-solving strategies to situations that include topics in different chapters. The first step is to identify the physical principles involved in the problem. The second step is to solve for the unknown using familiar problem-solving strategies. These are found throughout the text. Moreover, many worked examples show how to use problem-solving strategies for single topics. In this integrated concepts example, you can see how to apply them across several topics. You will find these techniques useful in applications of physics outside a physics course, such as in your profession, in other science disciplines, and in everyday life. The following problems will build your skills in the broad application of physical principles.

27.67 The 54.0 eV electron in Example 27.7 has a 0.167 nm wavelength. If such electrons are passed through a double slit and have their first maximum at an angle of 25.0°, what is the slit separation d?

27.68 An electron microscope produces electrons with a 2.00 pm wavelength. If these are passed through a 1.00 nm single slit, at what angle will the first diffraction minimum be found?

• 27.69 A certain heat lamp emits 200 W of mostly IR radiation averaging 1500 nm in wavelength. (a) What is the average photon energy in joules? (b) How many of these photons are required to increase the temperature of a person's shoulder by 2.0°C, assuming the affected mass is 4.0 kg with a specific heat of 0.83 kcal/kg·°C. Also assume no other significant heat transfer. (c) How long does this take?

• 27.70 On its high power setting, a microwave oven produces 900 W of 2560 MHz microwaves. (a) How many photons per second is this? (b) How many photons are required to increase the temperature of a 0.500 kg mass of Pasta Urone by 45.0°C, assuming a specific heat of 0.900 kcal/kg·°C? Neglect all other heat transfer. (c) How long must Urone wait for his pasta to be ready?

• 27.71 (a) Calculate the amount of microwave energy in joules needed to raise the temperature of 1.00 kg of soup from 20.0°C to 100°C. (b) What is the total momentum of all the microwave photons it takes to do this? (c) Calculate the velocity of a 1.00 kg mass with the same momentum. (d) What is the kinetic energy of this mass?

• 27.72 (a) What is γ for an electron emerging from the Stanford Linear Accelerator with a total energy of 50.0 GeV? (b) Find its momentum. (c) What is the electron's wavelength?

• 27.73 Repeat the previous problem for a proton having an energy of 1.00 TeV, produced by the Fermilab accelerator.

• 27.74 An electron microscope passes 1.00 pm wavelength electrons through a circular aperture 2.00 μm in diameter. What is the angle between two just resolvable point sources for this microscope?

• 27.75 Calculate the velocity of electrons that form the same pattern as 450 nm light when passed through a double slit.

• 27.76 (a) What is the separation between double slits that produces a second-order minimum at 45.0° for 650 nm light? (b) What slit separation is needed to produce the same pattern for 1.00 keV electrons?

• 27.77 Repeat Problem 27.76, but for 1.00 keV protons in part (b).

⁞ 27.78 Repeat Problem 27.17, but also: (c) Calculate the current of ejected electrons. (d) If the photoelectric material is electrically insulated and acts like a 2.00 pF capacitor, how long will current flow before the capacitor voltage stops it?

⁞ 27.79 One problem with x rays is that they are not sensed. To demonstrate why this might be, calculate the temperature increase of a researcher exposed in a few seconds to a nearly fatal accidental dose of x rays under the following conditions. The energy of the x-ray photons is 200 keV, and 4.00×10^{13} of them are absorbed per kilogram of tissue, the specific heat of which is 0.830 kcal/kg·°C. (Note that you need not to fear such an exposure from medical diagnostic x rays, unless you were x-rayed continuously for many hours.)

⁞ 27.80 A 1.00 fm photon has a wavelength short enough to detect some information about nuclei. (a) What is the photon momentum? (b) What is its energy in joules and MeV? (c) What is the (relativistic) velocity of an electron with the same momentum? (d) Calculate the electron's kinetic energy.

⁞ 27.81 The momentum of light is exactly reversed when reflected straight back from a mirror, assuming negligible recoil of the mirror. Thus the change in momentum is twice the photon momentum. Suppose light of intensity 1.00 kW/m² reflects from a mirror of area 2.00 m². (a) Calculate the energy reflected in 1.00 s. (b) What is the momentum imparted to the mirror? (c) Using the most general form of Newton's second law, what is the force on the mirror? (d) Does the assumption of no mirror recoil seem reasonable?

⁞ 27.82 Sunlight above the earth's atmosphere has an intensity of 1.30 kW/m². If this is reflected straight back from a mirror that has only a small recoil, the light's momentum is exactly reversed, giving the mirror twice the incident momentum. (a) Calculate the force per square meter of mirror. (b) Very low mass mirrors can be constructed in the near weightlessness of space, and attached to a spaceship to sail it. Once done, the average mass per square meter of the spaceship is 0.100 kg. Find the acceleration of the spaceship if all other forces are balanced. (c) How fast is it moving 24 hours later?

UNREASONABLE RESULTS

The following problems have results that are unreasonable because some premise is unreasonable or because certain of the premises are inconsistent with one another. Physical principles applied correctly then produce unreasonable results. The purpose of these problems is to give practice in assessing whether nature is being accurately described, and if it is not to trace the source of difficulty.

Note: Problem-solving strategies that can help you solve unreasonable results problems appear in several places, the most recent being Chapter 26. Consult the section of problems labeled *Unreasonable Results* at the end of that chapter or others.

27.83 (a) Assuming it is nonrelativistic, calculate the velocity of an electron with a 0.100 fm wavelength (small enough to detect details of a nucleus). (b) What is unreasonable about this result? (c) Which assumptions are unreasonable or inconsistent?

• 27.84 Red light having a wavelength of 700 nm is projected onto magnesium metal to which electrons are bound by 3.68 eV. (a) Use Equation 27.5 to calculate the kinetic energy of the ejected electrons. (b) What is unreasonable about this result? (c) Which assumptions are unreasonable or inconsistent?

• 27.85 (a) What is the binding energy of electrons to a material from which 4.00 eV electrons are ejected by 400 nm EM radiation? (b) What is unreasonable about this result? (c) Which assumptions are unreasonable or inconsistent?

⁞ 27.86 A car feels a small force due to the light it sends out from its headlights, equal to the momentum of the light divided by the time in which it is emitted. (a) Calculate the power of each headlight, if they exert a total force of 2.00×10^{-2} N backward on the car. (b) What is unreasonable about this result? (c) Which assumptions are unreasonable or inconsistent?

ATOMIC PHYSICS

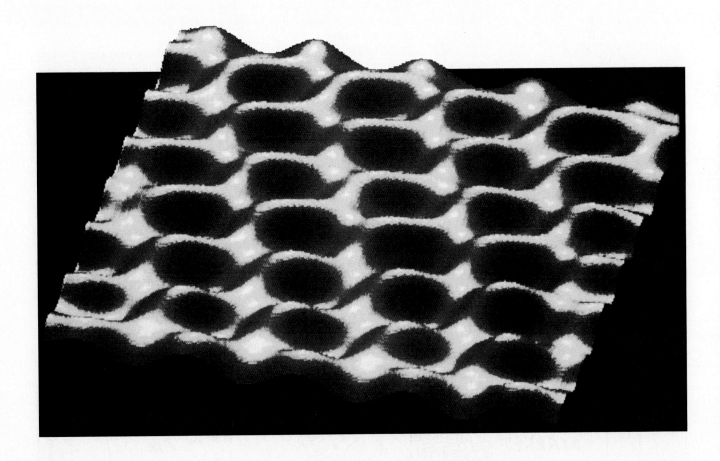

Individual carbon atoms in graphite, detected

by a scanning electron microscope.

From childhood on, we learn that atoms are a substructure of all things around us, from the air we breathe to the autumn leaves that blanket a forest trail. Invisible to the eye, the existence and properties of atoms are used to explain many phenomena—a theme found throughout this text. In this chapter, we discuss the discovery of atoms and their own substructures; we then apply quantum mechanics to the description of atoms, and their properties and interactions. Along the way, we will find, much like the scientists who made the original discoveries, that new concepts emerge with applications far beyond the boundaries of atomic physics.

28.1 DISCOVERY OF THE ATOM

How do we know that atoms are really there if we cannot see them with our eyes? A brief account of the progression from the proposal of atoms by the Greeks to the first direct evidence of their existence follows.

People have long speculated about the structure of matter and the existence of atoms. The earliest significant ideas to survive are due to the ancient Greeks in the fifth century B.C., especially the philosophers Leucippus and Democritus. They considered the question of whether a substance can be divided without limit into ever smaller pieces. Once the question is asked, only a few answers are possible. One is that infinitesimally small subdivision is possible. Another is what Democritus in particular believed—there is a smallest unit he called the atom, which cannot be further subdivided. We now know atoms themselves can be subdivided, but their identity is destroyed in the process, so that the Greeks were correct in a respect. The Greeks also felt that atoms were in constant motion, another correct notion.

The Greeks and others speculated about the properties of atoms, proposing that only a few types existed and that all matter was formed as various combinations of them. (See Figure 28.1.) The famous proposal that the basic elements were earth, air, fire, and water was brilliant, but incorrect. What the Greeks had done was to identify the most common examples of the four states of matter (solid, gas, plasma, and liquid), rather than the basic elements. More than 2000 years passed before observations could be made with equipment capable of revealing the true nature of atoms.

Over the centuries, discoveries were made regarding the properties of substances and their chemical reactions. Certain systematics were recognized, but similarities between common and rare elements resulted in efforts to transmute them (lead into gold, in particular) for financial gain. Secrecy was endemic. Alchemists discovered and rediscovered many facts but did not make them broadly available. As the Middle Ages ended, alchemy gradually faded, and the science of chemistry arose. It was no longer possible, nor considered desirable, to keep discoveries secret. Collective knowledge grew, and by the beginning of the 19th century, an important fact was well established—the masses of reactants in specific chemical reactions always have a particular mass ratio. This is very strong indirect evidence that there are basic units (atoms and molecules) that have these same mass ratios. The English chemist John Dalton (1766–1844) did much of this work,

FRANK & ERNEST® by Bob Thaves

Figure 28.1

with significant contributions by the Italian physicist Amedeo Avogadro (1776–1856). It was Avogadro who developed the idea of a fixed number of atoms and molecules in a mole, and this special number is called Avogadro's number in his honor.

Knowledge of the properties of elements and compounds grew, culminating in the mid-19th-century development of the periodic table of the elements by Dmitri Mendeleev (1834–1907), the great Russian chemist. Mendeleev proposed an ingenious array that highlighted the periodic nature of the properties of elements. Believing in the systematics of the periodic table, he also predicted the existence of then-unknown elements to complete it. Once these elements were discovered and determined to have properties predicted by Mendeleev, his periodic table became universally accepted.

Also during the 19th century, kinetic theory was developed. Kinetic theory is based on the existence of atoms and molecules in random thermal motion and provides a microscopic explanation of the gas laws, heat transfer, and thermodynamics (see Section 12.4 and many subsequent discussions in Chapters 12, 13, and 14). Kinetic theory works so well that it is another strong indication of the existence of atoms. But it is still indirect evidence—individual atoms and molecules had not been observed. There were heated debates about the validity of kinetic theory until direct evidence of atoms was obtained.

The first truly direct evidence is credited to Robert Brown, a Scottish botanist. In 1827 he noticed that tiny pollen grains suspended in still water moved about in complex paths. This can be observed with a microscope for any small particles in a fluid. The motion is caused by the random thermal motions of fluid molecules colliding with particles in the fluid, and it is now called **Brownian motion**. (See Figure 28.2.) Statistical fluctuations in the numbers of molecules striking the sides of a visible particle cause it to move first this way, then that. Although the molecules cannot be directly observed, their effects on the particle can be. By examining Brownian motion, the size of molecules can be calculated. The smaller and more numerous they are, the smaller the fluctuations in the numbers striking different sides.

It was Albert Einstein who, starting in his epochal year of 1905, published several papers that explained precisely how Brownian motion could be used to measure the size of atoms and molecules.* Their sizes were only approximately known to be 10^{-10} m, based on a comparison of latent heat of vaporization and surface tension made about 1805 by Thomas Young of double slit fame and the also famous astronomer and mathematician Simon Laplace.

Using Einstein's ideas, the French physicist Jean-Baptiste Perrin (1870–1942) carefully observed Brownian motion; not only did he confirm Einstein's theory, he also produced accurate sizes for atoms and molecules. Since molecular weights and densities of materials were well established, knowing atomic and molecular sizes allowed a precise value for Avogadro's number to be obtained. (If we know how big an atom is, we know how many fit into a certain volume.) Perrin also used these ideas to explain atomic and molecular agitation effects in sedimentation, and he received the 1926 Nobel prize for his achievements. Most scientists were already convinced of the existence of atoms, but the accurate observation and analysis of Brownian motion was conclusive—it was the first truly direct evidence.

A huge array of direct and indirect evidence for the existence of atoms now exists. For example, it has become possible to accelerate ions (much as electrons are accelerated in CRTs) and to detect them individually as well as measure their masses (see Section 21.11 for a discussion of mass spectrometers). Other devices that observe individual atoms, such as the tunneling electron microscope, will be discussed later. (See Figure 28.3.) All of our understanding of the properties of matter is based on and consistent with the atom. The atom's substructures, such as electron shells and the nucleus, are both interesting and important. The nucleus in turn has a substructure, as do the particles of which it is composed. These topics, and the question of whether there is a smallest basic structure to matter, will be explored in later parts of the text.

---

*In 1905 Einstein created special relativity, proposed photons as quanta of EM radiation, and produced a theory of Brownian motion that allowed the size of atoms to be determined. All this in his spare time, since he worked days as a patent examiner. Any one of these very basic works could have been the crowning achievement of an entire career—yet Einstein did even more in later years.

CONNECTIONS

Patterns and Systematics

The recognition and appreciation of patterns has enabled us to make many discoveries. The periodic table of elements was proposed as an organized summary of the elements long before all elements had been discovered, and it led to many other discoveries. We shall see in later chapters that patterns in the properties of subatomic particles led to the proposal of quarks as their underlying structure, an idea still bearing fruit.

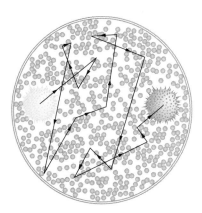

Figure 28.2 The position of a pollen grain in water, measured every few seconds under a microscope, exhibits Brownian motion. Brownian motion is due to fluctuations in the number of atoms and molecules colliding with a small mass, causing it to move about in complex paths. This is nearly direct evidence for the existence of atoms, providing a satisfactory alternative explanation cannot be found.

Figure 28.3 Individual atoms can be detected with devices such as the tunneling electron microscope that produced this image of individual gold atoms on a graphite substrate.

28.2 DISCOVERY OF THE PARTS OF THE ATOM: ELECTRONS AND NUCLEI

CONNECTIONS

Charges and Electromagnetic Forces

In previous discussions, we have noted that positive charge is associated with nuclei and negative charge with electrons. We have also covered many aspects of the electric and magnetic forces that affect charges. We will now explore the discovery of the electron and nucleus as substructures of the atom, and examine their contributions to the properties of atoms.

Just as atoms are a substructure of matter, electrons and nuclei are substructures of the atom. The experiments that were used to discover electrons and nuclei reveal some of the basic properties of atoms and can be readily understood using ideas such as electrostatic and magnetic force, already covered in previous chapters.

The Electron

Gas discharge tubes, such as that shown in Figure 28.4(a), glow when high voltage is applied to them. These tubes were precursors to today's neon lights, and they were first seriously studied starting in the 1860s, especially by Heinrich Geissler, a German inventor and glass blower. The English scientist William Crookes, among others, continued to study what for some time were called Crookes tubes. Electrons are freed from atoms and molecules in the rarefied gas inside the tube and are accelerated from the cathode (negative) to the anode (positive) by the high potential. Electrons colliding with gas atoms and molecules excite them, resulting in emission of EM radiation that makes the electrons' path visible as a ray that spreads and fades as it moves away from the cathode. Gas discharge tubes today are most commonly called **cathode ray tubes**, because the rays originate at the cathode. Crookes showed that the electrons carry momentum (they can make a small paddle wheel rotate). He also found that their normally straight path is bent by a magnet in the direction expected for a negative charge moving away from the cathode. These were the first direct indications of electrons and their charge.

The English physicist J. J. Thomson (1856–1940) improved and expanded the scope of experiments with gas discharge tubes. (See Figure 28.4(b).) He verified the negative charge of the rays with both magnetic and electric fields. Additionally, he collected the rays in a metal cup and found an excess of negative charge. Thomson was also able to measure the ratio of the charge of the electron to its mass, q_e/m_e—an important step to finding the actual values of both q_e and m_e. Figure 28.5(b) shows a cathode ray tube set up to produce a narrow beam that passes through crossed electric and magnetic fields, **E** and **B**. Those fields produce opposing forces on the electrons. As discussed for mass spectrometers in Section 21.11, if these opposing forces just balance, then the velocity of the charged particle is $v = E/B$. In this manner, Thomson determined the velocity of the electrons and then moved the beam up and down by adjusting the electric field.

To see how the amount of deflection is used to calculate q_e/m_e, note that the deflection is proportional to the electric force on the electron:

$$F = q_e E$$

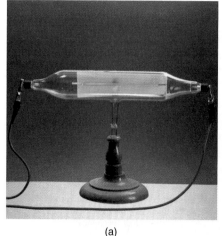

Figure 28.4 (a) A gas discharge tube glows when a high voltage is applied. Electrons emitted from the cathode are accelerated toward the anode; they excite atoms and molecules in the gas, which glow in response. Once called Geissler tubes and later Crookes tubes, they are now known as cathode ray tubes (CRTs) and are found in TVs, computer screens, and x-ray machines. When a magnetic field is applied, the beam bends in the direction expected for negative charge. (b) J. J. Thomson.

(a)   (b)

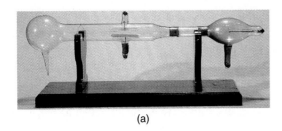

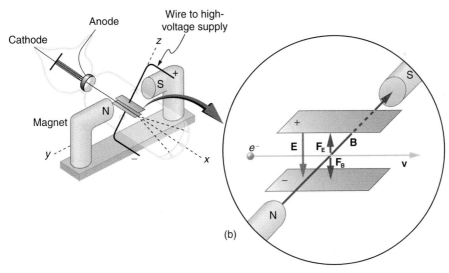

Figure 28.5 (a) Photograph of Thomson's CRT. (b) This schematic shows the electron beam passing in a CRT through crossed electric and magnetic fields and causing a phosphor to glow when striking the end of the tube. The beam can be deflected by varying the electric field, allowing q_e/m_e to be determined, as described in the text.

But the vertical deflection is also related to the electron's mass, since the electron's acceleration is

$$a = \frac{F}{m_e}$$

F is not known, since q_e was not yet known. Substituting the expression for electric force into the expression for acceleration yields

$$a = \frac{F}{m_e} = \frac{q_e E}{m_e}$$

Gathering terms, we have

$$\frac{q_e}{m_e} = \frac{a}{E}$$

The deflection is analyzed to get a, and E is known from the applied voltage and distance between the plates; thus, q_e/m_e can be determined. With the velocity known, another measurement of q_e/m_e can be obtained by bending the beam of electrons with the magnetic field. Since $F_{mag} = q_e vB = m_e a$, we have $q_e/m_e = a/vB$. Consistent results are obtained using magnetic deflection.

What is so important about q_e/m_e, the ratio of the electron's charge to its mass? The value obtained is

$$\frac{q_e}{m_e} = -1.76 \times 10^{11} \text{ C/kg} \qquad \text{(electron)} \qquad \textbf{(28.1a)}$$

This is a huge number, as Thomson realized, and it implies that the electron has a very small mass. It was known from electroplating that about 10^8 C/kg is needed to plate a material, a factor of about 1000 less than the charge per kilogram of electrons. Thomson went on to do the same experiment for positively charged hydrogen ions (now known to be bare protons) and found a charge per kilogram about 1000 times smaller than that for the electron, implying the proton is about 1000 times more massive than the electron. Today, we know more precisely that

$$\frac{q_p}{m_p} = 9.57 \times 10^7 \text{ C/kg} \qquad \text{(proton)} \qquad \textbf{(28.1b)}$$

where q_p is the charge of the proton and m_p is its mass. This ratio is 1836 times less charge per kilogram than for the electron. Since the charges of electrons and protons are equal in magnitude, this implies $m_p = 1836m_e$.

Thomson performed a variety of experiments using differing gases in discharge tubes and employing other methods, such as the photoelectric effect, for freeing electrons from atoms. He always found the same properties for the electron, proving it to be an independent particle. For his work, the important pieces of which he began to publish in 1897, Thomson was awarded the 1906 Nobel prize in physics. In retrospect, it is difficult to appreciate how astonishing it was to find that the atom has a substructure. Thomson himself said, "It was only when I was convinced that the experiment left no escape from it that I published my belief in the existence of bodies smaller than atoms."

Thomson attempted to measure the charge of individual electrons, but his method could only determine its charge to the order of magnitude expected.* An American physicist, Robert Millikan (1868–1953), decided to improve upon Thomson's experiment for measuring q_e and was eventually forced to try another approach, one that is now a classic experiment performed by every physics major. The Millikan oil drop experiment is shown in Figure 28.6.

In the Millikan oil drop experiment, fine drops of oil are sprayed from an atomizer. Some of these are charged by the process and can then be suspended between metal plates by a voltage between the plates. The weight of the drop is balanced by the electric force:

$$m_{\text{drop}}g = q_e E$$

The electric field is created by the applied voltage, so $E = V/d$, and V is adjusted to just balance the drop's weight. The drops can be seen as points of reflected light using a microscope, but they are too small to directly measure their size and mass. The mass of the drop is determined by observing how fast it falls when the voltage is turned off. Since air resistance is very significant for these submicroscopic drops, the more massive drops fall

---

*Since Faraday's experiments with electroplating in the 1830s, it had been known that about 100,000 C per mole was needed to plate singly ionized ions. Dividing this by the number of ions per mole (that is, by Avogadro's number), which was approximately known, the charge per ion was calculated to be about 1.6×10^{-19} C, close to the actual value.

(a)

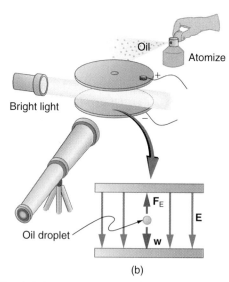
(b)

Figure 28.6 (a) Robert Millikan. (b) The Millikan oil drop experiment produced the first accurate direct measurement of the charge on electrons, one of the most fundamental constants in nature. Fine drops of oil become charged when sprayed. These are observed between metal plates with a potential applied to oppose gravity. The balance of gravity and electric force allows the calculation of the charge on a drop. The charge is found to be quantized in units of -1.6×10^{-19} C, thus determining directly the charge of the excess and missing electrons on the oil drops.

faster than the less massive, and sophisticated sedimentation calculations can reveal their mass. Oil is used rather than water, because it does not readily evaporate and so mass is nearly constant. Once the mass of the drop is known, the charge of the electron is given by rearranging the previous equation:

$$q_e = \frac{m_{\text{drop}}g}{E} = \frac{m_{\text{drop}}gd}{V}$$

where d is the separation of the plates and V is the voltage that holds the drop motionless. (The same drop can be observed for several hours to see that it really is motionless.) By 1913 Millikan had measured the charge of the electron q_e to an accuracy of 1%, and he improved this by a factor of 10 within a few years to a value of -1.60×10^{-19} C. He also observed that all charges were multiples of the basic electron charge and that sudden changes could occur in which electrons were added or removed from drops. For this very fundamental direct measurement of q_e and for his studies of the photoelectric effect, Millikan was awarded the 1923 Nobel prize in physics.

With the charge of the electron known and the charge to mass ratio known, the electron's mass can be calculated. It is

$$m_e = \frac{q_e}{q_e / m_e}$$

Substituting known values yields

$$m_e = \frac{-1.60 \times 10^{-19} \text{ C}}{-1.76 \times 10^{11} \text{ C/kg}}$$

or

$$m_e = 9.11 \times 10^{-31} \text{ kg} \qquad \text{(electron's mass)} \qquad \textbf{(28.2)}$$

The mass of the electron has been verified in many subsequent experiments and is now known to an accuracy of better than 1 part in one million.* It is an incredibly small mass and remains the smallest known mass of any particle that has mass. (Some particles, such as photons, are massless and cannot be brought to rest, but travel at the speed of light.) A similar calculation gives the masses of other particles, including the proton. To three digits, the mass of the proton is now known to be

$$m_p = 1.67 \times 10^{-27} \text{ kg} \qquad \text{(proton's mass)} \qquad \textbf{(28.3)}$$

which is nearly identical to the mass of a hydrogen atom. What Thomson and Millikan had done was to prove the existence of one substructure of atoms, the electron, and further to show that it had only a tiny fraction of the mass of an atom.† The nucleus of an atom contains most of its mass, and the nature of the nucleus was completely unanticipated.

The Nucleus

Here we examine the first direct evidence of the size and mass of the nucleus. In later chapters, we will examine many other aspects of nuclear physics, but the basic information on nuclear size and mass is so important to understanding the atom that we consider it here.

Nuclear radioactivity was discovered in 1896, and it was soon the subject of intense study by a number of the best scientists in the world. Among them was Ernest Rutherford, who made numerous fundamental discoveries. Rutherford used nuclear radiation to directly examine the size and mass of the atomic nucleus. The experiment he devised is shown in Figure 28.7. A radioactive source that emits alpha radiation was placed in a lead container with a hole in one side to produce a beam of alpha particles. A thin gold foil was placed in the beam, and the scattering of the alpha particles was observed by the glow they caused when they struck a phosphor.

---

*The preceding calculation produces a result of 9.09×10^{-31} kg due to the use of only three-digit figures in the calculation.

† Another important characteristic of quantum mechanics was also beginning to emerge. All electrons are identical to one another. The charge and mass of electrons are not average values; rather, they are unique values that all electrons must have. This is true of other fundamental entities at the submicroscopic level. All protons are identical to one another, and so on.

Figure 28.7 Rutherford's experiment gave direct evidence for the size and mass of the nucleus by scattering alpha particles from a thin gold foil. Alpha particles with energies of about 5 MeV are emitted from a radioactive source, are collimated into a beam, and fall upon the foil. The number that penetrate the foil or scatter to various angles indicates that gold nuclei are very small and contain nearly all of the gold atom's mass. This is particularly indicated by the alphas that scatter to very large angles, much like a soccer ball bouncing off a goalie's head.

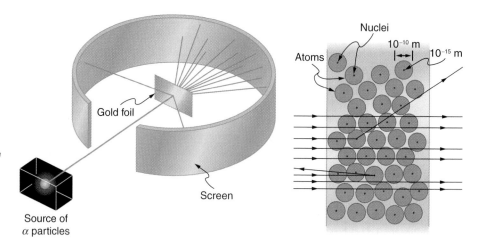

Alpha particles were known to be the doubly charged positive nuclei of helium atoms that had kinetic energies on the order of 5 MeV when emitted in nuclear decays. They interact with matter mostly via the Coulomb force, and the manner in which they scatter from nuclei can reveal nuclear size and mass. This is analogous to observing how a bowling ball is scattered by an object you cannot see directly. Because the alpha particle's energy is so large compared with the typical energies associated with atoms (MeV versus eV), you would expect the alpha particles to simply crash through a thin foil much like a supersonic bowling ball would crash through a few dozen rows of bowling pins. Instead, Rutherford and his collaborators found that alpha particles occasionally were scattered to large angles, some even back in the direction from which they came. Detailed analysis using conservation of momentum and energy—particularly of the small number that came straight back—implied that gold nuclei are very small compared with the size of a gold atom, contain almost all of the atom's mass, and are tightly bound. Since the gold nucleus is several times more massive than the alpha, a head-on collision would scatter the alpha straight back toward the source. The smaller the nucleus, the fewer that would hit head on.

Although the results of the experiment were published by his colleagues in 1909, it took Rutherford two years to convince himself of their meaning. Like Thomson before him, Rutherford was reluctant to accept such radical results. Nature on a small scale is so unlike our classical world that even those at the forefront of discovery are sometimes surprised. Rutherford later wrote: "It was almost as incredible as if you fired a 15-inch shell at a piece of tissue paper and it came back and hit you. On consideration, I realised that this scattering backwards . . . [meant] . . . the greatest part of the mass of the atom was concentrated in a tiny nucleus." In 1911 Rutherford published his analysis together with a proposed model of the atom. The size of the nucleus was determined to be about 10^{-15} m, or 100,000 times smaller than the atom! This implies a huge density, on the order of 10^{15} g/cm^3, vastly unlike any macroscopic matter. Also implied is the existence of previously unknown nuclear forces to counteract the huge repulsive Coulomb forces among the positive charges in the nucleus. Huge forces would also be consistent with the large energies emitted in nuclear radiation.

The small size of the nucleus also implies that the atom is mostly empty inside. In fact, in Rutherford's experiment most alphas went straight through the gold foil with very little scattering, since electrons have such small masses and since the atom was mostly empty with nothing for the alpha to hit. There were already hints of this at the time Rutherford performed his experiments, since energetic electrons had been observed to penetrate thin foils more easily than expected. Figure 28.8 shows a schematic of the atoms in a thin foil with circles representing the size of the atoms (about 10^{-10} m) and dots representing the nuclei. (The dots are not to scale—if they were, you would need a microscope to see them.) Most alphas miss the small nuclei and are only slightly scattered by electrons. Occasionally (about once in 8000

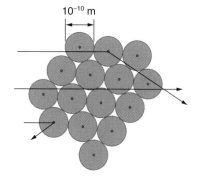

Figure 28.8 An expanded view of the atoms in the gold foil in Rutherford's experiment. Circles represent the atoms (about 10^{-10} m in diameter), while dots represent the nuclei (about 10^{-15} m in diameter). To be visible, the dots are much larger than to scale. Most alphas crash through relatively unaffected because of their high energy and the electron's small mass. Some, however, head straight toward a nucleus and are scattered straight back. Detailed analysis gives the size and mass of the nucleus.

times in Rutherford's experiment), an alpha hits a nucleus head-on and is scattered straight backward.

Based on the size and mass of the nucleus revealed by his experiment, as well as the mass of electrons, Rutherford proposed the **planetary model of the atom**. First mentioned in relationship to the separation of charges in static electricity in Section 17.1, the planetary model of the atom pictures low-mass electrons orbiting a large-mass nucleus. The size of the electron orbits are large compared with the size of the nucleus, with mostly vacuum inside the atom. This picture is analogous to how low-mass planets in our solar system orbit the large-mass sun at distances large compared with the size of the sun. In the atom, the attractive Coulomb force is analogous to gravitation in the planetary system. (See Figure 28.9.) Note that a model or mental picture is needed to explain experimental results, since the atom is too small to be directly observed with visible light.

Rutherford's planetary model of the atom was crucial to understanding the characteristics of atoms, and their interactions and energies, as we shall see in the next few sections. It also was an indication of how different nature is from the familiar classical world on the small, quantum mechanical, scale. The discovery of a substructure to all matter in the form of atoms and molecules was now being taken a step further to reveal a substructure to atoms that was simpler than the 92 elements then known. We have continued to search for deeper substructures, such as those inside the nucleus, with some success. In later chapters, we will follow this quest to the discussion of quarks and other elementary particles, and we will look at the direction the search seems now to be heading.

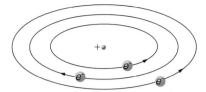

Figure 28.9 Rutherford's planetary model of the atom incorporates the characteristics of the nucleus, electrons, and the size of the atom. This model was the first to recognize the structure of atoms, in which low-mass electrons orbit a very small massive nucleus in orbits much larger than the nucleus. The atom is mostly empty and is analogous to our planetary system.

28.3 BOHR'S THEORY OF THE HYDROGEN ATOM

The great Danish physicist Niels Bohr (1885–1962) made immediate use of Rutherford's planetary model of the atom. (See Figure 28.10.) Bohr became convinced of its validity and spent part of 1912 at Rutherford's laboratory. In 1913, after returning to Copenhagen, he began publishing his theory of the simplest atom, hydrogen, based on the planetary model of the atom. For decades, many questions had been asked about atomic characteristics. From their sizes to their spectra, much was known about atoms, but little had been explained in terms of the laws of physics. Bohr's theory explained the atomic spectrum of hydrogen and established new and broadly applicable principles in quantum mechanics.

Mysteries of Atomic Spectra

As noted in Section 27.1, the energies of some small systems are quantized. Atomic and molecular spectra have been known to be discrete or quantized for about 200 years. (See Figure 28.11.) Maxwell and others had realized that there must be a connection between the spectrum of an atom and its structure, something like the resonant frequencies of musical instruments. But in spite of years of efforts by many great minds, no one had a workable theory.* Following Einstein's proposal of photons with quantized energies directly proportional to their wavelengths, it became even more evident that discrete spectra implied quantized energy states in atoms and molecules.

In some cases, it had been possible to devise formulas that described the spectra. As you might expect, the simplest atom—hydrogen with its single electron—has a relatively simple spectrum. Hydrogen's spectrum had been observed in the infrared (IR), visible, and ultraviolet (UV), and several series of spectral lines had been observed. (See Figure 28.12.) These series are named after early researchers who studied them in particular

Figure 28.10 Niels Bohr, Danish physicist, used the planetary model of the atom to explain the atomic spectrum and size of the hydrogen atom. His many contributions to the development of atomic physics and quantum mechanics, his personal influence on many students and colleagues, and his personal integrity, especially in the face of Nazi oppression, earned him a prominent place in history.

---

*It was a running joke that any theory of atomic and molecular spectra could be destroyed by throwing a book of data at it, so complex were the spectra.

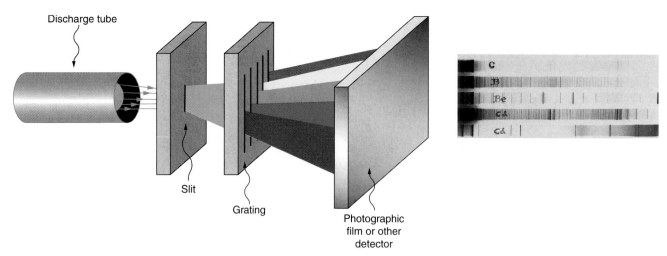

Figure 28.11 These emission spectra are clearly discrete, implying quantized energy states for the atoms and molecules creating them. Each is unique, providing a powerful and much used analytical tool, and many were well known for many years before they could be explained with physics. Also shown are a discharge tube, slit, and diffraction grating producing a line spectrum.

depth. The observed **hydrogen spectrum wavelengths** can be calculated using the following formula:

$$\frac{1}{\lambda} = R\left(\frac{1}{n_f^2} - \frac{1}{n_i^2}\right) \tag{28.4}$$

where λ is the wavelength of the emitted EM radiation and R is the **Rydberg constant**, determined by experiment to be

$$R = 1.097 \times 10^7/\text{m} \tag{28.5}$$

The constant n_f is a positive integer associated with a specific series. For the Lyman series, $n_f = 1$; for the Balmer series, $n_f = 2$; for the Paschen series, $n_f = 3$; and so on. The Lyman series is entirely in the UV, while part of the Balmer series is visible and the remainder UV. The Paschen series and all the rest are entirely IR. There are apparently an unlimited number of series, although they lie progressively farther into the infrared and become difficult to observe as n_f increases. The constant n_i is a positive integer, but it must be greater than n_f. Thus, for the Balmer series, $n_f = 2$ and $n_i = 3, 4, 5, 6, \ldots$. Note that n_i can approach infinity. While the formula in Equation 28.4 was just a recipe designed to fit data and was not based on physical principles, it did imply deeper meaning. Balmer first devised the formula for his series alone, and it was later found to describe all the other series by using different values of n_f. Bohr was the first to comprehend the deeper meaning.

Figure 28.12 A schematic of the hydrogen spectrum shows several series named for those who contributed most to their determination. Part of the Balmer series is in the visible spectrum, while the Lyman series is entirely in the UV, and the Paschen series and others are in the IR. Values of n_f and n_i are shown for some of the lines.

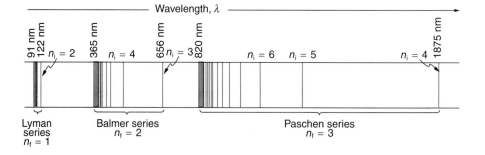

EXAMPLE 28.1 WAVELENGTH OF A HYDROGEN LINE

Calculate the wavelength of the second line in the Balmer series.

Strategy and Concept The Balmer series requires that $n_f = 2$. The first line in the series is taken to be for $n_i = 3$, and so the second would have $n_i = 4$.

Solution The calculation is a straightforward application of Equation 28.4. Entering the determined values for n_f and n_i yields

$$\frac{1}{\lambda} = R\left(\frac{1}{n_f^2} - \frac{1}{n_i^2}\right)$$

$$= (1.097 \times 10^7/\text{m})\left(\frac{1}{2^2} - \frac{1}{4^2}\right)$$

$$= 2.057 \times 10^6/\text{m}$$

Inverting to find λ gives

$$\lambda = \frac{1}{2.057 \times 10^6/\text{m}} = 486 \times 10^{-9}\text{ m}$$

$$= 486\text{ nm}$$

Discussion This is indeed the experimentally observed wavelength, corresponding to the second (blue-green) line in the Balmer series. More impressive is the fact that the same simple recipe predicts *all* of the hydrogen spectrum lines, including new ones observed in subsequent experiments. What is nature telling us?

Bohr's Solution for Hydrogen

Bohr was able to derive the formula for the hydrogen spectrum in Equation 28.4 using basic physics, the planetary model of the atom, and some important new proposals. His first proposal is that **the orbits of electrons in atoms are quantized**. That is, only certain orbits are allowed. Each orbit has a different energy, and electrons can move to a higher orbit by absorbing energy and drop to a lower orbit by emitting energy. If the orbits are quantized, the amount of energy absorbed or emitted is also quantized, producing discrete spectra. Photon absorption and emission are among the primary methods of transferring energy into and out of atoms. The energies of the photons are quantized, and their energy is explained as being equal to the change in energy of the electron when it moves from one orbit to another. In equation form, this is

$$\Delta E = hf = E_i - E_f \tag{28.6}$$

Here ΔE is the change in energy between the initial and final orbits, and hf is the energy of the absorbed or emitted photon. It is quite logical (that is, expected from our everyday experience) that energy is involved in changing orbits. A blast of energy is required for the space shuttle, for example, to climb to a higher orbit. What is not expected is that atomic orbits should be quantized. This is not observed for satellites or planets, which can have any orbit given the proper energy. (See Figure 28.13.)

Figure 28.14 shows an **energy level diagram**, a convenient way to display energy states. Energy is plotted vertically with the lowest, or ground state, at the bottom and with excited states above. Given the energies of the lines in an atomic spectrum, it is possible (although sometimes very difficult) to determine the energy levels of an atom. Energy level diagrams are used for many systems, including molecules and nuclei. A theory of the atom or other system must predict its energies based on the physics of the system.

Bohr was clever enough to find a way to calculate the electron orbital energies in hydrogen. This was an important first step that has been improved upon, but it is well worth repeating here, because it does correctly describe many characteristics of hydrogen. Assuming circular orbits, Bohr proposed that the **angular momentum of an electron in its orbit is quantized** and given by the formula

$$L = m_e v r_n = n\frac{h}{2\pi} \quad (n = 1, 2, 3, \ldots) \tag{28.7}$$

where L is the angular momentum, m_e is the electron's mass, r_n is the radius of the nth orbit, and h is Planck's constant.* At the time, Bohr himself did not know why angular

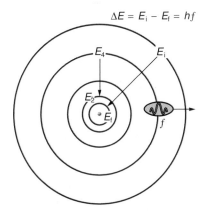

Figure 28.13 The planetary model of the atom, as modified by Bohr, has the orbits of the electrons quantized. Only certain orbits are allowed, explaining why atomic spectra are discrete (quantized). The energy carried away from an atom by a photon comes from the electron dropping from one allowed orbit to another and is thus quantized. This is likewise true for atomic absorption of photons.

---

*In Section 9.5 angular momentum was noted to be $L = I\omega$. For a small object at a radius r, $I = mr^2$ and $\omega = v/r$, so that $L = (mr^2)(v/r) = mvr$.

E

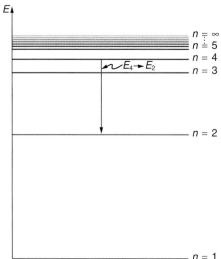

$n = \infty$
$n = 5$
$n = 4$
$\sim E_4 \rightarrow E_2$ $\quad n = 3$

$n = 2$

$n = 1$

Figure 28.14 An energy level diagram plots energy vertically and is useful in visualizing the energy states of a system and the transitions between them. This diagram is for the hydrogen atom, showing a transition between two orbits having energies E_4 and E_2.

momentum should be quantized (we will see why in the next section), but using this assumption he was able to calculate the energies in the hydrogen spectrum, something no one else had done.

From Bohr's assumptions, we will now derive a number of important properties of the hydrogen atom. We start by noting the centripetal force causing the electron to follow a circular path is supplied by the Coulomb force. [To be more general, we note that this analysis is valid for any single-electron atom. So if a nucleus has Z protons ($Z = 1$ for hydrogen, 2 for helium, etc.) and only one electron, that atom is called **hydrogen-like**. The spectra of hydrogen-like ions are similar to hydrogen, but shifted to higher energy by the greater attractive force between the electron and nucleus.] The magnitude of the centripetal force is $m_e v^2/r_n$, while the Coulomb force is $k(Zq_e)(q_e)/r_n^2$. Equating these,*

$$k\frac{Zq_e^2}{r_n^2} = \frac{m_e v^2}{r_n} \qquad \text{(Coulomb = centripetal)}$$

Angular momentum quantization is stated in Equation 28.7. We solve that equation for v, substitute it into the above, and rearrange the expression to obtain the radius of the orbit. This yields

$$r_n = \frac{n^2}{Z} a_B \qquad \text{(allowed orbits } n = 1, 2, 3, \ldots) \qquad \textbf{(28.8)}$$

where a_B is defined to be the **Bohr radius**, since for the lowest orbit ($n = 1$) and for hydrogen ($Z = 1$), $r_1 = a_B$. It is left as an end-of-chapter problem to show that the Bohr radius is

$$a_B = \frac{h^2}{4\pi^2 m_e k Z q_e^2} = 0.529 \times 10^{-10} \text{ m} \qquad \textbf{(28.9)}$$

These last two equations can be used to calculate the **radii of the allowed electron orbits in any hydrogen-like atom**. It is impressive that the formula gives the correct radius of hydrogen, which is measured experimentally to be very close to the Bohr radius. Equation 28.8 also tells us that the orbital radius is proportional to n^2, as illustrated in Figure 28.15.

To get the electron orbital energies, we start by noting that the electron energy is the sum of its kinetic and potential energy:

$$E_n = \text{KE} + \text{PE}$$

Kinetic energy is the familiar $\text{KE} = (1/2)m_e v^2$, assuming the electron is not moving at relativistic speeds. Potential energy for the electron is electrical, or $\text{PE} = q_e V$, where V is the potential due to the nucleus, which looks like a point charge. The nucleus has a positive charge Zq_e; thus, $V = kZq_e/r_n$, recalling Equation 18.8 for the potential due to a point charge. Since the electron's charge is negative, we see that $\text{PE} = -kZq_e^2/r_n$. Entering these expressions for KE and PE, we find

$$E_n = \frac{1}{2}m_e v^2 - k\frac{Zq_e^2}{r_n}$$

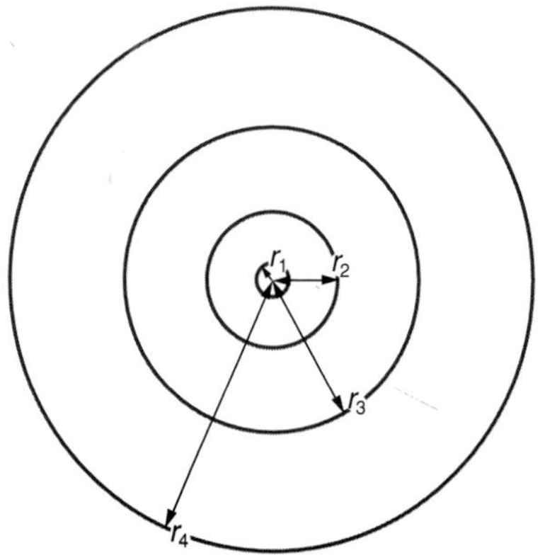

Now we substitute r_n from Equation 28.8 and v from Equation 28.7 into the above expression for energy. Algebraic manipulation yields

$$E_n = -\frac{Z^2}{n^2} E_0 \qquad (n = 1, 2, 3, \ldots) \qquad \textbf{(28.10)}$$

for the orbital **energies of hydrogen-like atoms**. E_0 is the **ground state energy** ($n = 1$) for hydrogen ($Z = 1$) and is given by

$$E_0 = \frac{2\pi^2 q_e^4 m_e k^2}{h^2} = 13.6 \text{ eV} \qquad \textbf{(28.11)}$$

Figure 28.15 The allowed electron orbits in hydrogen have the radii shown. These radii were first calculated by Bohr and are given by Equation 28.8. The lowest orbit has the experimentally verified diameter of a hydrogen atom.

---

*There is a tacit assumption here that the nucleus is so much more massive than the electron that it is stationary, and the electron orbits about it. This is consistent with the planetary model of the atom.

Thus, for hydrogen,

$$E_n = -\frac{13.6 \text{ eV}}{n^2} \qquad (n = 1, 2, 3, \ldots) \qquad \textbf{(28.12)}$$

Figure 28.16 shows an energy level diagram for hydrogen that also illustrates how the various spectral series for hydrogen are related to transitions between energy levels.

Electron total energies are negative, since the electron is bound to the nucleus, analogous to being in a hole without enough kinetic energy to escape. As n approaches infinity, the total energy becomes zero. This corresponds to a free electron with no kinetic energy, since r_n gets very large for large n, and the electric potential energy thus becomes zero. Thus 13.6 eV is needed to ionize hydrogen (to go from −13.6 eV to 0, or unbound), an experimentally verified number. Given more energy, the electron becomes unbound with some kinetic energy. For example, giving 15.0 eV to an electron in the ground state of hydrogen strips it from the atom and leaves it with 1.4 eV of kinetic energy.

Finally, let us consider the energy of a photon emitted in a downward transition, given by Equation 28.6 to be

$$\Delta E = hf = E_i - E_f$$

Substituting $E_n = -13.6 \text{ eV}/n^2$, we see that

$$hf = (13.6 \text{ eV})\left(\frac{1}{n_f^2} - \frac{1}{n_i^2}\right)$$

Dividing both sides of this equation by hc gives an expression for $1/\lambda$:

$$\frac{hf}{hc} = \frac{f}{c} = \frac{1}{\lambda} = \frac{13.6 \text{ eV}}{hc}\left(\frac{1}{n_f^2} - \frac{1}{n_i^2}\right)$$

It can be shown that $(13.6 \text{ eV})/hc = 1.097 \times 10^7/\text{m} = R$, Rydberg's constant. Thus we have used Bohr's assumptions to derive the formula

$$\frac{1}{\lambda} = R\left(\frac{1}{n_f^2} - \frac{1}{n_i^2}\right)$$

We see that Bohr's theory of the hydrogen atom answers the question as to why this previously known formula describes the hydrogen spectrum. It is because the energy levels are proportional to $1/n^2$, where n is a nonnegative integer. A downward transition releases energy, and so n_i must be greater than n_f. The various series are those where the transitions end on a certain level. For the Lyman series, $n_f = 1$—that is, all the transitions end in the ground state (see also Figure 28.16). For the Balmer series, $n_f = 2$, or all the transitions end in the first excited state; and so on. What was once a recipe is now based in physics, and something new is emerging—angular momentum is quantized.

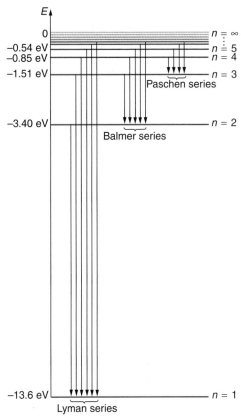

Figure 28.16 Energy level diagram for hydrogen showing the Lyman, Balmer, and Paschen series of transitions. The level energies are calculated using Equation 28.12, first derived by Bohr.

Triumphs and Limits of the Bohr Theory

Bohr did what no one had been able to do before. Not only did he explain the spectrum of hydrogen, he correctly calculated the size of the atom from basic physics. Some of his ideas are broadly applicable. Electron orbital energies are quantized in all atoms and molecules. Angular momentum is quantized. The electrons do not spiral into the nucleus, as expected classically (accelerated charges radiate, so that the electron orbits classically would decay quickly, and the electrons would sit on the nucleus—matter would collapse). These are major triumphs.

But there are limits to Bohr's theory. It cannot be applied to multielectron atoms, even one as simple as a two-electron helium atom. Bohr's model is what we call *semiclassical*. The orbits are quantized (nonclassical) but are assumed to be simple circular paths (classical). As quantum mechanics was developed, it became clear that there are no well-defined orbits; rather, there are clouds of probability. Bohr's theory also did not explain that some spectral lines are doublets (split into two) when examined closely. We shall examine many of these aspects of quantum mechanics in more detail, but it should be kept in mind that Bohr did not fail. Rather, he made very important steps along the path to greater knowledge and laid the foundation for all of atomic physics that has since evolved.

The Next Questions

Einstein once said it was important to keep asking the questions we eventually teach children not to ask. Why are the orbits quantized? Why is angular momentum quantized? You already know the key. Electrons are waves, as de Broglie later proposed. They can only exist where they interfere constructively, and the proper conditions are only met by certain orbits, as we shall see in the next section.

28.4 THE WAVE NATURE OF MATTER CAUSES QUANTIZATION

Following Bohr's initial work on the hydrogen atom, a decade was to pass before de Broglie proposed that matter has wave properties. The wave nature of matter was subsequently confirmed by observations of electron interference when scattered from crystals. (See Section 27.6 for details.) Electrons can only exist in locations where they interfere constructively. How does this affect electrons in atomic orbits? When an electron is bound to an atom, its wavelength must fit into a small space, something like a wave on a string. (See Figure 28.17.) Allowed orbits are those where an electron interferes with itself constructively. Not all orbits produce constructive interference. Thus only certain orbits are allowed—the orbits are quantized.

For a circular orbit, constructive interference occurs when the electron's wavelength fits neatly into the circumference, so that wave crests always align with crests and wave troughs align with troughs, as shown in Figure 28.17(b). More precisely, when an integral multiple of the electron's wavelength equals the circumference of the orbit, constructive interference is obtained. In equation form, the **condition for constructive interference and an allowed electron orbit** is

$$n\lambda_n = 2\pi r_n \qquad (n = 1, 2, 3, \ldots) \tag{28.13}$$

where λ_n is the electron's wavelength and r_n is the radius of that circular orbit. The de Broglie wavelength is $\lambda = h/p = h/mv$, and so here $\lambda_n = h/m_e v$. Substituting this into

Figure 28.17 (a) Waves on a string have a wavelength related to the length of the string, allowing them to interfere constructively. (b) If we imagine the string bent into a closed circle, we get a rough idea of how electrons in circular orbits can interfere constructively. (c) If the wavelength does not fit into the circumference, the electron interferes destructively; it cannot exist in such an orbit.

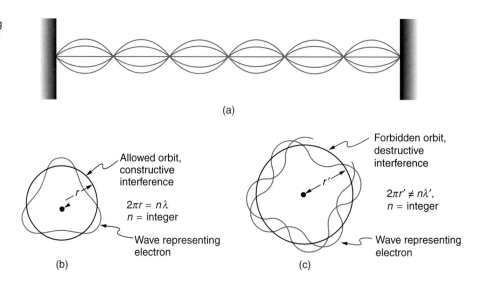

the condition for constructive interference produces an interesting result:

$$\frac{nh}{m_e v} = 2\pi r_n$$

Rearranging terms, and noting that $L = mvr$ for a circular orbit, we obtain the quantization of angular momentum as the condition for allowed orbits:

$$L = m_e v r_n = n\frac{h}{2\pi} \qquad (n = 1, 2, 3, \ldots)$$

This is what Bohr was forced to hypothesize as the rule for allowed orbits, as stated earlier in Equation 28.7. We now realize that it is the condition for constructive interference of an electron in a circular orbit. Figure 28.18 illustrates this for $n = 3$ and 4.

As discussed in Chapter 27, quantum mechanics flourished after de Broglie's insight about matter and photons having both wave and particle characteristics. Because of the wave character of matter, the idea of well-defined orbits gives way to a model in which there is a cloud of probability, consistent with Heisenberg's uncertainty principle. Figure 28.19 shows how this applies to the ground state of hydrogen. If you try to follow the electron in some well-defined orbit using a probe that has a small enough wavelength to get some details, you will instead knock the electron out of its orbit. Each measurement of the electron's position will find it to be in a definite location somewhere near the nucleus. Repeated measurements reveal a cloud of probability like that in the figure, with each speck the location determined by a single measurement. There is not a well-defined, circular orbit type of distribution. Nature again proves to be different on a small scale than on a macroscopic scale.

There are many examples in which the wave nature of matter causes quantization in bound systems such as the atom. Whenever a particle is confined (or bound) to a small space, its allowed wavelengths are those which fit into that space. This is true in blackbody radiators (atoms and molecules) as well as in atomic and molecular spectra. Various atoms and molecules will have different sets of electron orbits, depending on the size and complexity of the system. Other systems, such as protons bound in nuclei, will also have their own quantized states because of their size and the forces at play in them. The energy states of nuclei, like atoms, are quantized. When a system is large, such as a grain of sand, the tiny particle waves in it can fit in so many ways that it becomes impossible to see that the allowed states are discrete. Thus the correspondence principle is satisfied. As systems become large, they gradually look less grainy, and quantization becomes less evident. Unbound systems (small or not), such as an electron freed from an atom, do not have quantized energies, since their wavelengths are not constrained to fit in a certain space.

CONNECTIONS

Waves and Quantization

The wave nature of matter is responsible for the quantization of energy levels in bound systems. Only those states where matter interferes constructively exist, or are "allowed." Since there is a lowest orbit where this is possible in an atom, the electron cannot spiral into the nucleus. It cannot exist closer to or inside the nucleus. The wave nature of matter is what prevents matter from collapsing and gives atoms their sizes.

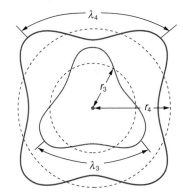

Figure 28.18 The third and fourth allowed circular orbits have three and four wavelengths, respectively, in their circumferences.

28.5 PATTERNS IN SPECTRA REVEAL MORE QUANTIZATION

High-resolution measurements of atomic and molecular spectra show that the lines are even more complex than they first appear. In this section, we will see that this complexity yielded important new information about electrons and their orbits in atoms.

In order to explore the substructure of atoms (and knowing that magnetic fields affect moving charges), the Dutch physicist Hendrik Lorentz (1853–1928) suggested that his student Pieter Zeeman (1865–1943) study how spectra might be affected by magnetic fields. What they found became known as the **Zeeman effect**—spectral lines are split by an external magnetic field, as shown in Figure 28.20. For their discoveries, Zeeman and Lorentz* shared the 1902 Nobel prize in physics.

Zeeman splitting is complex, like everything in atomic and molecular spectra. Some lines split into three lines, some into five, and so on. But one general feature is that the amount the split lines are separated is proportional to applied field strength, indicating an interaction with a moving charge. The splitting means that the quantized energy of an orbit is affected by an external magnetic field, causing the orbit to have several discrete

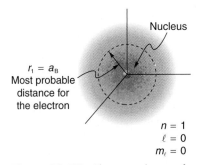

Figure 28.19 The ground state of a hydrogen atom has a probability cloud for the position of its electron. The probability of finding the electron is proportional to the darkness of the cloud. The electron can be closer or farther than the Bohr radius, but it is very unlikely to be a great distance from the nucleus.

---

*Lorentz also did trailblazing work on modern relativity that influenced Einstein.

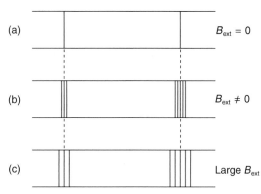

(a) $B_{ext} = 0$

(b) $B_{ext} \neq 0$

(c)   Large B_{ext}

Figure 28.20 The Zeeman effect is the splitting of spectral lines when a magnetic field is applied. The number of lines formed varies, but the spread is proportional to the strength of the applied field. (a) Two spectral lines with no external magnetic field. (b) The lines split when the field is applied. (c) The splitting is greater when a stronger field is applied.

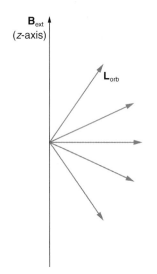

Figure 28.22 Only certain angles are allowed between the orbital angular momentum and an external magnetic field. This is implied by the fact that the Zeeman effect splits spectral lines into several discrete lines. Each line is associated with an angle between the external magnetic field and magnetic fields due to electrons and their orbits.

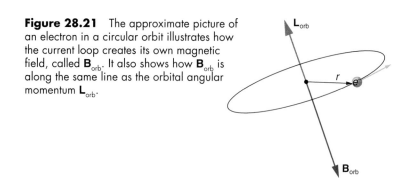

Figure 28.21 The approximate picture of an electron in a circular orbit illustrates how the current loop creates its own magnetic field, called **B**$_{orb}$. It also shows how **B**$_{orb}$ is along the same line as the orbital angular momentum **L**$_{orb}$.

energies instead of one. Even without an external magnetic field, very precise measurements showed that spectral lines are doublets (split into two), apparently by magnetic fields within the atom itself.

Bohr's theory of circular orbits is useful for visualizing how an electron's orbit is affected by a magnetic field. The circular orbit forms a current loop, which creates a magnetic field of its own. (See Figure 28.21.) Note that the **orbital magnetic field B**$_{orb}$ and the **orbital angular momentum L**$_{orb}$ are along the same line. The external magnetic field and the orbital magnetic field interact; a torque is exerted to align them. (See the discussions of torque on current loops and the fields created by current loops in Sections 21.8 and 21.9.) A torque rotating a system through some angle does work, so that there is energy associated with this interaction. Thus orbits at different angles to the external magnetic field have different energies. What is remarkable is that the energies are quantized—the magnetic field splits the spectral lines into several discrete lines. This means that only certain angles are allowed between the orbital angular momentum and the external field. (See Figure 28.22.)

We already know that the magnitude of angular momentum is quantized for electron orbits in atoms. The new insight is that the **direction of orbital angular momentum is also quantized**. The fact that orbital angular momentum can only have certain directions is called **space quantization**. Perhaps space itself is grainy and not all directions are possible. Like many aspects of quantum mechanics, this quantization of direction is totally unexpected. On the macroscopic scale, orbital angular momentum, such as that of the moon around the earth, can have any magnitude and be in any direction.

Detailed treatment of space quantization began to explain some of the complexities of atomic spectra, but certain patterns seemed to be caused by something else. As mentioned, spectral lines are actually closely spaced doublets, a characteristic called **fine structure**. (See Figure 28.23.) The doublet changes when a magnetic field is applied, implying that whatever causes the doublet interacts with a magnetic field. In 1925 Sem Goudsmit and George Uhlenbeck, two Dutch physicists who both later became U.S. citizens, successfully argued that electrons have properties analogous to a macroscopic charge spinning on its axis. Electrons, in fact, have an internal or intrinsic angular momentum called **intrinsic spin S**. Since electrons are charged, their intrinsic spin creates

Figure 28.23 Fine structure. (a) Upon close examination, spectral lines are doublets, even in the absence of an external magnetic field. The electron has an intrinsic magnetic field that interacts with its orbital magnetic field. (b) Fine structure splitting in hydrogen and deuterium.

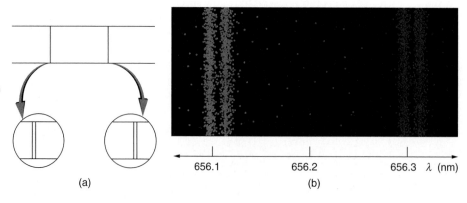

(a)   (b)

an **intrinsic magnetic field B**$_{int}$, which interacts with their orbital magnetic field **B**$_{orb}$. Furthermore, **electron intrinsic spin is quantized in magnitude and direction**, analogous to the situation for orbital angular momentum. The spin of the electron can only have one magnitude, and its direction can only be at one of two angles relative to a magnetic field. (See Figure 28.24.) Each spin direction has a different energy; hence, spectroscopic lines are split into two. Spectral doublets are now understood as being due to electron spin.

These two new insights—that the direction of angular momentum, whether orbital or spin, is quantized, and that electrons have intrinsic spin—help to explain many of the complexities of atomic and molecular spectra.

28.6 QUANTUM NUMBERS AND RULES

Physical characteristics that are quantized—such as energy, charge and angular momentum—are of such importance that names and symbols are given to them. The values of quantized entities are expressed in terms of **quantum numbers**, and the rules governing them are of the utmost importance in determining what nature is and does. This section covers some of the more important quantum numbers and rules—all of which apply far beyond the realm of atomic physics, where they were first discovered.

The *energy states of bound systems are quantized*, because the particle wavelength can only fit into the bounds of the system in certain ways. This was elaborated for the hydrogen atom, for which the allowed energies are expressed as $E_n \propto 1/n^2$, where $n = 1, 2, 3, \ldots$. We define n to be the **principal quantum number** that labels the basic states of a system. The lowest energy state has $n = 1$, the first excited state has $n = 2$, and so on. Thus the allowed values for the principal quantum number are

$$n = 1, 2, 3, \ldots \qquad (28.14)$$

This is more than just a numbering scheme, since the energy of the system, such as the hydrogen atom, can be expressed as some function of n, as can other characteristics (such as the orbital radii of the hydrogen atom).

The fact that the *magnitude of angular momentum is quantized* was first recognized by Bohr in relation to the hydrogen atom; it is now known to be true in general. With the development of quantum mechanics, it was found that the magnitude of angular momentum L can have only the values

$$L = \sqrt{\ell(\ell + 1)} \, \frac{h}{2\pi} \qquad (\ell = 0, 1, 2, \ldots, n - 1) \qquad (28.15)$$

where ℓ is defined to be the **angular momentum quantum number**. The rule for ℓ in atoms is given in the parentheses. Given n, the value of ℓ can be any integer from zero up to $n - 1$. For example, if $n = 4$, then ℓ can be 0, 1, 2, or 3.

Note that for $n = 1$, ℓ can only be zero. This means that the ground state angular momentum for hydrogen is actually zero, not $h/2\pi$ as Bohr proposed. The picture of circular orbits is not valid, because there would be angular momentum for any circular orbit. A more valid picture is the cloud of probability shown for the ground state of hydrogen in Figure 28.19. The electron actually spends time in and near the nucleus. The reason the electron does not remain in the nucleus is related to Heisenberg's uncertainty principle—the electron's energy would have to be much too large to be confined to the small space of the nucleus. (See the discussion in Section 27.7.)

Now the first excited state of hydrogen has $n = 2$, so that ℓ can be either 0 or 1, according to the rule in Equation 28.15. Similarly, for $n = 3$, ℓ can be 0, 1, or 2. It is often most convenient to state the value of ℓ, a simple integer, rather than calculating the value of L from Equation 28.15. For example, for $\ell = 2$, we see that $L = \sqrt{2(2 + 1)} \, h/2\pi = \sqrt{6} \, h/2\pi = 0.390h = 2.58 \times 10^{-34}$ J·s. It is much simpler to state $\ell = 2$.

As recognized in the Zeeman effect, the *direction of angular momentum is quantized*. We now know this is true in all circumstances. It is found that the component

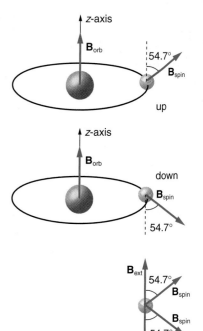

Figure 28.24 The intrinsic magnetic field **B**$_{int}$ of an electron is attributed to its spin **S**, roughly pictured to be due to its charge spinning on its axis. This is only a crude model, since electrons seem to have no size. The spin and intrinsic magnetic field of the electron can only make one of two angles with another magnetic field, such as that created by the electron's orbital motion. Space is quantized for spin as well as for orbital angular momentum.

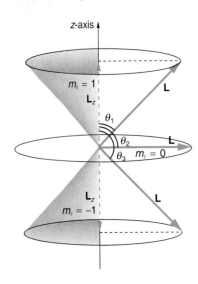

z-axis

$m_\ell = 1$
L_z
θ_1
L

θ_2
θ_3 $m_\ell = 0$
L

L_z
$m_\ell = -1$
L

Figure 28.25 The component of a given angular momentum along the z-axis (defined by the direction of a magnetic field) can only have certain values; these are shown here for $\ell = 1$, for which $m_\ell = -1$, 0, and +1. The direction of **L** is quantized in the sense that it can only have certain angles relative to the z-axis.

of angular momentum along one direction in space, usually called the z-axis,* can only have certain values L_z. The allowed values of L_z are

$$L_z = m_\ell \frac{h}{2\pi} \qquad (m_\ell = -\ell, -\ell + 1, \ldots, -1, 0, 1, \ldots, \ell - 1, \ell) \qquad \textbf{(28.16)}$$

where L_z is the **z-component of the angular momentum** and m_ℓ is the **angular momentum projection quantum number**. The rule in parentheses for the values of m_ℓ is that it can range from $-\ell$ to ℓ in steps of one. For example, if $\ell = 2$, then m_ℓ can have the five values $-2, -1, 0, 1,$ and 2. Each m_ℓ corresponds to a different energy in the presence of a magnetic field, so that they are related to the splitting of spectral lines into discrete parts, as discussed in the preceding section. If the z-component of angular momentum can only have certain values, then the angular momentum can only have certain directions, as illustrated in Figure 28.25.

EXAMPLE 28.2 WHAT ARE THE ALLOWED DIRECTIONS?

Calculate the angles that the angular momentum vector **L** can make with the z-axis for $\ell = 1$, as illustrated in Figure 28.25.

Strategy and Concept Figure 28.25 represents the vectors **L** and L_z as usual, with arrows proportional to their magnitudes and pointing in the correct directions. **L** and L_z form a right triangle, with **L** the hypotenuse and L_z the adjacent side. This means that the ratio of L_z to L is the cosine of the angle of interest. We can find L and L_z using Equations 28.15 and 28.16.

Solution We are given $\ell = 1$, so that m_ℓ can be +1, 0, or −1. Thus L has the value given by Equation 28.15:

$$L = \frac{\sqrt{1(1 + 1)}\,h}{2\pi} = \frac{\sqrt{2}\,h}{2\pi}$$

L_z can have three values given by Equation 28.16:

$$L_z = \frac{m_\ell h}{2\pi} = \begin{cases} \dfrac{h}{2\pi}, & m_\ell = +1 \\[2mm] 0, & m_\ell = 0 \\[2mm] -\dfrac{h}{2\pi}, & m_\ell = -1 \end{cases}$$

As can be seen in Figure 28.25, $\cos \theta = L_z/L$, and so for $m_\ell = +1$, we have

$$\cos \theta_1 = \frac{L_z}{L} = \frac{h/2\pi}{\sqrt{2}\,h/2\pi} = \frac{1}{\sqrt{2}} = 0.707$$

Thus,

$$\theta_1 = \cos^{-1}0.707 = 45.0°$$

Similarly, for $m_\ell = 0$, we find $\cos \theta_2 = 0$; thus,

$$\theta_2 = \cos^{-1}0 = 90.0°$$

And for $m_\ell = -1$,

$$\cos \theta_3 = \frac{L_z}{L} = \frac{-h/2\pi}{\sqrt{2}\,h/2\pi} = -\frac{1}{\sqrt{2}} = -0.707$$

so that

$$\theta_3 = \cos^{-1}(-0.707) = 135.0°$$

Discussion The angles are consistent with the figure. Only the angle relative to the z-axis is quantized. **L** can point in any direction as long as it makes the proper angle with the z-axis. Thus the angular momentum vectors lie on cones as illustrated. This behavior is not observed on the large scale. To see how the correspondence principle holds here, consider that the smallest angle (θ_1 in the example) is for the maximum value of m_ℓ, namely $m_\ell = \ell$. For that smallest angle, $\cos \theta = L_z/L = \ell/\sqrt{\ell(\ell + 1)}$, which approaches 1 as ℓ becomes very large. If $\cos \theta = 1$, then $\theta = 0°$. Furthermore, for large ℓ, there are many values of m_ℓ, so that all angles become possible as ℓ gets very large.

---

*The direction in space must be related to something physical, such as the direction of the magnetic field at that location. This is an aspect of relativity. Direction has no meaning if there is nothing that varies with direction, as does magnetic force.

Intrinsic Spin Angular Momentum Is Quantized in Magnitude and Direction

There are two more quantum numbers of immediate concern. (Both were first discovered for electrons in conjunction with fine structure in atomic spectra, as discussed in Section 28.5.) It is now well established that electrons and other fundamental particles have *intrinsic spin*, roughly analogous to a planet spinning on its axis. This spin is a fundamental characteristic of particles, and only one magnitude of intrinsic spin is allowed for a given type of particle. So intrinsic angular momentum is quantized independently of orbital angular momentum. Additionally, the direction of the spin is also quantized. It has been found that the **magnitude of the intrinsic** (internal) **spin angular momentum**, S, of an electron is given by

$$S = \sqrt{s(s + 1)}\,\frac{h}{2\pi} \qquad \left(s = \frac{1}{2} \text{ for electrons} \right) \qquad \textbf{(28.17)}$$

where s is defined to be the **spin quantum number**. This is very similar to the quantization of L given in Equation 28.15, except the only value allowed for s for electrons is 1/2.

The *direction of intrinsic spin is quantized*, just as is the direction of orbital angular momentum. The direction of spin angular momentum along one direction in space, again called the *z*-axis, can only have the values

$$S_z = m_s \frac{h}{2\pi} \qquad \left(m_s = -\frac{1}{2}, +\frac{1}{2} \right) \qquad \textbf{(28.18)}$$

for electrons. S_z is the **z-component of spin angular momentum** and m_s is the **spin projection quantum number**.* For electrons, s can only be 1/2, and m_s can be either $+1/2$ or $-1/2$. Spin projection $m_s = +1/2$ is referred to as *spin up*, whereas $m_s = -1/2$ is called *spin down*. These are illustrated in Figure 28.24.

To summarize: *For electrons in atoms*, the principal quantum number can have the values $n = 1, 2, 3, \ldots$. Once n is known, the values of the angular momentum quantum number are limited to $\ell = 0, 1, 2, \ldots, n - 1$. For a given value of ℓ, the angular momentum projection quantum number can only have the values $m_\ell = -\ell, -\ell + 1, \ldots, -1, 0, 1, \ldots, \ell - 1, \ell$. Electron spin is independent of n, ℓ, and m_ℓ, always having $s = 1/2$. The spin projection quantum number can have two values, $m_s = +1/2$ or $-1/2$. (See Table 28.1.)

The state of a system, such as the precise nature of an electron in an atom, is determined by its particular quantum numbers. For example, the ground state of hydrogen has $n = 1$, so that $\ell = 0$ and, thus, $m_\ell = 0$. Its cloud of probability was shown in Figure 28.19. Figure 28.26 shows several hydrogen states corresponding to different sets of quantum numbers. Note that these clouds of probability are the locations of electrons as determined by making repeated measurements—each measurement finds the electron in a definite location, with a greater chance of finding the electron in some places than others. With repeated measurements, the pattern of probability shown in the figure emerges. The

---

*For particles having other intrinsic spins, the rule in parentheses becomes that m_s can range from $-s$ to $+s$ in integral steps.

TABLE 28.1

ATOMIC QUANTUM NUMBERS		
Name	*Symbol*	*Allowed values*
Principal quantum number	n	$1, 2, 3, \ldots$
Angular momentum	ℓ	$0, 1, 2, \ldots, n - 1$
Angular momentum projection	m_ℓ	$-\ell, -\ell + 1, \ldots, -1, 0, 1, \ldots, \ell - 1, \ell$
		(or $0, \pm 1, \pm 2, \ldots, \pm\ell$)
Spin*	s	1/2 (electrons)
Spin projection	m_s	$-1/2, +1/2$

*The spin quantum number s is usually not stated, since it is always 1/2 for electrons.

Figure 28.26 Probability clouds for the electron in the ground state and several excited states of hydrogen. The nature of these states is determined by their sets of quantum numbers, here given as (n, ℓ, m_ℓ). The ground state is $(0, 0, 0)$; one of the possibilities for the second excited state is $(3, 2, 1)$.

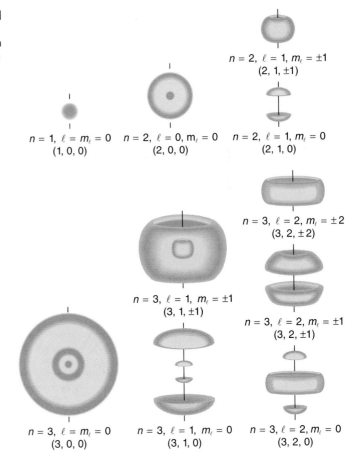

$n = 1, \ell = m_\ell = 0$
$(1, 0, 0)$

$n = 2, \ell = 0, m_\ell = 0$
$(2, 0, 0)$

$n = 2, \ell = 1, m_\ell = \pm 1$
$(2, 1, \pm 1)$

$n = 2, \ell = 1, m_\ell = 0$
$(2, 1, 0)$

$n = 3, \ell = 2, m_\ell = \pm 2$
$(3, 2, \pm 2)$

$n = 3, \ell = 1, m_\ell = \pm 1$
$(3, 1, \pm 1)$

$n = 3, \ell = 2, m_\ell = \pm 1$
$(3, 2, \pm 1)$

$n = 3, \ell = m_\ell = 0$
$(3, 0, 0)$

$n = 3, \ell = 1, m_\ell = 0$
$(3, 1, 0)$

$n = 3, \ell = 2, m_\ell = 0$
$(3, 2, 0)$

clouds of probability do not look like nor do they correspond to classical orbits. The uncertainty principle actually prevents us and nature from knowing how the electron gets from one place to another, and so an orbit really does not exist as such. Nature on a small scale is again much different from that on the large scale.

We will see that the quantum numbers discussed in this section are valid for a broad range of particles and other systems, such as nuclei. Some quantum numbers, such as intrinsic spin, are related to fundamental classifications of subatomic particles, and they obey laws that will give us further insight into the substructure of matter and its interactions.

28.7 THE PAULI EXCLUSION PRINCIPLE

Multiple-Electron Atoms

All atoms except hydrogen are multiple-electron atoms. The physical and chemical properties of elements are directly related to the number of electrons a neutral atom has. The periodic table of the elements, such as that at the end of the book, groups elements with similar properties into columns. This systematic organization is related to the number of electrons in a neutral atom, called the **atomic number** Z. We shall see in this section that the exclusion principle is key to the underlying explanations, and that it applies far beyond the realm of atomic physics.

In 1925, the Austrian physicist Wolfgang Pauli (see Figure 28.27) proposed the following rule:

Figure 28.27 The Austrian physicist Wolfgang Pauli (1900–1958) played a major role in the development of quantum mechanics. He proposed the exclusion principle; hypothesized the existence of an important particle, called the neutrino, before it was directly observed; made fundamental contributions to several areas of theoretical physics; and influenced many students who went on to do important work of their own.

> **No two electrons can have the same set of quantum numbers. That is, no two electrons can be in the same state.**

This statement is known as the **Pauli exclusion principle**, because it *excludes* electrons from being in the same state. The Pauli exclusion principle is extremely powerful and very broadly applicable. It applies to any identical particles with half-integral intrinsic spin—that is, having $s = 1/2, 3/2, \ldots$.* Thus no two electrons can have the same quantum numbers.

Let us examine how the exclusion principle applies to electrons in atoms. The quantum numbers involved were defined in the previous section—they are n, ℓ, m_ℓ, s, and m_s. Since s is always 1/2 for electrons, it is redundant to list s, and so we omit it and specify the state of an electron by a set of numbers (n, ℓ, m_ℓ, m_s). For example, the quantum numbers (2, 1, 0, −1/2) completely specify the state of an electron in an atom.

Since no two electrons can have the same quantum numbers, there are limits to how many of them can be in the same energy state. Note that n determines the energy state in the absence of a magnetic field. So we first choose n, and then we see how many electrons can be in this energy state or energy level. Consider the $n = 1$ level, for example. The only value ℓ can have is 0 (see Table 28.1 for a list of possible values once n is known), and thus m_ℓ can only be 0. The spin projection m_s can be either +1/2 or −1/2, and so there can be two electrons in the $n = 1$ state. One has quantum numbers (1, 0, 0, +1/2), and the other has (1, 0, 0, −1/2). Figure 28.28 illustrates that there can be one or two electrons having $n = 1$, but not three.

Shells and Subshells

Because of the Pauli exclusion principle, only hydrogen and helium can have all of their electrons in the $n = 1$ state. Lithium (see the periodic table) has three electrons, and so one must be in the $n = 2$ level. This leads to the concept of shells and shell filling. As we progress up in the number of electrons, we go from hydrogen to helium, lithium, beryllium, boron, and so on, and we see that there are limits to the number of electrons for each value of n. Higher ns correspond to higher energy, and they can allow more electrons because of the various combinations of ℓ, m_ℓ, and m_s that are possible. Each value of the principal quantum number n thus corresponds to an atomic **shell** into which a limited number of electrons can go.† Shells and the number of electrons in them determine the physical and chemical properties of atoms, since it is the outermost electrons that interact most with anything outside the atom.

The probability clouds of electrons with the lowest value of ℓ are closest to the nucleus and, thus, more tightly bound. Thus when shells fill they start with $\ell = 0$, progress to $\ell = 1$, and so on. Each value of ℓ thus corresponds to a **subshell**. Table 28.2 lists symbols traditionally used to denote shells and subshells.

To denote shells and subshells, we write $n\ell$ with a number for n and a letter for ℓ. For example, an electron in the $n = 1$ state must have $\ell = 0$, and it is denoted as a $1s$ electron. Two electrons in the $n = 1$ state is denoted as $1s^2$. Another example is an electron in the $n = 2$ state with $\ell = 1$, written as $2p$. The case of three electrons with these quantum numbers is written $2p^3$. This notation, called spectroscopic notation, is generalized as

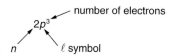

Counting the number of possible combinations of quantum numbers allowed by the exclusion principle, we can determine how many electrons it takes to fill each subshell and shell.

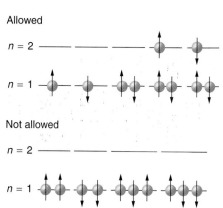

Figure 28.28 The Pauli exclusion principle explains why some configurations of electrons are allowed while others are not. Since electrons cannot have the same quantum numbers, a maximum of two can be in the $n = 1$ level, and a third electron must reside in the higher-energy $n = 2$ level. If there are two electrons in the $n = 1$ level, their spins must be in opposite directions. (More precisely, their spin projections must differ.)

TABLE 28.2

SHELL AND SUBSHELL SYMBOLS

Shell	Subshell	
n	ℓ	Symbol
1	0	s
2	1	p
3	2	d
4	3	f
5	4	g
⋮	5	h
	6*	i
	⋮	⋮

*It is unusual to deal with subshells having ℓ greater than 6, but when encountered they continue to be labeled in alphabetical order.

---

*Such particles belong to a family called fermions, which is discussed in Chapter 31. The exclusion principle does not apply to particles with integral intrinsic spin. Such integral spin particles, including photons, belong to a family called bosons, also discussed later.

† The shells are now labeled by $n = 1, 2, 3, \ldots$. At one time, it was common to call these the $K, L, M, \ldots$ shells, respectively, with the labeling starting with K (rather than A) since it was not known if that shell was the innermost.

EXAMPLE 28.3 HOW MANY ELECTRONS CAN BE IN THIS SHELL?

List all the possible sets of quantum numbers for the $n = 2$ shell, and determine the number of electrons that can be in the shell and each of its subshells.

Strategy Given $n = 2$ for the shell, the rules in Table 28.1 limit ℓ to be 0 or 1. The shell therefore has two subshells, labeled $2s$ and $2p$. Since the lowest ℓ subshell fills first, we start with the $2s$ subshell possibilities and then proceed with the $2p$ subshell.

Solution It is convenient to list the possible quantum numbers in a table.

n	ℓ	m_ℓ	m_s	Subshell	Total in subshell	Total in shell
2	0	0	+1/2	$2s$	2	
2	0	0	−1/2			
2	1	1	+1/2			
2	1	1	−1/2			
2	1	0	+1/2	$2p$	6	8
2	1	0	−1/2			
2	1	−1	+1/2			
2	1	−1	−1/2			

Discussion It is laborious to make a table like this every time we want to know how many electrons can be in a shell or subshell. There exist general rules that are easy to apply, as we shall now see.

The number of electrons that can be in a subshell depends entirely on the value of ℓ. Once ℓ is known, there are a fixed number of values of m_ℓ, each of which can have two values for m_s. First, since m_ℓ goes from $-\ell$ to ℓ in steps of 1, there are $2\ell + 1$ possibilities. This number is multiplied by 2, since each electron can be spin up or spin down. Thus the **maximum number of electrons that can be in a subshell** is

$$2(2\ell + 1) \tag{28.19}$$

For example, the $2s$ subshell in the preceding example has a maximum of 2 electrons in it, since $2(2\ell + 1) = 2(0 + 1) = 2$ for this subshell. Similarly, the $2p$ subshell has a maximum of 6 electrons, since $2(2\ell + 1) = 2(2 + 1) = 6$. For a shell, the maximum number is the sum of what can fit in the subshells. Some algebra (left as an end-of-chapter problem) shows that the **maximum number of electrons that can be in a shell** is

$$2n^2 \tag{28.20}$$

For example, for the first shell $n = 1$, and so $2n^2 = 2$. We have already seen that only two electrons can be in the $n = 1$ shell. Similarly, for the second shell $n = 2$, and so $2n^2 = 8$. As found in the preceding example, the total number of electrons in the $n = 2$ shell is 8.

EXAMPLE 28.4 SUBSHELLS AND TOTALS FOR $n = 3$

How many subshells are in the $n = 3$ shell? Identify each subshell, calculate the maximum number of electrons that will fit into each, and verify the total is $2n^2$.

Strategy Subshells are determined by the value of ℓ; thus, we first determine which ℓs are allowed, and then we apply Equation 28.19 to find the number of electrons in each subshell.

Solution Since $n = 3$, we know that ℓ can be 0, 1, or 2; thus, there are three possible subshells. In standard notation, they are labeled the $3s$, $3p$, and $3d$ subshells. We have already seen that 2 electrons can be in an s state, and 6 in a p state, but let us use Equation 28.19 to calculate the maximum number in each:

$3s$ has $\ell = 0$; thus, $2(2\ell + 1) = 2(0 + 1) = 2$
$3p$ has $\ell = 1$; thus, $2(2\ell + 1) = 2(2 + 1) = 6$
$3d$ has $\ell = 2$; thus, $2(2\ell + 1) = 2(4 + 1) = \underline{10}$
$$\text{Total} = 18$$
(in the $n = 3$ shell)

Equation 28.20 gives the maximum number in the $n = 3$ shell to be

Maximum number of electrons
$$= 2n^2 = 2(3)^2 = 2(9) = 18$$

Discussion The total number of electrons in the three possible subshells is thus the same as the formula $2n^2$ gives. In standard (spectroscopic) notation, a filled $n = 3$ shell is denoted as $3s^2 3p^6 3d^{10}$. Shells do not fill in a simple manner. Before the $n = 3$ shell is completely filled, for example, we begin to find electrons in the $n = 4$ shell.

Shell Filling and the Periodic Table

Table 28.3 shows electron configurations for the first 20 elements in the periodic table, starting with hydrogen and its single electron and ending with calcium. The Pauli exclusion principle determines the maximum number of electrons allowed in each shell and subshell. But the order in which the shells and subshells fill is complicated because of the large numbers of interactions between electrons.

Examining Table 28.3, you can see that as the number of electrons in an atom increases from 1 in hydrogen to 2 in helium and so on, the lowest-energy shell gets filled first—that is, the $n = 1$ shell fills first and then the $n = 2$ shell begins to fill. Within a shell, the subshells fill starting with the lowest ℓ, or with the s subshell, then the p, and so on, usually until all subshells are filled. The first exception to this occurs for potassium, where the $4s$ subshell begins to fill before any electrons go into the $3d$ subshell. The next exception is not shown in Table 28.3; it occurs for rubidium, where the $5s$ subshell starts to fill before the $4d$ subshell. The reason for these exceptions is that $\ell = 0$ electrons have probability clouds that penetrate closer to the nucleus and, thus, are more tightly bound (lower in energy).

The number of electrons in the outermost subshell determines the atom's chemical properties, since it is these electrons that are farthest from the nucleus and thus interact most with other atoms. If the outermost subshell can accept or give up an electron easily, then the atom will be highly reactive chemically. Each group in the periodic table is characterized by its outermost electron configuration. Perhaps the most familiar is Group 0, the noble gases. These gases are all characterized by a filled subshell that is particularly stable. This means that they have large ionization energies and do not readily give up an electron. Furthermore, if they were to accept an extra electron, it would be in a significantly higher level and thus loosely bound. Chemical reactions often involve sharing electrons. Noble gases can only be forced into unstable chemical compounds under high pressure and temperature.

Group VII contains the halogens, each of which has one less electron than a neighboring noble gas. Each halogen has 5 p electrons (a p^5 configuration), while the p subshell can hold 6 electrons. This means the halogens, such as fluorine and bromine, have one vacancy in their outermost subshell. They thus readily accept an extra electron (it becomes tightly bound, closing the shell as in noble gases) and are highly reactive chemically. The halogens are also likely to form singly negative ions, such as Cl⁻, fitting an extra electron into the vacancy in the outer subshell. In contrast, alkali metals, such as sodium and potassium, all have a single s electron in their outermost subshell (an s^1 configuration) and are members of Group I. These elements easily give up their extra electron and are thus highly reactive chemically. As you might expect, they also tend to form singly positive ions, such as Na⁺, by losing their loosely bound outermost electron. They are metals (conductors), because the loosely bound outer electron can move freely.

Of course, other groups are also of interest. Carbon, silicon, and germanium, for example, have similar chemistries and are in Group IV. Carbon in particular is extraordinary in its ability to form many types of bonds and be part of long chains, such as in organic molecules. The large group of transitional elements is characterized by filling of the d subshells. Heavier groups, such as the lanthanide series, are more complex—their shells do not fill in simple order. But the groups recognized by chemists such as Mendeleev have an explanation in the substructure of atoms.

TABLE 28.3

ELECTRON CONFIGURATIONS* OF ELEMENTS HYDROGEN THROUGH CALCIUM

Element	Number of electrons (Z)	Ground state configuration				
H	1	$1s^1$				
He	2	$1s^2$				
Li	3	$1s^2$	$2s^1$			
Be	4	"	$2s^2$			
B	5	"	$2s^2$	$2p^1$		
C	6	"	$2s^2$	$2p^2$		
N	7	"	$2s^2$	$2p^3$		
O	8	"	$2s^2$	$2p^4$		
F	9	"	$2s^2$	$2p^5$		
Ne	10	"	$2s^2$	$2p^6$		
Na	11	"	$2s^2$	$2p^6$	$3s^1$	
Mg	12	"	"	"	$3s^2$	
Al	13	"	"	"	$3s^2$	$3p^1$
Si	14	"	"	"	$3s^2$	$3p^2$
P	15	"	"	"	$3s^2$	$3p^3$
S	16	"	"	"	$3s^2$	$3p^4$
Cl	17	"	"	"	$3s^2$	$3p^5$
Ar	18	"	"	"	$3s^2$	$3p^6$
K	19	"	"	"	$3s^2$	$3p^6$ $4s^1$
Ca	20	"	"	"	"	" $4s^2$

*These are the ground state neutral atom configurations.

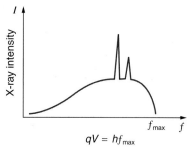

Figure 28.29 X-ray spectrum obtained when energetic electrons strike a material such as in the anode of a CRT. The smooth part of the spectrum is bremsstrahlung radiation, while the peaks are characteristic of the anode material. A different anode material would have characteristic x-ray peaks at different frequencies.

28.8 X RAYS: ATOMIC ORIGINS AND APPLICATIONS

Each type of atom (or element) has its own characteristic electromagnetic spectrum. X rays lie at the high end of an atom's spectrum and are characteristic of the atom as well. In this section, we explore characteristic x rays and some of the important applications of x rays.

We have previously discussed x rays, first as a part of the electromagnetic spectrum in Section 23.3, and again when discussing the properties of various types of photons in Section 27.3. Figure 27.10 in that section illustrated how an x-ray tube (a specialized CRT) creates x rays. Electrons boiled from a hot filament are accelerated with a high voltage, gaining significant kinetic energy and striking the anode. Electron energy deposited in the anode sometimes creates x-ray photons.

There are two processes by which x rays are created in the anode of an x-ray tube. In one process, the deceleration of electrons produces x rays, and these x rays are called *bremsstrahlung*, or braking radiation. The second process is atomic in nature and produces *characteristic x rays*, so called because they are characteristic of the anode material. The x-ray spectrum in Figure 28.29 is typical of what is created by an x-ray tube, showing a broad curve of bremsstrahlung radiation with characteristic x-ray peaks on it.

The spectrum in Figure 28.29 is collected over a period of time in which many electrons strike the anode, with a variety of possible outcomes for each hit. The broad range of x-ray energies in the bremsstrahlung radiation indicates that an incident electron's energy is not usually converted entirely into photon energy. The highest-energy x ray produced is one for which all of the electron's energy was converted to photon energy. Thus the accelerating voltage and the maximum x-ray energy are related by conservation of energy. Electric potential energy is converted to kinetic energy and then to photon energy, so that $E_{max} = hf_{max} = q_e V$. Units of electron volts are convenient. For example, a 100 kV accelerating voltage produces x-ray photons with a maximum energy of 100 keV.

Some electrons excite atoms in the anode. Part of the energy they deposit by collision with an atom results in one or more of the atom's electrons being knocked into a higher orbit or the atom being ionized. When the anode's atoms deexcite or recapture a missing electron, they emit characteristic electromagnetic radiation. The most energetic of these are created when an inner-shell vacancy is filled—that is, when an $n = 1$ or $n = 2$ shell electron has been excited to a higher level, and another electron falls into the vacant spot. A *characteristic x ray* (see Section 27.3) is *EM radiation emitted by an atom when an inner-shell vacancy is filled*. Figure 28.30 shows a representative energy level diagram that illustrates the labeling of characteristic x rays. X rays created when an electron falls into an $n = 1$ shell vacancy are called K_α when they come from the next higher level, an $n = 2$ to $n = 1$ transition.* A more energetic K_β x ray is created when an electron

---

*The labels $K, L, M, \ldots$ come from the older alphabetical labeling of shells starting with K rather than using the principal quantum numbers $1, 2, 3, \ldots$.

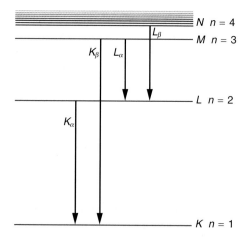

Figure 28.30 A characteristic x ray is emitted when an electron fills an inner-shell vacancy, as shown for several transitions in this approximate energy level diagram for a multiple-electron atom. Characteristic x rays are labeled according to the shell that had the vacancy and the shell from which the electron came. A K_α x ray, for example, is created when an electron fills the $n = 1$ shell vacancy coming from the $n = 2$ shell.

falls into an $n = 1$ shell vacancy from the $n = 3$ shell, an $n = 3$ to $n = 1$ transition. Similarly, when an electron falls into the $n = 2$ shell from the $n = 3$ shell, an L_α x ray is created. The energies of these x rays depend on the energies of electron states in the particular atom and, thus, are characteristic of that element.

EXAMPLE 28.5   CHARACTERISTIC X-RAY ENERGY

Calculate the approximate energy of a K_α x ray for a tungsten anode in an x-ray tube.

Strategy and Concept How do we calculate energies in a multiple-electron atom? In the case of characteristic x rays, the following approximate calculation is reasonable. Characteristic x rays are created when an inner-shell vacancy is filled. Inner-shell electrons are nearer the nucleus than others in an atom and, thus, feel little net effect from the others. This is similar to what happens inside a charged conductor, where its excess charge is distributed over the surface so that it creates no electric field inside. It is reasonable to assume the inner-shell electrons have hydrogen-like energies as given by Equation 28.10. As noted, a K_α x ray is created by an $n = 2$ to $n = 1$ transition. Since there are two electrons in a filled K shell, a vacancy would leave one electron, so that the effective charge would be $Z - 1$ rather than Z. For tungsten, $Z = 74$, so that the effective charge is 73.

Solution Equation 28.10 gives the energy for hydrogen-like atoms to be $E_n = -(Z^2/n^2)E_0$, where $E_0 = 13.6$ eV. As noted, the effective Z is 73. Now the K_α x-ray energy is given by

$$E_{K_\alpha} = \Delta E = E_i - E_f = E_2 - E_1$$

where

$$E_1 = -\frac{Z^2}{1^2} E_0 = -\frac{73^2}{1} \cdot 13.6 \text{ eV} = -72.5 \text{ keV}$$

and

$$E_2 = -\frac{Z^2}{2^2} E_0 = -\frac{73^2}{4} \cdot 13.6 \text{ eV} = -18.1 \text{ keV}$$

Thus,

$$E_{K_\alpha} = -18.1 \text{ keV} - (-72.5 \text{ keV}) = 54.4 \text{ keV}$$

Discussion This large photon energy is typical of characteristic x rays from heavy elements. It is large compared with other atomic emissions, because it is created when an inner-shell vacancy is filled, and inner-shell electrons are tightly bound. Characteristic x-ray energies become progressively larger for heavier elements, because their energy increases approximately as Z^2. Significant accelerating voltage is needed to create these inner-shell vacancies. In the case of tungsten, at least 72.5 kV is needed, because other shells are filled and you cannot simply bump one electron to a higher filled shell. Tungsten is a common anode material in x-ray tubes; so much of the energy of the impinging electrons is converted to heat that a high melting point material like tungsten is required.

Medical and Other Diagnostic Uses of X Rays

All of us can identify diagnostic uses of x-ray photons. Among these are the universal dental and medical x rays that have become an essential part of medical diagnostics. (See Figures 28.31 (b) and (c).) X rays are also used to inspect our luggage at airports (see Figure 28.31 (a)) and for early detection of cracks in crucial aircraft components. In fact, as with all common things,

Human & Medical Application

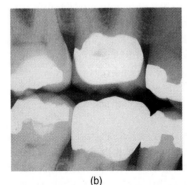
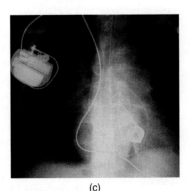

(a)   (b)   (c)

Figure 28.31   These x-ray images are shadows cast by the object exposed. The denser the material, the darker the shadow. (a) Contents of luggage can be examined. (b) Decay in the author's teeth can be detected and attributed to a taste for candy. (c) Many details such as artificial heart valves and a pacemaker can be seen in a person's chest.

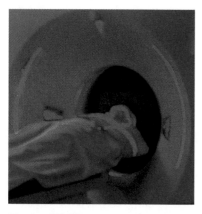

Figure 28.32 A patient being positioned in a CT scanner. A CT scanner passes x rays through a slice of the patient over a range of directions. The relative absorption of the x rays along different directions is computer analyzed to produce highly detailed images. Three-dimensional information can be obtained from multiple slices.

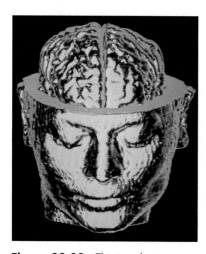

Figure 28.33 This is a three-dimensional image produced by analysis of several slices.

Figure 28.34 X-ray diffraction from a crystal produces this interference pattern. Analysis of the pattern yields information about the structure that created it.

new words have been created. An x ray is not only a noun meaning high-energy photon, it is also is an image produced by x rays, such as those shown in Figure 28.31, and it has been made into a familiar verb—to be x-rayed.

The most common x-ray images are simple shadows. Since x-ray photons have high energies, they penetrate materials that are opaque to visible light. The more energy an x-ray photon has, the more material it will penetrate. So an x-ray tube may be operated at 50.0 kV for a chest x ray, whereas it may need to be operated at 100 kV to examine a broken leg in a cast. The amount of penetration is related to the density of the material as well as to the energy of the photon. The denser the material, the fewer x-ray photons get through and the darker the shadow. Thus x rays excel at detecting breaks in bones and in imaging other physiological structures, such as some tumors, that differ in density from surrounding material. Because of their high photon energy, x rays create significant ionization in materials and damage cells in biological organisms. Modern uses minimize exposure to the patient and eliminate exposure to others. Biological effects of x rays will be explored in Chapter 30 along with other types of ionizing radiation such as those produced by nuclei.

While shadow images are sufficient in many applications, far more sophisticated images can be produced with modern technology. Figure 28.32 shows a computed tomography (CT) scanner. X rays are passed through a narrow section (called a slice) of the patient over a range of directions. Complex computer image processing of the relative absorption of the x rays along different directions produces a highly detailed image. Multiple images of different slices can also be computer analyzed to produce three-dimensional information, sometimes enhancing specific types of tissue, as shown in Figure 28.33.

X-Ray Diffraction and Crystallography

Since x-ray photons are very energetic, they have short wavelengths. For example, the 54.4 keV K_α x ray of Example 28.5 has a wavelength $\lambda = hc/E = (1240 \text{ eV} \cdot \text{nm})/(54.4 \text{ keV}) = 0.0228$ nm. Thus typical x-ray photons act like rays when they encounter macroscopic objects, like teeth, and produce sharp shadows. But since atoms are on the order of 0.1 nm in size, x rays can be used to detect the location, shape, and size of atoms and molecules. The process is called **x-ray diffraction**, because it involves the diffraction and interference of x rays to produce patterns that can be analyzed for information about the structures that scattered the x rays. Perhaps the most famous example of x-ray diffraction is the discovery of the double helix structure of DNA by James Watson and Francis Crick in 1953. Using x-ray diffraction data produced by Rosalind Franklin, they were the first to discern the structure of DNA that is so crucial to life. For this, Watson and Crick were awarded the 1962 Nobel prize in medicine.

Figure 28.34 shows a diffraction pattern created by scattering of x rays from a crystal. This process is known as x-ray crystallography because of the information it can yield about crystal structure, and it was the type of data Rosalind Franklin supplied to Watson and Crick for DNA. Not only do x rays confirm the size and shape of atoms, they give information on atomic arrangements in materials. For example, current research in high-temperature superconductors involves complex materials whose lattice arrangements are crucial to obtaining a superconducting material. These can be studied using x-ray crystallography.

Historically, scattering of x rays from crystals was used to *prove* that x rays are energetic EM waves. This was suspected from the time of the discovery of x rays in 1895, but it was not until 1912 that the German Max von Laue (1879–1960) convinced two of his colleagues to scatter x rays from crystals. If a diffraction pattern is obtained, he reasoned, then the x rays must be waves, and their wavelength could be determined. (The spacing of atoms in crystals was reasonably well known at the time, based on good values for Avogadro's number.) The experiments were convincing, and the 1914 Nobel prize in physics was given to von Laue for his suggestion leading to the proof that x rays are EM waves.

Certain other uses for x rays will be studied in later chapters. They are useful in the treatment of cancer because of the inhibiting effect they have on cell reproduction. X rays observed coming from outer space are useful in determining the nature of their sources, such as neutron stars and possibly black holes. Created in nuclear bomb explosions, x rays can also be used to detect clandestine atmospheric tests of these weapons. X rays can cause excitations of atoms, which then fluoresce (emitting characteristic EM radiation), making x-ray induced fluorescence a valuable analytical tool in a range of fields from art to archaeology. The atomic process of fluorescence is discussed in the next section.

28.9 APPLICATIONS OF ATOMIC EXCITATIONS AND DEEXCITATIONS

Many properties of matter are directly related to atomic energy levels and their associated excitations and deexcitations. The color of a rose, the output of a laser, and the transparency of air are but a few examples. (See Figure 28.35.) While it may not appear that glow-in-the-dark pajamas and lasers have much in common, they are in fact different applications of similar atomic deexcitations.

The color of a material is due to the ability of its atoms to absorb certain wavelengths while reflecting or reemitting others. A simple red material, for example, absorbs all visible wavelengths except red. This is because its atoms have levels separated by a variety of energies corresponding to all visible photon energies except red. (This may be due to atomic, molecular, or collective atomic effects that are complex.) Air is an interesting example. It is transparent to visible light, because there are few energy levels that visible photons can excite in air molecules and atoms. Visible light, thus, cannot be absorbed. Furthermore, visible light is only weakly scattered by air, because visible wavelengths are so much greater than the sizes of the air molecules and atoms. Light must pass through kilometers of air to scatter enough to cause red sunsets and blue skies.

Fluorescence and Phosphorescence

The ability of a material to emit various wavelengths of light is similarly related to its atomic energy levels. Figure 28.36 shows rocks illuminated by a UV lamp, sometimes called a black light. Each rock glows in a manner characteristic of its mineral composition. Black lights are also used to make retro posters glow.

The process is shown schematically in Figure 28.37; here an atom is excited to a level several steps above its ground state by the absorption of a relatively high energy UV photon. Once it is excited, the atom can deexcite in several ways, one of which is to reemit a photon of the same energy as excited it, a single step back to the ground state. All other paths of deexcitation involve smaller steps, in which lower-energy photons are emitted. Some of these may be in the visible range, such as for the rocks in Figure 28.36. **Fluorescence** is defined to be any process in which an atom or molecule is excited by one type of energy and deexcites by emission of a different form of energy.

Fluorescence can be induced by many types of energy input. Fluorescent paint, dyes, and even soap residues in clothes make colors seem brighter in sunlight by converting some UV into visible light. X rays can induce fluorescence, such as their use to make brighter visible images in x-ray fluoroscopy. Electric discharges can induce fluorescence,

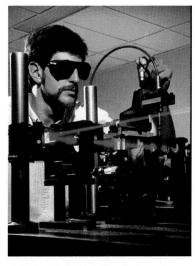

Figure 28.35 A laser is based on a particular type of atomic deexcitation.

Figure 28.36 Rocks glow in the visible spectrum when illuminated by an ultraviolet ("black") light. Emissions are characteristic of the mineral involved, since they are related to its energy levels. This is a colorful example of fluorescence in which excitation is induced by a different process than deexcitation.

Fifth	9.23 eV
Fourth	8.85 eV
Third	7.93 eV
Second	6.70 eV
First	4.89 eV
Ground state	0 eV
Hg	

Figure 28.37 This atom is excited to one of its higher levels by absorbing a UV photon. It can deexcite in a single step, reemitting a photon of the same energy, or in several steps. The process is called fluorescence if the atom deexcites in smaller steps, emitting energy different from that which excited it. Fluorescence can be induced by a variety of energy inputs, such as UV, x rays, and electrical discharge.

Figure 28.38 Atoms frozen in an excited state when this Chinese ceramic figure was fired can be stimulated to deexcite and emit EM radiation by heating a sample of the ceramic—a process called thermoluminescence. Since the states slowly deexcite over centuries, the amount of thermoluminescence decreases with age, making it possible to use this effect to date and authenticate antiquities.

as in so-called neon lights and in gas discharge tubes that produce atomic and molecular spectra. Common fluorescent lights use an electric discharge in mercury vapor to cause atomic emissions from mercury atoms. The inside of a fluorescent light is coated with a fluorescent material that emits visible light over a broad spectrum of wavelengths. By choosing an appropriate coating, fluorescent lights can be made more like sunlight or like the reddish glow of candlelight, depending on needs. Fluorescent lights are more efficient in converting electrical energy into visible light than incandescents, the blackbody radiation of which is primarily in the infrared due to temperature limitations.

Once excited, an atom or molecule will spontaneously deexcite quickly. (The electrons raised to higher levels are attracted to lower ones by the positive charge of the nucleus.) Spontaneous deexcitation is statistical in nature, with excited states having a mean lifetime of typically about 10^{-8} s. Some levels have significantly longer lifetimes, ranging up to milliseconds to minutes or even hours. (These levels are inhibited and slowed in deexciting because their quantum numbers differ greatly from those of available lower levels.) Although these level lifetimes are short in human terms, they are many orders of magnitude longer than typical and, thus, are said to be **metastable**, meaning relatively stable. **Phosphorescence** is the deexcitation of a metastable state. Glow-in-the-dark materials, such as luminous dials on some watches and clocks and on children's toys and pajamas, are made of phosphorescent substances. Visible light excites the atoms or molecules to metastable states that decay slowly, releasing the stored excitation energy. In some ceramics, atomic excitation energy can be frozen in after the ceramic has cooled from its firing. It is very slowly released, but the ceramic can be induced to phosphoresce by heating—a process called *thermoluminescence*. Since the release is slow, thermoluminescence can be used to date antiquities. The less emitted, the older the ceramic. (See Figure 28.38.)

Lasers

Lasers today are commonplace. Lasers are used to read bar codes at stores and in libraries, laser shows are staged for entertainment, laser printers produce high-quality images at relatively low cost, and lasers send prodigious numbers of telephone messages through optical fibers. Among other things, lasers are also employed in surveying, weapons guidance, tumor eradication, retinal welding, and for reading music CDs and computer CD-ROMs.

Why do lasers have so many varied applications? The answer is that lasers produce very pure wavelength EM radiation that is also very coherent—that is, the emitted photons are in phase. Laser output can, thus, be more precisely manipulated than incoherent mixed-wavelength EM radiation from other sources. The reason laser output is so pure and coherent is based on how it is produced, which in turn depends on a metastable state in the lasing material. Suppose a material had the energy levels shown in Figure 28.39. When energy is put into a large collection of these atoms, electrons are raised to all possible levels. Most return to the ground state in less than about 10^{-8} s, but those in the metastable state linger. This includes those electrons originally excited to the metastable state and those that fell into it from above. It is possible to get a majority of the atoms into the metastable state, a condition called a **population inversion**.

Once a population inversion is achieved, a very interesting thing can happen, as shown in Figure 28.40. An electron spontaneously falls from the metastable state, emitting a photon. This photon finds another atom in the metastable state and stimulates it to decay, emitting a second photon of *the same wavelength and in phase* with the first. **Stimulated emission**, in which an excited state is stimulated to decay, is most readily caused by a photon of the same energy that is necessary to excite the state; thus, the process can cascade as shown. Each photon can stimulate many atoms to decay, producing yet more identical photons that are in phase. The probability of absorption of a photon is the same as the probability of stimulated emission, and so a majority of atoms must be in the metastable state to produce energy.* The laser acts as a temporary energy

---

*Einstein (again Einstein, and back in 1917!) was one of the important contributors to the understanding of stimulated emission of radiation. Among other things, Einstein was the first to realize that stimulated emission and absorption are equally probable.

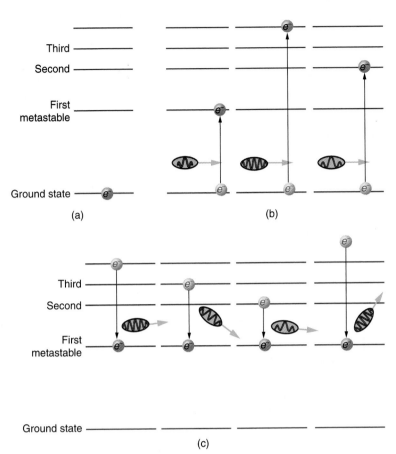

Figure 28.39 (a) Energy level diagram for an atom showing the first few states, one of which is metastable. (b) Massive energy input excites atoms to a variety of states. (c) Most states decay quickly, leaving electrons only in the metastable and ground state. If a majority are in the metastable state, a population inversion has been achieved.

storage device that subsequently produces a massive energy output of single-wavelength, in-phase photons.

The name **laser** is an acronym for **l**ight **a**mplification by **s**timulated **e**mission of **r**adiation, the process just described. The process was first proposed and developed in the early 1950s by Americans Charles Townes (1915–) and Arthur Schawlow (1921–) and others. The original devices were called masers, because they produced microwaves. Today the name laser is used for all such devices, although they have been developed to produce everything including microwave, infrared, visible, and ultraviolet radiation. Nobel prizes were awarded in 1964 to Townes and others for contributions toward the

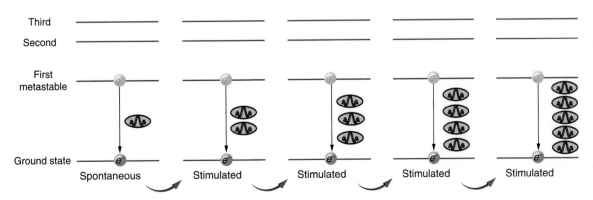

Figure 28.40 One atom in the metastable state spontaneously decays to a lower level, producing a photon that goes on to stimulate another atom to deexcite. The second photon has exactly the same energy and wavelength as the first and is in phase with it. Both go on to stimulate the emission of other photons. A population inversion is necessary for there to be a net production rather than net absorption of the photons.

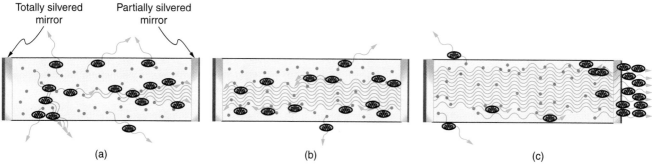

(a) (b) (c)

Figure 28.41 Typical laser construction has a method of pumping energy into the lasing material to produce a population inversion. (a) Spontaneous emission begins with some photons escaping and others stimulating further emissions. (b) and (c) Mirrors are used to enhance the probability of stimulated emission by passing photons through the material several times.

development of lasers, and in 1981 to Schawlow and others for pioneering laser applications. Figure 28.41 shows how a laser can be constructed to enhance the stimulated emission of radiation. Energy input can be from a flash tube, electrical discharge, or other sources, in a process sometimes called optical pumping. A large percentage of the original pumping energy is dissipated in other forms, but a population inversion must be achieved. Mirrors can be used to enhance stimulated emission by multiple passes of the radiation through the lasing material.

Lasers are constructed from many types of lasing materials, including gases, liquids, solids, and semiconductors. But all lasers are based on the existence of a metastable state or a phosphorescent material. Some lasers produce continuous output; others are pulsed in bursts as brief as 10^{-14} s. Some laser outputs are fantastically powerful—some greater than 10^{12} W—but, more commonly, lasers produce something on the order of 10^{-3} W. The helium-neon laser that produces a familiar red light is the most common of all. Figure 28.42 shows the energy levels of helium and neon, a pair of noble gases that work well together. An electrical discharge is passed through a helium-neon gas mixture in which the number of atoms of helium is ten times that of neon. The first excited state of helium is metastable and, thus, stores energy. This energy is easily transferred by collision to neon atoms, because they have an excited state at nearly the same energy as that in helium. That state in neon is also metastable, and this is the one that produces the laser output. (The most likely transition is to the nearby state, producing 1.96 eV photons, which have a wavelength of 633 nm and appear red.) A population inversion can be produced in neon, because there are so many more helium atoms and these put energy into the neon. Helium-neon lasers often have continuous output, because the population inversion can be maintained even while lasing occurs.

A medical application of lasers is shown in Figure 28.43. A detached retina can result in total loss of vision. Burns made by a laser focused to a small spot on the retina form scar tissue that can hold the retina in place, salvaging the patient's vision. Other

Medical Application

Figure 28.42 Energy levels in helium and neon. The common helium-neon laser pumps energy into the metastable states of both atoms. The gas mixture has about ten times more helium atoms than neon atoms. Excited helium atoms easily deexcite by transferring energy to neon in a collision. A population inversion in neon is achieved, allowing lasing by the neon to occur.

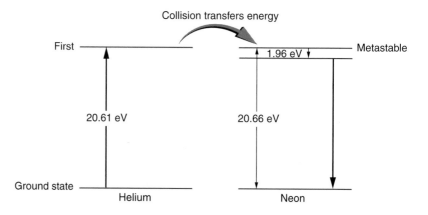

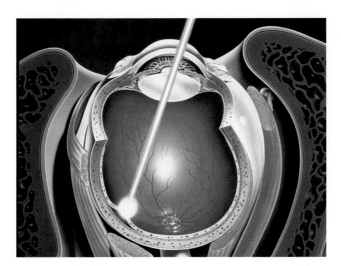

Figure 28.43   A detached retina is burned by a laser designed to focus to a small spot on the retina, the resulting scar tissue holding it in place. The lens of the eye is used to focus the light, as is the device bringing the laser output to the eye.

light sources cannot be focused as precisely as a laser's due to refractive dispersion of different wavelengths. Similarly, laser surgery in the form of cutting or burning away tissue is made more accurate because laser output can be very precisely focused and is preferentially absorbed because of its pure wavelength. Carbon dioxide lasers are convenient because their infrared output is strongly absorbed by tissue, which is jet black in the infrared.

The massive combination of lasers shown in Figure 28.44 is used to induce nuclear fusion, the energy source of the sun and hydrogen bombs. Since lasers can produce very high power in very brief pulses, they can be used to focus an enormous amount of energy on a small glass sphere containing fusion fuel. Not only does the energy heat the fuel to temperatures at which fusion can occur, it also compresses the fuel to great density, enhancing the probability of fusion. The compression is caused by the momentum of the impinging laser photons.

Music CDs are now so common that vinyl records are almost quaint antiquities. CDs store information digitally and have evolving applications in computer information storage because of their high capacity—one gigabyte and growing. An entire encyclopedia can be stored on a single CD. Figure 28.45 illustrates how the information is stored and read from the CD. Pits made in the CD by a laser can be tiny and very accurately spaced to record digital information. These are read by having an inexpensive solid state infrared laser beam scatter from pits as the CD spins, revealing their digital pattern and the information encoded upon them.

Figure 28.44   This system of lasers at Lawrence Livermore Laboratory is used to ignite nuclear fusion. A tremendous burst of energy is focused on a small fuel pellet, which is imploded to the high temperature and density needed to make the reaction go.

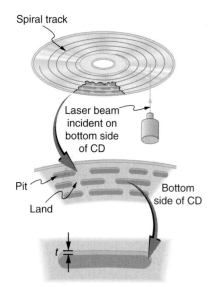

Spiral track

Laser beam incident on bottom side of CD

Pit

Land

Bottom side of CD

t

Figure 28.45 A CD has digital information stored in the form of laser-created pits on its surface. These in turn can be read by scattering laser output. Large information capacity is possible because of the precision of the laser.

(a)   (b)   (c)

Figure 28.46 (a) Credit cards commonly have holograms for logos, making them difficult to reproduce. (b, c) The same hologram viewed from two different angles has objects in different perspective in a true three-dimensional image. As in all holograms, you can see around and behind objects by changing your point of view.

Holograms, such as those in Figure 28.46, are true three-dimensional images recorded on film by lasers. Holograms are used for amusement, decoration on novelty items and magazine covers, security on credit cards and driver's licenses (a laser and other equipment is needed to reproduce them), and for serious three-dimensional information storage. You can see that a hologram is a true three-dimensional image, because objects change relative position in the image when viewed from different angles.

The name **hologram** means *entire picture* (from the Greek *holo*, as in holistic), because the image is three-dimensional. **Holography** is the process of producing holograms and, although they are recorded on photographic film, the process is quite different from normal photography. Holography uses light interference or wave optics, whereas normal photography uses geometric optics. Figure 28.47 shows one method of producing a hologram. Coherent light from a laser is split by a mirror, with part of the light illuminating the object. The remainder, called the reference beam, shines directly on a piece of film. Light scattered from the object interferes with the reference beam, producing constructive and destructive interference. As a result, the exposed film looks foggy, but close examination reveals a complicated interference pattern stored on it. Where the interference was constructive, the film (a negative actually) is darkened. Holography is sometimes called lensless photography, because it uses the wave characteristics of light as contrasted to normal photography, which uses geometric optics such as lenses.

Light falling on a hologram can form a three-dimensional image. The process is complicated in detail, but the basics can be understood as in Figure 28.48, in which a laser of the same type that exposed the film is now used to illuminate it. The myriad tiny exposed regions of the film are dark and block the light, while less exposed regions allow light to pass. The film thus acts much like a collection of diffraction gratings with various spacings. Light passing through the hologram is diffracted in various directions, producing both real and virtual images of the object used to expose the film. The interference pattern is the same as that produced by the object. Moving your eye to various places in the interference pattern gives you different perspectives, just as looking directly at the object would. The image thus looks like the object and is three-dimensional like the object.

The hologram illustrated in Figure 28.48 is a transmission hologram. Those viewed with reflected light, such as the white light holograms on Pog Slammers and credit cards, are reflection holograms and are more common. White light holograms often appear a little blurry with rainbow edges, because the diffraction patterns of various colors of light are at slightly different locations due to their different wavelengths. Further uses of holography include all types of three-dimensional information storage, such as of statues in museums and engineering studies of structures. Invented in the late 1940s by Dennis Gabor (1900–1970), who won the 1971 Noble prize in physics for his work, holography became far more practical with the development of the laser. Since lasers

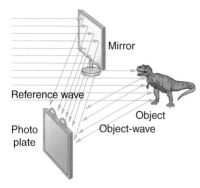

Mirror

Reference wave

Object

Photo plate

Object-wave

Figure 28.47 Production of a hologram. Pure-wavelength coherent light from a laser produces a well-defined interference pattern on a piece of film. The laser beam is split by a partially silvered mirror, with part of the light illuminating the object and the remainder shining directly on the film.

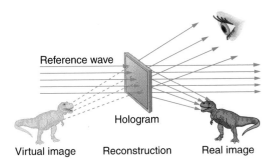

Figure 28.48 A transmission hologram is one that produces real and virtual images when a laser of the same type as exposed the hologram is passed through it. Diffraction from various parts of the film produces the same interference pattern as the object that was used to expose it.

Reference wave

Hologram

Virtual image   Reconstruction   Real image

produce coherent pure-wavelength light, their interference patterns are more pronounced. The precision is so great that it is even possible to record numerous holograms on a single piece of film by just changing the angle of the film for each successive image. This is how the holograms that move as you walk by them are produced—a kind of lensless movie.

SUMMARY

Atoms are the smallest unit of elements; atoms combine to form molecules, the smallest unit of compounds. The first direct observation of atoms was in **Brownian motion**, the statistical fluctuation of particles caused by collisions with atoms. Analysis of Brownian motion gave accurate sizes for atoms (average 10^{-10} m) and a precise value for Avogadro's number.

Atoms themselves have a substructure that is fundamental to their properties. Atoms are composed of negatively charged *electrons*, first proved to exist in **cathode ray tube** experiments, and a positively charged nucleus. All electrons are identical and have a charge to mass ratio of

$$\frac{q_e}{m_e} = -1.76 \times 10^{11} \text{ C/kg} \quad \text{(electron)} \quad \textbf{(28.1a)}$$

where q_e is the charge of an electron and m_e is its mass. This is an exceedingly large number, implying the mass of an electron is small. The positive charge in nuclei is carried by particles called *protons*, which have a charge to mass ratio of

$$\frac{q_p}{m_p} = 9.57 \times 10^7 \text{ C/kg} \quad \text{(proton)} \quad \textbf{(28.1b)}$$

where q_p is the charge of the proton and m_p is its mass. Since the charges of electrons and protons are identical in magnitude, the above two equations imply that the proton is about 1836 times as massive than the electron. Once the charge of an electron was measured, its mass could be calculated. The value has been confirmed by direct measurement to be

$$m_e = 9.11 \times 10^{-31} \text{ kg} \quad \text{(electron's mass)} \quad \textbf{(28.2)}$$

For the proton,

$$m_p = 1.67 \times 10^{-27} \text{ kg} \quad \text{(proton's mass)} \quad \textbf{(28.3)}$$

which implies most of the mass of an atom is found in its nucleus.

The **planetary model of the atom** pictures electrons orbiting the nucleus in the way that planets orbit the sun. Bohr used the planetary model to develop the first reasonable theory of hydrogen, the simplest atom. Atomic and molecular spectra are quantized, with hydrogen spectrum wavelengths given by the formula

$$\frac{1}{\lambda} = R\left(\frac{1}{n_f^2} - \frac{1}{n_i^2}\right) \quad \textbf{(28.4)}$$

where λ is the wavelength of the emitted EM radiation and R is the **Rydberg constant**, which has the value

$$R = 1.097 \times 10^7/\text{m} \quad \textbf{(28.5)}$$

The constants n_i and n_f are positive integers, and n_i must be greater than n_f.

Bohr correctly proposed that **the orbits of electrons in atoms are quantized**, with energy for transitions between orbits given by

$$\Delta E = hf = E_i - E_f \quad \textbf{(28.6)}$$

where ΔE is the change in energy between the initial and final orbits and hf is the energy of an absorbed or emitted photon. It is useful to plot orbit energies on a vertical graph called an **energy level diagram**. The allowed orbits are circular, Bohr proposed, and must have **quantized orbital angular momentum** given by

$$L = m_e v r_n = n\frac{h}{2\pi} \quad (n = 1, 2, 3, ...) \quad \textbf{(28.7)}$$

where L is the angular momentum, r_n is the radius of the nth orbit, and h is Planck's constant. For all one-electron (hydrogen-like) atoms, the radius of an orbit is given by

$$r_n = \frac{n^2}{Z} a_B \quad \text{(allowed orbits } n = 1, 2, 3, ...) \quad \textbf{(28.8)}$$

where Z is the **atomic number** of an element (its number of electrons when neutral) and a_B is defined to be the **Bohr radius**, which is

$$a_B = \frac{h^2}{4\pi^2 m_e k Z q_e^2} = 0.529 \times 10^{-10} \text{ m} \quad \textbf{(28.9)}$$

Furthermore, the **energies of hydrogen-like atoms** are given by

$$E_n = -\frac{Z^2}{n^2} E_0 \quad (n = 1, 2, 3, \ldots) \quad \textbf{(28.10)}$$

where E_0 is the **ground state energy** and is given by

$$E_0 = \frac{2\pi^2 q_e^4 m_e k^2}{h^2} = 13.6 \text{ eV} \quad \textbf{(28.11)}$$

Thus, for hydrogen,

$$E_n = -\frac{13.6 \text{ eV}}{n^2} \quad (n = 1, 2, 3, \ldots) \quad \textbf{(28.12)}$$

The Bohr theory gives accurate values for the energy levels in hydrogen-like atoms, but it has been improved upon in several respects.

Quantization of orbital energy is caused by the wave nature of matter. **Allowed orbits in atoms occur for constructive interference of electrons** in the orbit, requiring an integral number of wavelengths to fit in an orbit's circumference—that is,

$$n\lambda_n = 2\pi r_n \quad (n = 1, 2, 3, \ldots) \quad \textbf{(28.13)}$$

where λ_n is the electron's de Broglie wavelength. Because of the wave nature of electrons and the Heisenberg uncertainty principle, there are no well-defined orbits; rather, there are clouds of probability.

The **Zeeman effect**—the splitting of lines when a magnetic field is applied—is caused by other quantized entities in atoms. Both the **magnitude and direction of orbital angular momentum are quantized**. The same is true for the magnitude and direction of the **intrinsic spin** of electrons. **Quantum numbers** are used to express the allowed values of quantized entities. The **principal quantum number** n labels the basic states of a system and is given by

$$n = 1, 2, 3, \ldots \quad \textbf{(28.14)}$$

The magnitude of angular momentum is given by

$$L = \sqrt{\ell(\ell + 1)}\, \frac{h}{2\pi} \quad (\ell = 0, 1, 2, \ldots, n - 1) \quad \textbf{(28.15)}$$

where ℓ is the **angular momentum quantum number**. The direction of angular momentum is quantized, in that its component along an axis defined by a magnetic field, called the z-axis, is given by

$$L_z = m_\ell \frac{h}{2\pi} \quad (m_\ell = -\ell, -\ell + 1, \ldots, -1, 0, 1, \ldots, \ell - 1, \ell)$$

$$\textbf{(28.16)}$$

where L_z is the **z-component of the angular momentum** and m_ℓ is the **angular momentum projection quantum number**. Similarly, the electron's **intrinsic spin angular momentum** S is given by

$$S = \sqrt{s(s + 1)}\, \frac{h}{2\pi} \quad \left(s = \frac{1}{2} \text{ for electrons}\right) \quad \textbf{(28.17)}$$

where s is defined to be the **spin quantum number**. Finally, the direction of the electron's spin along the z-axis is given by

$$S_z = m_s \frac{h}{2\pi} \quad \left(m_s = -\frac{1}{2}, +\frac{1}{2}\right) \quad \textbf{(28.18)}$$

where S_z is the **z-component of spin angular momentum** and m_s is the **spin projection quantum number**. Spin projection $m_s = +1/2$ is referred to as *spin up*, whereas $m_s = -1/2$ is called *spin down*. Table 28.1 summarizes the atomic quantum numbers and their allowed values.

The state of a system is completely described by a complete set of quantum numbers. The **Pauli exclusion principle** places the following limit:

No two electrons can have the same set of quantum numbers. That is, no two electrons can be in the same state.

This exclusion limits the number of electrons in atomic shells and subshells. Each value of n corresponds to a **shell**, and each value of ℓ corresponds to a **subshell**. Table 28.2 gives the symbols for labeling shells and subshells. By counting the number of variations of quantum numbers, it is seen that the **maximum number of electrons that can be in a subshell** is

$$2(2\ell + 1) \quad \textbf{(28.19)}$$

and the **maximum number of electrons that can be in a shell** is

$$2n^2 \quad \textbf{(28.20)}$$

Table 28.3 and the accompanying text show electron configurations for the first 20 elements, illustrating how the shells and subshells fill and are related to the various groups in the periodic table of the elements.

X rays are high-frequency EM radiation created in two processes. The atomic process creates a *characteristic x ray*, which is defined to be the *EM radiation emitted by an atom when an inner-shell vacancy is filled*. X rays have many uses, including medical diagnostics and x-ray diffraction. Another important atomic process is **fluorescence**, defined to be any process in which an atom or molecule is excited by one type of energy and deexcites by emission of a different form or smaller amount of energy. Some states live much longer than others and are termed **metastable**. **Phosphorescence** is the deexcitation of a metastable state. **Lasers** produce coherent pure-wavelength EM radiation by **stimulated emission**, in which a metastable state is stimulated to decay. Lasing requires a **population inversion**, in which a majority of the atoms or molecules are in their metastable state.

CONCEPTUAL QUESTIONS

28.1 Name three different types of evidence for the existence of atoms.

28.2 Explain why patterns observed in the periodic table of the elements are evidence for the existence of atoms, and why

Brownian motion is a more direct type of evidence for their existence.

28.3 If atoms exist, why can't we see them with visible light?

28.4 What two pieces of evidence allowed the first calculation of m_e, the mass of the electron? (a) The ratios q_e/m_e and q_p/m_p. (b) The values of q_e and E_B. (c) The ratio q_e/m_e and q_e. Justify your response.

28.5 Is the claim that "nature abhors a vacuum" accurate in view of the known substructure of the atom? Explain.

28.6 How do the allowed orbits for electrons in atoms differ from the allowed orbits for planets around the sun? Explain how the correspondence principle applies here.

28.7 Atomic and molecular spectra are discrete. What does *discrete* mean, and how are discrete spectra related to the quantization of energy and electron orbits in atoms and molecules?

28.8 Hydrogen gas can only absorb EM radiation that has an energy corresponding to a transition in the atom, just as it can only emit these discrete energies. When a spectrum is taken of the solar corona, in which a broad range of EM wavelengths are passed through very hot hydrogen gas, the absorption spectrum shows all the features of the emission spectrum. But when such EM radiation passes through room temperature hydrogen gas, only the Lyman series is absorbed. Explain the difference.

28.9 Explain how Bohr's rule for the quantization of electron orbital angular momentum differs from the actual rule.

28.10 What is a hydrogen-like atom, and how are the energies and radii of its electron orbits related to those in hydrogen?

28.11 How is the de Broglie wavelength of electrons related to the quantization of their orbits in atoms and molecules?

28.12 What is the Zeeman effect, and what type of quantization was discovered because of this effect?

28.13 Define the quantum numbers n, ℓ, m_ℓ, s, and m_s.

28.14 For a given value of n, what are the allowed values of ℓ?

28.15 For a given value of ℓ, what are the allowed values of m_ℓ? What are the allowed values of m_ℓ for a given value of n? Give an example in each case.

28.16 List all the possible values of s and m_s for an electron. Are there particles for which these values are different? The same?

28.17 Identify the shell, subshell, and number of electrons for the following: (a) $2p^3$. (b) $4d^9$. (c) $3s^1$. (d) $5g^{16}$.

28.18 Which of the following are not allowed? State which rule is violated for any that are not allowed. (a) $1p^3$. (b) $2p^8$. (c) $3g^{11}$. (d) $4f^2$.

28.19 Radioactive strontium (Sr) is concentrated in the bones of those who ingest it. Why would you expect strontium to have a chemistry similar to calcium? And similar to what other elements?

28.20 Since the rare earths (lanthanum through lutetium) have chemistries similar to each other, would you expect the group of transuranic elements to have similar chemistries? Explain.

28.21 Explain why characteristic x rays are the most energetic in the EM spectrum of a given element.

28.22 Why does the energy of characteristic x rays become increasingly greater for heavier and heavier atoms?

28.23 Figure 28.29 shows an x-ray spectrum from a typical x-ray tube. Which features of the spectrum would change if the anode material were changed? Which features would change if the tube voltage were changed? Explain briefly.

28.24 Observers at a safe distance from an atmospheric test of a nuclear bomb feel its heat but receive none of its copious x rays. Why is air opaque to x rays but transparent to infrared?

28.25 Crystal lattices can be examined with x rays but not UV. Why?

28.26 CT scanners do not detect details smaller than about 0.5 mm. Is this limitation due to the wavelength of x rays? Explain.

28.27 What is the difference between fluorescence and phosphorescence?

28.28 Laser output is unique in two ways. What are they?

28.29 Why is a metastable state necessary for a laser material?

28.30 What is a population inversion, and why is it a necessary condition for lasing to occur?

28.31 How can you tell that a hologram is a true three-dimensional image and that those in 3-D movies are not?

PROBLEMS

Sections 28.1 and 28.2 The Atom's Size and Substructure

28.1 Using the given charge to mass ratios for electrons and protons, and knowing the magnitudes of their charges are equal, what is the ratio of the proton's mass to the electron's? (Note that since the charge to mass ratios are given to only three-digit accuracy, your answer may differ from the accepted ratio in the fourth digit.)

28.2 (a) Calculate the mass of a proton using the charge to mass ratio given for it in this chapter and its known charge. (b) How does your result compare with the proton mass given in this chapter?

28.3 If someone wanted to build a scale model of the atom with a nucleus 1.00 m in diameter, how far away would the nearest electron need to be?

28.4 An aspiring physicist wants to build a scale model of a hydrogen atom for her science fair project. If the atom is to be 1.00 m in diameter, how big should she try to make the nucleus?

28.5 (a) If 1.00 mol of singly charged ions has a total charge of approximately 100,000 C (as known from electroplating as long as 150 years ago), what is the approximate charge per ion? (b) What is the ratio of this charge to the known electron charge?

• **28.6** (a) If the solar system were on the same scale as a hydrogen atom, how far away would the nearest planet be from the sun, given the sun's average radius to be 7.0×10^8 m? (b) What is the ratio of this distance to the average distance of the earth from the sun (1.5×10^{11} m)?

⁞ **28.7** (a) Calculate approximately how many atoms could be in contact with a pollen grain by taking both to be cubes. Assume the atom is 10^{-10} m on a side and the pollen grain is 10^{-6} m on a side. (b) By what fraction could this number fluctuate if random thermal motions cause fluctuations that are approximately proportional to the square root of the number? (This gives some idea of how fine a measurement is needed to see Brownian motion.)

Section 28.3 Bohr's Theory of the Hydrogen Atom and the Hydrogen Spectrum

28.8 By calculating its wavelength, show that the first line in the Lyman series is UV radiation.

28.9 Find the wavelength of the third line in the Lyman series, and identify the type of EM radiation.

28.10 Look up the values of the quantities in Equation 28.9, and verify that the Bohr radius a_B is 0.529×10^{-10} m.

28.11 Verify that the ground state energy E_0 is 13.6 eV by using Equation 28.11 after looking up values for the quantities in the equation.

28.12 If a hydrogen atom has its electron in the $n = 4$ state, how much energy in eV is needed to ionize it?

28.13 A hydrogen atom in an excited state can be ionized with less energy than when it is in its ground state. What is n for a hydrogen atom if 0.85 eV of energy can ionize it?

28.14 Find the radius of a hydrogen atom in the $n = 2$ state according to Bohr's theory.

28.15 Show that $(13.6 \text{ eV})/hc = 1.097 \times 10^7/\text{m} = R$ (Rydberg's constant), as discussed in the text.

• **28.16** What is the smallest-wavelength line in the Balmer series? Is it in the visible part of the spectrum?

• **28.17** Show that the entire Paschen series is in the infrared part of the spectrum. To do this, you only need to calculate the shortest wavelength in the series.

• **28.18** Do the Balmer and Lyman series overlap? To answer this, calculate the shortest-wavelength Balmer line and the longest-wavelength Lyman line.

• **28.19** (a) Which line in the Balmer series is the first one in the UV part of the spectrum? (b) How many Balmer series lines are in the visible part of the spectrum? (c) How many are in the UV?

• **28.20** A wavelength of 4.653 μm is observed in a hydrogen spectrum for a transition that ends in the $n_f = 5$ level. What was n_i for the initial level of the electron?

• **28.21** A singly ionized helium ion has only one electron and is denoted He$^+$. What is the ion's radius in the ground state?

• **28.22** A beryllium ion with a single electron (denoted Be^{3+}) is in an excited state with radius the same as that of the ground state of hydrogen. (a) What is n for the Be^{3+} ion? (b) How much energy in eV is needed to ionize the ion from this excited state?

• **28.23** Atoms can be ionized by thermal collisions, such as at the high temperatures found in the solar corona. One such ion is C^{5+}, a carbon atom with only a single electron. (a) By what factor are the energies of its hydrogen-like levels greater than those of hydrogen? (b) What is the wavelength of the first line in this ion's Paschen series? (c) What type of EM radiation is this?

⁞ **28.24** (a) Calculate the radius of the orbit for the innermost electron in uranium, assuming it is relatively unaffected by the atom's other electrons. (b) What is the ratio of this orbital radius to the 7.5 fm radius of the uranium nucleus?

⁞ **28.25** Verify Equations 28.8 and 28.9 using the approach stated in the text. That is, equate the Coulomb and centripetal forces and then insert an expression for velocity from the condition for angular momentum quantization obtained from Equation 28.7.

Sections 28.4–28.6 Quantum Mechanics beyond Bohr's Theory of the Atom

28.26 If an atom has an electron in the $n = 5$ state with $m_\ell = 3$, what are the possible values of ℓ?

28.27 An atom has an electron with $m_\ell = 2$. What is the smallest value of n for this electron?

28.28 What are the possible values of m_ℓ for an electron in the $n = 4$ state?

28.29 What, if any, constraints does a value of $m_\ell = 1$ place on the other quantum numbers for an electron in an atom?

28.30 (a) Calculate the magnitude of the angular momentum for an $\ell = 1$ electron. (b) Compare your answer to the value Bohr proposed for the $n = 1$ state.

28.31 (a) What is the magnitude of the angular momentum for an $\ell = 1$ electron? (b) Calculate the magnitude of the electron's spin angular momentum. (c) What is the ratio of these angular momenta?

28.32 Repeat the preceding problem for $\ell = 3$.

• **28.33** (a) How many angles can **L** make with the z-axis for an $\ell = 2$ electron? (b) Calculate the value of the smallest angle.

• **28.34** What angles can the spin **S** of an electron make with the z-axis?

Section 28.7 The Pauli Exclusion Principle and Multiple-Electron Atoms

28.35 (a) How many electrons can be in the $n = 4$ shell? (b) What are its subshells, and how many electrons can be in each?

28.36 (a) What is the minimum value of ℓ for a subshell that has 11 electrons in it? (b) If this subshell is in the $n = 5$ shell, what is the spectroscopic notation for this atom?

28.37 (a) If one subshell of an atom has 9 electrons in it, what is the minimum value of ℓ? (b) What is the spectroscopic notation for this atom, if this subshell is part of the $n = 3$ shell?

28.38 (a) List all possible sets of quantum numbers (n, ℓ, m_ℓ, m_s) for the $n = 3$ shell, and determine the number of electrons that can be in the shell and each of its subshells. (b) Show that the number of electrons in the shell equals $2n^2$ and the number in each subshell is $2(2\ell + 1)$. (Do these by specific listing and counting, not by using equations.)

28.39 Which of the following spectroscopic notations are not allowed? (a) $5s^1$. (b) $1d^1$. (c) $4s^3$. (d) $3p^7$. (e) $5g^{15}$. State which rule is violated for each that is not allowed.

28.40 Which of the following spectroscopic notations are allowed (that is, which violate none of the rules regarding values of quantum numbers)? (a) $1s^1$. (b) $1d^3$. (c) $4s^2$. (d) $3p^7$. (e) $6h^{20}$.

⁞ **28.41** (a) Using the Pauli exclusion principle and the rules relating the allowed values of the quantum numbers (n, ℓ, m_ℓ, m_s), prove that the maximum number of electrons in a subshell is $2(2\ell + 1)$. (b) In a similar manner, prove that the maximum number of electrons in a shell is $2n^2$.

Section 28.8 X Rays

28.42 What is the shortest-wavelength x-ray radiation that can be generated in an x-ray tube with an applied voltage of 50.0 kV?

28.43 A color television tube also generates some x rays when its electron beam strikes the screen. What is the shortest wavelength of these x rays, if a 30.0 kV potential is used to accelerate the electrons? (Note that TVs have shielding to prevent these x rays from exposing viewers.)

28.44 An x-ray tube has an applied voltage of 100 kV. (a) What is the most energetic x-ray photon it can produce? Express your answer in electron volts and joules. (b) Find the wavelength of such an x ray.

• **28.45** What minimum accelerating voltage in an x-ray tube will produce 0.0200 nm wavelength radiation, such as might be useful for crystallography?

• **28.46** The maximum characteristic x-ray photon energy comes from the capture of a free electron into a K shell vacancy. What is this photon energy in keV for tungsten, assuming the free electron has no initial kinetic energy? (Hint: The answer appears in Example 28.5.)

• **28.47** Using the same approximation as in Example 28.5, calculate the energy of a K_β x ray for a tungsten anode in an x-ray tube.

• **28.48** What are the approximate energies of the K_α and K_β x rays for copper?

⁝ **28.49** Identify the element that has a K_α x-ray energy of 52.9 keV.

Section 28.9 Lasers and Other Applications

28.50 Figure 28.37 shows a simplified energy level diagram for a mercury atom. Fluorescent lights pass an electric discharge through mercury vapor, exciting atoms that then deexcite by emitting photons. Knowing that visible photons have energies from 1.63 to 3.26 eV, list the specific transitions between various levels that produce visible light. (For example, the transition from the fifth to the second excited state produces a $\Delta E = 9.23$ eV $- 6.70$ eV $= 2.53$ eV photon, which is visible.)

28.51 (a) Using the simplified energy level diagram in Figure 28.37, find the energy of the photon emitted when a mercury atom in its third excited state makes a transition to its second excited state. (b) What is its wavelength?

28.52 Figure 28.42 shows the energy level diagram for neon. (a) Verify that the energy of the photon emitted when neon goes from its metastable state to the one immediately below it has an energy of 1.96 eV. (b) Show that the wavelength of this radiation is 633 nm. (c) What wavelength is emitted when the neon then makes a direct transition to its ground state?

28.53 A helium-neon laser is pumped by electric discharge. What wavelength electromagnetic radiation would be needed to pump it? See Figure 28.42 for energy level information.

28.54 Ruby lasers have chromium atoms doped in an aluminum oxide crystal. The energy level diagram for chromium in a ruby is shown in Figure 28.49. What wavelength is emitted by a ruby laser?

• **28.55** (a) What energy photons can pump chromium atoms in a ruby laser from the ground state to its second and third excited states? (b) What are the wavelengths of these photons? Verify that they are in the visible part of the spectrum.

• **28.56** Some of the most powerful lasers are based on the energy levels of neodymium in solids, such as glass, as shown in Figure 28.50. (a) What average wavelength visible light can pump the neodymium into the levels above its metastable state? (b) Verify that the 1.17 eV transition produces 1.06 μm radiation.

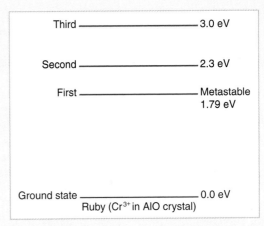

Figure 28.49 Chromium atoms in an aluminum oxide crystal have these energy levels, one of which is metastable. This is the basis of a ruby laser. Visible light can pump the atom into an excited state above the metastable state to achieve a population inversion. Problems 54 and 55.

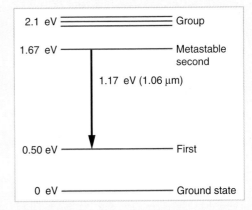

Figure 28.50 Neodymium atoms in glass have these energy levels, one of which is metastable. The group of levels above the metastable state is convenient for achieving a population inversion, since photons of many different energies can be absorbed. Problem 56.

⁝ **28.57** Not all transitions between atomic levels are equally likely. Those between dissimilar levels are inhibited. Consider the energy levels for mercury shown in simplified form in Figure 28.37. Suppose you examine a mercury spectrum and note that photons of energies 1.23, 4.89, and 6.70 eV are emitted, but there are no 1.81 or 3.04 eV photons. Which level is unlike the others?

INTEGRATED CONCEPTS

Physics is most interesting and most powerful when applied to general situations involving more than a narrow set of physical principles. The integration of concepts necessary to solve problems involving several physical principles also gives greater insight into the unity of physics. You may wish to refer to Chapters 6, 7, 9, 10, 13, 17, 18, 21, 25, and perhaps others, to solve the following problems.

Note: Problem-solving strategies and worked examples that can help you solve integrated concept problems appear in several places, the most recent being Chapter 27. Consult the section of problems labeled *Integrated Concepts* at the end of that chapter or others.

28.58 The electric and magnetic forces on an electron in the CRT in Figure 28.5 are supposed to be in opposite directions. Verify this by determining the direction of each force for the situation shown. Explain how you obtain the directions (that is, identify the rules used).

28.59 Estimate the density of a nucleus by calculating the density of a proton, taking it to be a sphere 1.2 fm in diameter. Compare your result with the value estimated in this chapter.

28.60 In a Millikan oil drop experiment using a setup like that in Figure 28.6, a 500 V potential difference is applied to plates separated by 2.50 cm. (a) What is the mass of an oil drop having two extra electrons that is suspended motionless by the field between the plates? (b) What is the diameter of the drop, assuming it is a sphere with the density of olive oil?

28.61 What double slit separation would produce a first-order maximum at 3.00° for 25.0 keV x rays? The small answer indicates that the wave character of x rays is best determined by having them interact with very small objects such as atoms and molecules.

28.62 (a) What is the distance between the slits of a diffraction grating that produces a first-order maximum for the first Balmer line at an angle of 20.0°? (b) At what angle will the fourth line of the Balmer series appear in first order? (c) At what angle will the second-order maximum be for the first line?

28.63 In a laboratory experiment designed to duplicate Thomson's determination of q_e/m_e, a beam of electrons having a velocity of 6.00×10^7 m/s enters a 5.00×10^{-3} T magnetic field. The beam moves perpendicular to the field in a path having a 6.8 cm radius of curvature. Determine q_e/m_e from these observations, and compare the result with the known value.

28.64 Particles called muons exist in cosmic rays and can be created in particle accelerators. Muons are very similar to electrons, having the same charge and spin, but they have a mass 207 times greater. When muons are captured by an atom, they orbit just like an electron but with a smaller radius, since the mass in Equation 28.9 is $207m_e$. (a) Calculate the radius of the $n = 1$ orbit for a muon in a uranium ion ($Z = 92$). (b) Compare this with the 7.5 fm radius of a uranium nucleus. Note that since the muon orbits inside the electrons, it falls into a hydrogen-like orbit. Since your answer is less than the radius of the nucleus, you can see that the photons emitted as the muon falls into its lowest orbit can give information about the nucleus.

28.65 Find the value of ℓ, the orbital angular momentum quantum number, for the moon around the earth. The extremely large

value obtained implies that it is impossible to tell the difference between adjacent quantized orbits for macroscopic objects.

28.66 Calculate the minimum amount of energy in joules needed to create a population inversion in a helium-neon laser containing 1.00×10^{-4} mol of neon.

28.67 A carbon dioxide laser used in surgery emits 10.6 μm infrared radiation. In 1.00 ms, this laser raised the temperature of 1.00 cm³ of flesh to 100°C and evaporated it. (a) How many photons are required? You may assume flesh has the same heat of vaporization as water. (b) What is the minimum power output during the flash?

28.68 Prove that the velocity of charged particles moving along a straight path through perpendicular electric and magnetic fields is $v = E/B$, as discussed in Section 28.2 and earlier in Section 21.11. Thus crossed electric and magnetic fields can be used as a velocity selector independent of the charge and mass of the particle involved.

UNREASONABLE RESULTS

The following problems have results that are unreasonable because some premise is unreasonable or because certain of the premises are inconsistent with one another. Physical principles applied correctly then produce unreasonable results. The purpose of these problems is to give practice in assessing whether nature is being accurately described, and if it is not to trace the source of difficulty.

Note: Problem-solving strategies that can help you solve unreasonable results problems appear in several places, the most recent being Chapter 26. Consult the section of problems labeled *Unreasonable Results* at the end of that chapter or others.

28.69 (a) What voltage must be applied to an x-ray tube to obtain 0.0100 fm wavelength x rays for use in exploring the details of nuclei? (b) What is unreasonable about this result? (c) Which assumptions are unreasonable or inconsistent?

28.70 (a) A student in a physics laboratory observes a hydrogen spectrum with a diffraction grating for the purpose of measuring the wavelengths of the emitted radiation. In the spectrum, she observes a yellow line and finds its wavelength to be 589 nm. Assuming this is part of the Balmer series, determine n_i, the principal quantum number of the initial state. (b) What is unreasonable about this result? (c) Which assumptions are unreasonable or inconsistent?

RADIOACTIVITY AND NUCLEAR PHYSICS

Visible evidence of nuclear decay is produced by high-level radiation

interacting with the water in this reactor.

There is an ongoing quest to find the substructures of matter. At one time, it was thought that atoms would be the ultimate substructure, but just when the first direct evidence of atoms was obtained, it became clear they have a substructure and a tiny nucleus. The nucleus itself has spectacular characteristics. For example, certain nuclei are unstable, and their decay emits radiation with energies millions of times greater than atomic energies. Some of the mysteries of nature, such as why the core of the earth remains molten and how the sun produces its energy, are explained by nuclear phenomena. The exploration of radioactivity and the nucleus revealed fundamental and previously unknown particles, forces, and conservation laws. That exploration has evolved into a search for further underlying structures, such as quarks. In this chapter the fundamentals of nuclear radioactivity and the nucleus are explored. The following two chapters explore the more important applications of nuclear physics and the basics of what we know about quarks and other substructures smaller than nuclei.

29.1 NUCLEAR RADIOACTIVITY

The discovery and study of nuclear radioactivity quickly revealed evidence of revolutionary new physics. Uses for nuclear radiation also emerged quickly—for example, people such as Rutherford used it to determine the size of the nucleus. (See Figure 29.1.) We therefore begin our study of nuclear physics with the discovery and basic features of nuclear radioactivity.

Discovery of Nuclear Radioactivity

In 1896 the French physicist Antoine Henri Becquerel (1852–1908) accidentally found that a uranium-rich mineral called pitchblende emits invisible, penetrating rays that can darken a photographic plate enclosed in an opaque envelope. The rays therefore carry energy; but amazingly, the pitchblende emits them continuously without any energy input. This is an apparent violation of the conservation of energy, one that we now understand to be due to the destruction of a small amount of mass, as related in Einstein's famous equation $E = mc^2$. (See the discussion in Section 26.6.) It was soon evident that Becquerel's rays originate in the nuclei of atoms, and have other unique characteristics in addition to violating a once well-established conservation law. The emission of these rays is called **nuclear radioactivity** or simply **radioactivity**. The rays themselves are called **nuclear radiation**. A nucleus that spontaneously destroys part of its mass to emit radiation is said to **decay** (a term also used to describe the emission of radiation by atoms in excited states). A substance or object that emits nuclear radiation is said to be **radioactive**.

Two types of evidence imply that Becquerel's rays originate deep in the heart (or nucleus) of an atom. First, the radiation is found to be associated with certain elements, such as uranium. Radiation does not vary with chemical state—that is, uranium is radioactive whether it is an element or in a compound. Additionally, radiation does not vary with temperature, pressure, or ionization state of the uranium atom. Since all of these things affect electrons in an atom, the radiation cannot come from electron transitions, as atomic spectra do. The huge energy emitted in each event is the second piece of evidence that the radiation cannot be atomic. Nuclear radiation has energies on the order of 10^6 eV per event, which is much greater than typical atomic energies (a few eV), such as observed in spectra and chemical reactions, and more than ten times as great as the most energetic characteristic x rays.

Becquerel did not vigorously pursue his discovery for very long, but by 1898 Marie Curie (1867–1934), then a graduate student married to the already well-known French physicist Pierre Curie (1859–1906), began her doctoral study of Becquerel's rays. She and her husband soon discovered two new radioactive elements, which she named *polonium* (after her native land) and *radium* (because it radiates). These two new elements filled holes in the periodic table and, furthermore, displayed much higher levels of radioactivity per gram of material than uranium. Over a period of four years, working under poor conditions and spending their own funds, the Curies processed more than a ton of uranium ore to isolate a gram of radium salt. Radium became highly sought after, because

Figure 29.1 The dial of this World War II aircraft instrument glows in the dark, because it is painted with radium-doped phosphorescent paint. It is a poignant reminder of the dual nature of radiation. Although radium paint dials are conveniently visible day or night, they emit radon, a radioactive gas that is hazardous and is not directly sensed.

it was about two million times as radioactive as uranium. The Curie's radium salt glowed visibly from the radiation that took its toll on them and other unaware researchers. Shortly after completion of her Ph.D., both Curies and Becquerel shared the 1903 Nobel prize in physics for their work on radioactivity. Pierre was killed in a horse cart accident in 1906, but Marie continued her study of radioactivity for nearly 30 more years. Awarded the 1911 Nobel prize in chemistry for her discovery of two new elements, she remains the only person to win Nobels in physics and chemistry. Marie's radioactive fingerprints on some pages of her notebooks can still expose film, and she suffered from radiation-induced lesions. She died of leukemia likely caused by radiation, but she was active in research until near her death in 1934. The following year, her daughter and son-in-law, Irene and Frederic Joliot-Curie, were awarded the Nobel prize in chemistry for their discovery of artificially induced radiation, adding to a remarkable family legacy.

Alpha, Beta, and Gamma

Research begun by people such as Ernest Rutherford soon after the discovery of nuclear radiation indicated that different types of rays are emitted. Eventually, three types were distinguished and named **alpha** (α), **beta** (β), and **gamma** (γ), because, like x rays, their identities were at first unknown. Figure 29.2 shows what happens if the rays are passed through a magnetic field. The γs are unaffected, while the αs and βs are deflected in opposite directions, indicating the αs are positive, the βs negative, and the γs chargeless. Rutherford used both magnetic and electric fields to show that αs have a positive charge twice the magnitude of an electron, or $+2|q_e|$. In the process, he found the α's charge to mass ratio to be several thousand times smaller than the electron's.* Later on, Rutherford collected αs from a radioactive source and passed an electric discharge through them, obtaining the spectrum of recently discovered helium gas. Among many important discoveries made by Rutherford and his collaborators was the proof that **α radiation is the emission of a helium nucleus**. Rutherford won the Nobel prize in chemistry in 1908 for his early work. He continued to make important contributions until his death in 1934.

Other researchers had already proved βs are negative and have the same mass and the same charge to mass ratio as the recently discovered electron. By 1902 it was

---

*Rutherford was J. J. Thomson's student, and he performed experiments similar to those in which Thomson demonstrated the existence of the electron. See Section 28.2.

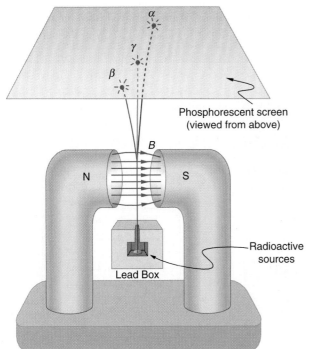

Figure 29.2 Alpha, beta, and gamma radiation are passed through a magnetic field on the way to a phosphorescent screen. The αs and βs bend in opposite directions, while the γs are unaffected, indicating a positive charge for αs, negative for βs, and neutral for γs. Consistent results are obtained with electric fields. Collection of the radiation offers further confirmation from the direct measurement of excess charge.

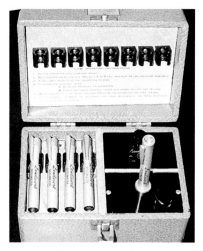

Figure 29.3 These dosimeters (literally, dose meters) are personal radiation monitors that detect the amount of radiation by the discharge of a rechargeable internal capacitor. The amount of discharge is related to the amount of ionizing radiation encountered, a measurement of dose. One dosimeter is shown in the charger. Its scale is read through an eyepiece on top.

recognized that **β radiation is the emission of an electron**. Although βs are electrons, they do not exist in the nucleus before it decays and are not ejected atomic electrons—the electron is created in the nucleus at the instant of decay.

Since γs are not affected by electric or magnetic fields, it is natural to think they might be photons. Evidence for this grew, but it was not until 1914 that this was proved by none other than Rutherford and collaborators. By scattering γ radiation from a crystal and observing interference, they demonstrated that **γ radiation is the emission of a high-energy photon by a nucleus**. In fact, γ radiation comes from the deexcitation of a nucleus, just as an x ray comes from the deexcitation of an atom. The names γ ray and x ray identify the source of the radiation. At the same energy, γ rays and x rays are otherwise identical.

Ionization and Range

Two of the most important characteristics of α, β, and γ radiation were recognized very early. All three types of nuclear radiation produce *ionization* in materials, but they penetrate different distances in materials—that is, they have different *ranges*. Let us examine both why they have these characteristics and what some of the consequences are.

Like x rays, nuclear radiation in the form of αs, βs, and γs has enough energy per event to ionize atoms and molecules in any material. The energy emitted in various nuclear decays ranges from a few keV to more than 10 MeV, and only a few eV are needed to produce ionization. The effects of x rays and nuclear radiation on biological tissue and other materials, such as solid state electronics, is directly related to the ionization they produce. All of them, for example, can damage electronics or kill cancer cells. Additionally, methods for detecting x rays and nuclear radiation are based on ionization, directly or indirectly. All of them can ionize the air between the plates of a capacitor, for example, causing it to discharge. This is the basis of inexpensive personal radiation monitors, such as pictured in Figure 29.3. There are other forms of nuclear radiation than α, β, and γ, and these also produce ionization with similar effects. We define **ionizing radiation** as any form of radiation that produces ionization whether nuclear in origin or not, since the effects and detection of the radiation are related to ionization.

The **range** of radiation is defined to be the distance it can travel through a material. Range is related to several factors, including the energy of the radiation, the material encountered, and the type of radiation. (See Figure 29.4.) The higher the *energy*, the greater the range, all other factors being the same. This makes good sense, since radiation loses its energy in materials primarily by producing ionization in them, and each ionization of an atom or molecule requires energy that is removed from the radiation. The amount of ionization is, thus, directly proportional to the energy of the particle of radiation, as is its range.

Figure 29.4 The range of radiation depends on its energy, the material it encounters, and the type of radiation. (a) Greater energy means greater range. (b) Radiation has a smaller range in materials with high electron density. (c) Alphas have the smallest range, betas have a greater range, and gammas penetrate the farthest.

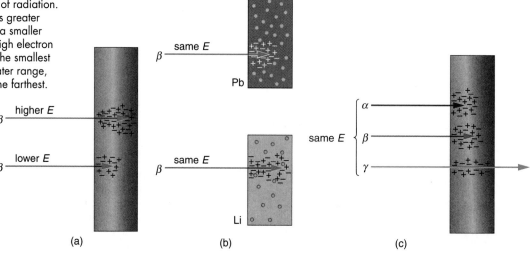

Radiation can be absorbed or shielded by materials, such as the lead aprons dentists drape on us when taking x rays. Lead is a particularly effective shield compared with other materials, such as plastic or air. How does the range of radiation depend on *material*? Ionizing radiation interacts best with charged particles in a material. Since electrons have small masses, they most readily absorb the energy of the radiation in collisions. The greater the density of a material and, in particular, the greater the density of electrons within a material, the smaller the range of radiation.

Different *types* of radiation have different ranges when compared at the same energy and in the same material. Alphas have the shortest range, betas penetrate farther, and gammas have the greatest range. This is directly related to the charge and speed of the particle or type of radiation. At a given energy, each α, β, or γ will produce the same number of ionizations in a material (each ionization requires a certain amount of energy on average). The more readily the particle produces ionization, the more quickly it will lose its energy. The effect of *charge* is as follows: The α has a charge of $+2|q_e|$, the β has a charge of $-|q_e|$, and the γ is chargeless. The electromagnetic force exerted by the α is thus twice as strong as that exerted by the β, and it is more likely to produce ionization. Although chargeless, the γ does interact weakly because it is an electromagnetic wave, but it is less likely to produce ionization in any encounter. More quantitatively, the impulse Δp given to a particle in the material is $\Delta p = F \Delta t$, where F is the force the α, β, or γ exerts over a time Δt. The smaller the charge, the smaller F is and the less momentum (and energy) lost.

The *speed* at which they travel is the other major factor affecting the range of αs, βs, and γs. The faster they move, the less time they spend in the vicinity of an atom or molecule, and the less likely they are to interact. (Quantitatively, Δt is small, and so $\Delta p = F \Delta t$ is small.) Since αs and βs are particles with mass (helium nuclei and electrons, respectively), their energy is kinetic, given classically by $(1/2)mv^2$. The mass of the β is thousands of times less than the α, so that βs must travel much faster than αs to have the same energy. Since βs move faster (most at relativistic speeds), they have less time to interact than αs. Gamma rays are photons, which must travel at the speed of light. They are even less likely to interact than a β, since they spend even less time near a given atom. The range of γs is thus greater than βs.

Alpha radiation from radioactive sources has a range much less than a millimeter of biological tissue, usually not enough to even penetrate the dead layers of skin on our bodies. The same α radiation can penetrate a few centimeters of air, so that mere distance from a source prevents α radiation from reaching us. This makes α radiation relatively safe external to the body. Typical β radiation can penetrate a few millimeters of tissue or about a meter of air. Beta radiation is thus hazardous even when not ingested. The range of βs in lead is about a millimeter, and so it is easy to store in radiation-proof containers. Gamma rays have a much greater range than either αs or βs. In fact, if a given thickness of material, like a lead brick, absorbs 90% of the γs, then a second lead brick will only absorb 90% of what got through the first. There is not a well-defined range that *no* γs penetrate; we can only cut down the number getting through. Typically, γs can penetrate many meters of air, go right through our bodies, and are effectively shielded (that is, reduced in intensity to acceptable levels) by many centimeters of lead. One benefit of γs is that they can be used as radioactive tracers. (See Figure 29.5.)

CONNECTIONS

Collisions

Conservation of energy and momentum often results in energy transfer to a less massive object in a collision. This was discussed in some detail in Chapter 7, for example.

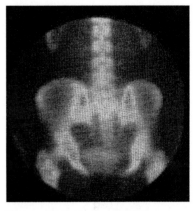

Figure 29.5 This image of the concentration of a radioactive tracer in a patient's body reveals where the most active bone cells are, an indication of bone cancer. A short-lived radioactive substance that locates itself selectively is given to the patient, and the radiation is measured with an external detector. The emitted γ radiation has a sufficent range to leave the body—the range of αs and βs is too small for them to be observed outside the patient.

Human & Medical Application

29.2 RADIATION DETECTION AND DETECTORS

It is well known that ionizing radiation affects us but does not trigger nerve impulses. Newspapers carry stories about unsuspecting victims of radiation poisoning who fall ill with radiation sickness, such as burns and blood count changes,* but who never felt the radiation directly. This makes the detection of radiation by instruments more than an important research tool. This section is a brief overview of radiation detection and some of its applications.

Human Application

---

*Biological effects of ionizing radiation are covered in the next chapter. These may be beneficial, as in cancer therapy, as well as detrimental.

Figure 29.6 (a) This radiograph was taken by placing the rock directly on a piece of film. (b) Film badges contain film similar to that used in dental x rays and is sandwiched between various absorbers to determine the penetrating ability of the radiation as well as the amount.

(a)   (b)

The first direct detection of radiation was Becquerel's fogged photographic plate. Photographic film is still the most common detector of ionizing radiation, being used routinely in medical and dental x rays. (See Section 28.8 for a discussion of medical and other diagnostic uses of x rays.) Nuclear radiation is also captured on film, such as seen in Figure 29.6(a). Film badge dosimeters are used to monitor radiation exposure to those who may encounter it in their profession. (See Figure 29.6(b).) This includes x-ray technicians, some nuclear power plant employees, as well as physicists at particle accelerator laboratories. The mechanism for film exposure by ionizing radiation is similar to that by photons. A quantum of energy interacts with the emulsion and alters it chemically, thus exposing the film. The quantum can be an α, β, or photon, provided it has more than the few eV of energy needed to induce the chemical change (as does all ionizing radiation). The process is not 100% efficient, since not all incident radiation interacts and not all interactions produce the chemical change. The amount of film darkening is related to exposure, but the darkening also depends on the type of radiation, so that absorbers and other devices must be used to obtain energy, charge, and particle identification information.

Another very common radiation detector is the Geiger tube, named after one of Rutherford's students who was partly responsible for its invention. The clicking and buzzing sound we hear in dramatizations and documentaries, as well as in our own physics labs, is usually an audio output of events detected by a Geiger counter. These relatively inexpensive radiation detectors are based on the simple and sturdy **Geiger tube**, shown schematically in Figure 29.7(b). A conducting cylinder with a wire along its axis is filled with an insulating gas so that a voltage applied between the cylinder and wire produces almost no current. Ionizing radiation passing through the tube creates free ion pairs that are attracted to the wire and cylinder, forming a current that is detected as a count. The name "count" implies that there is no information on the energy, charge, or type of radiation with a simple Geiger counter. They do not detect every particle, since some radiation can pass through without creating enough ionization to be detected. But Geiger

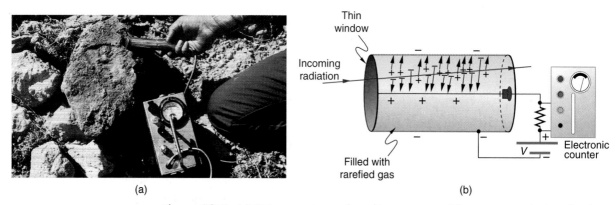

(a)   (b)

Figure 29.7 (a) Geiger counters such as this one are used for prompt monitoring of radiation levels, generally giving only relative intensity and not identifying the type or energy of the radiation. (b) Voltage applied between the cylinder and wire in a Geiger tube sweeps ions created by radiation passing through the gas-filled cylinder. This current is detected and registered as a count.

counters are very useful in producing a prompt output that reveals the existence and relative intensity of ionizing radiation.

Another radiation detection method records light produced when radiation interacts with materials. The energy of the radiation is sufficient to excite atoms in a material that may fluoresce, such as the phosphor used by Rutherford's group (described in Section 28.2). Materials called **scintillators** use a more complex collaborative process to convert radiation energy into light. Scintillators may be liquid or solid, and they can be very efficient. Their light output can provide information about the energy, charge, and type of radiation. Scintillator light flashes are very brief in duration, enabling the detection of huge numbers of particles in short periods of time. Scintillator detectors are used in a variety of research and diagnostic applications. Among these are the detection of the radiation from distant galaxies made on satellites, the analysis of radiation from a person indicating body burdens, and the detection of exotic particles in accelerator laboratories. (See Figure 29.8.)

Light from a scintillator is converted into electrical signals by devices such as the **photomultiplier** tube shown schematically in Figure 29.9. These tubes are based on the photoelectric effect, which is multiplied in stages into a cascade of electrons, hence the name photomultiplier. Light entering the photomultiplier strikes a metal plate, ejecting an electron that is attracted by a positive potential difference to the next plate, giving it enough energy to eject two or more electrons, and so on. The final output current can be made proportional to the energy of the light entering the tube, which is in turn proportional to the energy deposited in the scintillator. Very sophisticated information can be obtained with scintillators, including energy, charge, particle identification, direction of motion, and so on.

Solid state radiation detectors convert ionization produced in a semiconductor (like those found in computer chips) directly into an electrical signal. Semiconductors can be constructed that do not conduct current in one particular direction. When a voltage is applied in that direction, current flows only when ionization is created by radiation, similar to what happens in a Geiger tube. Moreover, the amount of current in a solid state detector is closely related to the energy deposited, and, since the detector is solid, it can have a high efficiency. As with scintillators, very sophisticated information can be obtained from solid state detectors.

Figure 29.8 This array of scintillators is a sophisticated particle detector at the Stanford Linear Accelerator.

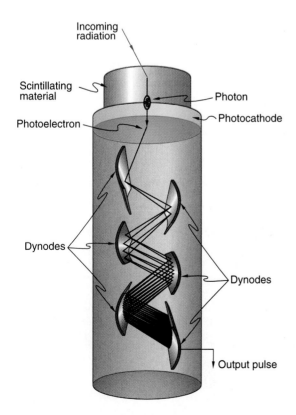

Figure 29.9 Photomultipliers use the photoelectric effect on the first dynode to convert the light output of a scintillator into an electrical signal. Each successive dynode has a more positive potential than the last and attracts the ejected electrons, giving them more energy. The number of electrons is thus multiplied at each dynode, resulting in an easily detected output current.

A variety of other radiation detectors are used. The interested reader can find a wealth of literature on them. Some of the stalwarts of the past, such as bubble chambers and cloud chambers, have been supplanted by modern innovations.

29.3 SUBSTRUCTURE OF THE NUCLEUS

What is inside the nucleus? Why are some stable while others decay? (See Figure 29.10.) Why are there different modes of decay (α, β, and γ)? Why are nuclear decay energies so large? Pursuing natural questions like these has led to far more fundamental discoveries than you might imagine.

We have already identified **protons** as the particles that carry positive charge in nuclei. But there are actually *two* types of particles in nuclei—the *proton* and the *neutron*, referred to collectively as **nucleons**, the constituents of nuclei. As its name implies, the **neutron** is a neutral particle ($q = 0$) that is nearly identical to the proton, having the same intrinsic spin and nearly the same mass. Table 29.1 compares the masses of protons, neutrons, and electrons. Note how close the proton and neutron masses are, but the neutron is slightly more massive once you look past the third digit. Both nucleons are much more massive than the electron. In fact, $m_p = 1836m_e$ (as noted in Chapter 28) and $m_n = 1839m_e$.

Table 29.1 also gives masses in mass units that are more convenient than kilograms on the atomic and nuclear scale. The first of these is the **unified atomic mass unit** (u), defined to be

$$1 \text{ u} = 1.6605 \times 10^{-27} \text{ kg} \tag{29.1a}$$

This unit is defined so that a neutral carbon atom has a mass of exactly 12 u. Masses are also expressed in units of MeV/c^2. These units are very convenient when considering the conversion of mass into energy (and vice versa), as is so prominent in nuclear processes. Using $E = mc^2$ and units of m in MeV/c^2, we find that c^2 cancels and E comes out conveniently in MeV. For example, if the rest mass of a proton is converted entirely into energy, then $E = mc^2 = (938.27 \text{ MeV}/c^2)\cdot c^2 = 938.27$ MeV. It is useful to note that 1 u of mass converted to energy produces 931.5 MeV, or

$$1 \text{ u} = 931.5 \text{ MeV}/c^2 \tag{29.1b}$$

All properties of a nucleus are determined by the number of protons and neutrons it has. A specific combination of protons and neutrons is called a **nuclide** and is a unique nucleus. The following notation is used to represent a particular nuclide:

$$^A_Z\text{X}_N$$

where the symbols A, X, Z, and N are defined as follows: The **number of protons in a nucleus** is the **atomic number** Z, as defined in Chapter 28. X **is the symbol for the element**, such as Ca for calcium. But once Z is known, the element is known; hence, Z and X are redundant.* For example, $Z = 20$ is always calcium, and calcium always has $Z = 20$. N **is the number of neutrons** in a nucleus. The symbol A is defined to be the number of nucleons or the **total number of protons and neutrons**,

$$A = N + Z \tag{29.2}$$

---

*See Section 28.7 for a discussion of how the number of positive charges or protons in the nucleus determines the number of electrons and chemistry of the element.

Figure 29.10 Why is the carbon in this coal stable (a) while the uranium in the rock slowly decays over billions of years (b)? Why is cesium in this wafer (c) even less stable than the uranium, decaying in far less than 1/1,000,000 the time? What is the reason uranium and cesium decay by different modes (α and β, respectively)?

(a)   (b)   (c)

TABLE 29.1

MASSES OF THE PROTON, NEUTRON, AND ELECTRON				
		Mass		
Particle	*Symbol*	*kg*	*u*	*MeV/c²*
Proton	p	1.67262×10^{-27}	1.007276	938.27
Neutron	n	1.67493×10^{-27}	1.008665	939.57
Electron	e	$9.1094 \ \times 10^{-31}$	0.00054858	0.511

where A is also called the **atomic mass**. This name for A is logical; the mass of an atom is nearly equal to the mass of its nucleus, since electrons have so little mass. The mass of the nucleus turns out to be nearly equal to the sum of the masses of the protons and neutrons in it, which is proportional to A. In this context, it is particularly convenient to express masses in units of u. Both protons and neutrons have masses close to 1 u, and so the mass of an atom is close to A u. For example, a carbon nucleus with six protons and six neutrons has $A = 12$, and its mass is 12 u. As noted above, the unified atomic mass unit is defined so that a neutral carbon atom (actually a ^{12}C atom) has a mass of *exactly* 12 u. Carbon was chosen as the standard partly because of its importance in chemistry. (See Appendix A.)

Let us look at a few examples of nuclides expressed in the $_Z^A$X$_N$ notation. The nucleus of the simplest atom, hydrogen, is a single proton, or $_1^1$H. (The zero for no neutrons is often omitted.) To check this symbol, refer to the periodic table—you see that the atomic number Z of hydrogen is 1. Since you are given that there are no neutrons, the atomic mass A is also 1. Suppose you are told that the helium nucleus or α particle has two protons and two neutrons. You can then see that it is written $_2^4$He$_2$. There is a scarce form of hydrogen called deuterium since its nucleus has one proton and one neutron and, hence, twice the mass of common hydrogen. The symbol for deuterium is, thus, $_1^2$H$_1$. An even rarer—and radioactive—form of hydrogen is called tritium since it has a single proton and two neutrons, and it is written $_1^3$H$_2$. These three varieties of hydrogen have nearly identical chemistries, but the nuclei differ greatly in mass, stability, and other characteristics. Nuclei (such as these of hydrogen) having the same Z and different Ns are defined to be **isotopes** of the same element.

There is some redundancy in the symbols A, X, Z, and N. If the element X is known, then Z can be found in a periodic table and is always the same for a given element. If both A and X are known, then N can also be determined (first find Z; then, $N = A - Z$). Thus the simpler notation for nuclides,

$$^A\text{X}$$

is sufficient and is most commonly used. For example, in this simpler notation, the three isotopes of hydrogen are ^{1}H, ^{2}H, and ^{3}H, while the α particle is ^{4}He. We read these backward, saying helium-4 for ^{4}He, or uranium-238 for ^{238}U. So for ^{238}U, should we need to know, we can determine that $Z = 92$ for uranium from the periodic table, and, thus, $N = 238 - 92 = 146$.

A variety of experiments indicate that a nucleus behaves something like a tightly packed ball of nucleons, as illustrated in Figure 29.11. These nucleons have large kinetic energies and, thus, move rapidly in very close contact. Nucleons can be separated by a large force, such as in a collision with another nucleus, but resist mightily being pushed closer together. The most compelling evidence that nucleons are closely packed in a nucleus is that the **radius of a nucleus**, r, is given approximately by

$$r = r_0 A^{1/3} \qquad (29.3)$$

where $r_0 = 1.2$ fm and A is the atomic mass of the nucleus.* Note that $r^3 \propto A$. Since many nuclei are spherical, and the volume of a sphere is $V = (4/3)\pi r^3$, we see that $V \propto A$—that is, the volume of a nucleus is proportional to the number of nucleons in it. This is what would happen if you pack nucleons so close there is no empty space between them.

○ Proton

○ Neutron

Figure 29.11 A model of the nucleus. Nucleons slide over one another in very close contact. They are held together by nuclear forces and resist both being pulled apart and pushed inside one another. The volume of the nucleus is the sum of the volumes of the nucleons in it, here shown in different colors to represent protons and neutrons.

---

*Noting that 1 fm (femtometer) = 10^{-15} m, we see that the nucleus is about the size determined by Rutherford as discussed in Section 28.2. Note also that much of the huge body of information about the radii of nuclei comes from scattering experiments.

EXAMPLE 29.1 HOW SMALL AND DENSE IS A NUCLEUS?

(a) Find the radius of an iron-56 nucleus. (b) Find its approximate density in kg/m^3, approximating the mass of ^{56}Fe to be 56 u.

Strategy and Concept (a) Finding the radius of ^{56}Fe is a straightforward application of Equation 29.3, given $A = 56$. (b) To find the approximate density, we assume the nucleus is spherical (this one actually is), calculate its volume using the radius found in part (a), and then find its density from $\rho = m/V$. Finally, we will need to convert density from units of u/fm^3 to kg/m^3.

Solution for (a) The radius of a nucleus is given by Equation 29.3 to be

$$r = r_0 A^{1/3}$$

Substituting the values for r_0 and A yields

$$r = (1.2 \text{ fm})(56)^{1/3} = (1.2 \text{ fm})(3.83)$$
$$= 4.6 \text{ fm}$$

Solution for (b) Density is defined to be $\rho = m/V$, which for a sphere of radius r is

$$\rho = \frac{m}{V} = \frac{m}{\frac{4}{3}\pi r^3}$$

Substituting known values gives

$$\rho = \frac{56 \text{ u}}{(1.33)(3.14)(4.6 \text{ fm})^3}$$
$$= 0.138 \text{ u/fm}^3$$

Converting to units of kg/m^3, we find

$$\rho = (0.138 \text{ u/fm}^3)(1.66 \times 10^{-27} \text{ kg/u})\left(\frac{1 \text{ fm}}{10^{-15} \text{ m}}\right)^3$$
$$= 2.3 \times 10^{17} \text{ kg/m}^3$$

Discussion and Implications (a) The radius of this medium-sized nucleus is found to be approximately 4.6 fm, and so its diameter is about 10 fm or 10^{-14} m. In our discussion of Rutherford's discovery of the nucleus, we noted that it is about 10^{-15} m in diameter (which it is for lighter nuclei), consistent with this result to an order of magnitude. The nucleus is much smaller in diameter than the typical atom, which has a diameter on the order of 10^{-10} m.

(b) The density found here is so large as to cause disbelief. It is consistent with earlier discussions we have had about the nucleus being very small and containing nearly all of the mass of the atom. Nuclear densities, such as found here, are about 2×10^{14} times greater than that of water, which has a density of "only" 10^3 kg/m^3. One cubic meter of nuclear matter, such as found in a neutron star, has the same mass as a cube of water 61 km on a side! The nucleus must be unlike anything we have ever seen, another unexpected and dramatic surprise from the realm of the very small.

Nuclear Forces and Stability

What holds a nucleus together? It is very small and its protons, being positive, exert tremendous repulsive forces on one another. (The Coulomb force increases as charges get closer, since it is proportional to $1/r^2$ even at the tiny distances found in nuclei.) The answer is that two previously unknown forces hold nucleons together and make it into a tightly packed ball of nucleons. These forces are called the **weak and strong nuclear forces** and were first discussed when the concept of force was introduced in Chapter 4. Table 4.2 compares basic properties of the four basic forces found in nature. Nuclear forces are very short ranged, meaning they fall to zero strength when nucleons are separated by only a few fm. But like glue, they are very attractive when the nucleons get close to one another. The strong nuclear force is about 100 times more attractive than the repulsive EM force, easily holding the nucleons together. Nuclear forces become extremely repulsive if the nucleons get too close, making nucleons strongly resist being pushed inside one another, something like ball bearings.*

The fact that nuclear forces are very strong is responsible for the very large energies emitted in nuclear decay. During decay the forces do work, producing energy, and since work is force times distance ($W = Fd \cos \theta$), a large force can result in a large emitted

---

*Nuclear forces are short ranged because their exchange particles have mass. See Section 27.7, where the role of the Heisenberg uncertainty principle in the exchange of particles was briefly discussed. In Chapter 31 we shall more fully explore the four forces, their exchange particles, and hopes for unification.

Figure 29.12 Simplified chart of the nuclides, a graph of *N* versus *Z* for known nuclides. The patterns of stable and unstable nuclides reveal characteristics of the nuclear forces. The dashed line is for *N* = *Z*. Numbers along diagonals are atomic masses *A*.

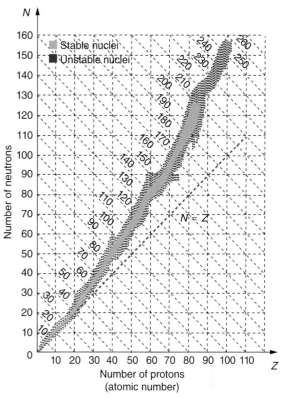

energy. In fact, we know that there are *two* distinct nuclear forces because of the different types of nuclear decay—the strong nuclear force is responsible for α decay, while the weak nuclear force is responsible for β decay.

The thousands of stable and unstable nuclei we have explored can be arranged in a table called the chart of the nuclides, a simplified version of which is shown in Figure 29.12. Nuclides are located on a plot of *N* versus *Z*. Examination of a detailed chart of the nuclides reveals patterns in the characteristics of nuclei, such as stability, abundance, and modes of decay, analogous to but more complex than the systematics in the periodic table of the elements.

In principle a nucleus can have any combination of protons and neutrons, but Figure 29.12 shows a definite pattern for those that are stable. For low-mass nuclei, there is a strong tendency for *N* and *Z* to be nearly equal. This means that the nuclear force is more attractive when *N* = *Z*. More detailed examination reveals greater stability when *N* and *Z* are even numbers—nuclear forces are more attractive when neutrons and protons are in pairs. For increasingly higher masses, there are progressively more neutrons than protons in stable nuclei. This is due to the ever-growing repulsion between protons. Since nuclear forces are short ranged, and the Coulomb force is long ranged, an excess of neutrons keeps the protons a little farther apart, reducing Coulomb repulsion. Decay modes of nuclides out of the region of stability consistently produce nuclides closer to the region of stability. There are more stable nuclei having certain numbers of protons and neutrons, called **magic numbers**. Magic numbers indicate a shell structure for the nucleus in which closed shells are more stable. Nuclear shell theory has been very successful in explaining nuclear energy levels, nuclear decay, and the greater stability of nuclei with closed shells. (See Figure 29.13.) We have been producing ever-heavier transuranic elements since the early 1940s, and we have now produced the element with *Z* = 110. There is a theoretical prediction of an island of stability at a shell closure expected at *Z* = 114.

29.4 NUCLEAR DECAY AND CONSERVATION LAWS

Nuclear decay has provided an amazing window into the realm of the very small. Nuclear decay gave the first indication of the connection between mass and energy, and it revealed the existence of two of the four basic forces in nature. In this section we explore the major modes of nuclear decay; and, like those who first explored them, we will discover evidence of previously unknown particles and conservation laws.

Some nuclides are stable, apparently living forever. Unstable nuclides decay (that is, they are radioactive), eventually producing a stable nuclide. We call the original nuclide the **parent** and its decay products the **daughters**. Some radioactive nuclides decay in a single step to a stable nucleus. For example, ^{60}Co is unstable and decays directly to ^{60}Ni, which is stable. Others, such as ^{238}U, decay to another unstable nuclide, resulting in a **decay series** in which each subsequent nuclide decays until a stable nuclide is finally produced. The decay series parented by ^{238}U is of particular interest, since it produces the radioactive isotopes ^{226}Ra and ^{210}Po that the Curies first discovered. (See Figure 29.14.) Notorious radon gas is also produced (^{222}Rn in the series), an increasingly recognized naturally occurring hazard. Since radon is a noble gas, it emanates from materials, such as soil, containing even trace amounts of ^{238}U, and can be ingested. The decay of radon and

Figure 29.13 The German-born American physicist Maria Goeppert Mayer (1906–1972) shared the 1963 Nobel prize in physics with J. Jensen for the creation of the nuclear shell model. This successful nuclear model has nucleons filling shells analogous to electron shells in atoms. It was inspired by patterns observed in nuclear properties.

Figure 29.14 The decay series produced by ^{238}U, the most common uranium isotope. Nuclides are graphed in the same manner as in the chart of nuclides. The mode of decay for each member of the series is shown, as well as the half-lives. Note that some nuclides decay by more than one mode. You can see why radium and polonium are found in uranium ore. A stable isotope of lead is the end product of the series.

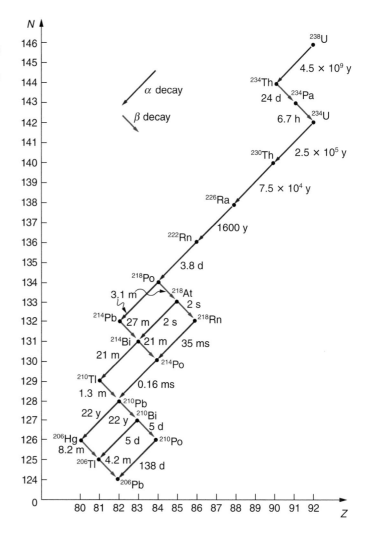

its daughters produces internal damage. The ^{238}U decay series ends with ^{206}Pb, a stable isotope of lead.

Note that the daughters of α decay shown in Figure 29.14 always have two less protons and two less neutrons than the parent. This seems reasonable, since we know that α decay is the emission of a ^{4}He nucleus, which has two protons and two neutrons. The daughters of β decay have one less neutron and one more proton than their parent. Beta decay is a little more subtle, as we shall see. No γ decays are shown in the figure, because they do not produce a daughter different than the parent.

Alpha Decay

In α decay, a ^{4}He nucleus simply breaks away from the parent nucleus, leaving a daughter with two fewer protons and two fewer neutrons than the parent. (See Figure 29.15.) One example of α decay is shown in Figure 29.14 for ^{238}U. Another nuclide that α decays is ^{239}Pu. The decay equations for these two nuclides are

$$^{238}\text{U} \rightarrow {}^{234}\text{Th} + {}^4\text{He}$$

and

$$^{239}\text{Pu} \rightarrow {}^{235}\text{U} + {}^4\text{He}$$

If you examine the periodic table of the elements, you will find that Th has $Z = 90$, two less than U, which has $Z = 92$. Similarly, in the second decay equation, we see that U has two less protons than Pu, which has $Z = 94$. The general rule for α decay is best written in the $^A_Z\text{X}_N$ format. If a certain nuclide is known to α decay (generally this information must be looked up in a table of isotopes, such as in Appendix B), its **α decay equation** is

Figure 29.15 Alpha decay is the separation of a ^{4}He nucleus from the parent. The daughter nucleus has two fewer protons and two fewer neutrons than the parent. This occurs spontaneously only if the daughter and ^{4}He nucleus have less total rest mass than the parent.

$$^A_Z\text{X}_N \rightarrow {}^{A-4}_{Z-2}\text{Y}_{N-2} + {}^4_2\text{He}_2 \qquad (\alpha \text{ decay}) \qquad \textbf{(29.4)}$$

where Y is the nuclide that has two fewer protons than X, such as Th having two fewer than U. So if you were told that ^{239}Pu α decays and asked to write the complete decay equation, you would first look up which element has two fewer protons (an atomic number two less) and find that this is uranium. Then since four nucleons have broken away from the original 239, its atomic mass would be 235.

It is instructive to examine conservation laws related to α decay. You can see from Equation 29.4 that total charge is conserved. Linear and angular momentum are conserved, too. Although conserved, angular momentum is not of great consequence in this type of decay, but conservation of linear momentum has interesting consequences. If the nucleus is at rest when it decays, its momentum is zero. In that case the fragments must fly in opposite directions with equal-magnitude momenta, so that total momentum remains zero. This results in the α particle carrying away most of the energy, as a bullet from a heavy rifle carries away most of the energy of the powder burned to shoot it. Total mass-energy is also conserved: the energy produced in the decay comes from the destruction of a fraction of the original mass. As discussed in Chapter 26, the general relationship is

$$E = (\Delta m)c^2 \tag{29.5}$$

Here E is a **nuclear reaction energy** (the reaction can be nuclear decay or any other reaction), and Δm is the difference in mass between initial and final products. When the final products have less total mass, Δm is positive, and the reaction releases energy (is exothermic). When the products have greater total mass, the reaction is endothermic (Δm is negative) and must be induced with an energy input. For α decay to be spontaneous, the decay products must have smaller mass than the parent.

EXAMPLE 29.2 ALPHA DECAY ENERGY FOUND FROM MASSES

Find the energy emitted in the α decay of ^{239}Pu.

Strategy Nuclear reaction energy, such as released in α decay, can be found using Equation 29.5. We must first find Δm, the difference in mass between the parent nucleus and the products of the decay. This is easily done using masses given in Appendix A.

Solution The decay equation was given above for ^{239}Pu; it is

$$^{239}\text{Pu} \rightarrow {}^{235}\text{U} + {}^4\text{He}$$

Thus the pertinent masses are those of ^{239}Pu, ^{235}U, and the α particle or ^{4}He, all of which are listed in Appendix A. The initial mass was $m(^{239}\text{Pu}) = 239.052157$ u. The final mass is the sum $m(^{235}\text{U}) + m(^4\text{He}) = 235.043924$ u $+ 4.002602$ u $= 239.046526$ u. Thus,

$$\Delta m = m(^{239}\text{Pu}) - [m(^{235}\text{U}) + m(^4\text{He})]$$

$$= 239.052157 \text{ u} - 239.046526 \text{ u}$$

$$= 0.005631 \text{ u}$$

Now we can find E by entering Δm into the equation:

$$E = (\Delta m)c^2 = (0.005631)(uc^2)$$

From Equation 29.1b we know 1 u = 931.5 MeV/c^2, and so

$$E = (0.005631)(931.5 \text{ MeV}/c^2)(c^2) = 5.25 \text{ MeV}$$

Discussion The energy released in this α decay is in the MeV range, about 10^6 as great as typical chemical reaction energies, consistent with many previous discussions. Most of this energy becomes kinetic energy of the α particle (or ^{4}He nucleus), which moves away at high speed. The energy carried away by the recoil of the ^{235}U nucleus is much smaller in order to conserve momentum. The ^{235}U nucleus can be left in an excited state to later emit photons (γ rays). This decay is spontaneous and releases energy, because the products have less mass than the parent nucleus. The question of why the products have less mass will be discussed in Section 29.6. Note that the masses given in Appendix A are atomic masses of neutral atoms, including their electrons. The mass of the electrons is the same before and after α decay, and so their masses subtract out when finding Δm. In this case, there are 94 electrons before and after the decay.

Beta Decay

There are actually *three* types of beta decay. The first discovered was "ordinary" beta decay and is called β^- decay or electron emission. The symbol β^- represents **an electron emitted in nuclear beta decay**. Cobalt-60 is a nuclide that β^- decays in the following manner:

$$^{60}\text{Co} \rightarrow {}^{60}\text{Ni} + \beta^- + \text{neutrino}$$

The **neutrino** is a particle emitted in beta decay that was unanticipated and is of fundamental importance. The neutrino was not even proposed in theory until more than

Figure 29.16 Enrico Fermi was nearly unique among 20th century physicists—he was accomplished as both an experimentalist and a theorist. His many contributions to theoretical physics included the identification of the weak nuclear force. The fermi (fm) is named for him, as are an entire class of subatomic particles (fermions), an element (Fermium), and a major research laboratory (Fermilab). His experimental works include studies of radioactivity, for which he won the 1938 Nobel prize in physics, and creation of the first nuclear chain reaction.

CONNECTIONS

Conservation Laws
New conservation laws were disclosed by patterns in nuclear decay processes. The search for other conservation laws has continued into particle physics. Conservation laws provide the broadest underlying unification principles in physics. They not only summarize our knowledge, they point the way to new discoveries.

β⁻ decay

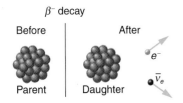

Figure 29.17 In β⁻ decay, the parent nucleus emits an electron and an antineutrino. The daughter nucleus has one more proton and one less neutron than its parent. Neutrinos interact so weakly that they are almost never directly observed, but they play a fundamental role in particle physics.

20 years after beta decay was known to involve electron emissions. Neutrinos are so difficult to detect that the first direct evidence of them was not obtained until 1953. Neutrinos are massless, chargeless, and do not feel the strong nuclear force. Traveling at the speed of light, they have little time to affect any nucleus they encounter. Being chargeless (and they are not EM waves), they do not interact through the EM force. They do interact via the relatively weak and very short ranged weak nuclear force. Consequently, neutrinos escape almost any detector and penetrate almost any shielding. But neutrinos do carry energy, angular momentum (they are fermions with half-integral spin), and linear momentum away from a beta decay. When accurate measurements of beta decay were made, it became apparent that energy, angular momentum, and linear momentum were not accounted for by the daughter nucleus and electron alone. Either a previously unsuspected particle was carrying them away, or three conservation laws were being violated. Wolfgang Pauli made a formal proposal for the existence of neutrinos in 1930. The Italian born American physicist Enrico Fermi (1901–1954) gave neutrinos their name, meaning little neutral ones, when he developed a sophisticated theory of beta decay. (See Figure 29.16.) Part of Fermi's theory was the identification of the weak nuclear force as being distinct from the strong nuclear force and in fact responsible for beta decay.

The neutrino also reveals a new conservation law. There are various families of particles, one of which is the electron family. The number of members of the electron family is constant in any process or any closed system. There are no members before the decay, and after there is an electron and a neutrino. Electrons are given an electron family number of +1. The neutrino in β^- decay is an **electron's antineutrino**, given the symbol $\bar{\nu}_e$, where ν is the Greek letter nu, and the subscript e means this neutrino is related to the electron. The bar indicates this is a particle of **antimatter**.* (All particles have antimatter counterparts that are nearly identical except that they have the opposite charge. Antimatter is almost entirely absent from normal matter, but it is found in nuclear decay and other nuclear and particle reactions.) The electron's antineutrino $\bar{\nu}_e$, being antimatter, has an electron family number of −1. The total is zero, before and after the decay. The new conservation law, obeyed in all circumstances, states that the **total electron family number is constant**. An electron cannot be created without also creating an antimatter family member. This law is analogous to conservation of charge in a situation where total charge is originally zero, and equal amounts of positive and negative charge must be created in a reaction to keep the total zero.

If a nuclide $^A_Z X_N$ is known to β^- decay, then its **β^- decay equation** is

$$^A_Z X_N \rightarrow \, ^A_{Z+1} Y_{N-1} + \beta^- + \bar{\nu}_e \qquad (\beta^- \text{ decay}) \qquad \textbf{(29.6)}$$

where Y is the nuclide having one more proton than X. (See Figure 29.17.) So if you know that a certain nuclide β^- decays, you can find the daughter nucleus by first looking up Z for the parent and then determining which element has atomic number $Z + 1$. In the example of the β^- decay of ^{60}Co given above, we see that $Z = 27$ for Co and $Z = 28$ is Ni. It is as if one of the neutrons in the parent nucleus decays into a proton, electron, and neutrino. In fact, neutrons outside of nuclei do just that—they live only an average of a few minutes and β^- decay in the following manner: $n \rightarrow p + \beta^- + \bar{\nu}_e$.

We see that charge is conserved in β^- decay, since the total charge is Z before and after the decay. For example, in ^{60}Co decay total charge is 27 before decay, since cobalt has $Z = 27$. After decay, the daughter nucleus is Ni, which has $Z = 28$, and there is an electron, so that the total charge is $28 - 1$ or 27, too. Angular momentum is conserved, but not obviously (you have to examine the spins and angular momenta of the final products in detail to verify this). Linear momentum is also conserved, again imparting most of the decay energy to the electron and the antineutrino, since they are of low and zero mass, respectively. Another new conservation law is obeyed here and elsewhere in nature. **The total number of nucleons A is conserved.** In ^{60}Co decay, for example, there are 60 nucleons before and after the decay. Note that total A is also conserved in α decay. Also note the total number of protons changes, as does the total number of neutrons, so that

---

*The families of particles and rules governing them are covered in Chapter 31, as are the characteristics of matter and antimatter.

total Z and total N are *not* conserved in β^- decay, as they are in α decay. Energy released in β^- decay can be calculated given the masses of the parent and products.

EXAMPLE 29.3 β^- DECAY ENERGY FROM MASSES

Find the energy emitted in the β^- decay of ^{60}Co.

Strategy and Concept As in the preceding example we must first find Δm, the difference in mass between the parent nucleus and the products of the decay, using masses given in Appendix A. Then the emitted energy is calculated as before using $E = (\Delta m)c^2$. The initial mass is just that of the parent nucleus, and the final mass is that of the daughter nucleus and the electron created in the decay. The neutrino is massless. But since the masses given in Appendix A are for neutral atoms, the daughter nucleus has one more electron than the parent, and so the extra electron mass is included in its atomic mass. Thus $\Delta m = m(^{60}\text{Co}) - m(^{60}\text{Ni})$.

Solution The β^- decay equation for ^{60}Co is

$$^{60}_{27}\text{Co}_{33} \rightarrow\ ^{60}_{28}\text{Ni}_{32} + \beta^- + \bar{\nu}_e$$

As noted,

$$\Delta m = m(^{60}\text{Co}) - m(^{60}\text{Ni})$$

Entering the masses found in Appendix A gives

$$\Delta m = 59.933820\ \text{u} - 59.930789\ \text{u} = 0.003031\ \text{u}$$

Thus,

$$E = (\Delta m)c^2 = (0.003031)(uc^2)$$

Using 1 u = 931.5 MeV/c^2, we obtain

$$E = (0.003031)(931.5\ \text{MeV}/c^2)(c^2) = 2.82\ \text{MeV}$$

Discussion and Implications Perhaps the most difficult thing about this example is convincing yourself that the β^- mass is included in the atomic mass of ^{60}Ni. Beyond that are other implications. Again the decay energy is in the MeV range. This energy is shared by all of the products of the decay. In many ^{60}Co decays, the daughter nucleus ^{60}Ni is left in an excited state and later emits photons (γ rays). Most of the remaining energy goes to the electron and neutrino, since the recoil kinetic energy of the daughter nucleus is small. One final note: the electron emitted in β^- decay is created in the nucleus at the time of decay.

The second type of beta decay is more exotic than the first. It is **β^+ decay**. Certain nuclides decay by the emission of a *positive* electron. This is **antielectron** or **positron decay**. (See Figure 29.18.) The antielectron is often represented by the symbol e^+, but in beta decay it is written as β^+ to indicate the antielectron was emitted in a nuclear decay. Antielectrons are the antimatter counterpart to electrons, being nearly identical, having the same mass, spin, and so on, but having a positive charge and an electron family number of -1. When a positron encounters an electron, there is a mutual annihilation in which all the mass of the antielectron-electron pair is converted into pure photon energy. (The reaction, $e^+ + e^- \rightarrow \gamma + \gamma$, conserves electron family number as well as all other conserved quantities.) If a nuclide $^A_Z X_N$ is known to β^+ decay, then its **β^+ decay equation** is

$$^A_Z X_N \rightarrow\ ^A_{Z-1} Y_{N+1} + \beta^+ + \nu_e \qquad (\beta^+\ \text{decay}) \qquad \text{(29.7)}$$

where Y is the nuclide having one less proton than X (to conserve charge) and ν_e is the symbol for the **electron's neutrino**, which has an electron family number of $+1$. Since an antimatter member of the electron family (the β^+) is created in the decay, a matter member of the family (here the ν_e) must also be created. Given, for example, that ^{22}Na β^+ decays, you can write its full decay equation by first finding that $Z = 11$ for ^{22}Na, so that the daughter nuclide will have $Z = 10$, the atomic number for neon. Thus the β^+ decay equation for ^{22}Na is

$$^{22}_{11}\text{Na}_{11} \rightarrow\ ^{22}_{10}\text{Ne}_{12} + \beta^+ + \nu_e$$

In β^+ decay, it is as if one of the protons in the parent nucleus decays into a neutron, a positron, and a neutrino. Protons do not do this outside of the nucleus, and so the decay is due to the complexities of the nuclear force. Note again that the total number of nucleons is constant in this and any other reaction. To find the energy emitted in β^+ decay, you must again count the number of electrons in the neutral atoms, since atomic masses are used. The daughter has one less electron than the parent, and one electron mass is created in the decay. Thus, in β^+ decay,

$$\Delta m = m(\text{parent}) - [m(\text{daughter}) + 2m_e]$$

since we use the masses of neutral atoms.

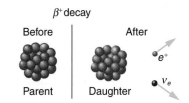

β^+ decay

Figure 29.18 β^+ decay is the emission of a positron that eventually finds an electron to annihilate, characteristically producing gammas in opposite directions.

Electron capture is the third type of beta decay. Here a nucleus captures an inner-shell electron and undergoes a nuclear reaction that has the same effect as β^+ decay. Electron capture is sometimes denoted by the letters EC. We know that electrons cannot reside in the nucleus, but this is a nuclear reaction that consumes the electron and only occurs spontaneously when the products have less mass than the parent plus the electron. If a nuclide $_Z^AX_N$ is known to undergo electron capture, then its **electron capture equation** is

$$_Z^AX_N + e^- \rightarrow \,_{Z-1}^{A}Y_{N+1} + \nu_e \qquad \text{(electron capture, or EC)} \qquad \text{(29.8)}$$

Any nuclide that can β^+ decay can also undergo electron capture (and often does both). The same conservation laws are obeyed for EC as for β^+ decay. It is good practice to confirm these for yourself.

All forms of beta decay occur because the parent nuclide is unstable and lies outside the region of stability in the chart of nuclides. (See Figure 29.12.) Those nuclides that have relatively more neutrons than those in the region of stability β^- decay to produce a daughter with fewer neutrons, producing a daughter nearer the region of stability. Similarly, those nuclides having relatively more protons than those in the region of stability β^+ decay or undergo electron capture to produce a daughter with fewer protons, nearer the region of stability.

Gamma Decay

Gamma decay is the simplest form of nuclear decay—it is the emission of energetic photons by nuclei left in an excited state by some earlier process. Protons and neutrons in an excited nucleus are in higher orbitals, and they fall to lower levels by photon emission (analogous to electrons in excited atoms). Nuclear excited states have lifetimes typically of only about 10^{-14} s, an indication of the great strength of the forces pulling the nucleons to lower states. The γ **decay equation** is simply

$$_Z^AX_N^* \rightarrow \,_Z^AX_N + \gamma_1 + \gamma_2 + \cdots \qquad (\gamma \text{ decay}) \qquad \text{(29.9)}$$

where the asterisk indicates the nucleus is in an excited state. There may be one or more γs emitted, depending on how the nuclide deexcites. In radioactive decay, γ emission is common and is preceded by α or β decay. For example, when ^{60}Co β^- decays, it most often leaves the daughter nucleus in an excited state, written ^{60}Ni*. Then the nickel nucleus quickly γ decays by the emission of two penetrating γs:

$$^{60}\text{Ni*} \rightarrow \,^{60}\text{Ni} + \gamma_1 + \gamma_2$$

These are called cobalt γ rays, although they come from nickel—they are used for cancer therapy, for example. It is again constructive to verify the conservation laws for gamma decay. Finally, since γ decay does not change the nuclide to another species, it is not prominently featured in charts of decay series, such as that in Figure 29.14.

Other types of nuclear decay have been found, but these are less common than α, β, and γ decay and do not reveal new conservation laws or forces in nature. Spontaneous fission is one of these and is covered in the next chapter because of its applications in nuclear power and weapons.

29.5 HALF-LIFE AND ACTIVITY

Unstable nuclei decay. But some nuclides are more unstable than others and decay faster. For example, radium and polonium, discovered by the Curies, decay faster than uranium. This means they have shorter lifetimes, producing a greater rate of decay. In this section we explore half-life and activity, the quantitative terms for lifetime and rate of decay.

Half-Life

Why use a term like half-life rather than lifetime? The answer can be found by examining Figure 29.19, which shows how the number of radioactive nuclei in a sample decreases with time. The **time in which half of the original nuclei decay** is defined to be the **half-life** $t_{1/2}$. Half of what remains decays in the next half-life. And half of that amount decays in the following half-life. So the number of radioactive nuclei goes from

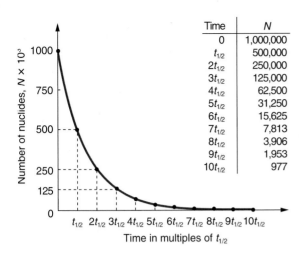

Time	N
0	1,000,000
$t_{1/2}$	500,000
$2t_{1/2}$	250,000
$3t_{1/2}$	125,000
$4t_{1/2}$	62,500
$5t_{1/2}$	31,250
$6t_{1/2}$	15,625
$7t_{1/2}$	7,813
$8t_{1/2}$	3,906
$9t_{1/2}$	1,953
$10t_{1/2}$	977

Figure 29.19 Radioactive decay reduces the number of radioactive nuclei over time. In one half-life $t_{1/2}$, the number decreases to half of its original value. Half of what remain decay in the next half-life, and half of those in the next, and so on. This is an exponential decay, as seen in the graph of the number of nuclei present as a function of time.

N to $N/2$ in one half-life, then to $N/4$ in the next, and to $N/8$ in the next, and so on. If N is a large number, then *many* half-lives (not just two) pass before all of the nuclei decay. Nuclear decay is an example of a purely statistical process. A more precise definition of half-life is that **each nucleus has a 50% chance of living for a time equal to one half-life $t_{1/2}$.** Thus if N is reasonably large, half of the original nuclei decay in a time of one half-life. If an individual nucleus makes it through that time, it still has a 50% chance of living another half-life. Even if it happens to make it through hundreds of half-lives, it still has a 50% chance of living for one more. The probability of decay is the same no matter when you start counting. This is like random coin flipping. The chance of heads is 50%, no matter what has happened before.

There is a tremendous range in the half-lives of various nuclides, being as short as 10^{-23} s for the most unstable, to more than 10^{16} y for the least unstable, or about 46 orders of magnitude! Nuclides with the shortest half-lives are those for which the nuclear forces are least attractive, some indication of how much the nuclear force can depend on the particular combination of neutrons and protons. The concept of half-life is applicable to other subatomic particles, as will be discussed in Chapter 31. It is also applicable to the decay of excited states in atoms and nuclei. All of these entities change form in a statistical manner with some *probability* of transformation occurring in a specific amount of time. Most notably, the probability is 50% in one half-life.

The following equation gives the quantitative relationship between the original number of nuclei present N_0 at time zero and the number N at a later time t:

$$N = N_0 e^{-\lambda t} \tag{29.10}$$

where $e = 2.71828\ldots$ is the base of the natural logarithm and λ is the **decay constant** for the nuclide. The shorter the half-life, the larger λ is, and the faster the exponential $e^{-\lambda t}$ decreases with time. The numerical relationship between the decay constant λ and the half-life $t_{1/2}$ is

$$\lambda = \frac{0.693}{t_{1/2}} \tag{29.11}$$

To see how the number of nuclei declines to half its original value in one half-life, let $t = t_{1/2}$ in the exponential in Equation 29.10. This gives $e^{-\lambda t} = e^{-0.693} = 0.500$, so that $N = 0.500N_0$. For integral numbers of half-lives, you can just divide the original number by 2 over and over again, rather than using the exponential relationship. For example, if ten half-lives have passed, we divide N by 2 ten times. This reduces it to $N/1024$. For an arbitrary time, not just a multiple of the half-life, the exponential relationship must be used.

Radioactive dating is a clever use of naturally occurring radioactivity. Its most famous application is **carbon-14 dating**. Carbon-14 has a half-life of 5730 years and is produced by a nuclear reaction induced when solar neutrons strike ^{14}N in the atmosphere. Radioactive carbon has the same chemistry as stable carbon, and so it mixes into the ecosphere, where it is consumed and becomes part of every living organism. Carbon-14 has

Figure 29.20 Part of the Shroud of Turin, which shows a remarkable negative image likeness of Jesus complete with evidence of crucifixion wounds. The shroud first surfaced in the 14th century and was only recently carbon-14 dated. It has not been determined how the image was placed on the material.

an abundance of 1.3 parts per trillion of normal carbon. Thus, if you know the number of carbon nuclei in an object (perhaps determined by mass and Avogadro's number), you multiply that number by 1.3×10^{-12} to find the number of ^{14}C nuclei in the object. When an organism dies, carbon exchange with the environment ceases, and ^{14}C is not replenished as it decays. By comparing the abundance of ^{14}C in an artifact, such as mummy wrappings, with the normal abundance in living tissue, it is possible to determine the artifact's age (or time since death). Carbon-14 dating can be used for biological tissue as old as 50 or 60 thousand years, but is most accurate for younger samples, since the abundance of ^{14}C nuclei in them is greater. Very old biological materials contain no ^{14}C at all. There are instances in which the date of an artifact can be determined by other means, such as historical knowledge or tree ring counting. These cross-references have confirmed the validity of carbon-14 dating and permitted us to calibrate the technique as well. Carbon-14 dating revolutionized parts of archaeology and is of such importance that it earned the 1960 Nobel prize in chemistry for its developer, the American chemist Willard Libby (1908–1980).

One of the most famous cases of carbon-14 dating involves the Shroud of Turin, a long piece of fabric purported to be the burial shroud of Jesus. (See Figure 29.20.) This relic was first displayed in Turin in 1354 and was denounced as a fraud at that time by a French bishop. Its remarkable negative image of an apparently crucified body resembles the then-accepted image of Jesus, and so the shroud was never disregarded completely and remained controversial over the centuries. Carbon-14 dating was not performed on the shroud until 1988, when the process had been refined to the point where only a small amount of material needed to be destroyed. Samples were tested at three independent laboratories, each being given four pieces of cloth, with only one unidentified piece from the shroud, to avoid prejudice. All three laboratories found samples of the shroud contain 92% of the ^{14}C found in living tissue, allowing the shroud to be dated.

EXAMPLE 29.4 HOW OLD IS THE SHROUD OF TURIN?

Calculate the age of the Shroud of Turin given that the amount of ^{14}C found in it is 92% of that in living tissue.

Strategy Knowing that 92% of the ^{14}C remains means that $N/N_0 = 0.92$. Thus Equation 29.10 can be used to find λt. We also know that the half-life of ^{14}C is 5730 y, and so once λt is known, we can use Equation 29.11 to find λ and then find t as requested. We are making the standard assumption that the decrease in ^{14}C is due to nuclear decay.

Solution Solving Equation 29.10 for N/N_0 gives

$$\frac{N}{N_0} = e^{-\lambda t}$$

Thus,

$$0.92 = e^{-\lambda t}$$

Taking the natural logarithm of both sides of the equation yields

$$\ln 0.92 = -\lambda t$$

so that

$$-0.0834 = -\lambda t$$

Rearranging to isolate t gives

$$t = \frac{0.0834}{\lambda}$$

Now, Equation 29.11 can be used to find λ for ^{14}C. Solving for λ and substituting the known half-life gives

$$\lambda = \frac{0.693}{t_{1/2}} = \frac{0.693}{5730 \text{ y}}$$

We enter this value into the previous equation to find t:

$$t = \frac{0.0834}{0.693/5730 \text{ y}} = 690 \text{ y}$$

Discussion This dates the material in the shroud to $1988 - 690 = $ A.D. 1300. Our calculation is only good to two digits, so that the year is rounded to 1300. The values obtained at the three independent laboratories gave a weighted average date of A.D. 1320 ± 60. The uncertainty is typical of carbon-14 dating and is due to the small amount of ^{14}C in living tissue, the amount of material available, and experimental uncertainties (reduced by having three independent measurements). It is meaningful that the date of the shroud is consistent with the first record of its existence and inconsistent with the lifetime of Jesus.

There are other forms of radioactive dating. Rocks, for example, can sometimes be dated based on the decay of ^{238}U. The decay series for ^{238}U ends with ^{206}Pb, so that the ratio of these nuclides in a rock is an indication of how long it has been since the rock solidified. The original composition of the rock, such as the absence of lead, must be known with some confidence. But as with carbon-14 dating, the technique can be verified by a consistent body of knowledge. Since ^{238}U has a half-life of 4.5×10^9 y, it is useful only for dating very old materials, showing, for example, that the oldest rocks on earth solidified about 3.5×10^9 y ago.

Activity, the Rate of Decay

What do we mean when we say a source is highly radioactive? Generally, this means the number of decays per unit time is very high. We define **activity** R to be the **rate of decay** expressed in decays per unit time. In equation form, this is

$$R = \frac{\Delta N}{\Delta t} \qquad (29.12)$$

where ΔN is the number of decays that occur in time Δt. The SI unit for activity is one decay per second and is given the name **becquerel** (Bq) in honor of the discoverer of radioactivity. That is,

$$1 \ \text{Bq} = 1 \ \text{decay/s} \qquad (29.13)$$

Activity R is often expressed in other units, such as decays per minute or decays per year. One of the most common units for activity is the **curie** (Ci), defined to be the activity of 1 g of ^{226}Ra, in honor of Marie Curie's work with radium. The official definition of the curie is

$$1 \ \text{Ci} = 3.70 \times 10^{10} \ \text{Bq} \qquad (29.14)$$

or 3.70×10^{10} decays per second. A curie is a large unit of activity, while a becquerel is a relatively small unit. In the United States, most radioactive sources, such as those used in medical diagnostics or in physics laboratories, are labeled in curies, typically being a few millicuries or less in activity. In countries that adhere more to SI units, labels often give activity in megabecquerel (MBq).

Intuitively, you expect the activity of a source to depend on two things: the amount of the radioactive substance present, and its half-life. The more nuclei you have, the more will decay per unit of time. The shorter the half-life, the more decays per unit time, for a given number of nuclei. So activity R should be proportional to the number of nuclei N and inversely proportional to their half-life $t_{1/2}$. In fact, your intuition is correct. It can be shown that the activity of a source is

$$R = \frac{0.693N}{t_{1/2}} \qquad (29.15)$$

where N is the number of radioactive nuclei present, having half-life $t_{1/2}$. This relationship is useful in a variety of calculations, as the next two examples illustrate.

EXAMPLE 29.5   HOW GREAT IS THE ^{14}C ACTIVITY IN LIVING TISSUE?

Calculate the activity due to ^{14}C in 1.00 kg of carbon found in a living organism. Express the activity in units of Bq and Ci.

Strategy To find the activity R using Equation 29.15, we must know N and $t_{1/2}$. The half-life of ^{14}C can be found in Appendix B, and was stated above as 5730 y. To find N, we first find the number of ^{12}C nuclei in 1.00 kg of carbon using the concept of a mole. As indicated above, we then multiply by 1.3×10^{-12} to get the number of ^{14}C nuclei in a living organism.

Solution One mole of carbon has a mass of 12.0 g, since it is nearly pure ^{12}C. (A mole has a mass equal to A found in the periodic table.) Thus the number of carbon nuclei in

a kilogram is

$$N(^{12}\text{C}) = \frac{6.02 \times 10^{23}}{12.0 \ \text{g}} \ 1000 \ \text{g} = 5.02 \times 10^{25}$$

So the number of ^{14}C nuclei is

$$N(^{14}\text{C}) = (5.02 \times 10^{25})(1.3 \times 10^{-12}) = 6.52 \times 10^{13}$$

Now the activity R is found using Equation 29.15:

$$R = \frac{0.693N}{t_{1/2}}$$

Entering known values gives

$$R = \frac{0.693(6.52 \times 10^{13})}{5730 \ \text{y}} = 7.89 \times 10^9 \text{/y}$$

(continued)

(continued)

or 7.89×10^9 decays per year. To convert this to Bq, we simply convert years to seconds. Thus,

$$R = 7.89 \times 10^9/\text{y} \cdot \frac{1.00 \text{ y}}{3.16 \times 10^7 \text{ s}} = 250 \text{ Bq}$$

or 250 decays per second. To express R in curies, we use the definition of the curie,

$$R = \frac{250 \text{ Bq}}{3.7 \times 10^{10} \text{ Bq/Ci}} = 6.75 \times 10^{-9} \text{ Ci}$$

Thus,

$$R = 6.75 \text{ nCi}$$

Discussion Our own bodies contain kilograms of carbon, and it is intriguing to think there are hundreds of ^{14}C

decays per second taking place in us. Carbon-14 and other naturally occurring radioactive substances in our bodies contribute to the background radiation we receive. In fact, sleeping with another person nightly significantly increases the amount of background radiation you receive. (Nevertheless, married men live several years longer on average than single men.) The small number of decays per second found for a kilogram of carbon in this example gives you some idea of how difficult it is to detect ^{14}C in a small sample of material. If there are 250 decays per second in a kilogram, then there are 0.25 decays per second in a gram of carbon in living tissue. To observe this, you must be able to distinguish decays from other forms of radiation, in order to reduce background noise. This becomes more difficult with old tissue, since it contains less ^{14}C, and for samples more than 50 thousand years old, it is impossible.

Human & Medical Application

Human-made (or artificial) radioactivity has been produced for decades and has many uses. Some of these include medical therapy for cancer, medical imaging and diagnostics, and food preservation by irradiation. Many applications as well as the biological effects of radiation are explored in Chapter 30, but it is clear that radiation is hazardous. A number of tragic examples of this exist, one of the most disastrous being the meltdown and fire at the Chernobyl reactor complex in the Ukraine. (See Figure 29.21.) Several radioactive isotopes were released in huge quantities, contaminating many thousands of square kilometers and directly affecting hundreds of thousands of people. The most significant releases were of ^{131}I, ^{90}Sr, ^{137}Cs, ^{239}Pu, ^{238}U, and ^{235}U.

EXAMPLE 29.6 WHAT MASS OF ^{137}Cs ESCAPED?

It is estimated that the Chernobyl disaster released 6.0 MCi of ^{137}Cs into the environment. Calculate the mass of ^{137}Cs released.

Strategy We can calculate the mass released using Avogadro's number and the concept of a mole if we can first find the number of nuclei N released. Since the activity R is given, and the half-life of ^{137}Cs is found in Appendix B to be 30.2 y, we can use Equation 29.15 to find N.

Solution Solving Equation 29.15 for N gives

$$N = \frac{Rt_{1/2}}{0.693}$$

Entering the given values yields

$$N = \frac{(6.0 \text{ MCi})(30.2 \text{ y})}{0.693}$$

Converting to units that cancel properly gives

$$N = \left[(6.0 \times 10^6 \text{ Ci})(3.7 \times 10^{10} \text{ Bq/Ci})(30.2 \text{ y}) \right.$$
$$\left. \times (3.16 \times 10^7 \text{ s/y}) \right] \div 0.693$$
$$= 3.1 \times 10^{26}$$

One mole of a nuclide ^AX has a mass of A grams, so that one mole of ^{137}Cs has a mass of 137 g. A mole has 6.02×10^{23} nuclei. Thus the mass of ^{137}Cs released was

$$m = \frac{137 \text{ g}}{6.02 \times 10^{23}} \cdot 3.1 \times 10^{26} = 70 \times 10^3 \text{ g}$$
$$= 70 \text{ kg}$$

Discussion While 70 kg of material is not a very large mass, it is extremely radioactive, since it only has a 30 year half-life. Six megacuries (6.0 MCi) is an extraordinary amount of activity but is only a fraction of what is produced in nuclear reactors. Similar amounts of the other isotopes were also released at Chernobyl. Although the chances of such a disaster may have seemed small, the consequences are extreme and required greater caution than was used. More will be said about safe reactor design in the next chapter, but it should be noted that Western reactors have a fundamentally safer design.

Figure 29.21 The Chernobyl reactor just after its meltdown and fire and after being encapsulated in a concrete sarcophagus. More than 100 people died soon after, and there will be thousands of deaths from radiation-induced cancer in the future. While the accident was due to a series of human errors, the cleanup efforts were heroic. Most of the immediate fatalities were firefighters and reactor personnel.

Activity R decreases in time, going to half its original value in one half-life, then to one-fourth its original value in the next half-life, and so on. Since $R = 0.693N/t_{1/2}$, the activity decreases as the number of nuclei decreases. The equation for R as a function of time is found by combining Equations 29.10 and 29.15, yielding

$$R = R_0 e^{-\lambda t} \qquad \textbf{(29.16)}$$

where R_0 is the activity at $t = 0$. For example, if a source originally has a 1.00 mCi activity, it declines to 0.500 mCi in one half-life, to 0.250 mCi in two half-lives, to 0.125 mCi in three half-lives, and so on. For times other than whole half-lives, Equation 29.16 must be used to find R.

29.6 BINDING ENERGY

The more tightly bound a system is, the stronger the forces that hold it together and the greater the energy required to pull it apart. We can therefore learn about nuclear forces by examining how tightly bound nuclei are. We define the **binding energy** (BE) **of a nucleus** to be the *energy required to completely disassemble it into separate protons and neutrons.* We can determine the BE of a nucleus from its rest mass. The two are connected through Einstein's famous relationship $E = (\Delta m)c^2$. A bound system has a *smaller* mass than its separate constituents; the more tightly bound, the smaller the mass.

Imagine pulling a nuclide apart as illustrated in Figure 29.22. Work done to overcome the nuclear forces holding the nucleus together puts energy into the system. By definition, the energy input equals the binding energy BE. The pieces are at rest when separated, and so the energy put into them increases their total rest mass compared with what it was when they were glued together as a nucleus. That mass increase is, thus, $\Delta m = \text{BE}/c^2$. When separated and at rest, the protons and neutrons have their normal rest masses, and so this implies the nucleus they came from had a mass smaller than the sum of its constituents. A nuclide ^{A}X has Z protons and N neutrons, so that the difference in mass is

$$\Delta m = (Zm_p + Nm_n) - m_{tot}$$

Thus,

$$\text{BE} = (\Delta m)c^2 = [(Zm_p + Nm_n) - m_{tot}]c^2$$

where m_{tot} is the mass of the nuclide ^{A}X, m_p is the mass of a proton, and m_n is the mass of a neutron. Traditionally, we deal with the masses of neutral atoms. To get atomic masses

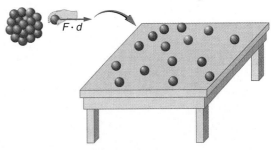

Figure 29.22 Work done to pull a nucleus apart into its constituent protons and neutrons increases the mass of the system. The work to disassemble the nucleus equals its binding energy BE. A bound system has less mass than the sum of its parts, especially noticeable in nuclei where forces and energies are very large.

THINGS GREAT AND SMALL

Nuclear Decay Helps Explain Earth's Hot Interior

A puzzle created by radioactive dating of rocks is resolved by radioactive heating of the earth's interior. This intriguing story is another example of how small-scale physics can explain large-scale phenomena.

Radioactive dating plays a role in determining the approximate age of the earth. As noted in Section 29.5, the oldest rocks on earth solidified about 3.5×10^9 y ago—a number determined by uranium-238 dating. These rocks could only have solidified once the surface of the earth had cooled sufficiently. The temperature of the earth at formation can be estimated based on gravitational potential energy of the assemblage of pieces being converted to thermal energy. Using heat transfer concepts discussed in Chapter 13, it is then possible to calculate how long it would take for the surface to cool to rock formation temperatures. The result is about 10^9 y. The first rocks formed have been solid for 3.5×10^9 y, so that the age of the earth is approximately 4.5×10^9 y. There is a large body of other types of evidence (both earth-bound and solar system characteristics are used) that supports this age. The puzzle is that, given its age and initial temperature, the center of the earth should be much cooler than it is today. (See Figure 29.A.)

We know from seismic waves produced by earthquakes that parts of the interior of the earth are liquid. Shear or transverse waves cannot travel through a liquid and are not transmitted through the earth's core. Yet compression or longitudinal waves can pass through a liquid and do go through the core. From this information, the temperature of the interior can be estimated. As noted, the interior should have cooled more from its initial temperature in the 4.5×10^9 y since its formation. In fact, it should have taken no more than about 10^9 y to cool to its present temperature. What is keeping it hot? The answer seems to be radioactive decay of primordial elements that were part of the material that formed the earth. (See the blowup in Figure 29.A.)

Nuclides such as ^{238}U and ^{40}K have half-lives similar to or longer than the age of the earth, and their decay still contributes energy to the interior. Some of the primordial radioactive nuclides have unstable decay products that also produce energy—^{238}U has a long decay chain of these. Furthermore, there were more of these early in the life of the earth, and thus the activity and energy contributed were greater then (perhaps by an order of magnitude). The amount of power created by these decays per cubic meter is very small. But since a huge volume of material lies deep below the surface, this relatively small amount of energy cannot quickly escape. That produced near the surface has much less distance to go to escape and has a negligible effect on surface temperatures.

A final effect of this trapped radiation merits mention. Alpha decay produces helium nuclei, which form helium atoms when they are stopped and capture electrons. Most of the helium on earth is obtained from wells and is thought to have been produced in this manner. Any helium in the atmosphere will escape in geologically short times because of its high thermal velocity.

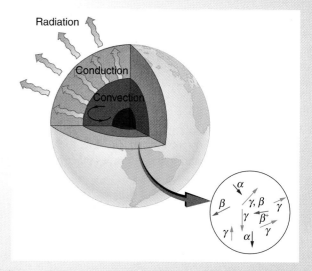

Figure 29.A The center of the earth cools by well-known heat transfer methods. Convection in the liquid regions and conduction move heat to the surface, where it radiates into cold dark space. Given the age of the earth and its initial temperature, it should have cooled to a lower temperature by now. The blowup shows that nuclear decay releases energy in the earth's interior. This energy has slowed the cooling process and is responsible for the interior still being molten.

into the last equation, we first add Z electrons to m_{tot}, which gives $m(^AX)$, the atomic mass of the nuclide. We then add Z electrons to the Z protons, which gives $Zm(^1H)$, or Z times the mass of a hydrogen atom.* Thus the binding energy of a nuclide ^AX is

$$\text{BE} = \{[Zm(^1H) + Nm_n] - m(^AX)\}c^2 \qquad (29.17)$$

The atomic masses can be found in Appendix A, most conveniently expressed in unified atomic mass units u ($1 \text{ u} = 931.5 \text{ MeV}/c^2$). BE is, thus, calculated from known atomic masses.

---

*Since m_{tot} is subtracted, this has the effect of adding and subtracting Z electrons from the equation, and the relationship is still valid.

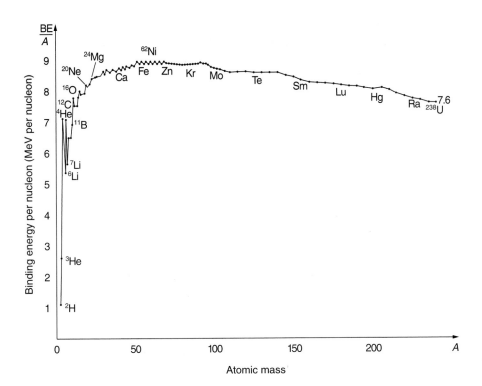

Figure 29.23 A graph of average binding energy per nucleon, BE/A, for stable nuclei. The most tightly bound nuclei are those with A near 60, where the attractive nuclear force has its greatest effect. At higher As, the Coulomb repulsion progressively reduces the binding energy per nucleon, because the nuclear force is short ranged. The spikes on the curve are very tightly bound nuclides and indicate shell closures.

What patterns and insights are gained from an examination of the binding energy of various nuclides? First, we find that BE is approximately proportional to the number of nucleons A in any nucleus. About twice as much energy is needed to pull apart a nucleus like ^{24}Mg compared with pulling apart ^{12}C, for example. To help us look at other effects, we divide BE by A* and consider the **binding energy per nucleon**, BE/A. The graph of BE/A in Figure 29.23 reveals some very interesting aspects of nuclei. We see that the binding energy per nucleon averages about 8 MeV, but is lower for both the lightest and heaviest nuclei. This overall trend, with nuclei having A about 60 having the greatest BE/A and, thus, being the most tightly bound, is due to the combined characteristics of the attractive nuclear forces and the repulsive Coulomb force. It is especially important to note two things—the strong nuclear force is about 100 times stronger than the Coulomb force, *and* the nuclear forces are short ranged. So for low-mass nuclei, the nuclear attraction dominates and each added nucleon forms bonds with all others, causing progressively heavier nuclei to have progressively greater BE/A. This continues up to A ≈ 60. Beyond an A of 60, new nucleons added to a nucleus will be too far from some others to feel their nuclear attraction. Added protons, however, feel the repulsion of all other protons, since the Coulomb force is long ranged. Coulomb repulsion grows for progressively heavier nuclei, but nuclear attraction remains about the same, and so BE/A becomes smaller. This is why stable nuclei heavier than A ≈ 40 have more neutrons than protons. Coulomb repulsion is reduced by having more neutrons to keep the protons farther apart. (See Figure 29.24.)

There are some noticeable spikes on the BE/A graph, which represent particularly tightly bound nuclei. These reveal further details of nuclear forces, such as confirming that closed-shell nuclei (those with magic numbers of protons or neutrons or both) are more tightly bound. The spikes also indicate that some nuclei with even numbers for Z and N, and with Z = N, are exceptionally tightly bound. This actually explains some of the cosmic abundances of the elements. The most common elements in the universe, as determined by observation of atomic spectra from outer space, are hydrogen, followed by ^{4}He, with much smaller amounts of ^{12}C and others. The most common are the most

Nucleons inside range feel nuclear force directly

Range of nuclear force

Figure 29.24 The nuclear force is attractive and stronger than the Coulomb force, but it is short ranged. In low-mass nuclei, each nucleon feels the nuclear attraction of all others. In larger nuclei, the range of the nuclear force, shown for a single nucleon, is smaller than the size of the nucleus, but the Coulomb repulsion from all protons reaches all others. If the nucleus is large enough, the Coulomb repulsion can add to overcome the nuclear attraction.

---

*If a quantity is proportional to A, then dividing it by A algebraically cancels this proportionality. Thus BE/A is related to other, more important, factors.

tightly bound, and these seem to have been synthesized in stars and supernova explosions, as will be explored in subsequent chapters. It is also no accident that one of the most tightly bound light nuclei is ^{4}He, emitted in α decay.

EXAMPLE 29.7 WHAT IS BE/A FOR AN α PARTICLE?

Calculate the binding energy per nucleon of ^{4}He, the α particle.

Strategy To find BE/A, we first find BE using Equation 29.17 and then divide by A. This is straightforward once we have looked up the appropriate atomic masses in Appendix A.

Solution The binding energy for a nucleus is given by Equation 29.17 to be

$$BE = \{[Zm(^1H) + Nm_n] - m(^AX)\}c^2$$

For ^{4}He, we have $Z = N = 2$; thus,

$$BE = \{[2m(^1H) + 2m_n] - m(^4He)\}c^2$$

Appendix A gives these masses as $m(^4He) = 4.002602$ u, $m(^1H) = 1.007825$ u, and $m_n = 1.008665$ u. Thus,

$$BE = (0.030378 \text{ u})c^2$$

Noting that 1 u = 931.5 MeV/c^2, we find

$$BE = (0.030378)(931.5 \text{ MeV}/c^2)c^2 = 28.3 \text{ MeV}$$

Since $A = 4$, we see that BE/A is this number divided by 4, or

$$\boxed{BE/A = 7.07 \text{ MeV/nucleon}}$$

Discussion This is a large binding energy per nucleon compared with those for other low-mass nuclei, which have BE/$A \approx 3$ MeV/nucleon. This indicates that ^{4}He is tightly bound compared with its neighbors on the chart of the nuclides. You can see the spike representing this value of BE/A for ^{4}He on the graph in Figure 29.23. This is why ^{4}He is stable. Since ^{4}He is tightly bound, it has less mass than other $A = 4$ nuclei and, therefore, cannot spontaneously decay into them. The large binding energy also helps to explain why some nuclei α decay. Smaller mass in the decay products can mean energy release, and such decays can be spontaneous. Furthermore, it can happen that two protons and two neutrons in a nucleus can randomly find themselves together, experience the exceptionally large nuclear force that binds this combination, and act as a ^{4}He within the nucleus, at least for a while. In some cases the ^{4}He escapes, and α decay has then taken place.

There is more to be learned from nuclear binding energies. The general trend in BE/A is fundamental to energy production in stars, and to fusion and fission energy sources on earth, for example. This is one of the applications of nuclear physics covered in Chapter 30. The abundance of elements on earth, in stars, and in the universe as a whole is related to the binding energy of nuclei and has implications for the continued expansion of the universe. These and other topics will be explored in Chapters 31 and 32.

PROBLEM-SOLVING STRATEGIES

FOR REACTION AND BINDING ENERGIES AND ACTIVITY CALCULATIONS IN NUCLEAR PHYSICS

Step 1. Identify exactly what needs to be determined in the problem (identify the unknowns). This will allow you to decide whether the energy of a decay or nuclear reaction is involved, for example, or whether the problem is primarily concerned with activity (rate of decay).

Step 2. Make a list of what is given or can be inferred from the problem as stated (identify the knowns).

Step 3. For reaction and binding energy problems, we use atomic rather than nuclear masses. Since the masses of neutral atoms are used, you must count the number of electrons involved. If these do not balance (such as in β^+ decay), then an energy adjustment of 0.511 MeV per electron must be made. Note also that atomic masses may not be given in a problem; they can be found in tables.

Step 4. For problems involving activity, the relationship of activity to half-life and the number of nuclei given in Equation 29.15 can be very useful. Because the number of nuclei is involved, you will also need to be familiar with moles and Avogadro's number.

Step 5. Perform the desired calculation; keep careful track of plus and minus signs as well as powers of 10.

Step 6. *Check the answer to see if it is reasonable: Does it make sense?* Compare your results with worked examples and other information in the text. (Heeding the advice in step 5 will also help you be certain of your result.) You must understand the problem conceptually to be able to determine if the numerical result is reasonable.

29.7* TUNNELING

Protons and neutrons are *bound* inside nuclei, meaning that energy must be supplied to break them away. The situation is analogous to a marble in a bowl that can roll around but lacks the energy to get over the rim. (See Figure 29.25.) If the marble could get over the rim, it would gain kinetic energy by rolling down the outside. But classically, if the marble does not have enough kinetic energy to get over the rim, it remains forever trapped in its well.

In a nucleus, the attractive nuclear potential is analogous to the bowl at the top of a volcano. (See Figure 29.26.) Protons and neutrons have kinetic energy, but it is about 8 MeV less than needed to get out. That is, they are bound by an average of 8 MeV per nucleon. For a nucleus, the hill outside the bowl is analogous to the repulsive Coulomb potential, such as for an α particle outside a positive nucleus. In α decay, two protons and two neutrons spontaneously break away as a ^{4}He unit. Yet the protons and neutrons do not have enough kinetic energy to get over the rim. So how does the α particle get out?

The answer was supplied in 1928 by the Russian physicist George Gamow (1904–1968), who later became an American, and independently by two American physicists. The α particle tunnels through a region of space it is forbidden to be in, and it comes out of the side of the nucleus. Like an electron making a transition between orbits around an atom, it gets from one point to another without ever having been in between. Figure 29.27 indicates how this works. A particle trapped in some region of space has a wavelength that must fit in such a way as to have constructive interference. (This is the criterion for electrons in their orbits about atoms, as discussed in Section 28.4.) At the boundary, the wave function (related to the probability cloud) of the particle does not become zero but decreases exponentially inside the barrier. The particle cannot be inside the barrier, because it does not have enough kinetic energy to get there. So we can never find the particle inside the barrier and, at first glance, that would seem to end the story. But at the outside of the barrier, the wave function again matches smoothly with what could be on the outside. There is not much left of the wave function, since it is decreasing exponentially inside the barrier, but it is not quite zero at the outside. The probability of finding a particle is related to the square of its wave function, and so there is a small probability of finding the particle outside the barrier. This process is called **barrier penetration** or **quantum mechanical tunneling**. This concept was developed in theory by J. Robert Oppenheimer and used by Gamow and others to describe α decay.

Good ideas explain more than one thing. In addition to qualitatively explaining how the four nucleons in an α particle can get out of the nucleus, the detailed theory also explains quantitatively the half-life of various nuclei that α decay. This description is what Gamow and other devised, and it works for α decay half-lives that vary by 17 orders of magnitude! Experiments have shown that the more energetic the α decay of a particular nuclide, the shorter its half-life. Tunneling explains this in the following manner. For the

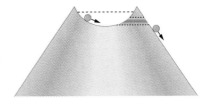

Figure 29.25 The marble in this bowl has enough kinetic energy to get to the altitude of the dashed line, but not enough to get over the rim, so that it is trapped forever. If it could find a tunnel through the barrier, it would escape, roll downhill, and gain kinetic energy.

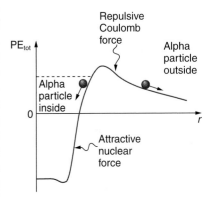

Figure 29.26 Nucleons experience a total potential as shown in this simplified graph. They are bound or trapped by the attractive nuclear force. An α particle outside the range of the nuclear force feels the repulsive Coulomb force. The α does not have enough kinetic energy to get over the rim, yet it does manage to get out by quantum mechanical tunneling.

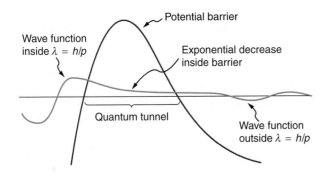

Figure 29.27 The wave function representing a quantum mechanical particle must vary smoothly. At the inside of the barrier, it does not abruptly become zero; rather, it decreases exponentially. At the outside of the barrier, the wave function is small but finite, and there it smoothly becomes sinusoidal. There is a small probability of finding the particle outside the barrier—it can *tunnel* through.

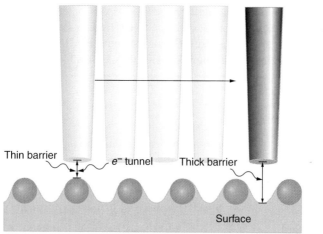

Figure 29.28 A scanning tunneling electron microscope can detect extremely small variations in dimensions, such as individual atoms. Electrons tunnel quantum mechanically between the probe and the sample. The probability of tunneling is extremely sensitive to barrier thickness, so that the electron current is a sensitive indication of surface features

decay to be more energetic, the nucleons must have more energy in the nucleus and rise a little closer to the rim. The barrier is therefore not as thick for more energetic decay, and the exponential decrease of the wave function inside the barrier is not as great. Thus the probability of finding the particle outside the barrier is greater, and the half-life is shorter.

Tunneling as an effect also occurs in quantum mechanical systems other than nuclei. Electrons trapped in solids can tunnel from one object to another if the barrier between the materials is thin enough. The process is the same in principle as described above for α decay. It is far more likely for a thin barrier than a thick one. Scanning tunneling electron microscopes function on this principle. The current of electrons between the probe and the sample tunnels through a barrier and is very sensitive to its thickness, allowing detection of individual atoms as shown in Figure 29.28. Tunneling also has applications in microelectronics across Josephson junctions that can act as very fast and small switches. It is one of the quirks of Nobel awards that Brian Josephson received a physics prize for his proposal of an application of tunneling while the original discoverers of the overall effect were not so honored.

SUMMARY

Some nuclei are **radioactive**—they spontaneously **decay**, destroying part of their mass and emitting energetic rays, a process called nuclear **radioactivity**. Nuclear radiation, like x rays, is **ionizing radiation**, because energy sufficient to ionize matter is emitted in each decay. The range (or distance traveled in a material) of ionizing radiation is directly related to the charge of the emitted particle and its speed, with greater-charge and lower-speed particles having the shortest ranges. Radiation detectors are based directly or indirectly upon the ionization created by radiation, as are the effects of radiation on living and inert materials.

Two particles, both called **nucleons**, are found inside nuclei. The two types of nucleons are **protons** and **neutrons**; they are very similar, except the proton is positively charged while the neutron is neutral. Some of their characteristics are given in Table 29.1 and compared with those of the electron. A mass unit convenient to atomic and nuclear processes is the

unified atomic mass unit (u), defined to be

$$1 \text{ u} = 1.6605 \times 10^{-27} \text{ kg} = 931.5 \text{ MeV}/c^2 \quad \textbf{(29.1)}$$

A **nuclide** is a specific combination of protons and neutrons, denoted by

$$_Z^A X_N \text{ or simply } ^A X$$

Z is the **number of protons** or **atomic number**, X is the **symbol for the element**, N is the **number of neutrons**, and A is the **atomic mass** or the **total number of protons and neutrons**,

$$A = N + Z \quad \textbf{(29.2)}$$

Nuclides having the same Z, but different Ns, are **isotopes** of the same element.

The **radius of a nucleus**, r, is approximately

$$r = r_0 A^{1/3} \quad \textbf{(29.3)}$$

where $r_0 = 1.2$ fm. Nuclear volumes are proportional to A. There are **two nuclear forces**, the **weak** and the **strong**.

Systematics in nuclear stability seen on the chart of the nuclides indicate that there are shell closures in nuclei for values of Z and N equal to the **magic numbers**, which correspond to highly stable nuclei.

When a **parent** nucleus decays, it produces a **daughter** nucleus following rules and conservation laws, some of which were revealed by radioactivity. There are three major types of nuclear decay, called **alpha** (α), **beta** (β), and **gamma** (γ). The α **decay equation** is

$$\,^A_Z X_N \rightarrow \,^{A-4}_{Z-2} Y_{N-2} + \,^4_2 He_2 \qquad (\alpha \text{ decay}) \qquad \textbf{(29.4)}$$

Nuclear decay (and all other spontaneous processes) emits energy E related to mass destroyed Δm by

$$E = (\Delta m)c^2 \qquad \textbf{(29.5)}$$

There are three forms of beta decay. The $\boldsymbol{\beta^-}$ **decay equation** is

$$\,^A_Z X_N \rightarrow \,^A_{Z+1} Y_{N-1} + \beta^- + \bar{\nu}_e \qquad (\beta^- \text{ decay}) \qquad \textbf{(29.6)}$$

The $\boldsymbol{\beta^+}$ **decay equation** is

$$\,^A_Z X_N \rightarrow \,^A_{Z-1} Y_{N+1} + \beta^+ + \nu_e \qquad (\beta^+ \text{ decay}) \qquad \textbf{(29.7)}$$

The **electron capture** equation is

$$\,^A_Z X_N + e^- \rightarrow \,^A_{Z-1} Y_{N+1} + \nu_e \qquad (\text{electron capture, or EC}) \qquad \textbf{(29.8)}$$

$\boldsymbol{\beta^-}$ is an electron, $\boldsymbol{\beta^+}$ is an **antielectron** or **positron**, ν_e represents an **electron's neutrino**, and $\bar{\nu}_e$ is an **electron's neutrino**. In addition to all previously known conservation laws, two new ones arise—**conservation of electron family number** and **conservation of the total number of nucleons**. The γ **decay equation** is

$$\,^A_Z X_N^* \rightarrow \,^A_Z X_N + \gamma_1 + \gamma_2 + \cdots \qquad (\gamma \text{ decay}) \qquad \textbf{(29.9)}$$

where a γ is a high-frequency photon originating in a nucleus.

Half-life $t_{1/2}$ is the **time in which there is a 50% chance that a nucleus will decay**. The number of nuclei N as a function of time is

$$N = N_0 e^{-\lambda t} \qquad \textbf{(29.10)}$$

where N_0 is the number present at $t = 0$, and λ is the **decay constant**, related to the half-life by

$$\lambda = \frac{0.693}{t_{1/2}} \qquad \textbf{(29.11)}$$

One of the applications of radioactive decay is **radioactive dating**, in which the age of a material is judged by the amount of decay that has taken place. The **rate of decay** is called the **activity** R:

$$R = \frac{\Delta N}{\Delta t} \qquad \textbf{(29.12)}$$

The SI unit for R is the **becquerel** (Bq), defined by

$$1 \text{ Bq} = 1 \text{ decay/s} \qquad \textbf{(29.13)}$$

R is also expressed in terms of **curies** (Ci), where

$$1 \text{ Ci} = 3.70 \times 10^{10} \text{ Bq} \qquad \textbf{(29.14)}$$

The activity R of a source is related to N and $t_{1/2}$ by

$$R = \frac{0.693 N}{t_{1/2}} \qquad \textbf{(29.15)}$$

Since N has an exponential behavior as in Equation 29.10, the activity also has an exponential behavior, given by

$$R = R_0 e^{-\lambda t} \qquad \textbf{(29.16)}$$

where R_0 is the activity at $t = 0$.

The **binding energy** (BE) of a nucleus is the energy needed to separate it into individual protons and neutrons. In terms of atomic masses,

$$\text{BE} = \{[Zm(^1H) + Nm_n] - m(^A X)\}c^2 \qquad \textbf{(29.17)}$$

where $m(^1H)$ is the mass of a hydrogen atom, $m(^A X)$ is the atomic mass of the nuclide, and m_n is the mass of a neutron. Patterns in the **binding energy per nucleon**, BE/A, reveal details of the nuclear force.

Tunneling is a quantum mechanical process of barrier penetration. The concept was first applied to explain α decay, but tunneling is found to occur in other quantum mechanical systems.

CONCEPTUAL QUESTIONS

29.1 What characteristics of radioactivity show it to be nuclear in origin and not atomic?

29.2 What is the source of the energy emitted in radioactive decay? Identify two old conservation laws, and describe how they were modified to take such processes into account.

29.3 Consider Figure 29.2. If an electric field is substituted for the magnetic field with positive charge instead of the north pole and negative charge instead of the south pole, in which directions will the α, β, and γ rays bend?

29.4 Explain how an α particle can have a larger range in air than a β particle of the same energy has in lead.

29.5 Arrange the following according to their ability to act as radiation shields, with the best first and worst last. (a) A solid material with low density composed of low-mass atoms. (b) A gas composed of high-mass atoms. (c) A gas composed of low-mass atoms. (d) A solid with high density composed of high-mass atoms. Explain your ordering in terms of how radiation loses its energy in matter.

29.6 Suppose the range for 5.0 MeV α radiation is known to be 2.0 mm in a certain material. Does this mean that every 5.0 MeV α that strikes this material travels 2.0 mm, or is the range an average value with some statistical fluctuations in the distances traveled? Explain.

29.7 What is the difference between γ rays and characteristic x rays? Is either necessarily more energetic than the other? Which can be the most energetic?

29.8 Ionizing radiation interacts with matter by scattering from electrons and nuclei in the substance. Based on conservation of momentum and energy, explain why electrons tend to absorb more energy than nuclei in these interactions.

29.9 Is it possible for light emitted by a scintillator to be too low in frequency to be used in a photomultiplier tube? Explain.

29.10 Identify two similarities and two differences between protons and neutrons.

29.11 The weak and strong nuclear forces are basic to the structure of matter. Why don't we experience them directly?

29.12 Define and make clear distinctions between the terms neutron, nucleon, nucleus, nuclide, and neutrino.

29.13 What are isotopes? Why do different isotopes of the same element have similar chemistries?

29.14 Star Trek fans have often heard the term "antimatter drive." Describe how you could use a magnetic field to trap antimatter, such as produced by nuclear decay, and later combine it with matter to produce energy. Be specific about the type of antimatter, the need for vacuum storage, and the fraction of matter converted into energy.

29.15 What conservation law requires an electron's neutrino to be produced in electron capture? Note that the electron no longer exists after it is captured by the nucleus.

29.16 Gravitational collapse during a supernova forces electrons to combine with protons to form neutrons. (A neutron star may be formed in the process.) What type of neutrinos should be produced in this reaction? Why?

29.17 Neutrinos are experimentally determined to have a mass close to zero, but with an uncertainty that allows for the possibility of a small mass. Huge numbers of neutrinos are created in a supernova at the same time as massive amounts of light are first produced. When the 1987A supernova occurred in the Magellanic Cloud, some 100,000 light years away, neutrinos from the explosion were observed at about the same time as the light from the blast. How could the relative arrival times of neutrinos and light be used to place limits on the mass of neutrinos?

29.18 What do the three types of beta decay have in common that is distinctly different from alpha decay?

29.19 If a 3×10^9 year old rock originally contained some ^{238}U, which has a half-life of 4.5×10^9 years, we expect to find some ^{238}U remaining in it. Why are ^{226}Ra, ^{222}Rn, and ^{210}Po also found in such a rock, even though they have much shorter half-lives (1600 years, 3.8 days, and 138 days, respectively)?

29.20 If dead biological tissue exchanges carbon with the environment, perhaps by being immersed in water with CO_2 in it, carbon-14 dating is affected. Would carbon-14 dating produce an age that is too old or too young? Explain.

29.21 Does the number of radioactive nuclei in a sample decrease to *exactly* half its original value in one half-life? Explain in terms of the statistical nature of radioactive decay.

29.22 We claim that radioactivity depends on the nucleus and not the atom or its chemical state. Why, then, is one kilogram of uranium more radioactive than one kilogram of uranium hexafluoride?

29.23 Explain how a bound system can have less mass than its components. Why is this not observed classically, say for a building made of bricks?

29.24 Spontaneous radioactive decay occurs only when the decay products have less mass than the parent, and it tends to produce a daughter that is more stable than the parent. Explain how this is related to the fact that more tightly bound nuclei are more stable. (Consider the binding energy per nucleon.)

29.25 To obtain the most precise value of BE from Equation 29.17, we should take into account the binding energy of the electrons in the neutral atoms. Will doing this produce a larger or smaller value for BE? Why is this effect usually negligible?

29.26 How does the finite range of the nuclear force relate to the fact that BE/A is greatest for A near 60?

29.27 Why is the number of neutrons greater than the number of protons in stable nuclei having A greater than about 40, and why is this effect more pronounced for the heaviest nuclei?

29.28* When a nucleus α decays, does the α particle move continuously from inside the nucleus to outside? That is, does it travel each point along an imaginary line from inside to out? Explain.

29.29* A physics student caught breaking conservation laws is imprisoned. She leans against the cell wall hoping to tunnel out quantum mechanically. Explain why her chances are negligible. (This is so in any classical situation.)

PROBLEMS

Sections 29.1–29.3 Radiation, Detectors, and the Nucleus

29.1 An energy of 30.0 eV is required to ionize a molecule of the gas inside a Geiger tube, thereby producing an ion pair. Suppose a particle of ionizing radiation deposits 0.500 MeV of energy in this Geiger tube. What maximum number of ion pairs can it create?

29.2 A particle of ionizing radiation creates 4000 ion pairs in the gas inside a Geiger tube as it passes through. What minimum energy was deposited, if 30.0 eV is required to create each ion pair?

29.3 Verify that a 2.3×10^{17} kg mass of water at normal density would make a cube 61 km on a side, as claimed in Example 29.1. (This mass at nuclear density would make a cube 1.0 m on a side.)

29.4 Find the length of a side of a cube having a mass of 1.0 kg and the density of nuclear matter, taking this to be 2.3×10^{17} kg/m^3.

29.5 What is the radius of an α particle?

29.6 Find the radius of a ^{238}Pu nucleus. ^{238}Pu is a manufactured nuclide that is used as a power source on some space probes.

29.7 (a) Calculate the radius of ^{58}Ni, one of the most tightly bound stable nuclei. (b) What is the ratio of the radius of ^{58}Ni to that of ^{258}Ha, one of the largest ever made? Note that the radius of the largest nucleus is still much smaller than the size of an atom.

29.8 The unified atomic mass unit is defined to be 1 u = 1.6605×10^{-27} kg. Verify that this amount of mass converted to energy yields 931.5 MeV. Note that you must use four-digit or better values for c and $|q_e|$.

• **29.9** What is the ratio of the velocity of a β ray to that of an α ray, if they have the same nonrelativistic kinetic energy? (A similar result is obtained in Problem 29.69 when the energy of the β is relativistic.)

• **29.10** If a 1.50 cm thick piece of lead can absorb 90.0% of the γ rays from a radioactive source, how many centimeters of lead are needed to absorb all but 0.100% of the γs?

• **29.11** The detail observable using a probe is limited by its wavelength. Calculate the energy of a γ-ray photon that has a wavelength of 1.00×10^{-16} m, small enough to detect details about one-tenth the size of a nucleon. Note that a photon having this energy is difficult to produce and interacts poorly with the nucleus, limiting the practicability of this probe.

⁝ **29.12** (a) Show that if you assume the average nucleus is spherical with a radius $r = r_0 A^{1/3}$, and with a mass of A u, then its density is independent of A. (b) Calculate that density in u/fm^3 and kg/m^3, and compare your results with those found in Example 29.1 for ^{56}Fe.

⁝ **29.13** When a large star supernovas, a neutron star can be formed by the gravitational collapse of its core. Electrons and protons in the core are combined to form neutrons in the neutron star remnant, which has approximately the density given for nuclear matter in Example 29.1. (a) Calculate the radius of a neutron star having a mass 1.3 times that of our sun. (b) Thinking of the star as a gigantic nucleus, find its atomic mass A from its radius.

Section 29.4 Nuclear Decay and Conservation Laws

In the following eight problems, write the complete decay equation for the given nuclide in the complete $_Z^A X_N$ notation. Refer to the periodic table for values of Z.

29.14 β^- decay of ^{3}H (tritium), a rare isotope of hydrogen used in some digital watch displays, and manufactured primarily for use in hydrogen bombs.

29.15 β^- decay of ^{40}K, a naturally occurring rare isotope of potassium responsible for some of our exposure to background radiation.

29.16 β^+ decay of ^{50}Mn.

29.17 β^+ decay of ^{52}Fe.

29.18 Electron capture by ^{7}Be.

29.19 Electron capture by ^{106}In.

29.20 α decay of ^{210}Po, the isotope of polonium in the decay series of ^{238}U that was discovered by the Curies. A favorite isotope in physics labs, since it has a short half-life and decays to a stable nuclide.

29.21 α decay of ^{226}Ra, another isotope in the decay series of ^{238}U, first recognized as a new element by the Curies. Poses special problems because its daughter is a radioactive noble gas.

In the following four problems, identify the parent nuclide and write the complete decay equation in the $_Z^A X_N$ notation. Refer to the periodic table for values of Z.

29.22 β^- decay producing ^{137}Ba. The parent nuclide is a major waste product of reactors and has chemistry similar to potassium and sodium, resulting in its concentration in your cells if ingested.

29.23 β^- decay producing ^{90}Y. The parent nuclide is a major waste product of reactors and has chemistry similar to calcium, so that it is concentrated in bones if ingested. (^{90}Y is also radioactive.)

29.24 α decay producing ^{228}Ra. The parent nuclide is nearly 100% of the natural element and is found in gas lantern mantles. (^{228}Ra is also radioactive.)

29.25 α decay producing ^{208}Pb. The parent nuclide is in the decay series produced by ^{232}Th, the only naturally occurring isotope of thorium.

29.26 When an electron and positron annihilate, both their masses are destroyed, creating two equal energy photons to preserve momentum. (a) Confirm that the annihilation equation $e^+ + e^- \rightarrow \gamma + \gamma$ conserves charge, electron family number, and total number of nucleons. To do this, identify the values of each before and after the annihilation. (b) Find the energy of each γ ray, assuming the electron and positron are initially nearly at rest.

• **29.27** Confirm that charge, electron family number, and the total number of nucleons are all conserved by the rule for α decay given in Equation 29.4. To do this, identify the values of each before and after the decay.

• **29.28** Confirm that charge, electron family number, and the total number of nucleons are all conserved by the rule for β^- decay given in Equation 29.6. To do this, identify the values of each before and after the decay.

• **29.29** Confirm that charge, electron family number, and the total number of nucleons are all conserved by the rule for β^+ decay given in Equation 29.7. To do this, identify the values of each before and after the decay.

• **29.30** Confirm that charge, electron family number, and the total number of nucleons are all conserved by the rule for electron capture given in Equation 29.8. To do this, identify the values of each before and after the capture.

• **29.31** An exotic and rare decay mode has been observed in which ^{222}Ra emits a ^{14}C nucleus. (a) The decay equation is ^{222}Ra $\rightarrow {}^A$X $+ {}^{14}$C. Identify the nuclide ^{A}X. (b) Find the energy emitted in the decay. The mass of ^{222}Ra is 222.015353 u.

• **29.32** (a) Write the complete α decay equation for ^{226}Ra. (b) Find the energy released in the decay.

• **29.33** (a) Write the complete α decay equation for ^{249}Cf. (b) Find the energy released in the decay.

• **29.34** (a) Write the complete β^- decay equation for the neutron. (b) Find the energy released in the decay.

• **29.35** (a) Write the complete β^- decay equation for ^{90}Sr, a major waste product of nuclear reactors. (b) Find the energy released in the decay.

• **29.36** Calculate the energy released in the β^+ decay of ^{22}Na, the equation for which is given in the text. The masses of ^{22}Na and ^{22}Ne are 21.994434 and 21.991383 u, respectively.

• **29.37** (a) Write the complete β^+ decay equation for ^{11}C. (b) Calculate the energy released in the decay. The masses of ^{11}C and ^{11}B are 11.011433 and 11.009305 u, respectively.

⁝ **29.38** (a) Calculate the energy released in the α decay of ^{238}U. (b) What fraction of the mass of a single ^{238}U is destroyed in the decay? The mass of ^{234}Th is 234.043593 u. (This mass loss is that of one nucleus in a slowly decaying substance, an indication that the missing mass is very difficult to observe in a macroscopic sample.)

⁝ **29.39** (a) Write the complete reaction equation for electron capture by ^{7}Be. (b) Calculate the energy released.

⁝ **29.40** (a) Write the complete reaction equation for electron capture by ^{15}O. (b) Calculate the energy released.

Section 29.5 Half-Life and Activity

Data from the appendices and from the periodic table may be needed for these problems.

29.41 An old campfire is uncovered during an archaeological dig. Its charcoal is found to contain less than 1/1000 the normal amount of ^{14}C. Estimate the minimum age of the charcoal, noting that $2^{10} = 1024$.

29.42 A ^{60}Co source is labeled 4.00 mCi, but its present activity is found to be 1.85×10^7 Bq. (a) What is the present activity in mCi? (b) How long ago did it actually have a 4.00 mCi activity?

29.43 (a) Calculate the activity R in curies of 1.00 g of ^{226}Ra. (b) Discuss why your answer is not exactly 1.00 Ci, given that the curie was originally supposed to be exactly the activity of a gram of radium.

29.44 Show the activity of the ^{14}C in 1.00 g of ^{12}C found in living tissue is 0.250 Bq.

29.45 Mantles for gas lanterns are made of thorium, because it forms an oxide that can survive being made incandescent for long periods of time. Natural thorium is almost 100% ^{232}Th, with a half-life of 1.405×10^{10} y. If the average lantern mantle contains 300 mg of thorium, what is its activity?

29.46 Cow's milk produced near nuclear reactors can be tested for as little as 1.00 pCi of ^{131}I per liter, to check for possible reactor leakage. What mass of ^{131}I has this activity?

29.47 (a) Natural potassium contains ^{40}K, which has a half-life of 1.277×10^9 y. What mass of ^{40}K in a person would have a decay rate of 4140 Bq? (b) What is the fraction of ^{40}K in natural potassium, given the person has 140 g in his body? (These numbers are typical for a 70 kg adult.)

29.48 There is more than one isotope of natural uranium. If a researcher isolates 1.00 mg of the relatively scarce ^{235}U and finds this mass to have an activity of 80.0 Bq, what is its half-life in years?

29.49 ^{50}V has one of the longest known radioactive half-lives. In a difficult experiment, a researcher found that the activity of 1.00 kg of ^{50}V is 1.75 Bq. What is the half-life in years?

• 29.50 You can sometimes find deep red crystal vases in antique stores, called uranium glass because their color was produced by doping the glass with uranium. Look up the natural isotopes of uranium and their half-lives, and calculate the activity of such a vase assuming it has 2.00 g of uranium in it. Neglect the activity of any daughter nuclides.

• 29.51 A tree falls in a forest. How many years must pass before the ^{14}C activity in 1.00 g of the tree's carbon drops to 1.00 decay per hour?

• 29.52 What fraction of the ^{40}K that was on earth when it formed 4.5×10^9 years ago is left today?

• 29.53 A 5000 Ci ^{60}Co source used for cancer therapy is considered too weak to be useful when its activity falls to 3500 Ci. How long after its manufacture does this happen?

• 29.54 Natural uranium is 0.7200% ^{235}U and 99.27% ^{238}U. What were the percentages of ^{235}U and ^{238}U in natural uranium when the earth formed 4.5×10^9 years ago?

⁚ 29.55 (a) The ^{210}Po source used in a physics laboratory is labeled as having an activity of 1.00 μCi on the date it was prepared. A student measures the radioactivity of this source with a Geiger counter and observes 1500 counts per minute.

She notices that the source was prepared 120 days before her lab. What fraction of the decays is she observing with her apparatus? (b) Identify some of the reasons that only a fraction of the αs emitted are observed by the detector.

⁚ 29.56 Armor-piercing shells with depleted uranium cores are fired by aircraft at tanks. (The high density of the uranium makes them effective.) The uranium is called depleted because it has had its ^{235}U removed for reactor use and is nearly pure ^{238}U. Depleted uranium has been erroneously called nonradioactive. To demonstrate this is wrong: (a) Calculate the activity of 60.0 g of pure ^{238}U. (b) Calculate the activity of 60.0 g of natural uranium, neglecting the ^{234}U and all daughter nuclides.

Section 29.6 Binding Energy

In the following five problems, calculate BE/A, the binding energy per nucleon, for the particular nuclide and compare it with the approximate value obtained from the graph in Figure 29.23.

29.57 ^{2}H. This loosely bound isotope of hydrogen, called deuterium or heavy hydrogen, is stable but relatively rare—it is 0.015% of natural hydrogen. Note that deuterium has $Z = N$, which should tend to make it more tightly bound, but both are odd numbers.

29.58 ^{56}Fe. Among the most tightly bound of all nuclides, it is more than 90% of natural iron. Note that ^{56}Fe has even numbers of both protons and neutrons.

29.59 ^{209}Bi. This is the heaviest stable nuclide, and its BE/A is low compared with medium-mass nuclides.

• 29.60 (a) ^{235}U. The rarer of the two most common uranium isotopes. (b) ^{238}U. Most of uranium is ^{238}U. Note that ^{238}U has even numbers of both protons and neutrons. Is the BE/A of ^{238}U significantly different from that of ^{235}U?

• 29.61 (a) ^{12}C. Stable and relatively tightly bound, this nuclide is most of natural carbon. (b) ^{14}C. Is the difference in BE/A between ^{12}C and ^{14}C significant? One is stable and common, and the other is unstable and rare.

• 29.62 The fact that BE/A is greatest for A near 60 implies that the range of the nuclear force is about the diameter of such nuclides. (a) Calculate the diameter of an $A = 60$ nucleus. (b) Compare BE/A for ^{58}Ni and ^{90}Sr. The first is one of the most tightly bound nuclides, while the second is larger and less tightly bound.

⁚ 29.63 The point of this problem is to show in three ways that the binding energy of the electron in a hydrogen atom is negligible compared with the masses of the proton and electron. (a) Calculate the mass equivalent in u of the 13.6 eV binding energy of an electron in a hydrogen atom, and compare this with the mass of the hydrogen atom obtained from Appendix A. (b) Subtract the mass of the proton given in Table 29.1 from the mass of the hydrogen atom given in Appendix A. You will find the difference is equal to the electron's mass to three digits, implying the binding energy is small in comparison. (c) Take the ratio of the binding energy of the electron (13.6 eV) to the energy equivalent of the electron's mass (0.511 MeV).

Section 29.7* Tunneling

⁚ 29.64* Derive an approximate relationship between the energy of α decay and half-life using the following data. It may be useful to graph the log of $t_{1/2}$ against E_α to find some straight-line relationship.

Energy and Half-Life for α Decay

Nuclide	E_α (MeV)	$t_{1/2}$
^{216}Ra	9.5	0.18 μs
^{194}Po	7.0	0.7 s
^{240}Cm	6.4	27 d
^{226}Ra	4.91	600 y
^{232}Th	4.1	1.4×10^{10} y

INTEGRATED CONCEPTS

The following problems involve concepts from this chapter and several others. Physics is most interesting when applied to general situations involving more than a narrow set of physical principles. For example, radiation carries energy, hence the relevance of Chapter 6, Work, Energy, and Power. The following topics are involved in some or all of the problems in this section:

Topics	Location
Work, energy, and power	Chapter 6
Linear momentum	Chapter 7
Heat and heat transfer	Chapter 13
Current and Ohm's law	Chapter 19
Magnetism	Chapter 21

PROBLEM-SOLVING STRATEGY

Step 1. *Identify which physical principles are involved.*

Step 2. *Solve the problem using strategies outlined in the text.*

The following worked example illustrates how this strategy is applied to an integrated concept problem.

EXAMPLE   ALPHA DECAY IN A SMOKE ALARM

The following topics are involved in this integrated concepts worked example:

Topics	Location
Activity (nuclear decay)	Chapter 29
Current	Chapter 19

Many smoke alarms use an α emitter that is blocked by smoke particles. Given the activity falling on the detector to be 1.00 μCi, find the electrical current it receives in this stream of positive particles.

Strategy Step 1 To solve an *integrated concept problem*, such as those following this example, we must first identify the physical principles involved and identify the chapters in which they are found. We are given the source *activity*, a topic of the present Chapter 29. We are asked to find *electric current*, a topic of Chapter 19.

Strategy Step 2 The following solution illustrates how specific problem-solving strategies are applied. These involve identifying knowns and unknowns, checking to see if the answer is reasonable, and so on.

Solution: Activity (Chapter 29) The activity is given in units of curies, defined in Equation 29.14. If the activity is converted to decays per second, then the current can be found by multiplying by the

charge per decay. The activity is

$$R = 1.00 \times 10^{-6} \text{ Ci} \cdot \frac{3.7 \times 10^{10} \text{ decays/s}}{\text{Ci}}$$

$$= 3.7 \times 10^4 \text{ decays/s}$$

Current (Chapter 19) The charge of each α particle is Z times the proton charge. Here $Z = 2$, and so the charge is 3.2×10^{-19} C. Current is defined to be charge per unit time. Thus current is found by converting decays per second to Coulombs per second:

$$I = (3.7 \times 10^4 \text{ decays/s})(3.2 \times 10^{-19} \text{ C/decay})$$

$$= 1.2 \times 10^{-14} \text{ A}$$

Discussion This tiny current can be detected because ionizing radiation is involved. It would otherwise be negligible.

This worked example illustrates how to apply problem-solving strategies to situations that include topics in different chapters. The first step is to identify the physical principles involved in the problem. The second step is to solve for the unknown using familiar problem-solving strategies. These are found throughout the text. Moreover, many worked examples show how to use problem-solving strategies for single topics. You will find these techniques useful in applications of physics outside a physics course, such as in your profession, in other science disciplines, and in everyday life. The following problems help build skills in the broad application of physical principles.

29.65 A 2.00 T magnetic field is applied perpendicular to the path of charged particles in a bubble chamber. What is the radius of curvature of the path of a 10.0 MeV proton in this field?

• **29.66** Suppose a particle of ionizing radiation deposits 1.00 MeV in the gas of a Geiger tube, all of which goes to creating ion pairs. Each ion pair requires 30.0 eV of energy. (a) The applied voltage sweeps the ions out of the gas in 1.00 μs. What is the current? (b) This current is smaller than the actual, since the applied voltage in the Geiger tube accelerates the separated ions, which then create other ion pairs in subsequent collisions. What is the current if this last effect multiplies the number of ion pairs by 900?

‡ **29.67** The Galileo space probe was launched on its long journey past several planets in 1989, with an ultimate goal of Jupiter. Its power source is 11.0 kg of ^{238}Pu, a by-product of nuclear weapons plutonium production. Electrical energy is generated thermoelectrically from the heat produced when the 5.59 MeV α particles emitted in each decay crash to a halt inside the plutonium and its shielding. The half-life of ^{238}Pu is 87.7 years. (a) What was the original activity of the ^{238}Pu in becquerel? (b) What power was emitted in kilowatts? (c) What power was emitted 5.00 y after launch? You may neglect any extra energy from daughter nuclides and any losses from escaping γ rays.

‡ **29.68** (a) Repeat Problem 29.2, and convert the energy to joules or calories. (b) If all of this energy is converted to thermal energy in the gas, what is its temperature increase, assuming 50.0 cm³ of ideal gas at 0.250 atm pressure? (The small answer is consistent with the fact that the energy is large on a quantum mechanical scale but small on a macroscopic scale.)

‡ **29.69** What is the ratio of the velocity of a 5.00 MeV β ray to that of an α particle with the same kinetic energy? This should confirm that βs travel much faster than αs even when relativity is taken into consideration. (See also Problem 29.9.)

⁞ 29.70 (a) What is the kinetic energy in MeV of a β ray that is traveling at $0.998c$? This gives some idea of how energetic a β ray must be to travel at nearly the same speed as a γ ray. (b) What is the velocity of the γ ray relative to the β ray?

• 29.71 Since the production of ions in radiation detectors is a statistical process, there is a spread in the number of ions created by a given amount of ionizing energy. In a purely statistical process, you can expect variations of one standard deviation to be the square root of the average number created, so that the spread is $\sqrt{N}/N$, where N is the number of ion pairs created. Compare this spread for two different detectors in which 1.00 MeV of ionization is created. In the first, a gas detector, 30.0 eV per ion pair is required. In the second, a solid state detector, 2.00 eV per ion pair is required.

⁞ 29.72 (a) Write the decay equation for the α decay of ^{235}U. (b) What energy is released in this decay? The mass of the daughter nuclide is 231.036298 u. (c) Assuming the residual nucleus is formed in its ground state, how much energy goes to the α particle?

UNREASONABLE RESULTS

The following problems have results that are unreasonable because some premise is unreasonable or because certain of the premises are inconsistent with one another. Physical principles applied correctly then produce unreasonable results. The purpose of these problems is to give practice in assessing whether nature is being accurately described, and if it is not to trace the source of difficulty.

PROBLEM-SOLVING STRATEGY

To determine if an answer is reasonable, and to determine the cause if it is not, do the following.

Step 1. Solve the problem using strategies as outlined in several places in the text. See the table of contents to locate problem-solving strategies. Use the format followed in the worked examples in this chapter to solve the problem as usual.

Step 2. Check to see if the answer is reasonable. Is it too large or too small, or does it have the wrong sign, improper units, . . .?

Step 3. If the answer is unreasonable, look for what specifically could cause the identified difficulty. Usually, the manner in which the answer is unreasonable is an indication of the difficulty.

29.73 A physicist scatters γ rays from a substance and sees evidence of a nucleus 7.5×10^{-13} m in radius. (a) Find the atomic mass of such a nucleus. (b) What is unreasonable about this result? (c) What is unreasonable about the assumption?

• 29.74 The relatively scarce naturally occurring calcium isotope ^{48}Ca has a half-life of about 2×10^{16} y. (a) A small sample of this isotope is labeled as having an activity of 1.0 Ci. What is the mass of the ^{48}Ca in the sample? (b) What is unreasonable about this result? (c) What assumption is responsible?

• 29.75 A particle physicist discovers a neutral particle with a mass of 2.02733 u that he assumes is two neutrons bound together. (a) Find the binding energy. (b) What is unreasonable about this result? (c) What assumptions are unreasonable or inconsistent?

• 29.76 A nuclear physicist finds 1.0 μg of ^{236}U in a piece of uranium ore and assumes it is primordial since its half-life is 2.3×10^7 y. (a) Calculate the amount of ^{236}U that would had to have been on earth when it formed 4.5×10^9 y ago for 1.0 μg to be left today. (b) What is unreasonable about this result? (c) What assumption is responsible?

• 29.77 A frazzled theoretical physicist reckons that all conservation laws are obeyed in the decay of a proton into a neutron, positron, and neutrino (as in β^+ decay of a nucleus) and sends a paper to a journal to announce the reaction as a possible end of the universe due to the spontaneous decay of protons. (a) What energy is released in this decay? (b) What is unreasonable about this result? (c) What assumption is responsible?

⁞ 29.78 (a) Repeat Problem 29.54 but include the 0.0055% natural abundance of ^{234}U with its 2.45×10^5 y half-life. (b) What is unreasonable about this result? (c) What assumption is responsible? (d) Where does the ^{234}U come from if it is not primordial?

⁞ 29.79 The manufacturer of a smoke alarm decides that the smallest current of α radiation he can detect is 1.00 μA. (a) Find the activity in curies of an α emitter that produces a 1.00 μA current of α particles. (b) What is unreasonable about this result? (c) What assumption is responsible?

APPLICATIONS OF NUCLEAR PHYSICS

Supernovas are the source of elements heavier than iron. Energy released

powers nucleosynthesis. Spectroscopic analysis of the ring of material ejected

by supernova 1987A shows evidence of heavy elements.

Figure 30.1 (a) This airport security gate inspects for explosives using neutron irradiation. Neutrons absorbed by nitrogen nuclei form a radioactive isotope that is easily identified by its γ-ray emissions and half-life. Nitrogen is a major component of explosives and can be difficult to detect with x rays. (b) This photograph shows the control panel for such a gate in operation.

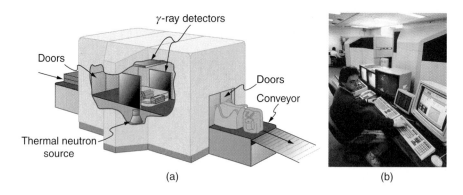

(a)

(b)

Applications of nuclear physics have become an integral part of modern life. From the bone scan that detects a cancer to the radioiodine treatment that cures another, nuclear radiation has diagnostic and therapeutic impact in medicine. From the fission power reactor to the hope of controlled fusion, nuclear energy is now commonplace and in our plans for the future. The destructive potential of nuclear weapons haunts us, as does the possibility of nuclear accidents. Not only has nuclear physics revealed secrets of nature, it has an inevitable impact based in its applications, intertwined as they are with human values. (See Figure 30.1.) Because of its potential for alleviation of suffering, and its power as an ultimate destructor of life, nuclear physics is often viewed with ambivalence. But it provides perhaps the best example that applications can be good or evil, while knowledge itself is neither.

30.1 MEDICAL IMAGING AND DIAGNOSTICS

Medical Application

A host of medical imaging techniques employ nuclear radiation. What makes nuclear radiation so useful? First, γ radiation can easily penetrate tissue, so that it is a useful probe. Second, nuclear radiation depends on the nuclide and not the chemical compound it is in, so that a radioactive nuclide can be put into a compound designed for specific purposes. The compound is said to be **tagged**. A tagged compound used for medical purposes is called a **radiopharmaceutical**. Radiation detectors external to the body can determine the location and concentration of a radiopharmaceutical to yield medically useful information. For example, certain drugs are concentrated in inflamed regions of the body, and this information can aid diagnosis and treatment. (See Figure 30.2.) Another application utilizes a radiopharmaceutical the body sends to bone cells, particularly those that are most active, to detect cancerous tumors or healing points. Images can be then produced of such bone scans.

Table 30.1 lists some medical diagnostic uses of radiopharmaceuticals, including isotopes and activities typically administered. Many organs can be imaged with a variety of nuclear isotopes. One common diagnostic employs iodine to image the thyroid, since iodine is concentrated in that organ. The most active thyroid cells, including cancerous cells, concentrate the most iodine and, therefore, emit the most radiation. Conversely, hypothyroidism is indicated by a lack of iodine uptake. Note that there is more than one isotope that can be used for many types of scans. Another common nuclear diagnostic is the thallium scan for the cardiovascular system, particularly used to evaluate blockages in the coronary arteries. The salt TlCl can be used, because it acts like NaCl and follows the blood. There are many other clever diagnostic applications of radiation that the interested reader can explore further.

Note that Table 30.1 lists many diagnostic uses for ^{99m}Tc, where the m stands for a metastable state of the technetium nucleus. Perhaps 80% of all radiopharmaceutical procedures employ ^{99m}Tc because of its many advantages. One is that the decay of its metastable state produces a single, easily identified 0.142 MeV γ ray. Additionally, the radiation dose to the patient is limited by the short 6.0 h half-life of ^{99m}Tc. And, although its half-life is short, it is easily and continuously produced on site. The basic process is neutron activation of molybdenum, which quickly β decays into ^{99m}Tc.

Figure 30.3 shows one of the simpler methods of imaging the concentration of nuclear activity, employing a device called an **Anger camera**. A piece of lead with holes

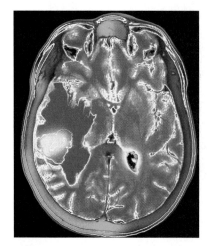

Figure 30.2 The radiopharmaceutical used to produce this image is concentrated in regions of tissue inflammation. In this case, a brain tumor and surrounding damaged brain tissue are evident.

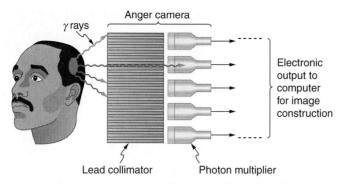

Figure 30.3 An Anger camera consists of a lead collimator and array of detectors that is scanned across a region of interest. A computer constructs an image from the detector output.

TABLE 30.1

DIAGNOSTIC USES OF RADIOPHARMACEUTICALS

Procedure, isotope	Typical activity (mCi)
Brain scan	
^{99m}Tc	7.5
^{113m}In	7.5
^{13}N (PET)	20
^{15}O (PET)	50
^{18}F (PET)	10
Lung scan	
^{99m}Tc	2
^{133}Xe	7.5
Cardiovascular blood pool	
^{131}I	0.2
^{99m}Tc	2
Cardiovascular arterial flow	
^{201}Tl	3
^{24}Na	7.5
Thyroid scan	
^{131}I	0.05
^{123}I	0.07
Liver scan	
^{198}Au (colloid)	0.1
^{99m}Tc (colloid)	2
Bone scan	
^{85}Sr	0.1
^{99m}Tc	10
Kidney scan	
^{197}Hg	0.1
^{99m}Tc	1.5

bored through it collimates γ rays emerging from the patient, allowing specific detectors to receive γ rays from specific directions only. The Anger camera is scanned across the region of interest, and computer analysis of detector signals produces an image such as the one in Figure 30.4.

Imaging techniques much like those in x-ray CT* scans are also used to form images with nuclear activity. Figure 30.5 shows a patient in a circular array of detectors that may be stationary or rotated, with detector output used by a computer to construct a detailed image. This technique is called **SPECT** for **single photon emission computed tomography**, or sometimes just SPET.

Images produced by β^+ emitters have become important in recent years. When the emitted positron (β^+) finds an electron, mutual annihilation occurs, producing two γ rays. These γs have identical 0.511 MeV energies (the energy supplied by the destruction of an electron or positron mass) and they move directly away from one another, allowing detectors to determine their point of origin accurately. (See Figure 30.6.) The system is called **PET** for **positron emission tomography**. It requires detectors on opposite sides to simultaneously detect photons of 0.511 MeV energy and utilizes computer imaging techniques similar to those in SPECT and CT scans. Examples of β^+ emitting isotopes used are ^{11}C, ^{13}N, ^{15}O, and ^{18}F. (See Figure 30.7.) The accuracy and sensitivity of PET scans make them useful for examining brain anatomy and function.

---

*CT stands for computed tomography. See Section 28.8.

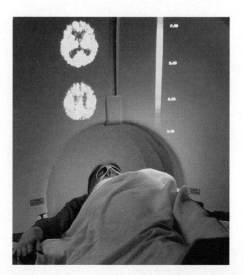

Figure 30.5 A SPECT, or single photon emission computed tomography system, uses a geometry much like a CT scanner to form an image of the concentration of a radiopharmaceutical.

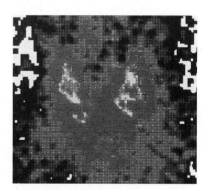

Figure 30.4 This thyroid scan was produced with an Anger camera detecting radiation from ^{123}I.

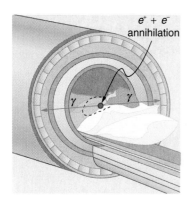

Figure 30.6 A positron emission tomography (PET) system takes advantage of the two identical γ-ray photons produced by positron-electron annihilation. These γs are emitted in opposite directions, so that their origin can be determined and computer analyzed to form an accurate image.

Figure 30.7 This PET scan shows brain activity in the optical centers of a subject.

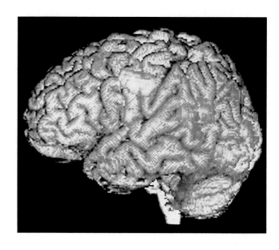

30.2 FUSION

While basking in the warmth of the summer sun, a student reads of the latest breakthrough in achieving sustained thermonuclear power and vaguely recalls the cold fusion controversy. The three are connected. The sun's energy is produced by fusion. (See Figure 30.8.) Thermonuclear power is the name given to the use of controlled fusion as an energy source. While progress is being made, high temperatures and attendant difficulties seem unavoidable. The cold fusion controversy centered around unsubstantiated claims of practical fusion power at room temperatures.

Nuclear fusion is a reaction in which two nuclei are combined, or *fused*, to form a larger nucleus. We know from Section 29.6 that all nuclei have less mass than the protons and neutrons that form them. The missing mass times c^2 equals the binding energy of the nucleus—the greater the binding energy, the greater the missing mass. We also know that BE/A, the binding energy per nucleon, is greatest for medium-mass nuclei. This means that if two low-mass nuclei can be fused together to form a larger nucleus, energy can be released. The larger nucleus has a greater binding energy and less mass per nucleon than the two that combined. Thus mass is destroyed in the fusion reaction, and energy is released. (See Figure 30.9.) On average, fusion of low-mass nuclei releases energy, but the details depend on the actual nuclides involved. There is ample fusion fuel on earth to supply human energy needs for millions of years if a practical generation method can be developed.

Figure 30.8 The sun's energy is produced by nuclear fusion.

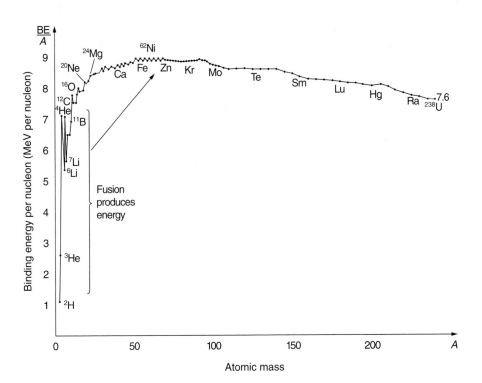

Figure 30.9 Fusion of light nuclei to form medium-mass nuclei destroys mass, because BE/A is greater for the product nuclei. The larger BE/A is, the less mass per nucleon, and so mass is destroyed and energy released in these fusion reactions.

The major obstruction to fusion is the Coulomb repulsion between nuclei. Since the attractive nuclear force that can fuse nuclei together is short ranged, the repulsion of like positive charges must be overcome to get nuclei close enough to induce fusion. Figure 30.10 shows an approximate graph of the potential energy between two nuclei as a function of the distance between their centers. The graph is analogous to a hill with a well in its center. A ball rolled from the right must have enough kinetic energy to get over the hump before it falls into the deeper well with a net gain in energy. So it is with fusion. If the nuclei are given enough kinetic energy to overcome the electrical potential energy due to repulsion, then they can combine, release energy, and fall into a deep well. One way to accomplish this is to heat fusion fuel to high temperatures so that the kinetic energy of thermal motion is sufficient to get the nuclei together.

You might think that in the core of our sun nuclei are coming into contact and fusing. But, in fact, temperatures on the order of 10^8 K are needed to actually get the nuclei in contact, exceeding the core temperature of the sun. Quantum mechanical tunneling is what makes fusion in the sun possible,* and tunneling is an important process in most other practical applications of fusion, too. Since the probability of tunneling is extremely sensitive to barrier height and width, increasing the temperature greatly increases the rate of fusion. The closer reactants get to one another, the more likely they are to fuse. (See Figure 30.11.) Thus most fusion in the sun and other stars takes place at their centers, where temperatures are highest. Moreover, high temperature is needed for thermonuclear power to be a practical source of energy.

The sun produces energy by fusing protons or hydrogen nuclei, ^{1}H (by far the sun's most abundant nuclide), into helium nuclei, ^{4}He. The principal sequence of fusion reactions form what is called the **proton-proton cycle**:

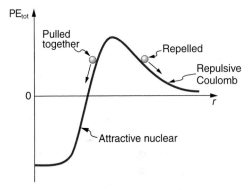

Figure 30.10 Potential energy between two light nuclei graphed as a function of distance between them. If the nuclei have enough kinetic energy to get over the Coulomb repulsion hump, they combine, release energy, and drop into a deep attractive well. Tunneling through the barrier is important in practice. The greater the kinetic energy and the higher the particles get up the barrier (or the lower the barrier), the more likely the tunneling.

$$
{}^1\mathrm{H} + {}^1\mathrm{H} \rightarrow {}^2\mathrm{H} + e^+ + \nu_e \qquad (0.42 \text{ MeV}) \qquad \textbf{(30.1a)}
$$

$$
{}^1\mathrm{H} + {}^2\mathrm{H} \rightarrow {}^3\mathrm{He} + \gamma \qquad (5.49 \text{ MeV}) \qquad \textbf{(30.1b)}
$$

$$
{}^3\mathrm{He} + {}^3\mathrm{He} \rightarrow {}^4\mathrm{He} + {}^1\mathrm{H} + {}^1\mathrm{H} \qquad (12.86 \text{ MeV}) \qquad \textbf{(30.1c)}
$$

---

*See the discussion of tunneling in Section 29.7. Here two nuclei combine by tunneling, whereas in α decay two nuclei split by tunneling.

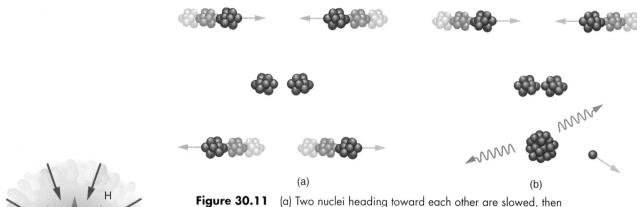

(a)

(b)

Figure 30.11 (a) Two nuclei heading toward each other are slowed, then stopped, and they fly away without touching or fusing. (b) At higher energy, the two nuclei approach close enough for fusion via tunneling. The probability of tunneling increases as they approach, but they do not have to touch for the reaction to occur.

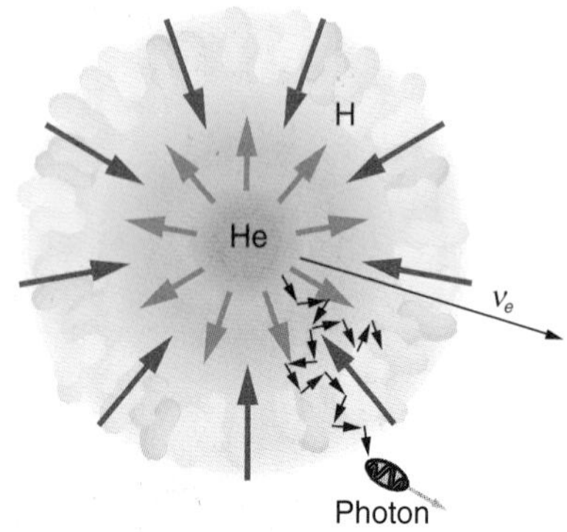

Figure 30.12 Nuclear fusion in the sun converts hydrogen nuclei into helium; fusion occurs primarily at the boundary of the helium core, where temperature is highest and sufficient hydrogen remains. Energy released diffuses slowly to the surface, with the exception of neutrinos, which escape immediately. Energy production remains stable because of negative feedback effects.

Figure 30.13 This large solar neutrino detector is buried underground to eliminate as much cosmic radiation as possible. In spite of its size and the huge flux of neutrinos that strike it, very few are detected each day since they interact so weakly. This, of course, is the same reason they escape the sun so readily.

where e^+ stands for a positron and ν_e is an electron's neutrino. (The energy in parentheses is that released by the reaction.) Note that the first two reactions must occur twice for the third to be possible, so that the cycle consumes six protons (^{1}H) but gives two back. Furthermore, the two positrons produced will find two electrons and annihilate to form four more γ rays, for a total of six. The overall effect of the cycle is thus

$$2e^- + 4\,^1\text{H} \rightarrow\,^4\text{He} + 2\nu_e + 6\gamma \quad (26.7 \text{ MeV}) \qquad \textbf{(30.2)}$$

where the 26.7 MeV includes the annihilation energy of the positrons and electrons and is distributed among all the reaction products. The solar interior is dense, and the reactions occur deep in the sun where temperatures are highest. It takes about 30,000 years for the energy to diffuse to the surface and radiate away. However, the neutrinos escape the sun in less than two seconds, carrying their energy with them, because they interact so weakly that the sun is transparent to them. Negative feedback in the sun acts as a thermostat to regulate overall energy output. For instance, if the interior of the sun becomes hotter than normal, the reaction rate increases, producing energy that expands the interior. This cools it and lowers the reaction rate. Conversely, if the interior becomes too cool, it contracts, increasing the temperature and reaction rate. (See Figure 30.12.) Stars like the sun are stable for billions of years, until a significant fraction of their hydrogen has been depleted. What happens then is discussed in Chapter 32.

Theories of the proton-proton cycle (and other energy-producing cycles in stars) were pioneered by the German-born, American physicist Hans Bethe, starting in 1938. He was awarded the 1967 Nobel prize in physics for this work, and he has made many other contributions to physics and society. Neutrinos produced in these cycles escape so readily that they provide us an excellent means to test these theories and study stellar interiors. Detectors have been constructed and operated for more than two decades now to measure solar neutrinos. (See Figure 30.13.) Although solar neutrinos are detected and neutrinos were observed from supernova 1987A, too few solar neutrinos are seen to be consistent with theories of solar energy production. This inconsistency has raised questions and motivated further research. It is possible that we do not understand the details of the fusion reactions as well as we think we do. Or it may be that the sun's energy production is temporarily turned down as part of its self-regulation. It is also possible that the neutrino detectors have unknown features that limit the fraction they detect or that we do not understand neutrino interactions.

The proton-proton cycle is not a practical source of energy on earth, in spite of the great abundance of hydrogen (^{1}H). The first reaction in Equation 30.1a has a very low

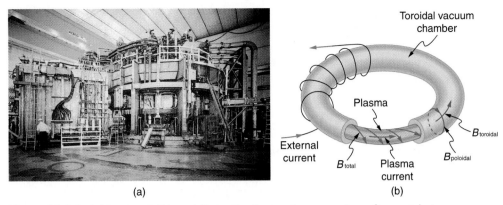

Figure 30.14 (a) Princeton's Tokamak Fusion Test Reactor is a magnetic confinement device that has nearly achieved break-even. (b) Schematic of tokamak's toroidal coil and confining field.

probability of occurring. (This is why our sun will last for about ten billion years.) But a number of other fusion reactions are easier to induce. Among them are

$$^2H + {}^2H \rightarrow {}^3H + {}^1H \qquad (4.03 \text{ MeV}) \qquad \textbf{(30.3a)}$$

$$^2H + {}^2H \rightarrow {}^3He + n \qquad (3.27 \text{ MeV}) \qquad \textbf{(30.3b)}$$

$$^2H + {}^3H \rightarrow {}^4He + n \qquad (17.59 \text{ MeV}) \qquad \textbf{(30.3c)}$$

$$^2H + {}^2H \rightarrow {}^4He + \gamma \qquad (23.85 \text{ MeV}) \qquad \textbf{(30.3d)}$$

Deuterium (^2H) is about 0.015% of natural hydrogen, and so there is an immense amount of it in sea water alone. In addition to an abundance of deuterium fuel, these fusion reactions produce large energies per reaction (in parentheses), but they do not produce much radioactive waste. Tritium (^3H) is radioactive, but it is consumed as a fuel (the reaction in Equation 30.3c), and the neutrons and γs can be shielded. The neutrons produced can also be used to create more energy and fuel in reactions like

$$n + {}^3He \rightarrow {}^4He + \gamma \quad (20.68 \text{ MeV})$$

and

$$n + {}^1H \rightarrow {}^2H + \gamma \quad (2.22 \text{ MeV})$$

Note that these last two reactions, and that in Equation 30.3d, put most of their energy output into the γ ray, and such energy is difficult to utilize.

The three keys to practical fusion energy generation are to achieve the temperatures necessary to make the reactions likely, to raise the density of the fuel, and to confine it long enough to produce large amounts of energy. These three factors—temperature, density, and time—complement one another, and so a deficiency in one can be compensated for by the others. **Ignition** is defined to occur when the reactions produce enough energy to be self-sustaining after external energy input is cut off. This goal, which must be reached before commercial plants can be a reality, has not been achieved. Another milestone, called **break-even**, occurs when the fusion power produced equals the heating power input. Break-even has nearly been reached and gives hope that ignition and commercial plants may become a reality in a few decades.

Two techniques have shown considerable promise. The first of these is called **magnetic confinement** and uses the property that charged particles have difficulty crossing magnetic field lines. The tokamak (first discussed in Section 21.5), shown in Figure 30.14, has shown particular promise. The tokamak's toroidal coil confines charged particles into a circular path with a helical twist due to the circulating ions themselves. In 1994 the Tokamak Fusion Test Reactor at Princeton came very close to break-even, and improvements continue.

The second promising technique aims massive lasers at tiny fuel pellets filled with a mixture of deuterium and tritium. (See Figure 30.15.) Huge power input heats the fuel,

Figure 30.15 The NOVA laser system at Lawrence Livermore Laboratory aims ten powerful lasers at a fuel pellet only a millimeter in diameter. The deuterium and tritium in the pellet are heated and compressed, causing fusion to occur.

evaporating the confining pellet and crushing the fuel to high density with the expanding hot plasma produced. This technique is called **inertial confinement**, because the fuel's inertia prevents it from escaping before significant fusion can take place. Higher densities have been reached than with tokamaks, but with smaller confinement times. Research on inertial confinement continues.

EXAMPLE 30.1 ENERGY AND POWER FROM FUSION

(a) Calculate the energy released by the fusion of a 1.00 kg mixture of deuterium and tritium, which produces helium. There are equal numbers of deuterium and tritium nuclei in the mixture. (b) If this takes place continuously over a period of a year, what is the average power output?

Strategy According to Equation 30.3c, the energy per reaction is 17.59 MeV. To find the total energy released, we must find the number of deuterium and tritium atoms in a kilogram. Deuterium has an atomic mass of 2 and tritium has an atomic mass of 3, for a total of about 5 g per mole of reactants or about 200 mol in 1.00 kg. To get a more precise figure, we will use the atomic masses from Appendix A. The power output is best expressed in watts, and so the energy output needs to be calculated in joules and then divided by the number of seconds in a year.

Solution for (a) The atomic mass of deuterium (^{2}H) is 2.014102 u, while that of tritium (^{3}H) is 3.016049 u, for a total of 5.030151 u per reaction. So a mole of reactants has a mass of 5.03 g, and in 1.00 kg there are (1000 g)/(5.03 g/mol) = 198.8 mol of reactants. The number of reactions that take place is therefore

(198.8 mol)(6.02 × 10^{23}/mol) = 1.20 × 10^{26} reactions

The total energy output is the number of reactions times the energy per reaction:

$$E = (1.20 \times 10^{26} \text{ reactions})(17.59 \text{ MeV/reaction})$$
$$\times (1.6 \times 10^{-13} \text{ J/MeV})$$
$$= 3.37 \times 10^{14} \text{ J}$$

Solution for (b) Power is energy per unit time. One year has 3.16 × 10^7 s; thus,

$$P = \frac{E}{t} = \frac{3.37 \times 10^{14} \text{ J}}{3.16 \times 10^7 \text{ s}}$$
$$= 1.07 \times 10^7 \text{ W} = 10.7 \text{ MW}$$

Discussion and Implications By now we expect nuclear processes to yield large amounts of energy, and we are not disappointed here. The energy output of 3.37 × 10^{14} J from fusing 1.00 kg of deuterium and tritium is equivalent to 2.6 million gallons of gasoline and about eight times the energy output of the bomb that destroyed Hiroshima. (See Table 6.1 for other important energies.) Yet the average backyard swimming pool has about 6 kg of deuterium in it, so that fuel is plentiful if it can be utilized in a controlled manner. The average power output over a year is more than 10 MW, impressive but a bit small for a commercial power plant. About 30 times this power output would allow generation of 100 MW of electricity, assuming a one-third efficiency in converting the fusion energy to electrical.

30.3 FISSION

Nuclear fission is a reaction in which a nucleus is split (or *fissured*). Controlled fission is a reality, whereas controlled fusion is a hope for the future. Hundreds of nuclear fission power plants around the world attest to the fact that, although its precise hazards are not known, controlled fission is practical and, at least in the short term, economical. (See Figure 30.16.)

Fission is the opposite of fusion and releases energy only if heavy nuclei are split. As discussed for fusion in the preceding section, energy is released if the products of a nuclear reaction have a greater binding energy per nucleon (BE/A) than the parent nuclei. Figure 30.9 shows that BE/A is greater for medium-mass nuclei than heavy nuclei, implying that when a heavy nucleus is split the products have less mass per nucleon, so that mass is destroyed and energy released in the reaction. The amount of energy per fission reaction can be large, even by nuclear standards. The graph in Figure 30.9 shows BE/A to be about 7.6 MeV/nucleon for the heaviest nuclei (A about 240), while BE/A is about 8.6 MeV/nucleon for nuclei having A about 120. Thus, if a heavy nucleus is split in half, then about 1 MeV per nucleon, or approximately 240 MeV per fission, is released. This is about 10 times the energy per fusion reaction, and about 100 times the energy of the average α, β, or γ decay.

Figure 30.16 Cows grazing near this nuclear power plant have no measurable radiation in them from the plant. A significant percentage of the world's electrical power is generated by controlled nuclear fission in such plants. The cooling towers are the most prominent feature and are not unique to nuclear power. The reactor is in the small domed building on the right.

EXAMPLE 30.2 ENERGY RELEASED BY FISSION

Calculate the energy released in the following fission reaction:

$$^{238}U \rightarrow ^{95}Sr + ^{140}Xe + 3n$$

given the atomic masses to be $m(^{238}U) = 238.050784$ u, $m(^{95}Sr) = 94.919388$ u, $m(^{140}Xe) = 139.921610$ u, and $m(n) = 1.008665$ u.

Strategy As always, the energy released is equal to the mass destroyed times c^2, and so we must find the difference in mass between the parent ^{238}U and the fission products.

Solution The products have a total mass

$$m_{products} = 94.919388 \text{ u} + 139.921610 \text{ u} + 3(1.008665 \text{ u})$$
$$= 237.866993 \text{ u}$$

The mass lost is the mass of ^{238}U minus $m_{products}$, or

$$\Delta m = 238.050784 \text{ u} - 237.866993 \text{ u} = 0.183791 \text{ u}$$

So the energy released is

$$E = (\Delta m)c^2$$
$$= (0.183791 \text{ u}) \frac{931.5 \text{ MeV}/c^2}{\text{u}} \cdot c^2 = 171 \text{ MeV}$$

Discussion and Implications A number of important things arise in this example. The 171 MeV energy released is large, but a little less than the 240 MeV estimated since this fission produces neutrons and does not split the nucleus into two equal parts. Fission of a given nuclide (like ^{238}U) does not always produce the same products. Fission is a statistical process in which an entire range of products are produced with various probabilities. Most fissions produce neutrons, although the number varies with each fission. This is an extremely important aspect of fission, because *neutrons can induce fission*, enabling self-sustaining chain reactions.

Spontaneous fission can occur, but this is usually not the most common decay mode for a given nuclide. For example, ^{238}U can spontaneously fission, but it decays mostly by α emission. Neutron-induced fission is crucial. (See Figure 30.17.) Being chargeless, even low-energy neutrons can strike a nucleus and be absorbed once they feel the attractive nuclear force. Large nuclei are described by a **liquid drop model** with surface tension and oscillation modes, because the large number of nucleons act something like atoms in a drop. The neutron is attracted and thus deposits energy, causing the nucleus to deform as a liquid drop might. If stretched enough, the nucleus narrows in the middle. The number of nucleons in contact and the strength of the nuclear force binding the nucleus together are reduced. Coulomb repulsion between the two ends then succeeds in fissioning the nucleus, which pops like a water drop into two large pieces and a few neutrons. **Neutron-induced fission** can be written as

$$n + {}^AX \rightarrow FF_1 + FF_2 + xn \qquad (30.4)$$

where FF_1 and FF_2 are the two daughter nuclei, called **fission fragments**, and x is the number of neutrons produced. Most of the energy released goes into the kinetic energy

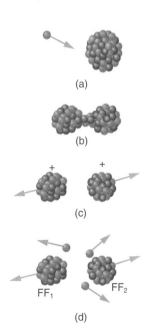

(a)

(b)

(c)

FF_1 FF_2

(d)

Figure 30.17 Neutron-induced fission. (a) Energy is put into this large nucleus when it absorbs a neutron. (b) Acting like a struck liquid drop, the nucleus deforms and begins to narrow in the middle. (c, d) Since fewer nucleons are in contact, the repulsive Coulomb force is able to break the nucleus in two parts with some neutrons also flying away.

Figure 30.18 A chain reaction can produce self-sustained fission if each fission produces enough neutrons to induce at least one more fission. This depends on several factors, including how many neutrons are produced in an average fission and how easy it is to make the particular type of nuclide fission.

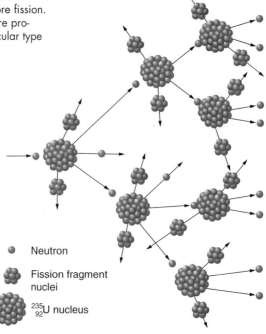

● Neutron

🔘 Fission fragment
    nuclei

⚫ $^{235}_{92}$U nucleus

of the fission fragments, with the remainder going into the neutrons and excited states of the fragments. Since neutrons can induce fission, a self-sustaining chain reaction is possible provided more than one neutron is produced on average—that is, if $x > 1$ in Equation 30.4. (See Figure 30.18.)

Not every neutron produced in a fission induces another fission. Some neutrons escape the fissionable material, while others interact with a nucleus without making it fission. We can enhance the number of fissions produced by neutrons by having a large amount of fissionable material. The minimum amount necessary for self-sustained fission of a given nuclide is called its **critical mass**. Some nuclides, such as ^{239}Pu, produce more neutrons per fission than others, such as ^{235}U. Additionally, some nuclides are easier to make fission than others. In particular, ^{235}U and ^{239}Pu are easier to fission than the much more abundant ^{238}U. Both factors affect critical mass, which is smallest for ^{239}Pu.

The reason ^{235}U and ^{239}Pu are easier to fission than ^{238}U is that the nuclear force is more attractive for an even number of neutrons in a nucleus than for an odd number. Consider that $^{235}_{92}$U$_{143}$ has 143 neutrons, and $^{239}_{94}$Pu$_{145}$ has 145 neutrons, whereas $^{238}_{92}$U$_{146}$ has 146. When a neutron encounters a nucleus with an odd number of neutrons, the nuclear force is more attractive, because the additional neutron will make the number even. About 2 MeV more energy is deposited in the resulting nucleus than would be the case if the number of neutrons was already even. This extra energy produces greater deformation, making fission more likely. Thus ^{235}U and ^{239}Pu are superior fission fuels. The isotope ^{235}U is only 0.72% of natural uranium, while ^{238}U is 99.27%, and ^{239}Pu does not exist in nature.

Most fission reactors utilize ^{235}U, which is separated from ^{238}U at some expense. The most common separation method is gas diffusion of uranium hexafluoride (UF_6) through membranes. Since ^{235}U has less mass than ^{238}U, its UF_6 molecules have higher average velocity at the same temperature and diffuse faster. Another interesting characteristic of ^{235}U is that it preferentially absorbs very slow moving neutrons (with energies a fraction of an eV), whereas fission reactions produce fast neutrons having energies on the order of an MeV. To make a self-sustained fission reactor with ^{235}U, it is thus necessary to slow down (thermalize) the neutrons. Water is very effective, since neutrons collide with protons in water molecules and lose energy. Figure 30.19 shows a schematic of one reactor design, called a pressurized water reactor.

Control rods containing nuclides that very strongly absorb neutrons are used to adjust neutron flux. To produce large power, reactors contain hundreds to thousands of critical masses, and the chain reaction easily becomes self-sustaining, a condition called **criticality**. Neutron flux must be carefully regulated to avoid an exponential increase in

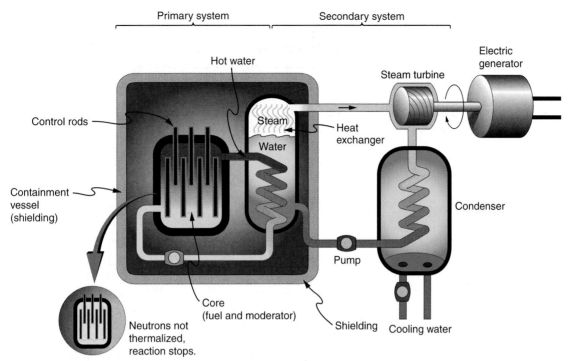

Figure 30.19 A pressurized water reactor is cleverly designed to control the fission of large amounts of ^{235}U, while using the heat to produce electrical energy. Control rods adjust neutron flux so that criticality is obtained, but not exceeded. Should the reactor overheat and boil the water away, the chain reaction terminates, because water is needed to thermalize the neutrons. This inherent safety feature can be overwhelmed in extreme circumstances.

fissions, called **supercriticality**. Control rods help to prevent overheating, perhaps even meltdown or explosive disassembly. Water used to thermalize neutrons, necessary to get them to induce fission in ^{235}U and achieve criticality, provides a negative feedback for temperature increases. Should the reactor overheat and boil the water to steam or be breached, the absence of water kills the chain reaction. Considerable heat, however, can still be generated by the reactor's radioactive fission products. Other safety features, thus, need to be incorporated, including auxiliary cooling water and pumps.

EXAMPLE 30.3 ENERGY FROM A KILOGRAM OF FISSION FUEL

Calculate the amount of energy produced by the fission of 1.00 kg of ^{235}U, given the average fission reaction of ^{235}U produces 200 MeV.

Strategy The total energy produced is the number of ^{235}U atoms times the given energy per ^{235}U fission. We must therefore find the number of ^{235}U atoms in 1.00 kg.

Solution The number of ^{235}U atoms in 1.00 kg is Avogadro's number times the number of moles. One mole of ^{235}U has a mass of 235.04 g; thus, there are (1000 g)/(235.04 g/mol) = 4.25 mol. The number of ^{235}U atoms is

$$(4.25 \text{ mol})(6.02 \times 10^{23} \text{ }^{235}\text{U/mol}) = 2.56 \times 10^{24} \text{ }^{235}\text{U}$$

So the total energy released is

$$E = (2.56 \times 10^{24} \text{ }^{235}\text{U}) \cdot \frac{200 \text{ MeV}}{^{235}\text{U}} \cdot \frac{1.60 \times 10^{-13} \text{ J}}{\text{MeV}}$$

$$= 8.20 \times 10^{13} \text{ J}$$

Discussion This is another impressively large energy, being equivalent to about 14,000 barrels of crude oil or 600,000 gallons of gasoline. But it is only one-fourth the energy produced by the fusion of a kilogram mixture of deuterium and tritium (see Example 30.1). Even though each fission reaction yields about ten times the energy of a fusion reaction, the energy per kilogram of fission fuel is less, because there are far fewer moles per kilogram of the heavy nuclides. Fission fuel is also much more scarce than fusion fuel, and less than 1% of uranium (the ^{235}U) is readily usable. The known safety problems of fission only add to our motivation to develop fusion as an energy source.

One nuclide already mentioned is ^{239}Pu, which has a 24,120 y half-life and does not exist in nature. Plutonium-239 is manufactured from ^{238}U in reactors, and it provides an opportunity to utilize the other 99% of natural uranium as an energy source. The following reaction sequence, called **breeding**, produces ^{239}Pu. Breeding begins with neutron capture by ^{238}U:

$$^{238}U + n \rightarrow {}^{239}U + \gamma$$

Uranium-239 then β^- decays:

$$^{239}U \rightarrow {}^{239}Np + \beta^- + \bar{\nu}_e \qquad (t_{1/2} = 23 \ \text{min})$$

Neptunium-239 also β^- decays:

$$^{239}Np \rightarrow {}^{239}Pu + \beta^- + \bar{\nu}_e \qquad (t_{1/2} = 2.4 \ \text{d})$$

Plutonium-239 builds up in reactor fuel at a rate that depends on the probability of neutron capture by ^{238}U (all reactor fuel contains more ^{238}U than ^{235}U). Reactors designed specifically to make plutonium are called **breeder reactors**. They seem to be inherently more hazardous than conventional reactors, but it remains unknown whether their hazards can be economically made acceptable. The four reactors at Chernobyl, including the one that was destroyed, are breeder reactors with a design significantly different from the pressurized water reactor illustrated above.

Plutonium-239 has advantages over ^{235}U as a reactor fuel—it produces more neutrons per fission on average, and it is somewhat easier for a thermal neutron to cause it to fission. This means ^{239}Pu has a particularly small critical mass, an advantage for nuclear weapons.

30.4 NUCLEAR WEAPONS

The world was in turmoil when fission was discovered in 1938. The discovery of fission, made by two German physicists, Otto Hahn and Fritz Strassman, was quickly verified by two Jewish refugees from Nazi Germany, Lise Meitner and her nephew Otto Frisch. Fermi, among others, soon found that not only did neutrons induce fission, more neutrons were produced during fission. The possibility of a self-sustained chain reaction was immediately recognized by leading scientists the world over. The enormous energy known to be in nuclei, but considered inaccessible, now seemed to be available large scale.

Within months after the announcement of the discovery of fission, Adolf Hitler banned the export of uranium from newly occupied Czechoslovakia. It seemed that the military value of uranium had been recognized in Nazi Germany, and that a serious effort to build a nuclear bomb had begun.

Alarmed scientists, many of them already affected by Nazi Germany, decided to take action. None was more famous or revered than Einstein. It was felt that his help was needed to get the American government to make a serious effort at nuclear weapons as a matter of survival. Leo Szilard, an escaped Hungarian physicist, took a draft of a letter to Einstein, who, although pacifistic, signed the final version. The letter was for President Franklin Roosevelt, warning of the German potential to build extremely powerful bombs of a new type. It was sent in August of 1939, just before the German invasion of Poland that marked the start of World War II.

It was not until December 6, 1941, the day before the Japanese attack on Pearl Harbor, that the United States made a massive commitment to building a nuclear bomb. The top secret Manhattan Project was a crash program aimed at beating the Germans. It was carried out in remote locations, such as Los Alamos, New Mexico, whenever possible, and eventually came to cost billions of dollars and employ the efforts of more than 100,000 people. J. Robert Oppenheimer (1904–1967), whose talent and ambitions made him ideal, was chosen to head the project. The first major step was made by the powerfully capable Fermi and his group in December 1942, when they achieved the first self-sustained nuclear reactor. This first "atomic pile" used carbon blocks to thermalize neutrons. It not only proved the chain reaction possible, it began the era of reactors and the breeding of plutonium. Glenn Seaborg, an American chemist and physicist, received the Nobel prize in physics in 1951 for discoveries of several transuranic elements, including plutonium. Carbon-moderated reactors are relatively inexpensive and simple in

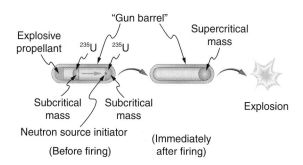

Figure 30.20 A gun-type fission bomb for ^{235}U utilizes two sub-critical masses forced together by explosive charges inside a cannon barrel. The energy yield depends on the amount of uranium and the time it can be held together before it disassembles itself.

design and are still used for breeding plutonium, such as at Chernobyl, where two such reactors remain in operation.

Plutonium was recognized as easier to fission with neutrons and, hence, a superior fission material very early in the Manhattan Project. Plutonium availability was uncertain, and so a uranium bomb was developed simultaneously. Figure 30.20 shows a gun-type bomb, which takes two subcritical uranium masses and blows them together. To get an appreciable yield, the critical mass must be held together by the explosive charges inside the cannon barrel for a few microseconds. Since the buildup of the uranium chain reaction is relatively slow, the device to hold the critical mass together can be relatively simple. Because the rate of spontaneous fission is low, a neutron source is triggered at the same time the critical mass is assembled.

Plutonium's special properties necessitated a more sophisticated critical mass assembly, shown schematically in Figure 30.21. A spherical mass of plutonium is surrounded by shape charges (high explosives that release most of their blast in one direction) that implode the plutonium, crushing it into a smaller volume to form a critical mass. The implosion technique is faster and more effective, because it compresses three-dimensionally rather than one-dimensionally as in the gun-type bomb. Again, a neutron source must be triggered at just the correct time to initiate the chain reaction.

Because of its complexity, the plutonium bomb needed to be tested before there was any attempt to use it. On July 16, 1945, the test named Trinity was conducted in the isolated Alamogordo Desert about 200 miles south of Los Alamos. (See Figure 30.22.) A new age had begun. The yield of this device was about 10 kilotons (kT), the equivalent of 5000 of the largest conventional bombs.

Although Germany surrendered on May 7, Japan had been steadfastly refusing to surrender for many months, forcing large casualties. Invasion plans by the Allies estimated a million casualties of our own and untold losses of Japanese lives. The bomb was viewed as a way to end the war. The first was a uranium bomb dropped on Hiroshima on

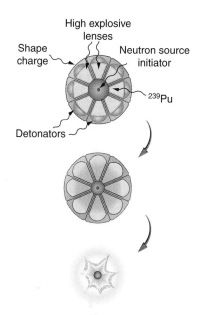

Figure 30.21 An implosion created by high explosives compresses a sphere of ^{239}Pu into a critical mass. The superior fissionability of plutonium has made it the universal bomb material.

Figure 30.22 Trinity test, the first nuclear bomb.

Figure 30.23 Destruction in Hiroshima.

August 6. Its yield of about 15 kT destroyed the city and killed an estimated 80,000 people, with 100,000 more being seriously injured. (See Figure 30.23.) The second was a plutonium bomb dropped on Nagasaki only three days later, on August 9. Its 20 kT yield killed at least 50,000 people, something less than Hiroshima because of the hilly terrain and the fact that it was a few kilometers off target. The Japanese were told that one bomb a week would be dropped until they unconditionally surrendered, which they did on August 14. In actuality, the United States had only enough plutonium for one more and as yet unassembled bomb.

Knowing that fusion produces several times more energy per kilogram of fuel than fission, some scientists pushed the idea of a fusion bomb starting very early on. Calling this bomb the Super, they realized that it could have another advantage over fission—high-energy neutrons would aid fusion, while they are ineffective in ^{239}Pu fission. Thus the fusion bomb could be virtually unlimited in energy release. The first was detonated by the United States on October 31, 1952, at Eniwetok Atoll with a yield of 10 megatons (MT), about 670 times that of the fission bomb that destroyed Hiroshima. The Soviets followed with a fusion device of their own in August 1953, and a weapons race, beyond the aims of this text to discuss, continued until the end of the Cold War.

Figure 30.24 shows a simple diagram of how a thermonuclear bomb is constructed. A fission bomb is exploded next to fusion fuel in the solid form of lithium deuteride. Before the shock wave blows it apart, γ rays heat and compress the fuel, and neutrons create tritium through the reaction $n + {}^6\text{Li} \rightarrow {}^3\text{H} + {}^4\text{He}$. Additional fusion and fission fuels are enclosed in a dense shell of ^{238}U. The shell reflects some of the neutrons back into the fuel to enhance its fusion, but at the high internal temperatures fast neutrons are created that also cause the plentiful and inexpensive ^{238}U to fission, part of what allows thermonuclear bombs to be so large.

The energy yield and the types of energy produced by nuclear bombs can be varied. Energy yields in current arsenals range from about 0.1 kT to 20 MT, although the Soviets once detonated a 67 MT device. Nuclear bombs differ from conventional explosives in more than size. Figure 30.25 shows the approximate fraction of energy output in various forms for conventional explosives and for two types of nuclear bombs. Nuclear bombs put a much larger fraction of their output into thermal energy than do conventional bombs, which tend to concentrate the energy in blast. Another difference is the immediate and residual radiation energy from nuclear weapons. This can be adjusted to put more energy into radiation (the so-called neutron bomb) so that the bomb can be used to irradiate advancing troops without killing friendly troops with blast and heat.

By 1988, the combined arsenals of the United States and the Soviet Union totaled about 50,000 nuclear warheads. In addition, the British, French, and Chinese each have several hundred bombs of various sizes, and a few other countries have a small number. Nuclear weapons are generally divided into two categories. Strategic nuclear weapons are

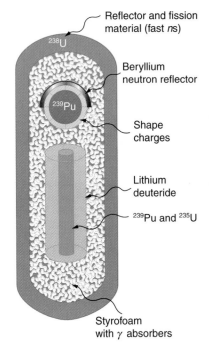

Reflector and fission material (fast ns)

^{238}U

Beryllium neutron reflector

^{239}Pu

Shape charges

Lithium deuteride

^{239}Pu and ^{235}U

Styrofoam with γ absorbers

Figure 30.24 This schematic of a fusion bomb (H-bomb) gives some idea of how the ^{239}Pu fission trigger is used to ignite fusion fuel. Neutrons and γ rays transmit energy to the fusion fuel, create tritium from deuterium, and heat and compress the fusion fuel. The outer shell of ^{238}U serves to reflect some neutrons back into the fuel, causing more fusion, and it boosts the energy output by fissioning itself when neutron energies become great enough.

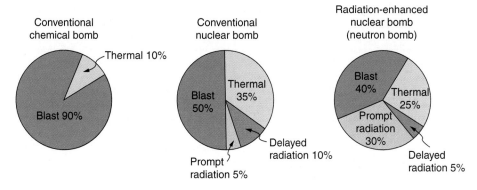

Figure 30.25 Approximate fractions of energy output by conventional and two types of nuclear weapons. In addition to yielding more energy than conventional weapons, nuclear bombs put a much larger fraction into thermal energy. This can be adjusted to enhance the radiation output to be more effective against troops. An enhanced radiation bomb is also called a neutron bomb.

those intended for military targets, such as bases and missile complexes, and moderate to large cities. There were about 20,000 strategic weapons in 1988. Tactical weapons are intended for use in smaller battles. Since the collapse of the Soviet Union and the end of the Cold War in 1989, most of the 30,000 tactical weapons (including Cruise missiles, artillery shells, land mines, torpedoes, depth charges, and backpacks) have been demobilized, and parts of the strategic weapons systems are being dismantled with warheads and missiles being disassembled. Many thousands will remain ready for use for the foreseeable future.

A few small countries have built or are capable of building nuclear bombs, as are some terrorist groups. Two things are needed—a minimum level of technical expertise and sufficient fissionable material. The first is easy. Fissionable material is controlled but is also available. There are international agreements and organizations that attempt to control nuclear proliferation, but it is increasingly difficult given the availability of fissionable material and the small amount needed for a crude bomb.

30.5 BIOLOGICAL EFFECTS OF IONIZING RADIATION

We hear many seemingly contradictory things about the biological effects of ionizing radiation. It can cause cancer, large exposures cause burns and hair loss, yet it is used to treat and even cure cancer. How do we go about understanding these effects? Once again there is an underlying simplicity in nature, even in complicated biological organisms. All of the effects of ionizing radiation on biological tissue can be understood by knowing that **ionizing radiation has two effects on cells**:

Human & Medical Application

1. Interference with cell reproduction
2. Destruction of the cell's function, or destruction of the cell itself

Since ionizing radiation interferes with cell reproduction, it has its greatest effect on cells that rapidly reproduce, including most types of cancer. Thus cancer cells are more sensitive to radiation than normal cells and can be more easily killed by it. Cancer is characterized by malfunction of cell reproduction, so that it can also be caused by ionizing radiation. Without contradiction, ionizing radiation can be both a cure and a cause.

To accurately discuss the biological effects of ionizing radiation, we need a radiation dose unit that is directly related to those effects. All effects of radiation are directly proportional to the amount of ionization created in the biological organism. The amount of ionization is in turn proportional to the amount of energy deposited. We therefore define a **radiation dose unit** called the **rad** (r), as 1/100 of a joule of ionizing energy deposited per kilogram of tissue. That is,

$$1 \text{ r} = 0.01 \text{ J/kg} \qquad \textbf{(30.5)}$$

For example, if a 50.0 kg person is exposed to ionizing radiation over her entire body and absorbs 1.00 J, then her whole body radiation dose is (1.00 J)/(50.0 kg) = 0.0200 J/kg = 2.00 rad. If the same 1.00 J of ionizing energy were absorbed in her 2.00 kg forearm alone, then the dose to the forearm would be (1.00 J)/(2.00 kg) = 0.500 J/kg = 50.0 rad, and the unaffected tissue would have a zero rad dose. When calculating radiation doses, you divide by the mass of affected tissue. You must specify the affected region

Figure 30.26 Ionization created in cells by α and γ radiation. Because of its shorter range, the ionization and damage created by the α is more concentrated and harder for the organism to repair. Thus the RBE for αs is greater than for γs, even though at the same energy they create the same amount of ionization.

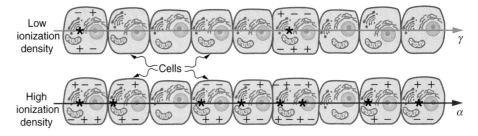

(whole body or forearm, for example) in addition to giving the numerical dose in rads. The SI unit for radiation dose is the **gray** (Gy), defined to be 1 Gy = 1 J/kg = 100 rad, but the rad is still commonly used. Although the energy per kilogram in 1 rad is small, it has significant effects since the energy causes ionization. The energy needed for a single ionization is a few eV, or less than 10^{-18} J. Thus 0.01 J of ionizing energy can create a huge number of ion pairs and have an effect at the cellular level.

The effects of ionizing radiation are directly proportional to the dose in rads, but they also depend on the type of radiation. That is, for a given dose in rads, the effects also depend on whether the radiation is α, β, γ, x ray, or some other type of ionizing radiation. In the discussion of the range of ionizing radiation (see Section 29.1), it was noted that energy is deposited in a series of ionizations and not in a single interaction. Each ion pair (or ionization) requires a certain amount of energy, so that the number of ion pairs is directly proportional to the amount of ionizing energy deposited. But if the range of the radiation is small, as it is for αs, then the ionization is more concentrated. (See Figure 30.26.) Concentrated damage is more difficult for biological organisms to repair than damage that is spread out, and so short-range particles have greater biological effects. The **relative biological effectiveness** (RBE) or **quality factor** (QF)* is given in Table 30.2 for several types of ionizing radiation— the effect of the radiation is directly proportional to RBE. A dose unit more closely related to effects in biological tissue is called the **roentgen equivalent man** (rem) and is defined to be the dose in rads multiplied by the relative biological effectiveness,

$$\text{rem} = \text{rad} \times \text{RBE} \qquad (30.6)$$

So if a person had a whole body dose of 2.00 rad of γ radiation, the dose in rem would be (2.00 rad)(1) = 2.00 rem whole body. If the person had a whole body dose of 2.00 rad of α radiation, then the dose in rem would be (2.00 rad)(20) = 40.0 rem whole body. The αs would have 20 times the effect on the person as the γs for the same energy deposited. The SI equivalent of the rem is the **sievert** (Sv), defined to be Sv = Gy $\times$ RBE, so that 1 Sv = 100 rem.

The RBEs given in Table 30.2 are approximate, but they yield certain insights. For example, the eyes are more sensitive to radiation, because the cells of the lens do not

---

*The greater the RBE or QF, the more effect the radiation has. Thus in cancer therapy the relative biological effectiveness is called the quality factor.

TABLE 30.2

RELATIVE BIOLOGICAL EFFECTIVENESS	
Type and energy of radiation	*RBE*
X rays	1
γ rays	1
β rays greater than 30 keV	1
β rays less than 30 keV	1.7
Neutrons, thermal to slow (<20 keV)	2–5
Neutrons, fast (1–10 MeV)	10 (body), 30 (eyes)
Protons (1–10 MeV)	10 (body), 30 (eyes)
α rays from radioactive decay	10–20
Heavy ions from accelerators	10–20

repair themselves. Neutrons cause more damage than γ rays, although both are neutral and have large ranges, because neutrons often cause secondary radiation when they are captured. Note that the RBEs are 1 for higher-energy βs, γs, and x rays, three of the most common types of radiation. For those types of radiation, the numerical values of the dose in rem and rad are identical. (1 rad of γ radiation is also 1 rem, for example.) For that reason, rads are still widely quoted rather than rem.

The large-scale effects of radiation on humans can be divided into two categories: immediate effects and long-term effects. Table 30.3 gives the immediate effects of whole body exposures received in less than one day. If the radiation exposure is spread out over more time, greater doses are needed to cause the effects listed. This is due to the body's ability to partially repair the damage. Any dose less than 10 rem is called a **low dose**, 10 to 100 rem is called a **moderate dose**, and anything greater than 100 rem is called a **high dose**. There is no known way to determine after the fact that a person has been exposed to less than 10 rem.

Immediate effects are explained by the effects of radiation on cells and the sensitivity of rapidly reproducing cells to radiation. The first clue that a person has been exposed to radiation is a change in blood count, which is not surprising since blood cells are the most rapidly reproducing in the body. At higher doses, nausea and hair loss are attributable to interference with cell reproduction. Cells in the lining of the digestive system are also rapidly reproducing, and their destruction causes nausea. When hair cells slow in their growth, the hair follicles thin and break off. High doses cause significant cell death in all systems, but the lowest doses that cause fatalities do so by weakening the immune system through loss of white blood cells.

The two known long-term effects of radiation are cancer and genetic defects. Both are directly attributable to the interference of radiation with cell reproduction. For high doses of radiation, the risk of cancer is reasonably well known from studies of exposed groups. (Hiroshima and Nagasaki survivors and a smaller number of people exposed by their occupation, such as radium dial painters, have been fully documented. Chernobyl victims will be studied for many decades, with some data already available.) The risk of a radiation-induced cancer for low and moderate doses is generally assumed to be proportional to the risk known for high doses. This is called the **linear hypothesis** and is prudent, but controversial. There is some evidence that, unlike the immediate effects of radiation, the long-term effects are cumulative and there is little self-repair. (This is analogous to the risk of skin cancer from UV exposure, which is known to be cumulative.) There is a latency period for the onset of radiation-induced cancer of about 2 years for leukemia and 15 years for most other forms. The person is at risk for at least 30 years after the latency period. Omitting many details, the overall risk of a radiation-induced cancer death per year per rem of exposure is about 10 in a million. Numerically, we can write this as $10/10^6 \cdot \text{rem} \cdot \text{y}$. If a person receives a dose of 1 rem, his risk each year of dying from radiation-induced cancer is 10 in a million and that risk continues for about 30 years. The lifetime risk is thus 300 in a million, or 0.03%. Since about 20% of all deaths are from cancer, the increase due to a 1 rem exposure is impossible to detect demographically. But 100 rem, which was the dose received by the average Hiroshima and

TABLE 30.3

IMMEDIATE EFFECTS OF RADIATION (ADULTS, WHOLE BODY, SINGLE EXPOSURE)

Dose in rem	Effect
0–10	No observable effect.
10–100	Slight to moderate decrease in white blood cell counts.
35, 50	Temporary sterility; 35 for women, 50 for men.
100–200	Significant reduction in blood cell counts, brief nausea and vomiting. Rarely fatal.
200–500	Nausea, vomiting, hair loss, severe blood damage, hemorrhage, fatalities.
450	LD50/30. Lethal to 50% so exposed in 30 days if untreated.
500–2000	Worst effects due to malfunction of small intestine and blood systems. Limited survival.
>2000	Fatal within hours due to collapse of central nervous system.

Nagasaki survivor, causes a 3% risk, which can be observed in the presence of a 20% normal incidence.

The incidence of genetic defects induced by radiation is about one-third that of cancer deaths, but is much more poorly known. The lifetime risk of a genetic defect due to a 1 rem exposure is about 100 in a million (or $3.3/10^6 \cdot$ rem$\cdot$y), but the normal incidence is 60,000 in a million. Evidence of such a small increase, tragic as it is, is nearly impossible to obtain. There is no evidence of increased genetic defects among the offspring of Hiroshima and Nagasaki survivors, for example. Animal studies do not seem to correlate well with effects on humans and are not very helpful. For both cancer and genetic defects, the approach to safety has been to use the linear hypothesis, which is likely to be an overestimate of the risks of low doses.

The risks are relatively small, and the average person is not exposed to large amounts of radiation. See Table 30.4 for average annual background radiation from natural and artificial sources. Cosmic rays are partially shielded by the atmosphere, and the doses range from 35 to 70 mrem/y in the United States, but the average is only 44 mrem/y since most of the population lives at low altitude. The contribution from ^{226}Ra averages 1 mrem/y, but this also varies with location, being as high as 350 mrem/y in areas where water and soil have high ^{226}Ra content. Medical and dental diagnostic exposures are mostly from x rays. It should be noted that x-ray doses tend to be localized and are becoming much smaller with improved techniques. The estimated exposure for ^{222}Rn is controversial. Recent studies indicate there is more radon in the home than had been realized, and it is speculated (and remains controversial) that radon may be responsible for 20% of lung cancers, being particularly hazardous to those who also smoke.

Radiation Protection

Laws regulate doses to which people can be exposed. The greatest whole body dose allowed is the occupational 5 rem/y (limited to 3 rem in any one quarter), rarely reached by medical and nuclear power workers. Higher doses are allowed for the hands. Much lower doses are permitted to the reproductive organs and the fetuses of pregnant women. Inadvertent doses to the public are limited to 1/10 of occupational doses, except for those caused by nuclear power, which cannot expose the public to more than 1/1000 the occupational limit (exceeded in the United States only near the Three Mile Island accident). Extensive monitoring with a variety of radiation detectors is performed to assure radiation safety.

TABLE 30.4

BACKGROUND RADIATION SOURCES AND AVERAGE DOSES (UNITED STATES)	
Source	*Dose (mrem/y)*
Natural radiation—external	
Cosmic rays	44
Soil, building materials	40
^{222}Rn (radon gas)	$\approx$200*
Natural radiation—internal	
^{14}C	1
^{40}K	16
^{226}Ra	1
Artificial radiation—external and internal	
Medical and dental	73
Weapons testing fallout	4
Occupational	1
Nuclear power to public	<0.003
Miscellaneous	2
Total (U.S. average)	*382*

*Approximate and subject to debate.

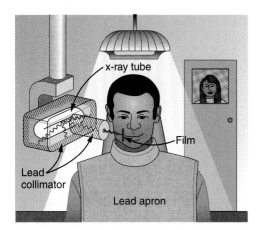

Figure 30.27 A lead apron is placed over the dental patient and shielding surrounds the x-ray tube to limit exposure to tissue other than that imaged. Fast films limit the time needed to obtain images, reducing exposure to imaged tissue. The technician stands a few meters away behind a lead-lined door with a lead glass window, reducing her occupational exposure.

To physically limit radiation doses, we use **shielding**, increase **distance** from a source, and limit **time of exposure**. Figure 30.27 illustrates how these are used to protect both the patient and the dental technician when an x ray is taken. Shielding absorbs radiation and can be provided by any material, including sufficient air. The greater the distance from the source, the more the radiation spreads out. The less time a person is exposed to a given source, the smaller the dose received. Doses from most medical diagnostics have decreased in recent years due to faster films, requiring less exposure time, for example.

PROBLEM-SOLVING STRATEGY
FOR DOSE CALCULATIONS

Step 1. Examine the situation to determine that a person is exposed to ionizing radiation.

Step 2. Identify exactly what needs to be determined in the problem (identify the unknowns). The most straightforward problems ask for a dose calculation.

Step 3. Make a list of what is given or can be inferred from the problem as stated (identify the knowns). Look in particular for information on the type of radiation, the energy per event, the activity, and the mass of tissue affected.

Step 4. For a dose calculation, you must determine the energy deposited. This may take one or more steps, depending on the given information.

Step 5. Divide the energy deposited by the mass of affected tissue. Use units of joules for energy and kilograms for mass. If a dose in rad is involved, use the definition that 1 r = 0.01 J/kg.

Step 6. If a dose in rem is involved, determine the RBE (QF) of the radiation. Recall that rem = rad × RBE.

Step 7. Check the answer to see if it is reasonable: Does it make sense? The dose should be consistent with the numbers given in the text for diagnostic, occupational, and therapeutic exposures, for example.

EXAMPLE 30.4 DOSE FROM INHALED PLUTONIUM

Calculate the dose in rem/y to the lungs of a weapons plant employee who inhales and retains an activity of 1.00 μCi of ^{239}Pu in an accident. The mass of affected lung tissue is 2.00 kg, the plutonium decays by emission of a 5.23 MeV α particle, and you may assume the higher value of the RBE for αs from Table 30.2.

Strategy Dose in rem is defined in Equations 30.5 and 30.6. The energy deposited is divided by the mass of tissue affected and then multiplied by the RBE. The latter two quantities are given, and so the main task in this example will be to find the energy deposited in one year. Since the activity of the source is given, we can calculate the number

(continued)

(continued)

of decays, multiply by the energy per decay, and convert MeV to joules to get the total energy.

Solution The activity $R = 1.00 \ \mu Ci = 3.70 \times 10^4 \ Bq = 3.70 \times 10^4$ decays/s. So the number of decays per year is obtained by multiplying by the number of seconds in a year:

$$(3.70 \times 10^4 \text{ decays/s})(3.16 \times 10^7 \text{ s})$$
$$= 1.17 \times 10^{12} \text{ decays}$$

Thus the ionizing energy deposited per year is

$$E = (1.17 \times 10^{12} \text{ decays})(5.23 \text{ MeV/decay})$$
$$\times \frac{1.6 \times 10^{-13} \text{ J}}{\text{MeV}} = 0.978 \text{ J}$$

Dividing by the mass of tissue affected gives

$$\frac{E}{\text{mass}} = \frac{0.978 \text{ J}}{2.00 \text{ kg}} = 0.489 \text{ J/kg}$$

One rad is 0.0100 J/kg, and so the dose in rads is

$$\text{dose in rad} = \frac{0.489 \text{ J/kg}}{0.010 \text{ J/kg/rad}} = 48.9 \text{ rad}$$

Now the dose in rem is

$$\boxed{\begin{array}{l} \text{dose in rem} = \text{rad} \times \text{RBE} \\ \qquad\qquad = (48.9 \text{ rad})(20) = 1.0 \times 10^3 \text{ rem} \end{array}}$$

Discussion First note that the dose is given to two digits, because the RBE is (at best) known only to two digits. By any standard, this yearly radiation dose is high and will have a devastating effect on the health of the worker. Worse yet, plutonium has a long radioactive half-life and is not readily eliminated by the body, and so it will remain in the lungs. Being an α emitter makes the effects 10 to 20 times worse than the same ionization produced by βs, γs, or x rays. An activity of 1.00 μCi is created by only 16 μg of ^{239}Pu (left as an end-of-chapter problem to verify), partly justifying claims that plutonium is the most toxic substance known. Its actual hazard depends on how likely it is to be spread out among a large population and ingested. The Chernobyl disaster's deadly legacy, for example, has nothing to do with the plutonium it put into the environment.

Figure 30.28 The therapeutic value of radiation was once touted for far more than its modern use in cancer therapy. Until the 1930s, a variety of "medicinal" uses were touted for radium and other radioactive elements, often with tragic results.

Medical Application

Risk versus Benefit

Medical doses of radiation are also limited. Diagnostic doses are generally low and becoming lower with improved techniques and faster films. With the possible exception of routine dental x rays, radiation is used diagnostically only when needed so that the low risk is justified by the benefit of diagnosis. Chest x rays give the lowest doses—about 10 mrem to the tissue affected, with less than 5% scattering into tissue not directly imaged. Other x-ray procedures range upward to about 1 rem per slice in a CT scan, and about 0.500 rem per dental x ray, again both only affecting the tissue imaged. Medical images with radiopharmaceuticals discussed in Section 30.1 give doses ranging from 0.100 to 5 rem, usually localized. One exception is the thyroid scan using ^{131}I. Because of its relatively long half-life, it exposes the thyroid to about 75 rem. The isotope ^{123}I is more difficult to produce, but its short half-life limits thyroid exposure to about 1.5 rem.

30.6 THERAPEUTIC USES OF IONIZING RADIATION

Therapeutic applications of ionizing radiation, called radiation therapy or **radiotherapy**, have existed since the discovery of x rays and nuclear radioactivity. Today radiotherapy is used almost exclusively for cancer therapy, where it saves thousands of lives and improves the quality of life and longevity of many it cannot save. Radiotherapy may be used alone or in combination with surgery and chemotherapy (drug treatment) depending on the type of cancer and the response of the patient. It is statistically well established that radiotherapy's beneficial effects far outweigh its long-term risks.

The earliest uses of ionizing radiation on humans were mostly harmful, with many on the level of snake oil. (See Figure 30.28.) Radium-doped cosmetics that glowed in the dark were used around World War I. As recently as the 1950s, radon mine tours were promoted as healthful and rejuvenating—those who toured were exposed but gained no benefits. Radium salts were sold as health elixirs for many years. The gruesome death of a wealthy industrialist, who became psychologically addicted to the brew, alerted the unsuspecting to the dangers of radium salt elixirs. Legislation in the 1950s finally ended most abuses.

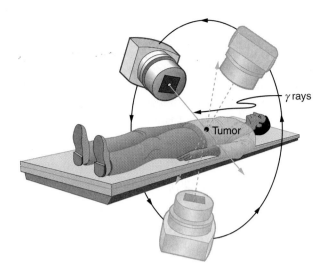

Figure 30.29 The source of radiation is rotated around the patient so that the common crossing point is in the tumor, concentrating the dose there. This geometric technique works for well-defined tumors.

Medically, a number of ailments *can* be cured with radiation. Acne responds to low-energy x rays, as does ringworm. Spinal vertebrae can be fused by damaging them with large x-ray doses, a technique used for many years to cure a condition called ankylosing spondilitis. These two radiotherapies are not used today, because effective alternative treatments exist, so that the risk of radiation-induced cancer is not worth the benefit (and may never have been).

Radiotherapy is effective against cancer, because cancer cells are rapidly reproducing and, consequently, more sensitive to radiation. The central problem in radiotherapy is to make the dose to cancer cells as high as possible while limiting the dose to normal cells. The ratio of abnormal cells killed to normal cells killed is called the **therapeutic ratio**, and all radiotherapy techniques are designed to enhance this ratio. The therapeutic ratio is best for the most rapidly reproducing cancers. But there is a trade-off in that rapidly reproducing cancer tumors are often oxygen poor. Ionizing radiation produces more toxic end products by knocking apart molecules where oxygen is plentiful. It is sometimes possible to oxygenate a tumor to aid the therapeutic ratio.

Radiation can be concentrated in cancerous tissue by a number of clever techniques. One of the most prevalent for well-defined tumors is a geometric technique shown in Figure 30.29. A narrow beam of radiation is passed through the patient from a variety of directions having a common crossing point in the tumor. This concentrates the dose in the tumor while spreading it out over a large volume of normal tissue. The external radiation can be x rays, ^{60}Co γ rays, or ionizing particle beams produced by accelerators. Accelerator-produced beams of neutrons, π-mesons, and heavy ions such as nitrogen nuclei have been employed, and these can be quite effective. These particles have larger QFs (RBEs) and sometimes can be better localized, producing a greater therapeutic ratio. But accelerator radiotherapy is much more expensive and less frequently employed than other forms.

Another form of radiotherapy utilizes chemically inert radioactive implants. Alpha emitters have the dual advantages of a large QF and a small range for better localization.

Radiopharmaceuticals are used for cancer therapy when they can be localized well enough to produce a favorable therapeutic ratio. Thyroid cancer is commonly treated utilizing radioactive iodine. Thyroid cells concentrate iodine, and cancerous thyroid cells are more aggressive in doing this. An ingenious use of radiopharmaceuticals in cancer therapy tags antibodies with radioisotopes. Antibodies produced by a patient to combat his cancer are extracted, cultured, loaded with a radioisotope, and then returned to the patient. The antibodies are concentrated almost entirely in the tissue they developed to fight, thus localizing the radiation in abnormal tissue. The therapeutic ratio can be quite high for short-range radiation. There is, however, a significant dose to organs that eliminate radiopharmaceuticals from the body, such as the liver, kidneys, and bladder. As with most radiotherapy, the technique is limited by the tolerable amount of damage to normal tissue. (See Figure 30.30.)

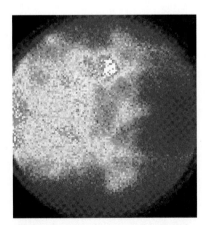

Figure 30.30 Radioimmunology attaches a radioisotope to antibodies produced against a cancer. This image shows how the radiation is localized in abnormal tissue by the fact that antibodies seek out the tissue they were produced to fight. Note that there is a significant amount of radiation in parts of the body that eliminate waste.

TABLE 30.5

CANCER RADIOTHERAPY	
	Typical dose (rem)
Lung	1000–2000
Hodgkin's disease	4000–4500
Skin	4000–5000
Ovarian	5000–7500
Breast	5000–8000+
Brain	
Neck	
Bone	8000+
Soft tissue	
Thyroid	

Table 30.5 lists typical therapeutic doses of radiation used against cancer. The doses are large, but not fatal because they are localized and spread out in time. Protocols for treatment vary with the type of cancer and the condition and response of the patient. Three to five 200 rem treatments per week for a period of several weeks is typical. Time between treatments allows the body to repair normal tissue. Abnormal tissue repairs itself, too, but not as completely, since damage is concentrated in it and it is more sensitive to radiation. Damage to normal tissue limits the doses. You will note that the greatest doses are given to tissue that is not rapidly reproducing, such as in the adult brain. Lung cancer, on the other end of the scale, cannot ordinarily be cured with radiation because of the sensitivity of lung tissue and blood to radiation. But radiotherapy for lung cancer does alleviate symptoms and prolong life and is therefore justified in some cases.

Finally, it is interesting to note that chemotherapy employs drugs that interfere with cell division and is, thus, effective against cancer. It also has many of the same side effects, such as nausea and hair loss, and risks, such as the inducement of another cancer.

30.7 FOOD IRRADIATION

Ionizing radiation is widely used to sterilize medical supplies, such as bandages, and consumer products, such as tampons and condoms. Worldwide it is also used to irradiate food, an application that promises to grow in the future. **Food irradiation** is the treatment of food with ionizing radiation. It is used to reduce pest infestation, and to delay spoilage and prevent illness caused by microorganisms. Food irradiation is controversial. Proponents see it as superior to pasteurization, preservatives, and insecticides, supplanting dangerous chemicals with a more effective process. Opponents see its safety as unproved, perhaps leaving worse toxic residues as well as presenting an environmental hazard at treatment sites. For developing countries, food irradiation might increase crop production by 25.0% or more, and reduce food spoilage by a similar amount. In the United States, food irradiation has been approved for many foods, but is used chiefly to treat spices and some fruits, such as strawberries.

Food irradiation exposes food to large doses of γ rays, x rays, or electron beams. These photons and electrons induce no nuclear reactions and thus create *no residual radioactivity*. (Some forms of ionizing radiation, such as neutron irradiation, cause residual radioactivity. These are not used for food irradiation.) The γ source is usually ^{60}Co or ^{137}Cs decay, the latter isotope being a major by-product of nuclear power. Cobalt-60 γ rays average 1.25 MeV, while those of ^{137}Cs are 0.660 MeV and are less penetrating. X rays used for food irradiation are created with voltages up to 5 MV and, thus, have photon energies up to 5 MeV. Electrons used for food irradiation are accelerated to energies up to 10 MeV. The higher the energy per particle, the more penetrating the radiation is and the more ionization it can create. Figure 30.31 shows a typical γ-irradiation plant.

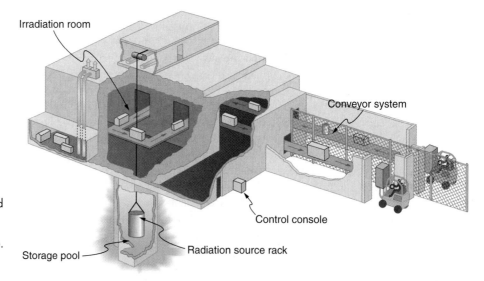

Figure 30.31 A food irradiation plant has a conveyor system to pass items through an intense radiation field behind thick shielding walls. The γ source is lowered into a deep pool of water for safe storage when not in use. Exposure times up to an hour expose food to doses up to 10^6 rad.

Irradiation room

Conveyor system

Control console

Storage pool

Radiation source rack

Because food irradiation seeks to destroy organisms such as insects and bacteria, much larger doses than those fatal to humans must be applied. Generally, the simpler the organism, the more radiation it can tolerate. (Cancer cells are a partial exception, because they are rapidly reproducing and, thus, more sensitive.) Current licensing allows up to 100,000 rads* to be applied to fresh fruits and vegetables, called a *low dose* in food irradiation. Such a dose is enough to discourage many microorganisms, but about 1,000,000 rad is needed to kill salmonella, and even more is needed to kill fungi and viruses. Doses greater than 1,000,000 rad are considered to be high doses in food irradiation and product sterilization.

The effectiveness of food irradiation varies with the type of food. Spices and many fruits and vegetables have dramatically longer shelf lives. These also show no degradation in taste and no loss of food value or vitamins. If not for the mandatory labeling, such foods subjected to low-level irradiation (up to 100,000 rad) could not be distinguished from untreated foods in quality. But some foods actually spoil faster after irradiation, particularly those with high water content like lettuce and peaches. Others, such as milk, are given a noticeably unpleasant taste. High-level irradiation produces significant and chemically measurable changes in foods. It produces about a 15% loss of nutrients and a 25% loss of vitamins, as well as some change in taste. Such losses are similar to those that occur in ordinary freezing and cooking.

How does food irradiation work? Ionization produces a random assortment of broken molecules and ions, some with unstable oxygen- or hydrogen-containing molecules known as **free radicals**. These undergo rapid chemical reactions, producing perhaps four or five thousand different compounds called **radiolytic products**, some of which make cell function impossible by breaking cell membranes, fracturing DNA, and so on. How safe is the food afterward? Critics argue that the radiolytic products present a lasting hazard, perhaps being carcinogenic. But in fact the safety of irradiated food is not precisely known. We do know that low-level food irradiation produces no compounds in amounts that can be measured chemically. This is not surprising, since trace amounts of several thousand compounds may be created. We also know that there have been no observable negative short-term effects on consumers. Long-term effects may show up if large numbers of people consume large quantities of irradiated food, but none have for the small amounts of irradiated food that are regularly consumed in this country and others. Finally, the hazard to consumers, if it exists, must be weighed against the benefits in food production and preservation. It must also be weighed against the very real hazards of existing insecticides and food preservatives.

---

*Units of rads are used, because rems are a human dose unit based on human response to radiation. Furthermore, the RBEs of γs, x rays, and electron beams are all 1, so that they should have similar effects for the same ionization produced.

SUMMARY

Radiopharmaceuticals are compounds used for medical imaging and therapeutics to which a radioactive substance has been attached, a process called **tagging**. Table 30.1 lists certain diagnostic uses of radiopharmaceuticals including the isotope and activity typically utilized. One common imaging device is the **Anger camera**, which consists of a lead collimator, radiation detectors, and an analysis computer. Tomography performed with γ-emitting radiopharmaceuticals is called **SPECT** for **single photon emission computed tomography**, and it has the advantages of x-ray CT scans coupled with organ- and function-specific drugs. **PET** or **positron emission tomography** is a similar technique that uses β^+ emitters and detects the annihilation γs, aiding source localization.

Nuclear fusion is a reaction in which two nuclei are combined to form a larger nucleus. It releases energy when light nuclei are fused to form medium-mass nuclei. Fusion is the source of energy in stars, with the **proton-proton cycle**,

$$^1H + {}^1H \rightarrow {}^2H + e^+ + \nu_e \qquad (0.42 \text{ MeV}) \qquad \textbf{(30.1a)}$$
$$^1H + {}^2H \rightarrow {}^3He + \gamma \qquad (5.49 \text{ MeV}) \qquad \textbf{(30.1b)}$$
$$^3He + {}^3He \rightarrow {}^4He + {}^1H + {}^1H \qquad (12.86 \text{ MeV}) \qquad \textbf{(30.1c)}$$

being the principal sequence of energy-producing reactions in our sun. The overall effect of the proton-proton cycle is

$$2\,e^- + 4\,{}^1H \rightarrow {}^4He + 2\nu_e + 6\gamma \qquad (26.7 \text{ MeV}) \quad \textbf{(30.2)}$$

where the 26.7 MeV includes the annihilation energy of the positrons emitted and annihilated. Attempts to utilize controlled fusion as an energy source on earth are related to deuterium and tritium, and the reactions

$$^2H + {}^2H \rightarrow {}^3H + {}^1H \qquad (4.03 \text{ MeV}) \qquad \textbf{(30.3a)}$$
$$^2H + {}^2H \rightarrow {}^3He + n \qquad (3.27 \text{ MeV}) \qquad \textbf{(30.3b)}$$
$$^2H + {}^3H \rightarrow {}^4He + n \qquad (17.59 \text{ MeV}) \qquad \textbf{(30.3c)}$$
$$^2H + {}^2H \rightarrow {}^4He + \gamma \qquad (23.85 \text{ MeV}) \qquad \textbf{(30.3d)}$$

play important roles. **Ignition** is the condition under which controlled fusion is self-sustaining; it has not been achieved. **Break-even**, in which the fusion energy output is as great as the external energy input, has nearly been achieved. **Magnetic confinement** and **inertial confinement** are the two methods being developed for heating fuel to high enough temperatures, at sufficient density, and for long enough times to achieve ignition. The first method uses magnetic fields and the second the momentum of impinging photons for confinement.

Nuclear fission is a reaction in which a nucleus is split. Fission releases energy when heavy nuclei are split into medium-mass nuclei, and it is broadly utilized. Self-sustained fission is possible, because **neutron-induced fission** also produces neutrons that can induce other fissions,

$$n + {}^A X \rightarrow FF_1 + FF_2 + xn \qquad (30.4)$$

where FF_1 and FF_2 are the two daughter nuclei, or **fission fragments**, and x is the number of neutrons produced. Some neutrons escape, and so a minimum mass, called the **critical mass**, that depends on the nuclide ${}^A X$, must be present to achieve **criticality**. More than a critical mass can produce **supercriticality**. The production of new or different isotopes (especially ${}^{239}Pu$) by nuclear transformation is called **breeding**, and reactors designed for this purpose are called **breeder reactors**.

There are two types of nuclear weapons—fission bombs use fission alone, whereas thermonuclear bombs use fission to ignite fusion. Both produce huge numbers of nuclear reactions in a very short time. Energy yields are measured in kilotons or megatons of equivalent conventional explosives and range from 0.1 kT to more than 20 MT. Nuclear bombs are characterized by far more thermal output and nuclear radiation output than conventional explosives.

The **biological effects of ionizing radiation** are due to two effects it has on cells: *interference with cell production*, and *destruction of cell function*. A **radiation dose unit** called the **rad** (r) is defined in terms of the ionizing energy deposited per kilogram of tissue:

$$1 \ r = 0.01 \ J/kg \qquad (30.5)$$

The SI unit for radiation dose is the **gray** (Gy), which is defined to be 1 Gy = 1 J/kg = 100 rad. To account for the effect of the type of particle creating the ionization, we use the **relative biological effectiveness** (RBE) or **quality factor** (QF) given in Table 30.2 and define a unit called the **roentgen equivalent man** (rem) as

$$rem = rad \times RBE \qquad (30.6)$$

Particles that have short ranges or create large ionization densities have RBEs greater than unity. The SI equivalent of the rem is the **sievert** (Sv), defined to be Sv = Gy × RBE, so that 1 Sv = 100 rem. Whole body, single-exposure doses of 10 rem or less are low doses, those of 10 to 100 rem are moderate, and those over 100 rem are high doses. Some immediate radiation effects are given in Table 30.3. Effects due to low doses are not observed, but their risk is assumed directly proportional to those of high doses, an assumption known as the **linear hypothesis**. Long-term effects are cancer deaths at the rate of $10/10^6 \cdot rem \cdot y$ and genetic defects at roughly one-third this rate. Background radiation doses and sources are given in Table 30.4. Radiation protection utilizes **shielding**, **distance**, and **time** to limit exposure.

Radiotherapy is the use of ionizing radiation to treat ailments, now limited to cancer therapy. The sensitivity of cancer cells to radiation enhances the ratio of cancer cells killed to normal cells killed, called the **therapeutic ratio**. Doses for various organs are limited by the tolerance of normal tissue for radiation and are given in Table 30.5. Treatment is localized in one region of the body and spread out in time.

CONCEPTUAL QUESTIONS

30.1 Why would an isotope that emits a single γ ray be easier to detect than one that emits several? (Think in terms of signal to noise.)

30.2 Why aren't α emitters used in medical diagnostics, unless the α decay is followed by γ emission?

30.3 In terms of radiation dose, what is the major difference between medical diagnostic uses of radiation and medical therapeutic uses?

30.4 One of the methods used to limit radiation dose to the patient in medical imaging is to employ isotopes with short half-lives. How might this limit dose?

30.5 Why does the fusion of light nuclei into heavier nuclei release energy?

30.6 Energy input is required to fuse medium-mass nuclei, such as iron or cobalt, into more massive nuclei. Explain why.

30.7 In considering potential fusion reactions, what is the advantage of the reaction ${}^2H + {}^3H \rightarrow {}^4He + n$ over the reaction ${}^2H + {}^2H \rightarrow {}^3He + n$?

30.8 Give reasons justifying the contention made in the text that energy from the fusion reaction ${}^2H + {}^2H \rightarrow {}^4He + \gamma$ is relatively difficult to capture and utilize.

30.9 Would you expect the sun to produce neutrinos or antineutrinos? What about fission reactors? Could these two sources be used to test for differences between neutrinos and antineutrinos?

30.10 Explain why the fission of heavy nuclei releases energy. Similarly, why is it that energy input is required to fission light nuclei?

30.11 Explain, in terms of conservation of momentum and energy, why collisions of neutrons with protons will thermalize neutrons better than collisions with oxygen.

30.12 The ruins of the Chernobyl reactor are enclosed in a huge concrete structure built around it after the accident. Some rain penetrates the building in winter, and radioactivity from the building increases. What does this imply is happening inside?

30.13 Since a uranium or plutonium nucleus fissions with a probability distribution over a wide range of pieces, would you expect more residual radioactivity from fission than fusion? Explain.

30.14 The core of a nuclear reactor generates large amounts of thermal energy from the decay of fission products even when the power-producing fission chain reaction is turned off. Would you expect this residual heat to be greatest after the reactor has run a long time or a short time? What if the reactor has been shut down for months? Years?

30.15 How can a nuclear reactor contain many critical masses and not go supercritical? What methods are used to control the fission in the reactor?

30.16 Why can heavy nuclei with odd numbers of neutrons be induced to fission with thermal neutrons, whereas those with even numbers of neutrons require more energy input to induce fission?

30.17 What are some of the reasons that plutonium rather than uranium is used in all fission bombs and as the trigger in all fusion bombs?

30.18 Use the laws of conservation of momentum and energy to explain how a shape charge can direct most of the energy released in an explosion in a specific direction. (Note that this is similar to the situation in guns and cannons—most of the energy goes into the bullet.)

30.19 How does the lithium deuteride in the thermonuclear bomb shown in Figure 30.24 supply tritium (^{3}H) as well as deuterium (^{2}H)?

30.20 Fallout from nuclear weapons tests in the atmosphere is now mainly ^{90}Sr and ^{137}Cs, which have 28.6 and 30.2 y half-lives, respectively. Atmospheric tests were terminated by most countries more than 30 years ago. It has been found that environmental activities of these two isotopes are decreasing faster than their half-lives. Why might this be?

30.21 Isotopes that emit α radiation are relatively safe outside the body and exceptionally hazardous inside. Yet those that emit γ radiation are hazardous outside and inside. Explain why.

30.22 Early researchers realized that radiation, such as from x rays, could redden the skin and limited their doses just enough to avoid this. Unfortunately, skin reddening starts at doses of about 100 rem. What long-term effects were they likely to suffer?

30.23 Why is it reasonable that the latency period for leukemia is significantly less than for other radiation-induced cancers?

30.24 Why is radon more closely associated with inducing lung cancer than other types of cancer?

30.25 The RBE for low-energy βs is 1.7, whereas that for higher-energy βs is only 1. Explain why, considering how the range of radiation depends on its energy.

30.26 Which methods of radiation protection were used in the device shown in Figure 30.32(a)? In the situation shown in Figure 30.32(b)?

30.27 What radioisotope might be a problem in homes built of cinder blocks made from uranium mine tailings? (This is true of homes and schools in certain regions near uranium mines.)

30.28 Radiotherapy is more likely to be used to treat cancer in elderly patients than in young ones. Explain why. Why is radiotherapy used to treat young people at all?

30.29 Are some types of cancer more sensitive to radiation than others? If so, what makes them more sensitive?

30.30 Does food irradiation leave the food radioactive? To what extent is the food altered chemically for low and high doses in food irradiation?

30.31 Compare a low dose of radiation to a human with a low dose of radiation used in food treatment.

30.32 Suppose one food irradiation plant uses a ^{137}Cs source while another uses an equal activity of ^{60}Co. Assuming equal fractions of the γ rays from the sources are absorbed, why is more time needed to get the same dose using the ^{137}Cs source?

(a)

(b)

Figure 30.32 (a) This x-ray fluorescence machine is one of thousands used in shoe stores to produce images of feet as a check on the fit of shoes. They are unshielded and remain on as long as feet are in them, producing doses *much* grater than medical images. Children were fascinated with them. These machines were used in shoe stores until laws preventing such unwarranted radiation exposure were enacted in the 1950s. (b) This nuclear plant employee is wearing protective clothing and holding a radiation monitor in preparation for working briefly in a high-radiation zone. Question 26

PROBLEMS

Note that in addition to the worked examples in this chapter, you may also find those in the previous chapter provide useful guidance in solving these problems.

Section 30.1 Medical Imaging with Nuclear Radiation

30.1 A neutron generator uses an α source, such as radium, to bombard beryllium, inducing the reaction $^4\text{He} + {}^9\text{Be} \rightarrow {}^{12}\text{C} + n$. Such neutron sources are called RaBe sources, or PuBe sources if they use plutonium to get the αs. Calculate the energy output of the reaction in MeV.

30.2 Neutrons from a source (perhaps the one discussed in the preceding problem) bombard natural molybdenum, which is 24% ^{98}Mo. What is the energy output of the reaction $^{98}\text{Mo} + n \rightarrow {}^{99}\text{Mo} + \gamma$? The mass of ^{98}Mo is given in Appendix A, and that of ^{99}Mo is 98.907711 u.

30.3 The purpose of producing ^{99}Mo (usually by neutron activation of natural molybdenum, as in the preceding problem) is to produce ^{99m}Tc. Using the rules found in Section 29.4, verify that the β^- decay of ^{99}Mo produces ^{99}Tc. (Most ^{99}Tc nuclei produced in this decay are left in a metastable excited state denoted ^{99m}Tc.)

30.4 (a) What energy is released in the β^- decay of ^{99}Mo, given its mass is 98.907711 u? (b) Is this sufficient energy to leave the residual technetium nucleus in its metastable state at 0.142 MeV excitation?

• **30.5** (a) Two annihilation γs in a PET scan originate at the same point and travel to detectors on either side of the patient. If the point of origin is 9.00 cm closer to one of the detectors, what is the difference in arrival times of the photons? (This could be used to give position information, but the time difference is small enough to make it difficult.) (b) How accurately would you need to be able to measure arrival time differences to get position resolution of 1.00 mm?

• **30.6** Table 30.1 indicates that 7.50 mCi of ^{99m}Tc is used in a brain scan. What mass of technetium is this?

• **30.7** The activities of ^{131}I and ^{123}I used in thyroid scans are given in Table 30.1 to be 50 and 70 μCi, respectively. Find and compare the masses of ^{131}I and ^{123}I in such scans, given their respective half-lives are 8.04 d and 13.2 h. Note that the masses are so small that the radioiodine is usually mixed with stable iodine as a carrier to ensure normal chemistry and distribution in the body.

‡ **30.8** (a) Neutron activation of sodium, which is 100% ^{23}Na, produces ^{24}Na, which is used in some heart scans (see Table 30.1). The equation for the reaction is $^{23}\text{Na} + n \rightarrow {}^{24}\text{Na} + \gamma$. Find its energy output, given the mass of ^{24}Na is 23.990962 u. (b) What mass of ^{24}Na produces the needed 5.0 mCi activity, given its half-life is 15.0 h?

Section 30.2 Fusion

30.9 Verify that the total number of nucleons, total charge, and electron family number are conserved for each of the fusion reactions in the proton-proton cycle in Equations 30.1a–c. (List the value of each of the conserved quantities before and after each of the reactions.)

• **30.10** Calculate the energy output in each of the fusion reactions in the proton-proton cycle, and verify the values given in Equations 30.1a–c.

• **30.11** Show that the total energy released in the proton-proton cycle is 26.7 MeV, considering the overall effect given in Equation 30.2 and being certain to include the annihilation energy.

• **30.12** Verify by listing the number of nucleons, total charge, and electron family number before and after the cycle that these quantities are conserved in the overall proton-proton cycle given in Equation 30.2.

30.13 The energy produced by the fusion of a 1.00 kg mixture of deuterium and tritium was found in Example 30.1. Approximately how many kilograms would be required to supply the annual energy use in the United States?

30.14 Calculate the time it takes a neutrino created at the center of the sun to escape.

30.15 Rare tritium fuel can be produced by the reaction $n + {}^2\text{H} \rightarrow {}^3\text{H} + \gamma$ (tritium is desired for fusion with plentiful deuterium). How much energy in MeV is released in this neutron capture?

• **30.16** Two fusion reactions mentioned in the text are

$$n + {}^3\text{He} \rightarrow {}^4\text{He} + \gamma$$

and

$$n + {}^1\text{H} \rightarrow {}^2\text{H} + \gamma$$

Both reactions release energy, but the second also creates more fuel. Confirm the energies produced in the reactions are 20.58 and 2.22 MeV, respectively. Comment on which product nuclide is most tightly bound, ^4He or ^2H.

• **30.17** (a) Calculate the number of grams of deuterium in an 80,000 L swimming pool, given deuterium is 0.0150% of natural hydrogen. (b) Find the energy released in joules if this deuterium is fused via the reaction $^2\text{H} + {}^2\text{H} \rightarrow {}^3\text{He} + n$ given in Equation 30.3b. (c) Could the neutrons be used to create more energy?

• **30.18** How many kilograms of water are needed to obtain the 198.8 mol of reactants in Example 30.1, assuming that deuterium is 0.01500% (by number) of natural hydrogen?

• **30.19** The power output of the sun is 4×10^{26} W. (a) If 90% of this is supplied by the proton-proton cycle, how many protons are consumed per second? (b) How many neutrinos per second should there be per square meter at the earth from this process? This huge number is indicative of how rarely a neutrino interacts, since large detectors observe very few per day.

• **30.20** Another set of reactions that result in the fusing of hydrogen into helium in the sun and especially in hotter stars is called the carbon cycle. It is

$$\begin{aligned}
{}^{12}\text{C} + {}^1\text{H} &\rightarrow {}^{13}\text{N} + \gamma \\
{}^{13}\text{N} &\rightarrow {}^{13}\text{C} + e^+ + \nu_e \\
{}^{13}\text{C} + {}^1\text{H} &\rightarrow {}^{14}\text{N} + \gamma \\
{}^{14}\text{N} + {}^1\text{H} &\rightarrow {}^{15}\text{O} + \gamma \\
{}^{15}\text{O} &\rightarrow {}^{15}\text{N} + e^+ + \nu_e \\
{}^{15}\text{N} + {}^1\text{H} &\rightarrow {}^{12}\text{C} + {}^4\text{He}
\end{aligned}$$

Write down the overall effect of the carbon cycle (as was done for the proton-proton cycle in Equation 30.2). Note the number of protons (^1H) required and assume that the positrons (e^+) annihilate electrons to form more γ rays.

• **30.21** Verify that the total number of nucleons, total charge, and electron family number are conserved for each of the fusion reactions in the carbon cycle given in Problem 30.20. (List the value of each of the conserved quantities before and after each of the reactions.)

: **30.22** (a) Find the total energy released in MeV in each carbon cycle (elaborated in Problem 30.20) including the annihilation energy. (b) How does this compare with the proton-proton cycle output?

Section 30.3 Fission

30.23 (a) Calculate the energy released in the neutron-induced fission (similar to the spontaneous fission in Example 30.2)

$$n + {}^{238}U \rightarrow {}^{96}Sr + {}^{140}Xe + 3n$$

given $m({}^{96}Sr) = 95.921750$ u. (b) This result is about 6 MeV greater than the spontaneous fission. Why? (c) Confirm that the total number of nucleons and total charge are conserved in this reaction.

30.24 (a) Calculate the energy released in the neutron-induced fission reaction

$$n + {}^{235}U \rightarrow {}^{92}Kr + {}^{142}Ba + 2n$$

given $m({}^{92}Kr) = 91.926269$ u and $m({}^{142}Ba) = 141.916361$ u. (b) Confirm that the total number of nucleons and total charge are conserved in this reaction.

30.25 (a) Calculate the energy released the neutron-induced fission reaction

$$n + {}^{239}Pu \rightarrow {}^{96}Sr + {}^{140}Ba + 4n$$

given $m({}^{96}Sr) = 95.921750$ u and $m({}^{140}Ba) = 139.910581$ u. (b) Confirm that the total number of nucleons and total charge are conserved in this reaction.

30.26 Confirm that each of the reactions listed for plutonium breeding just following Example 30.3 conserves the total number of nucleons, the total charge, and electron family number.

• **30.27** Breeding plutonium produces energy even before any plutonium is fissioned. Calculate the energy produced in each of the reactions listed for plutonium breeding just following Example 30.3. The pertinent masses are $m({}^{239}U) = 239.054289$ u, $m({}^{239}Np) = 239.052932$ u, and $m({}^{239}Pu) = 239.052157$ u.

• **30.28** The naturally occurring radioactive isotope ${}^{232}Th$ does not make good fission fuel, because it has an even number of neutrons, but it can be bred into a suitable fuel (much as ${}^{238}U$ is bred into ${}^{239}Pu$). (a) What are Z and N for ${}^{232}Th$? (b) Write the reaction equation for neutron capture by ${}^{232}Th$, and identify the nuclide ${}^{A}X$ produced in $n + {}^{232}Th \rightarrow {}^{A}X + \gamma$. (c) The product nucleus β^- decays, as does its daughter. Write the decay equations for each, and identify the final nucleus. (d) Confirm that the final nucleus has an odd number of neutrons, making it a better fission fuel. (e) Look up the half-life of the final nucleus to see if it lives long enough to be a useful fuel.

• **30.29** The electrical power output of a large nuclear reactor facility is 900 MW. It has a 35.0% efficiency in converting nuclear power to electrical. (a) What is the nuclear power output in megawatts? (b) How many ${}^{235}U$ nuclei fission each second, assuming the average fission produces 200 MeV? (c) What mass of ${}^{235}U$ is fissioned in one year of full-power operation?

• **30.30** A large power reactor that has been in operation for some months is turned off, but residual activity in the core still produces 150 MW of power. If the average energy per decay of the fission products is 1.00 MeV, what is the core activity in curies?

Section 30.4 Nuclear Weapons

In some of the following problems, the energy equivalents for bomb yields in kilotons and megatons are useful. They are 1 kT = 4.2×10^{12} J and 1 MT = 4.2×10^{15} J.

30.31 Find the mass converted into energy by a 12.0 kT bomb.

30.32 What mass is converted into energy by a 1.00 MT bomb?

30.33 Fusion bombs use neutrons from their fission trigger to create tritium fuel in the reaction $n + {}^{6}Li \rightarrow {}^{3}H + {}^{4}He$. What is the energy released by this reaction in MeV?

30.34 It is estimated that the total explosive yield of all the nuclear bombs in existence in 1988 was 12,000 MT. (a) Convert this amount of energy to kilowatt-hours, noting that 1 kW·h = 3.60×10^6 J. (b) What would the monetary value of this energy be if it could be converted to electricity costing 10 cents per kW·h?

• **30.35** A radiation-enhanced nuclear weapon (or neutron bomb) can have a smaller total yield and still produce more prompt radiation than a conventional nuclear bomb. This allows the use of neutron bombs to kill nearby advancing enemy forces with radiation without blowing up your own forces with the blast. For a 0.500 kT radiation-enhanced weapon and a 1.00 kT conventional nuclear bomb: (a) Compare the blast yields. (b) Compare the prompt radiation yields.

• **30.36** (a) How many ${}^{239}Pu$ nuclei must fission to produce a 20.0 kT yield, assuming 200 MeV per fission? (b) What is the mass of this much ${}^{239}Pu$?

• **30.37** Assume one-fourth of the yield of a typical 300 kT strategic bomb comes from fission reactions averaging 200 MeV and the remainder from fusion reactions averaging 20 MeV. (a) Calculate the number of fissions and the approximate mass of uranium and plutonium fissioned, taking the average atomic mass to be 238. (b) Find the number of fusions and calculate the approximate mass of fission fuel, assuming an average total atomic mass of the two nuclei in each reaction to be 5.

• **30.38** This problem gives some idea of the magnitude of the energy yield of a small tactical bomb. Assume that half the energy of a 1.00 kT nuclear depth charge set off under an aircraft carrier goes into lifting it out of the water—that is, into gravitational potential energy. How high is the carrier lifted if its mass is 90,000 tons?

• **30.39** Weapons tests in the atmosphere have deposited approximately 9 MCi of ${}^{90}Sr$ on the surface of the earth. Find the mass of this amount of ${}^{90}Sr$.

: **30.40** A 1.00 MT bomb exploded a few kilometers above the ground deposits 25.0% of its energy into radiant heat. (a) Find the calories per cm² at a distance of 10.0 km by assuming a uniform distribution over a spherical surface of that radius. (b) If this heat falls on a person's body, what temperature increase does it cause in the affected tissue, assuming it is absorbed in a layer 1.00 cm deep?

Sections 30.5–30.7 Biological Effects of Ionizing Radiation and Dose Calculations

30.41 What is the dose in rem for: (a) A 1.00 rad x ray? (b) 0.250 rad of neutron exposure to the eye? (c) 0.150 rad of α exposure?

30.42 Find the radiation dose in rad for: (a) A 10.0 rem fluoroscopic x-ray series. (b) 5.00 rem of skin exposure to an α emitter. (c) 16.0 mrem of β^- and γ rays from the ^{40}K in your body.

30.43 How many rad of exposure is needed to give a cancerous tumor a dose of 4000 rem if it is exposed to α activity?

30.44 What is the dose in rem in a cancer treatment that exposes the patient to 200 rad of γ rays?

30.45 One half the γs from ^{99m}Tc are absorbed by a 0.170 mm thickness of lead shielding. Half of the γs that pass through the first layer of lead are absorbed in a second layer of equal thickness. What thickness of lead will absorb all but one in 1000 of these γs?

30.46 A plumber at a nuclear power plant receives a whole body dose of 3 rem in 15 minutes while repairing a crucial valve. Find the radiation-induced yearly risk of death from cancer and the chance of genetic defect from this maximum allowable exposure.

30.47 In the 1980s the term picowave was used to describe food irradiation in order to overcome public resistance by playing on the well-known safety of microwave radiation. Find the energy in MeV of a photon having a wavelength of a picometer.

• **30.48** Find the mass of ^{239}Pu that has an activity of 1.00 μCi, and show that it is close to the value stated in Example 30.4.

• **30.49** Cisterns lined with radium salts were sold in the years around World War I. They were intended to be filled with water that dissolved some of the salt and would then be drunk daily for the benefits of radium ingestion. Suppose a person who does this ingests and concentrates 1.00 mg of ^{226}Ra in 8.00 kg of bone tissue, which absorbs all of the 4.80 MeV energy of the emitted α radiation. Calculate the yearly dose in rem to his bones. This calculation does not include the dose from radium daughters and assumes a balance between radium consumption and bodily elimination.

• **30.50** A beam of 168 MeV nitrogen nuclei is used for cancer therapy. If this beam is directed onto a 0.200 kg tumor and gives it a 200 rem dose, how many nitrogen nuclei were stopped? (Use an RBE of 20 for heavy ions.)

• **30.51** (a) If the average molecular mass of compounds in food is 50.0, how many molecules are there in 1.00 kg of food? (b) How many ion pairs are created in 1.00 kg of food, if it is exposed to 100,000 rad and it takes 30.0 eV to create an ion pair? (c) Find the ratio of ion pairs to molecules. (d) If these ion pairs recombine into a distribution of 2000 new compounds, how many parts per billion is each?

• **30.52** Calculate the dose in rem to the chest of a patient given an x ray under the following conditions. The x-ray beam intensity is 1.50 W/m^2, the area of the chest exposed is 0.0750 m^2, 35.0% of the x rays are absorbed in 20.0 kg of tissue, and the exposure time is 0.250 s.

• **30.53** A cancer patient is exposed to γ rays from a 5000 Ci ^{60}Co transillumination unit for 30.0 s. The γ rays are collimated in such a manner that only 1.00% of them strike the patient.

Of those, 20.0% are absorbed in a tumor having a mass of 1.50 kg. What is the dose in rem to the tumor, if the average γ energy per decay is 2.50 MeV? None of the βs from the decay reach the patient.

• **30.54** What is the mass of ^{60}Co in a cancer therapy transillumination unit containing 5000 Ci of ^{60}Co?

• **30.55** Large amounts of ^{65}Zn are produced in copper exposed to accelerator beams. While machining contaminated copper, a physicist ingests 50.0 μCi of ^{65}Zn. Each ^{65}Zn decay emits an average γ-ray energy of 0.550 MeV, 40.0% of which is absorbed in the scientist's 75.0 kg body. What dose in rem is caused by this in one day?

⁞ **30.56** Naturally occurring ^{40}K is listed as responsible for 16 mrem/y of background radiation in Table 30.4. Calculate the mass of ^{40}K that must be inside the 55 kg body of a woman to produce this dose. Each ^{40}K decay emits a 1.30 MeV β, 50% of which energy is absorbed inside the body.

⁞ **30.57** Background radiation due to ^{226}Ra averages only 1 mrem/y, but it can range upward depending on where a person lives. Find the mass of ^{226}Ra in the 80.0 kg body of a man who receives a dose of 250 mrem/y from it, noting that each ^{226}Ra decay emits a 4.80 MeV α particle. You may neglect dose due to daughters and assume a constant amount, evenly distributed due to balanced ingestion and bodily elimination.

⁞ **30.58** The annual radiation dose from ^{14}C in our bodies is 1 mrem/y. Each ^{14}C decay emits a β^- averaging 0.0750 MeV. Taking the fraction of ^{14}C to be 1.3×10^{-12} of normal ^{12}C, and assuming the body is 13% carbon, estimate the fraction of the decay energy absorbed. (The rest escapes, exposing those close to you.)

INTEGRATED CONCEPTS

Physics is most interesting and most powerful when applied to general situations involving more than a narrow set of physical principles. The integration of concepts necessary to solve problems involving several physical principles also gives greater insight into the unity of physics. You may wish to refer to Chapters 6, 7, 12, 13, 26, 27, and perhaps others, to solve the following problems.

Note: Problem-solving strategies and worked examples that can help you solve integrated concept problems appear in several places, the most recent being Chapter 29. Consult the section of problems labeled *Integrated Concepts* at the end of that chapter or others.

• **30.59** To show that the energy deposited by ionizing radiation is ordinarily small compared with macroscopic energies, calculate the temperature increase of a person who receives a 1.00 rad dose of radiation, assuming her specific heat is 0.840 kcal/kg·°C.

• **30.60** Calculate the temperature increase in food irradiated with a dose of 100,000 rad, assuming its specific heat is 0.900 kcal/kg·°C.

• **30.61** How many photons strike a patient being x-rayed, where an intensity of 1.50 W/m^2 illuminates 0.0750 m^2 of her body for 0.250 s? The energy of the x-ray photons is 100 keV.

• **30.62** One scheme to put nuclear weapons to nonmilitary use is to explode them underground in a geologically stable region and extract the geothermal energy for electricity production. There was a total yield of about 12,000 MT in the combined arsenals in 1988. If 1.00 MT per day could be converted to electricity with an efficiency of 10.0%: (a) What would the average electrical power output be? (b) How many years would the arsenal last at this rate? Assume the world's energy consumption rate is 20 times that of the United States.

⠸ 30.63 The laser system tested for inertial confinement can produce a 100 kJ pulse only 1.00 ns in duration. (a) What is the power output of the laser system during the brief pulse? (b) How many photons are in the pulse, given their wavelength is 1.06 μm? (c) What is the total momentum of all these photons? (d) How does the total photon momentum compare with that of a single 1.00 MeV deuterium nucleus?

⠸ 30.64 Find the amount of energy given to the ^{4}He nucleus and to the γ in the reaction $n + {}^3\text{He} \rightarrow {}^4\text{He} + \gamma$, using the conservation of momentum principle and taking the reactants to be initially at rest. This should confirm the contention that most of the energy goes to the γ ray.

⠸ 30.65 What temperature gas would have atoms moving fast enough to bring two ^{3}He nuclei into contact? Note that because both are moving, the average kinetic energy only needs to be half the electric potential energy of these doubly charged nuclei when just in contact with one another.

⠸ 30.66 Estimate the years that the deuterium fuel in the oceans could supply the energy needs of the world. Assume world energy consumption to be ten times that of the United States given in Table 6.1 and that the deuterium in the oceans could be converted to energy with an efficiency of 30.0%. You must estimate or look up the amount of water in the oceans and take the deuterium content to be 0.015% of natural hydrogen to find the mass of deuterium available. Note that Example 30.1 gives an approximate figure for the energy yield per kilogram of deuterium.

UNREASONABLE RESULTS

The following problems have results that are unreasonable because some premise is unreasonable or because certain of the premises are inconsistent with one another. Physical principles applied correctly then produce unreasonable results. The purpose of these problems is to give practice in assessing whether nature is being accurately described, and if it is not to trace the source of difficulty.

Note: Problem-solving strategies that can help you solve unreasonable results problems appear in several places, the most recent being Chapter 29. Consult the section of problems labeled *Unreasonable Results* at the end of that chapter or others.

30.67 A nuclear scan is to be taken of a patient in which 20,000,000 γ rays of energy 0.160 MeV will strike the patient with 75.0% being absorbed in 10.0 kg of tissue. (a) What is the radiation dose to the affected tissue? (b) What is unreasonable about this result? (c) What assumptions are unreasonable or inconsistent?

• 30.68 The activity label on a ^{60}Co γ ray calibration source (intended for calibrating γ-ray detector systems) is unclear, and the technician handling it assumes that the entire 10.0 g of the source is pure ^{60}Co. (a) What is the activity of 10.0 g of ^{60}Co? (b) What is unreasonable about this result? (c) What is unreasonable about the assumption?

• 30.69 An activity of 7.50 mCi of ^{99m}Tc is needed for a diagnostic scan. (a) What activity would the source have had to be when manufactured, if it was made 3 weeks before its use? (b) What is unreasonable about this result? (c) What assumptions are unreasonable or inconsistent?

This computer display shows the tracks of particles detected in an experiment

to probe the smallest structures of matter.

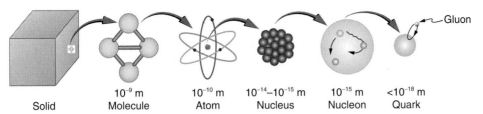

Figure 31.1 The properties of matter are based on substructures called molecules and atoms. Atoms in turn have the substructure of a nucleus with orbiting electrons, the interactions of which explain atomic properties. Protons and neutrons, the interactions of which explain the stability and abundance of elements, form the substructure of nuclei. Protons and neutrons themselves are not fundamental—they are composed of quarks. Like electrons and a few other particles, quarks may be the fundamental building blocks of all there is, lacking any further substructure. But the story is not complete, because quarks and electrons may have substructure smaller than details presently observable.

Following ideas remarkably similar to those of the ancient Greeks, we continue to look for smaller and smaller structure in nature, hoping ultimately to find and understand the most fundamental building blocks that exist. Atomic physics deals with the smallest units of elements and compounds. In its study, we found a relatively small number of atoms with systematic properties that explained a tremendous range of phenomena. Nuclear physics is concerned with the nuclei of atoms and their substructures. Here a smaller number of components—the proton and neutron—make up all nuclei. Exploring the systematics of their interactions has revealed even more about matter, forces, and energy. **Particle physics** deals with the substructures of atoms and nuclei and is particularly aimed at finding those truly fundamental particles that have no further substructure. Just as in atomic and nuclear physics, we have found a complex array of particles and properties with systematic characteristics analogous to the periodic table and the chart of nuclides. An underlying structure is apparent, and there is some reason to think that we *are* finding particles that have no substructure. Of course, we have been in similar situations before. Atoms were once thought to be the ultimate substructure, for example. Perhaps we will find deeper and deeper structures and never come to an ultimate substructure. We may never know. (See Figure 31.1.)

This chapter covers the basics of particle physics as we know it today. An amazing convergence of topics is evolving in particle physics. We find that particles are intimately related to forces, and that nature on the smallest scale may have its greatest influence on the large-scale character of the universe. It is an adventure exceeding the best science fiction because it is not only fantastic, it is real.

31.1 THE YUKAWA PARTICLE AND THE HEISENBERG UNCERTAINTY PRINCIPLE REVISITED

Particle physics as we know it today began with the ideas of Hideki Yukawa in 1935. Physicists had long been concerned with how forces are transmitted, finding the concept of a field to be very useful (such as electric and magnetic fields, for example). A field surrounds an object and carries the force the object exerts through space. Yukawa was interested in the strong nuclear force in particular and found an ingenious way to explain its short range. His idea is a blend of particles, forces, relativity, and quantum mechanics that is applicable to all forces. Yukawa proposed that force is transmitted by the exchange of particles (called carrier particles). The field consists of these carrier particles. On a macroscopic scale, such as the magnet you stick on a refrigerator, the number of particles is so large that the field seems continuous. But on a small scale, the quantum character is important.

Specifically for the strong nuclear force, Yukawa proposed that a previously unknown particle, now called a **pion**, is exchanged between nucleons, transmitting the force between them. Figure 31.2 illustrates how a pion would carry a force between a proton and a neutron. (Note that this concept was first discussed in Section 4.6 and illustrated

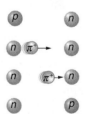

Figure 31.2 The strong nuclear force is transmitted between a proton and neutron by the creation and exchange of a pion. The pion is created through a temporary violation of conservation of mass-energy and travels from the proton to the neutron and is recaptured. It is not directly observable and is called a virtual particle. Note that the proton and neutron change identity in the process. The range of the force is limited by the fact that the pion can only exist for the short time allowed by the Heisenberg uncertainty principle. Yukawa used the finite range of the strong nuclear force to estimate the mass of the pion; the shorter the range, the larger the mass of the carrier particle.

there in Figure 4.17.) The pion has mass and can only be created by violating conservation of mass-energy. This is allowed by the Heisenberg uncertainty principle if it occurs for a sufficiently short period of time. As was discussed in Section 27.7, the Heisenberg uncertainty principle relates the uncertainties in energy ΔE and time Δt; by Equation 27.10,

$$\Delta E \, \Delta t \geq \frac{h}{4\pi}$$

where h is Planck's constant. So conservation of mass-energy can be violated by an amount ΔE for a time $\Delta t \approx h/(4\pi \Delta E)$, in which time no process can detect the violation. This allows the temporary creation of a particle of mass m, where $\Delta E = mc^2$. The larger the mass, the greater ΔE and the shorter the time it can exist. This means the range of the force is limited, because the particle can only travel a limited distance in a finite amount of time. In fact, the maximum distance is $d \approx c \, \Delta t$, where c is the speed of light. The pion must then be captured and, thus, cannot be directly observed because that would amount to a permanent violation of mass-energy conservation. Such particles (like the pion above) are called **virtual particles**, because they cannot be directly observed but their *effects* can be directly observed. Realizing all this, Yukawa used information about the range of the strong nuclear force to estimate the mass of the pion, the particle that carries it. His reasoning steps are approximately retraced in the following worked example.

EXAMPLE 31.1   ESTIMATING THE MASS OF A CARRIER PARTICLE CALLED A PION

Taking the range of the strong nuclear force to be about 1 fermi (10^{-15} m), calculate the approximate mass of the pion carrying the force, assuming it moves at nearly the speed of light.

Strategy and Concept The calculation is approximate because of the assumptions made about the range of the force and the speed of the pion, but also because a more accurate calculation would require the sophisticated mathematics of quantum mechanics. Here we use the Heisenberg uncertainty principle in the simple form stated above as developed in Chapter 27. First we must calculate the time Δt the pion exists, given the distance it travels at nearly the speed of light is about 1 fermi. Then the Heisenberg uncertainty principle can be solved for the energy ΔE, and from that the mass of the pion can be determined. We will use the units of MeV/c^2 for mass, which are convenient since we are often considering converting mass to energy and vice versa.

Solution The distance the pion travels is $d \approx c \, \Delta t$, and so the time it exists is approximately

$$\Delta t \approx \frac{d}{c} = \frac{10^{-15} \text{ m}}{3.0 \times 10^8 \text{ m/s}}$$

$$\approx 3.3 \times 10^{-24} \text{ s}$$

Now solving the Heisenberg uncertainty principle for ΔE gives

$$\Delta E \approx \frac{h}{4\pi \, \Delta t} = \frac{6.63 \times 10^{-34} \text{ J} \cdot \text{s}}{4\pi (3.3 \times 10^{-24} \text{ s})}$$

Solving this and converting the energy to MeV gives

$$\Delta E \approx 1.6 \times 10^{-11} \text{ J} \cdot \frac{1 \text{ MeV}}{1.6 \times 10^{-13} \text{ J}} = 100 \text{ MeV}$$

Mass is related to the energy by $\Delta E = mc^2$, so that the mass of the pion is $m = \Delta E/c^2$, or

$$m \approx 100 \text{ MeV}/c^2$$

Discussion This is about 200 times the mass of an electron and about one-tenth the mass of a nucleon. No such particles were known at the time Yukawa made his bold proposal.

Yukawa's proposal of particle exchange as the method of force transfer is intriguing. But how can we verify his proposal if we cannot observe the virtual pion directly? If sufficient energy is put into a nucleus, it might be possible to free the pion—that is, to create its mass from external energy input. This can be accomplished by collisions of energetic particles with nuclei, but energies greater than 100 MeV are required to conserve both energy and momentum. (See Figure 31.3.) In 1947, pions were observed in cosmic ray experiments in which nature supplies a small flux of high-energy protons that may collide with nuclei. Soon afterward, accelerators of sufficient energy were creating pions in the laboratory under controlled conditions. Three pions were discovered, two with charge and one neutral, and given the symbols π^+, π^-, and π^0. The masses of the π^+ and

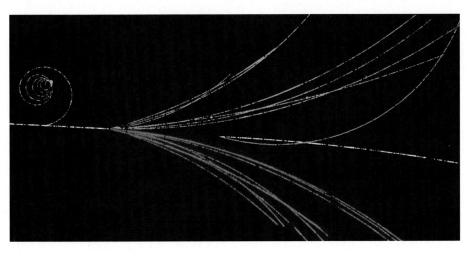

Figure 31.3 In this image, positive ions have red colors, and negative ions have blue colors. The tracks in this bubble chamber are in part due to a prolific production of pions in a nuclear collision. It is no accident that when enough energy is put into a nucleus the particles created from it are largely pions, the carriers of the strong nuclear force. (Bubble chambers are no longer in common use, as particle detector technology has advanced.)

π^- are identical, being equal to 139.6 MeV/c^2, while the π^0 has a mass of 135.0 MeV/c^2. These masses are close to the predicted value of 100 MeV/c^2 and, since they are intermediate between electron and nucleon masses, the particles are given the name **meson** (now an entire class of particles, as we shall see in Section 31.4).

Not only did the pions, or π-mesons as they are also called, have masses close to those predicted, they also had other predicted characteristics, such as zero intrinsic spin, and they do feel the strong nuclear force. Another previously unknown particle, now called the muon, was discovered in cosmic ray experiments in 1936 (one of its discoverers, Seth Neddermeyer, also originated the idea of implosion for plutonium bombs). Since the mass of a muon is about 106 MeV/c^2, it was at first thought to be the particle predicted by Yukawa. But it was soon realized that muons do not feel the strong nuclear force and could not be Yukawa's particle. Their role was unknown, causing the respected physicist I. I. Rabi to comment, "Who ordered that?" This remains a valid question today. We have discovered hundreds of subatomic particles; the roles of some are only partially understood. But there are patterns and relations to forces that have led to profound insights into nature's secrets.

31.2 THE FOUR BASIC FORCES

As first discussed in Section 4.6, and mentioned at various points in the text since then, there are only four distinct basic forces in all of nature. This is a remarkably small number considering the myriad phenomena they explain. Particle physics is intimately tied to these four forces. Certain fundamental particles, called carrier particles, carry these forces, and all particles can be classified according to which of the four forces they feel. Table 31.1 summarizes important characteristics of the four basic forces (it is very similar to Table 4.2).

TABLE 31.1

PROPERTIES OF THE FOUR BASIC FORCES				
Force	*Approximate relative strength*	*Range*	*+/−**	*Carrier particle*
Gravity	10^{-38}	∞	+ only	Graviton (conjectured)
Electromagnetic	10^{-2}	∞	+/−	Photon (observed)
Weak nuclear	10^{-13}	$<10^{-18}$ m	+/−	W^+, W^-, Z^0 (observed[†])
Strong nuclear	1	$<10^{-15}$ m	+/−	Gluons (conjectured[††])

*+, attractive; −, repulsive; +/−, both.
[†]Predicted by theory and first observed in 1983.
[††]Eight proposed—indirect evidence of existence. Underlie meson exchange.

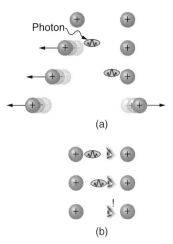

Figure 31.4 (a) The exchange of a virtual photon transmits the electromagnetic force between charges, just as virtual pion exchange carries the strong nuclear force between nucleons. (b) The photon cannot be directly observed in its passage, because this would disrupt it and alter the force. In this case it does not get to the other charge.

Figure 31.6 Richard Feynman was a brilliant physicist with a colorful personality. He made many fundamental contributions to QED and particle physics and has been the subject of several books. Feynman was fond of parlor tricks and puzzles and took a dim view of puffy authority. One particularly public example was his stinging criticism of NASA following his participation in the Challenger disaster investigation.

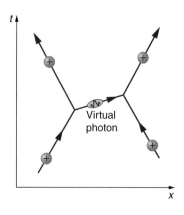

Figure 31.5 The Feynman diagram for the exchange of a virtual photon between two positive charges illustrates how the electromagnetic force is transmitted on a quantum mechanical scale. Time is graphed vertically and distance horizontally. The two positive charges are seen to be repelled by the photon exchange.

Although these four forces are distinct and differ greatly from one another under all but the most extreme circumstances, we can see some similarities among them. (In Section 31.6 we will discuss specifics about how the four forces may be different manifestations of a single unified force.) Perhaps the most important similarity among the forces is that they are all transmitted by the exchange of a carrier particle, exactly like what Yukawa had in mind for the strong nuclear force. Each carrier particle is a virtual particle—it cannot be directly observed while transmitting the force. Figure 31.4 shows the exchange of a virtual photon between two positive charges. It is evident that if the photon is observed, it is altered and the force is not transmitted.

Figure 31.5 shows a clever way of graphing the exchange of a virtual photon between two positive charges. This graph of time versus position is called a **Feynman diagram**, after the brilliant American physicist Richard Feynman (1918–1988) who developed it. (See Figure 31.6.) Figure 31.7 is a Feynman diagram for the exchange of a virtual pion between a proton and a neutron representing the same interaction as in Figure 31.2. Feynman diagrams are not only a useful tool for visualizing quantum interactions, they are also used to calculate details of interactions, such as their strengths and probability of occurring. Feynman was one of the theorists who developed the field of **quantum electrodynamics** (QED), which is the quantum mechanics of electromagnetism. QED has been spectacularly successful in describing electromagnetic interactions on the submicroscopic scale. Feynman was an inspiring teacher, had a colorful personality, and made a profound impact on generations of physicists. He shared the 1965 Nobel prize with Julian Schwinger and S. I. Tomonaga for work in QED with its deep implications for particle physics.

Why is it that particles called gluons are listed as the carrier particles for the strong nuclear force when we just found in the preceding section that pions carry that force? The answer is that pions are exchanged but they have a substructure, and as we explore it we will find the strong force is really related to the indirectly observed but more fundamental **gluons**. In fact, all the carrier particles are thought to be fundamental in the sense that they have no substructure. Another similarity among carrier particles is that they are all bosons, having integral intrinsic spins.

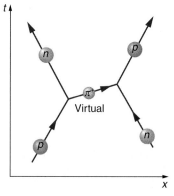

Figure 31.7 Feynman diagram for the exchange of a π^+ between a proton and a neutron, carrying the strong nuclear force between them. This diagram represents the situation shown more pictorially in Figure 31.2.

There is a relationship between the mass of the carrier particles and the range of the force. The photon is massless, and so it can be created without violating conservation of mass-energy and can travel an unlimited distance. Thus the electromagnetic force is infinite in range. This is also true for gravity. It is infinite in range because its carrier particle, the graviton, is massless. (Gravity is the most difficult of the four forces on a quantum scale, in that it affects the space and time in which the others act. But gravity is so weak that its effects are extremely difficult to observe quantum mechanically. We shall explore it further in Chapter 32.) The W^+, W^-, and Z^0 particles that carry the weak nuclear force are very massive, accounting for its very short range. In fact, the W^+, W^-, and Z^0 are about 1000 times more massive than pions, consistent with the fact that the range of the weak nuclear force is about 1/1000 that of the strong nuclear force. Gluons are actually massless, but since they act inside massive carrier particles like pions, the strong nuclear force is also short ranged.

The relative strengths of the forces given in Table 31.1 are those for the most common situations. When particles are brought very close together, the relative strengths change, and they may become identical at extremely close range. As we shall see in the section on the unification of forces, carrier particles may be altered by the energy required to bring particles very close together—in such a manner as to become identical.

31.3 ACCELERATORS CREATE MATTER FROM ENERGY

Before looking at all the particles we now know about, it is amusing to examine some of the machines that created them. The fundamental process in creating previously unknown particles is to accelerate known particles, such as protons or electrons, and direct a beam of them toward a target. Collisions with target nuclei provide a wealth of information, such as Rutherford obtained using energetic helium nuclei from natural α radiation. But if the energy of the incoming particles is great enough, new matter is sometimes created in the collision. The more energy input ΔE, the more matter m that can be created, since $m = \Delta E/c^2$. Limitations are placed on what can occur by known conservation laws, such as conservation of mass-energy, momentum, and charge. Even more interesting are the unknown limitations provided by nature. Some expected reactions do occur, while others do not, and still other unexpected reactions may appear. New laws are revealed, and the vast majority of what we know about particle physics has come from accelerator laboratories. It is the particle physicist's favorite indoor sport, partly inspired by theory.

Early Accelerators

One early accelerator is a relatively simple, large-scale version of the electron gun. The **Van de Graaff** (named for the Dutch physicist) you have likely seen in physics demonstrations (as in Figure 17.23) is a small version of the ones used for nuclear research since their invention for that purpose in 1932. (See Figure 31.8.) These machines are electrostatic, creating potentials as great as 50 MV, and are used to accelerate a variety of nuclei for a range of experiments. Energies produced by Van de Graaffs are insufficient to produce new particles, but they have been instrumental in exploring many aspects of the nucleus. Another, equally famous, early accelerator is the **cyclotron**, invented in 1930 by the American physicist E. O. Lawrence (1901–1958). (See Figure 31.9.) Cyclotrons use fixed-frequency alternating electric fields to accelerate particles. The particles spiral outward in a magnetic field, making increasingly larger radius orbits during acceleration. This clever arrangement allows for the successive addition of electric potential energy, and so greater particle energies are possible than in a Van de Graaff. Lawrence was involved in many early discoveries and in the promotion of physics programs in American universities. He was awarded the 1939 Nobel prize in physics for the cyclotron and nuclear activations, and he has an element and two major laboratories named for him.

A **synchrotron** is a version (actually several versions) of a cyclotron in which the frequency of the alternating voltage and the magnetic field strength are increased as the beam particles are accelerated. Particles are made to travel the same distance in a shorter time each cycle in fixed-radius orbits. A ring of magnets and accelerating

Figure 31.8 This Van de Graaff is part of the Free Electron Laser at the Thomas Jefferson National Accelerator Facility. Many modern accelerators are composed of several staged injectors and accelerators.

Figure 31.9 (a) Cyclotrons use a magnetic field to cause particles to have circular orbits. As the particles pass between the plates of the Ds, the voltage across the gap is oscillated to accelerate them twice in each orbit. (b) E. O. Lawrence.

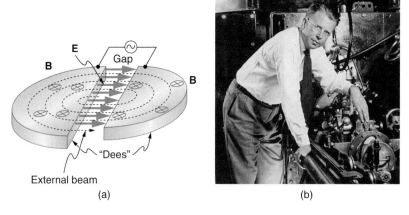

(a) (b)

tubes, such as shown in Figure 31.10, are the major components of synchrotrons. Accelerating voltages are synchronized with the particles to accelerate them, hence the name. Magnetic field strength is increased to keep the orbital radius constant as energy increases. High-energy particles require very large magnetic fields, and superconducting magnets are commonly employed. Still limited by achievable magnetic field strengths, synchrotrons must be very large at very high energies, since the radius of a high-energy particle's orbit is very large. Perhaps the major problem, other than the economics of scale in reaching very high energies, is the fact that charged particles radiate when accelerated, including when forced to follow a curved path in a magnetic field. This energy radiated from the beam must be replaced faster than it radiates away in order to continue to accelerate the particles (only a problem for electron accelerators). Radiation caused by a magnetic field accelerating a charged particle perpendicular to its velocity is called **synchrotron radiation** in honor of its importance in these machines. Synchrotron radiation has a characteristic spectrum and polarization and can be recognized in cosmic rays, implying large-scale magnetic fields acting on energetic charged particles in deep space. Synchrotron radiation produced by accelerators is sometimes used as a source of intense energetic electromagnetic radiation for research purposes.

Modern Behemoths and Colliding Beams

Physicists have built ever-larger machines, first to reduce the wavelength of the probe and obtain greater detail, then to put greater energy into collisions to create new particles. Each major energy increase brought new information, sometimes producing spectacular

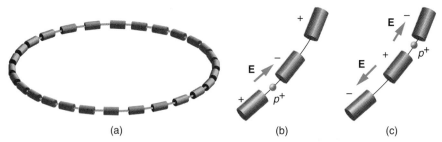

(a) (b) (c)

Figure 31.10 (a) A synchrotron has a ring of magnets and accelerating tubes. The frequency of the accelerating voltages is increased to cause the beam particles to travel the same distance in shorter times. The magnetic field must also be increased to keep each beam burst traveling in a fixed-radius path. Limits on magnetic field strength require these machines to be very large in order to accelerate particles to very high energies. (b) A positive particle is shown in the gap between accelerating tubes. (c) While the particle passes through the tube, the potentials are reversed so that there is another acceleration at the next gap. The frequency of the reversals must be varied as the particle is accelerated to achieve successive accelerations in each gap.

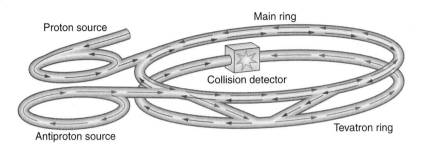

Proton source

Main ring

Collision detector

Antiproton source

Tevatron ring

Figure 31.11 This schematic shows the two rings of Fermilab's accelerator and the scheme for colliding protons and antiprotons.

progress, motivating the next step. One major innovation was driven by the desire to create more massive particles. Since momentum must be conserved in a collision, the particles created by a beam hitting a stationary target must recoil. This means that part of the energy input goes into recoil kinetic energy, significantly limiting the fraction of the beam energy that can be converted into new particles. One solution to this problem is to have head-on collisions between particles moving in opposite directions. **Colliding beams** are made to meet head-on at points where massive detectors are located. Since the total incoming momentum is zero, it is possible to create particles with momenta and kinetic energies near zero. Particles with masses equivalent to twice the beam energy can thus be created. Another innovation is to create the antimatter counterpart of the beam particle, which thus has the opposite charge and will circulate in the opposite direction in the same beam pipe. This has an extra advantage in that the initial quantum numbers in the collision mostly total zero, allowing for the creation of a wider range of particles. (See Figure 31.11.)

Detectors capable of finding the new particles in the spray of material that emerges from colliding beams are as impressive as the accelerators. Figure 31.12 shows a computer reconstruction of an event in a detector at the European Center for Nuclear Research (CERN) in Geneva. While the Fermilab Tevatron has proton and antiproton beam energies of about 1 TeV, so it can create particles up to 2 TeV/c^2, the Large Hadron Collider (LHC) at CERN achieves beam energies of 7 TeV, so that it has a 14 TeV collision energy. The now-canceled Superconducting Super Collider was being constructed in Texas with a design energy of 20 TeV to give 40 TeV collision energy. It was to be an oval 30 km in diameter. Its cost as well as the politics of international research funding doomed it.

In addition to the large synchrotrons that produce colliding beams of protons and antiprotons, there are two very large electron-positron accelerators. The oldest of these is a straight-line or **linear accelerator**, called the Stanford Linear Accelerator (SLAC), which has a beam energy of 50 GeV. (See Figure 31.13.) Positrons created by the accelerator are brought to the same energy and collided with electrons in

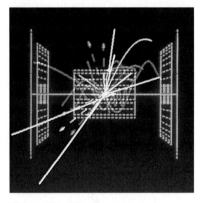

Figure 31.12 Computer reconstruction of an event observed by the detector.

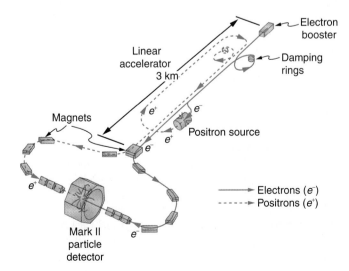

Electron booster

Linear accelerator 3 km

Damping rings

Magnets

e^+ e^-

Positron source

e^+

e^-

e^+

Mark II particle detector

e^-

→ Electrons (e^-)
--→ Positrons (e^+)

Figure 31.13 The Stanford Linear Accelerator is 3.2 km long and has the capability of colliding electron and positron beams. SLAC has also been used to probe nucleons by scattering extremely short wavelength electrons from them. This produced the first convincing evidence of a quark structure inside nucleons in an experiment analogous to those performed by Rutherford so long ago.

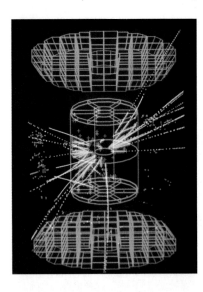

Figure 31.14 The Large Electron-Positron Collider at CERN actually straddles the border between Switzerland and France. This computer display reproduces the outline of the detector and a detected collision.

specially designed detectors. Linear accelerators use accelerating tubes similar to those in synchrotrons, but aligned in a straight line. This helps eliminate synchrotron radiation losses, which are particularly severe for electrons made to follow curved paths. CERN has an electron-positron collider appropriately called the Large Electron-Positron Collider (LEP) (see Figure 31.14), with beam energy of 46 GeV and collision energy of 92 GeV. It is 8.5 km in diameter, while the SLAC machine is 3.2 km long.

EXAMPLE 31.2 WHAT VOLTAGE DOES THIS ACCELERATOR NEED BETWEEN ACCELERATING TUBES?

A linear accelerator designed to produce a beam of 800 MeV protons has 2000 accelerating tubes. What average voltage must be applied between tubes (such as in the gaps in Figure 31.10) to achieve the desired energy?

Strategy The energy given to the proton in each gap between tubes is $PE_{elec} = qV$, where q is the proton's charge and V is the potential difference (voltage) across the gap. Since $q = q_e = 1.6 \times 10^{-19}$ C, and 1 eV = (1 V)(1.6 $\times$ 10^{-19} C), the proton gains 1 eV in energy for each volt across the gap it passes through. The AC voltage applied to the tubes is timed so that it adds to the energy in each gap, and so the effective voltage is the sum of the gap voltages and must equal 800 MV to produce an energy of 800 MeV.

Solution There are 2000 gaps and the sum of the voltages across them is 800 MV; thus,

$$V_{gap} = \frac{800 \text{ MV}}{2000} = 400 \text{ kV}$$

Discussion and Implications A voltage of this magnitude is not difficult to achieve in a vacuum. Much larger gap voltages would be required for higher energy, such as at the 50 GeV SLAC facility. Synchrotrons are aided by the circular path of the accelerated particles, which can orbit many times, effectively multiplying the number of accelerations by the number of orbits. This makes it possible to reach energies greater than 1 TeV.

31.4 PARTICLES, PATTERNS, AND CONSERVATION LAWS

In the early 1930s only a small number of subatomic particles were known to exist—the proton, neutron, electron, photon, and indirectly, the neutrino. Nature seemed relatively simple in some ways, but mysterious in others. Why, for example, should the particle that carries positive charge be almost 2000 times as massive as the one carrying negative charge? Why does a neutral particle like the neutron have a magnetic moment? Does this imply an internal structure with a distribution of moving charges? Why is it that the electron seems to have no size other than its wavelength, while the proton and neutron are about 1 fermi in size? So while the number of known particles was small and they explained a great deal of atomic and nuclear phenomena, there were glaring unexplained asymmetries and hints of further substructure.

Things soon became more complicated, both in theory and in the prediction and discovery of new particles. In 1928, the British physicist P. A. M. Dirac developed a highly successful relativistic quantum theory that laid the foundations of QED (quantum electrodynamics). (See Figure 31.15.) His theory, for example, explained electron spin and magnetic moment in a natural way. But Dirac's theory also predicted negative energy states for free electrons. By 1931, Dirac, along with Oppenheimer, realized this was a prediction of positive electrons (or positrons). In 1932, American physicist Carl Anderson discovered the positron in cosmic ray studies. The positron, or e^+, is the same particle as emitted in β^+ decay and was the first antimatter discovered. In 1935, Yukawa predicted pions as the carriers of the strong nuclear force, and they were eventually discovered. Muons were discovered in cosmic ray experiments in 1937, and they seemed to be heavy, unstable versions of electrons and positrons. After World War II, accelerators energetic enough to create these particles were built. Not only were predicted and known particles created, but many unexpected particles were observed. At first called elementary particles, their numbers proliferated to the dozens and then hundreds, and the term "particle zoo" became the physicist's lament at the lack of simplicity. But patterns were observed in the particle zoo that lead to simplifying ideas such as quarks, as we shall soon see.

Matter and Antimatter

The positron was only the first example of antimatter. Every particle in nature has an antimatter counterpart. Antimatter has charge opposite to that of matter (the positron is positive while the electron is negative, for example) but is nearly identical otherwise, having the same mass, intrinsic spin, half-life, and so on. When a particle and its antimatter counterpart interact, they annihilate one another, usually totally converting their masses to pure energy in the form of photons. (See Figure 31.16.) Neutral particles, such as neutrons, have neutral antimatter counterparts, which also annihilate when they interact. Certain neutral particles are their own antiparticle and live correspondingly short lives. For example, the neutral pion π^0 is its own antiparticle and has a half-life about 10^{-8} shorter than the π^+ and π^-, which are each other's antiparticles. Without exception, nature is symmetric—all particles have antimatter counterparts. Antiprotons and antineutrons were first created in accelerator experiments in 1956, for example, and the antiproton is negative. Antihydrogen atoms, consisting of an antiproton and antielectron, were observed in 1995 at CERN, too. Large-scale antimatter is possible in principle. At any rate, we now see that negative charge is associated with both low-mass (electrons) and high-mass particles (antiprotons) and the apparent asymmetry is not there. But this knowledge does raise another question—why is there such a predominance of matter and so little antimatter? Possible explanations emerge later in this and the next chapter.

Hadrons and Leptons

Particles can also be revealingly grouped according to what forces they feel. All particles (even those which are massless) are affected by gravity, since gravity affects the space and time in which particles exist. All charged particles are affected by the electromagnetic force, as are neutral particles that have an internal distribution of charge (such as the neutron with its magnetic moment). Special names are given to particles that feel the strong and weak nuclear forces. **Hadrons** are particles that feel the strong nuclear force, whereas **leptons** are particles that do not. The proton, neutron, and the pions are examples of hadrons. The electron, positron, muons, and neutrinos are examples of leptons, the name meaning low mass. Leptons do feel the weak nuclear force. In fact, all particles feel the weak nuclear force. This means that hadrons are distinguished by being able to feel both the strong and weak nuclear forces.

Table 31.2 lists characteristics of some of the most important subatomic particles, including the directly observed carrier particles for the electromagnetic and weak nuclear forces, all leptons, and some hadrons. Many hints about an underlying substructure emerge from an examination of these particle characteristics. First note that the carrier particles are called **gauge bosons**. First mentioned in Section 28.7, a **boson** is a particle with an integral intrinsic spin (such as $s = 0, 1, 2, \ldots$), whereas a **fermion** is a particle with a half-integral intrinsic spin ($s = 1/2, 3/2, \ldots$). Bosons do not obey the Pauli exclusion principle; fermions do. All the known and conjectured carrier particles are bosons. It is interesting to note that although the photon is its own antiparticle, it is stable and never decays. The reason for this is that photons are massless and travel at the speed of light. Time dilation approaches infinity as velocity approaches the speed of light. This implies that time does not pass for photons, and so they cannot self-annihilate.

All known leptons are listed in Table 31.2. There are only six (and their antiparticles), and they seem to be fundamental in that they have no apparent underlying structure.

Figure 31.15 P. A. M. Dirac's theory of relativistic quantum mechanics not only explained a great deal of what was known, it also predicted antimatter.

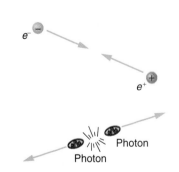

Figure 31.16 When a particle encounters its antiparticle, they annihilate, often producing pure energy in the form of photons. In this case an electron and a positron convert all of their mass into two identical energy γ rays, which move away in opposite directions to keep total momentum zero as it was before. Similar annihilations occur for other combinations of a particle with its antiparticle, sometimes producing yet more particles while obeying all conservation laws.

TABLE 31.2

SELECTED PARTICLE CHARACTERISTICS*

Category	Particle name	Symbol	Anti-particle	Rest mass (MeV/c²)	B	L_e	L_μ	L_τ	S	Lifetime† (s)
Gauge Bosons	Photon	γ	Self	0	0	0	0	0	0	Stable
	W	W^+	W^-	80.22×10^3	0	0	0	0	0	3×10^{-25}
	Z	Z^0	Self	91.19×10^3	0	0	0	0	0	3×10^{-25}
Leptons	Electron	e^-	e^+	0.511	0	±1	0	0	0	Stable
	Neutrino (e)	v_e	$\bar{v}_e$	0 (<7.0 eV)††	0	±1	0	0	0	Stable
	Muon	μ^-	μ^+	105.7	0	0	±1	0	0	2.20×10^{-6}
	Neutrino (μ)	v_μ	$\bar{v}_\mu$	0 (<0.27)	0	0	±1	0	0	Stable
	Tau	τ^-	τ^+	1777	0	0	0	±1	0	2.29×10^{-13}
	Neutrino (τ)	v_τ	$\bar{v}_\tau$	0 (<31)	0	0	0	±1	0	Stable
Hadrons (selected)										
Mesons	Pion	π^+	π^-	139.6	0	0	0	0	0	2.60×10^{-8}
		π^0	Self	135.0	0	0	0	0	0	0.84×10^{-16}
	Kaon	K^+	K^-	493.7	0	0	0	0	±1	1.24×10^{-8}
		K^0	$\bar{K}^0$	497.7	0	0	0	0	±1	0.89×10^{-10}
	Eta	η^0	Self	547.5	0	0	0	0	0	5×10^{-19}
(many other mesons known)										
Baryons	Proton	p	$\bar{p}$	938.3	±1	0	0	0	0	Stable†††
	Neutron	n	$\bar{n}$	939.6	±1	0	0	0	0	887
	Lambda	Λ^0	$\bar{\Lambda}^0$	1115.7	±1	0	0	0	∓1	2.63×10^{-10}
	Sigma	Σ^+	$\bar{\Sigma}^-$	1189.4	±1	0	0	0	∓1	0.80×10^{-10}
		Σ^0	$\bar{\Sigma}^0$	1192.6	±1	0	0	0	∓1	7.4×10^{-20}
		Σ^-	$\bar{\Sigma}^+$	1197.4	±1	0	0	0	∓1	1.48×10^{-10}
	Xi	Ξ^0	$\bar{\Xi}^0$	1314.9	±1	0	0	0	∓2	2.90×10^{-10}
		Ξ^-	Ξ^+	1321.3	±1	0	0	0	∓2	1.64×10^{-10}
	Omega	Ω^-	Ω^+	1672.5	±1	0	0	0	∓3	0.82×10^{-10}
(many other baryons known)										

*The lower of the ± or ∓ symbols are the values for antiparticles.
†Lifetimes are traditionally given as $t_{1/2}/0.693$ (which is $1/\lambda$, the inverse of the decay constant).
††Neutrino masses may be zero. Experimental upper limits are given in parentheses.
†††Experimental lower limit is $>5 \times 10^{32}$ y for proposed mode of decay.

Leptons have no discernible size other than their wavelength, so that we know they are pointlike down to about 10^{-18} m. But the leptons fall into three families, implying three conservation laws for three quantum numbers. One of these was known from β decay, where the existence of the electron's neutrino implied that a new quantum number, called the **electron family number** L_e, is conserved. In fact, experiments in particle physics imply that **conservation of total** L_e is universally obeyed. Thus, in β^- decay, an antielectron's neutrino $\bar{v}_e$ must be created with $L_e = -1$ when an electron with $L_e = +1$ is created, so that the total remains 0 as it was before decay.

Once the muon was discovered in cosmic rays, its decay mode was found to be

$$\mu^- \rightarrow e^- + \bar{v}_e + v_\mu$$

which implied another "family" and associated conservation principle. The particle v_μ is a muon's neutrino, and it is created to conserve **muon family number** L_μ. So muons are leptons with a family of their own, and **conservation of total** L_μ also seems to be universally obeyed.

More recently, a third lepton family was discovered when τ particles were created and observed to decay in a manner similar to muons. One principal decay mode is

$$\tau^- \rightarrow \mu^- + \bar{v}_\mu + v_\tau$$

where $\bar{v}_\mu$ is the muon's antineutrino and v_τ is the tau's neutrino. As you can see by adding the values for L_μ from the table, the total muon family number is zero before and after the decay. Similarly, the **tau family number** L_τ is 1 before and after the decay. All known decay modes for muons and taus conserve both the total muon family number and the total tau family number as well as the electron family number (first known from β decay). So **conservation of total L_τ** seems to be another universally obeyed law. While this may seem to be becoming ever more complicated, we shall find that the grouping of leptons into three families is mimicked by quarks and the force-carrying particles, which also fall into three families each. Additionally, these conservation laws seem to be universally obeyed, making them broad unifying principles.

Mesons and Baryons

Now note that the hadrons in Table 31.2 are divided into two subgroups, called mesons (originally for medium mass) and baryons (the name originally meaning large mass). The division between mesons and baryons is actually based on their observed decay modes and is not strictly associated with their masses. **Mesons** are hadrons that can decay to leptons and leave no hadrons. That is, mesons are not conserved in number. **Baryons** are hadrons that always decay to another baryon. A new physical quantity called **baryon number** B seems to always be conserved in nature and is listed for the various particles in Table 31.2. Mesons and leptons have $B = 0$, so that they can decay to other particles with $B = 0$. But baryons have $B = +1$ if they are matter, and $B = -1$ if they are antimatter. The **conservation of total baryon number** is a more general rule than first noted in nuclear physics, where it was observed that the total number of nucleons was always conserved in nuclear reactions and decays. That rule in nuclear physics is just one consequence of conservation of total baryon number.

Forces, Reactions, and Reaction Rates

The forces particles feel regulate how they interact with other particles. For example, pions feel the strong force and do not penetrate as far in matter as do muons, which do not feel the strong force. (This was the way those who discovered the muon knew it could not be the particle that carries the strong force—its penetration or range was too great for it to be feeling the strong force.) Similarly, reactions that create other particles, like cosmic rays interacting with nuclei in the atmosphere, have greater probability if they are caused by the strong force than if caused by the weak force. Such knowledge has been useful to physicists when analyzing the particles produced by various accelerators.

The forces felt by particles also govern how particles interact with themselves if they are unstable and decay. The stronger the force, the faster they decay and shorter their lifetime, for example. An example of a nuclear decay via the strong force is $^8Be \rightarrow \alpha + \alpha$, with a lifetime of about 10^{-16} s. The neutron is a good example of decay via the weak force. The process $n \rightarrow p + e^- + \bar{v}_e$ has a much longer lifetime of 887 s. The weak force causes this decay, as it does all β decay. An important clue that the weak force is responsible for β decay is the creation of leptons, such as e^- and $\bar{v}_e$. None would be created if the strong force was responsible, just as no leptons are created in the decay of ^8Be. The systematics of particle lifetimes is a little simpler than nuclear lifetimes when all the hundreds of particles are examined (not just the ones in Table 31.2). Particles that decay via the weak force have lifetimes mostly in the range of 10^{-6} to 10^{-12} s, whereas those that decay via the strong force have lifetimes mostly in the range of 10^{-16} to 10^{-23} s. Turning this around, if we measure the lifetime of a particle, we can tell if it decays via the weak or strong force.

Yet another quantum number emerges from decay lifetimes and patterns. Note that the particles Λ, Σ, Ξ, and Ω decay with lifetimes on the order of 10^{-10} s,* implying that their decay is caused by the weak force alone, although they are hadrons and feel the strong force. The decay modes of these particles also show patterns—in particular, certain decays that should be possible within all the known conservation laws do not occur.

---

*The exception is Σ^0, whose short lifetime is explained by its particular quark substructure.

Whenever something is possible in physics, it will happen. If something does not happen, it is forbidden by a rule. All this seemed strange to those studying these particles when they were first discovered, so that they named a new quantum number **strangeness**, given the symbol S in Table 31.2. The values of strangeness assigned to various particles are based on the decay systematics. It is found that **strangeness is conserved by the strong force**, which governs the production of most of these particles in accelerator experiments. However, **strangeness is *not conserved* by the weak force**. This conclusion is reached from the fact that particles that have long lifetimes decay via the weak force and do not conserve strangeness. All of this also has implications for the carrier particles, since they transmit forces and are thus involved in these decays.

EXAMPLE 31.3 CHECK ON QUANTUM NUMBERS IN TWO DECAYS

(a) The most common decay mode of the Ξ^- particle is $\Xi^- \rightarrow \Lambda^0 + \pi^-$. Using the quantum numbers in Table 31.2, show that strangeness changes by 1, baryon number and charge are conserved, and lepton family numbers are unaffected. (b) Is the decay $K^+ \rightarrow \mu^+ + \nu_\mu$ allowed, given the quantum numbers in Table 31.2?

Strategy and Concept In part (a), the conservation laws can be examined by adding the quantum numbers of the decay products and comparing them with the parent particle. In part (b), the same procedure can reveal if a conservation law is broken or not.

Solution for (a) Before the decay, the Ξ^- has strangeness $S = -2$. After the decay, the total strangeness is -1 for the Λ^0, plus 0 for the π^-. Thus total strangeness has gone from -2 to -1, for a change of $+1$. Baryon number for the Ξ^- is $B = +1$ before the decay, and after the decay the Λ^0 has $B = +1$ and the π^- has $B = 0$, so that the total baryon number remains $+1$. Charge is -1 before the decay, and the total charge after is also $0 - 1 = -1$. Lepton numbers for all the particles are zero, and so lepton numbers are conserved.

Discussion for (a) The Ξ^- decay is caused by the weak interaction, since strangeness changes, and it is consistent with the relatively long 1.64×10^{-10} s lifetime of the Ξ^-.

Solution for (b) The decay $K^+ \rightarrow \mu^+ + \nu_\mu$ is allowed if charge, baryon number, mass-energy, and lepton numbers are conserved. Strangeness can change due to the weak interaction. Charge is conserved at $+1 = +1 + 0$. Baryon number is conserved, since all particles have $B = 0$. Mass-energy is conserved in the sense that the K^+ has a greater mass than the products, so that the decay can be spontaneous. Lepton family numbers are conserved at 0 for the electron and tau family for all particles. The muon family number is $L_\mu = 0$ before and $L_\mu = -1 + 1 = 0$ after. Strangeness changes from $+1$ before to $0 + 0$ after, for an allowed change of 1. The decay is allowed by all these measures.

Discussion for (b) This decay is not only allowed by our reckoning, it is, in fact, the primary decay mode of the K^+ meson and is caused by the weak force, consistent with the long 1.24×10^{-8} s lifetime.

Figure 31.17 Murray Gell-Mann (b. 1929) proposed quarks as a substructure of hadrons in 1963 and was already known for his work on the concept of strangeness. Although quarks have never been directly observed, several predictions of the quark model were quickly confirmed, and their properties explain all known hadron characteristics. Gell-Mann was awarded the Nobel prize in 1969.

There are hundreds of particles, all hadrons, not listed in Table 31.2, most of which have shorter lifetimes. The systematics of those particle lifetimes, their production probabilities, and decay products are completely consistent with the conservation laws noted for lepton families, baryon number, and strangeness, but they also imply other quantum numbers and conservation laws. There are a finite, and in fact relatively small, number of these conserved quantities, however, implying a finite set of substructures. Additionally, some of these short-lived particles look very much like excited states of other particles, implying an internal structure. All of this jigsaw puzzle can be tied together and explained relatively simply by the existence of fundamental substructures. Leptons seem to be fundamental themselves. Hadrons seem to have a substructure called quarks. The next section explores the basics of the underlying quark building blocks.

31.5 QUARKS—IS THAT ALL THERE IS?

Quarks have been mentioned at various points in this text as fundamental building blocks and members of the exclusive club of truly elementary particles. Note that an elementary or **fundamental particle** has no substructure (it is not made of other particles) and has no finite size other than its wavelength. This does not mean that fundamental particles are stable—some decay, while others do not. Keep in mind that *all* leptons seem to be fundamental, whereas *no* hadrons are fundamental. There is strong evidence that **quarks**

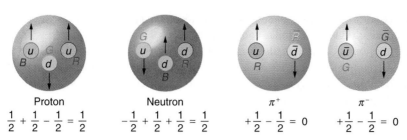

Figure 31.18 All baryons, such as the proton and neutron shown here, are composed of three quarks. All mesons, such as the pions shown here, are composed of a quark-antiquark pair. Arrows represent the spins of the quarks, which are also colored as we shall see. The colors are such that they must add to white for any possible combination of quarks.

Spin:

Proton: $\frac{1}{2} + \frac{1}{2} - \frac{1}{2} = \frac{1}{2}$

Neutron: $-\frac{1}{2} + \frac{1}{2} + \frac{1}{2} = \frac{1}{2}$

π^+: $+\frac{1}{2} - \frac{1}{2} = 0$

π^-: $+\frac{1}{2} - \frac{1}{2} = 0$

are the fundamental building blocks of hadrons. (See Figure 31.17.) Quarks are the second group of fundamental particles (leptons are the first). The third and perhaps final group of fundamental particles are the carrier particles for the four basic forces. Leptons, quarks, and carrier particles may be all there is. In this section we will discuss the quark substructure of hadrons and its relationship to forces as well as indicate some remaining questions and problems.

Conception Quarks were first proposed independently by American physicists Murray Gell-Mann and George Zweig in 1963. Their quaint name was taken by Gell-Mann from a James Joyce novel—Gell-Mann was also largely responsible for the concept and name of strangeness. (Whimsical names are common in particle physics, reflecting the less stuffy personalities of modern physicists.) Originally, three quark types—or **flavors**—were proposed to account for the then-known mesons and baryons. These quark flavors are named **up** (*u*), **down** (*d*), and **strange** (*s*). All quarks have half-integral spin and are thus fermions. All mesons have integral spin while all baryons have half-integral spin, and so mesons must be made of an even number of quarks, baryons an odd number. Figure 31.18 shows the quark substructure of the proton, neutron, and two pions. The most radical proposal by Gell-Mann and Zweig is the fractional charges of quarks, which are $\pm(2/3)q_e$ and $\pm(1/3)q_e$, while all directly observed particles have charges that are integral multiples of q_e. Table 31.3 lists characteristics of the six quark flavors now thought to exist. Discoveries made since 1963 have required extra quark flavors, which are divided into three families quite analogous to leptons.

How does it work? To get a feeling for how these quark substructures work, let us specifically examine the proton, neutron, and the two pions pictured in Figure 31.18 before moving on to more general considerations. First, the proton *p* is composed of the three quarks *uud*, so that its total charge is $+(2/3)q_e + (2/3)q_e - (1/3)q_e = q_e$, as expected. With the spins aligned as in the figure, the proton's intrinsic spin is $+(1/2) + (1/2) - (1/2) = (1/2)$, also as expected. Note that the spins of the up quarks are aligned, so that they would be in the same state except that they have different colors

TABLE 31.3

QUARKS AND ANTIQUARKS*

Name	Symbol	Anti-particle	Spin	Charge	B[†]	s	c	b	t	Mass (GeV/c²)[††]
Up	*u*	$\bar{u}$	1/2	$\pm(2/3)q_e$	$\pm 1/3$	0	0	0	0	0.005
Down	*d*	$\bar{d}$	1/2	$\mp(1/3)q_e$	$\pm 1/3$	0	0	0	0	0.008
Strange	*s*	$\bar{s}$	1/2	$\mp(1/3)q_e$	$\pm 1/3$	∓ 1	0	0	0	0.50
Charmed	*c*	$\bar{c}$	1/2	$\pm(2/3)q_e$	$\pm 1/3$	0	± 1	0	0	1.6
Bottom	*b*	$\bar{b}$	1/2	$\mp(1/3)q_e$	$\pm 1/3$	0	0	∓ 1	0	5
Top	*t*	$\bar{t}$	1/2	$\pm(2/3)q_e$	$\pm 1/3$	0	0	0	± 1	180

*The lower of the $\pm$ or $\mp$ symbols are the values for antiquarks.

[†]*B* is baryon number, *s* is strangeness, *c* is charm, *b* is bottomness, *t* is topness.

[††]Values are approximate, are not directly observable, and vary with model.

QUARK COMPOSITION OF SELECTED HADRONS*

Particle	Quark composition
Mesons	
π^+	$u\bar{d}$
π^-	$\bar{u}d$
π^0	$u\bar{u}$, $d\bar{d}$ mixture*
η^0	$u\bar{u}$, $d\bar{d}$ mixture*
K^0	$d\bar{s}$
$\bar{K}^0$	$\bar{d}s$
K^+	$u\bar{s}$
K^-	$\bar{u}s$
J/Ψ	$c\bar{c}$
Υ	$b\bar{b}$
Baryons†,††	
p	uud
n	udd
Δ^0	udd
Δ^+	uud
Δ^-	ddd
Δ^{++}	uuu
Λ^0	uds
Σ^0	uds
Σ^+	uus
Σ^-	dds
Ξ^0	uss
Ξ^-	dss
Ω^-	sss

*These two mesons are different mixtures, but each is its own antiparticle, as indicated by its quark composition.
†Antibaryons have the antiquarks of their counterparts. The antiproton $\bar{p}$ is $\bar{u}\bar{u}\bar{d}$, for example.
††Baryons composed of the same quarks are different states of the same particle. The Δ^+ is an excited state of the proton, for example.

CONNECTIONS

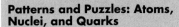

Patterns and Puzzles: Atoms, Nuclei, and Quarks

Patterns in the properties of atoms allowed the periodic table to be developed. From it, previously unknown elements were predicted and observed. Similarly, patterns were observed in the properties of nuclei, leading to the chart of nuclides and successful predictions of previously unknown nuclides. Now with particle physics, patterns imply a quark substructure that, if taken literally, predicts previously unknown particles. These have now been observed in another triumph of underlying unity.

(another quantum number to be elaborated upon a little later). Quarks obey the Pauli exclusion principle. Similar comments apply to the neutron n, which is composed of the three quarks udd. Note also that the neutron is made of charges that add to zero but move internally, producing its well-known magnetic moment. When the neutron β^- decays, it does so by changing the flavor of one of its quarks. Writing neutron β^- decay in terms of quarks,

$$n \rightarrow p + \beta^- + \bar{\nu}_e$$

becomes

$$udd \rightarrow uud + \beta^- + \bar{\nu}_e$$

We see that this is equivalent to a down quark changing flavor to become an up quark:

$$d \rightarrow u + \beta^- + \bar{\nu}_e$$

This is an example of the general fact that **the weak nuclear force can change the flavor of a quark**. By general, we mean that any quark can be converted into any other (change flavor) by the weak nuclear force. Not only can we get $d \rightarrow u$, we can also get $u \rightarrow d$. Furthermore, the strange quark can be changed by the weak force, too, making $s \rightarrow u$ and $s \rightarrow d$ possible. This explains the violation of conservation of strangeness by the weak force noted in the preceding section. Another general fact is that **the strong nuclear force cannot change the flavor of a quark**.

Again from Figure 31.18, we see that the π^+ meson (one of the three pions) is composed of an up quark plus an antidown quark, or $u\bar{d}$. Its total charge is thus $+(2/3)q_e + (1/3)q_e = q_e$, as expected. Its baryon number is 0, since it has a quark and an antiquark with baryon numbers $+(1/3) - (1/3) = 0$. The π^+ half-life is relatively long since, although it is composed of matter and antimatter, the quarks are different flavors and the weak force must cause the decay by changing the flavor of one into that of the other. The spins of the u and $\bar{d}$ quarks are antiparallel, making the pion have spin zero as observed experimentally. Finally, the π^- meson shown in Figure 31.18 is the antiparticle of the π^+ meson, and it is composed of the corresponding quark antiparticles. That is, the π^+ meson is $u\bar{d}$, while the π^- meson is $\bar{u}d$. These two pions annihilate each other quickly, because their constituent quarks are each other's antiparticles.

Two of the general rules for combining quarks to form hadrons are:

Baryons are composed of three quarks, and antibaryons are composed of three antiquarks.
Mesons are combinations of a quark and an antiquark.

One of the clever things about this scheme is that only integral charges result, even though the quarks have fractional charge.

All combinations are possible All quark combinations are possible. Table 31.4 lists some of these combinations. When Gell-Mann and Zweig proposed the original three quark flavors, particles corresponding to all combinations of those three had not been observed. The pattern was there, but it was incomplete—much as had been the case in the periodic table of the elements and the chart of nuclides. The Ω^- particle, in particular, had not been discovered but was predicted by quark theory. Its combination of three strange quarks, sss, gives it a strangeness of -3 (see Table 31.2) and other predictable characteristics, such as spin, charge, approximate mass, and lifetime. If the quark picture is complete, the Ω^- should exist. It was first observed in 1964 at the Brookhaven National Laboratory and had the predicted characteristics. (See Figure 31.19.) The discovery of the Ω^- was convincing indirect evidence for the existence of the three original quark flavors and boosted theoretical and experimental efforts to further explore particle physics in terms of quarks.

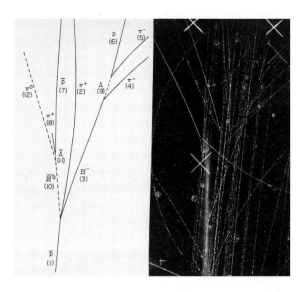

Figure 31.19 Discovery of the Ω^-. The important features of this famous bubble chamber photograph are highlighted on the left. It is a secondary reaction in which an accelerator-produced K^- collides with a proton via the strong force and conserves strangeness to produce the Ω^- with characteristics predicted by the quark model. As with other predictions of previously unobserved particles, this gave a tremendous boost to quark theory.

EXAMPLE 31.4 QUANTUM NUMBERS FROM QUARK COMPOSITION

Verify the quantum numbers given for the Ξ^0 particle in Table 31.2 by adding the quantum numbers for its quark composition as given in Table 31.4.

Strategy and Concept The composition of the Ξ^0 is given to be *uss* in Table 31.4. The quantum numbers for the constituent quarks are given in Table 31.3. We will not consider spin, because that is not given for the Ξ^0. But we can check on charge and the other quantum numbers given for the quarks.

Solution The total charge of *uss* is $+(2/3)q_e - (1/3)q_e - (1/3)q_e = 0$, which is correct for the Ξ^0. The baryon

number is $+(1/3) + (1/3) + (1/3) = 1$, also correct since the Ξ^0 is a matter baryon and has $B = 1$ listed in Table 31.2. Its strangeness is $S = 0 - 1 - 1 = -2$, also as expected from Table 31.2. Its charm, bottomness, and topness are 0, as are its lepton family numbers (it is not a lepton).

Discussion This procedure is similar to what the inventors of the quark hypothesis did when checking to see if their solution to the puzzle of particle patterns was correct. They also checked to see if all combinations were known, thereby predicting the previously unobserved Ω^- as the completion of a pattern.

What about direct evidence? At first, physicists expected that with sufficient energy we should be able to free quarks and observe them directly. This has not proved possible. There is still no direct observation of a fractional charge or any isolated quark. When large energies are put into collisions, other particles are created—but no quarks emerge. There is nearly direct evidence for quarks that is quite compelling. By 1967 experiments at SLAC scattering 20 GeV electrons from protons had produced results much like Rutherford had obtained for the nucleus nearly 60 years before. The SLAC scattering experiments showed unambiguously that there were three pointlike (meaning they had sizes considerably smaller than the probe's wavelength) charges inside the proton. (See Figure 31.20.) This evidence made all but the most skeptical admit that there was validity to the quark substructure of hadrons.

More recent and higher-energy experiments have produced jets of particles in collisions, highly suggestive of three quarks in a nucleon. (See Figure 31.21.) Since the quarks are very tightly bound, energy put into separating them pulls them only so far apart before it starts being converted into other particles. More energy produces more particles, not a separation of quarks. Conservation of momentum requires that the particles come out in jets along the three paths the quarks were being pulled. Note that there are only three jets, and that other characteristics of the particles are consistent with the three-quarks substructure.

Quarks have had their ups and downs The quark model actually lost some of its early popularity, not only because quarks were not directly observed, but also because there was something asymmetric about the three-quark model. The up and down

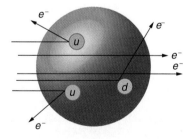

Proton

Figure 31.20 Scattering of high-energy electrons from protons at facilities like SLAC produces evidence of three pointlike charges consistent with proposed quark properties. This experiment is analogous to Rutherford's discovery of the small size of the nucleus by scattering α particles. High-energy electrons are used so that the probe wavelength is small enough to see details smaller than the proton.

Figure 31.21 This computer-generated analysis of a very high energy collision shows jets of particles. This is nearly direct evidence that quarks were moving in the directions of the jets before recombining.

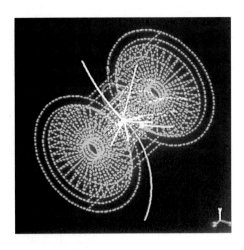

quarks seemed to compose normal matter (see Table 31.4), while the single strange quark explained strangeness. Why didn't it have a counterpart? A fourth quark flavor called **charm** (*c*) was proposed as the counterpart of the strange quark to make things symmetric—there would be two normal quarks (*u* and *d*) and two exotic quarks (*s* and *c*). Furthermore, at that time only four leptons were known, two normal and two exotic. It was attractive that there be four quarks and four leptons. The problem was that no known particles contained a charmed quark. Suddenly, in November of 1974, two groups (one headed by C. C. Ting at Brookhaven National Laboratory and the other by Burton Richter at SLAC) independently and nearly simultaneously discovered a new meson with characteristics that made it clear that its substructure is $c\bar{c}$. It was called *J* by one group and psi (Ψ) by the other and now is known as the *J/Ψ* meson. Numerous particles have since been discovered containing the charmed quark, consistent in every way with the quark model. The discovery of the *J/Ψ* meson had such a rejuvenating effect on quark theory that it is now called the November Revolution. Ting and Richter shared the 1976 Nobel prize.

History quickly repeated itself. In 1975 the tau (τ) was discovered, and a third family of leptons emerged (see Table 31.2).* Theorists quickly proposed two more quark flavors called **top** (*t*) or truth and **bottom** (*b*) or beauty to keep the number of quarks the same as the number of leptons. And in 1976 the upsilon (Υ) meson was discovered and shown to be composed of a bottom and an antibottom quark or $b\bar{b}$, quite analogous to the *J/Ψ* being $c\bar{c}$ (see Table 31.4). (Being a single flavor, these mesons are sometimes called bare charm and bare bottom and reveal the characteristics of their quarks most clearly.) Other mesons containing bottom quarks have since been observed. In 1995 two groups at Fermilab confirmed the top quark's existence, completing the picture of six quarks listed in Table 31.3. Each successive quark discovery—first *c*, then *b*, and finally *t*—has required higher energy because each has higher mass. Quark masses in Table 31.3 are only approximately known, because they are not directly observed. They must be inferred from the masses of the particles they combine to form.

What's color got to do with it?—A whiter shade of pale As mentioned and shown in Figure 31.18, quarks carry another quantum number we call **color**. Of course, it is not the color we sense with visible light, but its properties are analogous to those of three primary and three secondary colors. Specifically, a quark can have one of three color values we call **red** (*R*), **green** (*G*), and **blue** (*B*) in analogy to those primary visible colors. Antiquarks have three values we call **antired** or **cyan** ($\bar{R}$), **antigreen** or **magenta** ($\bar{G}$), and **antiblue** or **yellow** ($\bar{B}$) in analogy to those secondary visible colors. The reason for these names is that when certain visual colors are combined the eye sees white. The analogy of the colors combining to white is used to explain why baryons are made of three quarks, why mesons are a quark and an antiquark, and why we cannot isolate a single quark. The force between the quarks is such that their combined colors must produce white. This is illustrated

---

*Martin Perl, an American physicist, was awarded the 1995 Nobel prize for this discovery.

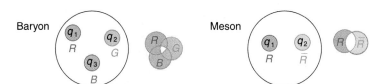

Baryon   Meson

Figure 31.22 The three quarks composing a baryon must be *RGB*, which adds to white. The quark and anti-quark composing a meson must be a color and anticolor, here $R\overline{R}$, also adding to white. The force between systems that have color is so great that they cannot be separated or exist as colored.

in Figure 31.22. A baryon must have one of each primary color or *RGB*, which produces white. A meson must have a primary color and its anticolor, also producing white.

Why must hadrons be white? The color scheme is intentionally devised to explain why baryons have three quarks and mesons have a quark and an antiquark. Quark color is thought of as being similar to charge, but with more values. An ion, by analogy, exerts much stronger forces than a neutral molecule. When the color of a combination of quarks is white, it is like a neutral atom. The forces a white particle exerts are like the polarization forces in molecules, but in hadrons these leftovers are the strong nuclear force! When a combination of quarks has color other than white, it exerts *extremely* large forces—even larger than the strong force—and perhaps cannot be stable or permanently separated. This is also part of the **theory of quark confinement**, which attempts to explain how quarks can exist and yet never be isolated or directly observed. Finally, an extra quantum number with three values (like those we assign to color) is necessary for quarks to obey the Pauli exclusion principle. Particles such as the Ω^-, which is composed of three strange quarks, *sss*, and the Δ^{++}, which is three up quarks, *uuu*, can exist because the quarks have different colors and do not have the same quantum numbers. At first viewed with some skepticism, color is very consistent with all observations and is now widely accepted. Quark theory including color is called **quantum chromodynamics** (QCD), also named by Gell-Mann.

The three families Fundamental particles are thought to be one of three types— leptons, quarks, or carrier particles. Each of those three types is further divided into three analogous families as illustrated in Figure 31.23. We have examined leptons and quarks in some detail. Each has six members (and their six antiparticles) divided into three analogous families. The first family is normal matter, of which most things are composed. The second is exotic, and the third more exotic and more massive than the second. The only stable particles are in the first family, which also has unstable members.

Always searching for symmetry and similarity, physicists have also divided the carrier particles into three families, omitting the graviton. Gravity is special among the four forces in that it affects the space and time in which the other forces exist and is proving most difficult to include in a Theory of Everything or TOE (to stub the pretension of such a theory). Gravity is thus often set apart. It is not certain that there is meaning in the groupings shown in Figure 31.23, but the analogies are tempting. In the past we have been able to make significant advances by looking for analogies and patterns, and this is an example of one under current scrutiny. There are connections between the families of leptons, in that the τ decays into the μ and the μ into the *e*. Similarly for quarks, the higher families eventually decay into the lowest, leaving only *u* and *d* quarks. We have long sought connections between the forces in nature. Since these are carried by particles, we will explore connections between gluons, $W^\pm$ and Z^0, and photons as part of the search for unification of forces discussed in the next section.

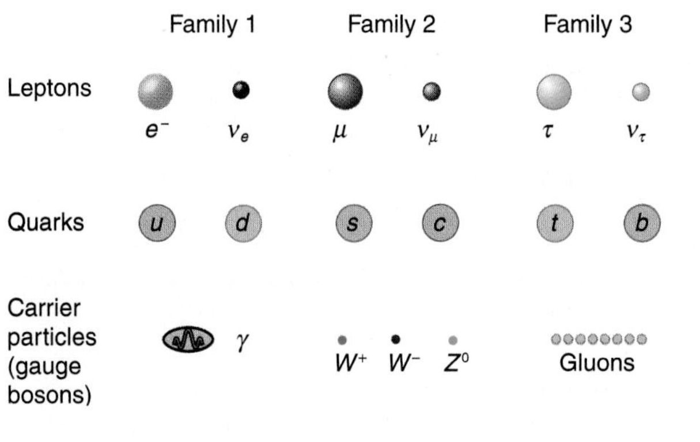

	Family 1		Family 2		Family 3	
Leptons	e^-	ν_e	μ	ν_μ	τ	ν_τ
Quarks	*u*	*d*	*s*	*c*	*t*	*b*
Carrier particles (gauge bosons)	γ		W^+ W^- Z^0		Gluons	

Figure 31.23 The three types of particles are leptons, quarks, and carrier particles. Each of those types is divided into three analogous families, with the graviton left out.

31.6 GUTs, THE UNIFICATION OF FORCES

Present quests to show that the four basic forces are different manifestations of a single unified force follow a long tradition. In the 19th century, the distinct electric and magnetic forces were shown to be intimately connected and are now collectively called the electromagnetic force. More recently, the weak nuclear force has been shown to be connected to the electromagnetic in a manner suggesting that all four forces may be unified. Certainly, there are similarities in how forces are transmitted by the exchange of carrier particles, and the carrier particles themselves (the gauge bosons in Table 31.2) are also similar in important ways. The analogy to the unification of electric and magnetic forces is quite good—the four forces are distinct under normal circumstances, but there are hints of connections even on the atomic scale, and there may be conditions under which the forces are intimately related and even indistinguishable. The search for a correct theory linking the forces, called **Grand Unified Theory** (GUT), is explored in this section in the realm of particle physics. The next chapter expands the story in making a connection with cosmology, on the opposite end of the distance scale.

Figure 31.24 is a Feynman diagram showing how the weak nuclear force is transmitted by the carrier particle Z^0, similar to the diagrams in Figures 31.5 and 31.7 for the electromagnetic and strong nuclear forces. In the 1960s a gauge theory,* called **electroweak theory**, was developed by Steven Weinberg, Sheldon Glashow, and Abdul Salam that proposed the weak and electromagnetic forces to be different manifestations of the same force. One of its predictions, in addition to describing both electromagnetic and weak force phenomena, was the existence of the W^+, W^-, and Z^0 carrier particles. Not only were three particles having spin 1 predicted, the mass of the W^+ and W^- were predicted to be 81 GeV/c^2, and that of the Z^0 to be 90 GeV/c^2. (Their masses had to be about 1000 times that of the pion, or about 100 GeV/c^2, since the range of the weak force is about 1000 times less than the strong force carried by virtual pions.) In 1983 these carrier particles were observed at CERN with the predicted characteristics, including masses having the predicted values. (See Table 31.2.) This was another triumph of particle theory and experimental effort, resulting in the 1984 Nobel prize to the experiment's group leaders Carlo Rubbia and Simon van der Meer. Theorists Weinberg, Glashow, and Salam had already been honored for other aspects of electroweak theory with the 1979 Nobel prize.

Although the weak nuclear force is very short ranged ($<10^{-18}$ m, as indicated in Table 31.1), its effects on atomic levels can be measured given the extreme precision of modern techniques. Since electrons spend some time in the nucleus, their energies are affected, and spectra can even indicate new aspects of the weak force, such as the possibility of other carrier particles. So systems many orders of magnitude larger than the range of the weak force supply evidence of electroweak unification in addition to evidence found at the particle scale.

Gluons (g) are the proposed carrier particles for the strong nuclear force, although they are not directly observed. Like quarks, gluons may be confined to systems having a total color of white. Less is known about gluons than the carriers of the weak and certainly of the electromagnetic force. QCD theory calls for eight gluons, all massless and all spin 1. Six of the gluons carry a color and an anticolor, while two do not carry color, as illustrated in Figure 31.25(a). There is indirect evidence of the existence of gluons in nucleons. When high-energy electrons are scattered from nucleons and evidence of quarks is seen, the momenta of the quarks are smaller than they would be if there were no gluons. That is, the gluons carrying force between quarks also carry some momentum, inferred by the already indirect quark momentum measurements. At any rate, the gluons carry color charge and can change the colors of quarks when exchanged, as seen in Figure 31.25(b). In that figure a red down quark interacts with a green strange quark by sending it a gluon. That gluon carries red away from the down quark and leaves it green, because it is an $R\overline{G}$

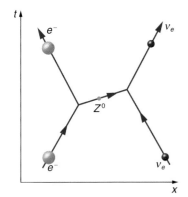

Figure 31.24 The exchange of a virtual Z^0 carries the weak nuclear force between an electron and a neutrino in this Feynman diagram. The Z^0 is one of the carrier particles for the weak nuclear force that has now been created in the laboratory with characteristics predicted by electroweak theory.

---

*Certain field theories, here related to the exchange of gauge bosons or carrier particles as the mechanism for transmission of force, are called gauge theories. See Table 31.2 for a partial list of gauge bosons. Figure 31.23 shows more gauge bosons.

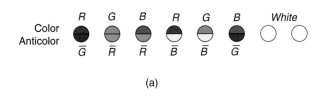

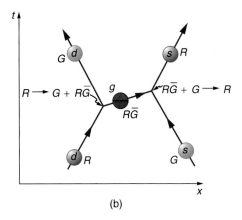

Figure 31.25   (a) The eight types of gluons that carry the strong nuclear force are divided into a group of six that carry color and a group of two that do not. (b) Exchange of gluons between quarks carries the strong force and may change the color of a quark.

(a)

(b)

(red-antigreen) gluon. (Taking antigreen away leaves you green.) Its antigreenness kills the green in the strange quark, and its redness turns the quark red.

The strong force is complicated, since observable particles that feel the strong force (hadrons) contain multiple quarks. Figure 31.26 shows the quark and gluon details of pion exchange between a proton and a neutron as illustrated earlier in Figures 31.2 and 31.7. The quarks within the proton and neutron move along together exchanging gluons, until the proton and neutron get close together. As the u quark leaves the proton, a gluon creates a pair of virtual particles, a d quark and a $\bar{d}$ antiquark. The d quark stays behind and the proton turns into a neutron, while the u and $\bar{d}$ move together as a π^+. (Table 31.4 confirms the $u\bar{d}$ composition for the π^+.) The $\bar{d}$ annihilates a d quark in the neutron, the u joins up, and the neutron becomes a proton. A pion has been exchanged and a force transmitted.

It is beyond the scope of this text to go into more detail on the types of quark and gluon interactions that underlie the observable particles, but the theory (quantum chromodynamics or QCD) is very self-consistent. So successful have QCD and the electroweak theory been that, taken together, they are called the **Standard Model**. Advances in knowledge are expected to modify, but not overthrow, the Standard Model of particle physics and forces.

How might forces be unified? They are definitely distinct under most circumstances, being carried by different particles and having greatly different strengths, for example. But experiments show that at very small distances, the strengths of the forces begin to become more similar. In fact, electroweak theory's prediction of the W^+, W^-,

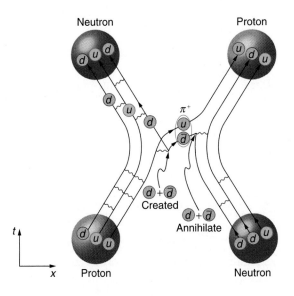

Figure 31.26   This Feynman diagram is the same interaction as shown in Figure 31.7, but it shows the quark and gluon details of the strong force interaction.

and Z^0 carrier particles was based on the strengths of the two forces being identical at very small distances. (See Figure 31.27.) As discussed for the creation of virtual particles for very short times, the small distances or short ranges correspond to the large masses of the carrier particles and the correspondingly large energies needed to create them. Thus the energy scale on the horizontal axis of Figure 31.27 corresponds to smaller and smaller distances, with 100 GeV corresponding to approximately 10^{-18} m, for example. At that distance, the strengths of the EM and weak forces are the same. To test physics at that distance, energies of about 100 GeV must be put into the system, and that is sufficient to create and release the W^+, W^-, and Z^0 carrier particles. At those and higher energies, the masses of the carrier particles becomes less and less relevant, and the Z^0 in particular becomes very similar to the massless, chargeless, spin 1 photon. In fact, there is enough energy when things are pushed to even smaller distances to transform the W^+, W^-, and Z^0 into massless carrier particles more similar to photons and gluons. These have not been observed experimentally, but there is a prediction of an associated particle called the **Higgs boson**. The mass of this particle is not predicted with nearly the certainty with which the W^+, W^-, and Z^0 particles were predicted, but it was hoped that the Higgs boson could be observed at the now-canceled Superconducting Super Collider (SSC). The existence of this more massive particle would give validity to the theory that the carrier particles become identical under certain circumstances.

The small distances and high energies at which the electroweak force becomes identical with the strong nuclear force are not reachable with any conceivable human-built accelerator. At energies of about 10^{14} GeV (16,000 J per particle), distances of about 10^{-30} m can be probed. Such energies are needed to test theory directly, but these are about 10^{10} higher than the proposed giant SSC would have had, and the distances are about 10^{-12} smaller than any structure we have direct knowledge of. This would be the realm of various GUTs, of which there are many since there is no constraining evidence at these energies and distances. Past experience has shown that any time you probe so many orders of magnitude further (here about 10^{12}), you find the unexpected. Even more extreme are the energies and distances at which gravity is thought to become unified with the other forces in a TOE. Most speculative and least constrained by experiment are TOEs, one of which is called **Superstring theory**. Superstrings are entities 10^{-35} m in scale that act like one-dimensional oscillating strings and are proposed to underlie all particles, forces, and space itself.

At the energy of GUTs, the carrier particles of the weak force would become massless and identical to gluons. If that happens, then both lepton and baryon conservation would be violated. We do not see such violations, because we do not encounter such energies. However, there is a tiny probability that at ordinary energies the virtual particles that violate conservation of baryon number may exist for extremely small amounts of time (corresponding to very small ranges). All GUTs thus predict that the proton should be unstable, but decaying with an extremely long lifetime of about 10^{31} y. The predicted decay mode is

$$p \rightarrow \pi^0 + e^+ \quad \text{(proposed proton decay)}$$

which violates both conservation of baryon number and electron family number. Although 10^{31} y is an extremely long time (about 10^{21} times the age of the universe), there are a lot of protons and detectors have been constructed to look for the proposed decay

Figure 31.27 The relative strengths of the four basic forces vary with distance and, hence, the energy needed to probe small distances. At ordinary energies (a few eV or less), the forces differ greatly as indicated in Table 31.1. But at energies available at accelerators, the weak and EM forces become identical, or unified. Unfortunately, the energies at which the strong and electroweak forces become the same are unreachable even in principle at any conceivable accelerator. The universe may provide a laboratory, and nature may show effects at ordinary energies that give us clues as to the validity of this graph.

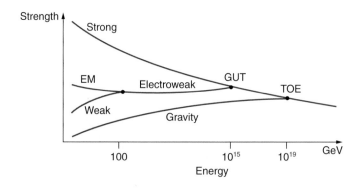

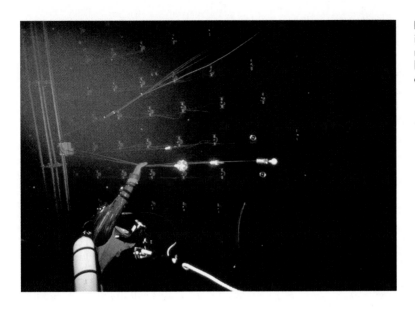

Figure 31.28 This gigantic proton decay detector is located deep beneath the earth to eliminate as many cosmic rays as possible. No proton decays have been observed in this or other such detectors, increasing the lower limit on its lifetime.

mode. (See Figure 31.28.) It is somewhat comforting that proton decay has not been detected, and its experimental lifetime is now greater than 5×10^{32} y. This does not prove GUTs wrong, but it does place greater constraints on the theories, benefiting theorists in many ways.

From looking increasingly inward at smaller and smaller details for direct evidence of electroweak theory and GUTs, we turn around and look to the universe for evidence of the unification of forces. In the 1920s, the expansion of the universe was discovered. Thinking backward in time, the universe must once have been very small, dense, and extremely hot. At a tiny fraction of a second after the fabled Big Bang, forces would have been unified and may have left their fingerprint on the existing universe. This is one of the most exciting forefronts of physics, the subject of the next and final chapter of this text.

SUMMARY

Particle physics is the study of and the quest for those truly fundamental particles having no substructure. Yukawa's idea of **virtual particle** exchange as the carrier of forces is crucial, with virtual particles being formed in temporary violation of the conservation of mass-energy as allowed by the Heisenberg uncertainty principle. The **four basic forces** and their carrier particles are summarized in Table 31.1. The **Feynman diagram** is a graph of time versus position and is a highly useful pictorial representation of particle processes. The theory of electromagnetism on the particle scale is called **quantum electrodynamics** (QED).

A variety of particle accelerators have been used to explore the nature of subatomic particles and to test predictions of particle theories. Modern accelerators used in particle physics are either large **synchrotrons** or **linear accelerators**. The use of **colliding beams** makes much greater energy available for the creation of particles, and collisions between matter and antimatter allow a greater range of final products.

All particles of matter have an **antimatter** counterpart that has the opposite charge and certain other quantum numbers as seen in Tables 31.2 and 31.3. These matter-antimatter pairs are otherwise very similar but will annihilate when

brought together. Known particles can be divided into three major groups—leptons, hadrons, and carrier particles (gauge bosons). **Leptons** do not feel the strong nuclear force and are further divided into three groups—electron family designated by **electron family number** L_e, which is conserved; muon family designated by **muon family number** L_μ, which is conserved; and tau family designated by **tau family number** L_τ, which is also conserved. **Hadrons** are particles that feel the strong nuclear force and are divided into **baryons**, the **baryon family number** B of which is conserved, and **mesons**.

Hadrons are thought to be composed of **quarks**, with baryons having three quarks and mesons having a quark and an antiquark. The characteristics of the six quarks and their antiquark counterparts are given in Table 31.3, and the quark compositions of certain hadrons are given in Table 31.4. Indirect evidence for quarks is very strong, explaining all known hadrons and their quantum numbers, such as **strangeness**, **charm**, **topness**, and **bottomness**. Quarks come in six **flavors** and three **colors** and occur only in combinations that produce white.

Fundamental particles have no further substructure, not even a size beyond their de Broglie wavelength. There are three types of fundamental particles—leptons, quarks, and carrier particles. Each type is divided into three analogous families as indicated in Figure 31.23.

Attempts to show unification of the four forces are called **Grand Unified Theories** (GUTs) and have been partially successful, with connections proven between EM and weak forces in **electroweak theory**. The strong force is carried by eight proposed particles called **gluons**, which are intimately connected to a quantum number called color—their governing theory is thus called **quantum chromodynamics** (QCD). Taken together, QCD and electroweak theory are widely accepted as the **Standard Model** of particle physics. Unification of the strong force is expected at such high energies that it cannot be directly tested, but it may have observable consequences in the as yet unobserved decay of the proton and topics to be discussed in the next chapter. Although unification of forces is generally anticipated, much remains to be done to prove its validity.

CONCEPTUAL QUESTIONS

31.1 The total energy in the beam of an accelerator is far greater than the energy of the individual beam particles. Why isn't this total energy available to create a single extremely massive particle?

31.2 Synchrotron radiation robs an accelerator beam of energy and is related to acceleration. Why would you expect the problem to be more severe for electron accelerators than proton accelerators?

31.3 What two major limitations prevent us from building high-energy accelerators that are physically small?

31.4 What are the advantages of colliding beam accelerators? What are the disadvantages?

31.5 Large quantities of antimatter isolated from normal matter should behave exactly like normal matter. An antiatom, for example, composed of positrons, antiprotons, and antineutrons should have the same atomic spectrum as its matter counterpart. Would you be able to tell it is antimatter by its emission of antiphotons? Explain briefly.

31.6 Massless particles are not only neutral, they are chargeless (unlike the neutron). Why is this?

31.7 Massless particles must travel at the speed of light, while others cannot reach this speed. Why are all massless particles stable? If evidence is found that neutrinos spontaneously decay into other particles, would this imply they have mass?

31.8 When a star erupts in a supernova explosion, huge numbers of electron neutrinos are formed in nuclear reactions. Such neutrinos from the 1987A supernova in the relatively nearby Magellanic Cloud were observed within hours of the initial brightening, indicating they traveled to earth at approximately the speed of light. Explain how this data can be used to set an upper limit on the mass of the neutrino, noting that if the mass is small the neutrinos could travel very close to the speed of light and have a reasonable energy (on the order of MeV).

31.9 Theorists have had spectacular success in predicting previously unknown particles. Considering past theoretical triumphs, why should we bother to perform experiments?

31.10 What lifetime do you expect for an antineutron isolated from normal matter?

31.11 Why does the η^0 meson have such a short lifetime compared with most other mesons?

31.12 (a) Is a hadron always a baryon? (b) Is a baryon always a hadron? (c) Can an unstable baryon decay into a meson, leaving no other baryon?

31.13 Explain how conservation of baryon number is responsible for conservation of total atomic mass (total number of nucleons) in nuclear decay and reactions.

31.14 The quark flavor change $d \to u$ takes place in β^- decay. Does this mean that the reverse quark flavor change $u \to d$ takes place in β^+ decay? Justify your response by writing the decay in terms of the quark constituents, noting that it looks as if a proton is converted into a neutron in β^+ decay.

31.15 Explain how the weak force can change strangeness by changing quark flavor.

31.16 Beta decay is caused by the weak force, as are all reactions in which strangeness changes. Does this imply that the weak force can change quark flavor? Explain.

31.17 Why is it easier to see the properties of the c, b, and t quarks in mesons having composition $c\bar{c}$, $b\bar{b}$, or $t\bar{t}$ rather than in baryons having a mixture of quarks, such as udb?

31.18 How can quarks, which are fermions, combine to form bosons? Why must an even number combine to form a boson? Give one example by stating the quark substructure of a boson.

31.19 What evidence is cited to support the contention that the gluon force between quarks is greater than the strong nuclear force between hadrons? How is this related to color? Is it also related to quark confinement?

31.20 Discuss how we know that π-mesons (π^+, π^-, and π^0) are not fundamental particles and are not the basic carriers of the strong force.

31.21 An antibaryon has three antiquarks with colors $\overline{R}\,\overline{G}\,\overline{B}$. What is its color?

31.22 Suppose leptons are created in a reaction. Does this imply the weak force is acting? (Consider β decay, for example.)

31.23 How can the lifetime of a particle indicate that its decay is caused by the strong nuclear force? How can a change in strangeness imply which force is responsible for a reaction? What does a change in quark flavor imply about the force that is responsible?

31.24 (a) Do all particles having strangeness also have at least one strange quark in them? (b) Do all hadrons with a strange quark also have nonzero strangeness?

31.25 The sigma zero particle decays mostly via the reaction $\Sigma^0 \to \Lambda^0 + \gamma$. Explain how this decay and the respective quark compositions imply that the Σ^0 is an excited state of the Λ^0.

31.26 What do the quark compositions and other quantum numbers imply about the relationships between the Δ^+ and the proton? The Δ^0 and the neutron?

31.27 Discuss the similarities and differences between the photon and the Z^0 in terms of particle properties, including forces felt.

31.28 Identify evidence for electroweak unification.

31.29 The quarks in a particle are confined, meaning individual quarks cannot be directly observed. Are gluons confined as well? Explain.

31.30 If a GUT is proven, and the four forces are unified, it will still be correct to say that the orbit of the moon is determined by the gravitational force. Explain why.

31.31 If the Higgs boson is discovered and found to have mass, will it be considered the ultimate carrier of the weak force? Explain your response.

31.32 Gluons and the photon are massless. Does this imply that the W^+, W^-, and Z^0 are the ultimate carriers of the weak force?

PROBLEMS

Throughout the text, the section to which a problem is most closely related has been identified. That practice is continued here, but since particle physics is an integration of many topics, the identification is not rigid and most of the following problems could be labeled as integrated concepts. You will therefore often find it necessary to refer to other parts of the text when solving the following problems.

Sections 31.1 and 31.2 Virtual Particles and the Four Forces

31.1 (a) Find the ratio of the strengths of the weak and electromagnetic forces under ordinary circumstances. (b) What does that ratio become under circumstances in which the forces are unified?

31.2 The ratio of the strong to the weak force and the ratio of the strong force to the electromagnetic force become 1 under circumstances where they are unified. What are the ratios of the strong force to those two forces under normal circumstances?

31.3 A virtual particle having an approximate mass of 10^{14} GeV/c^2 may be associated with the unification of the strong and electroweak forces. For what length of time could this virtual particle exist (in temporary violation of the conservation of mass-energy as allowed by the Heisenberg uncertainty principle)?

31.4 Calculate the mass in GeV/c^2 of a virtual carrier particle that has a range limited to 10^{-30} m by the Heisenberg uncertainty principle. Such a particle might be involved in the unification of the strong and electroweak forces.

31.5 Another component of the strong nuclear force is transmitted by the exchange of virtual K-mesons. Taking K-mesons to have an average mass of 495 MeV/c^2, what is the approximate range of this component of the strong force?

Section 31.3 Accelerators and Detectors

31.6 At full energy, protons in the 2.00 km diameter Fermilab synchrotron travel at nearly the speed of light, since their energy is about 1000 times their rest mass energy. (a) How long does it take for a proton to complete one trip around? (b) How many times per second will it pass through the target area?

31.7 Suppose a W^- created in a bubble chamber lives for 5.00×10^{-25} s. What distance does it move in this time if it is traveling at $0.900c$? Since this distance is too short to make a track, the presence of the W^- must be inferred from its decay products. Note that the time is longer than the given W^- lifetime, which can be due to the statistical nature of decay or to time dilation.

31.8 What length track does a π^+ traveling at $0.100c$ leave in a bubble chamber, if it is created there and lives for 2.60×10^{-8} s? (Those moving faster or living longer may escape the detector before decaying.)

31.9 The 3.20 km long SLAC produces a beam of 50.0 GeV electrons. If there are 15,000 accelerating tubes, what average voltage must be across the gaps between them to achieve this energy?

31.10 Because of energy loss due to synchrotron radiation in the LHC at CERN, only 5.00 MeV is added to the energy of each proton during each revolution around the main ring. How many revolutions are needed to produce 7.00 TeV (7000 GeV) protons, if they are injected with an initial energy of 8.00 GeV?

31.11 A proton and an antiproton collide head-on, with each having a kinetic energy of 7.00 TeV (such as in the LHC at CERN). How much collision energy is available, taking into account the annihilation of the two masses? (Note that this is not significantly greater than the extremely relativistic kinetic energy.)

31.12 When an electron and positron collide at the SLAC facility, they each have 50.0 GeV kinetic energies. What is the total collision energy available, taking into account the annihilation energy? Note that the annihilation energy is insignificant, because the electrons are highly relativistic.

Sections 31.4–31.6 Particles, Quarks, and Unified Forces

Tables 31.2–31.4 are crucial to many of the following problems. Units of MeV/c^2 and GeV/c^2 are particularly useful in calculating energy released in decays, because calculating $(\Delta m)c^2$ immediately produces units of MeV or GeV.

31.13 The π^0 is its own antiparticle and decays in the following manner: $\pi^0 \rightarrow \gamma + \gamma$. What is the energy of each γ ray if the π^0 is at rest when it decays?

31.14 The primary decay mode for the negative pion is $\pi^- \rightarrow \mu^- + \bar{\nu}_\mu$. What is the energy release in MeV in this decay?

31.15 The mass of a theoretical particle that may be associated with the unification of the electroweak and strong forces is 10^{14} GeV/c^2. (a) How many proton masses is this? (b) How many electron masses is this? (This indicates how extremely relativistic the accelerator would have to be in order to make the particle, and how large the relativistic quantity $\gamma = 1/\sqrt{1 - v^2/c^2}$ would have to be.)

31.16 The decay mode of the negative muon is $\mu^- \rightarrow e^- + \bar{\nu}_e + \nu_\mu$. (a) Find the energy released in MeV. (b) Verify that charge and lepton family numbers are conserved.

31.17 The decay mode of the positive tau is $\tau^+ \rightarrow \mu^+ + \nu_\mu + \bar{\nu}_\tau$. (a) What energy is released? (b) Verify that charge and lepton family numbers are conserved. (c) The τ^+ is the antiparticle of the τ^-. Verify that all the decay products of the τ^+ are the antiparticles of those in the decay of the τ^- given in the text.

• **31.18** The principal decay mode of the sigma zero is $\Sigma^0 \rightarrow \Lambda^0 + \gamma$. (a) What energy is released? (b) Considering the quark structure of the two baryons, does it appear that the Σ^0 is an excited state of the Λ^0? (c) Verify that strangeness, charge, and baryon number are conserved in the decay. (d) Considering the preceding and the short lifetime, can the weak force be responsible? State why or why not.

• **31.19** (a) What is the uncertainty in the energy released in the decay of a π^0 due to its short lifetime? (b) What fraction of the decay energy is this, noting that the decay mode is $\pi^0 \rightarrow \gamma + \gamma$ (so that all of the π^0 mass is destroyed)?

• **31.20** (a) What is the uncertainty in the energy released in the decay of a τ^- due to its short lifetime? (b) Is the uncertainty in this energy greater than or less than the uncertainty in the mass of the tau's neutrino? Discuss the source of the uncertainty.

• **31.21** (a) Verify from its quark composition that the Δ^+ particle could be an excited state of the proton. (b) There is a spread of about 100 MeV in the decay energy of the Δ^+, interpreted as uncertainty due to its short lifetime. What is its approximate lifetime? (c) Does its decay proceed via the strong or weak force?

• **31.22** Accelerators such as the Los Alamos Meson Physics Facility (LAMPF) produce secondary beams of pions by having an intense primary proton beam strike a target. Such "meson factories" have been used for many years to study the interaction of pions with nuclei and, hence, the strong nuclear force. One reaction that occurs is $\pi^+ + p \rightarrow \Delta^{++} \rightarrow \pi^+ + p$, where the Δ^{++} is a very short lived particle. The graph in Figure 31.29 shows the probability of this reaction as a function of energy. The width of the bump is the uncertainty in energy due to the short lifetime of the Δ^{++}. (a) Find this lifetime. (b) Verify from the quark composition of the particles that this reaction annihilates and then re-creates a d quark and a $\bar{d}$ antiquark, by writing the reaction and decay in terms of quarks. (c) Draw a Feynman diagram of the production and decay of the Δ^{++} showing the individual quarks involved.

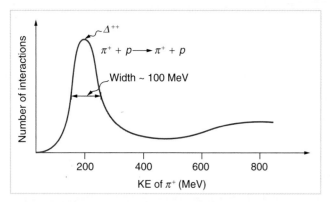

Figure 31.29 This graph shows the probability of an interaction between a π^+ and a proton as a function of energy. The bump is interpreted as a very short lived particle called a Δ^{++}. The approximately 100 MeV width of the bump is due to the short lifetime of the Δ^{++}. Problems 22 and 23.

• **31.23** The reaction $\pi^+ + p \rightarrow \Delta^{++}$ (described in the preceding problem) takes place via the strong force. (a) What is the baryon number of the Δ^{++} particle? (b) Draw a Feynman diagram of the reaction showing the individual quarks involved.

• **31.24** One of the decay modes of the omega minus is $\Omega^- \rightarrow \Xi^0 + \pi^-$. (a) What is the change in strangeness? (b) Verify that baryon number and charge are conserved, while lepton numbers are unaffected. (c) Write the equation in terms of the constituent quarks, indicating that the weak force is responsible.

• **31.25** Repeat Problem 31.24 for the decay mode $\Omega^- \rightarrow \Lambda^0 + K^-$.

• **31.26** One decay mode for the eta zero meson is $\eta^0 \rightarrow \gamma + \gamma$. (a) Find the energy released. (b) What is the uncertainty in the energy due to the short lifetime? (c) Write the decay in terms of the constituent quarks. (d) Verify that baryon number, lepton numbers, and charge are conserved.

• **31.27** One decay mode for the eta zero meson is $\eta^0 \rightarrow \pi^0 + \pi^0$. (a) Write the decay in terms of the quark constituents. (b) How much energy is released? (c) What is the ultimate release of energy, given the decay mode for the pi zero is $\pi^0 \rightarrow \gamma + \gamma$?

• **31.28** Is the decay $n \rightarrow e^+ + e^-$ possible considering the appropriate conservation laws? State why or why not.

• **31.29** Is the decay $\mu^- \rightarrow e^- + v_e + v_\mu$ possible considering the appropriate conservation laws? State why or why not.

• **31.30** (a) Is the decay $\Lambda^0 \rightarrow n + \pi^0$ possible considering the appropriate conservation laws? State why or why not. (b) Write the decay in terms of the quark constituents of the particles.

• **31.31** (a) Is the decay $\Sigma^- \rightarrow n + \pi^-$ possible considering the appropriate conservation laws? State why or why not. (b) Write the decay in terms of the quark constituents of the particles.

• **31.32** The only combination of quark colors that produces a white baryon is RGB. Identify all the color combinations that can produce a white meson.

• **31.33** (a) Three quarks form a baryon. How many combinations of the six known quarks are there if all combinations are possible? (b) This number is less than the number of known baryons. Explain why.

• **31.34** (a) Show that the conjectured decay of the proton, $p \rightarrow \pi^0 + e^+$, violates conservation of baryon number and conservation of lepton number. (b) What is the analogous decay process for the antiproton?

• **31.35** Verify the quantum numbers given for the Ω^+ in Table 31.2 by adding the quantum numbers for its quark constituents as inferred from Table 31.4.

• **31.36** Verify the quantum numbers given for the proton and neutron in Table 31.2 by adding the quantum numbers for their quark constituents as given in Table 31.4.

⁝ **31.37** (a) How much energy would be released if the proton did decay via the conjectured reaction $p \rightarrow \pi^0 + e^+$? (b) Given that the π^0 decays to two γs and that the e^+ will find an electron to annihilate, what total energy is ultimately produced in proton decay? (c) Why is this energy greater than the proton's total mass (converted to energy)?

⁝ **31.38** (a) Find the charge, baryon number, strangeness, charm, and bottomness of the J/Ψ particle from its quark composition. (b) Do the same for the Υ particle.

⁝ **31.39** There are particles called D-mesons. One of them is the D^+ meson, which has a single positive charge and a baryon number of zero, also the value of its strangeness, topness, and bottomness. It has a charm of +1. What is its quark configuration?

31.40 There are particles called bottom mesons or *B*-mesons. One of them is the B^- meson, which has a single negative charge; its baryon number is zero, as are its strangeness, charm, and topness. It has a bottomness of −1. What is its quark configuration?

31.41 (a) What particle has the quark composition $\bar{u}\bar{u}\bar{d}$? (b) What should its decay mode be?

31.42 (a) Show that all combinations of three quarks produce integral charges. Thus baryons must have integral charge. (b) Show that all combinations of a quark and an antiquark produce only integral charges. Thus mesons must have integral charge.

INTEGRATED CONCEPTS

The following problems involve greater integration of concepts than the preceding, although most topics in particle physics involve some integration of concepts.

Note: Problem-solving strategies and worked examples that can help you solve integrated concept problems appear in several places, the most recent being Chapter 29. Consult the section of problems labeled *Integrated Concepts* at the end of that chapter or others.

31.43 The intensity of cosmic ray radiation decreases rapidly with increasing energy, but there are occasionally extremely energetic cosmic rays that create a shower of radiation from all of the particles they create by striking a nucleus in the atmosphere. (See Figure 31.30.) Suppose a cosmic ray particle having an energy of 10^{10} GeV converts its energy into particles with masses averaging 200 MeV/c^2. (a) How many particles are created? (b) If the particles rain down on a 1.00 km² area, how many particles are there per square meter?

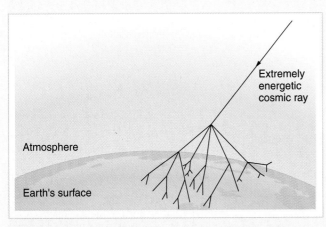

Figure 31.30 An extremely energetic cosmic ray creates a shower of particles on earth. The energy of these rare cosmic rays can approach a joule (about 10^{10} GeV), and after multiple collisions, huge numbers of particles are created from this energy. Cosmic ray showers have been observed to extend over many square kilometers. Problem 43.

31.44 Assuming conservation of momentum, what is the energy of each γ ray produced in the decay of a neutral at rest pion, in the reaction $\pi^0 \rightarrow \gamma + \gamma$?

31.45 What is the wavelength of a 50 GeV electron, such as produced at SLAC? This gives an idea of the limit to the detail it can probe.

31.46 (a) Calculate the relativistic quantity $\gamma = 1/\sqrt{1 - v^2/c^2}$ for 1.00 TeV protons produced at Fermilab. (b) If such a proton created a π^+ having the same speed, how long would its life be in the laboratory? (c) How far could it travel in this time?

31.47 The primary decay mode for the negative pion is $\pi^- \rightarrow \mu^- + \bar{\nu}_\mu$. (a) What is the energy release in MeV in this decay? (b) Using conservation of momentum, how much energy does each of the decay products receive, given the π^- is at rest when it decays? You may assume the muon's antineutrino is massless and has momentum $p = E/c$, just like a photon.

31.48 Plans for an accelerator that produces a secondary beam of *K*-mesons to scatter from nuclei, for the purpose of studying the strong force, call for them to have a kinetic energy of 500 MeV. (a) What would the relativistic quantity $\gamma = 1/\sqrt{1 - v^2/c^2}$ be for these particles? (b) How long would their average lifetime be in the laboratory? (c) How far could they travel in this time?

31.49 Suppose you are designing a proton decay experiment and you can detect 50% of the proton decays in a tank of water. (a) How many kilograms of water would you need to see one decay per month, assuming a lifetime of 10^{31} y? (b) How many cubic meters of water is this? (c) If the actual lifetime is 10^{33} y, how long would you have to wait on average to see a single proton decay?

31.50 Neutrinos are produced in huge amounts in supernovas. They were detected from the 1987A supernova in the Magellanic Cloud, which is about 120,000 light years away from earth (relatively close to our Milky Way galaxy). If neutrinos have a mass, they cannot travel at the speed of light, but if their mass is small, they can get close. (a) Suppose a neutrino with a 7 eV/c^2 mass has a kinetic energy of 700 keV. Find the relativistic quantity $\gamma = 1/\sqrt{1 - v^2/c^2}$ for it. (b) If the neutrino leaves the 1987A supernova at the same time as a photon and both travel to earth, how much sooner does the photon arrive? This is not a large time difference, given that it is impossible to know which neutrino left with which photon and the poor efficiency of the neutrino detectors. Thus the fact that neutrinos were observed within hours of the brightening of the supernova only places an upper limit on the neutrino's mass. (Hint: You may need to use a series expansion to find v for the neutrino, since its γ is so large.)

32) FRONTIERS OF PHYSICS

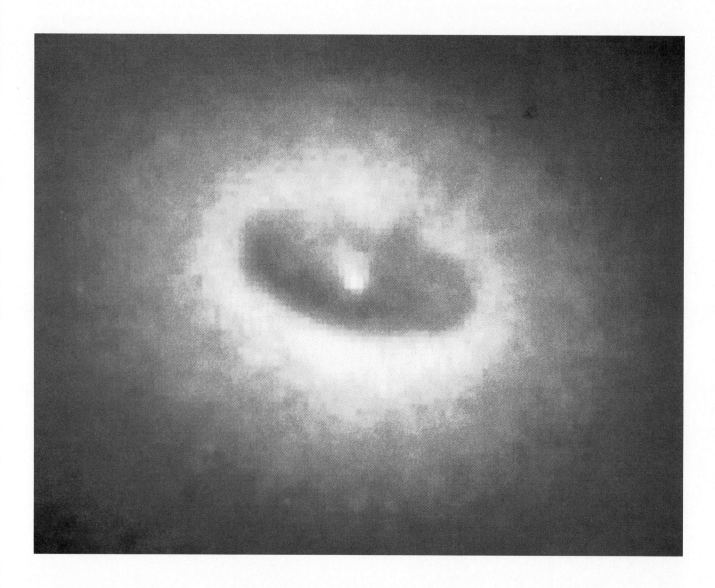

This galaxy is ejecting huge jets of matter, likely powered by an immensely massive black hole at its center.

Frontiers are exciting. There is mystery, surprise, adventure, and discovery. The satisfaction of finding the answer to a question is made keener by the fact that the answer always leads to a new question. The picture of nature becomes more complete, yet nature retains its sense of mystery and never loses its ability to awe us. The view of physics is beautiful looking both backward and forward in time. What marvelous patterns we have discovered. How clever nature seems in its rules and connections. How awesome. And we continue looking ever deeper and ever further, probing the basic structure of matter, energy, space, and time and wondering about the scope of the universe, its beginnings and future.

You are now in a wonderful position to explore the forefronts of physics, both the new discoveries and the unanswered questions. With the concepts, qualitative and quantitative, the problem-solving skills, the feeling for connections among topics, and all the rest you have mastered, you can more deeply appreciate and enjoy the brief treatments that follow. Years from now you will still enjoy the quest with an insight all the greater for your efforts. (See Figure 32.1.)

32.1 COSMOLOGY AND PARTICLE PHYSICS

Look at the sky some clear night when you are away from city lights. There you will see thousands of individual stars and a faint glowing background of millions more. The Milky Way, as it has been called since ancient times, is an arm of our galaxy of stars—the word *galaxy* coming from the Greek word *glaktik*, meaning milky. We know a great deal about our Milky Way galaxy and of the billions of other galaxies beyond its fringes. But they still provoke wonder and awe. And there are still many questions to be answered. Most remarkable when we view the universe on the large scale is that once again explanations of its character and evolution are tied to the very small. Particle physics and the questions being asked about the very small may also have their answers in the very large.

As has been noted in numerous Things Great and Small vignettes, this is not the first time the large has been explained by the small and vice versa. Newton realized that the nature of gravity on earth, such as pulls an apple to the ground, could explain the motion of the moon and planets so much farther away. Minute atoms and molecules explain the chemistry of substances on a much larger scale. Decays of tiny nuclei explain the hot interior of the earth. Fusion of nuclei likewise explains the energy of stars. Today, the patterns in particle physics seem to be explaining the evolution and character of the universe. And the nature of the universe has implications for unexplored regions of particle physics.

Cosmology is the study of the character and evolution of the universe. What are the major characteristics of the universe as we know them today? First, there are approximately 10^{11} galaxies in the observable part of the universe. An average galaxy contains

Figure 32.1 Take a moment to contemplate these clusters of galaxies, photographed by the Hubble Space Telescope. Trillions of stars linked by gravity in fantastic forms, glowing with light and showing evidence of undiscovered matter. What are they like, these myriad stars? How did they evolve? What can they tell us of matter, energy, space, and time?

more than 10^{11} stars, with our Milky Way galaxy being larger than average, both in its number of stars and its dimensions. Ours is a spiral-shaped galaxy with a diameter of about 100,000 light years and a thickness of about 2000 light years in the arms with a central bulge about 10,000 light years across. The sun lies about 30,000 light years from the center near the galactic plane. There are significant clouds of gas, and there is a halo of less dense regions of stars surrounding the main body. (See Figure 32.2.) Evidence strongly suggests the existence of a large amount of additional matter in galaxies that does not produce light—the mysterious dark matter we shall later discuss.

Distances are great even within our galaxy, being measured in light years (the distance traveled by light in one year). The average distance between galaxies is on the order of a million light years, but it varies greatly, with galaxies forming clusters such as shown in Figure 32.1. The Magellanic Clouds, for example, are small galaxies close to our own, some 160,000 light years from earth. The Andromeda galaxy is a large spiral like ours and lies 2 million light years away. It is just visible to the naked eye as an extended glow in the Andromeda constellation. Andromeda is the closest large galaxy in our local group, and we can see some individual stars in it with our larger telescopes. The most distant known galaxy is 14 billion light years from earth—a truly incredible distance. (See Figure 32.3.)

Consider the fact that the light we receive from these vast distances has been on its way to us for a long time. In fact, the time in years is the same as the distance in light years. For example, Andromeda galaxy is 2 million light years away, so that the light now reaching us left it 2 million years ago. If we could be there now, Andromeda would be different. Similarly, light from the most distant galaxy left it 14 billion years ago. We have an incredible view of the past when looking great distances. We can try to see if the universe was different then—if distant galaxies are more tightly packed or have younger looking stars, for example, than closer galaxies, then there has been an evolution in time. But the problem is that the uncertainties in our data are great. Cosmology is almost typified by these large uncertainties, so that we must be especially

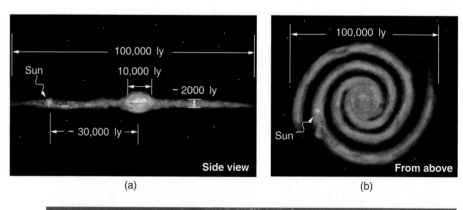

(a)   (b)

(c)

Figure 32.2 The Milky Way galaxy is typical of large spiral galaxies in its size, its shape, and the presence of gas and dust. We are fortunate to be in a location where we can see out of the galaxy and observe the vastly larger and fascinating universe around us. (a) Side view. (b) View from above. (c) A composite photograph of the Milky Way.

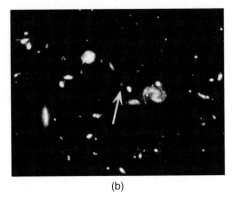

(a) (b)

Figure 32.3 (a) Andromeda is the closest large galaxy, at 2 million light years distance, and is very similar to our Milky Way. The blue regions harbor young and emerging stars, while dark streaks are vast clouds of gas and dust. A smaller satellite galaxy is clearly visible. (b) The arrow points to what may be the most distant known galaxy, estimated to be 14 billion light years from us. It exists in a much younger part of the universe.

cautious in drawing conclusions. One consequence is that there are more questions than answers, and so there are many competing theories. Another consequence is that any hard data produces a major result. Discoveries of some importance are being made on a regular basis, the hallmark of a field in its golden age.

Perhaps the most important characteristic of the universe is that all galaxies except those in our local cluster seem to be moving away from us at speeds proportional to their distance from our galaxy. It looks as if a gigantic explosion, universally called the **Big Bang**, threw matter out some billions of years ago. This amazing conclusion is based on the pioneering work of Edwin Hubble (1899–1953), the American astronomer. (See Figure 32.4(a).) In the 1920s Hubble first demonstrated conclusively that other galaxies, many previously called nebulae or clouds of stars, were outside our own. He then found that all but the closest galaxies have a red shift in their hydrogen spectra that is proportional to their distance. The explanation is that there is a **cosmological red shift** due to the expansion of space itself. The photon wavelength is stretched in transit from the source to the observer. Double the distance, and the red shift is doubled. While this cosmological red shift is often called a Doppler shift, it is not—space itself is expanding. As was illustrated back in Figure 3.29, it is not possible to determine where the center of expansion is, only to see that there is an expansion. (There may, in fact, be no center to the expansion.) All observers see themselves as stationary, with the others moving away from them. Hubble was directly responsible for discovering the universe was much larger than had previously been imagined and that it had this amazing characteristic of rapid expansion.

Universal expansion on the scale of galactic clusters (that is, galaxies at smaller distances are not uniformly receding from one another) is an integral part of modern cosmology. For galaxies farther away than about 20 Mly (20 million light years), the expansion is uniform with variations due to local motions of galaxies within clusters. A representative recession velocity v can be obtained from the simple formula

$$v = H_c d \tag{32.1}$$

where d is the distance to the galaxy and H_c is the **Hubble constant**. The Hubble constant is a central concept in cosmology. Its value is determined by taking the slope of a graph of velocity versus distance, obtained from red shift measurements, such as shown in Figure 32.4(b). We shall use an average value of $H_c = 20$ km/s/Mly, although researchers quote both higher and lower values because of disagreements on distance measurements and techniques. Thus Equation 32.1 is an average behavior for all but the closest galaxies. For example, a galaxy 100 Mly away (as determined by its size and brightness) typically moves away from us at a speed of $v = (20$ km/s/Mly$)(100$ Mly$) = 2000$ km/s. There can be variations in this speed due to so-called local motions or interactions with neighboring galaxies. Conversely, if a galaxy is found to be moving away from us at speed of 100,000 km/s based on its red shift, it is at a distance $d = v/H_c = (100,000$ km/s$)/(20$ km/s/Mly$) = 5000$ Mly $= 5$ Gly or 5×10^9 ly. This last calculation is approximate, because it assumes the expansion rate was the same 5 billion years ago as now.

Figure 32.5 shows how the recession of galaxies looks like the remnants of a gigantic explosion, the famous Big Bang. Extrapolating backward in time, the Big Bang would

(a)

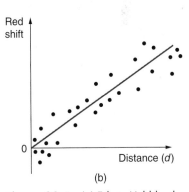

(b)

Figure 32.4 (a) Edwin Hubble, the great astronomer who not only discovered that galaxies were far outside the Milky Way, but also found that the universe is expanding as from some ancient explosion. Now called the Big Bang, this event is both a mystery and a source of knowledge. (b) This graph of red shift versus distance for galaxies shows a linear relationship, with larger red shifts at greater distances, implying an expanding universe. The slope gives the expansion rate, but uncertainties in distances make the value controversial.

Figure 32.5 Galaxies are flying apart from one another, with the more distant moving faster as if a primordial explosion expelled the matter from which they formed. The most distant known galaxies move nearly at the speed of light relative to us.

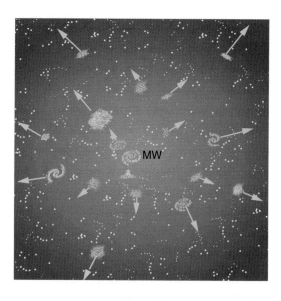

have occurred between 10 and 20 billion years ago* when all matter would have been at a point. Questions instantly arise. What caused the explosion? What happened before the Big Bang? Was there a before, or did time start then? Will the universe expand forever, or will gravity reverse it into a Big Crunch? And is there other evidence of the Big Bang besides the well-documented red shifts?

The answer to the last question is a resounding yes. The Russian-born American physicist George Gamow (1904–1968) was among the first to note that if there was a Big Bang, the remnants of the primordial fireball should still be evident and should be blackbody radiation. Since the radiation from this fireball has been traveling to us since shortly after the Big Bang, its wavelengths should be greatly stretched. If will look as if the fireball has cooled in the billions of years since the Big Bang. Gamow and collaborators predicted in the late 1940s that there should be blackbody radiation from the explosion filling space with a characteristic temperature of about 7 K. Such blackbody radiation would have its peak intensity in the microwave part of the spectrum. (See Figure 32.6.) In 1964 Arno Penzias and Robert Wilson, two American scientists working with Bell Telephone Laboratories on a low-noise radio antenna, detected the radiation and eventually recognized it for what it is. Figure 32.6(c) shows the spectrum of this microwave radiation that permeates space and is of cosmic origin. It is the most perfect blackbody spectrum known, and the temperature of the fireball remnant is determined from it to be 2.74 ± 0.002 K. The detection of what is now called the **cosmic microwave background** (CMBR) was so important (generally considered as important as Hubble's detection that the galactic red shift is proportional to distance) that virtually every scientist has accepted the expansion of the universe as fact. Penzias and Wilson shared the 1978 Nobel prize in physics for their discovery.

Matter versus antimatter We know from direct observation that antimatter is rare. The earth and the solar system are nearly pure matter. Space probes and cosmic rays give direct evidence—the landing of the Viking probes on Mars would have been spectacular explosions of mutual annihilation energy if Mars were antimatter. We also know that most of the universe is dominated by matter. This is proven by the lack of annihilation radiation coming to us from space, particularly the relative absence of 0.511 MeV γ rays created by the mutual annihilation of electrons and positrons. It seemed possible that there could be entire solar systems or galaxies made of antimatter in perfect symmetry with our matter-dominated systems. But the interactions between stars and galaxies

CONNECTIONS

Cosmology and Particle Physics

There are many connections of cosmology—by definition involving physics on the largest scale—with particle physics—by definition physics on the smallest scale. Among these are the dominance of matter over antimatter, the nearly perfect uniformity of the cosmic microwave background, and the mere existence of galaxies.

---

*This is based on calculating how long ago galaxies would have been on top of one another, or how long it would take to travel 1 Mly at a speed of 20 km/s as indicated by the Hubble constant. Uncertainties in distances and arguments among astronomers about methods of measuring them result in almost a factor of two variation in values claimed for the Hubble constant and, hence, a factor of two range in the estimated age of the universe. Recent results have reduced the discrepancies, and it is possible that a value good to 10% will be obtained in the next ten years.

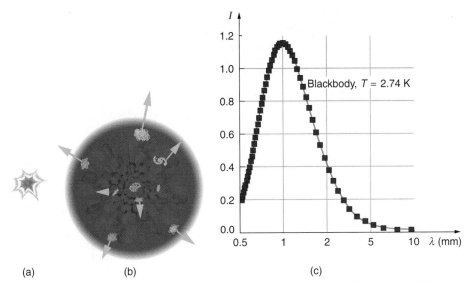

Figure 32.6 (a) The Big Bang is used to explain the present observed expansion of the universe. It was an incredibly energetic explosion some 10 to 20 billion years ago. (b) After expanding and cooling, galaxies form inside the now cold remnants of the primordial fireball. (c) The spectrum of cosmic microwave radiation is the most perfect blackbody spectrum ever detected. It is characteristic of a temperature of 2.74 K, the expansion-cooled temperature of the Big Bang's remnant. This radiation can be measured coming from any direction in space not obscured by some other source. It is compelling evidence of the creation of the universe in a gigantic explosion, already indicated by galactic red shifts.

would sometimes bring matter and antimatter together in large amounts. The annihilation radiation they would produce is simply not observed. Antimatter in nature is created in particle collisions and in β^+ decays, but only in small amounts that quickly annihilate, leaving almost pure matter surviving.

Particle physics seems symmetric in matter and antimatter. Why isn't the cosmos? The answer is that particle physics is not quite perfectly symmetric in this regard. The decay of one of the neutral K-mesons, for example, preferentially creates more matter than antimatter. This is caused by a fundamental small asymmetry in the basic forces. This small asymmetry produced slightly more matter than antimatter in the early universe. If there were only one part in 10^{12} more matter (a small asymmetry), the rest would annihilate pair for pair, leaving nearly pure matter to form the stars and galaxies we see today. So the vast number of stars we observe may be only a tiny remnant of the original matter created in the Big Bang. Here at last we see a very real and important asymmetry in nature. Rather than be disturbed by an asymmetry, most physicists are impressed by how small it is. Furthermore, if the universe were completely symmetric, the mutual annihilation would be more complete, leaving far less to form us and the universe we know.

How can something so old have so few wrinkles? Once the euphoria of discovering cosmic microwave background radiation (CMBR) subsided, a troubling aspect began to be recognized. True, the CMBR verified the Big Bang, had the correct temperature, and had a blackbody spectrum as expected. But the CMBR was too smooth—it looked identical in every direction. There should be hot and cool spots in the CMBR, nicknamed wrinkles, corresponding to dense and sparse regions of gas caused by turbulence or early quantum fluctuations. Over time, dense regions would contract under gravity and form stars and galaxies. Why aren't the fluctuations there? (This is a good example of an answer producing more questions.) Furthermore, galaxies are observed very far from us, so that they formed very long ago. The problem was to explain how galaxies could form so early and so quickly after the Big Bang if its remnant fingerprint is perfectly smooth. The answer is that if you look very closely, the CMBR is not perfectly smooth, only extremely smooth. (See Figure 32.7.)

A satellite called the Cosmic Background Explorer (COBE) carried an instrument that made very sensitive and accurate measurements of the CMBR. In April of 1992 there was extraordinary publicity of COBE's first results—there were small fluctuations in the

Figure 32.7 This map of the sky uses color to show fluctuations, or wrinkles, in the cosmic microwave background observed with the COBE satellite. Objects such as the Milky Way and other galaxies have been removed for clarity. Pink represents higher temperature and higher density, while blue is lower temperature and density. The fluctuations are small, only one part in 100,000, but these are still thought to be the cause of the eventual formation of galaxies.

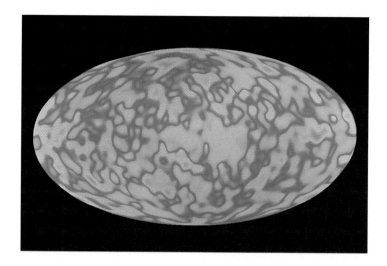

CMBR. These amount to temperature fluctuations of only 30 μK out of 2.7 K, around one part in 100,000. While this is not enough to make theorists comfortable with the apparently rapid evolution of galaxies after the Big Bang, it is enough to get them started on the path to a correct explanation. This data, shown in Figure 32.7, has since been refined and improved. The fluctuations are real, and astrophysicists are now trying to understand both how the wrinkles were created and how they evolve into galaxies.

Let us now examine the various stages of the overall evolution of the universe from the Big Bang to the present, illustrated in Figure 32.8. Note that scientific notation is used to encompass the many orders of magnitude in time, energy, temperature, and size of the universe. Going back in time, the two lines approach but do not cross (there is no zero on an exponential scale). Rather, they extend indefinitely in ever-smaller time intervals to some infinitesimal point.

Going back in time is equivalent to what would happen if expansion stopped and gravity pulled all the galaxies together, compressing and heating all matter. At a time long ago, the temperature and density were too high for stars and galaxies to exist. Before then, there was a time when the temperature was too great for atoms to exist. And farther back yet, there was a time when the temperature and density were so great that nuclei could not exist. Even farther back in time, the temperature was so high that average kinetic energy was great enough to create short-lived particles, and the density was high enough to make this likely. When we extrapolate back to the point of $W^{\pm}$ and Z^0 production (thermal energies reaching 1 TeV, or a temperature of about 10^{15} K), we reach the limits of what we know directly about particle physics. This is at a time about 10^{-12} s after the Big Bang. While 10^{-12} s may seem to be negligibly close to the instant of creation, it is not. There are important stages before this time that are tied to the unification of forces. At those stages, the universe was at extremely high energies and average particle separations were smaller than we can achieve with accelerators. What happened in the early stages before 10^{-12} s is crucial to all later stages and is possibly discerned by observing present conditions in the universe. One of these is the smoothness of the CMBR.

Names are given to early stages representing key conditions. The stage before 10^{-11} s back to 10^{-34} s is called the **electroweak epoch**, because the electromagnetic and weak forces become identical for energies above about 100 GeV. As discussed in Section 31.6, theorists expect that the strong force becomes identical to and thus unified with the electroweak force at energies of about 10^{14} GeV. The average particle energy would be this great at 10^{-34} s after the Big Bang, if there are no surprises in the unknown physics at energies above about 1 TeV. At the immense energy of 10^{14} GeV (corresponding to a temperature of about 10^{26} K), the $W^{\pm}$ and Z^0 carrier particles would be transformed into massless gauge bosons to accomplish the unification. Before 10^{-34} s back to about 10^{-43} s, we have Grand Unification in the **GUT epoch**, in which all forces except gravity are identical. At 10^{-43} s, the average energy reaches the immense 10^{19} GeV needed to unify gravity with the other forces in TOE, the Theory of Everything. Before that time is

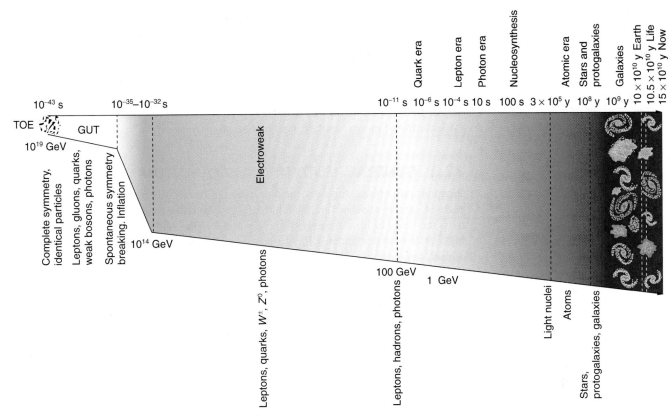

Figure 32.8 The evolution of the universe from the Big Bang onward is intimately tied to the laws of physics, especially those of particle physics at the earliest stages. The universe is relativistic throughout its history. Theories of the unification of forces at high energies may be verified by their shaping of the universe and its evolution.

the **TOE epoch**, but we have almost no idea as to the nature of the universe then, since we have no workable theory of quantum gravity. We call the hypothetical unified force **superforce**.

Now let us imagine starting at TOE and moving forward in time to see what type of universe is created from various events along the way. As temperatures and average energies decrease with expansion, the universe reaches the stage where average particle separations are large enough to see differences between the strong and electroweak forces (at about 10^{-35} s). After this time, the forces become distinct in almost all interactions—they are no longer unified or symmetric. This transition from GUT to electroweak is called **spontaneous symmetry breaking** to indicate that conditions spontaneously evolved to a point where the forces were no longer unified, breaking that symmetry. This is analogous to a phase transition in the universe, and a clever proposal by American physicist Alan Guth in the early 1980s ties it to the smoothness of the CMBR. Guth proposed that spontaneous symmetry breaking (like a phase transition during cooling of normal matter) released an immense amount of energy that caused the universe to expand extremely rapidly for the brief time from 10^{-35} s to about 10^{-32} s. This expansion may have been by an incredible factor of 10^{50} or more in the size of the universe and is, thus, called the **inflationary scenario**. One result of this inflation is that it would stretch the wrinkles in the universe nearly flat, leaving an extremely smooth CMBR. While speculative, there is as yet no other plausible explanation for the smoothness of the CMBR. Unless the CMBR is not really cosmic but local in origin, the distances between regions of similar temperatures are too great for any coordination to have caused them, since any coordination mechanism must travel at the speed of light. The wedding of particle physics and cosmology here is as intimate as it is in less speculative areas. There is little hope that we may be able to test the inflationary scenario directly, since it occurs at energies near 10^{14} GeV, vastly greater than the limits of modern accelerators. But the idea is so attractive that it is incorporated into most cosmological theories. Characteristics of the present universe may help us determine the validity of this intriguing idea.

It is important to note that if conditions such as those found in the early universe could be created in the laboratory, we would see the unification of forces directly today. The forces have not changed in time, but the average energy and separation of particles in the universe have. As discussed in Chapter 31, the four basic forces in nature are distinct under most circumstances found today. The early universe and its remnants provide evidence from times when they were unified under most circumstances.

32.2   GENERAL RELATIVITY AND QUANTUM GRAVITY

When we talk of black holes or the unification of forces, we are actually discussing aspects of general relativity and quantum gravity. We know from Chapter 26 that relativity is the study of how different observers measure the same event, particularly if they move relative to one another. Einstein's theory of **general relativity** describes all types of relative motion including accelerated motion and the effects of gravity. General relativity encompasses special relativity and classical relativity in situations where acceleration is zero and relative velocity is small compared with the speed of light. Many aspects of general relativity have been verified experimentally, some of which are better than science fiction in that they are bizarre but true. **Quantum gravity** is the theory that deals with particle exchange of gravitons as the mechanism for the force, and with extreme conditions where quantum mechanics and general relativity must both be used. A good theory of quantum gravity does not yet exist, but one will be needed to understand how all four forces may be unified. If we are successful, the theory of quantum gravity will encompass all others from classical physics to relativity to quantum mechanics—truly a Theory of Everything (TOE).

General Relativity

Einstein first considered the case of no observer acceleration when he developed the revolutionary special theory of relativity, publishing his first work on it in 1905. By 1916 he had laid the foundation of general relativity, again almost on his own. Much of what Einstein did to develop his ideas was to mentally analyze certain carefully and clearly defined situations—doing this is to perform a **thought experiment**. Figure 32.9 illustrates a thought experiment like the ones that convinced Einstein that light must fall in a gravitational field. Think about what a person feels in an elevator that is accelerated upward. It is identical to being in a stationary elevator in a gravitational field. Her feet are pressed against the floor, and objects released from her hand fall with identical accelerations. In fact, it is not possible, without looking outside, to know which is happening, acceleration upward or gravity. This led Einstein to correctly postulate that acceleration and gravity will produce identical effects in all situations. So if acceleration affects light, then gravity will, too. Figure 32.9 shows the effect of acceleration on a beam of light shone horizontally at one wall. Since the accelerated elevator moves up during the time light travels across the elevator, the beam of light strikes low, seeming to the person to bend down. (Normally a tiny effect, since the speed of light is so great.) The same effect must occur due to gravity, Einstein reasoned, since there is no way to tell the effects of

Figure 32.9   (a) A beam of light emerges from a flashlight in an upward-accelerating elevator. Since the elevator moves up during the time the light takes to reach the wall, the beam strikes lower than it would if the elevator were not accelerated. (b) Gravity has the same effect on the light, since it is not possible to tell whether the elevator is accelerating upward or acted upon by gravity.

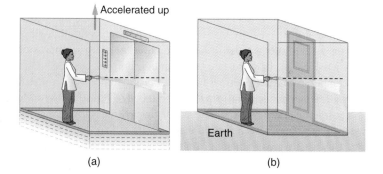

(a)   (b)

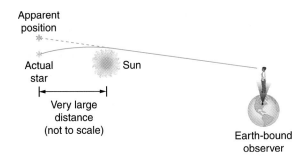

Figure 32.10 This schematic shows how light passing near a massive body like the sun is curved toward it. The light that reaches the earth seems then to come from different locations than the known positions of the originating stars. Not only was this effect observed, the amount of bending was precisely what Einstein predicted in his general theory of relativity.

gravity acting downward from acceleration of the elevator upward. Thus gravity affects the path of light, even though we think of gravity as acting between masses and photons are massless.

Of course, saying it is so does not make it so. Einstein's theory of general relativity got its first verification in 1919 when starlight passing near the sun was observed during a solar eclipse. (See Figure 32.10.) During an eclipse, the sky is darkened and we can briefly see stars. Those in a line of sight nearest the sun should have a shift in their apparent positions. Not only was this shift observed, but it agreed with Einstein's predictions well within experimental uncertainties. This discovery created a scientific and public sensation. Einstein was now a folk hero as well as a very great scientist. The bending of light by matter is equivalent to a bending of space itself, with light following the curve. This is another radical change in our concept of space, energy, matter, and time. It is also another connection between massless photons (which Einstein had correctly identified as particles of EM radiation) and particles that have mass—both are affected by gravity.

There are several current forefront efforts related to general relativity. One is the observation and analysis of gravitational lensing of light. Another is the search for definitive proof of the existence of black holes. Gravitational waves or moving wrinkles in space are being searched for. Theoretical efforts are also being aimed at the possibility of time travel and wormholes into other parts of space due to black holes.

Gravitational lensing As you can see in Figure 32.10, light is bent toward a mass, producing an effect much like a converging lens (large masses are needed to produce observable effects). On a galactic scale, the light from a distant galaxy could be "lensed" into several images when passing close by another galaxy on its way to earth. Einstein predicted this effect, but he considered it unlikely that we would ever observe it. A number of cases of this effect have now been observed; one is shown in Figure 32.11. This effect is a much larger scale verification of general relativity. But such gravitational lensing also is useful in verifying that red shift is proportional to distance. The red shift of the intervening galaxy is always less than that of the one being lensed, and each image of the lensed galaxy has the same red shift. This supplies more evidence that red shift is proportional to distance. Confidence that the multiple images are not different objects is

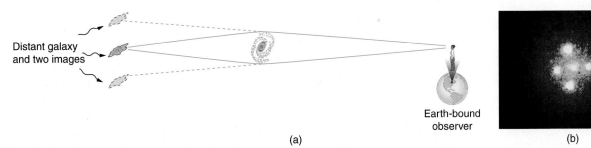

(a) (b)

Figure 32.11 (a) Light from a distant galaxy can travel different paths to the earth, because it is bent around an intermediary galaxy by gravity. This produces several images of the more distant galaxy. (b) The images around the central galaxy are produced by gravitational lensing. Each image has the same spectrum and a larger red shift than the intermediary.

bolstered by observations that if one varies in brightness over time, the others also vary in the same manner.

Black holes **Black holes** are objects having such large gravitational fields that things can fall in, but nothing, not even light, can escape. They are great fodder for science fiction, and we keep hearing conflicting reports as to their actual existence. Before addressing that question, let us briefly examine the concept of a black hole. Bodies, like the earth or sun, have what is called an **escape velocity**. If an object moves straight up from the body, starting at the escape velocity, it will just be able to escape the gravity of the body.* The greater the acceleration of gravity on the body, the greater the escape velocity. As long ago as the late 1700s, it was proposed that if the escape velocity is greater than the speed of light, then light cannot escape. Simon Laplace (1749–1827), the French astronomer and mathematician, even incorporated this idea of a dark star into his writings. But the idea was dropped after Young's double slit experiment showed light to be a wave. For some time, light was thought not to have particle characteristics and, thus, could not be acted upon by gravity. The idea of a black hole was very quickly reincarnated in 1916 after Einstein's theory of general relativity was published. It is now thought that black holes can form in the supernova collapse of a massive star, forming an object perhaps 10 km across and having a mass greater than our sun. It is interesting that several prominent physicists who worked on the concept, including Einstein, firmly believed that nature would find a way to prohibit such objects.

Black holes are difficult to observe directly, because they are small and no light comes directly from them. In fact, no light comes from inside the **event horizon**, which is defined to be at the distance from the object at which the escape velocity is exactly the speed of light. The radius of the event horizon is known as the **Schwarzschild radius R_s** and is given by

$$R_s = \frac{2GM}{c^2} \tag{32.2}$$

where G is the universal gravitational constant, M is the mass of the body, and c is the speed of light. The event horizon is the edge of the black hole and R_s is its radius (that is, the size of a black hole is twice R_s). Since G is small and c^2 is large, you can see that black holes are extremely small, only a few kilometers for masses a little greater than the sun's. The object itself is inside the event horizon. Physics near a black hole is fascinating. Gravity increases so rapidly as you approach a black hole that tidal effects tear matter apart, with matter closer to the hole being pulled in with much more force than that only slightly farther away. This can pull a companion star apart and heat inflowing gases to the point of producing x rays. (See Figure 32.12.) We have observed x rays from certain binary star systems that are consistent with such a picture. This is not quite proof of black holes, because the x rays could also be caused by matter falling onto a neutron star. These objects were first discovered in 1967 by British astrophysicists Jocelyn Bell and Anthony Hewish. **Neutron stars** are literally a star composed of neutrons. They are formed by the collapse of a star's core in a supernova, during which electrons and protons are forced together to form neutrons (the reverse of neutron β decay). Neutron stars are slightly larger than a black hole of the same mass and will not collapse further because of resistance by the strong force.

We also have evidence that supermassive black holes may exist at the cores of many galaxies, including the Milky Way. Such a black hole might have a mass millions or even billions times that of the sun, and it would probably have formed when matter first coalesced into a galaxy billions of years ago. Supporting this is the fact that very distant galaxies are more likely to have abnormally energetic cores. Some of the most distant galaxies, and hence the youngest, are known as **quasars** and emit as much or more energy than a normal galaxy but from a region less than a light year across. Quasar energy outputs may vary in times less than a year, so that the energy emitting region must be less than a light year across. The best explanation of quasars is that they are young galaxies with

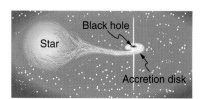

Figure 32.12 A black hole is shown pulling matter away from a companion star, forming a superheated accretion disk where x rays are emitted before the matter disappears forever into the hole. The infall energy also ejects some material, forming the two vertical spikes. (See also the photograph at the beginning of this chapter.) There are several x-ray emitting objects in space that are consistent with this picture and are likely to be black holes.

---

* Escape velocity occurs when kinetic energy at takeoff is just equal to gravitational potential energy at infinity. You may wish to refer to related concepts in Chapters 6 and 8.

a supermassive black hole forming at their core, and that they become less energetic over billions of years. In closer superactive galaxies, we observe tremendous amounts of energy being emitted from very small regions of space, consistent with stars falling into a black hole at the rate of one or more a month. The Hubble Space Telescope recently (1994) observed an accretion disk in the galaxy M87 rotating rapidly around a region of extreme energy emission. (See Figure 32.13.) A jet of material being ejected perpendicular to the plane of rotation gives further evidence of a supermassive black hole as the engine.

Gravitational waves   If a massive object distorts the space around it, like the foot of a water bug on the surface of a pond, then movement of the massive object should create waves in space like those on a pond. **Gravitational waves** are mass-created distortions in space that propagate at the speed of light and are predicted by general relativity. Since gravity is by far the weakest force, extreme conditions are needed to generate significant gravitational waves. Gravity near binary neutron star systems is so great that significant gravitational wave energy is radiated as the two neutron stars orbit one another. American astronomers Joseph Taylor and Russell Hulse measured changes in the orbit of such a binary neutron star system. They found its orbit to change precisely as predicted by general relativity, a strong indication of gravitational waves, and were awarded the 1993 Nobel prize. But direct detection of gravitational waves on earth would be conclusive. For many years, various attempts have been made to detect gravitational waves by observing vibrations induced in matter distorted by these waves. American physicist Joseph Weber pioneered this field in the 1960s, but no conclusive events have been observed. (No gravity wave detectors were in operation at the time of the 1987A supernova, unfortunately.) There are two ambitious systems of gravitational wave detectors under construction. The LIGO system has two laser interferometer detectors, one in the state of Washington and another in Louisiana. (See Figure 32.14.) The VIRGO facility in Italy has a single detector.

Quantum Gravity

Black holes radiate?   Quantum gravity is important in those situations where gravity is so extremely strong that it has effects on the quantum scale, where the other forces are ordinarily much stronger. The early universe was such a place, but black holes are another. The first significant connection between gravity and quantum effects was made by the Russian physicist Yakov Zel'dovich in 1971, and other significant advances followed from the British physicist Stephen Hawking (See Figure 32.15). These two showed that black holes could radiate away energy by quantum effects just outside the event horizon (nothing can escape from inside the event horizon). Black holes are, thus,

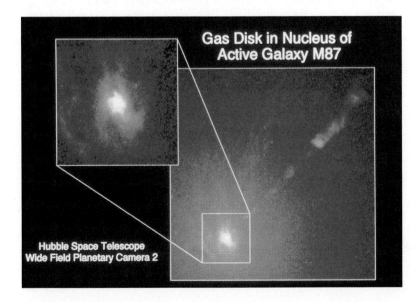

Figure 32.13 This Hubble Space Telescope photograph shows the extremely energetic core of the M87 galaxy. With the superior resolution of the orbiting telescope, it has been possible to observe the rotation of an accretion disk around the energy-producing object as well as mapping jets of material being ejected from the object. A supermassive black hole is consistent with these observations, but other possibilities are not quite eliminated.

Figure 32.14 Part of the LIGO gravitational wave detector under construction. Gravitational waves will cause extremely small vibrations in a mass in this detector, which will be detected by laser interferometer techniques. Such detection in coincidence with other detectors and with astronomical events, such as supernovas, would provide direct evidence of gravitational waves.

Figure 32.15 Stephen Hawking (b. 1942) has made many contributions to the theory of quantum gravity. Hawking is a long-time survivor of ALS and has produced popular books on general relativity, cosmology, and quantum gravity.

expected to radiate energy and shrink to nothing, although extremely slowly for most black holes. The mechanism is the creation of a particle-antiparticle pair from energy in the extremely strong gravitational field near the event horizon. One member of the pair falls into the hole and the other escapes, conserving momentum. (See Figure 32.16.) When a black hole loses energy and, hence, rest mass, its event horizon shrinks, creating an even greater gravitational field. This increases the rate of pair production so that the process grows exponentially until the black hole is nuclear in size. A final burst of particles and γ rays ensues. This is an extremely slow process for black holes about the mass of the sun (produced by supernovas) or larger ones (like those thought to be at galactic centers), taking on the order of 10^{67} years or longer! Smaller black holes would evaporate faster, but they are only speculated to exist as remnants of the Big Bang. Searches for characteristic γ-ray bursts have produced events attributable to more mundane objects like neutron stars accreting matter.

Wormholes and time travel The subject of time travel captures the imagination. Theoretical physicists, such as the American Kip Thorne, have treated the subject seriously, looking into the possibility that falling into a black hole could result in popping up in another time and place—a trip through a so-called wormhole. Time travel and wormholes appear in innumerable science fiction dramatizations, but the consensus is that time travel is not possible in theory. While still debated, it appears that quantum gravity effects inside a black hole prevent time travel due to the creation of particle pairs. Direct evidence is elusive.

The shortest time? Theoretical studies indicate that, at extremely high energies and correspondingly early in the universe, quantum fluctuations may make time intervals meaningful only down to some finite time limit. Early work indicated this might be the

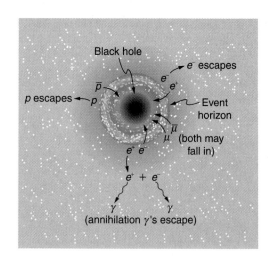

Figure 32.16 Gravity and quantum mechanics come into play when a black hole creates a particle-antiparticle pair from the energy in its gravitational field. One member of the pair falls into the hole while the other escapes, removing energy and shrinking the black hole. Searches are on for the characteristic energy.

case for times as long as 10^{-43} s, the time at which all forces were unified. If so, then it would be meaningless to consider the universe at times earlier than this. Subsequent studies indicate the crucial time may be as short as 10^{-95} s. But the point remains—quantum gravity seems to imply that there is no such thing as a vanishingly short time. Time may, in fact, be grainy with no meaning to time intervals shorter than some tiny but finite size.

The future of quantum gravity   Not only is quantum gravity in its infancy, no one knows how to get started on a theory of gravitons and unification of forces. The energies at which TOE should be valid are so high (at least 10^{19} GeV) and the necessary particle separation so small (less than 10^{-35} m) that only indirect evidence can provide clues. For some time, the common lament of theoretical physicists was one so familiar to struggling students—how do you even get started? But Hawking and others have made a start, and the approach most theorists are taking is called Superstring theory, the topic of the next section.

32.3 SUPERSTRINGS

Introduced in Section 31.6 (unification of forces), Superstring theory is an attempt to unify gravity with the other three forces and, thus, must contain quantum gravity. The main tenet of **Superstring theory** is that fundamental particles, including the graviton that carries the gravitational force, act like one-dimensional vibrating strings. Since gravity affects the time and space in which all else exists, Superstring theory is an attempt at a Theory of Everything (TOE). Each independent quantum number is thought of as a separate dimension in some superspace (analogous to the fact that the familiar dimensions of space are independent of one another) and is represented by a different type of Superstring. As the universe evolved after the Big Bang and forces became distinct (spontaneous symmetry breaking), some of the dimensions of superspace are imagined to have curled up and become unnoticed.

Forces are expected to be unified only at extremely high energies and at particle separations on the order of 10^{-35} m. This means that Superstrings must have dimensions or wavelengths of this size or smaller. Just as quantum gravity may imply that there are no time intervals shorter than some finite value, it also implies that there may be no sizes smaller than some tiny but finite value. That may be about 10^{-35} m. If so, and if Superstring theory can explain all it strives to, then the structures of Superstrings are at the lower limit to the smallest possible size and can have no further substructure. This would be the ultimate answer to the question the ancient Greeks considered. There is a finite lower limit to space.

Not only is Superstring theory in its infancy, it deals with dimensions about 17 orders of magnitude smaller than the 10^{-18} m details we have been able to observe directly. It is thus relatively unconstrained by experiment, and there are a host of theoretical possibilities to choose from. This has led theorists to make choices based subjectively (as always) on what is the most elegant theory with less hope than usual that experiment will guide them. It has also led to speculation of alternate universes, with their Big Bangs creating each new universe with a random set of rules. These speculations may not be tested even in principle, since an alternate universe is by definition unattainable. It is something like exploring a self-consistent field of mathematics, with its axioms and rules of logic, that is not consistent with nature. Such endeavors have often given insight to mathematicians and scientists alike and occasionally have been directly related to the description of new discoveries.

32.4 DARK MATTER AND CLOSURE

One of the most exciting problems in physics today is the fact that there is far more matter in the universe than we can see. The motion of stars in galaxies and the motions of galaxies in clusters imply that there is about ten times as much mass as in the luminous objects we can see. The indirectly observed nonluminous matter is called **dark matter**. Why is dark matter a problem? For one thing, we do not know what it is. It may well be

90% of all matter in the universe, yet there is a possibility it is of a completely unknown form—a stunning discovery if verified. Dark matter has implications for particle physics. It may be that neutrinos actually have small masses or that there are completely unknown types of particles. Dark matter also has implications for cosmology, since there may be enough dark matter to stop the expansion of the universe. That is another problem related to dark matter—we do not know how much there is. We keep finding evidence for more matter in the universe, and we have an idea of how much it would take to eventually stop the expansion of the universe, but whether there is enough is still unknown.

Evidence

The first clues that there is more matter than meets the eye came from the Swiss-born American astronomer Fritz Zwicky in the 1930s. Zwicky measured the velocities of stars orbiting the galaxy, using the Doppler shift of their spectra. (See Figure 32.17(a).) He found that velocity varied with distance from the center of the galaxy, as graphed in Figure 32.17(b). If the mass of the galaxy were concentrated in its center, as are its luminous stars, the velocities should decrease as the square root of the distance from the center. Instead, the velocity curve is almost flat, implying that there is a tremendous amount of matter in the galactic halo. While not immediately recognized for its significance, such measurements have now been made for many galaxies, with similar results. Furthermore, studies of galactic clusters have also indicated that galaxies have a mass distribution greater than that obtained from their brightness (proportional to the number of stars), which also extends into large halos surrounding the luminous parts of galaxies. Observations of other EM wavelengths, such as radio waves and x rays, have similarly confirmed the existence of dark matter. For example, x rays in the relatively dark space between galaxies indicates the presence of previously unobserved hot, ionized gas. (See Figure 32.17(c).)

Theoretical Yearnings for Closure

Is the universe open or closed? That is, is there enough matter to stop the expansion of the universe? If so, then the universe is closed. If not, the universe is open and will expand forever. (See Figure 32.18.)

Gravitational attraction between galaxies is slowing the expansion of the universe, but the amount of slowing is not directly known. In principle, we should be able to measure the average distance between galaxies at very large distances from us (hence very

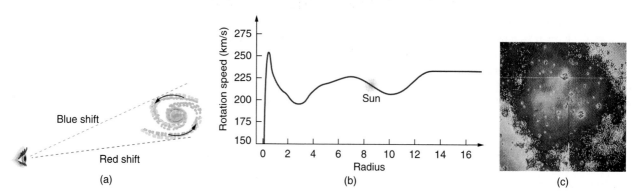

Figure 32.17 Evidence for dark matter. (a) We can measure the velocities of stars relative to their galaxies by observing the Doppler shift in emitted light, usually using the hydrogen spectrum. These measurements indicate the rotation of a spiral galaxy. (b) A graph of velocity versus distance from the galactic center shows that the velocity does not decrease as it would if the matter were concentrated in luminous stars. The flatness of the curve implies a massive galactic halo of dark matter extending beyond the visible stars. (c) This is a computer-generated image of x rays from a galactic cluster. The x rays indicate the presence of otherwise unseen hot clouds of ionized gas in the regions of space previously considered more empty.

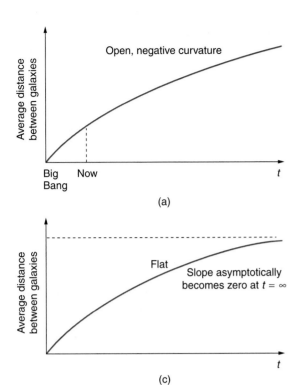

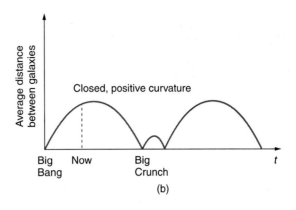

Figure 32.18 Is the universe open or closed? (a) In an open universe, the distance between galaxies increases without limit, although the expansion rate does slow. Space is said to be negatively curved, and the density of matter is less than the critical density. (b) In a closed universe, the distance between galaxies continues to increase until gravity slows and then reverses the expansion, resulting in a Big Crunch, the opposite of the Big Bang. It may be that the process is cyclical as shown, but perhaps with different rules in each cycle. Space is positively curved, since the density of matter is greater than the critical density. (c) A flat universe where the density of matter exactly equals the critical density and expansion halts asymptotically.

long ago) and see that they were once closer together than they are now. We should also be able to see that the expansion rate between distant galaxies was once greater than it is now. But uncertainties in distance measurements prevent us from determining how much the expansion rate of the universe has slowed in, say, 10 billion years. We can, however, calculate the amount of slowing based on the average density of matter we observe directly. Here we have a definite answer—there is far less visible matter than needed to stop expansion. The **critical density** ρ_c is defined to be the density needed to just halt universal expansion, as in Figure 32.18(c). It is estimated to be about

$$\rho_c \approx 10^{-26} \text{ kg/m}^3$$

But this estimate of ρ_c is only good to about a factor of two, due to uncertainties in the expansion rate of the universe. The critical density is equivalent to an average of only a few nucleons per cubic meter, remarkably small and indicative of how truly empty intergalactic space is. Luminous matter seems to account for roughly 0.5% to 2% of the critical density, far less than needed for closure. Taking into account the amount of dark matter we detect indirectly and all other types of indirectly observed normal matter, there is only 10% to 40% of what is needed for closure. But more is being found with every new part of the spectrum that is explored and with each new instrument that probes more accurately. So the question lingers. Will we eventually find enough matter to show the universe is closed?

Most theorists feel strongly that the universe should be just barely closed. Since matter can be thought to curve the space around it, we call an open universe that expands forever **negatively curved**. This means that you can in principle travel an unlimited distance in any direction. A universe that is closed and eventually contracts is called **positively curved**. This means that if you travel far enough in any direction, you will return to your starting point, analogous to circumnavigating the earth. And a universe that has exactly the critical density is called **flat** (zero curvature). Why do theorists feel the universe is flat? Flatness is part of the inflationary scenario that helps explain the flatness of the microwave background. In fact, since general relativity implies that matter creates the space in which it exists, there is a special symmetry to a flat universe. But, as noted, the actual condition is not yet known.

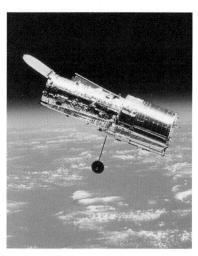

Figure 32.19 The Hubble Space Telescope is producing exciting data with its corrected optics and with the absence of atmospheric distortion. It has observed some MACHOs, disks of material around stars thought to precede planet formation, black hole candidates, and collisions of comets with Jupiter.

What *Is* the Dark Matter We See Indirectly?

There is virtually no doubt that dark matter exists, but what it is and how much there is are both still vigorously being studied. As always, we seek to explain new observations in terms of known principles. But as more discoveries are made, it is becoming more and more difficult to explain dark matter as a known type of matter. One of the possibilities for normal matter is being explored using the Hubble Space Telescope and employing the lensing effect of gravity on light. (See Figure 32.19.) Stars glow because of nuclear fusion in them, but planets are visible primarily by reflected light. Jupiter, for example, is too small to ignite fusion in its core and become a star, but we can see sunlight reflected from it since we are relatively close. If Jupiter orbited another star, we would not be able to see it directly. The question is open as to how many planets or other bodies smaller than about 1/1000 the mass of the sun are out there. If such bodies pass between us and a star, they will not block the star's light, being too small, but they will form a gravitational lens, as discussed in the preceding section. In a process called **microlensing**, light from the star is focused and the star appears to brighten in a characteristic manner. Searches for dark matter in this form are particularly interested in galactic halos because of the huge amount of mass that seems to be there. Such microlensing objects are, thus, called **massive compact halo objects** or **MACHOs**. To date, a few MACHOs have been observed, but not predominantly in galactic halos nor in the numbers needed to explain dark matter.

MACHOs are among the most conventional of unseen objects proposed to explain dark matter. Others being actively pursued are red dwarfs, which are small dim stars, but too few have been seen so far, even with the Hubble Telescope, to be of significance. Old remnants of stars called white dwarfs are also under consideration, since they contain about a solar mass but are as small as the earth and may dim to the point that we ordinarily do not observe them. While white dwarfs are known, old dim ones are not. Yet another possibility is the existence of large numbers of smaller than stellar mass black holes left from the Big Bang—here evidence is entirely absent.

There is a very real possibility that dark matter is composed of the known neutrinos, which may have small but finite masses. As discussed in Chapter 31, neutrinos are thought to be massless, but we only have upper limits on their masses rather than knowing they are exactly zero. So far, these upper limits come from difficult measurements of total energy emitted in the decays and reactions in which neutrinos are involved. There is an amusing possibility of proving that neutrinos have mass in a completely different way. In Section 31.4, we noted that there are three flavors of neutrinos (v_e, v_μ, and v_τ) and that the weak interaction could change quark flavor. It should also change neutrino flavor—that is, any type of neutrino could spontaneously change into any other, a process called **neutrino oscillations**. But this can only occur if neutrinos have mass. Why? Because if neutrinos are massless, they must travel at the speed of light and time will not pass for them, so that they cannot change without an interaction. Thus if we can observe neutrino oscillations, we will have proved that neutrinos have mass and may account for at least some of the dark matter in the universe.

Neutrino oscillations may also explain the low number of observed solar neutrinos. Detectors for observing solar neutrinos are specifically designed to detect electron neutrinos v_e produced in huge numbers by fusion in the sun. A large fraction of electron neutrinos v_e may be changing flavor to muon neutrinos v_μ on their way out of the sun, possibly enhanced by specific interactions, reducing the flux of electron neutrinos to observed levels. There is also a discrepancy in observations of neutrinos produced in cosmic ray showers (see Figure 31.30). While these showers of radiation produced by extremely energetic cosmic rays should contain twice as many v_μs as v_es, their numbers are nearly equal. This may be explained by neutrino oscillations from muon flavor to electron flavor. Massive neutrinos are particularly appealing possibility for explaining dark matter, since their existence is consistent with a large body of known information and explains more than dark matter. The question is not settled at this writing.

The most radical proposal to explain dark matter is that it consists of previously unknown leptons (sometimes obtusely referred to as nonbaryonic matter). These are called

Figure 32.20 Dark matter may shepherd normal matter gravitationally in space, as this stream moves leaves about. Dark matter may be invisible and even move through normal matter, as neutrinos penetrate us without small-scale effect.

weakly interacting massive particles, or **WIMPs**, and would also be chargeless, thus interacting negligibly with normal matter except through gravitation. One proposed group of WIMPs would have masses several orders of magnitude greater than nucleons and are sometimes called **neutralinos**. Others are called **axions** and would have masses about 10^{-10} of an electron mass. Both neutralinos and axions would be gravitationally attached to galaxies, but because they are chargeless and only feel the weak force, they would be in a halo rather than interacting and coalescing into spirals, and so on, like normal matter. (See Figure 32.20.) Some particle theorists have built WIMPs into their unified force theories and into the inflationary scenario of the evolution of the universe so popular today. These particles would have been produced in just the correct numbers to make the universe flat, shortly after the Big Bang. The proposal is radical in the sense that it invokes entirely new forms of matter, in fact *two* entirely new forms, in order to explain dark matter and other phenomena. WIMPs have the extra burden of automatically being almost impossible to observe directly, making them philosophically difficult to accept. This is somewhat analogous to quark confinement, which guarantees quarks are there but can never be seen directly. At any rate, before WIMPs are accepted as the best explanation, all other possibilities utilizing known phenomena will have to be shown inferior. Should that occur, we will be in the unanticipated position of admitting that all we know is only 10% of what there is. A far cry from the days when people firmly believed themselves to be not only the center of the universe but also the reason for its existence.

32.5 COMPLEXITY AND CHAOS

Much of what impresses us about physics is related to the underlying connections and basic simplicity of the laws we have discovered. The language of physics is precise and well defined because many of the basic systems we study are simple enough that we can perform controlled experiments and discover unambiguous relationships. Our most spectacular successes, such as the prediction of previously unobserved particles, come from the simple underlying patterns we have been able to recognize. But there are systems of interest to physicists that are inherently complex. The simple laws of physics apply, of course, but complex systems may reveal patterns that simple systems do not. The emerging field of **complexity** is devoted to the study of complex systems, including those outside the traditional bounds of physics. Of particular interest is the ability of complex systems to adapt and evolve.

What are some examples of complex adaptive systems? One is the primordial ocean. When the oceans first formed, they were a random mix of elements and compounds that obeyed the laws of physics and chemistry. In a relatively short geological time (about 500 million years), life had emerged. Laboratory simulations indicate that the emergence of life was far too fast to have come from random combinations of compounds, even if driven by lightning and heat. There must be an underlying ability of the complex system to organize itself, resulting in the self-replication we recognize as life. Living entities, even at the unicellular level, are highly organized and systematic. Systems of living organisms are themselves complex adaptive systems. The grandest of these evolved into the biological system we have today, leaving traces in the geological record of steps taken along the way. (See Figure 32.21.)

Complexity as a discipline examines complex systems, how they adapt and evolve, looking for similarities with other complex adaptive systems. Can, for example, parallels be drawn between biological evolution and the evolution of *economic systems*? Economic systems do emerge quickly, they show tendencies for self-organization, they are complex (in the number and types of transactions), and they adapt and evolve. Biological systems do all the same types of things. There are other examples of complex adaptive systems being studied for fundamental similarities. *Cultures* show signs of adaptation and evolution. The comparison of different cultural evolutions may bear fruit as well as comparisons to biological evolution. *Science* also is a complex system of human interactions, like culture and economics, that adapts to new information and political pressure, and evolves, usually becoming more organized rather than less. Those who study *creative thinking* also see parallels with complex systems. Humans sometimes organize almost random pieces of information, often subconsciously while doing other things, and come up with brilliant creative insights. The development of *language* is another complex adaptive system that may show similar tendencies. *Artificial intelligence* is an overt attempt to devise an adaptive system that will self-organize and evolve in the same manner as an intelligent living being learns. These are a few of the broad range of topics being studied by those who investigate complexity. There are now institutes, journals, and meetings, as well as popularizations of the emerging topic of complexity.

In traditional physics, the discipline of complexity may yield insights in certain areas. Thermodynamics treats systems on the average, while statistical mechanics deals in some detail with complex systems of atoms and molecules in random thermal motion. Yet there is organization, adaptation, and evolution in those complex systems. Nonequilibrium phenomena, such as heat transfer and phase changes, are characteristically complex in detail, and new approaches to them may evolve from complexity as a discipline. Crystal growth is another example of self-organization spontaneously emerging in a complex system. Alloys are also inherently complex mixtures that show certain simple characteristics implying some self-organization. The organization of iron atoms into magnetic domains as they cool is another. Perhaps insights into these difficult areas will emerge from complexity. But at the minimum, the discipline of complexity is another example of human effort to understand and organize the universe around us, partly rooted in the discipline of physics.

A predecessor to complexity, the topic of chaos has been widely publicized and become a discipline of its own. It is also based partly in physics and treats broad classes of

Figure 32.21

Figure 32.22 This image is related to the Mandelbrot set, a complex mathematical form that is chaotic. The patterns are infinitely fine as you look closer and closer, and they indicate order in the presence of chaos.

phenomena from many disciplines. **Chaos** is a word used to describe systems the outcomes of which are extremely sensitive to initial conditions. The orbit of the planet Pluto, for example, may be chaotic in that it can change tremendously due to small interactions with other planets. This makes its long-term behavior impossible to predict with precision, just as we cannot tell precisely where a decaying earth satellite will land or how many pieces it will break into. But the discipline of chaos has found ways to deal with such systems and has been applied to apparently unrelated systems. For example, the heartbeat of people with certain types of potentially lethal arrhythmias seems to be chaotic, and this knowledge may allow more sophisticated monitoring and recognition of the need for intervention.

Chaos is related to complexity. Some chaotic systems are also inherently complex; for example, vortices in a fluid as opposed to a double pendulum. Both are chaotic and not predictable in the same sense as other systems. But there can be organization in chaos. Among these are the beautiful fractal patterns such as in Figure 32.22. Some chaotic systems exhibit self-organization, a type of stable chaos. The orbits of the planets in our solar system, for example, may be chaotic (we are not certain yet). But they are definitely organized and systematic, with a simple formula describing the orbital radii of the first eight planets *and* the asteroid belt. Large-scale vortices in Jupiter's atmosphere are chaotic, but the Great Red Spot is a stable self-organization of rotational energy in that system. (See Figure 32.23.) The Great Red Spot has been in existence for at least 400 years and is a complex self-adaptive system.

The emerging field of complexity, like the now almost traditional field of chaos, is partly rooted in physics. Both attempt to see similar systematics in a very broad range of phenomena and, hence, generate better understanding of them. Time will tell what impact these fields have on more traditional areas of physics as well as the other disciplines they relate to.

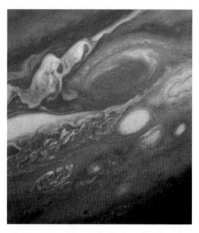

Figure 32.23 The Great Red Spot on Jupiter is an example of self-organization in a complex and chaotic system. Smaller vortices in Jupiter's atmosphere behave chaotically, but the triple-earth-size spot is self-organized and stable for at least hundreds of years.

32.6 HIGH-TEMPERATURE SUPERCONDUCTORS

Superconductors are materials with resistivity of zero. They are familiar to the general public because of their practical applications and have been mentioned at a number of points in the text (see Chapter 19, for example). Because the resistance of a piece of superconductor is zero, there are no heat losses for currents through them; they are used in magnets needing high currents, such as in MRI, and could cut energy losses in power transmission. But most superconductors must be cooled to temperatures only a few kelvin above absolute zero, a costly procedure limiting their practical applications. In the past decade, tremendous advances have been made in producing materials that become superconductors at relatively high temperatures. There is hope that room temperature superconductors may some day be manufactured.

Superconductivity was discovered accidentally in 1911 by the Dutch physicist H. Kamerlingh Onnes (1853–1926) when he used liquid helium to cool mercury. Onnes had been the first person to liquefy helium a few years earlier and was surprised to observe the resistivity of a mediocre conductor like mercury drop to zero at a temperature of 4.2 K. We define the temperature at which and below which a material becomes a superconductor to be its **critical temperature**, denoted T_c. (See Figure 32.24.) Progress in understanding how and why a material became a superconductor was relatively slow, with the first workable theory coming in 1957. Certain other elements were also found to become superconductors, but all had T_cs less than 10 K, which are expensive to maintain. Although Onnes received a Nobel prize in 1913, it was primarily for work with liquid helium.

In 1986 a breakthrough was announced—a ceramic compound was found to have an unprecedented T_c of 35 K. It looked as if much higher critical temperatures could be possible, and by early 1988 another ceramic (this of thallium, calcium, barium, copper, and

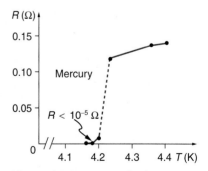

Figure 32.24 A graph of resistivity versus temperature for a superconductor shows a sharp transition to zero at the critical temperature T_c. High-temperature superconductors have verifiable T_cs greater than 125 K, well above the easily achieved 77 K temperature of liquid nitrogen.

Figure 32.25 One characteristic of a superconductor is that it excludes magnetic flux and, thus, repels other magnets. The small magnet levitated above a high-temperature superconductor, which is cooled by liquid nitrogen, gives evidence that the material is superconducting. When the material warms and becomes conducting, magnetic flux can penetrate it, and the magnet will rest upon it.

oxygen) had been found to have $T_c = 125$ K. (See Figure 32.25.) The economic potential of perfect conductors saving electric energy is immense for T_cs above 77 K, since that is the temperature of liquid nitrogen. Although liquid helium has a boiling point of 4 K and can be used to make materials superconducting, it costs about $5 per liter. Liquid nitrogen boils at 77 K, but only costs about $0.30 per liter. There was general euphoria at the discovery of these complex ceramic superconductors, but this soon subsided with the sobering difficulty of forming them into usable wires. The first commercial use of a high-temperature superconductor is in an electronic filter for cellular phones. High-temperature superconductors are used in experimental apparatus, and they are actively being researched, particularly in thin film applications.

The search is on for even higher T_c superconductors, many of complex and exotic copper oxide ceramics, sometimes including strontium, mercury, or yttrium as well as barium, calcium, and other elements. Room temperature (about 293 K) would be ideal, but any temperature close to room temperature is relatively cheap to produce and maintain. There are persistent reports of T_cs over 200 K and some in the vicinity of 270 K. Unfortunately, these observations are not routinely reproducible, with samples losing their superconducting nature once heated and recooled (cycled) a few times. (See Figure 32.26.) They are now called USOs for unidentified superconducting objects, out of frustration and the refusal of some

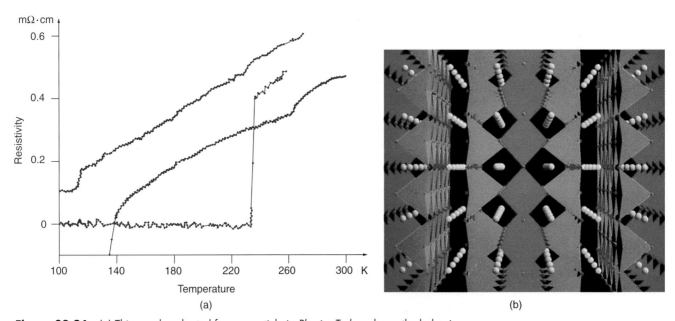

Figure 32.26 (a) This graph, adapted from an article in *Physics Today*, shows the behavior of a single sample of high-temperature superconductor in three different trials. In one case the sample exhibited a T_c of about 230 K, whereas in the others it did not become superconducting at all. The lack of reproducibility is typical of forefront experiments and prohibits definitive conclusions. (b) This colorful diagram shows the complex but systematic nature of the lattice structure of a high-temperature superconducting ceramic.

samples to show high T_c even though produced in the same manner as others. Reproducibility is crucial to discovery, and researchers are justifiably reluctant to claim the breakthrough they all seek. Time will tell whether USOs are real or an experimental quirk.

The theory of ordinary superconductors is difficult, involving quantum effects for widely separated electrons traveling through a material. Electrons couple in a manner that allows them to get through the material without losing energy to it, making it a superconductor. High-T_c superconductors are more difficult to understand theoretically, but theorists seem to be closing in on a workable theory. The difficulty of understanding how electrons can sneak through materials without losing energy in collisions is even greater at higher temperatures, where vibrating atoms should get in the way. Discoverers of high T_c may feel something analogous to what a politician once said upon an unexpected election victory—"I wonder what we did right?"

32.7 SOME QUESTIONS WE KNOW TO ASK

Throughout the text we have noted how essential it is to be curious and to ask questions in order to first understand what is known, and then to go a little farther. Some questions may go unanswered for centuries, others may not have answers, but some bear delicious fruit. Part of discovery is knowing which questions to ask. You have to know something before you can even phrase a decent question. As you may have noticed, the mere act of asking a question can give you the answer. The following questions are a sample of those physicists now know to ask, representative of the forefronts of physics. Although these questions are important, they will be replaced by others if answers are found to them. The fun continues.

On the Largest Scales

1. *Is the universe open or closed?* Theorists would like it to be just barely closed, but so far we have not discovered enough matter density to satisfy that desire. There is a connection to small-scale physics in the type and number of particles that may contribute to closing the universe.

2. *What is dark matter?* It is definitely there, but we really do not know what it is. Conventional possibilities are being ruled out, but one of them still may explain it. The answer could reveal whole new realms of physics.

3. *How do galaxies form?* They are there very early in the evolution of the universe, but the cosmic microwave background radiation is so wrinkle-free that it is difficult to understand how they evolved so quickly. Small fluctuations are visible in the CMBR and may yet explain galaxy formation.

4. *Do black holes exist?* Any confirmation must rule out all other possibilities, since direct observation is difficult. Evidence is building, but relativistic accretion disks would be definitive. The verified existence of any black hole will have far-reaching implications, as other forefront questions imply.

5. *What is the mechanism for the energy output of quasars?* These distant and extraordinarily energetic objects may be early stages of galactic evolution with a supermassive black hole devouring material. Connections are now being made with galaxies having energetic cores, and there is evidence consistent with less consuming supermassive black holes at the center of older galaxies.

6. *Where do the γ bursts come from?* We see bursts of γ rays coming from all directions in space, indicating the sources are very distant objects rather than something associated with our own galaxy. These γ bursts are not correlated with any known sources, but recently one was observed to occur simultaneously with other forms of radiation. This lends evidence for the possibility that they may be from binary neutron star interactions or black holes eating a companion neutron star.

On the Intermediate Scale

1. *How do phase transitions take place on the microscopic scale?* We know a lot about phase transitions, such as water freezing, but the details of how they occur molecule by molecule are not well understood. This may not seem like much, but similar questions about specific heat a century ago led to early quantum

mechanics. It is also an example of a complex adaptive system that may yield insights into other self-organizing systems.

2. *Is there a way to deal with nonlinear phenomena that reveals underlying connections?* Nonlinear phenomena lack a direct or linear proportionality that makes analysis and understanding a little easier. There are implications for nonlinear optics and broader topics such as chaos.

3. *How do high-T_c superconductors become resistanceless at such high temperatures?* Understanding how they work may help us to make them more practical or may simply result in other surprises, as unexpected as the discovery of superconductivity itself.

4. *There are magnetic effects in materials we do not understand—how do they work?* Although beyond the scope of this text, there is a great deal to learn in condensed matter physics (the physics of solids and liquids). We may find surprises analogous to lasing, the quantum Hall effect, and the quantization of magnetic flux. Complexity may play a role here, too.

On the Smallest Scale

1. *Are quarks and leptons fundamental, or do they have a substructure?* The higher-energy accelerators that are being constructed may supply some answers, but there will also be input from cosmology and other systematics.

2. *Why do leptons have integral charge while quarks have fractional charge?* If both are fundamental and analogous as thought, this question deserves an answer. It is obviously related to the previous question.

3. *Why are there three families of quarks and leptons?* First, does this imply some relationship? Second, why three and only three families?

4. *Are all forces truly equal (unified) under certain circumstances?* They don't have to be equal just because we want them to be. The answer may have to be indirectly obtained because of the extreme energy at which we think they are unified.

5. *Are there other fundamental forces?* There was a flurry of activity with claims of a fifth and even a sixth force a few years ago. Interest has subsided, since those forces have not been detected consistently. Moreover, the proposed forces have strengths similar to gravity, making them extraordinarily difficult to detect in the presence of stronger forces. But the question remains; and if there are no other forces, we need to ask why only four and why these four.

6. *Is the proton stable?* We have discussed this in some detail, but the question is related to fundamental aspects of the unification of forces. We may never know from experiment that the proton is stable, only that it is very long lived.

7. *Are there magnetic monopoles?* Many particle theories call for very massive individual north and south pole particles—magnetic monopoles. If they exist, why are they so different in mass and elusiveness from electric charges, and if they do not exist, why not?

8. *Do neutrinos have mass?* This may soon be answered, and the implications will be significant, as discussed in this chapter. There are effects on the closure of the universe and on the patterns in particle physics.

9. *Are lepton family numbers absolutely conserved?* This question has implications for the unification of forces, but we may be able to get an answer from double β decay. A rare form of nuclear decay, it is being studied for possible violations of lepton family number conservation.

10. *Is there an island of relatively stable nuclei around $Z = 114$?* With elements 110 and 111 now discovered, we may be close to an answer. We will learn more about the strong force, and there may be practical applications.

These lists of questions are not meant to be complete or consistently important—you can no doubt add to it yourself. There are also important questions in topics not broached in this text, such as certain particle symmetries, that are of current interest to physicists. The point is hopefully clear that no matter how much we learn, there always seems to be more to know. Although we are fortunate to have the hard-won wisdom of those who preceded us, we can look forward to new enlightenment, undoubtedly sprinkled with surprise.

SUMMARY

This chapter presents an overview of the most contemporary efforts and major areas of activity in physics, with an eye toward those most likely to produce important advances. Apologies are offered to those whose work is unintentionally slighted in this chapter.

Cosmology is the study of the character and evolution of the universe. The two most important features of the universe are the **cosmological red shifts of its galaxies being proportional to distance** and its **cosmic microwave background** (CMBR). Both support the notion that there was a gigantic explosion, known as the **Big Bang**, that created the universe. Galaxies farther away than our local group have, on average, a recessional velocity given by

$$v = H_c d \qquad (32.1)$$

where d is the distance to the galaxy and H_c is the **Hubble constant**, taken to have the average value $H_c = 20$ km/s/Mly.

Explanations of the large-scale characteristics of the universe are intimately tied to particle physics. The *dominance of matter over antimatter* and the *smoothness of the CMBR* are two that are tied to particle physics. The epochs of the universe are known back to very shortly after the Big Bang, based on known laws of physics. The earliest epochs are tied to the unification of forces, with the **electroweak epoch** being partially understood, the **GUT epoch** speculative, and the **TOE epoch** highly speculative since it involves an unknown single **superforce**. Part of the explanation for these epochs arises from present conditions in the universe. The transition from GUT to electroweak is called **spontaneous symmetry breaking**. It released energy that caused the **inflationary scenario**, which in turn explains the smoothness of the CMBR.

Einstein's theory of **general relativity** includes accelerated frames and, thus, encompasses special relativity and gravity. Created by use of careful **thought experiments**, it has been repeatedly verified by real experiment. One direct result of this behavior of nature is the **gravitational lensing** of light by massive objects, such as galaxies, also seen in the **microlensing** of light by smaller bodies in our galaxy. Another prediction is the existence of **black holes**, objects for which the escape velocity is greater than the speed of light and from which nothing can escape. The **event horizon** is the distance from the object at which the escape velocity equals the speed of light c. It is called the **Schwarzschild radius** R_s and is given by

$$R_s = \frac{2GM}{c^2} \qquad (32.2)$$

where G is the universal gravitational constant, and M is the mass of the body. Physics is unknown inside the event horizon, and the possibility of wormholes and time travel are being studied. Candidates for black holes may power the extremely energetic emissions of **quasars**, distant objects that seem to be early stages of galactic evolution. **Neutron stars** are stellar remnants, having the density of a nucleus, that hint that black holes could form from supernovas, too. **Gravitational waves** are wrinkles in space, predicted by general relativity but not yet observed, caused by changes in very massive objects.

Quantum gravity is an incompletely developed theory that strives to include general relativity, quantum mechanics, and unification of forces (thus, a TOE). One unconfirmed connection between general relativity and quantum mechanics is the prediction of characteristic radiation from just outside black holes. **Superstring theory** holds that fundamental particles are one-dimensional vibrations analogous to those on strings and is an attempt at a theory of quantum gravity.

Dark matter is nonluminous matter detected in and around galaxies and galactic clusters. It may be ten times the mass of luminous matter in the universe, and its amount may determine whether the universe is open or closed (expands forever or eventually stops). The determining factor is the **critical density** of the universe. An open universe is **negatively curved**, a closed universe is **positively curved**, whereas a universe with exactly the critical density is **flat**. Dark matter's composition is a major mystery, but it may be due to the unproven mass of neutrinos or a completely unknown type of leptonic matter. If neutrinos have mass, they will change families, a process known as **neutrino oscillations**, for which there is suggestive evidence.

Complexity is an emerging field, rooted primarily in physics, that considers complex adaptive systems and their evolution, including self-organization. It has applications in physics and many other disciplines, such as biological evolution. **Chaos** is a field that studies systems the *properties of which depend extremely sensitively on some variable*, and the evolution of which is impossible to predict. Chaotic systems may be simple or complex. Studies of chaos have led to methods for understanding and predicting certain chaotic behaviors.

High-temperature **superconductors** are materials that become superconducting at temperatures well above a few kelvin. The **critical temperature** T_c is the temperature below which a material is superconducting; some high-temperature superconductors have verified T_cs above 125 K, and there are reports of T_cs as high as 250 K.

CONCEPTUAL QUESTIONS

32.1 Explain why it only *appears* that we are at the center of expansion of the universe and why an observer in another galaxy would see the same relative motion of all but the closest galaxies away from her.

32.2 If there is no observable edge to the universe, can we determine where its center of expansion is? Explain.

32.3 If the universe is infinite, does it have a center? Discuss.

32.4 One cause of red shift in light is the source being in a high gravitational field. Discuss how this can be eliminated as the source of galactic red shifts, given that the shifts are proportional to distance and not to the size of the galaxy.

32.5 If another cause of red shift—such as light becoming "tired" from traveling long distances through empty space—is discovered, what effect would there be on cosmology?

32.6 Olbers's paradox poses an interesting question: If the universe is infinite, then any line of sight should eventually fall on a star's surface. Why then is the sky dark at night? Discuss the commonly accepted evolution of the universe as a solution to this paradox.

32.7 If the cosmic microwave background radiation (CMBR) is the remnant of the Big Bang's fireball, we expect to see hot and cold regions in it. What are two causes of these wrinkles in the CMBR? Are the observed temperature variations greater or less than originally expected?

32.8 The decay of one type of K-meson is cited as evidence that nature favors matter over antimatter. Since mesons are composed of a quark and an antiquark, is it surprising that they would preferentially decay to one type over another? Is this an asymmetry in nature? Is the predominance of matter over antimatter an asymmetry?

32.9 Distances to local galaxies are determined by measuring the brightness of stars, called Cepheid variables, that can be observed individually and that have absolute brightnesses at a standard distance that are well known. Explain how the measured brightness would vary with distance as compared with the absolute brightness.

32.10 Distances to remote galaxies are estimated based on their apparent type, which indicates the number of stars in the galaxy, and their measured brightness. Explain how the measured brightness would vary with distance. Would there be any correction necessary to compensate for the red shift of the galaxy (all distant galaxies have significant red shifts)? Discuss possible causes of uncertainties in these measurements.

32.11 Quantum gravity, if developed, would be an improvement on both general relativity and quantum mechanics, but more mathematically difficult. Under what circumstances would it be necessary to use quantum gravity? Similarly, under what circumstances could general relativity be used? Special relativity? Quantum mechanics? Classical physics?

32.12 Does observed gravitational lensing correspond to a converging or diverging lens? Explain briefly.

32.13 Suppose you measure the red shifts of all the images produced by gravitational lensing, such as in Figure 32.11(b). You find that the central image has a red shift less than the outer images, and those all have the same red shift. Discuss how this not only shows the images are of the same object, but also implies that the red shift is not affected by taking different paths through space. Does this imply that cosmological red shifts are not caused by traveling through space (light getting tired, perhaps)?

32.14 What are gravitational waves, and have they yet been observed either directly or indirectly?

32.15 Is the event horizon of a black hole the actual physical surface of the object?

32.16 Suppose black holes radiate their mass away and the lifetime of a black hole created by a supernova is about 10^{67} years. How does this lifetime compare with the accepted age of the universe? Is it surprising that we do not observe the predicted characteristic radiation?

32.17 If the smallest meaningful time interval is greater than zero, will the lines in Figure 32.8 ever meet?

32.18 Discuss the possibility that star velocities at the edges of galaxies being greater than expected is due to unknown properties of gravity rather than the existence of dark matter. Would, for example, this mean that gravity is greater or smaller than expected at large distances? Are there other tests that could be made of gravity at large distances, such as observing the motions of neighboring galaxies?

32.19 How does relativistic time dilation prohibit neutrino oscillations if they are massless?

32.20 If neutrino oscillations do occur, will they violate conservation of the various lepton family numbers (L_e, L_μ, and L_τ)? Will neutrino oscillations violate conservation of the total number of leptons?

32.21 Lacking direct evidence of WIMPs as dark matter, why must we eliminate all other possible explanations based on known forms of matter before we invoke their existence?

32.22 Must a complex system be adaptive to be of interest in the field of complexity? Give an example to support your answer.

32.23 State a necessary condition for a system to be chaotic.

32.24 What is critical temperature T_c? Do all materials have a critical temperature? Explain why or why not.

32.25 Explain how good thermal contact with liquid nitrogen can keep objects at a temperature of 77 K (liquid nitrogen's boiling point at atmospheric pressure).

32.26 Not only is liquid nitrogen a cheaper coolant than liquid helium, its boiling point is higher (77 K vs. 4.2 K). How does higher temperature help lower the cost of cooling a material? Explain in terms of the rate of heat transfer being related to the temperature difference between the sample and its surroundings.

32.27 For experimental evidence, particularly of previously unobserved phenomena, to be taken seriously it must be reproducible or of sufficiently high quality that a single observation is meaningful. Supernova 1987A is not reproducible. How do we know observations of it were valid? The fifth force is not broadly accepted. Is this due to lack of reproducibility or poor-quality experiments (or both)? Discuss why forefront experiments are more subject to observational problems than those involving established phenomena.

32.28 Discuss whether you think there are limits to what humans can understand about the laws of physics. Support your arguments.

PROBLEMS

Since the forefronts of physics represent an integration of many topics, most of the following problems could be labeled as integrated concepts. You will find it necessary to refer to other parts of the text when solving the following problems.

Note: Problem-solving strategies and worked examples that can help you solve integrated concept problems appear in several places, the most recent being Chapter 29. Consult the section of problems labeled *Integrated Concepts* at the end of that chapter or others.

32.1 Find the approximate mass of the luminous matter in the Milky Way galaxy, given it has approximately 10^{11} stars of average mass 1.5 times that of our sun.

32.2 Find the approximate mass of the dark and luminous matter in the Milky Way galaxy. Assume the luminous matter is due to approximately 10^{11} stars of average mass 1.5 times that of our sun, and take the dark matter to be 10 times as massive as the luminous.

32.3 (a) Estimate the mass of the luminous matter in the known universe, given there are 10^{11} galaxies, each containing 10^{11} stars of average mass 1.5 times that of our sun. (b) How many protons (the most abundant nuclide) are there in this mass? (c) Estimate the total number of particles in the observable universe by multiplying the answer to (b) by two, since there is an electron for each proton, and then by 10^9, since there are far more particles (such as photons and neutrinos) in space than in luminous matter.

32.4 Assume the average density of the universe is 0.1 of the critical density needed for closure. What is the average number of protons per cubic meter, assuming the universe is composed mostly of hydrogen?

32.5 On average, how far away are galaxies that are moving away from us at 2.0% of the speed of light?

32.6 If a galaxy is 500 Mly away from us, how fast do we expect it to be moving and in what direction?

32.7 (a) What is the approximate velocity relative to us of a galaxy near the edge of the known universe, some 10 Gly away? (b) What fraction of the speed of light is this? Note that we have observed galaxies moving away from us at greater than $0.9c$.

32.8 Our solar system orbits the center of the Milky Way galaxy. Assuming a circular orbit 30,000 ly in radius and an orbital speed of 250 km/s, how many years does it take for one revolution? Note that this is approximate, assuming constant speed and circular orbit, but it is representative of the time for our system and local stars to make one revolution around the galaxy.

32.9 If the dark matter in the Milky Way were composed entirely of MACHOs (evidence shows it is not), approximately how many would there have to be? Assume the average mass of a MACHO is 1/1000 that of the sun, and that dark matter has a mass 10 times that of the luminous Milky Way galaxy with its 10^{11} stars of average mass 1.5 times the sun's.

• 32.10 (a) Calculate the approximate age of the universe from the average value of the Hubble constant, $H_c = 20$ km/s/Mly. To do this, calculate the time it would take to travel 1 Mly at a constant expansion rate of 20 km/s. (b) If deceleration is taken into account, would the actual age of the universe be greater or less than that found here? Explain.

• 32.11 Assuming a circular orbit for the sun about the center of the Milky Way galaxy, calculate its orbital speed using the following information: The mass of the galaxy is equivalent to a single mass 1.5×10^{11} times that of the sun (or 3×10^{41} kg), located 30,000 ly away.

• 32.12 (a) What is the approximate force of gravity on a 70 kg person due to the Andromeda galaxy, assuming its total mass is 10^{13} that of our sun and acts like a single mass 2 Mly away? (b) What is the ratio of this force to the person's weight? Note that Andromeda is the closest large galaxy.

• 32.13 Andromeda galaxy is the closest large galaxy and is visible to the naked eye. Estimate its brightness relative to the sun, assuming it has a luminosity 10^{12} times that of the sun and lies 2 Mly away.

• 32.14 (a) A particle and its antiparticle are at rest relative to an observer and annihilate (completely destroying both masses), creating two γ rays of equal energy. What is the characteristic γ-ray energy you would look for if searching for evidence of proton-antiproton annihilation? (The fact that such radiation is rarely observed is evidence that there is very little antimatter in the universe.) (b) How does this compare with the 0.511 MeV energy associated with electron-positron annihilation?

• 32.15 The average particle energy needed to observe unification of forces is estimated to be 10^{19} GeV. (a) What is the rest mass in kilograms of a particle that has a rest mass of 10^{19} GeV/c^2? (b) How many times the mass of a hydrogen atom is this?

• 32.16 The characteristic length of entities in Superstring theory is approximately 10^{-35} m. (a) Find the energy in GeV of a photon of this wavelength. (b) Compare this with the average particle energy of 10^{19} GeV needed for unification of forces.

• 32.17 The critical mass density needed to just halt the expansion of the universe is approximately 10^{-26} kg/m³. (a) Convert this to (eV/c^2)/m³. (b) Find the number of neutrinos per cubic meter needed to close the universe if their average mass is 7 eV/c^2 and they have negligible kinetic energies.

• 32.18 The peak intensity of the CMBR occurs at a wavelength of 1.1 mm. (a) What is the energy in eV of a 1.1 mm photon? (b) There are approximately 10^9 photons for each massive particle in deep space. Calculate the energy of 10^9 such photons. (c) If the average massive particle in space has a mass half that of a proton, what energy would be created by converting its mass to energy? (d) Does this imply that space is "matter dominated"? Explain briefly.

• 32.19 Show that the velocity of a star orbiting its galaxy in a circular orbit is inversely proportional to the square root of its orbital radius, assuming the mass of the stars inside its orbit acts like a single mass at the center of the galaxy. You may use an equation from a previous chapter to support your conclusion, but you must justify its use and define all terms used.

• 32.20 (a) Use the Heisenberg uncertainty principle to calculate the uncertainty in energy for a corresponding time interval of 10^{-43} s. (b) Compare this energy with the 10^{19} GeV unification-of-forces energy, and discuss why they are similar.

• 32.21 What is the Schwarzschild radius of a black hole having a mass eight times that of our sun? Note that stars must be more massive than the sun to form black holes as a result of a supernova.

• 32.22 Black holes with masses smaller than those formed in supernovas may have been created in the Big Bang. Calculate the radius of one that has a mass equal to the earth's.

• 32.23 Supermassive black holes are thought to exist at the center of many galaxies. (a) What is the radius of such an object if it has a mass of 10^9 suns? (b) What is this radius in light years?

• 32.24 A section of superconducting wire carries a current of 100 A and requires 1.00 L of liquid nitrogen per hour to keep it below its critical temperature. For it to be economically advantageous to use a superconducting wire, the cost of cooling the wire must be less than the cost of energy lost to heat in the wire. Assume that the cost of liquid nitrogen is $0.30 per liter, and that electric energy costs $0.10 per kW·h. What is the resistance of a normal wire that costs as much in wasted electric energy as the cost of liquid nitrogen for the superconductor?

:32.25 (a) What Hubble constant corresponds to an approximate age of the universe of 10^{10} y? To get an approximate value, assume the expansion rate is constant and calculate the speed at which two galaxies must move apart to be separated by 1 Mly (present average galactic separation) in a time of 10^{10} y. (b) Similarly, what Hubble constant corresponds to a universe approximately 2×10^{10} y old?

:32.26 To get an idea of how empty deep space is on the average, perform the following calculations. (a) Find the volume our sun would occupy if it had an average density equal to the critical density of 10^{-26} kg/m^3 thought necessary to halt the expansion of the universe. (b) Find the radius of a sphere of this volume in light years. (c) What would this radius be if the density were that of luminous matter, which is approximately 5% that of the critical density? (d) Compare the radius found in part (c) with the 4 ly average separation of stars in the arms of the Milky Way.

:32.27 The core of a star collapses during a supernova, forming a neutron star. Angular momentum of the core is conserved, and so the neutron star spins rapidly. If the initial core radius is 5.0×10^5 km and it collapses to 10 km, find the neutron star's angular velocity in revolutions per second, given the core's was originally 1.0 revolution per 30 days.

:32.28 Using data from the previous problem, find the increase in rotational kinetic energy, given the core's mass is 1.3 times that of our sun. Where does this increase in kinetic energy come from?

:32.29 Distances to the nearest stars (up to 500 ly away) can be measured by a technique called parallax, as shown in Figure 32.27.

What are the angles θ_1 and θ_2 relative to the plane of the earth's orbit for a star 4.0 ly directly above the sun?

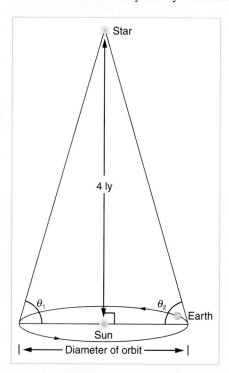

Figure 32.27 Distances to nearby stars are measured using triangulation, also called the parallax method. The angle of line of sight to the star is measured at intervals six months apart, and the distance is calculated by using the known diameter of the earth's orbit. This can be done for stars up to about 500 ly away. Problem 29.

EPILOGUE

If you have gained a greater sense of the wonder and beauty of the natural universe in using this book, then I have succeeded in my goals in writing it. The facts organized by laws, the connections underlying the complexity, the ability to make quantitative assessments—these make the universe all the more awe inspiring. What wonderful stories explain the workings of mass, energy, space, and time. Who would have thought that our explorations would lead us to the fantastic distances, to the remote times, and to the infinitesimal foundations we have discovered? And yet there seems no end to the story, no limit to what we can ask. How fortunate we are as humans to have the ability to understand the workings of things on earth, let alone to sense part of the grand scheme.

I hope that you will take with you a sense of how physics works and what wonders it holds. But most of all I hope you have gained the confidence that you have the tools to figure things out. Those tools we use in physics are useful elsewhere—it is sometimes said that the sign of an educated person is the ability to discover truth. So physics may or may not be of direct use to you, but its approach to scientific inquiry undoubtedly will be.

Knowledge only sharpens appreciation. With all we know of gravity, space, planetary orbits, and solar energy powering the winds of earth, we can still look at a photograph of the earth from space and be struck with how incredible it is.

APPENDIX A

Atomic Masses

ATOMIC NUMBER, Z	NAME	ATOMIC MASS NUMBER, A	SYMBOL	ATOMIC MASS* (u)	PERCENT ABUNDANCE OR DECAY MODE[†]	HALF-LIFE, $t_{1/2}$[†]
0	neutron	1	n	1.008 665	β^-	10.37 min
1	Hydrogen	1	^{1}H	1.007 825	99.985%	
	Deuterium	2	^{2}H or D	2.014 102	0.015%	
	Tritium	3	^{3}H or T	3.016 050	β^-	12.33 y
2	Helium	3	^{3}He	3.106 030	1.38×10^{-4}%	
		4	^{4}He	4.002 603	$\approx$100%	
3	Lithium	6	^{6}Li	6.015 121	7.5%	
		7	^{7}Li	7.016 003	92.5%	
4	Beryllium	7	^{7}Be	7.016 928	EC	53.29 d
		9	^{9}Be	9.012 182	100%	
5	Boron	10	^{10}B	10.012 937	19.9%	
		11	^{11}B	11.009 305	80.1%	
6	Carbon	11	^{11}C	11.011 432	EC, β^+	
		12	^{12}C	12.000 000	98.90%	
		13	^{13}C	13.003 355	1.10%	
		14	^{14}C	14.003 241	β^-	5730 y
7	Nitrogen	13	^{13}N	13.005 738	β^+	9.96 min
		14	^{14}N	14.003 074	99.63%	
		15	^{15}N	15.000 108	0.37%	
8	Oxygen	15	^{15}O	15.003 065	EC, β^+	122 s
		16	^{16}O	15.994 915	99.76%	
		18	^{18}O	17.999 160	0.200%	
9	Fluorine	18	^{18}F	18.000 937	EC, β^+	1.83 h
		19	^{19}F	18.998 403	100%	
10	Neon	20	^{20}Ne	19.992 435	90.51%	
		22	^{22}Ne	21.991 383	9.22%	
11	Sodium	22	^{22}Na	21.994 434	β^+	2.602 y
		23	^{23}Na	22.989 767	100%	
		24	^{24}Na	23.990 961	β^-	14.96 h
12	Magnesium	24	^{24}Mg	23.985 042	78.99%	
13	Aluminum	27	^{27}Al	26.981 539	100%	
14	Silicon	28	^{28}Si	27.976 927	92.23%	2.62 h
		31	^{31}Si	30.975 362	β^-	

* Atomic masses are for neutral atoms, including Z electrons.

† See Appendix B for details of the decay emissions of selected isotopes.

ATOMIC NUMBER, Z	NAME	ATOMIC MASS NUMBER, A	SYMBOL	ATOMIC MASS* (u)	PERCENT ABUNDANCE OR DECAY MODE†	HALF-LIFE, $t_{1/2}$†
15	Phosphorus	31	^{31}P	30.973 762	100%	
		32	^{32}P	31.973 907	β^-	14.28 d
16	Sulfur	32	^{32}S	31.972 070	95.02%	
		35	^{35}S	34.969 031	β^-	87.4 d
17	Chlorine	35	^{35}Cl	34.968 852	75.77%	
		37	^{37}Cl	36.965 903	24.23%	
18	Argon	40	^{40}Ar	39.962 384	99.60%	
19	Potassium	39	^{39}K	38.963 707	93.26%	
		40	^{40}K	39.963 999	0.0117%, EC, β^-	1.28×10^9 y
20	Calcium	40	^{40}Ca	39.962 591	96.94%	
21	Scandium	45	^{45}Sc	44.955 910	100%	
22	Titanium	48	^{48}Ti	47.947 947	73.8%	
23	Vanadium	51	^{51}V	50.943 962	99.75%	
24	Chromium	52	^{52}Cr	51.940 509	83.79%	
25	Manganese	55	^{55}Mn	54.938 047	100%	
26	Iron	56	^{56}Fe	55.934 939	91.72%	
27	Cobalt	59	^{59}Co	58.933 198	100%	
		60	^{60}Co	59.933 819	β^-	5.271 y
28	Nickel	58	^{58}Ni	57.935 346	68.27%	
		60	^{60}Ni	59.930 788	26.10%	
29	Copper	63	^{63}Cu	62.939 598	69.17%	
		65	^{65}Cu	64.927 793	30.83%	
30	Zinc	64	^{64}Zn	63.929 145	48.6%	
		66	^{66}Zn	65.926 034	27.9%	
31	Gallium	69	^{69}Ga	68.925 580	60.1%	
32	Germanium	72	^{72}Ge	71.922 079	27.4%	
		74	^{74}Ge	73.921 177	36.5%	
33	Arsenic	75	^{75}As	74.921 594	100%	
34	Selenium	80	^{80}Se	79.916 520	49.7%	
35	Bromine	79	^{79}Br	78.918 336	50.69%	
36	Krypton	84	^{84}Kr	83.911 507	57.0%	
37	Rubidium	85	^{85}Rb	84.911 794	72.17%	
38	Strontium	86	^{86}Sr	85.909 267	9.86%	
		88	^{88}Sr	87.905 619	82.58%	
		90	^{90}Sr	89.907 738	β^-	28.8 y
39	Yttrium	89	^{89}Y	88.905 849	100%	
		90	^{90}Y	89.907 152	β^-	64.1 h
40	Zirconium	90	^{90}Zr	89.904 703	51.45%	
41	Niobium	93	^{93}Nb	92.906 377	100%	
42	Molybdenum	98	^{98}Mo	97.905 406	24.13%	
43	Technetium	98	^{98}Tc	97.907 215	β^-	4.2×10^6 y
44	Ruthenium	102	^{102}Ru	101.904 348	31.6%	
45	Rhodium	103	^{103}Rh	102.905 500	100%	

ATOMIC NUMBER, Z	NAME	ATOMIC MASS NUMBER, A	SYMBOL	ATOMIC MASS* (u)	PERCENT ABUNDANCE OR DECAY MODE†	HALF-LIFE, $t_{1/2}$†
46	Palladium	106	^{106}Pd	105.903 478	27.33%	
47	Silver	107	^{107}Ag	106.905 092	51.84%	
		109	^{109}Ag	108.904 757	48.16%	
48	Cadmium	114	^{114}Cd	113.903 357	28.73%	
49	Indium	115	^{115}In	114.903 880	95.7%, β^-	4.4×10^{14} y
50	Tin	120	^{120}Sn	119.902 200	32.59%	
51	Antimony	121	^{121}Sb	120.903 821	57.3%	
52	Tellurium	130	^{130}Te	129.906 229	33.8%, β^-	2.5×10^{21} y
53	Iodine	127	^{127}I	126.904 473	100%	
		131	^{131}I	130.906 114	β^-	8.040 d
54	Xenon	132	^{132}Xe	131.904 144	26.9%	
		136	^{136}Xe	135.907 214	8.9%	
55	Cesium	133	^{133}Cs	132.905 429	100%	
		134	^{134}Cs	133.906 696	EC, β^-	2.06 y
56	Barium	137	^{137}Ba	136.905 812	11.23%	
		138	^{138}Ba	137.905 232	71.70%	
57	Lanthanum	139	^{139}La	138.906 346	99.91%	
58	Cerium	140	^{140}Ce	139.905 433	88.48%	
59	Praseodymium	141	^{141}Pr	140.907 647	100%	
60	Neodymium	142	^{142}Nd	141.907 719	27.13%	
61	Promethium	145	^{145}Pm	144.912 743	EC, α	17.7 y
62	Samarium	152	^{152}Sm	151.919 729	26.7%	
63	Europium	153	^{153}Eu	152.921 225	52.2%	
64	Gadolinium	158	^{158}Gd	157.924 099	24.84%	
65	Terbium	159	^{159}Tb	158.925 342	100%	
66	Dysprosium	164	^{164}Dy	163.929 171	28.2%	
67	Holmium	165	^{165}Ho	164.930 319	100%	
68	Erbium	166	^{166}Er	165.930 290	33.6%	
69	Thulium	169	^{169}Tm	168.934 212	100%	
70	Ytterbium	174	^{174}Yb	173.938 859	31.8%	
71	Lutecium	175	^{175}Lu	174.940 770	97.41%	
72	Hafnium	180	^{180}Hf	179.946 545	35.10%	
73	Tantalum	181	^{181}Ta	180.947 992	99.98%	
74	Tungsten	184	^{184}W	183.950 928	30.67%	
75	Rhenium	187	^{187}Re	186.955 744	62.6%, β^-	4.6×10^{10} y
76	Osmium	191	^{191}Os	190.960 920	β^-	15.4 d
		192	^{192}Os	191.961 467	41.0%	
77	Iridium	191	^{191}Ir	190.960 584	37.3%	
		193	^{193}Ir	192.962 917	62.7%	
78	Platinum	195	^{195}Pt	194.964 766	33.8%	
79	Gold	197	^{197}Au	196.966 543	100%	
		198	^{198}Au	197.968 217	β^-	2.696 d

ATOMIC NUMBER, Z	NAME	ATOMIC MASS NUMBER, A	SYMBOL	ATOMIC MASS* (u)	PERCENT ABUNDANCE OR DECAY MODE†	HALF-LIFE, $t_{1/2}$†
80	Mercury	199	^{199}Hg	198.968 253	16.87%	
		202	^{202}Hg	201.970 617	29.86%	
81	Thallium	205	^{205}Tl	204.974 401	70.48%	
82	Lead	206	^{206}Pb	205.974 440	24.1%	
		207	^{207}Pb	206.975 872	22.1%	
		208	^{208}Pb	207.976 627	52.4%	
		210	^{210}Pb	209.984 163	α, β^-	22.3 y
		211	^{211}Pb	210.988 735	β^-	36.1 min
		212	^{212}Pb	211.991 871	β^-	10.64 h
83	Bismuth	209	^{209}Bi	208.980 374	100%	
		211	^{211}Bi	210.987 255	α, β^-	2.14 min
84	Polonium	210	^{210}Po	209.982 848	α	138.38 d
85	Astatine	218	^{218}At	218.008 684	α, β^-	1.6 s
86	Radon	222	^{222}Rn	222.017 570	α	3.82 d
87	Francium	223	^{223}Fr	223.019 733	α, β^-	21.8 min
88	Radium	226	^{226}Ra	226.025 402	α	1.60×10^3 y
89	Actinium	227	^{227}Ac	227.027 750	α, β^-	21.8 y
90	Thorium	228	^{228}Th	228.028 715	α	1.91 y
		232	^{232}Th	232.038 054	100%, α	1.41×10^{10} y
91	Protactinium	231	^{231}Pa	231.035 880	α	3.28×10^4 y
92	Uranium	233	^{233}U	233.039 628	α	1.59×10^3 y
		235	^{235}U	235.043 924	0.720%, α	7.04×10^8 y
		236	^{236}U	236.045 562	α	2.34×10^7 y
		238	^{238}U	238.050 784	99.2745%, α	4.47×10^9 y
		239	^{239}U	239.054 289	β^-	23.5 min
93	Neptunium	239	^{239}Np	239.052 933	β^-	2.355 d
94	Plutonium	239	^{239}Pu	239.052 157	α	2.41×10^4 y
95	Americium	243	^{243}Am	243.061 375	α, fission	7.37×10^3 y
96	Curium	245	^{245}Cm	245.065 483	α	8.50×10^3 y
97	Berkelium	247	^{247}Bk	247.070 300	α	1.38×10^3 y
98	Californium	249	^{249}Cf	249.074 844	α	351 y
99	Einsteinium	254	^{254}Es	254.088 019	α, β^-	276 d
100	Fermium	253	^{253}Fm	253.085 173	EC, α	3.00 d
101	Mendelevium	255	^{255}Md	255.091 081	EC, α	27 min
102	Nobelium	255	^{255}No	255.093 260	EC, α	3.1 min
103	Lawrencium	257	^{257}Lr	257.099 480	EC, α	0.646 s
104	Rutherfordium	261	^{261}Rf	261.108 690	α	1.08 min
105	Dubnium	262	^{262}Db	262.113 760	α, fission	34 s
106	Seaborgium	263	^{263}Sg	263.118 6	α, fission	0.8 s
107	Bohrium	262	^{262}Bh	262.123 1	α	0.102 s
108	Hassium	264	^{264}Hs	264.128 5	α	0.08 ms
109	Meitnerium	266	^{266}Mt	266.137 8	α	3.4 ms

APPENDIX B

Selected Radioactive Isotopes

ISOTOPE	$t_{1/2}$	DECAY MODE(s)	ENERGY (MeV)	PERCENT		γ-RAY ENERGY (MeV)	PERCENT
^{3}H	12.33 y	β^-	0.0186	100%			
^{14}C	5730 y	β^-	0.156	100%			
^{13}N	9.96 min	β^+	1.20	100%			
^{22}Na	2.602 y	β^+	0.55	90%	γ	1.27	100%
^{32}P	14.28 d	β^-	1.71	100%			
^{35}S	87.4 d	β^-	0.167	100%			
^{36}Cl	3.00×10^5 y	β^-	0.710	100%			
^{40}K	1.28×10^9 y	β^-	1.31	89%			
^{43}K	22.3 h	β^-	0.827	87%	γs	0.373	87%
						0.618	87%
^{45}Ca	165 d	β^-	0.257	100%			
^{51}Cr	27.70 d	EC			γ	0.320	10%
^{52}Mn	5.59 d	β^+	3.69	28%	γs	1.33	28%
						1.43	28%
^{52}Fe	8.27 h	β^+	1.80	43%		0.169	43%
						0.378	43%
^{59}Fe	44.6 d	β^-s	0.273	45%	γs	1.10	57%
			0.466	55%		1.29	43%
^{60}Co	5.271 y	β^-	0.318	100%	γs	1.17	100%
						1.33	100%
^{65}Zn	244.1 d	EC			γ	1.12	51%
^{67}Ga	78.3 h	EC			γs	0.0933	70%
						0.185	35%
						0.300	19%
						others	
^{75}Se	118.5 d	EC			γs	0.121	20%
						0.136	65%
						0.265	68%
						0.280	20%
						others	
^{86}Rb	18.8 d	β^-s	0.69	9%	γ	1.08	9%
			1.77	91%			
^{85}Sr	64.8 d	EC			γ	0.514	100%

Decay modes are α, β^-, β^+, electron capture (EC), and isomeric transition (IT). EC results in the same daughter nucleus as would β^+ decay. IT is a transition from a metastable excited state. Energies for $\beta^{\pm}$ decays are the maxima; average energies are roughly one-half the maxima.

ISOTOPE	$t_{1/2}$	DECAY MODE(s)	ENERGY (MeV)	PERCENT		γ-RAY ENERGY (MeV)	PERCENT
^{90}Sr	28.8 y	β^-	0.546	100%			
^{90}Y	64.1 h	β^-	2.28	100%			
^{99m}Tc	6.02 h	IT			γ	0.142	100%
^{113m}In	99.5 min	IT			γ	0.392	100%
^{123}I	13.0 h	EC			γ	0.159	≈100%
^{131}I	8.040 d	β^-s	0.248	7%	γs	0.364	85%
			0.607	93%		others	
			others				
^{129}Cs	32.3 h	EC			γs	0.0400	35%
						0.372	32%
						0.411	25%
						others	
^{137}Cs	30.17 y	β^-s	0.511	95%	γ	0.662	95%
			1.17	5%			
^{140}Ba	12.79 d	β^-	1.035	≈100%	γs	0.030	25%
						0.044	65%
						0.537	24%
						others	
^{198}Au	2.696 d	β^-	1.161	≈100%	γ	0.412	≈100%
^{197}Hg	64.1 h	EC			γ	0.0733	100%
^{210}Po	138.38 d	α	5.41	100%			
^{226}Ra	1.60×10^3 y	αs	4.68	5%	γ	0.186	100%
			4.87	95%			
^{235}U	7.038×10^8 y	α	4.68	≈100%	γs	numerous	<0.400%
^{238}U	4.468×10^9 y	αs	4.22	23%	γ	0.050	23%
			4.27	77%			
^{237}Np	2.14×10^6 y	αs	numerous		γs	numerous	<0.250%
			4.96 (max.)				
^{239}Pu	2.41×10^4 y	αs	5.19	11%	γs	7.5×10^{-5}	73%
			5.23	15%		0.013	15%
			5.24	73%		0.052	10%
						others	
^{243}Am	7.37×10^3 y	αs	Max. 5.44		γs	0.075	
			5.37	88%		others	
			5.32	11%			
			others				

GLOSSARY OF KEY SYMBOLS AND NOTATION

In this glossary, key symbols and notation are briefly defined; the italic numbers in parentheses are the section and page numbers where each is first used.

-⊗-   alternating current (AC) voltage *(19.5, 488)*

$\overline{\text{any symbol}}$ average (indicated by a bar over a symbol—e.g., $\overline{v}$ is average velocity) *(2.2, 31)*

-⌒⌒-   inductor *(22.9, 579)*

°C   Celsius degree *(12.1, 297)*

°F   Fahrenheit degree *(12.1, 297)*

∥   parallel *(4.5, 92)*

⊥   perpendicular *(4.5, 92)*

∝   proportional to *(4.3, 84)*

±   plus or minus *(1.3, 12)*

$_0$ zero as a subscript denotes an initial value *(2.4, 36)*

α alpha rays *(29.1, 781)*

α angular acceleration *(9.1, 219)*

α temperature coefficient(s) of resistivity *(19.3, 485)*

β beta rays *(29.1, 781)*

β sound level *(16.3, 410)*

β volume coefficient of expansion *(12.2, 300)*

β^- electron emitted in nuclear beta decay *(29.4, 791)*

β^+ positron decay *(29.4, 793)*

γ elastic modulus or Young's modulus *(5.8, 134)*

γ gamma rays *(29.1, 781)*

γ surface tension *(10.8, 261)*

$\gamma = 1/\sqrt{1 - v^2/c^2}$ a constant used in relativity *(26.2, 696)*

Δ change in whatever quantity follows *(2.1, 29)*

δ uncertainty in whatever quantity follows *(1.3, 12)*

ΔE change in energy between the initial and final orbits of an electron in an atom *(28.3, 751)*

ΔE uncertainty in energy *(27.7, 733)*

Δm difference in mass between initial and final products *(29.4, 791)*

ΔN number of decays that occur *(29.5, 797)*

$\Delta \mathbf{p}$ change in momentum *(7.1, 174)*

Δp uncertainty in momentum *(27.7, 732)*

ΔPE_g change in gravitational potential energy *(6.3, 153)*

$\Delta \theta$ rotation angle *(8.1, 195)*

Δs distance traveled along a circular path *(8.1, 195)*

Δt uncertainty in time *(27.7, 733)*

Δt_0 proper time as measured by an observer at rest relative to the process *(26.2, 695)*

ΔV potential difference *(18.1, 457)*

Δx uncertainty in position *(27.7, 732)*

ε_0 permittivity of free space *(18.5, 466)*

η viscosity *(11.4, 282)*

θ angle between the force vector and the displacement vector *(6.1, 149)*

θ angle between two lines *(1.8*, 22)*

θ contact angle *(10.8, 262)*

θ direction of the resultant *(3.3, 62–63)*

θ_b Brewster's angle *(25.8, 680)*

θ_c critical angle *(24.4, 623)*

κ dielectric constant *(18.5, 467)*

λ decay constant of a nuclide *(29.5, 795)*

λ wavelength *(15.9, 392)*

λ_n wavelength in a medium *(25.1, 662)*

μ_0 permeability of free space *(23.4, 610)*

μ_k coefficient of kinetic friction *(4.5, 93–94)*

μ_s coefficient of static friction *(4.5, 93)*

ν_e electron's neutrino *(29.4, 793)*

π^+ positive pion *(31.1, 842)*

π^- negative pion *(31.1, 842)*

π^0 neutral pion *(31.1, 842)*

ρ density *(10.2, 247)*

ρ_c critical density, the density needed to just halt universal expansion *(32.4, 881)*

ρ_{fl} fluid density *(10.7, 257–258)*

$\overline{\rho}_{obj}$ average density of an object *(10.7, 257–258)*

ρ/ρ_w specific gravity *(10.7, 258)*

τ characteristic time constant for a resistance and inductance *(RL)* or resistance and capacitance *(RC)* circuit, *(22.10, 581)*

τ characteristic time for a resistor and capacitor *(RC)* circuit *(20.6, 524)*

τ torque *(5.2, 118–119)*

Υ upsilon meson *(31.5, 856)*

Φ magnetic flux *(22.1, 565)*

ϕ phase angle *(22.12, 587)*

Ω ohm (unit) *(19.2, 483)*

ω angular velocity *(8.1, 195)*

ω angular velocity *(22.6, 572)*

A   ampere (current unit) *(19.1, 479)*

A area *(15.11, 397)*

A cross-sectional area *(5.8, 134)*

A total number of nucleons *(29.3, 787)*

a acceleration *(2.3, 32)*

a_B Bohr radius *(28.3, 752)*

a_c centripetal acceleration *(8.3, 199)*

a_t tangential acceleration *(9.1, 220)*

AC   alternating current *(19.5, 488)*

AM   amplitude modulation *(23.3, 603)*

atm   atmosphere *(10.6, 255)*

B baryon number *(31.4, 851)*

B blue quark color *(31.5, 856)*

$\bar{B}$ antiblue (yellow) antiquark color *(31.5, 856)*

b quark flavor bottom or beauty *(31.5, 856)*

B bulk modulus *(5.8, 134)*

B magnetic field strength *(21.3, 539)*

$\mathbf{B}_{int}$ electron's intrinsic magnetic field *(28.5, 757)*

$\mathbf{B}_{orb}$ orbital magnetic field *(28.5, 756)*

BE   binding energy of a nucleus—it is the energy required to completely disassemble it into separate protons and neutrons *(29.6, 799)*

BE/A binding energy per nucleon *(29.6, 801)*

Bq   becquerel—one decay per second *(29.5, 797)*

C capacitance (amount of charge stored per volt) *(18.5, 465)*

C   coulomb (a fundamental SI unit of charge) *(1.2, 7)*

C_p total capacitance in parallel *(18.6, 470)*

C_s total capacitance in series *(18.6, 469)*

CG   center of gravity *(5.3, 122)*

CM   center of mass *(5.3, 121)*

c quark flavor charm *(31.5, 856)*

c specific heat *(13.2, 322)*

c speed of light *(23.3, 603)*

Cal   kilocalorie *(13.1, 320)*

cal   calorie *(13.1, 320)*

COP_{hp} heat pump's coefficient of performance *(14.5, 360)*

COP_{ref} coefficient of performance for refrigerators and air conditioners *(14.5, 361)*

$\cos\theta$ cosine *(1.8, 23)*

$\cot\theta$ cotangent *(1.8, 24)*

$\csc\theta$ cosecant *(1.8, 24)*

D diffusion constant *(11.7, 287)*

d displacement *(6.1, 149)*

d quark flavor down *(31.5, 853)*

dB   decibel *(16.3, 410)*

d_i distance of an image from the center of a lens *(24.6, 631)*

d_o distance of an object from the center of a lens *(24.6, 631)*

DC   direct current *(19.5, 488)*

E electric field strength *(17.4, 441)*

$\mathcal{E}$ emf (voltage) or Hall electromotive force *(21.6, 542)*

$\mathcal{E}$ or emf   electromotive force *(20.2, 512)*

E energy of a single photon *(27.3, 719)*

E nuclear reaction energy *(29.4, 791)*

E relativistic total energy *(26.6, 708)*

E total energy *(26.6, 704)*

E_0 ground state energy for hydrogen *(28.3, 752)*

E_0 rest energy *(26.6, 704)*

EC   electron capture *(29.4, 794)*

E_{cap} energy stored in a capacitor *(18.7, 472)*

Eff efficiency—the useful work output divided by the energy input *(6.8, 163)*

Eff_C Carnot efficiency *(14.4, 358)*

E_{in} energy consumed (food digested in humans) *(6.8, 163)*

E_{ind} energy stored in an inductor *(22.9, 580)*

E_{out} energy output *(6.8, 163)*

e emissivity of an object *(13.7, 336)*

e^+ antielectron or positron *(29.4, 793 & 30.2, 816)*

eV   electron volt *(18.1, 459)*

F   farad (unit of capacitance, a coulomb per volt) *(18.5, 465–466)*

$\mathcal{F}$ flow rate—volume per unit time flowing past a point *(11.1, 275)*

F   focal point of a lens *(24.6, 628)*

F force *(4.3, 84)*

F magnitude of a force *(5.2, 118–119)*

F restoring force *(15.1, 379)*

F_B buoyant force *(10.7, 256)*

F_c centripetal force *(8.3, 199)*

F_i force input *(5.6, 127)*

F_o force output *(5.6, 127)*

FM   frequency modulation *(23.3, 604)*

f focal length *(24.6, 628)*

f frequency *(15.2, 381)*

f_0 resonant frequency of a resistance, inductance, and capacitance (*RLC*) series circuit *(22.12, 586)*

f_0 threshold frequency for a particular material (photoelectric effect) *(27.2, 718)*

f_1 fundamental *(16.5, 418)*

f_2 first overtone *(16.5, 418)*

f_3 second overtone *(16.5, 418)*

f_B beat frequency *(16.7, 425)*

f_k magnitude of kinetic friction *(4.5, 93)*

f_s magnitude of static friction *(4.5, 93)*

G gravitational constant *(8.5, 204)*

G green quark color *(31.5, 856)*

$\bar{G}$ antigreen (magenta) antiquark color *(31.5, 856)*

g acceleration due to gravity *(2.6, 43)*

g gluons (carrier particles for strong nuclear force) *(31.6, 858)*

h change in vertical position *(6.3, 153)*

h height above some reference point *(11.2, 277)*

h maximum height of a projectile *(3.4, 67)*

h Planck's constant *(27.1, 716)*

hf photon energy *(27.2, 718)*

h_i height of the image *(24.6, 631)*

h_o height of the object *(24.6, 631)*

I electric current *(19.1, 479)*

I intensity *(15.11, 397)*

I intensity of a transmitted wave *(25.8, 678)*

I moment of inertia (also called rotational inertia) *(9.3, 225)*

I_0 intensity of a polarized wave before passing through a filter *(25.8, 678)*

I_{ave} average intensity for a continuous sinusoidal electro-magnetic wave *(23.4, 610)*

I_{rms} average current *(19.5, 489)*

J joule *(6.1, 149)*

J/Ψ J/psi meson *(31.5, 856)*

K kelvin *(12.1, 297)*

k Boltzmann constant *(12.3, 302)*

k force constant of a spring *(6.4, 155)*

K_α x rays created when an electron falls into an $n = 1$ shell vacancy from the $n = 3$ shell *(28.8, 764)*

K_β x rays created when an electron falls into an $n = 2$ shell vacancy from the $n = 3$ shell *(28.8, 764)*

kcal kilocalorie *(13.1, 320)*

KE translational kinetic energy *(6.2, 151)*

KE + PE mechanical energy *(6.4, 155)*

KE_e kinetic energy of an ejected electron *(27.2, 718)*

KE_{rel} relativistic kinetic energy *(26.6, 706)*

KE_{rot} rotational kinetic energy *(9.4, 229)*

$\overline{KE}$ thermal energy *(12.4, 307)*

kg kilogram (a fundamental SI unit of mass) *(1.2, 7)*

L angular momentum *(9.5, 232)*

L liter *(11.1, 275)*

L magnitude of angular momentum *(28.6, 757)*

L self-inductance *(22.9, 579)*

ℓ angular momentum quantum number *(28.6, 757)*

L_α x rays created when an electron falls into an $n = 2$ shell from the $n = 3$ shell *(28.8, 764)*

L_e electron total family number *(31.4, 850)*

L_μ muon family total number *(31.4, 850)*

L_τ tau family total number *(31.4, 851)*

L_f heat of fusion *(13.3, 325)*

L_f and L_v latent heat coefficients *(13.3, 325)*

$\mathbf{L}_{orb}$ orbital angular momentum *(28.5, 756)*

L_s heat of sublimation *(13.3, 328)*

L_v heat of vaporization *(13.3, 325)*

L_z z-component of the angular momentum *(28.6, 758)*

M angular magnification *(24.8, 643)*

M mutual inductance *(22.9, 578)*

m indicates metastable state *(30.1, 812)*

m magnification *(24.6, 631)*

m mass *(6.2, 151)*

m mass of an object as measured by a person at rest relative to the object *(26.5, 703)*

m meter (a fundamental SI unit of length) *(1.2, 7)*

m order of interference *(25.3, 665)*

m overall magnification (product of the individual magnifications) *(24.8, 641)*

$m(^AX)$ atomic mass of a nuclide *(29.6, 799)*

MA mechanical advantage *(5.6, 127)*

m_e magnification of the eyepiece *(24.8, 641)*

m_e mass of the electron *(28.3, 751)*

m_ℓ angular momentum projection quantum number *(28.6, 758)*

m_n mass of a neutron *(29.6, 799)*

m_o magnification of the objective lens *(24.8, 641)*

mol mole *(12.3, 303)*

m_p mass of a proton *(28.2, 746)*

m_s spin projection quantum number *(28.6, 759)*

N magnitude of the normal force *(4.5, 93)*

N newton *(4.3, 86)*

N normal force *(4.5, 91)*

N number of neutrons *(29.3, 787)*

n index of refraction *(24.3, 621)*

n number of free charges per unit volume *(19.1, 481)*

N_A Avogadro's number *(12.3, 303)*

N_r Reynolds number *(11.5, 285)*

net **F** net external force *(4.3, 84)*

N·m newton-meter (work-energy unit) *(6.1, 149)*

N·m newtons times meters (SI unit of torque) *(5.2, 119)*

OE other energy *(6.6, 159)*

P power *(6.7, 161)*

P power of a lens *(24.6, 629)*

P pressure *(10.3, 248)*

p momentum *(7.1, 174)*

p momentum magnitude *(7.1, 174)*

p relativistic momentum *(26.6, 708)*

$\mathbf{p}_{tot}$ total momentum *(7.3, 178)*

$\mathbf{p}'_{tot}$ total momentum some time later *(7.3, 178)*

P_{abs} absolute pressure *(10.6, 253)*

P_{atm} atmospheric pressure *(10.6, 253)*

P_{atm} standard atmospheric pressure *(10.4, 250)*

PE potential energy *(6.4, 154)*

PE_{el} elastic potential energy *(15.1, 380)*

PE_{elec} electric potential energy *(18.1, 457)*

PE_s potential energy of a spring *(6.4, 155)*

P_g gauge pressure *(10.6, 253)*

P_{in} power consumption or input *(6.8, 164)*

P_{out} useful power output going into useful work or a desired form of energy *(6.8, 164)*

Q latent heat *(13.3, 325)*

Q net heat transferred into a system *(14.1, 346)*

$+Q$　positive charge *(18.5, 465)*

$-Q$　negative charge *(18.5, 465)*

q　electron charge *(27.2, 717)*

q_p　charge of a proton *(28.2, 746)*

q　test charge *(17.4, 441)*

QF　quality factor (Note: same as RBE) *(30.5, 826)*

R　activity, the rate of decay *(29.5, 797)*

R　radius of curvature of a spherical mirror *(24.7, 637)*

R　red quark color *(31.5, 856)*

$\bar{R}$　antired (cyan) quark color *(31.5, 856)*

R　resistance *(19.2, 483)*

$\mathbf{R}$　resultant or total displacement *(3.3, 62)*

R　Rydberg constant *(28.3, 750)*

R　universal gas constant *(12.3, 304)*

r　distance from pivot point to the point where a force is applied *(5.2, 118–119)*

r　internal resistance *(20.2, 513)*

$r_\perp$　perpendicular lever arm *(5.2, 118–119)*

r　radius of a nucleus *(29.3, 787)*

r　radius of curvature *(8.1, 195)*

r　resistivity *(19.3, 484)*

r or rad　radiation dose unit *(30.5, 825)*

rem　roentgen equivalent man *(30.5, 826)*

rad　radian *(8.1, 195)*

RBE　relative biological effectiveness (Note: same as QF) *(30.5, 826)*

RC　resistor and capacitor circuit *(20.6, 523)*

rms　root mean square *(19.5, 489)*

r_n　radius of the nth H-atom orbit *(28.3, 751)*

R_p　total resistance of a parallel connection *(20.1, 508)*

R_s　total resistance of a series connection *(20.1, 507)*

R_s　Schwarzschild radius *(32.2, 876)*

S　entropy *(14.6, 362)*

$\mathbf{S}$　intrinsic spin (intrinsic angular momentum) *(28.5, 756)*

S　magnitude of the intrinsic (internal) spin angular momentum *(28.6, 759)*

S　shear modulus *(5.8, 134)*

S　strangeness quantum number *(31.4, 852)*

s　quark flavor strange *(31.5, 853)*

s　second (fundamental SI unit of time) *(1.2, 7)*

s　spin quantum number *(28.6, 759)*

$\mathbf{s}$　total displacement *(3.4, 64)*

sec θ　secant *(1.8, 24)*

sin θ　sine *(1.8, 23)*

s_z　z-component of spin angular momentum *(28.6, 759)*

T　period—time to complete one oscillation *(15.2, 381)*

T　temperature *(12.1, 297)*

T_c　critical temperature—temperature below which a material becomes a superconductor *(32.6, 885)*

T　tension *(4.5, 96)*

T　tesla (magnetic field strength B) *(21.4, 539)*

t　quark flavor top or truth *(31.5, 856)*

t　time *(2.2, 31)*

$t_{1/2}$　half-life—the time in which half of the original nuclei decay *(29.5, 794)*

tan θ　tangent *(1.8, 23)*

U　internal energy *(14.1, 346)*

u　quark flavor up *(31.5, 853)*

u　unified atomic mass unit *(29.3, 786)*

u　velocity of an object relative to an observer *(26.4, 702)*

u'　velocity relative to another observer *(26.4, 702)*

V　electric potential *(18.1, 457)*

V　terminal voltage *(20.2, 514)*

V　volt (unit) *(18.1, 457)*

V　volume *(10.2, 247)*

v　relative velocity between two observers *(26.4, 702)*

v　speed of light in a material *(24.3, 621)*

v　velocity *(2.2, 31)*

$\bar{v}$　average fluid velocity *(11.1, 275)*

$V_B - V_A$　change in potential *(18.1, 457)*

v_d　drift velocity *(19.1, 481)*

V_p　transformer input voltage *(22.7, 573)*

V_{rms}　rms voltage *(19.5, 489)*

V_s　transformer output voltage *(22.7, 573)*

$\mathbf{v}_{tot}$　total velocity *(3.5, 71)*

v_w　propagation speed of sound or other wave *(16.2, 407)*

v_w　wave velocity *(15.9, 391)*

W　work *(6.1, 149)*

W　net work done by a system *(14.1, 346)*

W　watt *(6.7, 161)*

w　weight *(4.5, 90)*

w_{fl}　weight of the fluid displaced by an object *(10.7, 256)*

W_c　total work done by all conservative forces *(6.5, 157)*

W_{nc}　total work done by all nonconservative forces *(6.5, 157)*

W_{out}　useful work output *(6.8, 163)*

X　amplitude *(15.3, 382)*

X　symbol for an element *(29.3, 787)*

$_Z^A X_N$　notation for a particular nuclide *(29.3, 787)*

x　deformation or displacement from equilibrium *(15.1, 379)*

x　displacement of a spring from its undeformed position *(6.4, 155)*

x　horizontal axis *(3.4, 64)*

X_C　capacitive reactance *(22.11, 583)*

X_L　inductive reactance *(22.11, 582)*

x_{rms}　root mean square diffusion distance *(11.7, 287)*

y　vertical axis *(3.4, 64)*

Z　atomic number (number of protons in a nucleus) *(29.3, 787)*

Z　impedance *(22.12, 585)*

ANSWERS

Answers to Odd-Numbered Problems

Chapter 1
1. Yes. $v/c = 2.50 \times 10^{-5} \ll 1\%$.
3. ~150 generations have passed.
5. 2×10^{31}
7. (a) 10 (b) 10^{10} (c) 10^{11}
9. ~50 atoms thick
11. (a) 10^{-3} (b) 10^{-3}
13. 5×10^{22} stars
15. 10^3 nerve impulses/s
17. (a) 97 km/h (b) 6.0×10^1 mi/h
19. 91.4 m
21. 1.85 m
23. 765 mi/h
25. (a) 29.9 km/s (b) 9.81×10^4 ft/s
27. 2.0 kg
29. (a) 85.5 to 94.5 km/h
    (b) 53.1 to 58.7 mi/h
31. (a) 7.6×10^7 beats (b) 7.57×10^7 beats
    (c) 7.574×10^7 beats
33. (a) 3 (b) 3 (c) 3
35. (a) 3.3% (b) 43.5 to 46.5 mi/h
37. 80 ± 3 beats/min
39. 2.8 h

Chapter 2
1. (a) 7 m (b) 7 m (c) +7 m
3. (a) 13 m (b) 9 m (c) +9 m
5. (a) 2.99×10^4 m/s (b) 0
7. (a) 6.61×10^{15} rev/s (b) 0
9. 1×10^7 y
11. 1×10^8 y
13. +6.00 m/s, −1.71 m/s, +4.04 m/s,
    +3.49 m/s
15. 4.29 m/s²
17. 10.8 m/s
19. 38.9 m/s
21. 502 m/s
23. (b) $a = 2.40$ m/s², $t = 12.0$ s, $v_0 = 0$
    (c) $x = 173$ m (d) $v = 28.8$ m/s
25. (b) $v_0 = 0$, $v = 30.0$ cm/s, $x - x_0 =$
    1.80 cm (c) $t = 0.120$ s
    (d) Reasonable since t is only a
    fraction of an entire heartbeat.
27. (a) 6.87 m/s² (b) 52.3 m
29. (a) 7.69×10^{-3} s (b) 8.45×10^3 m/s²
31. (a) −90.0 m/s² = $9.18g$
    (b) 6.67×10^{-3} s (c) $4.08g$
33. (a) 0.0933 s (b) −80.4 m/s²
35. (a) 0.307 m/s² (b) 0.132 m/s² (c) 5.45
    $\times 10^3$ km (d) 313 m/s (e) 484 m/s
37. 6.28 m, 10.1 m/s; 10.1 m, 5.20 m/s;
    11.5 m, 0.30 m/s; 10.4 m, −4.60 m/s
39. 4.95 m/s
41. (a) $v_0 = 13.0$ m/s, $y_0 = 0$,
    $a = -9.80$ m/s² (b) 8.62 m (c) 2.65 s
43. (a) 8.26 m (b) 0.717 s

45. (a) −70.0 m/s (b) 6.09 s
47. (a) −5.42 m/s (b) 5.33 m/s
    (c) 1.34×10^5 m/s² (d) 1.09×10^{-4} m
49. (a) ~115 m/s (b) ~5 m/s²
51. ~239 m/s
53. (a) ~1.4 m/s² (b) ~0.8 m/s²
55. (a) 0 (b) −4.0 m/s² at 5 s, 2.0 m/s² at
    15 s
57. 3.5 m/s², 2.2 m/s², 0.67 m/s²
59. (a) − 30.6 m/s² (b) Deceleration is
    too great. (c) Stopping distance is
    too short.

Chapter 3
1. (a) 480 m (b) 379 m, 18.4° east of
    north
3. North-component 3.2 km,
    east-component 3.8 km
5. 19.5 m, 4.65° south of west
7. (a) 26.6 m, 65.1° north of east
    (b) 26.6 m, 65.1° south of west
    (consistent)
9. 52.9 m, 90.1°
11. x-component 4.41 m/s,
    y-component 5.07 m/s
13. (a) 1.56 km (b) 120 m east
15. North-component 87.0 km,
    east-component 87.0 km
17. 30.8 m, 35.8° west of north
19. (a) 30.8 m, 54.2° south of west
    (b) 30.8 m, 54.2° north of east
21. 92.3 m, 53.7° south of west
23. 2.97 km, 22.2° west of south
25. 3.45 m/s, 3.94 m/s
27. x-component 6.35 m/s,
    y-component 2.20 m/s
29. 7 buses
31. (a) 18.9° (b) 71.1°, longer flight time
33. 128 m, 255 m, 128 m
35. (a) 15.2 m/s (b) 1.31 s (c) 2.11 m
37. 10.3 m/s, 73.1° below the horizontal
39. 20.1 m/s
41. 48.8° above the horizontal
43. $a = \tan\theta$, $b = -g/2(v_0\cos\theta)^2$
45. (a) 3.58×10^4 m, 45° south of east
    (b) 5.53 m/s, 45° south of east
    (c) 5.61×10^4 m, 45° south of east
47. (a) 0.70 m/s faster (b) Second runner
    wins (c) 4.17 m
49. 17.0 m/s, 22.1°
51. 220 m/s, 9.96° south of west
53. 6.68 m/s, 53.3° south of west
55. (a) 7.87 m/s, (b) 9.45° west of north
57. 1.72 m/s, 42.3° north of east
59. (a) 1.6×10^3 km (b) The distance is
    too great. (c) Air resistance and curva-

ture of the earth make the equation
inapplicable.
61. (a) 556 m/s (b) 278 m/s, 10.0° south
    of east (c) Both the total velocity and
    wind velocity are too great. (d) The
    large distance in the given amount of
    time is unreasonable.

Chapter 4
1. 265 N
3. 56.0 kg
5. 590 N
7. 12.0 m/s², friction not reduced
9. 1.26×10^3 N
11. 11.2 m/s²
13. (a) 4.41×10^5 N (b) 1.47×10^5 N
15. (a) 910 N (b) 1.11×10^3 N
17. (a) $F_1 = 16$ N, $F_2 = 11$ N
19. 10.2 m/s², 4.68° from the vertical
21. (a) -1.51×10^4 N (20.5w)
    (b) 2.79° above the horizontal
23. 352 lb
25. 779 N
27. (a) 1.00×10^3 N (b) 30.0 N
29. 1.53×10^{-3}
31. (a) 588 N (b) 1.96 m/s²
33. (a) 3.29 m/s² (b) 3.52 m/s²
    (c) 980 N, 945 N
39. (a) 4.20 m/s² (b) 2.74 m/s²
    (c) −0.195 m/s²
41. (a) 8.91×10^5 N (b) 2.98×10^5 N
43. (a) 588 N (b) 678 N
45. 12.9 N
47. (a) 1×10^{-13} (b) 1×10^{-11}
49. 1×10^2
51. (b) 1.42×10^3 N (c) 539 N
53. (b) 73.2 N
55. (a) 51.0 N (b) 0.720 m/s²
57. (a) 128 N (1.09w) (b) 118 N
    (1.00w) (c) 90.0 N (0.763w)
59. (a) 7.43 m/s (b) 2.97 m
61. (a) 4.20 m/s (b) 29.4 m/s²
    (c) 4.31×10^3 N
63. (a) 7.84×10^3 N (b) 9.80 m/s²
    (c) 31.9 m (d) 5.49×10^3 N, 6.86
    m/s², 45.6 m
65. (a) 4.47 m/s² (b) 5.60 s (c) 10.2 m/s²
    (d) 30.6 m
67. (a) 1.87×10^4 N (b) 1.67×10^4 N
    (c) 1.56×10^4 N (d) 19.4 m, 0 m/s
69. (a) −6.11 m/s (b) The brakes cannot
    reverse the direction of the car.
    (c) The deceleration time is too long
    for the given deceleration (the
    effective coefficient of friction is
    reasonable).

71. (a) 1860 N (2.53w) (b) The force is too great compared with the person's weight. (c) Either the increase in speed is too great or the given time is unreasonably short.

Chapter 5
1. 46.8 N·m
3. 233 N·m
5. 1.36 m
7. 1.43×10^3 N
9. 2.12×10^4 N
11. 4.51×10^5 m from the center of the sun
13. (a) 3.40 cm above the base (b) 2.44 cm above the base
15. $F_L = 16.3$ N and $F_R = 32.7$ N
17. (a) 0.167 (~1/6) (b) 2.0×10^4 N straight up
19. (a) 21.6 N (b) The force is horizontal (its vertical component is zero).
21. 629 N, 18.1° above the horizontal
23. (a) 18.5 (b) 29.1 N (c) 510 N
25. 15.6
27. 1.30×10^3 N
29. (a) 299 N (b) 897 N
31. (a) 172 N (b) Zero
33. 376 N
35. 1.72×10^3 N
37. (a) 25.0 N (b) 75.0 N
39. (a) 2.21×10^3 N (b) 2.94×10^3 N
41. 2.25×10^3 N
43. 1 mm
45. 2 cm
47. 0.536 mm
49. (a) 3.99×10^{-7} m (b) 9.67×10^{-8} m
51. 4×10^2 N/cm^2. This is about 36 atm, greater than a typical jar can withstand.
53. 1.44 cm
55. 1.16 cm
57. 0.22 mm
59. 6.0×10^2 m/s
61. (a) 2.33 m (b) This places the second child off the board (c) The first child must be closer to the pivot for the second child to be able to balance the board.

Chapter 6
1. 3.00 J $= 7.17 \times 10^{-4}$ kcal
3. (a) 5.92×10^5 J (b) -5.88×10^5 J (c) 0
5. 3.14×10^3 J
7. (a) −700 J (b) 0 (c) 700 J (d) 38.6 N (e) 0
9. $(9.34 \times 10^6$ J$)/(2.33 \times 10^9$ J$) = 1/250$ (truck/astronaut)
11. 1.1×10^{10} J
13. 2.82×10^3 N
15. 102 N
17. 2×10^3 bombs
19. (a) 1.96×10^{16} J (b) 0.52
21. (a) 80.9 J (b) 196 J
23. 3.92
25. (a) 22.7 m/s (b) 20.4 m/s (c) 3.83 m
27. (a) 26.2 m/s, 5.35 s (b) 26.3 m/s, 4.86 s
29. 24.8 m/s
31. 2×10^{-10}
33. (a) 40 (b) 8 million
35. $149
37. (a) 208 W (b) 141 s

39. (a) 3.20 s (b) 4.04 s
41. (a) 9.46×10^7 J (b) 2.54 y
43. 2.77×10^4 W
45. (a) 1.41 kW/m^2 (b) 28.9 km^2
47. 9.5 min, 69 flights of stairs
49. 641 W, 0.860 hp
51. 31 g
53. 14.3%
55. (a) 3.21×10^4 N (b) 2.35×10^3 N (c) 41.0 in part (a), 3.00 in part (b)
57. (a) 25.8 kcal (b) 599 W
59. (a) 144 J (b) 288 W
61. (a) 2.50×10^{12} J (b) 2.52% (c) 1.4×10^4 kg (14 metric tons)
63. (a) 294 N (b) 118 J (c) 49.0 W
65. 9.46 m/s
67. (a) 1.61×10^4 N (b) 3.22×10^5 J (c) 5.66 m/s (d) 4.00×10^3 J
69. (a) 4.20×10^3 kcal (b) 35.2 kcal/min (c) Greater than the highest output in Table 6.4 (26.5 kcal/min) (d) Impossible to maintain such a power output for 2 h.

Chapter 7
1. (a) 1.50×10^4 kg·m/s (b) 625 to 1 (c) 666 kg·m/s
3. 2.50 m/s
5. 9.00×10^3 N
7. (a) -2.40×10^3 N (b) No, the change in momentum and time interval are the same. Newton's third law states equal and opposite forces, so that both hands would experience the same magnitude force.
9. (a) −800 kg·m/s (b) −1.20 m/s
11. -4.69×10^5 N
13. 2.10×10^3 N
15. KE $= p^2/2m$
17. 0.122 m/s
19. 22.4 m/s
21. 24.8 m/s
23. (a) 1.02×10^{-6} m/s (b) -5.63×10^{20} J (almost all KE lost)
25. (a) −86.4 N (b) −0.389 J (c) 64.0%
27. (a) 8.06 m/s (b) −59.0 J (c) −7.88 m/s, −222 J
29. (a) 0.163 m/s (b) −81.6 J (c) −0.087° m/s, −81.5 J
31. 0.704 m/s, −2.25 m/s
33. (a) −4.58 m/s (b) 31.5 J (c) −0.491 m/s (d) 3.38 J
35. (a) 5.20 m/s, 3.00 m/s (direction 90° relative to incoming) (b) KE/KE′ = 1.00
37. (a) −2.26 m/s (b) 7.63×10^3 J
39. (a) 1.50×10^7 m/s; 5.36×10^5 m/s at −29.5° (b) 7.52×10^{-13} J
43. 39.2 m/s^2
45. 4.16×10^3 m/s
47. (a) 4.90×10^5 N (b) 488 m/s^2
49. 2.63×10^3 kg
51. (a) 73.7 m (b) 45.1 m/s (c) 369 m
53. (a) 21.4 m/s (b) 20 m/s (c) 3.0 m
55. (a) 44.3 m/s (b) 1.99 kg·m/s (c) 1.99×10^3 N (d) 2.5×10^{-4} m
57. 3.53 m/s^2
59. (a) 150 s (b) The time is unreasonably long. (c) A greater force is needed to stop the car in a reasonable time.

61. (a) −0.571 kg (b) The mass is negative. (c) The final speed of the combination cannot be greater than the initial speed of the incoming mass.

Chapter 8
1. 723 km
3. 0.556 rev/s, 3.49 rad/s
5. 4×10^{21} m
7. (a) 3.95×10^3 m/s^2 (b) 36.9% of the speed of sound
9. (a) 31.4 rad/s (b) 118 m/s^2 (c) 384 m/s^2
11. 0.313 rad/s
13. (a) 483 N (b) 17.4 N (c) 2.24, 0.0807
15. 4.14°
17. 8.03 m
19. (a) 0.250°/min (b) 0.469 solar diam/min
21. (b) 26.1°
23. (a) 3.33×10^{-5} m/s^2 (b) 5.93×10^{-3} m/s^2 (c) 178 to 1
25. (a) 274 m/s^2 (b) 28.0 to 1
27. 2.72×10^{-3} m/s^2
29. (a) 4.6×10^{-14} m/s^2 (b) 2.0×10^4 to 1
31. (a) 1.01×10^{13} m^3/s^2 (b) 3.21×10^5 m^3/s^2 (c) Jupiter more massive than Earth.
33. 1.98×10^{30} kg
35. 316 to 1
37. (a) 16.2 m/s (b) 0.234
39. (a) 14.8 m/s (b) 11.3 m (c) 7.35×10^4 J
41. (a) 1.84 (b) A coefficient of friction this much greater than 1 is unreasonable. (c) The assumed speed is too great for the tight curve.
43. (a) 2.94×10^{17} kg (b) 4.92×10^{-8} of the earth's mass (c) The mass of the mountain and its fraction of the earth's mass are too great. (d) The gravitational force assumed to be exerted by the mountain is too great.

Chapter 9
1. 0.368 rev/s
3. (a) 2.33 rad/s^2 (b) 0.417 rev
5. (a) −25.0 rad/s^2 (b) 28.7 rev (c) 3.80 s (d) 50.7 m (e) 26.6 m/s
7. (a) 0.338 s (b) 0.0404 rev (c) 0.313 s
9. (a) 50.4 N·m (b) 17.1 rad/s^2 (c) 17.0 rad/s^2
11. 3.96×10^{18} s (1.26×10^{11} y)
13. (a) 185 J (b) 0.0785 rev (c) 9.81 N
15. (a) 2.57×10^{29} J (b) 2.65×10^{33} J
17. (a) 128 rad/s (b) 20.0 m
19. (b) rolling to sliding time ratio = $(3/2)^{1/2} = 1.22$
21. 3.96×10^{18} s (1.26×10^{11} y)
23. (a) 2.66×10^{40} kg·m/s^2 (b) 7.08×10^{33} kg·m/s^2
25. 22.5 kg·m^2/s
27. 25.3 rev/min
29. (a) 1.74 rad/s (b) 22.6 J before, 1.57 J after (c) 1.50 kg·m/s before, 2.20 kg·m/s after
31. (a) 5.34×10^{-2} m/s (b) 0.916 rad/s (c) 28.1 J, 1.64 J

33. (a) 87.3 rad/s^2 (b) 8.29 m/s^2 (c) 1.04×10^7 m/s^2, 1.06×10^6 gs
35. (a) 29.7 m/s (b) 24 rad/s^2 (c) 27 rev
37. (a) 200 rad/s (b) This is about 32 rev/s, much too high. (c) Large angular acceleration for a long time is unreasonable.
39. (a) 1.84×10^3 rad/s (17,500 rpm) (b) Unreasonably large angular velocity. (c) Need a flywheel with a larger radius and greater mass.

Chapter 10

1. 1.610 cm^3
3. 2.58 g
5. 3.99 cm
7. 0.163 m
9. 1.2×10^3 kg/m^3
11. (a) 1×10^{18} kg/m^3 (b) About 20 km
13. 7.80×10^4 N/m^2
15. 0.760 m
17. 30.6 m
19. 6.11×10^3 N. There is an equal and opposite force exerted by the atmosphere on the other side of the paper.
21. 1.60×10^5 N/m^2, 16.0%
25. 136 N
27. Balloon: $P_g = 5.00$ cm H$_2$O, $P_{abs} = 1.035 \times 10^3$ cm H$_2$O. Jar: $P_g = -50.0$ mm Hg, $P_{abs} = 710$ mm Hg.
29. (a) 100 (b) 10.0 (c) Reduced by factor of 100
31. 1.49×10^4 N
33. (a) 10.3 m (b) 15 m
37. 0.917 or 91.7% submerged
39. 0.815 g/cm^3 = 815 kg/m^3
41. (a) 41.4 g (b) 41.4 cm^3 (c) 1.09 g/cm^3
43. (a) 39.5 g (b) 50 cm^3 (c) 0.79 g/cm^3. Ethyl alcohol.
45. 8.21 N
47. (a) 960 kg/m^3 (b) 6.34%
49. (a) 0.24 (b) 0.668 (c) Yes
53. The difference is 0.005%.
55. 1.56×10^{-3} N
57. 0.566 m
59. (a) 36.8 N/m (b) 2.94×10^3 N/m^2
61. (a) -1.46 cm (b) 0 cm
63. (a) 9.9 N/m^2 (b) 3.7 N/m^2 (c) 3.6 N/m^2
65. 0.39
67. -63.0 cm H$_2$O
69. (a) 3.81×10^3 N/m^2 (b) 28.7 mm Hg
71. 27.1 cm
73. (a) 13.6 cm H$_2$O (b) 76.5 cm H$_2$O
75. (a) 258 m (b) 2.4×10^{-2} m
77. (a) 1.09×10^3 atm (b) -5.0% (c) $+5.3\%$
79. (a) -8.82×10^5 N/m^2 (b) It is not possible to create a pressure more negative than -1.00 atm (-1.01×10^5 N/m^2). (c) It is unreasonable to assume you can raise water this high with suction.

Chapter 11

1. 2.10 cm^3/s
3. (a) 0.250 ft/s (b) 19.0 ft
5. (a) 22.2 h (b) 0.0535 s

7. (a) 3.14 m/s (b) 1.27×10^3 gal/min
9. 1.70 m/s
15. -153 mm Hg
17. 2.54×10^5 N
19. (a) 1.58×10^6 N/m^2 (b) 163 m
21. (a) 71.8 m/s (b) 257 m/s
23. 0.126 W
25. (a) 3.02×10^{-3} N (b) 1.03×10^{-3}
27. 1.60 cm^3/min
29. 1.52
31. 0.138 (N/m^2)·s. Olive oil.
33. (a) 1.62×10^4 N/m^2 (b) 0.111 cm^3/s (c) 10.6 cm
35. 2.21×10^6 N/m^2
37. $N_R = 1992 < 2000$ (to four digits, so it could be turbulent given the uncertainty in the data)
39. Hose: 3.51×10^4, not laminar. Nozzle: 1.27×10^5, not laminar.
41. 2.54, 2000
43. (a) ≥ 26.0 m (b) 2.68×10^{-6} N
45. 1.41×10^{-3} m
47. 125 s
49. (a) 23.7 atm, 344 lb/in^2 (b) The pressure is much too high. (c) The assumed flow rate is very high for a garden hose. (d) $N_R = 5.27 \times 10^6$, turbulent. Assumption of laminar flow not valid, so that actual pressure would be even higher.

Chapter 12

1. 102°F
3. 20.0°C and 25.6°C
5. 9890°F
7. (a) 22.2°C
9. 169.98 m (to five digits, since ΔL is small compared with L)
11. 5.4×10^{-6} m
13. 61.14 L
15. (a) 9.12 mL (b) 7.56 mL
17. 0.833 mm
21. 86.8 lb/in^2
23. (a) 9.85×10^{-2} atm (b) -0.902 atm
25. (a) 2.68×10^{19}/cm^3 (b) 2.68×10^7 atoms
27. 2.47×10^7 atoms
29. 6.86 atm
31. 57.9 mol
33. (a) 1.94 atm (b) 1.97 atm
35. (a) 3.73×10^{-17} N/m^2 (b) 6.02×10^{17} m^3 (c) 844 km
37. 176 m/s
39. 1.58×10^5 K
41. 3.09×10^9 K
43. 3.01 K
45. (a) 1.004 (b) 764 K
47. (a) 2.33×10^3 N/m^2 (from Table 12.4) (b) 2.30% (c) 1.43%
49. (a) 95°C (b) 90°C (both from Table 12.4)
51. 4.73×10^4 N/m^2 (from Table 12.4)
53. 3.07 g/m^3
55. 50.8 g/m^3
57. 50.9%
59. (a) 6.90×10^3 N/m^2 (b) 64.2%
61. 7.9 g/m^3
63. (a) 2.25 km (b) 647.4 K (c) No
65. (a) 4.76×10^5 N/m^2 (b) 9.70×10^6 N
67. (a) 4.40×10^4 °C (b) Unreasonably

high temperature (greater than the boiling point of aluminum). (c) A thermal expansion of 10% is much too large.
69. (a) 106% (b) Relative humidity cannot exceed 100% (c) The premise that the vapor density remains constant.

Chapter 13

1. 5.02×10^8 J
3. 3.07×10^3 J
5. 35.9 kcal
7. (a) 24.0 kcal (b) 87.8 kcal
9. (a) 591 kcal (b) 4.94×10^3 s
11. 10.8
13. (a) 3.08×10^9 J (b) 3.08×10^3 MW
15. 3.23 mW
17. (a) 148 kcal (b) 0.400 s, 3.19 s, 4.00 s, 21.6 s, 0.436 s
19. 33.0 g
21. 5.73 kcal/g
23. 0°C
25. -13.2°C
27. 44.0°C
29. (a) 1.57×10^4 kcal (b) 18.3 kW·h (c) 1.29×10^4 kcal
31. 617 W
33. 1.0 kW
35. 1×10^2 W
37. 2.59 kg
39. 10 m/s (from Table 13.4)
41. -8.80 kW
43. -543 W
45. (a) 40 W (b) 8.2×10^2 kcal
47. 35 to 1, window to wall
49. 10×10^3 K
51. 64 g
53. 77.5 kW
55. 293 W
57. (a) 1.31% (b) 20.5%
59. (a) 0.84 W (b) 31 d
61. 645°C
63. (a) -15.0 kW (b) 4.2 cm
65. 48.5°C
67. (a) 5.60 g/m^3 (b) 3.02 kcal (c) 13.6°C
69. 8.61 kg
71. (a) 2.27×10^{17} J (b) 24.0 y
73. (a) 0.117°C (b) 2.17×10^{-4}
75. (a) 36°C (b) Any temperature increase greater than about 3°C would be unreasonably large. In this case the final temperature of the person would rise to 73°C (163°F)! (c) The assumption of 95% heat retention is unreasonable.
77. (a) 1.46 kW (b) Very high power loss through a window. An electric heater of this power can keep an entire room warm. (c) The surface temperatures of the window do not differ by as great an amount as assumed. The inner surface will be warmer, and the outer surface will be cooler.

Chapter 14

1. 1.6×10^9 J
3. -9.30×10^8 J
5. (a) -2390 kcal (b) 5.00%

7. (a) 121 W (b) 2.09×10^6 J (c) 1.61×10^7 J (the motor produces 7.7 times the work)
9. 115 kcal
11. (a) 325 J (b) 325 J
13. 300 J
15. (a) 35.3 J (b) 118 N
17. 2.4 kJ
19. (a) 6.00% (b) 2.35×10^6 J
21. (a) 21.1% (b) 21.1 kJ
23. 7.25×10^{13} J (b) 1.38 (c) 5.25×10^{13} J
25. 5.12%
27. 403°C
29. (a) 244°C (b) 477°C
33. 3.61
35. 7.70
37. (a) 3.24 (b) 309 kcal (c) 3.59¢ (d) 1309 kcal
39. −2.60°C
43. 9.79×10^4 J/K
45. 8.01×10^5 J
47. (a) 1.04×10^{31} J/K (b) 3.28×10^{31} J
49. 199 J/K
51. (a) 2.47×10^{14} J (b) 1.60×10^{14} J (c) 2.86×10^{10} J/K (d) 8.31×10^{12} J
53. 2.0×10^{22} y
55. (a) 3.0×10^{29} (b) 23.5%
57. (a) -2.38×10^{-23} J/K (b) 5.6 times more likely
59. (b) 7 (c) 64 (d) 9.38% (e) 3.33 times more likely (20 to 6)
61. (a) 1.82×10^5 kg (b) 1.51 m³/s
63. (a) 0.0778 (b) 1.20×10^8 J
65. (a) 13.8 gal (b) 994 gal (c) a: 1.44×10^9 J b: 1.03×10^9 J
67. (a) 3.58% (b) 3.33×10^6 kcal (c) 212 m³/s
69. *W* is not equal to the difference of the heat input and output.
71. (a) 2.72×10^3 K (b) The temperature is much too high (iron melts at 1809 K, for example). (c) The assumed 80% efficiency is too high.

Chapter 15
1. 1.57×10^5 N/m
3. 7.54 cm
5. (a) 889 N/m (b) 133 N
7. (a) 6.53×10^3 N/m (b) Yes
9. 3.25×10^4 N/m
11. 400 N/m
13. 16.7 ms
15. 400 Hz
17. 1.50 kHz
19. (a) 3.95×10^6 N/m (b) 7.90×10^6 J
21. 1.94 s
23. 8.14 Hz
25. 0.389 kg
27. 6.21 cm
29. 2.01 s
31. 2.23 s
33. 2.99541 s (to six digits to show the effect)
35. (a) factor of 1.41 ($\sqrt{2}$) (b) 97.5% of old period
37. Slow by a factor of 2.45
39. 11:52:51 A.M. or 12:07:15 P.M.
41. 0.104 m/s

43. 0.300 J
45. 0.738 m/s
47. 30.0 m/s
49. 384 J
51. 141 J
53. 1.15 cm
55. (a) $\pm 4.90 \times 10^{-3}$ m (b) 5.00×10^5 J (c) 20.0 min
57. 11.3 m
59. 7.50
61. 16.0 m/s
63. 2.50 GHz
65. 34.0 cm
67. 4 Hz
69. 462 Hz, 4 Hz, 4 Hz
71. 21, 43, and 22 Hz
73. 0.250 mW
75. 7.07
77. 4.94 min
79. 16.0 d
81. 3.38×10^{-5} W/m²
83. 44.2 MW/m²
85. (a) 1.20032 m (to six digits because the fractional change is small) (b) Initial 2.19866 s, final 2.19895 s
87. (a) 11.3 m/s (b) 108 N (c) 21.6 m/s² = 2.20*g*. A person can tolerate this acceleration.
89. (a) 60.0 m/s or 216 km/h (b) Much too fast. (c) Ocean waves that crest 2.00 m apart do not arrive at such a high frequency as assumed (30 Hz is too high).
91. (a) 200 kHz (b) Well into the ultrasonic, so the frequency is too high. (c) The assumed wavelength is too short for audible sound.

Chapter 16
1. 0.288 m
3. 332 m/s
5. 343 m/s
7. 0.223
9. 7.70 m
11. 1.26 mW/m²
15. (a) 8.00×10^{-10} W/m² (b) 8.00×10^{-9} W/m²
17. 1.58×10^{-13} W/m²
19. 110 dB
21. 10^{-3} atm
23. 19.5 dB
25. 1.13×10^{-5} W
27. (a) 138 kHz (b) 1.77×10^3 Hz
29. 3.05 m/s
31. Ratio of frequencies = 1.030, so shift is easily perceived.
33. 0.990 m/s
35. (a) 44 Hz (b) 55 Hz (c) 33, 44, 55, 77, 88, and 132 Hz
37. 263.5 or 264.5 Hz
39. 96.0 Hz, 160 Hz, 224 Hz
41. 65.4 cm
43. 0.974 m
45. 3%
47. 7.35 kHz
49. 22.0°C
51. 1×10^9 m = 1×10^6 km
53. 498.5 or 501.5 Hz
55. 82 dB
57. 48, 9, 0, −7, and 20 dB

59. (a) 23 dB (b) 70 dB
61. Five factors of 10
63. (a) 2.00×10^{-10} W/m² (b) 2.00×10^{-13} W/m²
65. 2.51
67. 1.26
69. 3.16×10^3 W/m² is within the range used for diathermy (deep heating).
71. (a) 6.16 MHz (b) 12.5 cm
73. 1.30 μs
75. 19.2 cm/s
77. 4.62 cm/s
79. (a) 2.89 kJ (b) 2.75°C
81. (a) 25.2 kHz (b) An ultrasonic frequency (not audible). (c) Assumed length is too short.
83. (a) 0.345 m/s (b) Much too long a time for the 100 m dash. (c) The Doppler shift is unreasonably low (1001 Hz is too low).

Chapter 17
1. (a) 1.25×10^{10} (b) 3.13×10^{12}
3. −600 C
5. 1.03×10^{12}
7. 9.09×10^{-11}, electrons to protons
9. 1.48×10^8 C
11. 26.3 N
13. Separation decreased by a factor of 5.0
15. 3.45×10^{16} m/s²
17. 3.2 (square root of 10 to two digits)
19. 1.04×10^{-9} C
21. (a) 4.16×10^{42} (b) 1.24×10^{36} (c) Yes, they are consistent
23. 1.02×10^{-11}
25. (a) 0.859 m beyond negative charge on line connecting two charges (b) 0.104 m from lesser charge on line connecting two charges
27. 8.75×10^{-4} N (east)
29. 6.25 N/C
31. (a) 300 N/C (east) (b) 4.80×10^{-17} N (east)
33. 5.58×10^{-11} N/C (downward)
35. (a) -6.76×10^5 C (b) 2.63×10^{13} m/s² (upward) (c) 2.45×10^{-18} kg
37. The ratio of their charges is exactly 9 to 1.
47. (a) Field is in the direction of the electron's velocity. (b) 3.56×10^{-4} m (c) 1.42×10^{-10} s (d) -5.00×10^6 m/s (opposite its initial velocity)
49. (a) 1.33×10^7 m/s (b) 3.02×10^{-9} s
51. 2.15×10^{-8} C
53. (a) 3.78×10^{-14} N (b) -2.36×10^5 N/C
55. (a) 1.80×10^{11} m/s² (b) Unreasonably large acceleration for a macroscopic object. (c) The assumed charges are much too great (at least 103 as great as typical static electricity can be).

Chapter 18
1. 6.25×10^{-4} J
3. 600 J
5. 1.00×10^8 V
7. 5.00 mW
9. 3.00 V
11. −729 V
13. 47.0 kV

17. (a) 3.00 kV (b) 750 V
19. (a) 2.50×10^6 V; no (b) 1.7 mm (answer to two digits because maximum field strength is approximate)
21. 15 kV
23. (a) 800 keV (b) 25.0 km
25. 27.2 V
27. 90.0 m
29. 8.33×10^{-7} C
31. (a) 45.0 MV (b) 45.0 m (c) 132 MeV
41. 21.6 mC
43. 80.0 mC
45. 20.0 kV
47. 667 pF
49. 4.4 μF
51. 3.08 μF
53. (a) 4.43×10^{-12} F (b) 452 V (c) 4.52×10^{-7} J
55. (a) 6.38×10^{-5} J (b) 3.81×10^{-4} J
57. 2.79 μF
59. (a) 0.450 J (b) 225 s
61. 2.01×10^{-4} J
65. 1.00×10^5 K
67. (a) 2.00×10^9 J (b) 776 kg
69. 11.6°C. Yes, it is reasonable, since this does not bring the temperature close to the boiling point.
71. (a) 2.30×10^{-16} J (b) 1.11×10^7 K
73. (a) −3.00 μF (b) Capacitance cannot be negative. (c) A parallel connection always produces a greater capacitance, while here a smaller capacitance was assumed. This could only happen if the capacitors are connected in series.
75. (a) 1.44×10^{12} V (b) The voltage is unreasonably large. (c) It is not possible to store such a large charge on a metal sphere of this size.
77. (a) 14.2 kV (b) This voltage is more than 100 times the breakdown voltage of this material. (c) The assumed charge is unreasonably large and cannot be stored in a capacitor of these dimensions.

Chapter 19

1. 0.278 mA
3. 0.250 A
5. 1.50 ms
7. (a) 0.120 C (b) 7.50×10^{17} electrons
9. 96.3 s
11. (a) 7.81×10^{14} He^{++}/s (b) 4.00×10^3 s (c) 7.71×10^8 s
13. -1.13×10^{-4} m/s
15. 9.42×10^{13} electrons
17. 6.75 kΩ
19. 3.50 V
21. 0.200 mA
23. 0.322 Ω
25. 1.24 to 1, aluminum to copper
27. 276°C
29. Iron
31. (a) Gold (b) 110 Ω (last digit uncertain)
33. By a factor of 16
35. 40.0°C
37. 6.00 kW
39. 1.00 W
45. 1.50 kW

47. 480 V
49. 2.50 ms
51. 1.58 h
53. (a) 30.0 W (b) 7.50 MΩ
55. 16.0% decrease
57. (a) 3.33 Ω (b) $1080 (fourth digit uncertain)
59. 2.40 kW
61. 8.71% reduction
63. 25.5 W
65. (a) 1.39 ms (b) 4.17 ms (c) 8.33 ms
67. (a) 194 kW (b) 880 A
69. (a) 0.400 mA, no effect (b) 26.7 mA, muscular contraction for duration of shock (can't let go)
71. 1.20×10^5 Ω
73. (a) 1.00 Ω (b) 14.4 kW
75. (a) 10.6 g/min (b) 5.09 kg
77. 26.8 A
79. (a) 2.00×10^9 W (b) 1.50×10^3 m^3/s
81. (a) 410 kW 27.2 s (c) 0.734 m/s^2
83. (a) 4.7 Ω (total) (b) 3.00% decrease
85. Temperature increases 860°C. Very likely to be damaging.
87. (a) 5×10^5 °C (b) -3×10^5 °C (c) In part (a), the temperature is far above any reasonable temperature. In part (b), the temperature is far below absolute zero (impossible). (d) The assumption that the temperature coefficient of resistivity remains constant over such large changes is unreasonable.
89. (a) 10.0 kA (b) 1.00×10^{-4} Ω (c) 0.468 m (d) It is an unreasonably thick "wire." (e) The power loss assumed is too small, and the voltage should be much higher so that a lower current can carry the power.

Chapter 20

1. (a) 2.75 kΩ (b) 27.5 Ω
3. (a) 786 Ω (b) 20.3 Ω
5. 29.6 W
7. (a) 0.74 A (b) 0.742 A
9. (a) 60.8 W (b) 3.18 kW
13. (a) 6 (b) 9.24 V
15. (a) 1.34 V (b) 3.08 W (c) 2.68 W
17. (a) 18.0 V (b) 18.0 V
21. (a) 7.09 W (b) 7.99 W
23. (a) 0.617 A (b) 3.81 W (c) 17.9 Ω
35. 30.0 μA
37. 1.98 kΩ
39. 1.25×10^{-4} Ω
41. (a) 3.00 MΩ (b) 2.99 kΩ
43. (a) 1.58 mA (b) 1.5848 V (need five digits to see difference) (c) 0.99990 (need five digits to see difference from unity)
45. 15.0 μA
47. (a) 10.02 Ω (b) 0.9980, or a 0.20% decrease (c) 1.002, or a 0.20% increase
49. 1.540 V
51. 28.6 Ω
53. range 1.50 to 1.60 Ω
55. range 4.00 to 30.0 MΩ
57. (a) 2.50 μF (b) 2.00 s
59. 86.5%
61. (a) 20.0 s (b) 120 s (c) 16.0 ms

63. 17.3 ms
65. 3.33×10^{-3} Ω
67. (a) 4.99 s (b) 3.87°C
69. (a) −400 kΩ (b) Resistance cannot be negative. (c) Series resistance is assumed to be less than one of the resistors, and it must be greater than any of the resistors.
71. (a) −9.04 Ω (b) Resistance cannot be negative. (c) The combination of power delivered and resistance of the bulb requires a voltage greater than the emf of the battery, so that the internal resistance comes out to be negative.
73. (a) −20.0 Ω (b) Resistance cannot be negative. (c) A full-scale deflection for this galvanometer requires more than 0.5 mV.
75. (a) 10.0 kF (b) This is an enormous capacitance, which would be building-sized. (c) A much larger resistance is needed to obtain a long time constant. 0.100 Ω is quite small.

Chapter 21

7. 7.50×10^{-7} N
9. 3.01×10^{-5} T
11. 6.67×10^{-10} C (taking the earth's field to be 5.00×10^{-5} T)
13. 0.979 T
15. 4.80×10^{-19} C
17. (a) 400 kV/m (b) 4.00 kV
19. (a) 4.09×10^3 m/s (b) 7.83 km (c) 183 m (d) 4.27 m
21. 2.50 cm
23. 3.12 mV
25. 4.00 m/s
27. 75.0 nV
33. (a) 0.600 N/m (b) West
35. 50.0 N
37. (a) 196 A (b) 17.1 N
39. (a) The net **F** is down the page. The forces on the sides are equal in magnitude and opposite in direction and, hence, cancel. The magnitudes are the same because I, l, B, and θ are equal. The directions are opposite by RHR-1, because the currents are in opposite directions. (b) 1.00 N/T
41. (a) 389 N·m (b) 133 N·m
43. 1.50 T
45. (a) 64.2° (b) 30.0° (c) 5.74°
47. The 0.471 N·m torque is straight down, meaning the loop will rotate cw as viewed from directly above.
53. 1.02×10^{13} T
55. (a) 4.80×10^{-4} T (b) Zero
57. 39.8 A
59. 0.796 A
61. (a) To the lower left (b) $1/\pi$ (c) Into the page
63. 3.00×10^{-5} T to the right
65. 8.53 N repulsive
67. 400 A in the opposite direction
69. (a) 1.67×10^{-3} N/m (b) 3.33×10^{-3} N/m (c) repulsive
71. (a) Top wire: 2.65×10^{-4} N/m, 10.7° to left of up (b) Lower left wire: 3.61×10^{-4} N/m, 13.9° down from right

(c) Lower right wire: 3.46×10^{-4} N/m, 30.0° down from left

73. 9.09×10^{-7} N up

75. 60.2 cm

77. 1.02×10^3 N/m²; not a significant fraction of an atmosphere

79. 17×10^{-4} %/°C

81. 18.3 MHz

83. (a) Straight up (b) 6.00×10^{-4} N/m (c) 94.1 μm (d) 2.48 Ω/m, 49.4 V/m

85. (a) 571 C (b) Impossible to have such a large separated charge on such a small object. (c) The 1.00 N force is much too great to be realistic in the earth's field.

87. (a) 2.40×10^6 m/s (b) The speed is extremely high. (c) The assumption that you could generate such a voltage with a single wire in the earth's field is unreasonable.

89. (a) 25.0 kA (b) This current is extremely high for power lines. It implies 5.00 GW carried by the lines, much greater than practical. (c) The assumption of such a large field at this distance from power lines is unreasonable.

Chapter 22

1. Zero

3. (a) ccw (b) cw (c) No current induced

5. (a) 1 cw, 2 cw, 3 ccw (b) 1, 2, and 3 no current induced (c) 1 ccw, 2 ccw, 3 cw

9. 3.04 mV

11. 0.425 T

15. 0.630 V

17. 2.22 m/s

21. 24.2°

23. 764 rad/s

25. 15.7 kV

27. 0.477 T

29. (a) 5.89 V (b) At $t = 0$ (c) 0.393 s (d) 0.785 s

31. (a) 6.00 Ω (b) 3.33 A

33. 84.0 V

35. (a) 30.0 (b) 97.5 mA

37. (a) 20.0 mA (b) 2.40 W

39. 0.125 A

41. (a) 125 (b) 0.800 (c) 0.640, or 64.0%

43. 1.80 mH

45. 3.60 V

49. (a) 31.3 kV (b) 125 kJ (c) 1.56 MW (d) No, it is a very powerful magnet system.

51. (a) 0.139 mH (b) 3.99 V (c) Zero

53. 60.0 mH

55. 500 H

57. 50.0 Ω

59. 1.00×10^{-15} to 100 s

61. 95.0%

63. 24.6 ms

65. 531 Hz

67. 1.59 nF

69. (a) 2.55 A (b) 1.53 mA

71. 63.7 μH

73. (a) 21.1 mH (b) 8.00 Ω

75. (a) 3.18 mF (b) 16.7 Ω

77. (a) 532 Ω at 60 Hz, 40.1 Ω at 10 kHz

79. 1.13 kHz

81. 0.159 Hz to 5.03 GHz

83. 3.52 H

85. (a) 16.7 Ω (b) 3.71 Ω (c) 0.336 A at 120 Hz, 1.51 A at 5.00 kHz (d) 1.78 kHz (e) 2.24 A

87. (a) 0.150 (b) 81.4° (c) 0.281 W (d) 12.5 W

89. (a) 283 Ω (b) 18.8 nF (c) 589 W

91. 0.100 Ω

93. (a) 304 mA (b) 0.924 mW (c) 1.58 mT

95. (a) 200 H (b) 5.00°C

97. 251 V

99. (a) 1.94 m (b) 51.5 N/m (c) 97.0 J

101. (a) 6.11×10^5 rad/s (b) The angular velocity is extremely high. (c) The assumption that a voltage as great as 12.0 kV could be obtained is unreasonable.

103. (a) 2.50 GV (b) An extraordinarily high voltage. (c) It is not possible to shut off such a large current in such a large inductor in such an extremely short time.

Chapter 23

1. 2.999×10^8 m/s is obtained to four-digit accuracy.

3. 150 kV/m

5. 33.3 cm (900 MHz), 11.7 cm (2560 MHz)

7. 27.0 MHz

9. 2.00×10^{13} Hz

11. 0.600 m

13. 500 s = 8.33 min

15. (a) 5.00×10^6 m (b) 4.33×10^{-5} T

17. 1.50 MHz, AM band

19. 3.90×10^8 m

21. (a) 1.50×10^{11} m (b) 0.500 μs (c) 66.7 ns

23. 1.91 mW/m²

25. 0.236 μW/m²

27. (a) 89.2 cm (b) 27.4 V/m

29. (a) 333 T (b) 1.33×10^{19} W/m² (c) 41.7 kJ

33. (a) -3.5×10^2 W/m² (b) 88% (c) 1.7 μT

35. 13.5 pF

37. (a) 4.07 kW/m² (b) 1.75 kV/m (c) 5.84 μT (d) 2 min 19 s

39. (a) 183 ms (b) 1.23 MV/m

41. (a) 8.44×10^{-12} W (b) 5.65 mV

43. (a) 6.00×10^8 m/s (b) This is twice the speed of light, an impossibility. (c) The assumed combination of frequency and wavelength is not possible.

45. (a) 1.41×10^{-19} F (b) This capacitance is much too small for a macroscopic object. (c) Either the wavelength is too small or the inductance is too big (or both).

Chapter 24

1. Top 1.715 m from floor, bottom 0.825 m from floor. Height of mirror is 0.890 m, or precisely one-half the height of the person.

5. 2.25×10^8 m/s in water, 2.04×10^8 m/s in glycerine

7. 1.491, polystyrene

9. 1.28 s

11. 1.03 ns

13. 1.46, fused quartz

19. 66.3°

21. >1.414

23. 1.50, benzene

25. 41.9°

27. 46.5°, red; 46.0°, violet

29. (a) 0.043° (b) 1.33 m

31. 55.7°

33. 53.5°, red; 55.2°, violet

35. 5.00 to 12.5 D

37. −0.222 m

39. (a) 3.43 m (b) 0.800 by 1.20 m

41. (a) −1.35 m (on the same side of the lens as the mole) (b) +10.0 (c) 5.00 cm

43. 44.4 cm

45. (a) 6.60 cm (b) −0.333

47. +7.50 cm

49. (a) +6.67 (b) +20.0 (c) The magnification increases without limit (to infinity) as the object distance increases to the limit of the focal distance.

51. −0.933 mm

53. (b) The image distance approaches negative infinity (an unlimited large distance behind the lens on the same side as the object).

55. −0.333 D

57. +0.667 m

59. +1.60 m (concave)

61. +0.360 m (concave)

63. (a) +0.111 (b) −0.334 cm (c) −7.52 mm (convex)

65. (a) Infinite (b) Zero

69. (a) 4.00 (b) 400, 1600 (last digit uncertain)

71. <102.5 cm

73. +10.0 cm

75. (a) +18.3 cm (on the eyepiece side of the objective lens) (b) −60.0 (c) −11.3 cm (on the objective side of the eyepiece) (d) +6.67 (e) −400

77. (a) −25.0 (b) 9.55×10^{-3} ° (c) 0.239°

79. +50.3 D

81. 2.00 m

83. (a) +62.5 D (b) −0.250 mm (c) −0.0800 mm

85. 1.00 m

87. 10.6 cm

89. −5.00 D

91. 25.0 cm

93. −5.41 D

95. 26.8 cm

97. (a) −1.00 D (b) 24.5 cm

99. (a) 28.6 cm (b) +1.78 D

101. 6.82 kW/m²

103. (a) 0.898 (b) Implies a speed of light in the medium that is greater than the speed of light, an impossibility. (c) The refracted angle is too big relative to the incident angle.

Chapter 25

1. 1/1.333 = 0.750

3. 1.49, polystyrene

5. 0.877 glass to water
7. 0.997°
9. 0.290 µm
11. 577 nm
13. 62
15. 1.44 µm
17. (a) 20.3° (b) 4.98° (c) 5
19. 2.37 cm
21. 5.97°
23. 8.99×10^3
25. 0.707 nm
27. 589.1 and 589.6 nm
31. 11.8°, 12.5°, 14.1°, and 19.2°
33. 2.63×10^4; useful for UV but not for IR.
35. (b) 333 nm, UV (c) 6.58×10^3/cm
37. (a) 11.8° (b) 20.5°
39. (a) 0.693 µm (b) 612 nm
41. 600 nm
43. (a) 5.45° (b) 1.91 µm
45. (a) 1.00λ (b) 50.0λ (c) 1000λ
47. 5.00
49. 107 m
51. (a) 7.72×10^{-4} rad (b) 23.2 m
    (c) 590 km
53. (a) 2.24×10^{-4} rad (b) 5.81 km
    (c) 0.179 mm
55. 5.15 cm
57. 672 nm (red)
59. 128 nm
61. 195, 391, and 586 nm
63. 143 nm
65. (a) 0.442 mm (b) Material of slides irrelevant; path length difference giving rise to effect occurs in air.
67. 45.0°
69. 45.7 mW/m²
71. 90.0%
75. 48.8°
77. 41.2°
79. (a) $n = 1.92$ (not diamond) (b) 55.2°
81. 23.6°
83. 1.072 times longer
85. (a) 42.3 nm (b) Not a visible wavelength. (c) The combination of grating spacing and angle is inconsistent with the assumption of visible wavelengths.
87. (a) 0.820 cm (b) Unreasonably thick. (c) The assumption of thin film interference to destroy reflection is not reasonable with known materials.

Chapter 26

1. (a) 1.0328 (b) 1.155 (showing three digits difference from 1 in both cases)
3. 5.96×10^{-8} s
5. 6.23×10^{-8} s
7. $0.900c$
9. $0.140c$
11. (a) $0.745c$ (b) $0.99995c$ (to five digits to show effect)
13. (a) 2.01 km (b) 0.627 km
15. (a) 4.303 y (to four digits to show any effect) (b) 0.1434 y
17. (a) $0.909c$ (b) $0.400c$
19. $0.198c$
21. $0.991c$
23. $-0.696c$

27. (a) $0.9995c$ (b) 1.333×10^{11} y (c) 1.320×10^{11} y (all to four digits to show effects)
29. 4.09×10^{-19} kg·ṁ/s
31. (a) $3.000000015 \times 10^{13}$ kg·m/s (b) Ratio of relativistic to classical momenta = 1.000000005 (extra digits to show small effect)
33. 2.990×10^8 m/s
35. 939 MeV
37. (a) 0.333 g (b) 3.33×10^{-4}
39. 1.39×10^{20}
41. (a) 1.11×10^{16} kg (b) 1.11×10^{11} m³
43. $0.9024c$
45. $0.0511c$
47. (a) 1.02 MeV (1.64×10^{-13} J) (b) $0.0467c$ (c) $0.943c$
49. 1216 MeV
51. (a) 970 MeV (b) 243 MeV/c (c) No, $pc = .250 E$. $\gamma^2 - 1 = 0.0667$, so $pc << mc^2$.
53. (a) 51.2 GeV (b) 51.1 GeV
55. (a) 9.00×10^{16} J (b) 9.18×10^{11} kg
57. (a) 1583 y (b) 1583 (last digits uncertain in both (a) and (b))
59. (a) 3.62×10^{-10} (b) 775 tons of TNT
61. (a) .995 (b) γ cannot be less than 1. (c) Assumption that t is longer in moving ship is unreasonable.
63. (a) $10c$ (b) Speed cannot be greater than c. (c) Claim of $1.20c$ is unreasonable.

Chapter 27

1. (a) 0.070 eV (b) 14
3. (a) 2.21×10^{-34} J (b) 2.26×10^{34} (c) No
5. 555 nm (green)
7. 4.09 eV
9. 5.54 eV
11. 3.64 eV
13. 531 nm (green)
15. 8.44 µs
17. (a) 4.02×10^{15}/s (b) 0.533 mW
19. (a) 5.97×10^{-26} J (b) 3.73×10^{-7} eV
21. (a) 1.24 µm (b) 2.42×10^{14} Hz
23. (a) 0.0828 eV (b) 121 (c) 1.24 MeV (d) 1.24×10^5
27. 121 kV
29. 5.89×10^{26}/s
31. (a) 1.16×10^{32} photons/s (b) 9.23×10^{20} photons/s·m²
33. 3.04×10^{15} m (about 0.32 ly)
35. 1.66×10^{-32} kg·m/s
37. (a) 1.33 nm (b) 9.38 eV
39. (a) 2.65×10^{-28} kg·m/s (b) 291 m/s (c) electron 3.86×10^{-26} J, photon 7.96×10^{-20} J, ratio 2.06×10^6
41. (a) 1.32×10^{-13} m (b) 9.39 MeV (c) 47.0 keV
45. 8.09×10^{-11} m
47. 2.21×10^{-32} m/s
49. (a) 397 nm (b) 5.22×10^{-9} eV
51. 7.09×10^{-12} m
53. (a) 5.29 fm (b) 4.70×10^{-12} J (c) 29.4 MV
55. 3.70×10^{-12} m
57. (a) 57.9 m/s (b) 9.55×10^{-9} eV
59. 29.0 nm, 290 times greater
61. 1.10×10^{-13} eV

63. 3.23×10^{-23} s
65. 2.66×10^{-46} kg
67. 0.395 nm
69. (a) 1.33×10^{-19} J (b) 2.1×10^{23} (c) 140 s (last digit uncertain)
71. (a) 335 kJ (b) 1.12×10^{-3} kg·m/s (c) 1.12 mm/s (d) 6.23×10^{-7} J
73. (a) 1.06×10^3 (b) 5.33×10^{-16} kg·m/s (c) 1.24×10^{-18} m
75. 1.62 km/s
77. 2.57×10^{-12} m
79. 2.30×10^{-4} °C
81. (a) 2.00 kJ (b) 1.33×10^{-5} kg·m/s (c) 1.33×10^{-5} N (d) Yes
83. (a) 7.28×10^{12} m/s (b) This is thousands of times the speed of light (an impossibility). (c) The assumption that the electron is nonrelativistic is unreasonable at this wavelength.
85. (a) -0.892 eV (b) Binding energy cannot be negative (it would be unbound and fly away with this amount of energy). (c) It is unreasonable to assume the ejected electron has more energy than the incoming photon.

Chapter 28

1. 1839
3. 50 km
5. (a) 1.66×10^{-19} C (b) 1.04
7. (a) 6×10^8 (b) 4×10^{-5}
9. 97.2 nm (UV)
13. 4
17. 970 nm (IR)
19. (a) $n_i = 10$ (b) 7 (c) Infinite number
21. 0.265×10^{-10} m
23. (a) 36 (b) 52.1 nm (c) UV
27. 3
29. $1 \geq 1, n \geq 2$
31. (a) 1.49×10^{-34} J·s (b) 9.14×10^{-35} J·s (c) 1.63
33. (a) 5 (counting both positive and negative values) (b) 35.3°
35. (a) 32 (b) 2 in s, 6 in p, 10 in d, and 14 in f, for a total of 32
37. (a) 2 (b) $3d^9$
39. (b) $1d^1$ not allowed since $1 \geq n$ (c) $4s^3$ not allowed since $3 > 2(2l + 1)$ (d) $3d^9$ not allowed since $9 > 2(2l + 1)$
43. 4.14×10^{-11} m
45. 62.2 kV
47. 64.7 keV
49. Hf
51. (a) 1.23 eV (b) 1.01 µm
53. 60.0 nm
55. (a) 3.0 and 2.3 eV (b) 413 and 539 nm, respectively
57. The first level
59. 1.8×10^{15} g/cm³, 1.8 times greater than text estimate
61. 0.948 nm
63. 1.77×10^{11} C/kg (best to three digits is 1.76×10^{11} C/kg)
65. 2.63×10^{68}
67. (a) 1.34×10^{23} (b) 2.52 MW
69. (a) 1.24×10^{11} V (b) The voltage is extremely large compared with any practical value. (c) The assumption of such a short wavelength by this method is unreasonable.

Chapter 29

1. 1.67×10^4
5. 1.9×10^{-15} m
7. (a) 4.6×10^{-15} m (b) 1.6 to 1
9. 85.4 to 1
11. 12.4 GeV
13. (a) 18 km (b) 3.6×10^{57}
23. Parent is ^{90}Sr (you must provide complete decay equation)
25. Parent is ^{212}Po (you must provide complete decay equation)
27. $Z = Z - 2 + 2$, $A = A - 4 + 4$, electron family number $0 = 0 + 0$
29. $Z = Z - 1 + 1$, $A = A$, electron family number $0 = (-1) + (+1)$
31. (a) ^{208}Pb (b) 9.14×10^{-35} J·s (c) 1.63
33. (a) ^{249}Cf $\rightarrow$ ^{245}Cm $+$ ^{4}He (you must provide the values of Z and N) (b) 6.295 MeV
35. (a) ^{90}Sr $\rightarrow$ ^{90}Y $+ \beta^- + v_e$ (you must provide the values of Z and N) (b) 0.546 MeV
37. (a) ^{11}C $\rightarrow$ ^{11}B $+ \beta^+ + v_e$ (you must provide the values of Z and N) (b) 0.960 MeV
39. (b) ^{7}Be $+ e^- \rightarrow {^7}$Li $+ v_e$ (you must provide the values of Z and N) (c) 0.861 MeV
41. 57,000 y (last three digits uncertain)
43. (a) 0.986 Ci (b) The half-life of ^{226}Ra is now better known.
45. 1.22×10^3 Bq
47. (a) 16.0 mg (b) 0.0114% or 1.14 parts in 10^4
49. 1.51×10^{17} y
51. 7.9×10^4 y
53. 1.30 y
55. (a) 1.23×10^{-3} (b) Only part of the emitted radiation goes in the direction of the detector. Only a fraction of that causes a response in the detector. Some of the emitted radiation (mostly α particles) is absorbed within the source, some is absorbed by air before reaching the detector, and some does not penetrate the detector.
57. 1.112 MeV, consistent with graph
59. 7.848 MeV, consistent with graph
61. (a) 7.680 MeV, consistent with graph (b) 7.520 MeV, consistent with graph. Not significantly different from value for ^{12}C, but sufficiently lower to allow decay into another nuclide that is more tightly bound.
63. (a) 1.46×10^{-8} u (b) 0.00549 u (c) 2.66×10^{-5}
65. 22.9 cm
67. (a) 6.96×10^{15} Bq (b) 6.22 kW (c) 5.98 kW
69. 19.2 to 1
71. 5.48×10^{-3} spread in gas detector, 1.41×10^{-3} spread in solid state detector
73. (a) 2.4×10^8 (b) The greatest known values are about 260. This is extremely large. (c) The assumed radius is much too great.
75. (a) -9.315 MeV (b) Negative binding energy implies an unbound system.

(c) The assumption that it is two bound neutrons is incorrect.
77. (a) -1.805 MeV (b) Negative energy implies energy input is necessary to cause this reaction. (c) Although all conservation laws are obeyed, energy must be supplied, and so the assumption of spontaneous decay is incorrect.
79. (a) 84.5 Ci (b) An extemely large activity, many orders of magnitude greater than permitted for home use. (c) The assumption of 1.00 μA is unreasonably large. Other methods can detect much smaller decay rates.

Chapter 30

1. 5.700 MeV
3. ^{99}Mo $\rightarrow$ ^{99}Tc $+ \beta^- + v_e$ (you must provide the values of Z and N). Verify conservation of total A, charge, and electron family number by showing values before and after decay.
5. (a) 3.00×10^{-10} s (b) 3.33×10^{-12} s
7. ^{131}I, 4.03×10^{-10} g; ^{123}I, 3.63×10^{-11} g
9. (a) $A = 1 + 1 = 2$, $Z = 1 + 1 = 1 + 1$, efn $= 0 = 1 + (-1)$ (b) $A = 1 + 2 = 3$, $Z = 1 + 1 = 2$, efn $= 0 = 0$ (c) $A = 3 + 3 = 4 + 1 + 1$, $Z = 2 + 2 = 2 + 1 + 1$, efn $= 0 = 0$
13. 2×10^5 kg (about 200 tons)
15. 6.258 MeV
17. (a) 1.33 kg (b) 1.04×10^{14} J (c) Yes
19. (a) 3×10^{38}/s (b) 6×10^{14}/s·m²
21. (a) $A = 12 + 1 = 13$, $Z = 6 + 1 = 7$, efn $= 0 = 0$ (b) $A = 13 = 13$, $Z = 7 = 6 + 1$, efn $= 0 = -1 + 1$ (c) $A = 13 + 1 = 14$, $Z = 6 + 1 = 7$, efn $= 0 = 0$ (d) $A = 14 + 1 = 15$, $Z = 7 + 1 = 8$, efn $= 0 = 0$ (e) $A = 15 = 15$, $Z = 8 = 7 + 1$, efn $= 0 = -1 + 1$ (f) $A = 15 + 1 = 12 + 4$, $Z = 7 + 1 = 6 + 2$, efn $= 0 = 0$
23. (a) 177.1 MeV (b) Because the gain of an external neutron yields about 6 MeV, the average BE/A for heavy nuclei. (c) $A = 1 + 238 = 96 + 140 + 1 + 1 + 1$, $Z = 92 = 38 + 54$, efn $= 0 = 0$
25. (a) 180.6 MeV (b) $A = 1 + 239 = 96 + 140 + 1 + 1 + 1 + 1$, $Z = 94 = 38 + 56$, efn $= 0 = 0$
27. (a) 4.807 MeV (b) 0.753 MeV (c) 0.211 MeV
29. (a) 2.57×10^3 MW (b) 8.03×10^{19}/s (c) 991 kg
31. 0.560 g
33. 4.783 MeV
35. (a) Blast yield 2.1×10^{12} J to 8.4×10^{11} J, or 2.5 to 1, conventional to radiation enhanced (b) Prompt radiation yield 6.3×10^{11} J to 2.1×10^{11} J, or 3 to 1, radiation enhanced to conventional
37. (a) 9.84×10^{24} fissions, 3.89 kg (b) 3.0×10^{26} fissions, 2.0 kg
39. 7×10^4 g
41. (a) 1.00 rem (b) 7.50 rem (c) 3.00 rem
43. 200 rad
45. 1.06 mm
47. 1.24 MeV

49. 2.21×10^5 rem
51. (a) 1.20×10^{25} (b) 2.08×10^{20} (c) 1.74×10^{-5} (d) 8.67 parts per billion (for each of 2000 types)
53. 296 rem
55. 7.50 mrem (each day)
57. 1.13×10^{-8} g
59. 2.84×10^{-6} °C
61. 1.76×10^{12}
63. (a) 1.00×10^{14} W (b) 5.33×10^{23} (c) 3.33×10^{-4} kg·m/s (d) 1.13×10^{16} to 1, photon to deuterium
65. 6.43×10^9 K
67. (a) 3.84×10^{-6} rem (b) Dose is much too small to have any effect. (c) The assumed number of photons is much too small.
69. (a) 1.30×10^{23} Ci (b) An absurdly large activity. (c) The 6.01 h half-life is much too short to allow it to be manufactured 3 weeks ahead of its use.

Chapter 31

1. (a) 10^{-11} to 1, weak to EM (b) 1 to 1
3. 3×10^{-39} s
5. 2×10^{-16} m (0.2 fm)
7. 1.35×10^{-16} m (0.135 fm)
9. 3.33 MV
11. 14.0 TeV (14.002 TeV including the protons mass-energy)
13. 67.5 MeV
15. (a) 1×10^{14} (b) 1×10^{17}
17. (a) 1.67 GeV (b) $Z = 1 = 1 + 0 + 0$, τ fn $= -1 = 0 + 0 + (-1)$, μ fn $= 0 = -1 + 1 + 0$, efn $= 0 = 0 + 0 + 0$
19. (a) 3.9 eV (b) 2.9×10^{-8}
21. (a) The uud composition is the same as for a proton. (b) 3.3×10^{-24} s (c) Strong (short lifetime)
23. (a) 1
25. (a) $+1$ (b) $B = 1 = 1 + 0$, $Z = -1 = 0 + (-1)$, all lepton numbers are 0 before and after (c) $sss \rightarrow uds + \bar{u}s$
27. (a) $(u\bar{u} + d\bar{d}) \rightarrow (u\bar{u} + d\bar{d}) + (u\bar{u} + d\bar{d})$ (b) 278 MeV (c) 547.5 MeV
29. No, efn $= 0 = 1 + 1 + 0$, not conserved
31. (a) Yes. $Z = -1 = 0 + (-1)$, $B = 1 = 1 + 0$, all lepton family numbers are 0 before and after, spontaneous since mass greater before reaction (b) $dds \rightarrow udd + \bar{u}d$
33. (a) 216 (b) There are more baryons observed, because we have the 6 antiquarks and various mixtures of quarks (as for the π-meson) as well.
35. Ω^- ($\bar{s}\bar{s}\bar{s}$), $B = (-1/3) + (-1/3) + (-1/3) = -1$, lepton family numbers are all 0, $Z = 1/3 + 1/3 + 1/3 = 1$, $S = 1 + 1 + 1 = 3$
37. (a) 803 MeV (b) 938.78 MeV (c) The annihilation energy of an extra electron is obtained.
39. $c\bar{d}$
41. (a) The antiproton (b) $\bar{p} \rightarrow \pi^0 + e^-$
43. (a) 5×10^{10} (b) 5×10^4/m²
45. 2.5×10^{-17} m
47. (a) 33.9 MeV (b) Muon antineutrino 29.78 MeV, muon 4.12 MeV (kinetic energy)

49. (a) 7.2×10^5 kg (b) 7.2×10^2 m^3
(c) 100 months

Chapter 32
1. 3×10^{41} kg
3. (a) 3×10^{52} kg (b) 2×10^{79} (c) 4×10^{88}
5. 0.3 Gly
7. (a) 2.0×10^5 km/s (b) $0.67c$

9. 1.5×10^{15}
11. 2.7×10^5 m/s
13. 6×10^{-11} (an overestimate, since some of the light from Andromeda is blocked by gas and dust within that galaxy)
15. (a) 2×10^{-8} kg (b) 1×10^{19}
17. (a) 6×10^9 eV/c^2/m^3 (b) 8×10^8

21. 23.6 km
23. (a) 2.95×10^{12} m (b) 3.12×10^{-4} ly (about 5 light hours)
25. (a) (30 km/s)/Mly
(b) (15 km/s)/Mly
27. 965 rev/s (last digit uncertain)
29. 89.9999773° (many digits are needed to show difference from 90°)

INDEX

Boldface (**123**) indicates a definition; Italic (*123*) indicates a drawing or a photograph.

CREDITS

Credits are referenced by figure number.

About the Author   Chris Urone

Chapter 1   Tony Hallas/Photo Researchers, Inc.
1.1(a)   David Parker/Photo Researchers, Inc.
1.1(b)   Emilio Segre Visual Archive/American Institute of Physics
1.2(a)   David Young/Photo Edit
1.2(b)   Philippe Plailly/Photo Researchers, Inc.
1.3(a)   Audrey Ross/Bruce Coleman, Inc.
1.4(a)   Archive Photos
1.4(b)   Archive Photos
1.4(c)   Emilio Segre Visual Archive/American Institute for Physics
1.6(a)   Archive Photos
1.6(b)   Archive Photos
1.6(c)   Nordisk/Archive Photos
1.9   Courtesy of National Institute of Standards and Technology
1.12   Dr. P.P. Urone
1.13   Chronicle Features

Chapter 2   Kim Taylor/Bruce Coleman, Inc.
2.4   Christopher R. Harris/Uniphoto Picture Agency
2.8   David T. Overcash/Bruce Coleman, Inc.
2.23   Stephen Kline/Bruce Coleman, Inc.

Chapter 3   Tim Ribar/Uniphoto Picture Agency

Chapter 4   David Madison/Bruce Coleman, Inc.
4.15(a)   Francis Hogan, Electronic Publishing Services Inc., NYC
4.15(b)   Bruce Coleman, Inc.
4.18(a)   Tribune Media Services

Chapter 5   Uniphoto Picture Agency
5.29(a)   Dr. P.P. Urone
5.29(b)   Dr. P.P. Urone
5.29(c)   Dr. P.P. Urone
5.31   Anna E. Zuckerman/Photo Edit
5.35(a)   Columbia Pictures Corporation/Culver Pictures, Inc.

Chapter 6   Michael Dalton/Fundamental Photographs, New York
6.13   Bob Peterson/FPG International
6.14   Courtesy of NASA
6.16   Mark Burnett/Photo Researchers, Inc.
6.19   Walter Looss Jr., Inc./The Image Bank
6.20   Nell/Photo Researchers, Inc.

Chapter 7   Uniphoto Picture Agency
7.1   Ed Braverman/FPG International
7.2(b)   Tony Freeman/Photo Edit
7.A   S.P.L./Photo Researchers, Inc.
7.9   Richard Menga/Fundamental Photographs
7.12   Account/Phototake
7.14   Ron Lindsey/Pototake
7.16   Courtesy of Evan Jones/Dept. of Physics, Sierra College

Chapter 8   Joe Viesti/Viesti Associates, Inc.
8.14(a)   NASA/FPG International
8.14(e)   Jack Zehrt/FPG International
8.15   Francis Hogan, Electronic Publishing Services Inc., NYC
8.30(a)   Courtesy of NASA

Chapter 9   E.R. Degglreyer/Bruce Coleman, Inc.
9.1   Bob Burch/Bruce Coleman, Inc.
9.11   Bachmann/Photo Edit
9.20   Patti McConville/The Image Bank

Chapter 10   Peter Miller/The Image Bank
10.15(a)   A. Bignami/The Image Bank
10.15(b)   Joseph McNally/The Image Bank
10.15(c)   Comstock
10.18(a)   Courtesy of Seariver Maritime, Inc.
10.18(b)   Courtesy of Seariver Maritime, Inc.
10.20   D. Young Wolff/Photo Edit
10.22   Rudi Von Briel
10.28   Dr. P.P. Urone

Chapter 11   D. Young Wolff/Photo Edit
11.3   D. Young Wolff/Photo Edit
11.8(a)   Jeff Greene/Photo Edit
11.10   Thomas Wear/Comstock
11.11(c)   Comstock
11.13(c)   Dr. P.P. Urone

Chapter 12   Wolfgang Spunbarg/Photo Edit
12.3(a)   Richard Megna/Fundamental Photographs
12.3(b)   Photri, Inc.
12.4   NASA/Photo Researchers, Inc.
12.8   Robert Mathena/Fundamental Photographs
12.10   Francis Hogan, Electronic Publishing Services Inc., NYC
12.13(b)   A.D. Dowsett/Photo Researchers, Inc.
12.18   NASA/Bettmann Archive
12.23   Jack Grove/Photo Edit

Chapter 13   Comstock
13.6   D. Young Wolff/Photo Edit
13.9   Kristen Brochmann/Fundamental Photographs
13.10   Ron Thomas/FPG International
13.11(a)   Richard Megna/Fundamental Photographs
13.11(b)   Susan Van Stten /Photo Edit
13.18   Dennis MacDonald/Photo Edit
13.19   Anna E. Zuckerman/Photo Edit
13.20   Richard Hutchings/Photo Edit
13.21   Phil Jude/Photo Researchers, Inc.
13.22   Francis Hogan, Electronic Publishing Services Inc., NYC
13.24   NASA/Photo Researchers, Inc.
13.26   Courtesy of Bob Metcalf CSUS/Solar Cookers International, Sacramento, California

Chapter 14   Comstock
14.1   Photo Edit

Chapter 15   Vic Bider/Photo Edit
15.19   Photri-Microstock
15.23   Richard Megna/Fundamental Photographs
15.30   M. Richards/Photo Edit
15.33   Michael Newman/Photo Edit

Chapter 16   Comstock
16.1   Tribune Media Services
16.11   Dr. P.P. Urone

16.13(b)   Y. Arthus. Bertrand/Photo Researchers, Inc.
16.20(a)   Tony Freeman/Photo Edit
16.20(b)   Paul Silverman/Fundamental Photographs
16.21   James Lester/Photo Researchers, Inc.
16.30(b)   Dr. P.P. Urone
16.31   Menau Kulyk/Photo Researchers, Inc.
16.32   Mat Meadows/Photo Researchers, Inc.
16.35   Joe McDonald/Bruce Coleman, Inc.

Chapter 17   Fundamental Photographs
17.2(a)   Dr. P.P. Urone
17.2(b)   Robert Brenner/Photo Edit
17.10(d)   Richard Megna/Fundamental Photographs
17.23(a)   J. Tinning/Photo Researchers, Inc.
17.23(b)   Richard Megna/Fundamental Photographs
17.24   Argon National Laboratory/FPG International
17.28(a)   Julia Silva/The Image Bank
17.29(b)   Courtesy of Research Cortell, Inc.

Chapter 18   Mauritius GMBH/PhotoTake
18.5   Francis Hogan/Electronic Publishing Services Inc. NYC
18.18   Bonnie Rauch/Photo Researchers, Inc.

Chapter 19   Tony Freeman/Photo Edit
19.11   Michael Tamborrino/FPG International
19.12   Dr. P.P. Urone
19.16   Tony Freeman/Photo Edit
19.22   Kevin Cuff/FPG International
19.28   Han Reinhard/Bruce Coleman Inc.
19.32(b)   Patrica Peticolas/Fundamental Photographs

Chapter 20   Richard Sobol/Zuma Press
20.01   Michael Dalton/Fundamental Photographs
20.08(a)   Tony Freeman/Photo Edit
20.08(b)   Tony Freeman/Photo Edit
20.08(c)   Jack Pelkan/Fundamental Photographs
20.08(d)   Kristen Brochmann/Fundamental Photographs
20.11(a)   Petrified Colle/The Image Bank
20.11(b)   Dr. P.P. Urone
20.23   Michael Newman/Photo Edit
20.24   Dr. P.P. Urone
20.36   Bob and Clara Calhoun/Bruce Coleman Inc.
20.37(c)   Carl Vanderschul/FPG International

Chapter 21   Photo Researchers, Inc.
21.01   Richard Megna/Fundamental Photographs
21.07(a)   Richard Megna/Fundamental Photographs
21.07(b)   Richard Megna/Fundamental Photographs
21.15   Omikron/Photo Researchers, Inc.
21.17(a)   Dr. P.P. Urone
21.19(b)   World Perspectives/FPG International
21.21   Hank Morgan/Photo Researchers, Inc.
21.22(b)   Courtesy of Princeton Plasma Physics Laboratory
21.38(b)   Solar Telescope/Photri-Microstock
21.39(b)   Geoff Tompkinson/Photo Researchers, Inc.
21.41(a)   Takeshi Takahara/Photo Researchers, Inc.

PERIODIC TABLE OF THE ELEMENTS

1   Current ACS and IUPAC
IA   Preferred U.S.

1 IA	2 IIA	3 IIIB	4 IVB	5 VB	6 VIB	7 VIIB	8	9 VIIIB	10	11 IB	12 IIB	13 IIIA	14 IVA	15 VA	16 VIA	17 VIIA	18 VIIIA
1 H Hydrogen 1.008																	2 He Helium 4.003
3 Li Lithium 6.941	4 Be Beryllium 9.012											5 B Boron 10.81	6 C Carbon 12.01	7 N Nitrogen 14.01	8 O Oxygen 16.00	9 F Fluorine 19.00	10 Ne Neon 20.18
11 Na Sodium 22.99	12 Mg Magnesium 24.31											13 Al Aluminum 26.98	14 Si Silicon 28.09	15 P Phosphorous 30.97	16 S Sulfur 32.07	17 Cl Chlorine 35.45	18 Ar Argon 39.95
19 K Potassium 39.10	20 Ca Calcium 40.08	21 Sc Scandium 44.96	22 Ti Titanium 47.88	23 V Vanadium 50.94	24 Cr Chromium 52.00	25 Mn Manganese 54.94	26 Fe Iron 55.85	27 Co Cobalt 58.93	28 Ni Nickel 58.69	29 Cu Copper 63.55	30 Zn Zinc 65.39	31 Ga Gallium 69.72	32 Ge Germanium 72.61	33 As Arsenic 74.92	34 Se Selenium 78.96	35 Br Bromine 79.90	36 Kr Krypton 83.80
37 Rb Rubidium 85.47	38 Sr Strontium 87.62	39 Y Yttrium 88.91	40 Zr Zirconium 91.22	41 Nb Niobium 92.91	42 Mo Molybdenum 95.94	43 Tc Technetium (98)	44 Ru Ruthenium 101.1	45 Rh Rhodium 102.9	46 Pd Palladium 106.4	47 Ag Silver 107.9	48 Cd Cadmium 112.4	49 In Indium 114.8	50 Sn Tin 118.7	51 Sb Antimony 121.8	52 Te Tellurium 127.6	53 I Iodine 126.9	54 Xe Xenon 131.3
55 Cs Cesium 132.9	56 Ba Barium 137.33	71 Lu Lutetium 175.0	72 Hf Hafnium 178.5	73 Ta Tantalum 180.9	74 W Tungsten 183.8	75 Re Rhenium 186.2	76 Os Osmium 190.2	77 Ir Iridium 192.2	78 Pt Platinum 195.1	79 Au Gold 197.0	80 Hg Mercury 200.6	81 Tl Thallium 204.4	82 Pb Lead 207.2	83 Bi Bismuth 209.0	84 Po Polonium (209)	85 At Astatine (210)	86 Rn Radon (222)
87 Fr Francium (223)	88 Ra Radium 226.025	103 Lr Lawrencium (262)	104 Rf Rutherfordium (261)	105 Db Dubnium (262)	106 Sg Seaborgium (263)	107 Bh Bohrium (262)	108 Hs Hassium (265)	109 Mt Meitnerium (266)									

Names undecided (or not universally accepted)

* Lanthanides	57 La Lanthanum 138.9	58 Ce Cerium 140.1	59 Pr Praseodymium 140.9	60 Nd Neodymium 144.2	61 Pm Promethium (145)	62 Sm Samarium 150.4	63 Eu Europium 152.0	64 Gd Gadolinium 157.3	65 Tb Terbium 158.9	66 Dy Dysprosium 162.5	67 Ho Holmium 164.9	68 Er Erbium 167.3	69 Tm Thulium 168.9	70 Yb Ytterbium 173.0
◆ Actinides	89 Ac Actinium (227)	90 Th Thorium 232.0	91 Pa Protactinium 231.0	92 U Uranium 238.0	93 Np Neptunium (237)	94 Pu Plutonium (244)	95 Am Americium (243)	96 Cm Curium (247)	97 Bk Berkelium (247)	98 Cf Californium (251)	99 Es Einsteinium (254)	100 Fm Fermium (257)	101 Md Mendelevium (258)	102 No Nobelium (255)

— Metals
— Non-metals
— Semi-metals
— Artificially Prepared Metals

Atomic number —→ 1
Symbol —→ H
Name —→ Hydrogen
Atomic mass —→ 1.008

Note: The atomic masses shown in this table are the 1991 IUPAC atomic masses, which are based on carbon-12 = 12 u, rounded to four significant figures. Masses in parentheses are masses of the most stable isotope of radioactive elements.